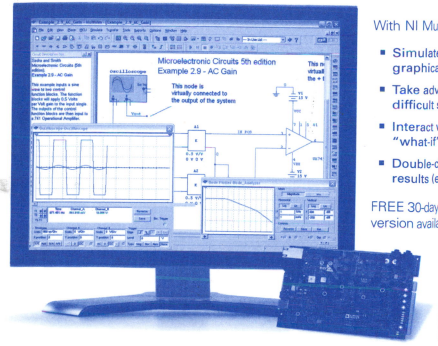

MICROELECTRONIC CIRCUITS

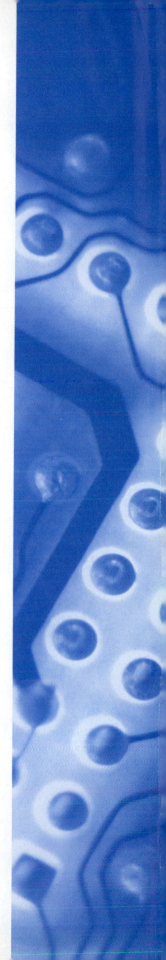

THE OXFORD SERIES IN ELECTRICAL AND COMPUTER ENGINEERING

Adel S. Sedra, Series Editor

Allen and Holberg, *CMOS Analog Circuit Design, 2nd Edition*
Bobrow, *Elementary Linear Circuit Analysis, 2nd Edition*
Bobrow, *Fundamentals of Electrical Engineering, 2nd Edition*
Burns and Roberts, *An Introduction to Mixed-Signal IC Test and Measurement*
Campbell, *The Science and Engineering of Microelectronic Fabrication, 2nd Edition*
Chen, *Digital Signal Processing*
Chen, *Linear System Theory and Design, 3rd Edition*
Chen, *Signals and Systems, 3rd Edition*
Comer, *Digital Logic and State Machine Design, 3rd Edition*
Comer, *Microprocessor-based System Design*
Cooper and McGillem, *Probabilistic Methods of Signal and System Analysis, 3rd Edition*
DeCarlo and Lin, *Linear Circuit Analysis, 2nd Edition*
Dimitrijev, *Understanding Semiconductor Devices*
Fortney, *Principles of Electronics: Analog & Digital*
Franco, *Electric Circuits Fundamentals*
Ghausi, *Electronic Devices and Circuits: Discrete and Integrated*
Guru and Hiziroglu, *Electric Machinery and Transformers, 3rd Edition*
Houts, *Signal Analysis in Linear Systems*
Jones, *Introduction to Optical Fiber Communication Systems*
Krein, *Elements of Power Electronics*
Kuo, *Digital Control Systems, 3rd Edition*
Lathi, *Linear Systems and Signals, 2nd Edition*
Lathi, *Modern Digital and Analog Communications Systems, 3rd Edition*
Lathi, *Signal Processing and Linear Systems*
Martin, *Digital Integrated Circuit Design*
Miner, *Lines and Electromagnetic Fields for Engineers*
Parhami, *Computer Arithmetic*
Roberts and Sedra, *SPICE, 2nd Edition*
Roulston, *An Introduction to the Physics of Semiconductor Devices*
Sadiku, *Elements of Electromagnetics, 3rd Edition*
Santina, Stubberud, and Hostetter, *Digital Control System Design, 2nd Edition*
Sarma, *Introduction to Electrical Engineering*
Schaumann and Van Valkenburg, *Design of Analog Filters*
Schwarz and Oldham, *Electrical Engineering: An Introduction, 2nd Edition*
Sedra and Smith, *Microelectronic Circuits, 5th Edition*
Stefani, Savant, Shahian, and Hostetter, *Design of Feedback Control Systems, 4th Edition*
Tsividis, *Operation and Modeling of the MOS Transistor, 2nd Edition*
Van Valkenburg, *Analog Filter Design*
Warner and Grung, *Semiconductor Device Electronics*
Wolovich, *Automatic Control Systems*
Yariv, *Optical Electronics in Modern Communications, 5th Edition*
Żak, *Systems and Control*

FIFTH EDITION

MICROELECTRONIC CIRCUITS

Adel S. Sedra
University of Waterloo

Kenneth C. Smith
University of Toronto

New York   Oxford

OXFORD UNIVERSITY PRESS

2004

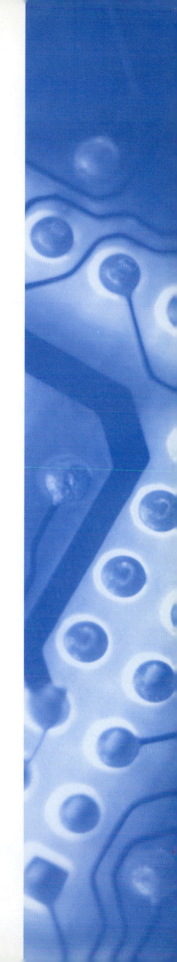

Oxford University Press, Inc., publishes works that further Oxford University's
objective of excellence in research, scholarship, and education.

Oxford New York
Auckland Cape Town Dar es Salaam Hong Kong Karachi
Kuala Lumpur Madrid Melbourne Mexico City Nairobi
New Delhi Shanghai Taipei Toronto

With offices in
Argentina Austria Brazil Chile Czech Republic France Greece
Guatemala Hungary Italy Japan Poland Portugal Singapore
South Korea Switzerland Thailand Turkey Ukraine Vietnam

Published by Oxford University Press, Inc.
198 Madison Avenue, New York, New York 10016
http://www.oup.com

Oxford is a registered trademark of Oxford University Press

ISBN 978-0-19-533883-6

Cover Illustration: The chip shown is an inside view of a mass-produced surface-micromachined gyroscope system, integrated on a 3mm by 3mm die, and using a standard 3-m 2-V BiCMOS process suited for the harsh automotive environment. This first single-chip gyroscopic sensor, in which micro-mechanical and electronic components are intimately entwined on the same chip, provides unprecedented performance through the use of a collection of precision-directed techniques, including emphasis on differential operation (both mechanically and electronically) bolstered by trimmable thin-film resistive components. This tiny, robust, low-power, angular-rate-to-voltage transducer, having a sensitivity of 12.5mV/°/s and resolution of 0.015°/s (or 50°/hour) has a myriad of applications—including automotive skid control and rollover detection, dead reckoning for GPS backup and robot motion control, and camera-field stabilization. The complete gyroscope package, weighing 1/3 gram with a volume of 1/6 cubic centimeter, uses 30mW from a 5-V supply. *Source:* John A. Geen, Steven J. Sherman, John F. Chang, Stephen R. Lewis; Single-chip surface micromachined integrated Gyroscope with 50°/h Allan deviation, IEEE Journal of Solid-State Circuits, vol. 37, pp. 1860–1866, December 2002. (Originally presented at ISSCC 2002.) Photographed by John Chang, provided by John Geen, both of Analog Devices, Micromachine Products Division, Cambridge, MA, USA.

Printing number: 9 8 7 6 5 4 3

Printed in the United States of America
on acid-free paper

CONDENSED TABLE OF CONTENTS

DETAILED TABLE OF CONTENTS

5 Bipolar Junction Transistors (BJTs) *377*

PREFACE

Microelectronic Circuits, fifth edition, is intended as a text for the core courses in electronic circuits taught to majors in electrical and computer engineering. It should also prove useful to engineers and other professionals wishing to update their knowledge through self-study.

As was the case with the first four editions, the objective of this book is to develop in the reader the ability to analyze and design electronic circuits, both analog and digital, discrete and integrated. While the application of integrated circuits is covered, emphasis is placed on transistor circuit design. This is done because of our belief that even if the majority of those studying the book were not to pursue a career in IC design, knowledge of what is inside the IC package would enable intelligent and innovative application of such chips. Furthermore, with the advances in VLSI technology and design methodology, IC design itself is becoming accessible to an increasing number of engineers.

PREREQUISITES

The prerequisite for studying the material in this book is a first course in circuit analysis. As a review, some linear circuits material is included here in appendixes: specifically, two-port network parameters in Appendix B; some useful network theorems in Appendix C; single-time-constant circuits in Appendix D; and *s*-domain analysis in Appendix E. No prior knowledge of physical electronics is assumed. All required device physics is included, and Appendix A provides a brief description of IC fabrication.

NEW TO THIS EDITION

Although the philosophy and pedagogical approach of the first four editions have been retained, several changes have been made to both organization and coverage.

1. The book has been reorganized into three parts. Part I: Devices and Basic Circuits, composed of the first five chapters, provides a coherent and reasonably comprehensive single-semester introductory course in electronics. Similarly, Part II: Analog and Digital Integrated Circuits (Chapters 6–10) presents a body of material suitable for a second one-semester course. Finally, four carefully chosen subjects are included in Part III: Selected Topics. These can be used as enhancements or substitutions for some of the material in earlier chapters, as resources for projects or thesis work, and/or as part of a third course.

2. Each chapter is organized so that the essential "must-cover" topics are placed first, and the more specialized material appears last. This allows considerable flexibility in teaching and learning from the book.

3. Chapter 4, MOSFETs, and Chapter 5, BJTs, have been completely rewritten, updated, and made completely independent of each other. The MOSFET chapter is placed first to reflect the fact that it is currently the most significant electronics device by a wide margin. However, if desired, the BJT can be covered first. Also, the identical structure of the two chapters makes teaching and learning about the second device easier and faster.

4. To make the first course comprehensive, both Chapters 4 and 5 include material on amplifier and digital-logic circuits. In addition, the frequency response of the basic common-source (common-emitter) amplifier is included. This is important for students who might not take a second course in electronics.

5. A new chapter on integrated-circuit (IC) amplifiers (Chapter 6) is added. It begins with a comprehensive comparison between the MOSFET and the BJT. Typical parameter values of devices produced by modern submicron fabrication processes are given and utilized in the examples, exercises, and end-of-chapter problems. The study of each amplifier configuration includes its frequency response. This should make the study of amplifier frequency response more interesting and somewhat easier.

6. The material on differential and multistage amplifiers in Chapter 7 has been rewritten to present the MOSFET differential pair first. Here also, the examples, exercises, and problems have been expanded and updated to utilize parameter values representative of modern submicron technologies.

7. Throughout the book, greater emphasis is placed on MOSFET circuits.

8. To make room for new material, some of the topics that have become less current, such as JFETs and TTL, or have remained highly specialized, such as GaAs devices and circuits, have been removed from the book. However, they are made available on the CD accompanying the book and on the book's website.

9. As a study aid and for easy reference, many summary tables have been added.

10. The review exercises, examples, and end-of-chapter problems have been updated and their numbers and variety increased.

11. The SPICE sections have been rewritten and the SPICE examples now utilize schematic entry. To enable further experimentation, the files for all SPICE examples are provided on the CD and website.

THE CD-ROM AND THE WEBSITE

A CD-ROM accompanies this book. It contains much useful supplementary information and material intended to enrich the student's learning experience. These include (1) A Student's Edition of OrCAD PSpice 9.2. (2) The input files for all the SPICE examples in this book. (3) A link to the book's website accessing PowerPoint slides of every figure in this book that students can print and carry to class to facilitate taking notes. (4) Bonus text material of specialized topics not covered in the current edition of the textbook. These include: JFETs, GaAs devices and circuits, and TTL circuits.

A website for the book has been set up (www.sedrasmith.org). Its content will change frequently to reflect new developments in the field. It features SPICE models and files for all PSpice examples, links to industrial and academic websites of interest, and a message center to communicate with the authors. There is also a link to the Higher Education Group of Oxford University Press so professors can receive complete text support.

EMPHASIS ON DESIGN

It has been our philosophy that circuit design is best taught by pointing out the various trade-offs available in selecting a circuit configuration and in selecting component values for a given configuration. The emphasis on design has been increased in this edition by including more design examples, exercise problems, and end-of-chapter problems. Those exercises and end-of-chapter problems that are considered "design-oriented" are indicated with a D. Also, the most valuable design aid, SPICE, is utilized throughout the book, as already outlined.

EXERCISES, END-OF-CHAPTER PROBLEMS, AND ADDITIONAL SOLVED PROBLEMS

Over 450 exercises are integrated throughout the text. The answer to each exercise is given below the exercise so students can check their understanding of the material as they read. Solving these exercises should enable the reader to gauge his or her grasp of the preceding material. In addition, more than 1370 end-of-chapter problems, about a third of which are new to this edition, are provided. The problems are keyed to the individual sections and their degree of difficulty is indicated by a rating system: difficult problems are marked with as asterisk (*); more difficult problems with two asterisks (**); and very difficult (and/or time consuming) problems with three asterisks (***). We must admit, however, that this classification is by no means exact. Our rating no doubt had depended to some degree on our thinking (and mood!) at the time a particular problem was created. Answers to about half the problems are given in Appendix H. Complete solutions for all exercises and problems are included in the *Instructor's Manual,* which is available from the publisher for those instructors who adopt the book.

As in the previous four editions, many examples are included. The examples, and indeed most of the problems and exercises, are based on real circuits and anticipate the applications encountered in designing real-life circuits. This edition continues the use of numbered solution steps in the figures for many examples, as an attempt to recreate the dynamics of the classroom.

A recurring request from many of the students who used earlier editions of the book has been for solved problems. To satisfy this need, a book of additional problems with solutions is available with this edition (see the list of available ancillaries later in this preface).

AN OUTLINE FOR THE READER

The book starts with an introduction to the basic concepts of electronics in Chapter 1. Signals, their frequency spectra, and their analog and digital forms are presented. Amplifiers are introduced as circuit building blocks and their various types and models are studied. The basic element of digital electronics, the digital logic inverter, is defined in terms of its voltage-transfer characteristic, and its various implementations using voltage and current switches are discussed. This chapter also establishes some of the terminology and conventions used throughout the text.

The next four chapters are devoted to the study of electronic devices and basic circuits and constitute the bulk of Part I of the text. Chapter 2 deals with operational amplifiers, their terminal characteristics, simple applications, and limitations. We have chosen to discuss the op amp as a circuit building block at this early stage simply because it is easy to deal with and because the student can experiment with op-amp circuits that perform nontrivial tasks with relative ease and with a sense of accomplishment. We have found this approach to be highly motivating to the student. We should point out, however, that part or all of this chapter can be skipped and studied at a later stage (for instance in conjunction with Chapter 7, Chapter 8, and/or Chapter 9) with no loss of continuity.

Chapter 3 is devoted to the study of the most fundamental electronic device, the *pn* junction diode. The diode terminal characteristics and its hierarchy of models and basic circuit applications are presented. To understand the physical operation of the diode, and indeed of the MOSFET and the BJT, a concise but substantial introduction to semiconductors and the *pn* junction is provided. This material is placed near the end of the chapter (Section 3.7) so that part or all of it can be skipped by those who have already had a course in physical electronics.

Chapters 4 and 5 deal with the two major electronic devices—the MOS field-effect transistor (MOSFET) and the bipolar junction transistor (BJT), respectively. The two chapters have an identical structure and are completely independent of each other and thus, can be covered in either order. Each chapter begins with a study of the device structure and its physical operation, leading to a description of its terminal characteristics. Then, to establish in the reader a high degree of familiarity with the operation of the transistor as a circuit element, a large number of examples are presented of dc circuits utilizing the device. The large-signal operation of the basic common-source (common-emitter) circuit is then studied and used to delineate the region over which the device can be used as a linear amplifier from those regions where it can be used as a switch. This makes clear the need for biasing the transistor and leads naturally to the study of biasing methods. At this point, the biasing methods used are mostly for discrete circuits, leaving the study of IC biasing to Chapter 6. Next, small-signal operation is studied and small-signal models are derived. This is followed by a study of the basic configurations of discrete-circuit amplifiers. The internal capacitive effects that limit the high-frequency operation of the transistor are then studied, and the high-frequency equivalent-circuit model is presented. This model is then used to determine the high-frequency response of a common-source (common-emitter) amplifier. As well, the low-frequency response resulting from the use of coupling and bypass capacitors is also presented. The basic digital-logic inverter circuit is then studied. Both chapters conclude with a study of the transistor models used in SPICE together with circuit-simulation examples using PSpice. This description should indicate that Chapters 4 and 5 contain the essential material for a first course in electronics.

Part II: Analog and Digital Integrated Circuits (Chapters 6–10) begins with a comprehensive compilation and comparison of the properties of the MOSFET and the BJT. The comparison is facilitated by the provision of typical parameter values of devices fabricated with modern process technologies. Following a study of biasing methods employed in IC amplifier design (Section 6.3), and some basic background material for the analysis of high-frequency amplifier response (Section 6.4), the various configurations of single-stage IC amplifiers are presented in a systematic manner. In each case, the MOS circuit is presented first. Some transistor–pair configurations that are usually treated as a single stage, such as the cascode and the Darlington circuits, are also studied. Each section includes a study of the high-frequency response of the particular amplifier configuration. Again, we believe that this "in-situ" study of frequency response is superior to the traditional approach of postponing all coverage of frequency response to a later chapter. As in other chapters, the more specialized material, including advanced current–mirror and current–source concepts, is placed in the second half of the chapter, allowing the reader to skip some of this material in a first reading. This chapter should provide an excellent preparation for an in-depth study of analog IC design.

The study of IC amplifiers is continued in Chapter 7 where the emphasis is on two major topics: differential amplifiers and multistage amplifiers. Here again, the MOSFET differential pair is treated first. Also, frequency response is discussed where needed, including in the two examples of multistage amplifiers.

Chapter 8 deals with the important topic of feedback. Practical circuit applications of negative feedback are presented. We also discuss the stability problem in feedback amplifiers and treat frequency compensation in some detail.

Chapter 9 integrates the material on analog IC design presented in the preceding three chapters and applies it to the analysis and design of two major analog IC functional blocks: op amps and data converters. Both CMOS and bipolar op amps are studied. The data-converter sections provide a bridge to the study of digital CMOS logic circuits in Chapter 10.

Chapter 10 builds on the introduction to CMOS logic circuits in Section 4.10 and includes a carefully selected set of topics on static and dynamic CMOS logic circuits that round out the study of analog and digital ICs in Part II.

The study of digital circuits is continued in the first of the four selected-topics chapters that comprise Part III. Specifically, Chapter 11 deals with memory and related circuits, such as latches, flip-flops, and monostable and stable multivibrators. As well, two somewhat specialized but significant digital circuit technologies are studied: emitter-coupled logic (ECL) and BiCMOS. The two digital chapters (10 and 11) together with the earlier material on digital circuits should prepare the reader well for a subsequent course on digital IC design or VLSI circuits.

The next two chapters of Part III, Chapters 12 and 13, are application or system oriented. Chapter 12 is devoted to the study of analog-filter design and tuned amplifiers. Chapter 13 presents a study of sinusoidal oscillators, waveform generators, and other nonlinear signal-processing circuits.

The last chapter of the book, Chapter 14, deals with various types of amplifier output stages. Thermal design is studied, and examples of IC power amplifiers are presented.

The eight appendixes contain much useful background and supplementary material. We wish to draw the reader's attention in particular to Appendix A, which provides a concise introduction to the important topic of IC fabrication technology including IC layout.

COURSE ORGANIZATION

The book contains sufficient material for a sequence of two single-semester courses (each of 40 to 50 lecture hours). The organization of the book provides considerable flexibility in course design. In the following, we suggest various possibilities for the two courses.

The First Course

The most obvious package for the first course consists of Chapters 1 through 5. However, if time is limited, some or all of the following sections can be postponed to the second course: 1.6, 1.7, 2.6, 2.7, 2.8, 3.6, 3.8, 4.8, 4.9, 4.10, 4.11, 5.8, 5.9, and 5.10. It is also quite possible to omit Chapter 2 altogether from this course. Also, it is possible to concentrate on the MOSFET (Chapter 4) and cover the BJT (Chapter 5) only partially and/or more quickly. Covering Chapter 5 thoroughly and Chapter 4 only partially and/or more quickly is also possible—but not recommended! An entirely analog first course is also possible by omitting Sections 1.7, 4.10, and 5.10. A digitally oriented first course is also possible. It would consist of the following sections; 1.1, 1.2, 1.3, 1.4, 1.7, 1.8, 3.1, 3.2, 3.3, 3.4, 3.7, 4.1, 4.2, 4.3, 4.4, 4.10, 4.12, 5.1, 5.2, 5.3, 5.4, 5.10, 5.11, all of Chapter 10, and selected topics from Chapter 11. Also, if time permits, some material from Chapter 2 on op amps would be beneficial.

The Second Course

An excellent place to begin the second course is Chapter 6 where Section 6.2 can serve as a review of the MOSFET and BJT characteristics. Ideally, the second course would cover

Chapters 6 through 10 (assuming, of course, that the first course covered Chapters 1 through 5). If time is short, either Chapter 10 can postponed to a subsequent course on digital circuits and/or some sections of Chapters 6–9 can be omitted. One possibility would be to de-emphasize bipolar circuits by omitting some or all of the bipolar sections in Chapters 6, 7, and 9. Another would be to reduce somewhat the coverage of feedback (Chapter 8). Also, data converters can be easily deleted from the second course. Still, for Chapter 9, perhaps only CMOS op amps need to be covered and the 741 deleted or postponed. It is also possible to replace some of the material from Chapters 6–10 by selected topics from Chapters 11–14. For instance, in an entirely analog second course, Chapter 10 can be replaced by a selection of topics from Chapters 13–14.

ANCILLARIES

A complete set of ancillary materials is available with this text to support your course.

For the Instructor

The *Instructor's Manual with Transparency Masters* provides complete worked solutions to all the exercises in each chapter and all the end-of-chapter problems in the text. It also contains 200 transparency masters that duplicate the figures in the text most often used in class.

A set of *Transparency Acetates* of the 200 most important figures in the book.

A *PowerPoint CD* with slides of every figure in the book and each corresponding caption.

For the Student and the Instructor

The *CD-ROM* included with every new copy of the textbook contains SPICE input files, a Student Edition of OrCAD PSpice 9.2 Lite Edition, a link to the website featuring PowerPoint slides of the book's illustrations, and bonus topics.

Laboratory Explorations for Microelectronic Circuits, 5th edition, by Kenneth C. Smith (KC), contains laboratory experiments and instructions for the major topics studied in the text.

KC's Problems and Solutions for Microelectronic Circuits, 5th edition, by Kenneth C. Smith (KC), contains hundreds of additional study problems with complete solutions, for students who want more practice.

SPICE, 2nd edition, by Gordon Roberts of McGill University and Adel Sedra, provides a detailed treatment of SPICE and its application in the analysis and design of circuits of the type studied in this book.

ACKNOWLEDGMENTS

Many of the changes in this fifth edition were made in response to feedback received from some of the instructors who adopted the fourth edition. We are grateful to all those who took the time to write to us. In addition, the following reviewers provided detailed commentary on the fourth edition and suggested many of the changes that we have incorporated in this revision. To all of them, we extend our sincere thanks: Maurice Aburdene, Bucknell University; Patrick L. Chapman, University of Illinois at Urbana-Champaign; Artice Davis, San Jose State University; Paul M. Furth, New Mexico State University; Roobik Gharabagi, St. Louis

University; Reza Hashemian, Northern Illinois University; Ward J. Helms, University of Washington; Hsiung Hsu, Ohio State University; Marian Kazimierczuk, Wright State University; Roger King, University of Toledo; Robert J. Krueger, University of Wisconsin-Milwaukee; Un-Ku Moon, Oregon State University; John A. Ringo, Washington State University; Zvi S. Roth, Florida Atlantic University; Mulukutla Sarma, Northeastern University; John Scalzo, Louisiana State University; Ali Sheikholeslami, University of Toronto; Pierre Schmidt, Florida International University; Charles Sullivan, Dartmouth College; Gregory M. Wierzba, Michigan State University; and Alex Zaslavsky, Brown University.

We are also grateful to the following colleagues and friends who have provided many helpful suggestions: Anthony Chan-Carusone, University of Toronto; Roman Genov, University of Toronto; David Johns, University of Toronto; Ken Martin, University of Toronto; Wai-Tung Ng, University of Toronto; Khoman Phang, University of Toronto; Gordon Roberts, McGill University; and Ali Sheikholeslami, University of Toronto.

We remain grateful to the reviewers of the four previous editions: Michael Bartz, University of Memphis; Roy H. Cornely, New Jersey Institute of Technology; Dale L. Critchlow, University of Vermont; Steven de Haas, California State University–Sacramento; Eby G. Friedman, University of Rochester; Rhett T. George, Jr., Duke University; Richard Hornsey, York University; Robert Irvine, California State University, Pamona; John Khoury, Columbia University; Steve Jantzi, Broadcom; Jacob B. Khurgin, The Johns Hopkins University; Joy Laskar, Georgia Institute of Technology; David Luke, University of New Brunswick; Bahram Nabet, Drexel University; Dipankar Nagchoudhuri, Indian Institute of Technology, Delhi, India; David Nairn, Analog Devices; Joseph H. Nevin, University of Cincinnati; Rabin Raut, Concordia University; Richard Schreier, Analog Devices; Dipankar Sengupta, Royal Melbourne Institute of Technology; Michael L. Simpson, University of Tennessee; Karl A. Spuhl, Washington University; Daniel van der Weide, University of Delaware.

A number of individuals made significant contributions to this edition. Anas Hamoui of the University of Toronto played a key role in shaping both the organization and content of this edition. In addition, he wrote the SPICE sections. Olivier Trescases of the University of Toronto preformed the SPICE simulations. Richard Schreier of Analog Devices helped us locate the excellent cover photo. Wai-Tung Ng of the University of Toronto completely rewrote Appendix A. Gordon Roberts of McGill University gave us permission to use some of the examples from the book *SPICE* by Roberts and Sedra. Mandana Amiri, Karen Kozma, Shahriar Mirabbasi, Roberto Rosales, Jim Somers of Sonora Designworks, and John Wilson all helped significantly in preparing the student and instructor support materials. Jennifer Rodrigues typed all the revisions with skill and good humor and assisted with many of the logistics. Laura Fujino assisted in the preparation of the index, and perhaps more importantly, in keeping one of us (KCS) focused. To all of these friends and colleagues we say thank you.

The authors would like to thank Cadence Design Systems, Inc., for allowing Oxford University Press to distribute OrCad Family Release 9.2 Lite Edition software with this book. We are grateful to John Geen from Analog Devices for providing the cover photo and to Tom McElwee (from TWM Research).

A large number of people at Oxford University Press contributed to the development of this edition and its various ancillaries. We would like to specifically mention Barbara Wasserman, Liza Murphy, Mary Beth Jarrad, Mac Hawkins, Barbara Brown, Cathleen Bennett, Celeste Alexander, Chris Critelli, Eve Siegel, Mary Hopkins, Jeanne Ambrosio, Trent Haywood, Jennifer Slomack, Ned Escobar, Debbie Agee, Valerie Alger, Susanne Arrington, Jim Brooks, Tina Bruce, Andrea Hill, Kathleen Kelly, Jennifer Lambert, Katrina Lewis, Earl Murphy, Sylvia Parrish, Terry Retchless, Diane Seagle, and Andre Turrentine.

We wish to extend special thanks to our Publisher at Oxford University Press, Chris Rogers. We are also grateful to Scott Burns, Marketing and Sales Director, for his many excellent and creative ideas and for his friendship. We received a great deal of support and advice from our previous editor and friend, Peter Gordon. After Peter's departure, the leadership of the project has been most ably assumed by Danielle Christensen, our current editor. Elyse Dubin, Director of Editorial, Design, and Production, played a pivotal role in ensuring that the book would receive the greatest possible attention in the various stages of design and production.

If there is a single person at Oxford University Press that is responsible for this book coming out on time and looking so good, it is our Managing Editor, Karen Shapiro: She has been simply great, and we are deeply indebted to her. We also wish to thank our families for their support and understanding.

Adel S. Sedra
Kenneth C. Smith

MICROELECTRONIC CIRCUITS

PART I

DEVICES AND BASIC CIRCUITS

INTRODUCTION

Part I, *Devices and Basic Circuits,* includes the most fundamental and essential topics for the study of electronic circuits. At the same time, it constitutes a complete package for a first course on the subject.

Besides silicon diodes and transistors, the basic electronic devices, the op amp is studied in Part I. Although not an electronic device in the most fundamental sense, the op amp is commercially available as an integrated circuit (IC) package and has well-defined terminal characteristics. Thus, despite the fact that the op amp's internal circuit is complex, typically incorporating 20 or more transistors, its almost-ideal terminal behavior makes it possible to treat the op amp as a circuit element and to use it in the design of powerful circuits, as we do in Chapter 2, without any knowledge of its internal construction. We should mention, however, that the study of op amps can be delayed to a later point, and Chapter 2 can be skipped with no loss of continuity.

The most basic silicon device is the diode. In addition to learning about diodes and a sample of their applications, Chapter 3 also introduces the general topic of device modeling for the purpose of circuit analysis and design. Also, Section 3.7 provides a substantial introduction to the physical operation of semiconductor devices. This subject is then continued in Section 4.1 for the MOSFET and in Section 5.1 for the BJT. Taken together, these three sections provide a physical background sufficient for the study of electronic circuits at the level presented in this book.

The heart of this book, and of any electronics course, is the study of the two transistor types in use today: the MOS field-effect transistor (MOSFET) in Chapter 4 and the bipolar junction transistor (BJT) in Chapter 5. These two chapters have been written to be completely independent of one another and thus can be studied in either desired order. Furthermore, the two chapters have the same structure, making it easier and faster to study the second device, as well as to draw comparisons between the two device types.

Chapter 1 provides both an introduction to the study of electronics and a number of important concepts for the study of amplifiers (Sections 1.4–1.6) and of digital circuits (Section 1.7).

Each of the five chapters concludes with a section on the use of SPICE simulation in circuit analysis and design. Of particular importance here are the device models employed by SPICE. Finally, note that as in most of the chapters of this book, the *must-know* material is placed near the beginning of a chapter while the *good-to-know* topics are placed in the latter part of the chapter. Some of this latter material can therefore be skipped in a first course and covered at a later time, when needed.

Introduction to Electronics

INTRODUCTION

The subject of this book is modern electronics, a field that has come to be known as **microelectronics. Microelectronics** refers to the integrated-circuit (IC) technology that at the time of this writing is capable of producing circuits that contain millions of components in a small piece of silicon (known as a **silicon chip**) whose area is on the order of 100 mm^2. One such microelectronic circuit, for example, is a complete digital computer, which accordingly is known as a **microcomputer** or, more generally, a **microprocessor.**

In this book we shall study electronic devices that can be used singly (in the design of **discrete circuits**) or as components of an **integrated-circuit (IC)** chip. We shall study the design and analysis of interconnections of these devices, which form discrete and integrated circuits of varying complexity and perform a wide variety of functions. We shall also learn about available IC chips and their application in the design of electronic systems.

The purpose of this first chapter is to introduce some basic concepts and terminology. In particular, we shall learn about signals and about one of the most important signal-processing functions electronic circuits are designed to perform, namely, signal amplification. We shall then look at models for linear amplifiers. These models will be employed in subsequent chapters in the design and analysis of actual amplifier circuits.

Whereas the amplifier is the basic element of analog circuits, the logic inverter plays this role in digital circuits. We shall therefore take a preliminary look at the digital inverter, its circuit function, and important characteristics.

In addition to motivating the study of electronics, this chapter serves as a bridge between the study of linear circuits and that of the subject of this book: the design and analysis of electronic circuits.

1.1 SIGNALS

Signals contain information about a variety of things and activities in our physical world. Examples abound: Information about the weather is contained in signals that represent the air temperature, pressure, wind speed, etc. The voice of a radio announcer reading the news into a microphone provides an acoustic signal that contains information about world affairs. To monitor the status of a nuclear reactor, instruments are used to measure a multitude of relevant parameters, each instrument producing a signal.

To extract required information from a set of signals, the observer (be it a human or a machine) invariably needs to **process** the signals in some predetermined manner. This **signal processing** is usually most conveniently performed by electronic systems. For this to be possible, however, the signal must first be converted into an electric signal, that is, a voltage or a current. This process is accomplished by devices known as **transducers.** A variety of transducers exist, each suitable for one of the various forms of physical signals. For instance, the sound waves generated by a human can be converted into electric signals using a microphone, which is in effect a pressure transducer. It is not our purpose here to study transducers; rather, we shall assume that the signals of interest already exist in the electrical domain and represent them by one of the two equivalent forms shown in Fig. 1.1. In Fig. 1.1(a) the signal is represented by a voltage source $v_s(t)$ having a source resistance R_s. In the alternate representation of Fig. 1.1(b) the signal is represented by a current source $i_s(t)$ having a source resistance R_s. Although the two representations are equivalent, that in Fig. 1.1(a) (known as the Thévenin form) is preferred when R_s is low. The representation of Fig. 1.1(b) (known as the Norton form) is preferred when R_s is high. The reader will come to appreciate this point later in this chapter when we study the different types of amplifiers. For the time being, it is important to be familiar with Thévenin's and Norton's theorems (for a brief review, see Appendix D) and to note that for the two representations in Fig. 1.1 to be equivalent, their parameters are related by

$$v_s(t) = R_s i_s(t)$$

From the discussion above, it should be apparent that a signal is a time-varying quantity that can be represented by a graph such as that shown in Fig. 1.2. In fact, the information content of the signal is represented by the changes in its magnitude as time progresses; that is, the information is contained in the "wiggles" in the signal waveform. In general, such waveforms are difficult to characterize mathematically. In other words, it is not easy to describe succinctly an arbitrary-looking waveform such as that of Fig. 1.2. Of course, such a

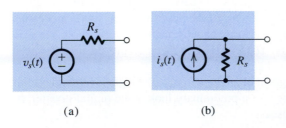

(a)   (b)

FIGURE 1.1 Two alternative representations of a signal source: **(a)** the Thévenin form, and **(b)** the Norton form.

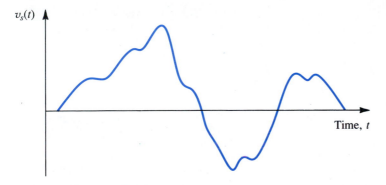

FIGURE 1.2   An arbitrary voltage signal $v_s(t)$.

description is of great importance for the purpose of designing appropriate signal-processing circuits that perform desired functions on the given signal.

EXERCISES

1.1   For the signal-source representations shown in Figs. 1.1(a) and 1.1(b), what are the open-circuit output voltages that would be observed? If, for each, the output terminals are short-circuited (i.e., wired together), what current would flow? For the representations to be equivalent, what must the relationship be between v_s, i_s, and R_s?

Ans. For (a), $v_{oc} = v_s(t)$; for (b), $v_{oc} = R_s i_s(t)$; for (a), $i_{sc} = v_s(t)/R_s$; for (b), $i_{sc} = i_s(t)$; for equivalency, $v_s(t) = R_s i_s(t)$

1.2   A signal source has an open-circuit voltage of 10 mV and a short-circuit current of 10 μA. What is the source resistance?

Ans. 1 kΩ

 ## 1.2   FREQUENCY SPECTRUM OF SIGNALS

An extremely useful characterization of a signal, and for that matter of any arbitrary function of time, is in terms of its **frequency spectrum.** Such a description of signals is obtained through the mathematical tools of **Fourier series** and **Fourier transform.**[1] We are not interested at this point in the details of these transformations; suffice it to say that they provide the means for representing a voltage signal $v_s(t)$ or a current signal $i_s(t)$ as the sum of sine-wave signals of different frequencies and amplitudes. This makes the sine wave a very important signal in the analysis, design, and testing of electronic circuits. Therefore, we shall briefly review the properties of the sinusoid.

Figure 1.3 shows a sine-wave voltage signal $v_a(t)$,

$$v_a(t) = V_a \sin \omega t \tag{1.1}$$

---

[1] The reader who has not yet studied these topics should not be alarmed. No detailed application of this material will be made until Chapter 6. Nevertheless, a general understanding of Section 1.2 should be very helpful when studying early parts of this book.

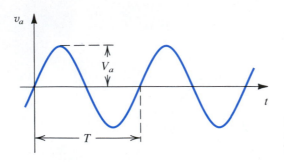

FIGURE 1.3 Sine-wave voltage signal of amplitude V_a and frequency $f = 1/T$ Hz. The angular frequency $\omega = 2\pi f$ rad/s.

where V_a denotes the peak value or amplitude in volts and ω denotes the angular frequency in radians per second; that is, $\omega = 2\pi f$ rad/s, where f is the frequency in hertz, $f = 1/T$ Hz, and T is the period in seconds.

The sine-wave signal is completely characterized by its peak value V_a, its frequency ω, and its phase with respect to an arbitrary reference time. In the case depicted in Fig. 1.3, the time origin has been chosen so that the phase angle is 0. It should be mentioned that it is common to express the amplitude of a sine-wave signal in terms of its root-mean-square (rms) value, which is equal to the peak value divided by $\sqrt{2}$. Thus the rms value of the sinusoid $v_a(t)$ of Fig. 1.3 is $V_a/\sqrt{2}$. For instance, when we speak of the wall power supply in our homes as being 120 V, we mean that it has a sine waveform of $120\sqrt{2}$ volts peak value.

Returning now to the representation of signals as the sum of sinusoids, we note that the Fourier series is utilized to accomplish this task for the special case when the signal is a periodic function of time. On the other hand, the Fourier transform is more general and can be used to obtain the frequency spectrum of a signal whose waveform is an arbitrary function of time.

The Fourier series allows us to express a given periodic function of time as the sum of an infinite number of sinusoids whose frequencies are harmonically related. For instance, the symmetrical square-wave signal in Fig. 1.4 can be expressed as

$$v(t) = \frac{4V}{\pi}(\sin \omega_0 t + \tfrac{1}{3} \sin 3\omega_0 t + \tfrac{1}{5} \sin 5\omega_0 t + \cdots) \qquad (1.2)$$

where V is the amplitude of the square wave and $\omega_0 = 2\pi/T$ (T is the period of the square wave) is called the **fundamental frequency.** Note that because the amplitudes of the harmonics progressively decrease, the infinite series can be truncated, with the truncated series providing an approximation to the square waveform.

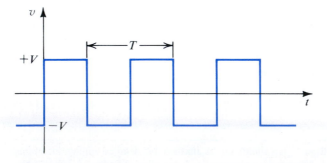

FIGURE 1.4 A symmetrical square-wave signal of amplitude V.

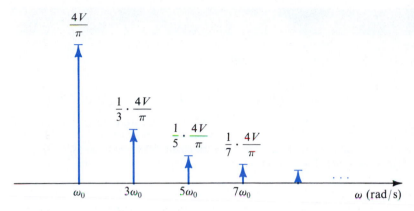

FIGURE 1.5 The frequency spectrum (also known as the line spectrum) of the periodic square wave of Fig. 1.4.

The sinusoidal components in the series of Eq. (1.2) constitute the frequency spectrum of the square-wave signal. Such a spectrum can be graphically represented as in Fig. 1.5, where the horizontal axis represents the angular frequency ω in radians per second.

The Fourier transform can be applied to a nonperiodic function of time, such as that depicted in Fig. 1.2, and provides its frequency spectrum as a continuous function of frequency, as indicated in Fig. 1.6. Unlike the case of periodic signals, where the spectrum consists of discrete frequencies (at ω_0 and its harmonics), the spectrum of a nonperiodic signal contains in general all possible frequencies. Nevertheless, the essential parts of the spectra of practical signals are usually confined to relatively short segments of the frequency (ω) axis—an observation that is very useful in the processing of such signals. For instance, the spectrum of audible sounds such as speech and music extends from about 20 Hz to about 20 kHz—a frequency range known as the **audio band.** Here we should note that although some musical tones have frequencies above 20 kHz, the human ear is incapable of hearing frequencies that are much above 20 kHz. As another example, analog video signals have their spectra in the range of 0 MHz to 4.5 MHz.

We conclude this section by noting that a signal can be represented either by the manner in which its waveform varies with time, as for the voltage signal $v_a(t)$ shown in Fig. 1.2, or in terms of its frequency spectrum, as in Fig. 1.6. The two alternative representations are known as the time-domain representation and the frequency-domain representation, respectively. The frequency-domain representation of $v_a(t)$ will be denoted by the symbol $V_a(\omega)$.

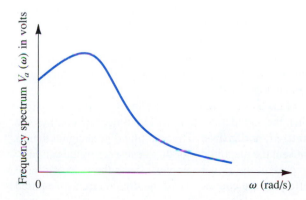

FIGURE 1.6 The frequency spectrum of an arbitrary waveform such as that in Fig. 1.2.

EXERCISES

1.3 Find the frequencies f and ω of a sine-wave signal with a period of 1 ms.

Ans. $f = 1000$ Hz; $\omega = 2\pi \times 10^3$ rad/s

1.4 What is the period T of sine waveforms characterized by frequencies of (a) $f = 60$ Hz? (b) $f = 10^{-3}$ Hz? (c) $f = 1$ MHz?

Ans. 16.7 ms; 1000 s; 1 μs

1.5 The UHF (Ultra High Frequency) television broadcast band begins with channel 14 and extends from 470 MHz to 806 MHz. If 6 MHz is allocated for each channel, how many channels can this band accommodate?

Ans. 56; channels 14 to 69

1.6 When the square-wave signal of Fig. 1.4, whose Fourier series is given in Eq. (1.2), is applied to a resistor, the total power dissipated may be calculated directly using the relationship $P = 1/T \int_0^T (v^2/R)\,dt$ or indirectly by summing the contribution of each of the harmonic components, that is, $P = P_1 + P_3 + P_5 + \cdots$, which may be found directly from rms values. Verify that the two approaches are equivalent. What fraction of the energy of a square wave is in its fundamental? In its first five harmonics? In its first seven? First nine? In what number of harmonics is 90% of the energy? (Note that in counting harmonics, the fundamental at ω_0 is the first, the one at $2\omega_0$ is the second, etc.)

Ans. 0.81; 0.93; 0.95; 0.96; 3

1.3 ANALOG AND DIGITAL SIGNALS

The voltage signal depicted in Fig. 1.2 is called an **analog signal.** The name derives from the fact that such a signal is analogous to the physical signal that it represents. The magnitude of an analog signal can take on any value; that is, the amplitude of an analog signal exhibits a continuous variation over its range of activity. The vast majority of signals in the world around us are analog. Electronic circuits that process such signals are known as **analog circuits.** A variety of analog circuits will be studied in this book.

An alternative form of signal representation is that of a sequence of numbers, each number representing the signal magnitude at an instant of time. The resulting signal is called a **digital signal.** To see how a signal can be represented in this form—that is, how signals can be converted from analog to digital form—consider Fig. 1.7(a). Here the curve represents a voltage signal, identical to that in Fig. 1.2. At equal intervals along the time axis we have marked the time instants t_0, t_1, t_2, and so on. At each of these time instants the magnitude of the signal is measured, a process known as **sampling.** Figure 1.7(b) shows a representation of the signal of Fig. 1.7(a) in terms of its samples. The signal of Fig. 1.7(b) is defined only at the sampling instants; it no longer is a continuous function of time, but rather, it is a **discrete-time signal.** However, since the magnitude of each sample can take any value in a continuous range, the signal in Fig. 1.7(b) is still an analog signal.

Now if we represent the magnitude of each of the signal samples in Fig. 1.7(b) by a number having a finite number of digits, then the signal amplitude will no longer be continuous; rather, it is said to be **quantized, discretized,** or **digitized.** The resulting digital signal then is simply a sequence of numbers that represent the magnitudes of the successive signal samples.

The choice of number system to represent the signal samples affects the type of digital signal produced and has a profound effect on the complexity of the digital circuits required

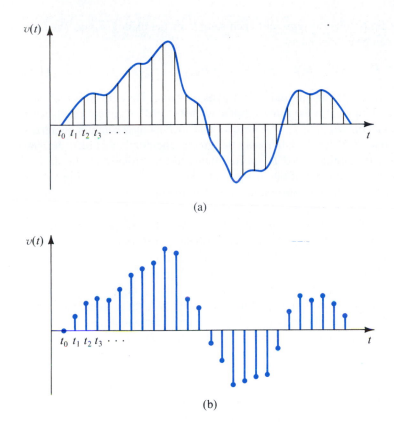

(a)

(b)

FIGURE 1.7 Sampling the continuous-time analog signal in **(a)** results in the discrete-time signal in **(b).**

to process the signals. It turns out that the **binary** number system results in the simplest possible digital signals and circuits. In a binary system, each digit in the number takes on one of only two possible values, denoted 0 and 1. Correspondingly, the digital signals in binary systems need have only two voltage levels, which can be labeled low and high. As an example, in some of the digital circuits studied in this book, the levels are 0 V and +5 V. Figure 1.8 shows the time variation of such a digital signal. Observe that the waveform is a pulse train with 0 V representing a 0 signal, or logic 0, and +5 V representing logic 1.

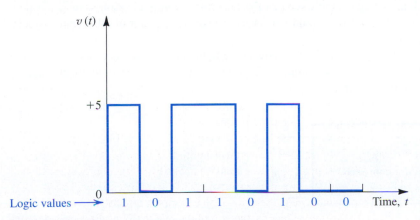

FIGURE 1.8 Variation of a particular binary digital signal with time.

If we use N binary dig*its* (bits) to represent each sample of the analog signal, then the digitized sample value can be expressed as

$$D = b_0 2^0 + b_1 2^1 + b_2 2^2 + \cdots + b_{N-1} 2^{N-1} \tag{1.3}$$

where b_0, b_1, ..., b_{N-1}, denote the N bits and have values of 0 or 1. Here bit b_0 is the **least significant bit** (LSB), and bit b_{N-1} is the **most significant bit** (MSB). Conventionally, this binary number is written as $b_{N-1} b_{N-2} \ldots b_0$. We observe that such a representation quantizes the analog sample into one of 2^N levels. Obviously the greater the number of bits (i.e., the larger the N), the closer the digital word D approximates the magnitude of the analog sample. That is, increasing the number of bits reduces the *quantization error* and increases the resolution of the analog-to-digital conversion. This improvement is, however, usually obtained at the expense of more complex and hence more costly circuit implementations. It is not our purpose here to delve into this topic any deeper; we merely want the reader to appreciate the nature of analog and digital signals. Nevertheless, it is an opportune time to introduce a very important circuit building block of modern electronic systems: the **analog-to-digital converter** (A/D or ADC) shown in block form in Fig. 1.9. The ADC accepts at its input the samples of an analog signal and provides for each input sample the corresponding N-bit digital representation (according to Eq. 1.3) at its N output terminals. Thus although the voltage at the input might be, say, 6.51 V, at each of the output terminals (say, at the ith terminal), the voltage will be either low (0 V) or high (5 V) if b_i is supposed to be 0 or 1, respectively. We shall study the ADC and its dual circuit the **digital-to-analog converter** (D/A or DAC) in Chapter 9.

Once the signal is in digital form, it can be processed using **digital circuits.** Of course digital circuits can deal also with signals that do not have an analog origin, such as the signals that represent the various instructions of a digital computer.

Since digital circuits deal exclusively with binary signals, their design is simpler than that of analog circuits. Furthermore, digital systems can be designed using a relatively few different kinds of digital circuit blocks. However, a large number (e.g., hundreds of thousands or even millions) of each of these blocks are usually needed. Thus the design of digital circuits poses its own set of challenges to the designer but provides reliable and economic implementations of a great variety of signal processing functions, some of which are not possible with analog circuits. At the present time, more and more of the signal processing functions are being performed digitally. Examples around us abound: from the digital watch and the calculator to digital audio systems and, more recently, digital television. Moreover, some longstanding analog systems such as the telephone communication system are now almost entirely digital. And we should not forget the most important of all digital systems, the digital computer.

The basic building blocks of digital systems are logic circuits and memory circuits. We shall study both in this book, beginning in Section 1.7 with the most fundamental digital circuit, the digital logic inverter.

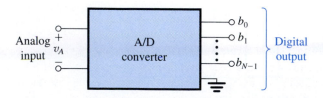

FIGURE 1.9 Block-diagram representation of the analog-to-digital converter (ADC).

One final remark: Although the digital processing of signals is at present all-pervasive, there remain many signal processing functions that are best performed by analog circuits. Indeed, many electronic systems include both analog and digital parts. It follows that a good electronics engineer must be proficient in the design of both analog and digital circuits, or **mixed-signal** or **mixed-mode** design as it is currently known. Such is the aim of this book.

EXERCISE

1.7 Consider a 4-bit digital word $D = b_3b_2b_1b_0$ (see Eq. 1.3) used to represent an analog signal v_A that varies between 0 V and +15 V.

(a) Give D corresponding to $v_A = 0$ V, 1 V, 2 V, and 15 V.

(b) What change in v_A causes a change from 0 to 1 in: (i) b_0, (ii) b_1, (iii) b_2, and (iv) b_3?

(c) If $v_A = 5.2$ V, what do you expect D to be? What is the resulting error in representation?

Ans. (a) 0000, 0001, 0010, 1111;   (b) +1 V, +2 V, +4 V, +8 V;   (c) 0101, −4%

 ## 1.4 AMPLIFIERS

In this section, we shall introduce a fundamental signal-processing function that is employed in some form in almost every electronic system, namely, signal amplification. We shall study the amplifier as a circuit building block, that is consider its external characteristics and leave the design of its internal circuit to later chapters.

1.4.1 Signal Amplification

From a conceptual point of view the simplest signal-processing task is that of **signal amplification.** The need for amplification arises because transducers provide signals that are said to be "weak," that is, in the microvolt (μV) or millivolt (mV) range and possessing little energy. Such signals are too small for reliable processing, and processing is much easier if the signal magnitude is made larger. The functional block that accomplishes this task is the **signal amplifier.**

It is appropriate at this point to discuss the need for **linearity** in amplifiers. When amplifying a signal, care must be exercised so that the information contained in the signal is not changed and no new information is introduced. Thus when feeding the signal shown in Fig. 1.2 to an amplifier, we want the output signal of the amplifier to be an exact replica of that at the input, except of course for having larger magnitude. In other words, the "wiggles" in the output waveform must be identical to those in the input waveform. Any change in waveform is considered to be **distortion** and is obviously undesirable.

An amplifier that preserves the details of the signal waveform is characterized by the relationship

$$v_o(t) = A v_i(t) \tag{1.4}$$

where v_i and v_o are the input and output signals, respectively, and A is a constant representing the magnitude of amplification, known as **amplifier gain.** Equation (1.4) is a linear relationship; hence the amplifier it describes is a **linear amplifier.** It should be easy to see that if the relationship between v_o and v_i contains higher powers of v_i, then the waveform of v_o will no longer be identical to that of v_i. The amplifier is then said to exhibit **nonlinear distortion.**

The amplifiers discussed so far are primarily intended to operate on very small input signals. Their purpose is to make the signal magnitude larger and therefore are thought of as **voltage amplifiers.** The **preamplifier** in the home stereo system is an example of a voltage amplifier. However, it usually does more than just amplify the signal; specifically, it per-forms some shaping of the frequency spectrum of the input signal. This topic, however, is beyond our need at this moment.

At this time we wish to mention another type of amplifier, namely, the **power amplifier.** Such an amplifier may provide only a modest amount of voltage gain but substantial current gain. Thus while absorbing little power from the input signal source to which it is con-nected, often a preamplifier, it delivers large amounts of power to its load. An example is found in the power amplifier of the home stereo system, whose purpose is to provide suffi-cient power to drive the loudspeaker, which is the amplifier load. Here we should note that the loudspeaker is the output transducer of the stereo system; it converts the electric output signal of the system into an acoustic signal. A further appreciation of the need for linearity can be acquired by reflecting on the power amplifier. A linear power amplifier causes both soft and loud music passages to be reproduced without distortion.

1.4.2 Amplifier Circuit Symbol

The signal amplifier is obviously a two-port network. Its function is conveniently repre-sented by the circuit symbol of Fig. 1.10(a). This symbol clearly distinguishes the input and output ports and indicates the direction of signal flow. Thus, in subsequent diagrams it will not be necessary to label the two ports "input" and "output." For generality we have shown the amplifier to have two input terminals that are distinct from the two output terminals. A more common situation is illustrated in Fig. 1.10(b), where a common terminal exists between the input and output ports of the amplifier. This common terminal is used as a ref-erence point and is called the **circuit ground.**

1.4.3 Voltage Gain

A linear amplifier accepts an input signal $v_I(t)$ and provides at the output, across a load resistance R_L (see Fig. 1.11(a)), an output signal $v_O(t)$ that is a magnified replica of $v_I(t)$. The **voltage gain** of the amplifier is defined by

$$\text{Voltage gain } (A_v) \equiv \frac{v_O}{v_I} \tag{1.5}$$

Fig. 1.11(b) shows the **transfer characteristic** of a linear amplifier. If we apply to the input of this amplifier a sinusoidal voltage of amplitude $\hat{V}$, we obtain at the output a sinusoid of amplitude $A_v\hat{V}$.

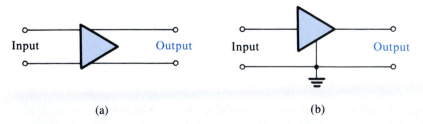

(a)	(b)

FIGURE 1.10 **(a)** Circuit symbol for amplifier. **(b)** An amplifier with a common terminal (ground) between the input and output ports.

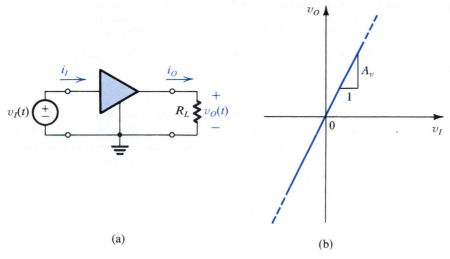

(a)

(b)

FIGURE 1.11 (a) A voltage amplifier fed with a signal $v_I(t)$ and connected to a load resistance R_L. (b) Transfer characteristic of a linear voltage amplifier with voltage gain A_v.

1.4.4 Power Gain and Current Gain

An amplifier increases the signal power, an important feature that distinguishes an amplifier from a transformer. In the case of a transformer, although the voltage delivered to the load could be greater than the voltage feeding the input side (the primary), the power delivered to the load (from the secondary side of the transformer) is less than or at most equal to the power supplied by the signal source. On the other hand, an amplifier provides the load with power greater than that obtained from the signal source. That is, amplifiers have power gain. The **power gain** of the amplifier in Fig. 1.11(a) is defined as

$$\text{Power gain } (A_p) \equiv \frac{\text{load power } (P_L)}{\text{input power } (P_I)} \tag{1.6}$$

$$= \frac{v_O i_O}{v_I i_I} \tag{1.7}$$

where i_O is the current that the amplifier delivers to the load (R_L), $i_O = v_O/R_L$, and i_I is the current the amplifier draws from the signal source. The **current gain** of the amplifier is defined as

$$\text{Current gain } (A_i) \equiv \frac{i_O}{i_I} \tag{1.8}$$

From Eqs. (1.5) to (1.8) we note that

$$A_p = A_v A_i \tag{1.9}$$

1.4.5 Expressing Gain in Decibels

The amplifier gains defined above are ratios of similarly dimensioned quantities. Thus they will be expressed either as dimensionless numbers or, for emphasis, as V/V for the voltage gain, A/A for the current gain, and W/W for the power gain. Alternatively, for a number of reasons, some of them historic, electronics engineers express amplifier gain with a logarithmic measure. Specifically the voltage gain A_v can be expressed as

$$\text{Voltage gain in decibels} = 20 \log|A_v| \qquad \text{dB}$$

and the current gain A_i can be expressed as

$$\text{Current gain in decibels} = 20 \log|A_i| \quad \text{dB}$$

Since power is related to voltage (or current) squared, the power gain A_p can be expressed in decibels as

$$\text{Power gain in decibels} = 10 \log A_p \quad \text{dB}$$

The absolute values of the voltage and current gains are used because in some cases A_v or A_i may be negative numbers. A negative gain A_v simply means that there is a 180° phase difference between input and output signals; it does not imply that the amplifier is **attenuating** the signal. On the other hand, an amplifier whose voltage gain is, say, −20 dB is in fact attenuating the input signal by a factor of 10 (i.e., $A_v = 0.1$ V/V).

1.4.6 The Amplifier Power Supplies

Since the power delivered to the load is greater than the power drawn from the signal source, the question arises as to the source of this additional power. The answer is found by observing that amplifiers need dc power supplies for their operation. These dc sources supply the extra power delivered to the load as well as any power that might be dissipated in the internal circuit of the amplifier (such power is converted to heat). In Fig. 1.11(a) we have not explicitly shown these dc sources.

Figure 1.12(a) shows an amplifier that requires two dc sources: one positive of value V_1 and one negative of value V_2. The amplifier has two terminals, labeled V^+ and V^-, for connection to the dc supplies. For the amplifier to operate, the terminal labeled V^+ has to be connected to the positive side of a dc source whose voltage is V_1 and whose negative side is connected to the circuit ground. Also, the terminal labeled V^- has to be connected to the negative side of a dc source whose voltage is V_2 and whose positive side is connected to the circuit ground. Now, if the current drawn from the positive supply is denoted I_1 and that from the negative supply is I_2 (see Fig. 1.12(a)), then the dc power delivered to the amplifier is

$$P_{\text{dc}} = V_1 I_1 + V_2 I_2$$

If the power dissipated in the amplifier circuit is denoted $P_{\text{dissipated}}$, the power-balance equation for the amplifier can be written as

$$P_{\text{dc}} + P_I = P_L + P_{\text{dissipated}}$$

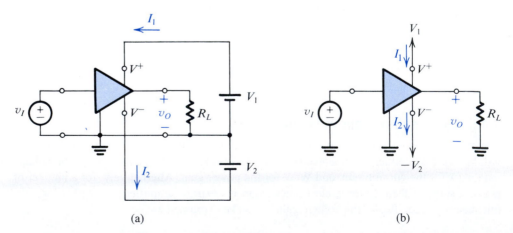

(a) (b)

FIGURE 1.12 An amplifier that requires two dc supplies (shown as batteries) for operation.

where P_I is the power drawn from the signal source and P_L is the power delivered to the load. Since the power drawn from the signal source is usually small, the amplifier **efficiency** is defined as

$$\eta \equiv \frac{P_L}{P_{dc}} \times 100 \tag{1.10}$$

The power efficiency is an important performance parameter for amplifiers that handle large amounts of power. Such amplifiers, called power amplifiers, are used, for example, as output amplifiers of stereo systems.

In order to simplify circuit diagrams, we shall adopt the convention illustrated in Fig. 1.12(b). Here the V^+ terminal is shown connected to an arrowhead pointing upward and the V^- terminal to an arrowhead pointing downward. The corresponding voltage is indicated next to each arrowhead. Note that in many cases we will not explicitly show the connections of the amplifier to the dc power sources. Finally, we note that some amplifiers require only one power supply.

EXAMPLE 1.1

Consider an amplifier operating from ±10-V power supplies. It is fed with a sinusoidal voltage having 1 V peak and delivers a sinusoidal voltage output of 9 V peak to a 1-kΩ load. The amplifier draws a current of 9.5 mA from each of its two power supplies. The input current of the amplifier is found to be sinusoidal with 0.1 mA peak. Find the voltage gain, the current gain, the power gain, the power drawn from the dc supplies, the power dissipated in the amplifier, and the amplifier efficiency.

Solution

$$A_v = \frac{9}{1} = 9 \text{ V/V}$$

or

$$A_v = 20 \log 9 \simeq 19.1 \text{ dB}$$

$$\hat{I}_o = \frac{9 \text{ V}}{1 \text{ k}\Omega} = 9 \text{ mA}$$

$$A_i = \frac{\hat{I}_o}{\hat{I}_i} = \frac{9}{0.1} = 90 \text{ A/A}$$

or

$$A_i = 20 \log 90 = 39.1 \text{ dB}$$

$$P_L = V_{o_{rms}} I_{o_{rms}} = \frac{9}{\sqrt{2}} \frac{9}{\sqrt{2}} = 40.5 \text{ mW}$$

$$P_I = V_{i_{rms}} I_{i_{rms}} = \frac{1}{\sqrt{2}} \frac{0.1}{\sqrt{2}} = 0.05 \text{ mW}$$

$$A_p = \frac{P_L}{P_I} = \frac{40.5}{0.05} = 810 \text{ W/W}$$

or

$$A_p = 10 \log 810 = 29.1 \text{ dB}$$

$$P_{\text{dc}} = 10 \times 9.5 + 10 \times 9.5 = 190 \text{ mW}$$

$$P_{\text{dissipated}} = P_{\text{dc}} + P_I - P_L$$

$$= 190 + 0.05 - 40.5 = 149.6 \text{ mW}$$

$$\eta = \frac{P_L}{P_{\text{dc}}} \times 100 = 21.3\%$$

From the above example we observe that the amplifier converts some of the dc power it draws from the power supplies to signal power that it delivers to the load.

1.4.7 Amplifier Saturation

Practically speaking, the amplifier transfer characteristic remains linear over only a limited range of input and output voltages. For an amplifier operated from two power supplies the output voltage cannot exceed a specified positive limit and cannot decrease below a specified negative limit. The resulting transfer characteristic is shown in Fig. 1.13, with the positive and

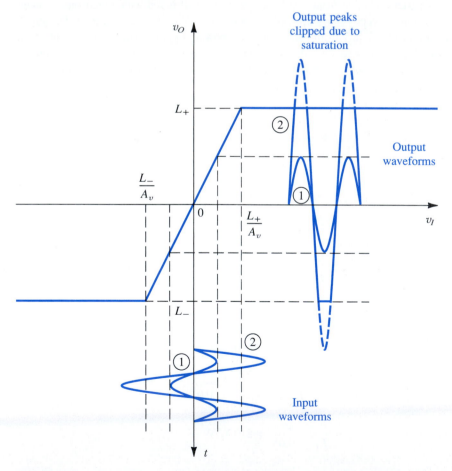

FIGURE 1.13 An amplifier transfer characteristic that is linear except for output saturation.

negative saturation levels denoted L_+ and L_-, respectively. Each of the two saturation levels is usually within a volt or so of the voltage of the corresponding power supply.

Obviously, in order to avoid distorting the output signal waveform, the input signal swing must be kept within the linear range of operation,

$$\frac{L_-}{A_v} \leq v_I \leq \frac{L_+}{A_v}$$

Figure 1.13 shows two input waveforms and the corresponding output waveforms. We note that the peaks of the larger waveform have been clipped off because of amplifier saturation.

1.4.8 Nonlinear Transfer Characteristics and Biasing

Except for the output saturation effect discussed above, the amplifier transfer characteristics have been assumed to be perfectly linear. In practical amplifiers the transfer characteristic may exhibit nonlinearities of various magnitudes, depending on how elaborate the amplifier circuit is and on how much effort has been expended in the design to ensure linear operation. Consider as an example the transfer characteristic depicted in Fig. 1.14. Such a characteristic is typical of simple amplifiers that are operated from a single (positive) power supply. The transfer characteristic is obviously nonlinear and, because of the single-supply operation, is not centered around the origin. Fortunately, a simple technique exists for obtaining linear amplification from an amplifier with such a nonlinear transfer characteristic.

The technique consists of first **biasing** the circuit to operate at a point near the middle of the transfer characteristic. This is achieved by applying a dc voltage V_I, as indicated in Fig. 1.14, where the operating point is labeled Q and the corresponding dc voltage at the output is V_O. The point Q is known as the **quiescent point,** the **dc bias point,** or simply the **operating point.** The time-varying signal to be amplified, $v_i(t)$, is then superimposed on the dc bias voltage V_I as indicated in Fig. 1.14. Now, as the total instantaneous input $v_I(t)$,

$$v_I(t) = V_I + v_i(t)$$

varies around V_I, the instantaneous operating point moves up and down the transfer curve around the dc operating point Q. In this way, one can determine the waveform of the total instantaneous output voltage $v_O(t)$. It can be seen that by keeping the amplitude of $v_i(t)$ sufficiently small, the instantaneous operating point can be confined to an almost linear segment of the transfer curve centered about Q. This in turn results in the time-varying portion of the output being proportional to $v_i(t)$; that is,

$$v_O(t) = V_O + v_o(t)$$

with

$$v_o(t) = A_v v_i(t)$$

where A_v is the slope of the almost linear segment of the transfer curve; that is,

$$A_v = \left.\frac{dv_O}{dv_I}\right|_{\text{at } Q}$$

In this manner, linear amplification is achieved. Of course, there is a limitation: The input signal must be kept sufficiently small. Increasing the amplitude of the input signal can cause

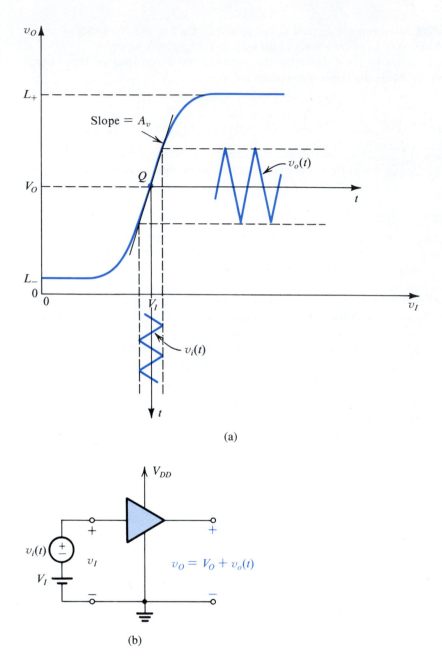

FIGURE 1.14 **(a)** An amplifier transfer characteristic that shows considerable nonlinearity. **(b)** To obtain linear operation the amplifier is biased as shown, and the signal amplitude is kept small. Observe that this amplifier is operated from a single power supply, V_{DD}.

the operation to be no longer restricted to an almost linear segment of the transfer curve. This in turn results in a distorted output signal waveform. Such nonlinear distortion is undesirable: The output signal contains additional spurious information that is not part of the input. We shall use this biasing technique and the associated **small-signal approximation** frequently in the design of transistor amplifiers.

EXAMPLE 1.2

A transistor amplifier has the transfer characteristic

$$v_O = 10 - 10^{-11} e^{40v_I} \tag{1.11}$$

which applies for $v_I \geq 0$ V and $v_O \geq 0.3$ V. Find the limits L_- and L_+ and the corresponding values of v_I. Also, find the value of the dc bias voltage V_I that results in $V_O = 5$ V and the voltage gain at the corresponding operating point.

Solution

The limit L_- is obviously 0.3 V. The corresponding value of v_I is obtained by substituting $v_O = 0.3$ V in Eq. (1.11); that is,

$$v_I = 0.690 \text{ V}$$

The limit L_+ is determined by $v_I = 0$ and is thus given by

$$L_+ = 10 - 10^{-11} \simeq 10 \text{ V}$$

To bias the device so that $V_O = 5$ V we require a dc input V_I whose value is obtained by substituting $v_O = 5$ V in Eq. (1.11) to find:

$$V_I = 0.673 \text{ V}$$

The gain at the operating point is obtained by evaluating the derivative dv_O/dv_I at $v_I = 0.673$ V. The result is

$$A_v = -200 \text{ V/V}$$

which indicates that this amplifier in an inverting one; that is, the output is 180° out of phase with the input. A sketch of the amplifier transfer characteristic (not to scale) is shown in Fig. 1.15, from which we observe the inverting nature of the amplifier.

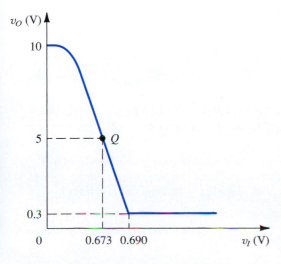

FIGURE 1.15 A sketch of the transfer characteristic of the amplifier of Example 1.2. Note that this amplifier is inverting (i.e., with a gain that is negative).

Once an amplifier is properly biased and the input signal is kept sufficiently small, the operation is assumed to be linear. We can then employ the techniques of linear circuit analysis to analyze the signal operation of the amplifier circuit. This is the topic of Sections 1.5 and 1.6.

1.4.9 Symbol Convention

At this point, we draw the reader's attention to the terminology used above and which we shall employ throughout the book. Total instantaneous quantities are denoted by a lowercase symbol with an uppercase subscript, for example, $i_A(t)$, $v_C(t)$. Direct-current (dc) quantities will be denoted by an uppercase symbol with an uppercase subscript, for example, I_A, V_C. Power-supply (dc) voltages are denoted by an uppercase V with a double-letter uppercase subscript, for example, V_{DD}. A similar notation is used for the dc current drawn from the power supply, for example, I_{DD}. Finally, incremental signal quantities will be denoted by a lowercase symbol with a lowercase subscript, for example, $i_a(t)$, $v_c(t)$. If the signal is a sine wave, then its amplitude is denoted by an uppercase letter with a lowercase subscript, for example, I_a, V_c. This notation is illustrated in Fig. 1.16.

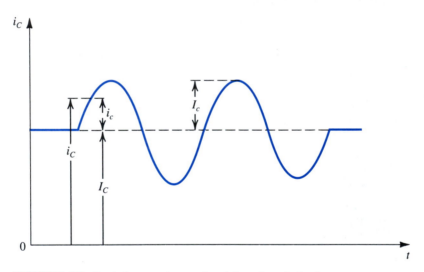

FIGURE 1.16 Symbol convention employed throughout the book.

EXERCISES

1.8 An amplifier has a voltage gain of 100 V/V and a current gain of 1000 A/A. Express the voltage and current gains in decibels and find the power gain.

Ans. 40 dB; 60 dB; 50 dB

1.9 An amplifier operating from a single 15-V supply provides a 12-V peak-to-peak sine-wave signal to a 1-kΩ load and draws negligible input current from the signal source. The dc current drawn from the 15-V supply is 8 mA. What is the power dissipated in the amplifier, and what is the amplifier efficiency?

Ans. 102 mW; 15%

1.10 The objective of this exercise is to investigate the limitation of the small-signal approximation. Consider the amplifier of Example 1.2 with a positive input signal of 1 mV superimposed on the dc bias voltage V_I. Find the corresponding signal at the output for two situations: (a) Assume the amplifier is linear around the operating point; that is, use the value of gain evaluated in Example 1.2. (b) Use the transfer characteristic of the amplifier. Repeat for input signals of 5 mV and 10 mV.

Ans. -0.2 V, -0.204 V; -1 V, -1.107 V; -2 V, -2.459 V

1.5 CIRCUIT MODELS FOR AMPLIFIERS

A good part of this book is concerned with the design of amplifier circuits using transistors of various types. Such circuits will vary in complexity from those using a single transistor to those with 20 or more devices. In order to be able to apply the resulting amplifier circuit as a building block in a system, one must be able to characterize, or **model,** its terminal behavior. In this section, we study simple but effective amplifier models. These models apply irrespective of the complexity of the internal circuit of the amplifier. The values of the model parameters can be found either by analyzing the amplifier circuit or by performing measurements at the amplifier terminals.

1.5.1 Voltage Amplifiers

Figure 1.17(a) shows a circuit model for the voltage amplifier. The model consists of a voltage-controlled voltage source having a gain factor A_{vo}, an input resistance R_i that accounts for the fact that the amplifier draws an input current from the signal source, and an output resistance R_o that accounts for the change in output voltage as the amplifier is called upon to

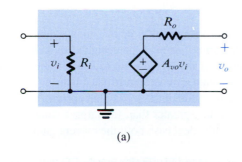

(a)

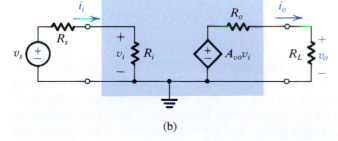

(b)

FIGURE 1.17 (a) Circuit model for the voltage amplifier. (b) The voltage amplifier with input signal source and load.

supply output current to a load. To be specific, we show in Fig. 1.17(b) the amplifier model fed with a signal voltage source v_s having a resistance R_s and connected at the output to a load resistance R_L. The nonzero output resistance R_o causes only a fraction of $A_{vo}v_i$ to appear across the output. Using the voltage-divider rule we obtain

$$v_o = A_{vo}v_i \frac{R_L}{R_L + R_o}$$

Thus the voltage gain is given by

$$A_v \equiv \frac{v_o}{v_i} = A_{vo} \frac{R_L}{R_L + R_o} \tag{1.12}$$

It follows that in order not to lose gain in coupling the amplifier output to a load, the output resistance R_o should be much smaller than the load resistance R_L. In other words, for a given R_L one must design the amplifier so that its R_o is much smaller than R_L. Furthermore, there are applications in which R_L is known to vary over a certain range. In order to keep the output voltage v_o as constant as possible, the amplifier is designed with R_o much smaller than the lowest value of R_L. An ideal voltage amplifier is one with $R_o = 0$. Equation (1.12) indicates also that for $R_L = \infty$, $A_v = A_{vo}$. Thus A_{vo} is the voltage gain of the unloaded amplifier, or the **open-circuit voltage gain.** It should also be clear that in specifying the voltage gain of an amplifier, one must also specify the value of load resistance at which this gain is measured or calculated. If a load resistance is not specified, it is normally assumed that the given voltage gain is the open-circuit gain A_{vo}.

The finite input resistance R_i introduces another voltage-divider action at the input, with the result that only a fraction of the source signal v_s actually reaches the input terminals of the amplifier; that is,

$$v_i = v_s \frac{R_i}{R_i + R_s} \tag{1.13}$$

It follows that in order not to lose a significant portion of the input signal in coupling the signal source to the amplifier input, the amplifier must be designed to have an input resistance R_i much greater than the resistance of the signal source, $R_i \gg R_s$. Furthermore, there are applications in which the source resistance is known to vary over a certain range. To minimize the effect of this variation on the value of the signal that appears at the input of the amplifier, the design ensures that R_i is much greater than the largest value of R_s. An ideal voltage amplifier is one with $R_i = \infty$. In this ideal case both the current gain and power gain become infinite.

The overall voltage gain (v_o/v_s) can be found by combining Eqs. (1.12) and (1.13),

$$\frac{v_o}{v_s} = A_{vo} \frac{R_i}{R_i + R_s} \frac{R_L}{R_L + R_o}$$

There are situations in which one is interested not in voltage gain but only in a significant power gain. For instance, the source signal can have a respectable voltage but a source resistance which is much greater than the load resistance. Connecting the source directly to the load would result in significant signal attenuation. In such a case, one requires an amplifier with a high input resistance (much greater than the source resistance) and a low output resistance (much smaller than the load resistance) but with a modest voltage gain (or even unity gain). Such an amplifier is referred to as a **buffer amplifier.** We shall encounter buffer amplifiers often throughout this book.

EXERCISES

1.11 A transducer characterized by a voltage of 1 V rms and a resistance of 1 MΩ is available to drive a 10-Ω load. If connected directly, what voltage and power levels result at the load? If a unity-gain (i.e., $A_{vo} = 1$) buffer amplifier with 1-MΩ input resistance and 10-Ω output resistance is interposed between source and load, what do the output voltage and power levels become? For the new arrangement find the voltage gain from source to load, and the power gain (both expressed in decibels).

Ans. 10 μV rms; 10^{-11} W; 0.25 V; 6.25 mW; −12 dB; 44 dB

1.12 The output voltage of a voltage amplifier has been found to decrease by 20% when a load resistance of 1 kΩ is connected. What is the value of the amplifier output resistance?

Ans. 250 Ω

1.13 An amplifier with a voltage gain of +40 dB, an input resistance of 10 kΩ, and an output resistance of 1 kΩ is used to drive a 1-kΩ load. What is the value of A_{vo}? Find the value of power gain in dB.

Ans. 100 V/V; 44 dB

1.5.2 Cascaded Amplifiers

To meet given amplifier specifications the need often arises to design the amplifier as a cascade of two or more stages. The stages are usually not identical; rather, each is designed to serve a specific purpose. For instance, the first stage is usually required to have a large input resistance, and the final stage in the cascade is usually designed to have a low output resistance. To illustrate the analysis and design of cascaded amplifiers, we consider a practical example.

EXAMPLE 1.3

Figure 1.18 depicts an amplifier composed of a cascade of three stages. The amplifier is fed by a signal source with a source resistance of 100 kΩ and delivers its output into a load resistance of 100 Ω. The first stage has a relatively high input resistance and a modest gain factor of 10. The second stage has a higher gain factor but lower input resistance. Finally, the last, or output, stage has unity gain but a low output resistance. We wish to evaluate the overall voltage gain, that is, v_L/v_s, the current gain, and the power gain.

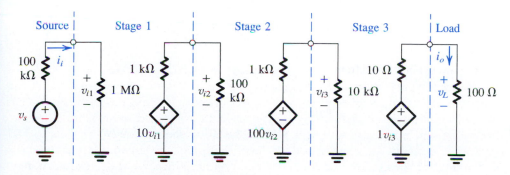

FIGURE 1.18   Three-stage amplifier for Example 1.3.

Solution

The fraction of source signal appearing at the input terminals of the amplifier is obtained using the voltage-divider rule at the input, as follows:

$$\frac{v_{i1}}{v_s} = \frac{1 \text{ M}\Omega}{1 \text{ M}\Omega + 100 \text{ k}\Omega} = 0.909 \text{ V/V}$$

The voltage gain of the first stage is obtained by considering the input resistance of the second stage to be the load of the first stage; that is,

$$A_{v1} \equiv \frac{v_{i2}}{v_{i1}} = 10\frac{100 \text{ k}\Omega}{100 \text{ k}\Omega + 1 \text{ k}\Omega} = 9.9 \text{ V/V}$$

Similarly, the voltage gain of the second stage is obtained by considering the input resistance of the third stage to be the load of the second stage,

$$A_{v2} \equiv \frac{v_{i3}}{v_{i2}} = 100\frac{10 \text{ k}\Omega}{10 \text{ k}\Omega + 1 \text{ k}\Omega} = 90.9 \text{ V/V}$$

Finally, the voltage gain of the output stage is as follows:

$$A_{v3} \equiv \frac{v_L}{v_{i3}} = 1\frac{100 \text{ }\Omega}{100 \text{ }\Omega + 10 \text{ }\Omega} = 0.909 \text{ V/V}$$

The total gain of the three stages in cascade can be now found from

$$A_v \equiv \frac{v_L}{v_{i1}} = A_{v1}A_{v2}A_{v3} = 818 \text{ V/V}$$

or 58.3 dB.

To find the voltage gain from source to load, we multiply A_v by the factor representing the loss of gain at the input; that is,

$$\frac{v_L}{v_s} = \frac{v_L}{v_{i1}}\frac{v_{i1}}{v_s} = A_v\frac{v_{i1}}{v_s}$$

$$= 818 \times 0.909 = 743.6 \text{ V/V}$$

or 57.4 dB.

The current gain is found as follows:

$$A_i \equiv \frac{i_o}{i_i} = \frac{v_L/100 \text{ }\Omega}{v_{i1}/1 \text{ M}\Omega}$$

$$= 10^4 \times A_v = 8.18 \times 10^6 \text{ A/A}$$

or 138.3 dB.

The power gain is found from

$$A_p \equiv \frac{P_L}{P_I} = \frac{v_L i_o}{v_{i1} i_i}$$

$$= A_v A_i = 818 \times 8.18 \times 10^6 = 66.9 \times 10^8 \text{ W/W}$$

or 98.3 dB. Note that

$$A_p(\text{dB}) = \tfrac{1}{2}[A_v(\text{dB}) + A_i(\text{dB})]$$

A few comments on the cascade amplifier in the above example are in order. To avoid losing signal strength at the amplifier input where the signal is usually very small, the first stage is designed to have a relatively large input resistance (1 MΩ), which is much larger than the source resistance. The trade-off appears to be a moderate voltage gain (10 V/V). The second stage does not need to have such a high input resistance; rather, here we need to realize the bulk of the required voltage gain. The third and final, or output, stage is not asked to provide any voltage gain; rather, it functions as a buffer amplifier, providing a relatively large input resistance and a low output resistance, much lower than R_L. It is this stage that enables connecting the amplifier to the 10-Ω load. These points can be made more concrete by solving the following exercises.

EXERCISES

1.14 What would the overall voltage gain of the cascade amplifier in Example 1.3 be without stage 3?

Ans. 81.8 V/V

1.15 For the cascade amplifier of Example 1.3, let v_s be 1 mV. Find v_{i1}, v_{i2}, v_{i3}, and v_L.

Ans. 0.91 mV; 9 mV; 818 mV; 744 mV

1.16 (a) Model the three-stage amplifier of Example 1.3 (without the source and load) using the voltage amplifier model. What are the values of R_i, A_{vo}, and R_o?

(b) If R_L varies in the range 10 Ω to 1000 Ω, find the corresponding range of the overall voltage gain, v_o/v_s.

Ans. 1 MΩ, 900 V/V, 10 Ω; 409 V/V to 810 V/V

1.5.3 Other Amplifier Types

In the design of an electronic system, the signal of interest—whether at the system input, at an intermediate stage, or at the output—can be either a voltage or a current. For instance, some transducers have very high output resistances and can be more appropriately modeled as current sources. Similarly, there are applications in which the output current rather than the voltage is of interest. Thus, although it is the most popular, the voltage amplifier considered above is just one of four possible amplifier types. The other three are the current amplifier, the transconductance amplifier, and the transresistance amplifier. Table 1.1 shows the four amplifier types, their circuit models, the definition of their gain parameters, and the ideal values of their input and output resistances.

1.5.4 Relationships Between the Four Amplifier Models

Although for a given amplifier a particular one of the four models in Table 1.1 is most preferable, *any of the four can be used to model the amplifier.* In fact, simple relationships can be derived to relate the parameters of the various models. For instance, the open-circuit voltage gain A_{vo} can be related to the short-circuit current gain A_{is} as follows: The open-circuit output voltage given by the voltage amplifier model of Table 1.1 is $A_{vo}v_i$. The current amplifier model in the same table gives an open-circuit output voltage of $A_{is}i_iR_o$. Equating these two values and noting that $i_i = v_i/R_i$ gives

$$A_{vo} = A_{is}\left(\frac{R_o}{R_i}\right) \tag{1.14}$$

TABLE 1.1 The Four Amplifier Types

Type	Circuit Model	Gain Parameter	Ideal Characteristics
Voltage Amplifier	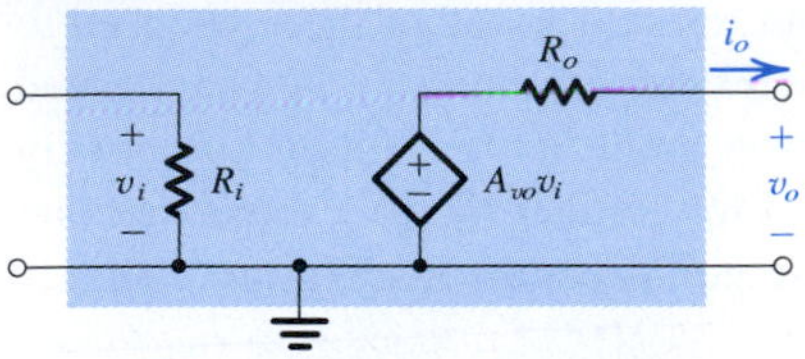	Open-Circuit Voltage Gain $$A_{vo} \equiv \left. \frac{v_o}{v_i} \right|_{i_o=0} \quad (V/V)$$	$R_i = \infty$ $R_o = 0$
Current Amplifier	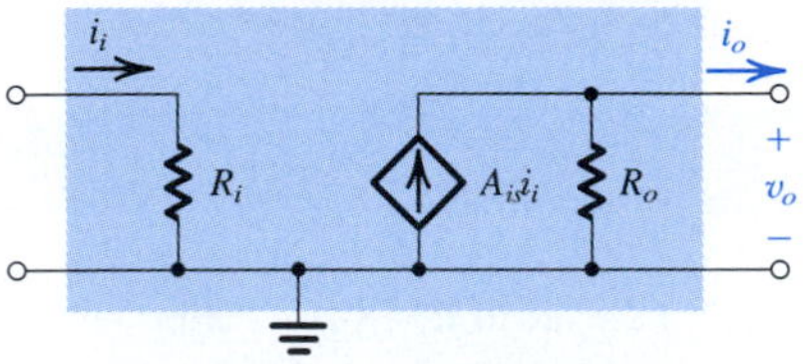	Short-Circuit Current Gain $$A_{is} \equiv \left. \frac{i_o}{i_i} \right|_{v_o=0} \quad (A/A)$$	$R_i = 0$ $R_o = \infty$
Transconductance Amplifier	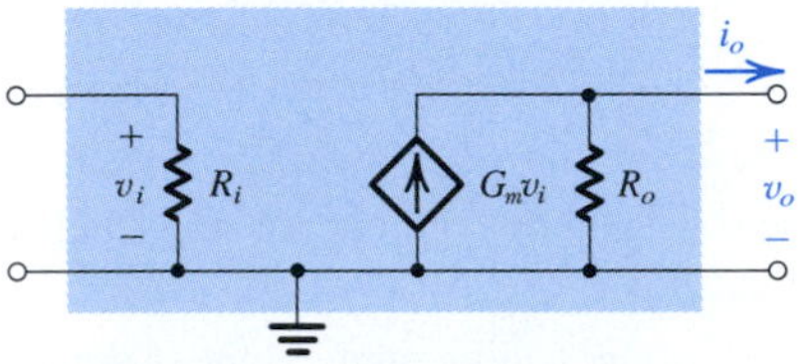	Short-Circuit Transconductance $$G_m \equiv \left. \frac{i_o}{v_i} \right|_{v_o=0} \quad (A/V)$$	$R_i = \infty$ $R_o = \infty$
Transresistance Amplifier	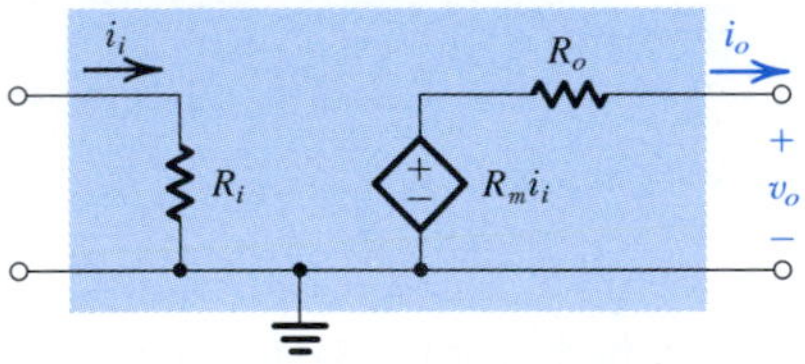	Open-Circuit Transresistance $$R_m \equiv \left. \frac{v_o}{i_i} \right|_{i_o=0} \quad (V/A)$$	$R_i = 0$ $R_o = 0$

Similarly, we can show that

$$A_{vo} = G_m R_o \tag{1.15}$$

and

$$A_{vo} = \frac{R_m}{R_i} \tag{1.16}$$

The expressions in Eqs. (1.14) to (1.16) can be used to relate any two of the gain parameters A_{vo}, A_{is}, G_m, and R_m.

From the amplifier circuit models given in Table 1.1, we observe that the input resistance R_i of the amplifier can be determined by applying an input voltage v_i and measuring (or calculating) the input current i_i; that is, $R_i = v_i/i_i$. The output resistance is found as the ratio of the open-circuit output voltage to the short-circuit output current. Alternatively, the output resistance can be found by eliminating the input signal source (then i_i and v_i will both be zero) and applying a voltage signal v_x to the output of the amplifier. If we denote the current drawn from v_x *into* the output terminals as i_x (note that i_x is opposite in direction to i_o), then $R_o = v_x/i_x$. Although these techniques are conceptually correct, in actual practice more refined methods are employed in measuring R_i and R_o.

The amplifier models considered above are **unilateral;** that is, signal flow is unidirectional, from input to output. Most real amplifiers show some reverse transmission, which is usually undesirable but must nonetheless be modeled. We shall not pursue this point further at this time except to mention that more complete models for linear two-port networks are given in Appendix B. Also, in Chapters 4 and 5, we will augment the models of Table 1.1 to take into account the nonunilateral nature of some transistor amplifiers.

EXAMPLE 1.4

The **bipolar junction transistor (BJT),** which will be studied in Chapter 5, is a three-terminal device that when dc biased and operated with small signals can be modeled by the linear circuit shown in Fig. 1.19(a). The three terminals are the **base (B)**, the **emitter (E)**, and the **collector (C).** The heart of the model is a transconductance amplifier represented by an input resistance between B and E (denoted r_π), a short-circuit transconductance g_m, and an output resistance r_o.

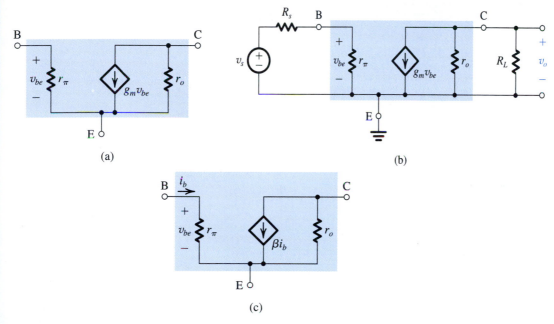

(a)

(b)

(c)

FIGURE 1.19 **(a)** Small-signal circuit model for a bipolar junction transistor (BJT). **(b)** The BJT connected as an amplifier with the emitter as a common terminal between input and output (called a common-emitter amplifier). **(c)** An alternative small-signal circuit model for the BJT.

(a) With the emitter used as a common terminal between input and output, Fig. 1.19(b) shows a transistor amplifier known as a **common-emitter** or **grounded-emitter** circuit. Derive an expression for the voltage gain v_o/v_s, and evaluate its magnitude for the case $R_s = 5$ kΩ, $r_\pi = 2.5$ kΩ, $g_m = 40$ mA/V, $r_o = 100$ kΩ, and $R_L = 5$ kΩ. What would the gain value be if the effect of r_o were neglected?

(b) An alternative model for the transistor in which a current amplifier rather than a transconductance amplifier is utilized is shown in Fig. 1.19(c). What must the short-circuit current-gain β be? Give both an expression and a value.

Solution

(a) Using the voltage-divider rule, we determine the fraction of input signal that appears at the amplifier input as

$$v_{be} = v_s \frac{r_\pi}{r_\pi + R_s} \tag{1.17}$$

Next we determine the output voltage v_o by multiplying the current $(g_m v_{be})$ by the resistance $(R_L \parallel r_o)$,

$$v_o = -g_m v_{be}(R_L \parallel r_o) \tag{1.18}$$

Substituting for v_{be} from Eq. (1.17) yields the voltage-gain expression

$$\frac{v_o}{v_s} = -\frac{r_\pi}{r_\pi + R_s} g_m(R_L \parallel r_o) \tag{1.19}$$

Observe that the gain is negative, indicating that this amplifier is inverting. For the given component values,

$$\frac{v_o}{v_s} = -\frac{2.5}{2.5 + 5} \times 40 \times (5 \parallel 100)$$

$$= -63.5 \text{ V/V}$$

Neglecting the effect of r_o, we obtain

$$\frac{v_o}{v_s} \simeq -\frac{2.5}{2.5 + 5} \times 40 \times 5$$

$$= -66.7 \text{ V/V}$$

which is quite close to the value obtained including r_o. This is not surprising since $r_o \gg R_L$. (b) For the model in Fig. 1.19(c) to be equivalent to that in Fig. 1.19(a),

$$\beta i_b = g_m v_{be}$$

But $i_b = v_{be}/r_\pi$; thus,

$$\beta = g_m r_\pi$$

For the values given,

$$\beta = 40 \text{ mA/V} \times 2.5 \text{ k}\Omega$$

$$= 100 \text{ A/A}$$

EXERCISES

1.17 Consider a current amplifier having the model shown in the second row of Table 1.1. Let the amplifier be fed with a signal current-source i_s having a resistance R_s, and let the output be connected to a load resistance R_L. Show that the overall current gain is given by

$$\frac{i_o}{i_s} = A_{is}\frac{R_s}{R_s + R_i}\frac{R_o}{R_o + R_L}$$

1.18 Consider the transconductance amplifier whose model is shown in the third row of Table 1.1. Let a voltage signal-source v_s with a source resistance R_s be connected to the input and a load resistance R_L be connected to the output. Show that the overall voltage-gain is given by

$$\frac{v_o}{v_s} = G_m\frac{R_i}{R_i + R_s}(R_o \parallel R_L)$$

1.19 Consider a transresistance amplifier having the model shown in the third row of Table 1.1. Let the amplifier be fed with a signal current-source i_s having a resistance R_s, and let the output be connected to a load resistance R_L. Show that the overall gain is given by

$$\frac{v_o}{i_s} = R_m \frac{R_s}{R_s + R_i} \frac{R_L}{R_L + R_o}$$

1.20 Find the input resistance between terminals B and G in the circuit shown in Fig. E1.20. The voltage v_x is a test voltage with the input resistance R_{in} defined as $R_{in} \equiv v_x/i_x$.

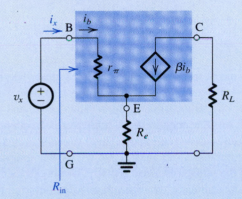

FIGURE E1.20

Ans. $R_{in} = r_\pi + (\beta + 1)R_e$

 ## 1.6 FREQUENCY RESPONSE OF AMPLIFIERS

From Section 1.2 we know that the input signal to an amplifier can always be expressed as the sum of sinusoidal signals. It follows that an important characterization of an amplifier is in terms of its response to input sinusoids of different frequencies. Such a characterization of amplifier performance is known as the amplifier frequency response.

1.6.1 Measuring the Amplifier Frequency Response

We shall introduce the subject of amplifier frequency response by showing how it can be measured. Figure 1.20 depicts a linear voltage amplifier fed at its input with a sine-wave signal of amplitude V_i and frequency ω. As the figure indicates, the signal measured at the

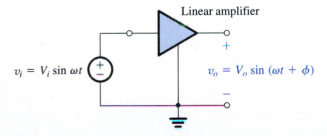

FIGURE 1.20 Measuring the frequency response of a linear amplifier. At the test frequency ω, the amplifier gain is characterized by its magnitude (V_o/V_i) and phase ϕ.

amplifier output also is sinusoidal with exactly the same frequency ω. This is an important point to note: *Whenever a sine-wave signal is applied to a linear circuit, the resulting output is sinusoidal with the same frequency as the input.* In fact, the sine wave is the only signal that does not change shape as it passes through a linear circuit. Observe, however, that the output sinusoid will in general have a different amplitude and will be shifted in phase relative to the input. The ratio of the amplitude of the output sinusoid (V_o) to the amplitude of the input sinusoid (V_i) is the magnitude of the amplifier gain (or transmission) at the test frequency ω. Also, the angle ϕ is the phase of the amplifier transmission at the test frequency ω. If we denote the **amplifier transmission,** or **transfer function** as it is more commonly known, by $T(\omega)$, then

$$|T(\omega)| = \frac{V_o}{V_i}$$
$$\angle T(\omega) = \phi$$

The response of the amplifier to a sinusoid of frequency ω is completely described by $|T(\omega)|$ and $\angle T(\omega)$. Now, to obtain the complete frequency response of the amplifier we simply change the frequency of the input sinusoid and measure the new value for $|T|$ and $\angle T$. The end result will be a table and/or graph of gain magnitude [$|T(\omega)|$] versus frequency and a table and/or graph of phase angle [$\angle T(\omega)$] versus frequency. These two plots together constitute the frequency response of the amplifier; the first is known as the **magnitude** or **amplitude response,** and the second is the **phase response.** Finally, we should mention that it is a common practice to express the magnitude of transmission in decibels and thus plot $20 \log |T(\omega)|$ versus frequency.

1.6.2 Amplifier Bandwidth

Figure 1.21 shows the magnitude response of an amplifier. It indicates that the gain is almost constant over a wide frequency range, roughly between ω_1 and ω_2. Signals whose frequencies are below ω_1 or above ω_2 will experience lower gain, with the gain decreasing as we move farther away from ω_1 and ω_2. The band of frequencies over which the gain of the amplifier is almost constant, to within a certain number of decibels (usually 3 dB), is called the **amplifier bandwidth.** Normally the amplifier is designed so that its bandwidth coincides with the spectrum of the signals it is required to amplify. If this were not the case, the amplifier would *distort* the frequency spectrum of the input signal, with different components of the input signal being amplified by different amounts.

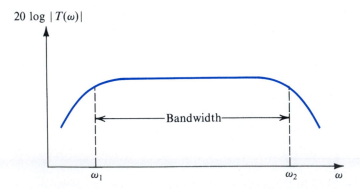

FIGURE 1.21 Typical magnitude response of an amplifier. $|T(\omega)|$ is the magnitude of the amplifier transfer function—that is, the ratio of the output $V_o(\omega)$ to the input $V_i(\omega)$.

1.6.3 Evaluating the Frequency Response of Amplifiers

Above, we described the method used to measure the frequency response of an amplifier. We now briefly discuss the method for analytically obtaining an expression for the frequency response. What we are about to say is just a preview of this important subject, whose detailed study starts in Chapter 4.

To evaluate the frequency response of an amplifier one has to analyze the amplifier equivalent circuit model, taking into account all reactive components.[2] Circuit analysis proceeds in the usual fashion but with inductances and capacitances represented by their reactances. An inductance L has a reactance or impedance $j\omega L$, and a capacitance C has a reactance or impedance $1/j\omega C$ or, equivalently, a susceptance or admittance $j\omega C$. Thus in a *frequency-domain* analysis we deal with impedances and/or admittances. The result of the analysis is the amplifier transfer function $T(\omega)$:

$$T(\omega) = \frac{V_o(\omega)}{V_i(\omega)}$$

where $V_i(\omega)$ and $V_o(\omega)$ denote the input and output signals, respectively. $T(\omega)$ is generally a complex function whose magnitude $|T(\omega)|$ gives the magnitude of transmission or the magnitude response of the amplifier. The phase of $T(\omega)$ gives the phase response of the amplifier.

In the analysis of a circuit to determine its frequency response, the algebraic manipulations can be considerably simplified by using the **complex frequency variable** s. In terms of s, the impedance of an inductance L is sL and that of a capacitance C is $1/sC$. Replacing the reactive elements with their impedances and performing standard circuit analysis, we obtain the transfer function $T(s)$ as

$$T(s) \equiv \frac{V_o(s)}{V_i(s)}$$

Subsequently, we replace s by $j\omega$ to determine the transfer function for **physical frequencies,** $T(j\omega)$. Note that $T(j\omega)$ is the same function we called $T(\omega)$ above;[3] the additional j is included in order to emphasize that $T(j\omega)$ is obtained from $T(s)$ by replacing s with $j\omega$.

1.6.4 Single-Time-Constant Networks

In analyzing amplifier circuits to determine their frequency response, one is greatly aided by knowledge of the frequency response characteristics of single-time-constant (STC) networks. An STC network is one that is composed of, or can be reduced to, one reactive component (inductance or capacitance) and one resistance. Examples are shown in Fig. 1.22. An STC network formed of an inductance L and a resistance R has a time constant $\tau = L/R$. The time constant τ of an STC network composed of a capacitance C and a resistance R is given by $\tau = CR$.

Appendix D presents a study of STC networks and their responses to sinusoidal, step, and pulse inputs. Knowledge of this material will be needed at various points throughout this book, and the reader will be encouraged to refer to the Appendix. At this point we need in particular the frequency response results; we will, in fact, briefly discuss this important topic, now.

---

[2] Note that in the models considered in previous sections no reactive components were included. These were simplified models and cannot be used alone to predict the amplifier frequency response.

[3] At this stage, we are using s simply as a shorthand for $j\omega$. We shall not require detailed knowledge of s-plane concepts until Chapter 6. A brief review of s-plane analysis is presented in Appendix E.

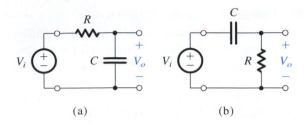

(a)   (b)

FIGURE 1.22 Two examples of STC networks: **(a)** a low-pass network and **(b)** a high-pass network.

TABLE 1.2 **Frequency Response of STC Networks**

	Low-Pass (LP)	High-Pass (HP)
Transfer Function $T(s)$	$\dfrac{K}{1+(s/\omega_0)}$	$\dfrac{Ks}{s+\omega_0}$
Transfer Function (for physical frequencies) $T(j\omega)$	$\dfrac{K}{1+j(\omega/\omega_0)}$	$\dfrac{K}{1-j(\omega_0/\omega)}$
Magnitude Response $\lvert T(j\omega)\rvert$	$\dfrac{\lvert K\rvert}{\sqrt{1+(\omega/\omega_0)^2}}$	$\dfrac{\lvert K\rvert}{\sqrt{1+(\omega_0/\omega)^2}}$
Phase Response $\angle T(j\omega)$	$-\tan^{-1}(\omega/\omega_0)$	$\tan^{-1}(\omega_0/\omega)$
Transmission at $\omega=0$ (dc)	K	0
Transmission at $\omega=\infty$	0	K
3-dB Frequency	$\omega_0=1/\tau;\ \tau\equiv$ time constant $\tau=CR$ or L/R	
Bode Plots	in Fig. 1.23	in Fig. 1.24

Most STC networks can be classified into two categories,[4] **low pass (LP)** and **high pass (HP),** with each of the two categories displaying distinctly different signal responses. As an example, the STC network shown in Fig. 1.22(a) is of the *low-pass* type and that in Fig. 1.22(b) is of the *high-pass* type. To see the reasoning behind this classification, observe that the transfer function of each of these two circuits can be expressed as a voltage-divider ratio, with the divider composed of a resistor and a capacitor. Now, recalling how the impedance of a capacitor varies with frequency $(Z = 1/j\omega C)$ it is easy to see that the transmission of the circuit in Fig. 1.22(a) will decrease with frequency and approach zero as ω approaches ∞. Thus the circuit of Fig. 1.22(a) acts as a **low-pass filter;**[5] it passes low-frequency sine-wave inputs with little or no attenuation (at $\omega = 0$, the transmission is unity) and attenuates high-frequency input sinusoids. The circuit of Fig. 1.22(b) does the opposite; its transmission is unity at $\omega = \infty$ and decreases as ω is reduced, reaching 0 for $\omega = 0$. The latter circuit, therefore, performs as a **high-pass filter.**

Table 1.2 provides a summary of the frequency response results for STC networks of both types.[6] Also, sketches of the magnitude and phase responses are given in Figs. 1.23 and 1.24.

---

[4] An important exception is the *all-pass* STC network studied in Chapter 11.

[5] A filter is a circuit that passes signals in a specified frequency band (the filter passband) and stops or severely attenuates (filters out) signals in another frequency band (the filter stopband). Filters will be studied in Chapter 12.

[6] The transfer functions in Table 1.2 are given in general form. For the circuits of Fig. 1.22, $K=1$ and $\omega_0 = 1/CR$.

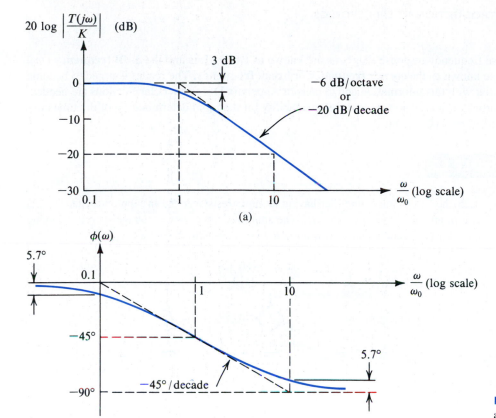

FIGURE 1.23 (a) Magnitude and (b) phase response of STC networks of the low-pass type.

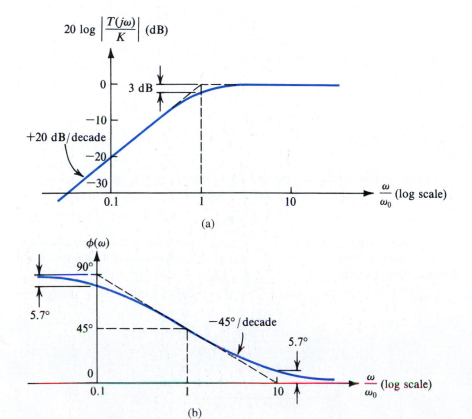

FIGURE 1.24 (a) Magnitude and (b) phase response of STC networks of the high-pass type.

These frequency response diagrams are known as **Bode plots** and the **3-dB frequency** (ω_0) is also known as the **corner frequency** or **break frequency.** The reader is urged to become familiar with this information and to consult Appendix D if further clarifications are needed. In particular, it is important to develop a facility for the rapid determination of the time constant τ of an STC circuit.

EXAMPLE 1.5

Figure 1.25 shows a voltage amplifier having an input resistance R_i, an input capacitance C_i, a gain factor μ, and an output resistance R_o. The amplifier is fed with a voltage source V_s having a source resistance R_s, and a load of resistance R_L is connected to the output.

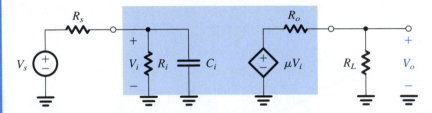

FIGURE 1.25 Circuit for Example 1.5.

(a) Derive an expression for the amplifier voltage gain V_o/V_s as a function of frequency. From this find expressions for the dc gain and the 3-dB frequency.

(b) Calculate the values of the dc gain, the 3-dB frequency, and the frequency at which the gain becomes 0 dB (i.e., unity) for the case $R_s = 20$ kΩ, $R_i = 100$ kΩ, $C_i = 60$ pF, $\mu = 144$ V/V, $R_o = 200$ Ω, and $R_L = 1$ kΩ.

(c) Find $v_o(t)$ for each of the following inputs:
  (i) $v_i = 0.1 \sin 10^2 t$, V
  (ii) $v_i = 0.1 \sin 10^5 t$, V
  (iii) $v_i = 0.1 \sin 10^6 t$, V
  (iv) $v_i = 0.1 \sin 10^8 t$, V

Solution

(a) Utilizing the voltage-divider rule, we can express V_i in terms of V_s as follows

$$V_i = V_s \frac{Z_i}{Z_i + R_s}$$

where Z_i is the amplifier input impedance. Since Z_i is composed of two parallel elements it is obviously easier to work in terms of $Y_i = 1/Z_i$. Toward that end we divide the numerator and denominator by Z_i, thus obtaining

$$V_i = V_s \frac{1}{1 + R_s Y_i}$$

$$= V_s \frac{1}{1 + R_s[(1/R_i) + sC_i]}$$

Thus,

$$\frac{V_i}{V_s} = \frac{1}{1 + (R_s/R_i) + sC_i R_s}$$

This expression can be put in the standard form for a low-pass STC network (see the top line of Table 1.2) by extracting $[1 + (R_s/R_i)]$ from the denominator; thus we have

$$\frac{V_i}{V_s} = \frac{1}{1 + (R_s/R_i)} \frac{1}{1 + sC_i[(R_sR_i)/(R_s + R_i)]} \tag{1.20}$$

At the output side of the amplifier we can use the voltage-divider rule to write

$$V_o = \mu V_i \frac{R_L}{R_L + R_o}$$

This equation can be combined with Eq. (1.20) to obtain the amplifier transfer function as

$$\frac{V_o}{V_s} = \mu \frac{1}{1 + (R_s/R_i)} \frac{1}{1 + (R_o/R_L)} \frac{1}{1 + sC_i[(R_sR_i)/(R_s + R_i)]} \tag{1.21}$$

We note that only the last factor in this expression is new (compared with the expression derived in the last section). This factor is a result of the input capacitance C_i, with the time constant being

$$\tau = C_i \frac{R_sR_i}{R_s + R_i} \tag{1.22}$$

$$= C_i(R_s//R_i)$$

We could have obtained this result by inspection: From Fig. 1.25 we see that the input circuit is an STC network and that its time constant can be found by reducing V_s to zero, with the result that the resistance seen by C_i is R_i in parallel with R_s. The transfer function in Eq. (1.21) is of the form $K/(1 + (s/\omega_0))$, which corresponds to a low-pass STC network. The dc gain is found as

$$K \equiv \frac{V_o}{V_s}(s = 0) = \mu \frac{1}{1 + (R_s/R_i)} \frac{1}{1 + (R_o/R_L)} \tag{1.23}$$

The 3-dB frequency ω_0 can be found from

$$\omega_0 = \frac{1}{\tau} = \frac{1}{C_i(R_s//R_i)} \tag{1.24}$$

Since the frequency response of this amplifier is of the low-pass STC type, the Bode plots for the gain magnitude and phase will take the form shown in Fig. 1.23, where K is given by Eq. (1.23) and ω_0 is given by Eq. (1.24).

(b) Substituting the numerical values given into Eq. (1.23) results in

$$K = 144 \frac{1}{1 + (20/100)} \frac{1}{1 + (200/1000)} = 100 \text{ V/V}$$

Thus the amplifier has a dc gain of 40 dB. Substituting the numerical values into Eq. (1.24) gives the 3-dB frequency

$$\omega_0 = \frac{1}{60 \text{ pF} \times (20 \text{ k}\Omega//100 \text{ k}\Omega)}$$

$$= \frac{1}{60 \times 10^{-12} \times (20 \times 100/(20 + 100)) \times 10^3} = 10^6 \text{ rad/s}$$

Thus,

$$f_0 = \frac{10^6}{2\pi} = 159.2 \text{ kHz}$$

Since the gain falls off at the rate of –20 dB/decade, starting at ω_0 (see Fig. 1.23a) the gain will reach 0 dB in two decades (a factor of 100); thus we have

$$\text{Unity-gain frequency} = 100 \times \omega_0 = 10^8 \text{ rad/s or } 15.92 \text{ MHz}$$

(c) To find $v_o(t)$ we need to determine the gain magnitude and phase at 10^2, 10^5, 10^6, and 10^8 rad/s. This can be done either approximately utilizing the Bode plots of Fig. 1.23 or exactly utilizing the expression for the amplifier transfer function,

$$T(j\omega) \equiv \frac{V_o}{V_s}(j\omega) = \frac{100}{1 + j(\omega/10^6)}$$

We shall do both:

(i) For $\omega = 10^2$ rad/s, which is $(\omega_0/10^4)$, the Bode plots of Fig. 1.23 suggest that $|T| \simeq K = 100$ and $\phi = 0°$. The transfer function expression gives $|T| \simeq 100$ and $\phi = -\tan^{-1} 10^{-4} \simeq 0°$. Thus,

$$v_o(t) = 10 \sin 10^2 t, \text{ V}$$

(ii) For $\omega = 10^5$ rad/s, which is $(\omega_0/10)$, the Bode plots of Fig. 1.23 suggest that $|T| \simeq K = 100$ and $\phi = -5.7°$. The transfer function expression gives $|T| = 99.5$ and $\phi = -\tan^{-1} 0.1 = -5.7°$. Thus,

$$v_o(t) = 9.95 \sin(10^5 t - 5.7°), \text{ V}$$

(iii) For $\omega = 10^6$ rad/s $= \omega_0$, $|T| = 100/\sqrt{2} = 70.7$ V/V or 37 dB and $\phi = -45°$. Thus,

$$v_o(t) = 7.07 \sin(10^6 t - 45°), \text{ V}$$

(iv) For $\omega = 10^8$ rad/s, which is $(100\omega_0)$, the Bode plots suggest that $|T| = 1$ and $\phi = -90°$. The transfer function expression gives

$$|T| \simeq 1 \quad \text{and} \quad \phi = -\tan^{-1} 100 = -89.4°,$$

Thus,

$$v_o(t) = 0.1 \sin(10^8 t - 89.4°), \text{ V}$$

1.6.5 Classification of Amplifiers Based on Frequency Response

Amplifiers can be classified based on the shape of their magnitude-response curve. Figure 1.26 shows typical frequency response curves for various amplifier types. In Fig. 1.26(a) the gain remains constant over a wide frequency range but falls off at low and high frequencies. This is a common type of frequency response found in audio amplifiers.

As will be shown in later chapters, **internal capacitances** in the device (a transistor) cause the falloff of gain at high frequencies, just as C_i did in the circuit of Example 1.5. On the other hand, the falloff of gain at low frequencies is usually caused by **coupling capacitors** used to connect one amplifier stage to another, as indicated in Fig. 1.27. This practice is usually adopted to simplify the design process of the different stages. The coupling capacitors are usually chosen quite large (a fraction of a microfarad to a few tens of microfarads) so that their reactance (impedance) is small at the frequencies of interest. Nevertheless, at sufficiently low frequencies the reactance of a coupling capacitor will become large enough to cause part of the signal being coupled to appear as a voltage drop across the coupling capacitor and thus not reach the subsequent stage. Coupling capacitors will thus cause loss of gain at low frequencies and cause the gain to be zero at dc. This is not at all surprising since from Fig. 1.27 we observe that the coupling capacitor, acting together with the input resistance of the subsequent stage, forms a

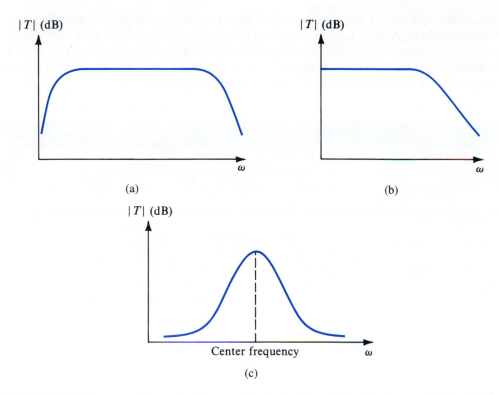

FIGURE 1.26 Frequency response for **(a)** a capacitively coupled amplifier, **(b)** a direct-coupled amplifier, and **(c)** a tuned or bandpass amplifier.

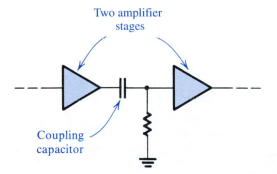

FIGURE 1.27 Use of a capacitor to couple amplifier stages.

high-pass STC circuit. It is the frequency response of this high-pass circuit that accounts for the shape of the amplifier frequency response in Fig. 1.26(a) at the low-frequency end.

There are many applications in which it is important that the amplifier maintain its gain at low frequencies down to dc. Furthermore, monolithic integrated-circuit (IC) technology does not allow the fabrication of large coupling capacitors. Thus IC amplifiers are usually designed as **directly coupled** or **dc amplifiers** (as opposed to **capacitively coupled** or **ac amplifiers**). Figure 1.26(b) shows the frequency response of a dc amplifier. Such a frequency response characterizes what is referred to as a **low-pass amplifier.**

In a number of applications, such as in the design of radio and TV receivers, the need arises for an amplifier whose frequency response peaks around a certain frequency (called the **center frequency**) and falls off on both sides of this frequency, as shown in Fig. 1.26(c).

Amplifiers with such a response are called **tuned amplifiers, bandpass amplifiers,** or **bandpass filters.** A tuned amplifier forms the heart of the front-end or tuner of a communication receiver; by adjusting its center frequency to coincide with the frequency of a desired communications channel (e.g., a radio station), the signal of this particular channel can be received while those of other channels are attenuated or filtered out.

EXERCISES

1.21 Consider a voltage amplifier having a frequency response of the low-pass STC type with a dc gain of 60 dB and a 3-dB frequency of 1000 Hz. Find the gain in dB at $f = 10$ Hz, 10 kHz, 100 kHz, and 1 MHz.

Ans. 60 dB; 40 dB; 20 dB; 0 dB

D1.22 Consider a transconductance amplifier having the model shown in Table 1.1 with $R_i = 5$ kΩ, $R_o = 50$ kΩ, and $G_m = 10$ mA/V. If the amplifier load consists of a resistance R_L in parallel with a capacitance C_L, convince yourself that the voltage transfer function realized, V_o/V_i, is of the low-pass STC type. What is the lowest value that R_L can have while a dc gain of at least 40 dB is obtained? With this value of R_L connected, find the highest value that C_L can have while a 3-dB bandwidth of at least 100 kHz is obtained.

Ans. 12.5 kΩ; 159.2 pF

D1.23 Consider the situation illustrated in Fig. 1.27. Let the output resistance of the first voltage amplifier be 1 kΩ and the input resistance of the second voltage amplifier (including the resistor shown) be 9 kΩ. The resulting equivalent circuit is shown in Fig. E1.23 where V_s and R_s are the output voltage and output resistance of the first amplifier, C is a coupling capacitor, and R_i is the input resistance of the second amplifier. Convince yourself that V_2/V_s is a high-pass STC function. What is the smallest value for C that will ensure that the 3-dB frequency is not higher than 100 Hz?

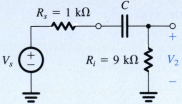

FIGURE E1.23

Ans. 0.16 μF

 ## 1.7 DIGITAL LOGIC INVERTERS[7]

The logic inverter is the most basic element in digital circuit design; it plays a role parallel to that of the amplifier in analog circuits. In this section we provide an introduction to the logic inverter.

1.7.1 Function of the Inverter

As its name implies, the logic inverter inverts the logic value of its input signal. Thus for a logic 0 input, the output will be a logic 1, and vice versa. In terms of voltage levels, consider

---

[7] If desired, study of this section can be postponed to just before study of the CMOS inverter (see Section 4.10).

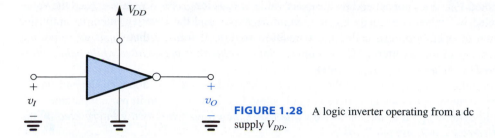

FIGURE 1.28 A logic inverter operating from a dc supply V_{DD}.

the inverter shown in block form in Fig. 1.28: When v_I is low (close to 0 V), the output v_O will be high (close to V_{DD}), and vice versa.

1.7.2 The Voltage Transfer Characteristic (VTC)

To quantify the operation of the inverter, we utilize its voltage transfer characteristic (VTC, as it is usually abbreviated). First we refer the reader to the amplifier considered in Example 1.2 whose transfer characteristic is sketched in Fig. 1.15. Observe that the transfer characteristic indicates that this inverting amplifier can be used as a logic inverter. Specifically, if the input is high ($v_I > 0.690$ V), v_O will be low at 0.3 V. On the other hand, if the input is low (close to 0 V), the output will be high (close to 10 V). Thus to use this amplifier as a logic inverter, we utilize its extreme regions of operation. This is exactly the opposite to its use as a signal amplifier, where it would be biased at the middle of the transfer characteristic and the signal kept sufficiently small so as to restrict operation to a short, almost linear, segment of the transfer curve. Digital applications, on the other hand, make use of the gross non-linearity exhibited by the VTC.

With these observations in mind, we show in Fig. 1.29 a possible VTC of a logic inverter. For simplicity, we are using three straight lines to approximate the VTC, which is usually a nonlinear curve such as that in Fig. 1.15. Observe that the output high level,

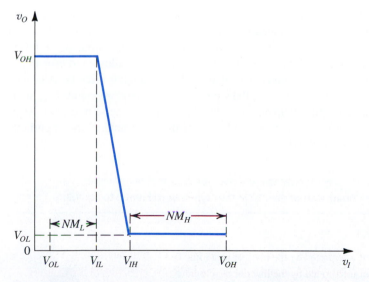

FIGURE 1.29 Voltage transfer characteristic of an inverter. The VTC is approximated by three straight-line segments. Note the four parameters of the VTC (V_{OH}, V_{OL}, V_{IL}, and V_{IH}) and their use in determining the noise margins (NM_H and NM_L).

denoted V_{OH}, does not depend on the exact value of v_I as long as v_I does not exceed the value labeled V_{IL}; when v_I exceeds V_{IL}, the output decreases and the inverter enters its amplifier region of operation, also called the **transition region.** It follows that V_{IL} is an important parameter of the inverter VTC: It is the *maximum value that v_I can have while being interpreted by the inverter as representing a logic 0.*

Similarly, we observe that the output low level, denoted V_{OL}, does not depend on the exact value of v_I as long as v_I does not fall below V_{IH}. Thus V_{IH} is an important parameter of the inverter VTC: It is the *minimum value that v_I can have while being interpreted by the inverter as representing a logic 1.*

1.7.3 Noise Margins

The insensitivity of the inverter output to the exact value of v_I within allowed regions is a great advantage that digital circuits have over analog circuits. To quantify this insensitivity property, consider the situation that occurs often in a digital system where an inverter (or a logic gate based on the inverter circuit) is driving another similar inverter. If the output of the driving inverter is high at V_{OH}, we see that we have a "margin of safety" equal to the difference between V_{OH} and V_{IH} (see Fig. 1.29). In other words, if for some reason a disturbing signal (called "electric noise," or simply **noise**) is superimposed on the output of the driving inverter, the driven inverter would not be "bothered" so long as this noise does not decrease the voltage at its input below V_{IH}. Thus we can say that the inverter has a **noise margin for high input,** NM_H, of

$$NM_H = V_{OH} - V_{IH} \tag{1.25}$$

Similarly, if the output of the driving inverter is low at V_{OL}, the driven inverter will provide a high output even if noise corrupts the V_{OL} level at its input, raising it up to nearly V_{IL}. Thus we can say that the inverter exhibits a **noise margin for low input,** NM_L, of

$$NM_L = V_{IL} - V_{OL} \tag{1.26}$$

In summary, four parameters, V_{OH}, V_{OL}, V_{IH}, and V_{IL}, define the VTC of an inverter and determine its noise margins, which in turn measure the ability of the inverter to tolerate variations in the input signal levels. In this regard, observe that changes in the input signal level within the noise margins are *rejected* by the inverter. Thus noise is not allowed to propagate further through the system, a definite advantage of digital over analog circuits. Alternatively, we can think of the inverter as *restoring* the signal levels to standard values (V_{OL} and V_{OH}) even when it is presented with corrupted signal levels (within the noise margins). As a summary, useful for future reference, we present a listing of the definitions of the important parameters of the inverter VTC in Table 1.3.

TABLE 1.3   **Important Parameters of the VTC of the Logic Inverter (Refer to Fig. 1.29)**

V_{OL}: Output low level

V_{OH}: Output high level

V_{IL}: Maximum value of input interpreted by the inverter as a logic 0

V_{IH}: Minimum value of input interpreted by the inverter as a logic 1

NM_L: Noise margin for low input = $V_{IL} - V_{OL}$

NM_H: Noise margin for high input = $V_{OH} - V_{IH}$

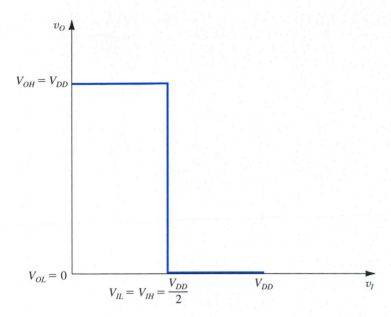

FIGURE 1.30 The VTC of an ideal inverter.

1.7.4 The Ideal VTC

The question naturally arises as to what constitutes an ideal VTC for an inverter. The answer follows directly from the preceding discussion: An ideal VTC is one that maximizes the noise margins and distributes them equally between the low and high input regions. Such a VTC is shown in Fig. 1.30 for an inverter operated from a dc supply V_{DD}. Observe that the output high level V_{OH} is at its maximum possible value of V_{DD}, and the output low level is at its minimum possible value of 0 V. Observe also that the threshold voltages V_{IL} and V_{IH} are equalized and placed at the middle of the power supply voltage ($V_{DD}/2$). Thus the width of the transition region between the high and low output regions has been reduced to zero. The transition region, though obviously very important for amplifier applications, is of no value in digital circuits. The ideal VTC exhibits a steep transition at the threshold voltage $V_{DD}/2$ with the gain in the transition region being infinite. The noise margins are now equal:

$$NM_H = NM_L = V_{DD}/2 \tag{1.27}$$

We will see in Chapter 4 that inverter circuits designed using the complementary metal-oxide-semiconductor (or CMOS) technology come very close to realizing the ideal VTC.

1.7.5 Inverter Implementation

Inverters are implemented using transistors (Chapters 4 and 5) operating as **voltage-controlled switches.** The simplest inverter implementation is shown in Fig. 1.31. The switch is controlled by the inverter input voltage v_I: When v_I is low, the switch will be open and $v_O = V_{DD}$ since no current flows through R. When v_I is high, the switch will be closed and, assuming an ideal switch, $v_O = 0$.

Transistor switches, however, as we will see in Chapters 4 and 5, are not perfect. Although their **off resistances** are very high and thus an open switch closely approximates

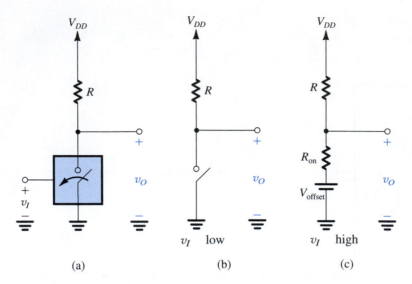

FIGURE 1.31 (a) The simplest implementation of a logic inverter using a voltage-controlled switch; (b) equivalent circuit when v_I is low; and (c) equivalent circuit when v_I is high. Note that the switch is assumed to close when v_I is high.

an open circuit, the "on" switch has a finite closure or **"on" resistance,** R_{on}. Furthermore, some switches (e.g., those implemented using bipolar transistors; see Chapter 5) exhibit in addition to R_{on} an **offset voltage,** V_{offset}. The result is that when v_I is high, the inverter has the equivalent circuit shown in Fig. 1.31(c), from which V_{OL} can be found.

More elaborate implementations of the logic inverter exist, and we show two of these in Figs. 1.32(a) and 1.33(a). The circuit in Fig. 1.32(a) utilizes a pair of **complementary switches,** the **"pull-up" (PU) switch** connects the output node to V_{DD}, and the **"pull-down" (PD) switch** connects the output node to ground. When v_I is low, the PU switch will be

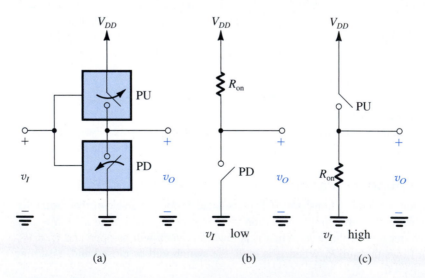

FIGURE 1.32 A more elaborate implementation of the logic inverter utilizing two complementary switches. This is the basis of the CMOS inverter studied in Section 4.10.

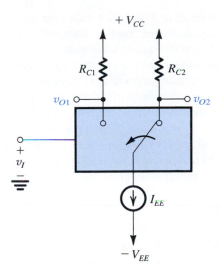

FIGURE 1.33 Another inverter implementation utilizing a double-throw switch to steer the constant current I_{EE} to R_{C1} (when v_I is high) or R_{C2} (when v_I is low). This is the basis of the emitter-coupled logic (ECL) studied in Chapters 7 and 11.

closed and the PD switch open, resulting in the equivalent circuit of Fig. 1.32(b). Observe that in this case R_{on} of PU connects the output to V_{DD}, thus establishing $V_{OH} = V_{DD}$. Also observe that no current flows and thus no power is dissipated in the circuit. Next, if v_I is raised to the logic 1 level, the PU switch will open while the PD switch will close, resulting in the equivalent circuit shown in Fig. 1.32(c). Here R_{on} of the PD switch connects the output to ground, thus establishing $V_{OL} = 0$. Here again no current flows, and no power is dissipated. The superiority of this implementation over that using the single pull-down switch and a resistor (known as a pull-up resistor) should be obvious. This circuit constitutes the basis of the CMOS inverter that we will study in Section 4.10. Note that we have not included off-set voltages in the equivalent circuits because MOS switches do not exhibit a voltage offset (Chapter 4).

Finally, consider the inverter implementation of Fig. 1.33. Here a double-throw switch is used to steer the constant current I_{EE} into one of two resistors connected to the positive supply V_{CC}. The reader is urged to show that if a high v_I results in the switch being connected to R_{C1}, then a logic inversion-function is realized at v_{O1}. Note that the output voltage is independent of the switch resistance. This *current-steering* or *current-mode* logic arrangement is the basis of the fastest available digital logic circuits, called emitter-coupled logic (ECL), introduced in Chapter 7 and studied in Chapter 11.

1.7.6 Power Dissipation

Digital systems are implemented using very large numbers of logic gates. For space and other economic considerations, it is desirable to implement the system with as few integrated-circuit (IC) chips as possible. It follows that one must pack as many logic gates as possible on an IC chip. At present, 100,000 gates or more can be fabricated on a single IC chip in what is known as **very-large-scale integration (VLSI).** To keep the power dissipated in the chip to acceptable limits (imposed by thermal considerations), the power dissipation per gate must be kept to a minimum. Indeed, a very important performance measure of the logic inverter is the power it dissipates.

The simple inverter of Fig. 1.31 obviously dissipates no power when v_I is low and the switch is open. In the other state, however, the power dissipation is approximately V_{DD}^2 / R and can be substantial. This power dissipation occurs even if the inverter is not switching and is thus known as **static power dissipation.** The inverter of Fig. 1.32 exhibits no static power dissipation, a definite advantage. Unfortunately, however, another component of power dissipation arises when a capacitance exists between the output node of the inverter and ground. This is always the case, for the devices that implement the switches have internal capacitances, the wires that connect the inverter output to other circuits have capacitance, and, of course, there is the input capacitance of whatever circuit the inverter is driving. Now, as the inverter is switched from one state to another, current must flow through the switch(es) to charge (and discharge) the load capacitance. These currents give rise to power dissipation in the switches, called **dynamic power dissipation.** In Chapter 4, we shall study dynamic power dissipation in the CMOS inverter, and we shall show that an inverter switched at a frequency f Hz exhibits a dynamic power dissipation

$$P_{\text{dynamic}} = fCV_{DD}^2 \qquad (1.28)$$

where C is the capacitance between the output node and ground and V_{DD} is the power-supply voltage. This result applies (approximately) to all inverter circuits.

1.7.7 Propagation Delay

Whereas the dynamic behavior of amplifiers is specified in terms of their frequency response, that of inverters is characterized in terms of the time delay between switching of v_I (from low to high or vice versa) and the corresponding change appearing at the output. Such a delay, called **propagation delay,** arises for two reasons: The transistors that implement the switches exhibit finite (nonzero) switching times, and the capacitance that is inevitably present between the inverter output node and ground needs to charge (or discharge, as the case may be) before the output reaches its required level of V_{OH} or V_{OL}. We shall analyze the inverter switching times in subsequent chapters. Such a study depends on a thorough familiarity with the time response of single-time-constant (STC) circuits. A review of this subject is presented in Appendix D. For our purposes here, we remind the reader of the key equation in determining the response to a step function:

Consider a step-function input applied to an STC network of either the low-pass or high-pass type, and let the network have a time constant τ. The output at any time t is given by

$$y(t) = Y_\infty - (Y_\infty - Y_{0+})e^{-t/\tau} \qquad (1.29)$$

where Y_∞ is the final value, that is, the value toward which the response is heading, and Y_{0+} is the value of the response immediately after $t = 0$. This equation states that the output at any time t is equal to the difference between the final value Y_∞ and a gap whose initial value is $Y_\infty - Y_{0+}$ and that is shrinking exponentially.

EXAMPLE 1.6

Consider the inverter of Fig. 1.31(a) with a capacitor $C = 10$ pF connected between the output and ground. Let $V_{DD} = 5$ V, $R = 1$ kΩ, $R_{\text{on}} = 100$ Ω, and $V_{\text{offset}} = 0.1$ V. If at $t = 0$, v_I goes low and neglecting the delay time of the switch, that is, assuming that it opens immediately, find the time for the output to reach $\frac{1}{2}(V_{OH} + V_{OL})$. The time to this 50% point on the output waveform is defined as the low-to-high propagation delay, t_{PLH}.

Solution

First we determine V_{OL}, which is the voltage at the output prior to $t = 0$. From the equivalent circuit in Fig. 1.31(b), we find

$$V_{OL} = V_{\text{offset}} + \frac{V_{DD} - V_{\text{offset}}}{R + R_{\text{on}}} R_{\text{on}}$$

$$= 0.1 + \frac{5 - 0.1}{1.1} \times 0.1 = 0.55 \text{ V}$$

Next, when the switch opens at $t = 0$, the circuit takes the form shown in Fig. 1.34(a). Since the voltage across the capacitor cannot change instantaneously, at $t = 0+$ the output will still be 0.55 V.

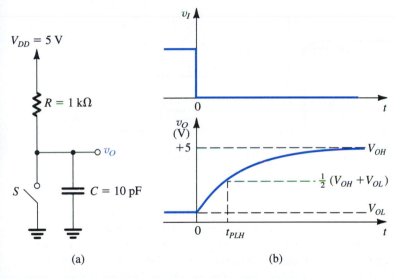

(a)   (b)

FIGURE 1.34   Example 1.6: **(a)** The inverter circuit after the switch opens (i.e., for $t \geq 0+$). **(b)** Waveforms of v_I and v_O. Observe that the switch is assumed to operate instantaneously. v_O rises exponentially, starting at V_{OL} and heading toward V_{OH}.

Then the capacitor charges through R, and v_O rises exponentially toward V_{DD}. The output waveform will be as shown in Fig. 1.34(b), and its equation can be obtained by substituting in Eq. (1.29), $v_O(\infty) = 5$ V and $v_O(0+) = 0.55$ V. Thus,

$$v_O(t) = 5 - (5 - 0.55)e^{-t/\tau}$$

where $\tau = CR$. To find t_{PLH}, we substitute

$$v_O(t_{PLH}) = \tfrac{1}{2}(V_{OH} + V_{OL})$$

$$= \tfrac{1}{2}(5 + 0.55)$$

The result is

$$t_{PLH} = 0.69\tau$$

$$= 0.69RC$$

$$= 0.69 \times 10^3 \times 10^{-11}$$

$$= 6.9 \text{ ns}$$

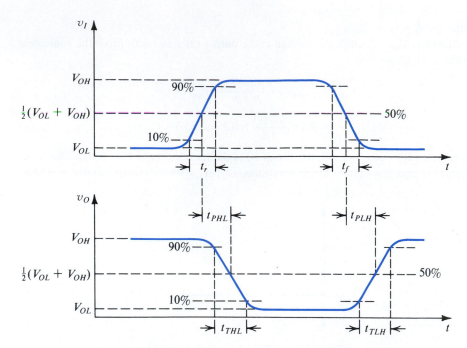

FIGURE 1.35 Definitions of propagation delays and transition times of the logic inverter.

We conclude this section by showing in Fig. 1.35 the formal definition of the propagation delay of an inverter. As shown, an input pulse with finite (nonzero) **rise and fall times** is applied. The inverted pulse at the output exhibits finite rise and fall times (labeled t_{TLH} and t_{THL}, where the subscript T denotes transition, LH denotes low-to-high, and HL denotes high-to-low). There is also a delay time between the input and output wave forms. The usual way to specify the propagation delay is to take the average of the high-to-low propagation delay, t_{PHL}, and the low-to-high propagation delay, t_{PLH}. As indicated, these delays are measured between the 50% points of the input and output waveforms. Also note that the transition times are specified using the 10% and 90% points of the output excursion $(V_{OH} - V_{OL})$.

EXERCISES

1.24 For the inverter in Fig. 1.31, let $V_{DD} = 5$ V, $R = 1$ kΩ, $R_{on} = 100$ Ω, $V_{offset} = 0.1$ V, $V_{IL} = 0.8$ V, and $V_{IH} = 1.2$ V. Find V_{OH}, V_{OL}, NM_H, and NM_L. Also find the average static power dissipation assuming that the inverter spends half the time in each of its two states.
Ans. 5 V; 0.55 V; 3.8 V; 0.25 V; 11.1 mW

1.25 Find the dynamic power dissipated in an inverter operated from a 5-V power supply. The inverter has a 2-pF capacitance load and is switched at 50 MHz.
Ans. 2.5 mW

1.8   CIRCUIT SIMULATION USING SPICE

The use of computer programs to simulate the operation of electronic circuits has become an essential step in the circuit-design process. This is especially the case for circuits that are to be fabricated in integrated-circuit form. However, even circuits that are assembled on a printed-circuit board using discrete components can and do benefit from circuit simulation. Circuit simulation enables the designer to verify that the design will meet specifications when actual components (with their many imperfections) are used, and it can also provide additional insight into circuit operation allowing the designer to fine-tune the final design prior to fabrication. However, notwithstanding the advantages of computer simulation, it is not a substitute for a thorough understanding of circuit operation. It should be performed only at a later stage in the design process and, most certainly, after a paper-and-pencil design has been done.

Among the various circuit-simulation programs available for the computer-aided numerical analysis of microelectronic circuits, **SPICE** (*S*imulation *P*rogram with *I*ntegrated *Ci*rcuit *E*mphasis) is generally regarded to be the most widely used. SPICE is an open-source program which has been under development by the University of California at Berkeley since the early 1970s. **PSpice** is a commercial personal-computer version of SPICE that is now commercially available from Cadence. Also available from Cadence is **PSpice A/D**— an advanced version of PSpice that can model the behavior and, hence, simulate circuits that process a mix of both analog and digital signals.[8] SPICE was originally a text-based program: The user had to describe the circuit to be simulated and the type of simulation to be performed using an input text file, called a **netlist.** The simulation results were also displayed as text. As an example of more recent developments, Cadence provides a graphical interface, called **OrCAD Capture CIS** (*C*omponent *I*nformation *S*ystem), for circuit-schematic entry and editing. Such graphical interface tools are referred to in the literature as **schematic entry, schematic editor,** or **schematic capture** tools. Furthermore, PSpice A/D includes a graphical postprocessor, called **Probe,** to numerically analyze and graphically display the results of the PSpice simulations. In this text, **"using PSpice"** or **"using SPICE"** loosely refers to using Capture CIS, PSpice A/D, and Probe to simulate a circuit and to numerically analyze and graphically display the simulation results.

An evaluation (student) version of Capture CIS and PSpice A/D are included on the CD accompanying this book. These correspond to the OrCAD Family Release 9.2 Lite Edition available from Cadence. Furthermore, the circuit diagrams entered in Capture CIS (called **Capture Schematics**) and the corresponding PSpice simulation files of all SPICE examples in this book can be found on the text's CD and website (www.sedrasmith.org). Access to these files will allow the reader to undertake further experimentation with these circuits, including investigating the effect of changing component values and operating conditions.

It is not our objective in this book to teach the reader *how* SPICE works nor the intricacies of using it effectively. This can be found in the SPICE books listed in Appendix F. Our objective in the sections of this book devoted to SPICE, usually the last section of each chapter, is twofold: to describe the models that are used by SPICE to represent the various electronic devices, and to illustrate how useful SPICE can be in investigating circuit operation.

---

[8] Such circuits are called mixed-signal circuits, and the simulation programs that can simulate such circuits are called mixed-signal simulators.

SUMMARY

- An electrical signal source can be represented in either the Thévenin form (a voltage source v_s in series with a source resistance R_s) or the Norton form (a current source i_s in parallel with a source resistance R_s). The Thévenin voltage v_s is the open-circuit voltage between the source terminals; equal to the Norton current i_s is equal to the short-circuit current between the source terminals. For the two representations to be equivalent, $v_s = R_s i_s$.

- The sine-wave signal is completely characterized by its peak value (or rms value which is the peak / $\sqrt{2}$), its frequency (ω in rad/s or f in Hz; $\omega = 2\pi f$ and $f = 1/T$ where T is the period in seconds), and its phase with respect to an arbitrary reference time.

- A signal can be represented either by its waveform versus time, or as the sum of sinusoids. The latter representation is known as the frequency spectrum of the signal.

- Analog signals have magnitudes that can assume any value. Electronic circuits that process analog signals are called analog circuits. Sampling the magnitude of an analog signal at discrete instants of time and representing each signal sample by a number, results in a digital signal. Digital signals are processed by digital circuits.

- The simplest digital signals are obtained when the binary system is used. An individual digital signal then assumes one of only two possible values: low and high (say, 0 V and +5 V), corresponding to logic 0 and logic 1, respectively.

- An analog-to-digital converter (ADC) provides at its output the digits of the binary number representing the analog signal sample applied to its input. The output digital signal can then be processed using digital circuits. Refer to Fig. 1.9 and Eq. 1.3.

- The transfer characteristic, v_O versus v_I, of a linear amplifier is a straight line with a slope equal to the voltage gain. Refer to Fig. 1.11.

- Amplifiers increase the signal power and thus require dc power supplies for their operation.

- The amplifier voltage gain can be expressed as a ratio A_v in V/V or in decibels, $20 \log|A_v|$, dB. Similarly, for current gain: A_i A/A or $20 \log|A_i|$, dB. For power gain: A_p W/W or $10 \log A_p$, dB.

- Linear amplification can be obtained from a device having a nonlinear transfer characteristic by employing dc biasing and keeping the input signal amplitude small. Refer to Fig. 1.14.

- Depending on the signal to be amplified (voltage or current) and on the desired form of output signal (voltage or current), there are four basic amplifier types: voltage, current, transconductance, and transresistance amplifiers. For the circuit models and ideal characteristics of these four amplifier types, refer to Table 1.1. A given amplifier can be modeled by any one of the four models, in which case their parameters are related by the formulas in Eqs. (1.14) to (1.16).

- A sinusoid is the only signal whose wave form is unchanged through a linear circuit. Sinusoidal signals are used to measure the frequency response of amplifiers.

- The transfer function $T(s) \equiv V_o(s)/V_i(s)$ of a voltage amplifier can be determined from circuit analysis. Substituting $s = j\omega$ gives $T(j\omega)$, whose magnitude $|T(j\omega)|$ is the magnitude response, and whose phase $\phi(\omega)$ is the phase response, of the amplifier.

- Amplifiers are classified according to the shape of their frequency response, $|T(j\omega)|$. Refer to Fig. 1.26.

- Single-time-constant (STC) networks are those networks that are composed of, or can be reduced to, one reactive component (L or C) and one resistance (R). The time constant τ is either L/R or CR.

- STC networks can be classified into two categories: low-pass (LP) and high-pass (HP). LP networks pass dc and low frequencies and attenuate high frequencies. The opposite is true for HP networks.

- The gain of an LP (HP) STC circuit drops by 3 dB below the zero-frequency (infinite-frequency) value at a frequency $\omega_0 = 1/\tau$. At high frequencies (low frequencies) the gain falls off at the rate of 6 dB/octave or 20 dB/decade. Refer to Table 1.2 on page 34 and Figs. (1.23) and (1.24). Further details are given in Appendix E.

- The digital logic inverter is the basic building block of digital circuits, just as the amplifier is the basic building block of analog circuits.

- The static operation of the inverter is described by its voltage transfer characteristic (VTC). The break-points of the transfer characteristic determine the inverter noise margins; refer to Fig. 1.29 and Table 1.3. In particular, note that $NM_H = V_{OH} - V_{IH}$ and $NM_L = V_{IL} - V_{OL}$.

- The inverter is implemented using transistors operating as voltage-controlled switches. The arrangement utilizing two switches operated in a complementary fashion results in a high-performance inverter. This is the basis for the CMOS inverter studied in Chapter 4.

- An important performance parameter of the inverter is the amount of power it dissipates. There are two components of power dissipation: static and dynamic. The first is a result of current flow in either the 0 or 1 state or both. The second occurs when the inverter is switched and has a capacitor load. Dynamic power dissipation is given approximately by fCV_{DD}^2.

■ Another very important performance parameter of the inverter is its propagation delay (see Fig. 1.35 for definitions).

PROBLEMS[1,2]

CIRCUIT BASICS

As a review of the basics of circuit analysis and in order for the readers to gauge their preparedness for the study of electronic circuits, this section presents a number of relevant circuit analysis problems. For a summary of Thévenin's and Norton's theorems, refer to Appendix D. The problems are grouped in appropriate categories.

RESISTORS AND OHM'S LAW

1.1 Ohm's law relates V, I, and R for a resistor. For each of the situations following, find the missing item:

(a) $R = 1\ \text{k}\Omega$, $V = 10\ \text{V}$
(b) $V = 10\ \text{V}$, $I = 1\ \text{mA}$
(c) $R = 10\ \text{k}\Omega$, $I = 10\ \text{mA}$
(d) $R = 100\ \Omega$, $V = 10\ \text{V}$

1.2 Measurements taken on various resistors are shown below. For each, calculate the power dissipated in the resistor and the power rating necessary for safe operation using standard components with power ratings of 1/8 W, 1/4 W, 1/2 W, 1 W, or 2 W:

(a) $1\ \text{k}\Omega$ conducting 30 mA
(b) $1\ \text{k}\Omega$ conducting 40 mA
(c) $10\ \text{k}\Omega$ conducting 3 mA
(d) $10\ \text{k}\Omega$ conducting 4 mA
(e) $1\ \text{k}\Omega$ dropping 20 V
(f) $1\ \text{k}\Omega$ dropping 11 V

1.3 Ohm's law and the power law for a resistor relate V, I, R, and P, making only two variables independent. For each pair identified below, find the other two:

(a) $R = 1\ \text{k}\Omega$, $I = 10\ \text{mA}$
(b) $V = 10\ \text{V}$, $I = 1\ \text{mA}$
(c) $V = 10\ \text{V}$, $P = 1\ \text{W}$
(d) $I = 10\ \text{mA}$, $P = 0.1\ \text{W}$
(e) $R = 1\ \text{k}\Omega$, $P = 1\ \text{W}$

COMBINING RESISTORS

1.4 You are given three resistors whose values are 10 kΩ, 20 kΩ, and 40 kΩ. How many different resistances can you

create using series and parallel combinations of these three? List them in value order, lowest first. Be thorough and organized. (*Hint:* In your search, first consider all parallel combinations, then consider series combinations, and then consider series-parallel combinations, of which there are two kinds).

1.5 In the analysis and test of electronic circuits, it is often useful to connect one resistor in parallel with another to obtain a nonstandard value, one which is smaller than the smaller of the two resistors. Often, particularly during circuit testing, one resistor is already installed, in which case the second, when connected in parallel, is said to "shunt" the first. If the original resistor is 10 kΩ, what is the value of the shunting resistor needed to reduce the combined value by 1%, 5%, 10%, and 50%? What is the result of shunting a 10-kΩ resistor by 1 MΩ? By 100 kΩ? By 10 kΩ?

VOLTAGE DIVIDERS

1.6 Figure P1.6(a) shows a two-resistor voltage divider. Its function is to generate a voltage V_O (smaller than the power-supply voltage V_{DD}) at its output node X. The circuit looking back at node X is equivalent to that shown in Fig. P1.6(b). Observe that this is the Thévenin equivalent of the voltage divider circuit. Find expressions for V_O and R_O.

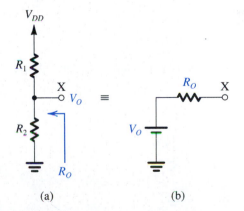

FIGURE P1.6

---

[1] Somewhat difficult problems are marked with an asterisk (*); more difficult problems are marked with two asterisks (**); and very difficult (and/or time-consuming) problems are marked with three asterisks (***).

[2] Design-oriented problems are marked with a D.

1.7 A two-resistor voltage divider employing a 3.3-kΩ and a 6.8-kΩ resistor is connected to a 9-V ground-referenced power supply to provide a relatively low voltage. Sketch the circuit. Assuming exact-valued resistors, what output voltage (measured to ground) and equivalent output resistance result? If the resistors used are not ideal but have a ±5% manufacturing tolerance, what are the extreme output voltages and resistances that can result?

1.8 You are given three resistors, each of 10 kΩ, and a 9-V battery whose negative terminal is connected to ground. With a voltage divider using some or all of your resistors, how many positive-voltage sources of magnitude less than 9 V can you design? List them in order, smallest first. What is the output resistance (i.e., the Thévenin resistance) of each?

D*1.9 Two resistors, with nominal values of 4.7 kΩ and 10 kΩ, are used in a voltage divider with a +15-V supply to create a nominal +10-V output. Assuming the resistor values to be exact, what is the actual output voltage produced? Which resistor must be shunted (paralleled) by what third resistor to create a voltage-divider output of 10.00 V? If an output resistance of exactly 3.33 kΩ is also required, what do you suggest? What should be done if the requirement is 10.00 V and 3.00 kΩ while still using the original 4.7-kΩ and 10-kΩ resistors?

CURRENT DIVIDERS

1.10 Current dividers play an important role in circuit design. Therefore it is important to develop a facility for dealing with current dividers in circuit analysis. Figure P1.10 shows a two-resistor current divider fed with an ideal current source *I*. Show that

$$I_1 = \frac{R_2}{R_1 + R_2}I$$

$$I_2 = \frac{R_1}{R_1 + R_2}I$$

and find the voltage *V* that develops across the current divider.

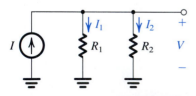

FIGURE P1.10

D1.11 Design a simple current divider that will reduce the current provided to a 1-kΩ load to 20% of that available from the source.

D1.12 A designer searches for a simple circuit to provide one-third of a signal current *I* to a load resistance *R*. Suggest a solution using one resistor. What must its value be? What is the input resistance of the resulting current divider? For a particular value *R*, the designer discovers that the otherwise-best-available resistor is 10% too high. Suggest two circuit topologies using one additional resistor that will solve this problem. What is the value of the resistor required? What is the input resistance of the current divider in each case?

D1.13 A particular electronic signal source generates currents in the range 0 mA to 1 mA under the condition that its load voltage not exceed 1 V. For loads causing more than 1 V to appear across the generator, the output current is no longer assured but will be reduced by some unknown amount. This circuit limitation, occurring, for example, at the peak of a signal sine wave, will lead to undesirable signal distortion that must be avoided. If a 10-kΩ load is to be connected, what must be done? What is the name of the circuit you must use? How many resistors are needed? What is (are) the(ir) value(s)?

THÉVENIN-EQUIVALENT CIRCUITS

1.14 For the circuit in Fig. P1.14, find the Thévenin equivalent circuit between terminals (a) 1 and 2, (b) 2 and 3, and (c) 1 and 3.

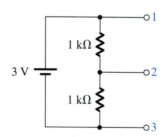

FIGURE P1.14

1.15 Through repeated application of Thévenin's theorem, find the Thévenin-equivalent of the circuit in Fig. P1.15 between node 4 and ground and hence find the current that flows through a load resistance of 1.5 kΩ connected between node 4 and ground.

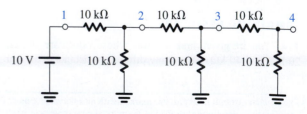

FIGURE P1.15

CIRCUIT ANALYSIS

1.16 For the circuit shown in Fig. P1.16, find the current in all resistors and the voltage (with respect to ground) at their common node using two methods:

(a) Current: Define branch currents I_1 and I_2 in R_1 and R_2, respectively; identify two equations; and solve them.
(b) Voltage: Define the node voltage V at the common node; identify a single equation; and solve it.

Which method do you prefer? Why?

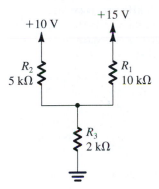

FIGURE P1.16

1.17 The circuit shown in Fig. P1.17 represents the equivalent circuit of an unbalanced bridge. It is required to calculate the current in the detector branch (R_5) and the voltage across it. Although this can be done using loop and node equations, a much easier approach is possible: Find the Thévenin equivalent of the circuit to the left of node 1 and the Thévenin equivalent of the circuit to the right of node 2. Then solve the resulting simplified circuit.

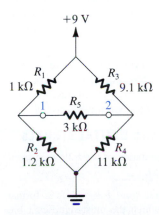

FIGURE P1.17

1.18 For the circuit in Fig. P1.18, find the equivalent resistance to ground, R_{eq}. To do this, apply a voltage V_x between terminal X and ground and find the current drawn from V_x. Note that you

can use particular special properties of the circuit to get the result directly! Now, if R_4 is raised to 1.2 kΩ, what does R_{eq} become?

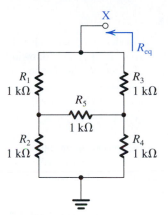

FIGURE P1.18

AC CIRCUITS

1.19 The periodicity of recurrent waveforms, such as sine waves or square waves, can be completely specified using only one of three possible parameters: radian frequency, ω, in radians per second (rad/s); (conventional) frequency, f, in Hertz (Hz); or period T, in seconds (s). As well, each of the parameters can be specified numerically in one of several ways: using letter prefixes associated with the basic units, using scientific notation, or using some combination of both. Thus, for example, a particular period may be specified as 100 ns, 0.1 μs, 10^{-1} μs, 10^5 ps, or 1×10^{-7} s. (For the definition of the various prefixes used in electronics, see Appendix H.) For each of the measures listed below, express the trio of terms in scientific notation associated with the basic unit (e.g., 10^{-7} s rather than 10^{-1} μs).

(a) $T = 10^{-4}$ ms
(b) $f = 1$ GHz
(c) $\omega = 6.28 \times 10^2$ rad/s
(d) $T = 10$ s
(e) $f = 60$ Hz
(f) $\omega = 1$ krad/s
(g) $f = 1900$ MHz

1.20 Find the complex impedance, Z, of each of the following basic circuit elements at 60 Hz, 100 kHz, and 1 GHz:

(a) $R = 1$ kΩ
(b) $C = 10$ nF
(c) $C = 2$ pF
(d) $L = 10$ mH
(e) $L = 1$ nH

1.21 Find the complex impedance at 10 kHz of the following networks:

(a) 1 kΩ in series with 10 nF
(b) 1 kΩ in parallel with 0.01 μF

(c) 100 kΩ in parallel with 100 pF
(d) 100 Ω in series with 10 mH

SECTION 1.1: SIGNALS

1.22 Any given signal source provides an open-circuit voltage, v_{oc}, and a short-circuit current i_{sc}. For the following sources, calculate the internal resistance, R_s; the Norton current, i_s; and the Thévenin voltage, v_s:

(a) $v_{oc} = 10$ V, $i_{sc} = 100$ μA
(b) $v_{oc} = 0.1$ V, $i_{sc} = 10$ μA

1.23 A particular signal source produces an output of 30 mV when loaded by a 100-kΩ resistor and 10 mV when loaded by a 10-kΩ resistor. Calculate the Thévenin voltage, Norton current, and source resistance.

1.24 A temperature sensor is specified to provide 2 mV/°C. When connected to a load resistance of 10 kΩ, the output voltage was measured to change by 10 mV, corresponding to a change in temperature of 10°C. What is the source resistance of the sensor?

1.25 Refer to the Thévenin and Norton representations of the signal source (Fig. 1.1). If the current supplied by the source is denoted i_o and the voltage appearing between the source output terminals is denoted v_o, sketch and clearly label v_o versus i_o for $0 \leq i_o \leq i_s$.

1.26 The connection of a signal source to an associated signal processor or amplifier generally involves some degree of signal loss as measured at the processor or amplifier input. Considering the two signal-source representations shown in Fig. 1.1, provide two sketches showing each signal-source representation connected to the input terminals (and corresponding input resistance) of a signal processor. What signal-processor input resistance will result in 90% of the open-circuit voltage being delivered to the processor? What input resistance will result in 90% of the short-circuit signal current entering the processor?

SECTION 1.2: FREQUENCY SPECTRUM OF SIGNALS

1.27 To familiarize yourself with typical values of angular frequency ω, conventional frequency f, and period T, complete the entries in the following table:

Case	ω (rad/s)	f (Hz)	T (s)
a		1×10^9	
b	1×10^9		
c			1×10^{-10}
d		60	
e	6.28×10^3		
f			1×10^{-6}

1.28 For the following peak or rms values of some important sine waves, calculate the corresponding other value:

(a) 117 V_{rms}, a household-power voltage in North America
(b) 33.9 V_{peak}, a somewhat common peak voltage in rectifier circuits
(c) 220 V_{rms}, a household-power voltage in parts of Europe
(d) 220 kV_{rms}, a high-voltage transmission-line voltage in North America

1.29 Give expressions for the sine-wave voltage signals having:

(a) 10-V peak amplitude and 10-kHz frequency
(b) 120-V rms and 60-Hz frequency
(c) 0.2-V peak-to-peak and 1000-rad/s frequency
(d) 100-mV peak and 1-ms period

1.30 Using the information provided by Eq. (1.2) in association with Fig. 1.4, characterize the signal represented by $v(t) = 1/2 + 2/\pi$ (sin $2000\pi t + \frac{1}{3}$ sin $6000\pi t + \frac{1}{5}$ sin $10,000\pi t + \cdots$). Sketch the waveform. What is its average value? Its peak-to-peak value? Its lowest value? Its highest value? Its frequency? Its period?

1.31 Measurements taken of a square-wave signal using a frequency-selective voltmeter (called a spectrum analyzer) show its spectrum to contain adjacent components (spectral lines) at 98 kHz and 126 kHz of amplitudes 63 mV and 49 mV, respectively. For this signal, what would direct measurement of the fundamental show its frequency and amplitude to be? What is the rms value of the fundamental? What are the peak-to-peak amplitude and period of the originating square wave?

1.32 What is the fundamental frequency of the highest-frequency square wave for which the fifth harmonic is barely audible by a relatively young listener? What is the fundamental frequency of the lowest-frequency square wave for which the fifth and some of the higher harmonics are directly heard? (Note that the psychoacoustic properties of human hearing allow a listener to sense the lower harmonics as well).

1.33 Find the amplitude of a symmetrical square wave of period T that provides the same power as a sine wave of peak amplitude $\hat{V}$ and the same frequency. Does this result depend on equality of the frequencies of the two waveforms?

SECTION 1.3: ANALOG AND DIGITAL SIGNALS

1.34 Give the binary representation of the following decimal numbers: 0, 5, 8, 25, and 57.

1.35 Consider a 4-bit digital word $b_3b_2b_1b_0$ in a format called signed-magnitude, in which the most-significant bit, b_3, is interpreted as a sign bit—0 for positive and 1 for negative values. List the values that can be represented by this scheme. What is peculiar about the representation of zero? For a particular analog-to-digital converter (ADC), each

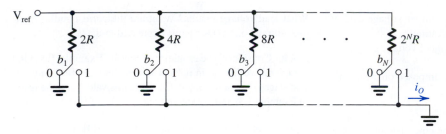

FIGURE P1.37

change in b_0 corresponds to a 0.5-V change in the analog input. What is the full range of the analog signal that can be represented? What signed-magnitude digital code results for an input of +2.5 V? For −3.0 V? For +2.7 V? For −2.8 V?

1.36 Consider an N-bit ADC whose analog input varies between 0 and V_{FS} (where the subscript FS denotes "full scale").

(a) Show that the least significant bit (LSB) corresponds to a change in the analog signal of $V_{FS}/(2^N - 1)$. This is the resolution of the converter.

(b) Convince yourself that the maximum error in the conversion (called the quantization error) is half the resolution; that is, the quantization error = $V_{FS}/2(2^N - 1)$.

(c) For V_{FS} = 10 V, how many bits are required to obtain a resolution of 5 mV or better? What is the actual resolution obtained? What is the resulting quantization error?

1.37 Figure P1.37 shows the circuit of an N-bit digital-to-analog converter (DAC). Each of the N bits of the digital word to be converted controls one of the switches. When the bit is 0, the switch is in the position labeled 0; when the bit is 1, the switch is in the position labeled 1. The analog output is the current i_O. V_{ref} is a constant reference voltage.

(a) Show that

$$i_O = \frac{V_{ref}}{R}\left(\frac{b_1}{2^1} + \frac{b_2}{2^2} + \cdots + \frac{b_N}{2^N}\right)$$

(b) Which bit is the LSB? Which is the MSB?
(c) For V_{ref} = 10 V, R = 5 kΩ, and N = 6, find the maximum value of i_O obtained. What is the change in i_O resulting from the LSB changing from 0 to 1?

1.38 In compact-disc (CD) audio technology, the audio signal is sampled at 44.1 kHz. Each sample is represented by 16 bits. What is the speed of this system in bits/second?

SECTION 1.4: AMPLIFIERS

1.39 Various amplifier and load combinations are measured as listed below using rms values. For each, find the voltage, current, and power gains (A_v, A_i, and A_p, respectively) both as ratios and in dB:

(a) v_I = 100 mV, i_I = 100 μA, v_O = 10 V, R_L = 100 Ω

(b) v_I = 10 μV, i_I = 100 nA, v_O = 2 V, R_L = 10 kΩ
(c) v_I = 1 V, i_I = 1 mA, v_O = 10 V, R_L = 10 Ω

1.40 An amplifier operating from ±3 V supplies provides a 2.2-V_{peak} sine wave across a 100-Ω load when provided with a 0.2-V_{peak} input from which 1.0 mA$_{peak}$ is drawn. The average current in each supply is measured to be 20 mA. Find the voltage gain, current gain, and power gain expressed as ratios and in dB as well as the supply power, amplifier dissipation, and amplifier efficiency.

1.41 An amplifier using balanced power supplies is known to saturate for signals extending within 1.2 V of either supply. For linear operation, its gain is 500 V/V. What is the rms value of the largest undistorted sine-wave output available, and input needed, with ±5-V supplies? With ±10-V supplies? With ±15-V supplies?

1.42 Symmetrically saturating amplifiers, operating in the so-called clipping mode, can be used to convert sine waves to pseudo-square waves. For an amplifier with a small-signal gain of 1000 and clipping levels of ±9 V, what peak value of input sinusoid is needed to produce an output whose extremes are just at the edge of clipping? Clipped 90% of the time? Clipped 99% of the time?

1.43 A particular amplifier operating from a single supply exhibits clipped peaks for signals intended to extend above 8 V and below 1.5 V. What is the peak value of the largest possible undistorted sine wave when this amplifier is biased at 4 V? At what bias point is the largest undistorted sine wave available?

D*1.44 An amplifier designed using a single metal-oxide-semiconductor (MOS) transistor has the transfer characteristic

$$v_O = 10 - 5(v_I - 2)^2$$

where v_I and v_O are in volts. This transfer characteristic applies for $2 \le v_I \le v_O + 2$ and v_O positive. At the limits of this region the amplifier saturates.

(a) Sketch and clearly label the transfer characteristic. What are the saturation levels L_+ and L_- and the corresponding values of v_I?

1

(b) Bias the amplifier to obtain a dc output voltage of 5 V. What value of input dc voltage V_I is required?

(c) Calculate the value of the small-signal voltage gain at the bias point.

(d) If a sinusoidal input signal is superimposed on the dc bias voltage V_I, that is,

$$v_I = V_I + V_i \cos \omega t$$

find the resulting v_O. Using the trigonometric identity $\cos^2 \theta = \frac{1}{2} + \frac{1}{2} \cos 2\theta$, express v_O as the sum of a dc component, a signal component with frequency ω, and a sinusoidal component with frequency 2ω. The latter component is undesirable and is a result of the nonlinear transfer characteristic of the amplifier. If it is required to limit the ratio of the second-harmonic component to the fundamental component to 1% (this ratio is known as the second-harmonic distortion), what is the corresponding upper limit on V_i? What output amplitude results?

SECTION 1.5: CIRCUIT MODELS FOR AMPLIFIERS

1.45 Consider the voltage-amplifier circuit model shown in Fig. 1.17(b), in which $A_{vo} = 10$ V/V under the following conditions:

(a) $R_i = 10R_s$, $R_L = 10R_o$
(b) $R_i = R_s$, $R_L = R_o$
(c) $R_i = R_s/10$, $R_L = R_o/10$

Calculate the overall voltage gain v_o/v_s in each case, expressed both directly and in dB.

1.46 An amplifier with 40 dB of small-signal open-circuit voltage gain, an input resistance of 1 MΩ, and an output resistance of 10 Ω drives a load of 100 Ω. What voltage and power gains (expressed in dB) would you expect with the load connected? If the amplifier has a peak output-current limitation of 100 mA, what is the rms value of the largest sine-wave input for which an undistorted output is possible? What is the corresponding output power available?

1.47 A 10-mV signal source having an internal resistance of 100 kΩ is connected to an amplifier for which the input resistance is 10 kΩ, the open-circuit voltage gain is 1000 V/V, and the output resistance is 1 kΩ. The amplifier is connected in turn to a 100-Ω load. What overall voltage gain results as measured from the source internal voltage to the load? Where did all the gain go? What would the gain be if the source was connected directly to the load? What is the ratio of these two gains? This ratio is a useful measure of the benefit the amplifier brings.

1.48 A buffer amplifier with a gain of 1 V/V has an input resistance of 1 MΩ and an output resistance of 10 Ω. It is connected between a 1-V, 100-kΩ source and a 100-Ω load. What load voltage results? What are the corresponding voltage, current, and power gains expressed in dB?

1.49 Consider the cascade amplifier of Example 1.3. Find the overall voltage gain v_o/v_s obtained when the first and second stages are interchanged. Compare this value with the result in Example 1.3, and comment.

1.50 You are given two amplifiers, A and B, to connect in cascade between a 10-mV, 100-kΩ source and a 100-Ω load. The amplifiers have voltage gain, input resistance, and output resistance as follows: For A, 100 V/V, 10 kΩ, 10 kΩ, respectively; for B, 1 V/V, 100 kΩ, 100 kΩ, respectively. Your problem is to decide how the amplifiers should be connected. To proceed, evaluate the two possible connections between source S and load L, namely, SABL and SBAL. Find the voltage gain for each both as a ratio and in dB. Which amplifier arrangement is best?

D*1.51 A designer has available voltage amplifiers with an input resistance of 10 kΩ, an output resistance of 1 kΩ, and an open-circuit voltage gain of 10. The signal source has a 10 kΩ resistance and provides a 10-mV rms signal, and it is required to provide a signal of at least 2 V rms to a 1-kΩ load. How many amplifier stages are required? What is the output voltage actually obtained.

D*1.52 Design an amplifier that provides 0.5 W of signal power to a 100-Ω load resistance. The signal source provides a 30-mV rms signal and has a resistance of 0.5 MΩ. Three types of voltage amplifier stages are available:

(a) A high-input-resistance type with $R_i = 1$ MΩ, $A_{vo} = 10$, and $R_o = 10$ kΩ
(b) A high-gain type with $R_i = 10$ kΩ, $A_{vo} = 100$, and $R_o = 1$ kΩ
(c) A low-output-resistance type with $R_i = 10$ kΩ, $A_{vo} = 1$, and $R_o = 20$ Ω

Design a suitable amplifier using a combination of these stages. Your design should utilize the minimum number of stages and should ensure that the signal level is not reduced below 10 mV at any point in the amplifier chain. Find the load voltage and power output realized.

D*1.53 It is required to design a voltage amplifier to be driven from a signal source having a 10-mV peak amplitude and a source resistance of 10 kΩ to supply a peak output of 3 V across a 1-kΩ load.

(a) What is the required voltage gain from the source to the load?

(b) If the peak current available from the source is 0.1 μA, what is the smallest input resistance allowed? For the design with this value of R_i, find the overall current gain and power gain.

(c) If the amplifier power supply limits the peak value of the output open-circuit voltage to 5 V, what is the largest output resistance allowed?

(d) For the design with R_i as in (b) and R_o as in (c), what is the required value of open-circuit voltage gain $\left(\text{i.e., } \dfrac{v_o}{v_i}\bigg|_{R_L=\infty}\right)$ of the amplifier?

(e) If, as a possible design option, you are able to increase R_i to the nearest value of the form $1 \times 10^n \ \Omega$ and to decrease R_o to the nearest value of the form $1 \times 10^m \ \Omega$, find (i) the input resistance achievable; (ii) the output resistance achievable; and (iii) the open-circuit voltage gain now required to meet the specifications.

D1.54 A voltage amplifier with an input resistance of 10 kΩ, an output resistance of 200 Ω, and a gain of 1000 V/V is connected between a 100-kΩ source with an open-circuit voltage of 10 mV and a 100-Ω load. For this situation:

(a) What output voltage results?
(b) What is the voltage gain from source to load?
(c) What is the voltage gain from the amplifier input to the load?
(d) If the output voltage across the load is twice that needed and there are signs of internal amplifier overload, suggest the location and value of a single resistor that would produce the desired output. Choose an arrangement that would cause minimum disruption to an operating circuit. (*Hint:* Use parallel rather than series connections.)

1.55 A current amplifier for which $R_i = 1$ kΩ, $R_o = 10$ kΩ, and $A_{is} = 100$ A/A is to be connected between a 100-mV source with a resistance of 100 kΩ and a load of 1 kΩ. What are the values of current gain i_o/i_i, of voltage gain v_o/v_s, and of power gain expressed directly and in dB?

1.56 A transconductance amplifier with $R_i = 2$ kΩ, $G_m = 40$ mA/V, and $R_o = 20$ kΩ is fed with a voltage source having a source resistance of 2 kΩ and is loaded with a 1-kΩ resistance. Find the voltage gain realized.

D**1.57 A designer is required to provide, across a 10-kΩ load, the weighted sum, $v_O = 10v_1 + 20v_2$, of input signals v_1 and v_2, each having a source resistance of 10 kΩ. She has a number of transconductance amplifiers for which the input and output resistances are both 10 kΩ and $G_m = 20$ mA/V, together with a selection of suitable resistors. Sketch an appropriate amplifier topology with additional resistors selected to provide the desired result. (*Hint:* In your design, arrange to add currents.)

1.58 Figure P1.58 shows a transconductance amplifier whose output is *fed back* to its input. Find the input resistance R_{in} of the resulting one-port network. (*Hint:* Apply a test

voltage v_x between the two input terminals, and find the current i_x drawn from the source. Then, $R_{in} \equiv v_x / i_x$.)

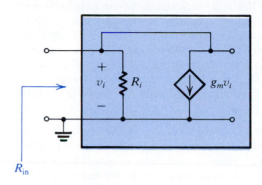

R_{in}

FIGURE P1.58

D1.59 It is required to design an amplifier to sense the open-circuit output voltage of a transducer and to provide a proportional voltage across a load resistor. The equivalent source resistance of the transducer is specified to vary in the range of 1 kΩ to 10 kΩ. Also, the load resistance varies in the range of 1 kΩ to 10 kΩ. The change in load voltage corresponding to the specified change in R_s should be 10% at most. Similarly, the change in load voltage corresponding to the specified change in R_L should be limited to 10%. Also, corresponding to a 10-mV transducer open-circuit output voltage, the amplifier should provide a minimum of 1 V across the load. What type of amplifier is required? Sketch its circuit model, and specify the values of its parameters. Specify appropriate values for R_i and R_o of the form $1 \times 10^m \ \Omega$.

D1.60 It is required to design an amplifier to sense the short-circuit output current of a transducer and to provide a proportional current through a load resistor. The equivalent source resistance of the transducer is specified to vary in the range of 1 kΩ to 10 kΩ. Similarly, the load resistance is known to vary over the range of 1 kΩ to 10 kΩ. The change in load current corresponding to the specified change in R_s is required to be limited to 10%. Similarly, the change in load current corresponding to the specified change in R_L should be 10% at most. Also, for a nominal short-circuit output current of the transducer of 10 μA, the amplifier is required to provide a minimum of 1 mA through the load. What type of amplifier is required? Sketch the circuit model of the amplifier, and specify values for its parameters. Select appropriate values for R_i and R_o in the form $1 \times 10^m \ \Omega$.

D1.61 It is required to design an amplifier to sense the open-circuit output voltage of a transducer and to provide a proportional current through a load resistor. The equivalent source resistance of the transducer is specified to vary in the range of 1 kΩ to 10 kΩ. Also, the load resistance is known to

1

vary in the range of 1 kΩ to 10 kΩ. The change in the current supplied to the load corresponding to the specified change in R_s is to be 10% at most. Similarly, the change in load current corresponding to the specified change in R_L is to be 10% at most. Also, for a nominal transducer open-circuit output voltage of 10 mV, the amplifier is required to provide a minimum of 1 mA current through the load. What type of amplifier is required? Sketch the amplifier circuit model, and specify values for its parameters. For R_i and R_o, specify values in the form 1×10^m Ω.

D1.62 It is required to design an amplifier to sense the short-circuit output current of a transducer and to provide a proportional voltage across a load resistor. The equivalent source resistance of the transducer is specified to vary in the range of 1 kΩ to 10 kΩ. Similarly, the load resistance is known to vary in the range of 1 kΩ to 10 kΩ. The change in load voltage corresponding to the specified change in R_s should be 10% at most. Similarly, the change in load voltage corresponding to the specified change in R_L is to be limited to 10%. Also, for a nominal transducer short-circuit output current of 10 μA, the amplifier is required to provide a minimum voltage across the load of 1 V. What type of amplifier is required? Sketch its circuit model, and specify the values of the model parameters. For R_i and R_o, specify appropriate values in the form 1×10^m Ω.

1.63 For the circuit in Fig. P1.63, show that

$$\frac{v_c}{v_b} = \frac{-\beta R_L}{r_\pi + (\beta + 1)R_E}$$

and

$$\frac{v_e}{v_b} = \frac{R_E}{R_E + [r_\pi/(\beta + 1)]}$$

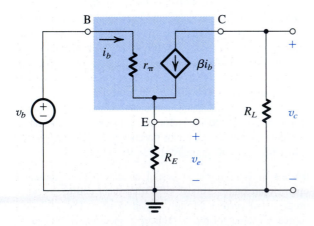

FIGURE P1.63

1.64 An amplifier with an input resistance of 10 kΩ, when driven by a current source of 1 μA and a source resistance of 100 kΩ, has a short-circuit output current of 10 mA and an open-circuit output voltage of 10 V. When driving a 4-kΩ load, what are the values of the voltage gain, current gain, and power gain expressed as ratios and in dB?

1.65 Figure P1.65(a) shows two transconductance amplifiers connected in a special configuration. Find v_o in terms of v_1 and v_2. Let $g_m = 100$ mA/V and $R = 5$ kΩ. If $v_1 = v_2 = 1$ V, find the value of v_o. Also, find v_o for the case $v_1 = 1.01$ V and $v_2 = 0.99$ V. (*Note:* This circuit is called a **differential amplifier** and is given the symbol shown in Fig. P1.65(b). A particular type of differential amplifier known as an **operational amplifier** will be studied in Chapter 2.)

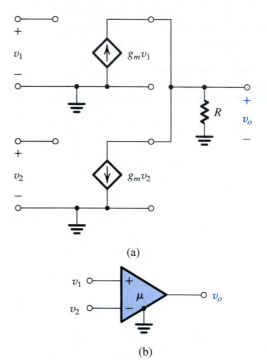

FIGURE P1.65

SECTION 1.6: FREQUENCY RESPONSE OF AMPLIFIERS

1.66 Using the voltage-divider rule, derive the transfer functions $T(s) \equiv V_o(s)/V_i(s)$ of the circuits shown in Fig. 1.22, and show that the transfer functions are of the form given at the top of Table 1.2.

1.67 Figure P1.67 shows a signal source connected to the input of an amplifier. Here R_s is the source resistance, and R_i and C_i are the input resistance and input capacitance, respectively, of the amplifier. Derive an expression for $V_i(s)/V_s(s)$, and show that it is of the low-pass STC type. Find the 3-dB frequency for the case $R_s = 20$ kΩ, $R_i = 80$ kΩ, and $C_i = 5$ pF.

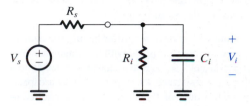

FIGURE P1.67

1.68 For the circuit shown in Fig. P1.68, find the transfer function $T(s) = V_o(s)/V_i(s)$, and arrange it in the appropriate standard form from Table 1.2. Is this a high-pass or a low-pass network? What is its transmission at very high frequencies? [Estimate this directly, as well as by letting $s \to \infty$ in your expression for $T(s)$.] What is the corner frequency ω_0? For $R_1 = 10$ kΩ, $R_2 = 40$ kΩ, and $C = 0.1$ μF, find f_0. What is the value of $|T(j\omega_0)|$?

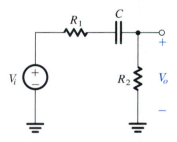

FIGURE P1.68

D1.69 It is required to couple a voltage source V_s with a resistance R_s to a load R_L via a capacitor C. Derive an expression for the transfer function from source to load (i.e., V_L/V_s), and show that it is of the high-pass STC type. For $R_s = 5$ kΩ and $R_L = 20$ kΩ, find the smallest coupling capacitor that will result in a 3-dB frequency no greater than 10 Hz.

1.70 Measurement of the frequency response of an amplifier yields the data in the following table:

f (Hz)	$|T|$ (dB)	$\angle T$ (°)
0	40	0
100	40	0
1000		
10^4	37	−45
10^5	20	
	0	

Provide plausible approximate values for the missing entries. Also, sketch and clearly label the magnitude frequency response (i.e., provide a Bode plot) for this amplifier.

1.71 Measurement of the frequency response of an amplifier yields the data in the following table:

f (Hz)	10	10^2	10^3	10^4	10^5	10^6	10^7
$|T|$ (dB)	0	20	37	40		37	20	0

Provide approximate plausible values for the missing table entries. Also, sketch and clearly label the magnitude frequency response (Bode plot) of this amplifier.

1.72 The unity-gain voltage amplifiers in the circuit of Fig. P1.72 have infinite input resistances and zero output resistances and thus function as perfect buffers. Convince yourself that the overall gain V_o/V_i will drop by 3 dB below the value at dc at the frequency for which the gain of each RC circuit is 1.0 dB down. What is that frequency in terms of CR?

1.73 An internal node of a high-frequency amplifier whose Thévenin-equivalent node resistance is 100 kΩ is accidentally shunted to ground by a capacitor (i.e., the node is connected to ground through a capacitor) through a manufacturing error. If the measured 3 dB bandwidth of the amplifier is reduced from the expected 6 MHz to 120 kHz, estimate the value of the shunting capacitor. If the original cutoff frequency can be attributed to a small parasitic capacitor at the same internal node (i.e., between the node and ground), what would you estimate it to be?

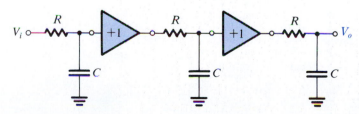

FIGURE P1.72

D*1.74 A designer wishing to lower the overall upper 3-dB frequency of a three-stage amplifier to 10 kHz considers shunting one of two nodes: Node A, between the output of the first stage and the input of the second stage, and Node B, between the output of the second stage and the input of the third stage, to ground with a small capacitor. While measuring the overall frequency response of the amplifier, she shunts a capacitor of 1 nF, first to node A and then to node B, lowering the 3-dB frequency from 2 MHz to 150 kHz and 15 kHz, respectively. If she knows that each amplifier stage has an input resistance of 100 kΩ, what output resistance must the driving stage have at node A? At node B? What capacitor value should she connect to which node to solve her design problem most economically?

D1.75 An amplifier with an input resistance of 100 kΩ and an output resistance of 1 kΩ is to be capacitor-coupled to a 10-kΩ source and a 1-kΩ load. Available capacitors have values only of the form 1×10^{-n} F. What are the values of the smallest capacitors needed to ensure that the corner frequency associated with each is less than 100 Hz? What actual corner frequencies result? For the situation in which the basic amplifier has an open-circuit voltage gain (A_{vo}) of 100 V/V, find an expression for $T(s) = V_o(s)/V_s(s)$.

*1.76 A voltage amplifier has the transfer function

$$A_v = \frac{100}{\left(1 + j\dfrac{f}{10^4}\right)\left(1 + \dfrac{10^2}{jf}\right)}$$

Using the Bode plots for low-pass and high-pass STC networks (Figs. 1.23 and 1.24), sketch a Bode plot for $|A_v|$. Give approximate values for the gain magnitude at $f = 10$ Hz, 10^2 Hz, 10^3 Hz, 10^4 Hz, 10^5 Hz, 10^6 Hz, and 10^7 Hz. Find the bandwidth of the amplifier (defined as the frequency range over which the gain remains within 3 dB of the maximum value).

*1.77 For the circuit shown in Fig. P1.77 first, evaluate $T_i(s) = V_i(s)/V_s(s)$ and the corresponding cutoff (corner) frequency. Second, evaluate $T_o(s) = V_o(s)/V_i(s)$ and the corresponding cutoff frequency. Put each of the transfer functions in the standard form (see Table 1.2), and combine them to form the overall transfer function, $T(s) = T_i(s) \times T_o(s)$. Provide a Bode magnitude plot for $|T(j\omega)|$. What is the bandwidth between 3-dB cutoff points?

D**1.78 A transconductance amplifier having the equivalent circuit shown in Table 1.1 is fed with a voltage source V_s having a source resistance R_s, and its output is connected to a load consisting of a resistance R_L in parallel with a capacitance C_L. For given values of R_s, R_L, and C_L, it is required to specify the values of the amplifier parameters R_i, G_m, and R_o to meet the following design constraints:

(a) At most, $x\%$ of the input signal is lost in coupling the signal source to the amplifier (i.e., $V_i \geq [1 - (x/100)]V_s$).
(b) The 3-dB frequency of the amplifier is equal to or greater than a specified value $f_{3\,dB}$.
(c) The dc gain V_o/V_s is equal to or greater than a specified value A_0.

Show that these constraints can be met by selecting

$$R_i \geq \left(\frac{100}{x} - 1\right)R_s$$

$$R_o \leq \frac{1}{2\pi f_{3\,dB}C_L - (1/R_L)}$$

$$G_m \geq \frac{A_0/[1 - (x/100)]}{(R_L \parallel R_o)}$$

Find R_i, R_o, and G_m for $R_s = 10$ kΩ, $x = 20\%$, $A_0 = 80$, $R_L = 10$ kΩ, $C_L = 10$ pF, and $f_{3\,dB} = 3$ MHz.

*1.79 Use the voltage-divider rule to find the transfer function $V_o(s)/V_i(s)$ of the circuit in Fig. P1.79. Show that the transfer function can be made independent of frequency if the condition $C_1R_1 = C_2R_2$ applies. Under this condition the circuit is called a **compensated attenuator** and is frequently

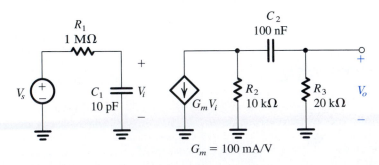

FIGURE P1.77

employed in the design of oscilloscope probes. Find the transmission of the compensated attenuator in terms of R_1 and R_2.

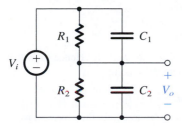

FIGURE P1.79

*1.80 An amplifier with a frequency response of the type shown in Fig. 1.21 is specified to have a phase shift of magnitude no greater than 11.4° over the amplifier bandwidth, which extends from 100 Hz to 1 kHz. It has been found that the gain falloff at the low-frequency end is determined by the response of a high-pass STC circuit and that at the high-frequency end it is determined by a low-pass STC circuit. What do you expect the corner frequencies of these two circuits to be? What is the drop in gain in decibels (relative to the maximum gain) at the two frequencies that define the amplifier bandwidth? What are the frequencies at which the drop in gain is 3 dB?

SECTION 1.7: DIGITAL LOGIC INVERTERS

1.81 A particular logic inverter is specified to have $V_{IL} = 1.3$ V, $V_{IH} = 1.7$ V, $V_{OL} = 0$ V, and $V_{OH} = 3.3$ V. Find the high and low noise margins, NM_H and NM_L.

1.82 The voltage-transfer characteristic of a particular logic inverter is modeled by three straight-line segments in the manner shown in Fig. 1.29. If $V_{IL} = 1.5$ V, $V_{IH} = 2.5$ V, $V_{OL} = 0.5$ V, and $V_{OH} = 4$ V, find:

(a) The noise margins
(b) The value of v_I at which $v_O = v_I$ (known as the **inverter threshold**)
(c) The voltage gain in the transition region

1.83 For a particular inverter design using a power supply V_{DD}, $V_{OL} = 0.1V_{DD}$, $V_{OH} = 0.8V_{DD}$, $V_{IL} = 0.4V_{DD}$, and $V_{IH} = 0.6V_{DD}$. What are the noise margins? What is the width of the transition region? For a minimum noise margin of 1 V, what value of V_{DD} is required?

1.84 A logic circuit family that used to be very popular is Transistor-Transistor Logic (TTL). The TTL logic gates and other building blocks are available commercially in small-scale integrated (SSI) and medium-scale-integrated (MSI) packages. Such packages can be assembled on printed-circuit boards to implement a digital system. The device data sheets provide the following specifications of the basic TTL inverter (of the SN7400 type):

Logic-1 input level required to ensure a logic-0 level at the output: MIN (minimum) 2 V
Logic-0 input level required to ensure a logic-1 level at the output: MAX (maximum) 0.8 V
Logic-1 output voltage: MIN 2.4 V, TYP (typical) 3.3 V
Logic-0 output voltage: TYP 0.22 V, MAX 0.4 V
Logic-0-level supply current: TYP 3 mA, MAX 5 mA
Logic-1-level supply current: TYP 1 mA, MAX 2 mA
Propagation delay time to logic-0 level (t_{PHL}): TYP 7 ns, MAX 15 ns
Propagation delay time to logic-1 level (t_{PLH}): TYP 11 ns, MAX 22 ns

(a) Find the worst-case values of the noise margins.
(b) Assuming that the inverter is in the 1-state 50% of the time and in the 0-state 50% of the time, find the average static power dissipation in a typical circuit. The power supply is 5 V.
(c) Assuming that the inverter drives a capacitance $C_L = 45$ pF and is switched at a 1-MHz rate, use the formula in Eq. (1.28) to estimate the dynamic power dissipation.
(d) Find the propagation delay t_P.

1.85 Consider an inverter implemented as in Fig. 1.31(a). Let $V_{DD} = 5$ V, $R = 2$ kΩ, $V_{offset} = 0.1$ V, $R_{on} = 200$ Ω, $V_{IL} = 1$ V, and $V_{IH} = 2$ V.

(a) Find V_{OL}, V_{OH}, NM_H, and NM_L.
(b) The inverter is driving N identical inverters. Each of these load inverters, or **fan-out** inverters as they are usually called, is specified to require an input current of 0.2 mA when the input voltage (of the fan-out inverter) is high and zero current when the input voltage is low. Noting that the input currents of the fan-out inverters will have to be supplied through R of the driving inverter, find the resulting value of V_{OH} and of NM_H as a function of the number of fan-out inverters N. Hence find the maximum value N can have while the inverter is still providing an NM_H value at least equal to its NM_L.
(c) Find the static power dissipation in the inverter in the two cases: (i) the output is low, and (ii) the output is high and driving the maximum fan-out found in (b).

1.86 A logic inverter is implemented using the arrangement of Fig. 1.32 with switches having $R_{on} = 1$ kΩ, $V_{DD} = 5$ V, and $V_{IL} = V_{IH} = V_{DD}/2$.

(a) Find V_{OL}, V_{OH}, NM_L, and NM_H.
(b) If v_I rises instantaneously from 0 V to +5 V and assuming the switches operate instantaneously—that is, at $t = 0$, PU opens, and PD closes—find an expression for $v_O(t)$ assuming that a capacitance C is connected between the output node and ground. Hence find the high-to-low propagation delay (t_{PHL}) for $C = 1$ pF. Also find t_{THL} (see Fig. 1.35).

(c) Repeat (b) for v_I falling instantaneously from +5 V to 0 V. Again assume that PD opens and PU closes instantaneously. Find an expression for $v_O(t)$, and hence find t_{PLH} and t_{TLH}.

1.87 For the current-mode inverter shown in Fig. 1.33, let $V_{CC} = 5$ V, $I_{EE} = 1$ mA, and $R_{C1} = R_{C2} = 2$ kΩ. Find V_{OL} and V_{OH}.

1.88 Consider a logic inverter of the type shown in Fig. 1.32. Let $V_{DD} = 5$ V, and let a 10-pF capacitance be connected between the output node and ground. If the inverter is switched at the rate of 100 MHz, use the expression in Eq. (1.28) to estimate the dynamic power dissipation. What is the average current drawn from the dc power supply?

D1.89** We wish to investigate the design of the inverter shown in Fig. 1.31(a). In particular we wish to determine the value for R. Selection of a suitable value for R is determined by two considerations: propagation delay, and power dissipation.

(a) Show that if v_I changes instantaneously from high to low and assuming that the switch opens instantaneously, the output voltage obtained across a load capacitance C will be

$$v_O(t) = V_{OH} - (V_{OH} - V_{OL})e^{-t/\tau_1}$$

where $\tau_1 = CR$. Hence show that the time required for $v_O(t)$ to reach the 50% point, $\frac{1}{2}(V_{OH} + V_{OL})$, is

$$t_{PLH} = 0.69CR$$

(b) Following a steady state, if v_I goes high and assuming that the switch closes immediately and has the equivalent circuit in Fig. 1.31, show that the output falls exponentially according to

$$v_O(t) = V_{OL} + (V_{OH} - V_{OL})e^{-t/\tau_2}$$

where $\tau_2 = C(R \| R_{on}) \cong CR_{on}$ for $R_{on} \ll R$. Hence show that the time for $v_O(t)$ to reach the 50% point is

$$t_{PHL} = 0.69CR_{on}$$

(c) Use the results of (a) and (b) to obtain the inverter propagation delay, defined as the average of t_{PLH} and t_{PHL} as

$$\tau_P \cong 0.35CR \text{ for } R_{on} \ll R$$

(d) Assuming that V_{offset} of the switch is much smaller than V_{DD}, show that for an inverter that spends half the time in the 0 state and half the time in the 1 state, the average static power dissipation is

$$P = \frac{1}{2}\frac{V_{DD}^2}{R}$$

(e) Now that the trade-offs in selecting R should be obvious, show that, for $V_{DD} = 5$ V and $C = 10$ pF, to obtain a propagation delay no greater than 10 ns and a power dissipation no greater than 10 mW, R should be in a specific range. Find that range and select an appropriate value for R. Then determine the resulting values of t_P and P.

Operational Amplifiers

INTRODUCTION

Having learned basic amplifier concepts and terminology, we are now ready to undertake the study of a circuit building block of universal importance: the operational amplifier (op amp). Op amps have been in use for a long time, their initial applications being primarily in the areas of analog computation and sophisticated instrumentation. Early op amps were constructed from discrete components (vacuum tubes and then transistors, and resistors), and their cost was prohibitively high (tens of dollars). In the mid-1960s the first integrated-circuit (IC) op amp was produced. This unit (the μA 709) was made up of a relatively large number of transistors and resistors all on the same silicon chip. Although its characteristics were poor (by today's standards) and its price was still quite high, its appearance signaled a new era in electronic circuit design. Electronics engineers started using op amps in large quantities, which caused their price to drop dramatically. They also demanded better-quality op amps. Semiconductor manufacturers responded quickly, and within the span of a few years, high-quality op amps became available at extremely low prices (tens of cents) from a large number of suppliers.

One of the reasons for the popularity of the op amp is its versatility. As we will shortly see, one can do almost anything with op amps! Equally important is the fact that the IC op amp has characteristics that closely approach the assumed ideal. This implies that it is quite easy to design circuits using the IC op amp. Also, op-amp circuits work at performance levels that are quite close to those predicted theoretically. It is for this reason that we are studying op amps at this early stage. It is expected that by the end of this chapter the reader should be able to design nontrivial circuits successfully using op amps.

As already implied, an IC op amp is made up of a large number (tens) of transistors, resistors, and (usually) one capacitor connected in a rather complex circuit. Since we have not yet studied transistor circuits, the circuit inside the op amp will not be discussed in this chapter. Rather, we will treat the op amp as a circuit building block and study its terminal characteristics and its applications. This approach is quite satisfactory in many op-amp applications. Nevertheless, for the more difficult and demanding applications it is quite useful to know what is inside the op-amp package. This topic will be studied in Chapter 9. Finally, it should be mentioned that more advanced applications of op amps will appear in later chapters.

2.1 THE IDEAL OP AMP

2.1.1 The Op-Amp Terminals

From a signal point-of-view the op amp has three terminals: two input terminals and one output terminal. Figure 2.1 shows the symbol we shall use to represent the op amp. Terminals 1 and 2 are input terminals, and terminal 3 is the output terminal. As explained in Section 1.4, amplifiers require dc power to operate. Most IC op amps require two dc power supplies, as shown in Fig. 2.2. Two terminals, 4 and 5, are brought out of the op-amp package and connected to a positive voltage V_{CC} and a negative voltage $-V_{EE}$, respectively. In Fig. 2.2(b)

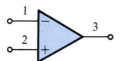

FIGURE 2.1 Circuit symbol for the op amp.

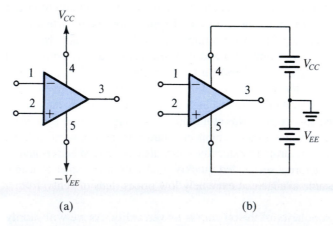

(a)   (b)

FIGURE 2.2 The op amp shown connected to dc power supplies.

we explicitly show the two dc power supplies as batteries with a common ground. It is interesting to note that the reference grounding point in op-amp circuits is just the common terminal of the two power supplies; that is, no terminal of the op-amp package is physically connected to ground. In what follows we will not explicitly show the op-amp power supplies.

In addition to the three signal terminals and the two power-supply terminals, an op amp may have other terminals for specific purposes. These other terminals can include terminals for frequency compensation and terminals for offset nulling; both functions will be explained in later sections.

2.1.2 Function and Characteristics of the Ideal Op Amp

We now consider the circuit function of the op amp. The op amp is designed to sense the difference between the voltage signals applied at its two input terminals (i.e., the quantity $v_2 - v_1$), multiply this by a number A, and cause the resulting voltage $A(v_2 - v_1)$ to appear at output terminal 3. Here it should be emphasized that when we talk about the voltage at a terminal we mean the voltage between that terminal and ground; thus v_1 means the voltage applied between terminal 1 and ground.

The ideal op amp is not supposed to draw any input current; that is, the signal current into terminal 1 and the signal current into terminal 2 are both zero. In other words, *the input impedance of an ideal op amp is supposed to be infinite.*

How about the output terminal 3? This terminal is supposed to act as the output terminal of an ideal voltage source. That is, the voltage between terminal 3 and ground will always be equal to $A(v_2 - v_1)$, independent of the current that may be drawn from terminal 3 into a load impedance. In other words, *the output impedance of an ideal op amp is supposed to be zero.*

Putting together all of the above, we arrive at the equivalent circuit model shown in Fig. 2.3. Note that the output is in phase with (has the same sign as) v_2 and is out of phase with (has the opposite sign of) v_1. For this reason, input terminal 1 is called the **inverting input terminal** and is distinguished by a "−" sign, while input terminal 2 is called the **non-inverting input terminal** and is distinguished by a "+" sign.

As can be seen from the above description, the op amp responds only to the *difference* signal $v_2 - v_1$ and hence ignores any signal *common* to both inputs. That is, if $v_1 = v_2 = 1$ V, then the output will—ideally—be zero. We call this property **common-mode rejection,** and we conclude that an ideal op amp has zero common-mode gain or, equivalently, infinite common-mode rejection. We will have more to say about this point later. For the time being note that the op amp is a **differential-input single-ended-output** amplifier, with the latter term referring to the fact that the output appears between terminal 3 and

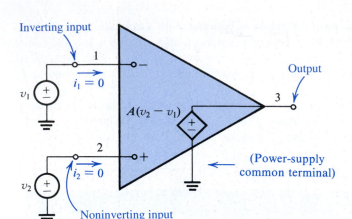

FIGURE 2.3 Equivalent circuit of the ideal op amp.

TABLE 2.1	Characteristics of the Ideal Op Amp

1. Infinite input impedance
2. Zero output impedance
3. Zero common-mode gain or, equivalently, infinite common-mode rejection
4. Infinite open-loop gain A
5. Infinite bandwidth

ground.[1] Furthermore, gain A is called the **differential gain,** for obvious reasons. Perhaps not so obvious is another name that we will attach to A: the **open-loop gain.** The reason for this name will become obvious later on when we "close the loop" around the op amp and define another gain, the closed-loop gain.

An important characteristic of op amps is that they are **direct-coupled** or **dc amplifiers,** where dc stands for direct-coupled (it could equally well stand for direct current, since a direct-coupled amplifier is one that amplifies signals whose frequency is as low as zero). The fact that op amps are direct-coupled devices will allow us to use them in many important applications. Unfortunately, though, the direct-coupling property can cause some serious practical problems, as will be discussed in a later section.

How about bandwidth? The ideal op amp has a gain A that remains constant down to zero frequency and up to infinite frequency. That is, ideal op amps will amplify signals of any frequency with equal gain, and are thus said to have *infinite bandwidth.*

We have discussed all of the properties of the ideal op amp except for one, which in fact is the most important. This has to do with the value of A. *The ideal op amp should have a gain A whose value is very large and ideally infinite.* One may justifiably ask: If the gain A is infinite, how are we going to use the op amp? The answer is very simple: In almost all applications the op amp will *not* be used alone in a so-called open-loop configuration. Rather, we will use other components to apply feedback to close the loop around the op amp, as will be illustrated in detail in Section 2.2.

For future reference, Table 2.1 lists the characteristics of the ideal op amp.

---

[1] Some op amps are designed to have differential outputs. This topic will be discussed in Chapter 9. In the current chapter we confine ourselves to single-ended-output op amps, which constitute the vast majority of commercially available op amps.

2.1.3 Differential and Common-Mode Signals

The differential input signal v_{Id} is simply the difference between the two input signals v_1 and v_2; that is,

$$v_{Id} = v_2 - v_1 \tag{2.1}$$

The common-mode input signal v_{Icm} is the average of the two input signals v_1 and v_2; namely,

$$v_{Icm} = \tfrac{1}{2}(v_1 + v_2) \tag{2.2}$$

Equations (2.1) and (2.2) can be used to express the input signals v_1 and v_2 in terms of their differential and common-mode components as follows:

$$v_1 = v_{Icm} - v_{Id}/2 \tag{2.3}$$

and

$$v_2 = v_{Icm} + v_{Id}/2 \tag{2.4}$$

These equations can in turn lead to the pictorial representation in Fig. 2.4.

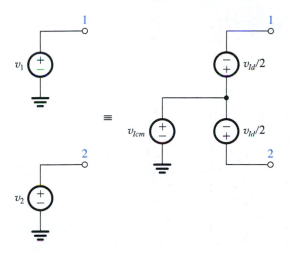

FIGURE 2.4 Representation of the signal sources v_1 and v_2 in terms of their differential and common-mode components.

EXERCISES

2.2 Consider an op amp that is ideal except that its open-loop gain $A = 10^3$. The op amp is used in a feedback circuit, and the voltages appearing at two of its three signal terminals are measured. In each of the following cases, use the measured values to find the expected value of the voltage at the third terminal. Also give the differential and common-mode input signals in each case. (a) $v_2 = 0$ V and $v_3 = 2$ V; (b) $v_2 = +5$ V and $v_3 = -10$ V; (c) $v_1 = 1.002$ V and $v_2 = 0.998$ V; (d) $v_1 = -3.6$ V and $v_3 = -3.6$ V.

Ans. (a) $v_1 = -0.002$ V, $v_{Id} = 2$ mV, $v_{Icm} = 1$ mV; (b) $v_1 = +5.01$ V, $v_{Id} = -10$ mV, $v_{Icm} = 5.005 \simeq 5$ V; (c) $v_3 = -4$ V, $v_{Id} = -4$ mV, $v_{Icm} = 1$ V; (d) $v_2 = -3.6036$ V, $v_{Id} = -3.6$ mV, $v_{Icm} \simeq -3.6$ V

2.3 The internal circuit of a particular op amp can be modeled by the circuit shown in Fig. E2.3. Express v_3 as a function of v_1 and v_2. For the case $G_m = 10$ mA/V, $R = 10$ kΩ, and $\mu = 100$, find the value of the open-loop gain A.

Ans. $v_3 = \mu G_m R(v_2 - v_1)$; $A = 10{,}000$ V/V or 80 dB

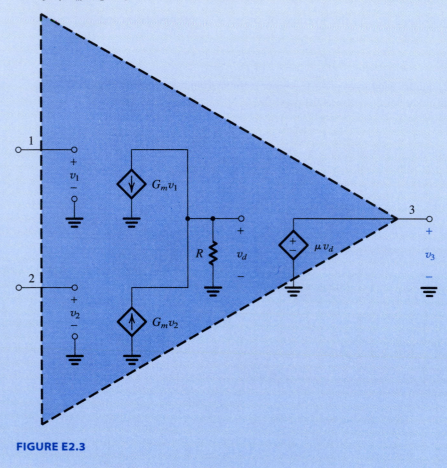

FIGURE E2.3

2.2 THE INVERTING CONFIGURATION

As mentioned above, op amps are not used alone; rather, the op amp is connected to passive components in a feedback circuit. There are two such basic circuit configurations employing an op amp and two resistors: the inverting configuration, which is studied in this section, and the noninverting configuration, which we shall study in the next section.

Figure 2.5 shows the inverting configuration. It consists of one op amp and two resistors R_1 and R_2. Resistor R_2 is connected from the output terminal of the op amp, terminal 3, *back* to the *inverting* or *negative* input terminal, terminal 1. We speak of R_2 as applying **negative feedback;** if R_2 were connected between terminals 3 and 2 we would have called this **positive feedback.** Note also that R_2 *closes the loop* around the op amp. In addition to adding R_2, we have grounded terminal 2 and connected a resistor R_1 between terminal 1 and an input signal source

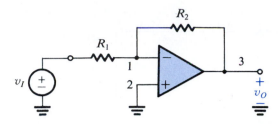

FIGURE 2.5 The inverting closed-loop configuration.

with a voltage v_I. The output of the overall circuit is taken at terminal 3 (i.e., between terminal 3 and ground). Terminal 3 is, of course, a convenient point to take the output, since the impedance level there is ideally zero. Thus the voltage v_O will not depend on the value of the current that might be supplied to a load impedance connected between terminal 3 and ground.

2.2.1 The Closed-Loop Gain

We now wish to analyze the circuit in Fig. 2.5 to determine the **closed-loop gain G,** defined as

$$G \equiv \frac{v_O}{v_I}$$

We will do so assuming the op amp to be ideal. Figure 2.6(a) shows the equivalent circuit, and the analysis proceeds as follows: The gain A is very large (ideally infinite). If we assume that the circuit is "working" and producing a finite output voltage at terminal 3, then the voltage between the op amp input terminals should be negligibly small and ideally zero. Specifically, if we call the output voltage v_O, then, by definition,

$$v_2 - v_1 = \frac{v_O}{A} = 0$$

It follows that the voltage at the inverting input terminal (v_1) is given by $v_1 = v_2$. That is, because the gain A approaches infinity, the voltage v_1 approaches and ideally equals v_2. We speak of this as the two input terminals "tracking each other in potential." We also speak of a "virtual short circuit" that exists between the two input terminals. Here the word *virtual* should be emphasized, and one should *not* make the mistake of physically shorting terminals 1 and 2 together while analyzing a circuit. A **virtual short circuit** means that whatever voltage is at 2 will automatically appear at 1 because of the infinite gain A. But terminal 2 happens to be connected to ground; thus $v_2 = 0$ and $v_1 = 0$. We speak of terminal 1 as being a **virtual ground**— that is, having zero voltage but not physically connected to ground.

Now that we have determined v_1 we are in a position to apply Ohm's law and find the current i_1 through R_1 (see Fig. 2.6) as follows:

$$i_1 = \frac{v_I - v_1}{R_1} = \frac{v_I - 0}{R_1} = \frac{v_I}{R_1}$$

Where will this current go? It cannot go into the op amp, since the ideal op amp has an infinite input impedance and hence draws zero current. It follows that i_1 will have to flow through R_2 to the low-impedance terminal 3. We can then apply Ohm's law to R_2 and determine v_O; that is,

$$v_O = v_1 - i_1 R_2$$

$$= 0 - \frac{v_I}{R_1} R_2$$

Thus,

$$\frac{v_O}{v_I} = -\frac{R_2}{R_1}$$

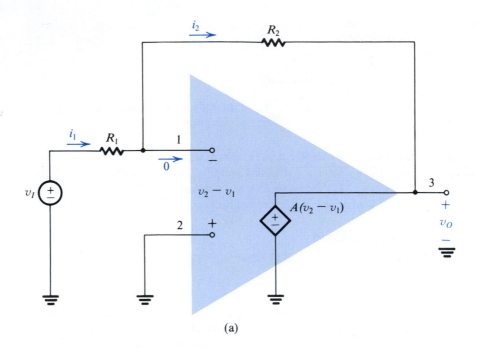

(a)

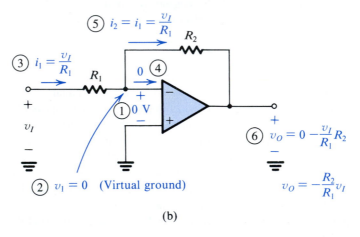

(b)

FIGURE 2.6 Analysis of the inverting configuration. The circled numbers indicate the order of the analysis steps.

which is the required closed-loop gain. Figure 2.6(b) illustrates these steps and indicates by the circled numbers the order in which the analysis is performed.

We thus see that the closed-loop gain is simply the ratio of the two resistances R_2 and R_1. The minus sign means that the closed-loop amplifier provides signal inversion. Thus if $R_2/R_1 = 10$ and we apply at the input (v_I) a sine-wave signal of 1 V peak-to-peak, then the output v_O will be a sine wave of 10 V peak-to-peak and phase-shifted 180° with respect to the input sine wave. Because of the minus sign associated with the closed-loop gain, this configuration is called the **inverting configuration.**

The fact that the closed-loop gain depends entirely on external passive components (resistors R_1 and R_2) is very significant. It means that we can make the closed-loop gain as accurate as we want by selecting passive components of appropriate accuracy. It also means that the closed-loop gain is (ideally) independent of the op-amp gain. This is a dramatic illustration of negative feedback: We started out with an amplifier having very large gain A, and through applying negative feedback we have obtained a closed-loop gain R_2/R_1 that is much smaller than A but is stable and predictable. That is, we are trading gain for accuracy.

2.2.2 Effect of Finite Open-Loop Gain

The points just made are more clearly illustrated by deriving an expression for the closed-loop gain under the assumption that the op-amp open-loop gain A is finite. Figure 2.7 shows the analysis. If we denote the output voltage v_O, then the voltage between the two input terminals of the op amp will be v_O/A. Since the positive input terminal is grounded, the voltage at the negative input terminal must be $-v_O/A$. The current i_1 through R_1 can now be found from

$$i_1 = \frac{v_I - (-v_O/A)}{R_1} = \frac{v_I + v_O/A}{R_1}$$

The infinite input impedance of the op amp forces the current i_1 to flow entirely through R_2. The output voltage v_O can thus be determined from

$$v_O = -\frac{v_O}{A} - i_1 R_2$$
$$= -\frac{v_O}{A} - \left(\frac{v_I + v_O/A}{R_1}\right) R_2$$

Collecting terms, the closed-loop gain G is found as

$$G \equiv \frac{v_O}{v_I} = \frac{-R_2/R_1}{1 + (1 + R_2/R_1)/A} \tag{2.5}$$

We note that as A approaches ∞, G approaches the ideal value of $-R_2/R_1$. Also, from Fig. 2.7 we see that as A approaches ∞, the voltage at the inverting input terminal approaches zero. This is the virtual-ground assumption we used in our earlier analysis when the op amp was assumed to be ideal. Finally, note that Eq. (2.5) in fact indicates that to minimize the dependence of the closed-loop gain G on the value of the open-loop gain A, we should make

$$1 + \frac{R_2}{R_1} \ll A$$

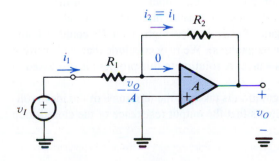

FIGURE 2.7 Analysis of the inverting configuration taking into account the finite open-loop gain of the op amp.

EXAMPLE 2.1

Consider the inverting configuration with $R_1 = 1$ kΩ and $R_2 = 100$ kΩ.

(a) Find the closed-loop gain for the cases $A = 10^3$, 10^4, and 10^5. In each case determine the percentage error in the magnitude of G relative to the ideal value of R_2/R_1 (obtained with $A = \infty$). Also determine the voltage v_1 that appears at the inverting input terminal when $v_I = 0.1$ V.

(b) If the open-loop gain A changes from 100,000 to 50,000 (i.e., drops by 50%), what is the corresponding percentage change in the magnitude of the closed-loop gain G?

Solution

(a) Substituting the given values in Eq. (2.5), we obtain the values given in the following table where the percentage error ε is defined as

$$\varepsilon \equiv \frac{|G| - (R_2/R_1)}{(R_2/R_1)} \times 100$$

The values of v_1 are obtained from $v_1 = -v_O/A = Gv_I/A$ with $v_I = 0.1$ V.

A	$|G|$	ε	v_1
10^3	90.83	-9.17%	-9.08 mV
10^4	99.00	-1.00%	-0.99 mV
10^5	99.90	-0.10%	-0.10 mV

(b) Using Eq. (2.5), we find that for $A = 50,000$, $|G| = 99.80$. Thus a -50% change in the open-loop gain results in a change of only -0.1% in the closed-loop gain!

2.2.3 Input and Output Resistances

Assuming an ideal op amp with infinite open-loop gain, the input resistance of the closed-loop inverting amplifier of Fig. 2.5 is simply equal to R_1. This can be seen from Fig. 2.6(b), where

$$R_i \equiv \frac{v_I}{i_1} = \frac{v_I}{v_I/R_1} = R_1$$

Now recall that in Section 1.5 we learned that the amplifier input resistance forms a voltage divider with the resistance of the source that feeds the amplifier. Thus, to avoid the loss of signal strength, voltage amplifiers are required to have high input resistance. In the case of the inverting op-amp configuration we are studying, to make R_i high we should select a high value for R_1. However, if the required gain R_2/R_1 is also high, then R_2 could become impractically large (e.g., greater than a few megaohms). We may conclude that the inverting configuration suffers from a low input resistance. A solution to this problem is discussed in Example 2.2 below.

Since the output of the inverting configuration is taken at the terminals of the ideal voltage source $A(v_2 - v_1)$ (see Fig. 2.6a), it follows that the output resistance of the closed-loop amplifier is zero.

EXAMPLE 2.2

Assuming the op amp to be ideal, derive an expression for the closed-loop gain v_O/v_I of the circuit shown in Fig. 2.8. Use this circuit to design an inverting amplifier with a gain of 100 and an input resistance of 1 MΩ. Assume that for practical reasons it is required not to use resistors greater than 1 MΩ. Compare your design with that based on the inverting configuration of Fig. 2.5.

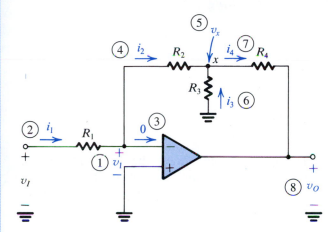

FIGURE 2.8   Circuit for Example 2.2. The circled numbers indicate the sequence of the steps in the analysis.

Solution

The analysis begins at the inverting input terminal of the op amp, where the voltage is

$$v_1 = \frac{-v_O}{A} = \frac{-v_O}{\infty} = 0$$

Here we have assumed that the circuit is "working" and producing a finite output voltage v_O. Knowing v_1, we can determine the current i_1 as follows:

$$i_1 = \frac{v_I - v_1}{R_1} = \frac{v_I - 0}{R_1} = \frac{v_I}{R_1}$$

Since zero current flows into the inverting input terminal, all of i_1 will flow through R_2, and thus

$$i_2 = i_1 = \frac{v_I}{R_1}$$

Now we can determine the voltage at node x:

$$v_x = v_1 - i_2 R_2 = 0 - \frac{v_I}{R_1} R_2 = -\frac{R_2}{R_1} v_I$$

This in turn enables us to find the current i_3:

$$i_3 = \frac{0 - v_x}{R_3} = \frac{R_2}{R_1 R_3} v_I$$

Next, a node equation at x yields i_4:

$$i_4 = i_2 + i_3 = \frac{v_I}{R_1} + \frac{R_2}{R_1 R_3} v_I$$

Finally, we can determine v_O from

$$v_O = v_x - i_4 R_4$$

$$= -\frac{R_2}{R_1} v_I - \left(\frac{v_I}{R_1} + \frac{R_2}{R_1 R_3} v_I \right) R_4$$

Thus the voltage gain is given by

$$\frac{v_O}{v_I} = -\left[\frac{R_2}{R_1} + \frac{R_4}{R_1} \left(1 + \frac{R_2}{R_3} \right) \right]$$

which can be written in the form

$$\frac{v_O}{v_I} = -\frac{R_2}{R_1} \left(1 + \frac{R_4}{R_2} + \frac{R_4}{R_3} \right)$$

Now, since an input resistance of 1 MΩ is required, we select $R_1 = 1$ MΩ. Then, with the limitation of using resistors no greater than 1 MΩ, the maximum value possible for the first factor in the gain expression is 1 and is obtained by selecting $R_2 = 1$ MΩ. To obtain a gain of -100, R_3 and R_4 must be selected so that the second factor in the gain expression is 100. If we select the maximum allowed (in this example) value of 1 MΩ for R_4, then the required value of R_3 can be calculated to be 10.2 kΩ. Thus this circuit utilizes three 1-MΩ resistors and a 10.2-kΩ resistor. In comparison, if the inverting configuration were used with $R_1 = 1$ MΩ we would have required a feedback resistor of 100 MΩ, an impractically large value!

Before leaving this example it is insightful to enquire into the mechanism by which the circuit is able to realize a large voltage gain without using large resistances in the feedback path. Toward that end, observe that because of the virtual ground at the inverting input terminal of the op amp, R_2 and R_3 are in effect in parallel. Thus, by making R_3 lower than R_2 by, say, a factor k (i.e., $R_3 = R_2/k$ where $k > 1$), R_3 is forced to carry a current k-times that in R_2. Thus, while $i_2 = i_1$, $i_3 = ki_1$ and $i_4 = (k + 1)i_1$. It is the current multiplication by a factor of $(k + 1)$ that enables a large voltage drop to develop across R_4 and hence a large v_O without using a large value for R_4. Notice also that the current through R_4 is independent of the value of R_4. It follows that the circuit can be used as a current amplifier as shown in Fig. 2.9.

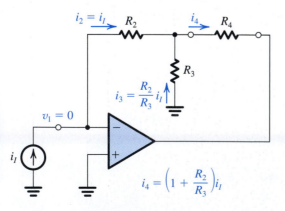

FIGURE 2.9 A current amplifier based on the circuit of Fig. 2.8. The amplifier delivers its output current to R_4. It has a current gain of $(1 + R_2/R_3)$, a zero input resistance, and an infinite output resistance. The load (R_4), however, must be floating (i.e., neither of its two terminals can be connected to ground).

EXERCISES

D2.4 Use the circuit of Fig. 2.5 to design an inverting amplifier having a gain of -10 and an input resistance of 100 kΩ. Give the values of R_1 and R_2.

Ans. $R_1 = 100$ kΩ; $R_2 = 1$ MΩ

2.5 The circuit shown in Fig. E2.5(a) can be used to implement a transresistance amplifier (see Table 1.1 in Section 1.5). Find the value of the input resistance R_i, the transresistance R_m, and the output resistance R_o of the transresistance amplifier. If the signal source shown in Fig. E2.5(b) is connected to the input of the transresistance amplifier, find its output voltage.

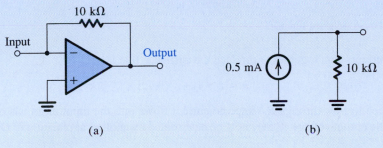

FIGURE E2.5

Ans. $R_i = 0$; $R_m = -10$ kΩ; $R_o = 0$; $v_o = -5$ V

2.6 For the circuit in Fig. E2.6 determine the values of v_1, i_1, i_2, v_O, i_L, and i_O. Also determine the voltage gain v_O/v_I, current gain i_L/i_I, and power gain P_O/P_I.

Ans. 0 V; 1 mA; 1 mA; -10 V; -10 mA; -11 mA; -10 V/V (20 dB), -10 A/A (20 dB); 100 W/W (20 dB)

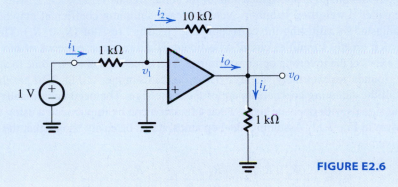

FIGURE E2.6

2.2.4 An Important Application—The Weighted Summer

A very important application of the inverting configuration is the weighted-summer circuit shown in Fig. 2.10. Here we have a resistance R_f in the negative-feedback path (as before), but we have a number of input signals $v_1, v_2, \ldots, v_n$ each applied to a corresponding resistor $R_1, R_2, \ldots, R_n$, which are connected to the inverting terminal of the op amp. From our previous discussion, the ideal op amp will have a virtual ground appearing at its negative input terminal. Ohm's law then tells us that the currents $i_1, i_2, \ldots, i_n$ are given by

$$i_1 = \frac{v_1}{R_1}, \qquad i_2 = \frac{v_2}{R_2}, \qquad \ldots, \qquad i_n = \frac{v_n}{R_n}$$

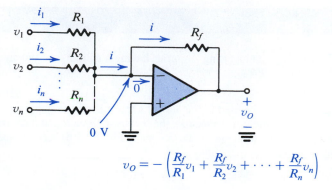

$$v_O = -\left(\frac{R_f}{R_1}v_1 + \frac{R_f}{R_2}v_2 + \cdots + \frac{R_f}{R_n}v_n\right)$$

FIGURE 2.10 A weighted summer.

All these currents sum together to produce the current i; that is,

$$i = i_1 + i_2 + \cdots + i_n \tag{2.6}$$

will be forced to flow through R_f (since no current flows into the input terminals of an ideal op amp). The output voltage v_O may now be determined by another application of Ohm's law,

$$v_O = 0 - iR_f = -iR_f$$

Thus,

$$v_O = -\left(\frac{R_f}{R_1}v_1 + \frac{R_f}{R_2}v_2 + \cdots + \frac{R_f}{R_n}v_n\right) \tag{2.7}$$

That is, the output voltage is a weighted sum of the input signals $v_1, v_2, \ldots, v_n$. This circuit is therefore called a **weighted summer.** Note that each summing coefficient may be independently adjusted by adjusting the corresponding "feed-in" resistor (R_1 to R_n). This nice property, which greatly simplifies circuit adjustment, is a direct consequence of the virtual ground that exists at the inverting op-amp terminal. As the reader will soon come to appreciate, virtual grounds are extremely "handy." The weighted summer of Fig. 2.10 has the constraint that all the summing coefficients are of the same sign. The need occasionally arises for summing signals with opposite signs. Such a function can be implemented using two op amps as shown in Fig. 2.11. Assuming ideal op amps, it can be easily shown that the output voltage is given by

$$v_O = v_1\left(\frac{R_a}{R_1}\right)\left(\frac{R_c}{R_b}\right) + v_2\left(\frac{R_a}{R_2}\right)\left(\frac{R_c}{R_b}\right) - v_3\left(\frac{R_c}{R_3}\right) - v_4\left(\frac{R_c}{R_4}\right) \tag{2.8}$$

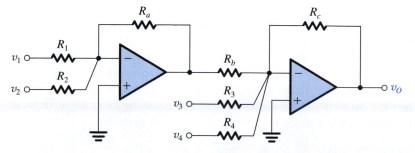

FIGURE 2.11 A weighted summer capable of implementing summing coefficients of both signs.

2.3 THE NONINVERTING CONFIGURATION

The second closed-loop configuration we shall study is shown in Fig. 2.12. Here the input signal v_I is applied directly to the positive input terminal of the op amp while one terminal of R_1 is connected to ground.

2.3.1 The Closed-Loop Gain

Analysis of the noninverting circuit to determine its closed-loop gain (v_O/v_I) is illustrated in Fig. 2.13. Notice that the order of the steps in the analysis is indicated by circled numbers.

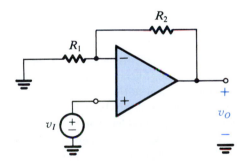

FIGURE 2.12 The noninverting configuration.

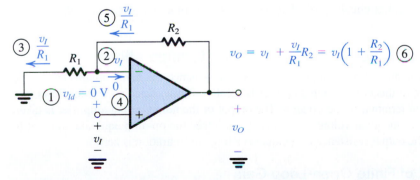

FIGURE 2.13 Analysis of the noninverting circuit. The sequence of the steps in the analysis is indicated by the circled numbers.

Assuming that the op amp is ideal with infinite gain, a virtual short circuit exists between its two input terminals. Hence the difference input signal is

$$v_{Id} = \frac{v_O}{A} = 0 \qquad \text{for } A = \infty$$

Thus the voltage at the inverting input terminal will be equal to that at the noninverting input terminal, which is the applied voltage v_I. The current through R_1 can then be determined as v_I/R_1. Because of the infinite input impedance of the op amp, this current will flow through R_2, as shown in Fig. 2.13. Now the output voltage can be determined from

$$v_O = v_I + \left(\frac{v_I}{R_1}\right)R_2$$

which yields

$$\frac{v_O}{v_I} = 1 + \frac{R_2}{R_1} \qquad\qquad (2.9)$$

Further insight into the operation of the noninverting configuration can be obtained by considering the following: Since the current into the op-amp inverting input is zero, the circuit composed of R_1 and R_2 acts in effect as a voltage divider feeding a fraction of the output voltage back to the inverting input terminal of the op amp; that is,

$$v_1 = v_O\left(\frac{R_1}{R_1 + R_2}\right) \qquad\qquad (2.10)$$

Then the infinite op-amp gain and the resulting virtual short circuit between the two input terminals of the op-amp forces this voltage to be equal to that applied at the positive input terminal; thus

$$v_O\left(\frac{R_1}{R_1 + R_2}\right) = v_I$$

which yields the gain expression given in Eq. (2.9).

This is an appropriate point to reflect further on the action of the negative feedback present in the noninverting circuit of Fig. 2.12. Let v_I increase. Such a change in v_I will cause v_{Id} to increase, and v_O will correspondingly increase as a result of the high (ideally infinite) gain of the op amp. However, a fraction of the increase in v_O will be fed back to the inverting input terminal of the op amp through the (R_1, R_2) voltage divider. The result of this feedback will be to counteract the increase in v_{Id}, driving v_{Id} back to zero, albeit at a higher value of v_O that corresponds to the increased value of v_I. This *degenerative* action of negative feedback gives it the alternative name **degenerative feedback.** Finally, note that the argument above applies equally well if v_I decreases. A formal and detailed study of feedback is presented in Chapter 8.

2.3.2 Characteristics of the Noninverting Configuration

The gain of the noninverting configuration is positive—hence the name *noninverting*. The input impedance of this closed-loop amplifier is ideally infinite, since no current flows into the positive input terminal of the op amp. The output of the noninverting amplifier is taken at the terminals of the ideal voltage source $A(v_2 - v_1)$ (see the op-amp equivalent circuit in Fig. 2.3), thus the output resistance of the noninverting configuration is zero.

2.3.3 Effect of Finite Open-Loop Gain

As we have done for the inverting configuration, we now consider the effect of the finite op-amp open-loop gain A on the gain of the noninverting configuration. Assuming the op amp to be ideal except for having a finite open-loop gain A, it can be shown that the closed-loop gain of the noninverting amplifier circuit of Fig. 2.12 is given by

$$G \equiv \frac{v_O}{v_I} = \frac{1 + (R_2/R_1)}{1 + \dfrac{1 + (R_2/R_1)}{A}} \tag{2.11}$$

Observe that the denominator is identical to that for the case of the inverting configuration (Eq. 2.5). This is no coincidence; it is a result of the fact that both the inverting and the non-inverting configurations have the same feedback loop, which can be readily seen if the input signal source is eliminated (i.e., short-circuited). The numerators, however, are different, for the numerator gives the ideal or nominal closed-loop gain ($-R_2/R_1$ for the inverting configuration, and $1 + R_2/R_1$ for the noninverting configuration). Finally, we note (with reassurance) that the gain expression in Eq. (2.11) reduces to the ideal value for $A = \infty$. In fact, it approximates the ideal value for

$$A \gg 1 + \frac{R_2}{R_1} \tag{2.12}$$

This is the same condition as in the inverting configuration, except that here the quantity on the right-hand side is the nominal closed-loop gain.

2.3.4 The Voltage Follower

The property of high input impedance is a very desirable feature of the noninverting configuration. It enables using this circuit as a buffer amplifier to connect a source with a high impedance to a low-impedance load. We have discussed the need for buffer amplifiers in Section 1.5. In many applications the buffer amplifier is not required to provide any voltage gain; rather, it is used mainly as an impedance transformer or a power amplifier. In such cases we may make $R_2 = 0$ and $R_1 = \infty$ to obtain the **unity-gain amplifier** shown in Fig. 2.14(a). This circuit is commonly referred to as a **voltage follower,** since the output "follows" the input. In the ideal case, $v_O = v_I$, $R_{\text{in}} = \infty$, $R_{\text{out}} = 0$, and the follower has the equivalent circuit shown in Fig. 2.14(b).

Since in the voltage-follower circuit the entire output is fed back to the inverting input, the circuit is said to have 100% negative feedback. The infinite gain of the op amp then acts to make $v_{Id} = 0$ and hence $v_O = v_I$. Observe that the circuit is elegant in its simplicity!

Since the noninverting configuration has a gain greater than or equal to unity, depending on the choice of R_2/R_1, some prefer to call it "a follower with gain."

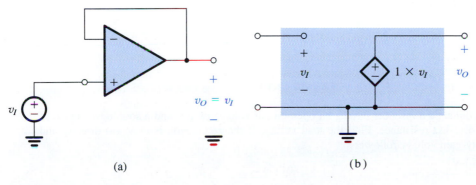

(a)   (b)

FIGURE 2.14 (a) The unity-gain buffer or follower amplifier. (b) Its equivalent circuit model.

EXERCISES

2.9 Use the superposition principle to find the output voltage of the circuit shown in Fig. E2.9.

Ans. $v_O = 6v_1 + 4v_2$

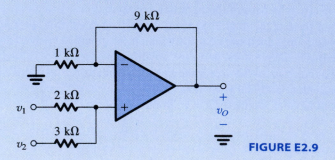

FIGURE E2.9

2.10 If in the circuit of Fig. E2.9 the 1-kΩ resistor is disconnected from ground and connected to a third signal source v_3, use superposition to determine v_O in terms of v_1, v_2, and v_3.

Ans. $v_O = 6v_1 + 4v_2 - 9v_3$

D2.11 Design a noninverting amplifier with a gain of 2. At the maximum output voltage of 10 V the current in the voltage divider is to be 10 μA.

Ans. $R_1 = R_2 = 0.5$ MΩ

2.12 (a) Show that if the op amp in the circuit of Fig. 2.12 has a finite open-loop gain A, then the closed-loop gain is given by Eq. (2.11). (b) For $R_1 = 1$ kΩ and $R_2 = 9$ kΩ find the percentage deviation ε of the closed-loop gain from the ideal value of $(1 + R_2/R_1)$ for the cases $A = 10^3$, 10^4, and 10^5. For $v_I = 1$ V, find in each case the voltage between the two input terminals of the op amp.

Ans. $\varepsilon = -1\%, -0.1\%, -0.01\%$; $v_2 - v_1 = 9.9$ mV, 1 mV, 0.1 mV

2.13 For the circuit in Fig. E2.13 find the values of i_I, v_1, i_1, i_2, v_O, i_L, and i_O. Also find the voltage gain v_O/v_I, the current gain i_L/i_I, and the power gain P_L/P_I.

Ans. 0; 1 V; 1 mA; 1 mA; 10 V; 10 mA; 11 mA; 10 V/V (20 dB); ∞; ∞.

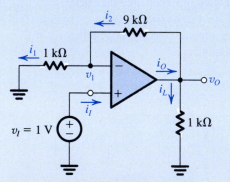

FIGURE E2.13

2.14 It is required to connect a transducer having an open-circuit voltage of 1 V and a source resistance of 1 MΩ to a load of 1-kΩ resistance. Find the load voltage if the connection is done (a) directly and (b) through a unity-gain voltage follower.

Ans. (a) 1 mV; (b) 1 V

2.4 DIFFERENCE AMPLIFIERS

Having studied the two basic configurations of op-amp circuits together with some of their direct applications, we are now ready to consider a somewhat more involved but very important application. Specifically, we shall study the use of op amps to design difference or differential amplifiers.[2] A difference amplifier is one that responds to the difference between the two signals applied at its input and ideally rejects signals that are common to the two inputs. The representation of signals in terms of their differential and common-mode components was given in Fig. 2.4. It is repeated here in Fig. 2.15 with slightly different symbols to serve as the input signals for the difference amplifiers we are about to design. Although ideally the difference amplifier will amplify only the differential input signal v_{Id} and reject completely the common-mode input signal v_{Icm}, practical circuits will have an output voltage v_O given by

$$v_O = A_d v_{Id} + A_{cm} v_{Icm} \tag{2.13}$$

where A_d denotes the amplifier differential gain and A_{cm} denotes its common-mode gain (ideally zero). The efficacy of a differential amplifier is measured by the degree of its rejection of common-mode signals in preference to differential signals. This is usually quantified by a measure known as the **common-mode rejection ratio (CMRR)**, defined as

$$\text{CMRR} = 20 \log \frac{|A_d|}{|A_{cm}|} \tag{2.14}$$

The need for difference amplifiers arises frequently in the design of electronic systems, especially those employed in instrumentation. As a common example, consider a transducer providing a small (e.g., 1 mV) signal between its two output terminals while each of the two wires leading from the transducer terminals to the measuring instrument may have a large interference signal (e.g., 1 V) relative to the circuit ground. The instrument front end obviously needs a difference amplifier.

Before we proceed any further we should address a question that the reader might have: The op amp is itself a difference amplifier; why not just use an op amp? The answer is that the very high (ideally infinite) gain of the op amp makes it impossible to use by itself.

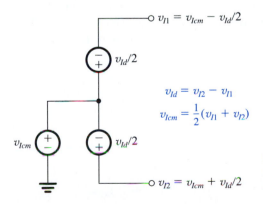

$$v_{I1} = v_{Icm} - v_{Id}/2$$
$$v_{Id}/2$$
$$v_{Id} = v_{I2} - v_{I1}$$
$$v_{Icm} = \frac{1}{2}(v_{I1} + v_{I2})$$
$$v_{Icm}$$
$$v_{Id}/2$$
$$v_{I2} = v_{Icm} + v_{Id}/2$$

FIGURE 2.15 Representing the input signals to a differential amplifier in terms of their differential and common-mode components.

[2] The terms *difference* and *differential* are usually used to describe somewhat different amplifier types. For our purposes at this point the distinction is not sufficiently significant. We will be more precise near the end of this section.

Rather, as we did before, we have to devise an appropriate feedback network to connect to the op amp to create a circuit whose closed-loop gain is finite, predictable, and stable.

2.4.1 A Single Op-Amp Difference Amplifier

Our first attempt at designing a difference amplifier is motivated by the observation that the gain of the noninverting amplifier configuration is positive, $(1 + R_2/R_1)$, while that of the inverting configuration is negative, $(-R_2/R_1)$. Combining the two configurations together is then a step in the right direction—namely, getting the difference between two input signals. Of course, we have to make the two gain magnitudes equal in order to reject common-mode signals. This, however, can be easily achieved by attenuating the positive input signal to reduce the gain of the positive path from $(1 + R_2/R_1)$ to (R_2/R_1). The resulting circuit would then look like that shown in Fig. 2.16, where the attenuation in the positive input path is achieved by the voltage divider (R_3, R_4). The proper ratio of this voltage divider can be determined from

$$\frac{R_4}{R_4 + R_3}\left(1 + \frac{R_2}{R_1}\right) = \frac{R_2}{R_1}$$

which can be put in the form

$$\frac{R_4}{R_4 + R_3} = \frac{R_2}{R_2 + R_1}$$

This condition is satisfied by selecting

$$\frac{R_4}{R_3} = \frac{R_2}{R_1} \tag{2.15}$$

This completes our work. However, we have perhaps proceeded a little too fast! Let's step back and verify that the circuit in Fig. 2.16 with R_3 and R_4 selected according to Eq. (2.15) does in fact function as a difference amplifier. Specifically, we wish to determine the output voltage v_O in terms of v_{I1} and a v_{I2}. Toward that end, we observe that the circuit is linear, and thus we can use superposition.

To apply superposition, we first reduce v_{I2} to zero—that is, ground the terminal to which v_{I2} is applied—and then find the corresponding output voltage, which will be due entirely to v_{I1}. We denote this output voltage v_{O1}. Its value may be found from the circuit in Fig. 2.17(a), which we recognize as that of the inverting configuration. The existence of R_3 and R_4 does not affect the gain expression, since no current flows through either of them. Thus,

$$v_{O1} = -\frac{R_2}{R_1}v_{I1}$$

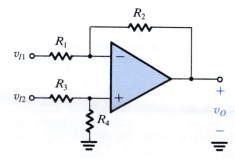

FIGURE 2.16 A difference amplifier.

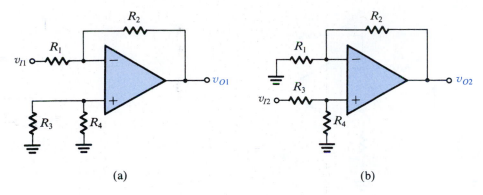

(a)	(b)

FIGURE 2.17 Application of superposition to the analysis of the circuit of Fig. 2.16.

Next, we reduce v_{I1} to zero and evaluate the corresponding output voltage v_{O2}. The circuit will now take the form shown in Fig. 2.17(b), which we recognize as the noninverting configuration with an additional voltage divider, made up of R_3 and R_4, connected to the input v_{I2}. The output voltage v_{O2} is therefore given by

$$v_{O2} = v_{I2}\frac{R_4}{R_3 + R_4}\left(1 + \frac{R_2}{R_1}\right) = \frac{R_2}{R_1}v_{I2}$$

where we have utilized Eq. (2.15).

The superposition principle tells us that the output voltage v_O is equal to the sum of v_{O1} and v_{O2}. Thus we have

$$v_O = \frac{R_2}{R_1}(v_{I2} - v_{I1}) = \frac{R_2}{R_1}v_{Id} \tag{2.16}$$

Thus, as expected, the circuit acts as a difference amplifier with a differential gain A_d of

$$A_d = \frac{R_2}{R_1} \tag{2.17}$$

Of course this is predicated on the op amp being ideal and furthermore on the selection of R_3 and R_4 so that their ratio matches that of R_1 and R_2 (Eq. 2.15). To make this matching requirement a little easier to satisfy, we usually select

$$R_3 = R_1 \quad \text{and} \quad R_4 = R_2$$

Let's next consider the circuit with only a common-mode signal applied at the input, as shown in Fig. 2.18. The figure also shows some of the analysis steps. Thus,

$$i_1 = \frac{1}{R_1}\left[v_{Icm} - \frac{R_4}{R_4 + R_3}v_{Icm}\right]$$

$$= v_{Icm}\frac{R_3}{R_4 + R_3}\frac{1}{R_1} \tag{2.18}$$

The output voltage can now be found from

$$v_O = \frac{R_4}{R_4 + R_3}v_{Icm} - i_2R_2$$

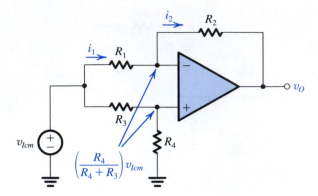

FIGURE 2.18 Analysis of the difference amplifier to determine its common-mode gain $A_{cm} \equiv v_O / v_{Icm}$.

Substituting $i_2 = i_1$ and for i_1 from Eq. (2.18),

$$v_O = \frac{R_4}{R_4 + R_3} v_{Icm} - \frac{R_2}{R_1} \frac{R_3}{R_4 + R_3} v_{Icm}$$

$$= \frac{R_4}{R_4 + R_3}\left(1 - \frac{R_2}{R_1} \frac{R_3}{R_4}\right) v_{Icm}$$

Thus,

$$A_{cm} \equiv \frac{v_O}{v_{Icm}} = \left(\frac{R_4}{R_4 + R_3}\right)\left(1 - \frac{R_2}{R_1} \frac{R_3}{R_4}\right) \tag{2.19}$$

For the design with the resistor ratios selected according to Eq. (2.15), we obtain

$$A_{cm} = 0$$

as expected. Note, however, that any mismatch in the resistance ratios can make A_{cm} nonzero, and hence CMRR finite.

In addition to rejecting common-mode signals, a difference amplifier is usually required to have a high input resistance. To find the input resistance between the two input terminals (i.e., the resistance seen by v_{Id}), called the **differential input resistance** R_{id}, consider Fig. 2.19. Here we have assumed that the resistors are selected so that

$$R_3 = R_1 \quad \text{and} \quad R_4 = R_2$$

Now

$$R_{id} \equiv \frac{v_{Id}}{i_I}$$

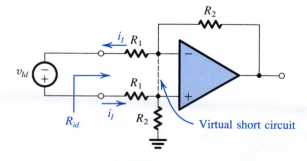

Virtual short circuit

FIGURE 2.19 Finding the input resistance of the difference amplifier for the case $R_3 = R_1$ and $R_4 = R_2$.

Since the two input terminals of the op amp track each other in potential, we may write a loop equation and obtain

$$v_{Id} = R_1 i_I + 0 + R_1 i_I$$

Thus,

$$R_{id} = 2R_1 \qquad\qquad (2.20)$$

Note that if the amplifier is required to have a large differential gain (R_2/R_1), then R_1 of necessity will be relatively small and the input resistance will be correspondingly low, a drawback of this circuit. Another drawback of the circuit is that it is not easy to vary the differential gain of the amplifier. Both of these drawbacks are overcome in the instrumentation amplifier discussed next.

EXERCISES

2.15 Consider the difference-amplifier circuit of Fig. 2.16 for the case $R_1 = R_3 = 2$ kΩ and $R_2 = R_4 = 200$ kΩ. (a) Find the value of the differential gain A_d. (b) Find the value of the differential input resistance R_{id} and the output resistance R_o. (c) If the resistors have 1% tolerance (i.e., each can be within ±1% of its nominal value), find the worst-case common-mode gain A_{cm} and the corresponding value of CMRR.
Ans. (a) 100 V/V (40 dB); (b) 4 kΩ, 0 Ω; (c) 0.04 V/V, 68 dB

D2.16 Find values for the resistances in the circuit of Fig. 2.16 so that the circuit behaves as a difference amplifier with an input resistance of 20 kΩ and a gain of 10.
Ans. $R_1 = R_3 = 10$ kΩ; $R_2 = R_4 = 100$ kΩ

2.4.2 A Superior Circuit—The Instrumentation Amplifier

The low-input-resistance problem of the difference amplifier of Fig. 2.16 can be solved by buffering the two input terminals using voltage followers; that is, a voltage follower of the type in Fig. 2.14 is connected between each input terminal and the corresponding input terminal of the difference amplifier. However, if we are going to use two additional op amps, we should ask the question: Can we get more from them than just impedance buffering? An obvious answer would be that we should try to get some voltage gain. It is especially interesting that we can achieve this without compromising the high input resistance simply by using followers-with-gain rather than unity-gain followers. Achieving some or indeed the bulk of the required gain in this new first stage of the differential amplifier eases the burden on the difference amplifier in the second stage, leaving it to its main task of implementing the differencing function and thus rejecting common-mode signals.

The resulting circuit is shown in Fig. 2.20(a). It consists of two stages. The first stage is formed by op amps A_1 and A_2 and their associated resistors, and the second stage is the by-now-familiar difference amplifier formed by op amp A_3 and its four associated resistors. Observe that as we set out to do, each of A_1 and A_2 is connected in the noninverting configuration and thus realizes a gain of $(1 + R_2/R_1)$. It follows that each of v_{I1} and v_{I2} is amplified by this factor, and the resulting amplified signals appear at the outputs of A_1 and A_2, respectively.

The difference amplifier in the second stage operates on the difference signal $(1 + R_2/R_1)(v_{I2} - v_{I1}) = (1 + R_2/R_1)v_{Id}$ and provides at its output

$$v_O = \frac{R_4}{R_3}\left(1 + \frac{R_2}{R_1}\right)v_{Id}$$

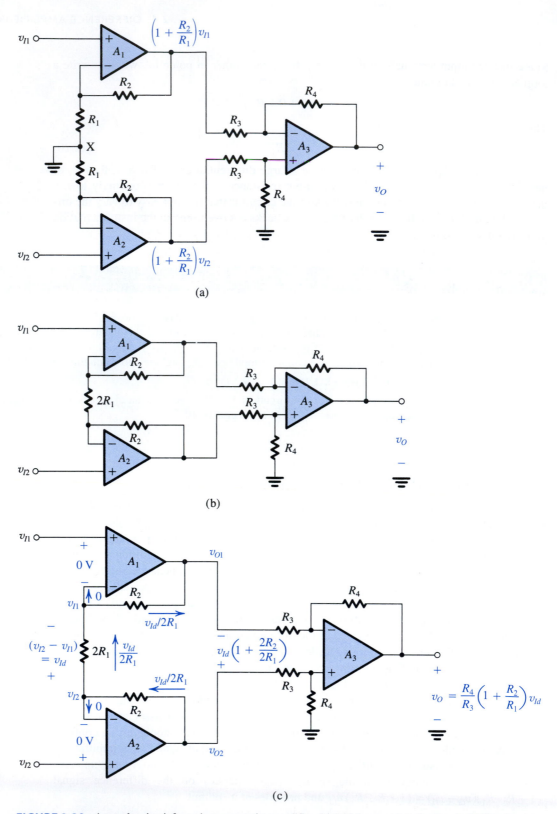

FIGURE 2.20 A popular circuit for an instrumentation amplifier: (a) Initial approach to the circuit; (b) The circuit in (a) with the connection between node X and ground removed and the two resistors R_1 and R_1 lumped together. This simple wiring change dramatically improves performance; (c) Analysis of the circuit in' (b) assuming ideal op amps.

Thus the differential gain realized is

$$A_d = \left(\frac{R_4}{R_3}\right)\left(1 + \frac{R_2}{R_1}\right) \tag{2.21}$$

The common-mode gain will be zero because of the differencing action of the second-stage amplifier.

The circuit in Fig. 2.20(a) has the advantage of very high (ideally infinite) input resistance and high differential gain. Also, provided that A_1 and A_2 and their corresponding resistors are matched, the two signal paths are symmetric—a definite advantage in the design of a differential amplifier. The circuit, however, has three major disadvantages:

1. The input common-mode signal v_{Icm} is amplified in the first stage by a gain equal to that experienced by the differential signal v_{Id}. This is a very serious issue, for it could result in the signals at the outputs of A_1 and A_3 being of such large magnitudes that the op amps saturate (more on op-amp saturation in Section 2.6). But even if the op amps do not saturate, the difference amplifier of the second stage will now have to deal with much larger common-mode signals, with the result that the CMRR of the overall amplifier will inevitably be reduced.

2. The two amplifier channels in the first stage have to be perfectly matched, otherwise a spurious signal may appear between their two outputs. Such a signal would get amplified by the difference amplifier in the second stage.

3. To vary the differential gain A_d, two resistors have to be varied simultaneously, say the two resistors labeled R_1. At each gain setting the two resistors have to be perfectly matched, a difficult task.

All three problems can be solved with a very simple wiring change: Simply disconnect the node between the two resistors labeled R_1, node X, from ground. The circuit with this small but functionally profound change is redrawn in Fig. 2.20(b), where we have lumped the two resistors (R_1 and R_1) together into a single resistor ($2R_1$).

Analysis of the circuit in Fig. 2.20(b), assuming ideal op amps, is straightforward, as is illustrated in Fig. 2.20(c). The key point is that the virtual short circuits at the inputs of op amps A_1 and A_2 cause the input voltages v_{I1} and v_{I2} to appear at the two terminals of resistor ($2R_1$). Thus the differential input voltage $v_{I2} - v_{I1} \equiv v_{Id}$ appears across $2R_1$ and causes a current $i = v_{Id}/2R_1$ to flow through $2R_1$ and the two resistors labeled R_2. This current in turn produces a voltage difference between the output terminals of A_1 and A_2 given by

$$v_{O2} - v_{O1} = \left(1 + \frac{2R_2}{2R_1}\right)v_{Id}$$

The difference amplifier formed by op amp A_3 and its associated resistors senses the voltage difference ($v_{O2} - v_{O1}$) and provides a proportional output voltage v_O:

$$v_O = \frac{R_4}{R_3}(v_{O2} - v_{O1})$$

$$= \frac{R_4}{R_3}\left(1 + \frac{R_2}{R_1}\right)v_{Id}$$

Thus the overall differential voltage gain is given by

$$A_d \equiv \frac{v_O}{v_{Id}} = \frac{R_4}{R_3}\left(1 + \frac{R_2}{R_1}\right) \tag{2.22}$$

Observe that proper differential operation does *not* depend on the matching of the two resistors labeled R_2. Indeed, if one of the two is of different value, say R_2', the expression for A_d becomes

$$A_d = \frac{R_4}{R_3}\left(1 + \frac{R_2 + R_2'}{2R_1}\right) \tag{2.23}$$

Consider next what happens when the two input terminals are connected together to a common-mode input voltage v_{Icm}. It is easy to see that an equal voltage appears at the negative input terminals of A_1 and A_2, causing the current through $2R_1$ to be zero. Thus there will be no current flowing in the R_2 resistors, and the voltages at the output terminals of A_1 and A_2 will be equal to the input (i.e., v_{Icm}). Thus the first stage no longer amplifies v_{Icm}; it simply propagates v_{Icm} to its two output terminals, where they are subtracted to produce a zero common-mode output by A_3. The difference amplifier in the second stage, however, now has a much improved situation at its input: The difference signal has been amplified by $(1 + R_2/R_1)$ while the common-mode voltage remained unchanged.

Finally, we observe from the expression in Eq. (2.22) that the gain can be varied by changing only one resistor, $2R_1$. We conclude that this is an excellent differential amplifier circuit and is widely employed as an instrumentation amplifier; that is, as the input amplifier used in a variety of electronic instruments.

EXAMPLE 2.3

Design the instrumentation amplifier circuit in Fig. 2.20(b) to provide a gain that can be varied over the range of 2 to 1000 utilizing a 100-kΩ variable resistance (a potentiometer, or "pot" for short).

Solution

It is usually preferable to obtain all the required gain in the first stage, leaving the second stage to perform the task of taking the difference between the outputs of the first stage and thereby rejecting the common-mode signal. In other words, the second stage is usually designed for a gain of 1. Adopting this approach, we select all the second-stage resistors to be equal to a practically convenient value, say 10 kΩ. The problem then reduces to designing the first stage to realize a gain adjustable over the range of 2 to 1000. Implementing $2R_1$ as the series combination of a fixed resistor R_{1f} and the variable resistor R_{1v} obtained using the 100-kΩ pot (Fig. 2.21), we can write

$$1 + \frac{2R_2}{R_{1f} + R_{1v}} = 2 \text{ to } 1000$$

Thus,

$$1 + \frac{2R_2}{R_{1f}} = 1000$$

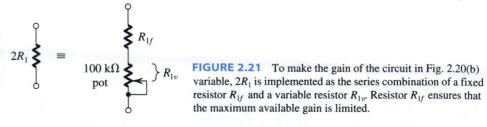

FIGURE 2.21 To make the gain of the circuit in Fig. 2.20(b) variable, $2R_1$ is implemented as the series combination of a fixed resistor R_{1f} and a variable resistor R_{1v}. Resistor R_{1f} ensures that the maximum available gain is limited.

and

$$1 + \frac{2R_2}{R_{1f} + 100 \text{ k}\Omega} = 2$$

These two equations yield $R_{1f} = 100.2 \ \Omega$ and $R_2 = 50.050$ kΩ. Other practical values may be selected; for instance, $R_{1f} = 100 \ \Omega$ and $R_2 = 49.9$ kΩ (both values are available as standard 1%-tolerance metal-film resistors; see Appendix G) results in a gain covering approximately the required range.

EXERCISE

2.17 Consider the instrumentation amplifier of Fig. 2.20(b) with a common-mode input voltage of +5 V (dc) and a differential input signal of 10-mV-peak sine wave. Let $(2R_1) = 1$ kΩ, $R_2 = 0.5$ MΩ, and $R_3 = R_4 = 10$ kΩ. Find the voltage at every node in the circuit.

Ans. $v_{I1} = 5 - 0.005 \sin \omega t$; $v_{I2} = 5 + 0.005 \sin \omega t$; v_- (op amp A_1) $= 5 - 0.005 \sin \omega t$; v_- (op amp A_2) = $5 + 0.005 \sin \omega t$; $v_{O1} = 5 - 5.005 \sin \omega t$; $v_{O2} = 5 + 5.005 \sin \omega t$; $v_-(A_3) = v_+(A_3) = 2.5 + 2.0025 \sin \omega t$; $v_O = 10.01 \sin \omega t$ (all in volts)

2.5 EFFECT OF FINITE OPEN-LOOP GAIN AND BANDWIDTH ON CIRCUIT PERFORMANCE

Above we defined the ideal op amp, and we presented a number of circuit applications of op amps. The analysis of these circuits assumed the op amps to be ideal. Although in many applications such an assumption is not a bad one, a circuit designer has to be thoroughly familiar with the characteristics of practical op amps and the effects of such characteristics on the performance of op-amp circuits. Only then will the designer be able to use the op amp intelligently, especially if the application at hand is not a straightforward one. The nonideal properties of op amps will, of course, limit the range of operation of the circuits analyzed in the previous examples.

In this and the two sections that follow, we consider some of the important nonideal properties of the op amp.[3] We do this by treating one parameter at a time, beginning in this section with the most serious op-amp nonidealities, its finite gain and limited bandwidth.

2.5.1 Frequency Dependence of the Open-Loop Gain

The differential open-loop gain of an op amp is not infinite; rather, it is finite and decreases with frequency. Figure 2.22 shows a plot for |A|, with the numbers typical of most commercially available general-purpose op amps (such as the 741-type op amp, which is available from many semiconductor manufacturers and whose internal circuit is studied in Chapter 9).

---

[3] We should note that real op amps have nonideal effects additional to those discussed in this chapter. These include finite (nonzero) common-mode gain or, equivalently, noninfinite CMRR, noninfinite input resistance, and nonzero output resistance. The effect of these, however, on the performance of most of the closed-loop circuits studied here is not very significant, and their study will be postponed to later chapters (in particular Chapters 8 and 9). Nevertheless, some of these nonideal characteristics will be modeled in Section 2.9 in the context of circuit simulation using SPICE.

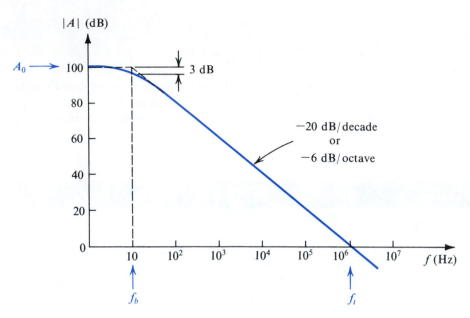

FIGURE 2.22 Open-loop gain of a typical general-purpose internally compensated op amp.

Note that although the gain is quite high at dc and low frequencies, it starts to fall off at a rather low frequency (10 Hz in our example). The uniform –20-dB/decade gain rolloff shown is typical of **internally compensated** op amps. These are units that have a network (usually a single capacitor) included within the same IC chip whose function is to cause the op-amp gain to have the single-time-constant (STC) low-pass response shown. This process of modifying the open-loop gain is termed **frequency compensation,** and its purpose is to ensure that op-amp circuits will be stable (as opposed to oscillatory). The subject of stability of op-amp circuits—or, more generally, of feedback amplifiers—will be studied in Chapter 8.

By analogy to the response of low-pass STC circuits (see Section 1.6 and, for more detail, Appendix D), the gain $A(s)$ of an internally compensated op amp may be expressed as

$$A(s) = \frac{A_0}{1 + s/\omega_b} \tag{2.24}$$

which for physical frequencies, $s = j\omega$, becomes

$$A(j\omega) = \frac{A_0}{1 + j\omega/\omega_b} \tag{2.25}$$

where A_0 denotes the dc gain and ω_b is the 3-dB frequency (corner frequency or "break" frequency). For the example shown in Fig. 2.22, $A_0 = 10^5$ and $\omega_b = 2\pi \times 10$ rad/s. For frequencies $\omega \gg \omega_b$ (about 10 times and higher) Eq. (2.25) may be approximated by

$$A(j\omega) \simeq \frac{A_0\omega_b}{j\omega} \tag{2.26}$$

Thus,

$$|A(j\omega)| = \frac{A_0\omega_b}{\omega} \tag{2.27}$$

from which it can be seen that the gain $|A|$ reaches unity (0 dB) at a frequency denoted by ω_t and given by

$$\omega_t = A_0 \omega_b \qquad (2.28)$$

Substituting in Eq. (2.26) gives

$$A(j\omega) \simeq \frac{\omega_t}{j\omega} \qquad (2.29)$$

The frequency $f_t = \omega_t/2\pi$ is usually specified on the data sheets of commercially available op amps and is known as the **unity-gain bandwidth.**[4] Also note that for $\omega \gg \omega_b$ the open-loop gain in Eq. (2.24) becomes

$$A(s) \simeq \frac{\omega_t}{s} \qquad (2.30)$$

The gain magnitude can be obtained from Eq. (2.29) as

$$|A(j\omega)| \simeq \frac{\omega_t}{\omega} = \frac{f_t}{f} \qquad (2.31)$$

Thus if f_t is known (10^6 Hz in our example), one can easily determine the magnitude of the op-amp gain at a given frequency f. Furthermore, observe that this relationship correlates with the Bode plot in Fig. 2.22. Specifically, for $f \gg f_b$, doubling f (an octave increase) results in halving the gain (a 6-dB reduction). Similarly, increasing f by a factor of 10 (a decade increase) results in reducing $|A|$ by a factor of 10 (20 dB).

As a matter of practical importance, we note that the production spread in the value of f_t between op-amp units of the same type is usually much smaller than that observed for A_0 and f_b. For this reason f_t is preferred as a specification parameter. Finally, it should be mentioned that an op amp having this uniform –6-dB/octave (or equivalently –20-dB/decade) gain rolloff is said to have a **single-pole model.** Also, since this single pole *dominates* the amplifier frequency response, it is called a **dominant pole.** For more on poles (and zeros), the reader may wish to consult Appendix E.

EXERCISE

2.18 An internally compensated op amp is specified to have an open-loop dc gain of 106 dB and a unity-gain bandwidth of 3 MHz. Find f_b and the open-loop gain (in dB) at f_b, 300 Hz, 3 kHz, 12 kHz, and 60 kHz.

Ans. 15 Hz; 103 dB; 80 dB; 60 dB; 48 dB; 34 dB

2.5.2 Frequency Response of Closed-Loop Amplifiers

We next consider the effect of limited op-amp gain and bandwidth on the closed-loop transfer functions of the two basic configurations: the inverting circuit of Fig. 2.5 and the noninverting circuit of Fig. 2.12. The closed-loop gain of the inverting amplifier, assuming a finite op-amp open-loop gain A, was derived in Section 2.2 and given in Eq. (2.5), which we repeat here as

$$\frac{V_o}{V_i} = \frac{-R_2/R_1}{1 + (1 + R_2/R_1)/A} \qquad (2.32)$$

---

[4] Since f_t is the product of the dc gain A_0 and the 3-dB bandwidth f_b (where $f_b = \omega_b/2\pi$), it is also known as the **gain–bandwidth product** (GB). The reader is cautioned, however, that in some amplifiers, the unity-gain frequency and the gain-bandwidth product are *not* equal.

Substituting for A from Eq. (2.24) gives

$$\frac{V_o(s)}{V_i(s)} = \frac{-R_2/R_1}{1 + \frac{1}{A_0}\left(1 + \frac{R_2}{R_1}\right) + \frac{s}{\omega_t/(1 + R_2/R_1)}} \tag{2.33}$$

For $A_0 \gg 1 + R_2/R_1$, which is usually the case,

$$\frac{V_o(s)}{V_i(s)} \simeq \frac{-R_2/R_1}{1 + \frac{s}{\omega_t/(1 + R_2/R_1)}} \tag{2.34}$$

which is of the same form as that for a low-pass STC network (see Table 1.2, page 34). Thus the inverting amplifier has an STC low-pass response with a dc gain of magnitude equal to R_2/R_1. The closed-loop gain rolls off at a uniform -20-dB/decade slope with a corner frequency (3-dB frequency) given by

$$\omega_{3\text{dB}} = \frac{\omega_t}{1 + R_2/R_1} \tag{2.35}$$

Similarly, analysis of the noninverting amplifier of Fig. 2.12, assuming a finite open-loop gain A, yields the closed-loop transfer function

$$\frac{V_o}{V_i} = \frac{1 + R_2/R_1}{1 + (1 + R_2/R_1)/A} \tag{2.36}$$

Substituting for A from Eq. (2.24) and making the approximation $A_0 \gg 1 + R_2/R_1$ results in

$$\frac{V_o(s)}{V_i(s)} \simeq \frac{1 + R_2/R_1}{1 + \frac{s}{\omega_t/(1 + R_2/R_1)}} \tag{2.37}$$

Thus the noninverting amplifier has an STC low-pass response with a dc gain of $(1 + R_2/R_1)$ and a 3-dB frequency given also by Eq. (2.35).

EXAMPLE 2.4

Consider an op amp with $f_t = 1$ MHz. Find the 3-dB frequency of closed-loop amplifiers with nominal gains of $+1000$, $+100$, $+10$, $+1$, -1, -10, -100, and -1000. Sketch the magnitude frequency response for the amplifiers with closed-loop gains of $+10$ and -10.

Solution

Using Eq. (2.35), we obtain the results given in the following table:

Closed-Loop Gain	R_2/R_1	$f_{3\text{dB}} = f_t/(1 + R_2/R_1)$
+1000	999	1 kHz
+100	99	10 kHz
+10	9	100 kHz
+1	0	1 MHz
−1	1	0.5 MHz
−10	10	90.9 kHz
−100	100	9.9 kHz
−1000	1000	≈1 kHz

Figure 2.23 shows the frequency response for the amplifier whose nominal dc gain is +10 (20 dB), and Fig. 2.24 shows the frequency response for the −10 (also 20 dB) case. An interesting observation follows from the table above: The unity-gain inverting amplifier has a 3-dB frequency of $f_t/2$ as compared to f_t for the unity-gain noninverting amplifier (the unity-gain voltage follower).

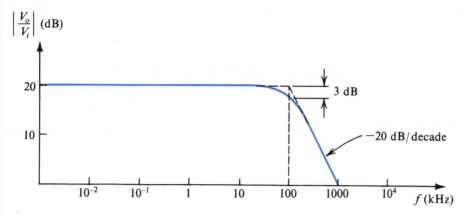

FIGURE 2.23 Frequency response of an amplifier with a nominal gain of +10 V/V.

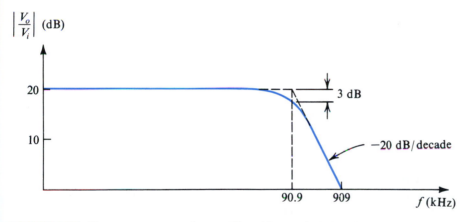

FIGURE 2.24 Frequency response of an amplifier with a nominal gain of −10 V/V.

The table in Example 2.4 above clearly illustrates the trade-off between gain and bandwidth: For a given op amp, the lower the closed-loop gain required, the wider the bandwidth achieved. Indeed, the noninverting configuration exhibits a constant **gain–bandwidth product** equal to f_t of the op amp. An interpretation of these results in terms of feedback theory will be given in Chapter 8.

EXERCISES

    2.19 An internally compensated op amp has a dc open-loop gain of 10^6 V/V and an ac open-loop gain of 40 dB at 10 kHz. Estimate its 3-dB frequency, its unity-gain frequency, its gain–bandwidth product, and its expected gain at 1 kHz.

    Ans. 1 Hz; 1 MHz; 1 MHz; 60 dB

2.20 An op amp having a 106-dB gain at dc and a single-pole frequency response with $f_t = 2$ MHz is used to design a noninverting amplifier with nominal dc gain of 100. Find the 3-dB frequency of the closed-loop gain.

Ans. 20 kHz

2.6 LARGE-SIGNAL OPERATION OF OP AMPS

In this section, we study the limitations on the performance of op-amp circuits when large output signals are present.

2.6.1 Output Voltage Saturation

Similar to all other amplifiers, op amps operate linearly over a limited range of output voltages. Specifically, the op-amp output saturates in the manner shown in Fig. 1.13 with L_+ and L_- within 1 V or so of the positive and negative power supplies, respectively. Thus, an op amp that is operating from ±15-V supplies will saturate when the output voltage reaches about +13 V in the positive direction and –13 V in the negative direction. For this particular op amp the **rated output voltage** is said to be ±13 V. To avoid clipping off the peaks of the output waveform, and the resulting waveform distortion, the input signal must be kept correspondingly small.

2.6.2 Output Current Limits

Another limitation on the operation of op amps is that their output current is limited to a specified maximum. For instance, the popular 741 op amp is specified to have a maximum output current of ±20 mA. Thus, in designing closed-loop circuits utilizing the 741, the designer has to ensure that under no condition will the op amp be required to supply an output current, in either direction, exceeding 20 mA. This, of course, has to include both the current in the feedback circuit as well as the current supplied to a load resistor. If the circuit requires a larger current, the op-amp output voltage will saturate at the level corresponding to the maximum allowed output current.

EXAMPLE 2.5

Consider the noninverting amplifier circuit shown in Fig. 2.25. As shown, the circuit is designed for a nominal gain $(1 + R_2/R_1) = 10$ V/V. It is fed with a low-frequency sine-wave signal of peak voltage V_p and is connected to a load resistor R_L. The op amp is specified to have output saturation voltages of ±13 V and output current limits of ±20 mA.

(a) For $V_p = 1$ V and $R_L = 1$ kΩ, specify the signal resulting at the output of the amplifier.

(b) For $V_p = 1.5$ V and $R_L = 1$ kΩ, specify the signal resulting at the output of the amplifier.

(c) For $R_L = 1$ kΩ, what is the maximum value of V_p for which an undistorted sine-wave output is obtained?

(d) For $V_p = 1$ V, what is the lowest value of R_L for which an undistorted sine-wave output is obtained?

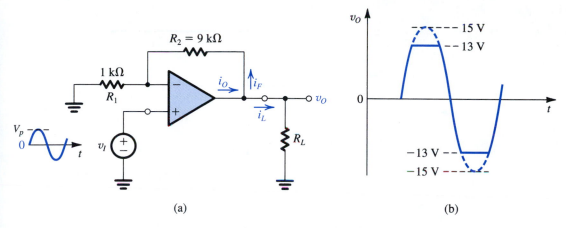

(a) (b)

FIGURE 2.25 **(a)** A noninverting amplifier with a nominal gain of 10 V/V designed using an op amp that saturates at ±13-V output voltage and has ±20-mA output current limits. **(b)** When the input sine wave has a peak of 1.5 V, the output is clipped off at ±13 V.

Solution

(a) For $V_p = 1$ V and $R_L = 1$ kΩ, the output will be a sine wave with peak value of 10 V. This is lower than output saturation levels of ±13 V, and thus the amplifier is not limited that way. Also, when the output is at its peak (10 V), the current in the load will be 10 V/1 kΩ = 10 mA, and the current in the feedback network will be 10 V/(9 + 1) kΩ = 1 mA, for a total op-amp output current of 11 mA, well under its limit of 20 mA.

(b) Now if V_p is increased to 1.5 V, ideally the output would be a sine wave of 15-V peak. The op amp, however, will saturate at ±13 V, thus clipping the sine-wave output at these levels. Let's next check on the op-amp output current: At 13-V output and $R_L = 1$ kΩ, $i_L = 13$ mA and $i_F = 1.3$ mA; thus $i_O = 14.3$ mA, again under the 20-mA limit. Thus the output will be a sine wave with its peaks clipped off at ±13 V, as shown in Fig. 2.25(b).

(c) For $R_L = 1$ kΩ, the maximum value of V_p for undistorted sine-wave output is 1.3 V. The output will be a 13-V peak sine wave, and the op-amp output current at the peaks will be 14.3 mA.

(d) For $V_p = 1$ V and R_L reduced, the lowest value possible for R_L while the output is remaining an undistorted sine wave of 10-V peak can be found from

$$i_{O\max} = 20 \text{ mA} = \frac{10 \text{ V}}{R_{L\min}} + \frac{10 \text{ V}}{9 \text{ k}\Omega + 1 \text{ k}\Omega}$$

which results in

$$R_{L\min} = 526 \text{ }\Omega$$

2.6.3 Slew Rate

Another phenomenon that can cause nonlinear distortion when large output signals are present is that of slew-rate limiting. This refers to the fact that there is a specific *maximum rate of change* possible at the output of a real op amp. This maximum is known as the slew rate (SR) of the op amp and is defined as

$$\text{SR} = \frac{dv_O}{dt}\bigg|_{\max} \tag{2.38}$$

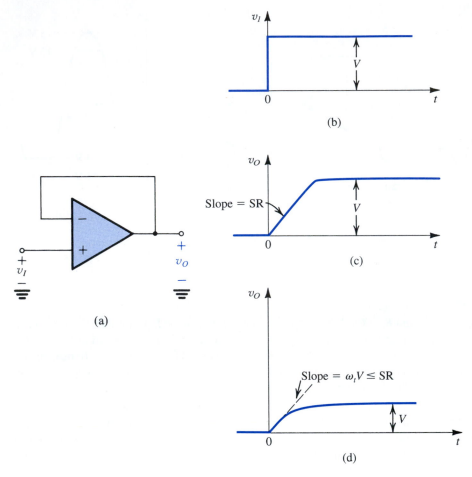

FIGURE 2.26 **(a)** Unity-gain follower. **(b)** Input step waveform. **(c)** Linearly rising output waveform obtained when the amplifier is slew-rate limited. **(d)** Exponentially rising output waveform obtained when V is sufficiently small so that the initial slope ($\omega_t V$) is smaller than or equal to SR.

and is usually specified on the op-amp data sheet in units of V/μs. It follows that if the input signal applied to an op-amp circuit is such that it demands an output response that is faster than the specified value of SR, the op amp will not comply. Rather, its output will change at the maximum possible rate, which is equal to its SR. As an example, consider an op amp connected in the unity-gain voltage-follower configuration shown in Fig. 2.26(a), and let the input signal be the step voltage shown in Fig. 2.26(b). The output of the op amp will not be able to rise instantaneously to the ideal value V; rather, the output will be the linear ramp of slope equal to SR, shown in Fig. 2.26(c). The amplifier is then said to be slewing, and its output is slew-rate limited.

In order to understand the origin of the slew-rate phenomenon, we need to know about the internal circuit of the op amp, and we will do so in Chapter 9. For the time being, however, it is sufficient to know about the phenomenon and to note that it is distinct from the finite op-amp bandwidth that limits the frequency response of the closed-loop amplifiers, studied in the previous section. The limited bandwidth is a linear phenomenon and does not result in a change in the shape of an input sinusoid; that is, it does not lead to nonlinear distortion. The slew-rate limitation, on the other hand, can cause nonlinear distortion to an input sinusoidal signal when its frequency and amplitude are such that the corresponding ideal output would require v_O to change at a rate greater than SR. This is the origin of another related op-amp specification, its full-power bandwidth, to be explained later.

Before leaving the example in Fig. 2.26, however, we should point out that if the step input voltage V is sufficiently small, the output can be the exponentially rising ramp shown in Fig. 2.26(d). Such an output would be expected from the follower if the only limitation on its dynamic performance is the finite op-amp bandwidth. Specifically, the transfer function of the follower can be found by substituting $R_1 = \infty$ and $R_2 = 0$ in Eq. (2.37) to obtain

$$\frac{V_o}{V_i} = \frac{1}{1 + s/\omega_t} \tag{2.39}$$

which is a low-pass STC response with a time constant $1/\omega_t$. Its step response would therefore be (see Appendix D)

$$v_O(t) = V(1 - e^{-\omega_t t}) \tag{2.40}$$

The initial slope of this exponentially rising function is $(\omega_t V)$. Thus, as long as V is sufficiently small so that $\omega_t V \le$ SR, the output will be as in Fig. 2.26(d).

EXERCISE

2.21 An op amp that has a slew rate of 1 V/μs and a unity-gain bandwidth f_t of 1 MHz is connected in the unity-gain follower configuration. Find the largest possible input voltage step for which the output waveform will still be given by the exponential ramp of Eq. (2.40). For this input voltage, what is the 10% to 90% rise time of the output waveform? If an input step 10 times as large is applied, find the 10% to 90% rise time of the output waveform.

Ans. 0.16 V; 0.35 μs; 1.28 μs

2.6.4 Full-Power Bandwidth

Op-amp slew-rate limiting can cause nonlinear distortion in sinusoidal waveforms. Consider once more the unity-gain follower with a sine wave input given by

$$v_I = \hat{V}_i \sin \omega t$$

The rate of change of this waveform is given by

$$\frac{dv_I}{dt} = \omega \hat{V}_i \cos \omega t$$

with a maximum value of $\omega \hat{V}_i$. This maximum occurs at the zero crossings of the input sinusoid. Now if $\omega \hat{V}_i$ exceeds the slew rate of the op amp, the output waveform will be distorted in the manner shown in Fig. 2.27. Observe that the output cannot keep up with the large rate of change of the sinusoid at its zero crossings, and the op amp slews.

The op-amp data sheets usually specify a frequency f_M called the **full-power bandwidth.** It is the frequency at which an output sinusoid with amplitude equal to the rated output voltage of the op amp begins to show distortion due to slew-rate limiting. If we denote the rated output voltage $V_{o\max}$, then f_M is related to SR as follows:

$$\omega_M V_{o\max} = \text{SR}$$

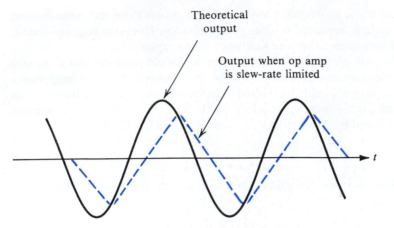

FIGURE 2.27 Effect of slew-rate limiting on output sinusoidal waveforms.

Thus,

$$f_M = \frac{SR}{2\pi V_{o\max}} \tag{2.41}$$

It should be obvious that output sinusoids of amplitudes smaller than $V_{o\max}$ will show slew-rate distortion at frequencies higher than ω_M. In fact, at a frequency ω higher than ω_M, the maximum amplitude of the undistorted output sinusoid is given by

$$V_o = V_{o\max}\left(\frac{\omega_M}{\omega}\right) \tag{2.42}$$

EXERCISE

2.22 An op amp has a rated output voltage of ±10 V and a slew rate of 1 V/μs. What is its full-power bandwidth? If an input sinusoid with frequency $f = 5f_M$ is applied to a unity-gain follower constructed using this op amp, what is the maximum possible amplitude that can be accommodated at the output without incurring SR distortion?

Ans. 15.9 kHz; 2 V (peak)

2.7 DC IMPERFECTIONS

2.7.1 Offset Voltage

Because op amps are direct-coupled devices with large gains at dc, they are prone to dc problems. The first such problem is the dc offset voltage. To understand this problem consider the following *conceptual* experiment: If the two input terminals of the op amp are tied together and connected to ground, it will be found that a finite dc voltage exists at the output. In fact, if the op amp has a high dc gain, the output will be at either the positive or negative saturation level. The op-amp output can be brought back to its ideal value of 0 V by connecting a dc voltage source of appropriate polarity and magnitude between the two input terminals of the op amp. This external source balances out the input offset voltage of the op amp. It follows that the **input offset voltage** (V_{OS}) must be of equal magnitude and of opposite polarity to the voltage we applied externally.

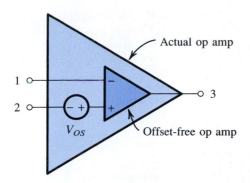

FIGURE 2.28 Circuit model for an op amp with input offset voltage V_{OS}.

The input offset voltage arises as a result of the unavoidable mismatches present in the input differential stage inside the op amp. In later chapters we shall study this topic in detail. Here, however, our concern is to investigate the effect of V_{OS} on the operation of closed-loop op-amp circuits. Toward that end, we note that general-purpose op amps exhibit V_{OS} in the range of 1 mV to 5 mV. Also, the value of V_{OS} depends on temperature. The op-amp data sheets usually specify typical and maximum values for V_{OS} at room temperature as well as the temperature coefficient of V_{OS} (usually in μV/°C). They do not, however, specify the polarity of V_{OS} because the component mismatches that give rise to V_{OS} are obviously not known *a priori;* different units of the same op-amp type may exhibit either a positive or a negative V_{OS}.

To analyze the effect of V_{OS} on the operation of op-amp circuits, we need a circuit model for the op amp with input offset voltage. Such a model is shown in Fig. 2.28. It consists of a dc source of value V_{OS} placed in series with the positive input lead of an offset-free op amp. The justification for this model follows from the description above.

EXERCISE

2.23 Use the model of Fig. 2.28 to sketch the transfer characteristic v_O versus v_{Id} ($v_O \equiv v_3$ and $v_{Id} \equiv v_2 - v_1$) of an op amp having $A_0 = 10^4$, output saturation levels of ± 10 V, and V_{OS} of +5 mV.

Ans. See Fig. E2.23.

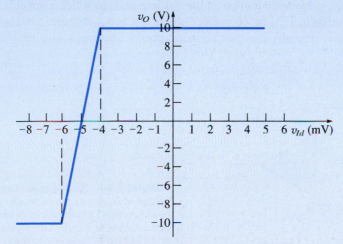

FIGURE E2.23 Transfer characteristic of an op amp with $V_{OS} = 5$ mV.

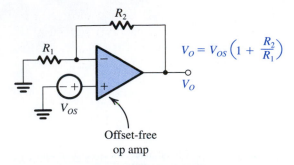

$$V_O = V_{OS}\left(1 + \frac{R_2}{R_1}\right)$$

FIGURE 2.29 Evaluating the output dc offset voltage due to V_{OS} in a closed-loop amplifier.

Analysis of op-amp circuits to determine the effect of the op-amp V_{OS} on their performance is straightforward: The input voltage signal source is short circuited and the op amp is replaced with the model of Fig. 2.28. (Eliminating the input signal, done to simplify matters, is based on the principle of superposition.) Following this procedure we find that both the inverting and the noninverting amplifier configurations result in the same circuit, that shown in Fig. 2.29, from which the output dc voltage due to V_{OS} is found to be

$$V_O = V_{OS}\left[1 + \frac{R_2}{R_1}\right] \qquad (2.43)$$

This output dc voltage can have a large magnitude. For instance, a noninverting amplifier with a closed-loop gain of 1000, when constructed from an op amp with a 5-mV input offset voltage, will have a dc output voltage of +5 V or −5 V (depending on the polarity of V_{OS}) rather than the ideal value of 0 V. Now, when an input signal is applied to the amplifier, the corresponding signal output will be superimposed on the 5-V dc. Obviously then, the allowable signal swing at the output will be reduced. Even worse, if the signal to be amplified is dc, we would not know whether the output is due to V_{OS} or to the signal!

Some op amps are provided with two additional terminals to which a specified circuit can be connected to trim to zero the output dc voltage due to V_{OS}. Figure 2.30 shows such an arrangement that is typically used with general-purpose op amps. A potentiometer is

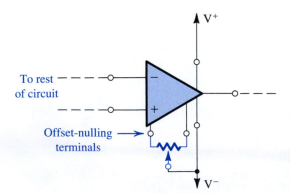

FIGURE 2.30 The output dc offset voltage of an op amp can be trimmed to zero by connecting a potentiometer to the two offset-nulling terminals. The wiper of the potentiometer is connected to the negative supply of the op amp.

connected between the offset-nulling terminals with the wiper of the potentiometer connected to the op-amp negative supply. Moving the potentiometer wiper introduces an imbalance that counteracts the asymmetry present in the internal op-amp circuitry and that gives rise to V_{OS}. We shall return to this point in the context of our study of the internal circuitry of op amps in Chapter 9. It should be noted, however, that even though the dc output offset can be trimmed to zero, the problem remains of the variation (or drift) of V_{OS} with temperature.

EXERCISE

2.24 Consider an inverting amplifier with a nominal gain of 1000 constructed from an op amp with an input offset voltage of 3 mV and with output saturation levels of ±10 V. (a) What is (approximately) the peak sine-wave input signal that can be applied without output clipping? (b) If the effect of V_{OS} is nulled at room temperature (25°C), how large an input can one now apply if: (i) the circuit is to operate at a constant temperature? (ii) the circuit is to operate at a temperature in the range 0°C to 75°C and the temperature coefficient of V_{OS} is 10 μV/°C?

Ans. (a) 7 mV; (b) 10 mV, 9.5 mV

One way to overcome the dc offset problem is by capacitively coupling the amplifier. This, however, will be possible only in applications where the closed-loop amplifier is not required to amplify dc or very low-frequency signals. Figure 2.31(a) shows a capacitively coupled amplifier. Because of its infinite impedance at dc, the coupling capacitor will cause the gain to be zero at dc. As a result the equivalent circuit for determining the dc output voltage resulting from the op-amp input offset voltage V_{OS} will be that shown in Fig. 2.31(b). Thus V_{OS} sees in effect a unity-gain voltage follower, and the dc output voltage V_O will be equal to V_{OS} rather than $V_{OS}(1 + R_2/R_1)$, which is the case without the coupling capacitor. As far as input signals are concerned, the coupling capacitor C forms together with R_1 an STC high-pass circuit with a corner frequency of $\omega_0 = 1/CR_1$. Thus the gain of the capacitively coupled amplifier will fall off at the low-frequency end [from a magnitude of $(1 + R_2/R_1)$ at high frequencies] and will be 3 dB down at ω_0.

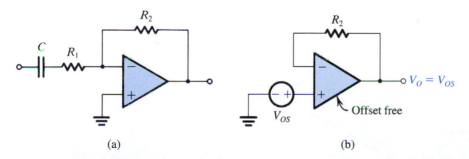

(a) (b)

FIGURE 2.31 **(a)** A capacitively coupled inverting amplifier, and **(b)** the equivalent circuit for determining its dc output offset voltage V_O.

2.25 Consider the same amplifier as in Exercise 2.24—that is, an inverting amplifier with a nominal gain of 1000 constructed from an op amp with an input offset voltage of 3 mV and with output saturation levels of ±10 V—except here let the amplifier be capacitively coupled as in Fig. 2.31(a). (a) What is the dc offset voltage at the output, and what (approximately) is the peak sine-wave signal that can be applied at the input without output clipping? Is there a need for offset trimming? (b) If $R_1 = 1$ kΩ and $R_2 = 1$ MΩ, find the value of the coupling capacitor C_1 that will ensure that the gain will be greater than 57 dB down to 100 Hz.

Ans. (a) 3 mV, 10 mV, no need for offset trimming; (b) 1.6 μF

2.7.2 Input Bias and Offset Currents

The second dc problem encountered in op amps is illustrated in Fig. 2.32. In order for the op amp to operate, its two input terminals have to be supplied with dc currents, termed the **input bias currents.** In Fig. 2.32 these two currents are represented by two current sources, I_{B1} and I_{B2}, connected to the two input terminals. It should be emphasized that the input bias currents are independent of the fact that a real op amp has finite though large input resistance (not shown in Fig. 2.32). The op-amp manufacturer usually specifies the average value of I_{B1} and I_{B2} as well as their expected difference. The average value I_B is called the **input bias current,**

$$I_B = \frac{I_{B1} + I_{B2}}{2}$$

and the difference is called the **input offset current** and is given by

$$I_{OS} = |I_{B1} - I_{B2}|$$

Typical values for general-purpose op amps that use bipolar transistors are $I_B = 100$ nA and $I_{OS} = 10$ nA. Op amps that utilize field-effect transistors in the input stage have a much smaller input bias current (of the order of picoamperes).

We now wish to find the dc output voltage of the closed-loop amplifier due to the input bias currents. To do this we ground the signal source and obtain the circuit shown in

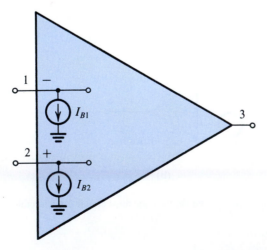

FIGURE 2.32 The op-amp input bias currents represented by two current sources I_{B1} and I_{B2}.

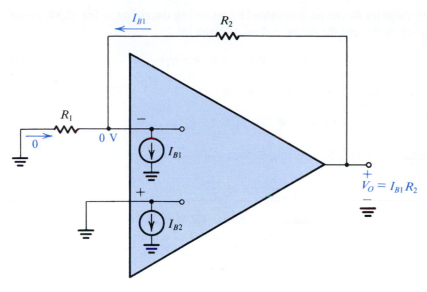

FIGURE 2.33 Analysis of the closed-loop amplifier, taking into account the input bias currents.

Fig. 2.33 for both the inverting and noninverting configurations. As shown in Fig. 2.33, the output dc voltage is given by

$$V_O = I_{B1}R_2 \simeq I_B R_2 \qquad (2.44)$$

This obviously places an upper limit on the value of R_2. Fortunately, however, a technique exists for reducing the value of the output dc voltage due to the input bias currents. The method consists of introducing a resistance R_3 in series with the noninverting input lead, as shown in Fig. 2.34. From a signal point of view, R_3 has a negligible effect (ideally no effect).

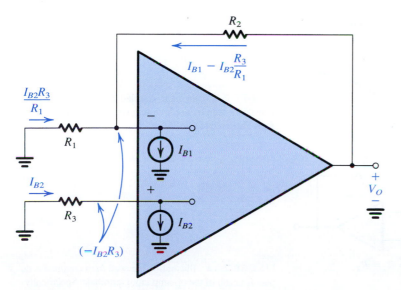

FIGURE 2.34 Reducing the effect of the input bias currents by introducing a resistor R_3.

The appropriate value for R_3 can be determined by analyzing the circuit in Fig. 2.34, where analysis details are shown and the output voltage is given by

$$V_O = -I_{B2}R_3 + R_2(I_{B1} - I_{B2}R_3/R_1) \qquad (2.45)$$

Consider first the case $I_{B1} = I_{B2} = I_B$, which results in

$$V_O = I_B[R_2 - R_3(1 + R_2/R_1)]$$

Thus we can reduce V_O to zero by selecting R_3 such that

$$R_3 = \frac{R_2}{1 + R_2/R_1} = \frac{R_1R_2}{R_1 + R_2} \qquad (2.46)$$

That is, R_3 should be made equal to the parallel equivalent of R_1 and R_2.

Having selected R_3 as above, let us evaluate the effect of a finite offset current I_{OS}. Let $I_{B1} = I_B + I_{OS}/2$ and $I_{B2} = I_B - I_{OS}/2$, and substitute in Eq. (2.45). The result is

$$V_O = I_{OS}R_2 \qquad (2.47)$$

which is usually about an order of magnitude smaller than the value obtained without R_3 (Eq. 2.44). We conclude that to minimize the effect of the input bias currents one should *place in the positive lead a resistance equal to the dc resistance seen by the inverting terminal.* We should emphasize the word *dc* in the last statement; note that if the amplifier is ac-coupled, we should select $R_3 = R_2$, as shown in Fig. 2.35.

While we are on the subject of ac-coupled amplifiers, we should note that one must always provide a continuous dc path between each of the input terminals of the op amp and ground. For this reason the ac-coupled noninverting amplifier of Fig. 2.36 will *not* work without the resistance R_3 to ground. Unfortunately, including R_3 lowers considerably the input resistance of the closed-loop amplifier.

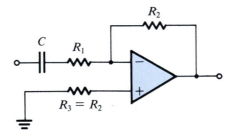

FIGURE 2.35 In an ac-coupled amplifier the dc resistance seen by the inverting terminal is R_2; hence R_3 is chosen equal to R_2.

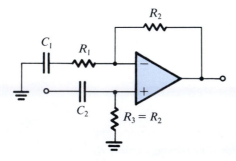

FIGURE 2.36 Illustrating the need for a continuous dc path for each of the op-amp input terminals. Specifically, note that the amplifier will *not* work without resistor R_3.

2.26 Consider an inverting amplifier circuit designed using an op amp and two resistors, $R_1 = 10\ \text{k}\Omega$ and $R_2 = 1\ \text{M}\Omega$. If the op amp is specified to have an input bias current of 100 nA and an input offset current of 10 nA, find the output dc offset voltage resulting and the value of a resistor R_3 to be placed in series with the positive input lead in order to minimize the output offset voltage. What is the new value of V_O?

Ans. 0.1 V; 9.9 kΩ ($\simeq$ 10 kΩ); 0.01 V

 ## 2.8 INTEGRATORS AND DIFFERENTIATORS

The op-amp circuit applications we have studied thus far utilized resistors in the op-amp feedback path and in connecting the signal source to the circuit, that is, in the feed-in path. As a result circuit operation has been (ideally) independent of frequency. The only exception has been the use of coupling capacitors in order to minimize the effect of the dc imperfections of op amps [e.g., the circuits in Figs. 2.31(a) and 2.36]. By allowing the use of capacitors together with resistors in the feedback and feed-in paths of op-amp circuits, we open the door to a very wide range of useful and exciting applications of the op amp. We begin our study of op-amp–*RC* circuits in this section by considering two basic applications, namely signal integrators and differentiators.

2.8.1 The Inverting Configuration with General Impedances

To begin with, consider the inverting closed-loop configuration with impedances $Z_1(s)$ and $Z_2(s)$ replacing resistors R_1 and R_2, respectively. The resulting circuit is shown in Fig. 2.37 and, for an ideal op amp, has the closed-loop gain or, more appropriately, the closed-loop transfer function

$$\frac{V_o(s)}{V_i(s)} = -\frac{Z_2(s)}{Z_1(s)} \tag{2.48}$$

As explained in Section 1.6, replacing s by $j\omega$ provides the transfer function for physical frequencies ω, that is, the transmission magnitude and phase for a sinusoidal input signal of frequency ω.

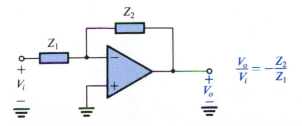

FIGURE 2.37 The inverting configuration with general impedances in the feedback and the feed-in paths.

EXAMPLE 2.6

For the circuit in Fig. 2.38, derive an expression for the transfer function $V_o(s)/V_i(s)$. Show that the transfer function is that of a low-pass STC circuit. By expressing the transfer function in the standard form shown in Table 1.2 on page 34, find the dc gain and the 3-dB frequency. Design the circuit to obtain a dc gain of 40 dB, a 3-dB frequency of 1 kHz, and an input resistance of 1 kΩ. At what frequency does the magnitude of transmission become unity? What is the phase angle at this frequency?

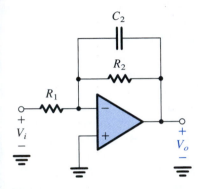

FIGURE 2.38 Circuit for Example 2.6.

Solution

To obtain the transfer function of the circuit in Fig. 2.38, we substitute in Eq. (2.48), $Z_1 = R_1$ and $Z_2 = R_2 \| (1/sC_2)$. Since Z_2 is the parallel connection of two components, it is more convenient to work in terms of Y_2; that is, we use the following alternative form of the transfer function:

$$\frac{V_o(s)}{V_i(s)} = -\frac{1}{Z_1(s)Y_2(s)}$$

and substitute $Z_1 = R_1$ and $Y_2(s) = (1/R_2) + sC_2$ to obtain

$$\frac{V_o(s)}{V_i(s)} = -\frac{1}{\dfrac{R_1}{R_2} + sC_2R_1}$$

This transfer function is of first order, has a finite dc gain (at $s = 0$, $V_o/V_i = -R_2/R_1$), and has zero gain at infinite frequency. Thus it is the transfer function of a low-pass STC network and can be expressed in the standard form of Table 1.2 as follows:

$$\frac{V_o(s)}{V_i(s)} = \frac{-R_2/R_1}{1 + sC_2R_2}$$

from which we find the dc gain K to be

$$K = -\frac{R_2}{R_1}$$

and the 3-dB frequency ω_0 as

$$\omega_0 = \frac{1}{C_2R_2}$$

We could have found all this from the circuit in Fig. 2.38 by inspection. Specifically, note that the capacitor behaves as an open circuit at dc; thus at dc the gain is simply $(-R_2/R_1)$. Furthermore, because there is a virtual ground at the inverting input terminal, the resistance seen by the capacitor is R_2, and thus the time constant of the STC network is C_2R_2.

Now to obtain a dc gain of 40 dB, that is, 100 V/V, we select $R_2/R_1 = 100$. For an input resistance of 1 kΩ, we select $R_1 = 1$ kΩ, and thus $R_2 = 100$ kΩ. Finally, for a 3-dB frequency $f_0 = 1$ kHz, we select C_2 from

$$2\pi \times 1 \times 10^3 = \frac{1}{C_2 \times 100 \times 10^3}$$

which yields $C_2 = 1.59$ nF.

The circuit has gain and phase Bode plots of the standard form in Fig. 1.23. As the gain falls off at the rate of –20 dB/decade, it will reach 0 dB in two decades, that is, at $f = 100f_0 = 100$ kHz. As Fig. 1.23(b) indicates, at such a frequency which is much greater than f_0, the phase is approximately –90°. To this, however, we must add the 180° arising from the inverting nature of the amplifier (i.e., the negative sign in the transfer function expression). Thus at 100 kHz, the total phase shift will be –270° or, equivalently, +90°.

2.8.2 The Inverting Integrator

By placing a capacitor in the feedback path (i.e., in place of Z_2 in Fig. 2.37) and a resistor at the input (in place of Z_1), we obtain the circuit of Fig. 2.39(a). We shall now show that this circuit realizes the mathematical operation of integration. Let the input be a time-varying function $v_I(t)$. The virtual ground at the inverting op-amp input causes $v_I(t)$ to appear in effect across R, and thus the current $i_1(t)$ will be $v_I(t)/R$. This current flows through the capacitor C, causing charge to accumulate on C. If we assume that the circuit begins operation at time $t = 0$, then at an arbitrary time t the current $i_1(t)$ will have deposited on C a charge equal to $\int_0^t i_1(t)\,dt$. Thus the capacitor voltage $v_C(t)$ will change by $\frac{1}{C}\int_0^t i_1(t)\,dt$. If the initial voltage on C (at $t = 0$) is denoted V_C, then

$$v_C(t) = V_C + \frac{1}{C}\int_0^t i_1(t)\,dt$$

Now the output voltage $v_O(t) = -v_C(t)$; thus,

$$v_O(t) = -\frac{1}{CR}\int_0^t v_I(t)\,dt - V_C \tag{2.49}$$

Thus the circuit provides an output voltage that is proportional to the time-integral of the input, with V_C being the initial condition of integration and CR the **integrator time-constant.** Note that, as expected, there is a negative sign attached to the output voltage, and thus this integrator circuit is said to be an inverting integrator. It is also known as a **Miller integrator** after an early worker in this area.

The operation of the integrator circuit can be described alternatively in the frequency domain by substituting $Z_1(s) = R$ and $Z_2(s) = 1/sC$ in Eq. (2.48) to obtain the transfer function

$$\frac{V_o(s)}{V_i(s)} = -\frac{1}{sCR} \tag{2.50}$$

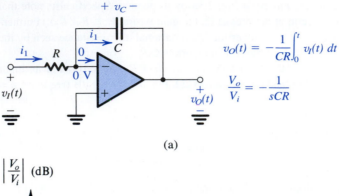

$$v_O(t) = -\frac{1}{CR}\int_0^t v_I(t)\, dt$$

$$\frac{V_o}{V_i} = -\frac{1}{sCR}$$

(a)

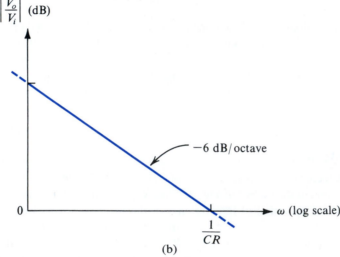

(b)

FIGURE 2.39 (a) The Miller or inverting integrator. (b) Frequency response of the integrator.

For physical frequencies, $s = j\omega$ and

$$\frac{V_o(j\omega)}{V_i(j\omega)} = -\frac{1}{j\omega CR} \tag{2.51}$$

Thus the integrator transfer function has magnitude

$$\left|\frac{V_o}{V_i}\right| = \frac{1}{\omega CR} \tag{2.52}$$

and phase

$$\phi = +90° \tag{2.53}$$

The Bode plot for the integrator magnitude response can be obtained by noting from Eq. (2.52) that as ω doubles (increases by an octave) the magnitude is halved (decreased by 6 dB). Thus the Bode plot is a straight line of slope –6 dB/octave (or, equivalently, –20 dB/decade). This line [shown in Fig. 2.39(b)] intercepts the 0-dB line at the frequency that makes $|V_o/V_i| = 1$, which from Eq. (2.52) is

$$\omega_{int} = \frac{1}{CR} \tag{2.54}$$

The frequency ω_{int} is known as the **integrator frequency** and is simply the inverse of the integrator time constant.

Comparison of the frequency response of the integrator to that of an STC low-pass network indicates that the integrator behaves as a low-pass filter with a corner frequency of zero. Observe also that at $\omega = 0$, the magnitude of the integrator transfer function is infinite. This indicates that at dc the op amp is operating with an open loop. This should also be obvious from the integrator circuit itself. Reference to Fig. 2.39(a) shows that the feedback element is a capacitor, and thus at dc, where the capacitor behaves as an open circuit, there is no negative feedback! This is a very significant observation and one that indicates a source of problems with the integrator circuit: Any tiny dc component in the input signal will theoretically produce an infinite output. Of course, no infinite output voltage results in practice; rather, the output of the amplifier saturates at a voltage close to the op-amp positive or negative power supply (L_+ or L_-), depending on the polarity of the input dc signal.

It should be clear from this discussion that the integrator circuit will suffer deleterious effects from the presence of the op-amp input dc offset voltage and current. To see the effect of the input dc offset voltage V_{OS}, consider the integrator circuit in Fig. 2.40, where for simplicity we have short-circuited the input signal source. Analysis of the circuit is straightforward and is shown in Fig. 2.40. Assuming for simplicity that at time $t = 0$ the voltage across the capacitor is zero, the output voltage as a function of time is given by

$$v_O = V_{OS} + \frac{V_{OS}}{CR} t \qquad (2.55)$$

Thus v_O increases linearly with time until the op amp saturates—clearly an unacceptable situation! As should be expected, the dc input offset current I_{OS} produces a similar problem. Figure 2.41 illustrates the situation. Observe that we have added a resistance R in the op-amp positive-input lead in order to keep the input bias current I_B from flowing through C.

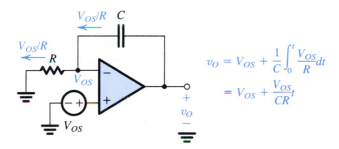

FIGURE 2.40   Determining the effect of the op-amp input offset voltage V_{OS} on the Miller integrator circuit. Note that since the output rises with time, the op amp eventually saturates.

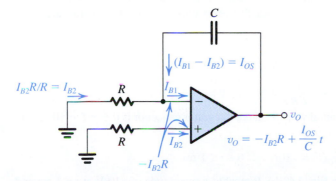

FIGURE 2.41   Effect of the op-amp input bias and offset currents on the performance of the Miller integrator circuit.

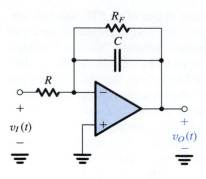

FIGURE 2.42 The Miller integrator with a large resistance R_F connected in parallel with C in order to provide negative feedback and hence finite gain at dc.

Nevertheless, the offset current I_{OS} will flow through C and cause v_O to ramp linearly with time until the op amp saturates.

The dc problem of the integrator circuit can be alleviated by connecting a resistor R_F across the integrator capacitor C, as shown in Fig. 2.42. Such a resistor provides a dc path through which the dc currents (V_{OS}/R) and I_{OS} can flow, with the result that v_O will now have a dc component $[V_{OS}(1 + R_F/R) + I_{OS}R_F]$ instead of rising linearly. To keep the dc offset at the output small, one would select a low value for R_F. Unfortunately, however, the lower the value of R_F, the less ideal the integrator circuit becomes. This is because R_F causes the frequency of the integrator pole to move from its ideal location at $\omega = 0$ to one determined by the corner frequency of the STC network (R_F, C). Specifically, the integrator transfer function becomes

$$\frac{V_o(s)}{V_i(s)} = -\frac{R_F/R}{1 + sCR_F}$$

as opposed to the ideal function of $-1/sCR$. The lower the value we select for R_F, the higher the corner frequency $(1/CR_F)$ will be and the more nonideal the integrator becomes. Thus selecting a value for R_F presents the designer with a trade-off between dc performance and signal performance. The effect of R_F on integrator performance is investigated further in the Example 2.7. Before doing so, however, observe that R_F closes the negative-feedback loop at dc and provides the integrator circuit with a finite dc gain of $-R_F/R$.

EXAMPLE 2.7

Find the output produced by a Miller integrator in response to an input pulse of 1-V height and 1-ms width [Fig. 2.43(a)]. Let $R = 10$ kΩ and $C = 10$ nF. If the integrator capacitor is shunted by a 1-MΩ resistor, how will the response be modified? The op amp is specified to saturate at ± 13 V.

Solution

In response to a 1-V, 1-ms input pulse, the integrator output will be

$$v_O(t) = -\frac{1}{CR}\int_0^t 1.dt, \qquad 0 \le t \le 1 \text{ ms}$$

where we have assumed that the initial voltage on the integrator capacitor is 0. For $C = 10$ nF and $R = 10$ kΩ, $CR = 0.1$ ms, and

$$v_O(t) = -10t, \qquad 0 \le t \le 1 \text{ ms}$$

which is the linear ramp shown in Fig. 2.43(b). It reaches a magnitude of -10 V at $t = 1$ ms and remains constant thereafter.

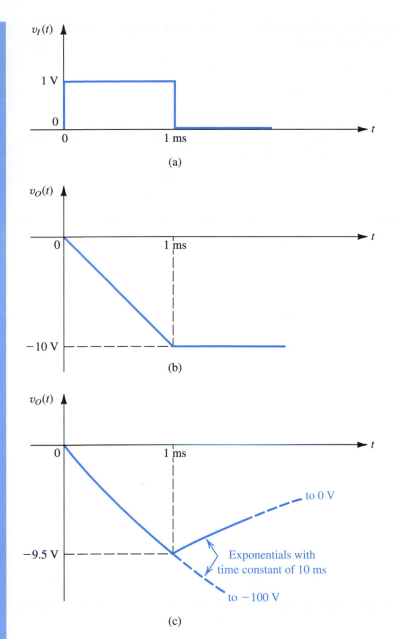

FIGURE 2.43 Waveforms for Example 2.7: **(a)** Input pulse. **(b)** Output linear ramp of ideal integrator with time constant of 0.1 ms. **(c)** Output exponential ramp with resistor R_F connected across integrator capacitor.

That the output is a linear ramp should also be obvious from the fact that the 1-V input pulse produces a $1\text{ V}/10\text{ k}\Omega = 0.1\text{ mA}$ constant current through the capacitor. This constant current $I = 0.1\text{ mA}$ supplies the capacitor with a charge It, and thus the capacitor voltage changes linearly as (It/C), resulting in $v_O = -(I/C)t$. It is worth remembering that charging a capacitor with a constant current produces a linear voltage across it.

Next consider the situation with resistor $R_F = 1\text{ M}\Omega$ connected across C. As before, the 1-V pulse will provide a constant current $I = 0.1\text{ mA}$. Now, however, this current is supplied to an

STC network composed of R_F in parallel with C. To find the output voltage, we use Eq. (1.29), which can be adapted to our case here as follows:

$$v_O(t) = v_O(\infty) - [v_O(\infty) - v_O(0+)]e^{-t/CR_F}$$

where $v_O(\infty)$ is the final value, obtained as

$$v_O(\infty) = -IR_F = -0.1 \times 10^{-3} \times 1 \times 10^6 = -100 \text{ V}$$

and $v_O(0+)$ is the initial value, which is zero. That is, the output will be an exponential heading toward -100 V with a time constant of $CR_F = 10 \times 10^{-9} \times 1 \times 10^6 = 10$ ms,

$$v_O(t) = -100(1 - e^{-t/10}), \qquad 0 \le t \le 1 \text{ ms}$$

Of course, the exponential will be interrupted at the end of the pulse, that is, at $t = 1$ ms, and the output will reach the value

$$v_O(1 \text{ ms}) = -100(1 - e^{-1/10}) = -9.5 \text{ V}$$

The output waveform is shown in Fig. 2.43(c), from which we see that including R_F causes the ramp to be slightly rounded such that the output reaches only -9.5 V, 0.5 V short of the ideal value of -10 V. Furthermore, for $t > 1$ ms, the capacitor discharges through R_F with the relatively long time-constant of 10 ms. Finally, we note that op amp saturation, specified to occur ±13 V, has no effect on the operation of this circuit.

The preceding example hints at an important application of integrators, namely, their use in providing triangular waveforms in response to square-wave inputs. This application is explored in Exercise 2.27. Integrators have many other applications, including their use in the design of filters (Chapter 12).

2.8.3 The Op-Amp Differentiator

Interchanging the location of the capacitor and the resistor of the integrator circuit results in the circuit in Fig. 2.44(a), which performs the mathematical function of differentiation. To see how this comes about, let the input be the time-varying function $v_I(t)$, and note that the virtual ground at the inverting input terminal of the op amp causes $v_I(t)$ to appear in effect across the capacitor C. Thus the current through C will be $C(dv_I/dt)$, and this current flows through the feedback resistor R providing at the op-amp output a voltage $v_O(t)$,

$$v_O(t) = -CR\frac{dv_I(t)}{dt} \tag{2.56}$$

The frequency-domain transfer function of the differentiator circuit can be found by substituting in Eq. (2.56), $Z_1(s) = 1/sC$ and $Z_2(s) = R$ to obtain

$$\frac{V_o(s)}{V_i(s)} = -sCR \tag{2.57}$$

which for physical frequencies $s = j\omega$ yields

$$\frac{V_o(j\omega)}{V_i(j\omega)} = -j\omega CR \tag{2.58}$$

Thus the transfer function has magnitude

$$\left|\frac{V_o}{V_i}\right| = \omega CR \tag{2.59}$$

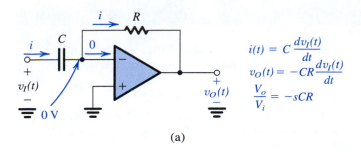

$$i(t) = C\frac{dv_I(t)}{dt}$$

$$v_O(t) = -CR\frac{dv_I(t)}{dt}$$

$$\frac{V_o}{V_i} = -sCR$$

(a)

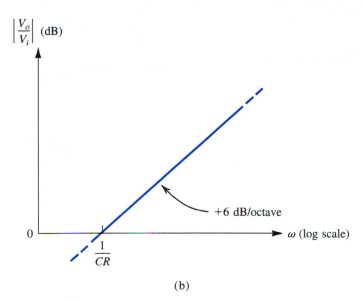

(b)

FIGURE 2.44 (a) A differentiator. (b) Frequency response of a differentiator with a time-constant CR.

and phase

$$\phi = -90° \tag{2.60}$$

The Bode plot of the magnitude response can be found from Eq. (2.59) by noting that for an octave increase in ω, the magnitude doubles (increases by 6 dB). Thus the plot is simply a straight line of slope +6 dB/octave (or, equivalently, +20 dB/decade) intersecting the 0-dB line (where $|V_o/V_i| = 1$) at $\omega = 1/CR$, where CR is the **differentiator time-constant** [see Fig. 2.44(b)].

The frequency response of the differentiator can be thought of as that of an STC highpass filter with a corner frequency at infinity (refer to Fig. 1.24). Finally, we should note that the very nature of a differentiator circuit causes it to be a "noise magnifier." This is due to the spike introduced at the output every time there is a sharp change in $v_I(t)$; such a change could be interference coupled electromagnetically ("picked-up") from adjacent signal sources. For this reason and because they suffer from stability problems (Chapter 8), differentiator circuits are generally avoided in practice. When the circuit of Fig. 2.44(a) is used, it is usually necessary to connect a small-valued resistor in series with the capacitor. This modification, unfortunately, turns the circuit into a nonideal differentiator.

EXERCISES

2.27 Consider a symmetrical square wave of 20-V peak-to-peak, 0 average, and 2-ms period applied to a Miller integrator. Find the value of the time constant CR such that the triangular waveform at the output has a 20-V peak-to-peak amplitude.

Ans. 0.5 ms

D2.28 Using an ideal op amp, design an inverting integrator with an input resistance of 10 kΩ and an integration time constant of 10^{-3} s. What is the gain magnitude and phase angle of this circuit at 10 rad/s and at 1 rad/s? What is the frequency at which the gain magnitude is unity?

Ans. $R = 10$ kΩ, $C = 0.1$ μF; at $\omega = 10$ rad/s: $|V_o / V_i| = 100$ V/V and $\phi = +90°$; at $\omega = 1$ rad/s: $|V_o / V_i| = 1{,}000$ V/V and $\phi = +90°$; 1000 rad/s

2.29 Consider a Miller integrator with a time constant of 1 ms and an input resistance of 10 kΩ. Let the op amp have $V_{OS} = 2$ mV and output saturation voltages of ± 12 V. (a) Assuming that when the power supply is turned on the capacitor voltage is zero, how long does it take for the amplifier to saturate? (b) Select the largest possible value for a feedback resistor R_F so that at least ± 10 V of output signal swing remains available. What is the corner frequency of the resulting STC network?

Ans. (a) 6 s; (b) 10 MΩ, 0.16 Hz

D2.30 Design a differentiator to have a time constant of 10^{-2} s and an input capacitance of 0.01 μF. What is the gain magnitude and phase of this circuit at 10 rad/s, and at 10^3 rad/s? In order to limit the high-frequency gain of the differentiator circuit to 100, a resistor is added in series with the capacitor. Find the required resistor value.

Ans. $C = 0.01$ μF; $R = 1$ MΩ; at $\omega = 10$ rad/s: $|V_o / V_i| = 0.1$ V/V and $\phi = -90°$; at $\omega = 1000$ rad/s: $|V_o / V_i| = 10$ V/V and $\phi = -90°$; 10 kΩ

2.9 THE SPICE OP-AMP MODEL AND SIMULATION EXAMPLES

As mentioned at the beginning of this chapter, the op amp is not a single electronic device, such as the junction diode or the MOS transistor, both of which we shall study later on; rather, it is a complex IC made up of a large number of electronic devices. Nevertheless, as we have seen in this chapter, the op amp can be treated and indeed effectively used as a circuit component or a circuit building block without the user needing to know the details of its internal circuitry. The user, however, needs to know the terminal characteristics of the op amp, such as its open-loop gain, its input resistance, its frequency response, etc. Furthermore, in designing circuits utilizing the op amp, it is useful to be able to represent the op amp with an equivalent circuit model. Indeed, we have already done this in this chapter, albeit with very simple equivalent circuit models suitable for hand analysis. Since we are now going to use computer simulation, the models we use can be more complex to account as fully as possible for the op amp's nonideal performance.

Op amp models that are based on their observed terminal characteristics are known as **macromodels.** These are to be distinguished from models that are obtained by modeling every device in the op amp's actual internal circuit. The latter type of model can become very complex and unwieldy, especially if one attempts to use it in the simulation of a circuit that utilizes a large number of op amps.

The goal of macromodeling of a circuit block (in our case here, the op amp) is to achieve a very close approximation to the actual performance of the op amp while using circuit model of significantly reduced complexity compared to the actual internal circuit. Advantages of

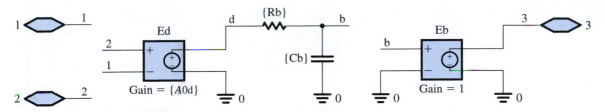

FIGURE 2.45 A linear macromodel used to model the finite gain and bandwidth of an internally compensated op amp.

using macromodels include: A macromodel can be developed on the basis of data-sheet specification, without having to know the details of the internal circuitry of the op amp. Moreover, macromodels allow the simulation of a circuit containing a number of op amps to be performed much faster.

2.9.1 Linear Macromodel

The Capture schematic[5] of a linear macromodel for an internally compensated op amp with finite gain and bandwidth is shown in Fig. 2.45. In this equivalent-circuit model, the gain constant A_{0d} of the voltage-controlled voltage source E_d corresponds to the differential gain of the op amp at dc. Resistor R_b and capacitor C_b form an STC filter with a corner frequency

$$f_b = \frac{1}{2\pi R_b C_b} \tag{2.61}$$

The low-pass response of this filter is used to model the frequency response of the internally compensated op amp. The values of R_b and C_b used in the macromodel are chosen such that f_b corresponds to the 3-dB frequency of the op amp being modeled. This is done by arbitrarily selecting a value for either R_b or C_b (the selected value does not need to be a practical one) and then using Eq. (2.61) to compute the other value. In Fig. 2.45, the voltage-controlled voltage source E_b with a gain constant of unity is used as a buffer to isolate the low-pass filter from any load at the op-amp output. Thus any op-amp loading will not affect the frequency response of the filter and hence that of the op amp.

The linear macromodel in Fig. 2.45 can be further expanded to account for other op-amp nonidealities. For example, the equivalent-circuit model in Fig. 2.46 can be used to model an internally compensated op amp while accounting for the following op-amp nonidealities:

1. **Input Offset Voltage (V_{OS}).** The dc voltage source V_{OS} models the op-amp input offset voltage.

2. **Input Bias Current (I_B) and Input Offset Current (I_{OS}).** The dc current sources I_{B1} and I_{B2} model the input bias current at each input terminal of the op amp, with

$$I_{B1} = I_B + \frac{I_{OS}}{2} \quad \text{and} \quad I_{B2} = I_B - \frac{I_{OS}}{2}$$

where I_B and I_{OS} are, respectively, the input bias current and the input offset current specified by the op-amp manufacturer.

---

[5] The reader is reminded that the Capture schematics and the corresponding PSpice simulation files of all SPICE examples in this book can be found on the text's CD, as well as on its website (www.sedrasmith.org).

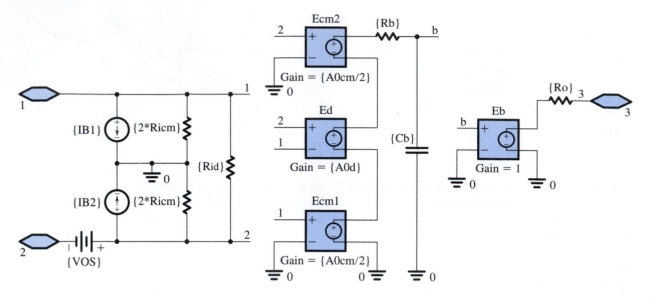

FIGURE 2.46 A comprehensive linear macromodel of an internally compensated op amp.

3. **Common-Mode Input Resistance (R_{icm}).** If the two input terminals of an op amp are tied together and the input resistance (to ground) is measured, the result is the common-mode input resistance R_{icm}. In the macromodel of Fig. 2.46, we have split R_{icm} into two equal parts ($2R_{icm}$), each connected between one of the input terminals and ground.

4. **Differential-Input Resistance (R_{id}).** The resistance seen between the two input terminals of an op amp is the differential input resistance R_{id}.

5. **Differential Gain at DC (A_{0d}) and Common-Mode Rejection Ratio (CMRR).** The output voltage of an op amp at dc can be expressed as

$$V_3 = A_{0d}(V_2 - V_1) + \frac{A_{0cm}}{2}(V_1 + V_2)$$

where A_{0d} and A_{0cm} are, respectively, the differential and common-mode gains of the op amp at dc. For an op amp with a finite CMRR,

$$A_{0cm} = A_{0d}/\text{CMRR} \qquad (2.62)$$

where CMRR is expressed in V/V (not in dB). Note that the CMRR value in Eq. (2.62) is that of the open-loop op amp while the CMRR in Eq. (2.14) is that of a particular closed-loop amplifier. In the macromodel of Fig. 2.46, the voltage-controlled voltage sources E_{cm1} and E_{cm2} with gain constants of $A_{0cm}/2$ account for the finite CMRR while source E_d models A_{0d}.

6. **Unity-Gain Frequency (f_t).** From Eq. (2.28), the 3-dB frequency f_b and the unity-gain frequency (or gain-bandwidth product) f_t of an internally compensated op amp with an STC frequency response are related through

$$f_b = \frac{f_t}{A_{0d}} \qquad (2.63)$$

As in Fig. 2.45, the finite op-amp bandwidth is accounted for in the macromodel of Fig. 2.46 by setting the corner frequency of the filter formed by resistor R_b and capacitor C_b (Eq. 2.61) to equal the 3-dB frequency of the op amp (Eq. 2.63). It should be noted that here we are assuming that the differential gain and the common-mode gain have the same frequency response (not always a valid assumption!).

7. **Output Resistance (R_o).** The resistance seen at the output terminal of an op amp is the output resistance R_o.

EXAMPLE 2.8

Performance of a Noninverting Amplifier

Consider an op amp with a differential input resistance of 2 MΩ, an input offset voltage of 1 mV, a dc gain of 100 dB, and an output resistance of 75 Ω. Assume the op amp is internally compensated and has an STC frequency response with a gain-bandwidth product of 1 MHz.

(a) Create a subcircuit model for this op amp in PSpice.

(b) Using this subcircuit, simulate the closed-loop noninverting amplifier in Fig. 2.12 with resistors $R_1 = 1$ kΩ and $R_2 = 100$ kΩ to find:

   (i) Its 3-dB bandwidth f_{3dB}.

   (ii) Its output offset voltage V_{OSout}.

   (iii) Its input resistance R_{in}.

   (iv) Its output resistance R_{out}.

(c) Simulate the step response of the closed-loop amplifier, and measure its rise time t_r. Verify that this time agrees with the 3-dB frequency measured above.

Solution

To model the op amp in PSpice, we use the equivalent circuit in Fig. 2.46 but with $R_{id} = 2$ MΩ, $R_{icm} = \infty$ (open circuit), $I_{B1} = I_{B2} = 0$ (open circuit), $V_{OS} = 1$ mV, $A_{0d} = 10^5$ V/V, $A_{0cm} = 0$ (short circuit), and $R_o = 75$ Ω. Furthermore, we set $C_b = 1$ μF and $R_b = 15.915$ kΩ to achieve an $f_t = 1$ MHz.

To measure the 3-dB frequency of the closed-loop amplifier, we apply a 1-V ac voltage at its input, perform an ac-analysis simulation in PSpice, and plot its output versus frequency. The output voltage, plotted in Fig. 2.47, corresponds to the gain of the amplifier because we chose an input voltage of 1 V. Thus, from Fig. 2.47, the closed-loop amplifier has a dc gain of $G_0 = 100.9$ V/V, and the frequency at which its gain drops to $G_0/\sqrt{2} = 71.35$ V/V is $f_{3dB} = 9.9$ kHz, which agrees with Eq. (2.28).

The input resistance R_{in} corresponds to the reciprocal of the current drawn out of the 1-V ac voltage source used in the above ac-analysis simulation at 0.1 Hz. (Theoretically, R_{in} is the small-signal input resistance at dc. However, ac-analysis simulations must start at frequencies greater than zero, so we use 0.1 Hz to approximate the dc point.) Accordingly, R_{in} is found to be 2 GΩ.

To measure R_{out}, we short-circuit the amplifier input to ground, inject a 1-A ac current at its output, and perform an ac-analysis simulation. R_{out} corresponds to the amplifier output voltage at 0.1 Hz and is found to be 76 mΩ. Although an ac test voltage source could equally well have been used to measure the output resistance in this case, it is a good practice to attach a current source rather than a voltage source between the output and ground. This is because an ac current source appears as an open circuit when the simulator computes the dc bias point of the circuit while an ac voltage source appears as a short circuit, which can erroneously force the dc output voltage to zero. For similar reasons, an ac test voltage source should be attached in series with the biasing dc voltage source for measuring the input resistance of a voltage amplifier.

A careful look at R_{in} and R_{out} of the closed-loop amplifier reveals that their values have, respectively, increased and decreased by a factor of about 1000 relative to the corresponding

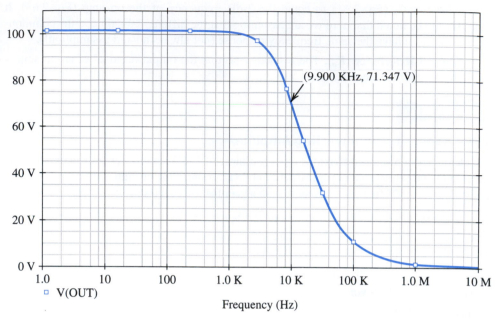

FIGURE 2.47   Frequency response of the closed-loop amplifier in Example 2.8.

resistances of the op amp. Such a large input resistance and small output resistance are indeed desirable characteristics for a voltage amplifier. This improvement in the small-signal resistances of the closed-loop amplifier is a direct consequence of applying negative feedback (through resistors R_1 and R_2) around the open-loop op amp. We will study negative feedback in Chapter 8, where we will also learn how the improvement factor (1000 in this case) corresponds to the ratio of the open-loop op-amp gain (10^5) to the closed-loop amplifier gain (100).

From Eqs. (2.37) and (2.35), the closed-loop amplifier has an STC low-pass response given by

$$\frac{V_o(s)}{V_i(s)} = \frac{G_0}{1 + \dfrac{s}{2\pi f_{3dB}}}$$

As described in Appendix D, the response of such an amplifier to an input step of height V_{step} is given by

$$v_O(t) = V_{final}(1 - e^{-t/\tau}) \tag{2.64}$$

where $V_{final} = G_0 V_{step}$ is the final output-voltage value (i.e., the voltage value toward which the output is heading) and $\tau = 1/(2\pi f_{3dB})$ is the time constant of the amplifier. If we define $t_{10\%}$ and $t_{90\%}$ to be the time it takes for the output waveform to rise to, respectively, 10% and 90% of V_{final}, then from Eq. (2.64), $t_{10\%} \approx 0.1\tau$ and $t_{90\%} \approx 2.3\tau$. Therefore, the rise time t_r of the amplifier can be expressed as

$$t_r = t_{90\%} - t_{10\%} = 2.2\tau = \frac{2.2}{2\pi f_{3dB}}$$

Therefore, if $f_{3dB} = 9.9$ kHz, then $t_r = 35.4$ μs. To simulate the step response of the closed-loop amplifier, we apply a step voltage at its input, using a piece-wise-linear (PWL) source (with a very short rise time); then perform a transient-analysis simulation, and measure the voltage at the output versus time. In our simulation, we applied a 1-V step input, plotted the output waveform in Fig. 2.48, and measured t_r to be 35.3 μs.

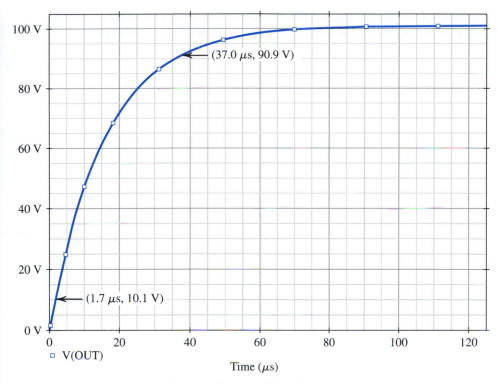

FIGURE 2.48 Step response of the closed-loop amplifier in Example 2.8.

The linear macromodels in Figs. 2.45 and 2.46 assume that the op-amp circuit is operating in its linear range, and do not account for its nonideal performance when large signals are present at the output. Therefore, nonlinear effects, such as output saturation and slew rate, are not modeled. This is why, in the step response of Fig. 2.48, we could see an output voltage of 100 V when we applied a 1-V step input. However, IC op amps are not capable of producing such large output voltages. Hence, a designer must be very careful when using these models.

It is important to point out that we also saw output voltages of 100 V or so in the ac analysis of Fig. 2.47, where for convenience we applied a 1-V ac input to measure the gain of the closed-loop amplifier. So, would we see such large output voltages if the op-amp macromodel accounted for nonlinear effects (particularly output saturation)? The answer is yes, because in an ac analysis PSpice uses a linear model for nonlinear devices with the linear-model parameters evaluated at a bias point. We will have more to say about this in subsequent chapters. Here, however, we must keep in mind that the voltage magnitudes encountered in an ac analysis may not be realistic. What is of importance to the designer in this case are the voltage and current ratios (e.g., the output-to-input voltage ratio as a measure of voltage gain).

2.9.2 Nonlinear Macromodel

The linear macromodel in Fig. 2.46 can be further expanded to account for the op-amp nonlinear performance. For example, the finite output voltage swing of the op amp can be modeled by placing limits on the output voltage of the voltage-controlled voltage source E_b. In PSpice this can be done using the ETABLE component in the analog-behavioral-modeling (ABM) library and setting the output voltage limits in the look-up table of this component. Further details on how to build nonlinear macromodels for the op amp can be found in the references on Spice simulation. In general, robust macromodels that account for the nonlinear effects in an IC are provided by the op-amp manufacturers. Most simulators include such macromodels for some of the popular off-the-shelf ICs in their libraries. For example, PSpice includes models for the μA741, the LF411, and the LM324 op amps.[6]

EXAMPLE 2.9

Characteristics of the 741 OP Amp

Consider the μA741 op amp whose macromodel is available in PSpice. Use PSpice to plot the open-loop gain and hence determine f_t. Also, investigate the SR limitation and the output saturation of this op amp.

Solution

Figure 2.49 shows the Capture schematic used to simulate the frequency response of the μA741 op amp. The μA741 part has seven terminals. Terminals 7 and 4 are, respectively, the positive and negative dc power-supply terminals of the op amp. 741-type op amps are typically operated from $\pm$15-V power supplies; therefore we connected the dc voltage sources $V_{CC} = +15$ V and $V_{EE} = -15$ V to terminals 7 and 4, respectively. Terminals 3 and 2 of the μA741 part correspond to the positive and negative input terminals, respectively of the op amp. In general, as outlined in Section 2.1.3, the op amp input signals are expressed as

$$v_{INP} = V_{CM} + \frac{V_d}{2}$$

$$v_{INN} = V_{CM} - \frac{V_d}{2}$$

where v_{INP} and v_{INN} are the signals at, respectively, the positive- and negative-input terminals of the op amp with V_{CM} being the common-mode input signal (which sets the dc bias voltage at the op amp input terminals) and V_d being the differential input signal to be amplified. The dc voltage source V_{CM} in Fig. 2.49 is used to set the common-mode input voltage. Typically, V_{CM} is set to the average of the dc power-supply voltages V_{CC} and V_{EE} to maximize the available input signal swing. Hence, we set $V_{CM} = 0$. The voltage source V_d in Fig. 2.49 is used to generate the differential input signal V_d. This signal is applied differentially to the op-amp input terminals using the voltage-controlled voltage sources E_p and E_n whose gain constants are set to 0.5.

Terminals 1 and 5 of part μA741 are the offset-nulling terminals of the op amp (as depicted in Fig. 2.30). However, a check of the PSpice netlist of this part (by selecting Edit → PSpice Model, in the Capture menus), reveals that these terminals are floating; therefore the offset-nulling characteristic of the op amp is not incorporated in this macromodel.

To measure f_t of the op-amp, we set the voltage of source V_d to be 1-V ac, perform an ac-analysis simulation in PSpice, and plot the output voltage versus frequency as shown in Fig. 2.50. Accordingly, the frequency at which the op-amp voltage gain drops to 0 dB is $f_t = 0.9$ MHz (which is close to the 1-MHz value reported in the data sheets for 741-type op amps).

To determine the slew rate of the μA741 op amp, we connect the op amp in a unity-gain configuration, as shown in Fig. 2.51, apply a large pulse signal at the input with very short rise and fall times to cause slew-rate limiting at the output, perform a transient-analysis simulation in PSpice, and plot the output voltage as shown in Fig. 2.52. The slope of the slew-rate limited output waveform corresponds to the slew-rate of the op amp and is found to be SR = 0.5 V/μs (which agrees with the value specified in the data sheets for 741-type op amps).

---

[6] The OrCAD 9.2 Lite Edition of PSpice, which is available on the CD accompanying this book, includes these models in its evaluation (EVAL) library.

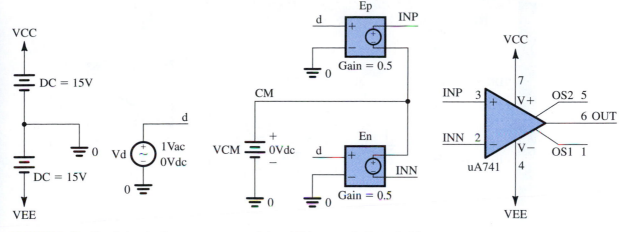

FIGURE 2.49 Simulating the frequency response of the μA741 op-amp in Example 2.9.

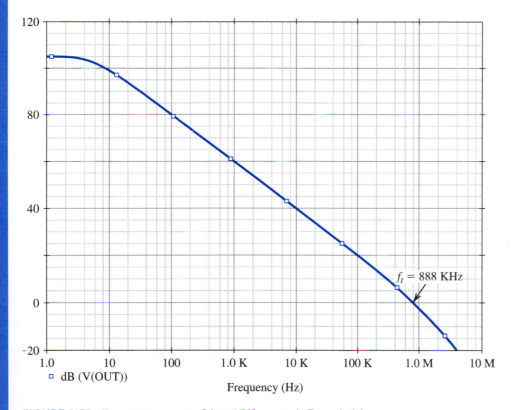

FIGURE 2.50 Frequency response of the μA741 op amp in Example 2.9.

To determine the maximum output voltage of the μA741 op amp, we set the dc voltage of the differential voltage source V_d in Fig. 2.49 to a large value, say +1 V, and perform a bias-point simulation in PSpice. The corresponding dc output voltage is the positive-output saturation voltage of the op amp. We repeat the simulation with the dc differential input voltage set to −1 V to find the negative-output saturation voltage. Accordingly, we find that the μA741 op amp has a maximum output voltage $V_{omax} = 14.8$ V.

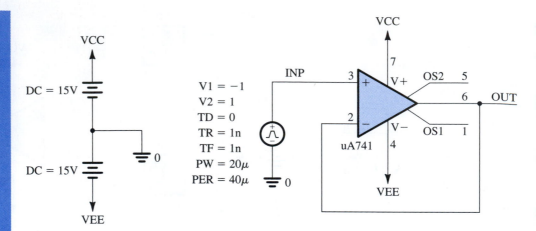

FIGURE 2.51 Circuit for determining the slew rate of the μA741 op amp in Example 2.9.

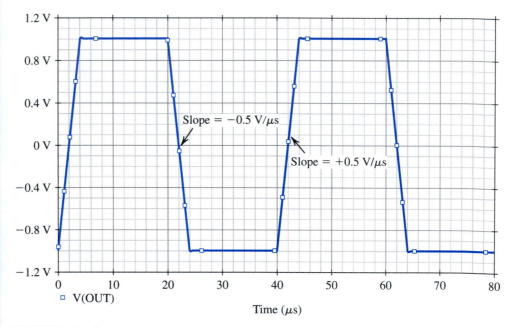

FIGURE 2.52 Square-wave response of the μA741 op amp connected in the unity-gain configuration shown in Fig. 2.51.

SUMMARY

■ The IC op amp is a versatile circuit building block. It is easy to apply, and the performance of op-amp circuits closely matches theoretical predictions.

■ The op-amp terminals are the inverting input terminal (1), the noninverting input terminal (2), the output terminal (3), the positive-supply terminal (V^+) to be connected to the

positive power supply, and the negative-supply terminal (V^-) to be connected to the negative supply. The common terminal of the two supplies is the circuit ground.

■ The ideal op amp responds only to the difference input signal, that is, $(v_2 - v_1)$; provides at the output, between terminal 3 and ground, a signal $A(v_2 - v_1)$, where A, the open-loop gain, is very large (10^4 to 10^6) and ideally infinite; and has an infinite input resistance and a zero output resistance.

■ Negative feedback is applied to an op amp by connecting a passive component between its output terminal and its inverting (negative) input terminal. Negative feedback causes the voltage between the two input terminals to become very small and ideally zero. Correspondingly, a virtual short circuit is said to exist between the two input terminals. If the positive input terminal is connected to ground, a virtual ground appears on the negative input terminal.

■ The two most important assumptions in the analysis of op-amp circuits, presuming negative feedback exists and the op amps are ideal, are: the two input terminals of the op amp are at the same voltage, and zero current flows into the op-amp input terminals.

■ With negative feedback applied and the loop closed, the closed-loop gain is almost entirely determined by external components: For the inverting configuration, $V_o/V_i = -R_2/R_1$; and for the noninverting configuration, $V_o/V_i = 1 + R_2/R_1$.

■ The noninverting closed-loop configuration features a very high input resistance. A special case is the unity-gain follower, frequently employed as a buffer amplifier to connect a high-resistance source to a low-resistance load.

■ For most internally compensated op amps, the open-loop gain falls off with frequency at a rate of −20 dB/decade, reaching unity at a frequency f_t (the unity-gain bandwidth). Frequency f_t is also known as the gain–bandwidth product of the op amp: $f_t = A_0 f_b$, where A_0 is the dc gain, and f_b is the 3-dB frequency of the open-loop gain. At any frequency $f(f \gg f_b)$, the op-amp gain $|A| \simeq f_t/f$.

■ For both the inverting and the noninverting closed-loop configurations, the 3-dB frequency is equal to $f_t/(1 + R_2/R_1)$.

■ The maximum rate at which the op-amp output voltage can change is called the slew rate. The slew rate, SR, is usually specified in V/μs. Op-amp slewing can result in nonlinear distortion of output signal waveforms.

■ The full-power bandwidth, f_M, is the maximum frequency at which an output sinusoid with an amplitude equal to the op-amp rated output voltage (V_{omax}) can be produced without distortion: $f_M = SR/2\pi V_{omax}$.

■ The input offset voltage, V_{OS}, is the magnitude of dc voltage that when applied between the op amp input terminals, with appropriate polarity, reduces the dc offset voltage at the output to zero.

■ The effect of V_{OS} on performance can be evaluated by including in the analysis a dc source V_{OS} in series with the op-amp positive input lead. For both the inverting and the noninverting configurations, V_{OS} results in a dc offset voltage at the output of $V_{OS}(1 + R_2/R_1)$.

■ Capacitively coupling an op amp reduces the dc offset voltage at the output considerably.

■ The average of the two dc currents, I_{B1} and I_{B2}, that flow in the input terminals of the op amp, is called the input bias current, I_B. In a closed-loop amplifier, I_B gives rise to a dc offset voltage at the output of magnitude $I_B R_2$. This voltage can be reduced to $I_{OS} R_2$ by connecting a resistance in series with the positive input terminal equal to the total dc resistance seen by the negative input terminal. I_{OS} is the input offset current; that is, $I_{OS} = |I_{B1} - I_{B2}|$.

■ Connecting a large resistance in parallel with the capacitor of an op-amp inverting integrator prevents op-amp saturation (due to the effect of V_{OS} and I_B).

PROBLEMS

SECTION 2.1: THE IDEAL OP AMP

2.1 What is the minimum number of pins required for a so-called dual-op-amp IC package, one containing two op amps? What is the number of pins required for a so-called quad-op-amp package, one containing four op amps?

2.2 The circuit of Fig. P2.2 uses an op amp that is ideal except for having a finite gain A. Measurements indicate $v_O = 4.0$ V when $v_I = 4.0$ V. What is the op amp gain A?

2.3 Measurement of a circuit incorporating what is thought to be an ideal op amp shows the voltage at the op amp output to be

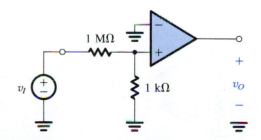

FIGURE P2.2

−2.000 V and that at the negative input to be −3.000 V. For the amplifier to be ideal, what would you expect the voltage at the positive input to be? If the measured voltage at the positive input is −3.020 V, what is likely to be the actual gain of the amplifier?

2.4 A set of experiments are run on an op amp that is ideal except for having a finite gain A. The results are tabulated below. Are the results consistent? If not, are they reasonable, in view of the possibility of experimental error? What do they show the gain to be? Using this value, predict values of the measurements that were accidentally omitted (the blank entries).

Experiment #	v_1	v_2	v_O
1	0.00	0.00	0.00
2	1.00	1.00	0.00
3		1.00	1.00
4	1.00	1.10	10.1
5	2.01	2.00	−0.99
6	1.99	2.00	1.00
7	5.10		−5.10

2.5 Refer to Exercise 2.3. This problem explores an alternative internal structure for the op amp. In particular, we wish to model the internal structure of a particular op amp using two transconductance amplifiers and one transresistance amplifier. Suggest an appropriate topology. For equal transconductances G_m and a transresistance R_m, find an expression for the open-loop gain A. For $G_m = 100$ mA/V and $R_m = 10^6$ Ω, what value of A results?

2.6 The two wires leading from the output terminals of a transducer pick up an interference signal that is a 60-Hz, 1-V sinusoid. The output signal of the transducer is sinusoidal of 10-mV amplitude and 1000-Hz frequency. Give expressions for v_{cm}, v_d, and the total signal between each wire and the system ground.

2.7 Nonideal (i.e., real) operational amplifiers respond to both the differential and common-mode components of their input signals (refer to Fig. 2.4 for signal representation). Thus the output voltage of the op amp can be expressed as

$$v_O = A_d v_{Id} + A_{cm} v_{Icm}$$

where A_d is the differential gain (referred to simply as A in the text) and A_{cm} is the common-mode gain (assumed to be zero in the text). The op amp's effectiveness in rejecting common-mode signals is measured by its CMRR, defined as

$$\text{CMRR} = 20 \log \left| \frac{A_d}{A_{cm}} \right|$$

Consider an op amp whose internal structure is of the type shown in Fig. E2.3 except for a mismatch ΔG_m between the transconductances of the two channels; that is,

$$G_{m1} = G_m - \tfrac{1}{2}\Delta G_m$$

$$G_{m2} = G_m + \tfrac{1}{2}\Delta G_m$$

Find expressions for A_d, A_{cm}, and CMRR. If A_d is 80 dB and the two transconductances are matched to within 0.1% of each other, calculate A_{cm} and CMRR.

SECTION 2.2: THE INVERTING CONFIGURATION

2.8 Assuming ideal op amps, find the voltage gain v_o / v_i and input resistance R_{in} of each of the circuits in Fig. P2.8.

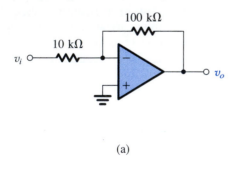

(a)

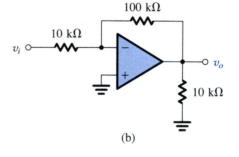

(b)

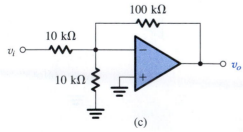

(c)

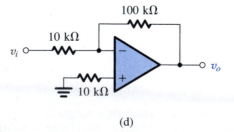

(d)

FIGURE P2.8

2.9 A particular inverting circuit uses an ideal op amp and two 10-kΩ resistors. What closed-loop gain would you expect? If a dc voltage of +5.00 V is applied at the input, what output result? If the 10-kΩ resistors are said to be "5% resistors," having values somewhere in the range (1 ± 0.05) times the nominal value, what range of outputs would you expect to actually measure for an input of precisely 5.00 V?

2.10 You are provided with an ideal op amp and three 10-kΩ resistors. Using series and parallel resistor combinations, how many different inverting-amplifier circuit topologies are possible? What is the largest (noninfinite) available voltage gain? What is the smallest (nonzero) available gain? What are the input resistances in these two cases?

2.11 For ideal op amps operating with the following feedback networks in the inverting configuration, what closed-loop gain results?

(a) $R_1 = 10$ kΩ, $R_2 = 10$ kΩ
(b) $R_1 = 10$ kΩ, $R_2 = 100$ kΩ
(c) $R_1 = 10$ kΩ, $R_2 = 1$ kΩ
(d) $R_1 = 100$ kΩ, $R_2 = 10$ MΩ
(e) $R_1 = 100$ kΩ, $R_2 = 1$ MΩ

D2.12 Using an ideal op amp, what are the values of the resistors R_1 and R_2 to be used to design amplifiers with the closed-loop gains listed below? In your designs, use at least one 10-kΩ resistor and another larger resistor.

(a) -1 V/V
(b) -2 V/V
(c) -0.5 V/V
(d) -100 V/V

D2.13 Design an inverting op-amp circuit for which the gain is -5 V/V and the total resistance used is 120 kΩ.

D2.14 Using the circuit of Fig. 2.5 and assuming an ideal op amp, design an inverting amplifier with a gain of 26 dB having the largest possible input resistance under the constraint of having to use resistors no larger than 10 MΩ. What is the input resistance of your design?

2.15 An ideal op amp connected as shown in Fig. 2.5 of the text with $R_1 = 10$ kΩ and $R_2 = 100$ kΩ. A symmetrical square-wave signal with levels of 0 V and 1 V is applied at the input. Sketch and clearly label the waveform of the resulting output voltage. What is its average value? What is its highest value? What is its lowest value?

2.16 For the circuit in Fig. P2.16, find the currents through all branches and the voltages at all nodes. Since the current supplied by the op amp is greater than the current drawn from the input signal source, where does the additional current come from?

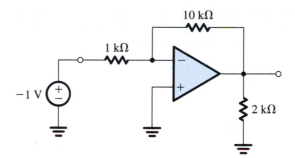

FIGURE P2.16

2.17 An inverting op amp circuit is fabricated with the resistors R_1 and R_2 having $x\%$ tolerance (i.e., the value of each resistance can deviate from the nominal value by as much as $\pm x\%$). What is the tolerance on the realized closed-loop gain? Assume the op amp to be ideal. If the nominal closed-loop gain is -100 V/V and $x = 5$, what is the range of gain values expected from such a circuit?

2.18 An ideal op amp with 5-kΩ and 15-kΩ resistors is used to create a +5-V supply from a -15-V reference. Sketch the circuit. What are the voltages at the ends of the 5-kΩ resistor? If these resistors are so-called 1% resistors, whose actual values are the range bounded by the nominal value $\pm 1\%$, what are the limits of the output voltage produced? If the -15-V supply can also vary by $\pm 1\%$, what is the range of the output voltages that might be found?

2.19 An inverting op-amp circuit for which the required gain is -50 V/V uses an op amp whose open-loop gain is only 200 V/V. If the larger resistor used is 100 kΩ, to what must the smaller be adjusted? With what resistor must a 2-kΩ resistor connected to the input be shunted to achieved this goal? (Note that a resistor R_a is said to be shunted by resistor R_b when R_b is placed in parallel with R_a.)

D2.20 (a) Design an inverting amplifier with a closed-loop gain of -100 V/V and an input resistance of 1 kΩ.
(b) If the op amp is known to have an open-loop gain of 1000 V/V, what do you expect the closed-loop gain of your circuit to be (assuming the resistors have precise values)?
(c) Give the value of a resistor you can place in parallel (shunt) with R_1 to restore the closed-loop gain to its nominal value. Use the closest standard 1% resistor value (see Appendix G).

2.21 An op amp with an open-loop gain of 1000 V/V is used in the inverting configuration. If in this application the output voltage ranges from -10 V to +10 V, what is the maximum voltage by which the "virtual ground node" departs from its ideal value?

2.22 The circuit in Fig. P2.22 is frequently used to provide an output voltage v_o proportional to an input signal current i_i. Derive expressions for the transresistance $R_m \equiv v_o/i_i$ and the input resistance $R_i \equiv v_i/i_i$ for the following cases:

(a) A is infinite.
(b) A is finite.

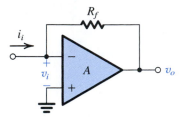

FIGURE P2.22

2.23 Derive an expression for the input resistance of the inverting amplifier of Fig. 2.5 taking into account the finite open-loop gain A of the op amp.

*2.24 For an inverting op amp with open-loop gain A and nominal closed-loop gain R_2/R_1, find the minimum value the gain A must have (in terms of R_2/R_1) for a gain error of 0.1%, 1%, and 10%. In each case, what value of resistor R_{1a} can be used to shunt R_1 to achieve the nominal result?

*2.25 Figure P2.25 shows an op amp that is ideal except for having a finite open-loop gain and is used to realize an inverting amplifier whose gain has a nominal magnitude $G = R_2/R_1$. To compensate for the gain reduction due to the finite A, a resistor R_c is shunted across R_1. Show that perfect compensation is achieved when R_c is selected according to

$$\frac{R_c}{R_1} = \frac{A - G}{1 + G}$$

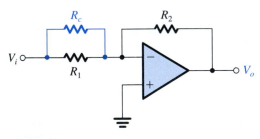

FIGURE P2.25

*2.26 Rearrange Eq. (2.5) to give the amplifier open-loop gain A required to realize a specified closed-loop gain $(G_{nominal} = -R_2/R_1)$ within a specified gain error ε,

$$\varepsilon \equiv \left| \frac{G - G_{nominal}}{G_{nominal}} \right|$$

For a closed-loop gain of -100 and a gain error of $\leq 10\%$, what is the minimum A required?

*2.27 Using Eq. (2.5), determine the value of A for which a reduction of A by $x\%$ results in a reduction in $|G|$ by $(x/k)\%$. Find the value of A required for the case in which the nominal closed-loop gain is 100, x is 50, and k is 100.

2.28 Consider the circuit in Fig. 2.8 with $R_1 = R_2 = R_4 = 1$ MΩ, and assume the op amp to be ideal. Find values for R_3 to obtain the following gains:

(a) -10 V/V
(b) -100 V/V
(c) -2 V/V

D2.29 An inverting op-amp circuit using an ideal op amp must be designed to have a gain of -1000 V/V using resistors no larger than 100 kΩ.

(a) For the simple two-resistor circuit, what input resistance would result?
(b) If the circuit in Fig. 2.8 is used with three resistors of maximum value, what input resistance results? What is the value of the smallest resistor needed?

2.30 The inverting circuit with the T network in the feedback is redrawn in Fig. P2.30 in a way that emphasizes the observation that R_2 and R_3 in effect are in parallel (because the ideal op amp forces a virtual ground at the inverting input terminal). Use this observation to derive an expression for the gain (v_O/v_I) by first finding (v_X/v_I) and (v_O/v_X).

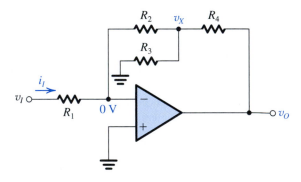

FIGURE P2.30

*2.31 The circuit in Fig. P2.31 can be considered an extension of the circuit in Fig. 2.8.

(a) Find the resistances looking into node 1, R_1; node 2, R_2; node 3, R_3; and node 4, R_4.
(b) Find the currents I_1, I_2, I_3, and I_4 in terms of the input current I.

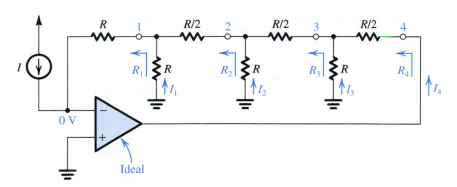

FIGURE P2.31

(c) Find the voltages at nodes 1, 2, 3, and 4, that is, V_1, V_2, V_3, and V_4 in terms of (IR).

2.32 The circuit in Fig. P2.32 utilizes an ideal op amp.

(a) Find I_1, I_2, I_3, and V_x.
(b) If V_O is not to be lower than -13 V, find the maximum allowed value for R_L.
(c) If R_L is varied in the range $100\ \Omega$ to $1\ k\Omega$, what is the corresponding change in I_L and in V_O?

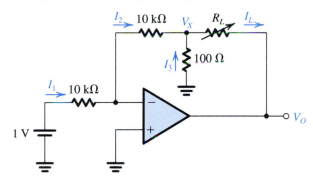

FIGURE P2.32

D2.33 Assuming the op amp to be ideal, it is required to design the circuit shown in Fig. P2.33 to implement a current amplifier with gain $i_L/i_I = 20$ A/A.

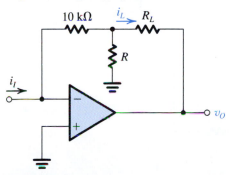

FIGURE P2.33

(a) Find the required value for R.
(b) If $R_L = 1\ k\Omega$ and the op amp operates in an ideal manner so long as v_O is in the range ± 12 V. What range of i_I is possible?
(c) What is the input resistance of the current amplifier? If the amplifier is fed with a current source having a current of 1 mA and a source resistance of $10\ k\Omega$, find i_L.

2.34 Figure P2.34 shows the inverting amplifier circuit of Fig. 2.8 redrawn to emphasize the fact that R_3 and R_4 can be thought of as a voltage divider connected across the output v_O and from which a fraction of the output voltage (that available at node A) is fed back through R_2. Assuming $R_2 \gg R_3$ and thus that the loading of the feedback network can be ignored, express v_A as a function of v_O. Now express v_A as a function of v_I. Use these two relationships to find the (approximate) relationship between v_O and v_I. With appropriate manipulation, compare it with the result obtained in Example 2.2. Show that the exact result can be obtained by noting that R_2 appears in effect across R_3 and, thus, that the voltage divider is composed of R_4 and $(R_3 \| R_2)$.

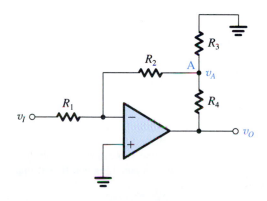

FIGURE P2.34

D2.35 Design the circuit shown in Fig. P2.35 to have an input resistance of $100\ k\Omega$ and a gain that can be varied from -1 V/V to -10 V/V using the 10-kΩ potentiometer R_4. What voltage gain results when the potentiometer is set exactly at its middle value?

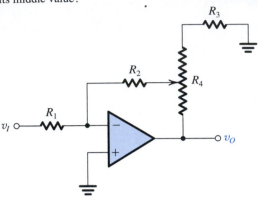

FIGURE P2.35

2.36 A weighted summer circuit using an ideal op amp has three inputs using 100-kΩ resistors and a feedback resistor of 50 kΩ. A signal v_1 is connected to two of the inputs while a signal v_2 is connected to the third. Express v_O in terms of v_1 and v_2. If $v_1 = 3$ V and $v_2 = -3$ V, what is v_O?

D2.37 Design an op amp circuit to provide an output $v_O = -[4v_1 + (v_2/3)]$. Choose relatively low values of resistors but ones for which the input current (from each input signal source) does not exceed 0.1 mA for 1-V input signals.

D2.38 Using the scheme illustrated in Fig. 2.10, design an op-amp circuit with inputs v_1, v_2, and v_3 whose output is $v_O = -(2v_1 + 4v_2 + 8v_3)$ using small resistors but no smaller than 10 kΩ.

D2.39 An ideal op amp is connected in the weighted summer configuration of Fig. 2.10. The feedback resistor $R_f = 10$ kΩ, and six 10-kΩ resistors are connected to the inverting input terminal of the op amp. Show, by sketching the various circuit configurations, how this basic circuit can be used to implement the following functions:

(a) $v_O = -(v_1 + 2v_2 + 3v_3)$
(b) $v_O = -(v_1 + v_2 + 2v_3 + 2v_4)$
(c) $v_O = -(v_1 + 5v_2)$
(d) $v_O = -6v_1$

In each case find the input resistance seen by each of the signal sources supplying v_1, v_2, v_3, and v_4. Suggest at least two additional summing functions that you can realize with this circuit. How would you realize a summing coefficient that is 0.5?

D2.40 Give a circuit, complete with component values, for a weighted summer that shifts the dc level of a sine-wave signal of 5 $\sin(\omega t)$ V from zero to -5 V. Assume that in addition to the sine-wave signal you have a dc reference voltage of 2 V available. Sketch the output signal waveform.

D2.41 Use two ideal op amps and resistors to implement the summing function.

$$v_O = v_1 + 2v_2 - 3v_3 - 4v_4$$

D*2.42 In an instrumentation system, there is a need to take the difference between two signals, one of $v_1 = 3 \sin(2\pi \times 60t) + 0.01 \sin(2\pi \times 1000t)$, volts and another of $v_2 = 3 \sin(2\pi \times 60t) - 0.01 \sin(2\pi \times 1000t)$ volts. Draw a circuit that finds the required difference using two op amps and mainly 10-kΩ resistors. Since it is desirable to amplify the 1000-Hz component in the process, arrange to provide an overall gain of 10 as well. The op amps available are ideal except that their output voltage swing is limited to ± 10 V.

***2.43** Figure P2.43 shows a circuit for a digital-to-analog converter (DAC). The circuit accepts a 4-bit input binary word $a_3a_2a_1a_0$, where a_0, a_1, a_2, and a_3 take the values of 0 or 1, and it provides an analog output voltage v_O proportional to the value of the digital input. Each of the bits of the input word controls the correspondingly numbered switch. For instance, if a_2 is 0 then switch S_2 connects the 20-kΩ resistor to ground, while if a_2 is 1 then S_2 connects the 20-kΩ resistor to the $+5$-V power supply. Show that v_O is given by

$$v_O = -\frac{R_f}{16}[2^0a_0 + 2^1a_1 + 2^2a_2 + 2^3a_3]$$

where R_f is in kΩ. Find the value of R_f so that v_O ranges from 0 to -12 volts.

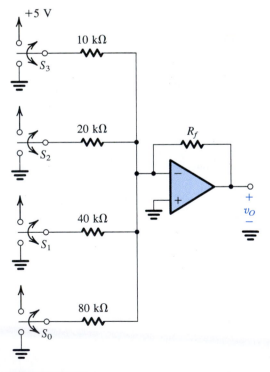

FIGURE P2.43

SECTION 2.3: THE NONINVERTING CONFIGURATION

D2.44 Using an ideal op amp to implement designs for the following closed-loop gains, what values of resistors (R_1, R_2) should be used? Where possible, use at least one 10-kΩ resistor as the smallest resistor in your design.

(a) +1 V/V
(b) +2 V/V
(c) +101 V/V
(d) +100 V/V

D2.45 Design a circuit based on the topology of the non-inverting amplifier to obtain a gain of +1.5 V/V, using only 10-kΩ resistors. Note that there are two possibilities. Which of these can be easily converted to have a gain of either +1.0 V/V or +2.0 V/V simply by short-circuiting a single resistor in each case?

D2.46 Figure P2.46 shows a circuit for an analog voltmeter of very high input resistance that uses an inexpensive moving-coil meter. The voltmeter measures the voltage V applied between the op amp's positive-input terminal and ground. Assuming that the moving coil produces full-scale deflection when the current passing through it is 100 μA, find the value of R such that full-scale reading is obtained when V is +10 V. Does the meter resistance shown affect the voltmeter calibration?

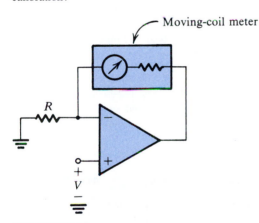

FIGURE P2.46

D*2.47 (a) Use superposition to show that the output of the circuit in Fig. P2.47 is given by

$$v_O = -\left[\frac{R_f}{R_{N1}}v_{N1} + \frac{R_f}{R_{N2}}v_{N2} + \cdots + \frac{R_f}{R_{Nn}}v_{Nn}\right]$$

$$+ \left[1 + \frac{R_f}{R_N}\right]\left[\frac{R_P}{R_{P1}}v_{P1} + \frac{R_P}{R_{P2}}v_{P2} + \cdots + \frac{R_P}{R_{Pn}}v_{Pn}\right]$$

where $R_N = R_{N1}//R_{N2}//\cdots//R_{Nn}$ and

$$R_P = R_{P1}//R_{P2}//\cdots//R_{Pn}//R_{P0}$$

(b) Design a circuit to obtain

$$v_O = -2v_{N1} + v_{P1} + 2v_{P2}$$

The smallest resistor used should be 10 kΩ.

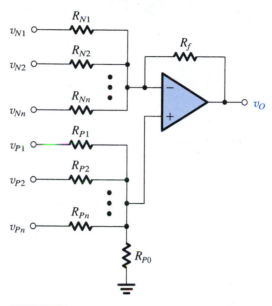

FIGURE P2.47

D2.48 Design a circuit, using one ideal op amp, whose output is $v_O = v_{I1} + 3v_{I2} - 2(v_{I3} + 3v_{I4})$. (*Hint:* Use a structure similar to that shown in general form in Fig. P2.47.)

2.49 Derive an expression for the voltage gain, v_O/v_I, of the circuit in Fig. P2.49.

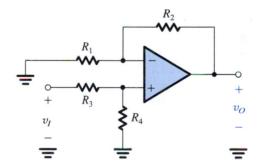

FIGURE P2.49

2.50 For the circuit in Fig. P2.50, use superposition to find v_O in terms of the input voltages v_1 and v_2. Assume an ideal op amp. For

$$v_1 = 10\sin(2\pi \times 60t) - 0.1\sin(2\pi \times 1000t), \text{ volts}$$

$$v_2 = 10\sin(2\pi \times 60t) + 0.1\sin(2\pi \times 1000t), \text{ volts}$$

find v_O.

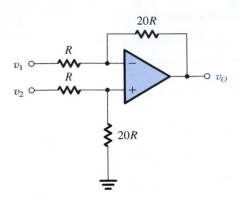

FIGURE P2.50

D2.51 The circuit shown in Fig. P2.51 utilizes a 10-kΩ potentiometer to realize an adjustable-gain amplifier. Derive an expression for the gain as a function of the potentiometer setting x. Assume the op amp to be ideal. What is the range of gains obtained? Show how to add a fixed resistor so that the gain range can be 1 to 21 V/V. What should the resistor value be?

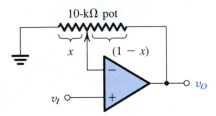

FIGURE P2.51

D2.52 Given the availability of resistors of value 1 kΩ and 10 kΩ only, design a circuit based on the noninverting configuration to realize a gain of +10 V/V.

2.53 It is required to connect a 10-V source with a source resistance of 100 kΩ to a 1-kΩ load. Find the voltage that will appear across the load if:

(a) The source is connected directly to the load.
(b) A unity-gain op-amp buffer is inserted between the source and the load.

In each case find the load current and the current supplied by the source. Where does the load current come from in case (b)?

2.54 Derive an expression for the gain of the voltage follower of Fig. 2.14 assuming the op amp to be ideal except for having a finite gain A. Calculate the value of the closed-loop gain for $A = 1000$, 100, and 10. In each case find the percentage error in gain magnitude from the nominal value of unity.

2.55 Complete the following table for feedback amplifiers created using one ideal op amp. Note that R_{in} signifies input resistance and R_1 and R_2 are feedback-network resistors as labelled in the inverting and noninverting configurations.

Case	Gain	R_{in}	R_1	R_2
a	−10 V/V	10 kΩ		
b	−1 V/V		100 kΩ	
c	−2 V/V			100 kΩ
d	+1 V/V	∞		
e	+2 V/V		10 kΩ	
f	+11 V/V			100 kΩ
g	−0.5 V/V	10 kΩ		

D2.56 A noninverting op-amp circuit with nominal gain of 10 V/V uses an op amp with open-loop gain of 50 V/V and a lowest-value resistor of 10 kΩ. What closed-loop gain actually results? With what value resistor can which resistor be shunted to achieve the nominal gain? If in the manufacturing process, an op amp of gain 100 V/V were used, what closed-loop gain would result in each case (the uncompensated one, and the compensated one)?

2.57 Using Eq. (2.11), show that if the reduction in the closed-loop gain G from the nominal value $G_0 = 1 + R_2/R_1$ is to be kept less than $x\%$ of G_0, then the open-loop gain of the op amp must exceed G_0 by at least a factor $F = (100/x) - 1 \simeq 100/x$. Find the required F for $x = 0.01, 0.1, 1$, and 10. Utilize these results to find for each value of x the minimum required open-loop gain to obtain closed-loop gains of 1, 10, 10^2, 10^3, and 10^4 V/V.

2.58 For each of the following combinations of op-amp open-loop gain A and nominal closed-loop gain G_0, calculate the actual closed-loop gain G that is achieved. Also, calculate the percentage by which $|G|$ falls short of the nominal gain magnitude $|G_0|$.

Case	G_0 (V/V)	A (V/V)
a	−1	10
b	+1	10
c	−1	100
d	+10	10
e	−10	100
f	−10	1000
g	+1	2

2.59 Figure P2.59 shows a circuit that provides an output voltage v_O whose value can be varied by turning the wiper of the 100-kΩ potentiometer. Find the range over which v_O can be varied. If the potentiometer is a "20-turn" device, find the change in v_O corresponding to each turn of the pot.

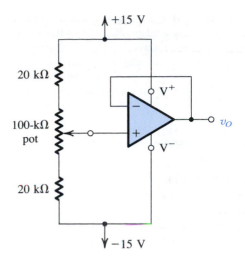

FIGURE P2.59

SECTION 2.4: DIFFERENCE AMPLIFIERS

2.60 Find the voltage gain v_O/v_{Id} for the difference amplifier of Fig. 2.16 for the case $R_1 = R_3 = 10\ \text{k}\Omega$ and $R_2 = R_4 = 100\ \text{k}\Omega$. What is the differential input resistance R_{id}? If the two key resistance ratios (R_2/R_1) and (R_4/R_3) are different from each other by 1%, what do you expect the common-mode gain A_{cm} to be? Also, find the CMRR in this case.

D2.61 Using the difference amplifier configuration of Fig. 2.16 and assuming an ideal op-amp, design the circuit to provide the following differential gains. In each case the differential input resistance should be 20 kΩ.

(a) 1 V/V
(b) 2 V/V
(c) 100 V/V
(d) 0.5 V/V

2.62 For the circuit shown in Fig. P2.62, express v_O as a function of v_1 and v_2. What is the input resistance seen by v_1 alone? By v_2 alone? By a source connected between the two input terminals? By a source connected to both input terminals simultaneously?

2.63 Consider the difference amplifier of Fig. 2.16 with the two input terminals connected together to an input common-mode signal source. For $R_2/R_1 = R_4/R_3$, show that the input common-mode resistance is $(R_3 + R_4) \| (R_1 + R_2)$.

2.64 Consider the circuit of Fig. 2.16, and let each of the v_{I1} and v_{I2} signal sources have a series resistance R_s. What condition must apply in addition to the condition in Eq. (2.15) in order for the amplifier to function as an ideal difference amplifier?

*2.65 For the difference amplifier shown in Fig. P2.62, let all the resistors be 100 k$\Omega \pm x\%$. Find an expression for the worst-case common-mode gain that results. Evaluate this for $x = 0.1$, 1, and 5. Also, evaluate the resulting CMRR in each case.

2.66 For the difference amplifier of Fig. 2.16, show that if each resistor has a tolerance of $\pm 100\ \varepsilon\%$ (i.e., for, say, a 5% resistor, $\varepsilon = 0.05$) then the worst-case CMRR is given approximately by

$$\text{CMRR} \cong 20 \log \left[\frac{K+1}{4\varepsilon} \right]$$

where K is the nominal (ideal) value of the ratios (R_2/R_1) and (R_4/R_3). Calculate the value of worst-case CMRR for an amplifier designed to have a differential gain of ideally 100 V/V, assuming that the op amp is ideal and that 1% resistors are used.

D*2.67 Design the difference amplifier circuit of Fig. 2.16 to realize a differential gain of 100, a differential input resistance of 20 kΩ, and a minimum CMRR of 80 dB. Assume the op amp to be ideal. Specify both the resistor values and their required tolerance (e.g., better than $x\%$).

*2.68 (a) Find A_d and A_{cm} for the difference amplifier circuit shown in Fig. P2.68.
(b) If the op amp is specified to operate properly so long as the common-mode voltage at its positive and negative inputs falls in the range ± 2.5 V, what is the corresponding limitation on the range of the input common-mode signal v_{Icm}? (This is known as the **common-mode range** of the differential amplifier).
(c) The circuit is modified by connecting a 10-kΩ resistor between node A and ground and another 10-kΩ resistor between node B and ground. What will now be the values of A_d, A_{cm}, and the input common-mode range?

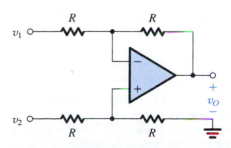

FIGURE P2.62

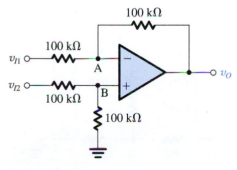

FIGURE P2.68

**2.69 To obtain a high-gain, high-input-resistance difference amplifier the circuit in Fig. P2.69 employs positive feedback, in addition to the negative feedback provided by the resistor R connected from the output to the negative input of the op amp. Specifically, a voltage divider (R_5, R_6) connected across the output feeds a fraction β of the output, that is, a voltage βv_O, back to the positive-input terminal of the op amp through a resistor R. Assume that R_5 and R_6 are much smaller than R so that the current through R is much lower than the current in the voltage divider, with the result that $\beta \cong R_6|(R_5 + R_6)$. Show that the differential gain is given by

$$A_d \equiv \frac{v_O}{v_{Id}} = \frac{1}{1 - \beta}$$

Design the circuit to obtain a differential gain of 10 V/V and differential input resistance of 2 MΩ. Select values for R, R_5, and R_6 such that $(R_5 + R_6) \le R/100$.

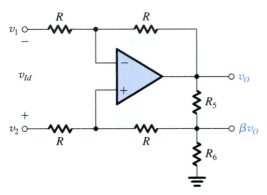

FIGURE P2.69

*2.70 Figure P2.70 shows a modified version of the difference amplifier. The modified circuit includes a resistor R_G, which can be used to vary the gain. Show that the differential voltage gain is given by

$$\frac{v_O}{v_{Id}} = -2\frac{R_2}{R_1}\left[1 + \frac{R_2}{R_G}\right]$$

(*Hint:* The virtual short circuit at the op amp input causes the current through the R_1 resistors to be $v_{Id}/2R_1$.)

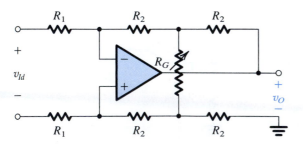

FIGURE P2.70

D*2.71 The circuit shown in Fig. P2.71 is a representation of a versatile, commercially available IC, the INA105, manufactured by Burr-Brown and known as a differential amplifier module. It consists of an op amp and precision, laser-trimmed, metal-film resistors. The circuit can be configured for a variety of applications by the appropriate connection of terminals A, B, C, D, and O.

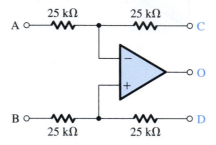

FIGURE P2.71

(a) Show how the circuit can be used to implement a difference amplifier of unity gain.
(b) Show how the circuit can be used to implement single-ended amplifiers with gains:

    (i) -1 V/V
    (ii) $+1$ V/V
    (iii) $+2$ V/V
    (iv) $+1/2$ V/V

Avoid leaving a terminal open-circuited, for such a terminal may act as an "antenna," picking up interference and noise through capacitive coupling. Rather, find a convenient node to connect such a terminal in a redundant way. When more than one circuit implementation is possible, comment on the relative merits of each, taking into account such considerations as dependence on component matching and input resistance.

2.72 Consider the instrumentation amplifier of Fig. 2.20(b) with a common-mode input voltage of $+3$ V (dc) and a differential input signal of 80-mV peak sine wave. Let $2R_1 = 1$ kΩ, $R_2 = 50$ kΩ, $R_3 = R_4 = 10$ kΩ. Find the voltage at every node in the circuit.

2.73 (a) Consider the instrumentation amplifier circuit of Fig. 2.20(a). If the op amps are ideal except that their outputs saturate at ±14 V, in the manner shown in Fig. 1.13, find the maximum allowed input common-mode signal for the case $R_1 = 1$ kΩ and $R_2 = 100$ kΩ.
(b) Repeat (a) for the circuit in Fig. 2.20(b), and comment on the difference between the two circuits.

2.74 (a) Expressing v_{I1} and v_{I2} in terms of differential and common-mode components, find v_{O1} and v_{O2} in the circuit in Fig. 2.20(a) and hence find their differential component $v_{O2} - v_{O1}$ and their common-mode component $\frac{1}{2}(v_{O1} + v_{O2})$. Now find the differential gain and the common-mode gain of the first stage of this instrumentation amplifier and hence the CMRR.

(b) Repeat for the circuit in Fig. 2.20(b), and comment on the difference between the two circuits.

2.75 For an instrumentation amplifier of the type shown in Fig. 2.20(b), a designer proposes to make $R_2 = R_3 = R_4 = 100$ kΩ, and $2R_1 = 10$ kΩ. For ideal components, what difference-mode gain, common-mode gain, and CMRR result? Reevaluate the worst-case values for these for the situation in which all resistors are specified as $\pm 1\%$ units. Repeat the latter analysis for the case in which $2R_1$ is reduced to 1 kΩ. What do you conclude about the importance of the relative difference gains of the first and second stages?

D2.76 Design the instrumentation-amplifier circuit of Fig. 2.20(b) to realize a differential gain, variable in the range 1 to 100, utilizing a 100-kΩ pot as variable resistor. (*Hint:* Design the second stage for a gain of 0.5.)

*2.77 The circuit shown in Fig. P2.77 is intended to supply a voltage to floating loads (those for which both terminals are ungrounded) while making greatest possible use of the available power supply.

(a) Assuming ideal op amps, sketch the voltage waveforms at nodes B and C for a 1-V peak-to-peak sine wave applied at A. Also sketch v_O.

(b) What is the voltage gain v_O / v_I?

(c) Assuming that the op amps operate from ± 15-V power supplies and that their output saturates at ± 14 V (in the manner

shown in Fig. 1.13), what is the largest sine wave output that can be accommodated? Specify both its peak-to-peak and rms values.

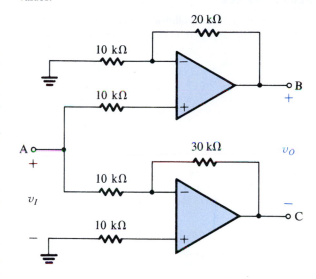

FIGURE P2.77

*2.78 The two circuits in Fig. P2.78 are intended to function as voltage-to-current converters; that is, they supply the load impedance Z_L with a current proportional to v_I and independent of the value of Z_L. Show that this is indeed the case, and find for each circuit i_O as a function of v_I. Comment on the differences between the two circuits.

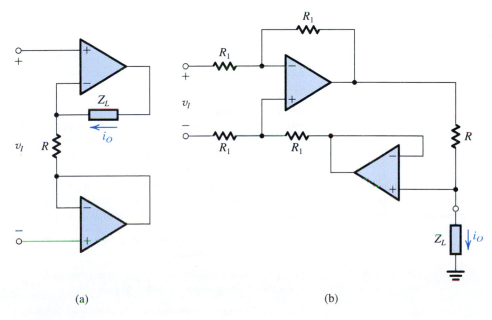

(a)   (b)

FIGURE P2.78

SECTION 2.5: EFFECT OF FINITE OPEN-LOOP GAIN AND BANDWIDTH ON CIRCUIT PERFORMANCE

2.79 The data in the following table apply to internally compensated op amps. Fill in the blank entries.

A_0	f_b (Hz)	f_t (Hz)
10^5	10^2	
10^6		10^6
	10^3	10^8
	10^{-1}	10^6
2×10^5	10	

2.80 A measurement of the open-loop gain of an internally compensated op amp at very low frequencies shows it to be 86 dB; at 100 kHz, this shows it is 40 dB. Estimate values for A_0, f_b, and f_t.

2.81 Measurements of the open-loop gain of a compensated op amp intended for high-frequency operation indicate that the gain is 5.1×10^3 at 100 kHz and 8.3×10^3 at 10 kHz. Estimate its 3-dB frequency, its unity-gain frequency, and its dc gain.

2.82 Measurements made on the internally compensated amplifiers listed below provide the dc gain and the frequency at which the gain has dropped by 20 dB. For each, what are the 3 dB and unity-gain frequencies?

(a) 3×10^5 V/V and 6×10^2 Hz
(b) 50×10^5 V/V and 10 Hz
(c) 1500 V/V and 0.1 MHz
(d) 100 V/V and 0.1 GHz
(e) 25 V/mV and 25 kHz

2.83 An inverting amplifier with nominal gain of -20 V/V employs an op amp having a dc gain of 10^4 and a unity-gain frequency of 10^6 Hz. What is the 3-dB frequency f_{3dB} of the closed-loop amplifier? What is its gain at $0.1 f_{3dB}$ and at $10 f_{3dB}$?

2.84 A particular op amp, characterized by a gain–bandwidth product of 20 MHz, is operated with a closed-loop gain of $+100$ V/V. What 3-dB bandwidth results? At what frequency does the closed-loop amplifier exhibit a $-6°$ phase shift? A $-84°$ phase shift?

2.85 Find the f_t required for internally compensated op amps to be used in the implementation of closed-loop amplifiers with the following nominal dc gains and 3-dB bandwidths:

(a) -100 V/V; 100 kHz
(b) $+100$ V/V; 100 kHz
(c) $+2$ V/V; 10 MHz
(d) -2 V/V; 10 MHz
(e) -1000 V/V; 20 kHz
(f) $+1$ V/V; 1 MHz
(g) -1 V/V; 1 MHz

2.86 A noninverting op-amp circuit with a gain of 100 V/V is found to have a 3-dB frequency of 8 kHz. For a particular system application, a bandwidth of 20 kHz is required. What is the highest gain available under these conditions?

2.87 Consider a unity-gain follower utilizing an internally compensated op amp with $f_t = 1$ MHz. What is the 3-dB frequency of the follower? At what frequency is the gain of the follower 1% below its low-frequency magnitude? If the input to the follower is a 1-V step, find the 10% to 90% rise time of the output voltage. (*Note:* The step response of STC low-pass networks is discussed in Appendix D.)

D*2.88 It is required to design a noninverting amplifier with a dc gain of 10. When a step voltage of 100 mV is applied at the input, it is required that the output be within 1% of its final value of 1 V in at most 100 ns. What must the f_t of the op amp be? (*Note:* The step response of STC low-pass networks is discussed in Appendix D.)

D*2.89 This problem illustrates the use of cascaded closed-loop amplifiers to obtain an overall bandwidth greater than can be achieved using a single-stage amplifier with the same overall gain.

(a) Show that cascading two identical amplifier stages, each having a low-pass STC frequency response with a 3-dB frequency f_1, results in an overall amplifier with a 3-dB frequency given by

$$f_{3dB} = \sqrt{\sqrt{2} - 1}\, f_1$$

(b) It is required to design a noninverting amplifier with a dc gain of 40 dB utilizing a single internally-compensated op amp with $f_t = 1$ MHz. What is the 3-dB frequency obtained?
(c) Redesign the amplifier of (b) by cascading two identical noninverting amplifiers each with a dc gain of 20 dB. What is the 3-dB frequency of the overall amplifier? Compare this to the value obtained in (b) above.

D2.90** A designer, wanting to achieve a stable gain of 100 V/V at 5 MHz, considers her choice of amplifier topologies. What unity-gain frequency would a single operational amplifier require to satisfy her need? Unfortunately, the best available amplifier has an f_t of 40 MHz. How many such amplifiers connected in a cascade of identical noninverting stages would she need to achieve her goal? What is the 3-dB frequency of each stage she can use? What is the overall 3-dB frequency?

2.91 Consider the use of an op amp with a unity-gain frequency f_t in the realization of

(a) an inverting amplifier with dc gain of magnitude K.
(b) a noninverting amplifier with a dc gain of K.

In each case find the 3-dB frequency and the gain-bandwidth product (GBP $\equiv$ |Gain| $\times f_{3dB}$). Comment on the results.

*2.92 Consider an inverting summer with two inputs V_1 and V_2 and with $V_o = -(V_1 + V_2)$. Find the 3-dB frequency of each of the gain functions V_o / V_1 and V_o / V_2 in terms of the op amp f_t. (*Hint:* In each case, the other input to the summer can be set to zero—an application of superposition.)

SECTION 2.6: LARGE-SIGNAL OPERATION OF OP AMPS

2.93 A particular op amp using ±15-V supplies operates linearly for outputs in the range −12 V to +12 V. If used in an inverting amplifier configuration of gain −100, what is the rms value of the largest possible sine wave that can be applied at the input without output clipping?

2.94 Consider an op amp connected in the inverting configuration to realize a closed-loop gain of −100 V/V utilizing resistors of 1 kΩ and 100 kΩ. A load resistance R_L is connected from the output to ground, and a low-frequency sine-wave signal of peak amplitude V_p is applied to the input. Let the op amp be ideal except that its output voltage saturates at ±10 V and its output current is limited to the range ±20 mA.

(a) For $R_L = 1$ kΩ, what is the maximum possible value of V_p while an undistorted output sinusoid is obtained?
(b) Repeat (a) for $R_L = 100$ Ω.
(c) If it is desired to obtain an output sinusoid of 10-V peak amplitude, what minimum value of R_L is allowed?

2.95 An op amp having a slew rate of 20 V/μs is to be used in the unity-gain follower configuration, with input pulses that rise from 0 to 3 V. What is the shortest pulse that can be used while ensuring full-amplitude output? For such a pulse, describe the output resulting.

*2.96 For operation with 10-V output pulses with the requirement that the sum of the rise and fall times should represent only 20% of the pulse width (at half amplitude), what is the slew-rate requirement for an op amp to handle pulses 2 μs wide? (*Note:* The rise and fall times of a pulse signal are usually measured between the 10%- and 90%-height points.)

2.97 What is the highest frequency of a triangle wave of 20-V peak-to-peak amplitude that can be reproduced by an op amp whose slew rate is 10 V/μs? For a sine wave of the same frequency, what is the maximum amplitude of output signal that remains undistorted?

2.98 For an amplifier having a slew rate of 60 V/μs, what is the highest frequency at which a 20-V peak-to-peak sine wave can be produced at the output?

D*2.99 In designing with op amps one has to check the limitations on the voltage and frequency ranges of operation of the closed-loop amplifier, imposed by the op amp finite bandwidth (f_t), slew rate (SR), and output saturation (V_{omax}).

This problem illustrates the point by considering the use of an op amp with $f_t = 2$ MHz, SR = 1 V/μs, and $V_{omax} = 10$ V in the design of a noninverting amplifier with a nominal gain of 10. Assume a sine-wave input with peak amplitude V_i.

(a) If $V_i = 0.5$ V, what is the maximum frequency before the output distorts?
(b) If $f = 20$ kHz, what is the maximum value of V_i before the output distorts?
(c) If $V_i = 50$ mV, what is the useful frequency range of operation?
(d) If $f = 5$ kHz, what is the useful input voltage range?

SECTION 2.7: DC IMPERFECTIONS

2.100 An op amp wired in the inverting configuration with the input grounded, having $R_2 = 100$ kΩ and $R_1 = 1$ kΩ, has an output dc voltage of −0.3 V. If the input bias current is known to be very small, find the input offset voltage.

2.101 A noninverting amplifier with a gain of 200 uses an op amp having an input offset voltage of ±2 mV. Find the output when the input is 0.01 sin ωt, volts.

2.102 A noninverting amplifier with a closed-loop gain of 1000 is designed using an op amp having an input offset voltage of 3 mV and output saturation levels of ±13 V. What is the maximum amplitude of the sine wave that can be applied at the input without the output clipping? If the amplifier is capacitively coupled in the manner indicated in Fig. 2.36, what would the maximum possible amplitude be?

2.103 An op amp connected in a closed-loop inverting configuration having a gain of 1000 V/V and using relatively small-valued resistors is measured with input grounded to have a dc output voltage of −1.4 V. What is its input offset voltage? Prepare an offset-voltage-source sketch resembling that in Fig. 2.28. Be careful of polarities.

2.104 A particular inverting amplifier with nominal gain of −100 V/V uses an imperfect op amp in conjunction with 100-kΩ and 10-MΩ resistors. The output voltage is found to be +9.31 V when measured with the input open and +9.09 V with the input grounded.

(a) What is the bias current of this amplifier? In what direction does it flow?
(b) Estimate the value of the input offset voltage.
(c) A 10-MΩ resistor is connected between the positive-input terminal and ground. With the input left floating (disconnected), the output dc voltage is measured to be −0.8 V. Estimate the input offset current.

D*2.105 A noninverting amplifier with a gain of +10 V/V using 100 kΩ as the feedback resistor operates from a 5-kΩ source. For an amplifier offset voltage of 0 mV, but with a bias current of 1 μA and an offset current of 0.1 μA, what range of outputs would you expect? Indicate where you would add an additional resistor to compensate for the bias currents. What does the range of possible outputs then become? A designer wishes to use this amplifier with a 15-kΩ source. In order to compensate for the bias current in this case, what resistor would you use? And where?

D2.106 The circuit of Fig. 2.36 is used to create an ac-coupled noninverting amplifier with a gain of 200 V/V using resistors no larger than 100 kΩ. What values of R_1, R_2, and R_3 should be used? For a break frequency due to C_1 at 100 Hz, and that due to C_2 at 10 Hz, what values of C_1 and C_2 are needed?

*2.107 Consider the difference amplifier circuit in Fig. 2.16. Let $R_1 = R_3 = 10$ kΩ and $R_2 = R_4 = 1$ MΩ. If the op amp has $V_{OS} = 4$ mV, $I_B = 0.3$ μA, and $I_{OS} = 50$ nA, find the worst-case (largest) dc offset voltage at the output.

*2.108 The circuit shown in Fig. P2.108 uses an op amp having a $\pm$4-mV offset. What is its output offset voltage? What does the output offset become with the input ac coupled through a capacitor C? If, instead, the 1-kΩ resistor is capacitively coupled to ground, what does the output offset become?

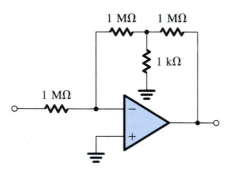

FIGURE P2.108

2.109 Using offset-nulling facilities provided for the op amp, a closed-loop amplifier with gain of +1000 is adjusted at 25°C to produce zero output with the input grounded. If the input offset-voltage drift of the op amp is specified to be 10 μV/°C, what output would you expect at 0°C and at 75°C? While nothing can be said separately about the polarity of the output offset at either 0 or 75°C, what would you expect their relative polarities to be?

2.110 An op amp is connected in a closed loop with gain of +100 utilizing a feedback resistor of 1 MΩ.

(a) If the input bias current is 100 nA, what output voltage results with the input grounded?

(b) If the input offset voltage is $\pm$1 mV and the input bias current as in (a), what is the largest possible output that can be observed with the input grounded?

(c) If bias-current compensation is used, what is the value of the required resistor? If the offset current is no more than one-tenth the bias current, what is the resulting output offset voltage (due to offset current alone)?

(d) With bias-current compensation as in (c) in place what is the largest dc voltage at the output due to the combined effect of offset voltage and offset current?

*2.111 An op amp intended for operation with a closed-loop gain of −100 V/V uses feedback resistors of 10 kΩ and 1 MΩ with a bias-current-compensation resistor R_3. What should the value of R_3 be? With input grounded, the output offset voltage is found to be +0.21 V. Estimate the input offset current assuming zero input offset voltage. If the input offset voltage can be as large as 1 mV of unknown polarity, what range of offset current is possible? What current injected into, or extracted from, the nongrounded end of R_3 would reduce the op amp output voltage to zero? For available $\pm$15-V supplies, what resistor and supply voltage would you use?

SECTION 2.8: INTEGRATORS AND DIFFERENTIATORS

2.112 A Miller integrator incorporates an ideal op amp, a resistor R of 100 kΩ, and a capacitor C of 10 nF. A sine-wave signal is applied to its input.

(a) At what frequency (in Hz) are the input and output signals equal in amplitude?

(b) At that frequency how does the phase of the output sine wave relate to that of the input?

(c) If the frequency is lowered by a factor of 10 from that found in (a), by what factor does the output voltage change, and in what direction (smaller or larger)?

(d) What is the phase relation between the input and output in situation (c)?

D2.113 Design a Miller integrator with a time constant of one second and an input resistance of 100 kΩ. For a dc voltage of −1 volt applied at the input at time 0, at which moment $v_O = -10$ V, how long does it take the output to reach 0 V? +10 V?

2.114 An op-amp-based inverting integrator is measured at 1 kHz to have a voltage gain of −100 V/V. At what frequency is its gain reduced to −1 V/V? What is the integrator time constant?

D2.115 Design a Miller integrator that has a unity-gain frequency of 1 krad/s and an input resistance of 100 kΩ. Sketch the output you would expect for the situation in which, with output initially at 0 V, a 2-V 2-ms pulse is applied to the input. Characterize the output that results when a sine wave $2 \sin 1000t$ is applied to the input?

D2.116 Design a Miller integrator whose input resistance is 20 kΩ and unity-gain frequency is 10 kHz. What components are needed? For long-term stability, a feedback resistor is introduced across the capacitor, which limits the dc gain to 40 dB. What is its value? What is the associated lower 3-dB frequency? Sketch and label the output which results with a 0.1-ms, 1-V positive-input pulse (initially at 0 V) with (a) no dc stabilization (but with the output initially at 0 V) and (b) the feedback resistor connected.

*2.117 A Miller integrator whose input and output voltages are initially zero and whose time constant is 1 ms is driven by the signal shown in Fig. P2.117. Sketch and label the output waveform that results. Indicate what happens if the input levels are ±2 V, with the time constant the same (1 ms) and with the time constant raised to 2 ms.

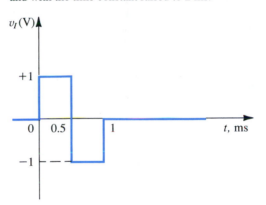

FIGURE P2.117

2.118 Consider a Miller integrator having a time constant of 1 ms, and whose output is initially zero, when fed with a string of pulses of 10-μs duration and 1-V amplitude rising from 0 V (see Fig. P2.118). Sketch and label the output waveform resulting. How many pulses are required for an output voltage change of 1 V?

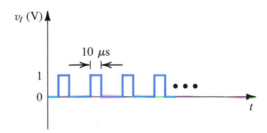

FIGURE P2.118

D2.119 Figure P2.119 shows a circuit that performs a low-pass STC function. Such a circuit is known as a first-order low-pass active filter. Derive the transfer function and show that the dc gain is $(-R_2/R_1)$ and the 3-dB frequency

$\omega_0 = 1/CR_2$. Design the circuit to obtain an input resistance of 1 kΩ, a dc gain of 20 dB, and a 3-dB frequency of 4 kHz. At what frequency does the magnitude of the transfer function reduce to unity?

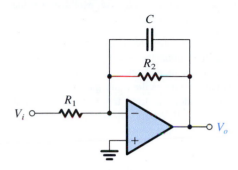

FIGURE P2.119

2.120 A Miller integrator with $R = 10$ kΩ and $C = 10$ nF is implemented using an op amp with $V_{OS} = 3$ mV, $I_B = 0.1$ μA, and $I_{OS} = 10$ nA. To provide a finite dc gain, a 1-MΩ resistor is connected across the capacitor.

(a) To compensate for the effect of I_B, a resistor is connected in series with the positive-input terminal of the op amp. What should its value be?
(b) With the resistor of (a) in place, find the worst-case dc output voltage of the integrator when the input is grounded.

2.121 A differentiator utilizes an ideal op amp, a 10-kΩ resistor, and a 0.01-μF capacitor. What is the frequency f_0 (in Hz) at which its input and output sine-wave signals have equal magnitude? What is the output signal for a 1-V peak-to-peak sine-wave input with frequency equal to $10f_0$?

2.122 An op-amp differentiator with 1-ms time constant is driven by the rate-controlled step shown in Fig. P2.122. Assuming v_O to be zero initially, sketch and label its waveform.

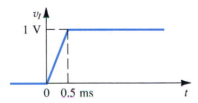

FIGURE P2.122

*2.123 An op-amp differentiator, employing the circuit shown in Fig. 2.44(a), has $R = 10$ kΩ and $C = 0.1$ μF. When a triangle wave of ±1-V peak amplitude at 1 kHz is applied to the input, what form of output results? What is its frequency? What is its peak amplitude? What is its average value? What value of R is needed to cause the output to have a 10-V peak amplitude? When a 1-V peak sine wave at 1 kHz is applied to

2

the (original) circuit, what output waveform is produced? What is its peak amplitude? Calculate this three ways: First, use the second formula in Fig. 2.44(a) directly; second, use the third formula in Fig. 2.44(a); third, use the maximum slope of the input sine wave. In each case, establish a value for the peak output voltage and its location.

2.124 Using an ideal op amp, design a differentiation circuit for which the time constant is 10^{-3} s using a 10-nF capacitor. What are the gains and phase shifts found for this circuit at one-tenth and 10 times the unity-gain frequency? A series input resistor is added to limit the gain magnitude at high frequencies to 100 V/V. What is the associated 3-dB frequency? What gain and phase shift result at 10 times the unity-gain frequency?

D2.125 Figure P2.125 shows a circuit that performs the high-pass single-time-constant function. Such a circuit is known as a first-order high-pass active filter. Derive the transfer function and show that the high-frequency gain is $(-R_2/R_1)$ and the 3-dB frequency $\omega_0 = 1/CR_1$. Design the circuit to obtain a high-frequency input resistance of 10 kΩ, a high-frequency gain of 40 dB, and a 3-dB frequency of 1000 Hz. At what frequency does the magnitude of the transfer function reduce to unity?

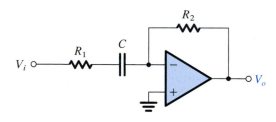

FIGURE P2.125

D**2.126** Derive the transfer function of the circuit in Fig. P2.126 (for an ideal op amp) and show that it can be written in the form

$$\frac{V_o}{V_i} = \frac{-R_2/R_1}{[1+(\omega_1/j\omega)][1+j(\omega/\omega_2)]}$$

where $\omega_1 = 1/C_1R_1$ and $\omega_2 = 1/C_2R_2$. Assuming that the circuit is designed such that $\omega_2 \gg \omega_1$, find approximate expressions for the transfer function in the following frequency regions:

(a) $\omega \ll \omega_1$
(b) $\omega_1 \ll \omega \ll \omega_2$
(c) $\omega \gg \omega_2$

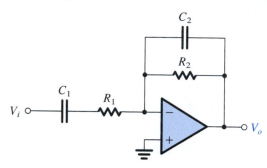

FIGURE P2.126

Use these approximations to sketch a Bode plot for the magnitude response. Observe that the circuit performs as an amplifier whose gain rolls off at the low-frequency end in the manner of a high-pass STC network, and at the high-frequency end in the manner of a low-pass STC network. Design the circuit to provide a gain of 60 dB in the "middle frequency range," a low-frequency 3-dB point at 100 Hz, a high-frequency 3-dB point at 10 kHz, and an input resistance (at $\omega \gg \omega_1$) of 1 kΩ.

Diodes

INTRODUCTION

In the previous chapter we dealt almost entirely with linear circuits; any nonlinearity, such as that introduced by amplifier output saturation, was considered a problem to be solved by the circuit designer. However, there are many other signal-processing functions that can be implemented only by nonlinear circuits. Examples include the generation of dc voltages from the ac power supply and the generation of signals of various waveforms (e.g., sinusoids, square waves, pulses, etc.). Also, digital logic and memory circuits constitute a special class of nonlinear circuits.

The simplest and most fundamental nonlinear circuit element is the diode. Just like a resistor, the diode has two terminals; but unlike the resistor, which has a linear (straight-line) relationship between the current flowing through it and the voltage appearing across it, the diode has a nonlinear i–v characteristic.

This chapter is concerned with the study of diodes. In order to understand the essence of the diode function, we begin with a fictitious element, the ideal diode. We then introduce the silicon junction diode, explain its terminal characteristics, and provide techniques for the analysis of diode circuits. The latter task involves the important subject of device modeling.

Our study of modeling the diode characteristics will lay the foundation for our study of modeling transistor operation in the next two chapters.

Of the many applications of diodes, their use in the design of rectifiers (which convert ac to dc) is the most common. Therefore we shall study rectifier circuits in some detail and briefly look at a number of other diode applications. Further nonlinear circuits that utilize diodes and other devices will be found throughout the book, but particularly in Chapter 13.

To understand the origin of the diode terminal characteristics, we consider its physical operation. Our study of the physical operation of the *pn* junction and of the basic concepts of semiconductor physics is intended to provide a foundation for understanding not only the characteristics of junction diodes but also those of the field-effect transistor, studied in the next chapter, and the bipolar junction transistor, studied in Chapter 5.

Although most of this chapter is concerned with the study of silicon *pn*-junction diodes, we briefly consider some specialized diode types, including the photodiode and the light-emitting diode. The chapter concludes with a description of the diode model utilized in the SPICE circuit-simulation program. We also present a design example that illustrates the use of SPICE simulation.

3.1 THE IDEAL DIODE

3.1.1 Current–Voltage Characteristic

The ideal diode may be considered the most fundamental nonlinear circuit element. It is a two-terminal device having the circuit symbol of Fig. 3.1(a) and the *i–v* characteristic shown in Fig. 3.1(b). The terminal characteristic of the ideal diode can be interpreted as follows: If a

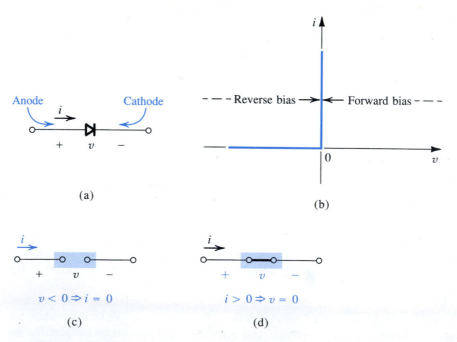

FIGURE 3.1 The ideal diode: **(a)** diode circuit symbol; **(b)** *i–v* characteristic; **(c)** equivalent circuit in the reverse direction; **(d)** equivalent circuit in the forward direction.

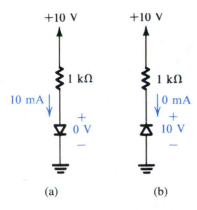

FIGURE 3.2 The two modes of operation of ideal diodes and the use of an external circuit to limit the forward current **(a)** and the reverse voltage **(b)**.

negative voltage (relative to the reference direction indicated in Fig. 3.1a) is applied to the diode, no current flows and the diode behaves as an open circuit (Fig. 3.1c). Diodes operated in this mode are said to be **reverse biased,** or operated in the reverse direction. An ideal diode has zero current when operated in the reverse direction and is said to be **cut off,** or simply **off.**

On the other hand, if a positive current (relative to the reference direction indicated in Fig. 3.1a) is applied to the ideal diode, zero voltage drop appears across the diode. In other words, the ideal diode behaves as a short circuit in the *forward* direction (Fig. 3.1d); it passes any current with zero voltage drop. A **forward-biased** diode is said to be **turned on,** or simply **on.**

From the above description it should be noted that the external circuit must be designed to limit the forward current through a conducting diode, and the reverse voltage across a cutoff diode, to predetermined values. Figure 3.2 shows two diode circuits that illustrate this point. In the circuit of Fig. 3.2(a) the diode is obviously conducting. Thus its voltage drop will be zero, and the current through it will be determined by the +10-V supply and the 1-kΩ resistor as 10 mA. The diode in the circuit of Fig. 3.2(b) is obviously cut off, and thus its current will be zero, which in turn means that the entire 10-V supply will appear as reverse bias across the diode.

The positive terminal of the diode is called the **anode** and the negative terminal the **cathode,** a carryover from the days of vacuum-tube diodes. The *i–v* characteristic of the ideal diode (conducting in one direction and not in the other) should explain the choice of its arrow-like circuit symbol.

As should be evident from the preceding description, the *i–v* characteristic of the ideal diode is highly nonlinear; although it consists of two straight-line segments, they are at 90° to one another. A nonlinear curve that consists of straight-line segments is said to be **piecewise linear.** If a device having a piecewise-linear characteristic is used in a particular application in such a way that the signal across its terminals swings along only one of the linear segments, then the device can be considered a linear circuit element as far as that particular circuit application is concerned. On the other hand, if signals swing past one or more of the break points in the characteristic, linear analysis is no longer possible.

3.1.2 A Simple Application: The Rectifier

A fundamental application of the diode, one that makes use of its severely nonlinear *i–v* curve, is the rectifier circuit shown in Fig. 3.3(a). The circuit consists of the series connection

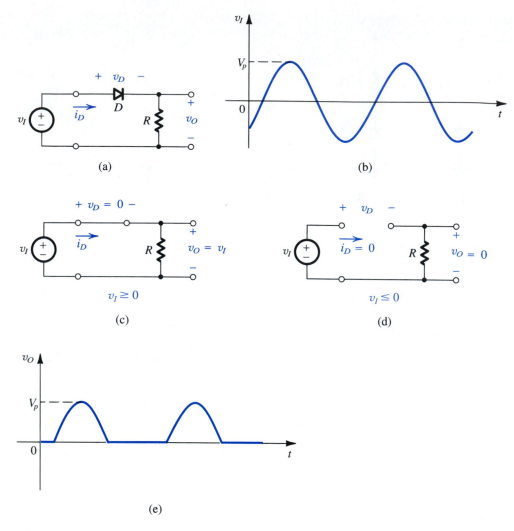

FIGURE 3.3 (a) Rectifier circuit. (b) Input waveform. (c) Equivalent circuit when $v_I \geq 0$. (d) Equivalent circuit when $v_I \leq 0$. (e) Output waveform.

of a diode D and a resistor R. Let the input voltage v_I be the sinusoid shown in Fig. 3.3(b), and assume the diode to be ideal. During the positive half-cycles of the input sinusoid, the positive v_I will cause current to flow through the diode in its forward direction. It follows that the diode voltage v_D will be very small—ideally zero. Thus the circuit will have the equivalent shown in Fig. 3.3(c), and the output voltage v_O will be equal to the input voltage v_I. On the other hand, during the negative half-cycles of v_I, the diode will not conduct. Thus the circuit will have the equivalent shown in Fig. 3.3(d), and v_O will be zero. Thus the output voltage will have the waveform shown in Fig. 3.3(e). Note that while v_I alternates in polarity and has a zero average value, v_O is unidirectional and has a finite average value or a *dc component*. Thus the circuit of Fig. 3.3(a) **rectifies** the signal and hence is called a **rectifier.** It can be used to generate dc from ac. We will study rectifier circuits in Section 3.5.

EXERCISES

3.1 For the circuit in Fig. 3.3(a), sketch the transfer characteristic v_O versus v_I.

Ans. See Fig. E3.1.

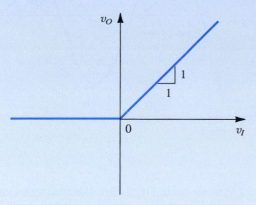

FIGURE E3.1

3.2 For the circuit in Fig. 3.3(a), sketch the waveform of v_D.

Ans. See Fig. E3.2.

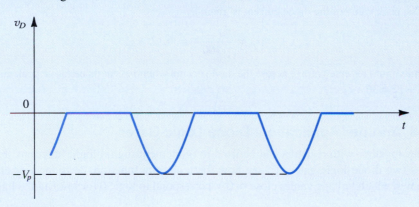

FIGURE E3.2

3.3 In the circuit of Fig. 3.3(a), let v_I have a peak value of 10 V and $R = 1$ kΩ. Find the peak value of i_D and the dc component of v_O.

Ans. 10 mA; 3.18 V

EXAMPLE 3.1

Figure 3.4(a) shows a circuit for charging a 12-V battery. If v_S is a sinusoid with 24-V peak amplitude, find the fraction of each cycle during which the diode conducts. Also, find the peak value of the diode current and the maximum reverse-bias voltage that appears across the diode.

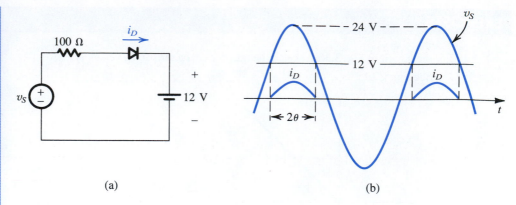

(a) (b)

FIGURE 3.4 Circuit and waveforms for Example 3.1.

Solution

The diode conducts when v_S exceeds 12 V, as shown in Fig. 3.4(b). The conduction angle is 2θ, where θ is given by

$$24 \cos\theta = 12$$

Thus $\theta = 60°$ and the conduction angle is 120°, or one-third of a cycle.

The peak value of the diode current is given by

$$I_d = \frac{24 - 12}{100} = 0.12 \text{ A}$$

The maximum reverse voltage across the diode occurs when v_S is at its negative peak and is equal to $24 + 12 = 36$ V.

3.1.3 Another Application: Diode Logic Gates

Diodes together with resistors can be used to implement digital logic functions. Figure 3.5 shows two diode logic gates. To see how these circuits function, consider a positive-logic system in which voltage values close to 0 V correspond to logic 0 (or low) and voltage values

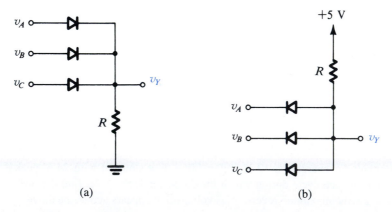

(a) (b)

FIGURE 3.5 Diode logic gates: **(a)** OR gate; **(b)** AND gate (in a positive-logic system).

close to +5 V correspond to logic 1 (or high). The circuit in Fig. 3.5(a) has three inputs, v_A, v_B, and v_C. It is easy to see that diodes connected to +5-V inputs will conduct, thus clamping the output v_Y to a value equal to +5 V. This positive voltage at the output will keep the diodes whose inputs are low (around 0 V) cut off. Thus the output will be high if one or more of the inputs are high. The circuit therefore implements the **logic OR function,** which in Boolean notation is expressed as

$$Y = A + B + C$$

Similarly, the reader is encouraged to show that using the same logic system mentioned above, the circuit of Fig. 3.5(b) implements the **logic AND function,**

$$Y = A \cdot B \cdot C$$

EXAMPLE 3.2

Assuming the diodes to be ideal, find the values of I and V in the circuits of Fig. 3.6.

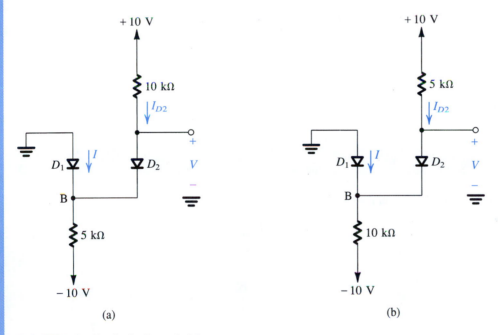

(a)   (b)

FIGURE 3.6 Circuits for Example 3.2.

Solution

In these circuits it might not be obvious at first sight whether none, one, or both diodes are conducting. In such a case, *we make a plausible assumption, proceed with the analysis, and then check whether we end up with a consistent solution.* For the circuit in Fig. 3.6(a), we shall assume that both diodes are conducting. It follows that $V_B = 0$ and $V = 0$. The current through D_2 can now be determined from

$$I_{D2} = \frac{10 - 0}{10} = 1 \text{ mA}$$

Writing a node equation at B,

$$I + 1 = \frac{0 - (-10)}{5}$$

results in $I = 1$ mA. Thus D_1 is conducting as originally assumed, and the final result is $I = 1$ mA and $V = 0$ V.

For the circuit in Fig. 3.6(b), if we assume that both diodes are conducting, then $V_B = 0$ and $V = 0$. The current in D_2 is obtained from

$$I_{D2} = \frac{10 - 0}{5} = 2 \text{ mA}$$

The node equation at B is

$$I + 2 = \frac{0 - (-10)}{10}$$

which yields $I = -1$ mA. Since this is not possible, our original assumption is not correct. We start again, assuming that D_1 is off and D_2 is on. The current I_{D2} is given by

$$I_{D2} = \frac{10 - (-10)}{15} = 1.33 \text{ mA}$$

and the voltage at node B is

$$V_B = -10 + 10 \times 1.33 = +3.3 \text{ V}$$

Thus D_1 is reverse biased as assumed, and the final result is $I = 0$ and $V = 3.3$ V.

EXERCISES

3.4 Find the values of I and V in the circuits shown in Fig. E3.4.

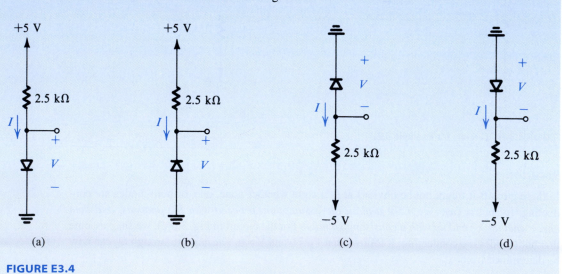

FIGURE E3.4

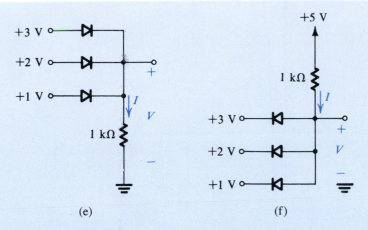

(e)   (f)

FIGURE E3.4   *(Continued)*

Ans. (a) 2 mA, 0 V; (b) 0 mA, 5 V; (c) 0 mA, 5 V; (d) 2 mA, 0 V; (e) 3 mA, +3 V; (f) 4 mA, +1 V.

3.5 Figure E3.5 shows a circuit for an ac voltmeter. It utilizes a moving-coil meter that gives a full-scale reading when the *average* current flowing through it is 1 mA. The moving-coil meter has a 50-Ω resistance.

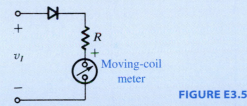

FIGURE E3.5

Find the value of R that results in the meter indicating a full-scale reading when the input sine-wave voltage v_I is 20 V peak-to-peak. (*Hint:* The average value of half-sine waves is V_p/π.)

Ans. 3.133 kΩ

3.2   TERMINAL CHARACTERISTICS OF JUNCTION DIODES

In this section we study the characteristics of real diodes—specifically, semiconductor junction diodes made of silicon. The physical processes that give rise to the diode terminal characteristics, and to the name "junction diode," will be studied in Section 3.7.

Figure 3.7 shows the i–v characteristic of a silicon junction diode. The same characteristic is shown in Fig. 3.8 with some scales expanded and others compressed to reveal details. Note that the scale changes have resulted in the apparent discontinuity at the origin.

As indicated, the characteristic curve consists of three distinct regions:

1. The forward-bias region, determined by $v > 0$

2. The reverse-bias region, determined by $v < 0$

3. The breakdown region, determined by $v < -V_{ZK}$

These three regions of operation are described in the following sections.

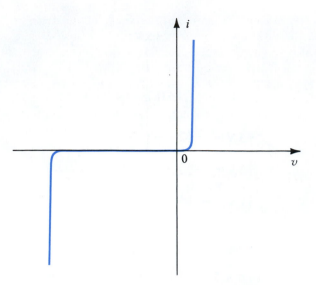

FIGURE 3.7 The i–v characteristic of a silicon junction diode.

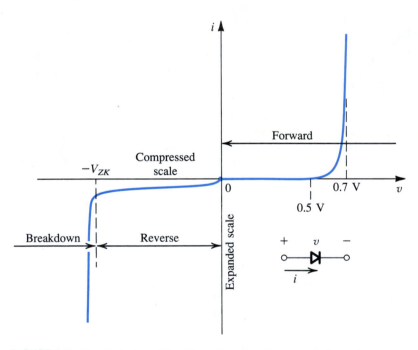

FIGURE 3.8 The diode i–v relationship with some scales expanded and others compressed in order to reveal details.

3.2.1 The Forward-Bias Region

The forward-bias—or simply forward—region of operation is entered when the terminal voltage v is positive. In the forward region the i–v relationship is closely approximated by

$$i = I_S(e^{v/nV_T} - 1) \tag{3.1}$$

In this equation I_S is a constant for a given diode at a given temperature. A formula for I_S in terms of the diode's physical parameters and temperature will be given in Section 3.7. The current I_S is usually called the **saturation current** (for reasons that will become apparent shortly). Another name for I_S, and one that we will occasionally use, is the **scale current.** This name arises from the fact that I_S is directly proportional to the cross-sectional area of the diode. Thus doubling of the junction area results in a diode with double the value of I_S and, as the diode equation indicates, double the value of current i for a given forward voltage v. For "small-signal" diodes, which are small-size diodes intended for low-power applications, I_S is on the order of 10^{-15} A. The value of I_S is, however, a very strong function of temperature. As a rule of thumb, I_S doubles in value for every 5°C rise in temperature.

The voltage V_T in Eq. (3.1) is a constant called the **thermal voltage** and is given by

$$V_T = \frac{kT}{q} \tag{3.2}$$

where

k = Boltzmann's constant = 1.38×10^{-23} joules/kelvin

T = the absolute temperature in kelvins = 273 + temperature in °C

q = the magnitude of electronic charge = 1.60×10^{-19} coulomb

At room temperature (20°C) the value of V_T is 25.2 mV. In rapid approximate circuit analysis we shall use $V_T \simeq 25$ mV at room temperature.[1]

In the diode equation the constant n has a value between 1 and 2, depending on the material and the physical structure of the diode. Diodes made using the standard integrated-circuit fabrication process exhibit $n = 1$ when operated under normal conditions.[2] Diodes available as discrete two-terminal components generally exhibit $n = 2$. In general, we shall assume $n = 1$ unless otherwise specified.

For appreciable current i in the forward direction, specifically for $i \gg I_S$, Eq. (3.1) can be approximated by the exponential relationship

$$i \simeq I_S e^{v/nV_T} \tag{3.3}$$

This relationship can be expressed alternatively in the logarithmic form

$$v = nV_T \ln \frac{i}{I_S} \tag{3.4}$$

where ln denotes the natural (base e) logarithm.

The exponential relationship of the current i to the voltage v holds over many decades of current (a span of as many as seven decades—i.e., a factor of 10^7—can be found). This is quite a remarkable property of junction diodes, one that is also found in bipolar junction transistors and that has been exploited in many interesting applications.

Let us consider the forward i–v relationship in Eq. (3.3) and evaluate the current I_1 corresponding to a diode voltage V_1:

$$I_1 = I_S e^{V_1/nV_T}$$

---

[1] A slightly higher ambient temperature (25°C or so) is usually assumed for electronic equipment operating inside a cabinet. At this temperature, $V_T \simeq 25.8$ mV. Nevertheless, for the sake of simplicity and to promote rapid circuit analysis, we shall use the more arithmetically convenient value of $V_T \simeq 25$ mV throughout this book.

[2] In an integrated circuit, diodes are usually obtained by connecting a bipolar junction transistor (BJT) as a two-terminal device, as will be seen in Chapter 5.

Similarly, if the voltage is V_2, the diode current I_2 will be

$$I_2 = I_S e^{V_2/nV_T}$$

These two equations can be combined to produce

$$\frac{I_2}{I_1} = e^{(V_2 - V_1)/nV_T}$$

which can be rewritten as

$$V_2 - V_1 = nV_T \ln \frac{I_2}{I_1}$$

or, in terms of base-10 logarithms,

$$V_2 - V_1 = 2.3nV_T \log \frac{I_2}{I_1} \qquad (3.5)$$

This equation simply states that for a decade (factor of 10) change in current, the diode voltage drop changes by $2.3nV_T$, which is approximately 60 mV for $n = 1$ and 120 mV for $n = 2$. This also suggests that the diode i–v relationship is most conveniently plotted on semilog paper. Using the vertical, linear axis for v and the horizontal, log axis for i, one obtains a straight line with a slope of $2.3nV_T$ per decade of current. Finally, it should be mentioned that not knowing the exact value of n (which can be obtained from a simple experiment), circuit designers use the convenient approximate number of 0.1 V/decade for the slope of the diode logarithmic characteristic.

A glance at the i–v characteristic in the forward region (Fig. 3.8) reveals that the current is negligibly small for v smaller than about 0.5 V. This value is usually referred to as the **cut-in voltage.** It should be emphasized, however, that this apparent threshold in the characteristic is simply a consequence of the exponential relationship. Another consequence of this relationship is the rapid increase of i. Thus, for a "fully conducting" diode, the voltage drop lies in a narrow range, approximately 0.6 V to 0.8 V. This gives rise to a simple "model" for the diode where it is assumed that a conducting diode has approximately a 0.7-V drop across it. Diodes with different current ratings (i.e., different areas and correspondingly different I_S) will exhibit the 0.7-V drop at different currents. For instance, a small-signal diode may be considered to have a 0.7-V drop at $i = 1$ mA, while a higher-power diode may have a 0.7-V drop at $i = 1$ A. We will study the topics of diode-circuit analysis and diode models in the next section.

EXAMPLE 3.3

A silicon diode said to be a 1-mA device displays a forward voltage of 0.7 V at a current of 1 mA. Evaluate the junction scaling constant I_S in the event that n is either 1 or 2. What scaling constants would apply for a 1-A diode of the same manufacture that conducts 1 A at 0.7 V?

Solution
Since

$$i = I_S e^{v/nV_T}$$

then

$$I_S = ie^{-v/nV_T}$$

For the 1-mA diode:

If $n = 1$: $I_S = 10^{-3}e^{-700/25} = 6.9 \times 10^{-16}$ A, or about 10^{-15} A

If $n = 2$: $I_S = 10^{-3}e^{-700/50} = 8.3 \times 10^{-10}$ A, or about 10^{-9} A

The diode conducting 1 A at 0.7 V corresponds to one-thousand 1-mA diodes in parallel with a total junction area 1000 times greater. Thus I_S is also 1000 times greater, being 1 pA and 1 μA, respectively for $n = 1$ and $n = 2$.

From this example it should be apparent that the value of n used can be quite important.

Since both I_S and V_T are functions of temperature, the forward i–v characteristic varies with temperature, as illustrated in Fig. 3.9. At a given constant diode current the voltage drop across the diode decreases by approximately 2 mV for every 1°C increase in temperature. The change in diode voltage with temperature has been exploited in the design of electronic thermometers.

EXERCISES

3.6 Consider a silicon diode with $n = 1.5$. Find the change in voltage if the current changes from 0.1 mA to 10 mA.

Ans. 172.5 mV

3.7 A silicon junction diode with $n = 1$ has $v = 0.7$ V at $i = 1$ mA. Find the voltage drop at $i = 0.1$ mA and $i = 10$ mA.

Ans. 0.64 V; 0.76 V

3.8 Using the fact that a silicon diode has $I_S = 10^{-14}$ A at 25°C and that I_S increases by 15% per °C rise in temperature, find the value of I_S at 125°C.

Ans. 1.17×10^{-8} A

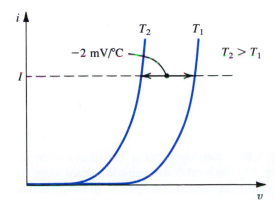

FIGURE 3.9 Illustrating the temperature dependence of the diode forward characteristic. At a constant current, the voltage drop decreases by approximately 2 mV for every 1°C increase in temperature.

3.2.2 The Reverse-Bias Region

The reverse-bias region of operation is entered when the diode voltage v is made negative. Equation (3.1) predicts that if v is negative and a few times larger than V_T (25 mV) in magnitude, the exponential term becomes negligibly small compared to unity, and the diode current becomes

$$i \simeq -I_S$$

That is, the current in the reverse direction is constant and equal to I_S. This constancy is the reason behind the term *saturation current*.

Real diodes exhibit reverse currents that, though quite small, are much larger than I_S. For instance, a small-signal diode whose I_S is on the order of 10^{-14} A to 10^{-15} A could show a reverse current on the order of 1 nA. The reverse current also increases somewhat with the increase in magnitude of the reverse voltage. Note that because of the very small magnitude of the current, these details are not clearly evident on the diode $i–v$ characteristic of Fig. 3.8.

A large part of the reverse current is due to leakage effects. These leakage currents are proportional to the junction area, just as I_S is. Their dependence on temperature, however, is different from that of I_S. Thus, whereas I_S doubles for every 5°C rise in temperature, the corresponding rule of thumb for the temperature dependence of the reverse current is that it doubles for every 10°C rise in temperature.

EXERCISE

3.9 The diode in the circuit of Fig. E3.9 is a large high-current device whose reverse leakage is reasonably independent of voltage. If $V = 1$ V at 20°C, find the value of V at 40°C and at 0°C.

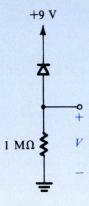

FIGURE E3.9

Ans. 4 V; 0.25 V

3.2.3 The Breakdown Region

The third distinct region of diode operation is the breakdown region, which can be easily identified on the diode $i–v$ characteristic in Fig. 3.8. The breakdown region is entered when the magnitude of the reverse voltage exceeds a threshold value that is specific to the particular diode, called the **breakdown voltage.** This is the voltage at the "knee" of the $i–v$ curve in

Fig. 3.8 and is denoted V_{ZK}, where the subscript Z stands for zener (to be explained shortly) and K denotes knee.

As can be seen from Fig. 3.8, in the breakdown region the reverse current increases rapidly, with the associated increase in voltage drop being very small. Diode breakdown is normally not destructive provided that the power dissipated in the diode is limited by external circuitry to a "safe" level. This safe value is normally specified on the device data sheets. It therefore is necessary to limit the reverse current in the breakdown region to a value consistent with the permissible power dissipation.

The fact that the diode $i–v$ characteristic in breakdown is almost a vertical line enables it to be used in voltage regulation. This subject will be studied in Section 3.5.

 ## 3.3 MODELING THE DIODE FORWARD CHARACTERISTIC

Having studied the diode terminal characteristics we are now ready to consider the analysis of circuits employing forward-conducting diodes. Figure 3.10 shows such a circuit. It consists of a dc source V_{DD}, a resistor R, and a diode. We wish to analyze this circuit to determine the diode voltage V_D and current I_D. Toward that end we consider developing a variety of models for the operation of the diode. We already know of two such models: the ideal-diode model, and the exponential model. In the following discussion we shall assess the suitability of these two models in various analysis situations. Also, we shall develop and comment on a number of other models. This material, besides being useful in the analysis and design of diode circuits, establishes a foundation for the modeling of transistor operation that we will study in the next two chapters.

3.3.1 The Exponential Model

The most accurate description of the diode operation in the forward region is provided by the exponential model. Unfortunately, however, its severely nonlinear nature makes this model the most difficult to use. To illustrate, let's analyze the circuit in Fig. 3.10 using the exponential diode model.

Assuming that V_{DD} is greater than 0.5 V or so, the diode current will be much greater than I_S, and we can represent the diode $i–v$ characteristic by the exponential relationship, resulting in

$$I_D = I_S e^{V_D/nV_T} \tag{3.6}$$

The other equation that governs circuit operation is obtained by writing a Kirchhoff loop equation, resulting in

$$I_D = \frac{V_{DD} - V_D}{R} \tag{3.7}$$

Assuming that the diode parameters I_S and n are known, Eqs. (3.6) and (3.7) are two equations in the two unknown quantities I_D and V_D. Two alternative ways for obtaining the solution are graphical analysis and iterative analysis.

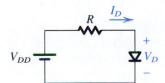

FIGURE 3.10 A simple circuit used to illustrate the analysis of circuits in which the diode is forward conducting.

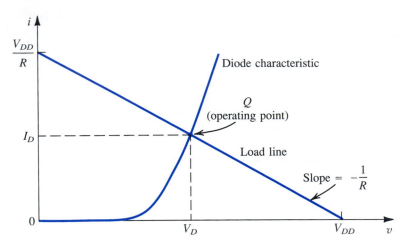

FIGURE 3.11 Graphical analysis of the circuit in Fig. 3.10 using the exponential diode model.

3.3.2 Graphical Analysis Using the Exponential Model

Graphical analysis is performed by plotting the relationships of Eqs. (3.6) and (3.7) on the i–v plane. The solution can then be obtained as the coordinates of the point of intersection of the two graphs. A sketch of the graphical construction is shown in Fig. 3.11. The curve represents the exponential diode equation (Eq. 3.6), and the straight line represents Eq. (3.7). Such a straight line is known as the **load line,** a name that will become more meaningful in later chapters. The load line intersects the diode curve at point Q, which represents the **operating point** of the circuit. Its coordinates give the values of I_D and V_D.

Graphical analysis aids in the visualization of circuit operation. However, the effort involved in performing such an analysis, particularly for complex circuits, is too great to be justified in practice.

3.3.3 Iterative Analysis Using the Exponential Model

Equations (3.6) and (3.7) can be solved using a simple iterative procedure, as illustrated in the following example.

EXAMPLE 3.4

Determine the current I_D and the diode voltage V_D for the circuit in Fig. 3.10 with $V_{DD} = 5$ V and $R = 1$ kΩ. Assume that the diode has a current of 1 mA at a voltage of 0.7 V and that its voltage drop changes by 0.1 V for every decade change in current.

Solution

To begin the iteration, we assume that $V_D = 0.7$ V and use Eq. (3.7) to determine the current,

$$I_D = \frac{V_{DD} - V_D}{R}$$

$$= \frac{5 - 0.7}{1} = 4.3 \text{ mA}$$

We then use the diode equation to obtain a better estimate for V_D. This can be done by employing Eq. (3.5), namely,

$$V_2 - V_1 = 2.3 n V_T \log \frac{I_2}{I_1}$$

For our case, $2.3 n V_T = 0.1$ V. Thus,

$$V_2 = V_1 + 0.1 \log \frac{I_2}{I_1}$$

Substituting $V_1 = 0.7$ V, $I_1 = 1$ mA, and $I_2 = 4.3$ mA results in $V_2 = 0.763$ V. Thus the results of the first iteration are $I_D = 4.3$ mA and $V_D = 0.763$ V. The second iteration proceeds in a similar manner:

$$I_D = \frac{5 - 0.763}{1} = 4.237 \text{ mA}$$

$$V_2 = 0.763 + 0.1 \log \left[\frac{4.237}{4.3} \right]$$

$$= 0.762 \text{ V}$$

Thus the second iteration yields $I_D = 4.237$ mA and $V_D = 0.762$ V. Since these values are not much different from the values obtained after the first iteration, no further iterations are necessary, and the solution is $I_D = 4.237$ mA and $V_D = 0.762$ V.

3.3.4 The Need for Rapid Analysis

The iterative analysis procedure utilized in the example above is simple and yields accurate results after two or three iterations. Nevertheless, there are situations in which the effort and time required are still greater than can be justified. Specifically, if one is doing a pencil-and-paper design of a relatively complex circuit, rapid circuit analysis is a necessity. Through quick analysis, the designer is able to evaluate various possibilities before deciding on a suitable circuit design. To speed up the analysis process one must be content with less precise results. This, however, is seldom a problem, because the more accurate analysis can be postponed until a final or almost-final design is obtained. Accurate analysis of the almost-final design can be performed with the aid of a computer circuit-analysis program such as SPICE (see Section 3.9). The results of such an analysis can then be used to further refine or "fine-tune" the design.

To speed up the analysis process, we must find simpler models for the diode forward characteristic.

3.3.5 The Piecewise-Linear Model

The analysis can be greatly simplified if we can find linear relationships to describe the diode terminal characteristics. An attempt in this direction is illustrated in Fig. 3.12, where the exponential curve is approximated by two straight lines, line A with zero slope and line B with a slope of $1/r_D$. It can be seen that for the particular case shown in Fig. 3.12, over the current range of 0.1 mA to 10 mA the voltages predicted by the straight-lines model shown differ from those predicted by the exponential model by less than 50 mV. Obviously the choice of these two straight lines is not unique; one can obtain a closer approximation by restricting the current range over which the approximation is required.

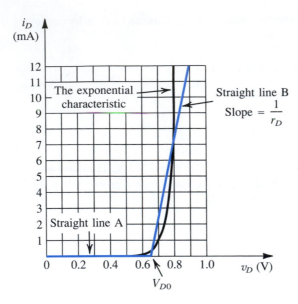

FIGURE 3.12 Approximating the diode forward characteristic with two straight lines: the piecewise-linear model.

The straight-lines (or piecewise-linear) model of Fig. 3.12 can be described by

$$i_D = 0, \quad v_D \leq V_{D0}$$
$$i_D = (v_D - V_{D0})/r_D, \quad v_D \geq V_{D0} \tag{3.8}$$

where V_{D0} is the intercept of line B on the voltage axis and r_D is the inverse of the slope of line B. For the particular example shown, $V_{D0} = 0.65$ V and $r_D = 20\ \Omega$.

The piecewise-linear model described by Eqs. (3.8) can be represented by the equivalent circuit shown in Fig. 3.13. Note that an ideal diode is included in this model to constrain i_D to flow in the forward direction only. This model is also known as the **battery-plus-resistance** model.

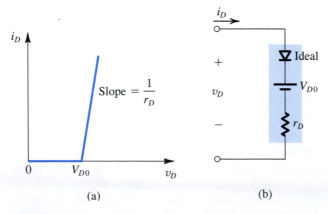

FIGURE 3.13 Piecewise-linear model of the diode forward characteristic and its equivalent circuit representation.

EXAMPLE 3.5

Repeat the problem in Example 3.4 utilizing the piecewise-linear model whose parameters are given in Fig. 3.12 ($V_{D0} = 0.65$ V, $r_D = 20$ Ω). Note that the characteristics depicted in this figure are those of the diode described in Example 3.4 (1 mA at 0.7 V and 0.1 V/decade).

Solution

Replacing the diode in the circuit of Fig. 3.10 with the equivalent circuit model of Fig. 3.13 results in the circuit in Fig. 3.14, from which we can write for the current I_D,

$$I_D = \frac{V_{DD} - V_{D0}}{R + r_D}$$

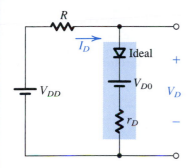

FIGURE 3.14 The circuit of Fig. 3.10 with the diode replaced with its piecewise-linear model of Fig. 3.13.

where the model parameters V_{D0} and r_D are seen from Fig. 3.12 to be $V_{D0} = 0.65$ V and $r_D = 20$ Ω. Thus,

$$I_D = \frac{5 - 0.65}{1 + 0.02} = 4.26 \text{ mA}$$

The diode voltage V_D can now be computed:

$$V_D = V_{D0} + I_D r_D$$
$$= 0.65 + 4.26 \times 0.02 = 0.735 \text{ V}$$

3.3.6 The Constant-Voltage-Drop Model

An even simpler model of the diode forward characteristics can be obtained if we use a vertical straight line to approximate the fast-rising part of the exponential curve, as shown in Fig. 3.15. The resulting model simply says that *a forward-conducting diode exhibits a constant voltage drop V_D*. The value of V_D is usually taken to be 0.7 V. Note that for the particular diode whose characteristics are depicted in Fig. 3.15, this model predicts the diode voltage to within ±0.1 V over the current range of 0.1 mA to 10 mA. The constant-voltage-drop model can be represented by the equivalent circuit shown in Fig. 3.16.

The constant-voltage-drop model is the one most frequently employed in the initial phases of analysis and design. This is especially true if at these stages one does not have detailed information about the diode characteristics, which is often the case.

Finally, note that if we employ the constant-voltage-drop model to solve the problem in Examples 3.4 and 3.5, we obtain

$$V_D = 0.7 \text{ V}$$

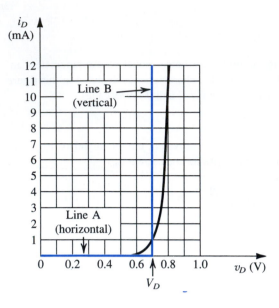

FIGURE 3.15 Development of the constant-voltage-drop model of the diode forward characteristics. A vertical straight line (**B**) is used to approximate the fast-rising exponential. Observe that this simple model predicts V_D to within ±0.1 V over the current range of 0.1 mA to 10 mA.

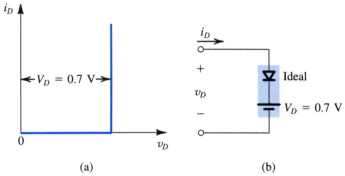

FIGURE 3.16 The constant-voltage-drop model of the diode forward characteristics and its equivalent-circuit representation.

and

$$I_D = \frac{V_{DD} - 0.7}{R}$$

$$= \frac{5 - 0.7}{1} = 4.3 \text{ mA}$$

which are not too different from the values obtained before with the more elaborate models.

3.3.7 The Ideal-Diode Model

In applications that involve voltages much greater than the diode voltage drop (0.6–0.8 V), we may neglect the diode voltage drop altogether while calculating the diode current. The result is the ideal-diode model, which we studied in Section 3.1. For the circuit in Examples 3.4 and 3.5 (i.e., Fig. 3.10 with $V_{DD} = 5$ V and $R = 1$ kΩ), utilization of the ideal-diode model leads to

$$V_D = 0 \text{ V}$$

$$I_D = \frac{5 - 0}{1} = 5 \text{ mA}$$

which for a very quick analysis would not be bad as a gross estimate. However, with almost no additional work, the 0.7-V-drop model yields much more realistic results. We note, however, that the greatest utility of the ideal-diode model is in determining which diodes are on and which are off in a multidiode circuit, such as those considered in Section 3.1.

EXERCISES

3.10 For the circuit in Fig. 3.10, find I_D and V_D for the case $V_{DD} = 5$ V and $R = 10$ kΩ. Assume that the diode has a voltage of 0.7 V at 1-mA current and that the voltage changes by 0.1 V/decade of current change. Use (a) iteration, (b) the piecewise-linear model with $V_{D0} = 0.65$ V and $r_D = 20$ Ω, (c) the constant-voltage-drop model with $V_D = 0.7$ V.

Ans. (a) 0.434 mA, 0.663 V; (b) 0.434 mA, 0.659 V; (c) 0.43 mA, 0.7 V

3.11 Consider a diode that is 100 times as large (in junction area) as that whose characteristics are displayed in Fig. 3.12. If we approximate the characteristics in a manner similar to that in Fig. 3.12 (but over a current range 100 times as large), how would the model parameters V_{D0} and r_D change?

Ans. V_{D0} does not change; r_D decreases by a factor of 100 to 0.2 Ω

D3.12 Design the circuit in Fig. E3.12 to provide an output voltage of 2.4 V. Assume that the diodes available have 0.7-V drop at 1 mA and that $\Delta V = 0.1$ V/decade change in current.

+10 V

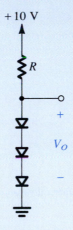

R

+

V_O

−

FIGURE E3.12

Ans. $R = 760$ Ω

3.13 Repeat Exercise 3.4 using the 0.7-V-drop model to obtain better estimates of I and V than those found in Exercise 3.4 (using the ideal-diode model).

Ans. (a) 1.72 mA, 0.7 V; (b) 0 mA, 5 V; (c) 0 mA, 5 V; (d) 1.72 mA, 0.7 V; (e) 2.3 mA, +2.3 V; (f) 3.3 mA, +1.7 V

3.3.8 The Small-Signal Model

There are applications in which a diode is biased to operate at a point on the forward i–v characteristic and a small ac signal is superimposed on the dc quantities. For this situation, we first have to determine the dc operating point (V_D and I_D) of the diode using one of the models discussed above. Most frequently, the 0.7-V-drop model is utilized. Then, for

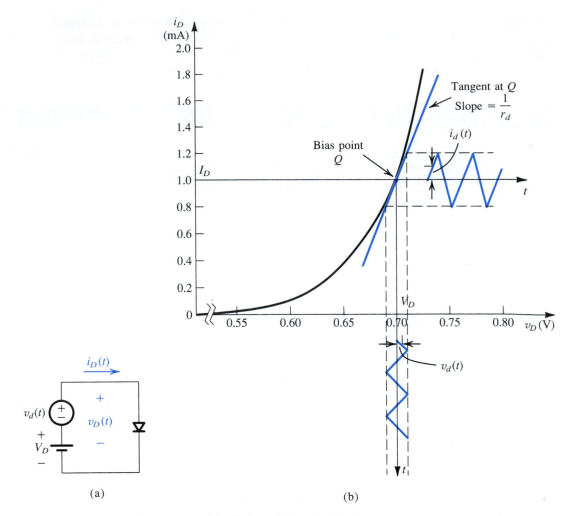

FIGURE 3.17 Development of the diode small-signal model. Note that the numerical values shown are for a diode with $n = 2$.

small-signal operation around the dc bias point, the diode is best modeled by a resistance equal to the inverse of the slope of the tangent to the exponential i–v characteristic at the bias point. The concept of biasing a nonlinear device and restricting signal excursion to a short, almost-linear segment of its characteristic around the bias point was introduced in Section 1.4 for two-port networks. In the following, we develop such a small-signal model for the junction diode and illustrate its application.

Consider the conceptual circuit in Fig. 3.17(a) and the corresponding graphical representation in Fig. 3.17(b). A dc voltage V_D, represented by a battery, is applied to the diode, and a time-varying signal $v_d(t)$, assumed (arbitrarily) to have a triangular waveform, is superimposed on the dc voltage V_D. In the absence of the signal $v_d(t)$ the diode voltage is equal to V_D, and correspondingly, the diode will conduct a dc current I_D given by

$$I_D = I_S e^{V_D/nV_T} \tag{3.9}$$

When the signal $v_d(t)$ is applied, the total instantaneous diode voltage $v_D(t)$ will be given by

$$v_D(t) = V_D + v_d(t) \tag{3.10}$$

Correspondingly, the total instantaneous diode current $i_D(t)$ will be

$$i_D(t) = I_S e^{v_D/nV_T} \tag{3.11}$$

Substituting for v_D from Eq. (3.10) gives

$$i_D(t) = I_S e^{(V_D+v_d)/nV_T}$$

which can be rewritten

$$i_D(t) = I_S e^{V_D/nV_T} e^{v_d/nV_T}$$

Using Eq. (3.9) we obtain

$$i_D(t) = I_D e^{v_d/nV_T} \tag{3.12}$$

Now if the amplitude of the signal $v_d(t)$ is kept sufficiently small such that

$$\frac{v_d}{nV_T} \ll 1 \tag{3.13}$$

then we may expand the exponential of Eq. (3.12) in a series and truncate the series after the first two terms to obtain the approximate expression

$$i_D(t) \simeq I_D \left(1 + \frac{v_d}{nV_T} \right) \tag{3.14}$$

This is the **small-signal approximation.** It is valid for signals whose amplitudes are smaller than about 10 mV for the case $n = 2$ and 5 mV for $n = 1$ (see Eq. 3.13 and recall that $V_T = 25$ mV).[3]

From Eq. (3.14) we have

$$i_D(t) = I_D + \frac{I_D}{nV_T} v_d \tag{3.15}$$

Thus, superimposed on the dc current I_D, we have a signal current component directly proportional to the signal voltage v_d. That is,

$$i_D = I_D + i_d \tag{3.16}$$

where

$$i_d = \frac{I_D}{nV_T} v_d \tag{3.17}$$

The quantity relating the signal current i_d to the signal voltage v_d has the dimensions of conductance, mhos (℧), and is called the **diode small-signal conductance.** The inverse of this parameter is the **diode small-signal resistance,** or **incremental resistance,** r_d,

$$r_d = \frac{nV_T}{I_D} \tag{3.18}$$

Note that the value of r_d is inversely proportional to the bias current I_D.

---

[3] For $n = 2$, $v_d/nV_T = 0.2$ with $v_d = 10$ mV. Thus the next term in the series expansion of the exponential will be $\frac{1}{2} \times 0.2^2 = 0.02$, a factor of 10 lower than the linear term we kept. A better approximation can be achieved by keeping v_d smaller. Also, note that for $n = 1$, v_d should be limited to, say, 5 mV.

Let us return to the graphical representation in Fig. 3.17(b). It is easy to see that using the small-signal approximation is equivalent to assuming that *the signal amplitude is sufficiently small such that the excursion along the i–v curve is limited to a short almost-linear segment.* The slope of this segment, which is equal to the slope of the tangent to the i–v curve at the operating point Q, is equal to the small-signal conductance. The reader is encouraged to prove that the slope of the i–v curve at $i = I_D$ is equal to I_D/nV_T, which is $1/r_d$; that is,

$$r_d = 1 \Big/ \left[\frac{\partial i_D}{\partial v_D} \right]_{i_D = I_D} \tag{3.19}$$

From the preceding we conclude that superimposed on the quantities V_D and I_D that define the dc bias point, or **quiescent point,** of the diode will be the small-signal quantities $v_d(t)$ and $i_d(t)$, which are related by the diode small-signal resistance r_d evaluated at the bias point (Eq. 3.18). Thus the small-signal analysis can be performed separately from the dc bias analysis, a great convenience that results from the linearization of the diode characteristics inherent in the small-signal approximation. Specifically, after the dc analysis is performed, the small-signal equivalent circuit is obtained by eliminating all dc sources (i.e., short-circuiting dc voltage sources and open-circuiting dc current sources) and replacing the diode by its small-signal resistance. The following example should illustrate the application of the small-signal model.

EXAMPLE 3.6

Consider the circuit shown in Fig. 3.18(a) for the case in which $R = 10$ kΩ. The power supply V^+ has a dc value of 10 V on which is superimposed a 60-Hz sinusoid of 1-V peak amplitude. (This "signal" component of the power-supply voltage is an imperfection in the power-supply design. It is known as the **power-supply ripple.** More on this later.) Calculate both the dc voltage of the diode and the amplitude of the sine-wave signal appearing across it. Assume the diode to have a 0.7-V drop at 1-mA current and $n = 2$.

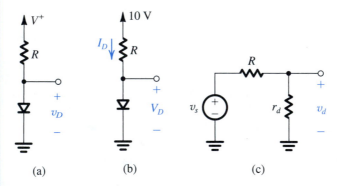

(a)   (b)   (c)

FIGURE 3.18   (**a**) Circuit for Example 3.6. (**b**) Circuit for calculating the dc operating point. (**c**) Small-signal equivalent circuit.

Solution

Considering dc quantities only, we assume $V_D \simeq 0.7$ V and calculate the diode dc current

$$I_D = \frac{10 - 0.7}{10} = 0.93 \text{ mA}$$

Since this value is very close to 1 mA, the diode voltage will be very close to the assumed value of 0.7 V. At this operating point, the diode incremental resistance r_d is

$$r_d = \frac{nV_T}{I_D} = \frac{2 \times 25}{0.93} = 53.8 \ \Omega$$

The signal voltage across the diode can be found from the small-signal equivalent circuit in Fig. 3.18(c). Here v_s denotes the 60-Hz 1-V peak sinusoidal component of V^+, and v_d is the corresponding signal across the diode. Using the voltage-divider rule provides the peak amplitude of v_d as follows:

$$v_d \ (\text{peak}) = \hat{V}_s \frac{r_d}{R + r_d}$$

$$= 1 \frac{0.0538}{10 + 0.0538} = 5.35 \ \text{mV}$$

Finally we note that since this value is quite small, our use of the small-signal model of the diode is justified.

3.3.9 Use of the Diode Forward Drop in Voltage Regulation

A further application of the diode small-signal model is found in a popular diode application, namely the use of diodes to create a regulated voltage. A voltage regulator is a circuit whose purpose is to provide a constant dc voltage between its output terminals. The output voltage is required to remain as constant as possible in spite of (a) changes in the load current drawn from the regulator output terminal and (b) changes in the dc power-supply voltage that feeds the regulator circuit. Since the forward voltage drop of the diode remains almost constant at approximately 0.7 V while the current through it varies by relatively large amounts, a forward-biased diode can make a simple voltage regulator. For instance, we have seen in Example 3.6 that while the 10-V dc supply voltage had a ripple of 2 V peak-to-peak (a ±10% variation), the corresponding ripple in the diode voltage was only about ±5.4 mV (a ±0.8% variation). Regulated voltages greater than 0.7 V can be obtained by connecting a number of diodes in series. For example, the use of three forward-biased diodes in series provides a voltage of about 2 V. One such circuit is investigated in the following example, which utilizes the diode small-signal model to quantify the efficacy of the voltage regulator that is realized.

EXAMPLE 3.7

Consider the circuit shown in Fig. 3.19. A string of three diodes is used to provide a constant voltage of about 2.1 V. We want to calculate the percentage change in this regulated voltage caused by (a) a ±10% change in the power-supply voltage and (b) connection of a 1-kΩ load resistance. Assume $n = 2$.

Solution

With no load, the nominal value of the current in the diode string is given by

$$I = \frac{10 - 2.1}{1} = 7.9 \ \text{mA}$$

Thus each diode will have an incremental resistance of

$$r_d = \frac{nV_T}{I}$$

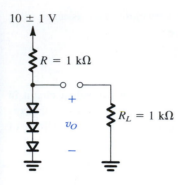

FIGURE 3.19 Circuit for Example 3.7.

Using $n = 2$ gives

$$r_d = \frac{2 \times 25}{7.9} = 6.3 \ \Omega$$

The three diodes in series will have a total incremental resistance of

$$r = 3r_d = 18.9 \ \Omega$$

This resistance, along with the resistance R, forms a voltage divider whose ratio can be used to calculate the change in output voltage due to a ±10% (i.e., ±1-V) change in supply voltage. Thus the peak-to-peak change in output voltage will be

$$\Delta v_O = 2 \frac{r}{r + R} = 2 \frac{0.0189}{0.0189 + 1} = 37.1 \ \text{mV}$$

That is, corresponding to the ±1-V (±10%) change in supply voltage, the output voltage will change by ±18.5 mV or ±0.9%. Since this implies a change of about ±6.2 mV per diode, our use of the small-signal model is justified.

When a load resistance of 1 kΩ is connected across the diode string, it draws a current of approximately 2.1 mA. Thus the current in the diodes decreases by 2.1 mA, resulting in a decrease in voltage across the diode string given by

$$\Delta v_O = -2.1 \times r = -2.1 \times 18.9 = -39.7 \ \text{mV}$$

Since this implies that the voltage across each diode decreases by about 13.2 mV, our use of the small-signal model is not entirely justified. Nevertheless, a detailed calculation of the voltage change using the exponential model results in $\Delta v_O = -35.5$ mV, which is not too different from the approximate value obtained using the incremental model.

EXERCISES

3.14 Find the value of the diode small-signal resistance r_d at bias currents of 0.1 mA, 1 mA, and 10 mA. Assume $n = 1$.

Ans. 250 Ω; 25 Ω; 2.5 Ω

3.15 Consider a diode with $n = 2$ biased at 1 mA. Find the change in current as a result of changing the voltage by (a) −20 mV, (b) −10 mV, (c) −5 mV, (d) +5 mV, (e) +10 mV, and (f) +20 mV. In each case, do the calculations (i) using the small-signal model and (ii) using the exponential model.

Ans. (a) −0.40, −0.33 mA; (b) −0.20, −0.18 mA; (c) −0.10, −0.10 mA; (d) +0.10, +0.11 mA; (e) +0.20, +0.22 mA; (f) +0.40, +0.49 mA

D3.16 Design the circuit of Fig. E3.16 so that $V_O = 3$ V when $I_L = 0$ and V_O changes by 40 mV per 1 mA of load current. Find the value of R and the junction area of each diode (assume all four diodes are identical) relative to a diode with 0.7-V drop at 1-mA current. Assume $n = 1$.

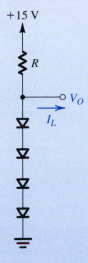

FIGURE E3.16

Ans. $R = 4.8$ kΩ; 0.34

3.3.10 Summary

As a summary of this important section on diode modeling, Table 3.1 lists the five diode models studied and provides pertinent comments regarding each. These comments are intended to aid in the selection of an appropriate model for a particular application. The question "which model?" is one that circuit designers face repeatedly, not just with diodes but with every circuit element. The problem is finding an appropriate compromise between accuracy and speed of analysis. One's ability to select appropriate device models improves with practice and experience.

TABLE 3.1   **Modeling the Diode Forward Characteristic**

Model	Graph	Equations	Circuit	Comments
Exponential		$i_D = I_S e^{v_D/nV_T}$ $v_D = 2.3 n V_T \log\left(\dfrac{i_D}{I_S}\right)$ $V_{D2} - V_{D1} = 2.3 n V_T \log\left(\dfrac{I_{D2}}{I_{D1}}\right)$ $2.3 n V_T = 60$ mV for $n = 1$ $2.3 n V_T = 120$ mV for $n = 2$		$I_S = 10^{-12}$ A to 10^{-15} A, depending on junction area $V_T \cong 25$ mV $n = 1$ to 2 Physically based and remarkably accurate model Useful when accurate analysis is needed

(Continued)

TABLE 3.1 *(Continued)*

Model	Graph	Equations	Circuit	Comments
Piecewise-linear (battery-plus-resistance)	Slope $= 1/r_D$; axes i_D vs v_D, intercept at V_{D0}	For $v_D \leq V_{D0}$: $i_D = 0$ For $v_D \geq V_{D0}$: $i_D = \dfrac{1}{r_D}(v_D - V_{D0})$	i_D, $+$, v_D, $-$; Ideal, V_{D0}, r_D	Choice of V_{D0} and r_D is determined by the current range over which the model is required. For the amount of work involved, not as useful as the constant-voltage-drop model. Used only infrequently.
Constant-voltage-drop (or the "0.7-V model")	axes i_D vs v_D, vertical at 0.7 V	For $i_D > 0$: $v_D = 0.7$ V	i_D, $+$, v_D, $-$; Ideal, 0.7 V	Easy to use and very popular for the quick, hand analysis that is essential in circuit design.
Ideal-diode	axes i_D vs v_D, vertical at 0	For $i_D > 0$: $v_D = 0$	i_D, $+$, v_D, $-$; Ideal	Good for determining which diodes are conducting and which are cutoff in a multiple-diode circuit. Good for obtaining very approximate values for diode currents, especially when the circuit voltages are much greater than V_D.
Small-signal	Slope $= 1/r_d$; axes i_D vs v_D, operating point I_D, V_D	For small signals superimposed on V_D and I_D: $i_d = v_d/r_d$ $r_d = nV_T/I_D$ (For $n = 1$, v_d is limited to 5 mV; for $n = 2$, 10 mV)	i_d, $+$, v_d, $-$; r_d	Useful for finding the signal component of the diode voltage (e.g., in the voltage-regulator application). Serves as the basis for small-signal modeling of transistors (Chapters 4 and 5).

3.4 OPERATION IN THE REVERSE BREAKDOWN REGION—ZENER DIODES

The very steep i–v curve that the diode exhibits in the breakdown region (Fig. 3.8) and the almost-constant voltage drop that this indicates suggest that diodes operating in the breakdown region can be used in the design of voltage regulators. From the previous section, the reader will recall that voltage regulators are circuits that provide constant dc output voltages in the face of changes in their load current and in the system power-supply voltage. This in fact turns out to be an important application of diodes operating in the reverse breakdown region, and special diodes are manufactured to operate specifically in the breakdown region. Such diodes are called **breakdown diodes** or, more commonly, **zener diodes,** after an early worker in the area.

Figure 3.20 shows the circuit symbol of the zener diode. In normal applications of zener diodes, current flows into the cathode, and the cathode is positive with respect to the anode. Thus I_Z and V_Z in Fig. 3.20 have positive values.

3.4.1 Specifying and Modeling the Zener Diode

Figure 3.21 shows details of the diode i–v characteristics in the breakdown region. We observe that for currents greater than the **knee current** I_{ZK} (specified on the data sheet of

FIGURE 3.20 Circuit symbol for a zener diode.

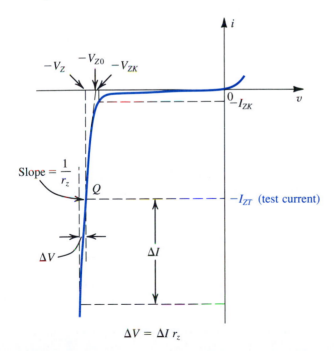

FIGURE 3.21 The diode i–v characteristic with the breakdown region shown in some detail.

the zener diode), the $i-v$ characteristic is almost a straight line. The manufacturer usually specifies the voltage across the zener diode V_Z at a specified test current, I_{ZT}. We have indicated these parameters in Fig. 3.21 as the coordinates of the point labeled Q. Thus a 6.8-V zener diode will exhibit at 6.8-V drop at a specified test current of, say, 10 mA. As the current through the zener deviates from I_{ZT}, the voltage across it will change, though only slightly. Figure 3.21 shows that corresponding to current change ΔI the zener voltage changes by ΔV, which is related to ΔI by

$$\Delta V = r_z \Delta I$$

where r_z is the inverse of the slope of the almost-linear $i-v$ curve at point Q. Resistance r_z is the **incremental resistance** of the zener diode at operating point Q. It is also known as the **dynamic resistance** of the zener, and its value is specified on the device data sheet. Typically, r_z is in the range of a few ohms to a few tens of ohms. Obviously, the lower the value of r_z is, the more constant the zener voltage remains as its current varies and thus the more ideal its performance becomes in the design of voltage regulators. In this regard, we observe from Fig. 3.21 that while r_z remains low and almost constant over a wide range of current, its value increases considerably in the vicinity of the knee. Therefore, as a general design guideline, one should avoid operating the zener in this low-current region.

Zener diodes are fabricated with voltages V_Z in the range of a few volts to a few hundred volts. In addition to specifying V_Z (at a particular current I_{ZT}), r_z, and I_{ZK}, the manufacturer also specifies the maximum power that the device can safely dissipate. Thus a 0.5-W, 6.8-V zener diode can operate safely at currents up to a maximum of about 70 mA.

The almost-linear $i-v$ characteristic of the zener diode suggests that the device can be modeled as indicated in Fig. 3.22. Here V_{Z0} denotes the point at which the straight line of slope $1/r_z$ intersects the voltage axis (refer to Fig. 3.21). Although V_{Z0} is shown to be slightly different from the knee voltage V_{ZK}, in practice their values are almost equal. The equivalent circuit model of Fig. 3.22 can be analytically described by

$$V_Z = V_{Z0} + r_z I_Z \tag{3.20}$$

and it applies for $I_Z > I_{ZK}$ and, obviously, $V_Z > V_{Z0}$.

3.4.2 Use of the Zener as a Shunt Regulator

We now illustrate, by way of an example, the use of zener diodes in the design of shunt regulators, so named because the regulator circuit appears in parallel (shunt) with the load.

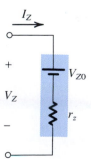

FIGURE 3.22 Model for the zener diode.

EXAMPLE 3.8

The 6.8-V zener diode in the circuit of Fig. 3.23(a) is specified to have $V_Z = 6.8$ V at $I_Z = 5$ mA, $r_z = 20\ \Omega$, and $I_{ZK} = 0.2$ mA. The supply voltage V^+ is nominally 10 V but can vary by ± 1 V.

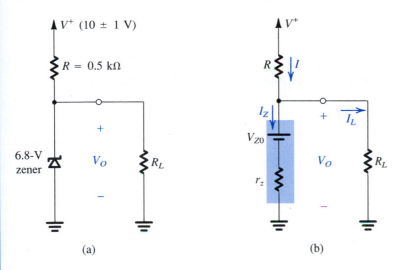

(a)   (b)

FIGURE 3.23   (a) Circuit for Example 3.8. (b) The circuit with the zener diode replaced with its equivalent circuit model.

(a) Find V_O with no load and with V^+ at its nominal value.

(b) Find the change in V_O resulting from the ± 1-V change in V^+. Note that $(\Delta V_O / \Delta V^+)$, usually expressed in mV/V, is known as **line regulation.**

(c) Find the change in V_O resulting from connecting a load resistance R_L that draws a current $I_L = 1$ mA, and hence find the **load regulation** $(\Delta V_O / \Delta I_L)$ in mV/mA.

(d) Find the change in V_O when $R_L = 2$ kΩ.

(e) Find the value of V_O when $R_L = 0.5$ kΩ.

(f) What is the minimum value of R_L for which the diode still operates in the breakdown region?

Solution

First we must determine the value of the parameter V_{Z0} of the zener diode model. Substituting $V_Z = 6.8$ V, $I_Z = 5$ mA, and $r_z = 20\ \Omega$ in Eq. (3.20) yields $V_{Z0} = 6.7$ V. Figure 3.23(b) shows the circuit with the zener diode replaced with its model.

(a) With no load connected, the current through the zener is given by

$$I_Z = I = \frac{V^+ - V_{Z0}}{R + r_z}$$

$$= \frac{10 - 6.7}{0.5 + 0.02} = 6.35 \text{ mA}$$

Thus,

$$V_O = V_{Z0} + I_Z r_z$$

$$= 6.7 + 6.35 \times 0.02 = 6.83 \text{ V}$$

(b) For a ±1-V change in V^+, the change in output voltage can be found from

$$\Delta V_O = \Delta V^+ \frac{r_z}{R + r_z}$$

$$= \pm 1 \times \frac{20}{500 + 20} = \pm 38.5 \text{ mV}$$

Thus,

$$\text{Line regulation} = 38.5 \text{ mV/V}$$

(c) When a load resistance R_L that draws a load current $I_L = 1$ mA is connected, the zener current will decrease by 1 mA. The corresponding change in zener voltage can be found from

$$\Delta V_O = r_z \Delta I_Z$$

$$= 20 \times -1 = -20 \text{ mV}$$

Thus the load regulation is

$$\text{Load regulation} \equiv \frac{\Delta V_O}{\Delta I_L} = -20 \text{ mV/mA}$$

(d) When a load resistance of 2 kΩ is connected, the load current will be approximately 6.8 V/2 kΩ = 3.4 mA. Thus the change in zener current will be $\Delta I_Z = -3.4$ mA, and the corresponding change in zener voltage (output voltage) will thus be

$$\Delta V_O = r_z \Delta I_Z$$

$$= 20 \times -3.4 = -68 \text{ mV}$$

This calculation, however, is approximate, because it neglects the change in the current I. A more accurate estimate of ΔV_O can be obtained by analyzing the circuit in Fig. 3.23(b). The result of such an analysis is $\Delta V_O = -70$ mV.

(e) An R_L of 0.5 kΩ would draw a load current of 6.8/0.5 = 13.6 mA. This is not possible, because the current I supplied through R is only 6.4 mA (for $V^+ = 10$ V). Therefore, the zener must be cut off. If this is indeed the case, then V_O is determined by the voltage divider formed by R_L and R (Fig. 3.23a),

$$V_O = V^+ \frac{R_L}{R + R_L}$$

$$= 10 \frac{0.5}{0.5 + 0.5} = 5 \text{ V}$$

Since this voltage is lower than the breakdown voltage of the zener, the diode is indeed no longer operating in the breakdown region.

(f) For the zener to be at the edge of the breakdown region, $I_Z = I_{ZK} = 0.2$ mA and $V_Z \simeq V_{ZK} \simeq$ 6.7 V. At this point the lowest (worst-case) current supplied through R is $(9 - 6.7)/0.5 = 4.6$ mA, and thus the load current is $4.6 - 0.2 = 4.4$ mA. The corresponding value of R_L is

$$R_L = \frac{6.7}{4.4} \simeq 1.5 \text{ kΩ}$$

3.4.3 Temperature Effects

The dependence of the zener voltage V_Z on temperature is specified in terms of its temperature coefficient TC, or **temco** as it is commonly known, which is usually expressed in mV/°C. The value of TC depends on the zener voltage, and for a given diode the TC varies with the operating current. Zener diodes whose V_Z are lower than about 5 V exhibit a negative TC. On the other hand, zeners with higher voltages exhibit a positive TC. The TC of a zener diode with a V_Z of about 5 V can be made zero by operating the diode at a specified current. Another commonly used technique for obtaining a reference voltage with low temperature coefficient is to connect a zener diode with a positive temperature coefficient of about 2 mV/°C in series with a forward-conducting diode. Since the forward-conducting diode has a voltage drop of ≈ 0.7 V and a TC of about −2 mV/°C, the series combination will provide a voltage of $(V_Z + 0.7)$ with a TC of about zero.

EXERCISES

3.17 A zener diode whose nominal voltage is 10 V at 10 mA has an incremental resistance of 50 Ω. What voltage do you expect if the diode current is halved? Doubled? What is the value of V_{Z0} in the zener model?

Ans. 9.75 V; 10.5 V; 9.5 V

D3.18 A zener diode exhibits a constant voltage of 5.6 V for currents greater than five times the knee current. I_{ZK} is specified to be 1 mA. The zener is to be used in the design of a shunt regulator fed from a 15-V supply. The load current varies over the range of 0 mA to 15 mA. Find a suitable value for the resistor R. What is the maximum power dissipation of the zener diode?

Ans. 470 Ω; 112 mW

3.19 A shunt regulator utilizes a zener diode whose voltage is 5.1 V at a current of 50 mA and whose incremental resistance is 7 Ω. The diode is fed from a supply of 15-V nominal voltage through a 200-Ω resistor. What is the output voltage at no load? Find the line regulation and the load regulation.

Ans. 5.1 V; 33.8 mV/V; −7 mV/mA

3.4.4 A Final Remark

Though simple and useful, zener diodes have lost a great deal of their popularity in recent years. They have been virtually replaced in voltage-regulator design by specially designed integrated circuits (ICs) that perform the voltage regulation function much more effectively and with greater flexibility than zener diodes.

3.5 RECTIFIER CIRCUITS

One of the most important applications of diodes is in the design of rectifier circuits. A diode rectifier forms an essential building block of the dc power supplies required to power electronic equipment. A block diagram of such a power supply is shown in Fig. 3.24. As indicated, the power supply is fed from the 120-V (rms) 60-Hz ac line, and it delivers a dc voltage V_O (usually in the range of 5–20 V) to an electronic circuit represented by the *load* block. The dc voltage V_O is required to be as constant as possible in spite of variations in the ac line voltage and in the current drawn by the load.

The first block in a dc power supply is the **power transformer.** It consists of two separate coils wound around an iron core that magnetically couples the two windings. The **primary winding,** having N_1 turns, is connected to the 120-V ac supply, and the **secondary winding,** having N_2 turns, is connected to the circuit of the dc power supply. Thus an ac voltage v_S

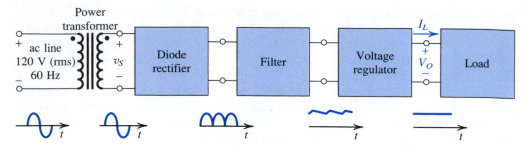

FIGURE 3.24   Block diagram of a dc power supply.

of $120(N_2/N_1)$ V (rms) develops between the two terminals of the secondary winding. By selecting an appropriate turns ratio (N_1/N_2) for the transformer, the designer can step the line voltage down to the value required to yield the particular dc voltage output of the supply. For instance, a secondary voltage of 8-V rms may be appropriate for a dc output of 5 V. This can be achieved with a 15:1 turns ratio.

In addition to providing the appropriate sinusoidal amplitude for the dc power supply, the power transformer provides electrical isolation between the electronic equipment and the power-line circuit. This isolation minimizes the risk of electric shock to the equipment user.

The diode rectifier converts the input sinusoid v_S to a unipolar output, which can have the pulsating waveform indicated in Fig. 3.24. Although this waveform has a nonzero average or a dc component, its pulsating nature makes it unsuitable as a dc source for electronic circuits, hence the need for a filter. The variations in the magnitude of the rectifier output are considerably reduced by the filter block in Fig. 3.24. In the following sections we shall study a number of rectifier circuits and a simple implementation of the output filter.

The output of the rectifier filter, though much more constant than without the filter, still contains a time-dependent component, known as **ripple.** To reduce the ripple and to stabilize the magnitude of the dc output voltage of the supply against variations caused by changes in load current, a voltage regulator is employed. Such a regulator can be implemented using the zener shunt regulator configuration studied in Section 3.4. Alternatively, and much more commonly at present, an integrated-circuit regulator can be used.

3.5.1 The Half-Wave Rectifier

The half-wave rectifier utilizes alternate half-cycles of the input sinusoid. Figure 3.25(a) shows the circuit of a half-wave rectifier. This circuit was analyzed in Section 3.1 (see Fig. 3.3) assuming an ideal diode. Using the more realistic battery-plus-resistance diode model, we obtain the equivalent circuit shown in Fig. 3.25(b), from which we can write

$$v_O = 0, \qquad v_S < V_{D0} \tag{3.21a}$$

$$v_O = \frac{R}{R+r_D}v_S - V_{D0}\frac{R}{R+r_D}, \qquad v_S \geq V_{D0} \tag{3.21b}$$

The transfer characteristic represented by these equations is sketched in Fig. 3.25(c). In many applications, $r_D \ll R$ and the second equation can be simplified to

$$v_O \simeq v_S - V_{D0} \tag{3.22}$$

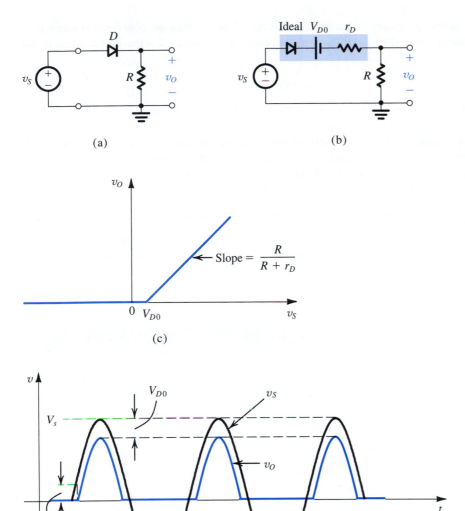

FIGURE 3.25 (a) Half-wave rectifier. (b) Equivalent circuit of the half-wave rectifier with the diode replaced with its battery-plus-resistance model. (c) Transfer characteristic of the rectifier circuit. (d) Input and output waveforms, assuming that $r_D \ll R$.

where $V_{D0} = 0.7$ V or 0.8 V. Figure 3.25(d) shows the output voltage obtained when the input v_S is a sinusoid.

In selecting diodes for rectifier design, two important parameters must be specified: the current-handling capability required of the diode, determined by the largest current the diode is expected to conduct, and the **peak inverse voltage** (PIV) that the diode must be able to withstand without breakdown, determined by the largest reverse voltage that is expected to appear across the diode. In the rectifier circuit of Fig. 3.25(a), we observe that when v_S is negative the diode will be cut off and v_O will be zero. It follows that the PIV is equal to the peak of v_S,

$$\text{PIV} = V_s$$

It is usually prudent, however, to select a diode that has a reverse breakdown voltage at least 50% greater than the expected PIV.

Before leaving the half-wave rectifier, the reader should note two points. First, it is possible to use the diode exponential characteristic to determine the exact transfer characteristic of the rectifier (see Problem 3.73). However, the amount of work involved is usually too great to be justified in practice. Of course, such an analysis can be easily done using a computer circuit-analysis program such as SPICE (see Section 3.9).

Second, whether we analyze the circuit accurately or not, it should be obvious that this circuit does not function properly when the input signal is small. For instance, this circuit cannot be used to rectify an input sinusoid of 100-mV amplitude. For such an application one resorts to a so-called precision rectifier, a circuit utilizing diodes in conjunction with op amps. One such circuit is presented in Section 3.5.5.

EXERCISE

3.20 For the half-wave rectifier circuit in Fig. 3.25(a), neglecting the effect of r_D, show the following: (a) For the half-cycles during which the diode conducts, conduction begins at an angle $\theta = \sin^{-1}(V_{D0}/V_s)$ and terminates at $(\pi - \theta)$, for a total conduction angle of $(\pi - 2\theta)$. (b) The average value (dc component) of v_O is $V_O \simeq (1/\pi)V_s - V_{D0}/2$. (c) The peak diode current is $(V_s - V_{D0})/R$.

Find numerical values for these quantities for the case of 12-V (rms) sinusoidal input, $V_{D0} \simeq 0.7$ V, and $R = 100\ \Omega$. Also, give the value for PIV.

Ans. (a) $\theta = 2.4°$, conduction angle = 175°; (b) 5.05 V; (c) 163 mA; 17 V

3.5.2 The Full-Wave Rectifier

The full-wave rectifier utilizes both halves of the input sinusoid. To provide a unipolar output, it inverts the negative halves of the sine wave. One possible implementation is shown in Fig. 3.26(a). Here the transformer secondary winding is **center-tapped** to provide two equal voltages v_S across the two halves of the secondary winding with the polarities indicated. Note that when the input line voltage (feeding the primary) is positive, both of the signals labeled v_S will be positive. In this case D_1 will conduct and D_2 will be reverse biased. The current through D_1 will flow through R and back to the center tap of the secondary. The circuit then behaves like a half-wave rectifier, and the output during the positive half cycles when D_1 conducts will be identical to that produced by the half-wave rectifier.

Now, during the negative half cycle of the ac line voltage, both of the voltages labeled v_S will be negative. Thus D_1 will be cut off while D_2 will conduct. The current conducted by D_2 will flow through R and back to the center tap. It follows that during the negative half-cycles while D_2 conducts, the circuit behaves again as a half-wave rectifier. The important point, however, is that the current through R always flows in the same direction, and thus v_O will be unipolar, as indicated in Fig. 3.26(c). The output waveform shown is obtained

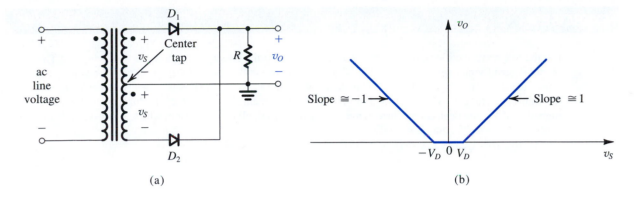

(a)   (b)

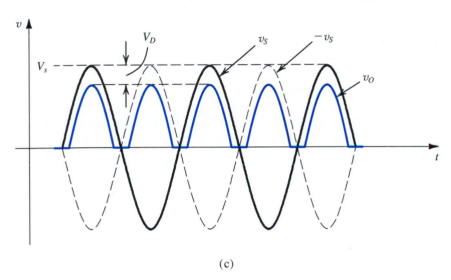

(c)

FIGURE 3.26 Full-wave rectifier utilizing a transformer with a center-tapped secondary winding: **(a)** circuit; **(b)** transfer characteristic assuming a constant-voltage-drop model for the diodes; **(c)** input and output waveforms.

by assuming that a conducting diode has a constant voltage drop V_D. Thus the transfer characteristic of the full-wave rectifier takes the shape shown in Fig. 3.26(b).

The full-wave rectifier obviously produces a more "energetic" waveform than that provided by the half-wave rectifier. In almost all rectifier applications, one opts for a full-wave type of some kind.

To find the PIV of the diodes in the full-wave rectifier circuit, consider the situation during the positive half-cycles. Diode D_1 is conducting, and D_2 is cut off. The voltage at the cathode of D_2 is v_O, and that at its anode is $-v_S$. Thus the reverse voltage across D_2 will be $(v_O + v_S)$, which will reach its maximum when v_O is at its peak value of $(V_s - V_D)$, and v_S is at its peak value of V_s; thus,

$$\text{PIV} = 2V_s - V_D$$

which is approximately twice that for the case of the half-wave rectifier.

3.21 For the full-wave rectifier circuit in Fig. 3.26(a), neglecting the effect of r_D, show the following: (a) The output is zero for an angle of $2 \sin^{-1}(V_D/V_s)$ centered around the zero-crossing points of the sine-wave input. (b) The average value (dc component) of v_O is $V_O \simeq (2/\pi)V_s - V_D$. (c) The peak current through each diode is $(V_s - V_D)/R$. Find the fraction (percentage) of each cycle during which $v_O > 0$, the value of V_O, the peak diode current, and the value of PIV, all for the case in which v_S is a 12-V (rms) sinusoid, $V_D \simeq 0.7$ V, and $R = 100$ Ω.

Ans. 97.4%; 10.1 V; 163 mA; 33.2 V

3.5.3 The Bridge Rectifier

An alternative implementation of the full-wave rectifier is shown in Fig. 3.27(a). The circuit, known as the bridge rectifier because of the similarity of its configuration to that of the Wheatstone bridge, does not require a center-tapped transformer, a distinct advantage over the full-wave rectifier circuit of Fig. 3.26. The bridge rectifier, however, requires four diodes as compared to two in the previous circuit. This is not much of a disadvantage, because diodes are inexpensive and one can buy a diode bridge in one package.

The bridge rectifier circuit operates as follows: During the positive half-cycles of the input voltage, v_S is positive, and thus current is conducted through diode D_1, resistor R, and

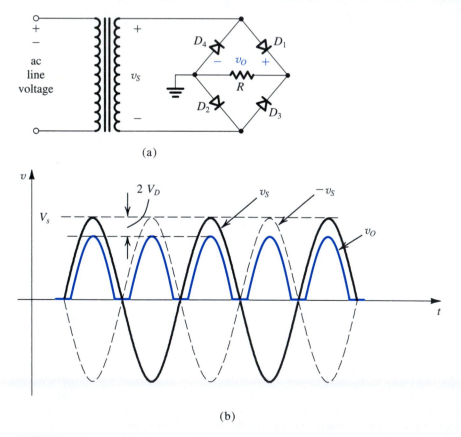

(a)

(b)

FIGURE 3.27 The bridge rectifier: **(a)** circuit; **(b)** input and output waveforms.

diode D_2. Meanwhile, diodes D_3 and D_4 will be reverse biased. Observe that there are two diodes in series in the conduction path, and thus v_O will be lower than v_S by two diode drops (compared to one drop in the circuit previously discussed). This is somewhat of a disadvantage of the bridge rectifier.

Next, consider the situation during the negative half-cycles of the input voltage. The secondary voltage v_S will be negative, and thus $-v_S$ will be positive, forcing current through D_3, R, and D_4. Meanwhile, diodes D_1 and D_2 will be reverse biased. The important point to note, though, is that during both half-cycles, current flows through R in the same direction (from right to left), and thus v_O will always be positive, as indicated in Fig. 3.27(b).

To determine the peak inverse voltage (PIV) of each diode, consider the circuit during the positive half-cycles. The reverse voltage across D_3 can be determined from the loop formed by D_3, R, and D_2 as

$$v_{D3} \text{ (reverse)} = v_O + v_{D2} \text{ (forward)}$$

Thus the maximum value of v_{D3} occurs at the peak of v_O and is given by

$$\text{PIV} = V_s - 2V_D + V_D = V_s - V_D$$

Observe that here the PIV is about half the value for the full-wave rectifier with a center-tapped transformer. This is another advantage of the bridge rectifier.

Yet one more advantage of the bridge rectifier circuit over that utilizing a center-tapped transformer is that only about half as many turns are required for the secondary winding of the transformer. Another way of looking at this point can be obtained by observing that each half of the secondary winding of the center-tapped transformer is utilized for only half the time. These advantages have made the bridge rectifier the most popular rectifier circuit configuration.

EXERCISE

3.22 For the bridge rectifier circuit of Fig. 3.27(a), use the constant-voltage-drop diode model to show that (a) the average (or dc component) of the output voltage is $V_O \simeq (2/\pi)V_s - 2V_D$ and (b) the peak diode current is $(V_s - 2V_D)/R$. Find numerical values for the quantities in (a) and (b) and the PIV for the case in which v_S is a 12-V (rms) sinusoid, $V_D \simeq 0.7$ V, and $R = 100 \, \Omega$.

Ans. 9.4 V; 156 mA; 16.3 V

3.5.4 The Rectifier with a Filter Capacitor—The Peak Rectifier

The pulsating nature of the output voltage produced by the rectifier circuits discussed above makes it unsuitable as a dc supply for electronic circuits. A simple way to reduce the variation of the output voltage is to place a capacitor across the load resistor. It will be shown that this **filter capacitor** serves to reduce substantially the variations in the rectifier output voltage.

To see how the rectifier circuit with a filter capacitor works, consider first the simple circuit shown in Fig. 3.28. Let the input v_I be a sinusoid with a peak value V_p, and assume the diode to be ideal. As v_I goes positive, the diode conducts and the capacitor is charged so that $v_O = v_I$. This situation continues until v_I reaches its peak value V_p. Beyond the peak, as v_I decreases the diode becomes reverse biased and the output voltage remains constant at the value V_p. In fact, theoretically speaking, the capacitor will retain its charge and hence its voltage indefinitely, because there is no way for the capacitor to discharge. Thus the circuit provides a dc voltage output equal to the peak of the input sine wave. This is a very encouraging result in view of our desire to produce a dc output.

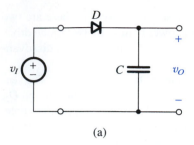

(a)

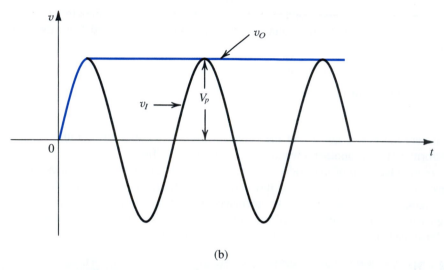

(b)

FIGURE 3.28 **(a)** A simple circuit used to illustrate the effect of a filter capacitor. **(b)** Input and output waveforms assuming an ideal diode. Note that the circuit provides a dc voltage equal to the peak of the input sine wave. The circuit is therefore known as a peak rectifier or a peak detector.

Next, we consider the more practical situation where a load resistance R is connected across the capacitor C, as depicted in Fig. 3.29(a). However, we will continue to assume the diode to be ideal. As before, for a sinusoidal input, the capacitor charges to the peak of the input V_p. Then the diode cuts off, and the capacitor discharges through the load resistance R. The capacitor discharge will continue for almost the entire cycle, until the time at which v_I exceeds the capacitor voltage. Then the diode turns on again and charges the capacitor up to the peak of v_I, and the process repeats itself. Observe that to keep the output voltage from decreasing too much during capacitor discharge, one selects a value for C so that the time constant CR is much greater than the discharge interval.

We are now ready to analyze the circuit in detail. Figure 3.29(b) shows the steady-state input and output voltage waveforms under the assumption that $CR \gg T$, where T is the period of the input sinusoid. The waveforms of the load current

$$i_L = v_O/R \tag{3.23}$$

and of the diode current (when it is conducting)

$$i_D = i_C + i_L \tag{3.24}$$

$$= C\frac{dv_I}{dt} + i_L \tag{3.25}$$

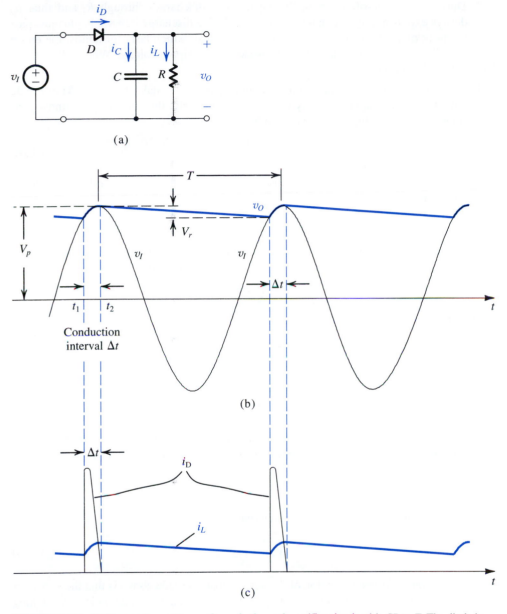

FIGURE 3.29 Voltage and current waveforms in the peak rectifier circuit with $CR \gg T$. The diode is assumed ideal.

are shown in Fig. 3.29(c). The following observations are in order:

1. The diode conducts for a brief interval, Δt, near the peak of the input sinusoid and supplies the capacitor with charge equal to that lost during the much longer discharge interval. The latter is approximately equal to the period T.

2. Assuming an ideal diode, the diode conduction begins at time t_1, at which the input v_I equals the exponentially decaying output v_O. Conduction stops at t_2 shortly after the peak of v_I; the exact value of t_2 can be determined by setting $i_D = 0$ in Eq. (3.25).

3. During the diode-off interval, the capacitor C discharges through R, and thus v_O decays exponentially with a time constant CR. The discharge interval begins just past the peak of v_I. At the end of the discharge interval, which lasts for almost the entire period T, $v_O = V_p - V_r$, where V_r is the peak-to-peak ripple voltage. When $CR \gg T$, the value of V_r is small.

4. When V_r is small, v_O is almost constant and equal to the peak value of v_I. Thus the dc output voltage is approximately equal to V_p. Similarly, the current i_L is almost constant, and its dc component I_L is given by

$$I_L = \frac{V_p}{R} \tag{3.26}$$

If desired, a more accurate expression for the output dc voltage can be obtained by taking the average of the extreme values of v_O,

$$V_O = V_p - \tfrac{1}{2}V_r \tag{3.27}$$

With these observations in hand, we now derive expressions for V_r and for the average and peak values of the diode current. During the diode-off interval, v_O can be expressed as

$$v_O = V_p e^{-t/CR}$$

At the end of the discharge interval we have

$$V_p - V_r \simeq V_p e^{-T/CR}$$

Now, since $CR \gg T$, we can use the approximation $e^{-T/CR} \simeq 1 - T/CR$ to obtain

$$V_r \simeq V_p \frac{T}{CR} \tag{3.28}$$

We observe that to keep V_r small we must select a capacitance C so that $CR \gg T$. The **ripple voltage** V_r in Eq. (3.28) can be expressed in terms of the frequency $f = 1/T$ as

$$V_r = \frac{V_p}{fCR} \tag{3.29a}$$

Using Eq. (3.26) we can express V_r by the alternate expression

$$V_r = \frac{I_L}{fC} \tag{3.29b}$$

Note that an alternative interpretation of the approximation made above is that the capacitor discharges by means of a constant current $I_L = V_p/R$. This approximation is valid as long as $V_r \ll V_p$.

Using Fig. 3.29(b) and assuming that diode conduction ceases almost at the peak of v_I, we can determine the **conduction interval** Δt from

$$V_p \cos(\omega \Delta t) = V_p - V_r$$

where $\omega = 2\pi f = 2\pi/T$ is the angular frequency of v_I. Since $(\omega \Delta t)$ is a small angle, we can employ the approximation $\cos(\omega \Delta t) \simeq 1 - \tfrac{1}{2}(\omega \Delta t)^2$ to obtain

$$\omega \Delta t \simeq \sqrt{2V_r/V_p} \tag{3.30}$$

We note that when $V_r \ll V_p$, the conduction angle $\omega \Delta t$ will be small, as assumed.

To determine the average diode current during conduction, i_{Dav}, we equate the charge that the diode supplies to the capacitor,

$$Q_{\text{supplied}} = i_{Cav} \Delta t$$

where from Eq. (3.24),

$$i_{Cav} = i_{Dav} - I_L$$

to the charge that the capacitor loses during the discharge interval,

$$Q_{\text{lost}} = CV_r$$

to obtain, using Eqs. (3.30) and (3.29a),

$$i_{Dav} = I_L(1 + \pi\sqrt{2V_p/V_r}) \tag{3.31}$$

Observe that when $V_r \ll V_p$, the average diode current during conduction is much greater than the dc load current. This is not surprising, since the diode conducts for a very short interval and must replenish the charge lost by the capacitor during the much longer interval in which it is discharged by I_L.

The peak value of the diode current, i_{Dmax}, can be determined by evaluating the expression in Eq. (3.25) at the onset of diode conduction—that is, at $t = t_1 = -\Delta t$ (where $t = 0$ is at the peak). Assuming that i_L is almost constant at the value given by Eq. (3.26), we obtain

$$i_{Dmax} = I_L(1 + 2\pi\sqrt{2V_p/V_r}) \tag{3.32}$$

From Eqs. (3.31) and (3.32), we see that for $V_r \ll V_p$, $i_{Dmax} \simeq 2i_{Dav}$, which correlates with the fact that the waveform of i_D is almost a right-angle triangle (see Fig. 3.29c).

EXAMPLE 3.9

Consider a peak rectifier fed by a 60-Hz sinusoid having a peak value $V_p = 100$ V. Let the load resistance $R = 10$ kΩ. Find the value of the capacitance C that will result in a peak-to-peak ripple of 2 V. Also, calculate the fraction of the cycle during which the diode is conducting and the average and peak values of the diode current.

Solution

From Eq. (3.29a) we obtain the value of C as

$$C = \frac{V_p}{V_r f R} = \frac{100}{2 \times 60 \times 10 \times 10^3} = 83.3 \ \mu\text{F}$$

The conduction angle $\omega \Delta t$ is found from Eq. (3.30) as

$$\omega \Delta t = \sqrt{2 \times 2/100} = 0.2 \text{ rad}$$

Thus the diode conducts for $(0.2/2\pi) \times 100 = 3.18\%$ of the cycle. The average diode current is obtained from Eq. (3.31), where $I_L = 100/10 = 10$ mA, as

$$i_{Dav} = 10(1 + \pi\sqrt{2 \times 100/2}) = 324 \text{ mA}$$

The peak diode current is found using Eq. (3.32),

$$i_{Dmax} = 10(1 + 2\pi\sqrt{2 \times 100/2}) = 638 \text{ mA}$$

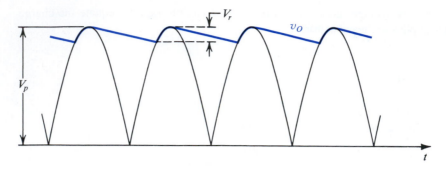

FIGURE 3.30 Waveforms in the full-wave peak rectifier.

The circuit of Fig. 3.29(a) is known as a half-wave **peak rectifier.** The full-wave rectifier circuits of Figs. 3.26(a) and 3.27(a) can be converted to peak rectifiers by including a capacitor across the load resistor. As in the half-wave case, the output dc voltage will be almost equal to the peak value of the input sine wave (Fig. 3.30). The ripple frequency, however, will be twice that of the input. The peak-to-peak ripple voltage, for this case, can be derived using a procedure identical to that above but with the discharge period T replaced by $T/2$, resulting in

$$V_r = \frac{V_p}{2fCR} \tag{3.33}$$

While the diode conduction interval, Δt, will still be given by Eq. (3.30), the average and peak currents in each of the diodes will be given by

$$i_{Dav} = I_L(1 + \pi\sqrt{V_p/2V_r}) \tag{3.34}$$

$$i_{Dmax} = I_L(1 + 2\pi\sqrt{V_p/2V_r}) \tag{3.35}$$

Comparing these expressions with the corresponding ones for the half-wave case, we note that for the same values of V_p, f, R, and V_r (and thus the same I_L), we need a capacitor half the size of that required in the half-wave rectifier. Also, the current in each diode in the full-wave rectifier is approximately half that which flows in the diode of the half-wave circuit.

The analysis above assumed ideal diodes. The accuracy of the results can be improved by taking the diode voltage drop into account. This can be easily done by replacing the peak voltage V_p to which the capacitor charges with $(V_p - V_D)$ for the half-wave circuit and the full-wave circuit using a center-tapped transformer and with $(V_p - 2V_D)$ for the bridge-rectifier case.

We conclude this section by noting that peak-rectifier circuits find application in signal-processing systems where it is required to detect the peak of an input signal. In such a case, the circuit is referred to as a **peak detector.** A particularly popular application of the peak detector is in the design of a demodulator for amplitude-modulated (AM) signals. We shall not discuss this application further here.

EXERCISES

3.23 Derive the expressions in Eqs. (3.33), (3.34), and (3.35).

D3.24 Consider a bridge-rectifier circuit with a filter capacitor C placed across the load resistor R for the case in which the transformer secondary delivers a sinusoid of 12 V (rms) having a 60-Hz frequency and assuming $V_D = 0.8$ V and a load resistance $R = 100\ \Omega$. Find the value of C that results in a ripple voltage no larger than 1 V peak-to-peak. What is the dc voltage at the output? Find the load current. Find the diodes' conduction angle. What is the average diode current? What is the peak reverse voltage across each diode? Specify the diode in terms of its peak current and its PIV.

Ans. 1281 μF; 15.4 V or (a better estimate) 14.9 V; 0.15 A; 0.36 rad (20.7°); 1.45 A; 2.74 A; 16.2 V. Thus select a diode with 3.5 A to 4 A peak current and a 20-V PIV rating.

3.5.5 Precision Half-Wave Rectifier—The Super Diode[4]

The rectifier circuits studied thus far suffer from having one or two diode drops in the signal paths. Thus these circuits work well only when the signal to be rectified is much larger than the voltage drop of a conducting diode (0.7 V or so). In such a case the details of the diode forward characteristics or the exact value of the diode voltage do not play a prominent role in determining circuit performance. This is indeed the case in the application of rectifier circuits in power-supply design. There are other applications, however, where the signal to be rectified is small (e.g., on the order of 100 mV or so) and thus clearly insufficient to turn on a diode. Also, in instrumentation applications, the need arises for rectifier circuits with very precise and predictable transfer characteristics. For these applications, a class of circuits has been developed utilizing op amps (Chapter 2) together with diodes to provide precision rectification. In the following discussion, we study one such circuit, leaving a more comprehensive study of op amp–diode circuits to Chapter 13.

Figure 3.31(a) shows a precision half-wave rectifier circuit consisting of a diode placed in the negative-feedback path of an op amp, with R being the rectifier load resistance. The op amp, of course, needs power supplies for its operation. For simplicity, these are not shown in the circuit diagram. The circuit works as follows: If v_I goes positive, the output voltage v_A of the op amp will go positive and the diode will conduct, thus establishing a closed feedback path between the op amp's output terminal and the negative input terminal.

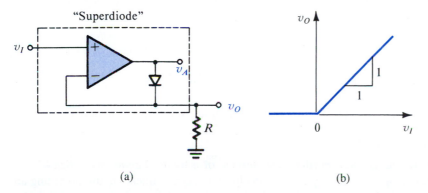

(a)   (b)

FIGURE 3.31 The "superdiode" precision half-wave rectifier and its almost-ideal transfer characteristic. Note that when $v_I > 0$ and the diode conducts, the op amp supplies the load current, and the source is conveniently buffered, an added advantage. Not shown are the op-amp power supplies.

---

[4] This section requires knowledge of operational amplifiers.

This negative-feedback path will cause a virtual short circuit to appear between the two input terminals. Thus the voltage at the negative input terminal, which is also the output voltage v_O, will equal (to within a few millivolts) that at the positive input terminal, which is the input voltage v_I,

$$v_O = v_I \qquad v_I \geq 0$$

Note that the offset voltage ($\simeq 0.6$ V) exhibited in the simple half-wave rectifier circuit of Fig. 3.25 is no longer present. For the op-amp circuit to start operation, v_I has to exceed only a negligibly small voltage equal to the diode drop divided by the op amp's open-loop gain. In other words, the straight-line transfer characteristic v_O–v_I almost passes through the origin. This makes this circuit suitable for applications involving very small signals.

Consider now the case when v_I goes negative. The op amp's output voltage v_A will tend to follow and go negative. This will reverse-bias the diode, and no current will flow through resistance R, causing v_O to remain equal to 0 V. Thus, for $v_I < 0$, $v_O = 0$. Since in this case the diode is off, the op amp will be operating in an open-loop fashion, and its output will be at the negative saturation level.

The transfer characteristic of this circuit will be that shown in Fig. 3.31(b), which is almost identical to the ideal characteristic of a half-wave rectifier. The nonideal diode characteristics have been almost completely masked by placing the diode in the negative-feedback path of an op amp. This is another dramatic application of negative feedback, a subject we will study formally in Chapter 8. The combination of diode and op amp, shown in the dotted box in Fig. 3.31(a), is appropriately referred to as a "superdiode."

EXERCISES

3.25 Consider the operational rectifier or superdiode circuit of Fig. 3.31(a), with $R = 1$ kΩ. For $v_I = 10$ mV, 1 V, and −1 V, what are the voltages that result at the rectifier output and at the output of the op amp? Assume that the op amp is ideal and that its output saturates at ±12 V. The diode has a 0.7-V drop at 1-mA current, and the voltage drop changes by 0.1 V per decade of current change.

Ans. 10 mV, 0.51 V; 1 V, 1.7 V; 0 V, −12 V

3.26 If the diode in the circuit of Fig. 3.31(a) is reversed, find the transfer characteristic v_O as a function of v_I.

Ans. $v_O = 0$ for $v_I \geq 0$; $v_O = v_I$ for $v_I \leq 0$

 ## 3.6 LIMITING AND CLAMPING CIRCUITS

In this section, we shall present additional nonlinear circuit applications of diodes.

3.6.1 Limiter Circuits

Figure 3.32 shows the general transfer characteristic of a limiter circuit. As indicated, for inputs in a certain range, $L_-/K \leq v_I \leq L_+/K$, the limiter acts as a linear circuit, providing an output proportional to the input, $v_O = Kv_I$. Although in general K can be greater than 1, the circuits discussed in this section have $K \leq 1$ and are known as passive limiters. (Examples of active limiters will be presented in Chapter 13.) If v_I exceeds the upper *threshold* (L_+/K), the output voltage is *limited* or clamped to the upper limiting level L_+. On the other hand, if v_I is reduced below the lower limiting threshold (L_-/K), the output voltage v_O is limited to the lower limiting level L_-.

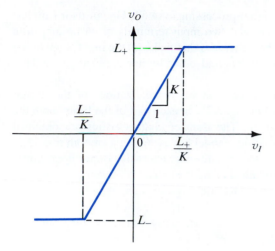

FIGURE 3.32 General transfer characteristic for a limiter circuit.

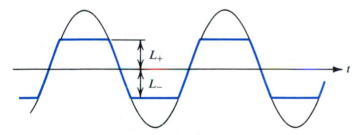

FIGURE 3.33 Applying a sine wave to a limiter can result in clipping off its two peaks.

The general transfer characteristic of Fig. 3.32 describes a **double limiter**—that is, a limiter that works on both the positive and negative peaks of an input waveform. **Single limiters,** of course, exist. Finally, note that if an input waveform such as that shown in Fig. 3.33 is fed to a double limiter, its two peaks will be *clipped off*. Limiters therefore are sometimes referred to as **clippers.**

The limiter whose characteristics are depicted in Fig. 3.32 is described as a **hard limiter. Soft limiting** is characterized by smoother transitions between the linear region and the saturation regions and a slope greater than zero in the saturation regions, as illustrated in Fig. 3.34. Depending on the application, either hard or soft limiting may be preferred.

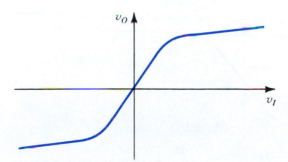

FIGURE 3.34 Soft limiting.

Limiters find application in a variety of signal-processing systems. One of their simplest applications is in limiting the voltage between the two input terminals of an op amp to a value lower than the breakdown voltage of the transistors that make up the input stage of the op-amp circuit. We will have more to say on this and other limiter applications at later points in this book.

Diodes can be combined with resistors to provide simple realizations of the limiter function. A number of examples are depicted in Fig. 3.35. In each part of the figure both the circuit and its transfer characteristic are given. The transfer characteristics are obtained using the constant-voltage-drop ($V_D = 0.7$ V) diode model but assuming a smooth transition between the linear and saturation regions of the transfer characteristic. Better approximations for the transfer characteristics can be obtained using the piecewise-linear diode model. If this is done, the saturation region of the characteristic acquires a slight slope (due to the effect of r_D).

The circuit in Fig. 3.35(a) is that of the half-wave rectifier except that here the output is taken across the diode. For $v_I < 0.5$ V, the diode is cut off, no current flows, and the voltage

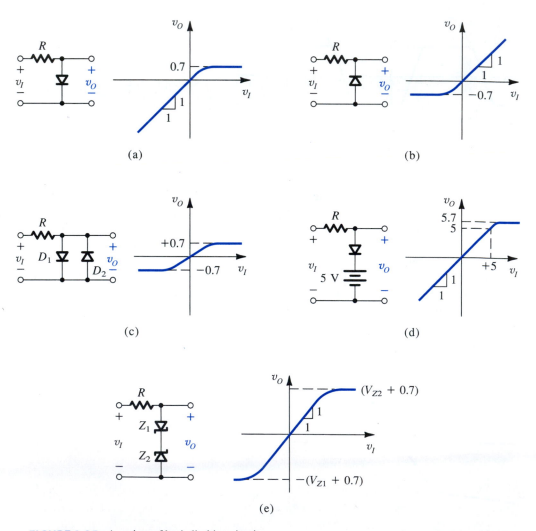

FIGURE 3.35 A variety of basic limiting circuits.

drop across R is zero; thus $v_O = v_I$. As v_I exceeds 0.5 V, the diode turns on, eventually limiting v_O to one diode drop (0.7 V). The circuit of Fig. 3.35(b) is similar to that in Fig. 3.35(a) except that the diode is reversed.

Double limiting can be implemented by placing two diodes of opposite polarity in parallel, as shown in Fig. 3.35(c). Here the linear region of the characteristic is obtained for $-0.5 \text{ V} \le v_I \le 0.5 \text{ V}$. For this range of v_I, both diodes are off and $v_O = v_I$. As v_I exceeds 0.5 V, D_1 turns on and eventually limits v_O to $+0.7$ V. Similarly, as v_I goes more negative than -0.5 V, D_2 turns on and eventually limits v_O to -0.7 V.

The thresholds and saturation levels of diode limiters can be controlled by using strings of diodes and/or by connecting a dc voltage in series with the diode(s). The latter idea is illustrated in Fig. 3.35(d). Finally, rather than strings of diodes, we may use two zener diodes in series, as shown in Fig. 3.35(e). In this circuit, limiting occurs in the positive direction at a voltage of $V_{Z2} + 0.7$, where 0.7 V represents the voltage drop across zener diode Z_1 when conducting in the *forward* direction. For negative inputs, Z_1 acts as a zener, while Z_2 conducts in the forward direction. It should be mentioned that pairs of zener diodes connected in series are available commercially for applications of this type under the name **double-anode zener.**

More flexible limiter circuits are possible if op amps are combined with diodes and resistors. Examples of such circuits are discussed in Chapter 13.

EXERCISE

3.27 Assuming the diodes to be ideal, describe the transfer characteristic of the circuit shown in Fig. E3.27.

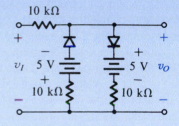

FIGURE E3.27

Ans. $v_O = v_I \quad$ for $-5 \le v_I \le +5$

$v_O = \frac{1}{2} v_I - 2.5 \quad$ for $v_I \le -5$

$v_O = \frac{1}{2} v_I + 2.5 \quad$ for $v_I \ge +5$

3.6.2 The Clamped Capacitor or DC Restorer

If in the basic peak-rectifier circuit the output is taken across the diode rather than across the capacitor, an interesting circuit with important applications results. The circuit, called a dc restorer, is shown in Fig. 3.36 fed with a square wave. Because of the polarity in which the diode is connected, the capacitor will charge to a voltage v_C with the polarity indicated in Fig. 3.36 and equal to the magnitude of the most negative peak of the input signal. Subsequently, the diode turns off and the capacitor retains its voltage indefinitely. If, for instance, the input square wave has the arbitrary levels -6 V and $+4$ V, then v_C will be equal to 6 V.

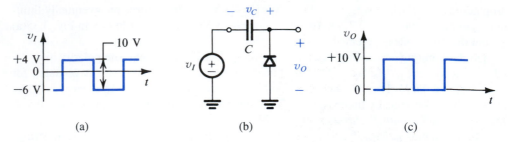

FIGURE 3.36 The clamped capacitor or dc restorer with a square-wave input and no load.

Now, since the output voltage v_O is given by

$$v_O = v_I + v_C$$

it follows that the output waveform will be identical to that of the input, except that it is shifted upward by v_C volts. In our example the output will thus be a square wave with levels of 0 V and +10 V.

Another way of visualizing the operation of the circuit in Fig. 3.36 is to note that because the diode is connected across the output with the polarity shown, it prevents the output voltage from going below 0 V (by conducting and charging up the capacitor, thus causing the output to rise to 0 V), but this connection will not constrain the positive excursion of v_O. The output waveform will therefore have its lowest peak *clamped* to 0 V, which is why the circuit is called a **clamped capacitor.** It should be obvious that reversing the diode polarity will provide an output waveform whose highest peak is clamped to 0 V. In either case, the output waveform will have a finite average value or dc component. This dc component is entirely unrelated to the average value of the input waveform. As an application, consider a pulse signal being transmitted through a capacitively coupled or ac-coupled system. The capacitive coupling will cause the pulse train to lose whatever dc component it originally had. Feeding the resulting pulse waveform to a clamping circuit provides it with a well-determined dc component, a process known as **dc restoration.** This is why the circuit is also called a **dc restorer.**

Restoring dc is useful because the dc component or average value of a pulse waveform is an effective measure of its duty cycle.[5] The duty cycle of a pulse waveform can be modulated (in a process called pulsewidth modulation) and made to carry information. In such a system, detection or demodulation could be achieved simply by feeding the received pulse waveform to a dc restorer and then using a simple RC low-pass filter to separate the average of the output waveform from the superimposed pulses.

When a load resistance R is connected across the diode in a clamping circuit, as shown in Fig. 3.37, the situation changes significantly. While the output is above ground, a net dc current must flow in R. Since at this time the diode is off, this current obviously comes from the capacitor, thus causing the capacitor to discharge and the output voltage to fall. This is shown in Fig. 3.37 for a square-wave input. During the interval t_0 to t_1, the output voltage falls exponentially with time constant CR. At t_1 the input decreases by V_a volts, and the output attempts to follow. This causes the diode to conduct heavily and to quickly charge the capacitor. At the end of the interval t_1 to t_2, the output voltage would normally be a few tenths of a volt negative (e.g., −0.5 V). Then, as the input rises by V_a volts (at t_2), the output follows, and the cycle repeats itself. In the steady state the charge lost by the capacitor during

---

[5] The duty cycle of a pulse waveform is the proportion of each cycle occupied by the pulse. In other words, it is the pulse width expressed as a fraction of the pulse period.

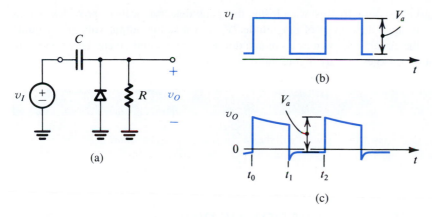

FIGURE 3.37   The clamped capacitor with a load resistance R.

the interval t_0 to t_1 is recovered during the interval t_1 to t_2. This charge equilibrium enables us to calculate the average diode current as well as the details of the output waveform.

3.6.3 The Voltage Doubler

Figure 3.38(a) shows a circuit composed of two sections in cascade: a clamp formed by C_1 and D_1, and a peak rectifier formed by D_2 and C_2. When excited by a sinusoid of amplitude V_p the clamping section provides the voltage waveform shown, assuming ideal diodes, in Fig. 3.38(b). Note that while the positive peaks are clamped to 0 V, the negative peak

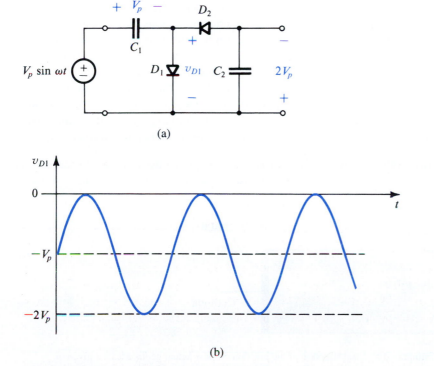

FIGURE 3.38   Voltage doubler: **(a)** circuit; **(b)** waveform of the voltage across D_1.

reaches $-2V_p$. In response to this waveform, the peak-detector section provides across capacitor C_2 a negative dc voltage of magnitude $2V_p$. Because the output voltage is double the input peak, the circuit is known as a voltage doubler. The technique can be extended to provide output dc voltages that are higher multiples of V_p.

EXERCISE

3.28 If the diode in the circuit of Fig. 3.36 is reversed, what will the dc component of v_O become?

Ans. -5 V

 ## 3.7 PHYSICAL OPERATION OF DIODES

Having studied the terminal characteristics and circuit applications of junction diodes, we will now briefly consider the physical processes that give rise to the observed terminal characteristics. The following treatment of device physics is somewhat simplified; nevertheless, it should provide sufficient background for a fuller understanding of diodes and for understanding the operation of transistors in the following two chapters.

3.7.1 Basic Semiconductor Concepts

The *pn* Junction   The semiconductor diode is basically a *pn* junction, as shown schematically in Fig. 3.39. As indicated, the *pn* junction consists of *p*-type semiconductor material (e.g., silicon) brought into close contact with *n*-type semiconductor material (also silicon). In actual practice, both the *p* and *n* regions are part of the same silicon crystal; that is, the *pn* junction is formed within a single silicon crystal by creating regions of different "dopings" (*p* and *n* regions). Appendix A provides a brief description of the process employed in the fabrication of *pn* junctions. As indicated in Fig. 3.39, external wire connections to the *p* and *n* regions (i.e., diode terminals) are made through metal (aluminum) contacts.

In addition to being essentially a diode, the *pn* junction is the basic element of bipolar junction transistors (BJTs) and plays an important role in the operation of field-effect transistors (FETs). Thus an understanding of the physical operation of *pn* junctions is important to the understanding of the operation and terminal characteristics both of diodes and transistors.

Intrinsic Silicon   Although either silicon or germanium can be used to manufacture semiconductor devices—indeed, earlier diodes and transistors were made of germanium—today's

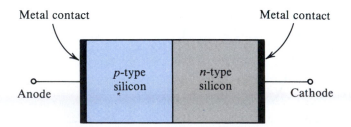

FIGURE 3.39 Simplified physical structure of the junction diode. (Actual geometries are given in Appendix A.)

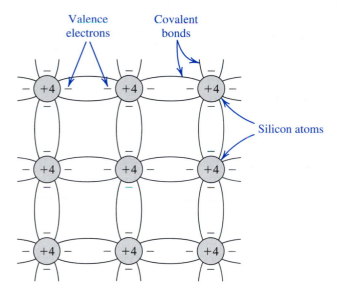

FIGURE 3.40 Two-dimensional representation of the silicon crystal. The circles represent the inner core of silicon atoms, with +4 indicating its positive charge of +4q, which is neutralized by the charge of the four valence electrons. Observe how the covalent bonds are formed by sharing of the valence electrons. At 0 K, all bonds are intact and no free electrons are available for current conduction.

integrated-circuit technology is based almost entirely on silicon. For this reason, we will deal mostly with silicon devices throughout this book.[6]

A crystal of pure or intrinsic silicon has a regular lattice structure where the atoms are held in their positions by bonds, called **covalent bonds,** formed by the four valence electrons associated with each silicon atom. Figure 3.40 shows a two-dimensional representation of such a structure. Observe that each atom shares each of its four valence electrons with a neighboring atom, with each pair of electrons forming a covalent bond. At sufficiently low temperatures, all covalent bonds are intact and no (or very few) **free electrons** are available to conduct electric current. However, at room temperature, some of the bonds are broken by thermal ionization and some electrons are freed. As shown in Fig. 3.41, when a covalent bond is broken, an electron leaves its parent atom; thus a positive charge, equal to the magnitude of the electron charge, is left with the parent atom. An electron from a neighboring atom may be attracted to this positive charge, leaving its parent atom. This action fills up the "hole" that existed in the ionized atom but creates a new hole in the other atom. This process may repeat itself, with the result that we effectively have a positively charged carrier, or **hole,** moving through the silicon crystal structure and being available to conduct electric current. The charge of a hole is equal in magnitude to the charge of an electron.

Thermal ionization results in free electrons and holes in equal numbers and hence equal concentrations. These free electrons and holes move randomly through the silicon crystal structure, and in the process some electrons may fill some of the holes. This process, called **recombination,** results in the disappearance of free electrons and holes. The recombination rate is proportional to the number of free electrons and holes, which, in turn, is determined by

---

[6] An exception is the subject of gallium arsenide (GaAs) circuits, which though not covered in this edition of the book, is studied in some detail in material provided on the text website and on the CD accompanying the text.

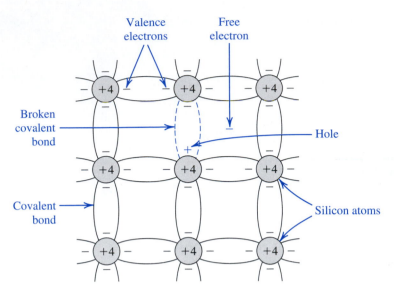

FIGURE 3.41 At room temperature, some of the covalent bonds are broken by thermal ionization. Each broken bond gives rise to a free electron and a hole, both of which become available for current conduction.

the ionization rate. The ionization rate is a strong function of temperature. In thermal equilibrium, the recombination rate is equal to the ionization or thermal-generation rate, and one can calculate the concentration of free electrons n, which is equal to the concentration of holes p,

$$n = p = n_i$$

where n_i denotes the concentration of free electrons or holes in intrinsic silicon at a given temperature. Study of semiconductor physics shows that at an absolute temperature T (in kelvins), the intrinsic concentration n_i (i.e., the number of free electrons and holes per cubic centimeter) can be found from

$$n_i^2 = BT^3 e^{-E_G/kT} \tag{3.36}$$

where B is a material-dependent parameter $= 5.4 \times 10^{31}$ for silicon, E_G is a parameter known as the bandgap energy $= 1.12$ electron volts (eV) for silicon, and k is Boltzmann's constant $= 8.62 \times 10^{-5}$ eV/K. Although we shall not make use of the bandgap energy in this circuit-focused introductory exposition, it is interesting to note that E_G represents the minimum energy required to break a covalent bond and thus generate an electron-hole pair. Substitution in Eq. (3.36) of the parameter values given shows that for intrinsic silicon at room temperature ($T \simeq 300$ K), $n_i \simeq 1.5 \times 10^{10}$ carriers/cm^3. To place this number in perspective, we note that the silicon crystal has about 5×10^{22} atoms/cm^3. Thus, at room temperature, only one of every billion atoms is ionized!

Finally, it should be mentioned that the reason that silicon is called a **semiconductor** is that its conductivity, which is determined by the number of charge carriers available to conduct electric current, is between that of conductors (e.g., metals) and that of insulators (e.g., glass).

Diffusion and Drift There are two mechanisms by which holes and electrons move through a silicon crystal—**diffusion** and **drift.** Diffusion is associated with random motion due to thermal agitation. In a piece of silicon with uniform concentrations of free electrons

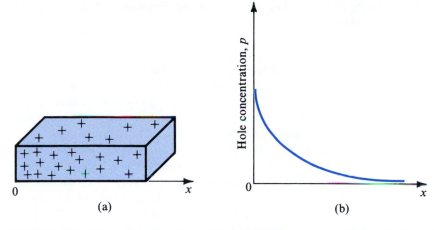

FIGURE 3.42 A bar of intrinsic silicon **(a)** in which the hole concentration profile shown in **(b)** has been created along the x-axis by some unspecified mechanism.

and holes, this random motion does not result in a net flow of charge (i.e., current). On the other hand, if by some mechanism the concentration of, say, free electrons is made higher in one part of the piece of silicon than in another, then electrons will diffuse from the region of high concentration to the region of low concentration. This diffusion process gives rise to a net flow of charge, or **diffusion current.** As an example, consider the bar of silicon shown in Fig. 3.42(a), in which the hole **concentration profile** shown in Fig. 3.42(b) has been created along the x-axis by some unspecified mechanism. The existence of such a concentration profile results in a hole diffusion current in the x direction, with the magnitude of the current at any point being proportional to the slope of the concentration curve, or the concentration gradient, at that point,

$$J_p = -qD_p\frac{dp}{dx} \tag{3.37}$$

where J_p is the current density (i.e., the current per unit area of the plane perpendicular to the x axis) in A/cm^2, q is the magnitude of electron charge $= 1.6 \times 10^{-19}$ C, and D_p is a constant called the **diffusion constant** or **diffusivity** of holes. Note that the gradient (dp/dx) is negative, resulting in a positive current in the x direction, as should be expected. In the case of electron diffusion resulting from an electron concentration gradient, a similar relationship applies, giving the electron-current density

$$J_n = qD_n\frac{dn}{dx} \tag{3.38}$$

where D_n is the diffusivity of electrons. Observe that a negative (dn/dx) gives rise to a negative current, a result of the convention that the positive direction of current is taken to be that of the flow of positive charge (and opposite to that of the flow of negative charge). For holes and electrons diffusing in intrinsic silicon, typical values for the diffusion constants are $D_p = 12$ cm^2/s and $D_n = 34$ cm^2/s.

The other mechanism for carrier motion in semiconductors is drift. **Carrier drift** occurs when an electric field is applied across a piece of silicon. Free electrons and holes are accelerated by the electric field and acquire a velocity component (superimposed on the velocity of their thermal motion) called **drift velocity.** If the electric field strength is denoted

E (in V/cm), the positively charged holes will drift in the direction of E and acquire a velocity v_{drift} (in cm/s) given by

$$v_{drift} = \mu_p E \tag{3.39}$$

where μ_p is a constant called the **mobility** of holes which has the units of $cm^2/V \cdot s$. For intrinsic silicon, μ_p is typically 480 $cm^2/V \cdot s$. The negatively charged electrons will drift in a direction opposite to that of the electric field, and their velocity is given by a relationship similar to that in Eq. (3.39), except that μ_p is replaced by μ_n, the electron mobility. For intrinsic silicon, μ_n is typically 1350 $cm^2/V \cdot s$, about 2.5 times greater than the hole mobility.

Consider now a silicon crystal having a hole density p and a free-electron density n subjected to an electric field E. The holes will drift in the same direction as E (call it the x direction) with a velocity $\mu_p E$. Thus we have a positive charge of density qp (coulomb/cm^3) moving in the x direction with velocity $\mu_p E$ (cm/s). It follows that in 1 second, a charge of $qp\mu_p EA$ (coulomb) will cross a plane of area A (cm^2) perpendicular to the x-axis. This is the current component caused by hole drift. Dividing by the area A gives the current density

$$J_{p-drift} = qp\mu_p E \tag{3.40a}$$

The free electrons will drift in the direction opposite to that of E. Thus we have a charge of density $(-qn)$ moving in the negative x direction, and thus it has a negative velocity $(-\mu_n E)$. The result is a positive current component with a density given by

$$J_{n-drift} = qn\mu_n E \tag{3.40b}$$

The total **drift current** density is obtained by combining Eqs. (3.40a) and (3.40b),

$$J_{drift} = q(p\mu_p + n\mu_n)E \tag{3.40c}$$

It should be noted that this is a form of Ohm's law with the resistivity ρ (in units of $\Omega \cdot cm$) given by

$$\rho = 1/[q(p\mu_p + n\mu_n)] \tag{3.41}$$

Finally, it is worth mentioning that a simple relationship, known as the **Einstein relationship,** exists between the carrier diffusivity and mobility,

$$\frac{D_n}{\mu_n} = \frac{D_p}{\mu_p} = V_T \tag{3.42}$$

where V_T is the thermal voltage that we have encountered before, in the diode i–v relationship (see Eq. 3.1). Recall that at room temperature, $V_T \simeq 25$ mV. The reader can easily check the validity of Eq. (3.42) by substituting the typical values given above for intrinsic silicon.

Doped Semiconductors   The intrinsic silicon crystal described above has equal concentrations of free electrons and holes generated by thermal ionization. These concentrations, denoted n_i, are strongly dependent on temperature. Doped semiconductors are materials in which carriers of one kind (electrons or holes) predominate. Doped silicon in which the majority of charge carriers are the *negatively* charged electrons is called **n type,** while silicon doped so that the majority of charge carriers are the *positively* charged holes is called **p type.**

Doping of a silicon crystal to turn it into n type or p type is achieved by introducing a small number of impurity atoms. For instance, introducing impurity atoms of a pentavalent element such as phosphorus results in n-type silicon, because the phosphorus atoms that replace some of the silicon atoms in the crystal structure have five valence electrons, four of

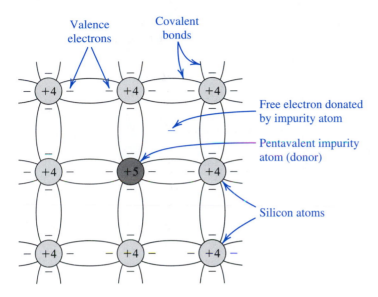

FIGURE 3.43 A silicon crystal doped by a pentavalent element. Each dopant atom donates a free electron and is thus called a donor. The doped semiconductor becomes *n* type.

which form bonds with the neighboring silicon atoms while the fifth becomes a free electron (Fig. 3.43). Thus each phosphorus atom *donates* a free electron to the silicon crystal, and the phosphorus impurity is called a **donor.** It should be clear, though, that no holes are generated by this process; hence the majority of charge carriers in the phosphorus-doped silicon will be electrons. In fact, if the concentration of donor atoms (phosphorus) is N_D, in thermal equilibrium the concentration of free electrons in the *n*-type silicon, n_{n0}, will be

$$n_{n0} \simeq N_D \qquad (3.43)$$

where the additional subscript 0 denotes thermal equilibrium. From semiconductor physics, it turns out that in thermal equilibrium the product of electron and hole concentrations remains constant; that is,

$$n_{n0}p_{n0} = n_i^2 \qquad (3.44)$$

Thus the concentration of holes, p_{n0}, that are generated by thermal ionization will be

$$p_{n0} \simeq \frac{n_i^2}{N_D} \qquad (3.45)$$

Since n_i is a function of temperature (Eq. 3.36), it follows that the concentration of the **minority** holes will be a function of temperature whereas that of the **majority** electrons is independent of temperature.

To produce a *p*-type semiconductor, silicon has to be doped with a trivalent impurity such as boron. Each of the impurity boron atoms *accepts* one electron from the silicon crystal, so that they may form covalent bonds in the lattice structure. Thus, as shown in Fig. 3.44, each boron atom gives rise to a hole, and the concentration of the majority holes in *p*-type silicon, under thermal equilibrium, is approximately equal to the concentration N_A of the **acceptor** (boron) impurity,

$$p_{p0} \simeq N_A \qquad (3.46)$$

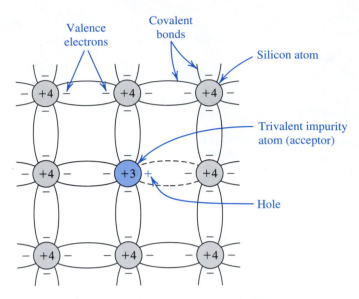

FIGURE 3.44 A silicon crystal doped with a trivalent impurity. Each dopant atom gives rise to a hole, and the semiconductor becomes *p* type.

In this *p*-type silicon, the concentration of the minority electrons, which are generated by thermal ionization, can be calculated using the fact that the product of carrier concentrations remains constant; thus,

$$n_{p0} \simeq \frac{n_i^2}{N_A} \tag{3.47}$$

It should be emphasized that a piece of *n*-type or *p*-type silicon is electrically neutral; the majority free carriers (electrons in *n*-type silicon and holes in *p*-type silicon) are neutralized by **bound charges** associated with the impurity atoms.

EXERCISES

3.29 Calculate the intrinsic carrier density n_i at 250 K, 300 K, and 350 K.

Ans. 1.5×10^8/cm³; 1.5×10^{10}/cm³; 4.18×10^{11}/cm³

3.30 Consider an *n*-type silicon in which the dopant concentration N_D is 10^{17}/cm³. Find the electron and hole concentrations at 250 K, 300 K, and 350 K. You may use the results of Exercise 3.29.

Ans. 10^{17}, 2.25×10^{-1}; 10^{17}, 2.25×10^3; 10^{17}, 1.75×10^6 (all per cm³)

3.31 Find the resistivity of (a) intrinsic silicon and (b) *p*-type silicon with $N_A = 10^{16}$/cm³. Use $n_i = 1.5 \times 10^{10}$/cm³, and assume that for intrinsic silicon $\mu_n = 1350$ cm²/V·s and $\mu_p = 480$ cm²/V·s and for the doped silicon $\mu_n = 1110$ cm²/V·s and $\mu_p = 400$ cm²/V·s. (Note that doping results in reduced carrier mobilities.)

Ans. (a) 2.28×10^5 Ω·cm; (b) 1.56 Ω·cm

3.7.2 The *pn* Junction Under Open-Circuit Conditions

Figure 3.45 shows a *pn* junction under open-circuit conditions—that is, the external terminals are left open. The "+" signs in the *p*-type material denote the majority holes. The charge

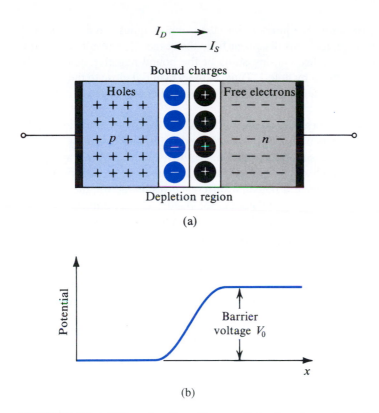

FIGURE 3.45 (a) The *pn* junction with no applied voltage (open-circuited terminals). (b) The potential distribution along an axis perpendicular to the junction.

of these holes is neutralized by an equal amount of bound negative charge associated with the acceptor atoms. For simplicity, these bound charges are not shown in the diagram. Also not shown are the minority electrons generated in the *p*-type material by thermal ionization.

In the *n*-type material the majority electrons are indicated by "–" signs. Here also, the bound positive charge, which neutralizes the charge of the majority electrons, is not shown in order to keep the diagram simple. The *n*-type material also contains minority holes generated by thermal ionization that are not shown in the diagram.

The Diffusion Current I_D Because the concentration of holes is high in the *p* region and low in the *n* region, holes diffuse across the junction from the *p* side to the *n* side; similarly, electrons diffuse across the junction from the *n* side to the *p* side. These two current components add together to form the diffusion current I_D, whose direction is from the *p* side to the *n* side, as indicated in Fig. 3.45.

The Depletion Region The holes that diffuse across the junction into the *n* region quickly recombine with some of the majority electrons present there and thus disappear from the scene. This recombination process results in the disappearance of some free electrons from the *n*-type material. Thus some of the bound positive charge will no longer be neutralized by free electrons, and this charge is said to have been **uncovered.** Since recombination takes place close to the junction, there will be a region close to the junction that is *depleted of free electrons* and contains uncovered bound positive charge, as indicated in Fig. 3.45.

The electrons that diffuse across the junction into the p region quickly recombine with some of the majority holes there, and thus disappear from the scene. This results also in the disappearance of some majority holes, causing some of the bound negative charge to be uncovered (i.e., no longer neutralized by holes). Thus, in the p material close to the junction, there will be a region *depleted of holes* and containing uncovered bound negative charge, as indicated in Fig. 3.45.

From the above it follows that a **carrier-depletion region** will exist on both sides of the junction, with the n side of this region positively charged and the p side negatively charged. This carrier-depletion region—or, simply, **depletion region**—is also called the **space-charge region.** The charges on both sides of the depletion region cause an electric field to be established across the region; hence a potential difference results across the depletion region, with the n side at a positive voltage relative to the p side, as shown in Fig. 3.45(b). Thus the resulting electric field opposes the diffusion of holes into the n region and electrons into the p region. In fact, the voltage drop across the depletion region acts as a **barrier** that has to be overcome for holes to diffuse into the n region and electrons to diffuse into the p region. The larger the barrier voltage, the smaller the number of carriers that will be able to overcome the barrier and hence the lower the magnitude of diffusion current. Thus the diffusion current I_D depends strongly on the voltage drop V_0 across the depletion region.

The Drift Current I_S and Equilibrium

In addition to the current component I_D due to majority-carrier diffusion, a component due to minority-carrier drift exists across the junction. Specifically, some of the thermally generated holes in the n material diffuse through the n material to the edge of the depletion region. There, they experience the electric field in the depletion region, which sweeps them across that region into the p side. Similarly, some of the minority thermally generated electrons in the p material diffuse to the edge of the depletion region and get swept by the electric field in the depletion region across that region into the n side. These two current components—electrons moved by drift from p to n and holes moved by drift from n to p—add together to form the drift current I_S, whose direction is from the n side to the p side of the junction, as indicated in Fig. 3.45. Since the current I_S is carried by thermally generated minority carriers, its value is strongly dependent on temperature; however, it is independent of the value of the depletion-layer voltage V_0.

Under open-circuit conditions (Fig. 3.45) no external current exists; thus the two opposite currents across the junction should be equal in magnitude:

$$I_D = I_S$$

This equilibrium condition is maintained by the barrier voltage V_0. Thus, if for some reason I_D exceeds I_S, then more bound charge will be uncovered on both sides of the junction, the depletion layer will widen, and the voltage across it (V_0) will increase. This in turn causes I_D to decrease until equilibrium is achieved with $I_D = I_S$. On the other hand, if I_S exceeds I_D, then the amount of uncovered charge will decrease, the depletion layer will narrow, and the voltage across it (V_0) will decrease. This causes I_D to increase until equilibrium is achieved with $I_D = I_S$.

The Junction Built-In Voltage

With no external voltage applied, the voltage V_0 across the pn junction can be shown to be given by

$$V_0 = V_T \ln\left(\frac{N_A N_D}{n_i^2}\right) \tag{3.48}$$

where N_A and N_D are the doping concentrations of the p side and n side of the junction, respectively. Thus V_0 depends both on doping concentrations and on temperature. It is known as the junction built-in voltage. Typically, for silicon at room temperature, V_0 is in the range of 0.6 V to 0.8 V.

When the *pn* junction terminals are left open-circuited, the voltage measured between them will be zero. That is, the voltage V_0 across the depletion region *does not* appear between the diode terminals. This is because of the contact voltages existing at the metal-semiconductor junctions at the diode terminals, which counter and exactly balance the barrier voltage. If this were not the case, we would have been able to draw energy from the isolated *pn* junction, which would clearly violate the principle of conservation of energy.

Width of the Depletion Region   From the above, it should be apparent that the depletion region exists in both the *p* and *n* materials and that equal amounts of charge exist on both sides. However, since usually the doping levels are not equal in the *p* and *n* materials, one can reason that the width of the depletion region will not be the same on the two sides. Rather, in order to uncover the same amount of charge, the depletion layer will extend deeper into the more lightly doped material. Specifically, if we denote the width of the depletion region in the *p* side by x_p and in the *n* side by x_n, this charge-equality condition can be stated as

$$qx_p A N_A = qx_n A N_D$$

where A is the cross-sectional area of the junction. This equation can be rearranged to yield

$$\frac{x_n}{x_p} = \frac{N_A}{N_D} \tag{3.49}$$

In actual practice, it is usual that one side of the junction is much more heavily doped than the other, with the result that the depletion region exists almost entirely on one side (the lightly doped side). Finally, from device physics, the width of the depletion region of an open-circuited junction is given by

$$W_{dep} = x_n + x_p = \sqrt{\frac{2\varepsilon_s}{q}\left(\frac{1}{N_A} + \frac{1}{N_D}\right)V_0} \tag{3.50}$$

where ε_s is the electrical permittivity of silicon = $11.7\varepsilon_0 = 1.04 \times 10^{-12}$ F/cm. Typically, W_{dep} is in the range of 0.1 μm to 1 μm.

EXERCISE

3.32 For a *pn* junction with $N_A = 10^{17}$/cm³ and $N_D = 10^{16}$/cm³, find, at $T = 300$ K, the built-in voltage, the width of the depletion region, and the distance it extends in the *p* side and in the *n* side of the junction. Use $n_i = 1.5 \times 10^{10}$/cm³.

Ans. 728 mV; 0.32 μm; 0.03 μm and 0.29 μm

3.7.3 The *pn* Junction Under Reverse-Bias Conditions

The behavior of the *pn* junction in the reverse direction is more easily explained on a microscopic scale if we consider exciting the junction with a constant-current source (rather than with a voltage source), as shown in Fig. 3.46. The current source I is obviously in the reverse direction. For the time being let the magnitude of I be less than I_S; if I is greater than I_S, breakdown will occur, as explained in Section 3.7.4.

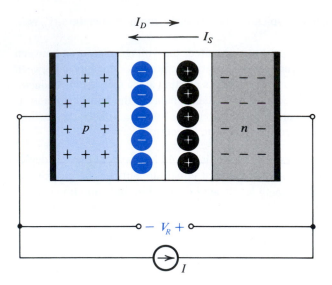

FIGURE 3.46 The *pn* junction excited by a constant-current source *I* in the reverse direction. To avoid breakdown, *I* is kept smaller than I_S. Note that the depletion layer widens and the barrier voltage increases by V_R volts, which appears between the terminals as a reverse voltage.

The current I will be carried by electrons flowing in the external circuit from the n material to the p material (i.e., in the direction opposite to that of I). This will cause electrons to leave the n material and holes to leave the p material. The free electrons leaving the n material cause the uncovered positive bound charge to increase. Similarly, the holes leaving the p material result in an increase in the uncovered negative bound charge. Thus the reverse current I will result in an increase in the width of, and the charge stored in, the depletion layer. This in turn will result in a higher voltage across the depletion region—that is, a greater barrier voltage—which causes the diffusion current I_D to decrease. The drift current I_S, being independent of the barrier voltage, will remain constant. Finally, equilibrium (steady state) will be reached when

$$I_S - I_D = I$$

In equilibrium, the increase in depletion-layer voltage, above the value of the built-in voltage V_0, will appear as an external voltage that can be measured between the diode terminals, with n being positive with respect to p. This voltage is denoted V_R in Fig. 3.46.

We can now consider exciting the pn junction by a reverse voltage V_R, where V_R is less than the breakdown voltage V_{ZK}. (Refer to Fig. 3.8 for the definition of V_{ZK}.) When the voltage V_R is first applied, a reverse current flows in the external circuit from p to n. This current causes the increase in width and charge of the depletion layer. Eventually the voltage across the depletion layer will increase by the magnitude of the external voltage V_R, at which time an equilibrium is reached with the external reverse current I equal to $(I_S - I_D)$. Note, however, that initially the external current can be much greater than I_S. The purpose of this initial transient is to *charge* the depletion layer and increase the voltage across it by V_R volts. Eventually, when a steady state is reached, I_D will be negligibly small, and the reverse current will be nearly equal to I_S.

The Depletion Capacitance   From the above we observe the analogy between the depletion layer of a pn junction and a capacitor. As the voltage across the pn junction changes,

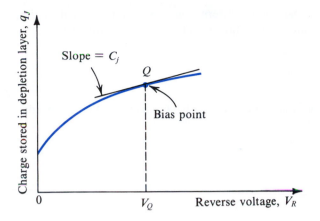

FIGURE 3.47   The charge stored on either side of the depletion layer as a function of the reverse voltage V_R.

the charge stored in the depletion layer changes accordingly. Figure 3.47 shows a sketch of typical charge-versus-external-voltage characteristic of a *pn* junction. Note that only the portion of the curve for the reverse-bias region is shown.

An expression for the depletion-layer stored charge q_J can be derived by finding the charge stored on either side of the junction (which charges are, of course, equal). Using the *n* side, we write

$$q_J = q_N = qN_D x_n A$$

where A is the cross-sectional area of the junction (in a plane perpendicular to the page). Next we use Eq. (3.49) to express x_n in terms of the depletion-layer width W_{dep} to obtain

$$q_J = q\frac{N_A N_D}{N_A + N_D}AW_{dep} \tag{3.51}$$

where W_{dep} can be found from Eq. (3.50) by replacing V_0 by the total voltage across the depletion region, $(V_0 + V_R)$,

$$W_{dep} = \sqrt{\left(\frac{2\varepsilon_s}{q}\right)\left(\frac{1}{N_A} + \frac{1}{N_D}\right)(V_0 + V_R)} \tag{3.52}$$

Combining Eqs. (3.51) and (3.52) yields the expression for the nonlinear q_J–V_R relationship depicted in Fig. 3.47. This relationship obviously does not represent a linear capacitor. However, a linear-capacitance approximation can be used if the device is biased and the signal swing around the bias point is small, as illustrated in Fig. 3.47. This is the technique we utilized in Section 1.4 to obtain linear amplification from an amplifier having a nonlinear transfer characteristic and in Section 3.3 to obtain a small-signal model for the diode in the forward-bias region. Under this small-signal approximation, the **depletion capacitance** (also called the **junction capacitance**) is simply the slope of the q_J–V_R curve at the bias point Q,

$$C_j = \frac{dq_J}{dV_R}\bigg|_{V_R = V_Q} \tag{3.53}$$

We can easily evaluate the derivative and find C_j. Alternatively, we can treat the depletion layer as a parallel-plate capacitor and obtain an identical expression for C_j using the familiar formula

$$C_j = \frac{\varepsilon_s A}{W_{dep}} \tag{3.54}$$

where W_{dep} is given in Eq. (3.52). The resulting expression for C_j can be written in the convenient form

$$C_j = \frac{C_{j0}}{\sqrt{1 + \dfrac{V_R}{V_0}}} \tag{3.55}$$

where C_{j0} is the value of C_j obtained for zero applied voltage,

$$C_{j0} = A\sqrt{\left(\frac{\varepsilon_s q}{2}\right)\left(\frac{N_A N_D}{N_A + N_D}\right)\left(\frac{1}{V_0}\right)} \tag{3.56}$$

The preceding analysis and the expression for C_j apply to junctions in which the carrier concentration is made to change abruptly at the junction boundary. A more general formula for C_j is

$$C_j = \frac{C_{j0}}{\left(1 + \dfrac{V_R}{V_0}\right)^m} \tag{3.57}$$

where m is a constant whose value depends on the manner in which the concentration changes from the p to the n side of the junction. It is called the **grading coefficient,** and its value ranges from $\frac{1}{3}$ to $\frac{1}{2}$.

To recap, as a reverse-bias voltage is applied to a pn junction, a transient occurs during which the depletion capacitance is charged to the new bias voltage. After the transient dies, the steady-state reverse current is simply equal to $I_S - I_D$. Usually I_D is very small when the diode is reverse-biased, and the reverse current is approximately equal to I_S. This, however, is only a theoretical model; one that does not apply very well. In actual fact, currents as high as few nanoamperes (10^{-9} A) flow in the reverse direction in devices for which I_S is on the order of 10^{-15} A. This large difference is due to leakage and other effects. Furthermore, the reverse current is dependent to a certain extent on the magnitude of the reverse voltage, contrary to the theoretical model which states that $I \simeq I_S$ independent of the value of the reverse voltage applied. Nevertheless, because of the very low currents involved, one is usually not interested in the details of the diode i–v characteristic in the reverse direction.

EXERCISE

3.33 For a pn junction with $N_A = 10^{17}$/cm^3 and $N_D = 10^{16}$/cm^3, operating at $T = 300$ K, find (a) the value of C_{j0} per unit junction area (μm^2 is a convenient unit here) and (b) the capacitance C_j at a reverse-bias voltage of 2 V, assuming a junction area of 2500 μm^2. Use $n_i = 1.5 \times 10^{10}$/cm^3, $m = \frac{1}{2}$, and the value of V_0 found in Exercise 3.32 ($V_0 = 0.728$ V).

Ans. (a) 0.32 fF/μm^2; (b) 0.41 pF

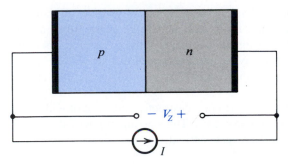

FIGURE 3.48 The *pn* junction excited by a reverse-current source *I*, where $I > I_S$. The junction breaks down, and a voltage V_Z, with the polarity indicated, develops across the junction.

3.7.4 The *pn* Junction in the Breakdown Region

In considering diode operation in the reverse-bias region in Section 3.7.3, it was assumed that the reverse-current source *I* (Fig. 3.46) was smaller than I_S or, equivalently, that the reverse voltage V_R was smaller than the breakdown voltage V_{ZK}. (Refer to Fig. 3.8 for the definition of V_{ZK}.) We now wish to consider the breakdown mechanisms in *pn* junctions and explain the reasons behind the almost-vertical line representing the *i–v* relationship in the breakdown region. For this purpose, let the *pn* junction be excited by a current source that causes a constant current *I* greater than I_S to flow in the reverse direction, as shown in Fig. 3.48. This current source will move holes from the *p* material through the external circuit[7] into the *n* material and electrons from the *n* material through the external circuit into the *p* material. This action results in more and more of the bound charge being uncovered; hence the depletion layer widens and the barrier voltage rises. This latter effect causes the diffusion current to decrease; eventually it will be reduced to almost zero. Nevertheless, this is not sufficient to reach a steady state, since *I* is greater than I_S. Therefore the process leading to the widening of the depletion layer continues until a sufficiently high junction voltage develops, at which point a new mechanism sets in to supply the charge carriers needed to support the current *I*. As will be now explained, this mechanism for supplying reverse currents in excess of I_S can take one of two forms depending on the *pn* junction material, structure, and so on.

The two possible breakdown mechanisms are the **zener effect** and the **avalanche effect.** If a *pn* junction breaks down with a breakdown voltage $V_Z < 5$ V, the breakdown mechanism is usually the zener effect. Avalanche breakdown occurs when V_Z is greater than approximately 7 V. For junctions that break down between 5 V and 7 V, the breakdown mechanism can be either the zener or the avalanche effect or a combination of the two.

Zener breakdown occurs when the electric field in the depletion layer increases to the point where it can break covalent bonds and generate electron-hole pairs. The electrons generated in this way will be swept by the electric field into the *n* side and the holes into the *p* side. Thus these electrons and holes constitute a reverse current across the junction that helps support the external current *I*. Once the zener effect starts, a large number of carriers can be generated, with a negligible increase in the junction voltage. Thus the reverse current in the breakdown region will be determined by the external circuit, while the reverse voltage appearing between the diode terminals will remain close to the rated breakdown voltage V_Z.

The other breakdown mechanism is avalanche breakdown, which occurs when the minority carriers that cross the depletion region under the influence of the electric field gain

---

[7] The current in the external circuit will, of course, be carried entirely by electrons.

sufficient kinetic energy to be able to break covalent bonds in atoms with which they collide. The carriers liberated by this process may have sufficiently high energy to be able to cause other carriers to be liberated in another ionizing collision. This process occurs in the fashion of an avalanche, with the result that many carriers are created that are able to support any value of reverse current, as determined by the external circuit, with a negligible change in the voltage drop across the junction.

As mentioned before, *pn* junction breakdown is not a destructive process, provided that the maximum specified power dissipation is not exceeded. This maximum power-dissipation rating in turn implies a maximum value for the reverse current.

3.7.5 The *pn* Junction Under Forward-Bias Conditions

We next consider operation of the *pn* junction in the forward-bias region. Again, it is easier to explain physical operation if we excite the junction by a constant-current source supplying a current I in the forward direction, as shown in Fig. 3.49. This causes majority carriers to be supplied to both sides of the junction by the external circuit: holes to the p material and electrons to the n material. These majority carriers will neutralize some of the uncovered bound charge, causing less charge to be stored in the depletion layer. Thus the depletion layer narrows and the depletion barrier voltage reduces. The reduction in barrier voltage enables more holes to cross the barrier from the p material into the n material and more electrons from the n side to cross into the p side. Thus the diffusion current I_D increases until equilibrium is achieved with $I_D - I_S = I$, the externally supplied forward current.

Let us now examine closely the current flow across the forward-biased *pn* junction in the steady state. The barrier voltage is now lower than V_0 by an amount V that appears between the diode terminals as a forward voltage drop (i.e., the anode of the diode will be more positive than the cathode by V volts). Owing to the decrease in the barrier voltage or, alternatively, because of the forward voltage drop V, holes are **injected** across the junction into the n region and electrons are injected across the junction into the p region. The holes injected into the n region will cause the minority-carrier concentration there, p_n, to exceed the thermal equilibrium value, p_{n0}. The *excess* concentration $(p_n - p_{n0})$ will be highest near the edge of

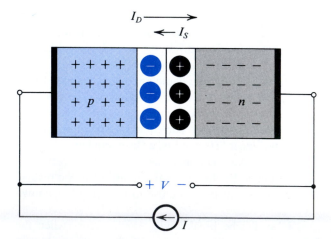

FIGURE 3.49 The *pn* junction excited by a constant-current source supplying a current I in the forward direction. The depletion layer narrows and the barrier voltage decreases by V volts, which appears as an external voltage in the forward direction.

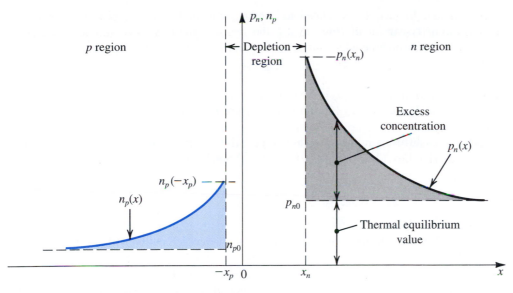

FIGURE 3.50 Minority-carrier distribution in a forward-biased *pn* junction. It is assumed that the *p* region is more heavily doped than the *n* region; $N_A \gg N_D$.

the depletion layer and will decrease (exponentially) as one moves away from the junction, eventually reaching zero. Figure 3.50 shows such a minority-carrier distribution.

In the steady state the concentration profile of **excess minority carriers** remains constant, and indeed it is such a distribution that gives rise to the increase of diffusion current I_D above the value I_S. This is because the distribution shown causes injected minority holes to diffuse away from the junction into the *n* region and disappear by recombination. To maintain equilibrium, an equal number of electrons will have to be supplied by the external circuit, thus replenishing the electron supply in the *n* material.

Similar statements can be made about the minority electrons in the *p* material. The diffusion current I_D is, of course, the sum of the electron and hole components.

The Current-Voltage Relationship   We now show how the diode *i–v* relationship of Eq. (3.1) arises. Toward that end, we consider in some detail the current component caused by the holes injected across the junction into the *n* region. An important result from semiconductor physics relates the concentration of minority carriers at the edge of the depletion region, denoted by $p_n(x_n)$ in Fig. 3.50, to the forward voltage V,

$$p_n(x_n) = p_{n0}e^{V/V_T} \tag{3.58}$$

This is known as the **law of the junction;** its proof is normally found in textbooks dealing with device physics.

The distribution of excess hole concentration in the *n* region, shown in Fig. 3.50, is an exponentially decaying function of distance and can be expressed as

$$p_n(x) = p_{n0} + [p_n(x_n) - p_{n0}]e^{-(x-x_n)/L_p} \tag{3.59}$$

where L_p is a constant that determines the steepness of the exponential decay. It is called the **diffusion length** of holes in the *n*-type silicon. The smaller the value of L_p, the faster the injected holes will recombine with majority electrons, resulting in a steeper decay of minority-carrier concentration. In fact, L_p is related to another semiconductor parameter known as the **excess-minority-carrier lifetime,** τ_p. It is the average time it takes for a hole injected into the n region to recombine with a majority electron. The relationship is

$$L_p = \sqrt{D_p \tau_p} \tag{3.60}$$

where, as mentioned before, D_p is the diffusion constant for holes in the n-type silicon. Typical values for L_p are 1 μm to 100 μm, and the corresponding values of τ_p are in the range of 1 ns to 10,000 ns.

The holes diffusing in the n region will give rise to a hole current whose density can be evaluated using Eqs. (3.37) and (3.59) with $p_n(x_n)$ obtained from Eq. (3.58),

$$J_p = q\frac{D_p}{L_p}p_{n0}(e^{V/V_T} - 1)e^{-(x - x_n)/L_p}$$

Observe that J_p is largest at the edge of the depletion region ($x = x_n$) and decays exponentially with distance. The decay is, of course, due to the recombination with the majority electrons. In the steady state, the majority carriers will have to be replenished, and thus electrons will be supplied from the external circuit to the n region at a rate that will keep the current constant at the value it has at $x = x_n$. Thus the current density due to hole injection is given by

$$J_p = q\frac{D_p}{L_p}p_{n0}(e^{V/V_T} - 1) \tag{3.61}$$

A similar analysis can be performed for the electrons injected across the junction into the p region resulting in the electron-current component J_n,

$$J_n = q\frac{D_n}{L_n}n_{p0}(e^{V/V_T} - 1) \tag{3.62}$$

where L_n is the diffusion length of electrons in the p region. Since J_p and J_n are in the same direction, they can be added and multiplied by the junction cross-sectional area A to obtain the total current I as

$$I = A\left(\frac{qD_p p_{n0}}{L_p} + \frac{qD_n n_{p0}}{L_n}\right)(e^{V/V_T} - 1)$$

Substituting for $p_{n0} = n_i^2/N_D$ and for $n_{p0} = n_i^2/N_A$, we can express I in the form

$$I = Aqn_i^2\left(\frac{D_p}{L_p N_D} + \frac{D_n}{L_n N_A}\right)(e^{V/V_T} - 1) \tag{3.63}$$

We recognize this as the diode equation where the saturation current I_S is given by

$$I_S = Aqn_i^2\left(\frac{D_p}{L_p N_D} + \frac{D_n}{L_n N_A}\right) \tag{3.64}$$

Observe that, as expected, I_S is directly proportional to the junction area A. Furthermore, I_S is proportional to n_i^2, which is a strong function of temperature (Eq. 3.36). Also, note that the exponential in Eq. (3.63) does not include the constant n; n is a "fix-up" parameter that is included to account for nonideal effects.

Diffusion Capacitance From the description of the operation of the pn junction in the forward region, we note that in the steady state a certain amount of excess minority-carrier charge is stored in each of the p and n bulk regions. If the terminal voltage changes, this charge will have to change before a new steady state is achieved. This charge-storage phenomenon gives rise to another capacitive effect, distinctly different from that due to charge storage in the depletion region.

To calculate the excess minority-carrier stored charge, refer to Fig. 3.50. The excess hole charge stored in the n region can be found from the shaded area under the exponential as follows:

$$Q_p = Aq \times \text{shaded area under the } p_n(x) \text{ exponential}$$
$$= Aq \times [p_n(x_n) - p_{n0}]L_p$$

Substituting for $p_n(x_n)$ from Eq. (3.58) and using Eq. (3.61) enables us to express Q_p as

$$Q_p = \frac{L_p^2}{D_p}I_p$$

where $I_p = AJ_p$ is the hole component of the current across the junction. Now, using Eq. (3.60), we can substitute for $L_p^2/D_p = \tau_p$, the hole lifetime, to obtain

$$Q_p = \tau_p I_p \qquad (3.65)$$

This attractive relationship says that the stored excess hole charge is proportional to both the hole current-component and the hole lifetime. A similar relationship can be developed for the electron charge stored in the p region,

$$Q_n = \tau_n I_n \qquad (3.66)$$

where τ_n is the electron lifetime in the p region. The total excess minority-carrier charge can be obtained by adding together Q_p and Q_n,

$$Q = \tau_p I_p + \tau_n I_n \qquad (3.67)$$

This charge can be expressed in terms of the diode current $I = I_p + I_n$ as

$$Q = \tau_T I \qquad (3.68)$$

where τ_T is called the **mean transit time** of the diode. Obviously, τ_T is related to τ_p and τ_n. Furthermore, in most practical devices, one side of the junction is much more heavily doped than the other. For instance, if $N_A \gg N_D$, one can show that $I_p \gg I_n$, $I \simeq I_p$, $Q_p \gg Q_n$, $Q \simeq Q_p$, and thus $\tau_T \simeq \tau_p$. This case is illustrated in Exercise 3.34 on page 208.

For small changes around a bias point, we can define the **small-signal diffusion capacitance** C_d as

$$C_d = \frac{dQ}{dV}$$

and can show that

$$C_d = \left(\frac{\tau_T}{V_T}\right)I \qquad (3.69)$$

where I is the diode current at the bias point. Note that C_d is directly proportional to the diode current I and is thus negligibly small when the diode is reverse biased. Also, note that to keep C_d small, the transit time τ_T must be made small, an important requirement for diodes intended for high-speed or high-frequency operation.

EXERCISE

3.34 A diode has $N_A = 10^{17}/\text{cm}^3$, $N_D = 10^{16}/\text{cm}^3$, $n_i = 1.5 \times 10^{10}/\text{cm}^3$, $L_p = 5$ μm, $L_n = 10$ μm, $A = 2500$ μm^2, D_p (in the n region) = 10 cm^2/V·s, and D_n (in the p region) = 18 cm^2/V·s. The diode is forward biased and conducting a current $I = 0.1$ mA. Calculate: (a) I_S; (b) the forward-bias voltage V; (c) the component of the current I due to hole injection and that due to electron injection across the junction; (d) τ_p and τ_n; (e) the excess hole charge in the n region Q_p, and the excess electron charge in the p region Q_n, and hence the total minority stored charge Q, as well as the transit time τ_T; and (f) the diffusion capacitance.

Ans. (a) 2×10^{-15} A; (b) 0.616 V; (c) 91.7 μA, 8.3 μA; (d) 25 ns, 55.6 ns; (e) 2.29 pC, 0.46 pC, 2.75 pC, 27.5 ns; (f) 110 pF

Junction Capacitance   The depletion-layer or junction capacitance under forward-bias conditions can be found by replacing V_R with $-V$ in Eq. (3.57). It turns out, however, that the accuracy of this relationship in the forward-bias region is rather poor. As an alternative, circuit designers use the following rule of thumb:

$$C_j \simeq 2C_{j0} \qquad (3.70)$$

3.7.6 Summary

For easy reference, Table 3.2 provides a listing of the important relationships that describe the physical operation of *pn* junctions.

TABLE 3.2   Summary of Important Equations for *pn*-Junction Operation

Quantity	Relationship	Values of Constants and Parameters (for Intrinsic *Si* at *T* = 300 K)
Carrier concentration in intrinsic silicon (/cm^3)	$n_i^2 = BT^3 e^{-E_G/kT}$	$B = 5.4 \times 10^{31}/(\text{K}^3\text{cm}^6)$ $E_G = 1.12$ eV $k = 8.62 \times 10^{-5}$ eV/K $n_i = 1.5 \times 10^{10}/\text{cm}^3$
Diffusion current density (A/cm^2)	$J_p = -qD_p\dfrac{dp}{dx}$ $J_n = qD_n\dfrac{dn}{dx}$	$q = 1.60 \times 10^{-19}$ coulomb $D_p = 12$ cm^2/s $D_n = 34$ cm^2/s
Drift current density (A/cm^2)	$J_{drift} = q(p\mu_p + n\mu_n)E$	$\mu_p = 480$ cm^2/V·s $\mu_n = 1350$ cm^2/V·s
Resistivity (Ω·cm)	$\rho = 1/[q(p\mu_p + n\mu_n)]$	μ_p and μ_n decrease with the increase in doping concentration
Relationship between mobility and diffusivity	$\dfrac{D_n}{\mu_n} = \dfrac{D_p}{\mu_p} = V_T$	$V_T = kT/q$ $\simeq 25.8$ mV
Carrier concentration in n-type silicon (/cm^3)	$n_{n0} \simeq N_D$ $p_{n0} = n_i^2/N_D$	

Quantity	Relationship	Values of Constants and Parameters (for Intrinsic *Si* at $T = 300$ K)
Carrier concentration in p-type silicon (/cm^3)	$p_{p0} \simeq N_A$ $n_{p0} = n_i^2/N_A$	
Junction built-in voltage (V)	$V_0 = V_T \ln\left(\dfrac{N_A N_D}{n_i^2}\right)$	
Width of depletion region (cm)	$\dfrac{x_n}{x_p} = \dfrac{N_A}{N_D}$ $W_{dep} = x_n + x_p$ $\qquad = \sqrt{\dfrac{2\varepsilon_s}{q}\left(\dfrac{1}{N_A} + \dfrac{1}{N_D}\right)(V_0 + V_R)}$	$\varepsilon_s = 11.7\varepsilon_0$ $\varepsilon_0 = 8.854 \times 10^{-14}$ F/cm
Charge stored in depletion layer (coulomb)	$q_J = q\dfrac{N_A N_D}{N_A + N_D} A W_{dep}$	
Depletion capacitance (F)	$C_j = \dfrac{\varepsilon_s A}{W_{dep}},\ C_{j0} = \dfrac{\varepsilon_s A}{W_{dep}\vert_{V_{R=0}}}$ $C_j = C_{j0}\Big/\left(1 + \dfrac{V_R}{V_0}\right)^m$ $C_j \simeq 2C_{j0}$ (for forward bias)	$m = \dfrac{1}{3}$ to $\dfrac{1}{2}$
Forward current (A)	$I = I_p + I_n$ $I_p = Aq n_i^2 \dfrac{D_p}{L_p N_D}(e^{V/V_T} - 1)$ $I_n = Aq n_i^2 \dfrac{D_n}{L_n N_A}(e^{V/V_T} - 1)$	
Saturation current (A)	$I_S = Aq n_i^2\left(\dfrac{D_p}{L_p N_D} + \dfrac{D_n}{L_n N_A}\right)$	
Minority-carrier lifetime (s)	$\tau_p = L_p^2/D_p \qquad \tau_n = L_n^2/D_n$	$L_p, L_n = 1\ \mu\text{m}$ to $100\ \mu\text{m}$ $\tau_p, \tau_n = 1$ ns to 10^4 ns
Minority-carrier charge storage (coulomb)	$Q_p = \tau_p I_p \qquad Q_n = \tau_n I_n$ $Q = Q_p + Q_n = \tau_T I$	
Diffusion capacitance (F)	$C_d = \left(\dfrac{\tau_T}{V_T}\right)I$	

 ## 3.8 SPECIAL DIODE TYPES[8]

In this section, we discuss briefly some important special types of diodes.

[8] This section can be skipped with no loss in continuity.

3.8.1 The Schottky-Barrier Diode (SBD)

The Schottky-barrier diode (SBD) is formed by bringing metal into contact with a moderately doped *n*-type semiconductor material. The resulting metal–semiconductor junction behaves like a diode, conducting current in one direction (from the metal anode to the semiconductor cathode) and acting as an open-circuit in the other, and is known as the Schottky-barrier diode or simply the Schottky diode. In fact, the current–voltage characteristic of the SBD is remarkably similar to that of a *pn*-junction diode, with two important exceptions:

1. In the SBD, current is conducted by majority carriers (electrons). Thus the SBD does not exhibit the minority-carrier charge-storage effects found in forward-biased *pn* junctions. As a result, Schottky diodes can be switched from on to off, and vice versa, much faster than is possible with *pn*-junction diodes.

2. The forward voltage drop of a conducting SBD is lower than that of a *pn*-junction diode. For example, an SBD made of silicon exhibits a forward voltage drop of 0.3 V to 0.5 V, compared to the 0.6 V to 0.8 V found in silicon *pn*-junction diodes. SBDs can also be made of gallium arsenide (GaAs) and, in fact, play an important role in the design of GaAs circuits.[9] Gallium-arsenide SBDs exhibit forward voltage drops of about 0.7 V.

Apart from GaAs circuits, Schottky diodes find application in the design of a special form of bipolar-transistor logic circuits, known as Schottky-TTL, where TTL stands for transistor-transistor logic.

Before leaving the subject of Schottky-barrier diodes, it is important to note that not every metal–semiconductor contact is a diode. In fact, metal is commonly deposited on the semiconductor surface in order to make terminals for the semiconductor devices and to connect different devices in an integrated-circuit chip. Such metal–semiconductor contacts are known as **ohmic contacts** to distinguish them from the rectifying contacts that result in SBDs. Ohmic contacts are usually made by depositing metal on very heavily doped (and thus low-resistivity) semiconductor regions.

3.8.2 Varactors

Earlier, we learned that reverse-biased *pn* junctions exhibit a charge-storage effect that is modeled with the depletion-layer or junction capacitance C_j. As Eq. (3.57) indicates, C_j is a function of the reverse-bias voltage V_R. This dependence turns out to be useful in a number of applications, such as the automatic tuning of radio receivers. Special diodes are therefore fabricated to be used as voltage-variable capacitors known as varactors. These devices are optimized to make the capacitance a strong function of voltage by arranging that the grading coefficient *m* is 3 or 4.

3.8.3 Photodiodes

If a reverse-biased *pn* junction is illuminated—that is, exposed to incident light—the photons impacting the junction cause covalent bonds to break, and thus electron-hole pairs are generated in the depletion layer. The electric field in the depletion region then sweeps the liberated electrons to the *n* side and the holes to the *p* side, giving rise to a reverse current across the junction. This current, known as photocurrent, is proportional to the intensity of

---

[9] The CD accompanying the text and the text's website contain material on GaAs circuits.

the incident light. Such a diode, called a photodiode, can be used to convert light signals into electrical signals.

Photodiodes are usually fabricated using a compound semiconductor[10] such as gallium arsenide. The photodiode is an important component of a growing family of circuits known as **optoelectronics** or **photonics.** As the name implies, such circuits utilize an optimum combination of electronics and optics for signal processing, storage, and transmission. Usually, electronics is the preferred means for signal processing, whereas optics is most suited for transmission and storage. Examples include fiber-optic transmission of telephone and television signals and the use of optical storage in CD-ROM computer disks. Optical transmission provides very wide bandwidths and low signal attenuation. Optical storage allows vast amounts of data to be stored reliably in a small space.

Finally, we should note that without reverse bias, the illuminated photodiode functions as a **solar cell.** Usually fabricated from low-cost silicon, a solar cell converts light to electrical energy.

3.8.4 Light-Emitting Diodes (LEDs)

The light-emitting diode (LED) performs the inverse of the function of the photodiode; it converts a forward current into light. The reader will recall that in a forward-biased *pn* junction, minority carriers are injected across the junction and diffuse into the *p* and *n* regions. The diffusing minority carriers then recombine with the majority carriers. Such recombination can be made to give rise to light emission. This can be done by fabricating the *pn* junction using a semiconductor of the type known as direct-bandgap materials. Gallium arsenide belongs to this group and can thus be used to fabricate light-emitting diodes.

The light emitted by an LED is proportional to the number of recombinations that take place, which in turn is proportional to the forward current in the diode.

LEDs are very popular devices. They find application in the design of numerous types of displays, including the displays of laboratory instruments such as digital voltmeters. They can be made to produce light in a variety of colors. Furthermore, LEDs can be designed so as to produce coherent light with a very narrow bandwidth. The resulting device is a **laser diode.** Laser diodes find application in optical communication systems and in CD players, among other things.

Combining an LED with a photodiode in the same package results in a device known as an **optoisolator.** The LED converts an electrical signal applied to the optoisolator into light, which the photodiode detects and converts back to an electrical signal at the output of the optoisolator. Use of the optoisolator provides complete electrical isolation between the electrical circuit that is connected to the isolator's input and the circuit that is connected to its output. Such isolation can be useful in reducing the effect of electrical interference on signal transmission within a system, and thus optoisolators are frequently employed in the design of digital systems. They can also be used in the design of medical instruments to reduce the risk of electrical shock to patients.

Note that the optical coupling between an LED and photodiode need not be accomplished inside a small package. Indeed, it can be implemented over a long distance using an optical fiber, as is done in fiber-optic communication links.

---

[10] Whereas an elemental semiconductor, such as silicon, uses an element from column IV of the periodic table, a compound semiconductor uses a combination of elements from columns III and V or II and VI. For example, GaAs is formed of gallium (column III) and arsenic (column V) and is thus known as a III-V compound.

3.9 THE SPICE DIODE MODEL AND SIMULATION EXAMPLES

We conclude this chapter with a description of the model that SPICE uses for the diode. We will also illustrate the use of SPICE in the design of a dc power supply.

3.9.1 The Diode Model

To the designer, the value of simulation results is a direct function of the quality of the models used for the devices. The more faithfully the model represents the various characteristics of the device, the more accurately the simulation results will describe the operation of an actual fabricated circuit. In other words, to see the effect of various imperfections in device operation on circuit performance, these imperfections must be included in the device model used by the circuit simulator. These comments about device modeling obviously apply to all devices and not just to diodes.

The large-signal SPICE model for the diode is shown in Fig. 3.51. The static behavior is modeled by the exponential i–v relationship. The dynamic behavior is represented by the nonlinear capacitor C_D, which is the sum of the diffusion capacitance C_d and the junction capacitance C_j. The series resistance R_S represents the total resistance of the p and n regions on both sides of the junction. The value of this parasitic resistance is ideally zero, but it is typically in the range of a few ohms for small-signal diodes. For small-signal analysis, SPICE uses the diode incremental resistance r_d and the incremental values of C_d and C_j.

Table 3.3 provides a partial listing of the diode-model parameters used by SPICE, all of which should be familiar to the reader. But, having a good device model solves only half of the modeling problem; the other half is to determine appropriate values for the model parameters. This is by no means an easy task. The values of the model parameters are determined using a combination of characterization of the device-fabrication process and specific measurements performed on the actual manufactured devices. Semiconductor manufacturers expend enormous effort and money to extract the values of the model parameters for their devices. For discrete diodes, the values of the SPICE model parameters can be determined from the diode data sheets, supplemented if needed by key measurements. Circuit simulators (such as PSpice) include in their libraries the model parameters of some of the popular off-the-self components. For instance, in Example 3.10, we will use the commercially available 1N4148 pn-junction diode whose SPICE model parameters are available in PSpice.

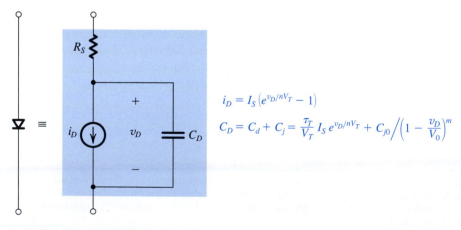

$$i_D = I_S \left(e^{v_D/nV_T} - 1 \right)$$

$$C_D = C_d + C_j = \frac{\tau_T}{V_T} I_S e^{v_D/nV_T} + C_{j0} \Big/ \left(1 - \frac{v_D}{V_0} \right)^m$$

FIGURE 3.51   The SPICE diode model.

TABLE 3.3 Parameters of the SPICE Diode Model (Partial Listing)

SPICE Parameter	Book Symbol	Description	Units
IS	I_S	Saturation current	A
N	n	Emission coefficient	
RS	R_S	Ohmic resistance	Ω
VJ	V_0	Built-in potential	V
CJ0	C_{j0}	Zero-bias depletion (junction) capacitance	F
M	m	Grading coefficient	
TT	τ_T	Transit time	s
BV	V_{ZK}	Breakdown voltage	V
IBV	I_{ZK}	Reverse current at V_{ZK}	A

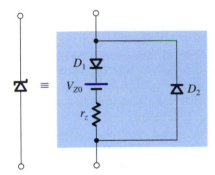

FIGURE 3.52 Equivalent-circuit model used to simulate the zener diode in SPICE. Diode D_1 is ideal and can be approximated in SPICE by using a very small value for n (say $n = 0.01$).

3.9.2 The Zener Diode Model

The diode model above does not adequately describe the operation of the diode in the breakdown region. Hence, it does not provide a satisfactory model for zener diodes. However, the equivalent-circuit model shown in Fig. 3.52 can be used to simulate a zener diode in SPICE. Here, diode D_1 is an ideal diode which can be approximated in SPICE by using a very small value for n (say $n = 0.01$). Diode D_2 is a regular diode that models the forward-bias region of the zener (for most applications, the parameters of D_2 are of little consequence).

EXAMPLE 3.10

DESIGN OF A DC POWER SUPPLY

In this example, we will design a dc power supply using the rectifier circuit whose Capture schematic[11] is shown in Fig. 3.53. This circuit consists of a full-wave diode rectifier, a filter

[11] The reader is reminded that the Capture schematics, and the corresponding PSpice simulation files, of all SPICE examples in this book can be found on the text's CD as well as on its website (www.sedrasmith.org). In these schematics (as shown in Fig. 3.53), we use variable parameters to enter the values of the various circuit components. This allows one to investigate the effect of changing component values by simply changing the corresponding parameter values.

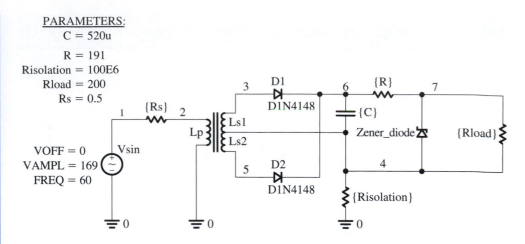

FIGURE 3.53 Capture schematic of the 5-V dc power supply in Example 3.10.

capacitor, and a zener voltage regulator. The only perhaps-puzzling component is $R_{\text{isolation}}$, the 100-MΩ resistor between the secondary winding of the transformer and ground. This resistor is included to provide dc continuity and thus "keep SPICE happy"; it has little effect on circuit operation.

Let it be required that the power supply (in Fig. 3.53) provide a nominal dc voltage of 5 V and be able to supply a load current I_{load} as large as 25 mA; that is, R_{load} can be as low as 200 Ω. The power supply is fed from a 120-V (rms) 60-Hz ac line. Note that in the PSpice schematic (Fig. 3.53), we use a sinusoidal voltage source with a 169-V peak amplitude to represent the 120-V rms supply (as 120-V rms = 169-V peak). Assume the availability of a 5.1-V zener diode having $r_z = 10$ Ω at $I_Z = 20$ mA (and thus $V_{Z0} = 4.9$ V), and that the required minimum current through the zener diode is $I_{Z\text{min}} = 5$ mA.

An approximate first-cut design can be obtained as follows: The 120-V (rms) supply is stepped down to provide 12-V (peak) sinusoids across each of the secondary windings using a 14:1 turns ratio for the center-tapped transformer. The choice of 12 V is a reasonable compromise between the need to allow for sufficient voltage (above the 5-V output) to operate the rectifier and the regulator, while keeping the PIV ratings of the diodes reasonably low. To determine a value for R, we can use the following expression:

$$R = \frac{V_{C\text{min}} - V_{Z0} - r_z I_{Z\text{min}}}{I_{Z\text{min}} + I_{L\text{max}}}$$

where an estimate for $V_{C\text{min}}$, the minimum voltage across the capacitor, can be obtained by subtracting a diode drop (say, 0.8 V) from 12 V and allowing for a ripple voltage across the capacitor of, say, $V_r = 0.5$ V. Thus, $V_{S\text{min}} = 10.7$ V. Furthermore, we note that $I_{L\text{max}} = 25$ mA and $I_{Z\text{min}} = 5$ mA, and that $V_{Z0} = 4.9$ V and $r_z = 10$ Ω. The result is that $R = 191$ Ω.

Next, we determine C using a restatement of Eq. (3.33) with V_p/R replaced by the current through the 191-Ω resistor. This current can be estimated by noting that the voltage across C varies from 10.7 to 11.2 V, and thus has an average value of 10.95 V. Furthermore, the desired voltage across the zener is 5 V. The result is $C = 520$ μF.

Now, with an approximate design in hand, we can proceed with the SPICE simulation. For the zener diode, we use the model of Fig. 3.52, and assume (arbitrarily) that D_1 has $I_S = 100$ pA

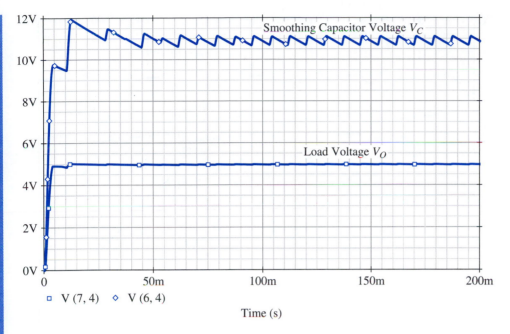

FIGURE 3.54 The voltage v_C across the smoothing capacitor C and the voltage v_O across the load resistor $R_{load} = 200\,\Omega$ in the 5-V power supply of Example 3.10.

and $n = 0.01$ while D_2 has $I_S = 100$ pA and $n = 1.7$. For the rectifier diodes, we use the commercially available 1N4148 type[12] (with $I_S = 2.682$ nA, $n = 1.836$, $R_S = 0.5664\,\Omega$, $V_0 = 0.5$ V, $C_{j0} = 4$ pF, $m = 0.333$, $\tau_T = 11.54$ ns, $V_{ZK} = 100$ V, $I_{ZK} = 100\,\mu A$).

In PSpice, we perform a transient analysis and plot the waveforms of both the voltage v_C across the smoothing capacitor C and the voltage v_O across the load resistor R_{load}. The simulation results for $R_{load} = 200\,\Omega$ ($I_{load} \cong 25$ mA) are presented in Fig. 3.54. Observe that v_C has an average of 10.85 V and a ripple of ± 0.21 V. Thus, $V_r = 0.42$ V, which is close to the 0.5-V value that we would expect from the chosen value of C. The output voltage v_O is very close to the required 5 V, with v_O varying between 4.957 V and 4.977 V for a ripple of only 20 mV. The variations of v_O with R_{load} is illustrated in Fig. 3.55 for $R_{load} = 500\,\Omega$, 250 Ω, 200 Ω, and 150 Ω. Accordingly, v_O remains close to the nominal value of 5 V for R_{load} as low as 200 Ω ($I_{load} \cong 25$ mA). For $R_{load} = 150\,\Omega$ (which implies $I_{load} \cong 33.3$ mA, greater than the maximum designed value), we see a significant drop in v_O (to about 4.8 V), as well as a large increase in the ripple voltage at the output (to about 190 mV). This is because the zener regulator is no longer operational; the zener has in fact cut off.

We conclude that the design meets the specifications, and we can stop here. Alternatively, we may consider fine-tuning the design using further runs of PSpice to help with the task. For instance, we could consider what happens if we use a lower value of C, and so on. We can also investigate other properties of the present design; for instance, the maximum current through each diode and ascertain whether this maximum is within the rating specified for the diode.

---

[12] The 1N4148 model is included in the evaluation (EVAL) library of PSpice (OrCad 9.2 Lite Edition), which is available on the CD accompanying this book.

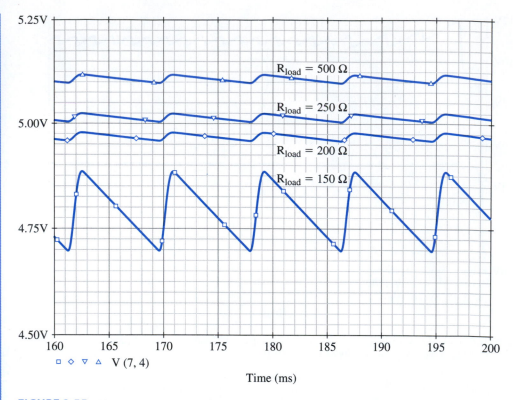

FIGURE 3.55 The output-voltage waveform from the 5-V power supply (in Example 3.10) for various load resistances: R_{load} = 500 Ω, 250 Ω, 200 Ω, and 150 Ω. The voltage regulation is lost at a load resistance of 150 Ω.

EXERCISE

3.35 Use PSpice to investigate the operation of the voltage doubler whose Capture schematic is shown in Fig. E3.35(a). Specifically, plot the transient behavior of the voltages v_2 and v_{out} when the input is a sinusoid of 10-V peak and 1-kHz frequency. Assume that the diodes are of the 1N4148 type (with I_S = 2.682 nA, n = 1.836, R_S = 0.5664 Ω, V_0 = 0.5 V, C_{j0} = 4 pF, m = 0.333, τ_T = 11.54 ns, V_{ZK} = 100 V, I_{ZK} = 100 μA).

Ans. The voltage waveforms are shown in Fig. E3.35(b).

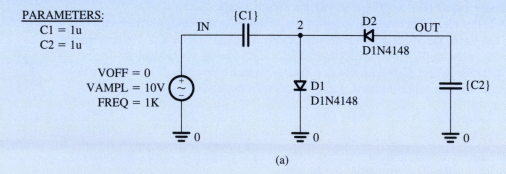

(a)

FIGURE E3.35 (a) Capture schematic of the voltage-doubler circuit (in Exercise 3.35).

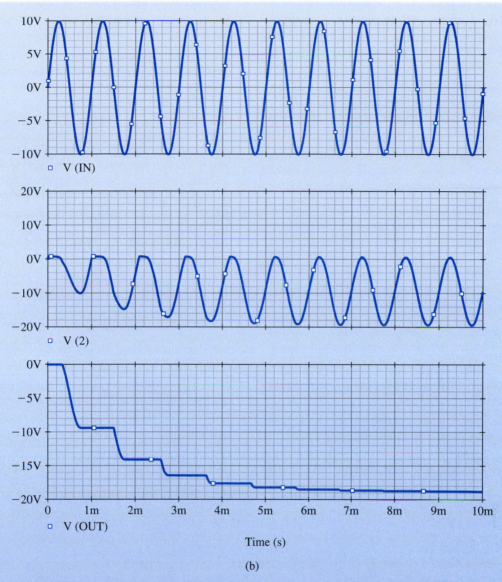

FIGURE E3.35 *(Continued)* **(b)** Various voltage waveforms in the voltage-doubler circuit. The top graph displays the input sine-wave voltage signal, the middle graph displays the voltage across diode D_1, and the bottom graph displays the voltage that appears at the output.

SUMMARY

■ In the forward direction, the ideal diode conducts any current forced by the external circuit while displaying a zero voltage drop. The ideal diode does not conduct in the reverse direction; any applied voltage appears as reverse bias across the diode.

■ The unidirectional-current-flow property makes the diode useful in the design of rectifier circuits.

■ The forward conduction of practical silicon diodes is accurately characterized by the relationship $i = I_S e^{v/nV_T}$.

■ A silicon diode conducts a negligible current until the forward voltage is at least 0.5 V. Then the current increases rapidly, with the voltage drop increasing by 60 mV to 120 mV (depending on the value of n) for every decade of current change.

■ In the reverse direction, a silicon diode conducts a current on the order of 10^{-9} A. This current is much greater than I_S and increases with the magnitude of reverse voltage.

■ Beyond a certain value of reverse voltage (that depends on the diode) breakdown occurs, and current increases rapidly with a small corresponding increase in voltage.

■ Diodes designed to operate in the breakdown region are called zener diodes. They are employed in the design of voltage regulators whose function is to provide a constant dc voltage that varies little with variations in power supply voltage and/or load current.

■ A hierarchy of diode models exists, with the selection of an appropriate model dictated by the application.

■ In many applications, a conducting diode is modeled as having a constant voltage drop, usually approximately 0.7 V.

■ A diode biased to operate at a dc current I_D has a small-signal resistance $r_d = nV_T/I_D$.

■ The silicon junction diode is basically a pn junction. Such a junction is formed in a single silicon crystal.

■ In p-type silicon there is an overabundance of holes (positively charged carriers), while in n-type silicon electrons are abundant.

■ A carrier-depletion region develops at the interface in a pn junction, with the n side positively charged and the p side negatively charged. The voltage difference resulting is called the barrier voltage.

■ A diffusion current I_D flows in the forward direction (carried by holes from the p side and electrons from the n side), and a current I_S flows in the reverse direction (carried by thermally generated minority carriers). In an open-circuited junction, $I_D = I_S$ and the barrier voltage is denoted V_0. V_0 is also called the junction built-in voltage.

■ Applying a reverse-bias voltage $|V|$ to a pn junction causes the depletion region to widen, and the barrier voltage increases to $(V_0 + |V|)$. The diffusion current decreases and a net reverse current of $(I_S - I_D)$ flows.

■ Applying a forward-bias voltage $|V|$ to a pn junction causes the depletion region to become narrower, and the barrier voltage decreases to $(V_0 - |V|)$. The diffusion current increases, and a net forward current of $(I_D - I_S)$ flows.

■ For a summary of the diode models in the forward region, refer to Table 3.1.

■ For a summary of the relationships that govern the physical operation of the pn junction, refer to Table 3.2.

PROBLEMS

SECTION 3.1: THE IDEAL DIODE

3.1 An AA flashlight cell, whose Thévenin equivalent is a voltage source of 1.5 V and a resistance of 1 Ω, is connected to the terminals of an ideal diode. Describe two possible situations that result. What are the diode current and terminal voltage when (a) the connection is between the diode cathode and the positive terminal of the battery and (b) the anode and the positive terminal are connected?

3.2 For the circuits shown in Fig. P3.2 using ideal diodes, find the values of the voltages and currents indicated.

3.3 For the circuits shown in Fig. P3.3 using ideal diodes, find the values of the labeled voltages and currents.

3.4 In each of the ideal-diode circuits shown in Fig. P3.4, v_I is a 1-kHz, 10-V peak sine wave. Sketch the waveform resulting at v_O. What are its positive and negative peak values?

3.5 The circuit shown in Fig. P3.5 is a model for a battery charger. Here v_I is a 10-V peak sine wave, D_1 and D_2 are ideal diodes, I is a 100-mA current source, and B is a 4.5-V battery. Sketch and label the waveform of the battery current i_B. What is its peak value? What is its average value? If the peak value

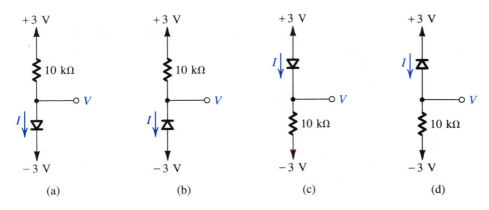

(a)   (b)   (c)   (d)

FIGURE P3.2

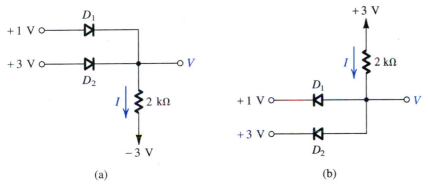

(a)   (b)

FIGURE P3.3

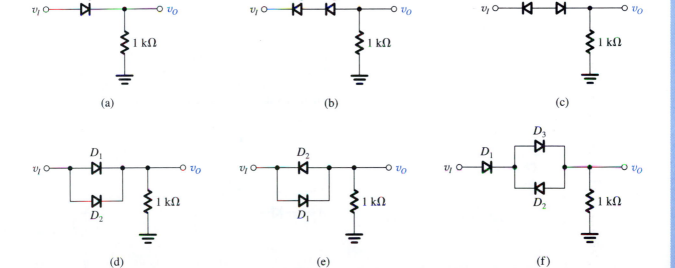

(a)   (b)   (c)

(d)   (e)   (f)

FIGURE P3.4 *(Continued)*

3

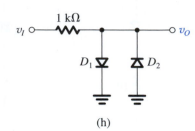

(g)

(h)

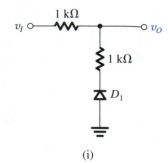

(i)

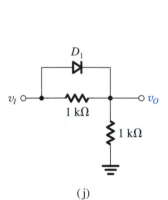

(j)

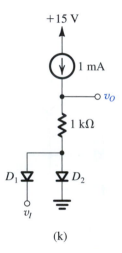

(k)

FIGURE P3.4 *(Continued)*

of v_I is reduced by 10%, what do the peak and average values of i_B become?

to denote the high value and "0" to denote the low value, prepare a table with four columns including all possible input combinations and the resulting values of X and Y. What logic function is X of A and B? What logic function is Y of A and B? For what values of A and B do X and Y have the same value? For what values of A and B do X and Y have opposite values?

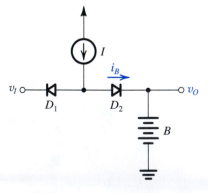

FIGURE P3.5

3.6 The circuits shown in Fig. P3.6 can function as logic gates for input voltages are either high or low. Using "1"

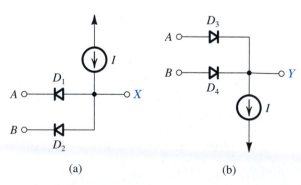

(a)

(b)

FIGURE P3.6

D3.7 For the logic gate of Fig. 3.5(a), assume ideal diodes and input voltage levels of 0 V and +5 V. Find a suitable value for R so that the current required from each of the input signal sources does not exceed 0.1 mA.

D3.8 Repeat Problem 3.7 for the logic gate of Fig. 3.5(b).

3.9 Assuming that the diodes in the circuits of Fig. P3.9 are ideal, find the values of the labeled voltages and currents.

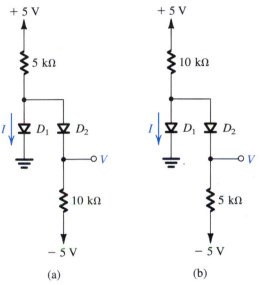

FIGURE P3.9

3.10 Assuming that the diodes in the circuits of Fig. P3.10 are ideal, utilize Thévenin's theorem to simplify the circuits and thus find the values of the labeled currents and voltages.

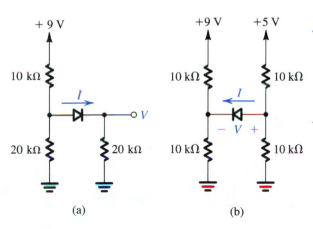

FIGURE P3.10

D3.11 For the rectifier circuit of Fig. 3.3(a), let the input sine wave have 120-V rms value and assume the diode to be ideal. Select a suitable value for R so that the peak diode current does not exceed 50 mA. What is the greatest reverse voltage that will appear across the diode?

3.12 Consider the rectifier circuit of Fig. 3.3 in the event that the input source v_I has a source resistance R_s. For the case $R_s = R$ and assuming the diode to be ideal, sketch and clearly label the transfer characteristic v_O versus v_I.

3.13 A square wave of 6-V peak-to-peak amplitude and zero average is applied to a circuit resembling that in Fig. 3.3(a) and employing a 100-Ω resistor. What is the peak output voltage that results? What is the average output voltage that results? What is the peak diode current? What is the average diode current? What is the maximum reverse voltage across the diode?

3.14 Repeat Problem 3.13 for the situation in which the average voltage of the square wave is 2 V while its peak-to-peak value remains at 6 V.

*D3.15 Design a battery-charging circuit, resembling that in Fig. 3.4 and using an ideal diode, in which current flows to the 12-V battery 20% of the time and has an average value of 100 mA. What peak-to-peak sine-wave voltage is required? What resistance is required? What peak diode current flows? What peak reverse voltage does the diode endure? If resistors can be specified to only one significant digit and the peak-to-peak voltage only to the nearest volt, what design would you choose to guarantee the required charging current? What fraction of the cycle does diode current flow? What is the average diode current? What is the peak diode current? What peak reverse voltage does the diode endure?

3.16 The circuit of Fig. P3.16 can be used in a signalling system using one wire plus a common ground return. At any moment, the input has one of three values: +3 V, 0 V, −3 V. What is the status of the lamps for each input value? (Note that the lamps can be located apart from each other and that there may be several of each type of connection, all on one wire!).

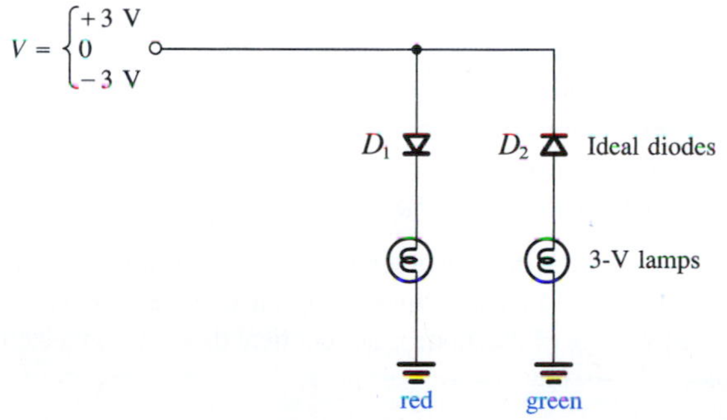

$$V = \begin{cases} +3 \text{ V} \\ 0 \\ -3 \text{ V} \end{cases}$$

D_1 D_2 Ideal diodes

3-V lamps

red   green

FIGURE P3.16

3

SECTION 3.2: TERMINAL CHARACTERISTICS OF JUNCTION DIODES

3.17 Calculate the value of the thermal voltage, V_T, at $-40°C$, $0°C$, $+40°C$, and $+150°C$. At what temperature is V_T exactly 25 mV?

3.18 At what forward voltage does a diode for which $n = 2$ conduct a current equal to $1000I_S$? In terms of I_S, what current flows in the same diode when its forward voltage is 0.7 V?

3.19 A diode for which the forward voltage drop is 0.7 V at 1.0 mA and for which $n = 1$ is operated at 0.5 V. What is the value of the current?

3.20 A particular diode, for which $n = 1$, is found to conduct 5 mA with a junction voltage of 0.7 V. What is its saturation current I_S? What current will flow in this diode if the junction voltage is raised to 0.71 V? To 0.8 V? If the junction voltage is lowered to 0.69 V? To 0.6 V? What change in junction voltage will increase the diode current by a factor of 10?

3.21 The following measurements are taken on particular junction diodes to which V is the terminal voltage and I is the diode current. For each diode, estimate values of I_S and the terminal voltage at 1% of the measured current for $n = 1$ and for $n = 2$. Use $V_T = 25$ mV in your computations.

(a) $V = 0.700$ V at $I = 1.00$ A
(b) $V = 0.650$ V at $I = 1.00$ mA
(c) $V = 0.650$ V at $I = 10$ μA
(d) $V = 0.700$ V at $I = 10$ mA

3.22 Listed below are the results of measurements taken on several different junction diodes. For each diode, the data provided are the diode current I, the corresponding diode voltage V, and the diode voltage at a current $I/10$. In each case, estimate I_S, n, and the diode voltage at $10I$.

(a) 10.0 mA, 700 mV, 600 mV
(b) 1.0 mA, 700 mV, 600 mV
(c) 10 A, 800 mV, 700 mV
(d) 1 mA, 700 mV, 580 mV
(e) 10 μA, 700 mV, 640 mV

3.23 The circuit in Fig. P3.23 utilizes three identical diodes having $n = 1$ and $I_S = 10^{-14}$ A. Find the value of the current I required to obtain an output voltage $V_O = 2$ V. If a current of 1 mA is drawn away from the output terminal by a load, what is the change in output voltage?

3.24 A junction diode is operated in a circuit in which it is supplied with a constant current I. What is the effect on the forward voltage of the diode if an identical diode is connected in parallel? Assume $n = 1$.

3.25 In the circuit shown in Fig. P3.25, both diodes have $n = 1$, but D_1 has 10 times the junction area of D_2. What value

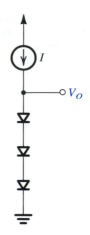

FIGURE P3.23

of V results? To obtain a value for V of 50 mV, what current I_2 is needed?

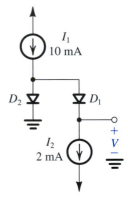

FIGURE P3.25

3.26 For the circuit shown in Fig. P3.26, both diodes are identical, conducting 10 mA at 0.7 V and 100 mA at 0.8 V. Find the value of R for which $V = 80$ mV.

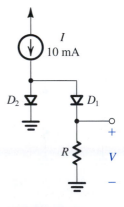

FIGURE P3.26

3.27 Several diodes having a range of sizes, but all with $n = 1$, are measured at various temperatures and junction currents as noted below. For each, estimate the diode voltage at 1 mA and 25°C.

(a) 620 mV at 10 μA and 0°C
(b) 790 mV at 1 A and 50°C
(c) 590 mV at 100 μA and 100°C
(d) 850 mV at 10 mA and −50°C
(e) 700 mV at 100 mA and 75°C

*3.28 In the circuit shown in Fig. P3.28, D_1 is a large-area high-current diode whose reverse leakage is high and independent of applied voltage while D_2 is a much smaller, low-current diode for which $n = 1$. At an ambient temperature of 20°C, resistor R_1 is adjusted to make $V_{R1} = V_2 = 520$ mV. Subsequent measurement indicates that R_1 is 520 kΩ. What do you expect the voltages V_{R1} and V_2 to become at 0°C and at 40°C?

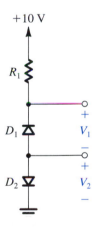

FIGURE P3.28

3.29 When a 15-A current is applied to a particular diode, it is found that the junction voltage immediately becomes 700 mV. However, as the power being dissipated in the diode raises its temperature, it is found that the voltage decreases and eventually reaches 580 mV. What is the apparent rise in junction temperature? What is the power dissipated in the diode in its final state? What is the temperature rise per watt of power dissipation? (This is called the thermal resistance.)

*3.30 A designer of an instrument that must operate over a wide supply-voltage range, noting that a diode's junction-voltage drop is relatively independent of junction current, considers the use of a large diode to establish a small relatively constant voltage. A power diode, for which the nominal current at 0.8 V is 10 A, is available. Furthermore, the designer has reason to believe that $n = 2$. For the available

current source, which varies from 0.5 mA to 1.5 mA, what junction voltage might be expected? What additional voltage change might be expected for a temperature variation of ±25°C?

*3.31 As an alternative to the idea suggested in Problem 3.30, the designer considers a second approach to producing a relatively constant small voltage from a variable current supply: It relies on the ability to make quite accurate copies of any small current that is available (using a process called current mirroring). The designer proposes to use this idea to supply two diodes of different junction areas with the same current and to measure their junction-voltage difference. Two types of diodes are available; for a forward voltage of 700 mV, one conducts 0.1 mA while the other conducts 1 A. Now, for identical currents in the range of 0.5 mA to 1.5 mA supplied to each, what range of difference voltages result? What is the effect of a temperature change of ±25°C on this arrangement? Assume $n = 1$.

SECTION 3.3: MODELING THE DIODE FORWARD CHARACTERISTIC

*3.32 Consider the graphical analysis of the diode circuit of Fig. 3.10 with $V_{DD} = 1$ V, $R = 1$ kΩ, and a diode having $I_S = 10^{-15}$ A and $n = 1$. Calculate a small number of points on the diode characteristic in the vicinity of where you expect the load line to intersect it, and use a graphical process to refine your estimate of diode current. What value of diode current and voltage do you find? Analytically, find the voltage corresponding to your estimate of current. By how much does it differ from the graphically estimated value?

3.33 Use the iterative-analysis procedure to determine the diode current and voltage in the circuit of Fig. 3.10 for $V_{DD} = 1$ V, $R = 1$ kΩ, and a diode having $I_S = 10^{-15}$ A and $n = 1$.

3.34 A "1-mA diode" (i.e., one that has $v_D = 0.7$ V at $i_D = 1$ mA) is connected in series with a 200-Ω resistor to a 1.0-V supply.

(a) Provide a rough estimate of the diode current you would expect.
(b) If the diode is characterized by $n = 2$, estimate the diode current more closely using iterative analysis.

3.35 A collection of circuits, which are variants of that shown in Fig. 3.10, are listed below. For each diode used, the measured junction current I_0 at junction voltage V_0 is provided, along with the change of junction voltage ΔV measured when the current is increased 10-fold. For each circuit, find the diode current I_D and diode voltage V_D that result, using the diode exponential equation and iteration. (Hint: To reduce your workload, notice the very special relation between the circuit and diode parameters in many—but not all—cases. Finally, note that using such relationships, or approximations to them, can often make your first pass at a circuit design much easier and faster!)

Circuit	V_{DD} (V)	R (kΩ)	I_0 (mA)	V_0 (mV)	ΔV (mV)
a	10.0	9.3	1.0	700	100
b	3.0	2.3	1.0	700	100
c	2.0	2.0	10	700	100
d	2.0	2.0	1.0	700	100
e	1.0	0.30	10	700	100
f	1.0	0.30	10	700	60
g	1.0	0.30	10	700	120
h	0.5	30	10	700	100

D3.36 Assuming the availability of diodes for which $v_D = 0.7$ V at $i_D = 1$ mA and $n = 1$, design a circuit that utilizes four diodes connected in series, in series with a resistor R connected to a 10-V power supply. The voltage across the string of diodes is to be 3.0 V.

3.37 Find the parameters of a piecewise-linear model of a diode for which $v_D = 0.7$ V at $i_D = 1$ mA and $n = 2$. The model is to fit exactly at 1 mA and 10 mA. Calculate the error in millivolts in predicting v_D using the piecewise-linear model at $i_D = 0.5$, 5, and 14 mA.

3.38 Using a copy of the diode curve presented in Fig. 3.12, approximate the diode characteristic using a straight line that exactly matches the diode characteristic at both 10 mA and 1 mA. What is the slope? What is r_D? What is V_{D0}?

3.39 On a copy of the diode characteristics presented in Fig. 3.12, draw a load line corresponding to an external circuit consisting of a 0.9-V voltage source and a 100-Ω resistor. What are the values of diode drop and loop current you estimate using:

(a) the actual diode characteristics?
(b) the two-segment model shown?

3.40 For the diodes characterized below, find r_D and V_{D0}, the elements of the battery-plus-resistor model for which the straight line intersects the diode exponential characteristic at 0.1× and 10× the specified diode current.

(a) $V_D = 0.7$ V at $I_D = 1$ mA and $n = 1$
(b) $V_D = 0.7$ V at $I_D = 1$ A and $n = 1$
(c) $V_D = 0.7$ V at $I_D = 10$ μA and $n = 1$

3.41 The diode whose characteristic curve is shown in Fig. 3.15 is to be operated at 10 mA. What would likely be a suitable voltage choice for an appropriate constant-voltage-drop model?

3.42 A diode operates in a series circuit with R and V. A designer, considering using a constant-voltage model, is uncertain whether to use 0.7 V or 0.6 V for V_D. For what value of V is the difference in the calculated values of current only 1%? For $V = 2$ V and $R = 1$ kΩ, what two currents would result from the use of the two values of V_D? What is their percentage difference?

D3.43 A designer has a relatively large number of diodes for which a current of 20 mA flows at 0.7 V and the 0.1-V/ decade approximation is relatively good. Using a 10-mA current source, the designer wishes to create a reference voltage of 1.25 V. Suggest a combination of series and parallel diodes that will do the job as well as possible. How many diodes are needed? What voltage is actually achieved?

3.44 Consider the half-wave rectifier circuit of Fig. 3.3(a) with $R = 1$ kΩ and the diode having the characteristics and the piecewise-linear model shown in Fig. 3.12 ($V_{D0} = 0.65$ V, $r_D = 20$ Ω). Analyze the rectifier circuit using the piecewise-linear model for the diode, and thus find the output voltage v_O as a function of v_I. Sketch the transfer characteristic v_O versus v_I for $0 \le v_I \le 10$ V. For v_I being a sinusoid with 10 V peak amplitude, sketch and clearly label the waveform of v_O.

3.45 Solve the problems in Example 3.2 using the constant-voltage-drop ($V_D = 0.7$ V) diode model.

3.46 For the circuits shown in Fig. P3.2, using the constant-voltage-drop ($V_D = 0.7$ V) diode model, find the voltages and currents indicated.

3.47 For the circuits shown in Fig. P3.3, using the constant-voltage-drop ($V_D = 0.7$ V) diode model, find the voltages and currents indicated.

3.48 For the circuits in Fig. P3.9, using the constant-voltage-drop ($V_D = 0.7$ V) diode model, find the values of the labeled currents and voltages.

3.49 For the circuits in Fig. P3.10, utilize Thévenin's theorem to simplify the circuits and find the values of the labeled currents and voltages. Assume that conducting diodes can be represented by the constant-voltage-drop model ($V_D = 0.7$ V).

D3.50 Repeat Problem 3.11, representing the diode by its constant-voltage-drop ($V_D = 0.7$ V) model. How different is the resulting design?

3.51 Repeat the problem in Example 3.1 assuming that the diode has 10 times the area of the device whose characteristics and piecewise-linear model are displayed in Fig. 3.12. Represent the diode by its piecewise-linear model ($v_D = 0.65 + 2i_D$).

3.52 The small-signal model is said to be valid for voltage variations of about 10 mV. To what percentage current change does this correspond (consider both positive and negative signals) for:

(a) $n = 1$?
(b) $n = 2$?

For each case, what is the maximum allowable voltage signal (positive or negative) if the current change is to be limited to 10%?

3.53 In a particular circuit application, ten "20-mA diodes" (a 20-mA diode is a diode that provides a 0.7-V drop when the current through it is 20 mA) connected in parallel operate at a total current of 0.1 A. For the diodes closely matched, with $n = 1$, what current flows in each? What is the corresponding small-signal resistance of each diode and of the combination? Compare this with the incremental resistance of a single diode conducting 0.1 A. If each of the 20-mA diodes has a series resistance of 0.2 Ω associated with the wire bonds to the junction, what is the equivalent resistance of the 10 parallel-connected diodes? What connection resistance would a single diode need in order to be totally equivalent? (*Note:* This is why the parallel connection of real diodes can often be used to advantage.)

3.54 In the circuit shown in Fig. P3.54, I is a dc current and v_s is a sinusoidal signal. Capacitors C_1 and C_2 are very large; their function is to couple the signal to and from the diode but block the dc current from flowing into the signal source or the load (not shown). Use the diode small-signal model to show that the signal component of the output voltage is

$$v_o = v_s \frac{nV_T}{nV_T + IR_s}$$

If $v_s = 10$ mV, find v_o for $I = 1$ mA, 0.1 mA, and 1 μA. Let $R_s = 1$ kΩ and $n = 2$. At what value of I does v_o become one-half of v_s? Note that this circuit functions as a signal attenuator with the attenuation factor controlled by the value of the dc current I.

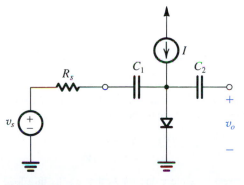

FIGURE P3.54

3.55 In the attenuator circuit of Fig. P3.54, let $R_s = 10$ kΩ. The diode is a 1-mA device; that is, it exhibits a voltage drop of 0.7 V at a dc current of 1 mA and has $n = 1$. For small input signals, what value of current I is needed for $v_o/v_s = 0.50$? 0.10? 0.01? 0.001? In each case, what is the largest input signal that can be used while ensuring that the signal component of the diode current is limited to $\pm$10% of its dc current? What output signals correspond?

3.56 In the capacitor-coupled attenuator circuit shown in Fig. P3.56, I is a dc current that varies from 0 mA to 1 mA, D_1 and D_2 are diodes with $n = 1$, and C_1 and C_2 are large coupling capacitors. For very small input signals, find the values of the ratio v_o/v_i for I equal to:

(a) 0 μA
(b) 1 μA
(c) 10 μA
(d) 100 μA
(e) 500 μA
(f) 600 μA
(g) 900 μA
(h) 990 μA
(i) 1 mA

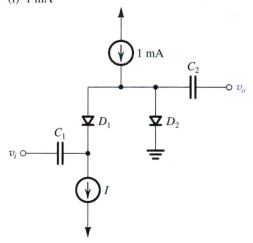

FIGURE P3.56

For the current in each diode in excess of 10 μA, what is the largest input signal for which the critical diode current remains within 10% of its dc value?

*3.57 In the circuit shown in Fig. P3.57, diodes D_1 through D_4 are identical. Each has $n = 1$ and is a "1-mA diode"; that is, it exhibits a voltage drop of 0.7 V at a 1-mA current.

(a) For small input signals (e.g., 10 mV peak), find values of the small-signal transmission v_o/v_i for various values of I: 0 μA, 1 μA, 10 μA, 100 μA, 1 mA, and 10 mA.

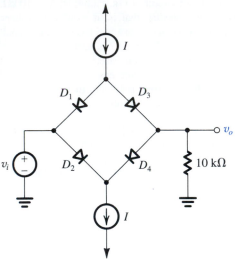

FIGURE P3.57

(b) For a forward-conducting diode, what is the largest signal-voltage magnitude that it can support while the corresponding signal current is limited to 10% of the dc bias current. Now, for the circuit in Fig. P3.57, for 10-mV peak input, what is the smallest value of I for which the diode currents remain within $\pm10\%$ of their dc value?

(c) For $I = 1$ mA, what is the largest possible output signal for which the diode currents deviate by at most 10% of their dc values? What is the corresponding peak input?

*3.58 In the circuit shown in Fig. P3.58, I is a dc current and v_i is a sinusoidal signal with small amplitude (less than 10 mV) and a frequency of 100 kHz. Representing the diode by its small-signal resistance r_d, which is a function of I, sketch the circuit for determining the sinusoidal output voltage V_o, and thus find the phase shift between V_i and V_o. Find the value of I that will provide a phase shift of $-45°$, and find the range of phase shift achieved as I is varied over the range of 0.1 to 10 times this value. Assume $n = 1$.

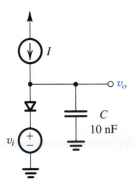

FIGURE P3.58

*3.59 Consider the voltage-regulator circuit shown in Fig. P3.59. The value of R is selected to obtain an output voltage V_O (across the diode) of 0.7 V.

(a) Use the diode small-signal model to show that the change in output voltage corresponding to a change of 1 V in V^+ is

$$\frac{\Delta V_O}{\Delta V^+} = \frac{nV_T}{V^+ + nV_T - 0.7}$$

This quantity is known as the line regulation and is usually expressed in mV/V.

(b) Generalize the expression above for the case of m diodes connected in series and the value of R adjusted so that the voltage across each diode is 0.7 V (and $V_O = 0.7m$ V).

(c) Calculate the value of line regulation for the case $V^+ = 10$ V (nominally) and (i) $m = 1$ and (ii) $m = 3$. Use $n = 2$.

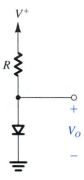

FIGURE P3.59

*D3.60 Consider the voltage-regulator circuit shown in Fig P3.59 under the condition that a load current I_L is drawn from the output terminal.

(a) If the value of I_L is sufficiently small so that the corresponding change in regulator output voltage ΔV_O is small enough to justify using the diode small-signal model, show that

$$\frac{\Delta V_O}{I_L} = -(r_d \mathbin{/\mkern-5mu/} R)$$

This quantity is known as the load regulation and is usually expressed in mV/mA.

(b) If the value of R is selected such that at no load the voltage across the diode is 0.7 V and the diode current is I_D, show that the expression derived in (a) becomes

$$\frac{\Delta V_O}{I_L} = -\frac{nV_T}{I_D}\frac{V^+ - 0.7}{V^+ - 0.7 + nV_T}$$

Select the lowest possible value for I_D that results in a load regulation ≤ 5 mV/mA. Assume $n = 2$. If V^+ is nominally 10 V, what value of R is required? Also, specify the diode required.

(c) Generalize the expression derived in (b) for the case of m diodes connected in series and R adjusted to obtain $V_O = 0.7m$ V at no load.

D3.61 Design a diode voltage regulator to supply 1.5 V to a 150-Ω load. Use two diodes specified to have a 0.7-V drop at a current of 10 mA and $n = 1$. The diodes are to be connected to a +5-V supply through a resistor R. Specify the value for R. What is the diode current with the load connected? What is the increase resulting in the output voltage when the load is disconnected? What change results if the load resistance is reduced to 100 Ω? To 75 Ω? To 50 Ω?

*D3.62 A voltage regulator consisting of two diodes in series fed with a constant-current source is used as a replacement for a single carbon-zinc cell (battery) of nominal voltage 1.5 V. The regulator load current varies from 2 mA to 7 mA. Constant-current supplies of 5 mA, 10 mA, and 15 mA are available. Which would you choose, and why? What change in output voltage would result when the load current varies over its full range? Assume that the diodes have $n = 2$.

*3.63 A particular design of a voltage regulator is shown in Fig. P3.63. Diodes D_1 and D_2 are 10-mA units; that is, each has a voltage drop of 0.7 V at a current of 10 mA. Each has $n = 1$.

(a) What is the regulator output voltage V_O with the 150-Ω load connected?
(b) Find V_O with no load.
(c) With the load connected, to what value can the 5-V supply be lowered while maintaining the loaded output voltage within 0.1 V of its nominal value?
(d) What does the loaded output voltage become when the 5-V supply is raised by the same amount as the drop found in (c)?
(e) For the range of changes explored in (c) and (d), by what percentage does the output voltage change for each percentage change of supply voltage in the worst case?

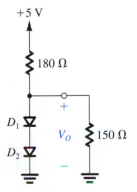

FIGURE P3.63

SECTION 3.4: OPERATION IN THE REVERSE BREAKDOWN REGION—ZENER DIODES

3.64 Partial specifications of a collection of zener diodes are provided below. Identify the missing parameter, and estimate its value. Note from Fig. 3.21 that $V_{ZK} \cong V_{Z0}$.

(a) $V_Z = 10.0$ V, $V_{ZK} = 9.6$ V, and $I_{ZT} = 50$ mA
(b) $I_{ZT} = 10$ mA, $V_Z = 9.1$ V, and $r_z = 30$ Ω
(c) $r_z = 2$ Ω, $V_Z = 6.8$ V, and $V_{ZK} = 6.6$ V
(d) $V_Z = 18$ V, $I_{ZT} = 5$ mA, and $V_{ZK} = 17.2$ V
(e) $I_{ZT} = 200$ mA, $V_Z = 7.5$ V, and $r_z = 1.5$ Ω

Assuming that the power rating of a breakdown diode is established at about twice the specified zener current (I_{ZT}), what is the power rating of each of the diodes described above?

D3.65 A designer requires a shunt regulator of approximately 20 V. Two kinds of zener diodes are available: 6.8-V devices with r_z of 10 Ω and 5.1-V devices with r_z of 30 Ω. For the two major choices possible, find the load regulation. In this calculation neglect the effect of the regulator resistance R.

3.66 A shunt regulator utilizing a zener diode with an incremental resistance of 5 Ω is fed through an 82-Ω resistor. If the raw supply changes by 1.3 V, what is the corresponding change in the regulated output voltage?

3.67 A 9.1-V zener diode exhibits its nominal voltage at a test current of 28 mA. At this current the incremental resistance is specified as 5 Ω. Find V_{Z0} of the zener model. Find the zener voltage at a current of 10 mA and at 100 mA.

D3.68 Design a 7.5-V zener regulator circuit using a 7.5-V zener specified at 12 mA. The zener has an incremental resistance $r_z = 30$ Ω and a knee current of 0.5 mA. The regulator operates from a 10-V supply and has a 1.2-kΩ load. What is the value of R you have chosen? What is the regulator output voltage when the supply is 10% high? Is 10% low? What is the output voltage when both the supply is 10% high and the load is removed? What is the smallest possible load resistor that can be used while the zener operates at a current no lower than the knee current while the supply is 10% low?

*D3.69 Provide two designs of shunt regulators utilizing the 1N5235 zener diode, which is specified as follows: $V_Z = 6.8$ V and $r_z = 5$ Ω for $I_Z = 20$ mA; at $I_Z = 0.25$ mA (nearer the knee), $r_z = 750$ Ω. For both designs, the supply voltage is nominally 9 V and varies by ±1 V. For the first design, assume that the availability of supply current is not a problem, and thus operate the diode at 20 mA. For the second design, assume that the current from the raw supply is limited, and therefore you are forced to operate the diode at 0.25 mA. For the purpose of these initial designs, assume no load. For each design find the value of R and the line regulation.

3

*D3.70 A zener shunt regulator employs a 9.1-V zener diode for which $V_Z = 9.1$ V at $I_Z = 9$ mA, with $r_z = 30$ Ω and $I_{ZK} = 0.3$ mA. The available supply voltage of 15 V can vary as much as $\pm10\%$. For this diode, what is the value of V_{Z0}? For a nominal load resistance R_L of 1 kΩ and a nominal zener current of 10 mA, what current must flow in the supply resistor R? For the nominal value of supply voltage, select a value for resistor R, specified to one significant digit, to provide at least that current. What nominal output voltage results? For a $\pm10\%$ change in the supply voltage, what variation in output voltage results? If the load current is reduced by 50%, what increase in V_O results? What is the smallest value of load resistance that can be tolerated while maintaining regulation when the supply voltage is low? What is the lowest possible output voltage that results? Calculate values for the line regulation and for the load regulation for this circuit using the numerical results obtained in this problem.

*D3.71 It is required to design a zener shunt regulator to provide a regulated voltage of about 10 V. The available 10-V, 1-W zener of type 1N4740 is specified to have a 10-V drop at a test current of 25 mA. At this current its r_z is 7 Ω. The raw supply available has a nominal value of 20 V but can vary by as much as $\pm25\%$. The regulator is required to supply a load current of 0 mA to 20 mA. Design for a minimum zener current of 5 mA.

(a) Find V_{Z0}.
(b) Calculate the required value of R.
(c) Find the line regulation. What is the change in V_O expressed as a percentage, corresponding to the $\pm25\%$ change in V_S?
(d) Find the load regulation. By what percentage does V_O change from the no-load to the full-load condition?
(e) What is the maximum current that the zener in your design is required to conduct? What is the zener power dissipation under this condition?

SECTION 3.5: RECTIFIER CIRCUITS

3.72 Consider the half-wave rectifier circuit of Fig. 3.25(a) with the diode reversed. Let v_S be a sinusoid with 15-V peak amplitude, and let $R = 1.5$ kΩ. Use the constant-voltage-drop diode model with $V_D = 0.7$ V.

(a) Sketch the transfer characteristic.
(b) Sketch the waveform of v_O.
(c) Find the average value of v_O.
(d) Find the peak current in the diode.
(e) Find the PIV of the diode.

3.73 Using the exponential diode characteristic, show that for v_S and v_O both greater than zero, the circuit of Fig. 3.25(a) has the transfer characteristic

$$v_O = v_S - v_D \text{ (at } i_D = 1 \text{ mA)} - nV_T \ln (v_O/R)$$

where v_S and v_O are in volts and R is in kilohms.

3.74 Consider a half-wave rectifier circuit with a triangular-wave input of 5-V peak-to-peak amplitude and zero average and with $R = 1$ kΩ. Assume that the diode can be represented by the piecewise-linear model with $V_{D0} = 0.65$ V and $r_D = 20$ Ω. Find the average value of v_O.

3.75 For a half-wave rectifier circuit with $R = 1$ kΩ, utilizing a diode whose voltage drop is 0.7 V at a current of 1 mA and exhibiting a 0.1-V change per decade of current variation, find the values of the input voltage to the rectifier corresponding to $v_O = 0.1$ V, 0.5 V, 1 V, 2 V, 5 V, and 10 V. Plot the rectifier transfer characteristic.

3.76 A half-wave rectifier circuit with a 1-kΩ load operates from a 120-V (rms) 60-Hz household supply through a 10-to-1 step-down transformer. It uses a silicon diode that can be modeled to have a 0.7-V drop for any current. What is the peak voltage of the rectified output? For what fraction of the cycle does the diode conduct? What is the average output voltage? What is the average current in the load?

3.77 A full-wave rectifier circuit with a 1-kΩ load operates from a 120-V (rms) 60-Hz household supply through a 5-to-1 transformer having a center-tapped secondary winding. It uses two silicon diodes that can be modeled to have a 0.7-V drop for all currents. What is the peak voltage of the rectified output? For what fraction of a cycle does each diode conduct? What is the average output voltage? What is the average current in the load?

3.78 A full-wave bridge rectifier circuit with a 1-kΩ load operates from a 120-V (rms) 60-Hz household supply through a 10-to-1 step-down transformer having a single secondary winding. It uses four diodes, each of which can be modeled to have a 0.7-V drop for any current. What is the peak value of the rectified voltage across the load? For what fraction of a cycle does each diode conduct? What is the average voltage across the load? What is the average current through the load?

D3.79 It is required to design a full-wave rectifier circuit using the circuit of Fig. 3.26 to provide an average output voltage of:

(a) 10 V
(b) 100 V

In each case find the required turns ratio of the transformer. Assume that a conducting diode has a voltage drop of 0.7 V. The ac line voltage is 120 V rms.

D3.80 Repeat Problem 3.79 for the bridge rectifier circuit of Fig. 3.27.

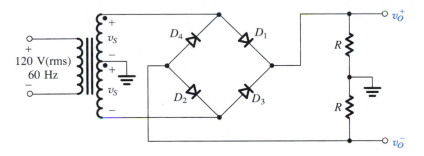

FIGURE P3.82

D3.81 Consider the full-wave rectifier in Fig. 3.26 when the transformer turns ratio is such that the voltage across the entire secondary winding is 24 V rms. If the input ac line voltage (120 V rms) fluctuates by as much as ±10%, find the required PIV of the diodes. (Remember to use a factor of safety in your design.)

*3.82 The circuit in Fig. P3.82 implements a complementary-output rectifier. Sketch and clearly label the waveforms of v_O^+ and v_O^-. Assume a 0.7-V drop across each conducting diode. If the magnitude of the average of each output is to be 15 V, find the required amplitude of the sine wave across the entire secondary winding. What is the PIV of each diode?

3.83 Augment the rectifier circuit of Problem 3.76 with a capacitor chosen to provide a peak-to-peak ripple voltage of (i) 10% of the peak output and (ii) 1% of the peak output. In each case:

(a) What average output voltage results?
(b) What fraction of the cycle does the diode conduct?
(c) What is the average diode current?
(d) What is the peak diode current?

3.84 Repeat Problem 3.83 for the rectifier in Problem 3.77.

3.85 Repeat Problem 3.83 for the rectifier in Problem 3.78.

*D3.86 It is required to use a peak rectifier to design a dc power supply that provides an average dc output voltage of 15 V on which a maximum of ±1-V ripple is allowed. The rectifier feeds a load of 150 Ω. The rectifier is fed from the line voltage (120 V rms, 60 Hz) through a transformer. The diodes available have 0.7-V drop when conducting. If the designer opts for the half-wave circuit:

(a) Specify the rms voltage that must appear across the transformer secondary.
(b) Find the required value of the filter capacitor.

(c) Find the maximum reverse voltage that will appear across the diode, and specify the PIV rating of the diode.
(d) Calculate the average current through the diode during conduction.
(e) Calculate the peak diode current.

*D3.87 Repeat Problem 3.86 for the case in which the designer opts for a full-wave circuit utilizing a center-tapped transformer.

*D3.88 Repeat Problem 3.86 for the case in which the designer opts for a full-wave bridge rectifier circuit.

*D3.89 Consider a half-wave peak rectifier fed with a voltage v_S having a triangular waveform with 20-V peak-to-peak amplitude, zero average, and 1-kHz frequency. Assume that the diode has a 0.7-V drop when conducting. Let the load resistance $R = 100\ \Omega$ and the filter capacitor $C = 100\ \mu F$. Find the average dc output voltage, the time interval during which the diode conducts, the average diode current during conduction, and the maximum diode current.

*D3.90 Consider the circuit in Fig. P3.82 with two equal filter capacitors placed across the load resistors R. Assume that the diodes available exhibit a 0.7-V drop when conducting. Design the circuit to provide ±15-V dc output voltages with a peak-to-peak ripple no greater than 1 V. Each supply should be capable of providing 200 mA dc current to its load resistor R. Completely specify the capacitors, diodes and the transformer.

3.91 The op amp in the precision rectifier circuit of Fig. P3.91 is ideal with output saturation levels of ±12 V. Assume that when conducting the diode exhibits a constant voltage drop of 0.7 V. Find v_-, v_O, and v_A for:

(a) $v_I = +1$ V
(b) $v_I = +2$ V
(c) $v_I = -1$ V
(d) $v_I = -2$ V

3

Also, find the average output voltage obtained when v_I is a symmetrical square wave of 1-kHz frequency, 5-V amplitude, and zero average.

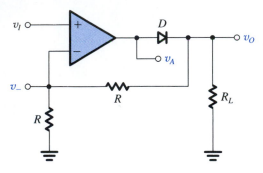

FIGURE P3.91

3.92 The op amp in the circuit of Fig. P3.92 is ideal with output saturation levels of ±12 V. The diodes exhibit a constant 0.7-V drop when conducting. Find v_-, v_A, and v_O for:

(a) $v_I = +1$ V
(b) $v_I = +2$ V
(c) $v_I = -1$ V
(d) $v_I = -2$ V

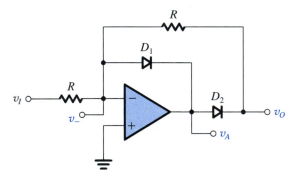

FIGURE P3.92

SECTION 3.6: LIMITING AND CLAMPING CIRCUITS

3.93 Sketch the transfer characteristic v_O versus v_I for the limiter circuits shown in Fig. P3.93. All diodes begin conducting at a forward voltage drop of 0.5 V and have voltage drops of 0.7 V when fully conducting.

3.94 Repeat Problem 3.93 assuming that the diodes are modeled with the piecewise-linear model with $V_{D0} = 0.65$ V and $r_D = 20$ Ω.

3.95 The circuits in Fig. P3.93(a) and (d) are connected as follows: The two input terminals are tied together, and the

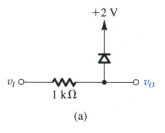

(a)

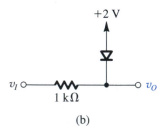

(b)

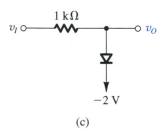

(c)

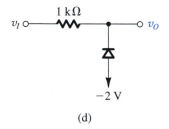

(d)

FIGURE P3.93

two output terminals are tied together. Sketch the transfer characteristic of the circuit resulting, assuming that the cut-in voltage of the diodes is 0.5 V and their voltage drop when fully conducting is 0.7 V.

3.96 Repeat Problem 3.95 for the two circuits in Fig. P3.93(a) and (b) connected together as follows: The two input terminals are tied together, and the two output terminals are tied together.

*3.97 Sketch and clearly label the transfer characteristic of the circuit in Fig. P3.97 for -20 V $\leq v_I \leq +20$ V. Assume that the diodes can be represented by a piecewise-linear model with $V_{D0} = 0.65$ V and $r_D = 20$ Ω. Assuming that the specified zener voltage (8.2 V) is measured at a current of 10 mA and that $r_z = 20\ \Omega$, represent the zener by a piecewise-linear model.

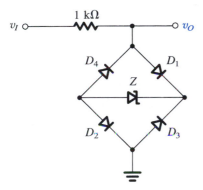

FIGURE P3.97

*3.98 Plot the transfer characteristic of the circuit in Fig. P3.98 by evaluating v_I corresponding to $v_O = 0.5$ V, 0.6 V, 0.7 V, 0.8 V, 0 V, −0.5 V, −0.6 V, −0.7 V, and −0.8 V. Assume that the diodes are 1-mA units (i.e., have 0.7-V drops at 1-mA currents) having a 0.1-V/decade logarithmic characteristic. Characterize the circuit as a hard or soft limiter. What is the value of K? Estimate L_+ and L_-.

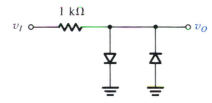

FIGURE P3.98

D3.99 Design limiter circuits using only diodes and 10-kΩ resistors to provide an output signal limited to the range:

(a) −0.7 V and above
(b) −2.1 V and above
(c) ±1.4 V

Assume that each diode has a 0.7-V drop when conducting.

D3.100 Design a two-sided limiting circuit using a resistor, two diodes, and two power supplies to feed a 1-kΩ load with nominal limiting levels of ±3 V. Use diodes modeled by a constant 0.7 V. In the nonlimiting region, the circuit voltage gain should be at least 0.95 V/V.

*3.101 Reconsider Problem 3.100 with diodes modeled by a 0.5-V offset and a resistor consistent with 10-mA conduction at 0.7 V. Sketch and quantify the output voltage for inputs of ±10 V.

*3.102 In the circuit shown in Fig. P3.102, the diodes exhibit a 0.7-V drop at 0.1 mA with a 0.1 V/decade characteristic.

For inputs over the range of ±5 V, provide a calibrated sketch of the voltages at outputs B and C. For a 5-V peak, 100-Hz sinusoid applied at A, sketch the signals at nodes B and C.

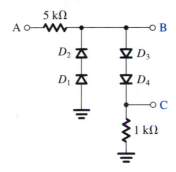

FIGURE P3.102

**3.103 Sketch and label the transfer characteristic of the circuit shown in Fig. P3.103 over a ±10-V range of input signals. All diodes are 1-mA units (i.e., each exhibits a 0.7-V drop at a current of 1 mA) with $n = 1$. What are the slopes of the characteristic at the extreme ±10-V levels?

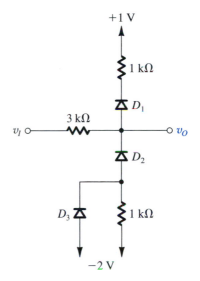

FIGURE P3.103

3.104 A clamped capacitor using an ideal diode with cathode grounded is supplied with a sine wave of 10-V rms. What is the average (dc) value of the resulting output?

**3.105 For the circuits in Fig. P3.105, each utilizing an ideal diode (or diodes), sketch the output for the input shown. Label the most positive and most negative output levels. Assume $CR \gg T$.

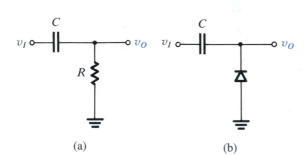

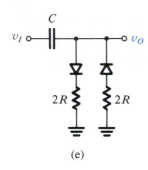

(a)

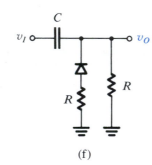

(b)

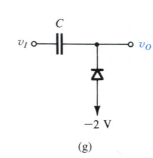

(c)

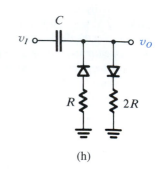

(d)

(e)

(f)

(g)

(h)

FIGURE P3.105

SECTION 3.7: PHYSICAL OPERATION OF DIODES

Note: If in the following problems the need arises for the values of particular parameters or physical constants that are not stated, please consult Table 3.1.

3.106 Find values of the intrinsic carrier concentration n_i for silicon at $-70°C$, $0°C$, $20°C$, $100°C$, and $125°C$. At each temperature, what fraction of the atoms is ionized? Recall that a silicon crystal has approximately 5×10^{22} atoms/cm^3.

3.107 A young designer, aiming to develop intuition concerning conducting paths within an integrated circuit, examines the end-to-end resistance of a connecting bar 10 μm long, 3 μm wide, and 1 μm thick, made of various materials. The designer considers:

(a) intrinsic silicon
(b) n-doped silicon with $N_D = 10^{16}$/cm^3
(c) n-doped silicon with $N_D = 10^{18}$/cm^3
(d) p-doped silicon with $N_A = 10^{10}$/cm^3
(e) aluminum with resistivity of 2.8 $\mu\Omega \cdot$ cm

Find the resistance in each case. For intrinsic silicon, use the data in Table 3.2. For doped silicon, assume $\mu_n \cong 2.5\mu_p = 1200$ cm^2/V·s. (Recall that $R = \rho L / A$.)

3.108 Holes are being steadily injected into a region of n-type silicon (connected to other devices, the details of which are not important for this question). In the steady state, the excess-hole concentration profile shown in Fig. P3.108 is established in the n-type silicon region. Here "excess" means

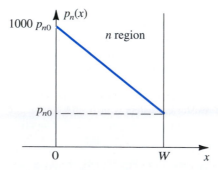

FIGURE P3.108

over and above the concentration p_{n0}. If $N_D = 10^{16}/\text{cm}^3$, $n_i = 1.5 \times 10^{10}/\text{cm}^3$, and $W = 5$ μm, find the density of the current that will flow in the x direction.

3.109 Contrast the electron and hole drift velocities through a 10-μm layer of intrinsic silicon across which a voltage of 5 V is imposed. Let $\mu_n = 1350$ cm^2/V·s and $\mu_p = 480$ cm^2/V·s.

3.110 Find the current flow in a silicon bar of 10-μm length having a 5-μm × 4-μm cross-section and having free-electron and hole densities of $10^5/\text{cm}^3$ and $10^{15}/\text{cm}^3$, respectively, with 1 V applied end-to-end. Use $\mu_n = 1200$ cm^2/V·s and $\mu_p = 500$ cm^2/V·s.

3.111 In a 10-μm long bar of donor-doped silicon, what donor concentration is needed to realize a current density of 1 mA/μm^2 in response to an applied voltage of 1 V. (*Note:* Although the carrier mobilities change with doping concentration [see the table associated with Problem 3.113], as a first approximation you may assume μ_n to be constant and use the value for intrinsic silicon, 1350 cm^2/V·s.)

3.112 In a phosphorous-doped silicon layer with impurity concentration of $10^{16}/\text{cm}^3$, find the hole and electron concentration at 25°C and 125°C.

3.113 Both the carrier mobility and diffusivity decrease as the doping concentration of silicon is increased. The following table provides a few data points for μ_n and μ_p versus doping concentration. Use the Einstein relationship to obtain the corresponding value for D_n and D_p.

Doping Concentration	μ_n cm^2/V·s	μ_p cm^2/V·s	D_n cm^2/s	D_p cm^2/s
Intrinsic	1350	480		
10^{16}	1100	400		
10^{17}	700	260		
10^{18}	360	150		

3.114 Calculate the built-in voltage of a junction in which the p and n regions are doped equally with 10^{16} atoms/cm^3. Assume $n_i \simeq 10^{10}/\text{cm}^3$. With no external voltage applied, what is the width of the depletion region, and how far does it extend into the p and n regions? If the cross-sectional area of the junction is 100 μm^2, find the magnitude of the charge stored on either side of the junction, and calculate the junction capacitance C_j.

3.115 If, for a particular junction, the acceptor concentration is $10^{16}/\text{cm}^3$ and the donor concentration is $10^{15}/\text{cm}^3$, find the junction built-in voltage. Assume $n_i \simeq 10^{10}/\text{cm}^3$. Also, find the width of the depletion region (W_{dep}) and its

extent in each of the p and n regions when the junction is reverse biased with $V_R = 5$ V. At this value of reverse bias, calculate the magnitude of the charge stored on either side of the junction. Assume the junction area is 400 μm^2. Also, calculate C_j.

3.116 Estimate the total charge stored in a 0.1-μm depletion layer on one side of a 10-μm × 10-μm junction. The doping concentration on that side of the junction is $10^{16}/\text{cm}^3$.

3.117 Combine Eqs. (3.51) and (3.52) to find q_J in terms of V_R. Differentiate this expression to find an expression for the junction capacitance C_j. Show that the expression you found is the same as the result obtained using Eq. (3.54) in conjunction with Eq. (3.52).

3.118 For a particular junction for which $C_{j0} = 0.6$ pF, $V_0 = 0.75$ V, and $m = 1/3$, find the capacitance at reverse-bias voltages of 1 V and 10 V.

3.119 An avalanche-breakdown diode, for which the breakdown voltage is 12 V, has a rated power dissipation of 0.25 W. What continuous operating current will raise the dissipation to half the maximum value? If breakdown occurs for only 10 ms in every 20 ms, what average breakdown current is allowed?

3.120 In a forward-biased pn junction show that the ratio of the current component due to hole injection across the junction to the component due to electron injection is given by

$$\frac{I_p}{I_n} = \frac{D_p}{D_n} \frac{L_n}{L_p} \frac{N_A}{N_D}$$

Evaluate this ratio for the case $N_A = 10^{18}/\text{cm}^3$, $N_D = 10^{16}/\text{cm}^3$, $L_p = 5$ μm, $L_n = 10$ μm, $D_p = 10$ cm^2/s, $D_n = 20$ cm^2/s, and hence find I_p and I_n for the case in which the diode is conducting a forward current $I = 1$ mA.

3.121 A p^+-n diode is one in which the doping concentration in the p region is much greater than that in the n region. In such a diode, the forward current is mostly due to hole injection across the junction. Show that

$$I \simeq I_p = A q n_i^2 \frac{D_p}{L_p N_D} (e^{V/V_T} - 1)$$

For the specific case in which $N_D = 5 \times 10^{16}/\text{cm}^3$, $D_p = 10$ cm^2/s, $\tau_p = 0.1$ μs, and $A = 10^4$ μm^2, find I_S and the voltage V obtained when $I = 0.2$ mA. Assume operation at 300 K where $n_i = 1.5 \times 10^{10}/\text{cm}^3$. Also, calculate the excess minority-carrier charge and the value of the diffusion capacitance at $I = 0.2$ mA.

**3.122 A short-base diode is one where the widths of the p and n regions are much smaller than L_n and L_p, respectively. As a result, the excess minority-carrier distribution in each region is a straight line rather than the exponentials shown in Fig. 3.50.

(a) For the short-base diode, sketch a figure corresponding to Fig. 3.50, and assume, as in Fig. 3.50, that $N_A \gg N_D$.

(b) Following a derivation similar to that given on page 205–206, show that if the widths of the p and n regions are denoted W_p and W_n then

$$I = A q n_i^2 \left[\frac{D_p}{(W_n - x_n)N_D} + \frac{D_n}{(W_p - x_p)N_A} \right](e^{V/V_T} - 1)$$

and

$$Q_p = \frac{1}{2} \frac{(W_n - x_n)^2}{D_p} I_p$$

$$\simeq \frac{1}{2} \frac{W_n^2}{D_p} I_p, \qquad \text{for } W_n \gg x_n$$

(c) Also, assuming $Q \simeq Q_p$, $I \simeq I_p$, show that

$$C_d = \frac{\tau_T}{V_T} I$$

where

$$\tau_T = \frac{1}{2} \frac{W_n^2}{D_p}$$

(d) If a designer wishes to limit C_d to 8 pF at $I = 1$ mA, what should W_n be? Assume $D_p = 10$ cm^2/s.

CHAPTER 4

MOS Field-Effect Transistors (MOSFETs)

INTRODUCTION

Having studied the junction diode, which is the most basic two-terminal semiconductor device, we now turn our attention to three-terminal semiconductor devices. Three-terminal devices are far more useful than two-terminal ones because they can be used in a multitude of applications, ranging from signal amplification to digital logic and memory. The basic principle involved is the use of the voltage between two terminals to control the current flowing in the third terminal. In this way a three-terminal device can be used to realize a controlled source, which as we have learned in Chapter 1 is the basis for amplifier design. Also, in the extreme, the control signal can be used to cause the current in the third terminal to change from zero to a large value, thus allowing the device to act as a switch. As we also learned in Chapter 1, the switch is the basis for the realization of the logic inverter, the basic element of digital circuits.

There are two major types of three-terminal semiconductor device: the metal-oxide-semiconductor field-effect transistor (MOSFET), which is studied in this chapter, and the bipolar junction transistor (BJT), which we shall study in Chapter 5. Although each of the two transistor types offers unique features and areas of application, the MOSFET has become by far the most widely used electronic device, especially in the design of integrated circuits (ICs), which are circuits fabricated on a single silicon chip.

Compared to BJTs, MOSFETs can be made quite small (i.e., requiring a small area on the silicon IC chip), and their manufacturing process is relatively simple (see Appendix A). Also, their operation requires comparatively little power. Furthermore, circuit designers have found ingenious ways to implement digital and analog functions utilizing MOSFETs almost exclusively (i.e., with very few or no resistors). All of these properties have made it possible to pack large numbers of MOSFETs (>200 million!) on a single IC chip to implement very sophisticated, very-large-scale-integrated (VLSI) circuits such as those for memory and micro-processors. Analog circuits such as amplifiers and filters are also implemented in MOS technology, albeit in smaller less-dense chips. Also, both analog and digital functions are increasingly being implemented on the same IC chip, in what is known as mixed-signal design.

The objective of this chapter is to develop in the reader a high degree of familiarity with the MOSFET: its physical structure and operation, terminal characteristics, circuit models, and basic circuit applications, both as an amplifier and a digital logic inverter. Although dis-crete MOS transistors exist, and the material studied in this chapter will enable the reader to design discrete MOS circuits, our study of the MOSFET is strongly influenced by the fact that most of its applications are in integrated-circuit design. The design of IC analog and digital MOS circuits occupies a large proportion of the remainder of this book.

4.1 DEVICE STRUCTURE AND PHYSICAL OPERATION

The enhancement-type MOSFET is the most widely used field-effect transistor. In this sec-tion, we shall study its structure and physical operation. This will lead to the current-voltage characteristics of the device, studied in the next section.

4.1.1 Device Structure

Figure 4.1, shows the physical structure of the n-channel enhancement-type MOSFET. The meaning of the names "enhancement" and "n-channel" will become apparent shortly. The transistor is fabricated on a p-type substrate, which is a single-crystal silicon wafer that pro-vides physical support for the device (and for the entire circuit in the case of an integrated circuit). Two heavily doped n-type regions, indicated in the figure as the n^+ **source**[1] and the n^+ **drain** regions, are created in the substrate. A thin layer of silicon dioxide (SiO_2) of thick-ness t_{ox} (typically 2–50 nm),[2] which is an excellent electrical insulator, is grown on the sur-face of the substrate, covering the area between the source and drain regions. Metal is deposited on top of the oxide layer to form the **gate electrode** of the device. Metal contacts are also made to the source region, the drain region, and the substrate, also known as the

---

[1] The notation n^+ indicates heavily doped n-type silicon. Conversely, n^- is used to denote lightly doped n-type silicon. Similar notation applies for p-type silicon.

[2] A nanometer (nm) is 10^{-9} m or 0.001 μm. A micrometer (μm), or micron, is 10^{-6} m. Sometimes the oxide thickness is expressed in angstroms. An angstrom (Å) is 10^{-1} nm, or 10^{-10} m.

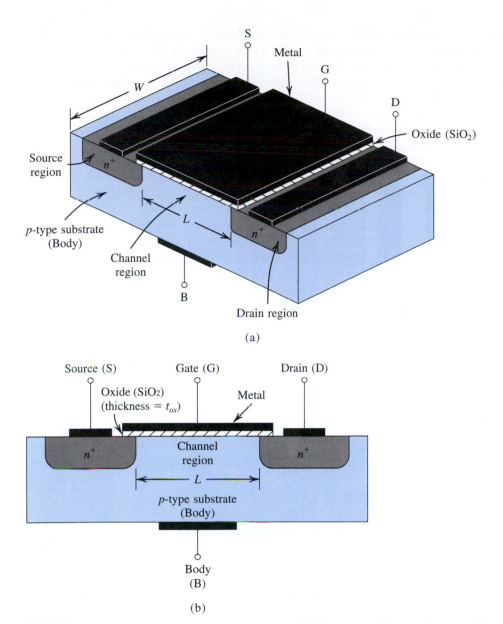

FIGURE 4.1 Physical structure of the enhancement-type NMOS transistor: **(a)** perspective view; **(b)** cross-section. Typically $L = 0.1$ to 3 μm, $W = 0.2$ to 100 μm, and the thickness of the oxide layer (t_{ox}) is in the range of 2 to 50 nm.

body.[3] Thus four terminals are brought out: the gate terminal (G), the source terminal (S), the drain terminal (D), and the substrate or body terminal (B).

At this point it should be clear that the name of the device (metal-oxide-semiconductor FET) is derived from its physical structure. The name, however, has become a general one and is

---

[3] In Fig. 4.1, the contact to the body is shown on the bottom of the device. This will prove helpful later in explaining a phenomenon known as the "body effect." It is important to note, however, that in actual ICs, contact to the body is made at a location on the top of the device.

used also for FETs that do not use metal for the gate electrode. In fact, most modern MOSFETs are fabricated using a process known as silicon-gate technology, in which a certain type of silicon, called polysilicon, is used to form the gate electrode (see Appendix A). Our description of MOSFET operation and characteristics applies irrespective of the type of gate electrode.

Another name for the MOSFET is the **insulated-gate FET** or **IGFET.** This name also arises from the physical structure of the device, emphasizing the fact that the gate electrode is electrically insulated from the device body (by the oxide layer). It is this insulation that causes the current in the gate terminal to be extremely small (of the order of 10^{-15} A).

Observe that the substrate forms *pn* junctions with the source and drain regions. In normal operation these *pn* junctions are kept reverse-biased at all times. Since the drain will be at a positive voltage relative to the source, the two *pn* junctions can be effectively cut off by simply connecting the substrate terminal to the source terminal. We shall assume this to be the case in the following description of MOSFET operation. Thus, here, the substrate will be considered as having no effect on device operation, and the MOSFET will be treated as a three-terminal device, with the terminals being the gate (G), the source (S), and the drain (D). It will be shown that a voltage applied to the gate controls current flow between source and drain. This current will flow in the longitudinal direction from drain to source in the region labeled "channel region." Note that this region has a length L and a width W, two important parameters of the MOSFET. Typically, L is in the range of 0.1 μm to 3 μm, and W is in the range of 0.2 μm to 100 μm. Finally, note that the MOSFET is a symmetrical device; thus its source and drain can be interchanged with no change in device characteristics.

4.1.2 Operation with No Gate Voltage

With no bias voltage applied to the gate, two back-to-back diodes exist in series between drain and source. One diode is formed by the *pn* junction between the n^+ drain region and the *p*-type substrate, and the other diode is formed by the *pn* junction between the *p*-type substrate and the n^+ source region. These back-to-back diodes prevent current conduction from drain to source when a voltage v_{DS} is applied. In fact, the path between drain and source has a very high resistance (of the order of 10^{12} Ω).

4.1.3 Creating a Channel for Current Flow

Consider next the situation depicted in Fig. 4.2. Here we have grounded the source and the drain and applied a positive voltage to the gate. Since the source is grounded, the gate voltage appears in effect between gate and source and thus is denoted v_{GS}. The positive voltage on the gate causes, in the first instance, the free holes (which are positively charged) to be repelled from the region of the substrate under the gate (the channel region). These holes are pushed downward into the substrate, leaving behind a carrier-depletion region. The depletion region is populated by the bound negative charge associated with the acceptor atoms. These charges are "uncovered" because the neutralizing holes have been pushed downward into the substrate.

As well, the positive gate voltage attracts electrons from the n^+ source and drain regions (where they are in abundance) into the channel region. When a sufficient number of electrons accumulate near the surface of the substrate under the gate, an *n* region is in effect created, connecting the source and drain regions, as indicated in Fig. 4.2. Now if a voltage is applied between drain and source, current flows through this induced *n* region, carried by the mobile electrons. The *induced n* region thus forms a **channel** for current flow from drain to source and *is aptly called so.* Correspondingly, the MOSFET of Fig. 4.2 is called an **n-channel MOSFET** or, alternatively, an **NMOS transistor.** Note that an *n*-channel MOSFET is formed in a *p*-type substrate: The channel is created by *inverting* the substrate surface from *p* type to *n* type. Hence the induced channel is also called an **inversion layer.**

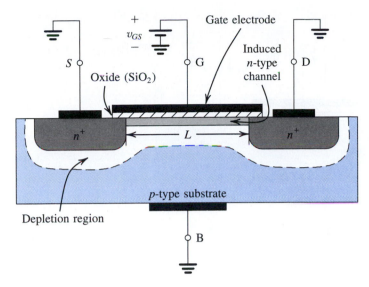

FIGURE 4.2 The enhancement-type NMOS transistor with a positive voltage applied to the gate. An *n* channel is induced at the top of the substrate beneath the gate.

The value of v_{GS} at which a sufficient number of mobile electrons accumulate in the channel region to form a conducting channel is called the **threshold voltage** and is denoted V_t.[4] Obviously, V_t for an *n*-channel FET is positive. The value of V_t is controlled during device fabrication and typically lies in the range of 0.5 V to 1.0 V.

The gate and the channel region of the MOSFET form a parallel-plate capacitor, with the oxide layer acting as the capacitor dielectric. The positive gate voltage causes positive charge to accumulate on the top plate of the capacitor (the gate electrode). The corresponding negative charge on the bottom plate is formed by the electrons in the induced channel. An electric field thus develops in the vertical direction. It is this field that controls the amount of charge in the channel, and thus it determines the channel conductivity and, in turn, the current that will flow through the channel when a voltage v_{DS} is applied.

4.1.4 Applying a Small v_{DS}

Having induced a channel, we now apply a positive voltage v_{DS} between drain and source, as shown in Fig. 4.3. We first consider the case where v_{DS} is small (i.e., 50 mV or so). The voltage v_{DS} causes a current i_D to flow through the induced *n* channel. Current is carried by free electrons traveling from source to drain (hence the names source and drain). By convention, the direction of current flow is opposite to that of the flow of negative charge. Thus the current in the channel, i_D, will be from drain to source, as indicated in Fig. 4.3. The magnitude of i_D depends on the density of electrons in the channel, which in turn depends on the magnitude of v_{GS}. Specifically, for $v_{GS} = V_t$ the channel is just induced and the current conducted is still negligibly small. As v_{GS} exceeds V_t, more electrons are attracted into the channel. We may visualize the increase in charge carriers in the channel as an increase in the channel depth. The result is a channel of increased conductance or, equivalently, reduced resistance. In fact, the conductance of the channel is proportional to the **excess gate voltage** ($v_{GS} - V_t$), also

---

[4] Some texts use V_T to denote the threshold voltage. We use V_t to avoid confusion with the thermal voltage V_T.

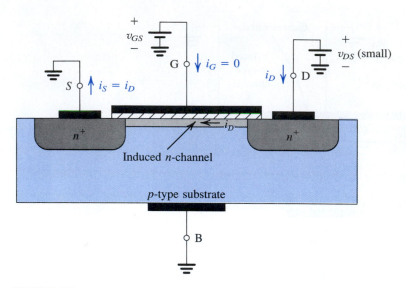

FIGURE 4.3 An NMOS transistor with $v_{GS} > V_t$ and with a small v_{DS} applied. The device acts as a resistance whose value is determined by v_{GS}. Specifically, the channel conductance is proportional to $v_{GS} - V_t$, and thus i_D is proportional to $(v_{GS} - V_t)v_{DS}$. Note that the depletion region is not shown (for simplicity).

known as the **effective voltage** or the **overdrive voltage.** It follows that the current i_D will be proportional to $v_{GS} - V_t$ and, of course, to the voltage v_{DS} that causes i_D to flow.

Figure 4.4 shows a sketch of i_D versus v_{DS} for various values of v_{GS}. We observe that the MOSFET is operating as a linear resistance whose value is controlled by v_{GS}. The resistance is infinite for $v_{GS} \leq V_t$, and its value decreases as v_{GS} exceeds V_t.

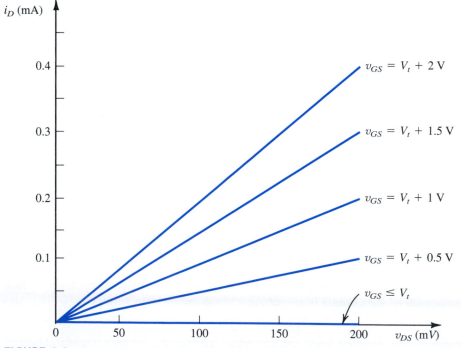

FIGURE 4.4 The i_D–v_{DS} characteristics of the MOSFET in Fig. 4.3 when the voltage applied between drain and source, v_{DS}, is kept small. The device operates as a linear resistor whose value is controlled by v_{GS}.

The description above indicates that for the MOSFET to conduct, a channel has to be induced. Then, increasing v_{GS} above the threshold voltage V_t enhances the channel, hence the names **enhancement-mode operation** and **enhancement-type MOSFET.** Finally, we note that the current that leaves the source terminal (i_S) is equal to the current that enters the drain terminal (i_D), and the gate current $i_G = 0$.

4.1.5 Operation as v_{DS} Is Increased

We next consider the situation as v_{DS} is increased. For this purpose let v_{GS} be held constant at a value greater than V_t. Refer to Fig. 4.5, and note that v_{DS} appears as a voltage drop across the length of the channel. That is, as we travel along the channel from source to drain, the voltage (measured relative to the source) increases from 0 to v_{DS}. Thus the voltage between the gate and points along the channel decreases from v_{GS} at the source end to $v_{GS} - v_{DS}$ at the drain end. Since the channel depth depends on this voltage, we find that the channel is no longer of uniform depth; rather, the channel will take the tapered form shown in Fig. 4.5, being deepest at the source end and shallowest at the drain end. As v_{DS} is increased, the channel becomes more tapered and its resistance increases correspondingly. Thus the i_D–v_{DS} curve does not continue as a straight line but bends as shown in Fig. 4.6. Eventually, when v_{DS} is increased to the value that reduces the voltage between gate and

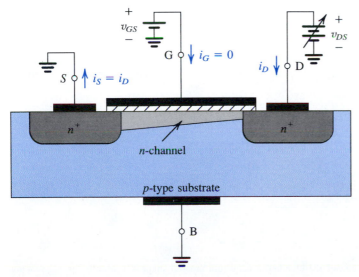

FIGURE 4.5 Operation of the enhancement NMOS transistor as v_{DS} is increased. The induced channel acquires a tapered shape, and its resistance increases as v_{DS} is increased. Here, v_{GS} is kept constant at a value $> V_t$.

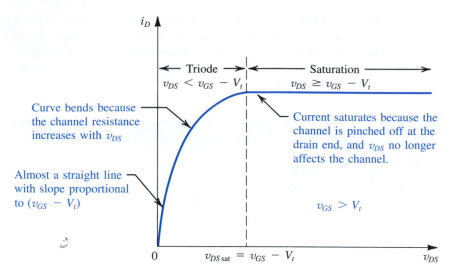

FIGURE 4.6 The drain current i_D versus the drain-to-source voltage v_{DS} for an enhancement-type NMOS transistor operated with $v_{GS} > V_t$.

channel at the drain end to V_t—that is, $v_{GD} = V_t$ or $v_{GS} - v_{DS} = V_t$ or $v_{DS} = v_{GS} - V_t$—the channel depth at the drain end decreases to almost zero, and the channel is said to be **pinched off.** Increasing v_{DS} beyond this value has little effect (theoretically, no effect) on the channel shape, and the current through the channel remains constant at the value reached for $v_{DS} = v_{GS} - V_t$. The drain current thus **saturates** at this value, and the MOSFET is said to have entered the **saturation region** of operation. The voltage v_{DS} at which saturation occurs is denoted v_{DSsat},

$$v_{DSsat} = v_{GS} - V_t \qquad (4.1)$$

Obviously, for every value of $v_{GS} \geq V_t$, there is a corresponding value of v_{DSsat}. The device operates in the saturation region if $v_{DS} \geq v_{DSsat}$. The region of the i_D–v_{DS} characteristic obtained for $v_{DS} < v_{DSsat}$ is called the **triode region,** a carryover from the days of vacuum-tube devices whose operation a FET resembles.

To help further in visualizing the effect of v_{DS}, we show in Fig. 4.7 sketches of the channel as v_{DS} is increased while v_{GS} is kept constant. Theoretically, any increase in v_{DS} above

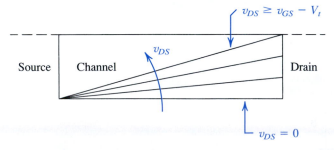

FIGURE 4.7 Increasing v_{DS} causes the channel to acquire a tapered shape. Eventually, as v_{DS} reaches $v_{GS} - V_t$, the channel is pinched off at the drain end. Increasing v_{DS} above $v_{GS} - V_t$ has little effect (theoretically, no effect) on the channel's shape.

v_{DSsat} (which is equal to $v_{GS} - V_t$) has no effect on the channel shape and simply appears across the depletion region surrounding the channel and the n^+ drain region.

4.1.6 Derivation of the i_D–v_{DS} Relationship

The description of physical operation presented above can be used to derive an expression for the i_D–v_{DS} relationship depicted in Fig. 4.6. Toward that end, assume that a voltage v_{GS} is applied between gate and source with $v_{GS} > V_t$ to induce a channel. Also, assume that a voltage v_{DS} is applied between drain and source. First, we shall consider operation in the triode region, for which the channel must be continuous and thus v_{GD} must be greater than V_t or, equivalently, $v_{DS} < v_{GS} - V_t$. In this case the channel will have the tapered shape shown in Fig. 4.8.

The reader will recall that in the MOSFET, the gate and the channel region form a parallel-plate capacitor for which the oxide layer serves as a dielectric. If the capacitance per unit gate area is denoted C_{ox} and the thickness of the oxide layer is t_{ox}, then

$$C_{ox} = \frac{\varepsilon_{ox}}{t_{ox}} \tag{4.2}$$

where ε_{ox} is the permittivity of the silicon oxide,

$$\varepsilon_{ox} = 3.9\varepsilon_0 = 3.9 \times 8.854 \times 10^{-12} = 3.45 \times 10^{-11} \text{ F/m}$$

The oxide thickness t_{ox} is determined by the process technology used to fabricate the MOSFET. As an example, for $t_{ox} = 10$ nm, $C_{ox} = 3.45 \times 10^{-3}$ F/m^2, or 3.45 fF/μm^2 as it is usually expressed.

Now refer to Fig. 4.8 and consider the infinitesimal strip of the gate at distance x from the source. The capacitance of this strip is $C_{ox}W\,dx$. To find the charge stored on this infinitesimal strip of the gate capacitance, we multiply the capacitance by the *effective voltage*

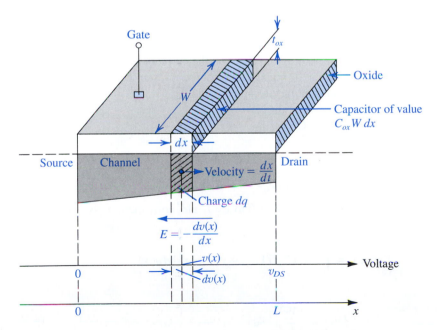

FIGURE 4.8 Derivation of the i_D–v_{DS} characteristic of the NMOS transistor.

between the gate and the channel at point x, where the effective voltage is the voltage that is responsible for inducing the channel at point x and is thus $[v_{GS} - v(x) - V_t]$ where $v(x)$ is the voltage in the channel at point x. It follows that the electron charge dq in the infinitesimal portion of the channel at point x is

$$dq = -C_{ox}(W\,dx)[v_{GS} - v(x) - V_t] \tag{4.3}$$

where the leading negative sign accounts for the fact that dq is a negative charge.

The voltage v_{DS} produces an electric field along the channel in the negative x direction. At point x this field can be expressed as

$$E(x) = -\frac{dv(x)}{dx}$$

The electric field $E(x)$ causes the electron charge dq to *drift* toward the drain with a velocity dx/dt,

$$\frac{dx}{dt} = -\mu_n E(x) = \mu_n \frac{dv(x)}{dx} \tag{4.4}$$

where μ_n is the mobility of electrons in the channel (called surface mobility). It is a physical parameter whose value depends on the fabrication process technology. The resulting drift current i can be obtained as follows:

$$i = \frac{dq}{dt}$$

$$= \frac{dq}{dx}\frac{dx}{dt}$$

Substituting for the charge-per-unit-length dq/dx from Eq. (4.3), and for the electron drift velocity dx/dt from Eq. (4.4), results in

$$i = -\mu_n C_{ox} W[v_{GS} - v(x) - V_t]\frac{dv(x)}{dx}$$

Although evaluated at a particular point in the channel, the current i must be constant at all points along the channel. Thus i must be equal to the source-to-drain current. Since we are interested in the drain-to-source current i_D, we can find it as

$$i_D = -i = \mu_n C_{ox} W[v_{GS} - v(x) - V_t]\frac{dv(x)}{dx}$$

which can be rearranged in the form

$$i_D\,dx = \mu_n C_{ox} W[v_{GS} - V_t - v(x)]\,dv(x)$$

Integrating both sides of this equation from $x = 0$ to $x = L$ and, correspondingly, for $v(0) = 0$ to $v(L) = v_{DS}$,

$$\int_0^L i_D\,dx = \int_0^{v_{DS}} \mu_n C_{ox} W[v_{GS} - V_t - v(x)]\,dv(x)$$

gives

$$i_D = (\mu_n C_{ox})\left(\frac{W}{L}\right)\left[(v_{GS} - V_t)v_{DS} - \frac{1}{2}v_{DS}^2\right] \tag{4.5}$$

This is the expression for the i_D–v_{DS} characteristic in the triode region. The value of the current at the edge of the triode region or, equivalently, at the beginning of the saturation region can be obtained by substituting $v_{DS} = v_{GS} - V_t$, resulting in

$$i_D = \frac{1}{2}(\mu_n C_{ox})\left(\frac{W}{L}\right)(v_{GS} - V_t)^2 \tag{4.6}$$

This is the expression for the i_D–v_{DS} characteristic in the saturation region; it simply gives the saturation value of i_D corresponding to the given v_{GS}. (Recall that in saturation i_D remains constant for a given v_{GS} as v_{DS} is varied.)

In the expressions in Eqs. (4.5) and (4.6), $\mu_n C_{ox}$ is a constant determined by the process technology used to fabricate the n-channel MOSFET. It is known as the **process transconductance parameter,** for as we shall see shortly, it determines the value of the MOSFET transconductance, is denoted k'_n, and has the dimensions of A/V^2:

$$k'_n = \mu_n C_{ox} \tag{4.7}$$

Of course, the i_D–v_{DS} expressions in Eqs. (4.5) and (4.6) can be written in terms of k'_n as follows:

$$i_D = k'_n \frac{W}{L}\left[(v_{GS} - V_t)v_{DS} - \frac{1}{2}v_{DS}^2\right] \quad \text{(Triode region)} \tag{4.5a}$$

$$i_D = \frac{1}{2}k'_n \frac{W}{L}(v_{GS} - V_t)^2 \quad \text{(Saturation region)} \tag{4.6a}$$

In this book we will use the forms with $(\mu_n C_{ox})$ and with k'_n interchangeably.

From Eqs. (4.5) and (4.6) we see that the drain current is proportional to the ratio of the channel width W to the channel length L, known as the **aspect ratio** of the MOSFET. The values of W and L can be selected by the circuit designer to obtain the desired i–v characteristics. For a given fabrication process, however, there is a minimum channel length, L_{min}. In fact, the minimum channel length that is possible with a given fabrication process is used to characterize the process and is being continually reduced as technology advances. For instance, at the time of this writing (2003) the state-of-the-art in MOS technology is a 0.13-μm process, meaning that for this process the minimum channel length possible is 0.13 μm. There also is a minimum value for the channel width W. For instance, for the 0.13-μm process just mentioned, W_{min} is 0.16 μm. Finally, we should note that the oxide thickness t_{ox} scales down with L_{min}. Thus, for a 1.5-μm technology, t_{ox} is 25 nm, but the modern 0.13-μm technology mentioned above has $t_{ox} = 2$ nm.

EXAMPLE 4.1

Consider a process technology for which $L_{min} = 0.4\ \mu$m, $t_{ox} = 8$ nm, $\mu_n = 450$ cm^2/V·s, and $V_t = 0.7$ V.

(a) Find C_{ox} and k'_n.

(b) For a MOSFET with $W/L = 8\ \mu$m/0.8 μm, calculate the values of V_{GS} and V_{DSmin} needed to operate the transistor in the saturation region with a dc current $I_D = 100\ \mu$A.

(c) For the device in (b), find the value of V_{GS} required to cause the device to operate as a 1000-Ω resistor for very small v_{DS}.

Solution

(a)
$$C_{ox} = \frac{\varepsilon_{ox}}{t_{ox}} = \frac{3.45 \times 10^{-11}}{8 \times 10^{-9}} = 4.32 \times 10^{-3} \, \text{F/m}^2$$

$$= 4.32 \, \text{fF/} \mu \text{m}^2$$

$$k'_n = \mu_n C_{ox} = 450 \, (\text{cm}^2/\text{V}\cdot\text{s}) \times 4.32 \, (\text{fF/} \mu \text{m}^2)$$

$$= 450 \times 10^8 \, (\mu \text{m}^2/\text{V}\cdot\text{s}) \times 4.32 \times 10^{-15} \, (\text{F/} \mu \text{m}^2)$$

$$= 194 \times 10^{-6} \, (\text{F/V}\cdot\text{s})$$

$$= 194 \, \mu \text{A/V}^2$$

(b) For operation in the saturation region,

$$i_D = \frac{1}{2} k'_n \frac{W}{L} (v_{GS} - V_t)^2$$

Thus,

$$100 = \frac{1}{2} \times 194 \times \frac{8}{0.8} (V_{GS} - 0.7)^2$$

which results in

$$V_{GS} - 0.7 = 0.32 \, \text{V}$$

or

$$V_{GS} = 1.02 \, \text{V}$$

and

$$V_{DS\text{min}} = V_{GS} - V_t = 0.32 \, \text{V}$$

(c) For the MOSFET in the triode region with v_{DS} very small,

$$i_D \cong k'_n \frac{W}{L} (v_{GS} - V_t) v_{DS}$$

from which the drain-to-source resistance r_{DS} can be found as

$$r_{DS} \equiv \frac{v_{DS}}{i_D} \bigg|_{\text{small } v_{DS}}$$

$$= 1 \bigg/ \left[k'_n \frac{W}{L} (V_{GS} - V_t) \right]$$

Thus

$$1000 = \frac{1}{194 \times 10^{-6} \times 10 (V_{GS} - 0.7)}$$

which yields

$$V_{GS} - 0.7 = 0.52 \, \text{V}$$

Thus,

$$V_{GS} = 1.22 \, \text{V}$$

EXERCISES

4.2 For a 0.8-μm process technology for which $t_{ox} = 15$ nm and $\mu_n = 550$ cm^2/V·s, find C_{ox}, k'_n, and the overdrive voltage $V_{OV} \equiv V_{GS} - V_t$ required to operate a transistor having $W/L = 20$ in saturation with $I_D = 0.2$ mA. What is the minimum value of V_{DS} needed?

Ans. 2.3 fF/μm^2; 127 μA/V^2; 0.40 V; 0.40 V

4.3 Use the expression for operation in the triode region to show that an n-channel MOSFET operated with an overdrive voltage $V_{OV} \equiv V_{GS} - V_t$ and having a small V_{DS} across it behaves approximately as a linear resistance r_{DS},

$$r_{DS} = 1 \Big/ \left[k'_n \frac{W}{L} V_{OV} \right]$$

Calculate the value of r_{DS} obtained for a device having $k'_n = 100$ μA/V^2 and $W/L = 10$ when operated with an overdrive voltage of 0.5 V.

Ans. 2 kΩ

4.1.7 The *p*-Channel MOSFET

A *p*-channel enhancement-type MOSFET (PMOS transistor), fabricated on an *n*-type substrate with p^+ regions for the drain and source, has holes as charge carriers. The device operates in the same manner as the *n*-channel device except that v_{GS} and v_{DS} are negative and the threshold voltage V_t is negative. Also, the current i_D enters the source terminal and leaves through the drain terminal.

PMOS technology originally dominated MOS manufacturing. However, because NMOS devices can be made smaller and thus operate faster, and because NMOS historically required lower supply voltages than PMOS, NMOS technology has virtually replaced PMOS. Nevertheless, it is important to be familiar with the PMOS transistor for two reasons: PMOS devices are still available for discrete-circuit design, and more importantly, both PMOS and NMOS transistors are utilized in **complementary MOS** or **CMOS** circuits, which is currently the dominant MOS technology.

4.1.8 Complementary MOS or CMOS

As the name implies, complementary MOS technology employs MOS transistors of both polarities. Although CMOS circuits are somewhat more difficult to fabricate than NMOS, the availability of complementary devices makes possible many powerful circuit-design possibilities. Indeed, at the present time CMOS is the most widely used of all the IC technologies. This statement applies to both analog and digital circuits. CMOS technology has virtually replaced designs based on NMOS transistors alone. Furthermore, at the time of this writing (2003), CMOS technology has taken over many applications that just a few years ago were possible only with bipolar devices. Throughout this book, we will study many CMOS circuit techniques.

Figure 4.9 shows a cross-section of a CMOS chip illustrating how the PMOS and NMOS transistors are fabricated. Observe that while the NMOS transistor is implemented directly in the *p*-type substrate, the PMOS transistor is fabricated in a specially created *n* region, known as an ***n* well.** The two devices are isolated from each other by a thick region of oxide that functions as an insulator. Not shown on the diagram are the connections made to the *p*-type body and to the *n* well. The latter connection serves as the body terminal for the PMOS transistor.

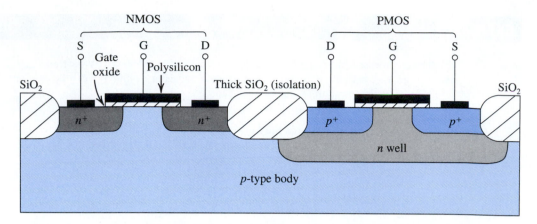

FIGURE 4.9 Cross-section of a CMOS integrated circuit. Note that the PMOS transistor is formed in a separate n-type region, known as an n well. Another arrangement is also possible in which an n-type body is used and the n device is formed in a p well. Not shown are the connections made to the p-type body and to the n well; the latter functions as the body terminal for the p-channel device.

4.1.9 Operating the MOS Transistor in the Subthreshold Region

The above description of the n-channel MOSFET operation implies that for $v_{GS} < V_t$, no current flows and the device is cut off. This is not entirely true, for it has been found that for values of v_{GS} smaller than but close to V_t, a small drain current flows. In this **subthreshold region** of operation the drain current is exponentially related to v_{GS}, much like the i_C–v_{BE} relationship of a BJT, as will be shown in the next chapter.

Although in most applications the MOS transistor is operated with $v_{GS} > V_t$, there are special, but a growing number of, applications that make use of subthreshold operation. In this book, we will not consider subthreshold operation any further and refer the reader to the references listed in Appendix F.

 ## 4.2 CURRENT–VOLTAGE CHARACTERISTICS

Building on the physical foundation established in the previous section for the operation of the enhancement MOS transistor, we present in this section its complete current-voltage characteristics. These characteristics can be measured at dc or at low frequencies and thus are called static characteristics. The dynamic effects that limit the operation of the MOSFET at high frequencies and high switching speeds will be discussed in Section 4.8.

4.2.1 Circuit Symbol

Figure 4.10(a) shows the circuit symbol for the n-channel enhancement-type MOSFET. Observe that the spacing between the two vertical lines that represent the gate and the channel indicates the fact that the gate electrode is insulated from the body of the device. The polarity of the p-type substrate (body) and the n channel is indicated by the arrowhead on the line representing the body (B). This arrowhead also indicates the polarity of the transistor, namely, that it is an n-channel device.

Although the MOSFET is a symmetrical device, it is often useful in circuit design to designate one terminal as the source and the other as the drain (without having to write S and D beside the terminals). This objective is achieved in the modified circuit symbol shown in Fig. 4.10(b). Here an arrowhead is placed on the source terminal, thus distinguishing it from

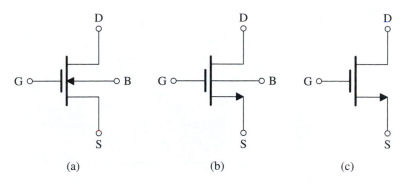

FIGURE 4.10 **(a)** Circuit symbol for the *n*-channel enhancement-type MOSFET. **(b)** Modified circuit symbol with an arrowhead on the source terminal to distinguish it from the drain and to indicate device polarity (i.e., *n* channel). **(c)** Simplified circuit symbol to be used when the source is connected to the body or when the effect of the body on device operation is unimportant.

the drain terminal. The arrowhead points in the normal direction of current flow and thus indicates the polarity of the device (i.e., *n* channel). Observe that in the modified symbol, there is no need to show the arrowhead on the body line. Although the circuit symbol of Fig. 4.10(b) clearly distinguishes the source from the drain, in practice it is the polarity of the voltage impressed across the device that determines source and drain; *the drain is always positive relative to the source in an n-channel FET*.

In applications where the source is connected to the body of the device, a further simplification of the circuit symbol is possible, as indicated in Fig. 4.10(c). This symbol is also used in applications when the effect of the body on circuit operation is not important, as will be seen later.

4.2.2 The i_D–v_{DS} Characteristics

Figure 4.11(a) shows an *n*-channel enhancement-type MOSFET with voltages v_{GS} and v_{DS} applied and with the normal directions of current flow indicated. This conceptual circuit can be used to measure the i_D–v_{DS} characteristics, which are a family of curves, each measured at a constant v_{GS}. From the study of physical operation in the previous section, we expect each of the i_D–v_{DS} curves to have the shape shown in Fig. 4.6. This indeed is the case, as is evident from Fig. 4.11(b), which shows a typical set of i_D–v_{DS} characteristics. A thorough understanding of the MOSFET terminal characteristics is essential for the reader who intends to design MOS circuits.

The characteristic curves in Fig. 4.11(b) indicate that there are three distinct regions of operation: the **cutoff region,** the **triode region,** and the **saturation region.** The saturation region is used if the FET is to operate as an amplifier. For operation as a switch, the cutoff and triode regions are utilized. The device is cut off when $v_{GS} < V_t$. To operate the MOSFET in the triode region we must first induce a channel,

$$v_{GS} \geq V_t \quad \text{(Induced channel)} \tag{4.8}$$

and then keep v_{DS} small enough so that the channel remains continuous. This is achieved by ensuring that the gate-to-drain voltage is

$$v_{GD} > V_t \quad \text{(Continuous channel)} \tag{4.9}$$

This condition can be stated explicitly in terms of v_{DS} by writing $v_{GD} = v_{GS} + v_{SD} = v_{GS} - v_{DS}$; thus,

$$v_{GS} - v_{DS} > V_t$$

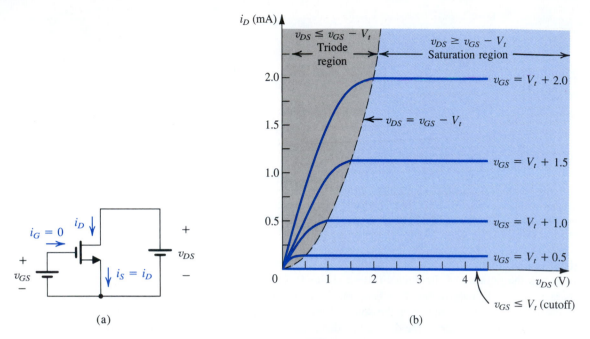

FIGURE 4.11 (a) An *n*-channel enhancement-type MOSFET with v_{GS} and v_{DS} applied and with the normal directions of current flow indicated. (b) The i_D–v_{DS} characteristics for a device with $k'_n (W/L) = 1.0$ mA/V^2.

which can be rearranged to yield

$$v_{DS} < v_{GS} - V_t \quad \text{(Continuous channel)} \tag{4.10}$$

Either Eq. (4.9) or Eq. (4.10) can be used to ascertain triode-region operation. In words, *the n-channel enhancement-type MOSFET operates in the triode region when v_{GS} is greater than V_t and the drain voltage is lower than the gate voltage by at least V_t volts.*

In the triode region, the i_D–v_{DS} characteristics can be described by the relationship of Eq. (4.5), which we repeat here,

$$i_D = k'_n \frac{W}{L} \left[(v_{GS} - V_t) v_{DS} - \frac{1}{2} v_{DS}^2 \right] \tag{4.11}$$

where $k'_n = \mu_n C_{ox}$ is the process transconductance parameter; its value is determined by the fabrication technology. If v_{DS} is sufficiently small so that we can neglect the v_{DS}^2 term in Eq. (4.11), we obtain for the i_D–v_{DS} characteristics near the origin the relationship

$$i_D \simeq k'_n \frac{W}{L} (v_{GS} - V_t) v_{DS} \tag{4.12}$$

This linear relationship represents the operation of the MOS transistor as a linear resistance r_{DS} whose value is controlled by v_{GS}. Specifically, for v_{GS} set to a value V_{GS}, r_{DS} is given by

$$r_{DS} \equiv \frac{v_{DS}}{i_D} \bigg|_{\substack{v_{DS} \text{ small} \\ v_{GS} = V_{GS}}} = \left[k'_n \frac{W}{L} (V_{GS} - V_t) \right]^{-1} \tag{4.13}$$

We discussed this region of operation in the previous section (refer to Fig. 4.4). It is also useful to express r_{DS} in terms of the **gate-to-source overdrive voltage,**

$$V_{OV} \equiv V_{GS} - V_t \tag{4.14}$$

as

$$r_{DS} = 1 \bigg/ \left[k_n' \left(\frac{W}{L} \right) V_{OV} \right] \tag{4.15}$$

Finally, we urge the reader to show that the approximation involved in writing Eq. (4.12) is based on the assumption that $v_{DS} \ll 2V_{OV}$.

To operate the MOSFET in the saturation region, a channel must be induced,

$$v_{GS} \geq V_t \quad \text{(Induced channel)} \tag{4.16}$$

and pinched off at the drain end by raising v_{DS} to a value that results in the gate-to-drain voltage falling below V_t,

$$v_{GD} \leq V_t \quad \text{(Pinched-off channel)} \tag{4.17}$$

This condition can be expressed explicitly in terms of v_{DS} as

$$v_{DS} \geq v_{GS} - V_t \quad \text{(Pinched-off channel)} \tag{4.18}$$

In words, *the n-channel enhancement-type MOSFET operates in the saturation region when v_{GS} is greater than V_t and the drain voltage does not fall below the gate voltage by more than V_t volts.*

The boundary between the triode region and the saturation region is characterized by

$$v_{DS} = v_{GS} - V_t \quad \text{(Boundary)} \tag{4.19}$$

Substituting this value of v_{DS} into Eq. (4.11) gives the saturation value of the current i_D as

$$i_D = \frac{1}{2} k_n' \frac{W}{L} (v_{GS} - V_t)^2 \tag{4.20}$$

Thus in saturation the MOSFET provides a drain current whose value is independent of the drain voltage v_{DS} and is determined by the gate voltage v_{GS} according to the square-law relationship in Eq. (4.20), a sketch of which is shown in Fig. 4.12. Since the drain current is independent of the drain voltage, the saturated MOSFET behaves as an ideal current source whose value is controlled by v_{GS} according to the nonlinear relationship in Eq. (4.20). Figure 4.13 shows a circuit representation of this view of MOSFET operation in the saturation region. Note that this is a **large-signal equivalent-circuit model.**

Referring back to the i_D–v_{DS} characteristics in Fig. 4.11(b), we note that the boundary between the triode and the saturation regions is shown as a broken-line curve. Since this curve is characterized by $v_{DS} = v_{GS} - V_t$, its equation can be found by substituting for $v_{GS} - V_t$ by v_{DS} in either the triode-region equation (Eq. 4.11) or the saturation-region equation (Eq. 4.20). The result is

$$i_D = \frac{1}{2} k_n' \frac{W}{L} v_{DS}^2 \tag{4.21}$$

It should be noted that the characteristics depicted in Figs. 4.4, 4.11, and 4.12 are for a MOSFET with $k_n'(W/L) = 1.0$ mA/V^2 and $V_t = 1$ V.

Finally, the chart in Fig. 4.14 shows the relative levels of the terminal voltages of the enhancement-type NMOS transistor for operation in the triode region and in the saturation region.

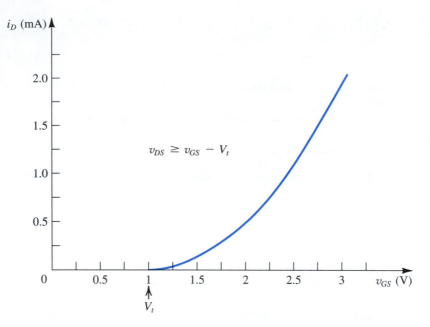

FIGURE 4.12 The i_D–v_{GS} characteristic for an enhancement-type NMOS transistor in saturation ($V_t = 1$ V, $k_n' W/L = 1.0$ mA/V^2).

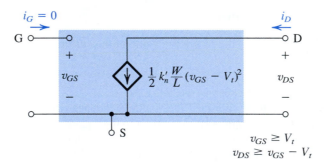

FIGURE 4.13 Large-signal equivalent-circuit model of an n-channel MOSFET operating in the saturation region.

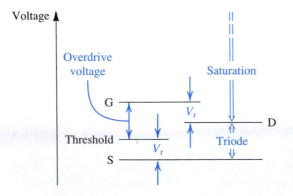

FIGURE 4.14 The relative levels of the terminal voltages of the enhancement NMOS transistor for operation in the triode region and in the saturation region.

EXERCISES

4.4 An enhancement-type NMOS transistor with $V_t = 0.7$ V has its source terminal grounded and a 1.5-V dc applied to the gate. In what region does the device operate for (a) $V_D = +0.5$ V? (b) $V_D = 0.9$ V? (c) $V_D = 3$ V?

Ans. (a) Triode; (b) Saturation; (c) Saturation

4.5 If the NMOS device in Exercise 4.4 has $\mu_n C_{ox} = 100 \ \mu\text{A/V}^2$, $W = 10 \ \mu\text{m}$, and $L = 1 \ \mu\text{m}$, find the value of drain current that results in each of the three cases (a), (b), and (c) specified in Exercise 4.4.

Ans. (a) 275 μA; (b) 320 μA; (c) 320 μA

4.6 An enhancement-type NMOS transistor with $V_t = 0.7$ V conducts a current $i_D = 100 \ \mu$A when $v_{GS} = v_{DS} = 1.2$ V. Find the value of i_D for $v_{GS} = 1.5$ V and $v_{DS} = 3$ V. Also, calculate the value of the drain-to-source resistance r_{DS} for small v_{DS} and $v_{GS} = 3.2$ V.

Ans. 256 μA; 500 Ω

4.2.3 Finite Output Resistance in Saturation

Equation (4.12) and the corresponding large-signal equivalent circuit in Fig. 4.13 indicate that in saturation, i_D is independent of v_{DS}. Thus a change Δv_{DS} in the drain-to-source voltage causes a zero change in i_D, which implies that the incremental resistance looking into the drain of a saturated MOSFET is infinite. This, however, is an idealization based on the premise that once the channel is pinched off at the drain end, further increases in v_{DS} have no effect on the channel's shape. But, in practice, increasing v_{DS} beyond v_{DSsat} does affect the channel somewhat. Specifically, as v_{DS} is increased, the channel pinch-off point is moved slightly away from the drain, toward the source. This is illustrated in Fig. 4.15, from which we note that the voltage across the channel remains constant at $v_{GS} - V_t = v_{DSsat}$, and the additional voltage applied to the drain appears as a voltage drop across the narrow depletion region between the end of the channel and the drain region. This voltage accelerates the electrons that reach the drain end of the channel and sweeps them across the depletion region into the drain. Note, however, that (with depletion-layer widening) the channel length is in effect reduced, from L to $L - \Delta L$, a phenomenon known as **channel-length modulation.** Now, since i_D is inversely proportional to the channel length (Eq. 4.20), i_D increases with v_{DS}.

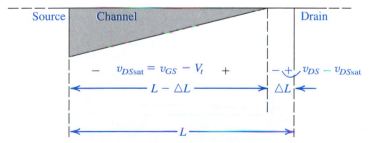

FIGURE 4.15 Increasing v_{DS} beyond v_{DSsat} causes the channel pinch-off point to move slightly away from the drain, thus reducing the effective channel length (by ΔL).

To account for the dependence of i_D on v_{DS} in saturation, we replace L in Eq. (4.20) with $L - \Delta L$ to obtain

$$i_D = \frac{1}{2} k_n' \frac{W}{L - \Delta L} (v_{GS} - V_t)^2$$

$$= \frac{1}{2} k_n' \frac{W}{L} \frac{1}{1 - (\Delta L / L)} (v_{GS} - V_t)^2$$

$$\cong \frac{1}{2} k_n' \frac{W}{L} \left(1 + \frac{\Delta L}{L}\right) (v_{GS} - V_t)^2$$

where we have assumed that $(\Delta L / L) \ll 1$. Now, if we assume that ΔL is proportional to v_{DS},

$$\Delta L = \lambda' v_{DS}$$

where λ' is a process-technology parameter with the dimensions of μm/V, we obtain for i_D,

$$i_D = \frac{1}{2} k_n' \frac{W}{L} \left(1 + \frac{\lambda'}{L} v_{DS}\right) (v_{GS} - V_t)^2$$

Usually, λ'/L is denoted λ,

$$\lambda = \frac{\lambda'}{L}$$

It follows that λ is a process-technology parameter with the dimensions of V^{-1} and that, for a given process, λ is inversely proportional to the length selected for the channel. In terms of λ, the expression for i_D becomes

$$i_D = \frac{1}{2} k_n' \frac{W}{L} (v_{GS} - V_t)^2 (1 + \lambda v_{DS}) \tag{4.22}$$

A typical set of i_D–v_{DS} characteristics showing the effect of channel-length modulation is displayed in Fig. 4.16. The observed linear dependence of i_D on v_{DS} in the saturation region is represented in Eq. (4.22) by the factor $(1 + \lambda v_{DS})$. From Fig. 4.16 we observe that when the straight-line i_D–v_{DS} characteristics are extrapolated they intercept the v_{DS}-axis at the point $v_{DS} = -V_A$, where V_A is a positive voltage. Equation (4.22), however, indicates that $i_D = 0$

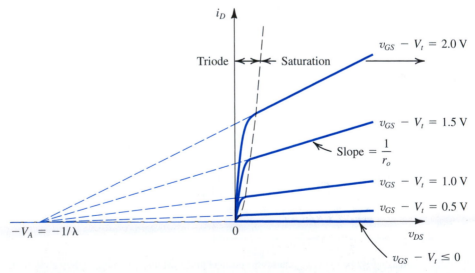

FIGURE 4.16 Effect of v_{DS} on i_D in the saturation region. The MOSFET parameter V_A depends on the process technology and, for a given process, is proportional to the channel length L.

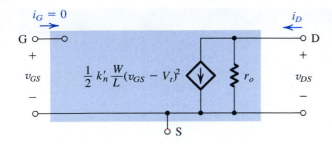

FIGURE 4.17 Large-signal equivalent circuit model of the *n*-channel MOSFET in saturation, incorporating the output resistance r_o. The output resistance models the linear dependence of i_D on v_{DS} and is given by Eq. (4.22).

at $v_{DS} = -1/\lambda$. It follows that

$$V_A = \frac{1}{\lambda}$$

and thus V_A is a process-technology parameter with the dimensions of V. For a given process, V_A is proportional to the channel length L that the designer selects for a MOSFET. Just as in the case of λ, we can isolate the dependence of V_A on L by expressing it as

$$V_A = V_A' L$$

where V_A' is entirely process-technology dependent with the dimensions of V/μm. Typically, V_A' falls in the range of 5 V/μm to 50 V/μm. The voltage V_A is usually referred to as the Early voltage, after J.M. Early, who discovered a similar phenomenon for the BJT (Chapter 5).

Equation (4.22) indicates that when channel-length modulation is taken into account, the saturation values of i_D depend on v_{DS}. Thus, for a given v_{GS}, a change Δv_{DS} yields a corresponding change Δi_D in the drain current i_D. It follows that the output resistance of the current source representing i_D in saturation is no longer infinite. Defining the output resistance r_o as[5]

$$r_o \equiv \left[\frac{\partial i_D}{\partial v_{DS}}\right]^{-1}_{v_{GS}\text{ constant}} \tag{4.23}$$

and using Eq. (4.22) results in

$$r_o = \left[\lambda \frac{k_n'}{2}\frac{W}{L}(V_{GS} - V_t)^2\right]^{-1} \tag{4.24}$$

which can be written as

$$r_o = \frac{1}{\lambda I_D} \tag{4.25}$$

or, equivalently,

$$r_o = \frac{V_A}{I_D} \tag{4.26}$$

where I_D is the drain current *without* channel-length modulation taken into account; that is,

$$I_D = \frac{1}{2}k_n'\frac{W}{L}(V_{GS} - V_t)^2$$

Thus the output resistance is inversely proportional to the drain current. Finally, we show in Fig. 4.17 the large-signal equivalent circuit model incorporating r_o.

---

[5] In this book we use r_o to denote the output resistance in saturation, and r_{DS} to denote the drain-to-source resistance in the triode region, for small v_{DS}.

EXERCISE

4.7 An NMOS transistor is fabricated in a 0.4-μm process having $\mu_n C_{ox} = 200$ μA/V^2 and $V'_A = 50$ V/μm of channel length. If $L = 0.8$ μm and $W = 16$ μm, find V_A and λ. Find the value of I_D that results when the device is operated with an overdrive voltage $V_{OV} = 0.5$ V and $V_{DS} = 1$ V. Also, find the value of r_o at this operating point. If V_{DS} is increased by 2 V, what is the corresponding change in I_D?

Ans. 40 V; 0.025 V^{-1}; 0.51 mA; 80 kΩ; 0.025 mA

4.2.4 Characteristics of the *p*-Channel MOSFET

The circuit symbol for the *p*-channel enhancement-type MOSFET is shown in Fig. 4.18(a). Figure 4.18(b) shows a modified circuit symbol in which an arrowhead pointing in the normal direction of current flow is included on the source terminal. For the case where the source is connected to the substrate, the simplified symbol of Fig. 4.18(c) is usually used. The voltage and current polarities for normal operation are indicated in Fig. 4.18(d). Recall that for the *p*-channel device the threshold voltage V_t is negative. To induce a channel we apply a gate voltage that is more negative than V_t,

$$v_{GS} \le V_t \quad \text{(Induced channel)} \tag{4.27}$$

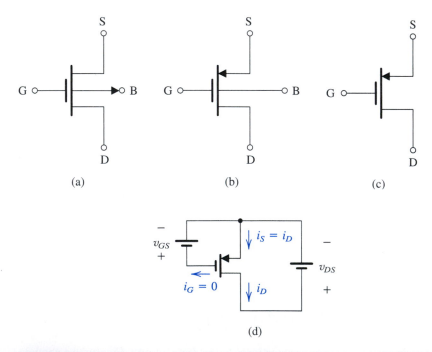

FIGURE 4.18 (a) Circuit symbol for the *p*-channel enhancement-type MOSFET. (b) Modified symbol with an arrowhead on the source lead. (c) Simplified circuit symbol for the case where the source is connected to the body. (d) The MOSFET with voltages applied and the directions of current flow indicated. Note that v_{GS} and v_{DS} are negative and i_D flows out of the drain terminal.

or, equivalently,

$$v_{SG} \geq |V_t|$$

and apply a drain voltage that is more negative than the source voltage (i.e., v_{DS} is negative or, equivalently, v_{SD} is positive). The current i_D flows out of the drain terminal, as indicated in the figure. To operate in the triode region, v_{DS} must satisfy

$$v_{DS} \geq v_{GS} - V_t \quad \text{(Continuous channel)} \tag{4.28}$$

that is, the drain voltage must be higher than the gate voltage by at least $|V_t|$. The current i_D is given by the same equation as for NMOS, Eq. (4.11), except for replacing k_n' with k_p',

$$i_D = k_p' \frac{W}{L} \left[(v_{GS} - V_t) v_{DS} - \frac{1}{2} v_{DS}^2 \right] \tag{4.29}$$

where v_{GS}, V_t, and v_{DS} are negative and the transconductance parameter k_p' is given by

$$k_p' = \mu_p C_{ox} \tag{4.30}$$

where μ_p is the mobility of holes in the induced p channel. Typically, $\mu_p = 0.25$ to $0.5\mu_n$ and is process-technology dependent.

To operate in saturation, v_{DS} must satisfy the relationship

$$v_{DS} \leq v_{GS} - V_t \quad \text{(Pinched-off channel)} \tag{4.31}$$

that is, the drain voltage must be lower than (gate voltage + $|V_t|$). The current i_D is given by the same equation used for NMOS, Eq. (4.22), again with k_n' replaced with k_p',

$$i_D = \frac{1}{2} k_p' \frac{W}{L} (v_{GS} - V_t)^2 (1 + \lambda v_{DS}) \tag{4.32}$$

where v_{GS}, V_t, λ, and v_{DS} are all negative. We should note, however, that in evaluating r_o using Eqs. (4.24) through (4.26), the magnitudes of λ and V_A should be used.

To recap, to turn a PMOS transistor on, the gate voltage has to be made lower than that of the source by at least $|V_t|$. To operate in the triode region, the drain voltage has to exceed that of the gate by at least $|V_t|$; otherwise, the PMOS operates in saturation.

Finally, the chart in Fig. 4.19 provides a pictorial representation of these operating conditions.

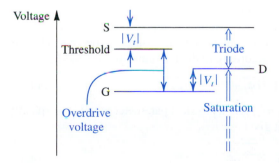

FIGURE 4.19 The relative levels of the terminal voltages of the enhancement-type PMOS transistor for operation in the triode region and in the saturation region.

EXERCISE

4.8 The PMOS transistor shown in Fig. E4.8 has $V_t = -1$ V, $k'_p = 60$ μA/V^2, and $W/L = 10$. (a) Find the range of V_G for which the transistor conducts. (b) In terms of V_G, find the range of V_D for which the transistor operates in the triode region. (c) In terms of V_G, find the range of V_D for which the transistor operates in saturation. (d) Neglecting channel-length modulation (i.e., assuming $\lambda = 0$), find the values of $|V_{OV}|$ and V_G and the corresponding range of V_D to operate the transistor in the saturation mode with $I_D = 75$ μA. (e) If $\lambda = -0.02$ V^{-1}, find the value of r_o corresponding to the overdrive voltage determined in (d). (f) For $\lambda = -0.02$ V^{-1} and for the value of V_{OV} determined in (d), find I_D at $V_D = +3$ V and at $V_D = 0$ V; hence, calculate the value of the apparent output resistance in saturation. Compare to the value found in (e).

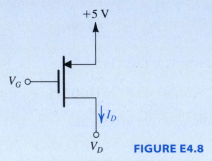

FIGURE E4.8

Ans. (a) $V_G \leq +4$ V; (b) $V_D \geq V_G + 1$; (c) $V_D \leq V_G + 1$; (d) 0.5 V, 3.5 V, ≤ 4.5 V; (e) 0.67 MΩ; (f) 78 μA, 82.5 μA, 0.67 MΩ (same).

4.2.5 The Role of the Substrate—The Body Effect

In many applications the source terminal is connected to the substrate (or body) terminal B, which results in the *pn* junction between the substrate and the induced channel (see Fig. 4.5) having a constant zero (cutoff) bias. In such a case the substrate does not play any role in circuit operation and its existence can be ignored altogether.

In integrated circuits, however, the substrate is usually common to many MOS transistors. In order to maintain the cutoff condition for all the substrate-to-channel junctions, the substrate is usually connected to the most negative power supply in an NMOS circuit (the most positive in a PMOS circuit). The resulting reverse-bias voltage between source and body (V_{SB} in an *n*-channel device) will have an effect on device operation. To appreciate this fact, consider an NMOS transistor and let its substrate be made negative relative to the source. The reverse bias voltage will widen the depletion region (refer to Fig. 4.2). This in turn reduces the channel depth. To return the channel to its former state, v_{GS} has to be increased.

The effect of V_{SB} on the channel can be most conveniently represented as a change in the threshold voltage V_t. Specifically, it has been shown that increasing the reverse substrate bias voltage V_{SB} results in an increase in V_t according to the relationship

$$V_t = V_{t0} + \gamma[\sqrt{2\phi_f + V_{SB}} - \sqrt{2\phi_f}] \qquad (4.33)$$

where V_{t0} is the threshold voltage for $V_{SB} = 0$; ϕ_f is a physical parameter with $(2\phi_f)$ typically 0.6 V; γ is a fabrication-process parameter given by

$$\gamma = \frac{\sqrt{2qN_A\varepsilon_s}}{C_{ox}} \qquad (4.34)$$

where q is the electron charge (1.6×10^{-19} C), N_A is the doping concentration of the p-type substrate, and ε_s is the permittivity of silicon ($11.7\varepsilon_0 = 11.7 \times 8.854 \times 10^{-14} = 1.04 \times 10^{-12}$ F/cm). The parameter γ has the dimension of $\sqrt{V}$ and is typically 0.4 $V^{1/2}$. Finally, note that Eq. (4.33) applies equally well for p-channel devices with V_{SB} replaced by the reverse bias of the substrate, V_{BS} (or, alternatively, replace V_{SB} by $|V_{SB}|$) and note that γ is negative. In evaluating γ, N_A must be replaced with N_D, the doping concentration of the n well in which the PMOS is formed. For p-channel devices, $2\phi_f$ is typically 0.75 V, and γ is typically -0.5 $V^{1/2}$.

Equation (4.33) indicates that an incremental change in V_{SB} gives rise to an incremental change in V_t, which in turn results in an incremental change in i_D even though v_{GS} might have been kept constant. It follows that the body voltage controls i_D; thus the body acts as another gate for the MOSFET, a phenomenon known as the **body effect.** Here we note that the parameter γ is known as the **body-effect parameter.** The body effect can cause considerable degradation in circuit performance, as will be shown in Chapter 6.

EXERCISE

4.9 An NMOS transistor has $V_{t0} = 0.8$ V, $2\phi_f = 0.7$ V, and $\gamma = 0.4$ $V^{1/2}$. Find V_t when $V_{SB} = 3$ V.

Ans. 1.23 V

4.2.6 Temperature Effects

Both V_t and k' are temperature sensitive. The magnitude of V_t decreases by about 2 mV for every 1°C rise in temperature. This decrease in $|V_t|$ gives rise to a corresponding increase in drain current as temperature is increased. However, because k' decreases with temperature and its effect is a dominant one, the overall observed effect of a temperature increase is a *decrease* in drain current. This very interesting result is put to use in applying the MOSFET in power circuits (Chapter 14).

4.2.7 Breakdown and Input Protection

As the voltage on the drain is increased, a value is reached at which the pn junction between the drain region and substrate suffers avalanche breakdown (see Section 3.7.4). This breakdown usually occurs at voltages of 20 V to 150 V and results in a somewhat rapid increase in current (known as a **weak avalanche**).

Another breakdown effect that occurs at lower voltages (about 20 V) in modern devices is called **punch-through.** It occurs in devices with relatively short channels when the drain voltage is increased to the point that the depletion region surrounding the drain region extends through the channel to the source. The drain current then increases rapidly. Normally, punch-through does not result in permanent damage to the device.

Yet another kind of breakdown occurs when the gate-to-source voltage exceeds about 30 V. This is the breakdown of the gate oxide and results in permanent damage to the device. Although 30 V may seem high, it must be remembered that the MOSFET has a very high input resistance, and a very small input capacitance, and thus small amounts of static charge accumulating on the gate capacitor can cause its breakdown voltage to be exceeded.

To prevent the accumulation of static charge on the gate capacitor of a MOSFET, gate-protection devices are usually included at the input terminals of MOS integrated circuits. The protection mechanism invariably makes use of clamping diodes.

4.2.8 Summary

For easy reference we present in Table 4.1 a summary of the current-voltage relationships for enhancement-type MOSFETs.

TABLE 4.1   **Summary of the MOSFET Current-Voltage Characteristics**

NMOS Transistor

Symbol:

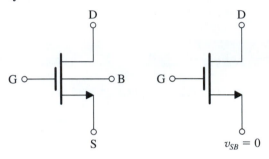

Overdrive voltage:

$$v_{OV} = v_{GS} - V_t$$

$$v_{GS} = V_t + v_{OV}$$

Operation in the *triode* region:

- Conditions:

    (1) $v_{GS} \geq V_t \quad \Leftrightarrow \quad v_{OV} \geq 0$

    (2) $v_{GD} \geq V_t \quad \Leftrightarrow \quad v_{DS} \leq v_{GS} - V_t \quad \Leftrightarrow \quad v_{DS} \leq v_{OV}$

- i-v Characteristics:

$$i_D = \mu_n C_{ox} \frac{W}{L} \left[(v_{GS} - V_t) v_{DS} - \frac{1}{2} v_{DS}^2 \right]$$

- For $v_{DS} \ll 2(v_{GS} - V_t) \quad \Leftrightarrow \quad v_{DS} \ll 2v_{OV}$

$$r_{DS} \equiv \frac{v_{DS}}{i_D} = 1 \bigg/ \left[\mu_n C_{ox} \frac{W}{L} (v_{GS} - V_t) \right]$$

Operation in the *saturation* region:

- Conditions:

    (1) $v_{GS} \geq V_t \quad \Leftrightarrow \quad v_{OV} \geq 0$

    (2) $v_{GD} \leq V_t \quad \Leftrightarrow \quad v_{DS} \geq v_{GS} - V_t \quad \Leftrightarrow \quad v_{DS} \geq v_{OV}$

- i-v Characteristics:

$$i_D = \frac{1}{2} \mu_n C_{ox} \frac{W}{L} (v_{GS} - V_t)^2 (1 + \lambda v_{DS})$$

- Large-signal equivalent circuit model:

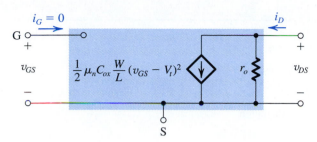

$$r_o = \left[\lambda \frac{1}{2} \mu_n C_{ox} \frac{W}{L} (V_{GS} - V_t)^2 \right]^{-1} = \frac{V_A}{I_D}$$

where

$$I_D = \frac{1}{2} \mu_n C_{ox} \frac{W}{L} (V_{GS} - V_t)^2$$

Threshold voltage:

$$V_t = V_{t0} + \gamma(\sqrt{2\phi_f + |V_{SB}|} - \sqrt{2\phi_f})$$

Process parameters:

$$
\begin{aligned}
C_{ox} &= \varepsilon_{ox}/t_{ox} & (\text{F/m}^2) \\
k_n' &= \mu_n C_{ox} & (\text{A/V}^2) \\
V_A' &= (V_A/L) & (\text{V/m}) \\
\lambda &= (1/V_A) & (\text{V}^{-1}) \\
\gamma &= \sqrt{2qN_A\varepsilon_s}/C_{ox} & (\text{V}^{1/2})
\end{aligned}
$$

Constants:

$$
\begin{aligned}
\varepsilon_0 &= 8.854 \times 10^{-12} \text{ F/m} \\
\varepsilon_{ox} &= 3.9\varepsilon_0 = 3.45 \times 10^{-11} \text{ F/m} \\
\varepsilon_s &= 11.7\varepsilon_0 = 1.04 \times 10^{-10} \text{ F/m} \\
q &= 1.602 \times 10^{-19} \text{ C}
\end{aligned}
$$

PMOS Transistor

Symbol:

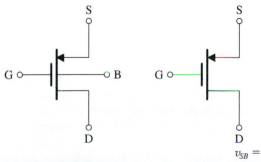

Overdrive voltage:

$$v_{OV} = v_{GS} - V_t$$
$$v_{SG} = |V_t| + |v_{OV}|$$

(Continued)

TABLE 4.1 *(Continued)*

i-v **Characteristics:**

Same relationships as for NMOS transistors except:

- Replace μ_n, k'_n, and N_A with μ_p, k'_p, and N_D, respectively.

- V_t, V_{t0}, V_A, λ, and γ are negative.

- Conditions for operation in the **triode** region:

  (1) $v_{GS} \leq V_t \iff v_{OV} \leq 0 \iff v_{SG} \geq |V_t|$

  (2) $v_{DG} \geq |V_t| \iff v_{DS} \geq v_{GS} - V_t \iff v_{SD} \leq |v_{OV}|$

- Conditions for operation in the **saturation** region:

  (1) $v_{GS} \leq V_t \iff v_{OV} \leq 0 \iff v_{SG} \geq |V_t|$

  (2) $v_{DG} \leq |V_t| \iff v_{DS} \leq v_{GS} - V_t \iff v_{SD} \geq |v_{OV}|$

- Large-signal equivalent circuit model:

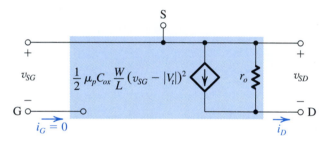

$$r_o = \left[|\lambda| \frac{1}{2} \mu_p C_{ox} \frac{W}{L} (V_{SG} - |V_t|)^2 \right]^{-1} = \frac{|V_A|}{I_D}$$

where

$$I_D = \frac{1}{2} \mu_p C_{ox} \frac{W}{L} (V_{SG} - |V_t|^2)$$

4.3 MOSFET CIRCUITS AT DC

Having studied the current-voltage characteristics of MOSFETs, we now consider circuits in which only dc voltages and currents are of concern. Specifically, we shall present a series of design and analysis examples of MOSFET circuits at dc. The objective is to instill in the reader a familiarity with the device and the ability to perform MOSFET circuit analysis both rapidly and effectively.

In the following examples, to keep matters simple and thus focus attention on the essence of MOSFET circuit operation, we will generally neglect channel-length modulation; that is, we will assume $\lambda = 0$. We will find it convenient to work in terms of the overdrive voltage; $V_{OV} = V_{GS} - V_t$. Recall that for NMOS, V_t and V_{OV} are positive while, for PMOS, V_t and V_{OV} are negative. For PMOS the reader may prefer to write $V_{SG} = |V_{GS}| = |V_t| + |V_{OV}|$.

EXAMPLE 4.2

Design the circuit of Fig. 4.20 so that the transistor operates at $I_D = 0.4$ mA and $V_D = +0.5$ V. The NMOS transistor has $V_t = 0.7$ V, $\mu_n C_{ox} = 100\ \mu\text{A/V}^2$, $L = 1\ \mu\text{m}$, and $W = 32\ \mu\text{m}$. Neglect the channel-length modulation effect (i.e., assume that $\lambda = 0$).

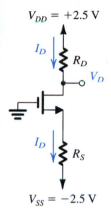

$V_{DD} = +2.5$ V

I_D

R_D

V_D

I_D

R_S

$V_{SS} = -2.5$ V

FIGURE 4.20 Circuit for Example 4.2.

Solution

Since $V_D = 0.5$ V is greater than V_G, this means the NMOS transistor is operating in the saturation region, and we use the saturation-region expression of i_D to determine the required value of V_{GS},

$$I_D = \frac{1}{2}\mu_n C_{ox}\frac{W}{L}(V_{GS} - V_t)^2$$

Substituting $V_{GS} - V_t = V_{OV}$, $I_D = 0.4$ mA $= 400\ \mu\text{A}$, $\mu_n C_{ox} = 100\ \mu\text{A/V}^2$, and $W/L = 32/1$ gives

$$400 = \frac{1}{2} \times 100 \times \frac{32}{1}V_{OV}^2$$

which results in

$$V_{OV} = 0.5\ \text{V}$$

Thus,

$$V_{GS} = V_t + V_{OV} = 0.7 + 0.5 = 1.2\ \text{V}$$

Referring to Fig. 4.20, we note that the gate is at ground potential. Thus the source must be at -1.2 V, and the required value of R_S can be determined from

$$R_S = \frac{V_S - V_{SS}}{I_D}$$

$$= \frac{-1.2 - (-2.5)}{0.4} = 3.25\ \text{k}\Omega$$

To establish a dc voltage of $+0.5$ V at the drain, we must select R_D as follows:

$$R_D = \frac{V_{DD} - V_D}{I_D}$$

$$= \frac{2.5 - 0.5}{0.4} = 5\ \text{k}\Omega$$

D4.10 Redesign the circuit of Fig. 4.20 for the following case: $V_{DD} = -V_{SS} = 2.5$ V, $V_t = 1$ V, $\mu_n C_{ox} = 60$ μA/V^2, $W/L = 120$ μm/3 μm, $I_D = 0.3$ mA, and $V_D = +0.4$ V.
   Ans. $R_S = 3.3$ kΩ; $R_D = 7$ kΩ

EXAMPLE 4.3

Design the circuit in Fig. 4.21 to obtain a current I_D of 80 μA. Find the value required for R, and find the dc voltage V_D. Let the NMOS transistor have $V_t = 0.6$ V, $\mu_n C_{ox} = 200$ μA/V^2, $L = 0.8$ μm, and $W = 4$ μm. Neglect the channel-length modulation effect (i.e., assume $\lambda = 0$).

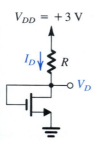

$V_{DD} = +3$ V

I_D ↓ R

V_D

FIGURE 4.21 Circuit for Example 4.3.

Solution

Because $V_{DG} = 0$, $V_D = V_G$ and the FET is operating in the saturation region. Thus,

$$I_D = \frac{1}{2}\mu_n C_{ox}\frac{W}{L}(V_{GS} - V_t)^2$$

$$= \frac{1}{2}\mu_n C_{ox}\frac{W}{L}V_{OV}^2$$

from which we obtain V_{OV} as

$$V_{OV} = \sqrt{\frac{2I_D}{\mu_n C_{ox}(W/L)}}$$

$$= \sqrt{\frac{2 \times 80}{200 \times (4/0.8)}} = 0.4 \text{ V}$$

Thus,

$$V_{GS} = V_t + V_{OV} = 0.6 + 0.4 = 1 \text{ V}$$

and the drain voltage will be

$$V_D = V_G = +1 \text{ V}$$

The required value for R can be found as follows:

$$R = \frac{V_{DD} - V_D}{I_D}$$

$$= \frac{3 - 1}{0.080} = 25 \text{ k}\Omega$$

EXERCISES

D4.11 Redesign the circuit in Example 4.3 to double the value of I_D without changing V_D. Give new values for W/L and R.

Ans. $W/L = 10$, say 8 μm/0.8 μm; $R = 12.5$ kΩ

4.12 Consider the circuit of Fig. 4.21, which is designed in Example 4.3 (to which you should refer before solving this problem). Let the voltage V_D be applied to the gate of another transistor Q_2, as shown in Fig. E4.12. Assume that Q_2 is identical to Q_1. Find the drain current and voltage of Q_2. (Assume $\lambda = 0$.)

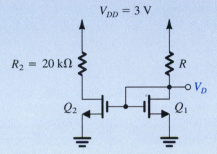

FIGURE E4.12

Ans. 80 μA; +1.4 V

EXAMPLE 4.4

Design the circuit in Fig. 4.22 to establish a drain voltage of 0.1 V. What is the effective resistance between drain and source at this operating point? Let $V_t = 1$ V and $k_n'(W/L) = 1$ mA/V^2.

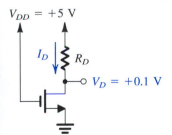

FIGURE 4.22 Circuit for Example 4.4.

Solution

Since the drain voltage is lower than the gate voltage by 4.9 V and $V_t = 1$ V, the MOSFET is operating in the triode region. Thus the current I_D is given by

$$I_D = k_n'\frac{W}{L}\left[(V_{GS} - V_t)V_{DS} - \frac{1}{2}V_{DS}^2\right]$$

$$I_D = 1 \times \left[(5-1) \times 0.1 - \frac{1}{2} \times 0.01\right]$$

$$= 0.395 \text{ mA}$$

The required value for R_D can be found as follows:

$$R_D = \frac{V_{DD} - V_D}{I_D}$$

$$= \frac{5 - 0.1}{0.395} = 12.4 \text{ k}\Omega$$

In a practical discrete-circuit design problem one selects the closest standard value available for, say, 5% resistors—in this case, 12 kΩ; see Appendix G. Since the transistor is operating in the triode region with a small V_{DS}, the effective drain-to-source resistance can be determined as follows:

$$r_{DS} = \frac{V_{DS}}{I_D}$$

$$= \frac{0.1}{0.395} = 253 \ \Omega$$

EXERCISE

4.13 If in the circuit of Example 4.4 the value of R_D is doubled, find approximate values for I_D and V_D.
Ans. 0.2 mA; 0.05 V

EXAMPLE 4.5

Analyze the circuit shown in Fig. 4.23(a) to determine the voltages at all nodes and the currents through all branches. Let $V_t = 1$ V and $k'_n(W/L) = 1$ mA/V^2. Neglect the channel-length modulation effect (i.e., assume $\lambda = 0$).

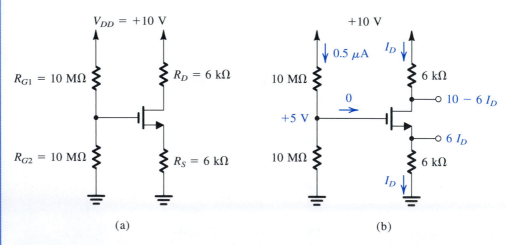

(a)   (b)

FIGURE 4.23 (a) Circuit for Example 4.5. (b) The circuit with some of the analysis details shown.

Solution

Since the gate current is zero, the voltage at the gate is simply determined by the voltage divider formed by the two 10-MΩ resistors,

$$V_G = V_{DD}\frac{R_{G2}}{R_{G2} + R_{G1}} = 10 \times \frac{10}{10 + 10} = +5 \text{ V}$$

With this positive voltage at the gate, the NMOS transistor will be turned on. We do not know, however, whether the transistor will be operating in the saturation region or in the triode region. We shall assume saturation-region operation, solve the problem, and then check the validity of our assumption. Obviously, if our assumption turns out not to be valid, we will have to solve the problem again for triode-region operation.

Refer to Fig. 4.23(b). Since the voltage at the gate is 5 V and the voltage at the source is I_D (mA) $\times 6$ (kΩ) $= 6I_D$ (V), we have

$$V_{GS} = 5 - 6I_D$$

Thus I_D is given by

$$I_D = \frac{1}{2}k'_n \frac{W}{L}(V_{GS} - V_t)^2$$

$$= \frac{1}{2} \times 1 \times (5 - 6I_D - 1)^2$$

which results in the following quadratic equation in I_D:

$$18I_D^2 - 25I_D + 8 = 0$$

This equation yields two values for I_D: 0.89 mA and 0.5 mA. The first value results in a source voltage of $6 \times 0.89 = 5.34$, which is greater than the gate voltage and does not make physical sense as it would imply that the NMOS transistor is cut off. Thus,

$$I_D = 0.5 \text{ mA}$$
$$V_S = 0.5 \times 6 = +3 \text{ V}$$
$$V_{GS} = 5 - 3 = 2 \text{ V}$$
$$V_D = 10 - 6 \times 0.5 = +7 \text{ V}$$

Since $V_D > V_G - V_t$, the transistor is operating in saturation, as initially assumed.

EXERCISES

4.14 For the circuit of Fig. 4.23, what is the largest value that R_D can have while the transistor remains in the saturation mode?

Ans. 12 kΩ

D4.15 Redesign the circuit of Fig. 4.23 for the following requirements: $V_{DD} = +5$ V, $I_D = 0.32$ mA, $V_S = 1.6$ V, $V_D = 3.4$ V, with a 1-μA current through the voltage divider R_{G1}, R_{G2}. Assume the same MOSFET as in Example 4.5.

Ans. $R_{G1} = 1.6$ MΩ; $R_{G2} = 3.4$ MΩ, $R_S = R_D = 5$ kΩ

EXAMPLE 4.6

Design the circuit of Fig. 4.24 so that the transistor operates in saturation with $I_D = 0.5$ mA and $V_D = +3$ V. Let the enhancement-type PMOS transistor have $V_t = -1$ V and $k_p'(W/L) = 1$ mA/V^2. Assume $\lambda = 0$. What is the largest value that R_D can have while maintaining saturation-region operation?

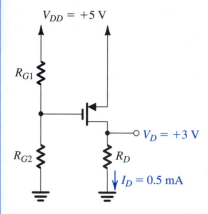

$V_{DD} = +5$ V

R_{G1}

$V_D = +3$ V

R_{G2} R_D

$I_D = 0.5$ mA

FIGURE 4.24 Circuit for Example 4.6.

Solution

Since the MOSFET is to be in saturation, we can write

$$I_D = \frac{1}{2} k_p' \frac{W}{L} (V_{GS} - V_t)^2$$

$$= \frac{1}{2} k_p' \frac{W}{L} V_{OV}^2$$

Substituting $I_D = 0.5$ mA and $k_p' W/L = 1$ mA/V^2 and recalling that for a PMOS transistor V_{OV} is negative, we obtain

$$V_{OV} = -1 \text{ V}$$

and

$$V_{GS} = V_t + V_{OV} = -1 - 1 = -2 \text{ V}$$

Since the source is at +5 V, the gate voltage must be set to +3 V. This can be achieved by the appropriate selection of the values of R_{G1} and R_{G2}. A possible selection is $R_{G1} = 2$ MΩ and $R_{G2} = 3$ MΩ.

The value of R_D can be found from

$$R_D = \frac{V_D}{I_D} = \frac{3}{0.5} = 6 \text{ k}\Omega$$

Saturation-mode operation will be maintained up to the point that V_D exceeds V_G by $|V_t|$; that is, until

$$V_{D_{max}} = 3 + 1 = 4 \text{ V}$$

This value of drain voltage is obtained with R_D given by

$$R_D = \frac{4}{0.5} = 8 \text{ k}\Omega$$

EXAMPLE 4.7

The NMOS and PMOS transistors in the circuit of Fig. 4.25(a) are matched with $k_n'(W_n/L_n) = k_p'(W_p/L_p) = 1 \text{ mA/V}^2$ and $V_{tn} = -V_{tp} = 1$ V. Assuming $\lambda = 0$ for both devices, find the drain currents i_{DN} and i_{DP}, as well as the voltage v_O, for $v_I = 0$ V, +2.5 V, and −2.5 V.

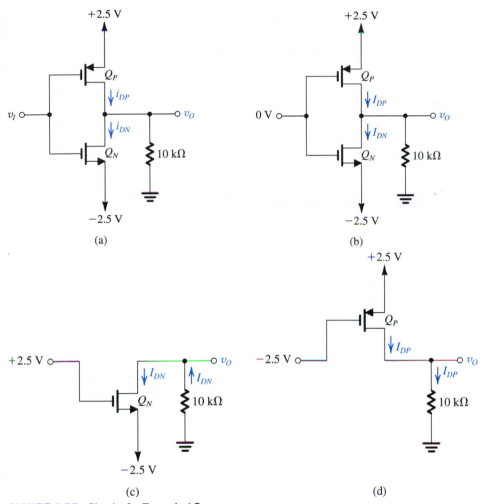

FIGURE 4.25 Circuits for Example 4.7.

Solution

Figure 4.25(b) shows the circuit for the case $v_I = 0$ V. We note that since Q_N and Q_P are perfectly matched and are operating at equal $|V_{GS}|$ (2.5 V), the circuit is symmetrical, which dictates that $v_O = 0$ V. Thus both Q_N and Q_P are operating with $|V_{DG}| = 0$ and, hence, in saturation. The drain currents can now be found from

$$I_{DP} = I_{DN} = \tfrac{1}{2} \times 1 \times (2.5 - 1)^2$$

$$= 1.125 \text{ mA}$$

Next, we consider the circuit with $v_I = +2.5$ V. Transistor Q_P will have a V_{GS} of zero and thus will be cut off, reducing the circuit to that shown in Fig. 4.25(c). We note that v_O will be negative, and thus v_{GD} will be greater than V_t, causing Q_N to operate in the triode region. For simplicity we shall assume that v_{DS} is small and thus use

$$I_{DN} \cong k'_n(W_n/L_n)(V_{GS} - V_t)V_{DS}$$
$$= 1[2.5 - (-2.5) - 1][v_O - (-2.5)]$$

From the circuit diagram shown in Fig. 4.25(c), we can also write

$$I_{DN}\,(\text{mA}) = \frac{0 - v_O}{10\,(\text{k}\Omega)}$$

These two equations can be solved simultaneously to yield

$$I_{DN} = 0.244\,\text{mA} \qquad v_O = -2.44\,\text{V}$$

Note that $V_{DS} = -2.44 - (-2.5) = 0.06$ V, which is small as assumed.

Finally, the situation for the case $v_I = -2.5$ V [Fig. 4.25(d)] will be the exact complement of the case $v_I = +2.5$ V: Transistor Q_N will be off. Thus $I_{DN} = 0$, Q_P will be operating in the triode region with $I_{DP} = 0.244$ mA and $v_O = +2.44$ V.

EXERCISE

4.16 The NMOS and PMOS transistors in the circuit of Fig. E4.16 are matched with $k'_n(W_n/L_n) = k'_p(W_p/L_p) = 1$ mA/V^2 and $V_{tn} = -V_{tp} = 1$ V. Assuming $\lambda = 0$ for both devices, find the drain currents i_{DN} and i_{DP} and the voltage v_O for $v_I = 0$ V, +2.5 V, and -2.5 V.

Ans. $v_I = 0$ V: 0 mA, 0 mA, 0 V; $v_I = +2.5$ V: 0.104 mA, 0 mA, 1.04 V; $v_I = -2.5$ V: 0 mA, 0.104 mA, -1.04 V

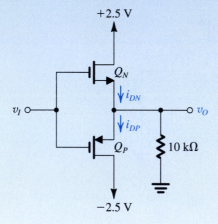

FIGURE E4.16

4.4 THE MOSFET AS AN AMPLIFIER AND AS A SWITCH

In this section we begin our study of the use of MOSFETs in the design of amplifier circuits.[6] The basis for this important MOSFET application is that when operated in the saturation region, the MOSFET acts as a voltage-controlled current source: Changes in the gate-to-source voltage

---

[6] An introduction to amplifiers from an external-terminals point of view was presented in Chapter 1 (Sections 1.4 and 1.5), and it would be helpful for readers who are not familiar with basic amplifier concepts to review some of this material before proceeding with the study of MOS amplifiers.

v_{GS} give rise to changes in the drain current i_D. Thus the saturated MOSFET can be used to implement a *transconductance amplifier* (see Section 1.5). However, since we are interested in linear amplification—that is, in amplifiers whose output signal (in this case, the drain current i_D) is linearly related to their input signal (in this case, the gate-to-source voltage v_{GS})—we will have to find a way around the highly nonlinear (square-law) relationship of i_D to v_{GS}.

The technique we will utilize to obtain linear amplification from a fundamentally non-linear device is that of **dc biasing** the MOSFET to operate at a certain appropriate V_{GS} and a corresponding I_D and then superimposing the voltage signal to be amplified, v_{gs}, on the dc bias voltage V_{GS}. By keeping the signal v_{gs} "small," the resulting change in drain current, i_d, can be made nearly proportional to v_{gs}. This technique was introduced in a general way in Section 1.4 and was applied in the case of the diode in Section 3.3.8. However, before considering the small-signal operation of the MOSFET amplifier, we will look at the "big picture": We will study the total or large-signal operation of a MOSFET amplifier. We will do this by deriving the voltage transfer characteristic of a commonly used MOSFET amplifier circuit. From the voltage transfer characteristic we will be able to clearly see the region over which the transistor can be biased to operate as a small-signal amplifier as well as those regions where it can be operated as a switch (i.e., being either fully "on" or fully "off"). MOS switches find application in both analog and digital circuits.

4.4.1 Large-Signal Operation—The Transfer Characteristic

Figure 4.26(a) shows the basic structure (skeleton) of the most commonly used MOSFET amplifier, the **common-source (CS)** circuit. The name common-source or **grounded-source**

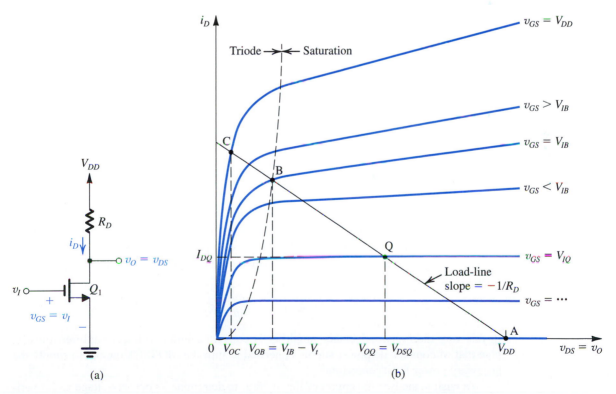

FIGURE 4.26 **(a)** Basic structure of the common-source amplifier. **(b)** Graphical construction to determine the transfer characteristic of the amplifier in (a).

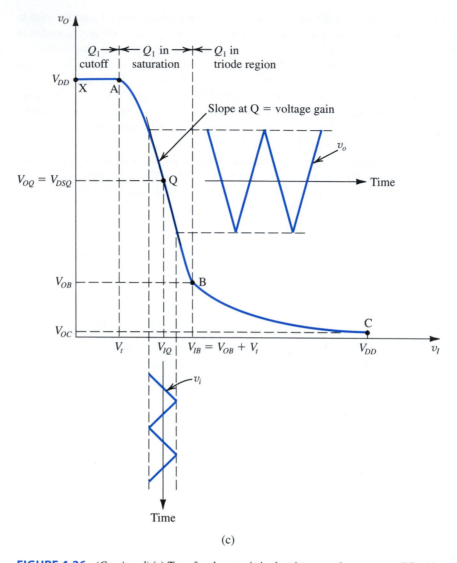

(c)

FIGURE 4.26 *(Continued)* **(c)** Transfer characteristic showing operation as an amplifier biased at point Q.

circuit arises because when the circuit is viewed as a two-port network, the grounded source terminal is common to both the input port, between gate and source, and the output port, between drain and source. Note that although the basic control action of the MOSFET is that changes in v_{GS} (here, changes in v_I as $v_{GS} = v_I$) give rise to changes in i_D, we are using a resistor R_D to obtain an output voltage v_O,

$$v_O = v_{DS} = V_{DD} - R_D i_D \qquad (4.35)$$

In this way the transconductance amplifier is converted into a voltage amplifier. Finally, note that of course a dc power supply is needed to turn the MOSFET on and to supply the necessary power for its operation.

We wish to analyze the circuit of Fig. 4.26(a) to determine its output voltage v_O for various values of its input voltage v_I, that is, to determine the **voltage transfer characteristic** of the CS amplifier. For this purpose, we will assume v_I to be in the range of 0 to V_{DD}. To obtain greater insight into the operation of the circuit, we will derive its transfer characteristic in two ways: graphically and analytically.

4.4.2 Graphical Derivation of the Transfer Characteristic

The operation of the common-source circuit is governed by the MOSFET's i_D–v_{DS} characteristics and by the relationship between i_D and v_{DS} imposed by connecting the drain to the power supply V_{DD} via resistor R_D, namely

$$v_{DS} = V_{DD} - R_D i_D \tag{4.36}$$

or, equivalently,

$$i_D = \frac{V_{DD}}{R_D} - \frac{1}{R_D} v_{DS} \tag{4.37}$$

Figure 4.26(b) shows a sketch of the MOSFET's i_D–v_{DS} characteristic curves superimposed on which is a straight line representing the i_D–v_{DS} relationship of Eq. (4.37). Observe that the straight line intersects the v_{DS}-axis at V_{DD} [since from Eq. (4.36) $v_{DS} = V_{DD}$ at $i_D = 0$] and has a slope of $-1/R_D$. Since R_D is usually thought of as the **load resistor** of the amplifier (i.e., the resistor across which the amplifier provides its output voltage), the straight line in Fig. 4.26(b) is known as the **load line.**

The graphical construction of Fig. 4.26(b) can now be used to determine v_O (equal to v_{DS}) for each given value of v_I ($v_{GS} = v_I$). Specifically, for any given value of v_I, we locate the corresponding i_D–v_{DS} curve and find v_O from the point of intersection of this curve with the load line.

Qualitatively, the circuit works as follows: Since $v_{GS} = v_I$, we see that for $v_I < V_t$, the transistor will be cut off, i_D will be zero, and $v_O = v_{DS} = V_{DD}$. Operation will be at the point labeled A. As v_I exceeds V_t, the transistor turns on, i_D increases, and v_O decreases. Since v_O will initially be high, the transistor will be operating in the saturation region. This corresponds to points along the segment of the load line from A to B. We have identified a particular point in this region of operation and labeled it Q. It is obtained for $V_{GS} = V_{IQ}$ and has the coordinates $V_{OQ} = V_{DSQ}$ and I_{DQ}.

Saturation-region operation continues until v_O decreases to the point that it is below v_I by V_t volts. At this point, $v_{DS} = v_{GS} - V_t$, and the MOSFET enters its triode region of operation. This is indicated in Fig. 4.26(b) by point B, which is at the intersection of the load line and the broken-line curve that defines the boundary between the saturation and the triode regions. Point B is defined by

$$V_{OB} = V_{IB} - V_t$$

For $v_I > V_{IB}$, the transistor is driven deeper into the triode region. Note that because the characteristic curves in the triode region are bunched together, the output voltage decreases slowly towards zero. Here we have identified a particular operating point C obtained for $v_I = V_{DD}$. The corresponding output voltage V_{OC} will usually be very small. This point-by-point determination of the transfer characteristic results in the transfer curve shown in Fig. 4.26(c). Observe that we have delineated its three distinct segments, each corresponding to one of the three regions of operation of MOSFET Q_1. We have also labeled the critical points of the transfer curve in correspondence with the points in Fig. 4.26(b).

4.4.3 Operation as a Switch

When the MOSFET is used as a switch, it is operated at the extreme points of the transfer curve. Specifically, the device is turned off by keeping $v_I < V_t$, resulting in operation somewhere on the segment XA with $v_O = V_{DD}$. The switch is turned on by applying a voltage close to V_{DD}, resulting in operation close to point C with v_O very small (at C, $v_O = V_{OC}$). At this juncture we observe that the transfer curve of Fig. 4.26(c) is of the form presented in Section 1.7 for the digital logic inverter. Indeed, the common-source MOS circuit can be used as a logic inverter with the "low" voltage level close to 0 V and the "high" level close to V_{DD}. More elaborate MOS logic inverters are studied in Section 4.10.

4.4.4 Operation as a Linear Amplifier

To operate the MOSFET as an amplifier we make use of the saturation-mode segment of the transfer curve. The device is biased at a point located somewhere close to the middle of the curve; point Q is a good example of an appropriate bias point. The dc bias point is also called the **quiescent point,** which is the reason for labeling it Q. The voltage signal to be amplified v_i is then superimposed on the dc voltage V_{IQ} as shown in Fig. 4.26(c). By keeping v_i sufficiently small to restrict operation to an almost linear segment of the transfer curve, the resulting output voltage signal v_o will be proportional to v_i. That is, the amplifier will be very nearly linear, and v_o will have the same waveform as v_i except that it will be larger by a factor equal to the voltage gain of the amplifier at Q, A_v, where

$$A_v \equiv \frac{dv_O}{dv_I}\bigg|_{v_I = V_{IQ}} \tag{4.38}$$

Thus the voltage gain is equal to the slope of the transfer curve at the bias point Q. Observe that the slope is negative, and thus the basic CS amplifier is inverting. This should be also evident from the waveforms of v_i and v_o shown in Fig. 4.26(c). It should be obvious that if the amplitude of the input signal v_i is increased, the output signal will become distorted since operation will no longer be restricted to an almost linear segment of the transfer curve.

We shall return to the small-signal operation of the MOSFET in Section 4.6. For the time being, however, we wish to make an important observation about selecting an appropriate location for the bias point Q. Since the output signal will be superimposed on the dc voltage at the drain V_{OQ} or V_{DSQ}, it is important that V_{DSQ} be of such value to allow for the required output signal swing. That is, V_{DSQ} should be lower than V_{DD} by a sufficient amount and higher than V_{OB} by a sufficient amount to allow for the required positive and negative output signal swing, respectively. If V_{DSQ} is too close to V_{DD}, the positive peaks of the output signals might "bump" into V_{DD} and would be clipped off, because the MOSFET would turn off for part of the cycle. We speak of this situation as the circuit not having sufficient "headroom." Similarly, if V_{DSQ} is too close to the boundary of the triode region, the MOSFET would enter the triode region for the part of the cycle near the negative peaks, resulting in a distorted output signal. We speak of this situation as the circuit not having sufficient "legroom." Finally, it is important to note that although we made our comments on the selection of bias-point location in the context of a given transfer curve, the circuit designer also has to decide on a value for R_D, which of course determines the transfer curve. It is therefore more appropriate when considering the location of the bias point Q to do so with reference to the i_D–v_{DS} plane. This point is further illustrated by the sketch in Fig. 4.27.

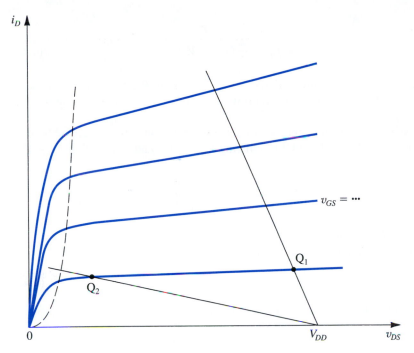

FIGURE 4.27 Two load lines and corresponding bias points. Bias point Q_1 does not leave sufficient room for positive signal swing at the drain (too close to V_{DD}). Bias point Q_2 is too close to the boundary of the triode region and might not allow for sufficient negative signal swing.

4.4.5 Analytical Expressions for the Transfer Characteristic

The i–v relationships that describe the MOSFET operation in the three regions—cutoff, saturation, and triode—can be easily used to derive analytical expressions for the three segments of the transfer characteristic in Fig. 4.26(a).

The Cutoff-Region Segment, XA   Here, $v_I \leq V_t$, and $v_O = V_{DD}$.

The Saturation-Region Segment, AQB   Here, $v_I \geq V_t$, and $v_O \geq v_I - V_t$. Neglecting channel-length modulation and substituting for i_D from

$$i_D = \frac{1}{2}(\mu_n C_{ox})\left(\frac{W}{L}\right)(v_I - V_t)^2$$

into

$$v_O = V_{DD} - R_D i_D$$

gives

$$v_O = V_{DD} - \frac{1}{2}R_D \mu_n C_{ox}\frac{W}{L}(v_I - V_t)^2 \tag{4.39}$$

We can use this relationship to derive an expression for the incremental voltage gain A_v at a bias point Q at which $v_I = V_{IQ}$ as follows:

$$A_v \equiv \left.\frac{dv_O}{dv_I}\right|_{v_I = V_{IQ}}$$

Thus,

$$A_v = -R_D \mu_n C_{ox} \frac{W}{L}(V_{IQ} - V_t) \tag{4.40}$$

Observe that the voltage gain is proportional to the values of R_D, the transconductance parameter $k_n' = \mu_n C_{ox}$, the transistor aspect ratio W/L, and the overdrive voltage at the bias point $V_{OV} = V_{IQ} - V_t$.

Another simple and very useful expression for the voltage gain can be obtained by substituting $v_I = V_{IQ}$ and $v_O = V_{OQ}$ in Eq. (4.39), utilizing Eq. (4.40), and substituting $V_{IQ} - V_t = V_{OV}$. The result is

$$A_v = -\frac{2(V_{DD} - V_{OQ})}{V_{OV}} = -\frac{2V_{RD}}{V_{OV}} \tag{4.41}$$

where V_{RD} is the dc voltage across the drain resistor R_D; that is, $V_{RD} = V_{DD} - V_{OQ}$.

The end point of the saturation-region segment is characterized by

$$V_{OB} = V_{IB} - V_t \tag{4.42}$$

Thus its coordinates can be determined by substituting $v_O = V_{OB}$ and $v_I = V_{IB}$ in Eq. (4.39) and solving the resulting equation simultaneously with Eq. (4.42).

The Triode-Region Segment, BC Here, $v_I \geq V_t$, and $v_O \leq v_I - V_t$. Substituting for i_D by the triode-region expression

$$i_D = \mu_n C_{ox} \frac{W}{L}\left[(v_I - V_t)v_O - \frac{1}{2}v_O^2\right]$$

into

$$v_O = V_{DD} - R_D i_D$$

gives

$$v_O = V_{DD} - R_D \mu_n C_{ox} \frac{W}{L}\left[(v_I - V_t)v_O - \frac{1}{2}v_O^2\right]$$

The portion of this segment for which v_O is small is given approximately by

$$v_O \cong V_{DD} - R_D \mu_n C_{ox} \frac{W}{L}(v_I - V_t)v_O$$

which reduces to

$$v_O = V_{DD}\Big/\left[1 + R_D \mu_n C_{ox} \frac{W}{L}(v_I - V_t)\right] \tag{4.43}$$

We can use the expression for r_{DS}, the drain-to-source resistance near the origin of the i_D–v_{DS} plane (Eq. 4.13),

$$r_{DS} = 1\Big/\left[\mu_n C_{ox} \frac{W}{L}(v_I - V_t)\right]$$

together with Eq. (4.43) to obtain

$$v_O = V_{DD}\frac{r_{DS}}{r_{DS} + R_D} \tag{4.44}$$

which makes intuitive sense: For small v_O, the MOSFET operates as a resistance r_{DS} (whose value is determined by v_I), which forms with R_D a voltage divider across V_{DD}. Usually, $r_{DS} \ll R_D$, and Eq. (4.44) reduces to

$$v_O \cong V_{DD}\frac{r_{DS}}{R_D} \tag{4.45}$$

EXAMPLE 4.8

To make the above analysis more concrete we consider a numerical example. Specifically, consider the CS circuit of Fig. 4.26(a) for the case $k_n'\,(W/L) = 1\ \text{mA/V}^2$, $V_t = 1\ \text{V}$, $R_D = 18\ \text{k}\Omega$, and $V_{DD} = 10\ \text{V}$.

Solution

First, we determine the coordinates of important points on the transfer curve.

(a) Point X:

$$v_I = 0\ \text{V}, \qquad v_O = 10\ \text{V}$$

(b) Point A:

$$v_I = 1\ \text{V}, \qquad v_O = 10\ \text{V}$$

(c) Point B: Substituting

$$v_I = V_{IB} = V_{OB} + V_t$$
$$= V_{OB} + 1$$

and $v_O = V_{OB}$ in Eq. (4.39) results in

$$9V_{OB}^2 + V_{OB} - 10 = 0$$

which has two roots, only one of which makes physical sense, namely,

$$V_{OB} = 1\ \text{V}$$

Correspondingly,

$$V_{IB} = 1 + 1 = 2\ \text{V}$$

(d) Point C: From Eq. (4.43) we find

$$V_{OC} = \frac{10}{1 + 18 \times 1 \times (10 - 1)} = 0.061\ \text{V}$$

which is very small, justifying our use of the approximate expression in Eq. (4.43).

Next, we bias the amplifier to operate at an appropriate point on the saturation-region segment. Since this segment extends from $v_O = 1\ \text{V}$ to $10\ \text{V}$, we choose to operate at $V_{OQ} = 4\ \text{V}$. This point allows for reasonable signal swing in both directions and provides a higher voltage gain than available at the middle of the range (i.e., at $V_{OQ} = 5.5\ \text{V}$). To operate at an output dc voltage

of 4 V, the dc drain current must be

$$I_D = \frac{V_{DD} - V_{OQ}}{R_D} = \frac{10 - 4}{18} = 0.333 \text{ mA}$$

We can find the required overdrive voltage V_{OV} from

$$I_D = \frac{1}{2} k_n' \frac{W}{L} V_{OV}^2$$

$$V_{OV} = \sqrt{\frac{2 \times 0.333}{1}} = 0.816 \text{ V}$$

Thus, we must operate the MOSFET at a dc gate-to-source voltage

$$V_{GSQ} = V_t + V_{OV} = 1.816 \text{ V}$$

The voltage-gain of the amplifier at this bias point can be found from Eq. (4.40) as

$$A_v = -18 \times 1 \times (1.816 - 1)$$

$$= -14.7 \text{ V/V}$$

To gain insight into the operation of the amplifier we apply an input signal v_i of, say, 150 mV peak-to-peak amplitude, of, say, triangular waveform. Figure 4.28(a) shows such a signal superimposed on the dc bias voltage $V_{GSQ} = 1.816$ V. As shown, v_{GS} varies linearly between 1.741 V and 1.891 V around the bias value of 1.816 V. Correspondingly, i_D will be

$$\text{At } v_{GS} = 1.741 \text{ V}, \ i_D = \tfrac{1}{2} \times 1 \times (1.741 - 1)^2 = 0.275 \text{ mA}$$

$$\text{At } v_{GS} = 1.816 \text{ V}, \ i_D = \tfrac{1}{2} \times 1 \times (1.816 - 1)^2 = 0.333 \text{ mA}$$

$$\text{At } v_{GS} = 1.891 \text{ V}, \ i_D = \tfrac{1}{2} \times 1 \times (1.891 - 1)^2 = 0.397 \text{ mA}$$

Note that the negative increment in i_D is (0.333 – 0.275) = 0.058 mA while the positive increment is (0.397 – 0.333) = 0.064 mA, which are slightly different, indicating that the segment of the i_D–v_{GS} curve (or, equivalently, of the v_O–v_I curve) is not perfectly linear, as should be expected. The output voltage will vary around the bias value $V_{OQ} = 4$ V and will have the following extremities:

$$\text{At } v_{GS} = 1.741 \text{ V}, \ i_D = 0.275 \text{ mA, and } v_O = 10 - 0.275 \times 18 = 5.05 \text{ V}$$

$$\text{At } v_{GS} = 1.891 \text{ V}, \ i_D = 0.397 \text{ mA, and } v_O = 10 - 0.397 \times 18 = 2.85 \text{ V}$$

Thus, while the positive increment is 1.05 V, the negative excursion is slightly larger at 1.15 V, again a result of the nonlinear transfer characteristic. The nonlinear distortion of v_O can be reduced by reducing the amplitude of the input signal.

Further insight into the operation of this amplifier can be gained by considering its graphical analysis shown in Fig. 4.28(b). Observe that as v_{GS} varies, because of v_i, the **instantaneous operating point** moves along the load line, being at the intersection of the load line and the i_D–v_{DS} curve corresponding to the instantaneous value of v_{GS}.

We note that by biasing the transistor at a quiescent point in the middle of the saturation region, we ensure that the instantaneous operating point always remains in the saturation region, and thus nonlinear distortion is minimized. Finally, we note that in this example we carried out our calculations to three decimal digits, simply to illustrate the concepts involved. In practice, this degree of precision is not justified for approximate manual analysis.

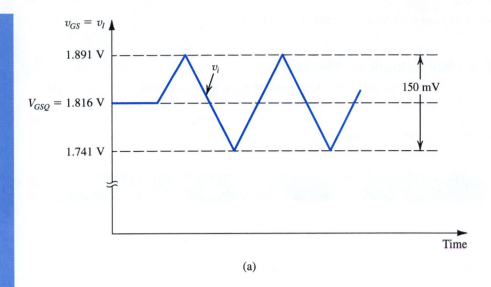

(a)

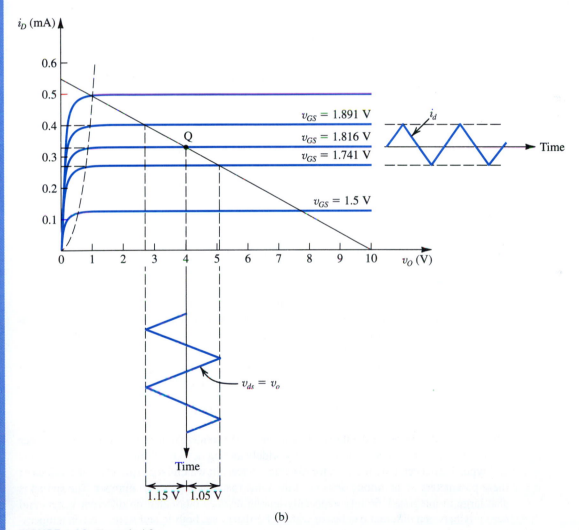

(b)

FIGURE 4.28 Example 4.8.

4.4.6 A Final Remark on Biasing

In the above example, the MOSFET was assumed to be biased at a constant v_{GS} of 1.816 V. Although it is possible to generate a constant bias voltage using an appropriate voltage-divider network across the power supply V_{DD} or across another reference voltage that may be available in the system, fixing the value of v_{GS} is not a good biasing technique. In the next section we will explain why this is so and present superior biasing schemes.

EXERCISES

4.17 For the circuit studied in Example 4.8 above and with reference to the transfer characteristic sketched in Fig. 4.26(c): (a) Give the values of V_{IQ}, V_{IB}, V_{OQ}, and V_{OB}. (b) Use the values in (a) to determine the largest allowable value of the negative peak of the output signal and the magnitude of the corresponding positive peak of the input signal. Disregard distortion caused by the square-law MOSFET characteristic. (c) Repeat (b) for the positive-output peak and the corresponding negative-input peak. (d) From the results of (b) and (c), what is the maximum amplitude of a sine wave that can be applied at the input and the corresponding output amplitude. What value of gain do these amplitudes imply? Why is it different from the 14.7 V/V found in Example 4.8?

Ans. (a) 1.816 V, 2 V, 4 V, 1 V; (b) 3 V, 0.184 V; (c) 6 V, 0.816 V; (d) 0.184 V, 3 V, 16.3 V/V, because of the nonlinear transfer characteristic.

4.18 Derive the voltage-gain expression in Eq. (4.41). Use the expression to verify the gain value found in Example 4.8.

4.5 BIASING IN MOS AMPLIFIER CIRCUITS

As mentioned in the previous section, an essential step in the design of a MOSFET amplifier circuit is the establishment of an appropriate dc operating point for the transistor. This is the step known as biasing or bias design. An appropriate dc operating point or bias point is characterized by a stable and predictable dc drain current I_D and by a dc drain-to-source voltage V_{DS} that ensures operation in the saturation region for all expected input-signal levels.

4.5.1 Biasing by Fixing V_{GS}

The most straightforward approach to biasing a MOSFET is to fix its gate-to-source voltage V_{GS} to the value required to provide the desired I_D. This voltage value can be derived from the power supply voltage V_{DD} through the use of an appropriate voltage divider. Alternatively, it can be derived from another suitable reference voltage that might be available in the system. Independent of how the voltage V_{GS} may be generated, this is not a good approach to biasing a MOSFET. To understand the reason for this statement, recall that

$$I_D = \frac{1}{2}\mu_n C_{ox}\frac{W}{L}(V_{GS} - V_t)^2$$

and note that the values of the threshold voltage V_t, the oxide-capacitance C_{ox}, and (to a lesser extent) the transistor aspect ratio W/L vary widely among devices of supposedly the same size and type. This is certainly the case for discrete devices, in which large spreads in the values of these parameters occur among devices of the same manufacturer's part number. The spread is also large in integrated circuits, especially among devices fabricated on different wafers and certainly between different batches of wafers. Furthermore, both V_t and μ_n depend on temperature, with the result that if we fix the value of V_{GS}, the drain current I_D becomes very much temperature dependent.

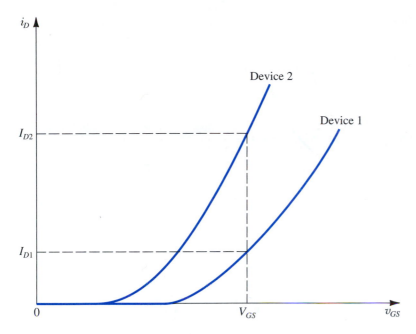

FIGURE 4.29 The use of fixed bias (constant V_{GS}) can result in a large variability in the value of I_D. Devices 1 and 2 represent extremes among units of the same type.

To emphasize the point that biasing by fixing V_{GS} is not a good technique, we show in Fig. 4.29 two i_D–v_{GS} characteristic curves representing extreme values in a batch of MOSFETs of the same type. Observe that for the fixed value of V_{GS}, the resultant spread in the values of the drain current can be substantial.

4.5.2 Biasing by Fixing V_G and Connecting a Resistance in the Source

An excellent biasing technique for discrete MOSFET circuits consists of fixing the dc voltage at the gate, V_G, and connecting a resistance in the source lead, as shown in Fig. 4.30(a). For this circuit we can write

$$V_G = V_{GS} + R_S I_D \qquad (4.46)$$

Now, if V_G is much greater than V_{GS}, I_D will be mostly determined by the values of V_G and R_S. However, even if V_G is not much larger than V_{GS}, resistor R_S provides *negative feedback*, which acts to stabilize the value of the bias current I_D. To see how this comes about consider the case when I_D increases for whatever reason. Equation (4.46) indicates that since V_G is constant, V_{GS} will have to decrease. This in turn results in a decrease in I_D, a change that is opposite to that initially assumed. Thus the action of R_S works to keep I_D as constant as possible. This negative feedback action of R_S gives it the name **degeneration resistance,** a name that we will appreciate much better at a later point in this text.

Figure 4.30(b) provides a graphical illustration of the effectiveness of this biasing scheme. Here we show the i_D–v_{GS} characteristics for two devices that represent the extremes of a batch of MOSFETs. Superimposed on the device characteristics is a straight line that represents the constraint imposed by the bias circuit—namely, Eq. (4.46). The intersection of this straight line with the i_D–v_{GS} characteristic curve provides the coordinates (I_D and V_{GS}) of the bias point. Observe that compared to the case of fixed V_{GS}, here the variability obtained in I_D is much smaller. Also, note that the variability decreases as V_G and R_S are made larger (providing a bias line that is less steep).

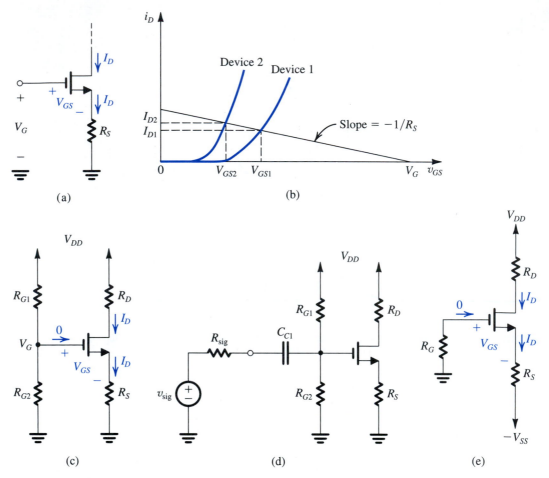

FIGURE 4.30 Biasing using a fixed voltage at the gate, V_G, and a resistance in the source lead, R_S: **(a)** basic arrangement; **(b)** reduced variability in I_D; **(c)** practical implementation using a single supply; **(d)** coupling of a signal source to the gate using a capacitor C_{C1}; **(e)** practical implementation using two supplies.

Two possible practical discrete implementations of this bias scheme are shown in Fig. 4.30(c) and (e). The circuit in Fig. 4.30(c) utilizes one power-supply V_{DD} and derives V_G through a voltage divider (R_{G1}, R_{G2}). Since $I_G = 0$, R_{G1} and R_{G2} can be selected to be very large (in the MΩ range), allowing the MOSFET to present a large input resistance to a signal source that may be connected to the gate through a coupling capacitor, as shown in Fig. 4.30(d). Here capacitor C_{C1} blocks dc and thus allows us to couple the signal v_{sig} to the amplifier input without disturbing the MOSFET dc bias point. The value of C_{C1} should be selected sufficiently large so that it approximates a short circuit at all signal frequencies of interest. We shall study capacitively coupled MOSFET amplifiers, which are suitable only in discrete circuit design, in Section 4.7. Finally, note that in the circuit of Fig. 4.30(c), resistor R_D is selected to be as large as possible to obtain high gain but small enough to allow for the desired signal swing at the drain while keeping the MOSFET in saturation at all times.

When two power supplies are available, as is often the case, the somewhat simpler bias arrangement of Fig. 4.30(e) can be utilized. This circuit is an implementation of Eq. (4.46), with V_G replaced by V_{SS}. Resistor R_G establishes a dc ground at the gate and presents a high input resistance to a signal source that may be connected to the gate through a coupling capacitor.

EXAMPLE 4.9

It is required to design the circuit of Fig. 4.30(c) to establish a dc drain current $I_D = 0.5$ mA. The MOSFET is specified to have $V_t = 1$ V and $k'_n W/L = 1$ mA/V^2. For simplicity, neglect the channel-length modulation effect (i.e., assume $\lambda = 0$). Use a power-supply $V_{DD} = 15$ V. Calculate the percentage change in the value of I_D obtained when the MOSFET is replaced with another unit having the same $k'_n W/L$ but $V_t = 1.5$ V.

Solution

As a rule of thumb for designing this classical biasing circuit, we choose R_D and R_S to provide one-third of the power-supply voltage V_{DD} as a drop across each of R_D, the transistor (i.e., V_{DS}) and R_S. For $V_{DD} = 15$ V, this choice makes $V_D = +10$ V and $V_S = +5$ V. Now, since I_D is required to be 0.5 mA, we can find the values of R_D and R_S as follows:

$$R_D = \frac{V_{DD} - V_D}{I_D} = \frac{15 - 10}{0.5} = 10 \text{ k}\Omega$$

$$R_S = \frac{V_S}{R_S} = \frac{5}{0.5} = 10 \text{ k}\Omega$$

The required value of V_{GS} can be determined by first calculating the overdrive voltage V_{OV} from

$$I_D = \tfrac{1}{2} k'_n (W/L) V_{OV}^2$$

$$0.5 = \tfrac{1}{2} \times 1 \times V_{OV}^2$$

which yields $V_{OV} = 1$ V, and thus,

$$V_{GS} = V_t + V_{OV} = 1 + 1 = 2 \text{ V}$$

Now, since $V_S = +5$ V, V_G must be

$$V_G = V_S + V_{GS} = 5 + 2 = 7 \text{ V}$$

To establish this voltage at the gate we may select $R_{G1} = 8$ MΩ and $R_{G2} = 7$ MΩ. The final circuit is shown in Fig. 4.31. Observe that the dc voltage at the drain (+10 V) allows for a positive signal swing of +5 V (i.e., up to V_{DD}) and a negative signal swing of −4 V [i.e., down to $(V_G - V_t)$].

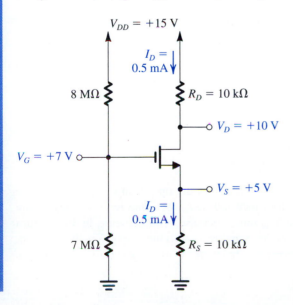

FIGURE 4.31 Circuit for Example 4.9.

If the NMOS transistor is replaced with another having $V_t = 1.5$ V, the new value of I_D can be found as follows:

$$I_D = \tfrac{1}{2} \times 1 \times (V_{GS} - 1.5)^2 \tag{4.47}$$

$$V_G = V_{GS} + I_D R_S$$

$$7 = V_{GS} + 10 I_D \tag{4.48}$$

Solving Eqs. (4.47) and (4.48) together yields

$$I_D = 0.455 \text{ mA}$$

Thus the change in I_D is

$$\Delta I_D = 0.455 - 0.5 = -0.045 \text{ mA}$$

which is $\dfrac{-0.045}{0.5} \times 100 = -9\%$ change.

EXERCISES

4.19 Consider the MOSFET in Example 4.9 when fixed-V_{GS} bias is used. Find the required value of V_{GS} to establish a dc bias current $I_D = 0.5$ mA. Recall that the device parameters are $V_t = 1$ V, $k_n' W/L = 1$ mA/V^2, and $\lambda = 0$. What is the percentage change in I_D obtained when the transistor is replaced with another having $V_t = 1.5$ V?

Ans. $V_{GS} = 2$ V; -75%

D4.20 Design the circuit of Fig. 4.30(e) to operate at a dc drain current of 0.5 mA and $V_D = +2$ V. Let $V_t = 1$ V, $k_n' W/L = 1$ mA/V^2, $\lambda = 0$, $V_{DD} = V_{SS} = 5$ V. Use standard 5% resistor values (see Appendix G), and give the resulting values of I_D, V_D, and V_S.

Ans. $R_D = R_S = 6.2$ kΩ; $I_D = 0.49$ mA, $V_S = -1.96$ V, and $V_D = +1.96$ V. R_G can be selected in the range of 1 MΩ to 10 MΩ.

4.5.3 Biasing Using a Drain-to-Gate Feedback Resistor

A simple and effective discrete-circuit biasing arrangement utilizing a feedback resistor connected between the drain and the gate is shown in Fig. 4.32. Here the large feedback resistance R_G (usually in the MΩ range) forces the dc voltage at the gate to be equal to that at the drain (because $I_G = 0$). Thus we can write

$$V_{GS} = V_{DS} = V_{DD} - R_D I_D$$

which can be rewritten in the form

$$V_{DD} = V_{GS} + R_D I_D \tag{4.49}$$

which is identical in form to Eq. (4.46), which describes the operation of the bias scheme discussed above [that in Fig. 4.30(a)]. Thus, here too, if I_D for some reason changes, say increases, then Eq. (4.49) indicates that V_{GS} must decrease. The decrease in V_{GS} in turn causes a decrease in I_D, a change that is opposite in direction to the one originally assumed. Thus the negative feedback or degeneration provided by R_G works to keep the value of I_D as constant as possible.

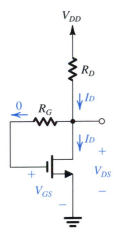

FIGURE 4.32 Biasing the MOSFET using a large drain-to-gate feedback resistance, R_G.

The circuit of Fig. 4.32 can be utilized as a CS amplifier by applying the input voltage signal to the gate via a coupling capacitor so as not to disturb the dc bias conditions already established. The amplified output signal at the drain can be coupled to another part of the circuit, again via a capacitor. We shall consider such a CS amplifier circuit in Section 4.6. There we will learn that this circuit has the drawback of a rather limited output voltage signal swing.

EXERCISE

D4.21 It is required to design the circuit in Fig. 4.32 to operate at a dc drain current of 0.5 mA. Assume $V_{DD} = +5$ V, $k_n' W/L = 1$ mA/V^2, $V_t = 1$ V, and $\lambda = 0$. Use a standard 5% resistance value for R_D, and give the actual values obtained for I_D and V_D.
Ans. $R_D = 6.2$ kΩ; $I_D \cong 0.49$ mA; $V_D \cong 1.96$ V

4.5.4 Biasing Using a Constant-Current Source

The most effective scheme for biasing a MOSFET amplifier is that using a constant-current source. Figure 4.33(a) shows such an arrangement applied to a discrete MOSFET. Here R_G (usually in the MΩ range) establishes a dc ground at the gate and presents a large resistance to an input signal source that can be capacitively coupled to the gate. Resistor R_D establishes an appropriate dc voltage at the drain to allow for the required output signal swing while ensuring that the transistor always remains in the saturation region.

A circuit for implementing the constant-current source I is shown in Fig. 4.33(b). The heart of the circuit is transistor Q_1, whose drain is shorted to its gate and thus is operating in the saturation region, such that

$$I_{D1} = \frac{1}{2}k_n'\left(\frac{W}{L}\right)_1 (V_{GS} - V_t)^2 \tag{4.50}$$

where we have neglected channel-length modulation (i.e., assumed $\lambda = 0$). The drain current of Q_1 is supplied by V_{DD} through resistor R. Since the gate currents are zero,

$$I_{D1} = I_{REF} = \frac{V_{DD} + V_{SS} - V_{GS}}{R} \tag{4.51}$$

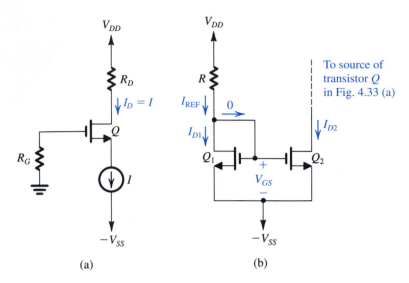

FIGURE 4.33 (a) Biasing the MOSFET using a constant-current source I. (b) Implementation of the constant-current source I using a current mirror.

where the current through R is considered to be the *reference current* of the current source and is denoted I_{REF}. Given the parameter values of Q_1 and a desired value for I_{REF}, Eqs. (4.50) and (4.51) can be used to determine the value of R. Now consider transistor Q_2: It has the same V_{GS} as Q_1; thus if we assume that it is operating in saturation, its drain current, which is the desired current I of the current source, will be

$$I = I_{D2} = \frac{1}{2}k_n'\left(\frac{W}{L}\right)_2(V_{GS} - V_t)^2 \tag{4.52}$$

where we have neglected channel-length modulation. Equations (4.51) and (4.52) enable us to relate the current I to the reference current I_{REF},

$$I = I_{REF}\frac{(W/L)_2}{(W/L)_1} \tag{4.53}$$

Thus I is related to I_{REF} by the ratio of the aspect ratios of Q_1 and Q_2. This circuit, known as a **current mirror,** is very popular in the design of IC MOS amplifiers and will be studied in great detail in Chapter 6.

EXERCISE

D4.22 Using two transistors Q_1 and Q_2 having equal lengths but widths related by $W_2/W_1 = 5$, design the circuit of Fig. 4.33(b) to obtain $I = 0.5$ mA. Let $V_{DD} = -V_{SS} = 5$ V, $k_n'(W/L)_1 = 0.8$ mA/V^2, $V_t = 1$ V, and $\lambda = 0$. Find the required value for R. What is the voltage at the gates of Q_1 and Q_2? What is the lowest voltage allowed at the drain of Q_2 while Q_2 remains in the saturation region?
Ans. 85 kΩ; −3.5 V; −4.5 V

4.5.5 A Final Remark

The bias circuits studied in this section are intended for discrete-circuit applications. The only exception is the current mirror circuit of Fig. 4.33(b) which, as mentioned above, is extensively used in IC design. Bias arrangements for IC MOS amplifiers will be studied in Chapter 6.

 ## 4.6 SMALL-SIGNAL OPERATION AND MODELS

In our study of the large-signal operation of the common-source MOSFET amplifier in Section 4.4 we learned that linear amplification can be obtained by biasing the MOSFET to operate in the saturation region and by keeping the input signal small. Having studied methods for biasing the MOS transistor in the previous section, we now turn our attention to exploring small-signal operation in some detail. For this purpose we utilize the conceptual common-source amplifier circuit shown in Fig. 4.34. Here the MOS transistor is biased by applying a dc voltage V_{GS}, a clearly impractical arrangement but one that is simple and useful for our purposes. The input signal to be amplified, v_{gs}, is shown superimposed on the dc bias voltage V_{GS}. The output voltage is taken at the drain.

4.6.1 The DC Bias Point

The dc bias current I_D can be found by setting the signal v_{gs} to zero; thus,

$$I_D = \frac{1}{2} k_n' \frac{W}{L} (V_{GS} - V_t)^2 \tag{4.54}$$

where we have neglected channel-length modulation (i.e., we have assumed $\lambda = 0$). The dc voltage at the drain, V_{DS} or simply V_D (since S is grounded), will be

$$V_D = V_{DD} - R_D I_D \tag{4.55}$$

To ensure saturation-region operation, we must have

$$V_D > V_{GS} - V_t$$

Furthermore, since the total voltage at the drain will have a signal component superimposed on V_D, V_D has to be sufficiently greater than $(V_{GS} - V_t)$ to allow for the required signal swing.

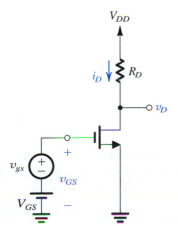

FIGURE 4.34 Conceptual circuit utilized to study the operation of the MOSFET as a small-signal amplifier.

4.6.2 The Signal Current in the Drain Terminal

Next, consider the situation with the input signal v_{gs} applied. The total instantaneous gate-to-source voltage will be

$$v_{GS} = V_{GS} + v_{gs} \tag{4.56}$$

resulting in a total instantaneous drain current i_D,

$$
\begin{aligned}
i_D &= \frac{1}{2} k_n' \frac{W}{L} (V_{GS} + v_{gs} - V_t)^2 \\
&= \frac{1}{2} k_n' \frac{W}{L} (V_{GS} - V_t)^2 + k_n' \frac{W}{L} (V_{GS} - V_t) v_{gs} + \frac{1}{2} k_n' \frac{W}{L} v_{gs}^2
\end{aligned}
\tag{4.57}
$$

The first term on the right-hand side of Eq. (4.57) can be recognized as the dc bias current I_D (Eq. 4.54). The second term represents a current component that is directly proportional to the input signal v_{gs}. The third term is a current component that is proportional to the square of the input signal. This last component is undesirable because it represents *nonlinear distortion*. To reduce the nonlinear distortion introduced by the MOSFET, the input signal should be kept small so that

$$\frac{1}{2} k_n' \frac{W}{L} v_{gs}^2 \ll k_n' \frac{W}{L} (V_{GS} - V_t) v_{gs}$$

resulting in

$$v_{gs} \ll 2(V_{GS} - V_t) \tag{4.58}$$

or, equivalently,

$$v_{gs} \ll 2 V_{OV} \tag{4.59}$$

where V_{OV} is the overdrive voltage at which the transistor is operating.

If this **small-signal condition** is satisfied, we may neglect the last term in Eq. (4.57) and express i_D as

$$i_D \simeq I_D + i_d \tag{4.60}$$

where

$$i_d = k_n' \frac{W}{L} (V_{GS} - V_t) v_{gs}$$

The parameter that relates i_d and v_{gs} is the MOSFET **transconductance** g_m,

$$g_m \equiv \frac{i_d}{v_{gs}} = k_n' \frac{W}{L} (V_{GS} - V_t) \tag{4.61}$$

or in terms of the overdrive voltage V_{OV},

$$g_m = k_n' \frac{W}{L} V_{OV} \tag{4.62}$$

Figure 4.35 presents a graphical interpretation of the small-signal operation of the enhancement MOSFET amplifier. Note that g_m is equal to the slope of the i_D–v_{GS} characteristic at the bias point,

$$g_m \equiv \frac{\partial i_D}{\partial v_{GS}} \bigg|_{v_{GS} = V_{GS}} \tag{4.63}$$

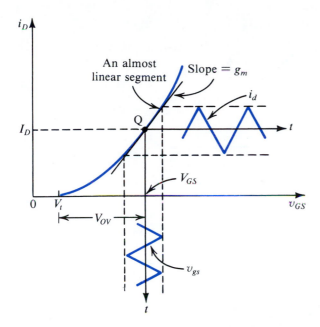

FIGURE 4.35   Small-signal operation of the enhancement MOSFET amplifier.

This is the formal definition of g_m, which can be shown to yield the expressions given in Eqs. (4.61) and (4.62).

4.6.3   The Voltage Gain

Returning to the circuit of Fig. 4.34, we can express the total instantaneous drain voltage v_D as follows:

$$v_D = V_{DD} - R_D i_D$$

Under the small-signal condition, we have

$$v_D = V_{DD} - R_D(I_D + i_d)$$

which can be rewritten as

$$v_D = V_D - R_D i_d$$

Thus the signal component of the drain voltage is

$$v_d = -i_d R_D = -g_m v_{gs} R_D \tag{4.64}$$

which indicates that the voltage gain is given by

$$A_v \equiv \frac{v_d}{v_{gs}} = -g_m R_D \tag{4.65}$$

The minus sign in Eq. (4.65) indicates that the output signal v_d is 180° out of phase with respect to the input signal v_{gs}. This is illustrated in Fig. 4.36, which shows v_{GS} and v_D. The input signal is assumed to have a triangular waveform with an amplitude much smaller than $2(V_{GS} - V_t)$, the small-signal condition in Eq. (4.58), to ensure linear operation. For operation in the saturation region at all times, the minimum value of v_D should not fall below the corresponding value of v_G by more than V_t. Also, the maximum value of v_D should be

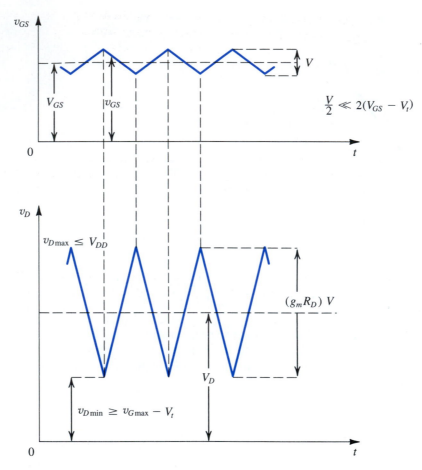

FIGURE 4.36 Total instantaneous voltages v_{GS} and v_D for the circuit in Fig. 4.34.

smaller than V_{DD}; otherwise the FET will enter the cutoff region and the peaks of the output signal waveform will be clipped off.

Finally, we note that by substituting for g_m from Eq. (4.61) the voltage gain expression in Eq. (4.65) becomes identical to that derived in Section 4.4—namely, Eq. (4.40).

4.6.4 Separating the DC Analysis and the Signal Analysis

From the preceding analysis, we see that under the small-signal approximation, signal quantities are superimposed on dc quantities. For instance, the total drain current i_D equals the dc current I_D plus the signal current i_d, the total drain voltage $v_D = V_D + v_d$, and so on. It follows that the analysis and design can be greatly simplified by separating dc or bias calculations from small-signal calculations. That is, once a stable dc operating point has been established and all dc quantities calculated, we may then perform signal analysis ignoring dc quantities.

4.6.5 Small-Signal Equivalent-Circuit Models

From a signal point of view the FET behaves as a voltage-controlled current source. It accepts a signal v_{gs} between gate and source and provides a current $g_m v_{gs}$ at the drain terminal. The input resistance of this controlled source is very high—ideally, infinite. The output resistance—that is, the resistance looking into the drain—also is high, and we have assumed

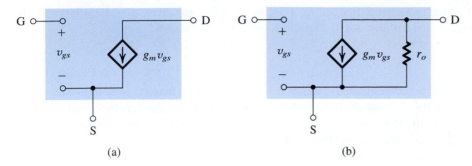

(a)   (b)

FIGURE 4.37 Small-signal models for the MOSFET: **(a)** neglecting the dependence of i_D on v_{DS} in saturation (the channel-length modulation effect); and **(b)** including the effect of channel-length modulation, modeled by output resistance $r_o = |V_A|/I_D$.

it to be infinite thus far. Putting all of this together, we arrive at the circuit in Fig. 4.37(a), which represents the small-signal operation of the MOSFET and is thus a **small-signal model** or a **small-signal equivalent circuit.**

In the analysis of a MOSFET amplifier circuit, the transistor can be replaced by the equivalent circuit model shown in Fig. 4.37(a). The rest of the circuit remains unchanged except that *ideal constant dc voltage sources are replaced by short circuits.* This is a result of the fact that the voltage across an ideal constant dc voltage source does not change, and thus there will always be a zero voltage signal across a constant dc voltage source. A dual statement applies for constant dc current sources; namely, the signal current of an ideal constant dc current source will always be zero, and thus *an ideal constant dc current source can be replaced by an open-circuit* in the small-signal equivalent circuit of the amplifier. The circuit resulting can then be used to perform any required signal analysis, such as calculating voltage gain.

The most serious shortcoming of the small-signal model of Fig. 4.37(a) is that it assumes the drain current in saturation is independent of the drain voltage. From our study of the MOSFET characteristics in saturation, we know that the drain current does in fact depend on v_{DS} in a linear manner. Such dependence was modeled by a finite resistance r_o between drain and source, whose value was given by Eq. (4.26) in Section 4.2.3, which we repeat here as

$$r_o = \frac{|V_A|}{I_D} \tag{4.66}$$

where $V_A = 1/\lambda$ is a MOSFET parameter that either is specified or can be measured. It should be recalled that for a given process technology, V_A is proportional to the MOSFET channel length. The current I_D is the value of the dc drain current without the channel-length modulation taken into account; that is,

$$I_D = \frac{1}{2}k_n'\frac{W}{L}V_{OV}^2 \tag{4.67}$$

Typically, r_o is in the range of 10 kΩ to 1000 kΩ. It follows that the accuracy of the small-signal model can be improved by including r_o in parallel with the controlled source, as shown in Fig. 4.37(b).

It is important to note that the small-signal model parameters g_m and r_o depend on the dc bias point of the MOSFET.

Returning to the amplifier of Fig. 4.34, we find that replacing the MOSFET with the small-signal model of Fig. 4.37(b) results in the voltage-gain expression

$$A_v = \frac{v_d}{v_{gs}} = -g_m(R_D /\!/ r_o) \qquad (4.68)$$

Thus the finite output resistance r_o results in a reduction in the magnitude of the voltage gain.

Although the analysis above is performed on an NMOS transistor, the results, and the equivalent circuit models of Fig. 4.37, apply equally well to PMOS devices, except for using $|V_{GS}|$, $|V_t|$, $|V_{OV}|$, and $|V_A|$ and replacing k'_n with k'_p.

4.6.6 The Transconductance g_m

We shall now take a closer look at the MOSFET transconductance given by Eq. (4.61), which we repeat here as

$$g_m = k'_n(W/L)(V_{GS} - V_t) = k'_n(W/L)V_{OV} \qquad (4.69)$$

This relationship indicates that g_m is proportional to the process transconductance parameter $k'_n = \mu_n C_{ox}$ and to the W/L ratio of the MOS transistor; hence to obtain relatively large transconductance the device must be short and wide. We also observe that for a given device the transconductance is proportional to the overdrive voltage, $V_{OV} = V_{GS} - V_t$, the amount by which the bias voltage V_{GS} exceeds the threshold voltage V_t. Note, however, that increasing g_m by biasing the device at a larger V_{GS} has the disadvantage of reducing the allowable voltage signal swing at the drain.

Another useful expression for g_m can be obtained by substituting for $(V_{GS} - V_t)$ in Eq. (4.69) by $\sqrt{2I_D/(k'_n(W/L))}$ [from Eq. (4.53)]:

$$g_m = \sqrt{2k'_n}\sqrt{W/L}\sqrt{I_D} \qquad (4.70)$$

This expression shows that

1. For a given MOSFET, g_m is proportional to the square root of the dc bias current.

2. At a given bias current, g_m is proportional to $\sqrt{W/L}$.

In contrast, the transconductance of the bipolar junction transistor (BJT) studied in Chapter 5 is proportional to the bias current and is independent of the physical size and geometry of the device.

To gain some insight into the values of g_m obtained in MOSFETs consider an integrated-circuit device operating at $I_D = 0.5$ mA and having $k'_n = 120$ μA/V^2. Equation (4.70) shows that for $W/L = 1$, $g_m = 0.35$ mA/V, whereas a device for which $W/L = 100$ has $g_m = 3.5$ mA/V. In contrast, a BJT operating at a collector current of 0.5 mA has $g_m = 20$ mA/V.

Yet another useful expression for g_m of the MOSFET can be obtained by substituting for $k'_n(W/L)$ in Eq. (4.69) by $2I_D/(V_{GS} - V_t)^2$:

$$g_m = \frac{2I_D}{V_{GS} - V_t} = \frac{2I_D}{V_{OV}} \qquad (4.71)$$

In summary, there are three different relationships for determining g_m—Eqs. (4.69), (4.70), and (4.71)—and there are three design parameters—(W/L), V_{OV}, and I_D, any two of which can be chosen independently. That is, the designer may choose to operate the MOSFET with a certain overdrive voltage V_{OV} and at a particular current I_D; the required W/L ratio can then be found and the resulting g_m determined.

EXAMPLE 4.10

Figure 4.38(a) shows a discrete common-source MOSFET amplifier utilizing the drain-to-gate feedback biasing arrangement. The input signal v_i is coupled to the gate via a large capacitor, and the output signal at the drain is coupled to the load resistance R_L via another large capacitor. We wish to analyze this amplifier circuit to determine its small-signal voltage gain, its input resistance, and the largest allowable input signal. The transistor has $V_t = 1.5$ V, $k_n'(W/L) = 0.25$ mA/V^2, and $V_A = 50$ V. Assume the coupling capacitors to be sufficiently large so as to act as short circuits at the signal frequencies of interest.

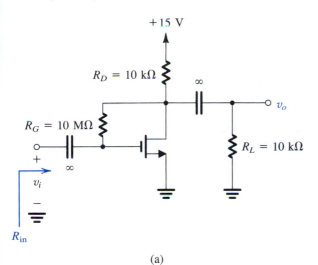

(a)

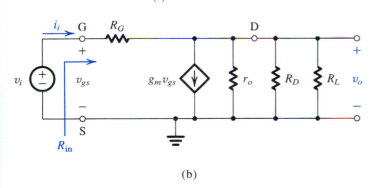

(b)

FIGURE 4.38 Example 4.10: **(a)** amplifier circuit; **(b)** equivalent-circuit model.

Solution

We first evaluate the dc operating point as follows:

$$I_D = \tfrac{1}{2} \times 0.25 (V_{GS} - 1.5)^2 \tag{4.72}$$

where, for simplicity, we have neglected the channel-length modulation effect. Since the dc gate current is zero, there will be no dc voltage drop across R_G; thus $V_{GS} = V_D$, which, when substituted in Eq. (4.72), yields

$$I_D = 0.125 (V_D - 1.5)^2 \tag{4.73}$$

Also,

$$V_D = 15 - R_D I_D = 15 - 10 I_D \tag{4.74}$$

Solving Eqs. (4.73) and (4.74) together gives

$$I_D = 1.06 \text{ mA} \quad \text{and} \quad V_D = 4.4 \text{ V}$$

(Note that the other solution to the quadratic equation is not physically meaningful.)

The value of g_m is given by

$$g_m = k_n' \frac{W}{L} (V_{GS} - V_t)$$

$$= 0.25(4.4 - 1.5) = 0.725 \text{ mA/V}$$

The output resistance r_o is given by

$$r_o = \frac{V_A}{I_D} = \frac{50}{1.06} = 47 \text{ k}\Omega$$

Figure 4.38(b) shows the small-signal equivalent circuit of the amplifier, where we observe that the coupling capacitors have been replaced with short circuits and the dc supply has been replaced with a short circuit to ground. Since R_G is very large (10 MΩ), the current through it can be neglected compared to that of the controlled source $g_m v_{gs}$, enabling us to write for the output voltage

$$v_o \simeq -g_m v_{gs} (R_D /\!/ R_L /\!/ r_o)$$

Since $v_{gs} = v_i$, the voltage gain is

$$A_v = \frac{v_o}{v_i} = -g_m (R_D /\!/ R_L /\!/ r_o)$$

$$= -0.725(10 /\!/ 10 /\!/ 47) = -3.3 \text{ V/V}$$

To evaluate the input resistance R_{in}, we note that the input current i_i is given by

$$i_i = (v_i - v_o)/R_G$$

$$= \frac{v_i}{R_G} \left(1 - \frac{v_o}{v_i} \right)$$

$$= \frac{v_i}{R_G}[1 - (-3.3)] = \frac{4.3 v_i}{R_G}$$

Thus,

$$R_{in} \equiv \frac{v_i}{i_i} = \frac{R_G}{4.3} = \frac{10}{4.3} = 2.33 \text{ M}\Omega$$

The largest allowable input signal $\hat{v}_i$ is determined by the need to keep the MOSFET in saturation at all times; that is,

$$v_{DS} \geq v_{GS} - V_t$$

Enforcing this condition, with equality, at the point v_{GS} is maximum and v_{DS} is correspondingly minimum, we write

$$v_{DSmin} = v_{GSmax} - V_t$$

$$V_{DS} - |A_v|\hat{v}_i = V_{GS} + \hat{v}_i - V_t$$

$$4.4 - 3.3\hat{v}_i = 4.4 + \hat{v}_i - 1.5$$

which results in

$$\hat{v}_i = 0.34 \text{ V}$$

Note that in the negative direction, this input signal amplitude results in $v_{GSmin} = 4.4 - 0.34 = 4.06$ V, which is larger than V_t, and thus the transistor remains conducting. Thus, as we have surmised, the limitation on input signal amplitude is posed by the upper-end considerations, and the maximum allowable input signal peak is 0.34 V.

4.6.7 The T Equivalent-Circuit Model

Through a simple circuit transformation it is possible to develop an alternative equivalent-circuit model for the MOSFET. The development of such a model, known as the T model, is illustrated in Fig. 4.39. Figure 4.39(a) shows the equivalent circuit studied above without r_o. In Fig. 4.39(b) we have added a second $g_m v_{gs}$ current source in series with the original controlled source. This addition obviously does not change the terminal currents and is thus allowed. The newly created circuit node, labeled X, is joined to the gate terminal G in Fig. 4.39(c). Observe that the gate current does not change—that is, it remains equal to zero—and thus this connection does not alter the terminal characteristics. We now note that

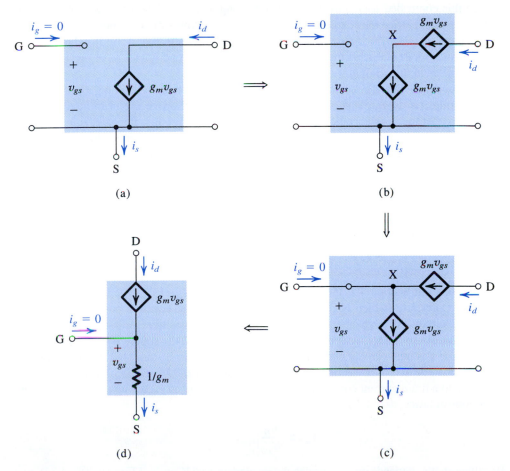

FIGURE 4.39 Development of the T equivalent-circuit model for the MOSFET. For simplicity, r_o has been omitted but can be added between D and S in the T model of (**d**).

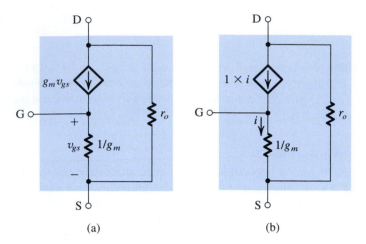

FIGURE 4.40 (a) The T model of the MOSFET augmented with the drain-to-source resistance r_o. (b) An alternative representation of the T model.

we have a controlled current source $g_m v_{gs}$ connected across its control voltage v_{gs}. We can replace this controlled source by a resistance as long as this resistance draws an equal current as the source. (See the source-absorption theorem in Appendix C.) Thus the value of the resistance is $v_{gs}/g_m v_{gs} = 1/g_m$. This replacement is shown in Fig. 4.39(d), which depicts the alternative model. Observe that i_g is still zero, $i_d = g_m v_{gs}$, and $i_s = v_{gs}/(1/g_m) = g_m v_{gs}$, all the same as in the original model in Fig. 4.39(a).

The model of Fig. 4.39(d) shows that the resistance between gate and source looking into the source is $1/g_m$. This observation and the T model prove useful in many applications. Note that the resistance between gate and source, looking into the gate, is infinite.

In developing the T model we did not include r_o. If desired, this can be done by incorporating in the circuit of Fig. 4.39(d) a resistance r_o between drain and source, as shown in Fig. 4.40(a). An alternative representation of the T model in which the voltage-controlled current source is replaced with a current-controlled current source is shown in Fig. 4.40(b).

Finally, we should note that in order to distinguish the model of Fig. 4.37(b) from the equivalent T model, the former is sometimes referred to as the **hybrid-π model,** a carryover from the bipolar transistor literature. The origin of this name will be explained in the next chapter.

4.6.8 Modeling the Body Effect

As mentioned in Section 4.2, the body effect occurs in a MOSFET when the source is not tied to the substrate (which is always connected to the most-negative power supply in the integrated circuit for n-channel devices and to the most-positive for p-channel devices). Thus the substrate (body) will be at signal ground, but since the source is not, a signal voltage v_{bs} develops between the body (B) and the source (S). In Section 4.2, it was mentioned that the substrate acts as a "second gate" or a **backgate** for the MOSFET. Thus the signal v_{bs} gives rise to a drain-current component, which we shall write as $g_{mb} v_{bs}$, where g_{mb} is the **body transconductance,** defined as

$$g_{mb} \equiv \frac{\partial i_D}{\partial v_{BS}}\bigg|_{\substack{v_{GS}\,=\,\text{constant} \\ v_{DS}\,=\,\text{constant}}} \tag{4.75}$$

Recalling that i_D depends on v_{BS} through the dependence of V_t on V_{BS}, Eqs. (4.20), (4.33), and (4.61) can be used to obtain

$$g_{mb} = \chi g_m \tag{4.76}$$

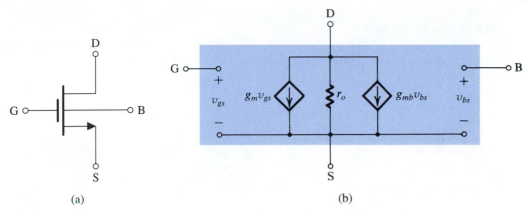

(a)   (b)

FIGURE 4.41 Small-signal equivalent-circuit model of a MOSFET in which the source is not connected to the body.

where

$$\chi \equiv \frac{\partial V_t}{\partial V_{SB}} = \frac{\gamma}{2\sqrt{2\phi_f + V_{SB}}} \qquad (4.77)$$

Typically the value of χ lies in the range 0.1 to 0.3.

Figure 4.41 shows the MOSFET model augmented to include the controlled source $g_{mb}v_{bs}$ that models the body effect. This is the model to be used whenever the source is not connected to the substrate.

Finally, although the analysis above was performed on a NMOS transistor, the results and the equivalent circuit of Fig. 4.41 apply equally well to PMOS transistors, except for using $|V_{GS}|$, $|V_t|$, $|V_{OV}|$, $|V_A|$, $|V_{SB}|$, $|\gamma|$, and $|\lambda|$ and replacing k'_n with k'_p.

4.6.9 Summary

We conclude this section by presenting in Table 4.2 a summary of the formulas for calculating the values of the small-signal MOSFET parameters. Observe that for g_m we have three different formulas, each providing the circuit designer with insight regarding design choices. We shall make frequent comments on these in later sections and chapters.

TABLE 4.2   Small-Signal Equivalent-Circuit Models for the MOSFET

Small-Signal Parameters

NMOS transistors:

- Transconductance:

$$g_m = \mu_n C_{ox} \frac{W}{L} V_{OV} = \sqrt{2\mu_n C_{ox} \frac{W}{L} I_D} = \frac{2I_D}{V_{OV}}$$

- Output resistance:

$$r_o = V_A / I_D = 1/\lambda I_D$$

- Body transconductance:

$$g_{mb} = \chi g_m = \frac{\gamma}{2\sqrt{2\phi_f + V_{SB}}} g_m$$

PMOS transistors:

Same formulas as for NMOS *except* using $|V_{OV}|$, $|V_A|$, $|\lambda|$, $|\gamma|$, $|V_{SB}|$, and $|\chi|$ and replacing μ_n with μ_p.

(Continued)

TABLE 4.2 *(Continued)*

Small-Signal Equivalent Circuit Models when $|V_{SB}| = 0$ (i.e., No Body Effect)

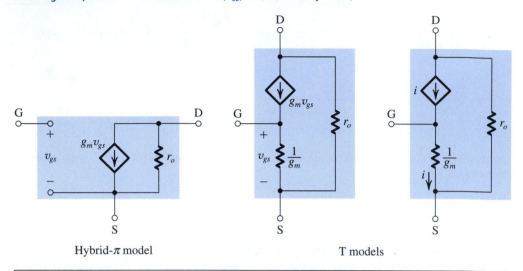

Hybrid-π model   T models

Small-Signal Circuit Model when $|V_{SB}| \neq 0$ (i.e., Including the Body Effect)

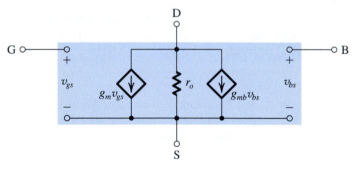

Hybrid-π model

EXERCISES

4.23 For the amplifier in Fig. 4.34, let $V_{DD} = 5$ V, $R_D = 10$ kΩ, $V_t = 1$ V, $= 20$ μA/V^2, $W/L = 20$, $V_{GS} = 2$ V, and $\lambda = 0$. (a) Find the dc current I_D and the dc voltage V_D. (b) Find g_m. (c) Find the voltage gain. (d) If $v_{gs} = 0.2 \sin \omega t$ volts, find v_d assuming that the small-signal approximation holds. What are the minimum and maximum values of v_D? (e) Use Eq. (4.57) to determine the various components of i_D. Using the identity ($\sin^2 \omega t = \cos 2\omega t$), show that there is a slight shift in I_D (by how much?) and that there is a second-harmonic component (i.e., a component with frequency 2ω). Express the amplitude of the second-harmonic component as a percentage of the amplitude of the fundamental. (This value is known as the second-harmonic distortion.)

Ans. (a) 200 μA, 3 V; (b) 0.4 mA/V; (c) -4 V/V; (d) $v_d = -0.8 \sin \omega t$ volts, 2.2 V, 3.8 V; (e) $i_D = (204 + 80 \sin \omega t - 4 \cos 2\omega t)$ μA, 5%

4.24 An NMOS transistor has $\mu_n C_{ox} = 60\ \mu\text{A/V}^2$, $W/L = 40$, $V_t = 1$ V, and $V_A = 15$ V. Find g_m and r_o when (a) the bias voltage $V_{GS} = 1.5$ V, and when (b) the bias current $I_D = 0.5$ mA.

Ans. (a) 1.2 mA/V, 50 kΩ; (b) 1.55 mA/V, 30 kΩ

4.25 A MOSFET is to operate at $I_D = 0.1$ mA and is to have $g_m = 1$ mA/V. If $k'_n = 50\ \mu\text{A/V}^2$, find the required W/L ratio and the overdrive voltage.

Ans. 100, 0.2 V

4.26 For a fabrication process for which $\mu_p \simeq 0.4\mu_n$, find the ratio of the width of a PMOS transistor to the width of an NMOS transistor so that the two devices have equal g_m for the same bias conditions. The two devices have equal channel lengths.

Ans. 2.5

4.27 For an NMOS transistor with $2\phi_f = 0.6$ V, $\gamma = 0.5$ V$^{1/2}$, and $V_{SB} = 4$ V, find $\chi \equiv g_{mb}/g_m$.

Ans. 0.12

4.28 A PMOS transistor has $V_t = -1$ V, $k'_p = 60\ \mu\text{A/V}^2$, and $W/L = 16\ \mu\text{m}/0.8\ \mu\text{m}$. Find I_D and g_m when the device is biased at $V_{GS} = -1.6$ V. Also, find the value of r_o if λ (at $L = 1\ \mu\text{m}$) $= -0.04$ V^{-1}.

Ans. 216 μA; 0.72 mA/V; 92.6 kΩ

4.29 Use the formulas in Table 4.2 to derive an expression for $(g_m r_o)$ in terms of V_A and V_{OV}. As we shall see in Chapter 6, this is an important transistor parameter and is known as the intrinsic gain. Evaluate the value for $g_m r_o$ for an NMOS transistor fabricated in a 0.8-μm CMOS process for which $V'_A = 12.5$ V/μm of channel length. Let the device have minimum channel length and be operated at an overdrive voltage of 0.2 V.

Ans. $g_m r_o = 2V_A/V_{OV}$; 100 V/V

 ## 4.7 SINGLE-STAGE MOS AMPLIFIERS

Having studied MOS amplifier biasing (Section 4.5) and the small-signal operation and models of the MOSFET amplifier (Section 4.6), we are now ready to consider the various configurations utilized in the design of MOS amplifiers. In this section we shall do this for the case of discrete MOS amplifiers, leaving the study of integrated-circuit (IC) MOS amplifiers to Chapter 6. Beside being useful in their own right, discrete MOS amplifiers are somewhat easier to understand than their IC counterparts for two main reasons: The separation between dc and signal quantities is more obvious in discrete circuits, and discrete circuits utilize resistors as amplifier loads. In contrast, as we shall see in Chapter 6, IC MOS amplifiers employ constant-current sources as amplifier loads, with these being implemented using additional MOSFETs and resulting in more complicated circuits. Thus the circuits studied in this section should provide us with both an introduction to the subject of MOS amplifier configurations and a solid base on which to build during our study of IC MOS amplifiers in Chapter 6.

Since in discrete circuits the MOSFET source is usually tied to the substrate, the body effect will be absent. Therefore in this section we shall *not* take the body effect into account. Also, in some circuits we will neglect r_o in order to keep the analysis simple and focus our attention at this early stage on the salient features of the amplifier configurations studied.

4.7.1 The Basic Structure

Figure 4.42 shows the basic circuit we shall utilize to implement the various configurations of discrete-circuit MOS amplifiers. Among the various schemes for biasing discrete MOS

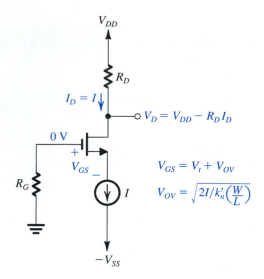

FIGURE 4.42 Basic structure of the circuit used to realize single-stage discrete-circuit MOS amplifier configurations.

amplifiers (Section 4.5) we have selected, for both its effectiveness and its simplicity, the one employing constant-current biasing. Figure 4.42 indicates the dc current and the dc voltages resulting at various nodes.

EXERCISE

4.30 Consider the circuit of Fig. 4.42 for the case $V_{DD} = V_{SS} = 10$ V, $I = 0.5$ mA, $R_G = 4.7$ MΩ, $R_D = 15$ kΩ, $V_t = 1.5$ V, and $k_n'(W/L) = 1$ mA/V^2. Find V_{OV}, V_{GS}, V_G, V_S, and V_D. Also, calculate the values of g_m and r_o, assuming that $V_A = 75$ V. What is the maximum possible signal swing at the drain for which the MOSFET remains in saturation?

Ans. See Fig. E4.30; without taking into account the signal swing at the gate, the drain can swing to −1.5 V, a negative signal swing of 4 V.

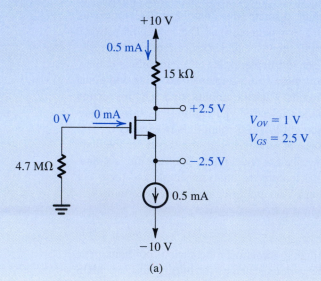

(a)

FIGURE E4.30

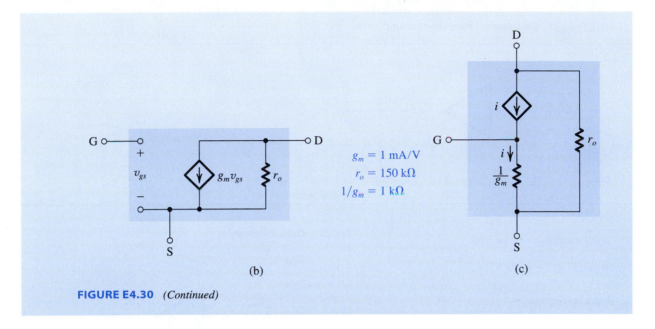

$$g_m = 1 \text{ mA/V}$$
$$r_o = 150 \text{ k}\Omega$$
$$1/g_m = 1 \text{ k}\Omega$$

(b) (c)

FIGURE E4.30 *(Continued)*

4.7.2 Characterizing Amplifiers

As we begin our study of MOS amplifier circuits, it is important to know how to characterize the performance of amplifiers as circuit building blocks. An introduction to this subject was presented in Section 1.5. However, the material of Section 1.5 was limited to **unilateral amplifiers.** A number of the amplifier circuits we shall study in this book, though none in this chapter, are not unilateral; that is, they have internal feedback that may cause their input resistance to depend on the load resistance. Similarly, internal feedback may cause the output resistance to depend on the value of the resistance of the signal source feeding the amplifier. To accommodate **nonunilateral amplifiers,** we present, in Table 4.3, a general set of parameters and equivalent circuits that we will employ in characterizing and comparing transistor amplifiers. A number of remarks are in order:

1. The amplifier is shown fed with a signal source having an open-circuit voltage v_{sig} and an internal resistance R_{sig}. These can be the parameters of an actual signal source or the Thévenin equivalent of the output circuit of another amplifier stage preceding the one under study in a cascade amplifier. Similarly, R_L can be an actual load resistance or the input resistance of a succeeding amplifier stage in a cascade amplifier.

2. Parameters R_i, R_o, A_{vo}, A_{is}, and G_m pertain to the *amplifier proper;* that is, they do *not* depend on the values of R_{sig} and R_L. By contrast, R_{in}, R_{out}, A_v, A_i, G_{vo}, and G_v may depend on one or both of R_{sig} and R_L. Also, observe the relationships of related pairs of these parameters; for instance, $R_i = R_{in}|_{R_L=\infty}$, and $R_o = R_{out}|_{R_{sig}=0}$.

3. As mentioned above, for nonunilateral amplifiers, R_{in} may depend on R_L, and R_{out} may depend on R_{sig}. Although none of the amplifiers studied in this chapter are of this type, we shall encounter nonunilateral MOSFET amplifiers in Chapter 6 and beyond. No such dependencies exist for unilateral amplifiers, for which $R_{in} = R_i$ and $R_{out} = R_o$.

4. The *loading* of the amplifier on the signal source is determined by the input resistance R_{in}. The value of R_{in} determines the current i_i that the amplifier draws from the signal source. It also determines the proportion of the signal v_{sig} that appears at the input of the amplifier proper (i.e., v_i).

TABLE 4.3 Characteristic Parameters of Amplifiers

Circuit

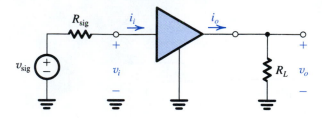

Definitions

■ Input resistance with no load:

$$R_i \equiv \frac{v_i}{i_i}\bigg|_{R_L=\infty}$$

■ Input resistance:

$$R_{in} \equiv \frac{v_i}{i_i}$$

■ Open-circuit voltage gain:

$$A_{vo} \equiv \frac{v_o}{v_i}\bigg|_{R_L=\infty}$$

■ Voltage gain:

$$A_v \equiv \frac{v_o}{v_i}$$

■ Short-circuit current gain:

$$A_{is} \equiv \frac{i_o}{i_i}\bigg|_{R_L=0}$$

■ Current gain:

$$A_i \equiv \frac{i_o}{i_i}$$

■ Short-circuit transconductance:

$$G_m \equiv \frac{i_o}{v_i}\bigg|_{R_L=0}$$

■ Output resistance of amplifier proper:

$$R_o \equiv \frac{v_x}{i_x}\bigg|_{v_i=0}$$

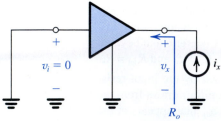

■ Output resistance:

$$R_{out} \equiv \frac{v_x}{i_x}\bigg|_{v_{sig}=0}$$

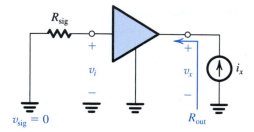

■ Open-circuit overall voltage gain:

$$G_{vo} \equiv \frac{v_o}{v_{sig}}\bigg|_{R_L=\infty}$$

■ Overall voltage gain:

$$G_v \equiv \frac{v_o}{v_{sig}}$$

Equivalent Circuits

A:

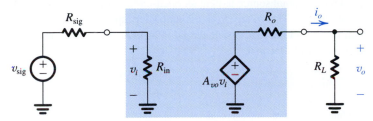

B:

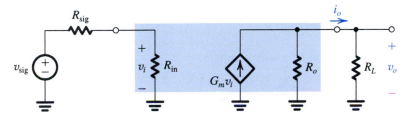

C:

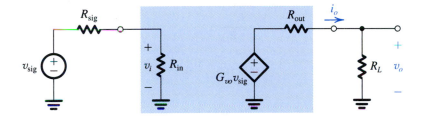

Relationships

$$\frac{v_i}{v_{\text{sig}}} = \frac{R_{\text{in}}}{R_{\text{in}} + R_{\text{sig}}}$$

$$A_v = A_{vo}\frac{R_L}{R_L + R_o}$$

$$A_{vo} = G_m R_o$$

$$G_v = \frac{R_{\text{in}}}{R_{\text{in}} + R_{\text{sig}}} A_{vo}\frac{R_L}{R_L + R_o}$$

$$G_{vo} = \frac{R_i}{R_i + R_{\text{sig}}} A_{vo}$$

$$G_v = G_{vo}\frac{R_L}{R_L + R_{\text{out}}}$$

5. When evaluating the gain A_v from the open-circuit value A_{vo}, R_o is the output resistance to use. This is because A_v is based on feeding the amplifier with an ideal voltage signal v_i. This should be evident from Equivalent Circuit A in Table 4.3. On the other hand, if we are evaluating the overall voltage gain G_v from its open-circuit value G_{vo}, the output resistance to use is R_{out}. This is because G_v is based on feeding the amplifier with v_{sig}, which has an internal resistance R_{sig}. This should be evident from Equivalent Circuit C in Table 4.3.

6. We urge the reader to carefully examine and reflect on the definitions and the six relationships presented in Table 4.3. Example 4.11 should help in this regard.

EXAMPLE 4.11

A transistor amplifier is fed with a signal source having an open-circuit voltage v_{sig} of 10 mV and an internal resistance R_{sig} of 100 kΩ. The voltage v_i at the amplifier input and the output voltage v_o are measured both without and with a load resistance $R_L = 10$ kΩ connected to the amplifier output. The measured results are as follows:

	v_i (mV)	v_o (mV)
Without R_L	9	90
With R_L connected	8	70

Find all the amplifier parameters.

Solution

First, we use the data obtained for $R_L = \infty$ to determine

$$A_{vo} = \frac{90}{9} = 10 \text{ V/V}$$

and

$$G_{vo} = \frac{90}{10} = 9 \text{ V/V}$$

Now, since

$$G_{vo} = \frac{R_i}{R_i + R_{sig}} A_{vo}$$

$$9 = \frac{R_i}{R_i + 100} \times 10$$

which gives

$$R_i = 900 \text{ kΩ}$$

Next, we use the data obtained when $R_L = 10$ kΩ is connected to the amplifier output to determine

$$A_v = \frac{70}{8} = 8.75 \text{ V/V}$$

and

$$G_v = \frac{70}{10} = 7 \text{ V/V}$$

The values of A_v and A_{vo} can be used to determine R_o as follows:

$$A_v = A_{vo}\frac{R_L}{R_L + R_o}$$

$$8.75 = 10\frac{10}{10 + R_o}$$

which gives

$$R_o = 1.43 \text{ kΩ}$$

Similarly, we use the values of G_v and G_{vo} to determine R_{out} from

$$G_v = G_{vo}\frac{R_L}{R_L + R_{out}}$$

$$7 = 9\frac{10}{10 + R_{out}}$$

resulting in

$$R_{out} = 2.86 \text{ k}\Omega$$

The value of R_{in} can be determined from

$$\frac{v_i}{v_{sig}} = \frac{R_{in}}{R_{in} + R_{sig}}$$

Thus,

$$\frac{8}{10} = \frac{R_{in}}{R_{in} + 100}$$

which yields

$$R_{in} = 400 \text{ k}\Omega$$

The short-circuit transconductance G_m can be found as follows:

$$G_m = \frac{A_{vo}}{R_o} = \frac{10}{1.43} = 7 \text{ mA/V}$$

and the current gain A_i can be determined as follows:

$$A_i = \frac{v_o/R_L}{v_i/R_{in}} = \frac{v_o}{v_i}\frac{R_{in}}{R_L}$$

$$= A_v\frac{R_{in}}{R_L} = 8.75 \times \frac{400}{10} = 350 \text{ A/A}$$

Finally, we determine the short-circuit current gain A_{is} as follows. From Equivalent Circuit A in Table 4.3, the short-circuit output current is

$$i_{osc} = A_{vo}v_i/R_o$$

However, to determine v_i we need to know the value of R_{in} obtained with $R_L = 0$. Toward that end, note that from Equivalent Circuit C, the output short-circuit current can be found as

$$i_{osc} = G_{vo}v_{sig}/R_{out}$$

Now, equating the two expressions for i_{osc} and substituting for G_{vo} by

$$G_{vo} = \frac{R_i}{R_i + R_{sig}}A_{vo}$$

and for v_i from

$$v_i = v_{sig}\frac{R_{in}|_{R_L=0}}{R_{in}|_{R_L=0} + R_{sig}}$$

results in

$$R_{in}|_{R_L=0} = R_{sig}\Bigg/\left[\left(1 + \frac{R_{sig}}{R_i}\right)\left(\frac{R_{out}}{R_o}\right) - 1\right]$$

$$= 81.8 \text{ k}\Omega$$

We now can use

$$i_{osc} = A_{vo}i_i R_{in}|_{R_L=0}/R_o$$

to obtain

$$A_{is} \equiv \frac{i_{osc}}{i_i} = 10 \times 81.8/1.43 = 572 \text{ A/A}$$

4.7.3 The Common-Source (CS) Amplifier

The common-source (CS) or grounded-source configuration is the most widely used of all MOSFET amplifier circuits. A common-source amplifier realized using the circuit of Fig. 4.42 is shown in Fig. 4.43(a). Observe that to establish a **signal ground,** or an **ac ground** as it is sometimes called, at the source, we have connected a large capacitor, C_S, between the source and ground. This capacitor, usually in the μF range, is required to provide a very small impedance (ideally, zero impedance; i.e., in effect, a short circuit) at all signal frequencies of interest. In this way, the signal current passes through C_S to ground and thus *bypasses* the output resistance of current source I (and any other circuit component that might be connected to the MOSFET source); hence, C_S is called a **bypass capacitor.** Obviously, the lower the signal frequency, the less effective the bypass capacitor becomes. This issue will be studied in Section 4.9. For our purposes here we shall assume that C_S is acting as a perfect short circuit and thus is establishing a zero signal voltage at the MOSFET source.

In order not to disturb the dc bias current and voltages, the signal to be amplified, shown as voltage source v_{sig} with an internal resistance R_{sig}, is connected to the gate through a large capacitor C_{C1}. Capacitor C_{C1}, known as a **coupling capacitor,** is required to act as a perfect short circuit at all signal frequencies of interest while blocking dc. Here again, we note that as the signal frequency is lowered, the impedance of C_{C1} (i.e., $1/j\omega C_{C1}$) will increase and its effectiveness as a coupling capacitor will be correspondingly reduced. This problem too will be considered in Section 4.9 when the dependence of the amplifier operation on frequency is studied. For our purposes here we shall assume C_{C1} is acting as a perfect short circuit as far as the signal is concerned. Before leaving C_{C1}, we should point out that in situations where the signal source can provide an appropriate dc path to ground, the gate can be connected directly to the signal source and both R_G and C_{C1} can be dispensed with.

The voltage signal resulting at the drain is coupled to the load resistance R_L via another coupling capacitor C_{C2}. We shall assume that C_{C2} acts as a perfect short circuit at all signal frequencies of interest and thus that the output voltage $v_o = v_d$. Note that R_L can be either an actual load resistor, to which the amplifier is required to provide its output voltage signal, or it can be the input resistance of another amplifier stage in cases where more than one stage of amplification is needed. (We will study multistage amplifiers in Chapter 7.)

To determine the terminal characteristics of the CS amplifier—that is, its input resistance, voltage gain, and output resistance—we replace the MOSFET with its small-signal model. The resulting circuit is shown in Fig. 4.43(b). At the outset we observe that this amplifier is unilateral. Therefore R_{in} does not depend on R_L, and thus $R_{in} = R_i$. Also, R_{out} will not depend on R_{sig}, and thus $R_{out} = R_o$. Analysis of this circuit is straightforward and proceeds in a step-by-step manner, from the signal source to the amplifier load. At the input

$$i_g = 0$$

$$R_{in} = R_G \tag{4.78}$$

$$v_i = v_{sig}\frac{R_{in}}{R_{in} + R_{sig}} = v_{sig}\frac{R_G}{R_G + R_{sig}} \tag{4.79}$$

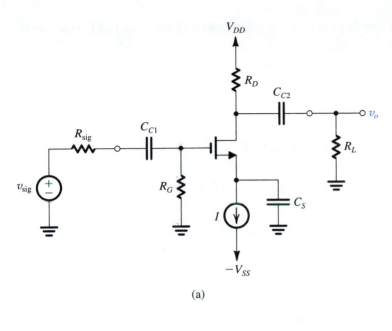

(a)

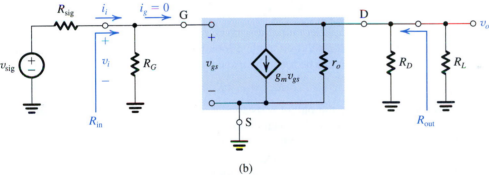

(b)

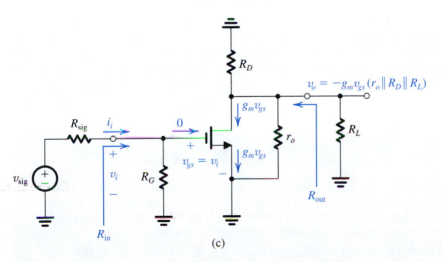

(c)

FIGURE 4.43 **(a)** Common-source amplifier based on the circuit of Fig. 4.42. **(b)** Equivalent circuit of the amplifier for small-signal analysis. **(c)** Small-signal analysis performed directly on the amplifier circuit with the MOSFET model implicitly utilized.

Usually R_G is selected very large (e.g., in the MΩ range) with the result that in many applications $R_G \gg R_{sig}$ and

$$v_i \cong v_{sig}$$

Now

$$v_{gs} = v_i$$

and

$$v_o = -g_m v_{gs}(r_o \parallel R_D \parallel R_L)$$

Thus the voltage gain A_v is

$$A_v = -g_m(r_o \parallel R_D \parallel R_L) \tag{4.80}$$

and the open-circuit voltage gain A_{vo} is

$$A_{vo} = -g_m(r_o \parallel R_D) \tag{4.81}$$

The overall voltage gain from the signal-source to the load will be

$$G_v = \frac{R_{in}}{R_{in} + R_{sig}} A_v$$

$$= -\frac{R_G}{R_G + R_{sig}} g_m(r_o \parallel R_D \parallel R_L) \tag{4.82}$$

Finally, to determine the amplifier output resistance R_{out} we set v_{sig} to 0; that is, we replace the signal generator v_{sig} with a short circuit and look back into the output terminal, as indicated in Fig. 4.43. The result can be found by inspection as

$$R_{out} = r_o \parallel R_D \tag{4.83}$$

As we have seen, including the output resistance r_o in the analysis of the CS amplifier is straightforward: Since r_o appears between drain and source, it in effect appears in parallel with R_D. Since it is usually the case that $r_o \gg R_D$, the effect of r_o will be a slight decrease in the voltage gain and a decrease in R_{out}—the latter being a beneficial effect!

Although small-signal equivalent circuit models provide a systematic process for the analysis of any amplifier circuit, the effort involved in drawing the equivalent circuit is sometimes not justified. That is, in simple situations and after a lot of practice, one can perform the small-signal analysis directly on the original circuit. In such a situation, the small-signal MOSFET model is employed implicitly rather than explicitly. In order to get the reader started in this direction, we show in Fig. 4.43(c) the small-signal analysis of the CS amplifier performed on a somewhat simplified version of the circuit. We urge the reader to examine this analysis and to correlate it with the analysis using the equivalent circuit of Fig. 4.43(b).

EXERCISE

4.32 Consider a CS amplifier based on the circuit analyzed in Exercise 4.30. Specifically, refer to the results of that exercise shown in Fig. E4.30. Find R_{in}, A_{vo}, and R_{out}, both without and with r_o taken into account. Then calculate the overall voltage gain G_v, with r_o taken into account, for the case $R_{sig} = 100$ kΩ and $R_L = 15$ kΩ. If v_{sig} is a 0.4-V peak-to-peak sinusoid, what output signal v_o results?

Ans. Without r_o: $R_{in} = 4.7$ MΩ, $A_{vo} = -15$ V/V, and $R_{out} = 15$ kΩ; with r_o: $R_{in} = 4.7$ MΩ, $A_{vo} = -13.6$ V/V, and $R_{out} = 13.6$ kΩ; $G_v = -7$ V/V; v_o is a 2.8-V peak-to-peak sinusoid superimposed on a dc drain voltage of +2.5 V.

We conclude our study of the CS amplifier by noting that it has a very high input resistance, a moderately high voltage gain, and a relatively high output resistance.

4.7.4 The Common-Source Amplifier with a Source Resistance

It is often beneficial to insert a resistance R_S in the source lead of the common-source amplifier, as shown in Fig. 4.44(a). The corresponding small-signal equivalent circuit is shown in

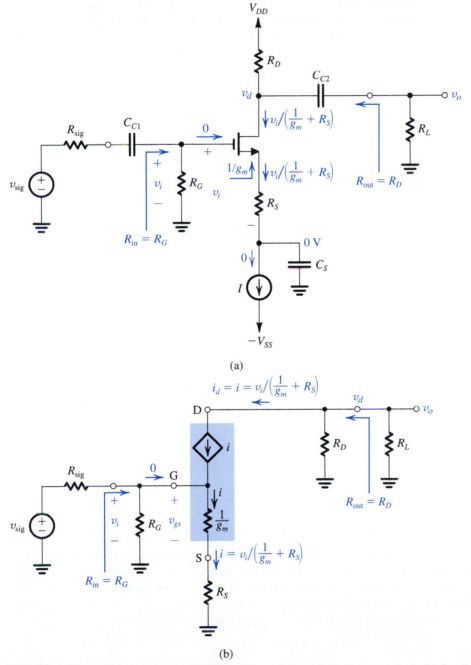

(a)

(b)

FIGURE 4.44 (a) Common-source amplifier with a resistance R_S in the source lead. (b) Small-signal equivalent circuit with r_o neglected.

Fig. 4.44(b) where we note that the transistor has been replaced by its T equivalent-circuit model. The T model is used in preference to the π model because it makes the analysis in this case somewhat simpler. In general, whenever a resistance is connected in the source lead, as for instance in the source-follower circuit we shall consider shortly, the T model is preferred: The source resistance then simply appears in series with the resistance $1/g_m$, which represents the resistance between source and gate, looking into the source.

It should be noted that we have not included r_o in the equivalent-circuit model. Including r_o would complicate the analysis considerably; r_o would connect the output node of the amplifier to the input side and thus would make the amplifier nonunilateral. Fortunately, it turns out that the effect of r_o on the operation of this discrete-circuit amplifier is not important. This can be verified using SPICE simulation (Section 4.12). This is *not* the case, however, for the integrated-circuit version of the circuit where r_o plays a major role and must be taken into account in the analysis and design of the circuit, which we shall do in Chapter 6.

From Fig. 4.44(b) we see that as in the case of the CS amplifier,

$$R_{\text{in}} = R_i = R_G \tag{4.84}$$

and thus,

$$v_i = v_{\text{sig}}\frac{R_G}{R_G + R_{\text{sig}}} \tag{4.85}$$

Unlike the CS circuit, however, here v_{gs} is only a fraction of v_i. It can be determined from the voltage divider composed of $1/g_m$ and R_S that appears across the amplifier input as follows:

$$v_{gs} = v_i\frac{\dfrac{1}{g_m}}{\dfrac{1}{g_m} + R_S} = \frac{v_i}{1 + g_m R_S} \tag{4.86}$$

Thus we can use the value of R_S to control the magnitude of the signal v_{gs} and thus ensure that v_{gs} does not become too large and cause unacceptably high nonlinear distortion. (Recall the constraint on v_{gs} given by Eq. 4.59). This is the first benefit of including resistor R_S. Other benefits will be encountered in later sections and chapters. For instance, we will show by SPICE simulation in Section 4.12 that R_S causes the useful bandwidth of the amplifier to be extended. The mechanism by which R_S causes such improvements in amplifier performance is that of negative feedback. Unfortunately, the price paid for these improvements is a reduction in voltage gain, as we shall now show.

The current i_d is equal to the current i flowing in the source lead; thus,

$$i_d = i = \frac{v_i}{\dfrac{1}{g_m} + R_S} = \frac{g_m v_i}{1 + g_m R_S} \tag{4.87}$$

Thus including R_S reduces i_d by the factor $(1 + g_m R_S)$, which is hardly surprising since this is the factor relating v_{gs} to v_i and the MOSFET produces $i_d = g_m v_{gs}$. Equation (4.87) indicates also that the effect of R_S can be thought of as reducing the effective g_m by the factor $(1 + g_m R_S)$.

The output voltage can now be found from

$$v_o = -i_d(R_D \parallel R_L)$$

$$= -\frac{g_m(R_D \parallel R_L)}{1 + g_m R_S}v_i$$

Thus the voltage gain is

$$A_v = -\frac{g_m(R_D \parallel R_L)}{1 + g_m R_S} \tag{4.88}$$

and setting $R_L = \infty$ gives

$$A_{vo} = -\frac{g_m R_D}{1 + g_m R_S} \tag{4.89}$$

The overall voltage gain G_v is

$$G_v = -\frac{R_G}{R_G + R_{sig}} \frac{g_m(R_D \parallel R_L)}{1 + g_m R_S} \tag{4.90}$$

Comparing Eqs. (4.88), (4.89), and (4.90) with their counterparts without R_S indicates that including R_S results in a gain reduction by the factor $(1 + g_m R_S)$. In Chapter 8 we shall study negative feedback in some detail. There we will learn that this factor is called the **amount of feedback** and that it determines both the magnitude of performance improvements and, as a trade-off, the reduction in gain. At this point, we should recall that in Section 4.5 we saw that a resistance R_S in the source lead increases dc bias stability; that is, R_S reduces the variability in I_D. The action of R_S that reduces the variability of I_D is exactly the same action we are observing here: R_S in the circuit of Fig. 4.44 is reducing i_d, which is, after all, just a variation in I_D. Because of its action in reducing the gain, R_S is called **source degeneration resistance.**

Another useful interpretation of the gain expression in Eq. (4.88) is that *the gain from gate to drain is simply the ratio of the total resistance in the drain, $(R_D \parallel R_L)$, to the total resistance in the source, $[(1/g_m) + R_S]$.*

Finally, we wish to direct the reader's attention to the small-signal analysis that is performed and indicated directly on the circuit in Fig. 4.44(a). Again, with some practice, the reader should be able to dispense, in simple situations, with the extra work involved in drawing a complete equivalent circuit model and use the MOSFET model implicitly. This also has the added advantage of providing greater insight regarding circuit operation and, furthermore, reduces the probability of making manipulation errors in circuit analysis.

EXERCISE

4.33 In Exercise 4.32 we applied an input signal of 0.4 V peak-to-peak, which resulted in an output signal of the CS amplifier of 2.8 V peak-to-peak. Assume that for some reason we now have an input signal three times as large as before (i.e., 1.2 V p-p) and that we wish to modify the circuit to keep the output signal level unchanged. What value should we use for R_S?

Ans. 2.15 kΩ

4.7.5 The Common-Gate (CG) Amplifier

By establishing a signal ground on the MOSFET gate terminal, a circuit configuration aptly named **common-gate (CG)** or **grounded-gate amplifier** is obtained. The input signal is applied to the source, and the output is taken at the drain, with the gate forming a common terminal between the input and output ports. Figure 4.45(a) shows a CG amplifier obtained from the circuit of Fig. 4.42. Observe that since both the dc and ac voltages at the gate are to be zero, we have connected the gate directly to ground, thus eliminating resistor R_G altogether. Coupling capacitors C_{C1} and C_{C2} perform similar functions to those in the CS circuit.

The small-signal equivalent circuit model of the CG amplifier is shown in Fig. 4.45(b). Since resistor R_{sig} appears directly in series with the MOSFET source lead we have selected the T model for the transistor. Either model, of course, can be used and yields identical

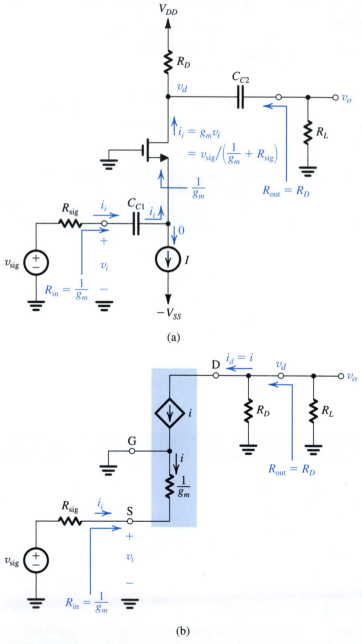

(a)

(b)

FIGURE 4.45 **(a)** A common-gate amplifier based on the circuit of Fig. 4.42. **(b)** A small-signal equivalent circuit of the amplifier in (a).

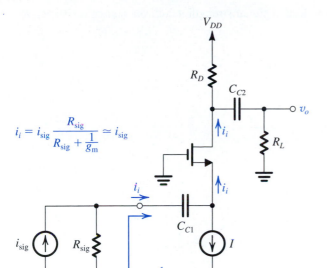

$$i_i = i_{sig} \frac{R_{sig}}{R_{sig} + \frac{1}{g_m}} \simeq i_{sig}$$

$$R_{in} = \frac{1}{g_m}$$

(c)

FIGURE 4.45 *(Continued)* **(c)** The common-gate amplifier fed with a current-signal input.

results; however, the T model is more convenient in this case. Observe also that we have not included r_o. Including r_o here would complicate the analysis considerably, for it would appear between the output and input of the amplifier. We will consider the effect of r_o when we study the IC form of the CG amplifier in Chapter 6.

From inspection of the equivalent-circuit model in Fig. 4.45(b) we see that the input resistance is

$$R_{in} = \frac{1}{g_m} \tag{4.91}$$

This should have been expected since we are looking into the source terminal of the MOSFET and the gate is grounded.[7] Furthermore, since the circuit is unilateral, R_{in} is independent of R_L, and $R_{in} = R_i$. Since g_m is of the order of 1 mA/V, the input resistance of the CG amplifier can be relatively low (of the order of 1 kΩ) and certainly much lower than in the case of the CS amplifier. It follows that significant loss of signal strength can occur in coupling the signal to the input of the CG amplifier, since

$$v_i = v_{sig} \frac{R_{in}}{R_{in} + R_{sig}} \tag{4.92}$$

Thus,

$$v_i = v_{sig} \frac{\frac{1}{g_m}}{\frac{1}{g_m} + R_{sig}} = v_{sig} \frac{1}{1 + g_m R_{sig}} \tag{4.93}$$

---

[7] As we will see in Chapter 6, when r_o is taken into account, R_{in} depends on R_D and R_L and can be quite different from $1/g_m$.

from which we see that to keep the loss in signal strength small, the source resistance R_{sig} should be small,

$$R_{sig} \ll \frac{1}{g_m}$$

The current i_i is given by

$$i_i = \frac{v_i}{R_{in}} = \frac{v_i}{1/g_m} = g_m v_i$$

and the drain current i_d is

$$i_d = i = -i_i = -g_m v_i$$

Thus the output voltage can be found as

$$v_o = v_d = -i_d(R_D \| R_L) = g_m(R_D \| R_L)v_i$$

resulting in the voltage gain

$$A_v = g_m(R_D \| R_L) \tag{4.94}$$

from which the open-circuit voltage gain can be found as

$$A_{vo} = g_m R_D \tag{4.95}$$

The overall voltage gain can be obtained as follows:

$$G_v = \frac{R_{in}}{R_{in} + R_{sig}} A_v = \frac{\dfrac{1}{g_m}}{\dfrac{1}{g_m} + R_{sig}} A_v = \frac{A_v}{1 + g_m R_{sig}} \tag{4.96a}$$

resulting in

$$G_v = \frac{g_m(R_D \| R_L)}{1 + g_m R_{sig}} \tag{4.96b}$$

Finally, the output resistance is found by inspection to be

$$R_{out} = R_o = R_D \tag{4.97}$$

Comparing these expressions with those for the common-source amplifier we make the following observations:

1. Unlike the CS amplifier, which is inverting, the CG amplifier is noninverting. This, however, is seldom a significant consideration.

2. While the CS amplifier has a very high input resistance, the input resistance of the CG amplifier is low.

3. While the A_v values of both CS and CG amplifiers are nearly identical, the overall voltage gain of the CG amplifier is smaller by the factor $1 + g_m R_{sig}$ (Eq. 4.96b), which is due to the low input resistance of the CG circuit.

The observations above do not show any particular advantage for the CG circuit; to explore this circuit further we take a closer look at its operation. Figure 4.45(c) shows the CG amplifier fed with a signal current-source i_{sig} having an internal resistance R_{sig}. This can, of course, be the Norton equivalent of the signal source used in Fig. 4.45(a). Now, using

$R_{in} = 1/g_m$ and the current-divider rule we can find the fraction of i_{sig} that flows into the MOSFET source, i_i,

$$i_i = i_{sig} \frac{R_{sig}}{R_{sig} + R_{in}} = i_{sig} \frac{R_{sig}}{R_{sig} + \dfrac{1}{g_m}} \qquad (4.98)$$

Normally, $R_{sig} \gg 1/g_m$, and

$$i_i \cong i_{sig} \qquad (4.98a)$$

Thus we see that the circuit presents a relatively low input resistance $1/g_m$ to the input signal-current source, resulting in very little signal-current attenuation at the input. The MOSFET then reproduces this current in the drain terminal at a much higher output resistance. The circuit thus acts in effect as a **unity-gain current amplifier** or a **current follower.** This view of the operation of the common-gate amplifier has resulted in its most popular application, in a configuration known as the **cascode circuit,** which we shall study in Chapter 6.

Another area of application of the CG amplifier makes use of its superior high-frequency performance, as compared to that of the CS stage (Section 4.9). We shall study wideband amplifier circuits in Chapter 6. Here we should note that the low input-resistance of the CG amplifier can be an advantage in some very-high-frequency applications where the input signal connection can be thought of as a *transmission line* and the $1/g_m$ input resistance of the CG amplifier can be made to function as the *termination resistance* of the transmission line (see Problem 4.86).

EXERCISE

4.34 Consider a CG amplifier designed using the circuit of Fig. 4.42, which is analyzed in Exercise 4.30 with the analysis results displayed in Fig. E4.30. Note that $g_m = 1$ mA/V and $R_D = 15$ kΩ. Find R_{in}, R_{out}, A_{vo}, A_v, and G_v for $R_L = 15$ kΩ and $R_{sig} = 50$ Ω. What will the overall voltage gain become for $R_{sig} = 1$ kΩ? 10 kΩ? 100 kΩ?

Ans. 1 kΩ, 15 kΩ, +15 V/V, +7.5 V/V; +6.85 V/V; +3.75 V/V; 0.68 V/V ; 0.07 V/V

4.7.6 The Common-Drain or Source-Follower Amplifier

The last single-stage MOSFET amplifier configuration we shall study is that obtained by establishing a signal ground at the drain and using it as a terminal common to the input port, between gate and drain, and the output port, between source and drain. By analogy to the CS and CG amplifier configurations, this circuit is called **common-drain** or **grounded-drain amplifier.** However, it is known more popularly as the source follower, for a reason that will become apparent shortly.

Figure 4.46(a) shows a common-drain amplifier based on the circuit of Fig. 4.42. Since the drain is to function as a signal ground, there is no need for resistor R_D, and it has therefore been eliminated. The input signal is coupled via capacitor C_{C1} to the MOSFET gate, and the output signal at the MOSFET source is coupled via capacitor C_{C2} to a load resistor R_L.

Since R_L is in effect connected in series with the source terminal of the transistor (current source I acts as an open circuit as far as signals are concerned), it is more convenient to use the MOSFET's T model. The resulting small-signal equivalent circuit of the common-drain

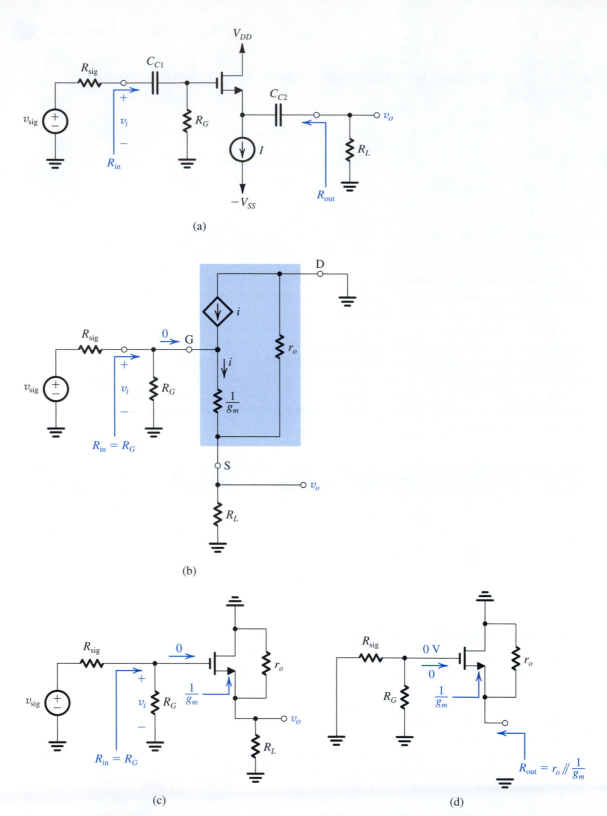

FIGURE 4.46 (a) A common-drain or source-follower amplifier. (b) Small-signal equivalent-circuit model. (c) Small-signal analysis performed directly on the circuit. (d) Circuit for determining the output resistance R_{out} of the source follower.

amplifier is shown in Fig. 4.46(b). Analysis of this circuit is straightforward and proceeds as follows: The input resistance R_{in} is given by

$$R_{\text{in}} = R_G \tag{4.99}$$

Thus,

$$v_i = v_{\text{sig}}\frac{R_{\text{in}}}{R_{\text{in}} + R_{\text{sig}}} = v_{\text{sig}}\frac{R_G}{R_G + R_{\text{sig}}} \tag{4.100}$$

Usually R_G is selected to be much larger than R_{sig} with the result that

$$v_i \cong v_{\text{sig}}$$

To proceed with the analysis, it is important to note that r_o appears in effect in parallel with R_L, with the result that between the gate and ground we have a resistance $(1/g_m)$ in series with $(R_L \parallel r_o)$. The signal v_i appears across this total resistance. Thus we may use the voltage-divider rule to determine v_o as

$$v_o = v_i\frac{R_L \parallel r_o}{(R_L \parallel r_o) + \dfrac{1}{g_m}} \tag{4.101}$$

from which the voltage gain A_v is obtained as

$$A_v = \frac{R_L \parallel r_o}{(R_L \parallel r_o) + \dfrac{1}{g_m}} \tag{4.102}$$

and the open-circuit voltage gain A_{vo} as

$$A_{vo} = \frac{r_o}{r_o + \dfrac{1}{g_m}} \tag{4.103}$$

Normally $r_o \gg 1/g_m$, causing the open-circuit voltage gain from gate to source, A_{vo} in Eq. (4.103), to become nearly unity. Thus the voltage at the source follows that at the gate, giving the circuit its popular name of **source follower.** Also, in many discrete-circuit applications, $r_o \gg R_L$, which enables Eq. (4.102) to be approximated by

$$A_v \cong \frac{R_L}{R_L + \dfrac{1}{g_m}} \tag{4.102a}$$

The overall voltage gain G_v can be found by combining Eqs. (4.100) and (4.102), with the result that

$$G_v = \frac{R_G}{R_G + R_{\text{sig}}}\frac{R_L \parallel r_o}{(R_L \parallel r_o) + \dfrac{1}{g_m}} \tag{4.104}$$

which approaches unity for $R_G \gg R_{\text{sig}}$, $r_o \gg 1/g_m$, and $r_o \gg R_L$.

To emphasize the fact that it is usually faster to perform the small-signal analysis directly on the circuit diagram with the MOSFET small-signal model utilized only implicitly, we show such as analysis in Fig. 4.46(c). Once again, observe that to separate the intrinsic action of the MOSFET from the Early effect, we have extracted the output resistance r_o and shown it separately.

The circuit for determining the output resistance R_{out} is shown in Fig. 4.46(d). Because the gate voltage is now zero, looking back into the source we see between the source and ground a resistance $1/g_m$ in parallel with r_o; thus,

$$R_{\text{out}} = \frac{1}{g_m} \parallel r_o \tag{4.105}$$

Normally, $r_o \gg 1/g_m$, reducing R_{out} to

$$R_{\text{out}} \cong \frac{1}{g_m} \tag{4.106}$$

which indicates that R_{out} will be moderately low.

We observe that although the source-follower circuit has a large amount of internal feedback (as we will find out in Chapter 8), its R_{in} is independent of R_L (and thus $R_i = R_{\text{in}}$) and its R_{out} is independent of R_{sig} (and thus $R_o = R_{\text{out}}$). The reason for this, however, is the zero gate current.

In conclusion, the source follower features a very high input resistance, a relatively low output resistance, and a voltage gain that is less than but close to unity. It finds application in situations in which we need to connect a voltage-signal source that is providing a signal of reasonable magnitude but has a very high internal resistance to a much smaller load resistance—that is, as a unity-gain voltage buffer amplifier. The need for such amplifiers was discussed in Section 1.5. The source follower is also used as the output stage in a multistage amplifier, where its function is to equip the overall amplifier with a low output resistance, thus enabling it to supply relatively large load currents without loss of gain (i.e., with little reduction of output signal level.) The design of output stages is studied in Chapter 14.

EXERCISE

4.35 Consider a source follower such as that in Fig. 4.46(a) designed on the basis of the circuit of Fig. 4.42, the results of whose analysis are displayed in Fig. E4.30. Specifically, note that $g_m = 1$ mA/V and $r_o = 150$ kΩ. Let $R_{\text{sig}} = 1$ MΩ and $R_L = 15$ kΩ. (a) Find R_{in}, A_{vo}, A_v, and R_{out} without and with r_o taken into account. (b) Find the overall small-signal voltage gain G_v with r_o taken into account.

Ans. (a) $R_{\text{in}} = 4.7$ MΩ; $A_{vo} = 1$ V/V (without r_o), 0.993 V/V (with r_o); $A_v = 0.938$ (without r_o), 0.932 V (with r_o); $R_{\text{out}} = 1$ kΩ (without r_o), 0.993 kΩ (with r_o); (b) 0.768 V/V

4.7.7 Summary and Comparisons

For easy reference we present in Table 4.4 a summary of the characteristics of the various configurations of discrete single-stage MOSFET amplifiers. In addition to the remarks already made throughout this section on the relative merits of the various configurations, the results displayed in Table 4.4 enable us to make the following concluding points:

1. The CS configuration is the best suited for obtaining the bulk of the gain required in an amplifier. Depending on the magnitude of the gain required, either a single CS stage or a cascade of two or three CS stages can be used.

2. Including a resistor R_S in the source lead of the CS stage provides a number of improvements in its performance, as will be seen in later chapters, at the expense of reduced gain.

TABLE 4.4   **Characteristics of Single-Stage Discrete MOS Amplifiers**

Common-Source

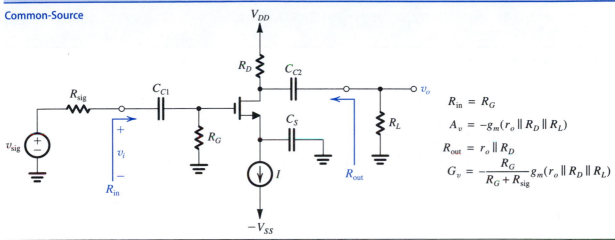

$$R_{\text{in}} = R_G$$

$$A_v = -g_m(r_o \parallel R_D \parallel R_L)$$

$$R_{\text{out}} = r_o \parallel R_D$$

$$G_v = -\frac{R_G}{R_G + R_{\text{sig}}} g_m(r_o \parallel R_D \parallel R_L)$$

Common-Source with Source Resistance

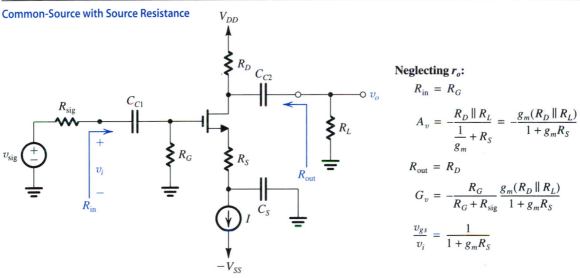

Neglecting r_o:

$$R_{\text{in}} = R_G$$

$$A_v = -\frac{R_D \parallel R_L}{\dfrac{1}{g_m} + R_S} = -\frac{g_m(R_D \parallel R_L)}{1 + g_m R_S}$$

$$R_{\text{out}} = R_D$$

$$G_v = -\frac{R_G}{R_G + R_{\text{sig}}} \frac{g_m(R_D \parallel R_L)}{1 + g_m R_S}$$

$$\frac{v_{gs}}{v_i} = \frac{1}{1 + g_m R_S}$$

Common Gate

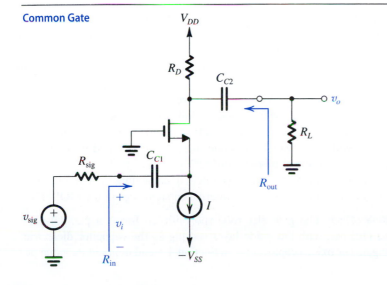

Neglecting r_o:

$$R_{\text{in}} = \frac{1}{g_m}$$

$$A_v = g_m(R_D \parallel R_L)$$

$$R_{\text{out}} = R_D$$

$$G_v = \frac{1}{1 + g_m R_{\text{sig}}} g_m(R_D \parallel R_L)$$

(Continued)

319

TABLE 4.4 *(Continued)*

Common-Drain or Source Follower

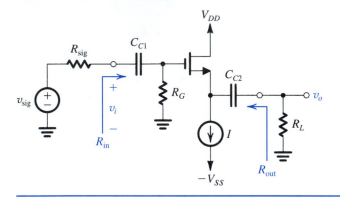

$$R_{in} = R_G$$

$$A_v = \frac{r_o \parallel R_L}{(r_o \parallel R_L) + \dfrac{1}{g_m}}$$

$$R_{out} = r_o \parallel \frac{1}{g_m} \cong \frac{1}{g_m}$$

$$G_v = \frac{R_G}{R_G + R_{sig}} \frac{r_o \parallel R_L}{(r_o \parallel R_L) + \dfrac{1}{g_m}}$$

3. The low input resistance of the CG amplifier makes it useful only in specific applications. These include voltage amplifiers that do not require a high input resistance and that take advantage of the excellent high-frequency performance of the CG configuration (see Chapter 6) and as a unity-gain current amplifier or current follower. This latter application gives rise to the most popular application of the common-gate configuration, the cascode amplifier (see Chapter 6).

4. The source follower finds application as a voltage buffer for connecting a high-resistance source to a low-resistance load and as the output stage in a multistage amplifier.

4.8 THE MOSFET INTERNAL CAPACITANCES AND HIGH-FREQUENCY MODEL

From our study of the physical operation of the MOSFET in Section 4.1, we know that the device has internal capacitances. In fact, we used one of these, the gate-to-channel capacitance, in our derivation of the MOSFET i–v characteristics. We did, however, implicitly assume that the steady-state charges on these capacitances are acquired instantaneously. In other words, we did not account for the finite time required to charge and discharge the various internal capacitances. As a result, the device models we derived, such as the small-signal model, do not include any capacitances. The use of these models would predict constant amplifier gains independent of frequency. We know, however, that this is (unfortunately) not the case; in fact, the gain of every MOSFET amplifier falls off at some high frequency. Similarly, the MOSFET digital logic inverter exhibits a finite nonzero propagation delay. To be able to predict these results, the MOSFET model must be augmented by including internal capacitances. This is the subject of this section.

To visualize the physical origin of the various internal capacitances, the reader is referred to Fig. 4.1. There are basically two types of internal capacitances in the MOSFET:

1. The gate capacitive effect: The gate electrode (polysilicon) forms a parallel-plate capacitor with the channel, with the oxide layer serving as the capacitor dielectric. We discussed the gate (or oxide) capacitance in Section 4.1 and denoted its value per unit area as C_{ox}.

2. The source-body and drain-body depletion-layer capacitances: These are the capacitances of the reverse-biased *pn* junctions formed by the n^+ source region (also called the **source diffusion**) and the *p*-type substrate and by the n^+ drain region (the **drain diffusion**) and the substrate. Evaluation of these capacitances will utilize the material studied in Chapter 3.

These two capacitive effects can be modeled by including capacitances in the MOSFET model between its four terminals, G, D, S, and B. There will be five capacitances in total: C_{gs}, C_{gd}, C_{gb}, C_{sb}, and C_{db}, where the subscripts indicate the location of the capacitances in the model. In the following, we show how the values of the five model capacitances can be determined. We will do so by considering each of the two capacitive effects separately.

4.8.1 The Gate Capacitive Effect

The gate capacitive effect can be modeled by the three capacitances C_{gs}, C_{gd}, and C_{gb}. The values of these capacitances can be determined as follows:

1. When the MOSFET is operating in the triode region at small v_{DS}, the channel will be of uniform depth. The gate-channel capacitance will be $WL C_{ox}$ and can be modeled by dividing it equally between the source and drain ends; thus,

$$C_{gs} = C_{gd} = \tfrac{1}{2} WL C_{ox} \quad \text{(triode region)} \qquad (4.107)$$

This is obviously an approximation (as all modeling is) but works well for triode-region operation even when v_{DS} is not small.

2. When the MOSFET operates in saturation, the channel has a tapered shape and is pinched off at or near the drain end. It can be shown that the gate-to-channel capacitance in this case is approximately $\tfrac{2}{3} WL C_{ox}$ and can be modeled by assigning this entire amount to C_{gs}, and a zero amount to C_{gd} (because the channel is pinched off at the drain); thus,

$$\left. \begin{aligned} C_{gs} &= \tfrac{2}{3} WL C_{ox} \\ C_{gd} &= 0 \end{aligned} \right\} \quad \text{(saturation region)} \qquad \begin{aligned} (4.108) \\ (4.109) \end{aligned}$$

3. When the MOSFET is cut off, the channel disappears, and thus $C_{gs} = C_{gd} = 0$. However, we can (after some rather complex reasoning) model the gate capacitive effect by assigning a capacitance $WL C_{ox}$ to the gate-body model capacitance; thus,

$$\left. \begin{aligned} C_{gs} &= C_{gd} = 0 \\ C_{gb} &= WL C_{ox} \end{aligned} \right\} \quad \text{(cutoff)} \qquad \begin{aligned} (4.110) \\ (4.111) \end{aligned}$$

4. There is an additional small capacitive component that should be added to C_{gs} and C_{gd} in all the preceding formulas. This is the capacitance that results from the fact that the source and drain diffusions extend slightly under the gate oxide (refer to Fig. 4.1). If the *overlap* length is denoted L_{ov}, we see that the **overlap capacitance** component is

$$C_{ov} = WL_{ov} C_{ox} \qquad (4.112)$$

Typically, $L_{ov} = 0.05$ to $0.1L$.

4.8.2 The Junction Capacitances

The depletion-layer capacitances of the two reverse-biased *pn* junctions formed between each of the source and the drain diffusions and the body can be determined using the formula developed in Section 3.7.3 (Eq. 3.56). Thus, for the source diffusion, we have the source-body capacitance, C_{sb},

$$C_{sb} = \frac{C_{sb0}}{\sqrt{1 + \dfrac{V_{SB}}{V_0}}} \tag{4.113}$$

where C_{sb0} is the value of C_{sb} at zero body-source bias, V_{SB} is the magnitude of the reverse-bias voltage, and V_0 is the junction built-in voltage (0.6 V to 0.8 V). Similarly, for the drain diffusion, we have the drain-body capacitance C_{db},

$$C_{db} = \frac{C_{db0}}{\sqrt{1 + \dfrac{V_{DB}}{V_0}}} \tag{4.114}$$

where C_{db0} is the capacitance value at zero reverse-bias voltage and V_{DB} is the magnitude of this reverse-bias voltage. Note that we have assumed that for both junctions, the grading coefficient $m = \frac{1}{2}$.

It should be noted also that each of these junction capacitances includes a component arising from the bottom side of the diffusion and a component arising from the *side walls* of the diffusion. In this regard, observe that each diffusion has three side walls that are in contact with the substrate and thus contribute to the junction capacitance (the fourth wall is in contact with the channel). In more advanced MOSFET modeling, the two components of each of the junction capacitances are calculated separately.

The formulas for the junction capacitances in Eqs. (4.113) and (4.114) assume small-signal operation. These formulas, however, can be modified to obtain approximate average values for the capacitances when the transistor is operating under large-signal conditions such as in logic circuits. Finally, typical values for the various capacitances exhibited by an *n*-channel MOSFET in a relatively modern (0.5 *μ*m) CMOS process are given in the following exercise.

EXERCISE

4.36 For an *n*-channel MOSFET with $t_{ox} = 10$ nm, $L = 1.0$ *μ*m, $W = 10$ *μ*m, $L_{ov} = 0.05$ *μ*m, $C_{sb0} = C_{db0} = 10$ fF, $V_0 = 0.6$ V, $V_{SB} = 1$ V, and $V_{DS} = 2$ V, calculate the following capacitances when the transistor is operating in saturation: C_{ox}, C_{ov}, C_{gs}, C_{gd}, C_{sb}, and C_{db}. (*Note:* You may consult Table 4.1 for values of the physical constants.)
Ans. 3.45 fF/μm^2; 1.72 fF; 24.7 fF; 1.72 fF; 6.1 fF; 4.1 fF

4.8.3 The High-Frequency MOSFET Model

Figure 4.47(a) shows the small-signal model of the MOSFET, including the four capacitances C_{gs}, C_{gd}, C_{sb}, and C_{db}. This model can be used to predict the high-frequency response of MOSFET amplifiers. It is, however, quite complex for manual analysis, and its use is

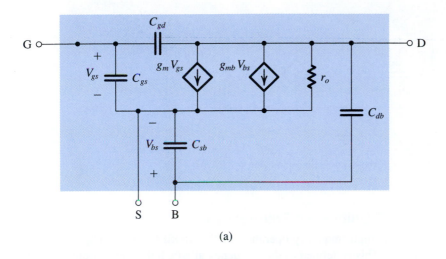

(a)

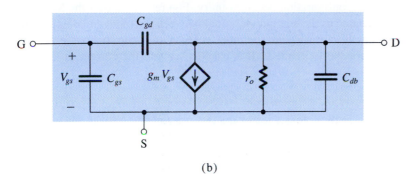

(b)

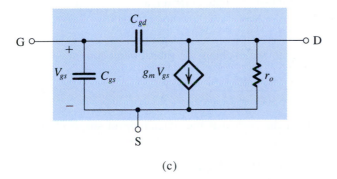

(c)

FIGURE 4.47 (a) High-frequency equivalent circuit model for the MOSFET. (b) The equivalent circuit for the case in which the source is connected to the substrate (body). (c) The equivalent circuit model of (b) with C_{db} neglected (to simplify analysis).

limited to computer simulation using, for example, SPICE. Fortunately, for the case when the source is connected to the body, the model simplifies considerably, as shown in Fig. 4.47(b). In this model, C_{gd}, although small, plays a significant role in determining the high-frequency response of amplifiers (Section 4.9) and thus must be kept in the model. Capacitance C_{db}, on the other hand, can usually be neglected, resulting in significant simplification of manual analysis. The resulting circuit is shown in Fig. 4.47(c).

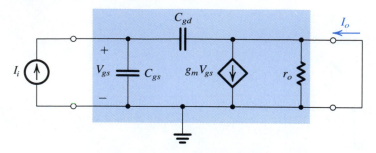

FIGURE 4.48 Determining the short-circuit current gain I_o/I_i.

4.8.4 The MOSFET Unity-Gain Frequency (f_T)

A figure of merit for the high-frequency operation of the MOSFET as an amplifier is the unity-gain frequency, f_T. This is defined as the frequency at which the short-circuit current-gain of the common-source configuration becomes unity. Figure 4.48 shows the MOSFET hybrid-π model with the source as the common terminal between the input and output ports. To determine the short-circuit current gain, the input is fed with a current-source signal I_i and the output terminals are short-circuited.[8] It is easy to see that the current in the short circuit is given by

$$I_o = g_m V_{gs} - s C_{gd} V_{gs}$$

Recalling that C_{gd} is small, at the frequencies of interest, the second term in this equation can be neglected,

$$I_o \simeq g_m V_{gs} \tag{4.115}$$

From Fig. 4.48, we can express V_{gs} in terms of the input current I_i as

$$V_{gs} = I_i / s(C_{gs} + C_{gd}) \tag{4.116}$$

Equations (4.115) and (4.116) can be combined to obtain the short-circuit current gain,

$$\frac{I_o}{I_i} = \frac{g_m}{s(C_{gs} + C_{gd})} \tag{4.117}$$

For physical frequencies $s = j\omega$, it can be seen that the magnitude of the current gain becomes unity at the frequency

$$\omega_T = g_m / (C_{gs} + C_{gd})$$

Thus the unity-gain frequency $f_T = \omega_T/2\pi$ is

$$f_T = \frac{g_m}{2\pi(C_{gs} + C_{gd})} \tag{4.118}$$

Since f_T is proportional to g_m and inversely proportional to the FET internal capacitances, the higher the value of f_T, the more effective the FET becomes as an amplifier. Substituting for g_m using Eq. (4.70), we can express f_T in terms of the bias current I_D (see Problem 4.92).

---

[8] Note that since we are now dealing with quantities (currents, in this case) that are functions of frequency, or, equivalently, the Laplace variable s, we are using capital letters with lowercase subscripts for our symbols. This conforms to the symbol notation introduced in Chapter 1.

Alternatively, we can substitute for g_m from Eq. (4.69) to express f_T in terms of the overdrive voltage V_{OV} (see Problem 4.93). Both expressions yield additional insight into the high-frequency operation of the MOSFET.

Typically, f_T ranges from about 100 MHz for the older technologies (e.g., a 5-μm CMOS process) to many GHz for newer high-speed technologies (e.g., a 0.13-μm CMOS process).

EXERCISE

4.37 Calculate f_T for the n-channel MOSFET whose capacitances were found in Exercise 4.36. Assume operation at 100 μA, and that $k_n' = 160$ μA/V^2.

Ans. 3.7 GHz

4.8.5 Summary

We conclude this section by presenting a summary in Table 4.5.

TABLE 4.5   The MOSFET High-Frequency Model

Model

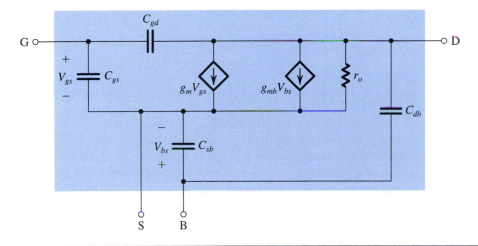

Model Parameters

$$g_m = \mu_n C_{ox} \frac{W}{L} V_{OV} = \sqrt{2\mu_n C_{ox} \frac{W}{L} I_D} = \frac{2I_D}{V_{OV}}$$

$$g_{mb} = \chi g_m = \frac{\gamma}{2\sqrt{2\phi_f + V_{SB}}} g_m$$

$$r_o = V_A / I_D$$

$$C_{gs} = \frac{2}{3} W L C_{ox} + W L_{ov} C_{ox}$$

$$C_{gd} = W L_{ov} C_{ox}$$

$$C_{sb} = \frac{C_{sb0}}{\sqrt{1 + \frac{V_{SB}}{V_0}}}$$

$$C_{db} = \frac{C_{db0}}{\sqrt{1 + \frac{V_{DB}}{V_0}}}$$

$$f_T = \frac{g_m}{2\pi(C_{gs} + C_{gd})}$$

4.9 FREQUENCY RESPONSE OF THE CS AMPLIFIER[9]

In this section we study the dependence of the gain of the MOSFET common-source amplifier of Fig. 4.49(a) on the frequency of the input signal. Before we begin, however, a note on terminology is in order: Since we will be dealing with voltages and currents that are functions of frequency or, more generally, the complex-frequency variable s, we will use uppercase letters with lowercase subscripts to represent them (e.g., V_{gs}, V_d, V_o).

4.9.1 The Three Frequency Bands

When the circuit of Fig. 4.49(a) was studied in Section 4.7.3, it was assumed that the coupling capacitors C_{C1} and C_{C2} and the bypass capacitor C_S were acting as perfect short circuits at all signal frequencies of interest. We also neglected the internal capacitances of the MOSFET; that is, C_{gs} and C_{gd} of the MOSFET high-frequency model shown in Fig. 4.47(c) were assumed to be sufficiently small to act as open circuits at all signal frequencies of interest. As a result of ignoring all capacitive effects, the gain expressions derived in Section 4.7.3 were independent of frequency. In reality, however, this situation applies over only a limited, though normally wide, band of frequencies. This is illustrated in Fig. 4.49(b), which shows a sketch of the magnitude of the overall voltage gain, $|G_v|$, of the CS amplifier versus frequency. We observe that the gain is almost constant over a wide frequency band, called the **midband.** The value of the midband gain A_M corresponds to the overall voltage gain G_v that we derived in Section 4.7.2, namely,

$$A_M \equiv \frac{V_o}{V_{\text{sig}}} = -\frac{R_G}{R_G + R_{\text{sig}}} g_m (r_o \parallel R_D \parallel R_L) \tag{4.119}$$

Figure 4.49(b) shows that the gain falls off at signal frequencies below and above the midband. The gain falloff in the **low-frequency band** is due to the fact that even though C_{C1}, C_{C2}, and C_S are large capacitors (in the μF range), as the signal frequency is reduced, their impedances increase, and they no longer behave as short circuits. On the other hand, the gain falls off in the **high-frequency band** as a result of C_{gs} and C_{gd}, which though very small (in the pF or fraction of pF range for discrete devices and much lower for IC devices), their impedances at high frequencies decrease and thus can no longer be considered as open circuits. It is our objective in this section to study the mechanisms by which these two sets of capacitances affect the amplifier gain in the low-frequency and the high-frequency bands. In this way, we will be able to determine the frequencies f_H and f_L, which define the extent of the midband, as shown in Fig. 4.49(b).

The midband is obviously the useful frequency band of the amplifier. Usually, f_L and f_H are the frequencies at which the gain drops by 3 dB below its value at midband. The amplifier **bandwidth** or **3-dB bandwidth** is defined as the difference between the lower (f_L) and the upper or higher (f_H) 3-dB frequencies,

$$BW \equiv f_H - f_L \tag{4.120}$$

and since, usually, $f_L \ll f_H$,

$$BW \cong f_H \tag{4.121}$$

[9] We strongly urge the reader to review Section 1.6 before proceeding with the study of this section.

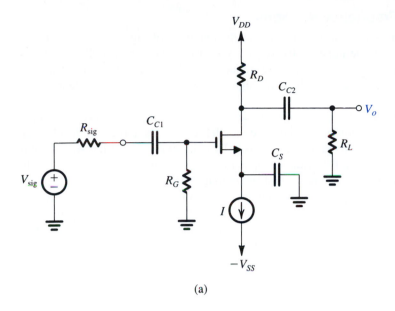

(a)

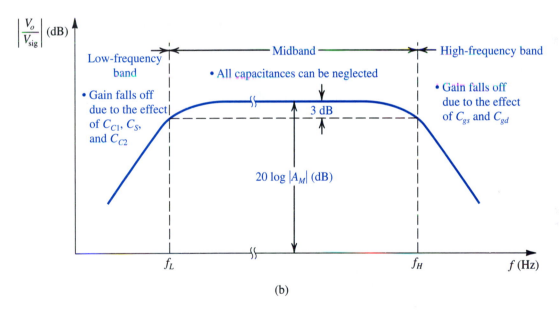

(b)

FIGURE 4.49 **(a)** Capacitively coupled common-source amplifier. **(b)** A sketch of the frequency response of the amplifier in (a) delineating the three frequency bands of interest.

A figure-of-merit for the amplifier is its **gain–bandwidth product,** which is defined as

$$GB \equiv |A_M|BW \qquad (4.122)$$

It will be shown at a later stage that in amplifier design it is usually possible to trade-off gain for bandwidth. One way to accomplish this, for instance, is by adding a source degeneration resistance R_S, as we have done in Section 4.7.4.

4.9.2 The High-Frequency Response

To determine the gain, or the transfer function, of the amplifier of Fig. 4.49(a) at high frequencies, and particularly the upper 3-dB frequency f_H, we replace the MOSFET with its high-frequency model of Fig. 4.47(c). At these frequencies, C_{C1}, C_{C2}, and C_S will be behaving as perfect short circuits. The result is the high-frequency amplifier equivalent circuit shown in Fig. 4.50(a).

The equivalent circuit of Fig. 4.50(a) can be simplified by utilizing the Thévenin theorem at the input side and by combining the three parallel resistances at the output side. The resulting simplified circuit is shown in Fig. 4.50(b). This circuit can be further simplified if we can find a way to deal with the bridging capacitor C_{gd} that connects the output node to the input side. Toward that end, consider first the output node. It can be seen that the load current is $(g_m V_{gs} - I_{gd})$, where $(g_m V_{gs})$ is the output current of the transistor and I_{gd} is the current supplied through the very small capacitance C_{gd}. At frequencies in the vicinity of f_H, which defines the edge of the midband, it is reasonable to assume that I_{gd} is still much smaller than $(g_m V_{gs})$, with the result that V_o can be given approximately by

$$V_o \cong -(g_m V_{gs})R'_L = -g_m R'_L V_{gs} \qquad (4.123)$$

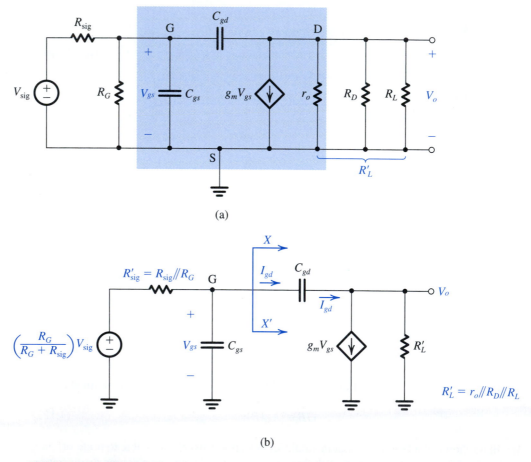

FIGURE 4.50 Determining the high-frequency response of the CS amplifier: **(a)** equivalent circuit; **(b)** the circuit of (a) simplified at the input and the output;

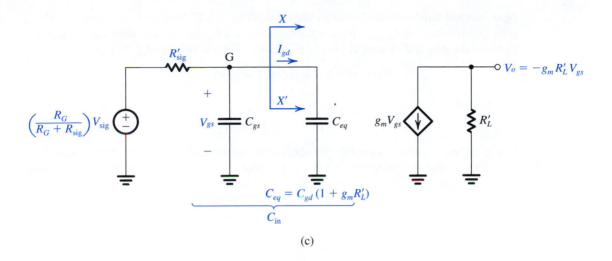

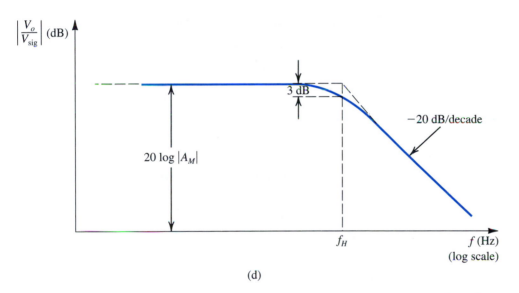

(d)

FIGURE 4.50 *(Continued)* **(c)** the equivalent circuit with C_{gd} replaced at the input side with the equivalent capacitance C_{eq}; **(d)** the frequency response plot, which is that of a low-pass single-time-constant circuit.

where

$$R_L' = r_o \parallel R_D \parallel R_L$$

Since $V_o = V_{ds}$, Eq. (4.123) indicates that the gain from gate to drain is $-g_m R_L'$, the same value as in the midband. The current I_{gd} can now be found as

$$I_{gd} = sC_{gd}(V_{gs} - V_o)$$

$$= sC_{gd}[V_{gs} - (-g_m R_L' V_{gs})]$$

$$= sC_{gd}(1 + g_m R_L')V_{gs}$$

Now, the left-hand side of the circuit in Fig. 4.50(b), at XX', knows of the existence of C_{gd} only through the current I_{gd}. Therefore, we can replace C_{gd} by an equivalent capacitance C_{eq} between the gate and ground as long as C_{eq} draws a current equal to I_{gd}. That is,

$$sC_{eq}V_{gs} = sC_{gd}(1 + g_m R_L')V_{gs}$$

which results in

$$C_{eq} = C_{gd}(1 + g_m R_L') \tag{4.124}$$

Using C_{eq} enables us to simplify the equivalent circuit at the input side to that shown in Fig. 4.50(c). We recognize the circuit of Fig. 4.50(c) as a single-time-constant (STC) circuit of the low-pass type (Section 1.6 and Appendix D). Reference to Table 1.2 enables us to express the output voltage V_{gs} of the STC circuit in the form

$$V_{gs} = \left(\frac{R_G}{R_G + R_{sig}} V_{sig}\right) \frac{1}{1 + \dfrac{s}{\omega_0}} \tag{4.125}$$

where ω_0 is the corner frequency or the break frequency of the STC circuit,

$$\omega_0 = 1/C_{in}R_{sig}' \tag{4.126}$$

with

$$C_{in} = C_{gs} + C_{eq} = C_{gs} + C_{gd}(1 + g_m R_L') \tag{4.127}$$

and

$$R_{sig}' = R_{sig} \parallel R_G \tag{4.128}$$

Combining Eqs. (4.123) and (4.125) results in the following expression for the high-frequency gain of the CS amplifier,

$$\frac{V_o}{V_{sig}} = -\left(\frac{R_G}{R_G + R_{sig}}\right)(g_m R_L') \frac{1}{1 + \dfrac{s}{\omega_0}} \tag{4.129}$$

which can be expressed in the form

$$\frac{V_o}{V_{sig}} = \frac{A_M}{1 + \dfrac{s}{\omega_H}} \tag{4.130}$$

where the midband gain A_M is given by Eq. (4.119) and ω_H is the upper 3-dB frequency,

$$\omega_H = \omega_0 = \frac{1}{C_{in}R_{sig}'} \tag{4.131}$$

and

$$f_H = \frac{\omega_H}{2\pi} = \frac{1}{2\pi C_{in}R_{sig}'} \tag{4.132}$$

We thus see that the high-frequency response will be that of a low-pass STC network with a 3-dB frequency f_H determined by the time constant $C_{in}R'_{sig}$. Figure 4.50(d) shows a sketch of the magnitude of the high-frequency gain.

Before leaving this section we wish to make a number of observations:

1. The upper 3-dB frequency is determined by the interaction of $R'_{sig} = R_{sig} \parallel R_G$ and $C_{in} = C_{gs} + C_{gd}(1 + g_m R'_L)$. Since the bias resistance R_G is usually very large, it can be neglected, resulting in $R'_{sig} \cong R_{sig}$, the resistance of the signal source. It follows that a large value of R_{sig} will cause f_H to be lowered.

2. The total input capacitance C_{in} is usually dominated by C_{eq}, which in turn is made large by the multiplication effect that C_{gd} undergoes. Thus, although C_{gd} is usually a very small capacitance, its effect on the amplifier frequency response can be very significant as a result of its multiplication by the factor $(1 + g_m R'_L)$, which is approximately equal to the midband gain of the amplifier.

3. The multiplication effect that C_{gd} undergoes comes about because it is connected between two nodes whose voltages are related by a large negative gain $(-g_m R'_L)$. This effect is known as the **Miller effect,** and $(1 + g_m R'_L)$ is known as the **Miller multiplier.** It is the Miller effect that causes the CS amplifier to have a large total input capacitance C_{in} and hence a low f_H.

4. To extend the high-frequency response of a MOSFET amplifier, we have to find configurations in which the Miller effect is absent or at least reduced. We shall return to this subject at great length in Chapter 6.

5. The above analysis, resulting in an STC or a single-pole response, is a simplified one. Specifically, it is based on neglecting I_{gd} relative to $g_m V_{gs}$, an assumption that applies well at frequencies not too much higher than f_H. A more exact analysis of the circuit in Fig. 4.50(a) will be carried out in Chapter 6. The results above, however, are more than sufficient for our current needs.

EXAMPLE 4.12

Find the midband gain A_M and the upper 3-dB frequency f_H of a CS amplifier fed with a signal source having an internal resistance $R_{sig} = 100$ kΩ. The amplifier has $R_G = 4.7$ MΩ, $R_D = R_L = 15$ kΩ, $g_m = 1$ mA/V, $r_o = 150$ kΩ, $C_{gs} = 1$ pF, and $C_{gd} = 0.4$ pF.

Solution

$$A_M = -\frac{R_G}{R_G + R_{sig}} g_m R'_L$$

where

$$R'_L = r_o \parallel R_D \parallel R_L = 150 \parallel 15 \parallel 15 = 7.14 \text{ k}\Omega.$$

$$g_m R'_L = 1 \times 7.14 = 7.14 \text{ V/V}$$

Thus,

$$A_M = -\frac{4.7}{4.7 + 0.1} \times 7.14 = -7 \text{ V/V}$$

The equivalent capacitance, C_{eq}, is found as

$$C_{eq} = (1 + g_m R_L')C_{gd}$$

$$= (1 + 7.14) \times 0.4 = 3.26 \text{ pF}$$

The total input capacitance C_{in} can be now obtained as

$$C_{in} = C_{gs} + C_{eq} = 1 + 3.26 = 4.26 \text{ pF}$$

The upper 3-dB frequency f_H is found from

$$f_H = \frac{1}{2\pi C_{in}(R_{sig} \parallel R_G)}$$

$$= \frac{1}{2\pi \times 4.26 \times 10^{-12}(0.1 \parallel 4.7) \times 10^6}$$

$$= 382 \text{ kHz}$$

EXERCISES

4.38 For the CS amplifier specified in Example 4.12, find the values of A_M and f_H that result when the signal-source resistance is reduced to 10 kΩ.

Ans. −7.12 V/V; 3.7 MHz

4.39 If it is possible to replace the MOSFET used in the amplifier in Example 4.12 with another having the same C_{gs} but a smaller C_{gd}, what is the maximum value that its C_{gd} can be in order to obtain an f_H of at least 1 MHz?

Ans. 0.08 pF

4.9.3 The Low-Frequency Response

To determine the low-frequency gain or transfer function of the common-source amplifier, we show in Fig. 4.51(a) the circuit with the dc sources eliminated (current source I open-circuited and voltage source V_{DD} short-circuited). We shall perform the small-signal analysis directly on this circuit. However, we will ignore r_o. This is done in order to keep the analysis simple and thus focus attention on significant issues. The effect of r_o on the low-frequency operation of this amplifier is minor, as can be verified by a SPICE simulation (Section 4.12).

The analysis begins at the signal generator by finding the fraction of V_{sig} that appears at the transistor gate,

$$V_g = V_{sig} \frac{R_G}{R_G + \dfrac{1}{sC_{C1}} + R_{sig}}$$

which can be written in the alternate form

$$V_g = V_{sig} \frac{R_G}{R_G + R_{sig}} \frac{s}{s + \dfrac{1}{C_{C1}(R_G + R_{sig})}} \tag{4.133}$$

Thus we see that the expression for the signal transmission from signal generator to amplifier input has acquired a frequency-dependent factor. From our study of frequency response

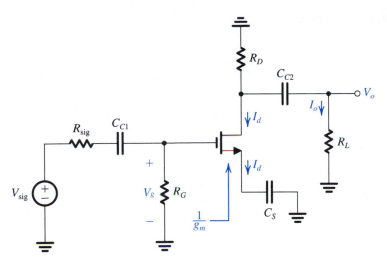

FIGURE 4.51 Analysis of the CS amplifier to determine its low-frequency transfer function. For simplicity, r_o is neglected.

in Section 1.6 (see also Appendix D), we recognize this factor as the transfer function of an STC network of the high-pass type with a break or corner frequency $\omega_0 = 1/C_{C1}(R_G + R_{\text{sig}})$. Thus the effect of the coupling capacitor C_{C1} is to introduce a high-pass STC response with a break frequency that we shall denote ω_{P1},

$$\omega_{P1} = \omega_0 = \frac{1}{C_{C1}(R_G + R_{\text{sig}})} \tag{4.134}$$

Continuing with the analysis, we next determine the drain current I_d by dividing V_g by the total impedance in the source circuit which is $[(1/g_m) + (1/sC_S)]$ to obtain

$$I_d = \frac{V_g}{\dfrac{1}{g_m} + \dfrac{1}{sC_S}}$$

which can be written in the alternate form

$$I_d = g_m V_g \frac{s}{s + \dfrac{g_m}{C_S}} \tag{4.135}$$

We observe that C_S introduces a frequency-dependent factor, which is also of the STC high-pass type. Thus the amplifier acquires another break frequency,

$$\omega_{P2} = \frac{g_m}{C_S} \tag{4.136}$$

To complete the analysis, we find V_o by first using the current-divider rule to determine the fraction of I_d that flows through R_L,

$$I_o = -I_d \frac{R_D}{R_D + \dfrac{1}{sC_{C2}} + R_L}$$

and then multiplying I_o by R_L to obtain

$$V_o = I_o R_L = -I_d \frac{R_D R_L}{R_D + R_L} \frac{s}{s + \dfrac{1}{C_{C2}(R_D + R_L)}} \tag{4.137}$$

from which we see that C_{C2} introduces a third STC high-pass factor, giving the amplifier a third break frequency at

$$\omega_{P3} = \frac{1}{C_{C2}(R_D + R_L)} \tag{4.138}$$

The overall low-frequency transfer function of the amplifier can be found by combining Eqs. (4.133), (4.135), and (4.137) and replacing the break frequencies by their symbols from Eqs. (4.134), (4.136), and (4.138),

$$\frac{V_o}{V_{\text{sig}}} = -\left(\frac{R_G}{R_G + R_{\text{sig}}}\right)[g_m(R_D \| R_L)]\left(\frac{s}{s + \omega_{P1}}\right)\left(\frac{s}{s + \omega_{P2}}\right)\left(\frac{s}{s + \omega_{P3}}\right) \tag{4.139}$$

The low-frequency magnitude response can be obtained from Eq. (4.139) by replacing s by $j\omega$ and finding $|V_o/V_{\text{sig}}|$. In many cases, however, one of the three break frequencies can be much higher than the other two, say by a factor greater than 4. In such a case, it is this highest-frequency break point that will determine the lower 3-dB frequency, f_L, and we do not have to do any additional hand analysis. For instance, because the expression for ω_{P2} includes g_m (Eq. 4.136), ω_{P2} is usually higher than ω_{P1} and ω_{P3}. If ω_{P2} is sufficiently separated from ω_{P1} and ω_{P3}, then

$$f_L \cong f_{P2}$$

which means that in such a case, the bypass capacitor determines the low end of the midband. Figure 4.52 shows a sketch of the low-frequency gain of a CS amplifier in which the three break frequencies are sufficiently separated so that their effects appear distinct. Observe that at each break frequency, the slope of the asymptotes to the gain function increases by 20 dB/decade. Readers familiar with poles and zeros will recognize f_{P1}, f_{P2}, and f_{P3} as the frequencies of the three real **low-frequency poles** of the amplifier. We will use poles and zeros and related s-plane concepts later on in Chapter 6 and beyond.

Before leaving this section, it is essential that the reader be able to quickly find the time-constant and hence the break frequency associated with each of the three capacitors. The procedure is simple:

1. Reduce V_{sig} to zero.

2. Consider each capacitor separately; that is, assume that the other two capacitors are acting as perfect short circuits.

3. For each capacitor, find the total resistance seen between its terminals. This is the resistance that determines the time constant associated with this capacitor.

The reader is encouraged to apply this procedure to C_{C1}, C_S, and C_{C2} and thus see that Eqs. (4.134), (4.136), and (4.138) can be written by inspection.

Selecting Values for the Coupling and Bypass Capacitors We now address the design issue of selecting appropriate values for C_{C1}, C_S, and C_{C2}. The design objective is to place the lower 3-dB frequency f_L at a specified value while minimizing the capacitor values.

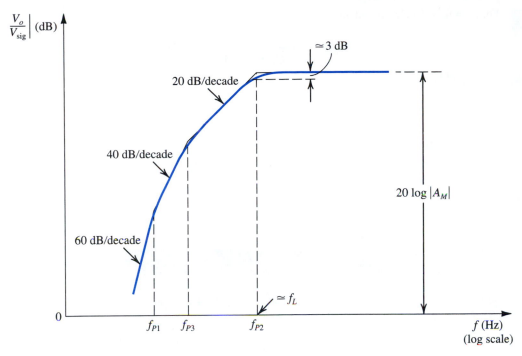

FIGURE 4.52 Sketch of the low-frequency magnitude response of a CS amplifier for which the three break frequencies are sufficiently separated for their effects to appear distinct.

Since as mentioned above C_S results in the highest of the three break frequencies, the total capacitance is minimized by selecting C_S so that its break frequency $f_{P2} = f_L$. We then decide on the location of the other two break frequencies, say 5 to 10 times lower than the frequency of the dominant one, f_{P2}. However, the values selected for f_{P1} and f_{P3} should not be too low, for that would require larger values for C_{C1} and C_{C2} than may be necessary. The design procedure will be illustrated by an example.

EXAMPLE 4.13

We wish to select appropriate values for the coupling capacitors C_{C1} and C_{C2} and the bypass capacitor C_S for the CS amplifier whose high-frequency response was analyzed in Example 4.12. The amplifier has $R_G = 4.7 \text{ M}\Omega$, $R_D = R_L = 15 \text{ k}\Omega$, $R_{\text{sig}} = 100 \text{ k}\Omega$, and $g_m = 1 \text{ mA/V}$. It is required to have f_L at 100 Hz and that the nearest break frequency be at least a decade lower.

Solution

We select C_S so that

$$f_{P2} = \frac{1}{2\pi(C_S/g_m)} = f_L$$

Thus,

$$C_S = \frac{g_m}{2\pi f_L} = \frac{1 \times 10^{-3}}{2\pi \times 100} = 1.6 \ \mu\text{F}$$

For $f_{P1} = f_{P3} = 10$ Hz, we obtain

$$10 = \frac{1}{2\pi C_{C1}(0.1 + 4.7) \times 10^6}$$

which yields

$$C_{C1} = 3.3 \text{ nF}$$

and

$$10 = \frac{1}{2\pi C_{C2}(15 + 15) \times 10^3}$$

which results in

$$C_{C2} = 0.53 \ \mu\text{F}$$

EXERCISE

4.40 A CS amplifier has $C_{C1} = C_S = C_{C2} = 1 \ \mu\text{F}$, $R_G = 10$ MΩ, $R_{\text{sig}} = 100$ kΩ, $g_m = 2$ mA/V, $R_D = R_L = 10$ kΩ. Find A_M, f_{P1}, f_{P2}, f_{P3}, and f_L.

Ans. −9.9 V/V; 0.016 Hz; 318.3 Hz; 8 Hz; 318.3 Hz

4.9.4 A Final Remark

The frequency response of the other amplifier configurations will be studied in Chapter 6.

 ## 4.10 THE CMOS DIGITAL LOGIC INVERTER

Complementary MOS or CMOS logic circuits have been available as standard packages for use in conventional digital system design since the early 1970s. Such packages contain logic gates and other digital system building blocks with the number of gates per package ranging from a few (small-scale integrated or SSI circuits) to few tens (medium-scale integrated or MSI circuits).

In the late 1970s, as the era of large- and very-large-scale integration (LSI and VLSI; hundreds to hundreds of thousands of gates per chip) began, circuits using only n-channel MOS transistors, known as NMOS, became the fabrication technology of choice. Indeed, early VLSI circuits, such as the early microprocessors, employed NMOS technology. Although at that time the design flexibility and other advantages that CMOS offers were known, the CMOS technology available then was too complex to produce such high-density VLSI chips economically. However, as advances in processing technology were made, this state of affairs changed radically. In fact, CMOS technology has now completely replaced NMOS at all levels of integration, in both analog and digital applications.

For any IC technology used in digital circuit design, the basic circuit element is the logic inverter.[10] Once the operation and characteristics of the inverter circuit are thoroughly

---

[10] A study of the digital logic inverter as a circuit building block was presented in Section 1.7. A review of this material before proceeding with the current section should prove helpful.

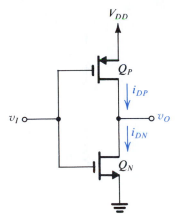

FIGURE 4.53 The CMOS inverter.

understood, the results can be extended to the design of logic gates and other more complex circuits. In this section we provide such a study for the CMOS inverter. Our study of the CMOS inverter and logic circuits will continue in Chapter 10.

The basic CMOS inverter is shown in Fig. 4.53. It utilizes two matched enhancement-type MOSFETs: one, Q_N, with an n channel and the other, Q_P, with a p channel. The body of each device is connected to its source and thus no body effect arises. As will be seen shortly, the CMOS circuit realizes the conceptual inverter implementation studied in Chapter 1 (Fig. 1.32), where a pair of switches are operated in a complementary fashion by the input voltage v_I.

4.10.1 Circuit Operation

We first consider the two extreme cases: when v_I is at logic-0 level, which is approximately 0 V; and when v_I is at logic-1 level, which is approximately V_{DD} volts. In both cases, for ease of exposition we shall consider the n-channel device Q_N to be the driving transistor and the p-channel device Q_P to be the load. However, since the circuit is completely symmetric, this assumption is obviously arbitrary, and the reverse would lead to identical results.

Figure 4.54 illustrates the case when $v_I = V_{DD}$, showing the i_D–v_{DS} characteristic curve for Q_N with $v_{GSN} = V_{DD}$. (Note that $i_D = i$ and $v_{DSN} = v_O$.) Superimposed on the Q_N character-istic curve is the load curve, which is the i_D–v_{SD} curve of Q_P for the case $v_{SGP} = 0$ V. Since $v_{SGP} < |V_t|$, the load curve will be a horizontal straight line at almost zero current level. The operating point will be at the intersection of the two curves, where we note that the output voltage is nearly zero (typically less than 10 mV) and the current through the two devices is also nearly zero. This means that the power dissipation in the circuit is very small (typically a fraction of a microwatt). Note, however, that although Q_N is operating at nearly zero current and zero drain-source voltage (i.e., near the origin of the i_D–v_{DS} plane), the operating point is on a steep segment of the i_D–v_{DS} characteristic curve. Thus Q_N provides a low-resistance path between the output terminal and ground, with the resistance obtained using Eq. (4.13) as

$$r_{DSN} = 1 \Big/ \left[k_n' \left(\frac{W}{L} \right)_n (V_{DD} - V_{tn}) \right] \qquad (4.140)$$

Figure 4.54(c) shows the equivalent circuit of the inverter when the input is high. This cir-cuit confirms that $v_O \equiv V_{OL} = 0$ V and that the power dissipation in the inverter is zero.

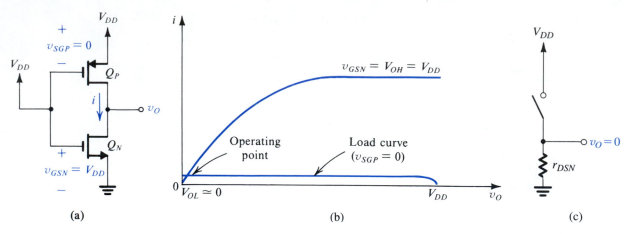

FIGURE 4.54 Operation of the CMOS inverter when v_I is high: **(a)** circuit with $v_I = V_{DD}$ (logic-1 level, or V_{OH}); **(b)** graphical construction to determine the operating point; **(c)** equivalent circuit.

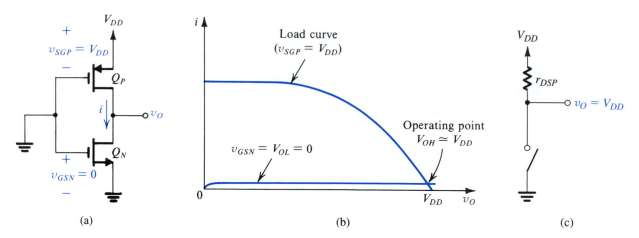

FIGURE 4.55 Operation of the CMOS inverter when v_I is low: **(a)** circuit with $v_I = 0$ V (logic-0 level, or V_{OL}); **(b)** graphical construction to determine the operating point; **(c)** equivalent circuit.

The other extreme case, when $v_I = 0$ V, is illustrated in Fig. 4.55. In this case Q_N is operating at $v_{GSN} = 0$; hence its i_D–v_{DS} characteristic is almost a horizontal straight line at zero current level. The load curve is the i_D–v_{SD} characteristic of the p-channel device with $v_{SGP} = V_{DD}$. As shown, at the operating point the output voltage is almost equal to V_{DD} (typically less than 10 mV below V_{DD}), and the current in the two devices is still nearly zero. Thus the power dissipation in the circuit is very small in both extreme states.

Figure 4.55(c) shows the equivalent circuit of the inverter when the input is low. Here we see that Q_P provides a low-resistance path between the output terminal and the dc supply V_{DD}, with the resistance given by

$$r_{DSP} = 1 \Big/ \left[k'_p \left(\frac{W}{L} \right)_p (V_{DD} - |V_{tp}|) \right] \tag{4.141}$$

The equivalent circuit confirms that in this case $v_O \equiv V_{OH} = V_{DD}$ and that the power dissipation in the inverter is zero.

It should be noted, however, that in spite of the fact that the quiescent current is zero, the load-driving capability of the CMOS inverter is high. For instance, with the input high, as in the circuit of Fig. 4.54, transistor Q_N can sink a relatively large load current. This current can quickly discharge the load capacitance, as will be seen shortly. Because of its action in sinking load current and thus pulling the output voltage down toward ground, transistor Q_N is known as the "pull-down" device. Similarly, with the input low, as in the circuit of Fig. 4.55, transistor Q_P can source a relatively large load current. This current can quickly charge up a load capacitance, thus pulling the output voltage up toward V_{DD}. Hence, Q_P is known as the "pull-up" device. The reader will recall that we used this terminology in connection with the conceptual inverter circuit of Fig. 1.32.

From the above, we conclude that the basic CMOS logic inverter behaves as an ideal inverter. In summary:

1. The output voltage levels are 0 and V_{DD}, and thus the signal swing is the maximum possible. This, coupled with the fact that the inverter can be designed to provide a symmetrical voltage-transfer characteristic, results in wide noise margins.

2. The static power dissipation in the inverter is zero (neglecting the dissipation due to leakage currents) in both of its states. (Recall that the static power dissipation is so named so as to distinguish it from the dynamic power dissipation arising from the repeated switching of the inverter, as will be discussed shortly.)

3. A low-resistance path exists between the output terminal and ground (in the low-output state) or V_{DD} (in the high-output state). These low-resistance paths ensure that the output voltage is 0 or V_{DD} independent of the exact values of the (W/L) ratios or other device parameters. Furthermore, the low output resistance makes the inverter less sensitive to the effects of noise and other disturbances.

4. The active pull-up and pull-down devices provide the inverter with high output-driving capability in both directions. As will be seen, this speeds up the operation considerably.

5. The input resistance of the inverter is infinite (because $I_G = 0$). Thus the inverter can drive an arbitrarily large number of similar inverters with no loss in signal level. Of course, each additional inverter increases the load capacitance on the driving inverter and slows down the operation. Shortly, we will consider the inverter switching times.

4.10.2 The Voltage Transfer Characteristic

The complete voltage-transfer characteristic (VTC) of the CMOS inverter can be obtained by repeating the graphical procedure, used above in the two extreme cases, for all intermediate values of v_I. In the following, we shall calculate the critical points of the resulting voltage transfer curve. For this we need the i–v relationships of Q_N and Q_P. For Q_N,

$$i_{DN} = k_n' \left(\frac{W}{L}\right)_n \left[(v_I - V_{tn})v_O - \frac{1}{2}v_O^2\right] \qquad \text{for } v_O \le v_I - V_{tn} \qquad (4.142)$$

and

$$i_{DN} = \frac{1}{2}k_n' \left(\frac{W}{L}\right)_n (v_I - V_{tn})^2 \qquad \text{for } v_O \ge v_I - V_{tn} \qquad (4.143)$$

For Q_P,

$$i_{DP} = k_p' \left(\frac{W}{L}\right)_p \left[(V_{DD} - v_I - |V_{tp}|)(V_{DD} - v_O) - \frac{1}{2}(V_{DD} - v_O)^2 \right]$$

$$\text{for } v_O \geq v_I + |V_{tp}| \qquad (4.144)$$

and

$$i_{DP} = \frac{1}{2}k_p' \left(\frac{W}{L}\right)_p (V_{DD} - v_I - |V_{tp}|)^2 \qquad \text{for } v_O \leq v_I + |V_{tp}| \qquad (4.145)$$

The CMOS inverter is usually designed to have $V_{tn} = |V_{tp}| = V_t$, and $k_n'(W/L)_n = k_p'(W/L)_p$. It should be noted that since μ_p is 0.3 to 0.5 times the value of μ_n, to make $k'(W/L)$ of the two devices equal, the width of the p-channel device is made two to three times that of the n-channel device. More specifically, the two devices are designed to have equal lengths, with widths related by

$$\frac{W_p}{W_n} = \frac{\mu_n}{\mu_p}$$

This will result in $k_n'(W/L)_n = k_p'(W/L)_p$, and the inverter will have a symmetric transfer characteristic and equal current-driving capability in both directions (pull-up and pull-down).

With Q_N and Q_P matched, the CMOS inverter has the voltage transfer characteristic shown in Fig. 4.56. As indicated, the transfer characteristic has five distinct segments corresponding to different combinations of modes of operation of Q_N and Q_P. The vertical

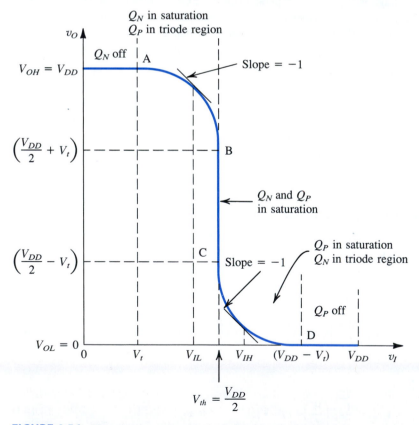

FIGURE 4.56 The voltage transfer characteristic of the CMOS inverter.

segment BC is obtained when both Q_N and Q_P are operating in the saturation region. Because we are neglecting the finite output resistance in saturation, the inverter gain in this region is infinite. From symmetry, this vertical segment occurs at $v_I = V_{DD}/2$ and is bounded by $v_O(\text{B}) = V_{DD}/2 + V_t$ and $v_O(\text{C}) = V_{DD}/2 - V_t$.

The reader will recall from Section 1.7 that in addition to V_{OL} and V_{OH}, two other points on the transfer curve determine the noise margins of the inverter. These are the maximum permitted logic-0 or "low" level at the input, V_{IL}, and the minimum permitted logic-1 or "high" level at the input, V_{IH}. These are formally defined as the two points on the transfer curve at which the incremental gain is unity (i.e., the slope is -1 V/V).

To determine V_{IH}, we note that Q_N is in the triode region, and thus its current is given by Eq. (4.142), while Q_P is in saturation and its current is given by Eq. (4.145). Equating i_{DN} and i_{DP}, and assuming matched devices, gives

$$(v_I - V_t)v_O - \tfrac{1}{2}v_O^2 = \tfrac{1}{2}(V_{DD} - v_I - V_t)^2 \tag{4.146}$$

Differentiating both sides relative to v_I results in

$$(v_I - V_t)\frac{dv_O}{dv_I} + v_O - v_O\frac{dv_O}{dv_I} = -(V_{DD} - v_I - V_t)$$

in which we substitute $v_I = V_{IH}$ and $dv_O/dv_I = -1$ to obtain

$$v_O = V_{IH} - \frac{V_{DD}}{2} \tag{4.147}$$

Substituting $v_I = V_{IH}$ and for v_O from Eq. (4.147) in Eq. (4.146) gives

$$V_{IH} = \tfrac{1}{8}(5V_{DD} - 2V_t) \tag{4.148}$$

V_{IL} can be determined in a manner similar to that used to find V_{IH}. Alternatively, we can use the symmetry relationship

$$V_{IH} - \frac{V_{DD}}{2} = \frac{V_{DD}}{2} - V_{IL}$$

together with V_{IH} from Eq. (4.148) to obtain

$$V_{IL} = \tfrac{1}{8}(3V_{DD} + 2V_t) \tag{4.149}$$

The noise margins can now be determined as follows:

$$
\begin{aligned}
NM_H &= V_{OH} - V_{IH} \\
&= V_{DD} - \tfrac{1}{8}(5V_{DD} - 2V_t) \\
&= \tfrac{1}{8}(3V_{DD} + 2V_t)
\end{aligned}
\tag{4.150}
$$

$$
\begin{aligned}
NM_L &= V_{IL} - V_{OL} \\
&= \tfrac{1}{8}(3V_{DD} + 2V_t) - 0 \\
&= \tfrac{1}{8}(3V_{DD} + 2V_t)
\end{aligned}
\tag{4.151}
$$

As expected, the symmetry of the voltage transfer characteristic results in equal noise margins. Of course, if Q_N and Q_P are not matched, the voltage transfer characteristic will no longer be symmetric, and the noise margins will not be equal (see Problem 4.107).

EXERCISES

4.41 For a CMOS inverter with matched MOSFETs having $V_t = 1$ V, find V_{IL}, V_{IH}, and the noise margins if $V_{DD} = 5$ V.

Ans. 2.1 V; 2.9 V; 2.1 V

4.42 Consider a CMOS inverter with $V_{tn} = |V_{tp}| = 2$ V, $(W/L)_n = 20$, $(W/L)_p = 40$, $\mu_n C_{ox} = 2\mu_p C_{ox} = 20$ μA/V^2, and $V_{DD} = 10$ V. For $v_I = V_{DD}$, find the maximum current that the inverter can sink while v_O remains ≤ 0.5 V.

Ans. 1.55 mA

4.43 An inverter fabricated in a 1.2-μm CMOS technology uses the minimum possible channel lengths (i.e., $L_n = L_p = 1.2$ μm). If $W_n = 1.8$ μm, find the value of W_p that would result in Q_N and Q_P being matched. For this technology, $k'_n = 80$ μA/V^2, $k'_p = 27$ μA/V^2, $V_{tn} = 0.8$ V, and $V_{DD} = 5$ V. Also, calculate the value of the output resistance of the inverter when $v_O = V_{OL}$.

Ans. 5.4 μm; 2 kΩ

4.44 Show that the threshold voltage V_{th} of a CMOS inverter (see Fig. 4.56) is given by

$$V_{th} = \frac{r(V_{DD} - |V_{tp}|) + V_{tn}}{1 + r}$$

where

$$r = \sqrt{\frac{k'_p (W/L)_p}{k'_n (W/L)_n}}$$

4.10.3 Dynamic Operation

As explained in Section 1.7, the speed of operation of a digital system (e.g., a computer) is determined by the **propagation delay** of the logic gates used to construct the system. Since the inverter is the basic logic gate of any digital IC technology, the propagation delay of the inverter is a fundamental parameter in characterizing the technology. In the following, we analyze the switching operation of the CMOS inverter to determine its propagation delay. Figure 4.57(a) shows the inverter with a capacitor C between the output node and ground. Here C represents the sum of the appropriate internal capacitances of the MOSFETs Q_N and Q_P, the capacitance of the interconnect wire between the inverter output node and the input(s) of the other logic gates the inverter is driving, and the total input capacitance of these load (or fan-out) gates. We assume that the inverter is driven by the ideal pulse (zero rise and fall times) shown in Fig. 4.57(b). Since the circuit is symmetric (assuming matched MOSFETs), the rise and fall times of the output waveform should be equal. It is sufficient, therefore, to consider either the turn-on or the turn-off process. In the following, we consider the first.

Figure 4.57(c) shows the trajectory of the operating point obtained when the input pulse goes from $V_{OL} = 0$ to $V_{OH} = V_{DD}$ at time $t = 0$. Just prior to the leading edge of the input pulse (that is, at $t = 0-$) the output voltage equals V_{DD} and capacitor C is charged to this voltage. At $t = 0$, v_I rises to V_{DD}, causing Q_P to turn off immediately. From then on, the circuit is equivalent to that shown in Fig. 4.57(d) with the initial value of $v_O = V_{DD}$. Thus the operating point at $t = 0+$ is point E, at which it can be seen that Q_N will be in the saturation region and conducting a large current. As C discharges, the current of Q_N remains constant until $v_O = V_{DD} - V_t$ (point F). Denoting this portion of the discharge interval t_{PHL1} (where the subscript

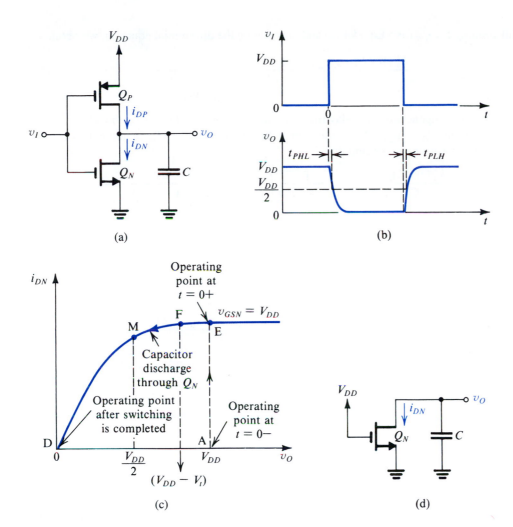

FIGURE 4.57 Dynamic operation of a capacitively loaded CMOS inverter: **(a)** circuit; **(b)** input and output waveforms; **(c)** trajectory of the operating point as the input goes high and C discharges through Q_N; **(d)** equivalent circuit during the capacitor discharge.

HL indicates the high-to-low output transition), we can write

$$t_{PHL1} = \frac{C[V_{DD} - (V_{DD} - V_t)]}{\frac{1}{2}k_n'\left(\frac{W}{L}\right)_n (V_{DD} - V_t)^2}$$

$$= \frac{CV_t}{\frac{1}{2}k_n'\left(\frac{W}{L}\right)_n (V_{DD} - V_t)^2} \qquad (4.152)$$

Beyond point F, transistor Q_N operates in the triode region, and thus its current is given by Eq. (4.142). This portion of the discharge interval can be described by

$$i_{DN}\, dt = -C\, dv_O$$

Substituting for i_{DN} from Eq. (4.142) and rearranging the differential equation, we obtain

$$\frac{k'_n (W/L)_n}{C} dt = \frac{1}{(V_{DD} - V_t)} \frac{dv_O}{\dfrac{1}{2(V_{DD} - V_t)} v_O^2 - v_O} \tag{4.153}$$

To find the component of the delay time t_{PHL} during which v_O decreases from $(V_{DD} - V_t)$ to the 50% point, $v_O = V_{DD}/2$, we integrate both sides of Eq. (4.153). Denoting this component of delay time t_{PHL2}, we find that

$$\frac{k'_n (W/L)_n}{C} t_{PHL2} = \frac{1}{(V_{DD} - V_t)} \int_{v_O = V_{DD} - V_t}^{v_O = V_{DD}/2} \frac{dv_O}{\dfrac{1}{2(V_{DD} - V_t)} v_O^2 - v_O} \tag{4.154}$$

Using the fact that

$$\int \frac{dx}{ax^2 - x} = \ln \left(1 - \frac{1}{ax} \right)$$

enables us to evaluate the integral in Eq. (4.154) and thus obtain

$$t_{PHL2} = \frac{C}{k'_n (W/L)_n (V_{DD} - V_t)} \ln \left(\frac{3V_{DD} - 4V_t}{V_{DD}} \right) \tag{4.155}$$

The two components of t_{PHL} in Eqs. (4.152) and (4.155) can be added to obtain

$$t_{PHL} = \frac{2C}{k'_n (W/L)_n (V_{DD} - V_t)} \left[\frac{V_t}{V_{DD} - V_t} + \frac{1}{2} \ln \left(\frac{3V_{DD} - 4V_t}{V_{DD}} \right) \right] \tag{4.156}$$

For the usual case of $V_t \simeq 0.2 V_{DD}$, this equation reduces to

$$t_{PHL} = \frac{1.6C}{k'_n (W/L)_n V_{DD}} \tag{4.157}$$

Similar analysis of the turn-off process yields an expression for t_{PLH} identical to that in Eq. (4.157) except for $k'_n(W/L)_n$ replaced with $k'_p(W/L)_p$. The propagation delay t_P is the average of t_{PHL} and t_{PLH}. From Eq. (4.157), we note that to obtain lower propagation delays and hence faster operation, C should be minimized, a higher process transconductance parameter k' should be utilized, the transistor W/L ratio should be increased, and the power-supply voltage V_{DD} should be increased. There are, of course, design trade-offs and physical limits involved in making choices for these parameter values. This subject, however, is too advanced for our present needs.

EXERCISES

4.45 A CMOS inverter in a VLSI circuit operating from a 5-V supply has $(W/L)_n = 10\ \mu\text{m}/5\ \mu\text{m}$, $(W/L)_p = 20\ \mu\text{m}/5\ \mu\text{m}$, $V_{tn} = |V_{tp}| = 1$ V, $\mu_n C_{ox} = 2\mu_p C_{ox} = 20\ \mu\text{A/V}^2$. If the total effective load capacitance is 0.1 pF, find t_{PHL}, t_{PLH}, and t_P.

Ans. 0.8 ns; 0.8 ns; 0.8 ns

4.46 For the CMOS inverter of Exercise 4.42, which is intended for SSI and MSI circuit applications, find t_P if the load capacitance is 15 pF.

Ans. 6 ns

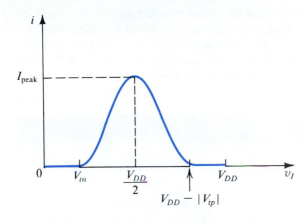

FIGURE 4.58 The current in the CMOS inverter versus the input voltage.

4.10.4 Current Flow and Power Dissipation

As the CMOS inverter is switched, current flows through the series connection of Q_N and Q_P. Figure 4.58 shows the inverter current as a function of v_I. We note that the current peaks at the switching threshold, $V_{th} = v_I = v_O = V_{DD}/2$. This current gives rise to dynamic power dissipation in the CMOS inverter. However, a more significant component of dynamic power dissipation results from the current that flows in Q_N and Q_P when the inverter is loaded by a capacitor C.

An expression for this latter component can be derived as follows: Consider once more the circuit in Fig. 4.57(a). At $t = 0-$, $v_O = V_{DD}$ and the energy stored on the capacitor is $\frac{1}{2}CV_{DD}^2$. At $t = 0$, v_I goes high to V_{DD}, Q_P turns off, and Q_N turns on. Transistor Q_N then discharges the capacitor, and at the end of the discharge interval, the capacitor voltage is reduced to zero. Thus during the discharge interval, energy of $\frac{1}{2}CV_{DD}^2$ is removed from C and dissipated in Q_N. Next consider the other half of the cycle when v_I goes low to zero. Transistor Q_N turns off, and Q_P conducts and charges the capacitor. Let the instantaneous current supplied by Q_P to C be denoted i. This current is, of course, coming from the power supply V_{DD}. Thus the energy drawn from the supply during the charging period will be $\int V_{DD}i\,dt = V_{DD}\int i\,dt = V_{DD}Q$, where Q is the charge supplied to the capacitor; that is, $Q = CV_{DD}$. Thus the energy drawn from the supply during the charging interval is CV_{DD}^2. At the end of the charging interval, the capacitor voltage will be V_{DD}, and thus the energy stored in it will be $\frac{1}{2}CV_{DD}^2$. It follows that during the charging interval, half of the energy drawn from the supply, $\frac{1}{2}CV_{DD}^2$, is dissipated in Q_P.

From the above, we see that in every cycle, $\frac{1}{2}CV_{DD}^2$ of energy is dissipated in Q_N and $\frac{1}{2}CV_{DD}^2$ dissipated in Q_P, for a total energy dissipation in the inverter of CV_{DD}^2. Now if the inverter is switched at the rate of f cycles per second, the dynamic power dissipation in it will be

$$P_D = fCV_{DD}^2 \tag{4.158}$$

Observe that the frequency of operation is related to the propagation delay: The lower the propagation delay, the higher the frequency at which the circuit can be operated and, according to Eq. (4.158), the higher the power dissipation in the circuit. A figure of merit or a quality measure of the particular circuit technology is the **delay-power product** (*DP*),

$$DP = P_D t_P \tag{4.159}$$

The delay-power product tends to be a constant for a particular digital circuit technology and can be used to compare different technologies. Obviously the lower the value of DP the more effective is the technology. The delay-power product has the units of joules, and is in effect a measure of the energy dissipated per cycle of operation. Thus for CMOS where most of the power dissipation is dynamic, we can take DP as simply CV_{DD}^2.

EXERCISES

4.47 For the inverter specified in Exercise 4.42, find the peak current drawn from V_{DD} during switching.

Ans. 1.8 mA

4.48 Let the inverter specified in Exercise 4.42 be loaded by a 15-pF capacitance. Find the dynamic power dissipation that results when the inverter is switched at a frequency of 2 MHz. What is the average current drawn from the power supply?

Ans. 3 mW; 0.3 mA

4.49 Consider a CMOS VLSI chip having 100,000 gates fabricated in a 1.2-μm CMOS technology. Let the load capacitance per gate be 30 fF. If the chip is operated from a 5-V supply and is switched at a rate of 100 MHz, find (a) the power dissipation per gate and (b) the total power dissipated in the chip assuming that only 30% of the gates are switched at any one time.

Ans. 75 μW; 2.25 W

4.10.5 Summary

In this section, we have provided an introduction to CMOS digital circuits. For convenient reference, Table 4.6 provides a summary of the important characteristics of the inverter. We shall return to this subject in Chapter 10, where a variety of CMOS logic circuits are studied.

4.11 THE DEPLETION-TYPE MOSFET

In this section we briefly discuss another type of MOSFET, the depletion-type MOSFET. Its structure is similar to that of the enhancement-type MOSFET with one important difference: The depletion MOSFET has a physically implanted channel. Thus an n-channel depletion-type MOSFET has an n-type silicon region connecting the n^+ source and the n^+ drain regions at the top of the p-type substrate. Thus if a voltage v_{DS} is applied between drain and source, a current i_D flows for $v_{GS} = 0$. In other words, there is no need to induce a channel, unlike the case of the enhancement MOSFET.

The channel depth and hence its conductivity can be controlled by v_{GS} in exactly the same manner as in the enhancement-type device. Applying a positive v_{GS} enhances the channel by attracting more electrons into it. Here, however, we also can apply a negative v_{GS}, which causes electrons to be repelled from the channel, and thus the channel becomes shallower and its conductivity decreases. The negative v_{GS} is said to **deplete** the channel of its charge carriers, and this mode of operation (negative v_{GS}) is called **depletion mode.** As the magnitude of v_{GS} is increased in the negative direction, a value is reached at which the channel is completely depleted of charge carriers and i_D is reduced to zero even though v_{DS}

TABLE 4.6	Summary of Important Characteristics of the CMOS Logic Inverter

Gate Output Resistance

- When v_O is low (current sinking) (Fig. 4.54):

$$r_{DSN} = 1 \Big/ \left[k_n' \left(\frac{W}{L} \right)_n (V_{DD} - V_{tn}) \right]$$

- When v_O is high (current sourcing) (Fig. 4.55):

$$r_{DSP} = 1 \Big/ \left[k_p' \left(\frac{W}{L} \right)_p (V_{DD} - |V_{tp}|) \right]$$

Gate Threshold Voltage

Point on VTC at which $v_O = v_I$:

$$V_{th} = \frac{r(V_{DD} - |V_{tp}|) + V_{tn}}{1 + r}$$

where

$$r = \sqrt{\frac{k_p'(W/L)_p}{k_n'(W/L)_n}}$$

Switching Current and Power Dissipation (Fig. 4.58)

$$I_{\text{peak}} = \frac{1}{2} k_n' \left(\frac{W}{L} \right)_n \left(\frac{V_{DD}}{2} - V_{tn} \right)^2$$

$$P_D = fCV_{DD}^2$$

Noise Margins (Fig. 4.56)

For matched devices, that is, $\mu_n \left(\dfrac{W}{L} \right)_n = \mu_p \left(\dfrac{W}{L} \right)_p$:

$$V_{th} = V_{DD}/2$$

$$V_{IL} = \tfrac{1}{8}(3V_{DD} + 2V_t)$$

$$V_{IH} = \tfrac{1}{8}(5V_{DD} - 2V_t)$$

$$NM_H = NM_L = \tfrac{1}{8}(3V_{DD} + 2V_t)$$

Propagation Delay (Fig. 4.57)

For $V_t \cong 0.2V_{DD}$:

$$t_{PHL} \cong \frac{1.6C}{k_n'(W/L)_n V_{DD}}$$

$$t_{PLH} \cong \frac{1.6C}{k_p'(W/L)_p V_{DD}}$$

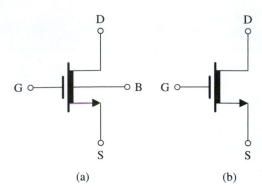

FIGURE 4.59 **(a)** Circuit symbol for the n-channel depletion-type MOSFET. **(b)** Simplified circuit symbol applicable for the case the substrate (B) is connected to the source (S).

may be still applied. This negative value of v_{GS} is the threshold voltage of the n-channel depletion-type MOSFET.

The description above suggests (correctly) that a depletion-type MOSFET can be operated in the enhancement mode by applying a positive v_{GS} and in the depletion mode by applying a negative v_{GS}. The i_D–v_{DS} characteristics are similar to those for the enhancement device except that V_t of the n-channel depletion device is negative.

Figure 4.59(a) shows the circuit symbol for the n-channel depletion-type MOSFET. This symbol differs from that of the enhancement-type device in only one respect: There is a shaded area next to the vertical line representing the channel, signifying that a physical channel exists. When the body (B) is connected to the source (S), the simplified symbol shown in Fig. 4.59(b) can be used.

The i_D–v_{DS} characteristics of a depletion-type n-channel MOSFET for which $V_t = -4$ V and $k_n'(W/L) = 2$ mA/V^2 are sketched in Fig. 4.60(b). (These numbers are typical of discrete devices.) Although these characteristics do not show the dependence of i_D on v_{DS} in saturation, such dependence exists and is identical to the case of the enhancement-type device. Observe that because the threshold voltage V_t is negative, *the depletion NMOS will operate in the triode region as long as the drain voltage does not exceed the gate voltage by more than* $|V_t|$. *For it to operate in saturation, the drain voltage must be greater than the gate voltage by at least* $|V_t|$ *volts.* The chart in Fig. 4.61 shows the relative levels of the terminal voltages of the depletion NMOS transistor for the two regions of operation.

Figure 4.60(c) shows the i_D–v_{GS} characteristics in saturation, indicating both the depletion and enhancement modes of operation.

The current-voltage characteristics of the depletion-type MOSFET are described by the equations identical to those for the enhancement device except that, for an n-channel depletion device, V_t is negative.

A special parameter for the depletion MOSFET is the value of drain current obtained in saturation with $v_{GS} = 0$. This is denoted I_{DSS} and is indicated in Fig. 4.60(b) and (c). It can be shown that

$$I_{DSS} = \frac{1}{2} k_n' \frac{W}{L} V_t^2 \tag{4.160}$$

Depletion-type MOSFETs can be fabricated on the same IC chip as enhancement-type devices, resulting in circuits with improved characteristics, as will be shown in a later chapter.

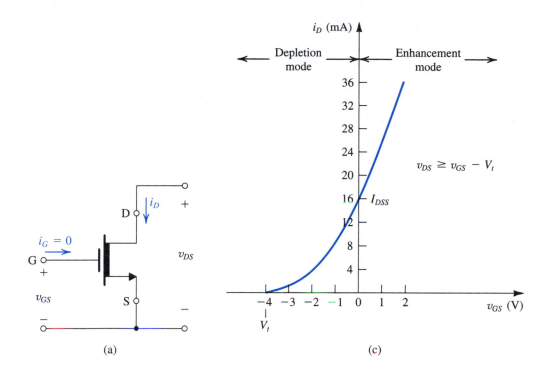

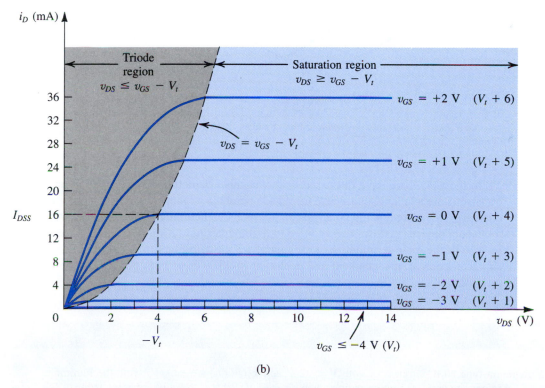

FIGURE 4.60 The current-voltage characteristics of a depletion-type n-channel MOSFET for which $V_t = -4$ V and $k_n'(W/L) = 2$ mA/V^2: **(a)** transistor with current and voltage polarities indicated; **(b)** the i_D–v_{DS} characteristics; **(c)** the i_D–v_{GS} characteristic in saturation.

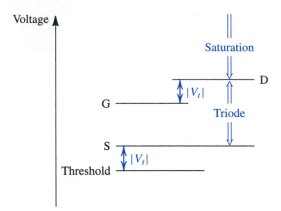

FIGURE 4.61 The relative levels of terminal voltages of a depletion-type NMOS transistor for operation in the triode and the saturation regions. The case shown is for operation in the enhancement mode (v_{GS} is positive).

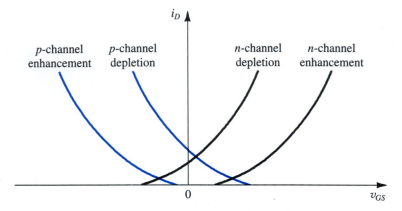

FIGURE 4.62 Sketches of the i_D–v_{GS} characteristics for MOSFETs of enhancement and depletion types, of both polarities (operating in saturation). Note that the characteristic curves intersect the v_{GS} axis at V_t. Also note that for generality somewhat different values of $|V_t|$ are shown for n-channel and p-channel devices.

In the above, we have discussed only n-channel depletion devices. Depletion PMOS transistors are available in discrete form and operate in a manner similar to their n-channel counterparts except that the polarities of all voltages (including V_t) are reversed. Also, in a p-channel device, i_D flows from source to drain, entering the source terminal and leaving by way of the drain terminal. As a summary, we show in Fig. 4.62 sketches of the i_D–v_{GS} characteristics of enhancement and depletion MOSFETs of both polarities (operating in saturation).

EXERCISES

4.50 For a depletion-type NMOS transistor with $V_t = -2$ V and $k_n'(W/L) = 2$ mA/V^2, find the minimum v_{DS} required to operate in the saturation region when $v_{GS} = +1$ V. What is the corresponding value of i_D?

Ans. 3 V; 9 mA

4.51 The depletion-type MOSFET in Fig. E4.51 has $k_n'(W/L) = 4$ mA/V^2 and $V_t = -2$ V. What is the value of I_{DSS}? Neglecting the effect of v_{DS} on i_D in the saturation region, find the voltage that will appear at the source terminal.

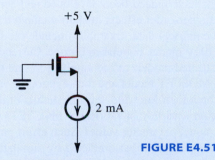

FIGURE E4.51

Ans. 8 mA; +1 V

4.52 Find i as a function of v for the circuit in Fig. E4.52. Neglect the effect of v_{DS} on i_D in the saturation region.

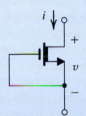

FIGURE E4.52

Ans. $i = k_n' \dfrac{W}{L}\left[-V_t v - \dfrac{1}{2}v^2\right]$, for $v \leq -V_t$; $i = \dfrac{1}{2}k_n'\dfrac{W}{L}V_t^2$, for $v \geq -V_t$

4.12 THE SPICE MOSFET MODEL AND SIMULATION EXAMPLE

We conclude this chapter with a discussion of the models that SPICE uses to simulate the MOSFET. We will also illustrate the use of SPICE in the simulation of the CS amplifier circuit.

4.12.1 MOSFET Models

To simulate the operation of a MOSFET circuit, a simulator requires a mathematical model to represent the characteristics of the MOSFET. The model we have derived in this chapter to represent the MOSFET is a simplified or first-order model. This model, called the **square-law model** because of the quadratic i–v relationship in saturation, works well for transistors with relatively *long* channels. However, for devices with *short* channels, especially submicron transistors, many physical effects that we have neglected come into play, with the result that the derived first-order model no longer accurately represents the actual operation of the MOSFET.

The simple square-law model is useful for understanding the basic operation of the MOSFET as a circuit element and is indeed used to obtain approximate pencil-and-paper circuit designs. However, more elaborate models, which account for short-channel effects, are required to be able to predict the performance of integrated circuits with a certain degree of precision prior to fabrication. Such models have indeed been developed and continue to be refined to more accurately represent the higher-order effects in short-channel transistors through a mix of physical relationships and empirical data. Examples include the Berkeley short-channel IGFET model (BSIM) and the EKV model, popular in Europe. Currently, semiconductor manufacturers rely on such sophisticated models to accurately represent the fabrication process. These manufacturers select a MOSFET model and then extract the values for the corresponding model parameters using both their knowledge of the details of the fabrication process and extensive measurements on a variety of fabricated MOSFETs. A great deal of effort is expended on extracting the model parameter values. Such effort pays off in fabricated circuits exhibiting performance very close to that predicted by simulation, thus reducing the need for costly redesign.

Although it is beyond the scope of this book to delve into the subject of MOSFET modeling and short-channel effects, it is important that the reader be aware of the limitations of the square-law model and of the availability of more accurate but, unfortunately, more complex MOSFET models. In fact, the power of computer simulation is more apparent when one has to use these complex device models in the analysis and design of integrated circuits.

SPICE-based simulators, like PSpice, provide the user with a choice of MOSFET models. The corresponding SPICE model parameters (whose values are provided by the semiconductor manufacturer) include a parameter, called LEVEL, which selects the MOSFET model to be used by the simulator. Although the value of this parameter is not always indicative of the accuracy, nor of the complexity of the corresponding MOSFET model, LEVEL = 1 corresponds to the simplest first-order model (called the Shichman-Hodges model) which is based on the square-law MOSFET equations presented in this chapter. For simplicity, we will use this model to illustrate the description of the MOSFET model parameters in SPICE and to simulate the example circuit in PSpice. However, the reader is again reminded of the need to use a more sophisticated model than the level-1 model to accurately predict the circuit performance, especially for submicron transistors.

4.12.2 MOSFET Model Parameters

Table 4.7 provides a listing of some of the MOSFET model parameters used in the Level-1 model of SPICE. The reader should already be familiar with these parameters, except for a few, which are described next.

MOSFET Diode Parameters   For the two reverse-biased diodes formed between each of the source and drain diffusion regions and the body (see Fig. 4.1) the saturation-current density is modeled in SPICE by the parameter JS. Furthermore, based on the parameters specified in Table 4.7, SPICE will calculate the depletion-layer (junction) capacitances discussed in Section 4.8.2 as

$$C_{db} = \frac{\text{CJ}}{\left(1 + \dfrac{V_{DB}}{\text{PB}}\right)^{\text{MJ}}} \text{AD} + \frac{\text{CJSW}}{\left(1 + \dfrac{V_{DB}}{\text{PB}}\right)^{\text{MJSW}}} \text{PD} \qquad (4.161)$$

$$C_{sb} = \frac{\text{CJ}}{\left(1 + \dfrac{V_{SB}}{\text{PB}}\right)^{\text{MJ}}} \text{AS} + \frac{\text{CJSW}}{\left(1 + \dfrac{V_{SB}}{\text{PB}}\right)^{\text{MJSW}}} \text{PS} \qquad (4.162)$$

TABLE 4.7 Parameters of the SPICE Level-1 MOSFET Model (Partial Listing)

SPICE Parameter	Book Symbol	Description	Units
		Basic Model Parameters	
LEVEL		MOSFET model selector	
TOX	t_{ox}	Gate-oxide thickness	m
COX	C_{ox}	Gate-oxide capacitance, per unit area	F/m²
UO	μ	Carrier mobility	cm²/V·s
KP	k'	Process transconductance parameter	A/V²
LAMBDA	λ	Channel-length modulation coefficient	V⁻¹
		Threshold Voltage Parameters	
VTO	V_{t0}	Zero-bias threshold voltage	V
GAMMA	γ	Body-effect parameter	V¹ᐟ²
NSUB	N_A, N_D	Substrate doping	cm⁻³
PHI	$2\Phi_f$	Surface inversion potential	V
		MOSFET Diode Parameters	
JS		Body-junction saturation-current density	A/m²
CJ		Zero-bias body-junction capacitance, per unit area over the drain/source region	F/m²
MJ		Grading coefficient, for area component	
CJSW		Zero-bias body-junction capacitance, per unit length along the sidewall (periphery) of the drain/source region	F/m
MJSW		Grading coefficient, for sidewall component	
PB	V_0	Body-junction built-in potential	V
		MOSFET Dimension Parameters	
LD	L_{ov}	Lateral diffusion into the channel from the source/drain diffusion regions	m
WD		Sideways diffusion into the channel from the body along the width	m
		MOS Gate-Capacitance Parameters	
CGBO		Gate-body overlap capacitance, per unit channel length	F/m
CGDO	C_{ov}/W	Gate-drain overlap capacitance, per unit channel width	F/m
CGSO	C_{ov}/W	Gate-source overlap capacitance, per unit channel width	F/m

where AD and AS are the areas while PD and PS are the perimeters of, respectively, the drain and source regions of the MOSFET. The first capacitance term in Eqs. (4.161) and (4.162) represents the depletion-layer (junction) capacitance over the bottom plate of the drain and source regions. The second capacitance term accounts for the depletion-layer capacitance along the sidewall (periphery) of these regions. Both terms are expressed using the formula developed in Section 3.7.3 (Eq. 3.56). The values of AD, AS, PD, and PS must be specified by the user based on the dimensions of the device being used.

MOSFET Dimension and Gate-Capacitance Parameters In a fabricated MOSFET, the effective channel length L_{eff} is shorter than the nominal (or drawn) channel length L (as specified by the designer) because the source and drain diffusion regions extend slightly under the gate oxide during fabrication. Furthermore, the effective channel width W_{eff} of the MOSFET is shorter than the nominal or drawn channel width W because of the sideways diffusion into the channel from the body along the width. Based on the parameters specified in Table 4.7,

$$L_{eff} = L - 2LD \tag{4.163}$$

$$W_{eff} = W - 2WD \tag{4.164}$$

In a manner analogous to using L_{ov} to denote LD, we will use the symbol W_{ov} to denote WD. Consequently, as indicated in Section 4.8.1, the gate-source capacitance C_{gs} and the gate-drain capacitance C_{gd} must be increased by an overlap component of, respectively,

$$C_{gs,ov} = W\,\mathrm{CGSO} \tag{4.165}$$

and

$$C_{gd,ov} = W\,\mathrm{CGDO} \tag{4.166}$$

Similarly, the gate-body capacitance C_{gb} must be increased by an overlap component of

$$C_{gb,ov} = L\,\mathrm{CGBO} \tag{4.167}$$

The reader may have observed that there is a built-in redundancy in specifying the MOSFET model parameters in SPICE. For example, the user may specify the value of KP for a MOSFET or, alternatively, specify TOX and UO and let SPICE compute KP as UO TOX. Similarly, GAMMA can be directly specified, or the physical parameters that enable SPICE to determine it can be specified (e.g., NSUB). In any case, *the user-specified values will always take precedence over (i.e., override) those values calculated by SPICE.* As another example, note that the user has the option of either directly specifying the overlap capacitances CGBO, CGDO, and CGSO or letting SPICE compute them as CGDO = CGSO = LD COX and CGBO = WD COX.

Table 4.8 provides typical values for the Level-1 MOSFET model parameters of a modern 0.5-μm CMOS technology and, for comparison, those of an old (even obsolete) 5-μm CMOS technology. The corresponding values for the minimum channel length L_{min}, minimum channel width W_{min}, and the maximum supply voltage $(V_{DD} + |V_{SS}|)_{max}$ are as follows:

Technology	L_{min}	W_{min}	$(V_{DD} + |V_{SS}|)_{max}$
5-μm CMOS	5 μm	12.5 μm	10 V
0.5-μm CMOS	0.5 μm	1.25 μm	3.3 V

Because of the thinner gate oxide in modern CMOS technologies, the maximum supply voltage must be reduced to ensure that the MOSFET terminal voltages do not cause a breakdown of the oxide dielectric under the gate. The shrinking supply voltage is one of the most challenging design aspects of analog integrated circuits in advanced CMOS technologies. From Table 4.8, the reader may have observed some other trends in CMOS processes. For example, as L_{min} is reduced, the channel-length modulation effect becomes more pronounced and, hence, the value of λ increases. This results in MOSFETs having smaller output resistance r_o and, therefore, smaller "intrinsic gains" (Chapter 6). Another example is the decrease in surface mobility μ in modern CMOS technologies and the corresponding increase in the ratio of μ_n/μ_p, from 2 to

TABLE 4.8 Values of the Level-1 MOSFET Model Parameters for Two CMOS Technologies[1]

	5-µm CMOS Process		0.5-µm CMOS Process	
	NMOS	PMOS	NMOS	PMOS
LEVEL	1	1	1	1
TOX	85e-9	85e-9	9.5e-9	9.5e-9
UO	750	250	460	115
LAMBDA	0.01	0.03	0.1	0.2
GAMMA	1.4	0.65	0.5	0.45
VTO	1	−1	0.7	−0.8
PHI	0.7	0.65	0.8	0.75
LD	0.7e-6	0.6e-6	0.08e-6	0.09e-6
JS	1e-6	1e-6	10e-9	5e-9
CJ	0.4e-3	0.18e-3	0.57e-3	0.93e-3
MJ	0.5	0.5	0.5	0.5
CJSW	0.8e-9	0.6e-9	0.12e-9	0.17e-9
MJSW	0.5	0.5	0.4	0.35
PB	0.7	0.7	0.9	0.9
CGBO	0.2e-9	0.2e-9	0.38e-9	0.38e-9
CGDO	0.4e-9	0.4e-9	0.4e-9	0.35e-9
CGSO	0.4e-9	0.4e-9	0.4e-9	0.35e-9

[1] In PSpice, we have created MOSFET parts corresponding to the above models. Readers can find these parts in the SEDRA.olb library, which is available on the CD accompanying this book as well as on-line at www.sedrasmith.org. The NMOS and PMOS parts for the 0.5-µm CMOS technology are labelled NMOS0P5_BODY and PMOS0P5_BODY, respectively. The NMOS and PMOS parts for the 5-µm CMOS technology are labelled NMOS5P0_BODY and PMOS5P0_BODY, respectively. Furthermore, parts NMOS0P5 and PMOS0P5 are created to correspond to, respectively, part NMOS0P5_BODY with its body connected to net 0 and part PMOS0P5_BODY with its body connected to net V_{DD}.

close to 5. The impact of this and other trends on the design of integrated circuits in advanced CMOS technologies are discussed in Chapter 6 (see in particular Section 6.2).

When simulating a MOSFET circuit, the user needs to specify both the values of the model parameters and the dimensions of each MOSFET in the circuit being simulated. At least, the channel length L and width W must be specified. The areas AD and AS and the perimeters PD and PS need to be specified for SPICE to model the body-junction capacitances (otherwise, zero capacitances would be assumed). The exact values of these geometry parameters depend on the actual layout of the device (Appendix A). However, to estimate these dimensions, we will assume that a metal contact is to be made to each of the source and drain regions of the MOSFET. For this purpose, typically, these diffusion regions must be extended *past* the end of the channel (i.e., in the L-direction in Fig. 4.1) by at least $2.75 L_{min}$. Thus, the minimum area and perimeter of a drain/source diffusion region with a contact are, respectively,

$$AD = AS = 2.75 L_{min} W \qquad (4.168)$$

and

$$PD = PS = 2 \times 2.75 L_{min} + W \qquad (4.169)$$

Unless otherwise specified, we will use Eqs. (4.168) and (4.169) to estimate the dimensions of the drain/source regions in our examples.

Finally, we note that SPICE computes *the values for the parameters of the MOSFET small-signal model based on the dc operating point (bias point)*. These are then used by SPICE to perform the small-signal analysis (ac analysis).

EXAMPLE 4.14

THE CS AMPLIFIER

In this example, we will use PSpice to compute the frequency response of the CS amplifier whose Capture schematic is shown in Fig. 4.63.[11] Observe that the MOSFET has its source and body connected in order to cancel the body effect. We will assume a 0.5-μm CMOS technology for the MOSFET and use the SPICE level-1 model parameters listed in Table 4.8. We will also assume a signal-source resistance $R_{sig} = 10$ kΩ, a load resistance $R_L = 50$ kΩ, and bypass and coupling capacitors of 10 μF. The targeted specifications for this CS amplifier are a midband gain $A_M = 10$ V/V and a maximum power consumption $P = 1.5$ mW. As should always be the case with computer simulation, we will begin with an approximate pencil-and-paper design. We will then use PSpice to fine-tune our design, and to investigate the performance of the final design. In this way, maximum advantage and insight can be obtained from simulation.

With a 3.3-V power supply, the drain current of the MOSFET must be limited to $I_D = P/V_{DD} = 1.5$ mW/3.3 V $= 0.45$ mA to meet the power consumption specification. Choosing $V_{OV} = 0.3$ V (a typical value in low-voltage designs) and $V_{DS} = V_{DD}/3$ (to achieve a large signal swing at the output), the MOSFET can now be sized as

$$\frac{W}{L_{eff}} = \frac{I_D}{\frac{1}{2}k'_n V_{OV}^2 (1 + \lambda V_{DS})} = \frac{0.45 \times 10^{-3}}{\frac{1}{2}(170.1 \times 10^{-6})(0.3)^2 [1 + 0.1(1.1)]} \cong 53 \qquad (4.170)$$

where $k'_n = \mu_n C_{ox} = 170.1$ μA/V^2 (from Table 4.8). Here, L_{eff} rather than L is used to more accurately compute I_D. The effect of using W_{eff} rather than W is much less important because typically $W \gg W_{ov}$. Thus, choosing $L = 0.6$ μm results in $L_{eff} = L - 2L_{ov} = 0.44$ μm and $W = 23.3$ μm. Note that we chose L slightly larger than L_{min}. This is a common practice in the design of analog ICs to minimize the effects of fabrication nonidealities on the actual value of L. As we will study in later chapters, this is particularly important when the circuit performance depends on the matching between the dimensions of two or more MOSFETs (e.g., in the current-mirror circuits we will study in Chapter 6).

Next, R_D is calculated based on the desired voltage gain:

$$|A_v| = g_m(R_D \| R_L \| r_o) = 10 \text{ V/V} \Rightarrow R_D \approx 4.2 \text{ k}\Omega \qquad (4.171)$$

where $g_m = 3.0$ mA/V and $r_o = 22.2$ kΩ. Hence, the output bias voltage is $V_O = V_{DD} - I_D R_D = 1.39$ V. An $R_S = (V_O - V_{DD}/3)/I_D = 630$ Ω is needed to bias the MOSFET at a $V_{DS} = V_{DD}/3$. Finally, resistors $R_{G1} = 2$ MΩ and $R_{G2} = 1.3$ MΩ are chosen to set the gate bias voltage at $V_G = I_D R_S + V_{OV} + V_{tn} \approx 1.29$ V. Using large values for these gate resistors ensures that both their power consumption and the loading effect on the input signal source are negligible. Note that we neglected the body effect in the expression for V_G to simplify our hand calculations.

We will now use PSpice to verify our design and investigate the performance of the CS amplifier. We begin by performing a bias-point simulation to verify that the MOSFET is properly

---

[11] The reader is reminded that the Capture schematics and the corresponding PSpice simulation files of all SPICE examples in this book can be found on the text's CD as well as on its website (www.sedrasmith.org). In these schematics (as shown in Fig. 4.63), we used variable parameters to enter the values of the various circuit components, including the dimensions of the MOSFET. This will allow the reader to investigate the effect of changing component values by simply changing the corresponding parameter values.

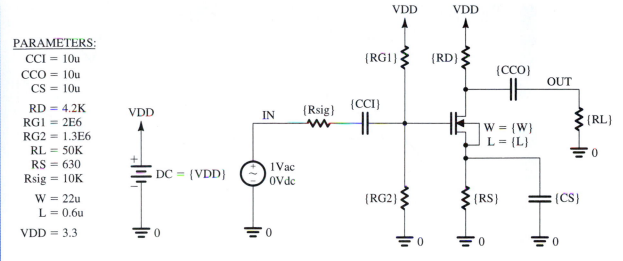

FIGURE 4.63   Capture schematic of the CS amplifier in Example 4.14.

biased in the saturation region and that the dc voltages and currents are within the desired specifications. Based on this simulation, we have decreased the value of W to 22 μm to limit I_D to about 0.45 mA. Next, to measure the midband gain A_M and the 3-dB frequencies f_L and f_H, we apply a 1-V ac voltage at the input, perform an ac-analysis simulation, and plot the output voltage magnitude (in dB) versus frequency as shown in Fig. 4.64. This corresponds to the magnitude response of the CS amplifier because we chose a 1-V input signal.[12] Accordingly, the midband gain is $A_M =$ 9.55 V/V and the 3-dB bandwidth is $BW = f_H - f_L \approx 122.1$ MHz. Figure 4.64 further shows that the gain begins to fall off at about 300 Hz but flattens out again at about 10 Hz. This flattening in the gain at low frequencies is due to a real transmission zero[13] introduced in the transfer function of the amplifier by R_S together with C_S. This zero occurs at a frequency $f_Z = 1/(2\pi R_S C_S) =$ 25.3 Hz, which is typically between the break frequencies f_{P2} and f_{P3} derived in Section 4.9.3 (Fig. 4.52). So, let us now verify this phenomenon by resimulating the CS amplifier with a $C_S = 0$ (i.e., removing C_S) in order to move f_Z to infinity and remove its effect. The corresponding frequency response is plotted also in Fig. 4.64. As expected, with $C_S = 0$, we do not observe any flattening in the low-frequency response of the amplifier, which now looks similar to that in Fig. 4.52. However, because the CS amplifier now includes a source resistor R_S, A_M has dropped by a factor of 2.6. This factor is approximately equal to $(1 + g_m R_S)$, as expected from our study of the CS amplifier with a source-degeneration resistance in Section 4.7.4. Note that the bandwidth BW has increased by approximately the same factor as the drop in gain A_M. As we will learn in Chapter 8 when we study negative feedback, the source-degeneration resistor R_S provides negative feedback, which allows us to trade off gain for wider bandwidth.

---

[12] The reader should not be alarmed about the use of a such a large signal amplitude. Recall (Section 2.9.1) that in a small-signal (ac) simulation, SPICE first finds the small-signal equivalent circuit at the bias point and then analyzes this linear circuit. Such ac analysis can, of course, be done with any ac signal amplitude. However, a 1-V ac input is convenient to use as the resulting ac output corresponds to the voltage gain of the circuit.

[13] Readers who have not yet studied poles and zeros can either refer to Appendix E or skip these few sentences.

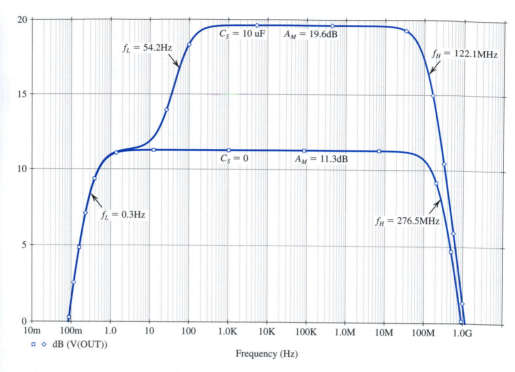

FIGURE 4.64 Frequency response of the CS amplifier in Example 4.14 with $C_S = 10 \, \mu$F and $C_S = 0$ (i.e., C_S removed).

To conclude this example, we will demonstrate the improved bias stability achieved when a source resistor R_S is used (see the discussion in Section 4.5.2). Specifically, we will change (in the MOSFET level-1 model for part NMOS0P5) the value of the zero-bias threshold voltage parameter VT0 by ±15% and perform a bias-point simulation in PSpice. Table 4.9 shows the corresponding variations in I_D and V_O for the case in which $R_S = 630 \, \Omega$. For the case without source degeneration, we use an $R_S = 0$ in the schematic of Fig. 4.63. Furthermore, to obtain the same I_D and V_O in both cases (for the nominal threshold voltage $V_{t0} = 0.7$ V), we use an $R_{G2} = 0.88$ MΩ to reduce V_G to around $V_{OV} + V_{tn} = 1$ V. The corresponding variations in the bias point are shown in Table 4.9. Accordingly, we see that the source degeneration resistor makes the bias point of the CS amplifier less sensitive to changes in the threshold voltage. In fact, the reader can show for the values displayed in Table 4.9 that the variation in bias current ($\Delta I / I$) is reduced by approximately the same factor, $(1 + g_m R_S)$. However, unless a large bypass capacitor C_S is used, this reduced sensitivity comes at the expense of a reduction in the midband gain (as we observed in this example when we simulated the frequency response of the CS amplifier with a $C_S = 0$).

TABLE 4.9   Variations in the Bias Point with the MOSFET Threshold Voltage

	$R_S = 630 \, \Omega$		$R_S = 0$	
V_{tn0}	I_D (mA)	V_O (V)	I_D (mA)	V_O (V)
0.60	0.56	0.962	0.71	0.33
0.7	0.46	1.39	0.45	1.40
0.81	0.36	1.81	0.21	2.40

SUMMARY

- The enhancement-type MOSFET is currently the most widely used semiconductor device. It is the basis of CMOS technology, which is the most popular IC fabrication technology at this time. CMOS provides both n-channel (NMOS) and p-channel (PMOS) transistors, which increases design flexibility. The minimum MOSFET channel length achievable with a given CMOS process is used to characterize the process. This figure has been continually reduced and is currently about 0.1 μm.

- The current-voltage characteristics of the MOSFET are presented in Section 4.2 and are summarized in Table 4.1.

- Techniques for analyzing MOSFET circuits at dc are illustrated in Section 4.3 via a number of examples.

- The large-signal operation of the basic common-source (CS) resistively loaded MOSFET is studied in Section 4.4. The voltage transfer characteristic is derived, both graphically and analytically, and is used to show the three regions of operation: cutoff and triode, which are useful for the application of the MOSFET as a switch and as a digital logic inverter; and saturation, which is the region for amplifier operation. To obtain linear amplification, the transistor is biased to operate somewhere near the middle of the saturation region, and the signal is superimposed on the dc bias V_{GS} and kept small. The small-signal gain is equal to the slope of the transfer characteristic at the bias point (see Fig. 4.26).

- A key step in the design of transistor amplifiers is to bias the transistor to operate at an appropriate point in the saturation region. A good bias design ensures that the parameters of the bias point, I_D, V_{OV}, and V_{DS}, are predictable and stable, and do not vary by a large amount when the transistor is replaced by another of the same type. A variety of biasing methods suitable for discrete-circuit design are presented in Section 4.5.

- The small-signal operation of the MOSFET as well as circuit models that represent it are covered in Section 4.6. A summary of the relationships for determining the values of MOSFET model parameters is provided in Table 4.2.

- Grounding one of the three terminals of the MOSFET results in a two-port network with the grounded terminal serving as a common terminal between the input and output ports. Accordingly, there are three basic MOSFET amplifier configurations: the CS configuration, which is the most widely used; the common-gate (CG) configuration, which has special applications and is particularly useful at

high frequencies; and the common-drain or source-follower configuration, which is employed as a voltage buffer or as the output stage of a multistage amplifier. Refer to the summary at the end of Section 4.7, and in particular to Table 4.4, which provides a summary and a comparison of the attributes of the various single-stage MOSFET amplifier configurations.

- For the MOSFET high-frequency model and the formulas for determining the model parameters, refer to Table 4.5.

- The internal capacitances of the MOSFET cause the gain of MOS amplifiers to fall off at high frequencies. Also, the coupling and bypass capacitors that are used in discrete MOS amplifiers cause the gain to fall off at low frequencies. The frequency band over which both sets of capacitors can be neglected, and hence over which the gain is constant, is known as the midband. The amplifier frequency response is characterized by the midband gain A_M and the lower and upper 3-dB frequencies f_L and f_H, respectively, and the bandwidth is $(f_H - f_L)$.

- Analysis of the frequency response of the common source amplifier (Section 4.9) shows that its high-frequency response is determined by the interaction of the total input capacitance C_{in} and the effective resistance of the signal source, R'_{sig}; $f_H = 1/2\pi C_{in}R'_{sig}$. The input capacitance $C_{in} = C_{gs} + (1 + g_m R'_L)C_{gd}$, which can be dominated by the second term. Thus, while C_{gd} is small, its effect can be very significant because it is multiplied by a factor approximately equal to the midband gain. This is the Miller effect.

- The CMOS digital logic inverter provides a near-ideal implementation of the logic inversion function. Its characteristics are studied in Section 4.10 and summarized in Table 4.6.

- The depletion-type MOSFET has an implanted channel and thus can be operated in either the depletion or enhancement modes. It is characterized by the same equations used for the enhancement device except for having a negative V_t (positive V_t for depletion PMOS transistors).

- Although there is no substitute for pencil-and-paper circuit design employing simplified device models, computer simulation using SPICE with more elaborate, and hence more precise, models is essential for checking and fine-tuning the design before fabrication.

- Our study of MOSFET amplifiers continues in Chapter 6 and that of digital CMOS circuits in Chapter 10.

PROBLEMS

SECTION 4.1: DEVICE STRUCTURE AND PHYSICAL OPERATION

4.1 MOS technology is used to fabricate a capacitor, utilizing the gate metallization and the substrate as the capacitor electrodes. Find the area required per 1-pF capacitance for oxide thickness ranging from 5 nm to 40 nm. For a square plate capacitor of 10 pF, what maximum dimensions are needed?

4.2 A particular MOSFET using the same gate structure and channel length as the transistor whose i_D–v_{DS} characteristics are shown in Fig. 4.4 has a channel width that is 10 times greater. How should the vertical axis be relabelled to represent this change? Find the new constant of proportionality relating i_D and $(v_{GS} - V_t)v_{DS}$. What is the range of drain-to-source resistance, r_{DS}, corresponding to an overdrive voltage $(v_{GS} - V_t)$ ranging from 0.5 V to 2 V?

4.3 With the knowledge that $\mu_p \approx 0.4\mu_n$, what must be the relative width of n-channel and p-channel devices if they are to have equal drain currents when operated in the saturation mode with overdrive voltages of the same magnitude?

4.4 An n-channel device has $k'_n = 50$ μA/V^2, $V_t = 0.8$ V, and $W/L = 20$. The device is to operate as a switch for small v_{DS}, utilizing a control voltage v_{GS} in the range 0 V to 5 V. Find the switch closure resistance, r_{DS}, and closure voltage, V_{DS}, obtained when $v_{GS} = 5$ V and $i_D = 1$ mA. Recalling that $\mu_p \approx 0.4\mu_n$, what must W/L be for a p-channel device that provides the same performance as the n-channel device in this application?

4.5 An n-channel MOS device in a technology for which oxide thickness is 20 nm, minimum gate length is 1 μm, $k'_n = 100$ μA/V^2, and $V_t = 0.8$ V operates in the triode region, with small v_{DS} and with the gate–source voltage in the range 0 V to +5 V. What device width is needed to ensure that the minimum available resistance is 1 kΩ?

4.6 Consider a CMOS process for which $L_{min} = 0.8$ μm, $t_{ox} = 15$ nm, $\mu_n = 550$ cm^2/V·s, and $V_t = 0.7$ V.

(a) Find C_{ox} and k'_n.
(b) For an NMOS transistor with $W/L = 16$ μm/0.8 μm, calculate the values of V_{OV}, V_{GS}, and V_{DSmin} needed to operate the transistor in the saturation region with a dc current $I_D = 100$ μA.
(c) For the device in (b), find the value of V_{OV} and V_{GS} required to cause the device to operate as a 1000-Ω resistor for very small v_{DS}.

4.7 Consider an n-channel MOSFET with $t_{ox} = 20$ nm, $\mu_n = 650$ cm^2/V·s, $V_t = 0.8$ V, and $W/L = 10$. Find the drain current in the following cases:

(a) $v_{GS} = 5$ V and $v_{DS} = 1$ V
(b) $v_{GS} = 2$ V and $v_{DS} = 1.2$ V
(c) $v_{GS} = 5$ V and $v_{DS} = 0.2$ V
(d) $v_{GS} = v_{DS} = 5$ V

SECTION 4.2: CURRENT-VOLTAGE CHARACTERISTICS

4.8 Consider an NMOS transistor that is identical to, except for having half the width of, the transistor whose i_D–v_{DS} characteristics are shown in Fig. 4.11(b). How should the vertical axis be relabeled so that the characteristics correspond to the narrower device? If the narrower device is operated in saturation with an overdrive voltage of 1.5 V, what value of i_D results?

4.9 Explain why the graphs in Fig. 4.11(b) do not change as V_t is changed. Can you devise a more general (i.e., V_t independent) representation of the characteristics presented in Fig. 4.12?

4.10 For the transistor whose i_D–v_{GS} characteristics are depicted in Fig. 4.12, sketch i_D versus the overdrive voltage $v_{OV} \equiv v_{GS} - V_t$ for $v_{DS} \geq v_{OV}$. What is the advantage of this graph over that in Fig. 4.12? Sketch, on the same diagram, the graph for a device that is identical except for having half the width.

4.11 An NMOS transistor having $V_t = 1$ V is operated in the triode region with v_{DS} small. With $V_{GS} = 1.5$ V, it is found to have a resistance r_{DS} of 1 kΩ. What value of V_{GS} is required to obtain $r_{DS} = 200$ Ω? Find the corresponding resistance values obtained with a device having twice the value of W.

4.12 A particular enhancement MOSFET for which $V_t = 1$ V and $k'_n(W/L) = 0.1$ mA/V^2 is to be operated in the saturation region. If i_D is to be 0.2 mA, find the required v_{GS} and the minimum required v_{DS}. Repeat for $i_D = 0.8$ mA.

4.13 A particular n-channel enhancement MOSFET is measured to have a drain current of 4 mA at $V_{GS} = V_{DS} = 5$ V and of 1 mA at $V_{GS} = V_{DS} = 3$ V. What are the values of $k'_n(W/L)$ and V_t for this device?

D4.14 For a particular IC-fabrication process, the transconductance parameter $k'_n = 50$ μA/V^2, and $V_t = 1$ V. In an application in which $v_{GS} = v_{DS} = V_{supply} = 5$ V, a drain current of 0.8 mA is required of a device of minimum length of 2 μm. What value of channel width must the design use?

4.15 An NMOS transistor, operating in the linear-resistance region with $v_{DS} = 0.1$ V, is found to conduct 60 μA for $v_{GS} = 2$ V and 160 μA for $v_{GS} = 4$ V. What is the apparent value of threshold voltage V_t? If $k'_n = 50$ μA/V^2, what is the device W/L ratio? What current would you expect to flow with $v_{GS} = 3$ V and $v_{DS} = 0.15$ V? If the device is operated at $v_{GS} = 3$ V, at what value of v_{DS} will the drain end of the MOSFET channel just reach pinch off, and what is the corresponding drain current?

4.16 For an NMOS transistor, for which $V_t = 0.8$ V, operating with v_{GS} in the range of 1.5 V to 4 V, what is the largest value of v_{DS} for which the channel remains continuous?

4.17 An NMOS transistor, fabricated with $W = 100\ \mu m$ and $L = 5\ \mu m$ in a technology for which $k'_n = 50\ \mu A/V^2$ and $V_t = 1$ V, is to be operated at very low values of v_{DS} as a linear resistor. For v_{GS} varying from 1.1 V to 11 V, what range of resistor values can be obtained? What is the available range if

(a) the device width is halved?
(b) the device length is halved?
(c) both the width and length are halved?

4.18 When the drain and gate of a MOSFET are connected together, a two-terminal device known as a "diode-connected transistor" results. Figure P4.18 shows such devices obtained from MOS transistors of both polarities. Show that

(a) the i–v relationship is given by

$$i = \frac{1}{2} k' \frac{W}{L} (v - |V_t|)^2$$

(b) the incremental resistance r for a device biased to operate at $v = |V_t| + V_{OV}$ is given by

$$r \equiv 1 \Big/ \left[\frac{\partial i}{\partial v}\right] = 1 \Big/ \left(k' \frac{W}{L} V_{OV}\right)$$

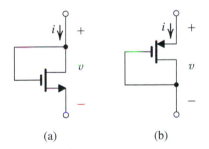

(a)   (b)

FIGURE P4.18

4.19 For a particular MOSFET operating in the saturation region at a constant v_{GS}, i_D is found to be 2 mA for $v_{DS} = 4$ V and 2.2 mA for $v_{DS} = 8$ V. What values of r_o, V_A, and λ correspond?

4.20 A particular MOSFET has $V_A = 50$ V. For operation at 0.1 mA and 1 mA, what are the expected output resistances? In each case, for a change in v_{DS} of 1 V, what percentage change in drain current would you expect?

D4.21 In a particular IC design in which the standard channel length is 2 μm, an NMOS device with W/L of 5 operating at

100 μA is found to have an output resistance of 0.5 MΩ, about $\frac{1}{4}$ of that needed. What dimensional change can be made to solve the problem? What is the new device length? The new device width? The new W/L ratio? What is V_A for the standard device in this IC? The new device?

D4.22 For a particular n-channel MOS technology, in which the minimum channel length is 1 μm, the associated value of λ is 0.02 V^{-1}. If a particular device for which L is 3 μm operates at $v_{DS} = 1$ V with a drain current of 80 μA, what does the drain current become if v_{DS} is raised to 5 V? What percentage change does this represent? What can be done to reduce the percentage by a factor of 2?

4.23 An NMOS transistor is fabricated in a 0.8-μm process having $k'_n = 130\ \mu A/V^2$ and $V'_A = 20$ V/μm of channel length. If $L = 1.6\ \mu m$ and $W = 16\ \mu m$, find V_A and λ. Find the value of I_D that results when the device is operated with an overdrive voltage of 0.5 V and $V_{DS} = 2$ V. Also, find the value of r_o at this operating point. If V_{DS} is increased by 1 V, what is the corresponding change in I_D?

4.24 Complete the missing entries in the following table, which describes characteristics of suitably biased NMOS transistors:

MOS	1	2	3	4
λ (V^{-1})		0.01		
V_A (V)	50			200
I_D (mA)	5		0.1	
r_o (kΩ)		30	100	1000

4.25 An NMOS transistor with $\lambda = 0.01$ V^{-1} is operating at a dc current $I_D = 1$ mA. If the channel length is doubled, find the new values of λ, V_A, I_D, and r_o for each of the following two cases:

(a) V_{GS} and V_{DS} are fixed.
(b) I_D and V_{DS} are fixed.

4.26 An enhancement PMOS transistor has $k'_p(W/L) = 80\ \mu A/V^2$, $V_t = -1.5$ V, and $\lambda = -0.02$ V^{-1}. The gate is connected to ground and the source to +5 V. Find the drain current for $v_D = +4$ V, +1.5 V, 0 V, and −5 V.

4.27 A p-channel transistor for which $|V_t| = 1$ V and $|V_A| = 50$ V operates in saturation with $|v_{GS}| = 3$ V, $|v_{DS}| = 4$ V, and $i_D = 3$ mA. Find corresponding signed values for v_{GS}, v_{SG}, v_{DS}, v_{SD}, V_t, V_A, λ, and $k'_p(W/L)$.

4.28 In a technology for which the gate-oxide thickness is 20 nm, find the value of N_A for which $\gamma = 0.5$ V$^{1/2}$. If the doping level is maintained but the gate oxide thickness is increased to 100 nm, what does γ become? If γ is to be kept constant at 0.5 V$^{1/2}$, to what value must the doping level be changed?

4

4.29 In a particular application, an n-channel MOSFET operates with V_{SB} in the range 0 V to 4 V. If V_{t0} is nominally 1.0 V, find the range of V_t that results if $\gamma = 0.5$ V$^{1/2}$ and $2\phi_f = 0.6$ V. If the gate oxide thickness is increased by a factor of 4, what does the threshold voltage become?

4.30 A p-channel transistor operates in saturation with its source voltage 3 V lower than its substrate. For $\gamma = 0.5$ V$^{1/2}$, $2\phi_f = 0.75$ V, and $V_{t0} = -0.7$ V, find V_t.

***4.31** (a) Using the expression for i_D in saturation and neglecting the channel-length modulation effect (i.e., let $\lambda = 0$), derive an expression for the per unit change in i_D per °C $[(\partial i_D/i_D)/\partial T]$ in terms of the per unit change in k_n' per °C $[(\partial k_n'/k_n')/\partial T]$ the temperature coefficient of V_t in V/°C $(\partial V_t/\partial T)$, and V_{GS} and V_t.
(b) If V_t decreases by 2 mV for every °C rise in temperature, find the temperature coefficient of k_n' that results in i_D decreasing by 0.2%/°C when the NMOS transistor with $V_t = 1$ V is operated at $V_{GS} = 5$ V.

***4.32** Various NMOS and PMOS transistors are measured in operation, as shown in the table at the bottom of the page. For each transistor, find the value of $\mu C_{ox} W/L$ and V_t that apply and complete the table, with V in volts, I in μA, and $\mu C_{ox} W/L$ in μA/V^2.

***4.33** All the transistors in the circuits shown in Fig. P4.33 have the same values of $|V_t|$, k', W/L, and λ. Moreover, λ is negligibly small. All operate in saturation at $I_D = I$ and $|V_{GS}| = |V_{DS}| = 3$ V. Find the voltages V_1, V_2, V_3, and V_4. If $|V_t| = 1$ V and $I = 2$ mA, how large a resistor can be inserted in series with each drain connection while maintaining saturation? What is the largest resistor that can be placed in series with each gate? If the current source I requires at least 2 V between its terminals to operate properly, what is the largest resistor that can be placed in series with each MOSFET source while ensuring saturated-mode operation of each transistor at $I_D = I$? In the latter limiting situation, what do V_1, V_2, V_3, and V_4 become?

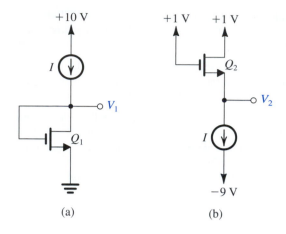

(a)   (b)

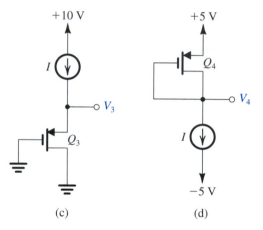

(c)   (d)

FIGURE P4.33

SECTION 4.3: MOSFET CIRCUITS AT DC

D4.34 Design the circuit of Fig. 4.20 to establish a drain current of 1 mA and a drain voltage of 0 V. The MOSFET has $V_t = 1$ V, $\mu_n C_{ox} = 60$ μA/V^2, $L = 3$ μm, and $W = 100$ μm.

Case	Transistor	V_S	V_G	V_D	I_D	Type	Mode	$\mu C_{ox} W/L$	V_t
a	1	0	2	5	100				
	1	0	3	5	400				
b	2	5	3	−4.5	50				
	2	5	2	−0.5	450				
c	3	5	3	4	200				
	3	5	2	0	800				
d	4	−2	0	0	72				
	4	−4	0	−3	270				

D4.35 Consider the circuit of Fig. E4.12. Let Q_1 and Q_2 have $V_t = 0.6$ V, $\mu_n C_{ox} = 200$ μA/V^2, $L_1 = L_2 = 0.8$ μm, $W_1 = 8$ μm, and $\lambda = 0$.

(a) Find the value of R required to establish a current of 0.2 mA in Q_1.
(b) Find W_2 and a new value for R_2 so that Q_2 operates in the saturation region with a current of 0.5 mA and a drain voltage of 1 V.

D4.36 The PMOS transistor in the circuit of Fig. P4.36 has $V_t = -0.7$ V, $\mu_p C_{ox} = 60$ μA/V^2, $L = 0.8$ μm, and $\lambda = 0$. Find the values required for W and R in order to establish a drain current of 115 μA and a voltage V_D of 3.5 V.

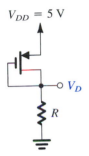

$V_{DD} = 5$ V

V_D

R

FIGURE P4.36

D4.37 The NMOS transistors in the circuit of Fig. P4.37 have $V_t = 1$ V, $\mu_n C_{ox} = 120$ μA/V^2, $\lambda = 0$, and $L_1 = L_2 = 1$ μm. Find the required values of gate width for each of Q_1 and Q_2, and the value of R, to obtain the voltage and current values indicated.

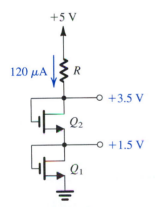

+5 V

120 μA R

o +3.5 V

Q_2

o +1.5 V

Q_1

FIGURE P4.37

D4.38 The NMOS transistors in the circuit of Fig. P4.38 have $V_t = 1$ V, $\mu_n C_{ox} = 120$ μA/V^2, $\lambda = 0$, and $L_1 = L_2 = L_3 = 1$ μm.

Find the required values of gate width for each of Q_1, Q_2, and Q_3 to obtain the voltage and current values indicated.

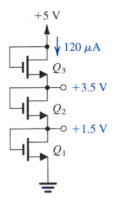

+5 V

120 μA

Q_3

o +3.5 V

Q_2

o +1.5 V

Q_1

FIGURE P4.38

4.39 Consider the circuit of Fig. 4.23(a). In Example 4.5 it was found that when $V_t = 1$ V and $k'_n(W/L) = 1$ mA/V^2, the drain current is 0.5 mA and the drain voltage is +7 V. If the transistor is replaced with another having $V_t = 2$ V and $k'_n(W/L) = 2$ mA/V^2, find the new values of I_D and V_D. Comment on how tolerant (or intolerant) the circuit is to changes in device parameters.

D4.40 Using an enhancement-type PMOS transistor with $V_t = -1.5$ V, $k'_p(W/L) = 1$ mA/V^2, and $\lambda = 0$, design a circuit that resembles that in Fig. 4.23(a). Using a 10-V supply design for a gate voltage of +6 V, a drain current of 0.5 mA, and a drain voltage of +5 V. Find the values of R_S and R_D.

4.41 The MOSFET in Fig. P4.41 has $V_t = 1$ V, $k'_n = 100$ μA/V^2, and $\lambda = 0$. Find the required values of W/L and of R so that when $v_I = V_{DD} = +5$ V, $r_{DS} = 50$ Ω, and $v_O = 50$ mV.

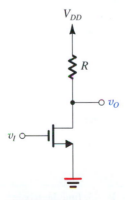

V_{DD}

R

v_O

v_I

FIGURE P4.41

4

4.42 In the circuits shown in Fig. P4.42, transistors are characterized by $|V_t| = 2$ V, $k'W/L = 1$ mA/V^2, and $\lambda = 0$.

(a) Find the labelled voltages V_1 through V_7.
(b) In each of the circuits, replace the current source with a resistor. Select the resistor value to yield a current as close to that of the current source as possible, while using resistors specified in the 1% table provided in Appendix G. Find the new values of V_1 to V_7.

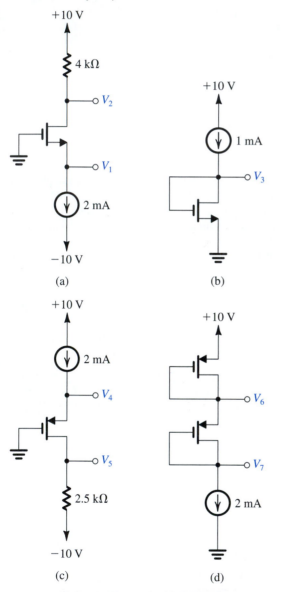

FIGURE P4.42

4.43 For each of the circuits in Fig. P4.43, find the labeled node voltages. For all transistors, $k'_n(W/L) = 0.4$ mA/V^2, $V_t = 1$ V, and $\lambda = 0$.

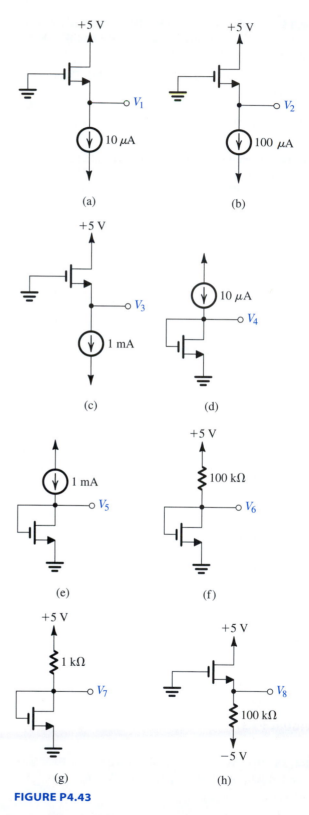

FIGURE P4.43

4.44 For each of the circuits shown in Fig. P4.44, find the labeled node voltages. The NMOS transistors have $V_t = 1$ V and $k'_n W/L = 2$ mA/V^2. Assume $\lambda = 0$.

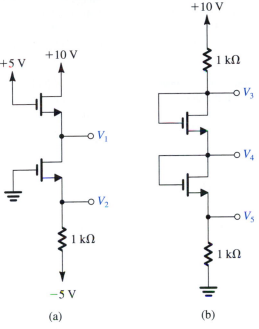

+10 V

+5 V +10 V

1 kΩ

V_3

V_1

V_4

V_2

V_5

1 kΩ

1 kΩ

−5 V

(a) (b)

FIGURE P4.44

*4.45 For the PMOS transistor in the circuit shown in Fig. P4.45, $k'_p = 8$ μA/V^2, $W/L = 25$, and $|V_{tp}| = 1$ V. For $I = 100$ μA, find the voltages V_{SD} and V_{SG} for $R = 0$, 10 kΩ, 30 kΩ, and 100 kΩ. For what value of R is $V_{SD} = V_{SG}$? $V_{SD} = V_{SG}/2$? $V_{SD} = V_{SG}/10$?

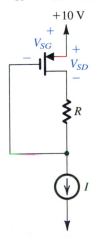

+10 V

+
V_{SG}
−

+
V_{SD}
−

R

I

FIGURE P4.45

4.46 For the circuits in Fig. P4.46, $\mu_n C_{ox} = 2.5$ $\mu_p C_{ox} = 20$ μA/V^2, $|V_t| = 1$ V, $\lambda = 0$, $\gamma = 0$, $L = 10$ μm, and $W = 30$ μm, unless otherwise specified. Find the labeled currents and voltages.

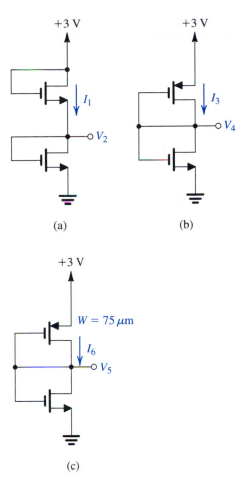

+3 V +3 V

I_1 I_3

V_2 V_4

(a) (b)

+3 V

$W = 75$ μm

I_6

V_5

(c)

FIGURE P4.46

*4.47 For the devices in the circuits of Fig. P4.47, $|V_t| = 1$ V, $\lambda = 0$, $\gamma = 0$, $\mu_n C_{ox} = 50$ μA/V^2, $L = 1$ μm, and $W = 10$ μm. Find V_2 and I_2. How do these values change if Q_3 and Q_4 are made to have $W = 100$ μm?

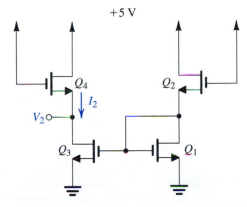

+5 V

Q_4 Q_2

I_2

V_2

Q_3 Q_1

FIGURE P4.47

4

4.48 In the circuit of Fig. P4.48, transistors Q_1 and Q_2 have $V_t = 1$ V, and the process transconductance parameter $k'_n = 100$ μA/V^2. Assuming $\lambda = 0$, find V_1, V_2, and V_3 for each of the following cases:

(a) $(W/L)_1 = (W/L)_2 = 20$
(b) $(W/L)_1 = 1.5(W/L)_2 = 20$

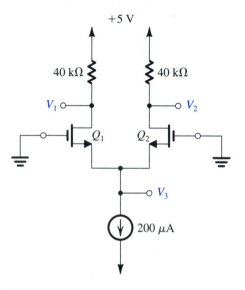

FIGURE P4.48

SECTION 4.4: THE MOSFET AS AN AMPLIFIER AND AS A SWITCH

4.49 Consider the CS amplifier of Fig. 4.26(a) for the case $V_{DD} = 5$ V, $R_D = 24$ kΩ, $k'_n(W/L) = 1$ mA/V^2, and $V_t = 1$ V.

(a) Find the coordinates of the two end points of the saturation-region segment of the amplifier transfer characteristic, that is, points A and B on the sketch of Fig. 4.26(c).
(b) If the amplifier is biased to operate with an overdrive voltage V_{OV} of 0.5 V, find the coordinates of the bias point Q_1 on the transfer characteristic. Also, find the value of I_D and of the incremental gain A_v at the bias point.
(c) For the situation in (b), and disregarding the distortion caused by the MOSFET's square-law characteristic, what is the largest amplitude of a sine-wave voltage signal that can be applied at the input while the transistor remains in saturation? What is the amplitude of the output voltage signal that results? What gain value does the combination of these amplitudes imply? By what percentage is this gain value different from the incremental gain value calculated above? Why is there a difference?

*4.50 We wish to investigate the operation of the CS amplifier circuit studied in Example 4.8 for various bias conditions, that is, for bias at various points along the saturation-region segment of the transfer characteristic. Prepare a table

giving the values of I_D (mA), V_{OV} (V), $V_{GS} = V_{IQ}$ (V), A_v (V/V), the magnitude of the largest allowable positive-output signal v_o^+ (V), and the magnitude of the largest allowable negative-output signal v_o^- (V) for values of $V_{DS} = V_{OQ}$ in the range of 1 V to 10 V, in increments of 1 V (i.e., there should be table rows for $V_{DS} = 1$ V, 2 V, 3 V, ..., 10 V). Note that v_o^+ is determined by the MOSFET entering cutoff and v_o^- by the MOSFET entering the triode region.

4.51 Various measurements are made on an NMOS amplifier for which the drain resistor R_D is 20 kΩ. First, dc measurements show the voltage across the drain resistor, V_{RD}, to be 2 V and the gate-to-source bias voltage to be 1.2 V. Then, ac measurements with small signals show the voltage gain to be -10 V/V. What is the value of V_t for this transistor? If the process transconductance parameter k'_n is 50 μA/V^2, what is the MOSFET's W/L?

*D4.52 Refer to the expression for the incremental voltage gain in Eq. (4.41). Various design considerations place a lower limit on the value of the overdrive voltage V_{OV}. For our purposes here, let this lower limit be 0.2 V. Also, assume that $V_{DD} = 5$ V.

(a) Without allowing any room for output voltage swing, what is the maximum voltage gain achievable?
(b) If we are required to allow for an output voltage swing of ± 0.5 V, what dc bias voltage should be established at the drain to obtain maximum gain? What gain value is achievable? What input signal results in a ± 0.5-V output swing?
(c) For the situation in (b), find W/L of the transistor to establish a dc drain current of 100 μA. For the given process technology, $k'_n = 100$ μA/V^2.
(d) Find the required value of R_D.

4.53 The expression for the incremental voltage gain A_v given in Eq. (4.41) can be written in as

$$A_v = -\frac{2(V_{DD} - V_{DS})}{V_{OV}}$$

where V_{DS} is the bias voltage at the drain (called V_{OQ} in the text). This expression indicates that for given values of V_{DD} and V_{OV}, the gain magnitude can be increased by biasing the transistor at a lower V_{DS}. This, however, reduces the allowable output signal swing in the negative direction. Assuming linear operation around the bias point, show that the largest possible negative output signal peak $\hat{v}_o$ that is achievable while the transistor remains saturated is

$$\hat{v}_o = (V_{DS} - V_{OV}) \Big/ \left(1 + \frac{1}{|A_v|}\right)$$

For $V_{DD} = 5$ V and $V_{OV} = 0.5$ V, provide a table of values for A_v, $\hat{v}_o$, and the corresponding $\hat{v}_i$ for $V_{DS} = 1$ V, 1.5 V, 2 V, and 2.5 V. If $k'_n W/L = 1$ mA/V^2, find I_D and R_D for the design for which $V_{DS} = 1$ V.

4.54 Figure P4.54 shows a CS amplifier in which the load resistor R_D has been replaced with another NMOS transistor Q_2 connected as a two-terminal device. Note that because v_{DG} of Q_2 is zero, it will be operating in saturation at all times, even when $v_I = 0$ and $i_{D2} = i_{D1} = 0$. Note also that the two transistors conduct equal drain currents. Using $i_{D1} = i_{D2}$, show that for the range of v_I over which Q_1 is operating in saturation, that is, for

$$V_{t1} \leq v_I \leq v_O + V_{t1}$$

the output voltage will be given by

$$v_O = V_{DD} - V_t + \sqrt{\frac{(W/L)_1}{(W/L)_2}} V_t - \sqrt{\frac{(W/L)_1}{(W/L)_2}} v_I$$

where we have assumed $V_{t1} = V_{t2} = V_t$. Thus the circuit functions as a linear amplifier, even for large input signals. For $(W/L)_1 = (50 \ \mu m/0.5 \ \mu m)$ and $(W/L)_2 = (5 \ \mu m/0.5 \ \mu m)$, find the voltage gain.

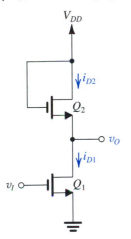

FIGURE P4.54

SECTION 4.5: BIASING IN MOS AMPLIFIER CIRCUITS

D4.55 Consider the classical biasing scheme shown in Fig. 4.30(c), using a 15-V supply. For the MOSFET, $V_t = 1.2$ V, $\lambda = 0$, $k'_n = 80 \ \mu A/V^2$, $W = 240 \ \mu m$, and $L = 6 \ \mu m$. Arrange that the drain current is 2 mA, with about one-third of the supply voltage across each of R_S and R_D. Use 22 MΩ for the larger of R_{G1} and R_{G2}. What are the values of R_{G1}, R_{G2}, R_S, and R_D that you have chosen? Specify them to two significant digits. For your design, how far is the drain voltage from the edge of saturation?

D4.56 Using the circuit topology displayed in Fig. 4.30(e), arrange to bias the NMOS transistor at $I_D = 2$ mA with V_D midway between cutoff and the beginning of triode operation. The available supplies are ±15 V. For the NMOS transistor, $V_t = 0.8$ V, $\lambda = 0$, $k'_n = 50 \ \mu A/V^2$, $W = 200 \ \mu m$, and $L = 4 \ \mu m$. Use a gate-bias resistor of 10 MΩ. Specify R_S and R_D to two significant digits.

*D4.57 In an electronic instrument using the biasing scheme shown in Fig. 4.30(c), a manufacturing error reduces R_S to zero. Let $V_{DD} = 12$ V, $R_{G1} = 5.6$ MΩ, and $R_{G2} = 2.2$ MΩ. What is the value of V_G created? If supplier specifications allow $k'_n(W/L)$ to vary from 220 to 380 $\mu A/V^2$ and V_t to vary from 1.3 to 2.4 V, what are the extreme values of I_D that may result? What value of R_S should have been installed to limit the maximum value of I_D to 0.15 mA? Choose an appropriate standard 5% resistor value (refer to Appendix G). What extreme values of current now result?

4.58 An enhancement NMOS transistor is connected in the bias circuit of Fig. 4.30(c), with $V_G = 4$ V and $R_S = 1$ kΩ. The transistor has $V_t = 2$ V and $k'_n(W/L) = 2$ mA/V^2. What bias current results? If a transistor for which $k'_n(W/L)$ is 50% higher is used, what is the resulting percentage increase in I_D?

4.59 The bias circuit of Fig. 4.30(c) is used in a design with $V_G = 5$ V and $R_S = 1$ kΩ. For an enhancement MOSFET with $k'_n(W/L) = 2$ mA/V^2, the source voltage was measured and found to be 2 V. What must V_t be for this device? If a device for which V_t is 0.5 V less is used, what does V_S become? What bias current results?

D4.60 Design the circuit of Fig. 4.30(e) for an enhancement MOSFET having $V_t = 2$ V and $k'_n(W/L) = 2$ mA/V^2. Let $V_{DD} = V_{SS} = 10$ V. Design for a dc bias current of 1 mA and for the largest possible voltage gain (and thus the largest possible R_D) consistent with allowing a 2-V peak-to-peak voltage swing at the drain. Assume that the signal voltage on the source terminal of the FET is zero.

D4.61 Design the circuit in Fig. P4.61 so that the transistor operates in saturation with V_D biased 1 V from the edge of the triode region, with $I_D = 1$ mA and $V_D = 3$ V, for each of the following two devices (use a 10-μA current in the voltage divider):

(a) $|V_t| = 1$ V and $k'_p W/L = 0.5$ mA/V^2
(b) $|V_t| = 2$ V and $k'_p W/L = 1.25$ mA/V^2

For each case, specify the values of $V_G, V_D, V_S, R_1, R_2, R_S$, and R_D.

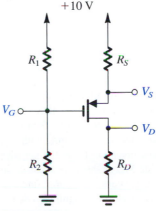

FIGURE P4.61

4

**D4.62 A very useful way to characterize the stability of the bias current I_D is to evaluate the sensitivity of I_D relative to a particular transistor parameter whose variability might be large. The sensitivity of I_D relative to the MOSFET parameter $K \equiv \frac{1}{2}k'(W/L)$ is defined as

$$S_K^{I_D} \equiv \frac{\delta I_D/I_D}{\delta K/K} = \frac{\delta I_D}{\delta K}\frac{K}{I_D}$$

and its value, when multiplied by the variability (or tolerance) of K, provides the corresponding expected variability of I_D. The purpose of this problem is to investigate the use of the sensitivity function in the design of the bias circuit of Fig. 4.30(e).

(a) Show that for V_t constant,

$$S_K^{I_D} = 1/(1 + 2\sqrt{KI_D}R_S)$$

(b) For a MOSFET having $K = 100\ \mu\text{A/V}^2$ with a variability of $\pm10\%$ and $V_t = 1$ V, find the value of R_S that would result in $I_D = 100\ \mu\text{A}$ with a variability of $\pm1\%$. Also, find V_{GS} and the required value of V_{SS}.

(c) If the available supply $V_{SS} = 5$ V, find the value of R_S for $I_D = 100\ \mu\text{A}$. Evaluate the sensitivity function, and give the expected variability of I_D in this case.

4.63 For the circuit in Fig. 4.33(a) with $I = 1$ mA, $R_G = 0$, $R_D = 5\ \text{k}\Omega$, and $V_{DD} = 10$ V, consider the behavior in each of the following two cases. In each case, find the voltages V_S, V_D, and V_{DS} that result.

(a) $V_t = 1$ V and $k_n' W/L = 0.5$ mA/V^2
(b) $V_t = 2$ V and $k_n' W/L = 1.25$ mA/V^2

4.64 In the circuit of Fig. 4.32, let $R_G = 10\ \text{M}\Omega$, $R_D = 10\ \text{k}\Omega$, and $V_{DD} = 10$ V. For each of the following two transistors, find the voltages V_D and V_G.

(a) $V_t = 1$ V and $k_n' W/L = 0.5$ mA/V^2
(b) $V_t = 2$ V and $k_n' W/L = 1.25$ mA/V^2

D4.65 Using the feedback bias arrangement shown in Fig. 4.32 with a 9-V supply and an NMOS device for which $V_t = 1$ V and $k_n'(W/L) = 0.4$ mA/V^2, find R_D to establish a drain current of 0.2 mA. If resistor values are limited to those on the 5% resistor scale (see Appendix G), what value would you choose? What values of current and V_D result?

D4.66 Figure P4.66 shows a variation of the feedback-bias circuit of Fig. 4.32. Using a 6-V supply with an NMOS transistor for which $V_t = 1.2$ V, $k_n' W/L = 3.2$ mA/V^2 and $\lambda = 0$, provide a design which biases the transistor at $I_D = 2$ mA, with V_{DS} large enough to allow saturation operation for a 2-V negative signal swing at the drain. Use 22 MΩ as the largest resistor in the feedback-bias network. What values of R_D, R_{G1}, and R_{G2} have you chosen? Specify all resistors to two significant digits.

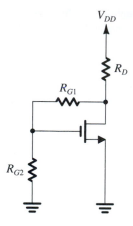

FIGURE P4.66

SECTION 4.6: SMALL-SIGNAL OPERATION AND MODELS

*4.67 This problem investigates the nonlinear distortion introduced by a MOSFET amplifier. Let the signal v_{gs} be a sine wave with amplitude V_{gs}, and substitute $v_{gs} = V_{gs}\sin\omega t$ in Eq. (4.57). Using the trigonometric identity $\sin^2\theta = \frac{1}{2} - \frac{1}{2}\cos2\theta$, show that the ratio of the signal at frequency 2ω to that at frequency ω, expressed as a percentage (known as the second-harmonic distortion) is

$$\text{Second-harmonic distortion} = \frac{1}{4}\frac{V_{gs}}{V_{OV}} \times 100$$

If in a particular application V_{gs} is 10 mV, find the minimum overdrive voltage at which the transistor should be operated so that the second-harmonic distortion is kept to less than 1%.

4.68 Consider an NMOS transistor having $k_n' W/L = 2$ mA/V^2. Let the transistor be biased at $V_{OV} = 1$ V. For operation in saturation, what dc bias current I_D results? If a +0.1-V signal is superimposed on V_{GS}, find the corresponding increment in collector current by evaluating the total collector current i_D and subtracting the dc bias current I_D. Repeat for a −0.1-V signal. Use these results to estimate g_m of the FET at this bias point. Compare with the value of g_m obtained using Eq. (4.62).

4.69 Consider the FET amplifier of Fig. 4.34 for the case $V_t = 2$ V, $k_n'(W/L) = 1$ mA/V^2, $V_{GS} = 4$ V, $V_{DD} = 10$ V, and $R_D = 3.6\ \text{k}\Omega$.

(a) Find the dc quantities I_D and V_D.
(b) Calculate the value of g_m at the bias point.
(c) Calculate the value of the voltage gain.
(d) If the MOSFET has $\lambda = 0.01$ V^{-1}, find r_o at the bias point and calculate the voltage gain.

*D4.70 An NMOS amplifier is to be designed to provide a 0.50-V peak output signal across a 50-kΩ load that can be used as a drain resistor. If a gain of at least 5 V/V is needed, what g_m is required? Using a dc supply of 3 V, what values of I_D and V_{OV} would you choose? What W/L ratio is required if $\mu_n C_{ox} = 100 \ \mu\text{A/V}^2$? If $V_t = 0.8$ V, find V_{GS}.

*D4.71 In this problem we investigate an optimum design of the CS amplifier circuit of Fig. 4.34. First, use the voltage gain expression $A_v = -g_m R_D$ together with Eq. (4.71) for g_m to show that

$$A_v = -\frac{2 I_D R_D}{V_{OV}} = -\frac{2(V_{DD} - V_D)}{V_{OV}}$$

which is the expression we obtained in Section 4.4 (Eq. 4.41). Next, let the maximum positive input signal be $\hat{v}_i$. To keep the second-harmonic distortion to an acceptable level, we bias the MOSFET to operate at an overdrive voltage $V_{OV} \gg \hat{v}_i$. Let $V_{OV} = m\hat{v}_i$. Now, to maximize the voltage gain $|A_v|$, we design for the lowest possible V_D. Show that the minimum V_D that is consistent with allowing a negative signal voltage swing at the drain of $|A_v|\hat{v}_i$ while maintaining saturation-mode operation is given by

$$V_D = \frac{V_{OV} + \hat{v}_i + 2 V_{DD}(\hat{v}_i / V_{OV})}{1 + 2(\hat{v}_i / V_{OV})}$$

Now, find V_{OV}, V_D, A_v, and $\hat{v}_o$ for the case $V_{DD} = 3$ V, $\hat{v}_i = 20$ mV, and $m = 10$. If it is desired to operate this transistor at $I_D = 100 \ \mu\text{A}$, find the values of R_D and W/L, assuming that for this process technology $k'_n = 100 \ \mu\text{A/V}^2$.

4.72 In the table below, for enhancement MOS transistors operating under a variety of conditions, complete as many entries as possible. Although some data is not available, it is always possible to calculate g_m using one of Eqs. (4.69), (4.70) or (4.71). In the table, current is in mA, voltage in V, and dimensions in μm. Assume $\mu_n = 500 \ \text{cm}^2/\text{Vs}$, $\mu_p = 250 \ \text{cm}^2/\text{Vs}$, and $C_{ox} = 0.4 \ \text{fF}/\mu\text{m}^2$.

4.73 An NMOS technology has $\mu_n C_{ox} = 50 \ \mu\text{A/V}^2$ and $V_t = 0.7$ V. For a transistor with $L = 1 \ \mu$m, find the value of W that results in $g_m = 1$ mA/V at $I_D = 0.5$ mA. Also, find the required V_{GS}.

4.74 For the NMOS amplifier in Fig. P4.74, replace the transistor with its T equivalent circuit of Fig. 4.39(d). Derive expressions for the voltage gains v_s/v_i and v_d/v_i.

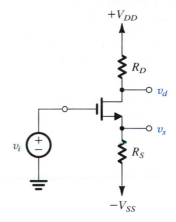

FIGURE P4.74

4.75 In the circuit of Fig. P4.75, the NMOS transistor has $|V_t| = 0.9$ V and $V_A = 50$ V and operates with $V_D = 2$ V. What is the voltage gain v_o/v_i? What do V_D and the gain become for I increased to 1 mA?

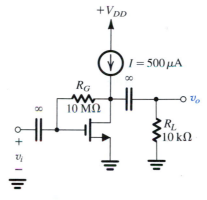

FIGURE P4.75

Case	Type	I_D	$|V_{GS}|$	$|V_t|$	$|V_{OV}|$	W	L	W/L	$k'(W/L)$	g_m
a	N	1	3	2			1			
b	N	1		0.7	0.5	50				
c	N	10		2			1			
d	N	0.5			0.5					
e	N	0.1				10	2			
f	N		1.8	0.8		40	4			
g	P	1			2			25		
h	P		3	1					500	
i	P	10				4000	2			
j	P	10			4					
k	P				1	30	3			
l	P	0.1			5				8	

4.76 For a 0.8-μm CMOS fabrication process: $V_{tn} = 0.8$ V, $V_{tp} = -0.9$ V, $\mu_n C_{ox} = 90$ μA/V^2, $\mu_p C_{ox} = 30$ μA/V^2, $C_{ox} = 1.9$ fF/μm^2, $\phi_f = 0.34$ V, $\gamma = 0.5$ V$^{1/2}$, V_A (n-channel devices) = $8L$ (μm), and $|V_A|$ (p-channel devices) = $12L$ (μm). Find the small-signal model parameters (g_m, r_o, and g_{mb}) for both an NMOS and a PMOS transistor having $W/L = 20$ μm/ 2 μm and operating at $I_D = 100$ μA with $|V_{SB}| = 1$V. Also, find the overdrive voltage at which each device must be operating.

4.77 Figure P4.77 shows a discrete-circuit CS amplifier employing the classical biasing scheme studied in Section 4.5. The input signal v_{sig} is coupled to the gate through a very large capacitor (shown as infinite). The transistor source is connected to ground at signal frequencies via a very large capacitor (shown as infinite). The output voltage signal that develops at the drain is coupled to a load resistance via a very large capacitor (shown as infinite).

(a) If the transistor has $V_t = 1$ V, and $k'_n W/L = 2$ mA/V^2, verify that the bias circuit establishes $V_{GS} = 2$ V, $I_D = 1$ mA, and $V_D = +7.5$ V. That is, assume these values, and verify that they are consistent with the values of the circuit components and the device parameters.
(b) Find g_m and r_o if $V_A = 100$ V.
(c) Draw a complete small-signal equivalent circuit for the amplifier assuming all capacitors behave as short circuits at signal frequencies.
(d) Find R_{in}, v_{gs}/v_{sig}, v_o/v_{gs}, and v_o/v_{sig}.

4.78 The fundamental relationship that describes MOSFET operation is the parabolic relationship between V_{OV} and i_D,

$$i_D = \frac{1}{2}k'_n \frac{W}{L} v_{OV}^2$$

Sketch this parabolic curve together with the tangent at a point whose coordinates are (V_{OV}, I_D). The slope of this tangent is g_m at this bias point. Show that this tangent intersects the v_{OV}-axis at $V_{OV}/2$ and thus that $g_m = 2I_D/V_{OV}$.

SECTION 4.7: SINGLE-STAGE MOS AMPLIFIERS

4.79 Calculate the overall voltage gain G_v of a common-source amplifier for which $g_m = 2$ mA/V, $r_o = 50$ kΩ, $R_D = 10$ kΩ, and $R_G = 10$ MΩ. The amplifier is fed from a signal source with a Thévenin resistance of 0.5 MΩ, and the amplifier output is coupled to a load resistance of 20 kΩ.

D4.80 This problem investigates a redesign of the common-source amplifier of Exercise 4.32 whose bias design was done in Exercise 4.30 and shown in Fig. E4.30. Please refer to these two exercises.

(a) The open-circuit voltage gain of the CS amplifier can be written as

$$A_{vo} = -\frac{2(V_{DD} - V_D)}{V_{OV}}$$

Verify that this expression yields the results in Exercise 4.32 (i.e., $A_{vo} = -15$ V/V).
(b) A_{vo} can be doubled by reducing V_{OV} by a factor of 2, (i.e., from 1 V to 0.5 V) while V_D is kept unchanged. What corresponding values for I_D, R_D, g_m, and r_o apply?
(c) Find A_{vo} and R_{out} with r_{o1} taken into account.
(d) For the same value of signal-generator resistance $R_{sig} = 100$ kΩ, the same value of gate-bias resistance $R_G = 4.8$ MΩ, and the same value of load resistance $R_L = 15$ kΩ, evaluate the new value of overall voltage gain G_v with r_o taken into account.
(e) Compare your results to those obtained in Exercises 4.30 and 4.32, and comment.

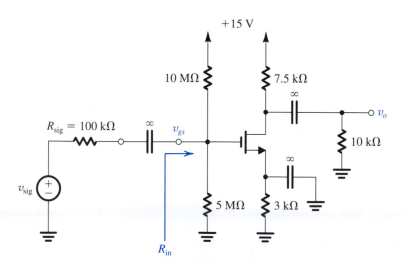

FIGURE P4.77

4.81 A common-gate amplifier using an n-channel enhancement MOS transistor for which $g_m = 5$ mA/V has a 5-kΩ drain resistance (R_D) and a 2-kΩ load resistance (R_L). The amplifier is driven by a voltage source having a 200-Ω resistance. What is the input resistance of the amplifier? What is the overall voltage gain G_v? If the circuit allows a bias-current increase by a factor of 4 while maintaining linear operation, what do the input resistance and voltage gain become?

4.82 A CS amplifier using an NMOS transistor biased in the manner of Fig. 4.43 for which $g_m = 2$ mA/V is found to have an overall voltage gain G_v of -16 V/V. What value should a resistance R_S inserted in the source lead have to reduce the voltage gain by a factor of 4?

4.83 The overall voltage gain of the amplifier of Fig. 4.44(a) was measured with a resistance R_S of 1 kΩ in place and found to be -10 V/V. When R_S is shorted, but the circuit operation remained linear the gain doubled. What must g_m be? What value of R_S is needed to obtain an overall voltage gain of -8 V/V?

4.84 Careful measurements performed on the source follower of Fig. 4.46(a) show that the open-circuit voltage gain is 0.98 V/V. Also, when R_L is connected and its value is varied, it is found that the gain is halved for $R_L = 500$ Ω. If the amplifier remained linear throughout this measurement, what must the values of g_m and r_o be?

4.85 The source follower of Fig. 4.46(a) uses a MOSFET biased to have $g_m = 5$ mA/V and $r_o = 20$ kΩ. Find the open-circuit voltage gain A_{vo} and the output resistance. What will the gain become when a 1-kΩ load resistance (R_L) is connected?

4.86 Figure P4.86 shows a scheme for coupling and amplifying a high-frequency pulse signal. The circuit utilizes two MOSFETs whose bias details are not shown and a 50-Ω coaxial cable. Transistor Q_1 operates as a CS amplifier and Q_2 as a CG amplifier. For proper operation, transistor Q_2 is required to present a 50-Ω resistance to the cable. This situation is known as "proper termination" of the cable and ensures that there will be no signal reflection coming back on the cable. When the cable is properly terminated, its input resistance is 50 Ω. What must g_{m2} be? If Q_1 is biased at the same point as Q_2, what is the amplitude of the current pulses in the drain of Q_1? What is the amplitude of the voltage pulses at the drain of Q_1? What value of R_D is required to provide 1-V pulses at the drain of Q_2?

*D4.87 The MOSFET in the circuit of Fig. P4.87 has $V_t = 1$ V, $k'_n W/L = 0.8$ mA/V^2, and $V_A = 40$ V.

(a) Find the values of R_S, R_D, and R_G so that $I_D = 0.1$ mA, the largest possible value for R_D is used while a maximum signal swing at the drain of ± 1 V is possible, and the input resistance at the gate is 10 MΩ.
(b) Find the values of g_m and r_o at the bias point.
(c) If terminal Z is grounded, terminal X is connected to a signal source having a resistance of 1 MΩ, and terminal Y is connected to a load resistance of 40 kΩ, find the voltage gain from signal source to load.
(d) If terminal Y is grounded, find the voltage gain from X to Z with Z open-circuited. What is the output resistance of the source follower?
(e) If terminal X is grounded and terminal Z is connected to a current source delivering a signal current of 10 μA and having a resistance of 100 kΩ, find the voltage signal that can be measured at Y. For simplicity, neglect the effect of r_o.

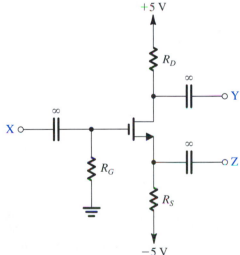

FIGURE P4.87

*4.88 (a) The NMOS transistor in the source-follower circuit of Fig. P4.88(a) has $g_m = 5$ mA/V and a large r_o. Find the open-circuit voltage gain and the output resistance.

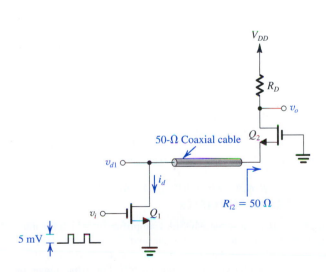

FIGURE P4.86

4

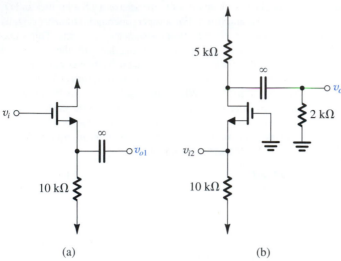

FIGURE P4.88

(b) The NMOS transistor in the common-gate amplifier of Fig. P4.88(b) has $g_m = 5$ mA/V and a large r_o. Find the input resistance and the voltage gain.

(c) If the output of the source follower in (a) is connected to the input of the common-gate amplifier in (b), use the results of (a) and (b) to obtain the overall voltage gain v_o/v_i.

*4.89 In this problem we investigate the large-signal operation of the source follower of Fig. 4.46(a). Specifically, consider the situation when negative input signals are applied. Let the negative signal voltage at the output be $-V$. The current in R_L will flow away from ground and will have a value of V/R_L. This current will subtract from the bias current I, resulting in a transistor current of $(I - V/R_L)$. One can use this current value to determine v_{GS}. Now the signal at the transistor source terminal will be $-V$, superimposed on the dc voltage, which is $-V_{GS}$ (corresponding to a drain current of I). We can thus find the signal voltage at the gate v_i. For the circuit analyzed in Exercise 4.34, find v_i for $v_o = -1$ V, -5 V, -6 V, and -7 V. At each point find the voltage gain v_o/v_i and compare to the small-signal value found in Exercise 4.34. What is the largest possible negative-output signal?

SECTION 4.8: THE MOSFET INTERNAL CAPACITANCES AND HIGH-FREQUENCY MODEL

4.90 Refer to the MOSFET high-frequency model in Fig. 4.47(a). Evaluate the model parameters for an NMOS transistor operating at $I_D = 100$ μA, $V_{SB} = 1$ V, and $V_{DS} = 1.5$ V. The MOSFET has $W = 20$ μm, $L = 1$ μm, $t_{ox} = 8$ nm, $\mu_n = 450$ cm²/Vs, $\gamma = 0.5$ V$^{1/2}$, $2\phi_f = 0.65$ V, $\lambda = 0.05$ V^{-1}, $V_0 = 0.7$ V, $C_{sb0} = C_{db0} = 15$ fF, and $L_{ov} = 0.05$ μm. (Recall that $g_{mb} = \chi g_m$, where $\chi = \gamma/(2\sqrt{2\phi_f + V_{SB}})$.)

4.91 Find f_T for a MOSFET operating at $I_D = 100$ μA and $V_{OV} = 0.25$ V. The MOSFET has $C_{gs} = 20$ fF and $C_{gd} = 5$ fF.

4.92 Starting from the definition of f_T for a MOSFET,

$$f_T = \frac{g_m}{2\pi(C_{gs} + C_{gd})}$$

and making the approximation that $C_{gs} \gg C_{gd}$ and that the overlap component of C_{gs} is negligibly small, show that

$$f_T \simeq \frac{1.5}{\pi L}\sqrt{\frac{\mu_n I_D}{2 C_{ox} WL}}$$

Thus note that to obtain a high f_T from a given device it must be operated at a high current. Also note that faster operation is obtained from smaller devices.

4.93 Starting from the expression for the MOSFET unity-gain frequency,

$$f_T = \frac{g_m}{2\pi(C_{gs} + C_{gd})}$$

and making the approximation that $C_{gs} \gg C_{gd}$ and that the overlap component of C_{gs} is negligibly small, show that for an n-channel device

$$f_T = \frac{3\mu_n V_{OV}}{4\pi L^2}$$

Observe that for a given device, f_T can be increased by operating the MOSFET at a higher overdrive voltage. Evaluate f_T for devices with $L = 1.0$ μm operated at overdrive voltages of 0.25 V and 0.5 V. Use $\mu_n = 450$ cm²/Vs.

SECTION 4.9: FREQUENCY RESPONSE OF THE CS AMPLIFIER

4.94 In a particular MOSFET amplifier for which the mid-band voltage gain between gate and drain is -27 V/V, the NMOS transistor has $C_{gs} = 0.3$ pF and $C_{gd} = 0.1$ pF. What input capacitance would you expect? For what range of signal-source resistances can you expect the 3-dB frequency to exceed 10 MHz? Neglect the effect of R_G.

D4.95 In a FET amplifier, such as that in Fig. 4.49(a), the resistance of the source $R_{sig} = 100$ kΩ, amplifier input resistance (which is due to the biasing network) $R_{in} = 100$ kΩ, $C_{gs} = 1$ pF, $C_{gd} = 0.2$ pF, $g_m = 3$ mA/V, $r_o = 50$ kΩ, $R_D = 8$ kΩ, and $R_L = 10$ kΩ. Determine the expected 3-dB cutoff frequency f_H and the midband gain. In evaluating ways to double f_H, a designer considers the alternatives of changing either R_{out} or R_{in}. To raise f_H as described, what separate change in each would be required? What midband voltage gain results in each case?

4.96 A discrete MOSFET common-source amplifier has $R_{in} = 2$ MΩ, $g_m = 4$ mA/V, $r_o = 100$ kΩ, $R_D = 10$ kΩ, $C_{gs} = 2$ pF, and $C_{gd} = 0.5$ pF. The amplifier is fed from a voltage source with an internal resistance of 500 kΩ and is connected to a 10-kΩ load. Find:

(a) the overall midband gain A_M
(b) the upper 3-dB frequency f_H

4.97 The analysis of the high-frequency response of the common-source amplifier, presented in the text, is based on the assumption that the resistance of the signal source, R_{sig}, is large and, thus, that its interaction with the input capacitance C_{in} produces the "dominant pole" that determines the upper 3-dB frequency f_H. There are situations, however, when the CS amplifier is fed with a very low R_{sig}. To investigate the high-frequency response of the amplifier in such a case, Fig. P4.97 shows the equivalent circuit when the CS amplifier is fed with an ideal voltage source V_{sig} having $R_{sig} = 0$. Note that C_L denotes the total capacitance at the output node. By writing a node equation at the output, show that the transfer function V_o/V_{sig} is given by

$$\frac{V_o}{V_{sig}} = -g_m R'_L \frac{1 - s(C_{gd}/g_m)}{1 + s(C_L + C_{gd})R'_L}$$

At frequencies $\omega \ll (g_m/C_{gd})$, the s term in the numerator can be neglected. In such case, what is the upper 3-dB frequency resulting? Compute the values of A_M and f_H for the case: $C_{gd} = 0.5$ pF, $C_L = 2$ pF, $g_m = 4$ mA/V, and $R'_L = 5$ kΩ.

D4.98 Consider the common-source amplifier of Fig. 4.49(a). For a situation in which $R_{sig} = 1$ MΩ and $R_G = 1$ MΩ, what

value of C_{C1} must be chosen to place the corresponding break frequency at 10 Hz? What value would you choose if available capacitors are specified to only one significant digit and the break frequency is not to exceed 10 Hz? What is the break frequency, f_{P1}, obtained with your choice? If a designer wishes to lower this by raising R_G, what is the most that he or she can expect if available resistors are limited to 10 times those now used?

D4.99 The amplifier in Fig. P4.99 is biased to operate at $I_D = 1$ mA and $g_m = 1$ mA/V. Neglecting r_o, find the midband gain. Find the value of C_S that places f_L at 10 Hz.

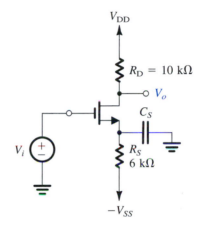

FIGURE P4.99

4.100 Consider the amplifier of Fig. 4.49(a). Let $R_D = 15$ kΩ, $r_o = 150$ kΩ, and $R_L = 10$ kΩ. Find the value of C_{C2}, specified to one significant digit, to ensure that the associated break frequency is at, or below, 10 Hz. If a higher-power design results in doubling I_D, with both R_D and r_o reduced by a factor of 2, what does the corner frequency (due to C_{C2}) become? For increasingly higher-power designs, what is the highest corner frequency that can be associated with C_{C2}.

4.101 The NMOS transistor in the discrete CS amplifier circuit of Fig. P4.101 is biased to have $g_m = 1$ mA/V. Find A_M, f_{P1}, f_{P2}, f_{P3}, and f_L.

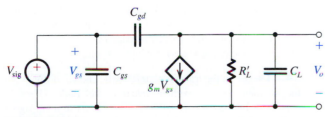

FIGURE P4.97

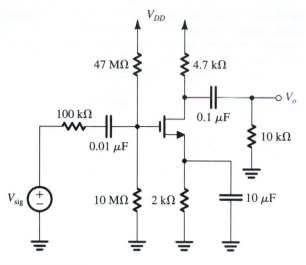

4.102 The NMOS transistor in the discrete CS amplifier circuit of Fig. P4.101 is biased to have $g_m = 1$ mA/V and $r_o = 100$ kΩ. Find A_M. If $C_{gs} = 1$ pF and $C_{gd} = 0.2$ pF, find f_H.

D4.103 Consider the low-frequency response of the CS amplifier of Fig. 4.49(a). Let $R_{sig} = 0.5$ MΩ, $R_G = 2$ MΩ, $g_m = 3$ mA/V, $R_D = 20$ kΩ, and $R_L = 10$ kΩ. Find A_M. Also, design the coupling and bypass capacitors to locate the three low-frequency poles at 50 Hz, 10 Hz, and 3 Hz. Use a minimum total capacitance, with capacitors specified only to a single significant digit. What value of f_L results?

4.104 Figure P4.104 shows a MOS amplifier whose bias design and midband analysis were performed in Example 4.10. Specifically, the MOSFET is biased at $I_D = 1.06$ mA and has $g_m = 0.725$ mA/V and $r_o = 47$ kΩ. The midband analysis showed that $V_o/V_i = -3.3$ V/V and $R_{in} = 2.33$ MΩ. Select appropriate values for the two capacitors so that the low-frequency response is dominated by a pole at 10 Hz with the other pole

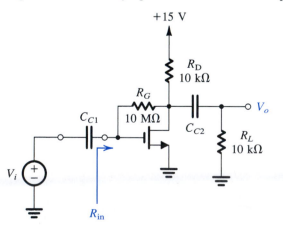

at least a decade lower. (*Hint:* In determining the pole due to C_{C2}, resistance R_G can be neglected.)

SECTION 4.10: THE CMOS DIGITAL LOGIC INVERTER

4.105 For a digital logic inverter fabricated in a 0.8-μm CMOS technology for which $k'_n = 120 \ \mu A/V^2$, $k'_p = 60 \ \mu A/V^2$, $V_{tn} = |V_{tp}| = 0.7$ V, $V_{DD} = 3$ V, $L_n = L_p = 0.8 \ \mu m$, $W_n = 1.2 \ \mu m$, and $W_p = 2.4 \ \mu m$, find:

(a) the output resistance for $v_O = V_{OL}$ and for $v_O = V_{OH}$
(b) the maximum current that the inverter can sink or source while the output remains within 0.1 V of ground or V_{DD}, respectively
(c) V_{IH}, V_{IL}, NM_H, and NM_L

4.106 For the technology specified in Problem 4.105, investigate how the threshold voltage of the inverter, V_{th}, varies with the degree of matching of the NMOS and PMOS devices. Use the formula given in Exercise 4.44, and find V_{th} for the cases $(W/L)_p = (W/L)_n$, $(W/L)_p = 2(W/L)_n$ (the matched case), and $(W/L)_p = 4(W/L)_n$.

4.107 For an inverter designed with equal-sized NMOS and PMOS transistors and fabricated in the technology specified in Problem 4.105 above, find V_{IL} and V_{IH} and hence the noise margins.

4.108 Repeat Exercise 4.41 for $V_{DD} = 10$ V and 15 V.

4.109 Repeat Exercise 4.42 for $V_t = 0.5$ V, 1.5 V, and 2 V.

4.110 For a technology in which $V_{tn} = 0.2V_{DD}$, show that the maximum current that the CMOS inverter can sink while its low output level does not exceed $0.1V_{DD}$ is $0.075k'_n(W/L)_n V_{DD}^2$. For $V_{DD} = 3$ V, $k'_n = 120 \ \mu A/V^2$, and $L_n = 0.8 \ \mu m$, find the required transistor width to obtain a current of 1 mA.

4.111 For the inverter specified in Problem 4.105, find the peak current drawn from the 3-V supply during switching.

4.112 For the inverter specified in Problem 4.105, find the value of t_{PHL} when the inverter is loaded with a capacitance $C = 0.05$ pF. Use both Eq. (4.156) and the approximate expression in Eq. (4.157), and compare the results.

4.113 Consider an inverter fabricated in the CMOS technology specified in Problem 4.105 and having $L_n = L_p = 0.8 \ \mu m$ and $(W/L)_p = 2(W/L)_n$. It is required to limit the propagation delay to 60 ps when the inverter is loaded with 0.05-pF capacitance. Find the required device widths, W_n and W_p.

*4.114 (a) In the transfer characteristic shown in Fig. 4.56, the segment BC is vertical because the Early effect is neglected. Taking the Early effect into account, use small-signal analysis to show that the slope of the transfer characteristic at

$v_I = v_O = V_{DD}/2$ is

$$\frac{-2\,|V_A|}{(V_{DD}/2) - V_t}$$

where V_A is the Early voltage for Q_N and Q_P. Assume Q_N and Q_P to be matched.

(b) A CMOS inverter with devices having $k_n'(W/L)_n = k_p'(W/L)_p$ is biased by connecting a resistor $R_G = 10\ M\Omega$ between input and output. What is the dc voltage at input and output? What is the small-signal voltage gain and input resistance of the resulting amplifier? Assume the inverter to have the characteristics specified in Problem 4.105 with $|V_A| = 50\ V$.

SECTION 4.11: THE DEPLETION-TYPE MOSFET

4.115 A depletion-type n-channel MOSFET with $k_n'W/L = 2\ mA/V^2$ and $V_t = -3\ V$ has its source and gate grounded. Find the region of operation and the drain current for $v_D = 0.1\ V$, 1 V, 3 V, and 5 V. Neglect the channel-length-modulation effect.

4.116 For a particular depletion-mode NMOS device, $V_t = -2\ V$, $k_n'W/L = 200\ \mu A/V^2$, and $\lambda = 0.02\ V^{-1}$. When operated at $v_{GS} = 0$, what is the drain current that flows for $v_{DS} = 1\ V$, 2 V, 3 V, and 10 V? What does each of these currents become if the device width is doubled with L the same? With L also doubled?

*4.117 Neglecting the channel-length-modulation effect show that for the depletion-type NMOS transistor of Fig. P4.117 the i–v relationship is given by

$$i = \tfrac{1}{2}k_n'(W/L)(v^2 - 2V_t v), \quad \text{for } v \geq V_t$$

$$i = -\tfrac{1}{2}k_n'(W/L)V_t^2, \quad \text{for } v \leq V_t$$

(Recall that V_t is negative.) Sketch the i–v relationship for the case: $V_t = -2\ V$ and $k_n'(W/L) = 2\ mA/V^2$.

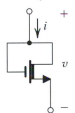

FIGURE P4.117

4.118 For the circuit analyzed in Exercise 4.51 (refer to Fig. E4.51), what does the voltage at the source become when the drain voltage is lowered to +1 V?

4.119 A depletion-type NMOS transistor operating in the saturation region with $v_{DS} = 5\ V$ conducts a drain current of 1 mA at $v_{GS} = -1\ V$, and 9 mA at $v_{GS} = +1\ V$. Find I_{DSS} and V_t. Assume $\lambda = 0$.

D4.120 Consider the circuit shown in Fig. P4.120 in which Q_1 with R_1 establishes the bias current for Q_2. R_2 has no effect

on the bias of Q_2 but provides an interesting function. R_3 acts as load resistor in the drain of Q_2. Assume that Q_1 and Q_2 are fabricated together (as a matched pair, or as part of an IC) and are identical. For each depletion NMOS, $I_{DSS} = 4\ mA$ and $|V_t| = 2\ V$. The voltage at the input is some value, say 0 V, that keeps Q_1 in saturation. What is the value of $k_n'(W/L)$ for these transistors?

Now, design R_1 so that $I_{D1} = I_{D2} = 1\ mA$. Make $R_2 = R_1$. Choose R_3 so that $v_E = 6\ V$. For $v_A = 0\ V$, what is the voltage v_C? Check what the voltage v_C is when $v_A = \pm 1\ V$. Notice the interesting behavior, namely, that node C follows node A. This circuit can be called a source follower, but it is a special one, one with zero offset! Note also that R_2 is not essential, since node B also follows node A but with a positive offset. In many applications, R_2 is short-circuited. Now, recognize that as the voltage on A rises, Q_2 will eventually enter the triode region. At what value of v_A does this occur? Also, as v_A lowers, Q_1 will enter its triode region. At what value of v_A? (Note that between these two values of v_A is the linear signal range of both v_A and v_C.)

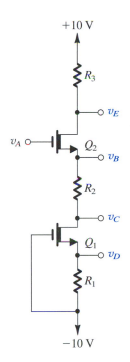

FIGURE P4.120

GENERAL PROBLEMS:

**4.121 The circuits shown in Fig. P4.121 employ negative feedback, a subject we shall study in detail in Chapter 8. Assume that each transistor is sized and biased so that $g_m = 1\ mA/V$ and $r_o = 100\ k\Omega$. Otherwise, ignore all dc biasing detail and concentrate on small-signal operation resulting in response to the input signal v_{sig}. For $R_L = 10\ k\Omega$, $R_1 = 500\ k\Omega$, and $R_2 = 1$ MΩ, find the overall voltage gain v_o/v_{sig} and the input resistance R_{in} for each circuit. Neglect the body effect. Do these circuits remind you of op-amp circuits? Comment.

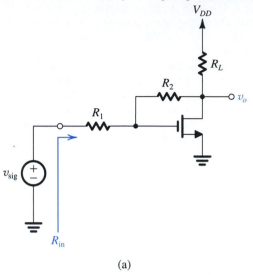

(a)

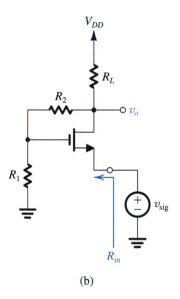

(b)

FIGURE P4.121

4.122 For the two circuits in Problem 4.121 (shown in Fig. P4.121), we wish to consider their dc bias design. Since v_{sig} has a zero dc component, we short circuit its generator.

For NMOS transistors with $V_t = 0.6$ V, find V_{OV}, $k'_n(W/L)$, and V_A to bias each device at $I_D = 0.1$ mA and to obtain the values of g_m and r_o specified in Problem 4.121; namely, $g_m = 1$ mA/V and $r_o = 100$ kΩ. For $R_1 = 0.5$ MΩ, $R_2 = 1$ MΩ, and $R_L = 10$ kΩ, find the required value of V_{DD}.

**4.123 In the amplifier shown in Fig. P4.123, transistors having $V_t = 0.6$ V and $V_A = 20$ V are operated at $V_{GS} = 0.8$ V using the appropriate choice of W/L ratio. In a particular application, Q_1 is to be sized to operate at 10 μA, while Q_2 is intended to operate at 1 mA. For $R_L = 2$ kΩ, the (R_1, R_2) network sized to consume only 1% of the current in R_L, v_{sig}, having zero dc component, and $I_1 = 10$ μA, find the values of R_1 and R_2 that satisfy all the requirements. (*Hint*: V_O must be +2 V.) What is the voltage gain v_o/v_i? Using a result from a theorem known as Miller's theorem (Chapter 6), find the input resistance R_{in} as $R_2/(1 - v_o/v_i)$. Now, calculate the value of the overall voltage gain v_o/v_{sig}. Does this result remind you of the inverting configuration of the op amp? Comment. How would you modify the circuit at the input using an additional resistor and a very large capacitor to raise the gain v_o/v_{sig} to −5 V/V? Neglect the body effect.

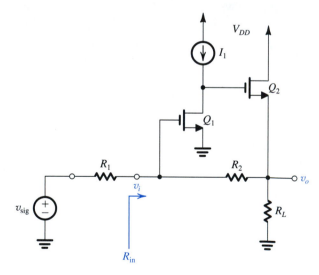

FIGURE P4.123

4.124 Consider the bias design of the circuit of Problem 4.123 (shown in Fig. P4.123). For $k'_n = 200$ μA/V^2 and $V_{DD} = 3.3$ V, find $(W/L)_1$ and $(W/L)_2$ to obtain the operating conditions specified in Problem 4.123.

CHAPTER 5

Bipolar Junction Transistors (BJTs)

INTRODUCTION

In this chapter, we study the other major three-terminal device: the bipolar junction transistor (BJT). The presentation of the material in this chapter parallels but does not rely on that for the MOSFET in Chapter 4; thus, if desired, the BJT can be studied before the MOSFET.

Three-terminal devices are far more useful than two-terminal ones, such as the diodes studied in Chapter 3, because they can be used in a multitude of applications, ranging from signal amplification to the design of digital logic and memory circuits. The basic principle involved is the use of the voltage between two terminals to control the current flowing in the third terminal. In this way, a three-terminal device can be used to realize a controlled source, which as we learned in Chapter 1 is the basis for amplifier design. Also, in the extreme, the control signal can be used to cause the current in the third terminal to change from zero to a large value, thus allowing the device to act as a switch. As we learned also in

Chapter 1, the switch is the basis for the realization of the logic inverter, the basic element of digital circuits.

The invention of the BJT in 1948 at the Bell Telephone Laboratories ushered in the era of solid-state circuits, which led to electronics changing the way we work, play, and indeed, live. The invention of the BJT also eventually led to the dominance of information technology and the emergence of the knowledge-based economy.

The bipolar transistor enjoyed nearly three decades as the device of choice in the design of both discrete and integrated circuits. Although the MOSFET had been known very early on, it was not until the 1970s and 1980s that it became a serious competitor to the BJT. At the time of this writing (2003), the MOSFET is undoubtedly the most widely used electronic device, and CMOS technology is the technology of choice in the design of integrated circuits. Nevertheless, the BJT remains a significant device that excels in certain applications. For instance, the reliability of BJT circuits under severe environmental conditions makes them the dominant device in automotive electronics, an important and still-growing area.

The BJT remains popular in discrete-circuit design, in which a very wide selection of BJT types are available to the designer. Here we should mention that the characteristics of the bipolar transistor are so well understood that one is able to design transistor circuits whose performance is remarkably predictable and quite insensitive to variations in device parameters.

The BJT is still the preferred device in very demanding analog circuit applications, both integrated and discrete. This is especially true in very-high-frequency applications, such as radio-frequency (RF) circuits for wireless systems. A very-high-speed digital logic-circuit family based on bipolar transistors, namely emitter-coupled logic, is still in use. Finally, bipolar transistors can be combined with MOSFETs to create innovative circuits that take advantage of the high-input-impedance and low-power operation of MOSFETs and the very-high-frequency operation and high-current-driving capability of bipolar transistors. The resulting technology is known as BiMOS or BiCMOS, and it is finding increasingly larger areas of application (see Chapters 6, 7, 9, and 11).

In this chapter, we shall start with a simple description of the physical operation of the BJT. Though simple, this physical description provides considerable insight regarding the performance of the transistor as a circuit element. We then quickly move from describing current flow in terms of electrons and holes to a study of the transistor terminal characteristics. Circuit models for transistor operation in different modes will be developed and utilized in the analysis and design of transistor circuits. The main objective of this chapter is to develop in the reader a high degree of familiarity with the BJT. Thus, by the end of the chapter, the reader should be able to perform rapid first-order analysis of transistor circuits and to design single-stage transistor amplifiers and simple logic inverters.

5.1 DEVICE STRUCTURE AND PHYSICAL OPERATION

5.1.1 Simplified Structure and Modes of Operation

Figure 5.1 shows a simplified structure for the BJT. A practical transistor structure will be shown later (see also Appendix A, which deals with fabrication technology).

As shown in Fig. 5.1, the BJT consists of three semiconductor regions: the emitter region (*n* type), the base region (*p* type), and the collector region (*n* type). Such a transistor is called an *npn* transistor. Another transistor, a dual of the *npn* as shown in Fig. 5.2, has a *p*-type emitter, an *n*-type base, and a *p*-type collector, and is appropriately called a *pnp* transistor.

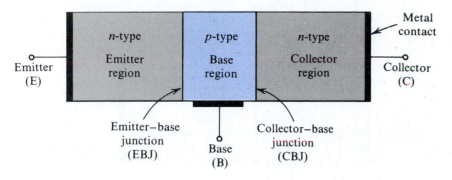

FIGURE 5.1 A simplified structure of the *npn* transistor.

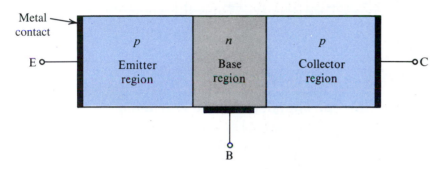

FIGURE 5.2 A simplified structure of the *pnp* transistor.

A terminal is connected to each of the three semiconductor regions of the transistor, with the terminals labeled **emitter** (E), **base** (B), and **collector** (C).

The transistor consists of two *pn* junctions, the **emitter–base junction** (EBJ) and the **collector–base junction** (CBJ). Depending on the bias condition (forward or reverse) of each of these junctions, different modes of operation of the BJT are obtained, as shown in Table 5.1.

The **active mode,** which is also called forward active mode, is the one used if the transistor is to operate as an amplifier. Switching applications (e.g., logic circuits) utilize both the **cutoff mode** and the **saturation mode.** The **reverse active** (or inverse active) mode has very limited application but is conceptually important.

As we will see shortly, charge carriers of both polarities—that is, electrons and holes—participate in the current-conduction process in a bipolar transistor, which is the reason for the name *bipolar*.

TABLE 5.1   BJT Modes of Operation

Mode	EBJ	CBJ
Cutoff	Reverse	Reverse
Active	Forward	Reverse
Reverse active	Reverse	Forward
Saturation	Forward	Forward

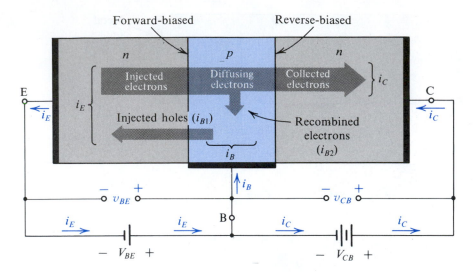

FIGURE 5.3 Current flow in an *npn* transistor biased to operate in the active mode. (Reverse current components due to drift of thermally generated minority carriers are not shown.)

5.1.2 Operation of the *npn* Transistor in the Active Mode

Let us start by considering the physical operation of the transistor in the active mode.[1] This situation is illustrated in Fig. 5.3 for the *npn* transistor. Two external voltage sources (shown as batteries) are used to establish the required bias conditions for active-mode operation. The voltage V_{BE} causes the *p*-type base to be higher in potential than the *n*-type emitter, thus forward-biasing the emitter–base junction. The collector–base voltage V_{CB} causes the *n*-type collector to be at a higher potential than the *p*-type base, thus reverse-biasing the collector–base junction.

Current Flow In the following description of current flow only diffusion-current components are considered. Drift currents, due to thermally generated minority carriers, are usually very small and can be neglected. Nevertheless, we will have more to say about these reverse-current components at a later stage.

The forward bias on the emitter–base junction will cause current to flow across this junction. Current will consist of two components: electrons injected from the emitter into the base, and holes injected from the base into the emitter. As will become apparent shortly, it is highly desirable to have the first component (electrons from emitter to base) at a much higher level than the second component (holes from base to emitter). This can be accomplished by fabricating the device with a heavily doped emitter and a lightly doped base; that is, the device is designed to have a high density of electrons in the emitter and a low density of holes in the base.

The current that flows across the emitter–base junction will constitute the emitter current i_E, as indicated in Fig. 5.3. The direction of i_E is "out of" the emitter lead, which is in the direction of the hole current and opposite to the direction of the electron current, with the emitter current i_E being equal to the sum of these two components. However, since the electron component is much larger than the hole component, the emitter current will be dominated by the electron component.

---

[1] The material in this section assumes that the reader is familiar with the operation of the *pn* junction under forward-bias conditions (Section 3.7.5).

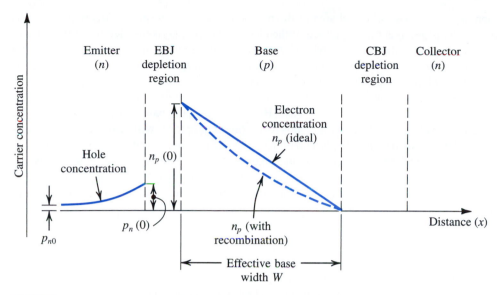

FIGURE 5.4   Profiles of minority-carrier concentrations in the base and in the emitter of an *npn* transistor operating in the active mode: $v_{BE} > 0$ and $v_{CB} \geq 0$.

Let us now consider the electrons injected from the emitter into the base. These electrons will be **minority carriers** in the *p*-type base region. Because the base is usually very thin, in the steady state the excess minority-carrier (electron) concentration in the base will have an almost-straight-line profile, as indicated by the solid straight line in Fig. 5.4. The electron concentration will be highest [denoted by $n_p(0)$] at the emitter side and lowest (zero) at the collector side.[2] As in the case of any forward-biased *pn* junction (Section 3.7.5), the concentration $n_p(0)$ will be proportional to e^{v_{BE}/V_T},

$$n_p(0) = n_{p0} e^{v_{BE}/V_T} \tag{5.1}$$

where n_{p0} is the thermal-equilibrium value of the minority-carrier (electron) concentration in the base region, v_{BE} is the forward base–emitter bias voltage, and V_T is the thermal voltage, which is equal to approximately 25 mV at room temperature. The reason for the zero concentration at the collector side of the base is that the positive collector voltage v_{CB} causes the electrons at that end to be swept across the CBJ depletion region.

The tapered minority-carrier concentration profile (Fig. 5.4) causes the electrons injected into the base to diffuse through the base region toward the collector. This electron diffusion current I_n is directly proportional to the slope of the straight-line concentration profile,

$$
\begin{aligned}
I_n &= A_E q D_n \frac{dn_p(x)}{dx} \\
&= A_E q D_n \left(-\frac{n_p(0)}{W} \right)
\end{aligned}
\tag{5.2}
$$

[2] This minority-carrier distribution in the base results from the boundary conditions imposed by the two junctions. It is not an exponentially decaying distribution, which would result if the base region were infinitely thick. Rather, the thin base causes the distribution to decay linearly. Furthermore, the reverse bias on the collector–base junction causes the electron concentration at the collector side of the base to be zero.

where A_E is the cross-sectional area of the base–emitter junction (in the direction perpendicular to the page), q is the magnitude of the electron charge, D_n is the electron diffusivity in the base, and W is the effective width of the base. Observe that the negative slope of the minority-carrier concentration results in a negative current I_n across the base; that is, I_n flows from right to left (in the negative direction of x).

Some of the electrons that are diffusing through the base region will combine with holes, which are the majority carriers in the base. However, since the base is usually very thin, the proportion of electrons "lost" through this recombination process will be quite small. Nevertheless, the recombination in the base region causes the excess minority-carrier concentration profile to deviate from a straight line and take the slightly concave shape indicated by the broken line in Fig. 5.4. The slope of the concentration profile at the EBJ is slightly higher than that at the CBJ, with the difference accounting for the small number of electrons lost in the base region through recombination.

The Collector Current   From the description above we see that most of the diffusing electrons will reach the boundary of the collector–base depletion region. Because the collector is more positive than the base (by v_{CB} volts), these successful electrons will be swept across the CBJ depletion region into the collector. They will thus get "collected" to constitute the collector current i_C. Thus $i_C = I_n$, which will yield a negative value for i_C, indicating that i_C flows in the negative direction of the x axis (i.e., from right to left). Since we will take this to be the positive direction of i_C, we can drop the negative sign in Eq. (5.2). Doing this and substituting for $n_p(0)$ from Eq. (5.1), we can thus express the collector current i_C as

$$i_C = I_S e^{v_{BE}/V_T} \qquad (5.3)$$

where the **saturation current** I_S is given by

$$I_S = A_E q D_n n_{p0}/W$$

Substituting $n_{p0} = n_i^2/N_A$, where n_i is the intrinsic carrier density and N_A is the doping concentration in the base, we can express I_S as

$$I_S = \frac{A_E q D_n n_i^2}{N_A W} \qquad (5.4)$$

An important observation to make here is that the magnitude of i_C is independent of v_{CB}. That is, as long as the collector is positive with respect to the base, the electrons that reach the collector side of the base region will be swept into the collector and register as collector current.

The saturation current I_S is inversely proportional to the base width W and is directly proportional to the area of the EBJ. Typically I_S is in the range of 10^{-12} A to 10^{-18} A (depending on the size of the device). Because I_S is proportional to n_i^2, it is a strong function of temperature, approximately doubling for every 5°C rise in temperature. (For the dependence of n_i^2 on temperature, refer to Eq. 3.37.)

Since I_S is directly proportional to the junction area (i.e., the device size), it will also be referred to as the **scale current.** Two transistors that are identical except that one has an EBJ area, say, twice that of the other will have saturation currents with that same ratio (i.e., 2). Thus for the same value of v_{BE} the larger device will have a collector current twice that in the smaller device. This concept is frequently employed in integrated-circuit design.

The Base Current   The base current i_B is composed of two components. The first component i_{B1} is due to the holes injected from the base region into the emitter region. This current component is proportional to e^{v_{BE}/V_T},

$$i_{B1} = \frac{A_E q D_p n_i^2}{N_D L_p} e^{v_{BE}/V_T} \tag{5.5}$$

where D_p is the hole diffusivity in the emitter, L_p is the hole diffusion length in the emitter, and N_D is the doping concentration of the emitter.

The second component of base current, i_{B2}, is due to holes that have to be supplied by the external circuit in order to replace the holes lost from the base through the recombination process. An expression for i_{B2} can be found by noting that if the average time for a minority electron to recombine with a majority hole in the base is denoted τ_b (called **minority-carrier lifetime**), then in τ_b seconds the minority-carrier charge in the base, Q_n, recombines with holes. Of course in the steady state, Q_n is replenished by electron injection from the emitter. To replenish the holes, the current i_{B2} must supply the base with a positive charge equal to Q_n every τ_b seconds,

$$i_{B2} = \frac{Q_n}{\tau_b} \tag{5.6}$$

The minority-carrier charge stored in the base region, Q_n, can be found by reference to Fig. 5.4. Specifically, Q_n is represented by the area of the triangle under the straight-line distribution in the base, thus

$$Q_n = A_E q \times \tfrac{1}{2} n_p(0) W$$

Substituting for $n_p(0)$ from Eq. (5.1) and replacing n_{p0} by n_i^2/N_A gives

$$Q_n = \frac{A_E q W n_i^2}{2 N_A} e^{v_{BE}/V_T} \tag{5.7}$$

which can be substituted in Eq. (5.6) to obtain

$$i_{B2} = \frac{1}{2} \frac{A_E q W n_i^2}{\tau_b N_A} e^{v_{BE}/V_T} \tag{5.8}$$

Combining Eqs. (5.5) and (5.8) and utilizing Eq. (5.4), we obtain for the total base current i_B the expression

$$i_B = I_S \left(\frac{D_p}{D_n} \frac{N_A}{N_D} \frac{W}{L_p} + \frac{1}{2} \frac{W^2}{D_n \tau_b} \right) e^{v_{BE}/V_T} \tag{5.9}$$

Comparing Eqs. (5.3) and (5.9), we see that i_B can be expressed as a fraction of i_C as follows:

$$i_B = \frac{i_C}{\beta} \tag{5.10}$$

That is,

$$i_B = \left(\frac{I_S}{\beta} \right) e^{v_{BE}/V_T} \tag{5.11}$$

where β is given by

$$\beta = 1 \bigg/ \left(\frac{D_p}{D_n} \frac{N_A}{N_D} \frac{W}{L_p} + \frac{1}{2} \frac{W^2}{D_n \tau_b} \right) \tag{5.12}$$

from which we see that β is a constant for a particular transistor. For modern *npn* transistors, β is in the range 50 to 200, but it can be as high as 1000 for special devices. For reasons that will become clear later, the constant β is called the **common-emitter current gain.**

Equation (5.12) indicates that the value of β is highly influenced by two factors: the width of the base region, W, and the relative dopings of the base region and the emitter region, (N_A/N_D). To obtain a high β (which is highly desirable since β represents a gain parameter) the base should be thin (W small) and lightly doped and the emitter heavily doped (making N_A/N_D small). Finally, we note that the discussion thus far assumes an idealized situation, where β is a constant for a given transistor.

The Emitter Current   Since the current that enters a transistor must leave it, it can be seen from Fig. 5.3 that the emitter current i_E is equal to the sum of the collector current i_C and the base current i_B; that is,

$$i_E = i_C + i_B \tag{5.13}$$

Use of Eqs. (5.10) and (5.13) gives

$$i_E = \frac{\beta + 1}{\beta} i_C \tag{5.14}$$

That is,

$$i_E = \frac{\beta + 1}{\beta} I_S e^{v_{BE}/V_T} \tag{5.15}$$

Alternatively, we can express Eq. (5.14) in the form

$$i_C = \alpha i_E \tag{5.16}$$

where the constant α is related to β by

$$\alpha = \frac{\beta}{\beta + 1} \tag{5.17}$$

Thus the emitter current in Eq. (5.15) can be written

$$i_E = (I_S/\alpha) e^{v_{BE}/V_T} \tag{5.18}$$

Finally, we can use Eq. (5.17) to express β in terms of α; that is,

$$\beta = \frac{\alpha}{1 - \alpha} \tag{5.19}$$

It can be seen from Eq. (5.17) that α is a constant (for a particular transistor) that is less than but very close to unity. For instance, if $\beta = 100$, then $\alpha \simeq 0.99$. Equation (5.19) reveals an important fact: Small changes in α correspond to very large changes in β. This mathematical observation manifests itself physically, with the result that transistors of the same type may have widely different values of β. For reasons that will become apparent later, α is called the **common-base current gain.**

Finally, we should note that because α and β characterize the operation of the BJT in the "forward-active" mode (as opposed to the "reverse-active" mode, which we shall discuss shortly), they are often denoted α_F and β_F. We shall use α and α_F interchangeably and, similarly, β and β_F.

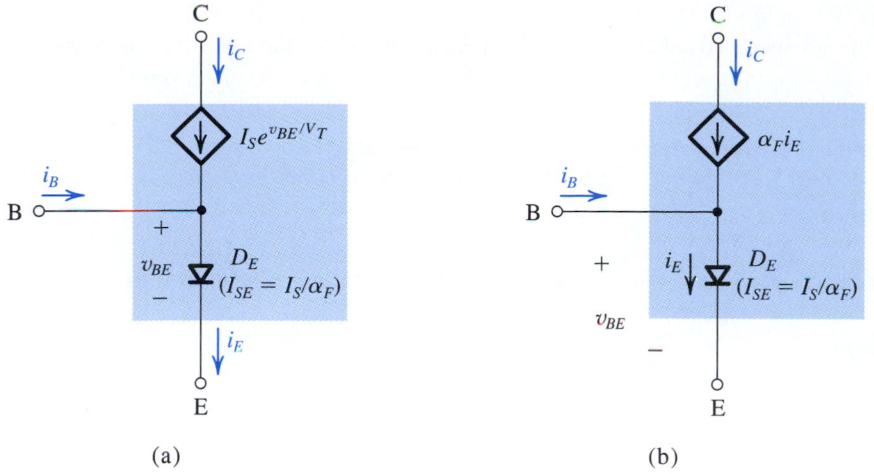

FIGURE 5.5 Large-signal equivalent-circuit models of the *npn* BJT operating in the forward active mode.

Recapitulation and Equivalent-Circuit Models We have presented a first-order model for the operation of the *npn* transistor in the active (or "forward" active) mode. Basically, the forward-bias voltage v_{BE} causes an exponentially related current i_C to flow in the collector terminal. The collector current i_C is independent of the value of the collector voltage as long as the collector–base junction remains reverse-biased; that is, $v_{CB} \geq 0$. Thus in the active mode the collector terminal behaves as an ideal constant-current source where the value of the current is determined by v_{BE}. The base current i_B is a factor $1/\beta_F$ of the collector current, and the emitter current is equal to the sum of the collector and base currents. Since i_B is much smaller than i_C (i.e., $\beta_F \gg 1$), $i_E \simeq i_C$. More precisely, the collector current is a fraction α_F of the emitter current, with α_F smaller than, but close to, unity.

This first-order model of transistor operation in the forward active mode can be represented by the equivalent circuit shown in Fig. 5.5(a). Here diode D_E has a scale current I_{SE} equal to (I_S/α_F) and thus provides a current i_E related to v_{BE} according to Eq. (5.18). The current of the controlled source, which is equal to the collector current, is controlled by v_{BE} according to the exponential relationship indicated, a restatement of Eq. (5.3). This model is in essence a nonlinear voltage-controlled current source. It can be converted to the current-controlled current-source model shown in Fig. 5.5(b) by expressing the current of the controlled source as $\alpha_F i_E$. Note that this model is also nonlinear because of the exponential relationship of the current i_E through diode D_E and the voltage v_{BE}. From this model we observe that if the transistor is used as a two-port network with the input port between E and B and the output port between C and B (i.e., with B as a common terminal), then the current gain observed is equal to α_F. Thus α_F is called the common-base current gain.

EXERCISES

5.1 Consider an *npn* transistor with $v_{BE} = 0.7$ V at $i_C = 1$ mA. Find v_{BE} at $i_C = 0.1$ mA and 10 mA.
Ans. 0.64 V; 0.76 V

5.2 Transistors of a certain type are specified to have β values in the range 50 to 150. Find the range of their α values.

Ans. 0.980 to 0.993

5.3 Measurement of an *npn* BJT in a particular circuit shows the base current to be 14.46 μA, the emitter current to be 1.460 mA, and the base–emitter voltage to be 0.7 V. For these conditions, calculate α, β, and I_S.

Ans. 0.99; 100; 10^{-15} A

5.4 Calculate β for two transistors for which $\alpha = 0.99$ and 0.98. For collector currents of 10 mA, find the base current of each transistor.

Ans. 99; 49; 0.1 mA; 0.2 mA

5.1.3 Structure of Actual Transistors

Figure 5.6 shows a more realistic (but still simplified) cross-section of an *npn* BJT. Note that the collector virtually surrounds the emitter region, thus making it difficult for the electrons injected into the thin base to escape being collected. In this way, the resulting α_F is close to unity and β_F is large. Also, observe that the device is not symmetrical. For more detail on the physical structure of actual devices, the reader is referred to Appendix A.

The fact that the BJT structure is not symmetrical means that if the emitter and collector are interchanged and the transistor is operated in the reverse active mode, the resulting values of α and β, denoted α_R and β_R, will be different from the forward active mode values, α_F and β_F. Furthermore, because the structure is optimized for forward mode operation, α_R and β_R will be much lower than their forward mode counterparts. Of course, α_R and β_R are related by equations identical to those that relate α_F and β_F. Typically, α_R is in the range of 0.01 to 0.5, and the corresponding range of β_R is 0.01 to 1.

The structure in Fig. 5.6 indicates also that the CBJ has a much larger area than the EBJ. It follows that if the transistor is operated in the reverse active mode (i.e., with the CBJ forward biased and the EBJ reverse biased) and the operation is modeled in the manner of Fig. 5.5(b), we obtain the model shown in Fig. 5.7. Here diode D_C represents the collector-base junction and has a scale current I_{SC} that is much larger than the scale current I_{SE} of diode D_E. The two scale currents have, of course, the same ratio as the areas of the corresponding junctions. Furthermore, a simple and elegant formula relates the scale currents I_{SE}, I_{SC}, and I_S and the current gains α_F and α_R, namely

$$\alpha_F I_{SE} = \alpha_R I_{SC} = I_S \tag{5.20}$$

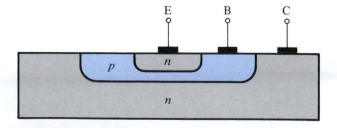

FIGURE 5.6 Cross-section of an *npn* BJT.

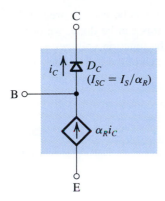

FIGURE 5.7   Model for the *npn* transistor when operated in the reverse active mode (i.e., with the CBJ forward biased and the EBJ reverse biased).

The large scale current I_{SC} has the effect that for the same current, the CBJ exhibits a lower voltage drop when forward biased than the forward voltage drop of the EBJ, V_{BE}. This point will have implications for the BJT's operation in the saturation mode.

EXERCISE

5.5   A particular transistor is said to have $\alpha_F \cong 1$ and $\alpha_R = 0.01$. Its emitter scale current (I_{SE}) is approximately 10^{-15} A. What is its collector scale current (I_{SC})? What is the size of the collector junction relative to the emitter junction? What is the value of β_R?

Ans. 10^{-13} A; 100 times larger; 0.01

5.1.4   The Ebers-Moll (EM) Model

The model of Fig. 5.5(a) can be combined with that of Fig. 5.7 to obtain the circuit model shown in Fig. 5.8. Note that we have relabelled the currents through D_E and D_C, and the corresponding control currents of the controlled sources, as i_{DE} and i_{DC}. Ebers and Moll, two

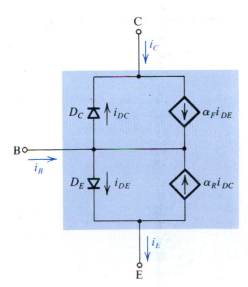

FIGURE 5.8   The Ebers-Moll (EM) model of the *npn* transistor.

early workers in the area, have shown that this composite model can be used to predict the operation of the BJT *in all of its possible modes*. To see how this can be done, we derive expressions for the terminal currents i_E, i_C, and i_B in terms of the junction voltages v_{BE} and v_{BC}. Toward that end, we write an expression for the current at each of the three nodes of the model in Fig. 5.8 as follows:

$$i_E = i_{DE} - \alpha_R i_{DC} \tag{5.21}$$

$$i_C = -i_{DC} + \alpha_F i_{DE} \tag{5.22}$$

$$i_B = (1 - \alpha_F)i_{DE} + (1 - \alpha_R)i_{DC} \tag{5.23}$$

Then we use the diode equation to express i_{DE} and i_{DC} as

$$i_{DE} = I_{SE}(e^{v_{BE}/V_T} - 1) \tag{5.24}$$

and

$$i_{DC} = I_{SC}(e^{v_{BC}/V_T} - 1) \tag{5.25}$$

Substituting for i_{DE} and i_{DC} in Eqs. (5.21), (5.22), and (5.23) and using the relationship in Eq. (5.20) yield the required expressions:

$$i_E = \left(\frac{I_S}{\alpha_F}\right)(e^{v_{BE}/V_T} - 1) - I_S(e^{v_{BC}/V_T} - 1) \tag{5.26}$$

$$i_C = I_S(e^{v_{BE}/V_T} - 1) - \left(\frac{I_S}{\alpha_R}\right)(e^{v_{BC}/V_T} - 1) \tag{5.27}$$

$$i_B = \left(\frac{I_S}{\beta_F}\right)(e^{v_{BE}/V_T} - 1) + \left(\frac{I_S}{\beta_R}\right)(e^{v_{BC}/V_T} - 1) \tag{5.28}$$

where

$$\beta_F = \frac{\alpha_F}{1 - \alpha_F} \tag{5.29}$$

and

$$\beta_R = \frac{\alpha_R}{1 - \alpha_R} \tag{5.30}$$

As a first application of the EM model, we shall use it to predict the terminal currents of a transistor operating in the forward active mode. Here v_{BE} is positive and in the range of 0.6 V to 0.8 V, and v_{BC} is negative. One can easily see that terms containing e^{v_{BC}/V_T} will be negligibly small and can be neglected to obtain

$$i_E \cong \left(\frac{I_S}{\alpha_F}\right)e^{v_{BE}/V_T} + I_S\left(1 - \frac{1}{\alpha_F}\right) \tag{5.31}$$

$$i_C \cong I_S e^{v_{BE}/V_T} + I_S\left(\frac{1}{\alpha_R} - 1\right) \tag{5.32}$$

$$i_B \cong \left(\frac{I_S}{\beta_F}\right)e^{v_{BE}/V_T} - I_S\left(\frac{1}{\beta_F} + \frac{1}{\beta_R}\right) \tag{5.33}$$

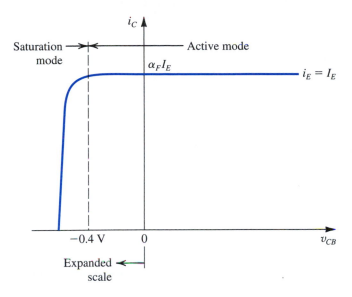

FIGURE 5.9 The i_C–v_{CB} characteristic of an *npn* transistor fed with a constant emitter current I_E. The transistor enters the saturation mode of operation for $v_{CB} < -0.4$ V, and the collector current diminishes.

In each of these three equations, one can normally neglect the second term on the right-hand side. This results in the familiar current-voltage relationships we derived earlier, namely, Eqs. (5.18), (5.3), and (5.11), respectively.

Thus far, we have stated the condition for forward active mode operation as $v_{CB} \geq 0$ to ensure that the CBJ is reverse biased. In actual fact, however, a *pn* junction does not become effectively forward biased until the forward voltage across it exceeds approximately 0.5 V. It follows that one can maintain active mode operation of an *npn* transistor for negative v_{CB} down to approximately -0.4 V or so. This is illustrated in Fig. 5.9, which shows a sketch of i_C versus v_{CB} for an *npn* transistor operated with a constant-emitter current I_E. Observe that i_C remains constant at $\alpha_F I_E$ for v_{CB} going negative to approximately -0.4 V. Below this value of v_{CB}, the CBJ begins to conduct sufficiently that the transistor leaves the active mode and enters the saturation mode of operation, where i_C decreases. We shall study BJT saturation next. For now, however, note that we can use the EM equations to verify that the terms containing e^{v_{BC}/V_T} remain negligibly small for v_{BC} as high as 0.4 V.

EXERCISE

5.6 For a BJT with $\alpha_F = 0.99$, $\alpha_R = 0.02$, and $I_S = 10^{-15}$ A, calculate the second term on the right-hand side of each of Eqs. (5.31), (5.32), and (5.33) to verify that they can be ignored. Then calculate i_E, i_C, and i_B for $v_{BE} = 0.7$ V.

Ans. -10^{-17} A; 49×10^{-15} A; -3×10^{-17} A; 1.461 mA; 1.446 mA; 0.0145 mA

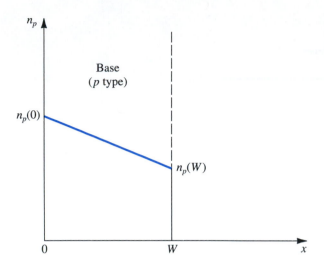

FIGURE 5.10 Concentration profile of the minority carriers (electrons) in the base of an *npn* transistor operating in the saturation mode.

5.1.5 Operation in the Saturation[3] Mode

Figure 5.9 indicates that as v_{CB} is lowered below approximately 0.4 V, the BJT enters the saturation mode of operation. Ideally, v_{CB} has no effect on the collector current in the active mode, but the situation changes dramatically in saturation: Increasing v_{CB} in the negative direction—that is, increasing the forward-bias voltage of the CBJ—reduces i_C. To see this analytically, consider the Ebers-Moll expression for i_C in Eq. (5.27) and, for simplicity, neglect the terms not involving exponentials to obtain

$$i_C = I_S e^{v_{BE}/V_T} - \left(\frac{I_S}{\alpha_R}\right) e^{v_{BC}/V_T} \tag{5.34}$$

The first term on the right-hand side is a result of the forward-biased EBJ, and the second term is a result of the forward-biased CBJ. The second term starts to play a role when v_{BC} exceeds approximately 0.4 V or so. As v_{BC} is increased, this term becomes larger and subtracts from the first term, causing i_C to reduce, eventually reaching zero. Of course, one can operate the saturated transistor at any value of i_C lower than $\alpha_F I_E$. We will have more to say about saturation-mode operation in subsequent sections. Here, however, it is instructive to examine the minority-carrier concentration profile in the base of the saturated transistor, as shown in Fig. 5.10. Observe that because the CBJ is now forward biased, the electron concentration at the collector edge of the base is no longer zero; rather, it is a value proportional to e^{v_{BC}/V_T}. Also note that the slope of the concentration profile is reduced in correspondence with the reduction in i_C.

---

[3] Saturation in a BJT means something completely different from that in a MOSFET. The saturation mode of operation of the BJT is analogous to the triode region of operation of the MOSFET. On the other hand, the saturation region of operation of the MOSFET corresponds to the active mode of BJT operation.

EXERCISE

5.7 (a) Use the EM expressions in Eqs. (5.26) and (5.27) to show that the i_C–v_{CB} relationship sketched in Fig. 5.9 can be described by $i_C = \alpha_F I_E + I_S[\alpha_F - (1/\alpha_R)]e^{v_{BC}/V_T}$. Neglect all terms not containing exponentials.

(b) For the case $I_S = 10^{-15}$ A, $I_E = 1$ mA, $\alpha_F \cong 1$, and $\alpha_R = 0.01$, find i_C for $v_{BC} = -1$ V, $+0.4$ V, $+0.5$ V, $+0.54$ V, and $+0.57$ V. Also find the value of v_{BC} at which $i_C = 0$.

(c) At the value of v_{BC} that makes i_C zero, what do you think i_B should be? Verify using Eq. (5.28).

Ans. (b) 1 mA; 1 mA; 0.95 mA; 0.76 mA; 0.20 mA; 576 mV; (c) 1 mA

5.1.6 The *pnp* Transistor

The *pnp* transistor operates in a manner similar to that of the *npn* device described above. Figure 5.11 shows a *pnp* transistor biased to operate in the active mode. Here the voltage V_{EB} causes the *p*-type emitter to be higher in potential than the *n*-type base, thus forward-biasing the base–emitter junction. The collector–base junction is reverse-biased by the voltage V_{BC}, which keeps the *p*-type collector lower in potential than the *n*-type base.

Unlike the *npn* transistor, current in the *pnp* device is mainly conducted by holes injected from the emitter into the base as a result of the forward-bias voltage V_{EB}. Since the component of emitter current contributed by electrons injected from base to emitter is kept small by using a lightly doped base, most of the emitter current will be due to holes. The electrons injected from base to emitter give rise to the first component of base current, i_{B1}. Also, a number of the holes injected into the base will recombine with the majority carriers in the base (electrons) and will thus be lost. The disappearing base electrons will have to be replaced from the external circuit, giving rise to the second component of base current, i_{B2}. The holes that succeed in reaching the boundary of the depletion region of the collector–base junction will be attracted by the negative voltage on the collector. Thus these holes will be swept across the depletion region into the collector and appear as collector current.

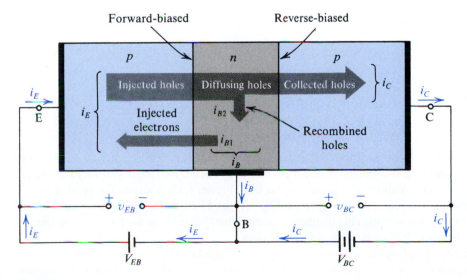

FIGURE 5.11 Current flow in a *pnp* transistor biased to operate in the active mode.

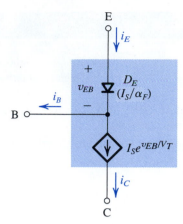

FIGURE 5.12 Large-signal model for the *pnp* transistor operating in the active mode.

It can easily be seen from the above description that the current-voltage relationships of the *pnp* transistor will be identical to those of the *npn* transistor except that v_{BE} has to be replaced by v_{EB}. Also, the large-signal active-mode operation of the *pnp* transistor can be modeled by the circuit depicted in Fig. 5.12. As in the *npn* case, another version of this equivalent circuit is possible in which the current source is replaced with a current-controlled current source $\alpha_F i_E$. Finally, we note that the *pnp* transistor can operate in the saturation mode in a manner analogous to that described for the *npn* device.

EXERCISES

5.8 Consider the model in Fig. 5.12 applied in the case of a *pnp* transistor whose base is grounded, the emitter is fed by a constant-current source that supplies a 2-mA current into the emitter terminal, and the collector is connected to a −10-V dc supply. Find the emitter voltage, the base current, and the collector current if for this transistor $\beta = 50$ and $I_S = 10^{-14}$ A.

Ans. 0.650 V; 39.2 μA; 1.96 mA

5.9 For a *pnp* transistor having $I_S = 10^{-11}$ A and $\beta = 100$, calculate v_{EB} for $i_C = 1.5$ A.

Ans. 0.643 V

 ## 5.2 CURRENT-VOLTAGE CHARACTERISTICS

5.2.1 Circuit Symbols and Conventions

The physical structure used thus far to explain transistor operation is rather cumbersome to employ in drawing the schematic of a multitransistor circuit. Fortunately, a very descriptive and convenient circuit symbol exists for the BJT. Figure 5.13(a) shows the symbol for the *npn* transistor; the *pnp* symbol is given in Fig. 5.13(b). In both symbols the emitter is distinguished by an arrowhead. This distinction is important because, as we have seen in the last section, practical BJTs are not symmetric devices.

The polarity of the device—*npn* or *pnp*—is indicated by the direction of the arrowhead on the emitter. This arrowhead points in the direction of normal current flow in the emitter, which is also the forward direction of the base–emitter junction. Since we have adopted a

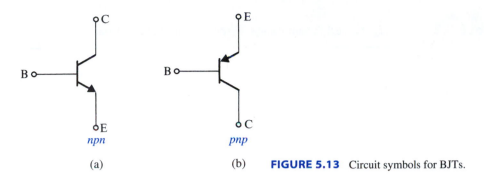

FIGURE 5.13 Circuit symbols for BJTs.

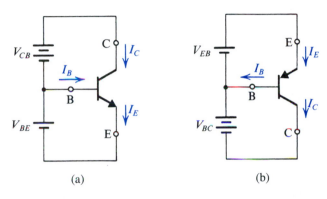

FIGURE 5.14 Voltage polarities and current flow in transistors biased in the active mode.

drawing convention by which currents flow from top to bottom, we will always draw *pnp* transistors in the manner shown in Fig. 5.13 (i.e., with their emitters on top).

Figure 5.14 shows *npn* and *pnp* transistors biased to operate in the active mode. It should be mentioned in passing that the biasing arrangement shown, utilizing two dc voltage sources, is not a usual one and is used here merely to illustrate operation. Practical biasing schemes will be presented in Section 5.5. Figure 5.14 also indicates the reference and actual directions of current flow throughout the transistor. Our convention will be to take the reference direction to coincide with the normal direction of current flow. Hence, normally, we should not encounter a negative value for i_E, i_B, or i_C.

The convenience of the circuit drawing convention that we have adopted should be obvious from Fig. 5.14. Note that currents flow from top to bottom and that voltages are higher at the top and lower at the bottom. The arrowhead on the emitter also implies the polarity of the emitter–base voltage that should be applied in order to forward bias the emitter–base junction. Just a glance at the circuit symbol of the *pnp* transistor, for example, indicates that we should make the emitter higher in voltage than the base (by v_{EB}) in order to cause current to flow into the emitter (downward). Note that the symbol v_{EB} means the voltage by which the emitter (E) is higher than the base (B). Thus for a *pnp* transistor operating in the active mode v_{EB} is positive, while in an *npn* transistor v_{BE} is positive.

From the discussion of Section 5.1 it follows that an *npn* transistor whose EBJ is forward biased will operate in the active mode *as long as the collector voltage does not fall below that of the base by more than approximately 0.4 V.* Otherwise, the transistor leaves the active mode and enters the saturation region of operation.

TABLE 5.2	Summary of the BJT Current-Voltage Relationships in the Active Mode

$$i_C = I_S e^{v_{BE}/V_T}$$

$$i_B = \frac{i_C}{\beta} = \left(\frac{I_S}{\beta}\right) e^{v_{BE}/V_T}$$

$$i_E = \frac{i_C}{\alpha} = \left(\frac{I_S}{\alpha}\right) e^{v_{BE}/V_T}$$

Note: For the *pnp* transistor, replace v_{BE} with v_{EB}.

$$i_C = \alpha i_E \qquad\qquad i_B = (1-\alpha)i_E = \frac{i_E}{\beta+1}$$

$$i_C = \beta i_B \qquad\qquad i_E = (\beta+1)i_B$$

$$\beta = \frac{\alpha}{1-\alpha} \qquad\qquad \alpha = \frac{\beta}{\beta+1}$$

$$V_T = \text{thermal voltage} = \frac{kT}{q} \cong 25 \text{ mV at room temperature}$$

In a parallel manner, the *pnp* transistor will operate in the active mode *if the EBJ is forward biased and the collector voltage is not allowed to rise above that of the base by more than 0.4 V or so.* Otherwise, the CBJ becomes forward biased, and the *pnp* transistor enters the saturation region of operation.

For easy reference, we present in Table 5.2 a summary of the BJT current-voltage relationships in the active mode of operation. Note that for simplicity we use α and β rather than α_F and β_F.

The Constant *n* In the diode equation (Chapter 3) we used a constant *n* in the exponential and mentioned that its value is between 1 and 2. For modern bipolar junction transistors the constant *n* is close to unity except in special cases: (1) at high currents (i.e., high relative to the normal current range of the particular transistor) the i_C–v_{BE} relationship exhibits a value for *n* that is close to 2, and (2) at low currents the i_B–v_{BE} relationship shows a value for *n* of approximately 2. Note that for our purposes we shall assume always that $n = 1$.

The Collector–Base Reverse Current (I_{CBO}) In our discussion of current flow in transistors we ignored the small reverse currents carried by thermally generated minority carriers. Although such currents can be safely neglected in modern transistors, the reverse current across the collector–base junction deserves some mention. This current, denoted I_{CBO}, is the reverse current flowing from collector to base with the emitter open-circuited (hence the subscript *O*). This current is usually in the nanoampere range, a value that is many times higher than its theoretically predicted value. As with the diode reverse current, I_{CBO} contains a substantial leakage component, and its value is dependent on v_{CB}. I_{CBO} depends strongly on temperature, approximately doubling for every 10°C rise.[4]

---

[4] The temperature coefficient of I_{CBO} is different from that of I_S because I_{CBO} contains a substantial leakage component.

EXAMPLE 5.1

The transistor in the circuit of Fig. 5.15(a) has $\beta = 100$ and exhibits a v_{BE} of 0.7 V at $i_C = 1$ mA. Design the circuit so that a current of 2 mA flows through the collector and a voltage of +5 V appears at the collector.

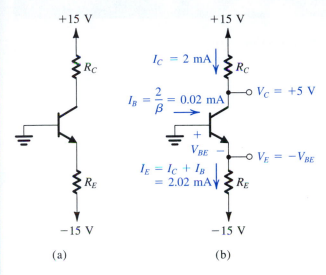

FIGURE 5.15 Circuit for Example 5.1.

Solution

Refer to Fig. 5.15(b). We note at the outset that since we are required to design for $V_C = +5$ V, the CBJ will be reverse biased and the BJT will be operating in the active mode. To obtain a voltage $V_C = +5$ V the voltage drop across R_C must be $15 - 5 = 10$ V. Now, since $I_C = 2$ mA, the value of R_C should be selected according to

$$R_C = \frac{10 \text{ V}}{2 \text{ mA}} = 5 \text{ k}\Omega$$

Since $v_{BE} = 0.7$ V at $i_C = 1$ mA, the value of v_{BE} at $i_C = 2$ mA is

$$V_{BE} = 0.7 + V_T \ln\left(\frac{2}{1}\right) = 0.717 \text{ V}$$

Since the base is at 0 V, the emitter voltage should be

$$V_E = -0.717 \text{ V}$$

For $\beta = 100$, $\alpha = 100/101 = 0.99$. Thus the emitter current should be

$$I_E = \frac{I_C}{\alpha} = \frac{2}{0.99} = 2.02 \text{ mA}$$

Now the value required for R_E can be determined from

$$R_E = \frac{V_E - (-15)}{I_E}$$

$$= \frac{-0.717 + 15}{2.02} = 7.07 \text{ k}\Omega$$

This completes the design. We should note, however, that the calculations above were made with a degree of accuracy that is usually neither necessary nor justified in practice in view, for instance, of the expected tolerances of component values. Nevertheless, we chose to do the design precisely in order to illustrate the various steps involved.

EXERCISES

5.10 In the circuit shown in Fig. E5.10, the voltage at the emitter was measured and found to be −0.7 V. If $\beta = 50$, find I_E, I_B, I_C, and V_C.

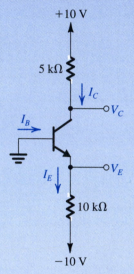

FIGURE E5.10

Ans. 0.93 mA; 18.2 μA; 0.91 mA; +5.45 V

5.11 In the circuit shown in Fig. E5.11, measurement indicates V_B to be +1.0 V and V_E to be +1.7 V. What are α and β for this transistor? What voltage V_C do you expect at the collector?

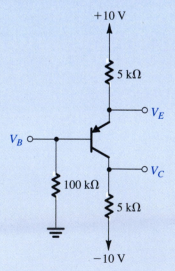

FIGURE E5.11

Ans. 0.994; 165; −1.75 V

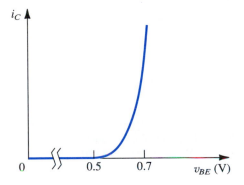

FIGURE 5.16   The i_C–v_{BE} characteristic for an *npn* transistor.

5.2.2 Graphical Representation of Transistor Characteristics

It is sometimes useful to describe the transistor i–v characteristics graphically. Figure 5.16 shows the i_C–v_{BE} characteristic, which is the exponential relationship

$$i_C = I_S e^{v_{BE}/V_T}$$

which is identical (except for the value of constant n) to the diode i–v relationship. The i_E–v_{BE} and i_B–v_{BE} characteristics are also exponential but with different scale currents: I_S/α for i_E, and I_S/β for i_B. Since the constant of the exponential characteristic, $1/V_T$, is quite high ($\simeq 40$), the curve rises very sharply. For v_{BE} smaller than about 0.5 V, the current is negligibly small.[5] Also, over most of the normal current range v_{BE} lies in the range of 0.6 V to 0.8 V. In performing rapid first-order dc calculations we normally will assume that $V_{BE} \simeq 0.7$ V, which is similar to the approach used in the analysis of diode circuits (Chapter 3). For a *pnp* transistor, the i_C–v_{EB} characteristic will look identical to that of Fig. 5.16 with v_{BE} replaced with v_{EB}.

As in silicon diodes, the voltage across the emitter–base junction decreases by about 2 mV for each rise of 1°C in temperature, provided that the junction is operating at a constant current. Figure 5.17 illustrates this temperature dependence by depicting i_C–v_{BE} curves at three different temperatures for an *npn* transistor.

The Common-Base Characteristics   One way to describe the operation of a bipolar transistor is to plot i_C versus v_{CB} for various values of i_E. We have already encountered one such graph, in Fig. 5.9, which we used to introduce the saturation mode of operation. A conceptual experimental setup for measuring such characteristics is shown in Fig. 5.18(a). Observe that in these measurements the base voltage is held constant, here at ground potential, and thus the base serves as a common terminal between the input and output ports. Consequently, the resulting set of characteristics, shown in Fig. 5.18(b), are known as common-base characteristics.

---

[5] The i_C–v_{BE} characteristic is the BJT's counterpart of the i_D–v_{GS} characteristic of the enhancement MOSFET. They share an important attribute: In both cases the voltage has to exceed a "threshold" for the device to conduct appreciably. In the case of the MOSFET, there is a formal threshold voltage, V_t, which lies typically in the range of 0.5 V to 1.0 V. For the BJT, there is an "apparent threshold" of approximately 0.5 V. The i_D–v_{GS} characteristic of the MOSFET is parabolic and thus is less steep than the i_C–v_{BE} characteristic of the BJT. This difference has a direct and significant implication on the value of transconductance g_m realized with each device.

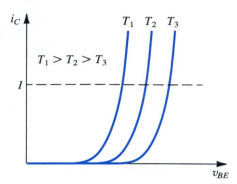

FIGURE 5.17 Effect of temperature on the i_C–v_{BE} characteristic. At a constant emitter current (broken line), v_{BE} changes by -2 mV/°C.

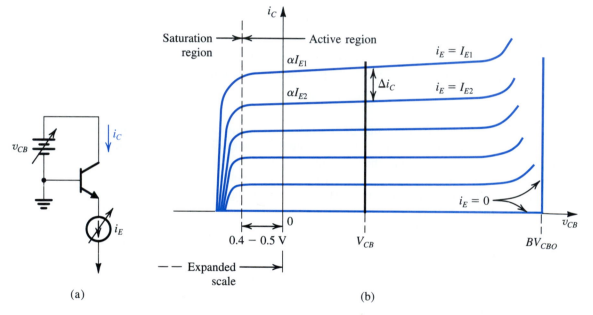

(a)

(b)

FIGURE 5.18 The i_C–v_{CB} characteristics of an *npn* transistor.

In the active region of operation, obtained for $v_{CB} \geq -0.4$ V or so, the i_C–v_{CB} curves deviate from our expectations in two ways. First, the curves are not horizontal straight lines but show a small positive slope, indicating that i_C depends slightly on v_{CB} in the active mode. We shall discuss this phenomenon shortly. Second, at relatively large values of v_{CB}, the collector current shows a rapid increase, which is a breakdown phenomenon that we will consider at a later stage.

As indicated in Fig. 5.18(b), each of the characteristic curves intersects the vertical axis at a current level equal to αI_E, where I_E is the constant emitter current at which the particular curve is measured. The resulting value of α is a **total** or **large-signal** α; that is, $\alpha = i_C/i_E$, where i_C and i_E denote total collector and emitter currents, respectively. Here we recall that α is appropriately called the common-base current gain. An **incremental** or **small-signal** α can be determined by measuring the change in i_C, Δi_C, obtained as a result of changing i_E by an increment Δi_E, $\alpha \equiv \Delta i_C/\Delta i_E$. This measurement is usually made at a constant v_{CB}, as indicated in Fig. 5.18(b). Usually, the values of incremental and total α differ slightly, but we shall not make a distinction between the two in this book.

Finally, turning to the saturation region, the Ebers-Moll equations can be used to obtain the following expression for the i_C–v_{CB} curve in the saturation region (for $i_E = I_E$),

$$i_C = \alpha_F I_E - I_S\left(\frac{1}{\alpha_R} - \alpha_F\right)e^{v_{BC}/V_T} \tag{5.35}$$

We can use this equation to determine the value of v_{BC} at which i_C is reduced to zero. Recalling that the CBJ is much larger than the EBJ, the forward-voltage drop v_{BC} will be smaller than v_{BE} resulting in a collector-emitter voltage, v_{CE}, of 0.1 V to 0.3 V in saturation.

EXERCISES

5.12 Consider a *pnp* transistor with $v_{EB} = 0.7$ V at $i_E = 1$ mA. Let the base be grounded, the emitter be fed by a 2-mA constant-current source, and the collector be connected to a –5-V supply through a 1-kΩ resistance. If the temperature increases by 30°C, find the changes in emitter and collector voltages. Neglect the effect of I_{CBO}.

Ans. –60 mV; 0 V.

5.13 Find the value of v_{CB} at which i_C of an *npn* transistor operated in the CB configuration with $I_E = 1$ mA is reduced (a) to half its active-mode value and (b) to zero. Assume $\alpha_F \cong 1$ and $\alpha_R = 0.1$. The value of V_{BE} was measured for $v_{CB} = 0$ [see measuring setup in Fig. 5.18(a)] and found to be 0.70 V. Repeat (a) and (b) for $\alpha_R = 0.01$.

Ans. –0.628 V; –0.645 V; –0.568 V; –0.585 V.

5.2.3 Dependence of i_C on the Collector Voltage—The Early Effect

When operated in the active region, practical BJTs show some dependence of the collector current on the collector voltage, with the result that their i_C–v_{CB} characteristics are not perfectly horizontal straight lines. To see this dependence more clearly, consider the conceptual circuit shown in Fig. 5.19(a). The transistor is connected in the **common-emitter configuration;** that is, here the emitter serves as a common terminal between the input and output ports. The voltage V_{BE} can be set to any desired value by adjusting the dc source connected between base and emitter. At each value of V_{BE}, the corresponding i_C–v_{CE} characteristic curve can be measured point-by-point by varying the dc source connected between collector and emitter and measuring the corresponding collector current. The result is the family of i_C–v_{CE} characteristic curves shown in Fig. 5.19(b) and known as **common-emitter characteristics.**

At low values of v_{CE}, as the collector voltage goes below that of the base by more than 0.4 V, the collector–base junction becomes forward biased and the transistor leaves the active mode and enters the saturation mode. We shall shortly look at the details of the i_C–v_{CE} curves in the saturation region. At this time, however, we wish to examine the characteristic curves in the active region in detail. We observe that the characteristic curves, though still straight lines, have finite slope. In fact, when extrapolated, the characteristic lines meet at a point on the negative v_{CE} axis, at $v_{CE} = -V_A$. The voltage V_A, a positive number, is a parameter for the particular BJT, with typical values in the range of 50 V to 100 V. It is called the **Early voltage,** after J. M. Early, the engineering scientist who first studied this phenomenon.

At a given value of v_{BE}, increasing v_{CE} increases the reverse-bias voltage on the collector–base junction and thus increases the width of the depletion region of this junction (refer to Fig. 5.3). This in turn results in a decrease in the **effective base width** W. Recalling that I_S is inversely proportional to W (Eq. 5.4), we see that I_S will increase and that i_C increases proportionally. This is the Early effect.

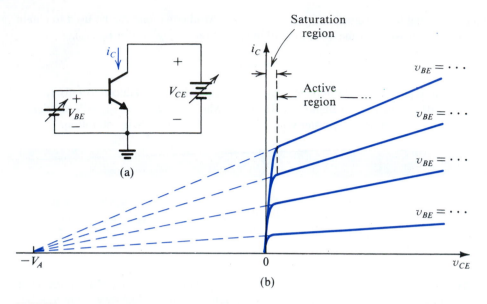

FIGURE 5.19 **(a)** Conceptual circuit for measuring the i_C–v_{CE} characteristics of the BJT. **(b)** The i_C–v_{CE} characteristics of a practical BJT.

The linear dependence of i_C on v_{CE} can be accounted for by assuming that I_S remains constant and including the factor $(1 + v_{CE}/V_A)$ in the equation for i_C as follows:

$$i_C = I_S e^{v_{BE}/V_T}\left(1 + \frac{v_{CE}}{V_A}\right) \tag{5.36}$$

The nonzero slope of the i_C–v_{CE} straight lines indicates that the **output resistance** looking into the collector is not infinite. Rather, it is finite and defined by

$$r_o \equiv \left[\frac{\partial i_C}{\partial v_{CE}}\bigg|_{v_{BE}=\text{constant}}\right]^{-1} \tag{5.37}$$

Using Eq. (5.36) we can show that

$$r_o = \frac{V_A + V_{CE}}{I_C} \tag{5.38}$$

where I_C and V_{CE} are the coordinates of the point at which the BJT is operating on the particular i_C–v_{CE} curve (i.e., the curve obtained for $v_{BE} = V_{BE}$). Alternatively, we can write

$$r_o = \frac{V_A}{I_C'} \tag{5.38a}$$

where I_C' is the value of the collector current with the Early effect neglected; that is,

$$I_C' = I_S e^{V_{BE}/V_T} \tag{5.38b}$$

It is rarely necessary to include the dependence of i_C on v_{CE} in dc bias design and analysis. However, the finite output resistance r_o can have a significant effect on the gain of transistor amplifiers, as will be seen in later sections and chapters.

The output resistance r_o can be included in the circuit model of the transistor. This is illustrated in Fig. 5.20, where we show large-signal circuit models of a common-emitter *npn* transistor operating in the active mode. Observe that diode D_B models the exponential dependence of i_B on v_{BE} and thus has a scale current $I_{SB} = I_S/\beta$. Also note that the two models

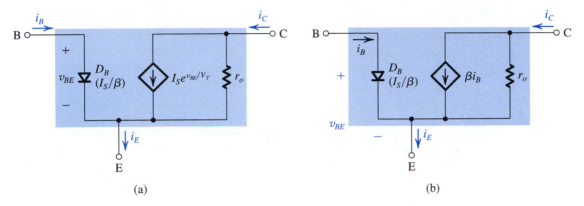

FIGURE 5.20 Large-signal equivalent-circuit models of an *npn* BJT operating in the active mode in the common-emitter configuration.

differ only in how the control function of the transistor is expressed: In the circuit of Fig. 5.20(a), voltage v_{BE} controls the collector current source, while in the circuit of Fig. 5.20(b), the base current i_B is the control parameter for the current source βi_B. Here we note that β represents the ideal current gain (i.e., when r_o is not present) of the common-emitter configuration, which is the reason for its name, the **common-emitter current gain.**

EXERCISES

5.14 Find the output resistance of a BJT for which $V_A = 100$ V at $I_C = 0.1$, 1, and 10 mA.

Ans. 1 MΩ; 100 kΩ; 10 kΩ

5.15 Consider the circuit in Fig. 5.19(a). At $V_{CE} = 1$ V, V_{BE} is adjusted to yield a collector current of 1 mA. Then, while V_{BE} is kept constant, V_{CE} is raised to 11 V. Find the new value of I_C. For this transistor, $V_A = 100$ V.

Ans. 1.1 mA

5.2.4 The Common-Emitter Characteristics

An alternative way of expressing the transistor common-emitter characteristics is illustrated in Fig. 5.21. Here the base current i_B rather than the base–emitter voltage v_{BE} is used as a parameter. That is, each i_C–v_{CE} curve is measured with the base fed with a constant current I_B. The resulting characteristics look similar to those in Fig. 5.19 except that here we show the breakdown phenomenon, which we shall discuss shortly. We should also mention that although it is not obvious from the graphs, the slope of the curves in the active region of operation differs from the corresponding slope in Fig. 5.19. This, however, is a rather subtle point and beyond our interest at this moment.

The Common-Emitter Current Gain β An important transistor parameter is the common-emitter current gain β_F or simply β. Thus far we have defined β as the ratio of the total current in the collector to the total current in the base, and we have assumed that β is constant for a given transistor, independent of the operating conditions. In the following we examine those two points in some detail.

Consider a transistor operating in the active region at the point labeled Q in Fig. 5.21, that is, at a collector current I_{CQ}, a base current I_{BQ}, and a collector-emitter voltage V_{CEQ}. The

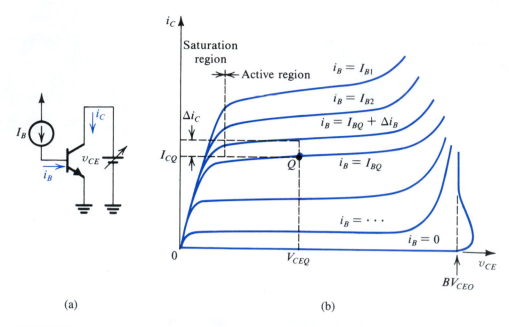

(a)

(b)

FIGURE 5.21 Common-emitter characteristics. Note that the horizontal scale is expanded around the origin to show the saturation region in some detail.

ratio of the collector current to the base current is the **large-signal or dc** β,

$$\beta_{dc} \equiv \frac{I_{CQ}}{I_{BQ}} \tag{5.39}$$

which is the β we have been using in our description of transistor operation. It is commonly referred to on the manufacturer's data sheets as h_{FE}, a symbol that comes from the use of the hybrid, or h, two-port parameters to characterize transistor operation (see Appendix B). One can define another β based on incremental or small-signal quantities. Referring to Fig. 5.21 we see that while keeping v_{CE} constant at the value V_{CEQ}, changing i_B from I_{BQ} to $(I_{BQ} + \Delta i_B)$ results in i_C increasing from I_{CQ} to $(I_{CQ} + \Delta i_C)$. Thus we can define the **incremental or ac** β, β_{ac}, as

$$\beta_{ac} = \frac{\Delta i_C}{\Delta i_B}\bigg|_{v_{CE}=\text{constant}} \tag{5.40}$$

The magnitudes of β_{ac} and β_{dc} differ, typically by approximately 10% to 20%. In this book we shall not normally make a distinction between the two. Finally, we should mention that the small-signal β or β_{ac} is also known by the alternate symbol h_{fe}. Because the small-signal β or h_{fe} is defined and measured at a constant v_{CE}—that is, with a zero signal component between collector and emitter—it is known as the **short-circuit common-emitter current gain.**

The value of β depends on the current at which the transistor is operating, and the relationship takes the form shown in Fig. 5.22. The physical processes that give rise to this relationship are beyond the scope of this book. Figure 5.22 also shows the temperature dependence of β.

The Saturation Voltage V_{CEsat} and Saturation Resistance R_{CEsat} An expanded view of the common-emitter characteristics in the saturation region is shown in Fig. 5.23. The fact that the curves are "bunched" together in the saturation region implies that the incremental β is lower there than in the active region. A possible operating point in the saturation region is that labeled X. It is characterized by a base current I_B, a collector current I_{Csat}, and a collector–emitter voltage V_{CEsat}. Note that $I_{Csat} < \beta_F I_B$. Since the value of I_{Csat} is established

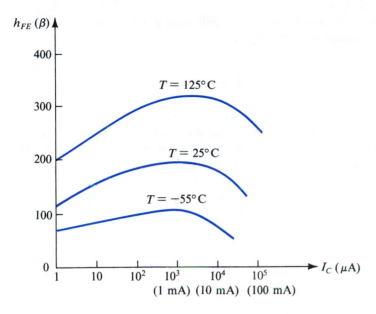

FIGURE 5.22 Typical dependence of β on I_C and on temperature in a modern integrated-circuit *npn* silicon transistor intended for operation around 1 mA.

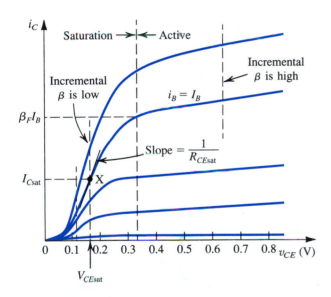

FIGURE 5.23 An expanded view of the common-emitter characteristics in the saturation region.

by the circuit designer, a saturated transistor is said to be operating at a **forced** β given by

$$\beta_{\text{forced}} \equiv \frac{I_{C\text{sat}}}{I_B} \tag{5.41}$$

Thus,

$$\beta_{\text{forced}} < \beta_F \tag{5.42}$$

The ratio of β_F to β_{forced} is known as the **overdrive factor.** The greater the overdrive factor, the deeper the transistor is driven into saturation and the lower $V_{CE\text{sat}}$ becomes.

The i_C–v_{CE} curves in saturation are rather steep, indicating that the saturated transistor exhibits a low collector-to-emitter resistance $R_{CE\text{sat}}$,

$$R_{CE\text{sat}} \equiv \left. \frac{\partial v_{CE}}{\partial i_C} \right|_{\substack{i_B=I_B \\ i_C=I_{C\text{sat}}}} \tag{5.43}$$

Typically, $R_{CE\text{sat}}$ ranges from a few ohms to a few tens of ohms.

Figure 5.24(b) shows one of the i_C–v_{CE} characteristic curves of the saturated transistor shown in Fig. 5.24(a). It is interesting to note that the curve intersects the v_{CE} axis at $V_T \ln (1/\alpha_R)$, a value common to all the i_C–v_{CE} curves. We have also shown in Fig. 5.24(b) the tangent at operating point X of slope $1/R_{CE\text{sat}}$. When extrapolated, the tangent intersects the v_{CE}-axis at a voltage $V_{CE\text{off}}$, typically approximately 0.1 V. It follows that the i_C–v_{CE} characteristic of a saturated transistor can be approximately represented by the equivalent circuit shown in Fig. 5.24(c). At the collector side, the transistor is represented by a resistance $R_{CE\text{sat}}$ in

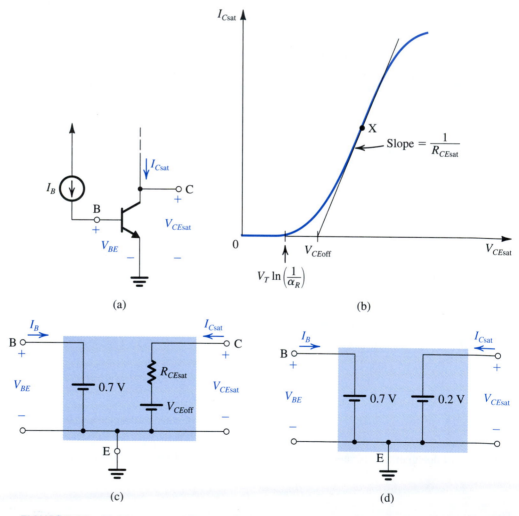

FIGURE 5.24 (a) An *npn* transistor operated in saturation mode with a constant base current I_B. (b) The i_C–v_{CE} characteristic curve corresponding to $i_B = I_B$. The curve can be approximated by a straight line of slope $1/R_{CE\text{sat}}$. (c) Equivalent-circuit representation of the saturated transistor. (d) A simplified equivalent-circuit model of the saturated transistor.

series with a battery V_{CEoff}. Thus the saturation voltage V_{CEsat} can be found from

$$V_{CEsat} = V_{CEoff} + I_{Csat}R_{CEsat} \tag{5.44}$$

Typically, V_{CEsat} falls in the range of 0.1 V to 0.3 V. For many applications the even simpler model shown in Fig. 5.24(d) suffices. The offset voltage of a saturated transistor, though small, makes the BJT less attractive as a switch than the MOSFET, whose i_D–v_{DS} characteristics go right through the origin of the i_D–v_{DS} plane.

It is interesting and instructive to use the Ebers-Moll model to derive analytical expressions for the characteristics of the saturated transistor. Toward that end we use Eqs. (5.28) and (5.27), substitute $i_B = I_B$, and neglect the small terms that do not include exponentials; thus,

$$I_B = \frac{I_S}{\beta_F}e^{v_{BE}/V_T} + \frac{I_S}{\beta_R}e^{v_{BC}/V_T} \tag{5.45}$$

$$i_C = I_S e^{v_{BE}/V_T} - \frac{I_S}{\alpha_R}e^{v_{BC}/V_T} \tag{5.46}$$

Dividing Eq. (5.46) by Eq. (5.45) and writing $v_{BE} = v_{BC} + v_{CE}$ enables us to express i_C in the form

$$i_C = (\beta_F I_B)\left(\frac{e^{v_{CE}/V_T} - \dfrac{1}{\alpha_R}}{e^{v_{CE}/V_T} + \dfrac{\beta_F}{\beta_R}}\right) \tag{5.47}$$

This is the equation of the i_C–v_{CE} characteristic curve obtained when the base is driven with a constant current I_B. Figure 5.25 shows a typical plot of the normalized collector current $i_C/(\beta_F I_B)$,

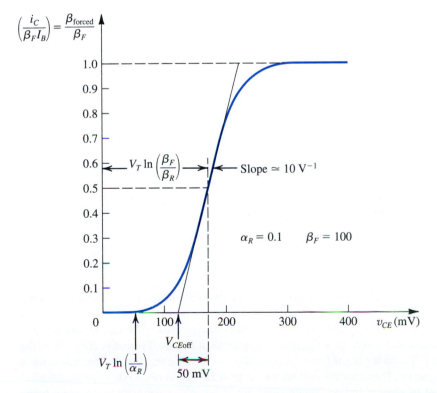

FIGURE 5.25 Plot of the normalized i_C versus v_{CE} for an *npn* transistor with $\beta_F = 100$ and $\alpha_R = 0.1$. This is a plot of Eq. (5.47), which is derived using the Ebers-Moll model.

which is equal to $(\beta_{\text{forced}}/\beta_F)$, versus v_{CE}. As shown, the curve can be approximated by a straight line coincident with the tangent at the point $\beta_{\text{forced}}/\beta_F = 0.5$. It can be shown that this tangent has a slope of approximately 10 V^{-1}, independent of the transistor parameters. Thus,

$$R_{CE\text{sat}} = 1/10\beta_F I_B \tag{5.48}$$

Other important parameters of the normalized plot are indicated in Fig. 5.25. Finally, we can obtain an expression for $V_{CE\text{sat}}$ by substituting $i_C = I_{C\text{sat}} = \beta_{\text{forced}} I_B$ and $v_{CE} = V_{CE\text{sat}}$ in Eq. (5.47),

$$V_{CE\text{sat}} = V_T \ln \frac{1 + (\beta_{\text{forced}} + 1)/\beta_R}{1 - (\beta_{\text{forced}}/\beta_F)} \tag{5.49}$$

EXERCISES

5.16 An *npn* transistor characterized by $\beta_F = 100$ and $\alpha_R = 0.1$ is operated in saturation with a constant base current of 0.1 mA and a forced β of 10. Find the values of V_{CE} at $i_C = 0$, $R_{CE\text{sat}}$, and $V_{CE\text{off}}$. Use the latter two figures to obtain an approximate value for $V_{CE\text{sat}}$ [i.e., using the equivalent circuit model of Fig. 5.24(c)]. Find a more accurate value for $V_{CE\text{sat}}$ using Eq. (5.49), and compare results. Repeat for a β_{forced} of 20.

Ans. 58 mV; 10 Ω; 120 mV; 130 mV and 118 mV; 140 mV and 137 mV

5.17 Measurements made on a BJT operated in saturation with a constant base-current drive provide the following data: at $i_C = 5$ mA, $v_{CE} = 170$ mV; at $i_C = 2$ mA, $v_{CE} = 110$ mV. What are the values of the offset voltage $V_{CE\text{off}}$ and saturation resistance $R_{CE\text{sat}}$ in this situation?

Ans. 70 mV; 20 Ω

5.2.5 Transistor Breakdown

The maximum voltages that can be applied to a BJT are limited by the EBJ and CBJ breakdown effects that follow the avalanche multiplication mechanism described in Section 3.7.4. Consider first the common-base configuration. The i_C–v_{CB} characteristics in Fig. 5.18(b) indicate that for $i_E = 0$ (i.e., with the emitter open-circuited) the collector–base junction breaks down at a voltage denoted by BV_{CBO}. For $i_E > 0$, breakdown occurs at voltages smaller than BV_{CBO}. Typically, BV_{CBO} is greater than 50 V.

Next consider the common-emitter characteristics of Fig. 5.21, which show breakdown occurring at a voltage BV_{CEO}. Here, although breakdown is still of the avalanche type, the effects on the characteristics are more complex than in the common-base configuration. We will not explain these in detail; it is sufficient to point out that typically BV_{CEO} is about half BV_{CBO}. On transistor data sheets, BV_{CEO} is sometimes referred to as the **sustaining voltage** LV_{CEO}.

Breakdown of the CBJ in either the common-base or common-emitter configuration is not destructive as long as the power dissipation in the device is kept within safe limits. This, however, is not the case with the breakdown of the emitter–base junction. The EBJ breaks down in an avalanche manner at a voltage BV_{EBO} much smaller than BV_{CBO}. Typically, BV_{EBO} is in the range of 6 V to 8 V, and the breakdown is destructive in the sense that the β of the transistor is permanently reduced. This does not prevent use of the EBJ as a zener diode to generate reference voltages in IC design. In such applications one is not concerned with the β-degradation effect. A circuit arrangement to prevent EBJ breakdown in IC amplifiers will be discussed in Chapter 9. Transistor breakdown and the maximum allowable power dissipation are important parameters in the design of power amplifiers (Chapter 14).

EXERCISE

5.18 What is the output voltage of the circuit in Fig. E5.18 if the transistor $BV_{BCO} = 70$ V?

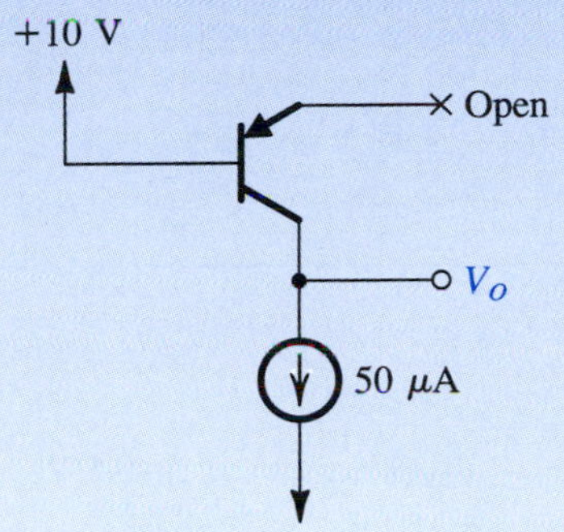

+10 V

× Open

○ V_O

50 μA

FIGURE E5.18

Ans. −60 V

5.2.6 Summary

We conclude our study of the current-voltage characteristics of the BJT with a summary of important results in Table 5.3.

5.3 THE BJT AS AN AMPLIFIER AND AS A SWITCH

Having studied the terminal characteristics of the BJT, we are now ready to consider its two major areas of application: as a signal amplifier,[6] and as a digital-circuit switch. The basis for the amplifier application is the fact that when the BJT is operated in the active mode, it acts as a voltage-controlled current source: Changes in the base–emitter voltage v_{BE} give rise to changes in the collector current i_C. Thus in the active mode the BJT can be used to implement a transconductance amplifier (see Section 1.5). Voltage amplification can be obtained simply by passing the collector current through a resistance R_C, as will be seen shortly.

[6] An introduction to amplifiers from an external-terminals point of view is presented in Sections 1.4 and 1.5. It would be helpful for readers who are not familiar with basic amplifier concepts to review this material before proceeding with the study of BJT amplifiers.

TABLE 5.3 **Summary of the BJT Current-Voltage Characteristics**

Circuit Symbol and Directions of Current Flow	*npn* Transistor	*pnp* Transistor
	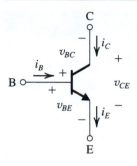	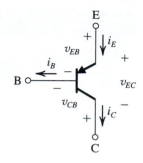

Operation in the Active Mode (for Amplifier Application)

Conditions:

1. EBJ Forward Biased

$v_{BE} > V_{BEon}$; $V_{BEon} \cong 0.5$ V

Typically, $v_{BE} = 0.7$ V

$v_{EB} > V_{EBon}$; $V_{EBon} \cong 0.5$ V

Typically, $v_{EB} = 0.7$ V

2. CBJ Reversed Biased

$v_{BC} \leq V_{BCon}$; $V_{BCon} \cong 0.4$ V

$\Rightarrow v_{CE} \geq 0.3$ V

$v_{CB} \leq V_{CBon}$; $V_{CBon} \cong 0.4$ V

$\Rightarrow v_{EC} \geq 0.3$ V

Current-Voltage Relationships

▪ $i_C = I_S e^{v_{BE}/V_T}$

▪ $i_C = I_S e^{v_{EB}/V_T}$

▪ $i_B = i_C/\beta \quad \Leftrightarrow \quad i_C = \beta i_B$

▪ $i_E = i_C/\alpha \quad \Leftrightarrow \quad i_C = \alpha i_E$

▪ $\beta = \dfrac{\alpha}{1-\alpha} \quad \Leftrightarrow \quad \alpha = \dfrac{\beta}{\beta+1}$

Large-Signal Equivalent-Circuit Model (Including the Early Effect)

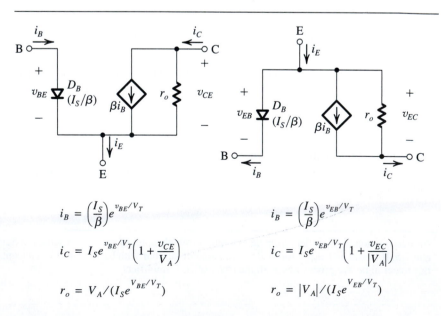

$i_B = \left(\dfrac{I_S}{\beta}\right)e^{v_{BE}/V_T}$

$i_C = I_S e^{v_{BE}/V_T}\left(1 + \dfrac{v_{CE}}{V_A}\right)$

$r_o = V_A/(I_S e^{v_{BE}/V_T})$

$i_B = \left(\dfrac{I_S}{\beta}\right)e^{v_{EB}/V_T}$

$i_C = I_S e^{v_{EB}/V_T}\left(1 + \dfrac{v_{EC}}{|V_A|}\right)$

$r_o = |V_A|/(I_S e^{v_{EB}/V_T})$

Ebers-Moll Model

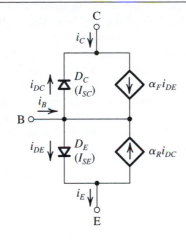
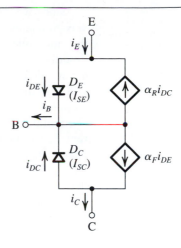

$$i_{DE} = I_{SE}(e^{v_{BE}/V_T} - 1)$$
$$i_{DC} = I_{SC}(e^{v_{BC}/V_T} - 1)$$

$$i_{DE} = I_{SE}(e^{v_{EB}/V_T} - 1)$$
$$i_{DC} = I_{SC}(e^{v_{CB}/V_T} - 1)$$

$$\alpha_F I_{SE} = \alpha_R I_{SC} = I_S$$

$$\frac{I_{SC}}{I_{SE}} = \frac{\alpha_F}{\alpha_R} = \frac{\text{CBJ Area}}{\text{EBJ Area}}$$

Operation in the Saturation Mode

Conditions:

1. EBJ Forward-Biased $v_{BE} > V_{BEon}; \; V_{BEon} \cong 0.5 \text{ V}$ $v_{EB} > V_{EBon}; \; V_{EBon} \cong 0.5 \text{ V}$

     Typically, $v_{BE} = 0.7\text{--}0.8 \text{ V}$ Typically, $v_{EB} = 0.7\text{--}0.8 \text{ V}$

2. CBJ Forward-Biased $v_{BC} \geq V_{BCon}; \; V_{BCon} \cong 0.4 \text{ V}$ $v_{CB} \geq V_{CBon}; \; V_{CBon} \cong 0.4 \text{ V}$

     Typically, $v_{BC} = 0.5\text{--}0.6 \text{ V}$ Typically, $v_{CB} = 0.5\text{--}0.6 \text{ V}$

       $\Rightarrow v_{CE} = V_{CEsat} = 0.1\text{--}0.2 \text{ V}$ $\Rightarrow v_{EC} = V_{ECsat} = 0.1\text{--}0.2 \text{ V}$

Currents

$$I_{Csat} = \beta_{forced} I_B$$

$$\beta_{forced} \leq \beta_F, \quad \frac{\beta_F}{\beta_{forced}} = \text{Overdrive factor}$$

Equivalent Circuits

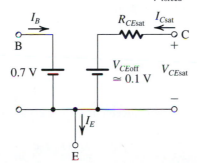
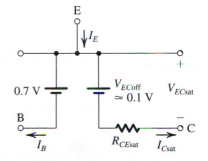

$$|V_{CEsat}| = V_T \ln \left[\frac{1 + (\beta_{forced} + 1)/\beta_F}{1 - \beta_{forced}/\beta_F} \right]$$

$$\text{For} \quad \beta_{forced} = \beta_F/2: \quad R_{CEsat} = 1/10\beta_F I_B$$

Since we are particularly interested in linear amplification, we will have to devise a way to achieve it in the face of the highly nonlinear behavior of the transistor, namely, that the collector current i_C is exponentially related to v_{BE}. We will use the approach described in general terms in Section 1.4. Specifically, we will **bias** the transistor to operate at a dc base–emitter voltage V_{BE} and a corresponding dc collector current I_C. Then we will superimpose the signal to be amplified, v_{be}, on the dc voltage V_{BE}. By keeping the amplitude of the signal v_{be} small, we will be able to constrain the transistor to operate on a short, almost linear segment of the i_C–v_{BE} characteristic; thus, the change in collector current, i_c, will be linearly related to v_{be}. We will study the small-signal operation of the BJT later in this section and in greater detail in Section 5.5. First, however, we will look at the "big picture": We will study the total or large-signal operation of a BJT amplifier. From the transfer characteristic of the circuit, we will be able to see clearly the region over which the circuit can be operated as a linear amplifier. We also will be able to see how the BJT can be employed as a switch.

5.3.1 Large-Signal Operation—The Transfer Characteristic

Figure 5.26(a) shows the basic structure (a skeleton) of the most commonly used BJT amplifier, the **grounded-emitter** or **common-emitter (CE)** circuit. The total input voltage v_I (bias + signal) is applied between base and emitter; that is, $v_{BE} = v_I$. The total output voltage v_O (bias + signal) is taken between collector and ground; that is, $v_O = v_{CE}$. Resistor R_C has two functions: to establish a desired dc bias voltage at the collector, and to convert the collector signal current i_c to an output voltage, v_{ce} or v_o. The supply voltage V_{CC} is needed to bias the BJT as well as to supply the power needed for the operation of the amplifier.

Figure 5.26(b) shows the voltage transfer characteristic of the CE circuit of Fig. 5.26(a). To understand how this characteristic arises, we first express v_O as

$$v_O = v_{CE} = V_{CC} - R_C i_C \tag{5.50}$$

Next, we observe that since $v_{BE} = v_I$, the transistor will be effectively cutoff for $v_I < 0.5$ V or so. Thus, for the range $0 < v_I < 0.5$ V, i_C will be negligibly small, and v_O will be equal to the supply voltage V_{CC} (segment XY of the transfer curve).

As v_I is increased above 0.5 V, the transistor begins to conduct, and i_C increases. From Eq. (5.50), we see that v_O decreases. However, since initially v_O will be large, the BJT will be operating in the active mode, which gives rise to the sharply descending segment YZ of the voltage transfer curve. The equation for this segment can be obtained by substituting in Eq. (5.50) the active-mode expression for i_C, namely,

$$i_C \cong I_S e^{v_{BE}/V_T}$$
$$= I_S e^{v_I/V_T}$$

where we have, for simplicity, neglected the Early effect. Thus we obtain

$$v_O = V_{CC} - R_C I_S e^{v_I/V_T} \tag{5.51}$$

We observe that the exponential term in this equation gives rise to the steep slope of the YZ segment of the transfer curve. Active-mode operation ends when the collector voltage (v_O or v_{CE}) falls by 0.4 V or so below that of the base (v_I or v_{BE}). At this point, the CBJ turns on, and the transistor enters the saturation region. This is indicated by point Z on the transfer curve.

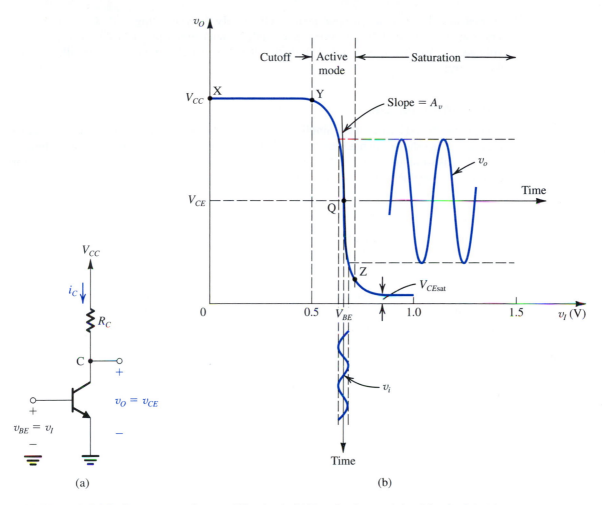

FIGURE 5.26 **(a)** Basic common-emitter amplifier circuit. **(b)** Transfer characteristic of the circuit in (a). The amplifier is biased at a point Q, and a small voltage signal v_i is superimposed on the dc bias voltage V_{BE}. The resulting output signal v_o appears superimposed on the dc collector voltage V_{CE}. The amplitude of v_o is larger than that of v_i by the voltage gain A_v.

Observe that a further increase in v_{BE} causes v_{CE} to decrease only slightly: In the saturation region, $v_{CE} = V_{CEsat}$, which falls in the narrow range of 0.1 V to 0.2 V. It is the almost-constant V_{CEsat} that gives this region of BJT operation the name *saturation*. The collector current will also remain nearly constant at the value I_{Csat},

$$I_{Csat} = \frac{V_{CC} - V_{CEsat}}{R_C} \tag{5.52}$$

We recall from our study of the saturation region of operation in the previous section that the saturated BJT exhibits a very small resistance R_{CEsat} between its collector and emitter. Thus, when saturated, the transistor in Fig. 5.26 provides a low-resistance path between the collector node C and ground and hence can be thought of as a closed switch. On the other hand, when the BJT is cut off, it conducts negligibly small (ideally zero) current and thus acts as an open switch, effectively disconnecting node C from ground. The status of the switch (i.e., open or closed) is determined by the value of the control voltage v_{BE}. Very shortly, we will show that the BJT switch can also be controlled by the base current.

5.3.2 Amplifier Gain

To operate the BJT as a linear amplifier, it must be biased at a point in the active region. Figure 5.26(b) shows such a bias point, labeled Q (for **quiescent point**), and characterized by a dc base–emitter voltage V_{BE} and a dc collector–emitter voltage V_{CE}. If the collector current at this value of V_{BE} is denoted I_C, that is,

$$I_C = I_S e^{V_{BE}/V_T} \tag{5.53}$$

then from the circuit in Fig. 5.26(a) we can write

$$V_{CE} = V_{CC} - R_C I_C \tag{5.54}$$

Now, if the signal to be amplified, v_i, is superimposed on V_{BE} and kept sufficiently small, as indicated in Fig. 5.26(b), the instantaneous operating point will be constrained to a relatively short, almost-linear segment of the transfer curve around the bias point Q. The slope of this linear segment will be equal to the slope of the tangent to the transfer curve at Q. This slope is the voltage gain of the amplifier for small-input signals around Q. An expression for the small-signal gain A_v can be found by differentiating the expression in Eq. (5.51) and evaluating the derivative at point Q; that is, for $v_I = V_{BE}$,

$$A_v \equiv \left. \frac{dv_O}{dv_I} \right|_{v_I = V_{BE}} \tag{5.55}$$

Thus,

$$A_v = -\frac{1}{V_T} I_S e^{V_{BE}/V_T} R_C$$

Now, using Eq. (5.53) we can express A_v in compact form:

$$A_v = -\frac{I_C R_C}{V_T} = -\frac{V_{RC}}{V_T} \tag{5.56}$$

where V_{RC} is the dc voltage drop across R_C,

$$V_{RC} = V_{CC} - V_{CE} \tag{5.57}$$

Observe that the CE amplifier is inverting; that is, the output signal is 180° out of phase relative to the input signal. The simple expression in Eq. (5.56) indicates that the voltage gain of the common-emitter amplifier is the ratio of the dc voltage drop across R_C to the thermal voltage V_T ($\cong 25$ mV at room temperature). It follows that to maximize the voltage gain we should use as large a voltage drop across R_C as possible. For a given value of V_{CC}, Eq. (5.57) indicates that to increase V_{RC} we have to operate at a lower V_{CE}. However, reference to Fig. 5.26(b) shows that a lower V_{CE} means a bias point Q close to the end of the active-region segment, which might not leave sufficient room for the negative-output signal swing without the amplifier entering the saturation region. If this happens, the negative peaks of the waveform of v_o will be flattened. Indeed, it is the need to allow sufficient room for output signal swing that determines the most effective placement of the bias point Q on the active-region segment, YZ, of the transfer curve. Placing Q too high on this segment not only results in reduced gain (because V_{RC} is lower) but could possibly limit the available range of positive signal swing. At the positive end, the limitation is imposed by the BJT cutting off, in which event the positive-output peaks would be clipped off at a level equal to V_{CC}. Finally, it is useful to note that the theoretical maximum gain A_v is obtained by biasing the BJT at the edge of saturation, which of course would not leave any room for negative signal swing. The resulting gain is given by

$$A_v = -\frac{V_{CC} - V_{CEsat}}{V_T} \tag{5.58}$$

Thus,

$$A_{vmax} \cong -\frac{V_{CC}}{V_T} \tag{5.59}$$

Although the gain can be increased by using a larger supply voltage, other considerations come into play when determining an appropriate value for V_{CC}. In fact, the trend has been toward using lower and lower supply voltages, currently approaching 1 V or so. At such low supply voltages, large gain values can be obtained by replacing the resistance R_C with a constant-current source, as will be seen in Chapter 6.

EXAMPLE 5.2

Consider a common-emitter circuit using a BJT having $I_S = 10^{-15}$ A, a collector resistance $R_C = 6.8$ kΩ, and a power supply $V_{CC} = 10$ V.

(a) Determine the value of the bias voltage V_{BE} required to operate the transistor at $V_{CE} = 3.2$ V. What is the corresponding value of I_C?

(b) Find the voltage gain A_v at this bias point. If an input sine-wave signal of 5-mV peak amplitude is superimposed on V_{BE}, find the amplitude of the output sine-wave signal (assume linear operation).

(c) Find the positive increment in v_{BE} (above V_{BE}) that drives the transistor to the edge of saturation, where $v_{CE} = 0.3$ V.

(d) Find the negative increment in v_{BE} that drives the transistor to within 1% of cutoff (i.e., to $v_O = 0.99V_{CC}$).

Solution

(a)

$$I_C = \frac{V_{CC} - V_{CE}}{R_C}$$

$$= \frac{10 - 3.2}{6.8} = 1 \text{ mA}$$

The value of V_{BE} can be determined from

$$1 \times 10^{-3} = 10^{-15} e^{V_{BE}/V_T}$$

which results in

$$V_{BE} = 690.8 \text{ mV}$$

(b)

$$A_v = -\frac{V_{CC} - V_{CE}}{V_T}$$

$$= -\frac{10 - 3.2}{0.025} = -272 \text{ V/V}$$

$$\hat{V}_o = 272 \times 0.005 = 1.36 \text{ V}$$

(c) For $v_{CE} = 0.3$ V,

$$i_C = \frac{10 - 0.3}{6.8} = 1.617 \text{ mA}$$

To increase i_C from 1 mA to 1.617 mA, v_{BE} must be increased by

$$\Delta v_{BE} = V_T \ln\left(\frac{1.617}{1}\right)$$

$$= 12 \text{ mV}$$

(d) For $v_O = 0.99 V_{CC} = 9.9$ V,

$$i_C = \frac{10 - 9.9}{6.8} = 0.0147 \text{ mA}$$

To decrease i_C from 1 mA to 0.0147 mA, v_{BE} must change by

$$\Delta v_{BE} = V_T \ln\left(\frac{0.0147}{1}\right)$$

$$= -105.5 \text{ mV}$$

EXERCISE

5.19 For the situation described in Example 5.2, while keeping I_C unchanged at 1 mA, find the value of R_C that will result in a voltage gain of -320 V/V. What is the largest negative signal swing allowed at the output (assume that v_{CE} is not to decrease below 0.3 V)? What (approximately) is the corresponding input signal amplitude? (Assume linear operation).

Ans. 8 kΩ; 1.7 V; 5.3 mV

5.3.3 Graphical Analysis

Although formal graphical methods are of little practical value in the analysis and design of most transistor circuits, it is illustrative to portray graphically the operation of a simple transistor amplifier circuit. Consider the circuit of Fig. 5.27, which is similar to the circuit we have been studying except for an added resistance in the base lead, R_B. A graphical analysis of the operation of this circuit can be performed as follows: First, we have to determine the dc bias point. Toward that end we set $v_i = 0$ and use the technique illustrated in Fig. 5.28 to determine the dc base current I_B. We next move to the $i_C–v_{CE}$ characteristics, shown in Fig. 5.29. We know that the operating point will lie on the $i_C–v_{CE}$ curve corresponding to the value of base current we have determined (the curve for $i_B = I_B$). Where it lies on the curve will be determined by the collector circuit. Specifically, the collector circuit imposes the constraint

$$v_{CE} = V_{CC} - i_C R_C$$

which can be rewritten as

$$i_C = \frac{V_{CC}}{R_C} - \frac{1}{R_C} v_{CE}$$

which represents a linear relationship between v_{CE} and i_C. This relationship can be represented by a straight line, as shown in Fig. 5.29. Since R_C can be considered the amplifier load,

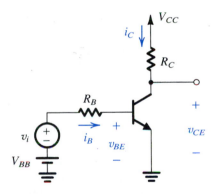

FIGURE 5.27 Circuit whose operation is to be analyzed graphically.

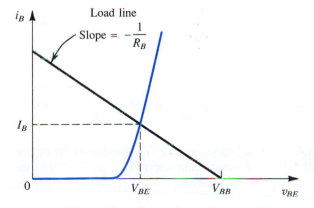

FIGURE 5.28 Graphical construction for the determination of the dc base current in the circuit of Fig. 5.27.

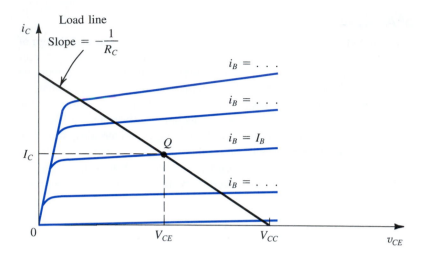

FIGURE 5.29 Graphical construction for determining the dc collector current I_C and the collector-to-emitter voltage V_{CE} in the circuit of Fig. 5.27.

the straight line of slope $-1/R_C$ is known as the load line.[7] The dc bias point, or quiescent point, Q will be at the intersection of the load line and the i_C–v_{CE} curve corresponding to the base current I_B. The coordinates of point Q give the dc collector current I_C and the dc collector-to-emitter voltage V_{CE}. Observe that for amplifier operation, Q should be in the active region and furthermore should be located so as to allow for a reasonable signal swing as the input signal v_i is applied. This will become clearer shortly.

The situation when v_i is applied is illustrated in Fig. 5.30. Consider first Fig. 5.30(a), which shows a signal v_i having a triangular waveform being superimposed on the dc voltage V_{BB}. Corresponding to each instantaneous value of $V_{BB} + v_i(t)$, one can draw a straight line with slope $-1/R_B$. Such an "instantaneous load line" intersects the i_B–v_{BE} curve at a point whose coordinates give the total instantaneous values of i_B and v_{BE} corresponding to the particular value of $V_{BB} + v_i(t)$. As an example, Fig. 5.30(a) shows the straight lines corresponding to $v_i = 0$, v_i at its positive peak, and v_i at its negative peak. Now, if the amplitude of v_i is sufficiently small so that the instantaneous operating point is confined to an almost-linear segment of the i_B–v_{BE} curve, then the resulting signals i_b and v_{be} will be triangular in waveform, as indicated in the figure. This, of course, is the small-signal approximation. In summary, the graphical construction in Fig. 5.30(a) can be used to determine the total instantaneous value of i_B corresponding to each value of v_i.

Next, we move to the i_C–v_{CE} characteristics of Fig. 5.30(b). The operating point will move along the load line of slope $-1/R_C$ as i_B goes through the instantaneous values determined from Fig. 5.30(a). For example, when v_i is at its positive peak, $i_B = i_{B2}$ (from Fig. 5.30(a)), and the instantaneous operating point in the i_C–v_{CE} plane will be at the intersection of the load line and the curve corresponding to $i_B = i_{B2}$. In this way, one can determine the waveforms of i_C and v_{CE} and hence of the signal components i_c and v_{ce}, as indicated in Fig. 5.30(b).

Effects of Bias-Point Location on Allowable Signal Swing The location of the dc bias point in the i_C–v_{CE} plane significantly affects the maximum allowable signal swing at the collector. Refer to Fig. 5.30(b) and observe that the positive peaks of v_{ce} cannot go beyond

---

[7] The term *load line* is also employed for the straight line in Fig. 5.28.

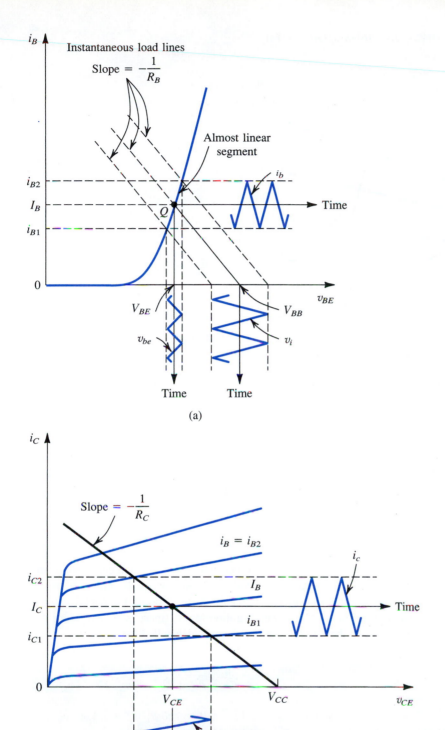

(a)

(b)

FIGURE 5.30 Graphical determination of the signal components v_{be}, i_b, i_c, and v_{ce} when a signal component v_i is superimposed on the dc voltage V_{BB} (see Fig. 5.27).

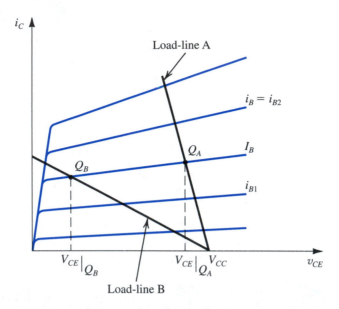

FIGURE 5.31 Effect of bias-point location on allowable signal swing: Load-line A results in bias point Q_A with a corresponding V_{CE} which is too close to V_{CC} and thus limits the positive swing of v_{CE}. At the other extreme, load-line B results in an operating point too close to the saturation region, thus limiting the negative swing of v_{CE}.

V_{CC}, otherwise the transistor enters the cutoff region. Similarly, the negative peaks of v_{ce} cannot extend below a few tenths of a volt (usually, 0.3 V), otherwise the transistor enters the saturation region. The location of the bias point in Fig. 5.30(b) allows for an approximately equal swing in each direction.

Next consider Fig. 5.31. Here we show load lines corresponding to two values of R_C. Line A corresponds to a low value of R_C and results in the operating point Q_A, where the value of V_{CE} is very close to V_{CC}. Thus the positive swing of v_{ce} will be severely limited; in this situation, it is said that there isn't sufficient "head room." On the other hand, line B, which corresponds to a large R_C, results in the bias point Q_B, whose V_{CE} is too low. Thus for line B, although there is ample room for the positive excursion of v_{ce} (there is a lot of head room), the negative signal swing is severely limited by the proximity to the saturation region (there is not sufficient "leg room"). A compromise between these two situations is obviously called for.

EXERCISE

5.20 Consider the circuit of Fig. 5.27 with $V_{BB} = 1.7$ V, $R_B = 100$ kΩ, $V_{CC} = 10$ V, and $R_C = 5$ kΩ. Let the transistor $\beta = 100$. The input signal v_i is a triangular wave of 0.4 V peak-to-peak. Refer to Fig. 5.30, and use the geometry of the graphical construction shown there to answer the following questions: (a) If $V_{BE} = 0.7$ V, find I_B. (b) Assuming operation on a straight line segment of the exponential i_B–v_{BE} curve, show that the inverse of its slope is V_T/I_B, and compute its value. (c) Find approximate values for the peak-to-peak amplitude of i_b and of v_{be}. (d) Assuming the i_C–v_{CE} curves to be horizontal (i.e., ignoring the Early effect), find I_C and V_{CE}. (e) Find the peak-to-peak amplitude of i_c and of v_{ce}. (f) What is the voltage gain of this amplifier?

Ans. (a) 10 μA; (b) 2.5 kΩ; (c) 4 μA, 10 mV; (d) 1 mA, 5 V; (e) 0.4 mA, 2 V; (f) –5 V/V

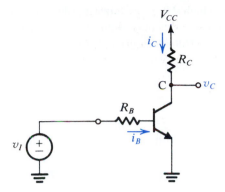

FIGURE 5.32 A simple circuit used to illustrate the different modes of operation of the BJT.

5.3.4 Operation as a Switch

To operate the BJT as a switch, we utilize the cutoff and the saturation modes of operation. To illustrate, consider once more the common-emitter circuit shown in Fig. 5.32 as the input v_I is varied. For v_I less than about 0.5 V, the transistor will be cut off; thus $i_B = 0$, $i_C = 0$, and $v_C = V_{CC}$. In this state, node C is disconnected from ground; the switch is in the open position.

To turn the transistor on, we have to increase v_I above 0.5 V. In fact, for appreciable currents to flow, v_{BE} should be about 0.7 V and v_I should be higher. The base current will be

$$i_B = \frac{v_I - V_{BE}}{R_B} \tag{5.60}$$

and the collector current will be

$$i_C = \beta i_B \tag{5.61}$$

which applies only when the device is in the active mode. This will be the case as long as the CBJ is not forward biased, that is, as long as $v_C > v_B - 0.4$ V, where v_C is given by

$$v_C = V_{CC} - R_C i_C \tag{5.62}$$

Obviously, as v_I is increased, i_B will increase (Eq. 5.60), i_C will correspondingly increase (Eq. 5.61), and v_C will decrease (Eq. 5.62). Eventually, v_C will become lower than v_B by 0.4 V, at which point the transistor leaves the active region and enters the saturation region. This **edge-of-saturation (EOS)** point is defined by

$$I_{C(EOS)} = \frac{V_{CC} - 0.3}{R_C} \tag{5.63}$$

where we have assumed that V_{BE} is approximately 0.7 V, and

$$I_{B(EOS)} = \frac{I_{C(EOS)}}{\beta} \tag{5.64}$$

The corresponding value of v_I required to drive the transistor to the edge-of-saturation can be found from

$$V_{I(EOS)} = I_{B(EOS)} R_B + V_{BE} \tag{5.65}$$

Increasing v_I above $V_{I(EOS)}$ increases the base current, which drives the transistor deeper into saturation. The collector-to-emitter voltage, however, decreases only slightly. As a reasonable approximation, we shall usually assume that for a saturated transistor, $V_{CEsat} \cong 0.2$ V. The collector current then remains nearly constant at I_{Csat},

$$I_{Csat} = \frac{V_{CC} - V_{CEsat}}{R_C} \tag{5.66}$$

Forcing more current into the base has very little effect on I_{Csat} and V_{CEsat}. In this state the switch is closed, with a low closure resistance R_{CEsat} and a small offset voltage V_{CEoff} (see Fig. 5.24c).

Finally, recall that in saturation one can force the transistor to operate at any desired β below the normal value; that is, the ratio of the collector current I_{Csat} to the base current can be set at will and is therefore called the forced β,

$$\beta_{forced} \equiv \frac{I_{Csat}}{I_B} \tag{5.67}$$

Also recall that the ratio of I_B to $I_{B(EOS)}$ is known as the overdrive factor.

EXAMPLE 5.3

The transistor in Fig. 5.33 is specified to have β in the range of 50 to 150. Find the value of R_B that results in saturation with an overdrive factor of at least 10.

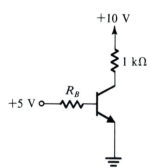

FIGURE 5.33   Circuit for Example 5.3.

Solution

When the transistor is saturated, the collector voltage will be

$$V_C = V_{CEsat} \cong 0.2 \text{ V}$$

Thus the collector current is given by

$$I_{Csat} = \frac{+10 - 0.2}{1} = 9.8 \text{ mA}$$

To saturate the transistor with the lowest β, we need to provide a base current of at least

$$I_{B(EOS)} = \frac{I_{Csat}}{\beta_{min}} = \frac{9.8}{50} = 0.196 \text{ mA}$$

For an overdrive factor of 10, the base current should be

$$I_B = 10 \times 0.196 = 1.96 \text{ mA}$$

Thus we require a value of R_B such that

$$\frac{+5 - 0.7}{R_B} = 1.96$$

$$R_B = \frac{4.3}{1.94} = 2.2 \text{ k}\Omega$$

EXERCISE

5.21 Consider the circuit in Fig. 5.32 for the case $V_{CC} = +5$ V, $v_I = +5$ V, $R_B = R_C = 1$ kΩ, and $\beta = 100$. Calculate the base current, the collector current, and the collector voltage. If the transistor is saturated, find β_{forced}. What value should R_B be raised to in order to bring the transistor to the edge of saturation?

Ans. 4.3 mA; 4.8 mA; 0.2 V; 1.1; 91.5 kΩ

 ## 5.4 BJT CIRCUITS AT DC

We are now ready to consider the analysis of BJT circuits to which only dc voltages are applied. In the following examples we will use the simple model in which, $|V_{BE}|$ of a conducting transistor is 0.7 V and $|V_{CE}|$ of a saturated transistor is 0.2 V, and we will neglect the Early effect. Better models can, of course, be used to obtain more accurate results. This, however, is usually achieved at the expense of speed of analysis, and more importantly, it could impede the circuit designer's ability to gain insight regarding circuit behavior. Accurate results using elaborate models can be obtained using circuit simulation with SPICE, as we shall see in Section 5.11. This is almost always done in the final stages of a design and certainly before circuit fabrication. Computer simulation, however, is not a substitute for quick pencil-and-paper circuit analysis, an essential ability that aspiring circuit designers must muster. The following series of examples is a step in that direction.

As will be seen, in analyzing a circuit the first question that one must answer is: In which mode is the transistor operating? In some cases, the answer will be obvious. In many cases, however, it will not. Needless to say, as the reader gains practice and experience in transistor circuit analysis and design, the answer will be obvious in a much larger proportion of problems. The answer, however, can always be determined by utilizing the following procedure:

Assume that the transistor is operating in the active mode, and proceed to determine the various voltages and currents that correspond. Then check for consistency of the results with the assumption of active-mode operation; that is, is v_{CB} of an *npn* transistor greater than −0.4 V (or v_{CB} of a *pnp* transistor lower than 0.4 V)? If the answer is yes, then our task is complete. If the answer is no, assume saturation-mode operation, and proceed to determine currents and voltages and then to check for consistency of the results with the assumption of saturation-mode operation. Here the test is usually to compute the ratio I_C/I_B and to verify that it is lower than the transistor β; i.e., $\beta_{forced} < \beta$. Since β for a given transistor varies over a wide range, one should use the lowest specified β for this test. Finally, note that the order of these two assumptions can be reversed.

Consider the circuit shown in Fig. 5.34(a), which is redrawn in Fig. 5.34(b) to remind the reader of the convention employed throughout this book for indicating connections to dc sources. We wish to analyze this circuit to determine all node voltages and branch currents. We will assume that β is specified to be 100.

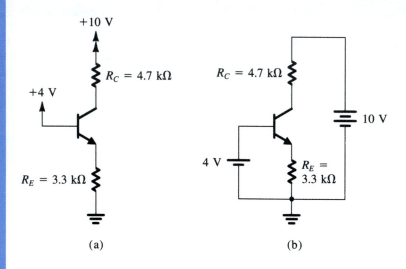

(a)   (b)

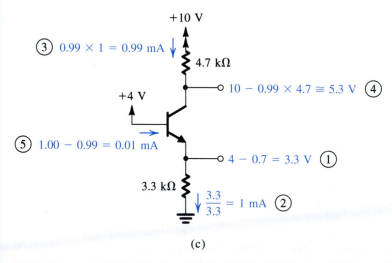

(c)

FIGURE 5.34 Analysis of the circuit for Example 5.4: **(a)** circuit; **(b)** circuit redrawn to remind the reader of the convention used in this book to show connections to the power supply; **(c)** analysis with the steps numbered.

Solution

Glancing at the circuit in Fig. 5.34(a), we note that the base is connected to +4 V and the emitter is connected to ground through a resistance R_E. It therefore is safe to conclude that the base–emitter junction will be forward biased. Assuming that this is the case and assuming that V_{BE} is approximately 0.7 V, it follows that the emitter voltage will be

$$V_E = 4 - V_{BE} \simeq 4 - 0.7 = 3.3 \text{ V}$$

We are now in an opportune position; we know the voltages at the two ends of R_E and thus can determine the current I_E through it,

$$I_E = \frac{V_E - 0}{R_E} = \frac{3.3}{3.3} = 1 \text{ mA}$$

Since the collector is connected through R_C to the +10-V power supply, it appears possible that the collector voltage will be higher than the base voltage, which is essential for active-mode operation. Assuming that this is the case, we can evaluate the collector current from

$$I_C = \alpha I_E$$

The value of α is obtained from

$$\alpha = \frac{\beta}{\beta + 1} = \frac{100}{101} \simeq 0.99$$

Thus I_C will be given by

$$I_C = 0.99 \times 1 = 0.99 \text{ mA}$$

We are now in a position to use Ohm's law to determine the collector voltage V_C,

$$V_C = 10 - I_C R_C = 10 - 0.99 \times 4.7 \simeq +5.3 \text{ V}$$

Since the base is at +4 V, the collector–base junction is reverse biased by 1.3 V, and the transistor is indeed in the active mode as assumed.

It remains only to determine the base current I_B, as follows:

$$I_B = \frac{I_E}{\beta + 1} = \frac{1}{101} \simeq 0.01 \text{ mA}$$

Before leaving this example we wish to emphasize strongly the value of carrying out the analysis directly on the circuit diagram. Only in this way will one be able to analyze complex circuits in a reasonable length of time. Figure 5.34(c) illustrates the above analysis on the circuit diagram, with the order of the analysis steps indicated by the circled numbers.

EXAMPLE 5.5

We wish to analyze the circuit of Fig. 5.35(a) to determine the voltages at all nodes and the currents through all branches. Note that this circuit is identical to that of Fig. 5.34 except that the voltage at the base is now +6 V. Assume that the transistor β is specified to be *at least* 50.

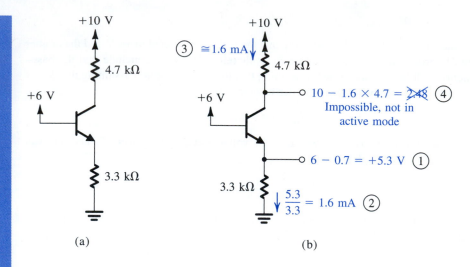

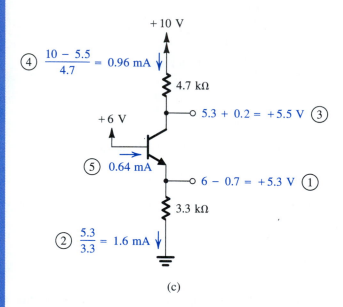

FIGURE 5.35 Analysis of the circuit for Example 5.5. Note that the circled numbers indicate the order of the analysis steps.

Solution

Assuming active-mode operation, we have

$$V_E = +6 - V_{BE} \simeq 6 - 0.7 = 5.3 \text{ V}$$

$$I_E = \frac{5.3}{3.3} = 1.6 \text{ mA}$$

$$V_C = +10 - 4.7 \times I_C \simeq 10 - 7.52 = 2.48 \text{ V}$$

The details of the analysis performed above are illustrated in Fig. 5.35(b).

Since the collector voltage calculated appears to be less than the base voltage by 3.52 V, it follows that our original assumption of active-mode operation is incorrect. In fact, the transistor has to be in the *saturation* mode. Assuming this to be the case, we have

$$V_E = +6 - 0.7 = +5.3 \text{ V}$$

$$I_E = \frac{V_E}{3.3} = \frac{5.3}{3.3} = 1.6 \text{ mA}$$

$$V_C = V_E + V_{CE\text{sat}} \simeq +5.3 + 0.2 = +5.5 \text{ V}$$

$$I_C = \frac{+10 - 5.5}{4.7} = 0.96 \text{ mA}$$

$$I_B = I_E - I_C = 1.6 - 0.96 = 0.64 \text{ mA}$$

Thus the transistor is operating at a forced β of

$$\beta_{\text{forced}} = \frac{I_C}{I_B} = \frac{0.96}{0.64} = 1.5$$

Since β_{forced} is less than the *minimum* specified value of β, the transistor is indeed saturated. We should emphasize here that in testing for saturation the minimum value of β should be used. By the same token, if we are designing a circuit in which a transistor is to be saturated, the design should be based on the minimum specified β. Obviously, if a transistor with this minimum β is saturated, then transistors with higher values of β will also be saturated. The details of the analysis are shown in Fig. 5.35(c), where the order of the steps used is indicated by the circled numbers.

EXAMPLE 5.6

We wish to analyze the circuit in Fig. 5.36(a) to determine the voltages at all nodes and the currents through all branches. Note that this circuit is identical to that considered in Examples 5.4 and 5.5 except that now the base voltage is zero.

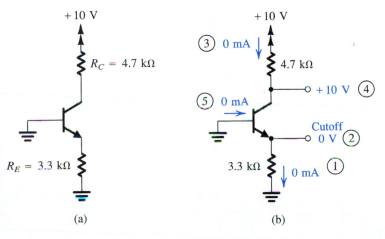

FIGURE 5.36 Example 5.6: **(a)** circuit; **(b)** analysis with the order of the analysis steps indicated by circled numbers.

Solution

Since the base is at zero volts and the emitter is connected to ground through R_E, the emitter–base junction cannot conduct and the emitter current is zero. Also, the collector–base junction cannot conduct since the n-type collector is connected through R_C to the positive power supply while the p-type base is at ground. It follows that the collector current will be zero. The base current will also have to be zero, and the transistor is in the *cutoff* mode of operation.

The emitter voltage will obviously be zero, while the collector voltage will be equal to +10 V, since the voltage drop across R_C is zero. Figure 5.36(b) shows the analysis details.

EXERCISES

D5.22 For the circuit in Fig. 5.34(a), find the highest voltage to which the base can be raised while the transistor remains in the active mode. Assume $\alpha \simeq 1$.
Ans. +4.7 V

D5.23 Redesign the circuit of Fig. 5.34(a) (i.e., find new values for R_E and R_C) to establish a collector current of 0.5 mA and a reverse-bias voltage on the collector–base junction of 2 V. Assume $\alpha \simeq 1$.
Ans. $R_E = 6.6$ kΩ; $R_C = 8$ kΩ

5.24 For the circuit in Fig. 5.35(a), find the value to which the base voltage should be changed to so that the transistor operates in saturation with a forced β of 5.
Ans. +5.18 V

EXAMPLE 5.7

We desire to analyze the circuit of Fig. 5.37(a) to determine the voltages at all nodes and the currents through all branches.

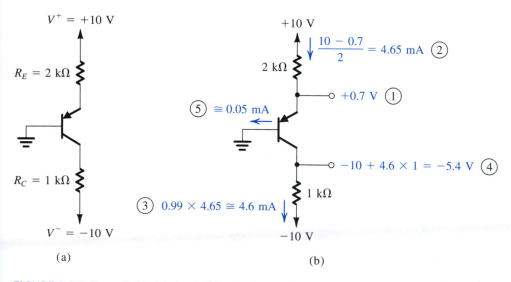

FIGURE 5.37 Example 5.7: **(a)** circuit; **(b)** analysis with the steps indicated by circled numbers.

Solution

The base of this *pnp* transistor is grounded, while the emitter is connected to a positive supply ($V^+ = +10$ V) through R_E. It follows that the emitter–base junction will be forward biased with

$$V_E = V_{EB} \simeq 0.7 \text{ V}$$

Thus the emitter current will be given by

$$I_E = \frac{V^+ - V_E}{R_E} = \frac{10 - 0.7}{2} = 4.65 \text{ mA}$$

Since the collector is connected to a negative supply (more negative than the base voltage) through R_C, it is *possible* that this transistor is operating in the active mode. Assuming this to be the case, we obtain

$$I_C = \alpha I_E$$

Since no value for β has been given, we shall assume $\beta = 100$, which results in $\alpha = 0.99$. Since large variations in β result in small differences in α, this assumption will not be critical as far as determining the value of I_C is concerned. Thus,

$$I_C = 0.99 \times 4.65 = 4.6 \text{ mA}$$

The collector voltage will be

$$V_C = V^- + I_C R_C$$
$$= -10 + 4.6 \times 1 = -5.4 \text{ V}$$

Thus the collector–base junction is reverse biased by 5.4 V, and the transistor is indeed in the active mode, which supports our original assumption.

It remains only to calculate the base current,

$$I_B = \frac{I_E}{\beta + 1} = \frac{4.65}{101} \simeq 0.05 \text{ mA}$$

Obviously, the value of β critically affects the base current. Note, however, that in this circuit the value of β will have no effect on the mode of operation of the transistor. Since β is generally an ill-specified parameter, this circuit represents a good design. As a rule, one should strive to *design the circuit such that its performance is as insensitive to the value of β as possible*. The analysis details are illustrated in Fig. 5.37(b).

EXERCISES

D5.25 For the circuit in Fig. 5.37(a), find the largest value to which R_C can be raised while the transistor remains in the active mode.
 Ans. 2.26 kΩ

D5.26 Redesign the circuit of Fig. 5.37(a) (i.e., find new values for R_E and R_C) to establish a collector current of 1 mA and a reverse bias on the collector–base junction of 4 V. Assume $\alpha \simeq 1$.
 Ans. $R_E = 9.3$ kΩ; $R_C = 6$ kΩ

EXAMPLE 5.8

We want to analyze the circuit in Fig. 5.38(a) to determine the voltages at all nodes and the currents in all branches. Assume $\beta = 100$.

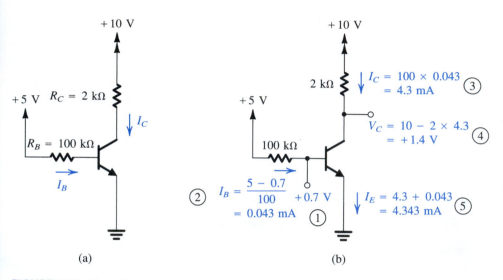

FIGURE 5.38 Example 5.8: **(a)** circuit; **(b)** analysis with the steps indicated by the circled numbers.

Solution

The base–emitter junction is clearly forward biased. Thus,

$$I_B = \frac{+5 - V_{BE}}{R_B} \simeq \frac{5 - 0.7}{100} = 0.043 \text{ mA}$$

Assume that the transistor is operating in the active mode. We now can write

$$I_C = \beta I_B = 100 \times 0.043 = 4.3 \text{ mA}$$

The collector voltage can now be determined as

$$V_C = +10 - I_C R_C = 10 - 4.3 \times 2 = +1.4 \text{ V}$$

Since the base voltage V_B is

$$V_B = V_{BE} \simeq +0.7 \text{ V}$$

it follows that the collector–base junction is reverse-biased by 0.7 V and the transistor is indeed in the active mode. The emitter current will be given by

$$I_E = (\beta + 1)I_B = 101 \times 0.043 \simeq 4.3 \text{ mA}$$

We note from this example that the collector and emitter currents depend critically on the value of β. In fact, if β were 10% higher, the transistor would leave the active mode and enter saturation. Therefore this clearly is a *bad* design. The analysis details are illustrated in Fig. 5.38(b).

D5.27 The circuit of Fig. 5.38(a) is to be fabricated using a transistor type whose β is specified to be in the range of 50 to 150. That is, individual units of this same transistor type can have β values anywhere in this range. Redesign the circuit by selecting a new value for R_C so that all fabricated circuits are guaranteed to be in the active mode. What is the range of collector voltages that the fabricated circuits may exhibit?

Ans. $R_C = 1.5$ kΩ; $V_C = 0.3$ V to 6.8 V

EXAMPLE 5.9

We want to analyze the circuit of Fig. 5.39 to determine the voltages at all nodes and the currents through all branches. The minimum value of β is specified to be 30.

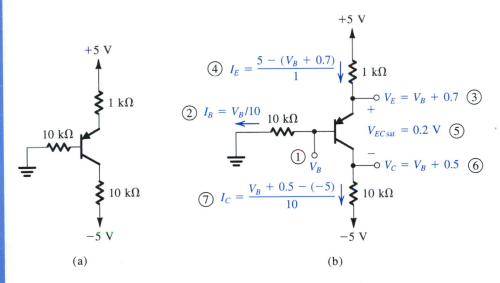

FIGURE 5.39 Example 5.9: **(a)** circuit; **(b)** analysis with steps numbered.

Solution

A quick glance at this circuit reveals that the transistor will be either active or saturated. Assuming active-mode operation and neglecting the base current, we see that the base voltage will be approximately zero volts, the emitter voltage will be approximately +0.7 V, and the emitter current will be approximately 4.3 mA. Since the maximum current that the collector can support while the transistor remains in the active mode is approximately 0.5 mA, it follows that the transistor is definitely saturated.

Assuming that the transistor is saturated and denoting the voltage at the base by V_B (refer to Fig. 5.39b), it follows that

$$V_E = V_B + V_{EB} \simeq V_B + 0.7$$

$$V_C = V_E - V_{EC\,sat} \simeq V_B + 0.7 - 0.2 = V_B + 0.5$$

$$I_E = \frac{+5 - V_E}{1} = \frac{5 - V_B - 0.7}{1} = 4.3 - V_B \text{ mA}$$

$$I_B = \frac{V_B}{10} = 0.1 V_B \text{ mA}$$

$$I_C = \frac{V_C - (-5)}{10} = \frac{V_B + 0.5 + 5}{10} = 0.1 V_B + 0.55 \text{ mA}$$

Using the relationship $I_E = I_B + I_C$, we obtain

$$4.3 - V_B = 0.1 V_B + 0.1 V_B + 0.55$$

which results in

$$V_B = \frac{3.75}{1.2} \simeq 3.13 \text{ V}$$

Substituting in the equations above, we obtain

$$V_E = 3.83 \text{ V}$$
$$V_C = 3.63 \text{ V}$$
$$I_E = 1.17 \text{ mA}$$
$$I_C = 0.86 \text{ mA}$$
$$I_B = 0.31 \text{ mA}$$

It is clear that the transistor is saturated, since the value of forced β is

$$\beta_{\text{forced}} = \frac{0.86}{0.31} \simeq 2.8$$

which is much smaller than the specified minimum β.

EXAMPLE 5.10

We want to analyze the circuit of Fig. 5.40(a) to determine the voltages at all nodes and the currents through all branches. Assume $\beta = 100$.

Solution

The first step in the analysis consists of simplifying the base circuit using Thévenin's theorem. The result is shown in Fig. 5.40(b), where

$$V_{BB} = +15 \frac{R_{B2}}{R_{B1} + R_{B2}} = 15 \frac{50}{100 + 50} = +5 \text{ V}$$

$$R_{BB} = (R_{B1} \mathbin{/\mkern-5mu/} R_{B2}) = (100 \mathbin{/\mkern-5mu/} 50) = 33.3 \text{ k}\Omega$$

To evaluate the base or the emitter current, we have to write a loop equation around the loop marked L in Fig. 5.40(b). Note, though, that the current through R_{BB} is different from the current through R_E. The loop equation will be

$$V_{BB} = I_B R_{BB} + V_{BE} + I_E R_E$$

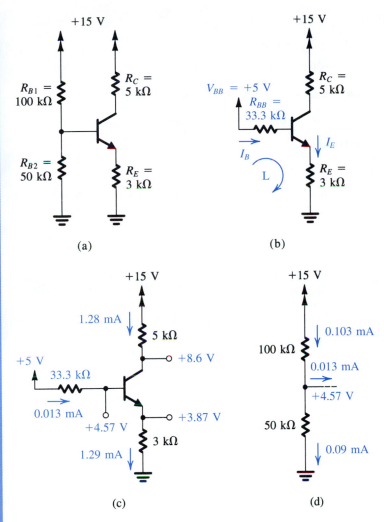

FIGURE 5.40 Circuits for Example 5.10.

Substituting for I_B by

$$I_B = \frac{I_E}{\beta + 1}$$

and rearranging the equation gives

$$I_E = \frac{V_{BB} - V_{BE}}{R_E + [R_{BB}/(\beta + 1)]}$$

For the numerical values given we have

$$I_E = \frac{5 - 0.7}{3 + (33.3/101)} = 1.29 \text{ mA}$$

The base current will be

$$I_B = \frac{1.29}{101} = 0.0128 \text{ mA}$$

The base voltage is given by

$$V_B = V_{BE} + I_E R_E$$

$$= 0.7 + 1.29 \times 3 = 4.57 \text{ V}$$

Assume active-mode operation. We can evaluate the collector current as

$$I_C = \alpha I_E = 0.99 \times 1.29 = 1.28 \text{ mA}$$

The collector voltage can now be evaluated as

$$V_C = +15 - I_C R_C = 15 - 1.28 \times 5 = 8.6 \text{ V}$$

It follows that the collector is higher in potential than the base by 4.03 V, which means that the transistor is in the active mode, as had been assumed. The results of the analysis are given in Figs. 5.40(c and d).

EXERCISE

5.28 If the transistor in the circuit of Fig. 5.40(a) is replaced with another having half the value of β (i.e., $\beta = 50$), find the new value of I_C, and express the change in I_C as a percentage.
Ans. $I_C = 1.15$ mA; -10%

EXAMPLE 5.11

We wish to analyze the circuit in Fig. 5.41(a) to determine the voltages at all nodes and the currents through all branches.

Solution

We first recognize that part of this circuit is identical to the circuit we analyzed in Example 5.10—namely, the circuit of Fig. 5.40(a). The difference, of course, is that in the new circuit we have an additional transistor Q_2 together with its associated resistors R_{E2} and R_{C2}. Assume that Q_1 is still in the active mode. The following values will be identical to those obtained in the previous example:

$$V_{B1} = +4.57 \text{ V} \qquad I_{E1} = 1.29 \text{ mA}$$

$$I_{B1} = 0.0128 \text{ mA} \qquad I_{C1} = 1.28 \text{ mA}$$

However, the collector voltage will be different than previously calculated since part of the collector current I_{C1} will flow in the base lead of Q_2 (I_{B2}). As a first approximation we may assume that I_{B2} is much smaller than I_{C1}; that is, we may assume that the current through R_{C1} is almost equal to I_{C1}. This will enable us to calculate V_{C1}:

$$V_{C1} \simeq +15 - I_{C1} R_{C1}$$

$$= 15 - 1.28 \times 5 = +8.6 \text{ V}$$

Thus Q_1 is in the active mode, as had been assumed.

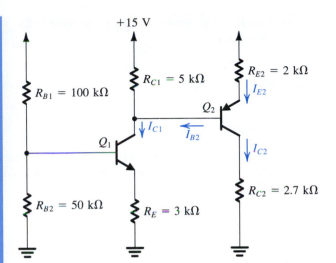

(a)

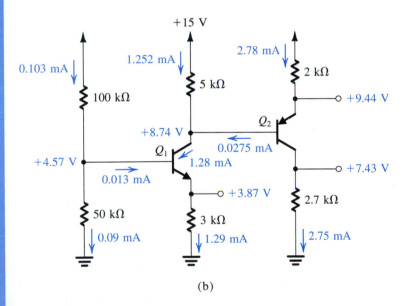

(b)

FIGURE 5.41   Circuits for Example 5.11.

As far as Q_2 is concerned, we note that its emitter is connected to +15 V through R_{E2}. It is therefore safe to assume that the emitter–base junction of Q_2 will be forward biased. Thus the emitter of Q_2 will be at a voltage V_{E2} given by

$$V_{E2} = V_{C1} + V_{EB}\big|_{Q_2} \simeq 8.6 + 0.7 = +9.3 \text{ V}$$

The emitter current of Q_2 may now be calculated as

$$I_{E2} = \frac{+15 - V_{E2}}{R_{E2}} = \frac{15 - 9.3}{2} = 2.85 \text{ mA}$$

Since the collector of Q_2 is returned to ground via R_{C2}, it is possible that Q_2 is operating in the active mode. Assume this to be the case. We now find I_{C2} as

$$I_{C2} = \alpha_2 I_{E2}$$
$$= 0.99 \times 2.85 = 2.82 \text{ mA} \quad (\text{assuming } \beta_2 = 100)$$

The collector voltage of Q_2 will be

$$V_{C2} = I_{C2} R_{C2} = 2.82 \times 2.7 = 7.62 \text{ V}$$

which is lower than V_{B2} by 0.98 V. Thus Q_2 is in the active mode, as assumed.

It is important at this stage to find the magnitude of the error incurred in our calculations by the assumption that I_{B2} is negligible. The value of I_{B2} is given by

$$I_{B2} = \frac{I_{E2}}{\beta_2 + 1} = \frac{2.85}{101} = 0.028 \text{ mA}$$

which is indeed much smaller than I_{C1} (1.28 mA). If desired, we can obtain more accurate results by iterating one more time, assuming I_{B2} to be 0.028 mA. The new values will be

$$\text{Current in } R_{C1} = I_{C1} - I_{B2} = 1.28 - 0.028 = 1.252 \text{ mA}$$

$$V_{C1} = 15 - 5 \times 1.252 = 8.74 \text{ V}$$

$$V_{E2} = 8.74 + 0.7 = 9.44 \text{ V}$$

$$I_{E2} = \frac{15 - 9.44}{2} = 2.78 \text{ mA}$$

$$I_{C2} = 0.99 \times 2.78 = 2.75 \text{ mA}$$

$$V_{C2} = 2.75 \times 2.7 = 7.43 \text{ V}$$

$$I_{B2} = \frac{2.78}{101} = 0.0275 \text{ mA}$$

Note that the new value of I_{B2} is very close to the value used in our iteration, and no further iterations are warranted. The final results are indicated in Fig. 5.41(b).

The reader justifiably might be wondering about the necessity for using an iterative scheme in solving a linear (or linearized) problem. Indeed, we can obtain the exact solution (if we can call anything we are doing with a first-order model exact!) by writing appropriate equations. The reader is encouraged to find this solution and then compare the results with those obtained above. It is important to emphasize, however, that in most such problems it is quite sufficient to obtain an approximate solution, provided that we can obtain it quickly and, of course, correctly.

In the above examples, we frequently used a precise value of α to calculate the collector current. Since $\alpha \simeq 1$, the error in such calculations will be very small if one assumes $\alpha = 1$ and $i_C = i_E$. Therefore, except in calculations that depend critically on the value of α (e.g., the calculation of base current), one usually assumes $\alpha \simeq 1$.

EXERCISES

5.29 For the circuit in Fig. 5.41, find the total current drawn from the power supply. Hence find the power dissipated in the circuit.
Ans. 4.135 mA; 62 mW

5.30 The circuit in Fig. E5.30 is to be connected to the circuit in Fig. 5.41(a) as indicated; specifically, the base of Q_3 is to be connected to the collector of Q_2. If Q_3 has $\beta = 100$, find the new value of V_{C2} and the values of V_{E3} and I_{C3}.

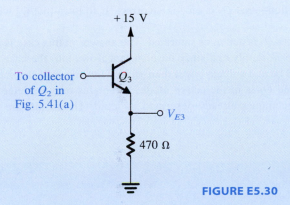

FIGURE E5.30

Ans. +7.06 V; +6.36 V; 13.4 mA

EXAMPLE 5.12

We desire to evaluate the voltages at all nodes and the currents through all branches in the circuit of Fig. 5.42(a). Assume $\beta = 100$.

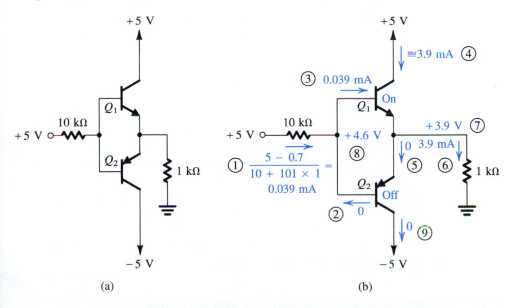

(a)

(b)

FIGURE 5.42 Example 5.12: **(a)** circuit; **(b)** analysis with the steps numbered.

Solution

By examining the circuit we conclude that the two transistors Q_1 and Q_2 cannot be simultaneously conducting. Thus if Q_1 is on, Q_2 will be off, and vice versa. Assume that Q_2 is on. It follows that current will flow from ground through the 1-kΩ load resistor into the emitter of Q_2. Thus the base of Q_2 will be at a negative voltage, and base current will be flowing out of the base through the 10-kΩ resistor and into the +5-V supply. This is impossible, since if the base is negative, current in the 10-kΩ resistor will have to flow into the base. Thus we conclude that our original assumption—that Q_2 is on—is incorrect. It follows that Q_2 will be off and Q_1 will be on.

The question now is whether Q_1 is active or saturated. The answer in this case is obvious. Since the base is fed with a +5-V supply and since base current flows into the base of Q_1, it follows that the base of Q_1 will be at a voltage lower than +5 V. Thus the collector–base junction of Q_1 is reverse biased and Q_1 is in the active mode. It remains only to determine the currents and voltages using techniques already described in detail. The results are given in Fig. 5.42(b).

EXERCISE

5.31 Solve the problem in Example 5.12 with the voltage feeding the bases changed to +10 V. Assume that $\beta_{min} = 30$, and find V_E, V_B, I_{C1}, and I_{C2}.
 Ans. +4.8 V; +5.5 V; 4.35 mA; 0

 ## 5.5 BIASING IN BJT AMPLIFIER CIRCUITS

The biasing problem is that of establishing a constant dc current in the collector of the BJT. This current has to be calculable, predictable, and insensitive to variations in temperature and to the large variations in the value of β encountered among transistors of the same type. Another important consideration in bias design is locating the dc bias point in the i_C–v_{CE} plane to allow for maximum output signal swing (see the discussion in Section 5.3.3). In this section, we shall deal with various approaches to solving the bias problem in transistor circuits designed with discrete devices. Bias methods for integrated-circuit design are presented in Chapter 6.

Before presenting the "good" biasing schemes, we should point out why two obvious arrangements are not good. First, attempting to bias the BJT by fixing the voltage V_{BE} by, for instance, using a voltage divider across the power supply V_{CC}, as shown in Fig. 5.43(a), is not a viable approach: The very sharp exponential relationship i_C–v_{BE} means that any small and inevitable differences in V_{BE} from the desired value will result in large differences in I_C and in V_{CE}. Second, biasing the BJT by establishing a constant current in the base, as shown in Fig. 5.43(b), where $I_B \cong (V_{CC} - 0.7)/R_B$, is also not a recommended approach. Here the typically large variations in the value of β among units of the same device type will result in correspondingly large variations in I_C and hence in V_{CE}.

5.5.1 The Classical Discrete-Circuit Bias Arrangement

Figure 5.44(a) shows the arrangement most commonly used for biasing a discrete-circuit transistor amplifier if only a single power supply is available. The technique consists of supplying the base of the transistor with a fraction of the supply voltage V_{CC} through the voltage divider R_1, R_2. In addition, a resistor R_E is connected to the emitter.

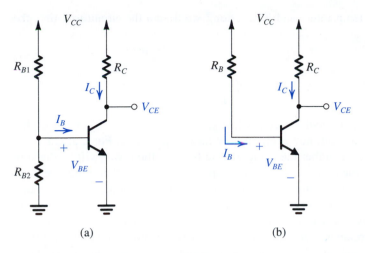

(a)   (b)

FIGURE 5.43 Two obvious schemes for biasing the BJT: **(a)** by fixing V_{BE}; **(b)** by fixing I_B. Both result in wide variations in I_C and hence in V_{CE} and therefore are considered to be "bad." Neither scheme is recommended.

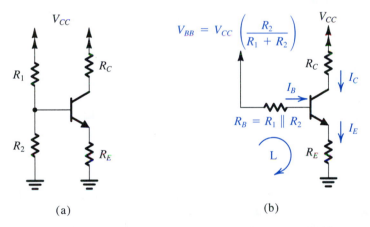

(a)   (b)

FIGURE 5.44 Classical biasing for BJTs using a single power supply: **(a)** circuit; **(b)** circuit with the voltage divider supplying the base replaced with its Thévenin equivalent.

Figure 5.44(b) shows the same circuit with the voltage-divider network replaced by its Thévenin equivalent,

$$V_{BB} = \frac{R_2}{R_1 + R_2} V_{CC} \tag{5.68}$$

$$R_B = \frac{R_1 R_2}{R_1 + R_2} \tag{5.69}$$

The current I_E can be determined by writing a Kirchhoff loop equation for the base–emitter–ground loop, labeled L, and substituting $I_B = I_E/(\beta + 1)$:

$$I_E = \frac{V_{BB} - V_{BE}}{R_E + R_B/(\beta + 1)} \tag{5.70}$$

To make I_E insensitive to temperature and β variation,[8] we design the circuit to satisfy the following two constraints:

$$V_{BB} \gg V_{BE} \tag{5.71}$$

$$R_E \gg \frac{R_B}{\beta + 1} \tag{5.72}$$

Condition (5.71) ensures that small variations in V_{BE} (≈ 0.7 V) will be swamped by the much larger V_{BB}. There is a limit, however, on how large V_{BB} can be: For a given value of the supply voltage V_{CC}, the higher the value we use for V_{BB}, the lower will be the sum of voltages across R_C and the collector–base junction (V_{CB}). On the other hand, we want the voltage across R_C to be large in order to obtain high voltage gain and large signal swing (before transistor cutoff). We also want V_{CB} (or V_{CE}) to be large to provide a large signal swing (before transistor saturation). Thus, as is the case in any design, we have a set of conflicting requirements, and the solution must be a compromise. As a rule of thumb, one designs for V_{BB} about $\frac{1}{3}V_{CC}$, V_{CB} (or V_{CE}) about $\frac{1}{3}V_{CC}$, and $I_C R_C$ about $\frac{1}{3}V_{CC}$.

Condition (5.72) makes I_E insensitive to variations in β and could be satisfied by selecting R_B small. This in turn is achieved by using low values for R_1 and R_2. Lower values for R_1 and R_2, however, will mean a higher current drain from the power supply, and will result in a lowering of the input resistance of the amplifier (if the input signal is coupled to the base), which is the trade-off involved in this part of the design. It should be noted that Condition (5.72) means that we want to make the base voltage independent of the value of β and determined solely by the voltage divider. This will obviously be satisfied if the current in the divider is made much larger than the base current. Typically one selects R_1 and R_2 such that their current is in the range of I_E to $0.1 I_E$.

Further insight regarding the mechanism by which the bias arrangement of Fig. 5.44(a) stabilizes the dc emitter (and hence collector) current is obtained by considering the feedback action provided by R_E. Consider that for some reason the emitter current increases. The voltage drop across R_E, and hence V_E will increase correspondingly. Now, if the base voltage is determined primarily by the voltage divider R_1, R_2, which is the case if R_B is small, it will remain constant, and the increase in V_E will result in a corresponding decrease in V_{BE}. This in turn reduces the collector (and emitter) current, a change opposite to that originally assumed. Thus R_E provides a *negative feedback* action that stabilizes the bias current. We shall study negative feedback formally in Chapter 8.

EXAMPLE 5.13

We wish to design the bias network of the amplifier in Fig. 5.44 to establish a current $I_E = 1$ mA using a power supply $V_{CC} = +12$ V. The transistor is specified to have a nominal β value of 100.

Solution

We shall follow the rule of thumb mentioned above and allocate one-third of the supply voltage to the voltage drop across R_2 and another one-third to the voltage drop across R_C, leaving one-third

---

[8] Bias design seeks to stabilize either I_E or I_C since $I_C = \alpha I_E$ and α varies very little. That is, a stable I_E will result in an equally stable I_C, and vice versa.

for possible signal swing at the collector. Thus,

$$V_B = +4 \text{ V}$$

$$V_E = 4 - V_{BE} \simeq 3.3 \text{ V}$$

and R_E is determined from

$$R_E = \frac{V_E}{I_E} = \frac{3.3}{1} = 3.3 \text{ k}\Omega$$

From the discussion above we select a voltage-divider current of $0.1I_E = 0.1 \times 1 = 0.1$ mA. Neglecting the base current, we find

$$R_1 + R_2 = \frac{12}{0.1} = 120 \text{ k}\Omega$$

and

$$\frac{R_2}{R_1 + R_2}V_{CC} = 4 \text{ V}$$

Thus $R_2 = 40$ kΩ and $R_1 = 80$ kΩ.

At this point, it is desirable to find a more accurate estimate for I_E, taking into account the nonzero base current. Using Eq. (5.70),

$$I_E = \frac{4 - 0.7}{3.3(\text{k}\Omega) + \dfrac{(80 \,//\, 40)(\text{k}\Omega)}{101}} = 0.93 \text{ mA}$$

This is quite a bit lower than the value we are aiming for of 1 mA. It is easy to see from the above equation that a simple way to restore I_E to its nominal value would be to reduce R_E from 3.3 kΩ by the magnitude of the second term in the denominator (0.267 kΩ). Thus a more suitable value for R_E in this case would be $R_E = 3$ kΩ, which results in $I_E = 1.01$ mA $\simeq 1$ mA.

It should be noted that if we are willing to draw a higher current from the power supply and to accept a lower input resistance for the amplifier, then we may use a voltage-divider current equal, say, to I_E (i.e., 1 mA), resulting in $R_1 = 8$ kΩ and $R_2 = 4$ kΩ. We shall refer to the circuit using these latter values as design 2, for which the actual value of I_E using the initial value of R_E of 3.3 kΩ will be

$$I_E = \frac{4 - 0.7}{3.3 + 0.027} = 0.99 \simeq 1 \text{ mA}$$

In this case, design 2, we need not change the value of R_E.

Finally, the value of R_C can be determined from

$$R_C = \frac{12 - V_C}{I_C}$$

Substituting $I_C = \alpha I_E = 0.99 \times 1 = 0.99$ mA $\simeq 1$ mA results, for both designs, in

$$R_C = \frac{12 - 8}{1} = 4 \text{ k}\Omega$$

5.32 For design 1 in Example 5.13, calculate the expected range of I_E if the transistor used has β in the range of 50 to 150. Express the range of I_E as a percentage of the nominal value ($I_E \simeq 1$ mA) obtained for $\beta = 100$. Repeat for design 2.

Ans. For design 1: 0.94 mA to 1.04 mA a 10% range; for design 2: 0.984 mA to 0.995 mA, a 1.1% range.

5.5.2 A Two-Power-Supply Version of the Classical Bias Arrangement

A somewhat simpler bias arrangement is possible if two power supplies are available, as shown in Fig. 5.45. Writing a loop equation for the loop labeled L gives

$$I_E = \frac{V_{EE} - V_{BE}}{R_E + R_B/(\beta + 1)} \qquad (5.73)$$

This equation is identical to Eq. (5.70) except for V_{EE} replacing V_{BB}. Thus the two constraints of Eqs. (5.71) and (5.72) apply here as well. Note that if the transistor is to be used with the base grounded (i.e., in the common-base configuration), then R_B can be eliminated altogether. On the other hand, if the input signal is to be coupled to the base, then R_B is needed. We shall study the various BJT amplifier configurations in Section 5.7.

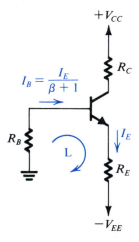

FIGURE 5.45 Biasing the BJT using two power supplies. Resistor R_B is needed only if the signal is to be capacitively coupled to the base. Otherwise, the base can be connected directly to ground, or to a grounded signal source, resulting in almost total β-independence of the bias current.

D5.33 The bias arrangement of Fig. 5.45 is to be used for a common-base amplifier. Design the circuit to establish a dc emitter current of 1 mA and provide the highest possible voltage gain while allowing for a maximum signal swing at the collector of ±2 V. Use +10-V and −5-V power supplies.

Ans. $R_B = 0$; $R_E = 4.3$ kΩ; $R_C = 8.4$ kΩ

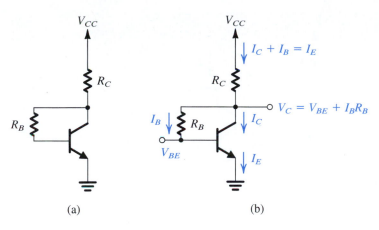

FIGURE 5.46 (a) A common-emitter transistor amplifier biased by a feedback resistor R_B. (b) Analysis of the circuit in (a).

5.5.3 Biasing Using a Collector-to-Base Feedback Resistor

Figure 5.46(a) shows a simple but effective alternative biasing arrangement suitable for common-emitter amplifiers. The circuit employs a resistor R_B connected between the collector and the base. Resistor R_B provides negative feedback, which helps to stabilize the bias point of the BJT. We shall study feedback formally in Chapter 8.

Analysis of the circuit is shown in Fig. 5.46(b), from which we can write

$$V_{CC} = I_E R_C + I_B R_B + V_{BE}$$

$$= I_E R_C + \frac{I_E}{\beta+1} R_B + V_{BE}$$

Thus the emitter bias current is given by

$$I_E = \frac{V_{CC} - V_{BE}}{R_C + R_B/(\beta+1)} \tag{5.74}$$

It is interesting to note that this equation is identical to Eq. (5.70), which governs the operation of the traditional bias circuit, except that V_{CC} replaces V_{BB} and R_C replaces R_E. It follow that to obtain a value of I_E that is insensitive to variation of β, we select $R_B/(\beta+1) \ll R_C$. Note, however, that the value of R_B determines the allowable signal swing at the collector since

$$V_{CB} = I_B R_B = I_E \frac{R_B}{\beta+1} \tag{5.75}$$

EXERCISE

D5.34 Design the circuit of Fig. 5.46 to obtain a dc emitter current of 1 mA and to ensure a ±2-V signal swing at the collector; that is, design for $V_{CE} = +2.3$ V. Let $V_{CC} = 10$ V and $\beta = 100$.
Ans. $R_B = 162$ kΩ; $R_C = 7.7$ kΩ. Note that if standard 5% resistor values are used (Appendix G) we select $R_B = 160$ kΩ and $R_C = 7.5$ kΩ. This results in $I_E = 1.02$ mA and $V_C = +2.3$ V.

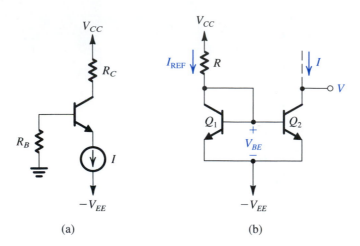

FIGURE 5.47 **(a)** A BJT biased using a constant-current source I. **(b)** Circuit for implementing the current source I.

5.5.4 Biasing Using a Constant-Current Source

The BJT can be biased using a constant-current source I as indicated in the circuit of Fig. 5.47(a). This circuit has the advantage that the emitter current is independent of the values of β and R_B. Thus R_B can be made large, enabling an increase in the input resistance at the base without adversely affecting bias stability. Further, current-source biasing leads to significant design simplification, as will become obvious in later sections and chapters.

A simple implementation of the constant-current source I is shown in Fig. 4.47(b). The circuit utilizes a pair of matched transistors Q_1 and Q_2, with Q_1 connected as a diode by shorting its collector to its base. If we assume that Q_1 and Q_2 have high β values, we can neglect their base currents. Thus the current through Q_1 will be approximately equal to I_{REF},

$$I_{\text{REF}} = \frac{V_{CC} - (-V_{EE}) - V_{BE}}{R} \qquad (5.76)$$

Now, since Q_1 and Q_2 have the same V_{BE}, their collector currents will be equal, resulting in

$$I = I_{\text{REF}} = \frac{V_{CC} + V_{EE} - V_{BE}}{R} \qquad (5.77)$$

Neglecting the Early effect in Q_2, the collector current will remain constant at the value given by this equation as long as Q_2 remains in the active region. This can be guaranteed by keeping the voltage at the collector, V, greater than that at the base $(-V_{EE} + V_{BE})$. The connection of Q_1 and Q_2 in Fig. 4.47(b) is known as a **current mirror.** We will study current mirrors in detail in Chapter 6.

EXERCISE

5.35 For the circuit in Fig. 5.47(a) with $V_{CC} = 10$ V, $I = 1$ mA, $\beta = 100$, $R_B = 100$ kΩ, and $R_C = 7.5$ kΩ, find the dc voltage at the base, the emitter, and the collector. For $V_{EE} = 10$ V, find the required value of R in order for the circuit of Fig. 5.47(b) to implement the current-source I.
Ans. -1 V; -1.7 V; $+2.6$ V; 19.3 kΩ

5.6 SMALL-SIGNAL OPERATION AND MODELS

Having learned how to bias the BJT to operate as an amplifier, we now take a closer look at the small-signal operation of the transistor. Toward that end, consider the *conceptual* circuit shown in Fig. 5.48(a). Here the base–emitter junction is forward biased by a dc voltage V_{BE} (battery). The reverse bias of the collector–base junction is established by connecting the collector to another power supply of voltage V_{CC} through a resistor R_C. The input signal to be amplified is represented by the voltage source v_{be} that is superimposed on V_{BE}.

We consider first the dc bias conditions by setting the signal v_{be} to zero. The circuit reduces to that in Fig. 5.48(b), and we can write the following relationships for the dc currents and voltages:

$$I_C = I_S e^{V_{BE}/V_T} \tag{5.78}$$

$$I_E = I_C / \alpha \tag{5.79}$$

$$I_B = I_C / \beta \tag{5.80}$$

$$V_C = V_{CE} = V_{CC} - I_C R_C \tag{5.81}$$

Obviously, for active-mode operation, V_C should be greater than $(V_B - 0.4)$ by an amount that allows for a reasonable signal swing at the collector.

5.6.1 The Collector Current and the Transconductance

If a signal v_{be} is applied as shown in Fig. 5.48(a), the total instantaneous base–emitter voltage v_{BE} becomes

$$v_{BE} = V_{BE} + v_{be}$$

Correspondingly, the collector current becomes

$$i_C = I_S e^{v_{BE}/V_T} = I_S e^{(V_{BE} + v_{be})/V_T}$$
$$= I_S e^{(V_{BE}/V_T)} e^{(v_{be}/V_T)}$$

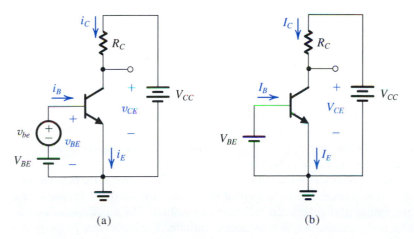

(a)   (b)

FIGURE 5.48 **(a)** Conceptual circuit to illustrate the operation of the transistor as an amplifier. **(b)** The circuit of (a) with the signal source v_{be} eliminated for dc (bias) analysis.

Use of Eq. (5.78) yields

$$i_C = I_C e^{v_{be}/V_T} \tag{5.82}$$

Now, if $v_{be} \ll V_T$, we may approximate Eq. (5.82) as

$$i_C \simeq I_C \left(1 + \frac{v_{be}}{V_T}\right) \tag{5.83}$$

Here we have expanded the exponential in Eq. (5.82) in a series and retained only the first two terms. This approximation, which is valid only for v_{be} less than approximately 10 mV, is referred to as the **small-signal approximation.** Under this approximation the total collector current is given by Eq. (5.83) and can be rewritten

$$i_C = I_C + \frac{I_C}{V_T} v_{be} \tag{5.84}$$

Thus the collector current is composed of the dc bias value I_C and a signal component i_c,

$$i_c = \frac{I_C}{V_T} v_{be} \tag{5.85}$$

This equation relates the signal current in the collector to the corresponding base–emitter signal voltage. It can be rewritten as

$$i_c = g_m v_{be} \tag{5.86}$$

where g_m is called the **transconductance,** and from Eq. (5.85), it is given by

$$g_m = \frac{I_C}{V_T} \tag{5.87}$$

We observe that the transconductance of the BJT is directly proportional to the collector bias current I_C. Thus to obtain a constant predictable value for g_m, we need a constant predictable I_C. Finally, we note that BJTs have relatively high transconductance (as compared to MOSFETs, which we studied in Chapter 4); for instance, at $I_C = 1$ mA, $g_m \simeq 40$ mA/V.

A graphical interpretation for g_m is given in Fig. 5.49, where it is shown that g_m is equal to the slope of the i_C–v_{BE} characteristic curve at $i_C = I_C$ (i.e., at the bias point Q). Thus,

$$g_m = \left.\frac{\partial i_C}{\partial v_{BE}}\right|_{i_C=I_C} \tag{5.88}$$

The small-signal approximation implies keeping the signal amplitude sufficiently small so that *operation is restricted to an almost-linear segment of the i_C–v_{BE} exponential curve.* Increasing the signal amplitude will result in the collector current having components non-linearly related to v_{be}. This, of course, is the same approximation that we discussed in the context of the amplifier transfer curve in Section 5.3.

The analysis above suggests that for small signals ($v_{be} \ll V_T$), the transistor behaves as a voltage-controlled current source. The input port of this controlled source is between base and emitter, and the output port is between collector and emitter. The transconductance of the controlled source is g_m, and the output resistance is infinite. The latter ideal property is a result of our first-order model of transistor operation in which the collector voltage has no effect on the collector current in the active mode. As we have seen in Section 5.2, practical

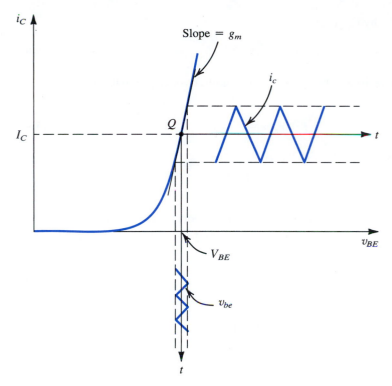

FIGURE 5.49 Linear operation of the transistor under the small-signal condition: A small signal v_{be} with a triangular waveform is superimposed on the dc voltage V_{BE}. It gives rise to a collector signal current i_c, also of triangular waveform, superimposed on the dc current I_C. Here, $i_c = g_m v_{be}$, where g_m is the slope of the i_C–v_{BE} curve at the bias point Q.

BJTs have finite output resistance because of the Early effect. The effect of the output resistance on amplifier performance will be considered later.

EXERCISE

5.36 Use Eq. (5.88) to derive the expression for g_m in Eq. (5.87).

5.6.2 The Base Current and the Input Resistance at the Base

To determine the resistance seen by v_{be}, we first evaluate the total base current i_B using Eq. (5.84), as follows:

$$i_B = \frac{i_C}{\beta} = \frac{I_C}{\beta} + \frac{1}{\beta}\frac{I_C}{V_T}v_{be}$$

Thus,

$$i_B = I_B + i_b \tag{5.89}$$

where I_B is equal to I_C/β and the signal component i_b is given by

$$i_b = \frac{1}{\beta}\frac{I_C}{V_T}v_{be} \tag{5.90}$$

Substituting for I_C/V_T by g_m gives

$$i_b = \frac{g_m}{\beta} v_{be} \qquad (5.91)$$

The small-signal input resistance between base and emitter, *looking into the base,* is denoted by r_π and is defined as

$$r_\pi \equiv \frac{v_{be}}{i_b} \qquad (5.92)$$

Using Eq. (5.91) gives

$$r_\pi = \frac{\beta}{g_m} \qquad (5.93)$$

Thus r_π is directly dependent on β and is inversely proportional to the bias current I_C. Substituting for g_m in Eq. (5.93) from Eq. (5.87) and replacing I_C/β by I_B gives an alternative expression for r_π,

$$r_\pi = \frac{V_T}{I_B} \qquad (5.94)$$

5.6.3 The Emitter Current and the Input Resistance at the Emitter

The total emitter current i_E can be determined from

$$i_E = \frac{i_C}{\alpha} = \frac{I_C}{\alpha} + \frac{i_c}{\alpha}$$

Thus,

$$i_E = I_E + i_e \qquad (5.95)$$

where I_E is equal to I_C/α and the signal current i_e is given by

$$i_e = \frac{i_c}{\alpha} = \frac{I_C}{\alpha V_T} v_{be} = \frac{I_E}{V_T} v_{be} \qquad (5.96)$$

If we denote the small-signal resistance between base and emitter, *looking into the emitter,* by r_e, it can be defined as

$$r_e \equiv \frac{v_{be}}{i_e} \qquad (5.97)$$

Using Eq. (5.96) we find that r_e, called the **emitter resistance,** is given by

$$r_e = \frac{V_T}{I_E} \qquad (5.98)$$

Comparison with Eq. (5.87) reveals that

$$r_e = \frac{\alpha}{g_m} \simeq \frac{1}{g_m} \qquad (5.99)$$

The relationship between r_π and r_e can be found by combining their respective definitions in Eqs. (5.92) and (5.97) as

$$v_{be} = i_b r_\pi = i_e r_e$$

Thus,

$$r_\pi = (i_e/i_b)r_e$$

which yields

$$r_\pi = (\beta + 1)r_e \qquad (5.100)$$

5.37 A BJT having $\beta = 100$ is biased at a dc collector current of 1 mA. Find the value of g_m, r_e, and r_π at the bias point.

Ans. 40 mA/V; 25 Ω; 2.5 kΩ

5.6.4 Voltage Gain

In the preceding section we have established only that the transistor senses the base–emitter signal v_{be} and causes a proportional current $g_m v_{be}$ to flow in the collector lead at a high (ideally infinite) impedance level. In this way the transistor is acting as a voltage-controlled current source. To obtain an output voltage signal, we may force this current to flow through a resistor, as is done in Fig. 5.48(a). Then the total collector voltage v_C will be

$$\begin{aligned} v_C &= V_{CC} - i_C R_C \\ &= V_{CC} - (I_C + i_c)R_C \\ &= (V_{CC} - I_C R_C) - i_c R_C \\ &= V_C - i_c R_C \end{aligned} \qquad (5.101)$$

Here the quantity V_C is the dc bias voltage at the collector, and the signal voltage is given by

$$\begin{aligned} v_c &= -i_c R_C = -g_m v_{be} R_C \\ &= (-g_m R_C) v_{be} \end{aligned} \qquad (5.102)$$

Thus the voltage gain of this amplifier A_v is

$$A_v \equiv \frac{v_c}{v_{be}} = -g_m R_C \qquad (5.103)$$

Here again we note that because g_m is directly proportional to the collector bias current, the gain will be as stable as the collector bias current is made. Substituting for g_m from Eq. (5.87) enables us to express the gain in the form

$$A_v = -\frac{I_C R_C}{V_T} \qquad (5.104)$$

which is identical to the expression we derived in Section 5.3 (Eq. 5.56).

5.38 In the circuit of Fig. 5.48(a), V_{BE} is adjusted to yield a dc collector current of 1 mA. Let $V_{CC} = 15$ V, $R_C = 10$ kΩ, and $\beta = 100$. Find the voltage gain v_c/v_{be}. If $v_{be} = 0.005 \sin \omega t$ volts, find $v_C(t)$ and $i_B(t)$.

Ans. -400 V/V; $5 - 2 \sin \omega t$ volts; $10 + 2 \sin \omega t$ μA

5.6.5 Separating the Signal and the DC Quantities

The analysis above indicates that every current and voltage in the amplifier circuit of Fig. 5.48(a) is composed of two components: a dc component and a signal component. For instance, $v_{BE} = V_{BE} + v_{be}$, $I_C = I_C + i_c$, and so on. The dc components are determined from the dc circuit given in Fig. 5.48(b) and from the relationships imposed by the transistor (Eqs. 5.78 through 5.81). On the other hand, a representation of the signal operation of the BJT can be obtained by eliminating the dc sources, as shown in Fig. 5.50. Observe that since the voltage of an ideal dc supply does not change, the signal voltage across it will be zero. For this reason we have replaced V_{CC} and V_{BE} with short circuits. Had the circuit contained ideal dc current sources, these would have been replaced by open circuits. Note, however, that the circuit of Fig. 5.50 is useful only in so far as it shows the various signal currents and voltages; it is *not* an actual amplifier circuit since the dc bias circuit is not shown.

Figure 5.50 also shows the expressions for the current increments (i_c, i_b, and i_e) obtained when a small signal v_{be} is applied. These relationships can be represented by a circuit. Such a circuit should have three terminals—C, B, and E—and should yield the same terminal currents indicated in Fig. 5.50. The resulting circuit is then *equivalent to the transistor as far as small-signal operation is concerned,* and thus it can be considered an equivalent small-signal circuit model.

5.6.6 The Hybrid-π Model

An equivalent circuit model for the BJT is shown in Fig. 5.51(a). This model represents the BJT as a voltage-controlled current source and explicitly includes the input resistance looking into

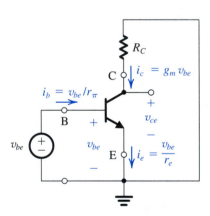

FIGURE 5.50 The amplifier circuit of Fig. 5.48(a) with the dc sources (V_{BE} and V_{CC}) eliminated (short circuited). Thus only the signal components are present. Note that this is a representation of the signal operation of the BJT and not an actual amplifier circuit.

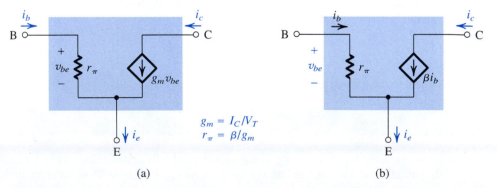

(a)   (b)

FIGURE 5.51 Two slightly different versions of the simplified hybrid-π model for the small-signal operation of the BJT. The equivalent circuit in **(a)** represents the BJT as a voltage-controlled current source (a transconductance amplifier), and that in **(b)** represents the BJT as a current-controlled current source (a current amplifier).

the base, r_π. The model obviously yields $i_c = g_m v_{be}$ and $i_b = v_{be}/r_\pi$. Not so obvious, however, is the fact that the model also yields the correct expression for i_e. This can be shown as follows: At the emitter node we have

$$i_e = \frac{v_{be}}{r_\pi} + g_m v_{be} = \frac{v_{be}}{r_\pi}(1 + g_m r_\pi)$$

$$= \frac{v_{be}}{r_\pi}(1 + \beta) = v_{be} \bigg/ \left(\frac{r_\pi}{1 + \beta}\right)$$

$$= v_{be}/r_e$$

A slightly different equivalent circuit model can be obtained by expressing the current of the controlled source ($g_m v_{be}$) in terms of the base current i_b as follows:

$$g_m v_{be} = g_m(i_b r_\pi)$$
$$= (g_m r_\pi) i_b = \beta i_b$$

This results in the alternative equivalent circuit model shown in Fig. 5.51(b). Here the transistor is represented as a current-controlled current source, with the control current being i_b.

The two models of Fig. 5.51 are simplified versions of what is known as the hybrid-π model. This is the most widely used model for the BJT.

It is important to note that the small-signal equivalent circuits of Fig. 5.51 model the operation of the BJT at a given bias point. This should be obvious from the fact that the model parameters g_m and r_π depend on the value of the dc bias current I_C, as indicated in Fig. 5.51. Finally, although the models have been developed for an *npn* transistor, they apply equally well to a *pnp* transistor *with no change of polarities*.

5.6.7 The T Model

Although the hybrid-π model (in one of its two variants shown in Fig. 5.51) can be used to carry out small-signal analysis of all transistor circuits, there are situations in which an alternative model, shown in Fig. 5.52, is much more convenient. This model, called the **T model,** is shown

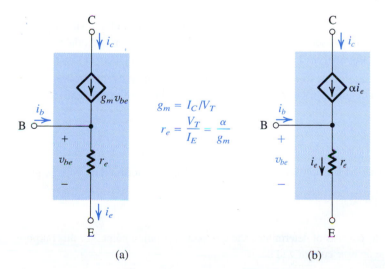

(a)   (b)

FIGURE 5.52 Two slightly different versions of what is known as the *T model* of the BJT. The circuit in (a) is a voltage-controlled current source representation and that in (b) is a current-controlled current source representation. These models explicitly show the emitter resistance r_e rather than the base resistance r_π featured in the hybrid-π model.

in two versions in Fig. 5.52. The model of Fig. 5.52(a) represents the BJT as a voltage-controlled current source with the control voltage being v_{be}. Here, however, the resistance between base and emitter, looking into the emitter, is explicitly shown. From Fig. 5.52(a) we see clearly that the model yields the correct expressions for i_c and i_e. For i_b we note that at the base node we have

$$i_b = \frac{v_{be}}{r_e} - g_m v_{be} = \frac{v_{be}}{r_e}(1 - g_m r_e)$$

$$= \frac{v_{be}}{r_e}(1 - \alpha) = \frac{v_{be}}{r_e}\left(1 - \frac{\beta}{\beta + 1}\right)$$

$$= \frac{v_{be}}{(\beta + 1)r_e} = \frac{v_{be}}{r_\pi}$$

as should be the case.

If in the model of Fig. 5.52(a) the current of the controlled source is expressed in terms of the emitter current as follows:

$$g_m v_{be} = g_m(i_e r_e)$$

$$= (g_m r_e)i_e = \alpha i_e$$

we obtain the alternative T model shown in Fig. 5.52(b). Here the BJT is represented as a current-controlled current source but with the control signal being i_e.

5.6.8 Application of the Small-Signal Equivalent Circuits

The availability of the small-signal BJT circuit models makes the analysis of transistor amplifier circuits a systematic process. The process consists of the following steps:

1. Determine the dc operating point of the BJT and in particular the dc collector current I_C.
2. Calculate the values of the small-signal model parameters: $g_m = I_C/V_T$, $r_\pi = \beta/g_m$, and $r_e = V_T/I_E = \alpha/g_m$.
3. Eliminate the dc sources by replacing each dc voltage source with a short circuit and each dc current source with an open circuit.
4. Replace the BJT with one of its small-signal equivalent circuit models. Although any one of the models can be used, one might be more convenient than the others for the particular circuit being analyzed. This point will be made clearer later in this chapter.
5. Analyze the resulting circuit to determine the required quantities (e.g., voltage gain, input resistance). The process will be illustrated by the following examples.

EXAMPLE 5.14

We wish to analyze the transistor amplifier shown in Fig. 5.53(a) to determine its voltage gain. Assume $\beta = 100$.

Solution

The first step in the analysis consists of determining the quiescent operating point. For this purpose we assume that $v_i = 0$. The dc base current will be

$$I_B = \frac{V_{BB} - V_{BE}}{R_{BB}}$$

$$\simeq \frac{3 - 0.7}{100} = 0.023 \text{ mA}$$

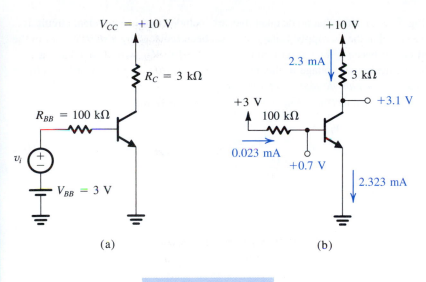

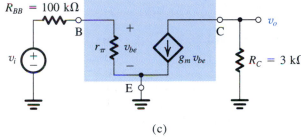

FIGURE 5.53 Example 5.14: **(a)** circuit; **(b)** dc analysis; **(c)** small-signal model.

The dc collector current will be

$$I_C = \beta I_B = 100 \times 0.023 = 2.3 \text{ mA}$$

The dc voltage at the collector will be

$$V_C = V_{CC} - I_C R_C$$
$$= +10 - 2.3 \times 3 = +3.1 \text{ V}$$

Since V_B at +0.7 V is less than V_C, it follows that in the quiescent condition the transistor will be operating in the active mode. The dc analysis is illustrated in Fig. 5.53(b).

Having determined the operating point, we may now proceed to determine the small-signal model parameters:

$$r_e = \frac{V_T}{I_E} = \frac{25 \text{ mV}}{(2.3/0.99) \text{ mA}} = 10.8 \text{ }\Omega$$

$$g_m = \frac{I_C}{V_T} = \frac{2.3 \text{ mA}}{25 \text{ mV}} = 92 \text{ mA/V}$$

$$r_\pi = \frac{\beta}{g_m} = \frac{100}{92} = 1.09 \text{ k}\Omega$$

To carry out the small-signal analysis it is equally convenient to employ either of the two hybrid-π equivalent circuit models of Fig. 5.51. Using the first results in the amplifier equivalent circuit given in Fig. 5.53(c). Note that no dc quantities are included in this equivalent circuit. It is most important to note that the dc supply voltage V_{CC} has been replaced by a *short circuit* in the signal equivalent circuit because the circuit terminal connected to V_{CC} will always have a constant voltage; that is, the signal voltage at this terminal will be zero. In other words, *a circuit terminal connected to a constant dc source can always be considered as a signal ground.*

Analysis of the equivalent circuit in Fig. 5.53(c) proceeds as follows:

$$v_{be} = v_i \frac{r_\pi}{r_\pi + R_{BB}}$$

$$= v_i \frac{1.09}{101.09} = 0.011 v_i \qquad (5.105)$$

The output voltage v_o is given by

$$v_o = -g_m v_{be} R_C$$

$$= -92 \times 0.011 v_i \times 3 = -3.04 v_i$$

Thus the voltage gain will be

$$A_v = \frac{v_o}{v_i} = -3.04 \text{ V/V} \qquad (5.106)$$

where the minus sign indicates a phase reversal.

EXAMPLE 5.15

To gain more insight into the operation of transistor amplifiers, we wish to consider the waveforms at various points in the circuit analyzed in the previous example. For this purpose assume that v_i has a triangular waveform. First determine the maximum amplitude that v_i is allowed to have. Then, with the amplitude of v_i set to this value, give the waveforms of $i_B(t)$, $v_{BE}(t)$, $i_C(t)$, and $v_C(t)$.

Solution

One constraint on signal amplitude is the small-signal approximation, which stipulates that v_{be} should not exceed about 10 mV. If we take the triangular waveform v_{be} to be 20 mV peak-to-peak and work backward, Eq. (5.105) can be used to determine the maximum possible peak of v_i,

$$\hat{V}_i = \frac{\hat{V}_{be}}{0.011} = \frac{10}{0.011} = 0.91 \text{ V}$$

To check whether or not the transistor remains in the active mode with v_i having a peak value $\hat{V}_i = 0.91$ V, we have to evaluate the collector voltage. The voltage at the collector will consist of a triangular wave v_c superimposed on the dc value $V_C = 3.1$ V. The peak voltage of the triangular waveform will be

$$\hat{V}_c = \hat{V}_i \times \text{gain} = 0.91 \times 3.04 = 2.77 \text{ V}$$

It follows that when the output swings negative, the collector voltage reaches a minimum of $3.1 - 2.77 = 0.33$ V, which is lower than the base voltage by less than 0.4 V. Thus the transistor will remain in the active mode with v_i having a peak value of 0.91 V. Nevertheless, we will use a somewhat lower value for $\hat{V}_i$ of approximately 0.8 V, as shown in Fig. 5.54(a), and complete the analysis of this problem. The signal current in the base will be triangular, with a peak value $\hat{I}_b$ of

$$\hat{I}_b = \frac{\hat{V}_i}{R_{BB} + r_\pi} = \frac{0.8}{100 + 1.09} = 0.008 \text{ mA}$$

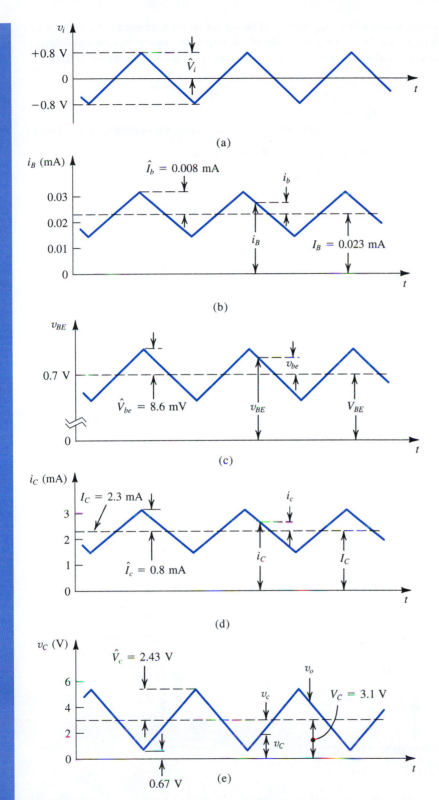

FIGURE 5.54 Signal waveforms in the circuit of Fig. 5.53.

This triangular-wave current will be superimposed on the quiescent base current I_B, as shown in Fig. 5.54(b). The base–emitter voltage will consist of a triangular-wave component superimposed on the dc V_{BE} that is approximately 0.7 V. The peak value of the triangular waveform will be

$$\hat{V}_{be} = \hat{V}_i \frac{r_\pi}{r_\pi + R_{BB}} = 0.8 \frac{1.09}{100 + 1.09} = 8.6 \text{ mV}$$

The total v_{BE} is sketched in Fig. 5.54(c).

The signal current in the collector will be triangular in waveform, with a peak value $\hat{I}_c$ given by

$$\hat{I}_c = \beta \hat{I}_b = 100 \times 0.008 = 0.8 \text{ mA}$$

This current will be superimposed on the quiescent collector current I_C (=2.3 mA), as shown in Fig. 5.54(d).

Finally, the signal voltage at the collector can be obtained by multiplying v_i by the voltage gain; that is,

$$\hat{V}_c = 3.04 \times 0.8 = 2.43 \text{ V}$$

Figure 5.54(e) shows a sketch of the total collector voltage v_C versus time. Note the phase reversal between the input signal v_i and the output signal v_c.

EXAMPLE 5.16

We need to analyze the circuit of Fig. 5.55(a) to determine the voltage gain and the signal waveforms at various points. The capacitor C is a coupling capacitor whose purpose is to couple the signal v_i to the emitter while blocking dc. In this way the dc bias established by V^+ and V^- together with R_E and R_C will not be disturbed when the signal v_i is connected. For the purpose of this example, C will be assumed to be very large and ideally infinite—that is, acting as a perfect short circuit at signal frequencies of interest. Similarly, another very large capacitor is used to couple the output signal v_o to other parts of the system.

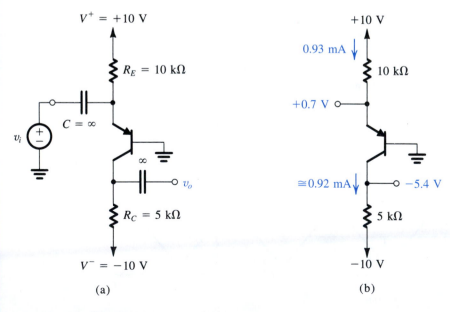

(a) (b)

FIGURE 5.55 Example 5.16: **(a)** circuit; **(b)** dc analysis;

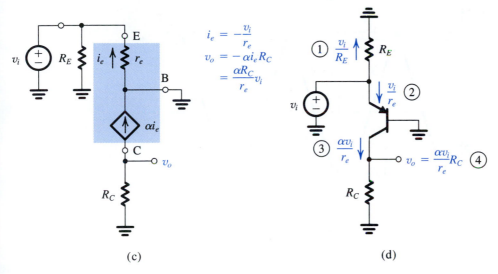

(c)

(d)

FIGURE 5.55 *(Continued)* **(c)** small-signal model; **(d)** small-signal analysis performed directly on the circuit.

Solution

We shall start by determining the dc operating point as follows (see Fig. 5.55b):

$$I_E = \frac{+10 - V_E}{R_E} \simeq \frac{+10 - 0.7}{10} = 0.93 \text{ mA}$$

Assuming $\beta = 100$, then $\alpha = 0.99$, and

$$I_C = 0.99 I_E = 0.92 \text{ mA}$$
$$V_C = -10 + I_C R_C$$
$$= -10 + 0.92 \times 5 = -5.4 \text{ V}$$

Thus the transistor is in the active mode. Furthermore, the collector signal can swing from −5.4 V to +0.4 V (which is 0.4 V above the base voltage) without the transistor going into saturation. However, a negative 5.8-V swing in the collector voltage will (theoretically) cause the minimum collector voltage to be −11.2 V, which is more negative than the power-supply voltage. It follows that if we attempt to apply an input that results in such an output signal, the transistor will cut off and the negative peaks of the output signal will be clipped off, as illustrated in Fig. 5.56. The waveform in Fig. 5.56, however, is shown to be linear (except for the clipped peaks); that is, the effect of the nonlinear $i_C - v_{BE}$ characteristic is not taken into account. This is not correct, since if we are driving the transistor into cutoff at the negative signal peaks, then we will surely be exceeding the small-signal limit, as will be shown later.

Let us now proceed to determine the small-signal voltage gain. Toward that end, we eliminate the dc sources and replace the BJT with its T equivalent circuit of Fig. 5.52(b). Note that because the base is grounded, the T model is somewhat more convenient than the hybrid-π model. Nevertheless, identical results can be obtained using the latter.

Figure 5.55(c) shows the resulting small-signal equivalent circuit of the amplifier. The model parameters are

$$\alpha = 0.99$$

$$r_e = \frac{V_T}{I_E} = \frac{25 \text{ mV}}{0.93 \text{ mA}} = 27 \ \Omega$$

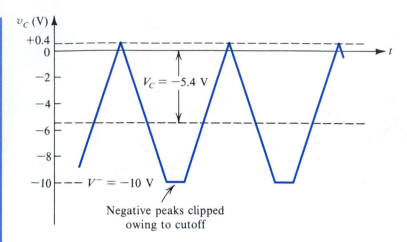

FIGURE 5.56 Distortion in output signal due to transistor cutoff. Note that it is assumed that no distortion due to the transistor nonlinear characteristics is occurring.

Analysis of the circuit in Fig. 5.55(c) to determine the output voltage v_o and hence the voltage gain v_o/v_i is straightforward and is given in the figure. The result is

$$A_v = \frac{v_o}{v_i} = 183.3 \text{ V/V}$$

Note that the voltage gain is positive, indicating that the output is in phase with the input signal. This property is due to the fact that the input signal is applied to the emitter rather than to the base, as was done in Example 5.14. We should emphasize that the positive gain has nothing to do with the fact that the transistor used in this example is of the *pnp* type.

Returning to the question of allowable signal magnitude, we observe from Fig. 5.55(c) that $v_{eb} = v_i$. Thus, if small-signal operation is desired (for linearity), then the peak of v_i should be limited to approximately 10 mV. With $\hat{V}_i$ set to this value, as shown for a sine-wave input in Fig. 5.57,

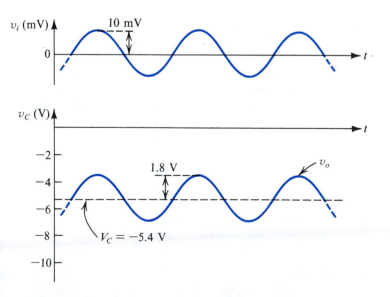

FIGURE 5.57 Input and output waveforms for the circuit of Fig. 5.55. Observe that this amplifier is noninverting, a property of the common-base configuration.

the peak amplitude at the collector, $\hat{V}_c$, will be

$$\hat{V}_c = 183.3 \times 0.01 = 1.833 \text{ V}$$

and the total instantaneous collector voltage $v_C(t)$ will be as depicted in Fig. 5.57.

EXERCISE

5.39 To increase the voltage gain of the amplifier analyzed in Example 5.16, the collector resistance R_C is increased to 7.5 kΩ. Find the new values of V_C, A_v, and the peak amplitude of the output sine wave corresponding to an input sine wave v_i of 10 mV peak.

Ans. −3.1 V; 275 V/V; 2.75 V

5.6.9 Performing Small-Signal Analysis Directly on the Circuit Diagram

In most cases one should explicitly replace each BJT with its small-signal model and analyze the resulting circuit, as we have done in the examples above. This systematic procedure is particularly recommended for beginning students. Experienced circuit designers, however, often perform a first-order analysis directly on the circuit. Figure 5.55(d) illustrates this process for the circuit we have just analyzed. The reader is urged to follow this direct analysis procedure (the steps are numbered). Observe that the equivalent circuit model is *implicitly* utilized; we are only saving the step of drawing the circuit with the BJT replaced with its model. Direct analysis, however, has an additional very important benefit: It provides insight regarding the signal transmission through the circuit. Such insight can prove invaluable in design, particularly at the stage of selecting a circuit configuration appropriate for a given application.

5.6.10 Augmenting the Small-Signal Models to Account for the Early Effect

The Early effect, discussed in Section 5.2, causes the collector current to depend not only on v_{BE} but also on v_{CE}. The dependence on v_{CE} can be modeled by assigning a finite output resistance to the controlled current-source in the hybrid-π model, as shown in Fig. 5.58. The output resistance r_o was defined in Eq. (5.37); its value is given by $r_o = (V_A + V_{CE})/I_C \simeq V_A/I_C$, where V_A is the Early voltage and V_{CE} and I_C are the coordinates of the dc bias point. Note that in the models of Fig. 5.58 we have renamed v_{be} as v_π, in order to conform with the literature.

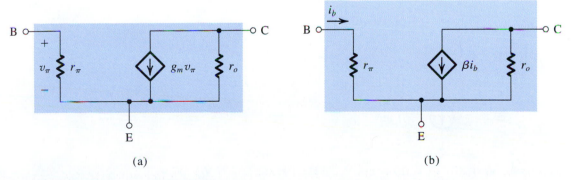

FIGURE 5.58 The hybrid-π small-signal model, in its two versions, with the resistance r_o included.

The question arises as to the effect of r_o on the operation of the transistor as an amplifier. In amplifier circuits in which the emitter is grounded (as in the circuit of Fig. 5.53), r_o simply appears in parallel with R_C. Thus, if we include r_o in the equivalent circuit of Fig. 5.53(c), for example, the output voltage v_o becomes

$$v_o = -g_m v_{be}(R_C \,//\, r_o)$$

Thus the gain will be somewhat reduced. Obviously if $r_o \gg R_C$, the reduction in gain will be negligible, and one can ignore the effect of r_o. In general, in such a configuration r_o can be neglected if it is greater than $10R_C$.

When the emitter of the transistor is not grounded, including r_o in the model can complicate the analysis. We will make comments regarding r_o and its inclusion or exclusion on frequent occasions throughout the book. We should also note that in integrated-circuit BJT amplifiers, r_o plays a dominant role, as will be seen in Chapter 6. Of course, if one is performing an accurate analysis of an almost-final design using computer-aided analysis, then r_o can be easily included (see Section 5.11).

Finally, it should be noted that either of the T models in Fig. 5.52 can be augmented to account for the Early effect by including r_o between collector and emitter.

5.6.11 Summary

The analysis and design of BJT amplifier circuits is greatly facilitated if the relationships between the various small-signal model parameters are at your fingertips. For easy reference, these are summarized in Table 5.4. Over time, however, we expect the reader to be able to recall these from memory.

EXERCISE

5.40 The transistor in Fig. E5.40 is biased with a constant current source $I = 1$ mA and has $\beta = 100$ and $V_A = 100$ V. (a) Find the dc voltages at the base, emitter, and collector. (b) Find g_m, r_π, and r_o. (c) If terminal Z is connected to ground, X to a signal source v_{sig} with a source resistance $R_{sig} = 2$ kΩ, and Y to an 8-kΩ load resistance, use the hybrid-π model of Fig. 5.58(a), to draw the small-signal equivalent circuit of the amplifier. (Note that the current source I should be replaced with an open circuit.) Calculate the overall voltage gain v_y/v_{sig}. If r_o is neglected what is the error in estimating the gain magnitude? (*Note:* An infinite capacitance is used to indicate that the capacitance is sufficiently large that it acts as a short circuit at all signal frequencies of interest. However, the capacitor still blocks dc.)

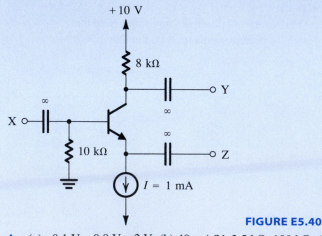

FIGURE E5.40

Ans. (a) −0.1 V, −0.8 V, +2 V; (b) 40 mA/V, 2.5 kΩ, 100 kΩ; (c) −77 V/V, +3.9%

TABLE 5.4   **Small-Signal Models of the BJT**

Hybrid-π Model

■ $(g_m v_\pi)$ Version   ■ (βi_b) Version

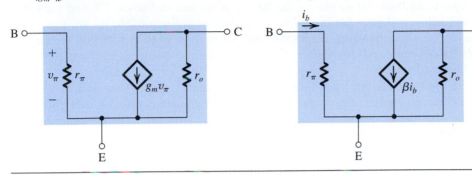

T Model

■ $(g_m v_\pi)$ Version   ■ (αi) Version

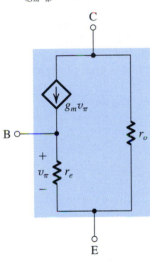 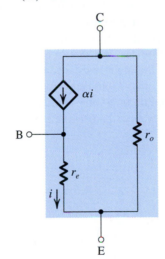

Model Parameters in Terms of DC Bias Currents

$$g_m = \frac{I_C}{V_T} \qquad r_e = \frac{V_T}{I_E} = \alpha\left(\frac{V_T}{I_C}\right) \qquad r_\pi = \frac{V_T}{I_B} = \beta\left(\frac{V_T}{I_C}\right) \qquad r_o = \frac{|V_A|}{I_C}$$

In Terms of g_m

$$r_e = \frac{\alpha}{g_m} \qquad r_\pi = \frac{\beta}{g_m}$$

In Terms of r_e

$$g_m = \frac{\alpha}{r_e} \qquad r_\pi = (\beta + 1)r_e \qquad g_m + \frac{1}{r_\pi} = \frac{1}{r_e}$$

Relationships Between α and β

$$\beta = \frac{\alpha}{1 - \alpha} \qquad \alpha = \frac{\beta}{\beta + 1} \qquad \beta + 1 = \frac{1}{1 - \alpha}$$

5.7 SINGLE-STAGE BJT AMPLIFIERS

We have studied the large-signal operation of BJT amplifiers in Section 5.3 and identified the region over which a properly biased transistor can be operated as a linear amplifier for small signals. Methods for dc biasing the BJT were studied in Section 5.5, and a detailed study of small-signal amplifier operation was presented in Section 5.6. We are now ready to consider practical transistor amplifiers, and we will do so in this section for circuits suitable for discrete-circuit fabrication. The design of integrated-circuit BJT amplifiers will be studied in Chapter 6.

There are basically three configurations for implementing single-stage BJT amplifiers: the common-emitter, the common-base, and the common-collector configurations. All three are studied below, utilizing the same basic structure with the same biasing arrangement.

5.7.1 The Basic Structure

Figure 5.59 shows the basic circuit that we shall utilize to implement the various configurations of BJT amplifiers. Among the various biasing schemes possible for discrete BJT amplifiers (Section 5.5), we have selected, for simplicity and effectiveness, the one employing constant-current biasing. Figure 5.59 indicates the dc currents in all branches and the dc voltages at all nodes. We should note that one would want to select a large value for R_B in order to keep the input resistance at the base large. However, we also want to limit the dc voltage drop across R_B and even more importantly the variability of this dc voltage resulting from the variation in β values among transistors of the same type. The dc voltage V_B determines the allowable signal swing at the collector.

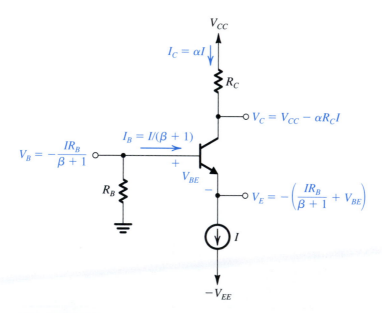

FIGURE 5.59 Basic structure of the circuit used to realize single-stage, discrete-circuit BJT amplifier configurations.

EXERCISE

5.41 Consider the circuit of Fig. 5.59 for the case: $V_{CC} = V_{EE} = 10$ V, $I = 1$ mA, $R_B = 100$ kΩ, $R_C = 8$ kΩ, and $\beta = 100$. Find all dc currents and voltages. What are the allowable signal swings at the collector in both directions? How do these values change as β is changed to 50? To 200? Find the values of the BJT small-signal parameters at the bias point (with $\beta = 100$). The Early voltage $V_A = 100$ V.

Ans. See Fig. E5.41. Signal swing: for $\beta = 100$, +8 V, −3.4 V; for $\beta = 50$, +8 V, −4.4 V; for $\beta = 200$, +8 V, −2.9 V.

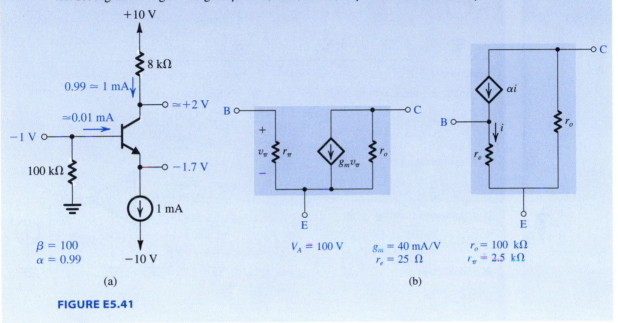

FIGURE E5.41

5.7.2 Characterizing BJT Amplifiers[9]

As we begin our study of BJT amplifier circuits, it is important to know how to characterize the performance of amplifiers as circuit building blocks. An introduction to this subject was presented in Section 1.5. However, the material of Section 1.5 was limited to **unilateral amplifiers.** A number of the amplifier circuits we shall study in this book are not unilateral; that is, they have internal feedback that may cause their input resistance to depend on the load resistance. Similarly, internal feedback may cause the output resistance to depend on the value of the resistance of the signal source feeding the amplifier. To accommodate **nonunilateral amplifiers,** we present in Table 5.5 a general set of parameters and equivalent circuits that we will employ in characterizing and comparing transistor amplifiers. A number of remarks are in order:

1. The amplifier in Table 5.5 is shown fed with a signal source having an open-circuit voltage v_{sig} and an internal resistance R_{sig}. These can be the parameters of an actual signal source or the Thévenin equivalent of the output circuit of another amplifier stage preceding the one under study in a cascade amplifier. Similarly, R_L can be an actual load resistance or the input resistance of a succeeding amplifier stage in a cascade amplifier.

---

[9] This section is identical to Section 4.7.2. Readers who studied Section 4.7.2 can skip this section.

TABLE 5.5 Characteristic Parameters of Amplifiers

Circuit

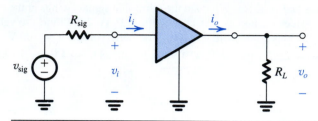

Definitions

- Input resistance with no load:

$$R_i \equiv \left. \frac{v_i}{i_i} \right|_{R_L = \infty}$$

- Input resistance:

$$R_{\text{in}} \equiv \frac{v_i}{i_i}$$

- Open-circuit voltage gain:

$$A_{vo} \equiv \left. \frac{v_o}{v_i} \right|_{R_L = \infty}$$

- Voltage gain:

$$A_v \equiv \frac{v_o}{v_i}$$

- Short-circuit current gain:

$$A_{is} \equiv \left. \frac{i_o}{i_i} \right|_{R_L = 0}$$

- Current gain:

$$A_i \equiv \frac{i_o}{i_i}$$

- Short-circuit transconductance:

$$G_m \equiv \left. \frac{i_o}{v_i} \right|_{R_L = 0}$$

- Output resistance of amplifier proper:

$$R_o \equiv \left. \frac{v_x}{i_x} \right|_{v_i = 0}$$

- Output resistance:

$$R_{\text{out}} \equiv \left. \frac{v_x}{i_x} \right|_{v_{\text{sig}} = 0}$$

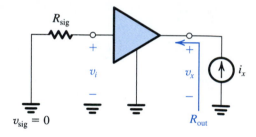

- Open-circuit overall voltage gain:

$$G_{vo} \equiv \left. \frac{v_o}{v_{\text{sig}}} \right|_{R_L = \infty}$$

- Overall voltage gain:

$$G_v \equiv \frac{v_o}{v_{\text{sig}}}$$

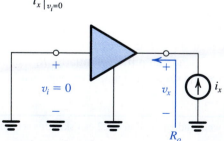

Equivalent Circuits

◼ A:

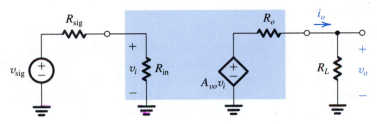

◼ B:

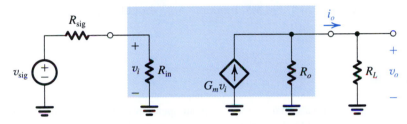

◼ C:

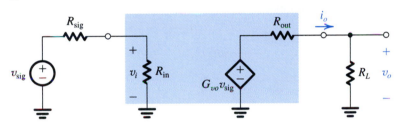

Relationships

◼ $\dfrac{v_i}{v_{sig}} = \dfrac{R_{in}}{R_{in} + R_{sig}}$

◼ $G_v = \dfrac{R_{in}}{R_{in} + R_{sig}} A_{vo} \dfrac{R_L}{R_L + R_o}$

◼ $A_v = A_{vo} \dfrac{R_L}{R_L + R_o}$

◼ $G_{vo} = \dfrac{R_i}{R_i + R_{sig}} A_{vo}$

◼ $A_{vo} = G_m R_o$

◼ $G_v = G_{vo} \dfrac{R_L}{R_L + R_{out}}$

2. Parameters R_i, R_o, A_{vo}, A_{is}, and G_m pertain to the *amplifier proper;* that is, they do *not* depend on the values of R_{sig} and R_L. By contrast, R_{in}, R_{out}, A_v, A_i, G_{vo}, and G_v may depend on one or both of R_{sig} and R_L. Also, observe the relationships of related pairs of these parameters; for instance, $R_i = R_{in}\big|_{R_L = \infty}$, and $R_o = R_{out}\big|_{R_{sig} = 0}$.

3. As mentioned above, for nonunilateral amplifiers, R_{in} may depend on R_L, and R_{out} may depend on R_{sig}. One such amplifier circuit is studied in Section 5.7.6. No such dependencies exist for unilateral amplifiers, for which $R_{in} = R_i$ and $R_{out} = R_o$.

4. The *loading* of the amplifier on the signal source is determined by the input resistance R_{in}. The value of R_{in} determines the current i_i that the amplifier draws from the signal source. It also determines the proportion of the signal v_{sig} that appears at the input of the amplifier proper, that is, v_i.

5. When evaluating the gain A_v from the open-circuit value A_{vo}, R_o is the output resistance to use. This is because A_v is based on feeding the amplifier with an ideal voltage signal v_i. This should be evident from Equivalent Circuit A in Table 5.5. On the other hand, if we are evaluating the overall voltage gain G_v from its open-circuit value G_{vo}, the output resistance to use is R_{out}. This is because G_v is based on feeding the amplifier with v_{sig}, which has an internal resistance R_{sig}. This should be evident from Equivalent Circuit C in Table 5.5.

6. We urge the reader to carefully examine and reflect on the definitions and the six relationships presented in Table 5.5. Example 5.17 should help in this regard.

EXAMPLE 5.17

A transistor amplifier is fed with a signal source having an open-circuit voltage v_{sig} of 10 mV and an internal resistance R_{sig} of 100 kΩ. The voltage v_i at the amplifier input and the output voltage v_o are measured both without and with a load resistance $R_L = 10$ kΩ connected to the amplifier output. The measured results are as follows:

	v_i (mV)	v_o (mV)
Without R_L	9	90
With R_L connected	8	70

Find all the amplifier parameters.

Solution

First, we use the data obtained for $R_L = \infty$ to determine

$$A_{vo} = \frac{90}{9} = 10 \text{ V/V}$$

and

$$G_{vo} = \frac{90}{10} = 9 \text{ V/V}$$

Now, since

$$G_{vo} = \frac{R_i}{R_i + R_{sig}} A_{vo}$$

$$9 = \frac{R_i}{R_i + 100} \times 10$$

which gives

$$R_i = 900 \text{ k}\Omega$$

Next, we use the data obtained when $R_L = 10$ kΩ is connected to the amplifier output to determine

$$A_v = \frac{70}{8} = 8.75 \text{ V/V}$$

and

$$G_v = \frac{70}{10} = 7 \text{ V/V}$$

The values of A_v and A_{vo} can be used to determine R_o as follows:

$$A_v = A_{vo} \frac{R_L}{R_L + R_o}$$

$$8.75 = 10 \frac{10}{10 + R_o}$$

which gives

$$R_o = 1.43 \text{ k}\Omega$$

Similarly, we use the values of G_v and G_{vo} to determine R_{out} from

$$G_v = G_{vo} \frac{R_L}{R_L + R_{\text{out}}}$$

$$7 = 9 \frac{10}{10 + R_{\text{out}}}$$

resulting in

$$R_{\text{out}} = 2.86 \text{ k}\Omega$$

The value of R_{in} can be determined from

$$\frac{v_i}{v_{\text{sig}}} = \frac{R_{\text{in}}}{R_{\text{in}} + R_{\text{sig}}}$$

Thus,

$$\frac{8}{10} = \frac{R_{\text{in}}}{R_{\text{in}} + 100}$$

which yields

$$R_{\text{in}} = 400 \text{ k}\Omega$$

The short-circuit transconductance G_m can be found as follows:

$$G_m = \frac{A_{vo}}{R_o} = \frac{10}{1.43} = 7 \text{ mA/V}$$

and the current gain A_i can be determined as follows:

$$A_i = \frac{v_o/R_L}{v_i/R_{in}} = \frac{v_o}{v_i}\frac{R_{in}}{R_L}$$

$$= A_v\frac{R_{in}}{R_L} = 8.75 \times \frac{400}{10} = 350 \text{ A/A}$$

Finally, we determine the short-circuit current gain A_{is} as follows. From Equivalent Circuit A, the short-circuit output current is

$$i_{osc} = A_{vo}v_i/R_o \qquad (5.107)$$

However, to determine v_i we need to know the value of R_{in} obtained with $R_L = 0$. Toward this end, note that from Equivalent Circuit C, the output short-circuit current can be found as

$$i_{osc} = G_{vo}v_{sig}/R_{out} \qquad (5.108)$$

Now, equating the two expressions for i_{osc} and substituting for G_{vo} by

$$G_{vo} = \frac{R_i}{R_i + R_{sig}}A_{vo}$$

and for v_i from

$$v_i = v_{sig}\frac{R_{in}|_{R_L=0}}{R_{in}|_{R_L=0} + R_{sig}}$$

results in

$$R_{in}|_{R_L=0} = R_{sig}\Big/\left[\left(1 + \frac{R_{sig}}{R_i}\right)\left(\frac{R_{out}}{R_o}\right) - 1\right]$$

$$= 81.8 \text{ k}\Omega$$

We now can use

$$i_{osc} = A_{vo}i_i R_{in}|_{R_L=0}/R_o$$

to obtain

$$A_{is} = \frac{i_{osc}}{i_i} = 10 \times 81.8/1.43 = 572 \text{ A/A}$$

EXERCISE

5.42 Refer to the amplifier of Example 5.17. (a) If R_{sig} is doubled, find the values for R_{in}, G_v, and R_{out}. (b) Repeat for R_L doubled (but R_{sig} unchanged; i.e., 100 kΩ). (c) Repeat for both R_{sig} and R_L doubled.
Ans. (a) 400 kΩ, 5.83 V/V, 4.03 kΩ; (b) 538 kΩ, 7.87 V/V, 2.86 kΩ; (c) 538 kΩ, 6.8 V/V, 4.03 kΩ

5.7.3 The Common-Emitter (CE) Amplifier

The CE configuration is the most widely used of all BJT amplifier circuits. Figure 5.60(a) shows a CE amplifier implemented using the circuit of Fig. 5.59. To establish a **signal ground** (or an **ac ground,** as it is sometimes called) at the emitter, a large capacitor C_E, usually in the μF or tens of μF range, is connected between emitter and ground. This capacitor is required to provide a very low impedance to ground (ideally, zero impedance; i.e., in effect, a short circuit) at all signal frequencies of interest. In this way, the emitter signal current passes through C_E to ground and thus *bypasses* the output resistance of the current source I (and any other circuit component that might be connected to the emitter); hence C_E is called a **bypass capacitor.** Obviously, the lower the signal frequency, the less effective the bypass capacitor becomes. This issue will be studied in Section 5.9. For our purposes here we shall assume that C_E is acting as a perfect short circuit and thus is establishing a zero signal voltage at the emitter.

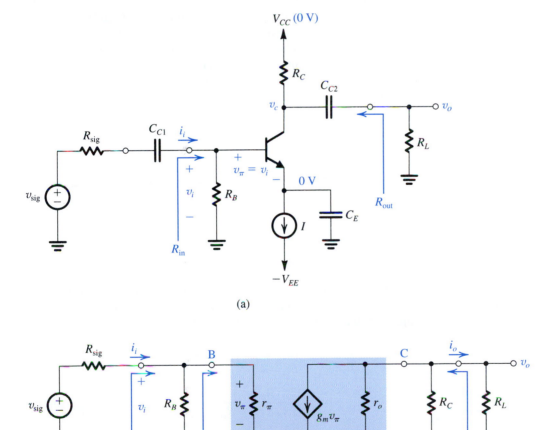

(a)

(b)

FIGURE 5.60 **(a)** A common-emitter amplifier using the structure of Fig. 5.59. **(b)** Equivalent circuit obtained by replacing the transistor with its hybrid-π model.

In order not to disturb the dc bias currents and voltages, the signal to be amplified, shown as a voltage source v_{sig} with an internal resistance R_{sig}, is connected to the base through a large capacitor C_{C1}. Capacitor C_{C1}, known as a **coupling capacitor,** is required to act as a perfect short circuit at all signal frequencies of interest while blocking dc. Here again we shall assume this to be the case and defer discussion of imperfect signal coupling, arising as a result of the rise of the impedance of C_{C1} at low frequencies, to Section 5.9. At this juncture, we should point out that in situations where the signal source can provide a dc path for the dc base current I_B without significantly changing the bias point we may connect the source directly to the base, thus dispensing with C_{C1} as well as R_B. Eliminating R_B has the added beneficial effect of raising the input resistance of the amplifier.

The voltage signal resulting at the collector, v_c, is coupled to the load resistance R_L via another coupling capacitor C_{C2}. We shall assume that C_{C2} also acts a perfect short circuit at all signal frequencies of interest; thus the output voltage $v_o = v_c$. Note that R_L can be an actual load resistor to which the amplifier is required to provide its output voltage signal, or it can be the input resistance of a subsequent amplifier stage in cases where more than one stage of amplification is needed. (We will study multistage amplifiers in Chapter 7).

To determine the terminal characteristics of the CE amplifier, that is, its input resistance, voltage gain, and output resistance, we replace the BJT with its hybrid-π small-signal model. The resulting small-signal equivalent circuit of the CE amplifier is shown in Fig. 5.60(b). We observe at the outset that this amplifier is unilateral and thus $R_{in} = R_i$ and $R_{out} = R_o$. Analysis of this circuit is straightforward and proceeds in a step-by-step manner, from the signal source to the amplifier load. At the amplifier input we have

$$R_{in} \equiv \frac{v_i}{i_i} = R_B \parallel R_{ib} \tag{5.109}$$

where R_{ib} is the input resistance looking into the base. Since the emitter is grounded,

$$R_{ib} = r_\pi \tag{5.110}$$

Normally, we select $R_B \gg r_\pi$, with the result that

$$R_{in} \cong r_\pi \tag{5.111}$$

Thus, we note that the input resistance of the CE amplifier will typically be a few kilohms, which can be thought of as low to moderate. The fraction of source signal v_{sig} that appears across the input terminals of the amplifier proper can be found from

$$v_i = v_{sig} \frac{R_{in}}{R_{in} + R_{sig}} \tag{5.112}$$

$$= v_{sig} \frac{(R_B \parallel r_\pi)}{(R_B \parallel r_\pi) + R_{sig}} \tag{5.113}$$

which for $R_B \gg r_\pi$ becomes

$$v_i \cong v_{sig} \frac{r_\pi}{r_\pi + R_{sig}} \tag{5.114}$$

Next we note that

$$v_\pi = v_i \tag{5.115}$$

At the output of the amplifier we have

$$v_o = -g_m v_\pi (r_o \parallel R_C \parallel R_L)$$

Replacing v_π by v_i we can write for the voltage gain of the amplifier proper; that is, the voltage gain from base to collector,

$$A_v = -g_m (r_o \parallel R_C \parallel R_L) \tag{5.116}$$

This equation simply says that the voltage gain from base to collector is found by multiplying g_m by the total resistance between collector and ground. The open-circuit voltage gain A_{vo} can be obtained by setting $R_L = \infty$ in Eq. (5.116); thus,

$$A_{vo} = -g_m (r_o \parallel R_C) \tag{5.117}$$

from which we note that the effect of r_o is simply to reduce the gain, usually only slightly since typically $r_o \gg R_C$, resulting in

$$A_{vo} \cong -g_m R_C \tag{5.118}$$

The output resistance R_{out} can be found from the equivalent circuit of Fig. 5.60(b) by looking back into the output terminal while short-circuiting the source v_{sig}. Since this will result in $v_\pi = 0$, we see that

$$R_{\text{out}} = R_C \parallel r_o \tag{5.119}$$

Thus r_o reduces the output resistance of the amplifier, again usually only slightly since typically $r_o \gg R_C$ and

$$R_{\text{out}} \cong R_C \tag{5.120}$$

Recalling that for this unilateral amplifier $R_o = R_{\text{out}}$, we can utilize A_{vo} and R_o to obtain the voltage gain A_v corresponding to any particular R_L,

$$A_v = A_{vo} \frac{R_L}{R_L + R_o}$$

The reader can easily verify that this approach does in fact lead to the expression for A_v in Eq. (5.116), which we have derived directly.

The overall voltage gain from source to load, G_v, can be obtained by multiplying (v_i / v_{sig}) from Eq. (5.113) by A_v from Eq. (5.116),

$$G_v = -\frac{(R_B \parallel r_\pi)}{(R_B \parallel r_\pi) + R_{\text{sig}}} g_m (r_o \parallel R_C \parallel R_L) \tag{5.121}$$

For the case $R_B \gg r_\pi$, this expression simplifies to

$$G_v \cong -\frac{\beta (R_C \parallel R_L \parallel r_o)}{r_\pi + R_{\text{sig}}} \tag{5.122}$$

From this expression we note that if $R_{\text{sig}} \gg r_\pi$, the overall gain will be highly dependent on β. This is not a desirable property since β varies considerably between units of the same transistor type. At the other extreme, if $R_{\text{sig}} \ll r_\pi$, we see that the expression for the overall voltage gain reduces to

$$G_v \cong -g_m(R_C \parallel R_L \parallel r_o) \tag{5.123}$$

which is the gain A_v; in other words, when R_{sig} is small, the overall voltage gain is almost equal to the gain of the CE circuit proper, which is independent of β. Typically a CE amplifier can realize a voltage gain on the order of a few hundred, which is very significant. It follows that the CE amplifier is used to realize the bulk of the voltage gain required in a usual amplifier design. Unfortunately, however, as we shall see in Section 5.9, the high-frequency response of the CE amplifier can be rather limited.

Before leaving the CE amplifier, we wish to evaluate its short-circuit current gain, A_{is}. This can be easily done by referring to the amplifier equivalent circuit in Fig. 5.60(b). When R_L is short circuited, the current through it will be equal to $-g_m v_\pi$,

$$i_{os} = -g_m v_\pi$$

Since v_π is related to i_i by

$$v_\pi = v_i = i_i R_{in}$$

the short-circuit current gain can be found as

$$A_{is} \equiv \frac{i_{os}}{i_i} = -g_m R_{in} \tag{5.124}$$

Substituting $R_{in} = R_B \parallel r_\pi$ we can see that in the case $R_B \gg r_\pi$, $|A_{is}|$ reduces to β, which is to be expected since β is, by definition, the short-circuit current gain of the common-emitter configuration.

In conclusion, the common-emitter configuration can provide large voltage and current gains, but R_{in} is relatively low and R_{out} is relatively high.

EXERCISE

5.43 Consider the CE amplifier of Fig. 5.60(a) when biased as in Exercise 5.41. In particular, refer to Fig. E5.41 for the bias currents and the values of the elements of the BJT model at the bias point. Evaluate R_{in} (without and with R_B taken into account), A_{vo} (without and with r_o taken into account), R_{out} (without and with r_o taken into account), and A_{is} (without and with R_B taken into account). For $R_L = 5$ kΩ, find A_v. If $R_{sig} = 5$ kΩ, find the overall voltage gain G_v. If the sine-wave v_π is to be limited to 5 mV peak, what is the maximum allowed peak amplitude of v_{sig} and the corresponding peak amplitude of v_o.

Ans. 2.5 kΩ, 2.4 kΩ; –320 V/V, –296 V/V; 8 kΩ, 7.4 kΩ; –100 A/A, –98 A/A; –119 V/V; –39 V/V; 15 mV; 0.6 V

5.7.4 The Common-Emitter Amplifier with an Emitter Resistance

Including a resistance in the signal path between emitter and ground, as shown in Fig. 5.61(a), can lead to significant changes in the amplifier characteristics. Thus such a resistor can be utilized by the designer as an effective design tool for tailoring the amplifier characteristics to fit the design requirements.

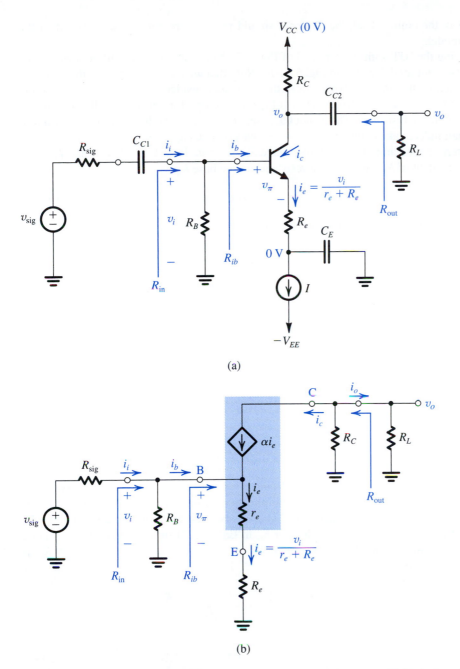

FIGURE 5.61 **(a)** A common-emitter amplifier with an emitter resistance R_e. **(b)** Equivalent circuit obtained by replacing the transistor with its T model.

Analysis of the circuit in Fig. 5.61(a) can be performed by replacing the BJT with one of its small-signal models. Although any one of the models of Figs. 5.51 and 5.52 can be used, the most convenient for this application is one of the two T models. This is because a resistance R_e in the emitter will appear in series with the emitter resistance r_e of the T model and can thus be added to it, simplifying the analysis considerably. In fact, whenever there is a resistance in the emitter lead, the T model should prove more convenient to use than the hybrid-π model.

Replacing the BJT with the T model of Fig. 5.52(b) results in the amplifier small-signal equivalent-circuit model shown in Fig. 5.61(b). Note that we have *not* included the BJT output resistance r_o; including r_o complicates the analysis considerably. Since for the discrete amplifier at hand it turns out that the effect of r_o on circuit performance is small, we shall not include it in the analysis here. This is not the case, however, for the IC version of this circuit, and we shall indeed take r_o into account in the analysis in Chapter 6.

To determine the amplifier input resistance R_{in}, we note from Fig. 5.61(b) that R_{in} is the parallel equivalent of R_B and the input resistance at the base R_{ib},

$$R_{\text{in}} = R_B \parallel R_{ib} \tag{5.125}$$

The input resistance at the base R_{ib} can be found from

$$R_{ib} \equiv \frac{v_i}{i_b}$$

where

$$i_b = (1 - \alpha)i_e = \frac{i_e}{\beta + 1}$$

and

$$i_e = \frac{v_i}{r_e + R_e} \tag{5.126}$$

Thus,

$$R_{ib} = (\beta + 1)(r_e + R_e) \tag{5.127}$$

This is a very important result. It says that *the input resistance looking into the base is $(\beta + 1)$ times the total resistance in the emitter*. Multiplication by the factor $(\beta + 1)$ is known as the **resistance-reflection rule.** The factor $(\beta + 1)$ arises because the base current is $1/(\beta + 1)$ times the emitter current. The expression for R_{ib} in Eq. (5.127) shows clearly that including a resistance R_e in the emitter can substantially increase R_{ib}. Indeed the value of R_{ib} is increased by the ratio

$$\frac{R_{ib}\,(\text{with } R_e \text{ included})}{R_{ib}\,(\text{without } R_e)} = \frac{(\beta + 1)(r_e + R_e)}{(\beta + 1)r_e}$$

$$= 1 + \frac{R_e}{r_e} \cong 1 + g_m R_e \tag{5.128}$$

Thus the circuit designer can use the value of R_e to control the value of R_{ib} and hence R_{in}. Of course, for this control to be effective, R_B must be much larger than R_{ib}; in other words, R_{ib} must dominate the input resistance.

To determine the voltage gain A_v, we see from Fig. 5.61(b) that

$$v_o = -i_c(R_C \parallel R_L)$$

$$= -\alpha i_e(R_C \parallel R_L)$$

Substituting for i_e from Eq. (5.126) gives

$$A_v \equiv \frac{v_o}{v_i} = -\frac{\alpha(R_C \parallel R_L)}{r_e + R_e} \tag{5.129}$$

Since $\alpha \cong 1$,

$$A_v \cong -\frac{R_C \parallel R_L}{r_e + R_e} \tag{5.130}$$

This simple relationship is very useful and is definitely worth remembering: *The voltage gain from base to collector is equal to the ratio of the total resistance in the collector to the total resistance in the emitter.* This statement is a general one and applies to any amplifier circuit. The open-circuit voltage gain A_{vo} can be found by setting $R_L = \infty$ in Eq. (5.129),

$$A_{vo} = -\frac{\alpha R_C}{r_e + R_e} \tag{5.131}$$

which can be expressed alternatively as

$$A_{vo} = -\frac{\alpha}{r_e} \frac{R_C}{1 + R_e / r_e}$$

$$A_{vo} = -\frac{g_m R_C}{1 + (R_e / r_e)} \cong -\frac{g_m R_C}{1 + g_m R_e} \tag{5.132}$$

Including R_e thus reduces the voltage gain by the factor $(1 + g_m R_e)$, which is the same factor by which R_{ib} is increased. This points out an interesting trade-off between gain and input resistance, a trade-off that the designer can exercise through the choice of an appropriate value for R_e.

The output resistance R_{out} can be found from the circuit in Fig. 5.61(b) by inspection:

$$R_{\text{out}} = R_C \tag{5.133}$$

At this point we should note that for this amplifier, $R_{\text{in}} = R_i$ and $R_{\text{out}} = R_o$.

The short-circuit current gain A_{is} can be found from the circuit in Fig. 5.61(b) as follows:

$$i_{os} = -\alpha i_e$$

$$i_i = v_i / R_{\text{in}}$$

Thus,

$$A_{is} = -\frac{\alpha R_{\text{in}} i_e}{v_i}$$

Substituting for i_e from Eq. (5.126) and for R_{in} from Eq. (5.125),

$$A_{is} = -\frac{\alpha(R_B \parallel R_{ib})}{r_e + R_e} \tag{5.134}$$

which for the case $R_B \gg R_{ib}$ reduces to

$$A_{is} = \frac{-\alpha(\beta + 1)(r_e + R_e)}{r_e + R_e} = -\beta$$

the same value as for the CE circuit.

The overall voltage gain from source to load can be obtained by multiplying A_v by (v_i/v_{sig}),

$$G_v = \frac{v_i}{v_{sig}} \cdot A_v = -\frac{R_{in}}{R_{sig} + R_{in}} \frac{\alpha(R_C \| R_L)}{r_e + R_e}$$

Substituting for R_{in} by $R_B \| R_{ib}$, assuming that $R_B \gg R_{ib}$, and substituting for R_{ib} from Eq. (5.127) results in

$$G_v \cong -\frac{\beta(R_C \| R_L)}{R_{sig} + (\beta + 1)(r_e + R_e)} \qquad (5.135)$$

We note that the gain is lower than that of the CE amplifier because of the additional term $(\beta + 1)R_e$ in the denominator. The gain, however, is less sensitive to the value of β, a desirable result.

Another important consequence of including the resistance R_e in the emitter is that it enables the amplifier to handle larger input signals without incurring nonlinear distortion. This is because only a fraction of the input signal at the base, v_i, appears between the base and the emitter. Specifically, from the circuit in Fig. 5.61(b), we see that

$$\frac{v_\pi}{v_i} = \frac{r_e}{r_e + R_e} \cong \frac{1}{1 + g_m R_e} \qquad (5.136)$$

Thus, for the same v_π, the signal at the input terminal of the amplifier, v_i, can be greater than for the CE amplifier by the factor $(1 + g_m R_e)$.

To summarize, including a resistance R_e in the emitter of the CE amplifier results in the following characteristics:

1. The input resistance R_{ib} is increased by the factor $(1 + g_m R_e)$.
2. The voltage gain from base to collector, A_v, is reduced by the factor $(1 + g_m R_e)$.
3. For the same nonlinear distortion, the input signal v_i can be increased by the factor $(1 + g_m R_e)$.
4. The overall voltage gain is less dependant on the value of β.
5. The high-frequency response is significantly improved (as we shall see in Chapter 6).

With the exception of gain reduction, these characteristics represent performance improvements. Indeed, the reduction in gain is the price paid for obtaining the other performance improvements. In many cases this is a good bargain; it is the underlying motive for the use of negative feedback. That the resistance R_e introduces negative feedback in the amplifier circuit can be seen by reference to Fig. 5.61(a): If for some reason the collector current increases, the emitter current also will increase, resulting in an increased voltage drop across R_e. Thus the emitter voltage rises, and the base–emitter voltage decreases. The latter effect causes the collector current to decrease, counteracting the initially assumed change, an indication of the presence of negative feedback. In Chapter 8, where we shall study negative feedback formally, we will find that the factor $(1 + g_m R_e)$ which appears repeatedly, is the "amount of negative feedback" introduced by R_e. Finally, we note that the negative feedback action of R_e gives it the name **emitter degeneration resistance.**

Before leaving this circuit we wish to point out that we have shown a number of the circuit analysis steps directly on the circuit diagram in Fig. 5.61(a). With practice, the reader should be able to do all of the small-signal analysis directly on the circuit diagram, thus dispensing with the task of drawing a complete small-signal equivalent-circuit model.

EXERCISE

5.44 Consider the emitter-degenerated CE circuit of Fig. 5.61 when biased as in Exercise 5.41. In particular refer to Fig. E5.41 for the bias currents and for the values of the elements of the BJT model at the bias point. Let the amplifier be fed from a source having $R_{sig} = 5$ kΩ, and let $R_L = 5$ kΩ. Find the value of R_e that results in R_{in} equal to four times the source resistance R_{sig}. For this value of R_e, find A_{vo}, R_{out}, A_v, G_v, and A_{is}. If v_π is to be limited to 5 mV, what is the maximum value v_{sig} can have with and without R_e included. Find the corresponding v_o.

Ans. 225 Ω; –32 V/V; 8 kΩ; –12.3 V/V; –9.8 V/V; –79.2 A/A; 62.5 mV; 15 mV; 0.6 V

5.7.5 The Common-Base (CB) Amplifier

By establishing a signal ground on the base terminal of the BJT, a circuit configuration aptly named common-base or **grounded-base amplifier** is obtained. The input signal is applied to the emitter, and the output is taken at the collector, with the base forming a common terminal between the input and output ports. Figure 5.62(a) shows a CB amplifier based on the circuit of Fig. 5.59. Observe that since both the dc and ac voltages at the base are zero, we have connected the base directly to ground, thus eliminating resistor R_B altogether. Coupling capacitors C_{C1} and C_{C2} perform similar functions to those in the CE circuit.

The small-signal equivalent circuit model of the amplifier is shown in Fig. 5.62(b). Since resistor R_{sig} appears in series with the emitter terminal, we have elected to use the T model for the transistor. Although the hybrid-π model would yield identical results, the T model is more convenient in this case. We have not included r_o. This is because including r_o would complicate the analysis considerably, for it would appear between the output and input of the amplifier. Fortunately, it turns out that the effect of r_o on the performance of a discrete CB amplifier is very small. We will consider the effect of r_o when we study the IC form of the CB amplifier in Chapter 6.

From inspection of the equivalent circuit model in Fig. 5.62(b), we see that the input resistance is

$$R_{in} = r_e \tag{5.137}$$

This should have been expected since we are looking into the emitter and the base is grounded. Typically r_e is a few ohms to a few tens of ohms; thus the CB amplifier has a low input resistance.

To determine the voltage gain, we write at the collector node

$$v_o = -\alpha i_e (R_C \parallel R_L)$$

and substitute for the emitter current from

$$i_e = -\frac{v_i}{r_e}$$

to obtain

$$A_v = \frac{v_o}{v_i} = \frac{\alpha}{r_e}(R_C \parallel R_L) = g_m(R_C \parallel R_L) \tag{5.138}$$

which except for its positive sign is identical to the expression for A_v for the CE amplifier.

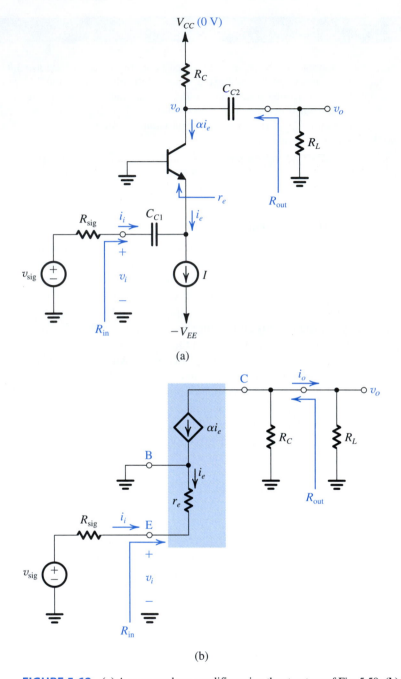

FIGURE 5.62 (a) A common-base amplifier using the structure of Fig. 5.59. (b) Equivalent circuit obtained by replacing the transistor with its T model.

The open-circuit voltage gain A_{vo} can be found from Eq. (5.138) by setting $R_L = \infty$:

$$A_{vo} = g_m R_C \qquad (5.139)$$

Again, this is identical to A_{vo} for the CE amplifier except that the CB amplifier is noninverting. The output resistance of the CB circuit can be found by inspection from the circuit in

Fig. 5.62(b) as

$$R_{\text{out}} = R_C$$

which is similar to the case of the CE amplifier. Here we should note that the CB amplifier with r_o neglected is unilateral, with the result that $R_{\text{in}} = R_i$ and $R_{\text{out}} = R_o$.

The short-circuit current gain A_{is} is given by

$$A_{is} = \frac{-\alpha i_e}{i_i} = \frac{-\alpha i_e}{-i_e} = \alpha \qquad (5.140)$$

which corresponds to our definition of α as the short-circuit current gain of the CB configuration.

Although the gain of the CB amplifier proper has the same magnitude as that of the CE amplifier, this is usually not the case as far as the overall voltage gain is concerned. The low input resistance of the CB amplifier can cause the input signal to be severely attenuated, specifically

$$\frac{v_i}{v_{\text{sig}}} = \frac{R_i}{R_{\text{sig}} + R_i} = \frac{r_e}{R_{\text{sig}} + r_e} \qquad (5.141)$$

from which we see that except for situations in which R_{sig} is on the order of r_e, the signal transmission factor v_i/v_{sig} can be very small. It is useful at this point to mention that one of the applications of the CB circuit is to amplify high-frequency signals that appear on a coaxial cable. To prevent signal reflection on the cable, the CB amplifier is required to have an input resistance equal to the characteristic resistance of the cable, which is usually in the range of 50 Ω to 75 Ω.

The overall voltage gain G_v of the CB amplifier can be obtained by multiplying the ratio v_i/v_{sig} of Eq. (5.141) by A_v from Eq. (5.138),

$$\begin{aligned} G_v &= \frac{r_e}{R_{\text{sig}} + r_e} g_m(R_C \parallel R_L) \\ &= \frac{\alpha(R_C \parallel R_L)}{R_{\text{sig}} + r_e} \end{aligned} \qquad (5.142)$$

Since $\alpha \cong 1$, we see that the overall voltage gain is simply the ratio of the total resistance in the collector circuit to the total resistance in the emitter circuit. We also note that the overall voltage gain is almost independent of the value of β (except through the small dependence of α on β), a desirable property. Observe that for R_{sig} of the same order as R_C and R_L, the gain will be very small.

In summary, the CB amplifier exhibits a very low input resistance (r_e), a short-circuit current gain that is nearly unity (α), an open-circuit voltage gain that is positive and equal in magnitude to that of the CE amplifier ($g_m R_C$), and like the CE amplifier, a relatively high output resistance (R_C). Because of its very low input resistance, the CB circuit *alone* is not attractive as a voltage amplifier except in specialized applications, such as the cable amplifier mentioned above. The CB amplifier has excellent high-frequency performance as well, which as we shall see in Chapter 6 makes it useful together with other circuits in the implementation of high-frequency amplifiers. Finally, a very significant application of the CB circuit is as a unity-gain current amplifier or **current buffer**: It accepts an input signal current at a low input resistance and delivers a nearly equal current at very high output resistance at the collector (the output resistance excluding R_C and neglecting r_o is infinite). We shall study such an application in the context of the IC version of the CB circuit in Chapter 6.

EXERCISES

5.45 Consider the CB amplifier of Fig. 5.62(a) when designed using the BJT and component values specified in Exercise 5.41. Specifically, refer to Fig. E5.41 for the bias quantities and the values of the components of the BJT small-signal model. Let $R_{\text{sig}} = R_L = 5$ kΩ. Find the values of R_{in}, A_{vo}, R_o, A_v, v_i/v_s, and G_v. To what value should R_{sig} be reduced to obtain an overall voltage gain equal to that found for the CE amplifier in Exercise 5.43, that is, −39 V/V?

Ans. 25 Ω; +320 V/V; 8 kΩ; +123 V/V; 0.005 V/V; 0.6 V/V; 54 Ω

D5.46 It is required to design a CB amplifier for a signal delivered by a 50-Ω coaxial cable. The amplifier is to provide a "proper termination" for the cable and to provide an overall voltage gain of +100 V/V. Specify the value of the bias current I_E and the total resistance in the collector circuit.

Ans. 0.5 mA; 10 kΩ

5.7.6 The Common-Collector (CC) Amplifier or Emitter Follower

The last of the basic BJT amplifier configurations is the common-collector (CC) circuit, a very important circuit that finds frequent application in the design of both small-signal and large-signal amplifiers (Chapter 14) and even in digital circuits (Chapter 11). The circuit is more commonly known by the alternate name *emitter follower,* the reason for which will shortly become apparent.

An emitter-follower circuit based on the structure of Fig. 5.59 is shown in Fig. 5.63(a). Observe that since the collector is to be at signal ground, we have eliminated the collector resistance R_C. The input signal is capacitively coupled to the base, and the output signal is capacitively coupled from the emitter to a load resistance R_L.

Since, as far as signals are concerned, resistance R_L is connected in series with the emitter, the T model of the BJT would be the more convenient one to use. Figure 5.63(b) shows the small-signal equivalent circuit of the emitter follower with the BJT replaced by its T model augmented to include r_o. Here it is relatively simple to take r_o into account, and we shall do so. Inspection of the circuit in Fig. 5.63(b) reveals that r_o appears in effect in parallel with R_L. Therefore the circuit is redrawn to emphasize this point, and indeed to simplify the analysis, in Fig. 5.63(c).

Unlike the CE and CB circuits we studied above, the emitter-follower circuit is *not unilateral;* that is, the input resistance depends on R_L, and the output resistance depends on R_{sig}. Care therefore must be exercised in characterizing the emitter follower. In the following we shall derive expressions for R_{in}, G_v, G_{vo}, and R_{out}. The expressions that we derive will shed light on the operation and characteristics of the emitter follower. More important than the actual expressions, however, are the methods we use to obtain them. It is in these that we hope the reader will become proficient.

Reference to Fig. 5.63(c) reveals that the BJT has a resistance $(r_o \| R_L)$ in series with the emitter resistance r_e. Thus application of the resistance reflection rule results in the equivalent circuit shown in Fig. 5.64(a). Recall that in reflecting resistances to the base side, we multiply all resistances in the emitter by $(\beta + 1)$, the ratio of i_e to i_b. In this way the voltages remain unchanged.

Inspection of the circuit in Fig. 5.64(a) shows that the input resistance at the base, R_{ib}, is

$$R_{ib} = (\beta + 1)[r_e + (r_o \| R_L)] \tag{5.143}$$

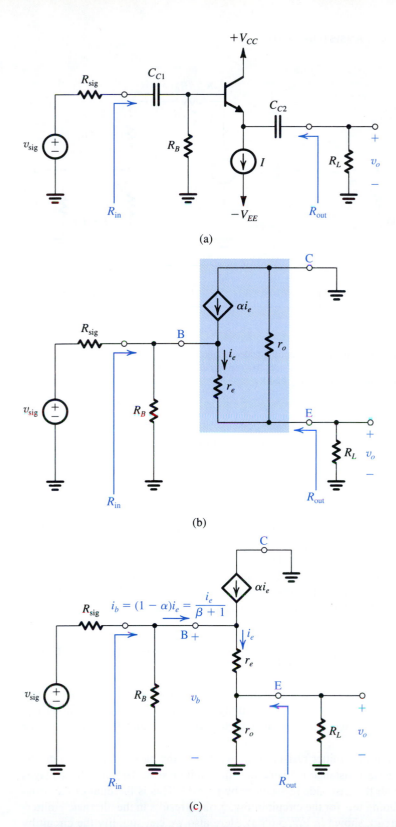

(a)

(b)

(c)

FIGURE 5.63 (a) An emitter-follower circuit based on the structure of Fig. 5.59. (b) Small-signal equivalent circuit of the emitter follower with the transistor replaced by its T model augmented with r_o. (c) The circuit in (b) redrawn to emphasize that r_o is in parallel with R_L. This simplifies the analysis considerably.

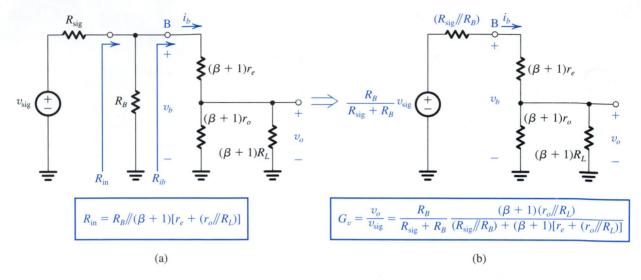

$$R_{in} = R_B // (\beta + 1)[r_e + (r_o // R_L)]$$

$$G_v = \frac{v_o}{v_{sig}} = \frac{R_B}{R_{sig} + R_B} \frac{(\beta + 1)(r_o // R_L)}{(R_{sig} // R_B) + (\beta + 1)[r_e + (r_o // R_L)]}$$

(a)   (b)

FIGURE 5.64 (a) An equivalent circuit of the emitter follower obtained from the circuit in Fig. 5.63(c) by reflecting all resistances in the emitter to the base side. (b) The circuit in (a) after application of Thévenin theorem to the input circuit composed of v_{sig}, R_{sig}, and R_B.

from which we see that the emitter follower acts to raise the resistance level of R_L (or $R_L \parallel r_o$ to be exact) by the factor $(\beta + 1)$ and presents to the source the increased resistance. The total input resistance of the follower is

$$R_{in} = R_B \parallel R_{ib}$$

from which we see that to realize the full effect of the increased R_{ib}, we have to choose as large a value for the bias resistance R_B as is practical (i.e., from a bias design point of view). Also, whenever possible, we should dispense with R_B altogether and connect the signal source directly to the base (in which case we also dispense with C_{C1}).

To find the overall voltage gain G_v, we first apply Thévenin theorem at the input side of the circuit in Fig. 5.64(a) to simplify it to the form shown in Fig. 5.64(b). From the latter circuit we see that v_o can be found by utilizing the voltage divider rule; thus,

$$G_v = \frac{R_B}{R_{sig} + R_B} \frac{(\beta + 1)(r_o \parallel R_L)}{(R_{sig} \parallel R_B) + (\beta + 1)[r_e + (r_o \parallel R_L)]} \tag{5.144}$$

We observe that the voltage gain is less than unity; however, for $R_B \gg R_{sig}$ and $(\beta + 1)[r_e + (r_o \parallel R_L)] \gg (R_{sig} \parallel R_B)$, it becomes very close to unity. Thus the voltage at the emitter (v_o) follows very closely the voltage at the input, which gives the circuit the name **emitter follower.**

Rather than reflecting the emitter resistance network into the base side, we can do the converse: Reflect the base resistance network into the emitter side. To keep the voltages unchanged, we divide all the base-side resistances by $(\beta + 1)$. This is the dual of the resistance reflection rule. Doing this for the circuit in Fig. 5.63(c) results in the alternate emitter-follower equivalent circuit, shown in Fig. 5.65(a). Here also we can simplify the circuit by applying Thévenin theorem at the input side, resulting in the circuit in Fig. 5.65(b). Inspection

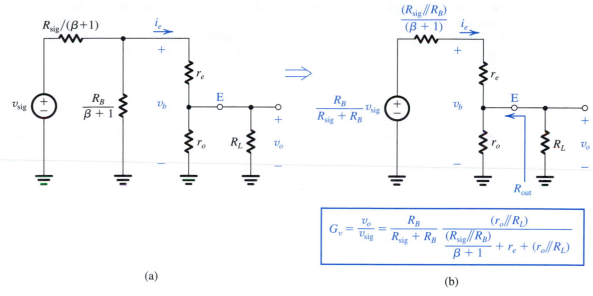

$$G_v = \frac{v_o}{v_{\text{sig}}} = \frac{R_B}{R_{\text{sig}} + R_B} \frac{(r_o /\!/ R_L)}{\dfrac{(R_{\text{sig}} /\!/ R_B)}{\beta + 1} + r_e + (r_o /\!/ R_L)}$$

(a)   (b)

FIGURE 5.65   **(a)** An alternate equivalent circuit of the emitter follower obtained by reflecting all base-circuit resistances to the emitter side. **(b)** The circuit in (a) after application of Thévenin theorem to the input circuit composed of v_{sig}, $R_{\text{sig}}/(\beta + 1)$, and $R_B/(\beta + 1)$.

of the latter reveals that the output voltage and hence v_o/v_{sig} can be found by a simple application of the voltage-divider rule, with the result that

$$G_v = \frac{R_B}{R_{\text{sig}} + R_B} \frac{(r_o \parallel R_L)}{\dfrac{R_{\text{sig}} \parallel R_B}{\beta + 1} + r_e + (r_o \parallel R_L)} \tag{5.145}$$

which, as expected, is identical to the expression in Eq. (5.144) except that both the numerator and denominator of the second factor on the right-hand side have been divided by $(\beta + 1)$. To gain further insight regarding the operation of the emitter follower, let's simplify this expression for the usual case of $R_B \gg R_{\text{sig}}$ and $r_o \gg R_L$. The result is

$$\frac{v_o}{v_{\text{sig}}} \cong \frac{R_L}{\dfrac{R_{\text{sig}}}{\beta + 1} + r_e + R_L} \tag{5.146}$$

which clearly indicates that the gain approaches unity when $R_{\text{sig}}/(\beta + 1)$ becomes much smaller than R_L or alternatively when $(\beta + 1)R_L$ becomes much larger than R_{sig}. This is the **buffering action** of the emitter follower, which derives from the fact that the circuit has a short-circuit current gain that is approximately equal to $(\beta + 1)$.

It is also useful to represent the output of the emitter follower by its Thévenin equivalent circuit. The open-circuit output voltage will be $G_{vo}v_{\text{sig}}$ where G_{vo} can be obtained from Eq. (5.145) by setting $R_L = \infty$,

$$G_{vo} = \frac{R_B}{R_{\text{sig}} + R_B} \frac{r_o}{\dfrac{R_{\text{sig}} \parallel R_B}{\beta + 1} + r_e + r_o} \tag{5.147}$$

Note that r_o usually is large and the second factor becomes almost unity. The first factor approaches unity for $R_B \gg R_{sig}$. The Thévenin resistance is the output resistance R_{out}. It can be determined by inspection of the circuit in Fig. 5.65(b): Reduce v_{sig} to zero, "grab hold" of the emitter terminal, and look back into the circuit. The result is

$$R_{out} = r_o \parallel \left(r_e + \frac{R_{sig} \parallel R_B}{\beta + 1} \right) \tag{5.148}$$

Usually r_o is much larger than the parallel component between the parentheses and can be neglected, leaving

$$R_{out} \cong r_e + \frac{R_{sig} \parallel R_B}{\beta + 1} \tag{5.149}$$

Thus the output resistance of the emitter follower is low, again a result of its impedance transformation or buffering action, which leads to the division of $(R_{sig} \parallel R_B)$ by $(\beta + 1)$. The Thévenin equivalent circuit of the emitter follower is shown together with the formulas for G_{vo} and R_{out} in Fig. 5.66. This circuit can be used to find v_o and hence G_v for any value of R_L.

In summary, the emitter follower exhibits a high input resistance, a low output resistance, a voltage gain that is smaller than but close to unity, and a relatively large current gain. It is therefore ideally suited for applications in which a high-resistance source is to be connected to a low-resistance load—namely, as the last stage or output stage in a multistage amplifier, where its purpose would be not to supply additional voltage gain but rather to give the cascade amplifier a low output resistance. We shall study the design of amplifier output stages in Chapter 14.

Before leaving the emitter follower, the question of the maximum allowed signal swing deserves comment. Since only a small fraction of the input signal appears between the base and the emitter, the emitter follower exhibits linear operation for a wide range of input-signal amplitude. There is, however, an absolute upper limit imposed on the value of the output-signal amplitude by transistor cutoff. To see how this comes about, consider the circuit of Fig. 5.63(a) when the input signal is a sine wave. As the input goes negative,

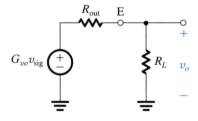

FIGURE 5.66 Thévenin equivalent circuit of the output of the emitter follower of Fig. 5.63(a). This circuit can be used to find v_o and hence the overall voltage gain v_o / v_{sig} for any desired R_L.

the output v_o will also go negative, and the current in R_L will be flowing from ground into the emitter terminal. The transistor will cut off when this current becomes equal to the bias current I. Thus the peak value of v_o can be found from

$$\frac{\hat{V}_o}{R_L} = I$$

or

$$\hat{V}_o = IR_L$$

The corresponding value of v_{sig} will be

$$\hat{V}_{\text{sig}} = \frac{IR_L}{G_v}$$

Increasing the amplitude of v_{sig} above this value results in the transistor becoming cut off and the negative peaks of the output-signal waveform being clipped off.

EXERCISE

5.47 The emitter follower in Fig. 5.63(a) is used to connect a source with $R_{\text{sig}} = 10$ kΩ to a load $R_L = 1$ kΩ. The transistor is biased at $I = 5$ mA, utilizes a resistance $R_B = 40$ kΩ, and has $\beta = 100$ and $V_A = 100$ V. Find R_{ib}, R_{in}, G_v, G_{vo}, and R_{out}. What is the largest peak amplitude of an output sinusoid that can be used without the transistor cutting off? If in order to limit nonlinear distortion the base–emitter signal voltage is limited to 10 mV peak, what is the corresponding amplitude at the output? What will the overall voltage gain become if R_L is changed to 2 kΩ? To 500 Ω?

Ans. 96.7 kΩ; 28.3 kΩ; 0.735 V/V; 0.8 V/V; 84 Ω; 5 V; 1.9 V; 0.768 V/V; 0.685 V/V

5.7.7 Summary and Comparisons

For easy reference and to enable comparisons, we present in Table 5.6 the formulas for determining the characteristic parameters of discrete single-stage BJT amplifiers. In addition to the remarks already made throughout this section about the characteristics and areas of applicability of the various configurations, we make the following concluding points:

1. The CE configuration is the best suited for realizing the bulk of the gain required in an amplifier. Depending on the magnitude of the gain required, either a single stage or a cascade of two or three stages can be used.

2. Including a resistor R_e in the emitter lead of the CE stage provides a number of performance improvements at the expense of gain reduction.

3. The low input resistance of the CB amplifier makes it useful only in specific applications. As we shall see in Chapter 6, it has a much better high-frequency response than the CE amplifier. This superiority will make it useful as a high-frequency amplifier, especially when combined with the CE circuit. We shall see one such combination in Chapter 6.

4. The emitter follower finds application as a voltage buffer for connecting a high-resistance source to a low-resistance load and as the output stage in a multistage amplifier.

Finally, we should point out that the Exercises in this section (except for that relating to the emitter follower) used the same component values to allow numerical comparisons.

TABLE 5.6 **Characteristics of Single-Stage Discrete BJT Amplifiers**

Common Emitter

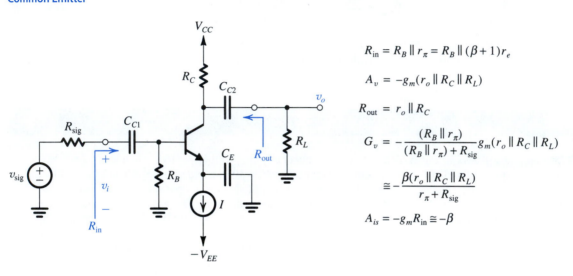

$$R_{\text{in}} = R_B \parallel r_\pi = R_B \parallel (\beta + 1) r_e$$

$$A_v = -g_m(r_o \parallel R_C \parallel R_L)$$

$$R_{\text{out}} = r_o \parallel R_C$$

$$G_v = -\frac{(R_B \parallel r_\pi)}{(R_B \parallel r_\pi) + R_{\text{sig}}} g_m(r_o \parallel R_C \parallel R_L)$$

$$\cong -\frac{\beta(r_o \parallel R_C \parallel R_L)}{r_\pi + R_{\text{sig}}}$$

$$A_{is} = -g_m R_{\text{in}} \cong -\beta$$

Common Emitter with Emitter Resistance

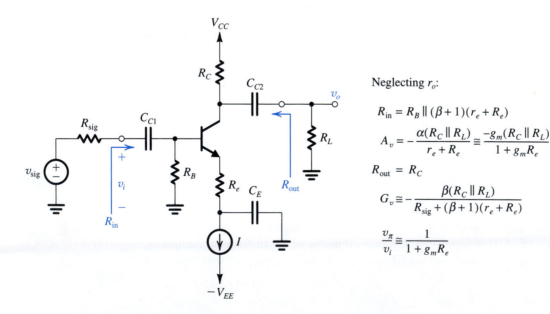

Neglecting r_o:

$$R_{\text{in}} = R_B \parallel (\beta + 1)(r_e + R_e)$$

$$A_v = -\frac{\alpha(R_C \parallel R_L)}{r_e + R_e} \cong \frac{-g_m(R_C \parallel R_L)}{1 + g_m R_e}$$

$$R_{\text{out}} = R_C$$

$$G_v \cong -\frac{\beta(R_C \parallel R_L)}{R_{\text{sig}} + (\beta + 1)(r_e + R_e)}$$

$$\frac{v_\pi}{v_i} \cong \frac{1}{1 + g_m R_e}$$

Common Base

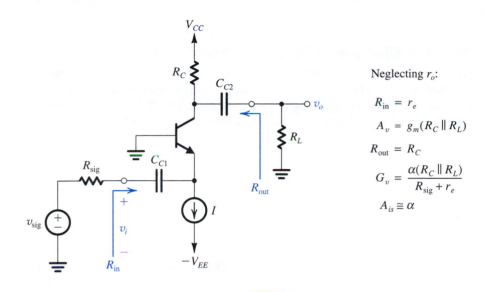

Neglecting r_o:

$$R_{in} = r_e$$

$$A_v = g_m(R_C \parallel R_L)$$

$$R_{out} = R_C$$

$$G_v = \frac{\alpha(R_C \parallel R_L)}{R_{sig} + r_e}$$

$$A_{is} \cong \alpha$$

Common Collector or Emitter Follower

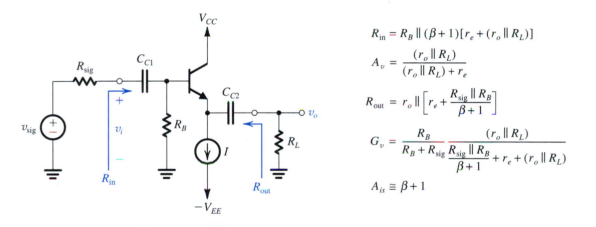

$$R_{in} = R_B \parallel (\beta + 1)[r_e + (r_o \parallel R_L)]$$

$$A_v = \frac{(r_o \parallel R_L)}{(r_o \parallel R_L) + r_e}$$

$$R_{out} = r_o \parallel \left[r_e + \frac{R_{sig} \parallel R_B}{\beta + 1} \right]$$

$$G_v = \frac{R_B}{R_B + R_{sig}} \frac{(r_o \parallel R_L)}{\frac{R_{sig} \parallel R_B}{\beta + 1} + r_e + (r_o \parallel R_L)}$$

$$A_{is} \cong \beta + 1$$

5.8 THE BJT INTERNAL CAPACITANCES AND HIGH-FREQUENCY MODEL

Thus far we have assumed transistor action to be instantaneous, and as a result the transistor models we have developed do not include any elements (i.e., capacitors or inductors) that would cause time or frequency dependence. Actual transistors, however, exhibit charge-storage phenomena that limit the speed and frequency of their operation. We have already encountered such effects in our study of the *pn* junction in Chapter 3 and learned that they can be modeled using capacitances. In the following we study the charge-storage effects that take place in the BJT and take them into account by adding capacitances to the hybrid-π

model. The resulting augmented BJT model will be able to predict the observed dependence of amplifier gain on frequency and the time delays that transistor switches and logic gates exhibit.

5.8.1 The Base-Charging or Diffusion Capacitance C_{de}

When the transistor is operating in the active or saturation modes, minority-carrier charge is stored in the base region. In fact, we have already derived an expression for this charge, Q_n, in the case of an *npn* transistor operating in the active mode (Eq. 5.7). Using the result in Eq. (5.7) together with Eqs. (5.3) and (5.4), we can express Q_n in terms of the collector current i_C as

$$Q_n = \frac{W^2}{2D_n} i_C = \tau_F i_C \tag{5.150}$$

where τ_F is a device constant,

$$\tau_F = \frac{W^2}{2D_n} \tag{5.151}$$

with the dimension of time. It is known as the **forward base-transit time** and represents the average time a charge carrier (electron) spends in crossing the base. Typically, τ_F is in the range of 10 ps to 100 ps. For operation in the reverse active mode, a corresponding constant τ_R applies and is many orders of magnitude larger than τ_F.

Equation (5.150) applies for large signals and, since i_C is exponentially related to v_{BE}, Q_n will similarly depend on v_{BE}. Thus this charge-storage mechanism represents a nonlinear capacitive effect. However, for small signals we can define the **small-signal diffusion capacitance C_{de}**,

$$C_{de} \equiv \frac{dQ_n}{dv_{BE}} \tag{5.152}$$

$$= \tau_F \frac{di_C}{dv_{BE}}$$

resulting in

$$C_{de} = \tau_F g_m = \tau_F \frac{I_C}{V_T} \tag{5.153}$$

5.8.2 The Base–Emitter Junction Capacitance C_{je}

Using the development in Chapter 3, and in particular Eq. (3.58), the base–emitter junction or depletion-layer capacitance C_{je} can be expressed as

$$C_{je} = \frac{C_{je0}}{\left(1 - \dfrac{V_{BE}}{V_{0e}}\right)^m} \tag{5.154}$$

where C_{je0} is the value of C_{je} at zero voltage, V_{0e} is the EBJ built-in voltage (typically, 0.9 V), and m is the grading coefficient of the EBJ junction (typically, 0.5). It turns out, however, that because the EBJ is forward biased in the active mode, Eq. (5.154) does not provide an accurate prediction of C_{je}. Alternatively, one typically uses an approximate value for C_{je},

$$C_{je} \cong 2C_{je0} \tag{5.155}$$

5.8.3 The Collector–Base Junction Capacitance C_μ

In active-mode operation, the CBJ is reverse biased, and its junction or **depletion capacitance**, usually denoted C_μ, can be found from

$$C_\mu = \frac{C_{\mu 0}}{\left(1 + \dfrac{V_{CB}}{V_{0c}}\right)^m} \tag{5.156}$$

where $C_{\mu 0}$ is the value of C_μ at zero voltage, V_{0c} is the CBJ built-in voltage (typically, 0.75 V), and m is its grading coefficient (typically, 0.2–0.5).

5.8.4 The High-Frequency Hybrid-π Model

Figure 5.67 shows the hybrid-π model of the BJT, including capacitive effects. Specifically, there are two capacitances: the emitter–base capacitance $C_\pi = C_{de} + C_{je}$ and the collector–base capacitance C_μ. Typically, C_π is in the range of a few picofarads to a few tens of picofarads, and C_μ is in the range of a fraction of a picofarad to a few picofarads. Note that we have also added a resistor r_x to model the resistance of the silicon material of the base region between the base terminal B and a fictitious internal, or intrinsic, base terminal B′ that is right under the emitter region (refer to Fig. 5.6). Typically, r_x is a few tens of ohms, and its value depends on the current level in a rather complicated manner. Since (usually) $r_x \ll r_\pi$, its effect is negligible at low frequencies. Its presence is felt, however, at high frequencies, as will become apparent later.

The values of the hybrid-π equivalent circuit parameters can be determined at a given bias point using the formulas presented in this chapter. They can also be found from the terminal measurements specified on the BJT data sheets. For computer simulation, SPICE uses the parameters of the given IC technology to evaluate the BJT model parameters (Section 5.11).

Before proceeding, a note on notation is in order. Since we are now dealing with voltages and currents that are functions of frequency, we have reverted to using symbols that are uppercase letters with lowercase subscripts (e.g., V_π, I_c). This conforms to the notation system used throughout this book.

5.8.5 The Cutoff Frequency

The transistor data sheets do not usually specify the value of C_π. Rather, the behavior of β (or h_{fe}) versus frequency is normally given. In order to determine C_π and C_μ we shall derive an expression for h_{fe}, the CE short-circuit current gain, as a function of frequency in terms of

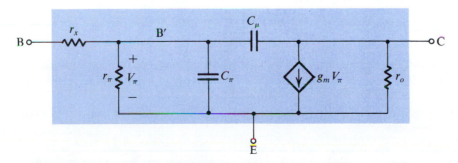

FIGURE 5.67 The high-frequency hybrid-π model.

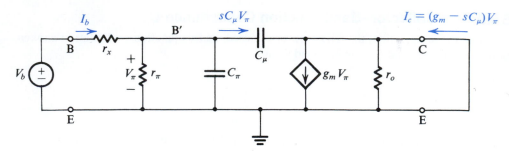

FIGURE 5.68 Circuit for deriving an expression for $h_{fe}(s) \equiv I_c / I_b$.

the hybrid-π components. For this purpose consider the circuit shown in Fig. 5.68, in which the collector is shorted to the emitter. A node equation at C provides the short-circuit collector current I_c as

$$I_c = (g_m - sC_\mu)V_\pi \qquad (5.157)$$

A relationship between V_π and I_b can be established by multiplying I_b by the impedance seen between B' and E:

$$V_\pi = I_b(r_\pi \,/\!/\, C_\pi \,/\!/\, C_\mu) = \frac{I_b}{1/r_\pi + sC_\pi + sC_\mu} \qquad (5.158)$$

Thus h_{fe} can be obtained by combining Eqs. (5.157) and (5.158):

$$h_{fe} \equiv \frac{I_c}{I_b} = \frac{g_m - sC_\mu}{1/r_\pi + s(C_\pi + C_\mu)}$$

At the frequencies for which this model is valid, $g_m \gg \omega C_\mu$; thus we can neglect the sC_μ term in the numerator and write

$$h_{fe} \simeq \frac{g_m r_\pi}{1 + s(C_\pi + C_\mu)r_\pi}$$

Thus,

$$h_{fe} = \frac{\beta_0}{1 + s(C_\pi + C_\mu)r_\pi} \qquad (5.159)$$

where β_0 is the low-frequency value of β. Thus h_{fe} has a single-pole (or STC) response[10] with a 3-dB frequency at $\omega = \omega_\beta$, where

$$\omega_\beta = \frac{1}{(C_\pi + C_\mu)r_\pi} \qquad (5.160)$$

Figure 5.69 shows a Bode plot for $|h_{fe}|$. From the -6-dB/octave slope it follows that the frequency at which $|h_{fe}|$ drops to unity, which is called the **unity-gain bandwidth** ω_T, is given by

$$\omega_T = \beta_0 \omega_\beta \qquad (5.161)$$

---

[10] The frequency response of single-time-constant (STC) networks was reviewed in Section 1.6. Also, a more detailed discussion of this important topic can be found in Appendix D.

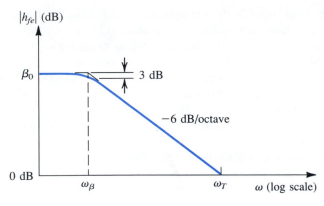

FIGURE 5.69 Bode plot for $\left|h_{fe}\right|$.

Thus,

$$\omega_T = \frac{g_m}{C_\pi + C_\mu} \tag{5.162}$$

and

$$f_T = \frac{g_m}{2\pi(C_\pi + C_\mu)} \tag{5.163}$$

The unity-gain bandwidth f_T is usually specified on the data sheets of a transistor. In some cases f_T is given as a function of I_C and V_{CE}. To see how f_T changes with I_C, recall that g_m is directly proportional to I_C, but only part of C_π (the diffusion capacitance C_{de}) is directly proportional to I_C. It follows that f_T decreases at low currents, as shown in Fig. 5.70. However, the decrease in f_T at high currents, also shown in Fig. 5.70, cannot be explained by this argument; rather it is due to the same phenomenon that causes β_0 to decrease at high currents. In the region where f_T is almost constant, C_π is dominated by the diffusion part.

Typically, f_T is in the range of 100 MHz to tens of GHz. The value of f_T can be used in Eq. (5.163) to determine $C_\pi + C_\mu$. The capacitance C_μ is usually determined separately by measuring the capacitance between base and collector at the desired reverse-bias voltage V_{CB}.

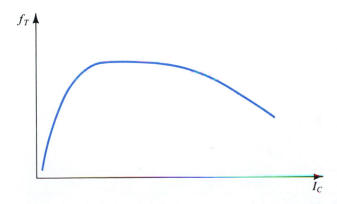

FIGURE 5.70 Variation of f_T with I_C.

Before leaving this section, we should mention that the hybrid-π model of Fig. 5.68 characterizes transistor operation fairly accurately up to a frequency of about $0.2f_T$. At higher frequencies one has to add other parasitic elements to the model as well as refine the model to account for the fact that the transistor is in fact a distributed-parameter network that we are trying to model with a lumped-component circuit. One such refinement consists of splitting r_x into a number of parts and replacing C_μ by a number of capacitors, each connected between the collector and one of the taps of r_x. This topic is beyond the scope of this book.

An important observation to make from the high-frequency model of Fig. 5.68 is that at frequencies above 5 to $10f_\beta$, one may ignore the resistance r_π. It can be seen then that r_x becomes the only resistive part of the input impedance at high frequencies. Thus r_x plays an important role in determining the frequency response of transistor circuits at high frequencies. It follows that an accurate determination of r_x should be made from a high-frequency measurement.

EXERCISES

5.48 Find C_{de}, C_{je}, C_π, C_μ, and f_T for a BJT operating at a dc collector current $I_C = 1$ mA and a CBJ reverse bias of 2 V. The device has $\tau_F = 20$ ps, $C_{je0} = 20$ fF, $C_{\mu0} = 20$ fF, $V_{0e} = 0.9$ V, $V_{0c} = 0.5$ V, and $m_{CBJ} = 0.33$.

Ans. 0.8 pF; 40 fF; 0.84 pF; 12 fF; 7.47 GHz

5.49 For a BJT operated at $I_C = 1$ mA, determine f_T and C_π if $C_\mu = 2$ pF and $|h_{fe}| = 10$ at 50 MHz.

Ans. 500 MHz; 10.7 pF

5.50 If C_π of the BJT in Exercise 5.49 includes a relatively constant depletion-layer capacitance of 2 pF, find f_T of the BJT when operated at $I_C = 0.1$ mA.

Ans. 130.7 MHz

5.8.6 Summary

For convenient reference, Table 5.7 provides a summary of the relationships used to determine the values of the parameters of the BJT high-frequency model.

TABLE 5.7   The BJT High-Frequency Model

$$g_m = I_C/V_T \qquad r_o = |V_A|/I_C \qquad r_\pi = \beta_0/g_m$$

$$C_\pi + C_\mu = \frac{g_m}{2\pi f_T} \qquad C_\pi = C_{de} + C_{je} \qquad C_{de} = \tau_F g_m \qquad C_{je} \cong 2C_{je0}$$

$$C_\mu = C_{jc0} \Big/ \left(1 + \frac{V_{CB}}{V_{0c}}\right)^m, \qquad m = 0.3\text{–}0.5$$

5.9 FREQUENCY RESPONSE OF THE COMMON-EMITTER AMPLIFIER

In this section we study the dependence of the gain of the BJT common-emitter amplifier of Fig. 5.71(a) on the frequency of the input signal.

5.9.1 The Three Frequency Bands

When the common-emitter amplifier circuit of Fig. 5.71(a) was studied in Section 5.7.3, it was assumed that the coupling capacitors C_{C1} and C_{C2} and the bypass capacitor C_E were

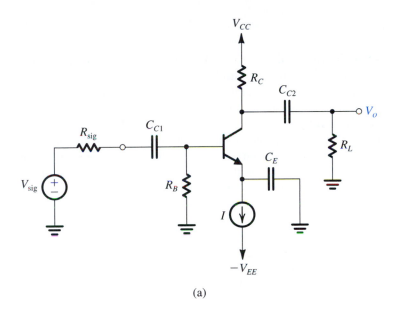

(a)

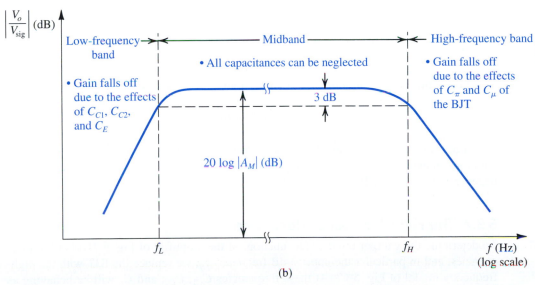

(b)

FIGURE 5.71 (a) Capacitively coupled common-emitter amplifier. (b) Sketch of the magnitude of the gain of the CE amplifier versus frequency. The graph delineates the three frequency bands relevant to frequency-response determination.

acting as perfect short circuits at all signal frequencies of interest. We also neglected the internal capacitances of the BJT. That is, C_π and C_μ of the BJT high-frequency model (Fig. 5.67) were assumed to be sufficiently small to act as open circuits at all signal frequencies of interest. As a result of ignoring all capacitive effects, the gain expressions derived in Section 5.7.3 were independent of frequency. In reality, however, this situation only applies over a limited, though usually wide, band of frequencies. This is illustrated in Fig. 5.71(b), which shows a sketch of the magnitude of the overall voltage gain, $|G_v|$, of the common-emitter amplifier versus frequency. We observe that the gain is almost constant over a wide frequency band, called the **midband.** The value of the midband gain A_M corresponds to the overall voltage gain G_v that we derived in Section 5.7.3, namely,

$$A_M = \frac{V_o}{V_{\text{sig}}} = -\frac{(R_B \| r_\pi)}{(R_B \| r_\pi) + R_{\text{sig}}} g_m (r_o \| R_C \| R_L) \tag{5.164}$$

Figure 5.71(b) shows that the gain falls off at signal frequencies below and above the midband. The gain falloff in the **low-frequency band** is due to the fact that even though C_{C1}, C_{C2}, and C_E are large capacitors (typically, in the μF range), as the signal frequency is reduced their impedances increase and they no longer behave as short circuits. On the other hand, the gain falls off in the **high-frequency band** as a result of C_{gs} and C_{gd}, which though very small (in the fraction of a pF to the pF range), their impedances at sufficiently high frequencies decrease; thus they can no longer be considered as open circuits. Our objective in this section is to study the mechanisms by which these two sets of capacitances affect the amplifier gain in the low-frequency and the high-frequency bands. In this way we will be able to determine the frequencies f_L and f_H, which define the extent of the midband, as shown in Fig. 5.71(b).

The midband is obviously the useful frequency band of the amplifier. Usually, f_L and f_H are the frequencies at which the gain drops by 3 dB below its value at midband; that is, at f_L and f_H, $|\text{gain}| = |A_M|/\sqrt{2}$. The amplifier **bandwidth** or 3-dB bandwidth is defined as the difference between the lower (f_L) and upper or higher (f_H) 3-dB frequencies:

$$BW \equiv f_H - f_L \tag{5.165}$$

Since usually $f_L \ll f_H$,

$$BW \cong f_H$$

A figure-of-merit for the amplifier is its gain–bandwidth product, defined as

$$GB = |A_M| BW \tag{5.166}$$

It will be shown at a later stage that in amplifier design, it is usually possible to trade off gain for bandwidth. One way of accomplishing this, for instance, is by including an emitter-degeneration resistance R_e, as we have done in Section 5.7.4.

5.9.2 The High-Frequency Response

To determine the gain, or the transfer function, of the amplifier of Fig. 5.71(a) at high frequencies, and in particular the upper 3-dB frequency f_H, we replace the BJT with the high-frequency model of Fig. 5.67. At these frequencies C_{C1}, C_{C2}, and C_E will be behaving as perfect short circuits. The result is the high-frequency amplifier equivalent circuit shown in Fig. 5.72(a).

The equivalent circuit of Fig. 5.72(a) can be simplified by utilizing Thévenin theorem at the input side and by combining the three parallel resistances at the output side. Specifically, the reader should be able to show that applying Thévenin theorem *twice* simplifies the resistive network at the input side to a signal generator V'_{sig} and a resistance R'_{sig},

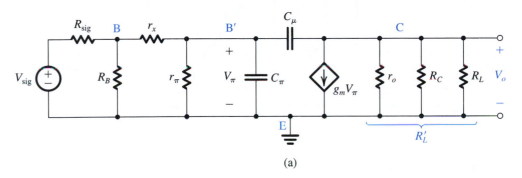

(a)

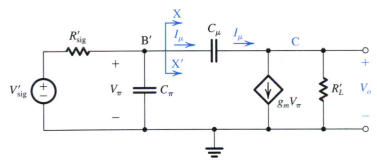

$$V'_{sig} = V_{sig} \frac{R_B}{R_B + R_{sig}} \frac{r_\pi}{r_\pi + r_x + (R_{sig}//R_B)} \qquad R'_L = r_o // R_C // R_L$$

$$R'_{sig} = r_\pi // [r_x + (R_B // R_{sig})]$$

(b)

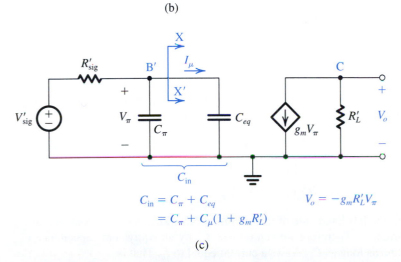

$$C_{in} = C_\pi + C_{eq} \qquad\qquad V_o = -g_m R'_L V_\pi$$
$$= C_\pi + C_\mu (1 + g_m R'_L)$$

(c)

FIGURE 5.72 Determining the high-frequency response of the CE amplifier: **(a)** equivalent circuit; **(b)** the circuit of (a) simplified at both the input side and the output side; **(c)** equivalent circuit with C_μ replaced at the input side with the equivalent capacitance C_{eq};

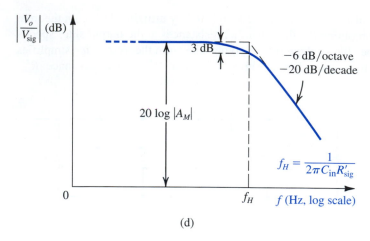

(d)

FIGURE 5.72 *(Continued)* **(d)** sketch of the frequency-response plot, which is that of a low-pass STC circuit.

where

$$V'_{\text{sig}} = V_{\text{sig}} \frac{R_B}{R_B + R_{\text{sig}}} \frac{r_\pi}{r_\pi + r_x + (R_{\text{sig}} \parallel R_B)} \tag{5.167}$$

$$R'_{\text{sig}} = r_\pi \parallel [r_x + (R_B \parallel R_{\text{sig}})] \tag{5.168}$$

Observe that R'_{sig} is the resistance seen looking back into the resistive network between nodes B′ and E.

The circuit in Fig. 5.72(b) can be simplified further if we can find a way to deal with the bridging capacitance C_μ that connects the output node to the "input" node, B′. Toward that end consider first the output node. It can be seen that the load current is $(g_m V_\pi - I_\mu)$, where $g_m V_\pi$ is the output current of the transistor and I_μ is the current supplied through the very small capacitance C_μ. In the vicinity of f_H, which is close to the edge of the midband, it is reasonable to assume that I_μ is still much smaller than $g_m V_\pi$, with the result that V_o can be given approximately by

$$V_o \cong -g_m V_\pi R'_L = -g_m R'_L V_\pi \tag{5.169}$$

Since $V_o = V_{ce}$, Eq. (5.169) indicates that the gain from B′ to C is $-g_m R'_L$, the same value as in the midband. The current I_μ can now be found from

$$
\begin{aligned}
I_\mu &= sC_\mu (V_\pi - V_o) \\
&= sC_\mu [V_\pi - (-g_m R'_L V_\pi)] \\
&= sC_\mu (1 + g_m R'_L) V_\pi
\end{aligned}
$$

Now, in Fig. 5.72(b), the left-hand-side of the circuit, at XX′, knows of the existence of C_μ only through the current I_μ. Therefore we can replace C_μ by an equivalent capacitance C_{eq} between B′ and ground as long as C_{eq} draws a current equal to I_μ. That is,

$$sC_{eq} V_\pi = I_\mu = sC_\mu (1 + g_m R'_L) V_\pi$$

which results in

$$C_{eq} = C_\mu(1 + g_m R_L') \tag{5.170}$$

Using C_{eq} enables us to simplify the equivalent circuit at the input side to that shown in Fig. 5.72(c), which we recognize as a single-time-constant (STC) network of the low-pass type (see Section 1.6 and Appendix D). Therefore we can express V_π in terms of V_{sig}' as

$$V_\pi = V_{sig}' \frac{1}{1 + s/\omega_0} \tag{5.171}$$

where ω_0 is the corner frequency of the STC network composed of C_{in} and R_{sig}',

$$\omega_0 = 1/C_{in}R_{sig}' \tag{5.172}$$

where C_{in} is the total input capacitance at B′,

$$C_{in} = C_\pi + C_{eq} = C_\pi + C_\mu(1 + g_m R_L') \tag{5.173}$$

and R_{sig}' is the effective source resistance, given by Eq. (5.168). Combining Eqs. (5.169), (5.171), and (5.167) give the voltage gain in the high-frequency band as

$$\frac{V_o}{V_{sig}} = -\left[\frac{R_B}{R_B + R_{sig}} \frac{r_\pi \cdot g_m R_L'}{r_\pi + r_x + (R_{sig} \| R_B)}\right]\left(\frac{1}{1 + \dfrac{s}{\omega_0}}\right) \tag{5.174}$$

The quantity between the square brackets of Eq. (5.174) is the midband gain, and except for the fact that here r_x is taken into account, this expression is the same as that in Eq. (5.164). Thus,

$$\frac{V_o}{V_{sig}} = \frac{A_M}{1 + \dfrac{s}{\omega_0}} \tag{5.175}$$

from which we deduce that the upper 3-dB frequency f_H must be

$$f_H = \frac{\omega_0}{2\pi} = \frac{1}{2\pi C_{in}R_{sig}'} \tag{5.176}$$

Thus we see that the high-frequency response will be that of a low-pass STC network with a 3-dB frequency f_H determined by the time constant $C_{in}R_{sig}'$. Fig. 5.72(d) shows a sketch of the magnitude of the high-frequency gain.

Before leaving this section we wish to make a number of observations:

1. The upper 3-dB frequency is determined by the interaction of R_{sig}' and C_{in}. If $R_B \gg R_{sig}$ and $r_x \ll R_{sig}$, then $R_{sig}' \cong R_{sig} \| r_\pi$. Thus the extent to which R_{sig} determines f_H depends on its value relative to r_π: If $R_{sig} \gg r_\pi$, then $R_{sig}' \cong r_\pi$; on the other hand, if R_{sig} is on the order of or smaller than r_π, then it has much greater influence on the value of f_H.

2. The input capacitance C_{in} is usually dominated by C_{eq}, which in turn is made large by the multiplication effect that C_μ undergoes. Thus, although C_μ is usually very small, its effect on the amplifier frequency response can be significant as a result of its

multiplication by the factor $(1 + g_m R'_L)$, which is approximately equal to the midband gain of the amplifier.

3. The multiplication effect that C_μ undergoes comes about because it is connected between two nodes (B' and C) whose voltages are related by a large negative gain $(-g_m R'_L)$. This effect is known as the **Miller effect,** and $(1 + g_m R'_L)$ is known as the **Miller multiplier.** It is the Miller effect that causes the CE amplifier to have a large input capacitance C_{in} and hence a low f_H.

4. To extend the high-frequency response of a BJT amplifier, we have to find configurations in which the Miller effect is absent or at least reduced. We shall return to this subject at great length in Chapter 6.

5. The above analysis, resulting in an STC or a single-pole response, is a simplified one. Specifically, it is based on neglecting I_μ relative to $g_m V_\pi$, an assumption that applies well at frequencies not too much higher than f_H. A more exact analysis of the circuit in Fig. 5.72(a) will be considered in Chapter 6. The results above, however, are more than sufficient for our current needs.

EXAMPLE 5.18

It is required to find the midband gain and the upper 3-dB frequency of the common-emitter amplifier of Fig. 5.71(a) for the following case: $V_{CC} = V_{EE} = 10$ V, $I = 1$ mA, $R_B = 100$ kΩ, $R_C = 8$ kΩ, $R_{\text{sig}} = 5$ kΩ, $R_L = 5$ kΩ, $\beta_0 = 100$, $V_A = 100$ V, $C_\mu = 1$ pF, $f_T = 800$ MHz, and $r_x = 50$ Ω.

Solution

The transistor is biased at $I_C \cong 1$ mA. Thus the values of its hybrid-π model parameters are

$$g_m = \frac{I_C}{V_T} = \frac{1 \text{ mA}}{25 \text{ mV}} = 40 \text{ mA/V}$$

$$r_\pi = \frac{\beta_0}{g_m} = \frac{100}{40 \text{ mA/V}} = 2.5 \text{ k}\Omega$$

$$r_o = \frac{V_A}{I_C} = \frac{100 \text{ V}}{1 \text{ mA}} = 100 \text{ k}\Omega$$

$$C_\pi + C_\mu = \frac{g_m}{\omega_T} = \frac{40 \times 10^{-3}}{2\pi \times 800 \times 10^6} = 8 \text{ pF}$$

$$C_\mu = 1 \text{ pF}$$

$$C_\pi = 7 \text{ pF}$$

$$r_x = 50 \text{ }\Omega$$

The midband voltage gain is

$$A_M = -\frac{R_B}{R_B + R_{\text{sig}}} \frac{r_\pi}{r_\pi + r_x + (R_B \parallel R_{\text{sig}})} g_m R'_L$$

where

$$R'_L = r_o \parallel R_C \parallel R_L$$

$$= (100 \parallel 8 \parallel 5) \text{ k}\Omega = 3 \text{ k}\Omega$$

Thus,

$$g_m R'_L = 40 \times 3 = 120 \text{ V/V}$$

and

$$A_M = -\frac{100}{100 + 5} \times \frac{2.5}{2.5 + 0.05 + (100 \parallel 5)} \times 120$$

$$= -39 \text{ V/V}$$

and

$$20 \log |A_M| = 32 \text{ dB}$$

To determine f_H we first find C_{in},

$$C_{in} = C_\pi + C_\mu(1 + g_m R'_L)$$

$$= 7 + 1(1 + 120) = 128 \text{ pF}$$

and the effective source resistance R'_{sig},

$$R'_{sig} = r_\pi \parallel [r_x + (R_B \parallel R_{sig})]$$

$$= 2.5 \parallel [0.05 + (100 \parallel 5)]$$

$$= 1.65 \text{ k}\Omega$$

Thus,

$$f_H = \frac{1}{2\pi C_{in} R'_{sig}} = \frac{1}{2\pi \times 128 \times 10^{-12} \times 1.65 \times 10^3} = 754 \text{ kHz}$$

5.9.3 The Low-Frequency Response

To determine the low-frequency gain (or transfer function) of the common-emitter amplifier circuit, we show in Fig. 5.73(a) the circuit with the dc sources eliminated (current source I open circuited and voltage source V_{CC} short circuited). We shall perform the small-signal analysis directly on this circuit. We will, of course, ignore C_π and C_μ since at such low frequencies their impedances will be very high and thus can be considered as open circuits. Also, to keep the analysis simple and thus focus attention on the mechanisms that limit the amplifier gain at low frequencies, we will neglect r_o. The reader can verify through SPICE simulation that the effect of r_o on the low-frequency amplifier gain is small. Finally, we shall also neglect r_x, which is usually much smaller than r_π, with which it appears in series.

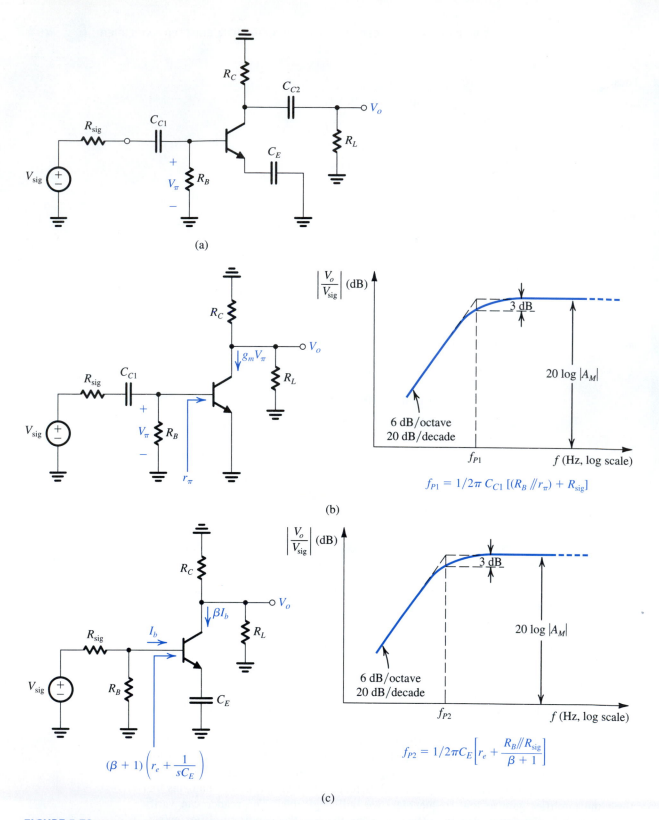

FIGURE 5.73 Analysis of the low-frequency response of the CE amplifier: **(a)** amplifier circuit with dc sources removed; **(b)** the effect of C_{C1} is determined with C_E and C_{C2} assumed to be acting as perfect short circuits; **(c)** the effect of C_E is determined with C_{C1} and C_{C2} assumed to be acting as perfect short circuits;

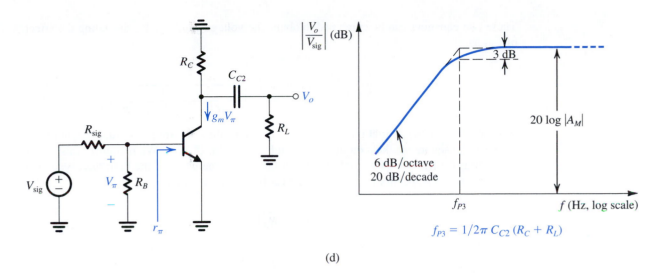

$$f_{P3} = 1/2\pi\, C_{C2}\,(R_C + R_L)$$

(d)

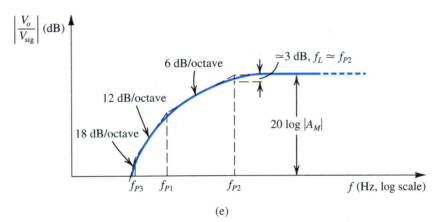

(e)

FIGURE 5.73 *(Continued)* **(d)** the effect of C_{C2} is determined with C_{C1} and C_E assumed to be acting as perfect short circuits; **(e)** sketch of the low-frequency gain under the assumptions that C_{C1}, C_E, and C_{C2} do not interact and that their break (or pole) frequencies are widely separated.

Our first cut at the analysis of the circuit in Fig. 5.72(a) is to consider the effect of the three capacitors C_{C1}, C_E, and C_{C2} *one at a time*. That is, when finding the effect of C_{C1}, we shall assume that C_E and C_{C2} are acting as perfect short circuits, and when considering C_E, we assume that C_{C1} and C_{C2} are perfect short circuits, and so on. This is obviously a major simplifying assumption—and one that might not be justified. However, it should serve as a first cut at the analysis enabling us to gain insight into the effect of these capacitances.

Figure 5.72(b) shows the circuit with C_E and C_{C2} replaced with short circuits. The voltage V_π at the base of the transistor can be written as

$$V_\pi = V_{sig} \frac{R_B \parallel r_\pi}{(R_B \parallel r_\pi) + R_{sig} + \dfrac{1}{sC_{C1}}}$$

and the output voltage is obtained as

$$V_o = -g_m V_\pi (R_C \parallel R_L)$$

These two equations can be combined to obtain the voltage gain V_o/V_{sig} including the effect of C_{C1} as

$$\frac{V_o}{V_{sig}} = -\frac{(R_B \parallel r_\pi)}{(R_B \parallel r_\pi) + R_{sig}} g_m(R_C \parallel R_L)\left[\frac{s}{s + \dfrac{1}{C_{C1}[(R_B \parallel r_\pi) + R_{sig}]}}\right] \tag{5.177}$$

from which we observe that the effect of C_{C1} is to introduce the frequency-dependent factor between the square brackets on the right-hand side of Eq. (5.177). We recognize this factor as the transfer fraction of a single-time-constant (STC) network of the high-pass type (see Section 1.6 and Appendix D) with a corner (or break) frequency ω_{P1},

$$\omega_{P1} = \frac{1}{C_{C1}[(R_B \parallel r_\pi) + R_{sig}]} \tag{5.178}$$

Note that $[(R_B \parallel r_\pi) + R_{sig}]$ is the *resistance seen between the terminals of C_{C1} when V_{sig} is set to zero.* The STC high-pass factor introduced by C_{C1} will cause the amplifier gain to roll off at low frequencies at the rate of 6 dB/octave (20 dB/decade) with a 3-dB frequency at $f_{P1} = \omega_{P1}/2\pi$, as indicated in Fig. 5.73(b).

Next, we consider the effect of C_E. For this purpose we assume that C_{C1} and C_{C2} are acting as perfect short circuits and thus obtain the circuit in Fig. 5.73(c). Reflecting r_e and C_E into the base circuit and utilizing Thévenin theorem enables us to obtain the base current as

$$I_b = V_{sig}\frac{R_B}{R_B + R_{sig}}\frac{1}{(R_B \parallel R_{sig}) + (\beta+1)\left(r_e + \dfrac{1}{sC_E}\right)}$$

The collector current can then be found as βI_b and the output voltage as

$$V_o = -\beta I_b(R_C \parallel R_L)$$

$$= -\frac{R_B}{R_B + R_{sig}}\frac{\beta(R_C \parallel R_L)}{(R_B \parallel R_{sig}) + (\beta+1)\left(r_e + \dfrac{1}{sC_E}\right)}V_{sig}$$

Thus the voltage gain including the effect of C_E can be expressed as

$$\frac{V_o}{V_{sig}} = -\frac{R_B}{R_B + R_{sig}}\frac{\beta(R_C \parallel R_L)}{(R_B \parallel R_{sig}) + (\beta+1)r_e}\frac{s}{s + \left[1/C_E\left(r_e + \dfrac{R_B \parallel R_{sig}}{\beta+1}\right)\right]} \tag{5.179}$$

We observe that C_E introduces the STC high-pass factor on the extreme right-hand side. Thus C_E causes the gain to fall off at low frequency at the rate of 6 dB/octave with a 3-dB frequency equal to the corner (or break) frequency of the high-pass STC factor; that is,

$$\omega_{P2} = \frac{1}{C_E\left[r_e + \dfrac{R_B \parallel R_{sig}}{\beta+1}\right]} \tag{5.180}$$

Observe that $[r_e + ((R_B \parallel R_{sig})/(\beta+1))]$ is the *resistance seen between the two terminals of C_E when V_{sig} is set to zero.* The effect of C_E on the amplifier frequency response is illustrated by the sketch in Fig. 5.73(c).

Finally, we consider the effect of C_{C2}. The circuit with C_{C1} and C_E assumed to be acting as perfect short circuits is shown in Fig. 5.73(d), for which we can write

$$V_\pi = V_{sig} \frac{R_B \| r_\pi}{(R_B \| r_\pi) + R_{sig}}$$

and

$$V_o = -g_m V_\pi \frac{R_C}{R_C + \dfrac{1}{sC_{C2}} + R_L} R_L$$

These two equations can be combined to obtain the low-frequency gain including the effect of C_{C2} as

$$\frac{V_o}{V_{sig}} = - \frac{R_B \| r_\pi}{(R_B \| r_\pi) + R_{sig}} g_m (R_C \| R_L) \left[\frac{s}{s + \dfrac{1}{C_{C2}(R_C + R_L)}} \right] \qquad (5.181)$$

We observe that C_{C2} introduces the frequency-dependent factor between the square brackets, which we recognize as the transfer function of a high-pass STC network with a break frequency ω_{P3},

$$\omega_{P3} = \frac{1}{C_{C2}(R_C + R_L)} \qquad (5.182)$$

Here we note that as expected, $(R_C + R_L)$ *is the resistance seen between the terminals of* C_{C2} *when* V_{sig} *is set to zero.* Thus capacitor C_{C2} causes the low-frequency gain of the amplifier to decrease at the rate of 6 dB/octave with a 3-dB frequency at $f_{P3} = \omega_{P3}/2\pi$, as illustrated by the sketch in Fig. 5.73(d).

Now that we have determined the effects of each of C_{C1}, C_E, and C_{C2} acting alone, the question becomes what will happen when all three are present at the same time. This question has two parts: First, what happens when all three capacitors are present but do not interact? The answer is that the amplifier low-frequency gain can be expressed as

$$\frac{V_o}{V_{sig}} = -A_M \left(\frac{s}{s + \omega_{P1}} \right) \left(\frac{s}{s + \omega_{P2}} \right) \left(\frac{s}{s + \omega_{P3}} \right) \qquad (5.183)$$

from which we see that it acquires three break frequencies at f_{P1}, f_{P2}, and f_{P3}, all in the low-frequency band. If the three frequencies are widely separated, their effects will be distinct, as indicated by the sketch in Fig. 5.73(e). The important point to note here is that the 3-dB frequency f_L *is determined by the highest of the three break frequencies.* This is usually the break frequency caused by the bypass capacitor C_E, simply because the resistance that it sees is usually quite small. Thus, even if one uses a large value for C_E, f_{P2} is usually the highest of the three break frequencies.

If f_{P1}, f_{P2}, and f_{P3} are close together, none of the three dominates, and to determine f_L, we have to evaluate $|V_o/V_{sig}|$ in Eq. (5.183) and calculate the frequency at which it drops to $|A_M|/\sqrt{2}$. The work involved in doing this, however, is usually too great and is rarely justified in practice, particularly because in any case, Eq. (5.183) is an approximation based on the assumption that the three capacitors do not interact. This leads to the second part of the question: What happens when all three capacitors are present and interact? We do know that C_{C1} and C_E usually interact and that their combined effect is two poles at frequencies that will differ somewhat from ω_{P1} and ω_{P2}. Of course, one can derive the overall transfer function taking this interaction into account and find more precisely the low-frequency response. This, however, will be too complicated to yield additional insight. As an alternative, for hand calculations we can obtain a reasonably good estimate for f_L using the following formula (which we will not derive here)[11]:

$$f_L \cong \frac{1}{2\pi}\left[\frac{1}{C_{C1}R_{C1}} + \frac{1}{C_E R_E} + \frac{1}{C_{C2}R_{C2}}\right] \tag{5.184}$$

or equivalently,

$$f_L = f_{P1} + f_{P2} + f_{P3} \tag{5.185}$$

where R_{C1}, R_E, and R_{C2} are the resistances seen by C_{C1}, C_E, and C_{C2}, respectively, when V_{sig} is set to zero and the other two capacitances are replaced with short circuits. Equations (5.184) and (5.185) provide insight regarding the relative contributions of the three capacitors to f_L. Finally, we note that a far more precise determination of the low-frequency gain and the 3-dB frequency f_L can be obtained using SPICE (Section 5.11).

Selecting Values for C_{C1}, C_E, and C_{C2} We now address the design issue of selecting appropriate values for C_{C1}, C_E, and C_{C2}. The design objective is to place the lower 3-dB frequency f_L at a specified location while minimizing the capacitor values. Since as mentioned above C_E usually sees the lowest of the three resistances, the total capacitance is minimized by selecting C_E so that its contribution to f_L is dominant. That is, by reference to Eq. (5.184), we may select C_E such that $1/(C_E R_E)$ is, say, 80% of $\omega_L = 2\pi f_L$, leaving each of the other capacitors to contribute 10% to the value of ω_L. Example 5.19 should help to illustrate this process.

EXAMPLE 5.19

We wish to select appropriate values for C_{C1}, C_{C2}, and C_E for the common-emitter amplifier whose high-frequency response was analyzed in Example 5.18. The amplifier has $R_B = 100$ kΩ, $R_C = 8$ kΩ, $R_L = 5$ kΩ, $R_{sig} = 5$ kΩ, $\beta_0 = 100$, $g_m = 40$ mA/V, and $r_\pi = 2.5$ kΩ. It is required to have $f_L = 100$ Hz.

Solution

We first determine the resistances seen by the three capacitors C_{C1}, C_E, and C_{C2} as follows:

$$R_{C1} = (R_B \| r_\pi) + R_{sig}$$

$$= (100 \| 2.5) + 5 = 7.44 \text{ kΩ}$$

$$R_E = r_e + \frac{R_B \| R_{sig}}{\beta + 1}$$

$$= 0.025 + \frac{100 \| 5}{101} = 0.072 \text{ kΩ} = 72 \text{ Ω}$$

$$R_{C2} = R_C + R_L = 8 + 5 = 13 \text{ kΩ}$$

Now, selecting C_E so that it contributes 80% of the value of ω_L gives

$$\frac{1}{C_E \times 72} = 0.8 \times 2\pi \times 100$$

$$C_E = 27.6 \ \mu\text{F}$$

---

[11] The interested reader can refer to Chapter 7 of the fourth edition of this book.

Next, if C_{C1} is to contribute 10% of f_L,

$$\frac{1}{C_{C1} \times 7.44 \times 10^3} = 0.1 \times 2\pi \times 100$$

$$C_{C1} = 2.1 \ \mu F$$

Similarly, if C_{C2} is to contribute 10% of f_L, its value should be selected as follows:

$$\frac{1}{C_{C2} \times 13 \times 10^3} = 0.1 \times 2\pi \times 100$$

$$C_{C2} = 1.2 \ \mu F$$

In practice, we would select the nearest standard values for the three capacitors while ensuring that $f_L \leq 100$ Hz.

EXERCISE

5.52 A common-emitter amplifier has $C_{C1} = C_E = C_{C2} = 1 \ \mu F$, $R_B = 100 \ k\Omega$, $R_{sig} = 5 \ k\Omega$, $g_m = 40 \ mA/V$, $r_\pi = 2.5 \ k\Omega$, $R_C = 8 \ k\Omega$, and $R_L = 5 \ k\Omega$. Assuming that the three capacitors do not interact, find f_{P1}, f_{P2}, and f_{P3}, and hence estimate f_L.

Ans. 21.4 Hz; 2.21 kHz; 12.2 Hz; since $f_{P2} \gg f_{P1}$ and f_{P3}, $f_L \cong f_{P2} = 2.21$ kHz; using Eq. (5.185), a somewhat better estimate for f_L is obtained as 2.24 kHz

5.9.4 A Final Remark

The frequency response of the other amplifier configurations will be studied in Chapter 6.

5.10 THE BASIC BJT DIGITAL LOGIC INVERTER

The most fundamental component of a digital system is the logic inverter. In Section 1.7, the logic inverter was studied at a conceptual level, and the realization of the inverter using voltage-controlled switches was presented. Having studied the BJT, we can now consider its application in the realization of a simple logic inverter. Such a circuit is shown in Fig. 5.74. The reader will note that we have already studied this circuit in some detail. In fact, we used it in Section 5.3.4 to illustrate the operation of the BJT as a switch. The operation of the circuit as a

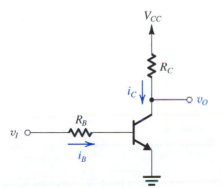

FIGURE 5.74 Basic BJT digital logic inverter.

logic inverter makes use of the cutoff and saturation modes. In very simple terms, if the input voltage v_I is "high," at a value close to the power-supply voltage V_{CC}, (representing a logic 1 in a positive-logic system) the transistor will be conducting and, with appropriate choice of values for R_B and R_C, saturated. Thus the output voltage will be $V_{CEsat} \cong 0.2$ V, representing a "low" logic level or logic 0 in a positive logic system. Conversely, if the input voltage is "low," at a value close to ground (e.g., V_{CEsat}), then the transistor will be cut off, i_C will be zero, and $v_O = V_{CC}$, which is "high" or logic 1.

The choice of cutoff and saturation as the two modes of operation of the BJT in this inverter circuit is motivated by the following two factors:

1. The power dissipation in the circuit is relatively low in both cutoff and saturation: In cutoff all currents are zero (except for very small leakage currents), and in saturation the voltage across the transistor is very small (V_{CEsat}).

2. The output voltage levels (V_{CC} and V_{CEsat}) are well defined. In contrast, if the transistor is operated in the active region, $v_O = V_{CC} - i_C R_C = V_{CC} - \beta i_B R_C$, which is highly dependent on the rather ill-controlled transistor parameter β.

5.10.1 The Voltage Transfer Characteristic

As mentioned in Section 1.7, the most useful characterization of an inverter circuit is in terms of its voltage transfer characteristic, v_O versus v_I. A sketch of the voltage transfer characteristic (VTC) of the inverter circuit of Fig. 5.74 is presented in Fig. 5.75. The transfer characteristic is approximated by three straight-line segments corresponding to the operation of the BJT in the cutoff, active, and saturation regions, as indicated. The actual transfer characteristic will obviously be a smooth curve but will closely follow the straight-line asymptotes indicated. We shall now compute the coordinates of the breakpoints of the transfer

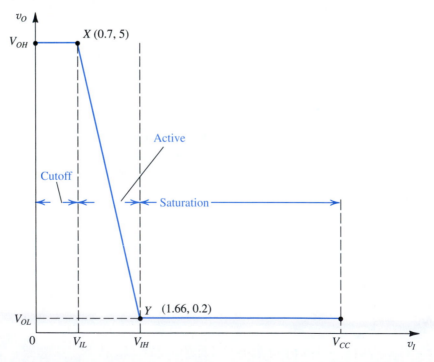

FIGURE 5.75 Sketch of the voltage transfer characteristic of the inverter circuit of Fig. 5.74 for the case $R_B = 10$ kΩ, $R_C = 1$ kΩ, $\beta = 50$, and $V_{CC} = 5$ V. For the calculation of the coordinates of X and Y, refer to the text.

characteristic of Fig. 5.75 for a representative case—$R_B = 10$ kΩ, $R_C = 1$ kΩ, $\beta = 50$, and $V_{CC} = 5$ V—as follows:

1. At $v_I = V_{OL} = V_{CEsat} = 0.2$ V, $v_O = V_{OH} = V_{CC} = 5$ V.

2. At $v_I = V_{IL}$, the transistor begins to turn on; thus,

$$V_{IL} \cong 0.7 \text{ V}$$

3. For $V_{IL} < v_I < V_{IH}$, the transistor is in the active region. It operates as an amplifier whose small-signal gain is

$$A_v \equiv \frac{v_o}{v_i} = -\beta \frac{R_C}{R_B + r_\pi}$$

    The gain depends on the value of r_π, which in turn is determined by the collector current and hence by the value of v_I. As the current through the transistor increases, r_π decreases and we can neglect r_π relative to R_B, thus simplifying the gain expression to

$$A_v \cong -\beta \frac{R_C}{R_B} = -50 \times \frac{1}{10} = -5 \text{ V/V}$$

4. At $v_I = V_{IH}$, the transistor enters the saturation region. Thus V_{IH} is the value of v_I that results in the transistor being at the edge of saturation,

$$I_B = \frac{(V_{CC} - V_{CEsat})/R_C}{\beta}$$

    For the values we are using, we obtain $I_B = 0.096$ mA, which can be used to find V_{IH},

$$V_{IH} = I_B R_B + V_{BE} = 1.66 \text{ V}$$

5. For $v_I = V_{OH} = 5$ V, the transistor will be deep into saturation with $v_O = V_{CEsat} \cong 0.2$ V, and

$$\beta_{\text{forced}} = \frac{(V_{CC} - V_{CEsat})/R_C}{(V_{OH} - V_{BE})/R_B}$$

$$= \frac{4.8}{0.43} = 11$$

6. The noise margins can now be computed using the formulas from Section 1.7,

$$NM_H = V_{OH} - V_{IH} = 5 - 1.66 = 3.34 \text{ V}$$
$$NM_L = V_{IL} - V_{OL} = 0.7 - 0.2 = 0.5 \text{ V}$$

    Obviously, the two noise margins are vastly different, making this inverter circuit less than ideal.

7. The gain in the transition region can be computed from the coordinates of the breakpoints X and Y,

$$\text{Voltage gain} = -\frac{5 - 0.2}{1.66 - 0.7} = -5 \text{ V/V}$$

    which is equal to the approximate value found above (the fact that it is exactly the same value is a coincidence).

5.10.2 Saturated Versus Nonsaturated BJT Digital Circuits

The inverter circuit just discussed belongs to the saturated variety of BJT digital circuits. A historically significant family of saturated BJT logic circuits is **transistor-transistor logic**

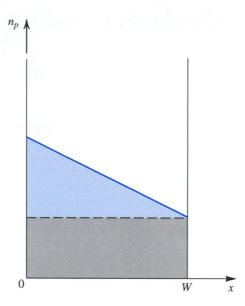

FIGURE 5.76 The minority-carrier charge stored in the base of a saturated transistor can be divided into two components: That in blue produces the gradient that gives rise to the diffusion current across the base, and that in gray results from driving the transistor deeper into saturation.

(TTL). Although some versions of TTL remains in use, saturated bipolar digital circuits generally are no longer the technology of choice in digital system design. This is because their speed of operation is severely limited by the relatively long time delay required to turn off a saturated transistor, as we will now explain, briefly.

In our study of BJT saturation in Section 5.1.5, we made use of the minority-carrier distribution in the base region (see Fig. 5.10). Such a distribution is shown in Fig. 5.76, where the minority carrier charge stored in the base has been divided into two components: The component represented by the blue triangle produces the gradient that gives rise to the diffusion current across the base; the other component, represented by the gray rectangle, causes the transistor to be driven deeper into saturation. The deeper the transistor is driven into saturation (i.e., the greater is the base overdrive factor), the greater the amount of the "gray" component of the stored charge will be. It is this "extra" stored base charge that represents a serious problem when it comes to turning off the transistor: Before the collector current can begin to decrease, all of the extra stored charge must first be removed. This adds a relatively large component to the turn-off time of a saturated transistor.

From the above we conclude that to achieve high operating speeds, the BJT should not be allowed to saturate. This is the case in **current-mode logic** in general and for the particular form called emitter-coupled logic (ECL), which will be studied briefly in Chapter 11. There, we will show why ECL is currently the highest-speed logic-circuit family available. It is based on the current-switching arrangement that was discussed conceptually in Section 1.7 (Fig. 1.33).

EXERCISE

5.53 Consider the inverter of Fig. 5.74 when v_I is low. Let the output be connected to the input terminals of N identical inverters. Convince yourself that the output level V_{OH} can be determined using the equivalent circuit shown in Fig. E5.53. Hence show that

$$V_{OH} = V_{CC} - R_C \frac{V_{CC} - V_{BE}}{R_C + R_B / N}$$

For $N = 5$, calculate V_{OH} using the component values of the example circuit discussed earlier (i.e., $R_B = 10$ kΩ, $R_C = 1$ kΩ, $V_{CC} = 5$ V). Note that this arrangement is historically important, as a precursor to the TTL logic form. It is called Resistor-Transistor Logic or RTL.

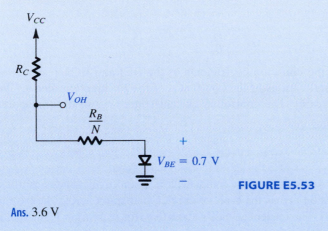

FIGURE E5.53

Ans. 3.6 V

5.11 THE SPICE BJT MODEL AND SIMULATION EXAMPLES

As we did in Chapter 4 for the MOSFET, we conclude this chapter with a discussion of the models that SPICE uses to simulate the BJT. We will also illustrate the use of SPICE in computing the dependence of β on the bias current and in simulating a CE amplifier.

5.11.1 The SPICE Ebers-Moll Model of the BJT

In Section 5.1.4, we studied the Ebers-Moll model of the BJT and showed a form of this model, known as the *injection* form, in Fig. 5.8. SPICE uses an equivalent form of the Ebers-Moll model, known as the *transport* form, which is shown in Fig. 5.77. Here, the currents of

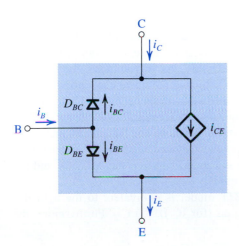

FIGURE 5.77 The transport form of the Ebers-Moll model for an *npn* BJT.

the base–emitter diode (D_{BE}) and the base–collector diode (D_{BC}) are given, respectively, by

$$i_{BE} = \frac{I_S}{\beta_F}(e^{v_{BE}/n_F V_T} - 1) \tag{5.186}$$

and

$$i_{BC} = \frac{I_S}{\beta_R}(e^{v_{BC}/n_R V_T} - 1) \tag{5.187}$$

where n_F and n_R are the emission coefficients of the BEJ and BCJ, respectively. These coefficients are generalizations of the constant n of the pn-junction diode. (We have so far assumed $n_F = n_R = 1$). The controlled current-source i_{CE} in the transport model is defined as

$$i_{CE} = I_S(e^{v_{BE}/n_F V_T} - e^{v_{BC}/n_R V_T}) \tag{5.188}$$

Observe that i_{CE} represents the current component of i_C and i_E that arises as a result of the minority-carrier diffusion across the base, or **carrier transport** across the base (hence the name transport model). The reader can easily show that, for $n_F = n_R = 1$, the relations

$$i_B = i_{BE} + i_{BC} \tag{5.189}$$

$$i_C = i_{CE} - i_{BC} \tag{5.190}$$

$$i_E = i_{CE} + i_{BE} \tag{5.191}$$

for the BJT currents in the transport model result in expressions identical to those derived in Eqs. (5.23), (5.26), and (5.27), respectively. Thus, the transport form (Fig. 5.77) of the Ebers-Moll model is exactly equivalent to its injection form (Fig. 5.8). Moreover, it has the advantage of being simpler, requiring only a single controlled source from collector to emitter. Hence, it is preferred for computer simulation.

The transport model can account for the Early effect (studied in Section 5.2.3) in a forward-biased BJT by including the factor $(1 - v_{BC}/V_A)$ in the expression for the transport current i_{CE} as follows:

$$i_{CE} = I_S(e^{v_{BE}/n_F V_T} - e^{v_{BC}/n_R V_T})\left(1 - \frac{v_{BC}}{V_A}\right) \tag{5.192}$$

Figure 5.78 shows the large-signal Ebers-Moll BJT model used in SPICE. It is based on the transport form of the Ebers-Moll model shown in Fig. 5.77. Here, resistors r_x, r_E, and r_C are added to represent the ohmic resistance of, respectively, the base, emitter, and collector regions. The dynamic operation of the BJT is modeled by two nonlinear capacitors, C_{BC} and C_{BE}. Each of these capacitors generally includes a diffusion component (i.e., C_{DC} and C_{DE}) and a depletion or junction component (i.e., C_{JC} and C_{JE}) to account for the charge-storage effects within the BJT (as described in Section 5.8). Furthermore, the BJT model includes a depletion junction capacitance C_{JS} to account for the collector–substrate junction in integrated-circuit BJTs, where a reverse-biased pn-junction is formed between the collector and the substrate (which is common to all components of the IC).

For small-signal (ac) analysis, the SPICE BJT model is equivalent to the hybrid-π model of Fig. 5.67, but augmented with r_E, r_C, and (for IC BJTs) C_{JS}. Furthermore, the

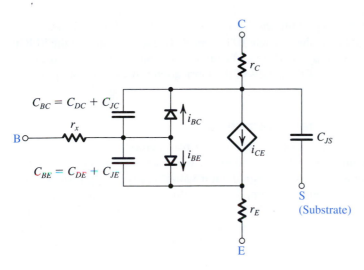

FIGURE 5.78   The SPICE large-signal Ebers-Moll model for an *npn* BJT.

model includes a large resistance r_μ between the base and collector (in parallel with C_μ) to account for the dependence of i_B on v_{CB}. This dependence can be noted from the CB characteristics of the BJT in Fig. 5.19(b), where i_C is observed to increase with v_{CB}: Since each i_C–v_{CB} curve in Fig. 5.19(b) is measured at a constant i_E, an increase in i_C with v_{CB} implies a corresponding decrease in i_B with v_{CB}. The resistance r_μ is very large, typically greater than $10\beta r_o$.

Although Fig. 5.77 shows the SPICE model for the *npn* BJT, the corresponding model for the *pnp* BJT can be obtained by reversing the direction of the currents and the polarity of the diodes and terminal voltages.

5.11.2 The SPICE Gummel-Poon Model of the BJT

The large-signal Ebers-Moll BJT model described in Section 5.11.1 lacks a representation of some second-order effects present in actual devices. One of the most important such effect is the variation of the current gains, β_F and β_R, with the current i_C. The Ebers-Moll model assumes β_F and β_R to be constant, thereby neglecting their current dependence (as depicted in Fig. 5.23). To account for this, and other second-order effects, SPICE uses a more accurate, yet more complex, BJT model called the Gummel-Poon model (named after Gummel and Poon, two pioneers in this field). This model is based on the relationship between the electrical terminal characteristics of a BJT and its base charge. It is beyond the scope of this book to delve into the model details. However, it is important for the reader to be aware of the existence of such a model.

In SPICE, the Gummel-Poon model automatically simplifies to the Ebers-Moll model when certain model parameters are not specified. Consequently, the BJT model to be used by SPICE need not be explicitly specified by the user (unlike the MOSFET case in which the model is specified by the LEVEL parameter). For discrete BJTs, the values of the SPICE model parameters can be determined from the data specified on the BJT data sheets, supplemented (if needed) by key measurements. For instance, in Example 5.20 (Section 5.11.4), we will use the Q2N3904 *npn* BJT (from Fairchild Semiconductor) whose SPICE model is available in PSpice. In fact, the PSpice library already includes the SPICE model parameters for many of the commercially available discrete BJTs. For IC BJTs, the values of the SPICE model parameters are determined by the IC manufacturer (using both measurements on the fabricated devices and knowledge of the details of the fabrication process) and are provided to the IC designers.

5.11.3 The SPICE BJT Model Parameters

Table 5.8 provides a listing of some of the BJT model parameters used in SPICE. The reader should be already familiar with these parameters. In the absence of a user-specified value for a particular parameter, SPICE uses a default value that typically results in the corresponding effect being ignored. For example, if no value is specified for the forward Early voltage VAF, SPICE assumes that VAF = ∞ and does not account for the Early effect. Although ignoring the forward Early voltage VAF can be a serious issue in some circuits, the same is not true, for example, for the value of the reverse Early voltage VAR.

5.11.4 The BJT Model Parameters BF and BR in SPICE

Before leaving the SPICE model, a comment on β is in order. SPICE interprets the user-specified model parameters BF and BR as the *ideal maximum* values of the forward and reverse dc current gains, repectively, versus the operating current. These parameters are not equal to the

TABLE 5.8 Parameters of the SPICE BJT Model (Partial Listing)

SPICE Parameter	Book Symbol	Description	Units
IS	I_S	Saturation current	A
BF		Ideal maximum forward current gain	
BR		Ideal maximum reverse current gain	
NF	n_F	Forward current emission coefficient	
NR	n_R	Reverse current emission coefficient	
VAF	V_A	Forward Early voltage	V
VAR		Reverse Early voltage	V
RB	r_x	Zero-bias base ohmic resistance	Ω
RC	r_C	Collector ohmic resistance	Ω
RE	r_E	Emitter ohmic resistance	Ω
TF	τ_F	Ideal forward transit time	s
TR	τ_R	Ideal reverse transit time	s
CJC	$C_{\mu 0}$	Zero-bias base-collector depletion (junction) capacitance	F
MJC	m_{BCJ}	Base-collector grading coefficient	
VJC	V_{0c}	Base-collector built-in potential	V
CJE	C_{je0}	Zero-bias base-emitter depletion (junction) capacitance	F
MJE	m_{BEJ}	Base-emitter grading coefficient	
VJE	V_{0e}	Base-emitter built-in potential	V
CJS		Zero-bias collector–substrate depletion (junction) capacitance	F
MJS		Collector–substrate grading coefficient	
VJS		Collector–substrate built-in potential	V

constant current-independent parameters β_F (β_{dc}) and β_R used in the Ebers-Moll model (and throughout this chapter) for the forward and reverse dc current gains of the BJT. SPICE uses a current-dependent model for β_F and β_R, and the user can specify other parameters (not shown in Table 5.8) for this model. Only when such parameters are not specified, and the Early effect is neglected, will SPICE assume that β_F and β_R are constant and equal to BF and BR, respectively. Furthermore, SPICE computes values for both β_{dc} and β_{ac}, the two parameters that we generally assume to be approximately equal. SPICE then uses β_{ac} to perform small-signal (ac) analysis.

EXAMPLE 5.20

DEPENDENCE OF β ON THE BIAS CURRENT

In this example, we use PSpice to simulate the dependence of β_{dc} on the collector bias current for the Q2N3904 discrete BJT (from Fairchild Semiconductor) whose model parameters are listed in Table 5.9 and are available in PSpice.[12] As shown in the Capture schematic[13] of Fig. 5.79, the V_{CE} of the BJT is fixed using a constant voltage source (in this example, $V_{CE} = 2$ V) and a dc current source I_B is applied at the base. To illustrate the dependence of β_{dc} on the collector current I_C, we perform a dc-analysis simulation in which the sweep variable is the current source I_B. The β_{dc} of the BJT, which corresponds to the ratio of the collector current I_C to the base current I_B, can then be plotted versus I_C using Probe (the graphical interface of PSpice), as shown in Fig. 5.80. We see that to operate at the maximum value of β_{dc} (i.e., $\beta_{dc} = 163$), at $V_{CE} = 2$ V, the BJT must be biased at an $I_C = 10$ mA. Since increasing the bias current of a transistor increases the power dissipation, it is clear from Fig. 5.80 that the choice of current I_C is a trade-off between the current gain β_{dc} and the power dissipation. Generally speaking, the optimum I_C depends on the application and technology in hand. For example, for the Q2N3904 BJT operating at $V_{CE} = 2$ V, decreasing I_C by a factor of 20 (from 10 mA to 0.5 mA) results in a drop in β_{dc} of about 25% (from 163 to 123).

TABLE 5.9	SPICE Model Parameters of the Q2N3904 Discrete BJT					
IS=6.734f	XTI=3	EG=1.11	VAF=74.03	BF=416.4	NE=1.259	ISE=6.734f
IKF=66.78m	XTB=1.5	BR=.7371	NC=2	ISC=0	IKR=0	RC=1
CJC=3.638p	MJC=.3085	VJC=.75	FC=.5	CJE=4.493p	MJE=.2593	VJE=.75
TR=239.5n	TF=301.2p	ITF=.4	VTF=4	XTF=2	RB=10	

---

[12] The Q2N3904 model is included in the evaluation (EVAL) library of PSpice (OrCad 9.2 Lite Edition) which is available on the CD accompanying this book.

[13] The reader is reminded that the Capture schematics and the corresponding PSpice simulation files of all SPICE examples in this book can be found on the text's CD as well as on its website (www.sedrasmith.org). In these schematics (as shown in Fig. 5.79), we use variable parameters to enter the values of the various circuit components. This allows one to investigate the effect of changing component values by simply changing the corresponding parameter values.

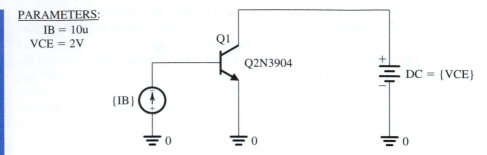

FIGURE 5.79 The PSpice testbench used to demonstrate the dependence of β_{dc} on the collector bias current I_C for the Q2N3904 discrete BJT (Example 5.20).

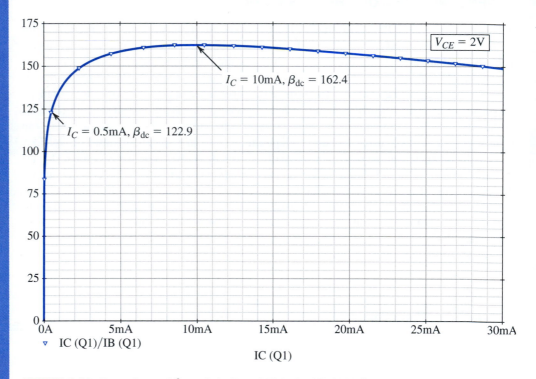

FIGURE 5.80 Dependence of β_{dc} on I_C (at $V_{CE} = 2$ V) in the Q2N3904 discrete BJT (Example 5.20).

EXAMPLE 5.21

THE CE AMPLIFIER WITH EMITTER RESISTANCE

In this example, we use PSpice to compute the frequency response of the CE amplifier and investigate its bias-point stability. A capture schematic of the CE amplifier is shown in Fig. 5.81. We will use part Q2N3904 for the BJT and a ±5-V power supply. We will also assume a signal-source resistor $R_{sig} = 10$ kΩ, a load resistor $R_L = 10$ kΩ, and bypass and coupling capacitors of

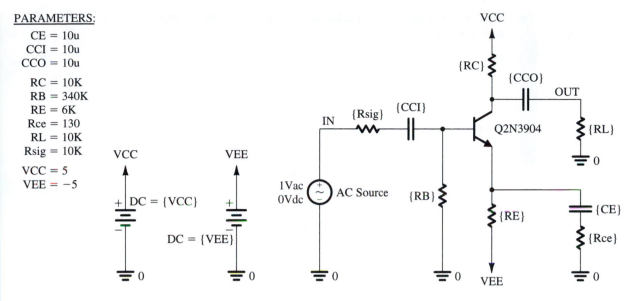

PARAMETERS:
    CE = 10u
    CCI = 10u
    CCO = 10u

    RC = 10K
    RB = 340K
    RE = 6K
    Rce = 130
    RL = 10K
    Rsig = 10K

    VCC = 5
    VEE = −5

FIGURE 5.81   Capture schematic of the CE amplifier in Example 5.21.

10 μF. To enable us to investigate the effect of including a resistance in the signal path of the emitter, a resistor R_{ce} is connected in series with the emitter bypass capacitor C_E. Note that the roles of R_E and R_{ce} are different. Resistor R_E is the **dc emitter degeneration resistor** because it appears in the dc path between the emitter and ground. It is therefore used to help stabilize the bias point for the amplifier. The equivalent resistance $R_e = R_E \parallel R_{ce}$ is the **small-signal emitter degeneration resistance** because it appears in the ac (small-signal) path between the emitter and ground and helps stabilize the gain of the amplifier. In this example, we will investigate the effects of both R_E and R_e on the performance of the CE amplifier. However, as should always be the case with computer simulation, we will begin with an approximate pencil-and-paper design. In this way, maximum advantage and insight can be obtained from simulation.

Based on the plot of β_{dc} versus I_C in Fig. 5.80, a collector bias current I_C of 0.5 mA is selected for the BJT, resulting in $\beta_{dc} = 123$. This choice of I_C is a reasonable compromise between power dissipation and current gain. Furthermore, a collector bias voltage V_C of 0 V (i.e., at the mid–supply rail) is selected to achieve a high signal swing at the amplifier output. For $V_{CE} = 2$ V, the result is that $V_E = -2$ V requires bias resistors with values

$$R_C = \frac{V_{CC} - V_C}{I_C} = 10 \text{ k}\Omega$$

and

$$R_E = \frac{V_E - V_{EE}}{I_C} = 6 \text{ k}\Omega$$

Assuming $V_{BE} = 0.7$ V and using $\beta_{dc} = 123$, we can determine

$$R_B = -\frac{V_B}{I_B} = -\frac{V_{BE} + V_E}{I_C/\beta_{dc}} = 320 \text{ k}\Omega$$

Next, the formulas of Section 5.7.4 can be used to determine the input resistance R_{in} and the midband voltage gain $|A_M|$ of the CE amplifier:

$$R_{\text{in}} = R_B \parallel (\beta_{\text{ac}} + 1)(r_e + R_e) \tag{5.193}$$

$$|A_M| = \left| -\frac{R_{\text{in}}}{R_{\text{sig}} + R_{\text{in}}} \times \frac{R_C \parallel R_L}{r_e + R_e} \right| \tag{5.194}$$

For simplicity, we will assume $\beta_{\text{ac}} \approx \beta_{\text{dc}} = 123$, resulting in

$$r_e = \left(\frac{\beta_{\text{ac}}}{\beta_{\text{ac}} + 1} \right) \left(\frac{V_T}{I_C} \right) = 49.6 \ \Omega$$

Thus, with no small-signal emitter degeneration (i.e., $R_{ce} = 0$), $R_{\text{in}} = 6.1$ kΩ and $|A_M| = 38.2$ V/V. Using Eq. (5.194) and assuming R_B is large enough to have a negligible effect on R_{in}, it can be shown that the emitter degeneration resistor R_e decreases the voltage gain $|A_M|$ by a factor of

$$\frac{1 + \dfrac{R_e}{r_e} + \dfrac{R_{\text{sig}}}{r_\pi}}{1 + \dfrac{R_{\text{sig}}}{r_\pi}}$$

Therefore, to limit the reduction in voltage gain to a factor of 2, we will select

$$R_e = r_e + \frac{R_{\text{sig}}}{\beta_{\text{ac}} + 1} \tag{5.195}$$

Thus, $R_{ce} \approx R_e = 130 \ \Omega$. Substituting this value in Eqs. (5.193) and (5.194) shows that R_{in} increases from 6.1 kΩ to 20.9 kΩ while $|A_M|$ drops from 38.2 V/V to 18.8 V/V.

We will now use PSpice to verify our design and investigate the performance of the CE amplifier. We begin by performing a bias-point simulation to verify that the BJT is properly biased in the active region and that the dc voltages and currents are within the desired specifications. Based on this simulation, we have increased the value of R_B to 340 kΩ in order to limit I_C to about 0.5 mA while using a standard 1% resistor value (Appendix G). Next, to measure the midband gain A_M and the 3-dB frequencies f_L and f_H, we apply a 1-V ac voltage at the input, perform an ac-analysis simulation, and plot the output voltage magnitude (in dB) versus frequency as shown in Fig. 5.82. This corresponds to the magnitude response of the CE amplifier because we chose a 1-V input signal.[14] Accordingly, with no emitter degeneration, the midband gain is $|A_M| = 38.5$ V/V = 31.7 dB and the 3-dB bandwidth is $BW = f_H - f_L = 145.7$ kHz. Using an $R_{ce} = 130 \ \Omega$ results in a drop in the midband gain $|A_M|$ by a factor of 2 (i.e., 6 dB). Interestingly, however, BW has now increased by approximately the same factor as the drop in $|A_M|$. As we will learn in Chapter 8 when we study negative feedback, the emitter-degeneration resistor R_{ce} provides negative feedback, which allows us to trade off gain for other desirable properties such as a larger input resistance, and a wider bandwidth.

To conclude this example, we will demonstrate the improved bias-point (or dc operating-point) stability achieved when an emitter resistor R_E is used (see the discussion in Section 5.5.1). Specifically, we will increase/decrease the value of the parameter BF (i.e., the ideal maximum

---

[14] The reader should not be alarmed about the use of such a large signal amplitude. Recall (Section 2.9.1) that, in a small-signal (ac) simulation, SPICE first finds the small-signal equivalent circuit at the dc bias point, and then analyzes this linear circuit. Such ac analysis can, of course, be done with any ac signal amplitude. However, a 1–V ac input is convenient to use as the resulting ac output corresponds to the voltage gain of the circuit.

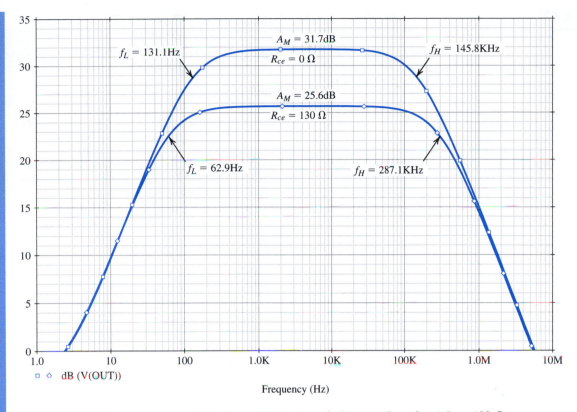

FIGURE 5.82 Frequency response of the CE amplifier in Example 5.21 with $R_{ce} = 0$ and $R_{ce} = 130$ Ω.

forward current gain) in the SPICE model for part Q2N3904 by a factor of 2 and perform a bias-point simulation. The corresponding change in BJT parameters (β_{dc} and β_{ac}) and bias-point (including I_C and V_{CE}) are presented in Table 5.10 for the case of $R_E = 6$ kΩ. Note that β_{ac} is not

TABLE 5.10   Variations in the Bias Point of the CE Amplifier with the SPICE Model-Parameter BF of BJT

BF (in SPICE)	$R_E = 6$ kΩ				$R_E = 0$			
	β_{ac}	β_{dc}	I_C (mA)	V_C (V)	β_{ac}	β_{dc}	I_C (mA)	V_C (V)
208	106	94.9	0.452	0.484	109	96.9	0.377	1.227
416.4 (nominal value)	143	123	0.494	0.062	148	127	0.494	0.060
832	173	144	0.518	−0.183	181	151	0.588	−0.878

equal to β_{dc} as we assumed, but is slightly larger. For the case without emitter degeneration, we will use $R_E = 0$ in the schematic of Fig. 5.81. Furthermore, to maintain the same I_C and V_C in both cases at the values obtained for nominal BF, we use $R_B = 1.12$ MΩ to limit I_C to approximately 0.5 mA. The corresponding variations in the BJT bias point are also shown in Table 5.10. Accordingly, we see that emitter degeneration makes the bias point of the CE amplifier much less sensitive to changes in β. However, unless a large bypass capacitor C_E is used, this reduced bias sensitivity comes at the expense of a reduction in the midband gain (as we observed in this example when we simulated the frequency response of the CE amplifier with an $R_e = 130$ Ω).

SUMMARY

■ Depending on the bias conditions on its two junctions, the BJT can operate in one of four possible modes: cutoff (both junctions reverse biased), active (the EBJ forward biased and the CBJ reverse biased), saturation (both junctions forward biased), and reverse-active (the EBJ reverse biased and the CBJ forward biased).

■ For amplifier applications, the BJT is operated in the active mode. Switching applications make use of the cutoff and saturation modes. The reverse-active mode of operation is of conceptual interest only.

■ A BJT operating in the active mode provides a collector current $i_C = I_S e^{|v_{BE}|/V_T}$. The base current $i_B = (i_C/\beta)$, and the emitter current $i_E = i_C + i_B$. Also, $i_C = \alpha i_E$, and thus $\beta = \alpha/(1-\alpha)$ and $\alpha = \beta/(\beta+1)$. See Table 5.2.

■ To ensure operation in the active mode, the collector voltage of an *npn* transistor must be kept higher than approximately 0.4 V below the base voltage. For a *pnp* transistor the collector voltage must be lower than approximately 0.4 V above the base voltage. Otherwise, the CBJ becomes forward biased, and the transistor enters the saturation region.

■ A convenient and intuitively appealing model for the large-signal operation of the BJT is the Ebers-Moll model shown in Fig. 5.8. A fundamental relationship between its parameters is $\alpha_F I_{SE} = \alpha_R I_{SC} = I_S$. While α_F is close to unity, α_R is very small (0.01–0.2), and β_R is correspondingly small. Use of the EM model enables expressing the terminal currents in terms of the voltages v_{BE} and v_{BC}. The resulting relationships are given in Eqs. (5.26) to (5.30).

■ In a saturated transistor, $|V_{CE\text{sat}}| \simeq 0.2$ V and $I_{C\text{sat}} = (V_{CC} - V_{CE\text{sat}})/R_C$. The ratio of $I_{C\text{sat}}$ to the base current is the forced β, which is lower than β. The collector-to-emitter resistance, $R_{CE\text{sat}}$, is small (few tens of ohms).

■ At a constant collector current, the magnitude of the base–emitter voltage decreases by about 2 mV for every 1°C rise in temperature.

■ With the emitter open-circuited ($i_E = 0$), the CBJ breaks down at a reverse voltage BV_{CBO} that is typically >50 V. For $i_E > 0$, the breakdown voltage is less than BV_{CBO}. In the common-emitter configuration the breakdown voltage specified is BV_{CEO}, which is about half BV_{CBO}. The emitter–base junction breaks down at a reverse bias of 6 V to 8 V. This breakdown usually has a permanent adverse effect on β.

■ A summary of the current-voltage characteristics and large-signal models of the BJTs in both the active and saturation modes of operation is presented in Table 5.3.

■ The dc analysis of transistor circuits is greatly simplified by assuming that $|V_{BE}| \simeq 0.7$ V.

■ To operate as a linear amplifier, the BJT is biased in the active region and the signal v_{be} is kept small ($v_{be} \ll V_T$).

■ For small signals, the BJT functions as a linear voltage-controlled current source with a transconductance $g_m = (I_C/V_T)$. The input resistance between base and emitter, looking into the base, is $r_\pi = \beta/g_m$. Simplified low-frequency equivalent-circuit models for the BJT are shown in Figs. 5.51 and 5.52. These models can be augmented by including the output resistance $r_o = |V_A|/I_C$ between the collector and the emitter. Table 5.4 provides a summary of the equations for determining the model parameters.

■ Bias design seeks to establish a dc collector current that is as independent of the value of β as possible.

■ In the common-emitter configuration, the emitter is at signal ground, the input signal is applied to the base, and the output is taken at the collector. A high voltage gain and a reasonably high input resistance are obtained, but the high-frequency response is limited.

■ The input resistance of the common-emitter amplifier can be increased by including an unbypassed resistance in the emitter lead. This emitter-degeneration resistance provides other performance improvements at the expense of reduced voltage gain.

■ In the common-base configuration, the base is at signal ground, the input signal is applied to the emitter, and the output is taken at the collector. A high voltage gain (from emitter to collector) and an excellent high-frequency response are obtained, but the input resistance is very low. The CB amplifier is useful as a current buffer.

■ In the emitter follower the collector is at signal ground, the input signal is applied to the base, and the output taken at the emitter. Although the voltage gain is less than unity, the input resistance is very high and the output resistance is very low. The circuit is useful as a voltage buffer.

■ Table 5.5 shows the parameters utilized to characterize amplifiers.

■ For a summary of the characteristics of discrete single-stage BJT amplifiers, refer to Table 5.6.

■ The high-frequency model of the BJT together with the formulas for determining its parameter values are shown in Table 5.7.

■ Analysis of the high-frequency gain of the CE amplifier in Section 5.9 shows that the gain rolls off at a slope of –6 dB/octave with the 3-dB frequency $f_H = 1/2\pi C_{\text{in}} R'_{\text{sig}}$. Here R'_{sig} is a modified value of R_{sig}, approximately equal to $R_{\text{sig}} \| r_\pi$, and $C_{\text{in}} = C_\pi + (1 + g_m R'_L) C_\mu$. The multiplication of C_μ by $(1 + g_m R'_L)$, known as the Miller effect, is the most significant factor limiting the high-frequency response of the CE amplifier.

■ For the analysis of the effect of C_{C1}, C_{C2}, and C_E on the low-frequency gain of the CE amplifier, refer to Section 5.9.3 and in particular to Fig. 5.73.

■ The basic BJT logic inverter utilizes the cutoff and saturation modes of transistor operation. A saturated transistor has a large amount of minority-carrier charge stored in its base region and is thus slow to turn off.

PROBLEMS

SECTION 5.1: DEVICE STRUCTURE AND PHYSICAL OPERATION

5.1 The terminal voltages of various *npn* transistors are measured during operation in their respective circuits with the following results:

Case	E	B	C	Mode
1	0	0.7	0.7	
2	0	0.8	0.1	
3	−0.7	0	0.7	
4	−0.7	0	−0.6	
5	0.7	0.7	0	
6	−2.7	−2.0	0	
7	0	0	5.0	
8	−0.10	5.0	5.0	

In this table, where the entries are in volts, 0 indicates the reference terminal to which the black (negative) probe of the voltmeter is connected. For each case, identify the mode of operation of the transistor.

5.2 An *npn* transistor has an emitter area of $10\ \mu m \times 10\ \mu m$. The doping concentrations are as follows: in the emitter $N_D = 10^{19}/cm^3$, in the base $N_A = 10^{17}/cm^3$, and in the collector $N_D = 10^{15}/cm^3$. The transistor is operating at $T = 300$ K, where $n_i = 1.5 \times 10^{10}/cm^3$. For electrons diffusing in the base, $L_n = 19\ \mu m$ and $D_n = 21.3\ cm^2/s$. For holes diffusing in the emitter, $L_p = 0.6\ \mu m$ and $D_p = 1.7\ cm^2/s$. Calculate I_S and β assuming that the base-width W is:

(a) $1\ \mu m$
(b) $2\ \mu m$
(c) $5\ \mu m$

For case (b), if $I_C = 1$ mA, find I_B, I_E, V_{BE}, and the minority-carrier charge stored in the base. (*Hint:* $\tau_b = L_n^2/D_n$. Recall that the electron charge $q = 1.6 \times 10^{-19}$ Coulomb.)

5.3 Two transistors, fabricated with the same technology but having different junction areas, when operated at a base-emitter voltage of 0.72 V, have collector currents of 0.2 mA and 12 mA. Find I_S for each device. What are the relative junction areas?

5.4 In a particular BJT, the base current is 7.5 μA, and the collector current is 400 μA. Find β and α for this device.

5.5 Find the values of β that correspond to α values of 0.5, 0.8, 0.9, 0.95, 0.99, 0.995, and 0.999.

5.6 Find the values of α that correspond to β values of 1, 2, 10, 20, 100, 200, 1000, and 2000.

5.7 Measurement of V_{BE} and two terminal currents taken on a number of *npn* transistors are tabulated below. For each, calculate the missing current value as well as α, β, and I_S as indicated by the table.

Transistor	a	b	c	d	e
V_{BE} (mV)	690	690	580	780	820
I_C (mA)	1.000	1.000		10.10	
I_B (μA)	50		7	120	1050
I_E (mA)		1.070	0.137		75.00
α					
β					
I_S					

5.8 Consider an *npn* transistor whose base–emitter drop is 0.76 V at a collector current of 10 mA. What current will it conduct at $v_{BE} = 0.70$ V? What is its base–emitter voltage for $i_C = 10\ \mu A$?

5.9 Show that for a transistor with α close to unity, if α changes by a small per-unit amount $(\Delta\alpha/\alpha)$ the corresponding per-unit change in β is given approximately by

$$\frac{\Delta\beta}{\beta} \simeq \beta\left(\frac{\Delta\alpha}{\alpha}\right)$$

5.10 An *npn* transistor of a type whose β is specified to range from 60 to 300 is connected in a circuit with emitter grounded, collector at +9 V, and a current of 50 μA injected into the base. Calculate the range of collector and emitter currents that can result. What is the maximum power dissipated in the transistor? (*Note:* Perhaps you can see why this is a bad way to establish the operating current in the collector of a BJT.)

5.11 A particular BJT when conducting a collector current of 10 mA is known to have $v_{BE} = 0.70$ V and $i_B = 100\ \mu A$. Use these data to create specific transistor models of the form shown in Figs. 5.5(a) and (b).

5.12 Using the *npn* transistor model of Fig. 5.5(b), consider the case of a transistor for which the base is connected to ground, the collector is connected to a 10-V dc source through a 2-kΩ resistor, and a 3-mA current source is connected to the emitter with the polarity so that current is drawn out of the emitter terminal. If $\beta = 100$ and $I_S = 10^{-15}$ A, find the voltages at the emitter and the collector and calculate the base current.

5.13 Consider an *npn* transistor for which $\beta_F = 100$, $\alpha_R = 0.1$, and $I_S = 10^{-15}$ A.

(a) If the transistor is operated in the forward active mode with $I_B = 10\ \mu A$ and $V_{CB} = 1$ V, find V_{BE}, I_C, and I_E.
(b) Now, operate the transistor in the reverse active mode with a forward-bias voltage V_{BC} equal to the value of V_{BE} found in (a) and with $V_{EB} = 1$ V. Find I_C, I_B, and I_E.

5.14 A transistor characterized by the Ebers-Moll model shown in Fig. 5.8 is operated with both emitter and collector grounded and a base current of 1 mA. If the collector junction is 10 times larger than the emitter junction and $\alpha_F \cong 1$, find i_C and i_E.

*5.15 (a) Use the Ebers-Moll expressions in Eqs. (5.26) and (5.27) to show that the i_C–v_{CB} relationship sketched in Fig. 5.9 can be described by

$$i_C = \alpha_F I_E - I_S \left(\frac{1}{\alpha_R} - \alpha_F \right) e^{v_{BC}/V_T}$$

(b) Calculate and sketch i_C–v_{CB} curves for a transistor for which $I_S = 10^{-15}$ A, $\alpha_F \cong 1$, and $\alpha_R = 0.1$. Sketch graphs for $I_E = 0.1$ mA, 0.5 mA, and 1 mA. For each, give the values of v_{BC}, v_{BE}, and v_{CE} for which (a) $i_C = 0.5\alpha_F I_E$ and (b) $i_C = 0$.

5.16 Consider the *pnp* large-signal model of Fig. 5.12 applied to a transistor having $I_S = 10^{-13}$ A and $\beta = 40$. If the

emitter is connected to ground, the base is connected to a current source that pulls 20 μA out of the base terminal, and the collector is connected to a negative supply of −10 V via a 10-kΩ resistor, find the collector voltage, the emitter current, and the base voltage.

5.17 A *pnp* transistor has $v_{EB} = 0.8$ V at a collector current of 1 A. What do you expect v_{EB} to become at $i_C = 10$ mA? At $i_C = 5$ A?

5.18 A *pnp* transistor modeled with the circuit in Fig. 5.12 is connected with its base at ground, collector at −1.5 V, and a 10-mA current injected into its emitter. If it is said to have $\beta = 10$, what are its base and collector currents? In which direction do they flow? If $I_S = 10^{-16}$ A, what voltage results at the emitter? What does the collector current become if a transistor with $\beta = 1000$ is substituted? (*Note:* The fact that the collector current changes by less than 10% for a large change of β illustrates that this is a good way to establish a specific collector current.)

5.19 A *pnp* power transistor operates with an emitter-to-collector voltage of 5 V, an emitter current of 10 A, and $V_{EB} = 0.85$ V. For $\beta = 15$, what base current is required? What is I_S for this transistor? Compare the emitter–base junction area of this transistor with that of a small-signal transistor that conducts $i_C = 1$ mA with $v_{EB} = 0.70$ V. How much larger is it?

SECTION 5.2: CURRENT-VOLTAGE CHARACTERISTICS

5.20 For the circuits in Fig. P5.20, assume that the transistors have very large β. Some measurements have been made

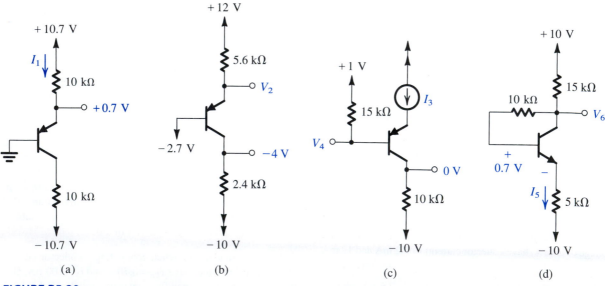

(a) (b) (c) (d)

FIGURE P5.20

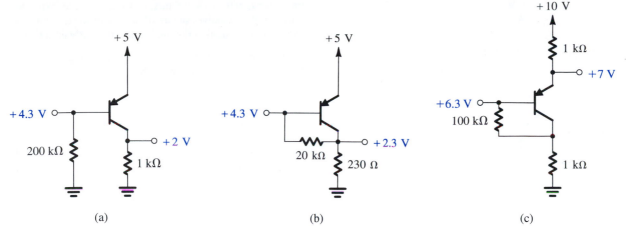

FIGURE P5.21

on these circuits, with the results indicated in the figure. Find the values of the other labeled voltages and currents.

5.21 Measurements on the circuits of Fig. P5.21 produce labeled voltages as indicated. Find the value of β for each transistor.

D5.22 Examination of the table of standard values for resistors with 5% tolerance in Appendix G reveals that the closest values to those found in the design of Example 5.1 are 5.1 kΩ and 6.8 kΩ. For these values use approximate calculations (e.g., $V_{BE} \simeq 0.7$ V and $\alpha \simeq 1$) to determine the values of collector current and collector voltage that are likely to result.

D5.23 Redesign the circuit in Example 5.1 to provide $V_C = +3$ V and $I_C = 5$ mA.

5.24 For each of the circuits shown in Fig. P5.24, find the emitter, base, and collector voltages and currents. Use $\beta = 30$, but assume $|V_{BE}| = 0.7$ V independent of current level.

5.25 Repeat Problem 5.24 using transistors for which $|V_{BE}| = 0.7$ V at $I_C = 1$ mA.

5.26 For the circuit shown in Fig. P5.26, measurement indicates that $V_B = -1.5$ V. Assuming $V_{BE} = 0.7$ V, calculate V_E, α, β, and V_C. If a transistor with $\beta = \infty$ is used, what values of V_B, V_E, and V_C result?

5

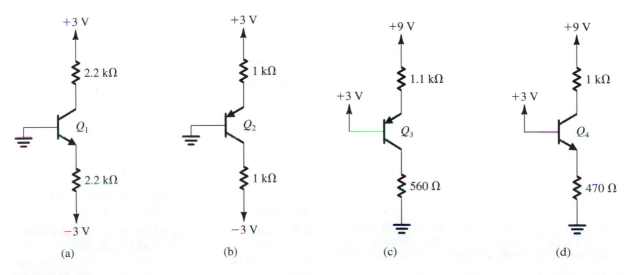

FIGURE P5.24

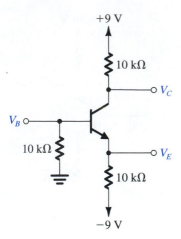

FIGURE P5.26

5.27 The current I_{CBO} of a small transistor is measured to be 20 nA at 25°C. If the temperature of the device is raised to 85°C, what do you expect I_{CBO} to become?

*5.28 Augment the model of the *npn* BJT shown in Fig. 5.20(a) by a current source representing I_{CBO}. Assume that r_o is very large and thus can be neglected. In terms of this addition, what do the terminal currents i_B, i_C, and i_E become? If the base lead is open-circuited while the emitter is connected to ground and the collector is connected to a positive supply, find the emitter and collector currents.

5.29 An *npn* transistor is accidentally connected with collector and emitter leads interchanged. The resulting currents in the normal emitter and base leads are 0.5 mA and 1 mA, respectively. What are the values of α_R and β_R?

5.30 A BJT whose emitter current is fixed at 1 mA has a base–emitter voltage of 0.69 V at 25°C. What base–emitter voltage would you expect at 0°C? At 100°C?

5.31 A particular *pnp* transistor operating at an emitter current of 0.5 mA at 20°C has an emitter–base voltage of 692 mV.

(a) What does v_{EB} become if the junction temperature rises to 50°C?
(b) If the transistor has $n = 1$ and is operated at a fixed emitter–base voltage of 700 mV, what emitter current flows at 20°C? At 50°C?

5.32 Consider a transistor for which the base–emitter voltage drop is 0.7 V at 10 mA. What current flows for $V_{BE} = 0.5$ V?

5.33 In Problem 5.32, the stated voltages are measured at 25°C. What values correspond at –25°C? At 125°C?

5.34 Use the Ebers-Moll expressions in Eqs. (5.26) and (5.27) to derive Eq. (5.35). Note that the emitter current is set to a constant value I_E. Ignore the terms not involving exponentials.

*5.35 Use Eq. (5.35) to plot the i_C–v_{CB} characteristics of an *npn* transistor having $\alpha_F \cong 1$, $\alpha_R = 0.1$, and $I_S = 10^{-15}$ A. Plot graphs for $I_E = 0.1$ mA, 0.5 mA, and 1 mA. Use an expanded scale for the negative values of v_{BC} in order to show the details of the saturation region. Neglect the Early effect.

*5.36 For the saturated transistor shown in Fig. P5.36, use the EM expressions to show that for $\alpha_F \cong 1$,

$$V_{CE\text{sat}} = V_T \ln\left(\frac{\dfrac{1}{\alpha_R} - \dfrac{I_{C\text{sat}}}{I_E}}{1 - \dfrac{I_{C\text{sat}}}{I_E}}\right)$$

For a BJT with $\alpha_R = 0.1$, evaluate $V_{CE\text{sat}}$ for $I_{C\text{sat}}/I_E = 0.9$, 0.5, 0.1, and 0.

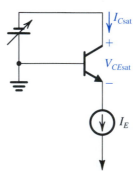

FIGURE P5.36

5.37 Use Eq. (5.36) to plot i_C versus v_{CE} for an *npn* transistor having $I_S = 10^{-15}$ A and $V_A = 100$ V. Provide curves for $v_{BE} = 0.65, 0.70, 0.72, 0.73$, and 0.74 volts. Show the characteristics for v_{CE} up to 15 V.

5.38 For a particular *npn* transistor operating at a v_{BE} of 670 mV and $I_C = 3$ mA, the i_C–v_{CE} characteristic has a slope of 3×10^{-5} ℧. To what value of output resistance does this correspond? What is the value of the Early voltage for this transistor? For operation at 30 mA, what would the output resistance become?

5.39 For a BJT having an Early voltage of 200 V, what is its output resistance at 1 mA? At 100 μA?

5.40 Measurements of the i_C–v_{CE} characteristic of a small-signal transistor operating at $v_{BE} = 720$ mV show that $i_C = 1.8$ mA at $v_{CE} = 2$ V and that $i_C = 2.4$ mA at $v_{CE} = 14$ V. What is the corresponding value of i_C near saturation? At what value of v_{CE} is $i_C = 2.0$ mA? What is the value of the Early voltage for this transistor? What is the output resistance that corresponds to operation at $v_{BE} = 720$ mV?

5.41 Give the *pnp* equivalent circuit models that correspond to those shown in Fig. 5.20 for the *npn* case.

5.42 A BJT operating at $i_B = 8$ μA and $i_C = 1.2$ mA undergoes a reduction in base current of 0.8 μA. It is found that when v_{CE} is held constant, the corresponding reduction in collector current is 0.1 mA. What are the values of h_{FE} and h_{fe} that apply? If the base current is increased from 8 μA to 10 μA and v_{CE} is increased from 8 V to 10 V, what collector current results? Assume $V_A = 100$ V.

5.43 For a transistor whose β characteristic is sketched in Fig. 5.22, estimate values of β at $-55°C$, $25°C$, and $125°C$ for $I_C = 100$ μA and 10 mA. For each current, estimate the temperature coefficient for temperatures above and below room temperature (four values needed).

5.44 Figure P5.44 shows a diode-connected *npn* transistor. Since $v_{CB} = 0$ results in active mode operation, the BJT will internally operate in the active mode; that is, its base and collector currents will be related by β_F. Use the EM equations to show that the diode-connected transistor has the $i–v$ characteristics,

$$i = \frac{I_S}{\alpha_F}(e^{v/V_T} - 1) \cong I_S e^{v/V_T}$$

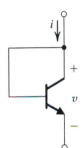

FIGURE P5.44

5.45 A BJT for which $\alpha_R = 0.2$ operates with a constant base current but with the collector open. What value of V_{CEsat} would you measure?

5.46 Find the saturation voltage V_{CEsat} and the saturation resistance R_{CEsat} of an *npn* BJT operated at a constant base current of 0.1 mA and a forced β of 20. The transistor has $\beta_F = 50$ and $\beta_R = 0.2$.

*5.47 Use Eq. (5.47) to show that the saturation resistance $R_{CEsat} \equiv \partial v_{CE}/\partial i_C$ of a transistor operated with a constant base current I_B is given by

$$R_{CEsat} = \frac{V_T}{\beta_F I_B} \frac{1}{x(1-x)}$$

where

$$x = \frac{I_{Csat}}{\beta_F I_B} = \frac{\beta_{forced}}{\beta_F}$$

Find R_{CEsat} for $\beta_{forced} = \beta_F/2$.

5.48 For a transistor for which $\beta_F = 70$ and $\beta_R = 0.7$, find an estimate of R_{CEsat} and V_{CEoff} for $I_B = 2$ mA by evaluating V_{CEsat} at $i_C = 3$ mA and at $i_C = 0.3$ mA (using Eq. 5.49). (*Note:* Because here we are modeling operation at a very low forced β, the value of R_{CEsat} will be much larger than that given by Eq. 5.48).

5.49 A transistor has $\beta_F = 150$ and the collector junction is 10 times larger than the emitter junction. Evaluate V_{CEsat} for $\beta_{forced}/\beta_F = 0.99$, 0.95, 0.9, 0.5, 0.1, 0.01, and 0.

5.50 A particular *npn* BJT with $v_{BE} = 720$ mV at $i_C = 600$ μA, and having $\beta = 150$, has a collector–base junction 20 times larger than the emitter–base junction.

(a) Find α_F, α_R, and β_R.
(b) For a collector current of 5 mA and nonsaturated operation, what is the base–emitter voltage and the base current?
(c) For the situation in (b) but with double the calculated base current, what is the value of forced β? What are the base–emitter and base–collector voltages? What are V_{CEsat} and R_{CEsat}?

*5.51 A BJT with fixed base current has $V_{CEsat} = 60$ mV with the emitter grounded and the collector open-circuited. When the collector is grounded and the emitter is open-circuited, V_{CEsat} becomes -1 mV. Estimate values for β_R and β_F for this transistor.

5.52 A BJT for which $I_B = 0.5$ mA has $V_{CEsat} = 140$ mV at $I_C = 10$ mA and $V_{CEsat} = 170$ mV at $I_C = 20$ mA. Estimate the values of its saturation resistance, R_{CEsat}, and its offset voltage, V_{CEoff}. Also, determine the values of β_F and β_R.

5.53 A BJT for which BV_{CBO} is 30 V is connected as shown in Fig. P5.53. What voltages would you measure on the collector, base, and emitter?

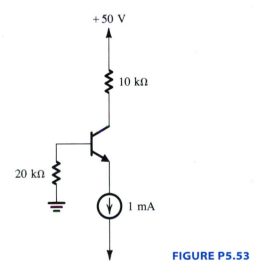

FIGURE P5.53

SECTION 5.3: THE BJT AS AN AMPLIFIER AND AS A SWITCH

5.54 A common-emitter amplifier circuit operated with $V_{CC} = +10$ V is biased at $V_{CE} = +1$ V. Find the voltage gain, the maximum allowed output negative swing without the transistor entering saturation, and the corresponding maximum input signal permitted.

5.55 For the common-emitter circuit in Fig. 5.26(a) with $V_{CC} = +10$ V and $R_C = 1$ kΩ, find V_{CE} and the voltage gain at the following dc collector bias currents: 1 mA, 2 mA, 5 mA, 8 mA, and 9 mA. For each, give the maximum possible positive- and negative-output signal swing as determined by the need to keep the transistor in the active region. Present your results in a table.

D5.56 Consider the CE amplifier circuit of Fig. 5.26(a) when operated with a dc supply $V_{CC} = +5$ V. It is required to find the point at which the transistor should be biased; that is, find the value of V_{CE} so that the output sine-wave signal v_{ce} resulting from an input sine-wave signal v_{be} of 5-mV peak amplitude has the maximum possible magnitude. What is the peak amplitude of the output sine wave and the value of the gain obtained? Assume linear operation around the bias point. (*Hint:* To obtain the maximum possible output amplitude for a given input, you need to bias the transistor as close to the edge of saturation as possible without entering saturation at any time, that is, without v_{CE} decreasing below 0.3 V.)

5.57 The transistor in the circuit of Fig. P5.57 is biased at a dc collector current of 0.5 mA. What is the voltage gain? (*Hint:* Use Thévenin theorem to convert the circuit to the form in Fig. 5.26a).

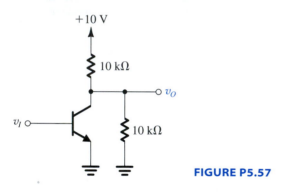

FIGURE P5.57

5.58 Sketch and label the voltage transfer characteristics of the *pnp* common-emitter amplifiers shown in Fig. P5.58.

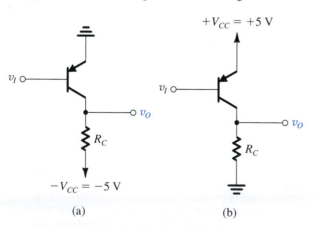

(a) (b)

FIGURE P5.58

*5.59 In deriving the expression for small-signal voltage gain A_v in Eq. (5.56) we neglected the Early effect. Derive this expression including the Early effect, by substituting

$$i_C = I_S e^{v_{BE}/V_T}\left(1 + \frac{v_{CE}}{V_A}\right)$$

in Eq. (5.50). Show that the gain expression changes to

$$A_v = \frac{-I_C R_C / V_T}{\left[1 + \dfrac{I_C R_C}{V_A + V_{CE}}\right]} = -\frac{(V_{CC} - V_{CE})/V_T}{\left[1 + \dfrac{V_{CC} - V_{CE}}{V_A + V_{CE}}\right]}$$

For the case $V_{CC} = 5$ V and $V_{CE} = 2.5$ V, what is the gain without and with the Early effect taken into account? Let $V_A = 100$ V.

5.60 When the common-emitter amplifier circuit of Fig. 5.26(a) is biased with a certain V_{BE}, the dc voltage at the collector is found to be +2 V. For $V_{CC} = +5$ V and $R_C = 1$ kΩ, find I_C and the small-signal voltage gain. For a change $\Delta v_{BE} = +5$ mV, calculate the resulting Δv_O. Calculate it two ways: by finding Δi_C using the transistor exponential characteristic and approximately using the small-signal voltage gain. Repeat for $\Delta v_{BE} = -5$ mV. Summarize your results in a table.

*5.61 Consider the common-emitter amplifier circuit of Fig. 5.26(a) when operated with a supply voltage $V_{CC} = +5$ V.

(a) What is the theoretical maximum voltage gain that this amplifier can provide?
(b) What value of V_{CE} must this amplifier be biased at to provide a voltage gain of −100 V/V?
(c) If the dc collector current I_C at the bias point in (b) is to be 0.5 mA, what value of R_C should be used?
(d) What is the value of V_{BE} required to provide the bias point mentioned above? Assume that the BJT has $I_S = 10^{-15}$ A.
(e) If a sine-wave signal v_{be} having a 5-mV peak amplitude is superimposed on V_{BE}, find the corresponding output voltage signal v_{ce} that will be superimposed on V_{CE} assuming linear operation around the bias point.
(f) Characterize the signal current i_c that will be superimposed on the dc bias current I_C.
(g) What is the value of the dc base current I_B at the bias point. Assume $\beta = 100$. Characterize the signal current i_b that will be superimposed on the base current I_B.
(h) Dividing the amplitude of v_{be} by the amplitude of i_b, evaluate the incremental (or small-signal) input resistance of the amplifier.
(i) Sketch and clearly label correlated graphs for v_{BE}, v_{CE}, i_C, and i_B. Note that each graph consists of a dc or average value and a superimposed sine wave. Be careful of the phase relationships of the sine waves.

5.62 The essence of transistor operation is that a change in v_{BE}, Δv_{BE}, produces a change in i_C, Δi_C. By keeping Δv_{BE} small, Δi_C is approximately linearly related to Δv_{BE}, $\Delta i_C = g_m \Delta v_{BE}$, where g_m is known as the transistor transconductance. By passing Δi_C through R_C, an output voltage signal Δv_O is obtained. Use the expression for the small-signal voltage gain in Eq. (5.56) to derive an expression for g_m. Find the value of g_m for a transistor biased at $I_C = 1$ mA.

5.63 Consider the characteristic curves shown in Fig. 5.29 with the following additional calibration data: Label, from the lowest colored line, $i_B = 1$ μA, 10 μA, 20 μA, 30 μA, and 40 μA. Assume the lines to be horizontal, and let $\beta = 100$. For $V_{CC} = 5$ V and $R_C = 1$ kΩ, what peak-to-peak collector voltage swing will result for i_B varying over the range 10 μA to 40 μA? If, at a new bias point (not the one shown in the figure) $V_{CE} = \frac{1}{2}V_{CC}$, find the value of I_C and I_B. If at this current $V_{BE} = 0.7$ V and if $R_B = 100$ kΩ, find the required value of V_{BB}.

*5.64 Sketch the i_C–v_{CE} characteristics of an *npn* transistor having $\beta = 100$ and $V_A = 100$ V. Sketch characteristic curves for $i_B = 20$ μA, 50 μA, 80 μA, and 100 μA. For the purpose of this sketch, assume that $i_C = \beta i_B$ at $v_{CE} = 0$. Also, sketch the load line obtained for $V_{CC} = 10$ V and $R_C = 1$ kΩ. If the dc bias current into the base is 50 μA, write the equation for the corresponding i_C–v_{CE} curve. Also, write the equation for the load line, and solve the two equations to obtain V_{CE} and I_C. If the input signal causes a sinusoidal signal of 30-μA peak amplitude to be superimposed on I_B, find the corresponding signal components of i_C and v_{CE}.

D5.65 For the circuit in Fig. P5.65 select a value for R_B so that the transistor saturates with an overdrive factor of 10. The BJT is specified to have a minimum β of 20 and $V_{CEsat} = 0.2$ V. What is the value of forced β achieved?

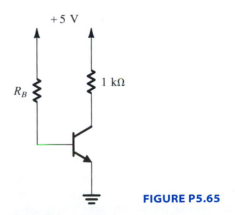

FIGURE P5.65

D5.66 For the circuit in Fig. P5.66 select a value for R_E so that the transistor saturates with a forced β of 10. Assume $V_{EB} = 0.7$ V and $V_{ECsat} = 0.2$ V.

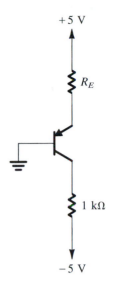

FIGURE P5.66

5.67 For each of the saturated circuits in Fig. P5.67, find i_B, i_C, and i_E. Use $|V_{BE}| = 0.7$ V and $|V_{CEsat}| = 0.2$ V.

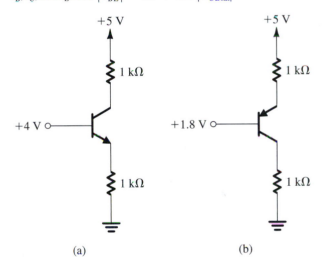

FIGURE P5.67

*5.68 Consider the operation of the circuit shown in Fig. P5.68 as v_B rises slowly from zero. For this transistor, assume $\beta = 50$, v_{BE} at which the transistor conducts is 0.5 V, v_{BE} when fully conducting is 0.7 V, saturation begins at $v_{BC} = 0.4$ V, and the transistor is deeply in saturation at $v_{BC} = 0.6$ V. Sketch and label v_E and v_C versus v_B. For what range of v_B is i_C essentially zero? What are the values of v_E, i_E, i_C, and v_C for $v_B = 1$ V and 3 V? For what value of v_B does saturation begin? What is i_B at this point? For $v_B = 4$ V and 6 V, what are the values of v_E, v_C, i_E, i_C, and i_B? Augment your sketch by adding a plot of i_B.

5

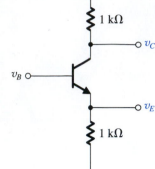

FIGURE P5.68

SECTION 5.4: BJT CIRCUITS AT DC

5.69 The transistor in the circuit of Fig. P5.69 has a very high β. Find V_E and V_C for V_B (a) +2 V, (b) +1 V, and (c) 0 V. Assume $V_{BE} \simeq 0.7$ V.

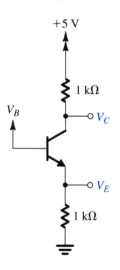

FIGURE P5.69

5.70 The transistor in the circuit of Fig. P5.69 has a very high β. Find the highest value of V_B for which the transistor still operates in the active mode. Also, find the value of V_B for which the transistor operates in saturation with a forced β of 1.

5.71 Consider the operation of the circuit shown in Fig. P5.71 for V_B at −1 V, 0 V, and +1 V. Assume that V_{BE} is 0.7 V for usual currents and that β is very high. What values of V_E and V_C result? At what value of V_B does the emitter current reduce to one-tenth of its value for $V_B = 0$ V? For what value of V_B is the transistor just at the edge of conduction? What

values of V_E and V_C correspond? For what value of V_B does the transistor reach saturation (when the base-to-collector junction reaches 0.5 V of forward bias)? What values of V_C and V_E correspond? Find the value of V_B for which the transistor operates in saturation with a forced β of 2.

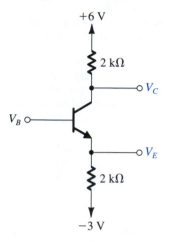

FIGURE P5.71

5.72 For the transistor shown in Fig. P5.72, assume $\alpha \cong 1$ and $v_{BE} = 0.5$ V at the edge of conduction. What are the values of V_E and V_C for $V_B = 0$ V? For what value of V_B does the transistor cut off? Saturate? In each case, what values of V_E and V_C result?

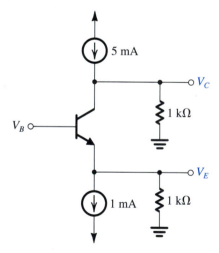

FIGURE P5.72

D5.73 Consider the circuit in Fig. P5.69 with the base voltage V_B obtained using a voltage divider across the 5-V supply. Assuming the transistor β to be very large (i.e., ignoring the base current), design the voltage divider to obtain $V_B = 2$ V. Design for a 0.2-mA current in the voltage divider. Now, if the BJT $\beta = 100$, analyze the circuit to determine the collector current and the collector voltage.

5.74 A single measurement indicates the emitter voltage of the transistor in the circuit of Fig. P5.74 to be 1.0 V. Under the assumption that $|V_{BE}| = 0.7$ V, what are V_B, I_B, I_E, I_C, V_C, β, and α? (*Note:* Isn't it surprising what a little measurement can lead to?)

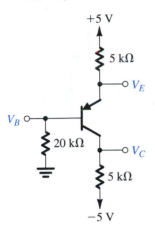

FIGURE P5.74

D5.75 Design a circuit using a *pnp* transistor for which $\alpha \cong$ 1 and $V_{EB} \cong 0.7$ V using two resistors connected appropriately to ±9 V so that $I_E = 2$ mA and $V_{BC} = 4.5$ V. What exact values of R_E and R_C would be needed? Now, consult a table of standard 5% resistor values (e.g., that provided in Appendix G) to select suitable practical values. What are the values of I_E and V_{BC} that result?

5.76 In the circuit shown in Fig. P5.76, the transistor has $\beta = 30$. Find the values of V_B, V_E, and V_C. If R_B is raised to 270 kΩ, what voltages result? With $R_B = 270$ kΩ, what value of β would return the voltages to the values first calculated?

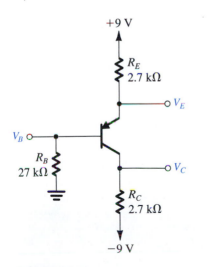

FIGURE P5.76

5.77 In the circuit shown in Fig. P5.76, the transistor has $\beta = 30$. Find the values of V_B, V_E, and V_C, and verify that the transistor is operating in the active mode. What is the largest value that R_C can have while the transistor remains in the active mode?

5.78 For the circuit in Fig. P5.78, find V_B, V_E, and V_C for $R_B = 100$ kΩ, 10 kΩ, and 1 kΩ. Let $\beta = 100$.

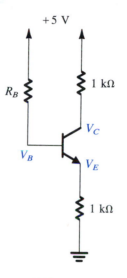

FIGURE P5.78

5.79 For the circuits in Fig. P5.79, find values for the labeled node voltages and branch currents. Assume β to be very high and $|V_{BE}| = 0.7$ V.

*5.80 Repeat the analysis of the circuits in Problem 5.79 using $\beta = 100$. Find all the labeled node voltages and branch currents. Assume $|V_{BE}| = 0.7$ V.

**D5.81 It is required to design the circuit in Fig. P5.81 so that a current of 1 mA is established in the emitter and a voltage of +5 V appears at the collector. The transistor type used has a nominal β of 100. However, the β value can be as low as 50 and as high as 150. Your design should ensure that the specified emitter current is obtained when $\beta = 100$ and that at the extreme values of β the emitter current does not change by more than 10% of its nominal value. Also, design for as large a value for R_B as possible. Give the values of R_B, R_E, and R_C to the nearest kilohm. What is the expected range of collector current and collector voltage corresponding to the full range of β values?

D5.82 The *pnp* transistor in the circuit of Fig. P5.82 has $\beta = 50$. Find the value for R_C to obtain $V_C = +5$ V. What happens if the transistor is replaced with another having $\beta = 100$?

5

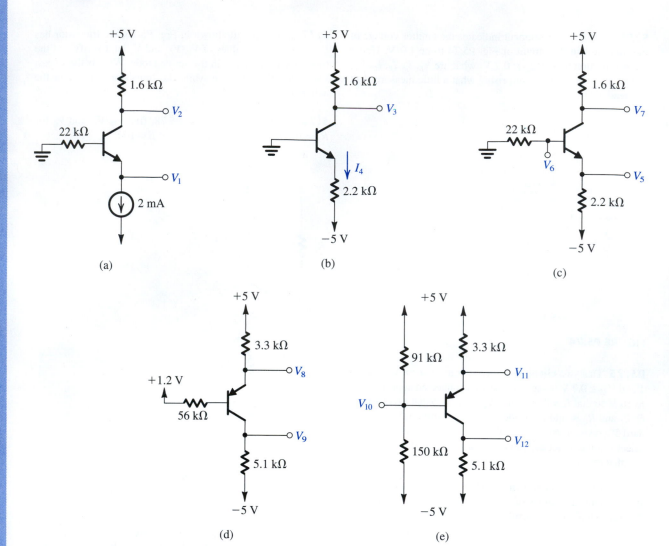

(a)

(b)

(c)

(d)

(e)

FIGURE P5.79

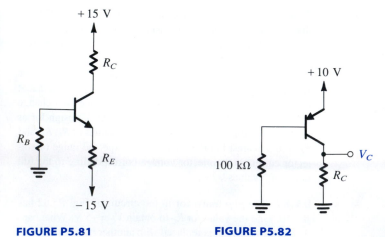

FIGURE P5.81

FIGURE P5.82

**5.83 Consider the circuit shown in Fig. P5.83. It resembles that in Fig. 5.41 but includes other features. First, note diodes D_1 and D_2 are included to make design (and analysis) easier and to provide temperature compensation for the emitter–base voltages of Q_1 and Q_2. Second, note resistor R, whose propose is provide negative feedback (more on this later in the book!). Using $|V_{BE}|$ and $V_D = 0.7$ V independent of current and $\beta = \infty$, find the voltages V_{B1}, V_{E1}, V_{C1}, V_{B2}, V_{E2}, and V_{C2}, initially with R open-circuited and then with R connected. Repeat for $\beta = 100$, initially with R open-circuited then connected.

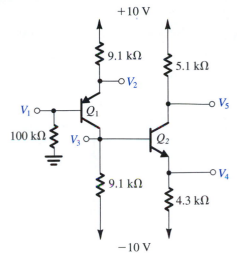

FIGURE P5.84

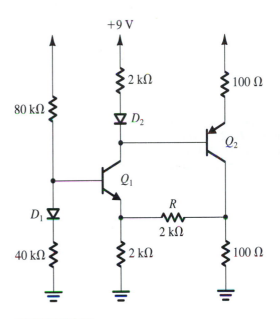

FIGURE P5.83

*5.84 For the circuit shown in Fig. P5.84, find the labeled node voltages for:

(a) $\beta = \infty$
(b) $\beta = 100$

**D5.85 Using $\beta = \infty$, design the circuit shown in Fig. P5.85 so that the bias currents in Q_1, Q_2, and Q_3 are 2 mA, 2 mA, and 4 mA, respectively, and $V_3 = 0$, $V_5 = -4$ V, and $V_7 = 2$ V. For each resistor, select the nearest standard value utilizing the table of standard values for 5% resistors in Appendix G. Now, for $\beta = 100$, find the values of V_3, V_4, V_5, V_6, and V_7.

5.86 For the circuit in Fig. P5.86, find V_B and V_E for $v_I = 0$ V, +3 V, −5 V, and −10 V. The BJTs have $\beta = 100$.

**5.87 Find approximate values for the collector voltages in the circuits of Fig. P5.87. Also, calculate forced β for each of the transistors. (*Hint:* Initially, assume all transistors are operating in saturation, and verify the assumption.)

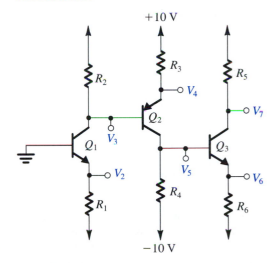

FIGURE P5.85

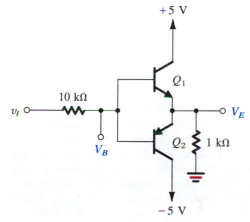

FIGURE P5.86

5

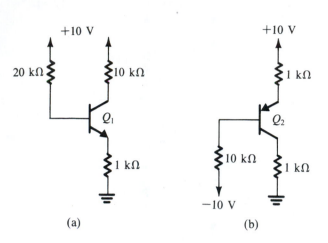

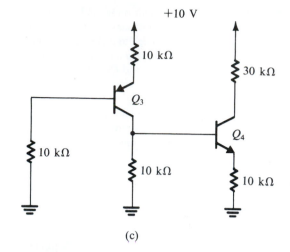

FIGURE P5.87

SECTION 5.5: BIASING IN BJT AMPLIFIER CIRCUITS

D5.88 For the circuit in Fig. 5.43(a), neglect the base current I_B in comparison with the current in the voltage divider. It is required to bias the transistor at $I_C = 1$ mA, which requires selecting R_{B1} and R_{B2} so that $V_{BE} = 0.690$ V. If $V_{CC} = 5$ V, what must the ratio R_{B1}/R_{B2} be? Now, if R_{B1} and R_{B2} are 1% resistors, that is, each can be in the range of 0.99 to 1.01 of its nominal value, what is the range obtained for V_{BE}? What is the corresponding range of I_C? If $R_C = 3$ kΩ, what is the range obtained for V_{CE}? Comment on the efficacy of this biasing arrangement.

D5.89 It is required to bias the transistor in the circuit of Fig. 5.43(b) at $I_C = 1$ mA. The transistor β is specified to be nominally 100, but it can fall in the range of 50 to 150. For $V_{CC} = +5$ V and $R_C = 3$ kΩ, find the required value of R_B to achieve $I_C = 1$ mA for the "nominal" transistor. What is the expected range for I_C and V_{CE}? Comment on the efficacy of this bias design.

D5.90 Consider the single-supply bias network shown in Fig. 5.44(a). Provide a design using a 9-V supply in which the supply voltage is equally split between R_C, V_{CE}, and R_E with a collector current of 3 mA. The transistor β is specified to have a minimum value of 90. Use a voltage-divider current of $I_E/10$, or slightly higher. Since a reasonable design should operate for the best transistors for which β is very high, do your initial design with $\beta = \infty$. Then choose suitable 5% resistors (see Appendix G), making the choice in a way that will result in a V_{BB} that is slightly higher than the ideal value. Specify the values you have chosen for R_E, R_C, R_1, and R_2. Now, find V_B, V_E, V_C, and I_C for your final design using $\beta = 90$.

D5.91 Repeat Problem 5.90, but use a voltage-divider current which is $I_E/2$. Check your design at $\beta = 90$. If you have the data available, find how low β can be while the value of I_C does not fall below that obtained with the design of Problem 5.90 for $\beta = 90$.

**D5.92 It is required to design the bias circuit of Fig. 5.44 for a BJT whose nominal $\beta = 100$.

(a) Find the largest ratio (R_B/R_E) that will guarantee I_E remain within $\pm 5\%$ of its nominal value for β as low as 50 and as high as 150.
(b) If the resistance ratio found in (a) is used, find an expression for the voltage $V_{BB} \equiv V_{CC} R_2/(R_1 + R_2)$ that will result in a voltage drop of $V_{CC}/3$ across R_E.
(c) For $V_{CC} = 10$ V, find the required values of R_1, R_2, and R_E to obtain $I_E = 2$ mA and to satisfy the requirement for stability of I_E in (a).
(d) Find R_C so that $V_{CE} = 3$ V for β equal to its nominal value.

Check your design by evaluating the resulting range of I_E.

*D5.93 Consider the two-supply bias arrangement shown in Fig. 5.45 using ± 3-V supplies. It is required to design the circuit so that $I_C = 3$ mA and V_C is placed midway between V_{CC} and V_E.

(a) For $\beta = \infty$, what values of R_E and R_C are required?
(b) If the BJT is specified to have a minimum β of 90, find the largest value for R_B consistent with the need to limit the voltage drop across it to one-tenth the voltage drop across R_E.
(c) What standard 5%-resistor values (see Appendix G) would you use for R_B, R_E, and R_C? In making your selection, use somewhat lower values in order to compensate for the low-β effects.
(d) For the values you selected in (c), find I_C, V_B, V_E, and V_C for $\beta = \infty$ and for $\beta = 90$.

*D5.94 Utilizing ±5-V power supplies, it is required to design a version of the circuit in Fig. 5.45 in which the signal will be coupled to the emitter and thus R_B can be set to zero. Find values for R_E and R_C so that a dc emitter current of 1 mA is obtained and so that the gain is maximized while allowing ±1 V of signal swing at the collector. If temperature increases from the nominal value of 25°C to 125°C, estimate the percentage change in collector bias current. In addition to the -2 mV/°C change in V_{BE}, assume that the transistor β changes over this temperature range from 50 to 150.

D5.95 Using a 5-V power supply, design a version of the circuit of Fig. 5.46 to provide a dc emitter current of 0.5 mA and to allow a ±1-V signal swing at the collector. The BJT has a nominal $\beta = 100$. Use standard 5%-resistor values (see Appendix G). If the actual BJT used has $\beta = 50$, what emitter current is obtained? Also, what is the allowable signal swing at the collector? Repeat for $\beta = 150$.

*D5.96 (a) Using a 3-V power supply, design the feedback bias circuit of Fig. 5.46 to provide $I_C = 3$ mA and $V_C = V_{CC}/2$ for $\beta = 90$.
(b) Select standard 5% resistor values, and reevaluate V_C and I_C for $\beta = 90$.
(c) Find V_C and I_C for $\beta = \infty$.
(d) To improve the situation that obtains when high-β transistors are used, we have to arrange for an additional current to flow through R_B. This can be achieved by connecting a resistor between base and emitter, as shown in Fig. P5.96. Design this circuit for $\beta = 90$. Use a current through R_{B2} equal to the base current. Now, what values of V_C and I_C result with $\beta = \infty$?

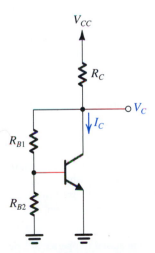

FIGURE P5.96

D5.97 A circuit that can provide a very large voltage gain for a high-resistance load is shown in Fig. P5.97. Find the

values of I and R_B to bias the BJT at $I_C = 3$ mA and $V_C = 1.5$ V. Let $\beta = 90$.

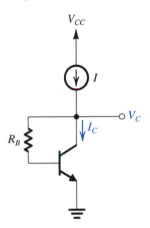

FIGURE P5.97

5.98 The circuit in Fig. P5.98 provides a constant current I_O as long as the circuit to which the collector is connected maintains the BJT in the active mode. Show that

$$I_O = \alpha \frac{V_{CC}[R_2/(R_1 + R_2)] - V_{BE}}{R_E + (R_1//R_2)/(\beta + 1)}$$

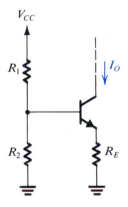

FIGURE P5.98

**D5.99 The current-bias circuit shown in Fig. P5.99 provides bias current to Q_1 that is independent of R_B and nearly independent of the value of β_1 (as long as Q_2 operates in the active mode). Prepare a design meeting the following specifications: Use ±5-V supplies; $I_{C1} = 0.1$ mA, $V_{RE} = 2$ V for $\beta = \infty$; the voltage across R_E decreases by at most 5% for $\beta = 50$; $V_{CE1} = 1.5$ V for $\beta = \infty$ and 2.5 V for $\beta = 50$. Use standard 5%-resistor values (see Appendix G). What values for R_1, R_2, R_E, R_B, and R_C do you choose? What values of I_{C1} and V_{CE1} result for $\beta = 50$, 100, and 200?

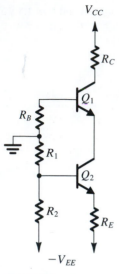

*D5.100 For the circuit in Fig. P5.100, assuming all transistors to be identical with β infinite, derive an expression for the output current I_O, and show that by selecting

$$R_1 = R_2$$

and keeping the current in each junction the same, the current I_O will be

$$I_O = \frac{\alpha V_{CC}}{2R_E}$$

which is independent of V_{BE}. What must the relationship of R_E to R_1 and R_2 be? For $V_{CC} = 10$ V and assuming $\alpha \simeq 1$ and $V_{BE} = 0.7$ V, design the circuit to obtain an output current of

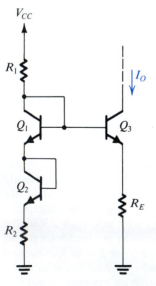

FIGURE P5.100

0.5 mA. What is the lowest voltage that can be applied to the collector of Q_3?

D5.101 For the circuit in Fig. P5.101 find the value of R that will result in $I_O \simeq 2$ mA. What is the largest voltage that can be applied to the collector? Assume $|V_{BE}| = 0.7$ V.

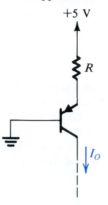

FIGURE P5.101

SECTION 5.6: SMALL-SIGNAL OPERATION AND MODELS

5.102 Consider a transistor biased to operate in the active mode at a dc collector current I_C. Calculate the collector signal current as a fraction of I_C (i.e., i_c/I_C) for input signals v_{be} of +1 mV, −1 mV, +2 mV, −2 mV, +5 mV, −5 mV, +8 mV, −8 mV, +10 mV, −10 mV, +12 mV, and −12 mV. In each case do the calculation two ways:

(a) using the exponential characteristic, and
(b) using the small-signal approximation.

Present your results in the form of a table that includes a column for the error introduced by the small-signal approximation. Comment on the range of validity of the small-signal approximation.

5.103 An *npn* BJT with grounded emitter is operated with $V_{BE} = 0.700$ V, at which the collector current is 1 mA. A 10-kΩ resistor connects the collector to a +15-V supply. What is the resulting collector voltage V_C? Now, if a signal applied to the base raises v_{BE} to 705 mV, find the resulting total collector current i_C and total collector voltage v_C using the exponential i_C–v_{BE} relationship. For this situation, what are v_{be} and v_c? Calculate the voltage gain v_c/v_{be}. Compare with the value obtained using the small-signal approximation, that is, $-g_m R_C$.

5.104 A transistor with $\beta = 120$ is biased to operate at a dc collector current of 1.2 mA. Find the values of g_m, r_π, and r_e. Repeat for a bias current of 120 μA.

5.105 A *pnp* BJT is biased to operate at $I_C = 2.0$ mA. What is the associated value of g_m? If $\beta = 50$, what is the value of the small-signal resistance seen looking into the emitter (r_e)? Into the base (r_π)? If the collector is connected to a 5-kΩ load, with a signal of 5-mV peak applied between base and emitter, what output signal voltage results?

D5.106 A designer wishes to create a BJT amplifier with a g_m of 50 mA/V and a base input resistance of 2000 Ω or more. What emitter-bias current should he choose? What is the minimum β he can tolerate for the transistor used?

5.107 A transistor operating with nominal g_m of 60 mA/V has a β that ranges from 50 to 200. Also, the bias circuit, being less than ideal, allows a ±20% variation in I_C. What are the extreme values found of the resistance looking into the base?

5.108 In the circuit of Fig. 5.48, V_{BE} is adjusted so that $V_C = 2$ V. If $V_{CC} = 5$ V, $R_C = 3$ kΩ, and a signal $v_{be} = 0.005 \sin \omega t$ volts is applied, find expressions for the total instantaneous quantities $i_C(t)$, $v_C(t)$, and $i_B(t)$. The transistor has $\beta = 100$. What is the voltage gain?

*D5.109 We wish to design the amplifier circuit of Fig. 5.48 under the constraint that V_{CC} is fixed. Let the input signal $v_{be} = \hat{V}_{be} \sin \omega t$, where $\hat{V}_{be}$ is the maximum value for acceptable linearity. For the design that results in the largest signal at the collector, without the BJT leaving the active region, show that

$$R_C I_C = (V_{CC} - 0.3 - \hat{V}_{be})\Big/\left(1 + \frac{\hat{V}_{be}}{V_T}\right)$$

and find an expression for the voltage gain obtained. For $V_{CC} = 5$ V and $\hat{V}_{be} = 5$ mV, find the dc voltage at the collector, the amplitude of the output voltage signal, and the voltage gain.

5.110 The following table summarizes some of the basic attributes of a number of BJTs of different types, operating as amplifiers under various conditions. Provide the missing entries.

5.111 A BJT is biased to operate in the active mode at a dc collector current of 1.0 mA. It has a β of 120. Give the four small-signal models (Figs. 5.51 and 5.52) of the BJT complete with the values of their parameters.

5.112 The transistor amplifier in Fig. P5.112 is biased with a current source I and has a very high β. Find the dc voltage at the collector, V_C. Also, find the value of g_m. Replace the transistor with the simplified hybrid-π model of Fig. 5.51(a) (note that the dc current source I should be replaced with an open circuit). Hence find the voltage gain v_c/v_i.

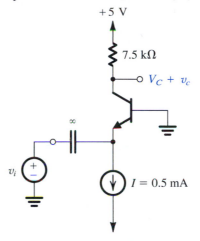

FIGURE P5.112

5.113 For the conceptual circuit shown in Fig. 5.50, $R_C = 2$ kΩ, $g_m = 50$ mA/V, and $\beta = 100$. If a peak-to-peak output voltage of 1 V is measured at the collector, what ac input voltage and current must be associated with the base?

5.114 A biased BJT operates as a grounded-emitter amplifier between a signal source, with a source resistance of 10 kΩ, connected to the base and a 10-kΩ load connected as a collector resistance R_C. In the corresponding model, g_m is 40 mA/V

Transistor	a	b	c	d	e	f	g
α	1.000					0.90	
β		100		∞			
I_C (mA)	1.00		1.00				
I_E (mA)		1.00				5	
I_B (mA)			0.020				1.10
g_m (mA/V)							700
r_e (Ω)				25	100		
r_π (Ω)					10.1 kΩ		

(*Note:* Isn't it remarkable how much two parameters can reveal?)

and r_π is 2.5 kΩ. Draw the complete amplifier model using the hybrid-π BJT equivalent circuit. Calculate the overall voltage gain (v_c/v_s). What is the value of BJT β implied by the values of the model parameters? To what value must β be increased to double the overall voltage gain?

5.115 For the circuit shown in Fig. P5.115, draw a complete small-signal equivalent circuit utilizing an appropriate T model for the BJT (use $\alpha = 0.99$). Your circuit should show the values of all components, including the model parameters. What is the input resistance R_{in}? Calculate the overall voltage gain (v_o/v_{sig}).

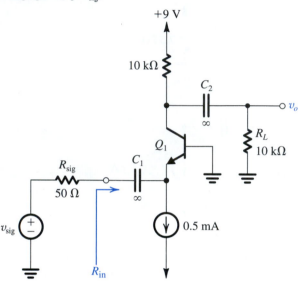

FIGURE P5.115

5.116 In the circuit shown in Fig. P5.116, the transistor has a β of 200. What is the dc voltage at the collector? Find the input resistances R_{ib} and R_{in} and the overall voltage gain

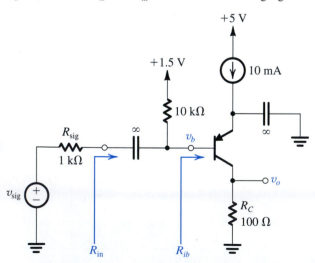

FIGURE P5.116

(v_o/v_{sig}). For an output signal of ±0.4 V, what values of v_{sig} and v_b are required?

5.117 Consider the augmented hybrid-π model shown in Fig. 5.58(a). Disregarding how biasing is to be done, what is the largest possible voltage gain available for a signal source connected directly to the base and a very-high-resistance load? Calculate the value of the maximum possible gain for $V_A = 25$ V and $V_A = 250$ V.

5.118 Reconsider the amplifier shown in Fig. 5.53 and analyzed in Example 5.14 under the condition that β is not well controlled. For what value of β does the circuit begin to saturate? We can conclude that large β is dangerous in this circuit. Now, consider the effect of reduced β, say, to $\beta = 25$. What values of r_e, g_m, and r_π result? What is the overall voltage gain? (*Note:* You can see that this circuit, using base-current control of bias, is very β-sensitive and usually *not recommended*.)

5.119 Reconsider the circuit shown in Fig. 5.55(a) under the condition that the signal source has an internal resistance of 100 Ω. What does the overall voltage gain become? What is the largest input signal voltage that can be used without output-signal clipping?

D5.120 Redesign the circuit of Fig. 5.55 by raising the resistor values by a factor n to increase the resistance seen by the input v_i to 75 Ω. What value of voltage gain results? Grounded-base circuits of this kind are used in systems such as cable TV, in which, for highest-quality signaling, load resistances need to be "matched" to the equivalent resistances of the interconnecting cables.

5.121 Using the BJT equivalent circuit model of Fig. 5.52(a), sketch the equivalent circuit of a transistor amplifier for which a resistance R_e is connected between the emitter and ground, the collector is grounded, and an input signal source v_b is connected between the base and ground. (It is assumed that the transistor is properly biased to operate in the active region.) Show that:

(a) the voltage gain between base and emitter, that is, v_e/v_b, is given by

$$\frac{v_e}{v_b} = \frac{R_e}{R_e + r_e}$$

(b) the input resistance,

$$R_{in} \equiv \frac{v_b}{i_b} = (\beta + 1)(R_e + r_e)$$

Find the numerical values for (v_e/v_b) and R_{in} for the case $R_e = 1$ kΩ, $\beta = 100$, and the emitter bias current $I_E = 1$ mA.

5.122 When the collector of a transistor is connected to its base, the transistor still operates (internally) in the active region because the collector–base junction is still in effect reverse biased. Use the simplified hybrid-π model to find the incremental (small-signal) resistance of the resulting two-terminal device (known as a diode-connected transistor.)

**D5.123 Design an amplifier using the configuration of Fig. 5.55(a). The power supplies available are ±10 V. The input signal source has a resistance of 100 Ω, and it is required that the amplifier input resistance match this value. (Note that $R_{in} = r_e \mathbin{/\mkern-5mu/} R_E \simeq r_e$.) The amplifier is to have the greatest possible voltage gain and the largest possible output signal but retain small-signal linear operation (i.e., the signal component across the base–emitter junction should be limited to no more than 10 mV). Find appropriate values for R_E and R_C. What is the value of voltage gain realized?

*5.124 The transistor in the circuit shown in Fig. P5.124 is biased to operate in the active mode. Assuming that β is very large, find the collector bias current I_C. Replace the transistor with the small-signal equivalent circuit model of Fig. 5.52(b) (remember to replace the dc power supply with a short circuit). Analyze the resulting amplifier equivalent circuit to show that

$$\frac{v_{o1}}{v_i} = \frac{R_E}{R_E + r_e}$$

$$\frac{v_{o2}}{v_i} = \frac{-\alpha R_C}{R_E + r_e}$$

Find the values of these voltage gains (for $\alpha \simeq 1$). Now, if the terminal labeled v_{o1} is connected to ground, what does the voltage gain v_{o2}/v_i become?

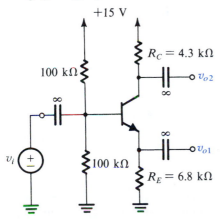

FIGURE P5.124

SECTION 5.7: SINGLE-STAGE BJT AMPLIFIERS

5.125 An amplifier is measured to have $R_i = 10$ kΩ, $A_{vo} = 100$ V/V, and $R_o = 100$ Ω. Also, when a load resistance R_L of 1 kΩ is connected between the output terminals, the input resistance is found to decrease to 8 kΩ. If the amplifier is fed with a signal source having an internal resistance of 2 kΩ, find G_m, A_v, G_{vo}, G_v, R_{out}, and A_i.

5.126 Figure P5.126 shows an alternative equivalent circuit for representing *any* linear two-port network including voltage

amplifiers. This non-unilateral equivalent circuit is based on the g-parameter two-port representation (see Appendix B).

(a) Using the values of R_i, A_{vo}, and R_o found in Example 5.17 together with the measured value of R_{in} of 400 kΩ obtained when a load R_L of 10 kΩ is connected to the output, determine the value of the feedback factor f.
(b) Now, use the equivalent circuit of Fig. P5.126 to determine the value of R_{out} obtained when the amplifier is fed with a signal generator having $R_{sig} = 100$ kΩ. Check your result against that found in Example 5.17.

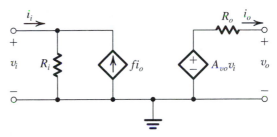

FIGURE P5.126

5.127 Refer to Table 5.5. By equating the expression for G_v obtained from Equivalent Circuit A to that obtained from Equivalent Circuit C with $G_{vo} = [R_i/(R_i + R_{sig})]A_{vo}$, show that

$$\frac{R_{in}}{R_i}\frac{R_{sig} + R_i}{R_{sig} + R_{in}} = \frac{R_L + R_o}{R_L + R_{out}}$$

Now, use this expression to:

(a) Show that for $R_L = \infty$, $R_{in} = R_i$.
(b) Show that for $R_{sig} = 0$, $R_{out} = R_o$.
(c) Find R_{out} when $R_{sig} = \infty$ (i.e., the amplifier input is open-circuited), and evaluate its value for the amplifier specified in Example 5.17.

5.128 A common-emitter amplifier of the type shown in Fig. 5.60(a) is biased to operate at $I_C = 0.2$ mA and has a collector resistance $R_C = 24$ kΩ. The transistor has $\beta = 100$ and a large V_A. The signal source is directly coupled to the base, and C_{C1} and R_B are eliminated. Find R_{in}, the voltage gain A_{vo}, and R_o. Use these results to determine the overall voltage gain when a 10-kΩ load resistor is connected to the collector and the source resistance $R_{sig} = 10$ kΩ.

5.129 Repeat Problem 5.128 with a 125-Ω resistance included in the signal path in the emitter. Furthermore, contrast the maximum amplitude of the input sine wave that can be applied with and without R_e assuming that to limit distortion the signal between base and emitter is not to exceed 5 mV.

5.130 For the common-emitter amplifier shown in Fig. P5.130, let $V_{CC} = 9$ V, $R_1 = 27$ kΩ, $R_2 = 15$ kΩ, $R_E = 1.2$ kΩ, and $R_C = 2.2$ kΩ. The transistor has $\beta = 100$ and $V_A = 100$ V.

Calculate the dc bias current I_E. If the amplifier operates between a source for which $R_{sig} = 10$ kΩ and a load of 2 kΩ, replace the transistor with its hybrid-π model, and find the values of R_{in}, the voltage gain v_o/v_{sig}, and the current gain i_o/i_i.

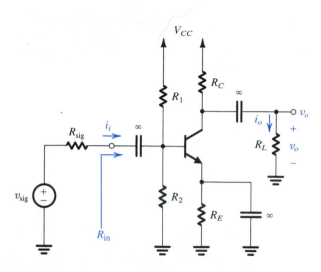

FIGURE P5.130

D5.131 Using the topology of Fig. P5.130, design an amplifier to operate between a 10-kΩ source and a 2-kΩ load with a gain v_o/v_{sig} of −8 V/V. The power supply available is 9 V. Use an emitter current of approximately 2 mA and a current of about one-tenth of that in the voltage divider that feeds the base, with the dc voltage at the base about one-third of the supply. The transistor available has $\beta = 100$ and $V_A = 100$ V. Use standard 5% resistor (see Appendix G).

5.132 A designer, having examined the situation described in Problem 5.130 and estimating the available gain to be approximately −8 V/V, wishes to explore the possibility of improvement by reducing the loading of the source by the amplifier input. As an experiment, the designer varies the resistance levels by a factor of approximately 3: R_1 to 82 kΩ, R_2 to 47 kΩ, R_E to 3.6 kΩ, and R_C to 6.8 kΩ (standard values of 5%-tolerance resistors). With $V_{CC} = 9$ V, $R_{sig} = 10$ kΩ, $R_L = 2$ kΩ, $\beta = 100$, and $V_A = 100$ V, what does the gain become? Comment.

D5.133 Consider the CE amplifier circuit of Fig. 5.60(a). It is required to design the circuit (i.e., find values for I, R_B, and R_C) to meet the following specifications:

(a) $R_{in} \cong 5$ kΩ.
(b) the dc voltage drop across R_B is approximately 0.5 V.
(c) the open-circuit voltage gain from base to collector is the maximum possible, consistent with the requirement that the collector voltage never falls by more than approximately 0.5 V

below the base voltage with the signal between base and emitter being as high as 5 mV.

Assume that v_{sig} is a sinusoidal source, the available supply $V_{CC} = 5$ V, and the transistor has $\beta = 100$ and a very large Early voltage. Use standard 5%-resistance values, and specify the value of I to one significant digit. What base-to-collector open-circuit voltage gain does your design provide? If $R_{sig} = R_L = 10$ kΩ, what is the overall voltage gain?

D5.134 In the circuit of Fig. P5.134, v_{sig} is a small sine-wave signal with zero average. The transistor β is 100.

(a) Find the value of R_E to establish a dc emitter current of about 0.5 mA.
(b) Find R_C to establish a dc collector voltage of about +5 V.
(c) For $R_L = 10$ kΩ and the transistor $r_o = 200$ kΩ, draw the small-signal equivalent circuit of the amplifier and determine its overall voltage gain.

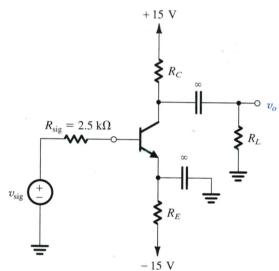

FIGURE P5.134

***5.135** The amplifier of Fig. P5.135 consists of two identical common-emitter amplifiers connected in cascade. Observe that the input resistance of the second stage, R_{in2}, constitutes the load resistance of the first stage.

(a) For $V_{CC} = 15$ V, $R_1 = 100$ kΩ, $R_2 = 47$ kΩ, $R_E = 3.9$ kΩ, $R_C = 6.8$ kΩ, and $\beta = 100$, determine the dc collector current and dc collector voltage of each transistor.
(b) Draw the small-signal equivalent circuit of the entire amplifier and give the values of all its components. Neglect r_{o1} and r_{o2}.
(c) Find R_{in1} and v_{b1}/v_{sig} for $R_{sig} = 5$ kΩ.
(d) Find R_{in2} and v_{b2}/v_{b1}.
(e) For $R_L = 2$ kΩ, find v_o/v_{b2}.
(f) Find the overall voltage gain v_o/v_{sig}.

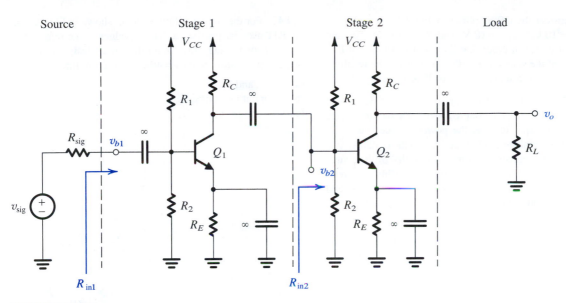

FIGURE P5.135

5.136 In the circuit of Fig. P5.136, v_{sig} is a small sine-wave signal. Find R_{in} and the gain v_o/v_{sig}. Assume $\beta = 100$. If the amplitude of the signal v_{be} is to be limited to 5 mV, what is the largest signal at the input? What is the corresponding signal at the output?

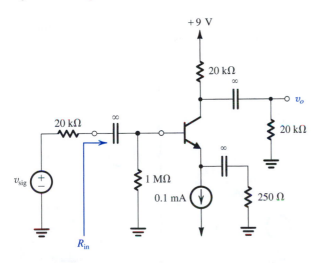

FIGURE P5.136

*5.137 The BJT in the circuit of Fig. P5.137 has $\beta = 100$.

(a) Find the dc collector current and the dc voltage at the collector.
(b) Replacing the transistor by its T model, draw the small-signal equivalent circuit of the amplifier. Analyze the resulting circuit to determine the voltage gain v_o/v_i.

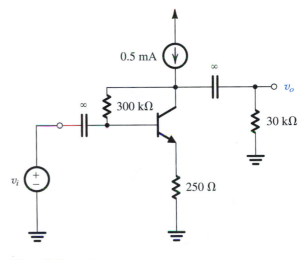

FIGURE P5.137

*5.138 Refer to the voltage-gain expression (in terms of transistor β) given in Eq. (5.135) for the CE amplifier with a resistance R_e in the emitter. Let the BJT be biased at an emitter current of 0.5 mA. The source resistance R_{sig} is 10 kΩ. The BJT β is specified to lie in the range of 50 to 150 with a nominal value of 100.

(a) What is the ratio of maximum to minimum voltage gain obtained without R_e?
(b) What value of R_e should be used to limit the ratio of maximum to minimum gain to 1.2?
(c) If the R_e found in (b) is used, by what factor is the gain reduced (compared to the case without R_e) for a BJT with a nominal β?

5

5.139 Consider the CB amplifier of Fig. 5.62(a) with $R_L = 10$ kΩ, $R_C = 10$ kΩ, $V_{CC} = 10$ V, and $R_{sig} = 100$ Ω. To what value must I be set in order that the input resistance at E is equal to that of the source (i.e., 100 Ω)? What is the resulting voltage gain from the source to the load? Assume $\alpha \simeq 1$.

**D5.140 Consider the CB amplifier of Fig. 5.62(a) with the collector voltage signal coupled to a 1-kΩ load resistance through a large capacitor. Let the power supplies be ±5 V. The source has a resistance of 50 Ω. Design the circuit so that the amplifier input resistance is matched to that of the source and the output signal swing is as large as possible with relatively low distortion (v_{be} limited to 10 mV). Find I and R_C and calculate the overall voltage gain obtained and the output signal swing. Assume $\alpha \simeq 1$.

5.141 For the circuit in Fig. P5.141, find the input resistance R_{in} and the voltage gain v_o/v_{sig}. Assume that the source provides a small signal v_{sig} and that $\beta = 100$.

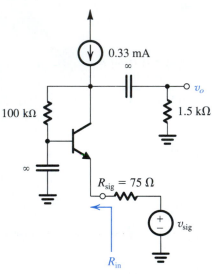

FIGURE P5.141

5.142 Consider the emitter follower of Fig. 5.63(a) for the case: $I = 1$ mA, $\beta = 100$, $V_A = 100$ V, $R_B = 100$ kΩ, $R_{sig} = 20$ kΩ, and $R_L = 1$ kΩ.

(a) Find R_{in}, v_b/v_{sig}, and v_o/v_{sig}.
(b) If v_{sig} is a sine-wave signal, to what value should its amplitude be limited in order that the transistor remains conducting at all times? For this amplitude, what is the corresponding amplitude across the base–emitter junction?
(c) If the signal amplitude across the base–emitter junction is to be limited to 10 mV, what is the corresponding amplitude of v_{sig} and of v_o?
(d) Find the open-circuit voltage gain v_o/v_{sig} and the output resistance. Use these values to determine the value of v_o/v_{sig} obtained with $R_L = 500$ Ω.

5.143 For the emitter-follower circuit shown in Fig. P5.143, the BJT used is specified to have β values in the range of 40 to 200 (a distressing situation for the circuit designer). For the two extreme values of β ($\beta = 40$ and $\beta = 200$), find:

(a) I_E, V_E, and V_B.
(b) the input resistance R_{in}.
(c) the voltage gain v_o/v_{sig}.

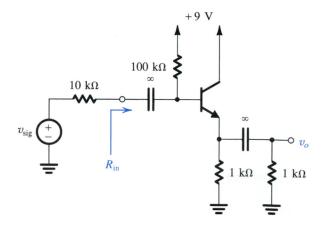

FIGURE P5.143

5.144 For the emitter follower in Fig. P5.144, the signal source is directly coupled to the transistor base. If the dc component of v_{sig} is zero, find the dc emitter current. Assume $\beta = 100$. Neglecting r_o, find R_{in}, the voltage gain v_o/v_{sig}, the current gain i_o/i_i, and the output resistance R_{out}.

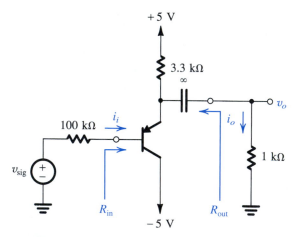

FIGURE P5.144

5.145 In the emitter follower of Fig. 5.63(a), the signal source is directly coupled to the base. Thus, C_{C1} and R_B are eliminated. The source has $R_{sig} = 10$ kΩ and a dc component of zero. The transistor has $\beta = 100$ and $V_A = 125$ V. The bias current $I = 2.5$ mA, and $V_{CC} = 3$ V. What is the output resistance of the follower? Find the gain v_o/v_{sig} with no load and with a load of 1 kΩ. With the 1-kΩ load connected, find the largest possible negative output signal. What is the largest possible positive output signal if operation is satisfactory up to the point that the base–collector junction is forward biased by 0.4 V?

5.146 The emitter follower of Fig. 5.63(a), when driven from a 10-kΩ source, was found to have an open-circuit voltage gain of 0.99 and an output resistance of 200 Ω. The output resistance increased to 300 Ω when the source resistance was increased to 20 kΩ. Find the overall voltage gain when the follower is driven by a 30-kΩ source and loaded by a 1-kΩ resistor. Assume r_o is very large.

5.147 For the circuit in Fig. P5.147, called a **bootstrapped follower:

(a) Find the dc emitter current and g_m, r_e, and r_π. Use $\beta = 100$.
(b) Replace the BJT with its T model (neglecting r_o), and analyze the circuit to determine the input resistance R_{in} and the voltage gain v_o/v_{sig}.
(c) Repeat (b) for the case when capacitor C_B is open-circuited. Compare the results with those obtained in (b) to find the advantages of bootstrapping.

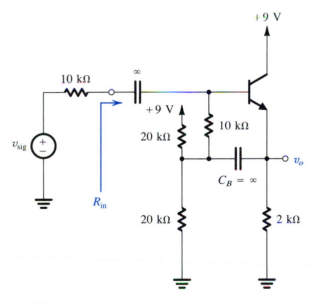

FIGURE P5.147

**5.148 For the follower circuit in Fig. P5.148 let transistor Q_1 have $\beta = 50$ and transistor Q_2 have $\beta = 100$, and neglect the effect of r_o. Use $V_{BE} = 0.7$ V.

(a) Find the dc emitter currents of Q_1 and Q_2. Also, find the dc voltages V_{B1} and V_{B2}.
(b) If a load resistance $R_L = 1$ kΩ is connected to the output terminal, find the voltage gain from the base to the emitter of

Q_2, v_o/v_{b2}, and find the input resistance R_{ib2} looking into the base of Q_2. (*Hint:* Consider Q_2 as an emitter follower fed by a voltage v_{b2} at its base.)
(c) Replacing Q_2 with its input resistance R_{ib2} found in (b), analyze the circuit of emitter follower Q_1 to determine its input resistance R_{in}, and the gain from its base to its emitter, v_{e1}/v_{b1}.
(d) If the circuit is fed with a source having a 100-kΩ resistance, find the transmission to the base of Q_1, v_{b1}/v_{sig}.
(e) Find the overall voltage gain v_o/v_{sig}.

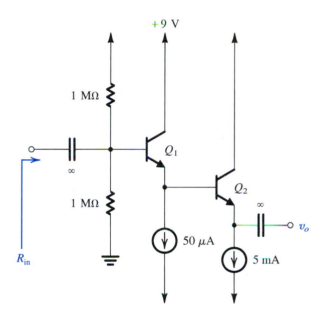

FIGURE P5.148

SECTION 5.8: THE BJT INTERNAL CAPACITANCES AND HIGH-FREQUENCY MODEL

5.149 An *npn* transistor is operated at $I_C = 0.5$ mA and $V_{CB} = 2$ V. It has $\beta_0 = 100$, $V_A = 50$ V, $\tau_F = 30$ ps, $C_{je0} = 20$ fF, $C_{\mu0} = 30$ fF, $V_{0c} = 0.75$ V, $m_{CBJ} = 0.5$, and $r_x = 100$ Ω. Sketch the complete hybrid-π model, and specify the values of all its components. Also, find f_T.

5.150 Measurement of h_{fe} of an *npn* transistor at 500 MHz shows that $|h_{fe}| = 2.5$ at $I_C = 0.2$ mA and 11.6 at $I_C = 1.0$ mA. Furthermore, C_μ was measured and found to be 0.05 pF. Find f_T at each of the two collector currents used. What must τ_F and C_{je} be?

5.151 A particular BJT operating at $I_C = 2$ mA has $C_\mu = 1$ pF, $C_\pi = 10$ pF, and $\beta = 150$. What are f_T and f_β for this situation?

5.152 For the transistor described in Problem 5.151, C_π includes a relatively constant depletion-layer capacitance of 2 pF.

Transistor	I_E (mA)	r_e (Ω)	g_m (mA/V)	r_π (kΩ)	β_0	f_T (MHz)	C_μ (pF)	C_π (pF)	f_β (MHz)
(a)	1				100	400	2		
(b)		25					2	10.7	4
(c)			2.525			400		13.84	
(d)	10				100	400	2		
(e)	0.1				100	100	2		
(f)	1				10	400	2		
(g)						800	1	9	80

If the device is operated at $I_C = 0.2$ mA, what does its f_T become?

5.153 A particular small-geometry BJT has f_T of 5 GHz and $C_\mu = 0.1$ pF when operated at $I_C = 0.5$ mA. What is C_π in this situation? Also, find g_m. For $\beta = 150$, find r_π and f_β.

5.154 For a BJT whose unity-gain bandwidth is 1 GHz and $\beta_0 = 200$, at what frequency does the magnitude of h_{fe} become 20? What is f_β?

*5.155 For a sufficiently high frequency, measurement of the complex input impedance of a BJT having (ac) grounded emitter and collector yields a real part approximating r_x. For what frequency, defined in terms of ω_β, is such an estimate of r_x good to within 10% under the condition that $r_x \le r_\pi/10$? Neglect C_μ.

*5.156 Complete the table entries above for transistors (a) through (g), under the conditions indicated. Neglect r_x.

SECTION 5.9: FREQUENCY RESPONSE OF THE COMMON-EMITTER AMPLIFIER

5.157 A designer wishes to investigate the effect of changing the bias current I on the midband gain and high-frequency response of the CE amplifier considered in Example 5.18. Let I be doubled to 2 mA, and assume that β_0 and f_T remain unchanged at 100 and 800 MHz, respectively. To keep the node voltages nearly unchanged, the designer reduces R_B and R_C by a factor of 2, to 50 kΩ and 4 kΩ, respectively. Assume $r_x = 50$ Ω, and recall that $V_A = 100$ V and that C_μ remains constant at 1 pF. As before, the amplifier is fed with a source having $R_{\text{sig}} = 5$ kΩ and feeds a load $R_L = 5$ kΩ. Find the new values of A_M, f_H, and the gain–bandwidth product, $|A_M| f_H$. Comment on the results. Note that the price paid for whatever improvement in performance is achieved is an increase in power. By what factor does the power dissipation increase?

*5.158 The purpose of this problem is to investigate the high-frequency response of the CE amplifier when it is fed with a relatively large source resistance R_{sig}. Refer to the amplifier in Fig. 5.71(a) and to its high-frequency equivalent-circuit model and the analysis shown in Fig. 5.72. Let $R_B \gg R_{\text{sig}}$, $r_x \ll R_{\text{sig}}$, $R_{\text{sig}} \gg r_\pi$, $g_m R_L' \gg 1$, and $g_m R_L' C_\mu \gg C_\pi$. Under these conditions, show that:

(a) the midband gain $A_M \cong -\beta(R_L'/R_{\text{sig}})$.
(b) the upper 3-dB frequency $f_H \cong 1/(2\pi C_\mu \beta R_L')$.
(c) the gain–bandwidth product $A_M f_H \cong 1/(2\pi C_\mu R_{\text{sig}})$.

Evaluate this approximate value of the gain–bandwidth product for the case $R_{\text{sig}} = 25$ kΩ and $C_\mu = 1$ pF. Now, if the transistor is biased at $I_C = 1$ mA and has $\beta = 100$, find the midband gain and f_H for the two cases $R_L' = 25$ kΩ and $R_L' = 2.5$ kΩ. On the same coordinates, sketch Bode plots for the gain magnitude versus frequency for the two cases. What f_H is obtained when the gain is unity? What value of R_L' corresponds?

5.159 Consider the common-emitter amplifier of Fig. P5.159 under the following conditions: $R_{\text{sig}} = 5$ kΩ, $R_1 = 33$ kΩ, $R_2 = 22$ kΩ, $R_E = 3.9$ kΩ, $R_C = 4.7$ kΩ, $R_L = 5.6$ kΩ, $V_{CC} = 5$ V. The dc emitter current can be shown to be $I_E \cong 0.3$ mA, at which $\beta_0 = 120$, $r_o = 300$ kΩ, and $r_x = 50$ Ω. Find the input resistance R_{in} and the midband gain A_M. If the transistor is specified to have $f_T = 700$ MHz and $C_\mu = 1$ pF, find the upper 3-dB frequency f_H.

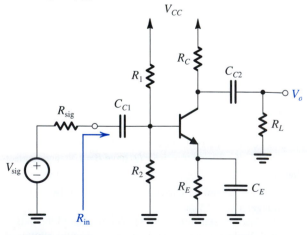

FIGURE P5.159

5.160 For a version of the CE amplifier circuit in Fig. P5.159, $R_{sig} = 10$ kΩ, $R_1 = 68$ kΩ, $R_2 = 27$ kΩ, $R_E = 2.2$ kΩ, $R_C = 4.7$ kΩ, and $R_L = 10$ kΩ. The collector current is 0.8 mA, $\beta = 200$, $f_T = 1$ GHz, and $C_\mu = 0.8$ pF. Neglecting the effect of r_x and r_o, find the midband voltage gain and the upper 3-dB frequency f_H.

*5.161 The amplifier shown in Fig. P5.161 has $R_{sig} = R_L = 1$ kΩ, $R_C = 1$ kΩ, $R_B = 47$ kΩ, $\beta = 100$, $C_\mu = 0.8$ pF, and $f_T = 600$ MHz.

(a) Find the dc collector current of the transistor.
(b) Find g_m and r_π.
(c) Neglecting r_o, find the midband voltage gain from base to collector (neglect the effect of R_B).
(d) Use the gain obtained in (c) to find the component of R_{in} that arises as a result of R_B. Hence find R_{in}.
(e) Find the overall gain at midband.
(f) Find C_{in}.
(g) Find f_H.

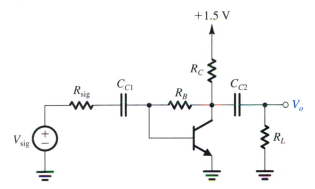

FIGURE P5.161

*5.162 Refer to Fig. P5.162. Utilizing the BJT high-frequency hybrid-π model with $r_x = 0$ and $r_o = \infty$, derive an expression for $Z_i(s)$ as a function of r_e and C_π. Find the frequency at which the impedance has a phase angle of 45° for the case in which the BJT has $f_T = 400$ MHz and the bias current is relatively high. What is the frequency when the bias current is reduced so that $C_\pi \simeq C_\mu$? Assume $\alpha = 1$.

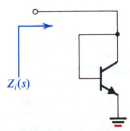

FIGURE P5.162

5.163 For the amplifier in Fig. P5.159, whose component values were specified in Problem 5.159, let $C_{C1} = C_{C2} = 1$ μF, and $C_E = 10$ μF. Find the break frequencies f_{P1}, f_{P2}, and f_{P3}

resulting from C_{C1}, C_E, and C_{C2}, respectively. Note that R_E has to be taken into account in evaluating f_{P2}. Hence, estimate the value of the lower 3-dB frequency f_L.

D5.164 For the amplifier described in Problem 5.163, design the coupling and bypass capacitors for a lower 3-dB frequency of 100 Hz. Design so that the contribution of each of C_{C1} and C_{C2} to determining f_L is only 5%.

5.165 Consider the circuit of Fig. P5.159. For $R_{sig} = 10$ kΩ, $R_B \equiv R_1 // R_2 = 10$ kΩ, $r_x = 100$ Ω, $r_\pi = 1$ kΩ, $\beta_0 = 100$, and $R_E = 1$ kΩ, what is the ratio C_E/C_{C1} that makes their contributions to the determination of f_L equal?

*D5.166 For the common-emitter amplifier of Fig. P5.166, neglect r_x and r_o, and assume the current source to be ideal.

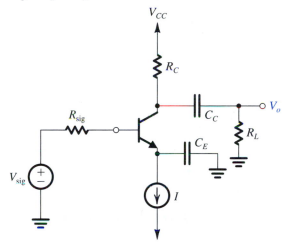

FIGURE P5.166

(a) Derive an expression for the midband gain.
(b) Derive expressions for the break frequencies caused by C_E and C_C.
(c) Give an expression for the amplifier voltage gain $A(s)$.
(d) For $R_{sig} = R_C = R_L = 10$ kΩ, $\beta = 100$, and $I = 1$ mA, find the value of the midband gain.
(e) Select values for C_E and C_C to place the two break frequencies a decade apart and to obtain a lower 3-dB frequency of 100 Hz while minimizing the total capacitance.
(f) Sketch a Bode plot for the gain magnitude, and estimate the frequency at which the gain becomes unity.
(g) Find the phase shift at 100 Hz.

5.167 The BJT common-emitter amplifier of Fig. P5.167 includes an emitter degeneration resistance R_e.

(a) Assuming $\alpha \cong 1$, neglecting r_x and r_o, and assuming the current source to be ideal, derive an expression for the small-signal voltage gain $A(s) \equiv V_o/V_{sig}$ that applies in the midband and the low frequency band. Hence find the midband gain A_M and the lower 3-dB frequency f_L.

5

(b) Show that including R_e reduces the magnitude of A_M by a certain factor. What is this factor?
(c) Show that including R_e reduces f_L by the same factor as in (b) and thus one can use R_e to trade-off gain for bandwidth.
(d) For $I = 1$ mA, $R_C = 10$ kΩ, and $C_E = 100$ μF, find $|A_M|$ and f_L with $R_e = 0$. Now find the value of R_e that lowers f_L by a factor of 5. What will the gain become?

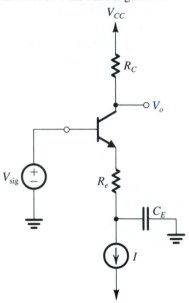

FIGURE P5.167

SECTION 5.10: THE BASIC BJT DIGITAL LOGIC INVERTER

5.168 Consider the inverter circuit in Fig. 5.74. In Exercise 5.53, the following expression is given for V_{OH} when the inverter is driving N identical inverters:

$$V_{OH} = V_{CC} - R_C \frac{V_{CC} - V_{BE}}{R_C + R_B/N}$$

For the same component values used in the analysis in the text (i.e., $V_{CC} = 5$ V, $R_C = 1$ kΩ, $R_B = 10$ kΩ, and $V_{BE} = 0.7$ V), find the maximum value of N that will still guarantee a high noise margin, NM_H, of at least 1 V. Assume $\beta = 50$ and $V_{CEsat} = 0.2$ V.

5.169 The purpose of this problem is to find the power dissipation of the inverter circuit of Fig. 5.74 in each of its two states. Assume that the component values are as given in the text (i.e., $V_{CC} = 5$ V, $R_C = 1$ kΩ, $R_B = 10$ kΩ, and $V_{BE} = 0.7$ V).

(a) With the input low at 0.2 V, the transistor is cut off. Let the inverter be driving 10 identical inverters. Find the total current supplied by the inverter and hence the power dissipated in R_C.

(b) With the input high and the transistor saturated, find the power dissipated in the inverter, neglecting the power dissipated in the base circuit.
(c) Use the results of (a) and (b) to find the average power dissipation in the inverter.

D5.170 Design a transistor inverter to operate from a 1.5-V supply. With the input connected to the 1.5-V supply through a resistor equal to R_C, the total power dissipated should be 1 mW, and forced β should be 10. Use $V_{BE} = 0.7$ V and $V_{CEsat} = 0.2$ V.

5.171 For the circuit in Fig. P5.171, consider the application of inputs of 5 V and 0.2 V to X and Y in any combination, and find the output voltage for each combination. Tabulate your results. How many input combinations are there? What happens when any input is high? What happens when both inputs are low? This is a logic gate that implements the NOR function: $Z = \overline{X + Y}$. (This logic-gate structure is called, historically, Resistor-Transistor Logic (RTL)).

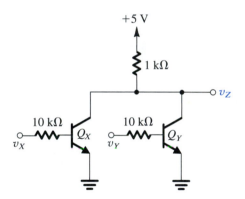

FIGURE P5.171

5.172 Consider the inverter of Fig. 5.74 with a load capacitor C connected between the output node and ground. We wish to find the contribution of C to the low-to-high delay time of the inverter, t_{PLH}. (For the formal definition of inverter delays, refer to Fig. 1.35.) Toward that end, assume that prior to $t = 0$, the transistor is on and saturated and $v_O = V_{OL} = V_{CEsat}$. Then, at $t = 0$, let the input fall to the low level, and assume that the transistor turns off instantaneously. Note that neglecting the turn-off time of a saturated transistor is an unrealistic assumption, but one that will help us concentrate on the effect of C. Now, with the transistor cut off, the capacitor will charge through R_C, and the output voltage will rise exponentially from $V_{OL} = V_{CEsat}$ to $V_{OH} = V_{CC}$. Find an expression for $v_O(t)$. Calculate the value of t_{PLH}, which in this case is the time for v_O to rise to $\frac{1}{2}(V_{OH} + V_{OL})$. Use $V_{CC} = 5$ V, $V_{CEsat} = 0.2$ V, $R_C = 1$ kΩ, and $C = 10$ pF. (*Hint:* The step response of RC circuits is reviewed in Section 1.7 and in greater detail in Appendix D.)

*5.173 Consider the inverter circuit of Fig. 5.74 with a load capacitor C connected between the output node and ground. We wish to find the contribution of C to the high-to-low delay time of the inverter, t_{PHL}. (For the formal definition of the inverter delays, refer to Fig. 1.35.) Toward that end, assume that prior to $t = 0$, the transistor is off and $v_O = V_{OH} = V_{CC}$. Then, at $t = 0$, let the input rise to the high level, and assume that the transistor turns on instantaneously. Note that neglecting the delay time of the transistor is unrealistic but will help us concentrate on the effect of the load capacitance C. Now, because C cannot discharge instantaneously, the transistor cannot saturate immediately. Rather, it will operate in the active mode, and its collector will supply a constant current of $\beta(V_{CC} - V_{BE})/R_B$. Find the Thévenin equivalent circuit for discharging the capacitor, and show that the voltage will fall exponentially, starting at V_{CC} and heading toward a large negative voltage of $[V_{CC} - \beta(V_{CC} - V_{BE})R_C/R_B]$. Find an expression for $v_O(t)$. This exponential discharge will stop when v_O reaches $V_{OL} = V_{CEsat}$ and the transistor saturates. Calculate the value of t_{PHL}, which in this case is the time for v_O to fall to $\frac{1}{2}(V_{OH} + V_{OL})$. Use $V_{CC} = 5$ V, $V_{CEsat} = 0.2$ V, $V_{BE} = 0.7$ V, $R_B = 10$ kΩ, $R_C = 1$ kΩ, $\beta = 50$, and $C = 10$ pF. If you have solved Problem 5.172, compare the value of t_{PHL} to that of t_{PLH} found there, and find the inverter delay, t_P. (*Hint:* The step response of RC circuits is reviewed in Section 1.7 and in greater detail in Appendix E).

5

PART II

ANALOG AND DIGITAL INTEGRATED CIRCUITS

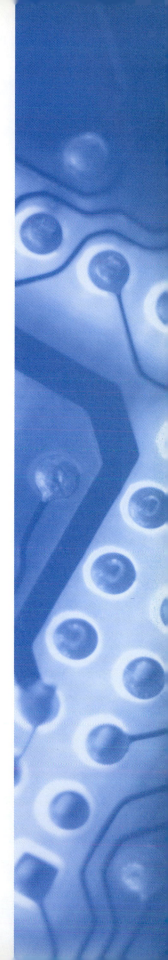

INTRODUCTION

Having studied the major electronic devices (the MOSFET and the BJT) and their basic circuit applications, we are now ready to consider the design of more complex analog and digital integrated circuits and systems. The five chapters of Part II are intended for this purpose. They provide a carefully selected set of topics suitable for a second course in electronics. Nevertheless, the flexibility inherent in this book should permit replacing some of the topics included with a selection from the special topics presented in Part III. As well, if desired, Chapter 10 on CMOS logic circuits can be studied at the beginning of the course.

Study of Part II assumes knowledge of MOSFET and BJT characteristics, models, and basic applications (Chapters 4 and 5). To review and consolidate this material and differences between the two devices, Section 6.2 with its three tables (6.1–6.3) is a *must* read. The remainder of Chapter 6 provides a systematic study of the circuit building blocks utilized in the design of analog ICs. In each case, both low-frequency and high-frequency operations are considered. Chapter 7 continues this study, concentrating on the most widely used configuration in analog IC design, the differential pair. It concludes with a section on multistage amplifiers. In both chapters, MOSFET circuits are presented first, simply because the MOSFET is now the device that is used in over 90% of integrated circuits. Bipolar transistor circuits are presented with the same depth but presented second and, on occasion, more briefly.

A formal study of the pivotal topic of feedback is presented in Chapter 8. Such a study is essential for the proper application of feedback in the design of amplifiers, to effect desirable properties such as more precise gain value, and to avoid problems such as instability. The analog material of Part II is integrated together in Chapter 9 in the study of op-amp circuits. Chapter 9 also presents an introduction to analog-to-digital and digital-to-analog converters, and thus acts as a bridge to the study of CMOS digital logic circuits in Chapter 10. Here again we concentrate on CMOS because it represents the technology in which the vast majority of digital systems are implemented.

The second course, based on Part II, is intended to prepare the reader for the practice of electronic design, and, if desired, to pursue more advanced courses on analog and digital IC design.

CHAPTER 6

Single-Stage Integrated-Circuit Amplifiers

INTRODUCTION

Having studied the two major transistor types, the MOSFET and the BJT, and their basic discrete-circuit amplifier configurations, we are now ready to begin the study of integrated-circuit amplifiers. This chapter and the next are devoted to the design of the basic building blocks of IC amplifiers.

In this chapter, we begin with a brief section on the design philosophy of integrated circuits, and how it differs from that of discrete circuits. Throughout this chapter, MOS and bipolar circuits are presented side-by-side, which allows a certain economy in presentation and, more importantly, provides an opportunity to compare and contrast the two circuit types. Toward that end, Section 6.2 provides a comprehensive comparison of the attributes of the two transistor types. This should serve both as a review as well as a guide to very interesting similarities and differences between the two devices.

Following the study of IC biasing, the various configurations of single-stage IC amplifiers are presented. This material builds on the study of basic discrete-amplifier configurations in Sections 4.7 and 5.7.

In addition to classical single-stage amplifiers, we also study some configurations that utilize two amplifying transistors. These "compound configurations" are usually treated as single-stage amplifiers (for reasons that will become clear later).

Current mirrors and current-source circuits play a major role in the design of IC amplifiers, where they serve both as biasing and load elements. For this reason, we return to the subject of current mirrors later in the chapter and consider some of their advanced (and indeed, ingenious) forms.

Although CMOS circuits are the most widely used at present, there are applications in which the addition of bipolar transistors can result in superior performance. Circuits that combine MOS and bipolar transistors, in a technology known as BiMOS or BiCMOS, are presented at appropriate locations throughout the chapter. The chapter concludes with SPICE simulation examples.

 ## 6.1 IC DESIGN PHILOSOPHY

Integrated-circuit fabrication technology (Appendix A) poses constraints on—and provides opportunities to—the circuit designer. Thus, while chip-area considerations dictate that large- and even moderate-value resistors are to be avoided, constant-current sources are readily available. Large capacitors, such as those we used in Sections 4.7 and 5.7 for signal coupling and bypass, are not available to be used, except perhaps as components external to the IC chip. Even then, the number of such capacitors has to be kept to a minimum; otherwise the number of chip terminals and hence its cost increase. Very small capacitors, in the picofarad and fraction of a picofarad range, however, are easy to fabricate in IC MOS technology and can be combined with MOS amplifiers and MOS switches to realize a wide range of signal processing functions, both analog (Chapter 12) and digital (Chapter 11).

As a general rule, in designing IC MOS circuits, one should strive to realize as many of the functions required as possible using MOS transistors only and, when needed, small MOS capacitors. MOS transistors can be sized; that is, their W and L values can be selected, to fit a wide range of design requirements. Also, arrays of transistors can be matched (or, more generally, made to have desired size ratios) to realize such useful circuit building blocks as current mirrors.

At this juncture, it is useful to mention that to pack a larger number of devices on the same IC chip, the trend has been to reduce the device dimensions. At the time of this writing (2003), CMOS process technologies capable of producing devices with a 0.1-μm minimum channel length are in use. Such small devices need to operate with dc voltage supplies close to 1 V. While low-voltage operation can help to reduce power dissipation, it poses a host of challenges to the circuit designer. For instance, such MOS transistors must be operated with overdrive voltages of only 0.2 V or so. In our study of MOS amplifiers, we will make frequent comments on such issues.

The MOS-amplifier circuits that we shall study will be designed almost entirely using MOSFETs of both polarities—that is, NMOS and PMOS—as are readily available in CMOS technology. As mentioned earlier, CMOS is currently the most widely used IC technology for both analog and digital as well as combined analog and digital (or mixed-signal) applications. Nevertheless, bipolar integrated circuits still offer many exciting opportunities to the analog design engineer. This is especially the case for general-propose circuit packages, such as high-quality op amps that are intended for assembly on printed-circuit (pc) boards (as opposed to being part of a system-on-chip). As well, bipolar circuits can provide much higher output currents and are favoured for certain applications, such as in the automotive industry, for their high reliability under severe environmental conditions. Finally, bipolar circuits can be combined with CMOS in innovative and exciting ways.

6.2 COMPARISON OF THE MOSFET AND THE BJT

In this section we present a comparison of the characteristics of the two major electronic devices: the MOSFET and the BJT. To facilitate this comparison, typical values for the important parameters of the two devices are first presented.

6.2.1 Typical Values of MOSFET Parameters

Typical values for the important parameters of NMOS and PMOS transistors fabricated in a number of CMOS processes are shown in Table 6.1. Each process is characterized by the minimum allowed channel length, $L_{\min}$; thus, for example, in a 0.18-μm process, the smallest transistor has a channel length $L = 0.18$ μm. The technologies presented in Table 6.1 are in descending order of channel length, with that having the shortest channel length being the most modern. Although the 0.8-μm process is now obsolete, its data are included to show trends in the values of various parameters. It should also be mentioned that although Table 6.1 stops at the 0.18-μm process, at the time of this writing (2003), a 0.13-μm fabrication process is commercially available and a 0.09-μm process is in the advanced stages of development. The 0.18-μm process, however, is currently the most popular and the one for which data are widely available. An important caution, however, is in order: The data presented in Table 6.1 do *not* pertain to any particular commercially available process. Accordingly, these generic data are not intended for use in an actual IC design; rather, they show trends and, as we shall see, help to illustrate design trade-offs as well as enable us to work out design examples and problems with parameter values that are as realistic as possible.

TABLE 6.1 Typical Values of CMOS Device Parameters

Parameter	0.8 μm		0.5 μm		0.25 μm		0.18 μm	
	NMOS	PMOS	NMOS	PMOS	NMOS	PMOS	NMOS	PMOS
t_{ox} (nm)	15	15	9	9	6	6	4	4
C_{ox} (fF/μm^2)	2.3	2.3	3.8	3.8	5.8	5.8	8.6	8.6
μ (cm^2/V·s)	550	250	500	180	460	160	450	100
μC_{ox} (μA/V^2)	127	58	190	68	267	93	387	86
V_{t0} (V)	0.7	−0.7	0.7	−0.8	0.43	−0.62	0.48	−0.45
V_{DD} (V)	5	5	3.3	3.3	2.5	2.5	1.8	1.8
$|V_A'|$ (V/μm)	25	20	20	10	5	6	5	6
C_{ov} (fF/μm)	0.2	0.2	0.4	0.4	0.3	0.3	0.37	0.33

As indicated in Table 6.1, the trend has been to reduce the minimum allowable channel length. This trend has been motivated by the desire to pack more transistors on a chip as well as to operate at higher speeds or, in analog terms, over wider bandwidths.

Observe that the oxide thickness, t_{ox}, scales down with the channel length, reaching 4 nm for the 0.18-μm process. Since the oxide capacitance C_{ox} is inversely proportional to t_{ox}, we see that C_{ox} increases as the technology scales down. The surface mobility μ decreases as the technology minimum-feature size is decreased, and μ_p decreases much faster than μ_n. As a result, the ratio of μ_p to μ_n has been decreasing with each generation of technology, falling from about 0.5 for older technologies to 0.2 or so for the newer ones. Despite the reduction of μ_n and μ_p, the transconductance parameters $k'_n = \mu_n C_{ox}$ and $k'_p = \mu_p C_{ox}$ have been steadily increasing. As a result, modern short-channel devices achieve required levels of bias currents at lower overdrive voltages. As well, they achieve higher transconductance, a major advantage.

Although the magnitudes of the threshold voltages V_{tn} and V_{tp} have been decreasing with L_{min} from about 0.7–0.8 V to 0.4–0.5 V, the reduction has not been as large as that of the power supply V_{DD}. The latter has been reduced dramatically, from 5 V for older technologies to 1.8 V for the 0.18-μm process. This reduction has been necessitated by the need to keep the electric fields in the smaller devices from reaching very high values. Another reason for reducing V_{DD} is to keep power dissipation as low as possible given that the IC chip now has a much larger number of transistors.[1]

The fact that in modern short-channel CMOS processes $|V_t|$ has become a much larger proportion of the power-supply voltage poses a serious challenge to the circuit design engineer. Recalling that $|V_{GS}| = |V_t| + |V_{OV}|$, where V_{OV} is the overdrive voltage, to keep $|V_{GS}|$ reasonably small, $|V_{OV}|$ for modern technologies is usually in the range of 0.2 V to 0.3 V. To appreciate this point further, recall that to operate a MOSFET in the saturation region, $|V_{DS}|$ must exceed $|V_{OV}|$; thus, to be able to have a number of devices stacked between the power-supply rails in a regime in which V_{DD} is only 1.8 V or lower, we need to keep $|V_{OV}|$ as low as possible. We will shortly see, however, that operating at a low $|V_{OV}|$ has some drawbacks.

Another significant though undesirable feature of modern submicron CMOS technologies is that the channel length modulation effect is very pronounced. As a result, V'_A has been steadily decreasing, which combined with the decreasing values of L has caused the Early voltage $V_A = V'_A L$ to become very small. Correspondingly, short-channel MOSFETs exhibit low output resistances.

From our study of the MOSFET high-frequency equivalent circuit model in the saturation mode in Section 4.8 and the high-frequency response of the common-source amplifier in Section 4.9, we know that two major MOSFET capacitances are C_{gs} and C_{gd}. While C_{gs} has an overlap component, C_{gd} is entirely an overlap capacitance. Both C_{gd} and the overlap component of C_{gs} are almost equal and are denoted C_{ov}. The last line of Table 6.1 provides the value of C_{ov} per micron of gate width. Although the normalized C_{ov} has been staying more or less constant with the reduction in L_{min}, we will shortly see that the shorter devices exhibit much higher operating speeds and wider amplifier bandwidths than the longer devices. Specifically, we will, for example, see that f_T for a 0.25-μm NMOS transistor can be as high as 10 GHz.

6.2.2 Typical Values of IC BJT Parameters

Table 6.2 provides typical values for the major parameters that characterize integrated-circuit bipolar transistors. Data are provided for devices fabricated in two different processes: the

---

[1] At the present time, chip power dissipation has become a very serious issue, with some of the recently reported ICs dissipating as much as 100 W. As a result, an important current area of research concerns what is termed "power-aware design."

TABLE 6.2	Typical Parameter Values for BJTs[1]			
	Standard High-Voltage Process		**Advanced Low-Voltage Process**	
Parameter	*npn*	Lateral *pnp*	*npn*	Lateral *pnp*
A_E (μm^2)	500	900	2	2
I_S (A)	5×10^{-15}	2×10^{-15}	6×10^{-18}	6×10^{-18}
β_0 (A/A)	200	50	100	50
V_A (V)	130	50	35	30
V_{CE0} (V)	50	60	8	18
τ_F	0.35 ns	30 ns	10 ps	650 ps
C_{je0}	1 pF	0.3 pF	5 fF	14 fF
$C_{\mu0}$	0.3 pF	1 pF	5 fF	15 fF
r_x (Ω)	200	300	400	200

[1]Adapted from Gray et al. (2000); see Bibliography.

standard, old process, known as the "high-voltage process"; and an advanced, modern process, referred to as a "low-voltage process." For each process we show the parameters of the standard *npn* transistor and those of a special type of *pnp* transistor known as a **lateral** (as opposed to **vertical** as in the *npn* case) **pnp** (see Appendix A). In this regard we should mention that a major drawback of standard bipolar integrated-circuit fabrication processes has been the lack of *pnp* transistors of a quality equal to that of the *npn* devices. Rather, there are a number of *pnp* implementations for which the lateral *pnp* is the most economical to fabricate. Unfortunately, however, as should be evident from Table 6.2, the lateral *pnp* has characteristics that are much inferior to those of the *npn*. Note in particular the lower value of β and the much larger value of the forward transit time τ_F that determines the emitter–base diffusion capacitance C_{de} and, hence, the transistor speed of operation. The data in Table 6.2 can be used to show that the unity-gain frequency of the lateral *pnp* is two orders of magnitude lower than that of the *npn* transistor fabricated in the same process. Another important difference between the lateral *pnp* and the corresponding *npn* transistor is the value of collector current at which their β values reach their maximums: For the high-voltage process, for example, this current is in the tens of microamperes range for the *pnp* and in the milliampere range for the *npn*. On the positive side, the problem of the lack of high-quality *pnp* transistors has spurred analog circuit designers to come up with highly innovative circuit topologies that either minimize the use of *pnp* transistors or minimize the dependence of circuit performance on that of the *pnp*. We shall encounter some of these ingenious circuits later in this book.

The dramatic reduction in device size achieved in the advanced low-voltage process should be evident from Table 6.2. As a result, the scale current I_S also has been reduced by about three orders of magnitude. Here we should note that the base width, W_B, achieved in the advanced process is on the order of 0.1 μm, as compared to a few microns in the standard high-voltage process. Note also the dramatic increase in speed; for the low-voltage *npn* transistor, $\tau_F = 10$ ps as opposed to 0.35 ns in the high-voltage process. As a result, f_T for the modern *npn* transistor is 10 GHz to 25 GHz, as compared to the 400 MHz to 600 MHz achieved in the high-voltage process. Although the Early voltage, V_A, for the modern process is lower than its value in the old high-voltage process, it is still reasonably high at 35 V. Another feature of the advanced process—and one that is not obvious from Table 6.2—is that β for the *npn* peaks at a collector current of 50 μA or so. Finally, note that as the name implies, *npn* transistors fabricated in the low-voltage process break down at collector-emitter voltages of 8 V, as compared to 50 V or so for the high-voltage process. Thus, while circuits designed with the standard high-voltage process utilize power supplies of ±15 V (e.g., in commercially available op amps of the 741 type), the total power-supply voltage utilized with modern bipolar devices is 5 V (or even 3.3 V to achieve compatibility with some of the submicron CMOS processes).

6.2.3 Comparison of Important Characteristics

Table 6.3 provides a compilation of the important characteristics of the NMOS and the *npn* transistors. The material is presented in a manner that facilitates comparison. In the following, we provide comments on the various items in Table 6.3. As well, a number of numerical examples and exercises are provided to illustrate how the wealth of information in Table 6.3 can be put to use. Before proceeding, note that the PMOS and the *pnp* transistors can be compared in a similar way.

TABLE 6.3   Comparison of the MOSFET and the BJT

	NMOS	*npn*
Circuit Symbol		
To Operate in the Active Mode, Two Conditions Have To Be Satisfied	(1) <u>Induce a channel:</u> $$v_{GS} \geq V_t, \quad V_t = 0.5 - 0.7 \text{ V}$$ Let $v_{GS} = V_t + v_{OV}$ (2) <u>Pinch-off channel at drain:</u> $$v_{GD} < V_t$$ or equivalently, $$v_{DS} \geq V_{OV}, \quad V_{OV} = 0.2 - 0.3 \text{ V}$$	(1) <u>Forward-bias EBJ:</u> $$v_{BE} \geq V_{BE\text{on}}, \quad V_{BE\text{on}} \cong 0.5 \text{ V}$$ (2) <u>Reverse-bias CBJ:</u> $$v_{BC} < V_{BC\text{on}}, \quad V_{BC\text{on}} \cong 0.4 \text{ V}$$ or equivalently, $$v_{CE} \geq 0.3 \text{ V}$$
Current-Voltage Characteristics in the Active Region	$$i_D = \frac{1}{2}\mu_n C_{ox}\frac{W}{L}(v_{GS}-V_t)^2\left(1+\frac{v_{DS}}{V_A}\right)$$ $$= \frac{1}{2}\mu_n C_{ox}\frac{W}{L}v_{OV}^2\left(1+\frac{v_{DS}}{V_A}\right)$$ $$i_G = 0$$	$$i_C = I_S e^{v_{BE}/V_T}\left(1+\frac{v_{CE}}{V_A}\right)$$ $$i_B = i_C/\beta$$
Low-Frequency Hybrid-π Model		

	NMOS	*npn*
Low-Frequency T Model		
Transconductance g_m	$g_m = I_D/(V_{OV}/2)$ $$g_m = (\mu_n C_{ox})\left(\frac{W}{L}\right)V_{OV}$$ $$g_m = \sqrt{2(\mu_n C_{ox})\left(\frac{W}{L}\right)I_D}$$	$g_m = I_C/V_T$
Output Resistance r_o	$$r_o = V_A/I_D = \frac{V_A' L}{I_D}$$	$r_o = V_A/I_C$
Intrinsic Gain $A_0 \equiv g_m r_o$	$A_0 = V_A/(V_{OV}/2)$ $$A_0 = \frac{2V_A' L}{V_{OV}}$$ $$A_0 = \frac{V_A' \sqrt{2\mu_n C_{ox} WL}}{\sqrt{I_D}}$$	$A_0 = V_A/V_T$
Input Resistance with Source (Emitter) Grounded	∞	$r_\pi = \beta/g_m$
High-Frequency Model		

(Continued)

TABLE 6.3 Comparison of the MOSFET and the BJT *(Continued)*

	NMOS	npn
Capacitances	$C_{gs} = \frac{2}{3}WLC_{ox} + WL_{ov}C_{ox}$ $C_{gd} = WL_{ov}C_{ox}$	$C_{\pi} = C_{de} + C_{je}$ $C_{de} = \tau_F g_m$ $C_{je} \cong 2C_{je0}$ $C_{\mu} = C_{\mu0} / \left[1 + \frac{V_{CB}}{V_{C0}}\right]^m$
Transition Frequency f_T	$f_T = \frac{g_m}{2\pi(C_{gs} + C_{gd})}$ For $C_{gs} \gg C_{gd}$ and $C_{gs} \cong \frac{2}{3}WLC_{ox}$, $f_T \cong \frac{1.5\mu_n V_{OV}}{2\pi L^2}$	$f_T = \frac{g_m}{2\pi(C_{\pi} + C_{\mu})}$ For $C_{\pi} \gg C_{\mu}$ and $C_{\pi} \cong C_{de}$, $f_T \cong \frac{2\mu_n V_T}{2\pi W_B^2}$
Design Parameters	$I_D, V_{OV}, L, \frac{W}{L}$	I_C, V_{BE}, A_E (or I_S)
Good Analog Switch?	Yes, because the device is symmetrical and thus the i_D–v_{DS} characteristics pass directly through the origin.	No, because the device is asymmetrical with an offset voltage V_{CEoff}.

Operating Conditions

At the outset, note that we shall use **active mode** or **active region** to denote both the active mode of operation of the BJT and the saturation-mode of operation of the MOSFET.

The conditions for operating in the active mode are very similar for the two devices: The explicit threshold V_t of the MOSFET has V_{BEon} as its implicit counterpart in the BJT. Furthermore, for modern processes, V_{BEon} and V_t are almost equal.

Also, pinching off the channel of the MOSFET at the drain end is very similar to reverse biasing the CBJ of the BJT. Note, however, that the asymmetry of the BJT results in V_{BCon} and V_{BEon} being unequal, while in the symmetrical MOSFET the operative threshold voltages at the source and the drain ends of the channel are identical (V_t). Finally, for both the MOSFET and the BJT to operate in the active mode, the voltage across the device (v_{DS}, v_{CE}) must be at least 0.2 V to 0.3 V.

Current-Voltage Characteristics

The square-law control characteristic, i_D–v_{GS}, in the MOSFET should be contrasted with the exponential control characteristic, i_C–v_{BE}, of the BJT. Obviously, the latter is a much more sensitive relationship, with the result that i_C can vary over a very wide range (five decades or more) within the same BJT. In the MOSFET, the range of i_D achieved in the same device is much more limited. To appreciate this point further, consider the parabolic relationship between i_D and v_{OV}, and recall from our discussion above that v_{OV} is usually kept in a narrow range (0.2 V to 0.4 V).

Next we consider the effect of the device dimensions on its current. For the bipolar transistor the control parameter is the area of the emitter–base junction (EBJ), A_E, which determines the scale current I_S. It can be varied over a relatively narrow range, such as 10 to 1. Thus, while the emitter area can be used to achieve current scaling in an IC (as we shall see in the next section in connection with the design of current mirrors) its narrow range of variation reduces its significance as a design parameter. This is particularly so if we compare A_E

with its counterpart in the MOSFET, the aspect ratio W/L. MOSFET devices can be designed with W/L ratios in a wide range, such as 0.1 to 100. As a result W/L is a very significant MOS design parameter. Like A_E, it is also used in current scaling, as we shall see in the next section. Combining the possible range of variation of v_{OV} and W/L, one can design MOS transistors to operate over an i_D range of four decades or so.

The channel-length modulation in the MOSFET and the base-width modulation in the BJT are similarly modeled and give rise to the dependence of $i_D(i_C)$ on $v_{DS}(v_{CE})$ and, hence, to the finite output resistance r_o in the active region. Two important differences, however, exist. In the BJT, V_A is solely a process-technology parameter and does not depend on the dimensions of the BJT. In the MOSFET, the situation is quite different: $V_A = V_A'L$, where V_A' is a process-technology parameter and L is the channel length used. Also, in modern submicron processes, V_A' is very low, resulting in V_A values much lower than the corresponding values for the BJT.

The last, and perhaps most important, difference between the current-voltage characteristics of the two devices concerns the input current into the control terminal: While the gate current of the MOSFET is practically zero and the input resistance looking into the gate is practically infinite, the BJT draws base current i_B that is proportional to the collector current; that is, $i_B = i_C/\beta$. The finite base current and the corresponding finite input resistance looking into the base is a definite disadvantage of the BJT in comparison to the MOSFET. Indeed, it is the infinite input resistance of the MOSFET that has made possible analog and digital circuit applications that are not feasible with the BJT. Examples include dynamic digital memory (Chapter 11) and switched-capacitor filters (Chapter 12).

EXAMPLE 6.1

(a) For an NMOS transistor with $W/L = 10$ fabricated in the 0.18-μm process whose data are given in Table 6.1, find the values of V_{OV} and V_{GS} required to operate the device at $I_D = 100\ \mu$A. Ignore channel-length modulation.

(b) Find V_{BE} for an *npn* transistor fabricated in the low-voltage process specified in Table 6.2 and operated at $I_C = 100\ \mu$A. Ignore base-width modulation.

Solution

(a)
$$I_D = \frac{1}{2}(\mu_n C_{ox})\left(\frac{W}{L}\right)V_{OV}^2$$

Substituting $I_D = 100\ \mu$A, $W/L = 10$, and, from Table 6.1, $\mu_n C_{ox} = 387\ \mu$A/V^2 results in

$$100 = \frac{1}{2} \times 387 \times 10 \times V_{OV}^2$$

$$V_{OV} = 0.23\ \text{V}$$

Thus,

$$V_{GS} = V_{tn} + V_{OV} = 0.48 + 0.23 = 0.71\ \text{V}$$

(b)
$$I_C = I_S e^{V_{BE}/V_T}$$

Substituting $I_C = 100\ \mu$A and, from Table 6.2, $I_S = 6 \times 10^{-18}$ A gives,

$$V_{BE} = 0.025 \ln\frac{100 \times 10^{-6}}{6 \times 10^{-18}} = 0.76\ \text{V}$$

Low-Frequency Small-Signal Models   The low-frequency models for the two devices are very similar except, of course, for the finite base current (finite β) of the BJT, which gives rise to r_π in the hybrid-π model and to the unequal currents in the emitter and collector in the T models ($\alpha < 1$). Here it is interesting to note that the low-frequency small-signal models become identical if one thinks of the MOSFET as a BJT with $\beta = \infty$ ($\alpha = 1$).

For both devices, the hybrid-π model indicates that the **open-circuit voltage gain** obtained from gate to drain (base to collector) with the source (emitter) grounded is $-g_m r_o$. It follows that $g_m r_o$ is the *maximum gain available from a single transistor* of either type. This important transistor parameter is given the name **intrinsic gain** and is denoted A_0. We will have more to say about the intrinsic gain shortly.

Although not included in the MOSFET low-frequency model shown in Table 6.3, the body effect can have a significant implication for the operation of the MOSFET as an amplifier. In simple terms, if the body (substrate) is not connected to the source, it can act as a second gate for the MOSFET. The voltage signal that develops between the body and the source, v_{bs}, gives rise to a drain current component $g_{mb}v_{bs}$, where the body transconductance g_{mb} is proportional to g_m; that is, $g_{mb} = \chi g_m$, where the factor χ is in the range of 0.1 to 0.2. We shall take the body effect into account in the study of IC MOS amplifiers in the succeeding sections. The body effect has no counterpart in the BJT.

The Transconductance   For the BJT, the transconductance g_m depends *only* on the dc collector current I_C. (Recall that V_T is a physical constant $\cong 0.025$ V at room temperature). It is interesting to observe that g_m does not depend on the geometry of the BJT, and its dependence on the EBJ area is only through the effect of the area on the total collector current I_C. Similarly, the dependence of g_m on V_{BE} is only through the fact that V_{BE} determines the total current in the collector. By contrast, g_m of the MOSFET depends on I_D, V_{OV}, and W/L. Therefore, we use three different (but equivalent) formulas to express g_m of the MOSFET.

The first formula given in Table 6.3 for the MOSFET's g_m is the most directly comparable with the formula for the BJT. It indicates that for the same operating current, g_m of the MOSFET is much smaller than that of the BJT. This is because $V_{OV}/2$ is in the range of 0.1 V to 0.2 V, which is four to eight times the corresponding term in the BJT's formula, namely V_T.

The second formula for the MOSFET's g_m indicates that for a given device (i.e., given W/L), g_m is proportional to V_{OV}. Thus a higher g_m is obtained by operating the MOSFET at a higher overdrive voltage. However, we should recall the limitations imposed on the magnitude of V_{OV} by the limited value of V_{DD}. Put differently, the need to obtain a reasonably high g_m constrains the designer's interest in reducing V_{OV}.

The third g_m formula shows that for a given transistor (i.e., given W/L), g_m is proportional to $\sqrt{I_D}$. This should be contrasted with the bipolar case, where g_m is directly proportional to I_C.

Output Resistance   The output resistance for both devices is determined by similar formulas, with r_o being the ratio of V_A to the bias current (I_D or I_C). Thus, for both transistors,

r_o is inversely proportional to the bias current. The difference in nature and magnitude of V_A between the two devices has already been discussed.

Intrinsic Gain   The intrinsic gain A_0 of the BJT is the ratio of V_A, which is solely a process parameter (35 V to 130 V), and V_T, which is a physical parameter (0.025 V at room temperature). Thus A_0 of a BJT is independent of the device junction area and of the operating current, and its value ranges from 1000 V/V to 5000 V/V. The situation in the MOSFET is very different: Table 6.3 provides three different (but equivalent) formulas for expressing the MOSFET's intrinsic gain. The first formula is the one most directly comparable to that of the BJT. Here, however, we note the following:

1. The quantity in the denominator is $V_{OV}/2$, which is a design parameter, and although it is becoming smaller in designs using short-channel technologies, it is still much larger than V_T. Furthermore, as we have seen earlier, there are reasons for selecting larger values for V_{OV}.

2. The numerator quantity V_A is both process- and device-dependent, and its value has been steadily decreasing.

   As a result, the intrinsic gain realized in a single MOSFET amplifier stage fabricated in a modern short-channel technology is only 20 V/V to 40 V/V, almost two orders of magnitude lower than that for a BJT.

   The third formula given for A_0 in Table 6.3 points out a very interesting fact: For a given process technology (V_A' and $\mu_n C_{ox}$) and a given device (W/L), the intrinsic gain is inversely proportional to $\sqrt{I_D}$. This is illustrated in Fig. 6.1, which shows a typical plot of A_0 versus the bias current I_D. The plot confirms that the gain increases as the bias current is lowered. The gain, however, levels off at very low currents. This is because the MOSFET enters the subthreshold region of operation (Section 4.1.9), where it becomes very much like a BJT with an exponential current-voltage characteristic. The intrinsic gain then becomes constant, just like that of a BJT. Note, however, that although a higher gain is achieved at lower bias currents, the price paid is a lower g_m and less ability to drive capacitive loads and thus a decrease in bandwidth. This point will be further illustrated shortly.

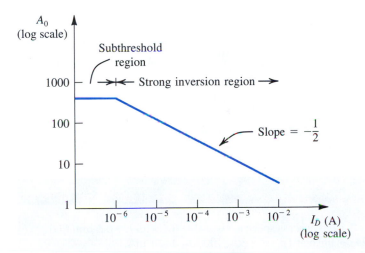

FIGURE 6.1   The intrinsic gain of the MOSFET versus bias current I_D. Outside the subthreshold region, this is a plot of $A_0 = V_A' \sqrt{2\mu_n C_{ox} WL/I_D}$ for the case: $\mu_n C_{ox} = 20\ \mu\text{A/V}^2$, $V_A' = 20\ \text{V/}\mu\text{m}$, $L = 2\ \mu\text{m}$, and $W = 20\ \mu\text{m}$.

EXAMPLE 6.2

We wish to compare the values of g_m, input resistance at the gate (base), r_o, and A_0 for an NMOS transistor fabricated in the 0.25-μm technology specified in Table 6.1 and an *npn* transistor fabricated in the low-voltage technology specified in Table 6.2. Assume both devices are operating at a drain (collector) current of 100 μA. For the MOSFET, let $L = 0.4$ μm and $W = 4$ μm, and specify the required V_{OV}.

Solution

For the NMOS transistor,

$$I_D = \frac{1}{2}(\mu_n C_{ox})\left(\frac{W}{L}\right)V_{OV}^2$$

$$100 = \frac{1}{2} \times 267 \times \frac{4}{0.4} \times V_{OV}^2$$

Thus,

$$V_{OV} = 0.27 \text{ V}$$

$$g_m = \sqrt{2(\mu_n C_{ox})\left(\frac{W}{L}\right)I_D}$$

$$= \sqrt{2 \times 267 \times 10 \times 100} = 0.73 \text{ mA/V}$$

$$R_{in} = \infty$$

$$r_o = \frac{V_A' L}{I_D} = \frac{5 \times 0.4}{0.1} = 20 \text{ k}\Omega$$

$$A_0 = g_m r_o = 0.73 \times 20 = 14.6 \text{ V/V}$$

For the *npn* transistor,

$$g_m = \frac{I_C}{V_T} = \frac{0.1 \text{ mA}}{0.025 \text{ V}} = 4 \text{ mA/V}$$

$$R_{in} = r_\pi = \beta_0/g_m = \frac{100}{4 \text{ mA/V}} = 25 \text{ k}\Omega$$

$$r_o = \frac{V_A}{I_C} = \frac{35}{0.1 \text{ mA}} = 350 \text{ k}\Omega$$

$$A_0 = g_m r_o = 4 \times 350 = 1400 \text{ V/V}$$

EXERCISE

6.2 For an NMOS transistor fabricated in the 0.5-μm process specified in Table 6.1 with $L = 0.5$ μm, find the transconductance and the intrinsic gain obtained at $I_D = 10$ μA, 100 μA, and 1 mA.
Ans. 0.2 mA/V, 200 V/V; 0.6 mA/V, 62 V/V; 2 mA/V, 20 V/V

High-Frequency Operation   The simplified high-frequency equivalent circuits for the MOSFET and the BJT are very similar, and so are the formulas for determining their unity-gain frequency (also called **transition frequency**) f_T. Recall that f_T is a measure of the *intrinsic* bandwidth of the transistor itself and does *not* take into account the effects of capacitive loads. We shall address the issue of capacitive loads shortly. For the time being, note the striking similarity between the approximate formulas given in Table 6.3 for the value of f_T of the two devices. In both cases f_T is inversely proportional to the square of the critical dimension of the device: the channel length for the MOSFET and the base width for the BJT. These formulas also clearly indicate that shorter-channel MOSFETs[2] and narrower-base BJTs are inherently capable of a wider bandwidth of operation. It is also important to note that while for the BJT the approximate expression for f_T indicates that it is entirely process determined, the corresponding expression for the MOSFET shows that f_T is proportional to the overdrive voltage V_{OV}. Thus we have conflicting requirements on V_{OV}: While a higher low-frequency gain is achieved by operating at a low V_{OV}, wider bandwidth requires an increase in V_{OV}. Therefore the selection of a value for V_{OV} involves, among other considerations, a trade-off between gain and bandwidth.

For *npn* transistors fabricated in the modern low-voltage process, f_T is in the range of 10 GHz to 20 GHz as compared to the 400 MHz to 600 MHz obtained with the standard high-voltage process. In the MOS case, NMOS transistors fabricated in a modern submicron technology, such as the 0.18-μm process, achieve f_T values in the range of 5 GHz to 15 GHz.

Before leaving the subject of high-frequency operation, let's look into the effect of a capacitive load on the bandwidth of the common-source (common-emitter) amplifier. For this purpose we shall assume that the frequencies of interest are much lower than f_T of the transistor. Hence we shall not take the transistor capacitances into account. Figure 6.2(a) shows a common-source amplifier with a capacitive load C_L. The voltage gain from gate to drain can be found as follows:

$$V_o = -g_m V_{gs}(r_o \| C_L)$$

$$= -g_m V_{gs} \frac{r_o \dfrac{1}{sC_L}}{r_o + \dfrac{1}{sC_L}}$$

$$A_v = \frac{V_o}{V_{gs}} = -\frac{g_m r_o}{1 + sC_L r_o} \tag{6.1}$$

Thus the gain has, as expected, a low-frequency value of $g_m r_o = A_0$ and a frequency response of the single-time-constant (STC) low-pass type with a break (pole) frequency at

$$\omega_P = \frac{1}{C_L r_o} \tag{6.2}$$

Obviously this pole is formed by r_o and C_L. A sketch of the magnitude of gain versus frequency is shown in Fig. 6.2(b). We observe that the gain crosses the 0-dB line at frequency ω_t,

$$\omega_t = A_0 \omega_P = (g_m r_o)\frac{1}{C_L r_o}$$

---

[2] Although the reason is beyond our capabilities at this stage, f_T of MOSFETs that have very short channels varies inversely with L rather than with L^2.

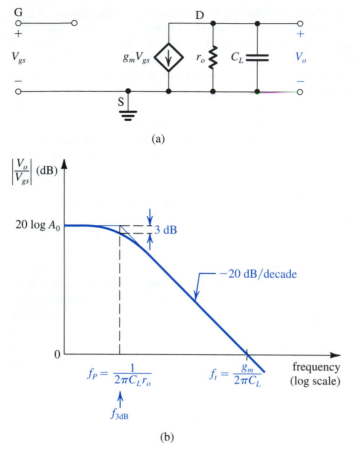

(a)

(b)

FIGURE 6.2 Frequency response of a CS amplifier loaded with a capacitance C_L and fed with an ideal voltage source. It is assumed that the transistor is operating at frequencies much lower than f_T, and thus the internal capacitances are not taken into account.

Thus,

$$\omega_t = \frac{g_m}{C_L} \tag{6.3}$$

That is, the **unity-gain frequency** or, equivalently, the **gain–bandwidth product**[3] ω_t is the ratio of g_m and C_L. We thus clearly see that for a given capacitive load C_L, a larger gain–bandwidth product is achieved by operating the MOSFET as a higher g_m. Identical analysis and conclusions apply to the case of the BJT. In each case, bandwidth increases as bias current is increased.

Design Parameters   For the BJT there are three design parameters—I_C, V_{BE}, and I_S (or, equivalently, the area of the emitter–base junction)—of which any two can be selected by the designer. However, since I_C is exponentially related to V_{BE} and is very sensitive to the value of V_{BE} (V_{BE} changes by only 60 mV for a factor of 10 change in I_C), I_C is much more useful than V_{BE} as a design parameter. As mentioned earlier, the utility of the EBJ area as a

---

[3] The unity-gain frequency and the gain–bandwidth product of an amplifier are the same when the frequency response is of the single-pole type; otherwise the two parameters may differ.

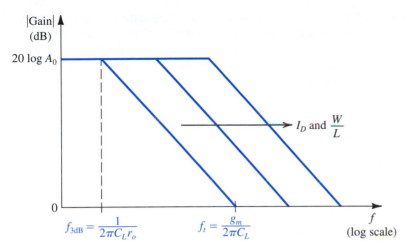

FIGURE 6.3 Increasing I_D or W/L increases the bandwidth of a MOSFET amplifier loaded by a constant capacitance C_L.

design parameter is rather limited because of the narrow range over which A_E can vary. It follows that for the BJT there is only one effective design parameter: the collector current I_C. Finally, note that we have not considered V_{CE} to be a design parameter, since its effect on I_C is only secondary. Of course, as we learned in Chapter 5, V_{CE} affects the output signal swing.

For the MOSFET there are four design parameters—I_D, V_{OV}, L, and W—of which any three can be selected by the designer. For analog circuit applications the trade-off in selecting a value for L is between the higher speeds of operation (wider amplifier bandwidth) obtained at lower values of L and the higher intrinsic gain obtained at larger values of L. Usually one selects an L of about 25% to 50% greater than $L_{\min}$.

The second design parameter is V_{OV}. We have already made numerous remarks about the effect of the value of V_{OV} on performance. Usually, for submicron technologies, V_{OV} is selected in the range of 0.2 V to 0.4 V.

Once values for L and V_{OV} are selected, the designer is left with the selection of the value of I_D or W (or, equivalently, W/L). For a given process and for the selected values of L and V_{OV}, I_D is proportional to W/L. It is important to note that the choice of I_D or, equivalently, of W/L has no bearing on the value of intrinsic gain A_0 and the transition frequency f_T. However, it affects the value of g_m and hence the gain–bandwidth product. Figure 6.3 illustrates this point by showing how the gain of a common-source amplifier operated at a constant V_{OV} varies with I_D (or, equivalently, W/L). Note that while the dc gain remains unchanged, increasing W/L and, correspondingly, I_D increases the bandwidth proportionally. This, however, assumes that the load capacitance C_L is not affected by the device size, an assumption that may not be entirely justified in some cases.

EXAMPLE 6.3

In this example we investigate the gain and the high-frequency response of an *npn* transistor and an NMOS transistor. For the *npn* transistor, assume that it is fabricated in the low-voltage process specified in Table 6.2, and assume that $C_\mu \cong C_{\mu 0}$. For $I_C = 10$ μA, 100 μA, and 1 mA, find g_m, r_o, A_0, C_{de}, C_{je}, C_π, C_μ, and f_T. Also, for each value of I_C, find the gain–bandwidth product f_t of a common-emitter amplifier loaded by a 1-pF capacitance, neglecting the internal capacitances of the transistor. For the NMOS transistor, assume that it is fabricated in the 0.25-μm CMOS process with $L = 0.4$ μm. Let the transistor be operated at $V_{OV} = 0.25$ V. Find W/L that is required to obtain $I_D = 10$ μA, 100 μA, and 1 mA. At each value of I_D, find g_m, r_o, A_0, C_{gs}, C_{gd}, and f_T. Also, for each value of I_D, determine the gain–bandwidth product f_t of a common-source amplifier loaded by a 1-pF capacitance, neglecting the internal capacitances of the transistor.

Solution

For the *npn* transistor,

$$g_m = \frac{I_C}{V_T} = \frac{I_C}{0.025} = 40 I_C \text{ A/V}$$

$$r_o = \frac{V_A}{I_C} = \frac{35}{I_C} \Omega$$

$$A_0 = \frac{V_A}{V_T} = \frac{35}{0.025} = 1400 \text{ V/V}$$

$$C_{de} = \tau_F g_m = 10 \times 10^{-12} \times 40 I_C = 0.4 \times 10^{-9} I_C \text{ F}$$

$$C_{je} \cong 2C_{je0} = 10 \text{ fF}$$

$$C_\pi = C_{de} + C_{je}$$

$$C_\mu \cong C_{\mu0} = 5 \text{ fF}$$

$$f_T = \frac{g_m}{2\pi(C_\pi + C_\mu)}$$

$$f_t = \frac{g_m}{2\pi C_L} = \frac{g_m}{2\pi \times 1 \times 10^{-12}}$$

We thus obtain the following results:

I_C	g_m (mA/V)	r_o (kΩ)	A_0 (V/V)	C_{de} (fF)	C_{je} (fF)	C_π (fF)	C_μ (fF)	f_T (GHz)	f_t (MHz)
10 μA	0.4	3500	1400	4	10	14	5	3.4	64
100 μA	4	350	1400	40	10	50	5	11.6	640
1 mA	40	35	1400	400	10	410	5	15.3	6400

For the NMOS transistor,

$$I_D = \frac{1}{2}\mu_n C_{ox}\frac{W}{L}V_{OV}^2$$

$$= \frac{1}{2} \times 267 \times \frac{W}{L} \times \frac{1}{16}$$

Thus,

$$\frac{W}{L} = 0.12 I_D$$

$$g_m = \frac{I_D}{V_{OV}/2} = \frac{I_D}{0.25/2} = 8 I_D \text{ A/V}$$

$$r_o = \frac{V_A' L}{I_D} = \frac{5 \times 0.4}{I_D} = \frac{2}{I_D} \Omega$$

$$A_0 = g_m r_o = 16 \text{ V/V}$$

$$C_{gs} = \tfrac{2}{3}WLC_{ox} + C_{ov} = \tfrac{2}{3}W \times 0.4 \times 5.8 + 0.6W$$

$$C_{gd} = C_{ov} = 0.6W$$

$$f_T = \frac{g_m}{2\pi(C_{gs} + C_{gd})}$$

$$f_t = \frac{g_m}{2\pi C_L}$$

We thus obtain the following results:

I_D	W/L	g_m (mA/V)	r_o (kΩ)	A_0 (V/V)	C_{gs} (fF)	C_{gd} (fF)	f_T (GHz)	f_t (MHz)
10 μA	1.2	0.08	200	16	1.03	0.29	9.7	12.7
100 μA	12	0.8	20	16	10.3	2.9	9.7	127
1 mA	120	8	2	16	103	29	9.7	1270

EXERCISE

6.3 Find I_D, g_m, r_o, A_0, C_{gs}, C_{gd}, and f_T for an NMOS transistor fabricated in the 0.5-μm CMOS technology specified in Table 6.1. Let $L = 0.5$ μm, $W = 5$ μm, and $V_{OV} = 0.3$ V.

Ans. 85.5 μA; 0.57 mA/V; 66.7 kΩ; 38 V/V; 8.3 fF; 2 fF; 8.8 GHz

6.2.4 Combining MOS and Bipolar Transistors—BiCMOS Circuits

From the discussion above it should be evident that the BJT has the advantage over the MOSFET of a much higher transconductance (g_m) at the same value of dc bias current. Thus, in addition to realizing much higher voltage gains per amplifier stage, bipolar transistor amplifiers have superior high-frequency performance compared to their MOS counterparts.

On the other hand, the practically infinite input resistance at the gate of a MOSFET makes it possible to design amplifiers with extremely high input resistances and an almost zero input bias current. Also, as mentioned earlier, the MOSFET provides an excellent implementation of a switch, a fact that has made CMOS technology capable of realizing a host of analog circuit functions that are not possible with bipolar transistors.

It can thus be seen that each of the two transistor types has its own distinct and unique advantages: Bipolar technology has been extremely useful in the design of very-high-quality general-purpose circuit building blocks, such as op amps. On the other hand, CMOS, with its very high packing density and its suitability for both digital and analog circuits, has become the technology of choice for the implementation of very-large-scale integrated circuits. Nevertheless, the performance of CMOS circuits can be improved if the designer has available (on the same chip) bipolar transistors that can be employed in functions that require their high g_m and excellent current-driving capability. A technology that allows the fabrication of high-quality bipolar transistors on the same chip as CMOS circuits is aptly called **BiCMOS.** At appropriate locations throughout this book we shall present interesting and useful BiCMOS circuit blocks.

6.2.5 Validity of the Square-Law MOSFET Model

We conclude this section with a comment on the validity of the simple square-law model we have been using to describe the operation of the MOS transistor. While this simple model works well for devices with relatively long channels (>1 μm) it does *not* provide an accurate representation of the operation of short-channel devices. This is because a number of physical phenomena come into play in these submicron devices, resulting in what are called **short-channel effects.** Although the study of short-channel effects is beyond the scope of this book, it should be mentioned that MOSFET models have been developed that take these effects into account. However, they are understandably quite complex and do not lend themselves to hand analysis of the type needed to develop insight into circuit operation. Rather, these models are suitable for computer simulation and are indeed used in SPICE (Section 6.13). For quick, manual analysis, however, we will continue to use the square-law model which is the basis for the comparison of Table 6.3.

6.3 IC BIASING—CURRENT SOURCES, CURRENT MIRRORS, AND CURRENT-STEERING CIRCUITS

Biasing in integrated-circuit design is based on the use of constant-current sources. On an IC chip with a number of amplifier stages, a constant dc current (called a **reference current**) is generated at one location and is then replicated at various other locations for biasing the various amplifier stages through a process known as **current steering.** This approach has the advantage that the effort expended on generating a predictable and stable reference current, usually utilizing a precision resistor external to the chip, need not be repeated for every amplifier stage. Furthermore, the bias currents of the various stages track each other in case of changes in power-supply voltage or in temperature.

In this section we study circuit building blocks and techniques employed in the bias design of IC amplifiers. These circuits are also utilized as amplifier load elements, as will be seen in Section 6.5 and beyond.

6.3.1 The Basic MOSFET Current Source

Figure 6.4 shows the circuit of a simple MOS constant-current source. The heart of the circuit is transistor Q_1, the drain of which is shorted to its gate,[4] thereby forcing it to operate in the saturation mode with

$$I_{D1} = \frac{1}{2}k_n'\left(\frac{W}{L}\right)_1 (V_{GS} - V_{tn})^2 \tag{6.4}$$

where we have neglected channel-length modulation. The drain current of Q_1 is supplied by V_{DD} through resistor R, which in most cases would be outside the IC chip. Since the gate currents are zero,

$$I_{D1} = I_{\text{REF}} = \frac{V_{DD} - V_{GS}}{R} \tag{6.5}$$

where the current through R is considered to be the reference current of the current source and is denoted I_{REF}. Equations (6.4) and (6.5) can be used to determine the value required for R.

---

[4] Such a transistor is said to be *diode connected.*

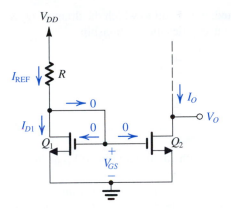

FIGURE 6.4 Circuit for a basic MOSFET constant-current source.

Now consider transistor Q_2: It has the same V_{GS} as Q_1; thus, if we assume that it is operating in saturation, its drain current, which is the output current I_O of the current source, will be

$$I_O = I_{D2} = \frac{1}{2} k_n' \left(\frac{W}{L} \right)_2 (V_{GS} - V_{tn})^2 \tag{6.6}$$

where we have neglected channel-length modulation. Equations (6.4) and (6.6) enable us to relate the output current I_O to the reference current I_{REF} as follows:

$$\frac{I_O}{I_{REF}} = \frac{(W/L)_2}{(W/L)_1} \tag{6.7}$$

This is a simple and attractive relationship: The special connection of Q_1 and Q_2 provides an output current I_O that is related to the reference current I_{REF} by the ratio of the aspect ratios of the transistors. In other words, the relationship between I_O and I_{REF} is solely determined by the geometries of the transistors. In the special case of identical transistors, $I_O = I_{REF}$, and the circuit simply replicates or mirrors the reference current in the output terminal. This has given the circuit composed of Q_1 and Q_2 the name **current mirror,** a name that is used irrespective of the ratio of device dimensions.

Figure 6.5 depicts the current mirror circuit with the input reference current shown as being supplied by a current source for both simplicity and generality. The **current gain** or **current transfer ratio** of the current mirror is given by Eq. (6.7).

Effect of V_O on I_O   In the description above for the operation of the current source of Fig. 6.4, we assumed Q_2 to be operating in saturation. This is obviously essential if Q_2 is to supply a

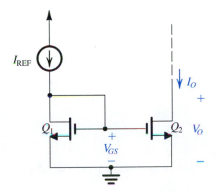

FIGURE 6.5 Basic MOSFET current mirror.

constant-current output. To ensure that Q_2 is saturated, the circuit to which the drain of Q_2 is to be connected must establish a drain voltage V_O that satisfies the relationship

$$V_O \geq V_{GS} - V_t \qquad (6.8)$$

or, equivalently, in terms of the overdrive voltage V_{OV} of Q_1 and Q_2,

$$V_O \geq V_{OV} \qquad (6.9)$$

In other words, the current source will operate properly with an output voltage V_O as low as V_{OV}, which is a few tenths of a volt.

Although thus far neglected, channel-length modulation can have a significant effect on the operation of the current source. Consider, for simplicity, the case of identical devices Q_1 and Q_2. The drain current of Q_2, I_O, will equal the current in Q_1, I_{REF}, at the value of V_O that causes the two devices to have the same V_{DS}, that is, at $V_O = V_{GS}$. As V_O is increased above this value, I_O will increase according to the incremental output resistance r_{o2} of Q_2. This is illustrated in Fig. 6.6, which shows I_O versus V_O. Observe that since Q_2 is operating at a constant V_{GS} (determined by passing I_{REF} through the matched device Q_1), the curve in Fig. 6.6 is simply the i_D–v_{DS} characteristic curve of Q_2 for v_{GS} equal to the particular value V_{GS}.

In summary, the current source of Fig. 6.4 and the current mirror of Fig. 6.5 have a finite output resistance R_o,

$$R_o \equiv \frac{\Delta V_O}{\Delta I_O} = r_{o2} = \frac{V_{A2}}{I_O} \qquad (6.10)$$

where I_O is given by Eq. (6.6) and V_{A2} is the Early voltage of Q_2. Also, recall that for a given process technology, V_A is proportional to the transistor channel length; thus, to obtain high output-resistance values, current sources are usually designed using transistors with relatively long channels. Finally, note that we can express the current I_O as

$$I_O = \frac{(W/L)_2}{(W/L)_1} I_{REF} \left(1 + \frac{V_O - V_{GS}}{V_{A2}}\right) \qquad (6.11)$$

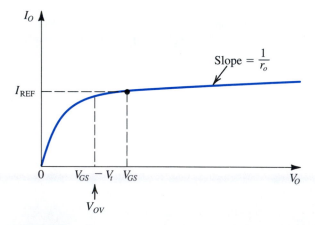

FIGURE 6.6 Output characteristic of the current source in Fig. 6.4 and the current mirror of Fig. 6.5 for the case Q_2 is matched to Q_1.

EXAMPLE 6.4

Given $V_{DD} = 3$ V and using $I_{REF} = 100$ μA, it is required to design the circuit of Fig. 6.4 to obtain an output current whose nominal value is 100 μA. Find R if Q_1 and Q_2 are matched and have channel lengths of 1 μm, channel widths of 10 μm, $V_t = 0.7$ V, and $k'_n = 200$ μA/V^2. What is the lowest possible value of V_O? Assuming that for this process technology the Early voltage $V'_A = 20$ V/μm, find the output resistance of the current source. Also, find the change in output current resulting from a +1-V change in V_O.

Solution

$$I_{D1} = I_{REF} = \frac{1}{2}k'_n\left(\frac{W}{L}\right)_1 V_{OV}^2$$

$$100 = \frac{1}{2} \times 200 \times 10 V_{OV}^2$$

Thus,

$$V_{OV} = 0.316 \text{ V}$$

and

$$V_{GS} = V_t + V_{OV} = 0.7 + 0.316 \cong 1 \text{ V}$$

$$R = \frac{V_{DD} - V_{GS}}{I_{REF}} = \frac{3-1}{0.1 \text{ mA}} = 20 \text{ k}\Omega$$

$$V_{O\min} = V_{OV} \cong 0.3 \text{ V}$$

For the transistors used, $L = 1$ μm. Thus,

$$V_A = 20 \times 1 = 20 \text{ V}$$

$$r_{o2} = \frac{20 \text{ V}}{100 \text{ }\mu\text{A}} = 0.2 \text{ M}\Omega$$

The output current will be 100 μA at $V_O = V_{GS} = 1$ V. If V_O changes by +1 V, the corresponding change in I_O will be

$$\Delta I_O = \frac{\Delta V_O}{r_{o2}} = \frac{1 \text{ V}}{0.2 \text{ M}\Omega} = 5 \text{ }\mu\text{A}$$

EXERCISE

D6.4 In the current source of Example 6.4, it is required to reduce the change in output current, ΔI_O, corresponding to a change in output voltage, ΔV_O, of 1 V to 1% of I_O. What should the dimensions of Q_1 and Q_2 be changed to? Assume that Q_1 and Q_2 are to remain matched.
Ans. $L = 5$ μm; $W = 50$ μm

6.3.2 MOS Current-Steering Circuits

As mentioned earlier, once a constant current is generated, it can be replicated to provide dc bias currents for the various amplifier stages in an IC. Current mirrors can obviously be used to implement this current-steering function. Figure 6.7 shows a simple current-steering circuit.

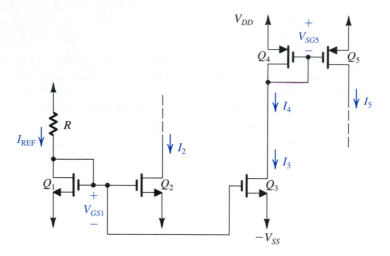

FIGURE 6.7 A current-steering circuit.

Here Q_1 together with R determine the reference current I_{REF}. Transistors Q_1, Q_2, and Q_3 form a two-output current mirror,

$$I_2 = I_{REF} \frac{(W/L)_2}{(W/L)_1} \tag{6.12}$$

$$I_3 = I_{REF} \frac{(W/L)_3}{(W/L)_1} \tag{6.13}$$

To ensure operation in the saturation region, the voltages at the drains of Q_2 and Q_3 are constrained as follows:

$$V_{D2}, V_{D3} \geq -V_{SS} + V_{GS1} - V_{tn} \tag{6.14}$$

or, equivalently,

$$V_{D2}, V_{D3} \geq -V_{SS} + V_{OV1} \tag{6.15}$$

where V_{OV1} is the overdrive voltage at which Q_1, Q_2, and Q_3 are operating. In other words, the drains of Q_2 and Q_3 will have to remain higher than $-V_{SS}$ by at least the overdrive voltage, which is usually a few tenths of a volt.

Continuing our discussion of the circuit in Fig. 6.7, we see that current I_3 is fed to the input side of a current mirror formed by PMOS transistors Q_4 and Q_5. This mirror provides

$$I_5 = I_4 \frac{(W/L)_5}{(W/L)_4} \tag{6.16}$$

where $I_4 = I_3$. To keep Q_5 in saturation, its drain voltage should be

$$V_{D5} \leq V_{DD} - |V_{OV5}| \tag{6.17}$$

where V_{OV5} is the overdrive voltage at which Q_5 is operating.

Finally, an important point to note is that while Q_2 *pulls* its current I_2 from a load (not shown in Fig. 6.7), Q_5 *pushes* its current I_5 into a load (not shown in Fig. 6.7). Thus Q_5 is appropriately called a **current source,** whereas Q_2 should more properly be called a **current sink.** In an IC, both current sources and current sinks are usually needed.

EXERCISE

6.5 For the circuit of Fig. 6.7, let $V_{DD} = V_{SS} = 1.5$ V, $V_{tn} = 0.6$ V, $V_{tp} = -0.6$ V, all channel lengths = 1 μm, $k'_n = 200$ μA/V^2, $k'_p = 80$ μA/V^2, and $\lambda = 0$. For $I_{REF} = 10$ μA, find the widths of all transistors to obtain $I_2 = 60$ μA, $I_3 = 20$ μA, and $I_5 = 80$ μA. It is further required that the voltage at the drain of Q_2 be allowed to go down to within 0.2 V of the negative supply and that the voltage at the drain of Q_5 be allowed to go up to within 0.2 V of the positive supply.

Ans. $W_1 = 2.5$ μm; $W_2 = 15$ μm; $W_3 = 5$ μm; $W_4 = 12.5$ μm; $W_5 = 50$ μm

---

6.3.3 BJT Circuits

The basic BJT current mirror is shown in Fig. 6.8. It works in a fashion very similar to that of the MOS mirror. However, there are two important differences: First, the nonzero base current of the BJT (or, equivalently, the finite β) causes an error in the current transfer ratio of the bipolar mirror. Second, the current transfer ratio is determined by the relative areas of the emitter–base junctions of Q_1 and Q_2.

Let us first consider the case when β is sufficiently high so that we can neglect the base currents. The reference current I_{REF} is passed through the diode-connected transistor Q_1 and thus establishes a corresponding voltage V_{BE}, which in turn is applied between base and emitter of Q_2. Now, if Q_2 is matched to Q_1 or, more specifically, if the EBJ area of Q_2 is the same as that of Q_1 and thus Q_2 has the same scale current I_S as Q_1, then the collector current of Q_2 will be equal to that of Q_1; that is,

$$I_O = I_{REF} \tag{6.18}$$

For this to happen, however, Q_2 must be operating in the active mode, which in turn is achieved so long as the collector voltage V_O is 0.3 V or so higher than that of the emitter.

To obtain a current transfer ratio other than unity, say m, we simply arrange that the area of the EBJ of Q_2 is m times that of Q_1. In this case,

$$I_O = mI_{REF} \tag{6.19}$$

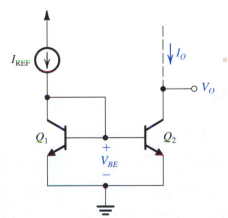

FIGURE 6.8 The basic BJT current mirror.

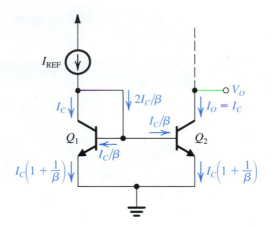

FIGURE 6.9 Analysis of the current mirror taking into account the finite β of the BJTs.

In general, the current transfer ratio is given by

$$\frac{I_O}{I_{REF}} = \frac{I_{S2}}{I_{S1}} = \frac{\text{Area of EBJ of } Q_2}{\text{Area of EBJ of } Q_1} \tag{6.20}$$

Alternatively, if the area ratio m is an integer, one can think of Q_2 as equivalent to m transistors, each matched to Q_1 and connected in parallel.

Next we consider the effect of finite transistor β on the current transfer ratio. The analysis for the case in which the current transfer ratio is nominally unity—that is, for the case in which Q_2 is matched to Q_1—is illustrated in Fig. 6.9. The key point here is that since Q_1 and Q_2 are matched and have the same V_{BE}, their collector currents will be equal. The rest of the analysis is straightforward. A node equation at the collector of Q_1 yields

$$I_{REF} = I_C + 2I_C/\beta = I_C\left(1 + \frac{2}{\beta}\right)$$

Finally, since $I_O = I_C$, the current transfer ratio can be found as

$$\frac{I_O}{I_{REF}} = \frac{I_C}{I_C\left(1 + \dfrac{2}{\beta}\right)} = \frac{1}{1 + \dfrac{2}{\beta}} \tag{6.21}$$

Note that as β approaches ∞, I_O/I_{REF} approaches the nominal value of unity. For typical values of β, however, the error in the current transfer ratio can be significant. For instance, $\beta = 100$ results in a 2% error in the current transfer ratio. Furthermore, the error due to the finite β increases as the nominal current transfer ratio is increased. The reader is encouraged to show that for a mirror with a nominal current transfer ratio m—that is, one in which $I_{S2} = mI_{S1}$—the actual current transfer ratio is given by

$$\frac{I_O}{I_{REF}} = \frac{m}{1 + \dfrac{m+1}{\beta}} \tag{6.22}$$

In common with the MOS current mirror, the BJT mirror has a finite output resistance R_o,

$$R_o \equiv \frac{\Delta V_O}{\Delta I_O} = r_{o2} = \frac{V_{A2}}{I_O} \tag{6.23}$$

where V_{A2} and r_{o2} are the Early voltage and the output resistance, respectively, of Q_2. Thus, even if we neglect the error due to finite β, the output current I_O will be at its nominal value only when Q_2 has the same V_{CE} as Q_1, namely at $V_O = V_{BE}$. As V_O is increased, I_O will correspondingly increase. Taking both the finite β and the finite R_o into account, we can express the output current of a BJT mirror with a nominal current transfer ratio m as

$$I_O = I_{REF} \left(\frac{m}{1 + \dfrac{m+1}{\beta}} \right) \left(1 + \frac{V_O - V_{BE}}{V_{A2}} \right) \tag{6.24}$$

where we note that the error term due to the Early effect is expressed so that it reduces to zero for $V_O = V_{BE}$.

A Simple Current Source   In a manner analogous to that in the MOS case, the basic BJT current mirror can be used to implement a simple current source, as shown in Fig. 6.10. Here the reference current is

$$I_{REF} = \frac{V_{CC} - V_{BE}}{R} \tag{6.25}$$

where V_{BE} is the base–emitter voltage corresponding to the desired value of output current I_O,

$$I_O = \frac{I_{REF}}{1 + (2/\beta)} \left(1 + \frac{V_O - V_{BE}}{V_A} \right) \tag{6.26}$$

The output resistance of this current source is r_o of Q_2,

$$R_o = r_{o2} \cong \frac{V_A}{I_O} \cong \frac{V_A}{I_{REF}} \tag{6.27}$$

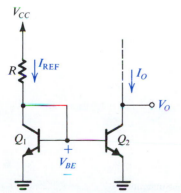

FIGURE 6.10   A simple BJT current source.

D6.7 Assuming the availability of BJTs with scale currents $I_S = 10^{-15}$ A, $\beta = 100$, and $V_A = 50$ V, design the current-source circuit of Fig. 6.10 to provide an output current $I_O = 0.5$ mA at $V_O = 2$ V. The power supply $V_{CC} = 5$ V. Give the values of I_{REF}, R, and V_{Omin}. Also, find I_O at $V_O = 5$ V.

Ans. 0.497 mA; 8.71 kΩ; 0.3 V; 0.53 mA

Current Steering   To generate bias currents for different amplifier stages in an IC, the current-steering approach described for MOS circuits can be applied in the bipolar case. As an example, consider the circuit shown in Fig. 6.11. The dc reference current I_{REF} is generated in the branch that consists of the diode-connected transistor Q_1, resistor R, and the diode-connected transistor Q_2:

$$I_{REF} = \frac{V_{CC} + V_{EE} - V_{EB1} - V_{BE2}}{R} \tag{6.28}$$

Now, for simplicity, assume that all the transistors have high β and thus that the base currents are negligibly small. We will also neglect the Early effect. The diode-connected transistor Q_1 forms a current mirror with Q_3; thus Q_3 will supply a constant current I_1 equal to I_{REF}. Transistor Q_3 can supply this current to any load as long as the voltage that develops at the collector does not exceed $(V_{CC} - 0.3$ V$)$; otherwise Q_3 would enter the saturation region.

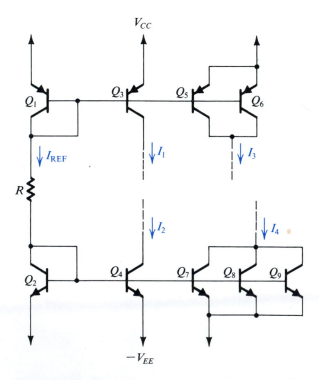

FIGURE 6.11   Generation of a number of constant currents of various magnitudes.

To generate a dc current twice the value of I_{REF}, two transistors, Q_5 and Q_6, each of which is matched to Q_1, are connected in parallel, and the combination forms a mirror with Q_1. Thus $I_3 = 2I_{REF}$. Note that the parallel combination of Q_5 and Q_6 is equivalent to a transistor with an EBJ area double that of Q_1, which is precisely what is done when this circuit is fabricated in IC form.

Transistor Q_4 forms a mirror with Q_2; thus Q_4 provides a constant current I_2 equal to I_{REF}. Note that while Q_3 *sources* its current to parts of the circuit whose voltage should not exceed ($V_{CC} - 0.3$ V), Q_4 *sinks* its current from parts of the circuit whose voltage should not decrease below $-V_{EE} + 0.3$ V. Finally, to generate a current three times I_{REF}, three transistors, Q_7, Q_8, and Q_9, each of which is matched to Q_2, are connected in parallel, and the combination is placed in a mirror configuration with Q_2. Again, in an IC implementation, Q_7, Q_8, and Q_9 would be replaced with a transistor having a junction area three times that of Q_2.

EXERCISE

6.8   Figure E6.8 shows an N-output current mirror. Assuming that all transistors are matched and have finite β and ignoring the effect of finite output resistances, show that

$$I_1 = I_2 = \cdots = I_N = \frac{1}{1 + (N+1)/\beta}$$

For $\beta = 100$, find the maximum number of outputs for an error not exceeding 10%.

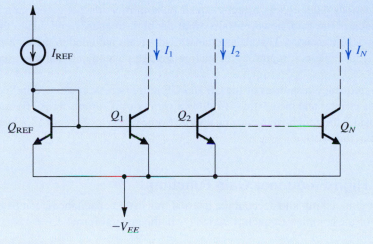

FIGURE E6.8

Ans. 9

 ## 6.4   HIGH-FREQUENCY RESPONSE—GENERAL CONSIDERATIONS

The amplifier circuits we shall study in this chapter and the next are intended for fabrication using IC technology. Therefore they do not employ bypass capacitors. Moreover, the various stages in an integrated-circuit cascade amplifier are *directly coupled*; that is, they do not utilize

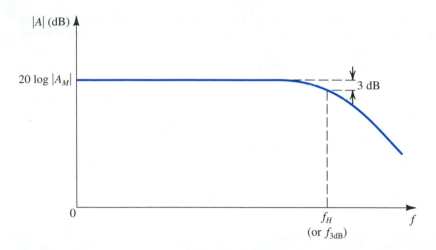

FIGURE 6.12 Frequency response of a direct-coupled (dc) amplifier. Observe that the gain does *not* fall off at low frequencies, and the midband gain A_M extends down to zero frequency.

large coupling capacitors, such as those we employed in Chapters 4 and 5.[5] The frequency response of these **direct-coupled or dc amplifiers** takes the general form shown in Fig. 6.12, from which we note that the gain remains constant at its midband value A_M down to zero frequency (dc). That is, compared to the capacitively coupled amplifiers that utilize bypass capacitors (Sections 4.9 and 5.9), direct-coupled IC amplifiers do not suffer gain reduction at low frequencies. The gain, however, falls off at the high-frequency end due to the internal capacitances of the transistor. These capacitances, which are included in the high-frequency device models in Table 6.3, represent the charge storage phenomena that take place inside the transistors.

The high-frequency responses of the CS and CE amplifiers were studied in Sections 4.9 and 5.9. In this chapter and the next, as we study a variety of IC-amplifier configurations, we shall also consider their high-frequency operation. Some of the tools needed for such a study are presented in this section.

6.4.1 The High-Frequency Gain Function

The amplifier gain, taking into account the internal transistor capacitances, can be expressed as a function of the complex-frequency variable s in the general form

$$A(s) = A_M F_H(s) \tag{6.29}$$

where A_M is the midband gain, which for the IC amplifiers we are studying here is equal to the low-frequency or dc gain. The value of A_M can be determined by analyzing the amplifier equivalent circuit while neglecting the effect of the transistor internal capacitances—that is, by assuming that they act as perfect open circuits. By taking these capacitances into account,

---

[5] In some cases there might be one or two off-chip coupling capacitors to connect the entire IC amplifier to a signal source and/or a load.

the gain acquires the factor $F_H(s)$, which can be expressed in terms of its poles and zeros,[6] which are usually real, as follows:

$$F_H(s) = \frac{(1+s/\omega_{Z1})(1+s/\omega_{Z2})\ldots(1+s/\omega_{Zn})}{(1+s/\omega_{P1})(1+s/\omega_{P2})\ldots(1+s/\omega_{Pn})} \tag{6.30}$$

where $\omega_{P1}, \omega_{P2}, \ldots, \omega_{Pn}$ are positive numbers representing the frequencies of the n real poles and $\omega_{Z1}, \omega_{Z2}, \ldots, \omega_{Zn}$ are positive, negative, or infinite numbers representing the frequencies of the n real transmission zeros. Note from Eq. (6.30) that, as should be expected, as s approaches 0, $F_H(s)$ approaches unity and the gain approaches A_M.

6.4.2 Determining the 3-dB Frequency f_H

The amplifier designer usually is particularly interested in the part of the high-frequency band that is close to the midband. This is because the designer needs to estimate—and if need be modify—the value of the upper 3-dB frequency f_H (or ω_H; $f_H = \omega_H/2\pi$). Toward that end it should be mentioned that in many cases the zeros are either at infinity or such high frequencies as to be of little significance to the determination of ω_H. If in addition one of the poles, say ω_{P1}, is of much lower frequency than any of the other poles, then this pole will have the greatest effect on the value of the amplifier ω_H. In other words, this pole will *dominate* the high-frequency response of the amplifier, and the amplifier is said to have a **dominant-pole response.** In such cases the function $F_H(s)$ can be approximated by

$$F_H(s) \cong \frac{1}{1+s/\omega_{P1}} \tag{6.31}$$

which is the transfer function of a first-order (or STC) low-pass network (Appendix D). It follows that if a dominant pole exists, then the determination of ω_H is greatly simplified;

$$\omega_H \cong \omega_{P1} \tag{6.32}$$

This is the situation we encountered in the case of the common-source amplifier analyzed in Section 4.9 and the common-emitter amplifier analyzed in Section 5.9. As a rule of thumb, *a dominant pole exists if the lowest-frequency pole is at least two octaves (a factor of 4) away from the nearest pole or zero.*

If a dominant pole does not exist, the 3-dB frequency ω_H can be determined from a plot of $|F_H(j\omega)|$. Alternatively, an approximate formula for ω_H can be derived as follows: Consider, for simplicity, the case of a circuit having two poles and two zeros in the high-frequency band; that is,

$$F_H(s) = \frac{(1+s/\omega_{Z1})(1+s/\omega_{Z2})}{(1+s/\omega_{P1})(1+s/\omega_{P2})} \tag{6.33}$$

Substituting $s = j\omega$ and taking the squared magnitude gives

$$|F_H(j\omega)|^2 = \frac{(1+\omega^2/\omega_{Z1}^2)(1+\omega^2/\omega_{Z2}^2)}{(1+\omega^2/\omega_{P1}^2)(1+\omega^2/\omega_{P2}^2)}$$

---

[6] At this point we assume that the reader is familiar with the subject of s-plane analysis and the notions of transfer-function poles and zeros as well as Bode plots. A brief review of this material is presented in Appendix E.

By definition, at $\omega = \omega_H$, $|F_H|^2 = \frac{1}{2}$; thus,

$$\frac{1}{2} = \frac{(1 + \omega_H^2/\omega_{Z1}^2)(1 + \omega_H^2/\omega_{Z2}^2)}{(1 + \omega_H^2/\omega_{P1}^2)(1 + \omega_H^2/\omega_{P2}^2)}$$

$$= \frac{1 + \omega_H^2\left(\dfrac{1}{\omega_{Z1}^2} + \dfrac{1}{\omega_{Z2}^2}\right) + \omega_H^4/\omega_{Z1}^2\omega_{Z2}^2}{1 + \omega_H^2\left(\dfrac{1}{\omega_{P1}^2} + \dfrac{1}{\omega_{P2}^2}\right) + \omega_H^4/\omega_{P1}^2\omega_{P2}^2} \tag{6.34}$$

Since ω_H is usually smaller than the frequencies of all the poles and zeros, we may neglect the terms containing ω_H^4 and solve for ω_H to obtain

$$\omega_H \cong 1\bigg/\sqrt{\frac{1}{\omega_{P1}^2} + \frac{1}{\omega_{P2}^2} - \frac{2}{\omega_{Z1}^2} - \frac{2}{\omega_{Z2}^2}} \tag{6.35}$$

This relationship can be extended to any number of poles and zeros as

$$\omega_H \cong 1\bigg/\sqrt{\left(\frac{1}{\omega_{P1}^2} + \frac{1}{\omega_{P2}^2} + \cdots\right) - 2\left(\frac{1}{\omega_{Z1}^2} + \frac{1}{\omega_{Z2}^2} + \cdots\right)} \tag{6.36}$$

Note that if one of the poles, say P_1, is dominant, then $\omega_{P1} \ll \omega_{P2}, \omega_{P3}, \ldots, \omega_{Z1}, \omega_{Z2}, \ldots$, and Eq. (6.36) reduces to Eq. (6.32).

EXAMPLE 6.5

The high-frequency response of an amplifier is characterized by the transfer function

$$F_H(s) = \frac{1 - s/10^5}{(1 + s/10^4)(1 + s/4 \times 10^4)}$$

Determine the 3-dB frequency approximately and exactly.

Solution

Noting that the lowest-frequency pole at 10^4 rad/s is two octaves lower than the second pole and a decade lower than the zero, we find that a dominant-pole situation almost exists and $\omega_H \simeq 10^4$ rad/s. A better estimate of ω_H can be obtained using Eq. (6.35), as follows:

$$\omega_H = 1\bigg/\sqrt{\frac{1}{10^8} + \frac{1}{16 \times 10^8} - \frac{2}{10^{10}}}$$

$$= 9800 \text{ rad/s}$$

The exact value of ω_H can be determined from the given transfer function as 9537 rad/s. Finally, we show in Fig. 6.13 a Bode plot and an exact plot for the given transfer function. Note that this is a plot of the high-frequency response of the amplifier normalized relative to its midband gain. That is, if the midband gain is, say, 100 dB, then the entire plot should be shifted upward by 100 dB.

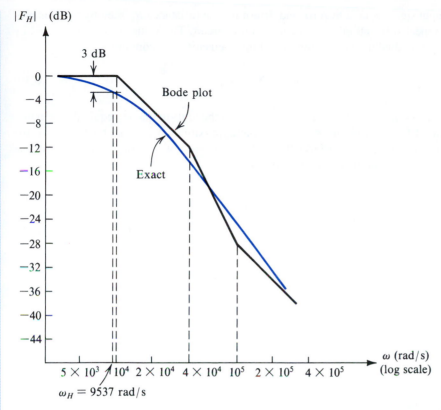

FIGURE 6.13 Normalized high-frequency response of the amplifier in Example 6.5.

6.4.3 Using Open-Circuit Time Constants for the Approximate Determination of f_H

If the poles and zeros of the amplifier transfer function can be determined easily, then we can determine f_H using the techniques above. In many cases, however, it is not a simple matter to determine the poles and zeros by quick hand analysis. In such cases an approximate value for f_H can be obtained using the following method.

Consider the function $F_H(s)$ (Eq. 6.30), which determines the high-frequency response of the amplifier. The numerator and denominator factors can be multiplied out and $F_H(s)$ expressed in the alternative form

$$F_H(s) = \frac{1 + a_1 s + a_2 s^2 + \cdots + a_n s^n}{1 + b_1 s + b_2 s^2 + \cdots + b_n s^n} \tag{6.37}$$

where the coefficients a and b are related to the frequencies of the zeros and poles, respectively. Specifically, the coefficient b_1 is given by

$$b_1 = \frac{1}{\omega_{P1}} + \frac{1}{\omega_{P2}} + \cdots + \frac{1}{\omega_{Pn}} \tag{6.38}$$

It can be shown [see Gray and Searle (1969)] that the value of b_1 can be obtained by considering the various capacitances in the high-frequency equivalent circuit one at a time while reducing all other capacitors to zero (or, equivalently, replacing them with open circuits). That is, to obtain the contribution of capacitance C_i we reduce all other capacitances to zero, reduce the input signal source to zero, and determine the resistance R_{io} seen by C_i. This process is then repeated for all other capacitors in the circuit. The value of b_1 is computed by summing the individual time constants, called **open-circuit time constants,**

$$b_1 = \sum_{i=1}^{n} C_i R_{io} \tag{6.39}$$

where we have assumed that there are n capacitors in the high-frequency equivalent circuit.

This method for determining b_1 is *exact;* the approximation comes about in using the value of b_1 to determine ω_H. Specifically, if the zeros are not dominant and if one of the poles, say P_1, is dominant, then from Eq. (6.38),

$$b_1 \simeq \frac{1}{\omega_{P1}} \tag{6.40}$$

But, also, the upper 3-dB frequency will be approximately equal to ω_{P1}, leading to the approximation

$$\omega_H \simeq \frac{1}{b_1} = \frac{1}{\left[\sum_i C_i R_{io}\right]} \tag{6.41}$$

Here it should be pointed out that in complex circuits we usually do not know whether or not a dominant pole exists. Nevertheless, using Eq. (6.41) to determine ω_H normally yields remarkably good results[7] even if a dominant pole does not exist. The method will be illustrated by an example.

EXAMPLE 6.6

Figure 6.14(a) shows the high-frequency equivalent circuit of a common-source MOSFET amplifier. The amplifier is fed with a signal generator V_{sig} having a resistance R_{sig}. Resistance R_{in} is due to the biasing network. Resistance R_L' is the parallel equivalent of the load resistance R_L, the drain bias resistance R_D, and the FET output resistance r_o. Capacitors C_{gs} and C_{gd} are the MOSFET internal capacitances. For $R_{sig} = 100$ kΩ, $R_{in} = 420$ kΩ, $C_{gs} = C_{gd} = 1$ pF, $g_m = 4$ mA/V, and $R_L' = 3.33$ kΩ, find the midband voltage gain, $A_M = V_o/V_{sig}$ and the upper 3-dB frequency, f_H.

Solution

The midband voltage gain is determined by assuming that the capacitors in the MOSFET model are perfect open circuits. This results in the midband equivalent circuit shown in Fig. 6.14(b), from which we find

$$A_M \equiv \frac{V_o}{V_{sig}} = -\frac{R_{in}}{R_{in} + R_{sig}}(g_m R_L')$$

$$= -\frac{420}{420 + 100} \times 4 \times 3.33 = -10.8 \text{ V/V}$$

We shall determine ω_H using the method of open-circuit time constants. The resistance R_{gs} seen by C_{gs} is found by setting $C_{gd} = 0$ and short-circuiting the signal generator V_{sig}. This results in the circuit

---

[7] The method of open-circuit time constants yields good results only when all the poles are real, as is the case in this chapter.

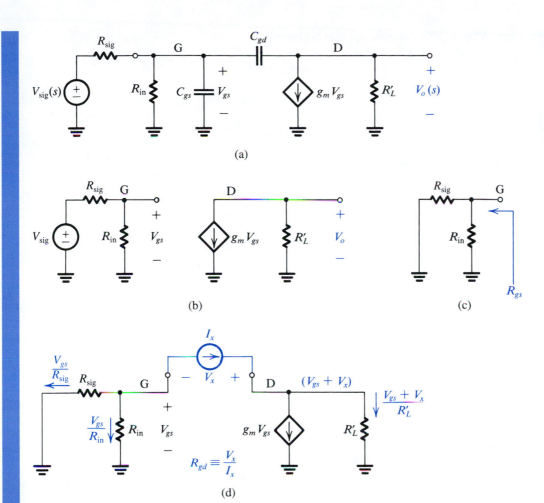

FIGURE 6.14 Circuits for Example 6.6: (**a**) high-frequency equivalent circuit of a MOSFET amplifier; (**b**) the equivalent circuit at midband frequencies; (**c**) circuit for determining the resistance seen by C_{gs}; and (**d**) circuit for determining the resistance seen by C_{gd}.

of Fig. 6.14(c), from which we find that

$$R_{gs} = R_{in}\|R_{sig} = 420 \text{ k}\Omega\|100 \text{ k}\Omega = 80.8 \text{ k}\Omega$$

Thus the open-circuit time constant of C_{gs} is

$$\tau_{gs} \equiv C_{gs}R_{gs} = 1 \times 10^{-12} \times 80.8 \times 10^3 = 80.8 \text{ ns}$$

The resistance R_{gd} seen by C_{gd} is found by setting $C_{gs} = 0$ and short-circuiting V_{sig}. The result is the circuit in Fig. 6.14(d), to which we apply a test current I_x. Writing a node equation at G gives

$$I_x = -\frac{V_{gs}}{R_{in}} - \frac{V_{gs}}{R_{sig}}$$

Thus,

$$V_{gs} = -I_x R' \tag{6.42}$$

where $R' = R_{in} \| R_{sig}$. A node equation at D provides

$$I_x = g_m V_{gs} + \frac{V_{gs} + V_x}{R_L'}$$

Substituting for V_{gs} from Eq. (6.42) and rearranging terms yields

$$R_{gd} \equiv \frac{V_x}{I_x} = R' + R_L' + g_m R_L' R' = 1.16 \text{ M}\Omega$$

Thus the open-circuit time constant of C_{gd} is

$$\tau_{gd} \equiv C_{gd} R_{gd}$$
$$= 1 \times 10^{-12} \times 1.16 \times 10^{6} = 1160 \text{ ns}$$

The upper 3-dB frequency ω_H can now be determined from

$$\omega_H \simeq \frac{1}{\tau_{gs} + \tau_{gd}}$$
$$= \frac{1}{(80.8 + 1160) \times 10^{-9}} = 806 \text{ krad/s}$$

Thus,

$$f_H = \frac{\omega_H}{2\pi} = 128.3 \text{ kHz}$$

The method of open-circuit time constants has an important advantage in that it tells the circuit designer which of the various capacitances is significant in determining the amplifier frequency response. Specifically, the relative contribution of the various capacitances to the effective time constant b_1 is immediately obvious. For instance, in the above example we see that C_{gd} is the dominant capacitance in determining f_H. We also note that, in effect to increase f_H either we use a MOSFET with smaller C_{gd} or, for a given MOSFET, we reduce R_{gd} by using a smaller R' or R_L'. If R' is fixed, then for a given MOSFET the only way to increase bandwidth is by reducing the load resistance. Unfortunately, this also decreases the midband gain. This is an example of the usual trade-off between gain and bandwidth, a common circumstance which was mentioned earlier.

6.4.4 Miller's Theorem

In our analysis of the high-frequency response of the common-source amplifier (Section 4.9), and of the common-emitter amplifier (Section 5.9), we employed a technique for replacing the bridging capacitance (C_{gs} or C_μ) by an equivalent input capacitance. This very useful and effective technique is based on a general theorem known as **Miller's theorem,** which we now present.

Consider the situation in Fig. 6.15(a). As part of a larger circuit that is not shown, we have isolated two circuit nodes, labeled 1 and 2, between which an impedance Z is connected. Nodes 1 and 2 are also connected to other parts of the circuit, as signified by the broken lines emanating from the two nodes. Furthermore, it is assumed that somehow it has been determined that the voltage at node 2 is related to that at node 1 by

$$V_2 = KV_1 \tag{6.43}$$

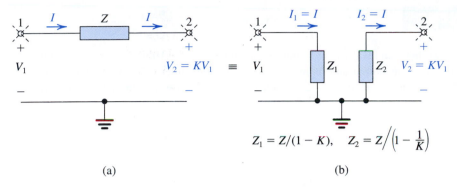

FIGURE 6.15 The Miller equivalent circuit.

In typical situations K is a gain factor that can be positive or negative and that has a magnitude usually larger than unity. This, however, is not an assumption for Miller's theorem.

Miller's theorem states that impedance Z can be replaced by two impedances: Z_1 connected between node 1 and ground and Z_2 connected between node 2 and ground, where

$$Z_1 = Z/(1 - K)$$ (6.44a)

and

$$Z_2 = Z \Big/ \left(1 - \frac{1}{K}\right)$$ (6.44b)

to obtain the equivalent circuit shown in Fig. 6.15(b).

The proof of Miller's theorem is achieved by deriving Eq. (6.44) as follows: In the original circuit of Fig. 6.15(a), the only way that node 1 "feels the existence" of impedance Z is through the current I that Z draws away from node 1. Therefore, to keep this current unchanged in the equivalent circuit, we must choose the value of Z_1 so that it draws an equal current,

$$I_1 = \frac{V_1}{Z_1} = I = \left(\frac{V_1 - KV_1}{Z}\right)$$

which yields the value of Z_1 in Eq. (6.44a). Similarly, to keep the current into node 2 unchanged, we must choose the value of Z_2 so that

$$I_2 = \frac{0 - V_2}{Z_2} = \frac{0 - KV_1}{Z_2} = I = \frac{V_1 - KV_1}{Z}$$

which yields the expression for Z_2 in Eq. (6.44b).

Although not highlighted, the Miller equivalent circuit derived above is valid only as long as the rest of the circuit remains unchanged; otherwise the ratio of V_2 to V_1 might change. It follows that the Miller equivalent circuit *cannot* be used directly to determine the output resistance of an amplifier. This is because in determining output resistances it is implicitly assumed that the source signal is reduced to zero and that a test-signal source (voltage or current) is applied to the output terminals—obviously a major change in the circuit, rendering the Miller equivalent circuit no longer valid.

EXAMPLE 6.7

Figure 6.16(a) shows an ideal voltage amplifier having a gain of −100 V/V with an impedance Z connected between its output and input terminals. Find the Miller equivalent circuit when Z is (a) a 1-MΩ resistance, and (b) a 1-pF capacitance. In each case, use the equivalent circuit to determine V_o/V_{sig}.

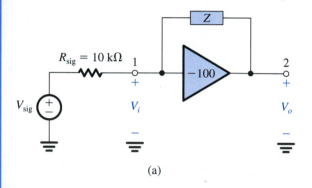

(a)

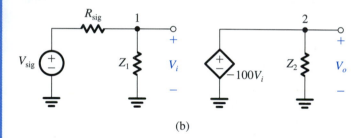

(b)

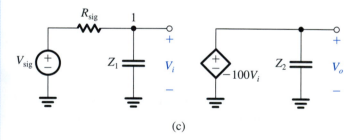

(c)

FIGURE 6.16 Circuit for Example 6.7.

Solution

(a) For $Z = 1$ MΩ, employing Miller's theorem results in the equivalent circuit in Fig. 6.16(b), where

$$Z_1 = \frac{Z}{1-K} = \frac{1000 \text{ k}\Omega}{1 + 100} = 9.9 \text{ k}\Omega$$

$$Z_2 = \frac{Z}{1 - \frac{1}{K}} = \frac{1 \text{ M}\Omega}{1 + \frac{1}{100}} = 0.99 \text{ M}\Omega$$

The voltage gain can be found as follows:

$$\frac{V_o}{V_{sig}} = \frac{V_o}{V_i} \frac{V_i}{V_{sig}} = -100 \times \frac{Z_1}{Z_1 + R_{sig}}$$

$$= -100 \times \frac{9.9}{9.9 + 10} = -49.7 \text{ V/V}$$

(b) For Z as a 1-pF capacitance—that is, $Z = 1/sC = 1/s \times 1 \times 10^{-12}$—applying Miller's theorem allows us to replace Z by Z_1 and Z_2, where

$$Z_1 = \frac{Z}{1-K} = \frac{1/sC}{1+100} = 1/s(101\,C)$$

$$Z_2 = \frac{Z}{1 - \frac{1}{K}} = \frac{1}{1.01} \frac{1}{sC} = \frac{1}{s(1.01\,C)}$$

It follows that Z_1 is a capacitance $101C = 101$ pF and that Z_2 is a capacitance $1.01C = 1.01$ pF. The resulting equivalent circuit is shown in Fig. 6.16(c), from which the voltage gain can be found as follows:

$$\frac{V_o}{V_{sig}} = \frac{V_o}{V_i} \frac{V_i}{V_{sig}} = -100 \frac{1/sC_1}{1/(sC_1) + R_{sig}}$$

$$= \frac{-100}{1 + sC_1 R_{sig}}$$

$$= \frac{-100}{1 + s \times 101 \times 1 \times 10^{-12} \times 10 \times 10^3}$$

$$= \frac{-100}{1 + s \times 1.01 \times 10^{-6}}$$

This is the transfer function of a first-order low-pass network with a dc gain of -100 and a 3-dB frequency f_{3dB} of

$$f_{3dB} = \frac{1}{2\pi \times 1.01 \times 10^{-6}} = 157.6 \text{ kHz}$$

From Example 6.7, we observe that the Miller replacement of a feedback or bridging resistance results, for a negative K, in a smaller resistance [by a factor $(1-K)$] at the input. If the feedback element is a capacitance, its value is multiplied by $(1-K)$ to obtain the equivalent capacitance at the input side. The multiplication of a feedback capacitance by $(1-K)$ is referred to as **Miller multiplication** or **Miller effect**. We have encountered the Miller effect in the analysis of the CS and CE amplifiers in Sections 4.9 and 5.9, respectively.

EXERCISES

6.9 A direct-coupled amplifier has a dc gain of 1000 V/V and an upper 3-dB frequency of 100 kHz. Find the transfer function and the gain–bandwidth product in hertz.

Ans. $\dfrac{1000}{1 + \dfrac{s}{2\pi \times 10^5}}$; 10^8 Hz

6.10 The high-frequency response of an amplifier is characterized by two zeros at $s = \infty$ and two poles at ω_{P1} and ω_{P2}. For $\omega_{P2} = k\omega_{P1}$, find the value of k that results in the exact value of ω_H being $0.9\omega_{P1}$. Repeat for $\omega_H = 0.99\omega_{P1}$.

Ans. 2.78; 9.88

6.11 For the amplifier described in Exercise 6.10, find the exact and approximate values (using Eq. 6.36) of ω_H (as a function of ω_{P1}) for the cases $k = 1, 2,$ and 4.

Ans. 0.64, 0.71; 0.84, 0.89; 0.95, 0.97

6.12 For the amplifier in Example 6.6, find the gain–bandwidth product in megahertz. Find the value of R'_L that will result in $f_H = 180$ kHz. Find the new values of the midband gain and of the gain–bandwidth product.

Ans. 1.39 MHz; 2.23 kΩ; −7.2 V/V; 1.30 MHz

6.13 Use Miller's theorem to investigate the performance of the inverting op-amp circuit shown in Fig. E6.13. Assume the op amp to be ideal except for having a finite differential gain, A. Without using any knowledge of op-amp circuit analysis, find R_{in}, V_i, V_o, and V_o/V_{sig}, for each of the following values of A: 10 V/V, 100 V/V, 1000 V/V, and 10,000 V/V. Assume $V_{sig} = 1$ V.

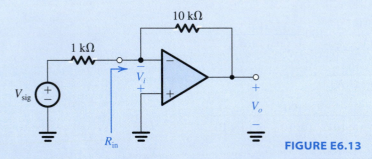

FIGURE E6.13

Ans.

A	R_{in}	V_i	V_o	V_o/V_{sig}
10 V/V	909 Ω	476 mV	−4.76 V	−4.76 V/V
100 V/V	99 Ω	90 mV	−9 V	−9 V/V
1000 V/V	9.99 Ω	9.9 mV	−9.9 V	−9.9 V/V
10,000 V/V	1 Ω	0.999 mV	−9.99 V	−9.99 V/V

6.5 THE COMMON-SOURCE AND COMMON-EMITTER AMPLIFIERS WITH ACTIVE LOADS

6.5.1 The Common-Source Circuit

Figure 6.17(a) shows the most basic IC MOS amplifier. It consists of a grounded-source MOS transistor with the drain resistor R_D replaced by a constant-current source I. As we shall see shortly, the current-source load can be implemented using a PMOS transistor and is therefore called an **active load,** and the CS amplifier of Fig. 6.17(a) is said to be **active-loaded.**

Before considering the small-signal operation of the active-loaded CS amplifier, a word on its dc bias design is in order. Obviously, Q_1 is biased at $I_D = I$, but what determines the dc voltages at the drain and at the gate? Usually, this circuit will be part of a larger circuit in which negative feedback is utilized to fix the values of V_{DS} and V_{GS}. We shall encounter

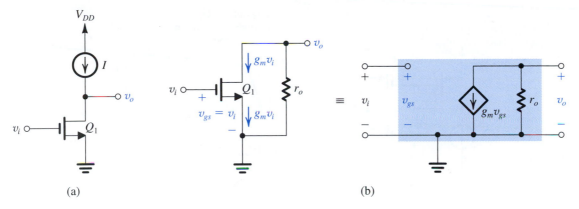

FIGURE 6.17 (a) Active-loaded common-source amplifier. (b) Small-signal analysis of the amplifier in (a), performed both directly on the circuit diagram and using the small-signal model explicitly.

examples of such circuits in later chapters. *For the time-being, however, we shall assume that the MOSFET is biased to operate in the saturation region.*

Small-signal analysis of the current-source-loaded CS amplifier is straightforward and is illustrated in Fig. 6.17(b). Here, along with the equivalent circuit model, we show the transistor with its r_o extracted and displayed separately and with the analysis performed directly on the circuit. From Fig. 6.17(b) we see that for this CS amplifier,[8]

$$R_i = \infty \tag{6.45a}$$

$$A_{vo} = -g_m r_o \tag{6.45b}$$

$$R_o = r_o \tag{6.45c}$$

We note that $|A_{vo}|$ in Eq. (6.45b) is the maximum voltage gain available from a common-source amplifier, namely the **intrinsic gain** of the MOSFET,

$$A_0 = g_m r_o \tag{6.46}$$

Recall that in Section 6.2 we discussed in some detail the intrinsic gain A_0 and presented in Table 6.3 formulas for its determination.

EXERCISE

6.14 Find A_0 for an NMOS transistor fabricated in a 0.4-μm CMOS process for which $k'_n = 200\ \mu\text{A/V}^2$ and $V'_A = 20\ \text{V}/\mu\text{m}$. The transistor has a 0.4-μm channel length and is operated with an overdrive voltage of 0.25 V. What must W be for the NMOS transistor to operate at $I_D = 100\ \mu$A? Also, find the values of g_m and r_o. Repeat for $L = 0.8\ \mu$m.

Ans. 64 V/V; 6.4 μm; 0.8 mA/V; 80 kΩ; 128 V/V; 12.8 μm; 0.8 mA/V; 160 kΩ

6.5.2 CMOS Implementation of the Common-Source Amplifier

A CMOS circuit implementation of the common-source amplifier is shown in Fig. 6.18(a). This circuit is based on that shown in Fig. 6.17(a) with the load current-source I implemented using transistor Q_2. The latter is the output transistor of the current mirror formed by Q_2 and Q_3 and fed with the bias current I_{REF}. We shall assume that Q_2 and Q_3 are matched; thus the

---

[8] For the definition of the parameters used to characterize amplifiers, the reader should consult Table 4.3.

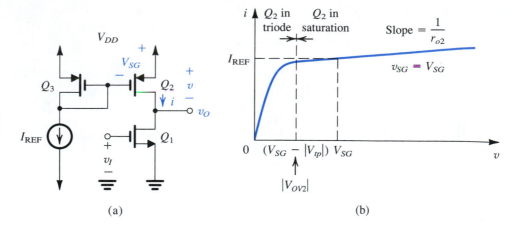

(a)

(b)

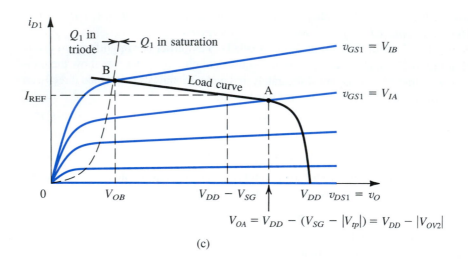

(c)

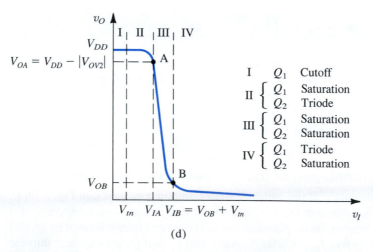

(d)

FIGURE 6.18 The CMOS common-source amplifier: **(a)** circuit; **(b)** i–v characteristic of the active-load Q_2; **(c)** graphical construction to determine the transfer characteristic; and **(d)** transfer characteristic.

i–v characteristic of the load device will be that shown in Fig. 6.18(b). This is simply the i_D–v_{SD} characteristic curve of the p-channel transistor Q_2 for a constant source-gate voltage V_{SG}. The value of V_{SG} is set by passing the reference bias current I_{REF} through Q_3. Observe that, as expected, Q_2 behaves as a current source when it operates in saturation, which in turn is obtained when $v = v_{SD}$ exceeds $(V_{SG} - |V_{tp}|)$, which is the magnitude of the overdrive voltage at which Q_2 and Q_3 are operating. When Q_2 is in saturation, it exhibits a finite incremental resistance r_{o2},

$$r_{o2} = \frac{|V_{A2}|}{I_{REF}} \tag{6.47}$$

where V_{A2} is the Early voltage of Q_2. In other words, the current-source load is not ideal but has a finite output resistance equal to the transistor r_o.

Before proceeding to determine the small-signal voltage gain of the amplifier, it is instructive to examine its transfer characteristic, v_O versus v_I. This can be determined using the graphical construction shown in Fig. 6.18(c). Here we have sketched the i_D–v_{DS} characteristics of the amplifying transistor Q_1 and superimposed the load curve on them. The latter is simply the i–v curve in Fig. 6.18(b) "flipped around" and shifted V_{DD} volts along the horizontal axis. Now, since $v_{GS1} = v_I$, each of the i_D–v_{DS} curves corresponds to a particular value of v_I. The intersection of each particular curve with the load curve gives the corresponding value of v_{DS1}, which is equal to v_O. Thus, in this way, we can obtain the v_O–v_I characteristic, point by point. The resulting transfer characteristic is sketched in Fig. 6.18(d). As indicated, it has four distinct segments, labeled I, II, III, and IV, each of which is obtained for one of the four combinations of the modes of operation of Q_1 and Q_2, which are also indicated in the diagram. Note also that we have labeled two important breakpoints on the transfer characteristic (A and B) in correspondence with the intersection points (A and B) in Fig. 6.18(c). We urge the reader to carefully study the transfer characteristic and its various details.

Not surprisingly, for amplifier operation segment III is the one of interest. Observe that in region III the transfer curve is almost linear and is very steep, indicating large voltage gain. In region III both the amplifying transistor Q_1 and the load transistor Q_2 are operating in saturation. The end points of region III are A and B: At A, defined by $v_O = V_{DD} - V_{OV2}$, Q_2 enters the triode region, and at B, defined by $v_O = v_I - V_{tn}$, Q_1 enters the triode region. When the amplifier is biased at a point in region III, the small-signal voltage gain can be determined by replacing Q_1 with its small-signal model and Q_2 with its output resistance, r_{o2}. The output resistance of Q_2 constitutes the load resistance of Q_1. The voltage gain A_v can be found by substituting the results from Eqs. (6.45) into

$$A_v \equiv \frac{v_o}{v_i} = A_{vo} \frac{R_L}{R_L + R_o} \tag{6.48}$$

to obtain

$$A_v = -(g_{m1} r_{o1}) \frac{r_{o2}}{r_{o2} + r_{o1}} = -g_{m1}(r_{o1} \parallel r_{o2}) \tag{6.49}$$

indicating that, as expected, A_v will be lower in magnitude than the intrinsic gain of Q_1, $g_{m1} r_{o1}$. For the case $r_{o2} = r_{o1}$, A_v will be $g_{m1} r_{o1}/2$. The result in Eq. (6.49) could, of course, have been obtained directly by multiplying $g_{m1} v_i$ by the total resistance between the output node and ground, $r_{o1} \parallel r_{o2}$.

The CMOS common-source amplifier can be designed to provide voltage gains of 15 to 100. It exhibits a very high input resistance; however, its output resistance is also high.

Two final comments need to be made before leaving the common-source amplifier:

1. The circuit is not affected by the body effect since the source terminals of both Q_1 and Q_2 are at signal ground.

2. The circuit is usually part of a larger amplifier circuit (as will be shown in Chapters 7 and 9), and negative feedback is utilized to ensure that the circuit in fact operates in region III of the amplifier transfer characteristic.

EXAMPLE 6.8

Consider the CMOS common-source amplifier in Fig. 6.18(a) for the case $V_{DD} = 3$ V, $V_{tn} = |V_{tp}| = 0.6$ V, $\mu_n C_{ox} = 200\ \mu\text{A}/\text{V}^2$, and $\mu_p C_{ox} = 65\ \mu\text{A}/\text{V}^2$. For all transistors, $L = 0.4\ \mu$m and $W = 4\ \mu$m. Also, $V_{An} = 20$ V, $|V_{Ap}| = 10$ V, and $I_{\text{REF}} = 100\ \mu$A. Find the small-signal voltage gain. Also, find the coordinates of the extremities of the amplifier region of the transfer characteristic—that is, points A and B.

Solution

$$g_{m1} = \sqrt{2k_n'\left(\frac{W}{L}\right)_1 I_{\text{REF}}}$$

$$= \sqrt{2 \times 200 \times \frac{4}{0.4} \times 100} = 0.63\ \text{mA/V}$$

$$r_{o1} = \frac{V_{An}}{I_{D1}} = \frac{20\ \text{V}}{0.1\ \text{mA}} = 200\ \text{k}\Omega$$

$$r_{o2} = \frac{V_{Ap}}{I_{D2}} = \frac{10\ \text{V}}{0.1\ \text{mA}} = 100\ \text{k}\Omega$$

Thus,

$$A_v = -g_{m1}(r_{o1} \parallel r_{o2})$$

$$= -0.63(\text{mA/V}) \times (200 \parallel 100)(\text{k}\Omega) = -42\ \text{V/V}$$

The extremities of the amplifier region of the transfer characteristic (region III) are found as follows (refer to Fig. 6.18): First, we determine V_{SG} of Q_2 and Q_3 corresponding to $I_D = I_{\text{REF}} = 100\ \mu$A using

$$I_D = \frac{1}{2} k_p'\left(\frac{W}{L}\right)_3 (V_{SG} - |V_{tp}|)^2\left(1 + \frac{V_{SD}}{|V_{Ap}|}\right)$$

Thus,

$$100 = \frac{1}{2} \times 65\left(\frac{4}{0.4}\right)|V_{OV3}|^2\left(1 + \frac{0.6 + |V_{OV3}|}{10}\right) \tag{6.50}$$

where $|V_{OV3}|$ is the magnitude of the overdrive voltage at which Q_3 and Q_2 are operating, and we have used the fact that, for Q_3, $V_{SD} = V_{SG}$. Equation (6.50) can be manipulated to the form

$$0.29 = |V_{OV3}|^2(1 + 0.09|V_{OV3}|)$$

which by trial-and-error yields

$$|V_{OV3}| = 0.53 \text{ V}$$

Thus,

$$V_{SG} = 0.6 + 0.53 = 1.13 \text{ V}$$

and

$$V_{OA} = V_{DD} - V_{OV3} = 2.47 \text{ V}$$

To find the corresponding value of v_I, V_{IA}, we derive an expression for v_O versus v_I in region III. Noting that in region III Q_1 and Q_2 are in saturation and obviously conduct equal currents, we can write

$$i_{D1} = i_{D2}$$

$$\frac{1}{2}k_n'\left(\frac{W}{L}\right)_1(v_I - V_{tn})^2\left(1 + \frac{v_O}{|V_{An}|}\right) = \frac{1}{2}k_p'\left(\frac{W}{L}\right)_2(V_{SG} - |V_{tp}|)^2\left(1 + \frac{V_{DD} - v_O}{|V_{Ap}|}\right)$$

Substituting numerical values, we obtain

$$8.55(v_I - 0.6)^2 = \frac{1 - 0.08 v_O}{1 + 0.05 v_O} \cong 1 - 0.13 v_O$$

which can be manipulated to the form

$$v_O = 7.69 - 65.77(v_I - 0.6)^2 \tag{6.51}$$

This is the equation of segment III of the transfer characteristic. Although it includes v_I^2, the reader should not be alarmed: Because region III is very narrow, v_I changes very little, and the characteristic is nearly linear. Substituting $v_O = 2.47$ V gives the corresponding value of v_I; that is, $V_{IA} = 0.88$ V. To determine the coordinates of B, we note that they are related by $V_{OB} = V_{IB} - V_{tn}$. Substituting in Eq. (6.51) and solving gives $V_{IB} = 0.93$ V and $V_{OB} = 0.33$ V. The width of the amplifier region is therefore

$$\Delta v_I = V_{IB} - V_{IA} = 0.05 \text{ V}$$

and the corresponding output range is

$$\Delta v_O = V_{OB} - V_{OA} = -2.14 \text{ V}$$

Thus the "large-signal" voltage gain is

$$\frac{\Delta v_O}{\Delta v_I} = -\frac{2.14}{0.05} = 42.8 \text{ V/V}$$

which is very close to the small-signal value of −42, indicating that segment III of the transfer characteristic is quite linear.

EXERCISE

6.15 A CMOS common-source amplifier fabricated in a 0.18-μm technology has $W/L = 7.2\ \mu m/0.36\ \mu m$ for all transistors, $k_n' = 387\ \mu A/V^2$, $k_p' = 86\ \mu A/V^2$, $I_{REF} = 100\ \mu A$, $V_{An}' = 5\ V/\mu m$, and $|V_{Ap}'| = 6\ V/\mu m$. Find g_{m1}, r_{o1}, r_{o2}, and the voltage gain.

Ans. 1.25 mA/V; 18 kΩ; 21.6 kΩ; −12.3 V/V

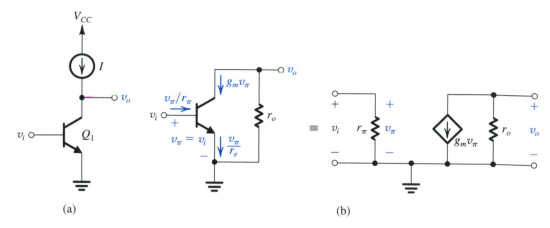

FIGURE 6.19 **(a)** Active-loaded common-emitter amplifier. **(b)** Small-signal analysis of the amplifier in (a), performed both directly on the circuit and using the hybrid-π model explicitly.

6.5.3 The Common-Emitter Circuit

The active-loaded common-emitter amplifier, shown in Fig. 6.19(a), is similar to the active-loaded common-source circuit studied above. Here also, the bias-stabilizing circuit is not shown. Small-signal analysis is similar to that for the MOS case and is illustrated in Fig. 6.19(b). The results are

$$R_i = r_\pi \tag{6.52a}$$

$$A_{vo} = -g_m r_o \tag{6.52b}$$

$$R_o = r_o \tag{6.52c}$$

which except for the rather low input resistance r_π are similar to the MOSFET case. Recall, however, from the comparison of Section 6.2 that the intrinsic gain $g_m r_o$ of the BJT is much higher than that for the MOSFET. This advantage, however, is counterbalanced by the practically infinite input resistance of the common-source amplifier. Further comparisons of the two amplifier types were presented in Section 6.2.

EXERCISE

6.16 Consider the active-loaded CE amplifier when the constant-current source I is implemented with a *pnp* transistor. Let $I = 0.1$ mA, $|V_A| = 50$ V (for both the *npn* and the *pnp* transistors), and $\beta = 100$. Find R_i, r_o (for each transistor), g_m, A_0, and the amplifier voltage gain.

Ans. 25 kΩ; 0.5 MΩ; 4 mA/V; 2000 V/V; −1000 V/V

6.6 HIGH-FREQUENCY RESPONSE OF THE CS AND CE AMPLIFIERS

We now consider the high-frequency response of the active-loaded common-source and common-emitter amplifiers. Figure 6.20 shows the high-frequency equivalent circuit of the common-source amplifier. This equivalent circuit applies equally well to the CE amplifier with a simple relabeling of components: C_{gs} would be replaced by C_π, C_{gd} by C_μ, and obviously V_{gs} by V_π.

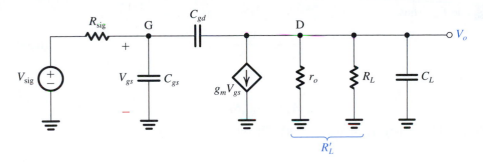

FIGURE 6.20 High-frequency equivalent-circuit model of the common-source amplifier. For the common-emitter amplifier, the values of V_{sig} and R_{sig} are modified to include the effects of r_π and r_x; C_{gs} is replaced by C_π, V_{gs} by V_π, and C_{gd} by C_μ.

The input-signal source is represented by V_{sig} and R_{sig}. In some cases, however, V_{sig} and R_{sig} would be modified values of the signal-source voltage and internal resistance, taking into account other resistive components such as a bias resistor R_G or R_B, the BJT resistances r_x and r_π, etc. We have seen examples of this kind of circuit simplification in Sections 4.9 and 5.9. The load resistance R_L represents the combination of an actual load resistance (if one is connected) and the output resistance of the current-source load. To avoid loss of gain, R_L is usually on the same order as r_o. We combine R_L with r_o and denote their parallel equivalent R'_L. The load capacitance C_L represents the total capacitance between drain (or collector) and ground; it includes the drain-to-body capacitance C_{db} (collector-to-substrate capacitance), the input capacitance of a succeeding amplifier stage, and in some cases, as we shall see in later chapters, a deliberately introduced capacitance. In IC MOS amplifiers, C_L can be relatively substantial.

6.6.1 Analysis Using Miller's Theorem

In situations when R_{sig} is relatively large and C_L is relatively small, Miller's theorem can be used to obtain a quick but approximate estimate of the 3-dB frequency f_H. We have already done this in Section 4.9 for the CS amplifier and in Section 5.9 for the CE amplifier. Therefore, here we will only state the results. Figure 6.21 shows the approximate equivalent circuit obtained for the CS case, from which we see that the amplifier has a dominant pole formed

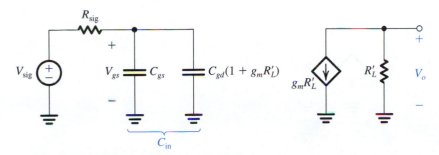

FIGURE 6.21 Approximate equivalent circuit obtained by applying Miller's theorem while neglecting C_L and the load current component supplied by C_{gd}. This model works reasonably well when R_{sig} is large and the amplifier high-frequency response is dominated by the pole formed by R_{sig} and C_{in}.

by R_{sig} and C_{in}. Thus,

$$\frac{V_o}{V_{sig}} \cong \frac{A_M}{1 + \dfrac{s}{\omega_H}} \tag{6.53}$$

where

$$A_M = -g_m R_L'$$

and the 3-dB frequency $f_H = \omega_H/2\pi$ is given by

$$f_H = \frac{1}{2\pi C_{in} R_{sig}} \tag{6.54}$$

where

$$C_{in} = C_{gs} + C_{gd}(1 + g_m R_L') \tag{6.55}$$

6.6.2 Analysis Using Open-Circuit Time Constants

The method of open-circuit time constants presented in Section 6.4.3 can be directly applied to the CS equivalent circuit of Fig. 6.20, as illustrated in Fig. 6.22, from which we see that the resistance seen by C_{gs}, $R_{gs} = R_{sig}$ and that seen by C_L is R_L'. The resistance R_{gd} seen by C_{gd} can be found by analyzing the circuit in Fig. 6.22(b) with the result that

$$R_{gd} = R_{sig}(1 + g_m R_L') + R_L' \tag{6.56}$$

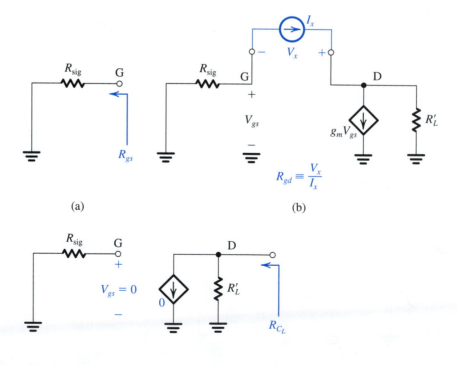

FIGURE 6.22 Application of the open-circuit time-constants method to the CS equivalent circuit of Fig. 6.20.

Thus the effective time-constant b_1 or τ_H can be found as

$$\begin{aligned}
\tau_H &= C_{gs}R_{gs} + C_{gd}R_{gd} + C_L R_{C_L} \\
&= C_{gs}R_{sig} + C_{gd}[R_{sig}(1 + g_m R'_L) + R'_L] + C_L R'_L
\end{aligned} \tag{6.57}$$

and the 3-dB frequency f_H is

$$f_H \cong \frac{1}{2\pi\tau_H} \tag{6.58}$$

For situations in which C_L is substantial, this approach yields a better estimate of f_H than that obtained using the Miller equivalence (simply because in the latter case we completely neglected C_L).

6.6.3 Exact Analysis

The approximate analysis presented above provides insight regarding the mechanism by which and the extent to which the various capacitances limit the high-frequency gain of the CS (and CE) amplifiers. Nevertheless, given that the circuit of Fig. 6.20 is relatively simple, it is instructive to also perform an exact analysis.[9] This is illustrated in Fig. 6.23. A node equation at the drain provides

$$sC_{gd}(V_{gs} - V_o) = g_m V_{gs} + \frac{V_o}{R'_L} + sC_L V_o$$

which can be manipulated to the form

$$V_{gs} = \frac{-V_o}{g_m R'_L} \frac{1 + s(C_L + C_{gd})R'_L}{1 - sC_{gd}/g_m} \tag{6.59}$$

A loop equation at the input yields

$$V_{sig} = I_i R_{sig} + V_{gs}$$

in which we can substitute for I_i from a node equation at G,

$$I_i = sC_{gs}V_{gs} + sC_{gd}(V_{gs} - V_o)$$

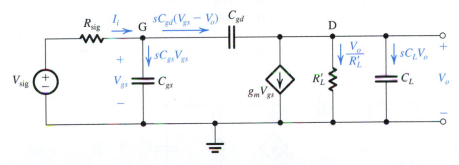

FIGURE 6.23 Analysis of the CS high-frequency equivalent circuit.

---

[9] "Exact" only in the sense that we are not making approximations in the circuit-analysis process. The reader is reminded, however, that the high-frequency model itself represents an approximation of the device performance.

to obtain

$$V_{\text{sig}} = V_{gs}[1 + s(C_{gs} + C_{gd})R_{\text{sig}}] - sC_{gd}R_{\text{sig}}V_o$$

We can now substitute in this equation for V_{gs} from Eq. (6.59) to obtain an equation in V_o and V_{sig} that can be arranged to yield the amplifier gain as

$$\frac{V_o}{V_{\text{sig}}} = \frac{-(g_m R_L')[1 - s(C_{gd}/g_m)]}{1 + s\{[C_{gs} + C_{gd}(1 + g_m R_L')]R_{\text{sig}} + (C_L + C_{gd})R_L'\} + s^2[(C_L + C_{gd})C_{gs} + C_L C_{gd}]R_{\text{sig}}R_L'}$$

$$(6.60)$$

The transfer function in Eq. (6.60) indicates that the amplifier has a second-order denominator, and hence two poles. Now, since the numerator is of the first order, it follows that one of the two transmission zeros is at infinite frequency. This is readily verifiable by noting that as s approaches ∞, (V_o/V_{sig}) approaches zero. The second zero is at

$$s = s_Z = \frac{g_m}{C_{gd}} \qquad (6.61)$$

That is, it is on the positive *real axis* of the s-plane and has a frequency ω_Z,

$$\omega_Z = g_m/C_{gd} \qquad (6.62)$$

Since g_m is usually large and C_{gd} is usually small, f_Z is normally a very high frequency and thus has negligible effect on the value of f_H.

It is useful at this point to show a simple method for finding the value of s at which $V_o = 0$—that is, s_Z. Figure 6.24 shows the circuit at $s = s_Z$. By definition, $V_o = 0$ and a node equation at D yields

$$s_Z C_{gd} V_{gs} = g_m V_{gs}$$

Now, since V_{gs} is *not* zero (why not?), we can divide both sides by V_{gs} to obtain

$$s_Z = \frac{g_m}{C_{gd}} \qquad (6.63)$$

Before considering the poles, we should note that in Eq. (6.60), as s goes toward zero, V_o/V_{sig} approaches the dc gain $(-g_m R_L')$, as should be the case. Let's now take a closer look at the denominator polynomial. First, we observe that the coefficient of the s term is equal to the effective time-constant τ_H obtained using the open-circuit time-constants method as given by Eq. (6.57). Again, this should have been expected since it is the basis for the open-circuit

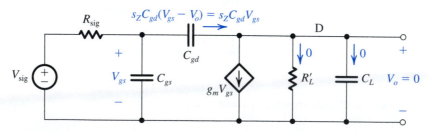

FIGURE 6.24 The CS circuit at $s = s_Z$. The output voltage $V_o = 0$, enabling us to determine s_Z from a node equation at D.

time-constants method (Section 6.4.3). Next, denoting the frequencies of the two poles ω_{P1} and ω_{P2}, we can express the denominator polynomial $D(s)$ as

$$D(s) = \left(1 + \frac{s}{\omega_{P1}}\right)\left(1 + \frac{s}{\omega_{P2}}\right)$$

$$= 1 + s\left(\frac{1}{\omega_{P1}} + \frac{1}{\omega_{P2}}\right) + \frac{s^2}{\omega_{P1}\omega_{P2}} \tag{6.64}$$

Now, if $\omega_{P2} \gg \omega_{P1}$—that is, the pole at ω_{P1} is dominant—we can approximate $D(s)$ as

$$D(s) \cong 1 + \frac{s}{\omega_{P1}} + \frac{s^2}{\omega_{P1}\omega_{P2}} \tag{6.65}$$

Equating the coefficients of the s term in denominator polynomial of Eq. (6.60) to that of the s term in Eq. (6.65) gives

$$\omega_{P1} \cong \frac{1}{[C_{gs} + C_{gd}(1 + g_m R'_L)]R_{sig} + (C_L + C_{gd})R'_L} \tag{6.66}$$

where the approximation is that involved in Eq. (6.65). Note that the expression in Eq. (6.66) is identical to the result obtained using open-circuit time constants and a little different from the result obtained using the Miller equivalence, the difference being the term $(C_L + C_{gd})R'_L$ related to the capacitance at the output, which was ignored in the original (simple) Miller derivation. Equating the coefficients of s^2 in Eqs. (6.60) and (6.65) and using Eq. (6.66) gives the frequency of the second pole:

$$\omega_{P2} = \frac{[C_{gs} + C_{gd}(1 + g_m R'_L)]R_{sig} + (C_L + C_{gd})R'_L}{[(C_L + C_{gd})C_{gs} + C_L C_{gd}]R'_L R_{sig}} \tag{6.67}$$

EXAMPLE 6.9

A CMOS common-source amplifier of the type shown in Fig. 6.18(a) has $W/L = 7.2 \ \mu m/0.36 \ \mu m$ for all transistors, $\mu_n C_{ox} = 387 \ \mu A/V^2$, $\mu_p C_{ox} = 86 \ \mu A/V^2$, $I_{REF} = 100 \ \mu A$, $V'_{An} = 5 \ V/\mu m$, and $|V'_{Ap}| = 6 \ V/\mu m$. For Q_1, $C_{gs} = 20$ fF, $C_{gd} = 5$ fF, $C_L = 25$ fF, and $R_{sig} = 10 \ k\Omega$. Assume that C_L includes all the capacitances introduced by Q_2 at the output node. Find f_H using both the Miller equivalence and the open-circuit time constants. Also, determine the exact values of f_{P1}, f_{P2}, and f_Z and hence provide another estimate for f_H.

Solution

$$I_D = I_{REF} = 100 \ \mu A = \frac{1}{2}\mu_n C_{ox}\left(\frac{W}{L}\right)V^2_{OV}$$

Thus,

$$100 = \frac{1}{2} \times 387 \times \left(\frac{7.2}{0.36}\right)V^2_{OV}$$

which results in

$$V_{OV} = 0.16 \ V$$

Thus,

$$g_m = \frac{I_D}{V_{OV}/2} = \frac{100 \ \mu A}{(0.16/2) \ V} = 1.25 \ mA/V$$

$$r_{o1} = \frac{V_{An}}{I_D} = \frac{5 \times 0.36}{0.1} = 18 \ k\Omega$$

$$r_{o2} = \frac{|V_{Ap}|}{I_D} = \frac{6 \times 0.36}{0.1} = 21.6 \ k\Omega$$

$$R'_L = r_{o1} \| r_{o2} = 18 \| 21.6 = 9.82 \ k\Omega$$

$$A_M = -g_m R'_L = -1.25 \times 9.82 = -12.3 \ V/V$$

Using the Miller equivalence:

$$C_{in} = C_{gs} + C_{gd}(1 + g_m R'_L)$$

$$= 20 + 5(1 + 12.3)$$

$$= 86.5 \ fF$$

$$f_H = \frac{1}{2\pi C_{in} R_{sig}}$$

$$= \frac{1}{2\pi \times 86.5 \times 10^{-15} \times 10 \times 10^3} = 184 \ MHz$$

Using the open-circuit time-constants method:

$$R_{gs} = R_{sig} = 10 \ k\Omega$$

$$R_{gd} = R_{sig}(1 + g_m R'_L) + R'_L$$

$$= 10(1 + 12.3) + 9.82 = 142.8 \ k\Omega$$

$$R_{C_L} = R'_L = 9.82 \ k\Omega$$

Thus,

$$\tau_{gs} = C_{gs} R_{gs} = 20 \times 10^{-15} \times 10 \times 10^3 = 200 \ ps$$

$$\tau_{gd} = C_{gd} R_{gd} = 5 \times 10^{-15} \times 142.8 \times 10^3 = 714 \ ps$$

$$\tau_{C_L} = C_L R_{C_L} = 25 \times 10^{-15} \times 9.82 \times 10^3 = 246 \ ps$$

which can be summed to obtain τ_H as

$$\tau_H = \tau_{gs} + \tau_{gd} + \tau_{C_L} = 1160 \ ps$$

from which we find the 3-dB frequency f_H,

$$f_H = \frac{1}{2\pi\tau_H} = \frac{1}{2\pi \times 1160 \times 10^{-12}} = 137 \ MHz$$

We note that this is about 25% lower than the estimate obtained using the Miller equivalence. The discrepancy is mostly a result of neglecting C_L in the Miller approach. Note that C_L here has a substantial magnitude and that its contribution to τ_H is significant (246 ps of the total 1160 ps, or 21%).

To determine the exact locations of the zero and the poles, we use the transfer function in Eq. (6.60). The frequency of the zero is given by Eq. (6.62):

$$f_Z = \frac{1}{2\pi} \frac{g_m}{C_{gd}} = \frac{1}{2\pi} \frac{1.25 \times 10^{-3}}{5 \times 10^{-15}} = 40 \text{ GHz}$$

The frequencies ω_{P1} and ω_{P2} are found as the roots of the equation obtained by equating the denominator polynomial of Eq. (6.60) to zero:

$$1 + 1.16 \times 10^{-9} s + 0.0712 \times 10^{-18} s^2 = 0$$

The result is

$$f_{P1} = 145.3 \text{ MHz}$$

and

$$f_{P2} = 2.45 \text{ GHz}$$

Since $f_Z, f_{P2} \gg f_{P1}$, a good estimate for f_H is

$$f_H \cong f_{P1} = 145.3 \text{ MHz}$$

Finally, we note that the estimate of f_{P1} obtained using Eq. (6.66) is about 5% lower than the exact value. Similarly, the estimate of f_H obtained using open-circuit time constants is 5% lower than the estimate found using the exact value of f_{P1}.

EXERCISES

6.17 For the CS amplifier in Example 6.9, using the value of f_H determined by the exact analysis, find the gain–bandwidth product. Also, convince yourself that this is the frequency at which the gain magnitude reduces to unity.

Ans. GBW = 1.79 GHz; since this is lower than f_{P2}, then f_t = 1.79 GHz

6.18 As a way to trade gain for bandwidth, the designer of the CS amplifier in Example 6.9 connects a load resistor at the output that results in halving the value of R_L'. Find the new values of $|A_M|$, f_H (using $f_H \cong f_{P1}$ of Eq. 6.66), and f_t.

Ans. 6.15 V/V; 226 MHz; 1.39 GHz

6.19 As another way to trade dc gain for bandwidth, the designer of the CS amplifier in Example 6.9 decides to operate the amplifying transistor at double the value of V_{OV} by increasing the bias current fourfold (i.e., to 400 μA). Find the new values of g_m, R_L', $|A_M|$, f_{P1}, f_H, and f_t. Use the approximate formula for f_{P1} given in Eq. (6.66).

Ans. 2.5 mA/V; 2.46 kΩ; 6.15 V/V; 252 MHz; 252 MHz; 1.55 GHz

6.6.4 Adapting the Formulas for the Case of the CE Amplifier

Adapting the formulas presented above to the case of the CE amplifier is straightforward. First, note from Fig. 6.25 how V_{sig} and R_{sig} are modified to take into account the effect of r_x and r_π,

$$V'_{sig} = V_{sig} \frac{r_\pi}{R_{sig} + r_x + r_\pi} \qquad (6.68)$$

$$R'_{sig} = r_\pi \| (R_{sig} + r_x) \qquad (6.69)$$

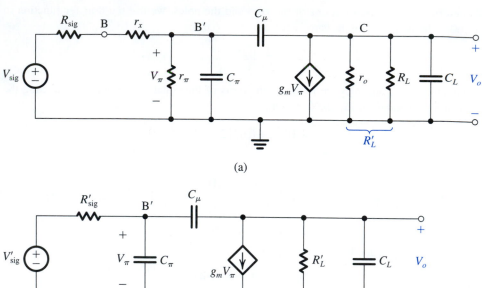

FIGURE 6.25 (a) High-frequency equivalent circuit of the common-emitter amplifier. (b) Equivalent circuit obtained after the Thévenin theorem is employed to simplify the resistive circuit at the input.

Thus the dc gain is now given by

$$A_M = -\frac{r_\pi}{R_{sig} + r_x + r_\pi}(g_m R_L')$$ (6.70)

Using Miller's theorem we obtain

$$C_{in} = C_\pi + C_\mu(1 + g_m R_L')$$ (6.71)

Correspondingly, the 3-dB frequency f_H can be estimated from

$$f_H \cong \frac{1}{2\pi C_{in} R_{sig}'}$$ (6.72)

Alternatively, using the method of open-circuit time constants yields

$$\tau_H = C_\pi R_\pi + C_\mu R_\mu + C_L C_{C_L}$$
$$= C_\pi R_{sig}' + C_\mu[(1 + g_m R_L')R_{sig}' + R_L'] + C_L R_L'$$ (6.73)

from which f_H can be estimated as

$$f_H \cong \frac{1}{2\pi \tau_H}$$ (6.74)

The exact analysis yields the following zero frequency:

$$f_Z = \frac{1}{2\pi}\frac{g_m}{C_\mu}$$ (6.75)

and, assuming that a dominant pole exists,

$$f_{P1} \cong \frac{1}{2\pi} \frac{1}{[C_\pi + C_\mu(1 + g_m R'_L)]R'_{sig} + (C_L + C_\mu)R'_L} \qquad (6.76)$$

$$f_{P2} \cong \frac{1}{2\pi} \frac{[C_\pi + C_\mu(1 + g_m R'_L)]R'_{sig} + (C_L + C_\mu)R'_L}{[C_\pi(C_L + C_\mu) + C_L C_\mu]R'_{sig}R'_L} \qquad (6.77)$$

For $f_Z, f_{P2} \gg f_{P1}$,

$$f_H \cong f_{P1}$$

EXERCISE

6.20 Consider a bipolar active-loaded CE amplifier having the load current-source implemented with a *pnp* transistor. Let the circuit be operating at a 1-mA bias current. The transistors are specified as follows: $\beta(npn) = 200$, $V_{An} = 130$ V, $|V_{Ap}| = 50$ V, $C_\pi = 16$ pF, $C_\mu = 0.3$ pF, $C_L = 5$ pF, and $r_x = 200\ \Omega$. The amplifier is fed with a signal source having a resistance of 36 kΩ. Determine: (a) A_M; (b) C_{in} and f_H using the Miller equivalence; (c) f_H using open-circuit time constants; (d) f_Z, f_{P1}, f_{P2}, and hence f_H (use the approximate expressions in Eqs. 6.76 and 6.77); and (e) f_t.
Ans. (a) −175 V/V; (b) 448 pF, 82.6 kHz; (c) 75.1 kHz; (d) 21.2 GHz, 75.1 kHz, 25.2 MHz, 75.1 kHz; (e) 13.1 MHz

---

6.6.5 The Situation When R_{sig} Is Low

There are applications in which the CS amplifier is fed with a low-resistance signal source. Obviously, in such a case, the high-frequency gain will no longer be limited by the interaction of the source resistance and the input capacitance. Rather, the high-frequency limitation happens at the amplifier output, as we shall now show.

Figure 6.26(a) shows the high-frequency equivalent circuit of the common-source amplifier in the limiting case when R_{sig} is zero. The voltage transfer function $V_o/V_{sig} = V_o/V_{gs}$ can be found by setting $R_{sig} = 0$ in Eq. (6.60). The result is

$$\frac{V_o}{V_{sig}} = \frac{(-g_m R'_L)[1 - s(C_{gd}/g_m)]}{1 + s(C_L + C_{gd})R'_L} \qquad (6.78)$$

Thus, while the dc gain and the frequency of the zero do not change, the high-frequency response is now determined by a pole formed by $C_L + C_{gd}$ together with R'_L. Thus the 3-dB frequency is now given by

$$f_H = \frac{1}{2\pi(C_L + C_{gd})R'_L} \qquad (6.79)$$

To see how this pole is formed, refer to Fig. 6.26(b), which shows the equivalent circuit with the input signal source reduced to zero. Observe that the circuit reduces to a capacitance $(C_L + C_{gd})$ in parallel with a resistance R'_L.

As we have seen above, the transfer-function zero is usually at a very high frequency and thus does not play a significant role in shaping the high-frequency response. The gain of the CS amplifier will therefore fall off at a rate of −6 dB/octave (−20 dB/decade) and reaches

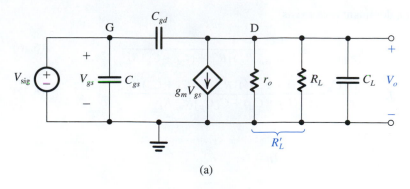

(a)

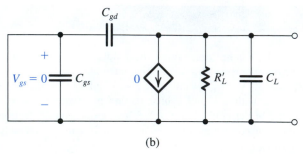

(b)

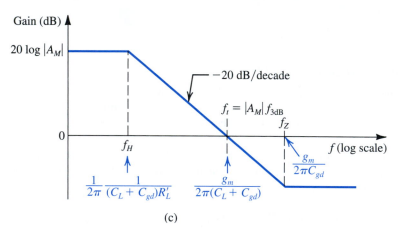

(c)

FIGURE 6.26 (a) High-frequency equivalent circuit of a CS amplifier fed with a signal source having a very low (effectively zero) resistance. (b) The circuit with V_{sig} reduced to zero. (c) Bode plot for the gain of the circuit in (a).

unity (0 dB) at a frequency f_t, which is equal to the **gain–bandwidth product,**

$$f_t = |A_M| f_H$$
$$= g_m R'_L \frac{1}{2\pi(C_L + C_{gd})R'_L}$$

Thus,

$$f_t = \frac{g_m}{2\pi(C_L + C_{gd})} \tag{6.80}$$

Figure 6.26(c) shows a sketch of the high-frequency gain of the CS amplifier.

EXAMPLE 6.10

Consider the CS amplifier specified in Example 6.9 when fed with a signal source having a negligible resistance (i.e., $R_{sig} = 0$). Find A_M, f_{3dB}, f_t, and f_Z. If the amplifying transistor is to be operated at twice the original overdrive voltage while W and L remain unchanged, what value of I_{REF} is needed? What are the new values of A_M, f_{3dB}, f_t, and f_Z?

Solution

In Example 6.9 we found that

$$A_M = -12.3 \text{ V/V}$$

The 3-dB frequency can be found using Eq. (6.79),

$$f_H = \frac{1}{2\pi(C_L + C_{gd})R_L'}$$

$$= \frac{1}{2\pi(25 + 5) \times 10^{-15} \times 9.82 \times 10^3}$$

$$= 540 \text{ MHz}$$

and the unity-gain frequency, which is equal to the gain–bandwidth product, can be determined as

$$f_t = |A_M|f_H = 12.3 \times 540 = 6.6 \text{ GHz}$$

The frequency of the zero is

$$f_Z = \frac{1}{2\pi}\frac{g_m}{C_{gd}}$$

$$= \frac{1}{2\pi}\frac{1.25 \times 10^{-3}}{5 \times 10^{-15}} \cong 40 \text{ GHz}$$

Now, to increase V_{OV} from 0.16 V to 0.32 V, I_D must be quadrupled by changing I_{REF} to

$$I_{REF} = 400 \text{ } \mu A$$

The new values of g_m, r_{o1}, r_{o2}, and R_L' can be found as follows:

$$g_m = \frac{I_D}{V_{OV}/2} = \frac{400}{0.32/2} = 2.5 \text{ mA/V}$$

$$r_{o1} = \frac{5 \times 0.36}{0.4 \text{ mA}} = 4.5 \text{ k}\Omega$$

$$r_{o2} = \frac{6 \times 0.36}{0.4 \text{ mA}} = 5.4 \text{ k}\Omega$$

$$R_L' = (4.5 \| 5.4) = 2.45 \text{ k}\Omega$$

Thus the new value of A_M becomes

$$A_M = -g_m R_L' = -2.5 \times 2.45 = -6.15 \text{ V/V}$$

That of f_H becomes

$$f_H = \frac{1}{2\pi(C_L + C_{gd})R'_L}$$

$$= \frac{1}{2\pi(25+5) \times 10^{-15} \times 2.45 \times 10^3}$$

$$= 2.16 \text{ GHz}$$

and the unity-gain frequency (i.e., the gain–bandwidth product) becomes

$$f_t = 6.15 \times 2.16 = 13.3 \text{ GHz}$$

We note that doubling V_{OV} results in reducing the dc gain by a factor of 2 and increasing the bandwidth by a factor of 4. Thus, the gain–bandwidth product is doubled—a good bargain!

EXERCISES

D6.21 For the CS amplifier considered in Example 6.10 operating at the original values of V_{OV} and I_D (i.e., $V_{OV} = 0.16$ V and $I_D = 100$ μA), find the value to which C_L should be increased to place f_t at 2 GHz.
  Ans. 94.4 fF

6.22 Show that the CS amplifier when fed with $R_{\text{sig}} = 0$ has a transfer-function zero whose frequency is related to f_t by

$$\frac{f_Z}{f_t} = 1 + \frac{C_L}{C_{gd}}$$

 ## 6.7 THE COMMON-GATE AND COMMON-BASE AMPLIFIERS WITH ACTIVE LOADS

6.7.1 The Common-Gate Amplifier

Figure 6.27(a) shows the basic IC MOS common-gate amplifier. The transistor has its gate grounded and its drain connected to an active load, shown as an ideal constant-current source I. The input signal source v_{sig} with a generator resistance R_s is connected to the source terminal.[10] Since the MOSFET source is *not* connected to the substrate, we show the substrate terminal, B, explicitly and indicate that it is connected to the lowest voltage in the circuit, in this case ground. Finally, observe that except for showing the current-source I, which determines the dc bias current I_D of the transistor, we have not shown any other bias detail. How the dc voltage V_{GS} will be established and how V_{DS} is determined are not of concern to us here. As mentioned before, however, bias stability is usually assured through the application of negative feedback to the larger circuit of which the CG amplifier is a part. For our purposes here, we shall assume that the MOSFET is operating in the saturation region and concentrate exclusively on its small-signal operation.

The Body Effect Since the substrate (i.e., body) is not connected to the source, the body effect plays a role in the operation of the common-gate amplifier. It turns out, however, that

---

[10] Rather than using R_{sig} to denote the resistance of the signal source, we use R_s since the resistance is in series with the source terminal of the MOSFET.

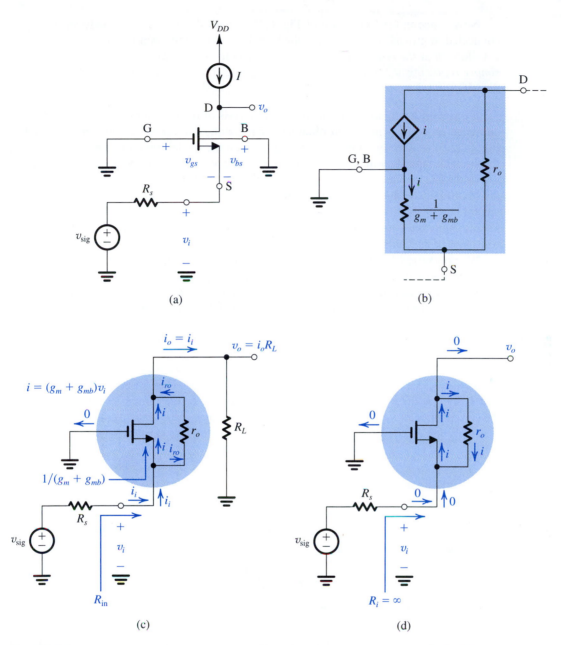

FIGURE 6.27 (a) Active-loaded common-gate amplifier. (b) MOSFET equivalent circuit for the CG case in which the body and gate terminals are connected to ground. (c) Small-signal analysis performed directly on the circuit diagram with the T model of (b) used implicitly. (d) Operation with the output open-circuited.

taking the body effect into account in the analysis of the CG circuit is a very simple matter. To see how this can be done, recall that the body terminal acts, in effect, as a second gate for the MOSFET. Thus, just as a signal voltage v_{gs} between the gate and the source gives rise to a drain current signal $g_m v_{gs}$, a signal voltage v_{bs} between the body and the source gives rise to a drain current signal $g_{mb} v_{bs}$. Thus the drain signal current becomes $(g_m v_{gs} + g_{mb} v_{bs})$, where the body transconductance g_{mb} is a small fraction χ of g_m; $g_{mb} = \chi g_m$ and $\chi = 0.1$ to 0.2.

Now, since in the CG circuit of Fig. 6.27(a) both the gate and the body terminals are connected to ground, $v_{bs} = v_{gs}$, and the signal current in the drain becomes $(g_m + g_{mb})v_{gs}$. It follows that *the body effect in the common-gate circuit can be fully accounted for by simply replacing g_m of the MOSFET by $(g_m + g_{mb})$*. As an example, Fig. 6.27(b) shows the MOSFET T model modified in this fashion.

Small-Signal Analysis   The small-signal analysis of the CG amplifier can be performed either on an equivalent circuit obtained by replacing the MOSFET with its T model of Fig. 6.27(b) or directly on the circuit diagram with the model used implicitly. We shall opt for the latter approach in order to gain greater insight into circuit operation. Figure 6.27(c) shows the CG circuit prepared for small-signal analysis. Note that we have "extracted" r_o of the MOSFET and shown it separately from the device. As well, we have indicated the resistance $1/(g_m + g_{mb})$, which appears in effect between gate and source looking into the source. Finally, note that a resistance R_L is shown at the output; it is assumed to include the output resistance of the current-source load I as well as any load resistance if one is connected.

We now proceed to analyze the circuit of Fig. 6.27(c) to determine the various parameters that characterize the CG amplifier. At this point we strongly urge the reader to consult Table 4.3 for a review of the definitions of amplifier characteristic parameters. This is especially useful here because the CG amplifier is *not* a unilateral circuit; the resistance r_o connects the output node to the input node, thus destroying unilateralism. As a result we should expect the amplifier input resistance R_{in} to depend on R_L and the output resistance R_{out} to depend on R_s.

Input Resistance   To determine the input resistance R_{in}, we must find a way to express i_i in terms of v_i. Inspection of the circuit in Fig. 6.27(c) reveals a key observation. The input current i_i splits at the source node into two components: the source current $i = (g_m + g_{mb})v_i$ and the current through r_o, i_{ro}. These two components combine at the drain to constitute the current i_o supplied to R_L; thus $i_o = i_i$ and $v_o = i_o R_L = i_i R_L$. Now we can write at the source node

$$i_i = (g_m + g_{mb})v_i + i_{ro} \tag{6.81}$$

and express i_{ro} as

$$i_{ro} = \frac{v_i - v_o}{r_o} = \frac{v_i - i_i R_L}{r_o} \tag{6.82}$$

Equations (6.81) and (6.82) can be combined to yield

$$i_i = \left(g_m + g_{mb} + \frac{1}{r_o}\right)v_i \Big/ \left(1 + \frac{R_L}{r_o}\right)$$

from which the input resistance R_{in} can be found as

$$R_{in} \equiv \frac{v_i}{i_i} = \frac{r_o + R_L}{1 + (g_m + g_{mb})r_o} \tag{6.83}$$

Observe that for $r_o = \infty$, R_{in} reduces to $1/(g_m + g_{mb})$, which is indeed the input resistance that we found for the discrete CG amplifier analyzed in Section 4.7.5 with r_o neglected (there we also neglected g_{mb}). When r_o is taken into account, this value of input resistance is obtained approximately only for $R_L = 0$. For the usual case of $R_L \cong r_o$, $R_i \cong 2/(g_m + g_{mb})$. Interestingly, for large values of R_L approaching infinity, $R_{in} = \infty$. This somewhat surprising result will be illustrated next.

Operation with $R_L = \infty$ Figure 6.27(d) shows the CG amplifier with R_L removed; that is, $R_L = \infty$ and the amplifier is operating with the output open-circuited. We immediately note that since $i_o = 0$, i_i must also be zero; the current i in the source terminal, $i = (g_m + g_{mb})v_i$, simply flows via the drain through r_o and back to the source node. It follows that the input resistance with no load, R_i, is infinite:

$$R_i = \infty$$

We can also use the circuit in Fig. 6.27(d) to determine the open-circuit voltage gain A_{vo} between the input (source) and output (drain) terminals as follows:

$$v_o = ir_o + v_i$$
$$= (g_m + g_{mb})r_o v_i + v_i \tag{6.84}$$

Thus,

$$A_{vo} = 1 + (g_m + g_{mb})r_o \tag{6.85}$$

This is a very important quantity that appears in almost all formulas that characterize the CG amplifier. We observe that A_{vo} differs from the intrinsic gain of the MOSFET in two minor respects: First, there is an additional term of unity, and second, g_{mb} is added to g_m. Typically A_{vo} is 10% to 20% larger than A_0.

We should also note that the gain of the CG circuit is positive. That is, unlike the CS amplifier, the CG amplifier is **noninverting.**

Utilizing Eqs. (6.83) and (6.85), we can express the input resistance of the CG amplifier in the compact and attractive form

$$R_{in} = \frac{r_o + R_L}{A_{vo}} \tag{6.86}$$

That is, the CG circuit divides the total resistance $(r_o + R_L)$ by the open-circuit voltage gain, which is approximately equal to the intrinsic gain of the MOSFET. Furthermore, since $A_{vo} \cong (g_m + g_{mb})r_o \cong A_0$, the expression for R_{in} can be simplified to

$$R_{in} \cong \frac{1}{g_m + g_{mb}} + \frac{R_L}{A_0} \tag{6.87}$$

This expression simply says that *taking r_o into account adds a component (R_L/A_0) to the input resistance.* This additional component becomes significant only when R_L is large.

Another interesting result follows directly from the fact that $i_i = 0$ in the circuit of Fig. 6.27(d): The voltage drop across R_s will be zero. Thus $v_i = v_{sig}$, and the open-circuit overall voltage gain, v_o/v_{sig}, will be equal to A_{vo},

$$G_{vo} = A_{vo} = 1 + (g_m + g_{mb})r_o \tag{6.88}$$

Voltage Gain The voltage gains A_v and G_v of the loaded CG amplifier of Fig. 6.27(c) can be obtained in a number of ways. The most direct approach is to make use, once more, of the fact that $i_o = i_i$ and express v_o as

$$v_o = i_o R_L = i_i R_L \tag{6.89}$$

The voltage v_i can be expressed in terms of i_i as

$$v_i = i_i R_{in} \tag{6.90}$$

Dividing Eq. (6.89) by Eq. (6.90) yields, for the voltage gain A_v,

$$A_v = \frac{v_o}{v_i} = \frac{R_L}{R_{in}} \tag{6.91}$$

Substituting for R_{in} from Eq. (6.86) provides

$$A_v = A_{vo}\frac{R_L}{R_L + r_o} \tag{6.92}$$

In a similar way we can derive an expression for the overall voltage gain, $G_v = v_o/v_{sig}$,

$$v_o = i_o R_L = i_i R_L$$
$$v_{sig} = i_i(R_s + R_{in})$$

Thus,

$$G_v = \frac{R_L}{R_s + R_{in}} \tag{6.93}$$

in which we can substitute for R_{in} from Eq. (6.86) to obtain

$$G_v = A_{vo}\frac{R_L}{R_L + r_o + A_{vo}R_s} \tag{6.94}$$

Recalling that $G_{vo} = A_{vo}$, we can express G_v as

$$G_v = G_{vo}\frac{R_L}{R_L + r_o + A_{vo}R_s} \tag{6.95}$$

Output Resistance   To complete our characterization of the CG amplifier, we find its output resistance. From the study of amplifier characterization in Section 4.7.2 (Table 4.3), we recall that there are two different output resistances: R_o, which is the output resistance when v_i is set to zero, and R_{out}, which is the output resistance when v_{sig} is set to zero. Both are illustrated in Fig. 6.28. Obviously R_o can be obtained from the expression for R_{out} by setting $R_s = 0$. It is important to be clear on the application of R_o and of R_{out}. Since R_o is the output resistance when the amplifier is fed with an ideal source v_i, it follows that it is the applicable output resistance for determining A_v from A_{vo},

$$A_v = A_{vo}\frac{R_L}{R_L + R_o} \tag{6.96}$$

On the other hand, R_{out} is the output resistance when the amplifier is fed with v_{sig} and its resistance R_s; thus it is the applicable output resistance for determining G_v from G_{vo},

$$G_v = G_{vo}\frac{R_L}{R_L + R_{out}} \tag{6.97}$$

Returning to the circuit in Fig. 6.28(a), we see by inspection that

$$R_o = r_o \tag{6.98}$$

A quick verification of this result is achieved by substituting $R_o = r_o$ in Eq. (6.96) and then observing that the resulting expression for A_v is identical to that in Eq. (6.92), which we derived directly from circuit analysis.

An expression for R_{out} can be derived using the circuit in Fig. 6.28(b) where a test voltage v_x is applied at the output. Our goal to find the current i_x drawn from v_x. Toward that

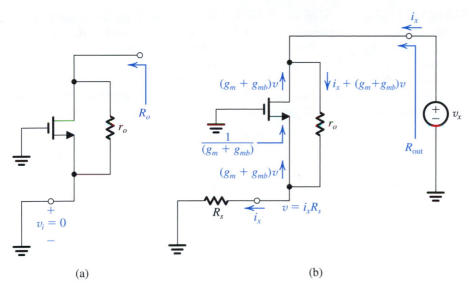

FIGURE 6.28   (a) The output resistance R_o is found by setting $v_i = 0$. (b) The output resistance R_{out} is obtained by setting $v_{\text{sig}} = 0$.

end note that the current through R_s is equal to i_x; thus we can express the voltage v at the MOSFET source as

$$v = i_x R_s \tag{6.99}$$

Utilizing the analysis indicated on the circuit diagram in Fig. 6.28(b), we can write for v_x

$$v_x = [i_x + (g_m + g_{mb})v]r_o + v \tag{6.100}$$

Equations (6.99) and (6.100) can be combined to eliminate v and obtain v_x in terms of i_x and hence $R_{\text{out}} \equiv v_x / i_x$,

$$R_{\text{out}} = r_o + [1 + (g_m + g_{mb})r_o]R_s \tag{6.101}$$

We recognize the term multiplying R_s as the open-circuit voltage gain A_{vo}; thus R_{out} can be expressed in an alternative, more compact form as

$$R_{\text{out}} = r_o + A_{vo}R_s \tag{6.102}$$

A quick verification of the formula for R_{out} in Eq. (6.102) can be obtained by substituting it in Eq. (6.97). The result will be seen to be identical to the gain expression in Eq. (6.95), which we derived by direct circuit analysis.

The expressions for R_{out} in Eqs. (6.101) and (6.102) are very useful results that we will employ frequently throughout the rest of this book. These formulas give the output resistance not only of the CG amplifier but also of a CS amplifier with a resistance R_s in the emitter. We will have more to say about this shortly. At this point, however, it is useful to interpret Eqs. (6.101) and (6.102). A first interpretation, immediately available from Eq. (6.102), is that the CG transistor increases the output resistance by adding to r_o a component $A_{vo}R_s$. In many cases the latter component would dominate, and one can think of the CG MOSFET as multiplying the resistance R_s in its source by A_{vo}, which is approximately equal to $g_m r_o$. Note that this action is the complement of what we saw earlier in regard to R_{in} where the MOSFET acts to divide R_L by A_{vo}. This impedance transformation action of the CG MOSFET is illustrated in Fig. 6.29 and is key to a number of applications of the CG circuit. One such application involves the use of the CG amplifier as a **current buffer.** Figure 6.30 shows an equivalent circuit that is suitable for such an application. The reader is urged to show that the **overall short-circuit current gain G_{is}** is given by

$$G_{is} = G_{vo} \frac{R_s}{R_{out}} \cong 1$$

The near-unity current gain together with the low input resistance and high output resistance are all characteristics of a good current buffer.

Yet another interpretation of the formula for R_{out} can be obtained by expressing Eq. (6.101) in the form

$$R_{out} = R_s + [1 + (g_m + g_{mb})R_s]r_o \qquad (6.103)$$

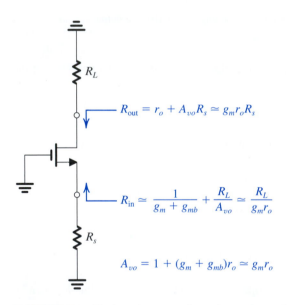

$$R_{out} = r_o + A_{vo}R_s \simeq g_m r_o R_s$$

$$R_{in} \simeq \frac{1}{g_m + g_{mb}} + \frac{R_L}{A_{vo}} \simeq \frac{R_L}{g_m r_o}$$

$$A_{vo} = 1 + (g_m + g_{mb})r_o \simeq g_m r_o$$

FIGURE 6.29 The impedance transformation property of the CG configuration.

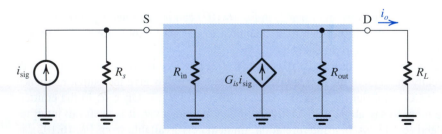

FIGURE 6.30 Equivalent circuit of the CG amplifier illustrating its application as a current buffer. R_{in} and R_{out} are given in Fig. 6.29, and $G_{is} = A_{vo}(R_s/R_{out}) \simeq 1$.

In this expression the second term often dominates, enabling the following approximation:

$$R_{\text{out}} \cong [1 + (g_m + g_{mb})R_s]r_o$$
$$\cong (1 + g_m R_s)r_o \qquad\qquad (6.104)$$

Thus placing a resistance R_s in the source lead results in multiplying the transistor output resistance r_o by a factor that we recognize from our discussion of the effect of source degeneration in Section 4.7.4. We will have more to say about Eq. (6.104) later.

High-Frequency Response Figure 6.31(a) shows the CG amplifier with the MOSFET internal capacitances C_{gs} and C_{gd} indicated. For generality, a capacitance C_L is included at the output node to represent the input capacitance of a succeeding amplifier stage. Capacitance C_L also includes the MOSFET capacitance C_{db}. Note the C_L appears in effect in parallel with C_{gd}; therefore, in the following discussion we will lump the two capacitances together.

It is important to note at the outset that each of the three capacitances in the circuit of Fig. 6.31(a) has a grounded node. Therefore none of the capacitances undergoes the Miller-multiplication effect observed in the CS stage. It follows that the CG circuit can be designed to have a much wider bandwidth than that of the CS circuit, especially when the resistance of the signal generator is large.

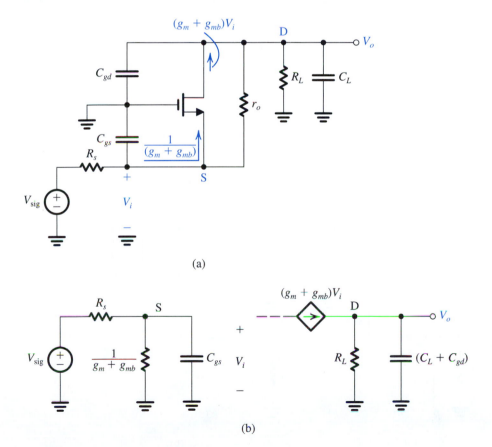

(a)

(b)

FIGURE 6.31 (a) The common-gate amplifier with the transistor internal capacitances shown. A load capacitance C_L is also included. (b) Equivalent circuit for the case in which r_o is neglected.

Analysis of the circuit in Fig. 6.31(a) is greatly simplified if r_o can be neglected. In such a case the input side is isolated from the output side, and the high-frequency equivalent circuit takes the form shown in Fig. 6.31(b). We immediately observe that there are two poles: one at the input side with a frequency f_{P1},

$$f_{P1} = \frac{1}{2\pi C_{gs}\left(R_s \parallel \dfrac{1}{g_m + g_{mb}}\right)} \tag{6.105}$$

and the other at the output side with a frequency f_{P2},

$$f_{P2} = \frac{1}{2\pi(C_{gd} + C_L)R_L} \tag{6.106}$$

The relative locations of the two poles will depend on the specific situation. However, f_{P2} is usually lower than f_{P1}; thus f_{P2} can be dominant. The important point to note is that both f_{P1} and f_{P2} are usually much higher than the frequency of the dominant input pole in the CS stage.

In situations when r_o has to be taken into account (because R_s and R_L are large), the method of open-circuit time constants can be employed to obtain an estimate for the 3-dB frequency f_H. Figure 6.32 shows the circuits for determining the resistances R_{gs} and R_{gd} seen by C_{gs} and $(C_{gd} + C_L)$, respectively. By inspection we obtain

$$R_{gs} = R_s \parallel R_{in} \tag{6.107}$$

and

$$R_{gd} = R_L \parallel R_{out} \tag{6.108}$$

which can be used to obtain f_H,

$$f_H = \frac{1}{2\pi[C_{gs}R_{gs} + (C_{gd} + C_L)R_{gd}]} \tag{6.109}$$

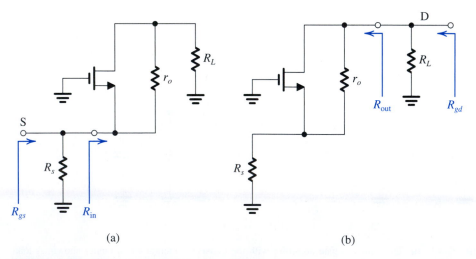

(a) (b)

FIGURE 6.32 Circuits for determining R_{gs} and R_{gd}.

EXAMPLE 6.11

Consider a common-gate amplifier specified as follows: $W/L = 7.2~\mu m/0.36~\mu m$, $\mu_n C_{ox} = 387~\mu A/V^2$, $r_o = 18~k\Omega$, $I_D = 100~\mu A$, $g_m = 1.25~mA/V$, $\chi = 0.2$, $R_s = 10~k\Omega$, $R_L = 100~k\Omega$, $C_{gs} = 20~fF$, $C_{gd} = 5~fF$, and $C_L = 0$. Find A_{vo}, R_{in}, R_{out}, G_v, G_{is}, G_i, and f_H.

Solution

$$g_m + g_{mb} = 1.25 + 0.2 \times 1.25 = 1.5~mA/V$$

$$A_{vo} = 1 + (g_m + g_{mb})r_o = 1 + 1.5 \times 18 = 28~V/V$$

$$R_{in} = \frac{r_o + R_L}{A_{vo}} = \frac{18 + 100}{28} = 4.2~k\Omega$$

$$R_{out} = r_o + A_{vo}R_s = 18 + 28 \times 10 = 298~k\Omega$$

$$G_v = G_{vo}\frac{R_L}{R_L + R_{out}} = A_{vo}\frac{R_L}{R_L + R_{out}} = 28\frac{100}{100 + 298} = 7~V/V$$

$$G_{is} = \frac{A_{vo}R_s}{R_{out}} = \frac{28 \times 10}{298} = 0.94~A/A$$

$$G_i = G_{is}\frac{R_{out}}{R_{out} + R_L} = 0.94\frac{298}{298 + 100} = 0.7~A/A$$

$$R_{gs} = R_s \| R_{in} = 10 \| 4.2 = 3~k\Omega$$

$$R_{gd} = R_L \| R_{out} = 100 \| 298 = 75~k\Omega$$

$$\tau_H = C_{gs}R_{gs} + C_{gd}R_{gd}$$

$$= 20 \times 3 + 5 \times 75$$

$$= 60 + 375 = 435~ps$$

$$f_H \cong \frac{1}{2\pi\tau_H} = \frac{1}{2\pi \times 435 \times 10^{-12}} = 366~MHz$$

We note that this circuit performs well as a current buffer, raising the resistance level from $R_{in} \cong 4~k\Omega$ to $R_{out} \cong 300~k\Omega$ and having an overall short-circuit current gain of 0.94 A/A. Because of the high output resistance, the amplifier bandwidth is determined primarily by the capacitance at the output node. Thus additional load capacitance can lower the bandwidth significantly.

EXERCISES

6.23 For the CG amplifier considered in Example 6.11, find the value of f_H when a capacitance $C_L = 5~fF$ is connected at the output.

Ans. 196 MHz

6.24 Repeat the problem in Example 6.11 for the case $R_s = 1~k\Omega$ and $R_L = 10~k\Omega$.

Ans. $A_{vo} = 28~V/V$; $R_{in} = 1~k\Omega$; $R_{out} = 46~k\Omega$; $G_v = 5~V/V$; $G_{is} = 0.61~A/A$; $G_i = 0.5~A/A$; $f_H = 2.61~GHz$

6.7.2 The Common-Base Amplifier

Analysis of the common-base amplifier parallels that of the common-gate circuit that we analyzed previously, with one major exception: The BJT has a finite β, and its base conducts signal current, which gives rise to the resistance r_π between base and emitter, looking into the base. Figure 6.33(a) shows the basic circuit for the active-loaded common-base

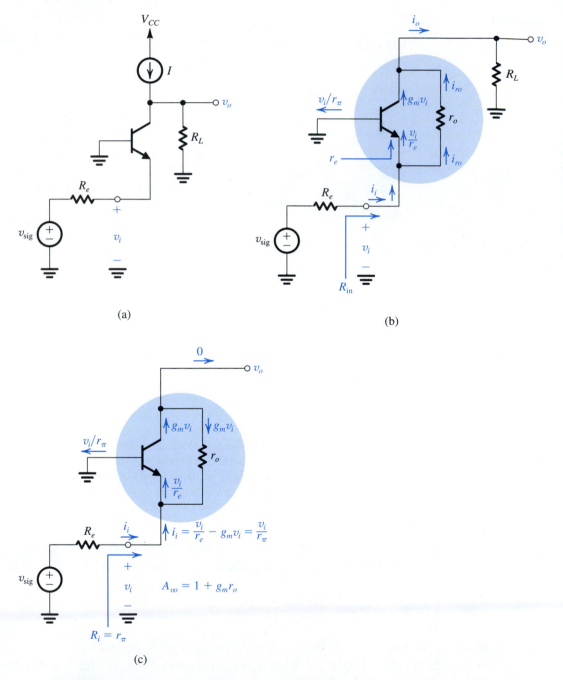

(a)

(b)

(c)

FIGURE 6.33 (a) Active-loaded common-base amplifier. (b) Small-signal analysis performed directly on the circuit diagram with the BJT T model used implicitly. (c) Small-signal analysis with the output open-circuited.

amplifier without the bias details. Note that resistance R_L represents the combination of a load resistance, if any, and the output resistance of the current source that realizes the active load I.

Figure 6.33(b) shows the small-signal analysis performed directly on the circuit with the T model of the BJT used implicitly. The analysis is very similar to that for the MOS case except that, as a result of the finite base current, v_i/r_π, the current i_o is related to i_i by

$$i_o = i_i - v_i/r_\pi \tag{6.110}$$

The reader can show that, neglecting r_x, the input resistance at the emitter R_{in} is given by

$$R_{in} = \frac{r_o + R_L}{1 + \dfrac{r_o}{r_e} + \dfrac{R_L}{(\beta + 1)r_e}} \tag{6.111}$$

We immediately observe that setting $\beta = \infty$ reduces this expression to that for the MOS case (Eq. 6.83) except that here $g_{mb} = 0$. Note that for $\beta = \infty$, $\alpha = 1$, and $r_e = \alpha/g_m = 1/g_m$.

With a slight approximation, the expression in Eq. (6.111) can be written as

$$R_{in} \cong r_e \frac{r_o + R_L}{r_o + R_L/(\beta + 1)} \tag{6.112}$$

Note that setting $r_o = \infty$ yields $R_{in} = r_e$, which is consistent with what we found in Section 5.7.5. Also, for $R_L = 0$, $R_{in} = r_e$. The value of R_{in} increases as R_L is raised, reaching a maximum of $(\beta + 1)r_e = r_\pi$ for $R_L = \infty$, that is, with the amplifier operating open-circuited (see Fig. 6.33c). For $R_L/(\beta + 1) \ll r_o$, Eq. (6.112) can be approximated as

$$R_{in} \cong r_e + \frac{R_L}{A_0} \tag{6.113}$$

where A_0 is the intrinsic gain $g_m r_o$. This equation is very similar to Eq. (6.87) in the MOSFET case.

The open-circuit voltage gain and input resistance can be easily found from the circuit in Fig. 6.33(c) as

$$A_{vo} = 1 + g_m r_o = 1 + A_0 \tag{6.114}$$

which is identical to Eq. (6.85) for the MOSFET except for the absence of g_{mb}. The input resistance with no load, R_i, is

$$R_i = r_\pi \tag{6.115}$$

as we have already found out from Eq. (6.112).

As in the MOSFET case, the output resistance R_o is given by

$$R_o = r_o \tag{6.116}$$

The output resistance including the source resistance R_e can be found by analysis of the circuit in Fig. 6.34 to be

$$R_{out} = r_o + (1 + g_m r_o)R_e' \tag{6.117a}$$

where $R_e' = R_e \parallel r_\pi$.

Note that the formula in Eq. (6.117a) is very similar to that for the MOS case, namely Eq. (6.101). However, there are two differences: First, g_{mb} is missing, and second, $R_e' = R_e \parallel r_\pi$

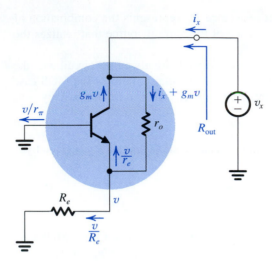

FIGURE 6.34 Analysis of the CB circuit to determine R_{out}. Observe that the current i_x that enters the transistor must equal the sum of the two currents v/r_π and v/R_e that leave the transistor; that is, $i_x = v/r_\pi + v/R_e$.

replaces R_s. The reason r_π appears in the BJT formula is the finite β of the BJT. The expression in Eq. (6.117a) can also be written in terms of the open-circuit voltage gain A_{vo} as

$$R_{\text{out}} = r_o + A_{vo}R_e' \tag{6.117b}$$

which is the BJT counterpart of the MOS expression in Eq. (6.102). Another useful form for R_{out} can be obtained from (6.117a),

$$R_{\text{out}} = R_e' + (1 + g_m R_e')r_o \tag{6.117c}$$

which is the BJT counterpart of the MOS expression in Eq. (6.103). In Eq. (6.117c) the second term is much larger than the first, resulting in the approximate expression

$$R_{\text{out}} \cong (1 + g_m R_e')r_o \tag{6.118}$$

which corresponds to Eq. (6.104) for the MOS case.

Equation (6.118) clearly indicates that the inclusion of an emitter resistance R_e increases the CB output resistance by the factor $(1 + g_m R_e')$. Thus, as R_e is increased from 0 to ∞, the output resistance increases from r_o to $(1 + g_m r_\pi)r_o = (1 + \beta)r_o \cong \beta r_o$. This upper limit on the value of R_{out}, dictated by the finite β of the BJT, has no counterpart in the MOS case and, as will be seen later, has important implications for circuit design. Finally, we note that for $R_e \ll r_\pi$, Eq. (6.118) can be approximated by

$$R_{\text{out}} \cong (1 + g_m R_e)r_o \tag{6.119}$$

A useful summary of the formulas for R_{in} and R_{out} is provided in Fig. 6.35.

The results above can be used to obtain the overall voltage gain G_v as

$$G_v = G_{vo}\frac{R_L}{R_L + R_{\text{out}}} \tag{6.120}$$

where

$$G_{vo} = \frac{R_i}{R_i + R_e}A_{vo} = \frac{r_\pi}{r_\pi + R_e}A_{vo} \tag{6.121}$$

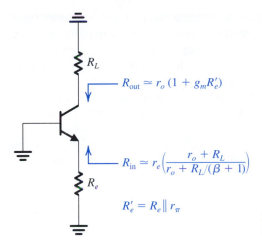

$$R_{\text{out}} \simeq r_o \left(1 + g_m R_e'\right)$$

$$R_{\text{in}} \simeq r_e \left(\frac{r_o + R_L}{r_o + R_L/(\beta + 1)}\right)$$

$$R_e' = R_e \| r_\pi$$

FIGURE 6.35 Input and output resistances of the CB amplifier.

The high-frequency response of the common-base circuit can be evaluated in a manner similar to that used for the MOSFET.

EXERCISE

6.25 Consider the CB amplifier of Fig. 6.33(a) for the case $I = 1$ mA, $\beta = 100$, $V_A = 100$ V, $R_L = 1$ MΩ, and $R_e = 1$ kΩ. Find R_{in}, A_{vo}, R_o, A_v, R_{out}, and G_v. Also, find v_o if v_{sig} is a 5-mV peak sine wave.

Ans. 250 Ω; 4001 V/V; 100 kΩ; 3637 V/V; 2.97 MΩ; 722 V/V; 3.61 V peak

---

6.7.3 A Concluding Remark

The common-gate and common-base circuits have open-circuit voltage gains A_{vo} almost equal to those of the common-source and common-emitter circuits. Their input resistance, however, is much smaller and their output resistance much larger than the corresponding values for the CS and CE amplifiers. These two properties, though not usually desirable in voltage amplifiers, make the CG and CB circuits suitable as current buffers. The absence of the Miller effect makes the high-frequency response of the CG and CB circuits far superior to that of the CS and CE amplifiers. The most significant application of the CG and CB circuits is in a configuration known as the *cascode amplifier,* which we shall study next.

6.8 THE CASCODE AMPLIFIER

By placing a common-gate (common-base) amplifier stage in cascade with a common-source (common-emitter) amplifier stage, a very useful and versatile amplifier circuit results. It is known as the **cascode configuration**[11] and has been in use for nearly three quarters of a century, obviously in a wide variety of technologies.

---

[11] The name *cascode* dates back to the days of vacuum tubes and is a shortened version of "*casc*aded cath*ode*" since, in the tube version, the output (anode) of the first tube feeds the cathode of the second.

The basic idea behind the cascode amplifier is to combine the high input resistance and large transconductance achieved in a common-source (common-emitter) amplifier with the current-buffering property and the superior high-frequency response of the common-gate (common-base) circuit. As will be seen shortly, the cascode amplifier can be designed to obtain a wider bandwidth but equal dc gain as compared to the common-source (common-emitter) amplifier. Alternatively, it can be designed to increase the dc gain while leaving the gain–bandwidth product unchanged. Of course, there is a continuum of possibilities between these two extremes.

Although the cascode amplifier is formed by cascading two amplifier stages, in many applications it is thought of and treated as a single-stage amplifier. Therefore it belongs in this chapter.

6.8.1 The MOS Cascode

Figure 6.36(a) shows the MOS cascode amplifier. Here transistor Q_1 is connected in the common-source configuration and provides its output to the input terminal (i.e., source) of transistor Q_2. Transistor Q_2 has a constant dc voltage, V_{BIAS}, applied to its gate. Thus the signal voltage at the gate of Q_2 is zero, and Q_2 is operating as a CG amplifier with a constant-current load, I. Obviously both Q_1 and Q_2 will be operating at dc drain currents equal to I. As in previous cases, feedback in the overall circuit that incorporates the cascode amplifier establishes an appropriate dc voltage at the gate of Q_1 so that its drain current is equal to I. Also, the value of V_{BIAS} has to be chosen so that both Q_1 and Q_2 operate in the saturation region at all times.

Small-Signal Analysis   We begin with a qualitative description of the operation of the cascode circuit. In response to the input signal voltage v_i, the common-source transistor Q_1 conducts a current signal $g_{m1}v_i$ in its drain terminal and feeds it to the source terminal of the common-gate transistor Q_2, called the **cascode transistor.** Transistor Q_2 passes the signal current $g_{m1}v_i$ on to its drain, where it is supplied to a load resistance R_L (not shown in Fig. 6.36) at a very high output resistance, R_{out}. The cascode transistor Q_2 acts in effect as a buffer, presenting a low input resistance to the drain of Q_1 and providing a high resistance at the amplifier output.

Next we analyze the cascode amplifier circuit to determine its characteristic parameters. Toward that end Fig. 6.36(b) shows the cascode circuit prepared for small-signal analysis and with a resistance R_L shown at the output. R_L is assumed to include the output resistance of current source I as well as an actual load resistance, if any. The diagram also indicates various input and output resistances obtained using the results of the analysis of the CS and CG amplifiers in previous sections. Note in particular that the CS transistor Q_1 provides the cascode amplifier with an infinite input resistance. Also, at the drain of Q_1 looking "downward," we see the output resistance of the CS transistor Q_1, r_{o1}. Looking "upward," we see the input resistance of the CG transistor Q_2,

$$R_{in2} = \frac{1}{g_{m2} + g_{mb2}} + \frac{R_L}{A_{vo2}} \tag{6.122}$$

where

$$A_{vo2} = 1 + (g_{m2} + g_{mb2})r_{o2} \tag{6.123}$$

Thus the total resistance between the drain of Q_1 and ground is

$$R_{d1} = r_{o1} \, \| \left[\frac{1}{g_{m2} + g_{mb2}} + \frac{R_L}{A_{vo2}} \right] \tag{6.124}$$

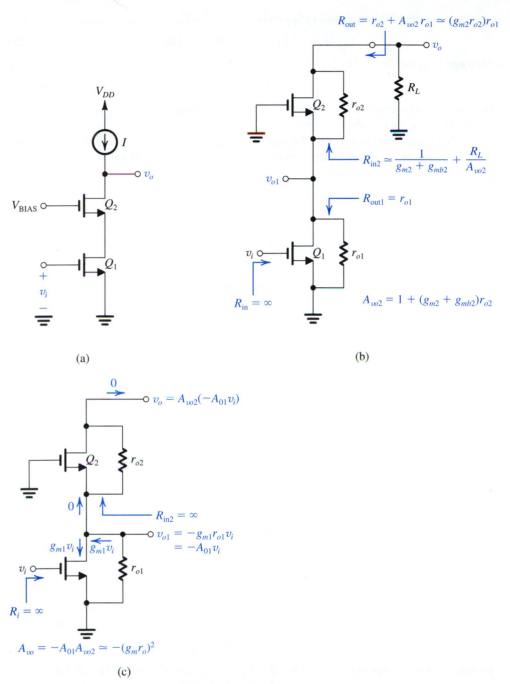

FIGURE 6.36 (a) The MOS cascode amplifier. (b) The circuit prepared for small-signal analysis with various input and output resistances indicated. (c) The cascode with the output open-circuited.

Figure 6.36(b) also indicates that the output resistance of the cascode amplifier, R_{out}, is given by

$$R_{\text{out}} = r_{o2} + A_{vo2}r_{o1} \qquad (6.125)$$

which has been obtained using the formula in Eq. (6.102) and noting that the resistance R_s in the source of the CG transistor Q_2 is the output resistance r_{o1} of Q_1. Substituting for

A_{vo2} from Eq. (6.123) into Eq. (6.125) yields

$$R_{\text{out}} = r_{o2} + [1 + (g_{m2} + g_{mb2})r_{o2}]r_{o1} \tag{6.126}$$

which can be approximated as

$$R_{\text{out}} \cong (g_{m2}r_{o2})r_{o1} = A_0 r_{o1} \tag{6.127}$$

Thus the cascode transistor raises the level of output resistance by a factor equal to its intrinsic gain, from r_{o1} of the CS amplifier to $A_0 r_{o1}$.

Another observation to make on the cascode amplifier circuit in Fig. 6.36(b) is that when a signal source v_{sig} with an internal resistance R_{sig} is connected to the input, the infinite input resistance of the amplifier causes

$$v_i = v_{\text{sig}}$$

Thus,

$$G_v = A_v$$

Also, note that the amplifier is unilateral; thus,

$$R_o = R_{\text{out}}$$

The open-circuit voltage gain A_{vo} of the cascode amplifier can be easily determined from the circuit in Fig. 6.36(c), which shows the amplifier operating with the output open-circuited. Since R_{in2} will be infinite, the gain of the CS stage Q_1 will be

$$\frac{v_{o1}}{v_i} = -g_{m1}r_{o1} = -A_{01}$$

The signal v_{o1} will be amplified by the open-circuit voltage gain A_{vo2} of the CG transistor Q_2 to obtain

$$v_o = A_{vo2}v_{o1}$$

Thus,

$$A_{vo} = -A_{01}A_{vo2} \tag{6.128}$$

$$\cong -A_{01}A_{02}$$

which for the usual case of equal intrinsic gains becomes

$$A_{vo} = -A_0^2 = -(g_m r_o)^2 \tag{6.129}$$

We conclude that cascoding increases the magnitude of the open-circuit voltage gain from A_0 of the CS amplifier to A_0^2.

We are now in a position to derive an expression for the **short-circuit transconductance** G_m of the cascode amplifier. From the definitions and the equivalent circuits in Table 4.3,

$$A_{vo} = -G_m R_o$$

Substituting for A_{vo} from Eq. (6.128) and for $R_o = R_{\text{out}}$ from Eq. (6.125) gives, for G_m,

$$G_m = \frac{A_{01}A_{vo2}}{r_{o2} + A_{vo2}r_{o1}}$$

$$= \frac{g_{m1}r_{o1}[1 + (g_{m2} + g_{mb2})r_{o2}]}{r_{o2} + [1 + (g_{m2} + g_{mb2})r_{o2}]r_{o1}} \tag{6.130}$$

$$\cong g_{m1}$$

which confirms the value obtained earlier in the qualitative analysis.

The operation of the cascode amplifier should now be apparent: In response to v_i the CS transistor provides a drain current $g_{m1}v_i$, which the CG transistor passes on to R_L and, in the process, increases the output resistance by A_0. It is the increase in R_{out} to $A_0 r_o$ that increases the open-circuit voltage gain to $(g_m)(A_0 r_o) = A_0^2$. Figure 6.37 provides a useful summary of the operation: Two output equivalent circuits are shown in Fig. 6.37(a) and (b), and an equivalent circuit for determining the voltage gain of the CS stage Q_1 is presented in Fig. 6.37(c). The voltage gain A_v can be found from either of the two equivalent circuits in Fig. 6.37(a) and (b). Using that in Fig. 6.37(a) gives

$$A_v = -A_0^2 \frac{R_L}{R_L + A_0 r_o} \tag{6.131}$$

We immediately see that if we are to realize the large gain of which the cascode is capable, resistance R_L should be large. At the very least, R_L should be of the order of $A_0 r_o$. For $R_L = A_0 r_o$, $A_v = -A_0^2/2$.

The gain of the CS stage is important because its value determines the Miller effect in that stage. From the equivalent circuit in Fig. 6.37(c), neglecting g_{mb},

$$\frac{v_{o1}}{v_i} = -g_m \left[r_o \parallel \left(\frac{1}{g_m} + \frac{R_L}{A_0} \right) \right] \tag{6.132}$$

For $R_L = A_0 r_o$,

$$\frac{v_{o1}}{v_i} = -g_m \left[r_o \parallel \left(\frac{1}{g_m} + r_o \right) \right]$$

$$\cong -\frac{1}{2} g_m r_o = -\frac{1}{2} A_0 \tag{6.133}$$

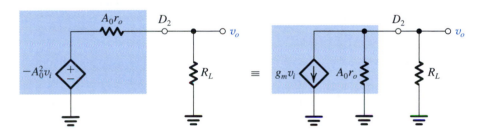

(a)

(b)

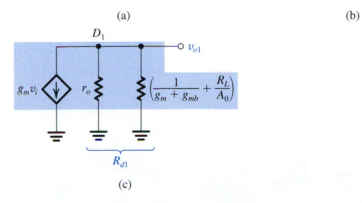

(c)

FIGURE 6.37 (**a** and **b**) Two equivalent circuits for the output of the cascode amplifier. Either circuit can be used to determine the gain $A_v = v_o/v_i$, which is equal to G_v because $R_{in} = \infty$ and thus $v_i = v_{sig}$. (**c**) Equivalent circuit for determining the voltage gain of the CS stage, Q_1.

Thus we see that when R_L is large and the cascode amplifier is realizing a substantial gain, a good part of the gain is obtained in the CS stage. This is not good news considering the Miller effect, as we shall see shortly. To keep the gain of the CS stage relatively low, R_L has to be lowered. For instance, for $R_L = r_o$, Eq. (6.132) indicates that

$$\frac{v_{o1}}{v_i} = -g_m\left[r_o \parallel \left(\frac{1}{g_m} + \frac{1}{g_m}\right)\right]$$

$$\cong -2 \text{ V/V}$$

Unfortunately, however, in this case the dc gain of the cascode is drastically reduced, as can be seen by substituting $R_L = r_o$ in Eq. (6.131),

$$A_v = -A_0^2 \frac{r_o}{r_o + A_0 r_o} \cong -A_0 \tag{6.134}$$

That is, the gain of the cascode becomes equal to that realized in a single CS stage! Does this mean that the cascode configuration (in this case) is not useful? Not at all, as we shall now see.

6.8.2 Frequency Response of the MOS Cascode

Figure 6.38 shows the cascode amplifier with all transistor internal capacitances indicated. Also included is a capacitance C_L at the output node to represent the combination of C_{db2}, the input capacitance of a succeeding amplifier stage (if any), and a load capacitance (if any). Note that C_{db1} and C_{gs2} appear in parallel, and we shall combine them in the following analysis. Similarly, C_L and C_{gd2} appear in parallel and will be combined.

The easiest and, in fact, quite insightful approach to determining the 3-dB frequency f_H is to employ the open-circuit time-constants method. We shall do so and, in the process, utilize the formulas derived in Sections 6.6.2 and 6.7.1 for the various resistances:

1. Capacitance C_{gs1} sees a resistance R_{sig}.
2. Capacitance C_{gd1} sees a resistance R_{gd1}, which can be obtained by adapting the formula in Eq. (6.56) to

$$R_{gd1} = (1 + g_{m1}R_{d1})R_{sig} + R_{d1} \tag{6.135}$$

where R_{d1}, the total resistance at D_1, is given by Eq. (6.124).

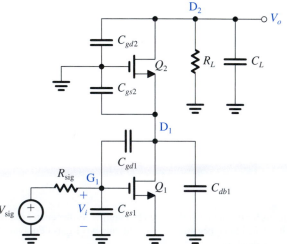

FIGURE 6.38 The cascode circuit with the various transistor capacitances indicated.

3. Capacitance $(C_{db1} + C_{gs2})$ sees a resistance R_{d1}.

4. Capacitance $(C_L + C_{gd2})$ sees a resistance $(R_L \parallel R_{out})$.

With the resistances determined, the effective time constant τ_H can be computed as

$$\tau_H = C_{gs1}R_{sig} + C_{gd1}[(1 + g_{m1}R_{d1})R_{sig} + R_{d1}]$$
$$+ (C_{db1} + C_{gs2})R_{d1} + (C_L + C_{gd2})(R_L \parallel R_{out}) \qquad (6.136)$$

and the 3-dB frequency f_H as

$$f_H \cong \frac{1}{2\pi\tau_H}$$

To gain insight regarding what limits the high-frequency gain of the MOS cascode amplifier, we rewrite Eq. (6.136) in the form

$$\tau_H = R_{sig}[C_{gs1} + C_{gd1}(1 + g_{m1}R_{d1})] + R_{d1}(C_{gd1} + C_{db1} + C_{gs2})$$
$$+ (R_L \parallel R_{out})(C_L + C_{gd2}) \qquad (6.137)$$

In the case of a large R_{sig}, the first term can dominate, especially if the Miller multiplier $(1 + g_{m1}R_{d1})$ is large. This in turn happens when the load resistance R_L is large (on the order of $A_0 r_o$), causing R_{in2} to be large and requiring the first stage, Q_1, to provide a large proportion of the gain. It follows that when R_{sig} is large, to extend the bandwidth we have to lower R_L to the order of r_o. This in turn lowers R_{in2} and hence R_{d1} and renders the Miller effect insignificant. Note, however, that the dc gain of the cascode will then be A_0. Thus, while the dc gain will be the same as (or a little higher than) that achieved in a CS amplifier, the bandwidth will be greater.

In the case when R_{sig} is small, the Miller effect in Q_1 will not be of concern. A large value of R_L (on the order of $A_0 r_o$) can then be used to realize the large dc gain possible with a cascode amplifier—that is, a dc gain on the order of A_0^2. Equation (6.137) indicates that in this case the third term will usually be dominant. To pursue this point a little further, consider the case $R_{sig} = 0$, and assume that the middle term is much smaller than the third term. It follows that

$$\tau_H \cong (C_L + C_{gd2})(R_L \parallel R_{out})$$

and the 3-dB frequency becomes

$$f_H = \frac{1}{2\pi(C_L + C_{gd2})(R_L \parallel R_{out})} \qquad (6.138)$$

which is of the same form as the formula for the CS amplifier with $R_{sig} = 0$ (Eq. 6.79). Here, however, $(R_L \parallel R_{out})$ is larger that R_L' by a factor of about A_0. Thus the f_H of the cascode will be lower than that of the CS amplifier by the same factor A_0. Figure 6.39 shows a sketch of the frequency response of the cascode and of the corresponding common-source amplifier. We observe that in this case cascoding increases the dc gain by a factor A_0 while keeping the unity-gain frequency unchanged at

$$f_t \cong \frac{1}{2\pi}\frac{g_m}{C_L + C_{gd2}} \qquad (6.139)$$

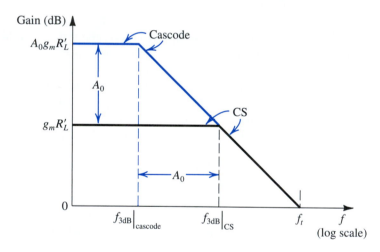

	Common Source	Cascode
Circuit	$R'_L = R_L \| r_o$	
DC Gain	$-g_m R'_L$	$-A_0 g_m R'_L$
f_{3dB}	$\dfrac{1}{2\pi(C_L + C_{gd})R'_L}$	$\dfrac{1}{2\pi(C_L + C_{gd})A_0 R'_L}$
f_t	$\dfrac{g_m}{2\pi(C_L + C_{gd})}$	$\dfrac{g_m}{2\pi(C_L + C_{gd})}$

FIGURE 6.39 Effect of cascoding on gain and bandwidth in the case $R_{sig} = 0$. Cascoding can increase the dc gain by the factor A_0 while keeping the unity-gain frequency constant. Note that to achieve the high gain, the load resistance must be increased by the factor A_0.

EXAMPLE 6.12

This example illustrates the advantages of cascoding by comparing the performance of a cascode amplifier with that of a common-source amplifier in two cases:

(a) The resistance of the signal source is significant, $R_{sig} = 10 \text{ k}\Omega$.

(b) R_{sig} is negligibly small.

Assume all MOSFETs have W/L of $7.2 \ \mu\text{m}/0.36 \ \mu\text{m}$ and are operating at $I_D = 100 \ \mu\text{A}$, $g_m = 1.25 \text{ mA/V}$, $\chi = 0.2$, $r_o = 20 \text{ k}\Omega$, $C_{gs} = 20 \text{ fF}$, $C_{gd} = 5 \text{ fF}$, $C_{db} = 5 \text{ fF}$, and C_L (excluding C_{db}) $= 5 \text{ fF}$. For case (a), let $R_L = r_o = 20 \text{ k}\Omega$ for both amplifiers. For case (b), let $R_L = r_o = 20 \text{ k}\Omega$ for the CS amplifier and $R_L = R_{out}$ for the cascode amplifier. For all cases, determine A_v, f_H, and f_t.

Solution

(a) For the CS amplifier:

$$A_0 = g_m r_o = 1.25 \times 20 = 25 \text{ V/V}$$

$$A_v = -g_m(R_L \parallel r_o) = -g_m(r_o \parallel r_o)$$

$$= -\tfrac{1}{2}A_0 = -12.5 \text{ V/V}$$

$$\tau_H = C_{gs}R_{\text{sig}} + C_{gd}[(1 + g_m R'_L)R_{\text{sig}} + R'_L] + (C_L + C_{db})R'_L$$

where

$$R'_L = r_o \parallel R_L = r_o \parallel r_o = 10 \text{ k}\Omega$$

$$\tau_H = 20 \times 10 + 5[(1 + 12.5)10 + 10] + (5 + 5)10$$

$$= 200 + 725 + 100 = 1025 \text{ ps}$$

Thus,

$$f_H = \frac{1}{2\pi \times 1025 \times 10^{-12}} = 155 \text{ MHz}$$

$$f_t = |A_v| f_H = 12.5 \times 155 = 1.94 \text{ GHz}$$

For the cascode amplifier:

$$A_{01} = g_{m1} r_{o1} = 1.25 \times 20 = 25 \text{ V/V}$$

$$A_{vo2} = 1 + (g_{m2} + g_{mb2})r_{o2} = 1 + (1.25 + 0.2 \times 1.25) \times 20$$

$$= 1 + 1.5 \times 20 = 31 \text{ V/V}$$

$$R_{\text{out1}} = r_{o1} = 20 \text{ k}\Omega$$

$$R_{\text{in2}} = \frac{1}{g_{m2} + g_{mb2}} + \frac{R_L}{A_{vo2}} = \frac{1}{1.5} + \frac{20}{31} = 1.3 \text{ k}\Omega$$

$$R_{d1} = R_{\text{out1}} \parallel R_{\text{in2}} = 20 \parallel 1.3 = 1.22 \text{ k}\Omega$$

$$R_{\text{out}} = r_{o2} + A_{vo2} r_{o1} = 20 + 31 \times 20 = 640 \text{ k}\Omega$$

$$\frac{v_{o1}}{v_i} = -g_{m1}R_{d1} = -1.25 \times 1.22 = -1.5 \text{ V/V}$$

$$A_v = A_{vo}\frac{R_L}{R_L + R_{\text{out}}} = -25 \times 31 \times \frac{20}{640 + 20} = -23.5 \text{ V/V}$$

$$\tau_H = R_{\text{sig}}[C_{gs1} + C_{gd1}(1 + g_{m1}R_{d1})] + R_{d1}(C_{gd1} + C_{db1} + C_{gs2})$$

$$\qquad + (R_L \parallel R_{\text{out}})(C_L + C_{db2} + C_{gd2})$$

$$\tau_H = 10[20 + 5(1 + 1.5)] + 1.22(5 + 5 + 20) + (20 \parallel 640)(5 + 5 + 5)$$

$$\qquad = 325 + 36.6 + 290.9$$

$$\qquad = 653 \text{ ps}$$

$$f_H = \frac{1}{2\pi \times 653 \times 10^{-12}} = 244 \text{ MHz}$$

$$f_t = 23.5 \times 244 = 5.73 \text{ GHz}$$

Thus cascoding has increased f_t by a factor of about 3.

(b) For the CS amplifier:

$$A_v = -12.5 \text{ V/V}$$

$$\tau_H = (C_{gd} + C_L + C_{db})R'_L$$

$$= (5 + 5 + 5)10 = 150 \text{ ps}$$

$$f_H = \frac{1}{2\pi \times 150 \times 10^{-12}} = 1.06 \text{ GHz}$$

$$f_t = 12.5 \times 1.06 = 13.3 \text{ GHz}$$

For the cascode amplifier:

$$A_v = A_{vo}\frac{R_L}{R_L + R_{out}}$$

$$= -25 \times 31 \times \frac{640}{640 + 640} = -388 \text{ V/V}$$

$$R_{in2} = \frac{1}{g_{m2} + g_{mb2}} + \frac{R_L}{A_{vo2}} = \frac{1}{1.5} + \frac{640}{31}$$

$$= 21.3 \text{ k}\Omega$$

$$R_{d1} = 21.3 \parallel 20 = 10.3 \text{ k}\Omega$$

$$\tau_H = R_{d1}(C_{gd1} + C_{db1} + C_{gs2}) + (R_L \parallel R_{out})(C_L + C_{gd2} + C_{db2})$$

$$= 10.3(5 + 5 + 20) + (640 \parallel 640)(5 + 5 + 5)$$

$$= 309 + 4800 = 5109 \text{ ps}$$

$$f_H = \frac{1}{2\pi \times 5109 \times 10^{-12}} = 31.2 \text{ MHz}$$

$$f_t = 388 \times 31.2 = 12.1 \text{ GHz}$$

Thus cascoding increases the dc gain from 12.5 to 388 V/V. The unity-gain frequency (i.e., gain–bandwidth product), however, remains nearly constant.

EXERCISES

6.26 What is the minimum value of V_{BIAS} required for a cascode amplifier operating at $I = 100\ \mu A$? Let $\mu_n C_{ox} = 300\ \mu A/V^2$, $W/L = 10$, and $V_{tn} = 0.6$ V.

Ans. 1.12 V

6.27 Consider a cascode amplifier operating at a bias current $I = 100\ \mu A$ and for which all transistors have $W/L = 5\ \mu m/0.5\ \mu m$, $V'_A = 20$ V/μm, $\mu_n C_{ox} = 190\ \mu A/V^2$, $\chi = 0.2$, $C_{gd} = 2$ fF, and $C_{db} = 3$ fF. For $R_{sig} = 0$, $R_L = R_{out}$, and $C_L = 5$ fF (excluding C_{db}), find A_{01}, A_{vo2}, A_{vo}, R_{out1}, R_{in2}, R_{d1}, R_{out}, A_v, f_t, and f_H. (*Hint:* Use the approximate formula for f_t in Eq. 6.139, but remember to add C_{db}.)

Ans. 62 V/V; 75 V/V; –4650 V/V; 100 kΩ; 103 kΩ; 50.7 kΩ; 7.6 MΩ; –2325 V/V; 9.8 GHz; 4.2 MHz

6.8.3 The BJT Cascode

Figure 6.40(a) shows the BJT cascode amplifier. The circuit is very similar to the MOS cascode, and the small-signal analysis follows in a similar fashion, as indicated in Fig. 6.40(b). Here we have shown the various input and output resistances. Observe that unlike the MOSFET cascode, which has an infinite input resistance, the BJT cascode has an input resistance of $r_{\pi 1}$ (neglecting r_x). The formula for R_{in2} is the one we found in the analysis

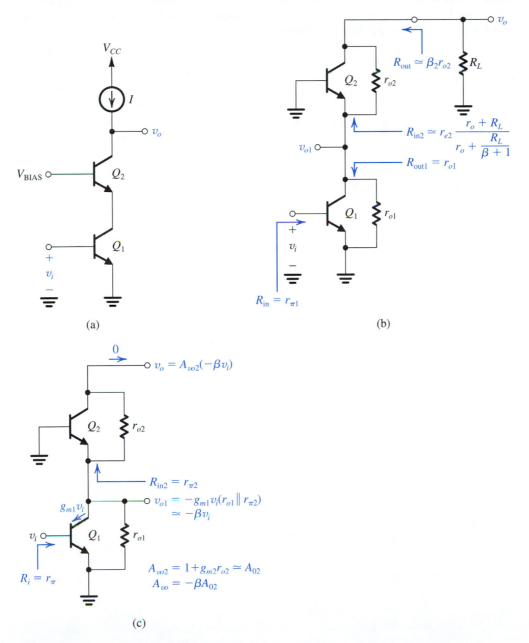

FIGURE 6.40 (a) The BJT cascode amplifier. (b) The circuit prepared for small-signal analysis with various input and output resistances indicated. Note that r_x is neglected. (c) The cascode with the output open-circuited.

of the common-base circuit (Eq. 6.112). The output resistance $R_{out} = \beta_2 r_{o2}$ is found by substituting $R_e = r_{o1}$ in Eq. (6.119) and making the approximation that $g_m r_o \gg \beta$. Recall that βr_o is the largest output resistance that a CB transistor can provide.

The open-circuit voltage gain A_{vo} and the no-load input resistance R_i can be found from the circuit in Fig. 6.40(c), in which the output is open-circuited. Observe that $R_{in2} = r_{\pi2}$, which is usually much smaller than r_{o1}. As a result the total resistance between the collector of Q_1 and ground is approximately $r_{\pi2}$; thus the voltage gain realized in the CE transistor Q_1 is $-g_{m1} r_{\pi2} = -\beta$. Recalling that the open-circuit voltage gain of a CB amplifier is $(1 + g_m r_o) \cong A_0$, we see that the voltage gain A_{vo} is

$$A_{vo} = -\beta A_0 \qquad (6.140)$$

Putting all of these results together we obtain for the BJT cascode amplifier the equivalent circuit shown in Fig. 6.41(a). We note that compared to the common-emitter amplifier, cascoding increases both the open-circuit voltage gain and the output resistance by a factor equal to the transistor β. This should be contrasted with the factor A_0 encountered in the MOS cascode. The equivalent circuit can be easily converted to the transconductance form shown in Fig. 6.41(b). It shows that the short-circuit transconductance G_m of the cascode amplifier is equal to the transconductance g_m of the BJTs. This should have been expected since Q_1 provides a current $g_{m1} v_i$ to the emitter of the cascode transistor Q_2, which in turn passes the current on (assuming $\alpha_2 \cong 1$) to its collector and to the load resistance R_L. In the process the cascode transistor raises the resistance level from r_o at the collector of Q_1 to βr_o at the collector of Q_2. This is the by-now-familiar current-buffering action of the common-base transistor.

The voltage gain of the CE transistor Q_1 can be determined from the equivalent circuit in Fig. 6.41(c). The resistance between the collector of Q_1 and ground is the parallel equivalent

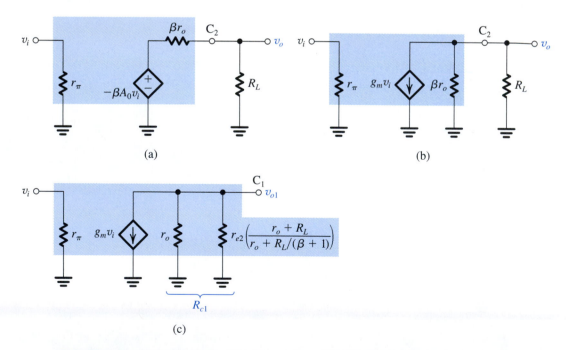

FIGURE 6.41 (a) Equivalent circuit for the cascode amplifier in terms of the open-circuit voltage gain $A_{vo} = -\beta A_0$. (b) Equivalent circuit in terms of the overall short-circuit transconductance $G_m \simeq g_m$. (c) Equivalent circuit for determining the gain of the CE stage, Q_1.

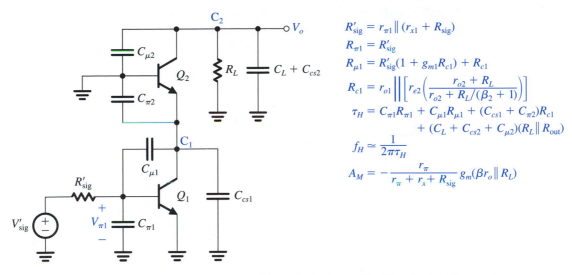

FIGURE 6.42 Determining the frequency response of the BJT cascode amplifier. Note that in addition to the BJT capacitances C_π and C_μ, the capacitance between the collector and the substrate C_{cs} for each transistor are also included.

of the output resistance of Q_1, r_o, and the input resistance of the CB transistor, Q_2, namely R_{in2}. Note that for $R_L \ll r_o$ the latter reduces to r_e, as expected. However, R_{in2} increases as R_L is increased. Of particular interest is the value of R_{in2} obtained for $R_L = \beta r_o$, namely $R_{in2} \cong r_\pi / 2$. It follows that for this value of R_L the CE stage has a voltage gain of $-\beta/2$.

Finally, we present in Fig. 6.42 the circuit and the formulas for determining the high-frequency response of the bipolar cascode. The analysis parallels that studied in the MOSFET case.

EXERCISE

6.28 The objective of this exercise is to evaluate the effect of cascoding on the performance of the CE amplifier of Exercise 6.20. The specifications are as follows: $I = 1$ mA, $\beta = 200$, $r_o = 130$ kΩ, $C_\pi = 16$ pF, $C_\mu = 0.3$ pF, $r_x = 200$ Ω, $C_{cs1} = C_{cs2} = 0$, $C_L = 5$ pF, $R_{sig} = 36$ kΩ, $R_L = 50$ kΩ. Find R_{in}, A_0, R_{out1}, R_{in2}, R_{out}, A_M, f_H and f_t. Compare A_M, f_H, and f_t with the corresponding values obtained in Exercise 6.20 for the CE amplifier. What should C_L be reduced to in order to have $f_H = 1$ MHz?

Ans. 5.2 kΩ; 5200 V/V; 130 kΩ; 35 Ω; 26 MΩ; −238 V/V; 469 kHz; 111.6 MHz. $|A_M|$ has increased from 175 V/V to 238 V/V; f_H has increased from 75 kHz to 469 kHz; f_t has increased from 13.1 MHz to 111.6 MHz. C_L must be reduced to 1.6 pF.

6.8.4 A Cascode Current Source

As mentioned above, to realize the high voltage gain of which the cascode amplifier is capable, the load resistance R_L must be at least on the order of $A_0 r_o$ for the MOSFET cascode or βr_o for the bipolar cascode. Recall, however, that R_L includes the output resistance of the circuit that implements the current-source load I. It follows that the current-source must have an output resistance that is at least $A_0 r_o$ for the MOS case (βr_o for the BJT case). This rules out using the simple current-source circuits of Section 6.2 since their output resistances are

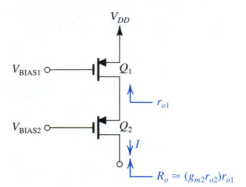

FIGURE 6.43 A cascode current-source.

equal to r_o. Fortunately, there is a conceptually simple and effective solution—namely, applying the cascoding principle to the current-source implementation. The idea is illustrated in Fig. 6.43, where Q_1 is the current-source transistor and Q_2 is the cascode transistor. The dc voltage V_{BIAS1} is chosen so that Q_1 provides the required value of I. V_{BIAS2} is chosen to keep Q_2 and Q_1 in saturation at all times. While the resistance looking into the drain of Q_1 is r_{o1}, the cascode transistor Q_2 multiplies this resistance by $(g_{m2}r_{o2})$ and provides an output resistance for the current source given approximately by

$$R_o \cong (g_{m2}r_{o2})r_{o1} \tag{6.141}$$

A similar arrangement can be used in the bipolar case. We will study a greater variety of current sources and current mirrors with improved performance in Section 6.12.

6.8.5 Double Cascoding

The essence of the operation of the MOS cascode is that the CG cascode transistor Q_2 multiplies the resistance in its source, which is r_o of the CS transistor Q_1, by its intrinsic gain A_{02} to provide an output resistance $A_{02}r_{o1}$. It follows that we can increase the output resistance further by adding another level of cascoding, as illustrated in Fig. 6.44. Here another CG

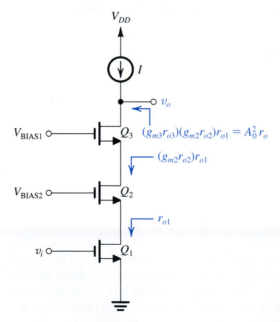

FIGURE 6.44 Double cascoding.

transistor Q_3 is added, with the result that the output resistance is increased by the factor A_{03}. Thus the output resistance of this double-cascode amplifier is $A_0^2 r_o$. Note that an additional bias voltage has to be generated for the additional cascode transistor Q_3.

A drawback of double cascoding is that an additional transistor is now stacked between the power supply rails. Furthermore, since we are now dealing with output resistances on the order of $A_0^2 r_o$, the current source I will also need to be implemented using a double cascode, which adds yet one more transistor to the stack. The difficulty posed by stacking additional transistors is appreciated by recalling that in modern CMOS process technologies V_{DD} is only a little more than 1 V.

Finally, note that since the largest output resistance possible in a bipolar cascode is βr_o, adding another level of cascoding does not provide any advantage.

6.8.6 The Folded Cascode

To avoid the problem of stacking a large number of transistors across a low-voltage power supply, one can use a PMOS transistor for the cascode device, as shown in Fig. 6.45. Here, as before, the NMOS transistor Q_1 is operating in the CS configuration, but the CG stage is implemented using the PMOS transistor Q_2. An additional current-source I_2 is needed to bias Q_2 and provide it with its active load. Note that Q_1 is now operating at a bias current of $(I_1 - I_2)$. Finally, a dc voltage V_{BIAS} is needed to provide an appropriate dc level for the gate of the cascode transistor Q_2. Its value has to be selected so that Q_2 and Q_1 operate in the saturation region.

The small-signal operation of the circuit in Fig. 6.45 is similar to that of the NMOS cascode. The difference here is that the signal current $g_m v_i$ is *folded down* and made to flow into the source terminal of Q_2, which gives the circuit the name **folded cascode.**[12] The folded cascode is a very popular building block in CMOS amplifiers.

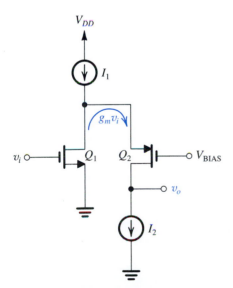

FIGURE 6.45 The folded cascode.

---

[12] The circuit itself can be thought of as having been folded. In this same vein, the regular cascode is sometimes referred to as a **telescopic cascode** because the stacking of transistors resembles the extension of a telescope.

EXERCISE

6.29 Consider the folded-cascode amplifier of Fig. 6.45 for the case: $V_{DD} = 1.8$ V, $k'_p = \frac{1}{4}k'_n$, and $V_{tn} = -V_{tp} = 0.5$ V. To operate Q_1 and Q_2 at equal bias currents I, $I_1 = 2I$ and $I_2 = I$. While current source I_1 is implemented using the simple circuit studied in Section 6.2, current source I_2 is realized using a cascoded circuit (i.e., the NMOS version of the circuit in Fig. 6.43). The transistor W/L ratios are selected so that each operates at an overdrive voltage of 0.2 V.

(a) What must the relationship of $(W/L)_2$ to $(W/L)_1$ be?

(b) What is the minimum dc voltage required for the proper operation of current-source I_1? Now, if a 0.1-V peak-to-peak signal swing is to be allowed at the drain of Q_1, what is the highest dc bias voltage that can be used at that node?

(c) What is the value of V_{SG} of Q_2, and hence what is the largest value to which V_{BIAS} can be set?

(d) What is the minimum dc voltage required for the proper operation of current-source I_2?

(e) Given the results of (c) and (d), what is the allowable range of signal swing at the output?

Ans. (a) $(W/L)_2 = 4(W/L)_1$; (b) 0.2 V, 1.55 V; (c) 0.7 V, 0.85 V; (d) 0.4 V; (e) 0.4 V to 1.35 V

6.8.7 BiCMOS Cascodes

As mentioned before, if the technology permits, the circuit designer can combine bipolar and MOS transistors in circuit configurations that take advantage of the unique features of each. As an example, Fig. 6.46 shows two possibilities for the BiCMOS implementation of the cascode amplifier. In the circuit of Fig. 6.46(a) a MOSFET is used for the input device, thus providing the cascode with an infinite input resistance. On the other hand, a bipolar transistor is used for the cascode device, thus providing a larger output resistance than is

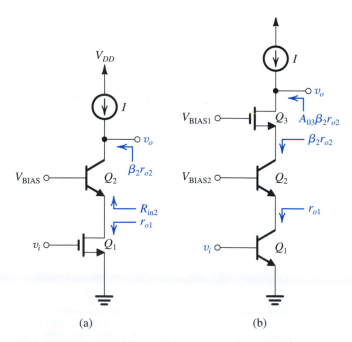

(a)   (b)

FIGURE 6.46 BiCMOS cascodes.

possible with a MOSFET cascode. This is because β of the BJT is usually larger than A_0 of the MOSFET and, more importantly, because r_o of the BJT is much larger than r_o of modern submicron MOSFETs. Also, the bipolar CB transistor provides a lower input resistance R_{in2} than is usually obtained with a CG transistor, especially when R_L is low. The result is a lower total resistance between the drain of Q_1 and ground and hence a reduced Miller effect in Q_1.

The circuit in Fig. 6.46(b) utilizes a MOSFET to implement the second level of cascoding in a bipolar cascode amplifier. The need for a MOSFET stems from the fact that while the maximum possible output resistance obtained with a BJT is βr_o, there is no such limit with the MOSFET, and indeed, Q_3 raises the output resistance by the factor A_{03}.

EXERCISE

6.30 For $I = 100\ \mu A$ find G_m, R_{out}, and the open-circuit voltage gain A_{vo} of the BiCMOS cascode amplifiers in Fig. 6.46. For the BJTs, $V_A = 50$ V and $\beta = 100$. For the MOSFETs, $V_A = 5$ V, $\mu_n C_{ox} = 200\ \mu A/V^2$, and $W/L = 25$.

Ans. For the circuit in Fig. 6.46(a): 1 mA/V, 50 MΩ, -5×10^4 V/V; for the circuit in Fig. 6.46(b): 4 mA/V, 2500 MΩ, -10^7 V/V.

 ## 6.9 THE CS AND CE AMPLIFIERS WITH SOURCE (EMITTER) DEGENERATION

Inserting a relatively small resistance (i.e., a small multiple of $1/g_m$) in the source of a CS amplifier (the emitter of a common-emitter amplifier) introduces negative feedback into the amplifier stage. As a result this resistance provides the circuit designer with an additional parameter that can be effectively utilized to obtain certain desirable properties as a trade-off for the gain reduction that source (emitter) degeneration causes. We have already seen some of this in Sections 4.7 and 5.7. In this section we consider source and emitter degeneration in IC amplifiers where r_o and g_{mb} have to be taken into account. We also demonstrate the use of source (emitter) degeneration to extend the amplifier bandwidth.

6.9.1 The CS Amplifier with a Source Resistance

Figure 6.47(a) shows an active-loaded CS amplifier with a source resistance R_s. Note that a signal v_{bs} will develop between body and source, and hence the body effect should be taken into account in the analysis. The circuit, prepared for small-signal analysis and with a resistance R_L shown at the output, is presented in Fig. 6.47(b). To determine the output resistance R_{out}, we reduce v_i to zero, which makes the circuit identical to that of a CG amplifier. Therefore we can obtain R_{out} by using Eq. (6.101) as

$$R_{out} = r_o + [1 + (g_m + g_{mb})r_o]R_s \qquad (6.142)$$

which for the usual situation $(g_m + g_{mb})r_o \gg 1$ reduces to

$$R_{out} \cong r_o[1 + (g_m + g_{mb})R_s] \qquad (6.143)$$

The open-circuit voltage gain can be found from the circuit in Fig. 6.47(c). Noting that the current in R_s must be zero, the voltage at the source, v_s, will be zero and thus $v_{gs} = v_i$ and $v_{bs} = 0$, resulting in

$$i = g_m v_{gs}$$

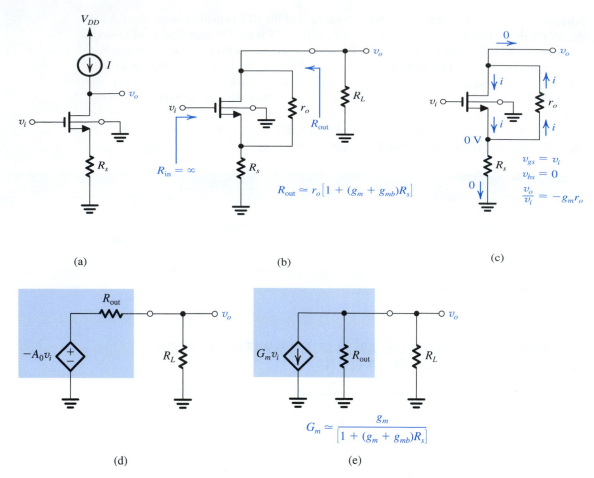

FIGURE 6.47 (a) A CS amplifier with a source-degeneration resistance R_s. (b) Circuit for small-signal analysis. (c) Circuit with the output open to determine A_{vo}. (d) Output equivalent circuit. (e) Another output equivalent circuit in terms of G_m.

and

$$v_o = -ir_o = -g_m r_o v_{gs} = -g_m r_o v_i$$

Thus,

$$A_{vo} = -g_m r_o = -A_0$$

In other words, *the resistance R_s has no effect on A_{vo}!*

Utilizing $A_{vo} = -A_0$ and R_{out} from Eq. (6.143) enables us to obtain the amplifier output equivalent circuit shown in Fig. 6.47(d). An alternative equivalent circuit in terms of the short-circuit transconductance G_m is shown in Fig. 6.47(e), where G_m can be found from

$$G_m = \frac{|A_{vo}|}{R_{out}} = \frac{g_m r_o}{r_o[1 + (g_m + g_{mb})R_s]}$$

Thus,

$$G_m = \frac{g_m}{1 + (g_m + g_{mb})R_s} \tag{6.144}$$

The effect of R_s is thus obvious: R_s *reduces the amplifier transconductance and increases its output resistance by the same factor:* $[1 + (g_m + g_{mb})R_s]$. We will find in Chapter 8 when we study negative feedback formally that this factor is the *amount of negative feedback* introduced by R_s.

The voltage gain A_v can be found as

$$A_v = -A_{vo}\frac{R_L}{R_L + R_{\text{out}}} \tag{6.145}$$

Thus, if R_L is kept unchanged, A_v will decrease, which is the price paid for the performance improvements obtained when R_s is introduced. One such improvement is in the linearity of the amplifier. This comes about because only a fraction v_{gs} of the input signal v_i now appears between gate and source. Derivation of an expression for v_{gs}/v_i is significantly complicated by the inclusion of r_o. The derivation should be done with the MOSFET equivalent-circuit model explicitly used. The result is

$$\frac{v_{gs}}{v_i} \cong \frac{1}{1 + (g_m + g_{mb})R_s}\frac{R_L \| R_{\text{out}}}{R_L \| r_o} \tag{6.146}$$

which for $r_o \gg R_L$ reduces to the familiar relationship

$$\frac{v_{gs}}{v_i} \cong \frac{1}{1 + (g_m + g_{mb})R_s} \tag{6.147}$$

Thus the value of R_s can be used to control the magnitude of v_{gs} so as to obtain the desired linearity—at the expense, of course, of gain reduction.

Frequency Response Another advantage of source degeneration is the ability to broaden the amplifier bandwidth. Figure 6.48(a) shows the amplifier with the internal capacitances C_{gs} and C_{gd} indicated. A capacitance C_L that *includes* the MOSFET capacitance C_{db} is also shown at the output. The method of open-circuit time constants can be employed to obtain an estimate of the 3-dB frequency f_H. Toward that end we show in Fig. 6.48(b) the circuit for determining R_{gd}, which is the resistance seen by C_{gd}. We observe that R_{gd} can be determined by simply adapting the formula in Eq. (6.56) to the case with source degeneration as follows:

$$R_{gd} = R_{\text{sig}}(1 + G_m R'_L) + R'_L \tag{6.148}$$

where

$$R'_L = R_L \| R_{\text{out}} \tag{6.149}$$

The formula for R_{C_L} can be seen to be simply

$$R_{C_L} = R_L \| R_{\text{out}} = R'_L \tag{6.150}$$

The formula for R_{gs} is the most difficult to derive, and the derivation should be performed with the hybrid-π model explicitly utilized. The result is

$$R_{gs} \cong \frac{R_{\text{sig}} + R_s}{1 + (g_m + g_{mb})R_s\left(\dfrac{r_o}{r_o + R_L}\right)} \tag{6.151}$$

When R_{sig} is relatively large, the frequency response will be dominated by the Miller multiplication of C_{gd}. Another way for saying this is that $C_{gd}R_{gd}$ will be the largest of the

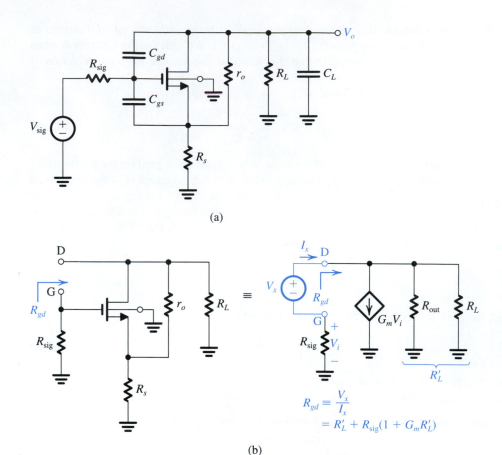

FIGURE 6.48 (a) The CS amplifier circuit, with a source resistance R_s, prepared for frequency-response analysis. (b) Determining the resistance R_{gd} seen by the capacitance C_{gd}.

three open-circuit time constants that make up τ_H,

$$\tau_H = C_{gs}R_{gs} + C_{gd}R_{gd} + C_L R_{C_L} \tag{6.152}$$

enabling us to approximate τ_H as

$$\tau_H \cong C_{gd}R_{gd} \tag{6.153}$$

and correspondingly to obtain f_H as

$$f_H \cong \frac{1}{2\pi C_{gd}R_{gd}} \tag{6.154}$$

Now, as R_s is increased, the gain magnitude, $|A_M| = G_m R_L'$, will decrease, causing R_{gd} to decrease (Eq. 6.148), which in turn causes f_H to increase (Eq. 6.154). To highlight the trade-off between gain and bandwidth that R_s affords the designer, let us simplify the expression for R_{gd} in Eq. (6.148) by assuming that $G_m R_L' \gg 1$ and $G_m R_{sig} \gg 1$,

$$R_{gd} \cong G_m R_L' R_{sig} = |A_M| R_{sig}$$

which can be substituted in Eq. (6.154) to obtain

$$f_H = \frac{1}{2\pi C_{gd} R_{sig} |A_M|} \tag{6.155}$$

which very clearly shows the gain–bandwidth trade-off. The gain–bandwidth product remains constant at

$$\text{Gain–bandwidth product, } f_t = |A_M| f_H = \frac{1}{2\pi C_{gd} R_{sig}} \tag{6.156}$$

In practice, however, the other capacitances will play a role in determining f_H, and f_t will decrease somewhat as R_s is increased.

EXERCISE

6.31 Consider a CS amplifier having $g_m = 2$ mA/V, $r_o = 20$ kΩ, $R_L = 20$ kΩ, $R_{sig} = 20$ kΩ, $C_{gs} = 20$ fF, $C_{gd} = 5$ fF, and $C_L = 5$ fF. (a) Find the voltage gain A_M and the 3-dB frequency f_H (using the method of open-circuit time constants) and hence the gain–bandwidth product. (b) Repeat (a) for the case in which a resistance R_s is connected in series with the source terminal with a value selected so that $(g_m + g_{mb})R_s = 2$.
Ans. (a) –20 V/V, 61.2 MHz, 1.22 GHz; (b) –10 V/V, 109.1 MHz, 1.1 GHz

6.9.2 The CE Amplifier with an Emitter Resistance

Emitter degeneration is even more useful in the CE amplifier than source degeneration is in the CS amplifier. This is because emitter degeneration increases the input resistance of the CE amplifier. The input resistance of the CS amplifier is, of course, practically infinite to start with. Figure 6.49(a) shows an active-loaded CE amplifier with an emitter resistance R_e, usually in the range of 1 to 5 times r_e. Figure 6.49(b) shows the circuit for determining the

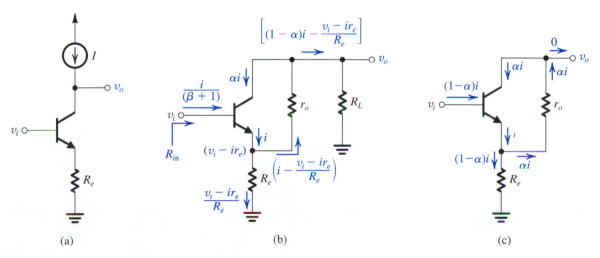

FIGURE 6.49 A CE amplifier with emitter degeneration: **(a)** circuit; **(b)** analysis to determine R_{in}; and **(c)** analysis to determine A_{vo}.

input resistance R_{in}, which due to the presence of r_o will depend on the value of R_L. With the aid of the analysis shown in Fig. 6.49(b), we can express the output voltage v_o as

$$v_o = \left[(1-\alpha)i - \frac{v_i - ir_e}{R_e}\right]R_L$$

Alternatively, we can express v_o as

$$v_o = (v_i - ir_e) - r_o\left[i - \frac{v_i - ir_e}{R_e}\right]$$

Equating these two expressions of v_o yields an equation in v_i and i, which can be rearranged to obtain

$$R_{\text{in}} = \frac{v_i}{i/(\beta+1)}$$

$$= (\beta+1)r_e + (\beta+1)R_e \frac{r_o + \dfrac{R_L}{\beta+1}}{r_o + R_L + R_e} \tag{6.157}$$

Usually R_L is on the order of r_o; thus $R_L/(\beta+1) \ll r_o$. Also, $R_e \ll r_o$. Taking account of these two conditions enables us to simplify the expression for R_{in} to

$$R_{\text{in}} \cong (\beta+1)r_e + (\beta+1)R_e \frac{1}{1 + R_L/r_o} \tag{6.158}$$

This expression indicates that the presence of r_o reduces the effect of R_e on increasing R_{in}. This is because r_o shunts away some of the current that would have flowed through R_e. For example, for $R_L = r_o$, $R_i = (\beta+1)(r_e + 0.5R_e)$.

To determine the open-circuit voltage gain A_{vo}, we utilize the circuit shown in Fig. 6.49(c). Analysis of this circuit is straightforward and can be shown to yield

$$A_{vo} \cong -g_m r_o \tag{6.159}$$

That is, the open-circuit voltage gain obtained with a relatively small R_e (i.e., on the order of r_e) remains very close to the value without R_e.

The output resistance R_o is identical to the value of R_{out} that we derived for the CB circuit (Eq. 6.118),

$$R_o \cong r_o(1 + g_m R_e') \tag{6.160}$$

where $R_e' = R_e \| r_\pi$. Since R_e is on the order of r_e, R_e is much smaller than r_π and $R_e' \cong R_e$. Thus,

$$R_o \cong r_o(1 + g_m R_e) \tag{6.161}$$

The expressions for R_{in}, A_{vo}, and R_o in Eqs. (6.158), (6.159), and (6.161), respectively, can be used to determine the overall voltage gain for given values of source resistance and load resistance. Finally, we should mention that A_{vo} and R_o can be used to find the effective short-circuit transconductance G_m of the emitter-degenerated CE amplifier as follows:

$$G_m = -\frac{A_{vo}}{R_o}$$

Thus,

$$G_m = \frac{g_m}{1 + g_m R_e} \tag{6.162}$$

which is identical to the expression we found for the discrete case in Section 5.7.

The high-frequency response of the CE amplifier with emitter degeneration can be found in a manner similar to that presented above for the CS amplifier.

In summary, including a relatively small resistance R_e (i.e., a small multiple of r_e) in the emitter of the active-loaded CE amplifier reduces its effective transconductance by the factor $(1 + g_m R_e)$ and increases its output resistance by the same factor, thus leaving the open-circuit voltage gain approximately unchanged. The input resistance R_{in} is increased by a factor that depends on R_L and that is somewhat lower than $(1 + g_m R_e)$. Also, including R_e reduces the severity of the Miller effect and correspondingly increases the amplifier bandwidth. Finally, an emitter-degeneration resistance R_e increases the linearity of the amplifier.

EXERCISE

6.32 Consider the active-loaded CE amplifier with emitter degeneration. Let $I = 1$ mA, $V_A = 100$ V, and $\beta = 100$. Find R_{in}, R_o, A_{vo}, G_m, and the overall voltage gain v_o/v_{sig} when $R_e = 75$ Ω, $R_{\text{sig}} = 5$ kΩ, and $R_L = 2r_o$.

Ans. 5 kΩ; 400 kΩ; −4000 V/V; 10 mA/V; −667 V/V

 ## 6.10 THE SOURCE AND EMITTER FOLLOWERS

The discrete-circuit source follower was presented in Section 4.7.6 and the discrete-circuit emitter follower in Section 5.7.6. In the following discussion we consider their IC versions, paying special attention to their high-frequency response.

6.10.1 The Source Follower

Figure 6.50(a) shows an IC source follower biased by a constant-current source I, which is usually implemented using an NMOS current mirror. The source follower would generally be part of a larger circuit that determines the dc voltage at the transistor gate. We will encounter such circuits in the following chapters. Here we note that v_i is the input signal appearing at the gate and that R_L represents the combination of a load resistance and the output resistance of the current-source I.

The low-frequency small-signal model of the source follower is shown in Fig. 6.50(b). Observe that r_o appears in parallel with R_L and thus can be combined with it. Also, the controlled current-source $g_{mb} v_{bs}$ feeds its current into the source terminal, where the voltage is $-v_{bs}$. Thus we can use the source-absorption theorem (Appendix C) to replace the current source with a resistance $1/g_{mb}$ between the source and ground, this can then be combined with R_L and r_o. With these two simplifications, the equivalent circuit takes the form shown in Fig. 6.50(c), where

$$R_L' = R_L \parallel r_o \parallel \frac{1}{g_{mb}} \tag{6.163}$$

We now can write for the output voltage v_o,

$$v_o = g_m v_{gs} R_L' \tag{6.164}$$

and for v_{gs},

$$v_{gs} = v_i - v_o \tag{6.165}$$

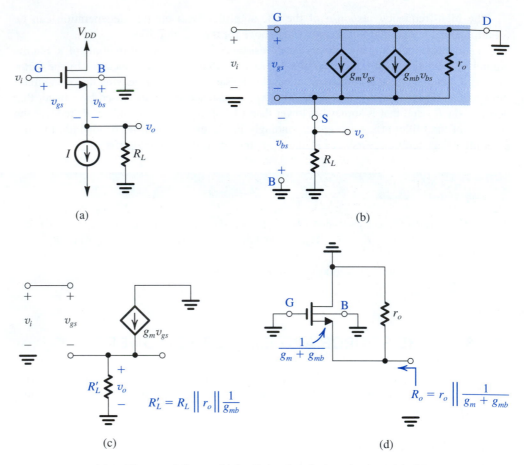

FIGURE 6.50 (a) An IC source follower. (b) Small-signal equivalent-circuit model of the source follower. (c) A simplified version of the equivalent circuit. (d) Determining the output resistance of the source follower.

Equations (6.164) and (6.165) can be combined to obtain the voltage gain

$$A_v \equiv \frac{v_o}{v_i} = \frac{g_m R'_L}{1 + g_m R'_L} \tag{6.166}$$

which, as expected, is less than unity. To obtain the open-circuit voltage gain, we set R_L in Eq. (6.163) to ∞, which reduces R'_L to $r_o \parallel (1/g_{mb})$. Substituting this value for R'_L in Eq. (6.166) gives

$$A_{vo} = \frac{g_m r_o}{1 + (g_m + g_{mb}) r_o} \tag{6.167}$$

which, for the usual case where $(g_m + g_{mb}) r_o \gg 1$, simplifies to

$$A_{vo} \cong \frac{g_m}{g_m + g_{mb}} = \frac{1}{1 + \chi} \tag{6.168}$$

Thus the highest value possible for the voltage gain of the source follower is limited to $1/(1 + \chi)$, which is typically 0.8 V/V to 0.9 V/V.

Finally, we can find the output resistance R_o of the source follower either using the equivalent circuit of Fig. 6.50(c) or by inspection of the circuit in Fig. 6.50(d) as

$$R_o = \frac{1}{g_m + g_{mb}} \parallel r_o \tag{6.169}$$

which can be approximated as

$$R_o \cong 1/[(1+\chi)g_m] \tag{6.170}$$

Similar to the discrete source follower, the IC source follower can be used as the output stage of a multistage amplifier to provide a low output resistance for driving low-impedance loads. It is also used to shift the dc level of the signal by an amount equal to V_{GS}.

6.10.2 Frequency Response of the Source Follower

A major advantage of the source follower is its excellent high-frequency response. This comes about because, as we shall now see, none of the internal capacitances suffers from the Miller effect. Figure 6.51(a) shows the high-frequency equivalent circuit of a source follower fed with a signal V_{sig} from a source having a resistance R_{sig}. In addition to the MOSFET capacitances C_{gs} and C_{gd}, a capacitance C_L is included between the output node and ground to account for the source-to-body capacitance C_{sb} as well as any actual load capacitance.

The simplifications performed above on the low-frequency equivalent circuit can be applied to the high-frequency model of Fig. 6.51(a) to obtain the equivalent circuit in Fig. 6.51(b), where R_L' is given by Eq. (6.163). Although one can derive an expression for the transfer function of this circuit, the resulting expression will be too complicated to yield insight regarding the role that each of the three capacitances plays. Rather, we shall first determine the location of the transmission zeros and then use the method of open-circuit time constants to estimate the 3-dB frequency, f_{3dB}.

Although there are three capacitances in the circuit of Fig. 6.51(b), the transfer function is of the second order. This is because the three capacitances form a continuous loop. To determine the location of the two transmission zeros, refer to the circuit in Fig. 6.51(b), and note that V_o is zero at the frequency at which C_L has a zero impedance and thus acts as a short circuit across the output, which is ω or $s = \infty$. Also, V_o will be zero at the value of s that causes the current into the impedance $R_L' \parallel C_L$ to be zero. Since this current is $(g_m + sC_{gs})V_{gs}$, the transmission zero will be at $s = s_Z$, where

$$s_Z = -\frac{g_m}{C_{gs}} \tag{6.171}$$

That is, the zero will be on the negative real-axis of the s-plane with a frequency

$$\omega_Z = \frac{g_m}{C_{gs}} \tag{6.172}$$

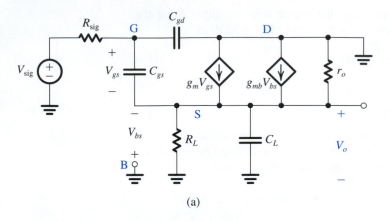

(a)

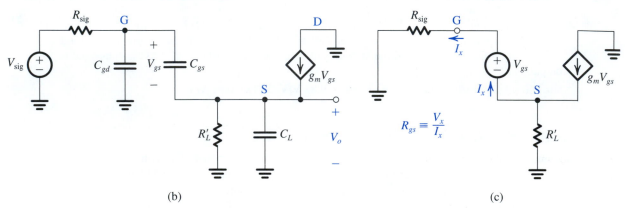

(b)   (c)

FIGURE 6.51 Analysis of the high-frequency response of the source follower: **(a)** Equivalent circuit; **(b)** simplified equivalent circuit; and **(c)** determining the resistance R_{gs} seen by C_{gs}.

Recalling that the MOSFET's $\omega_T = g_m/(C_{gs} + C_{gd})$ and that $C_{gd} \ll C_{gs}$, we see that ω_Z will be very close to ω_T,

$$f_Z \cong f_T \tag{6.173}$$

Next, we turn our attention to the poles. Specifically, we will find the resistance seen by each of three capacitances C_{gd}, C_{gs}, and C_L and then compute the time constant associated with each. With V_{sig} set to zero and C_{gs} and C_L assumed to be open circuited, we find by inspection that the resistance R_{gd} seen by C_{gd} is given by

$$R_{gd} = R_{sig} \tag{6.174}$$

This is intuitively obvious: Because of the ground at the drain terminal, the input capacitance of the source follower, in the absence of C_{gs} and C_L, is equal to C_{gd}. Thus, R_{sig} and C_{gd} form a high-frequency pole.

Next, we consider the effect of C_{gs}. The resistance R_{gs} seen by C_{gs} can be determined by straightforward analysis of the circuit in Fig. 6.51(c) to obtain

$$R_{gs} = \frac{R_{sig} + R'_L}{1 + g_m R'_L} \tag{6.175}$$

We note that the factor $(1 + g_m R'_L)$ in the denominator will result in reducing the effective resistance with which C_{gs} interacts. In the absence of the two other capacitances, C_{gs} together with R_{gs} introduce a pole with frequency $1/2\pi C_{gs} R_{gs}$.

Finally, it is easy to see from the circuit in Fig. 6.51(b) that C_L interacts with $R_L \parallel R_o$; that is,

$$R_{C_L} = R_L \parallel R_o$$

Usually, R_o (Eq. 6.169) is low. Thus R_{C_L} will be low, and the effect of C_L will be small. Nevertheless, all three time constants can be added to obtain τ_H and hence f_H,

$$f_H = \frac{1}{2\pi\tau_H} = 1/2\pi(C_{gd} R_{sig} + C_{gs} R_{gs} + C_L R_{C_L}) \tag{6.176}$$

EXERCISE

6.34 Consider a source follower specified as follows: $W/L = 7.2\ \mu m/0.36\ \mu m$, $I_D = 100\ \mu A$, $g_m = 1.25$ mA/V, $\chi = 0.2$, $r_o = 20$ kΩ, $R_{sig} = 20$ kΩ, $R_L = 10$ kΩ, $C_{gs} = 20$ fF, $C_{gd} = 5$ fF, and $C_L = 15$ fF. Find A_v, f_T, and f_Z. Also, find R_{gd}, R_{gs}, R_{C_L}, and hence the time constant associated with each of the three capacitances C_{gd}, C_{gs}, and C_L. Find τ_H and the percentage contribution to it from each of three capacitances. Find f_H.

Ans. 0.76 V/V; 8 GHz; 10 GHz; 20 kΩ; 5.45 kΩ; 0.61 kΩ; 100 ps; 109 ps; 9 ps; 218 ps; 46%; 50%; 4%; 730 MHz

6.10.3 The Emitter Follower

Figure 6.52(a) shows an emitter follower suitable for IC fabrication. It is biased by a constant-current source I. However, the circuit that sets the dc voltage at the base is not shown. The emitter follower is fed with a signal V_{sig} from a source with resistance R_{sig}. The resistance R_L, shown at the output, includes the output resistance of current source I as well as any actual load resistance.

Analysis of the emitter follower of Fig. 6.52(a) to determine its low-frequency gain, input resistance, and output resistance is identical to that performed on the capacitively coupled version in Section 5.7.6. Indeed, the formulas given in Table 5.6 can be easily adapted for the circuit in Fig. 6.52(a). Therefore we shall concentrate here on the analysis of the high-frequency response of the circuit.

Figure 6.52(b) shows the high-frequency equivalent circuit. Lumping r_o together with R_L and r_x together with R_{sig} and making a slight change in the way the circuit is drawn results in the simplified equivalent circuit shown in Fig. 6.52(c). We will follow a procedure for the analysis of this circuit similar to that used above for the source follower. Specifically, to obtain the location of the transmission zero, note that V_o will be zero at the frequency s_Z for which the current fed to R'_L is zero:

$$g_m V_\pi + \frac{V_\pi}{r_\pi} + s_Z C_\pi = 0$$

Thus,

$$s_Z = -\frac{g_m + (1/r_\pi)}{C_\pi} = -\frac{1}{C_\pi r_e} \tag{6.177}$$

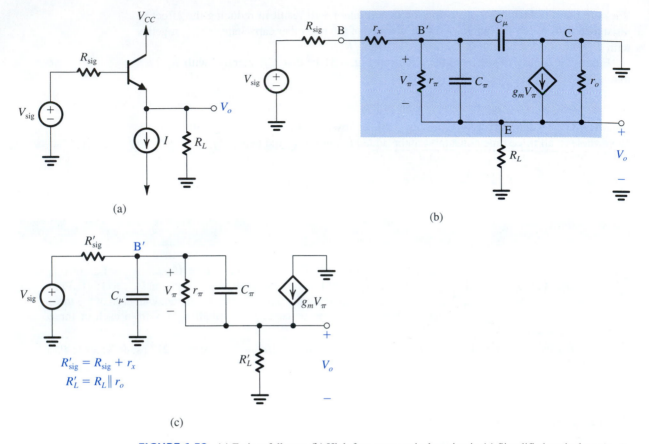

(a)

(b)

(c)

FIGURE 6.52 (a) Emitter follower. (b) High-frequency equivalent circuit. (c) Simplified equivalent circuit.

which is on the negative real-axis of the *s*-plane and has a frequency

$$\omega_Z = \frac{1}{C_\pi r_e} \tag{6.178}$$

This frequency is very close to the unity-gain frequency ω_T of the transistor. The other transmission zero is at $s = \infty$. This is because at this frequency, C_μ acts as a short circuit, making V_π zero, and hence V_o will be zero.

Next, we determine the resistances seen by C_μ and C_π. For C_μ the reader should be able to show that the resistance it sees, R_μ, is the parallel equivalent of R'_{sig} and the input resistance looking into B′; that is,

$$R_\mu = R'_{sig} \| [r_\pi + (\beta + 1)R'_L] \tag{6.179}$$

Equation (6.179) indicates that R_μ will be smaller than R'_{sig}, and since C_μ is usually very small, the time constant $C_\mu R_\mu$ will be correspondingly small.

The resistance R_π seen by C_π can be determined using an analysis similar to that employed for the determination of R_{gs} in the MOSFET case. The result is

$$R_\pi = \frac{R'_{sig} + R'_L}{1 + \dfrac{R'_{sig}}{r_\pi} + \dfrac{R'_L}{r_e}} \tag{6.180}$$

We observe that the term R_L'/r_e will usually make the denominator much greater than unity, thus rendering R_π rather low. Thus, the time constant $C_\pi R_\pi$ will be small. The end result is that the 3-dB frequency f_H of the emitter follower,

$$f_H = 1/2\pi[C_\mu R_\mu + C_\pi R_\pi] \qquad (6.181)$$

will usually be very high. We urge the reader to solve the following exercise to gain familiarity with typical values of the various parameters that determine f_H.

EXERCISE

6.35 For an emitter follower biased at $I_C = 1$ mA and having $R_{\text{sig}} = R_L = 1$ kΩ, $r_o = 100$ kΩ, $\beta = 100$, $C_\mu = 2$ pF, and $f_T = 400$ MHz, find the low-frequency gain, f_Z, R_μ, R_π, and f_H.
Ans. 0.965 V/V; 458 MHz; 1.09 kΩ; 51 Ω; 55 MHz

 ## 6.11 SOME USEFUL TRANSISTOR PAIRINGS

The cascode configuration studied in Section 6.8 combines CS and CG MOS transistors (CE and CB bipolar transistors) to great advantage. The key to the superior performance of the resulting combination is that the transistor pairing is done in a way that maximizes the advantages and minimizes the shortcomings of each of the two individual configurations. In this section we study a number of other such transistor pairings. In each case the transistor pair can be thought of as a compound device; thus the resulting amplifier may be considered as a single stage.

6.11.1 The CD–CS, CC–CE and CD–CE Configurations

Figure 6.53(a) shows an amplifier formed by cascading a common-drain (source-follower) transistor Q_1 with a common-source transistor Q_2. As should be expected, the voltage gain of the circuit will be a little lower than that of the CS amplifier. The advantage of this circuit

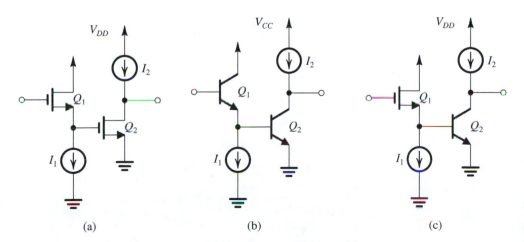

FIGURE 6.53 (a) CD–CS amplifier. (b) CC–CE amplifier. (c) CD–CE amplifier.

configuration, however, lies in its bandwidth, which is much wider than that obtained in a CS amplifier. To see how this comes about, note that the CS transistor Q_2 will still exhibit a Miller effect that results in a large input capacitance, C_{in2}, between its gate and ground. However, the resistance that this capacitance interacts with will be much lower than R_{sig}; the buffering action of the source follower causes a relatively low resistance, approximately equal to a $1/(g_{m1} + g_{mb1})$, to appear between the source of Q_1 and ground across C_{in2}.

The bipolar counterpart of the CD–CS circuit is shown in Fig. 6.53(b). Beside achieving a wider bandwidth than that obtained with a CE amplifier, the CC–CE configuration has an important additional advantage: The input resistance is increased by a factor equal to $(\beta_1 + 1)$. Finally, we show in Fig. 6.53(c) the BiCMOS version of this circuit type. Observe that Q_1 provides the amplifier with an infinite input resistance. Also, note that Q_2 provides the amplifier with a high g_m as compared to that obtained in the MOSFET circuit in Fig. 6.53(a) and hence high gain.

EXAMPLE 6.13

Consider a CC–CE amplifier such as that in Fig. 6.53(b) with the following specifications: $I_1 = I_2 = 1$ mA and identical transistors with $\beta = 100$, $f_T = 400$ MHz, and $C_\mu = 2$ pF. Let the amplifier be fed with a source V_{sig} having a resistance $R_{sig} = 4$ kΩ, and assume a load resistance of 4 kΩ. Find the voltage gain A_M, and estimate the 3-dB frequency, f_H. Compare the results with those obtained with a CE amplifier operating under the same conditions. For simplicity, neglect r_o and r_x.

Solution

At an emitter bias current of 1 mA, Q_1 and Q_2 have

$$g_m = 40 \text{ mA/V}$$

$$r_e = 25 \ \Omega$$

$$r_\pi = \frac{\beta}{g_m} = \frac{100}{40} = 2.5 \text{ k}\Omega$$

$$C_\pi + C_\mu = \frac{g_m}{\omega_T} = \frac{g_m}{2\pi f_T}$$

$$= \frac{40 \times 10^{-3}}{2\pi \times 400 \times 10^6} = 15.9 \text{ pF}$$

$$C_\mu = 2 \text{ pF}$$

$$C_\pi = 13.9 \text{ pF}$$

The voltage gain A_M can be determined from the circuit shown in Fig. 6.54(a) as follows:

$$R_{in2} = r_{\pi 2} = 2.5 \text{ k}\Omega$$

$$R_{in} = (\beta_1 + 1)(r_{e1} + R_{in2})$$

$$= 101(0.025 + 2.5) = 255 \text{ k}\Omega$$

$$\frac{V_{b1}}{V_{sig}} = \frac{R_{in}}{R_{in} + R_{sig}} = \frac{255}{255 + 4} = 0.98 \text{ V/V}$$

$$\frac{V_{b2}}{V_{b1}} = \frac{R_{in2}}{R_{in2} + r_{e1}} = \frac{2.5}{2.5 + 0.025} = 0.99 \text{ V/V}$$

$$\frac{V_o}{V_{b2}} = -g_{m2}R_L = -40 \times 4 = -160 \text{ V/V}$$

Thus,

$$A_M = \frac{V_o}{V_{sig}} = -160 \times 0.99 \times 0.98 = -155 \text{ V/V}$$

To determine f_H we use the method of open-circuit time constants. Figure 6.54(b) shows the circuit with V_{sig} set to zero and the four capacitances indicated. Capacitance $C_{\mu 1}$ sees a resistance $R_{\mu 1}$,

$$R_{\mu 1} = R_{sig} \| R_{in}$$
$$= 4 \| 255 = 3.94 \text{ k}\Omega$$

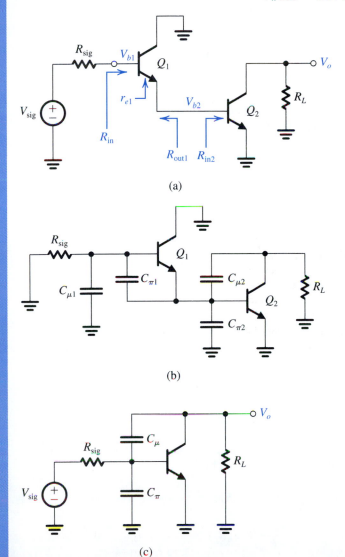

(a)

(b)

(c)

FIGURE 6.54 Circuits for Example 6.13: **(a)** The CC–CE circuit prepared for low-frequency small-signal analysis; **(b)** the circuit at high frequencies, with V_{sig} set to zero to enable determination of the open-circuit time constants; and **(c)** a CE amplifier for comparison.

To find the resistance $R_{\pi 1}$ seen by capacitance $C_{\pi 1}$ we refer to the analysis of the high-frequency response of the emitter follower in Section 6.10.3. Specifically, we adapt Eq. (6.180) to the situation here as follows:

$$R_{\pi 1} = \frac{R_{\text{sig}} + R_{\text{in2}}}{1 + \dfrac{R_{\text{sig}}}{r_{\pi 1}} + \dfrac{R_{\text{in2}}}{r_{e1}}}$$

$$= \frac{4000 + 2500}{1 + \dfrac{4000}{2500} + \dfrac{2500}{25}} = 63.4 \ \Omega$$

Capacitance $C_{\pi 2}$ sees a resistance $R_{\pi 2}$,

$$R_{\pi 2} = R_{\text{in2}} \ \| \ R_{\text{out1}}$$

$$= r_{\pi 2} \ \| \ \left[r_{e1} + \frac{R_{\text{sig}}}{\beta_1 + 1} \right]$$

$$= 2500 \ \| \ \left[25 + \frac{4000}{101} \right] = 63 \ \Omega$$

Capacitance $C_{\mu 2}$ sees a resistance $R_{\mu 2}$. To determine $R_{\mu 2}$ we refer to the analysis of the frequency response of the CE amplifier in Section 6.6 to obtain

$$R_{\mu 2} = (1 + g_{m2} R_L)(R_{\text{in2}} \ \| \ R_{\text{out1}}) + R_L$$

$$= (1 + 40 \times 4) \left[2500 \ \| \ \left(25 + \frac{4000}{101} \right) \right] + 4000$$

$$= 14{,}143 \ \Omega \cong 14.1 \ \text{k}\Omega$$

We now can determine τ_H from

$$\tau_H = C_{\mu 1} R_{\mu 1} + C_{\pi 1} R_{\pi 1} + C_{\mu 2} R_{\mu 2} + C_{\pi 2} R_{\pi 2}$$

$$= 2 \times 3.94 + 13.9 \times 0.0634 + 2 \times 14.1 + 13.9 \times 0.063$$

$$= 7.88 + 0.88 + 28.2 + 0.88 = 37.8 \ \text{ns}$$

We observe that $C_{\pi 1}$ and $C_{\pi 2}$ play a very minor role in determining the high-frequency response. As expected, $C_{\mu 2}$ through the Miller effect plays the most significant role. Also, $C_{\mu 1}$, which interacts directly with $(R_{\text{sig}} \ \| \ R_{\text{in}})$, also plays an important role. The 3-dB frequency f_H can be found as follows:

$$f_H = \frac{1}{2\pi \tau_H} = \frac{1}{2\pi \times 37.8 \times 10^{-9}} = 4.2 \ \text{MHz}$$

For comparison, we evaluate A_M and f_H of a CE amplifier operating under the same conditions. Refer to Fig. 6.54(c). The voltage gain A_M is given by

$$A_M = \frac{R_{\text{in}}}{R_{\text{in}} + R_{\text{sig}}} (-g_m R_L)$$

$$= \left(\frac{r_\pi}{r_\pi + R_{\text{sig}}} \right) (-g_m R_L)$$

$$= \left(\frac{2.5}{2.5 + 4} \right) (-40 \times 4)$$

$$= -61.5 \ \text{V/V}$$

$$R_\pi = r_\pi \| R_{sig} = 2.5 \| 4 = 1.54 \text{ k}\Omega$$

$$R_\mu = (1 + g_m R_L)(R_{sig} \| r_\pi) + R_L$$

$$= (1 + 40 \times 4)(4 \| 2.5) + 4$$

$$= 251.7 \text{ k}\Omega$$

Thus,

$$\tau_H = C_\pi R_\pi + C_\mu R_\mu$$

$$= 13.9 \times 1.54 + 2 \times 251.7$$

$$= 21.4 + 503.4 = 524.8 \text{ ns}$$

Observe the dominant role played by C_μ. The 3-dB frequency f_H is

$$f_H = \frac{1}{2\pi\tau_H} = \frac{1}{2\pi \times 524.8 \times 10^{-9}} = 303 \text{ kHz}$$

Thus, including the buffering transistor Q_1 increases the gain, $|A_M|$, from 61.5 V/V to 155 V/V—a factor of 2.5—and increases the bandwidth from 303 kHz to 4.2 MHz—a factor of 13.9! The gain–bandwidth product is increased from 18.63 MHz to 651 MHz—a factor of 35!

6.11.2 The Darlington Configuration

Figure 6.55(a) shows a popular BJT circuit known as the **Darlington configuration.** It can be thought of as a variation of the CC–CE circuit with the collector of Q_1 connected to that of Q_2. Alternatively, the Darlington pair can be thought of as a composite transistor with $\beta = \beta_1\beta_2$. It can therefore be used to implement a high-performance voltage follower, as illustrated in Fig. 6.55(b). Note that in this application the circuit can be considered as the cascade connection of two common-collector transistors (i.e., a CC–CC configuration).

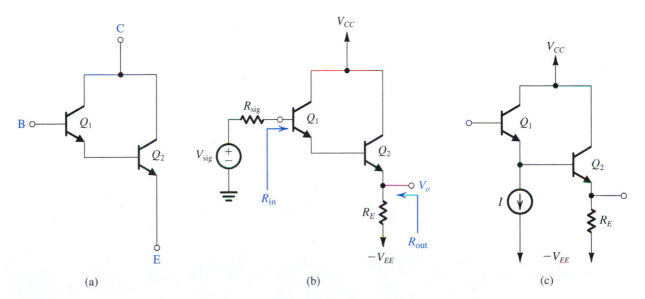

(a)   (b)   (c)

FIGURE 6.55 (a) The Darlington configuration; (b) voltage follower using the Darlington configuration; and (c) the Darlington follower with a bias current I applied to Q_1 to ensure that its β remains high.

Since the transistor β depends on the dc bias current, it is possible that Q_1 will be operating at a very low β, rendering the β-multiplication effect of the Darlington pair rather ineffective. A simple solution to this problem is to provide a bias current for Q_1, as shown in Fig. 6.55(c).

EXERCISE

6.36 For the Darlington voltage follower in Fig. 6.55(b), show that:

$$R_{in} = (\beta_1 + 1)[r_{e1} + (\beta_2 + 1)(r_{e2} + R_E)]$$

$$R_{out} = R_E \left\| \left[r_{e2} + \frac{r_{e1} + [R_{sig}/(\beta_1 + 1)]}{\beta_2 + 1} \right] \right.$$

$$\frac{V_o}{V_{sig}} = \frac{R_E}{R_E + r_{e2} + [r_{e1} + R_{sig}/(\beta_1 + 1)]/(\beta_2 + 1)}$$

Evaluate R_{in}, R_{out}, and V_o/V_{sig} for the case $I_{E2} = 5$ mA, $\beta_1 = \beta_2 = 100$, $R_E = 1$ kΩ, and $R_{sig} = 100$ kΩ.

Ans. 10.3 MΩ; 20 Ω; 0.98 V/V

6.11.3 The CC–CB and CD–CG Configurations

Cascading an emitter follower with a common-base amplifier, as shown in Fig. 6.56(a), results in a circuit with a low-frequency gain approximately equal to that of the CB but with the problem of the low input resistance of the CB solved by the buffering action of the CC stage. Since neither the CC nor the CB amplifier suffers from the Miller effect, the CC–CB

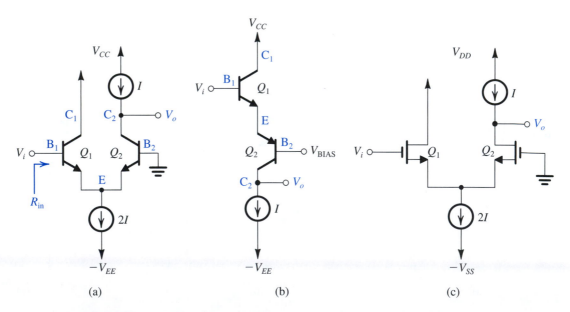

(a) (b) (c)

FIGURE 6.56 (a) A CC–CB amplifier. (b) Another version of the CC–CB circuit with Q_2 implemented using a *pnp* transistor. (c) The MOSFET version of the circuit in (a).

configuration has excellent high-frequency performance. Note that the biasing current sources shown in Fig. 6.56(a) ensure that each of Q_1 and Q_2 is operating at a bias current I. We are not showing, however, how the dc voltage at the base of Q_1 is set or the circuit that determines the dc voltage at the collector of Q_2. Both issues are usually looked after in the larger circuit of which the CC–CB amplifier is part.

An interesting version of the CC–CB configuration is shown in Fig. 6.56(b). Here the CB stage is implemented with a *pnp* transistor. Although only one current source is now needed, observe that we also need to establish an appropriate voltage at the base of Q_2. This circuit is part of the internal circuit of the popular 741 op amp, which will be studied in Chapter 9.

The MOSFET version of the circuit in Fig. 6.56(a) is the CD–CG amplifier shown in Fig. 6.56(c).

We now briefly analyze the circuit in Fig. 6.56(a) to determine its gain A_M and its high-frequency response. The analysis applies directly to the circuit in Fig. 6.56(b) and, with appropriate change of component and parameter names, to the MOSFET version in Fig. 6.56(c). For simplicity we shall neglect r_x and r_o of both transistors. The input resistance R_{in} is given by

$$R_{in} = (\beta_1 + 1)(r_{e1} + r_{e2}) \tag{6.182}$$

which for $r_{e1} = r_{e2} = r_e$ and $\beta_1 = \beta_2 = \beta$ becomes

$$R_{in} = 2r_\pi \tag{6.183}$$

If a load resistance R_L is connected at the output, the voltage gain V_o/V_i will be

$$\frac{V_o}{V_i} = \frac{\alpha_2 R_L}{r_{e1} + r_{e2}} = \frac{1}{2} g_m R_L \tag{6.184}$$

Now, if the amplifier is fed with a voltage signal V_{sig} from a source with a resistance R_{sig}, the overall voltage gain will be

$$\frac{V_o}{V_{sig}} = \frac{1}{2}\left(\frac{R_{in}}{R_{in} + R_{sig}}\right)(g_m R_L) \tag{6.185}$$

The high-frequency analysis is illustrated in Fig. 6.57(a). Here we have drawn the hybrid-π equivalent circuit for each of Q_1 and Q_2. Recalling that the two transistors are operating at equal bias currents, their corresponding model components will be equal (i.e., $r_{\pi1} = r_{\pi2}$, $C_{\pi1} = C_{\pi2}$, etc.). With this in mind the reader should be able to see that $V_{\pi1} = -V_{\pi2}$ and the horizontal line through the node labeled E in Fig. 6.57(a) can be deleted. Thus the circuit reduces to that in Fig. 6.57(b). This is a very attractive outcome because the circuit shows clearly the two poles that determine the high-frequency response: The pole at the input, with a frequency f_{P1}, is

$$f_{P1} = \frac{1}{2\pi\left(\dfrac{C_\pi}{2} + C_\mu\right)(R_{sig} \parallel 2r_\pi)} \tag{6.186}$$

and the pole at the output, with a frequency f_{P2}, is

$$f_{P2} = \frac{1}{2\pi C_\mu R_L} \tag{6.187}$$

This result is also intuitively obvious: The input impedance at B_1 of the circuit in Fig. 6.57(a) consists of the series connection of $r_{\pi1}$ and $r_{\pi2}$ in parallel with the series connection of $C_{\pi1}$ and $C_{\pi2}$. Then there is C_μ in parallel. At the output, we simply have R_L in parallel with C_μ.

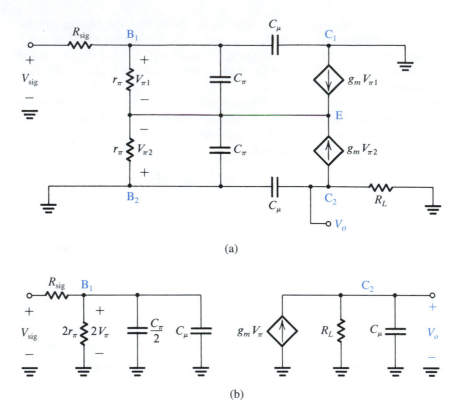

FIGURE 6.57 (a) Equivalent circuit for the amplifier in Fig. 6.56(a). (b) Simplified equivalent circuit. Note that the equivalent circuits in (a) and (b) also apply to the circuit shown in Fig. 6.56(b). In addition, they can be easily adapted for the MOSFET circuit in Fig. 6.56(c), with $2r_\pi$ eliminated, C_π replaced with C_{gs}, C_μ replaced with C_{gd}, and V_π replaced with V_{gs}.

Whether one of the two poles is dominant will depend on the relative values of R_{sig} and R_L. If the two poles are close to each other, then the 3-dB frequency f_H can be determined either by exact analysis—that is, finding the frequency at which the gain is down by 3 dB—or by using the approximate formula in Eq. (6.36),

$$f_H \cong 1 \Big/ \sqrt{\frac{1}{f_{P1}^2} + \frac{1}{f_{P2}^2}} \tag{6.188}$$

Finally, we note that the circuits in Fig. 6.56(a) and (c) are special forms of the differential amplifier, perhaps the most important circuit building block in analog IC design and the major topic of study in Chapter 7.

EXERCISE

6.37 For the CC–CB amplifier of Fig. 6.56(a), let $I = 0.5$ mA, $\beta = 100$, $C_\pi = 6$ pF, $C_\mu = 2$ pF, $R_{sig} = 10$ kΩ, and $R_L = 10$ kΩ. Find the low-frequency overall voltage gain A_M, the frequencies of the poles, and the 3-dB frequency f_H. Find f_H both exactly and using the approximate formula in Eq. (6.188).

Ans. 50 V/V; 6.4 MHz and 8 MHz; f_H by exact evaluation = 4.6 MHz; f_H using Eq. (6.188) = 5 MHz.

6.12 CURRENT-MIRROR CIRCUITS WITH IMPROVED PERFORMANCE

As we have seen throughout this chapter, current sources play a major role in the design of IC amplifiers: The constant-current source is used both in biasing and as active load. Simple forms of both MOS and bipolar current sources and, more generally, current mirrors were studied in Section 6.3. The need to improve the characteristics of the simple sources and mirrors has already been demonstrated. Specifically, two performance parameters need to be addressed: the *accuracy of the current transfer ratio of the mirror* and the *output resistance of the current source.*

The reader will recall from Section 6.3 that the accuracy of the current transfer ratio suffers particularly from the finite β of the BJT. The output resistance, which in the simple circuits is limited to r_o of the MOSFET and the BJT, also reduces accuracy and, much more seriously, severely limits the gain available from cascode amplifiers. In this section we study MOS and bipolar current mirrors with more accurate current transfer ratios and higher output resistances.

6.12.1 Cascode MOS Mirrors

A brief introduction to the use of the cascoding principle in the design of current sources was presented in Section 6.8.4. Figure 6.58 shows the basic cascode current mirror. Observe that in addition to the diode-connected transistor Q_1, which forms the basic mirror Q_1–Q_2, another diode-connected transistor, Q_4, is used to provide a suitable bias voltage for the gate of the cascode transistor Q_3. To determine the output resistance of the cascode mirror at the drain of Q_3, we set I_{REF} to zero. Also, since Q_1 and Q_4 have a relatively small incremental resistance, each of approximately $1/g_m$, the incremental voltages across them will be small, and we can assume that the gates of Q_3 and Q_2 are both grounded. Thus the output resistance R_o will be that of the CG transistor Q_3, which has a resistance r_{o1} in its source. Equation (6.101) can be adapted to obtain

$$R_o = r_{o3} + [1 + (g_{m3} + g_{mb3})r_{o3}]r_{o2} \qquad (6.189)$$

$$\cong g_{m3}r_{o3}r_{o2} \qquad (6.190)$$

Thus, as expected, cascoding raises the output resistance of the current source by the factor $g_{m3}r_{o3}$, which is the intrinsic gain of the cascode transistor.

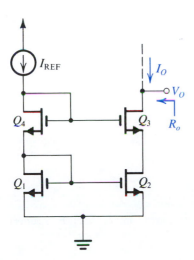

FIGURE 6.58 A cascode MOS current mirror.

A drawback of the cascode current mirror is that it consumes a relatively large portion of the steadily shrinking supply voltage V_{DD}. While the simple MOS mirror operates properly with a voltage as low as V_{OV} across its output transistor, the cascode circuit of Fig. 6.58 requires a minimum voltage of $V_t + 2V_{OV}$. This is because the gate of Q_3 is at $2V_{GS} = 2V_t + 2V_{OV}$. Thus the minimum voltage required across the output of the cascode mirror is 1 V or so. This obviously limits the signal swing at the output of the mirror (i.e., at the output of the amplifier that utilizes this current source as a load). In Chapter 9 we shall study a wide-swing cascode mirror.

6.38 For a cascode MOS mirror utilizing devices with $V_t = 0.5$ V, $\mu_n C_{ox} = 387$ μA/V^2, $V'_A = 5$ V/μm, $W/L = 3.6$ μm/0.36 μm, and $I_{REF} = 100$ μA, find the minimum dc voltage required at the output and the output resistance.

Ans. 0.95 V; 285 kΩ

6.12.2 A Bipolar Mirror with Base-Current Compensation

Figure 6.59 shows a bipolar current mirror with a current transfer ratio that is much less dependent on β than that of the simple current mirror. The reduced dependence on β is achieved by including transistor Q_3, the emitter of which supplies the base currents of Q_1 and Q_2. The sum of the base currents is then divided by $(\beta_3 + 1)$, resulting in a much smaller error current that has to be supplied by I_{REF}. Detailed analysis is shown on the circuit diagram; it is based on the assumption that Q_1 and Q_2 are matched and thus have equal collector currents, I_C. A node equation at the node labeled x gives

$$I_{REF} = I_C\left[1 + \frac{2}{\beta(\beta+1)}\right]$$

Since

$$I_O = I_C$$

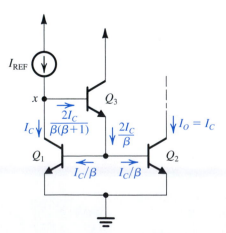

FIGURE 6.59 A current mirror with base-current compensation.

the current transfer ratio of the mirror will be

$$\frac{I_O}{I_{REF}} = \frac{1}{1 + 2/(\beta^2 + \beta)}$$

$$\cong \frac{1}{1 + 2/\beta^2} \tag{6.191}$$

which means that the error due to finite β has been reduced from $2/\beta$ in the simple mirror to $2/\beta^2$, a tremendous improvement. Unfortunately, however, the output resistance remains approximately equal to that of the simple mirror, namely r_o. Finally, note that if a reference current I_{REF} is not available, we simply connect node x to the power supply V_{CC} through a resistance R. The result is a reference current given by

$$I_{REF} = \frac{V_{CC} - V_{BE1} - V_{BE3}}{R} \tag{6.192}$$

6.12.3 The Wilson Current Mirror

A simple but ingenious modification of the basic bipolar mirror results in both reducing the β dependence and increasing the output resistance. The resulting circuit, known as the **Wilson mirror** after its inventor, George Wilson, an IC design engineer working for Tektronix, is shown in Fig. 6.60(a). The analysis to determine the effect of finite β on the current transfer

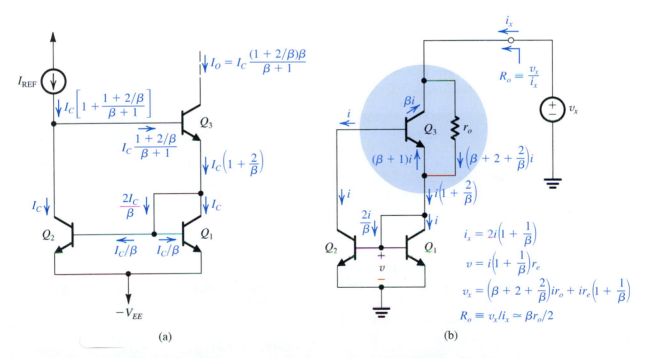

FIGURE 6.60 The Wilson bipolar current mirror: **(a)** circuit showing analysis to determine the current transfer ratio; and **(b)** determining the output resistance. Note that the current i_x that enters Q_3 must equal the sum of the currents that leave it, $2i$.

ratio is shown in Fig. 6.60(a), from which we can write

$$\frac{I_O}{I_{\text{REF}}} = \frac{I_C\left(1 + \frac{2}{\beta}\right)\beta\big/(\beta + 1)}{I_C\left[1 + \left(1 + \frac{2}{\beta}\right)\big/(\beta + 1)\right]}$$

$$= \frac{\beta + 2}{\beta + 1 + \dfrac{\beta + 2}{\beta}} = \frac{\beta + 2}{\beta + 2 + \dfrac{2}{\beta}}$$

$$= \frac{1}{1 + \dfrac{2}{\beta(\beta + 2)}}$$

$$\cong \frac{1}{1 + 2/\beta^2} \tag{6.193}$$

This analysis assumes that Q_1 and Q_2 conduct equal collector currents. There is, however, a slight problem with this assumption: The collector-to-emitter voltages of Q_1 and Q_2 are not equal, which introduces a current offset or a systematic error. The problem can be solved by adding a diode-connected transistor in series with the collector of Q_2, as we shall shortly show for the MOS version.

Analysis to determine the output resistance of the Wilson mirror is illustrated in Fig. 6.60(b), from which we see that

$$R_o = \beta r_o / 2 \tag{6.194}$$

Finally, we note that the Wilson mirror is preferred over the cascode circuit because the latter has the same dependence on β as the simple mirror. However, like the cascode mirror, the Wilson mirror requires an additional V_{BE} drop for its operation; that is, for proper operation we must allow for 1 V or so across the Wilson-mirror output.

EXERCISE

6.39 For $\beta = 100$ and $r_o = 100\ \text{k}\Omega$, contrast the Wilson mirror and the simple mirror by evaluating the transfer-ratio error due to finite β and the output resistance.
Ans. Transfer-ratio error: 0.02% for Wilson as opposed to 2% for the simple circuit; $R_o = 5\ \text{M}\Omega$ for Wilson compared to 100 kΩ for the simple circuit.

6.12.4 The Wilson MOS Mirror

Figure 6.61(a) shows the MOS version of the Wilson mirror. Obviously there is no β error to reduce here, and the advantage of the MOS Wilson lies in its enhanced output resistance. The analysis shown in Fig. 6.61(b) provides

$$R_o \cong r_{o3}(g_{m3}r_{o2} + 2)$$

$$\cong g_{m3}r_{o3}r_{o2}$$

where we have neglected, for simplicity, the body effect in Q_3. We observe that the output resistance is approximately the same as that achieved in the cascode circuit. Finally, to balance

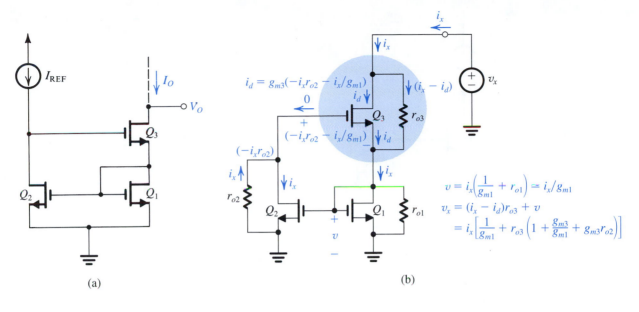

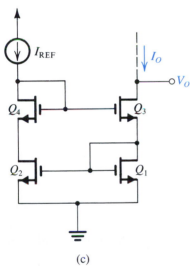

FIGURE 6.61 The Wilson MOS mirror: **(a)** circuit; **(b)** analysis to determine output resistance; and **(c)** modified circuit.

the two branches of the mirror and thus avoid the systematic current error resulting from the difference in V_{DS} between Q_1 and Q_2, the circuit can be modified as shown in Fig. 6.61(c).

6.12.5 The Widlar Current Source

Our final current-source circuit, known as the **Widlar current source,** is shown in Fig. 6.62. It differs from the basic current mirror circuit in an important way: A resistor R_E is included in the emitter lead of Q_2. Neglecting base currents we can write

$$V_{BE1} = V_T \ln\left(\frac{I_{\text{REF}}}{I_S}\right) \qquad (6.195)$$

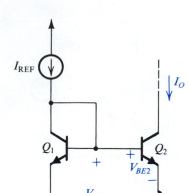

FIGURE 6.62 The Widlar current source.

and

$$V_{BE2} = V_T \ln\left(\frac{I_O}{I_S}\right) \tag{6.196}$$

where we have assumed that Q_1 and Q_2 are matched devices. Combining Eqs. (6.195) and (6.196) gives

$$V_{BE1} - V_{BE2} = V_T \ln\left(\frac{I_{REF}}{I_O}\right) \tag{6.197}$$

But from the circuit we see that

$$V_{BE1} = V_{BE2} + I_O R_E \tag{6.198}$$

Thus,

$$I_O R_E = V_T \ln\left(\frac{I_{REF}}{I_O}\right) \tag{6.199}$$

The design and advantages of the Widlar current source are illustrated in the following example.

EXAMPLE 6.14

Figure 6.63 shows two circuits for generating a constant current $I_O = 10\ \mu A$ which operate from a 10-V supply. Determine the values of the required resistors assuming that V_{BE} is 0.7 V at a current of 1 mA and neglecting the effect of finite β.

Solution

For the basic current-source circuit in Fig. 6.63(a) we choose a value for R_1 to result in $I_{REF} = 10\ \mu A$. At this current, the voltage drop across Q_1 will be

$$V_{BE1} = 0.7 + V_T \ln\left(\frac{10\ \mu A}{1\ mA}\right) = 0.58\ V$$

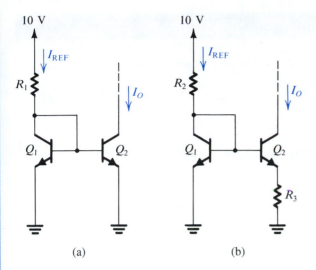

FIGURE 6.63 Circuits for Example 6.14.

Thus,

$$R_1 = \frac{10 - 0.58}{0.01} = 942 \text{ k}\Omega$$

For the Widlar circuit in Fig. 6.63(b) we must first decide on a suitable value for I_{REF}. If we select $I_{REF} = 1$ mA, then $V_{BE1} = 0.7$ V and R_2 is given by

$$R_2 = \frac{10 - 0.7}{1} = 9.3 \text{ k}\Omega$$

The value of R_3 can be determined using Eq. (6.199) as follows:

$$10 \times 10^{-6} R_3 = 0.025 \ln\left(\frac{1 \text{ mA}}{10 \text{ }\mu A}\right)$$

$$R_3 = 11.5 \text{ k}\Omega$$

From the above example we observe that using the Widlar circuit allows the generation of a small constant current using relatively small resistors. This is an important advantage that results in considerable savings in chip area. In fact the circuit of Fig. 6.63(a), requiring a 942-kΩ resistance, is totally impractical for implementation in IC form.

Another important characteristic of the Widlar current source is that its output resistance, is high. The increase in the output resistance, above that achieved in the basic current source, is due to the emitter degeneration resistance R_E. To determine the output resistance of Q_2, we assume that since the base of Q_2 is connected to ground via the small resistance r_e of Q_1, the incremental voltage at the base will be small. Thus we can use the formula derived in Section 6.7.2 for the CB amplifier, namely Eq. (6.118), and adapt it for our purposes here as follows:

$$R_o \cong [1 + g_m(R_E \| r_\pi)]r_o \tag{6.200}$$

Thus the output resistance is increased above r_o by a factor that can be significant.

6.13 SPICE SIMULATION EXAMPLES

We conclude this chapter by presenting two SPICE simulation examples. In the first example, we will use SPICE to investigate the operation of the CS amplifier circuit (studied in Section 6.5.2). In the second example, we will use SPICE to compare the high-frequency response of the CS amplifier (studied in Section 6.6) to that of the folded-cascode amplifier (studied in Section 6.8.6).

EXAMPLE 6.15

THE CMOS CS AMPLIFIER

In this example, we will use PSpice to compute the dc transfer characteristic of the CS amplifier whose Capture schematic is shown in Fig. 6.64. We will assume a 5-μm CMOS technology for the MOSFETs and use parts NMOS5P0 and PMOS5P0 whose SPICE level-1 parameters are listed in Table 4.8. To specify the dimensions of the MOSFETs in PSpice, we will use the multiplicative factor m together with the channel length L and the channel width W. The MOSFET parameter m,

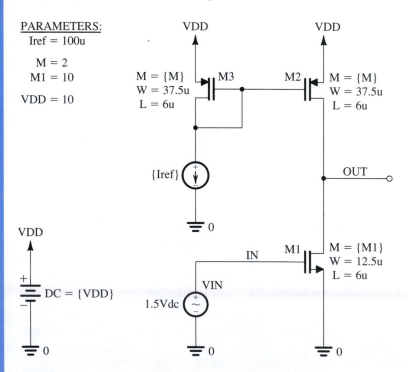

FIGURE 6.64 Capture schematic of the CS amplifier in Example 6.15.

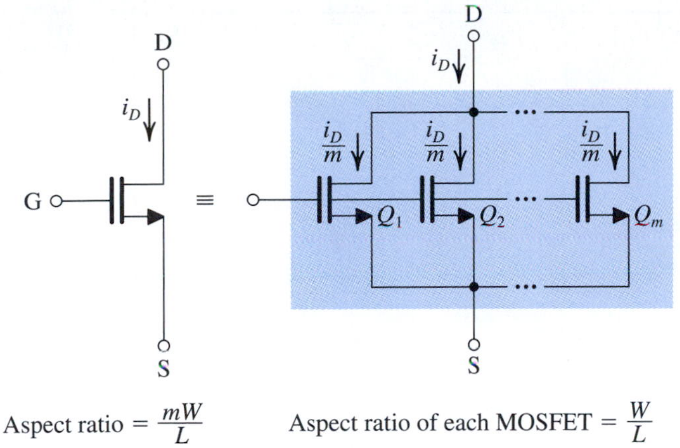

Aspect ratio $= \dfrac{mW}{L}$ Aspect ratio of each MOSFET $= \dfrac{W}{L}$

FIGURE 6.65 Transistor equivalency.

whose default value is 1, is used in SPICE to specify the number of MOSFETs connected in parallel. As depicted in Fig. 6.65, a wide transistor with channel length L and channel width $m \times W$ can be implemented using m narrower transistors in parallel, each having a channel length L and a channel width W. Thus, neglecting the channel-length modulation effect, the drain current of a MOSFET operating in the saturation region can be expressed as

$$I_D = \frac{1}{2}\mu C_{ox} m \frac{W}{L_{\text{eff}}} V_{OV}^2 \tag{6.201}$$

where L_{eff} rather than L is used to more accurately estimate the drain current (refer to Section 4.12.2).

The CS amplifier in Fig. 6.64 is designed for a bias current of 100 μA assuming a reference current $I_{\text{ref}} = 100$ μA and $V_{DD} = 10$ V. The current-mirror transistors M_2 and M_3 are sized for $V_{OV2} = V_{OV3} = 1$ V, while the input transistor M_1 is sized for $V_{OV1} = 0.5$ V. Note that a smaller overdrive voltage is selected for M_1 to achieve a larger voltage gain G_v for the CS amplifier, since

$$G_v = -g_{m1} R'_L = -g_{m1}(r_{o1} \parallel r_{o2}) = -\frac{2}{V_{OV1}}\left(\frac{V_{An} V_{Ap}}{V_{An} + V_{Ap}}\right) \tag{6.202}$$

where V_{An} and V_{Ap} are the magnitudes of the Early voltages of, respectively, the NMOS and PMOS transistors. Unit-size transistors are used with $W/L = 12.5$ μm/6 μm for the NMOS devices and $W/L = 37.5$ μm/6 μm for the PMOS devices. Thus, using Eq. (6.201) together with the 5-μm CMOS process parameters in Table 4.8, we find $m_1 = 10$ and $m_2 = m_3 = 2$ (rounded to the nearest integer). Furthermore, Eq. (6.202) gives $G_v = -100$ V/V.

To compute the dc transfer characteristic of the CS amplifier, we perform a dc analysis in PSpice with V_{IN} swept over the range 0 to V_{DD} and plot the corresponding output voltage V_{OUT}. Figure 6.66(a) shows the resulting transfer characteristic. The slope of this characteristic (i.e., $dV_{\text{OUT}}/dV_{\text{IN}}$) corresponds to the gain of the amplifier. The high-gain segment is clearly visible for V_{IN} around 1.5 V. This corresponds to an overdrive voltage for M_1 of $V_{OV1} = V_{\text{IN}} - V_{tn} = 0.5$ V, as desired. To examine the high-gain region more closely, we repeat the dc sweep for V_{IN} between 1.3 V and 1.7 V. The resulting transfer characteristic is plotted in Fig. 6.66 (b, middle curve). Using the Probe graphical interface of PSpice, we find that the linear region of this dc transfer characteristic is bounded approximately by $V_{\text{IN}} = 1.465$ V and $V_{\text{IN}} = 1.539$ V. The corresponding values of V_{OUT} are 8.838 V and 0.573 V. These results are close to the expected values. Specifically, transistors M_1 and M_2 will remain in the saturation region and, hence, the amplifier will operate in its linear region if $V_{OV1} \leq V_{\text{OUT}} \leq V_{DD} - V_{OV2}$ or 0.5 V $\leq V_{\text{OUT}} \leq 9$ V. From the results above,

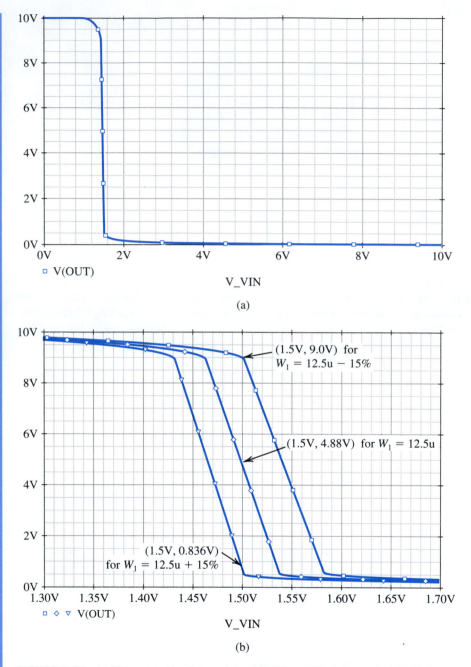

FIGURE 6.66 (a) Voltage transfer characteristic of the CS amplifier in Example 6.15. (b) Expanded view of the transfer characteristic in the high-gain region. Also shown are the transfer characteristics where process variations cause the width of transistor M_1 to change by +15% and −15% from its nominal value of $W_1 = 12.5$ μm.

the voltage gain G_v (i.e., the slope of the linear segment of the dc transfer characteristic) is approximately −112 V/V, which is reasonably close to the value obtained by hand analysis.

Note, from the dc transfer characteristic in Fig. 6.66(b), that for an input dc bias of $V_{IN} = 1.5$ V, the output dc bias is $V_{OUT} = 4.88$ V. This choice of V_{IN} maximizes the available signal swing at the out-put by setting V_{OUT} at the middle of the linear segment of the dc transfer characteristic.

However, because of the high resistance at the output node (or, equivalently, because of the high voltage gain), this value of V_{OUT} is highly sensitive to the effect of process and temperature variations on the characteristics of the transistors. To illustrate this point, consider what happens when the width of M_1 (i.e., W_1, which is normally 12.5 μm) changes by ±15%. The corresponding dc transfer characteristics are shown in Fig. 6.66(b). Accordingly, when $V_{IN} = 1.5$ V, V_{OUT} will drop to 0.84 V if W_1 increases by 15% and will rise to 9.0 V if W_1 decreases by 15%. In practical circuit implementations, this problem is circumvented by using negative feedback to accurately set the dc bias voltage at the output of the amplifier and, hence, to reduce the sensitivity of the circuit to process variations. The topic of negative feedback will be studied in Chapter 8.

EXAMPLE 6.16

FREQUENCY RESPONSE OF THE CS AND THE FOLDED-CASCODE AMPLIFIERS

In this example, we will use PSpice to compute the frequency response of both the CS and the folded-cascode amplifiers whose Capture schematics are shown in Figs. 6.67 and 6.69, respectively. We will assume that the dc bias levels at the output of the amplifiers are stabilized using negative feedback. However, before performing a small-signal analysis (an ac-analysis simulation) in SPICE to measure the frequency response, we will perform a dc analysis (a bias-point simulation) to verify that all MOSFETs are operating in the saturation region and, hence, ensure that the amplifier is operating in its linear region.

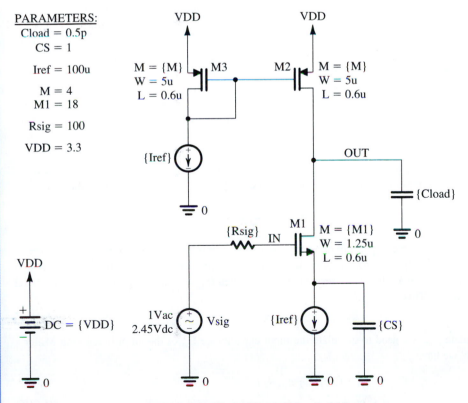

FIGURE 6.67 Capture schematic of the CS amplifier in Example 6.16.

In the following, we will assume a 0.5-μm CMOS technology for the MOSFETs and use parts NMOSOP5 and PMOSOP5 whose SPICE level-1 model parameters are listed in Table 4.8. To specify the dimensions of the MOSFETs in PSpice, we will use the multiplicative factor m, together with the channel length L and channel width W (as we did in Example 6.15).

THE CS AMPLIFIER

The CS amplifier circuit in Fig. 6.67 is identical to the one shown in Fig. 6.18, except that a current source is connected to the source of the input transistor M_1 to set its drain current I_{D1} independently of its drain voltage V_{D1}. Furthermore, in our PSpice simulations, we used an impractically large bypass capacitor C_S of 1 F. This sets the source of M_1 at approximately signal ground during the ac-analysis simulation. Accordingly, the CS amplifier circuits in Figs. 6.18 and 6.67 are equivalent for the purpose of frequency-response analysis. In Chapter 7, we will find out, in the context of studying the differential pair, how the goals of this biasing approach for the CS amplifier are realized in practical IC implementations.

The CS amplifier in Fig. 6.67 is designed assuming a reference current $I_{ref} = 100$ μA and $V_{DD} = 3.3$ V. The current-mirror transistors, M_2 and M_3, are sized for $V_{OV2} = V_{OV3} = 0.3$ V, while the input transistor M_1 is sized for $V_{OV1} = 0.15$ V. Unit-size transistors are used with $W/L = 1.25$ μm/ 0.6 μm for the NMOS devices and $W/L = 5$ μm/0.6 μm for the PMOS devices. Thus, using Eq. (6.201) together with the 0.5-μm CMOS process parameters in Table 4.8, we find $m_1 = 18$ and $m_2 = m_3 = 4$. Furthermore, Eq. (6.202) gives $G_v = -44.4$ V/V for the CS amplifier.

In the PSpice simulations of the CS amplifier in Fig. 6.67, the dc bias voltage of the signal source is set such that the voltage at the source terminal of M_1 is $V_{S1} = 1.3$ V. This requires the dc level of V_{sig} to be $V_{OV1} + V_{tn1} + V_{S1} = 2.45$ V because $V_{tn1} \cong 1$ V as a result of the body effect on M_1. The reasoning behind this choice of V_{S1} is that, in a practical circuit implementation, the current source that feeds the source of M_1 is realized using a cascode current mirror such as the one in Fig. 6.58. In this case, the minimum voltage required across the current source (i.e., the minimum V_{S1}) is $V_t + 2V_{OV} = 1.3$ V, assuming $V_{OV} = 0.3$ V for the current-mirror transistors.

A bias-point simulation is performed in PSpice to verify that all MOSFETs are biased in the saturation region. Next, to compute the frequency response of the amplifier, we set the ac voltage of the signal source to 1 V, perform an ac-analysis simulation, and plot the output voltage magnitude versus frequency. Figure 6.68(a) shows the resulting frequency response for $R_{sig} = 100$ Ω and $R_{sig} = 1$ MΩ. In both cases, a load capacitance of $C_{load} = 0.5$ pF is used. The corresponding values of the 3-dB frequency f_H of the amplifier are given in Table 6.4.

Observe that f_H drops when R_{sig} is increased. This is anticipated from our study of the high-frequency response of the CS amplifier in Section 6.6. Specifically, as R_{sig} increases, the pole

$$f_{p,\,in} = \frac{1}{2\pi}\frac{1}{R_{sig}C_{in}}$$

(6.203)

formed at the amplifier input will have an increasingly significant effect on the overall frequency response of the amplifier. As a result, the effective time constant τ_H in Eq. (6.57) increases and f_H decreases. When R_{sig} becomes very large, as it is when $R_{sig} = 1$ MΩ, a dominant pole is formed by R_{sig} and C_{in}. This results in

$$f_H \cong f_{p,\,in}$$

(6.204)

To estimate $f_{p,in}$, we need to calculate the input capacitance C_{in} of the amplifier. Using Miller's theorem, we have

$$C_{in} = C_{gs1} + C_{gd1}(1 + g_{m1}R'_L)$$
$$= (\tfrac{2}{3}m_1 W_1 L_1 C_{ox} + C_{gs,\,ov1}) + C_{gd,\,ov1}(1 + g_{m1}R'_L)$$

(6.205)

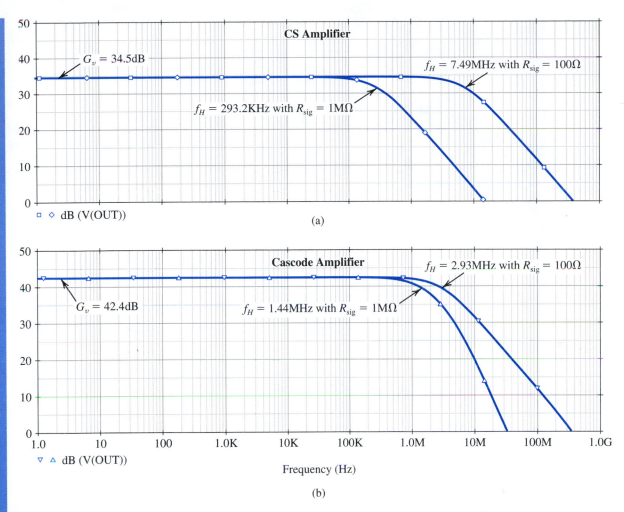

FIGURE 6.68 Frequency response of **(a)** the CS amplifier and **(b)** the folded-cascode amplifier in Example 6.16, with $R_{sig} = 100 \ \Omega$ and $R_{sig} = 1 \ M\Omega$.

	f_H	
TABLE 6.4 Dependence of the 3-dB Bandwidth f_H on R_{sig} for the CS and the Folded-Cascode Amplifiers in Example 6.16		
R_{sig}	**CS Amplifier**	**Folded-Cascode Amplifier**
100 Ω	7.49 MHz	2.93 MHz
1 MΩ	293.2 kHz	1.44 MHz

where

$$R'_L = r_{o1} \| r_{o2} \qquad (6.206)$$

Thus, C_{in} can be calculated using the values of C_{gs1} and C_{gd1} which are computed by PSpice and can be found in the output file of the bias-point simulation. Alternatively, C_{in} can be found using Eq. (6.205) with the values of the overlap capacitances $C_{gs, ov1}$ and $C_{gd, ov1}$ calculated using the

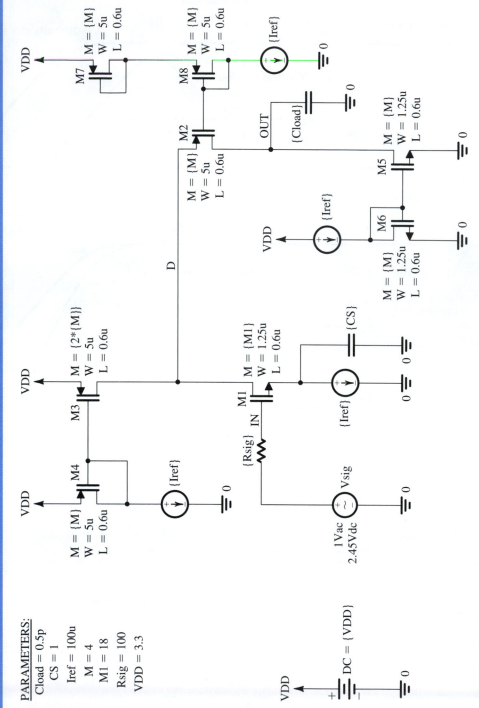

PARAMETERS:
Cload = 0.5p
CS = 1
Iref = 100u
M = 4
M1 = 18
Rsig = 100
VDD = 3.3

FIGURE 6.69 Capture schematic of the folded-cascode amplifier in Example 6.16.

process parameters in Table 4.8 (as described in Eqs. 4.170 and 4.171), that is:

$$C_{gs, ov1} = m_1 W_1 CGSO \tag{6.207}$$

$$C_{gd, ov1} = m_1 W_1 CGDO \tag{6.208}$$

This results in $C_{in} = 0.53$ pF when $|G_v| = g_{m1} R'_L = 53.2$ V/V. Accordingly, using Eqs. (6.203) and (6.204), $f_H = 300.3$ kHz when $R_{sig} = 1$ MΩ, which is close to the value computed by PSpice.

THE FOLDED-CASCODE AMPLIFIER

The folded-cascode amplifier circuit in Fig. 6.69 is equivalent to the one in Fig. 6.45, except that a current source is placed in the source of the input transistor M_1 (for the same dc-biasing purpose as in the case of the CS amplifier). Note that, in Fig. 6.69, the PMOS current mirror M_3–M_4 and the NMOS current mirror M_5–M_6 are used to realize, respectively, current sources I_1 and I_2 in the circuit of Fig. 6.45. Furthermore, the current transfer ratio of mirror M_3–M_4 is set to 2 (i.e., $m_3/m_4 = 2$). This results in $I_{D3} \cong 2I_{ref}$. Hence, transistor M_2 is biased at $I_{D2} = I_{D3} - I_{D1} = I_{ref}$. The gate bias voltage of transistor M_2 is generated using the diode-connected transistors M_7 and M_8. The size and drain current of these transistors are set equal to those of transistor M_2. Therefore, ignoring the body effect,

$$V_{G2} = V_{DD} - V_{SG7} - V_{SG8} \cong V_{DD} - 2(|V_{tp}| + |V_{OVp}|)$$

where V_{OVp} is the overdrive voltage of the PMOS transistors in the amplifier circuit. These transistors have the same overdrive voltage because their I_D/m is the same. Thus, such a biasing configuration results in $V_{SG2} = |V_{tp}| + |V_{OVp}|$ as desired, while setting $V_{SD3} = |V_{tp}| + |V_{OVp}|$ to improve the bias matching between M_3 and M_4.

The folded-cascode amplifier in Fig. 6.69 is designed assuming a reference current $I_{ref} = 100\ \mu A$ and $V_{DD} = 3.3$ V (similar to the case of the CS amplifier). All transistors are sized for an overdrive voltage of 0.3 V, except for the input transistor M_1, which is sized for $V_{OV1} = 0.15$ V. Thus, using Eq. (6.201), all the MOSFETs in the amplifier circuit are designed using $m = 4$, except for $m_1 = 18$.

The midband voltage gain of the folded-cascode amplifier in Fig. 6.69 can be expressed using Eq. (6.130) as

$$G_v = -g_{m1} R_{out} \tag{6.209}$$

where

$$R_{out} = R_{out2} \| R_{out5} \tag{6.210}$$

is the output resistance of the amplifier. Here, R_{out2} is the resistance seen looking into the drain of the cascode transistor M_2, while R_{out5} is the resistance seen looking into the drain of the current-mirror transistor M_5. Using Eq. (6.127), we have

$$R_{out2} \cong (g_{m2} r_{o2}) R_{s2} \tag{6.211}$$

where

$$R_{s2} = r_{o1} \| r_{o3} \tag{6.212}$$

is the effective resistance at the source of M_2. Furthermore,

$$R_{out5} = r_{o5} \tag{6.213}$$

Thus, for the folded-cascoded amplifier in Fig. 6.69,

$$R_{out} \cong r_{o5} \tag{6.214}$$

and

$$G_v \cong -g_{m1}r_{o5} = -2\frac{V_{An}}{V_{OV1}} \tag{6.215}$$

Using the 0.5-μm CMOS parameters, this gives $R_{out} = 100$ kΩ and $G_v = -133$ V/V. Therefore, R_{out} and, hence, $|G_v|$ of the folded-cascode amplifier in Fig. 6.69 are larger than those of the CS amplifier in Fig. 6.67 by a factor of 3.

Figure 6.68(b) shows the frequency response of the folded-cascode amplifier as computed by PSpice for the cases of $R_{sig} = 100$ Ω and $R_{sig} = 1$ MΩ. The corresponding values of the 3-dB frequency f_H of the amplifier are given in Table 6.4. Observe that, when R_{sig} is small, f_H of the folded-cascode amplifier is lower than that of the CS amplifier by a factor of approximately 2.6, approximately equal to the factor by which the gain is increased. This is because, when R_{sig} is small, the frequency response of both amplifiers is dominated by the pole formed at the output node, that is,

$$f_H \cong f_{p,\text{out}} = \frac{1}{2\pi}\frac{1}{R_{out}C_{out}} \tag{6.216}$$

Since the output resistance of the folded-cascode amplifier is larger than that of the CS amplifier (by a factor of approximately 3, as found through the hand analysis above) while their output capacitances are approximately equal, the folded-cascode amplifier has a lower f_H in this case.

On the other hand, when R_{sig} is large, f_H of the folded-cascode amplifier is much higher than that of CS amplifier. This is because, in this case, the effect of the pole at $f_{p,\text{in}}$ on the overall frequency response of the amplifier becomes significant. Since, due to the Miller effect, C_{in} of the CS amplifier is much larger than that of the folded-cascode amplifier its f_H is much lower in this case. To confirm this point, observe that C_{in} of the folded-cascode amplifier can be estimated by replacing R_L' in Eq. (6.205) with the total resistance R_{d1} between the drain of M_1 and ground. Here,

$$R_{d1} = r_{o1} \parallel r_{o3} \parallel R_{in2} \tag{6.217}$$

where R_{in2} is the input resistance of the common-gate transistor M_2 and can be obtained using an approximation of the relationship in Eq. (6.83) as

$$R_{in2} \cong \frac{r_{o2} + r_{o5}}{g_{m2}r_{o2}} \tag{6.218}$$

Thus,

$$R_{d1} \cong r_{o1} \parallel r_{o3} \parallel \frac{r_{o2} + r_{o5}}{g_{m2}r_{o2}} \cong \frac{2}{g_{m2}} \tag{6.219}$$

Therefore, R_{d1} is much smaller than R_L' in Eq. (6.206). Hence, C_{in} of the folded-cascode amplifier in Fig. 6.69 is indeed much smaller than that of the CS amplifier in Fig. 6.67. This confirms that the folded-cascode amplifier is much less impacted by the Miller effect and, therefore, can achieve a much higher f_H when R_{sig} is large.

The midband gain of the folded cascode amplifier can be significantly increased by replacing the current mirror M_5–M_6 with a current mirror having a larger output resistance, such as the cascode current mirror in Fig. 6.58 whose output resistance is approximately $g_m r_o^2$. In this case, however, R_{in2} and, hence R_{d1}, increase, causing an increased Miller effect and a corresponding reduction in f_H.

Finally, it is interesting to observe that the frequency response of the folded-cascode amplifier, shown in Fig. 6.68(b), drops beyond f_H at approximately –20 dB/decade when $R_{sig} = 100$ Ω and at approximately –40 dB/decade when $R_{sig} = 1$ MΩ. This is because, when R_{sig} is small, the frequency response is dominated by the pole at $f_{p,\text{out}}$. However, when R_{sig} is increased, $f_{p,\text{in}}$ is moved closer to $f_{p,\text{out}}$ and both poles, contribute to the gain falloff.

SUMMARY

■ Integrated-circuit fabrication technology offers the circuit designer many exciting opportunities, the most important of which is large numbers of inexpensive small-area MOS transistors. An overriding concern for IC designers, however, is the minimization of chip area or "silicon real estate." As a result, large-valued resistors and capacitors are virtually absent.

■ A review and comparison of the characteristics of the MOSFET and the BJT is presented in Section 6.2. Of particular interest is the summary provided in Table 6.3.

■ Biasing in integrated circuits utilizes current sources. Typically an accurate and stable reference current is generated and then replicated to provide bias currents for the various amplifier stages on the chip. The heart of the current-steering circuitry utilized to perform this function is the current mirror. The basic MOS and bipolar mirrors are studied in Section 6.3. Improved mirror circuits with more precise current transfer ratios, reduced dependence on the β value of the BJT, and higher output resistances are studied in Section 6.12.

■ IC amplifiers are usually direct-coupled; thus their midband gain A_M extends to zero frequency (dc). Their high-frequency response is limited by the transistor internal capacitances, mainly C_{gs} and C_{gd} in the MOSFET and C_π and C_μ in the BJT. There usually is also a capacitance C_L between the output node and ground. These capacitances cause the amplifier gain (or transfer function) to acquire a number of poles on the negative real-axis of the s-plane. In addition, there may be one transmission zero on the negative or positive real-axis, with the remaining transmission zeros at infinite frequency.

■ If the lowest-frequency pole is at least two octaves away from the nearest pole or zero, this pole, say at frequency f_{P1}, will play a dominant role in determining the high-frequency response, and the 3-dB frequency $f_H \cong f_{P1}$. If, on the other hand, none of the poles is dominant, an estimate of f_H can be obtained from

$$f_H = 1 \Big/ \sqrt{\frac{1}{f_{P1}^2} + \frac{1}{f_{P2}^2} + \cdots - 2\left(\frac{1}{f_{Z1}^2} + \frac{1}{f_{Z2}^2} + \cdots\right)}$$

■ If the poles and zeros cannot be easily determined, one can use the open-circuit time constants to obtain an estimate of f_H as follows:

$$f_H \cong 1/2\pi\tau_H$$

where

$$\tau_H = \sum_{i=1}^{n} C_i R_i$$

where C_i is a capacitance that determines the high-frequency response of the amplifier and R_i is the resistance that capacitance C_i "sees". To determine R_i, set V_{sig} and all capacitances to zero. Then apply a signal v_x between the terminals to which C_i was connected, determine the current i_x that the circuit draws from v_x, and calculate $R_i = v_x/i_x$.

■ Miller's theorem states that an impedance Z connected between two circuit nodes 1 and 2, whose voltages are related by $V_2 = KV_1$ can be replaced by two impedances: $Z_1 = Z/(1-K)$ between node 1 and ground and $Z_2 = Z/(1-(1/K))$ between node 2 and ground. Miller's theorem is very useful in the analysis of the high-frequency response of the CS and CE amplifiers.

■ IC amplifiers employ constant-current sources in place of the resistances $R_D(R_C)$ that connect the drain (collector) to the power supply. These active loads enable the realization of reasonably large voltage gains while using low-voltage supplies (as low as 1 V or so).

■ The largest voltage gain available from a CS or a CE amplifier is equal to the intrinsic gain of the transistor $A_0 = g_m r_o$, which for a BJT is 2000 to 4000 V/V and for a MOSFET is 20 to 100 V/V. Recall, however, that the CS amplifier has an infinite input resistance while the input resistance of the CE amplifier is limited by the finite β to r_π. Both CS and CE amplifiers have output resistances equal to the transistor r_o.

■ The high-frequency response of the CS amplifier is usually limited by the Miller multiplication of C_{gd}, which results in an input capacitance C_{in} of

$$C_{\text{in}} = C_{gs} + C_{gd}(1 + g_m R_L')$$

which interacts with the resistance R_{sig} of the signal source to form a dominant pole; thus $f_H \cong 1/2\pi C_{\text{in}} R_{\text{sig}}$. Alternatively, the method of open-circuit time constants can be used to obtain an estimate of f_H as $\cong 1/2\pi\tau_H$, where

$$\tau_H = C_{gs}R_{\text{sig}} + C_{gd}[R_{\text{sig}}(1 + g_m R_L') + R_L'] + C_L R_L'$$

■ Exact analysis of the high-frequency response of the CS amplifier yields the second-order transfer function given by Eq. (6.60), which can be used to determine the poles and zeros and hence f_H.

■ The high-frequency response of the CE amplifier can be found by adapting the CS equations as follows: Replace R_{sig} by $R_{\text{sig}}' = R_{\text{sig}} \| r_\pi$, C_{gs} by C_π, and C_{gd} by C_μ.

■ When the CS amplifier is fed with a low-resistance signal source, it has the frequency response shown in Fig. 6.26(c).

■ The CG and CB amplifiers act as current buffers. Their impedance transformation properties are displayed in Fig. 6.29 (CG) and Fig. 6.35 (CB).

■ The CG and CB amplifiers do *not* suffer from the Miller capacitance-multiplication effect and as a result have excellent high-frequency response. In situations when r_o can be neglected, the CG amplifier has two poles: one produced at the input node with a frequency $f_{P1} = 1/2\pi C_{gs}(R_s \parallel (1/(g_m + g_{mb})))$ and the other formed at the output node with a frequency $f_{P2} = 1/2\pi(C_L + C_{gd})R_L$. The 3-dB frequency f_H can be determined using f_{P1} and f_{P2}. When r_o is taken into account, an estimate of f_H can be obtained from

$$f_H = 1/2\pi[C_{gs}(R_s \parallel R_{in}) + (C_L + C_{gd})(R_L \parallel R_{out})]$$

■ In the cascode amplifier the CG (CB) transistor buffers the drain (collector) of Q_1 from the load. The result is a smaller signal at the drain (collector) of Q_1 and hence a reduced Miller effect and a wider bandwidth. Alternatively, we can think of the CG (CB) transistor as raising the output resistance and hence the open-circuit voltage gain by a factor of $g_{m2}r_{o2}$ (β_2 for the BJT). Refer to the output equivalent circuits shown in Fig. 6.37(a) and (b) for the MOS cascode and in Fig. 6.41(a) and (b) for the bipolar cascode.

■ A summary of the characteristics of the MOS cascode for the case of low signal-source resistance is provided in Fig. 6.39.

■ A summary of the frequency-response analysis of the BJT cascode is provided in Fig. 6.42.

■ Including a small resistance in the source (emitter) of a CS (CE) amplifier provides the designer with a tool to effect some performance improvements (e.g., wider bandwidth), in return for gain reduction (a trade-off characteristic of negative feedback).

■ The source and emitter followers are free of the negative effects of Miller capacitance multiplication and thus achieve very wide bandwidths.

■ Combining a source (or emitter) follower with a CS (or CE) amplifier results in amplifiers with gains equal to (or greater than) that of the CS (or CE) amplifier alone and, most importantly, wider bandwidth. The latter is a result of the buffering action of the input follower and the attendant reduction in the Miller effect that takes place at the input of the CS (or CE) stage.

■ Both the cascode and the Wilson MOS current mirrors achieve an increase in output resistance by a factor of $g_m r_o$. The Wilson bipolar mirror increases the output resistance by a factor of $\beta/2$ and dramatically reduces the current transfer-ratio error due to finite β.

PROBLEMS

SECTION 6.2: COMPARISON OF THE MOSFET AND THE BJT

6.1 Find the range of I_D obtained in a particular NMOS transistor as its overdrive voltage is increased from 0.15 V to 0.4 V. If the same current range is required in I_C of a BJT, what is the corresponding change in V_{BE}?

6.2 What range of I_C is obtained in an *npn* transistor as a result of changing the area of the emitter–base junction by a factor of 10 while keeping V_{BE} constant? If I_C is to be kept constant, by what amount must V_{BE} change?

6.3 For each of the CMOS technologies specified in Table 6.1, find the $|V_{OV}|$ and hence the $|V_{GS}|$ required to operate a device with a W/L of 10 at a drain current $I_D = 100\ \mu$A. Ignore channel-length modulation.

6.4 Consider NMOS and PMOS devices fabricated in the 0.25-μm process specified in Table 6.1. If both devices are to

operate at $|V_{OV}| = 0.25$ V and $I_D = 100\ \mu$A, what must their W/L ratios be?

6.5 Consider NMOS and PMOS transistors fabricated in the 0.25-μm process specified in Table 6.1. If the two devices are to be operated at equal drain currents, what must the ratio of $(W/L)_p$ to $(W/L)_n$ be to achieve equal values of g_m?

6.6 An NMOS transistor fabricated in the 0.18-μm CMOS process specified in Table 6.1 is operated at $V_{OV} = 0.2$ V. Find the required W/L and I_D to obtain a g_m of 10 mA/V. At what value of I_C must an *npn* transistor be operated to achieve this value of g_m?

6.7 For each of the CMOS process technologies specified in Table 6.1, find the g_m of an NMOS and a PMOS transistor with $W/L = 10$ operated at $I_D = 100\ \mu$A.

6.8 An NMOS transistor operated with an overdrive voltage of 0.25 V is required to have a g_m equal to that of an *npn* transistor operated at $I_C = 0.1$ mA. What must I_D be? What value of g_m is realized?

6.9 It is required to find the incremental (i.e., small-signal) resistance of each of the diode-connected transistors shown in Fig. P6.9. Assume that the dc bias current $I = 0.1$ mA. For the MOSFET, let $\mu_n C_{ox} = 200$ μA/V^2 and $W/L = 10$.

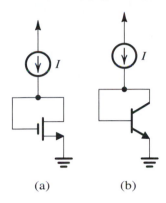

(a)   (b)

FIGURE P6.9

6.10 For an NMOS transistor with $L = 1$ μm fabricated in the 0.8-μm process specified in Table 6.1, find g_m, r_o, and A_0 if the device is operated with $V_{OV} = 0.5$ V and $I_D = 100$ μA. Also, find the required device width W.

6.11 For an NMOS transistor with $L = 0.3$ μm fabricated in the 0.18-μm process specified in Table 6.1, find g_m, r_o, and A_0 obtained when the device is operated at $I_D = 100$ μA with $V_{OV} = 0.2$ V. Also, find W.

6.12 Fill in the table below. For the BJT, let $\beta = 100$ and $V_A = 100$ V. For the MOSFET, let $\mu_n C_{ox} = 200$ μA/V^2, $W/L = 40$, and $V_A = 10$ V. Note that R_i refers to the input resistance at the control input terminal (gate, base) with the (source, emitter) grounded.

Device	BJT		MOSFET	
Bias Current	$I_C = 0.1$ mA	$I_C = 1$ mA	$I_D = 0.1$ mA	$I_D = 1$ mA
g_m (mA/V)				
r_o (kΩ)				
A_0 (V/V)				
R_i (kΩ)				

6.13 For an NMOS transistor fabricated in the 0.18-μm process specified in Table 6.1 with $L = 0.3$ μm and $W = 6$ μm, find the value of f_T obtained when the transistor is operated at $V_{OV} = 0.2$ V. Use both the formula in terms of C_{gs} and C_{gd} and the approximate formula. Why does the approximate formula overestimate f_T?

6.14 An NMOS transistor fabricated in the 0.18-μm process specified in Table 6.1 and having $L = 0.3$ μm and $W = 6$ μm is operated at $V_{OV} = 0.2$ V and used to drive a capacitive load of 100 fF. Find A_0, f_P (or f_{3dB}), and f_t. At what I_D value is the transistor operating? If it is required to double f_t, what must I_D become? What happens to A_0 and f_P in this case?

6.15 For an *npn* transistor fabricated in the high-voltage process specified in Table 6.2, evaluate f_T at $I_C = 10$ μA, 100 μA, and 1 mA. Assume $C_\mu \cong C_{\mu 0}$. Repeat for the low-voltage process.

6.16 Consider an NMOS transistor fabricated in the 0.8-μm process specified in Table 6.1. Let the transistor have $L = 1$ μm, and assume it is operated at $I_D = 100$ μA.

(a) For $V_{OV} = 0.25$ V, find W, g_m, r_o, A_0, C_{gs}, C_{gd}, and f_T.
(b) To what must V_{OV} be changed to double f_T? Find the new values of W, g_m, r_o, A_0, C_{gs}, and C_{gd}.

6.17 For a lateral *pnp* transistor fabricated in the high-voltage process specified in Table 6.2, find f_T if the device is operated at a collector bias current of 1 mA. Compare to the value obtained for a vertical *npn*.

6.18 Show that for a MOSFET the selection of L and V_{OV} determines A_0 and f_T. In other words, show that A_0 and f_T will not depend on I_D and W.

6.19 Consider an NMOS transistor fabricated in the 0.18-μm technology specified in Table 6.1. Let the transistor be operated at $V_{OV} = 0.2$ V. Find A_0 and f_T for $L = 0.2$ μm, 0.3 μm, and 0.4 μm.

D6.20 Consider an NMOS transistor fabricated in the 0.5-μm process specified in Table 6.1. Let $L = 0.5$ μm and $V_{OV} = 0.3$ V. If the MOSFET is connected as a common-source amplifier with a load capacitance $C_L = 1$ pF (as in Fig. 6.2a), find the required transistor width W and bias current I_D to obtain a unity-gain bandwidth of 100 MHz. Also, find A_0 and f_{3dB}.

SECTION 6.3: IC BIASING—CURRENT SOURCES, CURRENT MIRRORS, AND CURRENT-STEERING CIRCUITS

D6.21 For $V_{DD} = 1.8$ V and using $I_{REF} = 50$ μA, it is required to design the circuit of Fig. 6.4 to obtain an output current whose nominal value is 50 μA. Find R if Q_1 and Q_2 are matched with channel lengths of 0.5 μm, channel widths of 5 μm, $V_t = 0.5$ V, and $k'_n = 250$ μA/V^2. What is the lowest possible value of V_O? Assuming that for this process technology the Early voltage $V'_A = 20$ V/μm, find the output resistance of the current source. Also, find the change in output current resulting from a +1-V change in V_O.

D6.22 Using $V_{DD} = 1.8$ V and a pair of matched MOSFETs, design the current-source circuit of Fig. 6.4 to provide an output current of $100\text{-}\mu A$ nominal value. To simplify matters, assume that the nominal value of the output current is obtained at $V_O \cong V_{GS}$. It is further required that the circuit operate for V_O in the range of 0.25 V to V_{DD} and that the change in I_O over this range be limited to 5% of the nominal value of I_O. Find the required value of R and the device dimensions. For the fabrication-process technology utilized, $\mu_n C_{ox} = 250 \ \mu A/V^2$, $V'_A = 20 \ V/\mu m$, and $V_t = 0.6$ V.

6.23 Sketch the p-channel counterpart of the current-source circuit of Fig. 6.4. Note that while the circuit of Fig. 6.4 should more appropriately be called a current sink, the corresponding PMOS circuit is a current source. Let $V_{DD} = 1.8$ V, $|V_t| = 0.6$ V, Q_1 and Q_2 be matched, and $\mu_p C_{ox} = 100 \ \mu A/V^2$. Find the device (W/L) ratios and the value of the resistor that sets the value of I_{REF} so that a nominally $80\text{-}\mu A$ output current is obtained. The current source is required to operate for V_O as high as 1.6 V. Neglect channel-length modulation.

6.24 Consider the current-mirror circuit of Fig. 6.5 with two transistors having equal channel lengths but with Q_2 having a width four times that of Q_1. If I_{REF} is 20 μA and the transistors are operating at an overdrive voltage of 0.3 V, what I_O results? What is the minimum allowable value of V_O for proper operation of the current source? If $V_t = 0.5$ V, at what value of V_O will the nominal value of I_O be obtained? If V_O increases by 1 V, what is the corresponding increase in I_O? Let $V_A = 25$ V.

6.25 For the current-steering circuit of Fig. P6.25, find I_O in terms of I_{REF} and device (W/L) ratios.

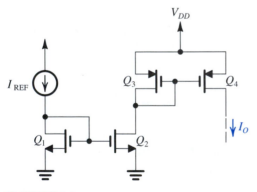

FIGURE P6.25

D6.26 The current-steering circuit of Fig. P6.26 is fabricated in a CMOS technology for which $\mu_n C_{ox} = 200 \ \mu A/V^2$, $\mu_p C_{ox} = 80 \ \mu A/V^2$, $V_{tn} = 0.6$ V, $V_{tp} = -0.6$ V, $V'_{An} = 10 \ V/\mu m$, and $|V'_{Ap}| = 12 \ V/\mu m$. If all devices have $L = 0.8 \ \mu m$, design the circuit so that $I_{REF} = 20 \ \mu A$, $I_2 = 100 \ \mu A$, $I_3 = I_4 = 20 \ \mu A$, and $I_5 = 50 \ \mu A$. Use the minimum possible device widths while achieving proper operation of the current source Q_2 for voltages at its drain as high at +1.3 V and proper operation of the current sink Q_5 with voltages at its drain as low as −1.3 V. Specify the widths of all devices and

the value of R. Find the output resistance of the current source Q_2 and the output resistance of the current sink Q_5.

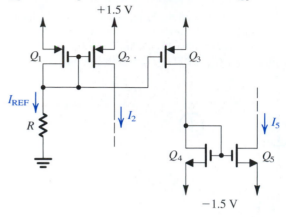

FIGURE P6.26

*6.27 A PMOS current mirror consists of three PMOS transistors, one diode-connected and two used as current outputs. All transistors have $|V_t| = 0.7$ V, $k'_p = 80 \ \mu A/V^2$, and $L = 1.0 \ \mu m$ but three different widths, namely 10 μm, 20 μm, and 40 μm. When the diode-connected transistor is supplied from a $100\text{-}\mu A$ source, how many different output currents are available? Repeat with two of the transistors diode-connected and the third used to provide current output. For each possible input-diode combination, give the values of the output currents and of the V_{SG} that results.

6.28 Although thus far we have focussed only on the application of current mirrors in dc biasing, they can also be used as signal-current amplifiers. One such application is illustrated in Fig. P6.28. Here Q_1 is a common-source amplifier fed with $v_I = V_{GS} + v_i$, where V_{GS} is the gate-to-source dc bias voltage of Q_1 and v_i is a small signal to be amplified. Find the signal component of the output voltage v_O and hence the small-signal voltage gain v_o / v_i.

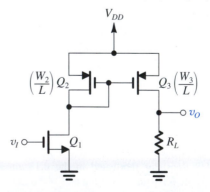

FIGURE P6.28

6.29 Consider the basic bipolar current mirror of Fig. 6.8 for the case in which Q_1 and Q_2 are identical devices having $I_S = 10^{-15}$ A.

(a) Assuming the transistor β is very high, find the range of V_{BE} and I_O corresponding to I_{REF} increasing from 10 μA to 10 mA. Assume that Q_2 remains in the active mode, and neglect the Early effect.

(b) Find the range of I_O corresponding to I_{REF} in the range of 10 μA to 10 mA, taking into account the finite β. Assume that β remains constant at 100 over the current range 0.1 mA to 5 mA but that at $I_C \cong 10$ mA, $\beta = 70$. Specify I_O corresponding to $I_{REF} = 10$ μA, 0.1 mA, 1 mA, and 10 mA. Note that β variation with current causes the current transfer ratio to vary with current.

6.30 Consider the basic BJT current mirror of Fig. 6.8 for the case in which Q_2 has m times the area of Q_1. Show that the current transfer ratio is given by Eq. (6.19). If β is specified to be a minimum of 80, what is the largest current transfer ratio possible while keeping the error introduced by the finite β limited to 5%?

6.31 Give the circuit for the *pnp* version of the basic current mirror of Fig. 6.8. If β of the *pnp* transistor is 20, what is the current gain (or transfer ratio) I_O/I_{REF}, neglecting the Early effect?

6.32 Consider the basic BJT current mirror of Fig. 6.8 when Q_1 and Q_2 are matched and $I_{REF} = 2$ mA. Neglecting the effect of finite β, find the change in I_O, both as an absolute value and as a percentage, corresponding to V_O changing from 1 V to 10 V. The Early voltage is 90 V.

D6.33 The current-source circuit of Fig. P6.33 utilizes a pair of matched *pnp* transistors having $I_S = 10^{-15}$A, $\beta = 50$, and $|V_A| = 50$ V. It is required to design the circuit to provide an output current $I_O = 1$ mA at $V_O = 2$ V. What values of I_{REF} and R are needed? What is the maximum allowed value of V_O while the current source continues to operate properly? What change occurs in I_O corresponding to V_O changing from the maximum positive value to -5 V?

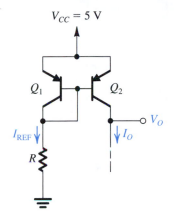

FIGURE P6.33

6.34 Find the voltages at all nodes and the currents through all branches in the circuit of Fig. P6.34. Assume $|V_{BE}| = 0.7$ V and $\beta = \infty$.

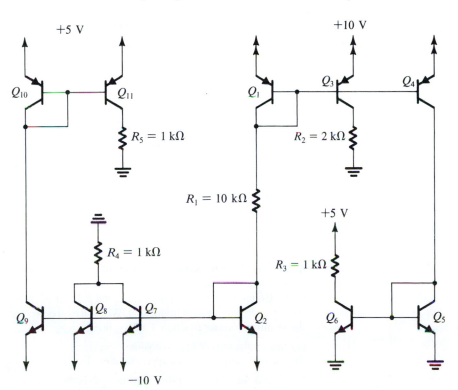

6.35 For the circuit in Fig. P6.35, let $|V_{BE}| = 0.7$ V and $\beta = \infty$. Find I, V_1, V_2, V_3, V_4, and V_5 for (a) $R = 10$ kΩ and (b) $R = 100$ kΩ.

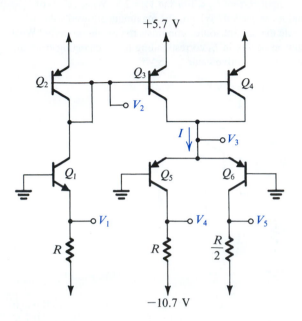

FIGURE P6.35

D6.36 Using the ideas embodied in Fig. 6.11, design a multiple-mirror circuit using power supplies of ±5 V to create source currents of 0.2 mA, 0.4 mA, and 0.8 mA and sink currents of 0.5 mA, 1 mA, and 2 mA. Assume that the BJTs have $V_{BE} \cong 0.7$ V and large β.

*6.37 The circuit shown in Fig. P6.37 is known as a **current conveyor.**

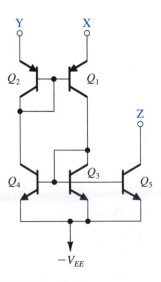

FIGURE P6.37

(a) Assuming that Y is connected to a voltage V, a current I is forced into X, and terminal Z is connected to a voltage that keeps Q_5 in the active region show that a current equal to I flows through terminal Y, that a voltage equal to V appears at terminal X, and that a current equal to I flows through terminal Z. Assume β to be large. Corresponding transistors are matched, and all transistors are operating in the active region. (b) With Y connected to ground, show that a virtual ground appears at X. Now, if X is connected to a +5-V supply through a 10-kΩ resistor, what current flows through Z?

SECTION 6.4: HIGH-FREQUENCY RESPONSE— GENERAL CONSIDERATIONS

6.38 A direct-coupled amplifier has a low-frequency gain of 40 dB, poles at 1 MHz and 10 MHz, a zero on the negative real-axis at 100 MHz, and another zero at infinite frequency. Express the amplifier gain function in the form of Eqs. (6.29) and (6.30), and sketch a Bode plot for the gain magnitude. What do you estimate the 3-dB frequency f_H to be?

6.39 An amplifier with a dc gain of 60 dB has a single-pole high-frequency response with a 3-dB frequency of 10 kHz.

(a) Give an expression for the gain function $A(s)$.
(b) Sketch Bode diagrams for the gain magnitude and phase.
(c) What is the gain–bandwidth product?
(d) What is the unity-gain frequency?
(e) If a change in the amplifier circuit causes its transfer function to acquire another pole at 100 kHz, sketch the resulting gain magnitude and specify the unity-gain frequency. Note that this is an example of an amplifier with a unity–gain bandwidth that is different from its gain–bandwidth product.

6.40 Consider an amplifier whose $F_H(s)$ is given by

$$F_H(s) = \cfrac{1}{\left(1 + \cfrac{s}{\omega_{P1}}\right)\left(1 + \cfrac{s}{\omega_{P2}}\right)}$$

with $\omega_{P1} < \omega_{P2}$. Find the ratio ω_{P2}/ω_{P1} for which the value of the 3-dB frequency ω_H calculated using the dominant-pole approximation differs from that calculated using the root-sum-of-squares formula (Eq. 6.36) by:

(a) 10%.
(b) 1%.

6.41 The high-frequency response of a direct-coupled amplifier having a dc gain of −100 V/V incorporates zeros at ∞ and 10^6 rad/s (one at each frequency) and poles at 10^5 rad/s and 10^7 rad/s (one at each frequency). Write an expression for the amplifier transfer function. Find ω_H using:

(a) the dominant-pole approximation.
(b) the root-sum-of-squares approximation (Eq. 6.36).

If a way is found to lower the frequency of the finite zero to 10^5 rad/s, what does the transfer function become? What is the 3-dB frequency of the resulting amplifier?

6.42 A direct-coupled amplifier has a dominant pole at 100 rad/s and three coincident poles at a much higher frequency. These nondominant poles cause the phase lag of the amplifier at high frequencies to exceed the 90° angle due to the dominant pole. It is required to limit the excess phase at $\omega = 10^6$ rad/s to 30° (i.e., to limit the total phase angle to $-120°$). Find the corresponding frequency of the nondominant poles.

D6.43 Refer to Example 6.6. Give an expression for ω_H in terms of C_{gs}, R' (note that $R' = R_{in} \| R_{sig}$), C_{gd}, R'_L, and g_m. If all component values except for the generator resistance R_{sig} are left unchanged, to what value must R_{sig} be reduced in order to raise f_H to 150 kHz?

6.44 In a particular amplifier design, two internal nodes having Thévenin equivalent resistances of 10 kΩ and 20 kΩ are expected to have node capacitances (to ground) of 5 pF and 2 pF, respectively, due to component and wiring capacitances. However, when the circuit is manufactured in a modular form, connections associated with each node add capacitances of 10 pF to each. What are the associated pole frequencies and overall 3-dB frequency in Hz for both the original and the manufactured designs?

6.45 A FET amplifier resembling that in Example 6.6, when operated at lower currents in a higher-impedance application, has $R_{sig} = 100$ kΩ, $R_{in} = 1.2$ MΩ, $g_m = 2$ mA/V, $R'_L = 12$ kΩ, and $C_{gs} = C_{gd} = 1$ pF. Find the midband voltage gain A_M and the 3-dB frequency f_H.

*6.46 Figure P6.46 shows the high-frequency equivalent circuit of a MOSFET amplifier with a resistance R_s connected in the source lead. The purpose of this problem is to show that the value of R_s can be used to control the gain and bandwidth of the amplifier, specifically to allow the designer to trade gain for increased bandwidth.

(a) Derive an expression for the low-frequency voltage gain (set C_{gs} and C_{gd} to zero).
(b) To be able to determine ω_H using the open-circuit time-constants method, derive expressions for R_{gs} and R_{gd}.
(c) Let $R_{sig} = 100$ kΩ, $g_m = 4$ mA/V, $R'_L = 5$ kΩ, and $C_{gs} = C_{gd} = 1$ pF. Use the expressions found in (a) and (b) to

determine the low-frequency gain and the 3-dB frequency f_H for three cases: $R_s = 0$ Ω, 100 Ω, and 250 Ω. In each case also evaluate the gain–bandwidth product.

6.47 A common-source MOS amplifier, whose equivalent circuit resembles that in Fig. 6.14(a), is to be evaluated for its high-frequency response. For this particular design, $R_{sig} = 1$ MΩ, $R_{in} = 5$ MΩ, $R'_L = 100$ kΩ, $C_{gs} = 0.2$ pF, $C_{gd} = 0.1$ pF, and $g_m = 0.3$ mA/V. Estimate the midband gain and the 3-dB frequency.

6.48 For a particular amplifier modeled by the circuit of Fig. 6.14(a), $g_m = 5$ mA/V, $R_{sig} = 150$ kΩ, $R_{in} = 0.65$ MΩ, $R'_L = 10$ kΩ, $C_{gs} = 2$ pF, and $C_{gd} = 0.5$ pF. There is also an output wiring capacitance of 3 pF. Find the corresponding midband voltage gain, the open-circuit time constants, and an estimate of the 3-dB frequency.

6.49 Consider the high-frequency response of an amplifier consisting of two identical stages, each with an input resistance of 10 kΩ and an output resistance of 2 kΩ. The two-stage amplifier is driven from a 5-kΩ source and drives a 1-kΩ load. Associated with each stage is a parasitic input capacitance (to ground) of 10 pF and a parasitic output capacitance (to ground) of 2 pF. Parasitic capacitances of 5 pF and 7 pF also are associated with the signal-source and load connections, respectively. For this arrangement, find the three poles and estimate the 3-dB frequency f_H.

6.50 Using the method of open-circuit time constants, a set of amplifiers are found to be characterized by the following time constants and/or frequencies. For each case, estimate the 3-dB cutoff frequency in rad/s and in Hz:

(a) 20 ns, 5 ns, 1 ns.
(b) 50 MHz, 200 MHz, 1 GHz.
(c) 50 Mrad/s, 200 Mrad/s, 1 Grad/s.
(d) 1 μs, 200 ns, 200 ns.
(e) 1 μs, 0.4 μs.
(f) 1 μs, 200 ns, 150 ns.
(g) 1 GHz, 2 GHz, 5 GHz, 5 GHz.

6

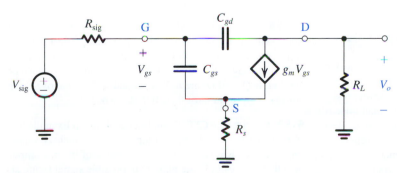

FIGURE P6.46

6.51 Consider an ideal voltage amplifier with a gain of 0.95 V/V and a resistance $R = 100$ kΩ connected in the feedback path—that is, between the output and input terminals. Use Miller's theorem to find the input resistance of this circuit.

6.52 An ideal voltage amplifier with a voltage gain of -1000 V/V has a 0.1-pF capacitance connected between its output and input terminals. What is the input capacitance of the amplifier? If the amplifier is fed from a voltage source V_{sig} having a resistance $R_{sig} = 1$ kΩ, find the transfer function V_o/V_{sig} as a function of the complex-frequency variable s and hence the 3-dB frequency f_H and the unity-gain frequency f_t.

6.53 The amplifiers listed below are characterized by the descriptor (A, C), where A is the voltage gain from input to output and C is an internal capacitor connected between input and output. For each, find the equivalent capacitances at the input and at the output as provided by the use of Miller's theorem:

(a) -1000 V/V, 1 pF.
(b) -10 V/V, 10 pF.
(c) -1 V/V, 10 pF.
(d) $+1$ V/V, 10 pF.
(e) $+10$ V/V, 10 pF.

Note that the input capacitances found in case (e) can be used to cancel the effect of other capacitance connected from input to ground. In (e), what capacitance can be cancelled?

6.54 Figure P6.54 shows an ideal voltage amplifier with a gain of $+2$ V/V (usually implemented with an op amp connected in the noninverting configuration) and a resistance R connected between output and input.

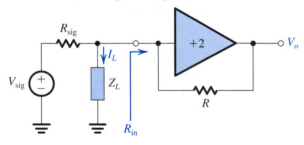

FIGURE P6.54

(a) Using Miller's theorem, show that the input resistance $R_{in} = -R$.
(b) Use Norton's theorem to replace V_{sig}, R_{sig}, and R_{in} with a signal current source and an equivalent parallel resistance. Show that by selecting $R_{sig} = R$ the equivalent parallel resistance becomes infinite and the current I_L into the load impedance Z_L becomes V_{sig}/R. The circuit then functions as an ideal voltage-controlled current source with an output current I_L.
(c) If Z_L is a capacitor C, find the transfer function V_o/V_{sig} and show it is that of an ideal noninverting integrator.

SECTION 6.5: THE COMMON-SOURCE AND COMMON-EMITTER AMPLIFIERS WITH ACTIVE LOADS

D6.55 Find the intrinsic gain of an NMOS transistor fabricated in a process for which $k'_n = 125$ μA/V^2 and $V'_A = 10$ V/μm. The transistor has a 1-μm channel length and is operated at $V_{OV} = 0.2$ V. If a 2-mA/V transconductance is required, what must I_D and W be?

6.56 An NMOS transistor fabricated in a certain process is found to have an intrinsic gain of 100 V/V when operated at I_D of 100 μA. Find the intrinsic gain for $I_D = 25$ μA and $I_D = 400$ μA. For each of these currents, find the ratio by which g_m changes from its value at $I_D = 100$ μA.

6.57 The NMOS transistor in the circuit of Fig. P6.57 has $V_t = 0.5$ V, $k'_n W/L = 2$ mA/V^2, and $V_A = 20$ V.

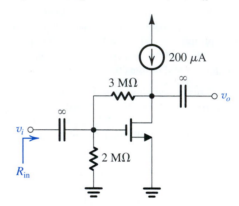

FIGURE P6.57

(a) Neglecting the dc current in the feedback network and the effect of r_o, find V_{GS} and V_{DS}. Now, find the dc current in the feedback network, and verify that you were justified in neglecting it.
(b) Find the small-signal voltage gain, v_o/v_i. What is the peak of the largest output sine-wave signal that is possible while the NMOS remains in saturation? What is the corresponding input signal?
(c) Find the small-signal input resistance R_{in}.

D6.58 Consider the CMOS amplifier of Fig. 6.18(a) when fabricated with a process for which $k'_n = 2.5k'_p = 250$ μA/V^2, $|V_t| = 0.6$ V, and $|V_A| = 10$ V. Find I_{REF} and $(W/L)_1$ to obtain a voltage gain of -40 V/V and an output resistance of 100 kΩ. If Q_2 and Q_3 are to be operated at the same overdrive voltage as Q_1, what must their W/L ratios be?

6.59 Consider the CMOS amplifier analyzed in Example 6.8. If v_I consists of a dc bias component on which is superimposed a sinusoidal signal, find the value of the dc component that will result in the maximum possible signal swing at the output with almost-linear operation. What is the amplitude of the output sinusoid resulting? (*Note:* In practice, the amplifier would have a feedback circuit that causes it to operate at a point near the middle of its linear region).

6.60 The power supply of the CMOS amplifier analyzed in Example 6.8 is increased to 5 V. What will the extent of the linear region at the output become?

6.61 Figure P6.61 shows an IC MOS amplifier formed by cascading two common-source stages. Assuming that $V_{An} = |V_{Ap}|$ and the biasing current-sources have output resistances equal to those of Q_1 and Q_2, find an expression for the overall voltage gain in terms of g_m and r_o of Q_1 and Q_2.

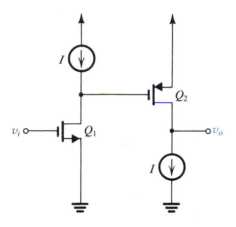

FIGURE P6.61

6.62 Consider the circuit shown in Fig. 6.18(a), using a 3.3-V supply and transistors for which $|V_t| = 0.8$ V and $L = 1$ μm. For Q_1, $k'_n = 100$ μA/V², $V_A = 100$ V, and $W = 20$ μm. For Q_2 and Q_3, $k'_p = 50$ μA/V² and $|V_A| = 50$ V. For Q_2, $W = 40$ μm. For Q_3, $W = 10$ μm.

(a) If Q_1 is to be biased at 100 μA, find I_{REF}. For simplicity, ignore the effect of V_A.
(b) What are the extreme values of v_O for which Q_1 and Q_2 just remain in saturation?
(c) What is the large-signal voltage gain?
(d) Find the slope of the transfer characteristic at $v_O = V_{DD}/2$.
(e) For operation as a small-signal amplifier around a bias point at $v_O = V_{DD}/2$, find the small-signal voltage gain and output resistance.

***6.63** The MOSFETs in the circuit of Fig. P6.63 are matched, having $k'_n(W/L)_1 = k'_p(W/L)_2 = 1$ mA/V² and $|V_t| = 0.5$ V. The resistance $R = 1$ MΩ.

(a) For G and D open, what are the drain currents I_{D1} and I_{D2}?
(b) For $r_o = \infty$, what is the voltage gain of the amplifier from G to D?

(c) For finite $r_o(|V_A| = 20$ V), what is the voltage gain from G to D and the input resistance at G?
(d) If G is driven (through a large coupling capacitor) from a source v_{sig} having a resistance of 100 kΩ, find the voltage gain v_d/v_{sig}.
(e) For what range of output signals do Q_1 and Q_2 remain in the saturation region?

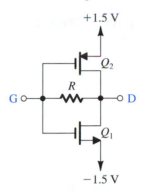

FIGURE P6.63

D6.64 Consider the active-loaded CE amplifier circuit of Fig. 6.19(a) for the case $I = 1$ mA, $\beta = 100$ and $V_A = 100$ V. Find R_i, A_{vo}, and R_o. If it is required to raise R_i by a factor of 4 by changing the bias current I, what value of I is required assuming that β remains unchanged? What are the new values of A_{vo} and R_o? If the amplifier is fed with a signal source having $R_{sig} = 5$ kΩ and is connected to a load of 100-kΩ resistance, find the overall voltage gain v_o/v_s in both cases.

6.65 Transistor Q_1 in the circuit of Fig. P6.65 is operating as a CE amplifier with an active load provided by transistor Q_2, which is the output transistor in a current mirror formed by Q_2 and Q_3. (Note that the biasing arrangement for Q_1 is *not* shown.)

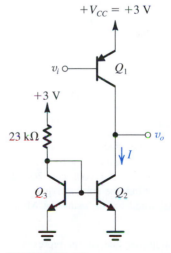

FIGURE P6.65

(a) Neglecting the finite base currents of Q_2 and Q_3 and assuming that their $V_{BE} \cong 0.7$ V and that Q_2 has five times the area of Q_3, find the value of I.

(b) If Q_1 and Q_2 are specified to have $|V_A| = 50$ V, find r_{o1} and r_{o2} and hence the total resistance at the collector of Q_1.

(c) Find $r_{\pi 1}$ and g_{m1} assuming that $\beta_1 = 50$.

(d) Find R_{in}, A_v, and R_o.

SECTION 6.6: HIGH-FREQUENCY RESPONSE OF THE CS AND CE AMPLIFIERS

6.66 A CS amplifier that can be represented by the equivalent circuit of Fig. 6.20 has $C_{gs} = 2$ pF, $C_{gd} = 0.1$ pF, $C_L = 1$ pF, $g_m = 5$ mA/V, and $R_{sig} = R'_L = 20$ kΩ. Find the midband gain A_M, the input capacitance C_{in} using the Miller equivalence, and hence an estimate of the 3-dB frequency f_H.

6.67 A CS amplifier that can be represented by the equivalent circuit of Fig. 6.20 has $C_{gs} = 2$ pF, $C_{gd} = 0.1$ pF, $C_L = 1$ pF, $g_m = 5$ mA/V, and $R_{sig} = R'_L = 20$ kΩ. Find the midband A_M gain, and estimate the 3-dB frequency f_H using the method of open-circuit time constants. Also, give the percentage contribution to τ_H by each of three capacitances. (Note that this is the same amplifier considered in Problem 6.66; if you have solved Problem 6.66, compare your results.)

6.68 A CS amplifier represented by the equivalent circuit of Fig. 6.20 has $C_{gs} = 2$ pF, $C_{gd} = 0.1$ pF, $C_L = 1$ pF, $g_m = 5$ mA/V, and $R_{sig} = R'_L = 20$ kΩ. Find the exact values of f_Z, f_{P1}, and f_{P2} using Eq. (6.60), and hence estimate f_H. Compare the values of f_{P1} and f_{P2} to the approximate values obtained using Eqs. (6.66) and (6.67). (Note that this is the same amplifier considered in Problems 6.66 and 6.67; if you have solved either or both of these problems, compare your results.)

6.69 A CS amplifier represented by the equivalent circuit of Fig. 6.20 has $C_{gs} = 2$ pF, $C_{gd} = 0.1$ pF, $C_L = 1$ pF, $g_m = 5$ mA/V, and $R_{sig} = R'_L = 20$ kΩ. It is required to find A_M, f_H, and the gain–bandwidth product for each of the following values of R'_L: 5 kΩ, 10 kΩ, and 20 kΩ. Use the approximate expression for f_{P1} in Eq. (6.66). However, in each case, also evaluate f_{P2} and f_Z to ensure that a dominant pole exists, and in each case, state whether the unity-gain frequency is equal to the gain–bandwidth product. Present your results in tabular form, and comment on the gain–bandwidth trade-off.

6.70 A common-emitter amplifier that can be represented by the equivalent circuit of Fig. 6.25(a) has $C_\pi = 10$ pF, $C_\mu = 0.5$ pF, $C_L = 2$ pF, $g_m = 20$ mA/V, $\beta = 100$, $r_x = 200$ Ω, $R'_L = 5$ kΩ, and $R_{sig} = 1$ kΩ. Find the midband gain A_M, and an estimate of the 3-dB frequency f_H using the Miller equivalence.

6.71 A common-emitter amplifier that can be represented by the equivalent circuit of Fig. 6.25(a) has $C_\pi = 10$ pF, $C_\mu =$

0.5 pF, $C_L = 2$ pF, $g_m = 20$ mA/V, $\beta = 100$, $r_x = 200$ Ω, $R'_L = 5$ kΩ, and $R_{sig} = 1$ kΩ. Find the midband gain A_M, the frequency of the zero f_Z, and the approximate values of the pole frequencies f_{P1} and f_{P2}. Hence, estimate the 3-dB frequency f_H. (Note that this is the same amplifier considered in Problem 6.70; if you have solved Problem 6.70, compare your results.)

*** 6.72** Refer to Fig. P6.72. Utilizing the BJT high-frequency hybrid-π model with $r_x = 0$ and $r_o = \infty$, derive an expression for $Z_i(s)$ as a function of r_e and C_π. Find the frequency at which the impedance has a phase angle of 45° for the case in which the BJT has $f_T = 400$ MHz and the bias current is relatively high. What does this frequency become if the bias current is reduced so that $C_\pi \cong C_\mu$? (Assume $\alpha = 1$).

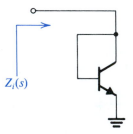

FIGURE P6.72

*** 6.73** For the current mirror in Fig. P6.73, derive an expression for the current transfer function $I_o(s)/I_i(s)$ taking into account the BJT internal capacitances and neglecting r_x and r_o. Assume the BJTs to be identical. Observe that a signal ground appears at the collector of Q_2. If the mirror is biased at 1 mA and the BJTs at this operating point are characterized by $f_T = 400$ MHz, $C_\mu = 2$ pF, and $\beta_0 = \infty$, find the frequencies of the pole and zero of the transfer function.

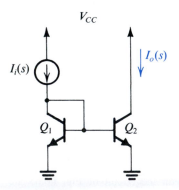

FIGURE P6.73

6.74 A CS amplifier modeled with the equivalent circuit of Fig. 6.26(a) is specified to have $C_{gs} = 2$ pF, $C_{gd} = 0.1$ pF, $g_m = 5$ mA/V, $C_L = 1$ pF, and $R'_L = 20$ kΩ. Find A_M, f_{3dB} and f_t.

*6.75 It is required to analyze the high-frequency response of the CMOS amplifier shown in Fig. P6.75. The dc bias current is 100 μA. For Q_1, $\mu_n C_{ox} = 90\ \mu$A/V^2, $V_A = 12.8$ V, $W/L = 100\ \mu$m/1.6 μm, $C_{gs} = 0.2$ pF, $C_{gd} = 0.015$ pF, and $C_{db} = 20$ fF. For Q_2, $C_{gd} = 0.015$ pF, $C_{db} = 36$ fF, and $|V_A| = 19.2$ V. Assume that the resistance of the input signal generator is negligibly small. Also, for simplicity assume that the signal voltage at the gate of Q_2 is zero. Find the low-frequency gain, the frequency of the pole, and the frequency of the zero.

(c) For $L = 0.5\ \mu$m, $W_2 = 25\ \mu$m, $f_T = 12$ GHz, and $\mu_n C_{ox} = 200\ \mu$A/V^2, design the circuit to obtain a gain of 3 V/V per stage. Bias the MOSFETs at $V_{OV} = 0.3$ V. Specify the required values of W_1 and I. What is the 3-dB frequency achieved?

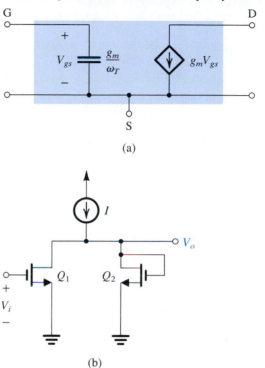

(a)

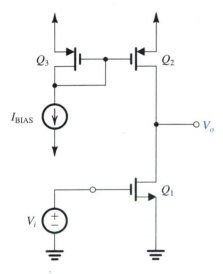

FIGURE P6.75

(b)

FIGURE P6.76

D**6.76 This problem investigates the use of MOSFETs in the design of wideband amplifiers (Steininger, 1990). Such amplifiers can be realized by cascading low-gain stages.

(a) Show that for the case $C_{gd} \ll C_{gs}$ and the gain of the common-source amplifier is low so that the Miller effect is negligible, the MOSFET can be modeled by the approximate equivalent circuit shown in Fig. P6.76(a), where ω_T is the unity-gain frequency of the MOSFET.

(b) Figure P6.76(b) shows an amplifier stage suitable for the realization of low gain and wide bandwidth. Transistors Q_1 and Q_2 have the same channel length L but different widths W_1 and W_2. They are biased at the same V_{GS} and have the same f_T. Use the MOSFET equivalent circuit of Fig. P6.76(a) to model this amplifier stage assuming that its output is connected to the input of an identical stage. Show that the voltage gain V_o / V_i is given by

$$\frac{V_o}{V_i} = -\frac{G_0}{1 + \dfrac{s}{\omega_T/(G_0 + 1)}}$$

where

$$G_0 = \frac{g_{m1}}{g_{m2}} = \frac{W_1}{W_2}$$

6.77 Consider an active-loaded common-emitter amplifier. Let the amplifier be fed with an ideal voltage source V_i, and neglect the effect of r_x. Assume that the bias current source has a very high resistance and that there is a capacitance C_L present between the output node and ground. This capacitance represents the sum of the input capacitance of the subsequent stage and the inevitable parasitic capacitance between collector and ground. Show that the voltage gain is given by

$$\frac{V_o}{V_i} = -g_m r_o \frac{1 - s(C_\mu/g_m)}{1 + s(C_L + C_\mu)r_o}$$

$$\simeq -\frac{g_m r_o}{1 + s(C_L + C_\mu)r_o}, \quad \text{for small } C_\mu$$

If the transistor is biased at $I_C = 200\ \mu$A and $V_A = 100$ V, $C_\mu = 0.2$ pF, and $C_L = 1$ pF, find the dc gain, the 3-dB frequency, and the frequency at which the gain reduces to unity. Sketch a Bode plot for the gain magnitude.

6.78 A common-source amplifier fed with a low-resistance signal source and operating with $g_m = 1$ mA/V has a unity-gain frequency of 2 GHz. What additional capacitance must be connected to the drain node to reduce f_t to 1 GHz?

SECTION 6.7: THE COMMON-GATE AND COMMON-BASE AMPLIFIERS WITH ACTIVE LOADS

6.79 Consider a CG amplifier for which $k_n' = 160 \ \mu A/V^2$, $\lambda = 0.1 \ V^{-1}$, $W/L = 50 \ \mu m/0.5 \ \mu m$, $\chi = 0.2$, $I = 0.5 \ mA$, and $R_L = R_s = r_o$. Find g_m, g_{mb}, r_o, R_o, A_{vo}, R_{out}, R_{in}, v_o/v_i, and v_o/v_{sig}. If the amplifier is instead fed with a current source i_{sig} having a source resistance R_s equal to r_o, find v_o/i_{sig} and i_o/i_{sig}, where i_o is the current through R_L.

6.80 Consider an NMOS CG amplifier for which the current-source load is implemented with a PMOS transistor having an output resistance r_o equal to that of the NMOS transistor. Design the circuit to obtain $v_o/v_i = 100 \ V/V$ and $R_{in} = 2 \ k\Omega$. Assume $|V_A| = 20 \ V$, $\chi = 0.2$, and $k_n' = 100 \ \mu A/V^2$. Specify I and W/L of the NMOS transistor.

6.81 Derive an expression for the overall short-circuit current gain of a CG amplifier, $G_{is} \equiv i_{osc}/i_{sig}$ in terms of A_{vo}, R_s, and r_o. Under what condition does G_{is} become close to unity? (*Hint:* Refer to the equivalent circuit in Fig. 6.30).

6.82 What is the approximate input resistance R_{in} of a CG amplifier loaded by a resistance $R_L = A_0 r_o$.

D6.83 The MOSFET current-source shown in Fig. P6.83 is required to deliver a dc current of 1 mA with $V_{GS} = 0.8 \ V$. If the MOSFET has $V_t = 0.55 \ V$, $V_A = 20 \ V$, and the body transconductance factor $\chi = 0.2$, find the value of R_s that results in a current-source output-resistance of 200 kΩ. Also, determine the required dc voltage V_{BIAS}.

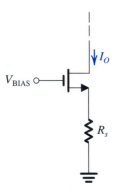

FIGURE P6.83

6.84 Figure P6.84 shows the CG amplifier with the output short-circuited. Use this circuit to obtain an expression for i_{osc} in terms of v_{sig}, and verify that this result is the same as that obtained using G_{vo} and R_{out} (i.e., using $i_{osc} = G_{vo}v_{sig}/R_{out}$).

6.85 In the common-gate amplifier circuit of Fig. P6.85, Q_2 and Q_3 are matched, $k_n'(W/L)_n = k_p'(W/L)_p = 4 \ mA/V^2$, and all transistors have $|V_t| = 0.8 \ V$ and $|V_A| = 20 \ V$.

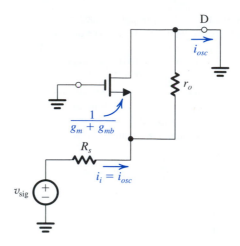

FIGURE P6.84

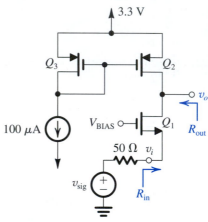

FIGURE P6.85

Transistor Q_1 has $\chi = 0.2$. The signal v_{sig} is a small sinusoidal signal with no dc component.

(a) Neglecting the effect of V_A, find the dc drain current of Q_1 and the required value of V_{BIAS}.
(b) Find the values of g_{m1} and g_{mb1} and of r_o for all transistors.
(c) Find the value of R_{in}.
(d) Find the value of R_{out}.
(e) Calculate the voltage gains v_o/v_i and v_o/v_{sig}.
(f) How large can v_{sig} be (peak-to-peak) while maintaining saturation-mode operation for Q_1 and Q_2?

6.86 A CG amplifier is specified to have $C_{gs} = 2 \ pF$, $C_{gd} = 0.1 \ pF$, $C_L = 2 \ pF$, $g_m = 5 \ mA/V$, $\chi = 0.2$, $R_{sig} = 1 \ k\Omega$, and $R_L' = 20 \ k\Omega$. Neglecting the effects of r_o, find the low-frequency gain v_o/v_{sig}, the frequencies of the poles f_{P1} and f_{P2}, and hence an estimate of the 3-dB frequency f_H.

6.87 For the CG amplifier considered in Problem 6.86, we wish to determine the low-frequency voltage gain v_o/v_{sig} and an estimate of f_H, this time taking into account the finite MOSFET r_o of 20 kΩ. If you have solved Problem 6.86, compare your results.

6.88 Use Fig. 6.33(b) together with Eq. (6.110) to derive the expression in Eq. (6.111).

6.89 Use Eq. (6.112) to explore the variation of the input resistance R_{in} with the load resistance R_L. Specifically, find R_{in} as a multiple of r_e for $R_L/r_o = 0, 1, 10, 100, 1000$, and ∞. Let $\beta = 100$. Present your results in tabular form.

6.90 Consider an active-loaded BJT connected in the common-base configuration with $I = 1$ mA. If the intrinsic gain of the BJT is 2000, what value of R_L causes the input resistance R_{in} to be double the value of r_e?

6.91 Use Fig. 6.34 to derive the expression in Eq. (6.117a).

6.92 Use Eq. (6.118) to explore the variation of the output resistance of the CB amplifier with the signal-generator resistance R_e. First, derive an expression for R_{out}/r_o as a function of β and m, where $m = R_e/r_e$. Then use this expression to generate a table for R_{out}/r_o versus R_e with entries for $R_e = r_e$, $2r_e$, $10r_e$, $(\beta/2)r_e$, βr_e, and $1000r_e$. Let $\beta = 100$.

6.93 As mentioned in the text, the CB amplifier functions as a current buffer. That is, when fed with a current signal, it passes it to the collector and supplies the output collector current at a high output resistance. Figure P6.93 shows a CB amplifier fed with a signal current i_{sig} having a source resistance $R_{\text{sig}} = 10$ kΩ. The BJT is specified to have $\beta = 100$ and $V_A = 50$ V. (Note that the bias arrangement is not shown.) The output at the collector is represented by its Norton equivalent circuit. Find the value of the current gain k and the output resistance R_{out}.

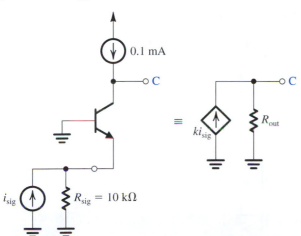

FIGURE P6.93

*6.94 Sketch the high-frequency equivalent circuit of a CB amplifier fed from a signal generator characterized by V_{sig}

and R_e and feeding a load resistance R_L in parallel with a capacitance C_L.

(a) Show that for $r_o = \infty$ the circuit can be separated into two parts: an input part that produces a pole at

$$f_{P1} = \frac{1}{2\pi C_\pi (R_e \parallel r_e)}$$

and an output part that forms a pole at

$$f_{P2} = \frac{1}{2\pi (C_\mu + C_L)R_L}$$

Note that these are the bipolar counterparts of the MOS expressions in Eqs. (6.105) and (6.106).

(b) Evaluate f_{P1} and f_{P2} and hence obtain an estimate for f_H for the case $C_\pi = 14$ pF, $C_\mu = 2$ pF, $C_L = 1$ pF, $I_C = 1$ mA, $R_{\text{sig}} = 1$ kΩ, and $R_L = 10$ kΩ. Also, find f_T of the transistor.

6.95 Adapt the expressions in Eqs. (6.107), (6.108), and (6.109) for the case of the CB amplifier.

6.96 For the constant-current source circuit shown in Fig. P6.96, find the collector current I and the output resistance. The BJT is specified to have $\beta = 100$ and $V_A = 100$ V. If the collector voltage undergoes a change of 10 V while the BJT remains in the active mode, what is the corresponding change in collector current?

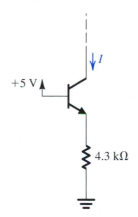

FIGURE P6.96

SECTION 6.8: THE CASCODE AMPLIFIER

6.97 For the cascode amplifier of Fig. 6.36(a), let Q_1 and Q_2 be identical with $V_t = 0.6$ V, $k'_n = 160$ μA/V^2, $\lambda = 0.05$ V^{-1}, $\chi = 0.2$, $W/L = 100$, and $V_{OV} = 0.2$ V.

(a) What must the bias current I be?
(b) Calculate the values of $g_{m1}, g_{m2}, g_{mb2}, r_{o1}, r_{o2}, A_0$, and A_{vo2}.
(c) Find the open-circuit voltage gain A_{vo}.
(d) Calculate the value of the effective short-circuit transconductance, G_m, of the cascode and the value of R_{out}.

6

(e) If the constant-current source I is implemented with a cascode circuit like that in Fig. 6.43 with an output resistance of 10 MΩ, find the voltage gain A_v.

(f) Ignoring the small signal swing at the input and at the drain of Q_1, find the lowest value that V_{BIAS} should have in order to operate Q_1 and Q_2 in saturation.

6.98 The cascode transistor can be thought of as providing a "shield" for the input transistor from the voltage variations at the output. To quantify this "shielding" property of the cascode, consider the situation in Fig. P6.98. Here we have grounded the input terminal (i.e., reduced v_i to zero), applied a small change v_x to the output node, and denoted the voltage change that results at the drain of Q_1 by v_y. By what factor is v_y smaller than v_x?

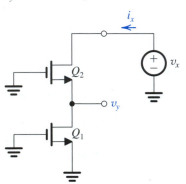

FIGURE P6.98

*6.99 In this problem we investigate whether, as an alternative to cascoding, we can simply increase the channel length L of the CS MOSFET. Specifically, we wish to compare the two circuits shown in Fig. P6.99(b) and (c). The circuit in Fig. P6.99(b) is a CS amplifier in which the channel length has been quadrupled relative to that of the original CS amplifier in Fig. P6.99(a) while the drain bias current has been kept constant.

(a) Show that for this circuit V_{OV} is double that of the original circuit, g_m is half that of the original circuit, and A_0 is double that of the original circuit.

(b) Compare these values to those of the cascode circuit in Fig. 6.99(c), which is operating at the same bias current and has the same minimum voltage requirement at the drain as in the circuit of Fig. P6.99(b).

*6.100 (a) Consider a CS amplifier having $C_{gd} = 0.2$ pF, $R_{sig} = R_L = 20$ kΩ, $g_m = 5$ mA/V, $C_{gs} = 2$ pF, C_L (including C_{db}) = 1 pF, $C_{db} = 0.2$ pF, and $r_o = 20$ kΩ. Find the low-frequency gain A_M, and estimate f_H using open-circuit time constants. Hence determine the gain–bandwidth product.

(b) If a CG stage is cascaded with the CS transistor in (a) to create a cascode amplifier, determine the new values of A_M, f_H, and gain–bandwidth product. Assume R_L remains unchanged and $\chi = 0.2$.

D6.101 It is required to design a cascode amplifier to provide a dc gain of 66 dB when driven with a low-resistance generator and utilizing NMOS transistors for which $V_A = 10$ V, $\mu_n C_{ox} = 200$ μA/V^2, $W/L = 10$, $C_{gd} = 0.1$ pF, and $C_L = 1$ pF. Assuming that $R_L = R_{out}$, determine the overdrive voltage and the drain current at which the MOSFETs should be operated. Neglect the body effect. Find the unity-gain frequency and the 3-dB frequency. If the cascode transistor is removed and R_L remains unchanged, what will the dc gain become? (*Hint:* The result is different than what can be inferred from Fig. 6.39. Be careful!)

6.102 Consider a bipolar cascode amplifier in which the current-source load is implemented with a circuit having an output resistance of βr_o. Let $\beta = 100$, $|V_A| = 100$ V, and $I = 0.1$ mA. Find R_{in}, G_m, R_{out}, and v_o/v_i. Also, find the gain of the CE stage.

6.103 Consider a bipolar cascode amplifier biased at a current of 1 mA. The transistors used have $\beta = 100$, $r_o = 100$ kΩ, $C_\pi = 14$ pF, $C_\mu = 2$ pF, $C_{cs} = 0$, and $r_x = 50$ Ω. The amplifier

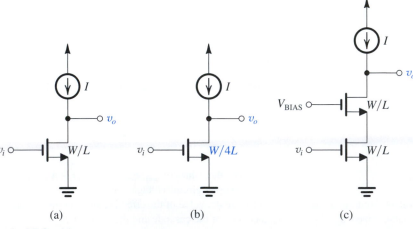

(a)   (b)   (c)

FIGURE P6.99

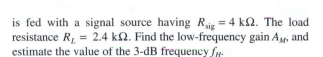

is fed with a signal source having $R_{sig} = 4$ kΩ. The load resistance $R_L = 2.4$ kΩ. Find the low-frequency gain A_M, and estimate the value of the 3-dB frequency f_H.

*6.104 In this problem we consider the frequency response of the bipolar cascode amplifier in the case that r_o can be neglected.

(a) Refer to the circuit in Fig. 6.42, and note that the total resistance between the collector of Q_1 and ground will be equal to r_{e2}, which is usually very small. It follows that the pole introduced at this node will typically be at a very high frequency and thus will have negligible effect on f_H. It also follows that at the frequencies of interest the gain from the base to the collector of Q_1 will be $-g_{m1}r_{e2} \cong -1$. Use this to find the capacitance at the input of Q_1 and hence show that the pole introduced at the input node will have a frequency

$$f_{P1} \cong \frac{1}{2\pi R'_{sig}(C_{\pi 1} + 2C_{\mu 1})}$$

Then show that the pole introduced at the output node will have a frequency

$$f_{P2} \cong \frac{1}{2\pi R_L(C_L + C_{cs2} + C_{\mu 2})}$$

(b) Evaluate f_{P1} and f_{P2}, and use the sum-of-the-squares formula to estimate f_H for the amplifier with $I = 1$ mA, $C_\pi = 5$ pF, $C_\mu = 1$ pF, $C_{cs} = C_L = 0$, $\beta = 100$, and $r_x = 0$ in the following two cases:

(i) $R_{sig} = 1$ kΩ.
(ii) $R_{sig} = 10$ kΩ.

D6.105 Design the circuit of Fig. 6.43 to provide an output current of 100 μA. Use $V_{DD} = 3.3$ V, and assume the PMOS transistors to have $\mu_p C_{ox} = 60$ μA/V^2, $V_{tp} = -0.8$ V, and $|V_A| = 5$ V. The current source is to have the widest possible signal swing at its output. Design for $V_{OV} = 0.2$ V, and specify the values of the transistor W/L ratios and of V_{BIAS1} and V_{BIAS2}. What is the highest allowable voltage at the output? What is the value of R_o?

6.106 Find the output resistance of a double-cascoded PMOS current source operating at $I_D = 0.2$ mA with each transistor having $V_{OV} = 0.25$ V. The PMOS transistors are specified to have $|V_A| = 5$ V.

*6.107 Figure P6.107 shows four possible realizations of the folded-cascode amplifier. Assume that for the BJTs $\beta = 200$ and $|V_A| = 100$ V and for the MOSFETs $k'W/L = 2$ mA/V^2,

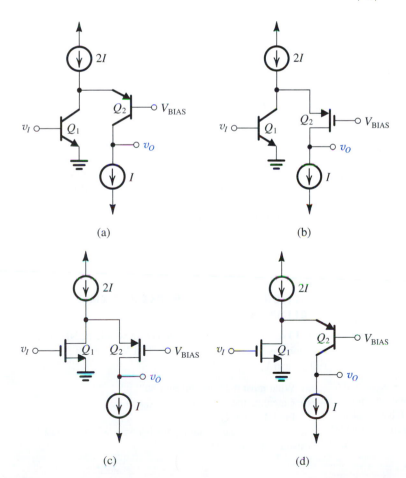

(a)

(b)

(c)

(d)

FIGURE P6.107

6

$|V_A| = 5$ V, and $|V_t| = 0.6$ V. Also, let $I = 100$ μA and $V_{\text{BIAS}} = +1$ V, and assume that the output resistance of current-source I is equal to the output resistance of its connected circuit. Current-source $2I$ should be assumed to be ideal. For each circuit, find:

(a) the bias current in Q_1.
(b) the voltage at the node between Q_1 and Q_2 (assume $|V_{BE}| = 0.7$ V).
(c) g_m and r_o for each device.
(d) the maximum allowable value of v_O.
(e) the input resistance.
(f) the output resistance.
(g) the voltage gain.

Does the current-source $2I$ have to be a sophisticated one? For this generator, what output resistance would reduce the overall gain by 1%?

SECTION 6.9: THE CS AND CE AMPLIFIERS WITH SOURCE (EMITTER) DEGENERATION

6.108 A common-source amplifier with $g_m = 2$ mA/V, $r_o = 50$ kΩ, $\chi = 0.2$, and $R_L = 50$ kΩ has a 500-Ω resistance connected in the source lead. Find R_{out}, A_{vo}, A_v, G_m, and the fraction of v_i that appears between gate and source.

D6.109 A common-source amplifier has $g_m = 2$ mA/V, $r_o = 50$ kΩ, $\chi = 0.2$, and $R_L = 50$ kΩ. Find the value of the resistance R_s that, when connected in the source, reduces the signal v_{gs} by a factor of 3 (i.e., with R_s connected $v_{gs}/v_i = \frac{1}{3}$). What is the corresponding value of voltage gain obtained?

6.110 A CS amplifier is specified to have $g_m = 5$ mA/V, $r_o = 40$ kΩ, $C_{gs} = 2$ pF, $C_{gd} = 0.1$ pF, $C_L = 1$ pF, $R_{sig} = 20$ kΩ, and $R_L = 40$ kΩ.

(a) Find the low-frequency gain A_M, and use open-circuit time constants to estimate the 3-dB frequency f_H. Hence determine the gain–bandwidth product.
(b) If a 500-Ω resistance is connected in the source lead, find the new values of $|A_M|$, f_H, and the gain–bandwidth product. Assume $g_{mb} = 1$ mA/V.

6.111 For the CS amplifier with a source-degeneration resistance R_s, show for $R_{sig} \gg R_s$ and $R_L = r_o$ that

$$\tau_H \cong \frac{C_{gs}R_{sig}}{1 + (k/2)} + C_{gd}R_{sig}\left(1 + \frac{A_0}{2+k}\right) + (C_L + C_{gd})r_o\left(\frac{1+k}{2+k}\right)$$

where $k \equiv (g_m + g_{mb})R_s$.

D*6.112 It is required to generate a table of $|A_M|$, f_H, and f_t versus $k \equiv (g_m + g_{mb})R_s$ for a CS amplifier with a source-degeneration resistance R_s. The table should have entries for $k = 0, 1, 2, \ldots, 15$. The amplifier is specified to have $g_m = 5$ mA/V, $g_{mb} = 1$ mA/V, $r_o = 40$ kΩ, $R_L = 40$ kΩ,

$R_{sig} = 20$ kΩ, $C_{gs} = 2$ pF, $C_{gd} = 0.1$ pF, and $C_L = 1$ pF. Use the formula for τ_H given in the statement for Problem 6.111. If $f_H = 2$ MHz is required, find the value needed for R_s and the corresponding value of $|A_M|$.

D6.113 (a) Use the approximate expression in Eq. (6.156) to determine the gain–bandwidth product of a CS amplifier with a source-degeneration resistance. Assume $C_{gd} = 0.1$ pF and $R_{sig} = 10$ kΩ.
(b) If a low-frequency gain of 20 V/V is required, what f_H corresponds?
(c) For $g_m = 5$ mA/V, $\chi = 0.2$, $A_0 = 100$ V/V, and $R_L = 20$ kΩ, find the required value of R_s.

6.114 A CE amplifier operating at a collector bias current of 0.5 mA has an emitter-degeneration resistance of 100 Ω. If $\beta = 100$, $V_A = 100$ V, and $R_L = r_o$, determine R_{in}, R_o, A_{vo}, G_m, A_v, and the overall voltage gain v_o/v_{sig} when $R_{sig} = 10$ kΩ.

*6.115 In this problem we investigate the effect of emitter degeneration on the frequency response of a common-emitter amplifier.

(a) Convince yourself that the MOSFET formulas in Equations (6.148) through (6.152) can be adapted to the BJT case to obtain the following:

$$R_\mu = [(R_{sig} + r_x) \| R_{in}](1 + G_m R_L') + R_L'$$
$$R_L' = R_L \| R_{out}$$
$$R_{C_L} = R_L \| R_{out} = R_L'$$
$$R_\pi = r_\pi \| \frac{R_{sig} + r_x + R_e}{1 + g_m R_e\left(\dfrac{r_o}{r_o + R_L}\right)}$$
$$\tau_H = C_\pi R_\pi + C_\mu R_\mu + C_L R_{C_L}$$

(b) Find A_M and f_H of a common-emitter amplifier having $C_\pi = 10$ pF, $C_\mu = 0.5$ pF, $C_L = 2$ pF, $g_m = 20$ mA/V, $\beta = 100$, $r_x = 200$ Ω, $r_o = 100$ kΩ, $R_L = 5.3$ kΩ, and $R_{sig} = 1$ kΩ for the following two cases:

(i) $R_e = 0$.
(ii) $R_e = 200$ Ω.

For simplicity, assume $R_{out} \cong R_o$.

SECTION 6.10: THE SOURCE AND EMITTER FOLLOWERS

6.116 Consider a source follower for which the NMOS transistor has $k'_n = 160$ μA/V^2, $\lambda = 0.05$ V^{-1}, $\chi = 0.2$, $W/L = 100$, and $V_{OV} = 0.5$ V.

(a) What must the bias current I be?
(b) Calculate the values of g_m, g_{mb}, and r_o.
(c) Find A_{vo} and R_o.
(d) What does the voltage gain become when a 1-kΩ load resistor is connected?

6.117 A source follower has $g_m = 5$ mA/V, $g_{mb} = 1$ mA/V, $r_o = 20$ kΩ, $R_{sig} = 20$ kΩ, $R_L = 20$ kΩ, $C_{gs} = 2$ pF, $C_{gd} = 0.1$ pF, and $C_L = 1$ pF. Find A_M, R_o, f_Z, and f_H. Also, find the percentage contribution of each of the three capacitances to the time-constant τ_H.

6.118 For the source follower, the term $C_L(R_L \| R_o)$ is usually very small and can be neglected in the determination of τ_H. If this is the case and, in addition, $R_{sig} \gg R_L'$, show that

$$f_H \cong 1/2\pi R_{sig}\left(C_{gd} + \frac{C_{gs}}{1 + g_m R_L'}\right)$$

where $R_L' = R_L \| r_o \| (1/g_{mb})$. For given values of C_{gd}, C_{gs}, and R_{sig}, f_H can be increased by reducing the term involving C_{gs}. This in turn can be done by increasing $g_m R_L'$. Note, however, that $g_m R_L'$ cannot exceed $1/\chi$. Why? What is the corresponding maximum for f_H? Calculate the value of the maximum f_H for the source follower specified in Problem 6.117.

6.119 For an emitter follower biased at $I_C = 5$ mA and having $R_{sig} = 10$ kΩ, $R_L = 1$ kΩ, $r_o = 20$ kΩ, $\beta = 100$, $C_\mu = 2$ pF, $r_x = 200$ Ω, and $f_T = 800$ MHz, find the low-frequency gain, f_Z, R_μ, R_π, and f_H.

6.120 For an emitter follower biased at $I_C = 1$mA and having $R_{sig} = R_L = 1$ kΩ, and using a transistor specified to have $f_T = 2$ GHz, $C_\mu = 0.1$ pF, $r_x = 100$ Ω, $\beta = 100$, and $V_A = 20$ V, evaluate the low-frequency gain A_M and the 3-dB frequency f_H.

*6.121 For the emitter follower shown in Fig. P6.121, find the low-frequency gain and the 3-dB frequency f_H for the following three cases:

(a) $R_{sig} = 1$ kΩ.
(b) $R_{sig} = 10$ kΩ.
(c) $R_{sig} = 100$ kΩ.

Let $\beta = 100$, $f_T = 400$ MHz, and $C_\mu = 2$ pF.

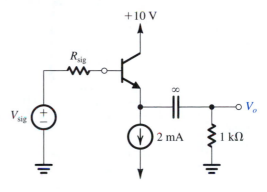

FIGURE P6.121

SECTION 6.11: SOME USEFUL TRANSISTOR PAIRINGS

D*6.122 The transistors in the circuit of Fig. P6.122 have $\beta_0 = 100$, $V_A = 100$ V, $C_\mu = 0.2$ pF, and $C_{je} = 0.8$ pF. At a bias current of 100 μA, $f_T = 400$ MHz. (Note that the bias details are not shown.)

(a) Find R_{in} and the midband gain.
(b) Find an estimate of the upper 3-dB frequency f_H. Which capacitor dominates? Which one is the second most significant?
(c) What are the effects of increasing the bias currents by a factor of 10?

D**6.123 Consider the BiCMOS amplifier shown in Fig. P6.123. The BJT has $|V_{BE}| = 0.7$ V, $\beta = 200$, $C_\mu = 0.8$ pF, and $f_T = 600$ MHz. The NMOS transistor has $V_t = 1$ V, $k_n' W/L = 2$ mA/V^2, and $C_{gs} = C_{gd} = 1$ pF.

(a) Consider the dc bias circuit. Neglect the base current of Q_2 in determining the current in Q_1, find the dc bias currents in Q_1 and Q_2, and show that they are approximately 100 μA and 1 mA, respectively.
(b) Evaluate the small-signal parameters of Q_1 and Q_2 at their bias points.
(c) Consider the circuit at midband frequencies. First, determine the small-signal voltage gain V_o/V_i. (Note that R_G can be

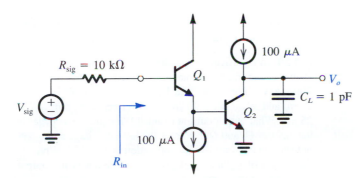

FIGURE P6.122

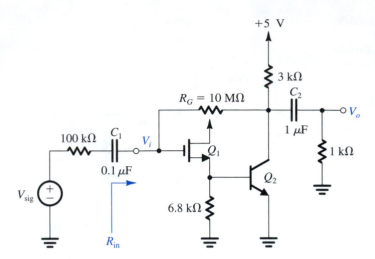

FIGURE P6.123

neglected in this process.) Then use Miller's theorem on R_G to determine the amplifier input resistance R_{in}. Finally, determine the overall voltage gain V_o/V_{sig}.

(d) Consider the circuit at low frequencies. Determine the frequency of the poles due to C_1 and C_2, and hence estimate the lower 3-dB frequency, f_L.

(e) Consider the circuit at higher frequencies. Use Miller's theorem to replace R_G with a resistance at the input. (The one at the output will be too large to matter.) Use open-circuit time constants to estimate f_H.

(f) To considerably reduce the effect of R_G on R_{in} and hence on amplifier performance, consider the effect of adding another 10-MΩ resistor in series with the existing one and placing a large bypass capacitor between their joint node and ground. What will R_{in}, A_M, and f_H become?

6.124 The BJTs in the Darlington follower of Fig. P6.124 have $\beta_0 = 100$. If the follower is fed with a source having a

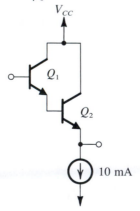

FIGURE P6.124

100-kΩ resistance and is loaded with 1 kΩ, find the input resistance and the output resistance (excluding the load). Also find the overall voltage gain, both open-circuited and with load.

6.125 For the amplifier in Fig. 6.56(a), let $I = 1$ mA, $\beta = 120$, $f_T = 700$ MHz, and $C_\mu = 0.5$ pF, and neglect r_x and r_o. Assume that a load resistance of 10 kΩ is connected to the output terminal. If the amplifier is fed with a signal V_{sig} having a source resistance $R_{sig} = 20$ kΩ, find A_M and f_H.

6.126 Consider the CD–CG amplifier of Fig. 6.56(c) for the case $g_m = 5$ mA/V, $C_{gs} = 2$ pF, $C_{gd} = 0.1$ pF, C_L (at the output node) $= 1$ pF, and $R_{sig} = R_L = 20$ kΩ. Neglecting r_o and the body effect, find A_M and f_H.

***6.127 In each of the six circuits in Fig. P6.127, let $\beta = 100$, $C_\mu = 2$ pF, and $f_T = 400$ MHz, and neglect r_x and r_o. Calculate the midband gain A_M and the 3-dB frequency f_H.

SECTION 6.12: CURRENT-MIRROR CIRCUITS WITH IMPROVED PERFORMANCE

6.128 For the cascode current mirror of Fig. 6.58 with $V_t = 0.5$ V, $k'_n W/L = 4$ mA/V^2, $V_A = 8$ V, $I_{REF} = 80$ μA, and $V_O = +5$ V, what value of I_O results? Specify the output resistance and the minimum allowable voltage at the output.

6.129 In a particular cascoded current mirror, such as that shown in Fig. 6.58, all transistors have $V_t = 0.6$ V, $\mu_n C_{ox} = 200$ μA/V^2, $L = 1$ μm, and $V_A = 20$ V. Width $W_1 = W_4 = 2$ μm, and $W_2 = W_3 = 40$ μm. The reference current I_{REF} is 25 μA. What output current results? What are the voltages at the gates of Q_2 and Q_3? What is the lowest voltage at the output for which current-source operation is possible? What are the values of g_m and r_o of Q_2 and Q_3? What is the output resistance of the mirror?

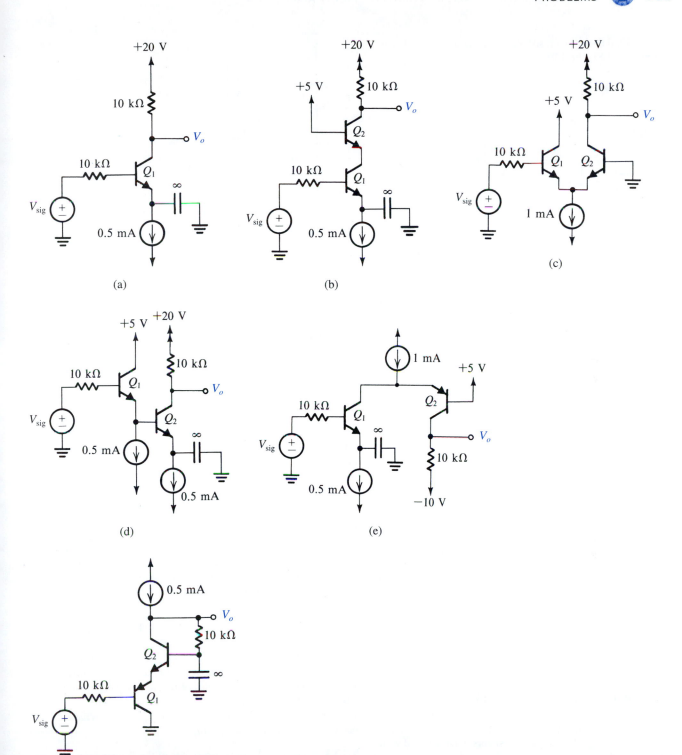

FIGURE P6.127

6.130 Find the output resistance of the double-cascode current mirror of Fig. P6.130.

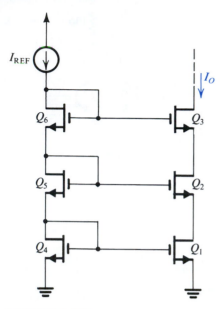

FIGURE P6.130

6.131 For the base-current-compensated mirror of Fig. 6.59, let the three transistors be matched and specified to have a collector current of 1 mA at $V_{BE} = 0.7$ V. For I_{REF} of 100 μA and assuming $\beta = 200$, what will the voltage at node x be? If I_{REF} is increased to 1 mA, what is the change in V_x? What is the value of I_O obtained with $V_O = V_x$ in both cases? Give the percentage difference between the actual and ideal value of I_O. What is the lowest voltage at the output for which proper current-source operation is maintained?

D6.132 Extend the current-mirror circuit of Fig. 6.59 to n outputs. What is the resulting current transfer ratio from the input to each output, I_O/I_{REF}? If the deviation from unity is to be kept at 0.1% or less, what is the maximum possible number of outputs for BJTs with $\beta = 100$?

*6.133 For the base-current-compensated mirror of Fig. 6.59, show that the incremental input resistance (seen by the reference current source) is approximately $2 V_T/I_{REF}$. Evaluate R_{in} for $I_{REF} = 100$ μA.

D*6.134 (a) The circuit in Fig. P6.134 is a modified version of the Wilson current mirror. Here the output transistor is "split" into two matched transistors, Q_3 and Q_4. Find I_{O1} and I_{O2} in terms of I_{REF}. Assume all transistors to be matched with current gain β.
(b) Use this idea to design a circuit that generates currents of 1 mA, 2 mA, and 4 mA using a reference current source of 7 mA. What are the actual values of the currents generated for $\beta = 50$?

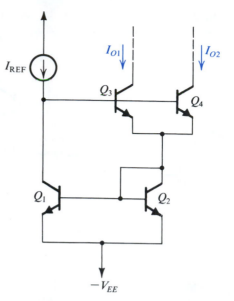

FIGURE P6.134

D6.135 Use the *pnp* version of the Wilson current mirror to design a 0.1-mA current source. The current source is required to operate with the voltage at its output terminal as low as −5 V. If the power supplies available are ±5 V, what is the highest voltage possible at the output terminal?

*6.136 For the Wilson current mirror of Fig. 6.60, show that the incremental input resistance seen by I_{REF} is approximately $2 V_T/I_{REF}$. (Neglect the Early effect in this derivation.) Evaluate R_{in} for $I_{REF} = 100$ μA.

6.137 Consider the Wilson current-mirror circuit of Fig. 6.60 when supplied with a reference current I_{REF} of 1 mA. What is the change in I_O corresponding to a change of +10 V in the voltage at the collector of Q_3? Give both the absolute value and the percentage change. Let $\beta = 100$, $V_A = 100$ V, and recall that the output resistance of the Wilson circuit is $\beta r_o/2$.

6.138 For the Wilson current mirror of Fig. 6.61(a), all transistors have $V_t = 0.6$ V, $\mu_n C_{ox} = 200$ μA/V^2, $L = 1$ μm, and $V_A = 20$ V. Width $W_1 = 2$ μm, and $W_2 = W_3 = 40$ μm. The reference current is 25 μA. What output current results? What are the voltages at the gates of Q_2 and Q_3? What is the lowest value of V_O for which current-source operation is possible? What are the values of g_m and r_o of Q_2 and Q_3? What is the output resistance of the mirror?

6.139 Show that the input resistance of the Wilson current mirror of Fig. 6.61(a) is approximately equal to $2/g_{m1}$ under the assumption that Q_2 and Q_3 are identical devices.

*6.140 A Wilson current mirror, such as that in Fig. 6.61(a), uses devices for which $V_t = 0.6$ V, $k'_n W/L = 2$ mA/V^2, and

$V_A = 20$ V. $I_{REF} = 100$ μA. What value of I_O results? If the circuit is modified to that in Fig. 6.61(c), what value of I_O results?

D*6.141 (a) Utilizing a reference current of 100 μA, design a Widlar current source to provide an output current of 10 μA. Let the BJTs have $v_{BE} = 0.7$ V at 1-mA current, and assume β to be high.

(b) If $\beta = 200$ and $V_A = 100$ V, find the value of the output resistance, and find the change in output current corresponding to a 5-V change in output voltage.

D6.142 Design three Widlar current sources, each having a 100-μA reference current: one with a current transfer ratio of 0.9, one with a ratio of 0.10, and one with a ratio of 0.01, all assuming high β. For each, find the output resistance, and contrast it with r_o of the basic unity-ratio source for which $R_E = 0$. Use $\beta = \infty$ and $V_A = 100$ V.

6.143 The BJT in the circuit of Fig. P6.143 has $V_{BE} = 0.7$ V, $\beta = 100$, and $V_A = 100$ V. Find R_o.

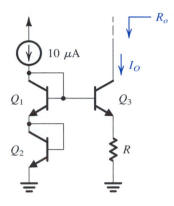

FIGURE P6.144

D*6.145 If the *pnp* transistor in the circuit of Fig. P6.145 is characterized by its exponential relationship with a scale current I_S, show that the dc current I is determined by $IR = V_T \ln(I/I_S)$. Assume Q_1 and Q_2 to be matched and Q_3, Q_4, and Q_5 to be matched. Find the value of R that yields a current $I = 10$ μA. For the BJT, $V_{EB} = 0.7$ V at $I_E = 1$ mA.

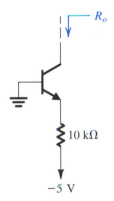

FIGURE P6.143

D6.144 (a) For the circuit in Fig. P6.144, assume BJTs with high β and $v_{BE} = 0.7$ V at 1 mA. Find the value of R that will result in $I_O = 10$ μA.

(b) For the design in (a), find R_o assuming $\beta = 100$ and $V_A = 100$ V.

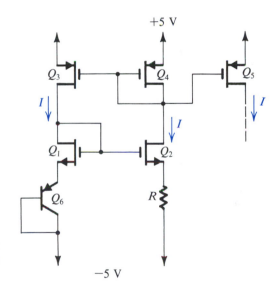

FIGURE P6.145

6

CHAPTER 7

Differential and Multistage Amplifiers

INTRODUCTION

The differential-pair or differential-amplifier configuration is the most widely used building block in analog integrated-circuit design. For instance, the input stage of every op amp is a differential amplifier. Also, the BJT differential amplifier is the basis of a very-high-speed logic circuit family, studied briefly in Chapter 11, called emitter-coupled logic (ECL).

Initially invented for use with vacuum tubes, the basic differential-amplifier configuration was subsequently implemented with discrete bipolar transistors. However, it was the advent of integrated circuits that has made the differential pair extremely popular in both bipolar and MOS technologies. There are two reasons why differential amplifiers are so well suited for IC fabrication: First, as we shall shortly see, the performance of the differential pair depends critically on the matching between the two sides of the circuit. Integrated-circuit fabrication is capable of providing matched devices whose parameters track over wide ranges of changes in environmental conditions. Second, by their very nature, differential amplifiers utilize more components (approaching twice as many) than single-ended circuits. Here again, the reader will recall from the discussion in Section 6.1 that a significant advantage of integrated-circuit technology is the availability of large numbers of transistors at relatively low cost.

We assume that the reader is familiar with the basic concept of a differential amplifier as presented in Section 2.1. Nevertheless it is worthwhile to answer the question: Why differential? Basically, there are two reasons for using differential in preference to single-ended amplifiers. First, differential circuits are much less sensitive to noise and interference than single-ended circuits. To appreciate this point, consider two wires carrying a small differential signal as the voltage difference between the two wires. Now, assume that there is an interference signal that is coupled to the two wires, either capacitively or inductively. As the two wires are physically close together, the interference voltages on the two wires (i.e., between each of the two wires and ground) will be equal. Since, in a differential system, only the difference signal between the two wires is sensed, it will contain no interference component!

The second reason for preferring differential amplifiers is that the differential configuration enables us to bias the amplifier and to couple amplifier stages together without the need for bypass and coupling capacitors such as those utilized in the design of discrete-circuit amplifiers (Sections 4.7 and 5.7). This is another reason why differential circuits are ideally suited for IC fabrication where large capacitors are impossible to fabricate economically.

The major topic of this chapter is the differential amplifier in both its MOS and bipolar implementations. As will be seen the design and analysis of differential amplifiers makes extensive use of the material on single-stage amplifiers presented in Chapter 6. We will follow the study of differential amplifiers with examples of multistage amplifiers, again in both MOS and bipolar technologies. The chapter concludes with two SPICE circuit simulation examples.

7.1 THE MOS DIFFERENTIAL PAIR

Figure 7.1 shows the basic MOS differential-pair configuration. It consists of two matched transistors, Q_1 and Q_2, whose sources are joined together and biased by a constant-current source I. The latter is usually implemented by a MOSFET circuit of the type studied in Sections 6.3 and 6.12. For the time being, we assume that the current source is ideal and that it has infinite output resistance. Although each drain is shown connected to the positive supply

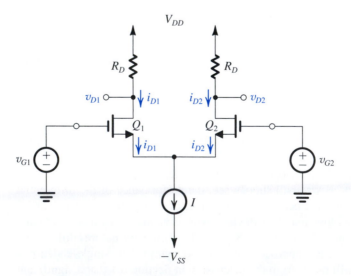

FIGURE 7.1 The basic MOS differential-pair configuration.

through a resistance R_D, in most cases active (current-source) loads are employed, as will be seen shortly. For the time being, however, we will explain the essence of the differential-pair operation utilizing simple resistive loads. Whatever type of load is used, it is essential that the MOSFETs not enter the triode region of operation.

7.1.1 Operation with a Common-Mode Input Voltage

To see how the differential pair works, consider first the case of the two gate terminals joined together and connected to a voltage v_{CM}, called the **common-mode voltage.** That is, as shown in Fig. 7.2, $v_{G1} = v_{G2} = v_{CM}$. Since Q_1 and Q_2 are matched, it follows from symmetry that the current I will divide equally between the two transistors. Thus, $i_{D1} = i_{D2} = I/2$, and the voltage at the sources, v_S, will be

$$v_S = v_{CM} - V_{GS} \tag{7.1}$$

where V_{GS} is the gate-to-source voltage corresponding to a drain current of $I/2$. Neglecting channel-length modulation, V_{GS} and $I/2$ are related by

$$\frac{I}{2} = \frac{1}{2} k'_n \frac{W}{L} (V_{GS} - V_t)^2 \tag{7.2}$$

or in terms of the overdrive voltage V_{OV},

$$V_{OV} = V_{GS} - V_t \tag{7.3}$$

$$\frac{I}{2} = \frac{1}{2} k'_n \frac{W}{L} V_{OV}^2 \tag{7.4}$$

$$V_{OV} = \sqrt{I/k'_n (W/L)} \tag{7.5}$$

The voltage at each drain will be

$$v_{D1} = v_{D2} = V_{DD} - \frac{I}{2} R_D \tag{7.6}$$

Thus, the difference in voltage between the two drains will be zero.

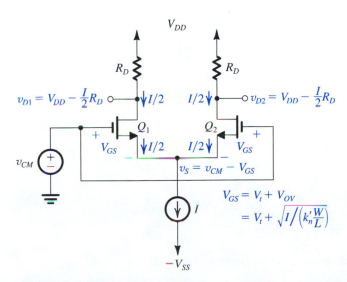

FIGURE 7.2 The MOS differential pair with a common-mode input voltage v_{CM}.

Now, let us vary the value of the common-mode voltage v_{CM}. Obviously, as long as Q_1 and Q_2 remain in the saturation region, the current I will divide equally between Q_1 and Q_2 and the voltages at the drains will not change. Thus the differential pair does not respond to (i.e., it *rejects*) common-mode input signals.

An important specification of a differential amplifier is its **input common-mode range.** This is the range of v_{CM} over which the differential pair operates properly. The highest value of v_{CM} is limited by the requirement that Q_1 and Q_2 remain in saturation, thus

$$v_{CMmax} = V_t + V_{DD} - \frac{I}{2}R_D \qquad (7.7)$$

The lowest value of v_{CM} is determined by the need to allow for a sufficient voltage across current source I for it to operate properly. If a voltage V_{CS} is needed across the current source, then

$$v_{CMmin} = -V_{SS} + V_{CS} + V_t + V_{OV} \qquad (7.8)$$

EXERCISE

7.1 For the MOS differential pair with a common-mode voltage v_{CM} applied, as shown in Fig. 7.2, let $V_{DD} = V_{SS} = 1.5$ V, $k_n'(W/L) = 4$ mA/V^2, $V_t = 0.5$ V, $I = 0.4$ mA, and $R_D = 2.5$ kΩ, and neglect channel-length modulation.

(a) Find V_{OV} and V_{GS} for each transistor.

(b) For $v_{CM} = 0$, find v_S, i_{D1}, i_{D2}, v_{D1}, and v_{D2}.

(c) Repeat (b) for $v_{CM} = +1$ V.

(d) Repeat (b) for $v_{CM} = -0.2$ V.

(e) What is the highest value of v_{CM} for which Q_1 and Q_2 remain in saturation?

(f) If current source I requires a minimum voltage of 0.4 V to operate properly, what is the lowest value allowed for v_S and hence for v_{CM}?

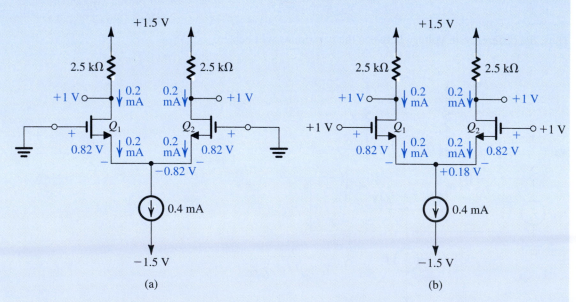

FIGURE 7.3 Circuits for Exercise 7.1. Effects of varying v_{CM} on the operation of the differential pair.

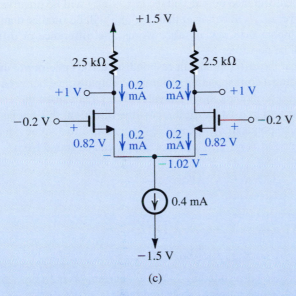

(c)

FIGURE 7.3 *(Continued)*

Ans. (a) 0.316 V, 0.82 V; (b) see Fig. 7.3(a); (c) see Fig. 7.3(b); (d) see Fig. 7.3(c) (It is assumed that 0.48 V is sufficient for the current-source to operate properly.); (e) +1.5 V; (f) −1.1 V, −0.28 V

7.1.2 Operation with a Differential Input Voltage

Next we apply a difference or differential input voltage by grounding the gate of Q_2 (i.e., setting $v_{G2} = 0$) and applying a signal v_{id} to the gate of Q_1, as shown in Fig. 7.4. It is easy to see that since $v_{id} = v_{GS1} - v_{GS2}$, if v_{id} is positive, v_{GS1} will be greater than v_{GS2} and hence i_{D1}

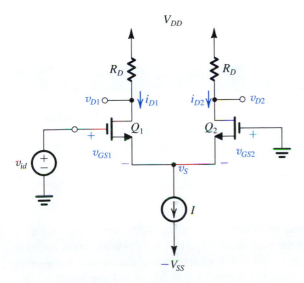

FIGURE 7.4 The MOS differential pair with a differential input signal v_{id} applied. With v_{id} positive: $v_{GS1} > v_{GS2}$, $i_{D1} > i_{D2}$, and $v_{D1} < v_{D2}$; thus $(v_{D2} - v_{D1})$ will be positive. With v_{id} negative: $v_{GS1} < v_{GS2}$, $i_{D1} < i_{D2}$, and $v_{D1} > v_{D2}$; thus $(v_{D2} - v_{D1})$ will be negative.

will be greater than i_{D2} and the difference output voltage $(v_{D2} - v_{D1})$ will be positive. On the other hand, when v_{id} is negative, v_{GS1} will be lower than v_{GS2}, i_{D1} will be smaller than i_{D2}, and correspondingly v_{D1} will be higher than v_{D2}; in other words, the difference or differential output voltage $(v_{D2} - v_{D1})$ will be negative.

From the above, we see that the differential pair responds to **difference-mode or differential input signals** by providing a corresponding differential output signal between the two drains. At this point, it is useful to inquire about the value of v_{id} that causes the entire bias current I to flow in one of the two transistors. In the positive direction, this happens when v_{GS1} reaches the value that corresponds to $i_{D1} = I$, and v_{GS2} is reduced to a value equal to the threshold voltage V_t, at which point $v_S = -V_t$. The value of v_{GS1} can be found from

$$I = \frac{1}{2}\left(k_n'\frac{W}{L}\right)(v_{GS1} - V_t)^2$$

as

$$v_{GS1} = V_t + \sqrt{2I/k_n'(W/L)}$$
$$= V_t + \sqrt{2}V_{OV} \tag{7.9}$$

where V_{OV} is the overdrive voltage corresponding to a drain current of $I/2$ (Eq. 7.5). Thus, the value of v_{id} at which the entire bias current I is steered into Q_1 is

$$v_{idmax} = v_{GS1} + v_S$$
$$= V_t + \sqrt{2}V_{OV} - V_t$$
$$= \sqrt{2}V_{OV} \tag{7.10}$$

if v_{id} is increased beyond $\sqrt{2}V_{OV}$, i_{D1} remains equal to I, v_{GS1} remains equal to $(V_t + \sqrt{2}V_{OV})$, and v_S rises correspondingly, thus keeping Q_2 off. In a similar manner we can show that in the negative direction, as v_{id} reaches $-\sqrt{2}V_{OV}$, Q_1 turns off and Q_2 conducts the entire bias current I. Thus the current I can be steered from one transistor to the other by varying v_{id} in the range

$$-\sqrt{2}V_{OV} \leq v_{id} \leq \sqrt{2}V_{OV}$$

which defines the range of differential-mode operation. Finally, observe that we have assumed that Q_1 and Q_2 remain in saturation even when one of them is conducting the entire current I.

EXERCISE

7.2 For the MOS differential pair specified in Exercise 7.1 find (a) the value of v_{id} that causes Q_1 to conduct the entire current I, and the corresponding values of v_{D1} and v_{D2}; (b) the value of v_{id} that causes Q_2 to conduct the entire current I, and the corresponding values of v_{D1} and v_{D2}; (c) the corresponding range of the differential output voltage $(v_{D2} - v_{D1})$.
Ans. (a) +0.45 V, 0.5 V, 1.5 V; (b) –0.45 V, 1.5 V, 0.5 V; (c) +1 V to –1 V

To use the differential pair as a linear amplifier, we keep the differential input signal v_{id} small. As a result, the current in one of the transistors (Q_1 when v_{id} is positive) will increase by an increment ΔI proportional to v_{id}, to $(I/2 + \Delta I)$. Simultaneously, the current in the other transistor will decrease by the same amount to become $(I/2 - \Delta I)$. A voltage signal $-\Delta I R_D$ develops at one of the drains and an opposite-polarity signal, $\Delta I R_D$, develops at the other drain. Thus the output voltage taken between the two drains will be $2\Delta I R_D$, which is proportional

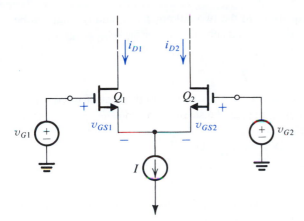

FIGURE 7.5 The MOSFET differential pair for the purpose of deriving the transfer characteristics, i_{D1} and i_{D2} versus $v_{id} = v_{G1} - v_{G2}$.

to the differential input signal v_{id}. The small-signal operation of the differential pair will be studied in detail in Section 7.2.

7.1.3 Large-Signal Operation

We shall now derive expressions for the drain currents i_{D1} and i_{D2} in terms of the input differential signal $v_{id} \equiv v_{G1} - v_{G2}$. Since these expressions do not depend on the details of the circuit to which the drains are connected we do not show these connections in Fig. 7.5; we simply assume that the circuit maintains Q_1 and Q_2 out of the triode region of operation at all times. The following derivation assumes that the differential pair is perfectly matched and neglects channel-length modulation ($\lambda = 0$) and the body effect.

To begin with, we express the drain currents of Q_1 and Q_2 as

$$i_{D1} = \frac{1}{2} k_n' \frac{W}{L} (v_{GS1} - V_t)^2 \tag{7.11}$$

$$i_{D2} = \frac{1}{2} k_n' \frac{W}{L} (v_{GS2} - V_t)^2 \tag{7.12}$$

Taking the square roots of both sides of each of Equations (7.11) and (7.12), we obtain

$$\sqrt{i_{D1}} = \sqrt{\frac{1}{2} k_n' \frac{W}{L}} (v_{GS1} - V_t) \tag{7.13}$$

$$\sqrt{i_{D2}} = \sqrt{\frac{1}{2} k_n' \frac{W}{L}} (v_{GS2} - V_t) \tag{7.14}$$

Subtracting Eq. (7.14) from Eq. (7.13) and substituting

$$v_{GS1} - v_{GS2} = v_{G1} - v_{G2} = v_{id} \tag{7.15}$$

results in

$$\sqrt{i_{D1}} - \sqrt{i_{D2}} = \sqrt{\frac{1}{2} k_n' \frac{W}{L}} v_{id} \tag{7.16}$$

The constant-current bias imposes the constraint

$$i_{D1} + i_{D2} = I \tag{7.17}$$

Equations (7.16) and (7.17) are two equations in the two unknowns i_{D1} and i_{D2} and can be solved as follows: Squaring both sides of (7.16) and substituting for $i_{D1} + i_{D2} = I$ gives

$$2\sqrt{i_{D1}i_{D2}} = I - \frac{1}{2}k'_n\frac{W}{L}v_{id}^2$$

Substituting for i_{D2} from Eq. (7.17) as $i_{D2} = I - i_{D1}$ and squaring both sides of the resulting equation provides a quadratic equation in i_{D1} that can be solved to yield

$$i_{D1} = \frac{I}{2} \pm \sqrt{k'_n\frac{W}{L}I}\left(\frac{v_{id}}{2}\right)\sqrt{1 - \frac{(v_{id}/2)^2}{I/k'_n\frac{W}{L}}}$$

Now since the increment in i_{D1} above the bias value of ($I/2$) must have the same polarity as v_{id}, only the root with the "+" sign in the second term is physically meaningful; thus

$$i_{D1} = \frac{I}{2} + \sqrt{k'_n\frac{W}{L}I}\left(\frac{v_{id}}{2}\right)\sqrt{1 - \frac{(v_{id}/2)^2}{I/k'_n\frac{W}{L}}} \tag{7.18}$$

The corresponding value of i_{D2} is found from $i_{D2} = I - i_{D1}$ as

$$i_{D2} = \frac{I}{2} - \sqrt{k'_n\frac{W}{L}I}\left(\frac{v_{id}}{2}\right)\sqrt{1 - \frac{(v_{id}/2)^2}{I/k'_n\frac{W}{L}}} \tag{7.19}$$

At the bias (quiescent) point, $v_{id} = 0$, leading to

$$i_{D1} = i_{D2} = \frac{I}{2} \tag{7.20}$$

Correspondingly,

$$v_{GS1} = v_{GS2} = V_{GS} \tag{7.21}$$

where

$$\frac{I}{2} = \frac{1}{2}k'_n\frac{W}{L}(V_{GS} - V_t)^2 = \frac{1}{2}k'_n\frac{W}{L}V_{OV}^2 \tag{7.22}$$

This relationship enables us to replace $k'_n(W/L)$ in Eqs. (7.18) and (7.19) with I/V_{OV}^2 to express i_{D1} and i_{D2} in the alternative form

$$i_{D1} = \frac{I}{2} + \left(\frac{I}{V_{OV}}\right)\left(\frac{v_{id}}{2}\right)\sqrt{1 - \left(\frac{v_{id}/2}{V_{OV}}\right)^2} \tag{7.23}$$

$$i_{D2} = \frac{I}{2} - \left(\frac{I}{V_{OV}}\right)\left(\frac{v_{id}}{2}\right)\sqrt{1 - \left(\frac{v_{id}/2}{V_{OV}}\right)^2} \tag{7.24}$$

These two equations describe the effect of applying a differential input signal v_{id} on the currents i_{D1} and i_{D2}. They can be used to obtain the normalized plots, i_{D1}/I and i_{D2}/I versus v_{id}/V_{OV}, shown in Fig. 7.6. Note that at $v_{id} = 0$, the two currents are equal to $I/2$. Making v_{id} positive causes i_{D1} to increase and i_{D2} to decrease by equal amounts so as to keep the sum constant, $i_{D1} + i_{D2} = I$. The current is steered entirely into Q_1 when v_{id} reaches the value $\sqrt{2}V_{OV}$, as we found out earlier. For v_{id} negative, identical statements can be made by interchanging i_{D1} and i_{D2}. In this case, $v_{id} = -\sqrt{2}V_{OV}$ steers the current entirely into Q_2.

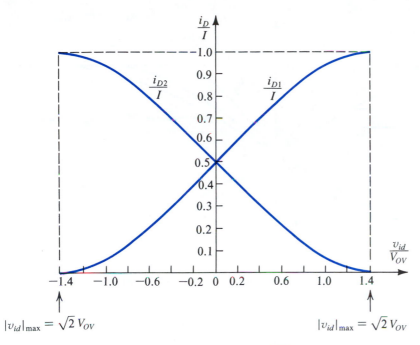

FIGURE 7.6 Normalized plots of the currents in a MOSFET differential pair. Note that V_{OV} is the overdrive voltage at which Q_1 and Q_2 operate when conducting drain currents equal to $I/2$.

The transfer characteristics of Eqs. (7.23) and (7.24) and Fig. 7.6 are obviously nonlinear. This is due to the term involving v_{id}^2. Since we are interested in obtaining linear amplification from the differential pair, we will strive to make this term as small as possible. For a given value of V_{OV}, the only thing we can do is keep $(v_{id}/2)$ much smaller than V_{OV} which is the condition for the small-signal approximation. It results in

$$i_{D1} \cong \frac{I}{2} + \left(\frac{I}{V_{OV}}\right)\left(\frac{v_{id}}{2}\right) \tag{7.25}$$

$$i_{D2} \cong \frac{I}{2} - \left(\frac{I}{V_{OV}}\right)\left(\frac{v_{id}}{2}\right) \tag{7.26}$$

which, as expected, indicate that i_{D1} increases by an increment i_d and i_{D2} decreases by the same amount, i_d, where i_d is proportional to the differential input signal v_{id},

$$i_d = \left(\frac{I}{V_{OV}}\right)\left(\frac{v_{id}}{2}\right) \tag{7.27}$$

Recalling from our study of the MOSFET in Chapter 4 and Section 6.2 (refer to Table 6.3), that a MOSFET biased at a current I_D has a transconductance $g_m = 2I_D/V_{OV}$, we recognize the factor (I/V_{OV}) in Eq. (7.27) as g_m of each of Q_1 and Q_2, which are biased at $I_D = I/2$. Now, why $v_{id}/2$? Simply because v_{id} divides equally between the two devices with $v_{gs1} = v_{id}/2$ and $v_{gs2} = -v_{id}/2$, which causes Q_1 to have a current increment i_d and Q_2 to have a current decrement i_d. We shall return to the small-signal operation of the MOS differential pair shortly. At this time, however, we wish to return to Eqs. (7.23) and (7.24) and note that linearity can be increased by increasing the overdrive voltage V_{OV} at which each of Q_1 and Q_2 is operating. This can be done by using smaller (W/L) ratios. The price paid for the increased linearity is a reduction in g_m and hence a reduction in gain. In this regard, we observe that the

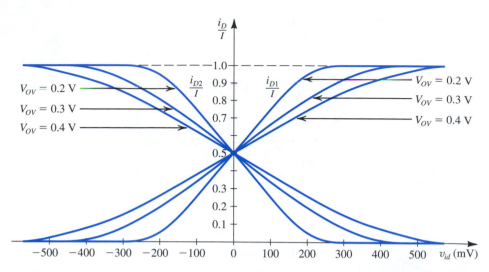

FIGURE 7.7 The linear range of operation of the MOS differential pair can be extended by operating the transistor at a higher value of V_{OV}.

normalized plot of Fig. 7.6, though compact, masks this design degree-of-freedom. Figure 7.7 shows plots of the transfer characteristics $i_{D1,2}/I$ versus v_{id} for various values of V_{OV}, assuming that the current I is kept constant. These graphs clearly illustrate the linearity–transconductance trade-off obtained by changing the value of V_{OV}: The linear range of operation can be extended by operating the MOSFETs at a higher V_{OV} (by using smaller W/L ratios) at the expense of reducing g_m and hence the gain. This trade-off is based on the assumption that the bias current I is kept constant. The bias current can, of course, be increased to obtain a higher g_m. The expense for doing this, however, is increased power dissipation, a serious limitation in IC design.

EXERCISE

7.3 A MOS differential pair is operated at a bias current I of 0.4 mA. If $\mu_n C_{ox} = 0.2$ mA/V^2, find the required values of W/L and the resulting g_m if the MOSFETs are operated at $V_{OV} = 0.2$, 0.3, and 0.4 V. For each value, give the maximum $|v_{id}|$ for which the term involving v_{id}^2 in Eqs. (7.23) and (7.24), namely $((v_{id}/2)/V_{OV})^2$, is limited to 0.1.

Ans.

V_{OV} (V)	0.2	0.3	0.4
W/L	50	22.2	12.5
g_m (mA/V)	2	1.33	1
$|v_{id}|_{max}$ (mV)	126	190	253

 ## 7.2 SMALL-SIGNAL OPERATION OF THE MOS DIFFERENTIAL PAIR

In this section we build on the understanding gained of the basic operation of the differential pair and consider in some detail its operation as a linear amplifier.

7.2.1 Differential Gain

Figure 7.8(a) shows the MOS differential amplifier with input voltages

$$v_{G1} = V_{CM} + \tfrac{1}{2}v_{id} \tag{7.28}$$

and

$$v_{G2} = V_{CM} - \tfrac{1}{2}v_{id} \tag{7.29}$$

Here, V_{CM} denotes a common-mode dc voltage within the input common-mode range of the differential amplifier. It is needed in order to set the dc voltage of the MOSFET gates.

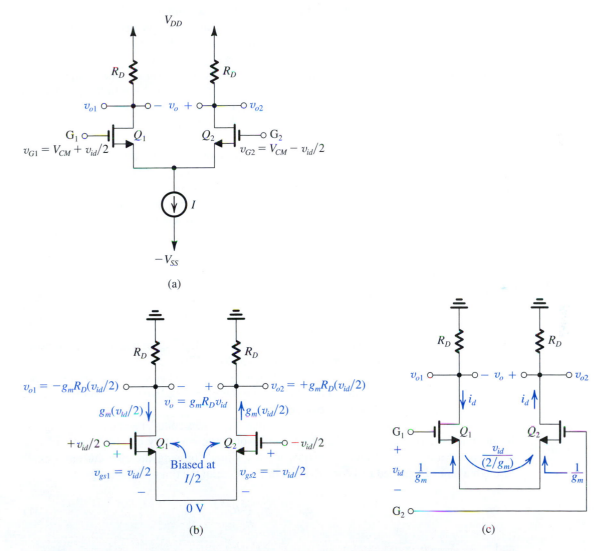

FIGURE 7.8   Small-signal analysis of the MOS differential amplifier: **(a)** The circuit with a common-mode voltage applied to set the dc bias voltage at the gates and with v_{id} applied in a complementary (or balanced) manner. **(b)** The circuit prepared for small-signal analysis. **(c)** An alternative way of looking at the small-signal operation of the circuit.

Typically V_{CM} is at the middle value of the power supply. Thus, for our case, where two complementary supplies are utilized, V_{CM} is typically 0 V.

The differential input signal v_{id} is applied in a **complementary (or balanced)** manner; that is, v_{G1} is increased by $v_{id}/2$ and v_{G2} is decreased by $v_{id}/2$. This would be the case, for instance, if the differential amplifier were fed from the output of another differential amplifier stage. Sometimes, however, the differential input is applied in a single-ended fashion, as we saw earlier in Fig. 7.4. The difference in the performance resulting is too subtle a point for our current needs.

As indicated in Fig. 7.8(a) the amplifier output can be taken either between one of the drains and ground or between the two drains. In the first case, the resulting **single-ended outputs** v_{o1} and v_{o2} will be riding on top of the dc voltages at the drains ($V_{DD} - \frac{I}{2}R_D$). This is not the case when the output is taken between the two drains; the resulting **differential** output v_o (having a 0 V dc component) will be entirely a signal component. We will see shortly that there are other significant advantages to taking the output voltage differentially.

Our objective now is to analyze the small-signal operation of the differential amplifier of Fig. 7.8(a) to determine its voltage gain in response to the differential input signal v_{id}. Toward that end we show in Fig. 7.8(b) the circuit with the power supplies removed and V_{CM} eliminated. For the time being we will neglect the effect of the MOSFET r_o, and as we have been doing since the beginning of this chapter, continue to neglect the body effect (i.e., continue to assume that $\chi = 0$). Finally note that each of Q_1 and Q_2 is biased at a dc current of $I/2$ and is operating at an overdrive voltage V_{OV}.

From the symmetry of the circuit as well as because of the balanced manner in which v_{id} is applied, we observe that the signal voltage at the joint source connection must be zero, acting as a sort of virtual ground. Thus Q_1 has a gate-to-source voltage signal $v_{gs1} = v_{id}/2$ and Q_2 has $v_{gs2} = -v_{id}/2$. Assuming $v_{id}/2 \ll V_{OV}$, the condition for the small-signal approximation, the changes resulting in the drain currents of Q_1 and Q_2 will be proportional to v_{gs1} and v_{gs2}, respectively. Thus Q_1 will have a drain current increment $g_m(v_{id}/2)$ and Q_2 will have a drain current decrement $g_m(v_{id}/2)$, where g_m denotes the equal transconductances of the two devices,

$$g_m = \frac{2I_D}{V_{OV}} = \frac{2(I/2)}{V_{OV}} = \frac{I}{V_{OV}} \qquad (7.30)$$

These results correspond to those obtained earlier using the large-signal transfer characteristics and imposing the small-signal condition, Eqs. (7.25) to (7.27).

It is useful at this point to observe again that a signal ground is established at the source terminals of the transistors without resorting to the use of a large bypass capacitor, clearly a major advantage of the differential-pair configuration.

The essence of differential-pair operation is that it provides complementary current signals in the drains; what we do with the resulting pair of complementary current signals is, in a sense, a separate issue. Here, of course, we are simply passing the two current signals through a pair of matched resistors, R_D, and thus obtaining the drain voltage signals

$$v_{o1} = -g_m \frac{v_{id}}{2} R_D \qquad (7.31)$$

and

$$v_{o2} = +g_m \frac{v_{id}}{2} R_D \qquad (7.32)$$

If the output is taken in a single-ended fashion, the resulting gain becomes

$$\frac{v_{o1}}{v_{id}} = -\frac{1}{2}g_m R_D \tag{7.33}$$

or

$$\frac{v_{o2}}{v_{id}} = \frac{1}{2}g_m R_D \tag{7.34}$$

Alternatively, if the output is taken differentially, the gain becomes

$$A_d \equiv \frac{v_{o2} - v_{o1}}{v_{id}} = g_m R_D \tag{7.35}$$

Thus, another advantage of taking the output differentially is an increase in gain by a factor of 2 (6 dB). It should be noted, however, that although differential outputs are preferred, a single-ended output is needed in some applications. We will have more to say about this later.

An alternative and useful way of viewing the operation of the differential pair in response to a differential input signal v_{id} is illustrated in Fig. 7.8(c). Here we are making use of the fact that the resistance between gate and source of a MOSFET, looking into the source, is $1/g_m$. As a result, between G_1 and G_2 we have a total resistance, in the source circuit, of $2/g_m$. It follows that we can obtain the current i_d simply by dividing v_{id} by $2/g_m$, as indicated in the figure.

Effect of the MOSFET's r_o Next we refine our analysis by considering the effect of the finite output resistance r_o of each of Q_1 and Q_2. As well, we make the realistic assumption that the bias current source I has a finite output resistance R_{SS}. The resulting differential-pair circuit, prepared for small-signal analysis, is shown in Fig. 7.9(a). Observe that the circuit remains perfectly symmetric, and as a result the voltage signal at the common source

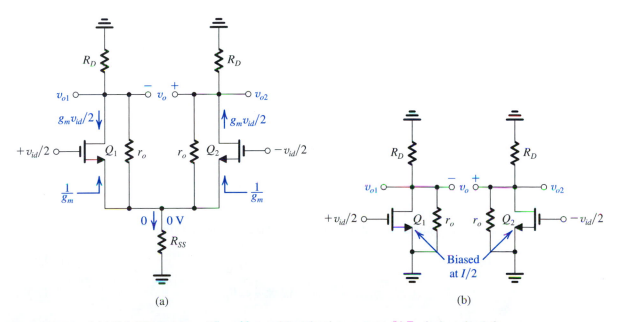

(a)   (b)

FIGURE 7.9 (a) MOS differential amplifier with r_o and R_{SS} taken into account. (b) Equivalent circuit for determining the differential gain. Each of the two halves of the differential amplifier circuit is a common-source amplifier, known as its differential "half-circuit."

connection will be zero. Thus the signal current through R_{SS} will be zero and R_{SS} plays no role in determining the differential gain.

The virtual ground on the common source connection enables us to obtain the equivalent circuit shown in Fig. 7.9(b). It consists of two identical common-source amplifiers, one fed with $+v_{id}/2$ and the other fed with $-v_{id}/2$. Obviously we need only one of the two circuits to perform any analysis we wish (including finding the frequency response, as we shall do shortly). Thus, either of the two common-source circuits is known as **the differential half-circuit.**

From the equivalent circuit in Fig. 7.9(b) we can write

$$v_{o1} = -g_m(R_D \parallel r_o)(v_{id}/2) \tag{7.36}$$

$$v_{o2} = g_m(R_D \parallel r_o)(v_{id}/2) \tag{7.37}$$

$$v_o = v_{o2} - v_{o1} = g_m(R_D \parallel r_o)v_{id} \tag{7.38}$$

EXERCISE

7.4 A MOS differential pair is operated at a total bias current of 0.8 mA, using transistors with a W/L ratio of 100, $\mu_n C_{ox} = 0.2$ mA/V^2, $V_A = 20$ V, and $R_D = 5$ kΩ. Find V_{OV}, g_m, r_o, and A_d.
Ans. 0.2 V; 4 mA/V; 50 kΩ; 18.2 V/V

7.2.2 Common-Mode Gain and Common-Mode Rejection Ratio (CMRR)

We next consider the operation of the MOS differential pair when a common-mode input signal v_{icm} is applied, as shown in Fig. 7.10(a). Here v_{icm} represents a disturbance or interference signal that is coupled somehow to both input terminals. Although not shown, the dc voltage of the input terminals must still be defined by a voltage V_{CM} as we have seen before.

The symmetry of the circuit enables us to break it into two identical halves, as shown in Fig. 7.10(b). Each of the two halves, known as a **CM half-circuit,** is a MOSFET biased at $I/2$ and having a source degeneration resistance $2R_{SS}$. Neglecting the effect of r_o, we can express the voltage gain of each of the two identical half-circuits as

$$\frac{v_{o1}}{v_{icm}} = \frac{v_{o2}}{v_{icm}} = -\frac{R_D}{\dfrac{1}{g_m} + 2R_{SS}} \tag{7.39}$$

Usually, $R_{SS} \gg 1/g_m$ enabling us to approximate Eq. (7.39) as

$$\frac{v_{o1}}{v_{icm}} = \frac{v_{o2}}{v_{icm}} \cong -\frac{R_D}{2R_{SS}} \tag{7.40}$$

Now, consider two cases:

(a) The output of the differential pair is taken single-endedly;

$$|A_{cm}| = \frac{R_D}{2R_{SS}} \tag{7.41}$$

$$|A_d| = \frac{1}{2}g_m R_D \tag{7.42}$$

Thus, the common-mode rejection ratio is given by

$$\text{CMRR} \equiv \left|\frac{A_d}{A_{cm}}\right| = g_m R_{SS} \tag{7.43}$$

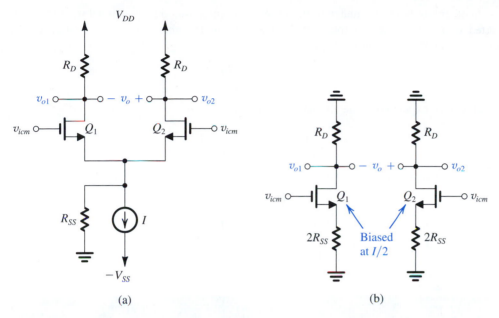

FIGURE 7.10 (a) The MOS differential amplifier with a common-mode input signal v_{icm}. (b) Equivalent circuit for determining the common-mode gain (with r_o ignored). Each half of the circuit is known as the "common-mode half-circuit."

(b) The output is taken differentially:

$$A_{cm} = \frac{v_{o2} - v_{o1}}{v_{icm}} = 0 \tag{7.44}$$

$$A_d = \frac{v_{o2} - v_{o1}}{v_{id}} = g_m R_D \tag{7.45}$$

Thus,

$$\text{CMRR} = \infty \tag{7.46}$$

Thus, even though R_{SS} is finite, taking the output differentially results in an infinite CMRR. However, this is true only when the circuit is perfectly matched.

Effect of R_D Mismatch on CMRR When the two drain resistances exhibit a mismatch of ΔR_D, as they inevitably do, the common-mode rejection ratio will be finite even if the output is taken differentially. To see how this comes about, consider the circuit in Fig. 7.10(b) for the case the load of Q_1 is R_D and that of Q_2 is $(R_D + \Delta R_D)$. The drain signal voltages arising from v_{icm} will be

$$v_{o1} \cong -\frac{R_D}{2R_{SS}} v_{icm} \tag{7.47}$$

$$v_{o2} \cong -\frac{R_D + \Delta R_D}{2R_{SS}} v_{icm} \tag{7.48}$$

Thus,

$$v_{o2} - v_{o1} = -\frac{\Delta R_D}{2R_{SS}} v_{icm} \tag{7.49}$$

In other words, the mismatch in R_D causes the common-mode input signal v_{icm} to be converted into a differential output signal; clearly an undesirable situation! Equation (7.49) indicates that the common-mode gain will be

$$A_{cm} = -\frac{\Delta R_D}{2R_{SS}} \tag{7.50}$$

which can be expressed in the alternative form

$$A_{cm} = -\frac{R_D}{2R_{SS}}\left(\frac{\Delta R_D}{R_D}\right) \tag{7.51}$$

Since the mismatch in R_D will have a negligible effect on the differential gain, we can write

$$A_d \cong -g_m R_D \tag{7.52}$$

and combine Eqs. (7.51) and (7.52) to obtain the CMRR resulting from a mismatch $(\Delta R_D/R_D)$ as

$$\text{CMRR} = \left|\frac{A_d}{A_{cm}}\right| = (2g_m R_{SS})\Big/\left(\frac{\Delta R_D}{R_D}\right) \tag{7.53}$$

EXERCISE

7.5 A MOS differential pair operated at a bias current of 0.8 mA employs transistors with $W/L = 100$ and $\mu_n C_{ox} = 0.2$ mA/V^2, using $R_D = 5$ kΩ, and $R_{SS} = 25$ kΩ.

(a) Find the differential gain, the common-mode gain, and the common-mode rejection ratio (in dB) if the output is taken single-endedly and the circuit is perfectly matched.

(b) Repeat (a) when the output is taken differentially.

(c) Repeat (a) when the output is taken differentially but the drain resistances have a 1% mismatch.

Ans. (a) 10 V/V, 0.1 V/V, 40 dB; (b) 20 V/V, 0 V/V, ∞ dB; (c) 20 V/V, 0.001 V/V, 86 dB

Effect of g_m Mismatch on CMRR Next we inquire into the effect of a mismatch between the values of the transconductance g_m of the two MOSFETs on the CMRR of the differential pair. Since the circuit is no longer matched, we cannot employ the common-mode half-circuit. Rather, we refer to the circuit shown in Fig. 7.11, and write

$$i_{d1} = g_{m1} v_{gs1} \tag{7.54}$$

$$i_{d2} = g_{m2} v_{gs2} \tag{7.55}$$

Since $v_{gs1} = v_{gs2}$ we can combine Eqs. (7.54) and (7.55) to obtain

$$\frac{i_{d1}}{i_{d2}} = \frac{g_{m1}}{g_{m2}} \tag{7.56}$$

The two drain currents sum together in R_{SS} to provide

$$v_s = (i_{d1} + i_{d2})R_{SS}$$

Thus

$$i_{d1} + i_{d2} = \frac{v_s}{R_{SS}} \tag{7.57}$$

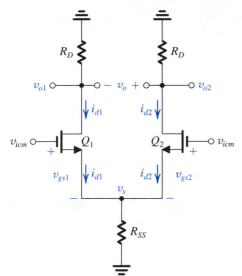

FIGURE 7.11 Analysis of the MOS differential amplifier to determine the common-mode gain resulting from a mismatch in the g_m values of Q_1 and Q_2.

Since Q_1 and Q_2 are in effect operating as source followers with a source resistance R_{SS} that is typically much larger than $1/g_m$,

$$v_s \cong v_{icm} \qquad (7.58)$$

enabling us to write Eq. (7.57) as

$$i_{d1} + i_{d2} \cong \frac{v_{icm}}{R_{SS}} \qquad (7.59)$$

We can now combine Eqs. (7.56) and (7.59) to obtain

$$i_{d1} = \frac{g_{m1} v_{icm}}{(g_{m1} + g_{m2}) R_{SS}} \qquad (7.60)$$

$$i_{d2} = \frac{g_{m2} v_{icm}}{(g_{m1} + g_{m2}) R_{SS}} \qquad (7.61)$$

If g_{m1} and g_{m2} exhibit a small mismatch Δg_m (i.e., $g_{m1} - g_{m2} = \Delta g_m$), we can assume that $g_{m1} + g_{m2} \cong 2g_m$, where g_m is the nominal value of g_{m1} and g_{m2}; thus

$$i_{d1} = \frac{g_{m1} v_{icm}}{2 g_m R_{SS}} \qquad (7.62)$$

and

$$i_{d2} = \frac{g_{m2} v_{icm}}{2 g_m R_{SS}} \qquad (7.63)$$

The differential output voltage can now be found as

$$v_{o2} - v_{o1} = -i_{d2} R_D + i_{d1} R_D$$

$$= R_D(i_{d1} - i_{d2}) = \frac{\Delta g_m R_D}{2 g_m R_{SS}} v_{icm}$$

from which the common-mode gain can be obtained as

$$A_{cm} = \left(\frac{R_D}{2R_{SS}}\right)\left(\frac{\Delta g_m}{g_m}\right) \tag{7.64}$$

Since the g_m mismatch will have a negligible effect on A_d,

$$A_d \cong -g_m R_D \tag{7.65}$$

and the CMRR resulting will be

$$\text{CMRR} \equiv \left|\frac{A_d}{A_{cm}}\right| = (2g_m R_{SS})\bigg/\left(\frac{\Delta g_m}{g_m}\right) \tag{7.66}$$

The similarity of this expression to that resulting from the R_D mismatch (Eq. 7.53) should be noted.

EXERCISE

7.6 For the MOS amplifier specified in Exercise 7.5 with the output taken differentially compute CMRR that results from a 1% mismatch in g_m.

Ans. 86 dB

 7.3 THE BJT DIFFERENTIAL PAIR

Figure 7.12 shows the basic BJT differential-pair configuration. It is very similar to the MOSFET circuit and consists of two matched transistors, Q_1 and Q_2, whose emitters are joined together and biased by a constant-current source I. The latter is usually implemented by a transistor circuit of the type studied in Sections 6.3 and 6.12. Although each collector is shown connected to the positive supply voltage V_{CC} through a resistance R_C, this connection is not essential to the operation of the differential pair—that is, in some applications the two collectors may be connected to other transistors rather than to resistive loads. It is essential, though, that the collector circuits be such that Q_1 and Q_2 never enter saturation.

7.3.1 Basic Operation

To see how the BJT differential pair works, consider first the case of the two bases joined together and connected to a common-mode voltage v_{CM}. That is, as shown in Fig. 7.13(a),

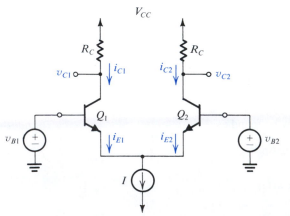

FIGURE 7.12 The basic BJT differential-pair configuration.

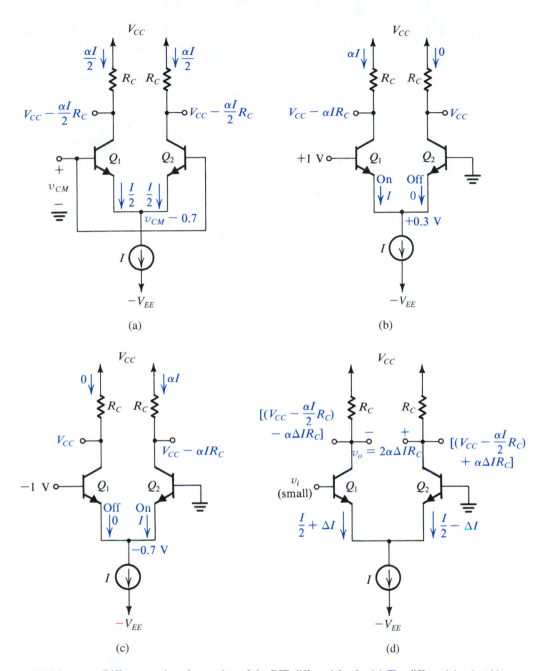

FIGURE 7.13 Different modes of operation of the BJT differential pair: **(a)** The differential pair with a common-mode input signal v_{CM}. **(b)** The differential pair with a "large" differential input signal. **(c)** The differential pair with a large differential input signal of polarity opposite to that in (b). **(d)** The differential pair with a small differential input signal v_i. Note that we have assumed the bias current source I to be ideal (i.e., it has an infinite output resistance) and thus I remains constant with the change in v_{CM}.

$v_{B1} = v_{B2} = v_{CM}$. Since Q_1 and Q_2 are matched, and assuming an ideal bias current source I with infinite output resistance, it follows that the current I will remain constant and from symmetry that I will divide equally between the two devices. Thus $i_{E1} = i_{E2} = I/2$, and the voltage at the emitters will be $v_{CM} - V_{BE}$, where V_{BE} is the base–emitter voltage (assumed in Fig 7.13a to be approximately 0.7 V) corresponding to an emitter current of $I/2$. The voltage at each collector will be $V_{CC} - \frac{1}{2}\alpha I R_C$, and the difference in voltage between the two collectors will be zero.

Now let us vary the value of the common-mode input signal v_{CM}. Obviously, as long as Q_1 and Q_2 remain in the active region the current I will still divide equally between Q_1 and Q_2, and the voltages at the collectors will not change. Thus the differential pair does not respond to (i.e., it *rejects*) common-mode input signals.

As another experiment, let the voltage v_{B2} be set to a constant value, say, zero (by grounding B_2), and let $v_{B1} = +1$ V (see Fig. 7.13b). With a bit of reasoning it can be seen that Q_1 will be on and conducting all of the current I and that Q_2 will be off. For Q_1 to be on (with $V_{BE1} = 0.7$ V), the emitter has to be at approximately $+0.3$ V, which keeps the EBJ of Q_2 reverse-biased. The collector voltages will be $v_{C1} = V_{CC} - \alpha I R_C$ and $v_{C2} = V_{CC}$.

Let us now change v_{B1} to -1 V (Fig. 7.13c). Again with some reasoning it can be seen that Q_1 will turn off, and Q_2 will carry all the current I. The common emitter will be at -0.7 V, which means that the EBJ of Q_1 will be reverse-biased by 0.3 V. The collector voltages will be $v_{C1} = V_{CC}$ and $v_{C2} = V_{CC} - \alpha I R_C$.

From the foregoing, we see that the differential pair certainly responds to large difference-mode (or differential) signals. In fact, with relatively small difference voltages we are able to steer the entire bias current from one side of the pair to the other. This current-steering property of the differential pair allows it to be used in logic circuits, as will be demonstrated in Chapter 11. Indeed, the reader can easily see that the differential pair implements the single-pole double-throw switch that we employed in the realization of the current-mode inverter of Fig. 1.33.

To use the BJT differential pair as a linear amplifier we apply a very small differential signal (a few millivolts), which will result in one of the transistors conducting a current of $I/2 + \Delta I$; the current in the other transistor will be $I/2 - \Delta I$, with ΔI being proportional to the difference input voltage (see Fig. 7.13d). The output voltage taken between the two collectors will be $2\alpha \Delta I R_C$, which is proportional to the differential input signal v_i. The small-signal operation of the differential pair will be studied next, in Section 7.3.

EXERCISE

7.7 Find v_E, v_{C1}, and v_{C2} in the circuit of Fig. E7.7. Assume that $|v_{BE}|$ of a conducting transistor is approximately 0.7 V and that $\alpha \simeq 1$.

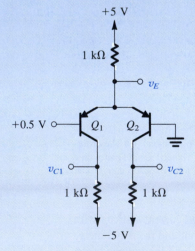

FIGURE E7.7

Ans. +0.7 V; −5 V; −0.7 V

7.3.2 Large-Signal Operation

We now present a general analysis of the BJT differential pair of Fig. 7.12. If we denote the voltage at the common emitter by v_E, the exponential relationship applied to each of the two transistors may be written

$$i_{E1} = \frac{I_S}{\alpha} e^{(v_{B1} - v_E)/V_T} \tag{7.67}$$

$$i_{E2} = \frac{I_S}{\alpha} e^{(v_{B2} - v_E)/V_T} \tag{7.68}$$

These two equations can be combined to obtain

$$\frac{i_{E1}}{i_{E2}} = e^{(v_{B1} - v_{B2})/V_T}$$

which can be manipulated to yield

$$\frac{i_{E1}}{i_{E1} + i_{E2}} = \frac{1}{1 + e^{(v_{B2} - v_{B1})/V_T}} \tag{7.69}$$

$$\frac{i_{E2}}{i_{E1} + i_{E2}} = \frac{1}{1 + e^{(v_{B1} - v_{B2})/V_T}} \tag{7.70}$$

The circuit imposes the additional constraint

$$i_{E1} + i_{E2} = I \tag{7.71}$$

Using Eq. (7.71) together with Eqs. (7.69) and (7.70) and substituting $v_{B1} - v_{B2} = v_{id}$ gives

$$i_{E1} = \frac{I}{1 + e^{-v_{id}/V_T}} \tag{7.72}$$

$$i_{E2} = \frac{I}{1 + e^{v_{id}/V_T}} \tag{7.73}$$

The collector currents i_{C1} and i_{C2} can be obtained simply by multiplying the emitter currents in Eqs. (7.72) and (7.73) by α, which is normally very close to unity.

The fundamental operation of the differential amplifier is illustrated by Eqs. (7.72) and (7.73). First, note that the amplifier responds only to the difference voltage v_{id}. That is, if $v_{B1} = v_{B2} = v_{CM}$, the current I divides equally between the two transistors irrespective of the value of the common-mode voltage v_{CM}. This is the essence of differential-amplifier operation, which also gives rise to its name.

Another important observation is that a relatively small difference voltage v_{id} will cause the current I to flow almost entirely in one of the two transistors. Figure 7.14 shows a plot of the two collector currents (assuming $\alpha \simeq 1$) as a function of the differential input signal. This is a normalized plot that can be used universally. Note that a difference voltage of about $4V_T$ ($\simeq 100$ mV) is sufficient to switch the current almost entirely to one side of the BJT pair. Note that this is much smaller than the corresponding voltage for the MOS pair, $\sqrt{2}V_{OV}$. The fact that such a small signal can switch the current from one side of the BJT differential pair to the other means that the BJT differential pair can be used as a fast current switch. Another reason for the high speed of operation of the differential device as a switch is that neither of the transistors saturates. The reader will recall from Chapter 5 that a saturated transistor stores charge in its base that must be removed before the device can turn off, generally a slow process that results in

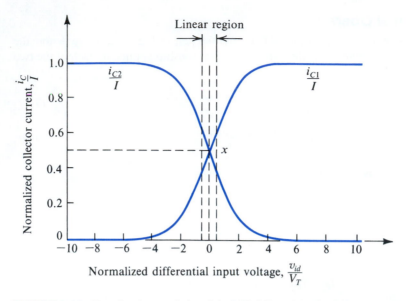

FIGURE 7.14 Transfer characteristics of the BJT differential pair of Fig. 7.12 assuming $\alpha \simeq 1$.

slow inverter. The absence of saturation[1] in the normal operation of the BJT differential pair makes the logic family based on it the fastest form of logic circuits available (see Chapter 11).

The nonlinear transfer characteristics of the differential pair, shown in Fig. 7.14, will not be utilized any further in this chapter. Rather, in the following we shall be interested specifically in the application of the differential pair as a small-signal amplifier. For this purpose the difference input signal is limited to less than about $V_T/2$ in order that we may operate on a linear segment of the characteristics around the midpoint x (in Fig. 7.14).

Before leaving the large-signal operation of the differential BJT pair, we wish to point out an effective technique frequently employed to extend the linear range of operation. It consists of including two equal resistances R_e in series with the emitters of Q_1 and Q_2, as shown in Fig. 7.15(a). The resulting transfer characteristics for three different values of R_e are sketched in Fig. 7.15(b). Observe that expansion of the linear range is obtained at the expense of reduced g_m (which is the slope of the transfer curve at $v_{id} = 0$) and hence reduced gain. This result should come as no surprise; R_e here is performing in exactly the same way as the emitter resistance R_e does in the CE amplifier with emitter degeneration (see Section 6.9.2). Finally, we also note that this linearization technique is in effect the bipolar counterpart of the technique employed for the MOS differential pair (Fig. 7.7). In the latter case, however, V_{OV} was varied by changing the transistors' W/L ratio, a design tool with no counterpart in the BJT.

EXERCISE

7.8 For the BJT differential pair of Fig. 7.12, find the value of input differential signal which is sufficient to cause $i_{E1} = 0.99I$.

Ans. 115 mV

---

[1] Recall that saturation of a BJT means something completely different from saturation of a MOSFET!

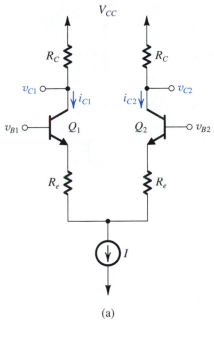

(a)

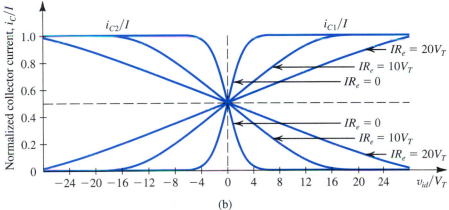

(b)

FIGURE 7.15 The transfer characteristics of the BJT differential pair (**a**) can be linearized (**b**) (i.e., the linear range of operation can be extended) by including resistances in the emitters.

7.3.3 Small-Signal Operation

In this section we shall study the application of the BJT differential pair in small-signal amplification. Figure 7.16 shows the BJT differential pair with a difference voltage signal v_{id} applied between the two bases. Implied is that the dc level at the input—that is, the common-mode input voltage—has been somehow established. For instance, one of the two input terminals can be grounded and v_{id} applied to the other input terminal. Alternatively, the differential amplifier may be fed from the output of another differential amplifier. In the latter case, the voltage at one of the input terminals will be $V_{CM} + v_{id}/2$ while that at the other input terminal will be $V_{CM} - v_{id}/2$. We will consider common-mode operation subsequently.

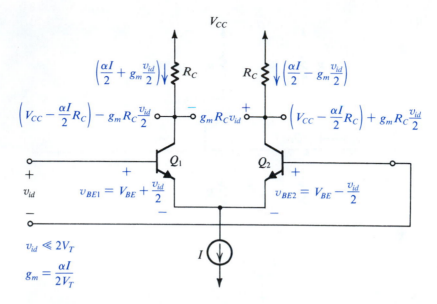

FIGURE 7.16 The currents and voltages in the differential amplifier when a small differential input signal v_{id} is applied.

The Collector Currents When v_{id} Is Applied For the circuit of Fig. 7.16, we may use Eqs. (7.72) and (7.73) to write

$$i_{C1} = \frac{\alpha I}{1 + e^{-v_{id}/V_T}} \tag{7.74}$$

$$i_{C2} = \frac{\alpha I}{1 + e^{v_{id}/V_T}} \tag{7.75}$$

Multiplying the numerator and the denominator of the right-hand side of Eq. (7.74) by $e^{v_{id}/2V_T}$ gives

$$i_{C1} = \frac{\alpha I e^{v_{id}/2V_T}}{e^{v_{id}/2V_T} + e^{-v_{id}/2V_T}} \tag{7.76}$$

Assume that $v_{id} \ll 2V_T$. We may thus expand the exponential $e^{(\pm v_{id}/2V_T)}$ in a series, and retain only the first two terms:

$$i_{C1} \simeq \frac{\alpha I (1 + v_{id}/2V_T)}{1 + v_{id}/2V_T + 1 - v_{id}/2V_T}$$

Thus

$$i_{C1} = \frac{\alpha I}{2} + \frac{\alpha I}{2V_T} \frac{v_{id}}{2} \tag{7.77}$$

Similar manipulations can be applied to Eq. (7.75) to obtain

$$i_{C2} = \frac{\alpha I}{2} - \frac{\alpha I}{2V_T} \frac{v_{id}}{2} \tag{7.78}$$

Equations (7.77) and (7.78) tell us that when $v_{id} = 0$, the bias current I divides equally between the two transistors of the pair. Thus each transistor is biased at an emitter current of

$I/2$. When a "small-signal" v_{id} is applied differentially (i.e., between the two bases), the collector current of Q_1 increases by an increment i_c and that of Q_2 decreases by an equal amount. This ensures that the sum of the total currents in Q_1 and Q_2 remains constant, as constrained by the current-source bias. The incremental (or signal) current component i_c is given by

$$i_c = \frac{\alpha I}{2V_T}\frac{v_{id}}{2} \tag{7.79}$$

Equation (7.79) has an easy interpretation. First, note from the symmetry of the circuit (Fig. 7.16) that the differential signal v_{id} should divide equally between the base–emitter junctions of the two transistors. Thus the total base–emitter voltages will be

$$v_{BE}|_{Q1} = V_{BE} + \frac{v_{id}}{2}$$

$$v_{BE}|_{Q2} = V_{BE} - \frac{v_{id}}{2}$$

where V_{BE} is the dc BE voltage corresponding to an emitter current of $I/2$. Therefore, the collector current of Q_1 will increase by $g_m v_{id}/2$ and the collector current of Q_2 will decrease by $g_m v_{id}/2$. Here g_m denotes the transconductance of Q_1 and of Q_2, which are equal and given by

$$g_m = \frac{I_C}{V_T} = \frac{\alpha I/2}{V_T} \tag{7.80}$$

Thus Eq. (7.79) simply states that $i_c = g_m v_{id}/2$.

An Alternative Viewpoint   There is an extremely useful alternative interpretation of the results above. Assume the current source I to be ideal. Its incremental resistance then will be infinite. Thus the voltage v_{id} appears across a total resistance of $2r_e$, where

$$r_e = \frac{V_T}{I_E} = \frac{V_T}{I/2} \tag{7.81}$$

Correspondingly there will be a signal current i_e, as illustrated in Fig. 7.17, given by

$$i_e = \frac{v_{id}}{2r_e} \tag{7.82}$$

Thus the collector of Q_1 will exhibit a current increment i_c and the collector of Q_2 will exhibit a current decrement i_c:

$$i_c = \alpha i_e = \frac{\alpha v_{id}}{2r_e} = g_m\frac{v_{id}}{2} \tag{7.83}$$

Note that in Fig. 7.17 we have shown signal quantities only. It is implied, of course, that each transistor is biased at an emitter current of $I/2$.

This method of analysis is particularly useful when resistances are included in the emitters, as shown in Fig. 7.18. For this circuit we have

$$i_e = \frac{v_{id}}{2r_e + 2R_e} \tag{7.84}$$

Input Differential Resistance   Unlike the MOS differential amplifier, which has an infinite input resistance, the bipolar differential pair exhibits a finite input resistance, a result of the finite β of the BJT.

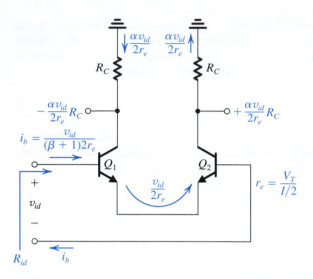

FIGURE 7.17 A simple technique for determining the signal currents in a differential amplifier excited by a differential voltage signal v_{id}; dc quantities are not shown.

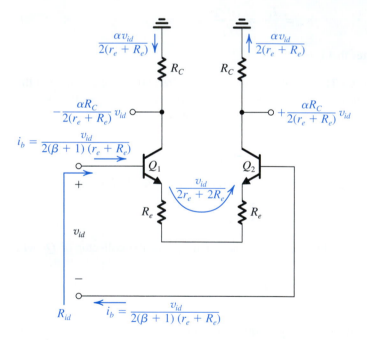

FIGURE 7.18 A differential amplifier with emitter resistances. Only signal quantities are shown (in color).

The input differential resistance is the resistance seen between the two bases; that is, it is the resistance seen by the differential input signal v_{id}. For the differential amplifier in Figs. 7.16 and 7.17 it can be seen that the base current of Q_1 shows an increment i_b and the base current of Q_2 shows an equal decrement,

$$i_b = \frac{i_e}{\beta + 1} = \frac{v_{id}/2r_e}{\beta + 1} \tag{7.85}$$

Thus the differential input resistance R_{id} is given by

$$R_{id} \equiv \frac{v_{id}}{i_b} = (\beta + 1)2r_e = 2r_\pi \tag{7.86}$$

This result is just a restatement of the familiar resistance-reflection rule; namely, *the resistance seen between the two bases is equal to the total resistance in the emitter circuit multiplied by* ($\beta + 1$). We can employ this rule to find the input differential resistance for the circuit in Fig. 7.18 as

$$R_{id} = (\beta + 1)(2r_e + 2R_e) \tag{7.87}$$

Differential Voltage Gain   We have established that for small difference input voltages ($v_{id} \ll 2V_T$; i.e., v_{id} smaller than about 20 mV) the collector currents are given by

$$i_{C1} = I_C + g_m \frac{v_{id}}{2} \tag{7.88}$$

$$i_{C2} = I_C - g_m \frac{v_{id}}{2} \tag{7.89}$$

where

$$I_C = \frac{\alpha I}{2} \tag{7.90}$$

Thus the total voltages at the collectors will be

$$v_{C1} = (V_{CC} - I_C R_C) - g_m R_C \frac{v_{id}}{2} \tag{7.91}$$

$$v_{C2} = (V_{CC} - I_C R_C) + g_m R_C \frac{v_{id}}{2} \tag{7.92}$$

The quantities in parentheses are simply the dc voltages at each of the two collectors.

As in the MOS case, the output voltage signal of a bipolar differential amplifier can be taken either *differentially* (i.e., between the two collectors) or *single-endedly* (i.e., between one collector and ground). If the output is taken differentially, then the differential gain (as opposed to the common-mode gain) of the differential amplifier will be

$$A_d = \frac{v_{c1} - v_{c2}}{v_d} = -g_m R_C \tag{7.93}$$

On the other hand, if we take the output single-endedly (say, between the collector of Q_1 and ground), then the differential gain will be given by

$$A_d = \frac{v_{c1}}{v_d} = -\frac{1}{2} g_m R_C \tag{7.94}$$

For the differential amplifier with resistances in the emitter leads (Fig. 7.18) the differential gain with the output is taken differentially is given by

$$A_d = -\frac{\alpha(2R_C)}{2r_e + 2R_e} \simeq -\frac{R_C}{r_e + R_e} \tag{7.95}$$

This equation is a familiar one: It states that *the voltage gain is equal to the ratio of the total resistance in the collector circuit* ($2R_C$) *to the total resistance in the emitter circuit* ($2r_e + 2R_e$).

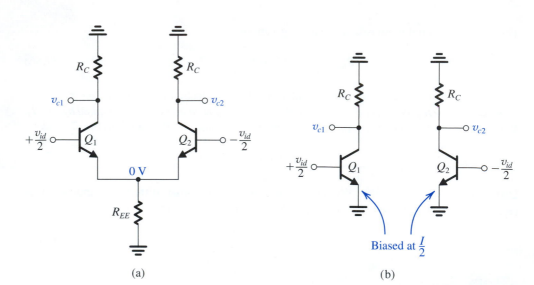

(a)   (b)

FIGURE 7.19 Equivalence of the BJT differential amplifier in **(a)** to the two common-emitter amplifiers in **(b).** This equivalence applies only for differential input signals. Either of the two common-emitter amplifiers in **(b)** can be used to find the differential gain, differential input resistance, frequency response, and so on, of the differential amplifier.

Equivalence of the Differential Amplifier to a Common-Emitter Amplifier

The analysis and results on the previous page are quite similar to those obtained in the case of a common-emitter amplifier stage. That the differential amplifier is in fact equivalent to a common-emitter amplifier is illustrated in Fig. 7.19. Figure 7.19(a) shows a differential amplifier fed by a differential signal v_{id} which is applied in a **complementary (push–pull or balanced)** manner. That is, while the base of Q_1 is raised by $v_{id}/2$, the base of Q_2 is lowered by $v_{id}/2$. We have also included the output resistance R_{EE} of the bias current source. From symmetry, it follows that the signal voltage at the emitters will be zero. Thus the circuit is equivalent to the two common-emitter amplifiers shown in Fig. 7.19(b), where each of the two transistors is biased at an emitter current of $I/2$. Note that the finite output resistance R_{EE} of the current source will have no effect on the operation. The equivalent circuit in Fig. 7.19(b) is valid for differential operation only.

In many applications the differential amplifier is not fed in a complementary fashion; rather, the input signal may be applied to one of the input terminals while the other terminal is grounded, as shown in Fig. 7.20. In this case the signal voltage at the emitters will not be zero, and thus the resistance R_{EE} will have an effect on the operation. Nevertheless, if R_{EE} is large ($R_{EE} \gg r_e$), as is usually the case,[2] then v_{id} will still divide equally (approximately) between the two junctions, as shown in Fig. 7.20. Thus the operation of the differential amplifier in this case will be almost identical to that in the case of symmetric feed, and the common-emitter equivalence can still be employed.

Since in Fig. 7.19 $v_{c2} = -v_{c1}$, the two common-emitter transistors in Fig. 7.19(b) yield similar results about the performance of the differential amplifier. Thus only one is needed to analyze the differential small-signal operation of the differential amplifier, and it is known as the **differential half-circuit.** If we take the common-emitter transistor fed with $+v_{id}/2$ as the differential half-circuit and replace the transistor with its low-frequency equivalent

---

[2] Note that R_{EE} appears in parallel with the much smaller r_e of Q_2.

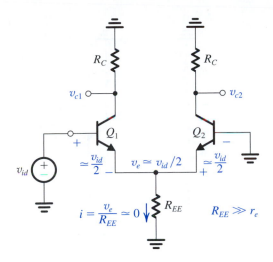

FIGURE 7.20 The differential amplifier fed in a single-ended fashion.

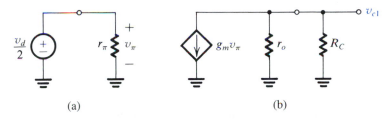

(a)   (b)

FIGURE 7.21 (a) The differential half-circuit and (b) its equivalent circuit model.

circuit model, the circuit in Fig. 7.21 results. In evaluating the model parameters r_π, g_m, and r_o, we must recall that the half-circuit is biased at $I/2$. The voltage gain of the differential amplifier (with the output taken differentially) is equal to the voltage gain of the half-circuit—that is, $v_{c1}/(v_{id}/2)$. Here, we note that including r_o will modify the gain expression in Eq. (7.93) to

$$A_d = -g_m(R_C \parallel r_o) \tag{7.96}$$

The input differential resistance of the differential amplifier is twice that of the half-circuit—that is, $2r_\pi$. Finally, we note that the differential half-circuit of the amplifier of Fig. 7.18 is a common-emitter transistor with a resistance R_e in the emitter lead.

Common-Mode Gain and CMRR Figure 7.22(a) shows a differential amplifier fed by a common-mode voltage signal v_{icm}. The resistance R_{EE} is the incremental output resistance of the bias current source. From symmetry it can be seen that the circuit is equivalent to that shown in Fig. 7.22(b), where each of the two transistors Q_1 and Q_2 is biased at an emitter current $I/2$ and has a resistance $2R_{EE}$ in its emitter lead. Thus the common-mode output voltage v_{c1} will be

$$v_{c1} = -v_{icm}\frac{\alpha R_C}{2R_{EE} + r_e} \simeq -v_{icm}\frac{\alpha R_C}{2R_{EE}} \tag{7.97}$$

At the other collector we have an equal common-mode signal v_{c2},

$$v_{c2} \simeq -v_{icm}\frac{\alpha R_C}{2R_{EE}} \tag{7.98}$$

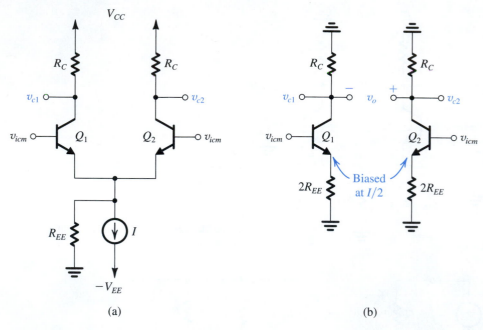

(a)   (b)

FIGURE 7.22 (a) The differential amplifier fed by a common-mode voltage signal v_{icm}. (b) Equivalent "half-circuits" for common-mode calculations.

Now, if the output is taken differentially, then the output common-mode voltage $v_o \equiv (v_{c1} - v_{c2})$ will be zero and the common-mode gain also will be zero. On the other hand, if the output is taken single-endedly, the common mode gain A_{cm} will be finite and given by[3]

$$A_{cm} = -\frac{\alpha R_C}{2R_{EE}} \qquad (7.99)$$

Since in this case the differential gain is

$$A_d = \tfrac{1}{2}g_m R_C \qquad (7.100)$$

the common-mode rejection ratio (CMRR) will be

$$\text{CMRR} = \left|\frac{A_d}{A_{cm}}\right| \simeq g_m R_{EE} \qquad (7.101)$$

Normally the CMRR is expressed in decibels,

$$\text{CMRR} = 20 \log \left|\frac{A_d}{A_{cm}}\right| \qquad (7.102)$$

Each of the circuits in Fig. 7.22(b) is called the **common-mode half-circuit.**

---

[3] The expressions in Eqs. (7.97) and (7.98) are obtained by neglecting r_o. A detailed derivation using the results of Section 6.4 shows that v_{c1}/v_{icm} and v_{c2}/v_{icm} are approximately

$$\frac{-\alpha R_C}{2R_{EE}}\left(1 - \frac{2R_{EE}}{\beta r_o}\right)$$

where it is assumed that $R_C \ll \beta r_o$ and $2R_{EE} \gg r_\pi$. This expression reduces to those in Eqs. (7.97) and (7.98) when $2R_{EE} \ll \beta r_o$.

The analysis on the facing page assumes that the circuit is perfectly symmetrical. However, practical circuits are not perfectly symmetrical, with the result that the common-mode gain will not be zero even if the output is taken differentially. To illustrate, consider the case of perfect symmetry except for a mismatch ΔR_C in the collector resistances. That is, let the collector of Q_1 have a load resistance R_C, and Q_2 have a load resistance $R_C + \Delta R_C$. It follows that

$$v_{c1} = -v_{icm}\frac{\alpha R_C}{2R_{EE} + r_e}$$

$$v_{c2} = -v_{icm}\frac{\alpha(R_C + \Delta R_C)}{2R_{EE} + r_e}$$

Thus the signal at the output due to the common-mode input signal will be

$$v_o = v_{c1} - v_{c2} = v_{icm}\frac{\alpha\Delta R_C}{2R_{EE} + r_e}$$

and the common-mode gain will be

$$A_{cm} = \frac{\alpha\Delta R_C}{2R_{EE} + r_e} \simeq \frac{\Delta R_C}{2R_{EE}}$$

This expression can be rewritten as

$$A_{cm} = \frac{R_C}{2R_{EE}}\frac{\Delta R_C}{R_C} \qquad (7.103)$$

Compare the common-mode gain in Eq. (7.103) with that for single-ended output in Eq. (7.99). We see that the common-mode gain is much smaller in the case of differential output. Therefore the input differential stage of an op amp, for example, is almost always a balanced one, with the output taken differentially. This ensures that the op amp will have the lowest possible common-mode gain or, equivalently, a high CMRR.

The input signals v_1 and v_2 to a differential amplifier usually contain a common-mode component, v_{icm},

$$v_{icm} \equiv \frac{v_1 + v_2}{2} \qquad (7.104)$$

and a differential component v_{id},

$$v_{id} \equiv v_1 - v_2 \qquad (7.105)$$

Thus the output signal will be given in general by

$$v_o = A_d(v_1 - v_2) + A_{cm}\left(\frac{v_1 + v_2}{2}\right) \qquad (7.106)$$

Input Common-Mode Resistance The definition of the **common-mode input resistance** R_{icm} is illustrated in Fig. 7.23(a). Figure 7.23(b) shows the equivalent common-mode half-circuit; its input resistance is $2R_{icm}$. The value of $2R_{icm}$ can be determined using the expression we derived in Section 6.9 for the input resistance of a CE amplifier with a resistance in the emitter. Specifically, we can use Eq. (6.157) and substitute $R_e = 2R_{EE}$ and $R_L = R_C$ to obtain for the case $R_C \ll r_o$ and $2R_{EE} \gg r_e$ the approximate expression

$$2R_{icm} \simeq (\beta + 1)(2R_{EE} \parallel r_o)$$

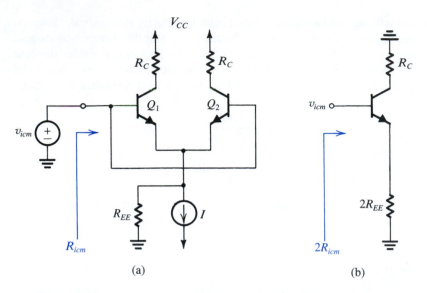

FIGURE 7.23 (a) Definition of the input common-mode resistance R_{icm}. (b) The equivalent common-mode half-circuit.

Thus,

$$R_{icm} \simeq (\beta + 1)\left(R_{EE} \parallel \frac{r_o}{2} \right) \tag{7.107}$$

Equation (7.107) indicates that since R_{EE} is typically of the order of r_o, R_{icm} will be very large.

EXAMPLE 7.1

The differential amplifier in Fig. 7.24 uses transistors with $\beta = 100$. Evaluate the following:

(a) The input differential resistance R_{id}.
(b) The overall differential voltage gain v_o/v_{sig} (neglect the effect of r_o).
(c) The worst-case common-mode gain if the two collector resistances are accurate to within $\pm 1\%$.
(d) The CMRR, in dB.
(e) The input common-mode resistance (assuming that the Early voltage $V_A = 100$ V).

Solution

(a) Each transistor is biased at an emitter current of 0.5 mA. Thus

$$r_{e1} = r_{e2} = \frac{V_T}{I_E} = \frac{25 \text{ mV}}{0.5 \text{ mA}} = 50 \ \Omega$$

The input differential resistance can now be found as

$$R_{id} = 2(\beta + 1)(r_e + R_E)$$
$$= 2 \times 101 \times (50 + 150) \cong 40 \text{ k}\Omega$$

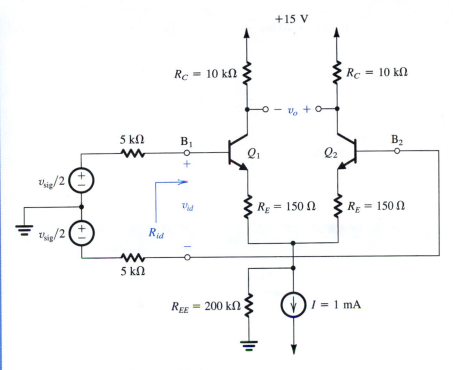

$+15$ V

$R_C = 10$ kΩ

$R_C = 10$ kΩ

$-\ v_o\ +$

5 kΩ

B_1

Q_1

Q_2

B_2

$v_{sig}/2$

v_{id}

$R_E = 150$ Ω

$R_E = 150$ Ω

R_{id}

$v_{sig}/2$

5 kΩ

$R_{EE} = 200$ kΩ

$I = 1$ mA

FIGURE 7.24 Circuit for Example 7.1.

(b) The voltage gain from the signal source to the bases of Q_1 and Q_2 is

$$\frac{v_{id}}{v_{sig}} = \frac{R_{id}}{R_{sig} + R_{id}}$$

$$= \frac{40}{5 + 5 + 40} = 0.8 \text{ V/V}$$

The voltage gain from the bases to the output is

$$\frac{v_o}{v_{id}} \cong \frac{\text{Total resistance in the collectors}}{\text{Total resistance in the emitters}}$$

$$= \frac{2R_C}{2(r_e + R_E)} = \frac{2 \times 10}{2(50 + 150) \times 10^{-3}} = 50 \text{ V/V}$$

The overall differential voltage gain can now be found as

$$A_d = \frac{v_o}{v_{sig}} = \frac{v_{id}}{v_{sig}} \frac{v_o}{v_{id}} = 0.8 \times 50 = 40 \text{ V/V}$$

(c) Using Eq. (7.103),

$$A_{cm} = \frac{R_C}{2R_{EE}} \frac{\Delta R_C}{R_C}$$

where $\Delta R_C = 0.02 R_C$ in the worst case. Thus,

$$A_{cm} = \frac{10}{2 \times 200} \times 0.02 = 5 \times 10^{-4} \text{ V/V}$$

(d)
$$\text{CMRR} = 20 \log \frac{A_d}{A_{cm}}$$

$$= 20 \log \frac{40}{5 \times 10^{-4}} = 98 \text{ dB}$$

(e)
$$r_o = \frac{V_A}{I/2} = \frac{100}{0.5} = 200 \text{ k}\Omega$$

Using Eq. (7.107),

$$R_{icm} = (\beta+1)\left(R_{EE} \| \frac{r_o}{2}\right)$$

$$= 101(200 \text{ k}\Omega \| 100 \text{ k}\Omega) = 6.7 \text{ M}\Omega$$

EXERCISE

7.9 For the circuit in Fig. 7.16, let $I = 1$ mA, $V_{CC} = 15$ V, $R_C = 10$ kΩ, with $\alpha = 1$, and let the input voltages be: $v_{B1} = 5 + 0.005 \sin 2\pi \times 1000t$, volts, and $v_{B2} = 5 - 0.005 \sin 2\pi \times 1000t$, volts. (a) If the BJTs are specified to have v_{BE} of 0.7 V at a collector current of 1 mA, find the voltage at the emitters. (*Hint:* Observe the symmetry of the circuit.) (b) Find g_m for each of the two transistors. (c) Find i_C for each of the two transistors. (d) Find v_C for each of the two transistors. (e) Find the voltage between the two collectors. (f) Find the gain experienced by the 1000-Hz signal.

Ans. (a) 4.317 V; (b) 20 mA/V; (c) $i_{C1} = 0.5 + 0.1 \sin 2\pi \times 1000t$, mA and $i_{C2} = 0.5 - 0.1 \sin 2\pi \times 1000t$, mA; (d) $v_{C1} = 10 - 1 \sin 2\pi \times 1000t$, V and $v_{C2} = 10 + 1 \sin 2\pi \times 1000t$, V; (e) $v_{C2} - v_{C1} = 2 \sin 2\pi \times 1000t$, V; (f) 200 V/V

7.4 OTHER NONIDEAL CHARACTERISTICS OF THE DIFFERENTIAL AMPLIFIER

7.4.1 Input Offset Voltage of the MOS Differential Pair

Consider the basic MOS differential amplifier with both inputs grounded, as shown in Fig. 7.25(a). If the two sides of the differential pair were perfectly matched (i.e., Q_1 and Q_2 identical and $R_{D1} = R_{D2} = R_D$), then current I would split equally between Q_1 and Q_2, and V_O would be zero. Practical circuits exhibit mismatches that result in a dc output voltage V_O even with both inputs grounded. We call V_O the **output dc offset voltage.** More commonly, we divide V_O by the differential gain of the amplifier, A_d, to obtain a quantity known as the **input offset voltage,** V_{OS},

$$V_{OS} = V_O/A_d \tag{7.108}$$

Obviously, if we apply a voltage $-V_{OS}$ between the input terminals of the differential amplifier, then the output voltage will be reduced to zero (see Fig. 7.25b). This observation gives rise to the usual definition of the input offset voltage. It should be noted, however, that since the offset voltage is a result of device mismatches, its polarity is not known a priori.

Three factors contribute to the dc offset voltage of the MOS differential pair: mismatch in load resistances, mismatch in W/L, and mismatch in V_t. We shall consider the three contributing factors one at a time.

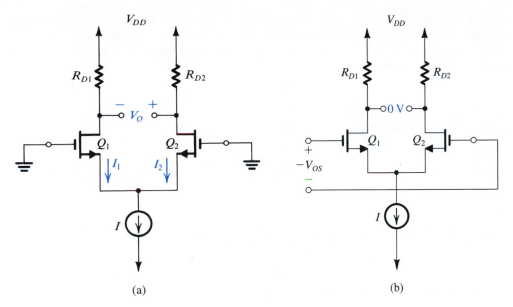

FIGURE 7.25 **(a)** The MOS differential pair with both inputs grounded. Owing to device and resistor mismatches, a finite dc output voltage V_O results. **(b)** Application of a voltage equal to the input offset voltage V_{OS} to the input terminals with opposite polarity reduces V_O to zero.

For the differential pair shown in Fig. 7.25(a) consider first the case where Q_1 and Q_2 are perfectly matched but R_{D1} and R_{D2} show a mismatch ΔR_D; that is,

$$R_{D1} = R_D + \frac{\Delta R_D}{2} \tag{7.109}$$

$$R_{D2} = R_D - \frac{\Delta R_D}{2} \tag{7.110}$$

Because Q_1 and Q_2 are matched, the current I will split equally between them. Nevertheless, because of the mismatch in load resistances, the output voltages V_{D1} and V_{D2} will be

$$V_{D1} = V_{DD} - \frac{I}{2}\left(R_D + \frac{\Delta R_D}{2}\right)$$

$$V_{D2} = V_{DD} - \frac{I}{2}\left(R_D - \frac{\Delta R_D}{2}\right)$$

Thus the differential output voltage V_O will be

$$V_O = V_{D2} - V_{D1}$$

$$= \left(\frac{I}{2}\right)\Delta R_D \tag{7.111}$$

The corresponding input offset voltage is obtained by dividing V_O by the gain $g_m R_D$ and substituting for g_m from Eq. (7.30). The result is

$$V_{OS} = \left(\frac{V_{OV}}{2}\right)\left(\frac{\Delta R_D}{R_D}\right) \tag{7.112}$$

Thus the offset voltage is directly proportional to V_{OV} and, of course, to $\Delta R_D/R_D$. As an example, consider a differential pair in which the two transistors are operating at an overdrive voltage of 0.2 V and each drain resistance is accurate to within ±1%. It follows that the worst-case resistor mismatch will be

$$\frac{\Delta R_D}{R_D} = 0.02$$

and the resulting input offset voltage will be

$$|V_{OS}| = 0.1 \times 0.02 = 2 \text{ mV}$$

Next, consider the effect of a mismatch in the W/L ratios of Q_1 and Q_2, expressed as

$$\left(\frac{W}{L}\right)_1 = \frac{W}{L} + \frac{1}{2}\Delta\left(\frac{W}{L}\right) \tag{7.113}$$

$$\left(\frac{W}{L}\right)_2 = \frac{W}{L} - \frac{1}{2}\Delta\left(\frac{W}{L}\right) \tag{7.114}$$

Such a mismatch causes the current I to no longer divide equally between Q_1 and Q_2. Rather, it can be shown that the currents I_1 and I_2 will be

$$I_1 = \frac{I}{2} + \frac{I}{2}\left(\frac{\Delta(W/L)}{2(W/L)}\right) \tag{7.115}$$

$$I_2 = \frac{I}{2} - \frac{I}{2}\left(\frac{\Delta(W/L)}{2(W/L)}\right) \tag{7.116}$$

Dividing the current increment

$$\frac{I}{2}\left(\frac{\Delta(W/L)}{2(W/L)}\right)$$

by g_m gives *half* the input offset voltage (due to the mismatch in W/L values). Thus

$$V_{OS} = \left(\frac{V_{OV}}{2}\right)\left(\frac{\Delta(W/L)}{(W/L)}\right) \tag{7.117}$$

Here again we note that V_{OS}, resulting from a (W/L) mismatch, is proportional to V_{OV} and, as expected, $\Delta(W/L)$.

Finally, we consider the effect of a mismatch ΔV_t between the two threshold voltages,

$$V_{t1} = V_t + \frac{\Delta V_t}{2} \tag{7.118}$$

$$V_{t2} = V_t - \frac{\Delta V_t}{2} \tag{7.119}$$

The current I_1 will be given by

$$I_1 = \frac{1}{2}k_n'\frac{W}{L}\left(V_{GS} - V_t - \frac{\Delta V_t}{2}\right)^2$$

$$= \frac{1}{2}k_n'\frac{W}{L}(V_{GS} - V_t)^2\left[1 - \frac{\Delta V_t}{2(V_{GS} - V_t)}\right]^2$$

which, for $\Delta V_t \ll 2(V_{GS} - V_t)$ [that is, $\Delta V_t \ll 2V_{OV}$], can be approximated as

$$I_1 \simeq \frac{1}{2}k_n'\frac{W}{L}(V_{GS} - V_t)^2\left(1 - \frac{\Delta V_t}{V_{GS} - V_t}\right)$$

Similarly,

$$I_2 \simeq \frac{1}{2}k_n' \frac{W}{L}(V_{GS} - V_t)^2 \left(1 + \frac{\Delta V_t}{V_{GS} - V_t}\right)$$

It follows that

$$\frac{1}{2}k_n' \frac{W}{L}(V_{GS} - V_t)^2 = \frac{I}{2}$$

and the current increment (decrement) in Q_2 (Q_1) is

$$\Delta I = \frac{I}{2} \frac{\Delta V_t}{V_{GS} - V_t} = \frac{I}{2} \frac{\Delta V_t}{V_{OV}}$$

Dividing ΔI by g_m gives *half* the input offset voltage (due to ΔV_t). Thus,

$$V_{OS} = \Delta V_t, \qquad (7.120)$$

a very logical result! For modern MOS technology ΔV_t can be easily as high as 2 mV. Finally, we note that since the three sources for offset voltage are not correlated, an estimate of the total input offset voltage can be found as

$$V_{OS} = \sqrt{\left(\frac{V_{OV}}{2}\frac{\Delta R_D}{R_D}\right)^2 + \left(\frac{V_{OV}}{2}\frac{\Delta(W/L)}{W/L}\right)^2 + (\Delta V_t)^2} \qquad (7.121)$$

EXERCISE

7.10 For the MOS differential pair specified in Exercise 7.4, find the three components of the input offset voltage. Let $\Delta R_D/R_D = 2\%$, $\Delta(W/L)/(W/L) = 2\%$, and $\Delta V_t = 2$ mV. Use Eq. (7.123) to obtain an estimate of the total V_{OS}.

Ans. 4 mV; 4 mV; 2 mV; 6 mV

7.4.2 Input Offset Voltage of the Bipolar Differential Pair

The offset voltage of the bipolar differential pair shown in Fig. 7.26(a) can be determined in a manner analogous to that used above for the MOS pair. Note, however, that in the bipolar case there is no analog to the V_t mismatch of the MOSFET pair. Here the output offset results from mismatches in the load resistances R_{C1} and R_{C2} and from junction area, β, and other mismatches in Q_1 and Q_2. Consider first the effect of the load mismatch. Let

$$R_{C1} = R_C + \frac{\Delta R_C}{2} \qquad (7.122)$$

$$R_{C2} = R_C - \frac{\Delta R_C}{2} \qquad (7.123)$$

and assume that Q_1 and Q_2 are perfectly matched. It follows that current I will divide equally between Q_1 and Q_2, and thus

$$V_{C1} = V_{CC} - \left(\frac{\alpha I}{2}\right)\left(R_C + \frac{\Delta R_C}{2}\right)$$

$$V_{C2} = V_{CC} - \left(\frac{\alpha I}{2}\right)\left(R_C - \frac{\Delta R_C}{2}\right)$$

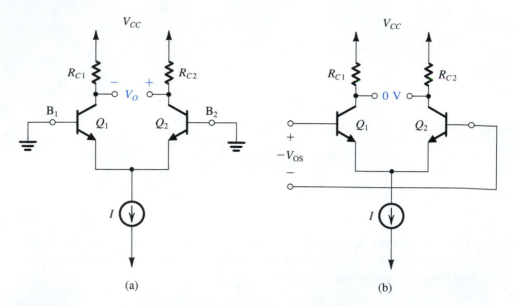

(a)   (b)

FIGURE 7.26 **(a)** The BJT differential pair with both inputs grounded. Device mismatches result in a finite dc output V_O. **(b)** Application of the input offset voltage $V_{OS} \equiv V_O/A_d$ to the input terminals with opposite polarity reduces V_O to zero.

Thus the output voltage will be

$$V_O = V_{C2} - V_{C1} = \alpha\left(\frac{I}{2}\right)(\Delta R_C)$$

and the input offset voltage will be

$$V_{OS} = \frac{\alpha(I/2)(\Delta R_C)}{A_d} \tag{7.124}$$

Substituting $A_d = g_m R_C$ and

$$g_m = \frac{\alpha I/2}{V_T}$$

gives

$$|V_{OS}| = V_T\left(\frac{\Delta R_C}{R_C}\right) \tag{7.125}$$

An important point to note is that in comparison to the corresponding expression for the MOS pair (Eq. 7.113) here the offset is proportional to V_T rather than $V_{OV}/2$. V_T at 25 mV is 4 to 10 times lower than $V_{OV}/2$. Hence bipolar differential pairs exhibit lower offsets than their MOS counterparts. As an example, consider the situation where the collector resistors are accurate to within $\pm 1\%$. Then the worst case mismatch will be

$$\frac{\Delta R_C}{R_C} = 0.02$$

and the resulting input offset voltage will be

$$|V_{OS}| = 25 \times 0.02 = 0.5 \text{ mV}$$

Next consider the effect of mismatches in transistors Q_1 and Q_2. In particular, let the transistors have a mismatch in their emitter–base junction areas. Such an area mismatch gives rise to a proportional mismatch in the scale currents I_S,

$$I_{S1} = I_S + \frac{\Delta I_S}{2} \tag{7.126}$$

$$I_{S2} = I_S - \frac{\Delta I_S}{2} \tag{7.127}$$

Refer to Fig. 7.26(a) and note that $V_{BE1} = V_{BE2}$. Thus, the current I will split between Q_1 and Q_2 in proportion to their I_S values, resulting in

$$I_{E1} = \frac{I}{2}\left(1 + \frac{\Delta I_S}{2I_S}\right) \tag{7.128}$$

$$I_{E2} = \frac{I}{2}\left(1 - \frac{\Delta I_S}{2I_S}\right) \tag{7.129}$$

It follows that the output offset voltage will be

$$V_O = \alpha\left(\frac{I}{2}\right)\left(\frac{\Delta I_S}{I_S}\right)R_C$$

and the corresponding input offset voltage will be

$$|V_{OS}| = V_T\left(\frac{\Delta I_S}{I_S}\right) \tag{7.130}$$

As an example, an area mismatch of 4% gives rise to $\Delta I_S/I_S = 0.04$ and an input offset voltage of 1 mV. Here again we note that the offset voltage is proportional to V_T rather than to the much larger V_{OV}, which determines the offset of the MOS pair due to $\Delta(W/L)$ mismatch.

Since the two contributions to the input offset voltage are not correlated, an estimate of the total input offset voltage can be found as

$$\begin{aligned} V_{OS} &= \sqrt{\left(V_T\frac{\Delta R_C}{R_C}\right)^2 + \left(V_T\frac{\Delta I_S}{I_S}\right)^2} \\ &= V_T\sqrt{\left(\frac{\Delta R_C}{R_C}\right)^2 + \left(\frac{\Delta I_S}{I_S}\right)^2} \end{aligned} \tag{7.131}$$

There are other possible sources for input offset voltage such as mismatches in the values of β and r_o. Some of these are investigated in the end-of-chapter problems. Finally, it should be noted that there is a popular scheme for compensating for the offset voltage. It involves introducing a deliberate mismatch in the values of the two collector resistances such that the differential output voltage is reduced to zero when both input terminals are grounded. Such an **offset-nulling** scheme is explored in Problem 7.57.

7.4.3 Input Bias and Offset Currents of the Bipolar Pair

In a perfectly symmetric differential pair the two input terminals carry equal dc currents; that is,

$$I_{B1} = I_{B2} = \frac{I/2}{\beta + 1} \tag{7.132}$$

This is the input bias current of the differential amplifier.

Mismatches in the amplifier circuit and most importantly a mismatch in β make the two input dc currents unequal. The resulting difference is the input offset current, I_{OS}, given as

$$I_{OS} = |I_{B1} - I_{B2}| \tag{7.133}$$

Let

$$\beta_1 = \beta + \frac{\Delta\beta}{2}$$

$$\beta_2 = \beta - \frac{\Delta\beta}{2}$$

then

$$I_{B1} = \frac{I}{2} \frac{1}{\beta + 1 + \Delta\beta/2} \simeq \frac{I}{2} \frac{1}{\beta + 1}\left(1 - \frac{\Delta\beta}{2\beta}\right) \tag{7.134}$$

$$I_{B2} = \frac{I}{2} \frac{1}{\beta + 1 - \Delta\beta/2} \simeq \frac{I}{2} \frac{1}{\beta + 1}\left(1 + \frac{\Delta\beta}{2\beta}\right) \tag{7.135}$$

$$I_{OS} = \frac{I}{2(\beta + 1)}\left(\frac{\Delta\beta}{\beta}\right) \tag{7.136}$$

Formally, the input bias current I_B is defined as follows:

$$I_B \equiv \frac{I_{B1} + I_{B2}}{2} = \frac{I}{2(\beta + 1)} \tag{7.137}$$

Thus

$$I_{OS} = I_B\left(\frac{\Delta\beta}{\beta}\right) \tag{7.138}$$

As an example, a 10% β mismatch results in an offset that is current one-tenth the value of the input bias current.

Finally note that obviously a great advantage of the MOS differential pair is that it does not suffer from a finite input bias current or from mismatches thereof!

7.4.4 Input Common-Mode Range

As mentioned earlier, the input common-mode range of a differential amplifier is the range of the input voltage v_{CM} over which the differential pair behaves as a linear amplifier for differential input signals. The upper limit of the common-mode range is determined by Q_1 and Q_2 leaving the active mode and entering the saturation mode of operation in the BJT case or the triode mode of operation in the MOS case. Thus, for the bipolar case the upper limit is approximately equal to 0.4 V above the dc collector voltage of Q_1 and Q_2. For the MOS case, the upper limit is equal to V_t volts above the voltage at the drains of Q_1 and Q_2. The lower limit is determined by the transistor that supplies the biasing current I leaving its active region of operation and thus no longer functioning as a constant-current source. Current-source circuits were studied in Sections 6.3 and 6.12.

7.4.5 A Concluding Remark

We conclude this section by noting that the definitions presented here are identical to those presented in Chapter 2 for op amps. In fact, as will be seen in Chapter 9, it is the input differential stage in an op-amp circuit that primarily determines the op-amp dc offset voltage, input bias and offset currents, and input common-mode range.

 ## 7.5 THE DIFFERENTIAL AMPLIFIER WITH ACTIVE LOAD

As we learned in Chapter 6, replacing the drain resistance R_D with a constant-current source results in a much higher voltage gain as well as savings in chip area. The same, of course, applies to the differential amplifier. In this section we study an ingenious circuit for implementing an active-loaded differential amplifier and at the same time converting the output from differential to single-ended. We shall study both the MOS and bipolar forms of this popular circuit.

7.5.1 Differential-to-Single-Ended Conversion

In the previous sections we found that taking the output of the differential amplifier as the voltage between the two drains (or collectors) results in double the value of the differential gain as well as a much reduced common-mode gain. In fact, the only reason a small fraction of an input common-mode signal appears between the differential output terminals is the mismatches inevitably present in the circuit. Thus if a multistage amplifier (such as an op amp) is to achieve a high CMRR, the output of its first stage must be taken differentially. Beyond the first stage, however, unless the system is fully differential, the signal is converted from differential to single-ended.

Figure 7.27 illustrates the simplest most-basic approach for differential-to-single-ended conversion. It consists of simply ignoring the drain current signal of Q_1 and eliminating its drain

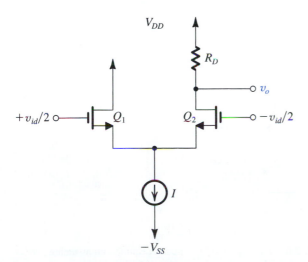

FIGURE 7.27 A simple but inefficient approach for differential to single-ended conversion.

resistor altogether, and taking the output between the drain of Q_2 and ground. The obvious draw-back of this scheme is that we lose a factor of 2 (or 6 dB) in gain as a result of "wasting" the drain signal current of Q_1. A much better approach would be to find a way of utilizing the drain-current signal of Q_1, and that is exactly what the circuit we are about to discuss accomplishes.

7.5.2 The Active-Loaded MOS Differential Pair

Figure 7.28(a) shows a MOS differential pair formed by transistors Q_1 and Q_2, loaded in a current mirror formed by transistors Q_3 and Q_4. To see how this circuit operates consider first the quiescent state with the two input terminals connected to a dc voltage equal to the common-mode equilibrium value, in this case 0 V, as shown in Fig. 7.28(b). Assuming per-fect matching, the bias current I divides equally between Q_1 and Q_2. The drain current of Q_1, $I/2$, is fed to the input transistor of the mirror, Q_3. Thus, a replica of this current is provided by the output transistor of the mirror, Q_4. Observe that at the output node the two currents $I/2$ balance each other out, leaving a zero current to flow out to the next stage or to a load (not shown). If Q_4 is perfectly matched to Q_3, its drain voltage will track the voltage at the drain

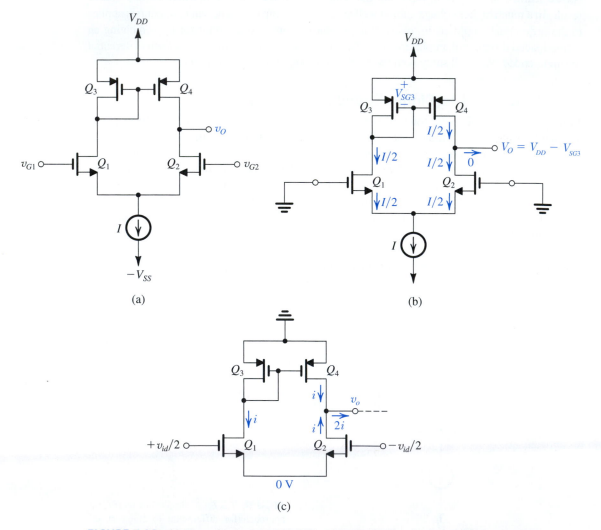

FIGURE 7.28 **(a)** The active-loaded MOS differential pair. **(b)** The circuit at equilibrium assuming perfect matching. **(c)** The circuit with a differential input signal applied, neglecting the r_o of all transistors.

of Q_3; thus in equilibrium the voltage at the output will be $V_{DD} - V_{SG3}$. It should be noted, however, that in practical implementations, there will always be mismatches, resulting in a net dc current at the output. In the absence of a load resistance, this current will flow into the output resistances of Q_2 and Q_4 and thus can cause a large deviation in the output voltage from the ideal value. Therefore, this circuit is always designed so that the dc bias voltage at the output node is defined by a feedback circuit rather than by simply relying on the matching of Q_4 and Q_3. We shall see how this is done later.

Next, consider the circuit with a differential input signal v_{id} applied to the input, as shown in Fig. 7.28(c). Since we are now investigating the small-signal operation of the circuit, we have removed the dc supplies (including the current source I). Also, for the time being let us ignore r_o of all transistors. As Fig. 7.28(c) shows, a virtual ground will develop at the common-source terminal of Q_1 and Q_2. Transistor Q_1 will conduct a drain signal current $i = g_{m1}v_{id}/2$, and transistor Q_2 will conduct an equal but opposite current i. The drain signal current i of Q_1 is fed to the input of the $Q_3 - Q_4$ mirror, which responds by providing a replica in the drain of Q_4. Now, at the output node we have two currents, each equal to i, which sum together to provide an output current $2i$. It is this factor of 2, which is a result of the current mirror action, that makes it possible to convert the signal to single-ended form (i.e., between the output node and ground) with no loss of gain! If a load resistance is connected to the output node, the current $2i$ flows through it and thus determines the output voltage v_o. In the absence of a load resistance, the output voltage is determined by the output current $2i$ and the output resistance of the circuit, as we shall shortly see.

7.5.3 Differential Gain of the Active-Loaded MOS Pair

As we have learned in Chapter 6, the output resistance r_o of the transistor plays a significant role in the operation of active-loaded amplifiers. Therefore, we shall now take r_o into account and derive an expression for the differential gain v_o/v_{id} of the active-loaded MOS differential pair. Unfortunately, because the circuit is *not* symmetrical we will not be able to use the differential half-circuit technique. Rather, we shall perform the derivation from first principles: We will first find the short-circuit transconductance G_m and the output resistance R_o. Then, the gain will be determined as $G_m R_o$.

Determining the Transconductance G_m   Figure 7.29(a) shows the circuit prepared for determining G_m. Note that we have short-circuited the output to ground in order to find G_m as i_o/v_{id}. Although the original circuit is not perfectly symmetrical, when the output is shorted to ground, the circuit becomes almost symmetrical. This is because the voltage between the drain of Q_1 and ground is very small. This in turn is due to the low resistance between that node and ground which is almost equal to $1/g_{m3}$. Thus, we can invoke symmetry and assume that a virtual ground will appear at the source of Q_1 and Q_2 and in this way obtain the equivalent circuit shown in Fig. 7.29(b). Here we have replaced the diode-connected transistor Q_3 by its equivalent resistance $[(1/g_{m3})\|r_{o3}]$. The voltage v_{g3} that develops at the common-gate line of the mirror can be found as

$$v_{g3} = -g_{m1}\left(\frac{v_{id}}{2}\right)\left(\frac{1}{g_{m3}} \| r_{o3} \| r_{o1}\right) \tag{7.139}$$

which for the usual case of r_{o1} and $r_{o3} \gg (1/g_{m3})$ reduces to

$$v_{g3} \cong -\left(\frac{g_{m1}}{g_{m3}}\right)\left(\frac{v_{id}}{2}\right) \tag{7.140}$$

This voltage controls the drain current of Q_4 resulting in a current of $g_{m4}v_{g3}$. Note that the ground at the output node causes the currents in r_{o2} and r_{o4} to be zero. Thus the output current

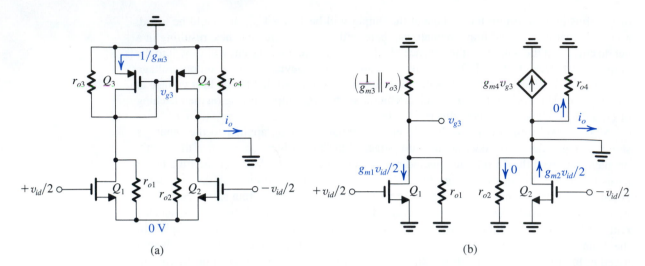

FIGURE 7.29 Determining the short-circuit transconductance $G_m \equiv i_o / v_{id}$ of the active-loaded MOS differential pair.

i_o will be

$$i_o = -g_{m4}v_{g3} + g_{m2}\left(\frac{v_{id}}{2}\right) \qquad (7.141)$$

Substituting for v_{g3} from (7.140) gives

$$i_o = g_{m1}\left(\frac{g_{m4}}{g_{m3}}\right)\left(\frac{v_{id}}{2}\right) + g_{m2}\left(\frac{v_{id}}{2}\right)$$

Now, since $g_{m3} = g_{m4}$ and $g_{m1} = g_{m2} = g_m$, the current i_o becomes

$$i_o = g_m v_{id}$$

from which G_m is found to be

$$G_m = g_m \qquad (7.142)$$

Thus the short-circuit transconductance of the circuit is equal to g_m of each of the two transistors of the differential pair. Here we should note that in the absence of the current–mirror action, G_m would be equal to $g_m/2$.

Determining the Output Resistance R_o Figure 7.30 shows the circuit for determining R_o. Observe that the current i that enters Q_2 must exit at its source. It then enters Q_1, exiting at the drain to feed the $Q_3 - Q_4$ mirror. Since for the diode-connected transistor Q_3, $1/g_{m3}$ is much smaller than r_{o3}, most of the current i will flow into the drain of Q_3. The mirror responds by providing an equal current i in the drain of Q_4. It now remains to determine the relationship between i and v_x. From Fig. 7.30 we see that

$$i = v_x / R_{o2} \qquad (7.143)$$

where R_{o2} is the output resistance of Q_2. Now, Q_2 is a CG transistor and has in its source lead the input resistance of Q_1. The latter is connected in the CG configuration with a small resistance in the drain (approximately equal to $1/g_{m3}$), thus its input resistance is approximately $1/g_{m1}$. We

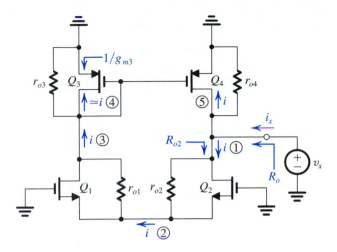

FIGURE 7.30   Circuit for determining R_o. The circled numbers indicate the order of the analysis steps.

can now use Eq. (6.101) to determine R_{o2} by substituting $g_{mb} = 0$ and $R_s = 1/g_{m1}$ to obtain

$$R_{o2} = r_{o2} + (1 + g_{m2}r_{o2})(1/g_{m1})$$

which for $g_{m1} = g_{m2} = g_m$ and $g_{m2}r_{o2} \gg 1$ yields

$$R_{o2} \cong 2r_{o2} \qquad (7.144)$$

Returning to the circuit in Fig. 7.30, we can write at the output node

$$i_x = i + i + \frac{v_x}{r_{o4}}$$

$$= 2i + \frac{v_x}{r_{o4}} = 2\frac{v_x}{R_{o2}} + \frac{v_x}{r_{o4}}$$

Substituting for R_{o2} from Eq. (7.144) we obtain

$$i_x = \frac{v_x}{r_{o2}} + \frac{v_x}{r_{o4}}$$

Thus,

$$R_o \equiv \frac{v_x}{i_x} = r_{o2} \| r_{o4} \qquad (7.145)$$

which is an intuitively appealing result.

Determining the Differential Gain   Equations (7.142) and (7.145) can be combined to obtain the differential gain A_d as

$$A_d \equiv \frac{v_o}{v_{id}} = G_m R_o = g_m(r_{o2} \| r_{o4}) \qquad (7.146)$$

For the case $r_{o2} = r_{o4} = r_o$,

$$A_d = \frac{1}{2}g_m r_o = \frac{A_0}{2} \qquad (7.147)$$

where A_0 is the intrinsic gain of the MOS transistor.

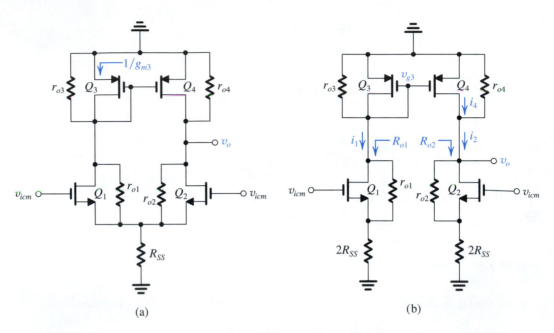

(a)   (b)

FIGURE 7.31 Analysis of the active-loaded MOS differential amplifier to determine its common-mode gain.

7.5.4 Common-Mode Gain and CMRR

Although its output is single-ended, the active-loaded MOS differential amplifier has a low common-mode gain and, correspondingly, a high CMRR. Figure 7.31(a) shows the circuit with v_{icm} applied and with the power supplies eliminated except, of course, for the output resistance R_{SS} of the bias-current source I. Although the circuit is not symmetrical and hence we cannot use the common-mode half-circuit, we can split R_{SS} equally between Q_1 and Q_2 as shown in Fig. 7.31b. It can now be seen that each of Q_1 and Q_2 is a CS transistor with a large source degeneration resistance $2R_{SS}$. We can use the formulas derived in Section 6.9.1 to determine the currents i_1 and i_2 that result from the application of an input signal v_{icm}. Alternatively, we observe that since $2R_{SS}$ is usually much larger than $1/g_m$ of each of Q_1 and Q_2, the signals at the source terminals will be approximately equal to v_{icm}. Also, the effect of r_{o1} and r_{o2} can be shown to be negligible. Thus, we can write

$$i_1 = i_2 \cong \frac{v_{icm}}{2R_{SS}} \tag{7.148}$$

The output resistance of each of Q_1 and Q_2 is given by Eq. (6.101) which for $R_s = 2R_{SS}$ and $g_{mb} = 0$ yields

$$R_{o1} = R_{o2} = r_o + 2R_{SS} + 2g_m r_o R_{SS} \tag{7.149}$$

where $r_{o1} = r_{o2} = r_o$ and $g_{m1} = g_{m2} = g_m$. Note that R_{o1} will be much greater than the parallel resistance introduced by Q_3, namely $(r_{o3} \parallel (1/g_{m3}))$. Similarly, R_{o2} will be much greater than r_{o4}. Thus, we can easily neglect R_{o1} and R_{o2} in finding the total resistance between each of the drain nodes and ground.

The current i_1 is passed through $((1/g_{m3}) \parallel r_{o3})$ and as a result produces a voltage v_{g3},

$$v_{g3} = -i_1\left(\frac{1}{g_{m3}} \parallel r_{o3}\right) \tag{7.150}$$

Transistor r_{o4} senses this voltage and hence provides a drain current i_4,

$$i_4 = -g_{m4}v_{g3}$$

$$= i_1 g_{m4}\left(\frac{1}{g_{m3}} \parallel r_{o3}\right) \tag{7.151}$$

Now, at the output node the current difference between i_4 and i_2 passes through r_{o4} (since $R_{o2} \gg r_{o4}$) to provide v_o,

$$v_o = (i_4 - i_2)r_{o4}$$

$$= \left[i_1 g_{m4}\left(\frac{1}{g_{m3}} \parallel r_{o3}\right) - i_2\right]r_{o4}$$

Substituting for i_1 and i_2 from Eq. (7.148) and setting $g_{m3} = g_{m4}$ we obtain after some straightforward manipulations

$$A_{cm} \equiv \frac{v_o}{v_{icm}} = -\frac{1}{2R_{SS}}\frac{r_{o4}}{1 + g_{m3}r_{o3}} \tag{7.152}$$

Usually, $g_{m3}r_{o3} \gg 1$ and $r_{o3} = r_{o4}$, yielding

$$A_{cm} \cong -\frac{1}{2g_{m3}R_{SS}} \tag{7.153}$$

Since R_{SS} is usually large, at least equal to r_o, A_{cm} will be small. The common-mode rejection ratio (CMRR) can now be obtained by utilizing Eqs. (7.146) and (7.153),

$$\text{CMRR} \equiv \frac{|A_d|}{|A_{cm}|} = [g_m(r_{o2} \parallel r_{o4})][2g_{m3}R_{SS}] \tag{7.154}$$

which for $r_{o2} = r_{o4} = r_o$ and $g_{m3} = g_m$ simplifies to

$$\text{CMRR} = (g_m r_o)(g_m R_{SS}) \tag{7.155}$$

We observe that to obtain a large CMRR we select an implementation of the biasing current source I that features a high output resistance. Such circuits include the cascode current source and the Wilson current source studied in Section 6.12.

EXERCISE

7.12 An active-loaded MOS differential amplifier of the type shown in Fig. 7.28(a) is specified as follows: $(W/L)_n = 100$, $(W/L)_p = 200$, $\mu_n C_{ox} = 2\mu_p C_{ox} = 0.2$ mA/V^2, $V_{An} = |V_{Ap}| = 20$ V, $I = 0.8$ mA, $R_{SS} = 25$ kΩ. Calculate G_m, R_o, A_d, $|A_{cm}|$, and CMRR.

Ans. 4 mA/V; 25 kΩ; 100 V/V; 0.005 V/V; 20,000 or 86 dB

7.5.5 The Bipolar Differential Pair with Active Load

The bipolar version of the active-loaded differential pair is shown in Fig. 7.32(a). The circuit structure and operation are very similar to those of its MOS counterpart except that here we have to contend with the effects of finite β and the resulting finite input resistance at the

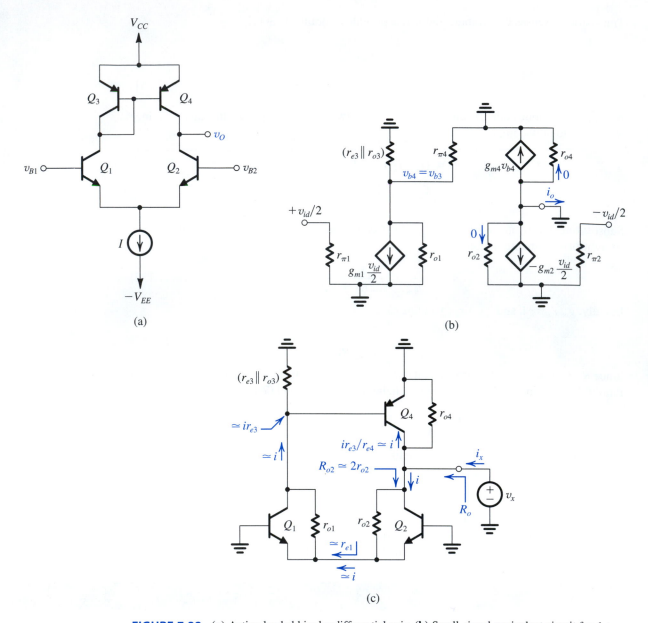

FIGURE 7.32 (a) Active-loaded bipolar differential pair. (b) Small-signal equivalent circuit for determining the transconductance $G_m \equiv i_o / v_{id}$. (c) Equivalent circuit for determining the output resistance $R_o \equiv v_x / i_x$.

base, r_π. For the time being, however, we shall ignore the effect of finite β on the dc bias of the four transistors and assume that in equilibrium all transistors are operating at a dc current of $I/2$.

Differential Gain   To obtain an expression for the differential gain, we apply an input differential signal v_{id} as shown in the equivalent circuit in Fig. 7.32(b). Note that the output is connected to ground in order to determine the overall short-circuit transconductance $G_m \equiv i_o / v_{id}$. Also, as in the MOS case, we have assumed that the circuit is sufficiently balanced so that a virtual ground develops on the common emitter terminal. This assumption is predicated on the fact that the voltage signal at the collector of Q_1 will be small as a result of the low resistance between that node and ground (approximately equal to r_{e3}). The voltage v_{b3} can be found from

$$v_{b3} = -g_{m1}\left(\frac{v_{id}}{2}\right)(r_{e3}\|r_{o3}\|r_{o1}\|r_{\pi4})$$

$$\cong -g_{m1}r_{e3}\left(\frac{v_{id}}{2}\right) \tag{7.156}$$

Since $v_{b4} = v_{b3}$, the collector current of Q_4 will be

$$g_{m4}v_{b4} = -g_{m4}g_{m1}r_{e3}\left(\frac{v_{id}}{2}\right) \tag{7.157}$$

The output current i_o can be found from a node equation at the output as

$$i_o = g_{m2}\left(\frac{v_{id}}{2}\right) - g_{m4}v_{b4} \tag{7.158}$$

Using Eq. (7.157) we obtain

$$i_o = g_{m2}\left(\frac{v_{id}}{2}\right) + g_{m4}g_{m1}r_{e3}\left(\frac{v_{id}}{2}\right) \tag{7.159}$$

Since all devices are operating at the same bias current, $g_{m1} = g_{m2} = g_{m4} = g_m$, where

$$g_m \cong \frac{I/2}{V_T} \tag{7.160}$$

and $r_{e3} = \alpha_3/g_{m3} = \alpha/g_m \cong 1/g_m$. Thus, for G_m, Eq. (7.159) yields

$$G_m = g_m \tag{7.161}$$

which is identical to the result found for the MOS pair.

Next we determine the output resistance of the amplifier utilizing the equivalent circuit shown in Fig. 7.32(c). We urge the reader to carefully examine this circuit and to note that the analysis is very similar to that for the MOS pair. The output resistance R_{o2} of transistor Q_2 can be found using Eq. (6.160) by noting that the resistance R_e in the emitter of Q_2 is approximately equal to r_{e1}, thus

$$R_{o2} = r_{o2}[1 + g_{m2}(r_{e1}\|r_{\pi2})]$$

$$\cong r_{o2}(1 + g_{m2}r_{e1})$$

$$\cong 2r_{o2} \tag{7.162}$$

where we made use of the fact that corresponding parameters of all four transistors are equal.

The current i can now be found as

$$i = \frac{v_x}{R_{o2}} = \frac{v_x}{2r_{o2}} \tag{7.163}$$

and the current i_x can be obtained from a node equation at the output as

$$i_x = 2i + \frac{v_x}{r_{o4}} = \frac{v_x}{r_{o2}} + \frac{v_x}{r_{o4}}$$

Thus,

$$R_o \equiv \frac{v_x}{i_x} = r_{o2}\|r_{o4} \tag{7.164}$$

This expression simply says that the output resistance of the amplifier is equal to the parallel equivalent of the output resistance of the differential pair and the output resistance of the current mirror; a result identical to that obtained for the MOS pair.

Equations (7.161) and (7.164) can now be combined to obtain the differential gain,

$$A_d \equiv \frac{v_o}{v_{id}} = G_m R_o = g_m(r_{o2} \| r_{o4}) \tag{7.165}$$

and since $r_{o2} = r_{o4} = r_o$, we can simplify Eq. (7.165) to

$$A_d = \tfrac{1}{2} g_m r_o \tag{7.166}$$

Although this expression is identical to that found for the MOS circuit, the gain here is much larger because $g_m r_o$ for the BJT is more than an order of magnitude greater than $g_m r_o$ of a MOSFET. The downside, however, lies in the low input resistance of BJT amplifiers. Indeed, the equivalent circuit of Fig. 7.32(b) indicates that, as expected, the differential input resistance of the differential amplifier is equal to $2r_\pi$,

$$R_{id} = 2r_\pi \tag{7.167}$$

in sharp contrast to the infinite input resistance of the MOS amplifier. Thus, while the voltage gain realized in an active-loaded amplifier stage is large, when a subsequent stage is connected to the output, its inevitably low input resistance will drastically reduce the overall voltage gain.

Common-Mode Gain and CMRR The common-mode gain A_{cm} and the common-mode rejection ratio (CMRR) can be found following a procedure identical to that utilized in the MOS case. Figure 7.33 shows the circuit prepared for common-mode signal analysis. The collector currents of Q_1 and Q_2 are given by

$$i_1 \cong i_2 \cong \frac{v_{icm}}{2R_{EE}} \tag{7.168}$$

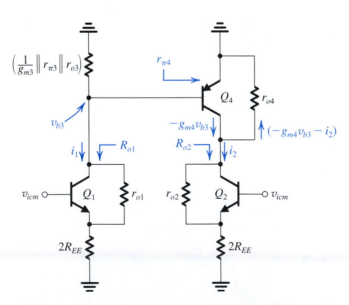

FIGURE 7.33 Analysis of the bipolar active-loaded differential amplifier to determine the common-mode gain.

It can be shown that the output resistances of Q_1 and Q_2, R_{o1} and R_{o2}, are very large and hence can be neglected. Then, the voltage v_{b3} at the common base connection of Q_3 and Q_4 can be found by multiplying i_1 by the total resistance between the common base node and ground as

$$v_{b3} = -i_1\left(\frac{1}{g_{m3}}\|r_{\pi3}\|r_{o3}\|r_{\pi4}\right) \tag{7.169}$$

In response to v_{b3} transistor Q_4 provides a collector current $g_{m4}v_{b3}$. At the output node we have

$$v_o = (-g_{m4}v_{b3} - i_2)r_{o4} \tag{7.170}$$

Substituting for v_{b3} from Eq. (7.169) and for i_1 and i_2 from Eq. (7.168) gives

$$A_{cm} \equiv \frac{v_o}{v_{icm}} = \frac{r_{o4}}{2R_{EE}}\left[g_{m4}\left(\frac{1}{g_{m3}}\|r_{\pi3}\|r_{o3}\|r_{\pi4}\right) - 1\right]$$

$$= -\frac{r_{o4}}{2R_{EE}}\frac{\dfrac{1}{r_{\pi3}} + \dfrac{1}{r_{\pi4}} + \dfrac{1}{r_{o3}}}{g_{m3} + \dfrac{1}{r_{\pi3}} + \dfrac{1}{r_{\pi4}} + \dfrac{1}{r_{o3}}} \tag{7.171}$$

where we have assumed $g_{m3} = g_{m4}$. Now, for $r_{\pi4} = r_{\pi3}$ and $r_{o3} \gg r_{\pi3}, r_{\pi4}$, Eq. (7.171) gives

$$A_{cm} \cong -\frac{r_{o4}}{2R_{EE}}\frac{\dfrac{2}{r_{\pi3}}}{g_{m3} + \dfrac{2}{r_{\pi3}}}$$

$$\cong -\frac{r_{o4}}{2R_{EE}}\frac{2}{\beta_3} = -\frac{r_{o4}}{\beta_3 R_{EE}} \tag{7.172}$$

Using A_d from Eq. (7.165) enables us to obtain the CMRR as

$$\text{CMRR} \equiv \frac{|A_d|}{|A_{cm}|} = g_m(r_{o2}\|r_{o4})\left(\frac{\beta_3 R_{EE}}{r_{o4}}\right) \tag{7.173}$$

For $r_{o2} = r_{o4} = r_o$,

$$\text{CMRR} = \tfrac{1}{2}\beta_3 g_m R_{EE} \tag{7.174}$$

from which we observe that to obtain a large CMRR, the circuit implementing the bias current source should have a large output resistance R_{EE}. This is possible with, say, a Wilson current mirror (Section 6.12.3).

EXERCISE

7.13 For the active-loaded BJT differential amplifier let $I = 0.8$ mA, $V_A = 100$ V, and $\beta = 160$. Find G_m, R_o, A_d, and R_{id}. If the bias current source is implemented with a simple *npn* current mirror, find R_{EE}, A_{cm}, and CMRR.

Ans. 16 mA/V; 125 kΩ; 2000 V/V; 20 kΩ; 125 kΩ; −0.0125 V/V; 160,000 or 104 dB

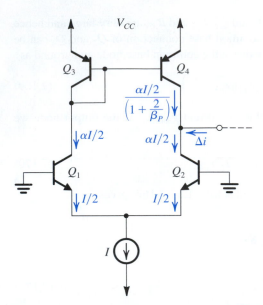

FIGURE 7.34 The active-loaded BJT differential pair suffers from a systematic input offset voltage resulting from the error in the current-transfer ratio of the current mirror.

Systematic Input Offset Voltage In addition to the random offset voltages that result from the mismatches inevitably present in the differential amplifier, the active-loaded bipolar differential pair suffers from a systematic offset voltage. This is due to the error in the current transfer ratio of the current-mirror load caused by the finite β of the *pnp* transistors that make up the mirror. To see how this comes about, refer to Fig. 7.34. Here the inputs are grounded and the transistors are assumed to be perfectly matched. Thus, the bias current I will divide equally between Q_1 and Q_2 with the result that their two collectors conduct equal currents of $\alpha I/2$. The collector current of Q_1 is fed to the input of the current mirror. From Section 6.3 we know that the current-transfer ratio of the mirror is

$$\frac{I_4}{I_3} = \frac{1}{1 + \dfrac{2}{\beta_P}} \tag{7.175}$$

where β_P is the value of β of the *pnp* transistors Q_3 and Q_4. Thus the collector current of Q_4 will be

$$I_4 = \frac{\alpha I/2}{1 + \dfrac{2}{\beta_P}} \tag{7.176}$$

which does not exactly balance the collector current of Q_2. It follows that the current difference Δi will flow into the output terminal of the amplifier with

$$
\begin{aligned}
\Delta i &= \frac{\alpha I}{2} - \frac{\alpha I/2}{1 + \dfrac{2}{\beta_P}} \\[2mm]
&= \frac{\alpha I}{2} \frac{2/\beta_P}{1 + \dfrac{2}{\beta_P}} \\[2mm]
&\cong \frac{\alpha I}{\beta_P} \tag{7.177}
\end{aligned}
$$

To reduce this output current to zero, an input voltage V_{OS} has to be applied with a value of

$$V_{OS} = -\frac{\Delta i}{G_m}$$

Substituting for Δi from Eq. (7.177) and for $G_m = g_m = (\alpha I/2)/V_T$, we obtain for the input offset voltage the expression

$$V_{OS} = -\frac{\alpha I/\beta_P}{\alpha I/2V_T} = -\frac{2V_T}{\beta_P} \tag{7.178}$$

As an example, for $\beta_P = 50$, $V_{OS} = -1$ mV. To reduce V_{OS}, an improved current mirror such as the Wilson circuit studied in Section 6.12 should be used. Such a circuit provides the added advantage of increased output resistance and hence voltage gain. However, to realize the full advantage of the higher output resistance of the active load, the output resistance of the differential pair should be raised by utilizing a cascode stage. Figure 7.35 shows such an arrangement: A folded cascode stage formed by *pnp* transistors Q_3 and Q_4 is utilized to raise the output resistance looking into the collector of Q_4 to $\beta_4 r_{o4}$. A Wilson mirror formed by transistors Q_5, Q_6, and Q_7 is used to implement the active load. From Section 6.12.3 we know that the output resistance of the Wilson mirror (i.e., looking into the collector of Q_5) is $\beta_5(r_{o5}/2)$. Thus the output resistance of the amplifier

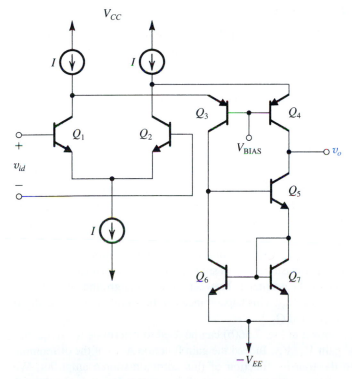

FIGURE 7.35 An active-loaded bipolar differential amplifier employing a folded cascode stage (Q_3 and Q_4) and a Wilson current mirror load (Q_5, Q_6, and Q_7).

is given by

$$R_o = \left[\beta_4 r_{o4} \parallel \beta_5 \frac{r_{o5}}{2} \right] \tag{7.179}$$

The transconductance G_m remains equal to g_m of Q_1 and Q_2. Thus the differential voltage gain becomes

$$A_d = g_m \left[\beta_4 r_{o4} \parallel \beta_5 \frac{r_{o5}}{2} \right] \tag{7.180}$$

which can be very large. Further examples of improved-performance differential amplifiers will be studied in Chapter 9.

EXERCISE

7.14 Find G_m and R_{o4}, R_{o5}, R_o, and A_d for the differential amplifier in Fig. 7.35 under the following conditions: $I = 1$ mA, $\beta_P = 50$, $\beta_N = 100$, and $V_A = 100$ V.

Ans. 20 mA/V; 10 MΩ; 10 MΩ; 5 MΩ; 10^5 V/V or 100 dB

7.6 FREQUENCY RESPONSE OF THE DIFFERENTIAL AMPLIFIER

In this section we study the frequency response of the differential amplifier. We will consider the variation with frequency of both the differential gain and the common-mode gain and hence of the CMRR. We will rely heavily on the study of frequency response of single-ended amplifiers presented in Chapter 6. Also, we will only consider MOS circuits; the bipolar case is a straightforward extension, as we saw on a number of occasions in Chapter 6.

7.6.1 Analysis of the Resistively Loaded MOS Amplifier

We begin with the basic, resistively loaded MOS differential pair shown in Fig. 7.36(a). Note that we have explicitly shown the transistor Q_S that supplies the bias current I. Although we are showing a dc bias voltage V_{BIAS} at its gate, usually Q_S is part of a current mirror. This detail, however, is of no consequence to our present needs. Most importantly, we are interested in the total impedance between node S and ground, Z_{SS}. As we shall shortly see, this impedance plays a significant role in determining the common-mode gain and the CMRR of the differential amplifier. Resistance R_{SS} is simply the output resistance of current source Q_S. Capacitance C_{SS} is the total capacitance between node S and ground and includes C_{db} and C_{gd} of Q_S, as well as C_{sb1}, and C_{sb2}. This capacitance can be significant, especially if wide transistors are used for Q_S, Q_1, and Q_2.

The differential half-circuit shown in Fig. 7.36(b) can be used to determine the frequency dependence of the differential gain V_o / V_{id}. Indeed the gain function $A_d(s)$ of the differential amplifier will be identical to the transfer function of this common-source amplifier. We studied the frequency response of the common-source amplifier at great length in Section 6.6 and will not repeat this material here.

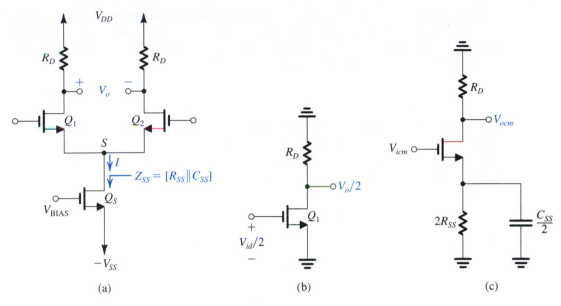

FIGURE 7.36 (a) A resistively loaded MOS differential pair with the transistor supplying the bias current explicitly shown. It is assumed that the total impedance between node S and ground, Z_{SS}, consists of a resistance R_{SS} in parallel with a capacitance C_{SS}. (b) Differential half-circuit. (c) Common-mode half-circuit.

EXERCISE

7.15 A MOSFET differential amplifier such as that in Fig. 7.36(a) is biased with a current $I = 0.8$ mA. The transistors have $W/L = 100$, $k'_n = 0.2$ mA/V^2, $V_A = 20$ V, $C_{gs} = 50$ fF, $C_{gd} = 10$ fF, and $C_{db} = 10$ fF. The drain resistors are 5 kΩ each. Also, there is a 100-fF capacitive load between each drain and ground.

(a) Find V_{OV} and g_m for each transistor.

(b) Find the differential gain A_d.

(c) If the input signal source has a small resistance R_{sig} and thus the frequency response is determined primarily by the output pole, estimate the 3-dB frequency f_H. (*Hint:* Refer to Section 6.6.5 and specifically to Eq. 6.79.)

(d) If, in a different situation, the amplifier is fed symmetrically with a signal source of 20 kΩ resistance (i.e., 10 kΩ in series with each gate terminal), use the open-circuit time-constants method to estimate f_H. (*Hint:* Refer to Section 6.6.2 and specifically to Eqs. 6.57 and 6.58.)

Ans. (a) 0.2 V, 4 mA/V; (b) 18.2 V/V; (c) 291 MHz; (d) 53.7 MHz

The common-mode half-circuit is shown in Fig. 7.36(c). Although this circuit has other capacitances, namely C_{gs}, C_{gd}, and C_{db} of the transistor in addition to other stray capacitances, we have chosen to show only $C_{SS}/2$. This is because $(C_{SS}/2)$ together with $(2R_{SS})$ forms a real-axis zero in the common-mode gain function at a frequency much lower than those of the other poles and zeros of the circuit. This zero then dominates the frequency dependence of A_{cm} and CMRR.

If the output of the differential amplifier is taken single-endedly, then the common-mode gain of interest is V_{ocm}/V_{icm}. More typically, the output is taken differentially. Nevertheless, as we have seen in Section 7.2, V_{ocm}/V_{icm} still plays a major role in determining the common-mode gain. To be specific, consider what happens when the output is taken differentially and there is a mismatch ΔR_D between the two drain resistances. The resulting common-mode gain was found in Section 7.2 to be (Eq. 7.51)

$$A_{cm} = -\left(\frac{R_D}{2R_{SS}}\right)\frac{\Delta R_D}{R_D} \tag{7.181}$$

which is simply the product of V_{ocm}/V_{icm} and the per-unit mismatch ($\Delta R_D/R_D$). Similar expressions can be found for the effects of other circuit mismatches. The important point to note is that the factor $R_D/(2R_{SS})$ is always present in these expressions. Thus, the frequency dependence of A_{cm} can be obtained by simply replacing R_{SS} by Z_{SS} in this factor. Doing so for the expression in Eq. (7.181) gives

$$\begin{aligned} A_{cm}(s) &= -\frac{R_D}{2Z_{SS}}\left(\frac{\Delta R_D}{R_D}\right) \\ &= -\frac{1}{2}R_D\left(\frac{\Delta R_D}{R_D}\right)Y_{SS} \\ &= -\frac{1}{2}R_D\left(\frac{\Delta R_D}{R_D}\right)\left(\frac{1}{R_{SS}} + sC_{SS}\right) \\ &= -\frac{R_D}{2R_{SS}}\left(\frac{\Delta R_D}{R_D}\right)(1 + sC_{SS}R_{SS}) \end{aligned} \tag{7.182}$$

from which we see that A_{cm} acquires a zero on the negative real-axis of the s-plane with frequency ω_Z,

$$\omega_Z = \frac{1}{C_{SS}R_{SS}} \tag{7.183}$$

or in hertz,

$$f_Z = \frac{\omega_Z}{2\pi} = \frac{1}{2\pi C_{SS}R_{SS}} \tag{7.184}$$

As mentioned above, usually f_Z is much lower than the frequencies of the other poles and zeros. As a result, the common-mode gain increases at the rate of +6 dB/octave (20 dB/decade) starting at a relatively low frequency, as indicated in Fig. 7.37(a). Of course, A_{cm} drops off at high frequencies because of the other poles of the common-mode half-circuit. It is, however, f_Z that is significant, for it is the frequency at which the CMRR of the differential amplifier begins to decrease, as indicated in Fig. 7.37(c). Note that if both A_d and A_{cm} are expressed and plotted in dB, then CMRR in dB is simply the difference between A_d and A_{cm}.

Although in the foregoing we considered only the common-mode gain resulting from an R_D mismatch, it should be obvious that the results apply to the common-mode gain resulting from any other mismatch. For instance, it applies equally well to the case of a g_m mismatch, modifying Eq. 7.64 by replacing R_{SS} by Z_{SS}, and so on.

Before leaving this section, it is interesting to point out an important trade-off found in the design of the current-source transistor Q_S: In order to operate this current source with a small V_{DS} (to conserve the already low V_{DD}), we desire to operate the transistor at a low overdrive voltage V_{OV}. For a given value of the current I, however, this means using a large W/L ratio (i.e., a wide transistor). This in turn increases C_{SS} and hence lowers f_Z with the result that the CMRR deteriorates (i.e., decreases) at a relatively low frequency. Thus there is a

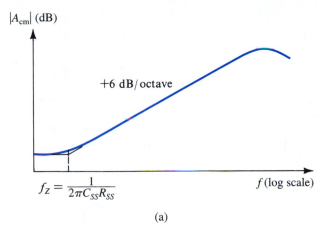

(a)

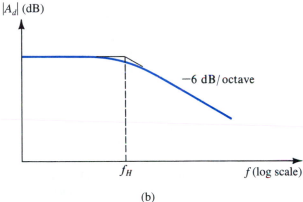

(b)

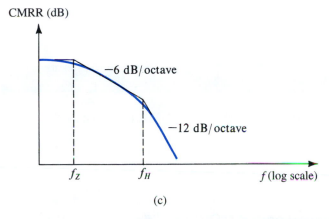

(c)

FIGURE 7.37 Variation of (**a**) common-mode gain, (**b**) differential gain, and (**c**) common-mode rejection ratio with frequency.

trade-off between the need to reduce the dc voltage across Q_S and the need to keep the CMRR reasonably high at higher frequencies.

To appreciate the need for high CMRR at higher frequencies, consider the situation illustrated in Fig. 7.38: We show two stages of a differential amplifier whose power-supply voltage V_{DD} is corrupted with high-frequency noise. Since the quiescent voltage at each of

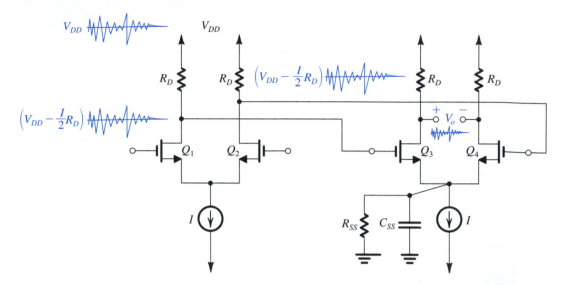

FIGURE 7.38 The second stage in a differential amplifier is relied on to suppress high-frequency noise injected by the power supply of the first stage, and therefore must maintain a high CMRR at higher frequencies.

the drains of Q_1 and Q_2 is $[V_{DD} - (I/2)R_D]$, we see that v_{D1} and v_{D2} will have the same high-frequency noise as V_{DD}. This high-frequency noise then constitutes a common-mode input signal to the second differential stage, formed by Q_3 and Q_4. If the second differential stage is perfectly matched, its differential output voltage V_o should be free of high-frequency noise. However, in practice there is no such thing as perfect matching and the second stage will have a finite common-mode gain. Furthermore, because of the zero formed by R_{SS} and C_{SS} of the second stage, the common-mode gain will increase with frequency, causing some of the noise to make its way to V_o. With careful design, this undesirable component of V_o can be kept small.

EXERCISE

7.16 The differential amplifier specified in Exercise 7.15 has $R_{SS} = 25$ kΩ and $C_{SS} = 0.4$ pF. Find the 3-dB frequency of the CMRR.
Ans. 15.9 MHz

7.6.2 Analysis of the Active-Loaded MOS Amplifier

We next consider the frequency response of the current-mirror-loaded MOS differential-pair circuit studied in Section 7.5. The circuit is shown in Fig. 7.39(a) with two capacitances indicated: C_m, which is the total capacitance at the input node of the current mirror, and C_L, which is the total capacitance at the output node. Capacitance C_m is mainly formed by C_{gs3} and C_{gs4} but also includes C_{gd1}, C_{db1}, and C_{db3},

$$C_m = C_{gd1} + C_{db1} + C_{db3} + C_{gs3} + C_{gs4} \tag{7.185}$$

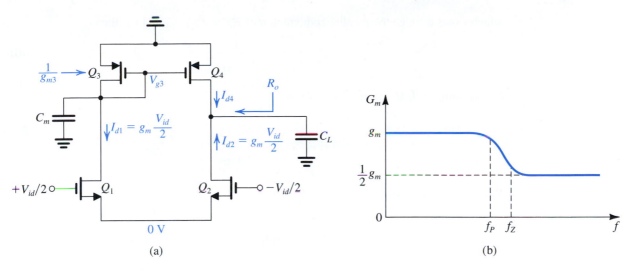

FIGURE 7.39 (a) Frequency–response analysis of the active-loaded MOS differential amplifier. (b) The overall transconductance G_m as a function of frequency.

Capacitance C_L includes C_{gd2}, C_{db2}, C_{db4}, C_{gd4} as well as an actual load capacitance and/or the input capacitance of a subsequent stage (C_x),

$$C_L = C_{gd2} + C_{db2} + C_{gd4} + C_{db4} + C_x \tag{7.186}$$

These two capacitances primarily determine the dependence of the differential gain of this amplifier on frequency.

As indicated in Fig. 7.39(a) the input differential signal V_{id} is applied in a balanced fashion. Transistor Q_1 will conduct a drain current signal of $g_m V_{id}/2$, which flows through the diode-connected transistor Q_3 and thus through the parallel combination of $(1/g_{m3})$ and C_m, where we have neglected the resistances r_{o1} and r_{o3} which are much larger than $(1/g_{m3})$, thus

$$V_{g3} = -\frac{g_m V_{id}/2}{g_{m3} + sC_m} \tag{7.187}$$

In response to V_{g3}, transistor Q_4 conducts a drain current I_{d4},

$$I_{d4} = -g_{m4} V_{g3} = \frac{g_{m4} g_m V_{id}/2}{g_{m3} + sC_m}$$

Since $g_{m3} = g_{m4}$, this equation reduces to

$$I_{d4} = \frac{g_m V_{id}/2}{1 + s\dfrac{C_m}{g_{m3}}} \tag{7.188}$$

Now, at the output node the total output current is

$$I_o = I_{d4} + I_{d2}$$

$$= \frac{g_m V_{id}/2}{1 + s\dfrac{C_m}{g_{m3}}} + g_m(V_{id}/2) \tag{7.189}$$

which flows through the parallel combination of $R_o = r_{o2} \| r_{o4}$ and C_L, thus

$$V_o = I_o \frac{1}{\dfrac{1}{R_o} + sC_L} \tag{7.190}$$

Substituting for I_o from Eq. (7.189) gives

$$V_o = g_m R_o \left(\frac{V_{id}}{2} \right) \left[1 + \frac{1}{1 + s\dfrac{C_m}{g_{m3}}} \right] \frac{1}{1 + sC_L R_o}$$

Which can be manipulated to yield

$$A_d(s) \equiv \frac{V_o}{V_{id}} = (g_m R_o) \left(\frac{1}{1 + sC_L R_o} \right) \frac{\left(1 + s\dfrac{C_m}{2g_{m3}} \right)}{1 + s\dfrac{C_m}{g_{m3}}} \tag{7.191}$$

We recognize the first factor on the right-hand side as the dc gain of the amplifier. The second factor indicates that C_L and R_o form a pole with frequency f_{P1},

$$f_{P1} = \frac{1}{2\pi C_L R_o} \tag{7.192}$$

This, of course, is an entirely expected result, and in fact this output pole is often dominant, especially when a large load capacitance is present. The third factor on the right-hand side of Eq. (7.191) indicates that the capacitance C_m at the input of the current mirror gives rise to a pole with frequency f_{P2},

$$f_{P2} = \frac{g_{m3}}{2\pi C_m} \tag{7.193}$$

and a zero with frequency f_Z,

$$f_Z = \frac{2g_{m3}}{2\pi C_m} \tag{7.194}$$

That is, the zero frequency is twice that of the pole. Since C_m is approximately $C_{gs3} + C_{gs4} = 2C_{gs3}$,

$$f_{P2} \cong \frac{g_{m3}}{2\pi(2C_{gs3})} \cong f_T/2 \tag{7.195}$$

and

$$f_Z \cong f_T \tag{7.196}$$

where f_T is the frequency at which the magnitude of the high-frequency current gain of the MOSFET becomes unity (see Sections 4.8 and 6.2). Thus, the **mirror pole and zero** occur at very high frequencies. Nevertheless, their effect can be significant.

It is interesting and useful to observe that the path of the signal current produced by Q_1 has a transfer function different from that of the signal current produced by Q_2. It is the first signal that encounters C_m and experiences the mirror pole. This observation leads to an interesting view of the effect of C_m on the overall transconductance G_m of the differential amplifier: As we learned in Section 7.5, at low frequencies I_{d1} is replicated by the mirror $Q_3 - Q_4$ in the collector of Q_4 as I_{d4}, which adds to I_{d2} to provide a factor-of-2 increase in G_m (thus making G_m equal to g_m, which is double the value available without the current mirror). Now, at high frequencies C_m acts as a short circuit causing V_{g3} to be zero and hence I_{d4} will be zero, reducing G_m to $g_m/2$. Thus, if the output is short-circuited to ground and the short-circuit transconductance G_m is plotted versus frequency, the plot will have the shape shown in Fig. 7.39(b).

EXAMPLE 7.2

Consider an active-loaded MOS differential amplifier of the type shown in Fig. 7.28(a). Assume that for all transistors, $W/L = 7.2\ \mu\text{m}/0.36\ \mu\text{m}$, $C_{gs} = 20$ fF, $C_{gd} = 5$ fF, and $C_{db} = 5$ fF. Also, let $\mu_n C_{ox} = 387\ \mu\text{A/V}^2$, $\mu_p C_{ox} = 86\ \mu\text{A/V}^2$, $V'_{An} = 5\ \text{V}/\mu\text{m}$, $|V'_{Ap}| = 6\ \text{V}/\mu\text{m}$. The bias current $I = 0.2$ mA, and the bias current source has an output resistance $R_{SS} = 25$ kΩ and an output capacitance $C_{SS} = 0.2$ pF. In addition to the capacitances introduced by the transistors at the output node, there is a capacitance C_x of 25 fF. It is required to determine the low-frequency values of A_d, A_{cm}, and CMRR. It is also required to find the poles and zero of A_d and the dominant pole of CMRR.

Solution

Since $I = 0.2$ mA, each of the four transistors is operating at a bias current of 100 μA. Thus, for Q_1 and Q_2,

$$100 = \frac{1}{2} \times 387 \times \frac{7.2}{0.36} \times V_{OV}^2$$

which leads to

$$V_{OV} = 0.16\ \text{V}$$

Thus,

$$g_m = g_{m1} = g_{m2} = \frac{2 \times 0.1}{0.16} = 1.25\ \text{mA/V}$$

$$r_{o1} = r_{o2} = \frac{5 \times 0.36}{0.1} = 18\ \text{k}\Omega$$

For Q_3 and Q_4 we have

$$100 = \frac{1}{2} \times 86 \times \frac{7.2}{0.36} \times V_{OV3,4}^2$$

Thus,

$$V_{OV3,4} = 0.34\ \text{V},$$

and

$$g_{m3} = g_{m4} = \frac{2 \times 0.1}{0.34} = 0.6\ \text{mA/V}$$

$$r_{o3} = r_{o4} = \frac{6 \times 0.36}{0.1} = 21.6\ \text{k}\Omega$$

The low-frequency value of the differential gain can be determined from

$$A_d = g_m(r_{o2} \| r_{o4})$$

$$= 1.25(18 \| 21.6) = 12.3\ \text{V/V}$$

The low-frequency value of the common-mode gain can be determined from Eq. (7.153) as

$$A_{cm} = -\frac{1}{2g_{m3}R_{SS}}$$

$$= -\frac{1}{2 \times 0.6 \times 25} = -0.033 \text{ V/V}$$

The low-frequency value of the CMRR can now be determined as

$$\text{CMRR} = \frac{|A_d|}{|A_{cm}|} = \frac{12.3}{0.033} = 369$$

or,

$$20 \log 369 = 51.3 \text{ dB}$$

To determine the poles and zero of A_d we first compute the values of the two pertinent capacitances C_m and C_L. Using Eq. (7.185),

$$C_m = C_{gd1} + C_{db1} + C_{db3} + C_{gs3} + C_{gs4}$$

$$= 5 + 5 + 5 + 20 + 20 = 55 \text{ fF}$$

Capacitance C_L is found using Eq. (7.186) as

$$C_L = C_{gd2} + C_{db2} + C_{gd4} + C_{db4} + C_x$$

$$= 5 + 5 + 5 + 5 + 25 = 45 \text{ fF}$$

Now, the poles and zero of A_d can be found from Eqs. (7.192) to (7.194) as

$$f_{P1} = \frac{1}{2\pi C_L R_o}$$

$$= \frac{1}{2\pi \times C_L(r_{o2} \| r_{o4})}$$

$$= \frac{1}{2\pi \times 45 \times 10^{-15}(18 \| 21.6)10^3}$$

$$= 360 \text{ MHz}$$

$$f_{P2} = \frac{g_{m3}}{2\pi C_m} = \frac{0.6 \times 10^{-3}}{2\pi \times 55 \times 10^{-15}} = 1.74 \text{ GHz}$$

$$f_Z = 2f_{P2} = 3.5 \text{ GHz}$$

Thus the dominant pole is that produced by C_L at the output node. As expected, the pole and zero of the mirror are at much higher frequencies.

The dominant pole of the CMRR is at the location of the common-mode-gain zero introduced by C_{SS} and R_{SS}, that is,

$$f_Z = \frac{1}{2\pi C_{SS}R_{SS}}$$

$$= \frac{1}{2\pi \times 0.2 \times 10^{-12} \times 25 \times 10^3}$$

$$= 31.8 \text{ MHz}$$

Thus, the CMRR begins to decrease at 31.8 MHz, which is much lower than f_{P1}.

7.17 A bipolar current-mirror-loaded differential amplifier is biased with a current source $I = 1$ mA. The transistors are specified to have $|V_A| = 100$ V. The total capacitance at the output node is 2 pF. Find the dc value and the frequency of the dominant high-frequency pole of the differential voltage gain.

Ans. 2000 V/V; 0.8 MHz

 ## 7.7 MULTISTAGE AMPLIFIERS

Practical transistor amplifiers usually consist of a number of stages connected in cascade. In addition to providing gain, the first (or input) stage is usually required to provide a high input resistance in order to avoid loss of signal level when the amplifier is fed from a high-resistance source. In a differential amplifier the input stage must also provide large common-mode rejection. The function of the middle stages of an amplifier cascade is to provide the bulk of the voltage gain. In addition, the middle stages provide such other functions as the conversion of the signal from differential mode to single-ended mode (unless, of course, the amplifier output also is differential) and the shifting of the dc level of the signal in order to allow the output signal to swing both positive and negative. These two functions and others will be illustrated later in this section and in greater detail in Chapter 9.

Finally, the main function of the last (or output) stage of an amplifier is to provide a low output resistance in order to avoid loss of gain when a low-valued load resistance is connected to the amplifier. Also, the output stage should be able to supply the current required by the load in an efficient manner—that is, without dissipating an unduly large amount of power in the output transistors. We have already studied one type of amplifier configuration suitable for implementing output stages, namely, the source follower and the emitter follower. It will be shown in Chapter 14 that the source and emitter followers are not optimum from the point of view of power efficiency and that other, more appropriate circuit configurations exist for output stages that are required to supply large amounts of output power. In fact, we will encounter some such output stages in the op-amp circuit examples studied in Chapter 9.

To illustrate the circuit structure and the method of analysis of multistage amplifiers, we will present two examples: a two-stage CMOS op amp and a four-stage bipolar op amp.

7.7.1 A Two-Stage CMOS Op Amp

Figure 7.40 shows a popular structure for CMOS op amps known as the **two-stage configuration.** The circuit utilizes two power supplies, which can range from ± 2.5 V for the 0.5-μm technology down to ± 0.9 V for the 0.18-μm technology. A reference bias current I_{REF} is generated either externally or using on-chip circuits. One such circuit will be discussed shortly. The current mirror formed by Q_8 and Q_5 supplies the differential pair $Q_1 - Q_2$ with bias current. The W/L ratio of Q_5 is selected to yield the desired value for the input-stage bias current I (or $I/2$ for each of Q_1 and Q_2). The input differential pair is actively loaded with the current mirror formed by Q_3 and Q_4. Thus the input stage is identical to that studied in Section 7.5 (except that here the differential pair is implemented with PMOS transistors and the current mirror with NMOS).

The second stage consists of Q_6, which is a common-source amplifier actively loaded with the current-source transistor Q_7. A capacitor C_C is included in the negative-feedback path of the second stage. Its function is to enhance the Miller effect already present in Q_6 (through the action of its C_{gd}) and thus provide the op amp with a dominant pole. By the careful placement of

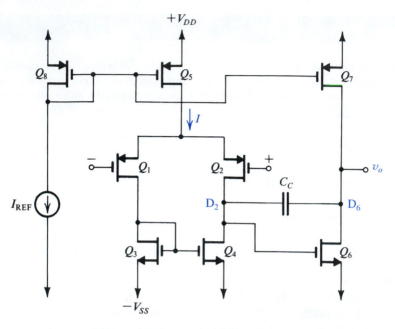

FIGURE 7.40 Two-stage CMOS op-amp configuration.

this pole, the op amp can be made to have a gain that decreases with frequency at the rate of −6 dB/octave, or, equivalently, −20 dB/decade down to unity gain or 0 dB. Op amps with such a gain function are guaranteed to operate in a stable fashion, as opposed to oscillating, with nearly all possible feedback connections. Such op amps are said to be *frequency compensated*. We shall study the subject of frequency compensation[4] in Chapters 8 and 9. Here, we will simply take C_C into account in the analysis of the frequency response of the circuit in Fig. 7.40.

A striking feature of the circuit in Fig. 7.40 is that it does not have a low-output-resistance stage. In fact, the output resistance of the circuit is equal to $(r_{o6} \parallel r_{o7})$ and is thus rather high. This circuit, therefore, is not suitable for driving low-impedance loads. Nevertheless, the circuit is very popular, and is used frequently for implementing op amps in VLSI circuits where the op amp needs to drive only a small capacitive load, for example, in switched-capacitor circuits (Chapter 12). The simplicity of the circuit results in an op amp of reasonably good quality realized in a very small chip area.

Voltage Gain The voltage gain of the first stage was found in Section 7.5 to be given by

$$A_1 = -g_{m1}(r_{o2} \parallel r_{o4}) \tag{7.197}$$

where g_{m1} is the transconductance of each of the transistors of the first stage, that is, Q_1 and Q_2.

The second stage is an actively loaded common-source amplifier whose low-frequency voltage gain is given by

$$A_2 = -g_{m6}(r_{o6} \parallel r_{o7}) \tag{7.198}$$

The dc open-loop gain of the op amp is the product of A_1 and A_2.

---

[4] Readers who have studied Chapter 2 will recall that commercially available op amps with this uniform gain rolloff of −20 dB/decade are said to be internally compensated. Here "internal" means that the frequency compensation network is internal to the package (i.e., on chip) and need not be supplied externally by the user. The μA 741 op amp is an example of an internally compensated op amp.

EXAMPLE 7.3

Consider the circuit in Fig. 7.40 with the following device geometries (in μm).

Transistor	Q_1	Q_2	Q_3	Q_4	Q_5	Q_6	Q_7	Q_8
W/L	20/0.8	20/0.8	5/0.8	5/0.8	40/0.8	10/0.8	40/0.8	40/0.8

Let $I_{REF} = 90\ \mu$A, $V_{tn} = 0.7$ V, $V_{tp} = -0.8$ V, $\mu_n C_{ox} = 160\ \mu$A/V^2, $\mu_p C_{ox} = 40\ \mu$A/V^2, $|V_A|$ (for all devices) $= 10$ V, $V_{DD} = V_{SS} = 2.5$ V. For all devices evaluate I_D, $|V_{OV}|$, $|V_{GS}|$, g_m, and r_o. Also find A_1, A_2, the dc open-loop voltage gain, the input common-mode range, and the output voltage range. Neglect the effect of V_A on bias current.

Solution

Refer to Fig. 7.40. Since Q_8 and Q_5 are matched, $I = I_{REF}$. Thus Q_1, Q_2, Q_3, and Q_4 each conducts a current equal to $I/2 = 45\ \mu$A. Since Q_7 is matched to Q_5 and Q_8, the current in Q_7 is equal to $I_{REF} = 90\ \mu$A. Finally, Q_6 conducts an equal current of $90\ \mu$A.

With I_D of each device known, we use

$$I_D = \tfrac{1}{2}(\mu C_{ox})(W/L)V_{OV}^2$$

to determine $|V_{OV}|$ for each transistor. Then we find $|V_{GS}|$ from $|V_{GS}| = |V_t| + |V_{OV}|$. The results are given in Table 7.1.

The transconductance of each device is determined from

$$g_m = 2I_D/|V_{OV}|$$

The value of r_o is determined from

$$r_o = |V_A|/I_D$$

The resulting values of g_m and r_o are given in Table 7.1.

The voltage gain of the first stage is determined from

$$A_1 = -g_{m1}(r_{o2} \| r_{o4})$$
$$= -0.3(222 \| 222) = -33.3\ \text{V/V}$$

The voltage gain of the second stage is determined from

$$A_2 = -g_{m6}(r_{o6} \| r_{o7})$$
$$= -0.6(111 \| 111) = -33.3\ \text{V/V}$$

TABLE 7.1

	Q_1	Q_2	Q_3	Q_4	Q_5	Q_6	Q_7	Q_8
I_D (μA)	45	45	45	45	90	90	90	90
$|V_{OV}|$ (V)	0.3	0.3	0.3	0.3	0.3	0.3	0.3	0.3
$|V_{GS}|$ (V)	1.1	1.1	1	1	1.1	1	1.1	1.1
g_m (mA/V)	0.3	0.3	0.3	0.3	0.6	0.6	0.6	0.6
r_o (kΩ)	222	222	222	222	111	111	111	111

Thus the overall dc open-loop gain is

$$A_0 = A_1 A_2 = (-33.3) \times (-33.3) = 1109 \text{ V/V}$$

or

$$20 \log 1109 = 61 \text{ dB}$$

The lower limit of the input common-mode range is the value of input voltage at which Q_1 and Q_2 leave the saturation region. This occurs when the input voltage falls below the voltage at the drain of Q_1 by $|V_{tp}|$ volts. Since the drain of Q_1 is at $-2.5 + 1 = -1.5$ V, then the lower limit of the input common-mode range is -2.3 V.

The upper limit of the input common-mode range is the value of input voltage at which Q_5 leaves the saturation region. Since for Q_5 to operate in saturation the voltage across it (i.e., V_{SD5}) should at least be equal to the overdrive voltage at which it is operating (i.e., 0.3 V), the highest voltage permitted at the drain of Q_5 should be $+2.2$ V. It follows that the highest value of v_{ICM} should be

$$v_{ICMmax} = 2.2 - 1.1 = 1.1 \text{ V}$$

The highest allowable output voltage is the value at which Q_7 leaves the saturation region, which is $V_{DD} - |V_{OV7}| = 2.5 - 0.3 = 2.2$ V. The lowest allowable output voltage is the value at which Q_6 leaves saturation, which is $-V_{SS} + V_{OV6} = -2.5 + 0.3 = -2.2$ V. Thus, the output voltage range is -2.2 V to $+2.2$ V.

Input Offset Voltage The device mismatches inevitably present in the input stage give rise to an input offset voltage. The components of this input offset voltage can be calculated using the methods developed in Section 7.4.1. Because device mismatches are random, the resulting offset voltage is referred to as **random offset.** This is to distinguish it from another type of input offset voltage that can be present even if all appropriate devices are perfectly matched. This predictable or **systematic offset** can be minimized by careful design. Although it occurs also in BJT op amps, and we have encountered it in Section 7.5.5, it is usually much more pronounced in CMOS op amps because their gain-per-stage is rather low.

To see how systematic offset can occur in the circuit of Fig. 7.40, let the two input terminals be grounded. If the input stage is perfectly balanced, then the voltage appearing at the drain of Q_4 will be equal to that at the drain of Q_3, which is $(-V_{SS} + V_{GS4})$. Now this is also the voltage that is fed to the gate of Q_6. In other words, a voltage equal to V_{GS4} appears between gate and source of Q_6. Thus the drain current of Q_6, I_6, will be related to the drain current of Q_4, which is equal to $I/2$, by the relationship

$$I_6 = \frac{(W/L)_6}{(W/L)_4}(I/2) \tag{7.199}$$

In order for no offset voltage to appear at the output, this current must be exactly equal to the current supplied by Q_7. The latter current is related to the current I of the parallel transistor Q_5 by

$$I_7 = \frac{(W/L)_7}{(W/L)_5}I \tag{7.200}$$

Now, the condition for making $I_6 = I_7$ can be found from Eqs. (7.199) and (7.200) as

$$\frac{(W/L)_6}{(W/L)_4} = 2\frac{(W/L)_7}{(W/L)_5} \tag{7.201}$$

If this condition is not met, a systematic offset will result. From the specification of the device geometries in Example 7.3, we can verify that condition (7.201) is satisfied, and, therefore, the op amp analyzed in that example should not exhibit a systematic input offset voltage.

EXERCISE

7.17 Consider the CMOS op amp of Fig. 7.40 when fabricated in a 0.8-μm CMOS technology for which $\mu_n C_{ox} = 3\mu_p C_{ox} = 90\ \mu\text{A/V}^2$, $|V_t| = 0.8$ V, and $V_{DD} = V_{SS} = 2.5$ V. For a particular design, $I = 100\ \mu\text{A}$, $(W/L)_1 = (W/L)_2 = (W/L)_5 = 200$, and $(W/L)_3 = (W/L)_4 = 100$.

(a) Find the (W/L) ratios of Q_6 and Q_7 so that $I_6 = 100\ \mu\text{A}$.

(b) Find the overdrive voltage, $|V_{OV}|$, at which each of Q_1, Q_2, and Q_6 is operating.

(c) Find g_m for Q_1, Q_2, and Q_6.

(d) If $|V_A| = 10$ V, find r_{o2}, r_{o4}, r_{o6}, and r_{o7}.

(e) Find the voltage gains A_1 and A_2, and the overall gain A.

Ans. (a) $(W/L)_6 = (W/L)_7 = 200$; (b) 0.129 V, 0.129 V, 0.105 V; (c) 0.775 mA/V, 0.775 mA/V, 1.90 mA/V; (d) 200 kΩ, 200 kΩ, 100 kΩ, 100 kΩ; (e) –77.5 V/V, –95 V/V, 7363 V/V

Frequency Response To determine the frequency response of the two-stage CMOS op amp of Fig. 7.40, consider its simplified small-signal equivalent circuit shown in Fig. 7.41. Here G_{m1} is the transconductance of the input stage ($G_{m1} = g_{m1} = g_{m2}$), R_1 is the output resistance of the first stage ($R_1 = r_{o2} \parallel r_{o4}$), and C_1 is the total capacitance at the interface between the first and second stages

$$C_1 = c_{gd4} + C_{db4} + C_{gd2} + C_{db2} + C_{gs6}$$

G_{m2} is the transconductance of the second stage ($G_{m2} = g_{m6}$), R_2 is the output resistance of the second stage ($R_2 = r_{o6} \parallel r_{o7}$), and C_2 is the total capacitance at the output node of the op amp

$$C_2 = C_{db6} + C_{db7} + C_{gd7} + C_L$$

where C_L is the load capacitance. Usually C_L is much larger than the transistor capacitances, with the result that C_2 is much larger than C_1. Finally, note that in the equivalent circuit of Fig. 7.41 we should have included C_{gd6} in parallel with C_C. Usually, however, $C_C \gg C_{gd6}$ which is the reason we have neglected C_{gd6}.

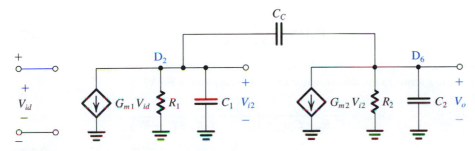

FIGURE 7.41 Equivalent circuit of the op amp in Fig. 7.40.

To determine V_o, analysis of the circuit in Fig. 7.41 proceeds as follows. Writing a node equation at node D_2 yields

$$G_{m1}V_{id} + \frac{V_{i2}}{R_1} + sC_1V_{i2} + sC_C(V_{i2} - V_o) = 0 \tag{7.202}$$

Writing a node equation at node D_6 yields

$$G_{m2}V_{i2} + \frac{V_o}{R_2} + sC_2V_o + sC_C(V_o - V_{i2}) = 0 \tag{7.203}$$

To eliminate V_{i2} and thus determine V_o in terms of V_{id}, we use Eq. (7.203) to express V_{i2} in terms of V_o and substitute the result into Eq. (7.202). After some straightforward manipulations we obtain the amplifier transfer function

$$\frac{V_o}{V_{id}} = \frac{G_{m1}(G_{m2} - sC_C)R_1R_2}{1 + s[C_1R_1 + C_2R_2 + C_C(G_{m2}R_1R_2 + R_1 + R_2)] + s^2[C_1C_2 + C_C(C_1 + C_2)]R_1R_2} \tag{7.204}$$

First we note that for $s = 0$ (i.e., dc), Eq. (7.204) gives $V_o/V_{id} = (G_{m1}R_1)(G_{m2}R_2)$, which is what we should have expected. Second, the transfer function in Eq. (7.204) indicates that the amplifier has a transmission zero at $s = s_Z$, which is determined from

$$G_{m2} - s_Z C_C = 0$$

Thus,

$$s_Z = \frac{G_{m2}}{C_C} \tag{7.205}$$

In other words, the zero is on the positive real axis with a frequency ω_Z of

$$\omega_Z = \frac{G_{m2}}{C_C} \tag{7.206}$$

Also, the amplifier has two poles that are the roots of the denominator polynomial of Eq. (7.204). If the frequencies of the two poles are denoted ω_{P1} and ω_{P2} then the denominator polynomial can be expressed as

$$D(s) = \left(1 + \frac{s}{\omega_{P1}}\right)\left(1 + \frac{s}{\omega_{P2}}\right) = 1 + s\left(\frac{1}{\omega_{P1}} + \frac{1}{\omega_{P2}}\right) + \frac{s^2}{\omega_{P1}\omega_{P2}}$$

Now if one of the poles, say that with frequency ω_{P1} is dominant, then $\omega_{P1} \ll \omega_{P2}$ and $D(s)$ can be approximated by

$$D(s) \cong 1 + \frac{s}{\omega_{P1}} + \frac{s^2}{\omega_{P1}\omega_{P2}} \tag{7.207}$$

The frequency of the dominant pole, ω_{P1}, can now be determined by equating the coefficients of the s terms in the denominator in Eq. (7.204) and in Eq. (7.207),

$$\omega_{P1} = \frac{1}{C_1R_1 + C_2R_2 + C_C(G_{m2}R_2R_1 + R_1 + R_2)}$$

$$= \frac{1}{R_1[C_1 + C_C(1 + G_{m2}R_2)] + R_2(C_2 + C_C)} \tag{7.208}$$

We recognize the first term in the denominator as arising at the interface between the first and second stages. Here, R_1, the output resistance of the first stage, is interacting with the total capacitance at the interface. The latter is the sum of C_1 and the Miller capacitance $C_C(1 + G_{m2}R_2)$, which results from connecting C_C in the negative-feedback path of the second stage whose gain is $G_{m2}R_2$. Now, since R_1 and R_2 are usually of comparable value, we see that the first term in the denominator will be much larger than the second and we can approximate ω_{P1} as

$$\omega_{P1} \cong \frac{1}{R_1[C_1 + C_C(1 + G_{m2}R_2)]}$$

A further approximation is possible because C_1 is usually much smaller than the Miller capacitance and $G_{m2}R_2 \gg 1$, thus

$$\omega_{P1} \cong \frac{1}{R_1 C_C G_{m2} R_2} \tag{7.209}$$

The frequency of the second, nondominant pole can be found by equating the coefficients of the s^2 terms in the denominator of Eq. (7.204) and in Eq. (7.207) and substituting for ω_{P1} from Eq. (7.209). The result is

$$\omega_{P2} = \frac{G_{m2}C_C}{C_1 C_2 + C_C(C_1 + C_2)}$$

Since $C_1 \ll C_2$ and $C_1 \ll C_C$, ω_{P2} can be approximated as

$$\omega_{P2} \cong \frac{G_{m2}}{C_2} \tag{7.210}$$

In order to provide the op amp with a uniform gain rolloff of -20 dB/decade down to 0 dB, the value of the compensation capacitor C_C is selected so that the resulting value of ω_{P1} (Eq. 7.209) when multiplied by the dc gain $(G_{m1}R_1G_{m2}R_2)$ results in a unity-gain frequency ω_t lower than ω_Z and ω_{P2}. Specifically

$$\omega_t = (G_{m1}R_1G_{m2}R_2)\omega_{P1}$$

$$\omega_t = \frac{G_{m1}}{C_C} \tag{7.211}$$

which must be lower than $\omega_Z = \dfrac{G_{m2}}{C_C}$ and $\omega_{P2} \cong \dfrac{G_{m2}}{C_2}$. We will have more to say about this point in Section 9.1.

EXERCISE

D7.19 Consider the frequency response of the op amp analyzed in Example 7.3. Let $C_1 = 0.1$ pF and $C_2 = 2$ pF. Find the value of C_C that results in $f_t = 10$ MHz and verify that f_t is lower than f_Z and f_{P2}.

Ans. $C_C = 4.8$ pF; $f_Z = 20$ MHz; $f_{P2} = 48$ MHz

A Bias Circuit That Stabilizes g_m   We conclude this section by presenting a bias circuit for the two-stage CMOS op amp. The circuit presented has the interesting and useful property of providing a bias current whose value is independent of both the supply voltage and the

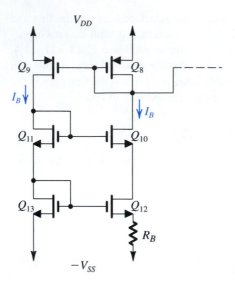

V_{DD}

Q_9 Q_8

I_B

I_B

Q_{11} Q_{10}

Q_{13} Q_{12}

R_B

$-V_{SS}$

FIGURE 7.42 Bias circuit for the CMOS op amp.

MOSFET threshold voltage. Furthermore, the transconductances of the transistors biased by this circuit have values that are determined only by a single resistor and the device dimensions.

The bias circuit is shown in Fig. 7.42. It consists of two deliberately mismatched transistors, Q_{12} and Q_{13}, with Q_{12} usually about four times wider than Q_{13} (Steininger, 1990; Johns and Martin, 1997). A resistor R_B is connected in series with the source of Q_{12}. Since, as will be shown, R_B determines both the bias current I_B and the transconductance g_{m12}, its value should be accurate and stable; in most applications, R_B would be an off-chip resistor. In order to minimize the channel-length modulation effect on Q_{12}, a cascode transistor Q_{10} and a matched diode-connected transistor Q_{11} to provide a bias voltage for Q_{10}, is included. Finally, a p-channel current mirror formed by a pair of matched devices, Q_8 and Q_9, replicates the current I_B back to Q_{11} and Q_{13}, as well as providing a bias line for Q_5 and Q_7 of the CMOS op-amp circuit of Fig. 7.40.[5]

The circuit operates as follows: The current mirror (Q_8, Q_9) causes Q_{13} to conduct a current equal to that in Q_{12}, that is, I_B. Thus,

$$I_B = \frac{1}{2}\mu_n C_{ox}\left(\frac{W}{L}\right)_{12}(V_{GS12} - V_t)^2 \tag{7.212}$$

and,

$$I_B = \frac{1}{2}\mu_n C_{ox}\left(\frac{W}{L}\right)_{13}(V_{GS13} - V_t)^2 \tag{7.213}$$

From the circuit, we see that the gate-source voltages of Q_{12} and Q_{13} are related by

$$V_{GS13} = V_{GS12} + I_B R_B$$

Subtracting V_t from both sides of this equation and using Eqs. (7.212) and (7.213) to replace $(V_{GS12} - V_t)$ and $(V_{GS13} - V_t)$ results in

$$\sqrt{\frac{2I_B}{\mu_n C_{ox}(W/L)_{13}}} = \sqrt{\frac{2I_B}{\mu_n C_{ox}(W/L)_{12}}} + I_B R_B \tag{7.214}$$

---

[5] We denote the bias current of this circuit by I_B. If this circuit is utilized to bias the CMOS op amp of Fig. 7.40, then I_B becomes the reference current I_{REF}.

This equation can be rearranged to yield

$$I_B = \frac{2}{\mu_n C_{ox}(W/L)_{12} R_B^2} \left(\sqrt{\frac{(W/L)_{12}}{(W/L)_{13}}} - 1 \right)^2 \qquad (7.215)$$

from which we observe that I_B is determined by the dimensions of Q_{12} and the value of R_B and by the ratio of the dimensions of Q_{12} and Q_{13}. Furthermore, Eq. (7.215) can be rearranged to the form

$$R_B = \frac{2}{\sqrt{2\mu_n C_{ox}(W/L)_{12} I_B}} \left(\sqrt{\frac{(W/L)_{12}}{(W/L)_{13}}} - 1 \right)$$

in which we recognize the factor $\sqrt{2\mu_n C_{ox}(W/L)_{12} I_B}$ as g_{m12}; thus,

$$g_{m12} = \frac{2}{R_B} \left(\sqrt{\frac{(W/L)_{12}}{(W/L)_{13}}} - 1 \right) \qquad (7.216)$$

This is a very interesting result: g_{m12} is determined solely by the value of R_B and the ratio of the dimensions of Q_{12} and Q_{13}. Furthermore, since g_m of a MOSFET is proportional to $\sqrt{I_D(W/L)}$, each transistor biased by the circuit of Fig. 7.42; that is, each transistor whose bias current is derived from I_B will have a g_m value that is a multiple of g_{m12}. Specifically, the ith n-channel MOSFET will have

$$g_{mi} = g_{m12} \sqrt{\frac{I_{Di}(W/L)_i}{I_B(W/L)_{12}}}$$

and the ith p-channel device will have

$$g_{mi} = g_{m12} \sqrt{\frac{\mu_p I_{Di}(W/L)_i}{\mu_n I_B(W/L)_{12}}}$$

Finally, it should be noted that the bias circuit of Fig. 7.42 employs positive feedback, and thus care should be exercised in its design to avoid unstable performance. Instability is avoided by making Q_{12} wider than Q_{13}, as has already been pointed out. Nevertheless, some form of instability may still occur; in fact, the circuit can operate in a stable state in which all currents are zero. To get it out of this state, current needs to be injected into one of its nodes, to "kick start" its operation. Feedback and stability will be studied in Chapter 8.

EXERCISES

7.20 Consider the bias circuit of Fig. 7.42 for the case: $(W/L)_8 = (W/L)_9 = (W/L)_{10} = (W/L)_{11} = (W/L)_{13} = 20$ and $(W/L)_{12} = 80$. Find the value of R_B that results in a bias current $I_B = 10$ μA. Also in a process technology having $\mu_n C_{ox} = 90$ μA/V^2, find the transconductance g_{m12}.

Ans. 5.27 kΩ; 0.379 mA/V

D7.21 Design the bias circuit of Fig. 7.42 to operate with the CMOS op amp of Example 7.3. Use Q_8 and Q_9 as identical devices with Q_8 having the dimensions given in Example 7.3. Transistors Q_{10}, Q_{11}, and Q_{13} are to be identical, with the same g_m as Q_8 and Q_9. Transistor Q_{12} is to be four times as wide as Q_{13}. Find the required value of R_B. What is the voltage drop across R_B? Also give the values of the dc voltages at the gates of Q_{12}, Q_{10}, and Q_8.

Ans. 1.67 kΩ; 150 mV; -1.5 V; -0.5 V; $+1.4$ V

7.7.2 A Bipolar Op Amp

Our second example of multistage amplifiers is the four-stage bipolar op amp shown in Fig. 7.43. The circuit consists of four stages. The input stage is **differential-in, differential-out** and consists of transistors Q_1 and Q_2, which are biased by current source Q_3. The second stage is also a differential-input amplifier, but its output is taken single-endedly at the collector of Q_5. This stage is formed by Q_4 and Q_5, which are biased by the current source Q_6. Note that the conversion from differential to single-ended as performed by the second stage results in a loss of gain by a factor of 2. A more elaborate method for accomplishing this conversion was studied in Section 7.5; it involves using a current mirror as an active load.

In addition to providing some voltage gain, the third stage, consisting of the *pnp* transistor Q_7, provides the essential function of *shifting the dc level* of the signal. Thus while the signal at the collector of Q_5 is not allowed to swing below the voltage at the base of Q_5 (+10 V), the signal at the collector of Q_7 can swing negatively (and positively, of course). From our study of op amps in Chapter 2 we know that the output terminal of the op amp should be capable of both positive and negative voltage swings. Therefore every op-amp circuit includes a **level-shifting** arrangement. Although the use of the complementary *pnp* transistor

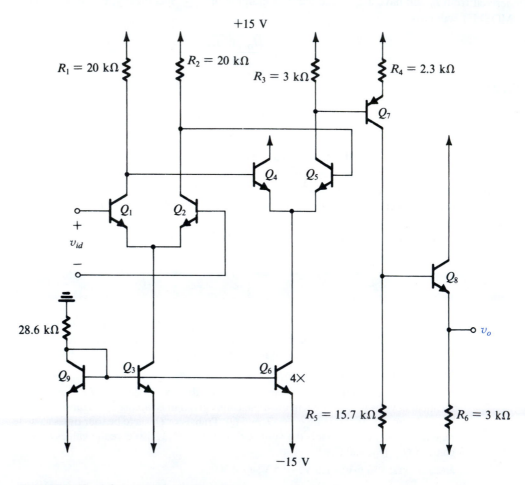

FIGURE 7.43 A four-stage bipolar op amp.

provides a simple solution to the level-shifting problem, other forms of level shifter exist, one of which will be discussed in Chapter 9. Furthermore, note that level-shifting is accomplished in the CMOS op amp we have been studying by using complementary devices for the two stages; that is, *p*-channel for the first stage and *n*-channel for the second stage.

The output stage of the op amp consists of emitter follower Q_8. As we know from our study of op amps in Chapter 2, the output operates ideally around zero volts. This and other features of the BJT op amp will be illustrated in Example 7.4.

EXAMPLE 7.4

In this example, we analyze the dc bias of the bipolar op-amp circuit of Fig. 7.43. Toward that end, Fig. 7.44 shows the circuit with the two input terminals connected to ground.

(a) Perform an approximate dc analysis (assuming $\beta \gg 1$, $|V_{BE}| \simeq 0.7$ V, and neglecting the Early effect) to calculate the dc currents and voltages everywhere in the circuit. Note that Q_6 has four times the area of each of Q_9 and Q_3.

(b) Calculate the quiescent power dissipation in this circuit.

(c) If transistors Q_1 and Q_2 have $\beta = 100$, calculate the input bias current of the op amp.

(d) What is the input common-mode range of this op amp?

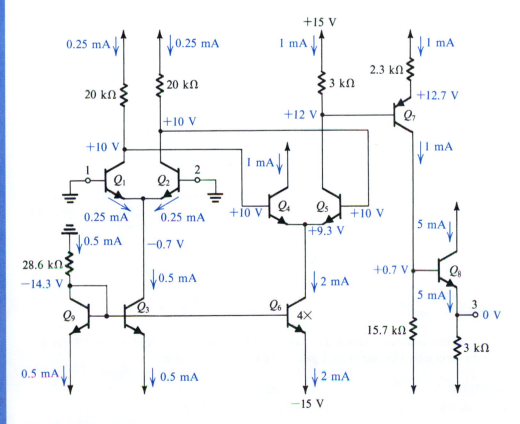

FIGURE 7.44 Circuit for Example 7.4.

Solution

(a) The values of all dc currents and voltages are indicated on the circuit diagram. These values were calculated by ignoring the base current of every transistor—that is, by assuming β to be very high. The analysis starts by determining the current through the diode-connected transistor Q_9 to be 0.5 mA. Then we see that transistor Q_3 conducts 0.5 mA and transistor Q_6 conducts 2 mA. The current-source transistor Q_3 feeds the differential pair (Q_1, Q_2) with 0.5 mA. Thus each of Q_1 and Q_2 will be biased at 0.25 mA. The collectors of Q_1 and Q_2 will be at $[+15 - 0.25 \times 20] = +10$ V.

Proceeding to the second differential stage formed by Q_4 and Q_5, we find the voltage at their emitters to be $[+10 - 0.7] = 9.3$ V. This differential pair is biased by the current-source transistor Q_6, which supplies a current of 2 mA; thus Q_4 and Q_5 will each be biased at 1 mA. We can now calculate the voltage at the collector of Q_5 as $[+15 - 1 \times 3] = +12$ V. This will cause the voltage at the emitter of the *pnp* transistor Q_7 to be $+12.7$ V, and the emitter current of Q_7 will be $(+15 - 12.7)/2.3 = 1$ mA.

The collector current of Q_7, 1 mA, causes the voltage at the collector to be $[-15 + 1 \times 15.7] = +0.7$ V. The emitter of Q_8 will be 0.7 V below the base; thus output terminal 3 will be at 0 V. Finally, the emitter current of Q_8 can be calculated to be $[0 - (-15)]/3 = 5$ mA.

(b) To calculate the power dissipated in the circuit in the quiescent state (i.e., with zero input signal) we simply evaluate the dc current that the circuit draws from each of the two power supplies. From the $+15$-V supply the dc current is $I^+ = 0.25 + 0.25 + 1 + 1 + 1 + 5 = 8.5$ mA. Thus the power supplied by the positive power supply is $P^+ = 15 \times 8.5 = 127.5$ mW. The -15-V supply provides a current I^- given by $I^- = 0.5 + 0.5 + 2 + 1 + 5 = 9$ mA. Thus the power provided by the negative supply is $P^- = 15 \times 9 = 135$ mW. Adding P^+ and P^- provides the total power dissipated in the circuit P_D: $P_D = P^+ + P^- = 262.5$ mW.

(c) The input bias current of the op amp is the average of the dc currents that flow in the two input terminals (i.e., in the bases of Q_1 and Q_2). These two currents are equal (because we have assumed matched devices); thus the bias current is given by

$$I_B = \frac{I_{E1}}{\beta + 1} \simeq 2.5 \ \mu\text{A}$$

(d) The upper limit on the input common-mode voltage is determined by the voltage at which Q_1 and Q_2 leave the active mode and enter saturation. This will happen if the input voltage exceeds the collector voltage, which is $+10$ V, by about 0.4 V. Thus the upper limit of the common-mode range is $+10.4$ V.

The lower limit of the input common-mode range is determined by the voltage at which Q_3 leaves the active mode and thus ceases to act as a constant-current source. This will happen if the collector voltage of Q_3 goes below the voltage at its base, which is -14.3 V, by more than 0.4 V. It follows that the input common-mode voltage should not go lower than $-14.7 + 0.7 = -14$ V. Thus the common-mode range is -14 V to $+10.4$ V.

EXAMPLE 7.5

Use the dc bias quantities evaluated in Example 7.4 to analyze the circuit in Fig. 7.43, to determine the input resistance, the voltage gain, and the output resistance.

Solution

The input differential resistance R_{id} is given by

$$R_{id} = r_{\pi1} + r_{\pi2}$$

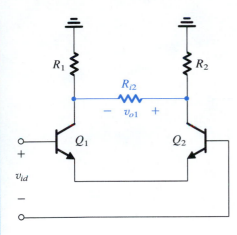

FIGURE 7.45 Equivalent circuit for calculating the gain of the input stage of the amplifier in Fig. 7.43.

Since Q_1 and Q_2 are each operating at an emitter current of 0.25 mA, it follows that

$$r_{e1} = r_{e2} = \frac{25}{0.25} = 100 \ \Omega$$

Assume $\beta = 100$; then

$$r_{\pi 1} = r_{\pi 2} = 101 \times 100 = 10.1 \text{ k}\Omega$$

Thus

$$R_{id} = 20.2 \text{ k}\Omega$$

To evaluate the gain of the first stage we first find the input resistance of the second stage, R_{i2},

$$R_{i2} = r_{\pi 4} + r_{\pi 5}$$

Q_4 and Q_5 are each operating at an emitter current of 1 mA; thus

$$r_{e4} = r_{e5} = 25 \ \Omega$$

$$r_{\pi 4} = r_{\pi 5} = 101 \times 25 = 2.525 \text{ k}\Omega$$

Thus $R_{i2} = 5.05$ kΩ. This resistance appears between the collectors of Q_1 and Q_2, as shown in Fig. 7.45. Thus the gain of the first stage will be

$$A_1 \equiv \frac{v_{o1}}{v_{id}} \simeq \frac{\text{Total resistance in collector circuit}}{\text{Total resistance in emitter circuit}}$$

$$= \frac{[R_{i2} \| (R_1 + R_2)]}{r_{e1} + r_{e2}}$$

$$= \frac{(5.05 \text{ k}\Omega \| 40 \text{ k}\Omega)}{200 \ \Omega} = 22.4 \text{ V/V}$$

Figure 7.46 shows an equivalent circuit for calculating the gain of the second stage. As indicated, the input voltage to the second stage is the output voltage of the first stage, v_{o1}. Also shown is the resistance R_{i3}, which is the input resistance of the third stage formed by Q_7. The value of R_{i3} can be found by multiplying the total resistance in the emitter of Q_7 by $(\beta + 1)$:

$$R_{i3} = (\beta + 1)(R_4 + r_{e7})$$

Since Q_7 is operating at an emitter current of 1 mA,

$$r_{e7} = \frac{25}{1} = 25 \ \Omega$$

$$R_{i3} = 101 \times 2.325 = 234.8 \text{ k}\Omega$$

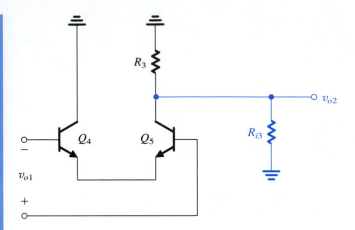

FIGURE 7.46 Equivalent circuit for calculating the gain of the second stage of the amplifier in Fig. 7.43.

We can now find the gain A_2 of the second stage as the ratio of the total resistance in the collector circuit to the total resistance in the emitter circuit:

$$A_2 \equiv \frac{v_{o2}}{v_{o1}} \simeq -\frac{(R_3 \| R_{i3})}{r_{e4} + r_{e5}}$$

$$= -\frac{(3 \text{ k}\Omega \| 234.8 \text{ k}\Omega)}{50 \ \Omega} = -59.2 \text{ V/V}$$

To obtain the gain of the third stage we refer to the equivalent circuit shown in Fig. 7.47, where R_{i4} is the input resistance of the output stage formed by Q_8. Using the resistance-reflection rule, we calculate the value of R_{i4} as

$$R_{i4} = (\beta + 1)(r_{e8} + R_6)$$

where

$$r_{e8} = \frac{25}{5} = 5 \ \Omega$$

$$R_{i4} = 101(5 + 3000) = 303.5 \text{ k}\Omega$$

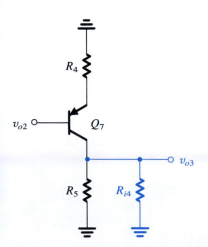

FIGURE 7.47 Equivalent circuit for evaluating the gain of the third stage in the amplifier circuit of Fig. 7.43.

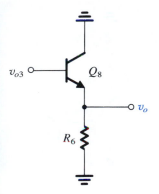

FIGURE 7.48 Equivalent circuit of the output stage of the amplifier circuit of Fig. 7.43.

The gain of the third stage is given by

$$A_3 \equiv \frac{v_{o3}}{v_{o2}} \simeq -\frac{(R_5 \parallel R_{i4})}{r_{e7} + R_4}$$

$$= -\frac{(15.7 \text{ k}\Omega \parallel 303.5 \text{ k}\Omega)}{2.325 \text{ k}\Omega} = -6.42 \text{ V/V}$$

Finally, to obtain the gain A_4 of the output stage we refer to the equivalent circuit in Fig. 7.48 and write

$$A_4 \equiv \frac{v_o}{v_{o3}} = \frac{R_6}{R_6 + r_{e8}}$$

$$= \frac{3000}{3000 + 5} = 0.998 \simeq 1$$

The overall voltage gain of the amplifier can then be obtained as follows:

$$\frac{v_o}{v_{id}} = A_1 A_2 A_3 A_4 = 8513 \text{ V/V}$$

or 78.6 dB.

To obtain the output resistance R_o we "grab hold" of the output terminal in Fig. 7.43 and look back into the circuit. By inspection we find

$$R_o = R_6 \parallel [r_{e8} + R_5/(\beta + 1)]$$

which gives

$$R_o = 152 \ \Omega$$

EXERCISE

7.22 Use the results of Example 7.5 to calculate the overall voltage gain of the amplifier in Fig. 7.43 when it is connected to a source having a resistance of 10 kΩ and a load of 1 kΩ.

Ans. 4943 V/V

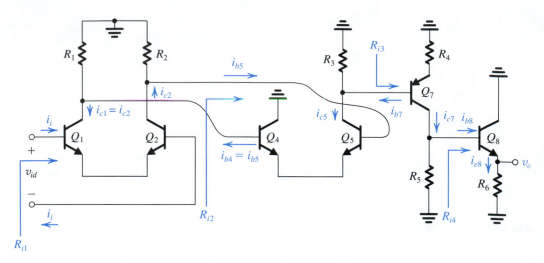

FIGURE 7.49 The circuit of the multistage amplifier of Fig. 7.43 prepared for small-signal analysis. Indicated are the signal currents throughout the amplifier and the input resistances of the four stages.

Analysis Using Current Gains There is an alternative method for the analysis of bipolar multistage amplifiers that can be somewhat easier to perform in some cases. The method makes use of current gains or more appropriately current transmission factors. In effect, one traces the transmission of the signal current throughout the amplifier cascade, evaluating all the current transmission factors in turn. We shall illustrate the method by using it to analyze the amplifier circuit of the preceding example.

Figure 7.49 shows the amplifier circuit prepared for small-signal analysis. We have indicated on the circuit diagram the signal currents through all the circuit branches. Also indicated are the input resistances of all four stages of the amplifier. These should be evaluated before commencing the following analysis.

The purpose of the analysis is to determine the overall voltage gain (v_o/v_{id}). Toward that end, we express v_o in terms of the signal current in the emitter of Q_8, i_{e8}, and v_{id} in terms of the input signal current i_i, as follows:

$$v_o = R_6 i_{e8}$$

$$v_{id} = R_{i1} i_i$$

Thus, the voltage gain can be expressed in terms of the current gain (i_{e8}/i_i) as

$$\frac{v_o}{v_{id}} = \frac{R_6}{R_{i1}} \frac{i_{e8}}{i_i}$$

Next, we expand the current gain (i_{e8}/i_i) in terms of the signal currents throughout the circuit as follows:

$$\frac{i_{e8}}{i_i} = \frac{i_{e8}}{i_{b8}} \times \frac{i_{b8}}{i_{c7}} \times \frac{i_{c7}}{i_{b7}} \times \frac{i_{b7}}{i_{c5}} \times \frac{i_{c5}}{i_{b5}} \times \frac{i_{b5}}{i_{c2}} \times \frac{i_{c2}}{i_i}$$

Each of the current-transmission factors on the right-hand side is either the current gain of a transistor or the ratio of a current divider. Thus, reference to Fig. 7.49 enables us to find these factors by inspection,

$$\frac{i_{e8}}{i_{b8}} = \beta_8 + 1$$

$$\frac{i_{b8}}{i_{c7}} = \frac{R_5}{R_5 + R_{i4}}$$

$$\frac{i_{c7}}{i_{b7}} = \beta_7$$

$$\frac{i_{b7}}{i_{c5}} = \frac{R_3}{R_3 + R_{i3}}$$

$$\frac{i_{c5}}{i_{b5}} = \beta_5$$

$$\frac{i_{b5}}{i_{c2}} = \frac{(R_1 + R_2)}{(R_1 + R_2) + R_{i2}}$$

$$\frac{i_{c2}}{i_i} = \beta_2$$

These ratios can be easily evaluated and their values used to determine the voltage gain.

With a little practice, it is possible to carry out such an analysis very quickly, forgoing explicitly labeling the signal currents on the circuit diagram. One simply "walks through" the circuit, from input to output, or vice versa, determining the current-transmission factors one at a time, in a chainlike fashion.

EXERCISE

7.23 Use the values of input resistance found in Example 7.5 to evaluate the seven current-transmission factors and hence the overall current gain and voltage gain.

Ans. The current-transmission factors in the order of their listing are 101, 0.0492, 100, 0.0126, 100, 0.8879, 100 A/A; the overall current gain is 55993 A/A; the voltage gain is 8256 V/V. This value differs slightly from that found in Example 7.5 because of the various approximations made in the example (e.g., $\alpha \cong 1$).

Frequency Response   The bipolar op-amp circuit of Fig. 7.43 is rather complex. Nevertheless, it is possible to obtain an approximate estimate of its high-frequency response. Figure 7.50(a) shows an approximate equivalent circuit for this purpose. Note that we have utilized the equivalent differential half-circuit concept, with Q_2 representing the input stage and Q_5 representing the second stage. We observe, of course, that the second stage is not symmetrical, and strictly speaking the equivalent half-circuit does not apply. Nevertheless, we use it as an approximation so as to obtain a quick pencil-and-paper estimate of the dominant high-frequency pole of the amplifier. More precise results can of course be obtained using computer simulation with SPICE (Section 7.8).

Examination of the equivalent circuit in Fig. 7.50(a) reveals that if the resistance of the source of signal V_i is small, the high-frequency limitation will not occur at the input but rather at the interface between the first and the second stages. This is because the total capacitance at node A will be high as a result of the Miller multiplication of $C_{\mu5}$. Also, the third stage, formed by transistor Q_7, should exhibit good high-frequency response, since Q_7 has a large emitter-degeneration resistance, R_3. The same is also true for the emitter-follower stage, Q_8.

To determine the frequency of the dominant pole that is formed at the interface between Q_2 and Q_5 we show in Fig. 7.50(b) the pertinent equivalent circuit. The total resistance

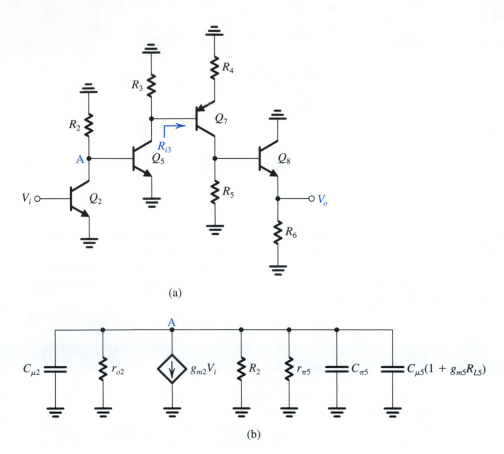

(a)

(b)

FIGURE 7.50 (a) Approximate equivalent circuit for determining the high-frequency response of the op amp of Fig. 7.43. (b) Equivalent circuit of the interface between the output of Q_2 and the input of Q_5.

between node A and ground can now be found as

$$R_{eq} = R_2 \parallel r_{o2} \parallel r_{\pi 5}$$

and the total capacitance is

$$C_{eq} = C_{\mu 2} + C_{\pi 5} + C_{\mu 5}(1 + g_{m5}R_{L5})$$

where

$$R_{L5} = R_3 \parallel r_{o5} \parallel R_{i3}$$

The frequency of the pole can be calculated from R_{eq} and C_{eq} as

$$f_P = \frac{1}{2\pi R_{eq} C_{eq}}$$

EXERCISE

7.24 Determine R_{eq}, C_{eq}, and f_P for the amplifier in Fig. 7.43 utilizing the facts that Q_2 is biased at 0.25 mA and Q_5 at 1 mA. Assume $\beta = 100$, $V_A = 100$ V, $f_T = 400$ MHz, and $C_\mu = 2$ pF. Assume $R_{L5} \cong R_3$.

Ans. 2.21 kΩ; 258 pF; 280 kHz

7.8 SPICE SIMULATION EXAMPLE

We conclude this chapter by presenting a SPICE simulation of the multistage differential amplifier whose dc bias was analyzed in Example 7.4 and whose small-signal performance was the subject of Example 7.5.

EXAMPLE 7.6

SPICE SIMULATION OF A MULTISTAGE DIFFERENTIAL AMPLIFIER

The Capture schematic of the multistage op-amp circuit analyzed in Examples 7.4 and 7.5 is shown in Fig. 7.51.[6] Observe the manner in which the differential signal input V_d and the common-mode input voltage V_{CM} are applied. Such an input bias configuration for an op-amp circuit was presented and used in Example 2.9. In the following simulations, we will use parts Q2N3904 and

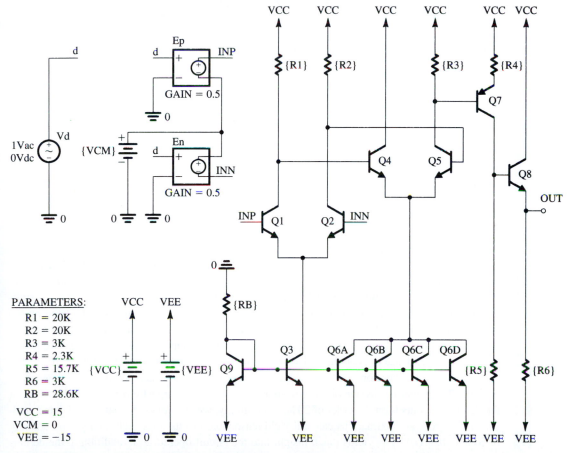

FIGURE 7.51 Capture schematic of the op-amp circuit in Example 7.6.

---

[6] This circuit cannot be simulated using the student evaluation version of PSpice (OrCAD 9.2 Lite Edition) that is included on the CD accompanying this book. This is because, in this free version of PSpice, circuit simulation is restricted to circuits with no more than 10 transistors.

TABLE 7.2 SPICE Model Parameters of the Q2N3904 and Q2N3906 Discrete BJTs

Q2N3904 Discrete BJT

IS=6.734f	XTI=3	EG=1.11	VAF=74.03	BF=416.4	NE=1.259	ISE=6.734f
IKF=66.78m	XTB=1.5	BR=.7371	NC=2	ISC=0	IKR=0	RC=1
CJC=3.638p	MJC=.3085	VJC=.75	FC=.5	CJE=4.493p	MJE=.2593	VJE=.75
TR=239.5n	TF=301.2p	ITF=.4	VTF=4	XTF=2	RB=10	

Q2N3906 Discrete BJT

IS=1.41f	XTI=3	EG=1.11	VAF=18.7	BF=180.7	NE=1.5	ISE=0
IKF=80m	XTB=1.5	BR=4.977	NC=2	ISC=0	IKR=0	RC=2.5
CJC=9.728p	MJC=.5776	VJC=.75	FC=.5	CJE=8.063p	MJE=.3677	VJE=.75
TR=33.42n	TF=179.3p	ITF=.4	VTF=4	XTF=6	RB=10	

TABLE 7.3 Dc Collector Currents of the Op-Amp Circuit in Fig. 7.51 as Computed by Hand Analysis (Example 7.4) and by PSpice

	Collector Currents (mA)		
Transistor	Hand Analysis (Example 7.4)	PSpice	Error (%)
Q_1	0.25	0.281	−11.0
Q_2	0.25	0.281	−11.0
Q_3	0.5	0.567	−11.8
Q_4	1.0	1.27	−21.3
Q_5	1.0	1.21	−17.4
Q_6	2.0	2.50	−20.0
Q_7	1.0	1.27	−21.3
Q_8	5.0	6.17	−18.9
Q_9	0.5	0.48	+4.2

Q2N3906 (from Fairchild Semiconductor) for the *npn* and *pnp* BJTs, respectively. The model parameters of these discrete BJTs are listed in Table 7.2 and are available in PSpice.

In PSpice, the common-mode input voltage V_{CM} of the op-amp circuit is set to 0 V (i.e., to the average of the dc power-supply voltages V_{CC} and V_{EE}) to maximize the available input signal swing. A bias-point simulation is performed to determine the dc operating point. Table 7.3 summarizes the values of the dc collector currents as computed by PSpice and as calculated by the hand analysis in Example 7.4. Recall that our hand analysis assumed β and the Early voltage V_A of the BJTs to be both infinite. However, our SPICE simulations in Example 5.21 (where we investigated the dependence of β on the collector current I_C) indicate that the Q2N3904 has $\beta \approx 125$ at $I_C = 0.25$ mA. Furthermore, its forward Early voltage (SPICE parameter VAF) is 74 V, as given in Table 7.2. Nevertheless, we observe from Table 7.3 that the largest error in the calculation of the dc bias currents is on the order of 20%. Accordingly, we can conclude that a quick hand analysis using gross approximations can still yield reasonable results for a preliminary estimate and, of course, hand analysis yields much insight into the circuit operation. In addition to the dc bias currents listed in Table 7.3, the bias-point simulation in PSpice shows that the output dc offset (i.e., V_{OUT} when $V_d = 0$) is 3.62 V and that the input bias current I_{B1} is 2.88 μA.

To compute the **large-signal differential transfer characteristic** of the op-amp circuit, we perform a dc-analysis simulation in PSpice with the differential voltage input V_d swept over the range $-V_{EE}$ to $+V_{CC}$, and we plot the corresponding output voltage V_{OUT}. Figure 7.52(a) shows the

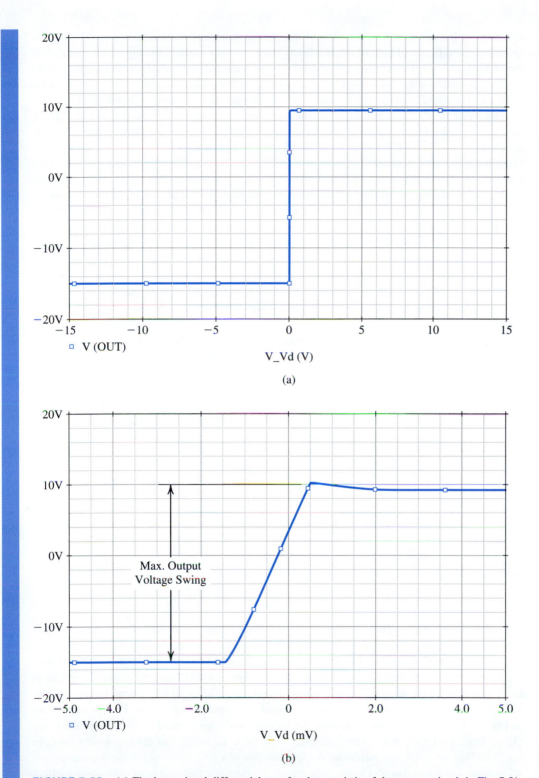

FIGURE 7.52 **(a)** The large-signal differential transfer characteristic of the op-amp circuit in Fig. 7.51. The common-mode input voltage V_{CM} is set to 0 V. **(b)** An expanded view of the transfer characteristic in the high-gain region.

resulting dc transfer characteristic. The slope of this characteristic (i.e., DV_{OUT}/DV_d) corresponds to the differential gain of the amplifier. Note that, as expected, the high-gain region is in the vicinity of $V_d = 0$ V. However, the resolution of the input-voltage axis is too gross to yield much information about the details of the high-gain region. Therefore, to examine this region more closely, the dc analysis is repeated with V_d swept over the range -5 mV to $+5$ mV at increments of 10μV. The resulting differential dc transfer characteristic is plotted in Fig. 7.52(b). Accordingly, the linear region of the large-signal differential characteristic is bounded approximately by $V_d = -1.5$ mV and $V_d = +0.5$ mV. Over this region, the output level changes from $V_{OUT} = -15$ V to about $V_{OUT} = +10$ V in a linear fashion. Thus, the output voltage swing for this amplifier is between -15 V and $+10$ V, a rather asymmetrical range. A rough estimate for the differential gain of this amplifier can be obtained from the boundaries of the linear region as $A_d = [10 - (-15)]$ V$/[0.5 - (-1.5)]$ mV $= 12.5 \times 10^3$ V/V. We also observe from Fig. 7.52(b) that $V_d \cong -260 \mu$V when $V_{OUT} = 0$. Therefore, the amplifier has an input offset voltage V_{OS} of $+260 \mu$V (by convention, the negative value of the x-axis intercept of the large-signal differential transfer characteristics). This corresponds to an output offset voltage of $A_d V_{OS} \cong$ $(12.5 \times 10^3)(260 \mu$V$) = 3.25$ V, which is close to the value found through the bias-point simulation. It should be emphasized that this offset voltage is inherent in the design and is not the result of component or device mismatches. Thus, it is usually referred to as a **systematic offset.**

Next, to compute the **frequency response** of the op-amp circuit and to measure its differential gain A_d and its 3-dB frequency f_H in PSpice, we set the differential input voltage V_d to be a 1-V ac signal (with 0-V dc level), perform an ac-analysis simulation, and plot the output voltage magnitude $|V_{OUT}|$ versus frequency. Figure 7.53(a) shows the resulting frequency response. Accordingly, $A_d = 13.96 \times 10^3$ V/V or 82.8 dB, and $f_H = 256.9$ kHz. Thus, this value of A_d is close to the value estimated using the large-signal differential transfer characteristic.

An approximate value of f_H can also be obtained using the expressions derived in Section 7.6.2. Specifically,

$$f_H \cong \frac{1}{2\pi R_{eq} C_{eq}} \tag{7.217}$$

where

$$C_{eq} = C_{\mu 2} + C_{\pi 5} + C_{\mu 5}[1 + g_{m5}(R_3 \parallel r_{o5} \parallel (r_{\pi 7} + (\beta + 1)R_4))]$$

and

$$R_{eq} = R_2 \parallel r_{o2} \parallel r_{\pi 5}$$

The values of the small-signal parameters as computed by PSpice can be found in the output file of a bias-point (or an ac-analysis) simulation. Using these values results in $C_{eq} = 338$ pF, $R_{eq} = 2.91$ kΩ, and $f_H = 161.7$ kHz. However, this approximate value of f_H is much smaller than the value computed by PSpice. The reason for this disagreement is that the foregoing expression for f_H was derived (in Section 7.6.2) using the equivalent differential half-circuit concept. However, the concept is accurate only when it is applied to a symmetrical circuit. The op-amp circuit in Fig. 7.51 is not symmetrical because the second gain stage formed by the differential pair Q_4–Q_5 has a load resistor R_3 in the collector of Q_5 only. To verify that the expression for f_H in Eq. (7.217) gives a close approximation for f_H in the case of a symmetric circuit, we insert a resistor R'_3 (whose size is equal to R_3) in the collector of Q_4. Note that this will have only a minor effect on the dc operating point. The op-amp circuit with Q_4 having a collector resistor R'_3 is then simulated in PSpice. Figure 7.53(b) shows the resulting frequency response of this symmetric op amp

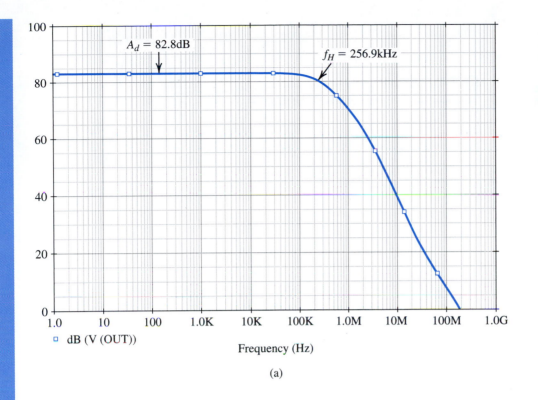

(a)

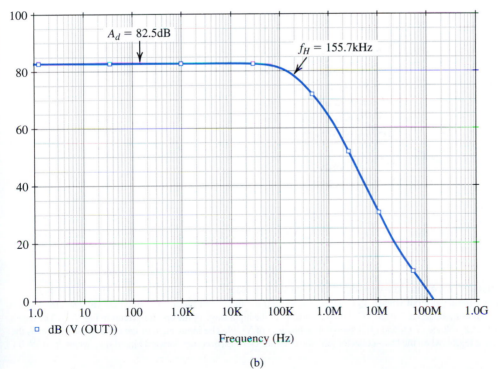

(b)

FIGURE 7.53 Frequency response of **(a)** the op-amp circuit in Fig. 7.51 and **(b)** the op-amp circuit in Fig. 7.51 but with a resistor $R'_3 = R_3$ inserted in the collector of Q_4 to make the op-amp circuit symmetrical.

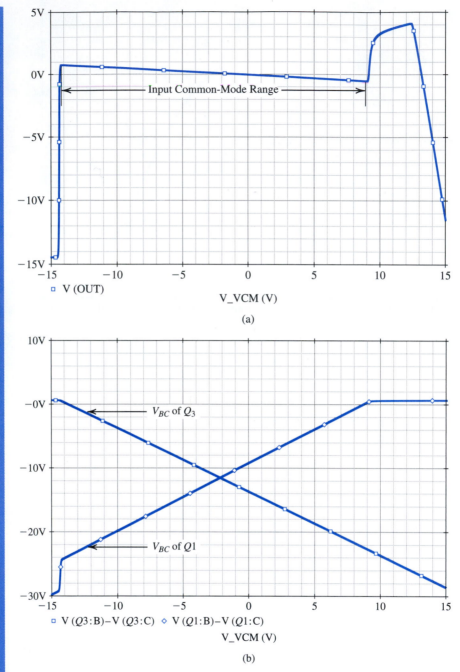

FIGURE 7.54 **(a)** The large-signal common-mode transfer characteristic of the op-amp circuit in Fig. 7.51. The differential input voltage V_d is set to $-V_{OS} = -260\ \mu\text{V}$ to prevent premature saturation. **(b)** The effect of the common-mode input voltage VCM on the linearity of the input stage of the op-amp circuit in Fig. 7.51. The base–collector voltage of Q_1 and Q_3 is shown as a function of VCM. The input stage of the op-amp circuit leaves the active region when the base–collector junction of either Q_1 or Q_3 becomes forward biased (i.e., when $\text{VBC} \geq 0$).

where $f_H = 155.7$ kHz. Accordingly, in the case of a perfectly symmetric op-amp circuit, the value of f_H in Eq. (7.217) closely approximates the value computed by PSpice. Comparing the frequency responses of the nonsymmetric (Fig. 7.53a) and the symmetric (Fig. 7.53b) op-amp circuits, we note that the 3-dB frequency of the op amp drops from 256.9 kHz to 155.7 kHz when resistor R_3' is inserted in the collector of Q_4 to make the op-amp circuit symmetrical. This is because, with a resistor R_3', the collector of Q_4 is no longer at signal ground and, hence, $C_{\mu 4}$ experiences the Miller effect. Consequently, the high-frequency response of the op-amp circuit is degraded.

Observe that in the preceding ac-analysis simulation, owing to the systematic offset inherent in the design, the op-amp circuit is operating at an output dc voltage of 3.62 V. However, in an actual circuit implementation (with $V_{CM} = 0$), negative feedback is employed (see Chapters 2 and 8) and the output dc voltage is stabilized at zero. Thus, the small-signal performance of the op-amp circuit can be more accurately simulated by biasing the circuit so as to force operation at this level of output voltage. This can be easily done by applying a differential dc input of $-V_{OS}$. Superimposed on this dc input, we can apply an ac signal to perform an ac-analysis simulation for the purpose of, for example, computing the differential gain and the 3-dB frequency.

Finally, to compute the input common-mode range of the op-amp circuit in Fig. 7.51, we perform a dc-analysis simulation in PSpice with the input common-mode voltage swept over the range $-V_{EE}$ to V_{CC}, while maintaining V_d constant at $-V_{OS}$ in order to cancel the output offset voltage (as discussed earlier) and, thus, prevent premature saturation of the BJTs. The corresponding output voltage V_{OUT} is plotted in Fig. 7.54(a). From this common-mode dc transfer characteristic we find that the amplifier behaves linearly over the V_{CM} range -14.1 V to $+8.9$ V, which is therefore the **input common-mode range.** In Example 7.4, we noted that the upper limit of this range is determined by Q_1 and Q_2 saturating, whereas the lower limit is determined by Q_3 saturating. To verify this assertion, we requested PSpice to plot the values of the collector–base voltages of these BJTs versus the input common-mode voltage V_{CM}. The results are shown in Fig. 7.54(b), from which we note that our assertion is indeed correct (recall that an *npn* BJT enters its saturation region when its base–collector junction becomes forward biased, i.e., $V_{BC} \geq 0$).

SUMMARY

- The differential-pair or differential-amplifier configuration is the most widely used building block in analog IC design. The input stage of every op amp is a differential amplifier.

- There are two reasons for preferring differential to single-ended amplifiers: Differential amplifiers are insensitive to interference, and they do not need bypass and coupling capacitors.

- For a MOS (bipolar) pair biased by a current source I, each device operates at a drain (collector, assuming $\alpha = 1$) current of $I/2$ and a corresponding overdrive voltage V_{OV} (no analog in bipolar). Each device has $g_m = I/V_{OV}$ ($\alpha I/2V_T$, for bipolar) and $r_o = |V_A|/(I/2)$.

■ With the two input terminals connected to a suitable dc voltage V_{CM}, the bias current I of a perfectly symmetrical differential pair divides equally between the two transistors of the pair, resulting in a zero voltage difference between the two drains (collectors). To steer the current completely to one side of the pair, a difference input voltage v_{id} of at least $\sqrt{2}V_{OV}$ ($4V_T$ for bipolar) is needed.

■ Superimposing a differential input signal v_{id} on the dc common-mode input voltage V_{CM} such that $v_{I1} = V_{CM} + v_{id}/2$ and $v_{I2} = V_{CM} - v_{id}/2$ causes a virtual signal ground to appear on the common source (emitter) connection. In response to v_{id}, the current in Q_1 increases by $g_m v_{id}/2$ and the current in Q_2 decreases by $g_m v_{id}/2$. Thus, voltage signals of $\pm g_m(R_D \parallel r_o)v_{id}/2$ develop at the two drains (collectors, with R_D replaced by R_C). If the output voltage is taken single-endedly, that is, between one of the drains (collectors) and ground, a differential gain of $\frac{1}{2}g_m(R_D \parallel r_o)$ is realized. Taking the output differentially, that is, between the two drains (collectors) the differential gain realized is twice as large: $g_m(R_D \parallel r_o)$.

■ The analysis of a differential amplifier to determine differential gain, differential input resistance, frequency response of differential gain, and so on is facilitated by employing the differential half-circuit, which is a common-source (common-emitter) transistor biased at $I/2$.

■ An input common-mode signal v_{icm} gives rise to drain (collector) voltage signals that are ideally equal and given by $-v_{icm}(R_D/2R_{SS})$ [$-v_{icm}(R_C/2R_{EE})$ for the bipolar pair], where R_{SS} (R_{EE}) is the output resistance of the current source that supplies the bias current I. When the output is taken single-endedly a common-mode gain of magnitude $|A_{cm}| = R_D/2R_{SS}$ ($R_C/2R_{EE}$ for the bipolar case) results. Taking the output differentially results in the perfectly matched case in zero A_{cm} (infinite CMRR). Mismatches between the two sides of the pair make A_{cm} finite even when the output is taken differentially: A mismatch ΔR_D causes $|A_{cm}| = (R_D/2R_{SS})(\Delta R_D/R_D)$; a mismatch Δg_m causes $|A_{cm}| = (R_D/2R_{SS})(\Delta g_m/g_m)$. Corresponding expressions apply for the bipolar pair.

■ While the input differential resistance R_{id} of the MOS pair is infinite, that for the bipolar pair is only $2r_\pi$ but can be increased to $2(\beta + 1)(r_e + R_e)$ by including resistances R_e in the two emitters. The latter action, however, lowers A_d.

■ Mismatches between the two sides of a differential pair result in a differential dc output voltage V_O even when the two input terminals are tied together and connected to a dc voltage V_{CM}. This signifies the presence of an input offset voltage $V_{OS} \equiv V_O/A_d$. In a MOS pair there are three main sources for V_{OS}:

$$\Delta R_D \Rightarrow V_{OS} = \frac{V_{OV}}{2}\frac{\Delta R_D}{R_D}$$

$$\Delta(W/L) \Rightarrow V_{OS} = \frac{V_{OV}}{2}\frac{\Delta(W/L)}{W/L}$$

$$\Delta V_t \Rightarrow V_{OS} = \Delta V_t$$

For the bipolar pair there are two main sources:

$$\Delta R_C \Rightarrow V_{OS} = V_T\frac{\Delta R_C}{R_C}$$

$$\Delta I_S \Rightarrow V_{OS} = V_T\frac{\Delta I_S}{I_S}$$

■ A popular circuit in both MOS and bipolar analog ICs is the current-mirror-loaded differential pair. It realizes a high differential gain $A_d = g_m(R_{o\,pair} \parallel R_{o\,mirror})$ and a low common-mode gain, $|A_{cm}| = \frac{1}{2}g_{m3}R_{SS}$ for the MOS circuit ($r_{o4}/\beta_3 R_{EE}$ for the bipolar circuit), as well as performing the differential-to-single-ended conversion with no loss of gain.

■ The common-mode gain of the differential amplifier exhibits a transmission zero caused by the finite output resistance and capacitance of the bias current source; $f_Z = \frac{1}{2\pi}C_{SS}R_{SS}$ ($\frac{1}{2\pi}C_{EE}R_{EE}$ for bipolar). Thus the CMRR has a pole at this relatively low frequency.

■ A multistage amplifier usually consists of three stages: an input stage having a high input resistance, a reasonably high gain, and, if differential, a high CMRR; an intermediate stage that realizes the bulk of the gain; and an output stage having a low output resistance. Many CMOS amplifiers serve to drive only small on-chip capacitive loads and hence do not need an output stage. In designing and analyzing a multistage amplifier; the loading effect of each stage on the one that precedes it must be taken into account.

PROBLEMS

SECTION 7.1: THE MOS DIFFERENTIAL PAIR

7.1 For an NMOS differential pair with a common-mode voltage v_{CM} applied, as shown in Fig. 7.2, let $V_{DD} = V_{SS} =$ 2.5 V, $k'_n W/L = 3$ mA/V^2, $V_{tn} = 0.7$ V, $I = 0.2$ mA, $R_D =$ 5 kΩ, and neglect channel-length modulation.

(a) Find V_{OV} and V_{GS} for each transistor.
(b) For $v_{CM} = 0$, find v_S, i_{D1}, i_{D2}, v_{D1}, and v_{D2}.
(c) Repeat (b) for $v_{CM} = +1$ V.
(d) Repeat (b) for $v_{CM} = -1$ V.
(e) What is the highest value of v_{CM} for which Q_1 and Q_2 remain in saturation?
(f) If current source I requires a minimum voltage of 0.3 V to operate properly, what is the lowest value allowed for v_S and hence for v_{CM}?

7.2 For the PMOS differential amplifier shown in Fig. P7.2 let $V_{tp} = -0.8$ V and $k'_p W/L = 3.5$ mA/V^2. Neglect channel-length modulation.

(a) For $v_{G1} = v_{G2} = 0$ V, find V_{OV} and V_{GS} for each of Q_1 and Q_2. Also find v_S, v_{D1}, and v_{D2}.
(b) If the current source requires a minimum voltage of 0.5 V, find the input common-mode range.

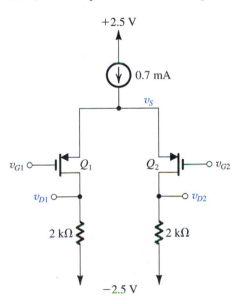

FIGURE P7.2

7.3 For the differential amplifier specified in Problem 7.1 let $v_{G2} = 0$ and $v_{G1} = v_{id}$. Find the value of v_{id} that corresponds to each of the following situations:

(a) $i_{D1} = i_{D2} = 0.1$ mA; (b) $i_{D1} = 0.15$ mA and $i_{D2} = 0.05$ mA; (c) $i_{D1} = 0.2$ mA and $i_{D2} = 0$ (Q_2 just cuts off); (d) $i_{D1} = 0.05$ mA

and $i_{D2} = 0.15$ mA; (e) $i_{D1} = 0$ mA (Q_1 just cuts off) and $i_{D2} =$ 0.2 mA. For each case, find v_S, v_{D1}, v_{D2}, and $(v_{D2} - v_{D1})$.

7.4 For the differential amplifier specified in Problem 7.2, let $v_{G2} = 0$ and $v_{G1} = v_{id}$. Find the range of v_{id} needed to steer the bias current from one side of the pair to the other. At each end of this range, give the value of the voltage at the common-source terminal and the drain voltages.

7.5 Consider the differential amplifier specified in Problem 7.1 with G_2 grounded and $v_{G1} = v_{id}$. Let v_{id} be adjusted to the value that causes $i_{D1} = 0.11$ mA and $i_{D2} = 0.09$ mA. Find the corresponding values of v_{GS2}, v_S, v_{GS1}, and hence v_{id}. What is the difference output voltage $v_{D2} - v_{D1}$? What is the voltage gain $(v_{D2} - v_{D1})/v_{id}$? What value of v_{id} results in $i_{D1} = 0.09$ mA and $i_{D2} = 0.11$ mA?

7.6 The table providing the answers to Exercise 7.3 shows that as the maximum input signal to be applied to the differential pair is increased, linearity is maintained at the same level by operating at a higher V_{OV}. If $|v_{id}|_{max}$ is to be 150 mV, use the data in the table to determine the required V_{OV} and the corresponding values of W/L and g_m.

7.7 Use Eq. (7.23) to show that if the term involving v_{id}^2 is to be kept to a maximum value of k then the maximum possible fractional change in the transistor current is given by

$$\left(\frac{\Delta I_{max}}{I/2}\right) = 2\sqrt{k(1-k)}$$

and the corresponding maximum value of v_{id} is given by

$$v_{id\,max} = 2\sqrt{k}V_{OV}$$

Evaluate both expressions for $k = 0.01, 0.1,$ and 0.2.

7.8 An NMOS differential amplifier utilizes a bias current of 200 μA. The devices have $V_t = 0.8$ V, $W = 100$ μm, and $L = 1.6$ μm, in a technology for which $\mu_n C_{ox} = 90$ μA/V^2. Find V_{GS}, g_m, and the value of v_{id} for full-current switching. To what value should the bias current be changed in order to double the value of v_{id} for full-current switching?

D7.9 Design the MOS differential amplifier of Fig. 7.5 to operate at $V_{OV} = 0.2$ V and to provide a transconductance g_m of 1 mA/V. Specify the W/L ratios and the bias current. The technology available provides $V_t = 0.8$ V and $\mu_n C_{ox} = 90$ μA/V^2.

7.10 Consider the NMOS differential pair illustrated in Fig. 7.5 under the conditions that $I = 100$ μA, using FETs for which $k'_n (W/L) = 400$ μA/V^2, and $V_t = 1$ V. What is the voltage on the common-source connection for $v_{G1} = v_{G2} = 0$? 2 V? What is the relation between the drain currents in each of these situations? Now for $v_{G2} = 0$ V, at what voltages must v_{G1} be placed to reduce i_{D2} by 10%? to increase i_{D2} by 10%? What is the differential voltage, $v_{id} = v_{G2} - v_{G1}$, for which the ratio of drain currents i_{D2}/i_{D1} is 1.0? 0.5? 0.9? 0.99? For the current ratio $i_{D1}/i_{D2} = 20.0$, what differential input is required?

SECTION 7.2: SMALL-SIGNAL OPERATION OF THE MOS DIFFERENTIAL PAIR

7.11 An NMOS differential amplifier is operated at a bias current I of 0.5 mA and has a W/L ratio of 50, $\mu_n C_{ox} = 250 \, \mu A/V^2$, $V_A = 10$ V, and $R_D = 4$ kΩ. Find V_{OV}, g_m, r_o, and A_d.

D7.12 It is required to design an NMOS differential amplifier to operate with a differential input voltage that can be as high as 0.2 V while keeping the nonlinear term under the square root in Eq. (7.23) to a maximum of 0.1. A transconductance g_m of 3 mA/V is needed. Find the required values of V_{OV}, I, and W/L. Assume that the technology available has $\mu_n C_{ox} = 100 \, \mu A/V^2$. What differential gain A_d results when $R_D = 5$ kΩ? Assume $\lambda = 0$. What is the resulting output signal corresponding to v_{id} at its maximum value?

D*7.13 Figure P7.13 shows a circuit for a differential amplifier with an active load. Here Q_1 and Q_2 form the differential pair while the current source transistors Q_4 and Q_5 form the active loads for Q_1 and Q_2, respectively. The dc

bias circuit that establishes an appropriate dc voltage at the drains of Q_1 and Q_2 is not shown. Note that the equivalent differential half-circuit is an active-loaded common-source transistor of the type studied in Section 6.5. It is required to design the circuit to meet the following specifications:

(a) Differential gain $A_d = 80$ V/V.
(b) $I_{REF} = I = 100 \, \mu A$.
(c) The dc voltage at the gates of Q_6 and Q_3 is +1.5 V.
(d) The dc voltage at the gates of Q_7, Q_4, and Q_5 is −1.5 V.

The technology available is specified as follows: $\mu_n C_{ox} = 3\mu_p C_{ox} = 90 \, \mu A/V^2$; $V_{tn} = |V_{tp}| = 0.7$ V, $V_{An} = |V_{Ap}| = 20$ V. Specify the required value of R and the W/L ratios for all transistors. Also specify I_D and $|V_{GS}|$ at which each transistor is operating. For dc bias calculations you may neglect channel-length modulation.

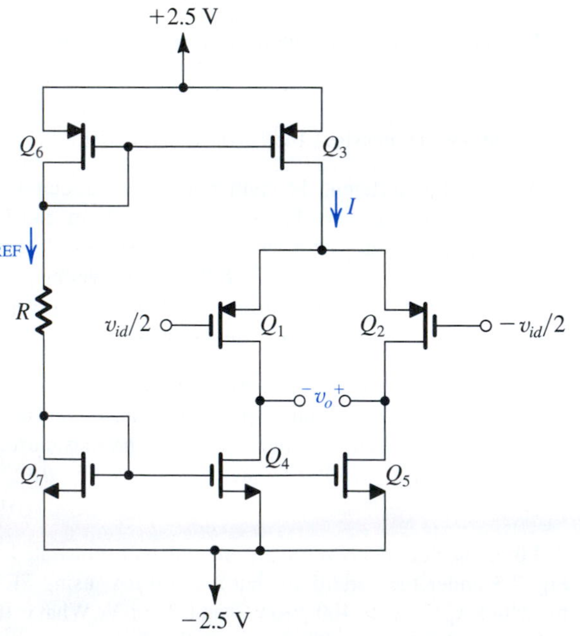

7.14 A design error has resulted in a gross mismatch in the circuit of Fig. P7.14. Specifically, Q_2 has twice the W/L ratio of Q_1. If v_{id} is a small sine-wave signal, find:

(a) I_{D1} and I_{D2}.
(b) V_{OV} for each of Q_1 and Q_2.
(c) The differential gain A_d in terms of R_D, I, and V_{OV}.

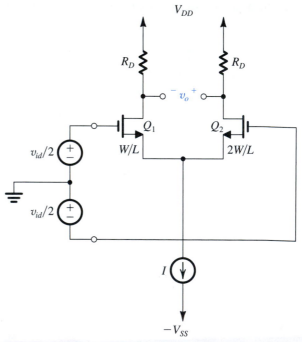

FIGURE P7.14

7.15 An NMOS differential pair is biased by a current source $I = 0.2$ mA having an output resistance $R_{SS} = 100$ kΩ. The amplifier has drain resistances $R_D = 10$ kΩ, using transistors with $k'_n W/L = 3$ mA/V^2, and r_o that is large.

FIGURE P7.13

(a) If the output is taken single-endedly, find $|A_d|$, $|A_{cm}|$, and CMRR.

(b) If the output is taken differentially and there is a 1% mismatch between the drain resistances, find $|A_d|$, $|A_{cm}|$, and CMRR.

7.16 For the differential amplifier shown in Fig. P7.2, let Q_1 and Q_2 have $k_p'(W/L) = 3.5$ mA/V^2, and assume that the bias current source has an output resistance of 30 kΩ. Find V_{OV}, g_m, $|A_d|$, $|A_{cm}|$, and the CMRR (in dB) obtained with the output taken differentially. The drain resistances are known to have a mismatch of 2%.

D*7.17 The differential amplifier in Fig. P7.17 utilizes a resistor R_{SS} to establish a 1-mA dc bias current. Note that this amplifier uses a single 5-V supply and thus needs a dc common-mode voltage V_{CM}. Transistors Q_1 and Q_2 have $k_n'W/L = 2.5$ mA/V^2, $V_t = 0.7$ V, and $\lambda = 0$.

(a) Find the required value of V_{CM}.

(b) Find the value of R_D that results in a differential gain A_d of 8 V/V.

(c) Determine the dc voltage at the drains.

(d) Determine the common-mode gain $\Delta V_{D1}/\Delta V_{CM}$. (*Hint:* You need to take $1/g_m$ into account.)

(e) Use the common-mode gain found in (d) to determine the change in V_{CM} that results in Q_1 and Q_2 entering the triode region.

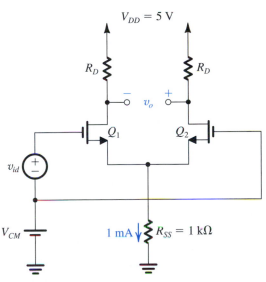

FIGURE P7.17

***7.18** The objective of this problem is to determine the common-mode gain and hence the CMRR of the differential pair arising from a simultaneous mismatch in g_m and in R_D.

(a) Refer to the circuit in Fig. 7.11 and let the two drain resistors be denoted R_{D1} and R_{D2} where $R_{D1} = R_D + (\Delta R_D/2)$ and

$R_{D2} = R_D - (\Delta R_D/2)$. Also let $g_{m1} = g_m + (\Delta g_m/2)$ and $g_{m2} = g_m - (\Delta g_m/2)$. Follow an analysis process similar to that used to derive Eq. (7.64) to show that

$$A_{cm} \cong \left(\frac{R_D}{2R_{SS}}\right)\left(\frac{\Delta g_m}{g_m} + \frac{\Delta R_D}{R_D}\right)$$

Note that this equation indicates that R_D can be deliberately varied to compensate for the initial variability in g_m and R_D, that is, to minimize A_{cm}.

(b) In a MOS differential amplifier for which $R_D = 5$ kΩ and $R_{SS} = 25$ kΩ, the common-mode gain is measured and found to be 0.002 V/V. Find the percentage change required in one of the two drain resistors so as to reduce A_{cm} to zero (or close to zero).

7.19 Recalling that g_m of a MOSFET is given by

$$g_m = k_n'\left(\frac{W}{L}\right)(V_{GS} - V_t)$$

we observe that there are two potential sources for a mismatch between the g_m values in a differential pair: a mismatch $\Delta(W/L)$ in the (W/L) values and a mismatch ΔV_t in the threshold voltage values. Hence show that

$$\frac{\Delta g_m}{g_m} = \frac{\Delta(W/L)}{W/L} + \frac{\Delta V_t}{V_{OV}}$$

Evaluate the worst-case fractional mismatch in g_m for a differential pair in which the (W/L) values have a tolerance of $\pm1\%$ and the largest mismatch in V_t is specified to be 5 mV. Assume that the pair is operating at $V_{OV} = 0.25$ V. If $R_D = 5$ kΩ and $R_{SS} = 25$ kΩ, find the worst-case value of A_{cm}. If the bias current $I = 1$ mA, find the corresponding worst-case CMRR.

SECTION 7.3: THE BJT DIFFERENTIAL PAIR

7.20 For the differential amplifier of Fig. 7.13(a) let $I = 1$ mA, $V_{CC} = 5$ V, $v_{CM} = -2$ V, $R_C = 3$ kΩ, and $\beta = 100$. Assume that the BJTs have $v_{BE} = 0.7$ V at $i_C = 1$ mA. Find the voltage at the emitters and at the outputs.

7.21 For the circuit of Fig. 7.13(b) with an input of +1 V as indicated, and with $I = 1$ mA, $V_{CC} = 5$ V, $R_C = 3$ kΩ, and $\beta = 100$, find the voltage at the emitters and the collector voltages. Assume that the BJTs have $v_{BE} = 0.7$ V at $i_C = 1$ mA.

7.22 Repeat Exercise 7.7 (page 706) for an input of −0.3 V.

7.23 For the BJT differential amplifier of Fig. 7.12 find the value of the input differential signal, $v_{id} \equiv v_{B1} - v_{B2}$, that causes $i_{E1} = 0.80I$.

D7.24 Consider the differential amplifier of Fig. 7.12 and let the BJT β be very large:

(a) What is the largest input common-mode signal that can be applied while the BJTs remain comfortably in the active region with $v_{CB} = 0$?

(b) If an input difference signal is applied that is large enough to steer the current entirely to one side of the pair, what is the change in voltage at each collector (from the condition for which $v_{id} = 0$)?

(c) If the available power supply V_{CC} is 5 V, what value of IR_C should you choose in order to allow a common-mode input signal of ±3 V?

(d) For the value of IR_C found in (c), select values for I and R_C. Use the largest possible value for I subject to the constraint that the base current of each transistor (when I divides equally) should not exceed 2 μA. Let $\beta = 100$.

7.25 To provide insight into the possibility of nonlinear distortion resulting from large differential input signals applied to the differential amplifier of Fig. 7.12, evaluate the normalized change in the current i_{E1}, $\Delta i_{E1}/I = (i_{E1} - (I/2))/I$, for differential input signals v_{id} of 5, 10, 20, 30, and 40 mV. Provide a tabulation of the ratio $((\Delta i_{E1}/I)/v_{id})$, which represents the proportional transconductance gain of the differential pair, versus v_{id}. Comment on the linearity of the differential pair as an amplifier.

D7.26 Design the circuit of Fig. 7.12 to provide a differential output voltage (i.e., one taken between the two collectors) of 1 V when the differential input signal is 10 mV. A current source of 2 mA and a positive supply of +10 V are available. What is the largest possible input common-mode voltage for which operation is as required? Assume $\alpha \simeq 1$.

D*7.27 One of the trade-offs available in the design of the basic differential amplifier circuit of Fig. 7.12 is between the value of the voltage gain and the range of common-mode input voltage. The purpose of this problem is to demonstrate this trade-off.

(a) Use Eqs. (7.72) and (7.73) to obtain i_{C1} and i_{C2} corresponding to a differential input signal of 5 mV (i.e., $v_{B1} - v_{B2} = 5$ mV). Assume β to be very high. Find the resulting voltage difference between the two collectors ($v_{C2} - v_{C1}$), and divide this value by 5 mV to obtain the voltage gain in terms of (IR_C).

(b) Find the maximum permitted value for v_{CM} (Fig. 7.13a) while the transistors remain comfortably in the active mode with $v_{CB} = 0$. Express this maximum in terms of V_{CC} and the gain, and hence show that for a given value of V_{CC}, the higher the gain achieved, the lower the common-mode range. Use this expression to find $v_{CM\,max}$ corresponding to a gain magnitude of 100, 200, 300, and 400 V/V. For each value, also give the required value of IR_C and the value of R_C for $I = 1$ mA.

*7.28 For the circuit in Fig. 7.12, assuming $\alpha = 1$ and $IR_C = 5$ V, use Eqs. (7.67) and (7.68) to find i_{C1} and i_{C2}, and hence determine $v_o = v_{C2} - v_{C1}$ for input differential signals $v_{id} \equiv v_{B1} - v_{B2}$ of 5 mV, 10 mV, 15 mV, 20 mV, 25 mV, 30 mV, 35 mV, and 40 mV. Plot v_o versus v_{id}, and hence comment on the amplifier linearity. As another way of visualizing linearity, determine the gain (v_o/v_{id}) versus v_{id}. Comment on the resulting graph.

7.29 In a differential amplifier using a 6-mA emitter bias current source the two BJTs are not matched. Rather, one has one-and-a-half times the emitter junction area of the other. For a differential input signal of zero volts, what do the collector currents become? What difference input is needed to equalize the collector currents? Assume $\alpha = 1$.

*7.30 Figure P7.30 shows a logic inverter based on the differential pair. Here, Q_1 and Q_2 form the differential pair, whereas Q_3 is an emitter follower that performs two functions: It shifts the level of the output voltage to make V_{OH} and V_{OL} centered on the reference voltage V_R, thus enabling one gate to drive another (this point will be explained in detail in Chapter 11), and it provides the inverter with a low output resistance. All transistors have $V_{BE} = 0.7$ V at $I_C = 1$ mA and have $\beta = 100$.

(a) For v_I sufficiently low that Q_1 is cut off, find the value of the output voltage v_O. This is V_{OH}.

(b) For v_I sufficiently high that Q_1 is carrying all the current I, find the output voltage v_O. This is V_{OL}.

(c) Determine the value of v_I that results in Q_1 conducting 1% of I. This can be taken as V_{IL}.

(d) Determine the value of v_I that results in Q_1 conducting 99% of I. This can be taken as V_{IH}.

(e) Sketch and clearly label the breakpoints of the inverter voltage transfer characteristic. Calculate the values of the noise margins NM_H and NM_L. Note the judicious choice of the value of the reference voltage V_R.

(For the definitions of the parameters that are used to characterize the inverter VTC, refer to Section 1.7.)

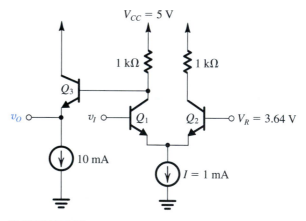

FIGURE P7.30

7.31 A BJT differential amplifier uses a 300-μA bias current. What is the value of g_m of each device? If β is 150, what is the differential input resistance?

D7.32 Design the basic BJT differential amplifier circuit of Fig. 7.16 to provide a differential input resistance of at least 10 kΩ and a differential voltage gain (with the output taken between the two collectors) of 200 V/V. The transistor β is specified to be at least 100. The available power supply is 10 V.

7.33 For a differential amplifier to which a total difference signal of 10 mV is applied, what is the equivalent signal to its corresponding CE half-circuit? If the emitter current source is 100 μA, what is r_e of the half-circuit? For a load resistance of 10 kΩ in each collector, what is the half-circuit gain? What magnitude of signal output voltage would you expect at each collector?

7.34 A BJT differential amplifier is biased from a 2-mA constant-current source and includes a 100-Ω resistor in each emitter. The collectors are connected to V_{CC} via 5-kΩ resistors. A differential input signal of 0.1 V is applied between the two bases.

(a) Find the signal current in the emitters (i_e) and the signal voltage v_{be} for each BJT.
(b) What is the total emitter current in each BJT?
(c) What is the signal voltage at each collector? Assume $\alpha = 1$.
(d) What is the voltage gain realized when the output is taken between the two collectors?

D7.35 Design a BJT differential amplifier to amplify a differential input signal of 0.2 V and provide a differential output signal of 4 V. To ensure adequate linearity, it is required to limit the signal amplitude across each base–emitter junction to a maximum of 5 mV. Another design requirement is that the differential input resistance be at least 80 kΩ. The BJTs available are specified to have $\beta \geq 200$. Give the circuit configuration and specify the values of all its components.

7.36 A particular differential amplifier operates from an emitter current source whose output resistance is 1 MΩ. What resistance is associated with each common-mode half-circuit? For collector resistors of 20 kΩ, what is the resulting common-mode gain for output taken (a) differentially, (b) single-endedly?

7.37 Find the voltage gain and the input resistance of the amplifier shown in Fig. P7.37 assuming $\beta = 100$.

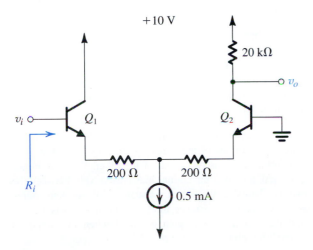

FIGURE P7.37

7.38 Find the voltage gain and input resistance of the amplifier in Fig. P7.38 assuming that $\beta = 100$.

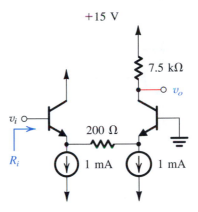

FIGURE P7.38

7.39 Derive an expression for the small-signal voltage gain v_o/v_i of the circuit shown in Fig. P7.39 in two different ways:

(a) as a differential amplifier
(b) as a cascade of a common-collector stage Q_1 and a common-base stage Q_2

Assume that the BJTs are matched and have a current gain α. Verify that both approaches lead to the same result.

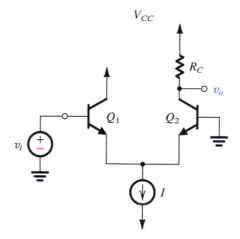

FIGURE P7.39

7.40 The differential amplifier circuit of Fig. P7.40 utilizes a resistor connected to the negative power supply to establish the bias current I.

(a) For $v_{B1} = v_{id}/2$ and $v_{B2} = -v_{id}/2$, where v_{id} is a small signal with zero average, find the magnitude of the differential gain, $|v_o/v_{id}|$.
(b) For $v_{B1} = v_{B2} = v_{icm}$, find the magnitude of the common-mode gain, $|v_o/v_{icm}|$.
(c) Calculate the CMRR.

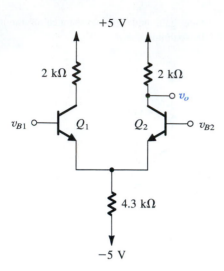

FIGURE P7.40

(d) If $v_{B1} = 0.1 \sin 2\pi \times 60t + 0.005 \sin 2\pi \times 1000t$ volts, $v_{B2} = 0.1 \sin 2\pi \times 60t - 0.005 \sin 2\pi \times 1000t$, volts, find v_o.

7.41 For the differential amplifier shown in Fig. P7.41, identify and sketch the differential half-circuit and the common-mode half-circuit. Find the differential gain, the differential input resistance, the common-mode gain, and the common-mode input resistance. For these transistors, $\beta = 100$ and $V_A = 100$ V.

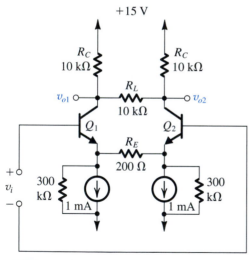

FIGURE P7.41

7.42 Consider the basic differential circuit in which the transistors have $\beta = 200$ and $V_A = 200$ V, with $I = 0.5$ mA, $R_{EE} = 1$ MΩ, and $R_C = 20$ kΩ. Find:

(a) the differential gain to a single-ended output
(b) the differential gain to a differential output
(c) the differential input resistance
(d) the common-mode gain to a single-ended output
(e) the common-mode gain to a differential output

7.43 In a differential-amplifier circuit resembling that shown in Fig. 7.23(a), the current generator represented by I and R_{EE} consists of a simple common-emitter transistor operating at 100 μA. For this transistor, and those used in the differential pair, $V_A = 200$ V and $\beta = 50$. What common-mode input resistance would apply?

D7.44 It is required to design a differential amplifier to provide the largest possible signal to a pair of 10-kΩ load resistances. The input differential signal is a sinusoid of 5-mV peak amplitude which is applied to one input terminal while the other input terminal is grounded. The power supply available is 10 V. To determine the bias current required, I, derive an expression for the total voltage at each of the collectors in terms of V_{CC} and I in the presence of the input signal. Then impose the condition that both transistors should remain well out of saturation with a minimum v_{CB} of approximately 0 V. Thus determine the required value of I. For this design, what differential gain is achieved? What is the amplitude of the signal voltage obtained between the two collectors? Assume $\alpha \cong 1$.

D*7.45 Design a BJT differential amplifier that provides two single-ended outputs (at the collectors). The amplifier is to have a differential gain (to each of the two outputs) of at least 100 V/V, a differential input resistance ≥ 10 kΩ, and a common-mode gain (to each of the two outputs) no greater than 0.1 V/V. Use a 2-mA current source for biasing. Give the complete circuit with component values and suitable power supplies that allow for ± 2 V swing at each collector. Specify the minimum value that the output resistance of the bias current source must have. The BJTs available have $\beta \geq 100$. What is the value of the input common-mode resistance when the bias source has the lowest acceptable resistance?

7.46 When the output of a BJT differential amplifier is taken differentially, its CMRR is found to be 40 dB higher than when the output is taken single-endedly. If the only source of common-mode gain when the output is taken differentially is the mismatch in collector resistances, what must this mismatch be (in percent)?

*7.47 In a particular BJT differential amplifier, a production error results in one of the transistors having an emitter–base junction area that is twice that of the other. With the inputs grounded, how will the emitter bias current split between the two transistors? If the output resistance of the current source is 1 MΩ and the resistance in each collector (R_C) is 12 kΩ, find the common-mode gain obtained when the output is taken differentially. Assume $\alpha \simeq 1$.

SECTION 7.4: OTHER NONIDEAL CHARACTERISTICS OF THE DIFFERENTIAL AMPLIFIER

D7.48 An NMOS differential pair is to be used in an amplifier whose drain resistors are 10 kΩ ± 1%. For the pair, $k'_n W/L = 4$ mA/V^2. A decision is to be made concerning the bias current I to be used, whether 200 μA or 400 μA. For differential output, contrast the differential gain and input offset voltage for the two possibilities.

D7.49 An NMOS amplifier, whose designed operating point is at $V_{OV} = 0.3$ V, is suspected to have a variability of V_t of ±5 mV, and of W/L and R_D (independently) of ±2%. What is the worst-case input offset voltage you would expect to find? What is the major contribution to this total offset? If you used a variation of one of the drain resistors to reduce the output offset to zero and thereby compensate for the uncertainties (including that of the other R_D), what percentage change from nominal would you require? If by selection you reduced the contribution of the worst cause of offset by a factor of 10, what change in R_D would be needed?

7.50 An NMOS differential pair operating at a bias current I of 100 μA uses transistors for which $k'_n = 100$ μA/V^2 and $W/L = 20$, with $V_t = 0.8$ V. Find the three components of input offset voltage under the conditions that $\Delta R_D/R_D = 5\%$, $\Delta(W/L)/(W/L) = 5\%$, and $\Delta V_t = 5$ mV. In the worst case, what might the total offset be? For the usual case of the three effects being independent, what is the offset likely to be? (*Hint:* For the latter situation, use a root-sum-of-squares computation.)

7.51 A differential amplifier using a 600-μA emitter bias source uses two well-matched transistors but collector load resistors that are mismatched by 10%. What input offset voltage is required to reduce the differential output voltage to zero?

7.52 A differential amplifier using a 600-μA emitter bias source uses two transistors whose scale currents I_S differ by 10%. If the two collector resistors are well matched, find the resulting input offset voltage.

7.53 Modify Eq. (7.125) for the case of a differential amplifier having a resistance R_E connected in the emitter of each transistor. Let the bias current source be I.

7.54 A differential amplifier uses two transistors whose β values are β_1 and β_2. If everything else is matched, show that the input offset voltage is approximately $V_T[(1/\beta_1) - (1/\beta_2)]$. Evaluate V_{OS} for $\beta_1 = 100$ and $\beta_2 = 200$. Assume the differential source resistance to be zero.

*7.55 A differential amplifier uses two transistors having V_A values of 100 V and 300 V. If everything else is matched, find the resulting input offset voltage. Assume that the two transistors are intended to be biased at a V_{CE} of about 10 V.

*7.56 A differential amplifier is fed in a balanced or push–pull manner with the source resistance in series with each base being R_s. Show that a mismatch ΔR_s between the values of the two source resistances gives rise to an input offset voltage of approximately $(I/2\beta)\Delta R_s$.

7.57 One approach to "offset correction" involves the adjustment of the values of R_{C1} and R_{C2} so as to reduce the differential output voltage to zero when both input terminals are grounded. This offset-nulling process can be accomplished by utilizing a potentiometer in the collector circuit, as shown in Fig. P7.57. We wish to find the potentiometer setting, represented by the fraction x of its value connected in series with R_{C1}, that is required for nulling the output offset voltage that results from:

(a) R_{C1} being 5% higher than nominal and R_{C2} 5% lower than nominal
(b) Q_1 having an area 10% larger than that of Q_2

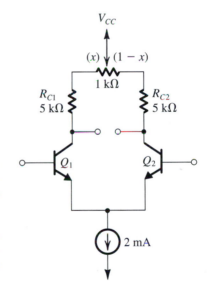

FIGURE P7.57

7.58 A differential amplifier for which the total emitter bias current is 600 μA uses transistors for which β is specified to lie between 80 and 200. What is the largest possible input bias current? The smallest possible input bias current? The largest possible input offset current?

*7.59 A BJT differential amplifier, operating at a bias current of 500 μA, employs collector resistors of 27 kΩ (each) connected to a +15-V supply. The emitter current source employs a BJT whose emitter voltage is −5 V. What are the positive and negative limits of the input common-mode range of the amplifier for differential signals of ≤20-mV peak amplitude, applied in a balanced or push–pull fashion?

**7.60 In a particular BJT differential amplifier, a production error results in one of the transistors having an emitter–base junction area twice that of the other. With both inputs grounded, find the current in each of the two transistors and hence the dc offset voltage at the output, assuming that the collector resistances are equal. Use small-signal analysis to find the input voltage that would restore current balance to the differential pair. Repeat using large-signal analysis and compare results. Also find the input bias and offset currents assuming $I = 0.1$ mA and $\beta_1 = \beta_2 = 100$.

D7.61 A large fraction of mass-produced differential-amplifier modules employing 20-kΩ collector resistors is found to have an input offset voltage ranging from +3 mV to −3 mV. If the gain of the input differential stage is 90 V/V, by what amount must one collector resistor be adjusted to reduce the input offset to zero? If an adjustment mechanism is devised that raises one collector resistor while correspondingly lowering the other, what resistance change is needed? Suggest a suitable circuit using the existing collector resistors and a potentiometer whose moving element is connected to V_{CC}. What value of potentiometer resistance (specified to 1 significant digit) is appropriate?

SECTION 7.5: THE DIFFERENTIAL AMPLIFIER WITH ACTIVE LOAD

D7.62 In an active-loaded differential amplifier of the form shown in Fig. 7.28(a), all transistors are characterized by $k'W/L = 3.2$ mA/V^2, and $|V_A| = 20$ V. Find the bias current I for which the gain $v_o/v_{id} = 80$ V/V.

7.63 In a version of the active-loaded MOS differential amplifier shown in Fig. 7.28(a), all transistors have $k'W/L = 0.2$ mA/V^2 and $|V_A| = 20$ V. For $V_{DD} = 5$ V, with the inputs near ground, and (a) $I = 100$ μA or (b) $I = 400$ μA, calculate the linear range of v_o, the g_m of Q_1 and Q_2, the output resistances of Q_2 and Q_4, the total output resistance, and the voltage gain.

7.64 Consider the active-loaded MOS differential amplifier of Fig. 7.28(a) in two cases:

(a) Current-source I is implemented with a simple current mirror.
(b) Current-source I is implemented with the modified Wilson current mirror shown in Fig. P7.64.

Recalling that for the simple mirror $R_{SS} = r_o|_{Q_5}$ and for the Wilson mirror $R_{SS} \cong g_{m7}r_{o7}r_{o5}$, and assuming that all transistors have the same $|V_A|$ and $k'W/L$, show that for case (a)

$$\text{CMRR} = 2\left(\frac{V_A}{V_{OV}}\right)^2$$

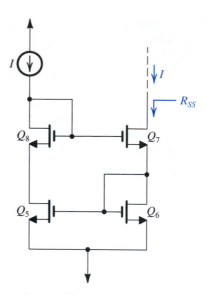

FIGURE P7.64

and for case (b)

$$\text{CMRR} = \sqrt{2}\left(\frac{V_A}{V_{OV}}\right)^3$$

where V_{OV} is the overdrive voltage that corresponds to a drain current of $I/2$. For $k'W/L = 10$ mA/V^2, $I = 1$ mA, and $|V_A| = 10$ V, find CMRR for both cases.

D*7.65 Consider an active-loaded differential amplifier such as that shown in Fig. 7.28(a) with the bias current source implemented with the modified Wilson mirror of Fig. P7.64 with $I = 100$ μA. The transistors have $|V_t| = 0.7$ V and $k'(W/L) = 800$ μA/V^2. What is the lowest value of the total power supply ($V_{DD} + V_{SS}$) that allows each transistor to operate with $|V_{DS}| \geq |V_{GS}|$?

*7.66 (a) Sketch the circuit of an active-loaded MOS differential amplifier in which the input transistors are cascoded, and a cascode current mirror is used for the load.
(b) Show that if all transistors are operated at an overdrive voltage V_{OV} and have equal Early voltages $|V_A|$, the gain is given by

$$A_d = 2(V_A/V_{OV})^2$$

Evaluate the gain for $V_{OV} = 0.25$ V and $V_A = 20$ V.

7.67 The differential amplifier in Fig. 7.32(a) is operated with $I = 100$ μA, with devices for which $V_A = 160$ V and $\beta = 100$. What differential input resistance, output resistance, equivalent transconductance, and open-circuit voltage gain would you expect? What will the voltage gain be if the input resistance of the subsequent stage is 100 kΩ?

D*7.68 Design the circuit of Fig. 7.32(a) using a basic current mirror to implement the current source I. It is required that the equivalent transconductance be 5 mA/V. Use ±5-V power supplies and BJTs that have $\beta = 150$ and $V_A = 100$ V. Give the complete circuit with component values and specify the differential input resistance R_{id}, the output resistance R_o, the open-circuit voltage gain A_d, the input bias current, the input common-mode range, and the common-mode input resistance.

D*7.69 Repeat the design of the amplifier specified in Problem 7.68 utilizing a Widlar current source [Fig. 6.62] to supply the bias current. Assume that the largest resistance available is 2 kΩ.

D7.70 Modify the design of the amplifier in Problem 7.68 by connecting emitter-degeneration resistances of values that result in $R_{id} = 100$ kΩ. What does A_d become?

7.71 An active-loaded bipolar differential amplifier such as that shown in Fig. 7.32(a) has $I = 0.5$ mA, $V_A = 120$ V, and $\beta = 150$. Find G_m, R_o, A_d, and R_{id}. If the bias-current source is implemented with a simple *npn* current mirror, find R_{EE}, A_{cm}, and CMRR. If the amplifier is fed differentially with a source having a total of 10 kΩ resistance (i.e., 5 kΩ in series with the base lead of each of Q_1 and Q_2), find the overall differential voltage gain.

*7.72 Consider the differential amplifier circuit of Fig. 7.32(a) with the two input terminals tied together and an input common-mode signal v_{icm} applied. Let the output resistance of the bias current source be denoted by R_{EE}, and let the β of the *pnp* transistors be denoted β_p. Assuming that β of the *npn* transistors is high, use the current transfer ratio of the mirror to show that there will be an output current of $v_{icm}/\beta_p R_{EE}$. Thus, show that the common-mode transconductance is $1/\beta_p R_{EE}$. Use this result together with the differential transconductance G_m (derived in the text) to find an alternative measure of the common-mode rejection. Observe that this result differs from the CMRR expression in Eq. (7.174) by a factor of 2, which is simply the ratio of the output resistance for common-mode inputs (r_{o4}) and the output resistance for differential inputs ($r_{o2} \parallel r_{o4}$).

*7.73 Repeat Problem 7.72 for the case in which the current mirror is replaced with a Wilson mirror. Show that in this case the output current will be $v_{icm}/\beta_p^2 R_{EE}$. Find the common-mode transconductance and the ratio G_{mcm}/G_m.

7.74 Figure P7.74 shows a differential cascode amplifier with an active load formed by a Wilson current mirror. Utilizing the expressions derived in Chapter 6 for the output resistance of a bipolar cascode and the output resistance of the Wilson mirror, and assuming all transistors to be identical, show that the differential voltage gain A_d is given by

$$A_d = \tfrac{1}{3}\beta g_m r_o$$

Evaluate A_d for the case $I = 0.4$ mA, $\beta = 100$, and $V_A = 120$ V.

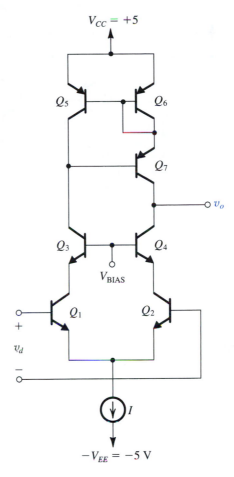

FIGURE P7.74

D7.75 Consider the bias design of the Wilson-loaded cascode differential amplifier shown in Fig. P7.74.

(a) What is the largest signal voltage possible at the output without Q_7 saturating? Assume that the CB junction conducts when the voltage across it exceeds 0.4 V.
(b) What should the dc bias voltage established at the output (by an arrangement not shown) be in order to allow for positive output signal swing of 1.5 V?
(c) What should the value of V_{BIAS} be in order to allow for a negative output signal swing of 1.5 V?
(d) What is the upper limit on the input common-mode voltage v_{CM}?

*7.76 Figure P7.76 shows a modified cascode differential amplifier. Here Q_3 and Q_4 are the cascode transistors. However, the manner in which Q_3 is connected with its base current feeding the current mirror Q_7–Q_8 results in very interesting input properties. Note that for simplicity the circuit is shown with the base of Q_2 grounded.

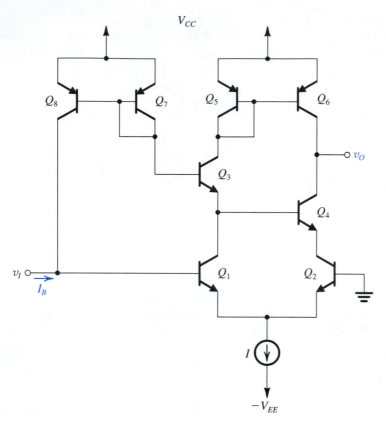

FIGURE P7.76

(a) With $v_I = 0$ V dc, find the input bias current I_B assuming all transistors have equal value of β. Compare the case without the Q_7–Q_8 connection.

(b) With $v_I = 0$ V (dc) $+ v_{id}$, find the input signal current i_b and hence the input differential resistance R_{id}. Compare with the case without the Q_7–Q_8 connection. (Observe that Q_4 arranges that the emitter currents of Q_1 and Q_2 are very nearly the same!)

7.77 Utilizing the expression for the current transfer ratio of the Wilson mirror derived in Section 6.12.3 (Eq. 6.193) derive an expression for the systematic offset voltage of a BJT differential amplifier that utilizes a *pnp* Wilson current mirror load. Evaluate V_{OS} for $\beta_P = 50$.

7.78 For the folded-cascode differential amplifier of Fig. 7.35, find the value of V_{BIAS} that results in the largest possible positive output swing, while keeping Q_3, Q_4, and the *pnp* transistors, that realize the current sources out of saturation. Assume $V_{CC} = V_{EE} = 5$ V. If the dc level at the output is 0 V, find the maximum allowable output signal swing. For $I = 0.4$ mA, $\beta_P = 50$, $\beta_N = 150$, and $V_A = 120$ V find G_m, R_{o4}, R_{o5}, R_o, and A_d.

7.79 For the BiCMOS differential amplifier in Fig. P7.79 let $V_{DD} = V_{SS} = 3$ V, $I = 0.4$ mA, $k_p'W/L = 6.4$ mA/V^2; $|V_A|$

for *p*-channel MOSFETs is 10 V, $|V_A|$ for *npn* transistors is 120 V. Find G_m, R_o, and A_d.

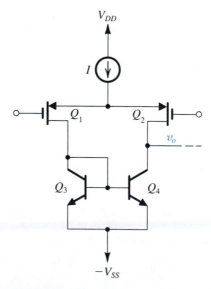

FIGURE P7.79

SECTION 7.6: FREQUENCY RESPONSE OF THE DIFFERENTIAL AMPLIFIER

7.80 A MOSFET differential amplifier such as that shown in Fig. 7.36(a) is biased with a current source $I = 200\ \mu A$. The transistors have $W/L = 25$, $k'_n = 128\ \mu A/V^2$, $V_A = 20$ V, $C_{gs} = 30$ fF, $C_{gd} = 5$ fF, and $C_{db} = 5$ fF. The drain resistors are 20 kΩ each. Also, there is a 90-fF capacitive load between each drain and ground.

(a) Find V_{OV} and g_m for each transistor.
(b) Find the differential gain A_d.
(c) If the input signal source has a small resistance R_{sig} and thus the frequency response is determined primarily by the output pole, estimate the 3-dB frequency f_H.
(d) If, in a different situation, the amplifier is fed symmetrically with a signal source of 40 kΩ resistance (i.e., 20 kΩ in series with each gate terminal), use the open-circuit time-constants method to estimate f_H.

7.81 The amplifier specified in Problem 7.80 has $R_{SS} = 100$ kΩ and $C_{SS} = 0.2$ pF. Find the 3-dB frequency of the CMRR.

7.82 A BJT differential amplifier operating with a 1-mA current source uses transistors for which $\beta = 100$, $f_T = 600$ MHz, $C_\mu = 0.5$ pF, and $r_x = 100\ \Omega$. Each of the collector resistances is 10 kΩ, and r_o is very large. The amplifier is fed in a symmetrical fashion with a source resistance of 10 kΩ in series with each of the two input terminals.

(a) Sketch the differential half-circuit and its high-frequency equivalent circuit.
(b) Determine the low-frequency value of the overall differential gain.
(c) Use Miller's theorem to determine the input capacitance and hence estimate the 3-dB frequency f_H and the gain–bandwidth product.

7.83 The differential amplifier circuit specified in Problem 7.82 is modified by including 100-Ω resistor in each of the emitters. Determine the low-frequency value of the overall differential voltage gain. Also, use the method of open-circuit time-constants to obtain an estimate for f_H. Toward that end, note that the resistance R_μ seen by C_μ is given by

$$R_\mu = [(R_{sig} + r_x)\ \|\ R_{in}](1 + G_m R_C) + R_C$$

where

$$R_{in} = (\beta + 1)(R_e + r_e)$$

$$G_m = \frac{g_m}{1 + g_m R_e}$$

The resistance R_π seen by C_π is given by

$$R_\pi = r_\pi\ \|\ \frac{R_{sig} + r_x + R_e}{1 + g_m R_e}$$

Also determine the gain–bandwidth product.

D7.84 It is required to increase the 3-dB frequency of the differential amplifier specified in Problem 7.82 to 1 MHz by adding an emitter resistance R_e. Use the open-circuit time-constants method to perform this design. Specifically, use the formulas for R_μ and R_π given in the statement for Problem 7.83 to determine the required value of the factor $(1 + g_m R_e)$ and hence find R_e. Make appropriate approximations to simplify the calculations. What does the dc gain become? Also determine the resulting gain–bandwidth product.

7.85 A current-mirror-loaded MOS differential amplifier is biased with a current source $I = 0.6$ mA. The two NMOS transistors of the differential pair are operating at $V_{OV} = 0.3$ V, and the PMOS devices of the mirror are operating at $V_{OV} = 0.5$ V. The Early voltage $V_{An} = |V_{Ap}| = 9$ V. The total capacitance at the input node of the mirror is 0.1 pF and that at the output node of the amplifier is 0.2 pF. Find the dc value and the frequencies of the poles and zero of the differential voltage gain.

7.86 A differential amplifier is biased by a current source having an output resistance of 1 MΩ and an output capacitance of 10 pF. The differential gain exhibits a dominant pole at 500 kHz. What are the poles of the CMRR?

7.87 For the differential amplifier specified in Problem 7.82, find the dc gain and f_H when the circuit is modified by eliminating the collector resistor of the left-hand-side transistor and the input signal is fed to the base of the left-hand-side transistor while the base of the other transistor in the pair is grounded. Let the source resistance be 20 kΩ and neglect r_x. (*Hint:* Refer to Fig. 6.57.)

7.88 Consider the circuit of Fig. P7.88 for the case: $I = 200\ \mu A$ and $V_{OV} = 0.25$ V, $R_{sig} = 200$ kΩ, $R_D = 50$ kΩ, $C_{gs} = C_{gd} = 1$ pF. Find the dc gain, the high-frequency poles, and an estimate of f_H.

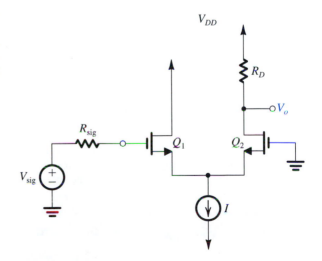

FIGURE P7.88

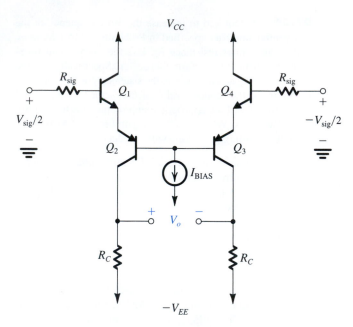

7.89 For the circuit in Fig. P7.89, let the bias be such that each transistor is operating at 100-μA collector current. Let the BJTs have $\beta = 200$, $f_T = 600$ MHz, and $C_\mu = 0.2$ pF, and neglect r_o and r_x. Also, $R_{sig} = R_C = 50$ kΩ. Find the low-frequency gain, the input differential resistance, the high-frequency poles, and an estimate of f_H.

SECTION 7.7: MULTISTAGE AMPLIFIERS

7.90 Consider the circuit in Fig. 7.40 with the device geometries (in μm) shown at the bottom of this page:

Let $I_{REF} = 225$ μA, $|V_t| = 0.75$ V for all devices, $\mu_n C_{ox} = 180$ μA/V^2, $\mu_p C_{ox} = 60$ μA/V^2, $|V_A| = 9$ V for all devices, $V_{DD} = V_{SS} = 1.5$ V. Determine the width of Q_6, W, that will ensure that the op amp will not have a systematic offset voltage. Then, for all devices evaluate I_D, $|V_{OV}|$, $|V_{GS}|$, g_m, and r_o. Provide your results in a table similar to Table 7.1. Also find A_1, A_2, the dc open-loop voltage gain, the input common-mode range, and the output voltage range. Neglect the effect of V_A on the bias current.

D*7.91 In a particular design of the CMOS op amp of Fig. 7.40 the designer wishes to investigate the effects of increasing the W/L ratio of both Q_1 and Q_2 by a factor of 4. Assuming that all other parameters are kept unchanged, refer to Example 7.3 to help you answer the following questions:

(a) Find the resulting change in $|V_{OV}|$ and in g_m of Q_1 and Q_2.
(b) What change results in the voltage gain of the input stage? In the overall voltage gain?
(c) What is the effect on the input offset voltages? (You might wish to refer to Section 7.4).
(d) If f_t is to be kept unchanged, how must C_C be changed?

7.92 Consider the amplifier of Fig. 7.40, whose parameters are specified in Example 7.3. If a manufacturing error results in the W/L ratio of Q_7 being 50/0.8, find the current that Q_7 will now conduct. Thus find the systematic offset voltage that will appear at the output. (Use the results of Example 7.3.) Assuming that the open-loop gain will remain approximately unchanged from the value found in Example 7.3, find the corresponding value of input offset voltage, V_{OS}.

7.93 Consider the input stage of the CMOS op amp in Fig. 7.40 with both inputs grounded. Assume that the two sides of the input stage are perfectly matched except that the threshold voltages of Q_3 and Q_4 have a mismatch ΔV_t. Show

Transistor	Q_1	Q_2	Q_3	Q_4	Q_5	Q_6	Q_7	Q_8
W/L	30/0.5	30/0.5	10/0.5	10/0.5	60/0.5	W/0.5	60/0.5	60/0.5

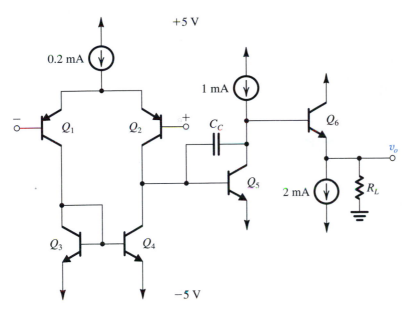

FIGURE P7.95

that a current $g_{m3}\Delta V_t$ appears at the output of the first stage. What is the corresponding input offset voltage? Evaluate this offset voltage for the circuit specified in Example 7.3 for $\Delta V_t = 2$ mV. (Use the results of Example 7.3.)

7.94 A CMOS op amp with the topology in Fig. 7.40 has $g_{m1} = g_{m2} = 1$ mA/V, $g_{m6} = 3$ mA/V, the total capacitance between node D_2 and ground = 0.2 pF, and the total capacitance between the output node and ground = 3 pF. Find the value of C_C that results in $f_t = 50$ MHz and verify that f_t is lower than f_Z and f_{P2}.

*7.95 Figure P7.95 shows a bipolar op-amp circuit that resembles the CMOS op amp of Fig. 7.40. Here, the input differential pair Q_1–Q_2 is loaded in a current mirror formed by Q_3 and Q_4. The second stage is formed by the current-source-loaded common-emitter transistor Q_5. Unlike the CMOS circuit, here there is an output stage formed by the emitter follower Q_6. Capacitor C_C is placed in the negative-feedback path of Q_5 and thus is Miller-multiplied by the gain of Q_5. The resulting large capacitance forms a dominant low-frequency pole with $r_{\pi5}$, thus providing the required uniform −20-dB/decade gain rolloff. All transistors have $\beta = 100$, $|V_{BE}| = 0.7$ V, and $r_o = \infty$.

(a) For inputs grounded and output held at 0 V (by negative feedback, not shown) find the emitter currents of all transistors.
(b) Calculate the dc gain of the amplifier with $R_L = 10$ kΩ.
(c) With R_L as in (b), find the value of C_C to obtain a 3-dB frequency of 100 Hz. What is the value of f_t that results?

D7.96 It is required to design the circuit of Fig. 7.42 to provide a bias current I_B of 225 μA with Q_8 and Q_9 as matched devices having $W/L = 60/0.5$. Transistors Q_{10}, Q_{11}, and Q_{13}

are to be identical and must have the same g_m as Q_8 and Q_9. Transistor Q_{12} is to be four times as wide as Q_{13}. Let $k'_n = 3k'_p = 180$ μA/V^2, and $V_{DD} = V_{SS} = 1.5$ V. Find the required value of R_B. What is the voltage drop across R_B? Also specify the W/L ratios of Q_{10}, Q_{11}, Q_{12}, and Q_{13} and give the expected dc voltages at the gates of Q_{12}, Q_{10}, and Q_8.

7.97 A BJT differential amplifier, biased to have $r_e = 50$ Ω and utilizing two 100-Ω emitter resistors and 5-kΩ loads, drives a second differential stage biased to have $r_e = 20$ Ω. All BJTs have $\beta = 120$. What is the voltage gain of the first stage? Also find the input resistance of the first stage, and the current gain from the input of the first stage to the collectors of the second stage.

7.98 In the multistage amplifier of Fig. 7.43, emitter resistors are to be introduced—100 Ω in the emitter lead of each of the first-stage transistors and 25 Ω for each of the second-stage transistors. What is the effect on input resistance, the voltage gain of the first stage, and the overall voltage gain? Use the bias values found in Example 7.4.

D7.99 Consider the circuit of Fig. 7.43 and its output resistance. Which resistor has the most effect on the output resistance? What should this resistor be changed to if the output resistance is to be reduced by a factor of 2? What will the amplifier gain become after this change? What other change can you make to restore the amplifier gain to approximately its prior value?

D*7.100 (a) If, in the multistage amplifier of Fig. 7.43, the resistor R_5 is replaced by a constant-current source $\simeq 1$ mA, such that the bias situation is essentially unaffected, what does the overall voltage gain of the amplifier become?

7

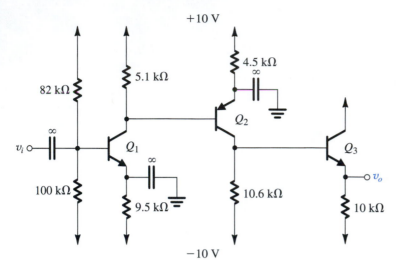

FIGURE P7.101

Assume that the output resistance of the current source is very high. Use the results of Example 7.5.
(b) With the modification suggested in (a), what is the effect of the change on output resistance? What is the overall gain of the amplifier when loaded by 100 Ω to ground? The original amplifier (before modification) has an output resistance of 152 Ω and a voltage gain of 8513 V/V. What is its gain when loaded by 100 Ω? Comment. Use $\beta = 100$.

*7.101 Figure P7.101 shows a three-stage amplifier in which the stages are directly coupled. The amplifier, however, utilizes bypass capacitors, and, as such, its frequency response falls off at low frequencies. For our purposes here, we shall assume that the capacitors are large enough to act as perfect short circuits at all signal frequencies of interest.

(a) Find the dc bias current in each of the three transistors. Also find the dc voltage at the output. Assume $|V_{BE}| = 0.7$ V, $\beta = 100$, and neglect the Early effect.
(b) Find the input resistance and the output resistance.
(c) Use the current-gain method to evaluate the voltage gain v_o/v_i.
(d) Find the frequency of the high-frequency pole formed at the interface between the first and the second stages. Assume that $C_{\mu2} = 2$ pF and $C_{\pi2} = 10$ pF.

D***7.102 For the circuit shown in Fig. P7.102, which uses a folded cascode involving transistor Q_3, all transistors have $|V_{BE}| = 0.7$ V for the currents involved, $V_A = 200$ V, and $\beta = 100$. The circuit is relatively conventional except for Q_5, which operates in a Class B mode (we will study this in Chapter 14) to provide an increased negative output swing for low-resistance loads.

(a) Perform a bias calculation assuming $|V_{BE}| = 0.7$ V, high β, $V_A = \infty$, $v_+ = v_- = 0$ V, and v_O is stabilized by feedback to

about 0 V. Find R so that the reference current I_{REF} is 100 μA. What are the voltages at all the labeled nodes?
(b) Provide in tabular form the bias currents in all transistors together with g_m and r_o for the signal transistors (Q_1, Q_2, Q_3, Q_4, and Q_5) and r_o for Q_C, Q_D, and Q_G.
(c) Now, using $\beta = 100$, find the voltage gain $v_o / (v_+ - v_-)$, and in the process, verify the polarity of the input terminals.
(d) Find the input and output resistances.
(e) Find the input common-mode range for linear operation.
(f) For no load, what is the range of available output voltages, assuming $|V_{CEsat}| = 0.3$ V?
(g) Now consider the situation with a load resistance connected from the output to ground. At the positive and negative limits of the output signal swing, find the smallest load resistance that can be driven if one or the other of Q_1 or Q_2 is allowed to cut off.

D***7.103 In the CMOS op amp shown in Fig. P7.103, all MOS devices have $|V_t| = 1$ V, $\mu_n C_{ox} = 2\mu_p C_{ox} = 40$ μA/V^2, $|V_A| = 50$ V, and $L = 5$ μm. Device widths are indicated on the diagram as multiples of W, where $W = 5$ μm.

(a) Design R to provide a 10-μA reference current.
(b) Assuming $v_O = 0$ V, as established by external feedback, perform a bias analysis, finding all the labeled node voltages, V_{GS} and I_D for all transistors.
(c) Provide in table form I_D, V_{GS}, g_m, and r_o for all devices.
(d) Calculate the voltage gain $v_o / (v_+ - v_-)$, the input resistance, and the output resistance.
(e) What is the input common-mode range?
(f) What is the output signal range for no load?
(g) For what load resistance connected to ground is the output negative voltage limited to −1 V before Q_7 begins to conduct?
(h) For a load resistance one-tenth of that found in (g), what is the output signal swing?

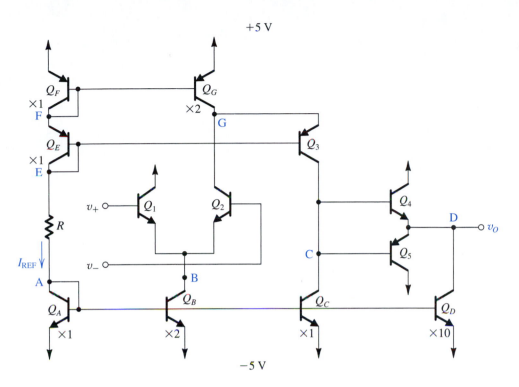

FIGURE P7.102

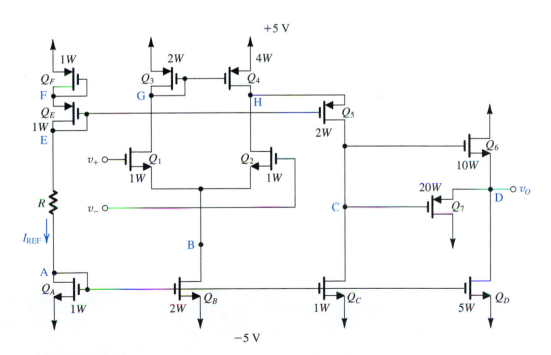

FIGURE P7.103

CHAPTER 8

Feedback

INTRODUCTION

Most physical systems incorporate some form of feedback. It is interesting to note, though, that the theory of negative feedback has been developed by electronics engineers. In his search for methods for the design of amplifiers with stable gain for use in telephone repeaters, Harold Black, an electronics engineer with the Western Electric Company, invented the feedback amplifier in 1928. Since then the technique has been so widely used that it is almost impossible to think of electronic circuits without some form of feedback, either implicit or explicit. Furthermore, the concept of feedback and its associated theory are currently used in areas other than engineering, such as in the modeling of biological systems.

Feedback can be either **negative (degenerative)** or **positive (regenerative).** In amplifier design, negative feedback is applied to effect one or more of the following properties:

1. *Desensitize the gain:* that is, make the value of the gain less sensitive to variations in the value of circuit components, such as might be caused by changes in temperature.

2. *Reduce nonlinear distortion:* that is, make the output proportional to the input (in other words, make the gain constant, independent of signal level).

3. *Reduce the effect of noise:* that is, minimize the contribution to the output of unwanted electric signals generated, either by the circuit components themselves, or by extraneous interference.

4. *Control the input and output impedances:* that is, raise or lower the input and output impedances by the selection of an appropriate feedback topology.

5. *Extend the bandwidth* of the amplifier.

All of the desirable properties above are obtained at the expense of a reduction in gain. It will be shown that the gain-reduction factor, called the **amount of feedback,** is the factor by which the circuit is desensitized, by which the input impedance of a voltage amplifier is increased, by which the bandwidth is extended, and so on. In short, *the basic idea of negative feedback is to trade off gain for other desirable properties.* This chapter is devoted to the study of negative-feedback amplifiers: their analysis, design, and characteristics.

Under certain conditions, the negative feedback in an amplifier can become positive and of such a magnitude as to cause oscillation. In fact, in Chapter 13 we will study the use of positive feedback in the design of oscillators and bistable circuits. Here, in this chapter, however, we are interested in the design of stable amplifiers. We shall therefore study the stability problem of negative-feedback amplifiers and their potential for oscillation.

It should not be implied, however, that positive feedback always leads to instability. In fact, positive feedback is quite useful in a number of nonregenerative applications, such as the design of active filters, which are studied in Chapter 12.

Before we begin our study of negative feedback, we wish to remind the reader that we have already encountered negative feedback in a number of applications. Almost all op-amp circuits employ negative feedback. Another popular application of negative feedback is the use of the emitter resistance R_E to stabilize the bias point of bipolar transistors and to increase the input resistance, bandwidth, and linearity of a BJT amplifier. In addition, the source follower and the emitter follower both employ a large amount of negative feedback. The question then arises about the need for a formal study of negative feedback. As will be appreciated by the end of this chapter, the formal study of feedback provides an invaluable tool for the analysis and design of electronic circuits. Also, the insight gained by thinking in terms of feedback can be extremely profitable.

8.1 THE GENERAL FEEDBACK STRUCTURE

Figure 8.1 shows the basic structure of a feedback amplifier. Rather than showing voltages and currents, Fig. 8.1 is a **signal-flow diagram,** where each of the quantities x can represent either a voltage or a current signal. The *open-loop* amplifier has a gain A; thus its output x_o is related to the input x_i by

$$x_o = Ax_i \tag{8.1}$$

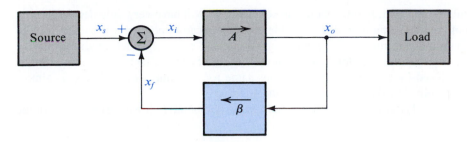

FIGURE 8.1 General structure of the feedback amplifier. This is a signal-flow diagram, and the quantities *x* represent either voltage or current signals.

The output x_o is fed to the load as well as to a feedback network, which produces a sample of the output. This sample x_f is related to x_o by the **feedback factor** β,

$$x_f = \beta x_o \tag{8.2}$$

The feedback signal x_f is *subtracted* from the source signal x_s, which is the input to the complete feedback amplifier,[1] to produce the signal x_i, which is the input to the basic amplifier,

$$x_i = x_s - x_f \tag{8.3}$$

Here we note that it is this subtraction that makes the feedback negative. In essence, negative feedback reduces the signal that appears at the input of the basic amplifier.

Implicit in the description above is that the source, the load, and the feedback network *do not* load the basic amplifier. That is, the gain A does not depend on any of these three networks. In practice this will not be the case, and we shall have to find a method for casting a real circuit into the ideal structure depicted in Fig. 8.1. Figure 8.1 also implies that the forward transmission occurs entirely through the basic amplifier and the reverse transmission occurs entirely through the feedback network.

The gain of the feedback amplifier can be obtained by combining Eqs. (8.1) through (8.3):

$$A_f \equiv \frac{x_o}{x_s} = \frac{A}{1 + A\beta} \tag{8.4}$$

The quantity $A\beta$ is called the **loop gain,** a name that follows from Fig. 8.1. For the feedback to be negative, the loop gain $A\beta$ should be positive; that is, the feedback signal x_f should have the same sign as x_s, thus resulting in a smaller difference signal x_i. Equation (8.4) indicates that for positive $A\beta$ the gain-with-feedback A_f will be smaller than the open-loop gain A by the quantity $1 + A\beta$, which is called the **amount of feedback.**

If, as is the case in many circuits, the loop gain $A\beta$ is large, $A\beta \gg 1$, then from Eq. (8.4) it follows that $A_f \simeq 1/\beta$, which is a very interesting result: *The gain of the feedback amplifier is almost entirely determined by the feedback network.* Since the feedback network usually consists of passive components, which usually can be chosen to be as accurate as one wishes, the advantage of negative feedback in obtaining accurate, predictable, and stable

---

[1] In earlier chapters, we used the subscript "sig" for quantities associated with the signal source (e.g., v_{sig} and R_{sig}). We did that to avoid confusion with the subscript "s," which is usually used with FETs to denote quantities associated with the source terminal of the transistor. At this point, however, it is expected that readers have become sufficiently familiar with the subject that the possibility of confusion is minimal. Therefore, we will revert to using the simpler subscript s for signal-source quantities.

gain should be apparent. In other words, the overall gain will have very little dependence on the gain of the basic amplifier, A, a desirable property because the gain A is usually a function of many manufacturing and application parameters, some of which might have wide tolerances. We have seen a dramatic illustration of all of these effects in op-amp circuits, where the **closed-loop gain** (which is another name for the gain-with-feedback) is almost entirely determined by the feedback elements.

Equations (8.1) through (8.3) can be combined to obtain the following expression for the feedback signal x_f:

$$x_f = \frac{A\beta}{1 + A\beta} x_s \tag{8.5}$$

Thus for $A\beta \gg 1$ we see that $x_f \simeq x_s$, which implies that the signal x_i at the input of the basic amplifier is reduced to almost zero. Thus if a large amount of negative feedback is employed, the feedback signal x_f becomes an almost identical replica of the input signal x_s. An outcome of this property is the tracking of the two input terminals of an op amp. The difference between x_s and x_f, which is x_i, is sometimes referred to as the "error signal." Accordingly, the input differencing circuit is often also called a **comparison circuit.** (It is also known as a **mixer.**) An expression for x_i can be easily determined as

$$x_i = \frac{1}{1 + A\beta} x_s \tag{8.6}$$

from which we can verify that for $A\beta \gg 1$, x_i becomes very small. Observe that negative feedback reduces the signal that appears at the input terminals of the basic amplifier by the amount of feedback, $(1 + A\beta)$.

EXERCISE

8.1 The noninverting op-amp configuration shown in Fig. E8.1 provides a direct implementation of the feedback loop of Fig. 8.1.

(a) Assume that the op amp has infinite input resistance and zero output resistance. Find an expression for the feedback factor β. (b) If the open-loop voltage gain $A = 10^4$, find R_2/R_1 to obtain a closed-loop voltage gain A_f of 10. (c) What is the amount of feedback in decibels? (d) If $V_s = 1$ V, find V_o, V_f, and V_i. (e) If A decreases by 20%, what is the corresponding decrease in A_f?

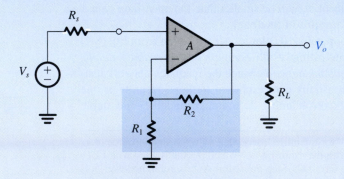

FIGURE E8.1

Ans. (a) $\beta = R_1/(R_1 + R_2)$; (b) 9.01; (c) 60 dB; (d) 10 V, 0.999 V, 0.001 V; (e) 0.02%

8.2 SOME PROPERTIES OF NEGATIVE FEEDBACK

The properties of negative feedback were mentioned in the Introduction. In the following, we shall consider some of these properties in more detail.

8.2.1 Gain Desensitivity

The effect of negative feedback on desensitizing the closed-loop gain was demonstrated in Exercise 8.1, where we saw that a 20% reduction in the gain of the basic amplifier gave rise to only a 0.02% reduction in the gain of the closed-loop amplifier. This sensitivity-reduction property can be analytically established as follows:

Assume that β is constant. Taking differentials of both sides of Eq. (8.4) results in

$$dA_f = \frac{dA}{(1 + A\beta)^2} \tag{8.7}$$

Dividing Eq. (8.7) by Eq. (8.4) yields

$$\frac{dA_f}{A_f} = \frac{1}{(1 + A\beta)} \frac{dA}{A} \tag{8.8}$$

which says that the percentage change in A_f (due to variations in some circuit parameter) is smaller than the percentage change in A by the amount of feedback. For this reason the amount of feedback, $1 + A\beta$, is also known as the **desensitivity factor.**

8.2.2 Bandwidth Extension

Consider an amplifier whose high-frequency response is characterized by a single pole. Its gain at mid and high frequencies can be expressed as

$$A(s) = \frac{A_M}{1 + s/\omega_H} \tag{8.9}$$

where A_M denotes the midband gain and ω_H is the upper 3-dB frequency. Application of negative feedback, with a frequency-independent factor β, around this amplifier results in a closed-loop gain $A_f(s)$ given by

$$A_f(s) = \frac{A(s)}{1 + \beta A(s)}$$

Substituting for $A(s)$ from Eq. (8.9) results, after a little manipulation, in

$$A_f(s) = \frac{A_M/(1 + A_M\beta)}{1 + s/\omega_H(1 + A_M\beta)} \tag{8.10}$$

Thus the feedback amplifier will have a midband gain of $A_M/(1 + A_M\beta)$ and an upper 3-dB frequency ω_{Hf} given by

$$\omega_{Hf} = \omega_H(1 + A_M\beta) \tag{8.11}$$

It follows that the upper 3-dB frequency is increased by a factor equal to the amount of feedback.

Similarly, it can be shown that if the open-loop gain is characterized by a dominant low-frequency pole giving rise to a lower 3-dB frequency ω_L, then the feedback amplifier will have a lower 3-dB frequency ω_{Lf},

$$\omega_{Lf} = \frac{\omega_L}{1 + A_M \beta} \tag{8.12}$$

Note that the amplifier bandwidth is increased by the same factor by which its midband gain is decreased, *maintaining the gain–bandwidth product at a constant value*.

EXERCISE

8.2 Consider the noninverting op-amp circuit of Exercise 8.1. Let the open-loop gain A have a low-frequency value of 10^4 and a uniform –6-dB/octave rolloff at high frequencies with a 3-dB frequency of 100 Hz. Find the low-frequency gain and the upper 3-dB frequency of a closed-loop amplifier with $R_1 = 1$ kΩ and $R_2 = 9$ kΩ.

Ans. 9.99 V/V; 100.1 kHz

8.2.3 Noise Reduction

Negative feedback can be employed to reduce the noise or interference in an amplifier or, more precisely, to increase the ratio of signal to noise. However, as we shall now explain, this noise-reduction process is possible only under certain conditions. Consider the situation illustrated in Fig. 8.2. Figure 8.2(a) shows an amplifier with gain A_1, an input signal V_s, and noise, or interference, V_n. It is assumed that for some reason this amplifier suffers from noise and that the noise can be assumed to be introduced at the input of the amplifier. The

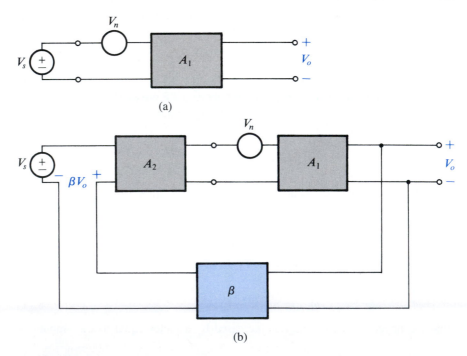

FIGURE 8.2 Illustrating the application of negative feedback to improve the signal-to-noise ratio in amplifiers.

signal-to-noise ratio for this amplifier is

$$S/N = V_s/V_n \tag{8.13}$$

Consider next the circuit in Fig. 8.2(b). Here we assume that it is possible to build another amplifier stage with gain A_2 that does not suffer from the noise problem. If this is the case, then we may precede our original amplifier A_1 by the *clean* amplifier A_2 and apply negative feedback around the overall cascade of such an amount as to keep the overall gain constant. The output voltage of the circuit in Fig. 8.2(b) can be found by superposition:

$$V_o = V_s \frac{A_1 A_2}{1 + A_1 A_2 \beta} + V_n \frac{A_1}{1 + A_1 A_2 \beta} \tag{8.14}$$

Thus the signal-to-noise ratio at the output becomes

$$\frac{S}{N} = \frac{V_s}{V_n} A_2 \tag{8.15}$$

which is A_2 times higher than in the original case.

We emphasize once more that the improvement in signal-to-noise ratio by the application of feedback is possible only if one can precede the noisy stage by a (relatively) noise-free stage. This situation, however, is not uncommon in practice. The best example is found in the output power-amplifier stage of an audio amplifier. Such a stage usually suffers from a problem known as **power-supply hum.** The problem arises because of the large currents that this stage draws from the power supply and the difficulty in providing adequate power-supply filtering inexpensively. The power-output stage is required to provide large power gain but little or no voltage gain. We may therefore precede the power-output stage by a small-signal amplifier that provides large voltage gain, and apply a large amount of negative feedback, thus restoring the voltage gain to its original value. Since the small-signal amplifier can be fed from another, less hefty (and hence better regulated) power supply, it will not suffer from the hum problem. The hum at the output will then be reduced by the amount of the voltage gain of this added **preamplifier.**

EXERCISE

8.3 Consider a power-output stage with voltage gain $A_1 = 1$, an input signal $V_s = 1$ V, and a hum V_n of 1 V. Assume that this power stage is preceded by a small-signal stage with gain $A_2 = 100$ V/V and that overall feedback with $\beta = 1$ is applied. If V_s and V_n remain unchanged, find the signal and noise voltages at the output and hence the improvement in S/N.

Ans. $\simeq 1$ V; $\simeq 0.01$ V; 100 (40 dB)

8.2.4 Reduction in Nonlinear Distortion

Curve (a) in Fig. 8.3 shows the transfer characteristic of an amplifier. As indicated, the characteristic is piecewise linear, with the voltage gain changing from 1000 to 100 and then to 0. This nonlinear transfer characteristic will result in this amplifier generating a large amount of nonlinear distortion.

The amplifier transfer characteristic can be considerably **linearized** (i.e., made less non-linear) through the application of negative feedback. That this is possible should not be too surprising, since we have already seen that negative feedback reduces the dependence of the overall closed-loop amplifier gain on the open-loop gain of the basic amplifier. Thus large

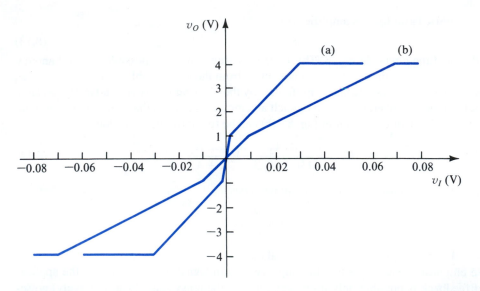

FIGURE 8.3 Illustrating the application of negative feedback to reduce the nonlinear distortion in amplifiers. Curve (a) shows the amplifier transfer characteristic without feedback. Curve (b) shows the characteristic with negative feedback ($\beta = 0.01$) applied.

changes in open-loop gain (1000 to 100 in this case) give rise to much smaller corresponding changes in the closed-loop gain.

To illustrate, let us apply negative feedback with $\beta = 0.01$ to the amplifier whose open-loop voltage transfer characteristic is depicted in Fig. 8.3. The resulting transfer characteristic of the closed-loop amplifier is shown in Fig. 8.3 as curve (b). Here the slope of the steepest segment is given by

$$A_{f1} = \frac{1000}{1 + 1000 \times 0.01} = 90.9$$

and the slope of the next segment is given by

$$A_{f2} = \frac{100}{1 + 100 \times 0.01} = 50$$

Thus the order-of-magnitude change in slope has been considerably reduced. The price paid, of course, is a reduction in voltage gain. Thus if the overall gain has to be restored, then a pre-amplifier should be added. This preamplifier should not present a severe nonlinear-distortion problem, since it will be dealing with smaller signals.

Finally, it should be noted that negative feedback can do nothing at all about amplifier saturation, since in saturation the gain is very small (almost zero) and hence the amount of feedback is also very small (almost zero).

8.3 THE FOUR BASIC FEEDBACK TOPOLOGIES

Based on the quantity to be amplified (voltage or current) and on the desired form of output (voltage or current), amplifiers can be classified into four categories. These categories were discussed in Chapter 1. In the following, we shall review this amplifier classification and point out the feedback topology appropriate in each case.

8.3.1 Voltage Amplifiers

Voltage amplifiers are intended to amplify an input voltage signal and provide an output voltage signal. The voltage amplifier is essentially a voltage-controlled voltage source. The input impedance is required to be high, and the output impedance is required to be low. Since the signal source is essentially a voltage source, it is convenient to represent it in terms of a Thévenin equivalent circuit. In a voltage amplifier the output quantity of interest is the output voltage. It follows that the feedback network should *sample* the output *voltage*. Also, because of the Thévenin representation of the source, the feedback signal x_f should be a *voltage* that can be *mixed* with the source voltage in *series*.

A suitable feedback topology for the voltage amplifier is the **voltage-mixing voltage-sampling** one shown in Fig. 8.4(a). Because of the series connection at the input and the parallel or shunt connection at the output, this feedback topology is also known as **series–shunt feedback.** As will be shown, this topology not only stabilizes the voltage gain but also results in a higher input resistance (intuitively, a result of the series connection at the input) and a lower output resistance (intuitively, a result of the parallel connection at the output), which are desirable properties for a voltage amplifier. The noninverting op-amp configuration of Fig. E8.1 is an example of series–shunt feedback.

8.3.2 Current Amplifiers

The input signal in a current amplifier is essentially a current, and thus the signal source is most conveniently represented by its Norton equivalent. The output quantity of interest is current; hence the feedback network should *sample* the output *current*. The feedback signal should be in *current* form so that it may be *mixed* in *shunt* with the source current. Thus the feedback topology suitable for a current amplifier is the **current-mixing current-sampling** topology, illustrated in Fig. 8.4(b). Because of the parallel (or shunt) connection at the input, and the series connection at the output, this feedback topology is also known as **shunt–series feedback.** As will be shown, this topology not only stabilizes the current gain but also results in a lower input resistance, and a higher output resistance, both desirable properties for a current amplifier.

An example of the shunt–series feedback topology is given in Fig. 8.5. Note that the bias details are not shown. Also note that the current being sampled is not the output current, but the equal current flowing from the source of Q_2. This use of a surrogate is done for circuit-design convenience and is quite usual in circuits involving current sampling.

The reference direction indicated in Fig. 8.5 for the feedback current I_f is such that it subtracts from I_s. This reference notation will be followed in all circuits in this chapter, since it is consistent with the notation used in the general feedback structure of Fig. 8.1. In all circuits, therefore, for the feedback to be negative, the loop gain $A\beta$ should be positive. The reader is urged to verify, through qualitative analysis, that in the circuit of Fig. 8.5, A is negative and β is negative.

It is of utmost importance to be able to ascertain qualitatively (and quickly) the feedback polarity (positive or negative). This can be done by "following the signal around the loop." For instance, let the current I_s in Fig. 8.5 increase. We see that the gate voltage of Q_1 will increase, and thus its drain current will also increase. This will cause the drain voltage of Q_1 (and the gate voltage of Q_2) to decrease, and thus the drain current of Q_2, I_o, will decrease. Thus the source current of Q_2, I_o, decreases. From the feedback network we see that if I_o decreases, then I_f (in the direction shown) will increase. The increase in I_f will subtract from I_s, causing a smaller increment to be seen by the amplifier. Hence the feedback is negative.

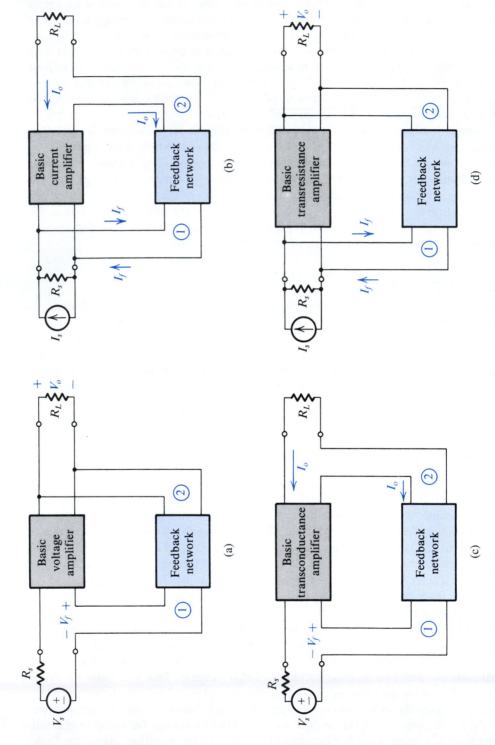

FIGURE 8.4 The four basic feedback topologies: **(a)** voltage-mixing voltage-sampling (series–shunt) topology; **(b)** current-mixing current-sampling (shunt–series) topology; **(c)** voltage-mixing current-sampling (series–series) topology; **(d)** current-mixing voltage-sampling (shunt–shunt) topology.

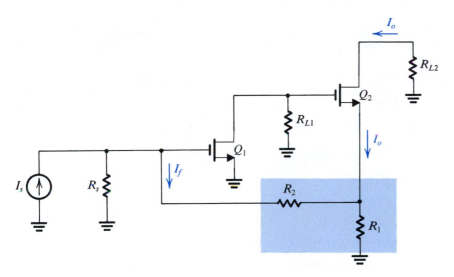

FIGURE 8.5 A transistor amplifier with shunt–series feedback. (Biasing not shown.)

8.3.3 Transconductance Amplifiers

In transconductance amplifiers the input signal is a voltage and the output signal is a current. It follows that the appropriate feedback topology is the **voltage-mixing current-sampling** topology, illustrated in Fig. 8.4(c). The presence of the series connection at both the input and the output gives this feedback topology the alternative name **series–series feedback.**

An example of this feedback topology is given in Fig. 8.6. Here, note that as in the circuit of Fig. 8.5 the current sampled is not the output current but the almost-equal emitter current of Q_3. In addition, the mixing loop is not a conventional one; it is not a simple series connection, since the feedback signal developed across R_{E1} is in the emitter circuit of Q_1, while the source is in the base circuit of Q_1. These two approximations are done for convenience of circuit design.

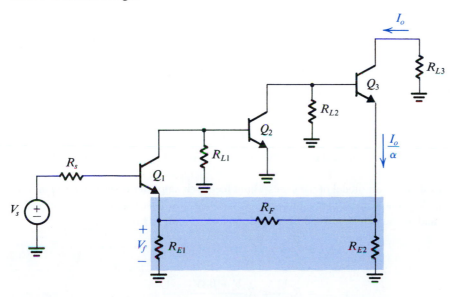

FIGURE 8.6 An example of the series–series feedback topology. (Biasing not shown.)

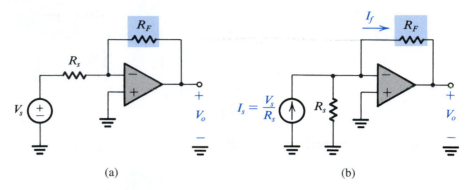

(a)   (b)

FIGURE 8.7 **(a)** The inverting op-amp configuration redrawn as **(b)** an example of shunt–shunt feedback.

8.3.4 Transresistance Amplifiers

In transresistance amplifiers the input signal is current and the output signal is voltage. It follows that the appropriate feedback topology is of the **current-mixing voltage-sampling** type, shown in Fig. 8.4(d). The presence of the parallel (or shunt) connection at both the input and the output makes this feedback topology also known as **shunt–shunt** feedback.

An example of this feedback topology is found in the inverting op-amp configuration of Fig. 8.7(a). The circuit is redrawn in Fig. 8.7(b) with the source converted to Norton's form.

8.4 THE SERIES–SHUNT FEEDBACK AMPLIFIER

8.4.1 The Ideal Situation

The ideal structure of the series–shunt feedback amplifier is shown in Fig. 8.8(a). It consists of a *unilateral* open-loop amplifier (the A circuit) and an ideal voltage-mixing voltage-sampling feedback network (the β circuit). The A circuit has an input resistance R_i, a voltage gain A, and an output resistance R_o. It is assumed that the source and load resistances have been included inside the A circuit (more on this point later). Furthermore, note that the β circuit does *not* load the A circuit; that is, connecting the β circuit does not change the value of A (defined as $A \equiv V_o/V_i$).

The circuit of Fig. 8.8(a) exactly follows the ideal feedback model of Fig. 8.1. Therefore the closed-loop voltage gain A_f is given by

$$A_f \equiv \frac{V_o}{V_s} = \frac{A}{1 + A\beta} \tag{8.16}$$

Note that A and β have reciprocal units. This in fact is always the case, resulting in a dimensionless loop gain $A\beta$.

The equivalent circuit model of the series–shunt feedback amplifier is shown in Fig. 8.8(b). Here R_{if} and R_{of} denote the *input and output resistances with feedback*. The relationship between R_{if} and R_i can be established by considering the circuit in Fig. 8.8(a):

$$R_{if} \equiv \frac{V_s}{I_i} = \frac{V_s}{V_i/R_i}$$

$$= R_i\frac{V_s}{V_i} = R_i\frac{V_i + \beta AV_i}{V_i}$$

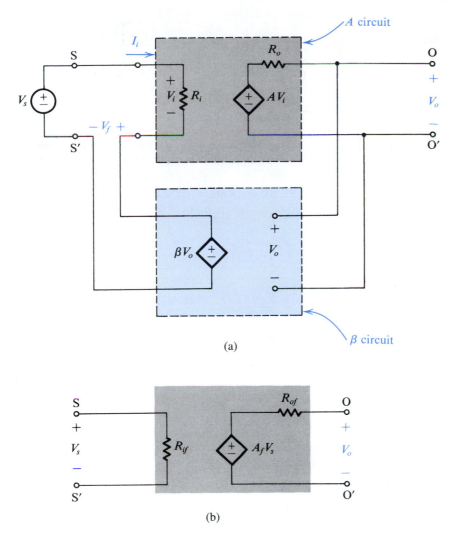

FIGURE 8.8 The series–shunt feedback amplifier: (**a**) ideal structure and (**b**) equivalent circuit.

Thus,

$$R_{if} = R_i(1 + A\beta) \tag{8.17}$$

That is, in this case *the negative feedback increases the input resistance by a factor equal to the amount of feedback.* Since the derivation above does not depend on the method of sampling (shunt or series), it follows that the relationship between R_{if} and R_i is a function *only* of the method of mixing. We shall discuss this point further in later sections.

Note, however, that this result is not surprising and is physically intuitive: Since the feedback voltage V_f subtracts from V_s, the voltage that appears across R_i—that is, V_i—becomes quite small $[V_i = V_s/(1 + A\beta)]$. Thus the input current I_i becomes correspondingly small and the resistance seen by V_s becomes large. Finally, it should be pointed out that Eq. (8.17) can be generalized to the form

$$Z_{if}(s) = Z_i(s)[1 + A(s)\beta(s)] \tag{8.18}$$

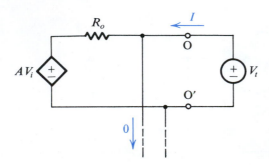

FIGURE 8.9 Measuring the output resistance of the feedback amplifier of Fig. 8.8(a): $R_{of} \equiv V_t / I$.

To find the output resistance, R_{of}, of the feedback amplifier in Fig. 8.8(a) we reduce V_s to zero and apply a test voltage V_t at the output, as shown in Fig. 8.9,

$$R_{of} \equiv \frac{V_t}{I}$$

From Fig. 8.9 we can write

$$I = \frac{V_t - AV_i}{R_o}$$

and since $V_s = 0$ it follows from Fig. 8.8(a) that

$$V_i = -V_f = -\beta V_o = -\beta V_t$$

Thus

$$I = \frac{V_t + A\beta V_t}{R_o}$$

leading to

$$R_{of} = \frac{R_o}{1 + A\beta} \tag{8.19}$$

That is, the *negative feedback in this case reduces the output resistance by a factor equal to the amount of feedback*. With a little thought one can see that the derivation of Eq. (8.19) does not depend on the method of mixing. Thus the relationship between R_{of} and R_o depends *only* on the method of sampling. Again, this result is not surprising and is physically intuitive: Since the feedback samples the output voltage V_o, it acts to stabilize the value of V_o: that is, to reduce changes in the value of V_o, including changes that might be brought about by changing the current drawn from the amplifier output terminals. This, in effect, means that voltage-sampling feedback reduces the output resistance. Finally, we note that Eq. (8.19) can be generalized to

$$Z_{of}(s) = \frac{Z_o(s)}{1 + A(s)\beta(s)} \tag{8.20}$$

8.4.2 The Practical Situation

In a practical series–shunt feedback amplifier, the feedback network will not be an ideal voltage-controlled voltage source. Rather, the feedback network is usually resistive and hence will load the basic amplifier and thus affect the values of A, R_i, and R_o. In addition, the source and load resistances will affect these three parameters. Thus the problem we have is as follows: Given a series–shunt feedback amplifier represented by the block diagram of Fig. 8.10(a), find the A circuit and the β circuit.

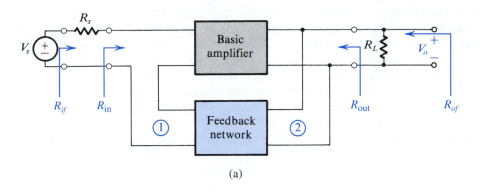

(a)

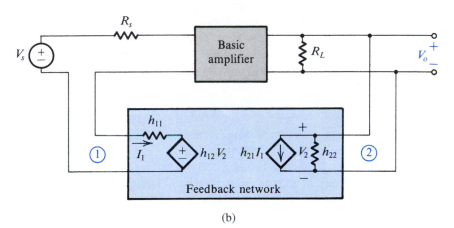

(b)

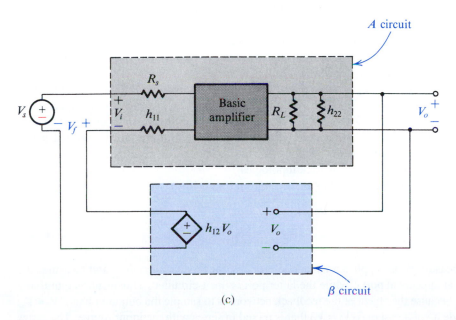

(c)

FIGURE 8.10 Derivation of the *A* circuit and *β* circuit for the series–shunt feedback amplifier. **(a)** Block diagram of a practical series–shunt feedback amplifier. **(b)** The circuit in (a) with the feedback network represented by its *h* parameters. **(c)** The circuit in (b) with h_{21} neglected.

Our problem essentially involves representing the amplifier of Fig. 8.10(a) by the ideal structure of Fig. 8.8(a). As a first step toward that end we observe that the source and load resistances should be lumped with the basic amplifier. This, together with representing the two-port feedback network in terms of its h parameters (see Appendix B), is illustrated in Fig. 8.10(b). The choice of h parameters is based on the fact that this is the only parameter set that represents the feedback network by a series network at port 1 and a parallel network at port 2. Such a representation is obviously convenient in view of the series connection at the input and the parallel connection at the output.

Examination of the circuit in Fig. 8.10(b) reveals that the current source $h_{21}I_1$ represents the forward transmission of the feedback network. Since the feedback network is usually passive, its forward transmission can be neglected in comparison to the much larger forward transmission of the basic amplifier. We will therefore assume that $|h_{21}|_{\substack{\text{feedback}\\\text{network}}} \ll |h_{21}|_{\substack{\text{basic}\\\text{amplifier}}}$ and thus omit the controlled source $h_{21}I_1$ altogether.

Compare the circuit of Fig. 8.10(b) (after eliminating the current source $h_{21}I_1$) with the ideal circuit of Fig. 8.8(a). We see that by including h_{11} and h_{22} with the basic amplifier we obtain the circuit shown in Fig. 8.10(c), which is very similar to the ideal circuit. Now, if the basic amplifier is unilateral (or almost unilateral), a situation that prevails when

$$|h_{12}|_{\substack{\text{basic}\\\text{amplifier}}} \ll |h_{12}|_{\substack{\text{feedback}\\\text{network}}} \tag{8.21}$$

then the circuit of Fig. 8.10(c) is equivalent (or approximately equivalent) to the ideal circuit. It follows then that the A circuit is obtained by augmenting the basic amplifier at the input with the source impedance R_s and the impedance h_{11} of the feedback network, and at the output with the load impedance R_L and the admittance h_{22} of the feedback network.

We conclude that the loading effect of the feedback network on the basic amplifier is represented by the components h_{11} and h_{22}. From the definitions of the h parameters in Appendix B we see that h_{11} is the impedance looking into port 1 of the feedback network with port 2 short-circuited. Since port 2 of the feedback network is connected in *shunt* with the output port of the amplifier, short-circuiting port 2 destroys the feedback. Similarly, h_{22} is the admittance looking into port 2 of the feedback network with port 1 open-circuited. Since port 1 of the feedback network is connected in *series* with the amplifier input, open-circuiting port 1 destroys the feedback.

These observations suggest a simple rule for finding the loading effects of the feedback network on the basic amplifier: The loading effect is found by looking into the appropriate port of the feedback network while the other port is open-circuited or short-circuited so as to destroy the feedback. If the connection is a shunt one, we short-circuit the port; if it is a series one, we open-circuit it. In Sections 8.5 and 8.6 it will be seen that this simple rule applies also to the other three feedback topologies.[2]

We next consider the determination of β. From Fig. 8.10(c), we see that β is equal to h_{12} of the feedback network,

$$\beta = h_{12} \equiv \left.\frac{V_1}{V_2}\right|_{I_1=0} \tag{8.22}$$

Thus to measure β, one applies a voltage to port 2 of the feedback network and measures the voltage that appears at port 1 while the latter port is open-circuited. This result is intuitively appealing because the object of the feedback network is to sample the output voltage ($V_2 = V_o$) and provide a voltage signal ($V_1 = V_f$) that is mixed in series with the input source. The series

---

[2] A simple rule to remember is: If the connection is *shunt*, *short* it; if *series*, *sever* it.

connection at the input suggests that (as in the case of finding the loading effects of the feedback network) β should be found with port 1 open-circuited.

8.4.3 Summary

A summary of the rules for finding the A circuit and β for a given series–shunt feedback amplifier of the form in Fig. 8.10(a) is given in Fig. 8.11. As for using the feedback formulas in Eqs. (8.17) and (8.19) to determine the input and output resistances, it is important to note that:

1. R_i and R_o are the input and output resistances, respectively, of the A circuit in Fig. 8.11(a).

2. R_{if} and R_{of} are the input and output resistances, respectively, of the feedback amplifier, *including* R_s and R_L (see Fig. 8.10a).

3. The actual input and output resistances of the feedback amplifier usually exclude R_s and R_L. These are denoted R_{in} and R_{out} in Fig. 8.10(a) and can be easily determined as

$$R_{\text{in}} = R_{if} - R_s \tag{8.23}$$

$$R_{\text{out}} = 1 \Big/ \left(\frac{1}{R_{of}} - \frac{1}{R_L} \right) \tag{8.24}$$

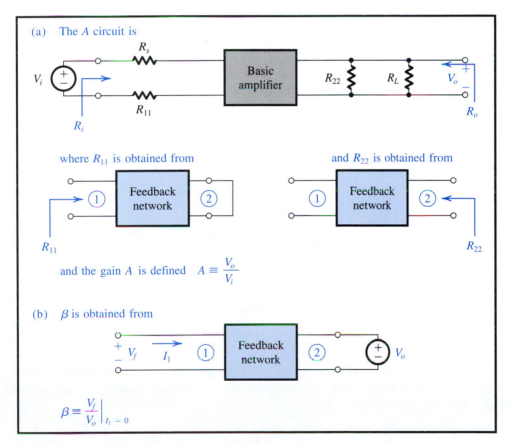

FIGURE 8.11 Summary of the rules for finding the A circuit and β for the voltage-mixing voltage-sampling case of Fig. 8.10(a).

EXAMPLE 8.1

Figure 8.12(a) shows an op amp connected in the noninverting configuration. The op amp has an open-loop gain μ, a differential input resistance R_{id}, and an output resistance r_o. Recall that in our analysis of op-amp circuits in Chapter 2, we neglected the effects of R_{id} (assumed it to be infinite) and of r_o (assumed it to be zero). Here we wish to use the feedback method to analyze the circuit taking both R_{id} and r_o into account. Find expressions for A, β, the closed-loop gain V_o/V_s, the input resistance R_{in} (see Fig. 8.12a), and the output resistance R_{out}. Also find numerical values, given $\mu = 10^4$, $R_{id} = 100$ kΩ, $r_o = 1$ kΩ, $R_L = 2$ kΩ, $R_1 = 1$ kΩ, $R_2 = 1$ MΩ, and $R_s = 10$ kΩ.

Solution

We observe that the feedback network consists of R_2 and R_1. This network samples the output voltage V_o and provides a voltage signal (across R_1) that is mixed in series with the input source V_s.

The A circuit can be easily obtained following the rules of Fig. 8.11, and is shown in Fig. 8.12(b). For this circuit we can write by inspection

$$A \equiv \frac{V_o}{V_i} = \mu \frac{[R_L \| (R_1 + R_2)]}{[R_L \| (R_1 + R_2)] + r_o} \frac{R_{id}}{R_{id} + R_s + (R_1 \| R_2)}$$

For the values given, we find that $A \simeq 6000$ V/V.

The circuit for obtaining β is shown in Fig. 8.12(c), from which we obtain

$$\beta \equiv \frac{V_f}{V_o} = \frac{R_1}{R_1 + R_2} \simeq 10^{-3} \text{ V/V}$$

The voltage gain with feedback is now obtained as

$$A_f \equiv \frac{V_o}{V_s} = \frac{A}{1 + A\beta} = \frac{6000}{7} = 857 \text{ V/V}$$

The input resistance R_{if} determined by the feedback equations is the resistance seen by the external source (see Fig. 8.12a), and is given by

$$R_{if} = R_i(1 + A\beta)$$

where R_i is the input resistance of the A circuit in Fig. 8.12(b):

$$R_i = R_s + R_{id} + (R_1 \| R_2)$$

For the values given, $R_i \simeq 111$ kΩ, resulting in

$$R_{if} = 111 \times 7 = 777 \text{ k}\Omega$$

This, however, is not the resistance asked for. What is required is R_{in}, indicated in Fig. 8.12(a). To obtain R_{in} we subtract R_s from R_{if}:

$$R_{in} = R_{if} - R_s$$

For the values given, $R_{in} = 739$ kΩ. The resistance R_{of} given by the feedback equations is the output resistance of the feedback amplifier, including the load resistance R_L, as indicated in Fig. 8.12(a). R_{of} is given by

$$R_{of} = \frac{R_o}{1 + A\beta}$$

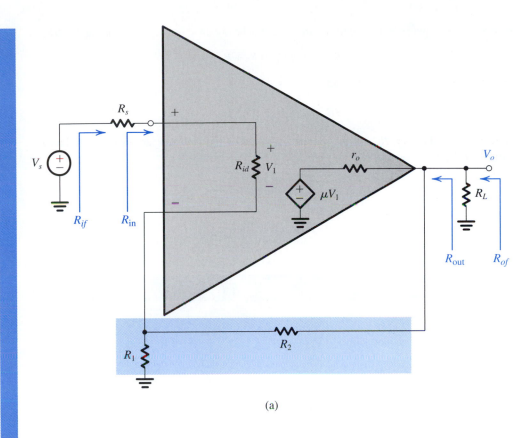

(a)

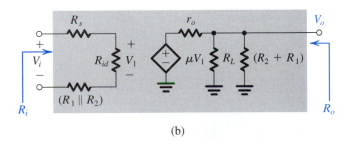

(b)

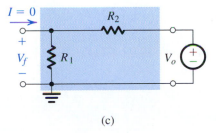

(c)

FIGURE 8.12 Circuits for Example 8.1.

where R_o is the output resistance of the A circuit. R_o can be obtained by inspection of Fig. 8.12(b) as

$$R_o = r_o \,\|\, R_L \,\|\, (R_2 + R_1)$$

For the values given, $R_o \simeq 667\ \Omega$, and

$$R_{of} = \frac{667}{7} = 95.3\ \Omega$$

The resistance asked for, R_{out}, is the output resistance of the feedback amplifier excluding R_L. From Fig. 8.12(a) we see that

$$R_{of} = R_{\text{out}} \| R_L$$

Thus

$$R_{\text{out}} \simeq 100\ \Omega$$

EXERCISES

8.4 If the op amp of Example 8.1 has a uniform –6-dB/octave high-frequency rolloff with $f_{3\text{dB}} = 1$ kHz, find the 3-dB frequency of the closed-loop gain V_o/V_s.

Ans. 7 kHz

8.5 The circuit shown in Fig. E8.5 consists of a differential stage followed by an emitter follower, with series–shunt feedback supplied by the resistors R_1 and R_2. Assuming that the dc component of V_s is zero, and that β of the BJTs is very high, find the dc operating current of each of the three transistors and show that the dc voltage at the output is approximately zero. Then find the values of A, β, $A_f \equiv V_o/V_s$, R_{in}, and R_{out}. Assume that the transistors have $\beta = 100$.

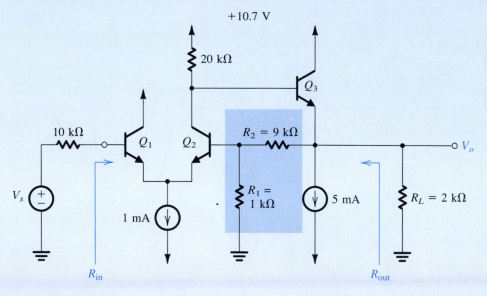

FIGURE E8.5

Ans. 85.7 V/V; 0.1 V/V; 8.96 V/V; 191 kΩ; 19.1 Ω

8.5 THE SERIES–SERIES FEEDBACK AMPLIFIER

8.5.1 The Ideal Case

As mentioned in Section 8.3, the series–series feedback topology stabilizes I_o/V_s and is therefore best suited for transconductance amplifiers. Figure 8.13(a) shows the ideal structure for the series–series feedback amplifier. It consists of a unilateral open-loop amplifier (the A circuit) and an ideal feedback network. Note that in this case A is a transconductance,

$$A \equiv \frac{I_o}{V_i} \qquad (8.25)$$

while β is a transresistance. Thus the loop gain $A\beta$ remains a dimensionless quantity, as it should always be.

In the ideal structure of Fig. 8.13(a), the load and source resistances have been absorbed inside the A circuit, and the β circuit does not load the A circuit. Thus the circuit follows the ideal feedback model of Fig. 8.1, and we can write

$$A_f \equiv \frac{I_o}{V_s} = \frac{A}{1 + A\beta} \qquad (8.26)$$

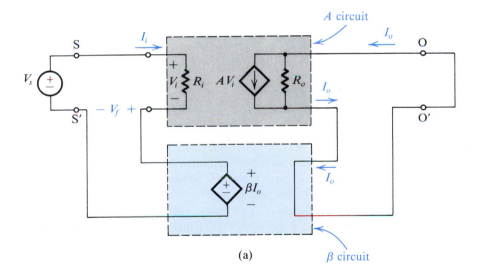

(a)

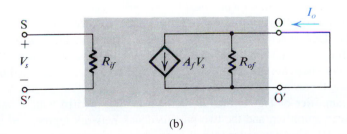

(b)

FIGURE 8.13 The series–series feedback amplifier: **(a)** ideal structure and **(b)** equivalent circuit.

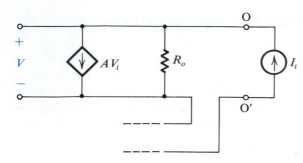

FIGURE 8.14 Measuring the output resistance R_{of} of the series–series feedback amplifier.

This transconductance-with-feedback is included in the equivalent circuit model of the feedback amplifier, shown in Fig. 8.13(b). In this model, R_{if} is the input resistance with feedback. Using an analysis similar to that in Section 8.4, we can show that

$$R_{if} = R_i(1 + A\beta) \qquad (8.27)$$

This relationship is identical to that obtained in the case of series–shunt feedback. This confirms our earlier observation that the relationship between R_{if} and R_i is a function only of the method of mixing. Voltage (or series) mixing therefore always increases the input resistance.

To find the output resistance R_{of} of the series–series feedback amplifier of Fig. 8.13(a) we reduce V_s to zero and break the output circuit to apply a test current I_t, as shown in Fig. 8.14:

$$R_{of} \equiv \frac{V}{I_t} \qquad (8.28)$$

In this case, $V_i = -V_f = -\beta I_o = -\beta I_t$. Thus for the circuit in Fig. 8.14 we obtain

$$V = (I_t - AV_i)R_o = (I_t + A\beta I_t)R_o$$

Hence

$$R_{of} = (1 + A\beta)R_o \qquad (8.29)$$

That is, in this case the negative feedback increases the output resistance. This should have been expected, since the negative feedback tries to make I_o constant in spite of changes in the output voltage, which means increased output resistance. This result also confirms our earlier observation: The relationship between R_{of} and R_o is a function only of the method of sampling. While voltage (shunt) sampling reduces the output resistance, current (series) sampling increases it.

8.5.2 The Practical Case

Figure 8.15(a) shows a block diagram for a practical series–series feedback amplifier. To be able to apply the feedback equations to this amplifier, we have to represent it by the ideal structure of Fig. 8.13(a). Our objective therefore is to devise a simple method for finding A and β. Observe the definition of the amplifier input resistance R_{in} and output resistance R_{out}. It is important to note that these are different from R_{if} and R_{of}, which are determined by the feedback equations, as will become clear shortly.

The series–series amplifier of Fig. 8.15(a) is redrawn in Fig. 8.15(b) with R_s and R_L shown closer to the basic amplifier, and the two-port feedback network represented by its z parameters (Appendix B). This parameter set has been chosen because it is the only one that provides a representation of the feedback network with a series circuit at the input and a

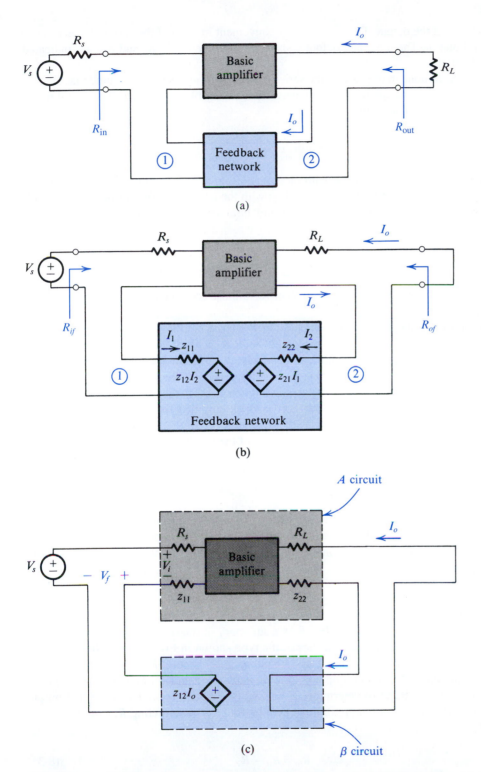

FIGURE 8.15 Derivation of the A circuit and the β circuit for series–series feedback amplifiers. **(a)** A series–series feedback amplifier. **(b)** The circuit of (a) with the feedback network represented by its z parameters. **(c)** A redrawing of the circuit in (b) with z_{21} neglected.

series circuit at the output. This is obviously convenient in view of the series connections at input and output. The input and output resistances with feedback, R_{if} and R_{of}, are indicated on the diagram.

As we have done in the case of the series–shunt amplifier, we shall assume that the forward transmission through the feedback network is negligible in comparison to that through the basic amplifier; that is, the condition

$$|z_{21}|_{\substack{\text{feedback} \\ \text{network}}} \ll |z_{21}|_{\substack{\text{basic} \\ \text{amplifier}}} \qquad (8.30)$$

is satisfied. We can then dispense with the voltage source $z_{21}I_1$ in Fig. 8.15(b). Doing this, and redrawing the circuit to include z_{11} and z_{22} with the basic amplifier, results in the circuit in Fig. 8.15(c). Now if the basic amplifier is unilateral (or almost unilateral), a situation that is obtained when

$$|z_{12}|_{\substack{\text{basic} \\ \text{amplifier}}} \ll |z_{12}|_{\substack{\text{feedback} \\ \text{network}}} \qquad (8.31)$$

then the circuit in Fig. 8.15(c) is equivalent (or almost equivalent) to the ideal circuit of Fig. 8.13(a).

It follows that the A circuit is composed of the basic amplifier augmented at the input with R_s and z_{11} and augmented at the output with R_L and z_{22}. Since z_{11} and z_{22} are the impedances looking into ports 1 and 2, respectively, of the feedback network with the other port open-circuited, we see that finding the loading effects of the feedback network on the basic amplifier follows the rule formulated in Section 8.4. That is, we look into one port of the feedback network while the other port is open-circuited or short-circuited so as to destroy the feedback (open if series and short if shunt).

From Fig. 8.15(c) we see that β is equal to z_{12} of the feedback network,

$$\beta = z_{12} \equiv \left. \frac{V_1}{I_2} \right|_{I_1=0} \qquad (8.32)$$

This result is intuitively appealing. Recall that in this case the feedback network samples the output current $[I_2 = I_o]$ and provides a voltage $[V_f = V_1]$ that is mixed in series with the input source. Again, the series connection at the input suggests that β is measured with port 1 open.

8.5.3 Summary

For future reference we present in Fig. 8.16 a summary of the rules for finding A and β for a given series–series feedback amplifier of the type shown in Fig. 8.15(a). Note that R_i is the input resistance of the A circuit, and its output resistance is R_o, which can be determined by breaking the output loop and looking between Y and Y'. R_i and R_o can be used in Eqs. (8.27) and (8.29) to determine R_{if} and R_{of} (see Fig. 8.15b). The input and output resistances of the feedback amplifier can then be found by subtracting R_s from R_{if} and R_L from R_{of},

$$R_{\text{in}} = R_{if} - R_s \qquad (8.33)$$

$$R_{\text{out}} = R'_{of} - R_L \qquad (8.34)$$

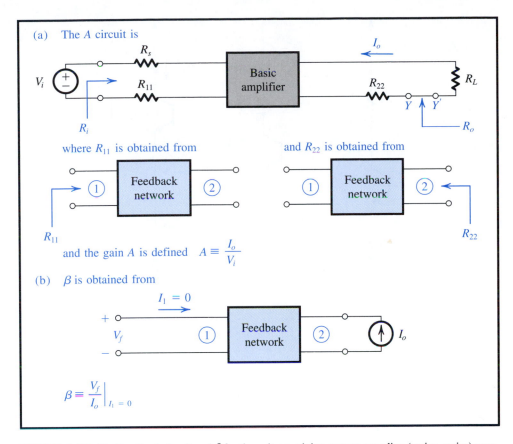

(a) The A circuit is

where R_{11} is obtained from   and R_{22} is obtained from

and the gain A is defined $A \equiv \dfrac{I_o}{V_i}$

(b) β is obtained from

$\beta \equiv \left. \dfrac{V_f}{I_o} \right|_{I_1 = 0}$

FIGURE 8.16 Finding the A circuit and β for the voltage-mixing current-sampling (series–series) case.

EXAMPLE 8.2

Because negative feedback extends the amplifier bandwidth, it is commonly used in the design of broadband amplifiers. One such amplifier is the MC1553. Part of the circuit of the MC1553 is shown in Fig. 8.17(a). The circuit shown (called a *feedback triple*) is composed of three gain stages with series–series feedback provided by the network composed of R_{E1}, R_F, and R_{E2}. Assume that the bias circuit, which is not shown, causes $I_{C1} = 0.6$ mA, $I_{C2} = 1$ mA, and $I_{C3} = 4$ mA. Using these values and assuming that $h_{fe} = 100$ and $r_o = \infty$, find the open-loop gain A, the feedback factor β, the closed-loop gain $A_f \equiv I_o/V_s$, the voltage gain V_o/V_s, the input resistance $R_{in} = R_{if}$, and the output resistance R_{of} (between nodes Y and Y', as indicated). Now, if r_o of Q_3 is 25 kΩ, estimate an approximate value of the output resistance R_{out}.

Solution

Employing the loading rules given in Fig. 8.16, we obtain the A circuit shown in Fig. 8.17(b). To find $A \equiv I_o/V_i$ we first determine the gain of the first stage. This can be written by inspection as

$$\frac{V_{c1}}{V_i} = \frac{-\alpha_1(R_{C1}//r_{\pi2})}{r_{e1} + [R_{E1}//(R_F + R_{E2})]}$$

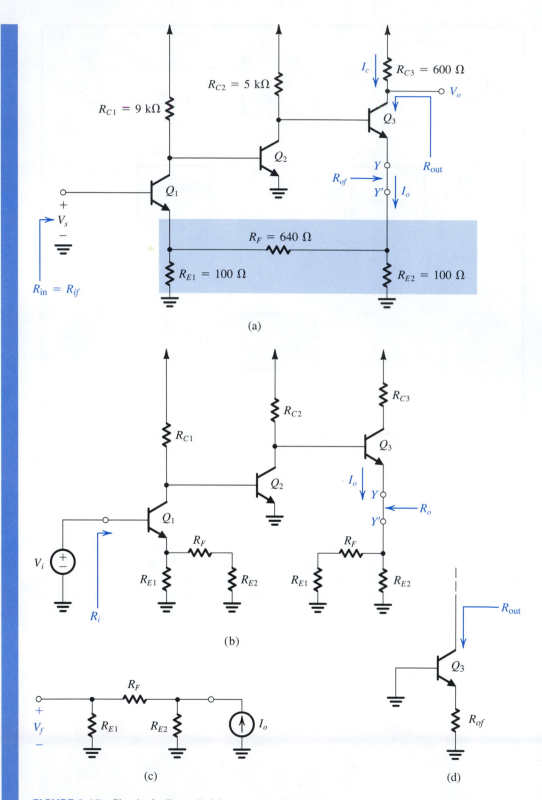

FIGURE 8.17 Circuits for Example 8.2.

Since Q_1 is biased at 0.6 mA, $r_{e1} = 41.7\ \Omega$. Transistor Q_2 is biased as 1 mA; thus $r_{\pi 2} = h_{fe}/g_{m2} = 100/40 = 2.5\ \text{k}\Omega$. Substituting these values together with $\alpha_1 = 0.99$, $R_{C1} = 9\ \text{k}\Omega$, $R_{E1} = 100\ \Omega$, $R_F = 640\ \Omega$, and $R_{E2} = 100\ \Omega$ results in

$$\frac{V_{c1}}{V_i} = -14.92\ \text{V/V}$$

Next, we determine the gain of the second stage, which can be written by inspection as (note that $V_{b2} = V_{c1}$)

$$\frac{V_{c2}}{V_{c1}} = -g_{m2}\{R_{C2}//(h_{fe}+1)[r_{e3} + (R_{E2}//(R_F + R_{E1}))]\}$$

Substituting $g_{m2} = 40\ \text{mA/V}$, $R_{C2} = 5\ \text{k}\Omega$, $h_{fe} = 100$, $r_{e3} = 25/4 = 6.25\ \Omega$, $R_{E2} = 100\ \Omega$, $R_F = 640\ \Omega$, and $R_{E1} = 100\ \Omega$, results in

$$\frac{V_{c2}}{V_{c1}} = -131.2\ \text{V/V}$$

Finally, for the third stage we can write by inspection

$$\frac{I_o}{V_{c2}} = \frac{I_{e3}}{V_{b3}} = \frac{1}{r_{e3} + (R_{E2}//(R_F + R_{E1}))}$$

$$= \frac{1}{6.25 + (100//740)} = 10.6\ \text{mA/V}$$

Combining the gains of the three stages results in

$$A \equiv \frac{I_o}{V_i} = -14.92 \times -131.2 \times 10.6 \times 10^{-3}$$

$$= 20.7\ \text{A/V}$$

The circuit for determining the feedback factor β is shown in Fig. 8.17(c), from which we find

$$\beta \equiv \frac{V_f}{I_o} = \frac{R_{E2}}{R_{E2} + R_F + R_{E1}} \times R_{E1}$$

$$= \frac{100}{100 + 640 + 100} \times 100 = 11.9\ \Omega$$

The closed-loop gain A_f can now be found from

$$A_f \equiv \frac{I_o}{V_s} = \frac{A}{1 + A\beta}$$

$$= \frac{20.7}{1 + 20.7 \times 11.9} = 83.7\ \text{mA/V}$$

The voltage gain is found from

$$\frac{V_o}{V_s} = \frac{-I_c R_{C3}}{V_s} \approx \frac{-I_o R_{C3}}{V_s} = -A_f R_{C3}$$

$$= -83.7 \times 10^{-3} \times 600 = -50.2\ \text{V/V}$$

The input resistance of the feedback amplifier is given by

$$R_{if} = R_i(1 + A\beta)$$

where R_i is the input resistance of the A circuit. The value of R_i can be found from the circuit in Fig. 8.17(b) as follows:

$$R_i = (h_{fe} + 1)[r_{e1} + (R_{E1}//(R_F + R_{E2}))]$$

$$= 13.65 \text{ k}\Omega$$

Thus,

$$R_{if} = 13.65(1 + 20.5 \times 11.9) = 3.34 \text{ M}\Omega$$

To find the output resistance R_o of the A circuit in Fig. 8.17(b), we break the circuit between Y and Y'. The resistance looking between these two nodes can be found to be

$$R_o = [R_{E2}//(R_F + R_{E1})] + r_{e3} + \frac{R_{C2}}{h_{fe} + 1}$$

which, for the values given, yields $R_o = 143.9 \ \Omega$. The output resistance R_{of} of the feedback amplifier can now be found as

$$R_{of} = R_o(1 + A\beta) = 143.9(1 + 20.7 \times 11.9) = 35.6 \text{ k}\Omega$$

Note that the feedback stabilizes the emitter current of Q_3, and thus the output resistance that is determined by the feedback formula is the resistance of the emitter loop (i.e., between Y and Y'), which we have just found, and not the resistance looking into the collector[3] of Q_3. This is because the output resistance r_o of Q_3 is in effect outside the feedback loop. We can, however, use the value of R_{of} to obtain an approximate value for R_{out}. To do this, we assume that the effect of the feedback is to place a resistance R_{of} (35.6 kΩ) in the emitter of Q_3, and find the output resistance from the equivalent circuit shown in Fig. 8.17(d). Using Eq. (6.117), R_{out} can be found as

$$R_{out} = r_o + (1 + g_{m3}r_o)(R_{of}//r_{\pi 3})$$

$$= 25 + (1 + 160 \times 25)(35.6//0.625) = 2.5 \text{ M}\Omega$$

Thus, the output resistance at the collector increases, but not by $(1 + A\beta)$.

8.6 Reconsider the circuit in Fig. 8.17(a), this time with the output voltage taken at the emitter of Q_3. In this case, the feedback can be considered to be of the voltage-mixing voltage-sampling type. Note, however, that the loop gain remains unchanged. Find the value of $A \equiv V_{e3}/V_i$ (from Fig. 8.17(b)), $A_f \equiv V_{e3}/V_s$, and the output resistance.
Ans. 1827 V/V; 7.4 V/V; 0.14 Ω

 ## 8.6 THE SHUNT–SHUNT AND SHUNT–SERIES FEEDBACK AMPLIFIERS

In this section we shall extend—without proof—the method of Sections 8.4 and 8.5 to the two remaining feedback topologies.

---

[3] This important point was first brought to the authors' attention by Gordon Roberts (see Roberts and Sedra, 1992).

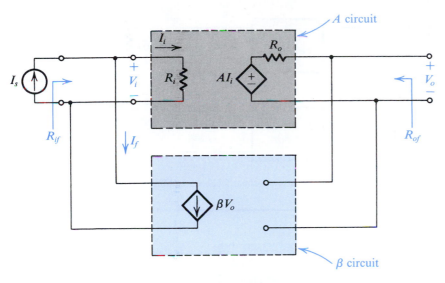

FIGURE 8.18 Ideal structure for the shunt–shunt feedback amplifier.

8.6.1 The Shunt–Shunt Configuration

Figure 8.18 shows the ideal structure for a shunt–shunt feedback amplifier. Here the A circuit has an input resistance R_i, a transresistance A, and an output resistance R_o. The β circuit is a voltage-controlled current source, and β is a transconductance. The closed-loop gain A_f is defined

$$A_f \equiv \frac{V_o}{I_s} \tag{8.35}$$

and is given by

$$A_f = \frac{A}{1 + A\beta}$$

The input resistance with feedback is given by

$$R_{if} = \frac{R_i}{1 + A\beta} \tag{8.36}$$

where we note that the shunt connection at the input results in a reduced input resistance. Also note that the resistance R_{if} is the resistance seen by the source I_s, and it includes any source resistance.

The output resistance with feedback is given by

$$R_{of} = \frac{R_o}{1 + A\beta} \tag{8.37}$$

where we note that the shunt connection at the output results in a reduced output resistance. This resistance includes any load resistance.

Given a practical shunt–shunt feedback amplifier having the block diagram of Fig. 8.19, we use the method given in Fig. 8.20 to obtain the A circuit and the circuit for determining β. As in Sections 8.4 and 8.5, the method of Fig. 8.20 assumes that the basic amplifier is almost unilateral and that the forward transmission through the feedback network is negligibly small.

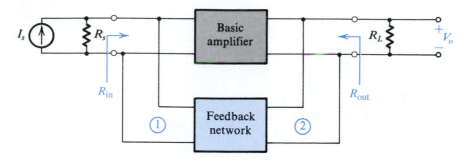

FIGURE 8.19 Block diagram for a practical shunt–shunt feedback amplifier.

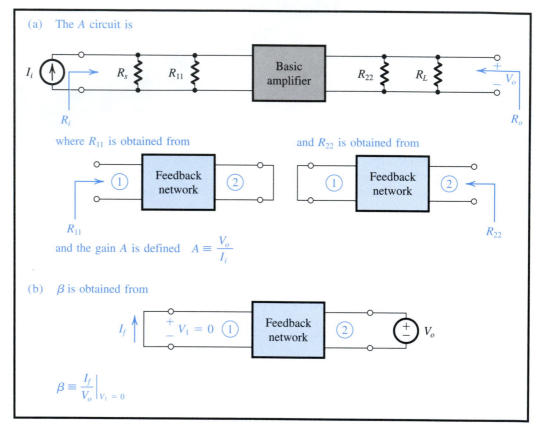

FIGURE 8.20 Finding the A circuit and β for the current-mixing voltage-sampling (shunt–shunt) feedback amplifier in Fig. 8.19.

The first assumption is justified when the reverse y parameters[4] of the basic amplifier and of the feedback network satisfy the condition

$$|y_{12}|_{\substack{\text{basic} \\ \text{amplifier}}} \ll |y_{12}|_{\substack{\text{feedback} \\ \text{network}}} \tag{8.38}$$

---

[4] Here, the y parameters (Appendix B) are used because this is the only two-port parameter set that provides a representation of the feedback network with a parallel circuit at the input and a parallel circuit at the output.

The second assumption is justified when the forward y parameters satisfy the condition

$$\left. |y_{21}| \right|_{\substack{\text{feedback} \\ \text{network}}} \ll \left. |y_{21}| \right|_{\substack{\text{basic} \\ \text{amplifier}}} \tag{8.39}$$

Finally, we note that once R_{if} and R_{of} have been determined using the feedback formulas (Eqs. 8.36 and 8.37), the input and output resistances of the amplifier proper (see definitions in Fig. 8.19) can be obtained as

$$R_{\text{in}} = 1 \bigg/ \left(\frac{1}{R_{if}} - \frac{1}{R_s} \right) \tag{8.40}$$

$$R_{\text{out}} = 1 \bigg/ \left(\frac{1}{R_{of}} - \frac{1}{R_L} \right) \tag{8.41}$$

EXAMPLE 8.3

We want to analyze the circuit of Fig. 8.21(a) to determine the small-signal voltage gain V_o/V_s, the input resistance R_{in}, and the output resistance $R_{\text{out}} = R_{of}$. The transistor has $\beta = 100$.

Solution

First we determine the transistor dc operating point. The dc analysis is illustrated in Fig. 8.21(b), from which we can write

$$V_C = 0.7 + (I_B + 0.07)47 = 3.99 + 47 I_B \quad \text{and} \quad \frac{12 - V_C}{4.7} = (\beta + 1) I_B + 0.07$$

These two equations can be solved to obtain $I_B \simeq 0.015$ mA, $I_C \simeq 1.5$ mA, and $V_C = 4.7$ V.

To carry out small-signal analysis we first recognize that the feedback is provided by R_f, which samples the output voltage V_o and feeds back a current that is mixed with the source current. Thus it is convenient to use the Norton source representation, as shown in Fig. 8.21(c). The A circuit can be easily obtained using the rules of Fig. 8.20, and it is shown in Fig. 8.21(d). For the A circuit we can write by inspection

$$V_\pi = I_i (R_s /\!/ R_f /\!/ r_\pi)$$

$$V_o = -g_m V_\pi (R_f /\!/ R_C)$$

Thus

$$A = \frac{V_o}{I_i} = -g_m (R_f /\!/ R_C)(R_s /\!/ R_f /\!/ r_\pi)$$

$$= -358.7 \text{ k}\Omega$$

The input and output resistances of the A circuit can be obtained from Fig. 8.21(d) as

$$R_i = R_s /\!/ R_f /\!/ r_\pi = 1.4 \text{ k}\Omega$$

$$R_o = R_C /\!/ R_f = 4.27 \text{ k}\Omega$$

The circuit for determining β is shown in Fig. 8.21(e), from which we obtain

$$\beta \equiv \frac{I_f}{V_o} = -\frac{1}{R_f} = -\frac{1}{47 \text{ k}\Omega}$$

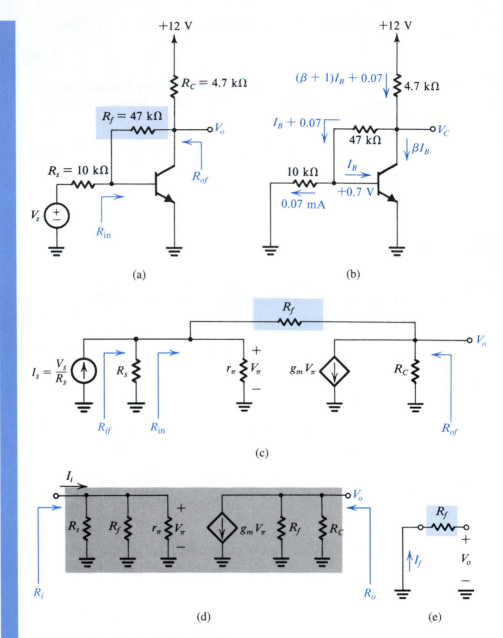

FIGURE 8.21 Circuits for Example 8.3.

Note that as usual the reference direction for I_f has been selected so that I_f subtracts from I_s. The resulting negative sign of β should cause no concern, since A is also negative, keeping the loop gain $A\beta$ positive, as it should be for the feedback to be negative.

We can now obtain A_f (for the circuit in Fig. 8.21c) as

$$A_f \equiv \frac{V_o}{I_s} = \frac{A}{1 + A\beta}$$

$$\frac{V_o}{I_s} = \frac{-358.7}{1 + 358.7/47} = \frac{-358.7}{8.63} = -41.6 \text{ k}\Omega$$

To find the voltage gain V_o/V_s we note that

$$V_s = I_s R_s$$

Thus

$$\frac{V_o}{V_s} = \frac{V_o}{I_s R_s} = \frac{-41.6}{10} \simeq -4.16 \text{ V/V}$$

The input resistance with feedback (see Fig. 8.21c) is given by

$$R_{if} = \frac{R_i}{1 + A\beta}$$

Thus

$$R_{if} = \frac{1.4}{8.63} = 162.2 \ \Omega$$

This is the resistance seen by the current source I_s in Fig. 8.21(c). To obtain the input resistance of the feedback amplifier excluding R_s (i.e., the required resistance R_{in}) we subtract $1/R_s$ from $1/R_{if}$ and invert the result; thus $R_{in} = 165 \ \Omega$. Finally, the amplifier output resistance R_{of} is evaluated using

$$R_{of} = \frac{R_o}{1 + A\beta} = \frac{4.27}{8.63} = 495 \ \Omega$$

8.6.2 An Important Note

The method we have been employing for the analysis of feedback amplifiers is predicated on two premises: Most of the forward transmission occurs in the basic amplifier, and most of the reverse transmission (feedback) occurs in the feedback network. For each of the three topologies considered thus far, these two assumptions were mathematically expressed as conditions on the relative magnitudes of the forward and reverse two-port parameters of the basic amplifier and the feedback network. Since the circuit considered in Example 8.3 is simple, we have a good opportunity to check the validity of these assumptions.

Reference to Fig. 8.21(d) indicates clearly that the basic amplifier is unilateral; thus *all* the reverse transmission takes place in the feedback network. The case with forward transmission, however, is not as clear, and we must evaluate the forward y parameters. For the A circuit in Fig. 8.21(d), $y_{21} = g_m$. For the feedback network it can be easily shown that $y_{21} = -1/R_f$. Thus for our analysis method to be valid we must have $g_m \gg 1/R_f$. For the numerical values in Example 8.3, $g_m = 60$ mA/V and $1/R_f = 0.02$ mA/V, indicating that this assumption is more than justified. Nevertheless, in designing feedback amplifiers, care should be taken in choosing component values to ensure that the two basic assumptions are valid.

8.6.3 The Shunt–Series Configuration

Figure 8.22 shows the ideal structure of the shunt–series feedback amplifier. It is a current amplifier whose gain with feedback is defined as

$$A_f \equiv \frac{I_o}{I_s} = \frac{A}{1 + A\beta} \tag{8.42}$$

The input resistance with feedback is the resistance seen by the current source I_s and is given by

$$R_{if} = \frac{R_i}{1 + A\beta} \tag{8.43}$$

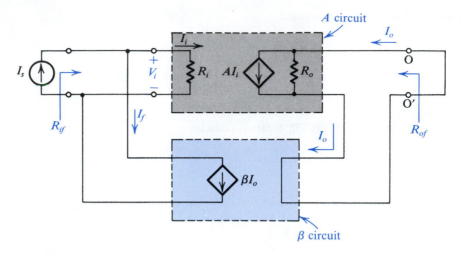

FIGURE 8.22 Ideal structure for the shunt–series feedback amplifier.

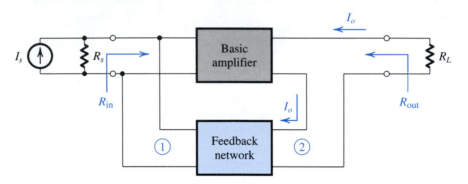

FIGURE 8.23 Block diagram for a practical shunt–series feedback amplifier.

Again we note that the shunt connection at the input reduces the input resistance. The output resistance with feedback is the resistance seen by breaking the output circuit, such as between O and O', and looking between the two terminals thus generated (i.e., between O and O'). This resistance, R_{of}, is given by

$$R_{of} = R_o(1 + A\beta) \tag{8.44}$$

where we note that the increase in output resistance is due to the current (series) sampling.

Given a practical shunt–series feedback amplifier, such as that represented by the block diagram of Fig. 8.23, we follow the method given in Fig. 8.24 in order to obtain A and β. Here again the analysis method is predicated on the assumption that most of the forward transmission occurs in the basic amplifier,[5]

$$|g_{21}|_{\substack{\text{feedback} \\ \text{network}}} \ll |g_{21}|_{\substack{\text{basic} \\ \text{amplifier}}} \tag{8.45}$$

---

[5] For this amplifier topology, the most convenient set of two-port parameters to use is the set of g parameters; it is the only set that provides a representation that is composed of a parallel circuit at the input and a series circuit at the output (see Appendix B).

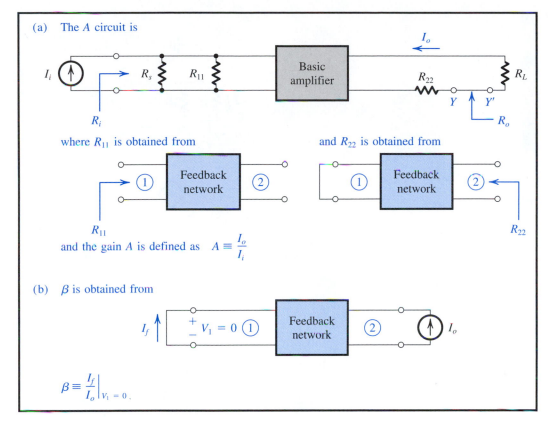

(a)   The A circuit is

where R_{11} is obtained from

and R_{22} is obtained from

and the gain A is defined as $A \equiv \dfrac{I_o}{I_i}$

(b) β is obtained from

$\beta \equiv \left.\dfrac{I_f}{I_o}\right|_{V_1 = 0}$.

FIGURE 8.24   Finding the A circuit and β for the current-mixing current-sampling (shunt–series) feedback amplifier of Fig. 8.23.

and that most of the reverse transmission takes place in the feedback network,

$$\left|g_{12}\right|_{\substack{\text{basic}\\\text{amplifier}}} \ll \left|g_{12}\right|_{\substack{\text{feedback}\\\text{network}}} \tag{8.46}$$

Finally, we note that once R_{if} and R_{of} have been determined using the feedback equations (Eqs. 8.43 and 8.44), the input and output resistances of the amplifier proper, R_{in} and R_{out} (Fig. 8.23), can be found as

$$R_{\text{in}} = 1 \Big/ \left(\frac{1}{R_{if}} - \frac{1}{R_s}\right) \tag{8.47}$$

$$R_{\text{out}} = R_{of} - R_L \tag{8.48}$$

EXAMPLE 8.4

Figure 8.25 shows a feedback circuit of the shunt–series type. Find $I_{\text{out}}/I_{\text{in}}$, R_{in}, and R_{out}. Assume the transistors to have $\beta = 100$ and $V_A = 75$ V.

Solution

We begin by determining the dc operating points. In this regard we note that the feedback signal is capacitively coupled; thus the feedback has no effect on dc bias. Neglecting the effect

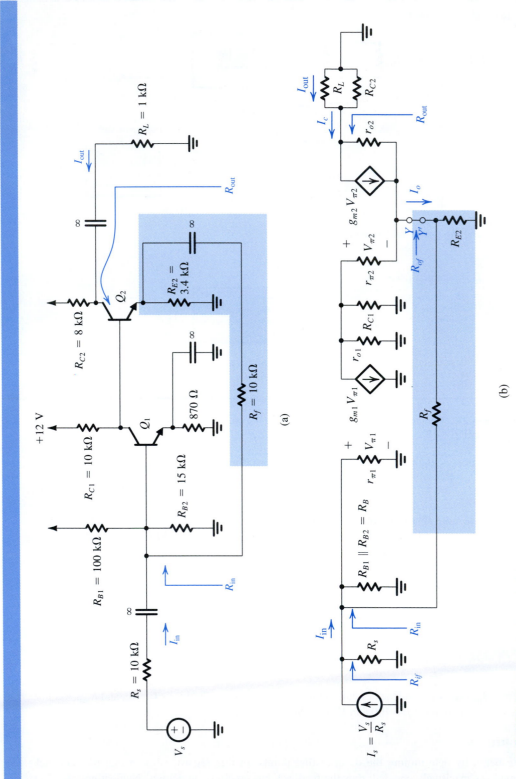

FIGURE 8.25 Circuits for Example 8.4.

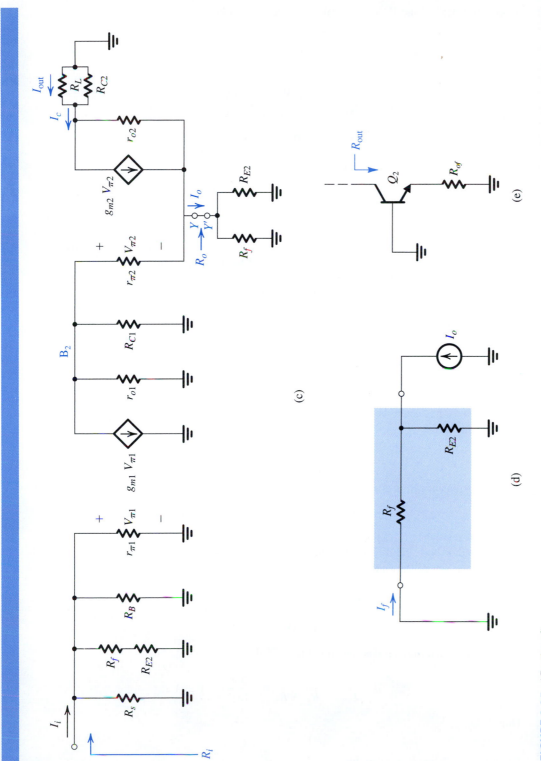

FIGURE 8.25 (*Continued*)

of finite transistor β and V_A, the dc analysis proceeds as follows:

$$V_{B1} \simeq 12 \frac{15}{100 + 15} = 1.57 \text{ V}$$

$$V_{E1} \simeq 1.57 - 0.7 = 0.87 \text{ V}$$

$$I_{E1} = 0.87/0.87 = 1 \text{ mA}$$

$$V_{C1} \simeq 12 - 10 \times 1 = 2 \text{ V}$$

$$V_{E2} \simeq 2 - 0.7 = 1.3 \text{ V}$$

$$I_{E2} \simeq 1.3/3.4 \simeq 0.4 \text{ mA}$$

$$V_{C2} \simeq 12 - 0.4 \times 8 = 8.8 \text{ V}$$

The amplifier equivalent circuit is shown in Fig. 8.25(b), from which we note that the feedback network is composed of R_{E2} and R_f. The feedback network samples the emitter current of Q_2, I_o, which is approximately equal to the collector current I_c. Also note that the required current gain, $I_{\text{out}}/I_{\text{in}}$, will be slightly different than the closed-loop current gain $A_f \equiv I_o/I_s$.

The A circuit is shown in Fig. 8.25(c), where we have obtained the loading effects of the feedback network using the rules of Fig. 8.24. For the A circuit we can write

$$V_{\pi 1} = I_i[R_s/\!/(R_{E2} + R_f)/\!/R_B/\!/r_{\pi 1}]$$

$$V_{b2} = -g_{m1}V_{\pi 1}\{r_{o1}/\!/R_{C1}/\!/[r_{\pi 2} + (\beta + 1)(R_{E2}/\!/R_f)]\}$$

$$I_o \simeq \frac{V_{b2}}{r_{e2} + (R_{E2}/\!/R_f)}$$

where we have neglected the effect of r_{o2}. These equations can be combined to obtain the open-loop current gain A,

$$A \equiv \frac{I_o}{I_i} \simeq -201.45 \text{ A/A}$$

The input resistance R_i is given by

$$R_i = R_s/\!/(R_{E2} + R_f)/\!/R_B/\!/r_{\pi 1} = 1.535 \text{ k}\Omega$$

The output resistance R_o is that found by looking into the output loop of the A circuit between nodes Y and Y' (see Fig. 8.25c) with the input excitation I_i set to zero. Neglecting the small effect of r_{o2} it can be shown that

$$R_o = (R_{E2}/\!/R_f) + r_{e2} + \frac{R_{C1}/\!/r_{o1}}{\beta + 1}$$

$$= 2.69 \text{ k}\Omega$$

The circuit for determining β is shown in Fig. 8.25(d), from which we find

$$\beta \equiv \frac{I_f}{I_o} = -\frac{R_{E2}}{R_{E2} + R_f} = -\frac{3.4}{13.4} = -0.254$$

Thus,

$$1 + A\beta = 52.1$$

The input resistance R_{if} is given by

$$R_{if} = \frac{R_i}{1 + A\beta} = 29.5 \ \Omega$$

The required input resistance R_{in} is given by (see Fig. 8.25b).

$$R_{in} = \frac{1}{1/R_{if} - 1/R_s} \simeq 29.5 \ \Omega$$

Since $R_{in} \simeq R_{if}$, it follows from Fig. 8.25(b) that $I_{in} \simeq I_s$. The current gain A_f is given by

$$A_f \equiv \frac{I_o}{I_s} = \frac{A}{1 + A\beta} = -3.87 \ \text{A/A}$$

Note that because $A\beta \gg 1$ the closed-loop gain is approximately equal to $1/\beta$.

Now, the required current gain is given by

$$\frac{I_{out}}{I_{in}} \simeq \frac{I_{out}}{I_s} = \frac{R_{C2}}{R_L + R_{C2}} \frac{I_c}{I_s} \simeq \frac{R_{C2}}{R_L + R_{C2}} \frac{I_o}{I_s}$$

Thus,

$$I_{out}/I_{in} = -3.44 \ \text{A/A}$$

The output resistance R_{of} is given by

$$R_{of} = R_o(1 + A\beta) \simeq 140.1 \ \text{k}\Omega$$

An estimate of the required output resistance R_{out} can be obtained using the technique employed in Example 8.2, namely, by considering that the effect of feedback is to place a resistance R_{of} in the emitter of Q_2 (see Fig. 8.25e). Thus, using Eq. (6.78), we can write

$$R_{out} = r_{o2}[1 + g_{m2}(r_{\pi2}//R_{of})]$$

Substituting, $r_{o2} = 75/0.4 = 187.5 \ \text{k}\Omega$, $g_{m2} = 16 \ \text{mA/V}$, $r_{\pi2} = 6.25 \ \text{k}\Omega$, and $R_{of} = 140.1 \ \text{k}\Omega$, results in

$$R_{out} = 18.1 \ \text{M}\Omega$$

Thus, while negative feedback considerably increases R_{out}, the increase is not by the factor $(1 + A\beta)$, simply because the feedback network samples the emitter current and not the collector current. Thus, in effect, the feedback network "does not know" about the existence of r_{o2}.

EXERCISE

8.7   Use the feedback method to find the voltage gain V_o/V_s, the input resistance R_{in}, and the output resistance R_{out} of the inverting op-amp configuration of Fig. E8.7. Let the op amp have open-loop gain $\mu = 10^4$ V/V, $R_{id} = 100 \ \text{k}\Omega$, and $r_o = 1 \ \text{k}\Omega$. (*Hint:* The feedback is of the shunt–shunt type.)

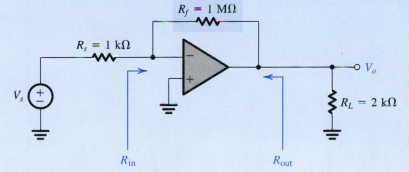

FIGURE E8.7

Ans. -870 V/V; $150 \ \Omega$; $92 \ \Omega$

TABLE 8.1 Summary of Relationships for the Four Feedback-Amplifier Topologies

Feedback Amplifier	x_i	x_o	x_f	x_s	A	β	A_f	Source Form	Loading of Feedback Network Is Obtained — At Input	Loading of Feedback Network Is Obtained — At Output	To Find β, Apply to Port 2 of Feedback Network	Z_{if}	Z_{of}	Refer to Figs.
Series–shunt (voltage amplifier)	V_i	V_o	V_f	V_s	$\dfrac{V_o}{V_i}$	$\dfrac{V_f}{V_o}$	$\dfrac{V_o}{V_s}$	Thévenin	By short-circuiting port 2 of feedback network	By open-circuiting port 1 of feedback network	a voltage, and find the open-circuit voltage at port 1	$Z_i(1+A\beta)$	$\dfrac{Z_o}{1+A\beta}$	8.4(a) 8.8 8.10 8.11
Shunt–series (current amplifier)	I_i	I_o	I_f	I_s	$\dfrac{I_o}{I_i}$	$\dfrac{I_f}{I_o}$	$\dfrac{I_o}{I_s}$	Norton	By open-circuiting port 2 of feedback network	By short-circuiting port 1 of feedback network	a current, and find the short-circuit current at port 1	$\dfrac{Z_i}{1+A\beta}$	$Z_o(1+A\beta)$	8.4(b) 8.22 8.23 8.24
Series–series (transconductance amplifier)	V_i	I_o	V_f	V_s	$\dfrac{I_o}{V_i}$	$\dfrac{V_f}{I_o}$	$\dfrac{I_o}{V_s}$	Thévenin	By open-circuiting port 2 of feedback network	By open-circuiting port 1 of feedback network	a current, and find the open-circuit voltage at port 1	$Z_i(1+A\beta)$	$Z_o(1+A\beta)$	8.4(c) 8.13 8.15 8.16
Shunt–shunt (transresistance amplifier)	I_i	V_o	I_f	I_s	$\dfrac{V_o}{I_i}$	$\dfrac{I_f}{V_o}$	$\dfrac{V_o}{I_s}$	Norton	By short-circuiting port 2 of feedback network	By short-circuiting port 1 of feedback network	a voltage, and find the short-circuit current at port 1	$\dfrac{Z_i}{1+A\beta}$	$\dfrac{Z_o}{1+A\beta}$	8.4(d) 8.18 8.19 8.20

8.6.4 Summary of Results

Table 8.1 provides a summary of the rules and relationships employed in the analysis of the four types of feedback amplifier.

 ## 8.7 DETERMINING THE LOOP GAIN

We have already seen that the loop gain $A\beta$ is a very important quantity that characterizes a feedback loop. Furthermore, in the following sections it will be shown that $A\beta$ determines whether the feedback amplifier is stable (as opposed to oscillatory). In this section, we shall describe an alternative approach to the determination of loop gain.

8.7.1 An Alternative Approach for Finding $A\beta$

Consider first the general feedback amplifier shown in Fig. 8.1. Let the external source x_s be set to zero. Open the feedback loop by breaking the connection of x_o to the feedback network and apply a test signal x_t. We see that the signal at the output of the feedback network is $x_f = \beta x_t$; that at the input of the basic amplifier is $x_i = -\beta x_t$; and the signal at the output of the amplifier, where the loop was broken, will be $x_o = -A\beta x_t$. It follows that the loop gain $A\beta$ is given by the negative of the ratio of the *returned* signal to the applied test signal; that is, $A\beta = -x_o/x_t$. It should also be obvious that this applies regardless of where the loop is broken.

However, in breaking the feedback loop of a practical amplifier circuit, we must ensure that the conditions that existed prior to breaking the loop do not change. This is achieved by terminating the loop where it is opened with an impedance equal to that seen before the loop was broken. To be specific, consider the conceptual feedback loop shown in Fig. 8.26(a). If we break the loop at XX′, and apply a test voltage V_t to the terminals thus created to the left of XX′, the terminals at the right of XX′ should be loaded with an impedance Z_t as shown in Fig. 8.26(b). The impedance Z_t is equal to that previously seen looking to the left of XX′. The loop gain $A\beta$ is then determined from

$$A\beta = -\frac{V_r}{V_t} \tag{8.49}$$

Finally, it should be noted that in some cases it may be convenient to determine $A\beta$ by applying a test current I_t and finding the returned current signal I_r. In this case, $A\beta = -I_r/I_t$.

An alternative equivalent method for determining $A\beta$ (see Rosenstark, 1986) that is usually convenient to employ especially in SPICE simulations is as follows: As before, the loop is broken at a convenient point. Then the open-circuit transfer function T_{oc} is determined as indicated in Fig. 8.26(c), and the short-circuit transfer function T_{sc} is determined as shown in Fig. 8.26(d). These two transfer functions are then combined to obtain the loop gain $A\beta$,

$$A\beta = -1 \Big/ \left(\frac{1}{T_{oc}} + \frac{1}{T_{sc}}\right) \tag{8.50}$$

This method is particularly useful when it is not easy to determine the termination impedance Z_t.

To illustrate the process of determining loop gain, we consider the feedback loop shown in Fig. 8.27(a). This feedback loop represents both the inverting and the noninverting op-amp

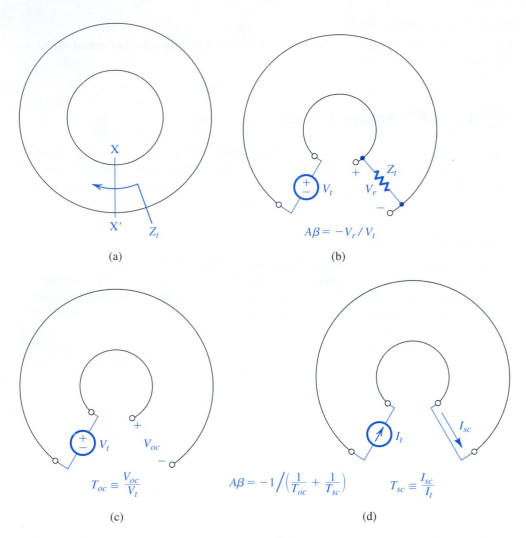

FIGURE 8.26 A conceptual feedback loop is broken at XX' and a test voltage V_t is applied. The impedance Z_t is equal to that previously seen looking to the left of XX'. The loop gain $A\beta = -V_r/V_t$, where V_r is the *returned* voltage. As an alternative, $A\beta$ can be determined by finding the open-circuit transfer function T_{oc}, as in (c), and the short-circuit transfer function T_{sc}, as in (d), and combining them as indicated.

configurations. Using a simple equivalent circuit model for the op amp we obtain the circuit of Fig. 8.27(b). Examination of this circuit reveals that a convenient place to break the loop is at the input terminals of the op amp. The loop, broken in this manner, is shown in Fig. 8.27(c) with a test signal V_t applied to the right-hand-side terminals and a resistance R_{id} terminating the left-hand-side terminals. The returned voltage V_r is found by inspection as

$$V_r = -\mu V_1 \frac{\{R_L//[R_2 + R_1//(R_{id} + R)]\}}{\{R_L//[R_2 + R_1//(R_{id} + R)]\} + r_o} \frac{[R_1//(R_{id} + R)]}{[R_1//(R_{id} + R)] + R_2} \frac{R_{id}}{R_{id} + R} \quad (8.51)$$

This equation can be used directly to find the loop gain $L = A\beta = -V_r/V_t = -V_r/V_1$.

Since the loop gain L is generally a function of frequency, it is usual to call it **loop transmission** and denote it by $L(s)$ or $L(j\omega)$.

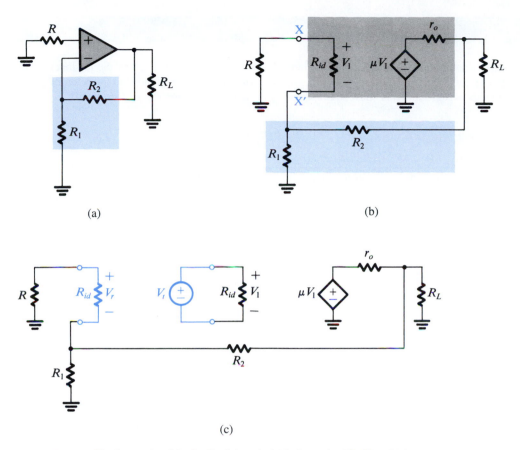

FIGURE 8.27   The loop gain of the feedback loop in (a) is determined in (b) and (c).

8.7.2 Equivalence of Circuits from a Feedback-Loop Point of View

From the study of circuit theory we know that the poles of a circuit are independent of the external excitation. In fact the poles, or the natural modes (which is a more appropriate name), are determined by setting the external excitation to zero. It follows that the poles of a feedback amplifier depend only on the feedback loop. This will be confirmed in a later section, where we show that the characteristic equation (whose roots are the poles) is completely determined by the loop gain. Thus, a given feedback loop may be used to generate a number of circuits having the same poles but different transmission zeros. The closed-loop gain and the transmission zeros depend on how and where the input signal is injected into the loop.

As an example consider the feedback loop of Fig. 8.27(a). This loop can be used to generate the noninverting op-amp circuit by feeding the input voltage signal to the terminal of R that is connected to ground; that is, we lift this terminal off ground and connect it to V_s. The same feedback loop can be used to generate the inverting op-amp circuit by feeding the input voltage signal to the terminal of R_1 that is connected to ground.

Recognition of the fact that two or more circuits are equivalent from a feedback-loop point of view is very useful because (as will be shown in Section 8.8) stability is a function of the loop. Thus one needs to perform the stability analysis only once for a given loop.

In Chapter 12 we shall employ the concept of loop equivalence in the synthesis of active filters.

EXERCISES

8.8 Consider the feedback amplifier whose equivalent circuit is shown in Fig. 8.25(b). Break the feedback loop at the input to transistor Q_1; that is, apply a test voltage V_t to the base of Q_1, and find the returned voltage that appears across $r_{\pi1}$. Thus show that the loop gain is given (neglecting r_{o2} to simplify the analysis) by:

$$A\beta = g_{m1}\{r_{o1}//R_{C1}//[r_{\pi2} + (h_{fe}+1)(R_{E2}//(R_f + (R_B//R_s//r_{\pi1})))]\}$$

$$\times \frac{R_{E2}//(R_f + (R_B//R_s//r_{\pi1}))}{R_{E2}//(R_f + (R_B//R_s//r_{\pi1})) + r_{e2}} \frac{R_s//R_B//r_{\pi1}}{(R_s//R_B//r_{\pi1}) + R_f}$$

Using the component values in Example 8.4, find the value of $A\beta$ and compare it with the value found in Example 8.4.

Ans. 49.3 versus 52.1

8.9 Find the numerical value of $A\beta$ for the amplifier in Exercise 8.7.

Ans. 6589 V/V

 ## 8.8 THE STABILITY PROBLEM

8.8.1 Transfer Function of the Feedback Amplifier

In a feedback amplifier such as that represented by the general structure of Fig. 8.1, the open-loop gain A is generally a function of frequency, and it should therefore be more accurately called the **open-loop transfer function,** $A(s)$. Also, we have been assuming for the most part that the feedback network is resistive and hence that the feedback factor β is constant, but this need not be always the case. We shall therefore assume that in the general case the **feedback transfer function** is $\beta(s)$. It follows that the **closed-loop transfer function** $A_f(s)$ is given by

$$A_f(s) = \frac{A(s)}{1 + A(s)\beta(s)} \tag{8.52}$$

To focus attention on the points central to our discussion in this section, we shall assume that the amplifier is direct-coupled with constant dc gain A_0 and with poles and zeros occurring in the high-frequency band. Also, for the time being let us assume that at low frequencies $\beta(s)$ reduces to a constant value. Thus at low frequencies the loop gain $A(s)\beta(s)$ becomes a constant, which should be a positive number; otherwise the feedback would not be negative. The question then is: What happens at higher frequencies?

For physical frequencies $s = j\omega$, Eq. (8.52) becomes

$$A_f(j\omega) = \frac{A(j\omega)}{1 + A(j\omega)\beta(j\omega)} \tag{8.53}$$

Thus the loop gain $A(j\omega)\beta(j\omega)$ is a complex number that can be represented by its magnitude and phase,

$$L(j\omega) \equiv A(j\omega)\beta(j\omega)$$
$$= |A(j\omega)\beta(j\omega)|e^{j\phi(\omega)} \tag{8.54}$$

It is the manner in which the loop gain varies with frequency that determines the stability or instability of the feedback amplifier. To appreciate this fact, consider the frequency at which the phase angle $\phi(\omega)$ becomes 180°. At this frequency, ω_{180}, the loop gain $A(j\omega)\beta(j\omega)$ will be a real number with a negative sign. Thus at this frequency the feedback will become positive. If at $\omega = \omega_{180}$ the magnitude of the loop gain is less than unity, then from Eq. (8.53) we see that the closed-loop gain $A_f(j\omega)$ will be greater than the open-loop gain $A(j\omega)$, since the denominator of Eq. (8.53) will be smaller than unity. Nevertheless, the feedback amplifier will be stable.

On the other hand, if at the frequency ω_{180} the magnitude of the loop gain is equal to unity, it follows from Eq. (8.53) that $A_f(j\omega)$ will be infinite. This means that the amplifier will have an output for zero input; this is by definition an **oscillator**. To visualize how this feedback loop may oscillate, consider the general loop of Fig. 8.1 with the external input x_s set to zero. Any disturbance in the circuit, such as the closure of the power-supply switch, will generate a signal $x_i(t)$ at the input to the amplifier. Such a noise signal usually contains a wide range of frequencies, and we shall now concentrate on the component with frequency $\omega = \omega_{180}$, that is, the signal $X_i \sin(\omega_{180}t)$. This input signal will result in a feedback signal given by

$$X_f = A(j\omega_{180})\beta(j\omega_{180})X_i = -X_i$$

Since X_f is further multiplied by -1 in the summer block at the input, we see that the feedback causes the signal X_i at the amplifier input to be *sustained*. That is, from this point on, there will be sinusoidal signals at the amplifier input and output of frequency ω_{180}. Thus the amplifier is said to oscillate at the frequency ω_{180}.

The question now is: What happens if at ω_{180} the magnitude of the loop gain is greater than unity? We shall answer this question, not in general, but for the restricted yet very important class of circuits in which we are interested here. The answer, which is not obvious from Eq. (8.53), is that the circuit will oscillate, and the oscillations will grow in amplitude until some nonlinearity (which is always present in some form) reduces the magnitude of the loop gain to exactly unity, at which point sustained oscillations will be obtained. This mechanism for starting oscillations by using positive feedback with a loop gain greater than unity, and then using a nonlinearity to reduce the loop gain to unity at the desired amplitude, will be exploited in the design of sinusoidal oscillators in Chapter 13. Our objective here is just the opposite: Now that we know how oscillations could occur in a negative-feedback amplifier, we wish to find methods to prevent their occurrence.

8.8.2 The Nyquist Plot

The Nyquist plot is a formalized approach for testing for stability based on the discussion above. It is simply a polar plot of loop gain with frequency used as a parameter. Figure 8.28 shows such a plot. Note that the radial distance is $|A\beta|$ and the angle is the phase angle ϕ. The solid-line plot is for positive frequencies. Since the loop gain—and for that matter any gain function of a physical network—has a magnitude that is an even function of frequency and a phase that is an odd function of frequency, the $A\beta$ plot for negative frequencies (shown in Fig. 8.28 as a broken line) can be drawn as a mirror image through the Re axis.

The Nyquist plot intersects the negative real axis at the frequency ω_{180}. Thus, if this intersection occurs to the left of the point $(-1, 0)$, we know that the magnitude of loop gain at this frequency is greater than unity and the amplifier will be unstable. On the other hand, if the intersection occurs to the right of the point $(-1, 0)$ the amplifier will be stable. It follows that if the Nyquist plot *encircles* the point $(-1, 0)$ then the amplifier will be

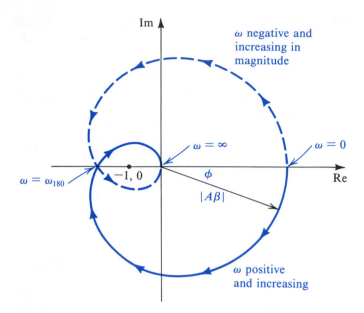

FIGURE 8.28 The Nyquist plot of an unstable amplifier.

unstable. It should be mentioned, however, that this statement is a simplified version of the **Nyquist criterion;** nevertheless, it applies to all the circuits in which we are interested. For the full theory behind the Nyquist method and for details of its application, consult Haykin (1970).

EXERCISE

8.10 Consider a feedback amplifier for which the open-loop transfer function $A(s)$ is given by

$$A(s) = \left(\frac{10}{1 + s/10^4} \right)^3$$

Let the feedback factor β be a constant independent of frequency. Find the frequency ω_{180} at which the phase shift is 180°. Then, show that the feedback amplifier will be stable if the feedback factor β is less than a critical value β_{cr} and unstable if $\beta \geq \beta_{cr}$, and find the value of β_{cr}.

Ans. $\omega_{180} = \sqrt{3} \times 10^4$ rad/s; $\beta_{cr} = 0.008$

 8.9 EFFECT OF FEEDBACK ON THE AMPLIFIER POLES

The amplifier frequency response and stability are determined directly by its poles. We shall therefore investigate the effect of feedback on the poles of the amplifier.[6]

---

[6] For a brief review of poles and zeros and related concepts, refer to Appendix E.

8.9.1 Stability and Pole Location

We shall begin by considering the relationship between stability and pole location. For an amplifier or any other system to be stable, its poles should lie in the left half of the s plane. A pair of complex-conjugate poles on the $j\omega$ axis gives rise to sustained sinusoidal oscillations. Poles in the right half of the s plane give rise to growing oscillations.

To verify the statement above, consider an amplifier with a pole pair at $s = \sigma_0 \pm j\omega_n$. If this amplifier is subjected to a disturbance, such as that caused by closure of the power-supply switch, its transient response will contain terms of the form

$$v(t) = e^{\sigma_0 t}[e^{+j\omega_n t} + e^{-j\omega_n t}] = 2e^{\sigma_0 t}\cos(\omega_n t) \qquad (8.55)$$

This is a sinusoidal signal with an envelope $e^{\sigma_0 t}$. Now if the poles are in the left half of the s plane, then σ_0 will be negative and the oscillations will decay exponentially toward zero, as shown in Fig. 8.29(a), indicating that the system is stable. If, on the other hand, the poles are in the right half-plane, then σ_0 will be positive, and the oscillations will grow exponentially

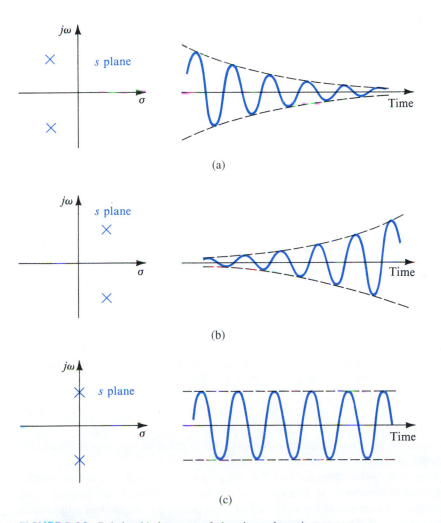

FIGURE 8.29 Relationship between pole location and transient response.

(until some nonlinearity limits their growth), as shown in Fig. 8.29(b). Finally, if the poles are on the $j\omega$ axis, then σ_0 will be zero and the oscillations will be sustained, as shown in Fig. 8.29(c).

Although the discussion above is in terms of complex-conjugate poles, it can be shown that the existence of any right-half-plane poles results in instability.

8.9.2 Poles of the Feedback Amplifier

From the closed-loop transfer function in Eq. (8.52), we see that the poles of the feedback amplifier are the zeros of $1 + A(s)\beta(s)$. That is, the feedback-amplifier poles are obtained by solving the equation

$$1 + A(s)\beta(s) = 0 \tag{8.56}$$

which is called the **characteristic equation** of the feedback loop. It should therefore be apparent that applying feedback to an amplifier changes its poles.

In the following, we shall consider how feedback affects the amplifier poles. For this purpose we shall assume that the open-loop amplifier has real poles and no finite zeros (i.e., all the zeros are at $s = \infty$). This will simplify the analysis and enable us to focus our attention on the fundamental concepts involved. We shall also assume that the feedback factor β is independent of frequency.

8.9.3 Amplifier with a Single-Pole Response

Consider first the case of an amplifier whose open-loop transfer function is characterized by a single pole:

$$A(s) = \frac{A_0}{1 + s/\omega_P} \tag{8.57}$$

The closed-loop transfer function is given by

$$A_f(s) = \frac{A_0/(1 + A_0\beta)}{1 + s/\omega_P(1 + A_0\beta)} \tag{8.58}$$

Thus the feedback moves the pole along the negative real axis to a frequency ω_{Pf},

$$\omega_{Pf} = \omega_P(1 + A_0\beta) \tag{8.59}$$

This process is illustrated in Fig. 8.30(a). Figure 8.30(b) shows Bode plots for $|A|$ and $|A_f|$. Note that while at low frequencies the difference between the two plots is $20 \log(1 + A_0\beta)$, the two curves coincide at high frequencies. One can show that this indeed is the case by approximating Eq. (8.58) for frequencies $\omega \gg \omega_P(1 + A_0\beta)$:

$$A_f(s) \simeq \frac{A_0\omega_P}{s} \simeq A(s) \tag{8.60}$$

Physically speaking, at such high frequencies the loop gain is much smaller than unity and the feedback is ineffective.

Figure 8.30(b) clearly illustrates the fact that applying negative feedback to an amplifier results in extending its bandwidth at the expense of a reduction in gain. Since the pole of the closed-loop amplifier never enters the right half of the s plane, the single-pole amplifier is stable for any value of β. Thus this amplifier is said to be **unconditionally stable.** This

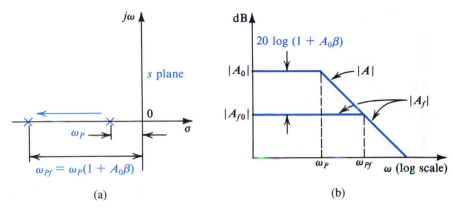

FIGURE 8.30 Effect of feedback on **(a)** the pole location and **(b)** the frequency response of an amplifier having a single-pole open-loop response.

result, however, is hardly surprising, since the phase lag associated with a single-pole response can never be greater than 90°. Thus the loop gain never achieves the 180° phase shift required for the feedback to become positive.

EXERCISE

8.11 An op amp having a single-pole rolloff at 100 Hz and a low-frequency gain of 10^5 is operated in a feedback loop with $\beta = 0.01$. What is the factor by which feedback shifts the pole? To what frequency? If β is changed to a value that results in a closed-loop gain of +1, to what frequency does the pole shift?

Ans. 1001; 100.1 kHz; 10 MHz

8.9.4 Amplifier with Two-Pole Response

Consider next an amplifier whose open-loop transfer function is characterized by two real-axis poles:

$$A(s) = \frac{A_0}{(1 + s/\omega_{P1})(1 + s/\omega_{P2})} \tag{8.61}$$

In this case, the closed-loop poles are obtained from $1 + A(s)\beta = 0$, which leads to

$$s^2 + s(\omega_{P1} + \omega_{P2}) + (1 + A_0\beta)\omega_{P1}\omega_{P2} = 0 \tag{8.62}$$

Thus the closed-loop poles are given by

$$s = -\tfrac{1}{2}(\omega_{P1} + \omega_{P2}) \pm \tfrac{1}{2}\sqrt{(\omega_{P1} + \omega_{P2})^2 - 4(1 + A_0\beta)\omega_{P1}\omega_{P2}} \tag{8.63}$$

From Eq. (8.63) we see that as the loop gain $A_0\beta$ is increased from zero, the poles are brought closer together. Then a value of loop gain is reached at which the poles become coincident. If the loop gain is further increased, the poles become complex conjugate and move along a vertical line. Figure 8.31 shows the locus of the poles for increasing loop gain. This plot is called a **root-locus diagram,** where "root" refers to the fact that the poles are the roots of the characteristic equation.

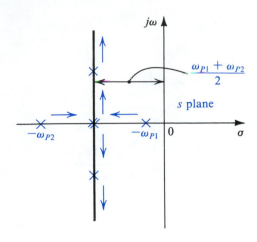

FIGURE 8.31 Root-locus diagram for a feedback amplifier whose open-loop transfer function has two real poles.

From the root-locus diagram of Fig. 8.31 we see that this feedback amplifier also is unconditionally stable. Again, this result should come as no surprise; the maximum phase shift of $A(s)$ in this case is 180° (90° per pole), but this value is reached at $\omega = \infty$. Thus there is no finite frequency at which the phase shift reaches 180°.

Another observation to make on the root-locus diagram of Fig. 8.31 is that the open-loop amplifier might have a dominant pole, but this is not necessarily the case for the closed-loop amplifier. The response of the closed-loop amplifier can, of course, always be plotted once the poles have been found from Eq. (8.63). As is the case with second-order responses generally, the closed-loop response can show a peak (see Chapter 12). To be more specific, the characteristic equation of a second-order network can be written in the standard form

$$s^2 + s\frac{\omega_0}{Q} + \omega_0^2 = 0 \tag{8.64}$$

where ω_0 is called the **pole frequency** and Q is called **pole Q factor.** The poles are complex if Q is greater than 0.5. A geometric interpretation for ω_0 and Q of a pair of complex-conjugate poles is given in Fig. 8.32, from which we note that ω_0 is the radial distance of the poles from the origin and that Q indicates the distance of the poles from the $j\omega$ axis. Poles on the $j\omega$ axis have $Q = \infty$.

By comparing Eqs. (8.62) and (8.64) we obtain the Q factor for the poles of the feedback amplifier as

$$Q = \frac{\sqrt{(1 + A_0\beta)\omega_{P1}\omega_{P2}}}{\omega_{P1} + \omega_{P2}} \tag{8.65}$$

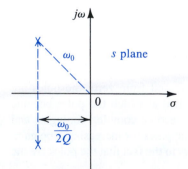

FIGURE 8.32 Definition of ω_0 and Q of a pair of complex-conjugate poles.

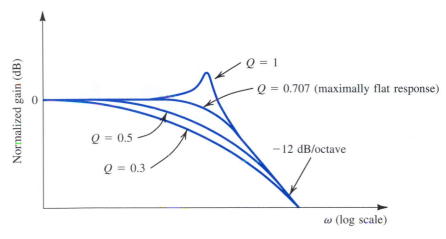

FIGURE 8.33 Normalized gain of a two-pole feedback amplifier for various values of Q. Note that Q is determined by the loop gain according to Eq. (8.65).

From the study of second-order network responses in Chapter 12, it will be seen that the response of the feedback amplifier under consideration shows no peaking for $Q \leq 0.707$. The boundary case corresponding to $Q = 0.707$ (poles at $45°$ angles) results in the **maximally flat** response. Figure 8.33 shows a number of possible responses obtained for various values of Q (or, correspondingly, various values of $A_0\beta$).

EXERCISE

8.12 An amplifier with a low-frequency gain of 100 and poles at 10^4 rad/s and 10^6 rad/s is incorporated in a negative-feedback loop with feedback factor β. For what value of β do the poles of the closed-loop amplifier coincide? What is the corresponding Q of the resulting second-order system? For what value of β is a maximally flat response achieved? What is the low-frequency closed-loop gain in the maximally flat case?

Ans. 0.245; 0.5; 0.5; 1.96 V/V

EXAMPLE 8.5

As an illustration of some of the ideas just discussed, we consider the positive-feedback circuit shown in Fig. 8.34(a). Find the loop transmission $L(s)$ and the characteristic equation. Sketch a root-locus diagram for varying K, and find the value of K that results in a maximally flat response, and the value of K that makes the circuit oscillate. Assume that the amplifier has infinite input impedance and zero output impedance.

Solution

To obtain the loop transmission, we short-circuit the signal source and break the loop at the amplifier input. We then apply a test voltage V_t and find the returned voltage V_r, as indicated in

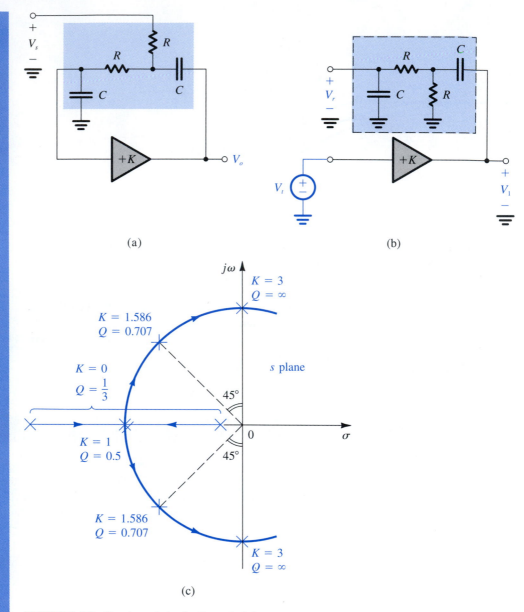

FIGURE 8.34 Circuits and plot for Example 8.5.

Fig. 8.34(b). The loop transmission $L(s) \equiv A(s)\beta(s)$ is given by

$$L(s) = -\frac{V_r}{V_t} = -KT(s) \tag{8.66}$$

where $T(s)$ is the transfer function of the two-port RC network shown inside the broken-line box in Fig. 8.34(b):

$$T(s) \equiv \frac{V_r}{V_1} = \frac{s(1/CR)}{s^2 + s(3/CR) + (1/CR)^2} \tag{8.67}$$

Thus,

$$L(s) = \frac{-s(K/CR)}{s^2 + s(3/CR) + (1/CR)^2} \tag{8.68}$$

The characteristic equation is

$$1 + L(s) = 0 \tag{8.69}$$

that is,

$$s^2 + s\frac{3}{CR} + \left(\frac{1}{CR}\right)^2 - s\frac{K}{CR} = 0$$

$$s^2 + s\frac{3-K}{CR} + \left(\frac{1}{CR}\right)^2 = 0 \tag{8.70}$$

By comparing this equation to the standard form of the second-order characteristic equation (Eq. 8.64) we see that the pole frequency ω_0 is given by

$$\omega_0 = \frac{1}{CR} \tag{8.71}$$

and the Q factor is

$$Q = \frac{1}{3-K} \tag{8.72}$$

Thus, for $K = 0$ the poles have $Q = \frac{1}{3}$ and are therefore located on the negative real axis. As K is increased the poles are brought closer together and eventually coincide ($Q = 0.5$, $K = 1$). Further increasing K results in the poles becoming complex and conjugate. The root locus is then a circle because the radial distance ω_0 remains constant (Eq. 8.71) independent of the value of K.

The maximally flat response is obtained when $Q = 0.707$, which results when $K = 1.586$. In this case the poles are at 45° angles, as indicated in Fig. 8.34(c). The poles cross the $j\omega$ axis into the right half of the s plane at the value of K that results in $Q = \infty$, that is, $K = 3$. Thus for $K \geq 3$ this circuit becomes unstable. This might appear to contradict our earlier conclusion that the feedback amplifier with a second-order response is unconditionally stable. Note, however, that the circuit in this example is quite different from the negative-feedback amplifier that we have been studying. Here we have an amplifier with a positive gain K and a feedback network whose transfer function $T(s)$ is frequency dependent. This feedback is in fact *positive,* and the circuit will oscillate at the frequency for which the phase of $T(j\omega)$ is zero.

Example 8.5 illustrates the use of feedback (positive feedback in this case) to move the poles of an RC network from their negative real-axis locations to complex-conjugate locations. One can accomplish the same task using negative feedback, as the root-locus diagram of Fig. 8.31 demonstrates. The process of pole control is the essence of *active-filter design,* as will be discussed in Chapter 12.

8.9.5 Amplifiers with Three or More Poles

Figure 8.35 shows the root-locus diagram for a feedback amplifier whose open-loop response is characterized by three poles. As indicated, increasing the loop gain from zero moves the highest-frequency pole outward while the two other poles are brought closer together. As $A_0\beta$ is increased further, the two poles become coincident and then become complex and conjugate. A value of $A_0\beta$ exists at which this pair of complex-conjugate poles enters the right half of the s plane, thus causing the amplifier to become unstable.

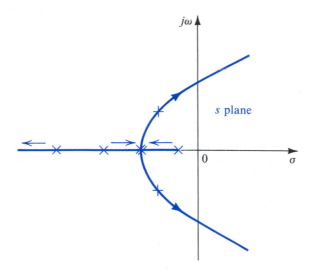

FIGURE 8.35 Root-locus diagram for an amplifier with three poles. The arrows indicate the pole movement as $A_0\beta$ is increased.

This result is not entirely unexpected, since an amplifier with three poles has a phase shift that reaches $-270°$ as ω approaches ∞. Thus there exists a finite frequency ω_{180}, at which the loop gain has $180°$ phase shift.

From the root-locus diagram of Fig. 8.35, we observe that one can always maintain amplifier stability by keeping the loop gain $A_0\beta$ smaller than the value corresponding to the poles entering the right half-plane. In terms of the Nyquist diagram, the critical value of $A_0\beta$ is that for which the diagram passes through the $(-1, 0)$ point. Reducing $A_0\beta$ below this value causes the Nyquist plot to shrink and thus intersect the negative real axis to the right of the $(-1, 0)$ point, indicating stable amplifier performance. On the other hand, increasing $A_0\beta$ above the critical value causes the Nyquist plot to expand, thus encircling the $(-1, 0)$ point and indicating unstable performance.

For a given open-loop gain A_0 the conclusions above can be stated in terms of the feedback factor β. That is, there exists a *maximum value* for β above which the feedback amplifier becomes unstable. Alternatively, we can state that there exists a *minimum value* for the closed-loop gain A_{f0} below which the amplifier becomes unstable. To obtain lower values of closed-loop gain one needs therefore to alter the loop transfer function $L(s)$. This is the process known as *frequency compensation*. We shall study the theory and techniques of frequency compensation in Section 8.11.

Before leaving this section we point out that construction of the root-locus diagram for amplifiers having three or more poles as well as finite zeros is an involved process for which a systematic procedure exists. However, such a procedure will not be presented here, and the interested reader should consult Haykin (1970). Although the root-locus diagram provides the amplifier designer with considerable insight, other, simpler techniques based on Bode plots can be effectively employed, as will be explained in Section 8.10.

EXERCISE

8.13 Consider a feedback amplifier for which the open-loop transfer function $A(s)$ is given by

$$A(s) = \left(\frac{10}{1 + s/10^4}\right)^3$$

Let the feedback factor β be frequency independent. Find the closed-loop poles as functions of β, and show that the root locus is that of Fig. E8.13. Also find the value of β at which the amplifier becomes unstable. (*Note:* This is the same amplifier that was considered in Exercise 8.10.)

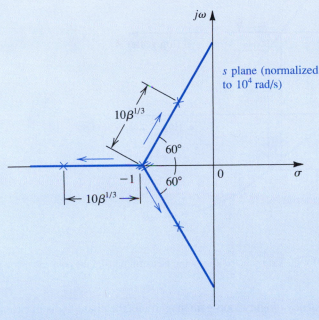

FIGURE E8.13

Ans. See Fig. E8.13; $\beta_{critical} = 0.008$

8.10 STABILITY STUDY USING BODE PLOTS

8.10.1 Gain and Phase Margins

From Sections 8.8 and 8.9 we know that one can determine whether a feedback amplifier is or is not stable by examining its loop gain $A\beta$ as a function of frequency. One of the simplest and most effective means for doing this is through the use of a Bode plot for $A\beta$, such as the one shown in Fig. 8.36. (Note that because the phase approaches $-360°$, the network examined is a fourth-order one.) The feedback amplifier whose loop gain is plotted in Fig. 8.36 will be stable, since at the frequency of $180°$ phase shift, ω_{180}, the magnitude of the loop gain is less than unity (negative dB). The difference between the value of $|A\beta|$ at ω_{180} and unity, called the **gain margin,** is usually expressed in decibels. The gain margin represents the amount by which the loop gain can be increased while stability is maintained. Feedback amplifiers are usually designed to have sufficient gain margin to allow for the inevitable changes in loop gain with temperature, time, and so on.

Another way to investigate the stability and to express its degree is to examine the Bode plot at the frequency for which $|A\beta| = 1$, which is the point at which the magnitude plot crosses the 0-dB line. If at this frequency the phase angle is less (in magnitude) than $180°$, then the amplifier is stable. This is the situation illustrated in Fig. 8.36. The difference between the phase angle at this frequency and $180°$ is termed the **phase margin.** On the

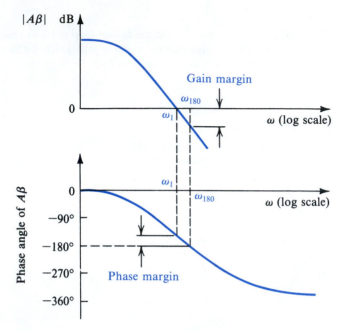

FIGURE 8.36 Bode plot for the loop gain $A\beta$ illustrating the definitions of the gain and phase margins.

other hand, if at the frequency of unity loop-gain magnitude, the phase lag is in excess of 180°, the amplifier will be unstable.

EXERCISE

8.14 Consider an op amp having a single-pole open-loop response with $A_0 = 10^5$ and $f_P = 10$ Hz. Let the op amp be ideal otherwise (infinite input impedance, zero output impedance, etc.). If this amplifier is connected in the noninverting configuration with a nominal low-frequency closed-loop gain of 100, find the frequency at which $|A\beta| = 1$. Also, find the phase margin.
Ans. 10^4 Hz; 90°

8.10.2 Effect of Phase Margin on Closed-Loop Response

Feedback amplifiers are normally designed with a phase margin of at least 45°. The amount of phase margin has a profound effect on the shape of the closed-loop gain response. To see this relationship, consider a feedback amplifier with a large low-frequency loop gain, $A_0\beta \gg 1$. It follows that the closed-loop gain at low frequencies is approximately $1/\beta$. Denoting the frequency at which the magnitude of loop gain is unity by ω_1 we have (refer to Fig. 8.36)

$$A(j\omega_1)\beta = 1 \times e^{-j\theta} \tag{8.73a}$$

where

$$\theta = 180° - \text{phase margin} \tag{8.73b}$$

At ω_1 the closed-loop gain is

$$A_f(j\omega_1) = \frac{A(j\omega_1)}{1 + A(j\omega_1)\beta} \tag{8.74}$$

Substituting from Eq. (8.73a) gives

$$A_f(j\omega_1) = \frac{(1/\beta)e^{-j\theta}}{1 + e^{-j\theta}} \tag{8.75}$$

Thus the magnitude of the gain at ω_1 is

$$|A_f(j\omega_1)| = \frac{1/\beta}{|1 + e^{-j\theta}|} \tag{8.76}$$

For a phase margin of $45°$, $\theta = 135°$; and we obtain

$$|A_f(j\omega_1)| = 1.3\frac{1}{\beta} \tag{8.77}$$

That is, the gain peaks by a factor of 1.3 above the low-frequency value of $1/\beta$. This peaking increases as the phase margin is reduced, eventually reaching ∞ when the phase margin is zero. Zero phase margin, of course, implies that the amplifier can sustain oscillations [poles on the $j\omega$ axis; Nyquist plot passing through $(-1, 0)$].

EXERCISE

8.15 Find the closed-loop gain at ω_1 relative to the low-frequency gain when the phase margin is $30°$, $60°$, and $90°$.
Ans. 1.93; 1; 0.707

8.10.3 An Alternative Approach for Investigating Stability

Investigating stability by constructing Bode plots for the loop gain $A\beta$ can be a tedious and time-consuming process, especially if we have to investigate the stability of a given amplifier for a variety of feedback networks. An alternative approach, which is much simpler, is to construct a Bode plot for the open-loop gain $A(j\omega)$ only. Assuming for the time being that β is independent of frequency, we can plot $20 \log(1/\beta)$ as a horizontal straight line on the same plane used for $20 \log|A|$. The difference between the two curves will be

$$20 \log|A(j\omega)| - 20 \log\frac{1}{\beta} = 20 \log|A\beta| \tag{8.78}$$

which is the loop gain (in dB). We may therefore study stability by examining the difference between the two plots. If we wish to evaluate stability for a different feedback factor we simply draw another horizontal straight line at the level $20 \log(1/\beta)$.

To illustrate, consider an amplifier whose open-loop transfer function is characterized by three poles. For simplicity let the three poles be widely separated—say, at 0.1 MHz, 1 MHz, and 10 MHz, as shown in Fig. 8.37. Note that because the poles are widely separated, the phase is approximately $-45°$ at the first pole frequency, $-135°$ at the second, and $-225°$ at the third. The frequency at which the phase of $A(j\omega)$ is $-180°$ lies on the -40-dB/decade segment, as indicated in Fig. 8.37.

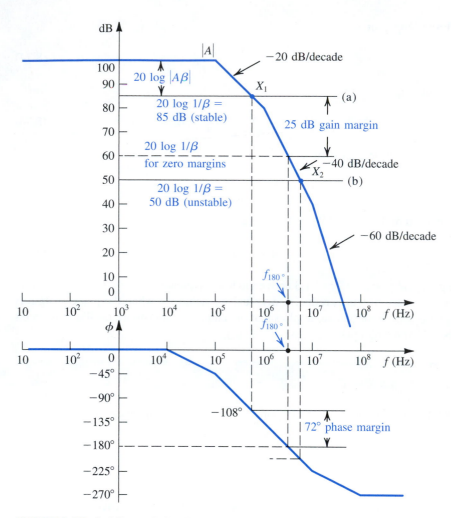

FIGURE 8.37 Stability analysis using Bode plot of |A|.

The open-loop gain of this amplifier can be expressed as

$$A = \frac{10^5}{(1 + jf/10^5)(1 + jf/10^6)(1 + jf/10^7)} \tag{8.79}$$

from which $|A|$ can be easily determined for any frequency f (in Hz), and the phase can be obtained as

$$\phi = -[\tan^{-1}(f/10^5) + \tan^{-1}(f/10^6) + \tan^{-1}(f/10^7)] \tag{8.80}$$

The magnitude and phase graphs shown in Fig. 8.37 are obtained using the method for constructing Bode plots (Appendix E). These graphs provide approximate values for important amplifier parameters, with more exact values obtainable from Eqs. (8.79) and (8.80). For example, the frequency f_{180} at which the phase angle is 180° can be found from Fig. 8.37 to be approximately 3.2×10^6 Hz. Using this value as a starting point, a more exact value can be found by trial and error using Eq. (8.80). The result is $f_{180} = 3.34 \times 10^6$ Hz. At this frequency, Eq. (8.79) gives a gain magnitude of 58.2 dB, which is reasonably close to the approximate value of 60 dB given by Fig. 8.37.

Consider next the straight line labeled (a) in Fig. 8.37. This line represents a feedback factor for which $20 \log(1/\beta) = 85$ dB, which corresponds to $\beta = 5.623 \times 10^{-5}$ and a closed-loop gain of 83.6 dB. Since the loop gain is the difference between the $|A|$ curve and the $1/\beta$ line, the point of intersection X_1 corresponds to the frequency at which $|A\beta| = 1$. Using the graphs of Fig. 8.37, this frequency can be found to be approximately 5.6×10^5 Hz. A more exact value of 4.936×10^5 can be obtained using the transfer-function equations. At this frequency the phase angle is approximately $-108°$. Thus the closed-loop amplifier, for which $20 \log(1/\beta) = 85$ dB, will be stable with a phase margin of $72°$. The gain margin can be easily obtained from Fig. 8.37; it is 25 dB.

Next, suppose that we wish to use this amplifier to obtain a closed-loop gain of 50-dB nominal value. Since $A_0 = 100$ dB, we see that $A_0\beta \gg 1$ and $20 \log(A_0\beta) \simeq 50$ dB, resulting in $20 \log(1/\beta) \simeq 50$ dB. To see whether this closed-loop amplifier is or is not stable, we draw line (b) in Fig. 8.37 with a height of 50 dB. This line intersects the open-loop gain curve at point X_2, where the corresponding phase is greater than $180°$. Thus the closed-loop amplifier with 50-dB gain will be unstable.

In fact, it can easily be seen from Fig. 8.37 that the *minimum* value of $20 \log(1/\beta)$ that can be used, with the resulting amplifier being stable, is 60 dB. In other words, the minimum value of stable closed-loop gain obtained with this amplifier is approximately 60 dB. At this value of gain, however, the amplifier may still oscillate, since no margin is left to allow for possible changes in gain.

Since the $180°$-phase point always occurs on the -40-dB/decade segment of the Bode plot for $|A|$, a rule of thumb to guarantee stability is as follows: *The closed-loop amplifier will be stable if the $20 \log(1/\beta)$ line intersects the $20 \log|A|$ curve at a point on the -20-dB/decade segment.* Following this rule ensures that a phase margin of at least $45°$ is obtained. For the example of Fig. 8.37, the rule implies that the maximum value of β is 10^{-4}, which corresponds to a closed-loop gain of approximately 80 dB.

The rule of thumb above can be generalized for the case in which β is a function of frequency. The general rule states that *at the intersection of $20 \log[1/|\beta(j\omega)|]$ and $20 \log|A(j\omega)|$ the difference of slopes* (called the **rate of closure**) *should not exceed 20 dB/decade.*

EXERCISE

8.16 Consider an op amp whose open-loop gain is identical to that of Fig. 8.37. Assume that the op amp is ideal otherwise. Let the op amp be connected as a differentiator. Use the rule of thumb above to show that for stable performance the differentiator time constant should be greater than 159 ms. [*Hint:* Recall that for a differentiator, the Bode plot for $1/|\beta(j\omega)|$ has a slope of $+20$ dB/decade and intersects the 0-dB line at $1/\tau$, where τ is the differentiator time constant.]

 ## 8.11 FREQUENCY COMPENSATION

In this section, we shall discuss methods for modifying the open-loop transfer function $A(s)$ of an amplifier having three or more poles so that the closed-loop amplifier is stable for any desired value of closed-loop gain.

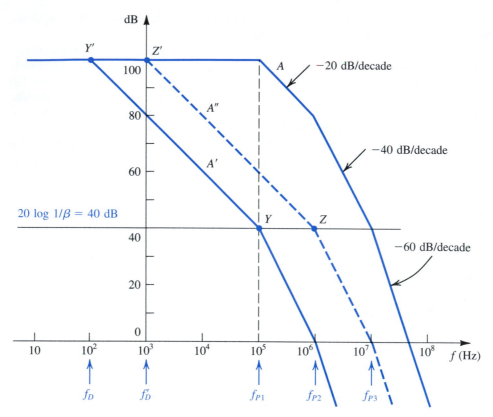

FIGURE 8.38 Frequency compensation for $\beta = 10^{-2}$. The response labeled A' is obtained by introducing an additional pole at f_D. The A'' response is obtained by moving the original low-frequency pole to f_D'.

8.11.1 Theory

The simplest method of frequency compensation consists of introducing a new pole in the function $A(s)$ at a sufficiently low frequency, f_D, such that the modified open-loop gain, $A'(s)$, intersects the $20 \log(1/|\beta|)$ curve with a slope difference of 20 dB/decade. As an example, let it be required to compensate the amplifier whose $A(s)$ is shown in Fig. 8.38 such that closed-loop amplifiers with β as high as 10^{-2} (i.e., closed-loop gains as low as approximately 40 dB) will be stable. First, we draw a horizontal straight line at the 40-dB level to represent $20 \log(1/\beta)$, as shown in Fig. 8.38. We then locate point Y on this line at the frequency of the first pole, f_{P1}. From Y we draw a line with −20-dB/decade slope and determine the point at which this line intersects the dc gain line, point Y'. This latter point gives the frequency f_D of the new pole that has to be introduced in the open-loop transfer function.

The compensated open-loop response $A'(s)$ is indicated in Fig. 8.38. It has four poles: at f_D, f_{P1}, f_{P2}, and f_{P3}. Thus $|A'|$ begins to roll off with a slope of −20 dB/decade at f_D. At f_{P1} the slope changes to −40 dB/decade, at f_{P2} it changes to −60 dB/decade, and so on. Since the $20 \log(1/\beta)$ line intersects the $20 \log|A'|$ curve at point Y on the −20-dB/decade segment, the closed-loop amplifier with this β value (or lower values) will be stable.

A serious disadvantage of this compensation method is that at most frequencies the open-loop gain has been drastically reduced. This means that at most frequencies the amount of feedback available will be small. Since all the advantages of negative feedback are directly proportional to the amount of feedback, the performance of the compensated amplifier has been impaired.

Careful examination of Fig. 8.38 shows that the gain $A'(s)$ is low because of the pole at f_{P1}. If we can somehow eliminate this pole, then—rather than locating point Y, drawing YY', and so on—we can start from point Z (at the frequency of the second pole) and draw the line ZZ'. This would result in the open-loop curve $A''(s)$, which shows considerably higher gain than $A'(s)$.

Although it is not possible to eliminate the pole at f_{P1}, it is usually possible to shift that pole from $f = f_{P1}$ to $f = f'_D$. This makes the pole dominant and eliminates the need for introducing an additional lower-frequency pole, as will be explained next.

8.11.2 Implementation

We shall now address the question of implementing the frequency-compensation scheme discussed above. The amplifier circuit normally consists of a number of cascaded gain stages, with each stage responsible for one or more of the transfer-function poles. Through manual and/or computer analysis of the circuit, one identifies which stage introduces each of the important poles f_{P1}, f_{P2}, and so on. For the purpose of our discussion, assume that the first pole f_{P1} is introduced at the interface between the two cascaded differential stages shown in Fig. 8.39(a). In Fig. 8.39(b) we show a simple small-signal model of the circuit at this interface. Current source I_x represents the output signal current of the $Q_1 - Q_2$ stage. Resistance R_x and capacitance C_x represent the total resistance and capacitance between the two nodes B and B'. It follows that the pole f_{P1} is given by

$$f_{P1} = \frac{1}{2\pi C_x R_x} \tag{8.81}$$

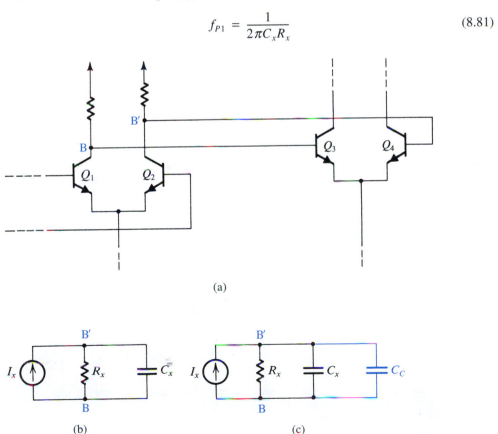

(a)

(b)   (c)

FIGURE 8.39 **(a)** Two cascaded gain stages of a multistage amplifier. **(b)** Equivalent circuit for the interface between the two stages in (a). **(c)** Same circuit as in (b) but with a compensating capacitor C_C added. Note that the analysis here applies equally well to MOS amplifiers.

Let us now connect the compensating capacitor C_C between nodes B and B′. This will result in the modified equivalent circuit shown in Fig. 8.39(c) from which we see that the pole introduced will no longer be at f_{P1}; rather, the pole can be at any desired lower frequency f'_D:

$$f'_D = \frac{1}{2\pi(C_x + C_C)R_x} \tag{8.82}$$

We thus conclude that one can select an appropriate value for C_C to shift the pole frequency from f_{P1} to the value f'_D determined by point Z' in Fig. 8.38.

At this juncture it should be pointed out that adding the capacitor C_C will usually result in changes in the location of the other poles (those at f_{P2} and f_{P3}). One might therefore need to calculate the new location of f_{P2} and perform a few iterations to arrive at the required value for C_C.

A disadvantage of this implementation method is that the required value of C_C is usually quite large. Thus if the amplifier to be compensated is an IC op amp, it will be difficult, and probably impossible, to include this compensating capacitor on the IC chip. (As pointed out in Chapter 6 and in Appendix A, the maximum practical size of a monolithic capacitor is about 100 pF.) An elegant solution to this problem is to connect the compensating capacitor in the feedback path of an amplifier stage. Because of the Miller effect, the compensating capacitance will be multiplied by the stage gain, resulting in a much larger effective capacitance. Furthermore, as explained later, another unexpected benefit accrues.

8.11.3 Miller Compensation and Pole Splitting

Figure 8.40(a) shows one gain stage in a multistage amplifier. For simplicity, the stage is shown as a common-emitter amplifier, but in practice it can be a more elaborate circuit. In the feedback path of this common-emitter stage we have placed a compensating capacitor C_f.

Figure 8.40(b) shows a simplified equivalent circuit of the gain stage of Fig. 8.40(a). Here R_1 and C_1 represent the total resistance and total capacitance between node B and ground. Similarly, R_2 and C_2 represent the total resistance and total capacitance between node C and ground. Furthermore, it is assumed that C_1 includes the Miller component due to capacitance C_μ, and C_2 includes the input capacitance of the succeeding amplifier stage. Finally, I_i represents the output signal current of the preceding stage.

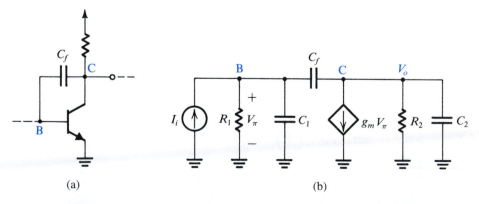

(a) (b)

FIGURE 8.40 **(a)** A gain stage in a multistage amplifier with a compensating capacitor connected in the feedback path and **(b)** an equivalent circuit. Note that although a BJT is shown, the analysis applies equally well to the MOSFET case.

In the absence of the compensating capacitor C_f, we can see from Fig. 8.40(b) that there are two poles—one at the input and one at the output. Let us assume that these two poles are f_{P1} and f_{P2} of Fig. 8.38; thus,

$$f_{P1} = \frac{1}{2\pi C_1 R_1} \qquad f_{P2} = \frac{1}{2\pi C_2 R_2} \qquad (8.83)$$

With C_f present, analysis of the circuit yields the transfer function

$$\frac{V_o}{I_i} = \frac{(sC_f - g_m)R_1 R_2}{1 + s[C_1 R_1 + C_2 R_2 + C_f(g_m R_1 R_2 + R_1 + R_2)] + s^2[C_1 C_2 + C_f(C_1 + C_2)]R_1 R_2} \qquad (8.84)$$

The zero is usually at a much higher frequency than the dominant pole, and we shall neglect its effect. The denominator polynomial $D(s)$ can be written in the form

$$D(s) = \left(1 + \frac{s}{\omega'_{P1}}\right)\left(1 + \frac{s}{\omega'_{P2}}\right) = 1 + s\left(\frac{1}{\omega'_{P1}} + \frac{1}{\omega'_{P2}}\right) + \frac{s^2}{\omega'_{P1}\omega'_{P2}} \qquad (8.85)$$

where ω'_{P1} and ω'_{P2} are the new frequencies of the two poles. Normally one of the poles will be dominant; $\omega'_{P1} \ll \omega'_{P2}$. Thus,

$$D(s) \simeq 1 + \frac{s}{\omega'_{P1}} + \frac{s^2}{\omega'_{P1}\omega'_{P2}} \qquad (8.86)$$

Equating the coefficients of s in the denominator of Eq. (8.84) and in Eq. (8.86) results in

$$\omega'_{P1} = \frac{1}{C_1 R_1 + C_2 R_2 + C_f(g_m R_1 R_2 + R_1 + R_2)}$$

which can be approximated by

$$\omega'_{P1} \simeq \frac{1}{g_m R_2 C_f R_1} \qquad (8.87)$$

To obtain ω'_{P2} we equate the coefficients of s^2 in the denominator of Eq. (8.84) and in Eq. (8.86) and use Eq. (8.87):

$$\omega'_{P2} \simeq \frac{g_m C_f}{C_1 C_2 + C_f(C_1 + C_2)} \qquad (8.88)$$

From Eqs. (8.87) and (8.88), we see that as C_f is increased, ω'_{P1} is reduced and ω'_{P2} is increased. This action is referred to as **pole splitting**. Note that the increase in ω'_{P2} is highly beneficial; it allows us to move point Z (see Fig. 8.38) further to the right, thus resulting in higher compensated open-loop gain. Finally, note from Eq. (8.87) that C_f is multiplied by the Miller-effect factor $g_m R_2$, thus resulting in a much larger capacitance, $g_m R_2 C_f$. In other words, the required value of C_f will be much smaller than that of C_C in Fig. 8.39.

EXAMPLE 8.6

Consider an op amp whose open-loop transfer function is identical to that shown in Fig. 8.37. We wish to compensate this op amp so that the closed-loop amplifier with resistive feedback is stable for any gain (i.e., for β up to unity). Assume that the op-amp circuit includes a stage such as that of Fig. 8.40 with $C_1 = 100$ pF, $C_2 = 5$ pF, and $g_m = 40$ mA/V, that the pole at f_{P1} is caused by the input circuit of that stage, and that the pole at f_{P2} is introduced by the output circuit. Find the value of the compensating capacitor for two cases: either if it is connected between the input node B and ground or in the feedback path of the transistor.

Solution

First we determine R_1 and R_2 from

$$f_{P1} = 0.1 \text{ MHz} = \frac{1}{2\pi C_1 R_1}$$

Thus,

$$R_1 = \frac{10^5}{2\pi} \ \Omega$$

$$f_{P2} = 1 \text{ MHz} = \frac{1}{2\pi C_2 R_2}$$

Thus,

$$R_2 = \frac{10^5}{\pi} \ \Omega$$

If a compensating capacitor C_C is connected across the input terminals of the transistor stage, then the frequency of the first pole changes from f_{P1} to f_D':

$$f_D' = \frac{1}{2\pi(C_1 + C_C)R_1}$$

The second pole remains unchanged. The required value for f_D' is determined by drawing a -20-dB/ decade line from the 1-MHz frequency point on the $20 \log(1/\beta) = 20 \log 1 = 0$ dB line. This line will intersect the 100-dB dc gain line at 10 Hz. Thus,

$$f_D' = 10 \text{ Hz} = \frac{1}{2\pi(C_1 + C_C)R_1}$$

which results in $C_C \simeq 1 \ \mu\text{F}$, which is quite large and certainly cannot be included on the IC chip.

Next, if a compensating capacitor C_f is connected in the feedback path of the transistor, then both poles change location to the values given by Eqs. (8.87) and (8.88):

$$f_{P1}' \simeq \frac{1}{2\pi g_m R_2 C_f R_1} \qquad f_{P2}' \simeq \frac{g_m C_f}{2\pi[C_1 C_2 + C_f(C_1 + C_2)]} \qquad (8.89)$$

To determine where we should locate the first pole, we need to know the value of f_{P2}'. As an approximation, let us assume that $C_f \gg C_2$, which enables us to obtain

$$f_{P2}' \simeq \frac{g_m}{2\pi(C_1 + C_2)} = 60.6 \text{ MHz}$$

Thus it appears that this pole will move to a frequency higher than f_{P3} (which is 10 MHz). Let us therefore assume that the second pole will be at f_{P3}. This requires that the first pole be located at

$$f_{P1}' = \frac{f_{P3}}{A_0} = \frac{10^7 \text{ Hz}}{10^5} = 100 \text{ Hz}$$

Thus,

$$f_{P1}' = 100 \text{ Hz} = \frac{1}{2\pi g_m R_2 C_f R_1}$$

which results in $C_f = 78.5$ pF. Although this value is indeed much greater than C_2, we can determine the location of the pole f_{P2}' from Eq. (8.89), which yields $f_{P2}' = 57.2$ MHz, confirming that this pole has indeed been moved past f_{P3}.

We conclude that using Miller compensation not only results in a much smaller compensating capacitor but, owing to pole splitting, also enables us to place the dominant pole a decade higher in frequency. This results in a wider bandwidth for the compensated op amp.

8.17 A multipole amplifier having a first pole at 1 MHz and an open-loop gain of 100 dB is to be compensated for closed-loop gains as low as 20 dB by the introduction of a new dominant pole. At what frequency must the new pole be placed?

Ans. 100 Hz

8.18 For the amplifier described in Exercise 8.17, rather than introducing a new dominant pole, we can use additional capacitance at the circuit node at which the first pole is formed to reduce the frequency of the first pole. If the frequency of the second pole is 10 MHz and if it remains unchanged while additional capacitance is introduced as mentioned, find the frequency to which the first pole must be lowered so that the resulting amplifier is stable for closed-loop gains as low as 20 dB. By what factor must the capacitance at the controlling node be increased?

Ans. 1000 Hz; 1000

 ## 8.12 SPICE SIMULATION EXAMPLE

We conclude this chapter by presenting an example that illustrates the use of SPICE in the analysis of feedback circuits.

EXAMPLE 8.7

DETERMINING THE LOOP GAIN USING SPICE

This example illustrates the use of SPICE to compute the loop gain $A\beta$. To be able to compare results, we shall use the same shunt–series feedback amplifier considered in Example 8.4 and redrawn in Fig. 8.41. This, however, does not limit the generality of the methods described.

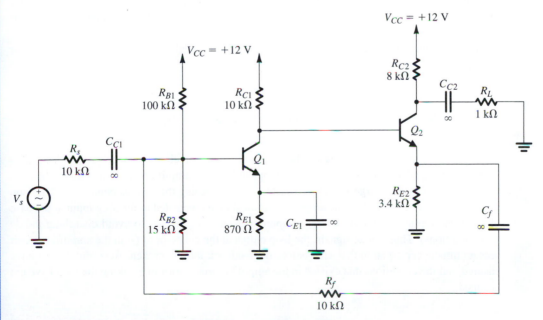

FIGURE 8.41 Circuit of the shunt–series feedback amplifier in Example 8.4.

To compute the loop gain, we set the input signal V_s to zero, and we choose to break the feedback loop between the collector of Q_1 and the base of Q_2. However, in breaking the feedback loop, we must ensure that the following two conditions that existed prior to breaking the feedback loop do not change: (1) the dc bias situation and (2) the ac signal termination.

To break the feedback loop without disturbing the dc bias conditions of the circuit, we insert a large inductor L_{break}, as shown in Fig. 8.42(a). Using a value of, say, $L_{break} = 1$ GH will ensure that the loop is opened for ac signals while keeping dc bias conditions unchanged.

To break the feedback loop without disturbing the signal termination conditions, we must load the loop output at the collector of Q_1 with a termination impedance Z_t whose value is equal to the impedance seen looking into the loop input at the base of Q_2. Furthermore, to avoid disturbing the dc bias conditions, Z_t must be connected to the collector of Q_1 via a large coupling capacitor. However, it is not always easy to determine the value of the termination impedance Z_t. So, we will describe two simulation methods to compute the loop gain without explicitly determining Z_t.

Method 1 *Using the open-circuit and short-circuit transfer functions*

As described in Section 8.7, the loop gain can be expressed as

$$A\beta = -1\Big/\left(\frac{1}{T_{oc}} + \frac{1}{T_{sc}}\right)$$

where T_{oc} is the open-circuit voltage transfer function and T_{sc} is the short-circuit current transfer function.

The circuit for determining T_{oc} is shown in Fig. 8.42(a). Here, an ac test signal voltage V_t is applied to the loop input at the base of Q_2 via a large coupling capacitor (having a value of, say, 1 kF) to avoid disturbing the dc bias conditions. Then,

$$T_{oc} = \frac{V_{oc}}{V_t}$$

where V_{oc} is the ac open-circuit output voltage at the collector of Q_1.

In the circuit for determining T_{sc} (Fig. 8.42b), an ac test signal current I_t is applied to the loop input at the base of Q_2. Note that a coupling capacitor is not needed in this case because the ac current source appears as an open circuit at dc, and, hence, does not disturb the dc bias conditions.

The loop output at the collector of Q_1 is ac short-circuited to ground via a large capacitor C_{to}. Then,

$$T_{sc} = \frac{I_{sc}}{I_t}$$

where I_{sc} is the ac short-circuit output current at the collector of Q_1.

Method 2 *Using a replica circuit*

As shown in Fig. 8.43, a replica of the feedback amplifier circuit can be simply used as a termination impedance. Here, the feedback loops of both the amplifier circuit and the replica circuit are broken using a large inductor L_{break} to avoid disturbing the dc bias conditions. The loop output at the collector of Q_1 in the amplifier circuit is then connected to the loop input at the base of Q_2 in the replica circuit via a large coupling capacitor C_{to} (again, to avoid disturbing the dc bias conditions). Thus, for ac signals, the loop output at the collector of Q_1 in the amplifier circuit sees an impedance equal to that seen before the feedback loop is broken. Accordingly, we have ensured that the conditions that existed in the amplifier circuit prior to breaking the loop have not changed.

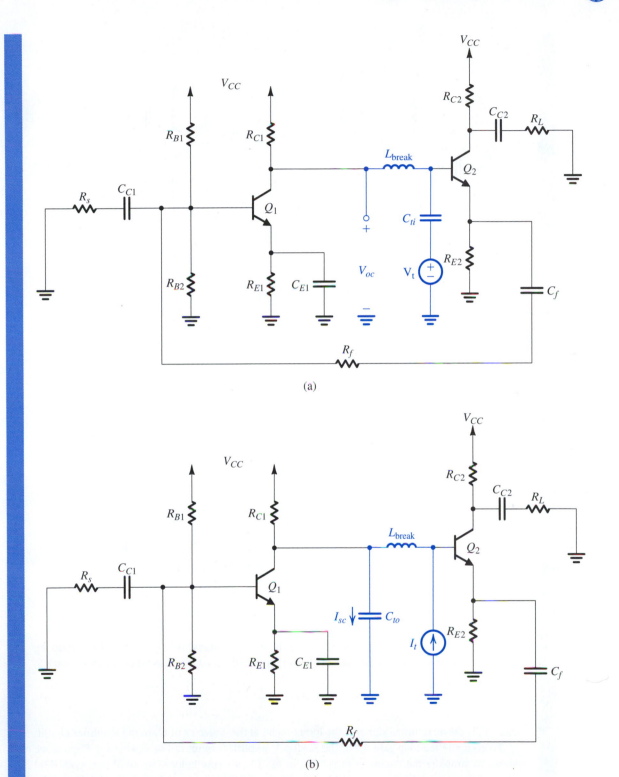

(a)

(b)

FIGURE 8.42 Circuits for simulating **(a)** the open-circuit voltage transfer function T_{oc} and **(b)** the short-circuit current transfer function T_{sc} of the feedback amplifier in Fig. 8.41 for the purpose of computing its loop gain.

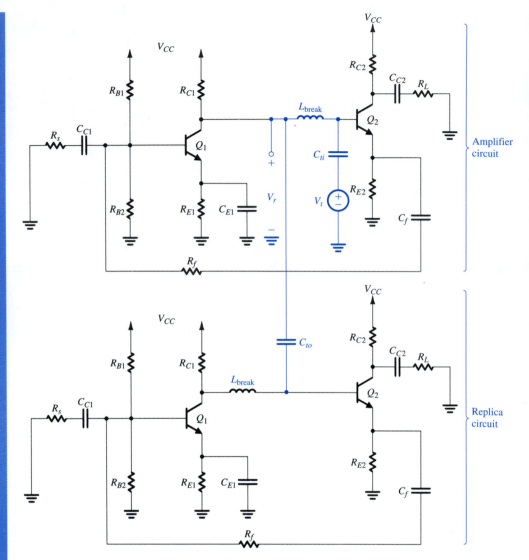

FIGURE 8.43 Circuit for simulating the loop gain of the feedback amplifier in Fig. 8.41 using the replica-circuit method.

Next, to determine the loop gain $A\beta$, we apply an ac test-signal voltage V_t via a large coupling capacitor C_{ti} to the loop input at the base of Q_2 in the amplifier circuit. Then, as described in Section 8.7,

$$A\beta = -\frac{V_r}{V_t}$$

where V_r is the ac returned signal at the loop output, at the collector of Q_1 in the amplifier circuit.

To compute the loop gain $A\beta$ of the feedback amplifier circuit in Fig. 8.41 using PSpice, we choose to simulate the circuit in Fig. 8.43. In the PSpice simulations, we used part Q2N3904 (whose SPICE model is given in Table 5.9) for the BJTs, and we set L_{break} to be 1 GH and the coupling and bypass capacitors to be 1 kF. The magnitude and phase of $A\beta$ are plotted in Fig. 8.44, from which we see that the feedback amplifier has a gain margin of 53.7 dB and a phase margin of 88.7°.

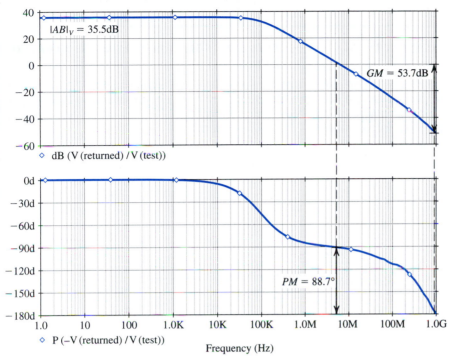

FIGURE 8.44 (a) Magnitude and (b) phase of the loop gain $A\beta$ of the feedback-amplifier circuit in Fig. 8.41.

SUMMARY

- Negative feedback is employed to make the amplifier gain less sensitive to component variations; to control input and output impedances; to extend bandwidth; to reduce nonlinear distortion; and to enhance signal-to-noise (and signal-to-interference) ratio.

- The advantages above are obtained at the expense of a reduction in gain and at the risk of the amplifier becoming unstable (that is, oscillating). The latter problem is solved by careful design.

- For each of the four basic types of amplifier, there is an appropriate feedback topology. The four topologies, together with their analysis procedure and their effects on input and output impedances, are summarized in Table 8.1 on page 830.

- The key feedback parameters are the loop gain ($A\beta$), which for negative feedback must be a positive dimensionless number, and the amount of feedback ($1 + A\beta$). The latter directly determines gain reduction, gain desensitivity, bandwidth extension, and changes in Z_i and Z_o.

- Since A and β are in general frequency dependent, the poles of the feedback amplifier are obtained by solving the characteristic equation $1 + A(s)\beta(s) = 0$.

- For the feedback amplifier to be stable, its poles must all be in the left half of the s plane.

- Stability is guaranteed if at the frequency for which the phase angle of $A\beta$ is 180° (i.e., ω_{180}), $|A\beta|$ is less than unity; the amount by which it is less than unity, expressed in decibels, is the gain margin. Alternatively, the amplifier is stable if, at the frequency at which $|A\beta| = 1$, the phase angle is less than 180°; the difference is the phase margin.

- The stability of a feedback amplifier can be analyzed by constructing a Bode plot for $|A|$ and superimposing on it a plot for $1/|\beta|$. Stability is guaranteed if the two plots intersect with a difference in slope no greater than 6 dB/octave.

- To make a given amplifier stable for a given feedback factor β, the open-loop frequency response is suitably modified by a process known as frequency compensation.

- A popular method for frequency compensation involves connecting a feedback capacitor across an inverting stage in the amplifier. This causes the pole formed at the input of the amplifier stage to shift to a lower frequency and thus become dominant, while the pole formed at the output of the amplifier stage is moved to a very high frequency and thus becomes unimportant. This process is known as pole splitting.

PROBLEMS

SECTION 8.1: THE GENERAL FEEDBACK STRUCTURE

8.1 A negative-feedback amplifier has a closed-loop gain $A_f = 100$ and an open-loop gain $A = 10^5$. What is the feedback factor β? If a manufacturing error results in a reduction of A to 10^3, what closed-loop gain results? What is the percentage change in A_f corresponding to this factor of 100 reduction in A?

8.2 Repeat Exercise 8.1, parts (b) through (e), for $A = 100$.

8.3 Repeat Exercise 8.1, parts (b) through (e), for $A_f = 10^3$. For part (d) use $V_s = 0.01$ V.

8.4 The noninverting buffer op-amp configuration shown in Fig. P8.4 provides a direct implementation of the feedback loop of Fig. 8.1. Assuming that the op amp has infinite input resistance and zero output resistance, what is β? If $A = 100$, what is the closed-loop voltage gain? What is the amount of feedback (in dB)? For $V_s = 1$ V, find V_o and V_i. If A decreases by 10%, what is the corresponding decrease in A_f?

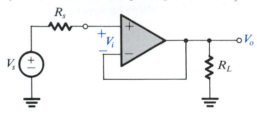

FIGURE P8.4

8.5 In a particular circuit represented by the block diagram of Fig. 8.1, a signal of 1 V from the source results in a difference signal of 10 mV being provided to the amplifying element A, and 10 V applied to the load. For this arrangement, identify the values of A and β that apply.

8.6 Find the open-loop gain, the loop gain, and the amount of feedback of a voltage amplifier for which A_f and $1/\beta$ differ by (a) 1%, (b) 5%, (c) 10%, (d) 50%.

8.7 In a particular amplifier design, the β network consists of a linear potentiometer for which β is 0.00 at one end, 1.00 at the other end, and 0.50 in the middle. As the potentiometer is adjusted, find the three values of closed-loop gain that result when the amplifier open-loop gain is (a) 1 V/V, (b) 10 V/V, (c) 100 V/V, (d) 10,000 V/V.

8.8 A newly constructed feedback amplifier undergoes a performance test with the following results: With the feedback connection removed, a source signal of 2 mV is required to provide a 10-V output to the load; with the feedback connected, a 10-V output requires a 200-mV source signal. For this amplifier, identify values of A, β, $A\beta$, the closed-loop gain, and the amount of feedback (in dB).

SECTION 8.2: SOME PROPERTIES OF NEGATIVE FEEDBACK

8.9 For the negative-feedback loop of Fig. 8.1, find the loop gain $A\beta$ for which the sensitivity of closed-loop gain to open-loop gain [i.e., $(dA_f/A_f)/(dA/A)$] is –20 dB. For what value of $A\beta$ does the sensitivity become 1/2?

D8.10 It is required to design an amplifier with a gain of 100 that is accurate to within ±1%. You have available amplifier stages with a gain of 1000 that is accurate to within ±30%. Provide a design that uses a number of these gain stages in cascade, with each stage employing negative feedback of an appropriate amount. Obviously, your design should use the lowest possible number of stages while meeting specification.

8.11 In a feedback amplifier for which $A = 10^4$ and $A_f = 10^3$, what is the gain-desensitivity factor? Find A_f exactly, and approximately using Eq. (8.8), in the two cases: (a) A drops by 10% and (b) A drops by 30%.

8.12 Consider an amplifier having a midband gain A_M and a low-frequency response characterized by a pole at $s = -\omega_L$ and a zero at $s = 0$. Let the amplifier be connected in a negative-feedback loop with a feedback factor β. Find an expression for the midband gain and the lower 3-dB frequency of the closed-loop amplifier. By what factor have both changed?

D*8.13 It is required to design an amplifier to have a nominal closed-loop gain of 10 V/V using a battery-operated amplifier whose gain reduces to half its normal full-battery value over the life of the battery. If only 2% drop in closed-loop gain is desired, what nominal open-loop amplifier gain must be used in the design? (Note that since the change in A is large, it is inaccurate to use differentials.) What value of β should be chosen? If component-value variation in the β network may produce as much as a ±1% variation in β, to what value must A be raised to ensure the required minimum gain?

8.14 A capacitively coupled amplifier has a midband gain of 100, a single high-frequency pole at 10 kHz, and a single low-frequency pole at 100 Hz. Negative feedback is employed so that the midband gain is reduced to 10. What are the upper and lower 3-dB frequencies of the closed-loop gain?

D8.15** It is required to design a dc amplifier with a low-frequency gain of 1000 and a 3-dB frequency of 0.5 MHz. You have available gain stages with a gain of 1000 but with a dominant high-frequency pole at 10 kHz. Provide a design that employs a number of such stages in cascade, each with negative feedback of an appropriate amount. Use identical stages. [*Hint:* When negative feedback of an amount $(1 + A\beta)$

is employed around a gain stage, its x-dB frequency is increased by the factor $(1 + A\beta)$.]

D8.16 Design a supply-ripple-reduced power amplifier, for which an output stage having a gain of 0.9 V/V and ± 1-V output supply ripple is used. A closed-loop gain of 10 V/V is desired. What is the gain of a low-ripple preamplifier needed to reduce the output ripple to ± 100 mV? To ± 10 mV? To ± 1 mV? For each case, specify the value required for the feedback factor β.

D8.17 Design a feedback amplifier that has a closed-loop gain of 100 V/V and is relatively insensitive to change in basic-amplifier gain. In particular, it should provide a reduction in A_f to 99 V/V for a reduction in A to one-tenth its nominal value. What is the required loop gain? What nominal value of A is required? What value of β should be used? What would the closed-loop gain become if A were increased tenfold? If A were made infinite?

D8.18 A feedback amplifier is to be designed using a feedback loop connected around a two-stage amplifier. The first stage is a direct-coupled small-signal amplifier with a high upper 3-dB frequency. The second stage is a power-output stage with a midband gain of 10 V/V and upper- and lower 3-dB frequencies of 8 kHz and 80 Hz, respectively. The feedback amplifier should have a midband gain of 100 V/V and an upper 3-dB frequency of 40 kHz. What is the required gain of the small-signal amplifier? What value of β should be used? What does the lower 3-dB frequency of the overall amplifier become?

*8.19 The complementary BJT follower shown in Fig. P8.19(a) has the approximate transfer characteristic shown in Fig. P8.19(b). Observe that for $-0.7\ V \le v_I \le +0.7\ V$, the output is zero. This "dead band" leads to crossover

distortion (see Section 14.3). Consider this follower driven by the output of a differential amplifier of gain 100 whose positive-input terminal is connected to the input signal source v_S and whose negative-input terminal is connected to the emitters of the follower. Sketch the transfer characteristic v_O versus v_S of the resulting feedback amplifier. What are the limits of the dead band and what are the gains outside the dead band?

D8.20 A particular amplifier has a nonlinear transfer characteristic that can be approximated as follows:

(a) For small input signals, $|v_I| \le 10$ mV, $v_O/v_I = 10^3$
(b) For intermediate input signals, $10\ \text{mV} \le |v_I| \le 50$ mV, $v_O/v_I = 10^2$
(c) For large input signals, $|v_I| \ge 50$ mV, the output saturates

If the amplifier is connected in a negative-feedback loop, find the feedback factor β that reduces the factor-of-10 change in gain (occurring at $|v_I| = 10$ mV) to only a 10% change. What is the transfer characteristic of the amplifier with feedback?

SECTION 8.3: THE FOUR BASIC FEEDBACK TOPOLOGIES

8.21 A series–shunt feedback amplifier representable by Fig. 8.4(a) and using an ideal basic voltage amplifier operates with $V_s = 100$ mV, $V_f = 95$ mV, and $V_o = 10$ V. What are the corresponding values of A and β? Include the correct units for each.

8.22 A shunt–series feedback amplifier representable by Fig. 8.4(b) and using an ideal basic current amplifier operates with $I_s = 100\ \mu$A, $I_f = 95\ \mu$A, and $I_o = 10$ mA. What are the corresponding values of A and β? Include the correct units for each:

8

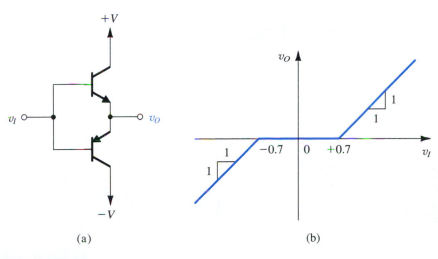

(a)   (b)

FIGURE P8.19

*8.23 Consider the shunt–series feedback amplifier of Fig. 8.5:

(a) For R_s, r_{o1}, and r_{o2} assumed very large, use direct circuit analysis (as opposed to feedback analysis) to show that the overall current gain is given by

$$A_f \equiv \frac{I_o}{I_s} = -\frac{R_1 + g_{m1}R_{L1}(R_1 + R_2)}{R_1 + \dfrac{1}{g_{m2}} + g_{m1}R_{L1}R_1}$$

and the input resistance is

$$R_{in} = R_1 + R_2 + A_f R_1$$

Hence, find approximate expressions for A_f and R_{in} for the case in which $g_{m1}R_{L1} \gg 1$ and $(1/g_{m2}) \ll R_1$.

(b) Evaluate A_f and R_{in}, exactly and approximately, for the case in which $g_{m1}R_{L1} = 100$, $R_1 = 10$ kΩ, $R_2 = 90$ kΩ, and $g_{m2} = 5$ mA/V.

(c) Since the negative feedback forces the input terminal of the amplifier toward ground, the value of β can be determined approximately as the current divider ratio of the (R_1, R_2) network. Find β and show that the approximate expression for A_f found above is simply $1/\beta$.

8.24 A series–series feedback circuit representable by Fig. 8.4(c) and using an ideal transconductance amplifier operates with $V_s = 100$ mV, $V_f = 95$ mV, and $I_o = 10$ mA. What are the corresponding values of A and β? Include the correct units for each.

8.25 A shunt–shunt feedback circuit representable by Fig. 8.4(d) and using an ideal transresistance amplifier operates with $I_s = 100$ μA, $I_f = 95$ μA, and $V_o = 10$ V. What are the corresponding values of A and β? Include the correct units for each.

*8.26 For each of the op-amp circuits shown in Fig. P8.26, identify the feedback topology and indicate the output variable being sampled and the feedback signal. In each case, assuming the op amp to be ideal, find an expression for β, and hence find A_f.

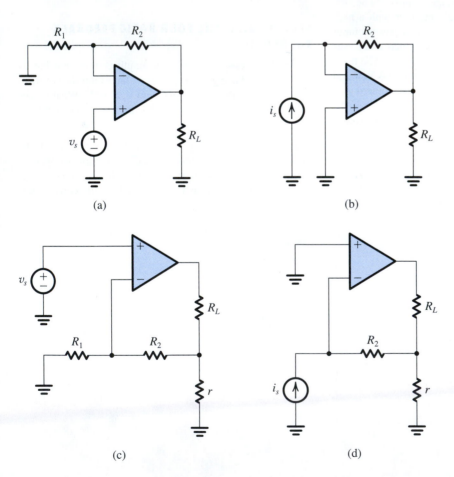

(a)

(b)

(c)

(d)

FIGURE P8.26

SECTION 8.4: THE SERIES–SHUNT FEEDBACK AMPLIFIER

8.27 A series–shunt feedback amplifier employs a basic amplifier with input and output resistances each of 1 kΩ and gain $A = 2000$ V/V. The feedback factor $\beta = 0.1$ V/V. Find the gain A_f, the input resistance R_{if}, and the output resistance R_{of} of the closed-loop amplifier.

8.28 For a particular amplifier connected in a feedback loop in which the output voltage is sampled, measurement of the output resistance before and after the loop is connected shows a change by a factor of 80. Is the resistance with feedback higher or lower? What is the value of the loop gain $A\beta$? If R_{of} is 100 Ω, what is R_o without feedback?

**8.29 A series–shunt feedback circuit employs a basic voltage amplifier that has a dc gain of 10^4 V/V and an STC frequency response with a unity-gain frequency of 1 MHz. The input resistance of the basic amplifier is 10 kΩ, and its output resistance is 1 kΩ. If the feedback factor $\beta = 0.1$ V/V, find the input impedance Z_{if} and the output impedance Z_{of} of the feedback amplifier. Give equivalent circuit representations of these impedances. Also find the value of each impedance at 10^3 Hz and at 10^5 Hz.

8.30 A series–shunt feedback amplifier utilizes the feedback circuit shown in Fig. P8.30.

(a) Find expressions for the h parameters of the feedback circuit (see Fig. 8.10b).
(b) If $R_1 = 1$ kΩ and $\beta = 0.01$, what are the values of all four h parameters? Give the units of each parameter.

(c) For the case $R_s = 1$ kΩ and $R_L = 1$ kΩ, sketch and label an equivalent circuit following the model in Fig. 8.10(c).

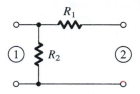

FIGURE P8.30

8.31 A feedback amplifier utilizing voltage sampling and employing a basic voltage amplifier with a gain of 100 V/V and an output resistance of 1000 Ω has a closed-loop output resistance of 100 Ω. What is the closed-loop gain? If the basic amplifier is used to implement a unity-gain voltage buffer, what output resistance do you expect?

*8.32 In the series–shunt amplifier shown in Fig. P8.32, the transistors operate at $V_{BE} \cong 0.7$ V with h_{FE} of 100 and an Early voltage that is very large.

(a) Derive expressions for A, β, R_i, and R_o.
(b) For $I_{B1} = 0.1$ mA, $I_{B2} = 1$ mA, $R_1 = 1$ kΩ, $R_2 = 10$ kΩ, $R_s = 100$ Ω, and $R_L = 1$ kΩ, find the dc bias voltages at the input and at the output, and find $A_f \equiv v_o / v_s$, R_{in}, and R_{out}.

D*8.33 Figure P8.33 shows a series–shunt amplifier with a feedback factor $\beta = 1$. The amplifier is designed so that $v_O = 0$ for $v_S = 0$, with small deviations in v_O from 0 V dc being minimized by the negative-feedback action. The technology utilized has $k_n' = 2k_p' = 120$ μA/V^2, $|V_t| = 0.7$ V, and $|V_A'| = 24$ V/μm.

8

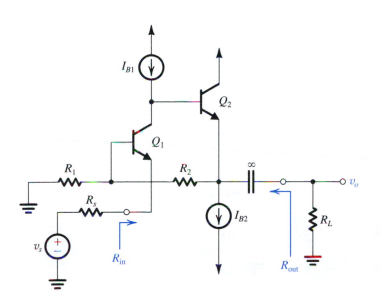

FIGURE P8.32

(a) With the feedback loop opened and the gate terminals of Q_1 and Q_2 grounded find the dc current and the overdrive voltage at which each of Q_1 to Q_5 is operating. Ignore the mismatch in I_D between Q_1 and Q_2 arising from their different drain voltages. Also find the dc voltage at the output.

(b) Find g_m and r_o of each of the five transistors.

(c) Find the values of A and R_o. Assume that the bias current sources are ideal.

(d) Find the gain-with-feedback, A_f, and the output resistance R_{out}.

(e) How would you modify the circuit to realize a closed-loop voltage gain of 5 V/V? What is the value of output resistance obtained?

**8.34 For the circuit in Fig. P8.34, $|V_t| = 1$ V, $k'W/L = 1$ mA/V^2, $h_{fe} = 100$, and the Early voltage magnitude for all

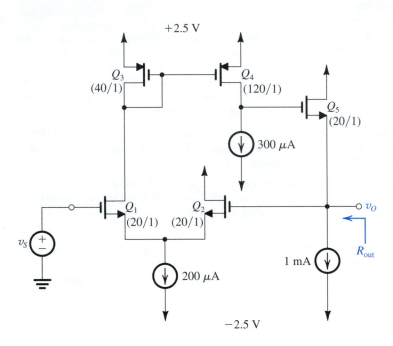

FIGURE P8.33

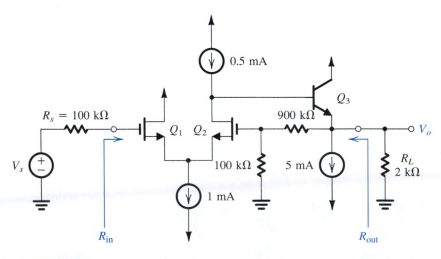

FIGURE P8.34

devices (including those that implement the current sources) is 100 V. The signal source V_s has a zero dc component. Find the dc voltage at the output and at the base of Q_3. Find the values of A, β, A_f, R_{in}, and R_{out}.

D*8.35 Figure P8.35 shows a series–shunt feedback amplifier without details of the bias circuit.

(a) Sketch the A circuit and the circuit for determining β.
(b) Show that if $A\beta$ is large then the closed-loop voltage gain is given approximately by

$$A_f \equiv \frac{V_o}{V_s} \simeq \frac{R_F + R_E}{R_E}$$

(c) If R_E is selected equal to 50 Ω, find R_F that will result in a closed-loop gain of approximately 25 V/V.
(d) If Q_1 is biased at 1 mA, Q_2 at 2 mA, and Q_3 at 5 mA, and assuming that the transistors have $h_{fe} = 100$, find approximate values for R_{C1} and R_{C2} to obtain gains from the stages of the A circuit as follows: a voltage gain of Q_1 of about -10 and a voltage gain of Q_2 of about -50.
(e) For your design, what is the closed-loop voltage gain realized?
(f) Calculate the input and output resistances of the closed-loop amplifier designed.

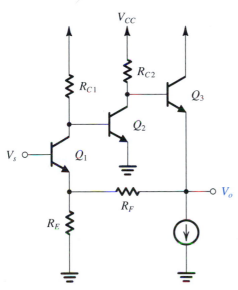

FIGURE P8.35

SECTION 8.5: THE SERIES–SERIES FEEDBACK AMPLIFIER

8.36 For the circuit in Fig. 8.17(a), find an approximate value for I_o/V_s assuming that the loop gain is large. Use it to determine the voltage gain V_o/V_s. Compare your results with the values found in Example 8.2.

8.37 A series–series feedback amplifier employs a transconductance amplifier having $G_m = 100$ mA/V, input resistance of 10 kΩ, and output resistance of 100 kΩ. The feedback network has $\beta = 0.1$ V/mA, an input resistance (with port 1 open-circuited) of 100 Ω, and an input resistance (with port 2 open-circuited) of 10 kΩ. The amplifier operates with a signal source having a resistance of 10 kΩ and with a load resistance of 10 kΩ. Find A_f, R_{in}, and R_{out}.

D*8.38 Figure P8.38 shows a circuit for a voltage-controlled current source employing series–series feedback through the resistor R_E. (The bias circuit for the transistor is not shown.) Show that if the loop gain $A\beta$ is large,

$$\frac{I_o}{V_s} \simeq \frac{1}{R_E}$$

Then find the value of R_E to obtain a circuit transconductance of 1 mA/V. If the voltage amplifier has a differential input resistance of 100 kΩ, a voltage gain of 100, and an output resistance of 1 kΩ, and if the transistor is biased at a current of 1 mA, and has h_{fe} of 100 and r_o of 100 kΩ, find the actual value of transconductance (I_o/V_s) realized. Use $R_s = 10$ kΩ. Also find the input resistance R_{in} and the output resistance R_{out}. For calculating R_{out}, recall that the output resistance of a BJT with an emitter resistance that is much larger than r_π is approximately $h_{fe} r_o$.

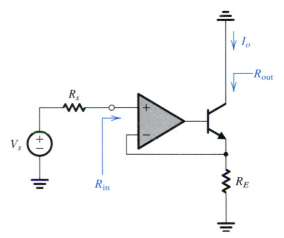

FIGURE P8.38

*8.39 Figure P8.39 shows a circuit for a voltage-to-current converter employing series–series feedback via resistor R_F. The MOSFETs have the dimensions shown and $\mu_n C_{ox} = 20$ μA/V^2, $|V_t| = 1$ V, and $|V_A| = 100$ V. What is the value of I_o/V_s obtained for large loop gain? Use feedback analysis to find a more exact value for I_o/V_s. Also, if the output voltage is taken at the source of Q_5, what closed-loop voltage gain is realized?

8.40 For the series–series feedback amplifier in Fig. P8.40, the op amp is characterized by an open-loop voltage gain μ,

8

+5 V

FIGURE P8.39

an input differential resistance $R_{id} = 10\ \text{k}\Omega$, and an output resistance $r_o = 100\ \Omega$. The amplifier supplies a current i_o to a load resistance $R_L = 1\ \text{k}\Omega$. The feedback network is composed of resistors $r = 100\ \Omega$, $R_2 = 10\ \text{k}\Omega$, and R_1. It is required to find the gain-with-feedback $A_f \equiv i_o / v_s$, the input resistance R_{in}, and the output resistance R_{out} for the following cases:

(a) $\mu = 10^5$ V/V and $R_1 = 100\ \Omega$
(b) $\mu = 10^4$ V/V and $R_1 = \infty$

Calculate this gain for the component values given on the circuit diagram, and compare the result with that found in Example 8.3. Find a new value for R_f to obtain a voltage gain of approximately −7.5 V/V.

8.42 The shunt–shunt feedback amplifier in Fig. P8.42 has $I = 1$ mA and $V_{GS} = 0.8$ V. The MOSFET has $V_t = 0.6$ V and $V_A = 30$ V. For $R_s = 10\ \text{k}\Omega$, $R_1 = 1\ \text{M}\Omega$, and $R_2 = 4.7\ \text{M}\Omega$, find the voltage gain v_o / v_s, the input resistance R_{in}, and the output resistance R_{out}.

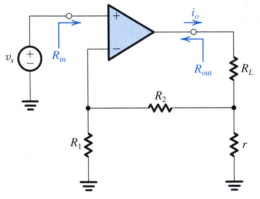

FIGURE P8.40

SECTION 8.6: THE SHUNT–SHUNT AND THE SHUNT–SERIES FEEDBACK AMPLIFIERS

D*8.41 For the amplifier topology shown in Fig. 8.21(a), show that for large loop gain,

$$\frac{V_o}{V_s} \simeq -\frac{R_f}{R_s}$$

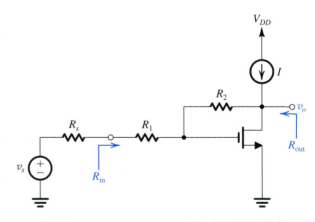

FIGURE P8.42

8.43 A transresistance amplifier having an open-circuit "gain" of 100 V/mA, an input resistance of 1 kΩ, and an output resistance of 1 kΩ is connected in a negative-feedback loop employing a shunt–shunt topology. The feedback network has an input resistance (with port 1 short-circuited) of

10 kΩ and an input resistance (with port 2 short-circuited) of 10 kΩ and provides a feedback factor $\beta = 0.1$ mA/V. The amplifier is fed with a current source having $R_s = 10$ kΩ, and a load resistance $R_L = 1$ kΩ is connected at the output. Find the transresistance A_f of the feedback amplifier, its input resistance R_{in}, and its output resistance R_{out}.

8.44 For the shunt–series feedback amplifier of Fig. P8.44, derive expressions for A, β, A_f, R_{in}, and R_{of} (the latter between the terminals labeled XX). Neglect r_o and the body effect. Evaluate all parameters for $g_{m1} = g_{m2} = 5$ mA/V, $R_D = 10$ kΩ, $R_S = 10$ kΩ, and $R_F = 90$ kΩ. Noting that R_{of} can be considered as a source-degeneration resistance for Q_2, find R_{out} for the case $r_{o2} = 20$ kΩ and neglecting the body effect. (*Hint:* A source-degeneration resistance R increases R_{out} by approximately $g_m R$.)

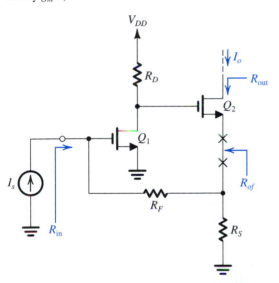

FIGURE P8.44

8.45 Reconsider the circuit in Fig. P8.44. Now let the drain of Q_2 be connected to V_{DD} and let the output be taken as the voltage at the source of Q_2. Now R_S should be considered as part of the A circuit, since the voltage V_o develops across it. Convince yourself that now the amplifier can be viewed as a shunt–shunt topology with the feedback network composed of R_F. Find expressions for A, β, A_f, R_{in}, and R_{out}, where R_{out} is the resistance looking back into the output terminal. Neglect r_o and the body effect. Find the values of all parameters for the case in which $g_{m1} = g_{m2} = 5$ mA/V, $R_D = 10$ kΩ, $R_S = 10$ kΩ, and $R_F = 90$ kΩ.

D8.46** (a) Show that for the circuit in Fig. P8.46(a) if the loop gain is large, the voltage gain V_o/V_s is given approximately by

$$\frac{V_o}{V_s} \simeq -\frac{R_f}{R_s}$$

(b) Using three cascaded stages of the type shown in Fig. P8.46(b) to implement the amplifier μ, design a feedback amplifier with a voltage gain of approximately -100 V/V. The amplifier is to operate between a source resistance $R_s = 10$ kΩ and a load resistance $R_L = 1$ kΩ. Calculate the actual value of V_o/V_s realized, the input resistance (excluding R_s), and the output resistance (excluding R_L). Assume that the BJTs have h_{fe} of 100. (*Note:* In practice, the three amplifier stages are not made identical, for stability reasons.)

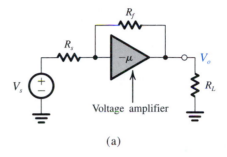

(a)

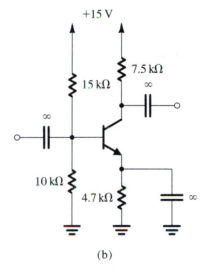

(b)

FIGURE P8.46

D8.47 Negative feedback is to be used to modify the characteristics of a particular amplifier for various purposes. Identify the feedback topology to be used if:

(a) Input resistance is to be lowered and output resistance raised.

(b) Both input and output resistances are to be raised.

(c) Both input and output resistances are to be lowered.

*8.48 For the circuit of Fig. P8.48, use the feedback method to find the voltage gain V_o/V_s, the input resistance R_{in}, and the output resistance R_{out}. The op amp has open-loop gain $\mu = 10^4$ V/V, $R_{id} = 100$ kΩ, and $r_o = 1$ kΩ.

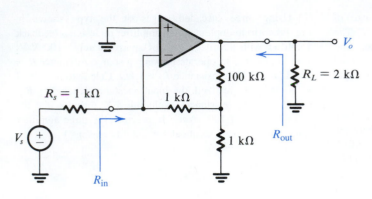

FIGURE P8.48

*8.49 Consider the amplifier of Fig. 8.25(a) to have its output at the emitter of the rightmost transistor Q_2. Use the technique for a shunt–shunt feedback amplifier to calculate (V_{out}/I_{in}) and R_{in}. Using this result, calculate I_{out}/I_{in}. Compare this with the results obtained in Example 8.4.

8.50 A current amplifier with a short-circuit current gain of 100 A/A, an input resistance of 1 kΩ, and an output resistance of 10 kΩ is connected in a negative-feedback loop employing the shunt–series topology. The feedback network provides a feedback factor $\beta = 0.1$ A/A. Lacking complete data about the situation, estimate the current gain, input resistance, and output resistance of the feedback amplifier.

*8.51 For the amplifier circuit in Fig. P8.51, assuming that V_s has a zero dc component, find the dc voltages at all nodes and the dc emitter currents of Q_1 and Q_2. Let the BJTs have $\beta = 100$. Use feedback analysis to find V_o/V_s and R_{in}.

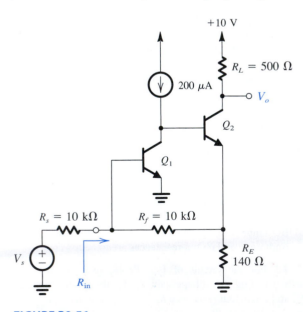

FIGURE P8.51

8.52 The feedback amplifier of Fig. P8.52 consists of a common-gate amplifier formed by Q_1 and R_D, and a feedback circuit formed by the capacitive divider (C_1, C_2) and the common-source transistor Q_f. Note that the bias circuit for Q_f is not shown. It is required to derive expressions for $A_f \equiv V_o/I_s$, R_{in}, and R_{out}. Assume that C_1 and C_2 are sufficiently small that their loading effect on the basic amplifier can be neglected. Also neglect r_o and the body effect. Find the values of A_f, R_{in}, and R_{out} for the case in which $g_{m1} = 5$ mA/V, $R_D = 10$ kΩ, $C_1 = 0.9$ pF, $C_2 = 0.1$ pF, and $g_{mf} = 1$ mA/V.

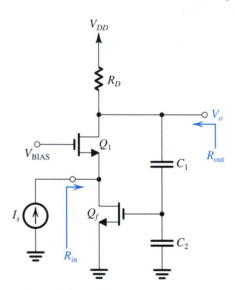

FIGURE P8.52

SECTION 8.7: DETERMINING THE LOOP GAIN

8.53 Determine the loop gain of the amplifier in Fig. P8.34 by breaking the loop at the gate of Q_2 and finding the returned voltage across the 100-kΩ resistor (while setting V_s to zero). The devices have $|V_t| = 1$ V, $k_n'W/L = 1$ mA/V^2, and $h_{fe} = 100$. The Early voltage magnitude for all devices (including those that implement the current sources) is 100 V. The signal source V_s has zero dc component. Determine the output resistance R_{out}.

8.54 It is required to determine the loop gain of the amplifier circuit shown in Fig. P8.35. The most convenient place to break the loop is at the base of Q_2. Thus, connect a resistance equal to $r_{\pi 2}$ between the collector of Q_1 and ground, apply a test voltage V_t to the base of Q_2, and determine the returned voltage at the collector of Q_1 (with V_s set to zero, of course). Show that

$$A\beta = \frac{g_{m2}R_{C2}(h_{fe3}+1)}{R_{C2}+(h_{fe3}+1)[r_{e3}+R_F+(R_E//r_{e1})]}$$
$$\times \frac{\alpha_1 R_E}{R_E+r_{e1}}(R_{C1}//r_{\pi 2})$$

8.55 Show that the loop gain of the amplifier circuit in Fig. P8.39 is

$$A\beta = g_{m1,2}(r_{o2}//r_{o4})\frac{R_F//r_{o5}}{(R_F//r_{o5})+1/g_{m5}}$$

where $g_{m1,2}$ is the g_m of each of Q_1 and Q_2.

8.56 Derive an expression for the loop gain of each of the four feedback circuits shown in Fig. P8.26. Assume that the op amp is modeled by an input resistance R_{id}, an open-circuit voltage gain μ, and an output resistance r_o.

*8.57 Find the loop gain of the feedback amplifier shown in Fig. P8.33 by breaking the loop at the gate of Q_2 (and, of course, setting $v_s = 0$). Use the values given in the statement of Problem 8.33. Determine the value of R_{out}.

8.58 For the feedback amplifier in Fig. P8.42, derive an expression for the loop gain by breaking the loop at the gate terminal of the MOSFET (and, of course, setting $v_s = 0$). Find the value of the loop gain for the component values given in Problem 8.42.

8.59 For the feedback amplifier in Fig. P8.44, set $I_s = 0$ and derive an expression for the loop gain by breaking the loop at the gate terminal of transistor Q_1.

8.60 For the feedback amplifier in Fig. P8.52, set $I_s = 0$ and derive an expression for the loop gain by breaking the loop at the gate terminal of transistor Q_f.

SECTION 8.8: THE STABILITY PROBLEM

8.61 An op amp designed to have a low-frequency gain of 10^5 and a high-frequency response dominated by a single pole at 100 rad/s, acquires, through a manufacturing error, a pair of additional poles at 10,000 rad/s. At what frequency does the total phase shift reach 180°? At this frequency, for what value of β, assumed to be frequency independent, does the loop gain reach a value of unity? What is the corresponding value of closed-loop gain at low frequencies?

**8.62 For the situation described in Problem 8.61, sketch Nyquist plots for $\beta = 1.0$ and 10^{-3}. (Plot for $\omega = 0$ rad/s, 100 rad/s, 10^3 rad/s, 10^4 rad/s, and ∞ rad/s.)

8.63 An op amp having a low-frequency gain of 10^3 and a single-pole rolloff at 10^4 rad/s is connected in a negative-feedback loop via a feedback network having a transmission k and a two-pole rolloff at 10^4 rad/s. Find the value of k above which the closed-loop amplifier becomes unstable.

8.64 Consider a feedback amplifier for which the open-loop gain $A(s)$ is given by

$$A(s) = \frac{1000}{(1+s/10^4)(1+s/10^5)^2}$$

If the feedback factor β is independent of frequency, find the frequency at which the phase shift is 180°, and find the critical value of β at which oscillation will commence.

SECTION 8.9: EFFECT OF FEEDBACK ON THE AMPLIFIER POLES

8.65 A dc amplifier having a single-pole response with pole frequency 10^4 Hz and unity-gain frequency of 10 MHz is operated in a loop whose frequency-independent feedback factor is 0.1. Find the low-frequency gain, the 3-dB frequency, and the unity-gain frequency of the closed-loop amplifier. By what factor does the pole shift?

*8.66 An amplifier having a low-frequency gain of 10^3 and poles at 10^4 Hz and 10^5 Hz is operated in a closed negative-feedback loop with a frequency-independent β.

(a) For what value of β do the closed-loop poles become coincident? At what frequency?
(b) What is the low-frequency gain corresponding to the situation in (a)? What is the value of the closed-loop gain at the frequency of the coincident poles?
(c) What is the value of Q corresponding to the situation in (a)?
(d) If β is increased by a factor of 10, what are the new pole locations? What is the corresponding pole Q?

D8.67 A dc amplifier has an open-loop gain of 1000 and two poles, a dominant one at 1 kHz and a high-frequency one whose location can be controlled. It is required to connect this amplifier in a negative-feedback loop that provides a dc closed-loop gain of 100 and a maximally flat response. Find the required value of β and the frequency at which the second pole should be placed.

8.68 Reconsider Example 8.5 with the circuit in Fig. 8.34 modified to incorporate a so-called tapered network, in which the components immediately adjacent to the amplifier input are raised in impedance to $C/10$ and $10\,R$. Find expressions for the resulting pole frequency ω_0 and Q factor. For what value of K do the poles coincide? For what value of K does the response become maximally flat? For what value of K does the circuit oscillate?

8.69 Three identical logic inverters, each of which can be characterized in its switching region as a linear amplifier having a gain $-K$ and a pole at 10^7 Hz, are connected in a ring. Regarding this as a negative-feedback loop with $\beta = 1$, find the minimum value of K for which the inverter ring must oscillate. What would the frequency of oscillation be for very small signal operation? [Note that in practice such a ring oscillator operates with relatively larger signal (logic levels) at a somewhat lower frequency.]

SECTION 8.10: STABILITY STUDY USING BODE PLOTS

8.70 Reconsider Exercise 8.14 for the case of the op amp wired as a unity-gain buffer. At what frequency is $|A\beta| = 1$? What is the corresponding phase margin?

8.71 Reconsider Exercise 8.14 for the case of a manufacturing error introducing a second pole at 10^4 Hz. What is now the frequency for which $|A\beta| = 1$? What is the corresponding phase margin? For what values of β is the phase margin $45°$ or more?

8.72 For what phase margin does the gain peaking have a value of 5%? Of 10%? Of 0.1 dB? Of 1 dB? (*Hint:* Use the result in Eq. 8.76.)

8.73 An amplifier has a dc gain of 10^5 and poles at 10^5 Hz, 3.16×10^5 Hz, and 10^6 Hz. Find the value of β, and the corresponding closed-loop gain, for which a phase margin of $45°$ is obtained.

8.74 A two-pole amplifier for which $A_0 = 10^3$ and having poles at 1 MHz and 10 MHz is to be connected as a differentiator. On the basis of the rate-of-closure rule, what is the smallest differentiator time constant for which operation is stable? What are the corresponding gain and phase margins?

*8.75 For the amplifier described by Fig. 8.37 and with frequency-independent feedback, what is the minimum closed-loop voltage gain that can be obtained for phase margins of $90°$ and $45°$?

SECTION 8.11: FREQUENCY COMPENSATION

D8.76 A multipole amplifier having a first pole at 2 MHz and a dc open-loop gain of 80 dB is to be compensated for closed-loop gains as low as unity by the introduction of a new dominant pole. At what frequency must the new pole be placed?

D8.77 For the amplifier described in Problem 8.76, rather than introducing a new dominant pole we can use additional capacitance at the circuit node at which the pole is formed to reduce the frequency of the first pole. If the frequency of the second pole is 10 MHz and if it remains unchanged while additional capacitance is introduced as mentioned, find the

frequency to which the first pole must be lowered so that the resulting amplifier is stable for closed-loop gains as low as unity. By what factor is the capacitance at the controlling node increased?

8.78 Contemplate the effects of pole splitting by considering Eqs. (8.87) and (8.88) under the conditions that $R_1 \simeq R_2 = R$, $C_2 \simeq C_1/10 = C$, $C_f \gg C$, and $g_m = 100/R$, by calculating ω_{P1}, ω_{P2}, and ω'_{P1}, ω'_{P2}.

D8.79 An op amp with open-loop voltage gain of 10^4 Hz and poles at 10^5 Hz, 10^6 Hz, and 10^7 Hz is to be compensated by the addition of a fourth dominant pole to operate stably with unity feedback ($\beta = 1$). What is the frequency of the required dominant pole? The compensation network is to consist of an RC low-pass network placed in the negative-feedback path of the op amp. The dc bias conditions are such that a 1-MΩ resistor can be tolerated in series with each of the negative and positive input terminals. What capacitor is required between the negative input and ground to implement the required fourth pole?

D*8.80 An op amp with an open-loop voltage gain of 80 dB and poles at 10^5 Hz, 10^6 Hz, and 2×10^6 Hz is to be compensated to be stable for unity β. Assume that the op amp incorporates an amplifier equivalent to that in Fig. 8.40, with $C_1 = 150$ pF, $C_2 = 5$ pF, and $g_m = 40$ mA/V, and that f_{P1} is caused by the input circuit and f_{P2} by the output circuit of this amplifier. Find the required value of the compensating Miller capacitance and the new frequency of the output pole.

**8.81 The op amp in the circuit of Fig. P8.81 has an open-loop gain of 10^5 and a single-pole rolloff with $\omega_{3dB} = 10$ rad/s.

(a) Sketch a Bode plot for the loop gain.
(b) Find the frequency at which $|A\beta| = 1$, and find the corresponding phase margin.
(c) Find the closed-loop transfer function, including its zero and poles. Sketch a pole-zero plot. Sketch the magnitude of the transfer function versus frequency, and label the important parameters on your sketch.

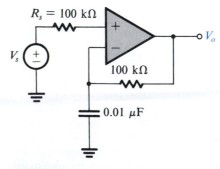

FIGURE P8.81

CHAPTER 9

Operational-Amplifier and Data-Converter Circuits

INTRODUCTION

Analog ICs include operational amplifiers, analog multipliers, analog-to-digital (A/D) and digital-to-analog (D/A) converters, phase-locked loops, and a variety of other, more specialized functional blocks. All these analog subsystems are constructed internally using the basic building blocks we have studied in earlier chapters, including single-stage amplifiers, differential pairs, current mirrors, and MOS switches.

In this chapter, we shall study the internal circuitry of the most important analog ICs, namely, operational amplifiers and data converters. The terminal characteristics and circuit applications of op amps have already been covered in Chapter 2. Here, our objective is to expose the reader to some of the ingenious techniques that have evolved over the years for combining elementary analog circuit building blocks to realize a complete op amp. We shall study both CMOS and bipolar op amps. The CMOS op-amp circuits considered find application in the design of analog and mixed-signal VLSI circuits. Because these op amps are usually designed with a specific application in mind, they can be optimized to meet a subset of the list of desired specifications, such as high dc gain, wide bandwidth, or large output-signal swing. In contrast, the bipolar op-amp circuit we shall study is of the *general-purpose* variety and therefore is designed to fit a wide range of specifications. As a result, its circuit represents a compromise between many performance parameters. This 741-type of op amp has been in existence for over 35 years. Nevertheless its internal circuit remains as relevant and interesting today as it ever was.

The material on data-converter circuits presented in this chapter should serve as a bridge between analog circuits, on which we have been concentrating in Chapters 6 to 8, and digital circuits whose study is undertaken in Chapters 10 and 11.

In addition to exposing the reader to some of the ideas that make analog IC design such an exciting topic, this chapter should serve to tie together many of the concepts and methods studied thus far.

9.1 THE TWO-STAGE CMOS OP AMP

The first op-amp circuit we shall study is the two-stage CMOS topology shown in Fig. 9.1. This simple but elegant circuit has become a classic and is used in a variety of forms in the design of VLSI systems. We have already studied this circuit in Section 7.7.1 as an example of a multistage CMOS amplifier. We urge the reader to review Section 7.7.1 before proceeding further. Here, our discussion will emphasize the performance characteristics of the circuit and the trade-offs involved in its design.

9.1.1 The Circuit

The circuit consists of two gain stages: The first stage is formed by the differential pair Q_1–Q_2 together with its current mirror load Q_3–Q_4. This differential-amplifier circuit, studied in detail in Section 7.5, provides a voltage gain that is typically in the range of 20 V/V to 60 V/V,

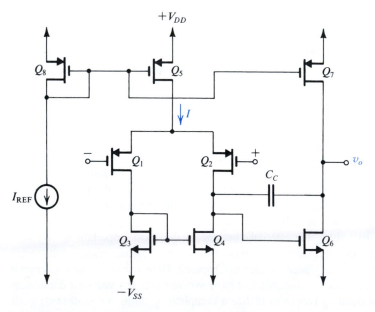

FIGURE 9.1 The basic two-stage CMOS op-amp configuration.

as well as performing conversion from differential to single-ended form while providing a reasonable common-mode rejection ratio (CMRR).

The differential pair is biased by current source Q_5, which is one of the two output transistors of the current mirror formed by Q_8, Q_5, and Q_7. The current mirror is fed by a reference current I_{REF}, which can be generated by simply connecting a precision resistor (external to the chip) to the negative supply voltage $-V_{SS}$ or to a more precise negative voltage reference if one is available in the same integrated circuit. Alternatively, for applications with more stringent requirements, I_{REF} can be generated using a circuit such as that studied in Section 7.7.1.

The second gain stage consists of the common-source transistor Q_6 and its current-source load Q_7. The second stage typically provides a gain of 50 V/V to 80 V/V. In addition, it takes part in the process of frequency compensating the op amp. From Section 8.11 the reader will recall that to guarantee that the op amp will operate in a stable fashion (as opposed to oscillating) when negative feedback of various amounts is applied, the open-loop gain is made to roll off with frequency at the uniform rate of −20 dB/decade. This in turn is achieved by introducing a pole at a relatively low frequency and arranging for it to dominate the frequency-response determination. In the circuit we are studying, this is implemented using a compensation capacitance C_C connected in the negative-feedback path of the second-stage amplifying transistor Q_6. As will be seen, C_C (together with the much smaller capacitance C_{gd6} across it) is Miller-multiplied by the gain of the second stage, and the resulting capacitance at the input of the second stage interacts with the total resistance there to provide the required dominant pole (more on this later).

Unless properly designed, the CMOS op-amp circuit of Fig. 9.1 can exhibit a **systematic output dc offset** voltage. This point was discussed in Section 7.7.1, where it was found that the dc offset can be eliminated by sizing the transistors so as to satisfy the following constraint:

$$\frac{(W/L)_6}{(W/L)_4} = 2\frac{(W/L)_7}{(W/L)_5} \tag{9.1}$$

9.1.2 Input Common-Mode Range and Output Swing

Refer to Fig. 9.1 and consider what happens when the two input terminals are tied together and connected to a voltage V_{ICM}. The lowest value of V_{ICM} has to be sufficiently large to keep Q_1 and Q_2 in saturation. Thus, the lowest value of V_{ICM} should not be lower than the voltage at the drain of Q_1 ($-V_{SS} + V_{GS3} = -V_{SS} + V_{tn} + V_{OV3}$) by more than $|V_{tp}|$, thus

$$V_{ICM} \geq -V_{SS} + V_{tn} + V_{OV3} - |V_{tp}| \tag{9.2}$$

The highest value of V_{ICM} should ensure that Q_5 remains in saturation; that is, the voltage across Q_5, V_{SD5}, should not decrease below $|V_{OV5}|$. Equivalently, the voltage at the drain of Q_5 should not go higher than $V_{DD} - |V_{OV5}|$. Thus the upper limit of V_{ICM} is

$$V_{ICM} \leq V_{DD} - |V_{OV5}| - V_{SG1}$$

or equivalently

$$V_{ICM} \leq V_{DD} - |V_{OV5}| - |V_{tp}| - |V_{OV1}| \tag{9.3}$$

The expressions in Eqs. (9.2) and (9.3) can be combined to express the input common-mode range as

$$-V_{SS} + V_{OV3} + V_{tn} - |V_{tp}| \leq V_{ICM} \leq V_{DD} - |V_{tp}| - |V_{OV1}| - |V_{OV5}| \tag{9.4}$$

As expected, the overdrive voltages, which are important design parameters, subtract from the dc supply voltages, thereby reducing the input common-mode range. It follows that from a V_{ICM} range point-of-view it is desirable to select the values of V_{OV} as low as possible.

The extent of the signal swing allowed at the output of the op amp is limited at the lower end by the need to keep Q_6 saturated and at the upper end by the need to keep Q_7 saturated, thus

$$-V_{SS} + V_{OV6} \leq v_O \leq V_{DD} - |V_{OV7}| \tag{9.5}$$

Here again we observe that to achieve a wide range for the output voltage swing we need to select values for $|V_{OV}|$ of Q_6 and Q_7 as low as possible. This requirement, however, is counteracted by the need to have a high transition frequency f_T for Q_6. From Table 6.3 and the corresponding discussion in Section 6.2.3, we know that f_T is proportional to V_{OV}; thus the high-frequency performance of a MOSFET improves with the increase of the overdrive voltage at which it is operated.

An important requirement of an op-amp circuit is that it be possible for its output terminal to be connected back to its negative input terminal so that a unity-gain amplifier is obtained. For such a connection to be possible, there must be a substantial overlap between the allowable range of v_O and the allowable range of V_{ICM}. This is usually the case in the CMOS amplifier circuit under study.

EXERCISE

9.1 For a particular design of the two-stage CMOS op amp of Fig. 9.1, ±1.65-V supplies are utilized and all transistors except for Q_6 and Q_7 are operated with overdrive voltages of 0.3-V magnitude; Q_6 and Q_7 use overdrive voltages of 0.5-V magnitude. The fabrication process employed provides $V_{tn} = |V_{tp}| = 0.5$ V. Find the input common-mode range and the range allowed for v_O.

Ans. −1.35 V to 0.55 V; −1.15 V to +1.15 V

9.1.3 Voltage Gain

To determine the voltage gain and the frequency response, consider a simplified equivalent circuit model for the small-signal operation of the CMOS amplifier (Fig. 9.2), where each of the two stages is modeled as a transconductance amplifier. As expected, the input resistance is practically infinite,

$$R_{in} = \infty$$

The first-stage transconductance G_{m1} is equal to the transconductance of each of Q_1 and Q_2 (see Section 7.5),

$$G_{m1} = g_{m1} = g_{m2} \tag{9.6}$$

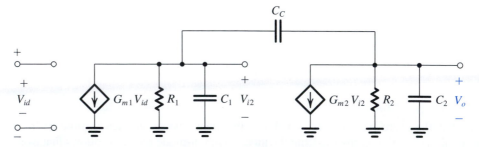

FIGURE 9.2 Small-signal equivalent circuit for the op amp in Fig. 9.1.

Since Q_1 and Q_2 are operated at equal bias currents ($I/2$) and equal overdrive voltages, $V_{OV1} = V_{OV2}$,

$$G_{m1} = \frac{2(I/2)}{V_{OV1}} = \frac{I}{V_{OV1}} \qquad (9.7)$$

Resistance R_1 represents the output resistance of the first stage, thus

$$R_1 = r_{o2} \| r_{o4} \qquad (9.8)$$

where

$$r_{o2} = \frac{|V_{A2}|}{I/2} \qquad (9.9)$$

and

$$r_{o4} = \frac{V_{A4}}{I/2} \qquad (9.10)$$

The dc gain of the first stage is thus

$$A_1 = -G_{m1}R_1 \qquad (9.11)$$

$$= -g_{m1}(r_{o2} \| r_{o4}) \qquad (9.12)$$

$$= -\frac{2}{V_{OV1}} \bigg/ \left[\frac{1}{|V_{A2}|} + \frac{1}{V_{A4}}\right] \qquad (9.13)$$

Observe that the magnitude of A_1 is increased by operating the differential-pair transistors, Q_1 and Q_2, at a low overdrive voltage, and by choosing a longer channel length to obtain larger Early voltages, $|V_A|$. Both actions, however, degrade the frequency response of the amplifier (see Table 6.3 and the corresponding discussion in Section 6.2.3).

Returning to the equivalent circuit in Fig. 9.2 and leaving the discussion of the various model capacitances until the next section, we note that the second-stage transconductance G_{m2} is given by

$$G_{m2} = g_{m6} = \frac{2I_{D6}}{V_{OV6}} \qquad (9.14)$$

Resistance R_2 represents the output resistance of the second stage, thus

$$R_2 = r_{o6} \| r_{o7} \qquad (9.15)$$

where

$$r_{o6} = \frac{V_{A6}}{I_{D6}} \qquad (9.16)$$

and

$$r_{o7} = \frac{|V_{A7}|}{I_{D7}} = \frac{|V_{A7}|}{I_{D6}} \qquad (9.17)$$

The voltage gain of the second stage can now be found as

$$A_2 = -G_{m2}R_2 \qquad (9.18)$$

$$= -g_{m6}(r_{o6} \| r_{o7}) \qquad (9.19)$$

$$= -\frac{2}{V_{OV6}} \bigg/ \left[\frac{1}{V_{A6}} + \frac{1}{|V_{A7}|}\right] \qquad (9.20)$$

Here again we observe that to increase the magnitude of A_2, Q_6 has to be operated at a low overdrive voltage, and the channel lengths of Q_6 and Q_7 should be made longer. Both these actions, however, would reduce the amplifier bandwidth, which presents the designer with an important trade-off.

The overall dc voltage gain can be found as the product $A_1 A_2$,

$$A_v = A_1 A_2$$

$$= G_{m1} R_1 G_{m2} R_2 \tag{9.21}$$

$$= g_{m1}(r_{o2} \parallel r_{o4}) g_{m6}(r_{o6} \parallel r_{o7}) \tag{9.22}$$

Note that A_v is of the order of $(g_m r_o)^2$. Thus the maximum value of A_v will be in the range of 500 V/V to 5000 V/V.

Finally, we note that the output resistance of the op amp is equal to the output resistance of the second stage,

$$R_o = r_{o6} \parallel r_{o7} \tag{9.23}$$

Hence R_o can be large (i.e., in the tens-of-kilohms range). Nevertheless, since on-chip CMOS op amps are rarely required to drive heavy loads, the large open-loop output resistance is usually not an important issue.

EXERCISE

9.2 The CMOS op amp of Fig. 9.1 is fabricated in a process for which $V'_{An} = |V'_{Ap}| = 20$ V/μm. Find A_1, A_2, and A_v if all devices are 1 μm long, $V_{OV1} = 0.2$ V, and $V_{OV6} = 0.5$ V. Also, find the op-amp output resistance obtained when the second stage is biased at 0.5 mA.
Ans. −100 V/V; −40 V/V; 4000 V/V; 20 kΩ

9.1.4 Frequency Response

Refer to the equivalent circuit in Fig. 9.2. Capacitance C_1 is the total capacitance between the output node of the first stage and ground, thus

$$C_1 = C_{gd2} + C_{db2} + C_{gd4} + C_{db4} + C_{gs6} \tag{9.24}$$

Capacitance C_2 represents the total capacitance between the output node of the op amp and ground and includes whatever load capacitance C_L that the amplifier is required to drive, thus

$$C_2 = C_{db6} + C_{db7} + C_{gd7} + C_L \tag{9.25}$$

Usually, C_L is larger than the transistor capacitances, with the result that C_2 becomes much larger than C_1. Finally, note that C_{gd6} should be shown in parallel with C_C but has been ignored because C_C is usually much larger.

The equivalent circuit of Fig. 9.2 was analyzed in detail in Section 7.7.1, where it was found that it has two poles and a positive real-axis zero with the following approximate frequencies:

$$f_{P1} \cong \frac{1}{2\pi R_1 G_{m2} R_2 C_C} \tag{9.26}$$

$$f_{P2} \cong \frac{G_{m2}}{2\pi C_2} \tag{9.27}$$

$$f_Z \cong \frac{G_{m2}}{2\pi C_C} \tag{9.28}$$

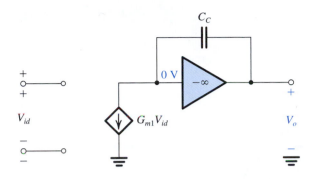

FIGURE 9.3 An approximate high-frequency equivalent circuit of the two-stage op amp. This circuit applies for frequencies $f \gg f_{P1}$.

Here, f_{P1} is the dominant pole formed by the interaction of Miller-multiplied C_C [i.e., $(1 + G_{m2}R_2)C_C \cong G_{m2}R_2C_C$] and R_1. To achieve the goal of a uniform –20 dB/decade gain rolloff down to 0 dB, the unity-gain frequency f_t,

$$f_t = |A_v| f_{P1} \tag{9.29}$$

$$= \frac{G_{m1}}{2\pi C_C} \tag{9.30}$$

must be lower than f_{P2} and f_Z, thus the design must satisfy the following two conditions

$$\frac{G_{m1}}{C_C} < \frac{G_{m2}}{C_2} \tag{9.31}$$

and

$$G_{m1} < G_{m2} \tag{9.32}$$

Simplified Equivalent Circuit   The uniform –20-dB/decade gain rolloff obtained at frequencies $f \gg f_{P1}$ suggests that at these frequencies, the op amp can be represented by the simplified equivalent circuit shown in Fig. 9.3. Observe that this attractive simplification is based on the assumption that the gain of the second stage, $|A_2|$, is large, and hence a virtual ground appears at the input terminal of the second stage. The second stage then effectively acts as an integrator that is fed with the output current signal of the first stage; $G_{m1}V_{id}$. Although derived for the CMOS amplifier, this simplified equivalent circuit is general and applies to a variety of two-stage op amps, including the first two stages of the 741-type bipolar op amp studied later in this chapter.

Phase Margin   The frequency compensation scheme utilized in the two-stage CMOS amplifier is of the pole-splitting type, studied in Section 8.11.3: It provides a dominant low-frequency pole with frequency f_{P1} and shifts the second pole beyond f_t. Figure 9.4 shows a representative Bode plot for the gain magnitude and phase. Note that at the unity-gain frequency f_t, the phase lag exceeds the 90° caused by the dominant pole at f_{P1}. This so-called excess phase shift is due to the second pole,

$$\phi_{P2} = -\tan^{-1}\left(\frac{f_t}{f_{P2}}\right) \tag{9.33}$$

and the right-half-plane zero,

$$\phi_Z = -\tan^{-1}\left(\frac{f_t}{f_Z}\right) \tag{9.34}$$

Thus the phase lag at $f = f_t$ will be

$$\phi_{\text{total}} = 90° + \tan^{-1}(f_t/f_{P2}) + \tan^{-1}(f_t/f_Z) \tag{9.35}$$

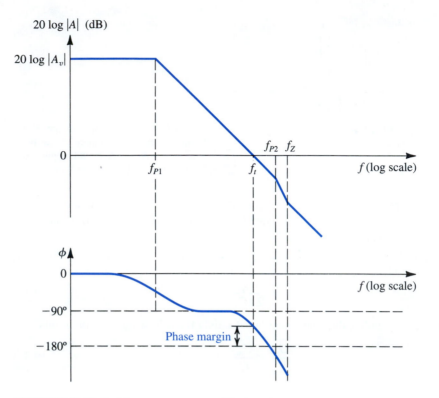

FIGURE 9.4 Typical frequency response of the two-stage op amp.

and thus the phase margin will be

$$\text{Phase margin} = 180° - \phi_{\text{total}}$$

$$= 90° - \tan^{-1}(f_t/f_{P2}) - \tan^{-1}(f_t/f_Z) \tag{9.36}$$

From our study of the stability of feedback amplifiers in Section 8.10.2, we know that the magnitude of the phase margin significantly affects the closed-loop gain. Therefore obtaining a desired minimum value of phase margin is usually a design requirement.

The problem of the additional phase lag provided by the zero has a rather simple and elegant solution: By including a resistance R in series with C_C, as shown in Fig. 9.5, the transmission zero can be moved to other less-harmful locations. To find the new location of the

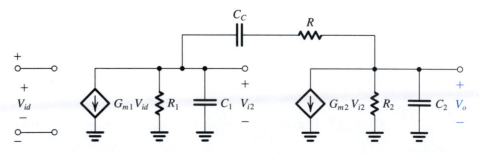

FIGURE 9.5 Small-signal equivalent circuit of the op amp in Fig. 9.1 with a resistance R included in series with C_C.

transmission zero, set $V_o = 0$. Then, the current through C_C will be $V_{i2} / (R + 1/sC_C)$, and a node equation at the output yields

$$\frac{V_{i2}}{R + \dfrac{1}{sC_C}} = G_{m2} V_{i2}$$

Thus the zero is now at

$$s = 1 \Big/ C_C \left(\frac{1}{G_{m2}} - R \right) \tag{9.37}$$

We observe that by selecting $R = 1/G_{m2}$, we can place the zero at infinite frequency. An even better choice would be to select R greater than $1/G_{m2}$, thus placing the zero at a negative real-axis location where the phase it introduces *adds* to the phase margin.

EXERCISE

9.3 A particular implementation of the CMOS amplifier of Figs. 9.1 and 9.2 provides $G_{m1} = 1$ mA/V, $G_{m2} = 2$ mA/V, $r_{o2} = r_{o4} = 100$ kΩ, $r_{o6} = r_{o7} = 40$ kΩ, and $C_2 = 1$ pF.

(a) Find the value of C_C that results in $f_t = 100$ MHz. What is the 3-dB frequency of the open-loop gain?

(b) Find the value of the resistance R that when placed in series with C_C causes the transmission zero to be located at infinite frequency.

(c) Find the frequency of the second pole and hence find the excess phase lag at $f = f_t$, introduced by the second pole, and the resulting phase margin assuming that the situation in (b) pertains.

Ans. 1.6 pF; 50 kHz; 500 Ω; 318 MHz; 17.4°; 72.6°

9.1.5 Slew Rate

The slew-rate limitation of op amps is discussed in Chapter 2. Here, we shall illustrate the origin of the slewing phenomenon in the context of the two-stage CMOS amplifier under study.

Consider the unity-gain follower of Fig. 9.6 with a step of say, 1 V applied at the input. Because of the amplifier dynamics, its output will not change in zero time. Thus, immediately after the input is applied, the entire value of the step will appear as a differential signal between the two input terminals. In all likelihood, such a large signal will exceed the voltage required to turn off one side of the pair ($\sqrt{2}V_{OV}$; see Fig. 7.6) and switch the entire bias

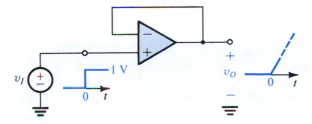

FIGURE 9.6 A unity-gain follower with a large step input. Since the output voltage cannot change immediately, a large differential voltage appears between the op-amp input terminals.

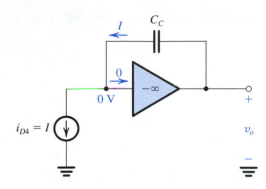

FIGURE 9.7 Model of the two-stage CMOS op amp of Fig. 9.1 when a large differential voltage is applied.

current I to the other side. Reference to Fig. 9.1 shows that for our example, Q_2 will turn off, and Q_1 will conduct the entire current I. Thus Q_4 will sink a current I that will be pulled from C_C, as shown in Fig. 9.7. Here, as we did in Fig. 9.3, we are modeling the second stage as an ideal integrator. We see that the output voltage will be a ramp with a slope of I/C_C:

$$v_o(t) = \frac{I}{C_C}t \tag{9.38}$$

Thus the slew rate, SR, is given by

$$SR = \frac{I}{C_C} \tag{9.39}$$

It should be pointed out, however, that this is a rather simplified model of the slewing process.

Relationship Between SR and f_t   A simple relationship exists between the unity-gain bandwidth f_t and the slew rate SR. This relationship can be found by combining Eqs. (9.30) and (9.39) and noting that $G_{m1} = g_{m1} = I/V_{OV1}$, to obtain

$$SR = 2\pi f_t V_{OV} \tag{9.40}$$

or equivalently,

$$SR = V_{OV}\omega_t \tag{9.41}$$

Thus, for a given ω_t, the slew rate is determined by the overdrive voltage at which the first-stage transistors are operated. A higher slew rate is obtained by operating Q_1 and Q_2 at a larger V_{OV}. Now, for a given bias current I, a larger V_{OV} is obtained if Q_1 and Q_2 are p-channel devices. This is an important reason for using p-channel rather than n-channel devices in the first stage of the CMOS op amp. Another reason is that it allows the second stage to employ an n-channel device. Now, since n-channel devices have greater transconductances than corresponding p-channel devices, G_{m2} will be high, resulting in a higher second-pole frequency and a correspondingly higher ω_t. However, the price paid for these improvements is a lower G_{m1} and hence a lower dc gain.

EXERCISE

9.4   Find SR for the CMOS op amp of Fig. 9.1 for the case $f_t = 100$ MHz and $V_{OV1} = 0.2$ V. If $C_C = 1.6$ pF, what must the bias current I be?

Ans. 126 V/μs; 200 μA

EXAMPLE 9.1

We conclude our study of the two-stage CMOS op amp with a design example. Let it be required to design the circuit to obtain a dc gain of 4000 V/V. Assume that the available fabrication technology is of the 0.5-μm type for which $V_{tn} = |V_{tp}| = 0.5$ V, $k'_n = 200$ μA/V^2, $k'_p = 80$ μA/V^2, $V'_{An} = |V'_{Ap}| = 20$ V/μm, and $V_{DD} = V_{SS} = 1.65$ V. To achieve a reasonable dc gain per stage, use $L = 1$ μm for all devices. Also, for simplicity, operate all devices at the same $|V_{OV}|$, in the range of 0.2 V to 0.4 V. Use $I = 200$ μA, and to obtain a higher G_{m2}, and hence a higher f_{P2}, use $I_{D6} = 0.5$ mA. Specify the W/L ratios for all transistors. Also give the values realized for the input common-mode range, the maximum possible output swing, R_{in} and R_o. If $C_1 = 0.2$ pF and $C_2 = 0.8$ pF, find the required values of C_C and the series resistance R to place the transmission zero at $s = \infty$ and to obtain the highest possible f_t consistent with a phase margin of 75°. Evaluate the values obtained for f_t and SR.

Solution

Using the voltage-gain expression in Eq. (9.22),

$$A_v = g_{m1}(r_{o2} \| r_{o4})g_{m6}(r_{o6} \| r_{o7})$$

$$= \frac{2(I/2)}{V_{OV}} \times \frac{1}{2} \times \frac{V_A}{(I/2)} \times \frac{2I_{D6}}{V_{OV}} \times \frac{1}{2} \times \frac{V_A}{I_{D6}}$$

$$= \left(\frac{V_A}{V_{OV}}\right)^2$$

To obtain $A_v = 4000$, given $V_A = 20$ V,

$$4000 = \frac{400}{V_{OV}^2}$$

$$V_{OV} = 0.316 \text{ V}$$

To obtain the required (W/L) ratios of Q_1 and Q_2,

$$I_{D1} = \frac{1}{2}k'_p\left(\frac{W}{L}\right)_1 V_{OV}^2$$

$$100 = \frac{1}{2} \times 80\left(\frac{W}{L}\right)_1 \times 0.316^2$$

Thus,

$$\left(\frac{W}{L}\right)_1 = \frac{25 \ \mu\text{m}}{1 \ \mu\text{m}}$$

and

$$\left(\frac{W}{L}\right)_2 = \frac{25 \ \mu\text{m}}{1 \ \mu\text{m}}$$

For Q_3 and Q_4 we write

$$100 = \frac{1}{2} \times 200\left(\frac{W}{L}\right)_3 \times 0.316^2$$

to obtain

$$\left(\frac{W}{L}\right)_3 = \left(\frac{W}{L}\right)_4 = \left(\frac{10\ \mu m}{1\ \mu m}\right)$$

For Q_5,

$$200 = \frac{1}{2} \times 80 \left(\frac{W}{L}\right)_5 \times 0.316^2$$

Thus,

$$\left(\frac{W}{L}\right)_5 = \left(\frac{50\ \mu m}{1\ \mu m}\right)$$

Since Q_7 is required to conduct 500 μA, its (W/L) ratio should be 2.5 times that of Q_5,

$$\left(\frac{W}{L}\right)_7 = 2.5 \left(\frac{W}{L}\right)_5 = \left(\frac{125\ \mu m}{1\ \mu m}\right)$$

For Q_6 we write

$$500 = \frac{1}{2} \times 200 \times \left(\frac{W}{L}\right)_6 \times 0.316^2$$

Thus,

$$\left(\frac{W}{L}\right)_6 = \frac{50\ \mu m}{1\ \mu m}$$

Finally, let's select $I_{REF} = 20\ \mu$A, thus

$$\left(\frac{W}{L}\right)_8 = 0.1 \left(\frac{W}{L}\right)_5 = \frac{5\ \mu m}{1\ \mu m}$$

The input common-mode range can be found using the expression in Eq. (9.4) as

$$-1.33\ \text{V} \leq V_{ICM} \leq 0.52\ \text{V}$$

The maximum signal swing allowable at the output is found using the expression in Eq. (9.5) as

$$-1.33\ \text{V} \leq v_O \leq 1.33\ \text{V}$$

The input resistance is practically infinite, and the output resistance is

$$R_o = r_{o6} \| r_{o7} = \frac{1}{2} \times \frac{20}{0.5} = 20\ \text{k}\Omega$$

To determine f_{P2} we use Eq. (9.27) and substitute for G_{m2},

$$G_{m2} = g_{m6} = \frac{2I_{D6}}{V_{OV}} = \frac{2 \times 0.5}{0.316} = 3.2\ \text{mA/V}$$

Thus,

$$f_{P2} = \frac{3.2 \times 10^{-3}}{2\pi \times 0.8 \times 10^{-12}} = 637\ \text{MHz}$$

To move the transmission zero to $s = \infty$, we select the value of R as

$$R = \frac{1}{G_{m2}} = \frac{1}{3.2 \times 10^{-3}} = 316\ \Omega$$

For a phase margin of 75°, the phase shift due to the second pole at $f = f_t$ must be 15°, that is,

$$\tan^{-1}\frac{f_t}{f_{P2}} = 15°$$

Thus,

$$f_t = 637 \times \tan 15° = 171\ \text{MHz}$$

The value of C_C can be found using Eq. (9.30),

$$C_C = \frac{G_{m1}}{2\pi f_t}$$

where

$$G_{m1} = g_{m1} = \frac{2 \times 100\ \mu\text{A}}{0.316\ \text{V}} = 0.63\ \text{mA/V}$$

Thus,

$$C_{C1} = \frac{0.63 \times 10^{-3}}{2\pi \times 171 \times 10^6} = 0.6\ \text{pF}$$

The value of SR can now be found using Eq. (9.40) as

$$SR = 2\pi \times 171 \times 10^6 \times 0.316$$

$$= 340\ \text{V/}\mu\text{s}$$

9.2 THE FOLDED-CASCODE CMOS OP AMP

In this section we study another type of CMOS op-amp circuit: the folded cascode. The circuit is based on the folded-cascode amplifier studied in Section 6.8.6. There, it was mentioned that although it is composed of a CS transistor and a CG transistor of opposite polarity, the folded-cascode configuration is generally considered to be a single-stage amplifier. Similarly, the op-amp circuit that is based on the cascode configuration is considered to be a single-stage op amp. Nevertheless, it can be designed to provide performance parameters that equal and in some respects exceed those of the two-stage topology studied in the preceding section. Indeed, the folded-cascode op-amp topology is currently as popular as the two-stage structure. Furthermore, the folded-cascode configuration can be used in conjunction with the two-stage structure to provide performance levels higher than those available from either circuit alone.

9.2.1 The Circuit

Figure 9.8 shows the structure of the CMOS folded-cascode op amp. Here, Q_1 and Q_2 form the input differential pair, and Q_3 and Q_4 are the cascode transistors. Recall that for differential input signals, each of Q_1 and Q_2 acts as a common-source amplifier. Also note that the gate terminals of Q_3 and Q_4 are connected to a constant dc voltage (V_{BIAS1}) and hence are

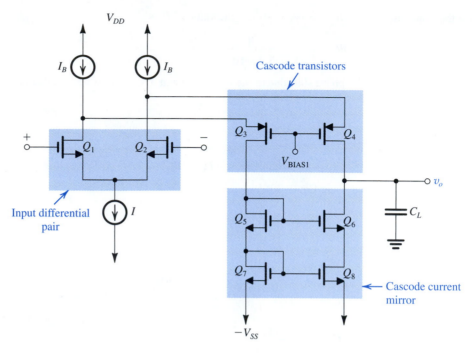

FIGURE 9.8 Structure of the folded-cascode CMOS op amp.

at signal ground. Thus, for differential input signals, each of the transistor pairs Q_1–Q_3 and Q_2–Q_4 acts as a folded-cascode amplifier, such as the one in Fig. 6.45. Note that the input differential pair is biased by a constant-current source I. Thus each of Q_1 and Q_2 is operating at a bias current $I/2$. A node equation at each of their drains shows that the bias current of each of Q_3 and Q_4 is $(I_B - I/2)$. Selecting $I_B = I$ forces all transistors to operate at the same bias current of $I/2$. For reasons that will be explained shortly, however, the value of I_B is usually made somewhat greater than I.

As we learned in Chapter 6, if the full advantage of the high output-resistance achieved through cascoding is to be realized, the output resistance of the current-source load must be equally high. This is the reason for using the cascode current mirror Q_5 to Q_8, in the circuit of Fig. 9.8. (This current-mirror circuit was studied in Section 6.12.1.) Finally, note that capacitance C_L denotes the total capacitance at the output node. It includes the internal transistor capacitances, an actual load capacitance (if any), and possibly an additional capacitance deliberately introduced for the purpose of frequency compensation. In many cases, however, the load capacitance will be sufficiently large, obviating the need to provide additional capacitance to achieve the desired frequency compensation. This topic will be discussed shortly. For the time being, we note that unlike the two-stage circuit, that requires the introduction of a separate compensation capacitor C_C, here the load capacitance contributes to frequency compensation.

A more complete circuit for the CMOS folded-cascode op amp is shown in Fig. 9.9. Here we show the two transistors Q_9 and Q_{10}, which provide the constant bias currents I_B, and transistor Q_{11}, which provides the constant current I utilized for biasing the differential pair. Observe that the details for generating the bias voltages V_{BIAS1}, V_{BIAS2}, and V_{BIAS3} are not shown. Nevertheless, we are interested in how the values of these voltages are to be selected. Toward that end, we evaluate the input common-mode range and the allowable output swing.

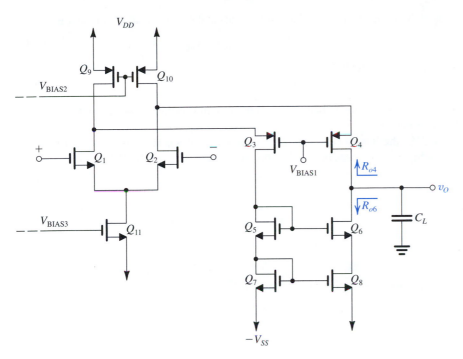

FIGURE 9.9 A more complete circuit for the folded-cascode CMOS amplifier of Fig. 9.8.

9.2.2 Input Common-Mode Range and the Output Voltage Swing

To find the input common-mode range, let the two input terminals be tied together and connected to a voltage V_{ICM}. The maximum value of V_{ICM} is limited by the requirement that Q_1 and Q_2 operate in saturation at all times. Thus V_{ICMmax} should be at most V_{tn} volts above the voltage at the drains of Q_1 and Q_2. The latter voltage is determined by V_{BIAS1} and must allow for a voltage drop across Q_9 and Q_{10} at least equal to their overdrive voltage, $|V_{OV9}| = |V_{OV10}|$. Assuming that Q_9 and Q_{10} are indeed operated at the edge of saturation, V_{ICMmax} will be

$$V_{ICMmax} = V_{DD} - |V_{OV9}| + V_{tn} \tag{9.42}$$

which can be larger than V_{DD}, a significant improvement over the case of the two-stage circuit. The value of V_{BIAS2} should be selected to yield the required value of I_B while operating Q_9 and Q_{10} at a small value of $|V_{OV}|$ (e.g., 0.2 V or so). The minimum value of V_{ICM} is the same as in the case of the two-stage circuit, namely

$$V_{ICMmin} = -V_{SS} + V_{OV11} + V_{OV1} + V_{tn} \tag{9.43}$$

The presence of the threshold voltage V_{tn} in this expression indicates that V_{ICMmin} is not sufficiently low. Later in this section we shall describe an ingenious technique for solving this problem. For the time being, note that the value of V_{BIAS3} should be selected to provide the required value of I while operating Q_{11} at a low overdrive voltage. Combining Eqs. (9.42) and (9.43) provides

$$-V_{SS} + V_{OV11} + V_{OV1} + V_{tn} \leq V_{ICM} \leq V_{DD} - |V_{OV9}| + V_{tn} \tag{9.44}$$

The upper end of the allowable range of v_O is determined by the need to maintain Q_{10} and Q_4 in saturation. Note that Q_{10} will operate in saturation as long as an overdrive voltage, $|V_{OV10}|$, appears across it. It follows that to maximize the allowable positive swing of v_O (and V_{ICMmax}), we should select the value of V_{BIAS1} so that Q_{10} operates at the edge of saturation, that is,

$$V_{BIAS1} = V_{DD} - |V_{OV10}| - V_{SG4} \qquad (9.45)$$

The upper limit of v_O will then be

$$v_{Omax} = V_{DD} - |V_{OV10}| - |V_{OV4}| \qquad (9.46)$$

which is two overdrive voltages below V_{DD}. The situation is not as good, however, at the other end: Since the voltage at the gate of Q_6 is $-V_{SS} + V_{GS7} + V_{GS5}$ or equivalently $-V_{SS} + V_{OV7} + V_{OV5} + 2V_{tn}$, the lowest possible v_O is obtained when Q_6 reaches the edge of saturation, namely, when v_O decreases below the voltage at the gate of Q_6 by V_{tn}, that is,

$$v_{Omin} = -V_{SS} + V_{OV7} + V_{OV5} + V_{tn} \qquad (9.47)$$

Note that this value is two overdrive voltages *plus* a threshold voltage above $-V_{SS}$. This is a drawback of utilizing the cascode mirror. The problem can be alleviated by using a modified mirror circuit, as we shall shortly see.

9.5 For a particular design of the folded-cascode op amp of Fig. 9.9, ±1.65-V supplies are utilized and all transistors are operated at overdrive voltages of 0.3-V magnitude. The fabrication process employed provides $V_{tn} = |V_{tp}| = 0.5$ V. Find the input common-mode range and the range allowed for v_O.

Ans. −0.55 V to +1.85 V; −0.55 V to +1.05 V.

9.2.3 Voltage Gain

The folded-cascode op amp is simply a transconductance amplifier with an infinite input resistance, a transconductance G_m and an output resistance R_o. G_m is equal to g_m of each of the two transistors of the differential pair,

$$G_m = g_{m1} = g_{m2} \qquad (9.48)$$

Thus,

$$G_m = \frac{2(I/2)}{V_{OV1}} = \frac{I}{V_{OV1}} \qquad (9.49)$$

The output resistance R_o is the parallel equivalent of the output resistance of the cascode amplifier and the output resistance of the cascode mirror, thus

$$R_o = R_{o4} \| R_{o6} \qquad (9.50)$$

Reference to Fig. 9.9 shows that the resistance R_{o4} is the output resistance of the CG transistor Q_4. The latter has a resistance $(r_{o2} \| r_{o10})$ in its source lead, thus

$$R_{o4} \cong (g_{m4}r_{o4})(r_{o2} \| r_{o10}) \qquad (9.51)$$

The resistance R_{o6} is the output resistance of the cascode mirror and is thus given by Eq. (6.141), thus

$$R_{o6} \cong g_{m6}r_{o6}r_{o8} \qquad (9.52)$$

Combining Eqs. (9.50) to (9.52) gives

$$R_o = [g_{m4}r_{o4}(r_{o2} \| r_{o10})] \| (g_{m6}r_{o6}r_{o8}) \qquad (9.53)$$

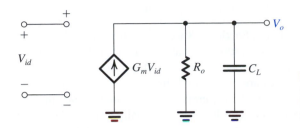

FIGURE 9.10 Small-signal equivalent circuit of the folded-cascode CMOS amplifier. Note that this circuit is in effect an operational transconductance amplifier (OTA).

The dc open-loop gain can now be found using G_m and R_o, as

$$A_v = G_m R_o \tag{9.54}$$

Thus,

$$A_v = g_{m1}\{[g_{m4} r_{o4}(r_{o2} \| r_{o10})] \| (g_{m6} r_{o6} r_{o8})\} \tag{9.55}$$

Figure 9.10 shows the equivalent circuit model including the load capacitance C_L, which we shall take into account shortly.

Because the folded-cascode op amp is a transconductance amplifier, it has been given the name **operational transconductance amplifier (OTA).** Its very high output resistance, which is of the order of $g_m r_o^2$ (see Eq. 9.53) is what makes it possible to realize a relatively high voltage gain in a single amplifier stage. However, such a high output resistance may be a cause of concern to the reader; after all, in Chapter 2, we stated that an ideal op amp has a zero output resistance! To alleviate this concern somewhat, let us find the closed-loop output resistance of a unity-gain follower formed by connecting the output terminal of the circuit of Fig. 9.9 back to the negative input terminal. Since this feedback is of the voltage sampling type, it reduces the output resistance by the factor $(1 + A\beta)$ where $A = A_v$ and $\beta = 1$, that is,

$$R_{of} = \frac{R_o}{1 + A_v} \cong \frac{R_o}{A_v} \tag{9.56}$$

Substituting for A_v from Eq. (9.54) gives

$$R_{of} \cong \frac{1}{G_m} \tag{9.57}$$

which is a general result that applies to any OTA to which 100% voltage feedback is applied. For our particular circuit, $G_m = g_{m1}$, thus

$$R_{of} = 1/g_{m1} \tag{9.58}$$

Since g_{m1} is of the order of 1 mA/V, R_{of} will be of the order of 1 kΩ. Although this is not very small, it is reasonable in view of the simplicity of the op-amp circuit as well as the fact that this type of op amp is not usually intended to drive low-valued resistive loads.

EXERCISE

9.6 The CMOS op amp of Figs. 9.8 and 9.9 is fabricated in a process for which $V_{An}' = |V_{Ap}'| = 20$ V/μm. If all devices have 1-μm channel length and are operated at equal overdrive voltages of 0.2-V magnitude, find the voltage gain obtained. If all devices are biased at 100 μA, what value of R_o is obtained?
Ans. 13,333 V/V; 13.3 MΩ.

9.2.4 Frequency Response

From our study of the cascode configuration in Section 6.8 we know that one of its advantages is its excellent high-frequency response. It has poles at the input, at the connection between the CS and CG transistors (i.e., at the source terminals of Q_3 and Q_4), and at the output terminal. Normally, the first two poles are at very high frequencies, especially when the resistance of the signal generator that feeds the differential pair is small. Since the primary purpose of CMOS op amps is to feed capacitive loads, C_L is usually large, and the pole at the output becomes dominant. Even if C_L is not large, we can increase it deliberately to give the op amp a dominant pole. From Fig. 9.10 we can write

$$\frac{V_o}{V_{id}} = \frac{G_m R_o}{1 + s C_L R_o} \qquad (9.59)$$

Thus, the dominant pole has a frequency f_P,

$$f_P = \frac{1}{2\pi C_L R_o} \qquad (9.60)$$

and the unity-gain frequency f_t will be

$$f_t = G_m R_o f_P = \frac{G_m}{2\pi C_L} \qquad (9.61)$$

From a design point-of-view, the value of C_L should be such that at $f = f_t$ the excess phase resulting from the nondominant poles is small enough to permit the required phase margin to be achieved. If C_L is not large enough to achieve this purpose, it can be augmented.

It is important to note the different effects of increasing the load capacitance on the operation of the two op-amp circuits we have studied. In the two-stage circuit, if C_L is increased, the frequency of the second pole decreases, the excess phase shift at $f = f_t$ increases, and the phase margin is reduced. Here, on the other hand, when C_L is increased, f_t decreases, but the phase margin increases. In other words, a heavier capacitive load decreases the bandwidth of the folded-cascode amplifier but does not impair its response (which happens when the phase margin decreases). Of course, if an increase in C_L is anticipated in the two-stage op-amp case, the designer can increase C_C, thus decreasing f_t and restoring the phase margin to its required value.

9.2.5 Slew Rate

As discussed in Section 9.1.5, slewing occurs when a large differential input signal is applied. Refer to Fig. 9.8 and consider the case when a large signal V_{id} is applied so that Q_2 cuts off and Q_1 conducts the entire bias current I. We see that Q_3 will now carry a current $(I_B - I)$, and Q_4 will conduct a current I_B. The current mirror will see an input current of $(I_B - I)$ through Q_5 and Q_7 and thus its output current in the drain of Q_6 will be $(I_B - I)$. It follows that at the output node the current that will flow into C_L will be $I_4 - I_6 = I_B - (I_B - I) = I$. Thus the output v_O will be a ramp with a slope of I/C_L which is the slew rate,

$$SR = \frac{I}{C_L} \qquad (9.62)$$

Note that the reason for selecting $I_B > I$ is to avoid turning off the current mirror completely; if the current mirror turns off, the output distortion increases. Typically, I_B is set 10% to 20% larger than I. Finally, Eqs. (9.61), (9.62), and (9.49), can be combined to obtain the following relationship between SR and f_t

$$SR = 2\pi f_t V_{OV1} \qquad (9.63)$$

which is identical to the corresponding relationship in the case of the two-stage design. Note, however, that this relationship applies only when $I_B > I$.

EXAMPLE 9.2

Consider a design of the folded-cascode op amp of Fig. 9.9 for which $I = 200\ \mu A$, $I_B = 250\ \mu A$, and $|V_{OV}|$ for all transistors is 0.25 V. Assume that the fabrication process provides $k'_n = 100\ \mu A/V^2$, $k'_p = 40\ \mu A/V^2$, $|V'_A| = 20\ V/\mu m$. $V_{DD} = V_{SS} = 2.5$ V, and $|V_t| = 0.75$ V. Let all transistors have $L = 1\ \mu m$ and assume that $C_L = 5$ pF. Find I_D, g_m, r_o, and W/L for all transistors. Find the allowable range of V_{ICM} and of the output voltage swing. Determine the values of A_v, f_t, f_P, and SR. What is the power dissipation of the op amp?

Solution

From the given values of I and I_B we can determine the drain current I_D for each transistor. The transconductance of each device is found using

$$g_m = \frac{2I_D}{V_{OV}} = \frac{2I_D}{0.25}$$

and the output resistance r_o from

$$r_o = \frac{|V_A|}{I_D} = \frac{20}{I_D}$$

The W/L ratio for each transistor is determined from

$$\left(\frac{W}{L}\right)_i = \frac{2I_{Di}}{k'V_{OV}^2}$$

The results are as follows:

	Q_1	Q_2	Q_3	Q_4	Q_5	Q_6	Q_7	Q_8	Q_9	Q_{10}	Q_{11}
I_D (μA)	100	100	150	150	150	150	150	150	250	250	200
g_m (mA/V)	0.8	0.8	1.2	1.2	1.2	1.2	1.2	1.2	2.0	2.0	1.6
r_o (kΩ)	200	200	133	133	133	133	133	133	80	80	100
W/L	32	32	120	120	48	48	48	48	200	200	64

Note that for all transistors,

$$g_m r_o = 160\ V/V$$

$$V_{GS} = 1.0\ V$$

Using the expression in Eq. (9.44), the input common-mode range is found to be

$$-1.25\ V \leq V_{ICM} \leq 3\ V$$

The output voltage swing is found using Eqs. (9.46) and (9.47) to be

$$-1.25\ V \leq v_O \leq 2\ V$$

To obtain the voltage gain, we first determine R_{o4} using Eq. (9.51) as

$$R_{o4} = 160(200 \| 80) = 9.14\ M\Omega$$

and R_{o6} using Eq. (9.52) as

$$R_{o6} = 21.28 \text{ M}\Omega$$

The output resistance R_o can then be found as

$$R_o = R_{o4} \| R_{o6} = 6.4 \text{ M}\Omega$$

and the voltage gain

$$A_v = G_m R_o = 0.8 \times 10^{-3} \times 6.4 \times 10^{6}$$

$$= 5120 \text{ V/V}$$

The unity-gain bandwidth is found using Eq. (9.61),

$$f_t = \frac{0.8 \times 10^{-3}}{2\pi \times 5 \times 10^{-12}} = 25.5 \text{ MHz}$$

Thus, the dominant-pole frequency must be

$$f_P = \frac{f_t}{A_v} = \frac{25.5 \text{ MHz}}{5120} = 5 \text{ kHz}$$

The slew rate can be determined using Eq. (9.62),

$$SR = \frac{I}{C_L} = \frac{200 \times 10^{-6}}{5 \times 10^{-12}} = 40 \text{V}/\mu\text{s}$$

Finally, to determine the power dissipation we note that the total current is 500 μA = 0.5 mA, and the total supply voltage is 5 V, thus

$$P_D = 5 \times 0.5 = 2.5 \text{ mW}$$

9.2.6 Increasing the Input Common-Mode Range: Rail-to-Rail Input Operation

In Section 9.2.2 we found that while the upper limit on the input common-mode range exceeds the supply voltage V_{DD}, the lower limit is significantly lower than V_{SS}. The opposite situation occurs if the input differential amplifier is made up of PMOS transistors. It follows that an NMOS and a PMOS differential pair placed in parallel would provide an input stage with a common-mode range that exceeds the power supply voltage in both directions. This is known as rail-to-rail input operation. Figure 9.11 shows such an arrangement. To keep the diagram simple, we have not shown the parallel connection of the two differential pairs: The two positive input terminals are to be connected together and the two negative input terminals are to be tied together. Transistors Q_5 and Q_6 are the cascode transistors for the Q_1–Q_2 pair, and transistors Q_7 and Q_8 are the cascode devices for the Q_3–Q_4 pair. The output voltage V_o is shown taken differentially between the drains of the cascode devices. To obtain a single-ended output, a differential-to-single-ended conversion circuit should be connected in cascade.

Figure 9.11 indicates by arrows the direction of the current increments that result from the application of a positive differential input signal V_{id}. Each of the current increments indicated is equal to $G_m(V_{id}/2)$ where $G_m = g_{m1} = g_{m2} = g_{m3} = g_{m4}$. Thus the total current feeding each of the two output nodes will be $G_m V_{id}$. Now, if the output resistance between each of the two nodes and ground is denoted R_o, the output voltage will be

$$V_o = 2G_m R_o V_{id} \tag{9.64}$$

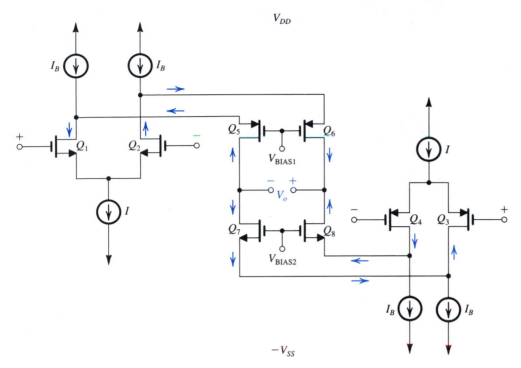

FIGURE 9.11 A folded-cascode op amp that employs two parallel complementary input stages to achieve rail-to-rail input common-mode operation. Note that the two "+" terminals are connected together and the two "−" terminals are connected together.

Thus, the voltage gain will be

$$A_v = 2G_m R_o \tag{9.65}$$

This, however, assumes that both differential pairs will be operating simultaneously. This in turn occurs only over a limited range of V_{ICM}. Over the remainder of the input common-mode range, only one of the two differential pairs will be operational, and the gain drops to half of the value in Eq. (9.65). This rail-to-rail folded-cascode structure is utilized in a commercially available op amp.[1]

EXERCISE

9.7 For the circuit in Fig. 9.11, assume that all transistors are operating at equal overdrive voltages of 0.3-V magnitude and have $|V_t| = 0.7$ V and that $V_{DD} = V_{SS} = 2.5$ V.

(a) Find the range over which the NMOS input stage operates.

(b) Find the range over which the PMOS input stage operates.

(c) Find the range over which both operate (the overlap range).

(d) Find the input common-mode range.

(Note that to operate properly, each of the current sources requires a minimum voltage of $|V_{OV}|$ across its terminals).

Ans. −1.2 V to +2.9 V; −2.9 V to +1.2 V, −1.2 V to +1.2 V; −2.9 V to +2.9 V

---

[1] The Texas Instruments OPA357.

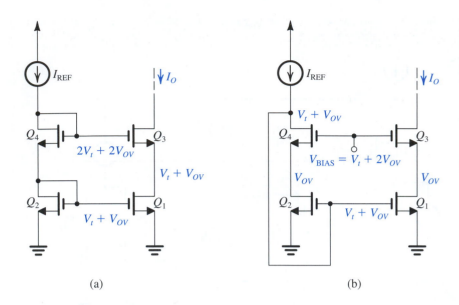

FIGURE 9.12 **(a)** Cascode current mirror with the voltages at all nodes indicated. Note that the minimum voltage allowed at the output is $V_t + V_{OV}$. **(b)** A modification of the cascode mirror that results in the reduction of the minimum output voltage to V_{OV}. This is the wide-swing current mirror.

9.2.7 Increasing the Output Voltage Range: The Wide-Swing Current Mirror

In Section 9.2.2 it was found that while the output voltage of the circuit of Fig. 9.9 can swing to within $2|V_{OV}|$ of V_{DD}, the cascode current mirror limits the negative swing to $[2|V_{OV}| + V_t]$ above $-V_{SS}$. In other words, the cascode mirror reduces the voltage swing by V_t volts. This point is further illustrated in Fig. 9.12(a), which shows a cascode mirror (with $V_{SS} = 0$, for simplicity) and indicates the voltages that result at the various nodes. Observe that because the voltage at the gate of Q_3 is $2V_t + 2V_{OV}$, the minimum voltage permitted at the output (while Q_3 remains saturated) is $V_t + 2V_{OV}$, hence the extra V_t. Also, observe that Q_1 is operating with a drain-to-source voltage $V_t + V_{OV}$, which is V_t volts greater than it needs to operate in saturation.

The observations above lead us to the conclusion that to permit the output voltage at the drain of Q_3 to swing as low as $2V_{OV}$, we must lower the voltage at the gate of Q_3 from $2V_t + 2V_{OV}$ to $V_t + 2V_{OV}$. This is exactly what is done in the modified mirror circuit in Fig. 9.12(b): The gate of Q_3 is now connected to a bias voltage $V_{BIAS} = V_t + 2V_{OV}$. Thus the output voltage can go down to $2V_{OV}$ with Q_3 still in saturation. Also, the voltage at the drain of Q_1 is now V_{OV} and thus Q_1 is operating at the edge of saturation. The same is true of Q_2 and thus the current tracking between Q_1 and Q_2 will be assured. Note, however, that we can no longer connect the gate of Q_2 to its drain. Rather, it is connected to the drain of Q_4. This establishes a voltage of $V_t + V_{OV}$ at the drain of Q_4 which is sufficient to operate Q_4 in saturation (as long as V_t is greater than V_{OV}, which is usually the case). This circuit is known as the **wide-swing current mirror.** Finally, note that Fig. 9.12(b) does not show the circuit for generating V_{BIAS}. There are a number of possible circuits to accomplish this task, one of which is explored in Exercise 9.8.

EXERCISE

9.8 Show that if transistor Q_5 in the circuit of Fig. E9.8 has a W/L ratio equal to one-quarter that of the transistors in the wide-swing current mirror of Fig. 9.12(b), and provided the same value of I_{REF} is utilized in both circuits, then the voltage generated, V_5 is $V_t + 2V_{OV}$, which is the value of V_{BIAS} needed for the gates of Q_3 and Q_4.

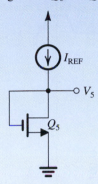

FIGURE E9.8

 ## 9.3 THE 741 OP-AMP CIRCUIT

Our study of BJT op amps is focused on the 741 op-amp circuit, which is shown in Fig. 9.13. Note that in keeping with the IC design philosophy the circuit uses a large number of transistors, but relatively few resistors, and only one capacitor. This philosophy is dictated by the economics (silicon area, ease of fabrication, quality of realizable components) of the fabrication of active and passive components in IC form (see Section 6.1 and Appendix A).

As is the case with most general-purpose IC op amps, the 741 requires two power supplies, $+V_{CC}$ and $-V_{EE}$. Normally, $V_{CC} = V_{EE} = 15$ V, but the circuit also operates satisfactorily with the power supplies reduced to much lower values (such as ±5 V). It is important to observe that no circuit node is connected to ground, the common terminal of the two supplies.

With a relatively large circuit such as that shown in Fig. 9.13, the first step in the analysis is the identification of its recognizable parts and their functions. This can be done as follows.

9.3.1 Bias Circuit

The reference bias current of the 741 circuit, I_{REF}, is generated in the branch at the extreme left of Fig. 9.13, consisting of the two diode-connected transistors Q_{11} and Q_{12} and the resistance R_5. Using a Widlar current source formed by Q_{11}, Q_{10}, and R_4, bias current for the first stage is generated in the collector of Q_{10}. Another current mirror formed by Q_8 and Q_9 takes part in biasing the first stage.

The reference bias current I_{REF} is used to provide two proportional currents in the collectors of Q_{13}. This double-collector *lateral*[2] *pnp* transistor can be thought of as two

---

[2] See Appendix A for a description of lateral *pnp* transistors. Also, their characteristics were discussed in Section 6.2.

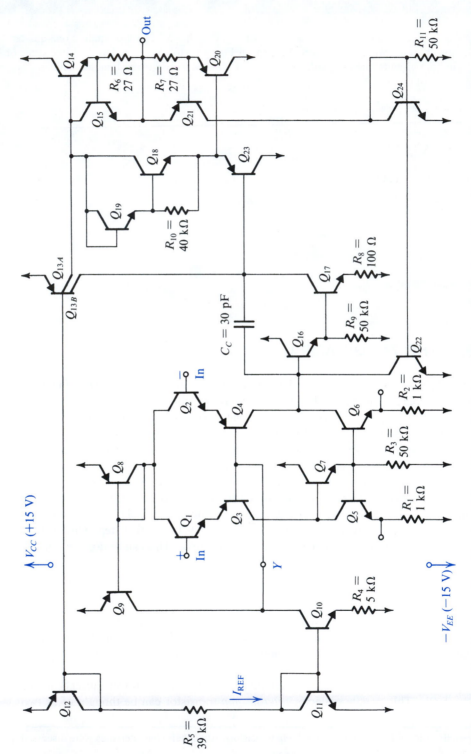

FIGURE 9.13 The 741 op-amp circuit. Q_{11}, Q_{12}, and R_5 generate a reference bias current, I_{REF}. Q_{10}, Q_9, and Q_8 bias the input stage, which is composed of Q_1 to Q_7. The second gain stage is composed of Q_{16} and Q_{17} with Q_{13B} acting as active load. The class AB output stage is formed by Q_{14} and Q_{20} with biasing devices Q_{13A}, Q_{18}, and Q_{19}, and an input buffer Q_{23}. Transistors Q_{15}, Q_{21}, Q_{24}, and Q_{22} serve to protect the amplifier against output short circuits and are normally cut off.

transistors whose base–emitter junctions are connected in parallel. Thus Q_{12} and Q_{13} form a two-output current mirror: One output, the collector of Q_{13B}, provides bias current for Q_{17}, and the other output, the collector of Q_{13A}, provides bias current for the output stage of the op amp.

Two more transistors, Q_{18} and Q_{19}, take part in the dc bias process. The purpose of Q_{18} and Q_{19} is to establish two V_{BE} drops between the bases of the output transistors Q_{14} and Q_{20}.

9.3.2 Short-Circuit Protection Circuitry

The 741 circuit includes a number of transistors that are normally off and conduct only in the event that one attempts to draw a large current from the op-amp output terminal. This happens, for example, if the output terminal is short-circuited to one of the two supplies. The short-circuit protection network consists of R_6, R_7, Q_{15}, Q_{21}, Q_{24}, R_{11}, and Q_{22}. In the following we shall assume that these transistors are off. Operation of the short-circuit protection network will be explained in Section 9.5.3.

9.3.3 The Input Stage

The 741 circuit consists of three stages: an input differential stage, an intermediate single-ended high-gain stage, and an output-buffering stage. The input stage consists of transistors Q_1 through Q_7, with biasing performed by Q_8, Q_9, and Q_{10}. Transistors Q_1 and Q_2 act as emitter followers, causing the input resistance to be high and delivering the differential input signal to the differential common-base amplifier formed by Q_3 and Q_4. Thus the input stage is the differential version of the common-collector common-base configuration discussed in Section 6.11.3.

Transistors Q_5, Q_6, and Q_7 and resistors R_1, R_2, and R_3 form the load circuit of the input stage. This is an elaborate current-mirror load circuit, which we will analyze in detail in Section 9.5.1. It will be shown that this load circuit not only provides a high-resistance load but also converts the signal from differential to single-ended form with no loss in gain or common-mode rejection. The output of the input stage is taken single-endedly at the collector of Q_6.

As mentioned in Section 7.7.2, every op-amp circuit includes a *level shifter* whose function is to shift the dc level of the signal so that the signal at the op-amp output can swing positive and negative. In the 741, level shifting is done in the first stage using the lateral *pnp* transistors Q_3 and Q_4. Although lateral *pnp* transistors have poor high-frequency performance, their use in the common-base configuration (which is known to have good high-frequency response) does not seriously impair the op-amp frequency response.

The use of the lateral *pnp* transistors Q_3 and Q_4 in the first stage results in an added advantage: protection of the input-stage transistors Q_1 and Q_2 against emitter–base junction breakdown. Since the emitter–base junction of an *npn* transistor breaks down at about 7 V of reverse bias (see Section 5.2.5), regular *npn* differential stages suffer such a breakdown if, say, the supply voltage is accidentally connected between the input terminals. Lateral *pnp* transistors, however, have high emitter–base breakdown voltages (about 50 V); and because they are connected in series with Q_1 and Q_2, they provide protection of the 741 input transistors, Q_1 and Q_2.

9.3.4 The Second Stage

The second or intermediate stage is composed of Q_{16}, Q_{17}, Q_{13B}, and the two resistors R_8 and R_9. Transistor Q_{16} acts as an emitter follower, thus giving the second stage a high input resistance. This minimizes the loading on the input stage and avoids loss of gain. Transistor Q_{17} acts as a common-emitter amplifier with a 100-Ω resistor in the emitter. Its load is composed of the high output resistance of the *pnp* current source Q_{13B} in parallel with the input resistance of the output stage (seen looking into the base of Q_{23}). Using a transistor current source as a load resistance (*active load*) enables one to obtain high gain without resorting to the use of large load resistances, which would occupy a large chip area and require large power-supply voltages.

The output of the second stage is taken at the collector of Q_{17}. Capacitor C_C is connected in the feedback path of the second stage to provide frequency compensation using the Miller compensation technique studied in Section 8.11. It will be shown in Section 9.5 that the relatively small capacitor C_C gives the 741 a dominant pole at about 4 Hz. Furthermore, pole splitting causes other poles to be shifted to much higher frequencies, giving the op amp a uniform −20-dB/decade gain rolloff with a unity-gain bandwidth of about 1 MHz. It should be pointed out that although C_C is small in value, the chip area that it occupies is about 13 times that of a standard *npn* transistor!

9.3.5 The Output Stage

The purpose of the output stage is to provide the amplifier with a low output resistance. In addition, the output stage should be able to supply relatively large load currents without dissipating an unduly large amount of power in the IC. The 741 uses an efficient output circuit known as a **class AB output stage.**

Output stages are studied in detail in Chapter 14, and the output stage of the 741 will be discussed in some detail in Section 9.4. For the time being we wish to point out the difference between the class AB output stage and the output stage we are familiar with, namely the emitter (or source) follower. Figure 9.14(a) shows an emitter follower biased with a constant-current source I. To keep the emitter-follower transistor conducting at all times and thus ensure the low output resistance it provides, the bias current I must be greater than the largest magnitude of load current i_L. This is known as **class A operation** and the emitter (source) follower is a **class A output stage.** The drawback of class A operation is the large power dissipated in the transistor.

The power dissipated in the output-stage can be reduced by arranging for the transistor to turn on only when an input signal is applied. For this to work, however, one needs two transistors, an *npn* to source output current and a *pnp* to sink output current. Such an arrangement is shown in Fig. 9.14(b). Observe that both transistors will be cut off when $v_I = 0$. In other words, the transistors are biased at a zero dc current. When v_I goes positive, Q_N conducts while Q_P remain off. When v_I goes negative the transistors reverse roles. This arrangement is known as **class B operation** and the circuit as a class B output stage.

Although efficient in terms of power dissipation, the class B circuit causes output-signal distortion, as illustrated in Fig. 9.14(c). This is a result of the fact that for $|v_I|$ less than about 0.5 V, neither of the transistors conducts and $v_O = 0$. This type of distortion is known as **crossover distortion.**

Crossover distortion can be reduced by biasing the output-stage transistors at a low current. This ensures that the output transistors Q_N and Q_P will remain conducting when v_I is small. As v_I increases, one of the two transistors conducts more, while the other shuts off, in a manner similar to that in the class B stage.

There are a number of ways for biasing the transistors of the class AB stage. Figure 9.14(d) shows one such approach utilizing two diode-connected transistors Q_1 and Q_2 with junction

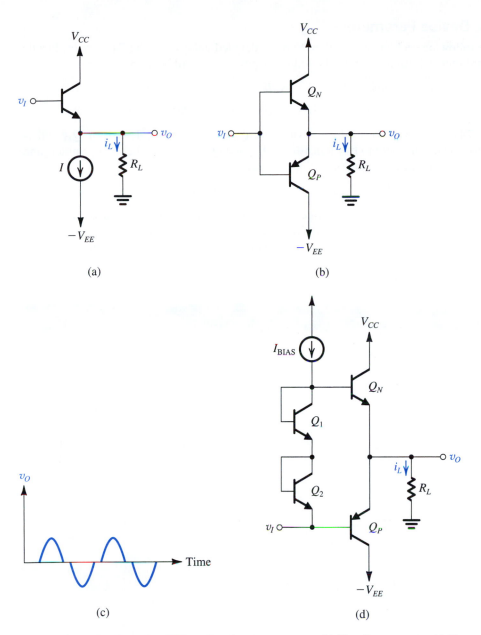

FIGURE 9.14 **(a)** The emitter follower is a class A output stage. **(b)** Class B output stage. **(c)** The output of a class B output stage fed with an input sinusoid. Observe the crossover distortion. **(d)** Class AB output stage.

areas much smaller than those of Q_N and Q_P. A somewhat more elaborate biasing network is utilized in the 741 output stage.

The output stage of the 741 consists of the complementary pair Q_{14} and Q_{20}, where Q_{20} is a *substrate pnp* (see Appendix A). Transistors Q_{18} and Q_{19} are fed by current source Q_{13A} and bias the output transistors Q_{14} and Q_{20}. Transistor Q_{23} (which is another substrate *pnp*) acts as an emitter follower, thus minimizing the loading effect of the output stage on the second stage.

9.3.6 Device Parameters

In the following sections we shall carry out a detailed analysis of the 741 circuit. For the standard *npn* and *pnp* transistors, the following parameters will be used:

$$npn: \quad I_S = 10^{-14}\text{A}, \ \beta = 200, \ V_A = 125 \text{ V}$$

$$pnp: \quad I_S = 10^{-14}\text{A}, \ \beta = 50, \ V_A = 50 \text{ V}$$

In the 741 circuit the nonstandard devices are Q_{13}, Q_{14}, and Q_{20}. Transistor Q_{13} will be assumed to be equivalent to two transistors, Q_{13A} and Q_{13B}, with parallel base–emitter junctions and having the following saturation currents:

$$I_{SA} = 0.25 \times 10^{-14}\text{A} \qquad I_{SB} = 0.75 \times 10^{-14}\text{A}$$

Transistors Q_{14} and Q_{20} will be assumed to each have an area three times that of a standard device. Output transistors usually have relatively large areas, to be able to supply large load currents and dissipate relatively large amounts of power with only a moderate increase in device temperature.

EXERCISES

9.9 For the standard *npn* transistor whose parameters are given in Section 9.3.6, find approximate values for the following parameters at $I_C = 1$ mA: V_{BE}, g_m, r_e, r_π, and r_o.

Ans. 633 mV; 40 mA/V; 25 Ω; 5 kΩ; 125 kΩ

9.10 For the circuit in Fig. E9.10, neglect base currents and use the exponential $i_C–v_{BE}$ relationship to show that

$$I_3 = I_1 \sqrt{\frac{I_{S3}I_{S4}}{I_{S1}I_{S2}}}$$

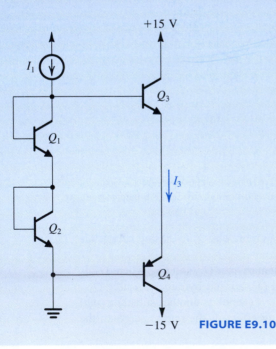

FIGURE E9.10

9.4 DC ANALYSIS OF THE 741

In this section, we shall carry out a dc analysis of the 741 circuit to determine the bias point of each device. For the dc analysis of an op-amp circuit the input terminals are grounded. Theoretically speaking, this should result in zero dc voltage at the output. However, because the op amp has very large gain, any slight approximation in the analysis will show that the output voltage is far from being zero and is close to either $+V_{CC}$ or $-V_{EE}$. In actual practice, an op amp left open-loop will have an output voltage saturated close to one of the two supplies. To overcome this problem in the dc analysis, it will be assumed that the op amp is connected in a negative-feedback loop that stabilizes the output dc voltage to zero volts.

9.4.1 Reference Bias Current

The reference bias current I_{REF} is generated in the branch composed of the two diode-connected transistors Q_{11} and Q_{12} and resistor R_5. With reference to Fig. 9.13, we can write

$$I_{REF} = \frac{V_{CC} - V_{EB12} - V_{BE11} - (-V_{EE})}{R_5}$$

For $V_{CC} = V_{EE} = 15$ V and $V_{BE11} = V_{EB12} \simeq 0.7$ V, we have $I_{REF} = 0.73$ mA.

9.4.2 Input-Stage Bias

Transistor Q_{11} is biased by I_{REF}, and the voltage developed across it is used to bias Q_{10}, which has a series emitter resistance R_4. This part of the circuit is redrawn in Fig. 9.15 and can be recognized as the Widlar current source studied in Section 6.12.5. From the circuit, and assuming β_{10} to be large, we have

$$V_{BE11} - V_{BE10} = I_{C10}R_4$$

Thus

$$V_T \ln \frac{I_{REF}}{I_{C10}} = I_{C10}R_4 \tag{9.66}$$

where it has been assumed that $I_{S10} = I_{S11}$. Substituting the known values for I_{REF} and R_4, this equation can be solved by trial and error to determine I_{C10}. For our case, the result is $I_{C10} = 19$ μA.

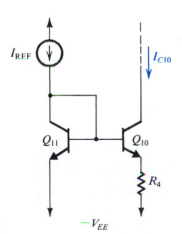

FIGURE 9.15 The Widlar current source.

D9.11 Design the Widlar current source of Fig. 9.15 to generate a current $I_{C10} = 10\ \mu A$ given that $I_{REF} = 1$ mA. If at a collector current of 1 mA, $V_{BE} = 0.7$ V, find V_{BE11} and V_{BE10}.

Ans. $R_4 = 11.5\ k\Omega$; $V_{BE11} = 0.7$ V; $V_{BE10} = 0.585$ V

Having determined I_{C10}, we proceed to determine the dc current in each of the input-stage transistors. Part of the input stage is redrawn in Fig. 9.16. From symmetry, we see that

$$I_{C1} = I_{C2}$$

Denote this current by I. We see that if the *npn* β is high, then

$$I_{E3} = I_{E4} \simeq I$$

and the base currents of Q_3 and Q_4 are equal, with a value of $I/(\beta_P + 1) \simeq I/\beta_P$, where β_P denotes β of the *pnp* devices.

The current mirror formed by Q_8 and Q_9 is fed by an input current of $2I$. Using the result in Eq. (6.21), we can express the output current of the mirror as

$$I_{C9} = \frac{2I}{1 + 2/\beta_P}$$

We can now write a node equation for node X in Fig. 9.16 and thus determine the value of I. If $\beta_P \gg 1$, then this node equation gives

$$2I \simeq I_{C10}$$

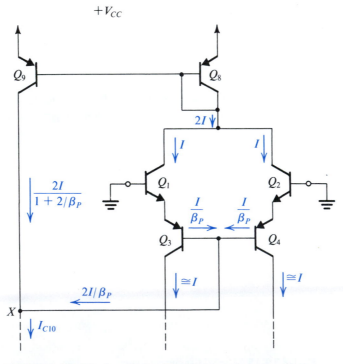

FIGURE 9.16 The dc analysis of the 741 input stage.

For the 741, $I_{C10} = 19$ μA; thus $I \simeq 9.5$ μA. We have thus determined that

$$I_{C1} = I_{C2} \simeq I_{C3} = I_{C4} = 9.5 \ \mu A$$

At this point, we should note that transistors Q_1 through Q_4, Q_8, and Q_9 form a **negative-feedback loop,** which works to stabilize the value of I at approximately $I_{C10}/2$. To appreciate this fact, assume that for some reason the current I in Q_1 and Q_2 increases. This will cause the current pulled from Q_8 to increase, and the output current of the Q_8–Q_9 mirror will correspondingly increase. However, since I_{C10} remains constant, node X forces the combined base currents of Q_3 and Q_4 to decrease. This in turn will cause the emitter currents of Q_3 and Q_4, and hence the collector currents of Q_1 and Q_2, to decrease. This is opposite in direction to the change originally assumed. Hence the feedback is negative, and it stabilizes the value of I.

Figure 9.17 shows the remainder of the 741 input stage. If we neglect the base current of Q_{16}, then

$$I_{C6} \simeq I$$

Similarly, neglecting the base current of Q_7 we obtain

$$I_{C5} \simeq I$$

The bias current of Q_7 can be determined from

$$I_{C7} \simeq I_{E7} = \frac{2I}{\beta_N} + \frac{V_{BE6} + IR_2}{R_3} \tag{9.67}$$

where β_N denotes β of the *npn* transistors. To determine V_{BE6} we use the transistor exponential relationship and write

$$V_{BE6} = V_T \ln \frac{I}{I_S}$$

Substituting $I_S = 10^{-14}$ A and $I = 9.5$ μA results in $V_{BE6} = 517$ mV. Then substituting in Eq. (9.67) yields $I_{C7} = 10.5$ μA. Note that the base current of Q_7 is indeed negligible in comparison to the value of I, as has been assumed.

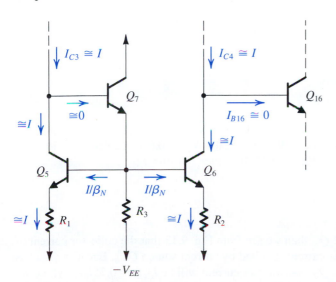

FIGURE 9.17 The dc analysis of the 741 input stage, continued.

9.4.3 Input Bias and Offset Currents

The **input bias current** of an op amp is defined (Chapters 2 and 7) as

$$I_B = \frac{I_{B1} + I_{B2}}{2}$$

For the 741 we obtain

$$I_B = \frac{I}{\beta_N}$$

Using $\beta_N = 200$, yields $I_B = 47.5$ nA. Note that this value is reasonably small and is typical of general-purpose op amps that use BJTs in the input stage. Much lower input bias currents (in the picoamp or femtoamp range) can be obtained using a FET input stage. Also, there exist techniques for reducing the input bias current of bipolar-input op amps.

Because of possible mismatches in the β values of Q_1 and Q_2, the input base currents will not be equal. Given the value of the β mismatch, one can use Eq. (7.137) to calculate the **input offset current,** defined as

$$I_{OS} = |I_{B1} - I_{B2}|$$

9.4.4 Input Offset Voltage

From Chapter 7 we know that the input offset voltage is determined primarily by mismatches between the two sides of the input stage. In the 741 op amp, the input offset voltage is due to mismatches between Q_1 and Q_2, between Q_3 and Q_4, between Q_5 and Q_6, and between R_1 and R_2. Evaluation of the components of V_{OS} corresponding to the various mismatches follows the method outlined in Section 7.4. Basically, we find the current that results at the output of the first stage due to the particular mismatch being considered. Then we find the differential input voltage that must be applied to reduce the output current to zero.

9.4.5 Input Common-Mode Range

The **input common-mode range** is the range of input common-mode voltages over which the input stage remains in the linear active mode. Refer to Fig. 9.13. We see that in the 741 circuit the input common-mode range is determined at the upper end by saturation of Q_1 and Q_2, and at the lower end by saturation of Q_3 and Q_4.

EXERCISE

9.12 Neglect the voltage drops across R_1 and R_2 and assume that $V_{CC} = V_{EE} = 15$ V. Show that the input common-mode range of the 741 is approximately -12.9 to $+14.7$ V. (Assume that $V_{BE} \simeq 0.6$ V and that to avoid saturation $V_{CB} \geq -0.3$ V for an *npn* transistor, and $V_{BC} \geq -0.3$ V for a *pnp* transistor.)

9.4.6 Second-Stage Bias

If we neglect the base current of Q_{23} then we see from Fig. 9.13 that the collector current of Q_{17} is approximately equal to the current supplied by current source Q_{13B}. Because Q_{13B} has a scale current 0.75 times that of Q_{12}, its collector current will be $I_{C13B} \simeq 0.75 I_{REF}$, where we have assumed that $\beta_P \gg 1$. Thus $I_{C13B} = 550$ μA and $I_{C17} \simeq 550$ μA. At this current level the

base–emitter voltage of Q_{17} is

$$V_{BE17} = V_T \ln \frac{I_{C17}}{I_S} = 618 \text{ mV}$$

The collector current of Q_{16} can be determined from

$$I_{C16} \simeq I_{E16} = I_{B17} + \frac{I_{E17}R_8 + V_{BE17}}{R_9}$$

This calculation yields $I_{C16} = 16.2 \ \mu A$. Note that the base current of Q_{16} will indeed be negligible compared to the input-stage bias I, as we have assumed.

9.4.7 Output-Stage Bias

Figure 9.18 shows the output stage of the 741 with the short-circuit-protection circuitry omitted. Current source Q_{13A} delivers a current of $0.25I_{REF}$ (because I_S of Q_{13A} is 0.25 times the I_S of Q_{12}) to the network composed of Q_{18}, Q_{19}, and R_{10}. If we neglect the base currents of Q_{14} and Q_{20}, then the emitter current of Q_{23} will also be equal to $0.25I_{REF}$. Thus

$$I_{C23} \simeq I_{E23} \simeq 0.25I_{REF} = 180 \ \mu A$$

Thus we see that the base current of Q_{23} is only $180/50 = 3.6 \ \mu A$, which is negligible compared to I_{C17}, as we have assumed.

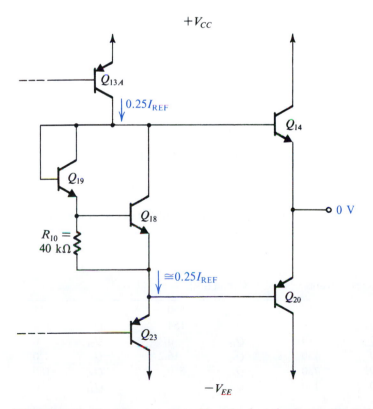

FIGURE 9.18 The 741 output stage without the short-circuit protection devices.

If we assume that V_{BE18} is approximately 0.6 V, we can determine the current in R_{10} as 15 μA. The emitter current of Q_{18} is therefore

$$I_{E18} = 180 - 15 = 165 \ \mu A$$

Also,

$$I_{C18} \simeq I_{E18} = 165 \ \mu A$$

At this value of current we find that $V_{BE18} = 588$ mV, which is quite close to the value assumed. The base current of Q_{18} is $165/200 = 0.8$ μA, which can be added to the current in R_{10} to determine the Q_{19} current as

$$I_{C19} \simeq I_{E19} = 15.8 \ \mu A$$

The voltage drop across the base–emitter junction of Q_{19} can now be determined as

$$V_{BE19} = V_T \ln \frac{I_{C19}}{I_S} = 530 \ \text{mV}$$

As mentioned in Section 9.3.5, the purpose of the Q_{18}–Q_{19} network is to establish two V_{BE} drops between the bases of the output transistors Q_{14} and Q_{20}. This voltage drop, V_{BB}, can be now calculated as

$$V_{BB} = V_{BE18} + V_{BE19} = 588 + 530 = 1.118 \ \text{V}$$

Since V_{BB} appears across the series combination of the base–emitter junctions of Q_{14} and Q_{20}, we can write

$$V_{BB} = V_T \ln \frac{I_{C14}}{I_{S14}} + V_T \ln \frac{I_{C20}}{I_{S20}}$$

Using the calculated value of V_{BB} and substituting $I_{S14} = I_{S20} = 3 \times 10^{-14}$ A, we determine the collector currents as

$$I_{C14} = I_{C20} = 154 \ \mu A$$

This is the small current at which the class AB output stage is biased.

9.4.8 Summary

For future reference, Table 9.1 provides a listing of the values of the collector bias currents of the 741 transistors.

TABLE 9.1	Dc Collector Currents of the 741 Circuit (μA)						
Q_1	9.5	Q_8	19	Q_{13B}	550	Q_{19}	15.8
Q_2	9.5	Q_9	19	Q_{14}	154	Q_{20}	154
Q_3	9.5	Q_{10}	19	Q_{15}	0	Q_{21}	0
Q_4	9.5	Q_{11}	730	Q_{16}	16.2	Q_{22}	0
Q_5	9.5	Q_{12}	730	Q_{17}	550	Q_{23}	180
Q_6	9.5	Q_{13A}	180	Q_{18}	165	Q_{24}	0
Q_7	10.5						

9.5 SMALL-SIGNAL ANALYSIS OF THE 741

9.5.1 The Input Stage

Figure 9.19 shows part of the 741 input stage for the purpose of performing small-signal analysis. Note that since the collectors of Q_1 and Q_2 are connected to a constant dc voltage, they are shown grounded. Also, the constant-current biasing of the bases of Q_3 and Q_4 is equivalent to having the common base terminal open-circuited.

The differential signal v_i applied between the input terminals effectively appears across four equal emitter resistances connected in series—those of Q_1, Q_2, Q_3, and Q_4. As a result, emitter signal currents flow as indicated in Fig. 9.19 with

$$i_e = \frac{v_i}{4r_e} \qquad (9.68)$$

where r_e denotes the emitter resistance of each of Q_1 through Q_4. Thus

$$r_e = \frac{V_T}{I} = \frac{25 \text{ mV}}{9.5 \ \mu A} = 2.63 \text{ k}\Omega$$

Thus the four transistors Q_1 through Q_4 supply the load circuit with a pair of complementary current signals αi_e, as indicated in Fig. 9.19.

The input differential resistance of the op amp can be obtained from Fig. 9.19 as

$$R_{id} = 4(\beta_N + 1)r_e \qquad (9.69)$$

For $\beta_N = 200$, we obtain $R_{id} = 2.1$ MΩ.

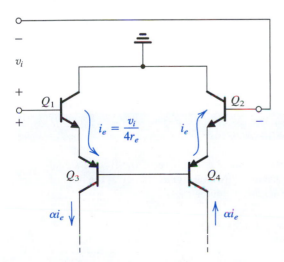

FIGURE 9.19 Small-signal analysis of the 741 input stage.

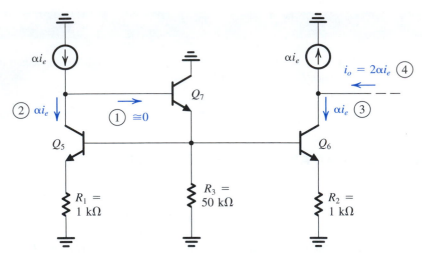

FIGURE 9.20 The load circuit of the input stage fed by the two complementary current signals generated by Q_1 through Q_4 in Fig. 9.19. Circled numbers indicate the order of the analysis steps.

Proceeding with the input-stage analysis, we show in Fig. 9.20 the load circuit fed with the complementary pair of current signals found earlier. Neglecting the signal current in the base of Q_7, we see that the collector signal current of Q_5 is approximately equal to the input current αi_e. Now, since Q_5 and Q_6 are identical and their bases are tied together, and since equal resistances are connected in their emitters, it follows that their collector signal currents must be equal. Thus the signal current in the collector of Q_6 is forced to be equal to αi_e. In other words, the load circuit functions as a current mirror.

Now consider the output node of the input stage. The output current i_o is given by

$$i_o = 2\alpha i_e \qquad (9.70)$$

The factor of 2 in this equation indicates that conversion from differential to single-ended is performed without losing half the signal. The trick, of course, is the use of the current mirror to invert one of the current signals and then add the result to the other current signal (see Section 7.5).

Equations (9.68) and (9.70) can be combined to obtain the transconductance of the input stage, G_{m1}:

$$G_{m1} \equiv \frac{i_o}{v_i} = \frac{\alpha}{2r_e} \qquad (9.71)$$

Substituting $r_e = 2.63$ kΩ and $\alpha \simeq 1$ yields $G_{m1} = 1/5.26$ mA/V.

EXERCISE

9.14 For the circuit in Fig. 9.20, find in terms of i_e: (a) the signal voltage at the base of Q_6; (b) the signal current in the emitter of Q_7; (c) the signal current in the base of Q_7; (d) the signal voltage at the base of Q_7; (e) the input resistance seen by the left-hand-side signal current source αi_e. (*Note:* For simplicity, assume that $I_{C7} \simeq I_{C5} = I_{C6}$.)
Ans. (a) 3.63 kΩ $\times i_e$; (b) $0.08i_e$; (c) $0.0004i_e$; (d) 3.84 kΩ $\times i_e$; (e) 3.84 kΩ

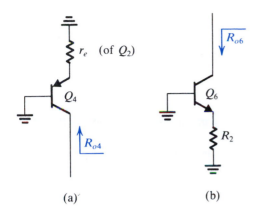

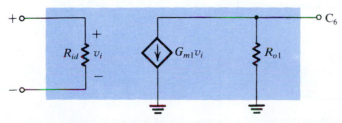

FIGURE 9.21 Simplified circuits for finding the two components of the output resistance R_{o1} of the first stage.

To complete our modeling of the 741 input stage we must find its output resistance R_{o1}. This is the resistance seen "looking back" into the collector terminal of Q_6 in Fig. 9.20. Thus R_{o1} is the parallel equivalent of the output resistance of the current source supplying the signal current αi_e and the output resistance of Q_6. The first component is the resistance looking into the collector of Q_4 in Fig. 9.19. Finding this resistance is considerably simplified if we assume that the common bases of Q_3 and Q_4 are at a *virtual ground*. This of course happens only when the input signal v_i is applied in a complementary fashion. Nevertheless, this assumption does not result in a large error.

Assuming that the base of Q_4 is at virtual ground, the resistance we are after is R_{o4}, indicated in Fig. 9.21(a). This is the output resistance of a common-base transistor that has a resistance (r_e of Q_2) in its emitter. To find R_{o4} we may use the following expression (Eq. 6.118):

$$R_o = r_o[1 + g_m(R_E//r_\pi)] \tag{9.72}$$

Substituting $R_E = r_e \equiv 2.63 \text{ k}\Omega$ and $r_o = V_A/I$, where $V_A = 50$ V and $I = 9.5$ μA (thus $r_o = 5.26$ MΩ), and neglecting r_π since it is ($\beta + 1$) times larger than R_E, results in $R_{o4} = 10.5$ MΩ.

The second component of the output resistance is that seen looking into the collector of Q_6 in Fig. 9.20. Although the base of Q_6 is not at signal ground, we shall assume that the signal voltage at the base is small enough to make this approximation valid. The circuit then takes the form shown in Fig. 9.21(b), and R_{o6} can be determined using Eq. (9.72) with $R_E = R_2$. Thus $R_{o6} \approx 18.2$ MΩ.

Finally, we combine R_{o4} and R_{o6} in parallel to obtain the output resistance of the input stage, R_{o1}, as $R_{o1} = 6.7$ MΩ.

Figure 9.22 shows the equivalent circuit that we have derived for the input stage.

FIGURE 9.22 Small-signal equivalent circuit for the input stage of the 741 op amp.

EXAMPLE 9.3

We wish to find the input offset voltage resulting from a 2% mismatch between the resistances R_1 and R_2 in Fig. 9.13.

Solution

Consider first the situation when both input terminals are grounded, and assume that $R_1 = R$ and $R_2 = R + \Delta R$, where $\Delta R/R = 0.02$. From Fig. 9.23 we see that while Q_5 still conducts a current equal to I, the current in Q_6 will be smaller by ΔI. The value of ΔI can be found from

$$V_{BE5} + IR = V_{BE6} + (I - \Delta I)(R + \Delta R)$$

Thus

$$V_{BE5} - V_{BE6} = I\Delta R - \Delta I(R + \Delta R) \tag{9.73}$$

The quantity on the left-hand side is in effect the change in V_{BE} due to a change in I_E of ΔI. We may therefore write

$$V_{BE5} - V_{BE6} \simeq \Delta I r_e \tag{9.74}$$

Equations (9.73) and (9.74) can be combined to obtain

$$\frac{\Delta I}{I} = \frac{\Delta R}{R + \Delta R + r_e} \tag{9.75}$$

Substituting $R = 1$ kΩ and $r_e = 2.63$ kΩ shows that a 2% mismatch between R_1 and R_2 gives rise to an output current $\Delta I = 5.5 \times 10^{-3}I$. To reduce this output current to zero we have to apply an input voltage V_{OS} given by

$$V_{OS} = \frac{\Delta I}{G_{m1}} = \frac{5.5 \times 10^{-3}I}{G_{m1}} \tag{9.76}$$

Substituting $I = 9.5$ μA and $G_{m1} = 1/5.26$ mA/V results in the offset voltage $V_{OS} \simeq 0.3$ mV.

It should be pointed out that the offset voltage calculated is only one component of the input offset voltage of the 741. Other components arise because of mismatches in transistor characteristics. The 741 offset voltage is specified to be typically 2 mV.

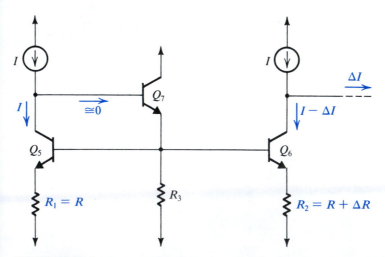

FIGURE 9.23 Input stage with both inputs grounded and a mismatch ΔR between R_1 and R_2.

EXERCISES

The purpose of this series of exercises is to determine the finite common-mode gain that results from a mismatch in the load circuit of the input stage of the 741 op amp. Figure E9.15 shows the input stage with an input common-mode signal v_{icm} applied and with a mismatch ΔR between the two resistances R_1 and R_2. Note that to simplify matters, we have opened the common-mode feedback loop and included a resistance R_o, which is the resistance seen looking to the left of node Y in the circuit of Fig. 9.13. Thus R_o is the parallel equivalent of R_{o9} (the output resistance of Q_9) and R_{o10} (the output resistance of Q_{10}).

9.15 Show that the current i (Fig. E9.15) is given approximately by

$$i = \frac{v_{icm}}{r_{e1} + r_{e3} + [2R_o/(\beta_P + 1)]}$$

9.16 Show that

$$i_o = -i\frac{\Delta R}{R + r_{e5} + \Delta R}$$

9.17 Using the results of Exercises 9.15 and 9.16, and assuming that $\Delta R \ll (R + r_e)$ and $R_o/(\beta_P + 1) \gg (r_{e1} + r_{e3})$, show that the common-mode transconductance G_{mcm} is given approximately by

$$G_{mcm} \equiv \frac{|i_o|}{v_{icm}} \simeq \frac{\beta_P}{2R_o}\frac{\Delta R}{R + r_{e5}}$$

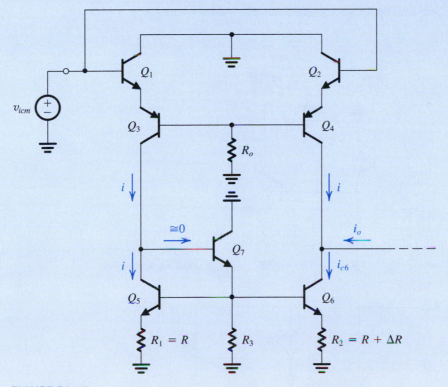

FIGURE E9.15

9.18 Refer to Fig. 9.13 and assume that the bases of Q_9 and Q_{10} are at approximately constant voltages (signal ground). Find R_{o9}, R_{o10} and hence R_o. Use $V_A = 125$ V for *npn* and 50 V for *pnp* transistors.

Ans. $R_{o9} = 2.63$ MΩ; $R_{o10} = 31.1$ MΩ; $R_o = 2.43$ MΩ

9.19 For $\beta_P = 50$ and $\Delta R/R = 0.02$, evaluate G_{mcm} obtained in Exercise 9.17.

Ans. 0.057 μA/V

9.20 Use the value of G_{mcm} obtained in Exercise 9.19 and the value of G_{m1} obtained from Eq. (9.71) to find the CMRR, that is the ratio of G_{m1} to G_{mcm}, expressed in decibels.

Ans. 70.5 dB

9.21 Noting that with the common-mode negative-feedback loop in place the common-mode gain will decrease by the amount of feedback, and noting that the loop gain is approximately equal to β_P (see Problem 9.26), find the CMRR with the feedback loop in place. ($\beta_P = 50$.)

Ans. 104.6 dB

9.5.2 The Second Stage

Figure 9.24 shows the 741 second stage prepared for small-signal analysis. In this section we shall analyze the second stage to determine the values of the parameters of the equivalent circuit shown in Fig. 9.25.

Input Resistance   The input resistance R_{i2} can be found by inspection to be

$$R_{i2} = (\beta_{16} + 1)[r_{e16} + R_9//(\beta_{17} + 1)(r_{e17} + R_8)] \tag{9.77}$$

Substituting the appropriate parameter values yields $R_{i2} \simeq 4$ MΩ.

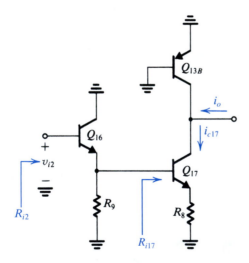

FIGURE 9.24   The 741 second stage prepared for small-signal analysis.

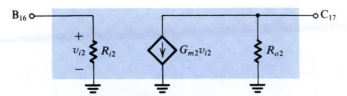

FIGURE 9.25   Small-signal equivalent circuit model of the second stage.

Transconductance From the equivalent circuit of Fig. 9.25, we see that the transconductance G_{m2} is the ratio of the *short-circuit output current* to the input voltage. Short-circuiting the output terminal of the second stage (Fig. 9.24) to ground makes the signal current through the output resistance of Q_{13B} zero, and the output short-circuit current becomes equal to the collector signal current of Q_{17} (i_{c17}). This latter current can be easily related to v_{i2} as follows:

$$i_{c17} = \frac{\alpha v_{b17}}{r_{e17} + R_8} \tag{9.78}$$

$$v_{b17} = v_{i2}\frac{(R_9//R_{i17})}{(R_9//R_{i17}) + r_{e16}} \tag{9.79}$$

$$R_{i17} = (\beta_{17} + 1)(r_{e17} + R_8) \tag{9.80}$$

These equations can be combined to obtain

$$G_{m2} \equiv \frac{i_{c17}}{v_{i2}} \tag{9.81}$$

which, for the 741 parameter values, is found to be $G_{m2} = 6.5$ mA/V.

Output Resistance To determine the output resistance R_{o2} of the second stage in Fig. 9.24, we ground the input terminal and find the resistance looking back into the output terminal. It follows that R_{o2} is given by

$$R_{o2} = (R_{o13B}//R_{o17}) \tag{9.82}$$

where R_{o13B} is the resistance looking into the collector of Q_{13B} while its base and emitter are connected to ground. It can be easily seen that

$$R_{o13B} = r_{o13B} \tag{9.83}$$

For the 741 component values we obtain $R_{o13B} = 90.9$ kΩ.

The second component in Eq. (9.82), R_{o17}, is the resistance seen looking into the collector of Q_{17}, as indicated in Fig. 9.26. Since the resistance between the base of Q_{17} and ground is relatively small, one can considerably simplify matters by assuming that the base is grounded. Doing this, we can use Eq. (9.72) to determine R_{o17}. For our case the result is $R_{o17} \cong 787$ kΩ. Combining R_{o13B} and R_{o17} in parallel yields $R_{o2} = 81$ kΩ.

Thévenin Equivalent Circuit The second-stage equivalent circuit can be converted to the Thévenin form, as shown in Fig. 9.27. Note that the stage open-circuit voltage gain is $-G_{m2}R_{o2}$.

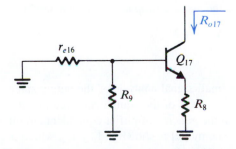

FIGURE 9.26 Definition of R_{o17}.

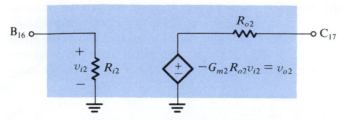

FIGURE 9.27 Thévenin form of the small-signal model of the second stage.

EXERCISES

9.22 Use Eq. (9.77) to show that $R_{i2} \simeq 4$ MΩ.

9.23 Use Eqs. (9.78) to (9.81) to verify that G_{m2} is 6.5 mA/V.

9.24 Verify that $R_{o2} \simeq 81$ kΩ.

9.25 Find the open-circuit voltage gain of the second stage of the 741.

Ans. −526.5 V/V

9.5.3 The Output Stage

The 741 output stage is shown in Fig. 9.28 without the short-circuit-protection circuitry. The stage is shown driven by the second-stage transistor Q_{17} and loaded with a 2-kΩ resistance. The circuit is of the AB class (Section 9.3.5), with the network composed of Q_{18}, Q_{19}, and R_{10} providing the bias of the output transistors Q_{14} and Q_{20}. The use of this network rather than two diode-connected transistors in series enables biasing the output transistors at a low current (0.15 mA) in spite of the fact that the output devices are three times as large as the standard devices. This is obtained by arranging that the current in Q_{19} is very small and thus its V_{BE} is also small. We analyzed the dc bias in Section 9.4.7.

Another feature of the 741 output stage worth noting is that the stage is driven by an emitter follower Q_{23}. As will be shown, this emitter follower provides added buffering, which makes the op-amp gain almost independent of the parameters of the output transistors.

Output Voltage Limits The maximum positive output voltage is limited by the saturation of current-source transistor Q_{13A}. Thus

$$v_{O\max} = V_{CC} - V_{CE\mathrm{sat}} - V_{BE14} \tag{9.84}$$

which is about 1 V below V_{CC}. The minimum output voltage (i.e., maximum negative amplitude) is limited by the saturation of Q_{17}. Neglecting the voltage drop across R_8, we obtain

$$v_{O\min} = -V_{EE} + V_{CE\mathrm{sat}} + V_{EB23} + V_{EB20} \tag{9.85}$$

which is about 1.5 V above $-V_{EE}$.

Small-Signal Model We shall now carry out a small-signal analysis of the output stage for the purpose of determining the values of the parameters of the equivalent circuit model shown in Fig. 9.29. Note that this model is based on the general amplifier equivalent circuit presented in Table 5.5 as "Equivalent Circuit C." The model is shown fed by v_{o2}, which is

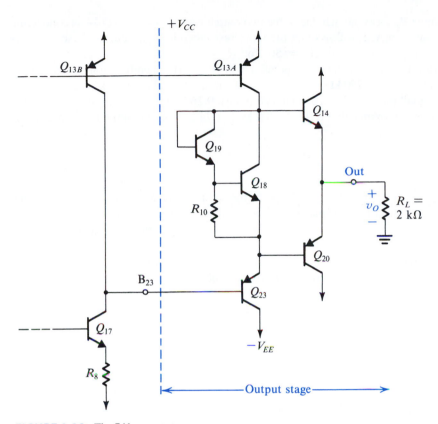

FIGURE 9.28 The 741 output stage.

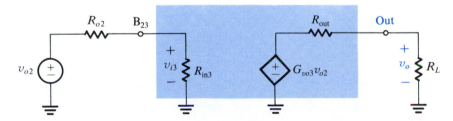

FIGURE 9.29 Model for the 741 output stage. This model is based on the amplifier equivalent circuit presented in Table 5.5 as "Equivalent Circuit C."

the open-circuit output voltage of the second stage. From Fig. 9.27, v_{o2} is given by

$$v_{o2} = -G_{m2}R_{o2}v_{i2} \tag{9.86}$$

where G_{m2} and R_{o2} were previously determined as $G_{m2} = 6.5$ mA/V and $R_{o2} = 81$ kΩ. Resistance R_{in3} is the input resistance of the output stage determined with the amplifier loaded with R_L. Although the effect of loading an amplifier stage on its input resistance is negligible in the input and second stages, this is not the case in general in an output stage. Defining R_{in3} in this manner (see Table 5.5) enables correct evaluation of the voltage gain of the second stage, A_2, as

$$A_2 \equiv \frac{v_{i3}}{v_{i2}} = -G_{m2}R_{o2}\frac{R_{in3}}{R_{in3} + R_{o2}} \tag{9.87}$$

To determine R_{in3}, assume that one of the two output transistors—say, Q_{20}—is conducting a current of, say, 5 mA. It follows that the input resistance looking into the base of Q_{20} is approximately $\beta_{20}R_L$. Assuming $\beta_{20} = 50$, for $R_L = 2$ kΩ the input resistance of Q_{20} is 100 kΩ. This resistance appears in parallel with the series combination of the output resistance of Q_{13A} ($r_{o13A} \simeq 280$ kΩ) and the resistance of the Q_{18}–Q_{19} network. The latter resistance is very small (about 160 Ω; see later: Exercise 9.26). Thus the total resistance in the emitter of Q_{23} is approximately (100 kΩ//280 kΩ) or 74 kΩ and the input resistance R_{in3} is given by

$$R_{in3} \simeq \beta_{23} \times 74 \text{ k}\Omega$$

which for $\beta_{23} = 50$ is $R_{in3} \simeq 3.7$ MΩ. Since $R_{o2} = 81$ kΩ, we see that $R_{in3} \gg R_{o2}$, and the value of R_{in3} will have little effect on the performance of the op amp. We can use the value obtained for R_{in3} to determine the gain of the second stage using Eq. (9.87) as $A_2 = -515$ V/V. The value of A_2 will be needed in Section 9.6 in connection with the frequency-response analysis.

Continuing with the determination of the equivalent circuit-model-parameters, we note from Fig. 9.29 that G_{vo3} is the **open-circuit overall voltage gain** of the output stage,

$$G_{vo3} = \left.\frac{v_o}{v_{o2}}\right|_{R_L=\infty} \tag{9.88}$$

With $R_L = \infty$, the gain of the emitter-follower output transistor (Q_{14} or Q_{20}) will be nearly unity. Also, with $R_L = \infty$ the resistance in the emitter of Q_{23} will be very large. This means that the gain of Q_{23} will be nearly unity and the input resistance of Q_{23} will be very large. We thus conclude that $G_{vo3} \simeq 1$.

Next, we shall find the value of the output resistance of the op amp, R_{out}. For this purpose refer to the circuit shown in Fig. 9.30. In accordance with the definition of R_{out}, the input source feeding the output stage is grounded, but its resistance (which is the output

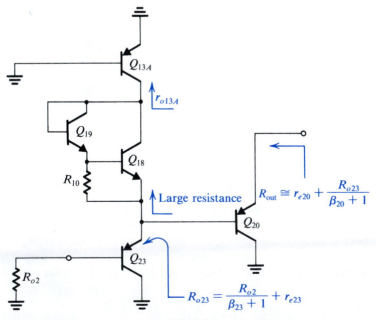

FIGURE 9.30 Circuit for finding the output resistance R_{out}.

resistance of the second stage, R_{o2}) is included. We have assumed that the output voltage v_O is negative, and thus Q_{20} is conducting most of the current; transistor Q_{14} has therefore been eliminated. The exact value of the output resistance will of course depend on which transistor (Q_{14} or Q_{20}) is conducting and on the value of load current. Nevertheless, we wish to find an estimate of R_{out}.

As indicated in Fig. 9.30, the resistance seen looking into the emitter of Q_{23} is

$$R_{o23} = \frac{R_{o2}}{\beta_{23} + 1} + r_{e23} \tag{9.89}$$

Substituting $R_{o2} = 81\ \text{k}\Omega$, $\beta_{23} = 50$, and $r_{e23} = 25/0.18 = 139\ \Omega$ yields $R_{o23} = 1.73\ \text{k}\Omega$. This resistance appears in parallel with the series combination of r_{o13A} and the resistance of the Q_{18}–Q_{19} network. Since r_{o13A} alone (0.28 MΩ) is much larger than R_{o23}, the effective resistance between the base of Q_{20} and ground is approximately equal to R_{o23}. Now we can find the output resistance R_{out} as

$$R_{\text{out}} = \frac{R_{o23}}{\beta_{20} + 1} + r_{e20} \tag{9.90}$$

For $\beta_{20} = 50$, the first component of R_{out} is 34 Ω. The second component depends critically on the value of output current. For an output current of 5 mA, r_{e20} is 5 Ω and R_{out} is 39 Ω. To this value we must add the resistance R_7 (27 Ω) (see Fig. 9.13), which is included for short-circuit protection. The output resistance of the 741 is specified to be typically 75 Ω.

EXERCISES

9.26 Using a simple (r_π, g_m) model for each of the two transistors Q_{18} and Q_{19} in Fig. E9.26, find the small-signal resistance between A and A'. (*Note:* From Table 9.1, $I_{C18} = 165\ \mu\text{A}$ and $I_{C19} \approx 16\ \mu\text{A}$.

Ans. 163 Ω

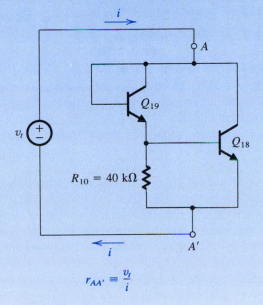

FIGURE E9.26

9.27 Figure E9.27 shows the circuit for determining the op-amp output resistance when v_O is positive and Q_{14} is conducting most of the current. Using the resistance of the Q_{18}–Q_{19} network calculated in Exercise 9.26 and neglecting the large output resistance of Q_{13A}, find R_{out} when Q_{14} is sourcing an output current of 5 mA.

Ans. 14.4 Ω

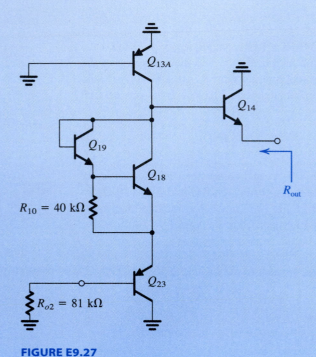

FIGURE E9.27

Output Short-Circuit Protection   If the op-amp output terminal is short-circuited to one of the power supplies, one of the two output transistors could conduct a large amount of current. Such a large current can result in sufficient heating to cause burnout of the IC (Chapter 14). To guard against this possibility, the 741 op amp is equipped with a special circuit for short-circuit protection. The function of this circuit is to limit the current in the output transistors in the event of a short circuit.

Refer to Fig. 9.13. Resistance R_6 together with transistor Q_{15} limits the current that would flow out of Q_{14} in the event of a short circuit. Specifically, if the current in the emitter of Q_{14} exceeds about 20 mA, the voltage drop across R_6 exceeds 540 mV, which turns Q_{15} on. As Q_{15} turns on, its collector robs some of the current supplied by Q_{13A}, thus reducing the base current of Q_{14}. This mechanism thus limits the maximum current that the op amp can source (i.e., supply from the output terminal in the outward direction) to about 20 mA.

Limiting of the maximum current that the op amp can sink, and hence the current through Q_{20}, is done by a mechanism similar to the one discussed above. The relevant circuit is composed of R_7, Q_{21}, Q_{24}, and Q_{22}. For the components shown, the current in the inward direction is limited also to about 20 mA.

9.6 GAIN, FREQUENCY RESPONSE, AND SLEW RATE OF THE 741

In this section we shall evaluate the overall small-signal voltage gain of the 741 op amp. We shall then consider the op amp's frequency response and its slew-rate limitation.

9.6.1 Small-Signal Gain

The overall small-signal gain can be easily found from the cascade of the equivalent circuits derived in the preceding sections for the three op-amp stages. This cascade is shown in Fig. 9.31, loaded with $R_L = 2$ kΩ, which is the typical value used in measuring and specifying the 741 data. The overall gain can be expressed as

$$\frac{v_o}{v_i} = \frac{v_{i2}}{v_i}\frac{v_{o2}}{v_{i2}}\frac{v_o}{v_{o2}} \tag{9.91}$$

$$= -G_{m1}(R_{o1}//R_{i2})(-G_{m2}R_{o2})G_{vo3}\frac{R_L}{R_L + R_{out}} \tag{9.92}$$

Using the values found earlier yields for the overall open-circuit voltage gain,

$$A_0 \equiv \frac{v_o}{v_i} = -476.1 \times (-526.5) \times 0.97 = 243,147 \text{ V/V} \tag{9.93}$$

$$\equiv 107.7 \text{ dB}$$

9.6.2 Frequency Response

The 741 is an internally compensated op amp. It employs the Miller compensation technique, studied in Section 8.11.3, to introduce a dominant low-frequency pole. Specifically, a 30-pF capacitor (C_C) is connected in the negative-feedback path of the second stage. An approximate estimate of the frequency of the dominant pole can be obtained as follows.

Using Miller's theorem (Section 6.4.4) the effective capacitance due to C_C between the base of Q_{16} and ground is (see Fig. 9.13)

$$C_{in} = C_C(1 + |A_2|) \tag{9.94}$$

where A_2 is the second-stage gain. Use of the value calculated for A_2 in Section 9.5.3, $A_2 = -515$, results in $C_{in} = 15,480$ pF. Since this capacitance is quite large, we shall neglect all other capacitances between the base of Q_{16} and signal ground. The total resistance between this node and ground is

$$R_t = (R_{o1}//R_{i2})$$

$$= (6.7 \text{ M}\Omega//4 \text{ M}\Omega) = 2.5 \text{ M}\Omega \tag{9.95}$$

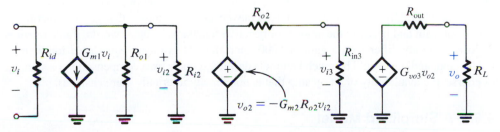

FIGURE 9.31 Cascading the small-signal equivalent circuits of the individual stages for the evaluation of the overall voltage gain.

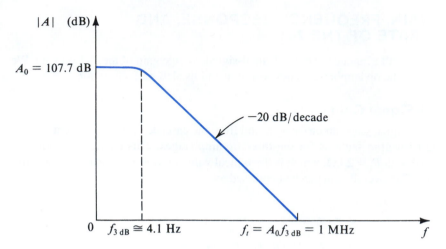

FIGURE 9.32   Bode plot for the 741 gain, neglecting nondominant poles.

Thus the dominant pole has a frequency f_P given by

$$f_P = \frac{1}{2\pi C_{in}R_t} = 4.1 \text{ Hz} \tag{9.96}$$

It should be noted that this approach is equivalent to using the approximate formula in Eq. (8.87).

As discussed in Section 8.11.3, Miller compensation provides an additional advantageous effect, namely pole splitting. As a result, the other poles of the circuit are moved to very high frequencies. This has been confirmed by computer-aided analysis [see Gray et al (2000)].

Assuming that all nondominant poles are at very high frequencies, the calculated values give rise to the Bode plot shown in Fig. 9.32 where $f_{3dB} = f_P$. The unity-gain bandwidth f_t can be calculated from

$$f_t = A_0 f_{3dB} \tag{9.97}$$

Thus,

$$f_t = 243{,}147 \times 4.1 \simeq 1 \text{ MHz} \tag{9.98}$$

Although this Bode plot implies that the phase shift at f_t is $-90°$ and thus that the phase margin is $90°$, in practice a phase margin of about $80°$ is obtained. The excess phase shift (about $10°$) is due to the nondominant poles. This phase margin is sufficient to provide stable operation of closed-loop amplifiers with any value of feedback factor β. This convenience of use of the internally compensated 741 is achieved at the expense of a great reduction in open-loop gain and hence in the amount of negative feedback. In other words, if one requires a closed-loop amplifier with a gain of 1000, then the 741 is *overcompensated* for such an application, and one would be much better off designing one's own compensation (assuming, of course, the availability of an op amp that is not already internally compensated).

9.6.3 A Simplified Model

Figure 9.33 shows a simplified model of the 741 op amp in which the high-gain second stage, with its feedback capacitance C_C, is modeled by an ideal integrator. In this model, the

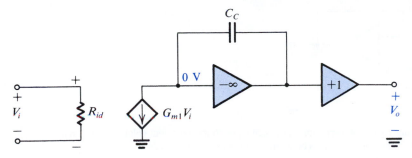

FIGURE 9.33 A simple model for the 741 based on modeling the second stage as an integrator.

gain of the second stage is assumed sufficiently large that a virtual ground appears at its input. For this reason the output resistance of the input stage and the input resistance of the second stage have been omitted. Furthermore, the output stage is assumed to be an ideal unity-gain follower. Except for the presence of the output stage, this model is identical to that which we used for the two-stage CMOS amplifier in Section 9.1.4 (Fig. 9.3).

Analysis of the model in Fig. 9.33 gives

$$A(s) \equiv \frac{V_o(s)}{V_i(s)} = \frac{G_{m1}}{sC_C} \tag{9.99}$$

Thus,

$$A(j\omega) = \frac{G_{m1}}{j\omega C_C} \tag{9.100}$$

and the magnitude of gain becomes unity at $\omega = \omega_t$, where

$$\omega_t = \frac{G_{m1}}{C_C} \tag{9.101}$$

Substituting $G_{m1} = 1/5.26$ mA/V and $C_C = 30$ pF yields

$$f_t = \frac{\omega_t}{2\pi} \simeq 1 \text{ MHz} \tag{9.102}$$

which is equal to the value calculated before. It should be pointed out, however, that this model is valid only at frequencies $f \gg f_{3dB}$. At such frequencies the gain falls off with a slope of −20 dB/decade, just like that of an integrator.

9.6.4 Slew Rate

The slew-rate limitation of op amps is discussed in Chapter 2. Here we shall illustrate the origin of the slewing phenomenon in the context of the 741 circuit.

Consider the unity-gain follower of Fig. 9.34 with a step of, say, 10 V applied at the input. Because of amplifier dynamics, its output will not change in zero time. Thus immediately after the input is applied, almost the entire value of the step will appear as a differential signal between the two input terminals. This large input voltage causes the input stage to be **overdriven,** and its small-signal model no longer applies. Rather, half the stage cuts off and the other half conducts all the current. Specifically, reference to Fig. 9.13 shows that a large positive differential input voltage causes Q_1 and Q_3 to conduct all the available bias current ($2I$) while Q_2 and Q_4 will be cut off. The current mirror Q_5, Q_6, and Q_7 will still function, and Q_6 will produce a collector current of $2I$.

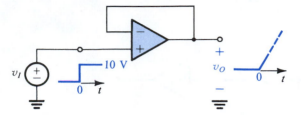

FIGURE 9.34 A unity-gain follower with a large step input. Since the output voltage cannot change instantaneously, a large differential voltage appears between the op-amp input terminals.

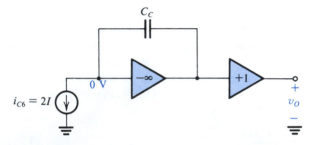

FIGURE 9.35 Model for the 741 op amp when a large positive differential signal is applied.

Using the observations above, and modeling the second stage as an ideal integrator, results in the model of Fig. 9.35. From this circuit we see that the output voltage will be a ramp with a slope of $2I/C_C$:

$$v_O(t) = \frac{2I}{C_C}t \qquad (9.103)$$

Thus the slew rate SR is given by

$$SR = \frac{2I}{C_C} \qquad (9.104)$$

For the 741, $I = 9.5\ \mu A$ and $C_C = 30$ pF, resulting in $SR = 0.63$ V/μs.

It should be pointed out that this is a rather simplified model of the slewing process. More detail can be found in Gray et al (2000).

EXERCISE

9.28 Use the value of the slew rate calculated above to find the full-power bandwidth f_M of the 741 op amp. Assume that the maximum output is ± 10 V.

Ans. 10 kHz

9.6.5 Relationship Between f_t and SR

A simple relationship exists between the unity-gain bandwidth f_t and the slew rate SR. This relationship is obtained from Eqs. (9.101) and (9.104) together with

$$G_{m1} = 2\frac{1}{4r_e}$$

where r_e is the emitter resistance of each of Q_1 through Q_4. Thus

$$r_e = \frac{V_T}{I}$$

and

$$G_{m1} = \frac{I}{2V_T} \tag{9.105}$$

Substituting in Eq. (9.101) results in

$$\omega_t = \frac{I}{2C_C V_T} \tag{9.106}$$

Substituting for I/C_C from Eq. (9.104) gives

$$\omega_t = \frac{SR}{4V_T} \tag{9.107}$$

which can be expressed in the alternative form

$$SR = 4V_T \omega_t \tag{9.108}$$

As a check, for the 741 we have

$$SR = 4 \times 25 \times 10^{-3} \times 2\pi \times 10^6 = 0.63 \text{ V}/\mu\text{s}$$

which is the result obtained previously. Observe that Eq. (9.108) is of the same form as Eq. (9.41), which applies to the two-stage CMOS op amp. Here, $4V_T$ replaces V_{OV}. Since, typically, V_{OV} will be two to three times the value of $4V_T$, a two-stage CMOS op amp with an f_t equal to that of the 741 exhibits a slew rate that is two to three times as large as that of the 741.

A general form for the relationship between SR and ω_t for an op amp with a structure similar to that of the 741 (including the two-stage CMOS circuit) is

$$SR = \omega_t/a$$

where a is the constant of proportionality relating the transconductance of the first stage G_{m1}, to the *total* bias current of the input differential stage. That is, for the 741 circuit $G_{m1} = a(2I)$, while for the CMOS circuit of Fig. 9.1, $G_{m1} = aI$.[3] For a given ω_t, a higher value of SR is obtained by making a smaller; that is, the total bias current is kept constant and G_{m1} is reduced. This is a viable technique for increasing slew rate. It is referred to as the G_m-reduction method (see Exercise 9.30).

EXERCISES

9.29 Consider the integrator model of the op amp in Fig. 9.33. Find the value of the resistor that, when connected across C_C, provides the correct value of the dc gain.
Ans. 1279 MΩ

D9.30 If a resistance R_E is included in each of the emitter leads of Q_3 and Q_4 show that $SR = 4(V_T + IR_E/2)\omega_t$. Hence find the value of R_E that would double the 741 slew rate while keeping ω_t and I unchanged. What are the new values of C_C, the dc gain, and the 3-dB frequency?
Ans. 5.26 kΩ; 15 pF; 101.7 dB (a 6-dB decrease); 8.2 Hz

---

[3] The difference is just a matter of notation; we used I to denote the total bias current of the input differential stage of the CMOS circuit, and we used $2I$ for the 741 case!

9.7 DATA CONVERTERS—AN INTRODUCTION

In this section we begin the study of another group of analog IC circuits of great importance; namely, data converters.

9.7.1 Digital Processing of Signals

Most physical signals, such as those obtained at transducer outputs, exist in analog form. Some of the processing required on these signals is most conveniently performed in an analog fashion. For instance, in instrumentation systems it is quite common to use a high-input-impedance, high-gain, high-CMRR differential amplifier right at the output of the transducer. This is usually followed by a filter whose purpose is to eliminate interference. However, further signal processing is usually required, which can range from simply obtaining a measurement of signal strength to performing some algebraic manipulations on this and related signals to obtain the value of a particular system parameter of interest, as is usually the case in systems intended to provide a complex control function. Another example of signal processing can be found in the common need for transmission of signals to a remote receiver.

Many such forms of signal processing can be performed by analog means. In earlier chapters we encountered circuits for implementing a number of such tasks. However, an attractive alternative exists: It is to convert, following some initial analog processing, the signal from analog to digital form and then use economical, accurate, and convenient digital ICs to perform **digital signal processing.** Such processing can in its simplest form provide us with a measure of the signal strength as an easy-to-read number (consider, e.g., the digital voltmeter). In more involved cases the digital signal processor can perform a variety of arithmetic and logic operations that implement a **filtering algorithm.** The resulting **digital filter** does many of the same tasks that an analog filter performs—namely, eliminate interference and noise. Yet another example of digital signal processing is found in digital communications systems, where signals are transmitted as a sequence of binary pulses, with the obvious advantage that corruption of the amplitudes of these pulses by noise is, to a large extent, of no consequence.

Once digital signal processing has been performed, we might be content to display the result in digital form, such as a printed list of numbers. Alternatively, we might require an analog output. Such is the case in a telecommunications system, where the usual output may be audible speech. If such an analog output is desired, then obviously we need to convert the digital signal back to an analog form.

It is not our purpose here to study the techniques of digital signal processing. Rather, we shall examine the interface circuits between the analog and digital domains. Specifically, we shall study the basic techniques and circuits employed to convert an analog signal to digital form (**analog-to-digital** or simply **A/D conversion**) and those used to convert a digital signal to analog form (**digital-to-analog** or simply **D/A conversion**). Digital circuits are studied in Chapters 10 and 11.

9.7.2 Sampling of Analog Signals

The principle underlying digital signal processing is that of **sampling** the analog signal. Figure 9.36 illustrates in a conceptual form the process of obtaining samples of an analog signal. The switch shown closes periodically under the control of a periodic pulse signal (clock). The closure time of the switch, τ, is relatively short, and the samples obtained are

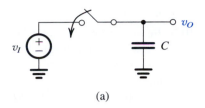

(a)

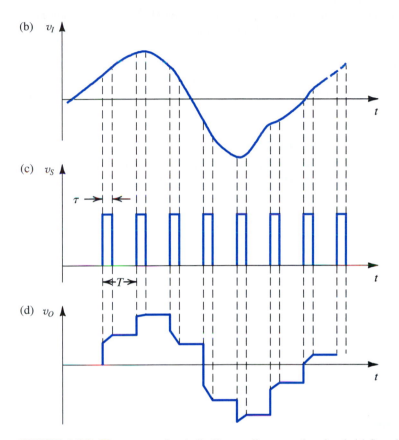

FIGURE 9.36 The process of periodically sampling an analog signal. **(a)** Sample-and-hold (S/H) circuit. The switch closes for a small part (τ seconds) of every clock period (T). **(b)** Input signal waveform. **(c)** Sampling signal (control signal for the switch). **(d)** Output signal (to be fed to A/D converter).

stored (held) on the capacitor. The circuit of Fig. 9.36 is known as a **sample-and-hold (S/H) circuit.** As indicated, the S/H circuit consists of an analog switch that can be implemented by a MOSFET transmission gate (Section 10.5), a storage capacitor, and (not shown) a buffer amplifier.

Between the sampling intervals—that is, during the *hold* intervals—the voltage level on the capacitor represents the signal samples we are after. Each of these voltage levels is then fed to the input of an A/D converter, which provides an N-bit binary number proportional to the value of signal sample.

The fact that we can do our processing on a limited number of samples of an analog signal while ignoring the analog-signal details between samples is based on the Shannon's sampling theorem [see Lathi (1965)].

9.7.3 Signal Quantization

Consider an analog signal whose values range from 0 to +10 V. Let us assume that we wish to convert this signal to digital form and that the required output is a 4-bit digital signal.[4] We know that a 4-bit binary number can represent 16 different values, 0 to 15; it follows that the *resolution* of our conversion will be 10 V/15 = $\frac{2}{3}$ V. Thus an analog signal of 0 V will be represented by 0000, $\frac{2}{3}$ V will be represented by 0001, 6 V will be represented by 1001, and 10 V will be represented by 1111.

All these sample numbers are multiples of the basic increment ($\frac{2}{3}$ V). A question now arises regarding the conversion of numbers that fall between these successive incremental levels. For instance, consider the case of a 6.2-V analog level. This falls between 18/3 and 20/3. However, since it is closer to 18/3 we treat it as if it were 6 V and *code* it as 1001. This process is called **quantization.** Obviously errors are inherent in this process; such errors are called quantization errors. Using more bits to represent (encode or, simply, code) an analog signal reduces quantization errors but requires more complex circuitry.

9.7.4 The A/D and D/A Converters as Functional Blocks

Figure 9.37 depicts the functional block representations of A/D and D/A converters. As indicated, the **A/D converter** (also called an **ADC**) accepts an analog sample v_A and produces an N-bit **digital word.** Conversely, the **D/A converter** (also called a **DAC**) accepts an n-bit digital word and produces an analog sample. The output samples of the D/A converter are often fed to a sample-and-hold circuit. At the output of the S/H circuit a staircase waveform, such as that in Fig. 9.38, is obtained. The staircase waveform can then be smoothed by a

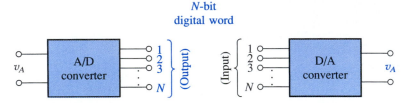

FIGURE 9.37 The A/D and D/A converters as circuit blocks.

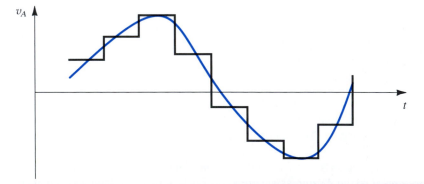

FIGURE 9.38 The analog samples at the output of a D/A converter are usually fed to a sample-and-hold circuit to obtain the staircase waveform shown. This waveform can then be filtered to obtain the smooth waveform, shown in color. The time delay usually introduced by the filter is not shown.

---

[4] *Bit* stands for *b*inary dig*it*.

low-pass filter, giving rise to the smooth curve shown in color in Fig. 9.38. In this way an analog output signal is reconstructed. Finally, note that the quantization error of an A/D converter is equivalent to $\pm\frac{1}{2}$ least significant bit (b_N).

EXERCISE

9.31 An analog signal in the range 0 to +10 V is to be converted to an 8-bit digital signal. What is the resolution of the conversion in volts? What is the digital representation of an input of 6 V? What is the representation of an input of 6.2 V? What is the error made in the quantization of 6.2 V in absolute terms and as a percentage of the input? As a percentage of full scale? What is the largest possible quantization error as a percentage of full scale?

Ans. 0.0392 V; 10011001; 10011110; –0.0064 V; –0.1%; –0.064%; 0.196%

9.8 D/A CONVERTER CIRCUITS

9.8.1 Basic Circuit Using Binary-Weighted Resistors

Figure 9.39 shows a simple circuit for an N-bit D/A converter. The circuit consists of a reference voltage V_{REF}, N binary-weighted resistors R, $2R$, $4R$, $8R$, . . . , $2^{N-1}R$, N single-pole double-throw switches S_1, S_2, . . . , S_N, and an op amp together with its feedback resistance $R_f = R/2$.

The switches are controlled by an N-bit digital input word D,

$$D = \frac{b_1}{2^1} + \frac{b_2}{2^2} + \cdots + \frac{b_N}{2^N} \tag{9.109}$$

where b_1, b_2, and so on are bit coefficients that are either 1 or 0. Note that the bit b_N is the **least significant bit (LSB)** and b_1 is the **most significant bit (MSB).** In the circuit in Fig. 9.39, b_1 controls switch S_1, b_2 controls S_2, and so on. When b_i is 0, switch S_i is in position 1, and when b_i is 1 switch S_i is in position 2.

Since position 1 of all switches is ground and position 2 is virtual ground, the current through each resistor remains constant. Each switch simply controls where its corresponding current goes: to ground (when the corresponding bit is 0) or to virtual ground (when the corresponding bit is 1). The currents flowing into the virtual ground add up, and the sum flows

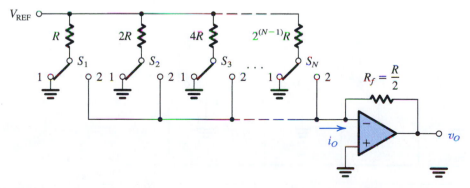

FIGURE 9.39 An N-bit D/A converter using a binary-weighted resistive ladder network.

through the feedback resistance R_f. The total current i_O is therefore given by

$$i_O = \frac{V_{REF}}{R}b_1 + \frac{V_{REF}}{2R}b_2 + \cdots + \frac{V_{REF}}{2^{N-1}R}b_N$$

$$= \frac{2V_{REF}}{R}\left(\frac{b_1}{2^1} + \frac{b_2}{2^2} + \cdots + \frac{b_N}{2^N}\right)$$

Thus,

$$i_O = \frac{2V_{REF}}{R}D \qquad (9.110)$$

and the output voltage v_O is given by

$$v_O = -i_O R_f = -V_{REF}D \qquad (9.111)$$

which is directly proportional to the digital word D, as desired.

It should be noted that the accuracy of the DAC depends critically on (1) the accuracy of V_{REF}, (2) the precision of the binary-weighted resistors, and (3) the perfection of the switches. Regarding the third point, we should emphasize that these switches handle analog signals; thus their perfection is of considerable interest. While the offset voltage and the finite on resistance are not of critical significance in a digital switch, these parameters are of immense importance in *analog switches*. The use of MOSFETs to implement analog switches will be discussed in Chapter 10. Also, we shall shortly see that in practical circuit implementations of the DAC, the binary-weighted currents are generated by current sources. In this case the analog switch can be realized using the differential-pair circuit, as will be shown shortly.

A disadvantage of the binary-weighted resistor network is that for a large number of bits ($N > 4$) the spread between the smallest and largest resistances becomes quite large. This implies difficulties in maintaining accuracy in resistor values. A more convenient scheme exists utilizing a resistive network called the *R-2R* ladder.

9.8.2 *R-2R* Ladders

Figure 9.40 shows the basic arrangement of a DAC using an *R-2R* ladder. Because of the small spread in resistance values, this network is usually preferred to the binary-weighted scheme discussed earlier, especially for $N > 4$. Operation of the *R-2R* ladder is straightforward. First, it can be shown, by starting from the right and working toward the left, that the

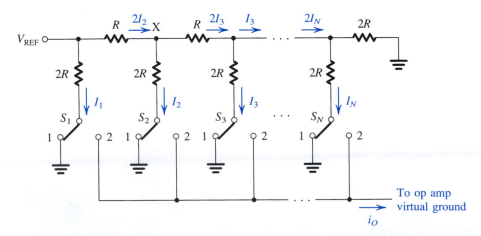

FIGURE 9.40 The basic circuit configuration of a DAC utilizing an *R-2R* ladder network.

resistance to the right of each ladder node, such as that labeled X, is equal to $2R$. Thus the current flowing to the right, away from each node, is equal to the current flowing downward to ground, and twice that current flows into the node from the left side. It follows that

$$I_1 = 2I_2 = 4I_3 = \cdots = 2^{N-1}I_N \tag{9.112}$$

Thus, as in the binary-weighted resistive network, the currents controlled by the switches are binary weighted. The output current i_O will therefore be given by

$$i_O = \frac{V_{REF}}{R}D \tag{9.113}$$

9.8.3 A Practical Circuit Implementation

A practical circuit implementation of the DAC utilizing an R-$2R$ ladder is shown in Fig. 9.41. The circuit utilizes BJTs to generate binary-weighted constant currents $I_1, I_2, \ldots, I_N$, which are switched between ground and virtual ground of an output summing op amp (not shown). We shall first show that the currents I_1 to I_N are indeed binary-weighted, with I_1 corresponding to the MSB and I_N corresponding to the LSB of the DAC.

Starting at the two rightmost transistors, Q_N and Q_t, we see that if they are matched, their emitter currents will be equal and are denoted (I_N/α). Transistor Q_t is included to provide proper termination of the R-$2R$ network. The voltage between the base line of the BJTs and node N will be

$$V_N = V_{BE_N} + \left(\frac{I_N}{\alpha}\right)(2R)$$

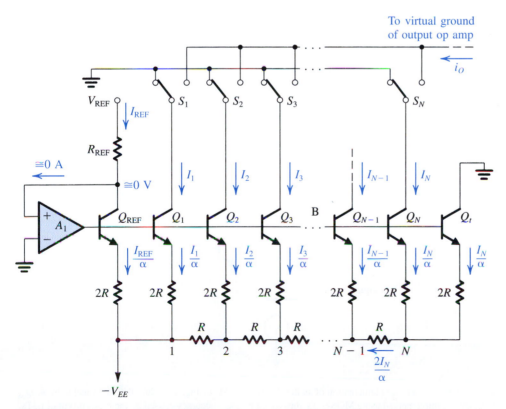

FIGURE 9.41 A practical circuit implementation of a DAC utilizing an R-$2R$ ladder network.

where V_{BE_N} is the base–emitter voltage of Q_N. Since the current flowing through the resistor R connected to node N is $(2I_N/\alpha)$, the voltage between node B and node $(N-1)$ will be

$$V_{N-1} = V_N + \left(\frac{2I_N}{\alpha}\right)R = V_{BE_N} + \frac{4I_N}{\alpha}R$$

Assuming, for the moment, that $V_{BE_{N-1}} = V_{BE_N}$, we see that a voltage of $(4I_N/\alpha)R$ appears across the resistance $2R$ in the emitter of Q_{N-1}. Thus Q_{N-1} will have an emitter current of $(2I_N/\alpha)$ and a collector current of $(2I_N)$, twice the current in Q_N. The two transistors will have equal V_{BE} drops if their junction areas are scaled in the same proportion as their currents, which is usually done in practice.

Proceeding in the manner above we can show that

$$I_1 = 2I_2 = 4I_3 = \cdots = 2^{N-1}I_N \tag{9.114}$$

under the assumption that the EBJ areas of Q_1 to Q_N are scaled in a binary-weighted fashion.

Next consider op amp A_1, which, together with the reference transistor Q_{REF}, forms a negative-feedback loop. (Convince yourself that the feedback is indeed negative.) A virtual ground appears at the collector of Q_{REF} forcing it to conduct a collector current $I_{REF} = V_{REF}/R_{REF}$ independent of whatever imperfections Q_{REF} might have. Now, if Q_{REF} and Q_1 are matched, their collector currents will be equal,

$$I_1 = I_{REF}$$

Thus, the binary-weighted currents are directly related to the reference current, independent of the exact values of V_{BE} and α. Also observe that op amp A_1 supplies the base currents of all the BJTs.

9.8.4 Current Switches

Each of the single-pole double-throw switches in the DAC circuit of Fig. 9.41 can be implemented by a circuit such as that shown in Fig. 9.42 for switch S_m. Here I_m denotes the current flowing in the collector of the mth-bit transistor. The circuit is a differential pair with the

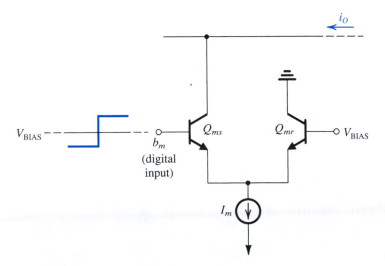

FIGURE 9.42 Circuit implementation of switch S_m in the DAC of Fig. 9.41. In a BiCMOS technology, Q_{ms} and Q_{mr} can be implemented using MOSFETs, thus avoiding the inaccuracy caused by the base current of BJTs.

base of the reference transistor Q_{mr} connected to a suitable dc voltage V_{BIAS}, and the digital signal representing the mth bit b_m applied to the base of the other transistor Q_{ms}. If the voltage representing b_m is higher than V_{BIAS} by a few hundred millivolts, Q_{ms} will turn on and Q_{mr} will turn off. The bit current I_m will flow through Q_{ms} and onto the output summing line. On the other hand, when b_m is low, Q_{ms} will be off and I_m will flow through Q_{mr} to ground.

The current switch of Fig. 9.42 is simple and features high-speed operation. It suffers, however, from the fact that part of the current I_m flows through the base of Q_{ms} and thus does not appear on the output summing line. More elaborate circuits for current switches can be found in Grebene (1984). Also, in a BiCMOS technology the differential-pair transistors Q_{ms} and Q_{mr} can be replaced with MOSFETs, thus eliminating the base current problem.

EXERCISES

9.32 What is the maximum resistor ratio required by a 12-bit D/A converter utilizing a binary-weighted resistor network?

Ans. 2048

9.33 If the input bias current of an op amp, used as the output summer in a 10-bit DAC, is to be no more than that equivalent to $\frac{1}{4}$ LSB, what is the maximum current required to flow in R_f for an op amp whose bias current is as great as 0.5 μA?

Ans. 2.046 mA

 ## 9.9 A/D CONVERTER CIRCUITS

There exist a number of A/D conversion techniques varying in complexity and speed. We shall discuss four different approaches: two simple, but slow, schemes, one complex (in terms of the amount of circuitry required) but extremely fast method, and, finally, a method particularly suited for MOS implementation.

9.9.1 The Feedback-Type Converter

Figure 9.43 shows a simple A/D converter that employs a comparator, an up/down counter, and a D/A converter. The comparator circuit provides an output that assumes one of two

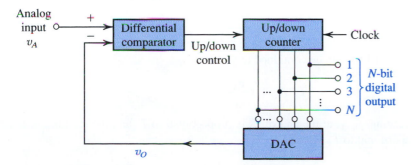

FIGURE 9.43 A simple feedback-type A/D converter.

distinct values: positive when the difference input signal is positive, and negative when the difference input signal is negative. We shall study comparator circuits in Chapter 13. An up/down counter is simply a counter that can count either up or down depending on the binary level applied at its up/down control terminal. Because the A/D converter of Fig. 9.43 employs a DAC in its feedback loop it is usually called a feedback-type A/D converter. It operates as follows: With a 0 count in the counter, the D/A converter output, v_O, will be zero and the output of the comparator will be high, instructing the counter to count the clock pulses in the up direction. As the count increases, the output of the DAC rises. The process continues until the DAC output reaches the value of the analog input signal, at which point the comparator switches and stops the counter. The counter output will then be the digital equivalent of the input analog voltage.

Operation of the converter of Fig. 9.43 is slow if it starts from zero. This converter however, tracks incremental changes in the input signal quite rapidly.

9.9.2 The Dual-Slope A/D Converter

A very popular high-resolution (12- to 14-bit) (but slow) A/D conversion scheme is illustrated in Fig. 9.44. To see how it operates, refer to Fig. 9.44 and assume that the analog input signal v_A is negative. Prior to the start of the conversion cycle, switch S_2 is closed, thus discharging capacitor C and setting $v_1 = 0$. The conversion cycle begins with opening S_2 and connecting the integrator input through switch S_1 to the analog input signal. Since v_A is negative, a current $I = v_A/R$ will flow through R in the direction away from the integrator. Thus v_1 rises linearly with a slope of $I/C = v_A/RC$, as indicated in Fig. 9.44(b). Simultaneously, the counter is enabled and it counts the pulses from a fixed-frequency clock. This phase of the conversion process continues for a fixed duration T_1. It ends when the counter has accumulated a fixed count denoted n_{REF}. Usually, for an N-bit converter, $n_{REF} = 2^N$. Denoting the peak voltage at the output of the integrator as V_{PEAK}, we can write with reference to Fig. 9.44(b)

$$\frac{V_{PEAK}}{T_1} = \frac{v_A}{RC} \tag{9.115}$$

At the end of this phase, the counter is reset to zero.

Phase II of the conversion begins at $t = T_1$ by connecting the integrator input through switch S_1 to the positive reference voltage V_{REF}. The current into the integrator reverses direction and is equal to V_{REF}/R. Thus v_1 decreases linearly with a slope of (V_{REF}/RC). Simultaneously the counter is enabled and it counts the pulses from the fixed-frequency clock. When v_1 reaches zero volts, the comparator signals the control logic to stop the counter. Denoting the duration of phase II by T_2, we can write, by reference to Fig. 9.44(b),

$$\frac{V_{PEAK}}{T_2} = \frac{V_{REF}}{RC} \tag{9.116}$$

Equations (9.115) and (9.116) can be combined to yield

$$T_2 = T_1 \left(\frac{v_A}{V_{REF}} \right) \tag{9.117}$$

Since the counter reading, n_{REF}, at the end of T_1 is proportional to T_1 and the reading, n, at the end of T_2 is proportional to T_2, we have

$$n = n_{REF} \left(\frac{v_A}{V_{REF}} \right) \tag{9.118}$$

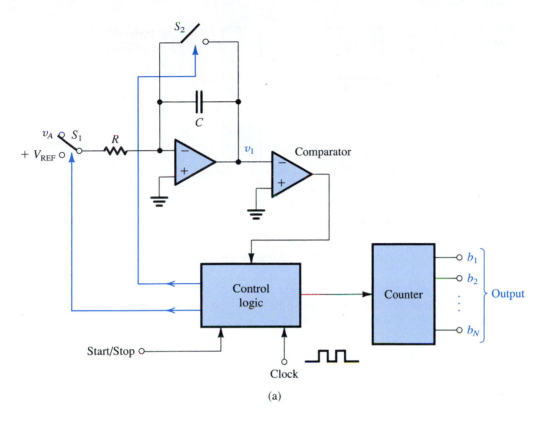

(a)

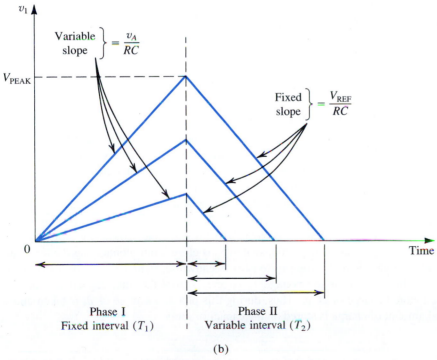

(b)

FIGURE 9.44 The dual-slope A/D conversion method. Note that v_A is assumed to be negative.

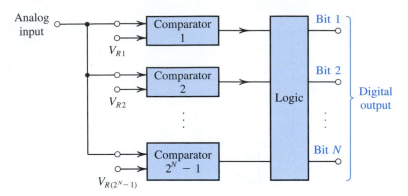

FIGURE 9.45 Parallel, simultaneous, or flash A/D conversion.

Thus the content of the counter,[5] n, at the end of the conversion process is the digital equivalent of v_A.

The dual-slope converter features high accuracy, since its performance is independent of the exact values of R and C. There exist many commercial implementations of the dual-slope method, some of which utilize CMOS technology.

9.9.3 The Parallel or Flash Converter

The fastest A/D conversion scheme is the simultaneous, parallel, or flash conversion process illustrated in Fig. 9.45. Conceptually, flash conversion is very simple. It utilizes $2^N - 1$ comparators to compare the input signal level with each of the $2^N - 1$ possible quantization levels. The outputs of the comparators are processed by an encoding-logic block to provide the N bits of the output digital word. Note that a complete conversion can be obtained within one clock cycle.

Although flash conversion is very fast, the price paid is a rather complex circuit implementation. Variations on the basic technique have been successfully employed in the design of IC converters.

9.9.4 The Charge-Redistribution Converter

The last A/D conversion technique that we shall discuss is particularly suited for CMOS implementation. As shown in Fig. 9.46, the circuit utilizes a binary-weighted capacitor array, a voltage comparator, and analog switches; control logic (not shown in Fig. 9.46) is also required. The circuit shown is for a 5-bit converter; capacitor C_T serves the purpose of terminating the capacitor array, making the total capacitance equal to the desired value of $2C$.

Operation of the converter can be divided into three distinct phases, as illustrated in Fig. 9.46. In the sample phase (Fig. 9.46a) switch S_B is closed, thus connecting the top plate of all capacitors to ground and setting v_O to zero. Meanwhile, switch S_A is connected to the analog input voltage v_A. Thus the voltage v_A appears across the total capacitance of $2C$, resulting in a stored charge of $2Cv_A$. Thus, during this phase, a sample of v_A is taken and a proportional amount of charge is stored on the capacitor array.

---

[5] Note that n is *not* a continuous function of v_A, as might be inferred from Eq. (9.118). Rather, n takes on discrete values corresponding to one of the 2^N quantized levels of v_A.

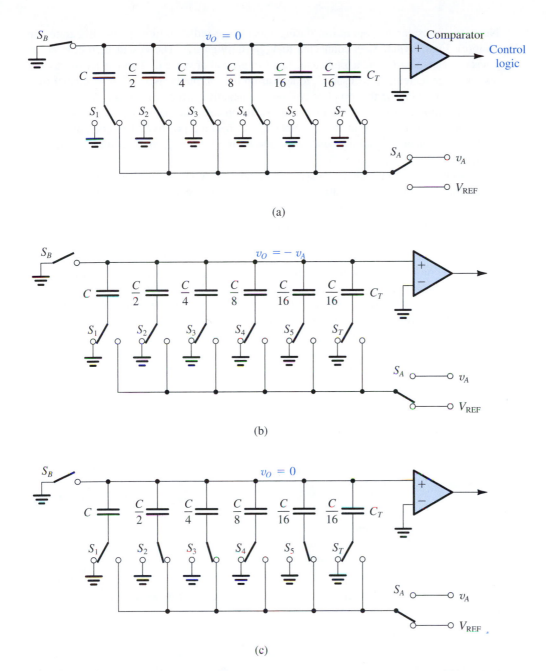

FIGURE 9.46 Charge-redistribution A/D converter suitable for CMOS implementation: **(a)** sample phase, **(b)** hold phase, and **(c)** charge-redistribution phase.

During the hold phase (Fig. 9.46b), switch S_B is opened and switches S_1 to S_5, and S_T are thrown to the ground side. Thus the top plate of the capacitor array is open-circuited while their bottom plates are connected to ground. Since no discharge path has been provided, the capacitor charges must remain constant, with the total equal to $2Cv_A$. It follows that the voltage at the top plate must become $-v_A$. Finally, note that during the hold phase, S_A is connected to V_{REF} in preparation for the charge-redistribution phase.

Next, we consider the operation during the charge-redistribution phase illustrated in Fig. 9.46(c). First, switch S_1 is connected to V_{REF} (through S_A). The circuit then consists of V_{REF}, a series capacitor C, and a total capacitance to ground of value C. This capacitive divider causes a voltage increment of $V_{REF}/2$ to appear on the top plates. Now, if v_A is greater than $V_{REF}/2$, the net voltage at the top plate will remain negative, which means that S_1 will be left in its new position as we move on to switch S_2. If, on the other hand, v_A was smaller than $V_{REF}/2$, then the net voltage at the top plate would become positive. The comparator will detect this situation and signal the control logic to return S_1 to its ground position and then to move on to S_2.

Next, switch S_2 is connected to V_{REF}, which causes a voltage increment of $V_{REF}/4$ to appear on the top plate. If the resulting voltage is still negative, S_2 is left in its new position; otherwise, S_2 is returned to its ground position. We then move on to switch S_3, and so on until all the bit switches S_1 to S_5 have been tried.

It can be seen that during the charge-redistribution phase the voltage on the top plate will be reduced incrementally to zero. The connection of the bit switches at the conclusion of this phase gives the output digital word; a switch connected to ground indicates a 0 value for the corresponding bit, whereas connection to V_{REF} indicates a 1. The particular switch configuration depicted in Fig. 9.46(c) is for $D = 01101$. Observe that at the end of the conversion process, all the charge is stored in the capacitors corresponding to 1 bits; the capacitors of the 0 bits have been discharged.

The accuracy of this A/D conversion method is independent of the value of stray capacitances from the bottom plate of the capacitors to ground. This is because the bottom plates are connected either to ground or to V_{REF}; thus the charge on the stray capacitances will not flow into the capacitor array. Also, because both the initial and the final voltages on the top plate are zero, the circuit is also insensitive to the stray capacitances between the top plates and ground.[6] The insensitivity to stray capacitances makes the charge-redistribution technique a reasonably accurate method capable of implementing A/D converters with as many as 10 bits.

EXERCISES

9.34 Consider the 5-bit charge-redistribution converter in Fig. 9.46 with $V_{REF} = 4$ V. What is the voltage increment appearing on the top plate when S_5 is switched? What is the full-scale voltage of this converter? If $v_A = 2.5$ V, which switches will be connected to V_{REF} at the end of conversion?

Ans. $\frac{1}{8}$ V; $\frac{31}{8}$ V; S_1 and S_3

9.35 Express the maximum quantization error of an N-bit A/D converter in terms of its least-significant bit (LSB) and in terms of its full-scale analog input V_{FS}.

Ans. $\pm\frac{1}{2}$ LSB; $V_{FS}/2(2^N - 1)$

 ## 9.10 SPICE SIMULATION EXAMPLE

We conclude this chapter with an example to illustrate the use of SPICE in the simulation of the two-stage CMOS op amp.

---

[6] More precisely, the final voltage can deviate from zero by as much as the analog equivalent of the LSB. Thus, the insensitivity to top-plate capacitance is not complete.

EXAMPLE 9.4

A TWO-STAGE CMOS OP AMP

In this example, we will use PSpice to aid in designing the frequency compensation of the two-stage CMOS circuit whose Capture schematic is shown in Fig. 9.47. PSpice will then be employed to determine the frequency response and the slew rate of the op amp. We will assume a 0.5-μm n-well CMOS technology for the MOSFETs and use the SPICE level-1 model parameters listed in Table 4.8. Observe that to eliminate the body effect and improve the matching between M_1 and M_2, the source terminals of the input PMOS transistors M_1 and M_2 are connected to their n well.

The op-amp circuit in Fig. 9.47 is designed using a reference current $I_{REF} = 90\ \mu A$, a supply voltage $V_{DD} = 3.3$ V, and a load capacitor $C_L = 1$ pF. Unit-size transistors with $W/L = 1.25\ \mu m/0.6\ \mu m$ are used for both the NMOS and PMOS devices. The transistors are sized for an overdrive voltage $V_{OV} = 0.3$ V. The corresponding multiplicative factors are given in Fig. 9.47.

In PSpice, the common-mode input voltage V_{CM} of the op-amp circuit is set to $V_{DD}/2 = 1.65$ V. A bias-point simulation is performed to determine the dc operating point. Using the values found in the simulation output file for the small-signal parameters of the MOSFETs, we obtain[7]

$$G_{m1} = 0.333\ \text{mA/V}$$
$$G_{m2} = 0.650\ \text{mA/V}$$
$$C_1 = 26.5\ \text{fF}$$
$$C_2 = 1.04\ \text{pF}$$

using Eqs. (9.7), (9.14), (9.24), and (9.25), respectively. Then, using Eq. (9.27), the frequency of the second, nondominant, pole can be found as

$$f_{P2} \simeq \frac{G_{m2}}{2\pi C_2} = 97.2\ \text{MHz}$$

In order to place the transmission zero, given by Eq. (9.37), at infinite frequency, we select

$$R = \frac{1}{G_{m2}} = 1.53\ \text{k}\Omega$$

Now, using Eq. (9.36), the phase margin of the op amp can be expressed as

$$\text{PM} = 90° - \tan^{-1}\left(\frac{f_t}{f_{P2}}\right) \tag{9.119}$$

where f_t is the unity-gain frequency, given in Eq. (9.30),

$$f_t = \frac{G_{m1}}{2\pi C_C} \tag{9.120}$$

Using Eqs. (9.119) and (9.120) we determine that compensation capacitors of $C_C = 0.78$ pF and $C_C = 2$ pF are required to achieve phase margins of PM = 55° and PM = 75°, respectively.

---

[7] Recall that G_{m1} and G_{m2} are the transconductances of, respectively, the first and second stages of the op amp. Capacitors C_1 and C_2 represent the total capacitance to ground at the output nodes of, respectively, the first and second stage of the op amp.

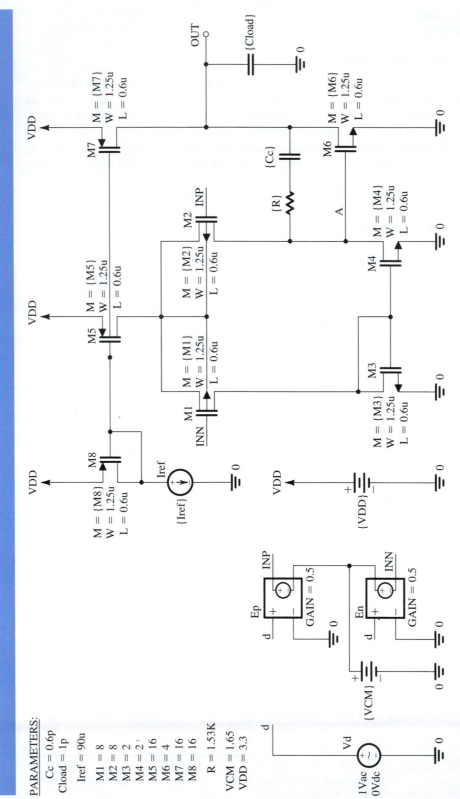

PARAMETERS:

Cc = 0.6p

Cload = 1p

Iref = 90u

M1 = 8
M2 = 8
M3 = 2
M4 = 2
M5 = 16
M6 = 4
M7 = 16
M8 = 16

R = 1.53K

VCM = 1.65
VDD = 3.3

FIGURE 9.47 Capture schematic of the two stage CMOS op amp in Example 9.4.

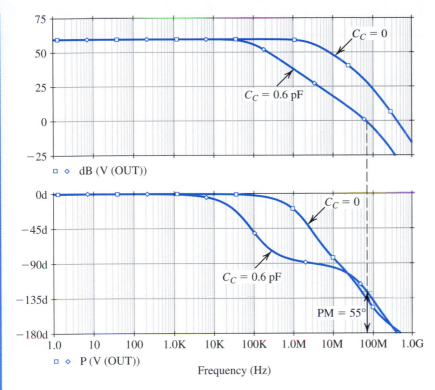

FIGURE 9.48 Magnitude and phase response of the op-amp circuit in Fig. 9.47: $R = 1.53$ kΩ, $C_C = 0$ (no frequency compensation), and $C_C = 0.6$ pF (PM = 55°).

Next, an ac-analysis simulation is performed in PSpice to compute the frequency response of the op amp and to verify the foregoing design values. It was found that, with $R = 1.53$ kΩ, we needed $C_C = 0.6$ pF and $C_C = 1.8$ pF to set PM = 55° and PM = 75°, respectively. We note that these values are reasonably close to those predicted by hand analysis. The corresponding frequency responses for the compensated op amp are plotted in Figs. 9.48 and 9.49. For comparison, we also show the frequency response of the uncompensated op amp ($C_C = 0$). Observe that, the unity gain frequency f_t drops from 70.2 MHz to 26.4 MHz as C_C is increased to improve PM (as anticipated from Eq. 9.120).

Rather than increasing the compensation capacitor C_C, the value of the series resistor R can be increased to improve the phase margin PM: For a given C_C, increasing R above $1/G_{m2}$ places the transmission zero at a negative real-axis location (Eq. 9.37), where the phase it introduces *adds* to the phase margin. Thus, PM can be improved without affecting f_t. To verify this point, we set C_C to 0.6 pF and simulate the op-amp circuit in PSpice for the cases of $R = 1.53$ kΩ and $R = 3.2$ kΩ. The corresponding frequency response is plotted in Fig. 9.50. Observe how f_t is approximately independent of R. However, by increasing R, PM is improved from 55° to 75°.

Increasing the PM is desirable because it reduces the overshoot in the step response of the op amp. To verify this point, we simulate in PSpice the step response of the op amp for PM = 55° and PM = 75°. To do that, we connect the op amp in a unity-gain configuration, apply a small (10-mV) pulse signal at the input with very short (1-ps) rise and fall times to emulate a step input, perform a transient-analysis simulation, and plot the output voltage as shown in Fig. 9.51. Observe that the overshoot in the step response drops from 15% to 1.4% when the phase margin is increased from 55° to 75°.

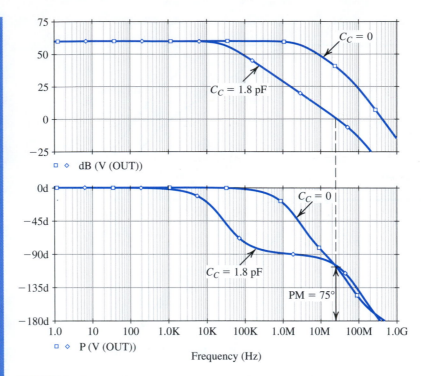

FIGURE 9.49 Magnitude and phase response of the op-amp circuit in Fig. 9.47: $R = 1.53$ kΩ, $C_C = 0$ (no frequency compensation), and $C_C = 1.8$ pF (PM = 75°).

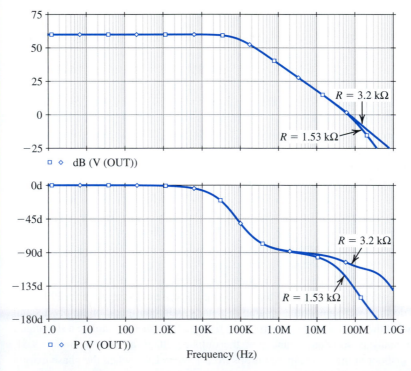

FIGURE 9.50 Magnitude and phase response of the op-amp circuit in Fig. 9.47: $C_C = 0.6$ pF, $R = 1.53$ kΩ (PM = 55°), and $R = 3.2$ kΩ (PM = 75°).

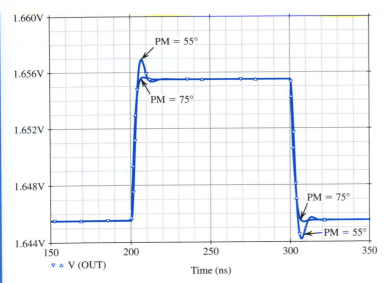

FIGURE 9.51 Small-signal step response (for a 10-mV step input) of the op-amp circuit in Fig. 9.47 connected in a unity-gain configuration: PM = 55° (C_C = 0.6 pF, R = 1.53 kΩ) and PM = 75° (C_C = 0.6 pF, R = 3.2 kΩ).

We conclude this example by computing *SR*, the slew rate of the op amp. From Eq. (9.40),

$$ SR = 2\pi f_t V_{OV} = \frac{G_{m1}}{C_C} V_{OV} = 166.5 \text{ V/}\mu s $$

when C_C = 0.6 pF. Next, to determine *SR* using PSpice (see Example 2.9), we again connect the op amp in a unity-gain configuration and perform a transient-analysis simulation. However, we now apply a large pulse signal (3.3 V) at the input to cause slew-rate limiting at the output. The corresponding output-voltage waveform is plotted in Fig. 9.52. The slope of the slew-rate-limited output waveform corresponds to the slew rate of the op amp and is found to be *SR* = 160 V/μs and 60 V/μs for the negative- and positive-going output, respectively. These results, with the unequal

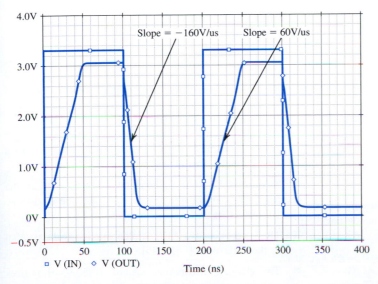

FIGURE 9.52 Large-signal step response (for a 3.3-V step-input) of the op-amp circuit in Fig. 9.47 connected in a unity-gain configuration. The slope of the rising and falling edges of the output waveform correspond to the slew rate of the op amp.

values of SR in the two directions, differ from those predicted by the simple model for the slew-rate limiting of the two-stage op-amp circuit (Section 9.1.5). The difference can perhaps be said to be a result of transistor M_4 entering the triode region and its output current (which is sourced through C_C) being correspondingly reduced. Of course, the availability of PSpice should enable the reader to explore this point further.

SUMMARY

- Most CMOS op amps are designed to operate as part of a VLSI circuit and thus are required to drive only small capacitive loads. Therefore, most do not have a low-output-resistance stage.

- There are basically two approaches to the design of CMOS op amps: a two-stage configuration, and a single-stage topology utilizing the folded-cascode circuit.

- In the two-stage CMOS op amp, approximately equal gains are realized in the two stages.

- The threshold mismatch ΔV_t together with the low transconductance of the input stage result in a larger input offset voltage for CMOS op amps than for bipolar units.

- Miller compensation is employed in the two-stage CMOS op amp, but a series resistor is required to place the transmission zero at either $s = \infty$ or on the negative real axis.

- CMOS op amps have higher slew rates than their bipolar counterparts with comparable f_t values.

- Use of the cascode configuration increases the gain of a CMOS amplifier stage by about two orders of magnitude, thus making possible a single-stage op amp.

- The dominant pole of the folded-cascode op amp is determined by the total capacitance at the output node, C_L. Increasing C_L improves the phase margin at the expense of reducing the bandwidth.

- By using two complementary input differential pairs in parallel, the input common-mode range can be extended to equal the entire power-supply voltage, providing so-called rail-to-rail operation at the input.

- The output voltage swing of the folded-cascode op amp can be extended by utilizing a wide-swing current mirror in place of the cascode mirror.

- The internal circuit of the 741 op amp embodies many of the design techniques employed in bipolar analog integrated circuits.

- The 741 circuit consists of an input differential stage, a high-gain single-ended second stage, and a class AB output stage. This structure is typical of modern BJT op amps and is known as the two-stage topology (not counting the output stage). It is the same structure used in the two-stage CMOS op amp of Section 9.1.

- To obtain low input offset voltage and current, and high CMRR, the 741 input stage is designed to be perfectly balanced. The CMRR is increased by common-mode feedback, which also stabilizes the dc operating point.

- To obtain high input resistance and low input bias current, the input stage of the 741 is operated at a very low current level.

- In the 741, output short-circuit protection is accomplished by turning on a transistor that takes away most of the base current drive of the output transistor.

- The use of Miller frequency compensation in the 741 circuit enables locating the dominant pole at a very low frequency, while utilizing a relatively small compensating capacitance.

- Two-stage op amps can be modeled as a transconductance amplifier feeding an ideal integrator with C_C as the integrating capacitor.

- The slew rate of a two-stage op amp is determined by the first-stage bias current and the frequency-compensation capacitor.

- A/D and D/A converters constitute an important group of analog ICs.

- A DAC consists of (a) a circuit that generates a reference current, (b) a circuit that assigns binary weights to the value of the reference current, (c) switches that, under the control of the bits of the input digital word, direct the proper combination of binary-weighted currents to an output summing node, and (d) an op amp that converts the current sum to an output voltage. The circuit of (b) can be implemented by either a binary-weighted resistive network or an R-$2R$ ladder.

- Two simple but slow implementations of the ADC are the feedback-type converter [Fig. 9.43] and the dual-slope converter [Fig. 9.44].

- The fastest possible ADC implementation is the parallel or flash converter [Fig. 9.45].

- The charge-redistribution method [Fig. 9.46] utilizes switched-capacitor techniques and is particularly suited for the implementation of ADCs in CMOS technology.

PROBLEMS

SECTION 9.1: THE TWO-STAGE CMOS OP AMP

9.1 A particular design of the two-stage CMOS operational amplifier of Fig. 9.1 utilizes ±2.5-V power supplies. All transistors are operated at overdrive voltages of 0.3-V magnitude. The process technology provides devices with $V_{tn} = |V_{tp}| = 0.7$ V. Find the input common-mode range, and the range allowed for v_O.

9.2 The CMOS op amp of Fig. 9.1 is fabricated in a process for which $V'_{An} = 25$ V/μm and $|V'_{Ap}| = 20$ V/μm. Find A_1, A_2, and A_v if all devices are 0.8-μm long and are operated at equal overdrive voltages of 0.25-V magnitude. Also, determine the op-amp output resistance obtained when the second stage is biased at 0.4 mA. What do you expect the output resistance of a unity-gain voltage amplifier to be, using this op amp?

D9.3 The CMOS op amp of Fig. 9.1 is fabricated in a process for which $|V'_A|$ for all devices is 10 V/μm. If all transistors have $L = 1$ μm and are operated at equal overdrive voltages, find the magnitude of the overdrive voltage required to obtain a dc open-loop gain of 2500 V/V.

9.4 This problem is identical to Problem 7.90.

Consider the circuit in Fig. 9.1 with the device geometries shown at the bottom of this page:

Let $I_{REF} = 225$ μA, $|V_t|$ for all devices = 0.75 V, $\mu_n C_{ox} = 180$ μA/V^2, $\mu_p C_{ox} = 60$ μA/V^2, $|V_A|$ for all devices = 9 V, $V_{DD} = V_{SS} = 1.5$ V. Determine the width of Q_6, W, that will ensure that the op amp will not have a systematic offset voltage. Then, for all devices, evaluate I_D, $|V_{OV}|$, $|V_{GS}|$, g_m, and r_o. Provide your results in a table. Also find A_1, A_2, the dc open-loop voltage gain, the input common-mode range, and the output voltage range. Neglect the effect of V_A on the bias currents.

D9.5 A particular implementation of the CMOS amplifier of Figs. 9.1 and 9.2 provides $G_{m1} = 0.3$ mA/V, $G_{m2} = 0.6$ mA/V, $r_{o2} = r_{o4} = 222$ kΩ, $r_{o6} = r_{o7} = 111$ kΩ, and $C_2 = 1$ pF.

(a) Find the frequency of the second pole, f_{P2}.
(b) Find the value of the resistance R which when placed in series with C_C causes the transmission zero to be located at $s = \infty$.

(c) With R in place, as in (b), find the value of C_C that results in the highest possible value of f_t while providing a phase margin of 80°. What value of f_t is realized? What is the corresponding frequency of the dominant pole?
(d) To what value should C_C be changed to double the value of f_t? At the new value of f_t, what is the phase shift introduced by the second pole? To reduce this excess phase shift to 10° and thus obtain an 80° phase margin, as before, what value should R be changed to?

D9.6 A two-stage CMOS op amp similar to that in Fig. 9.1 is found to have a capacitance between the output node and ground of 1 pF. If it is desired to have a unity-gain bandwidth f_t of 100 MHz with a phase margin of 75° what must g_{m6} be set to? Assume that a resistance R is connected in series with the frequency-compensation capacitor C_C and adjusted to place the transmission zero at infinity. What value should R have? If the first stage is operated at $|V_{OV}| = 0.2$ V, what is the value of slew rate obtained? If the first-stage bias current $I = 200$ μA, what is the required value of C_C?

D9.7 A CMOS op amp with the topology shown in Fig. 9.1 but with a resistance R included in series with C_C is designed to provide $G_{m1} = 1$ mA/V and $G_{m2} = 2$ mA/V.

(a) Find the value of C_C that results in $f_t = 100$ MHz.
(b) For $R = 500$ Ω, what is the maximum allowed value of C_2 for which a phase margin of at least 60° is obtained?

9.8 A two-stage CMOS op amp resembling that in Fig. 9.1 is found to have a slew rate of 60 V/μs and a unity-gain bandwidth f_t of 50 MHz.

(a) Estimate the value of the overdrive voltage at which the input-stage transistors are operating.
(b) If the first-stage bias current $I = 100$ μA, what value of C_C must be used?
(c) For a process for which $\mu_p C_{ox} = 50$ μA/V^2, what W/L ratio applies for Q_1 and Q_2?

D9.9 Sketch the circuit of a two-stage CMOS amplifier having the structure of Fig. 9.1 but utilizing NMOS transistors in the input stage (i.e., Q_1 and Q_2).

SECTION 9.2: THE FOLDED-CASCODE OP AMP

D9.10 If the circuit of Fig. 9.8 utilizes ±1.65-V power supplies and the power dissipation is to be limited to 1 mW, find

Transistor	Q_1	Q_2	Q_3	Q_4	Q_5	Q_6	Q_7	Q_8
W/L (μm/μm)	30/0.5	30/0.5	10/0.5	10/0.5	60/0.5	W/0.5	60/0.5	60/0.5

the values of I_B and I. To avoid turning off the current mirror during slewing, select I_B to be 20% larger than I.

D9.11 For the folded-cascode op amp utilizing power supplies of ±1.65 V, find the values of V_{BIAS1}, V_{BIAS2}, and V_{BIAS3} to maximize the allowable range of V_{ICM} and v_O. Assume that all transistors are operated at equal overdrive voltages of 0.2 V. Assume $|V_t|$ for all devices is 0.5 V. Specify the maximum range of V_{ICM} and of v_O.

D9.12 For the folded-cascode op-amp circuit of Figs. 9.8 and 9.9 with bias currents $I = 125\ \mu A$ and $I_B = 150\ \mu A$, and with all transistors operated at overdrive voltages of 0.2 V, find the W/L ratios for all devices. Assume that the technology available is characterized by $k'_n = 250\ \mu A/V^2$ and $k'_p = 90\ \mu A/V^2$.

D9.13 Consider the folded-cascode op amp when loaded with a 10-pF capacitance. What should the bias current I be to obtain a slew rate of at least 10 V/μs? If the input-stage transistors are operated at overdrive voltages of 0.2 V, what is the unity-gain bandwidth realized? If the two non-dominant poles have the same frequency of 25 MHz, what is the phase margin obtained? If it is required to have a phase margin of 75°, what must f_t be reduced to? By what amount should C_L be increased? What is the new value of SR?

9.14 Consider a design of the cascode op amp of Fig. 9.9 for which $I = 125\ \mu A$ and $I_B = 150\ \mu A$. Assume that all transistors are operated at $|V_{OV}| = 0.2$ V and that for all devices, $|V_A| = 10$ V. Find G_m, R_o, and A_v. Also, if the op amp is connected in the feedback configuration shown in Fig. P9.14, find the voltage gain and output resistance of the closed-loop amplifier.

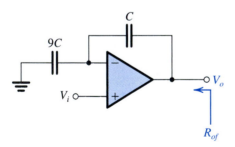

FIGURE P9.14

9.15 For the circuit in Fig. 9.11, assume that all transistors are operating at equal overdrive voltages of 0.2-V magnitude and have $|V_t| = 0.5$ V and that $V_{DD} = V_{SS} = 1.65$ V. Find (a) the range over which the NMOS input stage operates, (b) the range over which the PMOS input stage operates, (c) the range over which both operate (the overlap range), and (d) the input common-mode range.

9.16 A particular design of the wide-swing current mirror of Fig. 9.12(b) utilizes devices having $W/L = 25$ and $V_t = 0.5$ V. For $I_{REF} = 100\ \mu A$ give the voltages that you expect to appear at all nodes and specify the minimum voltage allowable at the output terminal. If V_A is specified to be 10 V, what is the output resistance of the mirror?

D9.17 It is required to design the folded-cascode circuit of Fig. 9.9 to provide voltage gain of 80 dB and a unity-gain frequency of 10 MHz when $C_L = 10$ pF. Design for $I_B = I$, and operate all devices at the same $|V_{OV}|$. Utilize transistors with 1-μm channel length for which $|V_A|$ is specified to be 20 V. Find the required overdrive voltages and bias currents. What slew rate is achieved? Also, for $k'_n = 2.5k'_p = 200\ \mu A/V^2$, specify the required width of each of the 11 transistors used.

D9.18 Sketch the circuit that is complementary to that in Fig. 9.9, that is, one that uses an input p-channel differential pair.

D9.19 For the folded-cascode circuit of Fig. 9.8, let the total capacitance to ground at each of the source nodes of Q_3 and Q_4 be denoted C_P. Show that the pole that arises at the interface between the first and second stages has a frequency $f_P = g_{m3}/2\pi C_P$. Now, if this is the only nondominant pole, what is the largest value that C_P can be (expressed as a fraction of C_L) while a phase margin of 75° is achieved? Assume that all transistors are operated at the same bias current and overdrive voltage.

SECTION 9.3: THE 741 OP-AMP CIRCUIT

9.20 In the 741 op-amp circuit of Fig. 9.13, Q_1, Q_2, Q_5, and Q_6 are biased at collector currents of 9.5 μA; Q_{16} is biased at a collector current of 16.2 μA; and Q_{17} is biased at a collector current of 550 μA. All these devices are of the "standard npn" type, having $I_S = 10^{-14}$ A, $\beta = 200$, and $V_A = 125$ V. For each of these transistors, find V_{BE}, g_m, r_e, r_π, and r_o. Provide your results in table form. (Note that these parameter values are utilized in the text in the analysis of the 741 circuit.)

D9.21 For the (mirror) bias circuit shown in Fig. E9.10 and the result verified in the associated Exercise, find I_1 for the case in which $I_{S3} = 3 \times 10^{-14}$ A, $I_{S4} = 6 \times 10^{-14}$ A, and $I_{S1} = I_{S2} = 10^{-14}$ A and for which a bias current $I_3 = 154\ \mu A$ is required.

9.22 Transistor Q_{13} in the circuit of Fig. 9.13 consists, in effect, of two transistors whose emitter–base junctions are connected in parallel and for which $I_{SA} = 0.25 \times 10^{-14}$ A, $I_{SB} = 0.75 \times 10^{-14}$ A, $\beta = 50$, and $V_A = 50$ V. For operation at a total emitter current of 0.73 mA, find values for the parameters V_{EB}, g_m, r_e, r_π, and r_o for the A and B devices.

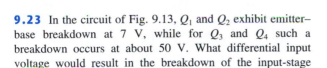

9.23 In the circuit of Fig. 9.13, Q_1 and Q_2 exhibit emitter–base breakdown at 7 V, while for Q_3 and Q_4 such a breakdown occurs at about 50 V. What differential input voltage would result in the breakdown of the input-stage transistors?

D*9.24 Figure P9.24 shows the CMOS version of the circuit in Fig. E9.10. Find the relationship between I_3 and I_1 in terms of k_1, k_2, k_3, and k_4 of the four transistors, assuming the threshold voltages of all devices to be equal in magnitude. Note that k denotes $\frac{1}{2}\mu C_{ox}W/L$. In the event that $k_1 = k_2$ and $k_3 = k_4 = 16k_1$, find the required value of I_1 to yield a bias current in Q_3 and Q_4 of 1.6 mA.

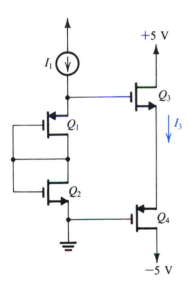

FIGURE P9.24

SECTION 9.4: DC ANALYSIS OF THE 741

D9.25 For the 741 circuit, estimate the input reference current I_{REF} in the event that ±5-V supplies are used. Find a more precise value assuming that for the two BJTs involved, $I_S = 10^{-14}$ A. What value of R_5 would be necessary to reestablish the same bias current for ±5-V supplies as exists for ±15 V in the original design?

***9.26** In the 741 circuit, consider the **common-mode feedback loop** comprising transistors Q_1, Q_2, Q_3, Q_4, Q_8, Q_9, and Q_{10}. We wish to find the loop gain. This can be conveniently done by breaking the loop between the common collector connection of Q_1 and Q_2, and the diode-connected transistor Q_8. Apply a test current signal I_t to Q_8 and find the returned current signal I_r in the combined collector connection of Q_1 and Q_2. Thus determine the loop gain. Assume that Q_9 and Q_{10} act as ideal current sources. If Q_3 and Q_4 have $\beta = 50$, find the amount of common-mode feedback in decibels.

D9.27 Design the Widlar current source of Fig. 9.15 to generate a current $I_{C10} = 20$ μA given that $I_{REF} = 0.5$ mA. If for the transistors, $I_S = 10^{-14}$ A, find V_{BE11} and V_{BE10}. Assume β to be high.

9.28 Consider the dc analysis of the 741 input stage shown in Fig. 9.16. For what value of β_P do the currents in Q_1 and Q_2 differ from the ideal value of $I_{C10}/2$ by 10%?

D9.29 Consider the dc analysis of the 741 input stage shown in Fig. 9.16 for the situation in which $I_{S9} = 2I_{S8}$. For $I_{C10} = 19$ μA and assuming β_P to be high, what does I become? Redesign the Widlar source to reestablish $I_{C1} = I_{C2} = 9.5$ μA.

9.30 For the mirror circuit shown in Fig. 9.17 with the bias and component values given in the text for the 741 circuit, what does the current in Q_6 become if R_2 is shorted?

D9.31 It is required to redesign the circuit of Fig. 9.17 by selecting a new value for R_3 so that when the base currents are *not* neglected, the collector currents of Q_5, Q_6, and Q_7 all become equal, assuming that the input current $I_{C3} = 9.4$ μA. Find the new value of R_3 and the three currents. Recall that $\beta_N = 200$.

9.32 Consider the input circuit of the 741 op amp of Fig. 9.13 when the emitter current of Q_8 is about 19 μA. If β of Q_1 is 150 and that of Q_2 is 200, find the input bias current I_B and the input offset current I_{OS} of the op amp.

9.33 For a particular application, consideration is being given to selecting 741 ICs for bias and offset currents limited to 40 nA and 4 nA, respectively. Assuming other aspects of the selected units to be normal, what minimum β_N and what β_N variation are implied?

9.34 A manufacturing problem in a 741 op amp causes the current transfer ratio of the mirror circuit that loads the input stage to become 0.9 A/A. For input devices (Q_1–Q_4) appropriately matched and with high β, and normally biased at 9.5 μA, what input offset voltage results?

D9.35 Consider the design of the second stage of the 741. What value of R_9 would be needed to reduce I_{C16} to 9.5 μA?

D9.36 Reconsider the 741 output stage as shown in Fig. 9.18, in which R_{10} is adjusted to make $I_{C19} = I_{C18}$. What is the new value of R_{10}? What values of I_{C14} and I_{C20} result?

D*9.37 An alternative approach to providing the voltage drop needed to bias the output transistors is the V_{BE}–multiplier circuit shown in Fig. P9.37. Design the circuit to provide a terminal voltage of 1.118 V (the same as in the 741 circuit). Base your design on half the current flowing through R_1, and assume that $I_S = 10^{-14}$ A and $\beta = 200$. What is the incremental

9

resistance between the two terminals of the V_{BE}–multiplier circuit?

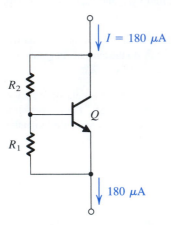

FIGURE P9.37

9.38 For the circuit of Fig. 9.13, what is the total current required from the power supplies when the op amp is operated in the linear mode, but with no load? Hence, estimate the quiescent power dissipation in the circuit. (*Hint:* Use the data given in Table 9.1.)

SECTION 9.5: SMALL-SIGNAL ANALYSIS OF THE 741

9.39 Consider the 741 input stage as modeled in Fig. 9.19, with two additional *npn* diode-connected transistors, Q_{1a} and Q_{2a}, connected between the present *npn* and *pnp* devices, one per side. Convince yourself that each of the additional devices will be biased at the same current as Q_1 to Q_4—that is, 9.5 μA. What does R_{id} become? What does G_{m1} become? What is the value of R_{o4} now? What is the output resistance of the first stage, R_{o1}? What is the new open-circuit voltage gain, $G_{m1}R_{o1}$? Compare these values with the original ones.

D9.40 What relatively simple change can be made to the mirror load of stage 1 to increase its output resistance, say by a factor of 2?

9.41 Repeat Exercise 9.14 with $R_1 = R_2$ replaced by 2-kΩ resistors.

9.42 In Example 9.3 we investigated the effect of a mismatch between R_1 and R_2 on the input offset voltage of the op amp. Conversely, R_1 and R_2 can be deliberately mismatched (using the circuit shown in Fig. P9.42, for example) to compensate for the op amp input offset voltage.

(a) Show that an input offset voltage V_{OS} can be compensated for (i.e., reduced to zero) by creating a relative mismatch

$\Delta R/R$ between R_1 and R_2,

$$\frac{\Delta R}{R} = \frac{V_{OS}}{2V_T}\frac{1 + r_e/R}{1 - V_{OS}/2V_T}$$

where r_e is the emitter resistance of each of Q_1 to Q_6, and R is the nominal value of R_1 and R_2. (*Hint:* Use Eq. 9.75)
(b) Find $\Delta R/R$ to trim a 5-mV offset to zero.
(c) What is the maximum offset voltage that can be trimmed this way (corresponding to R_2 completely shorted)?

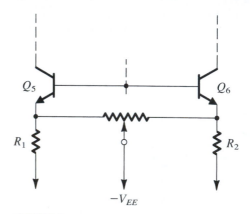

FIGURE P9.42

9.43 Through a processing imperfection, the β of Q_4 in Fig. 9.13 is reduced to 25, while the β of Q_3 remains at its regular value of 50. Find the input offset voltage that this mismatch introduces. (*Hint:* Follow the general procedure outlined in Example 9.3.)

9.44 Consider the circuit of Fig. 9.13 modified to include resistors R in series with the emitters of each of Q_8 and Q_9. What does the resistance looking into the collector of Q_9, R_{o9}, become? For what value of R does it equal R_{o10}? For this case, what does R_o looking to the left of node Y become?

*9.45 Refer to Fig. E9.15 and let $R_1 = R_2$. If Q_3 and Q_4 have a β mismatch such that for Q_3 the current gain is β_P and for Q_4 the current gain is $k\beta_P$, find i_o and G_{mcm}. For $R_o = 2.43$ MΩ, $\beta_P = 20$, $0.5 \le k \le 2$, G_{m1} (differential) $= 1/5.26$ kΩ, find the worst-case CMRR $\equiv G_{m1}/G_{mcm}$ (in dB) that results. Assume everything else is ideal.

*9.46 What is the effect on the differential gain of the 741 op amp of short-circuiting one, or the other, or both, of R_1 and R_2 in Fig. 9.13? (Refer to Fig. 9.20.) For simplicity, assume $\beta = \infty$.

*9.47 Figure P9.47 shows the equivalent common-mode half-circuit of the input stage of the 741. Here R_o is the resistance seen looking to the left of node Y in Fig. 9.13; its value is approximately 2.4 MΩ. Transistors Q_1 and Q_3 operate at a bias current of 9.5 μA. Find the input resistance of the common-mode half-circuit using $\beta_N = 200$, $\beta_P = 50$, and $V_A = 125$ V for *npn* and 50 V for *pnp* transistors. To find the common-mode input resistance of the 741 note that it has common-mode feedback that increases the input common-mode resistance. The loop gain is approximately equal to β_P. Find the value of R_{icm}.

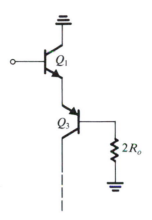

FIGURE P9.47

9.48 Consider a variation on the design of the 741 second stage in which $R_8 = 50$ Ω. What R_{i2} and G_{m2} correspond?

9.49 In the analysis of the 741 second stage, note that R_{o2} is affected most strongly by the low value of R_{o13B}. Consider the effect of placing appropriate resistors in the emitters of Q_{12}, Q_{13A}, and Q_{13B} on this value. What resistor in the emitter of Q_{13B} would be required to make R_{o13B} equal to R_{o17} and thus R_{o2} half as great? What resistors in each of the other emitters would be required?

9.50 For a 741 employing ±5-V supplies, $|V_{BE}| = 0.6$ V and $|V_{CEsat}| = 0.2$ V, find the output voltage limits that apply.

D9.51 Consider an alternative to the present 741 output stage in which Q_{23} is not used, that is, in which its base and emitter are joined. Reevaluate the reflection of $R_L = 2$ kΩ to the collector of Q_{17}. What does A_2 become?

9.52 Consider the positive current-limiting circuit involving Q_{13A}, Q_{15}, and R_6. Find the current in R_6 at which the collector current of Q_{15} equals the current available from Q_{13A} (180 μA) minus the base current of Q_{14}. (You need to perform a couple of iterations.)

D9.53 Consider the 741 sinking-current limit involving R_7, Q_{21}, Q_{24}, R_{11}, and Q_{22}. For what current through R_7 is

the current in Q_{22} equal to the maximum current available from the input stage (i.e., the current in Q_8)? What simple change would you make to reduce this current limit to 10 mA?

SECTION 9.6: GAIN, FREQUENCY RESPONSE, AND SLEW RATE OF THE 741

9.54 Using the data provided in Eq. (9.93) (alone) for the overall gain of the 741 with a 2-kΩ load, and realizing the significance of the factor 0.97 in relation to the load, calculate the open-circuit voltage gain, the output resistance, and the gain with a load of 200 Ω. What is the maximum output voltage available for such a load?

9.55 A 741 op amp has a phase margin of 80°. If the excess phase shift is due to a second single pole, what is the frequency of this pole?

9.56 A 741 op amp has a phase margin of 80°. If the op amp has nearly coincident second and third poles, what is their frequency?

D*9.57 For a modified 741 whose second pole is at 5 MHz, what dominant-pole frequency is required for 85° phase margin with a closed-loop gain of 100? Assuming C_C continues to control the dominant pole, what value of C_C would be required?

9.58 An internally compensated op amp having an f_t of 5 MHz and dc gain of 10^6 utilizes Miller compensation around an inverting amplifier stage with a gain of −1000. If space exists for at most a 50-pF capacitor, what resistance level must be reached at the input of the Miller amplifier for compensation to be possible?

9.59 Consider the integrator op-amp model shown in Fig. 9.33. For $G_{m1} = 10$ mA/V, $C_C = 50$ pF, and a resistance of 10^8 Ω shunting C_C, sketch and label a Bode plot for the magnitude of the open-loop gain. If G_{m1} is related to the first-stage bias current via Eq. (9.105), find the slew rate of this op amp.

9.60 For an amplifier with a slew rate of 10 V/μs, what is the full-power bandwidth for outputs of ±10 V? What unity-gain bandwidth, ω_t, would you expect if the topology was similar to that of the 741?

D*9.61 Figure P9.61 shows a circuit suitable for op-amp applications. For all transistors $\beta = 100$, $V_{BE} = 0.7$ V, and $r_o = \infty$.

(a) For inputs grounded and output held at 0 V (by negative feedback) find the emitter currents of all transistors.
(b) Calculate the gain of the amplifier with a load of 10 kΩ.

9

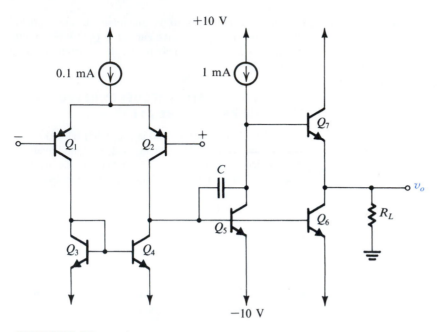

FIGURE P9.61

9

(c) With load as in (b) calculate the value of the capacitor C required for a 3-dB frequency of 1 kHz.

SECTION 9.7: DATA CONVERTERS—AN INTRODUCTION

9.62 An analog signal in the range 0 to +10 V is to be digitized with a quantization error of less than 1% of full scale. What is the number of bits required? What is the resolution of the conversion? If the range is to be extended to ±10 V with the same requirement, what is the number of bits required? For an extension to a range of 0 to +15 V, how many bits are required to provide the same resolution? What is the corresponding resolution and quantization error?

*9.63 Consider Fig. 9.38. On the staircase output of the S/H circuit sketch the output of a simple low-pass RC circuit with a time constant that is (a) one-third of the sampling interval and (b) equal to the sampling interval.

SECTION 9.8: D/A CONVERTER CIRCUITS

*9.64 Consider the DAC circuit of Fig. 9.39 for the cases $N = 2, 4,$ and 8. What is the tolerance, expressed as $\pm x\%$, to which the resistors should be selected to limit the resulting output error to the equivalent of $\pm\frac{1}{2}$ LSB?

9.65 The BJTs in the circuit of Fig. P9.65 have their base–emitter junction areas scaled in the ratios indicated. Find I_1 to I_4 in terms of I.

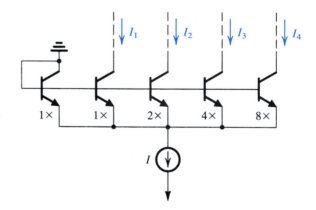

FIGURE P9.65

D9.66 A problem encountered in the DAC circuit of Fig. 9.41 is the large spread in transistor EBJ areas required when N is large. As an alternative arrangement, consider using the circuit in Fig. 9.41 for 4 bits only. Then, feed the current in the collector of the terminating transistor Q_t to the circuit of Fig. P9.65 (in place of the current source I), thus producing currents for 4 more bits. In this way, an 8-bit DAC can be implemented with a maximum spread in areas of 8. What is the total area of emitters needed in terms of the smallest device? Contrast this with the usual 8-bit circuit. Give the complete circuit of the converter thus realized.

D*9.67 The circuit in Fig. 9.41 can be used to multiply an analog signal by a digital one by feeding the analog signal to the V_{REF} terminal. In this case the D/A converter is called a **multiplying DAC** or MDAC. Given an input sine-wave signal of $0.1\sin\omega t$ volts, use the circuit of Fig. 9.41 together with an additional op amp to obtain $v_O = 10D\sin\omega t$, where D is the digital word given by Eq. (9.109) and $N = 4$. How many discrete sine-wave amplitudes are available at the output? What is the smallest? What is the largest? To what digital input does a 10-V peak-to-peak output correspond?

9.68 What is the input resistance seen by V_{REF} in the circuit of Fig. 9.41?

SECTION 9.9: A/D CONVERTER CIRCUITS

9.69 A 12-bit dual-slope ADC of the type illustrated in Fig. 9.44 utilizes a 1-MHz clock and has $V_{REF} = 10$ V. Its analog input voltage is in the range 0 to −10 V. The fixed interval T_1 is the time taken for the counter to accumulate a count of 2^N. What is the time required to convert an input voltage equal to the full-scale value? If the peak voltage reached at the output of the integrator is 10 V, what is the integrator time constant? If through aging, R increases by 2% and C decreases by 1%, what does V_{PEAK} become? Does the conversion accuracy change?

D9.70 The design of a 4-bit flash ADC such as that shown in Fig. 9.45 is being considered. How many comparators are required? For an input signal in the range of 0 to +10 V, what are the reference voltages needed? Show how they can be generated using a 10-V reference and several 1-kΩ resistors (how many?). If a comparison is possible in 50 ns and the associated logic requires 35 ns, what is the maximum possible conversion rate? Indicate the digital code you expect at the output of the comparators and at the output of the logic for an input of (a) 0 V, (b) +5.1 V, and (c) +10 V.

9

Digital CMOS Logic Circuits

INTRODUCTION

This chapter is concerned with the study of CMOS digital logic circuits. CMOS is by far the most popular technology for the implementation of digital systems. The small size, ease of fabrication, and low power dissipation of MOSFETs enable extremely high levels of integration of both logic and memory circuits. The latter will be studied in Chapter 11.

The chapter begins with an overview section whose objective is to place in proper perspective the material we shall study in this chapter and the next. Then, building on the study of the CMOS inverter in Section 4.10, we take a comprehensive look at its design and analysis. This material is then applied to the design of CMOS logic circuits and two other types of logic circuits (namely, pseudo-NMOS and pass-transistor logic) that are frequently employed in special applications, as supplements to CMOS.

To reduce the power dissipation even further, and simultaneously to increase performance (speed of operation), dynamic logic techniques are employed. This challenging topic is the subject of Section 10.6 and completes our study of logic circuits. The chapter concludes with a SPICE simulation example.

In summary, this chapter provides a reasonably comprehensive and in-depth treatment of CMOS digital integrated-circuit design, perhaps the most significant area (at least in terms of production volume and societal impact) of electronic circuits. To gain the most out of studying this chapter, the reader must be thoroughly familiar with the MOS transistor. Thus, a review of Chapter 4 is recommended, and a careful study of Section 4.10 is a must!

 ## 10.1 DIGITAL CIRCUIT DESIGN: AN OVERVIEW

In this section, we build on the introduction to digital circuits presented in Section 1.7 and provide an overview of the subject. We discuss the various technologies and logic-circuit families currently in use, consider the parameters employed to characterize the operation and performance of logic circuits, and finally mention the various styles for digital-system design.

10.1.1 Digital IC Technologies and Logic-Circuit Families

The chart in Figure 10.1 shows the major IC technologies and logic-circuit families that are currently in use. The concept of a logic-circuit family perhaps needs a few words of explanation. Members of each family are made with the same technology, have a similar circuit structure, and exhibit the same basic features. Each logic-circuit family offers a unique set of advantages and disadvantages. In the conventional style of designing systems, one selects an appropriate logic family (e.g., TTL, CMOS, or ECL) and attempts to implement as much of the system as possible using circuit modules (packages) that belong to this family. In this way, interconnection of the various packages is relatively straightforward. If, on the other hand, packages from more than one family are used, one has to design suitable *interface circuits*. The selection of a logic family is based on such considerations as logic flexibility, speed of operation, availability of complex functions, noise immunity, operating-temperature range, power dissipation, and cost. We will discuss some of these considerations in this chapter and the next. To begin with, we make some brief remarks on each of the four technologies listed in the chart of Fig. 10.1.

CMOS   Although shown as one of four possible technologies, this is not an indication of digital IC market share: CMOS technology is, by a large margin, the most dominant of all the IC technologies available for digital-circuit design. As mentioned earlier, CMOS has replaced NMOS, which was employed in the early days of VLSI (in the 1970s). There are a number of reasons for this development, the most important of which is the much lower power dissipation of CMOS circuits. CMOS has also replaced bipolar as the technology-of-choice in digital-system design, and has made possible levels of integration (or circuit-packing

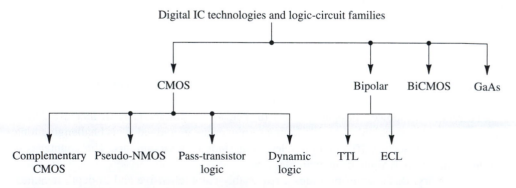

FIGURE 10.1 Digital IC technologies and logic-circuit families.

densities), and a range of applications, neither of which would have been possible with bipolar technology. Furthermore, CMOS continues to advance, whereas there appear to be few innovations at the present time in bipolar digital circuits. Some of the reasons for CMOS displacing bipolar technology in digital applications are as follows:

1. CMOS logic circuits dissipate much less power than bipolar logic circuits and thus one can pack more CMOS circuits on a chip than is possible with bipolar circuits. We will have a lot more to say about power dissipation in the following sections.

2. The high input impedance of the MOS transistor allows the designer to use charge storage as a means for the temporary storage of information in both logic and memory circuits. This technique cannot be used in bipolar circuits.

3. The feature size (i.e., minimum channel length) of the MOS transistor has decreased dramatically over the years, with some recently reported designs utilizing channel lengths as short as 0.06 μm. This permits very tight circuit packing and, correspondingly, very high levels of integration.

Of the various forms of CMOS, complementary CMOS circuits based on the inverter studied in Section 4.10 are the most widely used. They are available both as small-scale integrated (SSI) circuit packages (containing 1–10 logic gates) and medium-scale integrated (MSI) circuit packages (10–100 gates per chip) for assembling digital systems on printed-circuit boards. More significantly, complementary CMOS is used in VLSI logic (with millions of gates per chip) and memory-circuit design. In some applications, complementary CMOS is supplemented by one (or both) of two other MOS logic circuit forms. These are pseudo-NMOS, so-named because of the similarity of its structure to NMOS logic, and pass-transistor logic, both of which will be studied in this chapter.

A fourth type of CMOS logic circuit utilizes dynamic techniques to obtain faster circuit operation, while keeping the power dissipation very low. Dynamic CMOS logic represents an area of growing importance. Lastly, CMOS technology is used in the design of memory chips, as will be detailed in Chapter 11.

Bipolar Two logic-circuit families based on the bipolar junction transistor are in some use at present: TTL and ECL. Transistor–transistor logic (TTL or T^{2}L) was for many years the most widely used logic-circuit family. Its decline was precipitated by the advent of the VLSI era. TTL manufacturers, however, fought back with the introduction of low-power and high-speed versions. In these newer versions, the higher speeds of operation are made possible by preventing the BJT from saturating and thus avoiding the slow turnoff process of a saturated transistor. These nonsaturating versions of TTL utilize the Schottky diode discussed in Section 3.8 and are called Schottky TTL or variations of this name. Despite all these efforts, TTL is no longer a significant logic-circuit family and will not be studied in this book.

The other bipolar logic-circuit family in present use is emitter-coupled logic (ECL). It is based on the current-switch implementation of the inverter, discussed in Section 1.7. The basic element of ECL is the differential BJT pair studied in Chapter 7. Because ECL is basically a current-steering logic, and, correspondingly, also called **current-mode logic** (CML), in which saturation is avoided, very high speeds of operation are possible. Indeed, of all the commercially available logic-circuit families, ECL is the fastest. ECL is also used in VLSI circuit design when very high operating speeds are required and the designer is willing to accept higher power dissipation and increased silicon area. As such, ECL is considered an important specialty technology and will be briefly discussed in Chapter 11.

BiCMOS   BiCMOS combines the high operating speeds possible with BJTs (because of their inherently higher transconductance) with the low power dissipation and other excellent characteristics of CMOS. Like CMOS, BiCMOS allows for the implementation of both analog and digital circuits on the same chip. (See the discussion of analog BiCMOS circuits in Chapter 6.) At present, BiCMOS is used to great advantage in special applications, including memory chips, where its high performance as a high-speed capacitive-current driver justifies the more complex process technology it requires. A brief discussion of BiCMOS is provided in Chapter 11.

Gallium Arsenide (GaAs)   The high carrier mobility in GaAs results in very high speeds of operation. This has been demonstrated in a number of digital IC chips utilizing GaAs technology. It should be pointed out, however, that GaAs remains an "emerging technology," one that appears to have great potential but has not yet achieved such potential commercially. As such, it will not be studied in this book. Nevertheless, considerable material on GaAs devices and circuits, including digital circuits, can be found on the CD accompanying this book and on the book's website.

10.1.2 Logic-Circuit Characterization

The following parameters are usually used to characterize the operation and performance of a logic-circuit family.

Noise Margins   The static operation of a logic-circuit family is characterized by the voltage transfer characteristic (VTC) of its basic inverter. Figure 10.2 shows such a VTC and defines its four parameters; V_{OH}, V_{OL}, V_{IH}, and V_{IL}. Note that V_{IH} and V_{IL} are defined as the points at which the slope of the VTC is -1. Also indicated is the definition of the threshold voltage V_M, or V_{th} as we shall frequently call it, as the point at which $v_O = v_I$. Recall that we discussed the VTC in its generic form in Section 1.7, and have also seen actual VTCs: in Section 4.10 for the CMOS inverter, and in Section 5.10 for the BJT inverter.

The **robustness** of a logic-circuit family is determined by its ability to reject noise, and thus by the noise margins NH_H and NM_L,

$$NM_H \equiv V_{OH} - V_{IH} \tag{10.1}$$

$$NM_L \equiv V_{IL} - V_{OL} \tag{10.2}$$

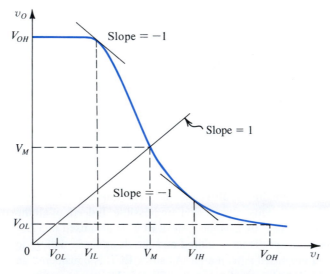

FIGURE 10.2   Typical voltage transfer characteristic (VTC) of a logic inverter, illustrating the definition of the critical points.

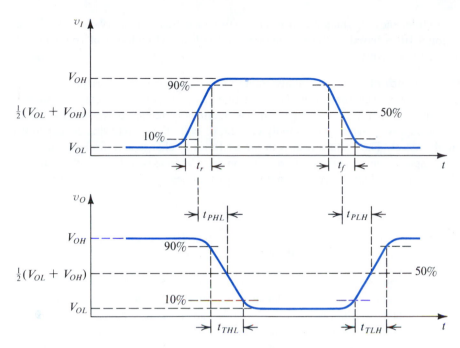

FIGURE 10.3 Definitions of propagation delays and switching times of the logic inverter.

An ideal inverter is one for which $NM_H = NM_L = V_{DD}/2$, where V_{DD} is the power-supply voltage. Further, for an ideal inverter, the threshold voltage $V_M = V_{DD}/2$.

Propagation Delay   The dynamic performance of a logic-circuit family is characterized by the propagation delay of its basic inverter. Figure 10.3 illustrates the definition of the low-to-high propagation delay (t_{PLH}) and the high-to-low propagation delay (t_{PHL}). The inverter propagation delay (t_P) is defined as the average of these two quantities:

$$t_P \equiv \tfrac{1}{2}(t_{PLH} + t_{PHL}) \tag{10.3}$$

Obviously, the shorter the propagation delay, the higher the speed at which the logic-circuit family can be operated.

Power Dissipation   Power dissipation is an important issue in digital-circuit design. The need to minimize the gate power dissipation is motivated by the desire to pack an ever-increasing number of gates on a chip, which in turn is motivated by space and economic considerations. In general, however, modern digital systems utilize large numbers of gates and memory cells, and thus to keep the total power requirement within reasonable bounds, the power dissipation per gate and per memory cell should be kept as low as possible. This is particularly the case for portable, battery-operated equipment such as cellular phones and personal digital assistants (PDAs).

There are two types of power dissipation in a logic gate: static and dynamic. Static power refers to the power that the gate dissipates in the absence of switching action. It results from the presence of a path in the gate circuit between the power supply and ground in one or both of its two states (i.e., with the output either low or high). Dynamic power, on the other hand, occurs only when the gate is switched: An inverter operated from a power supply V_{DD}, and driving a load capacitance C, dissipates dynamic power P_D,

$$P_D = fCV_{DD}^2 \tag{10.4}$$

where f is the frequency at which the inverter is being switched. The derivation of this formula (Section 4.10) is based on the assumption that the low and high output voltage levels are 0 and V_{DD}, respectively.

Delay–Power Product   One is usually interested in high-speed performance (low t_P) combined with low power dissipation. Unfortunately, these two requirements are often in conflict; generally, when designing a gate, if one attempts to reduce power dissipation by decreasing the supply voltage, or the supply current, or both, the current-driving capability of the gate decreases. This in turn results in longer times to charge and discharge the load and parasitic capacitances, and thus the propagation delay increases. It follows that a figure-of-merit for comparing logic-circuit technologies (or families) is the delay–power product, defined as

$$DP = P_D t_P \tag{10.5}$$

where P_D is the power dissipation of the gate. Note that DP has the units of joules. The lower the DP figure for a logic family, the more effective it is.

Silicon Area   An obvious objective in the design of digital VLSI circuits is the minimization of silicon area per logic gate. Smaller area requirement enables the fabrication of a larger number of gates per chip, which has economic and space advantages from a system-design standpoint. Area reduction occurs in three different ways: through advances in processing technology that enable the reduction of the minimum device size, through advances in circuit-design techniques, and through careful chip layout. In this book, our interest lies in circuit design, and we shall make frequent comments on the relationship between the circuit design and its silicon area. As a general rule, the simpler the circuit, the smaller the area required. As will be seen shortly, the circuit designer has to decide on device sizes. Choosing smaller devices has the obvious advantage of requiring smaller silicon area and at the same time reducing parasitic capacitances and thus increasing speed. Smaller devices, however, have lower current-driving capability, which tends to increase delay. Thus, as in all engineering design problems, there is a trade-off to be quantified and exercised in a manner that optimizes whatever aspect of the design is thought to be critical for the application at hand.

Fan-In and Fan-Out   The fan-in of a gate is the number of its inputs. Thus, a four-input NOR gate has a fan-in of 4. Fan-out is the maximum number of similar gates that a gate can drive while remaining within guaranteed specifications. As an example, we saw in Section 4.10 that increasing the fan-out of the BJT inverter reduces V_{OH} and hence NM_H. In this case, to keep NM_H above a certain minimum, the fan-out has to be limited to a calculable maximum value.

10.1.3 Styles for Digital System Design

The conventional approach to designing digital systems consists of assembling the system using standard IC packages of various levels of complexity (and hence integration). Many systems have been built this way using, for example, TTL SSI and MSI packages. The advent of VLSI, in addition to providing the system designer with more powerful off-the-shelf components such as microprocessors and memory chips, has made possible alternative design styles. One such alternative is to opt for implementing part or all of the system using one or more *custom VLSI* chips. However, custom IC design is usually economically justified only when the production volume is large (greater than about 100,000 parts).

An intermediate approach, known as *semicustom design,* utilizes *gate-array* chips. These are integrated circuits containing 100,000 or more unconnected logic gates. Their interconnection can be achieved by a final metallization step (performed at the IC fabrication facility) according to a pattern specified by the user to implement the user's particular functional need. A more recently available type of gate array, known as a *field-programmable gate array* (FPGA), can, as its name indicates, be programmed directly by the user. FPGAs provide a very convenient means for the digital-system designer to implement complex logic functions in VLSI form without having to incur either the cost or the "turnaround time" inherent in custom and, to a lesser extent, in semicustom IC design [see Brown and Rose (1996)].

10.1.4 Design Abstraction and Computer Aids

The design of very complex digital systems, whether on a single IC chip or using off-the-shelf components, is made possible by the use of different levels of design abstraction, and the use of a variety of computer aids. To appreciate the concept of design abstraction, consider the process of designing a digital system using off-the-shelf packages of logic gates. The designer consults data sheets (and books) to determine the input and output characteristics of the gates, their fan-in and fan-out limitations, and so on. In connecting the gates, the designer needs to adhere to a set of rules specified by the manufacturer in the data sheets. The designer does not need to consider, in a direct way, the circuit inside the gate package. In effect, the circuit has been abstracted in the form of a functional block that can be used as a component. This greatly simplifies system design. The digital-IC designer follows a similar process. Circuit blocks are designed, characterized, and stored in a library as *standard cells*. These cells can then be used by the IC designer to assemble a larger subsystem (e.g., an adder or a multiplier), which in turn is characterized and stored as a functional block to be used in the design of an even larger system (e.g., an entire processor).

At every level of design abstraction, the need arises for simulation and other computer programs that help make the design process as automated as possible. Whereas SPICE is employed in circuit simulation, other software tools are utilized at other levels and in other phases of the design process. Although digital-system design and design automation are outside the scope of this book, it is important that the reader appreciate the role of design abstraction and computer aids in digital design. They are what make it humanly possible to design a 100-million-transistor digital IC. Unfortunately, analog IC design does not lend itself to the same level of abstraction and automation. Each analog IC to a large extent has to be "handcrafted." As a result, the complexity and density of analog ICs remain much below what is possible in a digital IC.

Whatever approach or style is adopted in digital design, some familiarity with the various digital-circuit technologies and design techniques is essential. This chapter and the next aim to provide such a background.

 ## 10.2 DESIGN AND PERFORMANCE ANALYSIS OF THE CMOS INVERTER

The CMOS logic inverter was introduced and studied in Section 4.10, which we urge the reader to review before proceeding any further. In this section, we take a more comprehensive look at the inverter, investigating its performance and exploring the trade-offs available in its design. This material will serve as the foundation for the study of CMOS logic circuits in the following section.

10.2.1 Circuit Structure

The inverter circuit, shown in Fig. 10.4(a), consists of a pair of complementary MOSFETs switched by the input voltage v_I. Although not shown, the source of each device is connected

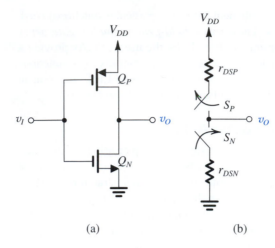

FIGURE 10.4 **(a)** The CMOS inverter and **(b)** its representation as a pair of switches operated in a complementary fashion.

(a)   (b)

to its body, thus eliminating the body effect. Usually, the threshold voltages V_{tn} and V_{tp} are equal in magnitude; that is, $V_{tn} = |V_{tp}| = V_t$, which is in the range of 0.2 V to 1 V, with values near the lower end of this range for modern process technologies having small feature size (e.g., with channel length of 0.5 to 0.1 μm or less).

The inverter circuit can be represented by a pair of switches operated in a complementary fashion, as shown in Fig. 10.4(b). As indicated, each switch is modeled by a finite on resistance, which is the source–drain resistance of the respective transistor, evaluated near $|v_{DS}| = 0$,

$$r_{DSN} = 1 \bigg/ \left[k_n' \left(\frac{W}{L} \right)_n (V_{DD} - V_t) \right] \tag{10.6}$$

$$r_{DSP} = 1 \bigg/ \left[k_p' \left(\frac{W}{L} \right)_p (V_{DD} - V_t) \right] \tag{10.7}$$

10.2.2 Static Operation

With $v_I = 0$, $v_O = V_{OH} = V_{DD}$, and the output node is connected to V_{DD} through the resistance r_{DSP} of the pull-up transistor Q_P. Similarly, with $v_I = V_{DD}$, $v_O = V_{OL} = 0$, and the output node is connected to ground through the resistance r_{DSN} of the pull-down transistor Q_N. Thus, in the steady state, no direct-current path exists between V_{DD} and ground, and the static-current and the static-power dissipation are both zero (leakage effects are usually negligibly small particularly for large-feature-size devices).

The voltage transfer characteristic of the inverter is shown in Fig. 10.5, from which it is confirmed that the output voltage levels are 0 and V_{DD}, and thus the output voltage swing is the maximum possible. The fact that V_{OL} and V_{OH} are independent of device dimensions makes CMOS very different from other forms of MOS logic.

The CMOS inverter can be made to switch at the midpoint of the logic swing, 0 to V_{DD}, that is, at $V_{DD}/2$, by appropriately *sizing the transistors*. Specifically, it can be shown that the switching threshold V_{th} (or V_M) is given by

$$V_{th} = \frac{V_{DD} - |V_{tp}| + \sqrt{k_n/k_p}\,V_{tn}}{1 + \sqrt{k_n/k_p}} \tag{10.8}$$

where $k_n = k_n'(W/L)_n$ and $k_p = k_p'(W/L)_p$, from which we see that for the typical case where $V_{tn} = |V_{tp}|$, $V_{th} = V_{DD}/2$ for $k_n = k_p$, that is,

$$k_n'(W/L)_n = k_p'(W/L)_p \tag{10.9}$$

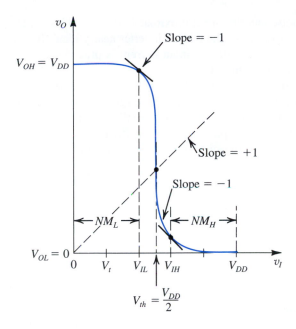

FIGURE 10.5 The voltage transfer characteristic (VTC) of the CMOS inverter when Q_N and Q_P are matched.

Thus a symmetrical transfer characteristic is obtained when the devices are designed to have equal transconductance parameters, a condition we refer to as *matching*. Since μ_n is two to four times larger than μ_p, matching is achieved by making $(W/L)_p$ two to four times (i.e., μ_n/μ_p times) $(W/L)_n$,

$$\left(\frac{W}{L}\right)_p = \frac{\mu_n}{\mu_p}\left(\frac{W}{L}\right)_n \tag{10.10}$$

Normally, the two devices have the same channel length, L, which is set at the minimum allowable for the given process technology. The minimum width of the NMOS transistor is usually one and a half to two times L, and the width of the PMOS transistor two to three times that. For example, for a 0.25-μm process for which $\mu_n/\mu_p = 3$, $L = 0.25$ μm, $(W/L)_n = 0.375$ μm$/0.25$ μm, and $(W/L)_p = 1.125$ μm$/0.25$ μm. As we shall discuss shortly, if the inverter is required to drive a relatively large capacitive load, the transistors are made wider. However, to conserve chip area, most of the inverters would have this "minimum size." For future purposes, we shall denote the (W/L) ratio of the NMOS transistor of this minimum-size inverter by n and the (W/L) ratio of the PMOS transistor by p. Since the inverter area can be represented by $W_n L_n + W_p L_p = (W_n + W_p)L$, the area of the minimum-size inverter is $(n + p)L^2$, and we can use the factor $(n + p)$ as a proxy for area. For the example cited earlier, $n = 1.5$, $p = 4.5$, and the area factor $n + p = 6$.

Besides placing the gate threshold at the center of the logic swing, matching the transconductance parameters of Q_N and Q_P provides the inverter with equal current-driving capability in both directions (pull-up and pull-down). Furthermore, and obviously related, it makes $r_{DSN} = r_{DSP}$. Thus an inverter with matched transistors will have equal propagation delays, t_{PLH} and t_{PHL}.

When the inverter threshold is at $V_{DD}/2$, the noise margins NM_H and NM_L are equalized, and their values are maximized, such that (Section 4.10):

$$NM_H = NM_L = \frac{3}{8}\left(V_{DD} + \frac{2}{3}V_t\right) \tag{10.11}$$

Since typically $V_t = 0.1$ to $0.2\, V_{DD}$, the noise margins are approximately $0.4\, V_{DD}$. This value, being close to half the power-supply voltage, makes the CMOS inverter nearly ideal from a noise-immunity standpoint. Further, since the inverter dc input current is practically zero, the noise margins are not dependent on the gate fan-out.

Although we have emphasized the advantages of matching Q_N and Q_P, there are occasions in which this scaling is not adopted. One might, for instance, forgo the advantages of matching in return for reducing chip area and simply make $(W/L)_p = (W/L)_n$. There are also instances in which a deliberate mismatch is used to place V_{th} at a specified value other than $V_{DD}/2$. Note that by making $k_n > k_p$, V_{th} moves closer to zero, whereas $k_p > k_n$ moves V_{th} closer to V_{DD}.

As a final comment on the inverter VTC, we note that the slope in the transition region, though large, is finite and is given by $-(g_{mN} + g_{mP})(r_{oN}//r_{oP})$.

10.2.3 Dynamic Operation

The propagation delay of the inverter is usually determined under the condition that it is driving an identical inverter. This situation is depicted in Fig. 10.6. We wish to analyze this circuit to determine the propagation delay of the inverter comprising Q_1 and Q_2, which is driven by a low-impedance source v_I, and is loaded by the inverter comprising Q_3 and Q_4. Indicated in the figure are the various transistor internal capacitances that are connected to the output node of the (Q_1, Q_2) inverter. Obviously, an exact pencil-and-paper analysis of this circuit will be too complicated to yield useful design insight, and a simplification of the circuit is in order. Specifically, we wish to replace all the capacitances attached to the inverter output node with a single capacitance C connected between the output node and ground. If we are able to do that, we can utilize the results of the transient analysis performed in Section 4.10. Toward that end, we note that during t_{PLH} or t_{PHL}, the output of the first inverter changes from 0 to $V_{DD}/2$ or from V_{DD} to $V_{DD}/2$, respectively. It follows that the second inverter remains in the same state during each of our analysis intervals. This observation will have an important bearing on our estimation of the equivalent input capacitance of the second inverter. Let's now consider the contribution of each of the capacitances in

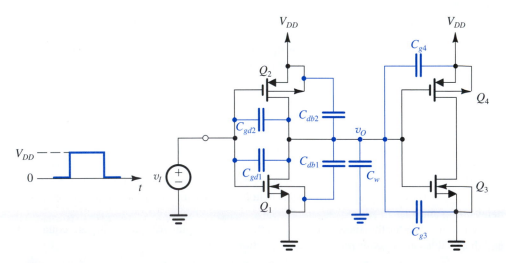

FIGURE 10.6 Circuit for analyzing the propagation delay of the inverter formed by Q_1 and Q_2, which is driving an identical inverter formed by Q_3 and Q_4.

Fig. 10.6 to the value of the equivalent load capacitance C:

1. The gate–drain overlap capacitance of Q_1, C_{gd1}, can be replaced by an equivalent capacitance between the output node and ground of $2C_{gd1}$. The factor 2 arises because of the Miller effect (Section 6.4.4). Specifically, note that as v_I goes high and v_O goes low by the same amount, the change in voltage across C_{gd1} is twice that amount. Thus the output node sees in effect twice the value of C_{gd1}. The same applies for the gate–drain overlap capacitance of Q_2, C_{gd2}, which can be replaced by a capacitance $2C_{gd2}$ between the output node and ground.

2. Each of the drain–body capacitances C_{db1} and C_{db2} has a terminal at a constant voltage. Thus for the purpose of our analysis here, C_{db1} and C_{db2} can be replaced with equal capacitances between the output node and ground. Note, however, that the formulas given in Section 4.8 for calculating C_{db1} and C_{db2} are small-signal relationships, whereas the analysis here is obviously a large-signal one. A technique has been developed for finding equivalent large-signal values for C_{db1} and C_{db2} [see Hodges and Jackson (1988) and Rabaey (2002)].

3. Since the second inverter does not switch states, we will assume that the input capacitances of Q_3 and Q_4 remain approximately constant and equal to the total gate capacitance ($WLC_{ox} + C_{gsov} + C_{gdov}$). That is, the input capacitance of the load inverter will be

$$C_{g3} + C_{g4} = (WL)_3 C_{ox} + (WL)_4 C_{ox} + C_{gsov3} + C_{gdov3} + C_{gsov4} + C_{gdov4}$$

4. The last component of C is the wiring capacitance C_w, which simply adds to the value of C.

Thus, the total value of C is given by

$$C = 2C_{gd1} + 2C_{gd2} + C_{db1} + C_{db2} + C_{g3} + C_{g4} + C_w \qquad (10.12)$$

Having determined an approximate value for the equivalent capacitance between the inverter output node and ground, we can utilize the circuits in Fig. 10.7 to determine t_{PHL} and t_{PLH}, respectively. Since the two circuits are similar, we need only consider one and apply the result directly to the other. Consider the circuit in Fig. 10.7(a), which applies when v_I goes high and Q_N discharges C from its initial voltage of V_{DD} to the final value of 0. The analysis is somewhat complicated by the fact that initially Q_N will be in the saturation mode and then, when v_O falls below $V_{DD} - V_t$, it will go into the triode region of operation. We have in fact performed this analysis in Section 4.10 and obtained the following approximate expression for t_{PHL}:

$$t_{PHL} = \frac{1.6C}{k'_n \left(\dfrac{W}{L}\right)_n V_{DD}} \qquad (10.13)$$

where we have assumed that $V_t \cong 0.2 V_{DD}$, which is typically the case.

There is an alternative, an approximate but simpler, method for analyzing the circuit in Fig. 10.7(a). It is based on computing an average value for the discharge current i_{DN} during the interval $t = 0$ to $t = t_{PHL}$. Specifically, at $t = 0$, Q_N will be saturated, and $i_{DN}(0)$ is given by

$$i_{DN}(0) = \frac{1}{2} k'_n \left(\frac{W}{L}\right)_n (V_{DD} - V_t)^2 \qquad (10.14)$$

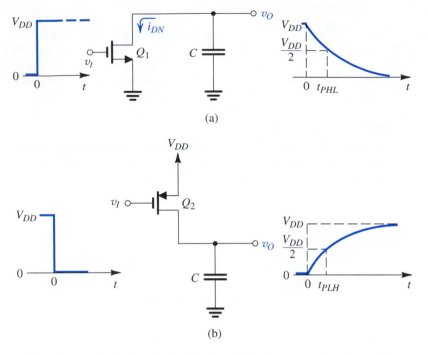

FIGURE 10.7 Equivalent circuits for determining the propagation delays **(a)** t_{PHL} and **(b)** t_{PLH} of the inverter.

At $t = t_{PHL}$, Q_N will be in the triode region, and $i_{DN}(t_{PHL})$ will be

$$i_{DN}(t_{PHL}) = k_n'\left(\frac{W}{L}\right)_n\left[(V_{DD} - V_t)\frac{V_{DD}}{2} - \frac{1}{2}\left(\frac{V_{DD}}{2}\right)^2\right] \qquad (10.15)$$

The average discharge current can then be found as

$$i_{DN}\big|_{av} = \tfrac{1}{2}[i_{DN}(0) + i_{DN}(t_{PHL})] \qquad (10.16)$$

and the discharge interval t_{PHL} computed from

$$t_{PHL} = \frac{C\,\Delta V}{i_{DN}\big|_{av}} = \frac{CV_{DD}/2}{i_{DN}\big|_{av}} \qquad (10.17)$$

Utilizing Eqs. (10.14) through (10.17) and substituting $V_t \cong 0.2\,V_{DD}$ gives

$$t_{PHL} \cong \frac{1.7C}{k_n'\left(\dfrac{W}{L}\right)_n V_{DD}} \qquad (10.18)$$

which yields a value very close to that obtained by the more precise formula of Eq. (10.13). Which formula to use is not very relevant, for we have already made many approximations. Indeed, our interest in these formulas is not in obtaining a precise value of t_{PHL} but in what they tell us about the effect of the various elements on determining the inverter delay. It is such insight that the circuit designer hopes to glean from manual analysis. Precise values for delay can be determined using computer simulation (Section 10.7).

An expression for the low-to-high inverter delay, t_{PLH}, can be written by analogy to the t_{PHL} expression in Eq. (10.17),

$$t_{PLH} \cong \frac{1.7C}{k_p'\left(\dfrac{W}{L}\right)_p V_{DD}} \tag{10.19}$$

Finally, the propagation delay t_P can be found as the average of t_{PHL} and t_{PLH},

$$t_P = \tfrac{1}{2}(t_{PHL} + t_{PLH})$$

Examination of the formulas in Eqs. (10.18) and (10.19) enables us to make a number of useful observations:

1. As expected, the two components of t_P can be equalized by selecting the (W/L) ratios to equalize k_n and k_p, that is, by matching Q_N and Q_P.

2. Since t_P is proportional to C, the designer should strive to reduce C. This is achieved by using the minimum possible channel length and by minimizing wiring and other parasitic capacitances. Careful layout of the chip can result in significant reduction in such capacitances and in the value of C_{db}.

3. Using a process technology with larger transconductance parameter k' can result in shorter propagation delays. Keep in mind, however, that for such processes C_{ox} is increased, and thus the value of C increases at the same time.

4. Using larger (W/L) ratios can result in a reduction in t_P. Care, however, should be exercised here also, since increasing the size of the devices increases the value of C, and thus the expected reduction in t_P might not materialize. Reducing t_P by increasing (W/L), however, is an effective strategy when C is dominated by components not directly related to the size of the driving device (such as wiring or fan-out devices).

5. A larger supply voltage V_{DD} results in a lower t_P. However, V_{DD} is determined by the process technology and thus is often not under the control of the designer. Furthermore, modern process technologies in which device sizes are reduced require lower V_{DD} (see Table 6.1). A motivating factor for lowering V_{DD} is the need to keep the dynamic power dissipation at acceptable levels, especially in very-high-density chips. We will have more to say on this point shortly.

These observations clearly illustrate the conflicting requirements and the trade-offs available in the design of a CMOS digital integrated circuit (and indeed in any engineering design problem).

10.2.4 Dynamic Power Dissipation

The negligible static power dissipation of CMOS has been a significant factor in its dominance as the technology of choice in implementing high-density VLSI circuits. However, as the number of gates per chip steadily increases, the dynamic power dissipation has become a serious issue. The dynamic power dissipated in the CMOS inverter is given by Eq. (10.4), which we repeat here as

$$P_D = f\, C V_{DD}^2 \tag{10.20}$$

where f is the frequency at which the gate is switched. It follows that minimizing C is an effective means for reducing dynamic-power dissipation. An even more effective strategy is the use of a lower power-supply voltage. As we have mentioned, new CMOS process technologies utilize V_{DD} values as low as 1 V. These newer chips, however, pack much more circuitry on the chip (as many as 100 million transistors) and operate at higher frequencies (microprocessor clock frequencies above 1 GHz are now available). The dynamic power dissipation of such high-density chips can be over 100 W.

EXAMPLE 10.1

Consider a CMOS inverter fabricated in a 0.25-μm process for which $C_{ox} = 6$ fF/μm^2, $\mu_n C_{ox} = 115$ μA/V^2, $\mu_p C_{ox} = 30$ μA/V^2, $V_{tn} = -V_{tp} = 0.4$ V, and $V_{DD} = 2.5$ V. The W/L ratio of Q_N is $0.375\,\mu$m$/0.25$ μm, and that for Q_P is 1.125 μm$/0.25$ μm. The gate–source and gate–drain overlap capacitances are specified to be 0.3 fF/μm of gate width. Further, the effective value of drain–body capacitances are $C_{dbn} = 1$ fF and $C_{dbp} = 1$ fF. The wiring capacitance $C_w = 0.2$ fF. Find t_{PHL}, t_{PLH}, and t_P.

Solution

First, we determine the value of the equivalent capacitance C using Eq. (10.12),

$$C = 2C_{gd1} + 2C_{gd2} + C_{db1} + C_{db2} + C_{g3} + C_{g4} + C_w$$

where

$$C_{gd1} = 0.3 \times W_n = 0.3 \times 0.375 = 0.1125 \text{ fF}$$

$$C_{gd2} = 0.3 \times W_p = 0.3 \times 1.125 = 0.3375 \text{ fF}$$

$$C_{db1} = 1 \text{ fF}$$

$$C_{db2} = 1 \text{ fF}$$

$$C_{g3} = 0.375 \times 0.25 \times 6 + 2 \times 0.3 \times 0.375 = 0.7875 \text{ fF}$$

$$C_{g4} = 1.125 \times 0.25 \times 6 + 2 \times 0.3 \times 1.125 = 2.3625 \text{ fF}$$

$$C_w = 0.2 \text{ fF}$$

Thus,

$$C = 2 \times 0.1125 + 2 \times 0.3375 + 1 + 1 + 0.7875 + 2.3625 + 0.2 = 6.25 \text{ fF}$$

Next, although we can use the formula in Eq. (10.18) to determine t_{PHL}, we shall take an alternative route. Specifically, we shall consider the discharge of C through Q_N and determine the average discharge current using Eqs. (10.14) through (10.16):

$$i_{DN}(0) = \frac{1}{2} k_n' \left(\frac{W}{L}\right)_n (V_{DD} - V_t)^2$$

$$= \frac{1}{2} \times 115 \left(\frac{0.375}{0.25}\right)(2.5 - 0.4)^2 = 380 \text{ } \mu\text{A}$$

$$i_{DN}(t_{PHL}) = k_n' \left(\frac{W}{L}\right)_n \left[(V_{DD} - V_t)\frac{V_{DD}}{2} - \frac{1}{2}\left(\frac{V_{DD}}{2}\right)^2\right]$$

$$= 115 \times \frac{0.375}{0.25}\left[(2.5 - 0.4)\frac{2.5}{2} - \frac{1}{2}\left(\frac{2.5}{2}\right)^2\right]$$

$$= 318 \text{ } \mu\text{A}$$

Thus

$$i_{DN}\big|_{\text{av}} = \frac{380 + 318}{2} = 349 \text{ } \mu\text{A}$$

and

$$t_{PHL} = \frac{C(V_{DD}/2)}{i_{DN}|_{av}} = \frac{6.25 \times 10^{-15} \times 1.25}{349 \times 10^{-6}} = 23.3 \text{ ps}$$

Since $W_p/W_n = 3$ and $\mu_n/\mu_p = 3.83$, the inverter is not perfectly matched. Therefore, we expect t_{PLH} to be greater than t_{PHL} by a factor of $3.83/3 = 1.3$, thus

$$t_{PLH} = 1.3 \times 23.3 = 30 \text{ ps}$$

and thus t_P will be

$$t_P = \tfrac{1}{2}(t_{PHL} + t_{PLH})$$
$$= \tfrac{1}{2}(23.3 + 30) = 26.5 \text{ ps}$$

EXERCISES

10.1 Consider the inverter specified in Example 10.1 when loaded with an additional 0.1-pF capacitance. What will the propagation delay become?

Ans. 437 ps

10.2 In an attempt to decrease the area of the inverter in Example 10.1, $(W/L)_p$ is made equal to $(W/L)_n$. What is the percentage reduction in area achieved? Find the new values of C, t_{PHL}, t_{PLH}, and t_P. Assume that C_{dbp} does not change significantly.

Ans. 50%; 4.225 fF; 15.8 ps; 20.5 ps; 18.1 ps

10.3 For the inverter of Example 10.1, find the dynamic power dissipation when clocked at a 500-MHz rate.

Ans. 19.5 μW

 ## 10.3 CMOS LOGIC-GATE CIRCUITS

In this section, we build on our knowledge of inverter design and consider the design of CMOS circuits that realize combinational-logic functions. In combinational circuits, the output at any time is a function only of the values of input signals at that time. Thus, these circuits do not have memory and do not employ feedback. Combinational-logic circuits are used in large quantities in a multitude of applications; indeed, every digital system contains large numbers of combinational-logic circuits.

10.3.1 Basic Structure

A CMOS logic circuit is in effect an extension, or a generalization, of the CMOS inverter: The inverter consists of an NMOS pull-down transistor, and a PMOS pull-up transistor, operated by the input voltage in a complementary fashion. The CMOS logic gate consists of two networks: the pull-down network (PDN) constructed of NMOS transistors, and the pull-up network (PUN) constructed of PMOS transistors (see Fig. 10.8). The two networks are operated by the input variables, in a complementary fashion. Thus, for the three-input gate represented in Fig. 10.8, the PDN will conduct for all input combinations that require a low output ($Y = 0$) and will then pull the output node down to ground, causing a zero voltage to

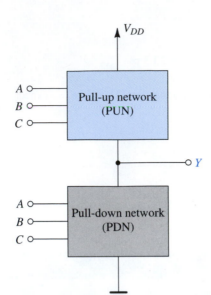

FIGURE 10.8 Representation of a three-input CMOS logic gate. The PUN comprises PMOS transistors, and the PDN comprises NMOS transistors.

appear at the output, $v_Y = 0$. Simultaneously, the PUN will be off, and no direct dc path will exist between V_{DD} and ground. On the other hand, all input combinations that call for a high output ($Y = 1$) will cause the PUN to conduct, and the PUN will then pull the output node up to V_{DD}, establishing an output voltage $v_Y = V_{DD}$. Simultaneously, the PDN will be cut off, and again, no dc current path between V_{DD} and ground will exist in the circuit.

Now, since the PDN comprises NMOS transistors, and since an NMOS transistor conducts when the signal at its gate is high, the PDN is activated (i.e., conducts) when the inputs are high. In a dual manner, the PUN comprises PMOS transistors, and a PMOS transistor conducts when the input signal at its gate is low; thus the PUN is activated when the inputs are low.

The PDN and the PUN each utilizes devices in parallel to form an OR function, and devices in series to form an AND function. Here, the OR and AND notation refer to current flow or conduction. Figure 10.9 shows examples of PDNs. For the circuit in Fig. 10.9(a), we observe that Q_A will conduct when A is high ($v_A = V_{DD}$) and will then pull the output node down to ground ($v_Y = 0$ V, $Y = 0$). Similarly, Q_B conducts and pulls Y down when B is high. Thus Y

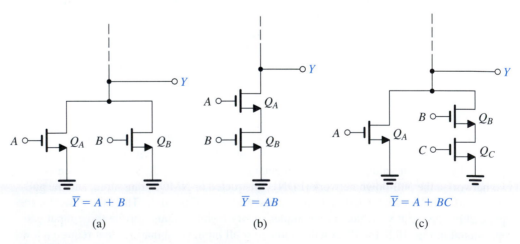

FIGURE 10.9 Examples of pull-down networks.

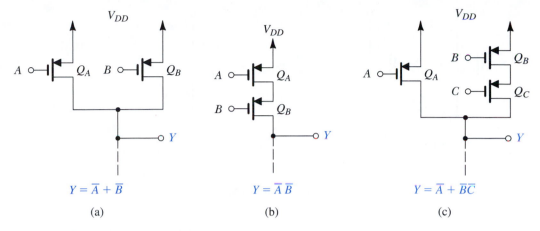

FIGURE 10.10 Examples of pull-up networks.

will be low when A is high *or* B is high, which can be expressed as

$$\bar{Y} = A + B$$

or equivalently

$$Y = \overline{A + B}$$

The PDN in Fig. 10.9(b) will conduct only when A and B are both high simultaneously. Thus Y will be low when A is high *and* B is high,

$$\bar{Y} = AB$$

or equivalently

$$Y = \overline{AB}$$

As a final example, the PDN in Fig. 10.9(c) will conduct and cause Y to be 0 when A is high *or* when B *and* C are both high, thus

$$\bar{Y} = A + BC$$

or equivalently

$$Y = \overline{A + BC}$$

Next consider the PUN examples shown in Fig. 10.10. The PUN in Fig. 10.10(a) will conduct and pull Y up to $V_{DD}(Y = 1)$ when A is low *or* B is low, thus

$$Y = \bar{A} + \bar{B}$$

The PUN in Fig. 10.10(b) will conduct and produce a high output ($v_Y = V_{DD}$, $Y = 1$) only when A *and* B are both low, thus

$$Y = \bar{A}\bar{B}$$

Finally, the PUN in Fig. 10.10(c) will conduct and cause Y to be high (logic 1) if A is low *or* if B *and* C are both low, thus

$$Y = \bar{A} + \bar{B}\bar{C}$$

Having developed an understanding and an appreciation of the structure and operation of PDNs and PUNs, we now consider complete CMOS gates. Before doing so, however, we wish to introduce alternative circuit symbols, that are almost universally used for MOS transistors by

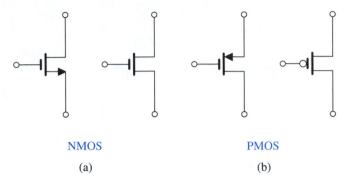

NMOS   PMOS

(a)   (b)

FIGURE 10.11 Usual and alternative circuit symbols for MOSFETs.

digital-circuit designers. Figure 10.11 shows our usual symbols (left) and the corresponding "digital" symbols (right). Observe that the symbol for the PMOS transistor with a circle at the gate terminal is intended to indicate that the signal at the gate has to be low for the device to be activated (i.e., to conduct). Thus, in terms of logic-circuit terminology, the gate terminal of the PMOS transistor is an *active low* input. Besides indicating this property of PMOS devices, the digital symbols omit any indication of which of the device terminals is the source and which is the drain. This should cause no difficulty at this stage of our study; simply remember that for an NMOS transistor, the drain is the terminal that is at the higher voltage (current flows from drain to source), and for a PMOS transistor the source is the terminal that is at the higher voltage (current flows from source to drain). To be consistent with the literature, we shall henceforth use these modified symbols for MOS transistors in logic applications, except in locations where our usual symbols help in understanding circuit operation.

10.3.2 The Two-Input NOR Gate

We first consider the CMOS gate that realizes the two-input NOR function

$$Y = \overline{A + B} = \overline{A}\,\overline{B} \tag{10.21}$$

We see that Y is to be low (PDN conducting) when A is high or B is high. Thus the PDN consists of two parallel NMOS devices with A and B as inputs (i.e., the circuit in Fig. 10.9a). For the PUN, we note from the second expression in Eq. (10.21) that Y is to be high when A and B are both low. Thus the PUN consists of two series PMOS devices with A and B as the inputs (i.e., the circuit in Fig. 10.10b). Putting the PDN and the PUN together gives the CMOS NOR gate shown in Fig. 10.12. Note that extension to a higher number of inputs is straightforward: For each additional input, an NMOS transistor is added in parallel with Q_{NA} and Q_{NB}, and a PMOS transistor is added in series with Q_{PA} and Q_{PB}.

10.3.3 The Two-Input NAND Gate

The two-input NAND function is described by the Boolean expression

$$Y = \overline{AB} = \overline{A} + \overline{B} \tag{10.22}$$

To synthesize the PDN, we consider the input combinations that require Y to be low: There is only one such combination, namely, A and B both high. Thus, the PDN simply comprises two NMOS transistors in series (such as the circuit in Fig. 10.9b). To synthesize the PUN, we consider the input combinations that result in Y being high. These are found from the

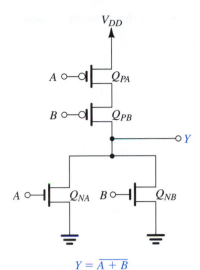

$$Y = \overline{A + B}$$

FIGURE 10.12 A two-input CMOS NOR gate.

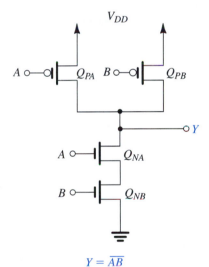

$$Y = \overline{AB}$$

FIGURE 10.13 A two-input CMOS NAND gate.

second expression in Eq. (10.22) as A low or B low. Thus, the PUN consists of two parallel PMOS transistors with A and B applied to their gates (such as the circuit in Fig. 10.10a). Putting the PDN and PUN together results in the CMOS NAND gate implementation shown in Fig. 10.13. Note that extension to a higher number of inputs is straightforward: For each additional input, we add an NMOS transistor in series with Q_{NA} and Q_{NB}, and a PMOS transistor in parallel with Q_{PA} and Q_{PB}.

10.3.4 A Complex Gate

Consider next the more complex logic function

$$Y = \overline{A(B + CD)} \tag{10.23}$$

Since $\overline{Y} = A(B + CD)$, we see that Y should be low for A high and simultaneously either B high or C and D both high, from which the PDN is directly obtained. To obtain the PUN, we need to express Y in terms of the complemented variables. We do this through repeated application of DeMorgan's law, as follows:

$$Y = \overline{A(B + CD)}$$

$$= \overline{A} + \overline{B + CD}$$

$$= \overline{A} + \overline{B} \ \overline{CD}$$

$$= \overline{A} + \overline{B}(\overline{C} + \overline{D}) \tag{10.24}$$

Thus, Y is high for A low or B low and either C or D low. The corresponding complete CMOS circuit will be as shown in Fig. 10.14.

10.3.5 Obtaining the PUN from the PDN and Vice Versa

From the CMOS gate circuits considered thus far (e.g., that in Fig. 10.14), we observe that the PDN and the PUN are dual networks: Where a series branch exists in one, a parallel branch exists in the other. Thus, we can obtain one from the other, a process that can be simpler than having to synthesize each separately from the Boolean expression of the function. For instance, in the circuit of Fig. 10.14, we found it relatively easy to obtain the PDN, simply because we already had $\overline{Y}$ in terms of the uncomplemented inputs. On the other hand, to obtain the PUN, we had to manipulate the given Boolean expression to express Y as a function of the complemented variables, the form convenient for synthesizing PUNs. Alternatively, we could have used this duality property to obtain the PUN from the PDN. The reader is urged to refer to Fig. 10.14 to convince herself that this is indeed possible.

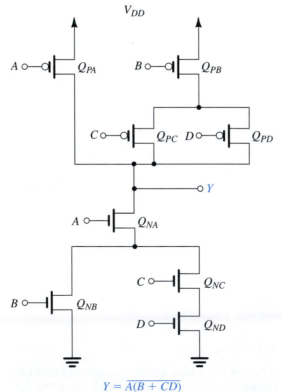

$$Y = \overline{A(B + CD)}$$

FIGURE 10.14 CMOS realization of a complex gate.

It should, however, be mentioned that at times it is not easy to obtain one of the two networks from the other using the duality property. For such cases, one has to resort to a more rigorous process, which is beyond the scope of this book [see Kang and Leblebici (1999)].

10.3.6 The Exclusive-OR Function

An important function that often arises in logic design is the exclusive-OR (XOR) function,

$$Y = A\bar{B} + \bar{A}B \tag{10.25}$$

We observe that since Y (rather than $\bar{Y}$) is given, it is easier to synthesize the PUN. We note, however, that unfortunately Y is not a function of the complemented variables only (as we would like it to be). Thus, we will need additional inverters. The PUN obtained directly from Eq. (10.25) is shown in Fig. 10.15(a). Note that the Q_1, Q_2 branch realizes the first term ($A\bar{B}$), whereas the Q_3, Q_4 branch realizes the second term ($\bar{A}B$). Note also the need for two additional inverters to generate $\bar{A}$ and $\bar{B}$.

As for synthesizing the PDN, we can obtain it as the dual network of the PUN in Fig. 10.15(a). Alternatively, we can develop an expression for $\bar{Y}$ and use it to synthesize the PDN. Leaving the first approach for the reader to do as an exercise, we shall utilize the direct synthesis approach. DeMorgan's law can be applied to the expression in Eq. (10.25) to obtain $\bar{Y}$ as

$$\bar{Y} = AB + \bar{A}\bar{B} \tag{10.26}$$

The corresponding PDN will be as in Fig. 10.15(b), which shows the CMOS realization of the exclusive-OR function except for the two additional inverters. Note that the exclusive-OR requires 12 transistors for its realization, a rather complex network. Later, in Section 10.5, we shall show a simpler realization of the XOR employing a different form of CMOS logic.

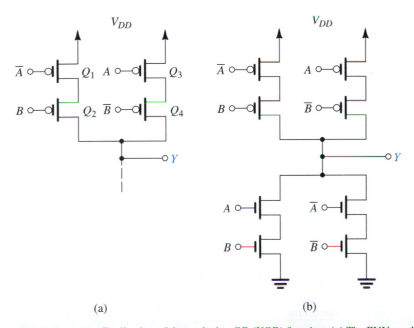

(a)   (b)

FIGURE 10.15 Realization of the exclusive-OR (XOR) function: **(a)** The PUN synthesized directly from the expression in Eq. (10.25). **(b)** The complete XOR realization utilizing the PUN in (a) and a PDN that is synthesized directly from the expression in Eq. (10.26). Note that two inverters (not shown) are needed to generate the complemented variables. Also note that in this XOR realization, the PDN and the PUN are not dual networks; however, a realization based on dual networks is possible (see Problem 10.27).

Another interesting observation follows from the circuit in Fig. 10.15(b). The PDN and the PUN here are *not* dual networks. Indeed, duality of the PDN and the PUN is not a necessary condition. Thus, although a dual of PDN (or PUN) can always be used for PUN (or PDN), the two networks are not necessarily duals.

10.3.7 Summary of the Synthesis Method

1. The PDN can be most directly synthesized by expressing $\bar{Y}$ as a function of the *uncomplemented* variables. If complemented variables appear in this expression, additional inverters will be required to generate them.

2. The PUN can be most directly synthesized by expressing Y as a function of the *complemented* variables and then applying the uncomplemented variables to the gates of the PMOS transistors. If uncomplemented variables appear in the expression, additional inverters will be needed.

3. The PDN can be obtained from the PUN (and vice versa) using the duality property.

10.3.8 Transistor Sizing

Once a CMOS gate circuit has been generated, the only significant step remaining in the design is to decide on W/L ratios for all devices. These ratios usually are selected to provide the gate with current-driving capability in both directions equal to that of the basic inverter. The reader will recall from Section 10.2 that for the basic inverter design, we denoted $(W/L)_n = n$ and $(W/L)_p = p$, where n is usually 1.5 to 2 and, for a matched design, $p = (\mu_n/\mu_p)n$. Thus, we wish to select individual W/L ratios for all transistors in a logic gate so that the PDN should be able to provide a capacitor discharge current *at least* equal to that of an NMOS transistor with $W/L = n$, and the PUN should be able to provide a charging current *at least* equal to that of a PMOS transistor with $W/L = p$. This will guarantee a *worst-case* gate delay equal to that of the basic inverter.[1]

In the preceding description, the idea of "worst case" should be emphasized. It means that in deciding on device sizing, we should find the input combinations that result in the lowest output current and then choose sizes that will make this current equal to that of the basic inverter. Before we consider examples, we need to address the issue of determining the current-driving capability of a circuit consisting of a number of MOS devices. In other words, we need to find the *equivalent W/L ratio* of a network of MOS transistors. Toward that end, we consider the parallel and series connection of MOSFETs and find the equivalent W/L ratios.

The derivation of the equivalent W/L ratio is based on the fact that the on resistance of a MOSFET is inversely proportional to W/L. Thus, if a number of MOSFETs having ratios of $(W/L)_1$, $(W/L)_2$, . . . are connected in series, the equivalent series resistance obtained by adding the on-resistances will be

$$
\begin{aligned}
R_{\text{series}} &= r_{DS1} + r_{DS2} + \cdots \\
&= \frac{\text{constant}}{(W/L)_1} + \frac{\text{constant}}{(W/L)_2} + \cdots \\
&= \text{constant}\left[\frac{1}{(W/L)_1} + \frac{1}{(W/L)_2} + \cdots\right] \\
&= \frac{\text{constant}}{(W/L)_{\text{eq}}}
\end{aligned}
$$

---

[1] This statement assumes that the total effective capacitance C of the logic gate is the same as that of the inverter. In actual practice, the value of C will be larger for a gate, especially as the fan-in is increased.

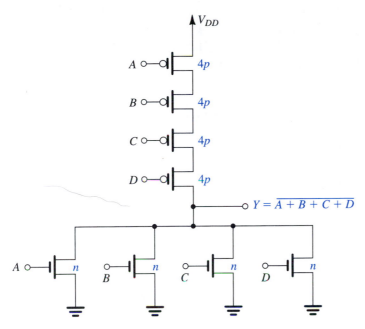

FIGURE 10.16 Proper transistor sizing for a four-input NOR gate. Note that n and p denote the (W/L) ratios of Q_N and Q_P, respectively, of the basic inverter.

resulting in the following expression for $(W/L)_{\text{eq}}$ for transistors connected in series:

$$(W/L)_{\text{eq}} = \frac{1}{\dfrac{1}{(W/L)_1} + \dfrac{1}{(W/L)_2} + \cdots} \tag{10.27}$$

Similarly, we can show that the parallel connection of transistors with W/L ratios of $(W/L)_1$, $(W/L)_2, \ldots$, results in an equivalent W/L of

$$(W/L)_{\text{eq}} = (W/L)_1 + (W/L)_2 + \cdots \tag{10.28}$$

As an example, two identical MOS transistors with individual W/L ratios of 4 result in an equivalent W/L of 2 when connected in series and of 8 when connected in parallel.

As an example of proper sizing, consider the four-input NOR in Fig. 10.16. Here, the worst case (the lowest current) for the PDN is obtained when only one of the NMOS transistors is conducting. We therefore select the W/L of each NMOS transistor to be equal to that of the NMOS transistor of the basic inverter, namely, n. For the PUN, however, the worst-case situation (and indeed the only case) is when all inputs are low and the four series PMOS transistors are conducting. Since the equivalent W/L will be one-quarter of that of each PMOS device, we should select the W/L ratio of each PMOS transistor to be four times that of Q_P of the basic inverter, that is, $4p$.

As another example, we show in Fig. 10.17 the proper sizing for a four-input NAND gate. Comparison of the NAND and NOR gates in Figs. 10.16 and 10.17 indicates that because p is usually two to three times n, the NOR gate will require much greater area than the NAND gate. For this reason, NAND gates are generally preferred for implementing combinational logic functions in CMOS.

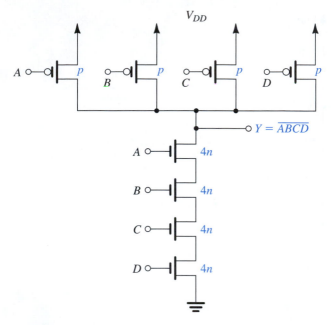

FIGURE 10.17 Proper transistor sizing for a four-input NAND gate. Note that n and p denote the (W/L) ratios of Q_N and Q_P, respectively, of the basic inverter.

EXAMPLE 10.2

Provide transistor W/L ratios for the logic circuit shown in Fig. 10.18. Assume that for the basic inverter $n = 1.5$ and $p = 5$ and that the channel length is 0.25 μm.

Solution

Refer to Fig. 10.18, and consider the PDN first. We note that the worst case occurs when Q_{NB} is on and either Q_{NC} or Q_{ND} is on. That is, in the worst case, we have two transistors in series. Therefore, we select each of Q_{NB}, Q_{NC}, and Q_{ND} to have twice the width of the n-channel device in the basic inverter, thus

$$Q_{NB}: \ W/L = 2n = 3 = 0.75/0.25$$

$$Q_{NC}: \ W/L = 2n = 3 = 0.75/0.25$$

$$Q_{ND}: \ W/L = 2n = 3 = 0.75/0.25$$

For transistor Q_{NA}, select W/L to be equal to that of the n-channel device in the basic inverter:

$$Q_{NA}: \ W/L = n = 1.5 = 0.375/0.25$$

Next, consider the PUN. Here, we see that in the worst case, we have three transistors in series: Q_{PA}, Q_{PC}, and Q_{PD}. Therefore, we select the W/L ratio of each of these to be three times that of Q_P in the basic inverter, that is, $3p$, thus

$$Q_{PA}: \ W/L = 3p = 15 = 3.75/0.25$$

$$Q_{PC}: \ W/L = 3p = 15 = 3.75/0.25$$

$$Q_{PD}: \ W/L = 3p = 15 = 3.75/0.25$$

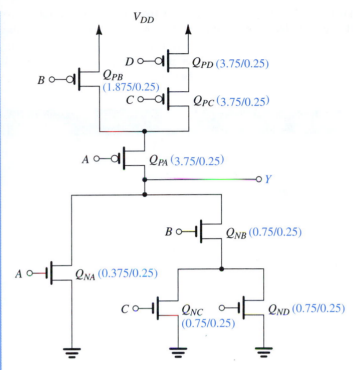

FIGURE 10.18 Circuit for Example 10.2.

Finally, the W/L ratio for Q_{PB} should be selected so that the equivalent W/L of the series connection of Q_{PB} and Q_{PA} should be equal to p. It follows that for Q_{PB} the ratio should be $1.5p$,

$$Q_{PB}: \ W/L = 1.5p = 7.5 = 1.875/0.25$$

Figure 10.18 shows the circuit with the transistor sizes indicated.

10.3.9 Effects of Fan-In and Fan-Out on Propagation Delay

Each additional input to a CMOS gate requires two additional transistors, one NMOS and one PMOS. This is in contrast to other forms of MOS logic, where each additional input requires only one additional transistor. The additional transistor in CMOS not only increases the chip area but also increases the total effective capacitance per gate and in turn increases the propagation delay. The size-scaling method described earlier compensates for some (but not all) of the increase in t_P. Specifically, by increasing device size, we are able to preserve the current-driving capability. However, the capacitance C increases because of both the increased number of inputs and the increase in device size. Thus t_P will still increase with fan-in, a fact that imposes a practical limit on the fan-in of, say, the NAND gate to about 4. If a higher number of inputs is required, then "clever" logic design should be adopted to realize the given Boolean function with gates of no more than four inputs. This would usually mean an increase in the number of cascaded stages and thus an increase in delay. However, such an increase in delay can be less than the increase due to the large fan-in (see Problem 10.36).

An increase in a gate's fan-out adds directly to its load capacitance and, thus, increases its propagation delay.

Thus although CMOS has many advantages, it does suffer from increased circuit complexity when the fan-in and fan-out are increased, and from the corresponding effects of this complexity on both chip area and propagation delay. In the following two sections, we shall study some simplified forms of CMOS logic that attempt to reduce this complexity, although at the expense of forgoing some of the advantages of basic CMOS.

EXERCISES

10.4 For a process technology with $L = 0.5 \ \mu m$, $n = 1.5$, $p = 6$, give the sizes of all transistors in (a) a four-input NOR and (b) a four-input NAND. Also, give the relative areas of the two gates.

Ans. (a) NMOS devices: $W/L = 0.75/0.5$, PMOS devices: $12/0.5$;

   (b) NMOS devices: $W/L = 3/0.5$, PMOS devices: $3/0.5$;

   NOR area/NAND area = 2.125

10.5 For the scaled NAND gate in Exercise 10.4, find the ratio of the maximum to minimum current available to (a) charge a load capacitance and (b) discharge a load capacitance.

Ans. (a) 4; (b) 1

 ## 10.4 PSEUDO-NMOS LOGIC CIRCUITS

As explained in Section 10.3, despite its many great advantages, CMOS suffers from increased area, and correspondingly increased capacitance and delay, as the logic gates become more complex. For this reason, designers of digital integrated circuits have been searching for forms of CMOS logic circuits that can be used to supplement the complementary-type circuits studied in Sections 10.2 and 10.3. These forms are not intended to replace complementary CMOS but rather to be used in special applications for special purposes. We shall examine two such CMOS logic styles in this and the following section.

10.4.1 The Pseudo-NMOS Inverter

Figure 10.19(a) shows a modified form of the CMOS inverter. Here, only Q_N is driven by the input voltage while the gate of Q_P is grounded, and Q_P acts as an active load for Q_N. Even before we examine the operation of this circuit in detail, an advantage over complementary CMOS is obvious: Each input must be connected to the gate of only one transistor or, alternatively, only one additional transistor (an NMOS) will be needed for each additional gate input. Thus the area and delay penalties arising from increased fan-in in a complementary CMOS gate will be reduced. This is indeed the motivation for exploring this modified inverter circuit.

The inverter circuit of Fig. 10.19(a) resembles other forms of NMOS logic that consist of a driver transistor (Q_N) and a load transistor (in this case, Q_P); hence the name pseudo-NMOS. For comparison purposes, we shall briefly mention two older forms of NMOS logic. The earliest form, popular in the mid-1970s, utilized an enhancement MOSFET for the load element, in a topology whose basic inverter is shown in Fig. 10.19(b). Enhancement-load NMOS logic circuits suffer from a relatively small logic swing, small noise margins, and high static power dissipation. For these reasons, this logic-circuit technology is now virtually obsolete. It was replaced in the late 1970s and early 1980s with depletion-load NMOS circuits, in which a depletion NMOS transistor with its gate connected to its source is used as the load element. The topology of the basic depletion-load inverter is shown in Fig. 10.19(c).

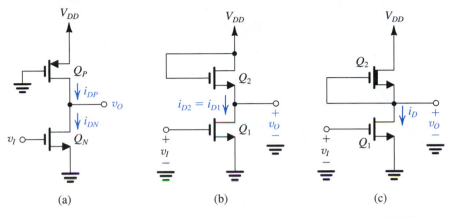

FIGURE 10.19 **(a)** The pseudo-NMOS logic inverter. **(b)** The enhancement-load NMOS inverter. **(c)** The depletion-load NMOS inverter.

It was initially expected that the depletion NMOS with $V_{GS} = 0$ would operate as a constant-current source and would thus provide an excellent load element.[2] However, it was quickly realized that the body effect in the depletion transistor causes its i–v characteristic to deviate considerably from that of a constant-current source. Nevertheless, depletion-load NMOS circuits feature significant improvements over their enhancement-load counterparts, enough to justify the extra processing step required to fabricate the depletion devices (namely, ion-implanting the channel). Although depletion-load NMOS has been virtually replaced by CMOS, one can still see some depletion-load circuits in specialized applications. We will not study depletion-load NMOS logic here (the interested reader can refer to the third edition of this book).

The pseudo-NMOS inverter that we are about to study is similar to depletion-load NMOS but with rather improved characteristics. It also has the advantage of being directly compatible with complementary CMOS circuits.

10.4.2 Static Characteristics

The static characteristics of the pseudo-NMOS inverter can be derived in a manner similar to that used for complementary CMOS. Toward that end, we note that the drain currents of Q_N and Q_P are given by

$$i_{DN} = \tfrac{1}{2}k_n(v_I - V_t)^2, \quad \text{for } v_O \geq v_I - V_t \quad \text{(saturation)} \tag{10.29}$$

$$i_{DN} = k_n[(v_I - V_t)v_O - \tfrac{1}{2}v_O^2], \quad \text{for } v_O \leq v_I - V_t \quad \text{(triode)} \tag{10.30}$$

$$i_{DP} = \tfrac{1}{2}k_p(V_{DD} - V_t)^2, \quad \text{for } v_O \leq V_t \quad \text{(saturation)} \tag{10.31}$$

$$i_{DP} = k_p[(V_{DD} - V_t)(V_{DD} - v_O) - \tfrac{1}{2}(V_{DD} - v_O)^2], \quad \text{for } v_O \geq V_t \quad \text{(triode)} \tag{10.32}$$

where we have assumed that $V_{tn} = -V_{tp} = V_t$, and have used $k_n = k_n'(W/L)_n$ and $k_p = k_p'(W/L)_p$ to simplify matters.

---

[2] A constant-current load provides a capacitor-charging current that does not diminish as v_O rises toward V_{DD}, as is the case with a resistive load. Thus the value of t_{PLH} obtained with a current-source load is significantly lower than that obtained with a resistive load (see Problem 10.38). Of course, a resistive load is simply out of the question because of the very large silicon area it would occupy (equivalent to that of thousands of transistors!).

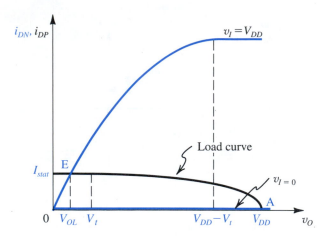

FIGURE 10.20 Graphical construction to determine the VTC of the inverter in Fig. 10.19.

To obtain the VTC of the inverter, we superimpose the load curve represented by Eqs. (10.31) and (10.32) on the i_D–v_{DS} characteristics of Q_N, which can be relabeled as i_{DN}–v_O and drawn for various values of $v_{GS} = v_I$. Such a graphical construction is shown in Fig. 10.20 where, to keep the diagram simple, we show the Q_N curves for only the two extreme values of v_I, namely, 0 and V_{DD}. Two observations follow:

1. The load curve represents a much lower saturation current (Eq. 10.31) than is represented by the corresponding curve for Q_N, namely, that for $v_I = V_{DD}$. This is a result of the fact that the pseudo-NMOS inverter is usually designed so that k_n is greater than k_p by a factor of 4 to 10. As we will show shortly, this inverter is of the so-called ratioed type,[3] and the ratio $r \equiv k_n/k_p$ determines all the breakpoints of the VTC, that is, V_{OL}, V_{IL}, V_{IH}, and so on, and thus determines the noise margins. Selection of a relatively high value for r reduces V_{OL} and widens the noise margins.

2. Although one tends to think of Q_P as acting as a constant-current source, it actually operates in saturation for only a small range of v_O, namely, $v_O \leq V_t$. For the remainder of the v_O range, Q_P operates in the triode region.

Consider first the two extreme cases of v_I: When $v_I = 0$, Q_N is cut off and Q_P is operating in the triode region, though with zero current and zero drain–source voltage. Thus the operating point is that labeled A in Fig. 10.20, where $v_O = V_{OH} = V_{DD}$, the static current is zero, and the static power dissipation is zero. When $v_I = V_{DD}$, the inverter will operate at the point labeled E in Fig. 10.20. Observe that unlike complementary CMOS, here V_{OL} is not zero, an obvious disadvantage. Another disadvantage is that the gate conducts current (I_{stat}) in the low-output state, and thus there will be static power dissipation ($P_D = I_{stat} \times V_{DD}$).

10.4.3 Derivation of the VTC

Figure 10.21 shows the VTC of the pseudo-NMOS inverter. As indicated, it has four distinct regions, labeled I through IV, corresponding to the different combinations of possible modes

---

[3] For the NMOS inverters, V_{OL} depends on the ratio of the transconductance parameters of the devices, that is, on the ratio $(k'(W/L))_{driver}/(k'(W/L))_{load}$. Such circuits are therefore known as *ratioed* logic circuits. Complementary CMOS logic circuits do not have such a dependency and can therefore be called *ratioless*.

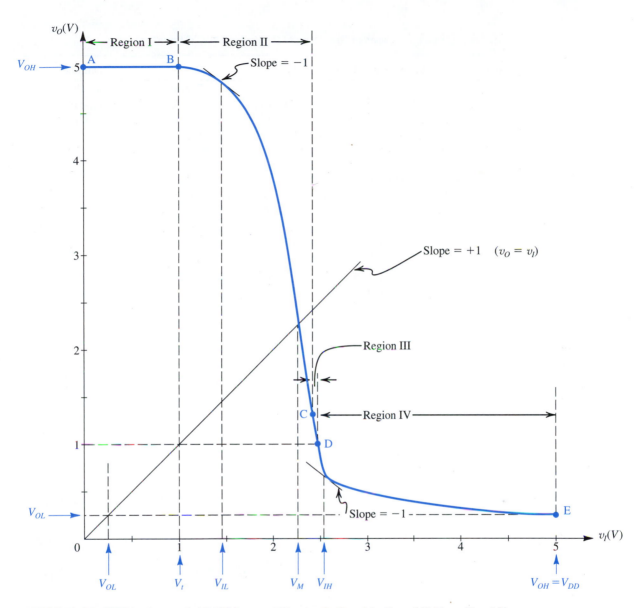

FIGURE 10.21 VTC for the pseudo-NMOS inverter. This curve is plotted for $V_{DD} = 5$ V, $V_{tn} = -V_{tp} = 1$ V, and $r = 9$.

of operation of Q_N and Q_P. The four regions, the corresponding transistor modes of operation, and the conditions that define the regions are listed in Table 10.1. We shall utilize the information in this table together with the device equations given in Eqs. (10.29) through (10.32) to derive expressions for the various segments of the VTC and in particular for the important parameters that characterize the static operation of the inverter.

■ **Region I (segment AB):**

$$v_O = V_{OH} = V_{DD} \qquad (10.33)$$

TABLE 10.1 Regions of Operation of the Pseudo-NMOS Inverter

Region	Segment of VTC	Q_N	Q_P	Condition
I	AB	Cutoff	Triode	$v_I < V_t$
II	BC	Saturation	Triode	$v_O \geq v_I - V_t$
III	CD	Triode	Triode	$V_t \leq v_O \leq v_I - V_t$
IV	DE	Triode	Saturation	$v_O \leq V_t$

- **Region II (segment BC):**
  Equating i_{DN} from Eq. (10.29) and i_{DP} from Eq. (10.32) together with substituting $k_n = rk_p$, and with some manipulations, we obtain

$$v_O = V_t + \sqrt{(V_{DD} - V_t)^2 - r(v_I - V_t)^2} \tag{10.34}$$

The value of V_{IL} can be obtained by differentiating this equation and substituting $\partial v_O / \partial v_I = -1$ and $v_I = V_{IL}$,

$$V_{IL} = V_t + \frac{V_{DD} - V_t}{\sqrt{r(r + 1)}} \tag{10.35}$$

The threshold voltage V_M (or V_{th}) is by definition the value of v_I for which $v_O = v_I$,

$$V_M = V_t + \frac{V_{DD} - V_t}{\sqrt{r + 1}} \tag{10.36}$$

Finally, the end of the region II segment (point C) can be found by substituting $v_O = v_I - V_t$ in Eq. (10.34), the condition for Q_N leaving saturation and entering the triode region.

- **Region III (segment CD)**
  This is a short segment that is not of great interest. Point D is characterized by $v_O = V_t$.

- **Region IV (segment DE)**
  Equating i_{DN} from Eq. (10.30) to i_{DP} from Eq. (10.31) and substituting $k_n = rk_p$ results in

$$v_O = (v_I - V_t) - \sqrt{(v_I - V_t)^2 - \frac{1}{r}(V_{DD} - V_t)^2} \tag{10.37}$$

The value of V_{IH} can be determined by differentiating this equation and setting $\partial v_O / \partial v_I = -1$ and $v_I = V_{IH}$,

$$V_{IH} = V_t + \frac{2}{\sqrt{3r}}(V_{DD} - V_t) \tag{10.38}$$

The value of V_{OL} can be found by substituting $v_I = V_{DD}$ into Eq. (10.37),

$$V_{OL} = (V_{DD} - V_t)\left[1 - \sqrt{1 - \frac{1}{r}}\right] \tag{10.39}$$

The static current conducted by the inverter in the low-output state is found from Eq. (10.31) as

$$I_{\text{stat}} = \tfrac{1}{2}k_p(V_{DD} - V_t)^2 \tag{10.40}$$

Finally, we can use Eqs. (10.35) and (10.39) to determine NM_L and Eqs. (10.33) and (10.38) to determine NM_H,

$$NM_L = V_t - (V_{DD} - V_t)\left[1 - \sqrt{1 - \frac{1}{r} - \frac{1}{\sqrt{r(r+1)}}}\right] \qquad (10.41)$$

$$NM_H = (V_{DD} - V_t)\left(1 - \frac{2}{\sqrt{3}r}\right) \qquad (10.42)$$

As a final observation, we note that since V_{DD} and V_t are determined by the process technology, the only design parameter for controlling the values of V_{OL} and the noise margins is the ratio r.

10.4.4 Dynamic Operation

Analysis of the inverter transient response to determine t_{PLH} with the inverter loaded by a capacitance C is identical to that of the complementary CMOS inverter. The capacitance will be charged by the current i_{DP}; we can determine an estimate for t_{PLH} by using the average value of i_{DP} over the range $v_O = 0$ to $v_O = V_{DD}/2$. The result is the following approximate expression (where we have assumed $V_t \cong 0.2 V_{DD}$):

$$t_{PLH} = \frac{1.7C}{k_p V_{DD}} \qquad (10.43)$$

The case for the capacitor discharge is somewhat different because the current i_{DP} has to be subtracted from i_{DN} to determine the discharge current. The result is the approximate expression,

$$t_{PHL} \cong \frac{1.7C}{k_n\left(1 - \frac{0.46}{r}\right)V_{DD}} \qquad (10.44)$$

which, for a large value of r, reduces to

$$t_{PHL} \cong \frac{1.7C}{k_n V_{DD}} \qquad (10.45)$$

Although these are identical formulas to those for the complementary CMOS inverter, the pseudo-NMOS inverter has a special problem: Since k_p is r times smaller than k_n, t_{PLH} will be r times larger than t_{PHL}. Thus the circuit exhibits an asymmetrical delay performance. Recall, however, that for gates with large fan-in, pseudo-NMOS requires fewer transistors and thus C can be smaller than in the corresponding complementary CMOS gate.

10.4.5 Design

The design involves selecting the ratio r and the (W/L) for one of the transistors. The value of (W/L) for the other device can then be obtained using r. The design parameters of interest are V_{OL}, NM_L, NM_H, I_{stat}, P_D, t_{PLH}, and t_{PHL}. Important design considerations are as follows:

1. The ratio r determines all the breakpoints of the VTC; the larger the value of r, the lower V_{OL} is (Eq. 10.39) and the wider the noise margins are (Eqs. 10.41 and 10.42). However, a larger r increases the asymmetry in the dynamic response and, for a given $(W/L)_p$, makes the gate larger. Thus selecting a value for r represents a compromise

between noise margins on the one hand and silicon area and t_P on the other. Usually, r is selected in the range 4 to 10.

2. Once r has been determined, a value for $(W/L)_p$ or $(W/L)_n$ can be selected and the other determined. Here, one would select a small $(W/L)_n$ to keep the gate area small and thus obtain a small value for C. Similarly, a small $(W/L)_p$ keeps I_{stat} and P_D low. On the other hand, one would want to select larger (W/L) ratios to obtain low t_P and thus fast response. For usual (high-speed) applications, $(W/L)_p$ is selected so that I_{stat} is in the range of 50 to 100 μA, which for $V_{DD} = 5$ V results in P_D in the range of 0.25 mW to 0.5 mW.

10.4.6 Gate Circuits

Except for the load device, the pseudo-NMOS gate circuit is identical to the PDN of the complementary CMOS gate. Four-input pseudo-NMOS NOR and NAND gates are shown in Fig. 10.22. Note that each requires five transistors compared to the eight used in complementary CMOS. In pseudo-NMOS, NOR gates are preferred over NAND gates since the former do not utilize transistors in series, and thus can be designed with minimum-size NMOS devices.

10.4.7 Concluding Remarks

Pseudo-NMOS is particularly suited for applications in which the output remains high most of the time. In such applications, the static power dissipation can be reasonably low (since the gate dissipates static power only in the low-output state). Further, the output transitions that matter would presumably be high-to-low ones where the propagation delay can be made as short as necessary. A particular application of this type can be found in the design of address decoders for memory chips (Section 11.5) and in read-only memories (Section 11.6).

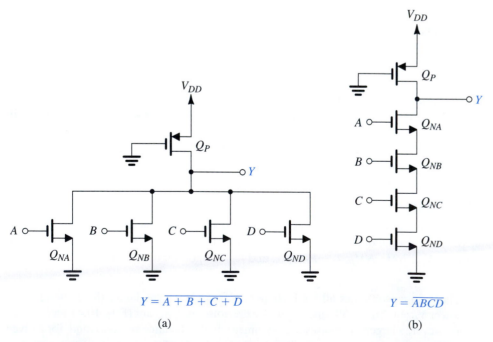

$$Y = \overline{A + B + C + D}$$

(a)

$$Y = \overline{ABCD}$$

(b)

FIGURE 10.22 NOR and NAND gates of the pseudo-NMOS type.

EXAMPLE 10.3

Consider a pseudo-NMOS inverter fabricated in the CMOS technology specified in Example 10.1 for which $\mu_n C_{ox} = 115$ $\mu A/V^2$, $\mu_p C_{ox} = 30$ $\mu A/V^2$, $V_{tn} = -V_{tp} = 0.4$ V, and $V_{DD} = 2.5$ V. Let the W/L ratio of Q_N be $(0.375\ \mu m / 0.25\ \mu m)$ and $r = 9$. Find:

(a) V_{OH}, V_{OL}, V_{IL}, V_{IH}, V_M, NM_H, and NM_L

(b) $(W/L)_p$

(c) I_{stat} and P_D

(d) t_{PLH}, t_{PHL}, and t_P, assuming a total capacitance at the inverter output of 7 fF

Solution

(a) $V_{OH} = V_{DD} = 2.5$ V

V_{OL} is determined from Eq. (10.39) as

$$V_{OL} = (2.5 - 0.4)[1 - \sqrt{1 - \tfrac{1}{9}}] = 0.12 \text{ V}$$

V_{IL} is determined from Eq. (10.35) as

$$V_{IL} = 0.4 + \frac{2.5 - 0.4}{\sqrt{9(9+1)}} = 0.62 \text{ V}$$

V_{IH} is determined from Eq. (10.38) as

$$V_{IH} = 0.4 + \frac{2}{\sqrt{3 \times 9}} \times (2.5 - 0.4) = 1.21 \text{ V}$$

V_M is determined from Eq. (10.36) as

$$V_M = 0.4 + \frac{2.5 - 0.4}{\sqrt{9+1}} = 1.06 \text{ V}$$

The noise margins can now be determined as

$$NM_H = V_{OH} - V_{IH} = 2.5 - 1.21 = 1.29 \text{ V}$$

$$NM_L = V_{IL} - V_{OL} = 0.62 - 0.12 = 0.50 \text{ V}$$

Observe that the noise margins are not equal and that NM_L is rather low.

(b) The (W/L) ratio of Q_P can be found from

$$\frac{\mu_n C_{ox}(W/L)_n}{\mu_p C_{ox}(W/L)_p} = 9$$

$$\frac{115 \times \dfrac{0.375}{0.25}}{30(W/L)_p} = 9$$

Thus

$$(W/L)_p = 0.64$$

(c) The dc current in the low-output state can be determined from Eq. (10.40), as

$$I_{stat} = \tfrac{1}{2} \times 30 \times 0.64(2.5 - 0.4)^2 = 42.3 \text{ } \mu A$$

The static power dissipation can now be found from

$$P_D = I_{\text{stat}} V_{DD}$$
$$= 42.3 \times 2.5 = 106 \ \mu\text{W}$$

(d) The low-to-high propagation delay can be found from Eq. (10.43), as

$$t_{PLH} = \frac{1.7 \times 7 \times 10^{-15}}{30 \times 10^{-6} \times 0.64 \times 2.5} = 0.25 \text{ ns}$$

The high-to-low propagation delay can be found from Eq. (10.45), as

$$t_{PHL} = \frac{1.7 \times 7 \times 10^{-15}}{115 \times 10^{-6} \times \dfrac{0.375}{0.25} \times 2.5} = 0.03 \text{ ns}$$

Now, the propagation delay can be determined, as

$$t_P = \tfrac{1}{2}(0.25 + 0.03) = 0.14 \text{ ns}$$

Although the propagation delay is considerably greater than that of the complementary CMOS inverter of Example 10.1, this is not an entirely fair comparison: Recall that the advantage of pseudo-NMOS occurs in gates with large fan-in, not in a single inverter.

EXERCISES

D10.6 While keeping r unchanged, redesign the inverter circuit of Example 10.3 to lower its static power dissipation to half the value found. Find the W/L ratios for the new design. Also find t_{PLH}, t_{PHL}, and t_P, assuming that C remains unchanged. Would the noise margins change?

Ans. $(W/L)_n = 1.5$; $(W/L)_p = 0.32$; 0.5 ns; 0.03 ns; 0.27 ns; no

D10.7 Redesign the inverter of Example 10.3 using $r = 4$. Find V_{OL} and the noise margins. If $(W/L)_n = 0.375 \ \mu\text{m}/0.25 \ \mu\text{m}$, find $(W/L)_p$, I_{stat}, P_D, t_{PLH}, t_{PHL}, and t_P. Assume $C = 7$ fF.

Ans. $V_{OL} = 0.28$ V; $NM_L = 0.59$ V; $NM_H = 0.89$ V; $(W/L)_p = 1.44$; $I_{\text{stat}} = 95.3 \ \mu\text{A}$; $P_D = 0.24$ mW; $t_{PLH} = 0.11$ ns; $t_{PHL} = 0.03$ ns; $t_P = 0.07$ ns

10.5 PASS-TRANSISTOR LOGIC CIRCUITS

A conceptually simple approach for implementing logic functions utilizes series and parallel combinations of switches that are controlled by input logic variables to connect the input and output nodes (see Fig. 10.23). Each of the switches can be implemented either by a single NMOS transistor (Fig. 10.24a) or by a pair of complementary MOS transistors connected in what is known as the **CMOS transmission-gate** configuration (Fig. 10.24b). The result is a simple form of logic circuit that is particularly suited for some special logic functions and is frequently used in conjunction with complementary CMOS logic to implement such functions efficiently.

Because this form of logic utilizes MOS transistors in the series path from input to output, to *pass* or block signal transmission, it is known as *pass-transistor logic* (PTL). As mentioned earlier, CMOS transmission gates are frequently employed to implement the switches, giving this logic-circuit form the alternative name, *transmission-gate logic*. The terms are used interchangeably independent of the actual implementation of the switches.

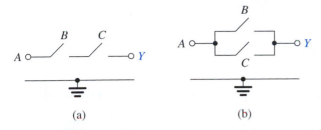

FIGURE 10.23 Conceptual pass-transistor logic gates. **(a)** Two switches, controlled by the input variables B and C, when connected in series in the path between the input node to which an input variable A is applied and the output node (with an implied load to ground) realize the function $Y = ABC$. **(b)** When the two switches are connected in parallel, the function realized is $Y = A(B + C)$.

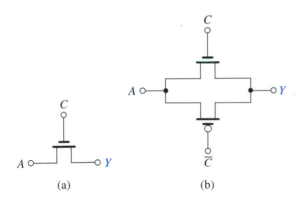

FIGURE 10.24 Two possible implementations of a voltage-controlled switch connecting nodes A and Y: **(a)** single NMOS transistor and **(b)** CMOS transmission gate.

Though conceptually simple, pass-transistor logic circuits have to be designed with care. In the following, we shall study the basic principles of PTL circuit design and present examples of its application.

10.5.1 An Essential Design Requirement

An essential requirement in the design of PTL circuits is ensuring that *every circuit node has at all times a low-resistance path to V_{DD} or ground*. To appreciate this point, consider the situation depicted in Fig. 10.25(a): A switch S_1 (usually part of a larger PTL network, not shown) is used to form the AND function of its controlling variable B and the variable A available at the output of a CMOS inverter. The output Y of the PTL circuit is shown connected to the input of another inverter. Obviously, if B is high, S_1 closes and $Y = A$. Node Y will then be connected either to V_{DD} (if A is high) through Q_2 or to ground (if A is low) through Q_1. But, what happens when B goes low and S_1 opens? Node Y will now become a high-impedance node. If initially, v_Y was zero, it will remain so. However, if initially, v_Y was high at V_{DD}, this voltage will be maintained by the charge on the parasitic capacitance C, but for only a time: The inevitable leakage currents will slowly discharge C, and v_Y will diminish correspondingly. In any case, the circuit can no longer be considered a static combinational logic circuit.

The problem can be easily solved by establishing for node Y a low-resistance path that is activated when B goes low, as shown in Fig. 10.25(b). Here, another switch, S_2, controlled by $\bar{B}$ is connected between Y and ground. When B goes low, S_2 closes and establishes a low-resistance path between Y and ground.

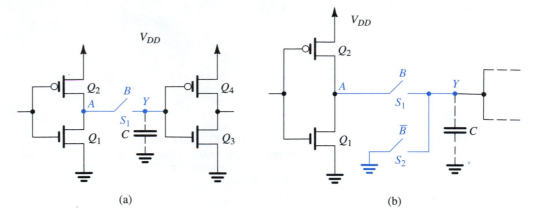

(a)   (b)

FIGURE 10.25 A basic design requirement of PTL circuits is that every node have, at all times, a low-resistance path to either ground or V_{DD}. Such a path does not exist in (a) when B is low and S_1 is open. It is provided in (b) through switch S_2.

10.5.2 Operation with NMOS Transistors as Switches

Implementing the switches in a PTL circuit with single NMOS transistors results in a simple circuit with small area and small node capacitances. These advantages, however, are obtained at the expense of serious shortcomings in both the static characteristics and the dynamic performance of the resulting circuits. To illustrate, consider the circuit shown in Fig. 10.26, where an NMOS transistor Q is used to implement a switch connecting an input node with voltage v_I and an output node. The total capacitance between the output node and ground is represented by capacitor C. The switch is shown in the closed state with the control signal applied to its gate being high at V_{DD}. We wish to analyze the operation of the circuit as the input voltage v_I goes high (to V_{DD}) at time $t = 0$. We assume that initially the output voltage v_O is zero and capacitor C is fully discharged.

When v_I goes high, the transistor operates in the saturation mode and delivers a current i_D to charge the capacitor,

$$i_D = \tfrac{1}{2}k_n(V_{DD} - v_O - V_t)^2 \tag{10.46}$$

where $k_n = k_n'(W/L)$, and V_t is determined by the body effect since the source is at a voltage v_O relative to the body, thus (see Eq. 4.33),

$$V_t = V_{t0} + \gamma(\sqrt{v_O + 2\phi_f} - \sqrt{2\phi_f}) \tag{10.47}$$

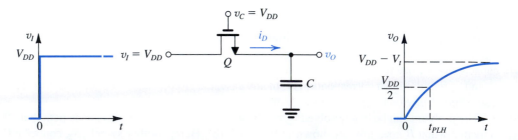

FIGURE 10.26 Operation of the NMOS transistor as a switch in the implementation of PTL circuits. This analysis is for the case with the switch closed (v_C is high) and the input going high ($v_I = V_{DD}$).

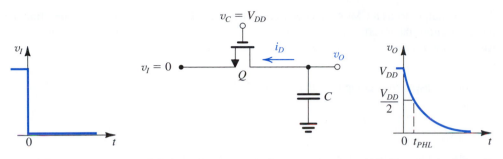

FIGURE 10.27 Operation of the NMOS switch as the input goes low ($v_I = 0$ V). Note that the drain of an NMOS transistor is always higher in voltage than the source; correspondingly, the drain and source terminals interchange roles in comparison to the circuit in Fig. 10.26.

Thus, initially (at $t = 0$), $V_t = V_{t0}$ and the current i_D is relatively large. However, as C charges up and v_O rises, V_t increases (Eq. 10.47) and i_D decreases. The latter effect is due to both the increase in v_O and in V_t. It follows that the process of charging the capacitor will be relatively slow. More seriously, observe from Eq. (10.46) that i_D reduces to zero when v_O reaches $(V_{DD} - V_t)$. Thus the high output voltage (V_{OH}) will not be equal to V_{DD}; rather, it will be lower by V_t, and to make matters worse, the value of V_t can be as high as 1.5 to 2 times V_{t0}!

In addition to reducing the gate noise immunity, the low value of V_{OH} (commonly referred to as a "poor 1") has another detrimental effect: Consider what happens when the output node is connected to the input of a complementary CMOS inverter (as was the case in Fig. 10.25). The low value of V_{OH} can cause Q_P of the load inverter to conduct. Thus the inverter will have a finite static current and static power dissipation.

The propagation delay t_{PLH} of the PTL gate of Fig. 10.26 can be determined as the time for v_O to reach $V_{DD}/2$. This can be calculated using techniques similar to those employed in the preceding sections, as will be illustrated shortly in an example.

Figure 10.27 shows the NMOS switch circuit when v_I is brought down to 0 V. We assume that initially $v_O = V_{DD}$. Thus at $t = 0+$, the transistor conducts and operates in the saturation region,

$$i_D = \tfrac{1}{2}k_n(V_{DD} - V_t)^2 \tag{10.48}$$

where we note that since the source is now at 0 V (note that the drain and source have interchanged roles), there will be no body effect, and V_t remains constant at V_{t0}. As C discharges, v_O decreases and the transistor enters the triode region at $v_O = V_{DD} - V_t$. Nevertheless, the capacitor discharge continues until C is fully discharged and $v_O = 0$. Thus, the NMOS transistor provides $V_{OL} = 0$, or a "good 0." Again, the propagation delay t_{PHL} can be determined using usual techniques, as illustrated by the following example.

EXAMPLE 10.4

Consider the NMOS transistor switch in the circuits of Figs. 10.26 and 10.27 to be fabricated in a technology for which $\mu_n C_{ox} = 50\ \mu A/V^2$, $\mu_p C_{ox} = 20\ \mu A/V^2$, $|V_{t0}| = 1$ V, $\gamma = 0.5\ V^{1/2}$, $2\phi_f = 0.6$ V, and $V_{DD} = 5$ V. Let the transistor be of the minimum size for this technology, namely, $4\ \mu m / 2\ \mu m$, and assume that the total capacitance between the output node and ground is $C = 50$ fF.

(a) For the case with v_I high (Fig. 10.26), find V_{OH}.

(b) If the output feeds a CMOS inverter whose $(W/L)_p = 2.5 (W/L)_n = 10 \ \mu m/2 \ \mu m$, find the static current of the inverter and its power dissipation when its input is at the value found in (a). Also find the inverter output voltage.

(c) Find t_{PLH}.

(d) For the case with v_I going low (Fig. 10.27), find t_{PHL}.

(e) Find t_P.

Solution

(a) Refer to Fig. 10.26. V_{OH} is the value of v_O at which Q stops conducting,

$$V_{DD} - V_{OH} - V_t = 0$$

thus,

$$V_{OH} = V_{DD} - V_t$$

where V_t is the value of the threshold voltage at a source–body reversed bias equal to V_{OH}. Using Eq. (10.47),

$$V_t = V_{t0} + \gamma(\sqrt{V_{OH} + 2\phi_f} - \sqrt{2\phi_f})$$
$$= V_{t0} + \gamma(\sqrt{V_{DD} - V_t + 2\phi_f} - \sqrt{2\phi_f})$$

Substituting $V_{t0} = 1$, $\gamma = 0.5$, $V_{DD} = 5$, and $2\phi_f = 0.6$, we obtain a quadratic equation in V_t whose solution yields

$$V_t = 1.6 \text{ V}$$

Thus,

$$V_{OH} = 3.4 \text{ V}$$

Note that this represents a significant loss in signal amplitude.

(b) The load inverter will have an input signal of 3.4 V. Thus, its Q_P will conduct a current of

$$i_{DP} = \tfrac{1}{2} \times 20 \times \tfrac{10}{2}(5 - 3.4 - 1)^2 = 18 \ \mu A$$

Thus, the static power dissipation of the inverter will be

$$P_D = V_{DD}i_{DP} = 5 \times 18 = 90 \ \mu W$$

The output voltage of the inverter can be found by noting that Q_N will be operating in the triode region. Equating its current to that of Q_P (i.e., 18 μA) enables us to determine the output voltage to be 0.08 V.

(c) To determine t_{PLH}, we need to find the current i_D at $t = 0$ (where $v_O = 0$, $V_t = V_{t0} = 1$ V) and at $t = t_{PLH}$ (where $v_O = 2.5$ V, V_t to be determined), as follows:

$$i_D(0) = \tfrac{1}{2} \times 50 \times \tfrac{4}{2} \times (5 - 1)^2 = 800 \ \mu A$$

$$V_t \text{ (at } v_O = 2.5 \text{ V)} = 1 + 0.5(\sqrt{2.5 + 0.6} - \sqrt{0.6}) = 1.49 \text{ V}$$

$$i_D(t_{PLH}) = \tfrac{1}{2} \times 50 \times \tfrac{4}{2}(5 - 2.5 - 1.49)^2 = 50 \ \mu A$$

We can now compute the average discharge current as

$$i_D|_{av} = \frac{800 + 50}{2} = 425 \ \mu A$$

and t_{PLH} can be found as

$$t_{PLH} = \frac{C(V_{DD}/2)}{i_D|_{av}}$$

$$= \frac{50 \times 10^{-15} \times 2.5}{425 \times 10^{-6}} = 0.29 \text{ ns}$$

(d) Refer to the circuit in Fig. 10.27. Observe that, here, V_t remains constant at $V_{t0} = 1$ V. The drain current at $t = 0$ is

$$i_D(0) = \frac{1}{2} \times 50 \times \frac{4}{2}(5-1)^2 = 800 \ \mu A$$

At $t = t_{PHL}$, Q will be operating in the triode region, and thus

$$i_D(t_{PHL}) = 50 \times \frac{4}{2}[(5-1) \times 2.5 - \frac{1}{2} \times 2.5^2]$$

$$= 690 \ \mu A$$

Thus, the average discharge current is given by

$$i_D|_{av} = \frac{1}{2}(800 + 690) = 740 \ \mu A$$

and t_{PHL} can be determined as

$$t_{PHL} = \frac{50 \times 10^{-15} \times 2.5}{740 \times 10^{-6}} = 0.17 \text{ ns}$$

(e) $t_P = \frac{1}{2}(t_{PLH} + t_{PHL}) = \frac{1}{2}(0.29 + 0.17) = 0.23 \text{ ns}$

Example 10.4 illustrates clearly the problem of signal-level loss and its deleterious effect on the operation of the succeeding CMOS inverter. Some rather ingenious techniques have been developed to restore the output level to V_{DD}. We shall briefly discuss two such techniques. One is circuit-based and the other is based on process technology.

The circuit-based approach is illustrated in Fig. 10.28. Here, Q_1 is a pass-transistor controlled by input B. The output node of the PTL network is connected to the input of a complementary inverter formed by Q_N and Q_P. A PMOS transistor Q_R, whose gate is controlled by the output voltage of the inverter, v_{O2}, has been added to the circuit. Observe that in the event that the output of the PTL gate, v_{O1}, is low (at ground), v_{O2} will be high (at V_{DD}), and Q_R will be off. On the other hand, if v_{O1} is high but not quite equal to V_{DD}, the output of the

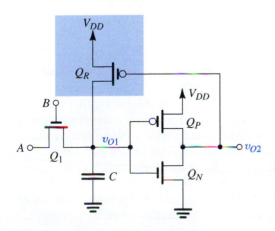

FIGURE 10.28 The use of transistor Q_R, connected in a feedback loop around the CMOS inverter, to restore the V_{OH} level, produced by Q_1, to V_{DD}.

inverter will be low (as it should be) and Q_R will turn on, supplying a current to charge C up to V_{DD}. This process will stop when $v_{O1} = V_{DD}$, that is, when the output voltage has been restored to its proper level. The "level-restoring" function performed by Q_R is frequently employed in MOS digital-circuit design. It should be noted that although the description of operation is relatively straightforward, the addition of Q_R closes a "positive-feedback" loop around the CMOS inverter, and thus operation is more involved than it appears, especially during transients. Selection of a W/L ratio for Q_R is also a somewhat involved process, although normally k_r is selected to be much lower than k_n (say a third or a fifth as large). Intuitively, this is appealing, for it implies that Q_R will not play a major role in circuit operation, apart from restoring the level of V_{OH} to V_{DD}, as explained [see Rabaey (1996)]. Transistor Q_R is said to be a "weak PMOS transistor."

The other technique for correcting for the loss of the high-output signal level (V_{OH}) is a technology-based solution. Specifically, recall that the loss in the value of V_{OH} is equal to V_{tn}. It follows that we can reduce the loss by using a lower value of V_{tn} for the NMOS switches, and we can eliminate the loss altogether by using devices for which $V_{tn} = 0$. These zero-threshold devices can be fabricated by using ion implantation to control the value of V_{tn} and are known as **natural devices.**

10.5.3 The Use of CMOS Transmission Gates as Switches

Great improvements in static and dynamic performance are obtained when the switches are implemented with CMOS transmission gates. The transmission gate utilizes a pair of complementary transistors connected in parallel. It acts as an excellent switch, providing bidirectional current flow, and it exhibits an on-resistance that remains almost constant for wide ranges of input voltage. These characteristics make the transmission gate not only an excellent switch in digital applications but also an excellent analog switch in such applications as data converters (Chapter 9) and switched-capacitor filters (Chapter 12).

Figure 10.29(a) shows the transmission-gate switch in the "on" position with the input, v_I, rising to V_{DD} at $t = 0$. Assuming, as before, that initially the output voltage is zero, we see that Q_N will be operating in saturation and providing a charging current of

$$i_{DN} = \tfrac{1}{2} k_n (V_{DD} - v_O - V_{tn})^2 \tag{10.49}$$

where, as in the case of the single NMOS switch, V_{tn} is determined by the body effect,

$$V_{tn} = V_{t0} + \gamma(\sqrt{v_O + 2\phi_f} - \sqrt{2\phi_f}) \tag{10.50}$$

Transistor Q_N will conduct a diminishing current that reduces to zero at $v_O = V_{DD} - V_{tn}$. Observe, however, that Q_P operates with $V_{SG} = V_{DD}$ and is initially in saturation,

$$i_{DP} = \tfrac{1}{2} k_p (V_{DD} - |V_{tp}|)^2 \tag{10.51}$$

where, since the body of Q_P is connected to V_{DD}, $|V_{tp}|$ remains constant at the value V_{t0}, assumed to be the same value as for the n-channel device. The total capacitor-charging current is the sum of i_{DN} and i_{DP}. Now, Q_P will enter the triode region at $v_O = |V_{tp}|$, but will continue to conduct until C is fully charged and $v_O = V_{OH} = V_{DD}$. Thus, the p-channel device will provide the gate with a "good 1." The value of t_{PLH} can be calculated using usual techniques, where we expect that as a result of the additional current available from the PMOS device, for the same value of C, t_{PLH} will be lower than in the case of the single NMOS switch. Note, however, that adding the PMOS transistor increases the value of C.

When v_I goes low, as shown in Fig. 10.29(b), Q_N and Q_P interchange roles. Analysis of the circuit in Fig. 10.29(b) will indicate that Q_P will cease conduction when v_O falls to $|V_{tp}|$,

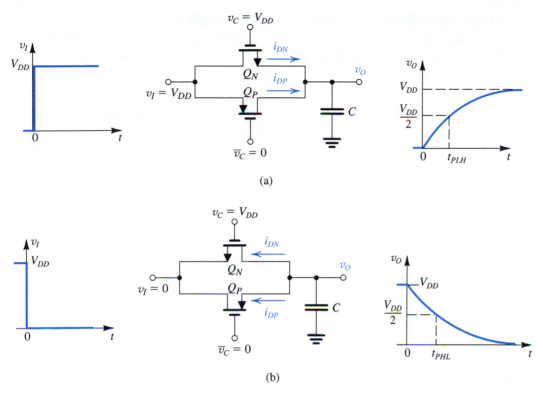

FIGURE 10.29 Operation of the transmission gate as a switch in PTL circuits with **(a)** v_I high and **(b)** v_I low.

where $|V_{tp}|$ is given by

$$|V_{tp}| = V_{t0} + \gamma[\sqrt{V_{DD} - v_O + 2\phi_f} - \sqrt{2\phi_f}] \quad (10.52)$$

Transistor Q_N, however, continues to conduct until C is fully discharged and $v_O = V_{OL} = 0$ V, a "good 0."

We conclude that transmission gates provide far superior performance, both static and dynamic, than is possible with single NMOS switches. The price paid is increased circuit complexity, area, and capacitance.

EXERCISE

10.8 The transmission gate of Figs. 10.29(a) and 10.29(b) is fabricated in a CMOS process technology for which $k'_n = 50$ μA/V^2, $k'_p = 20$ μA/V^2, $V_{tn} = |V_{tp}|$, $V_{t0} = 1$ V, $\gamma = 0.5$ V$^{1/2}$, $2\phi_f = 0.6$ V, and $V_{DD} = 5$ V. Let Q_N and Q_P be of the minimum size possible with this process technology, $(W/L)_n = (W/L)_p = 4$ μm/2 μm. The total capacitance at the output node is 70 fF. Utilize as many of the results of Example 10.4 as you need.

(a) For the situation in Fig. 10.29(a), find $i_{DN}(0)$, $i_{DP}(0)$, $i_{DN}(t_{PLH})$, $i_{DP}(t_{PLH})$, and t_{PLH}.

(b) For the situation depicted in Fig. 10.29(b), find $i_{DN}(0)$, $i_{DP}(0)$, $i_{DN}(t_{PHL})$, $i_{DP}(t_{PHL})$, and t_{PHL}. At what value of v_O will Q_P turn off?

(c) Find t_P.

Ans. (a) 800 μA, 320 μA, 50 μA, 275 μA, 0.24 ns; (b) 800 μA, 320 μA, 688 μA, 20 μA, 0.19 ns, 1.6 V; (c) 0.22 ns

10.5.4 Pass-Transistor Logic Circuit Examples

We conclude this section by showing examples of PTL logic circuits. Figure 10.30 shows a PTL realization of a two-to-one multiplexer: Depending on the logic value of C, either A or B is connected to the output Y. The circuit realizes the Boolean function

$$Y = CA + \overline{C}B$$

Our second example is an efficient realization of the exclusive-OR (XOR) function. The circuit, shown in Fig. 10.31, utilizes four transistors in the transmission gates and another four for the two inverters needed to generate the complements $\overline{A}$ and $\overline{B}$, for a total of eight transistors. Note that 12 transistors are needed in the realization with complementary CMOS.

Our final PTL example is the circuit shown in Fig. 10.32. It uses NMOS switches with low or zero threshold. Observe that both the input variables and their complements are

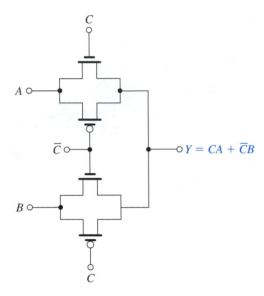

FIGURE 10.30 Realization of a two-to-one multiplexer using pass-transistor logic.

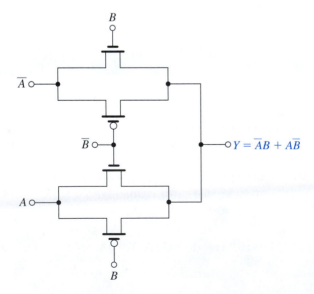

FIGURE 10.31 Realization of the XOR function using pass-transistor logic.

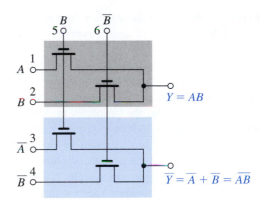

FIGURE 10.32 An example of a pass-transistor logic gate utilizing both the input variables and their complements. This type of circuit is therefore known as complementary pass-transistor logic or CPL. Note that both the output function and its complement are generated.

employed and that the circuit generates both the Boolean function and its complement. Thus this form of circuit is known as **complementary pass-transistor logic (CPL).** The circuit consists of two identical networks of pass transistors with the corresponding transistor gates controlled by the same signal (B and $\overline{B}$). The inputs to the PTL, however, are complemented: A and B for the first network, and $\overline{A}$ and $\overline{B}$ for the second. The circuit shown realizes both the AND and NAND functions.

EXERCISE

10.9 Consider the circuit in Fig. 10.32 with the input signals changed as follows. For each case, find Y and $\overline{Y}$:

(a) The signals at terminals 5 and 6 interchanged ($\overline{B}$ applied to 5 and B applied to 6). All the rest are the same.

(b) The signals at terminals 5 or 6 interchanged as in (a), and the signals at 2 and 4 changed to $\overline{A}$ and A, respectively. All the rest remain the same.

Ans. (a) $Y = A + B$, $\overline{Y} = \overline{A}\overline{B} = \overline{A + B}$ (i.e., OR–NOR); (b) $Y = A\overline{B} + \overline{A}B$, $\overline{Y} = \overline{A}\overline{B} + AB$ (i.e., XOR–XNOR)

10.5.5 A Final Remark

Although the use of zero-threshold devices solves the problem of the loss of signal levels when NMOS switches are used, the resulting circuits can be much more sensitive to noise and other effects, such as leakage currents resulting from subthreshold conduction.

10.6 DYNAMIC LOGIC CIRCUITS

The logic circuits that we have studied thus far are of the static type. In a static logic circuit, every node has, at all times, a low-resistance path to V_{DD} or ground. By the same token, the voltage of each node is well defined at all times, and no node is left floating. Static circuits do not need clocks (i.e., periodic timing signals) for their operation, although clocks may be present for other purposes. In contrast, the dynamic logic circuits we are about to discuss rely on the storage of signal voltages on parasitic capacitances at certain circuit nodes. Since charge will leak away with time, the circuits need to be *periodically refreshed;* thus the presence of a clock with a certain specified minimum frequency is essential.

To place dynamic-logic-circuit techniques into perspective, let's take stock of the various logic-circuit styles we have studied. Complementary CMOS excels in nearly every performance category: It is easy to design, has the maximum possible logic swing, is robust from a noise-immunity standpoint, dissipates no static power, and can be designed to provide equal low-to-high and high-to-low propagation delays. Its main disadvantage is the requirement of two transistors for each additional gate input, which for high fan-in gates can make the chip area large and increase the total capacitance and, correspondingly, the propagation delay and the dynamic power dissipation. Pseudo-NMOS reduces the number of required transistors at the expense of static power dissipation. Pass-transistor logic can result in simple small-area circuits but is limited to special applications and requires the use of complementary inverters to restore signal levels, especially when the switches are simple NMOS transistors. The dynamic logic techniques studied in this section maintain the low device count of pseudo-NMOS while reducing the static power dissipation to zero. As will be seen, this is achieved at the expense of more complex, and less robust, design.

10.6.1 Basic Principle

Figure 10.33(a) shows the basic dynamic-logic gate. It consists of a pull-down network (PDN) that realizes the logic function in exactly the same way as the PDN of a complementary CMOS gate or a pseudo-NMOS gate. Here, however, we have two switches in series that are periodically operated by the clock signal ϕ whose waveform is shown in Fig. 10.33(b). When ϕ is low, Q_p is turned on, and the circuit is said to be in the setup or **precharge phase.** When ϕ is high, Q_p is off and Q_e turns on, and the circuit is in the **evaluation phase.** Finally, note that C_L denotes the total capacitance between the output node and ground.

During precharge, Q_p conducts and charges capacitance C_L so that, at the end of the precharge interval, the voltage at Y is equal to V_{DD}. Also during precharge, the inputs A, B, and C are allowed to change and settle to their proper values. Observe that because Q_e is off, no path to ground exists.

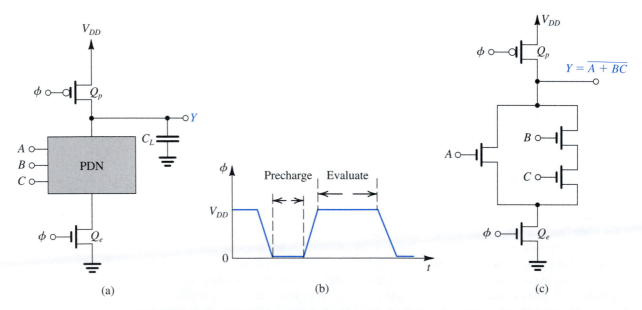

(a)   (b)   (c)

FIGURE 10.33 (a) Basic structure of dynamic-MOS logic circuits. (b) Waveform of the clock needed to operate the dynamic logic circuit. (c) An example circuit.

During the evaluation phase, Q_p is off and Q_e is turned on. Now, if the input combination is one that corresponds to a high output, the PDN does not conduct (just as in a complementary CMOS gate) and the output remains high at V_{DD}, thus $V_{OH} = V_{DD}$. Observe that no low-to-high propagation delay is required, thus $t_{PLH} = 0$. On the other hand, if the combination of inputs is one that corresponds to a low output, the appropriate NMOS transistors in the PDN will conduct and establish a path between the output node and ground through the on-transistor Q_e. Thus C_L will be discharged through the PDN, and the voltage at the output node will reduce to $V_{OL} = 0$ V. The high-to-low propagation delay t_{PHL} can be calculated in exactly the same way as for a complementary CMOS circuit except that here we have an additional transistor, Q_e, in the series path to ground. Although this will increase the delay slightly, the increase will be more than offset by the reduced capacitance at the output node as a result of the absence of the PUN.

As an example, we show in Fig. 10.33(c) the circuit that realizes the function $Y = \overline{A + BC}$. Sizing of the PDN transistors often follows the same procedure employed in the design of static CMOS. For Q_p, we select a W/L ratio large enough to ensure that C_L will be fully charged during the precharge interval. The size of Q_p, however, should be small so that the capacitance C_L will not be increased significantly. This is a ratioless form of MOS logic, where the output levels do not depend on the transistors' W/L ratios.

EXERCISES

Consider a four-input NAND gate realized in the dynamic logic form and fabricated in a CMOS process technology for which $\mu_n C_{ox} = 50$ μA/V^2, $\mu_p C_{ox} = 20$ μA/V^2, $V_{tn} = |V_{tp}| = 1$ V, $V_{DD} = 5$ V. To keep C_L small, minimum-size NMOS devices are used for which $W/L = 4$ μm/2 μm (this includes Q_e). The PMOS precharge transistor Q_p has a $W/L = 6$ μm/2 μm. The total capacitance C_L is found to be 30 fF.

10.10 Consider the precharge operation with the gate of Q_p falling to 0 V, and assume that at $t = 0$, C_L is fully discharged. We wish to calculate the rise time of the output voltage, defined as the time for v_Y to rise from 10% to 90% of the final value of 5 V. Find the current at $v_Y = 0.5$ V and the current at $v_Y = 4.5$ V, then compute an approximate value for t_r, $t_r = C_L(4.5 - 0.5)/I_{av}$, where I_{av} is the average value of the two currents.

Ans. 480 μA; 112 μA; 0.4 ns

10.11 Next, consider the computation of the high-to-low propagation delay t_{PHL}. Find the equivalent W/L ratio of the five NMOS transistors in series. Then, find the discharge current at $v_Y = 5$ V and at $v_Y = 2.5$ V. Finally, use the average of these two currents to compute an approximate value for t_{PHL}.

Ans. $(W/L)_{eq} = 0.4$; 160 μA; 138 μA; 0.5 ns

10.6.2 Nonideal Effects

We now briefly consider various sources of nonideal operation of dynamic logic circuits.

Noise Margins   Since, during the evaluation phase, the NMOS transistors begin to conduct for $v_I = V_{tn}$,

$$V_{IL} \cong V_{IH} \cong V_{tn}$$

and thus the noise margins will be

$$NM_L = V_{tn}$$
$$NM_H = V_{DD} - V_{tn}$$

Thus the noise margins are far from equal, and NM_L is rather low. Although NM_H is high, other nonideal effects reduce its value, as we shall shortly see. At this time, however, observe that the output node is a high-impedance node and thus will be susceptible to noise pickup and other disturbances.

Output Voltage Decay Due to Leakage Effects

In the absence of a path to ground through the PDN, the output voltage will ideally remain high at V_{DD}. This, however, is based on the assumption that the charge on C_L will remain intact. In practice, there will be leakage current that will cause C_L to slowly discharge and v_Y to decay. The principal source of leakage is the reverse current of the reverse-biased junction between the drain diffusion of transistors connected to the output node and the substrate. Such currents can be in the range of 10^{-12} A to 10^{-15} A, and they increase rapidly with temperature (approximately doubling for every $10°C$ rise in temperature). Thus the circuit can malfunction if the clock is operating at a very low frequency and the output node is not "refreshed" periodically. This exact same point will be encountered when we study dynamic memory cells in Chapter 11.

Charge Sharing

There is another and often more serious way for C_L to lose some of its charge and thus cause v_Y to fall significantly below V_{DD}. To see how this can happen, refer to Fig. 10.34(a), which shows only Q_1 and Q_2, the two top transistors of the PDN, together with the precharge transistor Q_p. Here, C_1 is the capacitance between the common node of Q_1 and Q_2 and ground. At the beginning of the evaluation phase, after Q_p has turned off and with C_L charged to V_{DD} (Fig. 10.34a), we assume that C_1 is initially discharged and that the inputs are such that at the gate of Q_1 we have a high signal, whereas at the gate of Q_2 the signal is low. We can easily see that Q_1 will turn on, and its drain current, i_{D1}, will flow as indicated.

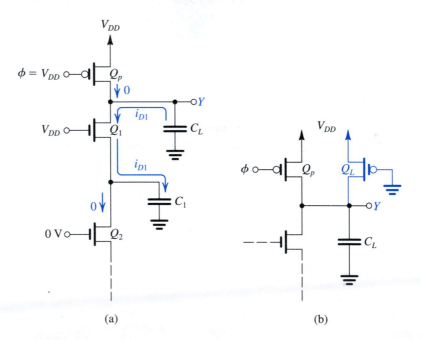

(a)   (b)

FIGURE 10.34 (a) Charge sharing. (b) Adding a permanently turned-on transistor Q_L solves the charge-sharing problem at the expense of static power dissipation.

Thus i_{D1} will discharge C_L and charge C_1. Although eventually i_{D1} will reduce to zero, C_L will have lost some of its charge, which will have been transferred to C_1. This phenomenon is known as charge sharing.

We shall not pursue the problem of charge sharing any further here, except to point out a couple of the techniques usually employed to minimize its effect. One approach involves adding a p-channel device that continuously conducts a small current to replenish the charge lost by C_L, as shown in Fig. 10.34(b). This arrangement should remind us of pseudo-NMOS. Indeed, adding this transistor will cause the gate to dissipate static power. On the positive side, however, the added transistor will lower the impedance level of the output node and make it less susceptible to noise as well as solving the leakage and charge-sharing problems. Another approach to solving the charge-sharing problem is to precharge the internal nodes, that is, to precharge capacitor C_1. The price paid in this case is increased circuit complexity and node capacitances.

Cascading Dynamic Logic Gates A serious problem arises if one attempts to cascade dynamic logic gates. Consider the situation depicted in Fig. 10.35, where two single-input dynamic gates are connected in cascade. During the precharge phase, C_{L1} and C_{L2} will be charged through Q_{p1} and Q_{p2}, respectively. Thus, at the end of the precharge interval, $v_{Y1} = V_{DD}$ and $v_{Y2} = V_{DD}$. Now consider what happens in the evaluation phase for the case of high input A. Obviously, the correct result will be Y_1 low ($v_{Y1} = 0$ V) and Y_2 high ($v_{Y2} = V_{DD}$). What happens, however, is somewhat different. As the evaluation phase begins, Q_1 turns on and C_{L1} begins to discharge. However, simultaneously, Q_2 turns on and C_{L2} also begins to discharge. Only when v_{Y1} drops below V_{tn} will Q_2 turn off. Unfortunately, however, by that time, C_{L2} will have lost a significant amount of its charge, and v_{Y2} will be less than the expected value of V_{DD}. (Here, it is important to note that in dynamic logic, once charge has been lost, it cannot be recovered.) This problem is sufficiently serious to make simple cascading an impractical proposition. As usual, however, the ingenuity of circuit designers has come to the rescue, and a number of schemes have been proposed to make cascading possible in dynamic-logic circuits. We shall discuss one such scheme after considering Exercise 10.12.

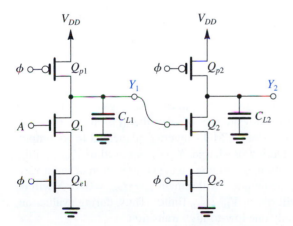

FIGURE 10.35 Two single-input dynamic logic gates connected in cascade. With the input A high, during the evaluation phase C_{L2} will partially discharge and the output at Y_2 will fall lower than V_{DD}, which can cause logic malfunction.

EXERCISE

10.12 To gain further insight into the cascading problem described, let us determine the decrease in the output voltage v_{Y2} for the circuit in Fig. 10.35. Specifically, consider the circuit as the evaluation phase begins: At $t = 0$, $v_{Y1} = v_{Y2} = V_{DD}$ and $v_\phi = v_A = V_{DD}$. Transistors Q_{p1} and Q_{p2} are cut off and can be removed from the equivalent circuit. Furthermore, for the purpose of this approximate analysis, we can replace the series combination of Q_1 and Q_{e1} with a single device having an appropriate W/L, and similarly for the combination of Q_2 and Q_{e2}. The result is the approximate equivalent circuit in Fig. E10.12. We are interested in the operation of this circuit in the interval Δt during which v_{Y1} falls from V_{DD} to V_t at which time Q_{eq2} turns off and C_{L2} stops discharging. Assume that the process technology has the parameter values specified in Example 10.4; that for all NMOS transistors in the circuit of Fig. 10.35, $W/L =$ 4 μm/2 μm, and that $C_{L1} = C_{L2} = 40$ fF.

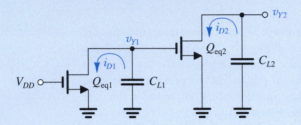

FIGURE E10.12

(a) Find $(W/L)_{eq1}$ and $(W/L)_{eq2}$.

(b) Find the values of i_{D1} at $v_{Y1} = V_{DD}$ and at $v_{Y1} = V_t$. Hence determine an average value for i_{D1}.

(c) Use the average value of i_{D1} found in (b) to determine an estimate for the interval Δt.

(d) Find the average value of i_{D2} during Δt. To simplify matters, take the average to be the value of i_{D2} obtained when the gate voltage v_{Y1} is midway through its excursion (i.e., $v_{Y1} = 3$ V). (*Hint:* Q_{eq2} will remain in saturation.)

(e) Use the value of Δt found in (c) together with the average value of i_{D2} determined in (d) to find an estimate of the reduction in v_{Y2} during Δt. Hence determine the final value of v_{Y2}.

Ans. (a) 1, 1; (b) 400 μA and 175 μA, for an average value of 288 μA; (c) 0.56 ns; (d) 100 μA; (e) $\Delta v_{Y2} =$ 1.4 V, thus v_{Y2} decreases to 3.6 V

10.6.3 Domino CMOS Logic

Domino CMOS logic is a form of dynamic logic that results in cascadable gates. Figure 10.36 shows the structure of the Domino CMOS logic gate. We observe that it is simply the basic dynamic-logic gate of Fig. 10.33(a) with a static CMOS inverter connected to its output. Operation of the gate is straightforward. During precharge, X will be raised to V_{DD}, and the gate output Y will be at 0 V. During evaluation, depending on the combination of input variables, either X will remain high and thus the output Y will remain low ($t_{PHL} = 0$) or X will be brought down to 0 V and the output Y will rise to V_{DD} (t_{PLH} finite). Thus, during evaluation, the output either remains low or makes only one low-to-high transition.

To see why Domino CMOS gates can be cascaded, consider the situation in Fig. 10.37(a), where we show two Domino gates connected in cascade. For simplicity, we show single-input gates. At the end of precharge, X_1 will be at V_{DD}, Y_1 will be at 0 V, X_2 will be at V_{DD}, and Y_2 will be at 0 V. As in the preceding case, assume A is high at the beginning of evaluation.

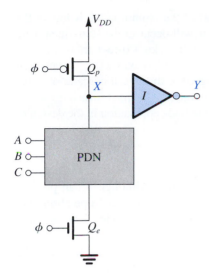

FIGURE 10.36 The Domino CMOS logic gate. The circuit consists of a dynamic-MOS logic gate with a static-CMOS inverter connected to the output. During evaluation, Y either will remain low (at 0 V) or will make one 0-to-1 transition (to V_{DD}).

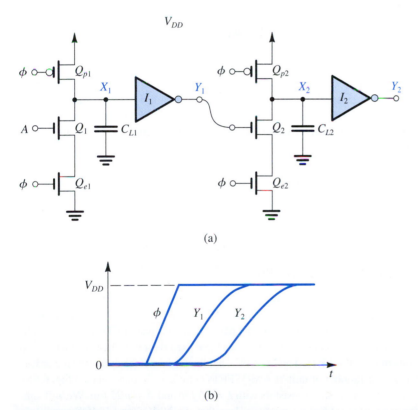

(a)

(b)

FIGURE 10.37 (a) Two single-input Domino CMOS logic gates connected in cascade. (b) Waveforms during the evaluation phase.

Thus, as ϕ goes up, capacitor C_{L1} will begin discharging, pulling X_1 down. Meanwhile, the low input at the gate of Q_2 keeps Q_2 off, and C_{L2} remains fully charged. When v_{X1} falls below the threshold voltage of inverter I_1, Y_1 will go up turning Q_2 on, which in turn begins to discharge C_{L2} and pulls X_2 low. Eventually, Y_2 rises to V_{DD}.

From this description, we see that because the output of the Domino gate is low at the beginning of evaluation, no premature capacitor discharge will occur in the subsequent gate in the cascade. As indicated in Fig. 10.37(b), output Y_1 will make a 0-to-1 transition t_{PLH} seconds after the rising edge of the clock. Subsequently, output Y_2 makes a 0-to-1 transition after another t_{PLH} interval. The propagation of the rising edge through a cascade of gates resembles contiguously placed dominoes falling over, each toppling the next, which is the origin of the name Domino CMOS logic. Domino CMOS logic finds application in the design of address decoders in memory chips, for example.

10.6.4 Concluding Remarks

Dynamic logic presents many challenges to the circuit designer. Although it can provide considerable reduction in the chip-area requirement, as well as high-speed operation, and zero (or little) static-power dissipation, the circuits are prone to many nonideal effects, some of which have been discussed here. It should also be remembered that dynamic-power dissipation is an important issue in dynamic logic. Another factor that should be considered is the "dead time" during precharge when the output of the circuit is not yet available.

10.7 SPICE SIMULATION EXAMPLE

We conclude this chapter with an example illustrating the use of SPICE in the analysis of CMOS digital circuits. To appreciate the need for SPICE, recall that throughout this chapter we have had to make many simplifying assumptions so that manual analysis can be made possible and also so that the results can be sufficiently simple to yield design insight. This is especially the case in the analysis of the dynamic operation of logic circuits. Computer-aided analysis using SPICE not only obviates the need to make approximations, thus providing accurate results, but it also allows the use of more precise MOSFET models. Such models, of course, are too complex to use in manual analysis.

EXAMPLE 10.5

OPERATION OF THE CMOS INVERTER

In this example, we will use PSpice to simulate the CMOS inverter whose Capture schematic is shown in Fig. 10.38. We will assume a 0.5-μm CMOS technology for the MOSFETs and use parts NMOS0P5 and PMOS0P5 whose level-1 model parameters are listed in Table 4.8. In addition to the channel length L and the channel width W, we have used the multiplicative factor m to specify the dimensions of the MOSFETs. The MOSFET parameter m, whose default value is 1, is used in SPICE to specify the number of unit-size MOSFETs connected in parallel (see Fig. 6.65). In our simulations, we will use unit-size transistors with $L = 0.5$ μm and $W = 1.25$ μm. We will simulate the inverter for two cases: (a) setting $m_p/m_n = 1$ so that the NMOS and PMOS transistors have equal widths, and (b) setting $m_p/m_n = \mu_n/\mu_p = 4$ so that the PMOS transistor is four times wider than the NMOS transistor (to compensate for the lower mobility in p-channel devices as compared with n-channel ones). Here, m_n and m_p are the multiplicative factors of, respectively, the NMOS and PMOS transistors of the inverter.

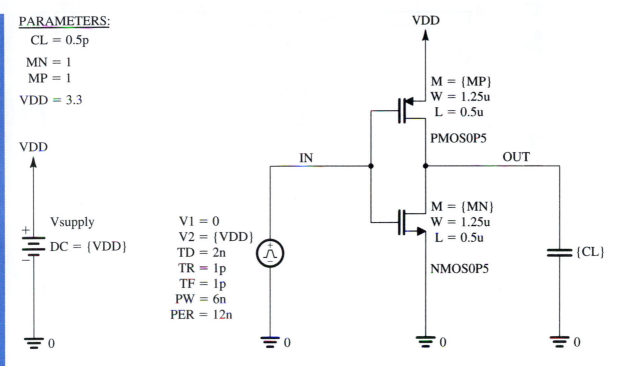

PARAMETERS:
  CL = 0.5p

MN = 1
MP = 1

VDD = 3.3

VDD

$+$ Vsupply
     DC = {VDD}
$-$

V1 = 0
V2 = {VDD}
TD = 2n
TR = 1p
TF = 1p
PW = 6n
PER = 12n

VDD

M = {MP}
W = 1.25u
L = 0.5u

PMOS0P5

IN   OUT

M = {MN}
W = 1.25u
L = 0.5u

NMOS0P5

{CL}

0   0   0   0

FIGURE 10.38 Capture schematic of the CMOS inverter in Example 10.5.

To compute both the voltage transfer characteristic (VTC) of the inverter and its supply current at various values of the input voltage V_{in}, we apply a dc voltage source at the input and perform a dc analysis with V_{in} swept over the range 0 to V_{DD}. The resulting VTC is plotted in Fig. 10.39. Note that the slope of the VTC in the switching region (where the NMOS and PMOS devices are both in saturation) is not infinite as predicted from the simple theory presented earlier (Section 4.10, Fig. 4.55). Rather, the nonzero value of λ causes the inverter gain to be finite. Using the derivative feature of Probe, we can find the two points on the VTC at which the inverter gain is unity (i.e., the VTC slope is -1 V/V) and, hence, determine V_{IL} and V_{IH}. Using the results given in Fig. 10.39, the corresponding noise margins are $NM_L = NM_H = 1.34$ V for the inverter with $m_p/m_n = 4$, while $NM_L = 0.975$ V and $NM_H = 1.74$ V for the inverter with $m_p/m_n = 1$. Observe that these results correlate reasonably well with the values obtained using the approximate formula in Eq. (10.8). Furthermore, note that, with $m_p/m_n = \mu_n/\mu_p = 4$, the NMOS and PMOS devices are closely matched and, hence, the two noise margins are equal.

The threshold voltage V_{th} of the CMOS inverter is defined as the input voltage v_{IN} that results in an identical output voltage v_{OUT}, that is,

$$V_{th} = v_{IN}\big|_{v_{OUT} = v_{IN}} \tag{10.53}$$

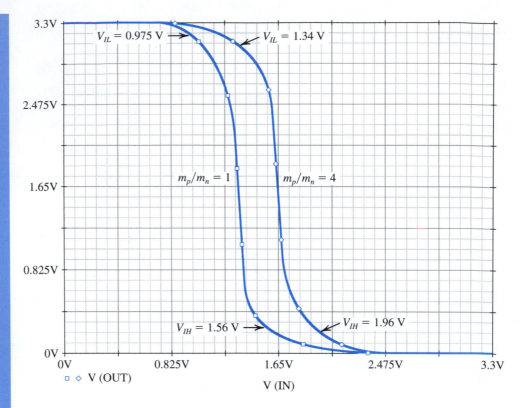

FIGURE 10.39 Input–output voltage transfer characteristic (VTC) of the CMOS inverter in Example 10.5 with $m_p/m_n = 1$ and $m_p/m_n = 4$.

Thus, as shown in Fig. 10.40, V_{th} is the intersection of the VTC with the straight line corresponding to $v_{OUT} = v_{IN}$ (this line can be simply generated in Probe by plotting v_{IN} versus v_{OUT}, as shown in Fig. 10.40). Note that $V_{th} \approx V_{DD}/2$ for the inverter with $m_p/m_n = 4$. Furthermore, decreasing m_p/m_n decreases V_{th} (see earlier: Exercise 4.44). Figure 10.40 also shows the inverter supply current versus v_{IN}. Observe that the location of the supply-current peak shifts with the threshold voltage.

To investigate the dynamic operation of the inverter with PSpice, we apply a pulse signal at the input (Fig. 10.38), perform a transient analysis, and plot the input and output waveforms as shown in Fig. 10.41. The rise and fall times of the pulse source are chosen to be very short. Note that increasing m_p/m_n from 1 to 4 decreases t_{PLH} (from 1.13 ns to 0.29 ns) because of the increased current available to charge C_L, with only a minor increase in t_{PHL} (from 0.33 ns to 0.34 ns). The two propagation delays, t_{PLH} and t_{PHL}, are not exactly equal when $m_p/m_n = 4$ because the NMOS and PMOS transistors are still not perfectly matched (e.g., $V_{tn} \neq |V_{tp}|$).

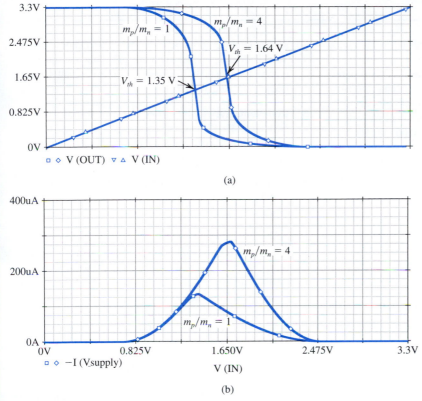

(a)

(b)

FIGURE 10.40 (a) Output voltage, and (b) supply current versus input voltage for the CMOS inverter in Example 10.5 with $m_p/m_n = 1$ and $m_p/m_n = 4$.

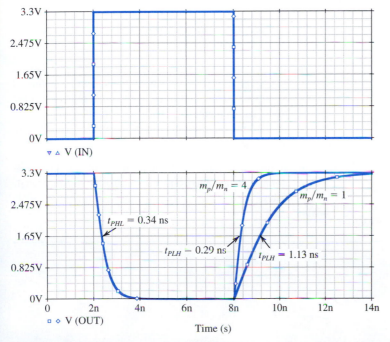

FIGURE 10.41 Transient response of the CMOS inverter in Example 10.5 with $m_p/m_n = 1$ and $m_p/m_n = 4$.

SUMMARY

- Although CMOS is one of four digital IC technologies currently in use (the others are bipolar, BiCMOS and GaAs), it is the most popular. This is due to its zero static-power dissipation and excellent static and dynamic characteristics. Further, advances in CMOS process technology have made possible the fabrication of MOS transistors with channel lengths as small as $0.06\ \mu$m. The high input impedance of MOS transistors allows the use of charge storage on capacitors as a means of realizing memory, a technique successfully exploited in both dynamic logic and dynamic memory.

- The CMOS inverter is usually designed using the minimum channel length for both the NMOS and PMOS transistors. The width of the NMOS transistor is usually 1.5 to 2 times L, and the width of the PMOS device is (μ_n/μ_p) times that. This latter (matching) condition ensures that the inverter will switch at $V_{DD}/2$ and gives equal current-driving capabilities in both directions and hence symmetrical propagation delays.

- A simple technique for determining the propagation delay of a logic gate is to determine the average current I_{av} available to charge (or discharge) a load capacitance C. Then, t_{PLH} (or t_{PHL}) is determined as $C(V_{DD}/2)/I_{av}$.

- A complementary CMOS logic gate consists of an NMOS pull-down network (PDN) and a PMOS pull-up network (PUN). The PDN conducts for every input combination that requires a low output. Since an NMOS transistor conducts when its input is high, the PDN is most directly synthesized from the expression for the low output ($\bar{Y}$) as a function of the uncomplemented inputs. In a complementary fashion, the PUN conducts for every input combination that corresponds to a high output. Since a PMOS conducts when its input is low, the PUN is most directly synthesized from the expression for a high output (Y) as a function of the complemented inputs.

- CMOS logic circuits are usually designed to provide equal current-driving capability in both directions. Furthermore, the worst-case value of the pull-up and pull-down currents are made equal to those of the basic (matched) inverter. Transistor sizing is based on this principle and makes use of the equivalent (W/L) ratios of series and parallel devices (Eqs. 10.27 and 10.28).

- Complementary CMOS logic utilizes two transistors, an NMOS and a PMOS, for each input variable. Thus the circuit complexity, silicon area, and parasitic capacitance all increase with fan-in.

- To reduce the device count, two other forms of static CMOS, namely, pseudo-NMOS and pass-transistor logic (PTL), are employed in special applications as supplements to complementary CMOS.

- Pseudo-NMOS utilizes the same PDN as in complementary CMOS logic but replaces the PUN with a single PMOS transistor whose gate is grounded. Unlike complementary CMOS, pseudo-NMOS is a ratioed form of logic in which V_{OL} is determined by the ratio r of k_n to k_p. Normally, r is selected in the range 4 to 10 and its value determines the noise margins.

- Pseudo-NMOS has the disadvantage of dissipating static power when the output of the logic gate is low. Static power can be eliminated by turning the PMOS load on for only a brief interval, known as the pre-charge interval, to charge the output node to V_{DD}. Then the inputs are applied, and depending on the input combination, the output node either remains high or is discharged through the PDN. This is the essence of dynamic logic.

- Pass-transistor logic utilizes either single NMOS transistors or CMOS transmission gates to implement a network of switches that are controlled by the input logic variables. Switches implemented by single NMOS transistors, though simple, result in the reduction of V_{OH} from V_{DD} to $V_{DD} - V_t$.

- A particular form of dynamic logic circuits, known as domino logic, allows the cascading of dynamic logic gates.

PROBLEMS

SECTION 10.1: DIGITAL CIRCUIT DESIGN: AN OVERVIEW

10.1 For a logic-circuit family employing a 3-V supply, suggest an ideal set of values for V_{th}, V_{IL}, V_{IH}, V_{OL}, V_{OH}, NM_L, NM_H. Also, sketch the VTC. What value of voltage gain in the transition region does your ideal specification imply?

10.2 For a particular logic-circuit family, the basic technology used provides an inherent limit to the small-signal low-frequency voltage gain of 50 V/V. If, with a 3.3-V supply, the values of V_{OL} and V_{OH} are ideal, but $V_{th} = 0.4 V_{DD}$, what are the best possible values of V_{IL} and V_{IH} that can be expected? What are the best possible noise margins you could expect? If the actual noise margins are only 7/10 of these values, what V_{IL} and V_{IH} result? What is the large-signal voltage gain [defined as $(V_{OH} - V_{OL})/(V_{IL} - V_{IH})$]. (*Hint:* Use straight-line approximations for the VTC.)

*10.3 A logic-circuit family intended for use in a digital-signal-processing application in a newly developed hearing aid can operate down to single-cell supply voltages of 1.2 V. If for its inverter, the output signals swing between 0 and V_{DD}, the "gain-of-one" points are separated by less than $1/3 V_{DD}$, and the noise margins are within 30% of one another, what ranges of values of V_{IL}, V_{IH}, V_{OL}, V_{OH}, NM_L, and NM_H can you expect for the lowest possible battery supply?

10.4 In a particular logic family, the standard inverter, when loaded by a similar circuit, has a propagation delay specified to be 1.2 ns:

(a) If the current available to charge a load capacitance is half as large as that available to discharge the capacitance, what do you expect t_{PLH} and t_{PHL} to be?
(b) If when an external capacitive load of 1 pF is added at the inverter output, its propagation delays increase by 70%, what do you estimate the normal combined capacitance of inverter output and input to be?
(c) If without the additional 1-pF load connected, the load inverter is removed and the propagation delays were observed to decrease by 40%, estimate the two components of the capacitance found in (b) that is, the component due to the inverter output and other associated parasitics, and the component due to the input of the load inverter?

10.5 In a particular logic family, operating with a 3.3-V supply, the basic inverter draws (from the supply) a current of 40 μA in one state and 0 μA in the other. When the inverter is switched at the rate of 100 MHz, the average supply current

becomes 150 μA. Estimate the equivalent capacitance at the output node of the inverter.

10.6 A collection of logic gates for which the static-power dissipation is zero, and the dynamic-power dissipation, as specified by Eq. (10.4), is 10 mW are operated at 50 MHz with a 5-V supply. By what fraction could the power dissipation be reduced if operation at 3.3 V were possible? If the frequency of operation is reduced by the same factor as the supply voltage (i.e., 3.3/5), what *additional* power can be saved?

D10.7 A logic-circuit family with zero static-power dissipation normally operates at $V_{DD} = 5$ V. To reduce its dynamic-power dissipation, which is specified by Eq. (10.4), operation at 3.3 V is considered. It is found, however, that the currents available to charge and discharge load capacitances also decrease. If current is (a) proportional to V_{DD}, or (b) proportional to V_{DD}^2, what reductions in maximum operating frequency do you expect in each case? What fractional change in delay–power product do you expect is each case?

D*10.8 Reconsider the situation described in Problem 10.7, for the situation in which a threshold relation exists such that the current depends on $(V_{DD} - V_t)$ rather than V_{DD} directly. Evaluate the change of current, propagation delay, operating frequency, dynamic power, and delay–power product as a result of decreasing V_{DD} from 5 V to 3.3 V. Assume that the currents are proportional to (a) $(V_{DD} - V_t)$, or (b) $(V_{DD} - V_t)^2$, for V_t equal to (i) 1 V and (ii) 0.5 V.

D*10.9 Consideration is being given to reducing by 10% all dimensions, including oxide thickness, of a silicon digital CMOS process. Recall that for a MOS device the available current is related to

$$i = \frac{1}{2}\mu C_{ox}\frac{W}{L}(V_{DD} - V_t)^2$$

where $C_{ox} = \varepsilon_{ox}/t_{ox}$. Also assume that the total effective capacitance that determines the propagation delay is divided about equally between MOS capacitances that are proportional to area and inversely proportional to oxide thickness, and reverse-bias junction capacitances that are proportional to area. Find the factors by which the following parameters change: chip area, current, effective capacitance, propagation delay, maximum operating frequency, dynamic power dissipation, delay–power product, and performance (in operations per unit area per second). If the supply voltage is also reduced by 10% (but V_t is not), what other changes result?

10

10.10 Consider an inverter for which t_{PLH}, t_{PHL}, t_{TLH}, and t_{THL} are 20 ns, 10 ns, 30 ns, and 15 ns, respectively. The rising and falling edges of the inverter output can be approximated by linear ramps. Two such inverters are connected in tandem and driven by an ideal input having zero rise and fall times. Calculate the time taken for the output voltage to complete 90% of its excursion for (a) a rising input and (b) a falling input. What is the propagation delay for the inverter?

10.11 A particular logic gate has t_{PLH} and t_{PHL} of 50 ns and 70 ns, respectively, and dissipates 1 mW with output low and 0.5 mW with output high. Calculate the corresponding delay–power product (under the assumption of a 50% duty-cycle signal and neglecting dynamic power dissipation).

SECTION 10.2: DESIGN AND PERFORMANCE ANALYSIS OF THE CMOS INVERTER

10.12 For a CMOS inverter operating from a 3.3-V supply in a technology for which $|V_t| = 0.8$ V, and $k'_n = 4k'_p = 180$ μA/V^2, evaluate the drain–source resistance associated with minimum-size transistors for which $W/L = 0.75$ μm/0.5 μm. For which ratio (W_p/W_n) will Q_N and Q_P, which have equal channel lengths, have equal resistances?

10.13 A CMOS inverter fabricated in the process specified in Problem 10.12 utilizes a p-channel device four times as wide as the n-channel device. If the V_{DD} supply is subject to very-high-frequency noise and there is an equivalent load capacitance of 1 pF, what is the 3-dB cutoff frequency embodied in each gate for this supply noise?

10.14 A CMOS inverter for which $k_n = 10k_p = 100\,\mu$A/V^2 and $V_t = 0.5$ V is connected as shown in Fig. P10.14 to a sinusoidal signal source having a Thévenin equivalent voltage of 0.1-V peak amplitude and resistance of 100 kΩ. What signal voltage appears at node A with $v_I = +1.5$ V? With $v_I = -1.5$ V?

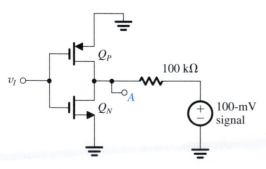

FIGURE P10.14

10.15 For a generalized CMOS inverter characterized by V_{tn}, V_{tp}, k_n, and k_p, derive the relation in Eq. (10.8) for V_{th}.

10.16 Use Eq. (10.8) to explore the variation of V_{th} with the ratio $r \equiv k_n/k_p$. Specifically, calculate V_{th} for the case $V_{tn} = |V_{tp}| = 0.5$ V and $V_{DD} = 2.5$ V for $r = 0.5$, 1, 1.5, 2, and 3. Note that V_{th} is not a strong function of r around the point $r = 1$.

D10.17 Design a "matched" inverter whose area is 15 μm^2 in a process for which the minimum length is 0.5 μm and $\mu_n/\mu_p = 3$. By what factor does the maximum output current available from this inverter exceed that of the minimum-size inverter for which the factor $n = 1.5$? What is the ratio of their areas? What is the ratio of their output resistances?

10.18 For a CMOS inverter having $k_n = k_p = 300$ μA/V^2, $V_{tn} = |V_{tp}| = 0.8$ V, $V_{DD} = 3.3$ V, and $\lambda_n = \lambda_p = 0.05$ V^{-1}, find V_{OH}, V_{IH}, V_{OL}, V_{IL}, NM_H, NM_L, V_{th}, and the voltage gain at the threshold point M. [Hint: The small-signal voltage gain is $-[(g_{mN} + g_{mP})(r_{oN}//r_{oP})]$.]

10.19 For a particular matched CMOS inverter, $k'_n = 75\,\mu$A/V^2, $(W/L)_n = 8\,\mu$m/0.8 μm, $\mu_n/\mu_p = 2.5$. The circuit has an equivalent output capacitance with two major components, one proportional to device width of 2-fF/μm width for each device, and the other fixed, at 50 fF. What total equivalent capacitance is associated with the output node? Calculate t_P using Eq. (10.13) for a supply of 3.3 V.

10.20 Use Eqs. (10.14) to (10.17) to derive an expression for t_{PHL} in which V_t is expressed as a fraction α of V_{DD} (i.e., $V_t = \alpha V_{DD}$). Find the value of the multiplier in the numerator of the expression, for α in the range 0.1 to 0.5 (e.g., for $\alpha = 0.2$ the multiplier is 1.7).

10.21 Find the propagation delay for a minimum-size inverter for which $k'_n = 3k'_p = 180$ μA/V^2 and $(W/L)_n = (W/L)_p = 0.75$ μm/0.5 μm, $V_{DD} = 3.3$ V, and the capacitance is roughly 2 fF/μm of device width plus 1 fF/device. What does t_P become if the design is changed to a matched one?

10.22 A CMOS microprocessor chip containing the equivalent of 1 million gates operates from a 5-V supply. The power dissipation is found to be 9 W when the chip is operating at 120 MHz, and 4.7 W when operating at 50 MHz. What is the power lost in the chip by some clock-independent mechanism, such as leakage and other static currents? If 70% of the gates are assumed to be active at any time, what is the average gate capacitance in such a design?

10.23 A matched CMOS inverter fabricated in a process for which $C_{ox} = 3.7$ fF/μm^2, $\mu_n C_{ox} = 180$ μA/V^2, $\mu_p C_{ox} = 45$ μA/V^2, $V_{tn} = -V_{tp} = 0.7$ V, and $V_{DD} = 3.3$ V, uses $W_n = 0.75$ μm and $L_n = L_p = 0.5$ μm. The gate–drain overlap capacitance and effective drain–body capacitance per micrometer of gate width are 0.4 fF and 1.0 fF, respectively. The wiring capacitance is $C_w = 2$ fF. Find t_{PLH}, t_{PHL}, and t_P. For how much additional capacitance load does the propagation delay increase by 50%?

10.24 Repeat Problem 10.23 for an inverter for which $(W/L)_n = (W/L)_p = 0.75 \, \mu m / 0.5 \, \mu m$. Find t_P and the dynamic power dissipation when the circuit is operated at a 250-MHz rate.

SECTION 10.3: CMOS LOGIC-GATE CIRCUITS

D10.25 Sketch a CMOS realization for the function $Y = \overline{A + B(C + D)}$.

D10.26 A CMOS logic gate is required to provide an output $Y = \overline{A}BC + A\overline{B}C + AB\overline{C}$. How many transistors does it need? Sketch a suitable PUN and PDN, obtaining each first independently, then one from the other using the dual-networks idea.

D10.27 Give two different realizations of the exclusive-OR function $Y = A\overline{B} + \overline{A}B$ in which the PDN and the PUN are dual networks.

D10.28 Sketch a CMOS logic circuit that realizes the function $Y = AB + \overline{A}\overline{B}$. This is called the equivalence or coincidence function.

D10.29 Sketch a CMOS logic circuit that realizes the function $Y = ABC + \overline{A}\overline{B}\overline{C}$.

D10.30 It is required to design a CMOS logic circuit that realizes a three-input even-parity checker. Specifically, the output Y is to be low when an even number (0 or 2) of the inputs A, B, and C are high.

(a) Give the Boolean function $\overline{Y}$.
(b) Sketch a PDN directly from the expression for $\overline{Y}$. Note that it requires 12 transistors in addition to those in the inverters.
(c) From inspection of the PDN circuit, reduce the number of transistors to 10.

(d) Find the PUN as a dual of the PDN in (c), and hence the complete realization.

D10.31 Give a CMOS logic circuit that realizes the function of three-input odd-parity checker. Specifically, the output is to be high when an odd number (1 or 3) of the inputs are high. Attempt a design with 10 transistors (not counting those in the inverters) in each of the PUN and the PDN.

D10.32 Design a CMOS full-adder circuit with inputs A, B, and C, and two outputs S and C_0 such that S is 1 if one or three inputs are 1, and C_0 is 1 if two or more inputs are 1.

D10.33 Consider the CMOS gate shown in Fig. 10.14. Specify W/L ratios for all transistors in terms of the ratios n and p of the basic inverter, such that the worst-case t_{PHL} and t_{PLH} of the gate are equal to those of the basic inverter.

D10.34 Find appropriate sizes for the transistors used in the exclusive-OR circuit of Fig. 10.15(b). Assume that the basic inverter has $(W/L)_n = 0.75 \, \mu m / 0.5 \, \mu m$ and $(W/L)_p = 3.0 \, \mu m / 0.5 \, \mu m$. What is the total area, including that of the required inverters?

10.35 Consider a four-input CMOS NAND gate for which the transient response is dominated by a fixed-size capacitance between the output node and ground. Compare the values of t_{PLH} and t_{PHL}, obtained when the devices are sized as in Fig. 10.17, to the values obtained when all n-channel devices have $W/L = n$ and all p-channel devices have $W/L = p$.

10.36 Figure P10.36 shows two approaches to realizing the OR function of six input variables. The circuit in Fig. P10.36(b), though it uses additional transistors, has in fact

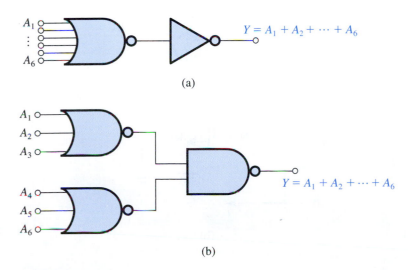

(a)

(b)

FIGURE P10.36

less total area and lower propagation delay because it uses NOR gates with lower fan-in. Assuming that the transistors in both circuits are properly sized to provide each gate with a current-driving capability equal to that of the basic matched inverter, find the number of transistors and the total area of each circuit. Assume the basic inverter to have a $(W/L)_n$ ratio of 1.2 μm/0.8 μm and a $(W/L)_p$ ratio of 3.6 μm/0.8 μm.

*10.37 Consider the two-input CMOS NOR gate of Fig. 10.12 whose transistors are properly sized so that the current-driving capability in each direction is equal to that of a matched inverter. For $|V_t| = 1$ V and $V_{DD} = 5$ V, find the gate threshold in the cases for which (a) input terminal A is connected to ground and (b) the two input terminals are tied together. Neglect the body effect in Q_{PB}.

SECTION 10.4: PSEUDO-NMOS LOGIC CIRCUITS

10.38 The purpose of this problem is to compare the value of t_{PLH} obtained with a resistive load (see Fig. P10.38a) to that obtained with a current–source load (see Fig. P10.38b). For a fair comparison, let the current source $I = V_{DD}/R_D$, which is the initial current available to charge the capacitor in the case of a resistive load. Find t_{PLH} for each case, and hence the percentage reduction obtained when a current–source load is used.

D*10.39 Design a pseudo-NMOS inverter that has equal positive and negative capacitive-driving output currents at $v_O = V_{DD}/4$ for use in a system with $V_{DD} = 5$ V, $|V_t| = 0.8$ V, $k'_n = 3k'_p = 75$ μA/V^2, and $(W/L)_n = 1.2$ μm/0.8 μm. What are the values of $(W/L)_p$, V_{IL}, V_{IH}, V_M, V_{OH}, V_{OL}, NM_H, and NM_L?

10.40 Consider a pseudo-NMOS inverter with $r = 2$, $(W/L)_n = 1.2$ μm/0.8 μm, $V_{DD} = 5$ V, $|V_t| = 0.8$ V, and $k'_n = 3k'_p = 75$ μA/V^2. Let the device capacitances per micrometer of device width be $C_{gs} = 1.5$ fF, $C_{gd} = 0.5$ fF, and $C_{db} = 2$ fF. Estimate the input and output capacitances and the values of t_{PLH}, t_{PHL}, and t_P obtained when the inverter is driving another identical inverter. Also find the corresponding values for a complementary CMOS inverter with a matched design.

*10.41 Use Eq. (10.41) to find the value of r for which NM_L is maximized. What is the corresponding value of NM_L?

D10.42 Design a pseudo-NMOS inverter that has $V_{OL} = 0.1$ V. Let $V_{DD} = 2.5$ V, $|V_t| = 0.4$ V, $k'_n = 4k'_p = 120$ μA/V^2, and $(W/L)_n = 0.375$ μm/0.25 μm. What is the value of $(W/L)_p$? Calculate the values of NM_L and the static power dissipation.

10.43 For what value of r does NM_H of a pseudo-NMOS inverter become zero? Prepare a table of NM_H versus r, for $r = 1$ to 16.

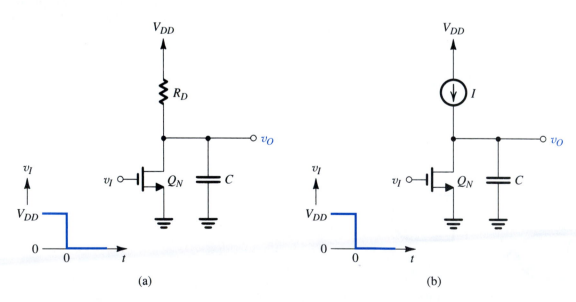

10.44 For a pseudo-NMOS inverter, what value of r results in $NM_L = NM_H$. Let $V_{DD} = 5$ V and $|V_t| = 0.8$ V. What is the resulting margin?

D*10.45 It is required to design a minimum-area pseudo-NMOS inverter with equal high and low noise margins using a 5-V supply and devices for which $|V_t| = 0.8$ V, $k'_n = 3k'_p = 75$ μA/V^2, and the minimum-size device has $(W/L) = 1.2$ μm/0.8 μm. Use $r = 2.72$ and show that $NM_L \approx NM_H$. Specify the values of $(W/L)_n$ and $(W/L)_p$. What is the power dissipated in this gate? What is the ratio of propagation delays for high and low transitions? For an external capacitive load of 1 pF, and neglecting the much smaller device capacitances, find t_{PLH}, t_{PHL}, and t_P. At what frequency of operation would the static and dynamic power levels be equal? Is this speed of operation possible in view of the t_P value you found? What is the ratio of dynamic power to static power at what you may assume is the maximum usable operating frequency [say, $1/(2t_{PLH} + 2t_{PHL})$]?

D10.46 Sketch a pseudo-NMOS realization of the function $Y = \overline{A + B(C + D)}$.

D10.47 Sketch a pseudo-NMOS realization of the exclusive-OR function $Y = A\overline{B} + \overline{A}B$.

D10.48 Consider a four-input pseudo-NMOS NOR gate in which the NMOS devices have $(W/L)_n = (1.8$ μm/1.2 μm$)$. It is required to find $(W/L)_p$ so that the worst-case value of V_{OL} is 0.2 V. Let $V_{DD} = 5$ V, $|V_t| = 0.8$ V, and $3k'_p = 75$ μA/V^2.

SECTION 10.5: PASS-TRANSISTOR LOGIC CIRCUITS

***10.49** A designer, beginning to experiment with the idea of pass-transistor logic, seizes upon what he sees as two good ideas:

(a) that a string of minimum-size single MOS transistors can do complex logic functions, but
(b) that there must always be a path between output and a supply terminal.

Correspondingly, he first considers two circuits (shown in Fig. P10.49). For each, express Y as a function of A and B. In each case, what can be said about general operation? About the logic levels at Y? About node X? Do either of these circuits look familiar? If in each case the terminal connected to V_{DD} is instead connected to the output of a CMOS inverter whose input is connected to a signal C, what does the function Y become?

10.50 Consider the circuits in Fig. P10.49 with all PMOS transistors replaced with NMOS, and all NMOS by PMOS,

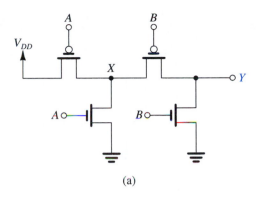

(a)

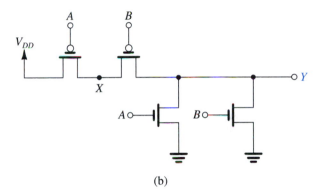

(b)

FIGURE P10.49

and with ground and V_{DD} connections interchanged. What do the output functions Y become?

***10.51** Is the circuit in Fig. P10.51 a satisfactory pass-transistor circuit? What are its deficiencies? What is Y as a function of A, B, C, D? What does the output become if the two V_{DD} connections are driven by a CMOS inverter with input E?

***10.52** An NMOS pass-transistor switch with $W/L = 1.2$ μm/0.8 μm, used in a 3.3-V system for which $V_{t0} = 0.8$ V, $\gamma = 0.5$ V$^{1/2}$, $2\phi_f = 0.6$ V, $\mu_n C_{ox} = 3\mu_p C_{ox} = 75$ μA/V^2, drives a 100-fF load capacitance at the input of a matched static inverter using $(W/L)_n = 1.2$ μm/0.8 μm. For the switch gate terminal at V_{DD}, evaluate the switch V_{OH} and V_{OL} for inputs at V_{DD} and 0 V, respectively. For this value of V_{OH}, what inverter static current results? Estimate t_{PLH} and t_{PHL} for this arrangement as measured from the input to the output of the switch itself.

***D10.53** The purpose of this problem is to design the level-restoring circuit of Fig. 10.28 and gain insight into its

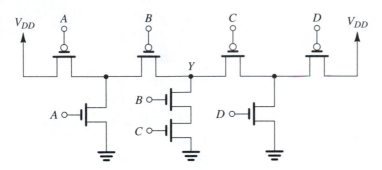

FIGURE P10.51

operation. Assume that $k'_n = 3k'_p = 75 \ \mu A/V^2$, $V_{DD} = 3.3$ V, $|V_{t0}| = 0.8$ V, $\gamma = 0.5$ V$^{1/2}$, $2\phi_f = 0.6$ V, $(W/L)_1 = (W/L)_n = 1.2 \ \mu m/0.8 \ \mu m$, $(W/L)_p = 3.6 \ \mu m/0.8 \ \mu m$, and $C = 20$ fF. Let $v_B = V_{DD}$.

(a) Consider first the situation with $v_A = V_{DD}$. Find the value of the voltage v_{O1} that causes v_{O2} to drop a threshold voltage below V_{DD}, that is, to 2.5 V so that Q_R turns on. At this value of v_{O1}, find V_t of Q_1. What is the capacitor-charging current available at this time? What is it at $v_{O1} = 0$? What is the average current available for charging C? Estimate t_{PLH} from the input to v_{O1}.

(b) Now, to determine a suitable W/L ratio for Q_R, consider the situation when v_A is brought down to 0 V and Q_1 conducts and begins to discharge C. The voltage v_{O1} will begin to drop. Meanwhile, v_{O2} is still low and Q_R is conducting. The current that Q_R conducts subtracts from the current of Q_1, reducing the current available to discharge C. Find the value of v_{O1} at which the inverter begins to switch. This is $V_{IH} = \frac{1}{8}(5V_{DD} - 2V_t)$. Then, find the current that Q_1 conducts at this value of v_{O1}. Choose W/L for Q_R so that the current it conducts is limited to one half the value of the current in Q_1. What is the W/L you have chosen? Estimate t_{PHL} as the time for v_{O1} to drop from V_{DD} to V_{IH}.

D10.54 (a) Use the idea embodied in the exclusive-OR realization in Fig. 10.31 to realize $\bar{Y} = AB + \bar{A}\bar{B}$. That is, find a realization for $\bar{Y}$ using two transmission gates.
(b) Now combine the circuit obtained in (a) with the circuit in Fig. 10.31 to obtain a realization of the function $Z = \bar{Y}C + Y\bar{C}$, where C is a third input. Sketch the complete 12-transistor circuit realization of Z. Note that Z is a three-input exclusive-OR.

*D10.55 Using the idea presented in Fig. 10.32, sketch a CPL circuit whose outputs are $Y = A\bar{B} + \bar{A}B$ and $\bar{Y} = AB + \bar{A}\bar{B}$.

D10.56 Extend the CPL idea in Fig. 10.32 to three variables to form $Z = ABC$ and $\bar{Z} = \overline{ABC} = \bar{A} + \bar{B} + \bar{C}$.

SECTION 10.6: DYNAMIC-LOGIC CIRCUITS

D10.57 Based on the basic dynamic-logic circuit of Fig. 10.33, sketch complete circuits for NOT, NAND, and NOR gates, the latter two with two inputs, and a circuit for which $\bar{Y} = AB + CD$.

10.58 In this and the following problem, we investigate the dynamic operation of a two-input NAND gate realized in the dynamic-logic form and fabricated in a CMOS process technology for which $k'_n = 3k'_p = 75 \ \mu A/V^2$, $V_{tn} = -V_{tp} = 0.8$ V, and $V_{DD} = 3$ V. To keep C_L small, minimum-size NMOS devices are used for which $W/L = 1.2 \ \mu m/0.8 \ \mu m$ (this includes Q_e). The PMOS precharge transistor Q_p has 2.4 $\mu m/0.8 \ \mu m$. The capacitance C_L is found to be 15 fF. Consider the precharge operation with the gate of Q_p at 0 V, and assume that at $t = 0$, C_L is fully discharged. We wish to calculate the rise time of the output voltage, defined as the time for v_Y to rise from 10% to 90% of the final value of 3 V. Find the current at $v_Y = 0.3$ V and the current at $v_Y = 2.7$ V, then compute an approximate value for t_r, $t_r = C_L(2.7 - 0.3)/I_{av}$, where I_{av} is the average value of the two currents.

10.59 For the gate specified in Problem 10.58, evaluate the high-to-low propagation delay, t_{PHL}. To obtain an approximate value of t_{PHL}, replace the three series NMOS transistors with an equivalent device and find the average discharge current.

*10.60 In this problem, we wish to calculate the reduction in the output voltage of a dynamic logic gate as a result of charge redistribution. Refer to the circuit in Fig. 10.34(a), and assume that at $t = 0-$, $v_Y = V_{DD}$, and $v_{C1} = 0$. At $t = 0$, ϕ goes high and Q_P turns off, and simultaneously the voltage at the gate of Q_1 goes high (to V_{DD}) turning Q_1 on. Transistor Q_1 will remain conducting until either the voltage at its source (v_{C1}) reaches $V_{DD} - V_{tn}$ or until $v_Y = v_{C1}$, whichever comes first. In both cases, the final value of v_Y can be found using charge conservation. For $V_{tn} = 1$ V, $V_{DD} = 5$ V, $C_L = 30$ fF, and neglecting the body effect in Q_1, find the drop in voltage at the output in the two cases: (a) $C_1 = 5$ fF and (b) $C_1 = 10$ fF (such that Q_1 remains in saturation during its entire conduction interval).

10.61 The leakage current in a dynamic-logic gate causes the capacitor C_L to discharge during the evaluation phase, even if the PDN is not conducting. For $C_L = 30$ fF, and $I_{leakage} = 10^{-12}$ A, find the longest allowable evaluate time if the decay in output voltage is to be limited to 0.5 V. If the precharge interval is much shorter than the maximum allowable evaluate time, find the minimum clocking frequency required.

10.62 For the four-input dynamic-logic NAND gate analyzed in Exercises 10.10 and 10.11, estimate the maximum clocking frequency allowed.

10

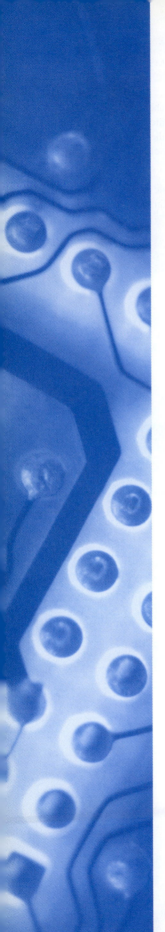

PART III

SELECTED TOPICS

INTRODUCTION

To round out our study of electronic circuits we have selected, from among the many possible somewhat-specialized topics, four to include in the third and final part of this book.

Chapter 11 deals with the important subject of digital memory. In addition, two advanced digital-circuit technologies—ECL and BiCMOS—are studied. The material in Chapter 11 follows naturally the study of logic circuits, presented in Chapter 10. Together, these two chapters should provide a preparation sufficient for advanced courses on digital electronics and VLSI design.

The subsequent two chapters, 12 and 13, have an applications or systems orientation: Chapter 12 deals with the design of filters, which are important building blocks of communications and instrumentation systems. Filter design is one of the rare areas of engineering for which a complete design theory exists, starting from specification and culminating in an actual working circuit. The material presented should allow the reader to perform such a complete design process.

In the design of electronic systems, the need usually arises for signals of various waveforms—sinusoidal, pulse, square-wave, etc. The generation of such signals is the subject of Chapter 13. It will be seen that some of the circuits utilized in waveform generation possess memory and are in fact the analog counterparts of the digital memory circuits studied in Chapter 11.

The material in Chapters 12 and 13 assumes knowledge of op amps (Chapter 2) and makes use of frequency response and related s-plane concepts (Chapter 6) and of feedback (Chapter 8).

The last of the four selected-topics chapters (Chapter 14) deals with the design of amplifiers that are required to deliver large amounts of load power; for example, the amplifier that drives the loudspeaker in a stereo system. As will be seen, the design of these high-power circuits is based on different considerations than those for small-signal amplifiers. Most of the material in Chapter 14 should be accessible to the reader who has studied Part I of this book.

Memory and Advanced Digital Circuits

INTRODUCTION

The logic circuits studied in Chapter 10 are called **combinational** (or *combinatorial*). Their output depends only on the present value of the input. Thus these circuits do *not* have memory. *Memory* is a very important part of digital systems. Its availability in digital computers allows for storing programs and data. Furthermore, it is important for temporary storage of the output produced by a combinational circuit for use at a later time in the operation of a digital system.

Logic circuits that incorporate memory are called **sequential circuits;** that is, their output depends not only on the present value of the input but also on the input's previous values. Such circuits require a timing generator (a *clock*) for their operation.

There are basically two approaches for providing memory to a digital circuit. The first relies on the application of positive feedback that, as will be seen shortly, can be arranged to provide a circuit with two stable states. Such a *bistable* circuit can then be used to store one bit of information: One stable state would correspond to a stored 0, and the other to a stored 1. A bistable circuit can remain in either state indefinitely, and thus belongs to the category of *static sequential circuits*. The other approach to realizing memory utilizes the storage of charge on a capacitor: When the capacitor is charged, it would be regarded as storing a 1; when it is discharged, it would be storing a 0. Since the inevitable leakage effects will cause the capacitor to discharge, such a form of memory requires the periodic recharging of the capacitor, a process known as *refresh*. Thus, like dynamic logic, memory based on charge storage is known as *dynamic memory* and the corresponding sequential circuits as *dynamic sequential circuits*.

In addition to the study of a variety of memory types and circuits in this chapter, we will also learn about two important digital-circuit technologies: Emitter-coupled logic (ECL), which utilizes bipolar transistors and achieves very high speeds of operation; and BiCMOS, which combines bipolar transistors and CMOS to great advantage.

11.1 LATCHES AND FLIP-FLOPS

In this section, we shall study the basic memory element, the latch, and consider a sampling of its applications. Both static and dynamic circuits will be considered.

11.1.1 The Latch

The basic memory element, the latch, is shown in Fig. 11.1(a). It consists of two cross-coupled logic inverters, G_1 and G_2. The inverters form a positive-feedback loop. To investigate the operation of the latch we break the feedback loop at the input of one of the inverters, say G_1, and apply an input signal, v_W, as shown in Fig. 11.1(b). Assuming that the input impedance of G_1 is large, breaking the feedback loop will not change the loop voltage transfer characteristic, which can be determined from the circuit of Fig. 11.1(b) by plotting v_Z versus v_W. This is the voltage transfer characteristic of two cascaded inverters and thus takes the shape shown in Fig. 11.1(c). Observe that the transfer characteristic consists of three segments, with the middle segment corresponding to the transition region of the inverters.

Also shown in Fig. 11.1(c) is a straight line with unity slope. This straight line represents the relationship $v_W = v_Z$ that is realized by reconnecting Z to W to close the feedback loop.

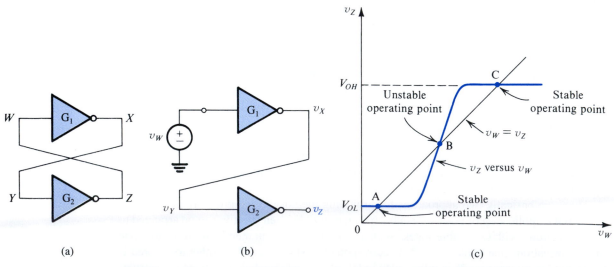

FIGURE 11.1 (a) Basic latch. (b) The latch with the feedback loop opened. (c) Determining the operating point(s) of the latch.

As indicated, the straight line intersects the loop transfer curve at three points, A, B, and C. Thus any of these three points can serve as the operating point for the latch. We shall now show that while points A and C are stable operating points in the sense that the circuit can remain at either indefinitely, point B is an unstable operating point; the latch cannot operate at B for any significant period of time.

The reason point B is unstable can be seen by considering the latch circuit in Fig. 11.1(a) to be operating at point B, and taking account of the electrical interference (or noise) that is inevitably present in any circuit. Let the voltage v_W increase by a small increment v_w. The voltage at X will increase (in magnitude) by a larger increment, equal to the product of v_w and the incremental gain of G_1 at point B. The resulting signal v_x is applied to G_2 and gives rise to an even larger signal at node Z. The voltage v_z is related to the original increment v_w by the loop gain at point B, which is the slope of the curve of v_Z versus v_W at point B. This gain is usually much greater than unity. Since v_z is coupled to the input of G_1, it will be further amplified by the loop gain. This regenerative process continues, shifting the operating point from B upward to point C. Since at C the loop gain is zero (or almost zero), no regeneration can take place.

In the description above, we assumed an initial positive voltage increment at W. Had we instead assumed a negative voltage increment, we would have seen that the operating point moves downward from B to A. Again, since at point A the slope of the transfer curve is zero (or almost zero), no regeneration can take place. In fact, for regeneration to occur the loop gain must be greater than unity, which is the case at point B.

The discussion above leads us to conclude that the latch has two stable operating points, A and C. At point C, v_W is high, v_X is low, v_Y is low, and v_Z is high. The reverse is true at point A. If we consider X and Z as the latch outputs, we see that in one of the stable states (say that corresponding to operating point A) v_X is high (at V_{OH}) and v_Z is low (at V_{OL}). In the other state (corresponding to operating point C) v_X is low (at V_{OL}) and v_Z is high (at V_{OH}). Thus the latch is a *bistable* circuit having two complementary outputs. The stable state in which the latch operates depends on the external excitation that forces it to the particular state. The latch then *memorizes* this external action by staying indefinitely in the acquired state. As a memory element the latch is capable of storing one bit of information. For instance, we can arbitrarily designate the state in which v_X is high and v_Z is low as corresponding to a stored logic 1. The other complementary state then is designated by a stored logic 0. Finally, it should be obvious that the latch circuit described is of the static variety.

It now remains to devise a mechanism by which the latch can be *triggered* to change state. The latch together with the triggering circuitry forms a *flip-flop*. This will be discussed next. Analog bistable circuits utilizing op amps will be presented in Chapter 13.

11.1.2 The SR Flip-Flop

The simplest type of flip-flop is the set/reset (SR) flip-flop shown in Fig. 11.2(a). It is formed by cross-coupling two NOR gates, and thus it incorporates a latch. The second input

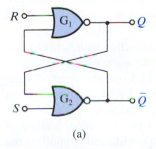

R	S	Q_{n+1}
0	0	Q_n
0	1	1
1	0	0
1	1	Not used

(a)   (b)

FIGURE 11.2 (a) The set/reset (SR) flip-flop and (b) its truth table.

of each NOR gate together serve as the trigger inputs of the flip-flop. These two inputs are labeled S (for set) and R (for reset). The outputs are labeled Q and $\overline{Q}$, emphasizing their complementarity. The flip-flop is considered to be set (i.e., storing a logic 1) when Q is high and $\overline{Q}$ is low. When the flip-flop is in the other state (Q low, $\overline{Q}$ high), it is considered to be reset (storing a logic 0).

In the rest or memory state (i.e., when we do not wish to change the state of the flip-flop), both the S and R inputs should be low. Consider the case when the flip-flop is storing a logic 0. Since Q will be low, both inputs to the NOR gate G_2 will be low. Its output will therefore be high. This high is applied to the input of G_1, causing its output Q to be low, satisfying the original assumption. To set the flip-flop we raise S to the logic-1 level while leaving R at 0. The 1 at the S terminal will force the output of G_2, $\overline{Q}$, to 0. Thus the two inputs to G_1 will be 0 and its output Q will go to 1. Now even if S returns to 0, the flip-flop remains in the newly acquired set state. Obviously, if we raise S to 1 again (with R remaining at 0) no change will occur. To reset the flip-flop we need to raise R to 1 while leaving $S = 0$. We can readily show that this forces the flip-flop into the reset state and that the flip-flop remains in this state even after R has returned to 0. It should be observed that the trigger signal merely starts the regenerative action of the positive-feedback loop of the latch.

Finally, we inquire into what happens if both S and R are simultaneously raised to 1. The two NOR gates will cause both Q and $\overline{Q}$ to become 0 (note that in this case the complementary labeling of these two variables is incorrect). However, if R and S return to the rest state ($R = S = 0$) simultaneously, the state of the flip-flop will be undefined. In other words, it will be impossible to predict the final state of the flip-flop. For this reason, this input combination is usually disallowed (i.e., not used). Note, however, that this situation arises only in the idealized case, when both R and S return to 0 precisely simultaneously. In actual practice one of the two will return to 0 first, and the final state will be determined by the input that remains high longest.

The operation of the flip-flop is summarized by the *truth table* in Fig. 11.2(b), where Q_n denotes the value of Q at time t_n just before the application of the R and S signals, and Q_{n+1} denotes the value of Q at time t_{n+1} after the application of the input signals.

Rather than using two NOR gates, one can also implement an SR flip-flop by cross-coupling two NAND gates in which case the set and reset functions are active when low and the inputs are correspondingly called $\overline{S}$ and $\overline{R}$.

11.1.3 CMOS Implementation of SR Flip-Flops

The SR flip-flop of Fig. 11.2 can be directly implemented in CMOS by simply replacing each of the NOR gates by its CMOS circuit realization. We encourage the reader to sketch the resulting circuit. Although the CMOS circuit thus obtained works well, it is somewhat complex. As an alternative, we consider a simplified circuit that furthermore implements additional logic. Specifically, Fig. 11.3 shows a *clocked* version of an SR flip-flop. Since the clock inputs form AND functions with the set and reset inputs, the flip-flop can be set or reset only when the clock ϕ is high. Observe that although the two cross-coupled inverters at the heart of the flip-flop are of the complementary CMOS type, only NMOS transistors are used for the set–reset circuitry. Nevertheless, since there is no conducting path between V_{DD} and ground (except during switching), the circuit does not dissipate any static power.

Except for the addition of clocking, the SR flip-flop of Fig. 11.3 operates in exactly the same way as its logic antecedent in Fig. 11.2: To illustrate, consider what happens when the flip-flop is in the reset state ($Q = 0$, $\overline{Q} = 1$, $v_Q = 0$, $v_{\overline{Q}} = V_{DD}$), and assume that we wish to set it. To do so, we arrange for a high (V_{DD}) signal to appear on the S input while R is held low at 0 V. Then, when the clock ϕ goes high, both Q_5 and Q_6 will conduct, pulling the

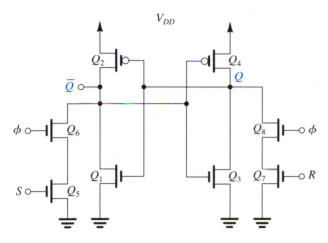

FIGURE 11.3 CMOS implementation of a clocked SR flip-flop. The clock signal is denoted by ϕ.

voltage $v_{\bar{Q}}$ down. If $v_{\bar{Q}}$ goes below the threshold of the (Q_3, Q_4) inverter, the inverter will switch states (or at least begin to switch states), and its output v_Q will rise. This increase in v_Q is fed back to the input of the (Q_1, Q_2) inverter, causing its output $v_{\bar{Q}}$ to go down even further; the regeneration process, characteristic of the positive-feedback latch, is now in progress.

The preceding description of flip-flop switching is predicated on two assumptions:

1. Transistors Q_5 and Q_6 supply sufficient current to pull the node $\bar{Q}$ down to a voltage at least slightly below the threshold of the (Q_3, Q_4) inverter. This is essential for the regenerative process to begin. Without this initial trigger, the flip-flop will fail to switch. In Example 11.1, we shall investigate the minimum W/L ratios that Q_5 and Q_6 must have to meet this requirement.

2. The set signal remains high for an interval long enough to cause regeneration to take over the switching process. An estimate of the minimum width required for the set pulse can be obtained as the sum of the interval during which $v_{\bar{Q}}$ is reduced from V_{DD} to $V_{DD}/2$, and the interval for the voltage v_Q to respond and rise to $V_{DD}/2$.

Finally, note that the symmetry of the circuit indicates that all the preceding remarks apply equally well to the reset process.

EXAMPLE 11.1

The CMOS SR flip-flop in Fig. 11.3 is fabricated in a process technology for which $\mu_n C_{ox} = 2.5\,\mu_p C_{ox} = 50\,\mu\text{A/V}^2$, $V_{tn} = |V_{tp}| = 1$ V, and $V_{DD} = 5$ V. The inverters have $(W/L)_n = 4\,\mu\text{m}/2\,\mu\text{m}$ and $(W/L)_p = 10\,\mu\text{m}/2\,\mu\text{m}$. The four NMOS transistors in the set–reset circuit have equal W/L ratios. Determine the minimum value required for this ratio to ensure that the flip-flop will switch.

Solution

Figure 11.4 shows the relevant portion of the circuit for our present purposes. Observe that since regeneration has not yet begun, we assume that $v_Q = 0$ and thus Q_2 will be conducting. The circuit is in effect a pseudo-NMOS gate, and our task is to select the W/L ratios for Q_5 and Q_6 so that V_{OL} of this inverter is lower than $V_{DD}/2$ (the threshold of the Q_3, Q_4 inverter whose Q_N and Q_P are

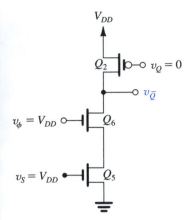

FIGURE 11.4 The relevant portion of the flip-flop circuit of Fig. 11.3 for determining the minimum W/L ratios of Q_5 and Q_6 needed to ensure that the flip-flop will switch.

matched). The minimum required W/L for Q_5 and Q_6 can be found by equating the current supplied by Q_5 and Q_6 to the current supplied by Q_2 at $v_{\bar{Q}} = V_{DD}/2$. To simplify matters, we assume that the series connection of Q_5 and Q_6 is approximately equivalent to a single transistor whose W/L is half the W/L of each of Q_5 and Q_6. Now, since at $v_{\bar{Q}} = V_{DD}/2$ both this equivalent transistor and Q_2 will be operating in the triode region, we can write

$$50 \times \frac{1}{2} \times \left(\frac{W}{L}\right)_5 \left[(5-1) \times \frac{5}{2} - \frac{1}{2} \times \left(\frac{5}{2}\right)^2\right] = 20 \times \frac{10}{2}\left[(5-1) \times \frac{5}{2} - \frac{1}{2} \times \left(\frac{5}{2}\right)^2\right]$$

which leads to

$$\left(\frac{W}{L}\right)_5 = 4 \quad \text{and} \quad \left(\frac{W}{L}\right)_6 = 4$$

Recalling that this is an absolute minimum value, we would in practice select a ratio of 5 or 6.

EXERCISES

11.1 Repeat Example 11.1 to determine the minimum required $(W/L)_5 = (W/L)_6$ so that switching is achieved when inputs S and ϕ are at $V_{DD}/2$.

Ans. 24.4

11.2 We wish to determine the minimum required width of the set pulse. Toward that end: (a) first consider the time for $v_{\bar{Q}}$ in the circuit of Fig. 11.4 to fall from V_{DD} to $V_{DD}/2$. Assume that the total capacitance between the $\bar{Q}$ node and ground is 50 fF. Determine the high-to-low propagation delay t_{PHL} by finding the average current available to discharge the capacitance over the voltage range V_{DD} to $V_{DD}/2$. Remember that Q_2 will be conducting a current that unfortunately reduces the current available to discharge C. Assume $(W/L)_5 = (W/L)_6 = 8$, and use the technology parameters given in Example 11.1. (b) Determine t_{PLH} for v_Q (Fig. 11.3) using the following formula:

$$t_{PLH} \cong \frac{1.7C}{k'_p \left(\dfrac{W}{L}\right)_p V_{DD}}$$

Assume a total node capacitance at Q of 50 fF. (c) What is the minimum width required of the set pulse?

Ans. (a) 0.11 ns; (b) 0.17 ns; (c) 0.28 ns

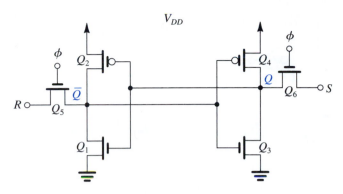

FIGURE 11.5 A simpler CMOS implementation of the clocked SR flip-flop. This circuit is popular as the basic cell in the design of static random-access memory (SRAM) chips.

11.1.4 A Simpler CMOS Implementation of the Clocked SR Flip-Flop

A simpler implementation of a clocked SR flip-flop is shown in Fig. 11.5. Here, pass-transistor logic is employed to implement the clocked set–reset functions. This circuit is very popular in the design of static random-access memory (SRAM) chips, where it is used as the basic memory cell (Section 11.4.1).

11.1.5 D Flip-Flop Circuits

A variety of flip-flop types exist and can be synthesized using logic gates. CMOS circuit implementations can be obtained by simply replacing the gates with their CMOS circuit realizations. This approach, however, usually results in rather complex circuits. In many cases, simpler circuits can be found by taking a circuit-design viewpoint, rather than a logic-design one. To illustrate this point, we shall consider the CMOS implementation of a very important type of flip-flop, the data, or D, flip-flop.

The D flip-flop is shown in block-diagram form in Fig. 11.6. It has two inputs, the data input D and the clock input ϕ. The complementary outputs are labeled Q and $\overline{Q}$. When the clock is low, the flip-flop is in the memory, or rest, state; signal changes on the D input line have no effect on the state of the flip-flop. As the clock goes high, the flip-flop acquires the logic level that existed on the D line just before the rising edge of the clock. Such a flip-flop is said to be **edge-triggered.** Some implementations of the D flip-flop include direct set and reset inputs that override the clocked operation just described.

A simple implementation of the D flip-flop is shown in Fig. 11.7. The circuit consists of two inverters connected in a positive-feedback loop, just as in the static latch of Fig. 11.1(a),

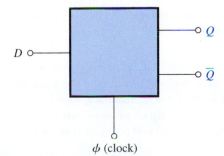

FIGURE 11.6 A block-diagram representation of the D flip-flop.

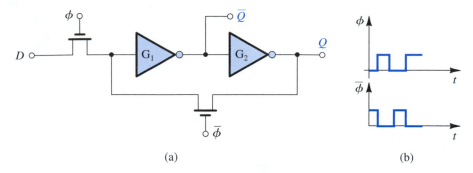

FIGURE 11.7 A simple implementation of the D flip-flop. The circuit in **(a)** utilizes the two-phase non-overlapping clock whose waveforms are shown in **(b)**.

except that here the loop is closed for only part of the time. Specifically, the loop is closed when the clock is low ($\phi = 0, \bar{\phi} = 1$). The input D is connected to the flip-flop through a switch that closes when the clock is high. Operation is straightforward: When ϕ is high, the loop is opened, and the input D is connected to the input of inverter G_1. The capacitance at the input node of G_1 is charged to the value of D, and the capacitance at the input node of G_2 is charged to the value of $\bar{D}$. Then, when the clock goes low, the input line is isolated from the flip-flop, the feedback loop is closed, and the latch acquires the state corresponding to the value of D just before ϕ went down, providing an output $Q = D$.

From the preceding, we observe that the circuit in Fig. 11.7 combines the positive-feedback technique of static bistable circuits and the charge-storage technique of dynamic circuits. It is important to note that the proper operation of this circuit, and of many circuits that use clocks, is predicated on the assumption that ϕ *and* $\bar{\phi}$ *will not be simultaneously high at any time*. This condition is defined by referring to the two clock phases as being *nonoverlapping*.

An inherent drawback of the D flip-flop implementation of Fig. 11.7 is that during ϕ, the output of the flip-flop simply follows the signal on the D input line. This can cause problems in certain logic design situations. The problem is solved very effectively by using the **master–slave** configuration shown in Fig. 11.8(a). Before discussing its circuit operation, we note that although the switches are shown implemented with single NMOS transistors, CMOS transmission gates are employed in many applications. We are simply using the single MOS transistor as a "shorthand notation" for a series switch.

The master–slave circuit consists of a pair of circuits of the type shown in Fig. 11.7, operated with alternate clock phases. Here, to emphasize that the two clock phases must be nonoverlapping, we denote them ϕ_1 and ϕ_2, and clearly show the nonoverlap interval in the waveforms of Fig. 11.8(b). Operation of the circuit is as follows:

1. When ϕ_1 is high and ϕ_2 is low, the input is connected to the master latch whose feedback loop is opened, while the slave latch is isolated. Thus, the output Q remains at the value stored previously in the slave latch whose loop is now closed. The node capacitances of the master latch are charged to the appropriate voltages corresponding to the present value of D.

2. When ϕ_1 goes low, the master latch is isolated from the input data line. Then, when ϕ_2 goes high, the feedback loop of the master latch is closed, locking in the value of D. Further, its output is connected to the slave latch whose feedback loop is now open. The node capacitances in the slave are appropriately charged so that when ϕ_1 goes high again the slave latch locks in the new value of D and provides it at the output, $Q = D$.

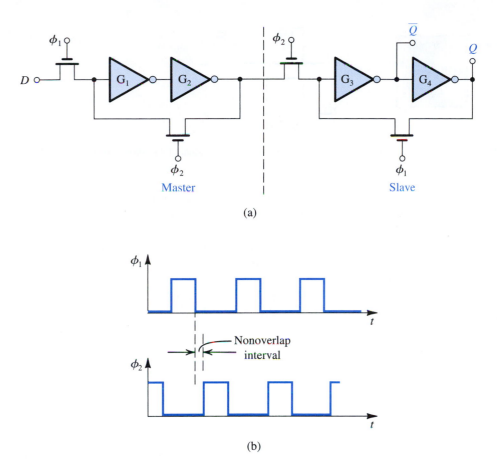

FIGURE 11.8 **(a)** A master–slave D flip-flop. The switches can be, and usually are, implemented with CMOS transmission gates. **(b)** Waveforms of the two-phase nonoverlapping clock required.

From this description, we note that at the positive transition of clock ϕ_2 the output Q adopts the value of D that existed on the D line at the end of the preceding clock phase, ϕ_1. This output value remains constant for one clock period. Finally, note that during the nonoverlap interval both latches have their feedback loops open and we are relying on the node capacitances to maintain most of their charge. It follows that the nonoverlap interval should be kept reasonably short (perhaps one-tenth or less of the clock period, and of the order of 1 ns or so in current practice).

 ## 11.2 MULTIVIBRATOR CIRCUITS

As mentioned before, the flip-flop has two stable states and is called a bistable multivibrator. There are two other types of multivibrator: monostable and astable. The **monostable multivibrator** has one stable state in which it can remain indefinitely. It has another *quasi-stable* state to which it can be triggered. The monostable multivibrator can remain in the quasi-stable state for a predetermined interval T, after which it automatically reverts to the stable state. In this way the monostable multivibrator generates an output pulse of duration T. This pulse duration is in no way related to the details of the triggering pulse, as is indicated schematically

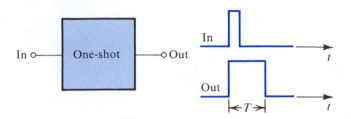

FIGURE 11.9 The monostable multivibrator (one-shot) as a functional block, shown to be triggered by a positive pulse. In addition, there are one shots that are triggered by a negative pulse.

in Fig. 11.9. The monostable multivibrator can therefore be used as a *pulse stretcher* or, more appropriately, a *pulse standardizer*. A monostable multivibrator is also referred to as a **one-shot.**

The **astable multivibrator** has no stable states. Rather, it has two quasi-stable states, and it remains in each for predetermined intervals T_1 and T_2. Thus after T_1 seconds in one of the quasi-stable states the astable switches to the other quasi-stable state and remains there for T_2 seconds, after which it reverts back to the original state, and so on. The astable multivibrator thus oscillates with a period $T = T_1 + T_2$ or a frequency $f = 1/T$, and it can be used to generate periodic pulses such as those required for clocking.

In Chapter 13 we will study astable and monostable multivibrator circuits that use op amps. In the following, we shall discuss monostable and astable circuits using logic gates. We also present an alternative, and very popular, oscillator circuit, the **ring oscillator.**

11.2.1 A CMOS Monostable Circuit

Figure 11.10 shows a simple and popular circuit for a monostable multivibrator. It is composed of two two-input CMOS NOR gates, G_1 and G_2, a capacitor of capacitance C, and a resistor of resistance R. The input source v_I supplies the triggering pulses for the monostable multivibrator.

Commercially available CMOS gates have a special arrangement of diodes connected at their input terminals, as indicated in Fig. 11.11(a). The purpose of these diodes is to prevent the input voltage signal from rising above the supply voltage V_{DD} (by more than one diode drop) and from falling below ground voltage (by more than one diode drop). These clamping diodes have an important effect on the operation of the monostable circuit. Specifically, we shall be interested in the effect of these diodes on the operation of the inverter-connected gate G_2. In this case, each pair of corresponding diodes appears in parallel, giving rise to the equivalent circuit in Fig. 11.11(b). While the diodes provide a low-resistance path to the power supply for voltages exceeding the power supply limits, the input current for intermediate voltages is essentially zero.

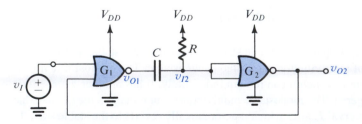

FIGURE 11.10 A monostable circuit using CMOS NOR gates. Signal source v_I supplies positive trigger pulses.

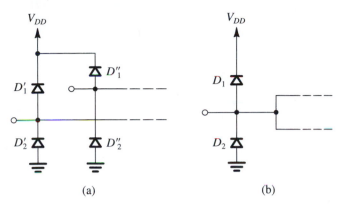

FIGURE 11.11 (a) Diodes at each input of a two-input CMOS gate. (b) Equivalent diode circuit when the two inputs of the gate are joined together. Note that the diodes are intended to protect the device gates from potentially destructive overvoltages due to static charge accumulation.

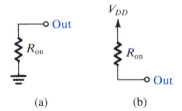

FIGURE 11.12 Output equivalent circuit of CMOS gate when the output is (a) low and (b) high.

To simplify matters we shall use the approximate output equivalent circuits of the gate, illustrated in Fig. 11.12. Figure 11.12(a) indicates that when the gate output is low, its output characteristics can be represented by a resistance R_{on} to ground, which is normally a few hundred ohms. In this state, current can flow from the external circuit into the output terminal of the gate; the gate is said to be *sinking* current. Similarly, the equivalent output circuit in Fig. 11.12(b) applies when the gate output is high. In this state, current can flow from V_{DD} through the output terminal of the gate into the external circuit; the gate is said to be *sourcing* current.

To see how the monostable circuit of Fig. 11.10 operates, consider the timing diagram given in Fig. 11.13. Here a short triggering pulse of duration τ is shown in Fig. 11.13(a). In the following we shall neglect the propagation delays through G_1 and G_2. These delays, however, set a lower limit on the pulse width τ, $\tau > (t_{P1} + t_{P2})$.

Consider first the stable state of the monostable circuit—that is, the state of the circuit before the trigger pulse is applied. The output of G_1 is high at V_{DD}, the capacitor is discharged, and the input voltage to G_2 is high at V_{DD}. Thus the output of G_2 is low, at ground voltage. This low voltage is fed back to G_1; since v_I also is low, the output of G_1 is high, as initially assumed.

Next consider what happens as the trigger pulse is applied. The output voltage of G_1 will go low. However, because G_1 will be sinking some current and because of its finite output resistance R_{on}, its output will not go all the way to 0 V. Rather, the output of G_1 drops by a value ΔV_1, which we shall shortly evaluate.

The drop ΔV_1 is coupled through C (which acts as a short circuit during the transient) to the input of G_2. Thus the input voltage of G_2 drops (from V_{DD}) by an identical amount ΔV_1. Here, we note that during the transient there will be an instantaneous current that flows from

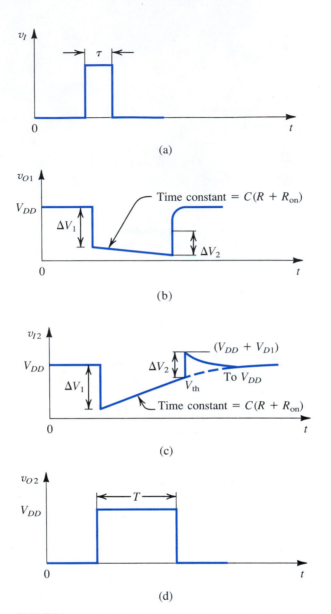

FIGURE 11.13 Timing diagram for the monostable circuit in Fig. 11.10.

V_{DD} through R and C and into the output terminal of G_1 to ground. We thus have a voltage divider formed by R and R_{on} (note that the instantaneous voltage across C is zero) from which we can determine ΔV_1 as

$$\Delta V_1 = V_{DD}\frac{R}{R+R_{on}} \tag{11.1}$$

Returning to G_2, we see that the drop of voltage at its input causes its output to go high (to V_{DD}). This signal keeps the output of G_1 low even after the triggering pulse has disappeared. The circuit is now in the quasi-stable state.

We next consider operation in the quasi-stable state. The current through R, C, and R_{on} causes C to charge, and the voltage v_{I2} rises exponentially toward V_{DD} with a time constant

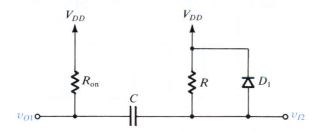

FIGURE 11.14 Circuit that applies during the discharge of C (at the end of the monostable pulse interval T).

$C(R + R_{on})$, as indicated in Fig. 11.13(c). The voltage v_{I2} will continue to rise until it reaches the value of the threshold voltage V_{th} of inverter G_2. At this time G_2 will switch and its output v_{O2} will go to 0 V, which will in turn cause G_1 to switch. The output of G_1 will attempt to rise to V_{DD}, but, as will become obvious shortly, its instantaneous rise will be limited to an amount ΔV_2. This rise in v_{O1} is coupled faithfully through C to the input of G_2. Thus the input of G_2 will rise by an equal amount ΔV_2. Note here that because of diode D_1, between the input of G_2 and V_{DD}, the voltage v_{I2} can rise only to $V_{DD} + V_{D1}$, where V_{D1} (approximately 0.7 V) is the drop across D_1. Thus from Fig. 11.13(c) we see that

$$\Delta V_2 = V_{DD} + V_{D1} - V_{th} \tag{11.2}$$

Thus it is diode D_1 that limits the size of the increment ΔV_2.

Because now v_{I2} is higher than V_{DD} (by V_{D1}), current will flow from the output of G_1 through C and then through the parallel combination of R and D_1. This current discharges C until v_{I2} drops to V_{DD} and v_{O1} rises to V_{DD}. The discharging circuit is depicted in Fig. 11.14, from which we note that the existence of the diode causes the discharging to be a nonlinear process. Although the details of the transient at the end of the pulse are not of immense interest, it is important to note that the monostable circuit should not be retriggered until the capacitor has been discharged, since otherwise the output obtained will not be the standard pulse, which the one-shot is intended to provide. The capacitor discharge interval is known as the *recovery time*.

An expression can be derived for the pulse interval T by referring to Fig. 11.13(c) and expressing $v_{I2}(t)$ as

$$v_{I2}(t) = V_{DD} - \Delta V_1 e^{-t/\tau_1}$$

where $\tau_1 = C(R + R_{on})$. Substituting for $t = T$ and $v_{I2}(T) = V_{th}$, and for ΔV_1 from Eq. (11.1) gives, after a little manipulation:

$$T = C(R + R_{on}) \ln\left(\frac{R}{R + R_{on}} \frac{V_{DD}}{V_{DD} - V_{th}}\right)$$

EXERCISES

11.3 For $V_{th} = V_{DD}/2$ and $R_{on} \ll R$, find an approximate expression for T.
Ans. $T = 0.69\,CR$

D11.4 If R_{on} is known to be less than 1 kΩ, use the approximation in Exercise 11.3 to design a one-shot that produces 10-μs pulses. Specify values for C and R. What is the maximum possible error in T due to neglecting R_{on} in the design?
Ans. Possible values are $C = 1$ nF, $R = 14.5$ kΩ; -3%

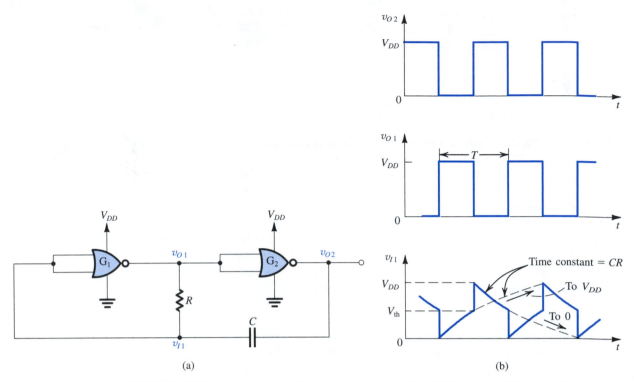

FIGURE 11.15 (a) A simple astable multivibrator circuit using CMOS gates. (b) Waveforms for the astable circuit in (a). The diodes at the gate input are assumed to be ideal and thus to limit the voltage v_{I1} to 0 and V_{DD}.

11.2.2 An Astable Circuit

Figure 11.15(a) shows a popular astable circuit composed of two inverter-connected NOR gates, a resistor, and a capacitor. We shall consider its operation, assuming that the NOR gates are of the CMOS family. However, to simplify matters we shall make some further approximations, neglecting the finite output resistance of the CMOS gate and assuming that the clamping diodes are ideal (thus have zero voltage drop when conducting).

With these simplifying assumptions, the waveforms of Fig. 11.15(b) are obtained. The reader is urged to consider the operation of this circuit in a step-by-step manner and verify that the waveforms shown indeed apply.[1]

EXERCISE

11.5 Using the waveforms in Fig. 11.15(b), derive an expression for the period T of the astable multivibrator of Fig. 11.15(a).

Ans. $T = CR \ln\left(\dfrac{V_{DD}}{V_{DD} - V_{th}} \dfrac{V_{DD}}{V_{th}}\right)$

---

[1] Practical circuits often use a large resistance in series with the input to G_1. This limits the effect of diode conduction and allows v_{I1} to rise to a voltage greater than V_{DD} and, as well, to fall below zero.

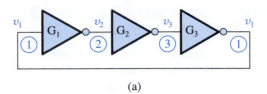

(a)

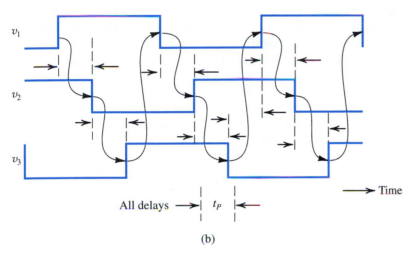

(b)

FIGURE 11.16 (a) A ring oscillator formed by connecting three inverters in cascade. (Normally at least five inverters are used.) (b) The resulting waveform. Observe that the circuit oscillates with frequency $1/6t_P$.

11.2.3 The Ring Oscillator

Another type of oscillator commonly used in digital circuits is the ring oscillator. It is formed by connecting an *odd* number of inverters in a loop. Although usually at least five inverters are used, we illustrate the principle of operation using a ring of three inverters, as shown in Fig. 11.16(a). Figure 11.16(b) shows the waveforms obtained at the outputs of the three inverters. These waveforms are idealized in the sense that their edges have zero rise and fall times. Nevertheless, they will serve to explain the circuit operation.

Observe that a rising edge at node 1 propagates through gates 1, 2, and 3 to return inverted after a delay of $3t_P$. This falling edge then propagates, and returns with the original (rising) polarity after another $3t_P$ interval. It follows that the circuit oscillates with a period of $6t_P$ or correspondingly with frequency $1/6t_P$. In general, a ring with N inverters (where N must be odd) will oscillate with period of $2Nt_P$ and frequency $1/2Nt_P$.

As a final remark, we note that the ring oscillator provides a relatively simple means for measuring the inverter propagation delay.

EXERCISE

11.6 Find the frequency of oscillation of a ring of five inverters if the inverter propagation delay is specified to be 1 ns.

Ans. 100 MHz

11.3 SEMICONDUCTOR MEMORIES: TYPES AND ARCHITECTURES

A computer system, whether a large machine or a microcomputer, requires memory for storing data and program instructions. Furthermore, within a given computer system there usually are various types of memory utilizing a variety of technologies and having different *access times*. Broadly speaking, computer memory can be divided into two types: **main memory** and **mass-storage** memory. The main memory is usually the most rapidly accessible memory and the one from which most, often all, instructions in programs are executed. The main memory is usually of the random-access type. A **random-access memory** (RAM) is one in which the time required for storing (writing) information and for retrieving (reading) information is independent of the physical location (within the memory) in which the information is stored.

Random-access memories should be contrasted with *serial* or *sequential* memories, such as disks and tapes, from which data are available only in the sequence in which the data were originally stored. Thus, in a serial memory the time to access particular information depends on the memory location in which the required information is stored, and the average access time is longer than the access time of random-access memory. In a computer system, serial memory is used for mass storage. Items not frequently accessed, such as large parts of the computer operating system, are usually stored in a *moving-surface memory* such as magnetic disk.

Another important classification of memory relates to whether it is a **read/write** or a **read-only memory.** Read/write (R/W) memory permits data to be stored and retrieved at comparable speeds. Computer systems require random-access read/write memory for data and program storage.

Read-only memories (ROM) permit reading at the same high speeds as R/W memories (or perhaps higher) but restrict the writing operation. ROMs can be used to store a microprocessor operating-system program. They are also employed in operations that require table lookup, such as finding the values of mathematical functions. A popular application of ROMs is their use in video game cartridges. It should be noted that read-only memory is usually of the random-access type. Nevertheless, in the digital circuit jargon, the acronym RAM usually refers to read/write, random-access memory, while ROM is used for read-only memory.

The regular structure of memory circuits has made them an ideal application for design of the circuits of the very-large-scale integrated (VLSI) type. Indeed, at any moment, memory chips represent the state of the art in packing density and hence integration level. Beginning with the introduction of the 1K-bit chip in 1970, memory-chip density has quadrupled about every 3 years. At the present time, chips containing 256M bits[2] are commercially available, while multigigabit memory chips are being tested in research and development laboratories. In this and the next two sections, we shall study some of the basic circuits employed in VLSI RAM chips. Read-only memory circuits are studied in Section 11.6.

11.3.1 Memory-Chip Organization

The bits on a memory chip are addressable either individually or in groups of 4 to 16. As an example, a 64M-bit chip in which all bits are individually addressable is said to be organized as 64M words $\times$ 1 bit (or simply 64M $\times$ 1). Such a chip needs a 26-bit address ($2^{26} = 67,108,864 = 64M$). On the other hand, the 64M-bit chip can be organized as 16M words $\times$ 4 bits

---

[2] The capacity of a memory chip to hold binary information as binary digits (or bits) is measured in K-bit and M-bit units, where 1K bit = 1024 bits and 1M bit = 1024 $\times$ 1024 = 1,048,576 bits. Thus a 64M-bit chip contains 67,108,864 bits of memory.

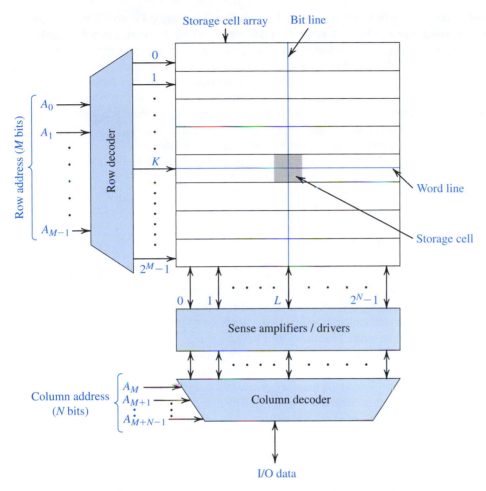

FIGURE 11.17 A 2^{M+N}-bit memory chip organized as an array of 2^M rows $\times$ 2^N columns.

(16M $\times$ 4), in which case a 24-bit address is required. For simplicity we shall assume in our subsequent discussion that all the bits on a memory chip are individually addressable.

The bulk of the memory chip consists of the cells in which the bits are stored. Each **memory cell** is an electronic circuit capable of storing one bit. We shall study memory-cell circuits in Section 11.4. For reasons that will become clear shortly, it is desirable to physically organize the storage cells on a chip in a square or a nearly square matrix. Figure 11.17 illustrates such an organization. The cell matrix has 2^M rows and 2^N columns, for a total storage capacity of 2^{M+N}. For example, a 1M-bit square matrix would have 1024 rows and 1024 columns ($M = N = 10$). Each cell in the array is connected to one of the 2^M row lines, known rather loosely, but universally, as **word lines,** and to one of the 2^N column lines, known as **digit lines** or, more commonly, **bit lines.** A particular cell is **selected** for reading or writing by activating its word line and its bit line.

Activating one of the 2^M word lines is performed by the **row decoder,** a combinational logic circuit that selects (raises the voltage of) the particular word line whose M-bit address is applied to the decoder input. The address bits are denoted $A_0, A_1, \ldots, A_{M-1}$. When the Kth word line is activated for, say, a read operation, all 2^N cells in row K will provide their contents to their respective bit lines. Thus, if the cell in column L (Fig. 11.17) is storing a 1, the voltage of bit-line number L will be raised, usually by a small voltage, say 0.1 V to 0.2 V. The readout voltage is small because the cell is small, a deliberate design decision, since the number of cells is very large. The small readout signal is applied to a sense amplifier connected to the bit line. As Fig. 11.17 indicates, there is a sense amplifier for every bit line. The sense amplifier provides a full-swing digital signal (from 0 to V_{DD}) at its output. This signal, together with the output signals from all the other cells in the selected row, is then delivered to the **column decoder.** The column decoder selects the signal of the particular column whose N-bit address is applied to the decoder input (the address bits are denoted $A_M, A_{M+1}, \ldots, A_{M+N-1}$) and causes this signal to appear on the chip input/output (I/O) data line.

A write operation proceeds in a similar manner: The data bit to be stored (1 or 0) is applied to the I/O line. The cell in which the data bit is to be stored is selected through the combination of its row address and its column address. The sense amplifier of the selected column acts as a **driver** to write the applied signal into the selected cell. Circuits for sense amplifiers and address decoders will be studied in Section 11.5.

Before leaving the topic of memory organization (or memory-chip architecture), we wish to mention a relatively recent innovation in organization dictated by the exponential increase in chip density. To appreciate the need for a change, note that as the number of cells in the array increases, the physical lengths of the word lines and the bit lines increase. This has occurred even though for each new generation of memory chips, the transistor size has decreased (currently, CMOS process technologies with 0.1–0.3 μm feature size are utilized). The net increase in word-line and bit-line lengths increases their total resistance and capacitance, and thus slows down their transient response. That is, as the lines lengthen, the exponential rise of the voltage of the word line becomes slower, and it takes longer for the cells to be activated. This problem has been solved by partitioning the memory chip into a number of blocks. Each of the blocks has an organization identical to that in Fig. 11.17. The row and column addresses are broadcast to all blocks, but the data selected come from only one of the blocks. Block selection is achieved by using an appropriate number of the address bits as a block address. Such an architecture can be thought of as three-dimensional: rows, columns, and blocks.

11.3.2 Memory-Chip Timing

The **memory access time** is the time between the initiation of a read operation and the appearance of the output data. The **memory cycle time** is the minimum time allowed between two consecutive memory operations. To be on the conservative side, a memory operation is usually taken to include both read and write (in the same location). MOS memories have access and cycle times in the range of a few to few hundred nanoseconds.

EXERCISES

11.7 A 4M-bit memory chip is partitioned into 32 blocks, with each block having 1024 rows and 128 columns. Give the number of bits required for the row address, column address, and block address.

Ans. 10; 7; 5

11.8 The word lines in a particular MOS memory chip are fabricated using polysilicon (see Appendix A). The resistance of each word line is estimated to be 5 kΩ, and the total capacitance between the line and ground is 2 pF. Find the time for the voltage on the word line to reach $V_{DD}/2$ assuming that the line is driven by a voltage V_{DD} provided by a low-impedance inverter. (*Note:* The line is actually a distributed network that we are approximating by a lumped circuit consisting of a single resistor and a single capacitor.)

Ans. 6.9 ns

11.4 RANDOM-ACCESS MEMORY (RAM) CELLS

As mentioned in Section 11.3, the major part of the memory chip is taken up by the storage cells. It follows that to be able to pack a large number of bits on a chip, it is imperative that the cell size be reduced to the smallest possible. The power dissipation per cell should be minimized also. Thus, many of the flip-flop circuits studied in Section 11.1 are too complex to be suitable for implementing the storage cells in a RAM chip.

There are basically two types of MOS RAM: static and dynamic. **Static RAMs** (called **SRAMs** for short) utilize static latches as the storage cells. Dynamic RAMs (called **DRAMs**), on the other hand, store the binary data on capacitors, resulting in further reduction in cell area, but at the expense of more complex read and write circuitry. In particular, while static RAMs can hold their stored data indefinitely, provided the power supply remains on, dynamic RAMs require *periodic refreshing* to regenerate the data stored on capacitors. This is because the storage capacitors will discharge, though slowly, as a result of the leakage currents inevitably present. By virtue of their smaller cell size, dynamic memory chips are usually four times as dense as their contemporary static chips. Both static and dynamic RAMs are *volatile;* that is, they require the continuous presence of a power supply. By contrast, most ROMs are of the nonvolatile type, as we shall see in Section 11.6. In the following sections, we shall study basic SRAM and DRAM storage cells.

11.4.1 Static Memory Cell

Figure 11.18 shows a typical static memory cell in CMOS technology. The circuit, which we encountered in Section 11.1, is a flip-flop comprising two cross-coupled inverters and two **access transistors,** Q_5 and Q_6. The access transistors are turned on when the word line is selected and its voltage raised to V_{DD}, and they connect the flip-flop to the column (bit or B) line and $\overline{\text{column}}$ ($\overline{\text{bit}}$ or $\overline{B}$) line. Note that both B and $\overline{B}$ lines are utilized. The access transistors act as transmission gates allowing bidirectional current flow between the flip-flop and the B and $\overline{B}$ lines.

The Read Operation   Consider first a read operation, and assume that the cell is storing a 1. In this case, Q will be high at V_{DD}, and $\overline{Q}$ will be low at 0 V. Before the read operation begins, the B and $\overline{B}$ lines are *precharged* to an intermediate voltage, between the low and high values,

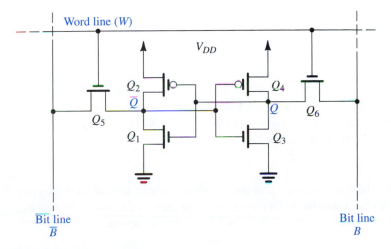

FIGURE 11.18   A CMOS SRAM memory cell.

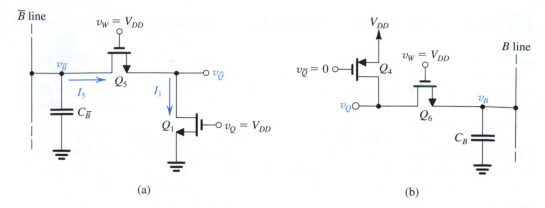

(a)　　　　　　　　　　　　　　　　　　　　(b)

FIGURE 11.19 Relevant parts of the SRAM cell circuit during a read operation when the cell is storing a logic 1. Note that initially $v_Q = V_{DD}$ and $v_{\bar{Q}} = 0$. Also note that the B and $\bar{B}$ lines are usually precharged to a voltage of about $V_{DD}/2$. However, in Example 11.2, it is assumed for simplicity that the precharge voltage is V_{DD}.

usually $V_{DD}/2$. (The circuit for precharging will be shown in Section 11.5 in conjunction with the sense amplifier.) When the word line is selected and Q_5 and Q_6 are turned on, we see that current will flow from V_{DD} through Q_4 and Q_6 and onto line B, charging the capacitance of line B, C_B. On the other side of the circuit, current will flow from the precharged $\bar{B}$ line through Q_5 and Q_1 to ground, thus discharging $C_{\bar{B}}$. It follows that the relevant parts of the circuit during a read operation are those shown in Fig. 11.19.

From this description, we note that during a read "1" operation, the voltage across C_B will rise and that across $C_{\bar{B}}$ will fall. Thus, a differential voltage $v_{B\bar{B}}$ develops between line B and line $\bar{B}$. Usually, only 0.2 V or so is required for the sense amplifier to detect the presence of a 1 in the cell. Observe that the cell must be designed so that the changes in v_Q and $v_{\bar{Q}}$ are small enough to prevent the flip-flop from changing state during readout. The read operation in an SRAM is *nondestructive*. Typically, each of the inverters is designed so that Q_N and Q_P are matched, thus placing the inverter threshold at $V_{DD}/2$. The access transistors are usually made two to three times wider than Q_N of the inverters.

EXAMPLE 11.2

The purpose of this example is to analyze the dynamic operation of the CMOS SRAM cell of Fig. 11.18. Assume that the cell is fabricated in a process technology for which $\mu_n C_{ox} = 50 \ \mu\text{A/V}^2$, $\mu_p C_{ox} = 20 \ \mu\text{A/V}^2$, $V_{tn0} = -V_{tp0} = 1$ V, $2\phi_f = 0.6$ V, $\gamma = 0.5$ V$^{1/2}$, and $V_{DD} = 5$ V. Let the cell transistors have $(W/L)_n = 4/2$, $(W/L)_p = 10/2$, and let the access transistors have $(W/L) = 10/2$. Assuming that the cell is storing a 1 and that the capacitance of each bit line is 1 pF, determine the time required to develop an output voltage of 0.2 V. To simplify the analysis, assume that the B and $\bar{B}$ lines are precharged to V_{DD}.

Solution

We note at the outset that the dynamic analysis of this circuit is complex, and we must therefore make a number of simplifying assumptions. Of course, a precise analysis can always be obtained using simulation. However, much insight can be gained from even an approximate paper-and-pencil analysis.

Refer to Fig. 11.19 and recall that initially $v_Q = V_{DD}$, $v_{\bar{Q}} = 0$, and $v_B = v_{\bar{B}} = V_{DD}$. We see immediately that the circuit in Fig. 11.19(b) will not be conducting, and thus v_B will remain constant at V_{DD}. Turning our attention then to the circuit in Fig. 11.19(a), we observe that since $v_{\bar{B}}$ will change by only 0.2 V (i.e., from 5 V to 4.8 V) during the readout process, transistor Q_5 will be operating in saturation, and thus $C_{\bar{B}}$ will be discharged with a constant current I_5. For transistor Q_1 to conduct, its drain voltage $v_{\bar{Q}}$ will have to rise. We hope, however, that this rise will not exceed the threshold of inverter (Q_3, Q_4), which is $V_{DD}/2$, since the p and n transistors in each inverter are matched. There will be a brief interval during which I_5 will charge the small parasitic capacitance between node $\bar{Q}$ and ground to a voltage $v_{\bar{Q}}$ sufficient to operate Q_1 in the triode mode at a current I_1 equal to I_5. The current I_1 can then be expressed as

$$I_1 = \mu_n C_{ox} \left(\frac{W}{L}\right)_1 \left[(V_{DD} - V_{t1})v_{\bar{Q}} - \frac{1}{2}v_{\bar{Q}}^2\right]$$

where we have assumed that v_Q will remain constant at V_{DD}. Since the source of Q_1 is at ground, $V_{t1} = 1$ V and,

$$I_1 = 50 \times \frac{4}{2}\left[(5-1)v_{\bar{Q}} - \frac{1}{2}v_{\bar{Q}}^2\right] \tag{11.3}$$

For Q_5 we can write

$$I_5 = \frac{1}{2}\mu_n C_{ox}\left(\frac{W}{L}\right)_5 (V_{DD} - v_{\bar{Q}} - V_{t5})^2$$

where the threshold voltage V_{t5} can be determined from

$$V_{t5} = 1 + 0.5(\sqrt{v_{\bar{Q}} + 0.6} - \sqrt{0.6}) \tag{11.4}$$

Since we do not yet know $v_{\bar{Q}}$, we need to solve by iteration. For a first iteration, we assume that $V_{t5} = 1$ V, thus I_5 will be

$$I_5 = \frac{1}{2} \times 50 \times \frac{10}{2}(5 - v_{\bar{Q}} - 1)^2 \tag{11.5}$$

Now equating I_1 from Eq. (11.3) to I_5 from Eq. (11.5) and solving for $v_{\bar{Q}}$ results in $v_{\bar{Q}} = 1.86$ V. As a second iteration, we use this value of $v_{\bar{Q}}$ in Eq. (11.4) to determine V_{t5}. The result is $V_{t5} = 1.4$ V. This value is then used in the expression for I_5 and the process repeated, with the result that $v_{\bar{Q}} = 1.6$ V. This is close enough to the original value, and no further iteration seems warranted. The current I_5 can now be determined, $I_5 = 0.5$ mA. Observe that $v_{\bar{Q}}$ is indeed less than $V_{DD}/2$, and thus the flip-flop will not switch state (a relief!). In fact, V_{IL} for this inverter is 2.125 V; thus the assumption that v_Q stays at V_{DD} is justified, although v_Q will change somewhat, a point we shall not pursue any further in this approximate analysis.

We can now determine the interval for a 0.2-V decrement to appear on the $\bar{B}$ line from

$$\Delta t = \frac{C_{\bar{B}}\Delta V}{I_5}$$

Thus,

$$\Delta t = \frac{1 \times 10^{-12} \times 0.2}{0.5 \times 10^{-3}} = 0.4 \text{ ns}$$

We should point out that Δt is only one component of the delay encountered in the read operation. Another significant component is due to the finite rise time of the voltage on the word line. Indeed, even the calculation of Δt is optimistic, since the word line will have only reached a voltage lower than V_{DD} when the process of discharging $C_{\bar{B}}$ takes place.

Another even more approximate (but faster) solution can be obtained by observing that in the circuit of Fig. 11.19(a), Q_1 and Q_5 have equal gate voltages (V_{DD}) and are connected in series. We may consider that they are approximately equivalent to a single transistor with a W/L ratio,

$$(W/L)_{eq} = \cfrac{1}{\cfrac{1}{(W/L)_1} + \cfrac{1}{(W/L)_5}} = \cfrac{1}{\cfrac{2}{4} + \cfrac{2}{10}} = \frac{10}{7}$$

The equivalent transistor will operate in saturation, thus its current I will be

$$I = \tfrac{1}{2} \times 50 \times \tfrac{10}{7}(5-1)^2 = 0.57 \text{ mA}$$

This is only 14% greater than the value found earlier. The voltage $v_{\bar{Q}}$ can be found by multiplying I by the approximate value of r_{DS} of Q_1 in the triode region,

$$r_{DS} = 1/[50 \times 10^{-6} \times \tfrac{4}{2} \times (5-1)] = 2.5 \text{ k}\Omega$$

Thus,

$$v_{\bar{Q}} = 0.57 \times 2.5 = 1.4 \text{ V}$$

Again, this is reasonably close to the value found earlier.

The Write Operation Next we consider the write operation. Assume that the cell is originally storing a 1 ($v_Q = V_{DD}$ and $v_{\bar{Q}} = 0$) and that we wish to write a 0. To do this, the B line is lowered to 0 V and the $\bar{B}$ line is raised to V_{DD}, and of course the cell is selected by raising the word line to V_{DD}. Figure 11.20 shows the relevant parts of the circuit during the interval in which node $\bar{Q}$ is being pulled up toward the threshold voltage $V_{DD}/2$ (Fig. 11.20a) and node Q is being pulled down toward $V_{DD}/2$ (Fig. 11.20b). Capacitors C_Q and $C_{\bar{Q}}$ are the parasitic capacitances at nodes Q and $\bar{Q}$, respectively. An approximate analysis can be performed on either circuit to determine the time required for toggling to take place. Note that the regenerative feedback that causes the flip-flop to switch will begin when *either* v_Q or $v_{\bar{Q}}$ reaches $V_{DD}/2$. When this happens, the positive feedback takes over, and the circuits in Fig. 11.20 no longer apply.

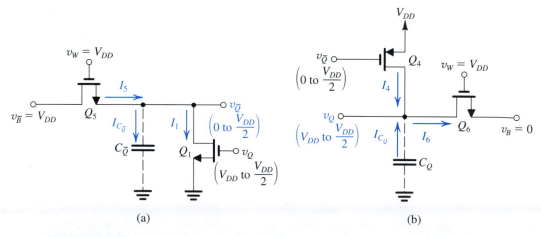

(a)   (b)

FIGURE 11.20 Relevant parts of the SRAM circuit during a write operation. Initially, the SRAM has a stored 1 and a 0 is being written. These equivalent circuits apply before switching takes place. **(a)** The circuit is pulling node $\bar{Q}$ up toward $V_{DD}/2$. **(b)** The circuit is pulling node Q down toward $V_{DD}/2$.

We shall briefly explain the operation of the circuits in Fig. 11.20, leaving the analysis for the reader to perform in Exercise 11.9 and Problems 11.23 and 11.24. Consider first the circuit in Fig. 11.20(a), and note that Q_5 will be operating in saturation. Initially, its source voltage will be 0, and thus its V_t will be equal to V_{t0}. Also initially, Q_1 will be off because its drain voltage is zero. The current I_5 will initially flow into $C_{\bar{Q}}$, charging it up, and thus $v_{\bar{Q}}$ will rise and Q_1 will conduct. Q_1 will be in the triode region and its current I_1 will subtract from I_5, reducing the current available for charging $C_{\bar{Q}}$. *Simultaneously*, as $v_{\bar{Q}}$ rises, V_{t5} will increase owing to the body effect, and I_5 will reduce. Another effect caused by the circuit in Fig. 11.20(b) is that v_Q will be falling from V_{DD} toward $V_{DD}/2$. This will cause a corresponding decrease in the current I_1. Despite all these complications, one can easily calculate an approximate average value for the capacitor charging current $I_{C_{\bar{Q}}}$ over the interval[3] beginning with ($v_Q = V_{DD}$, $v_{\bar{Q}} = 0$) and ending with ($v_Q = V_{DD}/2$, $v_{\bar{Q}} = V_{DD}/2$). We can then use this current value to determine the time for the voltage across $C_{\bar{Q}}$ to increase by $V_{DD}/2$.

The circuit in Fig. 11.20(b) operates in much the same fashion except that neither of the two transistors is susceptible to the body effect. Thus, this circuit will provide C_Q with a larger discharge current than the current provided by the circuit in Fig. 11.20(a) to charge $C_{\bar{Q}}$. The result will be that C_Q will discharge faster than $C_{\bar{Q}}$ will charge. In other words, v_Q will reach $V_{DD}/2$ before $v_{\bar{Q}}$ does. It follows that an estimate of this component of the write delay time can be obtained by considering only the circuit in Fig. 11.20(b).

Another component of write delay is that taken up by the switching action of the flip-flop. This can be approximated by the delay time of one inverter.

EXERCISE

11.9   Consider the circuit in Fig. 11.20(b), and assume that the device dimensions and process technology parameters are as specified in Example 11.2. We wish to determine the interval Δt required for C_Q to discharge, and its voltage to fall from V_{DD} to $V_{DD}/2$.

(a) At the beginning of interval Δt, find the values of I_4, I_6, and I_{C_Q}.

(b) At the end of interval Δt, find the values of I_4, I_6, and I_{C_Q}.

(c) Find an estimate of the average value of I_{C_Q} during interval Δt.

(d) If $C_Q = 50$ fF, estimate Δt.

Ans. (a) $I_4 = 0$, $I_6 = 2$ mA, $I_{C_Q} = 2$ mA; (b) $I_4 = 0.11$ mA, $I_6 = 1.72$ mA, $I_{C_Q} = 1.61$ mA; (c) $I_{C_Q}\big|_{av} = 1.8$ mA; (d) $\Delta t = 69.4$ ps

From the results of Exercise 11.9, we note that this component of write delay is much smaller than the corresponding component in the read operation. This is because in the write operation, only the small capacitance C_Q needs to be charged (or discharged), whereas in the read operation, we have to charge (or discharge) the much larger capacitances of the B or $\bar{B}$ lines. In the write operation, the B and $\bar{B}$ line capacitances are charged (and discharged) relatively quickly by the driver circuitry. The end result is that the delay time in the write operation is dominated by the word-line delay.

---

[3] Implicit in this statement is the assumption that both v_Q and $v_{\bar{Q}}$ will reach $V_{DD}/2$ simultaneously. As will be seen shortly, this is not the case. Nevertheless, it is a reasonable assumption to make for the purpose of obtaining an approximate estimate of the write delay time.

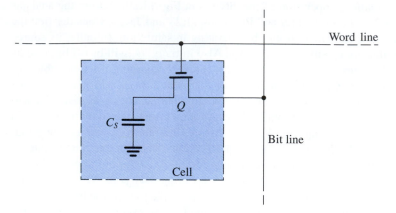

FIGURE 11.21 The one-transistor dynamic RAM cell.

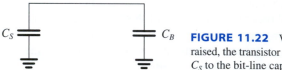

FIGURE 11.22 When the voltage of the selected word line is raised, the transistor conducts, thus connecting the storage capacitor C_S to the bit-line capacitance C_B.

11.4.2 Dynamic Memory Cell

Although a variety of DRAM storage cells have been proposed over the years, a particular cell, shown in Fig. 11.21, has become the industry standard. The cell consists of a single *n*-channel MOSFET, known as the *access* transistor, and a storage capacitor C_S. The cell is appropriately known as the **one-transistor cell.**[4] The gate of the transistor is connected to the word line, and its source (drain) is connected to the bit line. Observe that only one bit line is used in DRAMs, whereas in SRAMs both the bit and $\overline{\text{bit}}$ lines are utilized.

The DRAM cell stores its bit of information as charge on the cell capacitor C_S. When the cell is storing a 1, the capacitor is charged to $(V_{DD} - V_t)$; when a 0 is stored, the capacitor is discharged to a zero voltage.[5] Because of leakage effects, the capacitor charge will leak off, and hence the cell must be refreshed periodically. During *refresh,* the cell content is read and the data bit is rewritten, thus *restoring* the capacitor voltage to its proper value. The refresh operation must be performed every 5 ms to 10 ms.

Let us now consider the DRAM operation in more detail. As in the static RAM, the row decoder selects a particular row by raising the voltage of its word line. This causes all the access transistors in the selected row to become conductive, thereby connecting the storage capacitors of all the cells in the selected row to their respective bit lines. Thus the cell capacitor C_S is connected in parallel with the bit-line capacitance C_B, as indicated in Fig. 11.22. Here, it should be noted that C_S is typically 30 fF to 50 fF, whereas C_B is 30 to 50 times larger. Now, if the operation is a read, the bit line is precharged to $V_{DD}/2$. To find the change

---

[4] The name was originally used to distinguish this cell from earlier ones utilizing three transistors.

[5] The reason that the "1" level is less than V_{DD} by the magnitude of the threshold V_t is as follows: Consider a write-1 operation. The word line is at V_{DD} and the bit line is at V_{DD} and the transistor is conducting, charging C_S. The transistor will cease conduction when the voltage on C_S reaches $(V_{DD} - V_t)$, where V_t is higher than V_{t0} because of the body effect. We have analyzed this situation at length in Section 10.5 in connection with pass-transistor logic.

in the voltage on the bit line resulting from connecting a cell capacitor C_S to it, let the initial voltage on the cell capacitor be V_{CS} ($V_{CS} = V_{DD} - V_t$ when a 1 is stored, but $V_{CS} = 0$ V when a 0 is stored). Using charge conservation, we can write

$$C_S V_{CS} + C_B \frac{V_{DD}}{2} = (C_B + C_S)\left(\frac{V_{DD}}{2} + \Delta V\right)$$

from which we can obtain for ΔV

$$\Delta V = \frac{C_S}{C_B + C_S}\left(V_{CS} - \frac{V_{DD}}{2}\right) \tag{11.6}$$

and since $C_B \gg C_S$,

$$\Delta V \cong \frac{C_S}{C_B}\left(V_{CS} - \frac{V_{DD}}{2}\right) \tag{11.7}$$

Now, if the cell is storing a 1, $V_{CS} = V_{DD} - V_t$, and

$$\Delta V(1) \cong \frac{C_S}{C_B}\left(\frac{V_{DD}}{2} - V_t\right) \tag{11.8}$$

whereas if the cell is storing a 0, $V_{CS} = 0$, and

$$\Delta V(0) \cong -\frac{C_S}{C_B}\left(\frac{V_{DD}}{2}\right) \tag{11.9}$$

Since usually C_B is much greater than C_S, these readout voltages are very small. For example, for $C_B = 30\ C_S$, $V_{DD} = 5$ V, and $V_t = 1.5$ V, $\Delta V(0)$ will be about –83 mV, and $\Delta V(1)$ will be 33 mV. This is a best-case scenario, for the 1 level in the cell might very well be below $(V_{DD} - V_t)$. Furthermore, in modern memory chips, V_{DD} is 3.3 V or even lower. In any case, we see that a stored 1 in the cell results in a small positive increment in the bit-line voltage, whereas a stored zero results in a small negative increment. Observe also that the readout process is *destructive,* since the resulting voltage across C_S will no longer be $(V_{DD} - V_t)$ or 0.

The change of voltage on the bit line is detected and amplified by the column sense amplifier. The amplified signal is then impressed on the storage capacitor, thus restoring its signal to the proper level ($V_{DD} - V_t$ or 0). In this way, all the cells in the selected row are refreshed. Simultaneously, the signal at the output of the sense amplifier of the selected column is fed to the data-output line of the chip through the action of the column decoder.

The write operation proceeds similarly to the read operation, except that the data bit to be written, which is impressed on the data input line, is applied by the column decoder to the selected bit line. Thus, if the data bit to be written is a 1, the *B*-line voltage is raised to V_{DD} (i.e., C_B is charged to V_{DD}). When the access transistor of the particular cell is turned on, its capacitor C_S will be charged to $V_{DD} - V_t$; thus a 1 is written in the cell. Simultaneously, all the other cells in the selected row are simply refreshed.

Although the read and write operations result in automatic refreshing of all the cells in the selected row, provision must be made for the periodic refreshing of the entire memory every 5 to 10 ms, as specified for the particular chip. The refresh operation is carried out in a *burst mode,* one row at a time. During refresh, the chip will not be available for read or write operations. This is not a serious matter, however, since the interval required to refresh the entire chip is typically less than 2% of the time between refresh cycles. In other words, the memory chip is available for normal operation more than 98% of the time.

EXERCISES

11.10 In a particular dynamic memory chip, $C_S = 30$ fF, $C_B = 1$ pF, $V_{DD} = 5$ V, V_t (including the body effect) = 1.5 V, find the output readout voltage for a stored 1 and a stored 0. Recall that in a read operation, the bit lines are precharged to $V_{DD}/2$.

Ans. 30 mV; –75 mV

11.11 A 64M-bit DRAM chip fabricated in a 0.4-μm CMOS technology requires 2 μm^2 per cell. If the storage array is square, estimate its dimensions. Further, if the peripheral circuitry (e.g., sense amplifiers, decoders) add about 30% to the chip area, estimate the dimensions of the resulting chip.

Ans. 11.6 mm × 11.6 mm; 13.2 mm × 13.2 mm

11.5 SENSE AMPLIFIERS AND ADDRESS DECODERS

Having studied the circuits commonly used to implement the storage cells in SRAMs and DRAMs, we now consider some of the other important circuit blocks in a memory chip. The design of these circuits, commonly referred to as the memory *peripheral circuits,* presents exciting challenges and opportunities to integrated-circuit designers: Improving the performance of peripheral circuits can result in denser and faster memory chips that dissipate less power.

11.5.1 The Sense Amplifier

Next to the storage cells, the sense amplifier is the most critical component in a memory chip. Sense amplifiers are essential to the proper operation of DRAMs, and their use in SRAMs results in speed and area improvements.

A variety of sense-amplifier designs are in use, some of which closely resemble the active-load MOS differential amplifier studied in Chapter 7. Here, we describe a differential sense amplifier that employs positive feedback. Because the circuit is differential, it can be employed directly in SRAMs where the SRAM cell utilizes both the B and $\bar{B}$ lines. On the other hand, the one-transistor DRAM circuit we studied in Section 11.4.2 is a single-ended circuit, utilizing one bit line only. The DRAM circuit, however, can be made to resemble a differential signal source through the use of the "dummy cell" technique, which we shall discuss shortly. Therefore, we shall assume that the memory cell whose output is to be amplified develops a difference output voltage between the B and $\bar{B}$ lines. This signal, which can range between 30 mV and 500 mV depending on the memory type and cell design, will be applied to the input terminals of the sense amplifier. The sense amplifier in turn responds by providing a full-swing (0 to V_{DD}) signal at its output terminals. The particular amplifier circuit we shall discuss here has a rather unusual property: *Its output and input terminals are the same!*

A Sense Amplifier with Positive Feedback Figure 11.23 shows the sense amplifier together with some of the other column circuitry of a RAM chip. Note that the sense amplifier is nothing but the familiar latch formed by cross-coupling two CMOS inverters: One inverter is implemented by transistors Q_1 and Q_2, and the other by transistors Q_3 and Q_4. Transistors Q_5 and Q_6 act as switches that connect the sense amplifier to ground and V_{DD} only when data-sensing action is required. Otherwise, ϕ_s is low and the sense amplifier is turned off. This conserves power, an important consideration because usually there is one sense amplifier per column, resulting in *thousands of sense amplifiers per chip*. Note, again,

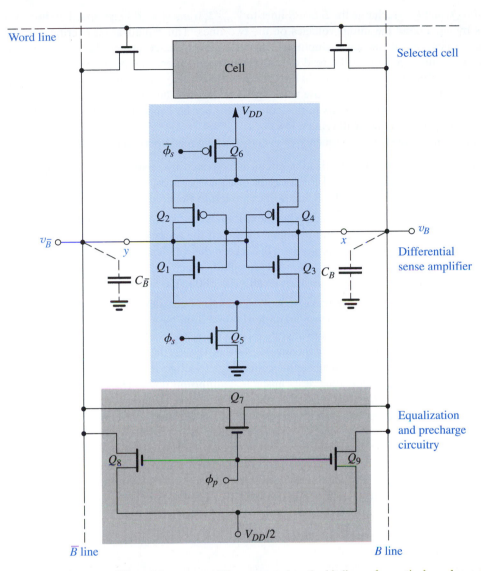

FIGURE 11.23 A differential sense amplifier connected to the bit lines of a particular column. This arrangement can be used directly for SRAMs (which utilize both the B and $\bar{B}$ lines). DRAMs can be turned into differential circuits by using the "dummy cell" arrangement shown in Fig. 11.25.

that terminals x and y are both the input and the output terminals of the amplifier. As indicated, these I/O terminals are connected to the B and $\bar{B}$ lines. The amplifier is required to detect a small signal appearing between B and $\bar{B}$, and to amplify it to provide a full-swing signal at B and $\bar{B}$. For instance, if during a read operation, the cell had a stored 1, then a small positive voltage will develop between B and $\bar{B}$, with v_B higher than $v_{\bar{B}}$. The amplifier will then cause v_B to rise to V_{DD} and $v_{\bar{B}}$ to fall to 0 V. This 1 output is then directed to the chip I/O pin by the column decoder (not shown) and at the same time is used to rewrite a 1 in the DRAM cell, thus performing the restore operation that is required because the DRAM readout process is destructive.

Figure 11.23 also shows the precharge and equalization circuit. Operation of this circuit is straightforward: When ϕ_p goes high prior to a read operation, all three transistors conduct.

While Q_8 and Q_9 precharge the $\overline{B}$ and B lines to $V_{DD}/2$, transistor Q_7 helps speed up this process by equalizing the initial voltages on the two lines. This equalization is critical to the proper operation of the sense amplifier. Any voltage difference present between B and $\overline{B}$ prior to commencement of the read operation can result in erroneous interpretation by the sense amplifier of its input signal. In Fig. 11.23, we show only one of the cells in this particular column, namely, the cell whose word line is activated. The cell can be either an SRAM or a DRAM cell. All other cells in this column will not be connected to the B and $\overline{B}$ lines (because their word lines will remain low).

Let us now consider the sequence of events during a read operation:

1. The precharge and equalization circuit is activated by raising the control signal ϕ_p. This will cause the B and $\overline{B}$ lines to be at equal voltages, equal to $V_{DD}/2$. The clock ϕ_p then goes low, and the B and $\overline{B}$ lines are left to float for a brief interval.

2. The word line goes up, connecting the cell to the B and $\overline{B}$ lines. A voltage then develops between B and $\overline{B}$, with v_B higher than $v_{\overline{B}}$ if the accessed cell is storing a 1, or v_B lower than $v_{\overline{B}}$ if the cell is storing a 0. To keep the cell design simple, and to facilitate operation at higher speeds, the readout signal, which the cell is required to provide between B and $\overline{B}$, is kept small (typically, 30–500 mV).

3. Once an adequate difference voltage signal has been developed between B and $\overline{B}$ by the storage cell, the sense amplifier is turned on by connecting it to ground and V_{DD} through Q_5 and Q_6, activated by raising the sense-control signal ϕ_s. Because initially the input terminals of the inverters are at $V_{DD}/2$, the inverters will be operating in their transition region where the gain is high (Section 10.2). It follows that initially the latch will be operating at its unstable equilibrium point. Thus, depending on the signal between the input terminals, the latch will quickly move to one of its two stable equilibrium points (refer to the description of the latch operation in Section 11.1). This is achieved by the regenerative action, inherent in positive feedback. Figure 11.24 clearly illustrates this point by showing the waveforms of the signal on the bit line for both a read-1 and a read-0 operation. Observe that once activated, the sense amplifier causes the small initial difference, $\Delta V(1)$ or $\Delta V(0)$, provided by the cell, to grow exponentially to either V_{DD} (for a read-1 operation) or 0 (for a read-0 operation).

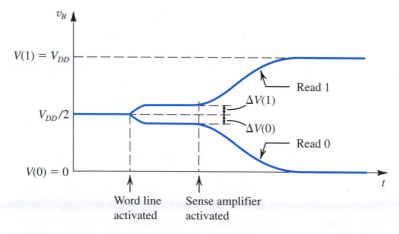

FIGURE 11.24 Waveforms of v_B before and after the activation of the sense amplifier. In a read-1 operation, the sense amplifier causes the initial small increment $\Delta V(1)$ to grow exponentially to V_{DD}. In a read-0 operation, the negative $\Delta V(0)$ grows to 0. Complementary signal waveforms develop on the $\overline{B}$ line.

The waveforms of the signal on the $\overline{B}$ line will be complementary to those shown in Fig. 11.24 for the B line. In the following, we quantify the process of exponential growth of v_B and $v_{\overline{B}}$.

A Closer Look at the Operation of the Sense Amplifier

Developing a precise expression for the output signal of the sense amplifier shown in Fig. 11.23 is a rather complex task requiring the use of large-signal (and thus nonlinear) models of the inverter voltage transfer characteristic, as well as taking the positive feedback into account. We will not do this here; rather, we shall consider the operation in a semiquantitative way.

Recall that at the time the sense amplifier is activated, each of its two inverters is operating in the transition region at $V_{DD}/2$. Thus, for small-signal operation, each inverter can be modeled using g_{mn} and g_{mp}, the transconductances of Q_N and Q_P, respectively, evaluated at an input bias of $V_{DD}/2$. Specifically, a small-signal v_i superimposed on $V_{DD}/2$ at the input of one of the inverters gives rise to an inverter output current signal of $(g_{mn} + g_{mp})v_i \equiv G_m v_i$. This output current is delivered to one of the capacitors, C_B or $C_{\overline{B}}$. The voltage thus developed across the capacitor is then fed back to the other inverter and is multiplied by its G_m, which gives rise to an output current feeding the other capacitor, and so on, in a regenerative process. The positive feedback in this loop will mean that the signal around the loop, and thus v_B and $v_{\overline{B}}$, will *rise or decay exponentially* (see Fig. 11.24) with a time constant of (C_B/G_m) [or $(C_{\overline{B}}/G_m)$, since we have been assuming $C_B = C_{\overline{B}}$]. Thus, for example, in a read-1 operation we obtain

$$v_B = \frac{V_{DD}}{2} + \Delta V(1)e^{(G_m/C_B)t} \qquad v_B \le V_{DD} \qquad (11.10)$$

whereas in a read-0 operation,

$$v_B = \frac{V_{DD}}{2} - \Delta V(0)e^{(G_m/C_B)t} \qquad (11.11)$$

Because these expressions have been derived assuming small-signal operation, they describe the exponential growth (decay) of v_B reasonably accurately only for values close to $V_{DD}/2$. Nevertheless, they can be used to obtain a reasonable estimate of the time required to develop a particular signal level on the bit line.

EXAMPLE 11.3

Consider the sense-amplifier circuit of Fig. 11.23 during the reading of a 1. Assume that the storage cell provides a voltage increment on the B line of $\Delta V(1) = 0.1$ V. If the NMOS devices in the amplifiers have $(W/L)_n = 12\ \mu\text{m}/4\ \mu\text{m}$ and the PMOS devices have $(W/L)_p = (30\ \mu\text{m}/4\ \mu\text{m})$, and assuming that the other parameters of the process technology are as specified in Example 11.2, find the time required for v_B to reach 4.5 V. Assume $C_B = 1$ pF.

Solution

First, we determine the transconductances g_{mn} and g_{mp}

$$g_{mn} = \mu_n C_{ox}\left(\frac{W}{L}\right)_n (V_{GS} - V_t)$$

$$= 50 \times \tfrac{12}{4}(2.5 - 1)$$

$$= 0.225 \text{ mA/V}$$

$$g_{mp} = \mu_p C_{ox} \left(\frac{W}{L}\right)_p (V_{GS} - |V_t|)$$

$$= 20 \times \tfrac{30}{4}(2.5 - 1) = 0.225 \text{ mA/V}$$

Thus, the inverter G_m is

$$G_m = g_{mn} + g_{mp} = 0.45 \text{ mA/V}$$

and the time constant τ for the exponential growth of v_B will be

$$\tau \equiv \frac{C}{G_m} = \frac{1 \times 10^{-12}}{0.45 \times 10^{-3}} = 2.22 \text{ ns}$$

Now, the time, Δt, for v_B to reach 4.5 V can be determined from

$$4.5 = 2.5 + 0.1 e^{\Delta t/2.22}$$

resulting in

$$\Delta t = 6.65 \text{ ns}$$

Obtaining Differential Operation in Dynamic RAMs The sense amplifier described earlier responds to difference signals appearing between the bit lines. Thus, it is capable of rejecting interference signals that are common to both lines, such as those caused by capacitive coupling from the word lines. For this *common-mode rejection* to be effective, great care has to be taken to match both sides of the amplifier, taking into account the circuits that feed each side. This is an important consideration in any attempt to make the inherently single-ended output of the DRAM cell appear differential. We shall now discuss an ingenious scheme for accomplishing this task. Although the technique has been around for many years (see the first edition of this book, published in 1982), it is still in use today. The method is illustrated in Fig. 11.25.

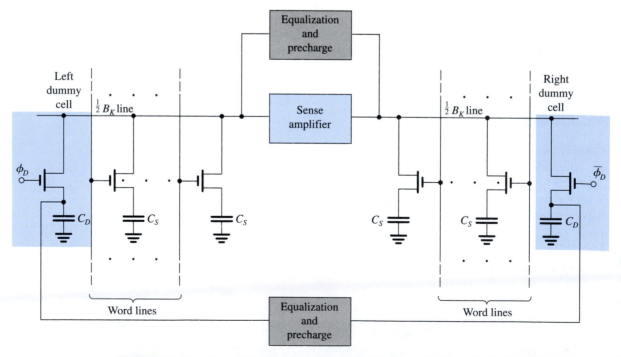

FIGURE 11.25 An arrangement for obtaining differential operation from the single-ended DRAM cell. Note the dummy cells at the far right and far left.

Basically, each bit line is split into two identical halves. Each half-line is connected to half the cells in the column and to an additional cell, known as a *dummy cell,* having a storage capacitor $C_D = C_S$. When a word line on the left side is selected for reading, the dummy cell on the right side (controlled by $\bar{\phi}_D$) is also selected, and vice versa; that is, when a word line on the right side is selected, the dummy cell on the left (controlled by ϕ_D) is also selected. In effect, then, the dummy cell serves as the other half of a differential DRAM cell. When the left-half bit line is in operation, the right-half bit line acts as its complement (or $\bar{B}$ line) and vice versa.

Operation of the circuit in Fig. 11.25 is as follows: The two halves of the line are precharged to $V_{DD}/2$ and their voltages are equalized. At the same time, the capacitors of the two dummy cells are precharged to $V_{DD}/2$. Then a word line is selected, and the dummy cell on the other side is enabled (with ϕ_D or $\bar{\phi}_D$ raised to V_{DD}). Thus the half-line connected to the selected cell will develop a voltage increment (around $V_{DD}/2$) of $\Delta V(1)$ or $\Delta V(0)$ depending on whether a 1 or a 0 is stored in the cell. Meanwhile, the other half of the line will have its voltage held equal to that of C_D (i.e., $V_{DD}/2$). The result is a differential signal of $\Delta V(1)$ or $\Delta V(0)$ that the sense amplifier detects and amplifies when it is enabled. As usual, by the end of the regenerative process, the amplifier will cause the voltage on one half of the line to become V_{DD} and that on the other half to become 0.

EXERCISES

11.12 It is required to reduce the time Δt of the sense-amplifier circuit in Example 11.3 to 4 ns by increasing g_m of the transistors (while retaining the matched design of each inverter). What must the (W/L) ratios of the *n*- and *p*-channel devices become?

Ans. $(W/L)_n = 5$; $(W/L)_p = 12.5$

11.13 If in the sense amplifier of Example 11.3, the signal available from the cell is only half as large (i.e., only 50 mV), what will Δt become?

Ans. 8.19 ns, an increase of 23%

11.5.2 The Row-Address Decoder

As described in Section 11.3, the row-address decoder is required to select one of the 2^M word lines in response to an *M*-bit address input. As an example, consider the case $M = 3$ and denote the three address bits A_0, A_1, and A_2, and the eight word lines $W_0, W_1, \ldots, W_7$. Conventionally, word line W_0 will be high when $A_0 = 0$, $A_1 = 0$, and $A_2 = 0$, thus we can express W_0 as a Boolean function of A_0, A_1, and A_2,

$$W_0 = \bar{A}_0\bar{A}_1\bar{A}_2 = \overline{A_0 + A_1 + A_2}$$

Thus the selection of W_0 can be accomplished by a three-input NOR gate whose three inputs are connected to A_0, A_1, and A_2 and whose output is connected to word line 0. Word line W_3 will be high when $A_0 = 1$, $A_1 = 1$, and $A_2 = 0$, thus

$$W_3 = A_0 A_1 \bar{A}_2 = \overline{\bar{A}_0 + \bar{A}_1 + A_2}$$

Thus the selection of W_3 can be realized by a three-input NOR gate whose three inputs are connected to $\bar{A}_0$, $\bar{A}_1$, and A_2, and whose output is connected to word line 3. We can thus see that this address decoder can be realized by eight three-input NOR gates. Each NOR gate is fed with the appropriate combination of address bits and their complements, corresponding to the word line to which its output is connected.

A simple approach to realizing these NOR functions is provided by the matrix structure shown in Fig. 11.26. The circuit shown is a dynamic one (Section 10.6). Attached to each row line is a p-channel device that is activated, prior to the decoding process, using the precharge control signal ϕ_P. During precharge (ϕ_P low), all the word lines are pulled high to V_{DD}. It is assumed that at this time the address input bits have not yet been applied and all the inputs are low; hence there is no need for the circuit to include the evaluation transistor utilized in dynamic logic gates. Then, the decoding operation begins when the address bits

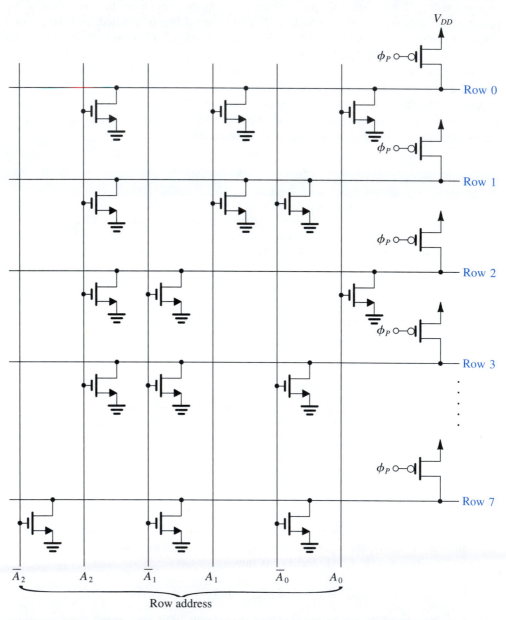

FIGURE 11.26 A NOR address decoder in array form. One out of eight lines (row lines) is selected using a 3-bit address.

and their complements are applied. Observe that the NMOS transistors are placed so that the word lines not selected will be discharged. For any input combination, only one word line will not be discharged, and thus its voltage remains high at V_{DD}. For instance, row 0 will be high only when $A_0 = 0$, $A_1 = 0$, and $A_2 = 0$; this is the only combination that will result in all three transistors connected to row 0 being cut off. Similarly, row 3 has transistors connected to $\overline{A}_0$, $\overline{A}_1$, and A_2, and thus it will be high when $A_0 = 1$, $A_1 = 1$, $A_2 = 0$, and so on. After the decoder outputs have stabilized, the output lines are connected to the word lines of the array, usually via clock-controlled transmission gates. This decoder is known as a NOR decoder. Observe that because of the precharge operation, the decoder circuit does not dissipate static power.

EXERCISE

11.14 How many transistors are needed for a NOR row decoder with an M-bit address?

Ans. $M2^M$ NMOS $+ 2^M$ PMOS $= 2^M(M + 1)$

11.5.3 The Column-Address Decoder

From the description in Section 11.3, the function of the column-address decoder is to connect one of the 2^N bit lines to the data I/O line of the chip. As such, it is a multiplexer and can be implemented using pass-transistor logic (Section 10.5) as shown in Fig. 11.27. Here, each bit line is connected to the data I/O line through an NMOS transistor. The gates of the pass transistors are controlled by 2^N lines, one of which is selected by a NOR decoder similar to that used for decoding the row address.

An alternative implementation of the column decoder that uses a smaller number of transistors (but at the expense of slower speed of operation) is shown in Fig. 11.28. This circuit, known as a *tree decoder,* has a simple structure of pass transistors. Unfortunately, since a relatively large number of transistors can exist in the signal path, the resistance of the bit lines increases, and the speed decreases correspondingly.

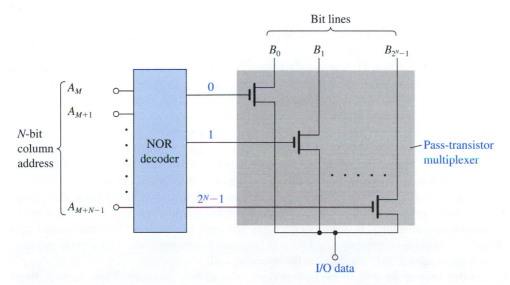

FIGURE 11.27 A column decoder realized by a combination of a NOR decoder and a pass-transistor multiplexer.

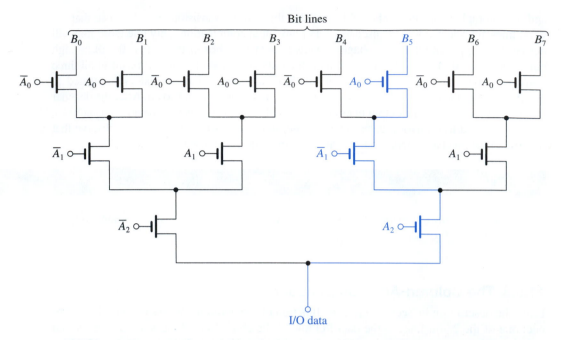

FIGURE 11.28 A tree column decoder. Note that the colored path shows the transistors that are conducting when $A_0 = 1$, $A_1 = 0$, and $A_2 = 1$, the address that results in connecting B_5 to the data line.

EXERCISE

11.15 How many transistors are needed for a tree decoder when there are 2^N bit lines?

Ans. $2(2^N - 1)$

 ## 11.6 READ-ONLY MEMORY (ROM)

As mentioned in Section 11.3, read-only memory (ROM) is memory that contains fixed data patterns. It is used in a variety of digital-system applications. Currently, a very popular application is the use of ROM in microprocessor systems to store the instructions of the system's basic operating program. ROM is particularly suited for such an application because it is nonvolatile; that is, it retains its contents when the power supply is switched off.

A ROM can be viewed as a combinational logic circuit for which the input is the collection of address bits of the ROM and the output is the set of data bits retrieved from the addressed location. This viewpoint leads to the application of ROMs in code conversion—that is, in changing the code of the signal from one system (say, binary) to another. Code conversion is employed, for instance, in secure communication systems, where the process is known as *scrambling*. It consists of feeding the code of the data to be transmitted to a ROM that provides corresponding bits in a (supposedly) secret code. The reverse process, which also uses a ROM, is applied at the receiving end.

In this section we will study various types of read-only memory. These include fixed ROM, which we refer to simply as ROM, programmable ROM (PROM), and erasable programmable ROM (EPROM).

11.6.1 A MOS ROM

Figure 11.29 shows a simplified 32-bit (or 8-word × 4-bit) MOS ROM. As indicated, the memory consists of an array of n-channel MOSFETs whose gates are connected to the word lines, whose sources are grounded, and whose drains are connected to the bit lines. Each bit line is connected to the power supply via a PMOS load transistor, in the manner of pseudo-NMOS logic (Section 10.4). An NMOS transistor exists in a particular cell if the cell is storing a 0; a cell storing a 1 has no MOSFET. This ROM can be thought of as 8 words of 4 bits each. The row decoder selects one of the 8 words by raising the voltage of the corresponding word line. The cell transistors connected to this word line will then conduct, thus pulling the voltage of the bit lines (to which transistors in the selected row are connected) down from V_{DD} to a voltage close to ground voltage (the logic-0 level). The bit lines that are connected to cells (of the selected word) without transistors (i.e., the cells that are storing 1s) will remain at the power-supply voltage (logic 1) because of the action of the pull-up PMOS load devices. In this way, the bits of the addressed word can be read.

A disadvantage of the ROM circuit in Fig. 11.29 is that it dissipates static power. Specifically, when a word is selected, the transistors in this particular row will conduct static current that is supplied by the PMOS load transistors. Static power dissipation can be eliminated by a simple change. Rather than grounding the gate terminals of the PMOS transistors, we can connect these transistors to a precharge line ϕ that is normally high. Just before a reading operation, ϕ is lowered and the bit lines are precharged to V_{DD} through the PMOS transistors. The precharge signal ϕ then goes high, and the word line is selected. The bit lines that have transistors in the selected word are then discharged, thus indicating stored zeros, whereas those lines for which no transistor is present remain at V_{DD}, indicating stored ones.

EXERCISE

11.16 The purpose of this exercise is to estimate the various delay times involved in the operation of a ROM. Consider the ROM in Fig. 11.29 with the gates of the PMOS devices disconnected from ground and connected to a precharge control signal ϕ. Let all the NMOS devices have $W/L = 6\ \mu m/2\ \mu m$ and all the PMOS devices have $W/L = 24\ \mu m/2\ \mu m$. Assume that $\mu_n C_{ox} = 50\ \mu A/V^2$, $\mu_p C_{ox} = 20\ \mu A/V^2$, $V_{tn} = -V_{tp} = 1$ V, and $V_{DD} = 5$ V.

(a) During the precharge interval, ϕ is lowered to 0 V. Estimate the time required to charge a bit line from 0 V to 5 V. Use as an average charging current the current supplied by a PMOS transistor at a bit-line voltage halfway through the 0-V to 5-V excursion (i.e., 2.5 V). The bit-line capacitance is 2 pF. Note that all NMOS transistors are cut off at this time.

(b) After completion of the precharge interval and the return of ϕ to V_{DD}, the row decoder raises the voltage of the selected word line. Because of the finite resistance and capacitance of the word line, the voltage rises exponentially toward V_{DD}. If the resistance of each of the polysilicon word lines is 3 kΩ and the capacitance between the word line and ground is 3 pF, what is the (10% to 90%) rise time of the word-line voltage? What is the voltage reached at the end of one time constant?

(c) We account for the exponential rise of the word-line voltage by approximating the word-line voltage by a step equal to the voltage reached in one time constant. Find the interval Δt required for an NMOS transistor to discharge the bit line and lower its voltage by 0.5 V. (It is assumed that the sense amplifier needs a 0.5-V change at its input to detect a low bit value.)

Ans. (a) 6.1 ns; (b) 19.8 ns, 3.16 V; (c) 2.9 ns

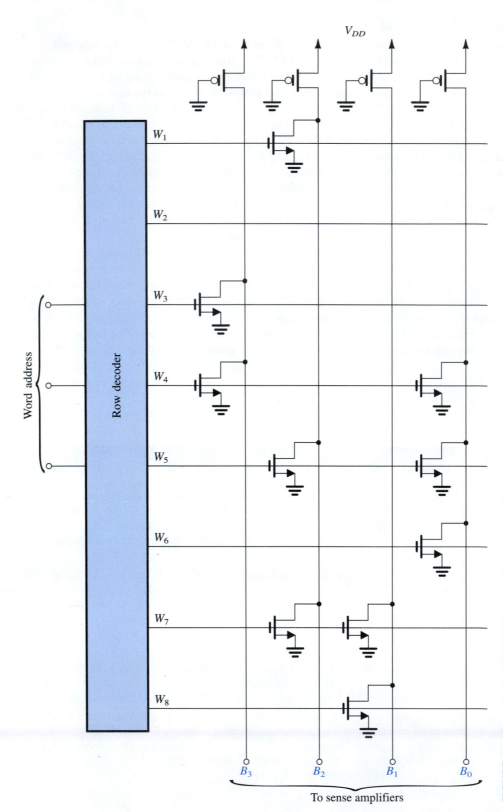

FIGURE 11.29 A simple MOS ROM organized as 8 words × 4 bits.

11.6.2 Mask-Programmable ROMs

The data stored in the ROMs discussed thus far is determined at the time of fabrication, according to the user's specifications. However, to avoid having to custom-design each ROM from scratch (which would be extremely costly), ROMs are manufactured using a process known as **mask programming.** As explained in Appendix A, integrated circuits are fabricated on a wafer of silicon using a sequence of processing steps that include photomasking, etching, and diffusion. In this way, a pattern of junctions and interconnections is created on the surface of the wafer. One of the final steps in the fabrication process consists of coating the surface of the wafer with a layer of aluminum and then selectively (using a mask) etching away portions of the aluminum, leaving aluminum only where interconnections are desired. This last step can be used to program (i.e., to store a desired pattern in) a ROM. For instance, if the ROM is made of MOS transistors as in Fig. 11.29, MOSFETs can be included at all bit locations, but only the gates of those transistors where 0s are to be stored are connected to the word lines; the gates of transistors where 1s are to be stored are not connected. This pattern is determined by the mask, which is produced according to the user's specifications.

The economic advantages of the mask programming process should be obvious: All ROMs are fabricated similarly; customization occurs only during one of the final steps in fabrication.

11.6.3 Programmable ROMs (PROMs and EPROMs)

PROMs are ROMs that can be programmed by the user, but only once. A typical arrangement employed in BJT PROMs involves using polysilicon fuses to connect the emitter of each BJT to the corresponding digit line. Depending on the desired content of a ROM cell, the fuse can be either left intact or blown out using a large current. The programming process is obviously irreversible.

An erasable programmable ROM, or EPROM, is a ROM that can be erased and reprogrammed as many times as the user wishes. It is therefore the most versatile type of read-only memory. It should be noted, however, that the process of erasure and reprogramming is time-consuming and is intended to be performed only infrequently.

State-of-the-art EPROMs use variants of the memory cell whose cross section is shown in Fig. 11.30(a). The cell is basically an enhancement-type *n*-channel MOSFET with two

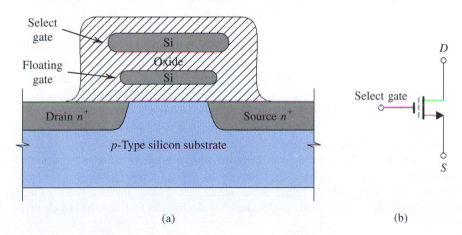

FIGURE 11.30 **(a)** Cross section and **(b)** circuit symbol of the floating-gate transistor used as an EPROM cell.

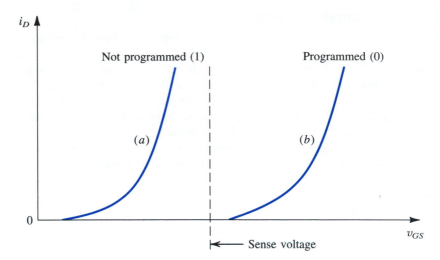

FIGURE 11.31 Illustrating the shift in the i_D–v_{GS} characteristic of a floating-gate transistor as a result of programming.

gates made of polysilicon material.[6] One of the gates is not electrically connected to any other part of the circuit; rather, it is left floating and is appropriately called a **floating gate.** The other gate, called a **select gate,** functions in the same manner as the gate of a regular enhancement MOSFET.

The MOS transistor of Fig. 11.30(a) is known as a **floating-gate transistor** and is given the circuit symbol shown in Fig. 11.30(b). In this symbol the broken line denotes the floating gate. The memory cell is known as the **stacked-gate cell.**

Let us now examine the operation of the floating-gate transistor. Before the cell is programmed (we will shortly explain what this means), no charge exists on the floating gate and the device operates as a regular n-channel enhancement MOSFET. It thus exhibits the i_D–v_{GS} characteristic shown as curve (a) in Fig. 11.31. Note that in this case the threshold voltage (V_t) is rather low. This state of the transistor is known as the **not-programmed state.** It is one of two states in which the floating-gate transistor can exist. Let us arbitrarily take the not-programmed state to represent a stored 1. That is, a floating-gate transistor whose i_D–v_{GS} characteristic is that shown as curve (a) in Fig. 11.31 will be said to be storing a 1.

To program the floating-gate transistor, a large voltage (16–20 V) is applied between its drain and source. Simultaneously, a large voltage (about 25 V) is applied to its select gate. Figure 11.32 shows the floating-gate MOSFET during programming. In the absence of any charge on the floating gate the device behaves as a regular n-channel enhancement MOSFET: An n-type inversion layer (channel) is created at the wafer surface as a result of the large positive voltage applied to the select gate. Because of the large positive voltage at the drain, the channel has a tapered shape.

The drain-to-source voltage accelerates electrons through the channel. As these electrons reach the drain end of the channel, they acquire large kinetic energy and are referred to as *hot electrons*. The large positive voltage on the select gate (greater than the drain voltage) establishes an electric field in the insulating oxide. This electric field attracts the hot electrons

---

[6] See Appendix A for a description of silicon-gate technology.

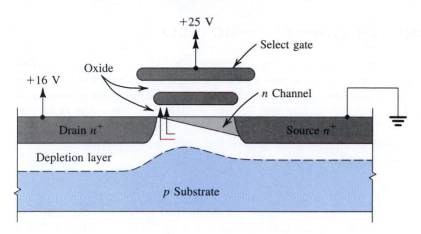

FIGURE 11.32 The floating-gate transistor during programming.

and accelerates them (through the oxide) toward the floating gate. In this way the floating gate is charged, and the charge that accumulates on it becomes trapped.

Fortunately, the process of charging the floating gate is self-limiting. The negative charge that accumulates on the floating gate reduces the strength of the electric field in the oxide to the point that it eventually becomes incapable of accelerating any more of the hot electrons.

Let us now inquire about the effect of the floating gate's negative charge on the operation of the transistor. The negative charge trapped on the floating gate will cause electrons to be repelled from the surface of the substrate. This implies that to form a channel, the positive voltage that has to be applied to the select gate will have to be greater than that required when the floating gate is not charged. In other words, the threshold voltage V_t of the programmed transistor will be higher than that of the not-programmed device. In fact, programming causes the i_D–v_{GS} characteristic to shift to the curve labeled (b) in Fig. 11.31. In this state, known as the *programmed state,* the cell is said to be storing a 0.

Once programmed, the floating-gate device retains its shifted i–v characteristic (curve b) even when the power supply is turned off. In fact, extrapolated experimental results indicate that the device can remain in the programmed state for as long as 100 years!

Reading the content of the stacked-gate cell is easy: A voltage V_{GS} somewhere between the low and high threshold values (see Fig. 11.31) is applied to the selected gate. While a programmed device (one that is storing a 0) will not conduct, a not-programmed device (one that is storing a 1) will conduct heavily.

To return the floating-gate MOSFET to its not-programmed state, the charge stored on the floating gate has to be returned to the substrate. This *erasure* process can be accomplished by illuminating the cell with ultraviolet light of the correct wavelength (2537 Å) for a specified duration. The ultraviolet light imparts sufficient photon energy to the trapped electrons to allow them to overcome the inherent energy barrier, and thus be transported through the oxide, back to the substrate. To allow this erasure process, the EPROM package contains a quartz window. Finally, it should be noted that the device is extremely durable, and can be erased and programmed many times.

A more versatile programmable ROM is the electrically erasable PROM (or EEPROM). As the name implies, an EEPROM can be erased and reprogrammed electrically without the need for ultraviolet illumination. EEPROMs utilize a variant of the floating-gate MOSFET. An important class of EEPROMs using a floating gate variant and implementing block erasure are referred to as Flash memories.

11.7 EMITTER-COUPLED LOGIC (ECL)

Emitter-coupled logic (ECL) is the fastest logic circuit family.[7] High speed is achieved by operating all transistors out of saturation, thus avoiding storage-time delays, and by keeping the logic signal swings relatively small (about 0.8 V or less), thus reducing the time required to charge and discharge the various load and parasitic capacitances. Saturation in ECL is avoided by using the BJT differential pair as a current switch.[8] The BJT differential pair was studied in Chapter 7, and we urge the reader to review the introduction given in Section 7.3 before proceeding with the study of ECL.

11.7.1 The Basic Principle

Emitter-coupled logic is based on the use of the current-steering switch introduced in Section 1.7. Such a switch can be most conveniently realized using the differential pair shown in Fig. 11.33. The pair is biased with a constant-current source I, and one side is connected to a reference voltage V_R. As shown in Section 7.3, the current I can be steered to either Q_1 or Q_2 under the control of the input signal v_I. Specifically, when v_I is greater than V_R by about $4V_T$ ($\cong 100$ mV), nearly all the current I is conducted by Q_1, and thus for $\alpha_1 \cong 1$, $v_{O1} = V_{CC} - IR_C$. Simultaneously, the current through Q_2 will be nearly zero, and thus $v_{O2} = V_{CC}$. Conversely, when v_I is lower than V_R by about $4V_T$, most of the current I will flow through Q_2 and the current through Q_1 will be nearly zero. Thus $v_{O1} = V_{CC}$ and $v_{O2} = V_{CC} - IR_C$.

The preceding description suggests that as a logic element, the differential pair realizes an inversion function at v_{O1} and simultaneously provides the complementary output signal at v_{O2}. The output logic levels are $V_{OH} = V_{CC}$ and $V_{OL} = V_{CC} - IR_C$, and thus the output logic

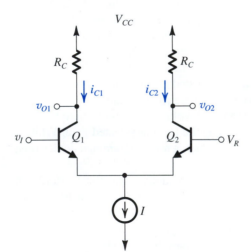

FIGURE 11.33 The basic element of ECL is the differential pair. Here, V_R is a reference voltage.

---

[7] Although higher speeds of operation can be obtained with gallium arsenide (GaAs) circuits, the latter are not available as off-the-shelf components for conventional digital system design. GaAs digital circuits are not covered in this book; however, a substantial amount of material on this subject can be found on the CD accompanying the book and on the website at www.sedrasmith.org.

[8] This is in sharp contrast to the technique utilized in a nonsaturating variant of transistor-transistor logic (TTL) known as Schottky TTL. There, a Schottky diode is placed across the CBJ junction to shunt away some of the base current and, owing to the low voltage drop of the Schottky diode, the CBJ from becoming forward biased.

swing is IR_C. A number of additional remarks can be made concerning this circuit:

1. The differential nature of the circuit makes it less susceptible to picked-up noise. In particular, an interfering signal will tend to affect both sides of the differential pair similarly and thus will not result in current switching. This is the common-mode rejection property of the differential pair (see Section 7.3).

2. The current drawn from the power supply remains constant during switching. Thus, unlike CMOS (and TTL), no supply current spikes occur in ECL, eliminating an important source of noise in digital circuits. This is a definite advantage, especially since ECL is usually designed to operate with small signal swings and has correspondingly low noise margins.

3. The output signal levels are both referenced to V_{CC} and thus can be made particularly stable by operating the circuit with $V_{CC} = 0$, in other words, by utilizing a negative power supply and connecting the V_{CC} line to ground. In this case, $V_{OH} = 0$ and $V_{OL} = -IR_C$.

4. Some means has to be provided to make the output signal levels compatible with those at the input so that one gate can drive another. As we shall see shortly, practical ECL gate circuits incorporate a level-shifting arrangement that serves to center the output signal levels on the value of V_R.

5. The availability of complementary outputs considerably simplifies logic design with ECL.

EXERCISE

11.17 For the circuit in Fig. 11.33, let $V_{CC} = 0$, $I = 4$ mA, $R_C = 220$ Ω, $V_R = -1.32$ V, and assume $\alpha \cong 1$. Determine V_{OH} and V_{OL}. By how much should the output levels be shifted so that the values of V_{OH} and V_{OL} become centered on V_R? What will the shifted values of V_{OH} and V_{OL} be?

Ans. 0; −0.88 V; −0.88 V; −0.88 V, −1.76 V

11.7.2 ECL Families

Currently there are two popular forms of commercially available ECL—namely, ECL 10K and ECL 100K. The ECL 100K series features gate delays of the order of 0.75 ns and dissipates about 40 mW/gate, for a delay–power product of 30 pJ. Although its power dissipation is relatively high, the 100K series provides the shortest available gate delay.

The ECL 10K series is slightly slower; it features a gate propagation delay of 2 ns and a power dissipation of 25 mW for a delay–power product of 50 pJ. Although the value of *DP* is higher than that obtained in the 100K series, the 10K series is easier to use. This is because the rise and fall times of the pulse signals are deliberately made longer, thus reducing signal coupling, or crosstalk, between adjacent signal lines. ECL 10K has an "edge speed" of about 3.5 ns, compared with the approximately 1 ns of ECL 100K. To give concreteness to our study of ECL, in the following we shall consider the popular ECL 10K in some detail. The same techniques, however, can be applied to other types of ECL.

In addition to its usage in small- and medium-scale integrated-circuit packages, ECL is also employed in large-scale and VLSI applications. A variant of ECL known as **current-mode logic** (CML) is utilized in VLSI applications [see Treadway (1989) and Wilson (1990)].

11.7.3 The Basic Gate Circuit

The basic gate circuit of the ECL 10K family is shown in Fig. 11.34. The circuit consists of three parts. The network composed of Q_1, D_1, D_2, R_1, R_2, and R_3 generates a reference voltage

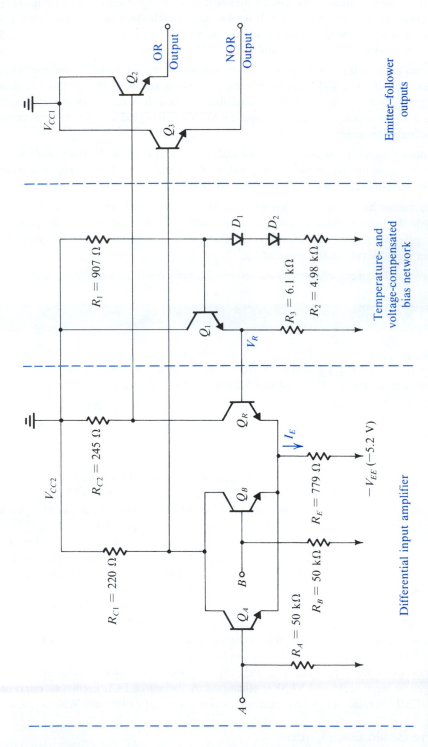

FIGURE 11.34 Basic circuit of the ECL 10K logic-gate family.

V_R whose value at room temperature is -1.32 V. As will be shown, the value of this reference voltage is made to change with temperature in a predetermined manner to keep the noise margins almost constant. Also, the reference voltage V_R is made relatively insensitive to variations in the power-supply voltage V_{EE}.

EXERCISE

11.18 Figure E11.18 shows the circuit that generates the reference voltage V_R. Assuming that the voltage drop across each of D_1, D_2, and the base–emitter junction of Q_1 is 0.75 V, calculate the value of V_R. Neglect the base current of Q_1.

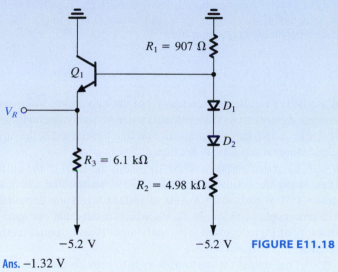

FIGURE E11.18

Ans. -1.32 V

The second part, and the heart of the gate, is the differential amplifier formed by Q_R and either Q_A or Q_B. This differential amplifier is biased not by a constant-current source, as was done in the circuit of Fig. 11.33, but with a resistance R_E connected to the negative supply $-V_{EE}$. Nevertheless, we will shortly show that the current in R_E remains approximately constant over the normal range of operation of the gate. One side of the differential amplifier consists of the reference transistor Q_R, whose base is connected to the reference voltage V_R. The other side consists of a number of transistors (two in the case shown), connected in parallel, with separated bases, each connected to a gate input. If the voltages applied to A and B are at the logic-0 level, which, as we will soon find out, is about 0.4 V below V_R, both Q_A and Q_B will be off and the current I_E in R_E will flow through the reference transistor Q_R. The resulting voltage drop across R_{C2} will cause the collector voltage of Q_R to be low.

On the other hand, when the voltage applied to A or B is at the logic-1 level, which, as we will show shortly, is about 0.4 V above V_R, transistor Q_A or Q_B, or both, will be on and Q_R will be off. Thus the current I_E will flow through Q_A or Q_B, or both, and an almost equal current flows through R_{C1}. The resulting voltage drop across R_{C1} will cause the collector voltage to drop. Meanwhile, since Q_R is off, its collector voltage rises. We thus see that the voltage at the collector of Q_R will be high if A or B, or both, is high, and thus at the collector of Q_R the OR logic function, $A + B$, is realized. On the other hand, the common collector of Q_A and Q_B will be high only when A and B are simultaneously low. Thus, at the common collector of Q_A and Q_B the logic function $\overline{A}\,\overline{B} = \overline{A+B}$ is realized. We therefore conclude that the two-input gate of Fig. 11.34 realizes the OR function and its complement, the NOR function. The availability of complementary outputs is an important advantage of ECL; it simplifies logic design and avoids the use of additional inverters with associated time delay.

It should be noted that the resistance connecting each of the gate input terminals to the negative supply enables the user to leave an unused input terminal open: An open input terminal will be *pulled down* to the negative supply voltage, and its associated transistor will be off.

EXERCISE

11.19 With input terminals A and B in Fig. 11.34 left open, find the current I_E through R_E. Also find the voltages at the collector of Q_R and at the common collector of the input transistors Q_A and Q_B. Use $V_R = -1.32$ V, V_{BE} of $Q_R \simeq 0.75$ V, and assume that β of Q_R is very high.
Ans. 4 mA; −1 V; 0 V

The third part of the ECL gate circuit is composed of the two emitter followers, Q_2 and Q_3. The emitter followers do not have on-chip loads, since in many applications of high-speed logic circuits the gate output drives a transmission line terminated at the other end, as indicated in Fig. 11.35. (More on this later in Section 11.7.6.)

The emitter followers have two purposes: First, they shift the level of the output signals by one V_{BE} drop. Thus, using the results of Exercise 11.19, we see that the output levels become approximately −1.75 V and −0.75 V. These shifted levels are centered approximately around the reference voltage ($V_R = -1.32$ V), which means that one gate can drive another. This compatibility of logic levels at input and output is an essential requirement in the design of gate circuits.

The second function of the output emitter followers is to provide the gate with low output resistances and with the large output currents required for charging load capacitances. Since these large transient currents can cause spikes on the power-supply line, the collectors of the emitter followers are connected to a power-supply terminal V_{CC1} separate from that of the differential amplifier and the reference-voltage circuit, V_{CC2}. Here we note that the supply current of the differential amplifier and the reference circuit remains almost constant. The use of separate power-supply terminals prevents the coupling of power-supply spikes from the output circuit to the gate circuit and thus lessens the likelihood of false gate switching. Both V_{CC1} and V_{CC2} are of course connected to the same system ground, external to the chip.

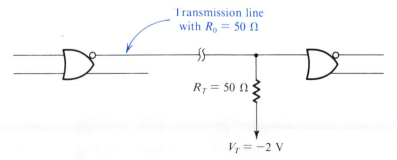

FIGURE 11.35 The proper way to connect high-speed logic gates such as ECL. Properly terminating the transmission line connecting the two gates eliminates the "ringing" that would otherwise corrupt the logic signals. (See Section 11.7.6.)

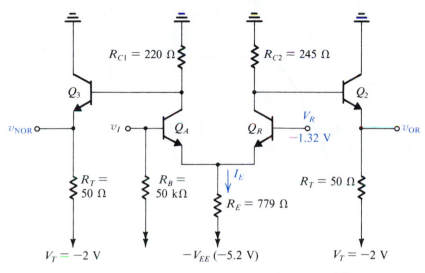

FIGURE 11.36 Simplified version of the ECL gate for the purpose of finding transfer characteristics.

11.7.4 Voltage Transfer Characteristics

Having provided a qualitative description of the operation of the ECL gate, we shall now derive its voltage transfer characteristics. This will be done under the conditions that the outputs are terminated in the manner indicated in Fig. 11.35. Assuming that the B input is low and thus Q_B is off, the circuit simplifies to that shown in Fig. 11.36. We wish to analyze this circuit to determine v_{OR} versus v_I and v_{NOR} versus v_I (where $v_I \equiv v_A$).

In the analysis to follow we shall make use of the exponential i_C–v_{BE} characteristic of the BJT. Since the BJTs used in ECL circuits have small areas (in order to have small capacitances and hence high f_T), their scale currents I_S are small. We will therefore assume that at an emitter current of 1 mA an ECL transistor has a V_{BE} drop of 0.75 V.

The OR Transfer Curve Figure 11.37 is a sketch of the OR transfer characteristic, v_{OR} versus v_I, with the parameters V_{OL}, V_{OH}, V_{IL}, and V_{IH} indicated. However, to simplify the calculation of V_{IL} and V_{IH}, we shall use an alternative to the unity-gain definition. Specifically, we shall assume that at point x, transistor Q_A is conducting 1% of I_E while Q_R is conducting 99% of I_E. The reverse will be assumed for point y. Thus at point x we have

$$\frac{I_E|_{Q_R}}{I_E|_{Q_A}} = 99$$

Using the exponential i_E–v_{BE} relationship, we obtain

$$V_{BE}|_{Q_R} - V_{BE}|_{Q_A} = V_T \ln 99 = 115 \text{ mV}$$

which gives

$$V_{IL} = -1.32 - 0.115 = -1.435 \text{ V}$$

Assuming Q_A and Q_R to be matched, we can write

$$V_{IH} - V_R = V_R - V_{IL}$$

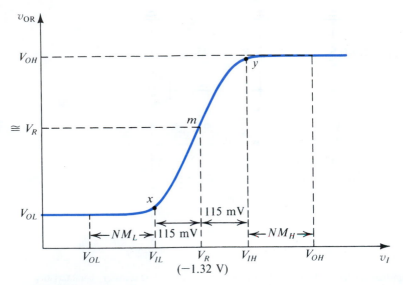

FIGURE 11.37 The OR transfer characteristic v_{OR} versus v_I, for the circuit in Fig. 11.36.

which can be used to find V_{IH} as

$$V_{IH} = -1.205 \text{ V}$$

To obtain V_{OL} we note that Q_A is off and Q_R carries the entire current I_E, given by

$$I_E = \frac{V_R - V_{BE}|_{Q_R} + V_{EE}}{R_E}$$

$$= \frac{-1.32 - 0.75 + 5.2}{0.779}$$

$$\simeq 4 \text{ mA}$$

(If we wish, we can iterate to determine a better estimate of $V_{BE}|_{Q_R}$ and hence of I_E.) Assuming that Q_R has a high β so that its $\alpha \simeq 1$, its collector current will be approximately 4 mA. If we neglect the base current of Q_2, we obtain for the collector voltage of Q_R

$$V_C|_{Q_R} \simeq -4 \times 0.245 = -0.98 \text{ V}$$

Thus a first approximation for the value of the output voltage V_{OL} is

$$V_{OL} = V_C|_{Q_R} - V_{BE}|_{Q_2}$$

$$\simeq -0.98 - 0.75 = -1.73 \text{ V}$$

We can use this value to find the emitter current of Q_2 and then iterate to determine a better estimate of its base–emitter voltage. The result is $V_{BE2} \simeq 0.79$ V and, correspondingly,

$$V_{OL} \simeq -1.77 \text{ V}$$

At this value of output voltage, Q_2 supplies a load current of about 4.6 mA.

To find the value of V_{OH} we assume that Q_R is completely cut off (because $v_I > V_{IH}$). Thus the circuit for determining V_{OH} simplifies to that in Fig. 11.38. Analysis of this circuit assuming $\beta_2 = 100$ results in $V_{BE2} \simeq 0.83$ V, $I_{E2} = 22.4$ mA, and

$$V_{OH} \simeq -0.88 \text{ V}$$

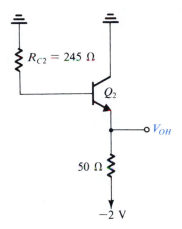

FIGURE 11.38 Circuit for determining V_{OH}.

Noise Margins The results of Exercise 11.20 indicate that the bias current I_E remains approximately constant. Also, the output voltage corresponding to $v_I = V_R$ is approximately equal to V_R. Notice further that this is also approximately the midpoint of the logic swing; specifically,

$$\frac{V_{OL} + V_{OH}}{2} = -1.325 \simeq V_R$$

Thus the output logic levels are centered around the midpoint of the input transition band. This is an ideal situation from the point of view of noise margins, and it is one of the reasons for selecting the rather arbitrary-looking numbers ($V_R = -1.32$ V and $V_{EE} = 5.2$ V) for reference and supply voltages.

The noise margins can now be evaluated as follows:

$$NM_H = V_{OH} - V_{IH} \qquad\qquad NM_L = V_{IL} - V_{OL}$$
$$= -0.88 - (-1.205) = 0.325 \text{ V} \qquad = -1.435 - (-1.77) = 0.335 \text{ V}$$

Note that these values are approximately equal.

The NOR Transfer Curve The NOR transfer characteristic, which is v_{NOR} versus v_I for the circuit in Fig. 11.36, is sketched in Fig. 11.39. The values of V_{IL} and V_{IH} are identical to those found earlier for the OR characteristic. To emphasize this we have labeled the threshold points x and y, the same letters used in Fig. 11.37.

For $v_I < V_{IL}$, Q_A is off and the output voltage v_{NOR} can be found by analyzing the circuit composed of R_{C1}, Q_3, and its 50-Ω termination. Except that R_{C1} is slightly smaller than R_{C2}, this circuit is identical to that in Fig. 11.38. Thus the output voltage will be only slightly

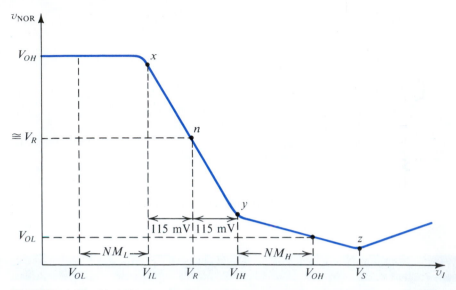

FIGURE 11.39 The NOR transfer characteristic, v_{NOR} versus v_I, for the circuit in Fig. 11.36.

greater than the value V_{OH} found earlier. In the sketch of Fig. 11.39 we have assumed that the output voltage is approximately equal to V_{OH}.

For $v_I > V_{IH}$, Q_A is on and is conducting the entire bias current. The circuit then simplifies to that in Fig. 11.40. This circuit can be easily analyzed to obtain v_{NOR} versus v_I for the range $v_I \geq V_{IH}$. A number of observations are in order. First, note that $v_I = V_{IH}$ results in an output voltage slightly higher than V_{OL}. This is because R_{C1} is smaller than R_{C2}. In fact, R_{C1} is chosen lower in value than R_{C2} so that with v_I equal to the normal logic-1 value (i.e., V_{OH}, which is approximately -0.88 V), the output will be equal to the V_{OL} value found earlier for the OR output.

Second, note that as v_I exceeds V_{IH}, transistor Q_A operates in the active mode and the circuit of Fig. 11.40 can be analyzed to find the gain of this amplifier, which is the slope of the segment yz of the transfer characteristic. At point z, transistor Q_A saturates. Further increments in v_I (beyond the point $v_I = V_S$) cause the collector voltage and hence v_{NOR} to increase.

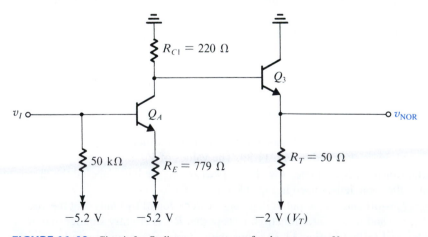

FIGURE 11.40 Circuit for finding v_{NOR} versus v_I for the range $v_I > V_{IH}$.

The slope of the segment of the transfer characteristic beyond point z, however, is not unity but is about 0.5 because as Q_A is driven deeper into saturation, a portion of the increment in v_I appears as an increment in the base–collector forward-bias voltage. The reader is urged to solve Exercise 11.21, which is concerned with the details of the NOR transfer characteristic.

EXERCISE

11.21 Consider the circuit in Fig. 11.40. (a) For $v_I = V_{IH} = -1.205$ V, find v_{NOR}. (b) For $v_I = V_{OH} = -0.88$ V, find v_{NOR}. (c) Find the slope of the transfer characteristic at the point $v_I = V_{OH} = -0.88$ V. (d) Find the value of v_I at which Q_A saturates (i.e., V_S). Assume that $V_{BE} = 0.75$ V at a current of 1 mA, $V_{CEsat} \simeq 0.3$ V, and $\beta = 100$.

Ans. (a) −1.70 V; (b) −1.79 V; (c) −0.24 V/V; (d) −0.58 V

Manufacturers' Specifications ECL manufacturers supply gate transfer characteristics of the form shown in Figs. 11.37 and 11.39. A manufacturer usually provides such curves measured at a number of temperatures. In addition, at each relevant temperature, worst-case values for the parameters V_{IL}, V_{IH}, V_{OL}, and V_{OH} are given. These worst-case values are specified with the inevitable component tolerances taken into account. As an example, Motorola specifies that for MECL 10,000 at 25°C the following worst-case values apply[9]:

$$V_{IL\max} = -1.475 \text{ V} \qquad V_{IH\min} = -1.105 \text{ V}$$

$$V_{OL\max} = -1.630 \text{ V} \qquad V_{OH\min} = -0.980 \text{ V}$$

These values can be used to determine worst-case noise margins,

$$NM_L = 0.155 \text{ V} \qquad NM_H = 0.125 \text{ V}$$

which are about half the *typical* values previously calculated.

For additional information on MECL specifications the interested reader is referred to the Motorola (1988, 1989) publications listed in the bibliography at the end of the book.

11.7.5 Fan-Out

When the input signal to an ECL gate is low, the input current is equal to the current that flows in the 50-kΩ pull-down resistor. Thus

$$I_{IL} = \frac{-1.77 + 5.2}{50} \simeq 69 \ \mu A$$

When the input is high, the input current is greater because of the base current of the input transistor. Thus, assuming a transistor β of 100, we obtain

$$I_{IH} = \frac{-0.88 + 5.2}{50} + \frac{4}{101} \simeq 126 \ \mu A$$

Both these current values are quite small, which, coupled with the very small output resistance of the ECL gate, ensures that little degradation of logic-signal levels results from the input currents of fan-out gates. It follows that the fan-out of ECL gates is not limited by

---

[9] MECL is the trade name used by Motorola for its ECL.

logic-level considerations but rather by the degradation of the circuit speed (rise and fall times). This latter effect is due to the capacitance that each fan-out gate presents to the driving gate (approximately 3 pF). Thus while the *dc fan-out* can be as high as 90 and thus does not represent a design problem, the *ac fan-out* is limited by considerations of circuit speed to 10 or so.

11.7.6 Speed of Operation and Signal Transmission

The speed of operation of a logic family is measured by the delay of its basic gate and by the rise and fall times of the output waveforms. Typical values of these parameters for ECL have already been given. Here we should note that because the output circuit is an emitter follower, the rise time of the output signal is shorter than its fall time, since on the rising edge of the output pulse the emitter follower functions and provides the output current required to charge up the load and parasitic capacitances. On the other hand, as the signal at the base of the emitter follower falls, the emitter follower cuts off, and the load capacitance discharges through the combination of load and pull-down resistances.

To take full advantage of the very high speed of operation possible with ECL, special attention should be paid to the method of interconnecting the various logic gates in a system. To appreciate this point, we shall briefly discuss the problem of signal transmission.

ECL deals with signals whose rise times may be 1 ns or even less, the time it takes for light to travel only 30 cm or so. For such signals a wire and its environment become a relatively complex circuit element along which signals propagate with finite speed (perhaps half the speed of light—i.e., 15 cm/ns). Unless special care is taken, energy that reaches the end of such a wire is not absorbed but rather returns as a *reflection* to the transmitting end, where (without special care) it may be re-reflected. The result of this process of reflection is what can be observed as **ringing,** a damped oscillatory excursion of the signal about its final value.

Unfortunately, ECL is particularly sensitive to ringing because the signal levels are so small. Thus it is important that transmission of signals be well controlled, and surplus energy absorbed, to prevent reflections. The accepted technique is to limit the nature of connecting wires in some way. One way is to insist that they be very short, where "short" is taken to mean with respect to the signal rise time. The reason for this is that if the wire connection is so short that reflections return while the input is still rising, the result becomes only a somewhat slowed and "bumpy" rising edge.

If, however, the reflection returns *after* the rising edge, it produces not simply a modification of the initiating edge but an *independent second event.* This is clearly bad! Thus the time taken for a signal to go from one end of a line and back is restricted to less than the rise time of the driving signal by some factor—say, 5. Thus for a signal with a 1-ns rise time and for propagation at the speed of light (30 cm/ns), a double path of only 0.2-ns equivalent length, or 6 cm, would be allowed, representing in the limit a wire only 3 cm from end to end.

Such is the restriction on ECL 100K. However, ECL 10K has an intentionally slower rise time of about 3.5 ns. Using the same rules, wires can accordingly be as long as about 10 cm for ECL 10K.

If greater lengths are needed, then transmission lines must be used. These are simply wires in a controlled environment in which the distance to a ground reference plane or second wire is highly controlled. Thus they might simply be twisted pairs of wires, one of which is grounded, or parallel ribbon wires, every second of which is grounded, or so-called microstrip lines on a printed-circuit board. The latter are simply copper strips of controlled geometry on one side of a thin printed-circuit board, the other side of which consists of a grounded plane.

Such transmission lines have a *characteristic impedance*, R_0, that ranges from a few tens of ohms to hundreds of ohms. Signals propagate on such lines somewhat more slowly than the speed of light, perhaps half as fast. When a transmission line is terminated at its receiving end in a resistance equal to its characteristic impedance, R_0, all the energy sent on the line is absorbed at the receiving end, and no reflections occur (since the termination acts as a limitless length of transmission line). Thus, signal integrity is maintained. Such transmission lines are said to be *properly terminated*. A properly terminated line appears at its sending end as a resistor of value R_0. The followers of ECL 10K with their open emitters and low output resistances (specified to be 7 Ω maximum) are ideally suited for driving transmission lines. ECL is also good as a line receiver. The simple gate with its high (50-kΩ) pull-down input resistor represents a very high resistance to the line. Thus a few such gates can be connected to a terminated line with little difficulty. Both of these ideas are represented in Fig. 11.35.

11.7.7 Power Dissipation

Because of the differential-amplifier nature of ECL, the gate current remains approximately constant and is simply steered from one side of the gate to the other depending on the input logic signals. Thus, the supply current and hence the gate power dissipation of unterminated ECL remain relatively constant independent of the logic state of the gate. It follows that no voltage spikes are introduced on the supply line. Such spikes can be a dangerous source of noise in a digital system. It follows that in ECL the need for supply-line bypassing is not as great as in, say, TTL. This is another advantage of ECL.

At this juncture we should reiterate a point we made earlier, namely, that although an ECL gate would operate with $V_{EE} = 0$ and $V_{CC} = +5.2$ V, the selection of $V_{EE} = -5.2$ V and $V_{CC} = 0$ V is recommended because in the circuit all signal levels are referenced to V_{CC}, and ground is certainly an excellent reference.

EXERCISE

11.22 For the ECL gate in Fig. 11.34, calculate an approximate value for the power dissipated in the circuit under the condition that all inputs are low and that the emitters of the output followers are left open. Assume that the reference circuit supplies four identical gates, and hence only a quarter of the power dissipated in the reference circuit should be attributed to a single gate.

Ans. 22.4 mW

11.7.8 Thermal Effects

In our analysis of the ECL gate of Fig. 11.34, we found that at room temperature the reference voltage V_R is -1.32 V. We have also shown that the midpoint of the output logic swing is approximately equal to this voltage, which is an ideal situation in that it results in equal high and low noise margins. In Example 11.4, we shall derive expressions for the temperature coefficients of the reference voltage and of the output low and high voltages. In this way, it will be shown that the midpoint of the output logic swing varies with temperature at the same rate as the reference voltage. As a result, although the magnitudes of the high and low noise margins change with temperature, their values remain equal. This is an added advantage of ECL and provides a demonstration of the high degree of design optimization of this gate circuit.

EXAMPLE 11.4

We wish to determine the temperature coefficient of the reference voltage V_R and of the midpoint between V_{OL} and V_{OH}.

Solution

To determine the temperature coefficient of V_R, consider the circuit in Fig. E11.18 and assume that the temperature changes by +1°C. Denoting the temperature coefficient of the diode and transistor voltage drops by δ, where $\delta \simeq -2$ mV/°C, we obtain the equivalent circuit shown in Fig. 11.41. In the latter circuit the changes in device voltage drops are considered as signals, and hence the power supply is shown as a signal ground.

In the circuit of Fig. 11.41 we have two signal generators, and we wish to analyze the circuit to determine ΔV_R, the change in V_R. We shall do so using the principle of superposition. Consider first the branch R_1, D_1, D_2, 2δ, and R_2, and neglect the signal base current of Q_1. The voltage signal at the base of Q_1 can be easily obtained from

$$v_{b1} = \frac{2\delta \times R_1}{R_1 + r_{d1} + r_{d2} + R_2}$$

where r_{d1} and r_{d2} denote the incremental resistances of diodes D_1 and D_2, respectively. The dc bias current through D_1 and D_2 is approximately 0.64 mA, and thus $r_{d1} = r_{d2} = 39.5\ \Omega$. Hence $v_{b1} \simeq 0.3\delta$. Since the gain of the emitter follower Q_1 is approximately unity, it follows that the component of ΔV_R due to the generator 2δ is approximately equal to v_{b1}, that is, $\Delta V_{R1} = 0.3\delta$.

Consider next the component of ΔV_R due to the generator δ. Reflection into the emitter circuit of the total resistance of the base circuit, $[R_1 \| (r_{d1} + r_{d2} + R_2)]$, by dividing it by $\beta + 1$ (with $\beta \simeq 100$) results in the following component of ΔV_R:

$$\Delta V_{R2} = -\frac{\delta \times R_3}{[R_B/(\beta + 1)] + r_{e1} + R_3}$$

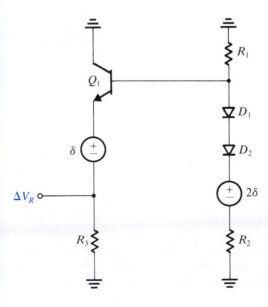

FIGURE 11.41 Equivalent circuit for determining the temperature coefficient of the reference voltage V_R.

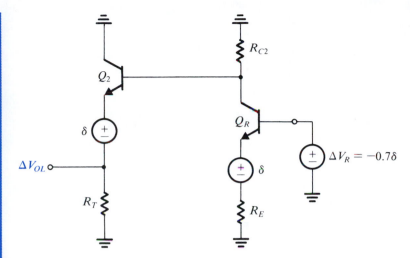

FIGURE 11.42 Equivalent circuit for determining the temperature coefficient of V_{OL}.

where R_B denotes the total resistance in the base circuit, and r_{e1} denotes the emitter resistance of Q_1 ($\simeq 40\ \Omega$). This calculation yields $\Delta V_{R2} \simeq -\delta$. Adding this value to that due to the generator 2δ gives $\Delta V_R \simeq -0.7\delta$. Thus for $\delta = -2$ mV/°C the temperature coefficient of V_R is +1.4 mV/°C.

We next consider the determination of the temperature coefficient of V_{OL}. The circuit on which to perform this analysis is shown in Fig. 11.42. Here we have three generators whose contributions can be considered separately and the resulting components of ΔV_{OL} summed. The result is

$$\Delta V_{OL} \simeq \Delta V_R \frac{-R_{C2}}{r_{eR}+R_E} \frac{R_T}{R_T+r_{e2}}$$

$$-\delta \frac{-R_{C2}}{r_{eR}+R_E} \frac{R_T}{R_T+r_{e2}}$$

$$-\delta \frac{R_T}{R_T+r_{e2}+R_{C2}/(\beta+1)}$$

Substituting the values given and those obtained throughout the analysis of this section, we find

$$\Delta V_{OL} \simeq -0.43\delta$$

The circuit for determining the temperature coefficient of V_{OH} is shown in Fig. 11.43, from which we obtain

$$\Delta V_{OH} = -\delta \frac{R_T}{R_T+r_{e2}+R_{C2}/(\beta+1)} \simeq -0.93\delta$$

We now can obtain the variation of the midpoint of the logic swing as

$$\frac{\Delta V_{OL}+\Delta V_{OH}}{2} = -0.68\delta$$

which is approximately equal to that of the reference voltage $V_R(-0.7\delta)$.

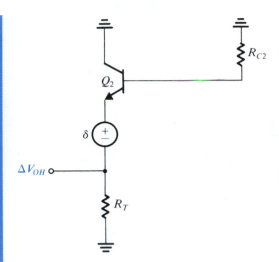

FIGURE 11.43 Equivalent circuit for determining the temperature coefficient of V_{OH}.

11.7.9 The Wired-OR Capability

The emitter-follower output stage of the ECL family allows an additional level of logic to be performed at very low cost by simply wiring the outputs of several gates in parallel. This is illustrated in Fig. 11.44, where the outputs of two gates are wired together. Note that the base–emitter diodes of the output followers realize an OR function: This **wired-OR** connection can be used to provide gates with high fan-in as well as to increase the flexibility of ECL in logic design.

11.7.10 Final Remarks

We have chosen to study ECL by focusing on a commercially available circuit family. As has been demonstrated, a great deal of design optimization has been applied to create a very-high-performance family of SSI and MSI logic circuits. As already mentioned, ECL and some of its variants are also used in VLSI circuit design. Applications include very-high-speed processors such as those used in supercomputers, as well as high-speed and high-frequency communication systems. When employed in VLSI design, current–source biasing is almost always utilized. Further, a variety of circuit configurations are employed [see Rabaey (1996)].

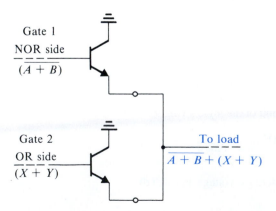

FIGURE 11.44 The wired-OR capability of ECL.

11.8 BiCMOS DIGITAL CIRCUITS

In this section, we provide an introduction to a VLSI circuit technology that is becoming increasingly popular, BiCMOS. As its name implies, BiCMOS technology combines *bi*polar and *CMOS* circuits on one IC chip. The aim is to combine the low power, high input impedance, and wide noise margins of CMOS with the high current-driving capability of bipolar transistors. Specifically, CMOS, although a nearly ideal logic-circuit technology in many respects, has a limited current-driving capability. This is not a serious problem when the CMOS gate has to drive a few other CMOS gates. It becomes a serious issue, however, when relatively large capacitive loads (e.g., greater than 0.5 pF or so) are present. In such cases, one has to either resort to the use of elaborate CMOS buffer circuits or face the usually unacceptable consequence of long propagation delays. On the other hand, we know that by virtue of its much larger transconductance, the BJT is capable of large output currents. We have seen a practical illustration of that in the emitter–follower output stage of ECL. Indeed, the high current-driving capability contributes to making ECL two to five times faster than CMOS (under equivalent conditions)—of course, at the expense of high power dissipation. In summary, then, BiCMOS seeks to combine the best of the CMOS and bipolar technologies to obtain a class of circuits that is particularly useful when output currents that are higher than possible with CMOS are needed. Furthermore, since BiCMOS technology is well suited for the implementation of high-performance analog circuits, it makes possible the realization of both analog and digital functions on the same IC chip, making the "system on a chip" an attainable goal. The price paid is a more complex, and hence more expensive (than CMOS) processing technology.

11.8.1 The BiCMOS Inverter

A variety of BiCMOS inverter circuits have been proposed and are in use. All of these are based on the use of *npn* transistors to increase the output current available from a CMOS inverter. This can be most simply achieved by cascading each of the Q_N and Q_P devices of the CMOS inverter with an *npn* transistor, as shown in Fig. 11.45(a). Observe that this circuit can be thought of as utilizing the pair of complementary composite MOS-BJT devices shown in Fig. 11.45(b). These composite devices[10] retain the high input impedance of the MOS transistor while in effect multiplying its rather low g_m by the β of the BJT. It is also useful to observe that the output stage formed by Q_1 and Q_2 has what is known as the **totem-pole configuration** utilized by TTL.[11]

The circuit of Fig. 11.45(a) operates as follows: When v_I is low, both Q_N and Q_2 are off while Q_P conducts and supplies Q_1 with base current, thus turning it on. Transistor Q_1 then provides a large output current to charge the load capacitance. The result is a very fast charging of the load capacitance and correspondingly a short low-to-high propagation delay, t_{PLH}. Transistor Q_1 turns off when v_O reaches a value about $V_{DD} - V_{BE1}$, and thus the output high level is lower than V_{DD}, a disadvantage. When v_I goes high, Q_P and Q_1 turn off, and Q_N turns on, providing its drain current into the base of Q_2. Transistor Q_2 then turns on and provides a large output current that quickly discharges the load capacitance. Here again the result is a short high-to-low propagation delay, t_{PHL}. On the negative side, Q_2 turns off when v_O reaches a value about V_{BE2}, and thus the output low level is greater than zero, a disadvantage.

---

[10] It is interesting to note that these composite devices were proposed as early as 1969 [see Lin et al. (1969)].

[11] Refer to the CD accompanying this book for a description of the basic TTL logic-gate circuit and its totem-pole output stage.

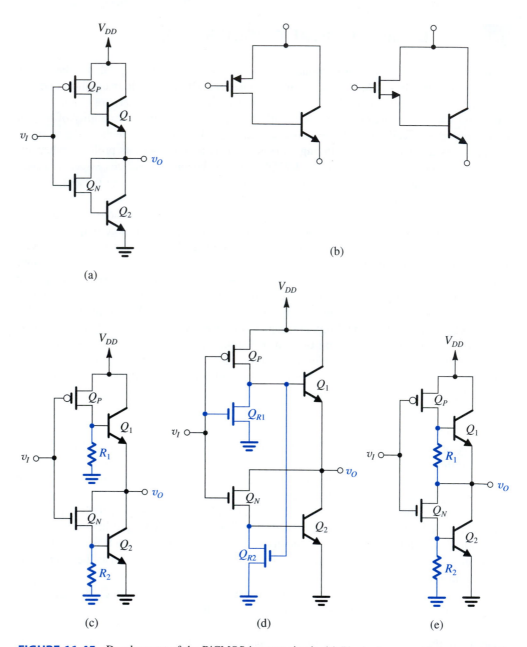

FIGURE 11.45 Development of the BiCMOS inverter circuit. **(a)** The basic concept is to use an additional bipolar transistor to increase the output current drive of each of Q_N and Q_P of the CMOS inverter. **(b)** The circuit in (a) can be thought of as utilizing these composite devices. **(c)** To reduce the turn-off times of Q_1 and Q_2, "bleeder resistors" R_1 and R_2 are added. **(d)** Implementation of the circuit in (c) using NMOS transistors to realize the resistors. **(e)** An improved version of the circuit in (c) obtained by connecting the lower end of R_1 to the output node.

Thus, while the circuit of Fig. 11.45(a) features large output currents and short propagation delays, it has the disadvantage of reduced logic swing, and, correspondingly, reduced noise margins. There is also another and perhaps more serious disadvantage, namely, the relatively long turn-off delays of Q_1 and Q_2 arising from the absence of circuit paths along which the base charge can be removed. This problem can be solved by adding a resistor between the base of each of Q_1 and Q_2 and ground, as shown in Fig. 11.45(c). Now when either Q_1 or Q_2 is turned off, its stored base charge is removed to ground through R_1 or R_2, respectively. Resistor R_2 provides an additional benefit: With v_I high, and after Q_2 cuts off, v_O continues to fall below V_{BE2}, and the output node is pulled to ground through the series path of Q_N and R_2. Thus R_2 functions as a pull-down resistor. The Q_N–R_2 path, however, is a high-impedance one with the result that pulling v_O to ground is a rather slow process. Incorporating the resistor R_1, however, is disadvantageous from a static power-dissipation standpoint: When v_I is low, a dc path exists between V_{DD} and ground through the conducting Q_P and R_1. Finally, it should be noted that R_1 and R_2 take some of the drain currents of Q_P and Q_N away from the bases of Q_1 and Q_2 and thus slightly reduce the gate output current available to charge and discharge the load capacitance.

Figure 11.45(d) shows the way in which R_1 and R_2 are usually implemented. As indicated, NMOS devices Q_{R1} and Q_{R2} are used to realize R_1 and R_2. As an added innovation, these two transistors are made to conduct only when needed. Thus, Q_{R1} will conduct only when v_I rises, at which time its drain current constitutes a reverse base current for Q_1, speeding up its turn-off. Similarly Q_{R2} will conduct only when v_I falls and Q_P conducts, pulling the gate of Q_{R2} high. The drain current of Q_{R2} then constitutes a reverse base current for Q_2, speeding up its turn-off.

As a final circuit for the BiCMOS inverter, we show the so-called R-circuit in Fig. 11.45(e). This circuit differs from that in Fig. 11.45(c) in only one respect: Rather than returning R_1 to ground, we have connected R_1 to the output node of the inverter. This simple change has two benefits. First, the problem of static power dissipation is now solved. Second, R_1 now functions as a pull-up resistor, pulling the output node voltage up to V_{DD} (through the conducting Q_P) after Q_1 has turned off. Thus, the R circuit in Fig. 11.45(e) does in fact have output levels very close to V_{DD} and ground.

As a final remark on the BiCMOS inverter, we note that the circuit is designed so that transistors Q_1 and Q_2 are never simultaneously conducting and neither is allowed to saturate. Unfortunately, sometimes the resistance of the collector region of the BJT in conjunction with large capacitive-charging currents causes saturation to occur. Specifically, at large output currents, the voltage developed across r_C (which can be of the order of 100 Ω) can lower the voltage at the intrinsic collector terminal and cause the CBJ to become forward biased. As the reader will recall, saturation is a harmful effect for two reasons: It limits the collector current to a value less than βI_B, and it slows down the transistor turn-off.

11.8.2 Dynamic Operation

A detailed analysis of the dynamic operation of the BiCMOS inverter circuit is a rather complex undertaking. Nevertheless, an estimate of its propagation delay can be obtained by considering only the time required to charge and discharge a load capacitance C. Such an approximation is justified when C is relatively large and thus its effect on inverter dynamics is dominant, in other words, when we are able to neglect the time required to charge the parasitic capacitances present at internal circuit nodes. Fortunately, this is usually the case in practice, for if the load capacitance is not large, one would use the simpler CMOS inverter. In fact, it has been shown [Embabi, Bellaouar, and Elmasry (1993)] that the speed advantage of BiCMOS (over CMOS) becomes evident only when the gate is required to drive a large fan-out or a large load capacitance. For instance, at a load capacitance of 50 fF to 100 fF, BiCMOS and CMOS typically feature equal delays. However, at a load capacitance of 1 pF, t_P of a BiCMOS inverter is 0.3 ns, whereas that of an otherwise comparable CMOS inverter is about 1 ns.

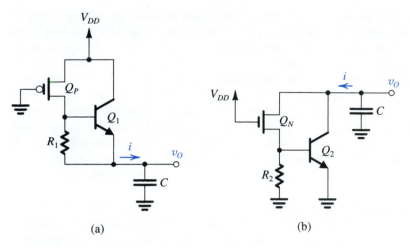

(a)   (b)

FIGURE 11.46 Equivalent circuits for charging and discharging a load capacitance C. Note that C includes all the capacitances present at the output node.

Finally, in Fig. 11.46, we show simplified equivalent circuits that can be employed in obtaining rough estimates of t_{PLH} and t_{PHL} of the R-type BiCMOS inverter (see Problem 11.55).

11.8.3 BiCMOS Logic Gates

In BiCMOS, the logic is performed by the CMOS part of the gate, with the bipolar portion simply functioning as an output stage. It follows that BiCMOS logic-gate circuits can be generated following the same approach used in CMOS. As an example, we show in Fig. 11.47 a BiCMOS two-input NAND gate.

As a final remark, we note that BiCMOS technology is applied in a variety of products including microprocessors, static RAMs, and gate arrays [see Alvarez (1993)].

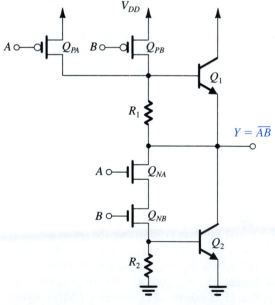

FIGURE 11.47 A BiCMOS two-input NAND gate.

11.9 SPICE SIMULATION EXAMPLE

We conclude this chapter by presenting an example that illustrates the use of SPICE in the analysis of bipolar digital circuits.

EXAMPLE 11.5

STATIC AND DYNAMIC OPERATION OF AN ECL GATE

In this example, we use PSpice to investigate the static and dynamic operation of the ECL gate (studied in Section 11.7) whose Capture schematic is shown in Fig. 11.48.

Having no access to the actual values for the SPICE model parameters of the BJTs utilized in commercially available ECL, we have selected parameter values representative of the technology utilized that, from our experience, would lead to reasonable agreement between simulation

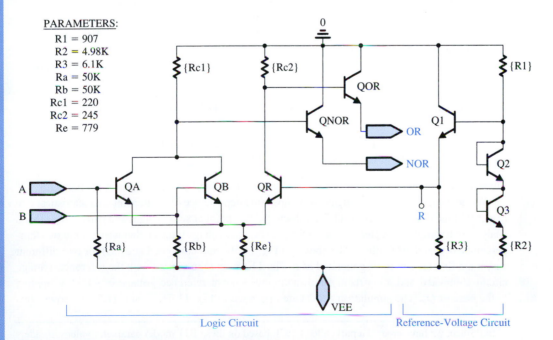

FIGURE 11.48 Capture schematic of the two-input ECL gate for Example 11.5.

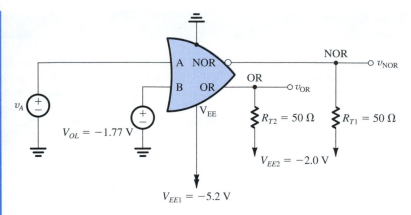

FIGURE 11.49 Circuit arrangement for computing the voltage transfer characteristics of the ECL gate in Fig. 11.48.

results and the measured performance data supplied by the manufacturer. It should be noted that this problem would not be encountered by an IC designer using SPICE as an aid; presumably, the designer would have full access to the proprietary process parameters and the corresponding device model parameters. In any case, for the simulations we conducted, we have utilized the following BJT model parameter values[12]: $I_S = 0.26$ fA, $\beta_F = 100$; $\beta_R = 1$, $\tau_F = 0.1$ ns, $C_{je} = 1$ pF, $C_{jc} = C_\mu = 1.5$ pF, and $|V_A| = 100$ V.

We use the circuit arrangement of Fig. 11.49 to compute the voltage transfer characteristics of the ECL gate, that is, v_{OR} and v_{NOR} versus v_A, where v_A is the input voltage at terminal A. For this investigation, the other input is deactivated by applying a voltage $v_B = V_{OL} = -1.77$ V. In PSpice, we perform a dc-analysis simulation with v_A swept over the range -2 V to 0 V in 10-mV increments and plot v_{OR} and v_{NOR} versus v_A. The simulation results are shown in Fig. 11.50. We immediately recognize the VTCs as those we have seen and (partially) verified by manual analysis in Section 11.7. The two transfer curves are symmetrical about an input voltage of -1.32 V. PSpice also determined that the voltage V_R at the base of the reference transistor Q_R has exactly this value (-1.32 V), which is also identical to the value we determined by hand analysis of the reference-voltage circuit.

Utilizing Probe (the graphical interface of PSpice), one can determine the values of the important parameters of the VTC, as follows:

OR output: $V_{OL} = -1.77$ V, $V_{OH} = -0.88$ V, $V_{IL} = -1.41$ V, and $V_{IH} = -1.22$ V; thus, $NM_H = 0.34$ V and $NM_L = 0.36$ V

NOR output: $V_{OL} = -1.78$ V, $V_{OH} = -0.88$ V, $V_{IL} = -1.41$ V, and $V_{IH} = -1.22$ V; thus, $NM_H = 0.34$ V and $NM_L = 0.37$ V

These values are remarkably close to those found by pencil-and-paper analysis in Section 11.6.

We next use PSpice to investigate the temperature dependence of the transfer characteristics. The reader will recall that in Section 11.7 we discussed this point at some length and carried out a hand analysis in Example 11.4. Here, we use PSpice to find the voltage transfer characteristics at two temperatures, 0°C and 70°C (the VTCs shown in Fig. 11.50 were computed at 27°C) for two different cases: the first case with V_R generated as in Fig. 11.48, and the second with the reference-voltage circuit eliminated and a constant, temperature-independent reference voltage of -1.32 V applied to the base of Q_R. The simulation results are displayed in Fig. 11.51. Figure 11.51(a) shows plots

---

[12] In PSpice, we have created a part called QECL based on these BJT model parameter values. Readers can find this part in the SEDRA.olb library which is available on the CD accompanying this book as well as on-line at www.sedrasmith.org.

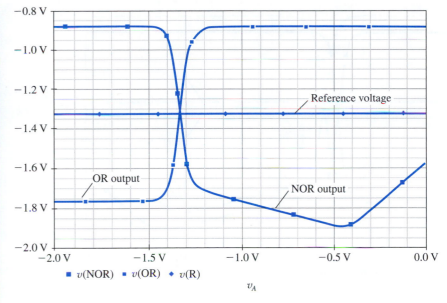

FIGURE 11.50 Voltage transfer characteristics of the OR and NOR outputs (see Fig. 11.49) for the ECL gate shown in Fig. 11.48. Also indicated is the reference voltage, $V_R = -1.32$ V.

of the transfer characteristics for the case in which the reference circuit is utilized and Fig. 11.51(b) shows plots for the case in which a constant reference voltage is employed. Figure 11.51(a) indicates that as the temperature is varied and V_R changes, the values of V_{OH} and V_{OL} also change but remain centered on V_R. In other words, the low and high noise margins remain nearly equal. As mentioned in Section 11.7 and demonstrated in the analysis of Example 11.4, this is the basic idea behind making V_R temperature dependent. When V_R is not temperature dependent, the symmetry of V_{OL} and V_{OH} around V_R is no longer maintained, as demonstrated in Fig. 11.51(b). Finally, we show in Table 11.1, some of the values obtained. Observe that for the temperature-compensated case, the

TABLE 11.1 PSpice-Computed Parameter Values of the ECL Gate, With and Without Temperature Compensation, at Two Different Temperatures

Temperature	Parameter	Temperature-Compensated		Not Temperature-Compensated	
		OR	NOR	OR	NOR
0°C	V_{OL}	−1.779 V	−1.799 V	−1.786 V	−1.799 V
	V_{OH}	−0.9142 V	−0.9092 V	−0.9142 V	−0.9092 V
	$V_{avg} = \dfrac{V_{OL} + V_{OH}}{2}$	−1.3466 V	−1.3541 V	−1.3501 V	−1.3541 V
	V_R	−1.345 V	−1.345 V	−1.32 V	−1.32 V
	$\lvert V_{avg} - V_R \rvert$	1.6 mV	9.1 mV	30.1 mV	34.1 mV
70°C	V_{OL}	−1.742 V	−1.759 V	−1.729 V	−1.759 V
	V_{OH}	−0.8338 V	−0.8285 V	−0.8338 V	−0.8285 V
	$V_{avg} = \dfrac{V_{OL} + V_{OH}}{2}$	−1.288 V	−1.294 V	−1.2814 V	−1.294 V
	V_R	−1.271 V	−1.271 V	−1.32 V	−1.32 V
	$\lvert V_{avg} - V_R \rvert$	17 mV	23 mV	38 mV	26.2 mV

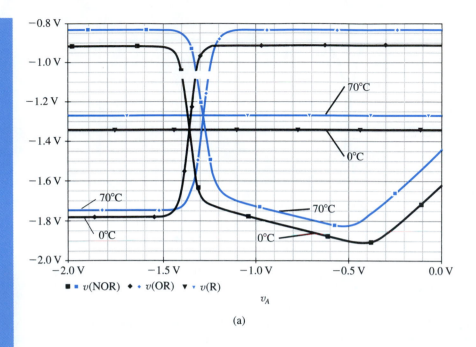

(a)

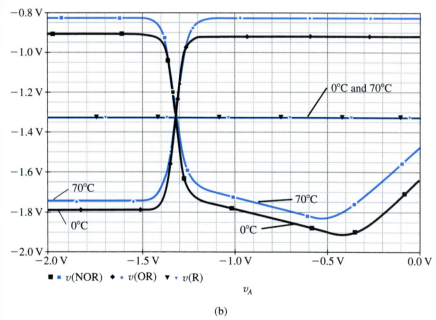

(b)

FIGURE 11.51 Comparing the voltage transfer characteristics of the OR and NOR outputs (see Fig. 11.49) of the ECL gate shown in Fig. 11.48, with the reference voltage V_R generated using: **(a)** the temperature-compensated bias network of Fig. 11.48. **(b)** a temperature-independent voltage source.

average value of V_{OL} and V_{OH} remains very close to V_R. The reader is encouraged to compare these results to those obtained in Example 11.4.

The dynamic operation of the ECL gate is investigated using the arrangement of Fig. 11.52. Here, two gates are connected by a 1.5-m coaxial cable having a characteristic impedance (Z_0) of

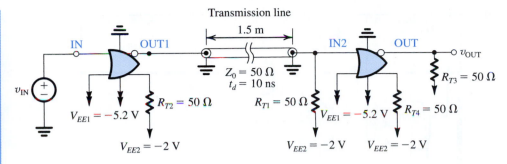

FIGURE 11.52 Circuit arrangement for investigating the dynamic operation of ECL. Two ECL gates (Fig. 11.48) are connected in cascade via a 1.5-m coaxial cable which has a characteristic impedance $Z_0 = 50\ \Omega$ and a propagation delay $t_d = 10$ ns. Resistor R_{T1} (50 Ω) provides proper termination for the coaxial cable.

50 Ω. The manufacturer specifies that signals propagate along this cable (when it is *properly terminated*) at about half the speed of light, or 15 cm/ns. Thus we would expect the 1.5-m cable we are using to introduce a delay t_d of 10 ns. Observe that in this circuit (Fig. 11.52), resistor R_{T1} provides the proper cable termination. The cable is assumed to be lossless and is modeled in PSpice using the *transmission line* element (the T part in the Analog library) with $Z_0 = 50\ \Omega$ and $t_d = 10$ ns. A voltage step, rising from -1.77 V to -0.884 V in 1 ns, is applied to the input of the first gate, and a transient analysis over a 30-ns interval is requested. Figure 11.53 shows plots of the waveforms of the input, the voltage at the output of the first gate, the voltage at the input of the second gate, and the output. Observe that despite the very high edge-speeds involved, the waveforms are reasonably clean and free of excessive ringing and reflections. This is particularly remarkable

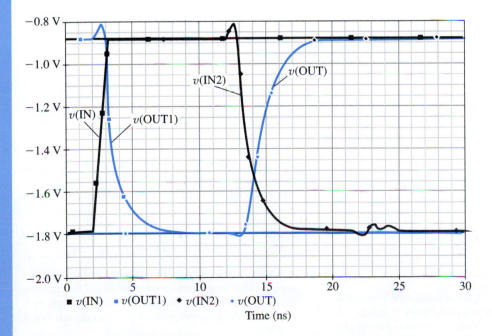

FIGURE 11.53 Transient response of a cascade of two ECL gates interconnected by a 1.5-m coaxial cable having a characteristic impedance of 50 Ω and a delay of 10 ns (see Fig. 11.52).

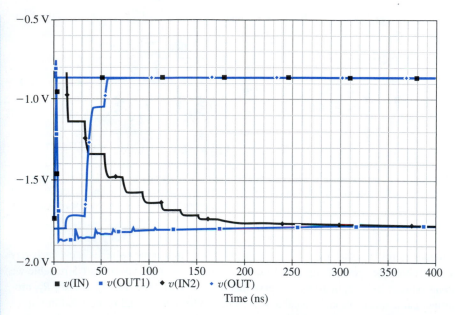

FIGURE 11.54 Transient response of a cascade of two ECL gates interconnected by a 1.5-m cable having a characteristic impedance of 300 Ω. The termination resistance R_{T1} (see Fig. 11.52) was kept unchanged at 50 Ω. Note the change in time scale of the plot.

because the signal is being transported over a relatively long distance. A detailed examination of the waveforms reveals that the delay along the cable is indeed 10 ns, and the delay of the second gate is about 1.06 ns.

Finally, to verify the need for properly terminating the transmission line, the dynamic analysis is repeated, this time with the 50-Ω coaxial cable replaced with a 300-Ω twisted-pair cable while keeping the termination resistance unchanged. The results are the slow rising and falling and long-delayed waveforms shown in Fig. 11.54. (Note the change of plotting scale.)

SUMMARY

- Flip-flops employ one or more latches. The basic static latch is a bistable circuit implemented using two inverters connected in a positive-feedback loop. The latch can remain in either stable state indefinitely.

- As an alternative to the positive-feedback approach, memory can be provided through the use of charge storage. A number of CMOS flip-flops are realized this way, including some master–slave D flip-flops.

- A monostable multivibrator has one stable state, in which it can remain indefinitely, and one quasi-stable state, which it enters upon triggering and in which it remains for a predetermined interval T. Monostable circuits can be used to generate a pulse signal of predetermined height and width.

- An astable multivibrator has no stable states. Rather, it has two quasi-stable states, between which it oscillates. The astable circuit, in its operation, is, in effect, a square-wave generator.

- A ring oscillator is implemented by connecting an odd number (N) of inverters in a loop, $f_{osc} = 1/2Nt_P$.

- A random-access memory (RAM) is one in which the time required for storing (writing) information and for retrieving (reading) information is independent of the physical location (within the memory) in which the information is stored.

- The major part of a memory chip consists of the cells in which the bits are stored and that are typically organized

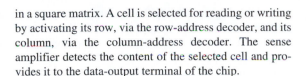

in a square matrix. A cell is selected for reading or writing by activating its row, via the row-address decoder, and its column, via the column-address decoder. The sense amplifier detects the content of the selected cell and provides it to the data-output terminal of the chip.

■ There are two kinds of MOS RAMs: static and dynamic. Static RAMs (SRAMs) employ flip-flops as the storage cells. In a dynamic RAM (DRAM), data is stored on a capacitor and thus must be periodically refreshed. DRAM chips provide the highest possible storage capacity for a given chip area.

■ Although sense amplifiers are utilized in SRAMs to speed up operation, they are essential in DRAMs. A typical sense amplifier is a differential circuit that employs positive feedback to obtain an output signal that grows exponentially toward either V_{DD} or 0.

■ Read-only memory (ROM) contains fixed data patterns that are stored at the time of fabrication and cannot be changed by the user. On the other hand, the contents of an erasable programmable ROM (EPROM) can be changed by the user. The erasure and reprogramming is a time-consuming process and is performed only infrequently.

■ Some EPROMS utilize floating-gate MOSFETs as the storage cells. The cell is programmed by applying a high voltage to the select gate. Erasure is achieved by illuminating the chip by ultraviolet light. Even more versatile, EEPROMs can be erased and reprogrammed electrically.

■ Emitter-coupled logic (ECL) is the fastest logic-circuit family. It achieves its high speed of operation by avoiding transistor saturation and by utilizing small logic-signal swings.

■ In ECL the input signals are used to steer a bias current between a reference transistor and an input transistor. The basic gate configuration is that of a differential amplifier.

■ There are two popular commercially available ECL types: ECL 10K, having $t_P = 2$ ns, $P_D = 25$ mW, and $DP = 50$ pJ; and ECL 100K, having $t_P = 0.75$ ns, $P_D = 40$ mW, and $DP = 30$ pJ. ECL 10K is easier to use because the rise and fall times of its signals are deliberately made long (about 3.5 ns).

■ Because of the very high operating speeds of ECL, care should be taken in connecting the output of one gate to the input of another. Transmission-line techniques are usually employed.

■ The design of the ECL gate is optimized so that the noise margins are equal and remain equal as temperature changes.

■ The ECL gate provides two complementary outputs, realizing the OR and NOR functions.

■ The outputs of ECL gates can be wired together to realize the OR function of the individual output variables.

■ BiCMOS combines the low-power and wide noise margins of CMOS with the high current-driving capability (and thus the short gate delays) of BJTs to obtain a technology that is capable of implementing very dense, low-power, high-speed VLSI circuits that can also include analog functions.

PROBLEMS

SECTION 11.1: LATCHES AND FLIP-FLOPS

11.1 Consider the clocked SR flip-flop of Fig. 11.3 for which a minimum-area design is required. Thus Q_1, Q_2, Q_3, and Q_4 are minimum-size devices for which $W/L = 2$ μm/1 μm. All the other devices should be sized equally so as to barely ensure regeneration. For this design, $V_{DD} = 5$ V, $|V_t| = 1$ V, and $k'_n = 2.5k'_p = 100$ μA/V^2. Calculate V_{th} for each of the internal inverters. Assuming that all the current from the conducting P device (say, Q_2) must be sustained for a moment at this voltage by current in (Q_5, Q_6) while both S and ϕ are high, find W/L of the equivalent transistor. What is the minimum W/L required for Q_5 and Q_6? Also find $W_5 = W_6$ for $L = 1$ μm. To guarantee operation and lower switching time, larger devices would normally be used.

11.2 For a flip-flop of the type shown in Fig. 11.3, determine the minimum width required of the set and reset pulse. Let Q_1, Q_2, Q_3, and Q_4 be minimum-size devices for which $W/L = 2$ μm/1 μm and all other devices have $W/L = 4$ μm/1 μm. $V_{DD} = 5$ V, $|V_t| = 1$ V, $k'_n = 2.5k'_p = 100$ μA/V^2, and the total capacitance at each of nodes Q and $\bar{Q}$ is 30 fF. (*Hint:* Follow the method outlined in Exercise 11.2.)

11.3 Consider another possibility for the circuit in Fig. 11.5: Relabel the R input as $\bar{S}$ and the S input as $\bar{R}$. Let $\bar{S}$ and $\bar{R}$ normally rest at relatively high voltages under the control of a relatively high-impedance source associated with "reading" the content of the flip-flop without changing its state. For "writing," that is, setting or resetting the flip-flop, $\bar{S}$ or $\bar{R}$ is brought low to 0 V with ϕ raised to V_{DD} to force $\bar{Q}$ or Q low to $V_{DD}/2$ at which point regeneration proceeds rapidly. For Q_1, Q_3, Q_5, and Q_6, all minimum size with $(W/L)_n = 2$, find $(W/L)_p$ so that $\overline{Q}$ can be lowered to 2.5 V in a 5-V system, when $\overline{S}$ is brought down to 0 V. Assume $|V_t| = 1$ V, $k'_n = 3\,k'_p = 75$ $\mu\text{A/V}^2$.

D11.4 The clocked SR flip-flop in Fig. 11.3 is not a fully complementary CMOS circuit. Sketch the fully complementary version by augmenting the circuit with the PUN corresponding to the PDN comprising Q_5, Q_6, Q_7, and Q_8. Note that the fully complementary circuit utilizes 12 transistors. Although the circuit is more complex, it switches faster.

D11.5 Sketch the complementary CMOS circuit implementation of the SR flip-flop of Fig. 11.2.

D11.6 Sketch the logic gate symbolic representation of an SR flip-flop using NAND gates. Give the truth table and describe the operation. Also sketch a CMOS circuit implementation.

**11.7 Consider the latch of Fig. 11.1 as implemented in CMOS technology. Let $\mu_n C_{ox} = 2\mu_p C_{ox} = 20$ $\mu\text{A/V}^2$, $W_p = 2W_n = 24$ μm, $L_p = L_n = 6$ μm, $|V_t| = 1$ V, and $V_{DD} = 5$ V.

(a) Plot the transfer characteristic of each inverter—that is, v_X versus v_W, and v_Z versus v_Y. Determine the output of each inverter at input voltages of 1, 1.5, 2, 2.25, 2.5, 2.75, 3, 3.5, 4, and 5 volts.
(b) Use the characteristics in (a) to determine the loop voltage-transfer curve of the latch—that is, v_Z versus v_W. Find the coordinates of points A, B, and C as defined in Fig. 11.1(c).
(c) If the finite output resistance of the saturated MOSFET is taken into account, with $|V_A| = 100$ V, find the slope of the loop transfer characteristic at point B. What is the approximate width of the transition region?

11.8 Two CMOS inverters operating from a 5-V supply have V_{IH} and V_{IL} of 2.42 and 2.00 V and corresponding outputs of 0.4 V and 4.6 V, respectively, and are connected as a latch. Approximating the corresponding transfer characteristic of each gate by a straight line between the threshold points, sketch the latch open-loop transfer characteristic. What are the coordinates of point B? What is the loop gain at B?

SECTION 11.2: MULTIVIBRATOR CIRCUITS

D11.9 For the monostable circuit of Fig. 11.11, use the approximate expression derived in Exercise 11.3 to find appropriate values for R and C so that $T = 1$ ms and the maximum error in the value obtained for T as a result of neglecting R_{on} in the design is 2%. Assume that R_{on} is limited to a maximum value of 1 kΩ.

11.10 Consider the monostable circuit of Fig. 11.10 under the condition that $R_{on} \ll R$. What does the expression for T

become? If V_{th} is nominally $0.5V_{DD}$ but can vary due to production variations in the range $0.4V_{DD}$ to $0.6V_{DD}$, find the corresponding variation in T expressed as a percentage of the nominal value.

*11.11 The waveforms for the monostable circuit of Fig. 11.10 are given in Fig. 11.13. Let $V_{DD} = 10$ V, $V_{th} = V_{DD}/2$, $R = 10$ kΩ, $C = 0.001$ μF, and $R_{on} = 200$ Ω. Find the values of T, ΔV_1, and ΔV_2. By how much does v_{O1} change during the quasi-stable state? What is the peak current that G_1 is required to sink? to source?

D11.12 Using the circuit of Fig. 11.10 design a monostable circuit with CMOS logic for which $R_{on} = 100$ Ω, $V_{DD} = 5$ V, and $V_{th} = 0.4$ V_{DD}. Use $C = 1$ μF to generate an output pulse of duration $T = 1$ s. What value of R should be used?

D11.13 (a) Use the expression given in Exercise 11.5 to find an expression for the frequency of oscillation f_0 for the astable multivibrator of Fig. 11.15 under the condition that $V_{th} = V_{DD}/2$.
(b) Find suitable values for R and C to obtain $f_0 = 100$ kHz.

11.14 Variations in manufacturing result in the CMOS gates used in implementing the astable circuit of Fig. 11.15 to have threshold voltages in the range $0.4V_{DD}$ to $0.6V_{DD}$ with $0.5V_{DD}$ being the nominal value. Express the expected corresponding variation in the value of f_0 (from nominal) as a percentage of the nominal value. (You may use the expression given in Exercise 11.5.)

*11.15 Consider a modification of the circuit of Fig. 11.15 in which a resistor equal to $10\,R$ is inserted between the common node of C and R and the input node of G_1. This resistor allows the voltage labeled v_{I1} to rise above V_{DD} and below ground. Sketch the resulting modified waveforms of v_{I1} and show that the period T is now given by

$$T = CR \ln\left[\frac{2V_{DD} - V_{th}}{V_{DD} - V_{th}} \cdot \frac{V_{DD} + V_{th}}{V_{th}}\right]$$

11.16 Consider a ring oscillator consisting of five inverters, each having $t_{PLH} = 60$ ns and $t_{PHL} = 40$ ns. Sketch one of the output waveforms, and specify its frequency and the percentage of the cycle during which the output is high.

11.17 A ring-of-eleven oscillator is found to oscillate at 20 MHz. Find the propagation delay of the inverter.

SECTION 11.3: SEMICONDUCTOR MEMORIES: TYPES AND ARCHITECTURES

11.18 A particular 1 M-bit square memory array has its peripheral circuits reorganized to allow for the readout of a 16-bit word. How many address bits will the new design need?

11.19 For the memory chip described in Problem 11.18, how many word lines must be supplied by the row decoder? How many sense amplifiers/drivers would a straightforward implementation require? If the chip power dissipation is 500 mW with a 5-V supply for continuous operation with a 200-ns cycle time, and that all the power loss is dynamic, estimate the total capacitance of all logic activated in any one cycle. If we assume that 90% of this power loss occurs in array access, and that the major capacitance contributor will be the bit line itself, calculate the capacitance per bit line and per bit for this design. If closer manufacturing control allows the memory array to operate at 3 V, how much larger a memory array can be designed in the same technology at about the same power level?

11.20 In a particular 1 G-bit memory of the dynamic type (called DRAM) under development by Samsung, using a 0.16-μm, 2-V technology, the cell array occupies about 50% of the area of the 21 mm $\times$ 31 mm chip. Estimate the cell area. If two cells form a square, estimate the cell dimensions.

11.21 An experimental 1.5-V, 1G-bit dynamic RAM (called DRAM) by Hitachi uses a 0.16-μm process with a cell size of $0.38 \times 0.76 \ \mu m^2$ in a $19 \times 38 \ mm^2$ chip. What fraction of the chip is occupied by the I/O connections, peripheral circuits, and interconnect?

11.22 A 256 M-bit RAM chip with a 16-bit readout employs a 16-block design with square cell arrays. How many address bits are needed for the block decoder, the row decoder, and the column decoder?

SECTION 11.4: RANDOM-ACCESS MEMORY (RAM) CELLS

D11.23 Consider the write operation of the SRAM cell of Fig. 11.18. Specifically, refer to relevant parts of the circuit, as depicted in Fig. 11.20. Let the process technology be characterized by $\mu_n/\mu_p = 2.5$, $\gamma = 0.5 \ V^{1/2}$, $|V_{t0}| = 0.8$ V, $2\phi_f = 0.6$ V, and $V_{DD} = 5$ V. Also let each of the two inverters be matched and $(W/L)_1 = (W/L)_3 = n$, where n denotes the W/L ratio of a minimum-size device.

(a) Using the circuit in Fig. 11.20(a), find the minimum required (W/L) of Q_5 (in terms of n) so that node $\bar{Q}$ can be pulled to $V_{DD}/2$, that is, at $v_{\bar{Q}} = 2.5$ V, $I_5 = I_1$.
(b) Using the circuit of Fig. 11.20(b), find the minimum required (W/L) ratio of Q_6 (in terms of n) so that node Q can be pulled down to $V_{DD}/2$, that is, at $v_Q = V_{DD}/2$, $I_6 = I_4$.
(c) Since Q_5 and Q_6 are designed to have equal W/L ratios, which of the two values found in (a) and (b) would you choose for a conservative design?
(d) For the value found in (c) and for $n = 2$, and $\mu_n C_{ox} = 50 \ \mu A/V^2$, determine the time for v_Q to reach $V_{DD}/2$. Let $C_Q = 50$ fF.

11.24 Consider the circuit in Fig. 11.20(a), and assume that the device dimensions and process technology parameters are as specified in Example 11.2. We wish to determine the interval Δt required for $C_{\bar{Q}}$ to charge, and its voltage to rise from 0 to $V_{DD}/2$.

(a) At the beginning of interval Δt, find the values of I_5, I_1, and $I_{C_{\bar{Q}}}$.
(b) At the end of interval Δt, find the values of I_5, I_1, and $I_{C_{\bar{Q}}}$.
(c) Find an estimate of the average value of $I_{C_{\bar{Q}}}$ during interval Δt.
(d) If $C_{\bar{Q}} = 50$ fF, estimate Δt. Compare this value to that found in Exercise 11.9 for v_Q to reach $V_{DD}/2$. Recalling that regeneration begins when either v_Q or $v_{\bar{Q}}$ reaches $V_{DD}/2$, what do you estimate the delay to be?

11.25 Reconsider the analysis of the read operation of the SRAM cell in Example 11.2. This time, assume that bit and $\overline{bit}$ lines are precharged to $V_{DD}/2$. Also consider the discharge of $C_{\bar{B}}$ [see Fig. 11.19(a)] to begin at the instant the voltage on the word line reaches $V_{DD}/2$. (Recall that the resistance and capacitance of the word line causes its voltage to rise relatively slowly toward V_{DD}.) Using an approach similar to that in Example 11.2, determine the read delay, defined at the time required to reduce the voltage of the $\bar{B}$ line by 0.2 V. Assume all technology and device parameters are those specified in Example 11.2.

11.26 For a particular DRAM design, the cell capacitance $C_S = 50$ fF, $V_{DD} = 5$ V, and V_t (including the body effect) = 1.4 V. Each cell represents a capacitive load on the bit line of 2 fF. The sense amplifier and other circuitry attached to the bit line has a 20-fF capacitance. What is the maximum number of cells that can be attached to a bit line while ensuring a minimum bit-line signal of 0.1 V? How many bits of row addressing can be used? If the sense-amplifier gain is increased by a factor of 5, how many word-line address bits can be accommodated?

11.27 For a DRAM available for regular use 98% of the time, having a row-to-column ratio of 2 to 1, a cycle time of 20 ns, and a refresh cycle of 8 ms, estimate the total memory capacity.

11.28 In a particular dynamic memory chip, $C_S = 25$ fF, the bit-line capacitance per cell is 1 fF and bit-line control circuitry involves 12 fF. For a 1 M-bit square array, what bit-line signals result when a stored 1 is read? when a stored 0 is read? Assume that $V_{DD} = 5$ V, and V_t (including the body effect) = 1.5 V. Recall that the bit lines are precharged to $V_{DD}/2$.

11.29 For a DRAM cell utilizing a capacitance of 20 fF, refresh is required within 10 ms. If a signal loss on the capacitor of 1 V can be tolerated, what is the largest acceptable leakage current present at the cell?

11

SECTION 11.5: SENSE AMPLIFIERS AND ADDRESS DECODERS

D11.30 Consider the operation of the differential sense amplifier of Fig. 11.23 following the rise of the sense control signal ϕ_s. Assume that a balanced differential signal of 0.1 V is established between the bit lines each of which has a 1 pF capacitance. For $V_{DD} = 3$ V, what is the value of G_m of each of the inverters in the amplifier required to cause the outputs to reach $0.1V_{DD}$ and $0.9V_{DD}$ (from initial values of $0.5V_{DD} + (0.1/2)$ and $0.5V_{DD} - (0.1/2)$ volts, respectively) in 2 ns? If for the matched inverters, $|V_t| = 0.8$ V and $k_n' = 3k_p' = 75$ μA/V^2, what are the device widths required? If the input signal is 0.2 V, what does the amplifier response time become?

11.31 A particular version of the regenerative sense amplifier of Fig. 11.23 in a 0.5-μm technology, uses transistors for which $|V_t| = 0.8$ V, $k_n' = 2.5k_p' = 100$ μA/V^2, $V_{DD} = 3.3$ V, with $(W/L)_n = 6$ μm/1.5 μm and $(W/L)_p = 15$ μm/1.5 μm. For each inverter, find the value of G_m. For a bit-line capacitance of 0.8 pF, and a delay until an output of $0.9V_{DD}$ is reached of 2 ns, find the initial difference-voltage required between the two bit lines. If the time can be relaxed by 1 ns, what input signal can be handled? With the increased delay time and with the input signal at the original level, by what percentage can the bit-line capacitance, and correspondingly the bit-line length, be increased? If the delay time required for the bit-line capacitances to charge by the constant current available from the storage cell, and thus develop the difference-voltage signal needed by the sense amplifier, was 5 ns, what does it increase to when longer lines are used?

D11.32 (a) For the sense amplifier of Fig. 11.23, show that the time required for the bit lines to reach $0.9V_{DD}$ and $0.1V_{DD}$ is given by $t_d = (C_B/G_m)\ln(0.8V_{DD}/\Delta V)$ where ΔV in the initial difference-voltage between the two bit lines.
(b) If the response time of the sense amplifier is to be reduced to one half the value of an original design, by what factor must the width of all transistors be increased?
(c) If for a particular design, $V_{DD} = 5$ V and $\Delta V = 0.2$ V, find the factor by which the width of all transistors must be increased so that ΔV is reduced by a factor of 4 while keeping t_d unchanged?

D11.33 It is required to design a sense amplifier of the type shown in Fig. 11.23 to operate with a DRAM using the dummy-cell technique illustrated in Fig. 11.25. The DRAM cell provides readout voltages of -100 mV when a 0 is stored and $+40$ mV when a 1 is stored. The sense amplifier is required to provide a differential output voltage of 2 V in at most 5 ns. Find the W/L ratios of the transistors in the amplifier inverters assuming that the processing technology is characterized by $k_n' = 2.5k_p' = 100$ μA/V^2, $|V_t| = 1$ V, and $V_{DD} = 5$ V. The capacitance of each half bit-line is 1 pF. What will be the amplifier response time when a 0 is read? When a 1 is read?

11.34 Consider a 512-row NOR decoder. To how many address bits does this correspond? How many output lines does it have? How many input lines does the NOR array require? How many NMOS and PMOS transistors does such a design need?

11.35 For the column decoder shown in Fig. 11.27, how many column-address bits are needed in a 256-K bit square array? How many NMOS pass transistors are needed in the multiplexer? How many NMOS transistors are needed in the NOR decoder? How many PMOS transistors? What is the total number of NMOS and PMOS transistors needed?

11.36 Consider the use of the tree column decoder shown in Fig. 11.28 for application with a square 256-K bit array. How many address bits are involved? How many levels of pass gates are used? How many pass transistors are there in total?

SECTION 11.6: READ-ONLY MEMORY (ROM)

11.37 Give the eight words stored in the ROM of Fig. 11.29.

D11.38 Design the bit pattern to be stored in a (16×4) ROM that provides the 4-bit product of two 2-bit variables. Give a circuit implementation of the ROM array using a form similar to that of Fig. 11.29.

11.39 Consider a dynamic version of the ROM in Fig. 11.29 in which the gates of the PMOS devices are connected to a precharge control signal ϕ. Let all the NMOS devices have $W/L = 3$ μm/1.2 μm and all the PMOS devices have $W/L = 12$ μm/1.2 μm. Assume $k_n' = 3k_p' = 90$ μA/V^2, $V_{tn} = -V_{tp} = 1$ V, and $V_{DD} = 5$ V.

(a) During the precharge interval, ϕ is lowered to 0 V. Estimate the time required to charge a bit line from 0 to 5 V. Use as an average charging current the current supplied by a PMOS transistor at a bit-line voltage half way through the 0 to 5 V excursion, i.e., 2.5 V. The bit-line capacitance is 1 pF. Note that all NMOS transistors are cut off at this time.
(b) After the precharge interval is completed and ϕ returns to V_{DD}, the row decoder raises the voltage of the selected word line. Because of the finite resistance and capacitance of the word line, the voltage rises exponentially toward V_{DD}. If the resistance of each of the polysilicon word lines is 5 kΩ and the capacitance between the word line and ground is 2 pF, what is the (10% to 90%) rise time of the word-line voltage? What is the voltage reached at the end of one time-constant?
(c) If we approximate the exponential rise of the word-line voltage by a step equal to the voltage reached in one time-constant, find the interval Δt required for an NMOS transistor to discharge the bit line and lower its voltage by 1 V.

SECTION 11.7: EMITTER-COUPLED LOGIC (ECL)

D11.40 For the ECL circuit in Fig. P11.40, the transistors exhibit V_{BE} of 0.75 V at an emitter current I and have very high β.

(a) Find V_{OH} and V_{OL}.
(b) For the input at B sufficiently negative for Q_B to be cut off, what voltage at A causes a current of $I/2$ to flow in Q_R?
(c) Repeat (b) for a current in Q_R of $0.99I$.
(d) Repeat (c) for a current in Q_R of $0.01I$.
(e) Use the results of (c) and (d) to specify V_{IL} and V_{IH}.
(f) Find NM_H and NM_L.
(g) Find the value of IR that makes the noise margins equal to the width of the transition region, $V_{IH} - V_{IL}$.
(h) Using the IR value obtained in (g), give numerical values for $V_{OH}, V_{OL}, V_{IH}, V_{IL},$ and V_R for this ECL gate.

*11.41 Three logic inverters are connected in a ring. Specifications for this family of gates indicates a typical propagation delay of 3 ns for high-to-low output transitions and 7 ns for low-to-high transitions. Assume that for some reason the input to one of the gates undergoes a low-to-high transition. By sketching the waveforms at the outputs of the three gates and keeping track of their relative positions, show that the circuit functions as an oscillator. What is the frequency of oscillation of this ring oscillator? In each cycle, how long is the output high? low?

*11.42 Following the idea of a ring oscillator introduced in Problem 11.41, consider an implementation using a ring of five ECL 100K inverters. Assume that the inverters have linearly rising and falling edges (and thus the waveforms are trapezoidal in shape). Let the 0 to 100% rise and fall times be equal to 1 ns. Also, let the propagation delay (for both transitions) be equal to 1 ns. Provide a labeled sketch of the five output signals, taking care that relevant phase information is provided. What is the frequency of oscillation?

*11.43 Using the logic and circuit flexibility of ECL indicated by Figs. 11.34 and 11.44, sketch an ECL logic circuit that realizes the exclusive OR function, $Y = \bar{A}B + A\bar{B}$.

*11.44 For the circuit in Fig. 11.36, whose transfer characteristic is shown in Fig. 11.37, calculate the incremental voltage gain from input to the OR output at points x, m, and y of the transfer characteristic. Assume $\beta = 100$. Use the results of Exercise 11.20, and let the output at x be -1.77 V and that at y be -0.88 V. Hint: Recall that x and y are defined by a 1%, 99% current split.

11.45 For the circuit in Fig. 11.36, whose transfer characteristic is shown in Fig. 11.37, find V_{IL} and V_{IH} if x and y are defined as the points at which

(a) 90% of the current I_E is switched.
(b) 99.9% of the current I_E is switched.

11.46 For the symmetrically loaded circuit of Fig. 11.36 and for typical output signal levels ($V_{OH} = -0.88$ V and $V_{OL} = -1.77$ V), calculate the power lost in both load resistors R_T and both output followers. What then is the total power

11

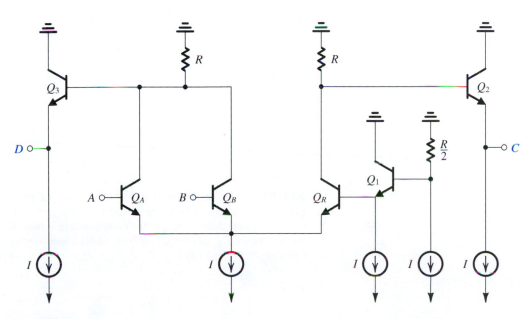

FIGURE P11.40

dissipation of a single ECL gate including its symmetrical output terminations?

11.47 Considering the circuit of Fig. 11.38, what is the value of β of Q_2, for which the high noise margin (NM_H) is reduced by 50%?

*11.48 Consider an ECL gate whose inverting output is terminated in a 50-Ω resistance connected to a -2-V supply. Let the total load capacitance be denoted C. As the input of the gate rises, the output emitter follower cuts off and the load capacitance C discharges through the 50-Ω load (until the emitter follower conducts again). Find the value of C that will result in a discharge time of 1 ns. Assume that the two output levels are -0.88 V and -1.77 V.

11.49 For signals whose rise and fall times are 3.5 ns, what length of unterminated gate-to-gate wire interconnect can be used if a ratio of rise time to return time of 5 to 1 is required? Assume the environment of the wire to be such that the signal propagates at two-thirds the speed of light (which is 30 cm/ns).

*11.50 For the circuit in Fig. P11.50 let the levels of the inputs A, B, C, and D be 0 and $+5$ V. For all inputs low at 0 V,

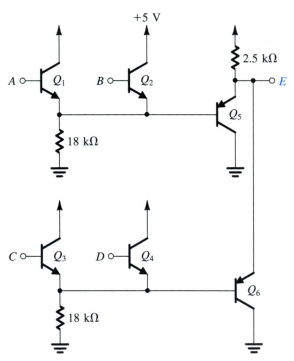

FIGURE P11.50

what is the voltage at E? If A and C are raised to $+5$ V, what is the voltage at E? Assume $\left| V_{BE} \right| = 0.7$ V and $\beta = 50$. Express E as a logic function of A, B, C, and D.

SECTION 11.8: BiCMOS DIGITAL CIRCUITS

11.51 Consider the conceptual BiCMOS circuit of Fig. 11.45(a), for the conditions that $V_{DD} = 5$ V, $\left| V_t \right| = 1$ V, $V_{BE} = 0.7$ V, $\beta = 100$, $k_n' = 2.5k_p' = 100\ \mu\text{A/V}^2$, and $(W/L)_n = 2\ \mu\text{m}/1\ \mu\text{m}$. For $v_I = v_O = V_{DD}/2$, find $(W/L)_p$ so that $I_{EQ_1} = I_{EQ_2}$. What is this totem-pole transient current?

11.52 Consider the conceptual BiCMOS circuit of Fig. 11.45(a) for the conditions stated in Problem 11.51. What is the threshold voltage of the inverter if both Q_N and Q_P have $W/L = 2\ \mu\text{m}/1\ \mu\text{m}$? What totem-pole current flows at v_I equal to the threshold voltage?

D11.53 Consider the choice of values for R_1 and R_2 in the circuit of Fig. 11.45(c). An important consideration in making this choice is that the loss of base drive current be limited. This loss becomes particularly acute when the current through Q_N and Q_P becomes small. This in turn happens near the end of the output signal swing when the associated MOS device is deeply in triode operation (say at $\left| v_{DS} \right| = \left| V_t \right|/3$). Determine values for R_1 and R_2 so that the loss in base current is limited to 50%. What is the ratio R_1/R_2? Repeat for a 20% loss in base drive.

11.54 For the circuit of Fig. 11.45(a) with parameters as in Problem 11.51 and with $(W/L)_p = (W/L)_n$, estimate the propagation delays t_{PLH}, t_{PHL} and t_P obtained for a load capacitance of 2 pF. Assume that the internal node capacitances do not contribute much to this result. Use average values for the capacitor charging and discharging currents.

11.55 Repeat Problem 11.54 for the circuit in Fig. 11.45(e) assuming that $R_1 = R_2 = 5$ kΩ.

D11.56 Consider the dynamic response of the NAND gate of Fig. 11.46 with a large external capacitive load. If the worst-case response is to be identical to that of the inverter of Fig. 11.45(e), how must the (W/L) ratios of Q_{NA}, Q_{NB}, Q_N, Q_{PA}, Q_{PB}, Q_P be related?

D11.57 Sketch the circuit of a BiCMOS two-input NOR gate. If when loaded with a large capacitance the gate is to have worst case delays equal to the corresponding values of the inverter of Fig. 11.45(e), find W/L of each transistor in terms of $(W/L)_n$ and $(W/L)_p$.

Filters and Tuned Amplifiers

INTRODUCTION

In this chapter, we study the design of an important building block of communications and instrumentation systems, the electronic filter. Filter design is one of the very few areas of engineering for which a complete design theory exists, starting from specification and ending with a circuit realization. A detailed study of filter design requires an entire book, and indeed such textbooks exist. In the limited space available here, we shall concentrate on a selection of topics that provide an introduction to the subject as well as a useful arsenal of filter circuits and design methods.

The oldest technology for realizing filters makes use of inductors and capacitors, and the resulting circuits are called **passive LC filters.** Such filters work well at high frequencies;

however, in low-frequency applications (dc to 100 kHz) the required inductors are large and physically bulky, and their characteristics are quite nonideal. Furthermore, such inductors are impossible to fabricate in monolithic form and are incompatible with any of the modern techniques for assembling electronic systems. Therefore, there has been considerable interest in finding filter realizations that do not require inductors. Of the various possible types of **inductorless filters,** we shall study **active-RC filters** and **switched-capacitor filters.**

Active-RC filters utilize op amps together with resistors and capacitors and are fabricated using discrete, hybrid thick-film, or hybrid thin-film technology. However, for large-volume production, such technologies do not yield the economies achieved by monolithic (IC) fabrication. At the present time, the most viable approach for realizing fully integrated monolithic filters is the switched-capacitor technique.

The last topic studied in this chapter is the tuned amplifier commonly employed in the design of radio and TV receivers. Although tuned amplifiers are in effect bandpass filters, they are studied separately because their design is based on somewhat different techniques.

12.1 FILTER TRANSMISSION, TYPES, AND SPECIFICATION

12.1.1 Filter Transmission

The filters we are about to study are linear circuits that can be represented by the general two-port network shown in Fig. 12.1. The filter **transfer function** $T(s)$ is the ratio of the output voltage $V_o(s)$ to the input voltage $V_i(s)$,

$$T(s) \equiv \frac{V_o(s)}{V_i(s)} \tag{12.1}$$

The filter **transmission** is found by evaluating $T(s)$ for physical frequencies, $s = j\omega$, and can be expressed in terms of its magnitude and phase as

$$T(j\omega) = |T(j\omega)| e^{j\phi(\omega)} \tag{12.2}$$

The magnitude of transmission is often expressed in decibels in terms of the **gain function**

$$G(\omega) \equiv 20 \log|T(j\omega)|, \, \text{dB} \tag{12.3}$$

or, alternatively, in terms of the **attenuation function**

$$A(\omega) \equiv -20 \log|T(j\omega)|, \, \text{dB} \tag{12.4}$$

A filter shapes the frequency spectrum of the input signal, $|V_i(j\omega)|$, according to the magnitude of the transfer function $|T(j\omega)|$, thus providing an output $V_o(j\omega)$ with a spectrum

$$|V_o(j\omega)| = |T(j\omega)||V_i(j\omega)| \tag{12.5}$$

Also, the phase characteristics of the signal are modified as it passes through the filter according to the filter phase function $\phi(\omega)$.

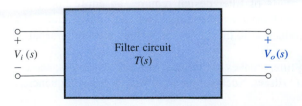

FIGURE 12.1 The filters studied in this chapter are linear circuits represented by the general two-port network shown. The filter transfer function $T(s) \equiv V_o(s)/V_i(s)$.

12.1.2 Filter Types

We are specifically interested here in filters that perform a **frequency-selection** function: **passing** signals whose frequency spectrum lies within a specified range, and **stopping** signals whose frequency spectrum falls outside this range. Such a filter has ideally a frequency band (or bands) over which the magnitude of transmission is unity (the filter **passband**) and a frequency band (or bands) over which the transmission is zero (the filter **stopband**). Figure 12.2 depicts the ideal transmission characteristics of the four major filter types: **low-pass** (LP) in Fig. 12.2(a), **high-pass** (HP) in Fig. 12.2(b), **bandpass** (BP) in Fig. 12.2(c), and **bandstop** (BS) or **band-reject** in Fig. 12.2(d). These idealized characteristics, by virtue of their vertical edges, are known as **brick-wall** responses.

12.1.3 Filter Specification

The filter-design process begins with the filter user specifying the transmission characteristics required of the filter. Such a specification cannot be of the form shown in Fig. 12.2 because physical circuits cannot realize these idealized characteristics. Figure 12.3 shows realistic specifications for the transmission characteristics of a low-pass filter. Observe that since a physical circuit cannot provide constant transmission at all passband frequencies, the specifications allow for deviation of the passband transmission from the ideal 0 dB, but places an upper bound, A_{max} (dB), on this deviation. Depending on the application, A_{max} typically ranges from 0.05 dB to 3 dB. Also, since a physical circuit cannot provide zero transmission at all stopband frequencies, the specifications in Fig. 12.3 allow for some transmission

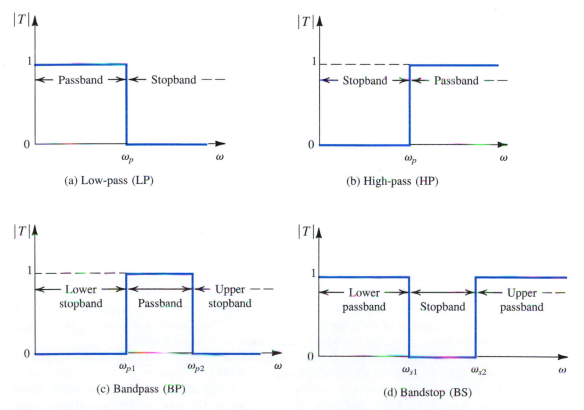

FIGURE 12.2 Ideal transmission characteristics of the four major filter types: **(a)** low-pass (LP), **(b)** high-pass (HP), **(c)** bandpass (BP), and **(d)** bandstop (BS).

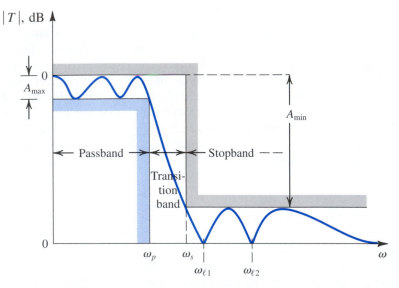

FIGURE 12.3 Specification of the transmission characteristics of a low-pass filter. The magnitude response of a filter that just meets specifications is also shown.

over the stopband. However, the specifications require the stopband signals to be attenuated by at least A_{min} (dB) relative to the passband signals. Depending on the filter application, A_{min} can range from 20 dB to 100 dB.

Since the transmission of a physical circuit cannot change abruptly at the edge of the passband, the specifications of Fig. 12.3 provide for a band of frequencies over which the attenuation increases from near 0 dB to A_{min}. This **transition band** extends from the passband edge ω_p to the stopband edge ω_s. The ratio ω_s/ω_p is usually used as a measure of the sharpness of the low-pass filter response and is called the **selectivity factor.** Finally, observe that for convenience the passband transmission is specified to be 0 dB. The final filter, however, can be given a passband gain, if desired, without changing its selectivity characteristics.

To summarize, the transmission of a low-pass filter is specified by four parameters:

1. The passband edge ω_p
2. The maximum allowed variation in passband transmission A_{max}
3. The stopband edge ω_s
4. The minimum required stopband attenuation A_{min}

The more tightly one specifies a filter—that is, lower A_{max}, higher A_{min}, and/or a selectivity ratio ω_s/ω_p closer to unity—the closer the response of the resulting filter will be to the ideal. However, the resulting filter circuit must be of higher order and thus more complex and expensive.

In addition to specifying the magnitude of transmission, there are applications in which the phase response of the filter is also of interest. The filter-design problem, however, is considerably complicated when both magnitude and phase are specified.

Once the filter specifications have been decided upon, the next step in the design is to find a transfer function whose magnitude meets the specification. To meet specification, the magnitude-response curve must lie in the unshaded area in Fig. 12.3. The curve shown in the figure is for a filter that *just* meets specifications. Observe that for this particular filter, the magnitude response *ripples* throughout the passband with the ripple peaks being all

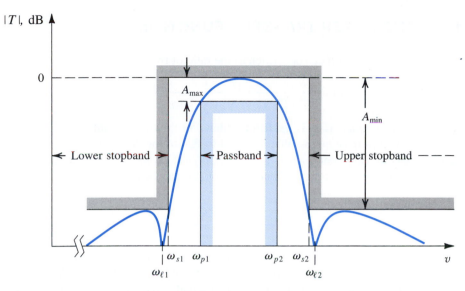

FIGURE 12.4 Transmission specifications for a bandpass filter. The magnitude response of a filter that just meets specifications is also shown. Note that this particular filter has a monotonically decreasing transmission in the passband on both sides of the peak frequency.

equal. Since the peak ripple is equal to A_{max} it is usual to refer to A_{max} as the **passband ripple** and to ω_p as the **ripple bandwidth.** The particular filter response shown ripples also in the stopband, again with the ripple peaks all equal and of such a value that the minimum stopband attenuation achieved is equal to the specified value, A_{min}. Thus this particular response is said to be **equiripple** in both the passband and the stopband.

The process of obtaining a transfer function that meets given specifications is known as **filter approximation.** Filter approximation is usually performed using computer programs (Snelgrove, 1982; Ouslis and Sedra, 1995) or filter design tables (Zverev, 1967). In simpler cases, filter approximation can be performed using closed-form expressions, as will be seen in Section 12.3.

Finally, Fig. 12.4 shows transmission specifications for a bandpass filter and the response of a filter that meets these specifications. For this example we have chosen an approximation function that does not ripple in the passband; rather, the transmission decreases monotonically on both sides of the center frequency, attaining the maximum allowable deviation at the two edges of the passband.

EXERCISES

12.1 Find approximate values of attenuation (in dB) corresponding to filter transmissions of: 1, 0.99, 0.9, 0.8, 0.7, 0.5, 0.1, 0.

Ans. 0, 0.1, 1, 2, 3, 6, 20, ∞ (dB)

12.2 If the magnitude of passband transmission is to remain constant to within ±5%, and if the stopband transmission is to be no greater than 1% of the passband transmission, find A_{max} and A_{min}.

Ans. 0.9 dB; 40 dB

12.2 THE FILTER TRANSFER FUNCTION

The filter transfer function $T(s)$ can be written as the ratio of two polynomials as

$$T(s) = \frac{a_M s^M + a_{M-1} s^{M-1} + \cdots + a_0}{s^N + b_{N-1} s^{N-1} + \cdots + b_0} \tag{12.6}$$

The degree of the denominator, N, is the **filter order.** For the filter circuit to be stable, the degree of the numerator must be less than or equal to that of the denominator; $M \leq N$. The numerator and denominator coefficients, $a_0, a_1, \ldots, a_M$ and $b_0, b_1, \ldots, b_{N-1}$, are real numbers. The polynomials in the numerator and denominator can be factored, and $T(s)$ can be expressed in the form

$$T(s) = \frac{a_M(s - z_1)(s - z_2) \cdots (s - z_M)}{(s - p_1)(s - p_2) \cdots (s - p_N)} \tag{12.7}$$

The numerator roots, $z_1, z_2, \ldots, z_M$, are the **transfer-function zeros,** or **transmission zeros;** and the denominator roots, $p_1, p_2, \ldots, p_N$, are the **transfer-function poles,** or the **natural modes.**[1] Each transmission zero or pole can be either a real or a complex number. Complex zeros and poles, however, must occur in conjugate pairs. Thus, if $-1 + j2$ happens to be a zero, then $-1 - j2$ also must be a zero.

Since in the filter stopband the transmission is required to be zero or small, the filter transmission zeros are usually placed on the $j\omega$ axis at stopband frequencies. This indeed is the case for the filter whose transmission function is sketched in Fig. 12.3. This particular filter can be seen to have infinite attenuation (zero transmission) at two stopband frequencies: ω_{l1} and ω_{l2}. The filter then must have transmission zeros at $s = +j\omega_{l1}$ and $s = +j\omega_{l2}$. However, since complex zeros occur in conjugate pairs, there must also be transmission zeros at $s = -j\omega_{l1}$ and $s = -j\omega_{l2}$. Thus the numerator polynomial of this filter will have the factors $(s + j\omega_{l1})(s - j\omega_{l1})(s + j\omega_{l2})(s - j\omega_{l2})$, which can be written as $(s^2 + \omega_{l1}^2)(s^2 + \omega_{l2}^2)$. For $s = j\omega$ (physical frequencies) the numerator becomes $(-\omega^2 + \omega_{l1}^2)(-\omega^2 + \omega_{l2}^2)$, which indeed is zero at $\omega = \omega_{l1}$ and $\omega = \omega_{l2}$.

Continuing with the example in Fig. 12.3, we observe that the transmission decreases toward $-\infty$ as ω approaches ∞. Thus the filter must have one or more transmission zeros at $s = \infty$. In general, the number of transmission zeros at $s = \infty$ is the difference between the degree of the numerator polynomial, M, and the degree of the denominator polynomial, N, of the transfer function in Eq. (12.6). This is because as s approaches ∞, $T(s)$ approaches a_M/s^{N-M} and thus is said to have $N - M$ zeros at $s = \infty$.

For a filter circuit to be stable, all its poles must lie in the left half of the s plane, and thus $p_1, p_2, \ldots, p_N$ must all have negative real parts. Figure 12.5 shows typical pole and zero locations for the low-pass filter whose transmission function is depicted in Fig. 12.3. We have assumed that this filter is of fifth order ($N = 5$). It has two pairs of complex-conjugate poles and one real-axis pole, for a total of five poles. All the poles lie in the vicinity of the passband, which is what gives the filter its high transmission at passband frequencies. The five transmission zeros are at $s = \pm j\omega_{l1}$, $s = \pm j\omega_{l2}$, and $s = \infty$. Thus, the transfer function for this filter is of the form

$$T(s) = \frac{a_4(s^2 + \omega_{l1}^2)(s^2 + \omega_{l2}^2)}{s^5 + b_4 s^4 + b_3 s^3 + b_2 s^2 + b_1 s + b_0} \tag{12.8}$$

---

[1] Throughout this chapter, we use the names *poles* and *natural modes* interchangeably.

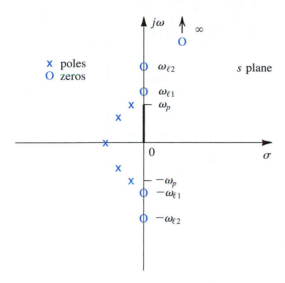

FIGURE 12.5 Pole–zero pattern for the low-pass filter whose transmission is sketched in Fig. 12.3. This is a fifth-order filter ($N = 5$).

As another example, consider the bandpass filter whose magnitude response is shown in Fig. 12.4. This filter has transmission zeros at $s = \pm j\omega_{l1}$ and $s = \pm j\omega_{l2}$. It also has one or more zeros at $s = 0$ and one or more zeros at $s = \infty$ (because the transmission decreases toward 0 as ω approaches 0 and ∞). Assuming that only one zero exists at each of $s = 0$ and $s = \infty$, the filter must be of sixth order, and its transfer function takes the form

$$T(s) = \frac{a_5 s(s^2 + \omega_{l1}^2)(s^2 + \omega_{l2}^2)}{s^6 + b_5 s^5 + \cdots + b_0} \tag{12.9}$$

A typical pole–zero plot for such a filter is shown in Fig. 12.6.

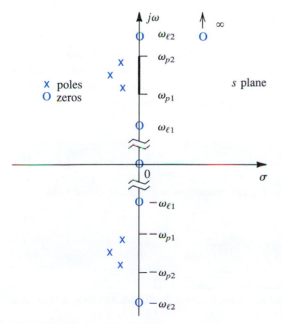

FIGURE 12.6 Pole–zero pattern for the bandpass filter whose transmission function is shown in Fig. 12.4. This is a sixth-order filter ($N = 6$).

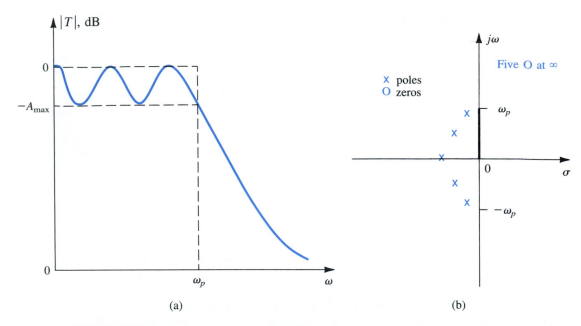

FIGURE 12.7 **(a)** Transmission characteristics of a fifth-order low-pass filter having all transmission zeros at infinity. **(b)** Pole–zero pattern for the filter in (a).

As a third and final example, consider the low-pass filter whose transmission function is depicted in Fig. 12.7(a). We observe that in this case there are no finite values of ω at which the attenuation is infinite (zero transmission). Thus it is possible that all the transmission zeros of this filter are at $s = \infty$. If this is the case, the filter transfer function takes the form

$$T(s) = \frac{a_0}{s^N + b_{N-1}s^{N-1} + \cdots + b_0} \tag{12.10}$$

Such a filter is known as an **all-pole filter.** Typical pole–zero locations for a fifth-order all-pole low-pass filter are shown in Fig. 12.7(b).

Almost all the filters studied in this chapter have all their transmission zeros on the $j\omega$ axis, in the filter stopband(s), including[2] $\omega = 0$ and $\omega = \infty$. Also, to obtain high selectivity, all the natural modes will be complex conjugate (except for the case of odd-order filters, where one natural mode must be on the real axis). Finally we note that the more selective the required filter response is, the higher its order must be, and the closer its natural modes are to the $j\omega$ axis.

EXERCISES

12.3 A second-order filter has its poles at $s = -(1/2) \pm j(\sqrt{3}/2)$. The transmission is zero at $\omega = 2$ rad/s and is unity at dc ($\omega = 0$). Find the transfer function.

Ans. $T(s) = \dfrac{1}{4}\,\dfrac{s^2 + 4}{s^2 + s + 1}$

---

[2] Obviously, a low-pass filter should *not* have a transmission zero at $\omega = 0$, and, similarly, a high-pass filter should not have a transmission zero at $\omega = \infty$.

12.4 A fourth-order filter has zero transmission at $\omega = 0$, $\omega = 2$ rad/s, and $\omega = \infty$. The natural modes are $-0.1 \pm j0.8$ and $-0.1 \pm j1.2$. Find $T(s)$.

Ans. $T(s) = \dfrac{a_3 s(s^2 + 4)}{(s^2 + 0.2s + 0.65)(s^2 + 0.2s + 1.45)}$

12.5 Find the transfer function $T(s)$ of a third-order all-pole low-pass filter whose poles are at a radial distance of 1 rad/s from the origin and whose complex poles are at $30°$ angles from the $j\omega$ axis. The dc gain is unity. Show that $|T(j\omega)| = 1/\sqrt{1 + \omega^6}$. Find ω_{3dB} and the attenuation at $\omega = 3$ rad/s.

Ans. $T(s) = 1/(s + 1)(s^2 + s + 1)$; 1 rad/s; 28.6 dB

 ## 12.3 BUTTERWORTH AND CHEBYSHEV FILTERS

In this section, we present two functions that are frequently used in approximating the transmission characteristics of low-pass filters. Closed-form expressions are available for the parameters of these functions, and thus one can use them in filter design without the need for computers or filter-design tables. Their utility, however, is limited to relatively simple applications.

Although in this section we discuss the design of low-pass filters only, the approximation functions presented can be applied to the design of other filter types through the use of frequency transformations [see Sedra and Brackett (1978)].

12.3.1 The Butterworth Filter

Figure 12.8 shows a sketch of the magnitude response of a Butterworth[3] filter. This filter exhibits a monotonically decreasing transmission with all the transmission zeros at $\omega = \infty$, making it an all-pole filter. The magnitude function for an Nth-order Butterworth filter with a passband edge ω_p is given by

$$|T(j\omega)| = \frac{1}{\sqrt{1 + \epsilon^2 \left(\dfrac{\omega}{\omega_p}\right)^{2N}}} \qquad (12.11)$$

At $\omega = \omega_p$,

$$|T(j\omega_p)| = \frac{1}{\sqrt{1 + \epsilon^2}} \qquad (12.12)$$

Thus, the parameter ϵ determines the maximum variation in passband transmission, A_{max}, according to

$$A_{max} = 20 \log \sqrt{1 + \epsilon^2} \qquad (12.13)$$

Conversely, given A_{max}, the value of ϵ can be determined from

$$\epsilon = \sqrt{10^{A_{max}/10} - 1} \qquad (12.14)$$

Observe that in the Butterworth response the maximum deviation in passband transmission (from the ideal value of unity) occurs at the passband edge only. It can be shown that the first

---

[3] The Butterworth filter approximation is named after S. Butterworth, a British engineer who in 1930 was among the first to employ it.

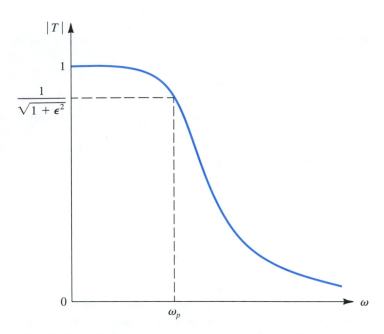

FIGURE 12.8 The magnitude response of a Butterworth filter.

$2N - 1$ derivatives of $|T|$ relative to ω are zero at $\omega = 0$ [see Van Valkenburg (1980)]. This property makes the Butterworth response very flat near $\omega = 0$ and gives the response the name **maximally flat** response. The degree of passband flatness increases as the order N is increased, as can be seen from Fig. 12.9. This figure indicates also that, as should be expected, as the order N is increased the filter response approaches the ideal brick-wall type of response.

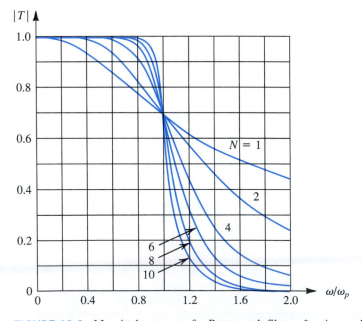

FIGURE 12.9 Magnitude response for Butterworth filters of various order with $\epsilon = 1$. Note that as the order increases, the response approaches the ideal brick-wall type of transmission.

At the edge of the stopband, $\omega = \omega_s$, the attenuation of the Butterworth filter is given by

$$A(\omega_s) = -20 \log[1/\sqrt{1 + \epsilon^2(\omega_s/\omega_p)^{2N}}]$$

$$= 10 \log[1 + \epsilon^2(\omega_s/\omega_p)^{2N}] \qquad (12.15)$$

This equation can be used to determine the filter order required, which is the lowest integer value of N that yields $A(\omega_s) \geq A_{\min}$.

The natural modes of an Nth-order Butterworth filter can be determined from the graphical construction shown in Fig. 12.10(a). Observe that the natural modes lie on a circle of

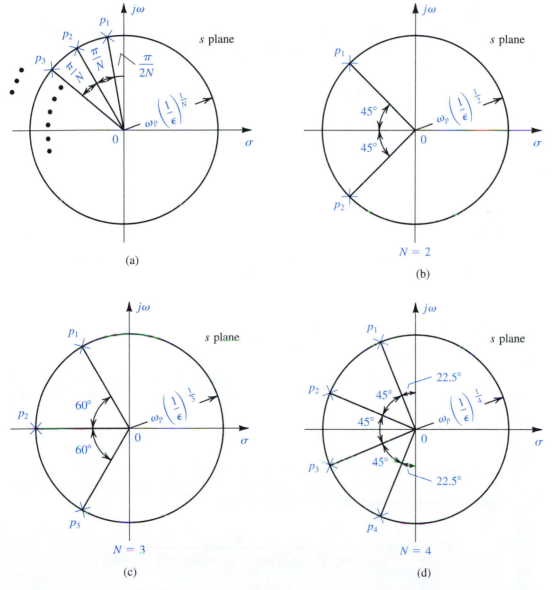

FIGURE 12.10 Graphical construction for determining the poles of a Butterworth filter of order N. All the poles lie in the left half of the s plane on a circle of radius $\omega_0 = \omega_p(1/\epsilon)^{1/N}$, where ϵ is the passband deviation parameter ($\epsilon = \sqrt{10^{A_{\max}/10} - 1}$): **(a)** the general case, **(b)** $N = 2$, **(c)** $N = 3$, and **(d)** $N = 4$.

radius $\omega_p(1/\epsilon)^{1/N}$ and are spaced by equal angles of π/N, with the first mode at an angle $\pi/2N$ from the $+j\omega$ axis. Since the natural modes all have equal radial distance from the origin they all have the same frequency $\omega_0 = \omega_p(1/\epsilon)^{1/N}$. Figure 12.10(b), (c), and (d) shows the natural modes of Butterworth filters of order $N = 2$, 3, and 4, respectively. Once the N natural modes $p_1, p_2, \ldots, p_N$ have been found, the transfer function can be written as

$$T(s) = \frac{K\omega_0^N}{(s - p_1)(s - p_2) \cdots (s - p_N)} \tag{12.16}$$

where K is a constant equal to the required dc gain of the filter.

To summarize, to find a Butterworth transfer function that meets transmission specifications of the form in Fig. 12.3 we perform the following procedure:

1. Determine ϵ from Eq. (12.14).

2. Use Eq. (12.15) to determine the required filter order as the lowest integer value of N that results in $A(\omega_s) \geq A_{min}$.

3. Use Fig. 12.10(a) to determine the N natural modes.

4. Use Eq. (12.16) to determine $T(s)$.

EXAMPLE 12.1

Find the Butterworth transfer function that meets the following low-pass filter specifications: $f_p = 10$ kHz, $A_{max} = 1$ dB, $f_s = 15$ kHz, $A_{min} = 25$ dB, dc gain = 1.

Solution

Substituting $A_{max} = 1$ dB into Eq. (12.14) yields $\epsilon = 0.5088$. Equation (12.15) is then used to determine the filter order by trying various values for N. We find that $N = 8$ yields $A(\omega_s) = 22.3$ dB and $N = 9$ gives 25.8 dB. We thus select $N = 9$.

Figure 12.11 shows the graphical construction for determining the poles. The poles all have the same frequency $\omega_0 = \omega_p(1/\epsilon)^{1/N} = 2\pi \times 10 \times 10^3(1/0.5088)^{1/9} = 6.773 \times 10^4$ rad/s. The first pole p_1 is given by

$$p_1 = \omega_0(-\cos 80° + j\sin 80°) = \omega_0(-0.1736 + j0.9848)$$

Combining p_1 with its complex conjugate p_9 yields the factor $(s^2 + s0.3472\omega_0 + \omega_0^2)$ in the denominator of the transfer function. The same can be done for the other complex poles, and the complete transfer function is obtained using Eq. (12.16),

$$T(s) = \frac{\omega_0^9}{(s + \omega_0)(s^2 + s1.8794\omega_0 + \omega_0^2)(s^2 + s1.5321\omega_0 + \omega_0^2)}$$

$$\times \frac{1}{(s^2 + s\omega_0 + \omega_0^2)(s^2 + s0.3472\omega_0 + \omega_0^2)} \tag{12.17}$$

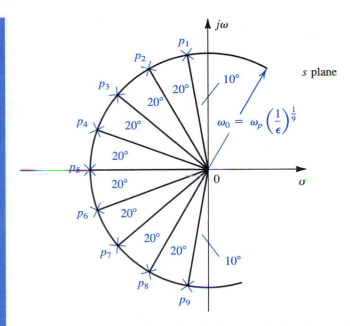

FIGURE 12.11 Poles of the ninth-order Butterworth filter of Example 12.1.

12.3.2 The Chebyshev Filter

Figure 12.12 shows representative transmission functions for Chebyshev[4] filters of even and odd order. The Chebyshev filter exhibits an equiripple response in the passband and a monotonically decreasing transmission in the stopband. While the odd-order filter has $|T(0)| = 1$, the even-order filter exhibits its maximum magnitude deviation at $\omega = 0$. In both cases the total number of passband maxima and minima equals the order of the filter, N. All the transmission zeros of the Chebyshev filter are at $\omega = \infty$, making it an all-pole filter.

The magnitude of the transfer function of an Nth-order Chebyshev filter with a passband edge (ripple bandwidth) ω_p is given by

$$|T(j\omega)| = \frac{1}{\sqrt{1 + \epsilon^2 \cos^2[N\cos^{-1}(\omega/\omega_p)]}} \quad \text{for } \omega \leq \omega_p \quad (12.18)$$

and

$$|T(j\omega)| = \frac{1}{\sqrt{1 + \epsilon^2 \cosh^2[N\cosh^{-1}(\omega/\omega_p)]}} \quad \text{for } \omega \geq \omega_p \quad (12.19)$$

At the passband edge, $\omega = \omega_p$, the magnitude function is given by

$$|T(j\omega_p)| = \frac{1}{\sqrt{1 + \epsilon^2}}$$

[4] Named after the Russian mathematician P. L. Chebyshev, who in 1899 used these functions in studying the construction of steam engines.

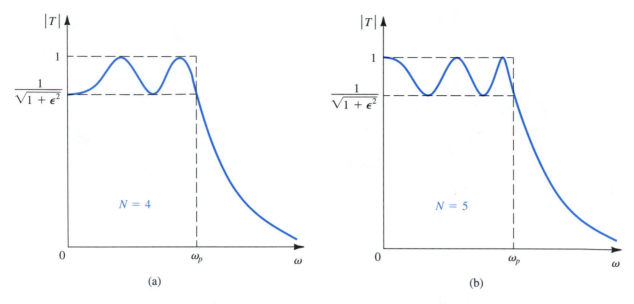

FIGURE 12.12 Sketches of the transmission characteristics of representative **(a)** even-order and **(b)** odd-order Chebyshev filters.

Thus, the parameter ϵ determines the passband ripple according to

$$A_{\max} = 10 \log(1 + \epsilon^2) \tag{12.20}$$

Conversely, given $A_{\max}$, the value of ϵ is determined from

$$\epsilon = \sqrt{10^{A_{\max}/10} - 1} \tag{12.21}$$

The attenuation achieved by the Chebyshev filter at the stopband edge ($\omega = \omega_s$) is found using Eq. (12.19) as

$$A(\omega_s) = 10 \log[1 + \epsilon^2 \cosh^2(N \cosh^{-1}(\omega_s/\omega_p))] \tag{12.22}$$

With the aid of a calculator this equation can be used to determine the order N required to obtain a specified $A_{\min}$ by finding the lowest integer value of N that yields $A(\omega_s) \geq A_{\min}$. As in the case of the Butterworth filter, increasing the order N of the Chebyshev filter causes its magnitude function to approach the ideal brick-wall low-pass response.

The poles of the Chebyshev filter are given by

$$p_k = -\omega_p \sin\left(\frac{2k-1}{N}\frac{\pi}{2}\right)\sinh\left(\frac{1}{N}\sinh^{-1}\frac{1}{\epsilon}\right)$$
$$+ j\omega_p \cos\left(\frac{2k-1}{N}\frac{\pi}{2}\right)\cosh\left(\frac{1}{N}\sinh^{-1}\frac{1}{\epsilon}\right) \qquad k = 1, 2, \ldots, N \tag{12.23}$$

Finally, the transfer function of the Chebyshev filter can be written as

$$T(s) = \frac{K\omega_p^N}{\epsilon 2^{N-1}(s - p_1)(s - p_2)\cdots(s - p_N)} \tag{12.24}$$

where K is the dc gain that the filter is required to have.

To summarize, given low-pass transmission specifications of the type shown in Fig. 12.3, the transfer function of a Chebyshev filter that meets these specifications can be found as follows:

1. Determine ϵ from Eq. (12.21).
2. Use Eq. (12.22) to determine the order required.
3. Determine the poles using Eq. (12.23).
4. Determine the transfer function using Eq. (12.24).

The Chebyshev filter provides a more efficient approximation than the Butterworth filter. Thus, for the same order and the same A_{max}, the Chebyshev filter provides greater stopband attenuation than the Butterworth filter. Alternatively, to meet identical specifications, one requires a lower order for the Chebyshev than for the Butterworth filter. This point will be illustrated by the following example.

EXAMPLE 12.2

Find the Chebyshev transfer function that meets the same low-pass filter specifications given in Example 12.1: namely, $f_p = 10$ kHz, $A_{max} = 1$ dB, $f_s = 15$ kHz, $A_{min} = 25$ dB, dc gain = 1.

Solution

Substituting $A_{max} = 1$ dB into Eq. (12.21) yields $\epsilon = 0.5088$. By trying various values for N in Eq. (12.22) we find that $N = 4$ yields $A(\omega_s) = 21.6$ dB and $N = 5$ provides 29.9 dB. We thus select $N = 5$. Recall that we required a ninth-order Butterworth filter to meet the same specifications in Example 12.1.

The poles are obtained by substituting in Eq. (12.23) as

$$p_1, p_5 = \omega_p(-0.0895 \pm j0.9901)$$

$$p_2, p_4 = \omega_p(-0.2342 \pm j0.6119)$$

$$p_5 = \omega_p(-0.2895)$$

The transfer function is obtained by substituting these values in Eq. (12.24) as

$$T(s) = \frac{\omega_p^5}{8.1408(s + 0.2895\,\omega_p)(s^2 + s0.4684\,\omega_p + 0.4293\,\omega_p^2)}$$

$$\times \frac{1}{s^2 + s0.1789\,\omega_p + 0.9883\,\omega_p^2} \qquad (12.25)$$

where $\omega_p = 2\pi \times 10^4$ rad/s.

EXERCISES

D12.6 Determine the order N of a Butterworth filter for which $A_{max} = 1$ dB, $\omega_s/\omega_p = 1.5$, and $A_{min} = 30$ dB. What is the actual value of minimum stopband attenuation realized? If A_{min} is to be exactly 30 dB, to what value can A_{max} be reduced?

Ans. $N = 11$; $A_{min} = 32.87$ dB; 0.54 dB

12.7 Find the natural modes and the transfer function of a Butterworth filter with $\omega_p = 1$ rad/s, $A_{max} = 3$ dB ($\epsilon \simeq 1$), and $N = 3$.

Ans. $-0.5 \pm j\sqrt{3}/2$ and -1; $T(s) = 1/(s+1)(s^2 + s + 1)$

12.8 Observe that Eq. (12.18) can be used to find the frequencies in the passband at which $|T|$ is at its peaks and at its valleys. (The peaks are reached when the $\cos^2[\ ]$ term is zero, and the valleys correspond to the $\cos^2[\ ]$ term equal to unity.) Find these frequencies for a fifth-order filter.

Ans. Peaks at $\omega = 0$, $0.59\omega_p$, and $0.95\omega_p$; the valleys at $\omega = 0.31\omega_p$ and $0.81\omega_p$

D12.9 Find the attenuation provided at $\omega = 2\omega_p$ by a seventh-order Chebyshev filter with a 0.5-dB passband ripple. If the passband ripple is allowed to increase to 1 dB, by how much does the stopband attenuation increase?

Ans. 64.9 dB; 3.3 dB

D12.10 It is required to design a low-pass filter having $f_p = 1$ kHz, $A_{max} = 1$ dB, $f_s = 1.5$ kHz, $A_{min} = 50$ dB. (a) Find the required order of a Chebyshev filter. What is the excess stopband attenuation obtained? (b) Repeat for a Butterworth filter.

Ans. (a) $N = 8$, 5 dB; (b) $N = 16$, 0.5 dB

12.4 FIRST-ORDER AND SECOND-ORDER FILTER FUNCTIONS

In this section, we shall study the simplest filter transfer functions, those of first and second order. These functions are useful in their own right in the design of simple filters. First- and second-order filters can also be cascaded to realize a high-order filter. Cascade design is in fact one of the most popular methods for the design of active filters (those utilizing op amps and RC circuits). Because the filter poles occur in complex-conjugate pairs, a high-order transfer function $T(s)$ is factored into the product of second-order functions. If $T(s)$ is odd, there will also be a first-order function in the factorization. Each of the second-order functions [and the first-order function when $T(s)$ is odd] is then realized using one of the op amp–RC circuits that will be studied in this chapter, and the resulting blocks are placed in cascade. If the output of each block is taken at the output terminal of an op amp where the impedance level is low (ideally zero), cascading does not change the transfer functions of the individual blocks. Thus the overall transfer function of the cascade is simply the product of the transfer functions of the individual blocks, which is the original $T(s)$.

12.4.1 First-Order Filters

The general first-order transfer function is given by

$$T(s) = \frac{a_1 s + a_0}{s + \omega_0} \tag{12.26}$$

This **bilinear transfer function** characterizes a first-order filter with a natural mode at $s = -\omega_0$, a transmission zero at $s = -a_0/a_1$, and a high-frequency gain that approaches a_1. The numerator coefficients, a_0 and a_1, determine the type of filter (e.g., low pass, high pass, etc.). Some special cases together with passive (RC) and active (op amp–RC) realizations are shown in Fig. 12.13. Note that the active realizations provide considerably more versatility than their passive counterparts; in many cases the gain can be set to a desired value, and some transfer-function parameters can be adjusted without affecting others. The output impedance of the active circuit is also very low, making cascading easily possible. The op amp, however, limits the high-frequency operation of the active circuits.

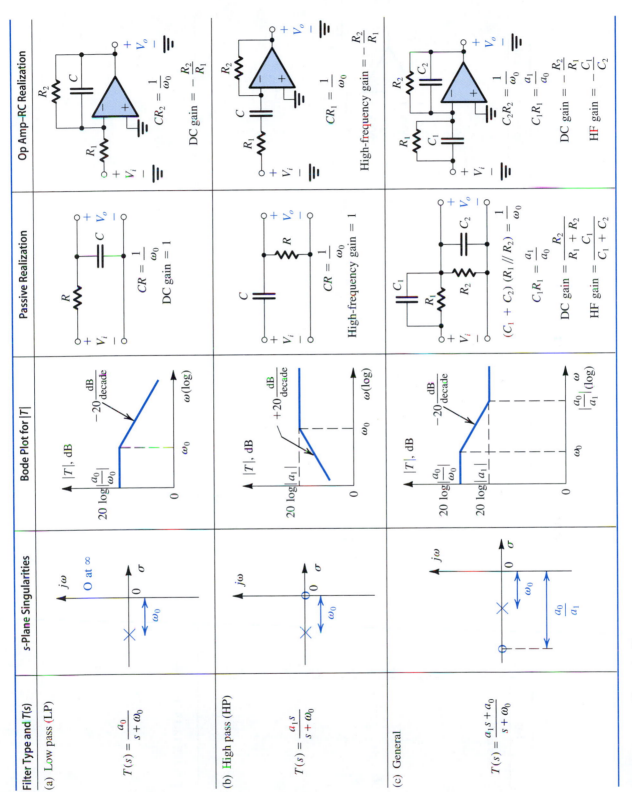

FIGURE 12.13 First-order filters.

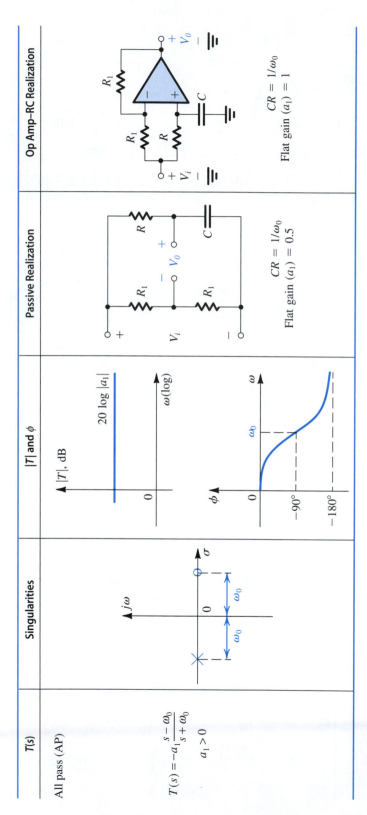

T(s)	Singularities	|T| and φ	Passive Realization	Op Amp–RC Realization

All pass (AP)

$$T(s) = -a_1 \frac{s - \omega_0}{s + \omega_0}$$

$$a_1 > 0$$

Passive Realization:

$CR = 1/\omega_0$
Flat gain $(a_1) = 0.5$

Op Amp–RC Realization:

$CR = 1/\omega_0$
Flat gain $(a_1) = 1$

FIGURE 12.14 First-order all-pass filter.

An important special case of the first-order filter function is the **all-pass filter** shown in Fig. 12.14. Here, the transmission zero and the natural mode are symmetrically located relative to the $j\omega$ axis. (They are said to display mirror-image symmetry with respect to the $j\omega$ axis.) Observe that although the transmission of the all-pass filter is (ideally) constant at all frequencies, its phase shows frequency selectivity. All-pass filters are used as phase shifters and in systems that require phase shaping (e.g., in the design of circuits called *delay equalizers,* which cause the overall time delay of a transmission system to be constant with frequency).

EXERCISES

D12.11 Using $R_1 = 10$ kΩ, design the op amp–RC circuit of Fig. 12.13(b) to realize a high-pass filter with a corner frequency of 10^4 rad/s and a high-frequency gain of 10.

Ans. $R_2 = 100$ kΩ; $C = 0.01$ μF

D12.12 Design the op amp–RC circuit of Fig. 12.14 to realize an all-pass filter with a 90° phase shift at 10^3 rad/s. Select suitable component values.

Ans. Possible choices: $R = R_1 = R_2 = 10$ kΩ; $C = 0.1$ μF

12.4.2 Second-Order Filter Functions

The general second-order (or **biquadratic**) filter transfer function is usually expressed in the standard form

$$T(s) = \frac{a_2 s^2 + a_1 s + a_0}{s^2 + (\omega_0/Q)s + \omega_0^2} \tag{12.27}$$

where ω_0 and Q determine the natural modes (poles) according to

$$p_1, p_2 = -\frac{\omega_0}{2Q} \pm j\omega_0\sqrt{1 - (1/4Q^2)} \tag{12.28}$$

We are usually interested in the case of complex-conjugate natural modes, obtained for $Q > 0.5$. Figure 12.15 shows the location of the pair of complex-conjugate poles in the s plane. Observe that the radial distance of the natural modes (from the origin) is equal to ω_0, which is known

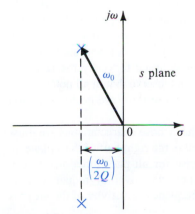

FIGURE 12.15 Definition of the parameters ω_0 and Q of a pair of complex-conjugate poles.

as the **pole frequency.** The parameter Q determines the distance of the poles from the $j\omega$ axis: the higher the value of Q, the closer the poles are to the $j\omega$ axis, and the more selective the filter response becomes. An infinite value for Q locates the poles on the $j\omega$ axis and can yield sustained oscillations in the circuit realization. A negative value of Q implies that the poles are in the right half of the s plane, which certainly produces oscillations. The parameter Q is called the **pole quality factor,** or simply, **pole Q.**

The transmission zeros of the second-order filter are determined by the numerator coefficients, a_0, a_1, and a_2. It follows that the numerator coefficients determine the type of second-order filter function (i.e., LP, HP, etc.). Seven special cases of interest are illustrated in Fig. 12.16. For each case we give the transfer function, the s-plane locations of the transfer-function singularities, and the magnitude response. Circuit realizations for the various second-order filter functions will be given in subsequent sections.

All seven special second-order filters have a pair of complex-conjugate natural modes characterized by a frequency ω_0 and a quality factor, Q.

In the low-pass (LP) case, shown in Fig. 12.16(a), the two transmission zeros are at $s = \infty$. The magnitude response can exhibit a peak with the details indicated. It can be shown that the peak occurs only for $Q > 1/\sqrt{2}$. The response obtained for $Q = 1/\sqrt{2}$ is the Butterworth, or maximally flat, response.

The high-pass (HP) function shown in Fig. 12.16(b) has both transmission zeros at $s = 0$ (dc). The magnitude response shows a peak for $Q > 1/\sqrt{2}$, with the details of the response as indicated. Observe the duality between the LP and HP responses.

Next consider the bandpass (BP) filter function shown in Fig. 12.16(c). Here, one transmission zero is at $s = 0$ (dc), and the other is at $s = \infty$. The magnitude response peaks at $\omega = \omega_0$. Thus the **center frequency** of the bandpass filter is equal to the pole frequency ω_0. The selectivity of the second-order bandpass filter is usually measured by its *3-dB bandwidth*. This is the difference between the two frequencies ω_1 and ω_2 at which the magnitude response is 3 dB below its maximum value (at ω_0). It can be shown that

$$\omega_1, \omega_2 = \omega_0 \sqrt{1 + (1/4Q^2)} \pm \frac{\omega_0}{2Q} \qquad (12.29)$$

Thus,

$$BW \equiv \omega_2 - \omega_1 = \omega_0/Q \qquad (12.30)$$

Observe that as Q increases, the bandwidth decreases and the bandpass filter becomes more selective.

If the transmission zeros are located on the $j\omega$ axis, at the complex-conjugate locations $\pm j\omega_n$, then the magnitude response exhibits zero transmission at $\omega = \omega_n$. Thus a **notch** in the magnitude response occurs at $\omega = \omega_n$, and ω_n is known as the **notch frequency.** Three cases of the second-order notch filter are possible: the regular notch, obtained when $\omega_n = \omega_0$ (Fig. 12.16d); the low-pass notch, obtained when $\omega_n > \omega_0$ (Fig. 12.16e); and the high-pass notch, obtained when $\omega_n < \omega_0$ (Fig. 12.16f). The reader is urged to verify the response details given in these figures (a rather tedious task, though!). Observe that in all notch cases, the transmission at dc and at $s = \infty$ is finite. This is so because there are no transmission zeros at either $s = 0$ or $s = \infty$.

The last special case of interest is the all-pass (AP) filter whose characteristics are illustrated in Fig. 12.16(g). Here the two transmission zeros are in the right half of the s plane, at the mirror-image locations of the poles. (This is the case for all-pass functions of any order.) The magnitude response of the all-pass function is constant over all frequencies; the **flat gain,** as it is called, is in our case equal to $|a_2|$. The frequency selectivity of the all-pass function is in its phase response.

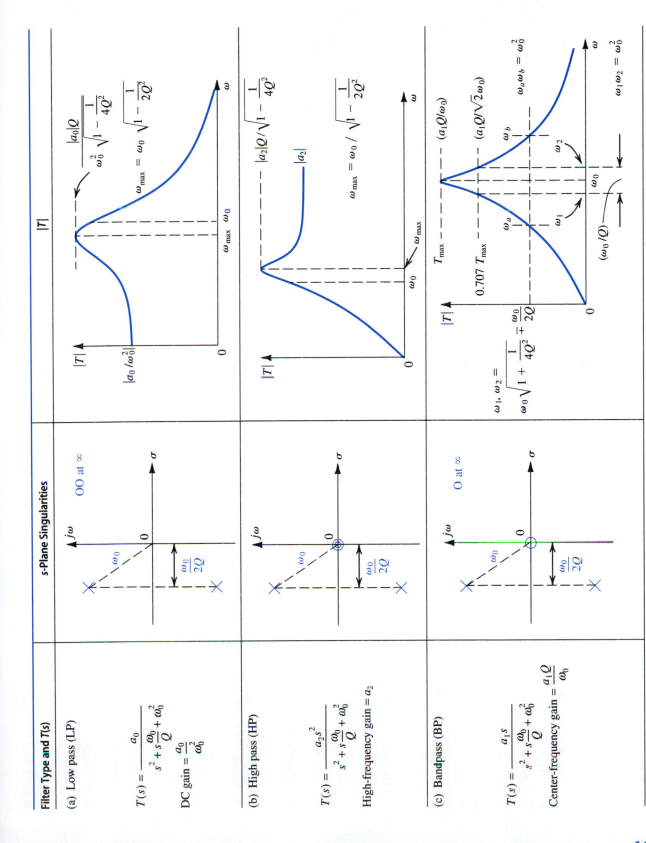

FIGURE 12.16 Second-order filtering functions.

Filter Type and $T(s)$	s-Plane Singularities	$|T|$

(d) Notch

$$T(s) = a_2 \frac{s^2 + \omega_0^2}{s^2 + s\frac{\omega_0}{Q} + \omega_0^2}$$

DC gain =
High-frequency gain = a_2

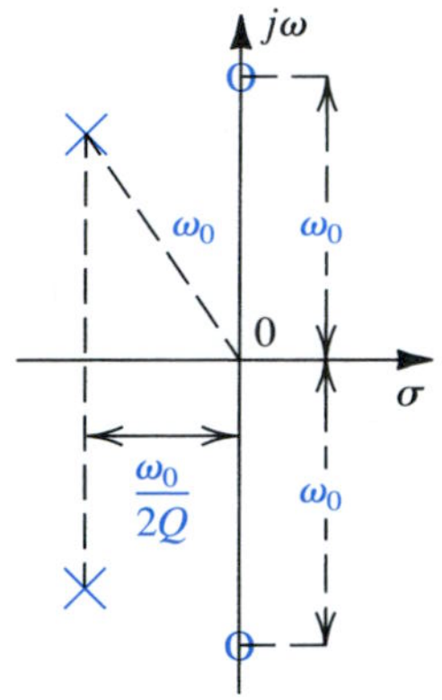

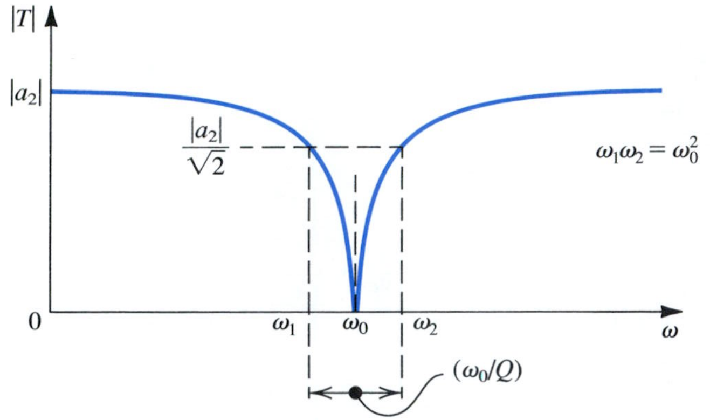

(e) Low-pass notch (LPN)

$$T(s) = a_2 \frac{s^2 + \omega_n^2}{s^2 + s\frac{\omega_0}{Q} + \omega_0^2}$$

$$\omega_n \geq \omega_0$$

DC gain = $a_2 \frac{\omega_n^2}{\omega_0^2}$

High-frequency gain = a_2

(f) High-pass notch (HPN)

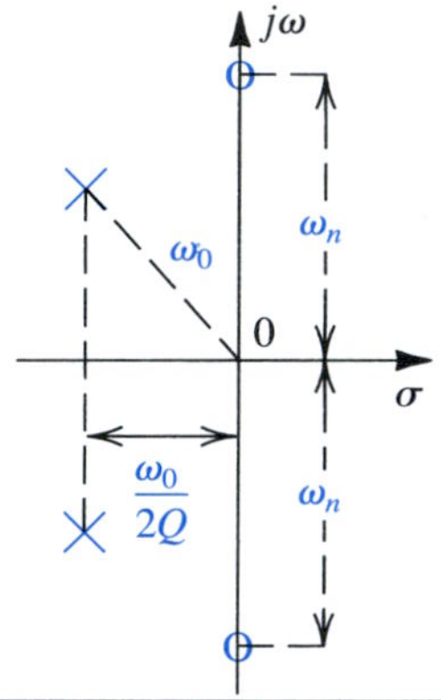

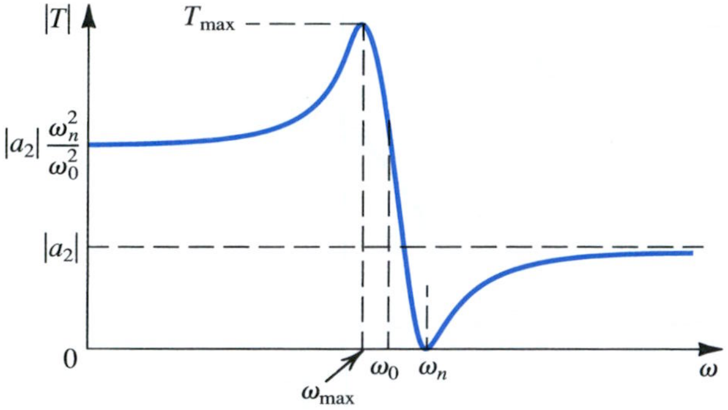

$$T(s) = a_2 \frac{s^2 + \omega_n^2}{s^2 + s\frac{\omega_0}{Q} + \omega_0^2}$$

$$\omega_n \leq \omega_0$$

DC gain = $a_2 \frac{\omega_n^2}{\omega_0^2}$

High-frequency gain = a_2

$$\omega_{max} = \omega_0 \sqrt{\frac{\sqrt{\left(\frac{\omega_n^2}{\omega_0^2}\right)\left(1 - \frac{1}{2Q^2}\right) - 1}}{\frac{\omega_n^2}{\omega_0^2}\frac{1}{\frac{1}{2Q^2} - 1}}}$$

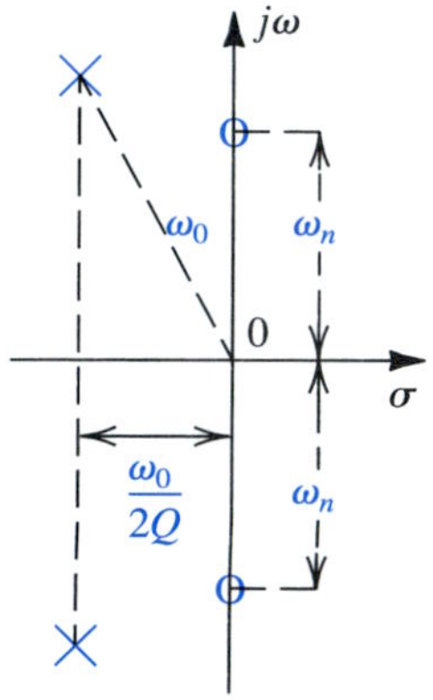

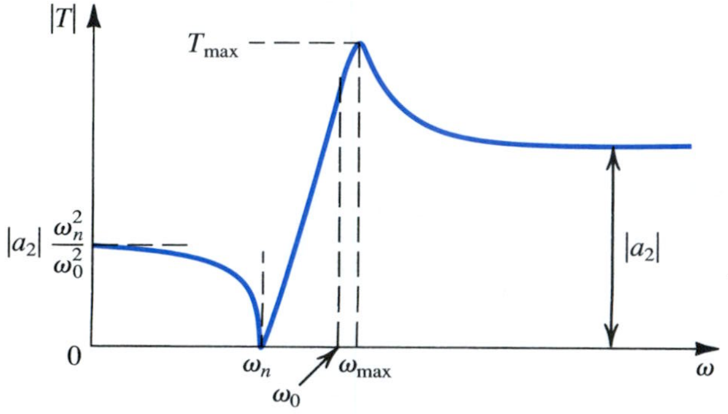

$$T_{max} = |a_2| \frac{|\omega_n^2 - \omega_{max}^2|}{\sqrt{(\omega_0^2 - \omega_{max}^2)^2 + \left(\frac{\omega_0}{Q}\right)^2 \omega_{max}^2}}$$

FIGURE 12.16 (Continued)

(g) All pass (AP)

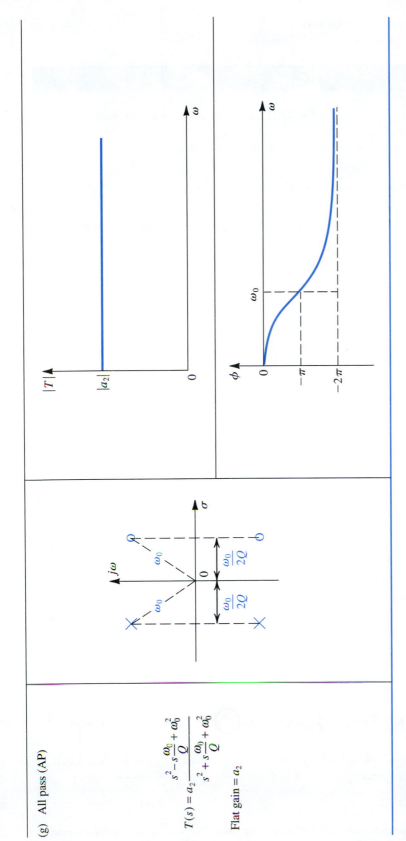

$$T(s) = a_2 \frac{s^2 - s\dfrac{\omega_0}{Q} + \omega_0^2}{s^2 + s\dfrac{\omega_0}{Q} + \omega_0^2}$$

Flat gain $= a_2$

FIGURE 12.16 *(Continued)*

EXERCISES

12.13 For a maximally flat second-order low-pass filter ($Q = 1/\sqrt{2}$), show that at $\omega = \omega_0$ the magnitude response is 3 dB below the value at dc.

12.14 Give the transfer function of a second-order bandpass filter with a center frequency of 10^5 rad/s, a center-frequency gain of 10, and a 3-dB bandwidth of 10^3 rad/s.

Ans. $T(s) = \dfrac{10^4 s}{s^2 + 10^3 s + 10^{10}}$

12.15 (a) For the second-order notch function with $\omega_n = \omega_0$, show that for the attenuation to be greater than A dB over a frequency band BW_a, the value of Q is given by

$$Q \le \frac{\omega_0}{BW_a\sqrt{10^{A/10} - 1}}$$

(*Hint:* First, show that any two frequencies, ω_1 and ω_2, at which $|T|$ is the same, are related by $\omega_1 \omega_2 = \omega_0^2$.) (b) Use the result of (a) to show that the 3-dB bandwidth is ω_0/Q, as indicated in Fig. 12.16(d).

12.16 Consider a low-pass notch with $\omega_0 = 1$ rad/s, $Q = 10$, $\omega_n = 1.2$ rad/s, and a dc gain of unity. Find the frequency and magnitude of the transmission peak. Also find the high-frequency transmission.

Ans. 0.986 rad/s; 3.17; 0.69

12.5 THE SECOND-ORDER LCR RESONATOR

In this section we shall study the second-order LCR resonator shown in Fig. 12.17(a). The use of this resonator to derive circuit realizations for the various second-order filter functions will be demonstrated. It will be shown in the next section that replacing the inductor L by a simulated inductance obtained using an op amp–RC circuit results in an op amp–RC resonator. The latter forms the basis of an important class of active-RC filters to be studied in Section 12.6.

12.5.1 The Resonator Natural Modes

The natural modes of the parallel resonance circuit of Fig. 12.17(a) can be determined by applying *an excitation that does not change the natural structure of the circuit*. Two possible ways of exciting the circuit are shown in Fig. 12.17(b) and (c). In Fig. 12.17(b) the resonator

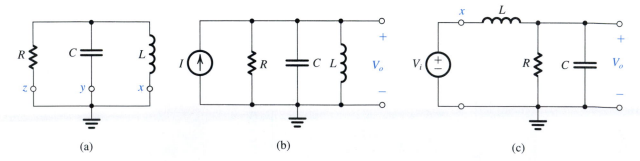

(a)   (b)   (c)

FIGURE 12.17 (**a**) The second-order parallel LCR resonator. (**b, c**) Two ways of exciting the resonator of (a) without changing its *natural structure:* resonator poles are those poles of V_o/I and V_o/V_i.

is excited with a current source I connected in parallel. Since, as far as the natural response of a circuit is concerned, an independent ideal current source is equivalent to an open circuit, the excitation of Fig. 12.17(b) does not alter the natural structure of the resonator. Thus the circuit in Fig. 12.17(b) can be used to determine the natural modes of the resonator by simply finding the poles of any response function. We can for instance take the voltage V_o across the resonator as the response and thus obtain the response function $V_o/I = Z$, where Z is the impedance of the parallel resonance circuit. It is obviously more convenient, however, to work in terms of the admittance Y; thus,

$$\frac{V_o}{I} = \frac{1}{Y} = \frac{1}{(1/sL) + sC + (1/R)}$$
$$= \frac{s/C}{s^2 + s(1/CR) + (1/LC)} \tag{12.31}$$

Equating the denominator to the standard form $[s^2 + s(\omega_0/Q) + \omega_0^2]$ leads to

$$\omega_0^2 = 1/LC \tag{12.32}$$

and

$$\omega_0/Q = 1/CR \tag{12.33}$$

Thus,

$$\omega_0 = 1/\sqrt{LC} \tag{12.34}$$

$$Q = \omega_0 CR \tag{12.35}$$

These expressions should be familiar to the reader from studies of parallel resonance circuits in introductory courses on circuit theory.

An alternative way of exciting the parallel LCR resonator for the purpose of determining its natural modes is shown in Fig. 12.17(c). Here, node x of inductor L has been disconnected from ground and connected to an ideal voltage source V_i. Now, since as far as the natural response of a circuit is concerned, an ideal independent voltage source is equivalent to a short circuit, the excitation of Fig. 12.17(c) does not alter the natural structure of the resonator. Thus we can use the circuit in Fig. 12.17(c) to determine the natural modes of the resonator. These are the poles of any response function. For instance, we can select V_o as the response variable and find the transfer function V_o/V_i. The reader can easily verify that this will lead to the natural modes determined earlier.

In a design problem, we will be given ω_0 and Q and will be asked to determine L, C, and R. Equations (12.34) and (12.35) are two equations in the three unknowns. The one available degree-of-freedom can be utilized to set the impedance level of the circuit to a value that results in practical component values.

12.5.2 Realization of Transmission Zeros

Having selected the component values of the LCR resonator to realize a given pair of complex-conjugate natural modes, we now consider the use of the resonator to realize a desired filter type (e.g., LP, HP, etc.). Specifically, we wish to find out where to inject the input voltage signal V_i so that the transfer function V_o/V_i is the desired one. Toward that end, note that in the resonator circuit in Fig. 12.17(a), any of the nodes labeled x, y, or z can be disconnected from ground and connected to V_i without altering the circuit's natural modes. When this is done the circuit takes the form of a voltage divider, as shown in Fig. 12.18(a). Thus the transfer function realized is

$$T(s) = \frac{V_o(s)}{V_i(s)} = \frac{Z_2(s)}{Z_1(s) + Z_2(s)} \tag{12.36}$$

We observe that *the transmission zeros are the values of s at which $Z_2(s)$ is zero, provided $Z_1(s)$ is not simultaneously zero, and the values of s at which $Z_1(s)$ is infinite, provided $Z_2(s)$ is not simultaneously infinite.* This statement makes physical sense: The output will be zero either when $Z_2(s)$ behaves as a short circuit or when $Z_1(s)$ behaves as an open circuit. If there is a value of s at which both Z_1 and Z_2 are zero, then V_o/V_i will be finite and no transmission zero is obtained. Similarly, if there is a value of s at which both Z_1 and Z_2 are infinite, then V_o/V_i will be finite and no transmission zero is realized.

12.5.3 Realization of the Low-Pass Function

Using the scheme just outlined we see that to realize a low-pass function, node x is disconnected from ground and connected to V_i, as shown in Fig. 12.18(b). The transmission zeros of this circuit will be at the value of s for which the series impedance becomes infinite (sL becomes infinite at $s = \infty$) and the value of s at which the shunt impedance becomes zero ($1/[sC + (1/R)]$ becomes zero at $s = \infty$). Thus this circuit has two transmission zeros at $s = \infty$, as an LP is supposed to. The transfer function can be written either by inspection or by using the voltage-divider rule. Following the latter approach, we obtain

$$T(s) \equiv \frac{V_o}{V_i} = \frac{Z_2}{Z_1 + Z_2} = \frac{Y_1}{Y_1 + Y_2} = \frac{1/sL}{(1/sL) + sC + (1/R)}$$

$$= \frac{1/LC}{s^2 + s(1/CR) + (1/LC)} \tag{12.37}$$

12.5.4 Realization of the High-Pass Function

To realize the second-order high-pass function, node y is disconnected from ground and connected to V_i, as shown in Fig. 12.18(c). Here the series capacitor introduces a transmission zero at $s = 0$ (dc), and the shunt inductor introduces another transmission zero at $s = 0$ (dc). Thus, by inspection, the transfer function may be written as

$$T(s) \equiv \frac{V_o}{V_i} = \frac{a_2 s^2}{s^2 + s(\omega_0/Q) + \omega_0^2} \tag{12.38}$$

where ω_0 and Q are the natural mode parameters given by Eqs. (12.34) and (12.35) and a_2 is the high-frequency transmission. The value of a_2 can be determined from the circuit by observing that as s approaches ∞, the capacitor approaches a short circuit and V_o approaches V_i, resulting in $a_2 = 1$.

12.5.5 Realization of the Bandpass Function

The bandpass function is realized by disconnecting node z from ground and connecting it to V_i, as shown in Fig. 12.18(d). Here the series impedance is resistive and thus does not introduce any transmission zeros. These are obtained as follows: One zero at $s = 0$ is

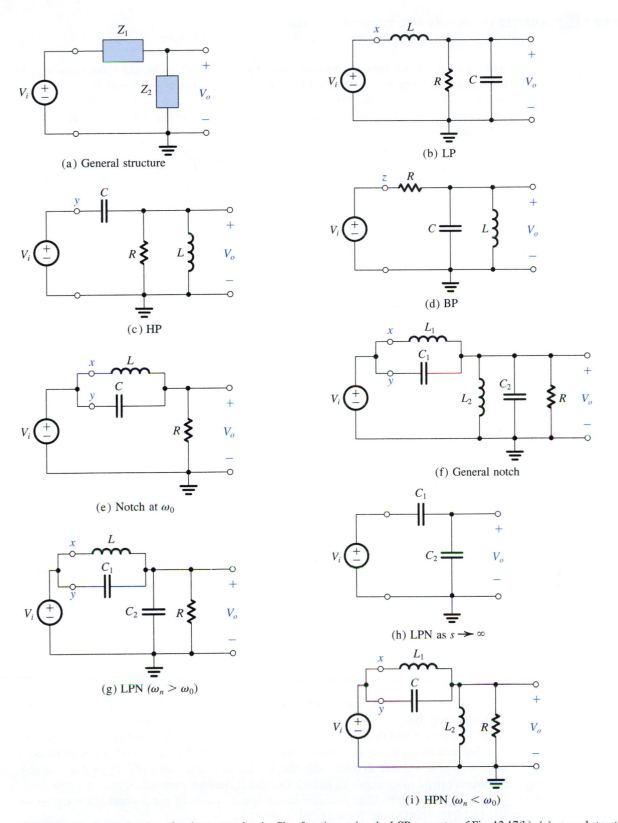

FIGURE 12.18 Realization of various second-order filter functions using the LCR resonator of Fig. 12.17(b): **(a)** general structure, **(b)** LP, **(c)** HP, **(d)** BP, **(e)** notch at ω_0, **(f)** general notch, **(g)** LPN ($\omega_n \geq \omega_0$), **(h)** LPN as $s \rightarrow \infty$, **(i)** HPN ($\omega_n < \omega_0$).

realized by the shunt inductor, and one zero at $s = \infty$ is realized by the shunt capacitor. At the center frequency ω_0, the parallel LC-tuned circuit exhibits an infinite impedance, and thus no current flows in the circuit. It follows that at $\omega = \omega_0$, $V_o = V_i$. In other words, the center-frequency gain of the bandpass filter is unity. Its transfer function can be obtained as follows:

$$T(s) = \frac{Y_R}{Y_R + Y_L + Y_C} = \frac{1/R}{(1/R) + (1/sL) + sC}$$
$$= \frac{s(1/CR)}{s^2 + s(1/CR) + (1/LC)} \tag{12.39}$$

12.5.6 Realization of the Notch Functions

To obtain a pair of transmission zeros on the $j\omega$ axis we use a parallel resonance circuit in the series arm, as shown in Fig. 12.18(e). Observe that this circuit is obtained by disconnecting both nodes x and y from ground and connecting them together to V_i. The impedance of the LC circuit becomes infinite at $\omega = \omega_0 = 1/\sqrt{LC}$, thus causing zero transmission at this frequency. The shunt impedance is resistive and thus does not introduce transmission zeros. It follows that the circuit in Fig. 12.18(e) will realize the notch transfer function

$$T(s) = a_2 \frac{s^2 + \omega_0^2}{s^2 + s(\omega_0/Q) + \omega_0^2} \tag{12.40}$$

The value of the high-frequency gain a_2 can be found from the circuit to be unity.

To obtain a notch-filter realization in which the notch frequency ω_n is arbitrarily placed relative to ω_0, we adopt a variation on the scheme above. We still use a parallel LC circuit in the series branch, as shown in Fig. 12.18(f) where L_1 and C_1 are selected so that

$$L_1 C_1 = 1/\omega_n^2 \tag{12.41}$$

Thus the $L_1 C_1$ tank circuit will introduce a pair of transmission zeros at $\pm j\omega_n$, provided the $L_2 C_2$ tank is not resonant at ω_n. Apart from this restriction, the values of L_2 and C_2 must be selected to ensure that the natural modes have not been altered; thus,

$$C_1 + C_2 = C \tag{12.42}$$

$$L_1 \| L_2 = L \tag{12.43}$$

In other words, when V_i is replaced by a short circuit, the circuit should reduce to the original LCR resonator. Another way of thinking about the circuit of Fig. 12.18(f) is that it is obtained from the original LCR resonator by lifting part of L and part of C off ground and connecting them to V_i.

It should be noted that in the circuit of Fig. 12.18(f), L_2 does *not* introduce a zero at $s = 0$ because at $s = 0$, the $L_1 C_1$ circuit also has a zero. In fact, at $s = 0$ the circuit reduces to an inductive voltage divider with the dc transmission being $L_2/(L_1 + L_2)$. Similar comments can be made about C_2 and the fact that it does *not* introduce a zero at $s = \infty$.

The LPN and HPN filter realizations are special cases of the general notch circuit of Fig. 12.18(f). Specifically, for the LPN,

$$\omega_n > \omega_0$$

and thus

$$L_1 C_1 < (L_1 \| L_2)(C_1 + C_2)$$

This condition can be satisfied with L_2 eliminated (i.e., $L_2 = \infty$ and $L_1 = L$), resulting in the LPN circuit in Fig. 12.18(g). The transfer function can be written by inspection as

$$T(s) \equiv \frac{V_o}{V_i} = a_2 \frac{s^2 + \omega_n^2}{s^2 + s(\omega_0/Q) + \omega_0^2} \qquad (12.44)$$

where $\omega_n^2 = 1/LC_1$, $\omega_0^2 = 1/L(C_1 + C_2)$, $\omega_0/Q = 1/CR$, and a_2 is the high-frequency gain. From the circuit we see that as $s \to \infty$, the circuit reduces to that in Fig. 12.18(h), for which

$$\frac{V_o}{V_i} = \frac{C_1}{C_1 + C_2}$$

Thus

$$a_2 = \frac{C_1}{C_1 + C_2} \qquad (12.45)$$

To obtain an HPN realization we start with the circuit of Fig. 12.18(f) and use the fact that $\omega_n < \omega_0$ to obtain

$$L_1 C_1 > (L_1 \| L_2)(C_1 + C_2)$$

which can be satisfied while selecting $C_2 = 0$ (i.e., $C_1 = C$). Thus we obtain the reduced circuit shown in Fig. 12.18(i). Observe that as $s \to \infty$, V_o approaches V_i and thus the high-frequency gain is unity. Thus, the transfer function can be expressed as

$$T(s) \equiv \frac{V_o}{V_i} = \frac{s^2 + (1/L_1 C)}{s^2 + s(1/CR) + [1/(L_1 \| L_2)C]} \qquad (12.46)$$

12.5.7 Realization of the All-Pass Function

The all-pass transfer function

$$T(s) = \frac{s^2 - s(\omega_0/Q) + \omega_0^2}{s^2 + s(\omega_0/Q) + \omega_0^2} \qquad (12.47)$$

can be written as

$$T(s) = 1 - \frac{s2(\omega_0/Q)}{s^2 + s(\omega_0/Q) + \omega_0^2} \qquad (12.48)$$

The second term on the right-hand side is a bandpass function with a center-frequency gain of 2. We already have a bandpass circuit (Fig. 12.18d) but with a center-frequency gain of unity. We shall therefore attempt an all-pass realization with a flat gain of 0.5, that is,

$$T(s) = 0.5 - \frac{s(\omega_0/Q)}{s^2 + s(\omega_0/Q) + \omega_0^2}$$

This function can be realized using a voltage divider with a transmission ratio of 0.5 together with the bandpass circuit of Fig. 12.18(d). To effect the subtraction, the output of the all-pass circuit is taken between the output terminal of the voltage divider and that of the

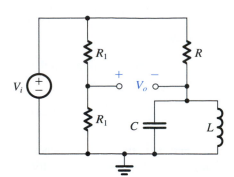

FIGURE 12.19 Realization of the second-order all-pass transfer function using a voltage divider and an LCR resonator.

bandpass filter, as shown in Fig. 12.19. Unfortunately this circuit has the disadvantage of lacking a common ground terminal between the input and the output. An op amp–RC realization of the all-pass function will be presented in the next section.

EXERCISES

12.17 Use the circuit of Fig. 12.18(b) to realize a second-order low-pass function of the maximally flat type with a 3-dB frequency of 100 kHz.

Ans. Selecting $R = 1$ kΩ, we obtain $C = 1125$ pF and $L = 2.25$ mH.

12.18 Use the circuit of Fig. 12.18(e) to design a notch filter to eliminate a bothersome power-supply hum at a 60-Hz frequency. The filter is to have a 3-dB bandwidth of 10 Hz (i.e., the attenuation is greater than 3 dB over a 10-Hz band around the 60-Hz center frequency; see Exercise 12.15 and Fig. 12.16d). Use $R = 10$ kΩ.

Ans. $C = 1.6$ μF and $L = 4.42$ H (Note the large inductor required. This is the reason passive filters are not practical in low-frequency applications.)

 ## 12.6 SECOND-ORDER ACTIVE FILTERS BASED ON INDUCTOR REPLACEMENT

In this section, we study a family of op amp–RC circuits that realize the various second-order filter functions. The circuits are based on an op amp–RC resonator obtained by replacing the inductor L in the LCR resonator with an op amp–RC circuit that has an inductive input impedance.

12.6.1 The Antoniou Inductance-Simulation Circuit

Over the years, many op amp–RC circuits have been proposed for simulating the operation of an inductor. Of these, one circuit invented by A. Antoniou[5] [see Antoniou (1969)] has proved to be the "best." By "best" we mean that the operation of the circuit is very tolerant of the nonideal properties of the op amps, in particular their finite gain and bandwidth. Figure 12.20(a) shows the Antoniou inductance-simulation circuit. If the circuit is fed at its input (node 1) with a voltage source V_1 and the input current is denoted I_1, then for ideal

---

[5] Andreas Antoniou is a Canadian academic, currently (2003) a member of the faculty of the University of Victoria, British Columbia.

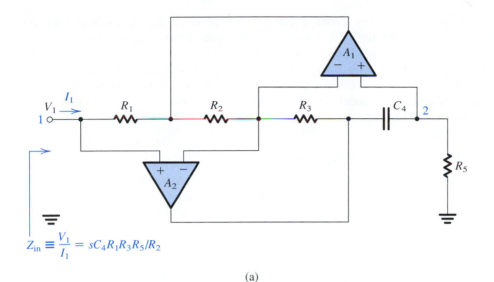

$$Z_{in} \equiv \frac{V_1}{I_1} = sC_4R_1R_3R_5/R_2$$

(a)

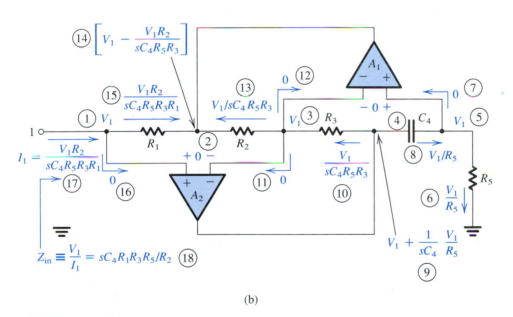

(b)

FIGURE 12.20 (a) The Antoniou inductance-simulation circuit. (b) Analysis of the circuit assuming ideal op amps. The order of the analysis steps is indicated by the circled numbers.

op amps the input impedance can be shown to be

$$Z_{in} \equiv V_1/I_1 = sC_4R_1R_3R_5/R_2 \qquad (12.49)$$

which is that of an inductance L given by

$$L = C_4R_1R_3R_5/R_2 \qquad (12.50)$$

Figure 12.20(b) shows the analysis of the circuit assuming that the op amps are ideal and thus that a virtual short circuit appears between the two input terminals of each op amp, and assuming also that the input currents of the op amps are zero. The analysis begins at node 1, which is assumed to be fed by a voltage source V_1, and proceeds step by step, with the order of the steps indicated by the circled numbers. The result of the analysis is the expression shown for the input current I_1 from which Z_{in} is found.

The design of this circuit is usually based on selecting $R_1 = R_2 = R_3 = R_5 = R$ and $C_4 = C$, which leads to $L = CR^2$. Convenient values are then selected for C and R to yield the desired inductance value L. More details on this circuit and the effect of the nonidealities of the op amps on its performance can be found in Sedra and Brackett (1978).

12.6.2 The Op Amp–RC Resonator

Figure 12.21(a) shows the LCR resonator we studied in detail in Section 12.5. Replacing the inductor L with a simulated inductance realized by the Antoniou circuit of Fig. 12.20(a) results in the op amp–RC resonator of Fig. 12.21(b). (Ignore for the moment the additional amplifier drawn with broken lines.) The circuit of Fig. 12.21(b) is a second-order resonator having a pole frequency

$$\omega_0 = 1/\sqrt{LC_6} = 1/\sqrt{C_4 C_6 R_1 R_3 R_5 / R_2} \tag{12.51}$$

where we have used the expression for L given in Eq. (12.50), and a pole Q factor,

$$Q = \omega_0 C_6 R_6 = R_6 \sqrt{\frac{C_6}{C_4} \frac{R_2}{R_1 R_3 R_5}} \tag{12.52}$$

Usually one selects $C_4 = C_6 = C$ and $R_1 = R_2 = R_3 = R_5 = R$, which results in

$$\omega_0 = 1/CR \tag{12.53}$$

$$Q = R_6/R \tag{12.54}$$

Thus, if we select a practically convenient value for C, we can use Eq. (12.53) to determine the value of R to realize a given ω_0, and then use Eq. (12.54) to determine the value of R_6 to realize a given Q.

12.6.3 Realization of the Various Filter Types

The op amp–RC resonator of Fig. 12.21(b) can be used to generate circuit realizations for the various second-order filter functions by following the approach described in detail in Section 12.5 in connection with the LCR resonator. Thus to obtain a bandpass function we disconnect node z from ground and connect it to the signal source V_i. A high-pass function is obtained by injecting V_i to node y. To realize a low-pass function using the LCR resonator, the inductor terminal x is disconnected from ground and connected to V_i. The corresponding node in the active resonator is the node at which R_5 is connected to ground,[6] labeled as node x in Fig. 12.21(b). A regular notch function ($\omega_n = \omega_0$) is obtained by feeding V_i to nodes x and y. In all cases the output can be taken as the voltage across the resonance circuit, V_r. However, this is not a convenient node to use as the filter output terminal because connecting a load there would change the filter characteristics. The problem can be solved easily by utilizing a buffer amplifier. This is the amplifier of gain K, drawn with broken lines in Fig. 12.21(b).

---

[6] This point might not be obvious! The reader, however, can show that when V_i is fed to this node the function V_r/V_i is indeed low pass.

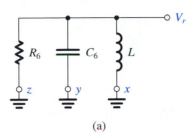

(a)

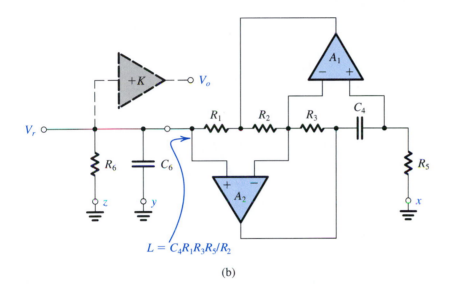

$$L = C_4 R_1 R_3 R_5 / R_2$$

(b)

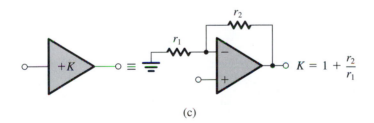

$$K = 1 + \frac{r_2}{r_1}$$

(c)

FIGURE 12.21 (a) An LCR resonator. (b) An op amp–RC resonator obtained by replacing the inductor L in the LCR resonator of (a) with a simulated inductance realized by the Antoniou circuit of Fig. 12.20(a). (c) Implementation of the buffer amplifier K.

Figure 12.21(c) shows how this amplifier can be simply implemented using an op amp connected in the noninverting configuration. Note that not only does the amplifier K buffer the output of the filter, but it also allows the designer to set the filter gain to any desired value by appropriately selecting the value of K.

Figure 12.22 shows the various second-order filter circuits obtained from the resonator of Fig. 12.21(b). The transfer functions and design equations for these circuits are given in

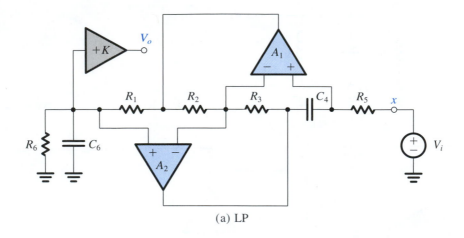

(a) LP

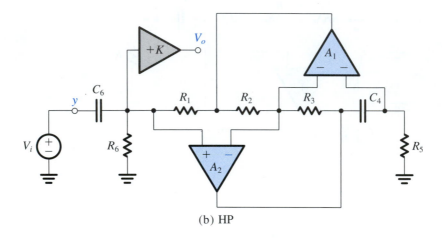

(b) HP

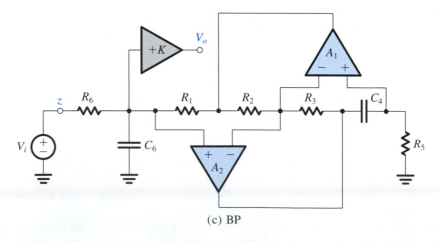

(c) BP

FIGURE 12.22 Realizations for the various second-order filter functions using the op amp–RC resonator of Fig. 12.21(b): **(a)** LP, **(b)** HP, **(c)** BP,

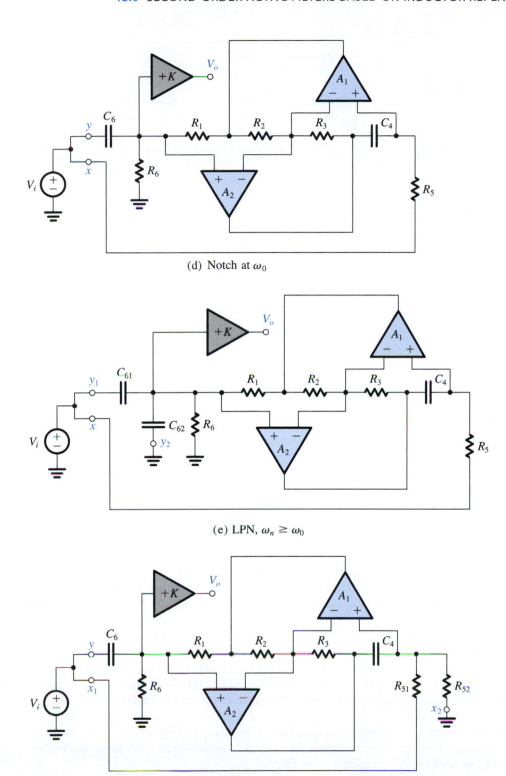

(d) Notch at ω_0

(e) LPN, $\omega_n \geq \omega_0$

(f) HPN, $\omega_n \leq \omega_0$

FIGURE 12.22 *(Continued)* **(d)** notch at ω_0, **(e)** LPN, $\omega_n \geq \omega_0$, **(f)** HPN, $\omega_n \leq \omega_0$, and

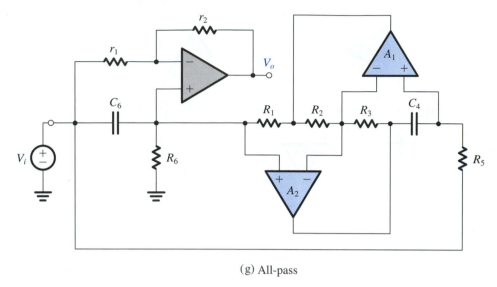

(g) All-pass

FIGURE 12.22 *(Continued)* **(g)** all pass. The circuits are based on the LCR circuits in Fig. 12.18. Design equations are given in Table 12.1.

Table 12.1. Note that the transfer functions can be written by analogy to those of the LCR resonator. We have already commented on the LP, HP, BP, and regular-notch circuits given in Fig. 12.22(a) to (d). The LPN and HPN circuits in Fig. 12.22(e) and (f) are obtained by direct analogy to their LCR counterparts in Fig. 12.18(g) and (i), respectively. The all-pass circuit in Fig. 12.22(g), however, deserves some explanation.

12.6.4 The All-Pass Circuit

An all-pass function with a flat gain of unity can be written as

$$AP = 1 - (BP \text{ with a center-frequency gain of } 2) \tag{12.55}$$

(see Eq. 12.48). Two circuits whose transfer functions are related in this fashion are said to be **complementary**.[7] Thus the all-pass circuit with unity flat gain is the complement of the bandpass circuit with a center-frequency gain of 2. A simple procedure exists for obtaining the complement of a given linear circuit: Disconnect all the circuit nodes that are connected to ground and connect them to V_i, and disconnect all the nodes that are connected to V_i and connect them to ground. That is, interchanging input and ground in a linear circuit generates a circuit whose transfer function is the complement of that of the original circuit.

Returning to the problem at hand, we first use the circuit of Fig. 12.22(c) to realize a BP with a gain of 2 by simply selecting $K = 2$ and implementing the buffer amplifier with the circuit of Fig. 12.21(c) with $r_1 = r_2$. We then interchange input and ground and thus obtain the all-pass circuit of Fig. 12.22(g).

Finally, in addition to being simple to design, the circuits in Fig. 12.22 exhibit excellent performance. They can be used on their own to realize second-order filter functions, or they can be cascaded to implement high-order filters.

---

[7] More about complementary circuits will be presented later in conjunction with Fig. 12.31.

TABLE 12.1 Design Data for the Circuits of Fig. 12.22

Circuit	Transfer Function and Other Parameters	Design Equations
Resonator Fig. 12.21(b)	$\omega_0 = 1/\sqrt{C_4 C_6 R_1 R_3 R_5 / R_2}$ $Q = R_6 \sqrt{\dfrac{C_6}{C_4} \dfrac{R_2}{R_1 R_3 R_5}}$	$C_4 = C_6 = C$ (practical value) $R_1 = R_2 = R_3 = R_5 = 1/\omega_0 C$ $R_6 = Q/\omega_0 C$
Low-pass (LP) Fig. 12.22(a)	$T(s) = \dfrac{K R_2 / C_4 C_6 R_1 R_3 R_5}{s^2 + s\dfrac{1}{C_6 R_6} + \dfrac{R_2}{C_4 C_6 R_1 R_3 R_5}}$	$K = $ DC gain
High-pass (HP) Fig. 12.22(b)	$T(s) = \dfrac{K s^2}{s^2 + s\dfrac{1}{C_6 R_6} + \dfrac{R_2}{C_4 C_6 R_1 R_3 R_5}}$	$K = $ High-frequency gain
Bandpass (BP) Fig. 12.22(c)	$T(s) = \dfrac{K s / C_6 R_6}{s^2 + s\dfrac{1}{C_6 R_6} + \dfrac{R_2}{C_4 C_6 R_1 R_3 R_5}}$	$K = $ Center-frequency gain
Regular notch (N) Fig. 12.22(d)	$T(s) = \dfrac{K[s^2 + (R_2 / C_4 C_6 R_1 R_3 R_5)]}{s^2 + s\dfrac{1}{C_6 R_6} + \dfrac{R_2}{C_4 C_6 R_1 R_3 R_5}}$	$K = $ Low- and high-frequency gain
Low-pass notch (LPN) Fig. 12.22(e)	$T(s) = K\dfrac{C_{61}}{C_{61} + C_{62}}$ $\times \dfrac{s^2 + (R_2 / C_4 C_{61} R_1 R_3 R_5)}{s^2 + s\dfrac{1}{(C_{61}+C_{62})R_6} + \dfrac{R_2}{C_4 (C_{61}+C_{62}) R_1 R_3 R_5}}$ $\omega_n = 1/\sqrt{C_4 C_{61} R_1 R_3 R_5 / R_2}$ $\omega_0 = 1/\sqrt{C_4 (C_{61}+C_{62}) R_1 R_3 R_5 / R_2}$ $Q = R_6 \sqrt{\dfrac{C_{61}+C_{62}}{C_4} \dfrac{R_2}{R_1 R_3 R_5}}$	$K = $ DC gain $C_{61} + C_{62} = C_6 = C$ $C_{61} = C(\omega_0 / \omega_n)^2$ $C_{62} = C - C_{61}$
High-pass notch (HPN) Fig. 12.22(f)	$T(s) = K\dfrac{s^2 + (R_2 / C_4 C_6 R_1 R_3 R_{51})}{s^2 + s\dfrac{1}{C_6 R_6} + \dfrac{R_2}{C_4 C_6 R_1 R_3}\left(\dfrac{1}{R_{51}} + \dfrac{1}{R_{52}}\right)}$ $\omega_n = 1/\sqrt{C_4 C_6 R_1 R_3 R_{51} / R_2}$ $\omega_0 = \sqrt{\dfrac{R_2}{C_4 C_6 R_1 R_3}\left(\dfrac{1}{R_{51}} + \dfrac{1}{R_{52}}\right)}$ $Q = R_6 \sqrt{\dfrac{C_6}{C_4} \dfrac{R_2}{R_1 R_3}\left(\dfrac{1}{R_{51}} + \dfrac{1}{R_{52}}\right)}$	$\dfrac{1}{R_{51}} + \dfrac{1}{R_{52}} = \dfrac{1}{R_5} = \omega_0 C$ $R_{51} = R_5(\omega_0 / \omega_n)^2$ $R_{52} = R_5 / [1 - (\omega_n / \omega_0)^2]$
All-pass (AP) Fig. 12.22(g)	$T(s) = \dfrac{s^2 - s\dfrac{1}{C_6 R_6}\dfrac{r_2}{r_1} + \dfrac{R_2}{C_4 C_6 R_1 R_3 R_5}}{s^2 + s\dfrac{1}{C_6 R_6} + \dfrac{R_2}{C_4 C_6 R_1 R_3 R_5}}$ $\omega_z = \omega_0 \quad Q_z = Q(r_1/r_2) \quad$ Flat gain $= 1$	$r_1 = r_2 = r$ (arbitrary) Adjust r_2 to make $Q_z = Q$

D12.19 Use the circuit of Fig. 12.22(c) to design a second-order bandpass filter with a center frequency of 10 kHz, a 3-dB bandwidth of 500 Hz, and a center-frequency gain of 10. Use $C = 1.2$ nF.

Ans. $R_1 = R_2 = R_3 = R_5 = 13.26$ kΩ; $R_6 = 265$ kΩ; $C_4 = C_6 = 1.2$ nF; $K = 10$, $r_1 = 10$ kΩ, $r_2 = 90$ kΩ

D12.20 Realize the Chebyshev filter of Example 12.2, whose transfer function is given in Eq. (12.25), as the cascade connection of three circuits: two of the type shown in Fig. 12.22(a) and one first-order op amp–RC circuit of the type shown in Fig. 12.13(a). Note that you can make the dc gain of all sections equal to unity. Do so. Use as many 10-kΩ resistors as possible.

Ans. First-order section: $R_1 = R_2 = 10$ kΩ, $C = 5.5$ nF; second-order section with $\omega_0 = 4.117 \times 10^4$ rad/s and $Q = 1.4$: $R_1 = R_2 = R_3 = R_5 = 10$ kΩ, $R_6 = 14$ kΩ, $C_4 = C_6 = 2.43$ nF, $r_1 = \infty$, $r_2 = 0$; second-order section with $\omega_0 = 6.246 \times 10^4$ rad/s and $Q = 5.56$: $R_1 = R_2 = R_3 = R_5 = 10$ kΩ, $R_6 = 55.6$ kΩ, $C_4 = C_6 = 1.6$ nF, $r_1 = \infty$, $r_2 = 0$

12.7 SECOND-ORDER ACTIVE FILTERS BASED ON THE TWO-INTEGRATOR-LOOP TOPOLOGY

In this section, we study another family of op amp–RC circuits that realize second-order filter functions. The circuits are based on the use of two integrators connected in cascade in an overall feedback loop and are thus known as two-integrator-loop circuits.

12.7.1 Derivation of the Two-Integrator-Loop Biquad

To derive the two-integrator-loop biquadratic circuit, or **biquad** as it is commonly known,[8] consider the second-order high-pass transfer function

$$\frac{V_{hp}}{V_i} = \frac{K s^2}{s^2 + s(\omega_0/Q) + \omega_0^2} \tag{12.56}$$

where K is the high-frequency gain. Cross-multiplying Eq. (12.56) and dividing both sides of the resulting equation by s^2 (to get all the terms involving s in the form $1/s$, which is the transfer function of an integrator) gives

$$V_{hp} + \frac{1}{Q}\left(\frac{\omega_0}{s}V_{hp}\right) + \left(\frac{\omega_0^2}{s^2}V_{hp}\right) = K V_i \tag{12.57}$$

In this equation we observe that the signal $(\omega_0/s)V_{hp}$ can be obtained by passing V_{hp} through an integrator with a time constant equal to $1/\omega_0$. Furthermore, passing the resulting signal through another identical integrator results in the third signal involving V_{hp} in Eq. (12.57)—namely, $(\omega_0^2/s^2)V_{hp}$. Figure 12.23(a) shows a block diagram for such a two-integrator arrangement. Note that in anticipation of the use of the inverting op-amp Miller integrator circuit to implement each integrator, the integrator blocks in Fig. 12.23(a) have been assigned negative signs.

The problem still remains, however, of how to form V_{hp}, the input signal feeding the two cascaded integrators. Toward that end, we rearrange Eq. (12.57), expressing V_{hp} in terms of its single- and double-integrated versions and of V_i as

$$V_{hp} = K V_i - \frac{1}{Q}\frac{\omega_0}{s}V_{hp} - \frac{\omega_0^2}{s^2}V_{hp} \tag{12.58}$$

---

[8] The name biquad stems from the fact that this circuit in its most general form is capable of realizing a biquadratic transfer function, that is, one that is the ratio of two quadratic polynomials.

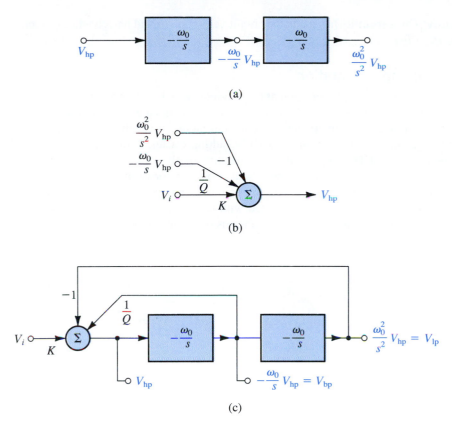

FIGURE 12.23 Derivation of a block diagram realization of the two-integrator-loop biquad.

which suggests that V_{hp} can be obtained by using the weighted summer of Fig. 12.23(b). Now it should be easy to see that a complete block diagram realization can be obtained by combining the integrator blocks of Fig. 12.23(a) with the summer block of Fig. 12.23(b), as shown in Fig. 12.23(c).

In the realization of Fig. 12.23(c), V_{hp}, obtained at the output of the summer, realizes the high-pass transfer function $T_{hp} \equiv V_{hp}/V_i$ of Eq. (12.56). The signal at the output of the first integrator is $-(\omega_0/s)V_{hp}$, which is a bandpass function,

$$\frac{(-\omega_0/s)V_{hp}}{V_i} = -\frac{K\omega_0 s}{s^2 + s(\omega_0/Q) + \omega_0^2} = T_{bp}(s) \qquad (12.59)$$

Therefore the signal at the output of the first integrator is labeled V_{bp}. Note that the center-frequency gain of the bandpass filter realized is equal to $-KQ$.

In a similar fashion, we can show that the transfer function realized at the output of the second integrator is the low-pass function,

$$\frac{(\omega_0^2/s^2)V_{hp}}{V_i} = \frac{K\omega_0^2}{s^2 + s(\omega_0/Q) + \omega_0^2} = T_{lp}(s) \qquad (12.60)$$

Thus the output of the second integrator is labeled V_{lp}. Note that the dc gain of the low-pass filter realized is equal to K.

We conclude that the two-integrator-loop biquad shown in block diagram form in Fig. 12.23(c) realizes the three basic second-order filtering functions, LP, BP, and HP, *simultaneously*. This versatility has made the circuit very popular and has given it the name *universal active filter*.

12.7.2 Circuit Implementation

To obtain an op-amp circuit implementation of the two-integrator-loop biquad of Fig. 12.23(c), we replace each integrator with a Miller integrator circuit having $CR = 1/\omega_0$, and we replace the summer block with an op-amp summing circuit that is capable of assigning both positive and negative weights to its inputs. The resulting circuit, known as the Kerwin–Huelsman–Newcomb or **KHN biquad** after its inventors, is shown in Fig. 12.24(a). Given values for ω_0, Q, and K, the design of the circuit is straightforward: We select suitably practical values for the components of the integrators C and R so that $CR = 1/\omega_0$. To determine the values of the resistors associated with the summer, we first use superposition to express the output of the summer V_{hp} in terms of its inputs, $V_{bp} = -(\omega_0/s)V_{hp}$ and $V_{lp} = (\omega_0^2/s^2)V_{hp}$, as

$$V_{hp} = \frac{R_3}{R_2 + R_3}\left(1 + \frac{R_f}{R_1}\right)V_i + \frac{R_2}{R_2 + R_3}\left(1 + \frac{R_f}{R_1}\right)\left(-\frac{\omega_0}{s}V_{hp}\right) - \frac{R_f}{R_1}\left(\frac{\omega_0^2}{s^2}V_{hp}\right) \qquad (12.61)$$

Equating the last right-hand-side terms of Eqs. (12.61) and (12.58) gives

$$R_f/R_1 = 1 \qquad (12.62)$$

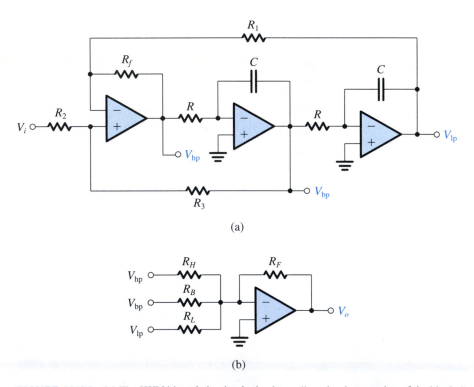

(a)

(b)

FIGURE 12.24   **(a)** The KHN biquad circuit, obtained as a direct implementation of the block diagram of Fig. 12.23(c). The three basic filtering functions, HP, BP, and LP, are simultaneously realized. **(b)** To obtain notch and all-pass functions, the three outputs are summed with appropriate weights using this op-amp summer.

which implies that we can select arbitrary but practically convenient equal values for R_1 and R_f. Then, equating the second-to-last terms on the right-hand side of Eqs. (12.61) and (12.58) and setting $R_1 = R_f$ yields the ratio R_3/R_2 required to realize a given Q as

$$R_3/R_2 = 2Q - 1 \qquad (12.63)$$

Thus an arbitrary but convenient value can be selected for either R_2 or R_3, and the value of the other resistance can be determined using Eq. (12.63). Finally, equating the coefficients of V_i in Eqs. (12.61) and (12.58) and substituting $R_f = R_1$ and for R_3/R_2 from Eq. (12.63) results in

$$K = 2 - (1/Q) \qquad (12.64)$$

Thus the gain parameter K is fixed to this value.

The KHN biquad can be used to realize notch and all-pass functions by summing weighted versions of the three outputs, LP, BP, and HP. Such an op-amp summer is shown in Fig. 12.24(b); for this summer we can write

$$
\begin{aligned}
V_o &= -\left(\frac{R_F}{R_H} V_{hp} + \frac{R_F}{R_B} V_{bp} + \frac{R_F}{R_L} V_{lp} \right) \\
&= -V_i \left(\frac{R_F}{R_H} T_{hp} + \frac{R_F}{R_B} T_{bp} + \frac{R_F}{R_L} T_{lp} \right)
\end{aligned}
\qquad (12.65)
$$

Substituting for T_{hp}, T_{bp}, and T_{lp} from Eqs. (12.56), (12.59), and (12.60), respectively, gives the overall transfer function

$$\frac{V_o}{V_i} = -K \frac{(R_F/R_H)s^2 - s(R_F/R_B)\omega_0 + (R_F/R_L)\omega_0^2}{s^2 + s(\omega_0/Q) + \omega_0^2} \qquad (12.66)$$

from which we can see that different transmission zeros can be obtained by the appropriate selection of the values of the summing resistors. For instance, a notch is obtained by selecting $R_B = \infty$ and

$$\frac{R_H}{R_L} = \left(\frac{\omega_n}{\omega_0} \right)^2 \qquad (12.67)$$

12.7.3 An Alternative Two-Integrator-Loop Biquad Circuit

An alternative two-integrator-loop biquad circuit in which all three op amps are used in a single-ended mode can be developed as follows: Rather than using the input summer to add signals with positive and negative coefficients, we can introduce an additional inverter, as shown in Fig. 12.25(a). Now all the coefficients of the summer have the same sign, and we can dispense with the summing amplifier altogether and perform the summation at the virtual-ground input of the first integrator. The resulting circuit is shown in Fig. 12.25(b), from which we observe that the high-pass function is no longer available! This is the price paid for obtaining a circuit that utilizes all op amps in a single-ended mode. The circuit of Fig. 12.25(b) is known as the **Tow–Thomas biquad,** after its originators.

Rather than using a fourth op amp to realize the finite transmission zeros required for the notch and all-pass functions, as was done with the KHN biquad, an economical *feedforward* scheme can be employed with the Tow–Thomas circuit. Specifically, the virtual ground available at the input of each of the three op amps in the Tow–Thomas circuit permits the input signal to be fed to all three op amps, as shown in Fig. 12.26. If V_o is taken at the output

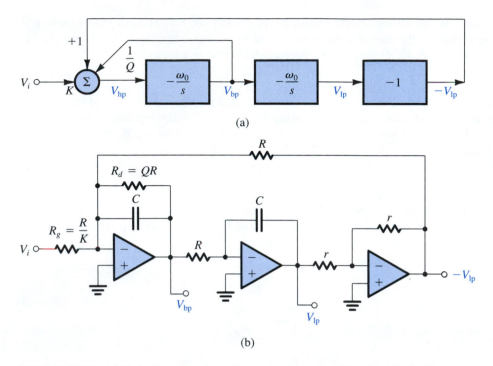

FIGURE 12.25 (**a**) Derivation of an alternative two-integrator-loop biquad in which all op amps are used in a single-ended fashion. (**b**) The resulting circuit, known as the Tow–Thomas biquad.

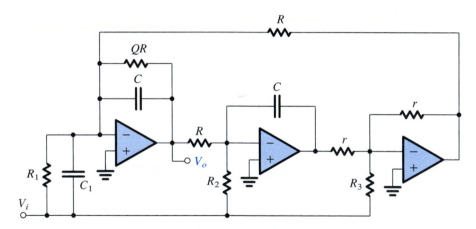

FIGURE 12.26 The Tow–Thomas biquad with feedforward. The transfer function of Eq. (12.68) is realized by feeding the input signal through appropriate components to the inputs of the three op amps. This circuit can realize all special second-order functions. The design equations are given in Table 12.2.

of the damped integrator, straightforward analysis yields the filter transfer function

$$\frac{V_o}{V_i} = -\frac{s^2\left(\dfrac{C_1}{C}\right) + s\dfrac{1}{C}\left(\dfrac{1}{R_1} - \dfrac{r}{RR_3}\right) + \dfrac{1}{C^2RR_2}}{s^2 + s\dfrac{1}{QCR} + \dfrac{1}{C^2R^2}} \tag{12.68}$$

which can be used to obtain the design data given in Table 12.2.

TABLE 12.2	Design Data for the Circuit in Fig. 12.26
All cases	C = arbitrary, $R = 1/\omega_0 C$, r = arbitrary
LP	$C_1 = 0$, $R_1 = \infty$, $R_2 = R/$dc gain, $R_3 = \infty$
Positive BP	$C_1 = 0$, $R_1 = \infty$, $R_2 = \infty$, $R_3 = Qr/$center-frequency gain
Negative BP	$C_1 = 0$, $R_1 = QR/$center-frequency gain, $R_2 = \infty$, $R_3 = \infty$
HP	$C_1 = C \times$ high-frequency gain, $R_1 = \infty$, $R_2 = \infty$, $R_3 = \infty$
Notch (all types)	$C_1 = C \times$ high-frequency gain, $R_1 = \infty$, $R_2 = R(\omega_0/\omega_n)^2/$high-frequency gain, $R_3 = \infty$
AP	$C_1 = C \times$ flat gain, $R_1 = \infty$, $R_2 = R/$gain, $R_3 = Qr/$gain

12.7.4 Final Remarks

Two-integrator-loop biquads are extremely versatile and easy to design. However, their performance is adversely affected by the finite bandwidth of the op amps. Special techniques exist for compensating the circuit for such effects [see the SPICE simulation in Section 12.12 and Sedra and Brackett (1978)].

EXERCISES

D12.21 Design the KHN circuit to realize a high-pass function with $f_0 = 10$ kHz and $Q = 2$. Choose $C = 1$ nF. What is the value of high-frequency gain obtained? What is the center-frequency gain of the bandpass function that is simultaneously available at the output of the first integrator?

Ans. $R = 15.9$ kΩ; $R_1 = R_f = R_2 = 10$ kΩ (arbitrary); $R_3 = 30$ kΩ; 1.5; 3

D12.22 Use the KHN circuit together with an output summing amplifier to design a low-pass notch filter with $f_0 = 5$ kHz, $f_n = 8$ kHz, $Q = 5$, and a dc gain of 3. Select $C = 1$ nF and $R_L = 10$ kΩ.

Ans. $R = 31.83$ kΩ; $R_1 = R_f = R_2 = 10$ kΩ (arbitrary); $R_3 = 90$ kΩ; $R_H = 25.6$ kΩ; $R_F = 16.7$ kΩ; $R_B = \infty$

D12.23 Use the Tow–Thomas biquad (Fig. 12.25b) to design a second-order bandpass filter with $f_0 = 10$ kHz, $Q = 20$, and unity center-frequency gain. If $R = 10$ kΩ, give the values of C, R_d, and R_g.

Ans. 1.59 nF; 200 kΩ; 200 kΩ

D12.24 Use the data of Table 12.2 to design the biquad circuit of Fig. 12.26 to realize an all-pass filter with $\omega_0 = 10^4$ rad/s, $Q = 5$, and flat gain = 1. Use $C = 10$ nF and $r = 10$ kΩ.

Ans. $R = 10$ kΩ; Q-determining resistor = 50 kΩ; $C_1 = 10$ nF; $R_1 = \infty$; $R_2 = 10$ kΩ; $R_3 = 50$ kΩ

12.8 SINGLE-AMPLIFIER BIQUADRATIC ACTIVE FILTERS

The op amp–RC biquadratic circuits studied in the two preceding sections provide good performance, are versatile, and are easy to design and to adjust (tune) after final assembly. Unfortunately, however, they are not economic in their use of op amps, requiring three or four amplifiers per second-order section. This can be a problem, especially in applications where power-supply current is to be conserved: for instance, in a battery-operated instrument. In this section we shall study a class of second-order filter circuits that requires only one op amp per biquad. These minimal realizations, however, suffer a greater dependence on the limited gain and bandwidth of the op amp and can also be more sensitive to the unavoidable tolerances in the values of resistors and capacitors than the multiple-op-amp biquads of the preceding sections. The **single-amplifier biquads** (SABs) are therefore limited to the less stringent filter specifications—for example, pole Q factors less than about 10.

The synthesis of SAB circuits is based on the use of feedback to move the poles of an RC circuit from the negative real axis, where they naturally lie, to the complex-conjugate locations required to provide selective filter response. The synthesis of SABs follows a two-step process:

1. Synthesis of a feedback loop that realizes a pair of complex-conjugate poles characterized by a frequency ω_0 and a Q factor Q.

2. Injecting the input signal in a way that realizes the desired transmission zeros.

12.8.1 Synthesis of the Feedback Loop

Consider the circuit shown in Fig. 12.27(a), which consists of a two-port RC network n placed in the negative-feedback path of an op amp. We shall assume that, except for having a finite gain A, the op amp is ideal. We shall denote by $t(s)$ the open-circuit voltage transfer function of the RC network n, where the definition of $t(s)$ is illustrated in Fig. 12.27(b). The transfer function $t(s)$ can in general be written as the ratio of two polynomials $N(s)$ and $D(s)$:

$$t(s) = \frac{N(s)}{D(s)}$$

The roots of $N(s)$ are the transmission zeros of the RC network, and the roots of $D(s)$ are its poles. Study of network theory shows that while the poles of an RC network are restricted to lie on the negative real axis, the zeros can in general lie anywhere in the s plane.

The loop gain $L(s)$ of the feedback circuit in Fig. 12.27(a) can be determined using the method of Section 8.7. It is simply the product of the op-amp gain A and the transfer function $t(s)$,

$$L(s) = At(s) = \frac{AN(s)}{D(s)} \tag{12.69}$$

Substituting for $L(s)$ into the characteristic equation

$$1 + L(s) = 0 \tag{12.70}$$

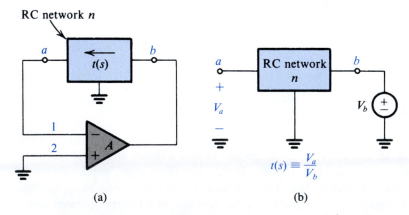

(a)   (b)

FIGURE 12.27 (a) Feedback loop obtained by placing a two-port RC network n in the feedback path of an op amp. (b) Definition of the open-circuit transfer function $t(s)$ of the RC network.

results in the poles s_P of the closed-loop circuit obtained as solutions to the equation

$$t(s_P) = -\frac{1}{A} \tag{12.71}$$

In the ideal case, $A = \infty$ and the poles are obtained from

$$N(s_P) = 0 \tag{12.72}$$

That is, *the filter poles are identical to the zeros of the RC network.*

Since our objective is to realize a pair of complex-conjugate poles, we should select an RC network that can have complex-conjugate transmission zeros. The simplest such networks are the bridged-T networks shown in Fig. 12.28 together with their transfer functions $t(s)$ from b to a, with a open-circuited. As an example, consider the circuit generated by placing the bridged-T network of Fig. 12.28(a) in the negative-feedback path of an op amp, as shown in Fig. 12.29.

$$t(s) = \frac{s^2 + s\left(\dfrac{1}{C_1} + \dfrac{1}{C_2}\right)\dfrac{1}{R_3} + \dfrac{1}{C_1 C_2 R_3 R_4}}{s^2 + s\left(\dfrac{1}{C_1 R_3} + \dfrac{1}{C_2 R_3} + \dfrac{1}{C_1 R_4}\right) + \dfrac{1}{C_1 C_2 R_3 R_4}}$$

(a)

$$t(s) = \frac{s^2 + s\left(\dfrac{1}{R_1} + \dfrac{1}{R_2}\right)\dfrac{1}{C_4} + \dfrac{1}{C_3 C_4 R_1 R_2}}{s^2 + s\left(\dfrac{1}{C_4 R_1} + \dfrac{1}{C_4 R_2} + \dfrac{1}{C_3 R_2}\right) + \dfrac{1}{C_3 C_4 R_1 R_2}}$$

(b)

FIGURE 12.28 Two RC networks (called bridged-T networks) that can have complex transmission zeros. The transfer functions given are from b to a, with a open-circuited.

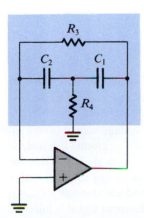

FIGURE 12.29 An active-filter feedback loop generated using the bridged-T network of Fig. 12.28(a).

The pole polynomial of the active-filter circuit will be equal to the numerator polynomial of the bridged-T network; thus,

$$s^2 + s\frac{\omega_0}{Q} + \omega_0^2 = s^2 + s\left(\frac{1}{C_1} + \frac{1}{C_2}\right)\frac{1}{R_3} + \frac{1}{C_1 C_2 R_3 R_4}$$

which enables us to obtain ω_0 and Q as

$$\omega_0 = \frac{1}{\sqrt{C_1 C_2 R_3 R_4}} \tag{12.73}$$

$$Q = \left[\frac{\sqrt{C_1 C_2 R_3 R_4}}{R_3}\left(\frac{1}{C_1} + \frac{1}{C_2}\right)\right]^{-1} \tag{12.74}$$

If we are designing this circuit, ω_0 and Q are given and Eqs. (12.73) and (12.74) can be used to determine C_1, C_2, R_3, and R_4. It follows that there are two degrees of freedom. Let us exhaust one of these by selecting $C_1 = C_2 = C$. Let us also denote $R_3 = R$ and $R_4 = R/m$. By substituting in Eqs. (12.73) and (12.74) and with some manipulation, we obtain

$$m = 4Q^2 \tag{12.75}$$

$$CR = \frac{2Q}{\omega_0} \tag{12.76}$$

Thus if we are given the value of Q, Eq. (12.75) can be used to determine the ratio of the two resistances R_3 and R_4. Then the given values of ω_0 and Q can be substituted in Eq. (12.76) to determine the time constant CR. There remains one degree of freedom—the value of C or R can be arbitrarily chosen. In an actual design, this value, which sets the *impedance level* of the circuit, should be chosen so that the resulting component values are practical.

EXERCISES

D12.25 Design the circuit of Fig. 12.29 to realize a pair of poles with $\omega_0 = 10^4$ rad/s and $Q = 1$. Select $C_1 = C_2 = 1$ nF.

Ans. $R_3 = 200$ kΩ; $R_4 = 50$ kΩ

12.26 For the circuit designed in Exercise 12.25, find the location of the poles of the RC network in the feedback loop.

Ans. -0.382×10^4 and -2.618×10^4 rad/s

12.8.2 Injecting the Input Signal

Having synthesized a feedback loop that realizes a given pair of poles, we now consider connecting the input signal source to the circuit. We wish to do this, of course, without altering the poles.

Since, for the purpose of finding the poles of a circuit, an ideal voltage source is equivalent to a short circuit, it follows that any circuit node that is connected to ground can instead be connected to the input voltage source without causing the poles to change. Thus the method of injecting the input voltage signal into the feedback loop is simply to disconnect a component (or several components) that is (are) connected to ground and connect it (them) to the input source. Depending on the component(s) through which the input signal is injected, different transmission zeros are obtained. This is, of course, the same method we used in Section 12.5 with the LCR resonator and in Section 12.6 with the biquads based on the LCR resonator.

As an example, consider the feedback loop of Fig. 12.29. Here we have two grounded nodes (one terminal of R_4 and the positive input terminal of the op amp) that can serve for injecting the input signal. Figure 12.30(a) shows the circuit with the input signal injected through part of the resistance R_4. Note that the two resistances R_4/α and $R_4/(1-\alpha)$ have a parallel equivalent of R_4.

Analysis of the circuit to determine its voltage transfer function $T(s) \equiv V_o(s)/V_i(s)$ is illustrated in Fig. 12.30(b). Note that we have assumed the op amp to be ideal, and have indicated the order of the analysis steps by the circled numbers. The final step, number 9,

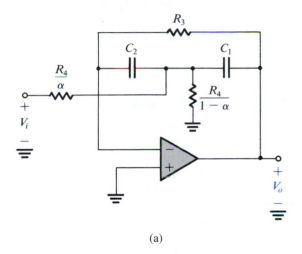

(a)

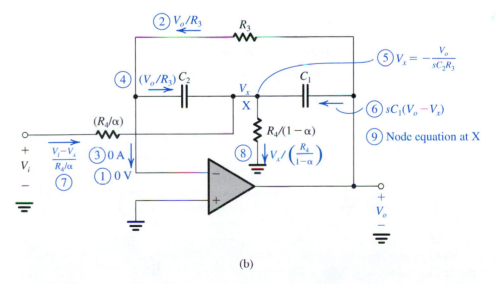

(b)

FIGURE 12.30 (a) The feedback loop of Fig. 12.29 with the input signal injected through part of resistance R_4. This circuit realizes the bandpass function. (b) Analysis of the circuit in (a) to determine its voltage transfer function $T(s)$ with the order of the analysis steps indicated by the circled numbers.

consists of writing a node equation at X and substituting for V_x by the value determined in step 5. The result is the transfer function

$$\frac{V_o}{V_i} = \frac{-s(\alpha/C_1 R_4)}{s^2 + s\left(\dfrac{1}{C_1} + \dfrac{1}{C_2}\right)\dfrac{1}{R_3} + \dfrac{1}{C_1 C_2 R_3 R_4}}$$

We recognize this as a bandpass function whose center-frequency gain can be controlled by the value of α. As expected, the denominator polynomial is identical to the numerator polynomial of $t(s)$ given in Fig. 12.28(a).

EXERCISE

12.27 Use the component values obtained in Exercise 12.25 to design the bandpass circuit of Fig. 12.30(a). Determine the values of (R_4/α) and $R_4/(1 - \alpha)$ to obtain a center-frequency gain of unity.

Ans. 100 kΩ; 100 kΩ

12.8.3 Generation of Equivalent Feedback Loops

The **complementary transformation** of feedback loops is based on the property of linear networks illustrated in Fig. 12.31 for the two-port (three-terminal) network n. In Fig. 12.31(a), terminal c is grounded and a signal V_b is applied to terminal b. The transfer function from b to a with c grounded is denoted t. Then, in Fig. 12.31(b), terminal b is grounded and the input signal is applied to terminal c. The transfer function from c to a with b grounded can be shown to be the complement of t—that is, $1 - t$. (Recall that we used this property in generating a circuit realization for the all-pass function in Section 12.6.)

Application of the complementary transformation to a feedback loop to generate an equivalent feedback loop is a two-step process:

1. Nodes of the feedback network and any of the op-amp inputs that are connected to ground should be disconnected from ground and connected to the op-amp output. Conversely, those nodes that were connected to the op-amp output should be now connected to ground. That is, we simply interchange the op-amp output terminal with ground.

2. The two input terminals of the op amp should be interchanged.

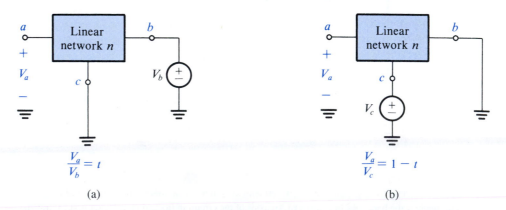

(a)   (b)

FIGURE 12.31 Interchanging input and ground results in the complement of the transfer function.

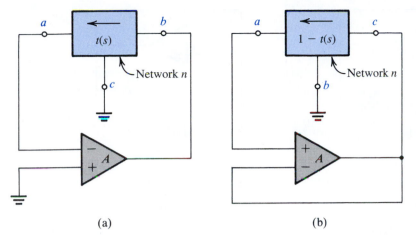

FIGURE 12.32 Application of the complementary transformation to the feedback loop in **(a)** results in the equivalent loop (same poles) shown in **(b)**.

The feedback loop generated by this transformation has the same characteristic equation, and hence the same poles, as the original loop.

To illustrate, we show in Fig. 12.32(a) the feedback loop formed by connecting a two-port RC network in the negative-feedback path of an op amp. Application of the complementary transformation to this loop results in the feedback loop of Fig. 12.32(b). Note that in the latter loop the op amp is used in the unity-gain follower configuration. We shall now show that the two loops of Fig. 12.32 are equivalent.

If the op amp has an open-loop gain A, the follower in the circuit of Fig. 12.32(b) will have a gain of $A/(A+1)$. This, together with the fact that the transfer function of network n from c to a is $1 - t$ (see Fig. 12.31), enables us to write for the circuit in Fig. 12.32(b) the characteristic equation

$$1 - \frac{A}{A+1}(1-t) = 0$$

This equation can be manipulated to the form

$$1 + At = 0$$

which is the characteristic equation of the loop in Fig. 12.32(a). As an example, consider the application of the complementary transformation to the feedback loop of Fig. 12.29: The feedback loop of Fig. 12.33(a) results. Injecting the input signal through C_1 results in the circuit in Fig. 12.33(b), which can be shown (by direct analysis) to realize a second-order high-pass function. This circuit is one of a family of SABs known as the Sallen-and-Key circuits, after their originators. The design of the circuit in Fig. 12.33(b) is based on Eqs. (12.73) through (12.76); namely, $R_3 = R$, $R_4 = R/4Q^2$, $C_1 = C_2 = C$, $CR = 2Q/\omega_0$, and the value of C is arbitrarily chosen to be practically convenient.

As another example, Fig. 12.34(a) shows the feedback loop generated by placing the two-port RC network of Fig. 12.28(b) in the negative-feedback path of an op amp. For an ideal op amp, this feedback loop realizes a pair of complex-conjugate natural modes having the same location as the zeros of $t(s)$ of the RC network. Thus, using the expression for $t(s)$

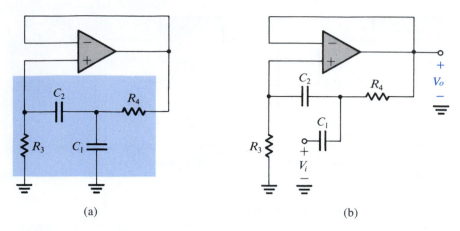

(a)　　　　　　　　　　　　　　　　(b)

FIGURE 12.33 **(a)** Feedback loop obtained by applying the complementary transformation to the loop in Fig. 12.29. **(b)** Injecting the input signal through C_1 realizes the high-pass function. This is one of the Sallen-and-Key family of circuits.

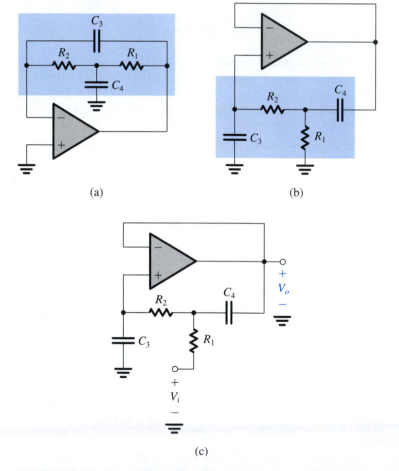

(a)　　　　　　　　　　　　　　　　(b)

(c)

FIGURE 12.34 **(a)** Feedback loop obtained by placing the bridged-T network of Fig. 12.28(b) in the negative-feedback path of an op amp. **(b)** Equivalent feedback loop generated by applying the complementary transformation to the loop in (a). **(c)** A low-pass filter obtained by injecting V_i through R_1 into the loop in (b).

given in Fig. 12.28(b), we can write for the active-filter poles

$$\omega_0 = 1/\sqrt{C_3 C_4 R_1 R_2} \tag{12.77}$$

$$Q = \left[\frac{\sqrt{C_3 C_4 R_1 R_2}}{C_4} \left(\frac{1}{R_1} + \frac{1}{R_2} \right) \right]^{-1} \tag{12.78}$$

Normally the design of this circuit is based on selecting $R_1 = R_2 = R$, $C_4 = C$, and $C_3 = C/m$. When substituted in Eqs. (12.77) and (12.78), these yield

$$m = 4Q^2 \tag{12.79}$$

$$CR = 2Q/\omega_0 \tag{12.80}$$

with the remaining degree of freedom (the value of C or R) left to the designer to choose.

Injecting the input signal to the C_4 terminal that is connected to ground can be shown to result in a bandpass realization. If, however, we apply the complementary transformation to the feedback loop in Fig. 12.34(a), we obtain the equivalent loop in Fig. 12.34(b). The loop equivalence means that the circuit of Fig. 12.34(b) has the same poles and thus the same ω_0 and Q and the same design equations (Eqs. 12.77 through 12.80). The new loop in Fig. 12.34(b) can be used to realize a low-pass function by injecting the input signal as shown in Fig. 12.34(c).

EXERCISES

12.28 Analyze the circuit in Fig. 12.34(c) to determine its transfer function $V_o(s)/V_i(s)$ and thus show that ω_0 and Q are indeed those in Eqs. (12.77) and (12.78). Also show that the dc gain is unity.

D12.29 Design the circuit in Fig. 12.34(c) to realize a low-pass filter with $f_0 = 4$ kHz and $Q = 1/\sqrt{2}$. Use 10-kΩ resistors.
Ans. $R_1 = R_2 = 10$ kΩ; $C_3 = 2.81$ nF; $C_4 = 5.63$ nF

12.9 SENSITIVITY

Because of the tolerances in component values and because of the finite op-amp gain, the response of the actual assembled filter will deviate from the ideal response. As a means for predicting such deviations, the filter designer employs the concept of **sensitivity**. Specifically, for second-order filters one is usually interested in finding how *sensitive* their poles are relative to variations (both initial tolerances and future drifts) in RC component values and amplifier gain. These sensitivities can be quantified using the **classical sensitivity function** S_x^y, defined as

$$S_x^y \equiv \lim_{\Delta x \to 0} \frac{\Delta y/y}{\Delta x/x} \tag{12.81}$$

Thus,

$$S_x^y = \frac{\partial y}{\partial x} \frac{x}{y} \tag{12.82}$$

Here, x denotes the value of a component (a resistor, a capacitor, or an amplifier gain) and y denotes a circuit parameter of interest (say, ω_0 or Q). For small changes

$$S_x^y \simeq \frac{\Delta y / y}{\Delta x / x} \tag{12.83}$$

Thus we can use the value of S_x^y to determine the per-unit change in y due to a given per-unit change in x. For instance, if the sensitivity of Q relative to a particular resistance R_1 is 5, then a 1% increase in R_1 results in a 5% increase in the value of Q.

EXAMPLE 12.3

For the feedback loop of Fig. 12.29, find the sensitivities of ω_0 and Q relative to all the passive components and the op-amp gain. Evaluate these sensitivities for the design considered in the preceding section for which $C_1 = C_2$.

Solution

To find the sensitivities with respect to the passive components, called **passive sensitivities,** we assume that the op-amp gain is infinite. In this case, ω_0 and Q are given by Eqs. (12.73) and (12.74). Thus for ω_0 we have

$$\omega_0 = \frac{1}{\sqrt{C_1 C_2 R_3 R_4}}$$

which can be used together with the sensitivity definition of Eq. (12.82) to obtain

$$S_{C_1}^{\omega_0} = S_{C_2}^{\omega_0} = S_{R_3}^{\omega_0} = S_{R_4}^{\omega_0} = -\tfrac{1}{2}$$

For Q we have

$$Q = \left[\sqrt{C_1 C_2 R_3 R_4} \left(\frac{1}{C_1} + \frac{1}{C_2} \right) \frac{1}{R_3} \right]^{-1}$$

to which we apply the sensitivity definition to obtain

$$S_{C_1}^Q = \frac{1}{2} \left(\sqrt{\frac{C_2}{C_1}} - \sqrt{\frac{C_1}{C_2}} \right) \left(\sqrt{\frac{C_2}{C_1}} + \sqrt{\frac{C_1}{C_2}} \right)^{-1}$$

For the design with $C_1 = C_2$ we see that $S_{C_1}^Q = 0$. Similarly, we can show that

$$S_{C_2}^Q = 0, \qquad S_{R_3}^Q = \tfrac{1}{2}, \qquad S_{R_4}^Q = -\tfrac{1}{2}$$

It is important to remember that the sensitivity expression should be derived *before* values corresponding to a particular design are substituted.

Next we consider the sensitivities relative to the amplifier gain. If we assume the op amp to have a finite gain A, the characteristic equation for the loop becomes

$$1 + At(s) = 0 \tag{12.84}$$

where $t(s)$ is given in Fig. 12.28(a). To simplify matters we can substitute for the passive components by their design values. This causes no errors in evaluating sensitivities, since we are now finding the sensitivity with respect to the amplifier gain. Using the design values obtained earlier—namely, $C_1 = C_2 = C$, $R_3 = R$, $R_4 = R/4Q^2$, and $CR = 2Q/\omega_0$—we get

$$t(s) = \frac{s^2 + s(\omega_0/Q) + \omega_0^2}{s^2 + s(\omega_0/Q)(2Q^2 + 1) + \omega_0^2} \tag{12.85}$$

where ω_0 and Q denote the nominal or design values of the pole frequency and Q factor. The actual values are obtained by substituting for $t(s)$ in Eq. (12.84):

$$s^2 + s\frac{\omega_0}{Q}(2Q^2 + 1) + \omega_0^2 + A\left(s^2 + s\frac{\omega_0}{Q} + \omega_0^2\right) = 0$$

Assuming the gain A to be real and dividing both sides by $A + 1$, we get

$$s^2 + s\frac{\omega_0}{Q}\left(1 + \frac{2Q^2}{A+1}\right) + \omega_0^2 = 0 \tag{12.86}$$

From this equation we see that the actual pole frequency, ω_{0a}, and the pole Q, Q_a, are

$$\omega_{0a} = \omega_0 \tag{12.87}$$

$$Q_a = \frac{Q}{1 + 2Q^2/(A+1)} \tag{12.88}$$

Thus

$$S_A^{\omega_{0a}} = 0$$

$$S_A^{Q_a} = \frac{A}{A+1}\frac{2Q^2/(A+1)}{1 + 2Q^2/(A+1)}$$

For $A \gg 2Q^2$ and $A \gg 1$ we obtain

$$S_A^{Q_a} \simeq \frac{2Q^2}{A}$$

It is usual to drop the subscript a in this expression and write

$$S_A^Q \simeq \frac{2Q^2}{A} \tag{12.89}$$

Note that if Q is high ($Q \ge 5$), its sensitivity relative to the amplifier gain can be quite high.[9]

12.9.1 A Concluding Remark

The results of Example 12.3 indicate a serious disadvantage of single-amplifier biquads—the sensitivity of Q relative to the amplifier gain is quite high. Although a technique exists for reducing S_A^Q in SABs [see Sedra et al. (1980)], this is done at the expense of increased passive sensitivities. Nevertheless, the resulting SABs are used extensively in many applications. However, for filters with Q factors greater than about 10, one usually opts for one of the multiamplifier biquads studied in Sections 12.6 and 12.7. For these circuits S_A^Q is proportional to Q, rather than to Q^2 as in the SAB case (Eq. 12.89).

---

[9] Because the open-loop gain A of op amps usually has wide tolerance, it is important to keep $S_A^{\omega_0}$ and S_A^Q very small.

12.30 In a particular filter utilizing the feedback loop of Fig. 12.29, with $C_1 = C_2$, find the expected percentage change in ω_0 and Q under the conditions that (a) R_3 is 2% high, (b) R_4 is 2% high, (c) both R_3 and R_4 are 2% high, and (d) both capacitors are 2% low and both resistors are 2% high.

Ans. (a) −1%, +1%; (b) −1%, −1%; (c) −2%, 0%; (d) 0%, 0%

12.10 SWITCHED-CAPACITOR FILTERS

The active-RC filter circuits presented above have two properties that make their production in monolithic IC form difficult, if not practically impossible; these are the need for large-valued capacitors and the requirement of accurate RC time constants. The search therefore has continued for a method of filter design that would lend itself more naturally to IC implementation. In this section we shall introduce one such method.

12.10.1 The Basic Principle

The switched-capacitor filter technique is based on the realization that a capacitor switched between two circuit nodes at a sufficiently high rate is equivalent to a resistor connecting these two nodes. To be specific, consider the active-RC integrator of Fig. 12.35(a). This is the familiar Miller integrator, which we used in the two-integrator-loop biquad in Section 12.7. In Fig. 12.35(b) we have replaced the input resistor R_1 by a grounded capacitor C_1

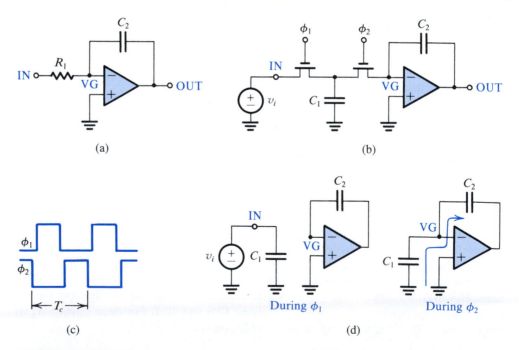

FIGURE 12.35 Basic principle of the switched-capacitor filter technique. **(a)** Active-RC integrator. **(b)** Switched-capacitor integrator. **(c)** Two-phase clock (nonoverlapping). **(d)** During ϕ_1, C_1 charges up to the current value of v_i and then, during ϕ_2, discharges into C_2.

together with two MOS transistors acting as switches. In some circuits, more elaborate switch configurations are used, but such details are beyond our present need.

The two MOS switches in Fig. 12.35(b) are driven by a *nonoverlapping* two-phase clock. Figure 12.35(c) shows the clock waveforms. We shall assume in this introductory exposition that the clock frequency f_c ($f_c = 1/T_c$) is much higher than the frequency of the signal being filtered. Thus during clock phase ϕ_1, when C_1 is connected across the input signal source v_i, the variations in the input signal are negligibly small. It follows that during ϕ_1 capacitor C_1 charges up to the voltage v_i,

$$q_{C1} = C_1 v_i$$

Then, during clock phase ϕ_2, capacitor C_1 is connected to the virtual-ground input of the op amp, as indicated in Fig. 12.35(d). Capacitor C_1 is thus forced to discharge, and its previous charge q_{C1} is transferred to C_2, in the direction indicated in Fig. 12.35(d).

From the description above we see that during each clock period T_c an amount of charge $q_{C1} = C_1 v_i$ is extracted from the input source and supplied to the integrator capacitor C_2. Thus the average current flowing between the input node (IN) and the virtual-ground node (VG) is

$$i_{av} = \frac{C_1 v_i}{T_c}$$

If T_c is sufficiently short, one can think of this process as almost continuous and define an equivalent resistance R_{eq} that is in effect present between nodes IN and VG:

$$R_{eq} \equiv v_i / i_{av}$$

Thus,

$$R_{eq} = T_c / C_1 \qquad (12.90)$$

Using R_{eq} we obtain an equivalent time constant for the integrator:

$$\text{Time constant} = C_2 R_{eq} = T_c \frac{C_2}{C_1} \qquad (12.91)$$

Thus the time constant that determines the frequency response of the filter is established by the clock period T_c and the capacitor ratio C_2/C_1. Both these parameters can be well controlled in an IC process. Specifically, note the dependence on capacitor ratios rather than on absolute values of capacitors. The accuracy of capacitor ratios in MOS technology can be controlled to within 0.1%.

Another point worth observing is that with a reasonable clocking frequency (such as 100 kHz) and not-too-large capacitor ratios (say, 10), one can obtain reasonably large time constants (such as 10^{-4} s) suitable for audio applications. Since capacitors typically occupy relatively large areas on the IC chip, one attempts to minimize their values. In this context, it is important to note that the ratio accuracies quoted earlier are obtainable with the smaller capacitor value as low as 0.1 pF.

12.10.2 Practical Circuits

The switched-capacitor (SC) circuit in Fig. 12.35(b) realizes an inverting integrator (note the direction of charge flow through C_2 in Fig. 12.35d). As we saw in Section 12.7, a two-integrator-loop active filter is composed of one inverting and one noninverting integrator.[10]

---

[10] In the two-integrator loop of Fig. 12.25(b) the noninverting integrator is realized by the cascade of a Miller integrator and an inverting amplifier.

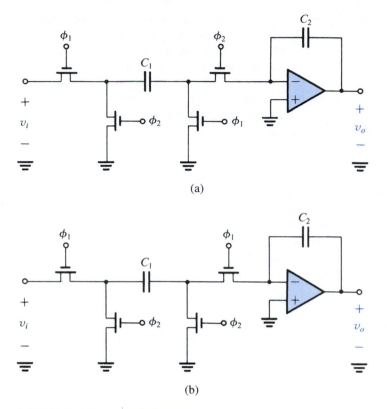

FIGURE 12.36 A pair of complementary stray-insensitive switched-capacitor integrators. **(a)** Noninverting switched-capacitor integrator. **(b)** Inverting switched-capacitor integrator.

To realize a switched-capacitor biquad filter we therefore need a pair of complementary switched-capacitor integrators. Figure 12.36(a) shows a noninverting, or positive, integrator circuit. The reader is urged to follow the operation of this circuit during the two clock phases and thus show that it operates in much the same way as the basic circuit of Fig. 12.35(b), except for a sign reversal.

In addition to realizing a noninverting integrator function, the circuit in Fig. 12.36(a) is insensitive to stray capacitances; however, we shall not explore this point any further. The interested reader is referred to Schaumann, Ghausi, and Laker (1990). By reversal of the clock phases on two of the switches, the circuit in Fig. 12.36(b) is obtained. This circuit realizes the inverting integrator function, like the circuit of Fig. 12.35(b), but is insensitive to stray capacitances (which the original circuit of Fig. 12.35b is not). The pair of complementary integrators of Fig. 12.36 has become the standard building block in the design of switched-capacitor filters.

Let us now consider the realization of a complete biquad circuit. Figure 12.37(a) shows the active-RC two-integrator-loop circuit studied earlier. By considering the cascade of integrator 2 and the inverter as a positive integrator, and then simply replacing each resistor by its switched-capacitor equivalent, we obtain the circuit in Fig. 12.37(b). Ignore the damping around the first integrator (i.e., the switched capacitor C_5) for the time being and note that the feedback loop indeed consists of one inverting and one noninverting integrator. Then note the phasing of the switched capacitor used for damping. Reversing the phases here would convert the feedback to positive and move the poles to the right half of the s plane.

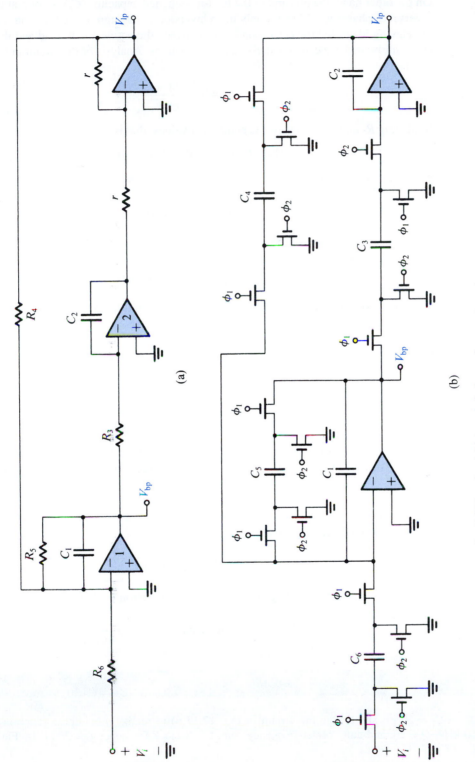

FIGURE 12.37 (a) A two-integrator-loop active-RC biquad and (b) its switched-capacitor counterpart.

On the other hand, the phasing of the feed-in switched capacitor (C_6) is not that important; a reversal of phases would result only in an inversion in the sign of the function realized.

Having identified the correspondences between the active-RC biquad and the switched-capacitor biquad, we can now derive design equations. Analysis of the circuit in Fig. 12.37(a) yields

$$\omega_0 = \frac{1}{\sqrt{C_1 C_2 R_3 R_4}} \tag{12.92}$$

Replacing R_2 and R_4 with their SC equivalent values, that is,

$$R_3 = T_c/C_3 \quad \text{and} \quad R_4 = T_c/C_4$$

gives ω_0 of the SC biquad as

$$\omega_0 = \frac{1}{T_c} \sqrt{\frac{C_3 C_4}{C_2 C_1}} \tag{12.93}$$

It is usual to select the time constants of the two integrators to be equal; that is,

$$\frac{T_c}{C_3} C_2 = \frac{T_c}{C_4} C_1 \tag{12.94}$$

If, further, we select the two integrating capacitors C_1 and C_2 to be equal,

$$C_1 = C_2 = C \tag{12.95}$$

then

$$C_3 = C_4 = KC \tag{12.96}$$

where from Eq. (12.93)

$$K = \omega_0 T_c \tag{12.97}$$

For the case of equal time constants, the Q factor of the circuit in Fig. 12.37(a) is given by R_5/R_4. Thus the Q factor of the corresponding SC circuit in Fig. 12.37(b) is given by

$$Q = \frac{T_c/C_5}{T_c/C_4} \tag{12.98}$$

Thus C_5 should be selected from

$$C_5 = \frac{C_4}{Q} = \frac{KC}{Q} = \omega_0 T_c \frac{C}{Q} \tag{12.99}$$

Finally, the center-frequency gain of the bandpass function is given by

$$\text{Center-frequency gain} = \frac{C_6}{C_5} = Q \frac{C_6}{\omega_0 T_c C} \tag{12.100}$$

EXERCISE

D12.31 Use $C_1 = C_2 = 20$ pF and design the circuit in Fig. 12.37(b) to realize a bandpass function with $f_0 = 10$ kHz, $Q = 20$, and unity center-frequency gain. Use a clock frequency $f_c = 200$ kHz. Find the values of C_3, C_4, C_5, and C_6.
Ans. 6.283 pF; 6.283 pF; 0.314 pF; 0.314 pF

12.10.3 A Final Remark

We have attempted to provide only an introduction to switched-capacitor filters. We have made many simplifying assumptions, the most important being the switched-capacitor–resistor equivalence (Eq. 12.90). This equivalence is correct only at $f_c = \infty$ and is approximately correct for $f_c \gg f$. Switched-capacitor filters are, in fact, sampled-data networks whose analysis and design can be carried out exactly using z-transform techniques. The interested reader is referred to the bibliography.

 ## 12.11 TUNED AMPLIFIERS

In this section, we study a special kind of frequency-selective network, the LC-tuned amplifier. Figure 12.38 shows the general shape of the frequency response of a tuned amplifier. The techniques discussed apply to amplifiers with center frequencies in the range of a few hundred kilohertz to a few hundred megahertz. Tuned amplifiers find application in the radio-frequency (RF) and intermediate-frequency (IF) sections of communications receivers and in a variety of other systems. It should be noted that the tuned-amplifier response of Fig. 12.38 is similar to that of the bandpass filter discussed in earlier sections.

As indicated in Fig. 12.38, the response is characterized by the center frequency ω_0, the 3-dB bandwidth B, and the *skirt selectivity,* which is usually measured as the ratio of the 30-dB bandwidth to the 3-dB bandwidth. In many applications, the 3-dB bandwidth is less than 5% of ω_0. This **narrow-band** property makes possible certain approximations that can simplify the design process, as will be explained later.

The tuned amplifiers studied in this section are small-signal voltage amplifiers in which the transistors operate in the "class A" mode that is, the transistors conduct at all times. Tuned power amplifiers based on class C and other switching modes of operation are not studied in this book. (For a discussion on the classification of amplifiers, refer to Section 14.1.)

12.11.1 The Basic Principle

The basic principle underlying the design of tuned amplifiers is the use of a parallel LCR circuit as the load, or at the input, of a BJT or a FET amplifier. This is illustrated in Fig. 12.39 with a MOSFET amplifier having a tuned-circuit load. For simplicity, the bias details are not included. Since this circuit uses a single tuned circuit, it is known as a **single-tuned amplifier.** The amplifier equivalent circuit is shown in Fig. 12.39(b). Here R denotes the

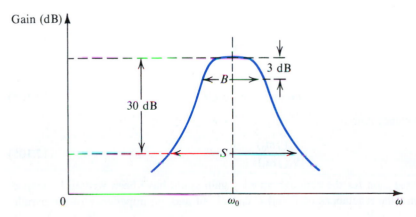

FIGURE 12.38 Frequency response of a tuned amplifier.

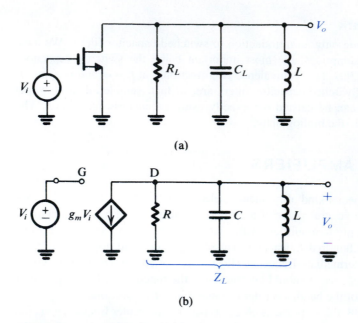

(a)

(b)

FIGURE 12.39 The basic principle of tuned amplifiers is illustrated using a MOSFET with a tuned-circuit load. Bias details are not shown.

parallel equivalent of R_L and the output resistance r_o of the FET, and C is the parallel equivalent of C_L and the FET output capacitance (usually very small). From the equivalent circuit we can write

$$V_o = \frac{-g_m V_i}{Y_L} = \frac{-g_m V_i}{sC + 1/R + 1/sL}$$

Thus the voltage gain can be expressed as

$$\frac{V_o}{V_i} = -\frac{g_m}{C} \frac{s}{s^2 + s(1/CR) + 1/LC} \tag{12.101}$$

which is a second-order bandpass function. Thus the tuned amplifier has a center frequency of

$$\omega_0 = 1/\sqrt{LC} \tag{12.102}$$

a 3-dB bandwidth of

$$B = \frac{1}{CR} \tag{12.103}$$

a Q factor of

$$Q \equiv \omega_0/B = \omega_0 CR \tag{12.104}$$

and a center-frequency gain of

$$\frac{V_o(j\omega_0)}{V_i(j\omega_0)} = -g_m R \tag{12.105}$$

Note that the expression for the center-frequency gain could have been written by inspection; At resonance the reactances of L and C cancel out and the impedance of the parallel LCR circuit reduces to R.

EXAMPLE 12.4

It is required to design a tuned amplifier of the type shown in Fig. 12.39, having $f_0 = 1$ MHz, 3-dB bandwidth = 10 kHz, and center-frequency gain = –10 V/V. The FET available has at the bias point $g_m = 5$ mA/V and $r_o = 10$ kΩ. The output capacitance is negligibly small. Determine the values of R_L, C_L, and L.

Solution

Center-frequency gain = $-10 = -5R$. Thus $R = 2$ kΩ. Since $R = R_L \| r_o$, then $R_L = 2.5$ kΩ.

$$B = 2\pi \times 10^4 = \frac{1}{CR}$$

Thus

$$C = \frac{1}{2\pi \times 10^4 \times 2 \times 10^3} = 7958 \text{ pF}$$

Since $\omega_0 = 2\pi \times 10^6 = 1/\sqrt{LC}$, we obtain

$$L = \frac{1}{4\pi^2 \times 10^{12} \times 7958 \times 10^{-12}} = 3.18 \ \mu\text{H}$$

12.11.2 Inductor Losses

The power loss in the inductor is usually represented by a series resistance r_s as shown in Fig. 12.40(a). However, rather than specifying the value of r_s, the usual practice is to specify the inductor Q factor at the frequency of interest,

$$Q_0 \equiv \frac{\omega_0 L}{r_s} \tag{12.106}$$

Typically, Q_0 is in the range of 50 to 200.

The analysis of a tuned amplifier is greatly simplified by representing the inductor loss by a parallel resistance R_p, as shown in Fig. 12.40(b). The relationship between R_p and Q_0 can be found by writing, for the admittance of the circuit in Fig. 12.40(a),

$$Y(j\omega_0) = \frac{1}{r_s + j\omega_0 L}$$

$$= \frac{1}{j\omega_0 L} \frac{1}{1 - j(1/Q_0)} = \frac{1}{j\omega_0 L} \frac{1 + j(1/Q_0)}{1 + (1/Q_0^2)}$$

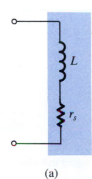
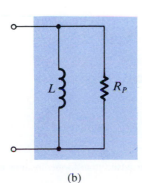

(a)   (b)   **FIGURE 12.40** Inductor equivalent circuits.

For $Q_0 \gg 1$,

$$Y(j\omega_0) \simeq \frac{1}{j\omega_0 L}\left(1 + j\frac{1}{Q_0}\right) \tag{12.107}$$

Equating this to the admittance of the circuit in Fig. 12.40(b) gives

$$Q_0 = \frac{R_p}{\omega_0 L} \tag{12.108}$$

or, equivalently,

$$R_p = \omega_0 L Q_0 \tag{12.109}$$

Finally, it should be noted that the coil Q factor poses an upper limit on the value of Q achieved by the tuned circuit.

EXERCISE

12.32 If the inductor in Example 12.4 has $Q_0 = 150$, find R_p and then find the value to which R_L should be changed to keep the overall Q, and hence the bandwidth, unchanged.

Ans. 3 kΩ; 15 kΩ

12.11.3 Use of Transformers

In many cases it is found that the required value of inductance is not practical, in the sense that coils with the required inductance might not be available with the required high values of Q_0. A simple solution is to use a transformer to effect an impedance change. Alternatively, a tapped coil, known as an **autotransformer,** can be used, as shown in Fig. 12.41. Provided the two parts of the inductor are tightly coupled, which can be achieved by winding on a ferrite core, the transformation relationships shown hold. The result is that the tuned circuit seen between terminals 1 and 1′ is equivalent to that in Fig. 12.39(b). For example, if a turns ratio $n = 3$ is used in the amplifier of Example 12.4, then a coil with inductance $L' = 9 \times 3.18 = 28.6 \ \mu$H and a capacitance $C' = 7958/9 = 884$ pF will be required. Both these values are more practical than the original ones.

In applications that involve coupling the output of a tuned amplifier to the input of another amplifier, the tapped coil can be used to raise the effective input resistance of the latter amplifier stage. In this way, one can avoid reduction of the overall Q. This point is illustrated in Fig. 12.42 and in the following exercises.

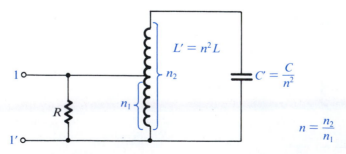

FIGURE 12.41 A tapped inductor is used as an impedance transformer to allow using a higher inductance, L', and a smaller capacitance, C'.

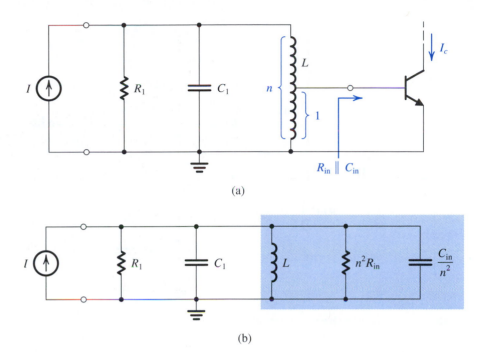

(a)

(b)

FIGURE 12.42 (a) The output of a tuned amplifier is coupled to the input of another amplifier via a tapped coil. (b) An equivalent circuit. Note that the use of a tapped coil increases the effective input impedance of the second amplifier stage.

EXERCISES

D12.33 Consider the circuit in Fig. 12.42(a), first without tapping the coil. Let $L = 5 \ \mu H$ and assume that R_1 is fixed at 1 kΩ. We wish to design a tuned amplifier with $f_0 = 455$ kHz and a 3-dB bandwidth of 10 kHz (this is the Intermediate Frequency (IF) amplifier of an AM radio). If the BJT has $R_{in} = 1$ kΩ and $C_{in} = 200$ pF, find the actual bandwidth obtained and the required value of C_1.

Ans. 13 kHz; 24.27 nF

D12.34 Since the bandwidth realized in Exercise 12.33 is greater than desired, find an alternative design utilizing a tapped coil as in Fig. 12.42(a). Find the value of n that allows the specifications to be just met. Also find the new required value of C_1 and the current gain I_c/I at resonance. Assume that at the bias point the BJT has $g_m = 40$ mA/V.

Ans. 1.36; 24.36 nF; 19.1 A/A

12.11.4 Amplifiers with Multiple Tuned Circuits

The selectivity achieved with the single-tuned circuit of Fig. 12.39 is not sufficient in many applications—for instance, in the IF amplifier of a radio or a TV receiver. Greater selectivity is obtained by using additional tuned stages. Figure 12.43 shows a BJT with tuned circuits at both the input and the output.[11] In this circuit the bias details are shown, from which we note that biasing is quite similar to the classical arrangement employed in low-frequency

---

[11] Note that because the input circuit is a parallel resonant circuit, an input current source (rather than voltage source) signal is utilized.

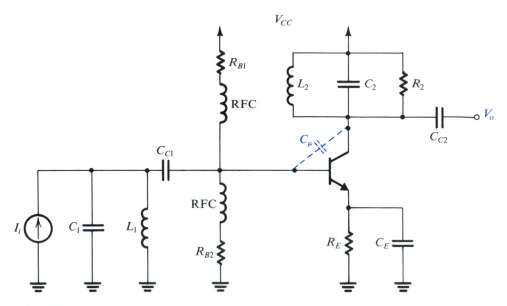

FIGURE 12.43 A BJT amplifier with tuned circuits at the input and the output.

discrete-circuit design. However, to avoid the loading effect of the bias resistors R_{B1} and R_{B2} on the input tuned circuit, a **radio-frequency choke** (RFC) is inserted in series with each resistor. Such chokes have high impedances at the frequencies of interest. The use of RFCs in biasing tuned RF amplifiers is common practice.

The analysis and design of the double-tuned amplifier of Fig. 12.43 is complicated by the Miller effect[12] due to capacitance C_μ. Since the load is not simply resistive, as was the case in the amplifiers studied in Section 6.4.4, the Miller impedance at the input will be complex. This reflected impedance will cause detuning of the input circuit as well as "skewing" of the response of the input circuit. Needless to say, the coupling introduced by C_μ makes tuning (or aligning) the amplifier quite difficult. Worse still, the capacitor C_μ can cause oscillations to occur [see Gray and Searle (1969) and Problem 12.75].

Methods exist for **neutralizing** the effect of C_μ, using additional circuits arranged to feed back a current equal and opposite to that through C_μ. An alternative, and preferred, approach is to use circuit configurations that do not suffer from the Miller effect. These are discussed later. Before leaving this section, however, we wish to point out that circuits of the type shown in Fig. 12.43 are usually designed utilizing the y-parameter model of the BJT (see Appendix B). This is done because here, in view of the fact that C_μ plays a significant role, the y-parameter model makes the analysis simpler (in comparison to that using the hybrid-π model). Also, the y parameters can easily be measured at the particular frequency of interest, ω_0. For narrow-band amplifiers, the assumption is usually made that the y parameters remain approximately constant over the passband.

12.11.5 The Cascode and the CC–CB Cascade

From our study of amplifier frequency response in Chapter 6 we know that two amplifier configurations do not suffer from the Miller effect. These are the cascode configuration and

---

[12] Here we use "Miller effect" to refer to the effect of the feedback capacitance C_μ in reflecting back an input impedance that is a function of the amplifier load impedance.

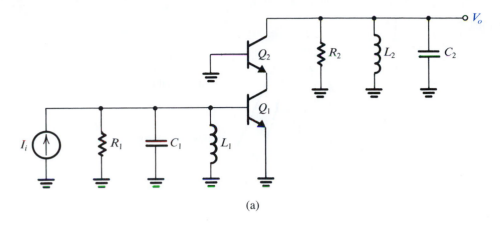

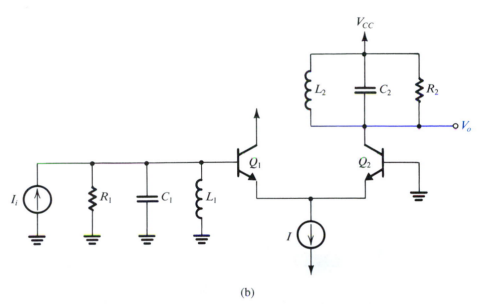

FIGURE 12.44 Two tuned-amplifier configurations that do not suffer from the Miller effect: **(a)** cascode and **(b)** common-collector common-base cascade. (Note that bias details of the cascode circuit are not shown.)

the common-collector common-base cascade. Figure 12.44 shows tuned amplifiers based on these two configurations. The CC–CB cascade is usually preferred in IC implementations because its differential structure makes it suitable for IC biasing techniques. (Note that the biasing details of the cascode circuit are not shown in Fig. 12.44(a). Biasing can be done using arrangements similar to those discussed in earlier chapters.)

12.11.6 Synchronous Tuning

In the design of a tuned amplifier with multiple tuned circuits the question of the frequency to which each circuit should be tuned arises. The objective, of course, is for the overall response to exhibit high passband flatness and skirt selectivity. To investigate this question, we shall assume that the overall response is the product of the individual responses: in other words, that the stages do not interact. This can easily be achieved using circuits such as those in Fig. 12.44.

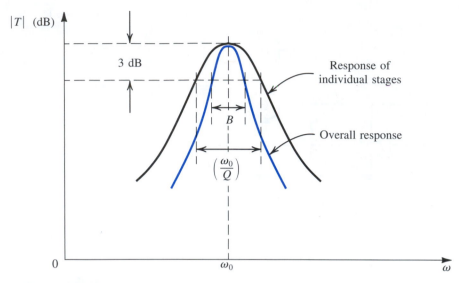

FIGURE 12.45 Frequency response of a synchronously tuned amplifier.

Consider first the case of N identical resonant circuits, known as the **synchronously tuned** case. Figure 12.45 shows the response of an individual stage and that of the cascade. Observe the bandwidth "shrinkage" of the overall response. The 3-dB bandwidth B of the overall amplifier is related to that of the individual tuned circuits, ω_0/Q, by (see Problem 12.77)

$$B = \frac{\omega_0}{Q}\sqrt{2^{1/N}-1} \qquad (12.110)$$

The factor $\sqrt{2^{1/N}-1}$ is known as the bandwidth-shrinkage factor. Given B and N, we can use Eq. (12.110) to determine the bandwidth required of the individual stages, ω_0/Q.

EXERCISE

D12.35 Consider the design of an IF amplifier for an FM radio receiver. Using two synchronously tuned stages with $f_0 = 10.7$ MHz, find the 3-dB bandwidth of each stage so that the overall bandwidth is 200 kHz. Using 3-μH inductors find C and R for each stage.
Ans. 310.8 kHz; 73.7 pF; 6.95 kΩ

12.11.7 Stagger-Tuning

A much better overall response is obtained by stagger-tuning the individual stages, as illustrated in Fig. 12.46. Stagger-tuned amplifiers are usually designed so that the overall response exhibits *maximal flatness* around the center frequency f_0. Such a response can be obtained by transforming the response of a maximally flat (Butterworth) low-pass filter up the frequency axis to ω_0. We show here how this can be done.

The transfer function of a second-order bandpass filter can be expressed in terms of its poles as

$$T(s) = \frac{a_1 s}{\left(s + \dfrac{\omega_0}{2Q} - j\omega_0\sqrt{1 - \dfrac{1}{4Q^2}}\right)\left(s + \dfrac{\omega_0}{2Q} + j\omega_0\sqrt{1 - \dfrac{1}{4Q^2}}\right)} \qquad (12.111)$$

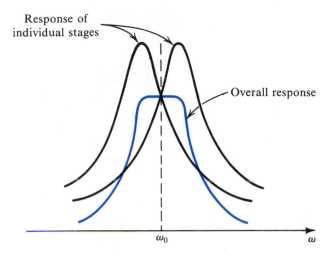

Response of individual stages

Overall response

ω_0

ω

FIGURE 12.46 Stagger-tuning the individual resonant circuits can result in an overall response with a passband flatter than that obtained with synchronous tuning (Fig. 12.45).

For a narrow-band filter, $Q \gg 1$, and for values of s in the neighborhood of $+j\omega_0$ (see Fig. 12.47b) the second factor in the denominator is approximately $(s + j\omega_0 \simeq 2s)$. Hence Eq. (12.111) can be approximated in the neighborhood of $j\omega_0$ by

$$T(s) \simeq \frac{a_1/2}{s + \omega_0/2Q - j\omega_0} = \frac{a_1/2}{(s - j\omega_0) + \omega_0/2Q} \qquad (12.112)$$

This is known as the **narrow-band approximation.**[13] Note that the magnitude response, for $s = j\omega$, has a peak value of $a_1 Q/\omega_0$ at $\omega = \omega_0$, as expected.

Now consider a first-order low-pass network with a single pole at $p = -\omega_0/2Q$ (we use p to denote the complex frequency variable for the low-pass filter). Its transfer function is

$$T(p) = \frac{K}{p + \omega_0/2Q} \qquad (12.113)$$

where K is a constant. Comparing Eqs. (12.112) and (12.113) we note that they are identical for $p = s - j\omega_0$ or, equivalently,

$$s = p + j\omega_0 \qquad (12.114)$$

This result implies that the response of the second-order bandpass filter *in the neighborhood of its center frequency $s = j\omega_0$* is identical to the response of a first-order low-pass filter with a pole at $(-\omega_0/2Q)$ *in the neighborhood of $p = 0$.* Thus the bandpass response can be obtained by shifting the pole of the low-pass prototype and adding the complex-conjugate pole, as illustrated in Fig. 12.47(b). This is called a **lowpass-to-bandpass transformation** for *narrow-band* filters.

The transformation $p = s - j\omega_0$ can be applied to low-pass filters of order greater than one. For instance, we can transform a maximally flat second-order low-pass filter ($Q = 1/\sqrt{2}$) to

---

[13] The bandpass response is *geometrically symmetrical* around the center frequency ω_0. That is, each pair of frequencies ω_1 and ω_2 at which the magnitude response is equal are related by $\omega_1\omega_2 = \omega_0^2$. For high Q, the symmetry becomes almost *arithmetic* for frequencies close to ω_0. That is, two frequencies with the same magnitude response are almost equally spaced from ω_0. The same is true for higher-order bandpass filters designed using the transformation presented in this section.

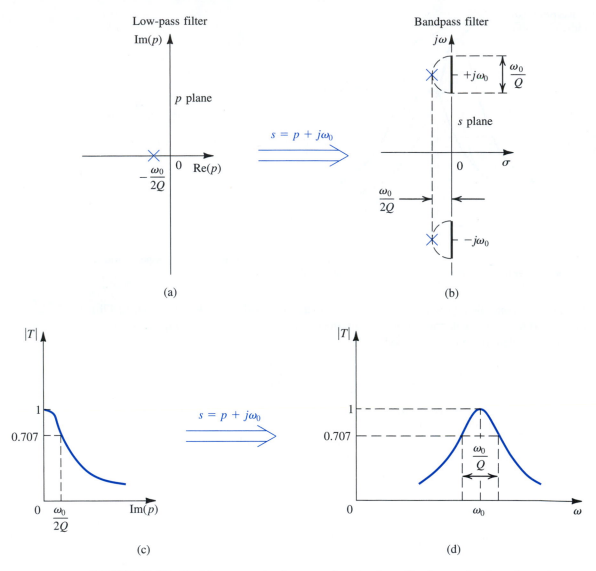

FIGURE 12.47 Obtaining a second-order narrow-band bandpass filter by transforming a first-order low-pass filter. **(a)** Pole of the first-order filter in the p plane. **(b)** Applying the transformation $s = p + j\omega_0$ and adding a complex-conjugate pole results in the poles of the second-order bandpass filter. **(c)** Magnitude response of the first-order low-pass filter. **(d)** Magnitude response of the second-order bandpass filter.

obtain a maximally flat bandpass filter. If the 3-dB bandwidth of the bandpass filter is to be B rad/s, then the low-pass filter should have a 3-dB frequency (and thus a pole frequency) of $(B/2)$ rad/s, as illustrated in Fig. 12.48. The resulting fourth-order bandpass filter will be a stagger-tuned one, with its two tuned circuits (refer to Fig. 12.48) having

$$\omega_{01} = \omega_0 + \frac{B}{2\sqrt{2}} \qquad B_1 = \frac{B}{\sqrt{2}} \qquad Q_1 \simeq \frac{\sqrt{2}\,\omega_0}{B} \qquad (12.115)$$

$$\omega_{02} = \omega_0 - \frac{B}{2\sqrt{2}} \qquad B_2 = \frac{B}{\sqrt{2}} \qquad Q_2 \simeq \frac{\sqrt{2}\,\omega_0}{B} \qquad (12.116)$$

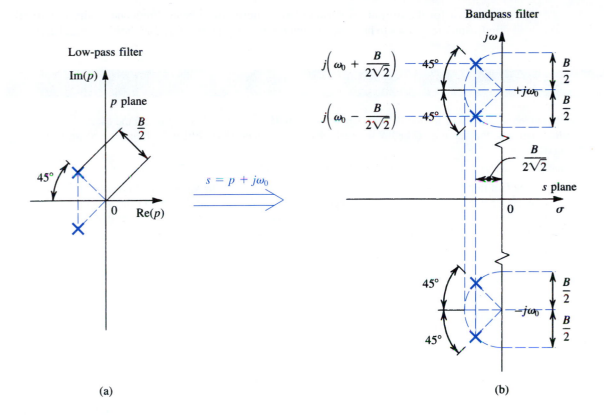

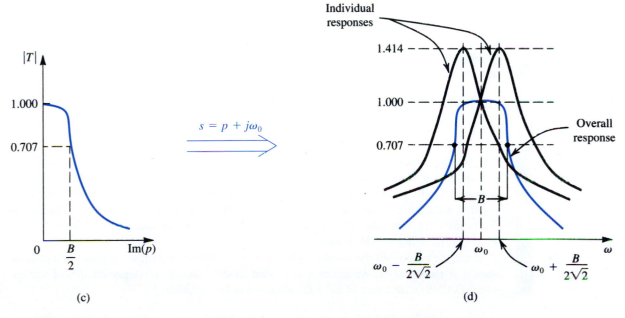

FIGURE 12.48 Obtaining the poles and the frequency response of a fourth-order stagger-tuned narrow-band bandpass amplifier by transforming a second-order low-pass maximally flat response.

Note that for the overall response to have a normalized center-frequency gain of unity, the individual responses to have equal center-frequency gains of $\sqrt{2}$, as shown in Fig. 12.48(d).

EXERCISES

D12.36 A stagger-tuned design for the IF amplifier specified in Exercise 12.35 is required. Find f_{01}, B_1, f_{02}, and B_2. Also give the value of C and R for each of the two stages. (Recall that 3-μH inductors are to be used.)

Ans. 10.77 MHz; 141.4 kHz; 10.63 MHz; 141.4 kHz; 72.8 pF; 15.5 kΩ; 74.7 pF; 15.1 kΩ

12.37 Using the fact that the voltage gain at resonance is proportional to the value of R, find the ratio of the gain at 10.7 MHz of the stagger-tuned amplifier designed in Exercise 12.36 and the synchronously tuned amplifier designed in Exercise 12.35. (*Hint:* For the stagger-tuned amplifier, note that the gain at ω_0 is equal to the product of the gains of the individual stages at their 3-dB frequencies.)

Ans. 2.42

 12.12 SPICE SIMULATION EXAMPLES

Circuit simulation is employed in filter design for at least three purposes: (1) to verify the correctness of the design using ideal components, (2) to investigate the effects of the nonideal characteristics of the op amps on the filter response, and (3) to determine the percentage of circuits fabricated with practical components, whose values have specified tolerance statistics, that meet the design specifications (this percentage is known as the yield). In this section, we present two examples that illustrate the use of SPICE for the first two purposes. The third area of computer-aided design, though very important, is a rather specialized topic, and is considered beyond the scope of this textbook.

EXAMPLE 12.5

VERIFICATION OF THE DESIGN OF A FIFTH-ORDER CHEBYSHEV FILTER

Our first example shows how SPICE can be utilized to verify the design of a fifth-order Chebyshev filter. Specifically, we simulate the operation of the circuit whose component values were obtained in Exercise 12.20. The complete circuit is shown in Fig. 12.49(a). It consists of a cascade of two second-order simulated-LCR resonators using the Antoniou circuit and a first-order op amp–RC circuit. Using PSpice, we would like to compare the magnitude of the filter response with that computed directly from its transfer function. Here, we note that PSpice can also be used to perform the latter task by using the Laplace transfer-function block in the analog-behavioral-modeling (ABM) library.

Since the purpose of the simulation is simply to verify the design, we assume ideal components. For the op amps, we utilize a near-ideal model, namely, a voltage-controlled voltage source (VCVS) with a gain of 10^6 V/V, as shown in Fig. 12.49(b).[14]

---

[14] SPICE models for the op amp are described in Section 2.9.

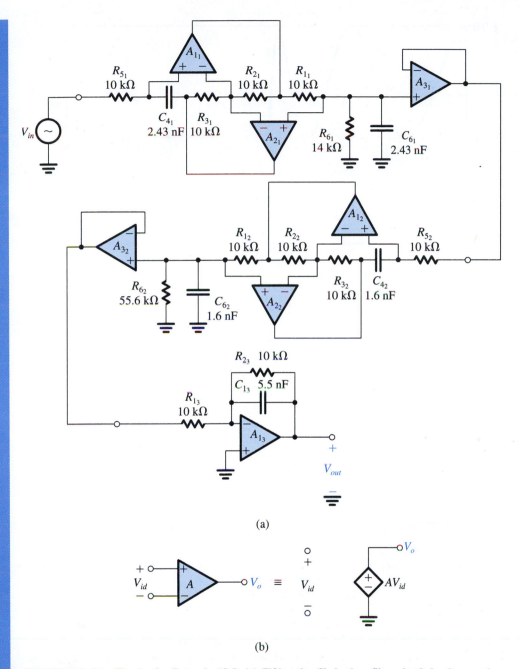

(a)

(b)

FIGURE 12.49 Circuits for Example 12.5. **(a)** Fifth-order Chebyshev filter circuit implemented as a cascade of two second-order simulated LCR resonator circuits and a single first-order op amp–RC circuit. **(b)** VCVS representation of an ideal op amp with gain A.

In SPICE, we apply a 1-V ac signal at the filter input, perform an ac-analysis simulation over the range 1 Hz to 20 kHz, and plot the output voltage magnitude versus frequency, as shown in Fig. 12.50. Both an expanded view of the passband and a view of the entire magnitude response are shown. These results are almost identical to those computed directly from the ideal transfer function, thereby verifying the correctness of the design.

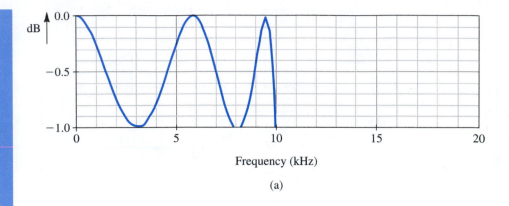

(a)

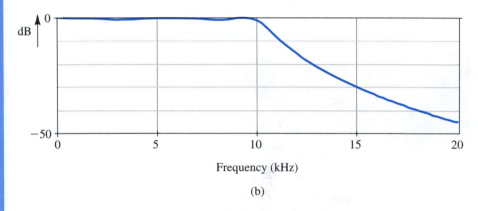

(b)

FIGURE 12.50 Magnitude response of the fifth-order lowpass filter circuit shown in Fig. 12.49: **(a)** an expanded view of the passband region; **(b)** a view of both the passband and stopband regions.

EXAMPLE 12.6

INVESTIGATING THE EFFECT OF FINITE OP-AMP BANDWIDTH ON THE OPERATION OF THE TWO-INTEGRATOR-LOOP FILTER

In this example, we investigate the effect of the finite bandwidth of practical op amps on the response of a two-integrator-loop bandpass filter utilizing the Tow–Thomas biquad circuit of Fig. 12.25(b). The circuit is designed to provide a bandpass response with $f_0 = 10$ kHz, $Q = 20$, and a unity center-frequency gain. The op amps are assumed to be of the 741 type. Specifically, we model the terminal behavior of the op amp with the single-time-constant linear network shown in Fig. 12.51. Since the analysis performed here is a small-signal (ac) analysis that ignores nonlinearities, no nonlinearities are included in this op-amp macromodel. (If the effects of op-amp nonlinearities are to be investigated, a transient analysis should be performed.) The following values are used for the parameters of the op-amp macromodel in Fig. 12.51:

$$R_{id} = 2 \text{ M}\Omega \qquad R_{icm} = 500 \text{ M}\Omega \qquad R_o = 75 \text{ }\Omega$$
$$G_m = 0.19 \text{ mA/V} \qquad R_b = 1.323 \times 10^9 \text{ }\Omega \qquad C_b = 30 \text{ pF}$$

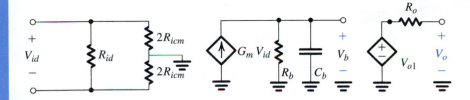

$$A_0 = G_m R_1 \qquad \omega_b = 1/R_b C_b$$

FIGURE 12.51 One-pole equivalent circuit macromodel of an op amp operated within its linear region.

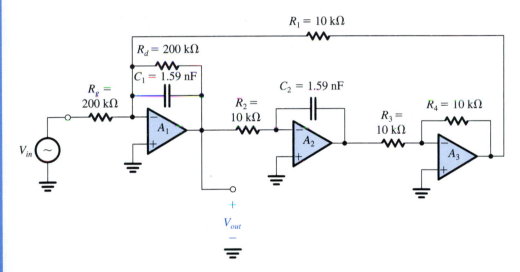

FIGURE 12.52 Circuit for Example 11.6. Second-order bandpass filter implemented with a Tow–Thomas biquad circuit having $f_0 = 10$ kHz, $Q = 20$, and unity center-frequency gain.

These values result in the specified input and output resistances of the 741-type op amp. Further, they provide a dc gain $A_0 = 2.52 \times 10^5$ V/V and a 3-dB frequency f_b of 4 Hz, again equal to the values specified for the 741. Note that the selection of the individual values of G_m, R_b, and C_b is immaterial as long as $G_m R_b = A_0$ and $C_b R_b = 1/2\pi f_b$.

The Tow–Thomas circuit simulated is shown in Fig. 12.52. The circuit is simulated in PSpice for two cases: (1) assuming 741-type op amps and using the linear macromodel in Fig. 12.51; and (2) assuming ideal op amps with dc gain of $A_0 = 10^6$ V/V and using the near-ideal model in Fig. 12.49. In both cases, we apply a 1-V ac signal at the filter input, perform an ac-analysis simulation over the range 8 kHz to 12 kHz, and plot the output-voltage magnitude versus frequency.

The simulation results are shown in Fig. 12.53, from which we observe the significant deviation between the response of the filter using the 741 op amp and that using the near-ideal op-amp model. Specifically, the response with practical op amps shows a deviation in the center

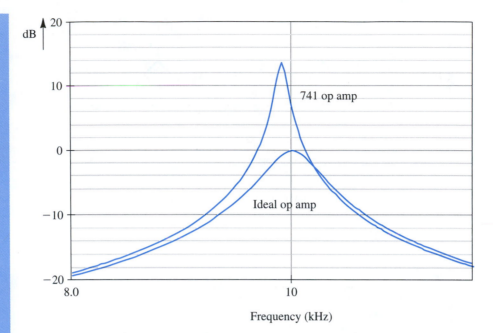

FIGURE 12.53 Comparing the magnitude response of the Tow–Thomas biquad circuit (shown in Fig. 12.52) constructed with 741-type op amps, with the ideal magnitude response. These results illustrate the effect of the finite dc gain and bandwidth of the 741 op amp on the frequency response of the Tow–Thomas biquad circuit.

frequency of about −100 Hz, and a reduction in the 3-dB bandwidth from 500 Hz to about 110 Hz. Thus, in effect, the filter Q factor has increased from the ideal value of 20 to about 90. This phenomenon, known as *Q-enhancement,* is predictable from an analysis of the two-integrator-loop biquad with the finite op-amp bandwidth taken into account [see Sedra and Brackett (1978)]. Such an analysis shows that Q-enhancement occurs as a res ult of the excess phase lag introduced by the finite op-amp bandwidth. The theory also shows that the Q-enhancement effect can be compensated for by introducing phase lead around the feedback loop. This can be accomplished by connecting a small capacitor, C_c, across resistor R_2. To investigate the potential of such a compensation technique, we repeat the PSpice simulation with various capacitance values. The results are displayed in Fig. 12.54(a). We observe that as the compensation capacitance is increased from 0 pF, both the filter Q and the resonance peak of the filter response move closer to the desired values. It is evident, however, that a compensation capacitance of 80 pF causes the response to deviate further from the ideal. Thus, optimum compensation is obtained with a capacitance value between 60 and 80 pF. Further experimentation using PSpice enabled us to determine that such an optimum is obtained with a compensation capacitance of 64 pF. The corresponding response is shown, together with the ideal response, in Fig. 12.54(b). We note that although the filter Q has been restored to its ideal value, there remains a deviation in the center frequency. We shall not pursue this matter any further here; our objective is not to present a detailed study of the design of two-integrator-loop biquads; rather, it is to illustrate the application of SPICE in investigating the nonideal performance of active-filter circuits, generally.

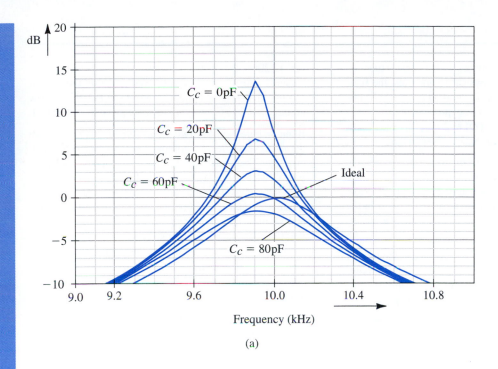

(a)

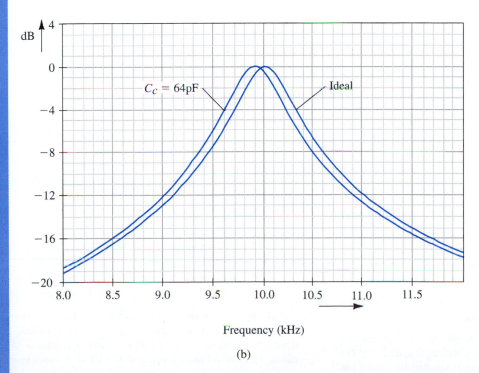

(b)

FIGURE 12.54 (a) Magnitude response of the Tow–Thomas biquad circuit with different values of compensation capacitance. For comparison, the ideal response is also shown. (b) Comparing the magnitude response of the Tow–Thomas biquad circuit using a 64-pF compensation capacitor and the ideal response.

SUMMARY

■ A filter is a linear two-port network with a transfer function $T(s) = V_o(s)/V_i(s)$. For physical frequencies, the filter transmission is expressed as $T(j\omega) = |T(j\omega)|e^{j\phi(\omega)}$. The magnitude of transmission can be expressed in decibels using either the gain function $G(\omega) \equiv 20\log|T|$ or the attenuation function $A(\omega) \equiv -20\log|T|$.

■ The transmission characteristics of a filter are specified in terms of the edges of the passband(s) and the stopband(s); the maximum allowed variation in passband transmission, A_{max} (dB); and the minimum attenuation required in the stopband, A_{min} (dB). In some applications, the phase characteristics are also specified.

■ The filter transfer function can be expressed as the ratio of two polynomials in s; the degree of the denominator polynomial, N, is the filter order. The N roots of the denominator polynomial are the poles (natural modes).

■ To obtain a highly selective response, the poles are complex and occur in conjugate pairs (except for one real pole when N is odd). The zeros are placed on the $j\omega$ axis in the stopband(s) including $\omega = 0$ and $\omega = \infty$.

■ The Butterworth filter approximation provides a low-pass response that is maximally flat at $\omega = 0$. The transmission decreases monotonically as ω increases, reaching 0 (infinite attenuation) at $\omega = \infty$, where all N transmission zeros lie. Eq. (12.11) gives $|T|$, where ϵ is given by Eq. (12.14) and the order N is determined using Eq. (12.15). The poles are found using the graphical construction of Fig. 12.10, and the transfer function is given by Eq. (12.16).

■ The Chebyshev filter approximation provides a low-pass response that is equiripple in the passband with the transmission decreasing monotonically in the stopband. All the transmission zeros are at $s = \infty$. Eq. (12.18) gives $|T|$ in the passband and Eq. (12.19) gives $|T|$ in the stopband, where ϵ is given by Eq. (12.21). The order N can be determined using Eq. (12.22). The poles are given by Eq. (12.23) and the transfer function by Eq. (12.24).

■ Figures 12.13 and 12.14 provide a summary of first-order filter functions and their realizations.

■ Figure 12.16 provides the characteristics of seven special second-order filtering functions.

■ The second-order LCR resonator of Fig. 12.17(a) realizes a pair of complex-conjugate poles with $\omega_0 = 1/\sqrt{LC}$ and $Q = \omega_0 CR$. This resonator can be used to realize the various special second-order filtering functions, as shown in Fig. 12.18.

■ By replacing the inductor of an LCR resonator with a simulated inductance obtained using the Antoniou circuit of

Fig. 12.20(a), the op amp–RC resonator of Fig. 12.21(b) is obtained. This resonator can be used to realize the various second-order filter functions as shown in Fig. 12.22. The design equations for these circuits are given in Table 12.1.

■ Biquads based on the two-integrator-loop topology are the most versatile and popular second-order filter realizations. There are two varieties: the KHN circuit of Fig. 12.24(a), which realizes the LP, BP, and HP functions simultaneously and can be combined with the output summing amplifier of Fig. 12.28(b) to realize the notch and all-pass functions; and the Tow–Thomas circuit of Fig. 12.25(b), which realizes the BP and LP functions simultaneously. Feedforward can be applied to the Tow–Thomas circuit to obtain the circuit of Fig. 12.26, which can be designed to realize any of the second-order functions (see Table 12.2).

■ Single-amplifier biquads (SABs) are obtained by placing a bridged-T network in the negative-feedback path of an op amp. If the op amp is ideal, the poles realized are at the same locations as the zeros of the RC network. The complementary transformation can be applied to the feedback loop to obtain another feedback loop having identical poles. Different transmission zeros are realized by feeding the input signal to circuit nodes that are connected to ground. SABs are economic in their use of op amps but are sensitive to the op-amp nonidealities and are thus limited to low-Q applications ($Q \leq 10$).

■ The classical sensitivity function

$$S_x^y = \frac{\partial y/y}{\partial x/x}$$

is a very useful tool in investigating how tolerant a filter circuit is to the unavoidable inaccuracies in component values and to the nonidealities of the op amps.

■ Switched-capacitor (SC) filters are based on the principle that a capacitor C, periodically switched between two circuit nodes at a high rate, f_c, is equivalent to a resistance $R = 1/Cf_c$ connecting the two circuit nodes. SC filters can be fabricated in monolithic form using CMOS IC technology.

■ Tuned amplifiers utilize LC-tuned circuits as loads, or at the input, of transistor amplifiers. They are used in the design of the RF tuner and the IF amplifier of communication receivers. The cascode and the CC–CB cascade configurations are frequently used in the design of tuned amplifiers. Stagger-tuning the individual tuned circuits results in a flatter passband response (in comparison to that obtained with all the tuned circuits synchronously tuned).

PROBLEMS

SECTION 12.1: FILTER TRANSMISSION, TYPES AND SPECIFICATION

12.1 The transfer function of a first-order low-pass filter (such as that realized by an RC circuit) can be expressed as $T(s) = \omega_0/(s + \omega_0)$, where ω_0 is the 3-dB frequency of the filter. Give in table form the values of $|T|$, ϕ, G, and A at $\omega = 0, 0.5\omega_0, \omega_0, 2\omega_0, 5\omega_0, 10\omega_0$, and $100\omega_0$.

*12.2 A filter has the transfer function $T(s) = 1/[(s + 1)(s^2 + s + 1)]$. Show that $|T| = \sqrt{1 + \omega^6}$ and find an expression for its phase response $\phi(\omega)$. Calculate the values of $|T|$ and ϕ for $\omega = 0.1, 1$, and 10 rad/s and then find the output corresponding to each of the following input signals:

(a) $2 \sin 0.1t$ (volts)
(b) $2 \sin t$ (volts)
(c) $2 \sin 10t$ (volts)

12.3 For the filter whose magnitude response is sketched (as the colored curve) in Fig. 12.3 find $|T|$ at $\omega = 0$, $\omega = \omega_p$, and $\omega = \omega_s$. $A_{max} = 0.5$ dB, and $A_{min} = 40$ dB.

D12.4 A low-pass filter is required to pass all signals within its passband, extending from 0 to 4 kHz, with a transmission variation of at most 10% (i.e., the ratio of the maximum to minimum transmission in the passband should not exceed 1.1). The transmission in the stopband, which extends from 5 kHz to ∞, should not exceed 0.1% of the maximum passband transmission. What are the values of A_{max}, A_{min}, and the selectivity factor for this filter?

12.5 A low-pass filter is specified to have $A_{max} = 1$ dB and $A_{min} = 10$ dB. It is found that these specifications can be just met with a single-time-constant RC circuit having a time constant of 1 s and a dc transmission of unity. What must ω_p and ω_s of this filter be? What is the selectivity factor?

12.6 Sketch transmission specifications for a high-pass filter having a passband defined by $f \geq 2$ kHz and a stopband defined by $f \leq 1$ kHz. $A_{max} = 0.5$ dB, and $A_{min} = 50$ dB.

12.7 Sketch transmission specifications for a bandstop filter that is required to pass signals over the bands $0 \leq f \leq 10$ kHz and 20 kHz $\leq f \leq \infty$ with A_{max} of 1 dB. The stopband extends from $f = 12$ kHz to $f = 16$ kHz, with a minimum required attenuation of 40 dB.

SECTION 12.2: THE FILTER TRANSFER FUNCTION

12.8 Consider a fifth-order filter whose poles are all at a radial distance from the origin of 10^3 rad/s. One pair of complex conjugate poles is at 18° angles from the $j\omega$ axis, and the other pair is at 54° angles. Give the transfer function in each of the following cases:

(a) The transmission zeros are all at $s = \infty$ and the dc gain is unity.
(b) The transmission zeros are all at $s = 0$ and the high-frequency gain is unity.

What type of filter results in each case?

12.9 A third-order low-pass filter has transmission zeros at $\omega = 2$ rad/s and $\omega = \infty$. Its natural modes are at $s = -1$ and $s = -0.5 \pm j0.8$. The dc gain is unity. Find $T(s)$.

12.10 Find the order N and the form of $T(s)$ of a bandpass filter having transmission zeros as follows: one at $\omega = 0$, one at $\omega = 10^3$ rad/s, one at 3×10^3 rad/s, one at 6×10^3 rad/s, and one at $\omega = \infty$. If this filter has a monotonically decreasing passband transmission with a peak at the center frequency of 2×10^3 rad/s, and equiripple response in the stopbands, sketch the shape of its $|T|$.

*12.11 Analyze the RLC network of Fig. P12.11 to determine its transfer function $V_o(s)/V_i(s)$ and hence its poles and zeros. (*Hint:* Begin the analysis at the output and work your way back to the input.)

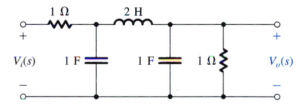

FIGURE P12.11

SECTION 12.3: BUTTERWORTH AND CHEBYSHEV FILTERS

D12.12 Determine the order N of the Butterworth filter for which $A_{max} = 1$ dB, $A_{min} \geq 20$ dB, and the selectivity ratio $\omega_s/\omega_p = 1.3$. What is the actual value of minimum stopband attenuation realized? If A_{min} is to be exactly 20 dB, to what value can A_{max} be reduced?

12.13 Calculate the value of attenuation obtained at a frequency 1.6 times the 3-dB frequency of a seventh-order Butterworth filter.

12.14 Find the natural modes of a Butterworth filter with a 1-dB bandwidth of 10^3 rad/s and $N = 5$.

D12.15 Design a Butterworth filter that meets the following low-pass specifications: $f_p = 10$ kHz, $A_{max} = 2$ dB, $f_s = 15$ kHz, and $A_{min} = 15$ dB. Find N, the natural modes, and $T(s)$. What is the attenuation provided at 20 kHz?

12

*12.16 Sketch $|T|$ for a seventh-order low-pass Chebyshev filter with $\omega_p = 1$ rad/s and $A_{max} = 1$ dB. Use Eq. (12.18) to determine the values of ω at which $|T| = 1$ and the values of ω at which $|T| = 1/\sqrt{1 + \epsilon^2}$. Indicate these values on your sketch. Use Eq. (12.19) to determine $|T|$ at $\omega = 2$ rad/s, and indicate this point on your sketch. For large values of ω, at what rate (in dB/octave) does the transmission decrease?

12.17 Contrast the attenuation provided by a fifth-order Chebyshev filter at $\omega_s = 2\omega_p$ to that provided by a Butterworth filter of equal order. For both, $A_{max} = 1$ dB. Sketch $|T|$ for both filters on the same axes.

D*12.18 It is required to design a low-pass filter to meet the following specifications: $f_p = 3.4$ kHz, $A_{max} = 1$ dB, $f_s = 4$ kHz, $A_{min} = 35$ dB.

(a) Find the required order of Chebyshev filter. What is the excess (above 35 dB) stopband attenuation obtained?
(b) Find the poles and the transfer function.

SECTION 12.4: FIRST-ORDER AND SECOND-ORDER FILTER FUNCTIONS

D12.19 Use the information displayed in Fig. 12.13 to design a first-order op amp–RC low-pass filter having a 3-dB frequency of 10 kHz, a dc gain magnitude of 10, and an input resistance of 10 kΩ.

D12.20 Use the information given in Fig. 12.13 to design a first-order op amp–RC high-pass filter with a 3-dB frequency of 100 Hz, a high-frequency input resistance of 100 kΩ, and a high-frequency gain magnitude of unity.

D*12.21 Use the information given in Fig. 12.13 to design a first-order op amp–RC spectrum-shaping network with a transmission zero frequency of 1 kHz, a pole frequency of 100 kHz, and a dc gain magnitude of unity. The low-frequency input resistance is to be 1 kΩ. What is the high-frequency gain that results? Sketch the magnitude of the transfer function versus frequency.

D*12.22 By cascading a first-order op amp–RC low-pass cir- cuit with a first-order op amp–RC high-pass circuit one can design a wideband bandpass filter. Provide such a design for the case in which the midband gain is 12 dB and the 3-dB bandwidth extends from 100 Hz to 10 kHz. Select appropriate component values under the constraint that no resistors higher than 100 kΩ are to be used, and that the input resistance is to be as high as possible.

D12.23 Derive $T(s)$ for the op amp–RC circuit in Fig. 12.14. We wish to use this circuit as a variable phase shifter by adjusting R. If the input signal frequency is 10^4 rad/s and if $C = 10$ nF, find the values of R required to obtain phase shifts of $-30°$, $-60°$, $-90°$, $-120°$, and $-150°$.

12.24 Show that by interchanging R and C in the op amp–RC circuit of Fig. 12.14, the resulting phase shift covers the range 0 to 180° (with 0° at high frequencies and 180° at low frequencies).

12.25 Use the information in Fig. 12.16(a) to obtain the transfer function of a second-order low-pass filter with $\omega_0 = 10^3$ rad/s, $Q = 1$, and dc gain = 1. At what frequency does $|T|$ peak? What is the peak transmission?

D***12.26 Use the information in Fig. 12.16(a) to obtain the transfer function of a second-order low-pass filter that just meets the specifications defined in Fig. 12.3 with $\omega_p = 1$ rad/s and $A_{max} = 3$ dB. Note that there are two possible solutions. For each, find ω_0 and Q. Also, if $\omega_s = 2$ rad/s, find the value of A_{min} obtained in each case.

D**12.27 Use two first-order op amp–RC all-pass circuits in cascade to design a circuit that provides a set of three-phase 60-Hz voltages, each separated by 120° and equal in magnitude, as shown in the phasor diagram of Fig. P12.27. These voltages simulate those used in three-phase power transmission systems. Use 1-μF capacitors.

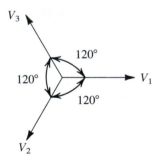

FIGURE P12.27

12.28 Use the information given in Fig. 12.16(b) to find the transfer function of a second-order high-pass filter with natural modes at $-0.5 \pm j\sqrt{3}/2$ and a high-frequency gain of unity.

D**12.29 (a) Show that $|T|$ of a second-order bandpass function is geometrically symmetrical around the center frequency ω_0. That is, the members of each pair of frequencies ω_1 and ω_2 for which $|T(j\omega_1)| = |T(j\omega_2)|$ are related by $\omega_1\omega_2 = \omega_0^2$.

(b) Find the transfer function of the second-order bandpass filter that meets specifications of the form in Fig. 12.4 where $\omega_{p1} = 8100$ rad/s, $\omega_{p2} = 10,000$ rad/s, and $A_{max} = 1$ dB. If $\omega_{s1} = 3000$ rad/s find A_{min} and ω_{s2}.

D*12.30 Use the result of Exercise 12.15 to find the transfer function of a notch filter that is required to eliminate a bothersome interference of 60-Hz frequency. Since the frequency of the interference is not stable, the filter should be designed to provide attenuation ≥ 20 dB over a 6-Hz band centered around 60 Hz. The dc transmission of the filter is to be unity.

12.31 Consider a second-order all-pass circuit in which errors in the component values result in the frequency of the zeros being slightly lower than that of the poles. Roughly sketch the expected $|T|$. Repeat for the case of the frequency of the zeros slightly higher than the frequency of the poles.

12.32 Consider a second-order all-pass filter in which errors in the component values result in the Q factor of the zeros being greater than the Q factor of the poles. Roughly sketch the expected $|T|$. Repeat for the case of the Q factor of the zeros lower than the Q factor of the poles.

SECTION 12.5: THE SECOND-ORDER LCR RESONATOR

D12.33 Design the LCR resonator of Fig. 12.17(a) to obtain natural modes with $\omega_0 = 10^4$ rad/s and $Q = 2$. Use $R = 10$ kΩ.

12.34 For the LCR resonator of Fig. 12.17(a) find the change in ω_0 that results from:

(a) increasing L by 1%
(b) increasing C by 1%
(c) increasing R by 1%

12.35 Derive an expression for $V_o(s)/V_i(s)$ of the high-pass circuit in Fig. 12.18(c).

D12.36 Use the circuit of Fig. 12.18(b) to design a low-pass filter with $\omega_0 = 10^5$ rad/s and $Q = 1/\sqrt{2}$. Utilize a 0.1-μF capacitor.

D12.37 Modify the bandpass circuit of Fig. 12.18(d) to change its center-frequency gain from 1 to 0.5 without changing ω_0 or Q.

12.38 Consider the LCR resonator of Fig. 12.17(a) with node x disconnected from ground and connected to an input signal source V_x, node y disconnected from ground and connected to another input signal source V_y, and node z disconnected from ground and connected to a third input signal source V_z. Use superposition to find the voltage that develops across the resonator, V_o, in terms of V_x, V_y, and V_z.

12.39 Consider the notch circuit shown in Fig. 12.18(i). For what ratio of L_1 to L_2 does the notch occur at $0.9\omega_0$? For this case, what is the magnitude of the transmission at frequencies $\ll \omega_0$? At frequencies $\gg \omega_0$?

SECTION 12.6: SECOND-ORDER ACTIVE FILTERS BASED ON INDUCTOR REPLACEMENT

D12.40 Design the circuit of Fig. 12.20 (utilizing suitable component values) to realize an inductance of (a) 10 H, (b) 1 H, and (c) 0.1 H.

*12.41 Starting from first principles and assuming ideal op amps, derive the transfer function of the circuit in Fig. 12.22(a).

D*12.42 It is required to design a fifth-order Butterworth filter having a 3-dB bandwidth of 10^4 rad/s and a unity dc gain. Use a cascade of two circuits of the type shown in Fig. 12.22(a) and a first-order op amp–RC circuit of the type shown in Fig. 12.13(a). Select appropriate component values.

D12.43 Design the circuit of Fig. 12.22(e) to realize an LPN function with $f_0 = 4$ kHz, $f_n = 5$ kHz, $Q = 10$, and a unity dc gain. Select $C_4 = 10$ nF.

D12.44 Design the all-pass circuit of Fig. 12.22(g) to provide a phase shift of 180° at $f = 1$ kHz and to have $Q = 1$. Use 1-nF capacitors.

12.45 Consider the Antoniou circuit of Fig. 12.20(a) with R_5 eliminated, a capacitor C_6 connected between node 1 and ground, and a voltage source V_2 connected to node 2. Show that the input impedance seen by V_2 is $R_2/s^2C_4C_6R_1R_3$. How does this impedance behave for physical frequencies ($s = j\omega$)? (This impedance is known as a **frequency-dependent negative resistance,** or FDNR.)

D12.46 Using the transfer function of the LPN filter, given in Table 12.1, derive the design equations also given.

D12.47 Using the transfer function of the HPN filter, given in Table 12.1, derive the design equations also given.

D**12.48 It is required to design a third-order low-pass filter whose $|T|$ is equiripple in both the passband and the stopband (in the manner shown in Fig. 12.3, except that the response shown is for $N = 5$). The filter passband extends from $\omega = 0$ to $\omega = 1$ rad/s and the passband transmission varies between 1 and 0.9. The stopband edge is at $\omega = 1.2$ rad/s. The following transfer function was obtained using filter design tables:

$$T(s) = \frac{0.4508(s^2 + 1.6996)}{(s + 0.7294)(s^2 + s0.2786 + 1.0504)}$$

The actual filter realized is to have $\omega_p = 10^4$ rad/s.

(a) Obtain the transfer function of the actual filter by replacing s by $s/10^4$.
(b) Realize this filter as the cascade connection of a first-order LP op amp–RC circuit of the type shown in Fig. 12.13(a) and a second-order LPN circuit of the type shown in Fig. 12.22(e). Each section is to have a dc gain of unity. Select appropriate component values. (*Note:* A filter with an equiripple response in both the passband and the stopband is known as an **elliptic filter.**)

SECTION 12.7: SECOND-ORDER ACTIVE FILTERS BASED ON THE TWO-INTEGRATOR-LOOP TOPOLOGY

D12.49 Design the KHN circuit of Fig. 12.24(a) to realize a bandpass filter with a center frequency of 1 kHz and a 3-dB bandwidth of 50 Hz. Use 10-nF capacitors. Give the complete circuit and specify all component values. What value of center-frequency gain is obtained?

12

D12.50 (a) Using the KHN biquad with the output summing amplifier of Fig. 12.24(b) show that an all-pass function is realized by selecting $R_L = R_H = R_B/Q$. Also show that the flat gain obtained is KR_F/R_H.
(b) Design the all-pass circuit to obtain $\omega_0 = 10^4$ rad/s, $Q = 2$, and flat gain $= 10$. Select appropriate component values.

D12.51 Consider a notch filter with $\omega_n = \omega_0$ realized using the KHN biquad with an output summing amplifier. If the summing resistors used have 1% tolerances, what is the worst-case percentage deviation between ω_n and ω_0?

D12.52 Design the circuit of Fig. 12.26 to realize a low-pass notch filter with $\omega_0 = 10^4$ rad/s, $Q = 10$, dc gain $= 1$, and $\omega_n = 1.2 \times 10^4$ rad/s. Use $C = 10$ nF and $r = 20$ kΩ.

D12.53 In the all-pass realization using the circuit of Fig. 12.26, which component(s) does one need to trim to adjust (a) only ω_z and (b) only Q_z?

D∗∗**12.54** Repeat Problem 12.48 using the Tow–Thomas biquad of Fig. 12.26 to realize the second-order section in the cascade.

SECTION 12.8: SINGLE-AMPLIFIER BIQUADRATIC ACTIVE FILTERS

D12.55 Design the circuit of Fig. 12.29 to realize a pair of poles with $\omega_0 = 10^4$ rad/s and $Q = 1/\sqrt{2}$. Use $C_1 = C_2 = 1$ nF.

12.56 Consider the bridged-T network of Fig. 12.28(a) with $R_1 = R_2 = R$ and $C_1 = C_2 = C$, and denote $CR = \tau$. Find the zeros and poles of the bridged-T network. If the network is placed in the negative-feedback path of an ideal infinite-gain op amp, as in Fig. 12.29, find the poles of the closed-loop amplifier.

∗**12.57** Consider the bridged-T network of Fig. 12.28(b) with $R_1 = R_2 = R$, $C_4 = C$, and $C_3 = C/16$. Let the network be placed in the negative-feedback path of an infinite-gain op amp and let C_4 be disconnected from ground and connected to the input signal source V_i. Analyze the resulting circuit to determine its transfer function $V_o(s)/V_i(s)$, where $V_o(s)$ is the voltage at the op-amp output. Show that the circuit realized is a bandpass filter and find its ω_0, Q, and the center-frequency gain.

D∗∗**12.58** Consider the bandpass circuit shown in Fig. 12.30. Let $C_1 = C_2 = C$, $R_3 = R$, $R_4 = R/4Q^2$, $CR = 2Q/\omega_0$, and $\alpha = 1$. Disconnect the positive input terminal of the op amp from ground and apply V_i through a voltage divider R_1, R_2 to the positive input terminal. Analyze the circuit to find its transfer function V_o/V_i. Find the voltage divider ratio $R_2/(R_1 + R_2)$ so that the circuit realizes (a) an all-pass function and (b) a notch function. Assume the op amp to be ideal.

D∗**12.59** Derive the transfer function of the circuit in Fig. 12.33(b) assuming the op amp to be ideal. Thus show that the circuit realizes a high-pass function. What is the high-

frequency gain of the circuit? Design the circuit for a maximally flat response with a 3-dB frequency of 10^3 rad/s. Use $C_1 = C_2 = 10$ nF. (*Hint:* For a maximally flat response, $Q = 1/\sqrt{2}$ and $\omega_{3dB} = \omega_0$.)

D∗**12.60** Design a fifth-order Butterworth low-pass filter with a 3-dB bandwidth of 5 kHz and a dc gain of unity using the cascade connection of two Sallen-and-Key circuits (Fig. 12.34c) and a first-order section (Fig. 12.13a). Use a 10-kΩ value for all resistors.

12.61 The process of obtaining the complement of a transfer function by interchanging input and ground, as illustrated in Fig. 12.31, applies to any general network (not just RC networks as shown). Show that if the network n is a bandpass with a center-frequency gain of unity, then the complement obtained is a notch. Verify this by using the RLC circuits of Fig. 12.18(d) and (e).

SECTION 12.9: SENSITIVITY

12.62 Evaluate the sensitivities of ω_0 and Q relative to R, L, and C of the bandpass circuit in Fig. 12.18(d).

∗**12.63** Verify the following sensitivity identities:

(a) If $y = uv$, then $S_x^y = S_x^u + S_x^v$.
(b) If $y = u/v$, then $S_x^y = S_x^u - S_x^v$.
(c) If $y = ku$, where k is a constant, then $S_x^y = S_x^u$.
(d) If $y = u^n$, where n is a constant, then $S_x^y = nS_x^u$.
(e) If $y = f_1(u)$ and $u = f_2(x)$, then $S_x^y = S_u^y \cdot S_x^u$.

∗**12.64** For the high-pass filter of Fig. 12.33(b), what are the sensitivities of ω_0 and Q to amplifier gain A?

∗**12.65** For the feedback loop of Fig. 12.34(a), use the expressions in Eqs. (12.77) and (12.78) to determine the sensitivities of ω_0 and Q relative to all passive components for the design in which $R_1 = R_2$.

12.66 For the op amp–RC resonator of Fig. 12.21(b), use the expressions for ω_0 and Q given in the top row of Table 12.1 to determine the sensitivities of ω_0 and Q to all resistors and capacitors.

SECTION 12.10: SWITCHED-CAPACITOR FILTERS

12.67 For the switched-capacitor input circuit of Fig. 12.35(b), in which a clock frequency of 100 kHz is used, what input resistances correspond to capacitance C_1 values of 1 pF and 10 pF?

12.68 For a dc voltage of 1 V applied to the input of the circuit of Fig. 12.35(b), in which C_1 is 1 pF, what charge is transferred for each cycle of the two-phase clock? For a 100-kHz clock, what is the average current drawn from the input source? For a feedback capacitance of 10 pF, what change would you expect in the output for each cycle of the clock? For an amplifier that saturates at ± 10 V and the feedback capacitor initially discharged, how many clock cycles would it take to saturate the amplifier? What is the average slope of the staircase output voltage produced?

D12.69 Repeat Exercise 12.31 for a clock frequency of 400 kHz.

D12.70 Repeat Exercise 12.31 for $Q = 40$.

D12.71 Design the circuit of Fig. 12.37(b) to realize, at the output of the second (noninverting) integrator, a maximally flat low-pass function with $\omega_{3dB} = 10^4$ rad/s and unity dc gain. Use a clock frequency $f_c = 100$ kHz and select $C_1 = C_2 = 10$ pF. Give the values of C_3, C_4, C_5, and C_6. (*Hint:* For a maximally flat response, $Q = 1/\sqrt{2}$ and $\omega_{3dB} = \omega_0$.)

SECTION 12.11: TUNED AMPLIFIERS

***12.72** A voltage signal source with a resistance $R_s = 10$ kΩ is connected to the input of a common-emitter BJT amplifier. Between base and emitter is connected a tuned circuit with $L = 1$ μH and $C = 200$ pF. The transistor is biased at 1 mA and has $\beta = 200$, $C_\pi = 10$ pF, and $C_\mu = 1$ pF. The transistor load is a resistance of 5 kΩ. Find ω_0, Q, the 3-dB bandwidth, and the center-frequency gain of this single-tuned amplifier.

12.73 A coil having an inductance of 10 μH is intended for applications around 1-MHz frequency. Its Q is specified to be 200. Find the equivalent parallel resistance R_p. What is the value of the capacitor required to produce resonance at 1 MHz? What additional parallel resistance is required to produce a 3-dB bandwidth of 10 kHz?

12.74 An inductance of 36 μH is resonated with a 1000-pF capacitor. If the inductor is tapped at one-third of its turns and a 1-kΩ resistor is connected across the one-third part, find f_0 and Q of the resonator.

***12.75** Consider a common-emitter transistor amplifier loaded with an inductance L. Ignoring r_o and r_x, show that for $\omega C_\mu \ll 1/\omega L$, the amplifier input admittance is given by

$$Y_{in} \simeq \left(\frac{1}{r_\pi} - \omega^2 C_\mu L g_m\right) + j\omega(C_\pi + C_\mu)$$

Note: The real part of the input admittance can be negative. This can lead to oscillations.

***12.76** (a) Substituting $s = j\omega$ in the transfer function $T(s)$ of a second-order bandpass filter (see Fig. 12.16c), find $|T(j\omega)|$. For ω in the vicinity of ω_0 [i.e., $\omega = \omega_0 + \delta\omega = \omega_0(1 + \delta\omega/\omega_0)$, where $\delta\omega/\omega_0 \ll 1$ so that $\omega^2 \simeq \omega_0^2(1 + 2\delta\omega/\omega_0)$], show that, for $Q \gg 1$,

$$|T(j\omega)| \simeq \frac{|T(j\omega_0)|}{\sqrt{1 + 4Q^2(\delta\omega/\omega_0)^2}}$$

(b) Use the result obtained in (a) to show that the 3-dB bandwidth B, of N synchronously tuned sections connected in cascade, is

$$B = (\omega_0/Q)\sqrt{2^{1/N} - 1}$$

***12.77** (a) Using the fact that for $Q \gg 1$ the second-order bandpass response in the neighborhood of ω_0 is the same as the response of a first-order low-pass with 3-dB frequency of $(\omega_0/2Q)$, show that the bandpass response at $\omega = \omega_0 + \delta\omega$, for $\delta\omega \ll \omega_0$, is given by

$$|T(j\omega)| \simeq \frac{|T(j\omega_0)|}{\sqrt{1 + 4Q^2(\delta\omega/\omega_0)^2}}$$

(b) Use the relationship derived in (a) together with Eq. (12.110) to show that a bandpass amplifier with a 3-dB bandwidth B, designed using N synchronously tuned stages, has an overall transfer function given by

$$|T(j\omega)|_{overall} = \frac{|T(j\omega_0)|_{overall}}{[1 + 4(2^{1/N} - 1)(\delta\omega/B)^2]^{N/2}}$$

(c) Use the relationship derived in (b) to find the attenuation (in decibels) obtained at a bandwidth $2B$ for $N = 1$ to 5. Also find the ratio of the 30-dB bandwidth to the 3-dB bandwidth for $N = 1$ to 5.

***12.78** This problem investigates the selectivity of maximally flat stagger-tuned amplifiers derived in the manner illustrated in Fig. 12.48.

(a) The low-pass maximally flat (Butterworth) filter having a 3-dB bandwidth $B/2$ and order N has the magnitude response

$$|T| = 1\Big/\sqrt{1 + \left(\frac{\Omega}{B/2}\right)^{2N}}$$

where $\Omega = \text{Im}(p)$ is the frequency in the low-pass domain. (This relationship can be obtained using the information provided in Section 12.3 on Butterworth filters.) Use this expression to obtain for the corresponding bandpass filter at $\omega = \omega_0 + \delta\omega$, where $\delta\omega \ll \omega_0$, the relationship

$$|T| = 1\Big/\sqrt{1 + \left(\frac{\delta\omega}{B/2}\right)^{2N}}$$

(b) Use the transfer function in (a) to find the attenuation (in decibels) obtained at a bandwidth of $2B$ for $N = 1$ to 5. Also find the ratio of the 30-dB bandwidth to the 3-dB bandwidth for $N = 1$ to 5.

***12.79** Consider a sixth-order stagger-tuned bandpass amplifier with center frequency ω_0 and 3-dB bandwidth B. The poles are to be obtained by shifting those of the third-order maximally flat low-pass filter, given in Fig. 12.10(c). For the three resonant circuits, find ω_0, the 3-dB bandwidth, and Q.

12

CHAPTER 13

Signal Generators and Waveform-Shaping Circuits

INTRODUCTION

In the design of electronic systems the need frequently arises for signals having prescribed standard waveforms, for example, sinusoidal, square, triangular, or pulse. Systems in which standard signals are required include computer and control systems where clock pulses are needed for, among other things, timing; communication systems where signals of a variety of waveforms are utilized as information carriers; and test and measurement systems where signals, again of a variety of waveforms, are employed for testing and characterizing electronic devices and circuits. In this chapter we study signal-generator circuits.

There are two distinctly different approaches for the generation of sinusoids, perhaps the most commonly used of the standard waveforms. The first approach, studied in Sections 13.1

to 13.3, employs a **positive-feedback loop** consisting of an amplifier and an RC or LC **frequency-selective network.** The amplitude of the generated sine waves is limited, or set, using a nonlinear mechanism, implemented either with a separate circuit or using the nonlinearities of the amplifying device itself. In spite of this, these circuits, which generate sine waves utilizing resonance phenomena, are known as **linear oscillators.** The name clearly distinguishes them from the circuits that generate sinusoids by way of the second approach. In these circuits, a sine wave is obtained by appropriately shaping a triangular waveform. We study waveform-shaping circuits in Section 13.9, following the study of triangular-waveform generators.

Circuits that generate square, triangular, pulse (etc.) waveforms, called **nonlinear oscillators** or **function generators,** employ circuit building blocks known as **multivibrators.** There are three types of multivibrator: the **bistable** (Section 13.4), the **astable** (Section 13.5), and the **monostable** (Section 13.6). The multivibrator circuits presented in this chapter employ op amps and are intended for precision analog applications. Multivibrator circuits using digital logic gates were studied in Chapter 11.

A general and versatile scheme for the generation of square and triangular waveforms is obtained by connecting a bistable multivibrator and an op-amp integrator in a feedback loop (Section 13.5). Similar results can be obtained using a commercially available versatile IC chip, the 555 timer (Section 13.7). The chapter includes also a study of precision circuits that implement the rectifier functions introduced in Chapter 3. The circuits studied here (Section 13.9), however, are intended for applications that demand precision, such as in instrumentation systems, including waveform generation. The chapter concludes with examples illustrating the use of SPICE in the simulation of oscillator circuits.

13.1 BASIC PRINCIPLES OF SINUSOIDAL OSCILLATORS

In this section, we study the basic principles of the design of linear sine-wave oscillators. In spite of the name *linear oscillator,* some form of nonlinearity has to be employed to provide control of the amplitude of the output sine wave. In fact, all oscillators are essentially nonlinear circuits. This complicates the task of analysis and design of oscillators; no longer is one able to apply transform (*s*-plane) methods directly. Nevertheless, techniques have been developed by which the design of sinusoidal oscillators can be performed in two steps: The first step is a linear one, and frequency-domain methods of feedback circuit analysis can be readily employed. Subsequently, a nonlinear mechanism for amplitude control can be provided.

13.1.1 The Oscillator Feedback Loop

The basic structure of a sinusoidal oscillator consists of an amplifier and a frequency-selective network connected in a positive-feedback loop, such as that shown in block diagram form in Fig. 13.1. Although in an actual oscillator circuit, no input signal will be present, we include an input signal here to help explain the principle of operation. It is important to note that unlike the negative-feedback loop of Fig. 8.1, here the feedback signal x_f is summed with a *positive* sign. Thus the gain-with-feedback is given by

$$A_f(s) = \frac{A(s)}{1 - A(s)\beta(s)} \tag{13.1}$$

where we note the negative sign in the denominator.

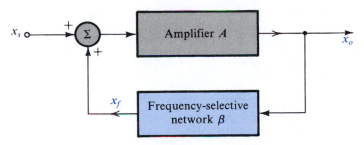

FIGURE 13.1 The basic structure of a sinusoidal oscillator. A positive-feedback loop is formed by an amplifier and a frequency-selective network. In an actual oscillator circuit, no input signal will be present; here an input signal x_s is employed to help explain the principle of operation.

According to the definition of loop gain in Chapter 8, the loop gain of the circuit in Fig. 13.1 is $-A(s)\beta(s)$. However, for our purposes here it is more convenient to drop the minus sign and define the loop gain $L(s)$ as

$$L(s) \equiv A(s)\beta(s) \qquad (13.2)$$

The characteristic equation thus becomes

$$1 - L(s) = 0 \qquad (13.3)$$

Note that this new definition of loop gain[1] corresponds directly to the actual gain seen around the feedback loop of Fig. 13.1.

13.1.2 The Oscillation Criterion

If at a specific frequency f_0 the loop gain $A\beta$ is equal to unity, it follows from Eq. (13.1) that A_f will be infinite. That is, at this frequency the circuit will have a finite output for zero input signal. Such a circuit is by definition an oscillator. Thus the condition for the feedback loop of Fig. 13.1 to provide sinusoidal oscillations of frequency ω_0 is

$$L(j\omega_0) \equiv A(j\omega_0)\beta(j\omega_0) = 1 \qquad (13.4)$$

That is, *at ω_0 the phase of the loop gain should be zero and the magnitude of the loop gain should be unity*. This is known as the **Barkhausen criterion.** Note that for the circuit to oscillate at one frequency, the oscillation criterion should be satisfied only at one frequency (i.e., ω_0); otherwise the resulting waveform will not be a simple sinusoid.

An intuitive feeling for the Barkhausen criterion can be gained by considering once more the feedback loop of Fig. 13.1. For this loop to *produce* and *sustain* an output x_o with no input applied ($x_s = 0$), the feedback signal x_f

$$x_f = \beta x_o$$

should be sufficiently large that when multiplied by A it produces x_o, that is,

$$A x_f = x_o$$

---

[1] For both the negative-feedback loop in Fig. 8.1 and the positive-feedback loop in Fig. 13.1, the loop gain $L = A\beta$. However, the negative sign with which the feedback signal is summed in the negative-feedback loop results in the characteristic equation being $1 + L = 0$. In the positive-feedback loop, the feedback signal is summed with a positive sign, thus resulting in the characteristic equation $1 - L = 0$.

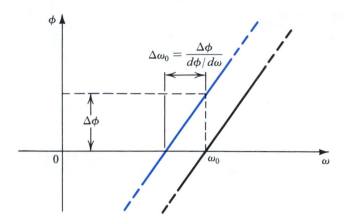

FIGURE 13.2 Dependence of the oscillator-frequency stability on the slope of the phase response. A steep phase response (i.e., large $d\phi/d\omega$) results in a small $\Delta\omega_0$ for a given change in phase $\Delta\phi$ (resulting from a change (due, for example, to temperature) in a circuit component).

that is,

$$A\beta x_o = x_o$$

which results in

$$A\beta = 1$$

It should be noted that the *frequency of oscillation* ω_0 is determined solely by the phase characteristics of the feedback loop; the loop oscillates at the frequency for which the phase is zero. It follows that the stability of the frequency of oscillation will be determined by the manner in which the phase $\phi(\omega)$ of the feedback loop varies with frequency. A "steep" function $\phi(\omega)$ will result in a more stable frequency. This can be seen if one imagines a change in phase $\Delta\phi$ due to a change in one of the circuit components. If $d\phi/d\omega$ is large, the resulting change in ω_0 will be small, as illustrated in Fig. 13.2.

An alternative approach to the study of oscillator circuits consists of examining the circuit poles, which are the roots of the **characteristic equation** (Eq. 13.3). For the circuit to produce **sustained oscillations** at a frequency ω_0 the characteristic equation has to have roots at $s = \pm j\omega_0$. Thus $1 - A(s)\beta(s)$ should have a factor of the form $s^2 + \omega_0^2$.

EXERCISE

13.1 Consider a sinusoidal oscillator formed of an amplifier with a gain of 2 and a second-order bandpass filter. Find the pole frequency and the center-frequency gain of the filter needed to produce sustained oscillations at 1 kHz.

Ans. 1 kHz; 0.5

13.1.3 Nonlinear Amplitude Control

The oscillation condition, the Barkhausen criterion, just discussed, guarantees sustained oscillations in a mathematical sense. It is well known, however, that the parameters of any physical system cannot be maintained constant for any length of time. In other words, suppose we work hard to make $A\beta = 1$ at $\omega = \omega_0$, and then the temperature changes and $A\beta$ becomes slightly less than unity. Obviously, oscillations will cease in this case. Conversely, if $A\beta$ exceeds unity, oscillations will grow in amplitude. We therefore need a mechanism for forcing $A\beta$ to remain equal to unity at the desired value of output amplitude. This task is accomplished by providing a nonlinear circuit for gain control.

Basically, the function of the gain-control mechanism is as follows: First, to ensure that oscillations will start, one designs the circuit such that $A\beta$ is slightly greater than unity. This corresponds to designing the circuit so that the poles are in the right half of the s plane. Thus as the power supply is turned on, oscillations will grow in amplitude. When the amplitude reaches the desired level, the nonlinear network comes into action and causes the loop gain to be reduced to exactly unity. In other words, the poles will be "pulled back" to the $j\omega$ axis. This action will cause the circuit to sustain oscillations at this desired amplitude. If, for some reason, the loop gain is reduced below unity, the amplitude of the sine wave will diminish. This will be detected by the nonlinear network, which will cause the loop gain to increase to exactly unity.

As will be seen, there are two basic approaches to the implementation of the nonlinear amplitude-stabilization mechanism. The first approach makes use of a limiter circuit (see Chapter 3). Oscillations are allowed to grow until the amplitude reaches the level to which the limiter is set. When the limiter comes into operation, the amplitude remains constant. Obviously, the limiter should be "soft" to minimize nonlinear distortion. Such distortion, however, is reduced by the filtering action of the frequency-selective network in the feedback loop. In fact, in one of the oscillator circuits studied in Section 13.2, the sine waves are hard limited, and the resulting square waves are applied to a bandpass filter present in the feedback loop. The "purity" of the output sine waves will be a function of the selectivity of this filter. That is, the higher the Q of the filter, the less the harmonic content of the sine-wave output.

The other mechanism for amplitude control utilizes an element whose resistance can be controlled by the amplitude of the output sinusoid. By placing this element in the feedback circuit so that its resistance determines the loop gain, the circuit can be designed to ensure that the loop gain reaches unity at the desired output amplitude. Diodes, or JFETs operated in the triode region,[2] are commonly employed to implement the controlled-resistance element.

13.1.4 A Popular Limiter Circuit for Amplitude Control

We conclude this section by presenting a limiter circuit that is frequently employed for the amplitude control of op-amp oscillators, as well as in a variety of other applications. The circuit is more precise and versatile than those presented in Chapter 3.

The limiter circuit is shown in Fig. 13.3(a), and its transfer characteristic is depicted in Fig. 13.3(b). To see how the transfer characteristic is obtained, consider first the case of a small (close to zero) input signal v_I and a small output voltage v_O, so that v_A is positive and v_B is negative. It can be easily seen that both diodes D_1 and D_2 will be off. Thus all of the input current v_I/R_1 flows through the feedback resistance R_f, and the output voltage is given by

$$v_O = -(R_f/R_1)v_I \qquad (13.5)$$

This is the linear portion of the limiter transfer characteristic in Fig. 13.3(b). We now can use superposition to find the voltages at nodes A and B in terms of $\pm V$ and v_O as

$$v_A = V\frac{R_3}{R_2 + R_3} + v_O\frac{R_2}{R_2 + R_3} \qquad (13.6)$$

$$v_B = -V\frac{R_4}{R_4 + R_5} + v_O\frac{R_5}{R_4 + R_5} \qquad (13.7)$$

---

[2] We have not studied JFETs in this book. However, the CD accompanying the book includes material on JFETs and JFET circuits. The same material can also be found on the book's website.

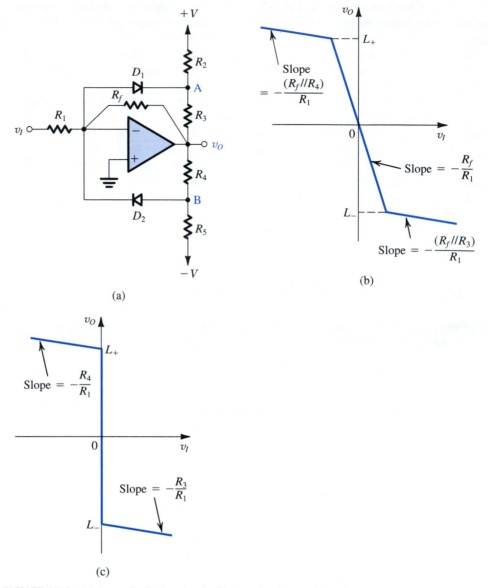

FIGURE 13.3 (a) A popular limiter circuit. (b) Transfer characteristic of the limiter circuit; L_- and L_+ are given by Eqs. (13.8) and (13.9), respectively. (c) When R_f is removed, the limiter turns into a comparator with the characteristic shown.

As v_I goes positive, v_O goes negative (Eq. 13.5), and we see from Eq. (13.7) that v_B will become more negative, thus keeping D_2 off. Equation (13.6) shows, however, that v_A becomes less positive. Then, if we continue to increase v_I, a negative value of v_O will be reached at which v_A becomes -0.7 V or so and diode D_1 conducts. If we use the constant-voltage-drop model for D_1 and denote the voltage drop V_D, the value of v_O at which D_1 conducts can be found from Eq. (13.6). This is the negative limiting level, which we denote L_-,

$$L_- = -V\frac{R_3}{R_2} - V_D\left(1 + \frac{R_3}{R_2}\right) \tag{13.8}$$

The corresponding value of v_I can be found by dividing L_- by the limiter gain $-R_f/R_1$. If v_I is increased beyond this value, more current is injected into D_1, and v_A remains at approximately $-V_D$. Thus the current through R_2 remains constant, and the additional diode current flows through R_3. Thus R_3 appears in effect in parallel with R_f, and the incremental gain (ignoring the diode resistance) is $-(R_f\|R_3)/R_1$. To make the slope of the transfer characteristic small in the limiting region, a low value should be selected for R_3.

The transfer characteristic for negative v_I can be found in a manner identical to that just employed. It can be easily seen that for negative v_I, diode D_2 plays an identical role to that played by diode D_1 for positive v_I. The positive limiting level L_+ can be found to be

$$L_+ = V\frac{R_4}{R_5} + V_D\left(1 + \frac{R_4}{R_5}\right) \tag{13.9}$$

and the slope of the transfer characteristic in the positive limiting region is $-(R_f\|R_4)/R_1$. We thus see that the circuit of Fig. 13.3(a) functions as a soft limiter, with the limiting levels L_+ and L_-, and the limiting gains, independently adjustable by the selection of appropriate resistor values.

Finally, we note that increasing R_f results in a higher gain in the linear region while keeping L_+ and L_- unchanged. In the limit, removing R_f altogether results in the transfer characteristic of Fig. 13.3(c), which is that of a comparator. That is, the circuit compares v_I with the comparator reference value of 0 V: $v_I > 0$ results in $v_O \simeq L_-$, and $v_I < 0$ yields $v_O \simeq L_+$.

EXERCISE

13.2 For the circuit of Fig. 13.3(a) with $V = 15$ V, $R_1 = 30$ kΩ, $R_f = 60$ kΩ, $R_2 = R_5 = 9$ kΩ, and $R_3 = R_4 = 3$ kΩ, find the limiting levels and the value of v_I at which the limiting levels are reached. Also determine the limiter gain and the slope of the transfer characteristic in the positive and negative limiting regions. Assume that $V_D = 0.7$ V.

Ans. ±5.93 V; ±2.97 V; −2; −0.095

 ## 13.2 OP AMP–RC OSCILLATOR CIRCUITS

In this section we shall study some practical oscillator circuits utilizing op amps and RC networks.

13.2.1 The Wien-Bridge Oscillator

One of the simplest oscillator circuits is based on the Wien bridge. Figure 13.4 shows a Wien-bridge oscillator without the nonlinear gain-control network. The circuit consists of an op amp connected in the noninverting configuration, with a closed-loop gain of $1 + R_2/R_1$. In the feedback path of this positive-gain amplifier an RC network is connected. The loop gain can be easily obtained by multiplying the transfer function $V_a(s)/V_o(s)$ of the feedback network by the amplifier gain,

$$L(s) = \left[1 + \frac{R_2}{R_1}\right]\frac{Z_p}{Z_p + Z_s}$$

Thus,

$$L(s) = \frac{1 + R_2/R_1}{3 + sCR + 1/sCR} \tag{13.10}$$

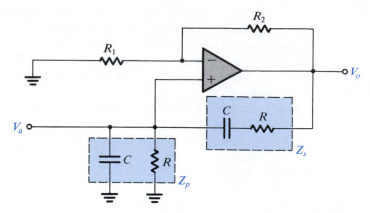

FIGURE 13.4 A Wien-bridge oscillator without amplitude stabilization.

Substituting $s = j\omega$ results in

$$L(j\omega) = \frac{1 + R_2/R_1}{3 + j(\omega CR - 1/\omega CR)} \tag{13.11}$$

The loop gain will be a real number (i.e., the phase will be zero) at one frequency given by

$$\omega_0 CR = \frac{1}{\omega_0 CR}$$

That is,

$$\omega_0 = 1/CR \tag{13.12}$$

To obtain sustained oscillations at this frequency, one should set the magnitude of the loop gain to unity. This can be achieved by selecting

$$R_2/R_1 = 2 \tag{13.13}$$

To ensure that oscillations will start, one chooses R_2/R_1 slightly greater than 2. The reader can easily verify that if $R_2/R_1 = 2 + \delta$, where δ is a small number, the roots of the characteristic equation $1 - L(s) = 0$ will be in the right half of the s plane.

The amplitude of oscillation can be determined and stabilized by using a nonlinear control network. Two different implementations of the amplitude-controlling function are shown in Figs. 13.5 and 13.6. The circuit in Fig. 13.5 employs a symmetrical feedback limiter of the type studied in Section 13.1.3. It is formed by diodes D_1 and D_2 together with resistors R_3, R_4, R_5, and R_6. The limiter operates in the following manner: At the positive peak of the output voltage v_O, the voltage at node b will exceed the voltage v_1 (which is about $\frac{1}{3}v_O$), and diode D_2 conducts. This will clamp the positive peak to a value determined by R_5, R_6, and the negative power supply. The value of the positive output peak can be calculated by setting $v_b = v_1 + V_{D2}$ and writing a node equation at node b while neglecting the current through D_2. Similarly, the negative peak of the output sine wave will be clamped to the value that causes diode D_1 to conduct. The value of the negative peak can be determined by setting $v_a = v_1 - V_{D1}$ and writing an equation at node a while neglecting the current through D_1. Finally, note that to obtain a symmetrical output waveform, R_3 is chosen equal to R_6, and R_4 equal to R_5.

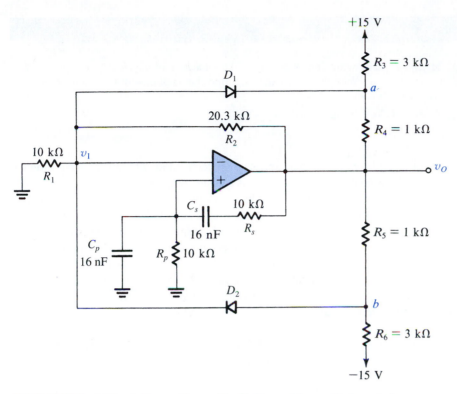

FIGURE 13.5 A Wien-bridge oscillator with a limiter used for amplitude control.

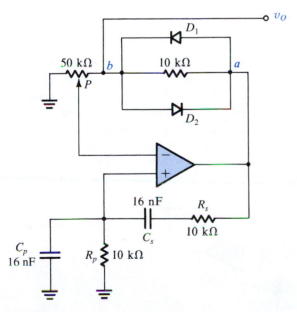

FIGURE 13.6 A Wien-bridge oscillator with an alternative method for amplitude stabilization.

The circuit of Fig. 13.6 employs an inexpensive implementation of the parameter-variation mechanism of amplitude control. Potentiometer P is adjusted until oscillations just start to grow. As the oscillations grow, the diodes start to conduct, causing the effective resistance between a and b to decrease. Equilibrium will be reached at the output amplitude that causes the loop gain to be exactly unity. The output amplitude can be varied by adjusting potentiometer P.

As indicated in Fig. 13.6, the output is taken at point b rather than at the op-amp output terminal because the signal at b has lower distortion than that at a. To appreciate this point, note that the voltage at b is proportional to the voltage at the op-amp input terminals and that the latter is a filtered (by the RC network) version of the voltage at node a. Node b, however, is a high-impedance node, and a buffer will be needed if a load is to be connected.

13.2.2 The Phase-Shift Oscillator

The basic structure of the phase-shift oscillator is shown in Fig. 13.7. It consists of a negative-gain amplifier ($-K$) with a three-section (third-order) RC ladder network in the feedback. The circuit will oscillate at the frequency for which the phase shift of the RC network is 180°. Only at this frequency will the total phase shift around the loop be 0° or 360°. Here we should note that the reason for using a three-section RC network is that three is the minimum number of sections (i.e., lowest order) that is capable of producing a 180° phase shift at a finite frequency.

For oscillations to be sustained, the value of K should be equal to the inverse of the magnitude of the RC network transfer function at the frequency of oscillation. However, to ensure that oscillations start, the value of K has to be chosen slightly higher than the value

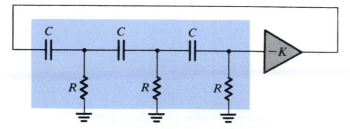

FIGURE 13.7 A phase-shift oscillator.

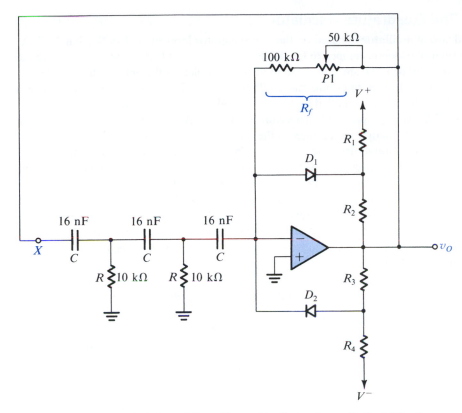

FIGURE 13.8 A practical phase-shift oscillator with a limiter for amplitude stabilization.

that satisfies the unity-loop-gain condition. Oscillations will then grow in magnitude until limited by some nonlinear control mechanism.

Figure 13.8 shows a practical phase-shift oscillator with a feedback limiter, consisting of diodes D_1 and D_2 and resistors R_1, R_2, R_3, and R_4 for amplitude stabilization. To start oscillations, R_f has to be made slightly greater than the minimum required value. Although the circuit stabilizes more rapidly, and provides sine waves with more stable amplitude, if R_f is made much larger than this minimum, the price paid is an increased output distortion.

EXERCISES

13.5 Consider the circuit of Fig. 13.8 *without* the limiter. Break the feedback loop at X and find the loop gain $A\beta \equiv V_o(j\omega)/V_x(j\omega)$. To do this, it is easier to start at the output and work backward, finding the various currents and voltages, and eventually V_x in terms of V_o.

Ans. $\dfrac{\omega^2 C^2 R R_f}{4 + j(3\omega C R - 1/\omega C R)}$

13.6 Use the expression derived in Exercise 13.5 to find the frequency of oscillation f_0, and the minimum required value of R_f for oscillations to start in the circuit of Fig. 13.8.

Ans. 574.3 Hz; 120 kΩ

13.2.3 The Quadrature Oscillator

The **quadrature oscillator** is based on the two-integrator loop studied in Section 12.7. As an active filter, the loop is damped to locate the poles in the left half of the s plane. Here, no such damping will be used, since we wish to locate the poles on the $j\omega$ axis to provide sustained oscillations. In fact, to ensure that oscillations start, the poles are initially located in the right half-plane and then "pulled back" by the nonlinear gain control.

Figure 13.9 shows a practical quadrature oscillator. Amplifier 1 is connected as an inverting Miller integrator with a limiter in the feedback for amplitude control. Amplifier 2 is connected as a noninverting integrator (thus replacing the cascade connection of the Miller integrator and the inverter in the two-integrator loop of Fig. 12.25b). To understand the operation of this noninverting integrator, consider the equivalent circuit shown in Fig. 13.9(b). Here, we have replaced the integrator input voltage v_{O1} and the series resistance $2R$ by the Norton equivalent composed of a current source $v_{O1}/2R$ and a parallel resistance $2R$. Now, since $v_{O2} = 2v$, where v is the voltage at the input of op amp 2, the current through R_f will be $(2v - v)/R_f = v/R_f$ in the direction from output to input. Thus R_f gives rise to a negative input resistance, $-R_f$, as indicated in the equivalent circuit of Fig. 13.9(b). Nominally, R_f is made equal to $2R$, and thus $-R_f$ cancels $2R$, and at the input we are left with a current source $v_{O1}/2R$ feeding a capacitor C. The result is that $v = \frac{1}{C}\int_0^t \frac{v_{O1}}{2R}dt$ and $v_{O2} = 2v = \frac{1}{CR}\int_0^t v_{O1}dt$. That is, for $R_f = 2R$, the circuit functions as a perfect noninverting integrator. If, however, R_f is made smaller than $2R$, a net negative resistance appears in parallel with C.

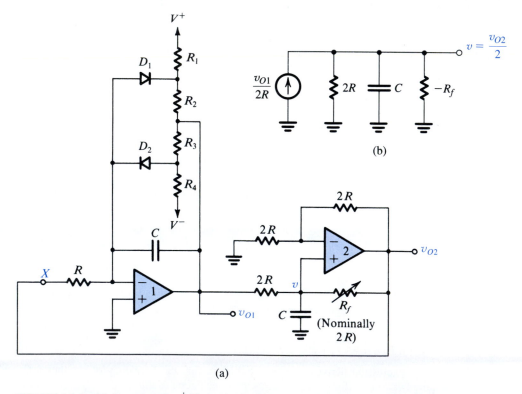

(a)

(b)

FIGURE 13.9 (a) A quadrature-oscillator circuit. (b) Equivalent circuit at the input of op amp 2.

Returning to the oscillator circuit in Fig. 13.9(a), we note that the resistance R_f in the positive-feedback path of op amp 2 is made variable, with a nominal value of $2R$. Decreasing the value of R_f moves the poles to the right half-plane (Problem 13.19) and ensures that the oscillations start. Too much positive feedback, although it results in better amplitude stability, also results in higher output distortion (because the limiter has to operate "harder"). In this regard, note that the output v_{O2} will be "purer" than v_{O1} because of the filtering action provided by the second integrator on the peak-limited output of the first integrator.

If we disregard the limiter and break the loop at X, the loop gain can be obtained as

$$L(s) \equiv \frac{V_{o2}}{V_x} = -\frac{1}{s^2 C^2 R^2} \tag{13.14}$$

Thus the loop will oscillate at frequency ω_0, given by

$$\omega_0 = \frac{1}{CR} \tag{13.15}$$

Finally, it should be pointed out that the name *quadrature oscillator* is used because the circuit provides two sinusoids with 90° phase difference. This should be obvious, since v_{O2} is the integral of v_{O1}. There are many applications for which quadrature sinusoids are required.

13.2.4 The Active-Filter-Tuned Oscillator

The last oscillator circuit that we shall discuss is quite simple both in principle and in design. Nevertheless, the approach is general and versatile and can result in high-quality (i.e., low-distortion) output sine waves. The basic principle is illustrated in Fig. 13.10. The circuit consists of a high-Q bandpass filter connected in a positive-feedback loop with a hard limiter. To understand how this circuit works, assume that oscillations have already started. The output of the bandpass filter will be a sine wave whose frequency is equal to the center frequency of the filter, f_0. The sine-wave signal v_1 is fed to the limiter, which produces at its output a square wave whose levels are determined by the limiting levels and whose

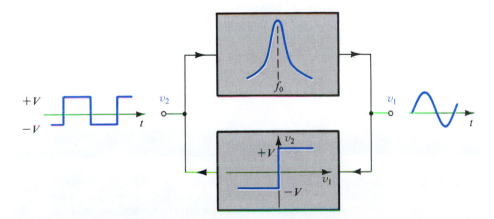

FIGURE 13.10 Block diagram of the active-filter-tuned oscillator.

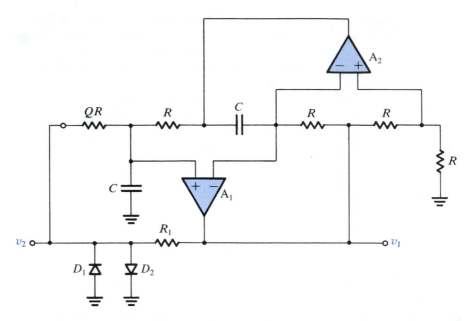

FIGURE 13.11 A practical implementation of the active-filter-tuned oscillator.

frequency is f_0. The square wave in turn is fed to the bandpass filter, which filters out the harmonics and provides a sinusoidal output v_1 at the fundamental frequency f_0. Obviously, the purity of the output sine wave will be a direct function of the selectivity (or Q factor) of the bandpass filter.

The simplicity of this approach to oscillator design should be apparent. We have independent control of frequency and amplitude as well as of distortion of the output sinusoid. Any filter circuit with positive gain can be used to implement the bandpass filter. The frequency stability of the oscillator will be directly determined by the frequency stability of the bandpass-filter circuit. Also, a variety of limiter circuits (see Chapter 3) with different degrees of sophistication can be used to implement the limiter block.

Figure 13.11 shows one possible implementation of the active-filter-tuned oscillator. This circuit uses a variation on the bandpass circuit based on the Antoniou inductance-simulation circuit (see Fig. 12.22c). Here resistor R_2 and capacitor C_4 are interchanged. This makes the output of the lower op amp directly proportional to (in fact, twice as large as) the voltage across the resonator, and we can therefore dispense with the buffer amplifier K. The limiter used is a very simple one consisting of a resistance R_1 and two diodes.

EXERCISE

13.7 Using $C = 16$ nF, find the value of R such that the circuit of Fig. 13.11 produces 1-kHz sine waves. If the diode drop is 0.7 V, find the peak-to-peak amplitude of the output sine wave. (*Hint:* A square wave with peak-to-peak amplitude of V volts has a fundamental component with $4V/\pi$ volts peak-to-peak amplitude.)

Ans. 10 kΩ; 3.6 V

13.2.5 A Final Remark

The op amp–RC oscillator circuits studied are useful for operation in the range 10 Hz to 100 kHz (or perhaps 1 MHz at most). Whereas the lower frequency limit is dictated by the size of passive components required, the upper limit is governed by the frequency-response and slew-rate limitations of op amps. For higher frequencies, circuits that employ transistors together with LC tuned circuits or crystals are frequently used.[3] These are discussed in Section 13.3.

13.3 LC AND CRYSTAL OSCILLATORS

Oscillators utilizing transistors (FETs or BJTs), with LC-tuned circuits or crystals as feedback elements, are used in the frequency range of 100 kHz to hundreds of megahertz. They exhibit higher Q than the RC types. However, LC oscillators are difficult to tune over wide ranges, and crystal oscillators operate at a single frequency.

13.3.1 LC-Tuned Oscillators

Figure 13.12 shows two commonly used configurations of LC-tuned oscillators. They are known as the **Colpitts oscillator** and the **Hartley oscillator.** Both utilize a parallel LC circuit connected between collector and base (or between drain and gate if a FET is used) with a fraction of the tuned-circuit voltage fed to the emitter (the source in a FET). This feedback is achieved by way of a capacitive divider in the Colpitts oscillator and by way of an inductive divider in the Hartley circuit. To focus attention on the oscillator's structure, the bias details are not shown. In both circuits, the resistor R models the combination of the losses of the inductors, the load resistance of the oscillator, and the output resistance of the transistor.

If the frequency of operation is sufficiently low that we can neglect the transistor capacitances, the frequency of oscillation will be determined by the resonance frequency of the parallel-tuned circuit (also known as a *tank circuit* because it behaves as a reservoir for

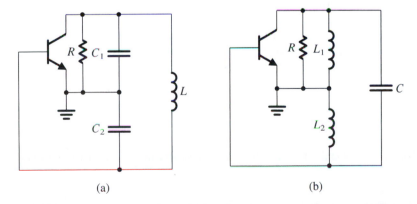

(a)   (b)

FIGURE 13.12 Two commonly used configurations of LC-tuned oscillators: **(a)** Colpitts and **(b)** Hartley.

---

[3] Of course, transistors can be used in place of the op amps in the circuits just studied. At higher frequencies, however, better results are obtained with LC-tuned circuits and crystals.

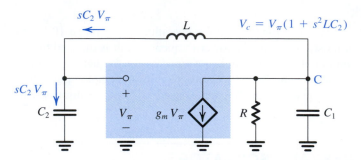

FIGURE 13.13 Equivalent circuit of the Colpitts oscillator of Fig. 13.12(a). To simplify the analysis, C_μ and r_π are neglected. We can consider C_π to be part of C_2, and we can include r_o in R.

energy storage). Thus for the Colpitts oscillator we have

$$\omega_0 = 1 \Big/ \sqrt{L\left(\frac{C_1 C_2}{C_1 + C_2}\right)} \tag{13.16}$$

and for the Hartley oscillator we have

$$\omega_0 = 1 \Big/ \sqrt{(L_1 + L_2)C} \tag{13.17}$$

The ratio L_1/L_2 or C_1/C_2 determines the feedback factor and thus must be adjusted in conjunction with the transistor gain to ensure that oscillations will start. To determine the oscillation condition for the Colpitts oscillator, we replace the transistor with its equivalent circuit, as shown in Fig. 13.13. To simplify the analysis we have neglected the transistor capacitance C_μ (C_{gd} for a FET). Capacitance C_π (C_{gs} for a FET), although not shown, can be considered to be a part of C_2. The input resistance r_π (infinite for a FET) has also been neglected, assuming that at the frequency of oscillation $r_\pi \gg (1/\omega C_2)$. Finally, as mentioned earlier, the resistance R includes r_o of the transistor.

To find the loop gain, we break the loop at the transistor base, apply an input voltage V_π, and find the returned voltage that appears across the input terminals of the transistor. We then equate the loop gain to unity. An alternative approach is to analyze the circuit and eliminate all current and voltage variables, and thus obtain one equation that governs circuit operation. Oscillations will start if this equation is satisfied. Thus the resulting equation will give us the conditions for oscillation.

A node equation at the transistor collector (node C) in the circuit of Fig. 13.13 yields

$$sC_2 V_\pi + g_m V_\pi + \left(\frac{1}{R} + sC_1\right)(1 + s^2 LC_2)V_\pi = 0$$

Since $V_\pi \neq 0$ (oscillations have started), it can be eliminated, and the equation can be rearranged in the form

$$s^3 LC_1 C_2 + s^2(LC_2/R) + s(C_1 + C_2) + \left(g_m + \frac{1}{R}\right) = 0 \tag{13.18}$$

Substituting $s = j\omega$ gives

$$\left(g_m + \frac{1}{R} - \frac{\omega^2 LC_2}{R}\right) + j[\omega(C_1 + C_2) - \omega^3 LC_1 C_2] = 0 \tag{13.19}$$

For oscillations to start, both the real and imaginary parts must be zero. Equating the imaginary part to zero gives the frequency of oscillation as

$$\omega_0 = 1 \Big/ \sqrt{L\left(\frac{C_1 C_2}{C_1 + C_2}\right)} \qquad (13.20)$$

which is the resonance frequency of the tank circuit, as anticipated.[4] Equating the real part to zero together with using Eq. (13.20) gives

$$C_2/C_1 = g_m R \qquad (13.21)$$

which has a simple physical interpretation: For sustained oscillations, the magnitude of the gain from base to collector ($g_m R$) must be equal to the inverse of the voltage ratio provided by the capacitive divider, which from Fig. 13.12(a) can be seen to be $v_{eb}/v_{ce} = C_1/C_2$. Of course, for oscillations to start, the loop gain must be made greater than unity, a condition that can be stated in the equivalent form

$$g_m R > C_2/C_1 \qquad (13.22)$$

As oscillations grow in amplitude, the transistor's nonlinear characteristics reduce the effective value of g_m and, correspondingly, reduce the loop gain to unity, thus sustaining the oscillations.

Analysis similar to the foregoing can be carried out for the Hartley circuit (see later-Exercise 13.8). At high frequencies, more accurate transistor models must be used. Alternatively, the *y* parameters of the transistor can be measured at the intended frequency ω_0, and the analysis can then be carried out using the *y*-parameter model (see Appendix B). This is usually simpler and more accurate, especially at frequencies above about 30% of the transistor f_T.

As an example of a practical LC oscillator we show in Fig. 13.14 the circuit of a Colpitts oscillator, complete with bias details. Here the radio-frequency choke (RFC) provides a high reactance at ω_0 but a low dc resistance.

Finally, a few words are in order on the mechanism that determines the amplitude of oscillations in the LC-tuned oscillators discussed above. Unlike the op-amp oscillators that incorporate special amplitude-control circuitry, LC-tuned oscillators utilize the nonlinear i_C–v_{BE} characteristics of the BJT (the i_D–v_{GS} characteristics of the FET) for amplitude control. Thus these LC-tuned oscillators are known as *self-limiting oscillators*. Specifically, as the oscillations grow in amplitude, the effective gain of the transistor is reduced below its small-signal value. Eventually, an amplitude is reached at which the effective gain is reduced to the point that the Barkhausen criterion is satisfied exactly. The amplitude then remains constant at this value.

Reliance on the nonlinear characteristics of the BJT (or the FET) implies that the collector (drain) current waveform will be nonlinearly distorted. Nevertheless, the output voltage signal will still be a sinusoid of high purity because of the filtering action of the LC tuned circuit. Detailed analysis of amplitude control, which makes use of nonlinear-circuit techniques, is beyond the scope of this book.

---

[4] If r_π is taken into account, the frequency of oscillation can be shown to shift slightly from the value given by Eq. (13.20).

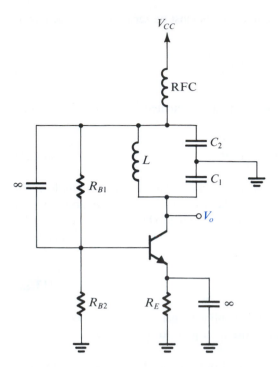

FIGURE 13.14 Complete circuit for a Colpitts oscillator.

EXERCISES

13.8 Show that for the Hartley oscillator of Fig. 13.12(b), the frequency of oscillation is given by Eq. (13.17) and that for oscillations to start $g_m R > (L_1/L_2)$.

D13.9 Using a BJT biased at $I_C = 1$ mA, design a Colpitts oscillator to operate at $\omega_0 = 10^6$ rad/s. Use $C_1 = 0.01$ μF, and assume that the coil available has a Q of 100 (this can be represented by a resistance in parallel with C_1 given by $Q/\omega_0 C_1$). Also assume that there is a load resistance at the collector of 2 kΩ and that for the BJT, $r_o = 100$ kΩ. Find C_2 and L.

Ans. 0.66 μF; 100 μH (a somewhat smaller C_2 would be used to allow oscillations to grow in amplitude)

13.3.2 Crystal Oscillators

A piezoelectric crystal, such as quartz, exhibits electromechanical-resonance characteristics that are very stable (with time and temperature) and highly selective (having very high Q factors). The circuit symbol of a crystal is shown in Fig. 13.15(a), and its equivalent circuit model is given in Fig. 13.15(b). The resonance properties are characterized by a large inductance L (as high as hundreds of henrys), a very small series capacitance C_s (as small as 0.0005 pF), a series resistance r representing a Q factor $\omega_0 L/r$ that can be as high as a few hundred thousand, and a parallel capacitance C_p (a few picofarads). Capacitor C_p represents the electrostatic capacitance between the two parallel plates of the crystal. Note that $C_p \gg C_s$.

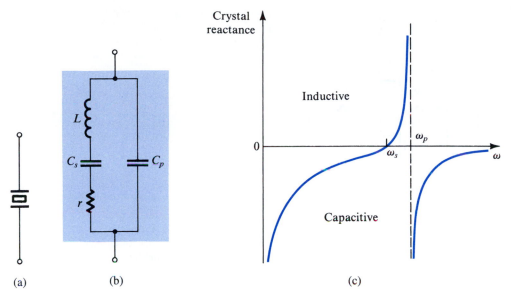

FIGURE 13.15 A piezoelectric crystal. **(a)** Circuit symbol. **(b)** Equivalent circuit. **(c)** Crystal reactance versus frequency [note that, neglecting the small resistance r, $Z_{\text{crystal}} = jX(\omega)$].

Since the Q factor is very high, we may neglect the resistance r and express the crystal impedance as

$$Z(s) = 1 \bigg/ \left[sC_p + \frac{1}{sL + 1/sC_s} \right]$$

which can be manipulated to the form

$$Z(s) = \frac{1}{sC_p} \frac{s^2 + (1/LC_s)}{s^2 + [(C_p + C_s)/LC_sC_p]} \tag{13.23}$$

From Eq. (13.23) and from Fig. 13.15(b) we see that the crystal has two resonance frequencies: a series resonance at ω_s

$$\omega_s = 1/\sqrt{LC_s} \tag{13.24}$$

and a parallel resonance at ω_p

$$\omega_p = 1 \bigg/ \sqrt{L\left(\frac{C_sC_p}{C_s + C_p}\right)} \tag{13.25}$$

Thus for $s = j\omega$ we can write

$$Z(j\omega) = -j\frac{1}{\omega C_p}\left(\frac{\omega^2 - \omega_s^2}{\omega^2 - \omega_p^2}\right) \tag{13.26}$$

From Eqs. (13.24) and (13.25) we note that $\omega_p > \omega_s$. However, since $C_p \gg C_s$, the two resonance frequencies are very close. Expressing $Z(j\omega) = jX(\omega)$, the crystal reactance $X(\omega)$ will

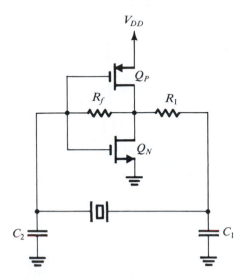

FIGURE 13.16 A Pierce crystal oscillator utilizing a CMOS inverter as an amplifier.

have the shape shown in Fig. 13.15(c). We observe that the crystal reactance is inductive over the very narrow frequency band between ω_s and ω_p. For a given crystal, this frequency band is well defined. Thus we may use the crystal to replace the inductor of the Colpitts oscillator (Fig. 13.12a). The resulting circuit will oscillate at the resonance frequency of the crystal inductance L with the series equivalent of C_s and $(C_p + C_1 C_2/(C_1 + C_2))$. Since C_s is much smaller than the three other capacitances, it will be dominant and

$$\omega_0 \simeq 1/\sqrt{LC_s} = \omega_s \qquad (13.27)$$

In addition to the basic Colpitts oscillator, a variety of configurations exist for crystal oscillators. Figure 13.16 shows a popular configuration (called the **Pierce oscillator**) utilizing a CMOS inverter (see Section 4.10) as amplifier. Resistor R_f determines a dc operating point in the high-gain region of the CMOS inverter. Resistor R_1 together with capacitor C_1 provides a low-pass filter that discourages the circuit from oscillating at a higher harmonic of the crystal frequency. Note that this circuit also is based on the Colpitts configuration.

The extremely stable resonance characteristics and the very high Q factors of quartz crystals result in oscillators with very accurate and stable frequencies. Crystals are available with resonance frequencies in the range of few kilohertz to hundreds of megahertz. Temperature coefficients of ω_0 of 1 or 2 parts per million (ppm) per degree Celsius are achievable. Unfortunately, however, crystal oscillators, being mechanical resonators, are fixed-frequency circuits.

EXERCISE

13.10 A 2-MHz quartz crystal is specified to have $L = 0.52$ H, $C_s = 0.012$ pF, $C_p = 4$ pF, and $r = 120\ \Omega$. Find f_s, f_p, and Q.
Ans. 2.015 MHz; 2.018 MHz; 55,000

13.4 BISTABLE MULTIVIBRATORS

In this section we begin the study of waveform-generating circuits of the other type—nonlinear oscillators or function generators. These devices make use of a special class of circuits known as **multivibrators.** As mentioned earlier, there are three types of multivibrator: bistable, monostable, and astable. This section is concerned with the first, the bistable multivibrator.[5]

As its name indicates, the **bistable multivibrator** has *two stable states*. The circuit can remain in either stable state indefinitely and moves to the other stable state only when appropriately *triggered*.

13.4.1 The Feedback Loop

Bistability can be obtained by connecting a dc amplifier in a positive-feedback loop having a loop gain greater than unity. Such a feedback loop is shown in Fig. 13.17; it consists of an op amp and a resistive voltage divider in the positive-feedback path. To see how bistability is obtained, consider operation with the positive input terminal of the op amp near ground potential. This is a reasonable starting point, since the circuit has no external excitation. Assume that the electrical noise that is inevitably present in every electronic circuit causes a small positive increment in the voltage v_+. This incremental signal will be amplified by the large open-loop gain A of the op amp, with the result that a much greater signal will appear in the op amp's output voltage v_O. The voltage divider (R_1, R_2) will feed a fraction $\beta \equiv R_1/(R_1 + R_2)$ of the output signal back to the positive input terminal of the op amp. If $A\beta$ is greater than unity, as is usually the case, the fed-back signal will be greater than the original increment in v_+. This *regenerative* process continues until eventually the op amp saturates with its output voltage at the positive saturation level, L_+. When this happens, the voltage at the positive input terminal, v_+, becomes $L_+R_1/(R_1 + R_2)$, which is positive and thus keeps the op amp in positive saturation. This is one of the two stable states of the circuit.

In the description above we assumed that when v_+ was near zero volts, a positive increment occurred in v_+. Had we assumed the equally probable situation of a negative increment, the op amp would have ended up saturated in the negative direction with $v_O = L_-$ and $v_+ = L_-R_1/(R_1 + R_2)$. This is the other stable state.

We thus conclude that the circuit of Fig. 13.17 has two stable states, one with the op amp in positive saturation and the other with the op amp in negative saturation. The circuit can exist in either of these two states indefinitely. We also note that the circuit cannot exist in the state for which $v_+ = 0$ and $v_O = 0$ for any length of time. This is a state of *unstable equilibrium* (also known as a **metastable state**); any disturbance, such as that caused by electrical noise,

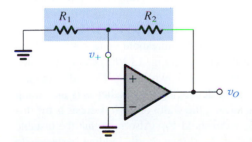

FIGURE 13.17 A positive-feedback loop capable of bistable operation.

---

[5] Digital implementations of multivibrators were presented in Chapter 11. Here, we are interested in implementations utilizing op amps.

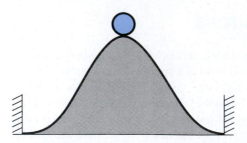

FIGURE 13.18 A physical analogy for the operation of the bistable circuit. The ball cannot remain at the top of the hill for any length of time (a state of unstable equilibrium or metastability); the inevitably present disturbance will cause the ball to fall to one side or the other, where it can remain indefinitely (the two stable states).

causes the bistable circuit to switch to one of its two stable states. This is in sharp contrast to the case when the feedback is negative, causing a virtual short circuit to appear between the op amp's input terminals and maintaining this virtual short circuit in the face of disturbances. A physical analogy for the operation of the bistable circuit is depicted in Fig. 13.18.

13.4.2 Transfer Characteristics of the Bistable Circuit

The question naturally arises as to how we can make the bistable circuit of Fig. 13.17 change state. To help answer this crucial question, we derive the transfer characteristics of the bistable. Reference to Fig. 13.17 indicates that either of the two circuit nodes that are connected to ground can serve as an input terminal. We investigate both possibilities.

Figure 13.19(a) shows the bistable circuit with a voltage v_I applied to the inverting input terminal of the op amp. To derive the transfer characteristic v_O–v_I, assume that v_O is at one of its two possible levels, say L_+, and thus $v_+ = \beta L_+$. Now as v_I is increased from 0 V we can see from the circuit that nothing happens until v_I reaches a value equal to v_+ (i.e., βL_+). As v_I begins to exceed this value, a net negative voltage develops between the input terminals of the op amp. This voltage is amplified by the open-loop gain of the op amp, and thus v_O goes negative. The voltage divider in turn causes v_+ to go negative, thus increasing the net negative input to the op amp and keeping the regenerative process going. This process culminates in the op amp saturating in the negative direction: that is, with $v_O = L_-$ and, correspondingly, $v_+ = \beta L_-$. It is easy to see that increasing v_I further has no effect on the acquired state of the bistable circuit. Figure 13.19(b) shows the transfer characteristic for increasing v_I. Observe that the characteristic is that of a comparator with a threshold voltage denoted V_{TH}, where $V_{TH} = \beta L_+$.

Next consider what happens as v_I is decreased. Since now $v_+ = \beta L_-$, we see that the circuit remains in the negative-saturation state until v_I goes negative to the point that it equals βL_-. As v_I goes below this value, a net positive voltage appears between the op amp's input terminals. This voltage is amplified by the op-amp gain and thus gives rise to a positive voltage at the op amp's output. The regenerative action of the positive-feedback loop then sets in and causes the circuit eventually to go to its positive-saturation state, in which $v_O = L_+$ and $v_+ = \beta L_+$. The transfer characteristic for decreasing v_I is shown in Fig. 13.19(c). Here again we observe that the characteristic is that of a comparator, but with a threshold voltage $V_{TL} = \beta L_-$.

The complete transfer characteristics, v_O–v_I, of the circuit in Fig. 13.19(a) can be obtained by combining the characteristics in Fig. 13.19(b) and (c), as shown in Fig. 13.19(d). As indicated, the circuit changes state at different values of v_I, depending on whether v_I is increasing or decreasing. Thus the circuit is said to exhibit *hysteresis;* the width of the hysteresis is the difference between the high threshold V_{TH} and the low threshold V_{TL}. Also note that the bistable circuit is in effect a comparator with hysteresis. As will be shown shortly, adding hysteresis to a comparator's characteristics can be very beneficial in certain applications. Finally, observe that because the bistable circuit of Fig. 13.19 switches from the positive state ($v_O = L_+$) to the negative state ($v_O = L_-$) as v_I is increased past the positive threshold V_{TH}, the circuit is said to be *inverting*. A bistable circuit with a *noninverting* transfer characteristic will be presented shortly.

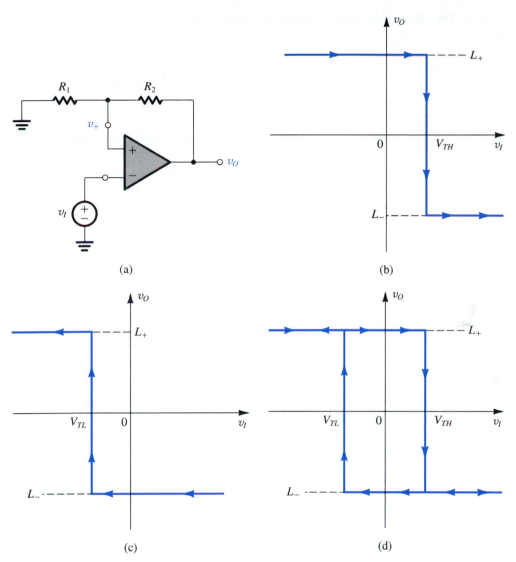

FIGURE 13.19 **(a)** The bistable circuit of Fig. 13.17 with the negative input terminal of the op amp disconnected from ground and connected to an input signal v_I. **(b)** The transfer characteristic of the circuit in (a) for increasing v_I. **(c)** The transfer characteristic for decreasing v_I. **(d)** The complete transfer characteristics.

13.4.3 Triggering the Bistable Circuit

Returning now to the question of how to make the bistable circuit change state, we observe from the transfer characteristics of Fig. 13.19(d) that if the circuit is in the L_+ state it can be switched to the L_- state by applying an input v_I of value greater than $V_{TH} \equiv \beta L_+$. Such an input causes a net negative voltage to appear between the input terminals of the op amp, which initiates the regenerative cycle that culminates in the circuit switching to the L_- stable state. Here it is important to note that the input v_I merely initiates or *triggers* regeneration. Thus we can remove v_I with no effect on the regeneration process. In other words, v_I can be simply a pulse of short duration. The input signal v_I is thus referred to as a **trigger signal,** or simply a **trigger.**

The characteristics of Fig. 13.19(d) indicate also that the bistable circuit can be switched to the positive state ($v_O = L_+$) by applying a negative trigger signal v_I of magnitude greater than that of the negative threshold V_{TL}.

13.4.4 The Bistable Circuit as a Memory Element

We observe from Fig. 13.19(d) that for input voltages in the range $V_{TL} < v_I < V_{TH}$, the output can be either L_+ or L_-, *depending on the state that the circuit is already in*. Thus, for this input range, the output is determined by the *previous* value of the trigger signal (the trigger signal that caused the circuit to be in its current state). Thus the circuit exhibits *memory*. Indeed, the bistable multivibrator is the basic memory element of digital systems, as we have seen in Chapter 11. Finally, note that in analog circuit applications, such as the ones of concern to us in this chapter, the bistable circuit is also known as a **Schmitt trigger.**

13.4.5 A Bistable Circuit with Noninverting Transfer Characteristics

The basic bistable feedback loop of Fig. 13.17 can be used to derive a circuit with noninverting transfer characteristics by applying the input signal v_I (the trigger signal) to the terminal of R_1 that is connected to ground. The resulting circuit is shown in Fig. 13.20(a). To obtain the transfer characteristics we first employ superposition to the linear circuit formed by R_1 and R_2, thus expressing v_+ in terms of v_I and v_O as

$$v_+ = v_I \frac{R_2}{R_1 + R_2} + v_O \frac{R_1}{R_1 + R_2} \tag{13.28}$$

From this equation we see that if the circuit is in the positive stable state with $v_O = L_+$, positive values for v_I will have no effect. To trigger the circuit into the L_- state, v_I must be made negative and of such a value as to make v_+ decrease below zero. Thus the low threshold V_{TL} can be found by substituting in Eq. (13.28) $v_O = L_+$, $v_+ = 0$, and $v_I = V_{TL}$. The result is

$$V_{TL} = -L_+(R_1/R_2) \tag{13.29}$$

Similarly, Eq. (13.28) indicates that when the circuit is in the negative-output state ($v_O = L_-$), negative values of v_I will make v_+ more negative with no effect on operation. To initiate the regeneration process that causes the circuit to switch to the positive state, v_+ must be made

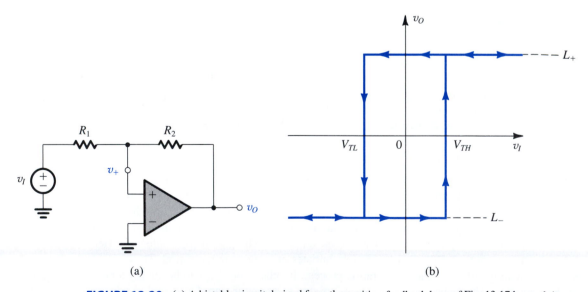

(a) (b)

FIGURE 13.20 **(a)** A bistable circuit derived from the positive-feedback loop of Fig. 13.17 by applying v_I through R_1. **(b)** The transfer characteristic of the circuit in (a) is noninverting. (Compare it to the inverting characteristic in Fig. 13.19d.)

to go slightly positive. The value of v_I that causes this to happen is the high threshold voltage V_{TH}, which can be found by substituting in Eq. (13.28) $v_O = L_-$ and $v_+ = 0$. The result is

$$V_{TH} = -L_-(R_1/R_2) \tag{13.30}$$

The complete transfer characteristic of the circuit of Fig. 13.20(a) is displayed in Fig. 13.20(b). Observe that a positive triggering signal v_I (of value greater than V_{TH}) causes the circuit to switch to the positive state (v_O goes from L_- to L_+). Thus the transfer characteristic of this circuit is noninverting.

13.4.6 Application of the Bistable Circuit as a Comparator

The comparator is an analog-circuit building block that is used in a variety of applications ranging from detecting the level of an input signal relative to a preset threshold value, to the design of analog-to-digital (A/D) converters (see Section 9.1). Although one normally thinks of the comparator as having a single threshold value (see Fig. 13.21a), it is useful in many applications to add hysteresis to the comparator characteristics. If this is done, the comparator exhibits two threshold values, V_{TL} and V_{TH}, symmetrically placed about the

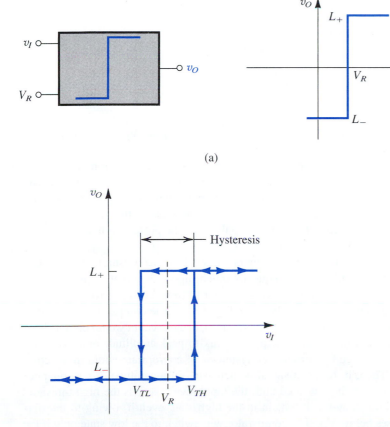

FIGURE 13.21 (a) Block diagram representation and transfer characteristic for a comparator having a reference, or threshold, voltage V_R. (b) Comparator characteristic with hysteresis.

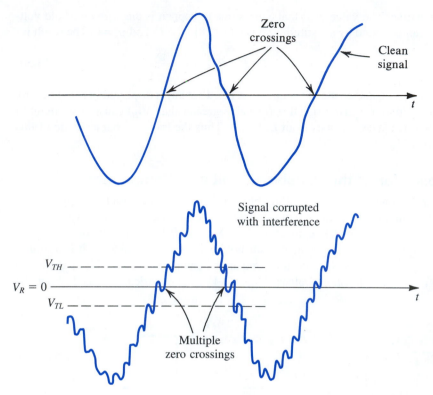

FIGURE 13.22 Illustrating the use of hysteresis in the comparator characteristics as a means of rejecting interference.

desired reference level, as indicated in Fig. 13.21(b). Usually V_{TH} and V_{TL} are separated by a small amount, say 100 mV.

To demonstrate the need for hysteresis we consider a common application of comparators. It is required to design a circuit that detects and counts the zero crossings of an arbitrary waveform. Such a function can be implemented using a comparator whose threshold is set to 0 V. The comparator provides a step change at its output every time a zero crossing occurs. Each step change can be used to generate a pulse, and the pulses are fed to a counter circuit.

Imagine now what happens if the signal being processed has—as it usually does have—interference superimposed on it, say of a frequency much higher than that of the signal. It follows that the signal might cross the zero axis a number of times around each of the zero-crossing points we are trying to detect, as shown in Fig. 13.22. The comparator would thus change state a number of times at each of the zero crossings, and our count would obviously be in error. However, if we have an idea of the expected peak-to-peak amplitude of the interference, the problem can be solved by introducing hysteresis of appropriate width in the comparator characteristics. Then, if the input signal is increasing in magnitude, the comparator with hysteresis will remain in the low state until the input level exceeds the high threshold V_{TH}. Subsequently the comparator will remain in the high state even if, owing to interference, the signal decreases below V_{TH}. The comparator will switch to the low state only if the input signal is decreased below the low threshold V_{TL}. The situation is illustrated in Fig. 13.22, from which we see that including hysteresis in the comparator characteristics provides an effective means for rejecting interference (thus providing another form of filtering).

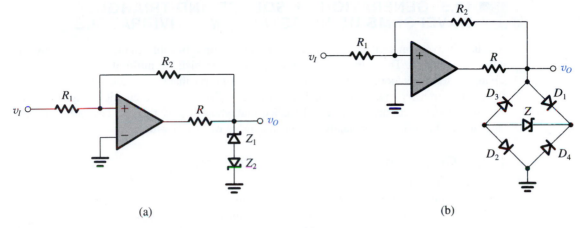

(a)   (b)

FIGURE 13.23 Limiter circuits are used to obtain more precise output levels for the bistable circuit. In both circuits the value of R should be chosen to yield the current required for the proper operation of the zener diodes. **(a)** For this circuit $L_+ = V_{Z_1} + V_D$ and $L_- = -(V_{Z_2} + V_D)$, where V_D is the forward diode drop. **(b)** For this circuit $L_+ = V_Z + V_{D_1} + V_{D_2}$ and $L_- = -(V_Z + V_{D_3} + V_{D_4})$.

13.4.7 Making the Output Levels More Precise

The output levels of the bistable circuit can be made more precise than the saturation voltages of the op amp are by cascading the op amp with a limiter circuit (see Section 3.6 for a discussion of limiter circuits). Two such arrangements are shown in Fig. 13.23.

EXERCISES

D13.11 The op amp in the bistable circuit of Fig. 13.19(a) has output saturation voltages of ± 13 V. Design the circuit to obtain threshold voltages of ± 5 V. For $R_1 = 10$ kΩ, find the value required for R_2.

Ans. 16 kΩ

D13.12 If the op amp in the circuit of Fig. 13.20(a) has ± 10-V output saturation levels, design the circuit to obtain ± 5-V thresholds. Give suitable component values.

Ans. Possible choice: $R_1 = 10$ kΩ and $R_2 = 20$ kΩ

13.13 Consider a bistable circuit with a noninverting transfer characteristic, and let $L_+ = -L_- = 10$ V and $V_{TH} = -V_{TL} = 5$ V. If v_I is a triangular wave with a 0-V average, a 10-V peak amplitude, and a 1-ms period, sketch the waveform of v_O. Find the time interval between the zero crossings of v_I and v_O.

Ans. v_O is a square wave with 0-V average, 10-V amplitude, and 1-ms period and is delayed by 125 μs relative to v_I

13.14 Consider an op amp having saturation levels of ± 12 V used without feedback, with the inverting input terminal connected to +3 V and the noninverting input terminal connected to v_I. Characterize its operation as a comparator. What are L_+, L_-, and V_R, as defined in Fig. 13.21(a)?

Ans. +12 V; −12 V; +3 V

13.15 In the circuit of Fig. 13.20(a) let $L_+ = -L_- = 10$ V and $R_1 = 1$ kΩ. Find a value for R_2 that gives a hysteresis of 100-mV width.

Ans. 200 kΩ

13.5 GENERATION OF SQUARE AND TRIANGULAR WAVEFORMS USING ASTABLE MULTIVIBRATORS

A square waveform can be generated by arranging for a bistable multivibrator to switch states periodically. This can be done by connecting the bistable multivibrator with an RC circuit in a feedback loop, as shown in Fig. 13.24(a). Observe that the bistable multivibrator has an inverting transfer characteristic and can thus be realized using the circuit of Fig. 13.19(a). This results in the circuit of Fig. 13.24(b). We shall show shortly that this circuit has no stable states and thus is appropriately named an **astable multivibrator.**

13.5.1 Operation of the Astable Multivibrator

To see how the astable multivibrator operates, refer to Fig. 13.24(b) and let the output of the bistable multivibrator be at one of its two possible levels, say L_+. Capacitor C will charge toward this level through resistor R. Thus the voltage across C, which is applied to the negative input terminal of the op amp and thus is denoted v_-, will rise exponentially toward L_+ with a time constant $\tau = CR$. Meanwhile, the voltage at the positive input terminal of the op amp is $v_+ = \beta L_+$. This situation will continue until the capacitor voltage reaches the positive threshold $V_{TH} = \beta L_+$ at which point the bistable multivibrator will switch to the other stable state in which $v_O = L_-$ and $v_+ = \beta L_-$. The capacitor will then start discharging, and its voltage, v_-, will decrease exponentially toward L_-. This new state will prevail until v_- reaches the negative threshold $V_{TL} = \beta L_-$, at which time the bistable multivibrator switches to the positive-output state, the capacitor begins to charge, and the cycle repeats itself.

From the preceding description we see that the astable circuit oscillates and produces a square waveform at the output of the op amp. This waveform, and the waveforms at the two input terminals of the op amp, are displayed in Fig. 13.24(c). The period T of the square

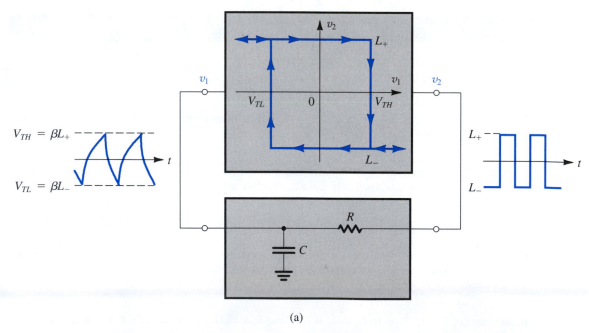

(a)

FIGURE 13.24 **(a)** Connecting a bistable multivibrator with inverting transfer characteristics in a feedback loop with an RC circuit results in a square-wave generator.

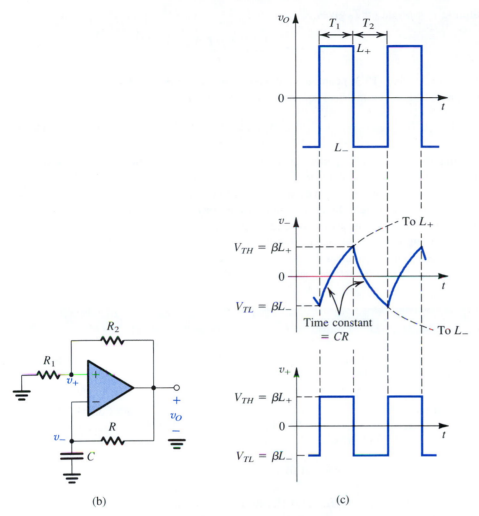

FIGURE 13.24 *(Continued)* **(b)** The circuit obtained when the bistable multivibrator is implemented with the circuit of Fig. 13.19(a). **(c)** Waveforms at various nodes of the circuit in (b). This circuit is called an astable multivibrator.

wave can be found as follows: During the charging interval T_1 the voltage v_- across the capacitor at any time t, with $t = 0$ at the beginning of T_1, is given by (see Appendix D)

$$v_- = L_+ - (L_+ - \beta L_-)e^{-t/\tau}$$

where $\tau = CR$. Substituting $v_- = \beta L_+$ at $t = T_1$ gives

$$T_1 = \tau \ln \frac{1 - \beta(L_-/L_+)}{1 - \beta} \tag{13.31}$$

Similarly, during the discharge interval T_2 the voltage v_- at any time t, with $t = 0$ at the beginning of T_2, is given by

$$v_- = L_- - (L_- - \beta L_+)e^{-t/\tau}$$

Substituting $v_- = \beta L_-$ at $t = T_2$ gives

$$T_2 = \tau \ln \frac{1 - \beta(L_+/L_-)}{1 - \beta} \tag{13.32}$$

Equations (13.31) and (13.32) can be combined to obtain the period $T = T_1 + T_2$. Normally, $L_+ = -L_-$, resulting in symmetrical square waves of period T given by

$$T = 2\tau \ln \frac{1 + \beta}{1 - \beta} \tag{13.33}$$

Note that this square-wave generator can be made to have variable frequency by switching different capacitors C (usually in decades) and by continuously adjusting R (to obtain continuous frequency control within each decade of frequency). Also, the waveform across C can be made almost triangular by using a small value for the parameter β. However, triangular waveforms of superior linearity can be easily generated using the scheme discussed next.

Before leaving this section, however, note that although the astable circuit has no stable states, it has two *quasi-stable* states and remains in each for a time interval determined by the time constant of the RC network and the thresholds of the bistable multivibrator.

EXERCISES

13.16 For the circuit in Fig. 13.24(b), let the op-amp saturation voltages be ±10 V, $R_1 = 100$ kΩ, $R_2 = R = 1$ MΩ, and $C = 0.01$ μF. Find the frequency of oscillation.

Ans. 274 Hz

13.17 Consider a modification of the circuit of Fig. 13.24(b) in which R_1 is replaced by a pair of diodes connected in parallel in opposite directions. For $L_+ = -L_- = 12$ V, $R_2 = R = 10$ kΩ, $C = 0.1$ μF, and the diode voltage a constant denoted V_D, find an expression for frequency as a function of V_D. If $V_D = 0.70$ V at 25°C with a *TC* of −2 mV/°C, find the frequency at 0°C, 25°C, 50°C, and 100°C. Note that the output of this circuit can be sent to a remotely connected frequency meter to provide a digital readout of temperature.

Ans. $f = 500/\ln[(12 + V_D)/(12 - V_D)]$ Hz; 3995 Hz, 4281 Hz, 4611 Hz, 5451 Hz

13.5.2 Generation of Triangular Waveforms

The exponential waveforms generated in the astable circuit of Fig. 13.24 can be changed to triangular by replacing the low-pass RC circuit with an integrator. (The integrator is, after all, a low-pass circuit with a corner frequency at dc.) The integrator causes linear charging and discharging of the capacitor, thus providing a triangular waveform. The resulting circuit is shown in Fig. 13.25(a). Observe that because the integrator is inverting, it is necessary to invert the characteristics of the bistable circuit. Thus the bistable circuit required here is of the noninverting type and can be implemented using the circuit of Fig. 13.2.

We now proceed to show how the feedback loop of Fig. 13.25(a) oscillates and generates a triangular waveform v_1 at the output of the integrator and a square waveform v_2 at the output of the bistable circuit: Let the output of the bistable circuit be at L_+. A current equal to L_+/R will flow into the resistor R and through capacitor C, causing the output of the integrator to *linearly* decrease with a slope of $-L_+/CR$, as shown in Fig. 13.25(c). This will continue until the integrator output reaches the lower threshold V_{TL} of the bistable circuit, at which point the bistable circuit will switch states, its output becoming negative and equal to

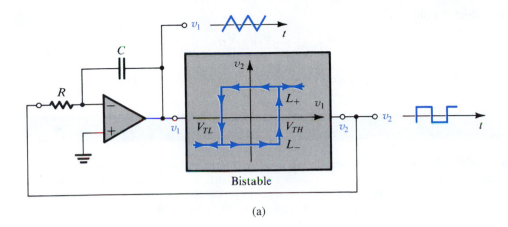

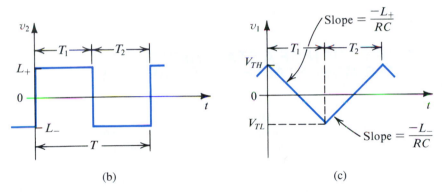

FIGURE 13.25 A general scheme for generating triangular and square waveforms.

L_-. At this moment the current through R and C will reverse direction, and its value will become equal to $|L_-|/R$. It follows that the integrator output will start to increase linearly with a positive slope equal to $|L_-|/CR$. This will continue until the integrator output voltage reaches the positive threshold of the bistable circuit, V_{TH}. At this point the bistable circuit switches, its output becomes positive (L_+), the current into the integrator reverses direction, and the output of the integrator starts to decrease linearly, beginning a new cycle.

From the discussion above it is relatively easy to derive an expression for the period T of the square and triangular waveforms. During the interval T_1 we have, from Fig. 13.25(c),

$$\frac{V_{TH} - V_{TL}}{T_1} = \frac{L_+}{CR}$$

from which we obtain

$$T_1 = CR\frac{V_{TH} - V_{TL}}{L_+} \tag{13.34}$$

Similarly, during T_2 we have

$$\frac{V_{TH} - V_{TL}}{T_2} = \frac{-L_-}{CR}$$

from which we obtain

$$T_2 = CR\frac{V_{TH} - V_{TL}}{-L_-} \qquad (13.35)$$

Thus to obtain symmetrical square waves we design the bistable circuit to have $L_+ = -L_-$.

EXERCISE

D13.18 Consider the circuit of Fig. 13.25(a) with the bistable circuit realized by the circuit in Fig. 13.20(a). If the op amps have saturation voltages of ± 10 V and if a capacitor $C = 0.01$ μF and a resistor $R_1 = 10$ kΩ are used, find the values of R and R_2 (note that R_1 and R_2 are associated with the bistable circuit of Fig. 13.20a) such that the frequency of oscillation is 1 kHz and the triangular waveform has a 10-V peak-to-peak amplitude.

Ans. 50 kΩ; 20 kΩ

13.6 GENERATION OF A STANDARDIZED PULSE—THE MONOSTABLE MULTIVIBRATOR

In some applications the need arises for a pulse of known height and width generated in response to a trigger signal. Because the width of the pulse is predictable, its trailing edge can be used for timing purposes—that is, to initiate a particular task at a specified time. Such a standardized pulse can be generated by the third type of multivibrator, the **monostable multivibrator.**

The monostable multivibrator has one stable state in which it can remain indefinitely. It also has a quasi-stable state to which it can be triggered and in which it stays for a predetermined interval equal to the desired width of the output pulse. When this interval expires, the monostable multivibrator returns to its stable state and remains there, awaiting another triggering signal. The action of the monostable multivibrator has given rise to its alternative name, the *one shot.*

Figure 13.26(a) shows an op-amp monostable circuit. We observe that this circuit is an augmented form of the astable circuit of Fig. 13.24(b). Specifically, a clamping diode D_1 is added across the capacitor C_1, and a trigger circuit composed of capacitor C_2, resistor R_4, and diode D_2 is connected to the noninverting input terminal of the op amp. The circuit operates as follows: In the stable state, which prevails in the absence of the triggering signal, the output of the op amp is at L_+ and diode D_1 is conducting through R_3 and thus clamping the voltage v_B to one diode drop above ground. We select R_4 much larger than R_1, so that diode D_2 will be conducting a very small current and the voltage v_C will be very closely determined by the voltage divider R_1, R_2. Thus $v_C = \beta L_+$, where $\beta = R_1/(R_1 + R_2)$. The stable state is maintained because βL_+ is greater than V_{D1}.

Now consider the application of a negative-going step at the trigger input and refer to the signal waveforms shown in Fig. 13.26(b). The negative triggering edge will be coupled to the cathode of diode D_2 via capacitor C_2, and thus D_2 conducts heavily and pulls node C down. If the trigger signal is of sufficient height to cause v_C to go below v_B, the op amp will see a net negative input voltage and its output will switch to L_-. This in turn will cause v_C to go negative to βL_-, keeping the op amp in its newly acquired state. Note that D_2 will then cut off, thus isolating the circuit from any further changes at the trigger input terminal.

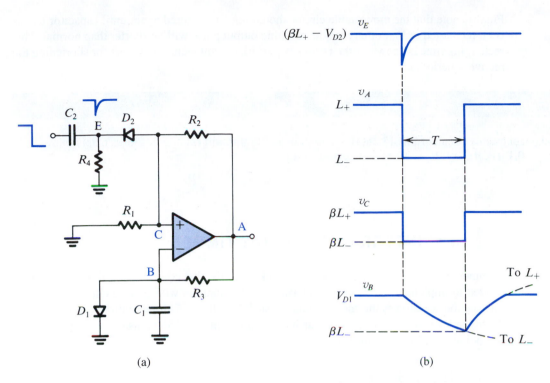

FIGURE 13.26 (a) An op-amp monostable circuit. (b) Signal waveforms in the circuit of (a).

The negative voltage at A causes D_1 to cut off, and C_1 begins to discharge exponentially toward L_- with a time constant $C_1 R_3$. The monostable multivibrator is now in its *quasi-stable state,* which will prevail until the declining v_B goes below the voltage at node C, which is βL_-. At this instant the op-amp output switches back to L_+ and the voltage at node C goes back to βL_+. Capacitor C_1 then charges toward L_+ until diode D_1 turns on and the circuit returns to its stable state.

From Fig. 13.26(b), we observe that a negative pulse is generated at the output during the quasi-stable state. The duration T of the output pulse is determined from the exponential waveform of v_B,

$$v_B(t) = L_- - (L_- - V_{D1})e^{-t/C_1 R_3}$$

by substituting $v_B(T) = \beta L_-$,

$$\beta L_- = L_- - (L_- - V_{D1})e^{-T/C_1 R_3}$$

which yields

$$T = C_1 R_3 \ln\left(\frac{V_{D1} - L_-}{\beta L_- - L_-}\right) \tag{13.36}$$

For $V_{D1} \ll |L_-|$, this equation can be approximated by

$$T \simeq C_1 R_3 \ln\left(\frac{1}{1 - \beta}\right) \tag{13.37}$$

Finally, note that the monostable circuit should not be triggered again until capacitor C_1 has been recharged to V_{D1}; otherwise the resulting output pulse will be shorter than normal. This recharging time is known as the **recovery period.** Circuit techniques exist for shortening the recovery period.

13.19 For the monostable circuit of Fig. 13.26(a) find the value of R_3 that will result in a 100-μs output pulse for $C_1 = 0.1 \ \mu$F, $\beta = 0.1$, $V_D = 0.7$ V, and $L_+ = -L_- = 12$ V.

Ans. 6171 Ω

13.7 INTEGRATED-CIRCUIT TIMERS

Commercially available integrated-circuit packages exist that contain the bulk of the circuitry needed to implement monostable and astable multivibrators with precise characteristics. In this section we discuss the most popular of such ICs, the **555 timer.** Introduced in 1972 by the Signetics Corporation as a bipolar integrated circuit, the 555 is also available in CMOS technology and from a number of manufacturers.

13.7.1 The 555 Circuit

Figure 13.27 shows a block-diagram representation of the 555 timer circuit [for the actual circuit, refer to Grebene (1984)]. The circuit consists of two comparators, an SR flip-flop,

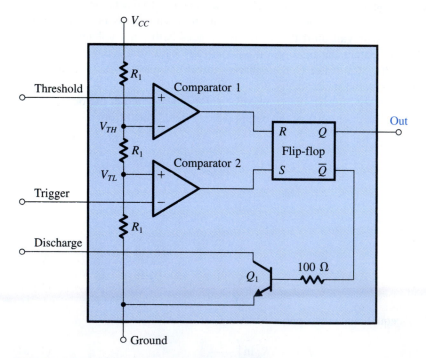

FIGURE 13.27 A block diagram representation of the internal circuit of the 555 integrated-circuit timer.

and a transistor Q_1 that operates as a switch. One power supply (V_{CC}) is required for operation, with the supply voltage typically 5 V. A resistive voltage divider, consisting of the three equal-valued resistors labeled R_1, is connected across V_{CC} and establishes the reference (threshold) voltages for the two comparators. These are $V_{TH} = \frac{2}{3}V_{CC}$ for comparator 1 and $V_{TL} = \frac{1}{3}V_{CC}$ for comparator 2.

We studied SR flip-flops in Chapter 11. For our purposes here we note that an SR flip-flop (also called a latch) is a bistable circuit having complementary outputs, denoted Q and $\bar{Q}$. In the *set* state, the output at Q is "high" (approximately equal to V_{CC}) and that at $\bar{Q}$ is "low" (approximately equal to 0 V). In the other stable state, termed the *reset* state, the output at Q is low and that at $\bar{Q}$ is high. The flip-flop is set by applying a high level (V_{CC}) to its set input terminal, labeled S. To reset the flip-flop, a high level is applied to the reset input terminal, labeled R. Note that the reset and set input terminals of the flip-flop in the 555 circuit are connected to the outputs of comparator 1 and comparator 2, respectively.

The positive-input terminal of comparator 1 is brought out to an external terminal of the 555 package, labeled Threshold. Similarly, the negative-input terminal of comparator 2 is connected to an external terminal labeled Trigger, and the collector of transistor Q_1 is connected to a terminal labeled Discharge. Finally, the Q output of the flip-flop is connected to the output terminal of the timer package, labeled Out.

13.7.2 Implementing a Monostable Multivibrator Using the 555 IC

Figure 13.28(a) shows a monostable multivibrator implemented using the 555 IC together with an external resistor R and an external capacitor C. In the stable state the flip-flop will be in the reset state, and thus its $\bar{Q}$ output will be high, turning on transistor Q_1. Transistor Q_1 will be saturated, and thus v_C will be close to 0 V, resulting in a low level at the output of comparator 1. The voltage at the trigger input terminal, labeled v_{trigger}, is kept high (greater than V_{TL}), and thus the output of comparator 2 also will be low. Finally, note that since the flip-flop is in the reset state, Q will be low and thus v_O will be close to 0 V.

To trigger the monostable multivibrator, a negative input pulse is applied to the trigger input terminal. As v_{trigger} goes below V_{TL}, the output of comparator 2 goes to the high level, thus setting the flip-flop. Output Q of the flip-flop goes high, and thus v_O goes high, and output $\bar{Q}$ goes low, turning off transistor Q_1. Capacitor C now begins to charge up through resistor R, and its voltage v_C rises exponentially toward V_{CC}, as shown in Fig. 13.28(b). The monostable multivibrator is now in its quasi-stable state. This state prevails until v_C reaches, and begins to exceed, the threshold of comparator 1, V_{TH}, at which time the output of comparator 1 goes high, resetting the flip-flop. Output $\bar{Q}$ of the flip-flop now goes high and turns on transistor Q_1. In turn, transistor Q_1 rapidly discharges capacitor C, causing v_C to go to 0 V. Also, when the flip-flop is reset its Q output goes low, and thus v_O goes back to 0 V. The monostable multivibrator is now back in its stable state and is ready to receive a new triggering pulse.

From the description above we see that the monostable multivibrator produces an output pulse v_O as indicated in Fig. 13.28(b). The width of the pulse, T, is the time interval that the monostable multivibrator spends in the quasi-stable state; it can be determined by reference to the waveforms in Fig. 13.28(b) as follows: Denoting the instant at which the trigger pulse is applied as $t = 0$, the exponential waveform of v_C can be expressed as

$$v_C = V_{CC}(1 - e^{-t/CR}) \tag{13.38}$$

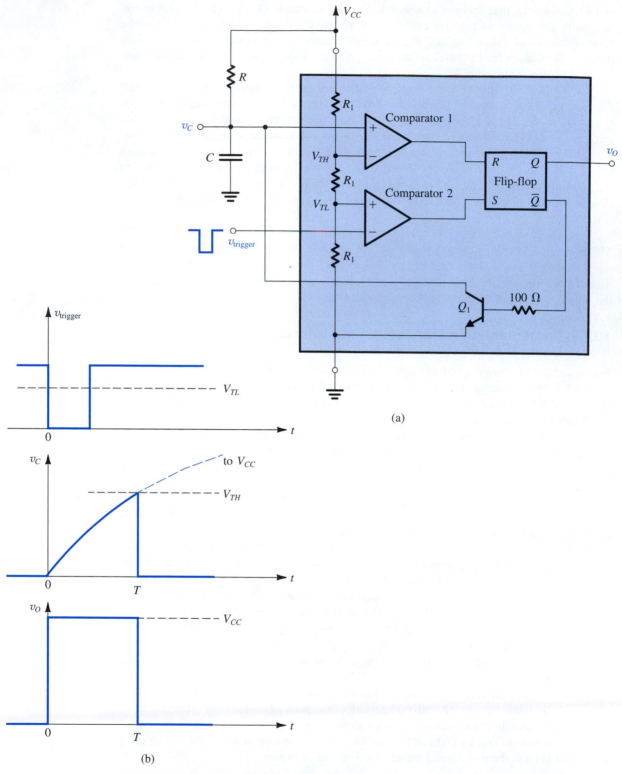

FIGURE 13.28 (a) The 555 timer connected to implement a monostable multivibrator. (b) Waveforms of the circuit in (a).

Substituting $v_C = V_{TH} = \frac{2}{3}V_{CC}$ at $t = T$ gives

$$T = CR \ln 3 \simeq 1.1 CR \tag{13.39}$$

Thus the pulse width is determined by the external components C and R, which can be selected to have values as precise as desired.

13.7.3 An Astable Multivibrator Using the 555 IC

Figure 13.29(a) shows the circuit of an astable multivibrator employing a 555 IC, two external resistors, R_A and R_B, and an external capacitor C. To see how the circuit operates refer to the waveforms depicted in Fig. 13.29(b). Assume that initially C is discharged and the flip-flop is set. Thus v_O is high and Q_1 is off. Capacitor C will charge up through the series combination of R_A and R_B, and the voltage across it, v_C, will rise exponentially toward V_{CC}. As v_C crosses the level equal to V_{TL}, the output of comparator 2 goes low. This, however, has no effect on the circuit operation, and the flip-flop remains set. Indeed, this state continues until v_C reaches and begins to exceed the threshold of comparator 1, V_{TH}. At this instant of time, the output of comparator 1 goes high and resets the flip-flop. Thus v_O goes low, $\bar{Q}$ goes high, and transistor Q_1 is turned on. The saturated transistor Q_1 causes a voltage of approximately zero volts to appear at the common node of R_A and R_B. Thus C begins to discharge through R_B and the collector of Q_1. The voltage v_C decreases exponentially with a time constant CR_B toward 0 V. When v_C reaches the threshold of comparator 2, V_{TL}, the output of comparator 2, goes high and sets the flip-flop. The output v_O then goes high, and $\bar{Q}$ goes low, turning off Q_1. Capacitor C begins to charge through the series equivalent of R_A and R_B, and its voltage rises exponentially toward V_{CC} with a time constant $C(R_A + R_B)$. This rise continues until v_C reaches V_{TH}, at which time the output of comparator 1 goes high, resetting the flip-flop, and the cycle continues.

From the description above we see that the circuit of Fig. 13.29(a) oscillates and produces a square waveform at the output. The frequency of oscillation can be determined as follows. Reference to Fig. 13.29(b) indicates that the output will be high during the interval T_H, in which v_C rises from V_{TL} to V_{TH}. The exponential rise of v_C can be described by

$$v_C = V_{CC} - (V_{CC} - V_{TL})e^{-t/C(R_A+R_B)} \tag{13.40}$$

where $t = 0$ is the instant at which the interval T_H begins. Substituting $v_C = V_{TH} = \frac{2}{3}V_{CC}$ at $t = T_H$ and $V_{TL} = \frac{1}{3}V_{CC}$ results in

$$T_H = C(R_A + R_B)\ln 2 \simeq 0.69\, C(R_A + R_B) \tag{13.41}$$

We also note from Fig. 13.29(b) that v_O will be low during the interval T_L, in which v_C falls from V_{TH} to V_{TL}. The exponential fall of v_C can be described by

$$v_C = V_{TH}e^{-t/CR_B} \tag{13.42}$$

where we have taken $t = 0$ as the beginning of the interval T_L. Substituting $v_C = V_{TL} = \frac{1}{3}V_{CC}$ at $t = T_L$ and $V_{TH} = \frac{2}{3}V_{CC}$ results in

$$T_L = CR_B \ln 2 \simeq 0.69\, CR_B \tag{13.43}$$

Equations (13.41) and (13.43) can be combined to obtain the period T of the output square wave as

$$T = T_H + T_L = 0.69\, C(R_A + 2R_B) \tag{13.44}$$

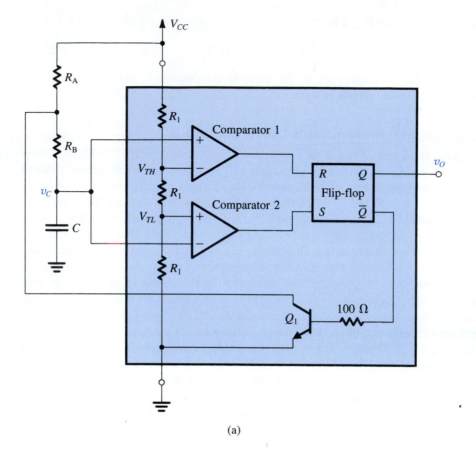

(a)

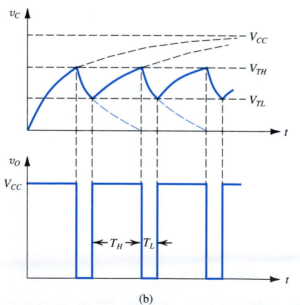

(b)

FIGURE 13.29 **(a)** The 555 timer connected to implement an astable multivibrator. **(b)** Waveforms of the circuit in (a).

Also, the **duty cycle** of the output square wave can be found from Eqs. (13.41) and (13.43):

$$\text{Duty cycle} \equiv \frac{T_H}{T_H + T_L} = \frac{R_A + R_B}{R_A + 2R_B} \qquad (13.45)$$

Note that the duty cycle will always be greater than 0.5 (50%); it approaches 0.5 if R_A is selected to be much smaller than R_B (unfortunately, at the expense of supply current).

EXERCISES

13.20 Using a 10-nF capacitor C, find the value of R that yields an output pulse of 100 μs in the monostable circuit of Fig. 13.28(a).

Ans. 9.1 kΩ

D13.21 For the circuit in Fig. 13.29(a), with a 1000-pF capacitor, and find the values of R_A and R_B that result in an oscillation frequency of 100 kHz and a duty cycle of 75%.

Ans. 7.2 kΩ, 3.6 kΩ

 ## 13.8 NONLINEAR WAVEFORM-SHAPING CIRCUITS

Diodes or transistors can be combined with resistors to synthesize two-port networks having arbitrary nonlinear transfer characteristics. Such two-port networks can be employed in **waveform shaping**—that is, changing the waveform of an input signal in a prescribed manner to produce a waveform of a desired shape at the output. In this section we illustrate this application by a concrete example: the **sine-wave shaper.** This is a circuit whose purpose is to change the waveform of an input triangular-wave signal to a sine wave. Though simple, the sine-wave shaper is a practical building block used extensively in function generators. This method of generating sine waves should be contrasted to that using linear oscillators (Sections 13.1–13.3). Although linear oscillators produce sine waves of high purity, they are not convenient at very low frequencies. Also, linear oscillators are in general more difficult to tune over wide frequency ranges. In the following we discuss two distinctly different techniques for designing sine-wave shapers.

13.8.1 The Breakpoint Method

In the breakpoint method the desired nonlinear transfer characteristic (in our case the sine function shown in Fig. 13.30) is implemented as a piecewise linear curve. Diodes are utilized as switches that turn on at the various breakpoints of the transfer characteristic, thus switching into the circuit additional resistors that cause the transfer characteristic to change slope.

Consider the circuit shown in Fig. 13.31(a). It consists of a chain of resistors connected across the entire symmetrical voltage supply $+V$, $-V$. The purpose of this voltage divider is to generate reference voltages that will serve to determine the breakpoints in the transfer characteristic. In our example these reference voltages are denoted $+V_2$, $+V_1$, $-V_1$, $-V_2$. Note that the entire circuit is symmetrical, driven by a symmetrical triangular wave and generating a symmetrical sine-wave output. The circuit approximates each quarter-cycle of the sine wave by three straight-line segments; the breakpoints between these segments are determined by the reference voltages V_1 and V_2.

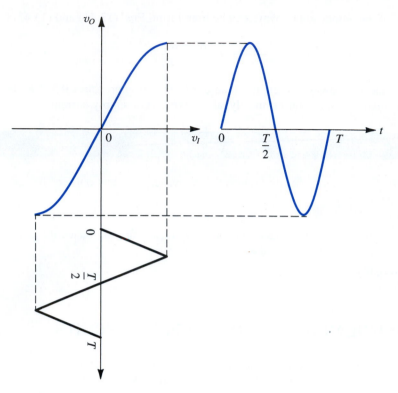

FIGURE 13.30 Using a nonlinear (sinusoidal) transfer characteristic to shape a triangular waveform into a sinusoid.

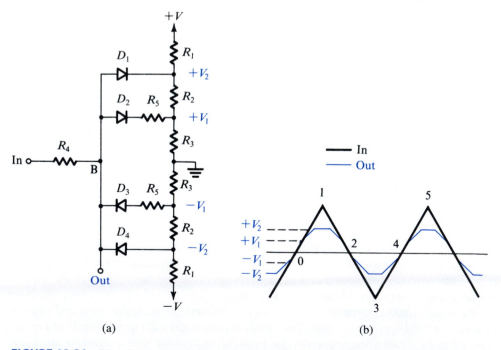

FIGURE 13.31 (a) A three-segment sine-wave shaper. (b) The input triangular waveform and the output approximately sinusoidal waveform.

The circuit works as follows: Let the input be the triangular wave shown in Fig. 13.31(b), and consider first the quarter-cycle defined by the two points labeled 0 and 1. When the input signal is less in magnitude than V_1, none of the diodes conducts. Thus zero current flows through R_4, and the output voltage at B will be equal to the input voltage. But as the input rises to V_1 and above, D_2 (assumed ideal) begins to conduct. Assuming that the conducting D_2 behaves as a short circuit, we see that, for $v_I > V_1$,

$$v_O = V_1 + (v_I - V_1)\frac{R_5}{R_4 + R_5}$$

This implies that as the input continues to rise above V_1 the output follows but with a reduced slope. This gives rise to the second segment in the output waveform, as shown in Fig. 13.31(b). Note that in developing the equation above we have assumed that the resistances in the voltage divider are low enough in value to cause the voltages V_1 and V_2 to be constant independent of the current coming from the input.

Next consider what happens as the voltage at point B reaches the second breakpoint determined by V_2. At this point, D_1 conducts, thus limiting the output v_O to V_2 (plus, of course, the voltage drop across D_1 if it is not assumed to be ideal). This gives rise to the third segment, which is flat, in the output waveform. The overall result is to "bend" the waveform and shape it into an approximation of the first quarter-cycle of a sine wave. Then, beyond the peak of the input triangular wave, as the input voltage decreases, the process unfolds, the output becoming progressively more like the input. Finally, when the input goes sufficiently negative, the process begins to repeat at $-V_1$ and $-V_2$ for the negative half-cycle.

Although the circuit is relatively simple, its performance is surprisingly good. A measure of goodness usually taken is to quantify the purity of the output sine wave by specifying the percentage **total harmonic distortion** (THD). This is the percentage ratio of the rms voltage of all harmonic components above the fundamental frequency (which is the frequency of the triangular wave) to the rms voltage of the fundamental (see also Chapter 14). Interestingly, one reason for the good performance of the diode shaper is the beneficial effects produced by the nonideal i–v characteristics of the diodes—that is, the exponential knee of the junction diode as it goes into forward conduction. The consequence is a relatively smooth transition from one line segment to the next.

Practical implementations of the breakpoint sine-wave shaper employ six to eight segments (compared with the three used in the example above). Also, transistors are usually employed to provide more versatility in the design, with the goal being increased precision and lower THD. [See Grebene (1984), pages 592–595.]

13.8.2 The Nonlinear-Amplification Method

The other method we discuss for the conversion of a triangular wave into a sine wave is based on feeding the triangular wave to the input of an amplifier having a nonlinear transfer characteristic that approximates the sine function. One such amplifier circuit consists of a differential pair with a resistance connected between the two emitters, as shown in Fig. 13.32. With appropriate choice of the values of the bias current I and the resistance R, the differential amplifier can be made to have a transfer characteristic that closely approximates that shown in Fig. 13.30. Observe that for small v_I the transfer characteristic of the circuit of Fig. 13.32 is almost linear, as a sine waveform is near its zero crossings. At large values of v_I the nonlinear characteristics of the BJTs reduce the gain of the amplifier and cause the transfer characteristic to bend, approximating the sine wave as it approaches its peak. [More details on this circuit can be found in Grebene (1984), pages 595–597.]

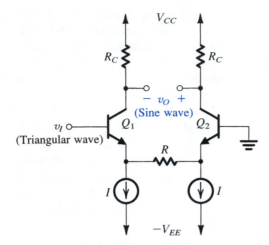

FIGURE 13.32 A differential pair with an emitter degeneration resistance used to implement a triangular-wave to sine-wave converter. Operation of the circuit can be graphically described by Fig. 13.30.

EXERCISES

D13.22 The circuit in Fig. E13.22 is required to provide a three-segment approximation to the nonlinear i–v characteristic, $i = 0.1v^2$, where v is the voltage in volts and i is the current in milliamperes. Find the values of R_1, R_2, and R_3 such that the approximation is perfect at $v = 2$ V, 4 V, and 8 V. Calculate the error in current value at $v = 3$ V, 5 V, 7 V, and 10 V. Assume ideal diodes.

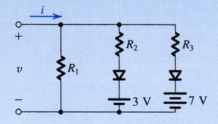

FIGURE E13.22

Ans. 5 kΩ, 1.25 kΩ, 1.25 kΩ; −0.3 mA, +0.1 mA, −0.3 mA, 0

13.23 A detailed analysis of the circuit in Fig. 13.32 shows that its optimum performance occurs when the values of I and R are selected so that $RI = 2.5V_T$, where V_T is the thermal voltage. For this design, the peak amplitude of the input triangular wave should be $6.6V_T$, and the corresponding sine wave across R has a peak value of $2.42V_T$. For $I = 0.25$ mA and $R_C = 10$ kΩ, find the peak amplitude of the sine-wave output v_O. Assume $\alpha \simeq 1$.

Ans. 4.84 V

13.9 PRECISION RECTIFIER CIRCUITS

Rectifier circuits were studied in Chapter 3, where the emphasis was on their application in power-supply design. In such applications the voltages being rectified are usually much greater than the diode voltage drop, rendering the exact value of the diode drop unimportant to the proper operation of the rectifier. Other applications exist, however, where this is not the case. For instance, in instrumentation applications, the signal to be rectified can be of a very small amplitude, say 0.1 V, making it impossible to employ the conventional rectifier circuits. Also, in instrumentation applications the need arises for rectifier circuits with very precise transfer characteristics.

In this section we study circuits that combine diodes and op amps to implement a variety of rectifier circuits with precise characteristics. Precision rectifiers, which can be considered a special class of wave-shaping circuits, find application in the design of instrumentation systems. An introduction to precision rectifiers was presented in Chapter 3. This material, however, is repeated here for the reader's convenience.

13.9.1 Precision Half-Wave Rectifier–The "Superdiode"

Figure 13.33(a) shows a precision half-wave-rectifier circuit consisting of a diode placed in the negative-feedback path of an op amp, with R being the rectifier load resistance. The circuit works as follows: If v_I goes positive, the output voltage v_A of the op amp will go positive and the diode will conduct, thus establishing a closed feedback path between the op amp's output terminal and the negative input terminal. This negative-feedback path will cause a virtual short circuit to appear between the two input terminals of the op amp. Thus the voltage at the negative input terminal, which is also the output voltage v_O, will equal (to within a few millivolts) that at the positive input terminal, which is the input voltage v_I,

$$v_O = v_I \quad v_I \geq 0$$

Note that the offset voltage ($\simeq 0.5$ V) exhibited in the simple half-wave rectifier circuit is no longer present. For the op-amp circuit to start operation, v_I has to exceed only a negligibly small voltage equal to the diode drop divided by the op amp's open-loop gain. In other words, the straight-line transfer characteristic v_O–v_I almost passes through the origin. This makes this circuit suitable for applications involving very small signals.

Consider now the case when v_I goes negative. The op amp's output voltage v_A will tend to follow and go negative. This will reverse-bias the diode, and no current will flow through resistance R, causing v_O to remain equal to 0 V. Thus for $v_I < 0$, $v_O = 0$. Since in this case the diode is off, the op amp will be operating in an open-loop fashion and its output will be at the negative saturation level.

The transfer characteristic of this circuit will be that shown in Fig. 13.33(b), which is almost identical to the ideal characteristic of a half-wave rectifier. The nonideal diode

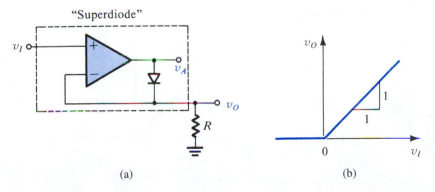

(a)   (b)

FIGURE 13.33 (a) The "superdiode" precision half-wave rectifier and (b) its almost ideal transfer characteristic. Note that when $v_I > 0$ and the diode conducts, the op amp supplies the load current, and the source is conveniently buffered, an added advantage.

characteristics have been almost completely masked by placing the diode in the negative-feedback path of an op amp. This is another dramatic application of negative feedback. The combination of diode and op amp, shown in the dashed box in Fig. 13.33(a), is appropriately referred to as a "superdiode."

As usual, though, not all is well. The circuit of Fig. 13.33 has some disadvantages: When v_I goes negative and $v_O = 0$, the entire magnitude of v_I appears between the two input terminals of the op amp. If this magnitude is greater than few volts, the op amp may be damaged unless it is equipped with what is called "overvoltage protection" (a feature that most modern IC op amps have). Another disadvantage is that when v_I is negative, the op amp will be saturated. Although not harmful to the op amp, saturation should usually be avoided, since getting the op amp out of the saturation region and back into its linear region of operation requires some time. This time delay will obviously slow down circuit operation and limit the frequency of operation of the superdiode half-wave-rectifier circuit.

13.9.2 An Alternative Circuit

An alternative precision rectifier circuit that does not suffer from the disadvantages mentioned above is shown in Fig. 13.34. The circuit operates in the following manner: For positive v_I, diode D_2 conducts and closes the negative-feedback loop around the op amp. A virtual ground therefore will appear at the inverting input terminal, and the op amp's output will be *clamped* at one diode drop below ground. This negative voltage will keep diode D_1 off, and no current will flow in the feedback resistance R_2. It follows that the rectifier output voltage will be zero.

As v_I goes negative, the voltage at the inverting input terminal will tend to go negative, causing the voltage at the op amp's output terminal to go positive. This will cause D_2 to be reverse-biased and hence cut off. Diode D_1, however, will conduct through R_2, thus establishing a negative-feedback path around the op amp and forcing a virtual ground to appear at the inverting input terminal. The current through the feedback resistance R_2 will be equal to the current through the input resistance R_1. Thus for $R_1 = R_2$ the output voltage v_O will be

$$v_O = -v_I \quad v_I \leq 0$$

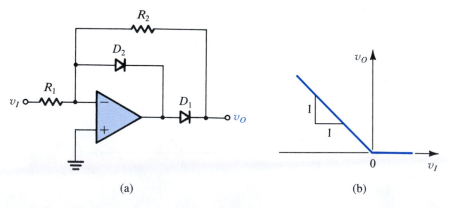

(a)   (b)

FIGURE 13.34 **(a)** An improved version of the precision half-wave rectifier: Diode D_2 is included to keep the feedback loop closed around the op amp during the off times of the rectifier diode D_1, thus preventing the op amp from saturating. **(b)** The transfer characteristic for $R_2 = R_1$.

The transfer characteristic of the circuit is shown in Fig. 13.34(b). Note that unlike the situation for the circuit shown in Fig. 13.33, here the slope of the characteristic can be set to any desired value, including unity, by selecting appropriate values for R_1 and R_2.

As mentioned before, the major advantage of the improved half-wave-rectifier circuit is that the feedback loop around the op amp remains closed at all times. Hence the op amp remains in its linear operating region, avoiding the possibility of saturation and the associated time delay required to "get out" of saturation. Diode D_2 "catches" the op-amp output voltage as it goes negative and clamps it to one diode drop below ground; hence D_2 is called a "catching diode."

13.9.3 An Application: Measuring AC Voltages

As one of the many possible applications of the precision rectifier circuits discussed in this section, consider the basic ac voltmeter circuit shown in Fig. 13.35. The circuit consists of a half-wave rectifier—formed by op amp A_1, diodes D_1 and D_2, and resistors R_1 and R_2—and a first-order low-pass filter—formed by op amp A_2, resistors R_3 and R_4, and capacitor C. For an input sinusoid having a peak amplitude V_p the output v_1 of the rectifier will consist of a half sine wave having a peak amplitude of $V_p R_2/R_1$. It can be shown using Fourier series analysis that the waveform of v_1 has an average value of $(V_p/\pi)(R_2/R_1)$ in addition to harmonics of the frequency ω of the input signal. To reduce the amplitudes of all these harmonics to negligible levels, the corner frequency of the low-pass filter should be chosen much smaller than the lowest expected frequency ω_{min} of the input sine wave. This leads to

$$\frac{1}{CR_4} \ll \omega_{min}$$

Then the output voltage v_2 will be mostly dc, with a value

$$V_2 = -\frac{V_p}{\pi}\frac{R_2 R_4}{R_1 R_3}$$

where R_4/R_3 is the dc gain of the low-pass filter. Note that this voltmeter essentially measures the average value of the negative parts of the input signal but can be calibrated to provide rms readings for input sinusoids.

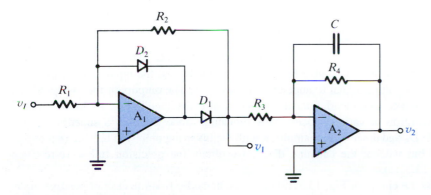

FIGURE 13.35 A simple ac voltmeter consisting of a precision half-wave rectifier followed by a first-order low-pass filter.

EXERCISES

13.24 Consider the operational rectifier or superdiode circuit of Fig. 13.33(a), with $R = 1$ kΩ. For $v_I = 10$ mV, 1 V, and −1 V, what are the voltages that result at the rectifier output and at the output of the op amp? Assume that the op amp is ideal and that its output saturates at ±12 V. The diode has a 0.7-V drop at 1-mA current, and the voltage drop changes by 0.1 V per decade of current change.

Ans. 10 mV, 0.51 V; 1 V, 1.7 V; 0 V, −12 V

13.25 If the diode in the circuit of Fig. 13.33(a) is reversed, what is the transfer characteristic v_O as a function of v_I?

Ans. $v_O = 0$ for $v_I \geq 0$; $v_O = v_I$ for $v_I \leq 0$

13.26 Consider the circuit in Fig. 13.34(a) with $R_1 = 1$ kΩ and $R_2 = 10$ kΩ. Find v_O and the voltage at the amplifier output for $v_I = +1$ V, −10 mV, and −1 V. Assume the op amp to be ideal with saturation voltages of ±12 V. The diodes have 0.7-V voltage drops at 1 mA, and the voltage drop changes by 0.1 V per decade of current change.

Ans. 0 V, −0.7 V; 0.1 V, 0.6 V; 10 V, 10.7 V

13.27 If the diodes in the circuit of Fig. 13.34(a) are reversed, what is the transfer characteristic v_O as a function of v_I?

Ans. $v_O = -(R_2/R_1)v_I$ for $v_I \geq 0$; $v_O = 0$ for $v_I \leq 0$

13.28 Find the transfer characteristic for the circuit in Fig. E13.28.

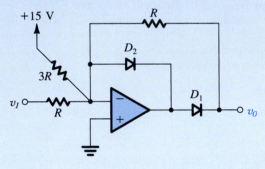

FIGURE E13.28

Ans. $v_O = 0$ for $v_I \geq -5$ V; $v_O = -v_I - 5$ for $v_I \leq -5$ V

13.9.4 Precision Full-Wave Rectifier

We now derive a circuit for a precision full-wave rectifier. From Chapter 3 we know that full-wave rectification is achieved by inverting the negative halves of the input-signal waveform and applying the resulting signal to another diode rectifier. The outputs of the two rectifiers are then joined to a common load. Such an arrangement is depicted in Fig. 13.36, which also shows the waveforms at various nodes. Now replacing diode D_A with a superdiode, and replacing diode D_B and the inverting amplifier with the inverting precision half-wave rectifier of Fig. 13.34 but without the catching diode, we obtain the precision full-wave-rectifier circuit of Fig. 13.37(a).

To see how the circuit of Fig. 13.37(a) operates, consider first the case of positive input at A. The output of A_2 will go positive, turning D_2 on, which will conduct through R_L and thus close the feedback loop around A_2. A virtual short circuit will thus be established between

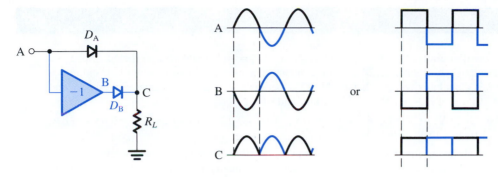

FIGURE 13.36 Principle of full-wave rectification.

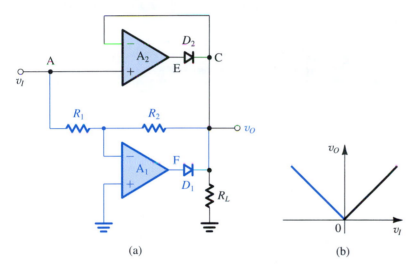

FIGURE 13.37 (a) Precision full-wave rectifier based on the conceptual circuit of Fig. 13.36. (b) Transfer characteristic of the circuit in (a).

the two input terminals of A_2, and the voltage at the negative-input terminal, which is the output voltage of the circuit, will become equal to the input. Thus no current will flow through R_1 and R_2, and the voltage at the inverting input of A_1 will be equal to the input and hence positive. Therefore the output terminal (F) of A_1 will go negative until A_1 saturates. This causes D_1 to be turned off.

Next consider what happens when A goes negative. The tendency for a negative voltage at the negative input of A_1 causes F to rise, making D_1 conduct to supply R_L and allowing the feedback loop around A_1 to be closed. Thus a virtual ground appears at the negative input of A_1, and the two equal resistances R_1 and R_2 force the voltage at C, which is the output voltage, to be equal to the negative of the input voltage at A and thus positive. The combination of positive voltage at C and negative voltage at A causes the output of A_2 to saturate in the negative direction, thus keeping D_2 off.

The overall result is perfect full-wave rectification, as represented by the transfer characteristic in Fig. 13.37(b). This precision is, of course, a result of placing the diodes in op-amp feedback loops, thus masking their nonidealities. This circuit is one of many possible precision full-wave-rectifier or **absolute-value circuits.** Another related implementation of this function is examined in Exercise 13.30.

EXERCISES

13.29 In the full-wave rectifier circuit of Fig. 13.37(a) let $R_1 = R_2 = R_L = 10$ kΩ and assume the op amps to be ideal except for output saturation at ±12 V. When conducting a current of 1 mA, each diode exhibits a voltage drop of 0.7 V, and this voltage changes by 0.1 V per decade of current change. Find v_O, v_E, and v_F corresponding to $v_I = +0.1$ V, +1 V, +10 V, −0.1 V, and −10 V.

Ans. +0.1 V, +0.6 V, −12 V; +1 V, +1.6 V, −12 V; +10 V, +10.7 V, −12 V; +0.1 V, −12 V, +0.63 V; +1 V, −12 V, +1.63 V; +10 V, −12 V, +10.73 V

D13.30 The block diagram shown in Fig. E13.30(a) gives another possible arrangement for implementing the absolute-value or full-wave-rectifier operation depicted symbolically in Fig. E13.30(b). The block diagram consists of two boxes: a half-wave rectifier, which can be implemented by the circuit in Fig. 13.34(a) after reversing both diodes, and a weighted inverting summer. Convince yourself that this block diagram does in fact realize the absolute-value operation. Then draw a complete circuit diagram, giving reasonable values for all resistors.

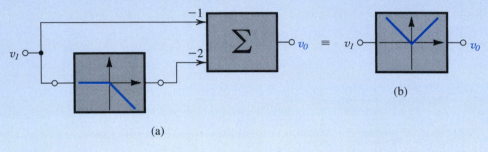

(a)

(b)

FIGURE E13.30

13.9.5 A Precision Bridge Rectifier for Instrumentation Applications

The bridge rectifier circuit studied in Chapter 3 can be combined with an op amp to provide useful precision circuits. One such arrangement is shown in Fig. 13.38. This circuit causes a current equal to $|v_A| / R$ to flow through the moving-coil meter M. Thus the meter provides

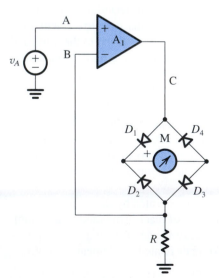

FIGURE 13.38 Use of the diode bridge in the design of an ac voltmeter.

a reading that is proportional to the average of the absolute value of the input voltage v_A. All the nonidealities of the meter and of the diodes are masked by placing the bridge circuit in the negative-feedback loop of the op amp. Observe that when v_A is positive, current flows from the op-amp output through D_1, M, D_3, and R. When v_A is negative, current flows into the op-amp output through R, D_2, M, and D_4. Thus the feedback loop remains closed for both polarities of v_A. The resulting virtual short circuit at the input terminals of the op amp causes a replica of v_A to appear across R. The circuit of Fig. 13.38 provides a relatively accurate high-input-impedance ac voltmeter using an inexpensive moving-coil meter.

EXERCISE

D13.31 In the circuit of Fig. 13.38, find the value of R that would cause the meter to provide a full-scale reading when the input voltage is a sine wave of 5 V rms. Let meter M have a 1-mA, 50-Ω movement (i.e., its resistance is 50 Ω, and it provides full-scale deflection when the average current through it is 1 mA). What are the approximate maximum and minimum voltages at the op amp's output? Assume that the diodes have constant 0.7-V drops when conducting.

Ans. 4.5 kΩ; +8.55 V; −8.55 V

13.9.6 Precision Peak Rectifiers

Including the diode of the peak rectifier studied in Chapter 3 inside the negative-feedback loop of an op amp, as shown in Fig. 13.39, results in a precision peak rectifier. The diode–op-amp combination will be recognized as the superdiode of Fig. 13.33(a). Operation of the circuit in Fig. 13.39 is quite straightforward. For v_I greater than the output voltage, the op amp will drive the diode on, thus closing the negative-feedback path and causing the op amp to act as a follower. The output voltage will therefore follow that of the input, with the op amp supplying the capacitor-charging current. This process continues until the input reaches its peak value. Beyond the positive peak, the op amp will see a negative voltage between its input terminals. Thus its output will go negative to the saturation level and the diode will turn off. Except for possible discharge through the load resistance, the capacitor will retain a voltage equal to the positive peak of the input. Inclusion of a load resistance is essential if the circuit is required to detect reductions in the magnitude of the positive peak.

13.9.7 A Buffered Precision Peak Detector

When the peak detector is required to hold the value of the peak for a long time, the capacitor should be buffered, as shown in the circuit of Fig. 13.40. Here op amp A_2, which should have high input impedance and low input bias current, is connected as a voltage follower. The remainder of the circuit is quite similar to the half-wave-rectifier circuit of Fig. 13.34.

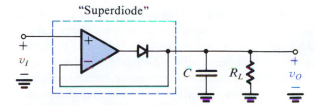

FIGURE 13.39 A precision peak rectifier obtained by placing the diode in the feedback loop of an op amp.

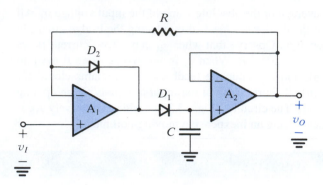

FIGURE 13.40 A buffered precision peak rectifier.

FIGURE 13.41 A precision clamping circuit.

While diode D_1 is the essential diode for the peak-rectification operation, diode D_2 acts as a catching diode to prevent negative saturation, and the associated delays, of op amp A_1. During the holding state, follower A_2 supplies D_2 with a small current through R. The output of op amp A_1 will then be clamped at one diode drop below the input voltage. Now if the input v_I increases above the value stored on C, which is equal to the output voltage v_O, op amp A_1 sees a net positive input that drives its output toward the positive saturation level, turning off diode D_2. Diode D_1 is then turned on and capacitor C is charged to the new positive peak of the input, after which time the circuit returns to the holding state. Finally, note that this circuit has a low-impedance output.

13.9.8 A Precision Clamping Circuit

By replacing the diode in the clamping circuit studied in Chapter 3 with a "superdiode," the precision clamp of Fig. 13.41 is obtained. Operation of this circuit should be self-explanatory.

13.10 SPICE SIMULATION EXAMPLES

The circuits studied in this chapter make use of the nonlinear operation of devices to perform a variety of tasks, such as the stabilization of the amplitude of a sine-wave oscillator and the shaping of a triangular waveform into a sinusoid. Although we have been able to devise simplified methods for the analysis and design of these circuits, a complete pencil-and-paper analysis is nearly impossible. The designer must therefore rely on computer simulation to obtain greater insight into detailed circuit operation, to experiment with different component values, and to optimize the design. In this section, we present two examples that illustrate the use of SPICE in the simulation of oscillator circuits.

EXAMPLE 13.1

WIEN-BRIDGE OSCILLATOR

For our first example, we shall simulate the operation of the Wien-bridge oscillator whose Capture schematic is shown in Fig. 13.42. The component values are selected to yield oscillations at 1 kHz. We would like to investigate the operation of the circuit for different settings of R_{1a} and R_{1b}, with $R_{1a} + R_{1b} = 50$ kΩ. Since oscillation just starts when $(R_2 + R_{1b})/R_{1a} = 2$ (see Exercise 13.4), that is, when $R_{1a} = 20$ kΩ and $R_{1b} = 30$ kΩ, we consider three possible settings: (a) $R_{1a} = 15$ kΩ, $R_{1b} = 35$ kΩ; (b) $R_{1a} = 18$ kΩ, $R_{1b} = 32$ kΩ; and (c) $R_{1a} = 25$ kΩ, $R_{1b} = 25$ kΩ. These settings correspond to loop gains of 1.33, 1.1, and 0.8, respectively.

In PSpice, a 741-type op amp and 1N4148-type diodes are used to simulate the circuit in Fig. 13.42.[6] A transient-analysis simulation is performed with the capacitor voltages initially set to zero. This demonstrates that the op-amp offset voltage is sufficient to cause the oscillations to start without the need for special start-up circuitry. Figure 13.43 shows the simulation results.

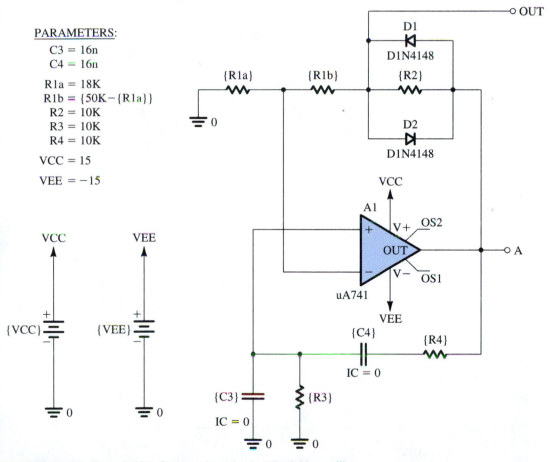

PARAMETERS:
  C3 = 16n
  C4 = 16n

  R1a = 18K
  R1b = {50K−{R1a}}
  R2 = 10K
  R3 = 10K
  R4 = 10K

  VCC = 15

  VEE = −15

FIGURE 13.42 Example 13.1: Capture schematic of a Wien-bridge oscillator.

[6] The SPICE models for the 741 op amp and the 1N4148 diode are available in PSpice. The 741 op amp was characterized in Example 2.9. The 1N4148 diode was used in Example 3.10.

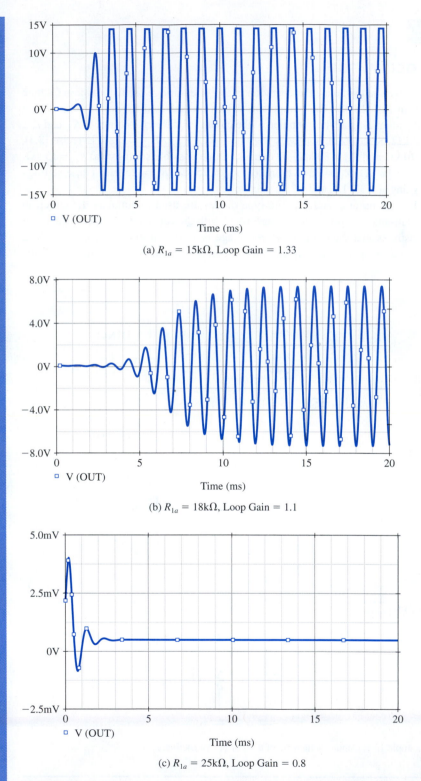

(a) $R_{1a} = 15\text{k}\Omega$, Loop Gain = 1.33

(b) $R_{1a} = 18\text{k}\Omega$, Loop Gain = 1.1

(c) $R_{1a} = 25\text{k}\Omega$, Loop Gain = 0.8

FIGURE 13.43 Start-up transient behavior of the Wien-bridge oscillator shown in Fig. 13.42 for various values of loop gain.

The graph in Fig. 13.43(a) shows the output waveform obtained for a loop gain of 1.33. Observe that although the oscillations grow and stabilize rapidly, the distortion is considerable. The output obtained for a loop gain of 1.1, shown in Fig. 13.43(b), is much less distorted. However, as expected, as the loop gain is reduced toward unity, it takes longer for the oscillations to build up and for the amplitude to stabilize. For this case, the frequency is 986.6 Hz, which is reasonably close to the design value of 1 kHz, and the amplitude is 7.37 V. Finally, for a loop gain of 0.8, the output shown in Fig. 13.43(c) confirms our expectation that sustained oscillations cannot be obtained when the loop gain is less than unity.

PSpice can be used to investigate the spectral purity of the output sine wave. This is achieved using the Fourier analysis facility. It is found that in the steady state the output for the case of a loop gain of 1.1 has a THD figure of 1.88%. When the oscillator output is taken at the op-amp output (voltage v_A), a THD of 2.57% is obtained, which as expected is higher than that for the voltage v_{OUT}, but not by very much. The output terminal of the op amp is of course a much more convenient place to take the output.

EXAMPLE 13.2

ACTIVE-FILTER-TUNED OSCILLATOR

In this example, we use PSpice to verify our contention that a superior op amp–oscillator can be realized using the active-filter-tuned circuit of Fig. 13.11. We also investigate the effect of changing the value of the filter Q factor on the spectral purity of the output sine wave.

Consider the circuit whose Capture schematic is shown in Fig. 13.44. For this circuit, the center frequency is 1 kHz, and the filter Q is 5 when $R_1 = 50$ kΩ and 20 when $R_1 = 200$ kΩ. As in the case of the Wien-bridge circuit in Example 13.1, 741-type op amps and 1N4148-type diodes are utilized. In PSpice a transient-analysis simulation is performed with the capacitor voltages initially set to zero. To be able to compute the Fourier components of the output, the analysis interval chosen must be long enough to allow the oscillator to reach a steady state. The time to reach a steady state is in turn determined by the value of the filter Q; the higher the Q, the longer it takes the output to settle. For $Q = 5$, it was determined, through a combination of approximate calculations and experimentation using PSpice, that 50 ms is a reasonable estimate for the analysis interval. For plotting purposes, we use 200 points per period of oscillation.

The results of the transient analysis are plotted in Fig. 13.45. The upper graph shows the sinusoidal waveform at the output of op amp A_1 (voltage v_1). The lower graph shows the waveform across the diode limiter (voltage v_2). The frequency of oscillation is found to be very close to the design value of 1 kHz. The amplitude of the sine wave is determined using Probe (the graphical interface of PSpice) to be 1.15 V (or 2.3 V p-p). Note that this is lower than the 3.6 V estimated in Exercise 13.7. The latter value, however, was based on an estimate of 0.7-V drop across each conducting diode in the limiter. The lower waveform in Fig. 13.45 indicates that the diode drop is closer to 0.5 V, for a 1-V peak-to-peak amplitude of the pseudo-square wave. We should therefore expect the peak-to-peak amplitude of the output sinusoid to be lower than 3.6 V by the same factor, and indeed it is approximately the case.

In PSpice, the Fourier analysis of the output sine wave indicates that THD = 1.61%. Repeating the simulation with Q increased to 20 (by increasing R_1 to 200 kΩ), we find that the value of THD is reduced to 1.01%. Thus, our expectations that the value of the filter Q can be used as an effective means for constrolling the THD of the output waveform are confirmed.

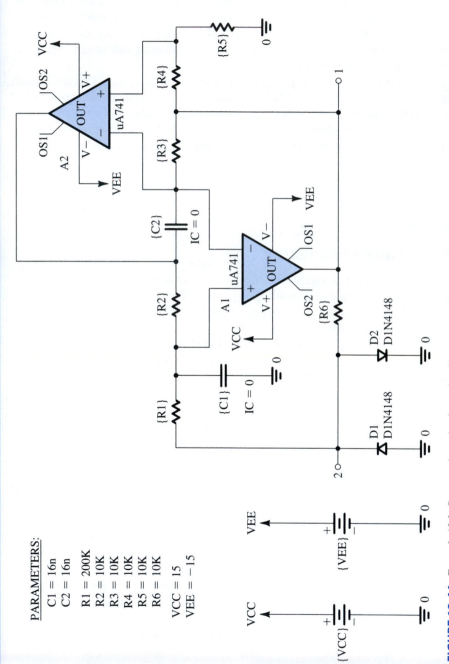

PARAMETERS:

C1 = 16n
C2 = 16n

R1 = 200K
R2 = 10K
R3 = 10K
R4 = 10K
R5 = 10K
R6 = 10K

VCC = 15
VEE = −15

FIGURE 13.44 Example 13.2: Capture schematic of an active-filter-tuned oscillator for which the Q of the filter is adjustable by changing R_1.

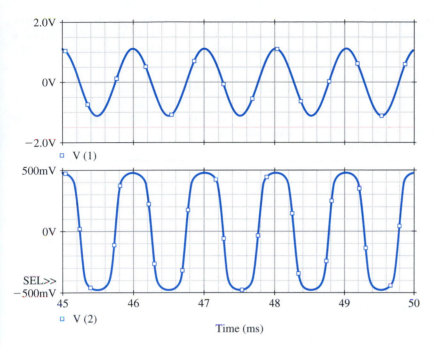

FIGURE 13.45 Output waveforms of the active-filter-tuned oscillator shown in Fig. 13.44 for $Q = 5$ ($R_1 = 50$ kΩ).

SUMMARY

- There are two distinctly different types of signal generator: linear oscillators, which utilize some form of resonance, and nonlinear oscillators or function generators, which employ a switching mechanism implemented with a multivibrator circuit.

- A linear oscillator can be realized by placing a frequency-selective network in the feedback path of an amplifier (an op amp or a transistor). The circuit will oscillate at the frequency at which the total phase shift around the loop is zero, provided that the magnitude of loop gain at this frequency is equal to, or greater than, unity.

- If in an oscillator the magnitude of loop gain is greater than unity, the amplitude will increase until a nonlinear amplitude-control mechanism is activated.

- The Wien-bridge oscillator, the phase-shift oscillator, the quadrature oscillator, and the active-filter-tuned oscillator are popular configurations for frequencies up to about 1 MHz. These circuits employ RC networks together with op amps or transistors. For higher frequencies, LC-tuned or crystal-tuned oscillators are utilized. A popular configuration is the Colpitts circuit.

- Crystal oscillators provide the highest possible frequency accuracy and stability.

- There are three types of multivibrator: bistable, monostable, and astable. Op-amp circuit implementations of multivibrators are useful in analog-circuit applications that require high precision. Implementations using digital logic gates are studied in Chapter 11.

- The bistable multivibrator has two stable states and can remain in either state indefinitely. It changes state when triggered. A comparator with hysteresis is bistable.

- A monostable multivibrator, also known as a one-shot multivibrator, has one stable state, in which it can remain indefinitely. When triggered, it goes into a quasistable state in which it remains for a predetermined interval, thus generating, at its output, a pulse of known width.

- An astable multivibrator has no stable state. It oscillates between two quasi-stable states, remaining in each for a predetermined interval. It thus generates a periodic waveform at the output.

■ A feedback loop consisting of an integrator and a bistable multivibrator can be used to generate triangular and square waveforms.

■ The 555 timer, a commercially available IC, can be used with external resistors and a capacitor to implement high-quality monostable and astable multivibrators.

■ A sine waveform can be generated by feeding a triangular waveform to a sine-wave shaper. A sine-wave shaper

can be implemented either by using diodes (or transistors) and resistors, or by using an amplifier having a nonlinear transfer characteristic that approximates the sine function.

■ Diodes can be combined with op amps to implement precision rectifier circuits in which negative feedback serves to mask the nonidealities of the diode characteristics.

PROBLEMS

SECTION 13.1: BASIC PRINCIPLES OF SINUSOIDAL OSCILLATORS

*13.1 Consider a sinusoidal oscillator consisting of an amplifier having a frequency-independent gain A (where A is positive) and a second-order bandpass filter with a pole frequency ω_0, a pole Q denoted Q, and a center-frequency gain K.

(a) Find the frequency of oscillation, and the condition that A and K must satisfy for sustained oscillation.
(b) Derive an expression for $d\phi/d\omega$, evaluated at $\omega = \omega_0$.
(c) Use the result of (b) to find an expression for the per-unit change in frequency of oscillation resulting from a phase-angle change of $\Delta\phi$, in the amplifier transfer function.

$$Hint: \frac{d}{dx}(\tan^{-1}y) = \frac{1}{1+y^2}\frac{dy}{dx}$$

13.2 For the oscillator described in Problem 13.1, show that, independent of the value of A and K, the poles of the circuit lie at a radial distance of ω_0. Find the value of AK that results in poles appearing (a) on the $j\omega$ axis, and (b) in the right-half of the s plane, at a horizontal distance from the $j\omega$ axis of $\omega_0/(2Q)$.

D13.3 Sketch a circuit for a sinusoidal oscillator formed by an op amp connected in the noninverting configuration and a bandpass filter implemented by an RLC resonator (such as that in Fig. 12.18d). What should the amplifier gain be to obtain sustained oscillation? What is the frequency of oscillation? Find the percentage change in ω_0, resulting from a change of +1% in the value of (a) L, (b) C, and (c) R.

13.4 An oscillator is formed by loading a transconductance amplifier having a positive gain with a parallel RLC circuit and connecting the output directly to the input (thus applying positive feedback with a factor $\beta = 1$). Let the transconductance amplifier have an input resistance of 10 kΩ and an output resistance of 10 kΩ. The LC resonator has $L = 10$ μH, $C = 1000$ pF, and $Q = 100$. For what value of

transconductance G_m will the circuit oscillate? At what frequency?

13.5 In a particular oscillator characterized by the structure of Fig. 13.1, the frequency-selective network exhibits a loss of 20 dB and a phase shift of 180° at ω_0. What is the minimum gain and the phase shift that the amplifier must have, for oscillation to begin?

D13.6 Consider the circuit of Fig. 13.3(a) with R_f removed to realize the comparator function. Find suitable values for all resistors so that the comparator output levels are ±6 V and the slope of the limiting characteristic is 0.1. Use power supply voltages of ±10 V and assume the voltage drop of a conducting diode to be 0.7 V.

D13.7 Consider the circuit of Fig. 13.3(a) with R_f removed to realize the comparator function. Sketch the transfer characteristic. Show that by connecting a dc source V_B to the virtual ground of the op amp through a resistor R_B, the transfer characteristic is shifted along the v_I axis to the point $v_I = -(R_1/R_B)V_B$. Utilizing available ±15-V dc supplies for ±V and for V_B, find suitable component values so that the limiting levels are ±5 V and the comparator threshold is at $v_I = +5$ V. Neglect the diode voltage drop (i.e., assume that $V_D = 0$). The input resistance of the comparator is to be 100 kΩ, and the slope in the limiting regions is to be ≤0.05 V/V. Use standard 5% resistors (see Appendix G).

13.8 Denoting the zener voltages of Z_1 and Z_2 by V_{Z1} and V_{Z2} and assuming that in the forward direction the voltage drop is approximately 0.7 V, sketch and clearly label the transfer characteristics v_O–v_I of the circuits in Fig. P13.8. Assume the op amps to be ideal.

SECTION 13.2: OP-AMP–RC OSCILLATOR CIRCUITS

13.9 For the Wien-bridge oscillator circuit in Fig. 13.4, show that the transfer function of the feedback network

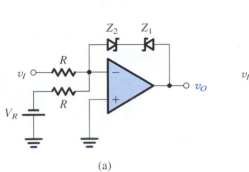

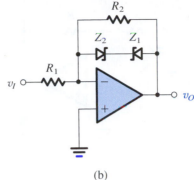

FIGURE P13.8

$[V_a(s)/V_o(s)]$ is that of a bandpass filter. Find ω_0 and Q of the poles, and find the center-frequency gain.

13.10 For the Wien-bridge oscillator of Fig. 13.4, let the closed-loop amplifier (formed by the op amp and the resistors R_1 and R_2) exhibit a phase shift of -0.1 rad in the neighborhood of $\omega = 1/CR$. Find the frequency at which oscillations can occur in this case, in terms of CR. (*Hint:* Use Eq. 13.11.)

13.11 For the Wien-bridge oscillator of Fig. 13.4, use the expression for loop gain in Eq. (13.10) to find the poles of the closed-loop system. Give the expression for the pole Q, and use it to show that to locate the poles in the right half of the s plane, R_2/R_1 must be selected to be greater than 2.

D*13.12 Reconsider Exercise 13.3 with R_3 and R_6 increased to reduce the output voltage. What values are required for a peak-to-peak output of 10 V? What results if R_3 and R_6 are open-circuited?

13.13 For the circuit in Fig. P13.13 find $L(s)$, $L(j\omega)$, the frequency for zero loop phase, and R_2/R_1 for oscillation.

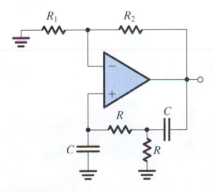

FIGURE P13.13

13.14 Repeat Problem 13.13 for the circuit in Fig. P13.14.

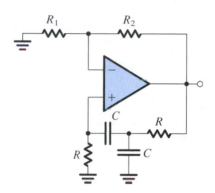

FIGURE P13.14

*13.15 Consider the circuit of Fig. 13.6 with the 50-kΩ potentiometer replaced by two fixed resistors: 10 kΩ between the op amp's negative input and ground, and 18 kΩ. Modeling each diode as a 0.65-V battery in series with a 100-Ω resistance, find the peak-to-peak amplitude of the output sinusoid.

D13.16** Redesign the circuit of Fig. 13.6 for operation at 10 kHz using the same values of resistance. If at 10 kHz the op amp provides an excess phase shift (lag) of 5.7°, what will be the frequency of oscillation? (Assume that the phase shift introduced by the op amp remains constant for frequencies around 10 kHz.) To restore operation to 10 kHz, what change must be made in the shunt resistor of the Wien bridge? Also, to what value must R_2/R_1 be changed?

*13.17 For the circuit of Fig. 13.8, connect an additional $R = 10$ kΩ resistor in series with the rightmost capacitor C. For this modification (and ignoring the amplitude stabilization circuitry) find the loop gain $A\beta$ by breaking the circuit at node X. Find R_f for oscillation to begin, and find f_0.

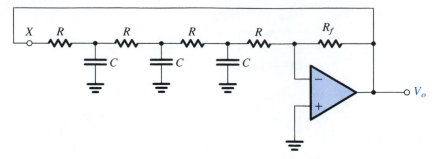

FIGURE P13.18

D13.18 For the circuit in Fig. P13.18, break the loop at node X and find the loop gain (working backward for simplicity to find V_x in terms of V_o). For $R = 10\ \text{k}\Omega$, find C and R_f to obtain sinusoidal oscillations at 10 kHz.

*13.19 Consider the quadrature-oscillator circuit of Fig. 13.9 without the limiter. Let the resistance R_f be equal to $2R/(1+\Delta)$, where $\Delta \ll 1$. Show that the poles of the characteristic equation are in the right-half s plane and given by $s \simeq (1/CR)[(\Delta/4) \pm j]$.

*13.20 Assuming that the diode-clipped waveform in Exercise 13.7 is nearly an ideal square wave and that the resonator Q is 20, provide an estimate of the distortion in the output sine wave by calculating the magnitude (relative to the fundamental) of

(a) the second harmonic
(b) the third harmonic
(c) the fifth harmonic
(d) the rms of harmonics to the tenth

Note that a square wave of amplitude V and frequency ω is represented by the series

$$\frac{4V}{\pi}\left(\cos \omega t - \frac{1}{3}\cos 3\omega t + \frac{1}{5}\cos 5\omega t - \frac{1}{7}\cos 7\omega t + \cdots\right)$$

SECTION 13.3: LC AND CRYSTAL OSCILLATORS

**13.21 Figure P13.21 shows four oscillator circuits of the Colpitts type, complete with bias detail. For each circuit, derive an equation governing circuit operation, and find the frequency of oscillation and the gain condition that ensures that oscillations start.

**13.22 Consider the oscillator circuit in Fig. P13.22, and assume for simplicity that $\beta = \infty$.

(a) Find the frequency of oscillation and the minimum value of R_C (in terms of the bias current I) for oscillation to start.

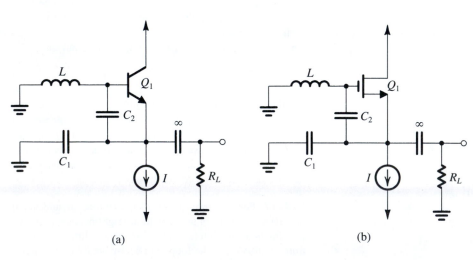

(a)   (b)

FIGURE P13.21

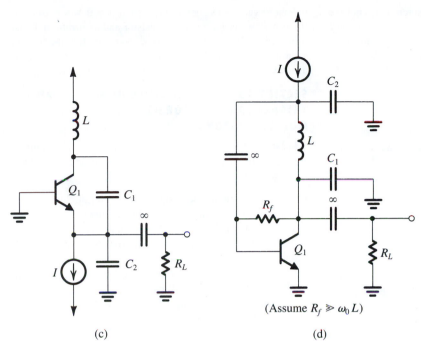

(c)

(Assume $R_f \gg \omega_0 L$)

(d)

FIGURE P13.21 (*Continued*)

(b) If R_C is selected equal to $(1/I)$ kΩ, where I is in milliamperes, convince yourself that oscillations will start. If oscillations grow to the point that V_o is large enough to turn the BJTs on and off, show that the voltage at the collector of Q_2 will be a square wave of 1 V peak to peak. Estimate the peak-to-peak amplitude of the output sine wave V_o.

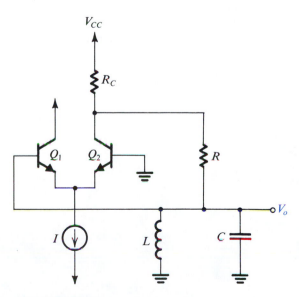

FIGURE P13.22

13.23 Consider the Pierce crystal oscillator of Fig. 13.16 with the crystal as specified in Exercise 13.10. Let C_1 be variable in the range 1 pF to 10 pF, and let C_2 be fixed at 10 pF. Find the range over which the oscillation frequency can be tuned. (*Hint:* Use the result in the statement leading to the expression in Eq. 13.27.)

SECTION 13.4: BISTABLE MULTIVIBRATORS

13.24 Consider the bistable circuit of Fig. 13.19(a) with the op amp's positive-input terminal connected to a positive-voltage source V through a resistor R_3.

(a) Derive expressions for the threshold voltages V_{TL} and V_{TH} in terms of the op amp's saturation levels L_+ and L_-, R_1, R_2, R_3, and V.
(b) Let $L_+ = -L_- = 13$ V, $V = 15$ V, and $R_1 = 10$ kΩ. Find the values of R_2 and R_3 that result in $V_{TL} = +4.9$ V and $V_{TH} = +5.1$ V.

13.25 Consider the bistable circuit of Fig. 13.20(a) with the op amp's negative-input terminal disconnected from ground and connected to a reference voltage V_R.

(a) Derive expressions for the threshold voltages V_{TL} and V_{TH} in terms of the op amp's saturation levels L_+ and L_-, R_1, R_2, and V_R.
(b) Let $L_+ = -L_- = V$ and $R_1 = 10$ kΩ. Find R_2 and V_R that result in threshold voltages of 0 and $V/10$.

13

13.26 For the circuit in Fig. P13.26, sketch and label the transfer characteristic v_O–v_I. The diodes are assumed to have a constant 0.7-V drop when conducting, and the op amp saturates at ±12 V. What is the maximum diode current?

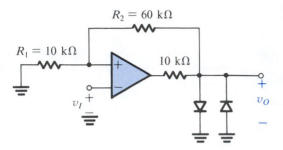

FIGURE P13.26

13.27 Consider the circuit of Fig. P13.26 with R_1 eliminated and R_2 short-circuited. Sketch and label the transfer characteristic v_O–v_I. Assume that the diodes have a constant 0.7-V drop when conducting and that the op amp saturates at ±12 V.

*13.28 Consider a bistable circuit having a noninverting transfer characteristic with $L_+ = -L_- = 12$ V, $V_{TL} = -1$ V, and $V_{TH} = +1$ V.

(a) For a 0.5-V-amplitude sine-wave input having zero average, what is the output?
(b) Describe the output if a sinusoid of frequency f and amplitude of 1.1 V is applied at the input. By how much can the average of this sinusoidal input shift before the output becomes a constant value?

D13.29 Design the circuit of Fig. 13.23(a) to realize a transfer characteristic with ±7.5-V output levels and ±7.5-V threshold values. Design so that when $v_I = 0$ V a current of 0.1 mA flows in the feedback resistor and a current of 1 mA flows through the zener diodes. Assume that the output saturation levels of the op amp are ±12 V. Specify the voltages of the zener diodes and give the values of all resistors.

SECTION 13.5: GENERATION OF SQUARE AND TRIANGULAR WAVEFORMS USING ASTABLE MULTIVIBRATORS

13.30 Find the frequency of oscillation of the circuit in Fig. 13.24(b) for the case $R_1 = 10$ kΩ, $R_2 = 16$ kΩ, $C = 10$ nF, and $R = 62$ kΩ.

D13.31 Augment the astable multivibrator circuit of Fig. 13.24(b) with an output limiter of the type shown in Fig. 13.23(b). Design the circuit to obtain an output square wave with 5-V amplitude and 1-kHz frequency using a 10-nF capacitor C. Use $\beta = 0.462$, and design for a current in the resistive divider approximately equal to the average current in the RC network over a half-cycle. Assuming ±13-V op-amp saturation voltages, arrange for the zener to operate at a current of 1 mA.

D13.32 Using the scheme of Fig. 13.25, design a circuit that provides square waves of 10 V peak to peak and triangular waves of 10 V peak to peak. The frequency is to be 1 kHz. Implement the bistable circuit with the circuit of Fig. 13.23(b). Use a 0.01-μF capacitor, and specify the values of all resistors and the required zener voltage. Design for a minimum zener current of 1 mA and for a maximum current in the resistive divider of 0.2 mA. Assume that the output saturation levels of the op amps are ±13 V.

D*13.33 The circuit of Fig. P13.33 consists of an inverting bistable multivibrator with an output limiter and a noninverting integrator. Using equal values for all resistors except R_7 and a 0.5-nF capacitor, design the circuit to obtain a square

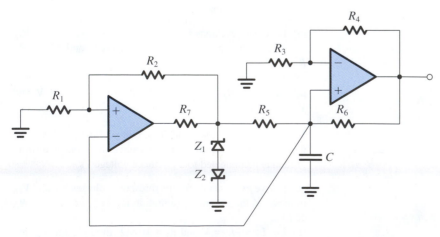

FIGURE P13.33

wave at the output of the bistable multivibrator of 15-V peak-to-peak amplitude and 10-kHz frequency. Sketch and label the waveform at the integrator output. Assuming ±13-V op-amp saturation levels, design for a minimum zener current of 1 mA. Specify the zener voltage required, and give the values of all resistors.

SECTION 13.6: GENERATION OF A STANDARDIZED PULSE—THE MONOSTABLE MULTIVIBRATOR

*13.34 Figure P13.34 shows a monostable multivibrator circuit. In the stable state, $v_O = L_+$, $v_A = 0$, and $v_B = -V_{ref}$. The circuit can be triggered by applying a positive input pulse of height greater than V_{ref}. For normal operation, $C_1R_1 \ll CR$. Show the resulting waveforms of v_O and v_A. Also, show that the pulse generated at the output will have a width T given by

$$T = CR \ln\left(\frac{L_+ - L_-}{V_{ref}}\right)$$

Note that this circuit has the interesting property that the pulse width can be controlled by changing V_{ref}.

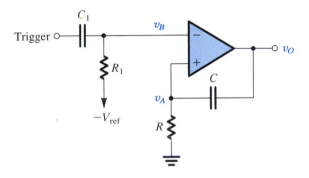

FIGURE P13.34

13.35 For the monostable circuit considered in Exercise 13.19, calculate the recovery time.

D*13.36 Using the circuit of Fig. 13.26, with a nearly ideal op amp for which the saturation levels are ±13 V, design a monostable multivibrator to provide a negative output pulse of 100-μs duration. Use capacitors of 0.1 nF and 1 nF. Wherever possible, choose resistors of 100 kΩ in your design. Diodes have a drop of 0.7 V. What is the minimum input step size that will ensure triggering? How long does the circuit take to recover to a state in which retriggering is possible with a normal output?

SECTION 13.7: INTEGRATED-CIRCUIT TIMERS

13.37 Consider the 555 circuit of Fig. 13.27 when the Threshold and the Trigger input terminals are joined together

and connected to an input voltage v_I. Verify that the transfer characteristic v_O–v_I is that of an inverting bistable circuit with thresholds $V_{TL} = \frac{1}{3}V_{CC}$ and $V_{TH} = \frac{2}{3}V_{CC}$ and output levels of 0 and V_{CC}.

13.38 (a) Using a 1-nF capacitor C in the circuit of Fig. 13.28(a), find the value of R that results in an output pulse of 10-μs duration.
(b) If the 555 timer used in (a) is powered with $V_{CC} = 15$ V, and assuming that V_{TH} can be varied externally (i.e., it need not remain equal to $\frac{2}{3}V_{CC}$), find its required value so that the pulse width is increased to 20 μs, with other conditions the same as in (a).

D13.39 Using a 680-pF capacitor, design the astable circuit of Fig. 13.29(a) to obtain a square wave with a 50-kHz frequency and a 75% duty cycle. Specify the values of R_A and R_B.

*13.40 The node in the 555 timer at which the voltage is V_{TH} (i.e., the inverting input terminal of comparator 1) is usually connected to an external terminal. This allows the user to change V_{TH} externally (i.e., V_{TH} no longer remains at $\frac{2}{3}V_{CC}$). Note, however, that whatever the value of V_{TH} becomes, V_{TL} always remains $\frac{1}{2}V_{TH}$.

(a) For the astable circuit of Fig. 13.29, rederive the expressions for T_H and T_L, expressing them in terms of V_{TH} and V_{TL}.
(b) For the case $C = 1$ nF, $R_A = 7.2$ kΩ, $R_B = 3.6$ kΩ, and $V_{CC} = 5$ V, find the frequency of oscillation and the duty cycle of the resulting square wave when no external voltage is applied to the terminal V_{TH}.
(c) For the design in (b), let a sine-wave signal of a much lower frequency than that found in (b) and of 1-V peak amplitude be capacitively coupled to the circuit node V_{TH}. This signal will cause V_{TH} to change around its quiescent value of $\frac{2}{3}V_{CC}$, and thus T_H will change correspondingly—a modulation process. Find T_H, and find the frequency of oscillation and the duty cycle at the two extreme values of V_{TH}.

SECTION 13.8: NONLINEAR WAVEFORM-SHAPING CIRCUITS

D*13.41 The two-diode circuit shown in Fig. P13.41 can provide a crude approximation to a sine-wave output when driven by a triangular waveform. To obtain a good approximation, we select the peak of the triangular waveform, V, so that the slope of the desired sine wave at the zero crossings is equal to that of the triangular wave. Also, the value of R is selected so that when v_I is at its peak the output voltage is equal to the desired peak of the sine wave. If the diodes exhibit a voltage drop of 0.7 V at 1-mA current, changing at the rate of 0.1 V per decade, find the values of V and R that will yield an approximation to a sine waveform of 0.7-V peak amplitude. Then find the angles θ (where $\theta = 90°$ when v_I is at its peak) at which the output of the circuit, in volts, is 0.7, 0.65, 0.6, 0.55, 0.5, 0.4, 0.3, 0.2, 0.1, and 0. Use the angle

13

values obtained to determine the values of the exact sine wave (i.e., 0.7 sin θ), and thus find the percentage error of this circuit as a sine shaper. Provide your results in tabular form.

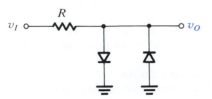

FIGURE P13.41

D13.42 Design a two-segment sine-wave shaper using a 10-kΩ input resistor, two diodes, and two clamping voltages. The circuit, fed by a 10-V peak-to-peak triangular wave, should limit the amplitude of the output signal via a 0.7-V diode to a value corresponding to that of a sine wave whose zero-crossing slope matches that of the triangle. What are the clamping voltages you have chosen?

13.43 Show that the output voltage of the circuit in Fig. P13.43 is given by

$$v_O = -n\, V_T \ln\left(\frac{v_I}{I_S R}\right) \quad v_I > 0$$

where I_S and n are the diode parameters and V_T is the thermal voltage. Since the output voltage is proportional to the logarithm of the input voltage, the circuit is known as a **logarithmic amplifier.** Such amplifiers find application in situations where it is desired to compress the signal range.

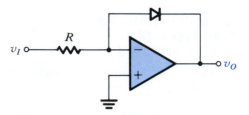

FIGURE P13.43

13.44 Verify that the circuit in Fig. P13.44 implements the transfer characteristic $v_O = v_1 v_2$ for v_1, $v_2 > 0$. Such a circuit is known as an analog multiplier. Check the circuit's performance for various combinations of input voltage of values, say, 0.5 V, 1 V, 2 V, and 3 V. Assume all diodes to be identical, with 700-mV drop at 1-mA current and $n = 2$. Note that a *squarer* can easily be produced using a single input (e.g., v_1) connected via a 0.5-kΩ resistor (rather than the 1-kΩ resistor shown).

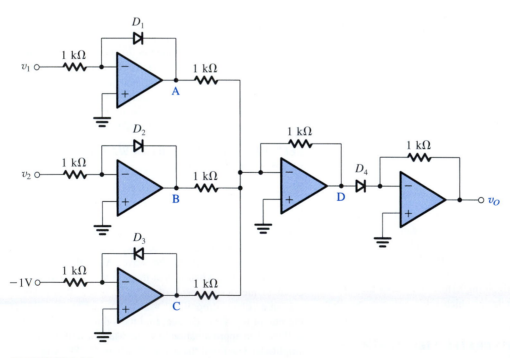

FIGURE P13.44

**13.45 Detailed analysis of the circuit in Fig. 13.32 shows that optimum performance (as a sine shaper) occurs when the values of I and R are selected so that $RI = 2.5V_T$, where V_T is the thermal voltage, and the peak amplitude of the input triangular wave is $6.6V_T$. If the output is taken across R (i.e., between the two emitters), find v_I corresponding to $v_O = 0.25V_T$, $0.5V_T$, V_T, $1.5V_T$, $2V_T$, $2.4V_T$, and $2.42V_T$. Plot v_O–v_I and compare to the ideal curve given by

$$v_O = 2.42\,V_T \sin\left(\frac{v_I}{6.6\,V_T} \times 90°\right)$$

SECTION 13.9: PRECISION RECTIFIER CIRCUITS

13.46 Two superdiode circuits connected to a common-load resistor and having the same input signal have their diodes reversed, one with cathode to the load, the other with anode to the load. For a sine-wave input of 10 V peak to peak, what is the output waveform? Note that each half-cycle of the load current is provided by a separate amplifier, and that while one amplifier supplies the load current, the other amplifier idles. This idea, called class-B operation (see Chapter 14), is important in the implementation of power amplifiers.

D13.47 The superdiode circuit of Fig. 13.33(a) can be made to have gain by connecting a resistor R_2 in place of the short circuit between the cathode of the diode and the negative-input terminal of the op amp, and a resistor R_1 between the negative-input terminal and ground. Design the circuit for a gain of 2. For a 10-V peak-to-peak input sine wave, what is the average output voltage resulting?

D13.48 Provide a design of the inverting precision rectifier shown in Fig. 13.34(a) in which the gain is −2 for negative inputs and zero otherwise, and the input resistance is 100 kΩ. What values of R_1 and R_2 do you choose?

D*13.49 Provide a design for a voltmeter circuit similar to the one in Fig. 13.35, which is intended to function at frequencies of 10 Hz and above. It should be calibrated for sine-wave input signals to provide an output of +10 V for an input of 1 V rms. The input resistance should be as high as possible. To extend the bandwidth of operation, keep the gain in the ac part of the circuit reasonably small. As well, the design should result in reduction of the size of the capacitor C required. The largest value of resistor available is 1 MΩ.

13.50 Plot the transfer characteristic of the circuit in Fig. P13.50.

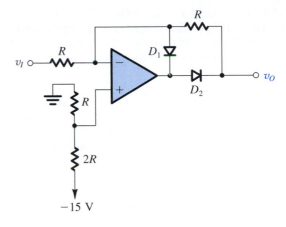

FIGURE P13.50

13.51 Plot the transfer characteristics v_{O1}–v_I and v_{O2}–v_I of the circuit in Fig. P13.51.

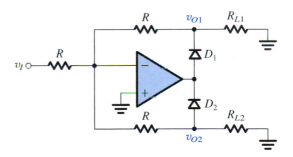

FIGURE P13.51

13.52 Sketch the transfer characteristics of the circuit in Fig. P13.52.

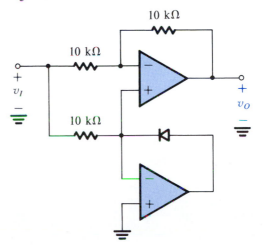

FIGURE P13.52

13

D13.53 A circuit related to that in Fig. 13.38 is to be used to provide a current proportional to v_A ($v_A \geq 0$) to a light-emitting diode (LED). The value of the current is to be independent of the diode's nonlinearities and variability. Indicate how this may be done easily.

*13.54 In the precision rectifier of Fig. 13.38, the resistor R is replaced by a capacitor C. What happens? For equivalent performance with a sine-wave input of 60-Hz frequency with $R = 1$ kΩ, what value of C should be used? What is the response of the modified circuit at 120 Hz? At 180 Hz? If the amplitude of v_A is kept fixed, what new function does this circuit perform? Now consider the effect of a waveform change on both circuits (the one with R and the one with C). For a triangular-wave input of 60-Hz frequency that produces an average meter current of 1 mA in the circuit with R, what does the average meter current become when R is replaced with the C whose value was just calculated?

*13.55 A positive-peak rectifier utilizing a fast op amp and a junction diode in a superdiode configuration, and a 10-μF capacitor initially uncharged, is driven by a series of 10-V pulses of 10-μs duration. If the maximum output current that the op amp can supply is 10 mA, what is the voltage on the capacitor following one pulse? Two pulses? Ten pulses? How many pulses are required to reach 0.5 V? 1.0 V? 2.0 V?

D13.56 Consider the buffered precision peak rectifier shown in Fig. 13.40 when connected to a triangular input of 1-V peak-to-peak amplitude and 1000-Hz frequency. It utilizes an op amp whose bias current (directed into A_2) is 10 nA and diodes whose reverse leakage current is 1 nA. What is the smallest capacitor that can be used to guarantee an output ripple less than 1%?

13

CHAPTER 14

Output Stages and Power Amplifiers

INTRODUCTION

An important function of the output stage is to provide the amplifier with a low output resistance so that it can deliver the output signal to the load without loss of gain. Since the output stage is the final stage of the amplifier, it usually deals with relatively large signals. Thus the small-signal approximations and models either are not applicable or must be used with care. Nevertheless, linearity remains a very important requirement. In fact, a measure of goodness of the design of the output stage is the **total harmonic distortion** (THD) it introduces. This is the rms value of the harmonic components of the output signal, excluding the fundamental, expressed as a percentage of the rms of the fundamental. A high-fidelity audio power amplifier features a THD of the order of a fraction of a percent.

The most challenging requirement in the design of the output stage is that it deliver the required amount of power to the load in an *efficient* manner. This implies that the power *dissipated* in the output-stage transistors must be as low as possible. This requirement stems mainly from the fact that the power dissipated in a transistor raises its internal **junction temperature,** and there is a maximum temperature (in the range of 150°C to 200°C for silicon devices) above which the transistor is destroyed. A high power-conversion efficiency also may be required to prolong the life of batteries employed in battery-powered circuits, to permit a smaller, lower-cost power supply, or to obviate the need for cooling fans.

We begin this chapter with a study of the various output-stage configurations employed in amplifiers that handle both low and high power. In this context, "high power" generally means greater than 1 W. We then consider the specific requirements of BJTs employed in the design of high-power output stages, called **power transistors.** Special attention will be paid to the thermal properties of such transistors.

A power amplifier is simply an amplifier with a high-power output stage. Examples of discrete- and integrated-circuit power amplifiers will be presented. Also included is a brief discussion of MOSFET structures that are currently finding application in power-circuit design. The chapter concludes with an example illustrating the use of SPICE simulation in the analysis and design of output stages.

 ## 14.1 CLASSIFICATION OF OUTPUT STAGES

Output stages are classified according to the collector current waveform that results when an input signal is applied. Figure 14.1 illustrates the classification for the case of a sinusoidal input signal. The class A stage, whose associated waveform is shown in Fig. 14.1(a), is biased at a current I_C greater than the amplitude of the signal current, $\hat{I}_c$. Thus the transistor in a class A stage conducts for the entire cycle of the input signal; that is, the conduction angle is 360°. In contrast, the class B stage, whose associated waveform is shown in Fig. 14.1(b), is biased at zero dc current. Thus a transistor in a class B stage conducts for only half the cycle of the input sine wave, resulting in a conduction angle of 180°. As will be seen later, the negative halves of the sinusoid will be supplied by another transistor that also operates in the class B mode and conducts during the alternate half-cycles.

An intermediate class between A and B, appropriately named class AB, involves biasing the transistor at a nonzero dc current much smaller than the peak current of the sine-wave signal. As a result, the transistor conducts for an interval slightly greater than half a cycle, as illustrated in Fig. 14.1(c). The resulting conduction angle is greater than 180° but much less than 360°. The class AB stage has another transistor that conducts for an interval slightly greater than that of the negative half-cycle, and the currents from the two transistors are combined in the load. It follows that, during the intervals near the zero crossings of the input sinusoid, both transistors conduct.

Figure 14.1(d) shows the collector-current waveform for a transistor operated as a class C amplifier. Observe that the transistor conducts for an interval shorter than that of a half-cycle; that is, the conduction angle is less than 180°. The result is the periodically pulsating current waveform shown. To obtain a sinusoidal output voltage, this current is passed through a parallel LC circuit, tuned to the frequency of the input sinusoid. The tuned circuit acts as a bandpass filter and provides an output voltage proportional to the amplitude of the fundamental component in the Fourier-series representation of the current waveform.

Class A, AB, and B amplifiers are studied in this chapter. They are employed as output stages of op amps and audio power amplifiers. In the latter application, class AB is the preferred choice, for reasons that will be explained in the following sections. Class C amplifiers are usually employed for radio-frequency (RF) power amplification (required, e.g., in mobile phones and radio and TV transmitters). The design of class C amplifiers is a rather specialized topic and is not included in this book.

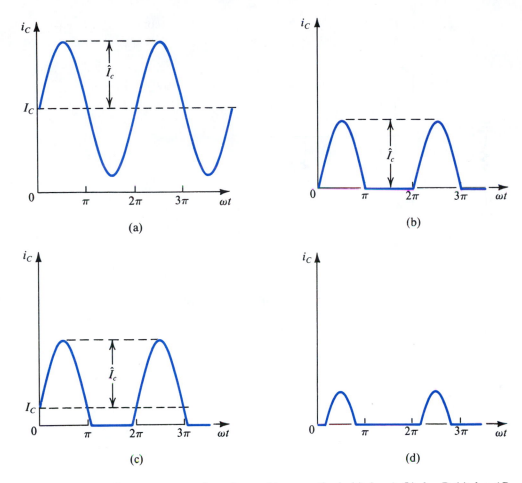

FIGURE 14.1 Collector current waveforms for transistors operating in (**a**) class A, (**b**) class B, (**c**) class AB, and (**d**) class C amplifier stages.

Although the BJT has been used to illustrate the definition of the various output-stage classes, the same classification applies to output stages implemented with MOSFETs. Furthermore, the classification above extends to amplifier stages other than those used at the output. In this regard, all the common-emitter, common-base, and common-collector amplifiers (and their FET counterparts) studied in earlier chapters fall into the class A category.

 ## 14.2 CLASS A OUTPUT STAGE

Because of its low output resistance, the emitter follower is the most popular class A output stage. We have already studied the emitter follower in Chapters 5 and 6; in the following we consider its large-signal operation.

14.2.1 Transfer Characteristic

Figure 14.2 shows an emitter follower Q_1 biased with a constant current I supplied by transistor Q_2. Since the emitter current $i_{E1} = I + i_L$, the bias current I must be greater than the

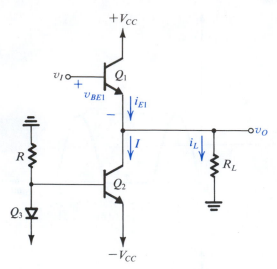

FIGURE 14.2 An emitter follower (Q_1) biased with a constant current I supplied by transistor Q_2.

largest negative load current; otherwise, Q_1 cuts off and class A operation will no longer be maintained.

The transfer characteristic of the emitter follower of Fig. 14.2 is described by

$$v_O = v_I - v_{BE1} \tag{14.1}$$

where v_{BE1} depends on the emitter current i_{E1} and thus on the load current i_L. If we neglect the relatively small changes in v_{BE1} (60 mV for every factor-of-10 change in emitter current), the linear transfer curve shown in Fig. 14.3 results. As indicated, the positive limit of the linear region is determined by the saturation of Q_1; thus

$$v_{Omax} = V_{CC} - V_{CE1sat} \tag{14.2}$$

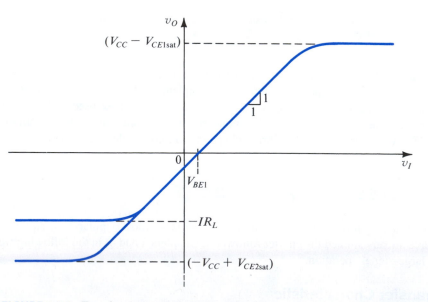

FIGURE 14.3 Transfer characteristic of the emitter follower in Fig. 14.2. This linear characteristic is obtained by neglecting the change in v_{BE1} with i_L. The maximum positive output is determined by the saturation of Q_1. In the negative direction, the limit of the linear region is determined either by Q_1 turning off or by Q_2 saturating, depending on the values of I and R_L.

In the negative direction, depending on the values of I and R_L, the limit of the linear region is determined either by Q_1 turning off,

$$v_{O\min} = -IR_L \tag{14.3}$$

or by Q_2 saturating,

$$v_{O\min} = -V_{CC} + V_{CE2\text{sat}} \tag{14.4}$$

The absolutely lowest output voltage is that given by Eq. (14.4) and is achieved provided the bias current I is greater than the magnitude of the corresponding load current,

$$I \geq \frac{|-V_{CC} + V_{CE2\text{sat}}|}{R_L} \tag{14.5}$$

14.2.2 Signal Waveforms

Consider the operation of the emitter-follower circuit of Fig. 14.2 for sine-wave input. Neglecting $V_{CE\text{sat}}$, we see that if the bias current I is properly selected, the output voltage can swing from $-V_{CC}$ to $+V_{CC}$ with the quiescent value being zero, as shown in Fig. 14.4(a). Figure 14.4(b) shows the corresponding waveform of $v_{CE1} = V_{CC} - v_O$. Now, assuming that the bias current I is selected to allow a maximum negative load current of V_{CC}/R_L, the collector current of Q_1 will have the waveform shown in Fig. 14.4(c). Finally, Fig. 14.4(d) shows the waveform of the **instantaneous power dissipation** in Q_1,

$$p_{D1} \equiv v_{CE1}i_{C1} \tag{14.6}$$

14.2.3 Power Dissipation

Figure 14.4(d) indicates that the maximum instantaneous power dissipation in Q_1 is $V_{CC}I$. This is equal to the quiescent power dissipation in Q_1. Thus the emitter-follower transistor dissipates the largest amount of power when $v_O = 0$. Since this condition (no input signal) can easily prevail for prolonged periods of time, transistor Q_1 must be able to withstand a continuous power dissipation of $V_{CC}I$.

The power dissipation in Q_1 depends on the value of R_L. Consider the extreme case of an output open circuit, that is, $R_L = \infty$. In this case, $i_{C1} = I$ is constant and the instantaneous power dissipation in Q_1 will depend on the instantaneous value of v_O. The maximum power dissipation will occur when $v_O = -V_{CC}$, for in this case v_{CE1} is a maximum of $2V_{CC}$ and $p_{D1} = 2V_{CC}I$. This condition, however, would not normally persist for a prolonged interval, so the

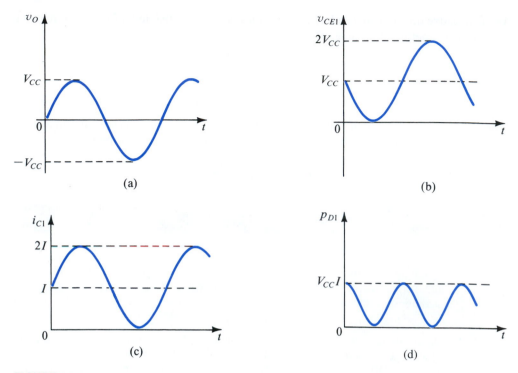

FIGURE 14.4 Maximum signal waveforms in the class A output stage of Fig. 14.2 under the condition $I = V_{CC}/R_L$ or, equivalently, $R_L = V_{CC}/I$.

design need not be that conservative. Observe that with an open-circuit load the average power dissipation in Q_1 is $V_{CC}I$. A far more dangerous situation occurs at the other extreme of R_L—specifically, $R_L = 0$. In the event of an output short circuit, a positive input voltage would theoretically result in an infinite load current. In practice, a very large current may flow through Q_1, and if the short-circuit condition persists, the resulting large power dissipation in Q_1 can raise its junction temperature beyond the specified maximum, causing Q_1 to burn up. To guard against such a situation, output stages are usually equipped with *short-circuit protection,* as will be explained later.

The power dissipation in Q_2 also must be taken into account in designing an emitter-follower output stage. Since Q_2 conducts a constant current I, and the maximum value of v_{CE2} is $2V_{CC}$, the maximum instantaneous power dissipation in Q_2 is $2V_{CC}I$. This maximum, however, occurs when $v_O = V_{CC}$, a condition that would not normally prevail for a prolonged period of time. A more significant quantity for design purposes is the average power dissipation in Q_2, which is $V_{CC}I$.

EXERCISE

14.3 Consider the emitter follower in Fig. 14.2 with $V_{CC} = 10$ V, $I = 100$ mA, and $R_L = 100$ Ω. Find the power dissipated in Q_1 and Q_2 under quiescent conditions ($v_O = 0$). For a sinusoidal output voltage of maximum possible amplitude (neglecting $V_{CE\text{sat}}$), find the average power dissipation in Q_1 and Q_2. Also find the load power.

Ans. 1 W, 1 W; 0.5 W, 1 W; 0.5 W

14.2.4 Power-Conversion Efficiency

The power-conversion efficiency of an output stage is defined as

$$\eta \equiv \frac{\text{Load power } (P_L)}{\text{Supply power } (P_S)} \tag{14.7}$$

For the emitter follower of Fig. 14.2, assuming that the output voltage is a sinusoid with the peak value $\hat{V}_o$, the average load power will be

$$P_L = \frac{(\hat{V}_o/\sqrt{2})^2}{R_L} = \frac{1}{2}\frac{\hat{V}_o^2}{R_L} \tag{14.8}$$

Since the current in Q_2 is constant (I), the power drawn from the negative supply[1] is $V_{CC}I$. The *average* current in Q_1 is equal to I, and thus the average power drawn from the positive supply is $V_{CC}I$. Thus the total average supply power is

$$P_S = 2V_{CC}I \tag{14.9}$$

Equations (14.8) and (14.9) can be combined to yield

$$\eta = \frac{1}{4}\frac{\hat{V}_o^2}{IR_LV_{CC}}$$

$$= \frac{1}{4}\left(\frac{\hat{V}_o}{IR_L}\right)\left(\frac{\hat{V}_o}{V_{CC}}\right) \tag{14.10}$$

Since $\hat{V}_o \leq V_{CC}$ and $\hat{V}_o \leq IR_L$, maximum efficiency is obtained when

$$\hat{V}_o = V_{CC} = IR_L \tag{14.11}$$

The maximum efficiency attainable is 25%. Because this is a rather low figure, the class A output stage is rarely used in high-power applications (>1 W). Note also that in practice the output voltage swing is limited to lower values to avoid transistor saturation and associated nonlinear distortion. Thus the efficiency achieved is usually in the 10% to 20% range.

EXERCISE

14.4   For the emitter follower of Fig. 14.2, let $V_{CC} = 10$ V, $I = 100$ mA, and $R_L = 100 \, \Omega$. If the output voltage is an 8-V-peak sinusoid, find the following: (a) the power delivered to the load; (b) the average power drawn from the supplies; (c) the power-conversion efficiency.
   Ignore the loss in Q_3 and R.
Ans. 0.32 W; 2 W; 16%

 ## 14.3 CLASS B OUTPUT STAGE

Figure 14.5 shows a class B output stage. It consists of a complementary pair of transistors (an *npn* and a *pnp*) connected in such a way that both cannot conduct simultaneously.

---

[1] This does *not* include the power drawn by the biasing resistor R and the diode-connected transistor Q_3.

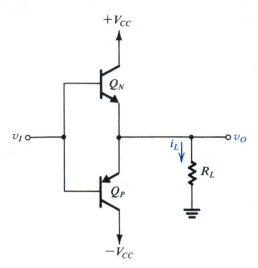

FIGURE 14.5 A class B output stage.

14.3.1 Circuit Operation

When the input voltage v_I is zero, both transistors are cut off and the output voltage v_O is zero. As v_I goes positive and exceeds about 0.5 V, Q_N conducts and operates as an emitter follower. In this case v_O follows v_I (i.e., $v_O = v_I - v_{BEN}$) and Q_N supplies the load current. Meanwhile, the emitter–base junction of Q_P will be reverse-biased by the V_{BE} of Q_N, which is approximately 0.7 V. Thus Q_P will be cut off.

If the input goes negative by more than about 0.5 V, Q_P turns on and acts as an emitter follower. Again v_O follows v_I (i.e., $v_O = v_I + v_{EBP}$), but in this case Q_P supplies the load current and Q_N will be cut off.

We conclude that the transistors in the class B stage of Fig. 14.5 are biased at zero current and conduct only when the input signal is present. The circuit operates in a **push–pull** fashion: Q_N *pushes* (sources) current into the load when v_I is positive, and Q_P *pulls* (sinks) current from the load when v_I is negative.

14.3.2 Transfer Characteristic

A sketch of the transfer characteristic of the class B stage is shown in Fig. 14.6. Note that there exists a range of v_I centered around zero where both transistors are cut off and v_O is zero. This **dead band** results in the **crossover distortion** illustrated in Fig. 14.7 for the case of an input sine wave. The effect of crossover distortion will be most pronounced when the amplitude of the input signal is small. Crossover distortion in audio power amplifiers gives rise to unpleasant sounds.

14.3.3 Power-Conversion Efficiency

To calculate the power-conversion efficiency, η, of the class B stage, we neglect the crossover distortion and consider the case of an output sinusoid of peak amplitude $\hat{V}_o$. The average load power will be

$$P_L = \frac{1}{2}\frac{\hat{V}_o^2}{R_L} \tag{14.12}$$

The current drawn from each supply will consist of half-sine waves of peak amplitude $(\hat{V}_o/R_L)$. Thus the average current drawn from each of the two power supplies will be

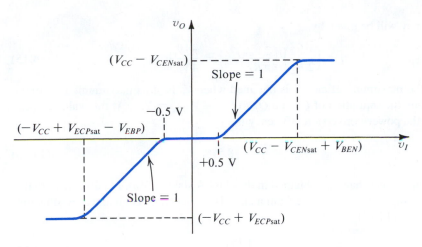

FIGURE 14.6 Transfer characteristic for the class B output stage in Fig. 14.5.

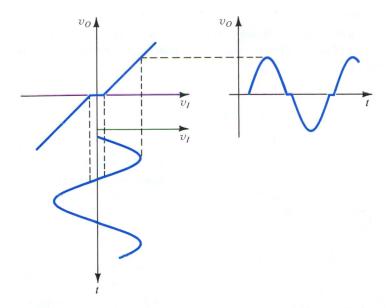

FIGURE 14.7 Illustrating how the dead band in the class B transfer characteristic results in crossover distortion.

$\hat{V}_o/\pi R_L$. It follows that the average power drawn from each of the two power supplies will be the same,

$$P_{S+} = P_{S-} = \frac{1}{\pi}\frac{\hat{V}_o}{R_L}V_{CC} \tag{14.13}$$

and the total supply power will be

$$P_S = \frac{2}{\pi}\frac{\hat{V}_o}{R_L}V_{CC} \tag{14.14}$$

Thus the efficiency will be given by

$$\eta = \left(\frac{1}{2}\frac{\hat{V}_o^2}{R_L}\right)\Big/\left(\frac{2}{\pi}\frac{\hat{V}_o}{R_L}V_{CC}\right) = \frac{\pi}{4}\frac{\hat{V}_o}{V_{CC}} \tag{14.15}$$

It follows that the maximum efficiency is obtained when $\hat{V}_o$ is at its maximum. This maximum is limited by the saturation of Q_N and Q_P to $V_{CC} - V_{CEsat} \simeq V_{CC}$. At this value of peak output voltage, the power-conversion efficiency is

$$\eta_{max} = \frac{\pi}{4} = 78.5\% \tag{14.16}$$

This value is much larger than that obtained in the class A stage (25%). Finally, we note that the maximum average power available from a class B output stage is obtained by substituting $\hat{V}_o = V_{CC}$ in Eq. (14.12),

$$P_{Lmax} = \frac{1}{2}\frac{V_{CC}^2}{R_L} \tag{14.17}$$

14.3.4 Power Dissipation

Unlike the class A stage, which dissipates maximum power under quiescent conditions ($v_O = 0$), the quiescent power dissipation of the class B stage is zero. When an input signal is applied, the *average* power dissipated in the class B stage is given by

$$P_D = P_S - P_L \tag{14.18}$$

Substituting for P_S from Eq. (14.14) and for P_L from Eq. (14.12) results in

$$P_D = \frac{2}{\pi}\frac{\hat{V}_o}{R_L}V_{CC} - \frac{1}{2}\frac{\hat{V}_o^2}{R_L} \tag{14.19}$$

From symmetry we see that half of P_D is dissipated in Q_N and the other half in Q_P. Thus Q_N and Q_P must be capable of safely dissipating $\frac{1}{2}P_D$ watts. Since P_D depends on $\hat{V}_o$, we must find the worst-case power dissipation, P_{Dmax}. Differentiating Eq. (14.19) with respect to $\hat{V}_o$ and equating the derivative to zero gives the value of $\hat{V}_o$ that results in maximum average power dissipation as

$$\hat{V}_o\big|_{P_{Dmax}} = \frac{2}{\pi}V_{CC} \tag{14.20}$$

Substituting this value in Eq. (14.19) gives

$$P_{Dmax} = \frac{2V_{CC}^2}{\pi^2 R_L} \tag{14.21}$$

Thus,

$$P_{DNmax} = P_{DPmax} = \frac{V_{CC}^2}{\pi^2 R_L} \tag{14.22}$$

At the point of maximum power dissipation the efficiency can be evaluated by substituting for $\hat{V}_o$ from Eq. (14.20) into Eq. (14.15); hence, $\eta = 50\%$.

Figure 14.8 shows a sketch of P_D (Eq. 14.19) versus the peak output voltage $\hat{V}_o$. Curves such as this are usually given on the data sheets of IC power amplifiers. (Usually, however, P_D is plotted versus P_L, as $P_L = \frac{1}{2}(\hat{V}_o^2/R_L)$, rather than $\hat{V}_o$). An interesting observation follows from Fig. 14.8: Increasing $\hat{V}_o$ beyond $2V_{CC}/\pi$ *decreases* the power dissipated in the class B

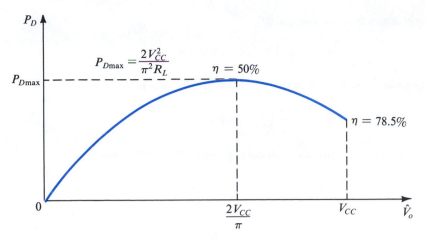

FIGURE 14.8 Power dissipation of the class B output stage versus amplitude of the output sinusoid.

stage while increasing the load power. The price paid is an increase in nonlinear distortion as a result of approaching the saturation region of operation of Q_N and Q_P. Transistor saturation flattens the peaks of the output sine waveform. Unfortunately, this type of distortion cannot be significantly reduced by the application of negative feedback (see Section 8.2), and thus transistor saturation should be avoided in applications requiring low THD.

EXAMPLE 14.1

It is required to design a class B output stage to deliver an average power of 20 W to an 8-Ω load. The power supply is to be selected such that V_{CC} is about 5 V greater than the peak output voltage. This avoids transistor saturation and the associated nonlinear distortion, and allows for including short-circuit protection circuitry. (The latter will be discussed in Section 14.7.) Determine the supply voltage required, the peak current drawn from each supply, the total supply power, and the power-conversion efficiency. Also determine the maximum power that each transistor must be able to dissipate safely.

Solution

Since

$$P_L = \frac{1}{2}\frac{\hat{V}_o^2}{R_L}$$

then

$$\hat{V}_o = \sqrt{2P_L R_L}$$

$$= \sqrt{2 \times 20 \times 8} = 17.9 \text{ V}$$

Therefore we select $V_{CC} = 23$ V.

The peak current drawn from each supply is

$$\hat{I}_o = \frac{\hat{V}_o}{R_L} = \frac{17.9}{8} = 2.24 \text{ A}$$

The average power drawn from each supply is

$$P_{S+} = P_{S-} = \frac{1}{\pi} \times 2.24 \times 23 = 16.4 \text{ W}$$

for a total supply power of 32.8 W. The power-conversion efficiency is

$$\eta = \frac{P_L}{P_S} = \frac{20}{32.8} \times 100 = 61\%$$

The maximum power dissipated in each transistor is given by Eq. (14.22); thus

$$P_{DN\text{max}} = P_{DP\text{max}} = \frac{V_{CC}^2}{\pi^2 R_L}$$

$$= \frac{(23)^2}{\pi^2 \times 8} = 6.7 \text{ W}$$

14.3.5 Reducing Crossover Distortion

The crossover distortion of a class B output stage can be reduced substantially by employing a high-gain op amp and overall negative feedback, as shown in Fig. 14.9. The ±0.7-V dead band is reduced to $\pm 0.7/A_0$ volt, where A_0 is the dc gain of the op amp. Nevertheless, the slew-rate limitation of the op amp will cause the alternate turning on and off of the output transistors to be noticeable, especially at high frequencies. A more practical method for reducing and almost eliminating crossover distortion is found in the class AB stage, which will be studied in the next section.

14.3.6 Single-Supply Operation

The class B stage can be operated from a single power supply, in which case the load is capacitively coupled, as shown in Fig. 14.10. Note that to make the formulas derived in Section 14.3.4 directly applicable, the single power supply is denoted $2V_{CC}$.

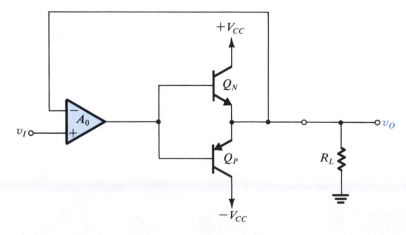

FIGURE 14.9 Class B circuit with an op amp connected in a negative-feedback loop to reduce crossover distortion.

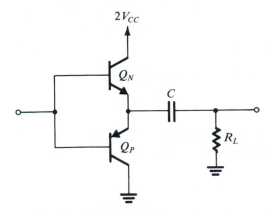

FIGURE 14.10 Class B output stage operated with a single power supply.

14.5 For the class B output stage of Fig. 14.5, let $V_{CC} = 6$ V and $R_L = 4\ \Omega$. If the output is a sinusoid with 4.5-V peak amplitude, find (a) the output power; (b) the average power drawn from each supply; (c) the power efficiency obtained at this output voltage; (d) the peak currents supplied by v_I, assuming that $\beta_N = \beta_P = 50$; (e) the maximum power that each transistor must be capable of dissipating safely.

Ans. (a) 2.53 W; (b) 2.15 W; (c) 59%; (d) 22.1 mA; (e) 0.91 W

 ## 14.4 CLASS AB OUTPUT STAGE

Crossover distortion can be virtually eliminated by biasing the complementary output transistors at a small nonzero current. The result is the class AB output stage shown in Fig. 14.11. A bias voltage V_{BB} is applied between the bases of Q_N and Q_P. For $v_I = 0$, $v_O = 0$, and a voltage $V_{BB}/2$ appears across the base–emitter junction of each of Q_N and Q_P. Assuming

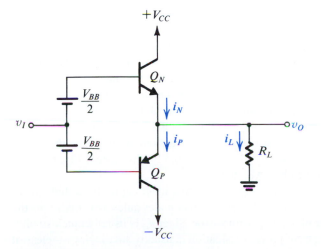

FIGURE 14.11 Class AB output stage. A bias voltage V_{BB} is applied between the bases of Q_N and Q_P, giving rise to a bias current I_Q given by Eq. (14.23). Thus, for small v_I, both transistors conduct and crossover distortion is almost completely eliminated.

matched devices,

$$i_N = i_P = I_Q = I_S e^{V_{BB}/2V_T} \tag{14.23}$$

The value of V_{BB} is selected to yield the required quiescent current I_Q.

14.4.1 Circuit Operation

When v_I goes positive by a certain amount, the voltage at the base of Q_N increases by the same amount and the output becomes positive at an almost equal value,

$$v_O = v_I + \frac{V_{BB}}{2} - v_{BEN} \tag{14.24}$$

The positive v_O causes a current i_L to flow through R_L, and thus i_N must increase; that is,

$$i_N = i_P + i_L \tag{14.25}$$

The increase in i_N will be accompanied by a corresponding increase in v_{BEN} (above the quiescent value of $V_{BB}/2$). However, since the voltage between the two bases remains constant at V_{BB}, the increase in v_{BEN} will result in an equal decrease in v_{EBP} and hence in i_P. The relationship between i_N and i_P can be derived as follows:

$$v_{BEN} + v_{EBP} = V_{BB}$$

$$V_T \ln \frac{i_N}{I_S} + V_T \ln \frac{i_P}{I_S} = 2V_T \ln \frac{I_Q}{I_S}$$

$$i_N i_P = I_Q^2 \tag{14.26}$$

Thus, as i_N increases, i_P decreases by the same ratio while the product remains constant. Equations (14.25) and (14.26) can be combined to yield i_N for a given i_L as the solution to the quadratic equation

$$i_N^2 - i_L i_N - I_Q^2 = 0 \tag{14.27}$$

From the equations above, we can see that for positive output voltages, the load current is supplied by Q_N, which acts as the output emitter follower. Meanwhile, Q_P will be conducting a current that decreases as v_O increases; for large v_O the current in Q_P can be ignored altogether.

For negative input voltages the opposite occurs: The load current will be supplied by Q_P, which acts as the output emitter follower, while Q_N conducts a current that gets smaller as v_I becomes more negative. Equation (14.26), relating i_N and i_P, holds for negative inputs as well.

We conclude that the class AB stage operates in much the same manner as the class B circuit, with one important exception: For small v_I, both transistors conduct, and as v_I is increased or decreased, one of the two transistors takes over the operation. Since the transition is a smooth one, crossover distortion will be almost totally eliminated. Figure 14.12 shows the transfer characteristic of the class AB stage.

The power relationships in the class AB stage are almost identical to those derived for the class B circuit in Section 14.3. The only difference is that under quiescent conditions the class AB circuit dissipates a power of $V_{CC}I_Q$ per transistor. Since I_Q is usually much smaller than the peak load current, the quiescent power dissipation is usually small. Nevertheless, it can be taken into account easily. Specifically, we can simply add the quiescent dissipation per transistor to its maximum power dissipation with an input signal applied, to obtain the total power dissipation that the transistor must be able to handle safely.

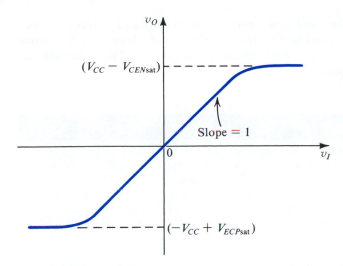

FIGURE 14.12 Transfer characteristic of the class AB stage in Fig. 14.11.

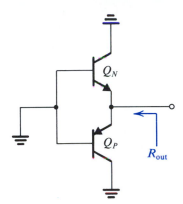

FIGURE 14.13 Determining the small-signal output resistance of the class AB circuit of Fig. 14.11.

14.4.2 Output Resistance

If we assume that the source supplying v_I is ideal, then the output resistance of the class AB stage can be determined from the circuit in Fig. 14.13 as

$$R_{\text{out}} = r_{eN} \parallel r_{eP} \tag{14.28}$$

where r_{eN} and r_{eP} are the small-signal emitter resistances of Q_N and Q_P, respectively. At a given input voltage, the currents i_N and i_P can be determined, and r_{eN} and r_{eP} are given by

$$r_{eN} = \frac{V_T}{i_N} \tag{14.29}$$

$$r_{eP} = \frac{V_T}{i_P} \tag{14.30}$$

Thus

$$R_{\text{out}} = \frac{V_T}{i_N} \parallel \frac{V_T}{i_P} = \frac{V_T}{i_P + i_N} \tag{14.31}$$

Since as i_N increases, i_P decreases, and vice versa, the output resistance remains approximately constant in the region around $v_I = 0$. This, in effect, is the reason for the virtual absence of crossover distortion. At larger load currents, either i_N or i_P will be significant, and R_{out} decreases as the load current increases.

EXERCISE

14.6 Consider a class AB circuit with $V_{CC} = 15$ V, $I_Q = 2$ mA, and $R_L = 100$ Ω. Determine V_{BB}. Construct a table giving i_L, i_N, i_P, v_{BEN}, v_{EBP}, v_I, v_O/v_I, R_{out}, and v_o/v_i versus v_O for $v_O = 0, 0.1, 0.2, 0.5, 1, 5, 10,$ $-0.1, -0.2, -0.5, -1, -5$ and -10 V. Note that v_O/v_I is the large-signal voltage gain and v_o/v_i is the incremental gain obtained as $R_L/(R_L + R_{out})$. The incremental gain is equal to the slope of the transfer curve at the operating point. Assume Q_N and Q_P to be matched, with $I_S = 10^{-13}$ A.

Ans. $V_{BB} = 1.186$ V

v_O (V)	i_L (mA)	i_N (mA)	i_P (mA)	v_{BEN} (V)	v_{EBP} (V)	v_I (V)	v_O/v_I	R_{out} (Ω)	v_o/v_i
+10.0	100	100.04	0.04	0.691	0.495	10.1	0.99	0.25	1.00
+5.0	50	50.08	0.08	0.673	0.513	5.08	0.98	0.50	1.00
+1.0	10	10.39	0.39	0.634	0.552	1.041	0.96	2.32	0.98
+0.5	5	5.70	0.70	0.619	0.567	0.526	0.95	4.03	0.96
+0.2	2	3.24	1.24	0.605	0.581	0.212	0.94	5.58	0.95
+0.1	1	2.56	1.56	0.599	0.587	0.106	0.94	6.07	0.94
0	0	2	2	0.593	0.593	0	—	6.25	0.94
−0.1	−1	1.56	2.56	0.587	0.599	−0.106	0.94	6.07	0.94
−0.2	−2	1.24	3.24	0.581	0.605	−0.212	0.94	5.58	0.95
−0.5	−5	0.70	5.70	0.567	0.619	−0.526	0.95	4.03	0.96
−1.0	−10	0.39	10.39	0.552	0.634	−1.041	0.96	2.32	0.98
−5.0	−50	0.08	50.08	0.513	0.673	−5.08	0.98	0.50	1.00
−10.0	−100	0.04	100.04	0.495	0.691	−10.1	0.99	0.25	1.00

14.5 BIASING THE CLASS AB CIRCUIT

In this section we discuss two approaches for generating the voltage V_{BB} required for biasing the class AB output stage.

14.5.1 Biasing Using Diodes

Figure 14.14 shows a class AB circuit in which the bias voltage V_{BB} is generated by passing a constant current I_{BIAS} through a pair of diodes, or diode-connected transistors, D_1 and D_2. In circuits that supply large amounts of power, the output transistors are large-geometry devices. The biasing diodes, however, need not be large devices, and thus the quiescent current I_Q established in Q_N and Q_P will be $I_Q = nI_{BIAS}$, where n is the ratio of the emitter–junction area of the output devices to the junction area of the biasing diodes. In other words, the saturation (or scale) current I_S of the output transistors is n times that of the biasing diodes. Area ratioing is simple to implement in integrated circuits but difficult to realize in discrete-circuit designs.

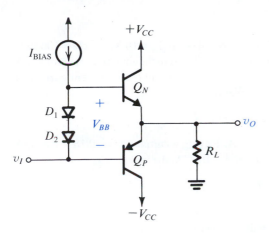

FIGURE 14.14 A class AB output stage utilizing diodes for biasing. If the junction area of the output devices, Q_N and Q_P, is n times that of the biasing devices D_1 and D_2, and a quiescent current $I_Q = nI_{BIAS}$ flows in the output devices.

When the output stage of Fig. 14.14 is sourcing current to the load, the base current of Q_N increases from I_Q/β_N (which is usually small) to approximately i_L/β_N. This base current drive must be supplied by the current source I_{BIAS}. It follows that I_{BIAS} must be greater than the maximum anticipated base drive for Q_N. This sets a lower limit on the value of I_{BIAS}. Now, since $I_Q = nI_{BIAS}$ and since I_Q is usually much smaller than the peak load current (<10%), we see that we cannot make n a large number. In other words, we cannot make the diodes much smaller than the output devices. This is a disadvantage of the diode biasing scheme.

From the discussion above we see that the current through the biasing diodes will decrease when the output stage is sourcing current to the load. Thus the bias voltage V_{BB} will also decrease, and the analysis of Section 14.4 must be modified to take this effect into account.

The diode biasing arrangement has an important advantage: It can provide thermal stabilization of the quiescent current in the output stage. To appreciate this point recall that the class AB output stage dissipates power under quiescent conditions. Power dissipation raises the internal temperature of the BJTs. From Chapter 5 we know that a rise in transistor temperature results in a decrease in its V_{BE} (approximately −2 mV/°C) if the collector current is held constant. Alternatively, if V_{BE} is held constant and the temperature increases, the collector current increases. The increase in collector current increases the power dissipation, which in turn increases the collector current. Thus a positive-feedback mechanism exists that can result in a phenomenon called **thermal runaway.** Unless checked, thermal runaway can lead to the ultimate destruction of the BJT. Diode biasing can be arranged to provide a compensating effect that can protect the output transistors against thermal runaway under quiescent conditions. Specifically, if the diodes are in close thermal contact with the output transistors, their temperature will increase by the same amount as that of Q_N and Q_P. Thus V_{BB} will decrease at the same rate as $V_{BEN} + V_{EBP}$, with the result that I_Q remains constant. Close thermal contact is easily achieved in IC fabrication. It is obtained in discrete circuits by mounting the bias diodes on the metal case of Q_N or Q_P.

EXAMPLE 14.2

Consider the class AB output stage under the conditions that $V_{CC} = 15$ V, $R_L = 100$ Ω, and the output is sinusoidal with a maximum amplitude of 10 V. Let Q_N and Q_P be matched with $I_S = 10^{-13}$ A and $\beta = 50$. Assume that the biasing diodes have one-third the junction area of the output devices. Find the value of I_{BIAS} that guarantees a minimum of 1 mA through the diodes at all times. Determine the quiescent current and the quiescent power dissipation in the output transistors (i.e., at $v_O = 0$). Also find V_{BB} for $v_O = 0$, +10 V, and −10 V.

Solution

The maximum current through Q_N is approximately equal to $i_{L\max} = 10\text{ V}/0.1\text{ k}\Omega = 100\text{ mA}$. Thus the maximum base current in Q_N is approximately 2 mA. To maintain a minimum of 1 mA through the diodes, we select $I_{BIAS} = 3$ mA. The area ratio of 3 yields a quiescent current of 9 mA through Q_N and Q_P. The quiescent power dissipation is

$$P_{DQ} = 2 \times 15 \times 9 = 270 \text{ mW}$$

For $v_O = 0$, the base current of Q_N is $9/51 \simeq 0.18$ mA, leaving a current of $3 - 0.18 = 2.82$ mA to flow through the diodes. Since the diodes have $I_S = \frac{1}{3} \times 10^{-13}$A, the voltage V_{BB} will be

$$V_{BB} = 2V_T \ln \frac{2.82 \text{ mA}}{I_S} = 1.26 \text{ V}$$

At $v_O = +10$ V, the current through the diodes will decrease to 1 mA, resulting in $V_{BB} \simeq 1.21$ V. At the other extreme of $v_O = -10$ V, Q_N will be conducting a very small current; thus its base current will be negligibly small and all of I_{BIAS} (3 mA) flows through the diodes, resulting in $V_{BB} \simeq 1.26$ V.

EXERCISES

14.7 For the circuit of Example 14.2, find i_N and i_P for $v_O = +10$ V and $v_O = -10$ V.

Ans. 100.1 mA, 0.1 mA; 0.8 mA, 100.8 mA

14.8 If the collector current of a transistor is held constant, its v_{BE} decreases by 2 mV for every 1°C rise in temperature. Alternatively, if v_{BE} is held constant, then i_C increases by approximately $g_m \times 2$ mV for every 1°C rise in temperature. For a device operating at $I_C = 10$ mA, find the change in collector current resulting from an increase in temperature of 5°C.

Ans. 4 mA

14.5.2 Biasing Using the V_{BE} Multiplier

An alternative biasing arrangement that provides the designer with considerably more flexibility in both discrete and integrated designs is shown in Fig. 14.15. The bias circuit consists of transistor Q_1 with a resistor R_1 connected between base and emitter and a feedback resistor R_2 connected between collector and base. The resulting two-terminal network is fed with a constant-current source I_{BIAS}. If we neglect the base current of Q_1, then R_1 and R_2 will carry the same current I_R, given by

$$I_R = \frac{V_{BE1}}{R_1} \tag{14.32}$$

and the voltage V_{BB} across the bias network will be

$$\begin{aligned} V_{BB} &= I_R(R_1 + R_2) \\ &= V_{BE1}\left(1 + \frac{R_2}{R_1}\right) \end{aligned} \tag{14.33}$$

Thus the circuit simply multiplies V_{BE1} by the factor $(1 + R_2/R_1)$ and is known as the "V_{BE} multiplier." The multiplication factor is obviously under the designer's control and can be

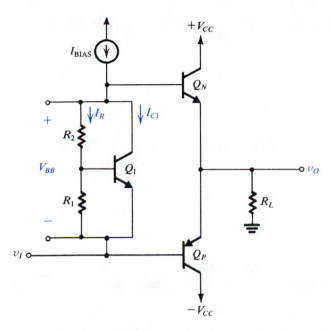

FIGURE 14.15 A class AB output stage utilizing a V_{BE} multiplier for biasing.

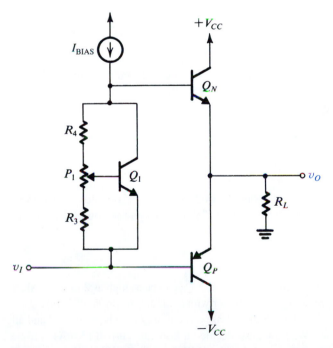

FIGURE 14.16 A discrete-circuit class AB output stage with a potentiometer used in the V_{BE} multiplier. The potentiometer is adjusted to yield the desired value of quiescent current in Q_N and Q_P.

used to establish the value of V_{BB} required to yield a desired quiescent current I_Q. In IC design it is relatively easy to control accurately the ratio of two resistances. In discrete-circuit design, a potentiometer can be used, as shown in Fig. 14.16, and is manually set to produce the desired value of I_Q.

The value of V_{BE1} in Eq. (14.33) is determined by the portion of I_{BIAS} that flows through the collector of Q_1; that is,

$$I_{C1} = I_{BIAS} - I_R \tag{14.34}$$

$$V_{BE1} = V_T \ln \frac{I_{C1}}{I_{S1}} \tag{14.35}$$

where we have neglected the base current of Q_N, which is normally small both under quiescent conditions and when the output voltage is swinging negative. However, for positive v_O, especially at and near its peak value, the base current of Q_N can become sizable and will reduce the current available for the V_{BE} multiplier. Nevertheless, since large changes in I_{C1} correspond to only small changes in V_{BE1}, the decrease in current will be mostly absorbed by Q_1, leaving I_R, and hence V_{BB}, almost constant.

EXERCISE

14.9 Consider a V_{BE} multiplier with $R_1 = R_2 = 1.2$ kΩ, utilizing a transistor that has $V_{BE} = 0.6$ V at $I_C = 1$ mA, and a very high β. (a) Find the value of the current I that should be supplied to the multiplier to obtain a terminal voltage of 1.2 V. (b) Find the value of I that will result in the terminal voltage changing (from the 1.2-V value) by +50 mV, +100 mV, +200 mV, –50 mV, –100 mV, –200 mV.

Ans. (a) 1.5 mA; (b) 3.24 mA, 7.93 mA, 55.18 mA, 0.85 mA, 0.59 mA, 0.43 mA

Like the diode biasing network, the V_{BE}–multiplier circuit can provide thermal stabilization of I_Q. This is especially true if $R_1 = R_2$ and Q_1 is in close thermal contact with the output transistors.

EXAMPLE 14.3

It is required to redesign the output stage of Example 14.2 utilizing a V_{BE} multiplier for biasing. Use a small-geometry transistor for Q_1 with $I_S = 10^{-14}$ A and design for a quiescent current $I_Q = 2$ mA.

Solution

Since the peak positive current is 100 mA, the base current of Q_N can be as high as 2 mA. We shall therefore select $I_{BIAS} = 3$ mA, thus providing the multiplier with a minimum current of 1 mA.

Under quiescent conditions ($v_O = 0$ and $i_L = 0$) the base current of Q_N can be neglected and all of I_{BIAS} flows through the multiplier. We now must decide on how this current (3 mA) is to be divided between I_{C1} and I_R. If we select I_R greater than 1 mA, the transistor will be almost cut off at the positive peak of v_O. Therefore, we shall select $I_R = 0.5$ mA, leaving 2.5 mA for I_{C1}.

To obtain a quiescent current of 2 mA in the output transistors, V_{BB} should be

$$V_{BB} = 2V_T \ln \frac{2 \times 10^{-3}}{10^{-13}} = 1.19 \text{ V}$$

We can now determine $R_1 + R_2$ as follows:

$$R_1 + R_2 = \frac{V_{BB}}{I_R} = \frac{1.19}{0.5} = 2.38 \text{ k}\Omega$$

At a collector current of 2.5 mA, Q_1 has

$$V_{BE1} = V_T \ln \frac{2.5 \times 10^{-3}}{10^{-14}} = 0.66 \text{ V}$$

The value of R_1 can now be determined as

$$R_1 = \frac{0.66}{0.5} = 1.32 \text{ k}\Omega$$

and R_2 as

$$R_2 = 2.38 - 1.32 = 1.06 \text{ k}\Omega$$

14.6 POWER BJTs

Transistors that are required to conduct currents in the ampere range and withstand power dissipation in the watts and tens-of-watts ranges differ in their physical structure, packaging, and specification from the small-signal transistors considered in earlier chapters. In this section we consider some of the important properties of power transistors, especially those aspects that pertain to the design of circuits of the type discussed earlier. There are, of course, other important applications of power transistors, such as their use as switching elements in power inverters and motor-control circuits. Such applications are not studied in this book.

14.6.1 Junction Temperature

Power transistors dissipate large amounts of power in their collector–base junctions. The dissipated power is converted into heat, which raises the junction temperature. However, the junction temperature T_J must not be allowed to exceed a specified maximum, $T_{J\text{max}}$; otherwise the transistor could suffer permanent damage. For silicon devices, $T_{J\text{max}}$ is in the range of 150°C to 200°C.

14.6.2 Thermal Resistance

Consider first the situation of a transistor operating in free air—that is, with no special arrangements for cooling. The heat dissipated in the transistor junction will be conducted away from the junction to the transistor case, and from the case to the surrounding environment. In a steady state in which the transistor is dissipating P_D watts, the temperature rise of the junction relative to the surrounding ambience can be expressed as

$$T_J - T_A = \theta_{JA} P_D \qquad (14.36)$$

where θ_{JA} is the **thermal resistance** between junction and ambience, having the units of degrees Celsius per watt. Note that θ_{JA} simply gives the rise in junction temperature over the ambient temperature for each watt of dissipated power. Since we wish to be able to dissipate large amounts of power without raising the junction temperature above $T_{J\text{max}}$, it is desirable to have, for the thermal resistance θ_{JA}, as small a value as possible. For operation in free air,

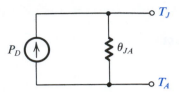

FIGURE 14.17 Electrical equivalent circuit of the thermal-conduction process; $T_J - T_A = P_D \theta_{JA}$.

θ_{JA} depends primarily on the type of case in which the transistor is packaged. The value of θ_{JA} is usually specified on the transistor data sheet.

Equation (14.36), which describes the thermal-conduction process, is analogous to Ohm's law, which describes the electrical-conduction process. In this analogy, power dissipation corresponds to current, temperature difference corresponds to voltage difference, and thermal resistance corresponds to electrical resistance. Thus, we may represent the thermal-conduction process by the electric circuit shown in Fig. 14.17.

14.6.3 Power Dissipation Versus Temperature

The transistor manufacturer usually specifies the maximum junction temperature $T_{J\max}$, the maximum power dissipation at a particular ambient temperature T_{A0} (usually, 25°C), and the thermal resistance θ_{JA}. In addition, a graph such as that shown in Fig. 14.18 is usually provided. The graph simply states that for operation at ambient temperatures below T_{A0}, the device can safely dissipate the rated value of P_{D0} watts. However, if the device is to be operated at higher ambient temperatures, the maximum allowable power dissipation must be **derated** according to the straight line shown in Fig. 14.18. The power-derating curve is a graphical representation of Eq. (14.36). Specifically, note that if the ambient temperature is T_{A0} and the power dissipation is at the maximum allowed (P_{D0}), then the junction temperature will be $T_{J\max}$. Substituting these quantities in Eq. (14.36) results in

$$\theta_{JA} = \frac{T_{J\max} - T_{A0}}{P_{D0}} \tag{14.37}$$

which is the inverse of the slope of the power-derating straight line. At an ambient temperature T_A, higher than T_{A0}, the maximum allowable power dissipation $P_{D\max}$ can be obtained from Eq. (14.36) by substituting $T_J = T_{J\max}$; thus,

$$P_{D\max} = \frac{T_{J\max} - T_A}{\theta_{JA}} \tag{14.38}$$

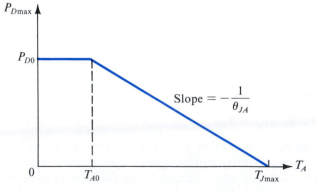

FIGURE 14.18 Maximum allowable power dissipation versus ambient temperature for a BJT operated in free air. This is known as a "power-derating" curve.

Observe that as T_A approaches $T_{J\max}$, the allowable power dissipation decreases; the lower thermal gradient limits the amount of heat that can be removed from the junction. In the extreme situation of $T_A = T_{J\max}$, no power can be dissipated because no heat can be removed from the junction.

EXAMPLE 14.4

A BJT is specified to have a maximum power dissipation P_{D0} of 2 W at an ambient temperature T_{A0} of 25°C, and a maximum junction temperature $T_{J\max}$ of 150°C. Find the following:

(a) The thermal resistance θ_{JA}.

(b) The maximum power that can be safely dissipated at an ambient temperature of 50°C.

(c) The junction temperature if the device is operating at $T_A = 25$°C and is dissipating 1 W.

Solution

(a) $\theta_{JA} = \dfrac{T_{J\max} - T_{A0}}{P_{D0}} = \dfrac{150 - 25}{2} = 62.5\,°\text{C/W}$

(b) $P_{D\max} = \dfrac{T_{J\max} - T_A}{\theta_{JA}} = \dfrac{150 - 50}{62.5} = 1.6\text{ W}$

(c) $T_J = T_A + \theta_{JA} P_D = 25 + 62.5 \times 1 = 87.5\,°\text{C}$

14.6.4 Transistor Case and Heat Sink

The thermal resistance between junction and ambience, θ_{JA}, can be expressed as

$$\theta_{JA} = \theta_{JC} + \theta_{CA} \tag{14.39}$$

where θ_{JC} is the thermal resistance between junction and transistor case (package) and θ_{CA} is the thermal resistance between case and ambience. For a given transistor, θ_{JC} is fixed by the device design and packaging. The device manufacturer can reduce θ_{JC} by encapsulating the device in a relatively large metal case and placing the collector (where most of the heat is dissipated) in direct contact with the case. Most high-power transistors are packaged in this fashion. Figure 14.19 shows a sketch of a typical package.

Although the circuit designer has no control over θ_{JC} (once a particular transistor has been selected), the designer can considerably reduce θ_{CA} below its free-air value (specified by the manufacturer as part of θ_{JA}). Reduction of θ_{CA} can be effected by providing means to facilitate heat transfer from case to ambience. A popular approach is to bolt the transistor to the chassis or to an extended metal surface. Such a metal surface then functions as a **heat sink.** Heat is easily conducted from the transistor case to the heat sink; that is, the thermal resistance θ_{CS} is usually very small. Also, heat is efficiently transferred (by convection and

FIGURE 14.19 The popular TO3 package for power transistors. The case is metal with a diameter of about 2.2 cm; the outside dimension of the "seating plane" is about 4 cm. The seating plane has two holes for screws to bolt it to a heat sink. The collector is electrically connected to the case. Therefore an electrically insulating but thermally conducting spacer is used between the transistor case and the "heat sink."

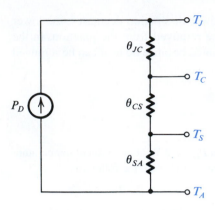

FIGURE 14.20 Electrical analog of the thermal conduction process when a heat sink is utilized.

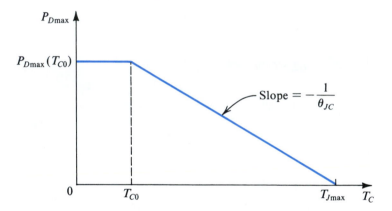

FIGURE 14.21 Maximum allowable power dissipation versus transistor-case temperature.

radiation) from the heat sink to the ambience, resulting in a low thermal resistance θ_{SA}. Thus, if a heat sink is utilized, the case-to-ambience thermal resistance given by

$$\theta_{CA} = \theta_{CS} + \theta_{SA} \qquad (14.40)$$

can be small because its two components can be made small by the choice of an appropriate heat sink.[2] For example, in very high-power applications the heat sink is usually equipped with fins that further facilitate cooling by radiation and convection.

The electrical analog of the thermal-conduction process when a heat sink is employed is shown in Fig. 14.20, from which we can write

$$T_J - T_A = P_D(\theta_{JC} + \theta_{CS} + \theta_{SA}) \qquad (14.41)$$

As well as specifying θ_{JC}, the device manufacturer usually supplies a derating curve for P_{Dmax} versus the case temperature, T_C. Such a curve is shown in Fig. 14.21. Note that the slope of the power-derating straight line is $-1/\theta_{JC}$. For a given transistor, the maximum

---

[2] As noted earlier, the metal case of a power transistor is electrically connected to the collector. Thus an electrically insulating material such as mica is usually placed between the metal case and the metal heat sink. Also, insulating bushings and washers are generally used in bolting the transistor to the heat sink.

power dissipation at a *case temperature* T_{C0} (usually 25°C) is much greater than that at an *ambient temperature* T_{A0} (usually 25°C). If the device can be maintained at a case temperature T_C, $T_{C0} \leq T_C \leq T_{Jmax}$, then the maximum safe power dissipation is obtained when $T_J = T_{Jmax}$,

$$P_{Dmax} = \frac{T_{Jmax} - T_C}{\theta_{JC}} \tag{14.42}$$

EXAMPLE 14.5

A BJT is specified to have $T_{Jmax} = 150°C$ and to be capable of dissipating maximum power as follows:

$$40 \text{ W at } T_C = 25°C$$
$$2 \text{ W at } T_A = 25°C$$

Above 25°C, the maximum power dissipation is to be derated linearly with $\theta_{JC} = 3.12°C/W$ and $\theta_{JA} = 62.5°C/W$. Find the following:

(a) The maximum power that can be dissipated safely by this transistor when operated in free air at $T_A = 50°C$.

(b) The maximum power that can be dissipated safely by this transistor when operated at an ambient temperature of 50°C, but with a heat sink for which $\theta_{CS} = 0.5°C/W$ and $\theta_{SA} = 4°C/W$. Find the temperature of the case and of the heat sink.

(c) The maximum power that can be dissipated safely if an *infinite heat sink* is used and $T_A = 50°C$.

Solution

(a)

$$P_{Dmax} = \frac{T_{Jmax} - T_A}{\theta_{JA}} = \frac{150 - 50}{62.5} = 1.6 \text{ W}$$

(b) With a heat sink, θ_{JA} becomes

$$\theta_{JA} = \theta_{JC} + \theta_{CS} + \theta_{SA}$$
$$= 3.12 + 0.5 + 4 = 7.62°C/W$$

Thus,

$$P_{Dmax} = \frac{150 - 50}{7.62} = 13.1 \text{ W}$$

Figure 14.22 shows the thermal equivalent circuit with the various temperatures indicated.

(c) An infinite heat sink, if it existed, would cause the case temperature T_C to equal the ambient temperature T_A. The infinite heat sink has $\theta_{CA} = 0$. Obviously, one cannot buy an infinite heat sink; nevertheless, this terminology is used by some manufacturers to describe the power-derating curve of Fig. 14.21. The abscissa is then labeled T_A and the curve is called "power dissipation versus ambient temperature with an infinite heat sink." For our example, with infinite heat sink,

$$P_{Dmax} = \frac{T_{Jmax} - T_A}{\theta_{JC}} = \frac{150 - 50}{3.12} = 32 \text{ W}$$

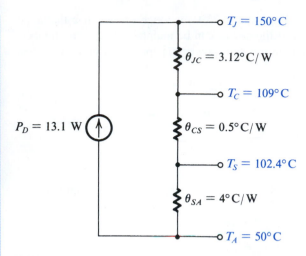

FIGURE 14.22 Thermal equivalent circuit for Example 14.5.

The advantage of using a heat sink is clearly evident from Example 14.5: With a heat sink, the maximum allowable power dissipation increases from 1.6 W to 13.1 W. Also note that although the transistor considered can be called a "40-W transistor," this level of power dissipation cannot be achieved in practice; it would require an infinite heat sink and an ambient temperature $T_A \leq 25°C$.

EXERCISE

14.10 The 2N6306 power transistor is specified to have $T_{Jmax} = 200°C$ and $P_{Dmax} = 125$ W for $T_C \leq 25°C$. For $T_C \geq 25°C$, $\theta_{JC} = 1.4°C/W$. If in a particular application this device is to dissipate 50 W and operate at an ambient temperature of 25°C, find the maximum thermal resistance of the heat sink that must be used (i.e., θ_{SA}). Assume $\theta_{CS} = 0.6°C/W$. What is the case temperature, T_C?

Ans. 1.5°C/W; 130°C

14.6.5 The BJT Safe Operating Area

In addition to specifying the maximum power dissipation at different case temperatures, power-transistor manufacturers usually provide a plot of the boundary of the safe operating area (SOA) in the $i_C - v_{CE}$ plane. The SOA specification takes the form illustrated by the sketch in Fig. 14.23; the following paragraph numbers correspond to the boundaries on the sketch.

1. The maximum allowable current I_{Cmax}. Exceeding this current on a continuous basis can result in melting the wires that bond the device to the package terminals.

2. The maximum power dissipation hyperbola. This is the locus of the points for which $v_{CE}i_C = P_{Dmax}$ (at T_{C0}). For temperatures $T_C > T_{C0}$, the power-derating curves described in Section 14.6.4 should be used to obtain the applicable P_{Dmax} and thus a correspondingly lower hyperbola. Although the operating point can be allowed to move temporarily above the hyperbola, the *average* power dissipation should not be allowed to exceed P_{Dmax}.

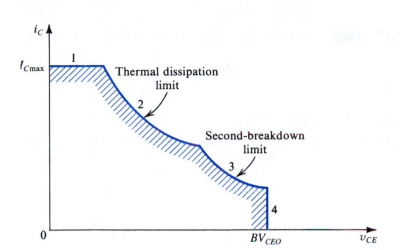

FIGURE 14.23 Safe operating area (SOA) of a BJT.

3. The **second-breakdown** limit. Second breakdown is a phenomenon that results because current flow across the emitter–base junction is not uniform. Rather, the current density is greatest near the periphery of the junction. This "current crowding" gives rise to increased localized power dissipation and hence temperature rise (at locations called hot spots). Since a temperature rise causes an increase in current, a localized form of thermal runaway can occur, leading to junction destruction.

4. The collector-to-emitter breakdown voltage, BV_{CEO}. The instantaneous value of v_{CE} should never be allowed to exceed BV_{CEO}; otherwise, avalanche breakdown of the collector–base junction may occur (see Section 5.2.5).

Finally, it should be mentioned that logarithmic scales are usually used for i_C and v_{CE}, leading to an SOA boundary that consists of straight lines.

14.6.6 Parameter Values of Power Transistors

Owing to their large geometry and high operating currents, power transistors display typical parameter values that can be quite different from those of small-signal transistors. The important differences are as follows:

1. At high currents, the exponential i_C–v_{BE} relationship exhibits a constant $n = 2$; that is, $i_C = I_S e^{v_{BE}/2V_T}$.

2. β is low, typically 30 to 80, but can be as low as 5. Here, it is important to note that β has a positive temperature coefficient.

3. At high currents, r_π becomes very small (a few ohms) and r_x becomes important (r_x is defined and explained in Section 5.8.4).

4. f_T is low (a few megahertz), C_μ is large (hundreds of picofarads), and C_π is even larger. (These parameters are defined and explained in Section 5.8).

5. I_{CBO} is large (a few tens of microamps) and, as usual, doubles for every 10°C rise in temperature.

6. BV_{CEO} is typically 50 to 100 V but can be as high as 500 V.

7. I_{Cmax} is typically in the ampere range but can be as high as 100 A.

14.7 VARIATIONS ON THE CLASS AB CONFIGURATION

In this section, we discuss a number of circuit improvements and protection techniques for the class AB output stage.

14.7.1 Use of Input Emitter Followers

Figure 14.24 shows a class AB circuit biased using transistors Q_1 and Q_2, which also function as emitter followers, thus providing the circuit with a high input resistance. In effect, the circuit functions as a unity-gain buffer amplifier. Since all four transistors are usually matched, the quiescent current ($v_I = 0$, $R_L = \infty$) in Q_3 and Q_4 is equal to that in Q_1 and Q_2. Resistors R_3 and R_4 are usually very small and are included to compensate for possible mismatches between Q_3 and Q_4 and to guard against the possibility of thermal runaway due to temperature differences between the input- and output-stage transistors. The latter point can be appreciated by noting that an increase in the current of, say, Q_3 causes an increase in the voltage drop across R_3 and a corresponding decrease in V_{BE3}. Thus R_3 provides negative feedback that helps stabilize the current through Q_3.

Because the circuit of Fig. 14.24 requires high-quality *pnp* transistors, it is not suitable for implementation in conventional monolithic IC technology. However, excellent results have been obtained with this circuit implemented in hybrid thick-film technology (Wong and Sherwin, 1979). This technology permits component trimming, for instance, to minimize the output offset voltage. The circuit can be used alone or together with an op amp to

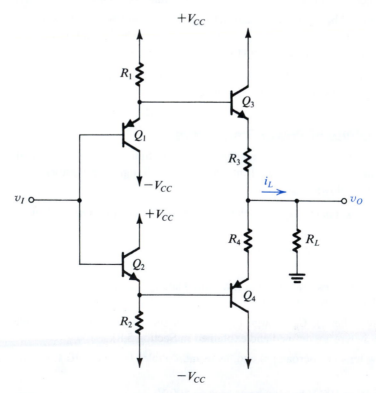

FIGURE 14.24 A class AB output stage with an input buffer. In addition to providing a high input resistance, the buffer transistors Q_1 and Q_2 bias the output transistors Q_3 and Q_4.

provide increased output driving capability. The latter application will be discussed in the next section.

14.7.2 Use of Compound Devices

To increase the current gain of the output-stage transistors, and thus reduce the required base current drive, the Darlington configuration shown in Fig. 14.25 is frequently used to replace the *npn* transistor of the class AB stage. The Darlington configuration (Section 6.11.2) is equivalent to a single *npn* transistor having $\beta \simeq \beta_1 \beta_2$, but almost twice the value of V_{BE}.

The Darlington configuration can be also used for *pnp* transistors, and this is indeed done in discrete-circuit design. In IC design, however, the lack of good-quality *pnp* transistors prompted the use of the alternative compound configuration shown in Fig. 14.26. This compound device is equivalent to a single *pnp* transistor having $\beta \simeq \beta_1 \beta_2$. When fabricated with standard IC technology, Q_1 is usually a lateral *pnp* having a low β ($\beta = 5 - 10$) and poor high-frequency response ($f_T \simeq 5$ MHz); see Appendix A. The compound device, although it has a relatively high equivalent β, still suffers from a poor high-frequency response. It also suffers from another problem: The feedback loop formed by Q_1 and Q_2 is prone to high-frequency oscillations (with frequency near f_T of the *pnp* device, i.e., about 5 MHz). Methods exist for preventing such oscillations. The subject of feedback-amplifier stability was studied in Chapter 8.

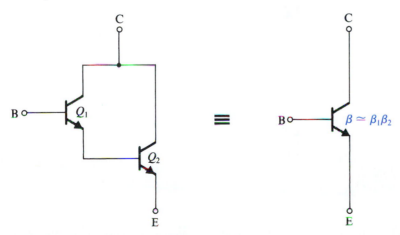

FIGURE 14.25 The Darlington configuration.

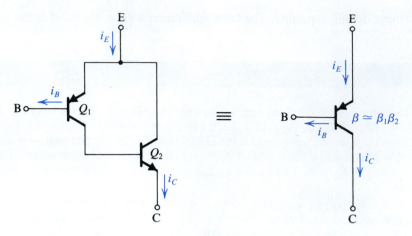

FIGURE 14.26 The compound-*pnp* configuration.

To illustrate the application of the Darlington configuration and of the compound *pnp*, we show in Fig. 14.27 an output stage utilizing both. Class AB biasing is achieved using a V_{BE} multiplier. Note that the Darlington *npn* adds one more V_{BE} drop, and thus the V_{BE} multiplier is required to provide a bias voltage of about 2 V. The design of this class AB stage is investigated in Problem 14.39.

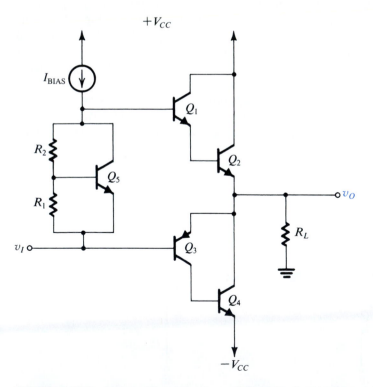

FIGURE 14.27 A class AB output stage utilizing a Darlington *npn* and a compound *pnp*. Biasing is obtained using a V_{BE} multiplier.

EXERCISE

14.12 (a) Refer to Fig. 14.26. Show that, for the composite *pnp* transistor,

$$i_B \simeq \frac{i_C}{\beta_N \beta_P}$$

and

$$i_E \simeq i_C$$

Hence show that

$$i_C \simeq \beta_N I_{SP} e^{v_{BE}/V_T}$$

and thus the transistor has an effective scale current

$$I_S = \beta_N I_{SP}$$

where I_{SP} is the saturation current of the *pnp* transistor Q_1.

(b) For $\beta_P = 20$, $\beta_N = 50$, $I_{SP} = 10^{-14}$ A, find the effective current gain of the compound device and its v_{EB} when $i_C = 100$ mA. Let $n = 1$.

Ans. (b) 1000; 0.651 V

14.7.3 Short-Circuit Protection

Figure 14.28 shows a class AB output stage equipped with protection against the effect of short-circuiting the output while the stage is sourcing current. The large current that flows through Q_1 in the event of a short circuit will develop a voltage drop across R_{E1} of sufficient

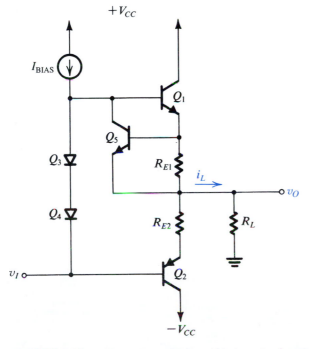

FIGURE 14.28 A class AB output stage with short-circuit protection. The protection circuit shown operates in the event of an output short circuit while v_O is positive.

value to turn Q_5 on. The collector of Q_5 will then conduct most of the current I_{BIAS}, robbing Q_1 of its base drive. The current through Q_1 will thus be reduced to a safe operating level.

This method of short-circuit protection is effective in ensuring device safety, but it has the disadvantage that under normal operation about 0.5 V drop might appear across each R_E. This means that the voltage swing at the output will be reduced by that much, in each direction. On the other hand, the inclusion of emitter resistors provides the additional benefit of protecting the output transistors against thermal runaway.

EXERCISE

D14.13 In the circuit of Fig. 14.28 let $I_{BIAS} = 2$ mA. Find the value of R_{E1} that causes Q_5 to turn on and absorb all 2 mA when the output current being sourced reaches 150 mA. For Q_5, $I_S = 10^{-14}$ A and $n = 1$. If the normal peak output current is 100 mA, find the voltage drop across R_{E1} and the collector current of Q_5.

Ans. 4.3 Ω; 430 mV; 0.3 μA

14.7.4 Thermal Shutdown

In addition to short-circuit protection, most IC power amplifiers are usually equipped with a circuit that senses the temperature of the chip and turns on a transistor in the event that the temperature exceeds a safe preset value. The turned-on transistor is connected in such a way that it absorbs the bias current of the amplifier, thus virtually shutting down its operation.

Figure 14.29 shows a thermal-shutdown circuit. Here, transistor Q_2 is normally off. As the chip temperature rises, the combination of the positive temperature coefficient of zener diode Z_1 and the negative temperature coefficient of V_{BE1} causes the voltage at the emitter of Q_1 to rise. This in turn raises the voltage at the base of Q_2 to the point at which Q_2 turns on.

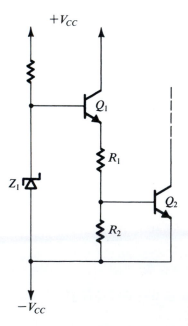

FIGURE 14.29 Thermal-shutdown circuit.

14.8 IC POWER AMPLIFIERS

A variety of IC power amplifiers are available. Most consist of a high-gain small-signal amplifier followed by a class AB output stage. Some have overall negative feedback already applied, resulting in a fixed closed-loop voltage gain. Others do not have on-chip feedback and are, in effect, op amps with large output-power capability. In fact, the output current-driving capability of any general-purpose op amp can be increased by cascading it with a class B or class AB output stage and applying overall negative feedback. The additional output stage can be either a discrete circuit or a hybrid IC such as the buffer discussed in the preceding section. In the following we discuss some power amplifier examples.

14.8.1 A Fixed-Gain IC Power Amplifier

Our first example is the LM380 (a product of National Semiconductor Corporation), which is a fixed-gain monolithic power amplifier. A simplified version of the internal circuit of the amplifier[3] is shown in Fig. 14.30. The circuit consists of an input differential amplifier utilizing Q_1 and Q_2 as emitter followers for input buffering, and Q_3 and Q_4 as a differential pair with an emitter resistor R_3. The two resistors R_4 and R_5 provide dc paths to ground for the base currents of Q_1 and Q_2, thus enabling the input signal source to be capacitively coupled to either of the two input terminals.

The differential amplifier transistors Q_3 and Q_4 are biased by two separate direct currents: Q_3 is biased by a current from the dc supply V_S through the diode-connected transistor Q_{10}, and resistor R_1; Q_4 is biased by a dc current from the output terminal through R_2. Under quiescent conditions (i.e., with no input signal applied) the two bias currents will be equal, and the current through and the voltage across R_3 will be zero. For the emitter current of Q_3 we can write

$$I_3 \simeq \frac{V_S - V_{EB10} - V_{EB3} - V_{EB1}}{R_1}$$

where we have neglected the small dc voltage drop across R_4. Assuming, for simplicity, all V_{EB} to be equal,

$$I_3 \simeq \frac{V_S - 3V_{EB}}{R_1} \tag{14.43}$$

For the emitter current of Q_4 we have

$$I_4 = \frac{V_O - V_{EB4} - V_{EB2}}{R_2}$$
$$\simeq \frac{V_O - 2V_{EB}}{R_2} \tag{14.44}$$

where V_O is the dc voltage at the output and we have neglected the small drop across R_5. Equating I_3 and I_4 and using the fact that $R_1 = 2R_2$ results in

$$V_O = \tfrac{1}{2}V_S + \tfrac{1}{2}V_{EB} \tag{14.45}$$

Thus the output is biased at approximately half the power-supply voltage, as desired for maximum output voltage swing. An important feature is the dc feedback from the output to

---

[3] The main objective of showing this circuit is to point out some interesting design features. The circuit is *not* a detailed schematic diagram of what is actually on the chip.

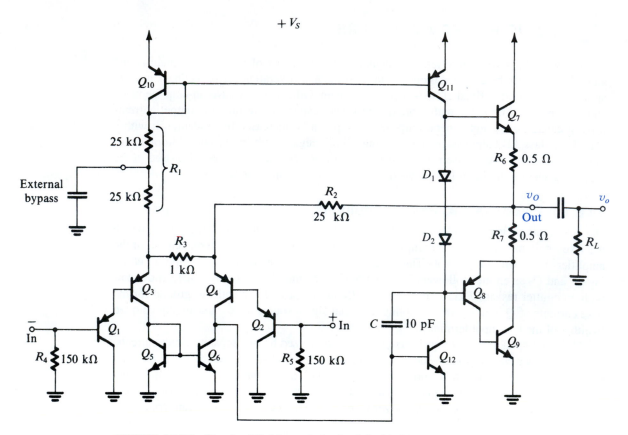

FIGURE 14.30 The simplified internal circuit of the LM380 IC power amplifier. (Courtesy National Semiconductor Corporation.)

the emitter of Q_4, through R_2. This dc feedback acts to stabilize the output dc bias voltage at the value in Eq. (14.45). Qualitatively, the dc feedback functions as follows: If for some reason V_O increases, a corresponding current increment will flow through R_2 and into the emitter of Q_4. Thus the collector current of Q_4 increases, resulting in a positive increment in the voltage at the base of Q_{12}. This, in turn, causes the collector current of Q_{12} to increase, thus bringing down the voltage at the base of Q_7 and hence V_O.

Continuing with the description of the circuit in Fig. 14.30, we observe that the differential amplifier (Q_3, Q_4) has a current mirror load composed of Q_5 and Q_6 (refer to Section 7.5.5 for a discussion of active loads). The single-ended output voltage signal of the first stage appears at the collector of Q_6 and thus is applied to the base of the second-stage common-emitter amplifier Q_{12}. Transistor Q_{12} is biased by the constant-current source Q_{11}, which also acts as its active load. In actual operation, however, the load of Q_{12} will be dominated by the reflected resistance due to R_L. Capacitor C provides frequency compensation (see Chapter 8).

The output stage is class AB, utilizing a compound *pnp* transistor (Q_8 and Q_9). Negative feedback is applied from the output to the emitter of Q_4 via resistor R_2. To find the closed-loop gain consider the small-signal equivalent circuit shown in Fig. 14.31. Here, we have replaced the second-stage common-emitter amplifier and the output stage with an inverting amplifier block with gain A. We shall assume that the amplifier A has high gain and high input resistance, and thus the input signal current into A is negligibly small. Under this assumption, Fig. 14.31 shows the analysis details with an input signal v_i applied to the inverting input

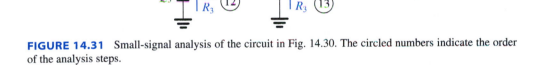

FIGURE 14.31 Small-signal analysis of the circuit in Fig. 14.30. The circled numbers indicate the order of the analysis steps.

terminal. The order of the analysis steps is indicated by the circled numbers. Note that since the input differential amplifier has a relatively large resistance, R_3, in the emitter circuit, most of the applied input voltage appears across R_3. In other words, the signal voltages across the emitter–base junctions of Q_1, Q_2, Q_3, and Q_4 are small in comparison to the voltage across R_3. Accordingly, the voltage gain can be found by writing a node equation at the collector of Q_6:

$$\frac{v_i}{R_3} + \frac{v_o}{R_2} + \frac{v_i}{R_3} = 0$$

which yields

$$\frac{v_o}{v_i} = -\frac{2R_2}{R_3} \approx -50 \text{ V/V}$$

EXERCISE

14.14 Denoting the total resistance between the collector of Q_6 and ground by R, show, using Fig. 14.31, that

$$\frac{v_o}{v_i} = \frac{-2R_2/R_3}{1 + (R_2/AR)}$$

which reduces to $(-2R_2/R_3)$ under the condition that $AR \gg R_2$.

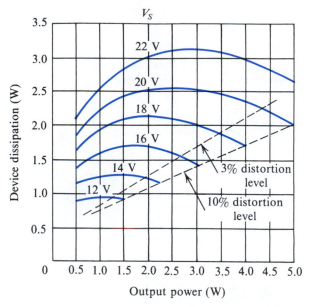

FIGURE 14.32 Power dissipation (P_D) versus output power (P_L) for the LM380 with $R_L = 8\ \Omega$. (Courtesy National Semiconductor Corporation.)

As was demonstrated in Chapter 8, one of the advantages of negative feedback is the reduction of nonlinear distortion. This is the case in the circuit of the LM380.

The LM380 is designed to operate from a single supply V_S in the range of 12 V to 22 V. The selection of supply voltage depends on the value of R_L and the required output power P_L. The manufacturer supplies curves for the device power dissipation versus output power for a given load resistance and various supply voltages. One such set of curves for $R_L = 8\ \Omega$ is shown in Fig. 14.32. Note the similarity to the class B power dissipation curve of Fig. 14.8. In fact, the reader can easily verify that the location and value of the peaks of the curves in Fig. 14.32 are accurately predicted by Eqs. (14.20) and (14.21), respectively (where $V_{CC} = \frac{1}{2}V_S$). The line labeled "3% distortion level" in Fig. 14.32 is the locus of the points on the various curves at which the distortion (THD) reaches 3%. A THD of 3% represents the onset of peak clipping due to output-transistor saturation.

The manufacturer also supplies curves for maximum power dissipation versus temperature (derating curves) similar to those discussed in Section 14.6 for discrete power transistors.

EXERCISES

14.15 The manufacturer specifies that for ambient temperatures below 25°C the LM380 can dissipate a maximum of 3.6 W. This is obtained under the condition that its dual-in-line package be soldered onto a printed-circuit board in close thermal contact with 6 square inches of 2-ounce copper foil. Above $T_A = 25°C$, the thermal resistance is $\theta_{JA} = 35°C/W$. T_{Jmax} is specified to be 150°C. Find the maximum power dissipation possible if the ambient temperature is to be 50°C.

Ans. 2.9 W

D14.16 It is required to use the LM380 to drive an 8-Ω loudspeaker. Use the curves of Fig. 14.32 to determine the maximum power supply possible while limiting the maximum power dissipation to the 2.9 W determined in Exercise 14.15. If for this application a 3% THD is allowed, find P_L and the peak-to-peak output voltage.

Ans. 20 V; 4.2 W; 16.4 V

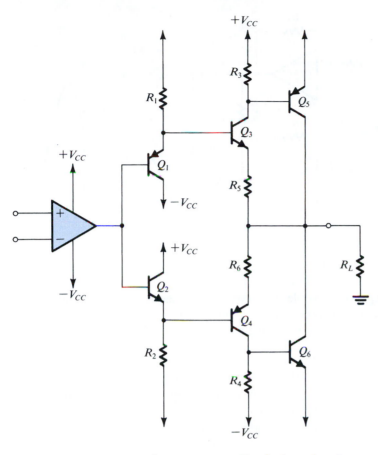

FIGURE 14.33 Structure of a power op amp. The circuit consists of an op amp followed by a class AB buffer similar to that discussed in Section 14.7.1. The output current capability of the buffer, consisting of Q_1, Q_2, Q_3, and Q_4, is further boosted by Q_5 and Q_6.

14.8.2 Power Op Amps

Figure 14.33 shows the general structure of a power op amp. It consists of a low-power op amp followed by a class AB buffer similar to that discussed in Section 14.7.1. The buffer consists of transistors Q_1, Q_2, Q_3, and Q_4, with bias resistors R_1 and R_2 and emitter degeneration resistors R_5 and R_6. The buffer supplies the required load current until the current increases to the point that the voltage drop across R_3 (in the current-sourcing mode) becomes sufficiently large to turn Q_5 on. Transistor Q_5 then supplies the additional load current required. In the current-sinking mode, Q_4 supplies the load current until sufficient voltage develops across R_4 to turn Q_6 on. Then, Q_6 sinks the additional load current. Thus the stage formed by Q_5 and Q_6 acts as a **current booster.** The power op amp is intended to be used with negative feedback in the usual closed-loop configurations. A circuit based on the structure of Fig. 14.33 is commercially available from National Semiconductor as LH0101. This op amp is capable of providing a continuous output current of 2 A, and with appropriate heat sinking can provide 40 W of output power (Wong and Johnson, 1981). The LH0101 is fabricated using hybrid thick-film technology.

14.8.3 The Bridge Amplifier

We conclude this section with a discussion of a circuit configuration that is popular in high-power applications. This is the bridge amplifier configuration shown in Fig. 14.34 utilizing

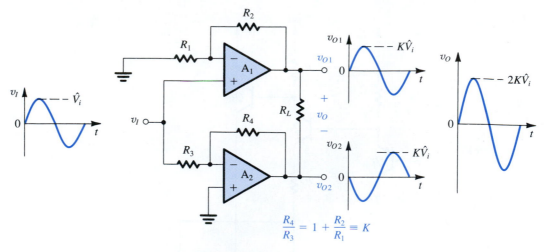

FIGURE 14.34 The bridge amplifier configuration.

two power op amps, A_1 and A_2. While A_1 is connected in the noninverting configuration with a gain $K = 1 + (R_2/R_1)$, A_2 is connected as an inverting amplifier with a gain of equal magnitude $K = R_4/R_3$. The load R_L is floating and is connected between the output terminals of the two op amps.

If v_I is a sinusoid with amplitude $\hat{V}_i$, the voltage swing at the output of each op amp will be $\pm K\hat{V}_i$, and that across the load will be $\pm 2K\hat{V}_i$. Thus, with op amps operated from ± 15-V supplies and capable of providing, say a ± 12-V output swing, an output swing of ± 24 V is obtained across the load of the bridge amplifier.

In designing bridge amplifiers, note should be taken of the fact that the peak current drawn from each op amp is $2K\hat{V}_i/R_L$. This effect can be taken into account by considering the load seen by each op amp (to ground) to be $R_L/2$.

EXERCISE

14.17 Consider the circuit of Fig. 14.34 with $R_1 = R_3 = 10$ kΩ, $R_2 = 5$ kΩ, $R_4 = 15$ kΩ, and $R_L = 8$ Ω. Find the voltage gain and the input resistance. The power supply used is ± 18 V. If v_I is a 20-V peak-to-peak sine wave, what is the peak-to-peak output voltage? What is the peak load current? What is the load power?

Ans. 3 V/V; 10 kΩ; 60 V; 3.75 A; 56.25 W

 ## 14.9 MOS POWER TRANSISTORS

Although, thus far in this chapter we have dealt exclusively with BJT circuits there exist MOS power transistors with specifications that are quite competitive with those of BJTs. In this section we consider the structure, characteristics, and application of power MOSFETs.

14.9.1 Structure of the Power MOSFET

The MOSFET structure studied in Chapter 4 (Fig. 4.1) is not suitable for high-power applications. To appreciate this fact, recall that the drain current of an n-channel MOSFET

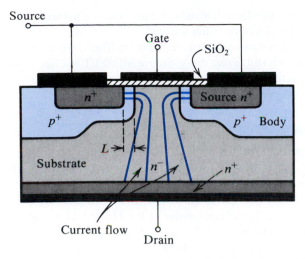

FIGURE 14.35 Double-diffused vertical MOS transistor (DMOS).

operating in the saturation region is given by

$$i_D = \frac{1}{2}\mu_n C_{ox}\left(\frac{W}{L}\right)(v_{GS} - V_t)^2 \tag{14.46}$$

It follows that to increase the current capability of the MOSFET, its width W should be made large and its channel length L should be made as small as possible. Unfortunately, however, reducing the channel length of the standard MOSFET structure results in a drastic reduction in its breakdown voltage. Specifically, the depletion region of the reverse-biased body-to-drain junction spreads into the short channel, resulting in breakdown at a relatively low voltage. Thus the resulting device would not be capable of handling the high voltages typical of power-transistor applications. For this reason, new structures had to be found for fabricating short-channel (1- to 2-μm) MOSFETs with high breakdown voltages.

At the present time the most popular structure for a power MOSFET is the double-diffused or DMOS transistor shown in Fig. 14.35. As indicated, the device is fabricated on a lightly doped n-type substrate with a heavily doped region at the bottom for the drain contact. Two diffusions[4] are employed, one to form the p-type body region and another to form the n-type source region.

The DMOS device operates as follows. Application of a positive gate voltage, v_{GS}, greater than the threshold voltage V_t, induces a lateral n channel in the p-type body region underneath the gate oxide. The resulting channel is short; its length is denoted L in Fig. 14.35. Current is then conducted by electrons from the source moving through the resulting short channel to the substrate and then vertically down the substrate to the drain. This should be contrasted with the lateral current flow in the standard small-signal MOSFET structure (Chapter 4).

Even though the DMOS transistor has a short channel, its breakdown voltage can be very high (as high as 600 V). This is because the depletion region between the substrate and the body extends mostly in the lightly doped substrate and does not spread into the channel. The result is a MOS transistor that simultaneously has a high current capability (50 A is possible)

---

[4] See Appendix A for a description of the IC fabrication process.

as well as the high breakdown voltage just mentioned. Finally, we note that the vertical structure of the device provides efficient utilization of the silicon area.

An earlier structure used for power MOS transistors deserves mention. This is the V-groove MOS device [see Severns (1984)]. Although still in use, the V-groove MOSFET has lost application ground to the vertical DMOS structure of Fig. 14.35, except possibly for high-frequency applications. Because of space limitations, we shall not describe the V-groove MOSFET.

14.9.2 Characteristics of Power MOSFETs

In spite of their radically different structure, power MOSFETs exhibit characteristics that are quite similar to those of the small-signal MOSFETs studied in Chapter 4. Important differences exist, however, and these are discussed next.

Power MOSFETs have threshold voltages in the range of 2 V to 4 V. In saturation, the drain current is related to v_{GS} by the square-law characteristic of Eq. (14.46). However, as shown in Fig. 14.36, the i_D–v_{GS} characteristic becomes linear for larger values of v_{GS}. The linear portion of the characteristic occurs as a result of the high electric field along the short channel, causing the velocity of charge carriers to reach an upper limit, a phenomenon known as **velocity saturation.** The drain current is then given by

$$i_D = \tfrac{1}{2} C_{ox} W U_{\text{sat}} (v_{GS} - V_t) \tag{14.47}$$

where U_{sat} is the saturated velocity value (5×10^6 cm/s for electrons in silicon). The linear i_D–v_{GS} relationship implies a constant g_m in the velocity-saturation region. It is interesting to note that g_m is proportional to W, which is usually large for power devices; thus power MOSFETs exhibit relatively high transconductance values.

The i_D–v_{GS} characteristic shown in Fig. 14.36 includes a segment labeled "subthreshold." Though of little significance for power devices, the subthreshold region of operation is of interest in very-low-power applications (see Section 4.1.9).

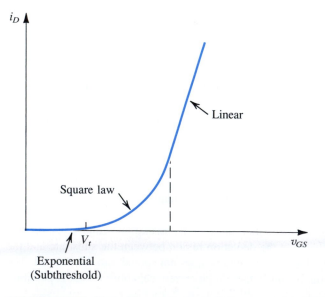

FIGURE 14.36 Typical i_D–v_{GS} characteristic for a power MOSFET.

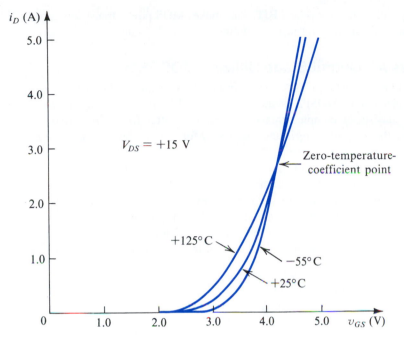

FIGURE 14.37 The i_D–v_{GS} characteristic curve of a power MOS transistor (IRF 630, Siliconix) at case temperatures of –55°C, +25°C, and +125°C. (Courtesy Siliconix Inc.)

14.9.3 Temperature Effects

Of considerable interest in the design of MOS power circuits is the variation of the MOSFET characteristics with temperature, illustrated in Fig. 14.37. Observe that there is a value of v_{GS} (in the range of 4 V to 6 V for most power MOSFETs) at which the temperature coefficient of i_D is zero. At higher values of v_{GS}, i_D exhibits a negative temperature coefficient. This is a significant property: It implies that a MOSFET operating beyond the zero-temperature-coefficient point does not suffer from the possibility of thermal runaway. This is *not* the case, however, at low currents (i.e., lower than the zero-temperature-coefficient point). In the (relatively) low-current region, the temperature coefficient of i_D is positive, and the power MOSFET can easily suffer thermal runaway (with unhappy consequences). Since class AB output stages are biased at low currents, means must be provided to guard against thermal runaway.

The reason for the positive temperature coefficient of i_D at low currents is that $v_{OV} = (v_{GS} - V_t)$ is relatively low, and the temperature dependence is dominated by the negative temperature coefficient of V_t (in the range of –3 mV/°C to –6 mV/°C) which causes v_{OV} to rise with temperature.

14.9.4 Comparison with BJTs

The power MOSFET does not suffer from second breakdown, which limits the safe operating area of BJTs. Also, power MOSFETs do not require the large dc base-drive currents of power BJTs. Note, however, that the driver stage in a MOS power amplifier should be capable of supplying sufficient current to charge and discharge the MOSFET's large and nonlinear input capacitance in the time allotted. Finally, the power MOSFET features, in general, a higher speed of operation than the power BJT. This makes MOS power transistors especially suited to switching applications—for instance, in motor-control circuits.

14.9.5 A Class AB Output Stage Utilizing MOSFETs

As an application of power MOSFETs, we show in Fig. 14.38 a class AB output stage utilizing a pair of complementary MOSFETs and employing BJTs for biasing and in the driver stage. The latter consists of complementary Darlington emitter followers formed by Q_1 through Q_4 and has the low output resistance necessary for driving the output MOSFETs at high speeds.

Of special interest in the circuit of Fig. 14.38 is the bias circuit utilizing two V_{BE} multipliers formed by Q_5 and Q_6 and their associated resistors. Transistor Q_6 is placed in direct thermal contact with the output transistors; this is achieved by simply mounting Q_6 on their common heat sink. Thus, by the appropriate choice of the V_{BE} multiplication factor of Q_6, the bias voltage V_{GG} (between the gates of the output transistors) can be made to decrease with temperature at the same rate as that of the sum of the threshold voltages $(V_{tN} + |V_{tP}|)$

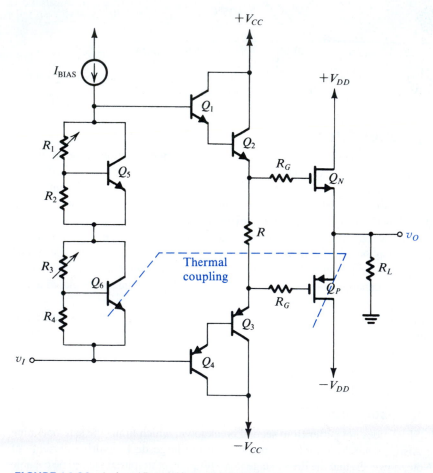

FIGURE 14.38 A class AB amplifier with MOS output transistors and BJT drivers. Resistor R_3 is adjusted to provide temperature compensation while R_1 is adjusted to yield the desired value of quiescent current in the output transistors. Resistors R_G are used to suppress parasitic oscillations at high frequencies. Typically, $R_G = 100\ \Omega$.

of the output MOSFETs. In this way the quiescent current of the output transistors can be stabilized against temperature variations.

Analytically, V_{GG} is given by

$$V_{GG} = \left(1 + \frac{R_3}{R_4}\right)V_{BE6} + \left(1 + \frac{R_1}{R_2}\right)V_{BE5} - 4V_{BE} \tag{14.48}$$

Since V_{BE6} is thermally coupled to the output devices while the other BJTs remain at constant temperature, we have

$$\frac{\partial V_{GG}}{\partial T} = \left(1 + \frac{R_3}{R_4}\right)\frac{\partial V_{BE6}}{\partial T} \tag{14.49}$$

which is the relationship needed to determine R_3/R_4 so that $\partial V_{GG}/\partial T = \partial(V_{tN} + |V_{tP}|)/\partial T$. The other V_{BE} multiplier is then adjusted to yield the value of V_{GG}, required for the desired quiescent current in Q_N and Q_P.

EXERCISES

14.18 For the circuit in Fig. 14.38, find the ratio R_3/R_4 that provides temperature stabilization of the quiescent current in Q_N and Q_P. Assume that $|V_t|$ changes at -3 mV/°C and that $\partial V_{BE}/\partial T = -2$ mV/°C.

Ans. 2

14.19 For the circuit in Fig. 14.38 assume that the BJTs have a nominal V_{BE} of 0.7 V and that the MOSFETs have $|V_t| = 3$ V and $\mu_n C_{ox}(W/L) = 2$ A/V^2. It is required to establish a quiescent current of 100 mA in the output stage and 20 mA in the driver stage. Find $|V_{GS}|$, V_{GG}, R, and R_1/R_2. Use the value of R_3/R_4 found in Exercise 14.18.

Ans. 3.32 V; 6.64 V; 332 Ω; 9.5

14.10 SPICE SIMULATION EXAMPLE

We conclude this chapter by presenting an example that illustrates the use of SPICE in the analysis of output circuits.

EXAMPLE 14.6

CLASS B OUTPUT STAGE

We investigate the operation of the class B output stage whose Capture schematic is shown in Fig. 14.39. For the power transistors, we use the discrete BJTs MJE243 and MJE253 (from ON Semiconductor)[5] which are rated for a maximum continuous collector current $I_{Cmax} = 4$ A and a maximum collector-emitter voltage of $V_{CEmax} = 100$ V. To permit comparison with the hand analysis performed in Example 14.1, we use, in the simulation, component and voltage values identical

---

[5] In PSpice, we have created BJT parts for these power transistors based on the values of the SPICE model parameters available on the data sheets available from ON Semiconductor. Readers can find these parts (labelled QMJE243 and QMJE253) in the SEDRA.olb library which is available on the CD accompanying this book as well as at www.sedrasmith.org.

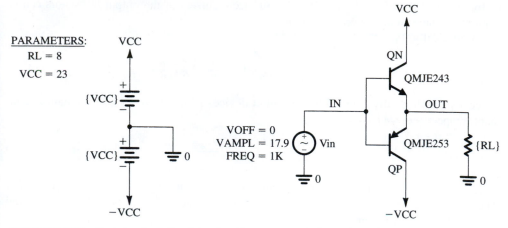

FIGURE 14.39 Capture schematic of the class B output stage in Example 14.6.

(or close) to those of the circuit designed in Example 14.1. Specifically, we use a load resistance of 8 Ω, an input sine-wave signal of 17.9-V peak and 1-kHz frequency, and 23-V power supplies. In PSpice, a transient-analysis simulation is performed over the interval 0 ms to 3 ms, and the waveforms of various node voltages and branch currents are plotted. In this example, Probe (the graphical interface of PSpice) is utilized to compute various power-dissipation values. Some of the resulting waveforms are displayed in Fig. 14.40. The upper and middle graphs show the load voltage and current, respectively. The peak voltage amplitude is 16.9 V, and the peak current

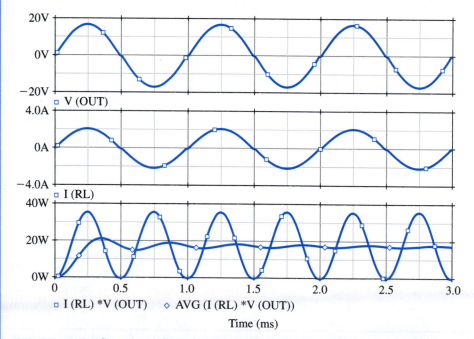

FIGURE 14.40 Several waveforms associated with the class B output stage (shown in Fig. 14.39) when excited by a 17.9-V, 1-kHz sinusoidal signal. The upper graph displays the voltage across the load resistance, the middle graph displays the load current, and the lower graph displays the instantaneous and average power dissipated by the load.

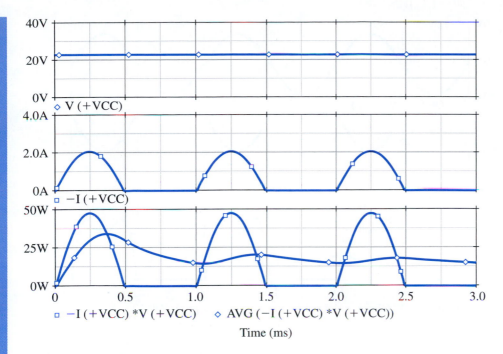

FIGURE 14.41 The voltage (upper graph), current (middle graph), and instantaneous and average power (bottom graph) supplied by the positive voltage supply ($+V_{CC}$) in the circuit of Fig. 14.39.

amplitude is 2.1 A. If one looks carefully, one can observe that both exhibit crossover distortion. The bottom graph displays the instantaneous and the average power dissipated in the load resistance as computed using Probe by multiplying the voltage and current values to obtain the instantaneous power, and taking a running average for the average load power P_L. The transient behavior of the average load power, which eventually settles into a quasiconstant steady state of about 17.6 W, is an artifact of the PSpice algorithm used to compute the running average of a waveform.

The upper two graphs of Fig. 14.41 show the voltage and current waveforms, respectively, of the positive supply, $+V_{CC}$. The bottom graph shows the instantaneous and average power supplied by $+V_{CC}$. Similar waveforms can be plotted for the negative supply, $-V_{CC}$. The average power provided by each supply is found to be about 15 W, for a total supply power P_S of 30 W. Thus, the power-conversion efficiency can be computed to be

$$\eta = P_L/P_S = \frac{17.6}{30} \times 100\% = 58.6\%$$

Figure 14.42 shows plots of the voltage, current, and power waveforms associated with transistor Q_P. Similar waveforms can be obtained for Q_N. As expected, the voltage waveform is a sinusoid, and the current waveform consists of half-sinusoids. The waveform of the instantaneous power, however, is rather unusual. It indicates the presence of some distortion as a result of driving the transistors rather hard. This can be verified by reducing the amplitude of the input signal. Specifically, by reducing the amplitude to about 17 V, the "dip" in the power waveform vanishes. The average power dissipated in each of Q_N and Q_P can be computed by Probe and are found to be approximately 6 W.

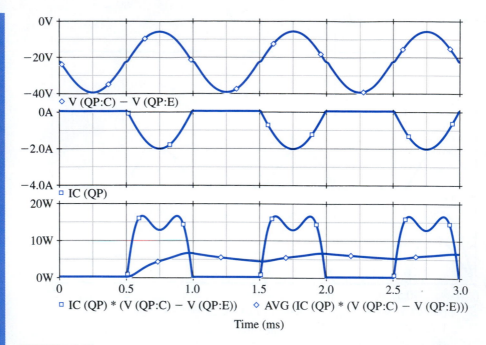

FIGURE 14.42 Waveforms of the voltage across, the current through, and the power dissipated in the *pnp* transistor Q_P of the output stage shown in Fig. 14.39.

TABLE 14.1	Various Power Terms Associated with the Class B Output Stage Shown in Fig. 14.39 as Computed by Hand and by PSpice Analysis			
Power/Efficiency	Equation	Hand Analysis (Example 14.1)	PSpice	Error %[1]
P_S	$\dfrac{2}{\pi}\dfrac{\hat{V}_o}{R_L}V_{CC}$	31.2 W	30.0 W	4
P_D	$\dfrac{2}{\pi}\dfrac{\hat{V}_o}{R_L}V_{CC} - \dfrac{1}{2}\dfrac{\hat{V}_o^2}{R_L}$	13.0 W	12.4 W	4.6
P_L	$\dfrac{1}{2}\dfrac{\hat{V}_o^2}{R_L}$	18.2 W	17.6 W	3.3
η	$\dfrac{P_L}{P_S} \times 100\%$	58.3%	58.6%	−0.5

[1] Relative percentage error between the values predicted by hand and by PSpice.

Table 14.1 provides a comparison of the results found from the PSpice simulation and the corresponding values obtained using hand analysis in Example 14.1. Observe that the two sets of results are quite close.

To investigate the crossover distortion further, we present in Fig. 14.43 a plot of the voltage transfer characteristic (VTC) of the class B output stage. This plot is obtained through a dc-analysis simulation with v_{IN} swept over the range −10 V to +10 V in 1.0-mV increments. Using

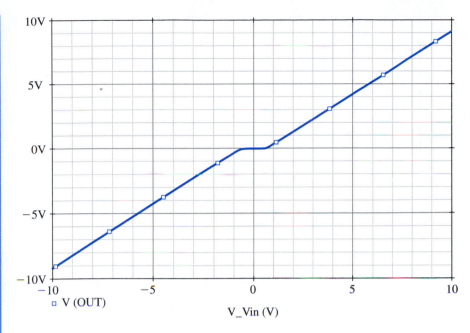

FIGURE 14.43   Transfer characteristic of the class B output stage of Fig. 14.39.

Probe we determine that the slope of the VTC is nearly unity and that the dead band extends from −0.60 to +0.58 V. The effect of the crossover distortion can be quantified by performing a Fourier analysis on the output voltage waveform in PSpice. This analysis decomposes the waveform generated through a transient analysis into its Fourier-series components. Further, PSpice computes the total harmonic distortion (THD) of the output waveform. The results obtained from the simulation output file are as follows:

```
FOURIER COMPONENTS OF TRANSIENT RESPONSE V(OUT)

DC COMPONENT = -1.525229E-02
```

HARMONIC NO	FREQUENCY (HZ)	FOURIER COMPONENT	NORMALIZED COMPONENT	PHASE (DEG)	NORMALIZED PHASE (DEG)
1	1.000E+03	1.674E+01	1.000E+00	-2.292E-03	0.000E+00
2	2.000E+03	9.088E-03	5.428E-04	9.044E+01	9.044E+01
3	3.000E+03	2.747E-01	1.641E-02	-1.799E+02	-1.799E+02
4	4.000E+03	4.074E-03	2.433E-04	9.035E+01	9.036E+01
5	5.000E+03	1.739E-01	1.039E-02	-1.799E+02	-1.799E+02
6	6.000E+03	5.833E-04	3.484E-05	9.159E+01	9.161E+01
7	7.000E+03	1.195E-01	7.140E-03	-1.800E+02	-1.799E+02
8	8.000E+03	5.750E-04	3.435E-05	9.128E+01	9.129E+01
9	9.000E+03	9.090E-02	5.429E-03	-1.800E+02	-1.799E+02
10	1.000E+04	3.243E-04	1.937E-05	9.120E+01	9.122E+01

```
TOTAL HARMONIC DISTORTION = 2.140017E+00 PERCENT
```

These Fourier components are used to plot the line spectrum shown in Fig. 14.44. We note that the output waveform is rather rich in odd harmonics and that the resulting THD is rather high (2.14%).

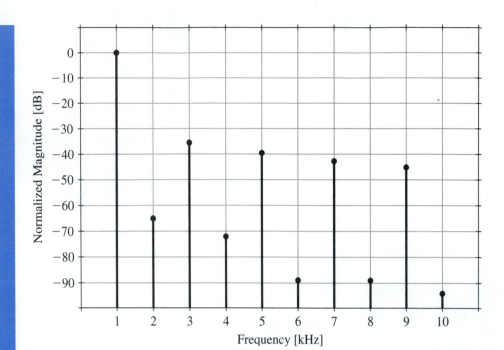

FIGURE 14.44 Fourier-series components of the output waveform of the class B output stage in Fig. 14.39.

SUMMARY

- Output stages are classified according to the transistor conduction angle: class A (360°), class AB (slightly more than 180°), class B (180°), and class C (less than 180°).

- The most common class A output stage is the emitter follower. It is biased at a current greater than the peak load current.

- The class A output stage dissipates its maximum power under quiescent conditions ($v_O = 0$). It achieves a maximum power-conversion efficiency of 25%.

- The class B stage is biased at zero current, and thus dissipates no power in quiescence.

- The class B stage can achieve a power conversion efficiency as high as 78.5%. It dissipates its maximum power for $\hat{V}_o = (2/\pi)V_{CC}$.

- The class B stage suffers from crossover distortion.

- The class AB output stage is biased at a small current; thus both transistors conduct for small input signals, and crossover distortion is virtually eliminated.

- Except for an additional small quiescent power dissipation, the power relationships of the class AB stage are similar to those in class B.

- To guard against the possibility of thermal runaway, the bias voltage of the class AB circuit is made to vary with temperature in the same manner as does V_{BE} of the output transistors.

- To facilitate the removal of heat from the silicon chip, power devices are usually mounted on heat sinks. The maximum power that can be safely dissipated in the device is given by

$$P_{D\text{max}} = \frac{T_{J\text{max}} - T_A}{\theta_{JC} + \theta_{CS} + \theta_{SA}}$$

where $T_{J\text{max}}$ and θ_{JC} are specified by the manufacturer, while θ_{CS} and θ_{SA} depend on the heat-sink design.

- Use of the Darlington configuration in the class AB output stage reduces the base-current drive requirement. In integrated circuits, the compound *pnp* configuration is commonly used.

■ Output stages are usually equipped with circuitry that, in the event of a short circuit, can turn on and limit the base-current drive, and hence the emitter current, of the output transistors.

■ IC power amplifiers consist of a small-signal voltage amplifier cascaded with a high-power output stage. Overall feedback is applied either on-chip or externally.

■ The bridge amplifier configuration provides, across a floating load, a peak-to-peak output voltage which is twice that possible from a single amplifier with a grounded load.

■ The DMOS transistor is a short-channel power device capable of both high-current and high-voltage operation.

■ The drain current of a power MOSFET exhibits a positive temperature coefficient at low currents, and thus the device can suffer thermal runaway. At high currents the temperature coefficient of i_D is negative.

PROBLEMS

SECTION 14.2: CLASS A OUTPUT STAGE

14.1 A class A emitter follower, biased using the circuit shown in Fig. 14.2, uses $V_{CC} = 5$ V, $R = R_L = 1$ kΩ, with all transistors (including Q_3) identical. Assume $V_{BE} = 0.7$ V, $V_{CEsat} = 0.3$ V, and β to be very large. For linear operation, what are the upper and lower limits of output voltage, and the corresponding inputs? How do these values change if the emitter–base junction area of Q_3 is made twice as big as that of Q_2? Half as big?

14.2 A source-follower circuit using NMOS transistors is constructed following the pattern shown in Fig. 14.2. All three transistors used are identical, with $V_t = 1$ V and $\mu_n C_{ox} W/L = 20$ mA/V^2; $V_{CC} = 5$ V, $R = R_L = 1$ kΩ. For linear operation, what are the upper and lower limits of the output voltage, and the corresponding inputs?

D14.3 Using the follower configuration shown in Fig. 14.2 with ±9-V supplies, provide a design capable of ±7-V outputs with a 1-kΩ load, using the smallest possible total supply current. You are provided with four identical, high-β BJTs and a resistor of your choice.

D14.4 An emitter follower using the circuit of Fig. 14.2, for which the output voltage range is ±5 V, is required using $V_{CC} = 10$ V. The circuit is to be designed such that the current variation in the emitter-follower transistor is no greater than a factor of 10, for load resistances as low as 100 Ω. What is the value of R required? Find the incremental voltage gain of the resulting follower at $v_O = +5$, 0, and -5 V, with a 100-Ω load. What is the percentage change in gain over this range of v_O?

*14.5 Consider the operation of the follower circuit of Fig. 14.2 for which $R_L = V_{CC}/I$, when driven by a square wave such that the output ranges from $+V_{CC}$ to $-V_{CC}$ (ignoring V_{CEsat}). For this situation, sketch the equivalent of Fig. 14.4 for v_O, i_{C1}, and p_{D1}. Repeat for a square-wave output that has peak levels of $\pm V_{CC}/2$. What is the average power dissipation in Q_1 in each case? Compare these results to those for sine waves of peak amplitude V_{CC} and $V_{CC}/2$, respectively.

14.6 Consider the situation described in Problem 14.5. For square-wave outputs having peak-to-peak values of $2V_{CC}$ and V_{CC}, and for sine waves of the same peak-to-peak values, find the average power loss in the current-source transistor Q_2.

14.7 Reconsider the situation described in Exercise 14.4 for variation in V_{CC}—specifically for $V_{CC} = 16$ V, 12 V, 10 V, and 8 V. Assume V_{CEsat} is nearly zero. What is the power-conversion efficiency in each case?

14.8 The BiCMOS follower shown in Fig. P14.8 uses devices for which $V_{BE} = 0.7$ V, $V_{CEsat} = 0.3$ V, $\mu_n C_{ox} W/L = 20$ mA/V^2,

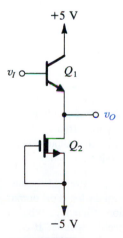

FIGURE P14.8

and $V_t = -2$ V. For linear operation, what is the range of output voltages obtained with $R_L = \infty$? With $R_L = 100$ Ω? What is the smallest load resistor allowed for which a 1-V peak sine-wave output is available? What is the corresponding power-conversion efficiency?

SECTION 14.3: CLASS B OUTPUT STAGE

14.9 Consider the circuit of a complementary-BJT class B output stage. For what amplitude of input signal does the crossover distortion represent a 10% loss in peak amplitude?

14.10 Consider the feedback configuration with a class B output stage shown in Fig. 14.9. Let the amplifier gain $A_0 = 100$ V/V. Derive an expression for v_O versus v_I, assuming that $|V_{BE}| = 0.7$ V. Sketch the transfer characteristic v_O versus v_I, and compare it with that without feedback.

14.11 Consider the class B output stage, using enhancement MOSFETs, shown in Fig. P14.11. Let the devices have $|V_t| = 1$ V and $\mu C_{ox}W/L = 200$ μA/V^2. With a 10-kHz sine-wave input of 5-V peak and a high value of load resistance, what peak output would you expect? What fraction of the sine-wave period does the crossover interval represent? For what value of load resistor is the peak output voltage reduced to half the input?

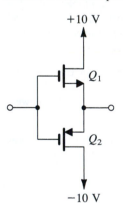

FIGURE P14.11

14.12 Consider the complementary-BJT class B output stage and neglect the effects of finite V_{BE} and V_{CEsat}. For ±10-V power supplies and a 100-Ω load resistance, what is the maximum sine-wave output power available? What supply power corresponds? What is the power-conversion efficiency? For output signals of half this amplitude, find the output power, the supply power, and the power-conversion efficiency.

D14.13 A class B output stage operates from ±5-V supplies. Assuming relatively ideal transistors, what is the output voltage for maximum power-conversion efficiency? What is the output voltage for maximum device dissipation? If each of the output devices is individually rated for 1-W dissipation, and a factor-of-2 safety margin is to be used, what is the smallest value of load resistance that can be tolerated, if operation is always at full output voltage? If operation is allowed at half the full output voltage, what is the smallest load permitted? What is the greatest possible output power available, in each case?

D14.14 A class B output stage is required to deliver an average power of 100 W into a 16-Ω load. The power supply should be 4 V greater than the corresponding peak sine-wave output voltage. Determine the power-supply voltage required (to the nearest volt in the appropriate direction), the peak current from each supply, the total supply power, and the power-conversion efficiency. Also, determine the maximum possible power dissipation in each transistor for a sine-wave input.

14.15 Consider the class B BJT output stage with a square-wave output voltage of amplitude $\hat{V}_o$ across a load R_L and employing power supplies ±V_{SS}. Neglecting the effects of finite V_{BE} and V_{CEsat}, determine the load power, the supply power, the power-conversion efficiency, the maximum attainable power-conversion efficiency and the corresponding value of $\hat{V}_o$, and the maximum available load power. Also find the value of $\hat{V}_o$ at which the power dissipation in the transistors reaches its peak, and the corresponding value of power-conversion efficiency.

SECTION 14.4: CLASS AB OUTPUT STAGE

D14.16 Design the quiescent current of a class AB BJT output stage so that the incremental voltage gain for v_I in the vicinity of the origin is in excess of 0.99 V/V for loads larger than 100 Ω. Assume that the BJTs have V_{BE} of 0.7 V at a current of 100 mA and determine the value of V_{BB} required.

D14.17 The design of a class AB MOS output stage is being considered. The available devices have $|V_t| = 1$ V and $\mu C_{ox}W/L = 200$ mA/V^2. What value of gate-to-gate bias voltage, V_{GG}, is required to reduce the incremental output resistance in the quiescent state to 10 Ω?

*14.18 A class AB output stage, resembling that in Fig. 14.11 but utilizing a single supply of +10 V and biased at $V_I = 6$ V, is capacitively coupled to a 100-Ω load. For transistors for which $|V_{BE}| = 0.7$ V at 1 mA and for a bias voltage $V_{BB} = 1.4$ V, what quiescent current results? For a step change in output from 0 to –1 V, what input step is required? Assuming transistor saturation voltages of zero, find the largest possible positive-going and negative-going steps at the output.

SECTION 14.5: BIASING THE CLASS AB CIRCUIT

D14.19 Consider the diode-biased class AB circuit of Fig. 14.14. For $I_{BIAS} = 100$ μA, find the relative size (n) that should be used for the output devices (in comparison to the biasing devices) to ensure an output resistance of 10 Ω or less.

D*14.20 A class AB output stage using a two-diode bias network as shown in Fig. 14.14 utilizes diodes having the same junction area as the output transistors. For $V_{CC} = 10$ V, $I_{BIAS} = 0.5$ mA, $R_L = 100$ Ω, $\beta_N = 50$, and $|V_{CEsat}| = 0$ V, what is the quiescent current? What are the largest possible positive and negative output signal levels? To achieve a positive peak output level equal to the negative peak level, what value of β_N is needed if I_{BIAS} is not changed? What value of I_{BIAS} is needed if β_N is held at 50? For this value, what does I_Q become?

**14.21 A class AB output stage using a two-diode bias network as shown in Fig. 14.14 utilizes diodes having the same junction area as the output transistors. At a room temperature of about 20°C the quiescent current is 1 mA and $|V_{BE}| = 0.6$ V. Through a manufacturing error, the thermal coupling between the output transistors and the biasing diode-connected transistors is omitted. After some output activity, the output devices heat up to 70°C while the biasing devices remain at 20°C. Thus while the V_{BE} of each device remains unchanged, the quiescent current in the output devices increases. To calculate the new current value, recall that there are two effects: I_S increases by about 14%/°C and $V_T = kT/q$ changes, where $T = (273° + \text{temperature in } °\text{C})$, and $V_T = 25$ mV only at 20°C. However, you may assume that β_N remains almost constant. This assumption is based on the fact that β increases with temperature but decreases with current (see Fig. 5.22). What is the new value of I_Q? If the power supply is ±20 V, what additional power is dissipated? If thermal runaway occurs, and the temperature of the output transistors increases by 10°C for every watt of additional power dissipation, what additional temperature rise and current increase result?

D14.22 Figure P14.22 shows a MOSFET class AB output stage. All transistors have $|V_t| = 1$ V and $k_1 = k_2 = nk_3 = nk_4$,

where $k = \mu C_{ox} W/L$ is the MOSFET transconductance parameter. Also, $k_3 = 2$ mA/V^2. For $I_{BIAS} = 100$ μA and $R_L = 1$ kΩ find the value of n that results in a small-signal gain of 0.99 for output voltages around zero. Find the corresponding value of I_Q.

D14.23 Repeat Example 14.3 for the situation in which the peak positive output current is 200 mA. Use the same general approach to safety margins. What are the values of R_1 and R_2 you have chosen?

**14.24 A V_{BE} multiplier is designed with equal resistances for nominal operation at a terminal current of 1 mA, with half the current flowing in the bias network. The initial design is based on $\beta = \infty$ and $V_{BE} = 0.7$ V at 1 mA.

(a) Find the required resistor values and the terminal voltage.
(b) Find the terminal voltage that results when the terminal current increases to 2 mA. Assume $\beta = \infty$.
(c) Repeat (b) for the case the terminal current becomes 10 mA.
(d) Repeat (c) using the more realistic value of $\beta = 100$.

SECTION 14.6: POWER BJTs

D14.25 A particular transistor having a thermal resistance $\theta_{JA} = 2°$C/W is operating at an ambient temperature of 30°C with a collector–emitter voltage of 20 V. If long life requires a maximum junction temperature of 130°C, what is the corresponding device power rating? What is the greatest average collector current that should be considered?

14.26 A particular transistor has a power rating at 25°C of 200 mW, and a maximum junction temperature of 150°C. What is its thermal resistance? What is its power rating when operated at an ambient temperature of 70°C? What is its junction temperature when dissipating 100 mW at an ambient temperature of 50°C?

14.27 A power transistor operating at an ambient temperature of 50°C, and an average emitter current of 3A, dissipates 30 W. If the thermal resistance of the transistor is known to be less than 3°C/W, what is the greatest junction temperature you would expect? If the transistor V_{BE} measured using a pulsed emitter current of 3 A at a junction temperature of 25°C is 0.80 V, what average V_{BE} would you expect under normal operating conditions? (Use a temperature coefficient of −2 mV/°C.)

14.28 For a particular application of the transistor specified in Example 14.4, extreme reliability is essential. To improve reliability the maximum junction temperature is to be limited to 100°C. What are the consequences of this decision for the conditions specified?

14.29 A power transistor is specified to have a maximum junction temperature of 130°C. When the device is operated at this junction temperature with a heat sink, the case temperature

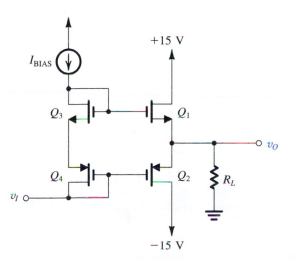

FIGURE P14.22

is found to be 90°C. The case is attached to the heat sink with a bond having a thermal resistance $\theta_{CS} = 0.5$°C/W and the thermal resistance of the heat sink $\theta_{SA} = 0.1$°C/W. If the ambient temperature is 30°C what is the power being dissipated in the device? What is the thermal resistance of the device, θ_{JC}, from junction to case?

14.30 A power transistor for which $T_{Jmax} = 180$°C can dissipate 50 W at a case temperature of 50°C. If it is connected to a heat sink using an insulating washer for which the thermal resistance is 0.6°C/W, what heat-sink temperature is necessary to ensure safe operation at 30 W? For an ambient temperature of 39°C, what heat-sink thermal resistance is required? If, for a particular extruded-aluminum-finned heat sink, the thermal resistance is still air is 4.5°C/W per centimeter of length, how long a heat sink is needed?

14.31 An *npn* power transistor operating at $I_C = 10$ A is found to have a base current of 0.5 A and an incremental base input resistance of 0.95 Ω. What value of r_x do you suspect? (At this high current density, $n = 2$.)

14.32 A base spreading resistance (r_x) of 0.8 Ω has been measured for an *npn* power transistor operating at $I_C = 5$ A, with a base–emitter voltage of 1.05 V and a base current of 190 mA. Assuming that $n = 2$ for high-current-density operation, what base–emitter voltage would you expect for operation at $I_C = 2$ A?

SECTION 14.7: VARIATIONS ON THE CLASS AB CONFIGURATION

14.33 Use the results given in the answer to Exercise 14.11 to determine the input current of the circuit in Fig. 14.24 for $v_I = 0$ and ±10 V with infinite and 100-Ω loads.

D*14.34** Consider the circuit of Fig. 14.24 in which Q_1 and Q_2 are matched, and Q_3 and Q_4 are matched but have three times the junction area of the others. For $V_{CC} = 10$ V, find values for resistors R_1 through R_4 which allow for a base current of at least 10 mA in Q_3 and Q_4 at $v_I = +5$ V (when a load demands it) with at most a 2-to-1 variation in currents in Q_1 and Q_2, and a no-load quiescent current of 40 mA in Q_3 and Q_4; $\beta_{1,2} \geq 150$, and $\beta_{3,4} \geq 50$. For input voltages around 0 V, estimate the output resistance of the overall follower driven by a source having zero resistance. For an input voltage of +1 V and a load resistance of 2 Ω, what output voltage results? Q_1 and Q_2 have $|V_{BE}|$ of 0.7 V at a current of 10 mA and exhibit a constant $n = 1$.

14.35 A circuit resembling that in Fig. 14.24 uses four matched transistors for which $|V_{BE}| = 0.7$ V at 10 mA, $n = 1$, and $\beta \geq 50$. Resistors R_1 and R_2 are replaced by 2-mA current sources, and $R_3 = R_4 = 0$. What quiescent current flows in the output transistors? What bias current flows in the bases of

the input transistors? Where does it flow? What is the net input current (the offset current) for a β mismatch of 10%? For a load resistance $R_L = 100$ Ω, what is the input resistance? What is the small-signal voltage gain?

14.36 Characterize a Darlington compound transistor formed from two *npn* BJTs for which $\beta \geq 50$, $V_{BE} = 0.7$ V at 1 mA, and $n = 1$. For operation at 10 mA, what values would you expect for β_{eq}, V_{BEeq}, $r_{\pi eq}$, and g_{meq}.

14.37 For the circuit in Fig. P14.37 in which the transistors have $V_{BE} = 0.7$ V and $\beta = 100$, find i_c, g_{meq}, v_o/v_i, and R_{in}.

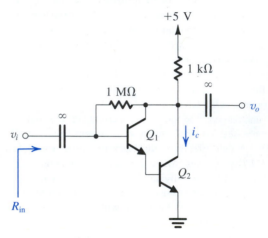

FIGURE P14.37

**14.38 The BJTs in the circuit of Fig. P14.38 have $\beta_P = 10$, $\beta_N = 100$, $|V_{BE}| = 0.7$ V, and $|V_A| = 100$ V.

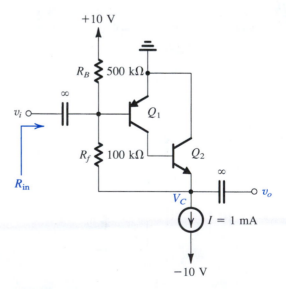

FIGURE P14.38

(a) Find the dc collector current of each transistor and the value of V_C.

(b) Replacing each BJT with its hybrid-π model, show that

$$\frac{v_o}{v_i} \simeq g_{m1}[r_{o1} \| \beta_N (r_{o2} \| R_f)]$$

(c) Find the values of v_o / v_i and R_{in}.

D****14.39** Consider the compound-transistor class AB output stage shown in Fig. 14.27 in which Q_2 and Q_4 are matched transistors with $V_{BE} = 0.7$ V at 10 mA and $\beta = 100$, Q_1 and Q_5 have $V_{BE} = 0.7$ V at 1-mA currents and $\beta = 100$, and Q_3 has $V_{EB} = 0.7$ V at a 1-mA current and $\beta = 10$. All transistors have $n = 1$. Design the circuit for a quiescent current of 2 mA in Q_2 and Q_4, I_{BIAS} that is 100 times the standby base current in Q_1, and a current in Q_5 that is nine times that in the associated resistors. Find the values of the input voltage required to produce outputs of ±10 V for a 1-kΩ load. Use V_{CC} of 15 V.

14.40 Repeat Exercise 14.13 for a design variation in which transistor Q_5 is increased in size by a factor of 10, all other conditions remaining the same.

14.41 Repeat Exercise 14.13 for a design in which the limiting output current and normal peak current are 50 mA and 33.3 mA, respectively.

D14.42 The circuit shown in Fig. P14.42 operates in a manner analogous to that in Fig. 14.28 to limit the output

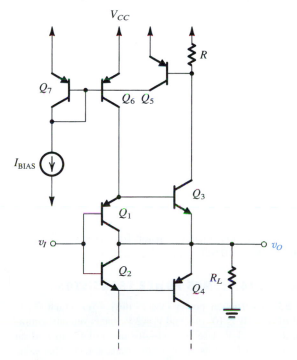

FIGURE P14.42

current from Q_3 in the event of a short circuit or other mishap. It has the advantage that the current-sensing resistor R does not appear directly at the output. Find the value of R that causes Q_5 to turn on and absorb all of $I_{BIAS} = 2$ mA, when the current being sourced reaches 150 mA. For Q_5, $I_S = 10^{-14}$ A and $n = 1$. If the normal peak output current is 100 mA, find the voltage drop across R and the collector current in Q_5.

D14.43 Consider the thermal shutdown circuit shown in Fig. 14.29. At 25°C, Z_1 is a 6.8-V zener diode with a TC of 2 mV/°C, and Q_1 and Q_2 are BJTs that display V_{BE} of 0.7 V at a current of 100 μA and have a TC of -2 mV/°C. Design the circuit so that at 125°C, a current of 100 μA flows in each of Q_1 and Q_2. What is the current in Q_2 at 25°C?

SECTION 14.8: IC POWER AMPLIFIERS

D14.44 In the power-amplifier circuit of Fig. 14.30 two resistors are important in controlling the overall voltage gain. Which are they? Which controls the gain alone? Which affects both the dc output level and the gain? A new design is being considered in which the output dc level is approximately $\frac{1}{3}V_S$ (rather than approximately $\frac{1}{2}V_S$) with a gain of 50 (as before). What changes are needed?

14.45 Consider the front end of the circuit in Fig. 14.30. For $V_S = 20$ V, calculate approximate values for the bias currents in Q_1 through Q_6. Assume $\beta_{npn} = 100$, $\beta_{pnp} = 20$, and $|V_{BE}| = 0.7$ V. Also find the dc voltage at the output.

***14.46** Assume that the output voltage of the circuit of Fig. 14.30 is at signal ground (and thus the signal feedback is deactivated) and find the differential and common-mode input resistances. For this purpose do not include R_4 and R_5. Let $V_S = 20$ V, $\beta_{npn} = 100$, and $\beta_{pnp} = 20$. Also find the transconductance from the input to the output of the first stage (at the connection of the collectors of Q_4 and Q_6 and the base of Q_{12}).

14.47 It is required to use the LM380 power amplifier to drive an 8-Ω loudspeaker while limiting the maximum possible device dissipation to 1.5 W. Use the graph of Fig. 14.32 to determine the maximum possible power-supply voltage that can be used. (Use only the given graphs; do not interpolate.) If the maximum allowed THD is to be 3%, what is the maximum possible load power? To deliver this power to the load what peak-to-peak output sinusoidal voltage is required?

14.48 Consider the LM380 amplifier. Assume that when the amplifier is operated with a 20-V supply, the transconductance of the first stage is 1.6 mA/V. Find the unity-gain

14

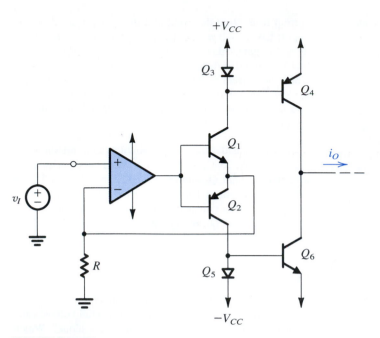

FIGURE P14.50

14

bandwidth (f_t). Since the closed-loop gain is approximately 50 V/V, find its 3-dB bandwidth.

D14.49 Consider the power-op-amp output stage shown in Fig. 14.33. Using a ±15-V supply, provide a design that provides an output of ±11 V or more, with currents up to ±20 mA provided primarily by Q_3 and Q_4 with a 10% contribution by Q_5 and Q_6, and peak output currents of 1 A at full output (+11 V). As the basis of an initial design, use $\beta = 50$ and $|V_{BE}| = 0.7$ V for all devices at all currents. Also use $R_5 = R_6 = 0$.

14.50 For the circuit in Fig. P14.50, assuming all transistors to have large β, show that $i_O = v_I/R$. [This voltage-to-current converter is an application of a versatile circuit building block known as the *current conveyor;* see Sedra and Roberts (1990)]. For $\beta = 100$, by what approximate percentage is i_O actually lower than this ideal value?

D14.51 For the bridge amplifier of Fig. 14.34, let $R_1 = R_3 = 10$ kΩ. Find R_2 and R_4 to obtain an overall gain of 10.

D14.52 An alternative bridge amplifier configuration, with high input resistance, is shown in Fig. P14.52. (Note the similarity of this circuit to the front end of the instrumentation amplifier circuit shown in Fig. 2.20(b).) What is the gain v_O/v_I? For op amps (using ±15-V supplies) that limit at ±13 V, what is the largest sine wave you can provide across R_L?

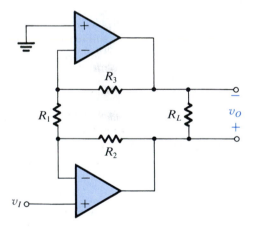

FIGURE P14.52

Using 1 kΩ as the smallest resistor, find resistor values that make $v_O/v_I = 10$ V/V.

SECTION 14.9: MOS POWER TRANSISTORS

14.53 A particular power DMOS device for which C_{ox} is 400 μF/m², W is 10^5 μm, and $V_t = 2$ V, enters velocity saturation at $v_{GS} = 5$ V. Use Eqs. (14.46) and (14.47) to find an expression for L and its value for this transistor. At what value of drain current does velocity saturation begin? For electrons in silicon, $U_{sat} = 5 \times 10^6$ cm/s and $\mu_n = 500$ cm^2/Vs. What is g_m for this device at high currents?

D14.54 Consider the design of the class AB amplifier of Fig. 14.38 under the following conditions: $|V_t| = 2$ V, $\mu C_{ox} W/L = 200$ mA/V^2, $|V_{BE}| = 0.7$ V, β is high, $I_{QN} = I_{QP} = I_R = 10$ mA, $I_{BIAS} = 100$ μA, $I_{Q5} = I_{Q6} = I_{BIAS}/2$, $R_2 = R_4$, the temperature coefficient of $V_{BE} = -2$ mV/°C, and the temperature coefficient of $V_t = -3$ mV/°C in the low-current region. Find the values of R, R_1, R_2, R_3, and R_4. Assume Q_6, Q_P, and Q_N to be thermally coupled. (R_G, used to suppress parasitic oscillation at high frequency, is usually 100 Ω or so.)

14

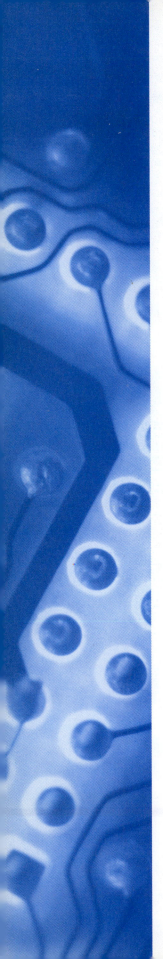

APPENDIXES

VLSI Fabrication Technology

INTRODUCTION

The purpose of this appendix is to familiarize the reader with VLSI (very-large-scale integrated circuit) fabrication technology. Brief explanations of standard silicon VLSI processing steps are given. The characteristics of devices available in **CMOS** and **BiCMOS** fabrication technologies are also presented. In particular, the aspects of IC (integrated-circuit) design that are distinct from discrete-circuit design will be discussed. To take proper advantage of the economics of integrated circuits, designers have had to overcome some serious device limitations (such as poor tolerances) while exploiting device advantages (such as good component matching). An understanding of device characteristics is therefore essential in designing good custom VLSIs or application-specific ICs (ASICs). This understanding is also very helpful when selecting commercially available ICs to implement a system design.

This appendix will consider only silicon-based technologies. Although gallium arsenide (GaAs) is also used to implement VLSI chips, silicon (Si) is by far the most popular material, featuring a wide range of cost–performance trade-offs. Recent development in SiGe and strained-silicon technologies will further strengthen the position of Si-based fabrication processes in the microelectronics industry in the coming years.

Silicon is an abundant element, which occurs naturally in the form of sand. It can be refined using well-established techniques of purification and crystal growth. Silicon also exhibits suitable physical properties for fabricating active devices with good electrical characteristics. Moreover, silicon can be easily oxidized to form an excellent insulator, SiO_2 (glass). This native oxide is useful for constructing capacitors and MOSFETs. It also serves as a diffusion barrier that can mask against the diffusion of unwanted impurities into nearby high-purity silicon material. This masking property of silicon oxide allows the electrical properties of silicon to be easily altered in predefined areas. Therefore, active and passive elements can be built on the same piece of material (or, substrate). The components can then be interconnected using metal layers (similar to those used in printed-circuit boards) to form a so-called monolithic IC, which is essentially a single piece of material (rock!).

 ## A.1 IC FABRICATION STEPS

The basic IC fabrication steps will be described in the following subsections. Some of these steps may be carried out many times, in different combinations and under different processing conditions during a complete fabrication run.

Silicon wafers

Silicon ingot

FIGURE A.1   Silicon ingot and wafer slices.

A.1.1 Wafer Preparation

The starting material for modern integrated circuits is very-high-purity silicon. The material is grown as a single-crystal ingot. It takes the form of a steel gray solid cylinder 10 cm to 30 cm in diameter (Fig. A.1) and can be 1 m to 2 m in length. This crystal is then sawed (like a loaf of bread) to produce circular wafers that are 400 μm to 600 μm thick (a micrometer or micron is a millionth of a meter). The surface of the wafer is then polished to a mirror finish using chemical and mechanical polishing (CMP) techniques. Semiconductor manufacturers usually purchase ready-made silicon wafers from a supplier and rarely start their process at the ingot stage.

The basic electrical and mechanical properties of the wafer depend on the orientation of the crystalline planes, as well as the concentration and type of impurities present. These variables are strictly controlled during crystal growth. Controlled amounts of impurities can be added to the pure silicon in a process known as doping. This allows the alteration of the electrical properties of the silicon, in particular its resistivity. It is also possible to control the conduction-carrier type, either holes (in p-type silicon) or electrons (in n-type silicon), that is responsible for electrical conduction. If a large number of impurity atoms is added, then the silicon is said to be heavily doped (e.g., concentration > 10^{18} atoms/cm^3). When designating the relative doping concentrations in semiconductor device structures, it is common to use + and − symbols. A heavily doped (low-resistivity) n-type silicon wafer would be referred to as $n+$ material, while a lightly doped region may be referred to as $n-$. The ability to control the type of impurity and the doping concentration in the silicon permits the formation of diodes, transistors, and resistors in flexible integrated-circuit form.

A.1.2 Oxidation

Oxidation refers to the chemical process of silicon reacting with oxygen to form silicon dioxide (SiO$_2$). To speed up the reaction, it is necessary to use special high-temperature (e.g., 1000–1200°C) ultraclean furnaces. To avoid the introduction of even small quantities of contaminants (which could significantly alter the electrical properties of the silicon), it is necessary to maintain a clean environment. This is true for all processing steps involved in the fabrication of an integrated circuit. Specially filtered air is circulated in the processing area, and all personnel must wear special lint-free clothing.

The oxygen used in the reaction can be introduced either as a high-purity gas (in a process referred to as a "dry oxidation") or as steam (for "wet oxidation"). In general, wet oxidation has a faster growth rate, but dry oxidation gives better electrical characteristics. In either case, the thermally grown oxide layer has excellent electrical insulation properties. The dielectric strength for SiO$_2$ is approximately 10^7 V/cm. It has a dielectric constant of about 3.9 and can be used to form excellent capacitors. As noted, silicon dioxide serves as an effective mask against many impurities, allowing the introduction of dopants into the silicon only in regions that are not covered with oxide. This masking property is one of the essential enablers of mass fabrication of VLSI devices.

Silicon dioxide is a thin transparent film, and the silicon surface is highly reflective. If white light is shone on an oxidized wafer, constructive and destructive interference will cause certain colors to be reflected. The wavelengths of the reflected light depend on the thickness of the oxide layer. In fact, by categorizing the color of the wafer surface, one can deduce the thickness of the oxide layer. The same principle is used by sophisticated optical inferometers to measure film thickness. On a processed wafer, there will be regions with different oxide thicknesses. The corresponding colors can be quite vivid, and thickness variations are immediately obvious when a finished wafer is viewed with the naked eye.

A.1.3 Diffusion

Diffusion is the process by which atoms move from a high-concentration region to a low-concentration region through the semiconductor crystal. The diffusion process is very much like a drop of ink dispersing through a glass of water except that it occurs much more slowly in solids. In fabrication, diffusion is a method by which to introduce impurity atoms (dopants) into silicon to change its resistivity. The rate at which dopants diffuse in silicon is a strong function of temperature. Thus, for speed, diffusion of impurities is usually carried out at high temperatures (1000–1200°C) to obtain the desired doping profile. When the wafer is cooled to room temperature, the impurities are essentially "frozen" in position. The diffusion process is performed in furnaces similar to those used for oxidation. The depth to which the impurities diffuse depends on both the temperature and the time allocated.

The most common impurities used as dopants are boron, phosphorus, and arsenic. Boron is a *p*-type dopant, while phosphorus and arsenic are *n*-type dopants. These dopants can be effectively masked by thin silicon dioxide layers. By diffusing boron into an *n*-type substrate, a *pn* junction (diode) is formed. If the doping concentration is heavy enough, the diffused layer can also be used as a conductor.

A.1.4 Ion Implantation

Ion implantation is another method used to introduce impurity atoms into the semiconductor crystal. An ion implanter produces ions of the desired dopant, accelerates them by an electric field, and allows them to strike the semiconductor surface. The ions become embedded in the crystal lattice. The depth of penetration is related to the energy of the ion beam, which can be controlled by the accelerating-field voltage. The quantity of ions implanted can be controlled by varying the beam current (flow of ions). Since both voltage and current can be accurately measured and controlled, ion implantation results in much more accurate and reproducible impurity profiles than can be obtained by diffusion. In addition, ion implantation can be performed at room temperature. Ion implantation normally is used when accurate control of the doping profile is essential for device operation.

A.1.5 Chemical-Vapor Deposition

Chemical-vapor deposition (CVD) is a process by which gases or vapors are chemically reacted, leading to the formation of solids on a substrate. CVD can be used to deposit various materials on a silicon substrate including SiO_2, Si_3N_4, and polysilicon. For instance, if silane gas and oxygen are allowed to react above a silicon substrate, the end product, silicon dioxide, will be deposited as a solid film on the silicon wafer surface. The properties of the CVD oxide layer are not as good as those of a thermally grown oxide, but such a layer is sufficient to act as an electrical insulator. The advantage of a CVD layer is that the oxide deposits at a fast rate and a low temperature (below 500°C).

If silane gas alone is used, then a silicon layer will be deposited on the wafer. If the reaction temperature is high enough (above 1000°C), the layer deposited will be a crystalline layer (assuming that there is an exposed crystalline silicon substrate). Such a layer is called an **epitaxial** layer, and the deposition process is referred to as **epitaxy**, rather than CVD. At lower temperatures, or if the substrate surface is not single-crystal silicon, the atoms will not be able to align in the same crystalline direction. Such a layer is called polycrystalline silicon (**poly Si**), since it consists of many small crystals of silicon whose crystalline axes are oriented in random directions. These layers are normally doped very heavily to form highly conductive regions that can be used for electrical interconnections.

A.1.6 Metallization

The purpose of metallization is to interconnect the various components (transistors, capacitors, etc.) to form the desired integrated circuit. Metallization involves the initial deposition of a metal over the entire surface of the silicon. The required interconnection pattern is then selectively etched. The metal layer is normally deposited via a sputtering process. A pure metal disk (e.g., 99.99% aluminum) is placed under an argon (Ar) ion gun inside a vacuum chamber. The wafers are also mounted inside the chamber above the target. The Ar ions will not react with the metal, since Ar is a noble gas. However, their ions are made to physically bombard the target and literally knock metal atoms out of it. These metal atoms will then coat all the surface inside the chamber, including the wafers. The thickness of the metal film can be controlled by the length of time for sputtering, which is normally in the range of 1 to 2 minutes.

A.1.7 Photolithography

The surface geometry of the various integrated-circuit components is defined photographically. First, the wafer surface is coated with a photosensitive layer (called photoresist) using a spin-on technique. After this, a photographic plate with drawn patterns (e.g., a quartz plate with a chromium pattern) will be used to selectively expose the photoresist under ultraviolet (UV) illumination. The exposed areas will become softened (for positive photoresist). The exposed layer can then be removed using a chemical developer, causing the mask pattern to appear on the wafer. Very fine surface geometries can be reproduced accurately by this technique. Photolithography requires some of the most expensive equipment in VLSI fabrication. Currently, we are already approaching the physical limits of the photolithographic process. Deep UV light or electron beam can be used to define patterns with resolution as fine as 50 nm. However, another technological breakthrough will be needed to achieve further geometry downscaling.

The patterned photoresist layer can be used as an effective masking layer to protect materials below from wet chemical etching or reactive ion etching. Correspondingly, silicon dioxide, silicon nitride, polysilicon, and metal layers can be selectively removed using the appropriate etching methods. After the etching step(s), the photoresist is stripped away, leaving behind a permanent pattern, an image of the photomask, on the wafer surface.

To make this process even more challenging, multiple masking layers (there can be more than 20 layers in advanced VLSI fabrication processes) must be aligned precisely on top of previous layers. This must be done with even greater precision than is associated with the minimum dimensions of the masking patterns. This requirement imposes very critical mechanical and optical constraints on the photolithography equipment.

A.1.8 Packaging

A finished silicon wafer may contain several hundred or more finished circuits or chips. Each chip may contain from 10 to 10^8 or more transistors in a rectangular shape, typically between 1 mm and 10 mm on a side. The circuits are first tested electrically (while still in wafer form) using an automatic probing station. Bad circuits are marked for later identification. The circuits are then separated from each other (by dicing), and the good circuits (called dies) are mounted in packages (or headers). Examples of such IC packages are given in Fig. A.2. Fine gold wires are normally used to connect the pins of the package to the metallization pattern on the die. Finally, the package is sealed using plastic or epoxy under vacuum or in an inert atmosphere.

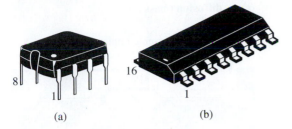

 ## A.2 VLSI PROCESSES

Integrated-circuit fabrication was originally dominated by bipolar technology. But, by the late 1970s metal-oxide-semiconductor (MOS) technology was perceived to be more promising for VLSI implementation, owing to its higher packing density and lower power consumption. Since the early 1980s, complementary MOS (CMOS) technology has grown prodigiously to almost completely dominate the VLSI scene, leaving bipolar technology to fill specialized functions such as digital and high-speed analog and RF circuits. CMOS technologies continue to evolve, and in the late 1980s, the incorporation of bipolar devices led to the emergence of high-performance bipolar-CMOS (Bi-CMOS) fabrication processes that provided the best of both technologies. However, BiCMOS processes are often very complicated and costly, since they require upward of 15 to 20 masking levels per implementation—by comparison, standard CMOS processes require only 10 to 12 masking levels.

The performance of CMOS and BiCMOS processes continues to improve, offering finer lithographic resolution. However, fundamental limitations on processing techniques and semiconductor properties have prompted the need to explore alternate materials. Silicon-germanium (SiGe) and strained-Si technologies have emerged as good compromises which improve performance while maintaining manufacturing compatibility (hence low cost) with existing silicon-based CMOS fabrication equipment.

In the subsections that follow, we will examine, in turn, three aspects of modern IC fabrication, namely; a typical CMOS process flow, the performance of the available components, and the inclusion of bipolar devices to form a BiCMOS process.

A.2.1 *n*-Well CMOS Process

Depending on the choice of starting material (substrate), CMOS processes can be identified as **n-well, p-well,** or **twin-well** processes, the latter being the most complicated but also the most flexible in the optimization of both the *n*- and *p*-channel devices. In addition, many advanced CMOS processes may make use of trench isolation, and silicon-on-insulator (SOI) technology, to reduce parasitic capacitance (to achieve higher speed) and to improve packing density.

For simplicity, an *n*-well CMOS process is chosen for discussion. Another benefit of this choice is that it can also be easily extended into a BiCMOS process. The typical process flow is as shown in Fig. A.3. A minimum of 7 masking layers is necessary. However, in practice most CMOS processes will also require additional layers such as *n* and *p* guards for better latchup immunity, a second polysilicon layer for capacitors, and multilayer metals for high-density interconnections. The inclusion of these layers would increase the total number of masking layers from 15 to 20.

The starting material for the *n*-well CMOS is a *p*-type substrate. The process begins with an *n*-well diffusion (Fig. A.3a). The *n* well is required wherever *p*-type MOSFETs are to be

(a) Define *n*-well diffusion (mask #1)

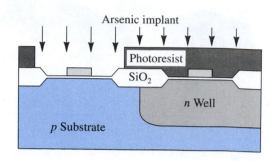

(b) Define active regions (mask #2)

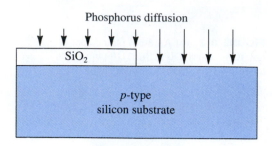

(c) LOCOS oxidation

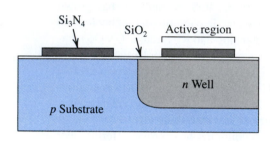

(d) Polysilicon gate (mask #3)

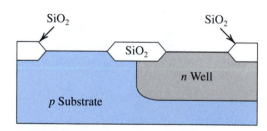

(e) *n*+ diffusion (mask #4)

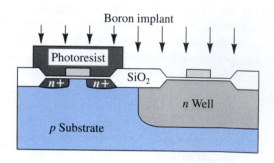

(f) *p*+ diffusion (mask #5)

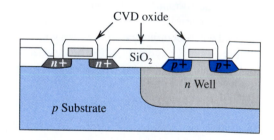

(g) Contact holes (mask #6)

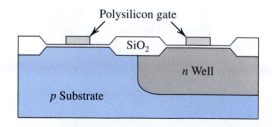

(h) Metallization (mask #7)

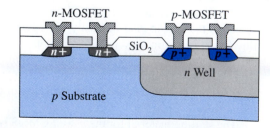

FIGURE A.3 A typical *n*-well CMOS process flow.

placed. A thick silicon dioxide layer is etched to expose the regions for *n*-well diffusion. The unexposed regions will be protected from the *n*-type phosphorus impurity. Phosphorus is usually used for deep diffusions because it has a large diffusion coefficient and can diffuse faster than arsenic into the substrate.

The second step is to define the active region (where transistors are to be placed) using a technique called **local oxidation** (LOCOS). A silicon nitride (Si_3N_4) layer is deposited and patterned relative to the previous *n*-well regions (Fig. A.3b). The nitride-covered regions will not be oxidized. After a long wet oxidation step, thick-field oxide will appear in regions between transistors (Fig. A.3c). This thick-field oxide is necessary for isolating the transistors. It also allow interconnection layers to be routed on top of the field oxide without inadvertently forming a conduction channel at the silicon surface.

The next step is the formation of the polysilicon gate (Fig. A.3d). This is one of the most critical steps in the CMOS process. The thin oxide layer in the active region is first removed using wet etching followed by the growth of a high-quality thin gate oxide. Current 0.13 μm and 0.18 μm processes routinely use oxide thicknesses as thin as 20 Å to 50 Å (1 angstrom = 10^{-8} cm). A polysilicon layer, usually arsenic doped (*n* type), is then deposited and patterned. The photolithography is most demanding in this step, since the finest resolution is required to produce the shortest possible MOS channel length.

The polysilicon gate is a self-aligned structure and is preferred over the older type of metal gate structure. A heavy arsenic implant can be used to form the *n*+ source and drain regions of the *n*-MOSFETs. The polysilicon gate also acts as a barrier for this implant to protect the channel region. A layer of photoresist can be used to block the regions where *p*-MOSFETs are to be formed (Fig. A.3e). The thick-field oxide stops the implant and prevents *n*+ regions from forming outside the active regions. A reversed photolithography step can be used to protect the *n*-MOSFETs during the *p*+ boron source and drain implant for the *p*-MOSFETs (Fig. A.3f). In both cases the separation between the source and drain diffusions—the channel length—is defined by the polysilicon gate mask alone, hence the self-alignment.

Before contact holes are opened, a thick layer of CVD oxide is deposited over the entire wafer. A photomask is used to define the contact window opening (Fig. A.3g) followed by a wet or dry oxide etch. A thin aluminum layer is then evaporated or sputtered onto the wafer. A final masking and etching step is used to pattern the interconnection (Fig. A.3h).

Not shown in the process flow is the final passivation step prior to packaging and wire bonding. A thick CVD oxide or pyrox glass is usually deposited on the wafer to serve as a protective layer.

A.2.2 Integrated Devices

Besides the obvious *n*- and *p*-channel MOSFETs, there are other devices that can be obtained by manipulating the masking layers. These include *pn* junction diodes, MOS capacitors, and resistors.

A.2.3 MOSFETs

The *n*-channel MOSFET is preferred over the *p*-MOSFET (Fig. A.4). The electron surface mobility of the *n*-channel device is two to four times higher than that for holes. Therefore, with the same device size (*W* and *L*), the *n*-MOSFET offers higher current drive (or lower on-resistance) as well as higher transconductance.

In an integrated-circuit design environment, MOSFETs are characterized by their threshold voltage and device sizes. Usually the *n*- and *p*-channel devices are designed to have threshold voltages of similar magnitude for a particular process. The transconductance can be adjusted

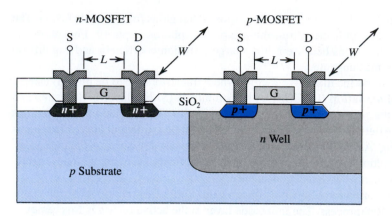

FIGURE A.4 Cross-sectional diagram of an *n*- and *p*-MOSFET.

by changing the device surface dimensions (*W* and *L*). This feature is not available for bipolar transistors; thus integrated MOSFET circuits are much more flexible in their design.

A.2.4 Resistors

Resistors in integrated form are not very precise. They can be made from various diffusion regions as shown in Fig. A.5. Different diffusion regions have different resistivity. The *n* well is usually used for medium-value resistors, while the *n*+ and *p*+ diffusions are useful for low-value resistors. The actual resistance value can be defined by changing the length and width of diffused regions. The tolerance of the resistor value is very poor (20–50%), but the matching of two similar resistor values is quite good (5%). Thus circuit designers should design circuits that exploit resistor matching and avoid designs that require a specific resistor value.

All diffused resistors are self-isolated by the reversed-biased *pn* junctions. However, a serious drawback for these resistors is that they are accompanied by a substantial parasitic junction capacitance, making them not very useful for high-frequency applications. The reversed-biased *pn* junctions also exhibit a JFET effect, leading to a variation in the resistance value as the applied voltage is changed (a large voltage coefficient is undesirable). Since the mobilities of carriers vary with temperature, diffused resistors also exhibit a significant temperature coefficient.

A more useful resistor can be fabricated using the polysilicon layer that is placed on top of the thick-field oxide. The thin polysilicon layer provides better surface area matching and

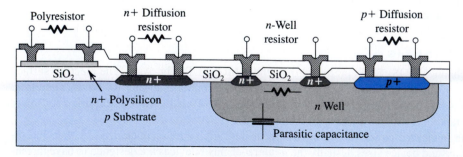

FIGURE A.5 Cross sections of resistors of various types available from a typical *n*-well CMOS process.

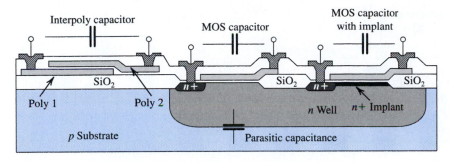

FIGURE A.6 Interpoly and MOS capacitors in an *n*-well CMOS process.

hence more accurate resistor ratios. Furthermore, the poly resistor is physically separated from the substrate, resulting in much lower parasitic capacitance and voltage coefficient.

A.2.5 Capacitors

Two types of capacitor structure are available in CMOS processes, MOS and interpoly capacitors (also MIM—metal-insulator-metal). The cross sections of these structures are as shown in Fig. A.6. The MOS gate capacitance, depicted in the center structure, is basically the gate-to-source capacitance of a MOSFET. The capacitance value is dependent on the gate area. The oxide thickness is the same as the gate oxide thickness in the MOSFETs. This capacitor exhibits a large voltage dependence. To eliminate this problem, an additional *n*+ implant is required to form the bottom plate of the capacitors, as shown in the structure on the right. Both these MOS capacitors are physically in contact with the substrate, resulting in a large parasitic *pn* junction capacitance at the bottom plate.

The interpoly capacitor exhibits near ideal characteristics but at the expense of the inclusion of a second polysilicon layer to the CMOS process. Since this capacitor is placed on top of the thick-field oxide, parasitic effects are kept to a minimum.

A third and less often used capacitor is the junction capacitor. Any *pn* junction under reversed bias produces a depletion region that acts as a dielectric between the *p* and the *n* regions. The capacitance is determined by geometry and doping levels and has a large voltage coefficient. This type of capacitor is often used as a variactor (variable capacitor) for tuning circuits. However, this capacitor works only with reversed-bias voltages.

For interpoly and MOS capacitors, the capacitance values can be controlled to within 1%. Practical capacitance values range from 0.5 pF to a few 10s of picofarads. The matching between similar-size capacitors can be within 0.1%. This property is extremely useful for designing precision analog CMOS circuits.

A.2.6 *pn* Junction Diodes

Whenever *n*-type and *p*-type diffusion regions are placed next to each other, a *pn* junction diode results. A useful structure is the *n*-well diode shown in Fig. A.7. The diode fabricated in an *n* well can provide a high breakdown voltage. This diode is essential for the input clamping circuits for protection against electrostatic discharge. The diode is also very useful as an on-chip temperature sensor by monitoring the variation of its forward voltage drop.

A.2.7 BiCMOS Process

An *npn* vertical bipolar transistor can be integrated into the *n*-well CMOS process with the addition of a *p*-base diffusion region (Fig. A.8). The characteristics of this device depend on

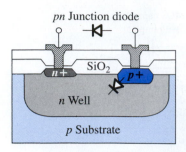

FIGURE A.7 A *pn* junction diode in an *n*-well CMOS process.

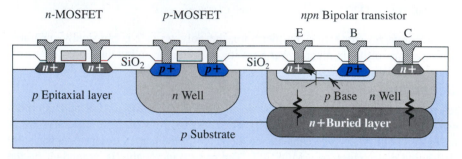

FIGURE A.8 Cross-sectional diagram of a BiCMOS process.

the base width and the emitter area. The base width is determined by the difference in junction depth between the *n*+ and the *p*-base diffusions. The emitter area is determined by the junction area of the *n*+ diffusion at the emitter. The *n* well serves as the collector for the *npn* transistor. Typically, the *npn* transistor has a β in the range of 50 to 100 and a cutoff frequency greater than 10 GHz.

Normally, an *n*+ buried layer is used to reduce the series resistance of the collector, since the *n* well has a very high resistivity. However, this would further complicate the process with the introduction of *p*-type epitaxy and one more masking step. Other variations on the bipolar transistor include the use of a polyemitter and self-aligned base contact to minimize parasitic effects.

A.2.8 Lateral *pnp* Transistor

The fact that most BiCMOS processes do not have optimized *pnp* transistors makes circuit design somewhat difficult. However, in noncritical situations, a parasitic lateral *pnp* transistor can be used (Fig. A.9).

In this case, the *n* well serves as the *n*-base region, with *p*+ diffusions as the emitter and the collector. The base width is determined by the separation between the two *p*+ diffusions. Since the doping profile is not optimized for the base–collector junctions, and because the

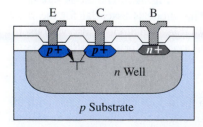

FIGURE A.9 A lateral *pnp* transistor.

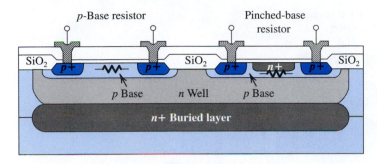

FIGURE A.10 *p*-Base and pinched *p*-base resistors.

base width is limited by the minimum photolithographic resolution, the performance of this device is not very good—typically, β of around 10, with a low cutoff frequency.

A.2.9 *p*-Base and Pinched-Base Resistors

With the additional *p*-base diffusion in the BiCMOS process, two additional resistor structures are available. The *p*-base diffusion can be used to form a straightforward *p*-base resistor as shown in Fig. A.10. Since the base region is usually of a relatively low doping level and with a moderate junction depth, it is suitable for medium-value resistors (a few kilohms). If a large resistor value is required, the pinched-base resistor can be used. In this structure, the *p*-base region is encroached by the *n*+ diffusion, restricting the conduction path. Resistor values in the range of 10 kΩ to 100 kΩ can be obtained. As with the diffusion resistors discussed earlier, these resistors exhibit poor tolerance and temperature coefficients but relatively good matching.

A.2.10 The SiGe BiCMOS Process

With the advent of wireless applications, the demand for high-performance, high-frequency RF integrated circuits is enjoying a tremendous growth. The fundamental limitations of physical material properties initially prevented silicon-based technology from competing with more expensive III–V compound technologies such GaAs. By incorporating a controlled amount (typically no more than 15 mole %) of germanium (Ge) into crystalline silicon (Si), the energy bandgap can be altered. The specific concentration profile of the Ge can be engineered in such a way that the energy bandgap can be gradually reduced from that in the pure Si region to a lower value in the SiGe region. This energy-bandgap reduction produces a built-in electrical field that can assist the movement of carriers, hence giving faster operating speed. Therefore, SiGe bipolar transistors can achieve significant higher cutoff frequency (e.g., in the 50–70 GHz range). Another benefit is that the SiGe process is compatible with existing Si-based fabrication technology, ensuring a very favorable cost/performance ratio.

To take advantage of the SiGe material characteristics, the basic bipolar transistor structure must also be modify to further reduce parasitic capacitance (for higher speed) and to improve the injection efficiency (for higher gain). A symmetrical bipolar device structure is shown in Fig. A.11. The device makes use of trench isolation to reduce the collector sidewall capacitance between the *n*-well/*n*+ buried layer and the *p* substrate. The emitter size and the *p*+ base contact size are defined by a self-aligned process to minimize the base–collector junction (Miller) capacitance. This type of device is called a heterojunction bipolar transistor (HBT), since the emitter–base junction is formed from two different types of material, polysilicon emitter and SiGe base. The injection efficiency is significantly better than a homojunction device (as in a

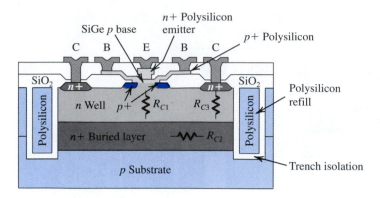

FIGURE A.11 Cross-sectional diagram of a symmetrical self-aligned *npn* SiGe heterojunction bipolar transistor (HBT).

conventional BJT). Coupled with the fact that base width is typically only around 50 nm, it is easy to achieve a current gain of more than 100. In addition, not shown in Fig. A.11, multiple layers of metallization can be used to further reduce the device size and interconnect resistance. All these device features are necessary to complement the speed performance of the SiGe material.

A.3 VLSI LAYOUT

Each designed circuit schematic must be transformed into a layout that consists of the geometric representation of the circuit components and interconnections. With the advent of computer-aided design (CAD) tools, many of the conversion steps from schematic to layout can be carried out semi- or fully automatically. However, any good mixed-signal IC designer must have practiced full-custom layout at one point or another. To illustrate such a procedure we will consider the layout of a CMOS inverter.

Similar to the requirement in a printed-circuit-board layout to reduce crossover paths, the circuit must first be "flattened" and redrawn to eliminate any interconnection crossovers. Each process is made up of a specific set of masking layers. In this case, 7 layers are used. Each layer is usually assigned a unique color and fill pattern for ease of identification on a computer screen or on a printed color plot. The layout begins with the placement of the transistors. For illustration purposes (Fig. A.12), the *p*- and *n*-MOSFETs are placed in arrangements similar to that of the schematic. In practice, the designer is free to choose the most area-efficient layout that she can identify. The MOSFETs are defined by the active areas overlapped by the "Poly 1" layer. The MOS channel length and width are defined by the width of the "Poly 1" strip and that of the active region, respectively. The *p*-MOSFET is enclosed in an *n* well. For more complex circuits, multiple *n* wells can be used for different groups of *p*-MOSFETs. The *n*-MOSFET is enclosed by the *n*+ diffusion mask to form the source and drain, while the *p*-MOSFET is enclosed by the *p*+ diffusion mask. Contact holes are placed in regions that require connection to the metal layer. Finally, the "Metal 1" layer completes the interconnections.

The corresponding cross-sectional diagram of the CMOS inverter, viewed along the AA′ plane is shown in Fig. A.13. The poly-Si gates for both transistor are connected together to form the input terminal, X. The drains of both transistors are tied together via "Metal 1" to form the output terminal, Y. The sources of the *n*- and *p*-MOSFETs are connected to GND and V_{DD}, respectively. Note that butting contacts consisting of side-by-side *n*+/*p*+ diffusions are used to tie the body potential of the *n*- and *p*-MOSFETs to the appropriate voltage levels.

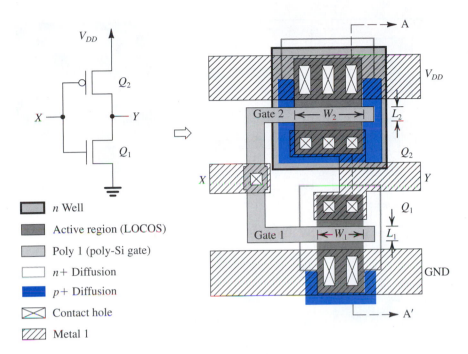

FIGURE A.12 A CMOS inverter schematic and its layout.

Legend:
- n Well
- Active region (LOCOS)
- Poly 1 (poly-Si gate)
- n+ Diffusion
- p+ Diffusion
- Contact hole
- Metal 1

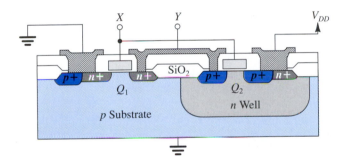

FIGURE A.13 The cross section along the plane AA′ of a CMOS inverter.

When the layout is completed, the circuit must be verified using appropriate CAD tools including circuit extractor, design-rule checker (DRC), and circuit simulator. Once these verifications have been satisfied, the design can be "taped out" to a mask-making facility. A pattern generator (PG machine) can then draw the geometries on a glass or quartz photoplate using electronically driven shutters. Layers are drawn one-by-one onto different photoplates. When these plates have been developed, clear and dark patterns depicting the geometries on the layout will appear. A set of the photoplates for the CMOS inverter example is shown in Fig. A.14. Depending on whether the drawn geometries are meant to be opened as windows or kept as patterns, the plates can be "positive" or "negative" images with **clear** or **dark** fields. Note that these layers must be processed in sequence. In the steps of this sequence, they must be aligned within very fine tolerances to form the transistors and interconnections. Naturally, the greater the number of layers, the more difficult it is to maintain the alignment. A process with more layers also requires better photolithography equipment and possibly results in lower yield. Hence, each additional mask will be reflected in an increase in the final cost of the IC chip.

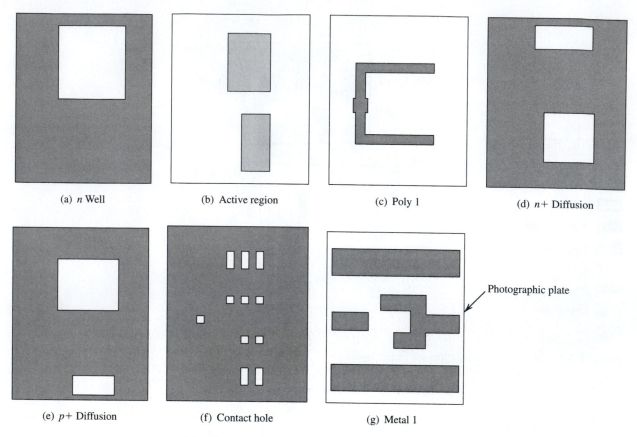

(a) *n* Well

(b) Active region

(c) Poly 1

(d) *n*+ Diffusion

(e) *p*+ Diffusion

(f) Contact hole

(g) Metal 1

Photographic plate

FIGURE A.14   A set of photomasks for the *n*-well CMOS inverter. Note that each layer requires a separate plate: **(a)**, **(d)**, **(e)**, and **(f)** are dark-field masks; **(b)**, **(c)**, and **(g)** are clear-field masks.

Summary

■ This appendix presented an overview of the various aspects of VLSI fabrication procedures. This includes component characteristics, process flows, and layouts. This is by no means a complete account of state-of-the-art VLSI technologies. Interested readers should consult reference textbooks on this subject for more detailed discussions.

Two-Port Network Parameters

INTRODUCTION

At various points throughout the text, we make use of some of the different possible ways to characterize linear two-port networks. A summary of this topic is presented in this appendix.

B.1 CHARACTERIZATION OF LINEAR TWO-PORT NETWORKS

A two-port network (Fig. B.1) has four port variables: V_1, I_1, V_2, and I_2. If the two-port network is linear, we can use two of the variables as excitation variables and the other two as response variables. For instance, the network can be excited by a voltage V_1 at port 1 and a voltage V_2 at port 2, and the two currents, I_1 and I_2, can be measured to represent the network response. In this case V_1 and V_2 are independent variables and I_1 and I_2 are dependent variables, and the network operation can be described by the two equations

$$I_1 = y_{11}V_1 + y_{12}V_2 \tag{B.1}$$

$$I_2 = y_{21}V_1 + y_{22}V_2 \tag{B.2}$$

Here, the four parameters y_{11}, y_{12}, y_{21}, and y_{22} are admittances, and their values completely characterize the linear two-port network.

Depending on which two of the four port variables are used to represent the network excitation, a different set of equations (and a correspondingly different set of parameters) is obtained for characterizing the network. We shall present the four parameter sets commonly used in electronics.

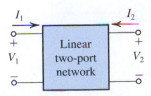

FIGURE B.1 The reference directions of the four port variables in a linear two-port network.

B.1.1 *y* Parameters

The short-circuit admittance (or *y*-parameter) characterization is based on exciting the network by V_1 and V_2, as shown in Fig. B.2(a). The describing equations are Eqs. (B.1) and (B.2). The four admittance parameters can be defined according to their roles in Eqs. (B.1) and (B.2).

Specifically, from Eq. (B.1) we see that y_{11} is defined as

$$y_{11} = \frac{I_1}{V_1}\bigg|_{V_2=0} \tag{B.3}$$

Thus y_{11} is the input admittance at port 1 with port 2 short-circuited. This definition is illustrated in Fig. B.2(b), which also provides a conceptual method for measuring the input short-circuit admittance y_{11}.

The definition of y_{12} can be obtained from Eq. (B.1) as

$$y_{12} = \frac{I_1}{V_2}\bigg|_{V_1=0} \tag{B.4}$$

Thus y_{12} represents transmission from port 2 to port 1. Since in amplifiers, port 1 represents the input port and port 2 the output port, y_{12} represents internal *feedback* in the network. Figure B.2(c) illustrates the definition and the method for measuring y_{12}.

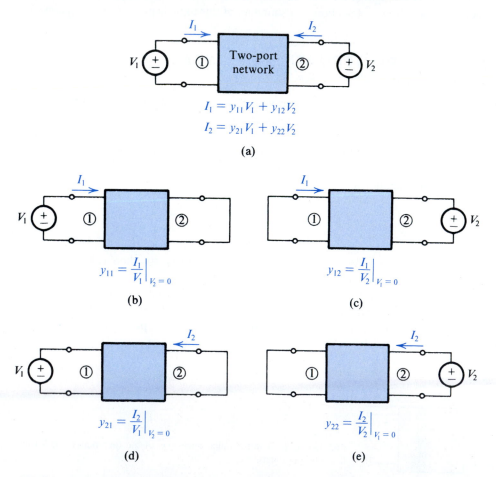

FIGURE B.2 Definition and conceptual measurement circuits for the *y* parameters.

The definition of y_{21} can be obtained from Eq. (B.2) as

$$y_{21} = \left.\frac{I_2}{V_1}\right|_{V_2=0} \tag{B.5}$$

Thus y_{21} represents transmission from port 1 to port 2. If port 1 is the input port and port 2 the output port of an amplifier, then y_{21} provides a measure of the forward gain or transmission. Figure B.2(d) illustrates the definition and the method for measuring y_{21}.

The parameter y_{22} can be defined, based on Eq. (B.2), as

$$y_{22} = \left.\frac{I_2}{V_2}\right|_{V_1=0} \tag{B.6}$$

Thus y_{22} is the admittance looking into port 2 while port 1 is short-circuited. For amplifiers, y_{22} is the output short-circuit admittance. Figure B.2(e) illustrates the definition and the method for measuring y_{22}.

B.1.2 z Parameters

The open-circuit impedance (or z-parameter) characterization of two-port networks is based on exciting the network by I_1 and I_2, as shown in Fig. B.3(a). The describing equations are

$$V_1 = z_{11}I_1 + z_{12}I_2 \tag{B.7}$$

$$V_2 = z_{21}I_1 + z_{22}I_2 \tag{B.8}$$

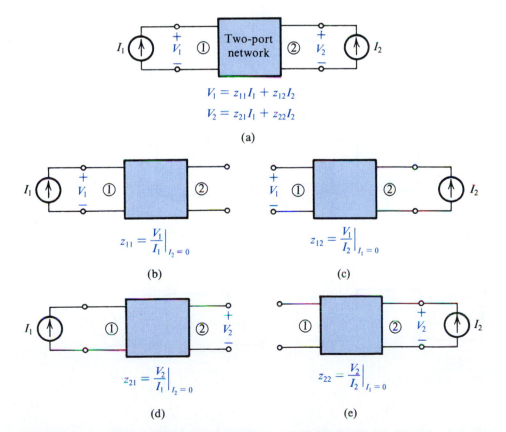

FIGURE B.3 Definition and conceptual measurement circuits for the z parameters.

Owing to the duality between the z- and y-parameter characterizations, we shall not give a detailed discussion of z parameters. The definition and method of measuring each of the four z parameters are given in Fig. B.3.

B.1.3 h Parameters

The hybrid (or h-parameter) characterization of two-port networks is based on exciting the network by I_1 and V_2, as shown in Fig. B.4(a) (note the reason behind the name *hybrid*). The describing equations are

$$V_1 = h_{11}I_1 + h_{12}V_2 \tag{B.9}$$

$$I_2 = h_{21}I_1 + h_{22}V_2 \tag{B.10}$$

from which the definition of the four h parameters can be obtained as

$$h_{11} = \left.\frac{V_1}{I_1}\right|_{V_2=0} \qquad h_{21} = \left.\frac{I_2}{I_1}\right|_{V_2=0}$$

$$h_{12} = \left.\frac{V_1}{V_2}\right|_{I_1=0} \qquad h_{22} = \left.\frac{I_2}{V_2}\right|_{I_1=0}$$

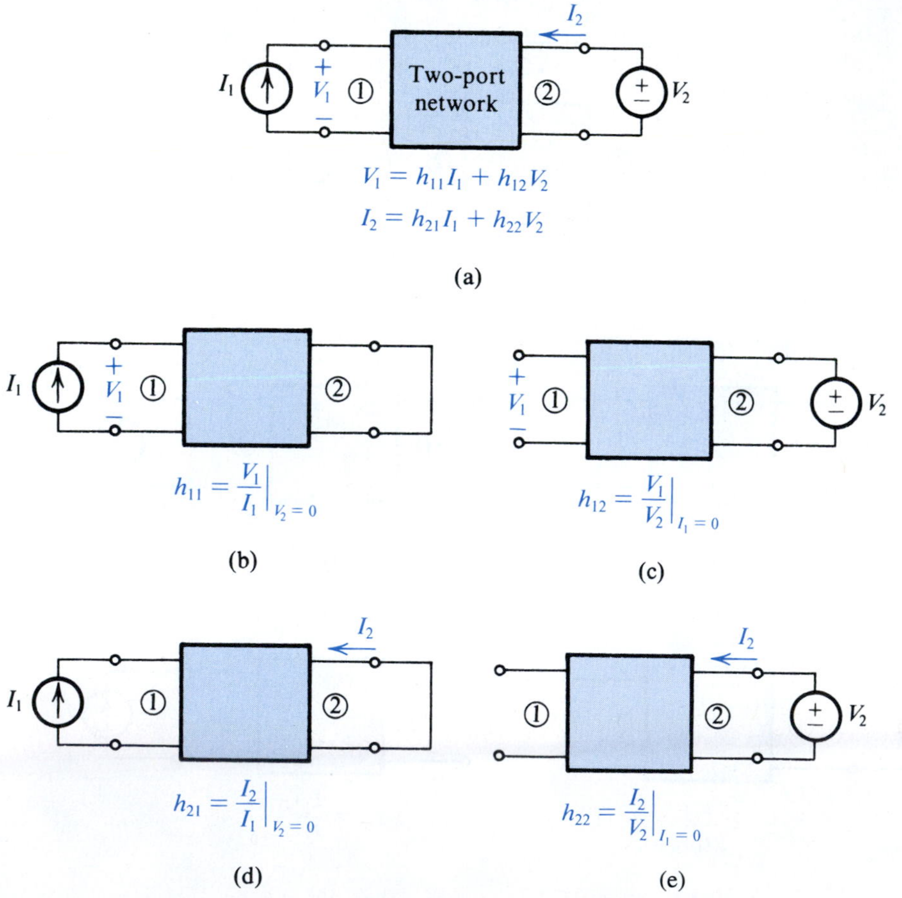

$$V_1 = h_{11}I_1 + h_{12}V_2$$
$$I_2 = h_{21}I_1 + h_{22}V_2$$

(a)

$$h_{11} = \left.\frac{V_1}{I_1}\right|_{V_2=0}$$

(b)

$$h_{12} = \left.\frac{V_1}{V_2}\right|_{I_1=0}$$

(c)

$$h_{21} = \left.\frac{I_2}{I_1}\right|_{V_2=0}$$

(d)

$$h_{22} = \left.\frac{I_2}{V_2}\right|_{I_1=0}$$

(e)

FIGURE B.4 Definition and conceptual measurement circuits for the h parameters.

Thus, h_{11} is the input impedance at port 1 with port 2 short-circuited. The parameter h_{12} represents the reverse or feedback voltage ratio of the network, measured with the input port open-circuited. The forward-transmission parameter h_{21} represents the current gain of the network with the output port short-circuited; for this reason, h_{21} is called the *short-circuit current gain*. Finally, h_{22} is the output admittance with the input port open-circuited.

The definitions and conceptual measuring setups of the h parameters are given in Fig. B.4.

B.1.4 *g* Parameters

The inverse-hybrid (or g-parameter) characterization of two-port networks is based on excitation of the network by V_1 and I_2, as shown in Fig. B.5(a). The describing equations are

$$I_1 = g_{11}V_1 + g_{12}I_2 \tag{B.11}$$

$$V_2 = g_{21}V_1 + g_{22}I_2 \tag{B.12}$$

The definitions and conceptual measuring setups are given in Fig. B.5.

B.1.5 Equivalent-Circuit Representation

A two-port network can be represented by an equivalent circuit based on the set of parameters used for its characterization. Figure B.6 shows four possible equivalent circuits corresponding

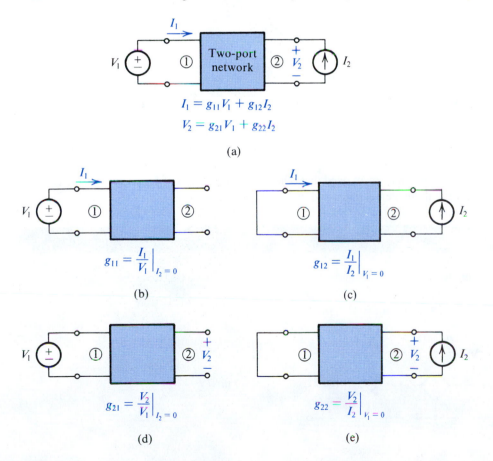

FIGURE B.5 Definition and conceptual measurement circuits for the g parameters.

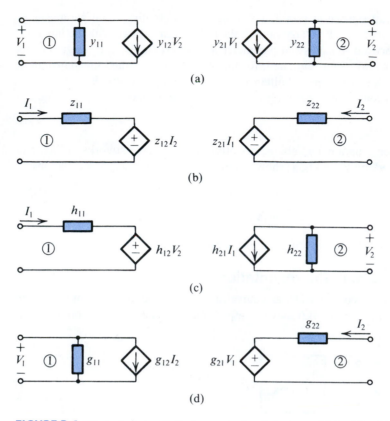

FIGURE B.6 Equivalent circuits for two-port networks in terms of (a) y, (b) z, (c) h, and (d) g parameters.

to the four parameter types just discussed. Each of these equivalent circuits is a direct pictorial representation of the corresponding two equations describing the network in terms of the particular parameter set.

Finally, it should be mentioned that other parameter sets exist for characterizing two-port networks, but these are not discussed or used in this book.

EXERCISE

B.1 Figure EB.1 shows the small-signal equivalent-circuit model of a transistor. Calculate the values of the h parameters.

Ans. $h_{11} \simeq 2.6 \text{ k}\Omega$; $h_{12} \simeq 2.5 \times 10^{-4}$; $h_{21} \simeq 100$; $h_{22} \simeq 2 \times 10^{-5}\ \Omega$

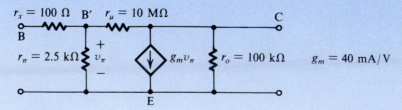

FIGURE EB.1

PROBLEMS

B.1 (a) An amplifier characterized by the h-parameter equivalent circuit of Fig. B.6(c) is fed with a source having a voltage V_s and a resistance R_s, and is loaded in a resistance R_L. Show that its voltage gain is given by

$$\frac{V_2}{V_s} = \frac{-h_{21}}{(h_{11} + R_s)(h_{22} + 1/R_L) - h_{12}h_{21}}$$

(b) Use the expression derived in (a) to find the voltage gain of the transistor in Exercise B.1 for $R_s = 1\text{ k}\Omega$ and $R_L = 10\text{ k}\Omega$.

B.2 The terminal properties of a two-port network are measured with the following results: With the output short-circuited and an input current of 0.01 mA, the output current is 1.0 mA and the input voltage is 26 mV. With the input open-circuited and a voltage of 10 V applied to the output,

the current in the output is 0.2 mA and the voltage measured at the input is 2.5 mV. Find values for the h parameters of this network.

B.3 Figure PB.3 shows the high-frequency equivalent circuit of a BJT. (For simplicity, r_x has been omitted.) Find the y parameters.

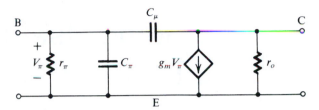

FIGURE PB.3

APPENDIX C

Some Useful Network Theorems

INTRODUCTION

In this appendix we review three network theorems that are useful in simplifying the analysis of electronic circuits: Thévenin's theorem, Norton's theorem, and the source-absorption theorem.

 ## C.1   THÉVENIN'S THEOREM

Thévenin's theorem is used to represent a part of a network by a voltage source V_t and a series impedance Z_t, as shown in Fig. C.1. Figure C.1(a) shows a network divided into two parts, A and B. In Fig. C.1(b) part A of the network has been replaced by its Thévenin equivalent: a voltage source V_t and a series impedance Z_t. Figure C.1(c) illustrates how V_t is to be determined: Simply open-circuit the two terminals of network A and measure (or calculate) the voltage that appears between these two terminals. To determine Z_t we reduce all external (i.e., independent) sources in network A to zero by short-circuiting voltage sources and open-circuiting current sources. The impedance Z_t will be equal to the input impedance of network A after this reduction has been performed, as illustrated in Fig. C.1(d).

 ## C.2   NORTON'S THEOREM

Norton's theorem is the *dual* of Thévenin's theorem. It is used to represent a part of a network by a current source I_n and a parallel impedance Z_n, as shown in Fig. C.2. Figure C.2(a) shows a network divided into two parts, A and B. In Fig. C.2(b) part A has been replaced by its Norton's equivalent: a current source I_n and a parallel impedance Z_n. The Norton's current source I_n can be measured (or calculated) as shown in Fig. C.2(c). The terminals of the network being reduced (network A) are shorted, and the current I_n will be equal simply to the short-circuit current. To determine the impedance Z_n we first reduce the external excitation in network A to zero: That is, we short-circuit independent voltage sources and open-circuit independent current sources. The impedance Z_n will be equal to the input impedance of network A after this source-elimination process has taken place. Thus the Norton impedance Z_n is equal to the Thévenin impedance Z_t. Finally, note that $I_n = V_t/Z$, where $Z = Z_n = Z_t$.

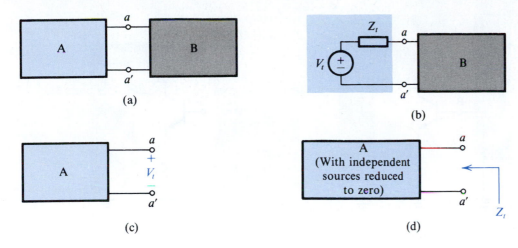

FIGURE C.1 Thévenin's theorem.

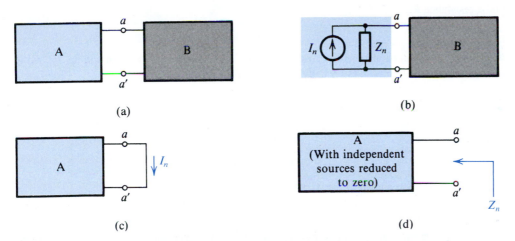

FIGURE C.2 Norton's theorem.

EXAMPLE C.1

Figure C.3(a) shows a bipolar junction transistor circuit. The transistor is a three-terminal device with the terminals labeled E (emitter), B (base), and C (collector). As shown, the base is connected to the dc power supply V^+ via the voltage divider composed of R_1 and R_2. The collector is connected to the dc supply V^+ through R_3 and to ground through R_4. To simplify the analysis we wish to apply Thévenin's theorem to reduce the circuit.

Solution

Thévenin's theorem can be used at the base side to reduce the network composed of V^+, R_1, and R_2 to a dc voltage source V_{BB},

$$V_{BB} = V^+ \frac{R_2}{R_1 + R_2}$$

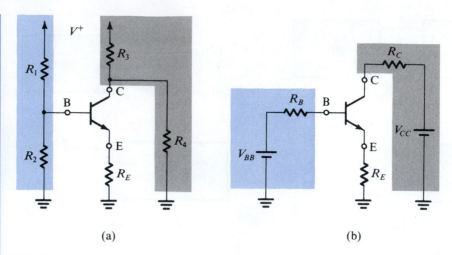

(a)	(b)

FIGURE C.3 Thévenin's theorem applied to simplify the circuit of **(a)** to that in **(b).** (See Example C.1.)

and a resistance R_B,

$$R_B = R_1 // R_2$$

where // denotes "in parallel with." At the collector side, Thévenin's theorem can be applied to reduce the network composed of V^+, R_3, and R_4 to a dc voltage source V_{CC},

$$V_{CC} = V^+ \frac{R_4}{R_3 + R_4}$$

and a resistance R_C,

$$R_C = R_3 // R_4$$

The reduced circuit is shown in Fig. C.3(b).

C.3 SOURCE-ABSORPTION THEOREM

Consider the situation shown in Fig. C.4. In the course of analyzing a network we find a controlled current source I_x appearing between two nodes whose voltage difference is the controlling voltage V_x. That is, $I_x = g_m V_x$ where g_m is a conductance. We can replace this controlled source by an impedance $Z_x = V_x / I_x = 1/g_m$, as shown in Fig. C.4, because the current drawn by this impedance will be equal to the current of the controlled source that we have replaced.

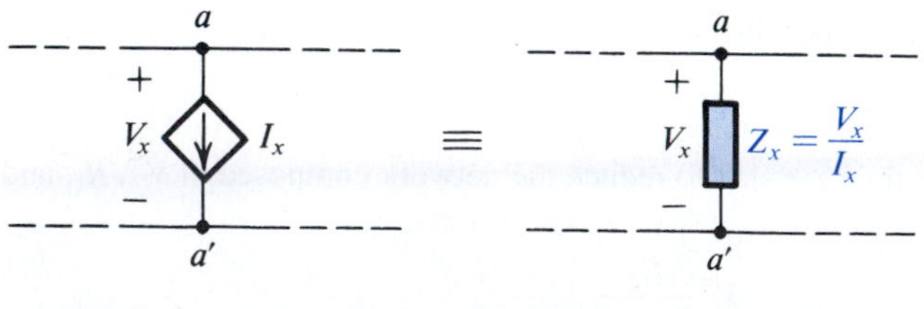

FIGURE C.4 The source-absorption theorem.

EXAMPLE C.2

Figure C.5(a) shows the small-signal equivalent-circuit model of a transistor. We want to find the resistance R_{in} "looking into" the emitter terminal E—that is, the resistance between the emitter and ground—with the base B and collector C grounded.

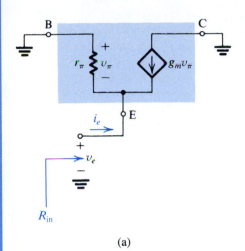

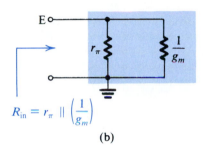

$$R_{in} = r_\pi \,\|\, \left(\frac{1}{g_m}\right)$$

(a)   (b)

FIGURE C.5   Circuit for Example C.2.

Solution

From Fig. C.5(a) we see that the voltage v_π will be equal to $-v_e$. Thus looking between E and ground we see a resistance r_π in parallel with a current source drawing a current $g_m v_e$ away from terminal E. The latter source can be replaced by a resistance $(1/g_m)$, resulting in the input resistance R_{in} given by

$$R_{in} = r_\pi \,\|\, (1/g_m)$$

as illustrated in Fig. C.5(b).

EXERCISES

C.1 A source is measured and found to have a 10-V open-circuit voltage and to provide 1 mA into a short circuit. Calculate its Thévenin and Norton equivalent source parameters.

Ans. $V_t = 10$ V; $Z_t = Z_n = 10$ kΩ; $I_n = 1$ mA

C.2 In the circuit shown in Fig. EC.2 the diode has a voltage drop $V_D \simeq 0.7$ V. Use Thévenin's theorem to simplify the circuit and hence calculate the diode current I_D.

Ans. 1 mA

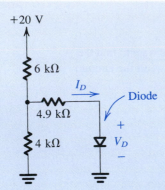

FIGURE EC.2

C.3 The two-terminal device M in the circuit of Fig. EC.3 has a current $I_M \simeq 1$ mA independent of the voltage V_M across it. Use Norton's theorem to simplify the circuit and hence calculate the voltage V_M.

Ans. 5 V

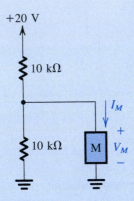

FIGURE EC.3

PROBLEMS

C.1 Consider the Thévenin equivalent circuit characterized by V_t and Z_t. Find the open-circuit voltage V_{oc} and the short-circuit current (i.e., the current that flows when the terminals are shorted together) I_{sc}. Express Z_t in terms of V_{oc} and I_{sc}.

C.2 Repeat Problem C.1 for a Norton equivalent circuit characterized by I_n and Z_n.

C.3 A voltage divider consists of a 9-kΩ resistor connected to +10 V and a resistor of 1 kΩ connected to ground. What is the Thévenin equivalent of this voltage divider?

What output voltage results if it is loaded with 1 kΩ? Calculate this two ways: directly and using your Thévenin equivalent.

C.4 Find the output voltage and output resistance of the circuit shown in Fig. PC.4 by considering a succession of Thévenin equivalent circuits.

C.5 Repeat Example C.2 with a resistance R_B connected between B and ground in Fig. C.5 (i.e., rather than directly grounding the base B as indicated in Fig. C.5).

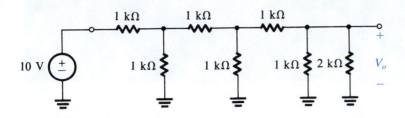

FIGURE PC.4

C.6 Figure PC.6(a) shows the circuit symbol of a device known as the *p*-channel junction field-effect transistor (JFET). As indicated, the JFET has three terminals. When the gate terminal G is connected to the source terminal S, the two-terminal device shown in Fig. PC.6(b) is obtained. Its *i*–*v* characteristic is given by

$$i = I_{DSS}\left[2\frac{v}{V_P} - \left(\frac{v}{V_P}\right)^2\right] \quad \text{for } v \le V_P$$

$$i = I_{DSS} \quad \text{for } v \ge V_P$$

where I_{DSS} and V_P are positive constants for the particular JFET. Now consider the circuit shown in Fig. PC.6(c) and let $V_P = 2$ V and $I_{DSS} = 2$ mA. For $V^+ = 10$ V show that the JFET is operating in the constant-current mode and find the voltage across it. What is the minimum value of V^+ for which this mode of operation is maintained? For $V^+ = 2$ V find the values of I and V.

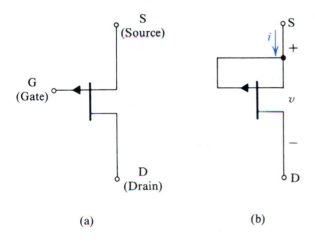

(a)

(b)

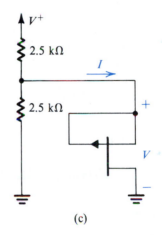

(c)

FIGURE PC.6

Single-Time-Constant Circuits

INTRODUCTION

Single-time-constant (STC) circuits are those circuits that are composed of, or can be reduced to, one reactive component (inductance or capacitance) and one resistance. An STC circuit formed of an inductance L and a resistance R has a time constant $\tau = L/R$. The time constant τ of an STC circuit composed of a capacitance C and a resistance R is given by $\tau = CR$.

Although STC circuits are quite simple, they play an important role in the design and analysis of linear and digital circuits. For instance, the analysis of an amplifier circuit can usually be reduced to the analysis of one or more STC circuits. For this reason, we will review in this appendix the process of evaluating the response of STC circuits to sinusoidal and other input signals such as step and pulse waveforms. The latter signal waveforms are encountered in some amplifier applications but are more important in switching circuits, including digital circuits.

D.1 EVALUATING THE TIME CONSTANT

The first step in the analysis of an STC circuit is to evaluate its time constant τ.

EXAMPLE D.1

Reduce the circuit in Fig. D.1(a) to an STC circuit, and find its time constant.

Solution

The reduction process is illustrated in Fig. D.1 and consists of repeated applications of Thévenin's theorem. From the final circuit (Fig. D.1c), we obtain the time constant as

$$\tau = C\{R_4 // [R_3 + (R_1 // R_2)]\}$$

D.1.1 Rapid Evaluation of τ

In many instances, it will be important to be able to evaluate rapidly the time constant τ of a given STC circuit. A simple method for accomplishing this goal consists first of reducing the excitation to zero; that is, if the excitation is by a voltage source, short it, and if by a current

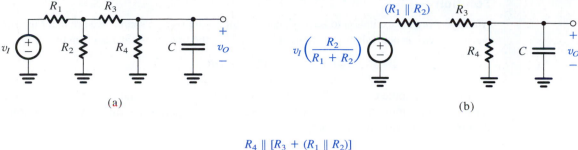

(a)   (b)

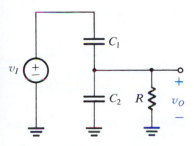

(c)

FIGURE D.1 The reduction of the circuit in **(a)** to the STC circuit in **(c)** through the repeated application of Thévenin's theorem.

source, open it. Then if the circuit has one reactive component and a number of resistances, "grab hold" of the two terminals of the reactive component (capacitance or inductance) and find the equivalent resistance R_{eq} seen by the component. The time constant is then either L/R_{eq} or CR_{eq}. As an example, in the circuit of Fig. D.1(a) we find that the capacitor C "sees" a resistance R_4 in parallel with the series combination of R_3 and (R_2 in parallel with R_1). Thus

$$R_{eq} = R_4 // [R_3 + (R_2 // R_1)]$$

and the time constant is CR_{eq}.

In some cases it may be found that the circuit has one resistance and a number of capacitances or inductances. In such a case the procedure should be inverted; that is, "grab hold" of the resistance terminals and find the equivalent capacitance C_{eq}, or equivalent inductance L_{eq}, seen by this resistance. The time constant is then found as $C_{eq}R$ or L_{eq}/R. This is illustrated in Example D.2.

EXAMPLE D.2

Find the time constant of the circuit in Fig. D.2.

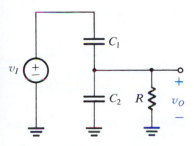

FIGURE D.2 Circuit for Example D.2.

Solution

After reducing the excitation to zero by short-circuiting the voltage source, we see that the resistance R "sees" an equivalent capacitance $C_1 + C_2$. Thus the time constant τ is given by

$$\tau = (C_1 + C_2)R$$

Finally, in some cases an STC circuit has more than one resistance and more than one capacitance (or more than one inductance). Such cases require some initial work to simplify the circuit, as illustrated by Example D.3.

EXAMPLE D.3

Here we show that the response of the circuit in Fig. D.3(a) can be obtained using the method of analysis of STC circuits.

Solution

The analysis steps are illustrated in Fig. D.3. In Fig. D.3(b) we show the circuit excited by two separate but equal voltage sources. The reader should convince himself or herself of the equivalence

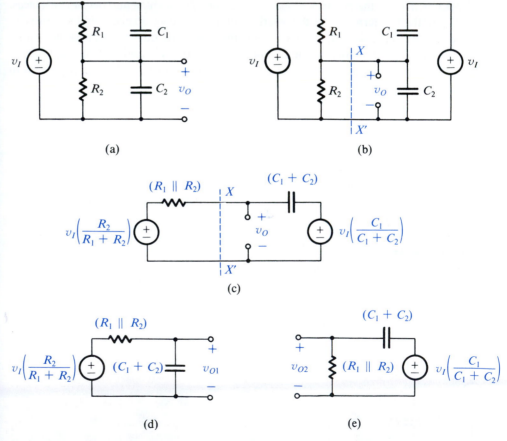

FIGURE D.3 The response of the circuit in **(a)** can be found by superposition, that is, by summing the responses of the circuits in **(d)** and **(e)**.

of the circuits in Fig. D.3(a) and D.3(b). The "trick" employed to obtain the arrangement in Fig. D.3(b) is a very useful one.

Application of Thévenin's theorem to the circuit to the left of the line XX' and then to the circuit to the right of that line result in the circuit of Fig. D.3(c). Since this is a linear circuit, the response may be obtained using the principle of superposition. Specifically, the output voltage v_O will be the sum of the two components v_{O1} and v_{O2}. The first component, v_{O1}, is the output due to the left-hand-side voltage source with the other voltage source reduced to zero. The circuit for calculating v_{O1} is shown in Fig. D.3(d). It is an STC circuit with a time constant given by

$$\tau = (C_1 + C_2)(R_1 // R_2)$$

Similarly, the second component v_{O2} is the output obtained with the left-hand-side voltage source reduced to zero. It can be calculated from the circuit of Fig. D.3(e), which is an STC circuit with the same time constant τ.

Finally, it should be observed that the fact that the circuit is an STC one can also be ascertained by setting the independent source v_I in Fig. D.3(a) to zero. Also, the time constant is then immediately obvious.

D.2 CLASSIFICATION OF STC CIRCUITS

STC circuits can be classified into two categories, *low-pass* (LP) and *high-pass* (HP) types, with each category displaying distinctly different signal responses. The task of finding whether an STC circuit is of LP or HP type may be accomplished in a number of ways, the simplest of which uses the frequency-domain response. Specifically, low-pass circuits pass dc (i.e., signals with zero frequency) and attenuate high frequencies, with the transmission being zero at $\omega = \infty$. Thus we can test for the circuit type either at $\omega = 0$ or at $\omega = \infty$. At $\omega = 0$ capacitors should be replaced by open circuits ($1/j\omega C = \infty$) and inductors should be replaced by short circuits ($j\omega L = 0$). Then if the output is zero, the circuit is of the high-pass type, while if the output is finite, the circuit is of the low-pass type. Alternatively, we may test at $\omega = \infty$ by replacing capacitors by short circuits ($1/j\omega C = 0$) and inductors by open circuits ($j\omega L = \infty$). Then if the output is finite, the circuit is of the HP type, whereas if the output is zero, the circuit is of the LP type. In Table D.1, which provides a summary of these results, s.c. stands for short circuit and o.c. for open circuit.

Figure D.4 shows examples of low-pass STC circuits, and Fig. D.5 shows examples of high-pass STC circuits. For each circuit we have indicated the input and output variables of interest. Note that a given circuit can be of either category, depending on the input and output variables. The reader is urged to verify, using the rules of Table D.1, that the circuits of Figs. D.4 and D.5 are correctly classified.

TABLE D.1	Rules for Finding the Type of STC Circuit		
Test At	Replace	Circuit Is LP If	Circuit Is HP If
$\omega = 0$	C by o.c. L by s.c.	output is finite	output is zero
$\omega = \infty$	C by s.c. L by o.c.	output is zero	output is finite

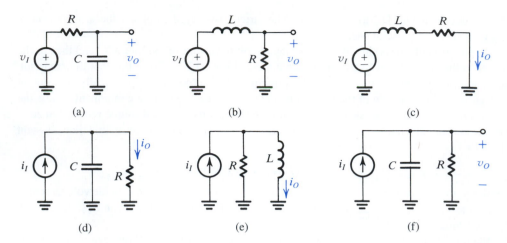

FIGURE D.4 STC circuits of the low-pass type.

FIGURE D.5 STC circuits of the high-pass type.

EXERCISES

D.1 Find the time constants for the circuits shown in Fig. ED.1.

Ans. (a) $\dfrac{(L_1//L_2)}{R}$; (b) $\dfrac{(L_1//L_2)}{(R_1//R_2)}$

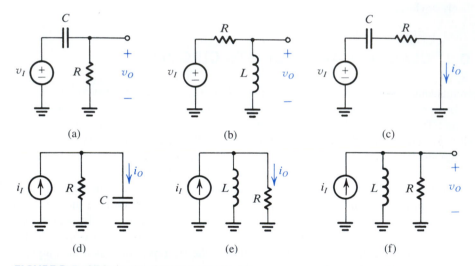

FIGURE ED.1

D.3 FREQUENCY RESPONSE OF STC CIRCUITS

D.3.1 Low-Pass Circuits

The transfer function $T(s)$ of an STC low-pass circuit always can be written in the form

$$T(s) = \frac{K}{1 + (s/\omega_0)} \tag{D.1}$$

which, for physical frequencies, where $s = j\omega$, becomes

$$T(j\omega) = \frac{K}{1 + j(\omega/\omega_0)} \tag{D.2}$$

where K is the magnitude of the transfer function at $\omega = 0$ (dc) and ω_0 is defined by

$$\omega_0 = 1/\tau$$

with τ being the time constant. Thus the magnitude response is given by

$$|T(j\omega)| = \frac{K}{\sqrt{1 + (\omega/\omega_0)^2}} \tag{D.3}$$

and the phase response is given by

$$\phi(\omega) = -\tan^{-1}(\omega/\omega_0) \tag{D.4}$$

Figure D.6 sketches the magnitude and phase responses for an STC low-pass circuit. The magnitude response shown in Fig. D.6(a) is simply a graph of the function in Eq. (D.3). The magnitude is normalized with respect to the dc gain K and is expressed in decibels; that is, the plot is for $20 \log|T(j\omega)/K|$, with a logarithmic scale used for the frequency axis. Furthermore, the frequency variable has been normalized with respect to ω_0. As shown, the magnitude curve is closely defined by two straight-line asymptotes. The low-frequency asymptote is a horizontal straight line at 0 dB. To find the slope of the high-frequency asymptote consider Eq. (D.3) and let $\omega/\omega_0 \gg 1$, resulting in

$$|T(j\omega)| \simeq K\frac{\omega_0}{\omega}$$

It follows that if ω doubles in value, the magnitude is halved. On a logarithmic frequency axis, doublings of ω represent equally spaced points, with each interval called an *octave*. Halving the magnitude function corresponds to a 6-dB reduction in transmission ($20 \log 0.5 = -6$ dB). Thus the slope of the high-frequency asymptote is -6 dB/octave. This can be equivalently expressed as -20 dB/decade, where "decade" indicates an increase in frequency by a factor of 10.

The two straight-line asymptotes of the magnitude–response curve meet at the "corner frequency" or "break frequency" ω_0. The difference between the actual magnitude–response curve and the asymptotic response is largest at the corner frequency, where its value is 3 dB.

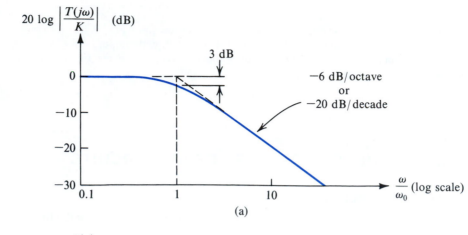

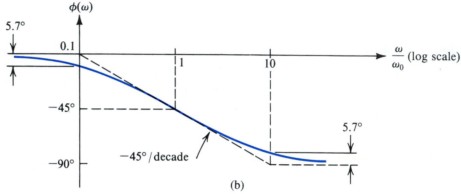

FIGURE D.6   (a) Magnitude and (b) phase response of STC circuits of the low-pass type.

To verify that this value is correct, simply substitute $\omega = \omega_0$ in Eq. (D.3) to obtain

$$|T(j\omega_0)| = K/\sqrt{2}$$

Thus at $\omega = \omega_0$ the gain drops by a factor of $\sqrt{2}$ relative to the dc gain, which corresponds to a 3-dB reduction in gain. The corner frequency ω_0 is appropriately referred to as the 3-dB frequency.

Similar to the magnitude response, the phase–response curve, shown in Fig. E.6(b), is closely defined by straight-line asymptotes. Note that at the corner frequency the phase is $-45°$, and that for $\omega \gg \omega_0$ the phase approaches $-90°$. Also note that the $-45°$/decade straight line approximates the phase function, with a maximum error of $5.7°$, over the frequency range $0.1\omega_0$ to $10\omega_0$.

EXAMPLE D.4

Consider the circuit shown in Fig. D.7(a), where an ideal voltage amplifier of gain $\mu = -100$ has a small (10-pF) capacitance connected in its feedback path. The amplifier is fed by a voltage source having a source resistance of 100 kΩ. Show that the frequency response V_o/V_s of this amplifier is equivalent to that of an STC circuit, and sketch the magnitude response.

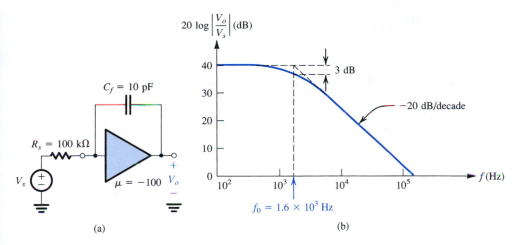

FIGURE D.7   (a) An amplifier circuit and (b) a sketch of the magnitude of its transfer function.

Solution

Direct analysis of the circuit in Fig. D.7(a) results in the transfer function

$$\frac{V_o}{V_s} = \frac{\mu}{1 + sRC_f(-\mu + 1)}$$

which can be seen to be that of a low-pass STC circuit with a dc gain $\mu = -100$ (or, equivalently, 40 dB) and a time constant $\tau = RC_f(-\mu + 1) = 100 \times 10^3 \times 10 \times 10^{-12} \times 101 \simeq 10^{-4}$ s, which corresponds to a frequency $\omega_0 = 1/\tau = 10^4$ rad/s. The magnitude response is sketched in Fig. D.7(b).

D.3.2 High-Pass Circuits

The transfer function $T(s)$ of an STC high-pass circuit always can be expressed in the form

$$T(s) = \frac{Ks}{s + \omega_0} \qquad \text{(D.5)}$$

which for physical frequencies $s = j\omega$ becomes

$$T(j\omega) = \frac{K}{1 - j\omega_0/\omega} \qquad \text{(D.6)}$$

where K denotes the gain as s or ω approaches infinity and ω_0 is the inverse of the time constant τ,

$$\omega_0 = 1/\tau$$

The magnitude response

$$|T(j\omega)| = \frac{K}{\sqrt{1 + (\omega_0/\omega)^2}} \qquad \text{(D.7)}$$

and the phase response

$$\phi(\omega) = \tan^{-1}(\omega_0/\omega) \qquad \text{(D.8)}$$

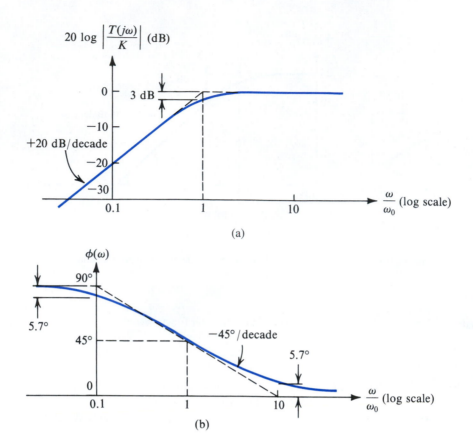

(a)

(b)

FIGURE D.8 (a) Magnitude and (b) phase response of STC circuits of the high-pass type.

are sketched in Fig. D.8. As in the low-pass case, the magnitude and phase curves are well defined by straight-line asymptotes. Because of the similarity (or, more appropriately, duality) with the low-pass case, no further explanation will be given.

EXERCISES

D.3 Find the dc transmission, the corner frequency f_0, and the transmission at $f = 2$ MHz for the low-pass STC circuit shown in Fig. ED.3.

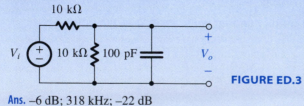

FIGURE ED.3

Ans. −6 dB; 318 kHz; −22 dB

D.4 Find the transfer function $T(s)$ of the circuit in Fig. D.2. What type of STC network is it?

Ans. $T(s) = \dfrac{C_1}{C_1 + C_2} \dfrac{s}{s + [1/(C_1 + C_2)R]}$; HP

D.5 For the situation discussed in Exercise D.4, if $R = 10$ kΩ, find the capacitor values that result in the circuit having a high-frequency transmission of 0.5 V/V and a corner frequency $\omega_0 = 10$ rad/s.

Ans. $C_1 = C_2 = 5$ μF

D.6 Find the high-frequency gain, the 3-dB frequency f_0, and the gain at $f = 1$ Hz of the capacitively coupled amplifier shown in Fig. ED.6. Assume the voltage amplifier to be ideal.

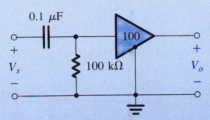

FIGURE ED.6

Ans. 40 dB; 15.9 Hz; 16 dB

D.4 STEP RESPONSE OF STC CIRCUITS

In this section we consider the response of STC circuits to the step-function signal shown in Fig. D.9. Knowledge of the step response enables rapid evaluation of the response to other switching-signal waveforms, such as pulses and square waves.

D.4.1 Low-Pass Circuits

In response to an input step signal of height S, a low-pass STC circuit (with a dc gain $K = 1$) produces the waveform shown in Fig. D.10. Note that while the input rises from 0 to S at $t = 0$, the output does not respond immediately to this transient and simply begins to rise exponentially toward the *final* dc value of the input, S. In the long term—that is, for $t \gg \tau$—the output approaches the dc value S, a manifestation of the fact that low-pass circuits faithfully pass dc.

The equation of the output waveform can be obtained from the expression

$$y(t) = Y_\infty - (Y_\infty - Y_{0+})e^{-t/\tau} \tag{D.9}$$

where Y_∞ denotes the *final* value or the value toward which the output is heading and Y_{0+} denotes the value of the output immediately after $t = 0$. This equation states that *the output at any time t is equal to the difference between the final value Y_∞ and a gap that has an initial value of $Y_\infty - Y_{0+}$ and is "shrinking" exponentially*. In our case, $Y_\infty = S$ and $Y_{0+} = 0$; thus,

$$y(t) = S(1 - e^{-t/\tau}) \tag{D.10}$$

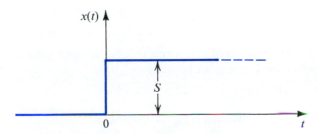

FIGURE D.9 A step-function signal of height S.

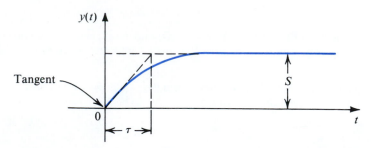

FIGURE D.10   The output $y(t)$ of a low-pass STC circuit excited by a step of height S.

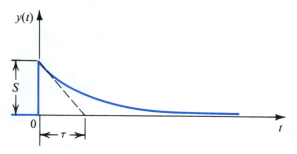

FIGURE D.11   The output $y(t)$ of a high-pass STC circuit excited by a step of height S.

The reader's attention is drawn to the slope of the tangent to $y(t)$ at $t = 0$, which is indicated in Fig. D.10.

D.4.2 High-Pass Circuits

The response of an STC high-pass circuit (with a high-frequency gain $K = 1$) to an input step of height S is shown in Fig. D.11. The high-pass circuit faithfully transmits the transient part of the input signal (the step change) but blocks the dc. Thus the output at $t = 0$ follows the input,

$$Y_{0+} = S$$

and then it decays toward zero,

$$Y_\infty = 0$$

Substituting for Y_{0+} and Y_∞ in Eq. (D.9) results in the output $y(t)$,

$$y(t) = Se^{-t/\tau} \tag{D.11}$$

The reader's attention is drawn to the slope of the tangent to $y(t)$ at $t = 0$, indicated in Fig. D.11.

EXAMPLE D.5

This example is a continuation of the problem considered in Example D.3. For an input v_I that is a 10-V step, find the condition under which the output v_O is a perfect step.

Solution

Following the analysis in Example D.3, which is illustrated in Fig. D.3, we have

$$v_{O1} = k_r[10(1 - e^{-t/\tau})]$$

where

$$k_r \equiv \frac{R_2}{R_1 + R_2}$$

and

$$v_{O2} = k_c(10e^{-t/\tau})$$

where

$$k_c \equiv \frac{C_1}{C_1 + C_2}$$

and

$$\tau = (C_1 + C_2)(R_1 /\!/ R_2)$$

Thus

$$v_O = v_{O1} + v_{O2}$$
$$= 10k_r + 10e^{-t/\tau}(k_c - k_r)$$

It follows that the output can be made a perfect step of height $10k_r$ volts if we arrange that

$$k_c = k_r$$

that is, if the resistive voltage-divider ratio is made equal to the capacitive voltage divider ratio.

This example illustrates an important technique, namely, that of the "compensated attenuator." An application of this technique is found in the design of the oscilloscope probe. The oscilloscope probe problem is investigated in Problem D.3.

EXERCISES

D.7 For the circuit of Fig. D.4(f) find v_O if i_I is a 3-mA step, $R = 1$ kΩ, and $C = 100$ pF.

Ans. $3(1 - e^{-10^7 t})$

D.8 In the circuit of Fig. D.5(f) find $v_O(t)$ if i_I is a 2-mA step, $R = 2$ kΩ, and $L = 10$ μH.

Ans. $4e^{-2 \times 10^8 t}$

D.9 The amplifier circuit of Fig. ED.6 is fed with a signal source that delivers a 20-mV step. If the source resistance is 100 kΩ, find the time constant τ and $v_O(t)$.

Ans. $\tau = 2 \times 10^{-2}$ s; $v_O(t) = 1 \times e^{-50t}$

D.10 For the circuit in Fig. D.2 with $C_1 = C_2 = 0.5$ μF, $R = 1$ MΩ, find $v_O(t)$ if $v_I(t)$ is a 10-V step.

Ans. $5e^{-t}$

D.11 Show that the area under the exponential of Fig. D.11 is equal to that of the rectangle of height S and width τ.

 ## D.5 PULSE RESPONSE OF STC CIRCUITS

Figure D.12 shows a pulse signal whose height is P and whose width is T. We wish to find the response of STC circuits to input signals of this form. Note at the outset that a pulse can be considered as the sum of two steps: a positive one of height P occurring at $t = 0$ and a

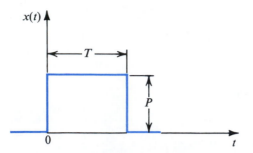

FIGURE D.12 A pulse signal with height P and width T.

negative one of height P occurring at $t = T$. Thus the response of a linear circuit to the pulse signal can be obtained by summing the responses to the two step signals.

D.5.1 Low-Pass Circuits

Figure D.13(a) shows the response of a low-pass STC circuit (having unity dc gain) to an input pulse of the form shown in Fig. D.12. In this case we have assumed that the time constant τ is in the same range as the pulse width T. As shown, the LP circuit does not respond immediately to the step change at the leading edge of the pulse; rather, the output starts to rise exponentially toward a final value of P. This exponential rise, however, will be stopped at time $t = T$, that is, at the trailing edge of the pulse when the input undergoes a negative step change. Again the output will respond by starting an exponential decay toward the final value of the input, which is zero. Finally, note that the area under the output waveform will be equal to the area under the input pulse waveform, since the LP circuit faithfully passes dc.

A low-pass effect usually occurs when a pulse signal from one part of an electronic system is connected to another. The low-pass circuit in this case is formed by the output resistance (Thévenin's equivalent resistance) of the system part from which the signal originates and the input capacitance of the system part to which the signal is fed. This unavoidable low-pass filter will cause distortion—of the type shown in Fig. D.13(a)—of the pulse signal. In a well-designed system such distortion is kept to a low value by arranging that the time constant τ be much smaller than the pulse width T. In this case the result will be a slight rounding of the pulse edges, as shown in Fig. D.13(b). Note, however, that the edges are still exponential.

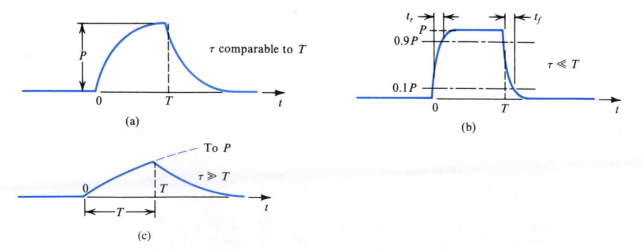

FIGURE D.13 Pulse responses of three STC low-pass circuits.

The distortion of a pulse signal by a parasitic (i.e., unwanted) low-pass circuit is measured by its *rise time* and *fall time*. The rise time is conventionally defined as the time taken by the amplitude to increase from 10% to 90% of the final value. Similarly, the fall time is the time during which the pulse amplitude falls from 90% to 10% of the maximum value. These definitions are illustrated in Fig. D.13(b). By use of the exponential equations of the rising and falling edges of the output waveform, it can be easily shown that

$$t_r = t_f \simeq 2.2\tau \qquad\qquad (D.12)$$

which can be also expressed in terms of $f_0 = \omega_0/2\pi = 1/2\pi\tau$ as

$$t_r = t_f \simeq \frac{0.35}{f_0} \qquad\qquad (D.13)$$

Finally, we note that the effect of the parasitic low-pass circuits that are always present in a system is to "slow down" the operation of the system: To keep the signal distortion within acceptable limits, one has to use a relatively long pulse width (for a given low-pass time constant).

The other extreme case—namely, when τ is much larger than T—is illustrated in Fig. D.13(c). As shown, the output waveform rises exponentially toward the level P. However, since $\tau \gg T$, the value reached at $t = T$ will be much smaller than P. At $t = T$ the output waveform starts its exponential decay toward zero. Note that in this case the output waveform bears little resemblance to the input pulse. Also note that because $\tau \gg T$ the portion of the exponential curve from $t = 0$ to $t = T$ is almost linear. Since the slope of this linear curve is proportional to the height of the input pulse, we see that the output waveform approximates the time integral of the input pulse. That is, a low-pass network with a large time constant approximates the operation of an *integrator*.

D.5.2 High-Pass Circuits

Figure D.14(a) shows the output of an STC HP circuit (with unity high-frequency gain) excited by the input pulse of Fig. D.12, assuming that τ and T are comparable in value. As shown, the step transition at the leading edge of the input pulse is faithfully reproduced at the output of the HP circuit. However, since the HP circuit blocks dc, the output waveform immediately starts an exponential decay toward zero. This decay process is stopped at $t = T$, when the negative step transition of the input occurs and the HP circuit faithfully reproduces it. Thus at $t = T$ the output waveform exhibits an *undershoot*. Then it starts an exponential decay toward zero. Finally, note that the area of the output waveform above the zero axis will be equal to that below the axis for a total average area of zero, consistent with the fact that HP circuits block dc.

In many applications an STC high-pass circuit is used to couple a pulse from one part of a system to another part. In such an application it is necessary to keep the distortion in the pulse shape as small as possible. This can be accomplished by selecting the time constant τ to be much longer than the pulse width T. If this is indeed the case, the loss in amplitude during the pulse period T will be very small, as shown in Fig. D.14(b). Nevertheless, the output waveform still swings negatively, and the area under the negative portion will be equal to that under the positive portion.

Consider the waveform in Fig. D.14(b). Since τ is much larger than T, it follows that the portion of the exponential curve from $t = 0$ to $t = T$ will be almost linear and that its slope will be equal to the slope of the exponential curve at $t = 0$, which is P/τ. We can use this value of

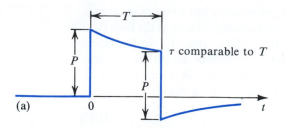

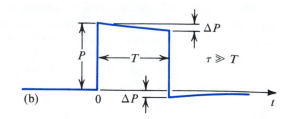

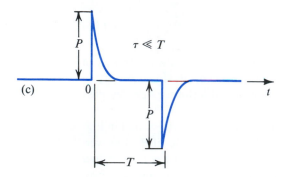

FIGURE D.14 Pulse responses of three STC high-pass circuits.

the slope to determine the loss in amplitude ΔP as

$$\Delta P \simeq \frac{P}{\tau} T \tag{D.14}$$

The distortion effect of the high-pass circuit on the input pulse is usually specified in terms of the per-unit or percentage loss in pulse height. This quantity is taken as an indication of the "sag" in the output pulse,

$$\text{Percentage sag} \equiv \frac{\Delta P}{P} \times 100 \tag{D.15}$$

Thus

$$\text{Percentage sag} = \frac{T}{\tau} \times 100 \tag{D.16}$$

Finally, note that the magnitude of the undershoot at $t = T$ is equal to ΔP.

The other extreme case—namely, $\tau \ll T$—is illustrated in Fig. D.14(c). In this case the exponential decay is quite rapid, resulting in the output becoming almost zero shortly beyond the leading edge of the pulse. At the trailing edge of the pulse the output swings negatively by an amount almost equal to the pulse height P. Then the waveform decays rapidly to zero. As seen from Fig. D.14(c), the output waveform bears no resemblance to the input pulse. It consists of two spikes: a positive one at the leading edge and a negative one at the trailing edge. Note that the output waveform is approximately equal to the time derivative of the input pulse. That is, for $\tau \ll T$ an STC high-pass circuit approximates a *differentiator*. However, the resulting differentiator is not an ideal one; an ideal differentiator would produce two impulses. Nevertheless, high-pass STC circuits with short time constants are employed in some applications to produce sharp pulses or spikes at the transitions of an input waveform.

EXERCISES

D.12 Find the rise and fall times of a 1-μs pulse after it has passed through a low-pass RC circuit with a corner frequency of 10 MHz.

Ans. 35 ns

D.13 Consider the pulse response of a low-pass STC circuit, as shown in Fig. D.13(c). If $\tau = 100T$, find the output voltage at $t = T$. Also, find the difference in the slope of the rising portion of the output waveform at $t = 0$ and $t = T$ (expressed as a percentage of the slope at $t = 0$).

Ans. $0.01P$; 1%

D.14 The output of an amplifier stage is connected to the input of another stage via a capacitance C. If the first stage has an output resistance of 10 kΩ, and the second stage has an input resistance of 40 kΩ, find the minimum value of C such that a 10-μs pulse exhibits less than 1% sag.

Ans. 0.02 μF

D.15 A high-pass STC circuit with a time constant of 100 μs is excited by a pulse of 1-V height and 100-μs width. Calculate the value of the undershoot in the output waveform.

Ans. 0.632 V

PROBLEMS

D.1 Consider the circuit of Fig. D.3(a) and the equivalent shown in (d) and (e). There, the output, $v_O = v_{O1} + v_{O2}$, is the sum of outputs of a low-pass and a high-pass circuit, each with the time constant $\tau = (C_1 + C_2)(R_1 // R_2)$. What is the condition that makes the contribution of the low-pass circuit at zero frequency equal to the contribution of the high-pass circuit at infinite frequency? Show that this condition can be expressed as $C_1 R_1 = C_2 R_2$. If this condition applies, sketch $|V_o / V_i|$ versus frequency for the case $R_1 = R_2$.

D.2 Use the voltage divider rule to find the transfer function $V_o(s)/V_i(s)$ of the circuit in Fig. D.3(a). Show that the transfer function can be made independent of frequency if the condition $C_1 R_1 = C_2 R_2$ applies. Under this condition the circuit is called a *compensated attenuator*. Find the transmission of the compensated attenuator in terms of R_1 and R_2.

D****D.3** The circuit of Fig. D.3(a) is used as a compensated attenuator (see Problems D.1 and D.2) for an oscilloscope probe. The objective is to reduce the signal voltage applied to the input amplifier of the oscilloscope, with the signal attenuation independent of frequency. The probe itself includes R_1 and C_1, while R_2 and C_2 model the oscilloscope input circuit. For an oscilloscope having an input resistance of 1 MΩ and an input capacitance of 30 pF, design a compensated "10-to-1

probe"—that is, a probe that attenuates the input signal by a factor of 10. Find the input impedance of the probe when connected to the oscilloscope, which is the impedance seen by v_I in Fig. D.3(a). Show that this impedance is 10 times higher than that of the oscilloscope itself. This is the great advantage of the 10:1 probe.

D.4 In the circuits of Figs. D.4 and D.5, let $L = 10$ mH, $C = 0.01$ μF, and $R = 1$ kΩ. At what frequency does a phase angle of 45° occur?

***D.5** Consider a voltage amplifier with an open-circuit voltage gain $A_{vo} = -100$ V/V, $R_o = 0$, $R_i = 10$ kΩ, and an input capacitance C_i (in parallel with R_i) of 10 pF. The amplifier has a feedback capacitance (a capacitance connected between output and input) $C_f = 1$ pF. The amplifier is fed with a voltage source V_s having a resistance $R_s = 10$ kΩ. Find the amplifier transfer function $V_o(s)/V_s(s)$ and sketch its magnitude response versus frequency (dB vs frequency) on a log axis.

D.6 For the circuit in Fig. PD.6 assume the voltage amplifier to be ideal. Derive the transfer function $V_o(s)/V_i(s)$. What type of STC response is this? For $C = 0.01$ μF and $R = 100$ kΩ, find the corner frequency.

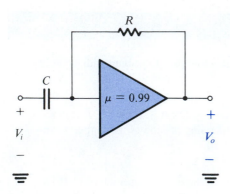

FIGURE PD.6

D.7 For the circuits of Figs. D.4(b) and D.5(b), find $v_O(t)$ if v_I is a 10-V step, $R = 1$ kΩ, and $L = 1$ mH.

D.8 Consider the exponential response of an STC low-pass circuit to a 10-V step input. In terms of the time constant τ, find the time taken for the output to reach 5 V, 9 V, 9.9 V, and 9.99 V.

D.9 The high-frequency response of an oscilloscope is specified to be like that of an STC LP circuit with a 100-MHz corner frequency. If this oscilloscope is used to display an ideal step waveform, what rise time (10% to 90%) would you expect to observe?

D.10 An oscilloscope whose step response is like that of a low-pass STC circuit has a rise time of t_s seconds. If an input signal having a rise time of t_w seconds is displayed, the waveform seen will have a rise time t_d seconds, which can be found using the empirical formula $t_d = \sqrt{t_s^2 + t_w^2}$. If $t_s = 35$ ns, what is the 3-dB frequency of the oscilloscope? What is the observed rise time for a waveform rising in 100 ns, 35 ns, and

10 ns? What is the actual rise time of a waveform whose displayed rise time is 49.5 ns?

D.11 A pulse of 10-ms width and 10-V amplitude is transmitted through a system characterized as having an STC high-pass response with a corner frequency of 10 Hz. What undershoot would you expect?

D.12 An RC differentiator having a time constant τ is used to implement a short-pulse detector. When a long pulse with $T \gg \tau$ is fed to the circuit, the positive and negative peak outputs are of equal magnitude. At what pulse width does the negative output peak differ from the positive one by 10%?

D.13 A high-pass STC circuit with a time constant of 1 ms is excited by a pulse of 10-V height and 1-ms width. Calculate the value of the undershoot in the output waveform. If an undershoot of 1 V or less is required, what is the time constant necessary?

DD.14 A capacitor C is used to couple the output of an amplifier stage to the input of the next stage. If the first stage has an output resistance of 2 kΩ and the second stage has an input resistance of 3 kΩ, find the value of C so that a 1-ms pulse exhibits less than 1% sag. What is the associated 3-dB frequency?

DD.15 An RC differentiator is used to convert a step voltage change V to a single pulse for a digital-logic application. The logic circuit that the differentiator drives distinguishes signals above $V/2$ as "high" and below $V/2$ as "low." What must the time constant of the circuit be to convert a step input into a pulse that will be interpreted as "high" for 10 μs?

DD.16 Consider the circuit in Fig. D.7(a) with $\mu = -100$, $C_f = 100$ pF, and the amplifier being ideal. Find the value of R so that the gain $|V_o/V_s|$ has a 3-dB frequency of 1 kHz.

s-Domain Analysis: Poles, Zeros, and Bode Plots

In analyzing the frequency response of an amplifier, most of the work involves finding the amplifier voltage gain as a function of the complex frequency s. In this s-domain analysis, a capacitance C is replaced by an admittance sC, or equivalently an impedance $1/sC$, and an inductance L is replaced by an impedance sL. Then, using usual circuit-analysis techniques, one derives the voltage transfer function $T(s) \equiv V_o(s)/V_i(s)$.

EXERCISE

E.1 Find the voltage transfer function $T(s) \equiv V_o(s)/V_i(s)$ for the STC network shown in Fig. EE.1.

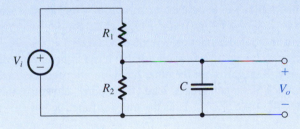

FIGURE EE.1

Ans. $T(s) = \dfrac{1/CR_1}{s + 1/C(R_1 \parallel R_2)}$

Once the transfer function $T(s)$ is obtained, it can be evaluated for **physical frequencies** by replacing s by $j\omega$. The resulting transfer function $T(j\omega)$ is in general a complex quantity whose magnitude gives the magnitude response (or transmission) and whose angle gives the phase response of the amplifier.

In many cases it will not be necessary to substitute $s = j\omega$ and evaluate $T(j\omega)$; rather, the form of $T(s)$ will reveal many useful facts about the circuit performance. In general, for all the circuits dealt with in this book, $T(s)$ can be expressed in the form

$$T(s) = \frac{a_m s^m + a_{m-1} s^{m-1} + \cdots + a_0}{s^n + b_{n-1} s^{n-1} + \cdots + b_0} \tag{E.1}$$

where the coefficients a and b are real numbers, and the order m of the numerator is smaller than or equal to the order n of the denominator; the latter is called the **order of the network.** Furthermore, for a **stable circuit**—that is, one that does not generate signals on its own—the denominator coefficients should be such that *the roots of the denominator polynomial all have negative real parts.* The problem of amplifier stability is studied in Chapter 8.

 ## E.1   POLES AND ZEROS

An alternate form for expressing $T(s)$ is

$$T(s) = a_m \frac{(s - Z_1)(s - Z_2) \cdots (s - Z_m)}{(s - P_1)(s - P_2) \cdots (s - P_n)} \tag{E.2}$$

where a_m is a multiplicative constant (the coefficient of s^m in the numerator), $Z_1, Z_2, \ldots, Z_m$ are the roots of the numerator polynomial, and $P_1, P_2, \ldots, P_n$ are the roots of the denominator polynomial. $Z_1, Z_2, \ldots, Z_m$ are called the **transfer-function zeros** or **transmission zeros,** and $P_1, P_2, \ldots, P_n$ are the **transfer-function poles** or the **natural modes** of the network. A transfer function is completely specified in terms of its poles and zeros together with the value of the multiplicative constant.

The poles and zeros can be either real or complex numbers. However, since the a and b coefficients are real numbers, the complex poles (or zeros) must occur in **conjugate pairs.** That is, if $5 + j3$ is a zero, then $5 - j3$ also must be a zero. A zero that is pure imaginary $(\pm j\omega_Z)$ causes the transfer function $T(j\omega)$ to be exactly zero at $\omega = \omega_Z$. This is because the numerator will have the factors $(s + j\omega_Z)(s - j\omega_Z) = (s^2 + \omega_Z^2)$, which for physical frequencies becomes $(-\omega^2 + \omega_Z^2)$, and thus the transfer fraction will be exactly zero at $\omega = \omega_Z$. Thus the "trap" one places at the input of a television set is a circuit that has a transmission zero at the particular interfering frequency. Real zeros, on the other hand, do not produce transmission nulls. Finally, note that for values of s much greater than all the poles and zeros, the transfer function in Eq. (E.1) becomes $T(s) \simeq a_m / s^{n-m}$. Thus the transfer function has $(n - m)$ zeros at $s = \infty$.

 ## E.2   FIRST-ORDER FUNCTIONS

Many of the transfer functions encountered in this book have real poles and zeros and can therefore be written as the product of first-order transfer functions of the general form

$$T(s) = \frac{a_1 s + a_0}{s + \omega_0} \tag{E.3}$$

where $-\omega_0$ is the location of the real pole. The quantity ω_0, called the **pole frequency,** is equal to the inverse of the time constant of this single-time-constant (STC) network (see Appendix D). The constants a_0 and a_1 determine the type of STC network. Specifically, we studied in Chapter 1 two types of STC networks, low pass and high pass. For the low-pass

first-order network we have

$$T(s) = \frac{a_0}{s + \omega_0} \tag{E.4}$$

In this case the dc gain is a_0/ω_0, and ω_0 is the corner or 3-dB frequency. Note that this transfer function has one zero at $s = \infty$. On the other hand, the first-order high-pass transfer function has a zero at dc and can be written as

$$T(s) = \frac{a_1 s}{s + \omega_0} \tag{E.5}$$

At this point the reader is strongly urged to review the material on STC networks and their frequency and pulse responses in Appendix D. Of specific interest are the plots of the magnitude and phase responses of the two special kinds of STC networks. Such plots can be employed to generate the magnitude and phase plots of a high-order transfer function, as explained below.

E.3 BODE PLOTS

A simple technique exists for obtaining an approximate plot of the magnitude and phase of a transfer function given its poles and zeros. The technique is particularly useful in the case of real poles and zeros. The method was developed by H. Bode, and the resulting diagrams are called **Bode plots.**

A transfer function of the form depicted in Eq. (E.2) consists of a product of factors of the form $s + a$, where such a factor appears on top if it corresponds to a zero and on the bottom if it corresponds to a pole. It follows that the magnitude response in decibels of the network can be obtained by summing together terms of the form $20 \log_{10} \sqrt{a^2 + \omega^2}$, and the phase response can be obtained by summing terms of the form $\tan^{-1}(\omega/a)$. In both cases the terms corresponding to poles are summed with negative signs. For convenience we can extract the constant a and write the typical magnitude term in the form $20 \log \sqrt{1 + (\omega/a)^2}$. On a plot of decibels versus log frequency this term gives rise to the curve and straight-line asymptotes shown in Fig. E.1. Here the low-frequency asymptote is a horizontal straight line

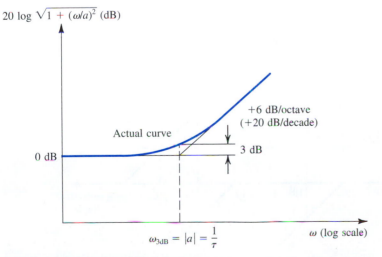

FIGURE E.1 Bode plot for the typical magnitude term. The curve shown applies for the case of a zero. For a pole, the high-frequency asymptote should be drawn with a −6-dB/octave slope.

at 0-dB level and the high-frequency asymptote is a straight line with a slope of 6 dB/octave or, equivalently, 20 dB/decade. The two asymptotes meet at the frequency $\omega = |a|$, which is called the **corner frequency.** As indicated, the actual magnitude plot differs slightly from the value given by the asymptotes; the maximum difference is 3 dB and occurs at the corner frequency.

For $a = 0$—that is, a pole or a zero at $s = 0$—the plot is simply a straight line of 6 dB/octave slope intersecting the 0-dB line at $\omega = 1$.

In summary, to obtain the Bode plot for the magnitude of a transfer function, the asymptotic plot for each pole and zero is first drawn. The slope of the high-frequency asymptote of the curve corresponding to a zero is +20 dB/decade, while that for a pole is −20 dB/decade. The various plots are then added together, and the overall curve is shifted vertically by an amount determined by the multiplicative constant of the transfer function.

EXAMPLE E.1

An amplifier has the voltage transfer function

$$T(s) = \frac{10s}{(1 + s/10^2)(1 + s/10^5)}$$

Find the poles and zeros and sketch the magnitude of the gain versus frequency. Find approximate values for the gain at $\omega = 10$, 10^3, and 10^6 rad/s.

Solution

The zeros are as follows: one at $s = 0$ and one at $s = \infty$. The poles are as follows: one at $s = -10^2$ rad/s and one at $s = -10^5$ rad/s.

Figure E.2 shows the asymptotic Bode plots of the different factors of the transfer function. Curve 1, which is a straight line intersecting the ω-axis at 1 rad/s and having a +20 dB/decade slope, corresponds to the s term (that is, the zero at $s = 0$) in the numerator. The pole at $s = -10^2$ results in curve 2, which consists of two asymptotes intersecting at $\omega = 10^2$. Similarly, the pole at $s = -10^5$ is represented by curve 3, where the intersection of the asymptotes is at $\omega = 10^5$. Finally, curve 4 represents the multiplicative constant of value 10.

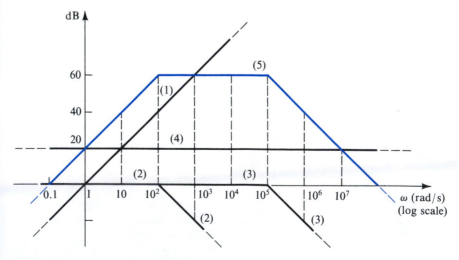

FIGURE E.2 Bode plots for Example E.1.

Adding the four curves results in the asymptotic Bode diagram of the amplifier gain (curve 5). Note that since the two poles are widely separated, the gain will be very close to 10^3 (60 dB) over the frequency range 10^2 to 10^5 rad/s. At the two corner frequencies (10^2 and 10^5 rad/s) the gain will be approximately 3 dB below the maximum of 60 dB. At the three specific frequencies, the values of the gain as obtained from the Bode plot and from exact evaluation of the transfer function are as follows:

ω	Approximate Gain	Exact Gain
10	40 dB	39.96 dB
10^3	60 dB	59.96 dB
10^6	40 dB	39.96 dB

We next consider the Bode phase plot. Figure E.3 shows a plot of the typical phase term $\tan^{-1}(\omega/a)$, assuming that a is negative. Also shown is an asymptotic straight-line approximation of the arctan function. The asymptotic plot consists of three straight lines. The first is horizontal at $\phi = 0$ and extends up to $\omega = 0.1|a|$. The second line has a slope of $-45°$/decade and extends from $\omega = 0.1|a|$ to $\omega = 10|a|$. The third line has a zero slope and a level of $\phi = -90°$. The complete phase response can be obtained by summing the asymptotic Bode plots of the phase of all poles and zeros.

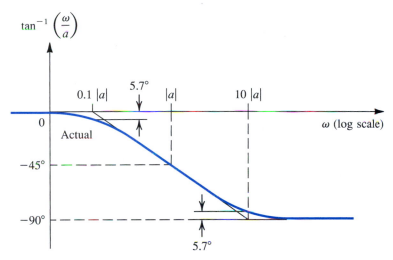

FIGURE E.3 Bode plot of the typical phase term $\tan^{-1}(\omega/a)$ when a is negative.

EXAMPLE E.2

Find the Bode plot for the phase of the transfer function of the amplifier considered in Example E.1.

Solution

The zero at $s = 0$ gives rise to a constant $+90°$ phase function represented by curve 1 in Fig. E.4. The pole at $s = -10^2$ gives rise to the phase function

$$\phi_1 = -\tan^{-1}\frac{\omega}{10^2}$$

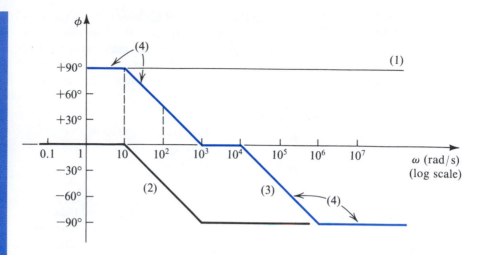

FIGURE E.4 Phase plots for Example E.2.

(the leading minus sign is due to the fact that this singularity is a pole). The asymptotic plot for this function is given by curve 2 in Fig. E.4. Similarly, the pole at $s = -10^5$ gives rise to the phase function

$$\phi_2 = -\tan^{-1} \frac{\omega}{10^5}$$

whose asymptotic plot is given by curve 3. The overall phase response (curve 4) is obtained by direct summation of the three plots. We see that at 100 rad/s, the amplifier phase leads by 45° and at 10^5 rad/s the phase lags by 45°.

E.4 AN IMPORTANT REMARK

For constructing Bode plots, it is most convenient to express the transfer-function factors in the form $(1 + s/a)$. The material of Figs. E.1 and E.2 and of the preceding two examples is then directly applicable.

PROBLEMS

E.1 Find the transfer function $T(s) = V_o(s)/V_i(s)$ of the circuit in Fig. PE.1. Is this an STC network? If so, of what type? For $C_1 = C_2 = 0.5\ \mu F$ and $R = 100\ k\Omega$, find the location of the pole(s) and zero(s), and sketch Bode plots for the magnitude response and the phase response.

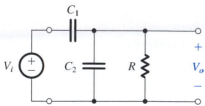

FIGURE PE.1

D*E.2 (a) Find the voltage transfer function $T(s) = V_o(s)/V_i(s)$, for the STC network shown in Fig. PE.2.

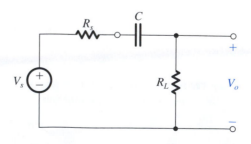

FIGURE PE.2

(b) In this circuit, capacitor C is used to couple the signal source V_s having a resistance R_s to a load R_L. For $R_s = 10$ kΩ, design the circuit, specifying the values of R_L and C to only one significant digit to meet the following requirements:

(i) The load resistance should be as small as possible.

(ii) The output signal should be at least 70% of the input at high frequencies.

(iii) The output should be at least 10% of the input at 10 Hz.

E.3 Two STC RC circuits, each with a pole at 100 rad/s and a maximum gain of unity, are connected in cascade with an intervening unity-gain buffer that ensures that they function separately. Characterize the possible combinations (of low-pass and high-pass circuits) by providing (i) the relevant transfer functions, (ii) the voltage gain at 10 rad/s, (iii) the voltage gain at 100 rad/s, and (iv) the voltage gain at 1000 rad/s.

E.4 Design the transfer function in Eq. (E.5) by specifying a_1 and ω_0 so that the gain is 10 V/V at high frequencies and 1 V/V at 10 Hz.

E.5 An amplifier has a low-pass STC frequency response. The magnitude of the gain is 20 dB at dc and 0 dB at 100 kHz. What is the corner frequency? At what frequency is the gain 19 dB? At what frequency is the phase $-6°$?

E.6 A transfer function has poles at (-5), $(-7 + j10)$, and (-20), and a zero at $(-1 - j20)$. Since this function represents an actual physical circuit, where must other poles and zeros be found?

E.7 An amplifier has a voltage transfer function $T(s) = 10^6 s/(s + 10)(s + 10^3)$. Convert this to the form convenient for constructing Bode plots [that is, place the denominator factors in the form $(1 + s/a)$]. Provide a Bode plot for the magnitude response, and use it to find approximate values for

the amplifier gain at 1, 10, 10^2, 10^3, 10^4, and 10^5 rad/s. What would the actual gain be at 10 rad/s? At 10^3 rad/s?

E.8 Find the Bode phase plot of the transfer function of the amplifier considered in Problem E.7. Estimate the phase angle at 1, 10, 10^2, 10^3, 10^4, and 10^5 rad/s. For comparison, calculate the actual phase at 1, 10, and 100 rad/s.

E.9 A transfer function has the following zeros and poles: one zero at $s = 0$ and one zero at $s = \infty$; one pole at $s = -100$ and one pole at $s = -10^6$. The magnitude of the transfer function at $\omega = 10^4$ rad/s is 100. Find the transfer function $T(s)$ and sketch a Bode plot for its magnitude.

E.10 Sketch Bode plots for the magnitude and phase of the transfer function

$$T(s) = \frac{10^4(1 + s/10^5)}{(1 + s/10^3)(1 + s/10^4)}$$

From your sketches, determine approximate values for the magnitude and phase at $\omega = 10^6$ rad/s. What are the exact values determined from the transfer function?

E.11 A particular amplifier has a voltage transfer function $T(s) = 10s^2/(1 + s/10)(1 + s/100)(1 + s/10^6)$. Find the poles and zeros. Sketch the magnitude of the gain in dB versus frequency on a logarithmic scale. Estimate the gain at 10^0, 10^3, 10^5, and 10^7 rad/s.

E.12 A direct-coupled differential amplifier has a differential gain of 100 V/V with poles at 10^6 and 10^8 rad/s, and a common-mode gain of 10^{-3} V/V with a zero at 10^4 rad/s and a pole at 10^8 rad/s. Sketch the Bode magnitude plots for the differential gain, the common-mode gain, and the CMRR. What is the CMRR at 10^7 rad/s? (*Hint:* Division of magnitudes corresponds to subtraction of logarithms.)

Bibliography

GENERAL TEXTBOOKS ON ELECTRONIC CIRCUITS

E.F. Angelo Jr., *Electronics: BJTs, FETs, and Microcircuits,* New York: McGraw-Hill, 1969.

S.B. Burns and P.R. Bond, *Principles of Electronic Circuits,* St. Paul: West, 1987.

M.S. Ghausi, *Electronic Devices and Circuits: Discrete and Integrated,* New York: Holt, Rinehart and Winston, 1985.

P.E. Gray and C.L. Searle, *Electronic Principles,* New York: Wiley, 1969.

A.R. Hambley, *Electronics, 2nd ed.,* Upper Saddle River, NJ: Prentice-Hall, 1999.

W.H. Hayt and G.W. Neudeck, *Electronic Circuit Analysis and Design, 2nd ed.,* Boston: Houghton Mifflin Co., 1984.

C.A. Holt, *Electronic Circuits,* New York: Wiley, 1978.

M.N. Horenstein, *Microelectronic Circuits and Devices, 2nd ed.,* Englewood Cliffs, NJ: Prentice-Hall, 1995.

R.T. Howe and C.G. Sodini, *Microelectronics—An Integrated Approach,* Englewood Cliffs, NJ: Prentice-Hall, 1997.

R.C. Jaeger and T.N. Blalock, *Microelectronic Circuit Design, 2nd ed.,* New York: McGraw-Hill, 2004.

N.R. Malik, *Electronic Circuits: Analysis, Simulation, and Design,* Englewood Cliffs, NJ: Prentice-Hall, 1995.

J. Millman and A. Grabel, *Microelectronics, 2nd ed.,* New York: McGraw-Hill, 1987.

D.A. Neamen, *Electronic Circuit Analysis and Design, 2nd ed.,* New York: McGraw-Hill, 2001.

M.H. Rashid, *Microelectronic Circuits: Analysis and Design,* Boston: PWS, 1999.

D.L. Schilling and C. Belove, *Electronic Circuits, 2nd ed.,* New York: McGraw-Hill, 1979.

R.A. Spencer and M.S. Ghausi, *Introduction to Electronic Circuit Design,* Upper Saddle River, NJ: Pearson Education Inc. (Prentice-Hall), 2003.

CIRCUIT AND SYSTEM ANALYSIS

L.S. Bobrow, *Elementary Linear Circuit Analysis, 2nd ed.,* New York: Holt, Rinehart and Winston, 1987.

A.M. Davis, *Linear Circuit Analysis,* Boston, MA: PWS Publishing Company, 1998.

S.S. Haykin, *Active Network Theory,* Reading, MA: Addison-Wesley, 1970.

W.H. Hayt, G.E. Kemmerly, and S.M. Durbin, *Engineering Circuit Analysis, 6th ed.,* New York: McGraw-Hill, 2003.

D. Irwin, *Basic Engineering Circuit Analysis, 7th ed.,* New York: Wiley, 2001.

B.P. Lathi, *Linear Systems and Signals,* New York: Oxford University Press, 1992.

J.W. Nilsson and S. Riedel, *Electronic Circuits, 6th ed.,* Revised Printing, Upper Saddle River, NJ: Prentice-Hall, 2001.

DEVICES AND IC FABRICATION

R.S.C. Cobbold, *Theory and Applications of Field Effect Transistors,* New York: Wiley, 1969.

I. Getreu, *Modeling the Bipolar Transistor,* Beaverton, OR: Tektronix, Inc., 1976.

R.S. Muller and T.I. Kamins, *Device Electronics for Integrated Circuits, 3rd ed.,* New York: Wiley, 2003.

J.D. Plummer, M.D. Deal, and P.B. Griffin, *Silicon VLSI Technology,* Upper Saddle River, NJ: Prentice-Hall, 2000.

D.L. Pulfrey and N.G. Tarr, *Introduction to Micro-electronic Devices,* Englewood Cliffs, NJ: Prentice-Hall, 1989.

C.L. Searle, A.R. Boothroyd, E.J. Angelo, Jr., P.E. Gray, and D.O. Pederson, *Elementary Circuit Properties of Transistors,* Vol. 3 of the SEEC Series, New York: Wiley, 1964.

B.G. Streetman and S. Banerjee, *Solid-State Electronic Devices, 5th ed.,* Upper Saddle River, NJ: Prentice-Hall, 2000.

Y. Tsividis, *Operation and Modeling of the MOS Transistor, 2nd ed.,* New York: Oxford University Press, 1999.

OPERATIONAL AMPLIFIERS

G.B. Clayton, *Experimenting with Operational Amplifiers,* London: Macmillan, 1975.

G.B. Clayton, *Operational Amplifiers, 2nd ed.,* London: Newnes-Butterworths, 1979.

S. Franco, *Design with Operational Amplifiers and Analog Integrated Circuits, 3rd ed.,* New York: McGraw-Hill, 2001.

J.G. Graeme, G.E. Tobey, and L.P. Huelsman, *Operational Amplifiers: Design and Applications,* New York: McGraw-Hill, 1971.

W. Jung, *IC Op Amp Cookbook,* Indianapolis: Howard Sams, 1974.

E.J. Kennedy, *Operational Amplifier Circuits: Theory and Applications,* New York: Holt, Rinehart and Winston, 1988.

J.K. Roberge, *Operational Amplifiers: Theory and Practice,* New York: Wiley, 1975.

J.L. Smith, *Modern Operational Circuit Design,* New York: Wiley-Interscience, 1971.

J.V. Wait, L.P. Huelsman, and G.A. Korn, *Introduction to Operational Amplifiers Theory and Applications,* New York: McGraw-Hill, 1975.

ANALOG CIRCUITS

P.E. Allen and D.R. Holberg, *CMOS Analog Circuit Design, 2nd ed.,* New York: Oxford University Press, 2002.

K. Bult, *Transistor-Level Analog IC Design.* Notes for a short course organized by Mead, Ecole Polytechnique Féderal De Lausanne, 2002.

R.L. Geiger, P.E. Allen, and N.R. Strader, *VLSI Design Techniques for Analog and Digital Circuits,* New York: McGraw-Hill, 1990.

P.R. Gray, P.J. Hurst, S.H. Lewis, and R.G. Meyer, *Analysis and Design of Analog Integrated Circuits, 4th ed.,* New York: Wiley, 2001.

A.B. Grebene, *Bipolar and MOS Analog Integrated Circuit Design,* New York: Wiley, 1984.

R. Gregorian and G.C. Temes, *Analog MOS Integrated Circuits for Signal Processing,* New York: Wiley, 1986.

IEEE Journal of Solid-State Circuits, a monthly publication of the IEEE.

D.A. Johns and K. Martin, *Analog Integrated Circuit Design,* New York: Wiley, 1997.

K. Laker and W. Sansen, *Design for Analog Integrated Circuits and Systems,* New York: McGraw-Hill, 1999.

H.S. Lee, "Analog Design," Chapter 8 in *BiCMOS Technology and Applications,* A.R. Alvarez, editor, Boston: Kluwer Academic Publishers, 1989.

B. Razavi, *Design of Analog CMOS Integrated Circuits,* New York: McGraw-Hill, 2001.

J.K. Roberge, *Operational Amplifiers: Theory and Practice,* New York: Wiley, 1975.

S. Rosenstark, *Feedback Amplifier Principles,* New York: Macmillan, 1986.

A.S. Sedra and G.W. Roberts, "Current Conveyor Theory and Practice," Chapter 3 in *Analogue IC Design: The Current-Mode Approach,* C. Toumazou, F.J. Lidgey, and D.G. Haigh, editors, London: Peter Peregrinus, 1990.

R. Severns, editor, *MOSPOWER Applications Handbook,* Santa Clara, CA: Siliconix, 1984.

Texas Instruments, Inc., *Power Transistor and TTL Integrated-Circuit Applications,* New York: McGraw-Hill, 1977.

S. Soclof, *Applications of Analog Integrated Circuits,* Englewood Cliffs, NJ: Prentice-Hall, 1985.

National Semiconductor Corporation, Audio/Radio Handbook, Santa Clara, CA: National Semiconductor Corporation, 1980.

J.M. Steininger, "Understanding wideband MOS transistors," *IEEE Circuits and Devices,* Vol. 6, No. 3, pp. 26–31, May 1990.

DIGITAL CIRCUITS

A.R. Alvarez, editor, *BiCMOS Technology and Applications, 2nd ed.,* Boston: Kluwer, 1993.

S.H.K Embabi, A. Bellaour, M.I. Elmasry, *Digital BiCMOS Integrated Circuit Design,* Boston: Kluwer, 1993.

M.I. Elmasry, editor, *Digital MOS Integrated Circuits,* New York: IEEE Press, 1981. Also, *Digital MOS Integrated Circuits II,* 1992.

D.A. Hodges and H.G. Jackson, *Analysis and Design of Digital Integrated Circuits, 2nd ed.,* New York: McGraw-Hill, 1988.

IEEE Journal of Solid-State Circuits, a monthly publication of the IEEE.

S.M. Kang and Y. Leblebici, *CMOS Digital Integrated Circuits, 3rd ed.,* New York: McGraw-Hill, 2003.

R. Littauer, *Pulse Electronics,* New York: McGraw-Hill, 1965.

K. Martin, *Digital Integrated Circuit Design,* New York: Oxford University Press, 2000.

J. Millman and H. Taub, *Pulse, Digital, and Switching Waveforms,* New York: McGraw-Hill, 1965.

Motorola, *MECL Device Data,* Phoenix, AZ: Motorola Semiconductor Products, Inc., 1989.

Motorola, *MECL System Design Handbook,* Phoenix, AZ: Motorola Semiconductor Products, Inc., 1988.

J.M. Rabaey, *Digital Integrated Circuits,* Englewood Cliffs, NJ: Prentice-Hall, 1996. Note: Also a 2nd ed., with A. Chandrakasan and B. Nikolic, appeared in 2003.

L. Strauss, *Wave Generation and Shaping, 2nd ed.,* New York: McGraw-Hill, 1970.

H. Taub and D. Schilling, *Digital Integrated Electronics,* New York: McGraw-Hill, 1977.

N. Weste and K. Eshraghian, *Principles of CMOS VLSI Design,* Reading, MA: Addison-Wesley, 1985 and 1993.

FILTERS AND TUNED AMPLIFIERS

P.E. Allen and E. Sanchez-Sinencio, *Switched-Capacitor Circuits,* New York: Van Nostrand Reinhold, 1984.

K.K. Clarke and D.T. Hess, *Communication Circuits: Analysis and Design,* Ch. 6, Reading, MA: Addison Wesley, 1971.

G. Daryanani, *Principles of Active Network Synthesis and Design,* New York: Wiley, 1976.

R. Gregorian and G.C. Temes, *Analog MOS Integrated Circuits for Signal Processing,* New York: Wiley-Interscience, 1986.

C. Ouslis and A. Sedra, "Designing custom filters," *IEEE Circuits and Devices,* May 1995, pp. 29–37.

S.K. Mitra and C.F. Kurth, editors, *Miniaturized and Integrated Filters,* New York: Wiley-Interscience, 1989.

R. Schaumann, M.S. Ghausi, and K.R. Laker, *Design of Analog Filters,* Englewood Cliffs, NJ: Prentice-Hall, 1990.

R. Schaumann, M. Soderstand, and K. Laker, editors, *Modern Active Filter Design,* New York: IEEE Press, 1981.

R. Schaumann and M.E. Van Valkenburg, *Design of Analog Filters,* New York: Oxford University Press, 2001.

A.S. Sedra, "Switched-capacitor filter synthesis," in *MOS VLSI Circuits for Telecommunications,* Y. Tsividis and P. Antognetti, editors, Englewood Cliffs, NJ: Prentice-Hall, 1985.

A.S. Sedra and P.O. Brackett, *Filter Theory and Design: Active and Passive,* Portland, OR: Matrix, 1978.

M.E. Van Valkenburg, *Analog Filter Design,* New York: Holt, Rinehart and Winston, 1981.

A.I. Zverev, *Handbook of Filter Synthesis,* New York: Wiley, 1967.

SPICE

M.E. Herniter, *Schematic Capture with Cadence PSpice, 2nd ed.,* NJ: Prentice-Hall, 2003.

G. Massobrio and P. Antognetti, *Semiconductor Device Modeling with SPICE, 2nd ed.,* New York: McGraw-Hill, 1993.

G.W. Roberts and A.S. Sedra, *SPICE,* New York: Oxford University Press, 1992 and 1997.

J.A. Svoboda, *PSpice for Linear Circuits,* New York: Wiley, 2002.

P.W. Tuinenga, *SPICE: A Guide To Circuit Simulation & Analysis Using PSpice, 2nd ed.,* NJ: Prentice-Hall, 1992.

Standard Resistance Values and Unit Prefixes

Discrete resistors are available only in standard values. Table G.1 provides the multipliers for the standard values of 5%-tolerance and 1%-tolerance resistors. Thus, in the kilohm

TABLE G.1 Standard Resistance Values

5% Resistor Values (kΩ)	1% Resistor Values (kΩ)			
	100–174	178–309	316–549	562–976
10	100	178	316	562
11	102	182	324	576
12	105	187	332	590
13	107	191	340	604
15	110	196	348	619
16	113	200	357	634
18	115	205	365	649
20	118	210	374	665
22	121	215	383	681
24	124	221	392	698
27	127	226	402	715
30	130	232	412	732
33	133	237	422	750
36	137	243	432	768
39	140	249	442	787
43	143	255	453	806
47	147	261	464	825
51	150	267	475	845
56	154	274	487	866
62	158	280	499	887
68	162	287	511	909
75	165	294	523	931
82	169	301	536	953
91	174	309	549	976

range of 5% resistors one finds resistances of 1.0, 1.1, 1.2, 1.3, 1.5, ... In the same range, one finds 1% resistors of kilohm values of 1.00, 1.02, 1.05, 1.07, 1.10, ...

Table G.2 provides the SI unit prefixes used in this book and in all modern works in English.

TABLE G.2 SI Unit Prefixes

Name	Symbol	Factor
femto	f	$\times 10^{-15}$
pico	p	$\times 10^{-12}$
nano	n	$\times 10^{-9}$
micro	μ	$\times 10^{-6}$
milli	m	$\times 10^{-3}$
kilo	k	$\times 10^{3}$
mega	M	$\times 10^{6}$
giga	G	$\times 10^{9}$
tera	T	$\times 10^{12}$
peta	P	$\times 10^{15}$

Answers to Selected Problems

CHAPTER 1

1.1 (a) 10 mA; (b) 10 kΩ; (c) 100 V; (d) 0.1 A   1.2 (a) 0.9 W, 1 W; (c) 0.09 W, 1/8 W; (f) 0.121 W, 1/8 W but preferably 1/4 W   1.4 17; 5.7, 6.7, 8.0, 8.6, 10, 13.3, 14.3, 17.1, 20, 23.3, 28, 30, 40, 46.7, 50, 60, 70 (all in kΩ)
1.7 2.94 V, 2.22 kΩ; 2.75 V to 3.14 V, 2.11 kΩ to 2.33 kΩ 1.9 10.2 V; shunt the 10-kΩ resistor a 157-kΩ resistor; add a series resistor of 200 Ω; shunt the 4.7-kΩ resistor with a 157 kΩ and the 10-kΩ resistor with 90 kΩ
1.11 Shunt the 1-kΩ resistor with a 250 Ω 1.13 Shunt R_L with a 1.1-kΩ resistor; current divider
1.15 0.77 V and 6.15 kΩ; 0.1 mA   1.17 1.88 μA; 5.64 V   1.19 (a) 10^{-7} s, 10^7 Hz, 6.28×10^7 Hz; (f) 10^3 rad/s,
1.59×10^2 Hz, 6.28×10^{-3} s   1.21 (a) $(1 - j1.59)$ kΩ; (c) $(71.72 - j45.04)$ kΩ 1.22 (b) 0.1 V, 10 μA, 10 kΩ
1.24 10 kΩ 1.28 (a) 165 V; (b) 24 V   1.30 0.5 V; 1 V; 0 V; 1 V; 1000 Hz; 10^{-3} s   1.32 4 kHz; 4 Hz
1.34 0, 101, 1000, 11001, 111001   1.36 (c) 11; 4.9 mV; 2.4 mV   1.38 7.056×10^5 bits per second
1.40 11 V/V or 20.8 dB; 22 A/A or 26.8 dB; 242 W/W or 23.8 dB; 120 mW; 95.8 mW; 20.2%
1.42 9 mV; 57.5 mV; 0.573 V   1.45 (a) 8.26 V/V or 18.3 dB; (b) 2.5 V/V or 8 dB; (c) 0.083 V/V or −21.6 dB
1.48 0.83 V; −1.6 dB; 79.2 dB; 38.8 dB   1.53 (a) 300 V/V; (b) 90 kΩ, 3×10^4 A/A, 9×10^6 W/W; (c) 667 Ω;
(d) 555.7 V/V; (e) 100 kΩ, 100 Ω, 363 V/V   1.59 A voltage amplifier; $R_i = 100$ kΩ, $R_o = 100$ s, $A_{v_o} = 121$ V/V
1.66 $s/(s + 1/CR)$ 1.69 0.64 μF   1.72 $0.51/CR$ 1.73 13.3 pF; 0.26 pF   1.76 20 dB; 37 dB; 40 dB; 37 dB;
20 dB; 0 dB; −20 dB; 9900 Hz   1.77 $1/(sC_1R_1 + 1)$; 16 kHz; $-G_m s(R_2//R_3)/(s + 1/(C_2(R_2 + R_3)))$; 53 Hz; 16 kHz
1.81 1.6 V, 1.3 V   1.82 (a) 1.5 V, 1 V; (b) 2.06 V; (c) −3.5 V   1.84 (a) 0.4 V, 0.4 V; (b) 8 mW; (c) 1.12 mW;
(d) 52.8 ns   1.85 (a) 0.545 V, 5 V, 3 V, 0.455 V; (b) 6; (c) 10.9 mW, 2.88 mW   1.88 25 mW, 5 mA

CHAPTER 2

2.2 1001 V/V   2.5 $A = G_m R_m = 100,000$ V/V   2.8 (a) −10 V/V, 10 kΩ; (b) −10 V/V, 10 kΩ; (c) −10 V/V, 10 kΩ;
(d) −10 V/V, 10 kΩ 2.11 (a) −1 V/V; (b) −10 V/V; (c) −0.1 V/V; (d) −100 V/V; (e) −10 V/V
2.12 $R_1 = 20$ kΩ, $R_2 = 100$ kΩ 2.14 $R_1 = 500$ kΩ, $R_2 = 10$ MΩ; 500 kΩ 2.17 $2x\%$; −110.5 to 90.5
2.18 0 V, 5 V; −4.9 V to −5.1 V   2.20 (a) $R_1 = 1$ kΩ, $R_2 = 100$ kΩ; (b) −90.8 V/V; (c) 8.9 kΩ 2.21 ±10 mV
2.23 $R_{in} = R_1 + R_2/(1 + A)$ 2.26 909 V/V   2.27 $A = (1 + R_2/R_1)(k - 1)/(1 - x/100)$; 2×10^4 V/V
2.29 100 Ω; 100 kΩ; ≈100 Ω 2.31 (a) R, R, R, R; (b) $I, 2I, 4I, 8I$; (c) $-IR, -2IR, -4IR, -8IR$ 2.33 (a) 0.53 kΩ;
(b) −0.4 to +0.4 mA; (c) 0 Ω, 20 mA   2.36 $v_O = v_1 - v_2/2$; −1.5 V   2.37 $R_{v_1} = 20$ kΩ; $R_{v_2} = 120$ kΩ; $R_f = 40$ kΩ
2.43 12.8 kΩ 2.46 $R = 100$ kΩ; No   2.50 $v_O = 10(v_2 - v_1)$; $v_O = 4 \sin(2\pi \times 1000t)$ 2.51 $v_O/v_I = 1/x$; +1 to +∞;
add 0.5 kΩ in series with grounded end of pot   2.53 (a) 0.099 V; 0.099 mA; 0.099 mA; (b) 10 V; 10 mA; 0 mA
2.54 $v_o/v_i = 1/(1 + 1/A)$; 0.999, −0.1%; 0.990, −1.0%; 0.909, −9.1%   2.56 8.33 V/V; Shunt R_1 with $R_{sh} = 36$ kΩ;
9.09 V/V; 11.1 V/V   2.59 −10.714 to +10.714 V; 1.07 V   2.62 $v_O = v_2 - v_1$; R; $2R$; $2R$; R 2.64 $R_1 = R_3$
2.66 68 dB   2.68 (a) 1, 0; (b) −5 V to +5 V; (c) 1, 0, −30 to +30 V   2.73 (a) −0.14 to +0.14 V; −14 to +14 V
2.76 $R_1 = 100$ kΩ pot plus 0.5 kΩ fixed; $R_2 = 50$ kΩ; $R_3 = 200$ kΩ; $R_4 = 100$ kΩ 2.77 (a) 3.0 V (peak-to-peak),

3.0 V (peak-to-peak) of opposite phase, 6.0 V (peak-to-peak); (b) 6 V/V; (c) 56 V (peak-to-peak), 19.8 V (rms)
2.80 86 dB; 500 Hz; 10 MHz   2.83 47.6 kHz; 19.9 V/V; 1.99 V/V   2.86 40 V/V   2.89 (a) $(\sqrt{2}-1)^{1/2} f_1$;
(b) 10 kHz; (c) 64.4 kHz, about six times greater   2.91 (a) $f_t/(1+K)$, $Kf_t/(1+K)$; (b) f_t/K, f_t; noninverting
preferred at low gains   2.92 For each, $f_{3dB} = f_t/3$ 2.99 (a) 31.8 kHz; (b) 0.795 V; (c) 0 to 200 kHz;
(d) 1 V peak   2.103 1.4 mV   2.105 42.5 to 57.5 mV; Add a 5-kΩ resistor in series with the positive input terminal;
±10 mV; add 5-kΩ resistor in series with the negative input terminal.   2.107 4.54 mV   2.110 (a) 100 mV;
(b) 0.2 V; (c) 10 kΩ, 10 mV; (d) 110 mV   2.114 100 kHz; 1.59 μs   2.118 100 pulses   2.119 $V_o(s)/V_i(s) = -(R_2/R_1)/(1+sR_2C)$; $R_1 = 1$ kΩ; $R_2 = 10$ kΩ; $C = 3.98$ nF; 39.8 kHz   2.121 1.59 kHz; 10 V (peak-to-peak)

CHAPTER 3

3.1 The diode can be reversed biased and thus no current would flow; or forward biased where current would flow;
(a) 0 A; 1.5 V; (b) 1.5 A; 0 V   3.2 (a) −3 V; 0.6 mA; (b) +3 V; 0 mA; (c) +3 V; 0.6 mA; (d) −3 V;
0 mA   3.5 100 mA; 35 mA; 100 mA; 33.3 mA   3.8 50 kΩ   3.9 (a) 0 V; 0.5 mA; (b) 1.67 V; 0 A
3.10 (a) 4.5 V; 0.225 mA; (b) 2 V; 0 A   3.13 3 V; 1.5 V; 30 mA; 15 mA   3.15 29.67 V; 3.75 Ω; 0.75 A;
26.83 V; 30 V; 3 Ω; 20.5%; 136 mA; 1 A; 27 V   3.16 red lights; neither lights; green lights   3.18 345 mV;
$1.2 \times 10^6 I_S$ 3.20 3.46×10^{-15} A; 7.46 mA; 273.2 mA; 3.35 mA; 91.65 μA; 57.6 mV   3.23 3.81 mA; −22.8 mV
3.26 57.1 Ω   3.27 (a) 678 mV; (b) 647 mV; (c) 814 mV; (d) 656 mV; (e) 662 mV   3.29 60°C; 8.7 W;
6.9°C/W   3.33 0.6638 V; 0.3362 mA   3.36 $R = 947$ Ω   3.37 0.687 V; 12.8 Ω; +28.1 mV; −29.5 mV; +34.2 mV
3.39 0.73 V; 1.7 mA; 0.7 V; 2 mA   3.41 0.8 V   3.45 0.86 mA; 0 V; 0 A; 3.6 V   3.46 (a) 0.53 mA; 2.3 V;
(b) 0 A; +3 V; (c) 0.53 mA; 2.3 V; (d) 0 A; −3 V   3.48 (a) 0.36 mA; 0 V; (b) 0 A; −1.9 V   3.52 (a) +49%
to −33%; (b) +22% to −18%; −2.6 to +2.4 mV ($n = 1$); −5.3 to +4.8 mV ($n = 2$)   3.56 (a) 0 V/V; (b) 0.001 V/V;
(c) 0.01 V/V; (d) 0.1 V/V; (e) 0.5 V/V; (f) 0.6 V/V; (g) 0.9 V/V; (h) 0.99 V/V; (i) 1 V/V; 2.5 mV (peak).
3.58 157 μA; −84.3° to −5.7°   3.62 15-mA supply; −10 mV/mA, for a total output change of −50 mV
3.65 −30 Ω; −120 Ω   3.67 8.96 V; 9.01 V; 9.46 V   3.70 8.83 V; 19.13 mA; 300 Ω; 9.14 V; ±0.01 V; +0.12 V;
578 Ω; 8.83 V; 90 mV/V; −27.3 mA/mA   3.76 16.27 V; 48.7%; 0.13; 5.06 V; 5.06 mA   3.77 16.27 V; 97.4%;
10.12 V; 10.12 mA   3.78 15.57 V; 94.7%; 9.4 V; 9.4 mA   3.81 56 V   3.83 (a) 166.7 μF; 15.4 V; 7.1%;
231 mA; 448 mA; (b) 1667 μF; 16.19 V; 2.2%; 735 mA; 1455 mA   3.85 (a) 83.3 μF; 14.79 V; 14.2%; 119 mA;
222 mA; (b) 833 μF; 15.49 V; 4.5%; 360 mA; 704 mA   3.87 (a) 23.6 V; (b) 444.4 μF; (c) 32.7 V; 49 V;
(d) 0.73 A; (e) 1.35 A   3.98 0.51 V; 0.7 V; 1.7 V; 10.8 V; 0 V; −0.51 V; −0.7 V; −1.7 V; −10.8 V; fairly hard; +1
3.104 14.14 V   3.106 2.75×10^5/cm³; 1.55×10^9/cm³; 8.76×10^9/cm³; 1.55×10^{12}/cm³; 4.79×10^{12}/cm³
3.113 34 cm²/s; 12 cm²/s; 28 cm²/s; 10 cm²/s; 18 cm²/s; 6 cm²/s; 9 cm²/s; 4 cm²/s   3.114 1.27 V; 0.57 μm;
0.28 μm; 45.6×10^{-15} C; 18.2 fF   3.116 16×10^{-15} C   3.121 0.72 fA; 0.684 V; 2×10^{-11} C; 800 pF

CHAPTER 4

4.3 $W_p/W_n = 2.5$ 4.4 238 Ω; 238 mV; 50   4.5 2.38 μm   4.7 (a) 4.15 mA; (b) 0.8 mA; 0.92 mA; 9.9 mA
4.11 3.5 V; 500 Ω; 100 Ω   4.12 3 V; 2 V; 5 V; 4 V   4.14 4 μm   4.16 0.7 V   4.17 100 Ω to 10 kΩ; (a) 200 Ω
to 20 kΩ; (b) 50 Ω to 5 kΩ; (c) 100 Ω to 10 kΩ   4.19 20 kΩ; 36 V; 0.028 V⁻¹   4.20 500 kΩ; 50 kΩ; 2%; 2%
4.22 82.13 μA; 2.7%; use $L = 6$ μm   4.26 240 μA; 524 μA; 539 μA; 588 μA   4.27 −3 V; +3 V; −4 V; +4 V;
−1 V; −50 V; −0.02 V⁻¹; 1.39 mA/V²   4.29 1 V to 1.69 V; 1 V to 3.7 V   4.31 (b) −0.3%/°C   4.34 $R_D = 5$ kΩ;
$R_S = 3$ kΩ   4.35 (a) 9.75 kΩ; (b) 20 μm; 4 kΩ   4.36 4.8 μm; 30.4 kΩ   4.37 8 μm; 2 μm; 12.5 kΩ   4.39 0.4 mA;
7.6 V   4.44 (a) 2.51 V; −2.79 V; (b) 7.56 V; 5 V; 2.44 V   4.46 (a) 7.5 μA; 1.5 V; (b) 4.8 μA; 1.4 V; (c) 1.5 V;
7.5 μA   4.48 (a) 1 V; 1 V; −1.32 V; (b) 0.2 V; 1.8 V; −1.35 V   4.51 0.8 V; 25   4.57 3.4 V; 110 μA to 838 μA;
8.2 kΩ; 40 μA to 0.15 mA   4.58 1 mA; 13%   4.59 1.59 V; 2.37 V; 2.37 mA   4.60 $R_D = 11$ kΩ; $R_S = 7$ kΩ
4.63 (a) −3 V; +5 V; 8 V; (b) −3.3 V; +5 V; +8.3 V   4.65 36 kΩ; 0.21 mA; 2 V   4.69 (a) 2 mA; 2.8 V;
(b) 2 mA/V; (c) −7.2 V/V; (d) 50 kΩ; −6.7 V/V   4.73 20 μm; 1.7 V   4.75 −8.3 V/V; 2.5 V; −10.8 V/V
4.76 NMOS: 0.42 mA/V, 160 kΩ, 0.08 mA/V, 0.5 V; PMOS: 0.245 mA/V, 240 kΩ, 0.05 mA/V, 0.8 V
4.79 −11.2 V/V   4.81 200 Ω; 3.57 V/V; 100 Ω; 4.76 V/V   4.85 0.99 V/V; 200 Ω; 0.83 V/V   4.91 5.1 GHz

4.93 2.7 GHz; 5.4 GHz 4.96 (a) −15.24 V/V; 33.1 kHz 4.99 −10 V/V, 18.6 μF 4.103 −16 V/V; C_{C1} = 20 nF, C_S = 10 μF; C_{C2} = 0.5 μF; 47.7 Hz 4.106 1.36 V; 1.5 V; 1.64 V 4.110 10 μm 4.114 (b) −125 V/V; 80 kΩ 4.115 0.59 mA, 5 mA; 9 mA; 9 mA 4.116 300 μA; 416 μA; 424 μA; 480 μA; 600 μA; 832 μA; 848 μA; 960 μA; 300 μA; 416 μA; 424 μA; 480 μA 4.118 +0.586 V

CHAPTER 5

5.1 active; saturation; active; saturation; inversed active; active; cutoff; cutoff 5.2 (a) 7.7×10^{-17} A, 368; (b) 3.8×10^{-17} A, 122; (c) 1.5×10^{-17} A, 24.2; 1.008 mA; 0.7 V; 0.96 pC 5.4 53.3; 0.982 5.6 0.5; 0.667; 0.909; 0.952; 0.991, 0.995, 0.999; 0.9995 5.8 0.907 mA; 0.587 V 5.10 3 to 15 mA; 3.05 to 15.05 mA; 135 mW 5.12 −0.718 V; 4.06 V; 0.03 mA 5.13 (a) 0.691 V, 1 mA, 1.01 mA; (b) −10.09 mA, 9.08 mA, −1.01 mA 5.16 −2 V; 0.82 mA; −0.57 V 5.18 0.91 mA; 9.09 mA; 0.803 V; 9.99 mA 5.20 (a) 1 mA; (b) −2 V; (c) 1 mA; 1 V; (d) 0.965 mA; 0.35 V 5.22 4.3 V; 2.1 mA 5.24 (a) −0.7 V, 0 V, 0.756 V, 1.05 mA, 0.034 mA, 1.02 mA; (b) 0.7 V, 0 V, −0.77 V, 2.3 mA, 0.074 mA, 2.23 mA; (c) 3.7 V, 3 V, 2.62 V, 4.82 mA, 0.155 mA, 4.66 mA; (d) 2.3 V, 3 V, 4.22 V, 4.89 mA, 0.158 mA, 4.73 mA 5.26 −2.2 V; 0.779; 3.53; 3.7 V; 0 V; −0.7 V; +0.7 V 5.29 1/3; 1/2 5.30 0.74 V; 0.54 V 5.32 3.35 μA 5.38 33.3 kΩ; 100 V; 3.3 kΩ 5.40 1.72 mA; 6 V; 34 V; 20 kΩ 5.42 150; 125; 1.474 mA 5.45 40.2 mV 5.52 3 Ω; 110 mV; 68.2; 0.11 5.54 −360 V/V; 0.7 V, 2 mV 5.57 −100 V/V 5.60 3 mA; −120 V/V; −0.66 V; −0.6 V; 0.54 V; 0.6 V 5.63 3 V; 2.5 mA; 25 μA; 3.2 V 5.65 1.8 kΩ; 2 5.67 (a) 1.8 mA, 1.5 mA, 3.3 mA; (b) 1.8 mA, 0.3 mA; 2.5 mA 5.69 (a) 1.3 V, 3.7 V; (b) 0.3 V, 4.7 V; (c) 0 V, +5 V 5.72 −0.7 V; +4.7 V; −0.5 V (−1 V; +5 V); +2.6 V (1.9 V, 2.6 V) 5.74 0.3 V; 15 μA; 0.8 mA; 0.785 mA; −1.075 V; 52.3; 0.98 5.79 (a) −0.7 V, 1.8 V; (b) 1.872 V, 1.955 mA; (c) −0.7 V, 0 V, 1.872 V; (d) 1.9 V, −0.209 V; (e) 1.224 V, 1.924 V, −0.246 V 5.82 1.08 kΩ; the transistor saturates. 5.112 1.25 V; 20 mA/V; 150 V/V 5.118 135; 41.8 Ω; 23 mA/V; 1.09 kΩ; −0.76 V/V 5.123 9.3 kΩ; 28.6 kΩ; 143 V/V 5.124 1 mA; 0.996 V/V; 0.63 V/V 5.146 0.7 V/V 5.147 (a) 1.73 mA, 68.5 mA/V, 14.5 Ω, 1.46 kΩ; (b) 148.2 kΩ, 0.93 V/V; (c) 18.21 kΩ, 0.64 V/V 5.150 1.25 GHz, 5.8 GHz, 2.47 ps, 0.95 pF 5.153 0.54 pF; 20 mA/V; 7.5 kΩ; 33.3 MΩ 5.168 19 5.169 2.15 mA; 4.62 mW; 24 mW; 14.3 mW 5.170 R_B = 11 kΩ; R_C = 2.2 kΩ

CHAPTER 6

6.4 12; 34 6.5 2.875 6.6 25.8; 1 mA; 0.25 mA 6.8 0.5 mA; 4 mA/V 6.10 0.4 mA/V; 250 kΩ; 100 V/V; 6.3 μm 6.13 16.7 GHz; 23.9 GHz; because the overlap capacitance is neglected. 6.14 15 V/V; 164.2 MHz; 2.5 GHz, 0.155 mA; quadrupled to 0.62 mA; 3.75 V/V; 656.8 MHz 6.17 5.3 MHz; 391 MHz 6.21 20 kΩ; 0.2 V; 200 kΩ; 5 μA 6.24 80 μA; 0.3 V; 0.8 V; 3.2 μA 6.27 4: 25, 50, 200, 400 μA; 3: 16.7, 40, 133 μA; 1.53 V 6.29 (a) 10 μA to 10 mA; 0.576 to 0.748 V 6.32 0.2 mA; 10% 6.35 (a) 2 mA, −0.7 V, 5 V, 0.7 V, −0.7 V, −5.7 V; (b) 0.2 mA, −0.7 V, 5 V, 0.7 V, 0.7 V, −0.7 V 6.37 0.5 mA 6.40 (a) 2.07; (b) 7.02 6.41 (a) 10^5 rad/s; (b) 1.01×10^5 rad/s; 10^7 rad/s 6.42 5.67×10^6 rad/s 6.44 2.5 MHz; 0.56 MHz 6.46 (a) $-g_m R_L/(1 + g_m R_s)$; (b) $R_{gs} = (R_{sig} + R_s)/(1 + g_m R_s)$, $R_{gd} = R_L + R_{sig} + (g_m R_L/(1 + g_m R_s))R_{sig}$; (c) For $R_s = 0$: $A_0 = -20$ V/V, ω_H = 453.5 krad/s, GBW = 9.07 Mrad/s; for $R_s = 100$ Ω: $A_0 = -14.3$ V/V, ω_H = 624.3 krad/s, GBW = 8.93 Mrad/s; for $R_s = 250$ Ω: $A_0 = -10$ V/V, ω_H = 865.7 krad/s, GBW = 8.66 Mrad/s 6.48 40.6 V/V; 243.8 ns; 3100 ns; 30 ns; 47.2 kHz 6.53 (a) −1000 V/V, C_i = 1.001 nF, C_o = 1.001 pF; (b) −10 V/V, C_i = 110 pF, C_o = 11 pF; (c) −1 V/V, C_i = 20 pF, C_o = 20 pF; (d) 1 V/V, C_i = 0 pF, C_o = 0 pF; (e) 10 V/V, C_i = −90 pF, C_o = 4 pF 6.59 0.905 V; 1.4 V 6.65 (a) 0.5 mA; (b) 100 kΩ, 100 kΩ, 50 kΩ; (c) 2.5 kΩ, 20 mA/V; (d) 2.5 kΩ, 50 kΩ, −1000 V/V 6.68 7.96 GHz; 611.5 kHz; 45.06 MHz; 611 kHz; 602.9 kHz; 45.7 MHz 6.71 −80.7 V/V; 6.37 GHz; 1.87 MHz; 86.8 MHz; 1.87 MHz 6.74 −100 V/V; 7.23 MHz; 723 MHz 6.78 80 fF 6.83 932.6 Ω; 1.73 V 6.86 17.1 V/V; 557 MHz; 3.79 MHz; 3.79 MHz 6.90 50 kΩ 6.93 0.97 A/A; 2.63 MΩ 6.98 $v_y/v_x = r_{o1}/\{r_{o2} + [1 + (g_{m2} + g_{mb2})r_{o2}]r_{o1}\} \simeq 1/g_{m2}r_{o2}$ 6.102 25 kΩ; 4 mA/V, 100 MΩ; -2×10^5 V/V; −50 V/V 6.108 110 kΩ; −100 V/V; −31.25 V/V; 0.91 mA/V; 0.45 V/V 6.116 (a) 2 mA; (b) 8 mA/V, 1.6 mA/V, 10 kΩ; (c) 0.82 V/V, 103 Ω; (d) 0.75 V/V 6.120 0.964 V/V; 544 MHz 6.122 (a) 2.51 MΩ, −3943 V/V; (b) 107.8 kHz, C_L dominates,

$C_{\mu 2}$ is the second most significant; f_H increases by a factor of 7, A_M remains unchanged   6.124 10.3 MΩ; 14.8 Ω; 1 V/V; 0.985 V/V   6.128 80 μA; 8 MΩ; 0.9 V   6.132 $1/(1 + (n + 1)/\beta^2)$; 9   6.133 0.5 kΩ   6.135 4.1 V 6.137 2 μA; 0.2%   6.141 (a) 5.76 kΩ; (b) 33 MΩ, 0.15 μA   6.143 11 MΩ   6.144 (a) 58.5 kΩ; (b) 200 MΩ

CHAPTER 7

7.8 1.19 V; 1.06 mA/V; 0.27 V; 800 μA   7.10 −1.5 V; +0.5 V; equal in both cases; 0.05 V; −0.05 V; 0.536 V 7.19 −2.68 V; 3.52 V; 3.52 V   7.20 −2.683 V; +3.515 V   7.22 −0.4 V   7.24 (a) $V_{CC} − (I/2)R_C$; (b) $−(I/2)R_C$, $+(I/2)R_C$; (c) 4 V; (d) 0.4 mA, 10 kΩ   7.27 (a) $20IR_C$ V/V; (b) $V_{CC} − 0.0275A_v$ 7.28 2.4 mA; 3.6 mA; 10.1 mV   7.29 $I_{C1} = 3.6$ mA, $I_{C2} = 2.4$ mA; 10.1 mV   7.30 (a) 4.14 V; (b) 3.15 V; (c) 3.525 V; (d) 3.755 V   7.32 1 mA; 10 kΩ   7.34 (a) 0.4 mA, 10 mV; (b) 1.40 mA, 0.60 mA; (c) −2.0 V, +2.0 V; (d) 40 V/V   7.37 40 V/V; 50 kΩ   7.38 30 V/V; ≃25 kΩ   7.41 26.7 V/V; 17.8 kΩ; 0.033 V/V; 15 kΩ 7.42 (a) 100 V/V; (b) 200 V/V; (c) 40.2 kΩ; (d) 0.1 V/V; (e) 0   7.44 1.8 mA; 360 V/V; 1.8 sin ωt V 7.45 $R_E = 25$ Ω; $R_C = 10$ kΩ; $R_o \geq 50$ kΩ; $R_{i_{cm}} = 5$ MΩ; ±12 V would do, ±15 V would be better.   7.46 2% mismatch, for example ±1% resistors   7.47 0.004 V/V   7.54 −125 μV   7.55 $V_{OS} = V_T((V_{CE}/V_{A1}) − (V_{CE}/V_{A2}))$ 7.57 (a) 0.25; (b) 0.225   7.60 $I/3$; $2I/3$; $R_C I/3$; 16.7 mV; 17.3 mV; 0.495 μA; 0.5 μA; 0.33 μA 7.98 R_5; reduce to 7.37 kΩ; 4104 V/V; reduce R_4 to 1.12 kΩ   7.99 $R_5 = 7.37$ kΩ; 4104 V/V; $R_4 = 1.11$ kΩ 7.100 173.1×10^3 V/V   7.101 (a) 1 mA; (b) 2.37 kΩ, 128 Ω; (c) 2.81×10^4 V/V

CHAPTER 8

8.1 9.99×10^{-3}; 90.99; −9%   8.3 (b) 1110; (c) 20 dB; (d) 10 V, 9 mV, 1 mV; (e) −2.44% 8.12 $A_{Mf} = A_M/(1 + A_M\beta)$; $W_{Lf} = W_L/(1 + A_M\beta)$; $1 + A_M\beta$ 8.14 100 kHz; 10 Hz   8.20 0.08; 12.34; 10.1 8.29 $10^4 + 10^7/(1 + jf/100)$; $10^{-3} + 1/(1 + jf/100)$; 1 MΩ; 14.1 kΩ; 10 Ω; 700 Ω   8.30 (a) $h_{11} = R_1R_2/(R_1 + R_2)$ Ω, $h_{12} = R_2/(R_1 + R_2)$ V/V, $h_{21} = − R_2/(R_1 + R_2)$ A/A, $h_{22} = 1/(R_1 + R_2)$ ℧; (b) $h_{11} = 10$ Ω, $h_{12} = 0.01$ V/V, $h_{21} = − 0.01$ A/V, $h_{22} = 0.99 \times 10^{-3}$ ℧   8.31 10 V/V; 9.9 Ω   8.34 0.0 V; 0.7 V; 31.3 V/V; 0.1 V/V; 7.6 V/V; ∞; 163 Ω 8.35 (b) $≃1 + (R_F/R_E)$; (c) 1.2 kΩ; (d) 1.75 kΩ, 628.1 Ω; (e) 23.8 V/V; (f) 154 kΩ, 0.53 Ω 8.37 7.52 mA/V; 110.8 kΩ; 433.4 kΩ   8.41 −4.7 V/V; 75 kΩ   8.47 (a) shunt-series; (b) series-series; (c) shunt-shunt   8.48 −5.66 V/V; 142 kΩ; 5.63 kΩ; 142.9 kΩ; −5.61 V/V; 5.96 kΩ   8.49 −9.83 kΩ; 29.7 Ω; −7.6 A/A   8.50 9.09 A/A; 90.9 Ω; 110 kΩ   8.53 3.13; 163 Ω   8.61 10^4 rad/s; $\beta = 0.002$; 500 V/V 8.63 $K < 0.008$ 8.65 9.9 V/V; 1.01 MHz; 10 MHz; 101   8.66 (a) 5.5×10^4 Hz, $\beta = 2.025 \times 10^{-3}$; (b) 330.6 V/ V; (c) 165.3 V/V, 1/2; (d) 1.33   8.68 $\omega_0 = 1/CR$; $Q = 1/(2.1 − K)$; 0.1; 0.686; $K = 2.1$ 8.69 $K \geq 2$; 17.3 MHz   8.70 1 MHz; 90°   8.72 56.87°; 54.07°; 59.24°; 52.93°   8.74 159.2 μs; 39.3°; 20 dB   8.76 200 Hz 8.77 10^3 Hz; 2000   8.78 1/10RC; 1/RC; $1/100RC_f$; 9.1/RC   8.79 10 Hz; 15.9 nF   8.80 58.8 pF; 38.8 MHz

CHAPTER 9

9.21 36.3 μA   9.22 0.625 V; for A, 7.3 mA/V, 134.3 Ω, 6.85 kΩ, 274 kΩ; for B, 21.9 mA/V, 44.7 Ω, 2.28 kΩ, 91.3 kΩ   9.27 616 mV; 535 mV; 4.02 kΩ   9.29 4.75 μA; 1.94 kΩ   9.31 56.5 kΩ; 9.353 μA   9.33 226 to 250; ±5%   9.36 6.37 kΩ; 270 μA   9.38 1.68 mA; 50.4 mW   9.40 Raise R'_1, R'_2 to 4.63 kΩ   9.43 0.96 mV 9.45 33.9 dB   9.48 3.10 MΩ; 9.38 mA/V   9.50 4.2 V to −3.6 V   9.52 21 mA   9.54 108 dB; 61.9 Ω; 105.6 dB; $|V_o| < 4$ V   9.56 11.4 MHz   9.58 637 kΩ   9.60 159 kHz; 15.9 MHz   9.62 six bits; 0.156 V; seven bits; seven bits; 0.117 V; 0.059 V   9.65 $I/16$; $I/8$; $I/4$; $I/2$ 9.67 Use op amp with $R/2$ input and $50R$ feedback to drive V_{ref}; 15 sine-wave amplitudes, from 0.625 V peak to 9.375 V peak; an output of 10 V (peak-to-peak) corresponds to a digital input of (1000).   9.69 8.19 ms; 4.096 ms; 9.9 V; no, stays the same!

CHAPTER 10

10.1 1.5 V; 1.5 V; 1.5 V; 0 V; 3 V; 1.5 V; 1.5 V; ∞   10.3 0.35 to 0.45 V; 0.75 to 0.85 V; 0 V; 1.2 V; 0.45 to 0.35 V; 0.35 to 0.45 V   10.4 (a) $t_{PLH} = 1.6$ ns, $t_{PHL} = 0.8$ ns; (b) $C = 1.43$ pF; (c) $C_o = 0.86$ pF, $C_i = 0.57$ pF

10.6 0.436; 1.48 mW   **10.7** Maximum operating frequency is reduced by a factor of (a) 0.66, (b) 0.44. DP decreases by a factor of 0.44 in both cases.   **10.9** Effect of changes in device dimensions is to change the performance parameters by the factors: 0.81, 1.11, 0.86, 0.77, 1.30, 1.11, 0.86, 1.60.   **10.14** 9.1 mV; 50 mV **10.19** 106 fF; 68.5 ps   **10.26** 24   **10.33** $p_A = p$; $p_B = p_C = p_D = 2p$; and $n_A = n_B = 2n$; $n_C = n_D = 2(2n) = 4n$. **10.35** With the proper sizing, t_{PHL} is one quarter the value obtained with the smaller-size devices; t_{PLH} is the same in both cases.   **10.38** (a) $0.69CR_D$; (b) $0.5CR_D$, for a 27.5% reduction   **10.39** 1.152; 1.76 V; 3.25 V; 2.70 V; +5.0 V; 0.58 V; 1.75 V; 1.18 V   **10.40** 2.4 fF; 10.5 fF; 63.5 ps; 41.2 ps; 52.4 ps; 9.6 fF; 24.0 fF; 72.5 ps; 72.5 ps; 72.5 ps   **10.41** $r \simeq 2$; NM_{Lmax} 1.28 V   **10.43** 1.33; 0.92 V   **10.53** (a) 1.62 V; 1.16 V; 15.3 μA; 351.6 μA; 183 μA; 177 ps   **10.60** 0.67 V; 1.25 V   **10.62** 1.1 GHz

CHAPTER 11

11.1 2.16 V; 0.93; 1.86   **11.3** 6   **11.11** 10.4 μs; 9.8 V; 5.7 V; $\approx$0.1 V; 21.5 mA; The source current can be as large as 21 mA (for $R_{on} = 200\ \Omega$), but is clearly limited by kp of G_1 to a much smaller value   **11.13** (a) $1.39CR$; (b) 10 kΩ; 721 pF   **11.14** 97.2%   **11.18** 16 bits   **11.19** 1024; 1024; 4000 pF; 225 pF; 220 fF/bit; 2.8 times **11.20** 0.3 μm^2; 0.39 μm $\times$ 0.78 μm   **11.21** 60%   **11.22** 4; 12; 28   **11.27** 32 Mbits   **11.29** 2 pA **11.30** 1.589 mA/V; 11.36 μm; 34.1 μm; 1.56 ns   **11.31** 0.68 mA/V; 0.48 V; 0.21 V; 50%; 7.5 ns   **11.32** (b) 2; (c) 1.46   **11.34** 9; 512; 18; 4608 NMOS and 512 PMOS transistors   **11.35** 9; 1024; 4608; 512; 5641; 521 **11.36** 262144; 9; 1022   **11.39** 2.42 ns; 22 ns, 3.16 V; 1.9 ns   **11.41** 33.3 MHz; high for 13 ns; low for 17 ns **11.44** 0.329 V/V; 8.94 V/V; 0.368 V/V   **11.45** (a) -1.375 V, -1.265 V; (b) -1.493 V, -1.147 V   **11.47** 21.2 **11.49** 7 cm   **11.51** $(W/L)_p = 5\ \mu$m/1 μm; 6.5 mA   **11.52** 2.32 V; 3.88 mA   **11.53** For R_1: 50%; 36.5 kΩ; 20%; 91.1 kΩ; for R_2: 50%; 6.70 kΩ; 20%; 16.7 kΩ; 50%; $R_1/R_2 = 5.45$; 20%; $R_1/R_2 = 5.45$ **11.54** 83.2 ps; 50.7 ps; 67.0 ps   **11.56** $(W/L)_{Q_{NA}} = (W/L)_{Q_{NB}} = 2(W/L)_{Q_N}$; $(W/L)_{Q_{PA}} = (W/L)_{Q_{PB}} = (W/L)_{Q_P}$

CHAPTER 12

12.1 1 V/V,   0°,   0   dB, 0   dB
0.894 V/V, $-26.6°$, -0.97 dB, 0.97 dB
0.707 V/V, $-45.0°$, -3.01 dB, 3.01 dB
0.447 V/V, $-63.4°$, -6.99 dB, 6.99 dB
0.196 V/V, $-78.7°$, -14.1 dB, 14.1 dB
0.100 V/V, $-84.3°$, -20.0 dB, 20.0 dB
0.010 V/V, $-89.4°$, -40.0 dB, 40.0 dB
12.3 1.000; 0.944; 0.010   **12.5** 0.509 rad/s; 3 rad/s; 5.90
12.8 $T(s) = 10^{15}/[(s + 10^3)(s^2 + 618s + 10^6)(s^2 + 1618s + 10^6)]$, low-pass;
$T(s) = s^5/[(s + 10^3)(s^2 + 618s + 10^6)(s^2 + 1618s + 10^6)]$, high-pass   **12.9** $T(s) = 0.2225\ (s^2 + 4)/[(s + 1)(s^2 + s + 0.89)]$
12.11 $T(s) = 0.5/[(s + 1)(s^2 + s + 1)]$; poles at $s = -1$, $-\frac{1}{2} \pm j\sqrt{3}/2$, 3 zeros at $s = \infty$ **12.13** 28.6 dB
12.15 $N = 5$; $f_0 = 10.55$ kHz, at $-108°$, $-144°$, $-180°$, $-216°$, $-252°$; $p_1 = -20.484 \times 10^3 + j63.043 \times 10^3$ (rad/s), $p_2 = -53.628 \times 10^3 + j38.963 \times 10^3$ (rad/s), $p_3 = -\omega_0 = -66.288 \times 10^3$ rad/s, $p_4 = -53.628 \times 10^3 - j38.963 \times 10^3$ (rad/s), $p_5 = -20.484 \times 10^3 - j63.043 \times 10^3$ (rad/s); $T(s) = \omega_0^5/[(s + \omega_0)(s^2 + 1.618\omega_0 s + \omega_0^2)(s^2 + 0.618\omega_0 s + \omega_0^2)]$; 27.8 dB   **12.19** $R_1 = 10$ kΩ; $R_2 = 100$ kΩ; $C = 159$ pF   **12.21** $R_1 = 1$ kΩ; $R_2 = 1$ kΩ; $C_1 = 0.159\ \mu$F; $C_2 = 1.59$ nF; High-frequency gain $= -100$ V/V   **12.23** $T(s) = (1 - RCs)/(1 + RCs)$; 2.68 k$\Omega$, 5.77 k$\Omega$, 10 k$\Omega$, 17.3 k$\Omega$, 37.3 k$\Omega$ **12.25** $T(s) = 10^6/(s^2 + 10^3s + 10^6)$; 707 rad/s; 1.16 V/V   **12.27** $R = 4.59$ kΩ; $R_1 = 10$ kΩ **12.28** $T(s) = s^2/(s^2 + s + 1)$ **12.30** $T(s) = (s^2 + 1.42 \times 10^5)/(s^2 + 375s + 1.42 \times 10^5)$ **12.33** $L = 0.5$ H; $C = 20$ nF **12.35** $V_o(s)/V_i(s) = s^2/(s^2 + s/RC + 1/LC)$ **12.37** Split R into two parts, leaving $2R$ in its place and adding $2R$ from the output to ground.   **12.39** $L_1/L_2 = 0.235$; $|T| = L_2/(L_1 + L_2)$; $|T| = 1$ **12.40** For all resistors $= 10$ kΩ, C_4 is (a) 0.1 μF, (b) 0.01 μF, (c) 1000 pF; For $R_5 = 100$ kΩ and $R_1 = R_2 = R_3 = 10$ kΩ, C_4 is (a) 0.01 μF, (b) 1000 pF, (c) 100 pF   **12.43** $R_1 = R_2 = R_3 = R_5 = 3979\ \Omega$; $R_6 = 39.79$ kΩ; $C_{61} = 6.4$ nF; $C_{62} = 3.6$ nF

12.44 $C_4 = C_6 = 1$ nF; $R_1 = R_2 = R_3 = R_5 = R_6 = r_1 = r_2 = 159$ kΩ   12.48 (a) $T(s) = 0.451 \times 10^4(s^2 + 1.70 \times 10^8)/$
$[(s + 0.729 \times 10^4)(s^2 + 0.279 \times 10^4 s + 1.05 \times 10^8)]$; (b) For LP section: $C = 10$ nF, $R_1 = R_2 = 13.7$ kΩ;
For LPN section: $C = 10$ nF, $R_1 = R_2 = R_3 = R_5 = 9.76$ kΩ, $R_6 = 35.9$ kΩ, $C_{61} = 6.18$ nF, $C_{62} = 3.82$ nF
12.49 $C = 10$ nF; $R = 15.9$ kΩ; $R_1 = R_f = 10$ kΩ; $R_2 = 10$ kΩ; $R_3 = 390$ kΩ; 39 V/V   12.51 ±1%   12.53 (a) For
only ω_z, change C_1 and r or R_3, or change R_2 and r or R_3; R_2 and R_3 preferred; (b) For only Q_z, change only r,
or only R_3 12.55 $R_3 = 141.4$ kΩ; $R_4 = 70.7$ kΩ   12.57 $T(s) = -(16s/RC)/[s^2 + 2s/RC + 16/(RC)^2]$; Bandpass;
$\omega_0 = 4/RC$; $Q = 2$; Center-frequency gain = 8 V/V   12.59 $T(s) = s^2/[s^2 + (C_1 + C_2)s/R_3C_1C_2 + 1/R_4R_3C_1C_2]$;
High-pass; High-frequency gain = 1 V/V; $R_3 = 141.4$ kΩ; $R_4 = 70.7$ kΩ   12.60 For first-order section: $C_1 = 3.18$ nF;
For one S and K section, the grounded and floating capacitors are, respectively, $C_2 = 984$ pF and $C_3 = 10.3$ nF;
For the other S and K section, corresponding capacitors are $C_4 = 2.57$ nF and $C_5 = 3.93$ nF, respectively.
12.62 Sensitivities of ω_0 to R, L, C are $0, -\frac{1}{2}, -\frac{1}{2}$ respectively, and of Q are $1, -\frac{1}{2}, \frac{1}{2}$ respectively.

CHAPTER 13

13.1 (a) $\omega = \omega_0$, $AK = 1$; (b) $d\phi/d\omega$ at $\omega = \omega_0$ is $-2Q/\omega_0$; (c) $\Delta\omega_0/\omega_0 = -\Delta\phi/2Q$.   13.3 To non-inverting input,
connect LC to ground and R to output; $A = 1 + R_2/R_1 \geq 1.0$; Use $R_1 = 10$ kΩ, $R_2 = 100$ Ω (say); $\omega_0 = 1/\sqrt{LC}$
(a) $-\frac{1}{2}\%$; (b) $-\frac{1}{2}\%$; (c) 0%.   13.5 Minimum gain is 20 dB; phase shift is 180°.   13.6 Use $R_2 = R_5 = 10$ kΩ;
$R_3 = R_4 = 5$ kΩ; $R_1 = 50$ kΩ   13.9 $V_a(s)/V_o(s) = (s/RC)/[s^2 + 3s/RC + 1/R^2C^2]$; with magnitude zero at $s = 0$,
$s = \infty$; $\omega_0 = 1/RC$; $Q = \frac{1}{3}$; Gain at $\omega_0 = \frac{1}{3}$.   13.10 $\omega = 1.16/CR$.   13.12 $R_3 = R_6 = 6.5$ kΩ; $v_O = 2.08$ V$_{(peak-to-peak)}$
13.13 $L(s) = (1 + R_2/R_1)(s/RC)/[s^2 + s3/RC + 1/R^2C^2]$; $L(j\omega) = (1 + R_2/R_1)/[3 - j(1/\omega RC - \omega RC)]$; $\omega = 1/RC$;
for oscillation, $R_2/R_1 = 2$.   13.15 20.3 V.   13.17 $A\beta(s) = -(R_f/R)/[1 + 6/RCs + 5/R^2C^2s^2 + 1/R^3C^3s^3]$;
$R_f = 29R$; $f_0 = 0.065/RC$ 13.21 For circuits (a), (b), (d), characteristic equation is: $C_1C_2Ls^3 + (C_2L/R_L)s^2 +$
$(C_1 + C_2)s + 1/R_L + g_m = 0$; $\omega_0 = [(C_1 + C_2)/C_1C_2L]^{1/2}$; $g_mR_L = C_2/C_1$; For circuit (c): $LC_1C_2s^3 + (C_1L/R_L)s^2 +$
$(C_1 + C_2)s + 1/R_L + g_m = 0$; $\omega_0 = [(C_1 + C_2)/C_1C_2L]^{1/2}$; $g_mR_L = C_1/C_2$.   13.23 From 2.01612 MHz to
2.01724 MHz.   13.25 (a) $V_{TL} = V_R(1 + R_1/R_2) - L_+R_1/R_2$, $V_{TH} = V_R(1 + R_1/R_2) - L_-R_1/R_2$; (b) $R_2 = 200$ kΩ,
$V_R = 0.0476$ V.   13.28 (a) Either +12 V or −12 V; (b) Symmetrical square wave of frequency f and amplitude ±12 V,
and lags the input by 65.4°. Maximum shift of average is 0.1 V.   13.29 $V_Z = 6.8$ V; $R_1 = R_2 = 37.5$ kΩ; $R = 4.1$ kΩ
13.31 $V_Z = 3.6$ V, $R_2 = 6.67$ kΩ; $R = 50$ kΩ; $R_1 = 24$ kΩ, $R_2 = 27$ kΩ.   13.33 $V_Z = 6.8$ V; $R_1 = R_2 = R_3 = R_4 = R_5 =$
$R_6 = 100$ kΩ; $R_7 = 5.0$ kΩ; Output is a symmetric triangle with half period of 50 μs and ±7.5 V peaks.
13.35 96 μs   13.36 $R_1 = R_2 = 100$ kΩ; $R_3 = 134.1$ kΩ; $R_4 = 470$ kΩ; 6.5 V; 61.8 μs.   13.38 (a) 9.1 kΩ; (b) 13.3 V
13.39 $R_A = 21.3$ kΩ; $R_B = 10.7$ kΩ   13.41 $V = 1.0996$ V; $R = 400$ Ω; Table rows, for v_O, θ, 0.7 sin θ, error % are:
0.70 V,   90°, 0.700 V,   0%;
0.65 V, 63.6°, 0.627 V, 3.7%;
0.60 V, 52.4°, 0.554 V, 8.2%;
0.55 V, 46.1°, 0.504 V, 9.1%;
0.50 V, 41.3°, 0.462 V, 8.3%;
0.40 V, 32.8°, 0.379 V, 5.6%;
0.30 V, 24.6°, 0.291 V, 3.1%;
0.20 V, 16.4°, 0.197 V, 1.5%;
0.10 V, 8.2°, 0.100 V,   0%;
0.00 V,   0°, 0.0 V,   0%.
13.42 ±2.5 V   13.45 Table rows; circuit v_O/V_T, circuit v_I/V_T, ideal v_O/v_T, and error as a % of ideal are:
0.250, 0.451, 0.259, −3.6%
0.500, 0.905, 0.517, −3.4%
1.000, 1.847, 1.030, −2.9%
1.500, 2.886, 1.535, −2.3%
2.000, 4.197, 2.035, −1.7%
2.400, 6.292, 2.413, −0.6%
2.420, 6.539, 2.420,   0.0%

13.47 $R_1 = R_2 = 10$ kΩ (say); 3.18 V   **13.49** $R_1 = 1$ MΩ; $R_2 = 1$ MΩ; $R_3 = 45$ kΩ; $R_4 = 1$ MΩ; $C = 0.16$ μF for a corner frequency of 1 Hz.   **13.53** Use op amp circuit with v_A connected to positive input, LED between output and negative input and resistor R between negative input and ground; $I_{LED} = v_A/R$.   **13.54** $i_M = C|dv/dt|$; $C = 2.65$ μF; $i_{M120} = 2i_{M60}$; $i_{M180} = 3i_{M60}$; Acts as a linear frequency meter for fixed input amplitude; with C, has a dependence on waveform rate of change; 1.272 mA.   **13.55** 10 mV, 20 mV, 100 mV; 50 pulses, 100 pulses, 200 pulses

CHAPTER 14

14.1 Upper limit (same in all cases): 4.7 V, 5.4 V; lower limits: −4.3 V, −3.6 V; −2.15 V, −1.45 V   **14.4** 152 Ω; 0.998 V/V; 0.996 V/V; 0.978 V/V; 2%   **14.6** $V_{CC}I$ **14.9** 5 V   **14.11** 4 V; 12.8%; 11.1 kΩ   **14.13** 5.0 V peak; 3.18 V peak; 3.425 Ω; 4.83 Ω; 3.65 W; 0.647 W   **14.15** $\hat{V}_o^2/R_L$; $V_{SS}\hat{V}_o/R_L$; $\hat{V}_o/V_{SS}$; 100%; V_{SS}; V_{SS}^2/R_L; $V_{SS}/2$; 50%   **14.17** 2.5 V   **14.19** 12.5   **14.21** 20.7 mA; 788 mW; 7.9°C; 37.6 mA   **14.23** 1.34 kΩ; 1.04 kΩ   **14.25** 50 W; 2.5 A   **14.27** 140°C; 0.57 V   **14.29** 100 W; 0.4°C/W   **14.31** 0.85 Ω   **14.33** 0 mA, 0 mA; 20 μA, 22.5 μA; −20 μA; −22.5 μA   **14.35** 1.96 mA; 38.4 μA; out of base 1 and into base 2; 3.4 μA; 277 kΩ; 0.94 V/V   **14.37** 0.033 mA; 66 mA/V; −66 V/V; 13.6 kΩ   **14.39** $R_1 = 300$ kΩ; $R_2 = 632$ kΩ; 9.48 V; −10.65 V   **14.41** 13 Ω; 433 mV; 0.33 μ   **14.43** $R_1 = 60$ kΩ; $R_2 = 5$ kΩ; 0.01 μ   **14.45** $I_{E1} = I_{E2} - 17$ μA; $I_{E3} = I_{E4} - 358$ μA; $I_{E5} - I_{E6} = 341$ μA; 10.5 V   **14.47** 14 V; 1.9 W; 11 V   **14.49** $R_3 = R_4 = 40$ Ω; $R_1 = R_2 = 2.2$ kΩ   **14.51** 40 kΩ; 50 kΩ   **14.53** $L = \mu_n(v_{GS} - V_t)/U_{sat}$; 3 μm; 3 A; 1 A/V

APPENDIX B

B.2 $h_{11} = 2.6$ kΩ; $h_{12} = 2.5 \times 10^{-4}$; $h_{21} = 100$; $h_{22} = 2 \times 10^{-5}$ ℧
B.3 $y_{11} = 1/r_\pi + s(C_\pi + C_\mu)$; $y_{12} = -sC_\mu$; $y_{21} = -sC_\mu + g_m$; $y_{22} = 1/r_o + sC_\mu$

APPENDIX C

C.1 $Z_t = V_{oc}/I_{sc}$ **C.3** 1 V, 0.90 kΩ; 0.526 V   **C.5** $R_{in} = (r_\pi + R_B)/(1 + g_m r_\pi)$

APPENDIX D

D.2 $V_o(s)/V_i(s) = R_2/(R_1 + R_2)$ **D.4** 10^5 rad/s   **D.6** HP; 10 rad/s   **D.7** $v_O(t) = 10(1 - e^{-t/10^{-6}})$; $v_o(t) = 10e^{-10^6 t}$
D.9 3.5 ns   **D.11** −4.67 V   **D.13** −6.32 V; 9.5 ms   **D.15** 14.4 μs

APPENDIX E

E.1 $V_o(s)/V_i(s) = RC_1 s/(1 + sR(C_1 + C_2))$; STC with $C_{eq} = C_1//C_2$; high-pass; zero at 0 Hz; pole at 1.59 Hz
E.5 10 kHz; 5.1 kHz; 1.05 kHz   **E.10** 0 dB, −90°; +0.04 dB, −95.0°

INDEX